ADVANCES IN ENERGY SCIENCE AND EQUIPMENT ENGINEERING II

PROCEEDINGS OF THE 2ND INTERNATIONAL CONFERENCE ON ENERGY EQUIPMENT SCIENCE AND ENGINEERING (ICEESE 2016), 12–14 NOVEMBER 2016, GUANGZHOU, CHINA

Advances in Energy Science and Equipment Engineering II

Editors

Shiquan Zhou
*China-EU Institute for Clean and Renewable Energy at Huazhong
University of Science and Technology, Wuhan, China*

Aragona Patty
*Institute of Applied Industrial Technology Division,
Portland Community College, Portland, OR, USA*

Shiming Chen
College of Civil Engineering, Tongji University, Shanghai, China

VOLUME 2

 **CRC Press**
Taylor & Francis Group
Boca Raton London New York

CRC Press is an imprint of the
Taylor & Francis Group, an **informa** business

A BALKEMA BOOK

Published by:
CRC Press/Balkema
P.O. Box 447, 2300 AK Leiden, The Netherlands
e-mail: Pub.NL@taylorandfrancis.com
www.crcpress.com – www.taylorandfrancis.com

First issued in paperback 2020

Typeset by V Publishing Solutions Pvt Ltd., Chennai, India

ISBN 13: 978-1-138-71798-5 (set of 2 volumes)
ISBN 13: 978-0-138-05682-4 (Vol 1)
ISBN 13: 978-0-367-73628-6 (Vol 2) (pbk)
ISBN 13: 978-1-138-05683-1 (Vol 2) (hbk)

Visit the Taylor & Francis Web site at
http://www.taylorandfrancis.com

and the CRC Press Web site at
http://www.crcpress.com

Advances in Energy Science and Equipment Engineering II – Zhou, Patty & Chen (Eds)
© 2017 Taylor & Francis Group, London, ISBN 978-1-138-71798-5

Table of contents

Mechanical engineering

Environmental and architectural engineering

VOLUME 2

Structural and materials science

Computer simulation & computer and electrical engineering

xix

Preface

The previous First International Conference on Energy Equipment Science and Engineering (ICEESE 2015) successfully took place on May 30-31, 2015 in Guangzhou, China. All accepted papers were published by CRC Press/Balkema (Taylor & Francis Group) and have been indexed by Ei Compendex and Scopus and CPCI.

The 2016 2nd International Conference on Energy Equipment Science and Engineering (ICEESE 2016) was held on November 12–14, 2016 in Guangzhou, China. ICEESE 2016 is bringing together innovative academics and industrial experts in the field of energy equipment science and engineering to a common forum. The primary goal of the conference is to promote research and developmental activities in energy equipment science and engineering and another goal is to promote scientific information interchange between researchers, developers, engineers, students, and practitioners working all around the world.

The conference will be held every year to make it an ideal platform for people to share views and experiences in energy equipment science and engineering and related areas. We invite original papers describing an idea or concept, addressing issues and problems, or focusing empirically on potential or realistic fields.

ICEESE 2016 has received about 1000 papers from 8 countries and regions. The papers originate from both academia and industry concentrating on the international flavor of this event in the topics of energy equipment science and engineering. Based on the peer review reports, 434 papers were accepted to be presented in ICEESE 2016 by the editors. All the accepted papers will be presented on the conference, mainly by oral presentations in 5 sessions: Energy and Environmental Engineering, Mechanical Engineering, Environmental and Architectural Engineering, Structural and Materials Science, Computer Simulation & Computer and Electrical Engineering.

We express our thanks to all the members of 2016 2nd International Conference on Energy Equipment Science and Engineering (ICEESE 2016). Thanks are also given to CRC Press/Balkema (Taylor & Francis Group) for producing this volume.

We hope that ICEESE 2016 has been successful and enjoyable to all participants.

The Organizing Committee of ICEESE 2016

Organizing committees

CONFERENCE CHAIRS

Prof. Shiquan Zhou, *Huazhong University of Science & Technology*
Prof. Aragona Patty, *Portland Community College, United States*

TECHNICAL PROGRAM COMMITTEE

Prof. Zhandeng Dong, *Chinese Academy for Environmental Planning(CAEP), China*
Dr. Qing Fu, *Sun Yat-sen University, China*
Prof. Hongwei Wang, *Hebei University, China*
Dr. Bo Cai, *Research Institute of Petroleum Exploration and Development-Langfang Branch, PetroChina, China*
Prof. Tao Zhu, *China University of Mining & Technology (Beijing), China*
Prof. Zicheng Zhou, *Mechanical Engineering Dept., Shanghai Urban Construction College, China*
A. Prof. Jinyou Dai, *College of Petroleum Engineering, China University of Petroleum-Beijing, China*
Assoc. Prof. Eng. Krzysztof Witkowski, *University of Zielona Gora, Poland*
Prof. Mohd Khairol Anuar Mohd Ariffin, *Universiti Putra Malaysia, Malaysia*
Prof. Mingming Mao, *Shandong University of Technology, China*
A. Prof. Yu-Ming Fei, *Chilhlee University of Technology, Taiwan*
Prof. Dr. Aidy Ali, *Department of Mechanical Engineering, Universiti Pertahanan Nasional, Malaysia*
A. Prof. Gajendra Sharma, *Kathmandu University, Department of Computer Science and Engineering, Nepal*
Dr. Harish Kumar Sahoo, *International Institute of Information Technology (IIIT), Bhubaneswar, India*
Prof. Alaimo Andrea, *Faculty of Engineering and Architecture, Italy*
Prof. Govind Sharan Dangayach, *Malaviya National Institute of Technology, India*
Prof. Antonio Gil, *Public University of Navarra, Spain*
Prof. Dzintra Atstaja, *BA School of Business and Finance, Latvia*
Dr. Asimananda Khandual, *Biju Pattanaik University of Technology, India*
Dr. Fábio Robereto Chavarette, *Department of Mathematics, Brasil*
Prof. Heyong He, *Fudan University, China*
Prof. Leonid Getsov, *The Polzunov Central Boiler and Turbine Inst, Russia*
Prof. Loganina, *Penza State University of Architecture and Construction, Russia*
Prof. Maroš Soldán, *Slovak Technical University, Slovak Republic*
Dr. Saima Shabbir, *Department of Materials Science and Engineering, Institute of Space Technology, Pakistan*
Prof. Saffi Mohamed, *Mohamed V University in Rabat (ESTS), Morocco*
Prof. Ahmad Rezaeian, *Isfahan University of Technology, Iranian/Canadian*
Prof. Yingkui Yang, *Hubei University, China*

Structural and materials science

Advances in Energy Science and Equipment Engineering II – Zhou, Patty & Chen (Eds)
© *2017 Taylor & Francis Group, London, ISBN 978-1-138-71798-5*

Thermal degradation kinetics and flame-retardant properties of acrylonitrile butadiene styrene/thermoplastic polyurethanes/halloysite nanotubes and nanocomposites

Huiwen Yu

Technology Development Center for Polymer Processing Engineering of Guangdong Colleges and Universities, Guangdong Industry Polytechnic, Guangzhou, China
Key Laboratory of Polymer Processing Engineering of the Ministry of Education, National Engineering Research Center of Novel Equipment for Polymer Processing, South China University of Technology, Guangzhou, China

Baiping Xu, Wenliu Zhuang & Meigui Wang

Technology Development Center for Polymer Processing Engineering of Guangdong Colleges and Universities, Guangdong Industry Polytechnic, Guangzhou, China

Hongwu Wu

Key Laboratory of Polymer Processing Engineering of the Ministry of Education, National Engineering Research Center of Novel Equipment for Polymer Processing, South China University of Technology, Guangzhou, China

ABSTRACT: In this paper, preparation of a novel halogen-free, flame-retardant Acrylonitrile Butadiene Styrene (ABS) nanocomposite is described. The synergistic effect was observed by adding Halloysite Nanotubes (HNTs) and Thermoplastic Polyurethanes (TPU) into an intumescent flame-retardant system consisting of ammonium polyphosphate. The Limiting Oxygen Index (LOI) test and the Thermogravimetric Analysis (TGA) were used to describe the fire behaviors, together with the thermal degradation kinetics. The experimental results showed that adding TPU into the flame-retardant ABS increased the oxygen index from 27.4% to 35.2%, carbon residue by 4%, and the intumescent formulations was V0 rated according to the UL-94 standard. The addition of only 1 wt% of HNTs could improve the thermal stability. A pyrolysis kinetics model was established by using the Coats–Redfern method, the kinetic equation of major degradation stages was derived, and the mechanism of flame was preliminarily speculated.

1 INTRODUCTION

In recent years, the use of Acrylonitrile–Butadiene–Styrene (ABS) is becoming more and more dramatic because of its chemical resistance, good mechanical properties, and processing advantages (S.F. Wang, 2002). However, its relatively high flammability, accompanied by the emission of smoke and melt dripping during combustion, restricted its further application in the related fields of building materials and electrical products (N.F. Attia, 2014). Consequently, how to improve its fire resistance and thermal stability is a bottleneck for extending the massive use of ABS to most applications.

The Intumescent Flame Retardant (IFR) system is the most important and commonly used system to improve the flame-retarding property of ABS. A desired fire-retarding property was achieved by incorporating the high concentrations of IFR additives into the ABS matrix during processing; however, the mechanical properties of the composites declined considerably (B.M. Le, 2000). So far, it is difficult to improve its flame-retardant property and thermal stability, meanwhile maintaining the mechanical properties of the composites as excellent as those of pure ABS matrix. To overcome this shortcoming, it was reported that some synergistic agents can be used with IFR to enhance the flame-retardant property and improve the mechanical properties of Montmorillonite Clay (S. Pack, 2010), Carbon Nanotubes5, and Nanomagnesium Hydroxide (C.M. Jiao, 2006). In addition, the major advantage of the nanocomposites, compared to other synergists, is that nanoadditives increased thermal stability and reduced cost simultaneously.

In this paper, Halloysite Nanotubes (HNTs) were employed as the synergistic agents to prepare the halogen-free, flame-retardant composite material, especially in order to improve the thermostability of the composites. The synergistic effect of HNTs was studied by the Limiting Oxygen Index (LOI), UL-94 tests, and thermogravimetric analysis (TGA). Then, the Coats–Redfern integral method was used to solve the pyrolysis kinetics model.

2 EXPERIMENT

2.1 Materials

Acrylonitrile Butadiene Styrene (ABS) was provided by Jinling Petroleum Chemical Company. Thermoplastic Polyurethanes (TPU) were acquired from BASF.

Halloysite Nanotubes (HNTs, produced in Hubei Province) had outer diameter of 20–70 nm, inner diameter of 10–30 nm, and length of 0.5–2 μm; the morphologies of HNTs are shown in Fig. 1. HNTs consisted of tubes of variable length. Ammonium Polyphosphate (APP) was supplied by Shanghai Aohui Industrail Co., Ltd. Pentaerythritol (PER) and ethylalcohol were obtained from Aladdin.

2.2 Preparation of flame-retardant ABS composites

HNTs were prepared by dissolving them in anhydrous alcohol; then they were dried and sieved. IFR consisted of APP and PER. The APP/PER (3/1 by weight) intumescent flame-retardant ABS/IFR/HNTs composites were melt-compounded using a Brabender mixer at 185°C for 8 min at a screw speed of 50 rpm. All formulations occupied 30% by weight of a total loading; when the nanoscale particulate additives were used as the synergistic agent, the IFR content was reduced to maintain the same total loading of the additives. All formulations are given in Table 1.

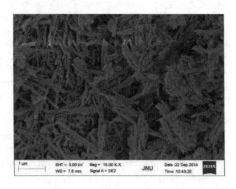

Figure 1. SEM micrograph of HNTs.

Table 1. Formulations of flame-retardant ABS composites.

Sample	ABS (g)	TPU (g)	HNTs (g)	APP (g)	PER (g)	Antioxidants (g)
ABS0	84	–	–	27	9	0.24
ABS1	63	21	–	27	9	0.24
ABS2	63	21	1.2	26.1	8.7	0.24
ABS3	63	21	2.4	25.2	8.4	0.24
ABS4	63	21	3.6	24.3	8.1	0.24
ABS5	63	21	4.8	23.4	7.8	0.24

2.3 Measurements

Thermogravimetric Analysis (TGA) was carried out using a thermal gravimetric analyzer in the temperature range of 32–700°C at a heating rate of 10°C/min under nitrogen atmosphere with a constant flow rate during analysis.

The Limiting Oxygen Indices (LOIs) were measured according to ASTM-2680, with the test using the sheets with dimensions of 80 mm × 10 mm × 4 mm. The apparatus used was an HC-2C oxygen index meter (Jiangning Analytical Instrument Factory, China).

Vertical burning tests (UL-94) were measured on a vertical burning test instrument (CZF-2-type, Jiangning Analytical Instrument Factory, China) according to ASTM D3801. The specimens used were of dimensions 130 mm × 13 mm × 3 mm.

Scanning Electronic Microscopy (SEM) was used to study the morphology, using a JSM-6330F field emission scanning electron microscope (Japan Electron Optics Laboratory CO., Ltd). The samples were fractured in liquid nitrogen and then covered with gold before being examined with a microscope.

3 RESULTS AND DISCUSSION

3.1 Limiting oxygen index

The UL-94 test results for various ABS composites are shown in Table 2.

It was found that the pure ABS used in this work has an LOI value of just 17.8, which was classified as "combustible." However, the LOI value of ABS0 with 30wt% IFR increased up to 27.4. In the case of ABS1, with the incorporation of TPU, the LOI value was increased up to 35.2, and the UL-94 was rated as V-0 in contrast to ABS0 as V-2. This was perhaps attributed to the fact that the ammonium polyphosphate was decomposed into polyphosphoric acid at high temperatures, and TPU was dehydrated into carbon to form the gunk to wrap the material surface, and then the oxygen was cut off. Compared with ABS1, ABS2 had an LOI

Table 2. TGA data of flame-retardant ABS composites.

Sample	ABS	ABS0	ABS1	ABS2	ABS3	ABS4	ABS5
LOI (%)	17.8	27.4	35.2	36.2	31.0	27.8	26.7
UL-94	Fail	V-2	V-0	V-0	V-1	V-2	Fail

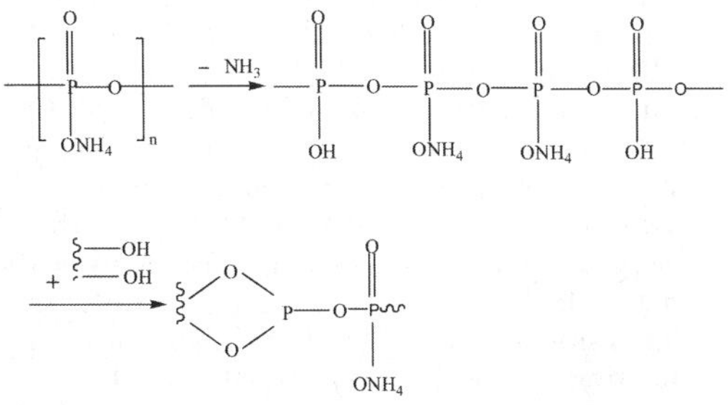

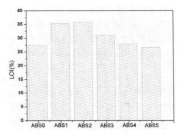

Figure 2. Limiting oxygen indices of different samples.

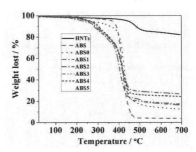

Scheme 1. Reaction of polyphosphoric acid with HNTs.

value of 36.2 with the enhancement of the thermal and mechanical properties, even when only 1 wt% HNT was used as the synergistic agent for ABS2. The first reason is that the surface of HNTs was rich of hydrate, and the heat that could be taken away during dehydration. The second reason is that the phosphate resulting from the pyrolysis of APP provided the phosphoric acid, which reacted with the high-density hydrogen provided by the HNTs to form polyol phosphates by esterification reaction, as shown in Scheme 1.

On the one hand, this process helped to increase char formation. On the other hand, the heat was taken away along with the loss of water and the formation of anhydrides. However, the LOI values of the other composites, say ABS3, ABS4, and ABS5, gradually decreased with the further increase of the relative HNT content, as shown in Figure 2. It may be due to the decrease of the flame-retardant content with the increase of HNTs, while the synergist content was fixed to a total loading of 30% by weight.

3.2 Thermogravimetric analysis

The thermogravimetric curves obtained for the above ABS composites under nitrogen atmosphere are shown in Figure 3. It is evident from the figure that ABS/TPU/IFR/HNTS nanocomposites indicated the three similar stages of decomposition. The first decomposition stage was perhaps related to the decomposition of APP into phosphoric acid and ammonia. The second decomposition could be attributed to the degradation of TPU and its

Figure 3. TGA thermograms for different samples.

Table 3. TGA data of flame-retardant ABS composites.

Sample	T_{max1} (°C)	T_{max2} (°C)	W (wt%/min)	Residue at 700°C (%)
HNTs	451.2	530.2	1.65	82.10
ABS	366.9	484.5	19.4	3.75
ABS0	374.5	459.5	17.2	12.46
ABS1	376.7	457.2	10.79	16.2
ABS2	376.8	457.1	9.18	26.65
ABS3	387.6	454.3	9.32	24.21
ABS4	387.6	454.8	9.41	24.15
ABS5	387.6	454.9	10.15	18.92

W is the decomposing rate at T_{max}.

additives. The third decomposition, which was the major degradation for all samples, occurred in the temperature range of 375–450°C, and this was responsible for the ABS degradation.

The decomposing rate at T_{max} and the residual amounts of all samples are summarized in Table 3. The degradation of all the samples at different temperatures depended on the compositions of the ABS/TPU/HNTs. It was observed that the addition of HNTs could enhance the thermal stability of the resin, especially when the loading of HNTs equaled 1 wt%, and the decomposing rate at T_{max} (wt%/min) of the composite had the smallest value of 9.18 wt%/min. However, with the further increase of HNTs, an upward tendency in the decomposing rate at T_{max} of ABS composite was found, which suggested an increase in the thermal stability.

It is seen from Table 3 that the residual amounts at 700°C increase because of the presence of HNTs. In general, for neat ABS, the amount of residue was only 3.7% when the temperature reached 700°C. However, the incorporation of 30 wt% IFR into the ABS resin led to a sharp increase in char yield during TGA analysis, where 12.46% char residue was observed in ABS0 even at 700°C. In addition, the thermal stability of the ABS composite was strongly affected by the presence of HNTs, resulting in an evident improvement in the carbon residue of ABS2 with the value of 26.65% at 700°C. When the content of HNTs was further increased up to 4%, the residue did not increase any more, but dropped down to 18.92%.

3.3 Thermal degradation kinetics

In order to determine whether there were different mechanisms among ABS0, ABS1, and ABS3, Coats–Redfern method was chosen, as it involved with the mechanisms of thermal degradation processes, as shown in Table 4.

For ABS0, $\alpha < 0.1$, this stage was mainly due to the loss of water and other small molecules. When $\alpha > 0.8$, a single model was not satisfied for the degradation of the material. There were further carbonizations of ABS composite material and the degradation of a small amount of unstable carbide residue. Hence, if we take $\alpha = 0.2$–0.8, we could find the activation energies calculated by C-R7 relationship and Friedman method matched, and the correlation coefficient approached 1.0, and the relative error was smaller at the same time. It seemed that the degradation mechanism of the ABS0 composite was due to diffusion, and the kinetic mechanism function is $F(\alpha) = \alpha^2$ ($0.2 \leq \alpha \leq 0.8$). In other words, the way of controlling the reaction process was plane symmetry. For ABS1 and ABS3, take $0.4 \leq \alpha \leq 0.6$, the activation energy calculated by C-R9 and Friedman method matched, the degradation mechanism was three-dimensional diffusion, and the kinetic mechanism function obeyed $F(\alpha) = \left[1 - (1 - \alpha)^{1/3}\right]^2$, where $\alpha < 0.4$ and $\alpha > 0.6$. With the addition of TPU, the degradation process of the composites became more complicated, where the pyrolysis process was accompanied by thermal decomposition of the intumescent flame-retardant behavior, and the single thermal degradation mechanism could not meet the degradation of TPU and ABS any more, which is due to the coexistence of various mechanisms.

Table 4. Activation energy values and correlation coefficients of ABS0, ABS1, and ABS3 with Coats–Redfern method.

Mechanism		ABS0 (kJ · mol⁻¹)			ABS1 (kJ · mol⁻¹)			ABS3 (kJ · mol⁻¹)		
		5 K/min	10 K/min	15 K/min	5 K/min	10 K/min	15 K/min	5 K/min	10 K/min	15 K/min
C-R1	E	136.98	153.30	147.58	92.48	92.85	89.69	79.64	83.76	75.36
	CR	0.9830	0.9900	0.9910	0.9972	0.9951	0.9932	0.9892	0.9912	0.9851
C-R2	E	145.89	163.15	156.92	98.90	99.29	95.94	85.23	89.63	80.70
	CR	0.9863	0.9919	0.9912	0.9963	0.9940	0.9918	0.9875	0.9897	0.9833
C-R3	E	165.14	184.42	177.10	112.54	113.01	109.24	97.14	102.12	92.09
	CR	0.9905	0.9943	0.9885	0.9942	0.9914	0.9890	0.9842	0.9865	0.9794
C-R4	E	76.91	86.45	82.74	50.72	50.84	48.90	43.05	45.47	40.36
	CR	0.9890	0.9934	0.9870	0.9930	0.9896	0.9864	0.9801	0.9832	0.9736
C-R5	E	47.50	53.79	51.29	30.11	30.12	28.79	25.03	26.53	23.12
	R	0.9872	0.9925	0.9852	0.9913	0.9869	0.9828	0.9741	0.9783	0.9648
C-R6	E	32.80	37.46	35.57	19.81	19.76	18.73	16.01	17.08	14.50
	CR	0.9849	0.9913	0.9828	0.9888	0.9832	0.9774	0.9650	0.9710	0.9505
C-R7	E	237.37	265.16	256.39	160.92	161.71	156.56	139.61	146.60	132.75
	CR	0.9706	0.9792	0.9886	0.9994	0.9983	0.9972	0.9947	0.9959	0.9920
C-R8	E	266.87	297.80	287.46	182.86	183.76	177.95	158.77	166.69	151.06
	CR	0.9798	0.9868	0.9914	0.9983	0.9968	0.9953	0.9922	0.9937	0.9890
C-R9	E	303.09	337.83	325.46	208.89	209.91	203.32	181.49	190.52	172.77
	CR	0.9874	0.9925	0.9918	0.9966	0.9945	0.9927	0.9889	0.9908	0.9853
C-R10	E	278.83	311.02	300.02	191.49	192.43	186.36	166.31	174.59	158.26
	CR	0.9829	0.9892	0.9920	0.9978	0.9960	0.9944	0.9911	0.9927	0.9877

4 CONCLUSIONS

1. The results showed that adding TPU into the flame-retardant ABS increased the oxygen index from 27.4% to 35.2%. Moreover, among all the specimens we tried, adding 1% HNTS was the preferred choice, and the intumescent formulations was V0 rated according to the UL-94 standard. However, with the further increase of HNTS, the flame-retardant property declined.
2. In a nitrogen environment, at the temperature rise rate of 10°C/min, TG results showed that the residue at 700°C reached the highest value for the addition amount of 1% of HNTs by weight. With further increase in the amount of HNTs, the flame-retardant property declined, so the reside amount reduced.
3. The main degradation kinetics mechanism was due to diffusion with the control of reaction process by plane symmetry. After the extra addition of TPU, the pyrolysis process became rather complicated; in the stage of $0.4 \leq \alpha \leq 0.6$, the degradation mechanism was attributed to three-dimensional diffusion with the kinetic mechanism function: $F(\alpha) = \left[1 - (1 - \alpha)^{1/3}\right]^2$.

ACKNOWLEDGMENTS

This work was supported by the National Natural Science Foundation of China (No. 11272093) and supported by Guangdong Province Higher Vocational Colleges & Schools Pearl River Scholar Funded Scheme (2012).

REFERENCES

Attia N.F., M.A. Hassan, M.A. Nourb and K.E. Geckeler:' Flame-retardant materials: synergistic effect of halloysite nanotubes on the flammability properties of acrylonitrile–butadiene–styrene composites'. *Polym. Int.,* 2014, 63(7), 1168–1173.

Jiao C.M., Z.Z. Wang, Z. Ye, Y. Hu and W.C. Fan: 'Flame Retardation of Ethylene Vinyl Acetate Copolymer Using Nano Magnesium Hydroxide and Nano Hydrotalcite', *J. Fire. Sci.,* 2006, 24(1), 47–64.

Le B.M., M. Bugajny, J. Lefebvre and S. Bourbigot: 'Use of polyurethanes as char-forming agents in polypropylene intumescent formulations', *Polym. Int.,* 2000, 49(10), 1115–1124.

Pack S., T. Kashiwagi, C.H. Cao, C.S. Korach, M. Lewin, and M.H. Rafailovich:' Role of Surface Interactions in the Synergizing Polymer/Clay Flame Retardant Properties', *Macromolecules,* 2010, 43(12), 5338–5351.

Peeterbroeck S., F. Laoutid, J.M. Taulemesse, F. Monteverde and J.M. Lopez-Cuesta:' Mechanical Properties and Flame-Retardant Behavior of Ethylene Vinyl Acetate/High-Density Polyethylene Coated Carbon Nanotube Nanocomposites', *Adv. Funct. Mater.,* 2007, 17(15), 2787–2791.

Wang S.F., Y. Hu, L Song, Z.Z. Wang, Z.Y. Chen and W.C. Fan: 'Preparation and thermal properties of ABS/montmorillonite nanocomposite', *Polym. degrad. Stab.,* 2002, 77(3), 423–426.

Advances in Energy Science and Equipment Engineering II – Zhou, Patty & Chen (Eds)
© 2017 Taylor & Francis Group, London, ISBN 978-1-138-71798-5

Application of nanoparticle hybrid nanofibers in biological sensing

Yuqi Guo
State Key Laboratory of Chemical Resource Engineering, Beijing University of Chemical Technology, Beijing, P.R. China

ABSTRACT: Nanoparticle hybrid nanofibers are powerful tools for various applications, including biological and chemical sensing, battery, hydrogen storage, environmental protection, and human tissue regeneration. The performance of the nanoparticle hybrid nanofibers is mainly limited by material properties. Recently, successful attempts have been made to add nanoparticles or nanotubes to polymer nanofibers by electrospinning technique, and nanocomposite materials were functionalized by the modification with nanomaterials. This paper focuses on carbon nanotubes and metal nanoparticle hybrid polymer nanofibers, which can be used to detect glucose, H_2O_2, H_2, H_2S, alcohol, and other substances.

1 INTRODUCTION

Development of electrospinning technology is a nanoengineering process for manufacturing nanofibers and woven films in a magnetic force field. However, the development of the nanofibers is mainly limited by two aspects. The first one is that the spinning requires ultrahigh voltage, while the control of the strong electrostatic field was far from mature, and also the properties of the nanomaterials, e.g., nanofibers, have not been understood with the limited characterization methods. With the development of nanotechnology, the electrostatic spinning technology has attracted more and more attention and has been widely used in various fields as well.

The electrostatic spinning technique possesses a well-defined demonstration in the laboratory scale, and is easy to control by the combination of spinning equipment. The typical electrospinning machine consists of a metal needle-shaped spinneret, injection pump, high-voltage power supply, plate electrode, collection plate, and collecting roller shaft. The structure of ES machine is shown in Figure 1. First, polymer spinning solution is loaded in the needle cylinder, and under the effect of static electricity, spinning solution sprays from the metal needle, which was collected on the collecting board or roller shaft. By doing this, we can achieve the diameter distribution of fibers from nanometer to micron level. In addition, during the spinning process, if changing the roll axis direction of motion into rotation and reciprocating movement along the axial direction, we will obtain a certain degree of orientation of the nanofibers, which stands a better chance to generate various structures of nanofibers. However, including the roll shaft will affect the nanofiber structure, and some underlying factors also have effects on the

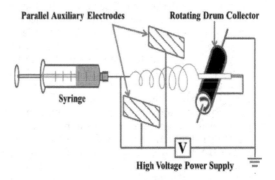

Figure 1. Structure of electrospinning machine.

structure of nanofiber, such as spinning solution viscosity, spinning operating conditions (voltage, needle cylinder speed, and needle diameter), and environmental conditions (humidity and temperature).

2 PREPARATION OF HYBRID NANOFIBERS BY USING ELECTROSPINNING

2.1 Polymer nanofibers

2.1.1 PAN
The performances of nanofibers obtained by electrospinning were better than those of the corresponding bulky materials, which was contributed by their dimensional characteristics: high diameter ratio, large specific surface area, and free volume. Nanofiber also has many different structures, such as single-oriented grid structure (Su Z, 2012), double-oriented grid structure (Su Z, 2012), strip structure (Sureeporn K, 2001), necklace structure

(Jin Y, 2010), and hollow structure (Dror Y, 2007) (Figure 2). These fiber structures can be controlled by electrospinning. Recently, studies have reported that the orientation of poly(ethylene oxide) (PEO) nanofibers (Su Z, 2012) can be controlled by electrospinning. In Figure 2, A and B show that the fiber structure can be obtained by controlling different conditions by electrospinning. The single orientation of the nanofiber structure can be obtained through using different density of nanofiber material, and the dual orientation of the nanofiber structure can be obtained by adjusting the spinning time, roller speed, and spinning voltage.

2.2 Nanoparticles hybrid nanofibers

2.2.1 CNTs hybrid nanofiber
Nanofiber doped with carbon nanotubes gained an orientated structure, assisted by the dispersion of Carbon Nanotubes (CNTs). Adding carbon nanotubes into conductive nanofibers can greatly improve the mechanical properties, electrical properties, and thermal properties of fiber material. In order to maximize the strength of composite fiber (CNTs/NFs), both the orientation and dispersion of carbon nanotubes in the fiber materials play a crucial role, while the dispersion and orientation of CNTs in the fiber required further investigation. For instance, electrospinning was used to prepare nanofiber film doped with carbon nanotubes to improve its conductivity (Maitra T, 2012; Min J K, 2010).

For the first time, Reneker's team found that by using electrospinning, the orientation of the carbon nanotubes in the nanofibers was higher (Figure 3 B)

than that of PAN (Ge J J, 2004). This finding indicates that not only the surface tension has effects on the orientation, but also the extruding that happens near the needle during electrospinning will affect the orientation of the carbon nanotubes. Because of the highly anisotropic orientation of carbon nanotubes in polymer nanofibers, PAN nanofibers doped with carbon nanotubes have better electrical conductivity, mechanical strength, and thermal stability. At the same time, Haddon's team used electrospinning technology to prepare SWCNT-reinforced composite polymer film (SWCNT-PS) (Zhao B, 2005), and the diameter of the nanofibers lies in the range of 50–100 nm. TEM shows that the orientation of a small cluster of SWCNT is parallel to the axial direction of the nanofiber.

2.2.2 Noble metal nanoparticle hybrid nanofibers
The study shows that the electrochemical catalytic properties of metal nanoparticles are related to their size and dispersion. The dispersion of metal nanoparticles plays a crucial role in the functional material. Materials doped with metal nanoparticles with a high degree of dispersion show a better electrochemical performance. If the metal nanoparticles coalescence, this phenomenon will reduce their catalytic effect and recycle time. Therefore, how to design and produce a high degree of dispersion and achieve a high catalytic performance of metal nanoparticle composite material is a challenge. Nowadays, researchers have used electrospinning to hybrid metal nanoparticles into the polymer nanofibers, which shows a good dispersion.

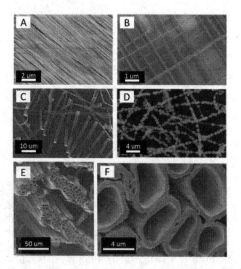

Figure 2. Structures of nanofiber: (A) single-oriented nanofibers; (B) double-oriented nanofibers; (C) strip nanofibers; (D) necklace beads; and (E, F) hollow nanofibers.

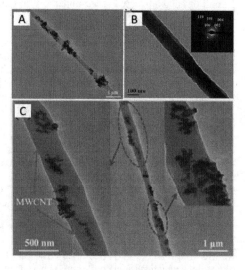

Figure 3. (A,C) Microstructure of Pt nanoparticles dispersed on the PVDF nanofibers; (B) High-magnification bright-field TEM images of electrospun MWNT/PAN composite nanofibers.

Panpan Zhang (Zhang P, 2014) used electrospinning to prepare composite PVDF nanofiber, which was doped with MCNTs and PtNPs. This material can be used as a biological membrane to detect glucose and H₂O₂; moreover, it possesses good stability and high sensitivity. The microstructure (Figure 3 A,C) under a Transmission Electron Microscope (TEM) indicates that the Pt nanoparticles dispersed evenly on the PVDF nanofibers, which indicates high dispersion effect and excellent electrochemical catalysis of the biofilm. Figure 2 shows that the dispersion of MCNTs on the composite fibers is also evenhanded, which improves the conductivity of the composite nanofibers.

3 APPLICATION IN BIOCHEMICAL SENSING

3.1 *Application in the detection*

3.1.1 *Sensing of alcohols*

For the first time, Schim modified Nylon 66 with a hydroxyl group of carbon nanotubes; furthermore, composites (Choi J, 2010) of nylon and carbon nanotube were obtained by electrospinning technique. By testing the reaction between the low molecular weight alcohol vapor and the composite (Figure 1), we find that the composite material has a high sensitivity to alcohols such as methanol, ethanol, and butanol. The mechanism of the detection of alcohols was that the hydrogen bonds formed between hydroxyl groups of ethanol vapor and hydroxyl groups of carbon nanotube/amino amide groups of nylon and this interaction led to the changes in resistance during the detection process. CNTs are P-type semiconductors under room temperature, and alcohol will provide electrons during the interaction of CNTs and alcohol, which has a great influence on the strength of hydrogen bonds. Their research indicated that the resistance of the composite increased gradually, during the process of detecting alcohol vapor. By using this CNT-modified nanofiber sensor detecting ethanol vapor, five rounds of reversible adsorption-release curves showed that the sensor had a reversible reaction for ethanol detection (Figure 4).

In addition, Yang's team developed a carbon fiber doped with nickel nanoparticles that can be attached to the electrode surface, which can be used to detect ethanol (Yang L, 2010). In order to produce this kind of sensor, a large number of spherical nanoparticles were embedded into carbon fibers, and then the modified carbon fibers were prepared by electrospinning technique. During the process of ethanol detection, the material shows a great current response and operation stability. The correction curve for ethanol (Figure 5) shows that

the relationship between the concentration and the response current is linear, when the concentration of ethanol is in the range of 0.25–87.5 mM.

3.1.2 *Sensing of glucose and H₂O₂*

The electrochemical biosensor based on enzyme modification has attracted much attention for its high sensitivity and specificity. In order to maximize the effect of the enzyme biosensor, the substrate material should have high specific surface area, good thermal stability, and chemical inertia, which does not swell and dissolve in aqueous solution. The biological fiber film that was obtained by electrospinning can meet the above requirements, and can be used as a base material as well. The sensor

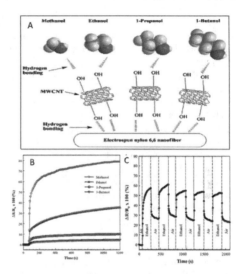

Figure 4. (A)Sensing mechanism of MWCNT–OH adsorbed electrospun Nylon 6,6 nanofibers to various low molecular weight alcohol vapors. (B) Response changes of pristine MWCNT and MWCNT–OH adsorbed electrospun Nylon 6,6 nanofibers to methanol vapor. (C) Response of MWCNT–OH adsorbed electrospun Nylon 6,6 nanofibers upon cyclic exposure to ethanol vapor at 25°C.

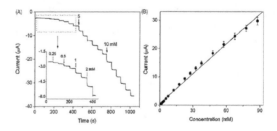

Figure 5. (A) Current–time responses of the NiCFP electrode upon successive addition of ethanol at different concentrations. (B) Calibration curve for ethanol detection.

845

obtained by electrospinning nanofibers possesses some merits; for instance, the design of this kind of sensor is quite flexible, high stability in the detection of both liquids and gases, high specific surface area, safe operation, and recyclability of materials.

Lee (2012) used the electrospinning technique to construct the carbon nanotube-modified PAN nanofibers and then used the glucose enzyme (GOD) to detect the glucose. By embedding GOD, the specific surface area and pore volume of the material become larger, which makes the attachment of more carbon nanotubes to the nanofibers possible. The effect of carbon nanotubes in the material is to enhance the overall conductivity of the porous carbon fiber, and the cyclic voltammetry curve (Figure 6) can be used to study the electrochemical performance of the material. The conductivity and the current peak intensity of the curve were increased upon the addition of carbon nanotubes into PAN nanofiber. When glucose is added in the environment and detected by electrochemical workstation, the glucose concentration in the environment is increased. After adding glucose every time, the current response time was as short as 7 s to yield a sensor with fast response.

Liu (2011) and other researchers have prepared a type of Pt nanoparticle hybrid carbon nanofibers by electrospinning, which can be applied in the detection of H_2O_2. During the detection process, CNF-PtNP-modified electrodes show a lower detection limit, fast response, and high sensitivity. In addition, under physiological conditions of pH value and the addition of ascorbic acid, acetaminophen, and uric acid, CNF-PtNP-modified electrode shows a high specificity in the detection of hydrogen peroxide (H_2O_2). Figure 7 A shows the current response of CNF-PtNP-modified electrodes under the continuous addition of different concentrations of H_2O_2 solution. Every time the injection of H_2O_2 solution showed a strong and rapid response, which is due to the high electrocatalytic ability of highly dispersed Pt metal particles. Figure 7B refers to the current response of the electrode, which is divided into eight times, with each time injected into the 0.1 mM H_2O_2 solution. The detection limit obtained from the electrodes was $0.6\mu M$, and the linear range was $1–800\mu M$ (R = 0.9991).

3.1.3 Sensing of H_2S gas

Wang (2011) prepared composite nanofibers doped with ZnO and Cu, by using electrospinning technique. At a high temperature of 230°C, this hybrid material for the sensor has a better detection limit to low concentration (1–10ppm) of the H_2S. Finally, the experimental results show that the detection of H_2S is greatly improved by using Cu to modify the composite nanofibers. The sensitivity of 6wt% Cu-modified composite nanofibers toward H_2S gas is high, and the detection sensitivity of 10ppm H_2S is better than that of the porous nanofibers prepared by ZnO. Figure 8 shows the response of a kind of gas sensor prepared by Cu-modified ZnO nanofibers under different concentrations of H_2S gas. It can be found that the response of the sensor for the sample intensified with the increase in H_2S gas concentration, and the response time and recovery time are 18 and 20 s, respectively.

3.1.4 Sensing of H_2

Fong (Zhang L, 2012) used PdNP-modified carbon nanofibers (nanobrous mat nano-felt (carbon)) by electrospinning to detect H_2. Figure 9 shows the response of different sensors in H_2 and vacuum environment. The experimental results show that the resistance of this kind of material will decrease when the material contacted with H_2, but the variation rate of the resistance will decrease with

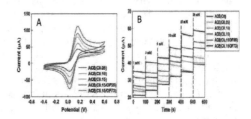

Figure 7. (A) Current response of CNF-PtNP-modified electrodes; (B) current response of the electrode.

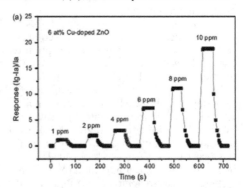

Figure 8. Response and recovery time of Cu-doped ZnO nanofiber sensors upon exposure to 1, 2, 4, 6, 8, and 10 ppm H_2S at the working temperature of 230°C.

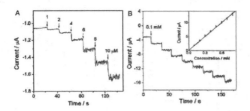

Figure 6. (A) Cyclic voltagram based on the effects of CNT additives and oxyfluorination; (B) Amperometric response of the glucose sensor at various glucose concentrations.

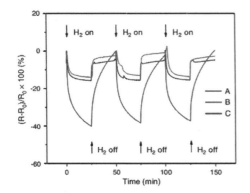

Figure 9. Response curves of different carbon nano-felts upon exposure to H2 at room temperature.

prolonged detection time. After isolating H_2, the resistance will return to the original value. This result also shows that the response of the sensor for H_2 is reversible. At the same time, experiment indicated that the detection sensitivity to H_2 was very high because the H_2 molecule, PdNPs, and carbon fibers have electronic contact and metastatic phenomenon. During the detection process, H_2 will react with PdNPs and generate palladium hydride, which has a positive effect on the reaction.

4 CONCLUSION

In the past decades, the preparation of functional nanomaterials by electrospinning technology has been described applied in various fields. It is indisputable in the future that, with the development of science and technology, electrostatic spinning technology will play a significant role in the preparation of nanostructured materials and high-sensitivity sensors. Therefore, this paper focus on how to use electrospinning technique to prepare novel composite fiber with metal nanoparticles and carbon nanotube hybrid nanofibers and the application of the electrochemical sensors and biosensors. However, there are still several challenges of electrospinning that need to be addressed. In order to control the size and shape of the product structure, we need to conduct more experiments and design more models to study. After modifying with carbon nanotubes and metal nanoparticles, nanofibers can also be used in other fields such as battery, hydrogen storage, environmental protection, and human tissue regeneration.

REFERENCES

Choi J, Park E J, Dong W P, et al. MWCNT–OH adsorbed electrospun nylon 6,6 nanofibers chemiresistor and their application in low molecular weight alcohol vapours sensing[J]. Synthetic Metals, 2010, 160(23):2664–2669.

Dror Y, Salalha W, Avrahami R, et al. One-Step Production of Polymeric Microtubes by Co-electrospinning[J]. Small, 2007, 3(6):1064–1073.

Ge J J, Hou H, Li Q, et al. Assembly of well-aligned multiwalled carbon nanotubes in confined polyacrylonitrile environments: electrospun composite nanofiber sheets.[J]. Journal of the American Chemical Society, 2004, 126(48):15754–61.

Ji S I, Yun J, Kim J G, et al. The effects of carbon nanotube addition and oxyfluorination on the glucose-sensing capabilities of glucose oxidase-coated carbon fiber electrodes[J]. Applied Surface Science, 2012, 258(7):2219–2225.

Jin Y, Yang D, Kang D, et al. Fabrication of necklace-like structures via electrospinning.[J]. Langmuir the Acs Journal of Surfaces & Colloids, 2010, 26(2):1186–90.

Liu Y, Wang D, Xu L, et al. A novel and simple route to prepare a Pt nanoparticle-loaded carbon nanofiber electrode for hydrogen peroxide sensing[J]. Biosensors & Bioelectronics, 2011, 26(11):4585–90.

Maitra T, Sharma S, Srivastava A, et al. Improved graphitization and electrical conductivity of suspended carbon nanofibers derived from carbon nanotube/polyacrylonitrile composites by directed electrospinning[J]. Carbon, 2012, 50(5):1753–1761.

Min J K, Lee J, Jung D, et al. Electrospun poly(vinyl alcohol) nanofibers incorporating PEGylated multiwall carbon nanotube[J]. Synthetic Metals, 2010, 160(13–14):1410–1414.

Su Z, Li J, Li Q, et al. Chain conformation, crystallization behavior, electrical and mechanical properties of electrospun polymer-carbon nanotube hybrid nanofibers with different orientations[J]. Carbon, 2012, 50(15):5605–5617.

Sureeporn K, Liu W, Reneker D H. Flat polymer ribbons and other shapes by electrospinning[J]. Journal of Polymer Science Part B Polymer Physics, 2001, 39(21):2598–2606.

Yang L, Lei Z, Guo Q, et al. Enzyme-free ethanol sensor based on electrospun nickel nanoparticle-loaded carbon fiber paste electrode[J]. AnalyticaChimicaActa, 2010, 663(2):153–157.

Zhang L, Wang X, Zhao Y, et al. Electrospun carbon nano-felt surface-attached with Pd nanoparticles for hydrogen sensing application[J]. Materials Letters, 2012, 68(1):133–136.

Zhang P, Zhao X, Xuan Z, et al. Electrospun Doping of Carbon Nanotubes and Platinum Nanoparticles into the β-Phase Polyvinylidene Difluoride Nanofibrous Membrane for Biosensor and Catalysis Applications[J]. Acs Applied Materials & Interfaces, 2014, 6(10):397–401(5).

Zhao B, Hu H, Yu A, et al. Synthesis and characterization of water soluble single-walled carbon nanotube graft copolymers.[J]. Journal of the American Chemical Society, 2005, 127(22):8197–203.

Zhao M, Wang X, Ning L, et al. Electrospun Cu-doped ZnO nanofibers for H 2 S sensing[J]. Sensors & Actuators B Chemical, 2011, 156(2):588–592.

Advances in Energy Science and Equipment Engineering II – Zhou, Patty & Chen (Eds)
© 2017 Taylor & Francis Group, London, ISBN 978-1-138-71798-5

Use of fine recycled aggregates for producing mortar—a preliminary study

Qin Li

Faculty of Chemical Engineering Yunnan Open University, Kunming, China

ABSTRACT: This paper presents a preliminary research on the use of fine recycled aggregates from an aggregate plant in Yunnan Province, China, to produce mortar. The fine recycled aggregates were used to replace the sand. The experiment was carried out in two stages. In the first stage, the effect of the amount of cement on the compressive strength of the produced mortar was investigated. An optimal amount of cement was determined according to the experimental data. In the second stage, the optimal amount of cement was determined from the first stage. In addition, lime and gypsum were incorporated into the mixture, and their effect on the compressive strength of the mortars was studied. On the basis of the experimental observations, it was found that there also exist optimal contents of lime and gypsum for the produced mortar. Below the optimal value, the compressive strength of the mortar increases with the lime and gypsum; however, when the optimal value exceeded, the mortar compressive strength decreases with the increase in the amounts of lime and gypsum.

1 INTRODUCTION

Recycling and reuse of waste concretes as aggregates for producing new concretes is beneficial and necessary from the viewpoint of environment preservation and efficient use of natural resources. In the past several decades, a large number of research activities on the material and structural properties of concrete containing coarse recycled aggregate crushed from waste concrete have been carried out (Xiao J, 2005). In the processing of producing recycled coarse aggregates from waste concrete, fine aggregates with maximum size of 5 mm, which constitute 25% by weight of the materials, are also generated (Hansen, T.C, 1992). Many experiments on the use of fine recycled aggregates for mixing new concretes have yielded less satisfactory results (Jongsung, 2011). The produced concrete generally has lower mechanical properties and weaker durability aspects, compared to concrete with natural fine aggregates (Ravindrarajah, 1987). Alternative use of the fine recycled aggregates is thus required.

The use of fine recycled aggregates for producing mortars seems to be a suitable option for more efficient use of recycled aggregates, as indicated from several experimental investigations (Braga M, 2012; Neno C, 2014). In the production of mortars with fine recycled aggregates, these materials are often used to partially or completely replace natural sand. It can be expected that increasing the amount of cement in the mortars can lead to an increase of the mortar compressive strength.

However, from the viewpoint of economy, it is not wise to use a large quantity of cement for achieving high compressive strength of mortar. A better solution might be the use of the less costly lime or gypsum for filling the voids in the fine recycled aggregates (S. Tsivilis, 2002).

This paper presents an experimental research on the synthesis of mortars with fine recycled aggregates produced from an aggregate plant in Yunnan Province, China. The investigation was focused on the influence of the cement content, lime, and gypsum on the compressive strength of the mortars.

2 MATERIALS

2.1 *Material for the experiment*

The materials used in the experiments were slag Portland cement produced by a local cement manufacturer with a 28-day characteristic compressive strength of 32.5 MPa; tap water from Kunming city; and fine recycled aggregate, which was produced by a local aggregate plant from the waste concretes demolished from some old structures in Kunming city. The physical and mechanical properties of the cement and the fine recycled aggregates are presented in Tables 1 and 2, respectively.

2.2 *Mixture proportions and test method*

The objective of this phase of experiment is to investigate the influence of the amount of cement

Table 1. Physical and mechanical properties of cement.

Density (g/cm³)	Specific surface area (cm²/g)	Setting time		Compressive strength (MPa)		Flexural strength (MPa)	
		Initial	Final	3 days	28 days	3 days	28 days
3.02	3460	2:13	5:25	22.1	46.5	5.4	8.9

Table 2. Physical characteristics of fine recycled aggregate.

Density (kg/m³)	Porosity (%)	Water content (%)	Specific surface area (cm²/g)
1980	36.55	3.08	279.34

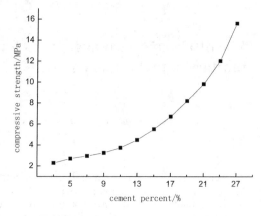

Figure 1. Variation of compressive strength of mortar with the percentage of cement.

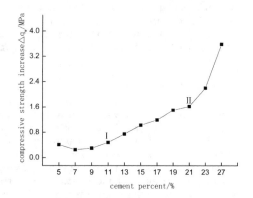

Figure 2. Relationship between the strength in cement of the stabilized debris and cement content.

on the behavior of the mortars with fine recycled aggregate. To this end, a total of 11 mixtures were produced. In the mixtures, the water/cement ratio was fixed to be 0.50. The percentages of the cement (to the sum of cement and fine recycled aggregate) were varied from 4% to 24% in a step of 2%.

All the mortars were mixed in a mixing machine using the following process: first, cement and fine recycled aggregates were mixed for 2 min, then water was added, and the mortars were mixed for another 5 min. Then, the mixtures were put in steel moulds (dimension: Φ50 mm × 50 mm) with three layers and were then vibrated for 1 min. The specimens were demolded after 24 h and stored in a curing room for 28 days before the test.

The experiments on the compressive strength of the mortars were carried out according to the Chinese Standard of *Experiment Methods of Materials Stabilized with Inorganic Binders for Highway Engineering* (JTJ057-94).

2.3 Test results and discussions

The variation of the compressive strength of the mortars (denoted as q_w) with the cement content is shown in Fig. 1. It can be seen from the figure that the mortar strength increases with the increase in the percentage of the cement.

Let Δq_w represent the increment of the compressive strength of two adjacent specimens in the terms of cement percentage, that is, $\Delta q_w = q_w - q_{w-2}$. Fig. 2 illustrates the influence of the cement percentage on the strength increment. From Fig. 1, it can be seen that the compressive strength variation can be divided into three sections. The transition point of the first section is 12%, whereas that of the second section is 25%. When the cement content is lower than or equal to 12%, the increment of the mortar compressive strength remains nearly constant with the increase of cement content. When the percentage of the cement varies from 12% to 25%, there is almost a linear increase of the increment for the compressive strength of the mortar. The compressive strength of the mortar increases dramatically with the cement content as the cement percentage exceeds 25%.

The experimental findings described above indicate different roles of the cement in increasing the compressive strength of the mortar. When the cement percentage is lower than 12%, the cement plays mainly a role for the compressive strength of the mortar due to its hydration. When the cement content exceeds 25%, the cement contributes to the mortar compressive strength in virtue of both a hydration and filling of the voids in the fine recycled aggregate. This means that a minimum cement percentage of 12% is required for achieving sufficient hydration.

3 EXPERIMENT PROGRAM—PHASE II

In this series of experiments, the cement percentage was fixed at 12% while the amounts of lime and gypsum were changed. The objective is to investigate the effect of the lime and the gypsum on the compressive strength of the mortar and to determine the optimal amount of these two materials.

3.1 Materials, mixtures, and test methods

Cement, water, and fine recycled aggregates used in this phase were the same as that in the first phase. The lime and gypsum were produced from local manufactures. The water/cement ratio was set at 0.50. As mentioned above, the cement content was set at 12%, which was thought to be the minimum amount for sufficient hydration for the fine recycled aggregates. The variation of lime and gypsum was the same, which was from 1% to 10% at a step of 1%. The preparation of the specimens and the test method were similar to those used in the last series of experiments.

3.2 Test results and discussions

Fig. 3 presents the experimental observations of the variation of the mortar compressive strength with the contents of lime and gypsum. It can be seen that with the increase of the content of lime or gypsum, the compressive strength of the mortar first increases to a maximum value and then decreases gradually. There exists a optimal value corresponding to the maximum compressive strength, which is 4% for both agents. The influence of the lime and

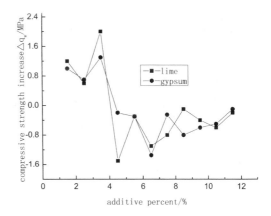

Figure 4. Relationship between compressive strength in the cement of the stabilized debris and loading agent content.

gypsum on the compressive strength increment, as shown in Fig. 4, implies similar results.

4 CONCLUSIONS AND REMARKS

This paper presents a study on the use of fine recycled aggregate to produce mortars. On the basis of the interesting test observations and within the investigated parameters in this study, the following conclusions can be drawn:

1. When only cement is used in the mortar, the compressive strength of the mortar increases with the cement content.
2. A minimum cement percentage of 12% is required to achieve sufficient hydration for the fine recycled aggregate with sufficient compressive strength.
3. With a cement percentage of 12%, lime and gypsum have a remarkable influence on the compressive strength of the mortar.
4. There exists an optimal value for the amount of lime and gypsum, which is 4% for both agents.

REFERENCES

Braga, M., De Brito, J., Veiga, R. Incorporation of fine concrete aggregates in mortars. *Construction and Building Materials*, 2012, Vol. 36, pp. 960–968.
Hansen, T.C. Recycling of Demolished Concrete and Masonry. London: E & FN SPON, 1992.
Jongsung, S., Cheolwoo, P., Compressive strength and resistance to chloride ion penetration and carbonate on of recycled aggregate concrete with varying

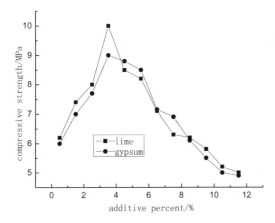

Figure 3. Relationship between compressive strength of the stabilized debris and loading agent content.

amount of fly ash and fine recycled aggregate. Waste Management, 2011, Vol. 31, No. 11, pp. 2352–2360.

Neno, C., De Brito, J., Veiga, R. Using fine recycled concrete aggregate for mortar production. *Material Research*, 2014, Vol. 17, No. 1, pp. 168–177.

Ravindrarajah, R.S., Tam, C.T. Recycling concrete as fine aggregate in concrete. *The International Journal of Cement Composites and Light Weight Concrete*, 1987, Vol. 9, No. 4, pp. 235–241.

Tsivilis S., E. Chaniotakis, G. Kakali, G. Batis. An analysis of the properties of Portland limestone cements and concrete [J]. Cement & Concrete Composites, 2002, Vol. 24, pp. 371–378.

Xiao, J., Li, J., Zhang, Ch. Mechanical properties of recycled aggregate concrete under uniaxial loading. *Cement and Concrete Research*, 2005, Vol. 35, No. 6, pp. 1187–1194.

Spectrophotometric determination of penicillamine in pharmaceutical samples by phosphorus molybdenum heteropoly acid

Xinrong Wen & Changqing Tu
College of Chemistry and Environment, Jiaying University, Meizhou, China

ABSTRACT: A novel spectrophotometric method for the determination of penicillamine by phosphorus molybdenum heteropoly acid is developed. The various effect factors on the determination of penicillamine by phosphorus molybdenum heteropoly acid spectrophotometry have been investigated in detail. The results show that in H_2SO_4 medium, $H_2PO_4^-$ reacts with $Mo_7O_{24}^{6-}$ to form phosphorus molybdenum heteropoly acid ($H_3P(Mo_3O_{10})_4$). Then $H_3P(Mo_3O_{10})_4$ is reduced by hydrosulfuryl(-SH) in penicillamine to form phosphorus molybdenum blue ($H_3P(Mo_3O_{10})_2(Mo_3O_9)_2$) which its maximum absorption wavelength is 719.2 nm. The content of penicillamine can be determined based on the absorbance of $H_3P(Mo_3O_{10})_2(Mo_3O_9)_2$. A good linear relationship is obtained between the absorbance and the concentration of penicillamine in the range of 4.016~120.5 µg·mL^{-1}, and the equation of the linear regression is $A = 0.0109+11.713C$ (mg·mL^{-1}), with a linear correlation coefficient of 0.9997. This proposed method has been successfully applied to the determination of penicillamine in pharmaceutical sample, and the results agree well with those obtained by pharmacopoeial method.

1 INTRODUCTION

Penicillamine (PA) is a sulfur-containing amino acid. Penicillamine is used in the treatment of rheumatoid arthritis, Wilson's disease, primary biliary cirrhosis, fibrotic lung disease, cystinuria and certain toxic metal poisoning. It is of great importance and significance for life science.

Some different methods have been reported for the determination of penicillamine, including spectro-Photometry (M. Skowron, 2011; M. Reza Hormozi-Nezhad, 2014; Corominas, B. Gómez-Taylor, 2005), fluorescence spectrophotometry (Wang Peng, 2015; Rasha Abdel-Aziz Shaalan, 2010), flow injection analysis (Martinović Anita, 2008; P. Viñas, 1990), voltammetry (J. B. Raoof, 2009; Ensafi, Ali A, 2010; Mazloum-Ardakani, Mohammad, 2010; Shahrokhian, Saeed, 2006), capillary electrophoresis (Yang Xiupei, 2007), HPLC method (Shu-Cai Liang, 2005), HPLC-UV method (Kuśmierek Krzysztof, 2009), HPLC-electrochemical method (Maria A. Saracino, 2013), LC-UV method (Kuśmierek, Krzysztof, 2007), chemiluminescence-LC method (Z. Zhang, 1997), kinetic potentiometric determination (Martinović Anita, 2009; Martinović Anita, 2007), electrochemical analysis (Fereshteh Chekin, 2012), etc. However, the fluorescence spectrophotometry, the chemiluminescence or the electrochemical method is less commonly available in the lab because of requiring special detectors. The process of determination of fow-injection spectrophotometry is relatively complex. HPLC method, HPLC-UV method, LC-UV method, etc are time-consuming, analytical cost high and requires complicated instrumentation. Capillary electrophoresis method requires demanding reaction conditions. Therefore, it is essential and significant to develop a simple, accurate, rapid and sensitive method for the determination of penicillamine in clinical analysis and drug quality control.

Spectrophotometry is a commonly used method of determination owing to its simplicity, rapidity and low-cost instrumentation. In this paper, we develop a novel spectrophotometric method for the determination of penicillamine by phosphorus molybdenum heteropoly acid. Under the optimum reaction conditions, in H_2SO_4 medium, $H_2PO_4^-$ reacts with $Mo_7O_{24}^{6-}$ to form phosphorus molybdenum heteropoly acid ($H_3P(Mo_3O_{10})_4$). Then $H_3P(Mo_3O_{10})_4$ is reduced by hydrosulfuryl(-SH) in penicillamine to form phosphorus molybdenum blue ($H_3P(Mo_3O_{10})_2(Mo_3O_9)_2$) which its maximum absorption wavelength is 719.2 nm. The content of penicillamine can be determined based on the absorbance of $H_3P(Mo_3O_{10})_2(Mo_3O_9)_2$. This proposed method has been successfully applied to the determination of penicillamine in pharmaceutical sample with satisfactory result. In contrast with previous methods (Wang Peng, 2015; Rasha Abdel-Aziz Shaalan, 2010; Martinović Anita, 2008; P. Viñas, 1990; J. B. Raoof, 2009; Ensafi, Ali A, 2010; Mazloum-Ardakani, Mohammad, 2010;

Shahrokhian, Saeed, 2006; Yang Xiupei, 2007; Shu-Cai Liang, 2005; Kuśmierek Krzysztof, 2009; Maria A. Saracino, 2013; Kuśmierek, Krzysztof, 2007; Z. Zhang, 1997; Martinović Anita, 2009; Martinović Anita, 2007; ereshteh Chekin, 2012), this method has the advantages of simply, rapidness, convenience, accuracy and so on.

2   EXPERIMENT

2.1   Equipment and reagents

A model 723S spectrophotometer (Shanghai Precision & Scientific Instrument Co., Ltd.) was used for photometric measurements. A model UV-2401 UV–visible spectrophotometer (The Shimadzu Corporation, japan) was used for scanning the absorption spectrum.

A stock of standard solution of $1.004 \ mg \cdot mL^{-1}$ penicillamine was prepared by dissolving 0.2008 g of penicillamine in 200 mL bidistilled water and was kept at 4°C, shieding from light. A $20.00 \ mg \cdot mL^{-1}$ KH_2PO_4 stock solution was prepared by dissolving 5.0000 g of KH_2PO_4 in bidistilled water and diluting it to 250 mL. A $0.1200 \ g \cdot mL^{-1}$ $(NH_4)_6Mo_7O_{24}$ stock solution was prepared by dissolving 63.7146 g of ammonium molybdate $((NH_4)_6Mo_7O_{24} \cdot 4H_2O)$ in bidistilled water, transferring into a 500 mL standard flask, diluting to the mark with bidistilled water. H_2SO_4 solution: $0.1840 \ mol \cdot L^{-1}$.

All reagents were of analytical reagent grade. Bidistilled water was used throughout.

2.2   Method

1.50 mL of $0.1200 \ g \cdot mL^{-1}$ $(NH_4)_6Mo_7O_{24}$ solution, 1.00 mL of $20.00 \ mg \cdot mL^{-1}$ KH_2PO_4, 0.20 mL of $0.1840 \ mol \cdot L^{-1}$ H_2SO_4 solution and 1.20 mL of $1.004 \ mg \cdot mL^{-1}$ penicillamine solution were added into a 25 mL volumetric flask, the solution was diluted to the mark with bidistilled water and mixed well. Aftering the mixture reacted for 35 min at 75°C in water both and cooled back to room temperature, the absorbance was measured at 719.2 nm against the reagent blank prepared in the same way except for penicillamine.

3   RESULTS AND DISCUSSIONS

3.1   Reaction mechanism

1. In H_2SO_4 Medium, $H_2PO_4^-$ reacts with $Mo_7O_{24}^{6-}$ to form phosphorus molybdenum heteropoly acid $(H_3P(Mo_3O_{10})_4)$:

$$7H_2PO_4^- + 12Mo_7O_{24}^{6-} + 79H^+$$
$$\rightarrow 7H_3P(Mo_3O_{10})_4 + 36H_2O$$

2. $H_3P(Mo_3O_{10})_4$ is reduced by hydrosulfuryl(-SH) in penicillamine to form phosphorus molybdenum blue $(H_3P(Mo_3O_{10})_2(Mo_3O_9)_2)$:

$$H_3P(Mo_3O_{10})_4 + 2R\text{-}SH \rightarrow H_3P(Mo_3O_{10})_2$$
$$(Mo_3O_9)_2 + RS\text{-}SR$$

According to the relationship between the amount of penicillamine and the amount of the formed phosphorus molybdenum blue in system, the content of penicillamine can be obtained based on the absorbance of the phosphorus molybdenum blue.

3.2   Absorption spectrum

According to the experimental method, in the range of 550~850 nm, the absorption spectrum of soluble phosphorus molybdenum blue formed from the reaction of $H_2PO_4^-$, $Mo_7O_{24}^{6-}$ and penicillamine is shown in Figure 1. It can be seen that the product of phosphorus molybdenum blue has an absorption peak at 719.2 nm. So, all the following measurements were carried out at 719.2 nm against reagent blank.

3.3   Effects of reaction temperature and time

In order to investigate the effect of different reaction temperature on absorbance, 3.00 mL of $0.1200 \ g \cdot mL^{-1}$ $(NH_4)_6Mo_7O_{24}$ solution, 2.00 mL of $20.00 \ mg \cdot mL^{-1}$ KH_2PO_4 solution, 0.50 mL of $0.1840 \ mol \cdot L^{-1}$ H_2SO_4 solution and 1.20 mL of $1.004 \ mg \cdot mL^{-1}$ penicillamine solution were applied to the proposed method. The absorbance of different reaction temperature (30, 35, 40, 45, 50, 55,

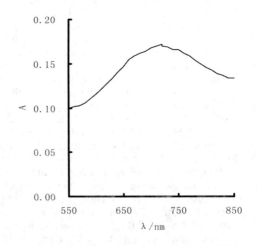

Figure 1.   Absorption spectrum MA: 2.50 mL; KH_2PO_4: 1.00 mL; H_2SO_4: 0.50 mL; PA: 0.50 mL; reaction temperature: 65°C; reaction time: 15 min.

60, 65, 70, 75, 80, 85, 90, 95, 100°C) was measured aftering the mixture reacted for 15 min. The results show that the absorbance reaches greater and remains constant when the temperature is 75~80°C. Hence, 75°C was selected for all further studies.

The absorbance of different reaction time (10, 15, 20, 25, 30, 35, 40, 45 min) was also measured at 75°C. It is found that the absorbance of solution reaches greater at 30 min and does not change when the reaction time is 30~45 min. This result clearly means that the reaction $H_2PO_4^-$-$Mo_7O_{24}^{6-}$-penicillamine was fast, and all penicillamine was completely oxidized and the formed phosphorus molybdenum blue ($H_3P(Mo_3O_{10})_2(Mo_3O_9)_2$) in the solution reached its maximum. Hence, 35 min was employed.

3.4 Effects of the dosage of H_2SO_4

Fixing the dosage of 0.1200 g·mL^{-1} $(NH_4)_6Mo_7O_{24}$ solution at 3.00 mL, 20.00 mg·mL^{-1} KH_2PO_4 solution at 2.00 mL, 1.004 mg·mL^{-1} penicillamine solution at 1.20 mL, reaction temperature and reaction time were 75°C and 35 min, the effect of the dosage of 0.1840 mol·L^{-1} H_2SO_4 on absorbance was studied (Figure 2). It can be seen from Figure 2 that as the amount of H_2SO_4 increased from 0.10 mL to 0.20 mL, the absorbance increased from 0.369 to 0.418, while the absorbance decreased when the amount of H_2SO_4 was above 0.20 mL. This result clearly demonstrated that the absorbance of the formed phosphorus molybdenum blue reaches its maximum when the dosage of H_2SO_4 solution is 0.20 mL. So, 0.20 mL of the dosage of H_2SO_4 has been selected for all further studies.

3.5 Effect of the dosage of KH_2PO_4

In order to study the effect of the dosage of 20.00 mg·mL^{-1} KH_2PO_4 on absorbance, 3.00 mL of 0.1200 g·mL^{-1} $(NH_4)_6Mo_7O_{24}$ solution, 0.20 mL of 0.1840 mol·L^{-1} H_2SO_4 solution and 1.20 mL of 1.004 mg·mL^{-1} penicillamine solution were added. The dosage of 20.00 mg·mL^{-1} KH_2PO_4 ranging from 0.20 to 1.80 mL was studied (Figure 3). It is found from Figure 3 that the absorbance reaches its maximum value when the dosage of KH_2PO_4 was 0.60 mL and remains constant in the range of 0.60 mL~1.20 mL. Hence, 1.00 mL of 20.00 mg·mL^{-1} KH_2PO_4 was chosen in the subsequent studies.

3.6 Effect of the dosage of ammonium molybdate

The dosage of ammonium molybdate is regarded as an important factor on the formation of phosphorus molybdenum blue. Keeping the dosage of 20.00 mg·mL^{-1} KH_2PO_4 solution at 1.00 mL, 0.1840 mol·L^{-1} H_2SO_4 solution at 0.20 mL, 1.004 mg·mL^{-1} penicillamine solution at 1.20 mL, reaction temperature and reaction time were 75°C and 35 min, the effect of ammonium molybdate on absorbance can be seen in Figure 4. As shown in Figure 4, the absorbance increase with the increase of ammonium molybdate dosage, the absorbance reaches its maximum value when the amount of ammonium molybdate is 1.00 mL, and it does not change any further with the increasing amount of ammonium molybdate. This result clearly indicated that the formed phosphorus molybdenum hetero-poly acid ($H_3P(Mo_3O_{10})_4$) in the solution reached a maximum, and all penicillamine was completely oxidized. Meanwhile, the formed phosphorus

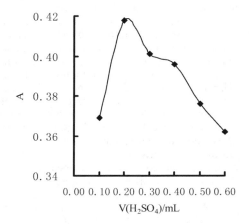

Figure 2. Effect of the dosage of H_2SO_4 MA: 3.00 mL; KH_2PO_4: 1.00 mL; PA: 1.20 mL; reaction temperature: 75°C; reaction time: 35 min.

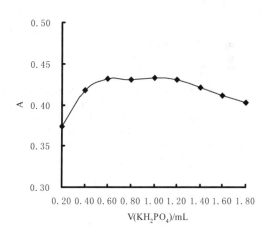

Figure 3. Effect of the dosage of KH_2PO_4 MA: 3.00 mL; H_2SO_4: 0.20 mL; PA: 1.20 mL; reaction temperature: 75°C; reaction time: 35 min.

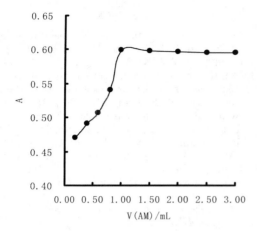

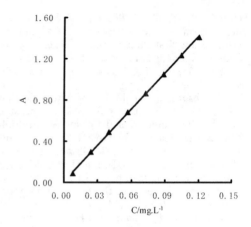

Figure 4. Effect of the dosage of ammonium molyb-date KH$_2$PO$_4$: 1.00 mL; H$_2$SO$_4$: 0.20 mL; PA: 1.20 mL; reaction temperature: 75°C; reaction time: 35 min.

Figure 5. Calibration curve MA: 1.50 mL; KH$_2$PO$_4$: 1.00 mL; H$_2$SO$_4$: 0.20 mL; reaction temperature: 75°C; reaction time: 35 min.

molybdenum blue (H$_3$P(Mo$_3$O$_{10}$)$_2$(Mo$_3$O$_9$)$_2$) rea-ched its maximum. So, 1.50 mL of ammonium molybdate has been selected.

3.7 Interference of coexisting components

A systematic study of the influence of excipients, aminoacids, carbohydrates and minerals on the determination of penicillamine were carried out. The tolerance levels are defined with an error of the deter-mination less than ±5%. A conclusion can be drawn from the following: 38.55 mg·mL^{-1} sucrose, lactose; 14.46 mg·mL^{-1} starch, L-glutamate; 4.819 mg·mL^{-1} glycine, serine, Na$^+$, Ca^{2+}, Mg^{2+}; 2.410 mg·mL^{-1} glucose, arginine, proline, K$^+$, SO$_4^{2-}$, NO$_3^-$, Cl$^-$, Br$^-$; 0.9638 mg·mL^{-1} Zn^{2+}, Co^{2+}, Mn^{2+}, Cu^{2+}, Cd^{2+}; 0.4819 mg·mL^{-1} Ni^{2+}, Pb^{2+}, Al^{3+}, Bi^{3+} do not affect the determination. But a certain amount of fructose affect the determination.

3.8 Calibration curve

According to the proposed procedures, a series of standard solutions of penicillamine was prepared. Ab-sorbance has been plotted as function of the concentration of penicillamine. A linear relation-ship between Absorbance (A) and the Concentra-tion (C) of penicillamine is obtained in the range of 4.016~120.5 μg·mL^{-1} (Fig. 5). The linear regres-sion equation is A = 0.0109 + 11.713C (mg·mL^{-1}), with a correlation coefficient of 0.9997.

3.9 Determination of penicillamine in pharmaceutical samples

Ten tablets of penicillamine weighed 2.8018 g after shucking sugar-coats off, then round and

blended. 1.5052 g powder of penicillamine was weighed precisely and dissolved in bidistilled water and was transferred into a 100 mL volumetric flask, the solution was diluted to the mark with bidistilled water and mixed well. After dry filtra-tion, 10.00 mL of filtrated solution was diluted to 100 mL with bidistilled water and mixed well. The solution was preserved at 4°C, shilding from light.

Based on the experimental method, the prepared solution of penicillamine sample was determined. The content of penicillamine in penicillamine tab-let can be obtained. Meanwhile, the recovery tests of standard addition were performed. The results obtained was compared with those obtained by pharmacopoeia method, as show in Table 1.

Table 1 shows that the content of penicillamine in penicillamine tablet is 124.5 mg·tablet^{-1} by this proposed method, and the content of penicillamine in penicillamine tablet is 129.5 mg·tablet^{-1} by phar-macopoeial method. Obviously, the result of this

Table 1. The determination result of penicillamine in penicillamine tablets n = 5.

Sample	Penicillamine tablet
Proposed method (mg·tablet^{-1})	124.5
RSD (%)	0.4
Pharmacopoeia method[21] (mg·tablet^{-1})	129.5
Added (μg/mL)	4.819
	8.032
Recovered (μg/mL)	4.922
	8.039
Recovery (%)	102.1
	100.1

proposed method agreed well with those obtained by pharmacopoeial method. It is indicated that the content of penicillamine in real pharmaceutical can be accurately determined by phosphorus molybdenum heteropoly acid spectrophotometry.

4 CONCLUSION

A novel spectrophotometric method for the determination of penicillamine is established by phosphorus molybdenum heteropoly acid. The proposed method has been successfully applied to the determination of penicillamine in pharmaceutical sample with satisfactory results, the recoveries are 100.1%~102.1%. This proposed method has the advantages of simply, rapidness, convenience, accuracy and so on. It is obvious that the determination of penicillamine by phosphorus molybdenum heteropoly acid spectrophotometry has certain practical significance and foreground of application.

REFERENCES

Corominas, B. Gómez-Taylor; Pferzschner, Julia; Icardo, M. Catalá; Zamora, L. Lahuerta; Martínez Calatayud, J. In situ generation of Co(II) by use of a solid-phase reactor in an FIA assembly for the spectrophotometric determination of penicillamine. Journal of Pharmaceutical & Biomedical Analysis. 2005, 39(1/2): 281–284.

Ensafi, Ali A; Khoddami, Elaheh; Rezaei, Behzad; Karimi-Maleh, Hassan. p-Aminophenol–multiwall carbon nanotubes–TiO$_2$ electrode as a sensor for simultaneous determination of penicillamine and uric acid. Colloids & Surfaces B: Biointerfaces. 2010, 81(1): 42–49.

Fereshteh Chekin, Jahan-Bakhsh Raoof, Samira Bagheri, Sharifah Bee Abd Hamid. Fabrication of Chitosan-Multiwall Carbon Nanotube Nanocomposite Containing Ferri/Ferrocyanide: Application for Simultaneous Detection of D-Penicillamine and Tryptophan. Journal of the Chinese Chemical Society. 2012, 59(11): 1461–1467.

Kuśmierek, Krzysztof; Bald, Edward. Simultaneous determination of tiopronin and d-penicillamine in human urine by liquid chromatography with ultraviolet detection. Analytica Chimica Acta. 2007, 590(1): 132–137.

Kuśmierek, Krzysztof; Chwatko, Grażyna; Głowacki, Rafał; Bald, Edward. Determination of endogenous thiols and thiol drugs in urine by HPLC with ultraviolet detection. Journal of Chromatography B: Analytical Technologies in the Biomedical & Life Sciences. 2009, 877(28): 3300–3308.

Maria A. Saracino, Cristina Cannistraci, Francesca Bugamelli, Emanuele Morganti, Iria Neri, Riccardo Balestri, Annalisa Patrizi, Maria A. Raggi. A novel HPLC-electrochemical detection approach for the determination of d-penicillamine in skin specimens. Talanta, 2013, 103: 355–360.

Martinović, Anita; Cerjan-Stefenović, Štefica; Radić, Njegomir. Flow Injection Analysis with Two Parallel Detectors: Potentiometric and Spectrophotometric Determination of Thiols and Ascorbic Acid in Mixture. Journal of Chemical Metrology. 2008, 2(1): 1–12.

Martinović, Anita; Radić, Njegomir. Kinetic Potentiometric Determination of Penicillamine and N-Acetyl-L-Cysteine Based on Reaction with Iodate. Acta Chimica Slovenica. 2009, 56(2): 503–506.

Martinović, Anita; Radić, Njegomir. Kinetic Potentiometric Determination of some Thiols with Iodide Ion-Sensitive Electrode. Analytical Letters. 2007, 40(15): 2851–2859.

Mazloum-Ardakani, Mohammad; Beitollahi, Hadi; Taleat, Zahra; Naeimi, Hossein; Taghavinia, Nima. Selective voltammetric determination of d-penicillamine in the presence of tryptophan at a modified carbon paste electrode incorporating TiO$_2$ nanoparticles and quinizarine. Journal of Electroanalytical Chemistry, 2010, 644(1): 1–6.

Raoof J.B., R. Ojani, M. Majidian, F. Chekin. Homogeneous electrocatalytic oxidation of d-penicillamine with ferrocyanide at a carbon paste electrode: application to voltammetric determination. Journal of Applied Electrochemistry, 2009, 39(6): 799–805.

Rasha Abdel-Aziz Shaalan. Improved spectrofluorimetric methods for determination of penicillamine in capsules. Central European Journal of Chemistry, 2010, 8(4): 892–898.

Reza Hormozi-Nezhad M., M. Azargun, N. Fahimi-Kashani. A colorimetric assay for d-Penicillamine in urine and plasma samples based on the aggregation of gold nanoparticles. Journal of the Iranian Chemical Society, 2014, 11(5): 1249–1255.

Shahrokhian, Saeed; Bozorgzadeh, Somayeh. Electrochemical oxidation of dopamine in the presence of sulfhydryl compounds: Application to the square-wave voltammetric detection of penicillamine and cysteine. Electrochimica Acta. 2006, 51(20): 4271–4276.

Shu-Cai Liang; Hong Wang; Zhi-Min Zhang; Hua-Shan Zhang. Determination of thiol by high-performance liquid chromatography and fluorescence detection with 5-methyl-(2-(m-iodoacetylaminophenyl) benzoxazole. Analytical & Bioanalytical Chemistry. 2005, 381(5): 1095–1100.

Skowron M., W. Ciesielski. Spectrophotometric determination of methimazole, D-penicillamine, captopril, and disulfiram in pure form and drug formulations. Journal of Analytical Chemistry, 2011, 66(8): 714–719.

The Pharmacopeial Committee of China. The Pharmacopoeia of the People's Republic of China (second-part); China Medical Science Press; Beijing, 2010, p428.

Viñas P., J.A. Sanchez-Prieto, M. Hernandez Cordoba. Flow injection analysis and batch procedures for the routine determination of N-penicillamine. Microchemical Journal, Volume 41, Issue 1, February 1990, Pages 2–9.

Wang, Peng; Li, Bang Lin; Li, Nian Bing; Luo, Hong Qun. A fluorescence detection of d-penicillamine based on Cu^{2+}-induced fluorescence quenching system of protein-stabilized gold nanoclusters. Spectrochimica

Acta Part A: Molecular & Biomolecular Spectroscopy. 2015, 135: 198–202.

Yang, Xiupei; Yuan, Hongyan; Wang, Chunling; Su, Xiaodong; Hu, Li; Xiao, Dan. Determination of penicillamine in pharmaceuticals and human plasma by capillary electrophoresis with in-column fiber optics light-emitting diode induced fluorescence detection.

Journal of Pharmaceutical & Biomedical Analysis. 2007, 45(2): 362–366.

Zhang Z., W.R.G. Baeyens, X. Zhang, Y. Zhao, G. Van Der Weken. Chemiluminescence detection coupled to liquid chromatography for the determination of penicillamine in human urine. Analytica Chimica Acta, 1997, 347(3): 325–332.

Advances in Energy Science and Equipment Engineering II – Zhou, Patty & Chen (Eds)
© 2017 Taylor & Francis Group, London, ISBN 978-1-138-71798-5

Study on the volatilization rate of metal tin and metal antimony under vacuum condition

Jibiao Han, Anxiang Wang, Yifu Li, Zhenghao Pu, Baoqiang Xu, Bin Yang & Yongnian Dai
National Engineering Laboratory for Vacuum Metallurgy, Kunming University of Science and Technology,
Kunming, Yunnan, China
State Key Laboratory of Complex Non-ferrous Metal Resources Clear Utilization, Kunming, Yunnan, China
Key Laboratory of Vacuum Metallurgy for Nonferrous Metal of Yunnan Province, Kunming, Yunnan, China

ABSTRACT: In order to study the separation law in vacuum distillation of tin-antimony alloy, the volatilization rate of metal tin and metal antimony in different temperature (1273 K–1673 K) and residual pressure (5 Pa–90 Pa) conditions was studied by differential gravimetric analysis. The experimental results showed, with the increase of temperature and the decrease of residual pressure, the volatilization rate of tin and antimony were increased and reached the maximum volatilization rate at 5 Pa. The critical pressure of metal tin and antimony is increased with the increase of temperature under the certain temperature conditions. The relationship between the maximum volatilization rate and the temperature of metal tin and antimony is obtained which can be used in vacuum distillation process. It is found that the maximum volatilization rate of antimony was much higher than tin. The effective separation of tin and antimony can be realized under the vacuum condition.

1 INTRODUCTION

Tin and antimony is Chinese traditional mineral resources, excessive exploitation leads to the increasing scarcity of mineral resources and the declining ore grade (Sun Hu et al, 2012). A large number of complex and diverse types of low grade ore are put into the practical production of non-ferrous metallurgy which will produce complex tin alloy. It is a common secondary resource in the process of metal recovery which produces the complex tin alloy. (Bin Yang et al, 2015; Bin Yang et al, 1998; Anxiang Wang et al, 2015).

Vacuum metallurgical technology has become the world's leading edge technology for the extraction of metallurgy by using a fully enclosed vacuum furnace in the production process, it also has the advantages of high metal recovery rate, small pollution, simple process flow, high utilization rate of resource, low production cost and friendly environment (Dachun Liu et al, 2006; Keqiang Qiu et al, 2003). But the results show that the separation effect of Sn-Sb alloy is not ideal. In industrial experiment of tin based alloy by vacuum distillation, the antimony removal limit is 0.1% which is difficult to be separated efficiently. In order to explore the volatilization behavior of Sn-Sb alloy in the process of vacuum distillation, in this paper, the relation between the Sn and Sb volatilization rate, temperature and residual pressure is studied.

Dai Yongnian (Yongnian Dai et al, 1994) explores the relation between the volatilization rate of metal antimony (873–923 K), metal tin (1173–1473 K), however, the temperature of vacuum distillation separation of Sn-Sb alloy is more than 1273 K. In this temperature range, the vacuum volatilization rules of Sn and Sb has not been reported.

Therefore, this study was based on the determination of the volatilization rate of Sn and Sb at different temperatures (1273–1673 K) and different residual pressure (5–90 Pa), the relation between the volatilization rate and residual pressure, the critical pressure and temperature, the maximum volatilization rate and temperature were obtained. From the experimental point of view, the volatilization rule of Sn and Sb was revealed, in order to guide the vacuum distillation separation process of Sn-Sb alloy.

2 EXPERIMENT

2.1 Raw materials

The raw materials for the experiment were pure tin (99.95%) and pure antimony (99.95%).

2.2 Experimental method

The volatilization rate of pure metals was determined under certain temperature, time and residual pressure condition. First, volatilization experiment was carried out to measure the amount of W_1 in the $t_1 + t_2 + t_3$ time range (t_1-Heating up time, t_2-Soaking time, t_3-Cooling time). The blank experiment was did to measure the volatilization W_2 in the $t_1 + t_3$ time range, eventually get the volatilize quantity (ΔW) of the sample in the t_2 time, $\Delta W = W_1 - W_2$. During the experiment the amount of raw material is much greater than the amount of volatile, there must still a large amount of raw material remaining, and the experiment of volatilization experiment and blank experiment were did at the same time, same temperature and cooling rate.

The calculation formula of the volatilization rate of metal can be expressed as:

$$\omega = \frac{\Delta W}{\Delta S \cdot t} \qquad (1)$$

ω is the metal volatilization rate, g/(cm² · s); W is the weight difference, g; ΔS is volatilization area of metal, cm²; t is volatilization time, S.

3 RESULTS AND DISCUSSION

3.1 *Relation between volatilization rate and residual pressure of metal tin and metal antimony*

Figure 1 is the relations between volatilization rate and the residual pressure of tin which is 200 g metal tin under six vacuum strength (5 Pa, 10 Pa, 30 Pa, 50 Pa, 70 Pa, 90 Pa) and five temperature (1273 K, 1373 K, 1473 K, 1573 K, 1673 K), the distillation time is 5 min, volatilization area is 78.5 cm².

As shown in Figure 1, with the decreased of residual pressure, the change of the volatilization rate of tin can be clearly divided into three areas at the same temperature: namely the straight line, the transition zone, the level area. Linear region

corresponds to "surface evaporation" and "boiling evaporation". The residual pressure decreased, the volatilization rate of tin rapidly increased in the linear region. The residual pressure from 90 Pa reduced to 50 Pa in 1673 K, the volatilization rate of the linear area of tin increased from 6.60×10^{-7} g • cm⁻² • s⁻¹ to 1.42×10^{-4} g • cm⁻² • s⁻¹ which increased by three orders of magnitude. This is due to the large amount of residual molecules reduce in the original system, metal vapor is reduced by resistance in the system which makes the vapor molecule volatilization rate increased. When the residual pressure is decreased from 50 Pa to 10 Pa, the reaction enters the transition zone. The volatilization rate of metal in the transition zone increases slowly with the decrease of the residual pressure which increased from 1.42×10^{-4} g • cm⁻² • s⁻¹ to 4.36×10^{-4} g • cm⁻² • s⁻¹. From the data, it can be seen that the change of the rate of volatilization is in the same order of magnitude, that is the change of the rate of metal tin volatilization is small. When the residual pressure is decreased from 10 Pa to 5 Pa, the reaction is entered into the horizontal zone, and the horizontal zone corresponds to the "molecular distillation". The volatilization rate of Sn in this area is changed from 4.36×10^{-4} g • cm⁻² • s⁻¹ to 4.30×10^{-4} g • cm⁻² • s⁻¹ which basically does not change. Less residual gas molecules in the system shows that gas molecules between the internal friction is negligible, volatilization rate is almost not affected by the change in the pressure of environment impact, at this time, the volatilization rate of metal tin has nothing to do with the residual pressure. The volatilization rate reached the maximum.

Figure 2 is the relations between volatilization rate and the residual pressure which is 200 g metal antimony under six vacuum strength (5 Pa, 10 Pa, 30 Pa, 50 Pa, 70 Pa, 90 Pa and five temperature (1273 K, 1373 K, 1473 K, 1573 K, 1673 K), and the distillation time is 5 min, volatilization area is 78.5 cm².

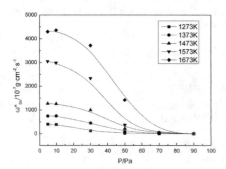

Figure 1.   The relations of the volatilization rate of Sn with the pressures at different temperatures.

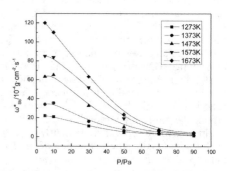

Figure 2.   The relations of the volatilization rate of Sb with the pressures at different temperatures.

As shown in Figure 2, the volatilization rate of metal antimony and the change trend of the metal tin volatilization rate are basically the same. This is consistent with the theoretical analysis of the volatilization rate and system pressure of pure metals. With the decrease of the residual pressure, the volatilization rate of antimony in the linear region increased rapidly. The residual pressure from 90 Pa reduced to 30 Pa in 1673 K, the volatilization rate of the linear area of tin increased from 4.31×10^{-4} g • cm^{-2} • s^{-1} to 6.33×10^{-3} g • cm^{-2} • s^{-1}. When the residual pressure is reduced from 30 Pa to 10 Pa, the reaction enters the transition zone, the volatilization rate of the metal antimony in the transition zone increases slowly from 6.33×10^{-3} g • cm^{-2} • s^{-1} to 1.10×10^{-2} g • cm^{-2} • s^{-1} with the decrease of the residual pressure. When the residual pressure is reduced from 10 Pa to 5 Pa, the volatilization rate increases from 1.10×10^{-2} g • cm^{-2} • s^{-1} to 1.20×10^{-2} g • cm^{-2} • s^{-1}. The volatilization rate of the metal is not changed, and the volatilization rate of the metal antimony is at the maximum this moment.

3.2 Relation between critical pressure and temperature of metal tin and metal antimony

The experiment results show that when the system residual pressure to a certain extent, continue to reduce the residual pressure has been unable to improve the volatilization rate of tin and antimony in the vacuum distillation process. As a turning point in Figure 1 and Figure 2 of relations between volatilization rate and residual pressure, the corresponding pressure of this turning point is called the critical pressure Pc. The critical pressure of metal tin and antimony can be obtained by plotting method, and the data can be found in Table 1 and Table 2.

It can be seen from the Table 1 and Table 2, with the increase of temperature, the critical pressure of tin and antimony were increased. For metal tin, the temperature increased from 1273 K to 1673 K, its critical pressure was increases from 12.7 Pa to 38.5 Pa. For metal antimony, the temperature increased from 1273 K to 1673 K, its critical pressure was increases from 7.8 Pa to 20.0 Pa.

3.3 Relation between maximum volatilization rate and temperature of metal tin and metal antimony

When the temperature is constant, the pressure is less than a certain value, the volatilization rate of metal tin, metal antimony reaches the maximum. The measured maximum volatilization rate of metal tin, metal antimony are shown in Table 3 at different temperatures.

The Table 3 shows that with the increase of temperature, the maximum volatilization rate of tin and antimony are increasing, because of the increase of temperature, the saturation vapor pressure of tin and antimony increases, the volatilization rate of tin and antimony increases. At 1673 K, the volatilization rate of metal tin reached 4.30×10^{-4} g · cm^{-2} · s^{-1}, the volatilization rate of metal antimony reached 1.20×10^{-2} g · cm^{-2} · s^{-1}. As can be seen from the data, the volatilization rate of antimony is much greater than the volatilization rate of tin which shows that antimony is more easily volatile in vacuum conditions.

By linear regression, the relation between the maximum volatilization rate and temperature of tin and antimony is shown in Figures 3 and 4. Linear regression equations are shown in formula 2 and formula 3.

$$\lg \omega^*_{Sn-max} = \frac{510.35}{T} - 3.5075 \tag{2}$$

$$\lg \omega^*_{Sb-max} = -\frac{2624}{T} - 0.0449 \tag{3}$$

Table 1. The critical pressures of tin.

T/K	1273	1373	1473	1573	1673
P_c/Pa	12.7	21.7	27.5	27.8	38.5

Table 2. The critical pressures of antimony.

T/K	1273	1373	1473	1573	1673
P_c/Pa	7.8	11.7	16.3	19.3	20.0

Table 3. The maximum volatilization rate of Sn and Sb from experimental measurement.

T/K	1273	1373	1473	1573	1673
ω^*_{Sn-max}/g·cm^{-2}·s^{-1}	4.04×10^{-5}	7.46×10^{-5}	1.27×10^{-4}	3.05×10^{-4}	4.30×10^{-4}
ω^*_{Sb-max}/g·cm^{-2}·s^{-1}	2.21×10^{-3}	3.41×10^{-3}	6.33×10^{-3}	8.50×10^{-3}	1.20×10^{-2}

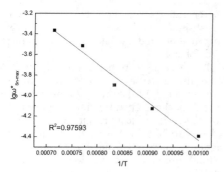

Figure 3. The relations of the maximum volatilization rate of Sn of with the temperature.

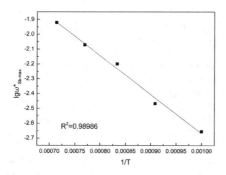

Figure 4. The relations of the maximum volatilization rate of Sb of with the temperature.

From the above content can be seen, the determination coefficient of formula 2 is $R^2 = 0.97596$, the determination coefficient of formula 3 is $R^2 = 0.98986$, the confidence is 95%. Therefore, the equation has application value which can be used to guide the process of separation of tin-antimony alloys in vacuum distillation.

4 CONCLUSION

1. With the decrease of the residual pressure, the volatilization of metal tin and antimony were divided into three regions. In the linear region, the volatilization rate increases rapidly, in the transition region, the volatilization rate gradually slows down, when the residual pressure is reduced to 5–10 Pa, the volatilization rate no longer changes and reaches to the maximum.
2. When the temperature is 1273 K–1473 K, the critical pressure of tin and antimony increases with the increase of temperature. Under a certain temperature condition, when the residual pressure is lower than the critical pressure, the volatilization rate of tin and antimony is not increased.
3. The relationship between the maximum volatilization rate and the temperature of metal tin and antimony is obtained, the equation has application value which can be used to guide the process of separation of tin-antimony alloys in vacuum distillation.

REFERENCES

Anxiang Wang, Bin Yang & Yifu Li (2015). Simulation of removal of Sb from Sn-Sb Alloy in vacuum with molecular interaction Volume Model. *J. Chinese journal of vaccum science and technology.* 35(4), 502–507.

Bin Yang, Lingxin Kong & Baoqiang Xu (2015). Recycling of metals from waste Sn-based alloys by vacuum separation. *J. Transactions of Nonferrous Metals Society of China.* (4), 1315–1324.

Bin Yang, Yongnian Dai & Guojing Zhang (1998). The separation of Sn-Sb Alloys by vaccum distillation. *J. Journal of Kunming University of Science and Technology.* 23(3), 104–107.

Dachun Liu, Yongnian Dai & Bin Yang (2006). Application of vacuum Metallurgical technology in extraction and purification of scattered metals. *J. Chinese journal of rare metal.* 30, 89–92.

Keqiang Qiu, Xuelin Yang & Lulu Zhang (2003). The new treatment technology of antimony-rich lead anode slime. *J. Gold.* 24(11), 37–39.

Lin Wang, Dachun Liu & Bin Yang (2010). Purification of crude nickel by vaccum distillation. *J. Chinese journal of vaccum science and technolugy.* 30(3), 283–287.

Sun Hu, Wang Jian-ping & Wang Yu-feng (2012). Status and suggestions of sustainable development to China's tin ores. *J. Resources and industries.* 14(4), 59–62.

Yongnian Dai, Dankui Xia & Yan Chen (1994). Evaporation of metals in vaccum. *J. Journal of Kunming institute of technology.* 19(6), 27–32.

Study on evaluating influence factors and establishing a kinetic model in synthesizing 1-decene

Sihan Wang
Daqing Petrochemical Research Center of Petrochina, Daqing, Heilongjiang, China
College of Chemistry and Chemical Engineering, Northeast Petroleum University, Daqing, Heilongjiang, China

Hongliang Huo, Lili Ma, Han Gao, Libo Wang, Tingting Xu, Zidong Wang & Wei Sun
Daqing Petrochemical Research Center of Petrochina, Daqing, Heilongjiang, China

ABSTRACT: As one kind of important alpha olefins, 1-decene has extensive application value. There is no technology of directional production of 1-decene in the world at present. In this paper, the self-made chromium catalyst was used in synthesizing 1-decene from ethylene oligomerization. The effects of reaction temperature, reaction pressure and the catalyst concentration on the catalytic activity were examined. We also established apparent reaction kinetic model of producing 1-decene in this paper. The results showed that the difference of the theoretical value and actual value was within 5% by applying this kinetic model, which played an important role on realizing effective control of the reaction.

1 INTRODUCTION

1-Decene is the primary raw material for the senior lube base oil (Li & Wu 2008). However, there is no international technology for oriented production of 1-decene, which exists only as byproduct at present. The selectivities of 1-decene in four advanced technologies (Chevron technology, B P Amoco process, SHOP process and Phillips process) were below 20% (Wang J. 2016).

There are many reports on the synthesis of alpha olefins by ethylene polymerization at home and abroad. But, it is mainly focused on the study of ethylene trimerization and tetramerization including the evaluation of influencing factors (Zhang et al. 2013, Yacoob et al. 2012 & Shaikh et al. 2012) and the exploration of mechanism et al (Agapie T, 2011). There are few scholars to study the synthesis of 1-decene (Ahamad et al. 2010, Ahamad & Alshehri 2013, Ahamad & Alshehri 2014). And the kinetic study on the highly selective synthesis of 1-decene is rarely reported.

In this paper, the effects of reaction temperature, reaction pressure and the catalyst concentration on the catalytic activity were evaluated based on the synthesizing 1-decene of high selectivity. In addition, the reaction kinetic model in synthetic process of 1-decene was established. The research will be significant in proving the inherent law of 1-decene in the ethylene oligomerization, achieving effective control of the reaction process and producing 1-decene of high selectivity.

2 EXPERIMENT

2.1 Materials

Ethylene was polymer-grade purchased from the Daqing Petrochemical Company Limited of China. Cyclohexane was industrial grade provided by the Fushun Petrochemical Company Limited (China). Cyclohexane was dryed by refluxing with NaH after being soaked with molecular sieve. The content of water in cyclohexane should be reduced to below 5 ug/g. Chromium catalyst, self-made.

2.2 Experimental methods

2.2.1 Evaluation of impact factors
The 500 mL of cyclohexane solvent was added in the 1L autoclave. The reactor was heated up to 130°C and stirred for 0.5 h. The solvent was released and the reactor was purged with hot nitrogen for 10 min. After that, the reactor was cooled down to a predetermined reaction temperature by circulating water. The 200 mL of cyclohexane and a certain amount of catalyst were added to the reactor Subsequently, ethylene was injected to the reactor with stirring. The reactor was raised to certain reaction condition. After a period time, the reaction rate of 1-decene was calculated.

2.2.2 Establishment of apparent kinetic equation
The key factors which affect the reaction rate were determined by evaluating the effects of reaction temperature, reaction pressure and

catalyst concentration on the reaction rate. And the key parameters in the rate equation were obtained based on the experimental data and the basic equation. Besides, the relative solubility, correlation coefficient and other parameters were calculated by using ASPEN software. The quantitative relationship between reaction rate and these three factors were determined. Finally, the apparent kinetic equation was established and validated.

3 RESULTS AND DISCUSSION

3.1 Influence of temperature on reaction rate

The ethylene oligomerization is exothermic reaction, and the reaction temperature plays a significant role on the process. The reaction temperature was studied with the consumption of ethylene as the reference value. The change of flow rate of ethylene with time at different temperature is shown in Figure 1.

As shown in Figure 1, the flow rate of ethylene with the time showed a trend of increasing first and then decreasing when the range of temperature is from 75 to 110°C. At lower temperature (below 95°C), the maximum reaction rate was achieved for a long time. However, at higher temperature (105~110°C), the maximum reaction rate was achieved very quickly. The reaction rate gradually decreased and the final reaction rate was consistent at higher temperature. The study found that the optimum reaction temperature was 95°C. In general, the ethylene polymerization occurs after ethylene molecules collide with each other in the catalytic system. The ethylene molecules get higher energy and the thermal motion is intensified at higher temperature. The reaction system could achieve the maximum reaction rate in a short time for increased collision frequency.

Then the reaction rate decreased gradually with the continuous consumption of active species in the reaction system.

3.2 Influence of pressure on reaction rate

Ethylene is the main raw material in the reaction. The variation of the reaction rate with time at different pressure directly exhibited the dynamic performance of the whole reaction. The result is shown in Figure 2.

As shown in Figure 2, at lower pressure (<2 MPa), the trend of the ethylene flow rate was very gentle with the prolonged reaction time. When the pressure was rised to 2 MPa, the reaction rate reached the maximum in a short time, and then decreased gradually. Under certain conditions, the reaction rate increases rapidly to the maximum value when the ethylene pressure increases, that is, the increase of the ethylene concentration in the liquid phase. After that, the reaction rate decrease with the consumption of the active center. It can be seen that ethylene pressure is the dominant factor, which has great influence on the reaction rate.

3.3 Influence of catalyst concentration on reaction rate

The change of flow rate of ethylene with time under different amount of catalyst was investigated. The result is shown in Figure 3.

As shown in Figure 3, the reaction rate gradually decreased with the extension of the reaction time when the concentration of the catalyst was in the lower range. It could be seen that the reaction was basically a concentration of the first order reaction when the dosage of the pre-catalyst was 0.1 mL. When the amount of catalyst was increased, the reaction rate increased first and then decreased with the extension of reaction time. On one hand,

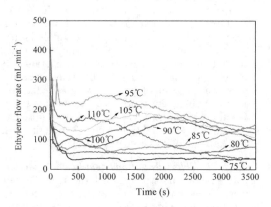

Figure 1.   Effect of temperature on reaction rate.

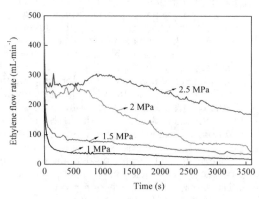

Figure 2.   Effect of pressure on reaction rate.

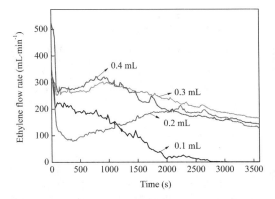

Figure 3. Effect of catalyst concentration on reaction rate.

the quantity of active center in the reaction system increases with the increasing of the concentration of the catalyst. The reaction rate was increased by increasing the collision probability of the ethylene molecule and the active center. On the other hand, the activation energy of the reaction was reduced by the catalyst, which greatly increased the percentage of active molecule per unit volume in the reactant molecule. Thus, the reaction rate was accelerated for increased probability of the effective collision.

4 FITTING OF THE APPARENT REACTION KINETIC EQUATION

According to the study, the ethylene concentration, active species concentration and reaction temperature were the main factors affecting the reaction rate. The fitting of the apparent reaction kinetic equation was mainly based on the following equations:

$$R = M / t \tag{1}$$

$$R = k(C_{ca})^n (C_e)^m \tag{2}$$

$$k = f(T) = A \times e^{-Ea/RT} \tag{3}$$

$$Ce = hPe \tag{4}$$

where R = average the reaction rate(g · h^{-1}); M = total ethylene consumption (g), t = reaction time (h), k = reaction rate constant, Cca = concentration of metal Cr(ug · L^{-1}), Ce = ethylene concentration in liquid phase (MPa), n = series of the catalyst concentration, m = series of ethylene concentration, A = pre-exponential factor, Ea = activation energy of reaction(J · mol^{-1}), R = gas state constant, T = reaction temperature (K), h = modified Herry coefficient, and Pe = gas phase partial

pressure of ethylene (MPa). The concentration of active center was difficult to determine. Thus, the catalyst dosage was used as catalyst concentration, which was assumed to be activated into the active center.

4.1 Determination of the reaction series of catalyst concentration

The equation (2) was the apparent kinetic equation of the ethylene oligomerization. Being taken to derivative on both sides, the equation (5) was obtained.

$$\log R = \log k = n \log C_{ca} + m \log C_e \tag{5}$$

Under certain reaction temperature and reaction pressure, the equation was simplified as,

$$\log R = n \log C_{ca} + b \tag{6}$$

The average reaction rate of the oligomer was measured under different concentration of the catalyst at fixed reaction temperature and pressure. The logarithm of the catalyst concentration and the logarithm of the reaction rate were plotted. The result is shown in Figure 4.

It was noted that the series of the catalyst concentration was 1.12 by linear regression of the other three points with removal of a point which represented a high concentration of the catalyst from Figure 4. Therefore, it was considered that the polymerization was the first order reaction of catalyst concentration.

4.2 Determination of the reaction series of ethylene concentration

Under certain reaction temperature and reaction pressure, the concentration of ethylene in

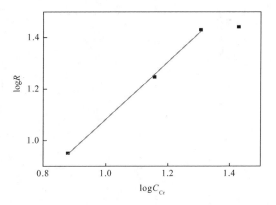

Figure 4. Relationship between $\log R$ and $\log C_{Cr}$.

liquid phase is the equilibrium concentration of ethylene pressure in gas phase. The solubility of ethylene in solution is basically in compliance with Herry's law, that is, the relationship between the solubility of ethylene and ethylene partial pressure is proportional under a certain reaction temperature. The ethylene concentration in liquid phase was replaced by ethylene partial pressure at fixed reaction temperature and catalyst concentration. The kinetic equation (7) was obtained.

$$R = k(C_{ca})^n (P_e)^m \tag{7}$$

The kinetic equation was taken to derivative on both sides, the equation (8) was obtained.

$$\log R = \log k + n \log C_{ca} + m \log P_e \tag{8}$$

Under a certain reaction temperature and catalyst concentration, the equation was simplified as,

$$\log R = m \log P_e + b \tag{9}$$

The average reaction rate was measured under different reaction pressure at fixed catalyst concentration and reaction temperature. The reaction pressure was converted to ethylene partial pressure by using ASPEN software. The logarithm of the partial pressure of ethylene and the logarithm of the reaction rate were plotted in the horizontal and vertical coordinates. The result is shown in Figure 5.

From Figure 5, the reaction series of the ethylene pressure was 1.96 by linear fitting of the above data. It could be obtained that the ethylene pressure played an important role on the reaction rate.

4.3 Determinations of the pre-exponential factor and the activation energy

The reaction rate constant is a function of the reaction temperature under certain catalyst concentration and the reaction pressure in the equation (3). Being taken to derivative on both sides, the equation was as follows,

$$\ln k = \ln A - \frac{E_a}{R} \times \frac{1}{T} \tag{10}$$

The average reaction rate of the oligomer was measured under different reaction temperature at fixed catalyst concentration and reaction pressure. The negative reciprocal of the reaction temperature and natural logarithm of the reaction rate constant were plotted in the horizontal and vertical coordinates. The result is shown in Figure 6.

The ratio of the activation energy to the gas state constant was −3062.28 and the natural logarithm of the pre-exponential factor was −6.54 by linear fitting of the above data from Figure 6. Hence, the activation energy of the ethylene oligomerization was −25000 J/mol and the pre-exponential factor was 0.00144.

4.4 Verification of apparent kinetic equation

The apparent kinetic equation of the ethylene oligomerization was as follows,

$$R = 0.00144 \cdot e^{(-25000/RT)} \cdot C_{Cr}^{1.12} \cdot C_e^{1.96} \tag{11}$$

In order to verify the accuracy of the apparent kinetic equation, the actual reaction rate was recorded during the ethylene oligomerization under different ethylene pressure and catalyst concentration at a reaction temperature of 95°C. Meanwhile, the theoretical reaction rate was calculated using

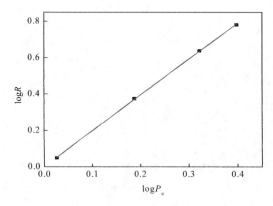

Figure 5.   The relationship between $\log R$ and $\log P_e$.

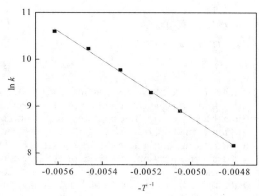

Figure 6.   Relationship between $\ln k$ and $(-T^{-1})$.

Table 1. Verification of apparent kinetic equation.

No.	$T/$ K	$C_{Cr}/$ ug·L^{-1}	$Pe/$ MPa	$Ce/$ g/g	$R_1/$ g·h^{-1}	$R_2/$ g·h^{-1}	R_2- $R_1/$%$
1	368	14.28	1.5	0.05	0.322	0.336	4.3
2	368	20.20	2.0	0.07	0.918	0.956	4.1
3	368	26.82	2.5	0.08	1.640	1.720	4.8

the apparent kinetic equation. The difference between two values is listed in Table 1.

From the data of Table 1, the difference between the actual reaction rate and the theoretical reaction rate was less than 5% under the catalytic system. It can theoretically verify that the difference between the actual catalytic activity and selectivity of 1-decene and the theoretical value was less than 5% in the ethylene oligomerization.

5 CONCLUSIONS

In this paper, the effects of reaction temperature, reaction pressure and catalyst concentration on the reaction rate were studied. The reaction rate increased with the reaction temperature and the optimum reaction temperature was 95°C. The reaction pressure had great influence on the reaction rate. When the reaction pressure was lower than 2 MPa, the reaction rate was stable. But the maximum of the reaction rate appeared in the short time when pressure was higher than 2 MPa. The higher the catalyst concentration was, the larger the maximum reaction rate was. In addition, the apparent kinetic model of synthesizing 1-decene was established. It was proved that the difference between actual reaction rate and theoretical value

which calculated by the apparent kinetic model was within 5%.

REFERENCES

Agapie, T. (2011) Selective ethylene oligomerization: recent advances in chromium catalysis and mechanistic investigations. *Coordination Chemistry Reviews*, 255(7), 861–880.

Ahamad, T., S. M. Alshehri, & S. F. Mapolie (2010) Synthesis characterization of polyamide metallodendrimers and their catalytic activities in ethylene oligomerization. *Catalysis Letters*, 138(3–4), 171–179.

Ahamad, T. & S. M. Alshehri (2013) Synthesis and characterization of polymer metal complexes and their catalytic activity in ethylene oligomerization. *Advances in Polymer Technology*, 32(3), 586–589.

Ahamad, T. & S. M. Alshehri (2014) Synthesis and characterization of first- and second-generation polyamide pyridylimine nickel dihalide metallodendrimers and their uses as catalysts for ethylene polymerization. *Polymer International*, 63(11), 1965–1973.

Li, J. L. & Y. R. Wu (2008). Production technology and application of 1-decene. Petrochemical Industry Technology, 15(1), 66–68.

Shaikh, Y., J. Gurnham, K. Albahily et al. (2012) Aminophosphine-based chromium catalysts for selective ethylene tetramerization. *Organometallics*, 31(21), 7427–7433.

Wang, J., H. L. Huo, L. L. Ma et al. (2016) Progress in catalytic systems of decene synthesis from ethylene oligomerization. *Chemistry Online*, 79(1), 31–36.

Yacoob, S., A. Khalid, S. Matthew et al. (2012) A highly selective ethylene tetramerization catalyst. *Angewandte Chemie International Edition*, 124(3), 1395–1398.

Zhang, J., X. Wang, X. J. Zhang et al. (2013) Switchable ethylene tri-/tetramerization with high activity: subtle effect presented by backbone-substituent of carbon-bridged diphosphine ligands. *American Chemical Society Catalysis*, 3(10), 2311–2317.

Advances in Energy Science and Equipment Engineering II – Zhou, Patty & Chen (Eds)
© 2017 Taylor & Francis Group, London, ISBN 978-1-138-71798-5

Effect of bleached shellac on the quality of the fluidized-bed drying process

Liqing Du & Xinghao Tu
Key Laboratory of Tropical Fruit Biology, South Subtropical Crops Research Institute, Chinese Academy of Tropical Agricultural Sciences, Ministry of Agriculture, Zhanjiang, Guangdong, China

Hong Zhang & Kun Li
Research Institute of Resources Insects, Chinese Academy of Forestry, Kunming, China

ABSTRACT: In this paper, bleached shellac was dried with fluidized bed, which was expected to dry rapidly at lower temperatures. Two main factors, including intake air temperature and airflow rate, which would influence the product color index and drying time, were considered in the phase of single-factor tests. By comparing and analyzing the quality of the products obtained by different drying methods, we found that the bleached shellac of fluidized-bed drying has high quality, which could reach the first grade of the Chinese National Standard. The results could provide valuable information for the industrialization of fluidized-bed drying process of bleached shellac.

1 INTRODUCTION

Shellac, a natural product from lac, has a considerable economic value as forest product and is an important industrial raw material. The current annual production of shellac in China is about 3000t, which is only half of the demand (6000t). Bleached shellac had a wide range of applications in military and food industry. For example, it could be used as a preservative in fruits or edible packaging film. The wet material of bleached shellac contains more water through physical and mechanical bonding, which is very sensitive to the heating temperature and time, which in turn could have a significant impact on its properties.

The moisture content of wet shellac is as high as 60–65%. Industrial production mainly uses natural drying way, although the content of moisture could reach 3% or less, but with longer drying time and less efficiency, and the products exhibit obvious color reversion phenomenon. Therefore, the shortcomings would directly affect the quality of products and large-scale industrial production of bleached shellac. After conducting a long-term study on the drying of bleached shellac, the author's research group has carried out research on the fixed-bed, vacuum, and microwave vacuum drying technology. Recently, we have found that the microwave vacuum technology for bleaching of shellac shows the best effect; however, because of the higher equipment cost and energy consumption, it is difficult to apply in practice. Therefore, it is necessary to develop a novel, economic method for drying bleached

shellac. When the fluidized bed, the material, and the drying medium had large contact areas, the material and the drying medium have large contact areas, thereby achieving uniform temperature distribution, better heat transfer effect, and short drying time, especially for granular, dry powder materials. In some countries, it was mainly used in milk powder and olive pomace powder as particles of fast drying to achieve good results, thereby realizing rapid drying at lower temperatures. In this paper, the process of fluidized-bed drying of bleached shellac, the effects of inlet air temperature, and the two main factors of the drying time of bleached shellac and color index were studied. The products made by different drying methods were compared and analyzed. The application of fluidized-bed drying technology in the drying process of bleached shellac provides certain theoretical guidance.

2 MATERIALS AND METHODS

The test material of wet bleached shellac (moisture content: 60–65%) was provided by Research Center of Engineering and Technology on Forest Resources with Characteristics, State Forestry Administration.

2.1 Fluidized-bed drying process of bleached shellac

Sampling method, according to GB/T8142–2008, as specified in wet material of bleached shellac was used to measure the moisture content of wet material with halogen moisture analyzer. Each time, 200 g of material (moisture content 63.1%) was placed in a fluidized bed vessel, and drying experiments were carried out under a certain temperature and airflow rate (the ambient temperature was 16°C and air humidity was 60%). The change of moisture content was obtained at every fixed time. By changing the airflow rate and temperature, until the water content was less than or equal to 3% at the end of bleached shellac drying experiment, the drying time was recorded. The color index of the sample after drying was measured, and the value of three repeated experiments were averaged.

2.2 Analytical test method of dry sample

A halogen moisture analyzer was used to calculate the mass loss per unit of time, and each sample was measured at 60°C for 2 h three times and the average value was found. Hot ethanol insolubles, water-soluble materials, color index, acidity, and other indicators were determined in accordance with the method "lac products test methods."

2.3 Data processing and analysis

Origin 8.0 software was adopted to process the test data and establish the regression equation and surface chart. This group of charts was used for the analysis and evaluation of the any two kinds of interaction effect and determination of the optimum operating conditions.

3 RESULT AND DISCUSSION

3.1 Fluidized-bed drying effect on the quality of the bleached lac

The effect of intake air temperature on bleached shellac color index and drying time is shown in Figure 1. It is evident from the figure that at higher inlet temperature, the index of bleached shellac is higher and the drying time is lower. When the inlet temperature was 38°C, the color index reaches 0.98; when the temperature continues to increase, the color index increases rapidly, but the dry time was not obvious, so the appropriate inlet temperature was 38°C.

The effect of airflow rate on the color index of bleached shellac and drying time is shown in Figure 2. From the figure, we can conclude that at high airflow rate, the color index of bleached shellac and drying time decreased. When the airflow rate was over 40 m³/h, the color index and drying time decreased slowly, but the energy consumption was large, and the energy use efficiency decreased significantly with the increase of flow rate. At a airflow rate of 40 m³/h, the color index was 0.96, which is the

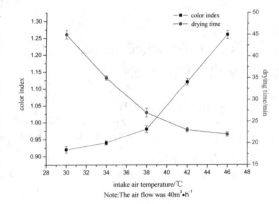

Figure 1. Effect of intake air temperature on the color index and drying time of bleached shellac.

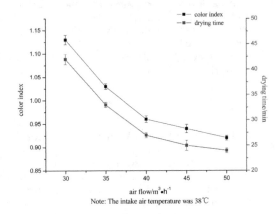

Figure 2. Effect of airflow rate on the color index and drying time of bleached shellac.

national standard for first-grade refined bleached shellac. Therefore, airflow rate of 40 m³/h used for drying of bleached shellac was appropriate.

3.2 Comparative analysis of product quality inspection results of different drying methods

The national standard and five different drying methods of bleached shellac are shown in Table 1. It is evident from the table that the quality of products obtained by fluidized-bed drying was better, and the indicators reached the technical requirements of GB/T 8140-2009. Compared with other drying methods, this method involves the shortest fluidized-bed drying time for 1 kg of materials (wet basis) and the consumption was the lowest.

Acid analysis of bleached shellac after the formation of particles contains mainly free water and bound water. The water particles between the bleached shellac particles can be easily dried. The internal bleaching of lac particles due to the

Table 1. Quality test results of bleached shellac by different drying methods.

Project	1	2	3	4	5	6
Color index	≤1.0	1.90	1.00	0.90	1.01	0.90
Hot ethanol insoluble substance/%	≤0.1	0.085	0.085	0.075	0.076	0.091
Volatile matter (moisture)/%	≤3	0.92	2.87	2.89	1.63	1.68
Water soluble substance/%	≤0.5	0.038	0.062	0.051	0.128	0.052
The acid value (with KOH)/ $(mg \cdot g^{-1})$	75~90	60.3	75.6	76.2	75.6	80.2
Drying time/min	–	45	510	30	1800	28
Energy consumption/ $(kJ \cdot kg^{-1})$	–	1624	1686	1763	2037	987

Note: No. 1 for the first-level standard of national standard of bleached shellac. No. 2 for fixed-bed drying, drying conditions for atmospheric pressure and 50°C. No. 3 for the vacuum drying, the drying conditions for 30 kPa, 45°C. No. 4 for microwave vacuum drying, drying conditions for 4.5 kPa, 600 W. No. 5 for vacuum freezing and drying, drying conditions for 18°C and 0.3 kPa. No. 6 for fluidized-bed drying, drying conditions for 36°C, 45 m³/h, power = power*time/quality.

precipitation formed a porous structure containing more water. Because of the capillary force combined with bleaching lac particles, it is difficult to remove the water by physical methods, thereby making the drying of bleached shellac significant. At the same time, lac is a natural resin with thermoplastic and thermosetting properties. In fixed-bed drying, vacuum drying, and microwave vacuum drying, the material layer was fixed relative to heat source, easy to soften the bleached shellac wet material heating surface, polymerize, cure to form a dense layer, and hinder its internal water discharge. On the contrary, the wet material of bleached shellac contains water for internal acid solution of higher concentration. During vacuum freeze-drying, this part of the electrolyte remains in the frozen state. Compared with other conventional drying methods, water-soluble substances are more, and because these electrolytes are soluble in ethanol, a high color index of the product results.

In the fluidized-bed drying of bleached shellac, wet material particles contribute to the moisture content in the lower air blowing drum, which suspend freely in moving fluid, wet material is mixed with each other, and the collision of particles and the hot air is effective, and the material temperature is uniform and stable. Therefore, the fluidized-bed drying of bleached shellac has obvious advantages.

3.3 Comparison of infrared spectra and Differential Scanning Calorimetry (DSC) curves of the fluidized-bed drying and vacuum freeze drying

Vacuum freeze drying is low-temperature drying, without heating process, and the quality and morphology of bleached shellac can be better preserved, so the products of vacuum freeze drying can be used as a contrast like other drying methods.

As shown in Figure 3, the infrared spectra prepared by the fluidized-bed drying and vacuum freeze drying methods of bleached shellac coincide. A broad-band peak was observed near 3420 cm⁻¹ in the infrared spectrum, the peak was O–H stretching vibration peak, two sharp peaks were observed at 2930 and 2857 cm⁻¹ with greater intensity, a C–H vibration peak was observed at 715 cm⁻¹, and a sharp absorption stretching vibration peak of C = O and stretching vibration peak of C = O aldehyde group were observed. The absorption peak of fingerprint region was attributed to –CH₂ and –CH₃ of asymmetric deformation vibration at 1467 cm⁻¹ and the symmetric deformation vibration of 1380 cm⁻¹ for C–CH₃. These features indicate that fluidized-bed drying retains the hydroxyl, carbonyl, and methyl characteristic absorption peaks of bleached shellac, without any change in their chemical structures. The fluidized-bed drying and vacuum freeze drying of the bleached lac DSC map is shown in Figure 4.

Figure 3. IR spectra of vacuum-freeze-dried and fluidized-bed-dried bleached shellac.

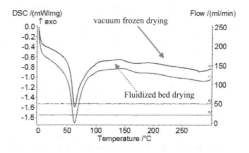

Figure 4. DSC of vacuum-freeze-dried and fluidized-bed-dried bleached shellac.

871

It is evident from the figure that the endothermic peaks and valleys were basically the same, and the softening points were 52.63 and 52.90°C, and the melting points were 62.53 and 62.50°C, and basically no changes have occurred.

4 CONCLUSION

When the bleached shellac was dried with fluidized bed, the air inlet temperature and airflow rate had different degrees of influence on the color index and drying time, and the air temperature has a great impact on the color index ($P < 0.0001$). Compared with the fixed-bed drying, vacuum drying, vacuum freeze drying, microwave vacuum drying, and fluidized-bed drying could be achieved at a relatively low temperature with fast drying, shortest drying time, and good color index, which are more suitable for industrial production.

The products of fluidized-bed drying and vacuum freeze drying, through analysis and comparison of IR spectrum and DSC curves, show that fluidized-bed drying method conforms to the national standards of the first grad, without any change in the chemical structure and condensation. The test results show that the fluidized bed device could be used in bleached shellac and quick drying, and the experimental data and conditions can provide a reference for the actual production.

It is important to note that this paper only focuses on fluidized-bed drying technology process conditions, without considering the influence of different weight and moisture contents. Further research will complement the tests, in order to expand the scope of application and model.

ACKNOWLEDGMENTS

The authors gratefully acknowledge the financial support from the National High Technology Research and Development Program of China (863 Program, 2014 AA021801) and the Special Fund Project for the Scientific Research of State Forest Public Welfare Industry, SFA, China (201204602).

When listing facts, use either the style tag list signs or the style tag list numbers.

REFERENCES

Akbari H, Karimi K, Lundin M, et al. Optimization of baker's yeast drying in industrial continuous fluidized bed dryer [J]. Food and Bioproducts Processing, 2012, 90(1): 52–57.

Alonso C P, Olivares J C, Ramirez A, et al. Moisture diffusion in allspice (Pimenta Dioical L. Merril) fruits during fluidized bed drying [J]. Journal of Food Processing and Preservation, 2011, 35(3): 308–312.

Chauhan O P, Raju P S, Singh A, et al. Shellac and aloe-gel-based surface coatings for maintaining keeping of apple slices quality [J]. Food Chemistry, 2011, 126(3): 961–966.

Coelho C, Nanabala R, Menager M, et al. Molecular changes during natural biopolymer ageing—The case of shellac [J]. Polymer Degradation and Stability, 2012, 97(6): 936–940.

Demers A M, Gosselin R, Simard J S, et al. In-line near infrared spectroscopy monitoring of pharmaceutical powder moisture in a fluidised bed dryer: An efficient methodology for chemometric model development [J]. The Canadian Journal of Chemical Engineering, 2012, 90(2): 299–303.

Farag Y, Leopold C S. Development of shellac-coated sustained release pellet formulations [J]. European Journal of Pharmaceutical Sciences, 2011, 42(4): 400–405.

GB/T 8140–2009 Bleached lac [S].

GB/T 8142–2008 Methods of sampling lac products [S].

GB/T 8143–2008 Methods of testing lac products [S].

Kozanoglua B, Martineza J, Alvareza S, et al. Influence of particle size on vacuum-fluidized bed drying [J]. Drying Technology, 2012, 30(2): 138–145.

Li Kai, Zhou Meicun, Zhang Hong, et al. Solubility and physico-chemical properties of lac resin and its sodium salt [J]. Food Science, 2010, 31(21): 159–164.

Liu Bin, Yuan Rubing, Zhang Qiang, et al. Experimental investigation on heat and mass transfer of fluidized drying of zymotic orange peel [J]. Transactions of the Chinese Society of Agricultural Engineering (Transactions of the CSAE), 2011, 27(7): 353–357.

Nazghelichia T, Aghbashloab M, Kianmehra M H, et al. Prediction of energy and exergy of carrot cubes in a fluidized bed dryer by artificial neural networks [J]. Drying Technology, 2011, 29(3): 295–307.

Yazdanpanah N, Langrishfast T A G. crystallization of lactose and milk powder in fluidized bed dryer/ crystallizer [J]. Dairy Science and Technology, 2011, 91(3): 323–340.

Yin Leichang, Wang Xiangyou, Yang Wen. Modeling of drying corn in pulsed fluidized bed and test validation[J]. Transactions of the Chinese Society of Agricultural Engineering (Transactions of the CSAE), 2007, 23(10): 251–255.

Yu Liansong, Zhang Hong, Zheng Hua, et al. Research on microwave-vacuum drying process of bleached shellac by response surface method [J]. Applied Chemical Industry, 2012, 41(1): 37–43. (in Chinese with English abstract)

Yu Liansong, Zhou Meicun, Zheng Hua, et al. Research of Conditions in Determining the Glass Transition Temperature of Lac Resin by Differential Scanning Calorimetry [J]. Materials Review B, 2014, 28(4):98–104,121. (in Chinese with English abstract)

Yu Liansong. Study on Drying Technology of Bleached Shellac [D]. Kunming: Kunming University of Science and Technology. 2010, 1–71. (in Chinese with English abstract)

Zhang Ruguo, Zhang Hong, Zhang Zhen, et al. Characterization of five natural resins and waxes by differential scanning calorimetry (DSC) [C] //Guilin, Advanced Materials Research. 2012, 418: 643–650.

Advances in Energy Science and Equipment Engineering II – Zhou, Patty & Chen (Eds)
© 2017 Taylor & Francis Group, London, ISBN 978-1-138-71798-5

Analysis of an aquaporin in *Gardenia jasminoides*

Lan Gao & Hao-Ming Li
School of Basic Courses, Guangdong Pharmaceutical University, Guangzhou Higher Education Mega Center, Guangzhou, P.R. China

ABSTRACT: Aquaporins (AQPs) are a family of membrane channels primarily responsible for conducting water across cellular membranes and maintenance of water balance. In addition to water, some AQPs can conduct other small molecules such as urea, ammonia, hydrogen peroxide, carbon dioxide, metalloids, and nitric oxide. In this paper, the structure of putative tonoplast aquaporin from *Gardenia jasminoides* (GjTIP) was analyzed. A phylogenetic analysis conducted with previously characterized aquaporins from other plant species indicates that the protein belongs to TIP 1 subfamily. A three-dimensional model structure of GjTIP was built on the basis of crystal structure of an ammonia-permeable AtTIP 2-1 from *Arabidopsis thaliana*. The model structure is displayed as a homo-tetramer. Each monomer has six transmembrane helices, and the narrowest pore in each monomer channel is the selectivity filter. The data suggest that the GjTIP has a tendency to be a mixed-function aquaporin, which might involve in water, urea, and hydrogen peroxide transport.

1 INTRODUCTION

Aquaporins are membrane protein channels that transport water along osmotic or hydrostatic pressure gradients. Aquaporins are widely distributed in most living species and share a common tetrameric fold with each monomer functioning as an individual water channel. The individual water channel has six Transmembrane (TM) helices separated by five loops. The helices in the membrane form a primary pore structure that substrates travel through. Two conserved regions are central for selective function. The first region is the two highly conserved asparagine–proline–alanine (NPA) motifs in Loops B and E, respectively; they embed into the membrane to form constriction selectivity for water (Forrest et al. 2007). The second region is known as the aromatic/arginine (ar/R) filter (SF), which is located near the extracellular pore entrance and forms the narrowest portion of the channel. It is formed from four residues; two from TM helices, 2 and 5, and two from Loops B and E, respectively (Maurel et al., 2008). Recently, Kirscht et al. (2016) have suggested an extended selectivity filter, including a fifth amino acid residue at position LC in Loop C. Additional conserved amino acids also influence the selectivity. Five key amino residues have been proposed to discriminate whether the AQPs transport water or glycerol by Froger et al. (1998). P1 is in Loop C, P2–3 within Loop E, and P4-5 within TM6.

High-plant aquaporins constitute a large protein family, including 30 to more than 70 homologs, compared with 13 AQP isoforms found in microbes and mammals (Agre & Kozono 2003). A total of 35 AQP isoforms have been found in Arabidopsis (Johanson et al. 2001), 71 in cotton (Park et al. 2001), and 36 in maize (Chaumont et al. 2001). On the basis of sequence similarity, plant AQPs fall into five subfamilies: the Plasma membrane Intrinsic Proteins (PIPs), the Tonoplast Intrinsic Proteins (TIPs), the Nodulin26-like Intrinsic Proteins (NIPs), the Small basic Intrinsic Proteins (SIPs), and the X Intrinsic Proteins (XIPs) (Johanson et al. 2001). Physiological and molecular studies proved that PIPs and TIPs play key functions in water homeostasis.

AQPs effectively mediate fast transmembrane transport of water during the plant growth and development processes; they were associated with cell growth, seed germination, stomata movement, and fruit development (Reuscher et al. 2016). AQPs are also important for the facilitated membrane diffusion of other uncharged solutes, such as glycerol (Biela et al. 1999), CO_2 (Otto et al. 2010), NH_3/NH_4^+ (Loqué et al. 2005), silicon (Mitani-Ueno et al. 2011), H_2O_2 (Bienert et al. 2007), boric acid (Schnurbusch et al. 2010), and urea (Kirscht et al. 2016). A number of plant AQPs are permeable to more than one substrate. For example, tobacco NtTIPa can conduct water, glycerol, and urea (Gerbeau et al. 1999), and Arabidopsis AtNIP 5-1 can conduct silicon, boron, and arsenic (Schnurbusch et al. 2010).

Crystal structures of aquaporins from bacteria, algae, yeast, plasmodium, plants, mammals, and human have been established. Fu et al. (2000), Savage et al. (2003), Törnroth-Horsefield et al.

(2006), and Kirscht et al. (2016) provided the molecular understanding of how aquaporins facilitate substrates flux across membranes.

Gardenia jasminoides originates in Asia and is cultivated for at least 1000 years. The fruit of G. jasminoides is used in Asian countries as a natural colorant and a traditional herbal medicine. Crocin, crocetin, and geniposide are the main secondary metabolites in the fruit, which exhibit a wide range of pharmacological activities (Hong et al. 2013). We have isolated a tonoplast intrinsic protein (GjTIP) gene from G. jasminoides (Gao & Guo 2013) and found that the transcript levels of GjTIP have increased during fruit maturation. In this paper, the structure of GjTIP was analyzed.

2 MATERIALS AND METHODS

2.1 *Alignment and phylogenetic analysis of AQPs*

The Aquaporin (GjTIP) from Gardenia jasminoides was aligned with aquaporins from other species using Clustal Omega program (http://www.ebi.ac.uk) and Blast (http://www.ncbi.nlm.nih.gov). A consensus tree was computed in MEGA4.

2.2 *Building a three-dimensional model*

A three-dimensional model was built using the Swiss-model Workspace (http://swissmodel.expasy.org). A homology model of GjTIP based on X-ray crystallography of Arabidopsis thaliana tonoplast aquaporin AtTIP 2-1 (PDB 5I32) was constructed.

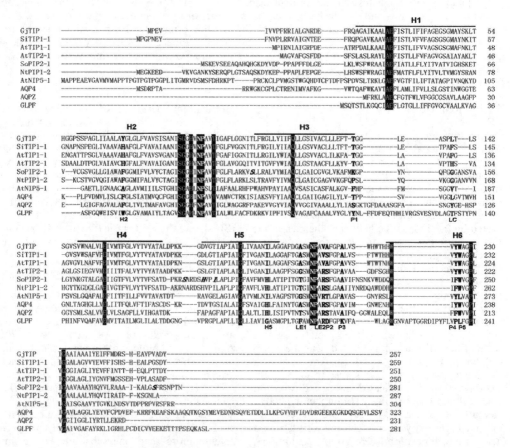

Figure 1. Sequence alignment of GjTIP with AQP amino acid sequences from other organism. Amino acids are numbered at the right of the sequence. The abbreviation and Genbank accession number are: GjTIP (*Gardenia jasminoides*, AEF59492), SiTIP 1-1 (*Sesamum indicum*, XP_011089965), AtTIP 1-1 (*Arabidopsis thaliana*, AAD31569), AtTIP 2-1 (BAB01264, PDB 5I32), AtNIP 5-1 (Q9SV84), SoPIP 2-1 (*Spinacia oleracea*, PDB:2B5F), NtPIP 1-2 (*Nicotiana tabacum*, NP_001312721), AQP4 (*Homo sapiens*, PDB 3GD8), AQPZ (*Escherichia coli*, P60844), GLPF (*Escherichia coli*, PDB 1LDI). The transmembrane domains are labeled H1-6. The Ar/R filter, the presumed fifth residue of filte LC, and Froger's position (P1–P5) are in bold black. Residues involved in gating by phosphorylation as well as gating by pH are indicated with italic bold black letters. The identical residues including NPA domain are shaded.

3 RESULTS AND DISCUSSION

The predicted amino acid sequence of GjTIP gene specifies a polypeptide with 257 amino acids. The pairwise sequence identities between 10 members of the AQPs in Figure 1 range from 73.8% (SiTIP 1-1) to 23.3% (GLPF). The highest degree of conservation is in the HB and HE that forms the aqueous pore and contains two NPA motifs. A phylogenetic analysis conducted with previously characterized aquaporins from other plant species indicates that the protein belongs to TIP 1 subfamily (not shown). In Table 1, the amino acid composition of the Froger's position and aromatic/arginine selectivity filters of these sequences were compared. It suggests that the GjTIP may transport water, urea, and H_2O_2.

The 3D model structure of GjTIP was constructed as a tetramer using SWISS-PDB software (Figure 2), and the X-ray crystallography at resolutions down to 1.18 Å of AtTIP 2-1 was used as template. The amino acid sequence of GjTIP (from 10 templates) and the amino acid sequence of GjTIP (from 10 to 246 residual range) have 48.6% amino acid identity with AtTIP 2-1 (from 26 to 250 residual range).

In the model, each of the monomers has typical six-membrane spanning helices and N- and C-terminal regions at the cytoplasmic vestibule. The two NPA motifs meet in the middle of the membrane. The ar/R selectivity filter is shown in black and the fifth selectivity filter residue L139 is shown in gray, which formed the narrowest part of the pore, near the vacuolar vestibule. AQP structures are shown in top view and periplasmic side view in Figure 2, including AtTIP 2-1 and GjTIP. The patterns of ar/R residues surrounding the pore regions are similar in these structures, and the sites of the fifth residue in the pore regions are equitable. It was suggested that the histidine at H2 and an aromatic residue at position LC is a common feature among ammonia-permeable AQPs (Kirscht 2016), which means that GjTIP was not an ammonia transporter (H2 is tyrosine, LC is leucine).

Table 1. Comparison of conserved amino acids involved in selectivity.

		GjTIP	AtTIP 1-1	AtTIP 2-1	Plant AQP
Ar/R filter	H2	Y	H	H	H/F/W
	H5	I	I	I	I/H/V
	LE1	A	A	G	A/T/G
	LE2	V	V	R	V/R
	LC	L	F	H	Undefined
Froger's position (P1–P5)	P1	T	T	T	T/Q/F
	P2	A	A	S	A/S
	P3	A	A	A	A
	P4	Y	Y	Y	Y/F
	P5	W	W	W	W/I
Permeation		Undefined	Water, urea, H_2O_2	Water, H_2O_2, urea, ammonia	Water, H_2O_2
Reference		Gao et al. 2013	Kirscht et al. 2016	Kirscht et al. 2016	Johanson et al. 2001

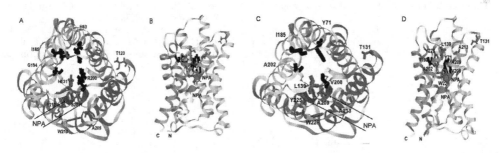

Figure 2. Three-dimensional structure of AQPs. (A) (C) Top view into the pore for monomeric AtTIP 2-1 (PDB 5I32) and monomeric GjTIP model, respectively. (B) (D) Viewed from the periplasmic side of AtTIP 2-1 and GjTIP model, respectively. The N- and C-termini faced the cytosol. NPA domains are shown as gray ribbons. Ar/R selectivity filter amino acids are shown as black sticks, the LC residues are shown as bold gray sticks (H131, L139), and Froger's positions are shown as gray sticks.

4 CONCLUSIONS

Plant cell extension is a developmental process. It requires a fast inflow of water, which ends up mainly in the vacuole. During rapid cell elongation, the up-regulated expression of aquaporin tonoplast genes aimed at increasing the water channel. A high abundance of aquaporins was detected in samples from fruits in the cell expansion stage. We predicted that the GjTIP might have water, urea, and H_2O_2 permeability of membranes during fruit maturation.

REFERENCES

Agre, P. & Kozono, D. (2003). Aquaporin water channels: molecular mechanisms for human diseases. *FEBS Lett* 555: 72–78.

Biela, A., Grote, K., Otto, B., Hoth, S., Hedrich, R. & Kaldenhoff, R. (1999). The Nicotiana tabacum plasma membrane aquaporin NtAQP1 is mercury-insensitive and permeable for glycerol. *Plant J* 18: 565–570.

Bienert, G.P., Møller, A.L., Kristiansen, K.A., Schulz, A., Møller, I.M. & Schjoerring, J.K. et al. (2007). Specific aquaporins facilitate the diffusion of hydrogen peroxide across membranes. *J. Biol. Chem* 282: 1183–1192.

Chaumont, F., Barrieu, F., Wojcik, E., Chrispeels, MJ. & Jung, R. (2001). Aquaporins constitute a large and highly divergent protein family in maize. *Plant Physiol* 125: 1206–1215.

Forrest, K.L. & Bhave, M. (2007). Major intrinsic proteins (MIPs) in plants: a complex gene family with major impacts on plant phenotype. *Funct Integr Genomics* 7: 263–289.

Froger, A., Tallur, B., Thomas, D. & Delamarche, C. (1998). Prediction of functional residues in water channels and related proteins. *Protein Sci* 7: 1458–1468.

Fu, D., Libson, A., Miercke, L.J., Weitzman, C., Nollert, P. & Krucinski, J. et al. (2000). Structure of a glycerol-conducting channel and the basis for its selectivity. *Science* 290: 481–486.

Gao, L. & Guo, Y.J. (2013). Isolation of a fruit ripening-related tonoplast aquaporin (GjTIP) gene from Gardenia jasminoides. *Physiol Mol Biol Plants* 19: 555–561.

Gerbeau, P., Guclu, J., Ripoche, P. & Maurel, C. (1999). Aquaporin Nt-TIPa can account for the high permeability of tobacco cell vacuolar membrane to small neutral solutes. *Plant J* 18: 577–587.

Hong, Y.J. & Yang, K.S. (2013). Anti-inflammatory activities of crocetin derivatives from processed Gardenia jasminoides. *Arch Pharm Res* 36: 933–940.

Johanson, U., Karlsson, M., Johansson, I., Gustavsson, S., Sjövall, S. & Fraysse, L. et al. (2001). The complete set of genes encoding major intrinsic proteins in Arabidopsis provides a frame work for a new nomenclature for major intrinsic proteins in plants. *Plant Physiol* 126: 1358–1369.

Kirscht, A., Kaptan, S.S., Bienert, G.P., Chaumont, F., Nissen, P. & de Groot B.L. et al. (2016). Crystal Structure of an Ammonia-Permeable Aquaporin. *PLoS Biol* 14: e1002411.

Loqué, D., Ludewig, U., Yuan, L. & von Wirén, N. (2005). Tonoplast intrinsic proteins AtTIP2;1 and AtTIP2;3 facilitate NH3 transport into the vacuole. *Plant Physiol* 137: 671–680.

Maurel, C., Verdoucq, L., Luu, D.T. & Santoni, V. (2008). Plant aquaporins: membrane channels with multiple integrated functions. *Annu Rev Plant Biol* 59: 595–624.

Mitani-Ueno, N., Yamaji, N., Zhao, F.J. & Ma, J.F. (2011). The aromatic/arginine selectivity filter of NIP aquaporins plays a critical role in substrate selectivity for silicon, boron, and arsenic. *J. Exp. Bot* 62: 4391–4398.

Otto, B., Uehlein, N., Sdorra, S., Fischer, M., Ayaz, M. & Belastegui-Macadam, X. et al. (2010). Aquaporin tetramer composition modifies the function of tobacco aquaporins. *J Biol Chem* 285: 31253–31260.

Park, W., Scheffler, B.E., Bauer, P.J. & Campbell, B.T. (2010). Identification of the family of aquaporin genes and their expression in upland cotton (Gossypium hirsutum L.). BMC *Plant Biol* 10: 142.

Reuscher, S., Fukao, Y., Morimoto, R., Otagaki, S., Oikawa, A. & Isuzugawa, K. et al. (2016). Quantitative Proteomics-Based Reconstruction and Identification of Metabolic Pathways. *Plant Cell Physiol* 57: 505–518.

Savage, D.F., Egea, P.F., Robles-Colmenares, Y., O'Connell, J.D. 3rd & Stroud, R.M. (2003). Architecture and selectivity in aquaporins: 2.5 Å X-ray structure of aquaporin Z. *PLoS Biol* 1: E72.

Schnurbusch, T., Hayes, J., Hrmova, M., Baumann, U., Ramesh, S.A. & Tyerman, S.D. et al. (2010). Boron toxicity tolerance in barley through reduced expression of the multifunctional aquaporin HvNIP2;1. *Plant Physiol* 153: 1706–1715.

Törnroth-Horsefield, S., Wang, Y., Hedfalk, K., Johanson, U., Karlsson, M. & Tajkhorshid, E. et al. (2006). Structural mechanism of plant aquaporin gating. *Nature* 439: 688–694.

Advances in Energy Science and Equipment Engineering II – Zhou, Patty & Chen (Eds)
© *2017 Taylor & Francis Group, London, ISBN 978-1-138-71798-5*

Isolation and analysis of a tropinone reductase gene from *Gardenia jasminoides*

Lan Gao & Hao-Ming Li

School of Basic Courses, Guangzhou Higher Education Mega Center, Guangdong Pharmaceutical University, Guangzhou, P.R. China

ABSTRACT: Tropinone Reductases (TRs) catalyze NADPH-dependent reductions of the 3-carbonyl group of their common substrate, tropinone, to hydroxy groups with different diastereomeric configurations. TR-I produces tropine and TR-II produces pseudotropine. The products of the tropine-forming tropinone reductase I, hyoscyamine and scopolamine, are important pharmaceutical and source compounds for the synthesis of derivatives. In this paper, a *Gardenia jasminoides* fruit cDNA library was constructed, and the GjTR cDNA was isolated from the cDNA library by sequencing method. The GjTR cDNA contains a predicted 825 bp open reading frame that encodes 274 amino acids. The bioinformatics analysis suggested that the GjTR protein has conserved Ser-Tyr-Lys triad active site, NADPH-binding motif, and tropinone-binding motif with previously characterized TR-I from other species. A three-dimensional dimer model of GjTR was built, whose structure is similar to that of TR-I from *Datura stramonium*. The results suggest that GjTR is a TR-I of *G. jasminoides*.

1 INTRODUCTION

Various tropane esters, such as hyoscyamine and scopolamine, are derived from tropine. They have medicinal applications as muscarinic receptor antagonists (Richter et al. 2005). The biosynthesis of hyoscyamine and scopolamine requires the first enzymatic steps of the general tropane alkaloid pathway. Reduction of tropinone to the stereoisomeric tropine and pseudotropine constitutes the diversion of tropine and calystegine ester alkaloid formation. TR-I (EC 1.1.1.206) produces tropine and TR-II (EC 1.1.1.236) produces pseudotropine (see Figure 1). Two tropinone reductases responsible for tropane alcohol formation have been isolated and characterized from root cultures of several species: *Datura stramonium* L (Nakajima et al. 1993, El Bazaoui et al. 2011), *Hyoscyamus niger* L (Dräger et al. 1988, Jaremicz et al. 2014), *Hyoscyamus senecionis* (Dehghan et al. 2013), *Anisodus acutangulus* (Cui et al. 2015), *Brugmansia arborea* (Qiang et al. 2016), *Withania somnifera* (Kushwaha et al. 2013), and *Atropa belladonna* (Dräger et al. 1994). Roots are the major organs of tropane alkaloid biosynthesis. In wild-type roots, tropine, as a product of tropinone reduction, is esterified to yield hyoscyamine and scopolamine (Rothe et al. 2001, Bedewitz et al. 2014).

Because of novel drugs based on natural alkaloids (Harvey et al. 2015) and in particular on the tropane skeleton, the importance of tropane

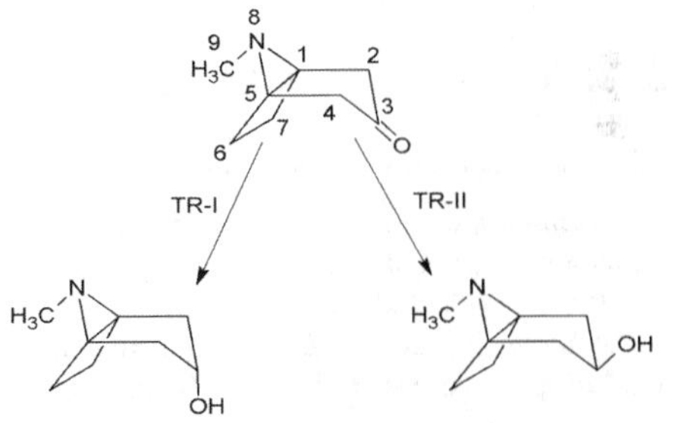

Figure 1. TR-I and TR-II catalyzes NADPH-dependent reduction of the 3-carbonyl group of tropinone to α or β-hydroxyl group and produces tropine or pseudotropine respectively.

alkaloids has increased. Muscarinic receptor antagonists, belladonna alkaloids in particular, have been used for a long time to treat a variety of clinical conditions, such as peptic ulcer, asthma, and Parkinson's disease (Grynkiewicz & Gadzikowska 2008). Structure activity relationship studies of muscarinic antagonists have indicated that such key functional groups as a nitrogen atom, an ester group, and a bulky hydrophobic portion were required in a molecule to attach antimuscarinic properties (Sampson et al. 2014).

The amino acid sequence of TR-I and TR-II share a high homology, and they are a member of the Short-chain Dehydrogenase/Reductase (SDR) family (Nakajima et al. 1998). Members of this family have a highly conserved Ser-Tyr-Lys triad sequence, which is thought to be essential for their catalytic activities (Rothe et al. 2001).

Crystallographic analysis of TR-I and TR-II from *Datura stramonium* indicated that the two structures are almost indistinguishable from each other, in both subunit folding and their association in dimers (Nakajima et al. 1998, Nakajima et al. 1999). The architectures of NADP(H) binding sites in both TRs are also similar, suggesting similar binding modes of NADP(H) in TR-II and TR-I. TR-II and TR-I presumably bind their substrates in different orientations toward the NADPH to catalyze the production of different tropine isomers (Nakajima et al. 1998).

Gardenia jasminoides originates in Asia and is cultivated for at least 1000 years. The fruit of *G. jasminoides* is used in Asian countries as a natural colorant and a traditional herbal medicine. Crocin, crocetin, and geniposide are the main secondary metabolites in the fruit, and they all exhibit a wide range of pharmacological activities (Pesaresi et al. 2011). In this paper, we identified and analyzed a Tropinone Reductases (GjTR) gene in *G. jasminoides*.

2 MATERIALS AND METHODS

2.1 *Plant and growth conditions*

G. jasminoides plants cultivated at Guangdong Pharmaceutical University were used as materials. Fruits were collected at development stage II, closed with yellowish green exocarp and orange mesocarp. The samples were stored at –80°C until use.

2.2 *CDNA library construction, EST sequencing, and cloning of GjTIP*

Total RNA was extracted from Gardenia fruit (stage II) using a modified CTAB (hexadecyl trimethyl ammonium bromide) based extraction protocol (Bekesiova et al. 1999). From total RNA, the cDNA library construction and amplification were performed following the users manual of the CreatorTM SMARTTM cDNA Library construction Kit (Clontech, USA). The SMART cDNAs were ligated into SfiI-digested pDNR-LIB vector and transformed into Escherichia coli strain DH5α. Colonies were randomly picked and inoculated to a separated PCR reaction solution. They were lysed by heating the mixed solutions at 95°C in a PTC-200 Thermocycler (MJ Research, USA) for 5 min. Then, PCR amplification was conducted with M13 primers provided by the CreatorTM SMARTTM cDNA Library Construction Kit. The amplified PCR products (ESTs, expressed sequence tags) were analyzed by 1.2% agarose gel electrophoresis. When the amplified PCR products were longer than 1000 bp, the isolated colonies were incubated and the ESTs were sequenced. A total of 40 ESTs were sequenced. After sequencing and analysis, the colony containing the predicted pDNR-LIB-GjTR was isolated.

3 RESULTS AND DISCUSSION

We identified novel TR homologs in *Gardenia jasminoides* (named GjTR) by exploiting the fruit cDNA library of *G. jasminoides*. The full-length GjTR cDNA (Genbank accession No. KM371229) was obtained. The cDNA contains a predicted 825 bp ORF that encodes 274 amino acids. The produced protein sequence of GjTR was compared to Genbank database, and the best homology was found to TR-like protein of Coffea canephora. The two proteins share 92% identical amino acids and compared GjTR to the TR proteins from *Nicotiana tomentosiformis* (identity 80%), *Anisodus acutangulus* (TR I:54.26%, TR II:54.87%), *Hyoscyamus niger* (TR I: 52.65%, TR II: 44.16%), *Datura stramonium* (TR I: 54.96%, TR II: 54.51%), and *Solanum tuberosum* (TR I: 55.04%, TR II: 55.23%). The GjTR secondary structure was predicted by Swiss-model software (Figure 2). Various species of TR have a common Ser-Tyr-Lys triad site. Tropinone binding residues and NADP binding residues are conserved for TR I or TR II, respectively, among which GjTR and NtoTR have some positions with substitutions resembling TR I or TR II. The main differences are at the N-terminal, carboxyl terminal, and the region between the sixth β-sheet and eighth α-helix, with GjTR having long N-terminal end, resembling TR I.

The 3D model structure of GjTR was predicted using SWISS-PDB software, the X-ray diffraction at resolutions down to 2.4 Å of *Datura stramonium* TR I complexes with NADP (PBD code 1 AE1) was used as template (Nakajima et al. 1998) (see Figure 3A), the TR I includes residues 16–205 and 219–273 of one subunit, designated chain A and residues 16–273 of the other Chain B, as well as two NADP+ and 98 water molecules per dimer. The amino acid sequence of GjTR has 54.96% amino acid identity with Dst TR I. The structure was successfully built as a homodimer (see Figure 3B), the fold of GjTR consists of eight α-helices flanked by a seven-stranded parallel β-sheet, which constitutes the Rossmann fold topology, indicating a high level of structural similarity to *D. stramonium* TR I. Both TR subunits consist of a core domain that includes most of the polypeptide and a small lobe that protrudes from the core.

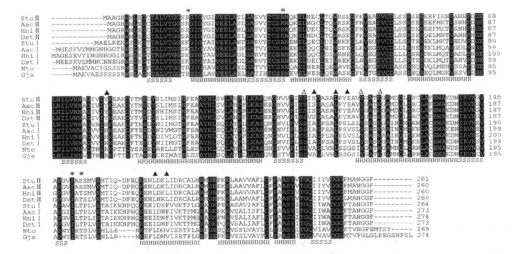

Figure 2. Alignment of GjTR with TR amino acid sequences from other organism. Amino acids are numbered at the right of the sequence. The abbreviation and Genbank accession number are: Gja (*Gardenia jasminoides*, AIX10937), Stu (*Solanum tuberosu.*, I CAC34420, II CAC19810), Aac (*Anisodus acutangulus*, I ACB71202, II ACB71203), Hni (*Hyoscyamus niger*, I BAA85844, II AAB09776), Dst (*Datura stramonium*, I AAA33281, PDB:1AE1, II PDB code: 1IPF), Nto (*Nicotiana tomentosiformis*, XP_009603003). Solid arrowheads indicate tropinone binding residues. Open arrowhead indicate active sites. Stars indicate NADP binding residues. Identical residues are shaded; Predicted alpha helices and beta sheets by Swiss-model of GjTR are marked with H and S, respectively.

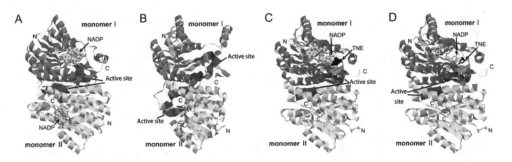

Figure 3. 3D model of GjTR. (A): Crystal structure of dimer TRI from *Datura stramonium* complex with NADP, PBD code: 1AE1. (B): Model structure of dimer GjTR (Genbank accession no. AIX10937), predited by Swiss-PDB software. (C) and (D): Crystal structure of dimer TRII from *Datura stramonium* complex with NADP and tropinone, PBD code: 1IPF. Shown in ribbon diagram. NADP shown in white spacefill (A and C) or stick diagram (D), ψ-tropinone (TNE) shown in black stick diagram. In all: Active site shown in grey.

In Dst TR I, NADP+ is located at the bottom of the cleft between the core domain and the small lobe. NADP+ is found in an extended conformation. The binding site of tropine is predicted in the crystal structure with resolution down to 2.5Å of *D. stramonium*. TR II combined with NADP. and ψ-tropine (PBD code 1IPF) (Nakajima et al. 1999) was shown (see Figure 3C). The complex was a dimer, including all residues except for the first N-terminal residue (2–273), and the structure is similar to that of the unliganded TR II and TR I. The deep cleft between the core domain and the small lobe is the binding site for tropinone.

In GjTR, the corresponding residues resemble the Dst TR I.

TR I and TR II contain a highly conserved Ser-Tyr-Lys triad at the active site. They are located near the hydroxyl group of ψ-tropine in the ternary complex of TR II, and similar positions are found in Dst TR-I and GjTR. A catalytic mechanism for tropinone reduction has been proposed, which involves stereospecific hydride transfer from the NADPH cofactor to the carbon atom of the keto group followed by donation of a proton to the oxygen of the resultant alkoxide ion intermediate from a base provided by the protein. The serine residue of the triad was suggested to be involved in binding of the substrate, reaction intermediates, and products (Reinhardt et al. 2014). The proposed base in TR II is the phenolic oxygen of Tyr159, with the amino group of Lys163 providing the positive charge to stabilize the transition state (Hashimoto & Yamada 1992). In GjTR, the corresponding residues are Ser154, K171, and Y167; in Dst TR I, the corresponding residues are S158, K175, and Y171.

4 CONCLUSIONS

The results indicated that GjTR has relatively conserved NADP-binding site, tropinone-binding position, and conserved Ser-Tyr-Lys triad active site. The structure comparison of the dimer of DstTR I and DstTR II with GjTR indicates a high level of structural similarity, and the results of the primary structure aliment suggest that GjTR is a TR I.

REFERENCES

Bedewitz, M.A., Góngora-Castillo, E., Uebler, J.B., Gonzales-Vigil, E., Wiegert-Rininger, K.E. & Childs, K.L. (2014). A root-expressed L-phenylalanine: 4-hydroxyphenylpyruvate aminotransferase is required for tropane alkaloid biosynthesis in Atropa belladonna. *Plant Cell* 26:3745–3762.

Bekesiova, I., Nap, J.P. & Mlynarova, L. (1999). Isolation of high quality DNA and RNA from leaves of the carnivorous plant Drosera rotundifolia. *Plant Mol. Biol. Rep* 17: 269–277.

Cui, L., Huang, F., Zhang, D., Lin, Y., Liao, P. & Zong, J. et al. (2015). Transcriptome exploration for further un derstanding of the tropane alkaloids biosynthesis in Anisodus acutangulus. *Mol Genet Genomics* 290: 1367–1377.

Dehghan, E., Shahriari Ahmadi, F., Ghotbi Ravandi, E., Reed, D.W., Covello, P.S. & Bahrami, A.R. (2013). An atypical pattern of accumulation of scopolamine and other tropane alkaloids and expression of alkaloid pathway genes in Hyoscyamus senecionis. *Plant Physiol Biochem* 70: 188–194.

Dräger, B., Hashimoto, T. & Yamada, Y. (1988). Purification and characterization of pseudotropine forming tropinone reductase from Hyoscyamus niger root cultures. *Agricultural and Biological Chemistry* 52: 2663–2667.

Dräger, B. & Schaal, A. (1994). Tropinone reduction in Atropa belladonna root cultures. *Phytochemistry* 35: 1441–1447.

El Bazaoui, A., Bellimam, M.A. & Soulaymani, A. (2011). Nine new tropane alkaloids from Datura stramonium L. identified by GC/MS. *Fitoterapia* 82: 193–197.

Grynkiewicz, G. & Gadzikowska, M. (2008). Tropane alkaloids as medicinally useful natural products and their synthetic derivatives as new drugs. Pharmacol Rep. 60: 439–463.

Hashimoto, T. & Yamada, Y. (1992). Two Tropinone Reductases with Distinct Stereospecificities from Cultured Roots of Hyoscyamus niger. *Plant Physiol* 100: 836–845.

Harvey, A.L., Edrada-Ebel, R. & Quinn. R.J. (2015). The re-emergence of natural products for drug discovery in the genomics era. *Nat Rev Drug Discov* 14: 111–129.

Jaremicz, Z., Luczkiewicz, M., Kokotkiewicz, A., Krolicka, A. & Sowinski, P. (2014). Production of tropane alkaloids in Hyoscyamus niger (black henbane) hairy roots grown in bubble-column and spray bioreactors. *Biotechnol Lett* 36: 843–853.

Kushwaha, A.K., Sangwan, N.S., Trivedi, P.K., Negi, A.S., Misra, L. & Sangwan, R.S. (2013). Tropine forming tropinone reductase gene from Withania somnifera (Ashwagandha): biochemical characteristics of the recombinant enzyme and novel physiological overtones of tissue-wide gene expression patterns. *PLoS One* 8: e74777.

Nakajima, K., Hashimoto, T. & Yamada, Y. (1993). Two tropinone reductases with different stereospecificities are short-chain dehydrogenases evolved from a common ancestor. *Proc. Natl. Acad. Sci. USA* 90: 9591–9595.

Nakajima, K., Yamashita, A., Akama, H., Nakatsu, T., Kato, H. & Hashimoto, T. et al. (1998). Crystal structures of two tropinone reductases: different reaction stereospecificities in the same protein fold. *Proc Natl Acad Sci USA* 95: 4876–4881.

Nakajima, K. & Hashimoto, T. (1999). Two tropinone reductases that catalyze opposite stereospecific reductions in tropane alkaloid biosynthesis are localized in plant root with different cell-specific patterns. *Plant Cell Physiol* 40: 1099–1107.

Pesaresi, P., Pribil, M., Wunder, T. & Leister, D. (2011). Dynamics of reversible protein phosphorylation in thylakoids of flowering plants: the roles of STN7, STN8 and TAP38. *Biochim. Biophys. Act*, 1807: 887–896.

Qiang, W., Xia, K., Zhang, Q., Zeng, J., Huang, Y. & Yang, C. (2016). Functional characterisation of a tropine-forming reductase gene from Brugmansia arborea, a woody plant species producing tropane alkaloids. *Phytochemistr* 127: 12–22.

Reinhardt, N., Fischer, J., Coppi, R., Blum, E., Brandt, W. & Dräger, B. (2014). Substrate flexibility and reaction specificity of tropinone reductase-like short-chain dehydrogenases. *Bioorg Chem* 53: 37–49.

Richter, U., Rothe, G., Fabian, A.K., Rahfeld, B. & Dräger, B. (2005). Overexpression of tropinone reductases alters alkaloid composition in Atropa belladonna root cultures. *J Exp Bot* 56: 645–652.

Rothe G, Garske U, Dräger B. (2001). Calystegines in root cultures of Atropa belladonna respond to sucrose, not to elicitation. *Plant Sci* 160: 1043–1053.

Sampson, D., Bricker, B., Zhu, X.Y., Peprah, K., Lamango, N.S. & Setola, V. et al. (2014). Further evaluation of the tropane analogs of haloperidol. *Bioorg Med Chem Lett* 24: 4294–4297.

Advances in Energy Science and Equipment Engineering II – Zhou, Patty & Chen (Eds)
© 2017 Taylor & Francis Group, London, ISBN 978-1-138-71798-5

One-step hydrotreatment of *Jatropha oil* in the production of bioaviation kerosene over bifunctional catalyst of Pt/SAPO-11

S.P. Zhang, Y.B. Chen, Y.J. Hao, Q. Wang & L. Su
*The China-Laos Joint Lab for Renewable Energy Utilization and Cooperative Development,
Yunnan Normal University, Kunming, Yunnan, China*

Q. Liu & J.C. Du
Kunming Institute of Precious Metals, Kunming, Yunnan, China

B.Y. Han
Kunming University of Science and Technology, China

ABSTRACT: In this paper, preparation of bioaviation kerosene by one-step catalytic hydrotreatment of *Jatropha oil* over Pt/SAPO-11 in a high-pressure fixed-bed microreactor is described. Furthermore, the effect of reaction conditions, such as temperature, pressure, hydrogen/oil ratio, and space velocity, on the conversion of Jatropha oil, the selectivity of C_8–C_{16} hydrocarbons, and the isomer ratio of the C_8–C_{16} alkanes are investigated. The reaction pathway for the hydrodeoxygenation of Jatropha oil is also discussed. The experimental results showed that temperature of 397°C, H_2/oil ratio of 1000, pressure of 5 MPa and LHSV of 1.2h^{-1} are the best reaction conditions. Under these conditions, the conversion of Jatropha oil was 99.45%, the selectivity of the C_8–C_{16} hydrocarbons was 44.67%, and the isomer ratio of the C_8–C_{16} alkanes was 25.18%. With the increase in the reaction temperature or the decrease in the H_2/oil ratio and reaction pressure, the deoxygenation via decarboxylation/decarbonylation was enhanced.

1 INTRODUCTION

Since the 21st century, because of the excessive consumption of fossil fuels, environmental pollution (Lee et al. 2009), and the deteriorating quality of crude oil (Huber et al. 2006), people have been actively seeking renewable energy and clean raw material to maintain our industrialized society. Biomass, as a sole carbon fossil energy resource (Yang et al. 2013), has broad prospects in the field of clean energy production. In recent years, several types of fuels have been developed by utilizing biomass, which is used directly or added proportionally into gasoline or diesel oil, to replace traditional fossil fuels (Ayhan 2007, Hahn-Hagerdal et al. 2006). Aviation kerosene is composed of hydrocarbons containing 8–16 carbons (Fei et al. 2012); the main components are C_8–C_{16} *iso*-alkanes, a small part of the cyclic alkanes, olefins, and aromatics (Wu et al. 2010). Production of aviation kerosene is much more difficult than that of other fuels because of the stringent international standards of aviation kerosene, particularly the energy density and flow properties (freezing point and viscosity) (Rye & Wilson 2010). Transforming biomass to C_8–C_{16} *iso*-alkanes, cyclic alkanes, olefins, and aromatics to prepare biological aviation kerosene is the best

way to replace the traditional aviation kerosene (Emma et al. 2009, Michelle et al. 2006).

At present, a variety of biological aviation kerosene production technologies have been developed worldwide, including Fischer–Tropsch synthesis technology, hydrotreating technology, and bio-synthetic hydrocarbon technology (Liu et al. 2012, Adesina et al. 1995). The most advanced method is the two-stage hydrogenation process (Dong et al. 2013), which produces an intermediate product from the hydrodeoxygenation of animal and vegetable oils, further cracking and isomerization of which results biological aviation kerosene (Brodzki et al. 1993, Duan et al. 2013, Gusmão et al. 1989). However, the two-step method has drawbacks such as high hydrogen consumption and complex operation.

In recent years, some studies have indicated that the noble metal Pd and Pt had a significant effect (Siswati et al. 2008, Siswati et al. 2009, Bartosz et al. 2010) in terms of catalytic hydrocracking, which provided the possibility for the one-step hydrogenation to produce biological aviation kerosene, causing great concern. Congxin Wang and others (Wang et al. 2013) explored the effects of the reaction conditions on the yield of *iso*-alkanes in the process of the one-step hydrotreatment of soybean over Pt/SAPO-11, and the results showed that temperature and airflow rate had significant impact on *iso*-alkanes yield. The United States Advanced Bio-energy Technology Co., Ltd (ABT) used low content of triglyceride coconut oil as raw material and ABT patent catalyst to produce biological aviation kerosene by one-step hydrogenation method (Xu 2013). These laid the foundation for the production of biological aviation kerosene by one-step method. However, according to the studies conducted by Congxin Wang Group, only a small amount of aviation kerosene components was produced, which needs to be improved.

In this paper, we studied the effects of temperature, pressure, hydrogen/oil ratio, space velocity, and other parameters on the conversion of Jatropha oil, the selectivity of C_8–C_{16} hydrocarbons, and the isomer ratio of the C_8–C_{16} alkanes in the production of biological aviation kerosene by one-step hydrogenation over bifunctional catalyst Pt/SAPO-11 using Jatropha oil as the feedstock. Meanwhile, we also analyzed the effect of reaction conditions on the decarboxylation/decarboxylation and hydrodeoxygenation pathways.

2 EXPERIMENT

2.1 *Material and reagents*

Jatropha oil, produced in Yunnan Shuang Bai, is analyzed by GC-MS; the relative content of fatty acid: $C_{16:0}$ (formula containing 16 carbons without C = C bonds), 17.65%; $C_{16:1}$, 0.91%; $C_{18:0}$, 7.58%; $C_{18:1}$, 40.18%; and $C_{18:2}$, 33.68%.

All chemical reagents used in this study are of analytical reagent (AR) grade. These include anhydrous sodium sulfate, methanol, concentrated sulfuric acid (West Long Chemical Co., Ltd.), petroleum ether, methylene chloride, ethanol (Tianjin Chemical Reagent Technology Co. sailboat company), and quartz sand (Tianjin Kermel chemical reagent Co., Ltd).

Pt/SAPO-11 catalyst: R&D group developed a certain amount of the active component Pt loaded on functionalized SAPO-11 zeolite by impregnation, high-temperature burning, cooling, and milling (Kabe et al. 2009, Moshe et al. 2015). The catalyst is activated for 6 h at 320°C/1 MPa before use.

2.2 *Instruments*

MRT-H00521JB high-pressure miniature fixed-bed catalytic reaction evaluation equipment (designed by R&D Group and manufactured by Yunnan Normal University), Clarus 680-Clarus SQ8T Gas-mass spectrometry (USA PerkinElmer, Co.), DHG-9203 A electric blast oven (Shanghai-Heng Science Instrument Co., Ltd.), HZQ-C double air bath temperature oscillator (Jintan Earth Automation

Instrument Factory), analytical balance (Shanghai Shun Yuheng Scientific Instruments Co., Ltd), 125 mL liquid separation funnel (Shanghai Wu Yi Glass Instrument Factory), and distiller (Friends of the Tianjin-Technology Co., Ltd).

2.3 *Methods*

2.3.1 *Characterization of catalyst*

XRD was tested by a Rigaku D/MAX-2000 automatic X-ray diffractometer. The experimental test conditions were as follows: radiation source, CuKα; wavelength, 0.15406 nm; tube current, 40 mA; tube voltage, 40kV (Du et al. 2015).

The catalyst surface area, pore volume, and average aperture were verified using a NOVA2000 e-type specific surface area and porosity analyzer obtained from the United States Contador company. The N_2 adsorption temperature was 77 K, and the surface area, pore volume, and average aperture were calculated using the BET specific instrument (Du et al. 2015).

2.3.2 *Preparation of biological aviation kerosene*

Biological aviation kerosene was prepared in high-pressure miniature fixed-bed catalytic reaction evaluation equipment. Catalyst (6 mL) was filled up in the reactor. Hydrogen gas at the desired pressure was injected into the reactor using a pressurizing pump. The reaction mixture was heated up to the reaction temperature. Then, the amount of oil intake was adjusted according to the airflow rate and volume. The flow rate of hydrogen flow was controlled by a mass flow meter according to the set value of the hydrogen/oil ratio. The obtained product was separated in a gas–liquid separator after condensed by a condenser. Then, the liquid was passed to an oil storage tank and finally, the reactor was emptied after alkaline cleaning and pickling emptying.

2.3.3 *Analysis*

The liquid product was encapsulated and inspected using methyl ester and its composition was determined by GC-MS. GC conditions: Column, CETM-5 (30 m × 0.25 mm × 0.25 μm); inlet temperature, 200°C; temperature program: initial temperature is 80°C, holding time: 2 min, heating to 280°C at a rate of 5°C/min, holding time: 8 min; and carrier gas (He) flow rate, 1.5 mL/min. MS conditions: electron bombardment (EI) ion source (electron energy, 70 eV; temperature, 250°C); transfer line temperature, 270°C; detection voltage, 0.9kV; mass scan range, m/z 32–500; and data acquisition time range, 1–50 min.

The conversion of Jatropha oil is defined as $X = [M_{(T0)} - M_{(TG)}]/M_{(T0)} \times 100\%$, where $M_{(T0)}$ and $M_{(TG)}$ are the mass percents of Jatropha oil before and after reaction, respectively; the selectivity of

C_8–C_{16} hydrocarbons is defined as $Y = \sum M(i)/[M_{(T0)} - M_{(TG)}] \times 100\%$, where $\sum M(i)$ is the total mass percent of C_8–C_{16} hydrocarbon components; the isomer ratio of the C_8–C_{16} alkanes is defined as $Z = \sum M(x)/[M_{(T0)} - M_{(TG)}] \times 100\%$, where $\sum M(x)$ is the total mass percent of *iso*-alkanes within the range C_8–C_{16}.

3 RESULTS AND DISCUSSION

3.1 *Catalyst characterization results*

Fig. 1 shows the XRD patterns of catalysts SAPO-11 and Pt/SAPO-11. It is evident from the figure that the modification of SAPO-11 had no obvious impact on the crystallinity of the carrier compared with that of SAPO-11. Table 1 shows the physicochemical properties of support and catalyst. It is evident from the table that the specific surface area, pore volume, and pore diameter of Pt/SAPO-11 catalyst were significantly decreased compared with those of the blank SAPO-11 carrier.

3.2 *Effects of catalyst on Jatropha oil hydrogenation*

The performance of catalytic hydrogenation of Jatropha oil for the production of biological aviation kerosene under various conditions and the composition of hydrocarbon fuel liquid product are shown in Table 2.

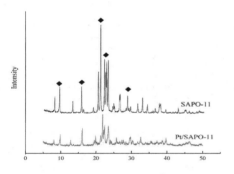

Figure 1. XRD patterns of SAPO-11 and Pt/SAPO-11 catalysts.

Table 1. Physicochemical properties of support and catalyst.

Support and catalyst	Surface area (m²/g)	Pore volume (cm³/g)	Average pore diameter (nm)
SAPO-11	259.83	0.26	4.07
Pt/SAPO-11	190.57	0.24	3.57

3.2.1 *Effect of temperature*

Table 2 shows the catalytic performance of one-step hydrotreatment of Jatropha oil under different reaction conditions. It is evident from the table that with the increase in the reaction temperature, the conversion of Jatropha oil first increased, then decreased, and kept at above 95%. Similarly, the selectivity of C_8–C_{16} hydrocarbons also increased first, then decreased, and reached the maximum value (34.28%) at 397°C; the *i/n* ratio quickly increased from 1.48 (397°C) to 2.09 (417°C), which indicated that the isomerization reaction was dynamic in the range of 397–417°C. Fig. 2 shows the *iso*-C_8–C_{16} yield and C_{17}/C_{18} ratio as a function of temperature. The yield of *iso*-C_8–C_{16} increased sharply from 357 to 397°C and then decreased slightly. This is due to the too high temperature, which increased the pyrolysis reaction and decreased the selectivity of the isomerization reaction (Liu et al. 2015). C_{17}/C_{18} ratio gradually increased from 1.03 (357°C) to 4.07 (417°C) and then decreased to 2.52 (437°C), which suggested that high temperature promotes the deoxygenation selectivity of Jatropha oil to decarboxylation/decarbonylation direction. Precisely, 397°C was the best reaction temperature.

3.2.2 *Effect of H_2/oil ratio*

It can be seen from Table 2 (3, 6–9) that the conversion of Jatropha oil first increased, then decreased with the increase in H_2/oil ratio, and reached the maximum value (99.16%) when the H_2/oil ratio was 1000. The selectivity of C_8–C_{16} hydrocarbons kept at about 35%, with H_2/oil ratio in the range of 1000–1400, and the yield of *iso*-C_8–C_{16} maintained at about 20% (Fig. 3), indicating that Pt/SAPO-11 catalyst had a higher cracking selectivity. The *i/n* ratio showed no significant change with the increase in the H_2/oil ratio, which suggested that this ratio had little effect on the isomerization of alkanes. C_{17}/C_{18} ratio significantly decreased with the increase in the H_2/oil ratio (Fig. 3), from 3.47 (H_2/oil, 800) to 1.39 (H_2/oil, 1600), which indicated that the low H_2/oil ratio favored the decarboxylation/decarbonylation of Jatropha oil. Precisely, 1000 was the best H_2/oil ratio.

3.2.3 *Effect of pressure*

It can be seen from Table 2 (3, 10–13) that the conversion of Jatropha oil was 99.16% and the selectivity of C_8–C_{16} hydrocarbons was 34.28% under 1.2h⁻¹, 397°C, 3 MPa, and H_2/oil = 1000. When the reaction pressure was increased to 5 MPa, the conversion of Jatropha oil and the selectivity of C_8–C_{16} hydrocarbons reached 99.45% and 44.67%, respectively. However, *i/n* ratio decreased with the increase in reaction pressure, which suggested that low pressure favored the isomerization reaction.

Table 2. Performance of the one-step hydrotreatment of Jatropha oil under different reaction conditions.

No.	Temperature/°C	Pressure/ MPa	H_2/oil ratio	LHSV/h^{-1}	Conversion /%	C_8–C_{16} hydrocarbons/%	i/n ratio (C_8–C_{16})
1	357	3	1000	1.2	99.11	20.89	1.49
2	377	3	1000	1.2	99.45	28.42	1.75
3	397	3	1000	1.2	99.16	34.28	1.48
4	417	3	1000	1.2	97.49	27.26	2.09
5	437	3	1000	1.2	94.31	26.2	1.73
6	397	3	800	1.2	95.51	28.22	1.68
7	397	3	1200	1.2	97.15	34.13	1.37
8	397	3	1400	1.2	93.84	35.8	1.32
9	397	3	1600	1.2	95.4	25.1	1.49
10	397	4	1000	1.2	97.63	40.91	1.5
11	397	5	1000	1.2	99.45	44.67	1.29
12	397	6	1000	1.2	95.52	34.18	1.02
13	397	7	1000	1.2	99.16	39.17	0.82
14	397	5	1000	0.8	90.84	29.04	1.25
15	397	5	1000	1.0	93.68	48.74	0.99
16	397	5	1000	1.4	97.35	30.47	2.54
17	397	5	1000	1.6	97.06	35.67	1.27

i/n ratio = ratio of isomers (i) to normal (n) alkanes.

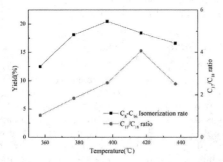

Figure 2. Yield of *iso*-alkanes and C_{17}/C_{18} ratio as a function of temperature.
Reaction conditions: P = 3 MPa, H_2/oil = 1000, LHSV = 1.2h^{-1}

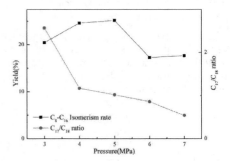

Figure 4. Yield of *iso*-alkanes and C_{17}/C_{18} ratio as a function of pressure.
Reaction conditions: T = 397°C, H_2/oil = 1000, LHSV = 1.2 h^{-1}.

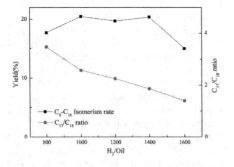

Figure 3. The *iso*-alkanes yield and the ratio of C_{17}/C_{18} as a function of H_2/oil ratio.
Reaction conditions: T = 397°C, P = 3 MPa, LHSV = 1.2h^{-1}

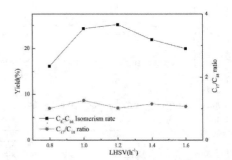

Figure 5. Yield of *iso*-alkanes yield and C_{17}/C_{18} ratio as a function of LHSV.
Reaction conditions: T = 397°C, P = 5 MPa, H_2/oil = 1000.

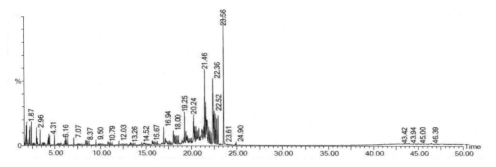

Figure 6. GC-MS spectra of the liquid product component.
Reaction conditions: $T = 397°C$, $P = 5$ MPa, H_2/oil = 1000, LHSV = 1.2 h^{-1}.

Fig. 4 shows the yield of iso-C_8–C_{16} yield and C_{17}/C_{18} ratio as a function of pressure. The isomerism rate of C_8–C_{16} alkanes increased from 20.47% at 3 MPa to 25.18% at 5 MPa and then decreased with the increase in reaction pressure; C_{17}/C_{18} ratio sharply decreased from 2.57 (3 MPa) to 1.17 (5 MPa) first, and then slightly decreased to 0.54 (7 MPa), which indicated that low pressure favored the decarboxylation/decarbonylation of Jatropha oil. Precisely, 5 MPa was the best reaction pressure.

3.2.4 Effect of LHSV

The results of the catalytic performance as a function of LHSV is presented in Table 2 (11, 14–17), which shows that the conversion of Jatropha oil reached the maximum value (99.45%) under 1.2h^{-1}, 397°C, 5 MPa, and H_2/oil = 1000. The selectivity of C_8–C_{16} hydrocarbons reached the maximum value (48.74%) when the LHSV was 1.0 h^{-1}, whereas when i/n ratio was only 0.99, a very low isomer-to-n-alkane ratio was not favorable for the isomerization reaction. The isomer ratio of the C_8–C_{16} alkanes first increased, then decreased with the increase of LHSV (Fig. 5), and reached the maximum value (25.18%) at 1.2 h^{-1}. C_{17}/C_{18} ratio was only slightly changed (remains at about 1.0) with the increase in LHSV, indicating that the ratio of decarboxylation/decarbonylation to hydrodeoxygenation did not change with the increase in LHSV. Precisely, 1.2 h^{-1} was the best reaction LHSV.

In conclusion, a higher conversion (99.45%) of Jatropha oil to aviation kerosene, 44.67% yield of C_8–C_{16} hydrocarbons, and 25.18% isomer ratio of the C_8–C_{16} alkanes were achieved at the temperature of 397°C, H_2/oil ratio of 1000, pressure of 5 MPa, and LHSV of 1.2 h^{-1}. The liquid product component is shown in Fig. 6.

4 CONCLUSIONS

In this paper, the effect of reaction conditions on the selectivity of C_8–C_{16} hydrocarbons and the yield of iso-C_8–C_{16} in the one-step hydrotreatment of Jatropha oil over bifunctional catalyst Pt/SAPO-11 is studied. The obtained liquid product has better aviation kerosene components at temperature 397°C, H_2/oil ratio 1000, pressure 5 MPa and LHSV 1.2 h^{-1}, resulting in 99.45% conversion of Jatropha oil, 44.67% yield of C_8–C_{16} hydrocarbons, and 25.18% isomer ratio of the C_8–C_{16} alkanes. In addition, the reaction conditions also affect the hydrodeoxygenation pathway of Jatropha oil. From the analysis of C_{17}/C_{18} ratio, it is evident that a high temperature, low H_2/oil ratio, and low pressure were favorable for the decarboxylation/decarbonylation of Jatropha oil.

ACKNOWLEDGMENT

This work was supported by the National Natural Science Foundation of China (No. 21266032).

REFERENCES

Adesina A.A., Hudgins R.R. & Silveston P.L. (1995). Fischer-Tropsch synthesis under periodic operation [J]. Catal. Today, 25(2), 127–144.

Ayhan D. (2007). Progress and recent trends in biofuels [J]. Prog. Energy & Comb. Sci., 33(1), 1–18.

Bartosz R., Päivi M. & Siswati L. (2010). Catalytic deoxygenation of tall oil fatty acids over a palladium-mesoporous carbon catalyst: a new source of biofuels [J].Topics Catal., 53(15–18), 1274–1277.

Brodzki D. & Djéga-Mariadassou G. (1993). Formation of alkanes, alkylcycloalkanes and alkylbenzenes during the catalytic hydrocracking of vegetable oils [J]. Fuel, 72(4), 543–549.

Dong P., Dong H. F. & Li J. Z. (2013). Present situation and development suggestion of preparation of bio-aviation kerosene by hydrogenation [J]. Petrochem. Technol. Appl., 31(6), 461–466.

Du J. C., Zhao Y. Y. & Xia W. Z. (2015). Effects of precious metals and supports on catalytic hydrodeoxygenation and hydrocracking performance over catalysts [J]. Rare metal mater. eng., 44 (9), 2210–2215.

Duan P. G., Jin B. B. & Xu Y. P. (2013). Thermo-chemical conversion of *Chlorella pyrenoidosa* to liquid biofuels [J]. *Bioresour. Technol.*, 133(4), 197–205.

Emma N., Kjell A. & Mikael H. (2009). Aviation fuel and future oil production scenarios [J]. *Energy Policy*, 37(10), 4003–4010.

Fei L., Harvind K. R. & Joshua H. (2012). Preparation of mesoporous silica-supported palladium catalysts for biofuel upgrades [J]. *J. Nanotechnol.*, 2012, 937–946.

Gusmão J., Brodzki D. & Djéga-Mariadassou G. (1989). Utilization of vegetable oils as an alternative source for diesel-type fuel: hydrocracking on reduced Ni/SiO₂ and sulphided Ni-Mo/γ-Al₂O₃ [J]. *Catal. Today*, 5(5), 533–544.

Hahn-Hagerdal B., Galbe M. & Gorwa-Grauslund M. F. (2006). Bio-ethanol-the fuel of tomorrow from the residues of today [J].*Trends in Biotechnol.*, 24(12), 549–56.

Huber G. W., Iborra S. & Corma A. (2006). Synthesis of transportation fuels from biomass: Chemistry, Catalysts, and Engineering [J]. *Chem. Rev.*, 106(9), 4044–4098.

Kabe T., Qian W. & Hirai Y. (2000). Hydrodesulfurization and hydrogenation reactions on noble metal catalysts: I. Elucidation of the behavior of sulfur on alumina-supported platinum and palladium using the 35S radioisotope Tracer method [J]. *J. Catal.*, 190(1), 191–198.

Lee D. S., Fahey D. W., Forster P. & Sausen R. (2009). Aviation and global climate change in the 21st century [J]. *Atmos. Environ.*, 43, 3520–3537.

Liu G. R., Yan B. B. & Chen G. Y. (2012). Review and prospect of the preparation technology of bio-aviation fuel [J]. *Biom. Chem. Eng.*, 46(3), 46–48.

Liu G. Z., Han L.J. & Shi Y. W. (2015). Selective hydroisomerization of medium fraction of Fischer-Tropsch Synthetic fuel over Pt/SAPO-11 for production of alternative Jet fuel [J]. *Petrochem. Technol.*, 44 (2), 144–149.

Michelle E. R., Harold W. G. & Sophie M. R. (2006). Characterization of microbial contamination in United States Air Force aviation fuel tanks[J]. *J. Industr. Microbiol. & Biotechnol.*, 33(1), 29–36.

Moshe R., Miron V. L. & Moti H. (2015). Conversion of vegetable oils on Pt/Al₂O₃/SAPO-11 to diesel and jet fuels containing aromatics [J]. *Fuel*, 161, 287–294.

Rye L., Blakey S. & Wilson C. W.(2010). Sustainability of supply or the planet: a review of potential drop-in alternative aviation fuels [J]. *Energy & Environ. Sci.*, 3(3), 17–27.

Siswati L., Irina S. & Anton T. (2008). Synthesis of biodiesel via deoxygenation of stearic acid over supported Pd/C catalyst [J].*Catal. Letters*, 54(122), 247–251.

Siswati L., Päivi M. & Irina S. (2009). Catalytic deoxygenation of stearic acid and palmitic acid in semibatch mode [J]. *Catal. Letters*, 130(1), 48–51.

Wang C. X., Liu Q. H. & Liu X. B. (2013). Influence of reaction conditions on one-step hydrotreatment of lipids in the production of *iso*-alkanes over Pt/SAPO-11[J].*Chin. J. Catal.*, 34(6), 1128–1138.

Wu W.G. & Huang J.K. (2010). Economic feasibility analysis of forest bio diesel raw material for leprosy tree planting [J]. *Chin. Rur. Econ.* 7, 56–63.

Xu T. (2013). Liquid biofuels: from fossil to biomass [M]. Beijing: Chemical Industry Press.

Yang C. Y., Nie R. F. & Fu J. (2013).Production of aviation fuel via catalytic hydrothermal decarboxylation of fatty acids in microalgae oil [J]. *Bioresour. Technol.*, 146(1), 569–573.

Advances in Energy Science and Equipment Engineering II – Zhou, Patty & Chen (Eds)
© 2017 Taylor & Francis Group, London, ISBN 978-1-138-71798-5

Effect of nanoparticle size on the breakdown strength of transformer oil-based Fe_3O_4 nanofluids

Y.Z. Lv & K. Yi
School of Energy, Power and Mechanical Engineering, North China Electric Power University, Beijing, China

X. Chen, B.L. Shan, C.R. Li & B. Qi
Beijing Key Laboratory of High Voltage and EMC, School of Electric and Electronic Engineering, North China Electric Power University, Beijing, China

ABSTRACT: In this study, monodisperse Fe_3O_4 nanoparticles with controlled size were prepared by a wet chemical method and used to synthesize transformer oil-based nanofluids. The dispersion stability of nanoparticles in nanofluids was examined by natural sedimentation and zeta potential measurement. The positive impulse breakdown strength of pure oil and three kinds of nanofluids was comparatively investigated. Experimental results show that the breakdown strength of nanofluids is significantly enhanced by the presence of Fe_3O_4 nanoparticles and the modification effect of nanoparticles is closely related to their size.

1 INTRODUCTION

Transformer oil as the most widely-used insulating liquid has been intensively investigated to enhance its dielectric strength and thermal property (C. Choi, 2008; V. Segal, 1998; C. Srinivasan, 2012; P. Kopcansky, 2005). Recently, various nanoparticles have been employed to improve the dielectric strength of transformer oil by dispersing them into the base oil (P. P. C. Sartoratto, 2005; M. Hanai, 2013; Y. F. Du, 2011; Y. Du, 2012; D. Mansour, 2012). It has been demonstrated that the breakdown strength of both mineral and vegetable transformer oil can be greatly increased with the presence of Fe_3O_4 nanoparticles (Sima, 2014; Bin Du, 2015). An electron-scavenging model is proposed and used to explain these phenomena based on modeling the electrodynamic process in Fe_3O_4 nanofluids (J. G. Hwang, 2010). The charge dynamics of Fe_3O_4 nanoparticle in transformer oil indicates that electron trapping of nanoparticles is the cause of higher electrical breakdown strength of nanofluids, which converts fast electrons from field ionization of oil into slow negatively charge carriers due to the electron charging of the nanoparticles. It is clarified that the modification effect of nanoparticles on the insulating property of transformer oil is mainly determined by basic physical property, size and concentration of nanoparticles (J. G. Hwang, 2010; J. C. Lee, 2012).

In the case of vegetable oil, size effect of Fe_3O_4 nanoparticles on its breakdown strength is studied (Bin Du, 2015). It was found that the calculated maximum electrical potential well depth for Fe_3O_4 nanoparticle is increased with the increasing of nanoparticle size from 10 nm, 20 nm to 30 nm. The increased electrical potential well depth in nanofluids could inhibit the free charge spread and enhance the breakdown strength of nanofluids. This is the reason why the nanoparticles with large size of 30 nm has the optimum modification effect on the breakdown performance of vegetable oil-based nanofluids. Both the electron-scavenging model and electrical potential well depth method are based on the polarization of Fe_3O_4 nanoparticles, which is greatly influenced by the physicochemical property of the liquid media (Jian Li, 2016). In comparison with the definite component of vegetable transformer oil, the composition of mineral transformer oil is quite complicated and diverse (Zou Ping, 2011). Therefore, it is of great necessity and importance to investigate size effect of Fe_3O_4 nanoparticles on the mineral transformer oil for further understanding the fundamental modification effect of nanoparticles.

In this paper, monodisperse Fe_3O_4 nanoparticles with different sizes were synthesized by a wet chemical method using oleic acid as surfactant and used to prepare well-dispersed mineral transformer oil-based nanofluids. The dispersion stability of nanofluids was measured and analyzed according to zeta potential of nanoparticles. Size effect of nanoparticles on the breakdown strength of nanofluids were measured and discussed.

2 EXPERIMENTS

2.1 *Preparation of nanoparticles and nanofluids*

In typical procedure, iron oleate precursor and oleic acid with a mole ratio of 1:1 were mixed in 50 mL octadecene by vigorous stirring under nitrogen protection to remove oxygen. Then, the resulting mixture was heated to 320°C with 1 h, 6 h, and 12 h, respectively. After cooling down to room temperature, the nanoparticles were subsequently centrifuged and washed several times with ethanol and cyclohexane.

Fe_3O_4 nanofluids were prepared by dispersing nanoparticles into the Mineral oil (25# Karamay) by stirring and ultrasonic treatment with the volume fractions of 0.2‰. All the nanofluids and the base oil were degassed at less than 1 kPa for 2 hours before measurement.

2.2 *Characterization*

The X-ray Diffraction (XRD) patterns were measured by a powder X-ray diffractometer equipped with a Cu-Kα radiation source at a step width of 0.02°. The morphology of the as-prepared powders was characterized by transmission electron microscopy (JEM-2100F). The dispersion stability of nanoparticles in the nanofluids was recorded by comparing the sediment of nanoparticles in the fresh and aged nanofluids after aging at room temperature for 15 days. A dynamic light scattering device (Malvern Nano ZS90) was used to determine the zeta potential of nanoparticles in the nanofluids.

Impulse breakdown voltage was tested according with IEC 60897 standard at room temperature. A 10 stages impulse generator was used to provide 1.2 μs/50 μs standard lightning impulse. The initial standing time was 5 minutes and the time interval between two successive shots was fixed at 1 minute. After one breakdown occurs, the oil sample and electrodes were changed. In this paper, 5 impulse breakdown tests were carried out for each oil sample.

3 RESULTS AND DISCUSSION

3.1 *Structure and morphology of nanoparticles*

Figure 1 shows the XRD pattern of Fe_3O_4 nanoparticles obtained by reacting for 12 h. The diffraction pattern and relative intensities of all diffraction peaks match well with those of JCPDS card no. 79-0419 for magnetite. The as-synthesized product can be easily indexed to the spinel structure of Fe_3O_4. The 2 theta values of 30.1°, 35.5°, 43.1°, 56.9°, and 62.6° are signatures of (220),

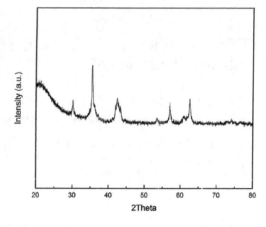

Figure 1. XRD patterns of as-prepared Fe_3O_4 nanoparticles.

(311), (400), (511), and (440) crystal face for Fe_3O_4, respectively.

The morphology of the as-prepared nanoparticles is depicted in Figure 2. The TEM image clearly demonstrates that the nanocrystals are highly uniform in particle-size distribution and well dispersed without any agglomeration. The average sizes of nanoparticles reacting for 1 h, 6 h and 12 h are 17.4 nm, 20.7 nm and 25.1 nm, respectively.

3.2 *Dispersion stability of nanofluids*

Natural sedimentation is the most immediate way to judge the dispersion stability of nanofluids. The images for three kinds of fresh Fe_3O_4 nanofluids are shown in Figure 3(a). The nanofluids are transparent with a color of pale yellow. There is no any visible agglomeration of the nanoparticles after aging for 15 days. These indicate that Fe_3O_4 nanoparticles are well dispersed in the nanofluids and exhibit a good dispersion stability.

To evaluate the dispersion stability of nanofluids, zeta potential measurement was also carried out for nanofluids. The zeta potential data of nanoparticles in the three kinds of nanofluids are 43.7 mV, 23.7 mV and 15.4 mV, respectively. It can be seen that the zeta potential is increased with the decreasing of nanoparticle size, which is well in agreement with the observation in vegetable oil-based nanofluids [12]. This result further demonstrates that the as-prepared nanofluids have a good dispersion stability.

3.3 *Breakdown strength of nanofluids*

The positive impulse breakdown voltages of the base oil and nanofluids are depicted in Figure 4.

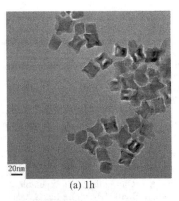

(a) 1h

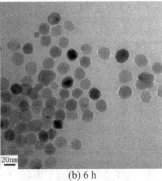

(b) 6 h

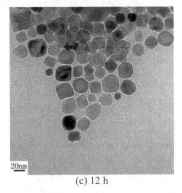

(c) 12 h

Figure 2. TEM images for as-prepared Fe_3O_4 nanoparticles.

It can be seen that the breakdown strength of nanofluids are greatly affected by the size of Fe_3O_4 nanoparticles. The nanofluid containing nanoparticles of 17.4 nm exhibits the highest breakdown voltage of 86.7 kV, which is enhanced by 10% compared to that of the base mineral oil. With the presence of nanoparticle of 20.7 nm, the breakdown voltage is only increased by about 5%. While the improvement is reduced to 1.3% for nanofluids containing nanoparticles of 25.1 nm. These results are consistent with the previous report, the increase

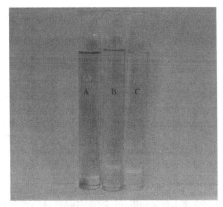

(a) freshly prepared nanofluids

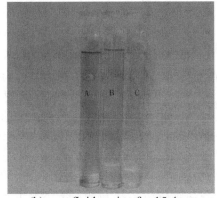

(b) nanofluids aging for 15 days

Figure 3. Dispersibility of three kinds of Fe_3O_4 nanoparticles in nanofluids (A) 17.4 nm, (B) 20.7 nm, (C) 25.1 nm.

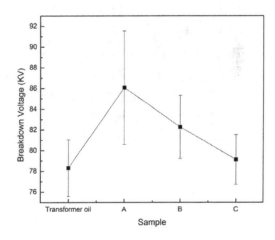

Figure 4. Positive impulse breakdown voltage for the base oil and nanofluids (A) 17.4 nm, (B) 20.7 nm, (C) 25.1 nm.

of nanoparticle size will weaken the modification effect on breakdown strength of nanofluids.

4 CONCLUSION

Monodisperse Fe_3O_4 nanoparticles with different sizes have been synthesized by a simple wet chemical method. The Fe_3O_4 nanofluids exhibit a good dispersion stability. The breakdown strength of nanofluids is improved by all three kinds of nanoparticles and significantly affected by the size of nanoparticles. The smaller nanoparticles in nanofluids, the higher the dispersion stability and breakdown strength of nanofluids. The enhancement of 10% in positive impulse breakdown strength is achieved by the presence of small size nanoparticles.

ACKNOWLEDGMENTS

The authors would like to thank National Natural Science Foundation of China for supporting this research under Contract No. 51337003, 51472084, and 51477052, and the Fundamental Research Funds for the Central Universities (JB2015019).

REFERENCES

Bin Du, Jian Li, Feipeng Wang, Wei Yao & Shuhan Yao (2015). Influence of Monodisperse Fe_3O_4 Nanoparticle Size on Electrical Properties of Vegetable Oil-Based Nanofluids. *Journal of Nanomaterials*. 560352.

Choi C., Yoo H.S. & Oh J.M. (2008). Preparation and heat transfer properties of nanoparticle-in-transformer oil dispersions as advanced energy efficient coolants. *Current Applied Physics 8(6)*.710–712.

Du Y., Lv Y., Li C., Chen M., Zhong Y., Zhou J., Li X. & Zhou Y. (2012). Effect of semiconductive nanoparticles on insulating performances of transformer oil. *IEEE Trans. Dielectr. Electr. Insul.* 770–776.

Du Y.F., Lv Y.Z., Li C.R., Chen M.T., Zhou J.Q., Li X.X., Zhou Y. & Zhong Y.X. (2011). Effect of electron shallow trap on breakdown performance of transformer oil-based nanofluids. *J. Appl. Phys.* 104104.

Du Y.F., Lv Y.Z., Wang F.C., Liu X.X. & Li C.R. (2011). Effect of TiO_2 nanoparticles on the breakdown strength of transformer oil. *in Conf. Rec. 2010 IEEE Int. Symp. Electrical Insulation (ISEI)*, 1–3.

Hanai M., Hosomi S., Kojima H., Hayakawa N. & Okubo H. (2013). Dependence of TiO_2 and ZnO nanoparticle concentration on electrical insulation characteristics of insulating oil. *In 2013 Annu. Rep. Conf. Electrical Insulation and Dielectric Phenomena (CEIDP)*. 780–783.

Hwang J.G., Zahn M., Osullivan F.M., Pettersson L.A.A., Hjortstam O. & Liu R. (2010). Effects of nanoparticle charging on streamer development in transformer oil-based nanofluids. *J. Appl. Phys.* 107.

Jian Li, Bin Du, Feipeng Wang, Wei Yao & Shuhan Yao. (2016). The effect of nanoparticle surfactant polarization on trapping depth of vegetable insulating oil-based nanofluids. *Physics Letters A*. 604–608.

Kopcansky P., Tomco L., Marton K., Koneracka M., Timko M. & Potocova I. (2005). The dc dielectric breakdown strength of magnetic fluids based on transformer oil. *J. Magn. Magn. Mater.* 415–418.

Lee, J.C. Lee W.-H., Lee S.-H. & Lee S. (2012). Positive and negative effects of dielectric breakdown in transformer oil based magnetic fluids. *Materials Research Bulletin*. 2984–2987.

Mansour D., Atiya E., Khattab R. & Azmy A. (2012). Effect of Titania Nanoparticles on the Dielectric Properties of Transformer Oil-Based Nanofluids. *IEEE Conf. Electr. Insul. Dielectr. Phenom. (CEIDP)*. 295–298.

Sartoratto P.P.C., Neto A.V.S., Lima E.C.D., Rodrigues de Sá A.L.C. & Morais P.C. (2005). Preparation and electrical properties of oil-based magnetic fluids. *J. Appl. Phys.* 917.

Segal V., Hjorstberg A., Rabinovich A., Nattrass D. & Raj K. (1998). AC and impulse breakdown strength of a colloidal fluid based on transformer oil and magnetite nanoparticles. *IEEE Int. Symp. Electrical Insulation (ISEI)*. 619–622.

Sima, Wen-Xia; Cao, Xue-Fei; Yang, Qing; Song, He & Shi, Jian (2014). Preparation of Three Transformer Oil-Based Nanofluids and Comparison of Their Impulse Breakdown Characteristics. *Nanoscience and Nanotechnology Letters*. 250–256.

Srinivasan C. & Saraswathi R. (2012). Nano-Oil with High Thermal Conductivity and Excellent Electrical Insulation Properties for Transformers. *Curr. Sci. 102(10)*. 1361–1363.

Zou, Ping; Li, Jian; Sun, Cai-Xin; Zhang, Zhao-Tao & Liao, Rui-Jin (2011). Dielectric Properties and Electrodynamic Process of Natural Ester-Based Insulating Nanofluid. *Mod. Phys. Lett. B 25*. 2021–2031.

Advances in Energy Science and Equipment Engineering II – Zhou, Patty & Chen (Eds)
© 2017 Taylor & Francis Group, London, ISBN 978-1-138-71798-5

Extracting caproic acid based on chemical complexation

Y.F. Peng, J. Tan & W.S. Deng
School of Chemical Engineering and Environment, Beijing Institute of Technology, Beijing, China

ABSTRACT: In this paper, extraction of caproic acid from dilute solution was investigated. Tributyl phosphate and trioctylamine (TOA) were selected as the solvents, while *n*-octanol and kerosene were selected as diluents, to extract caproic acid from water. The effects of the initial concentration of caproic acid, the concentration of trioctylamine, the phase ratio, and the temperature on the distribution coefficient of extraction equilibrium have been discussed. Based on these findings, a research plan for extracting caproic acid was designed using a multi-factor orthogonal test. The results showed that an extraction process based on chemical complexation shows a high capacity for dilute polar organic solutions. According to the orthogonal experimental results, the optimal scheme uses a TOA volume fraction of 30%, n-octanol volume fraction of 50%, and a temperature of 25°C.

1 INTRODUCTION

Caproic acid (hexanoic acid) is widely used in the synthesis of spices, as a medicine, flotation agent, and as an agent for thickening lubricating oil (Xue 2002). Caproic acid has a high boiling point. So its separation from dilute solution by distillation is extremely energy-intensive. Extraction of caproic acid is an alternative process that offers the advantages of high separation efficiency, large production capacity, and low energy consumption. Liquid-liquid extraction relies on the solute's distribution between two liquid phases (usually water and an organic solvent) to separate compounds (Tan 2011, Barta 2008, Tan 2013). In the extraction process, the phase ratio relies on the distribution coefficient between the aqueous phase and the organic phase (Tan et al. 2013). As is well known, physical extraction usually has a low distribution coefficient of extraction equilibrium for dilute polar organic solutions. Although increasing the polarity of the solvent can increase the value of the distribution coefficient, it can lead to an increase in the solvent's solubility in water, causing loss of the solvent. Complex extraction is a chemical extraction process that overcomes some of the drawbacks of physical extraction and thus has a wide range of industrial applications (Li et al. 2009). (Zhao et al. 1995) found that different ratios of trioctylamine, *n*-octanol, and kerosene were able to extract succinic acid from dilute solutions, with the best extraction provided by a mixture of 20% trioctylamine + 30% *n*-octanol + 50% kerosene. (Datta et al. 2013) studied that TOA dissolved in five different diluents extracted pyridine-3-carboxylic acid (NA) and pyridine-4-carboxylic acid (iNA) from

aqueous solution. Their study reported that for both of the above acids, the extraction ability of diluents with TOA varied in the order chloroform > decan-1-ol > isopropyl methyl ketone (MIBK) > methylbenzene > dodecane. (Li et al. 2002) investigated the extraction equilibrium for the extraction of maleic acid by TOA, reporting that TOA gave a high distribution coefficient in dilute aqueous maleic acid solutions. (Guan et al. 2001) investigated the extraction of succinic acid from its dilute aqueous solution, by mixtures of tributyl phosphate and trioctylamine (7301) as organic solvents, and toluene, *n*-octanol, MIBK, and kerosene as diluents. They found that mixtures of complex extraction agents could obtain satisfactory separation effects, especially the mixture of trioctylamine (7301) and *n*-octanol. Although many authors have investigated the extraction equilibria of carboxylic acids, there have been relatively few studies on caproic acid. There is a Lewis acid group contained in caproic acid, which provides the possibility in separation it with complex extraction. However, the selection of extraction solvent and the optimization of the operation conditions, e.g. temperature, concentrations of different chemicals contained in extraction agent, which are both important for extraction of caproic acid, have not been reported yet.

We report the results of investigations of the extraction of caproic acid from water by tributyl phosphate and TOA as solvents, and *n*-octanol and kero-sene as diluents. We investigated the effects of the initial caproic acid concentration, the trioctylamine concentration, the phase ratio, and the temperature on the distribution coefficient of the extraction equilibrium. With this data, we designed a research plan for investigating caproic acid extraction by the multi-factor orthogonal test.

2 EXPERIMENT SECTION

2.1 Materials

i. Caproic acid (>99%) was purchased from Tianjin Guangfu Fine Chemical Research Institute.
ii. Trioctylamine (>90%) was purchased from Shanghai Macklin Biochemical Co. Ltd.
iii. n-octanol (>98%) was purchased from Shanghai Zhanyun Chemical Co. Ltd.
iv. Kerosene was commercially available from Honsun Biological.
v. Tributyl phosphate was purchased from Tianjin Fuchen Chemical Reagents Factory.

All reagents were of AR grade.

2.2 Apparatus

A constant temperature oscillator (SHA-B) purchased from Changzhou Guohua Instrument Factory was used to control extraction temperature. An automatic potentiometric titrator (ZD-2) purchased from Shanghai Leici Instrument Factory was used to measure the concentration of caproic acid.

2.3 Procedure

During the experiments, the temperature was maintained at (25 ± 0.5) °C. Dilute caproic acid solutions, with different initial concentrations, were mixed with the complex extraction agent in a W/O ratio of 60:1 by volume and this mixture was put into constant temperature oscillator with water. To ensure that the extraction reached balance, the oscillation time was 90 minutes and the clarification time was 20 minutes. After the aqueous solution and organic phase were separated, the concentration of caproic acid in the water was analyzed by titration with standard sodium hydroxide solution. The concentration of caproic acid in the organic phase was calculated with conservation of mass. The relative deviation between calculated results and experimental results was less than 2%.

2.4 Choice of complex extraction agent

Caproic acid is a kind of Lewis acid, and thus requires a Lewis base for its complex extraction. Common Lewis-base complex extraction agents are high boiling species such as amines, phospholipids and phosphorus oxygen. The bond energy between these species and carboxylic acids is relatively low,

and thus regeneration of the extraction agent is facile. Phosphorus oxygen species, such as trialkyl phosphine oxide (TRPO), have the highest extraction capacity. However, their application in industrial extraction is limited by their high market value. In this paper, tributyl phosphate and trioctylamine (TOA) were selected as complex extraction agents.

Kerosene and n-octanol were added to the extraction agents as diluents, in order to improve the physical properties of the extraction mixture and to shorten the stratification time. The concentrations of caproic acid in the aqueous and organic phases were measured, and the equilibrium distribution coefficient was calculated as shown in Eq. (1). The results are shown in Table 1.

$$D = \frac{C_O}{C_W} = \frac{\text{Caproic acid concentration in organic phase}}{\text{Caproic acid concentration in aqueous phase}} \quad (1)$$

As shown in Table 1, the equilibrium distribution coefficient increased with the tributyl phosphate concentration. In most cases, kerosene gave a higher equilibrium distribution coefficient than n-octanol for the same tributyl phosphate concentration. Therefore, kerosene showed a higher capacity than n-octanol to act with tributyl phosphate to extract caproic acid from aqueous solution.

However, when using TOA, the distribution coefficient provided by n-octanol was much higher

Table 1. Distribution coefficient of caproic acid into different organic solvent (C_0 = 12.00 mmol/L).

Organic solvent	D mol/L/(mol/L)
20%TBP+80% kerosene	70.35±0.51
40%TBP+60% kerosene	139.58±1.18
60%TBP+40% kerosene	200.05±2.42
80%TBP+20% kerosene	253.84±2.28
20%TOA+80% kerosene	77.92±0.09
40%TOA+60% kerosene	118.96±0.84
60%TOA+40% kerosene	148.16±0.75
80%TOA+20% kerosene	163.99±0.76
20%TBP+80% n-octanol	118.39±0.34
40%TBP+60% n-octanol	132.89±1.26
60%TBP+40% n-octanol	164.33±1.52
80%TBP+20% n-octanol	217.25±2.32
20%TOA+80% n-octanol	371.11±3.13
40%TOA+60% n-octanol	655.80±0.11
60%TOA+40% n-octanol	587.30±1.46
80%TOA+20% n-octanol	396.17±0.72

than that of kerosene. This result shows that TOA is greatly influenced by the choice of diluent. When the TOA content was above 60%, the equilibrium distribution coefficients obtained were lower than for TOA contents of less than 60%. This decrease in distribution coefficient was probably due to the poor solubility of the complex formed by TOA and caproic acid. This also indicated that *n*-octanol plays an important role as a diluent for the TOA complex extraction process. At the same time, Table 1 indicated that TOA was a much more efficient extraction solvent since the distribution coefficient attained for TOA was much greater than for tributyl phosphate. Based on this finding, TOA was chosen for further investigation. From an economic point of view, in the system of TOA and n-octanol, much cheaper kerosene was added. Experimental results indicated that it could also give high distribution coefficient.

2.5 Effect of temperature on the equilibrium distribution coefficient

In the extraction process, the temperature is an important operating parameter. The change of temperature can shift the extraction balance so that the extraction equilibrium constant is changed. With an initial caproic acid concentration of 12.00 mmol/L, the temperature was varied between 25–55°C. The effect of this variation on the equilibrium distribution coefficient is shown in Figure 1. As shown in Figure 1, the plots of ln D versus 1/T were found to be linear, with R^2 equal to 0.994. Based on the linear fitting, we can obtain the enthalpy change (ΔH) of the extraction process. We calculated $\Delta H = -10.61$ kJ/mol. Since this enthalpy is negative, the extraction of caproic acid by TOA is exothermic, and thus the equilibrium distribution coefficient of complex extraction decreases with an

increase in temperature. Therefore, increasing the temperature is not only detrimental to the extraction operation but also increases the cost. Therefore, the optimum temperature for the extraction operation is 25°C.

2.6 Effect of complex agent volume fraction on the equilibrium distribution coefficient

The experiment was carried out at 25°C and atmospheric pressure. With the initial concentration of 15.00 mmol/L caproic acid as separate object, the effect of TOA volume fraction (10%-50%) on the equilibrium distribution coefficient for a certain volume fraction of n-octanol is shown in Figure 2. As shown in Figure 2, the equilibrium distribution coefficient gradually increases with increasing vol-

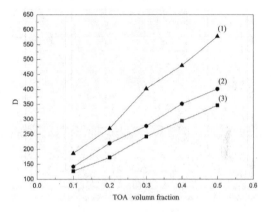

Figure 2. Effect of TOA volume fraction on the equilibrium distribution coefficient(1) organic phase: 50% volume fraction *n*-octanol in TOA (2) organic phase: 30% volume fraction n-octanol in TOA (3) organic phase: 20% volume fraction n-octanol in TOA.

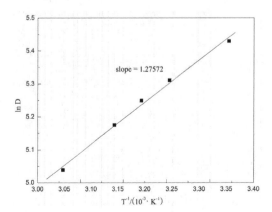

Figure 1. Effect of temperature on the equilibrium distribution coefficient.

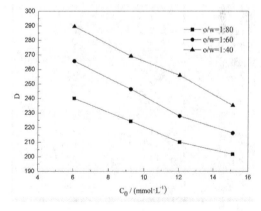

Figure 3. Effect of different initial concentration of caproic acid on the equilibrium distribution coefficient.

Table 2. Factors and levels of orthogonal test (Sun et al. 2009).

	Levels		
Factors	1	2	3
A: initial concentration of caproic acid (mol/L)	0.018	0.024	0.030
B: volume fraction of TOA (%)	10	20	30
C: volume fraction of n-octanol (%)	10	30	50
D: extraction temperature (°C)	25	35	45

ume fraction of TOA. Simultaneously, the equilibrium distribution coefficient gradually increases with increasing volume fraction of *n*-octanol. Thus, addition of *n*-octanol to the TOA improves the extraction of caproic acid.

2.7 Effect of initial concentration of caproic acid on the equilibrium distribution coefficient

The experiment was carried out at 25°C and atmospheric pressure. The initial caproic acid concentration was varied for three O/W phase ratios, and the

Table 3. L_{27} $(3)^4$ orthogonal test results.

NO.	A, initial concentration of caproic acid mol/L	B, volume fraction of TOA %	C, volume fraction of n-octanol %	D, extraction temperature °C	Distribution coefficient mol/L/(mol/L)
1	1	1	1	1	93.47±0.14
2	1	1	2	3	113.37±1.14
3	1	1	3	2	141.02±1.40
4	1	2	1	3	101.04±1.42
5	1	2	2	2	181.20±0.46
6	1	2	3	1	226.76±0.49
7	1	3	1	2	164.07±0.76
8	1	3	2	1	279.32±0.53
9	1	3	3	3	276.68±0.25
10	2	1	1	3	81.87±0.25
11	2	1	2	2	96.24±0.08
12	2	1	3	1	140.21±0.90
13	2	2	1	2	118.96±0.44
14	2	2	2	1	170.07±1.26
15	2	2	3	3	177.54±0.48
16	2	3	1	1	199.61±0.92
17	2	3	2	3	204.15±1.15
18	2	3	3	2	266.66±1.70
19	3	1	1	2	63.25±0.74
20	3	1	2	1	97.36±1.35
21	3	1	3	3	107.23±1.34
22	3	2	1	1	123.50±0.35
23	3	2	2	3	125.16±1.76
24	3	2	3	2	223.62±1.19
25	3	3	1	3	133.12±0.22
26	3	3	2	2	242.61±0.10
27	3	3	3	1	246.61±0.39
K_1	1576.94	934.02	1078.88	1576.90	4394.68
K_2	1455.30	1447.84	1509.48	1497.63	
K_3	1362.44	2012.82	1806.32	1320.15	
k_1	175.22	103.78	119.88	175.21	
k_2	161.70	160.87	167.72	166.40	
k_3	151.38	223.65	200.70	146.68	
R^*	23.83	119.87	80.83	28.53	

* refers to the result of extreme analysis.

observed equilibrium distribution coefficients, D, are shown in Figure 3.

As shown in Figure 3, increasing the initial concentration of caproic acid led to a decrease in the equilibrium distribution coefficient in each of the three tested O/W phase ratios. When the concentration of caproic acid increased from 6.00 mmol/L to 15.00 mmol/L, the equilibrium distribution coefficient decreased from 289.48 to 235.23 at the phase ratio of oil/water of 40:1. At the same time, with the increase in phase ratio of oil/water, the equilibrium distribution coefficient also decreased for the same initial concentration of caproic acid. Thus, the complex extraction method has advantages in separating lower concentration and phase ratios of caproic acid.

2.8 *The design of orthogonal test and results*

As our results show, it is not sufficient to consider only a single factor to determine the efficiency of the actual complex extraction. Therefore, we have varied the factors considered the most important for a desirable equilibrium distribution coefficient, i.e., the initial concentration of caproic acid, the volume fraction of TOA, the volume fraction of *n*-octanol, and the temperature. The result of orthogonal tests conducted to investigate these factors is listed in Table 2.

Each selected factor was tested using an orthogonal L_{27} $(3)^4$ test design. The statistical software SPSS 19.0 was used to analyze the results of the orthogonal tests, which are listed in Table 3. Although the maximum distribution coefficient obtained is 279.32, this extraction condition was not considered as the optimal scheme. By orthogonal analysis, the values of K, k, R are calculated by statistical software. The extremes of the four considered factors are 23.83, 119.87, 80.83, 28.53, so the order of influence of these factors on the distribution coefficient of caproic acid complex extraction is B > C > D > A. The maximum distribution coefficient is A1B3C3D1 (0.01800 mol/L, 30%, 50% and 25°C). According to the result of a variable table (data not shown), an extra experiment should be done. According to these test results, the maximum obtainable distribution coefficient is 503.48, greater than the value of 279.32 found earlier. Therefore, this scheme is the optimal scheme.

3 CONCLUSION

In this paper, the abilities of tributyl phosphate and trioctylamine to extract caproic acid from dilute aqueous solutions were investigated. Compared with tributyl phosphate, trioctylamine gave

much greater equilibrium distribution coefficients. According to the orthogonal experimental results and extreme analysis, the optimal scheme for extraction in the investigated factors used a TOA volume fraction of 30%, *n*-octanol volume fraction of 50%, and a temperature of 25 °C.

ACKNOWLEDGEMENT

We would like to acknowledge the support of the National Natural Science Foundation of China (21206081), Beijing Institute of Technology Research Fund Program for Young Scholars for this work.

REFERENCES

Barta H.J., Drumm, C. & Attarakih M.M. (2008). Process intensification with reactive extraction columns. J. Chemical Engineering Process, 47, 745–754.

Dipaloy Datta & Sushil Kumar (2013). Reactive Extraction of Pyridine Carboxylic Acids with N, N-Dioctyloctan-1-Amine:Experimental and Theoretical Studies. J. Separation Science and Technology, 48, 898–908.

Guan G.F., Ma X.L., Lu J.A., Yao H.Q. (2001). Extraction of succinic acid from its dilute solution based on chemical complexation. J. Journal of Nanjing University of Chemical Technology, 23, 33–37.

Li D.L., Liu Q., Chang Z.X. & Zhang Z.J. (2009). Development of extraction technique based on chemical complexation for polar organic diluents separation. J. Chemical Industry and Engineering Progress, 28, 13–18.

Li Z.Y., Qin W., Li F., Dai Y.Y. (2002). Extraction of maleic acid by reversible chemical complextion (I) extraction equilibrium. J. Journal of Chemical Industry and Engineering, 53, 729–732.

Sun Y.X., Li Y.J., Tong H.B, Yang X.D., Liu J.C. (2009). Optimization of extraction technology of the Anemone raddeana polysaccharides (ARP) by orthogonal test design and evaluation of its anti-tumor activity. J. Carbohydrate Polymers, 75, 575–579.

Tan J. Liu, Z.D., Lu Y.C., Xu J.H. & Luo G.S (2011). Process intensification of H_2O_2 extraction using gas-liquid-liquid microdispersion system. J. Separation and Purification Technology, 80, 225–234.

Tan J., Lu Y.C., Xu J.H. & Luo G.S. (2013). Modeling investigation of mass transfer of gas–liquid–liquid dispersion systems. J. Separation and Purification Technology, 108, 111–118.

Xue L.H., Li, Y.Z. & Huang, W.H. (2002). The application of capillary gas chromatography method to the synthesis of hexanoic acid. J. Journal of Jilin Institute of Chemical Technology, 19, 8–9.

Zhao H., Yang Y.Y. & Dai Y.Y. (1995). Extraction of Succinic acid from dilute solution by chemical complexation. J. Journal of Tsinghua University (Sci & Tech), 35, 43–48.

Advances in Energy Science and Equipment Engineering II – Zhou, Patty & Chen (Eds)
© 2017 Taylor & Francis Group, London, ISBN 978-1-138-71798-5

Research progresses on liquid propellant ignition and combustion performance

Lan Ma, Xuan-jun Wang, Yue Zhang & Bo-lin Zha
Xi'an Research Institute of High-tech, Xi'an, China

ABSTRACT: The new technical hot points, liquid propellant ignition and combustion process analysis, are being increasingly studied, mainly involving the experimental measurement and calculation of the ignition delay time, the ignition characteristics experiment and the characteristics of combustion flame temperature distribution related to the combustion mechanism and thermodynamics theory analysis. Especially, a laser-based diagnostic technique accurately measures the chemical delay time imbedded within the ignition delay time, which identifies several distinct regions within the classical ignition delay time for hypergolic reactions. Otherwise, the flame temperature estimations in the near-flame field using thin-filament pyrometry, which performed three representative experiments of the hypergolic ignition of MMH and RFNA, show the time-dependent spatially averaged or maximum flame temperatures at each filament height. The advanced studies from ignition and combustion performance analysis provide important data for quantitative comparison with hypergolic ignition models, oxidation transformation in the course of reservation, and future studies of alternative hypergolic propellants of practical interest.

1 INTRODUCTION

The liquid propellant ignition and combustion process in the rocket engine are compactly related to the energy performance of the propellant, showing the combustion kinetics mechanism and thermodynamics theory analysis through ignition delay time and characteristics of combustion flame temperature distribution. When the constituent of propellant are remodeled or transferred, the data of ignition and combustion process analysis would respond obviously.

Unsymmetrical dimethylhydrazine, a kind of methyl derivatives of HZ, having a great caloric value of combustion and high-density impulse, is widely used as the main body of liquid rocket and missile burners. UDMH can be slowly oxidized by little air or O_2 at room temperature. At the same time, with strong moisture absorption, it can react with CO_2 from the air, further reducing fuel quality of UDMH and declining the performance of the propellant energy. Therefore, it is more necessary to research in the ignition and combustion performance of the UDMH, which is long time reserved, in favor of controlling the propellant combustion stability and improving the reliability of the propulsion system.

2 RESEARCH IN THE IGNITION PROCESS

2.1 The ignition delay time

For the binary propellant ignition, the ignition delay time is the time interval during which the oxidizer and fuel come into contact to produce the flame, which often lasts 0–50 ms. Ignition processes include a series of physical and chemical processes. Physical delay time includes gas-phase or liquid-phase fuel preheating, diffusion, and mixing processes. Chemical delay time refers to the time in front of the formation of the flame, determined by the chemical reaction dynamics (Cho. K, 2010). A laser-based diagnostic technique, which measures ignition delay time, identifies several distinct regions within the classical ignition delay time for hypergolic reactions, as the first optical drop test laboratory technique to accurately measure the chemical delay time imbedded within the ignition delay time. We can finish it through lowering a droplet of fuel into a small pool of oxidizer contained within a quartz tube, as shown in Figure 1.

It should be pointed out that the determination results of ignition delay time vary with experimental conditions, such as open cup method and drop method. Otherwise, using commercial software such as CHEMKIN4.1, the ignition delay time can

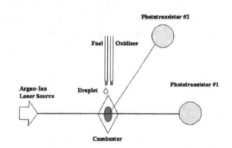

Figure 1. Phototransistor placement.

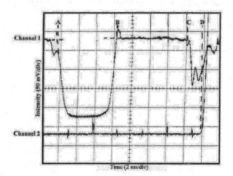

Figure 2.   Typical result of drop test for UMDH/RFNA.

be calculated from the reaction activation energy (Hong-bin Hu, 2011) so that it can calculate the concentration of reactants and temperature change with time and carry out the numerical simulation study on the combustion reaction process (Zhao Wei, 2012). And the model prediction accuracy depends on the chemical reaction mechanism, as shown in Figure 2.

2.2   *Experiment study on the ignition characteristics*

For liquid rocket engine, the main ignition characteristic parameters include combustion chamber ignition and starting (i.e., the time t_{90} when the combustion chamber pressure calibrates 90%), shutdown features (i.e., the time t_{10} when the interior pressure drops to calibrate 10%), and the combustion efficiency of the combustion chamber η_c.

Hang drop device and high-speed photography system are the important instrumentality for observing the influence of the interaction on the ignition process of the liquid droplets, organizing the quantitative test on average in droplet group, including ignition temperature, ignition delay and the environment temperature, and the relationship between droplet center spacing. The researching groups of Nanjing University of Science and Technology (Wang Yu-qiang, 2009; Li Ming, 2008; Pan Yu-zhu, 2013), Xi'an Institute of Aerospace Power (Meng-zheng Zhang, 2012), National University of Defense Technology (Chen Yao-heng, 2008), and Liberation Army Equipment Institute (Wan-sheng Nie, 2013), all acquired some evolvement in the liquid and the gel fuel droplet combustion process.

3   COMBUSTION PERFORMANCE RESEARCH

3.1   *Infrared imaging temperature measuring*

Ordinarily, the liquid propellant combustion flame temperature is commonly measured by thermocouple by inserting or surface pasting method. To insert the thermocouple, researchers need to punch on the surface of the object to be tested, which is generally not allowed, whereas surface-pasting-type thermocouple often runs off in the engine experiment due to vibration environment and other factors, so the measurement error is also very big.

However, the essence of the infrared thermal radiation temperature measurement is to determine the temperature by receiving radiation energy of the object. The sensitive elements of infrared thermal imager accept thermal radiation emitted by objects, convert to electrical signals, and feed certain temperature information into a circuit. It can simultaneously measure all the temperature information in visual range. Significantly, using the color index table to indicate the temperature, infrared thermal imager corresponds each temperature to one color within the selected scope and then shows the thermal image on the screen (Qian-qian Zhu, 2013). Its measuring principle by the Planck radiation law mathematical description is given by formula (1). Blackbody radiation and their spectral distribution are related to the temperature and wavelength. For the general object, introducing monochromatic emissivity $\varepsilon_\lambda = \varepsilon(\lambda, T)$, we have:

$$E(\lambda, T) = \varepsilon(\lambda, T) \cdot E_b(\lambda, T).$$
$$E_b(\lambda, T) = \frac{c_1}{\lambda^5 \left[e^{c2/\lambda T -1} \right]} \qquad (1)$$

3.2   *Infrared monitoring liquid propellant flame*

At Purdue University, researchers (Erik M. Dambach, 2010) studied the dynamic characteristics of the propellant combustion process, monitoring the flame radiation intensity online, by infrared imaging method based on the thin-filament pyrometry. They designed and developed airtight reaction vessel, which mix oxidant and burners with a certain proportion, and observed the combustion flame thermal radiation strength online with infrared camera, including the flame shape and duration and the relation between the flame along the height and temperature. The same set of methyl hydrazine nitrate/red fuming nitric acid combustion process of infrared image at different times are recorded as shown in Figure 3.

Thin-Filament Pyrometry (TFP) technology (Erik M. Dambach, 2012), developed in the late 1980s, is a well-established technique for estimating temperature profiles in laminar flames. This diagnostic technique typically involves placing a thin Silicon Carbide (SiC) fiber in a flame, measuring the radiation intensity emitted from the fiber, and calibrating the intensity measurements to determine a temperature profile along the fiber. β-SiC fiber has high strength and heat resistance (about

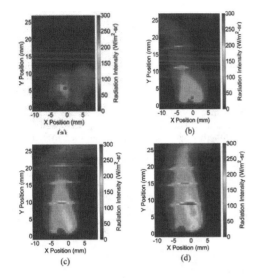

Figure 3. MMH/RFNA combustion infrared image at different time (a: 30.5 ms; b: 141.0 ms; c: 307.5 ms; d: 418.5 ms).

up to 2400 k), and can go through any hot gas, so it can produce the information along the profile of the temperature, reducing the disturbance to the fluid in the greatest degree.

Goss, Vilimpoc, and coworkers (1989) first developed and applied thin-filament pyrometry to obtain temperature estimates in an unsteady laminar diffusion flame with a spatial and temporal resolution of approximately 200 μm and 1 ms, respectively. Since the initial development of this technique, it has been applied extensively to laminar diffusion (Blunck et al., 2009; Maun et al., 2007), partially premixed, and countered flow flames. In addition, thin-filament pyrometry studies of turbulent premixed flames and microgravity droplet combustion have been considered.

3.3 Combustion mechanism and thermodynamics theory analysis

According to the energy theory, the calorific value of the propellant depends on the difference of heat between the propellant and the products of combustion. Therefore, the propellant having higher calorific value can be confirmed through thermal calculation, which can connect compound structure with energy.

At present, there are mainly two problems about the combustion numerical simulation, one is the too great amount of calculation due to the complexity of the combustion reactions, whose particular model often includes hundreds or thousands of reactions, so the flow equations coupling with chemical reaction kinetics model engender a huge amount of calculation. This calculation is more complex when the numerical simulation should be structured for the turbulence of premixed reaction process in engineering, even beyond the existing computing power. The other problem is the stability of the calculation. Because of the particularity of the chemical reaction process, the concentrations and residence time of different components have more difference on the order of magnitude, especially the reactions with some intermediate component. The different elementary reaction in the detailed chemical reaction kinetics, whose characteristic timescale persists from $10^0 S$ to $10^{-10} S$, causes the strong rigidity of differential equation and influences the stable convergence of the calculation as the greater difficulty. Therefore, the reaction kinetics mechanism must be simplified to a certain extent so as to reduce the amount of calculation and increase calculation efficiency. In the process of numerical simulation, researchers are concerned about some important parameters characterizing by combustion process, such as the ignition delay time, the flame propagation speed, and the concentration change of some major components with its effect on the reaction process. Under the condition of little error introduced by these important parameters and the detailed mechanism, we can significantly reduce the amount of calculation and calculation time. This is the target of simplification when the simplified mechanism is used for calculation. According this, researchers from Shanghai Jiao Tong University (Liu He, 2012) and Zhe Jiang University (Zhou Hua, 2012) successfully predigested and optimized the chemical kinetics mechanism of hydrocarbon fuels and the boron propellant combustion products.

4 CONCLUSION

In conclusion, with the new technology such as thin-filament pyrometry and a laser-based diagnostic technique, the liquid propellant ignition and combustion process analysis can be enhanced, which has practical significance for energy performance analysis of liquid propellant.

REFERENCES

Chen Yao-heng. 2008. Gel propellant single droplet combustion theory and experimental research. *National University of Defense Technology*: Chang Sha.
Cho. K, Pourpoint. T, and Heister, S, 2010. Ignition of the Advanced Hypergolic Propellants. *The 46th AIAA/ASME/SAE/ASEE be Propulsion Conference*: 25–29, Nashville, TN.
Erik M. Dambach, Brent a. Rankin. 2012 Temperature Estimations in the Near—Flame Field Resulting from Hypergolic Ignition Using Thin—Filament Pyrometry.

Journal of Purdue University. Combustion Science and Technology: 184:2, 205–223.

Erik M. Dambach. 2010. Ignition of Hypergolic Propellants Ph. D. *Dissertation, School of Aeronautics and Astronautics, Purdue University:* West Lafayette, IN.

Hong-bin Hu, 2011. Non-equilibrium plasma fuel of low calorific value gas fuel experimental research.: *Institute of Engineering Thermal Physics of Chinese Academy of Sciences*, Beijing.

Li Ming. 2008. Liquid propellant electric ignition characteristics research. *Nan Jing University of Science and Technology*: Nan Jing.

Liu He. 2012. The combustion reaction mechanism based on sensitivity analysis and genetic algorithm to simplify and optimize.*Shang Hai Jiao Tong University*: Shanghai.

Meng-zheng Zhang, Zhi-chao Hao, Zhang Mei. 2012, Carbon partial ignition and combustion characteristics of dimethyl hydrazine. *Journal of combustion science and technology*: 17 (4): 315–318.

Pan Yu-zhu. 2013. HAN based liquid propellant high pressure combustion characteristic of experiment research and numerical simulation. *Nan Jing University of Science and Technology*.: Nan Jing.

Qian-qian Zhu. 2013. Conclusion ceramic porous media combustion temperature and flame surface movement characteristics experimental study.: *Zhe Jiang University*: Hang Zhou.

Wang Yu-qiang. 2009. Liquid propellant liquid fog sequence pulse electric ignition characteristics research. *Nan Jing University of Science and Technology*: Nan Jing.

Wan-sheng Nie, He Bo, Ling-yu Su. 2013. Organic gel characteristics and influencing factors of partial dimethyl hydrazine droplet ignition and combustion experimental research. *Journal of experimental fluid mechanics*: 27 (4): 23 to 30.

Zhao Wei, 2012. Composite bottom row of propellant ignition combustion characteristic research.: *Nan Jing Unive-rsity of Science and Technology*: Nan Jing

Zhou Hua. 2012. Boron propellant combustion products once the ignition combustion characteristic research.: *Zhe Jiang University*: Hang Zhou.

Advances in Energy Science and Equipment Engineering II – Zhou, Patty & Chen (Eds)
© 2017 Taylor & Francis Group, London, ISBN 978-1-138-71798-5

Design and application of modified particle analyzer on fire-fighting foam

X.C. Fu, J.J. Xia, Z.M. Bao, T. Chen, X.Z. Zhang, Y. Chen, R.J. Wang, C. Hu & L.S. Jing
Tianjin Fire Research Institute of MPS, Tianjin, China

ABSTRACT: The purpose of this study is to recognize the foam boundary using a modified particle analyzer. First, the foam graphs were captured by a stereomicroscope. Because of the aided light, the foam's brightness increased. Second, the traditional particle analyzer was modified to improve the identification effect. Then, more foam could be recognized automatically by introducing Canny operator. Third, recognizing test was carried out using modified particle analyzer. The variation of different parameters with time was compared and analyzed. The results show that this modified particle analyzer could be used to characterize foam parameters, including foam particle size distribution, average diameter, aspect ratio, and foam number per unit area.

1 INTRODUCTION

Chemical foams are formed by trapping pockets of gas in a liquid. In most foam, the volume of gas is large. The films of liquid membrane could separate the regions of gas one-by-one. The foam is collected by a receiver for observation as shown in Figure 1, and many reports show that foam parameters change with time (Xia J J, 2013).

The micrographs of foam structure could be captured by a camera, video recorder, or microscope. Some researchers studied the structural parameters of foam using micrographs. However, manual counting is too difficult, as hundreds of foam particles can be present in a micrograph. In addition, other foam parameters, such as average diameter, aspect ratio, foam number per unit area, are difficult to obtain too. Therefore, a special analyzer on foam should be designed to increase the recognizing efficiency.

Figure 1. Fire-fighting foam in a receiver.

2 FOAM GRAPH

2.1 Stereomicroscope

The micrograph of foam could be captured by a microscope (Yang F, 2016). The average diameter of foam particles was about 1–500 μm. Therefore, the maximum magnification should be from $50 \times$ to $100 \times$. Then, the diameter of the foam could reach 0.5 mm, which is in the range of human eye's resolution. In addition, this diameter could be enlarged further in a computer monitor. Therefore, the micrographs of foam could be captured by a stereomicroscope with maximum magnification from $50 \times$ to $100 \times$.

Stereomicroscope is an optical microscope variant designed for observation, typically using light reflected from the surface of an object rather than transmitted through the sample (Sawamura S, 2012). There are always two lights: the up light and back light, as shown near the objective table in Figure 2.

Micrographs could be captured using a stereomicroscope with suitable parameters including brightness, contrast, color balance, gain, and gamma level. However, the foams in this micrograph were still difficult to recognize because of the low brightness as shown in Figure 3. Therefore, suitable aided light was required to increase the recognition efficiency.

2.2 Aided light

To solve the problem of low brightness, an aided light was introduced to obtain micrographs with higher brightness, as shown in Figure 4. The two

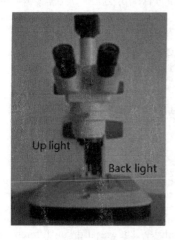

Figure 2.   Stereomicroscope with up light and back light.

Figure 3.   Micrograph obtained using only up light.

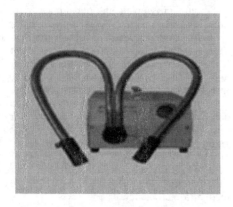

Figure 4.   Aided light with two light lenses.

light lenses will be fixed near the up light of stereomicroscope to increase brightness, and then the foam boundaries in the micrograph could be recognized easily.

Figure 5.   Micrograph obtained using aided light.

The micrograph obtained using aided light is shown in Figure 5. Compared with Figure 4, the foam boundaries can be recognized more easily.

3   DESIGN OF MODIFIED PARTICLE ANALYZER

3.1   *Traditional particle analyzer*

Particle size analyzers could recognize and measure the particle size automatically in a short time. The equipment could be introduced to calculate parameters such as light scattering, sedimentation, and laser diffraction. The particle boundaries could be recognized easily. Then, the size distribution could be obtained automatically, which could characterize the material's discrete situation. Although foams were not solid particles, their boundary was similar to that of solid particles, which could be recognized and counted, as shown in Figure 6.

3.2   *Introduction of edge detector*

Even though the foam's boundary obtained using aided light was clear, it was still difficult to recognize the foam automatically in micrograph using traditional particle analyzer. The foam's boundaries are not as clear as those of solid particles. Therefore, suitable edge detector should be introduced to modify the particle analyzer such as Kirsch edge detector, Laplace edge detector, and Canny edge detector.

The Kirsch detector is a nonlinear edge detector that can find the maximum edge strength in a few predetermined directions. The operator takes a single kernel mask and rotates it in 45° increments through all directions. The recognition result obtained using particle analyzer by introducing Kirsch detector was not very well, as shown in Figure 7, in which not all the foams could be recognized.

Laplace detector is a differential edge detection operator given by the divergence of the gradient of a function on Euclidean space. In a Cartesian

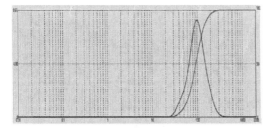

Figure 6. Micrograph obtained using aided light. (X-axis: particle size diameter; Y-axis: distribution percentage and cumulative percentage).

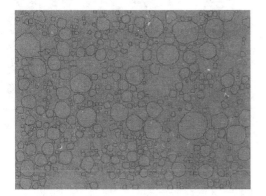

Figure 7. Recognition result using particle analyzer by introducing Kirsch detector.

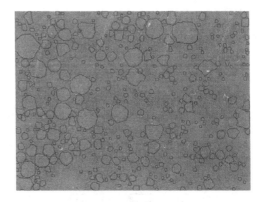

Figure 8. Recognition result using particle analyzer by introducing Laplace detector.

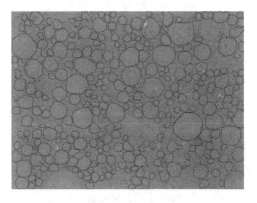

Figure 9. Recognition result using particle analyzer by introducing Canny detector.

coordinate system, the Laplacian was given by the sum of second partial derivatives of the function with respect to each independent variable. The recognition result obtained using particle analyzer by introducing Laplace detector was not very well, as shown in Figure 8, in which the foam boundaries were not recognized accurately.

The Canny edge detector is an edge detection operator that uses a multistage algorithm to detect a wide range of edges in graphs. The optimal function in Canny detector was described by the sum of four exponential terms, but it could be approximated by the first derivative of a Gaussian. The recognition result obtained using particle analyzer by introducing Canny detector was better, as shown in Figure 9, in which more foams with accurate boundary were recognized. Therefore, Canny detector is more suitable to recognize foam's boundary.

3.3 Modified particle analyzer with Canny detector

The graph was obtained by stereomicroscope with aided light. Then, the modified particle analyzer with Canny detector would recognize the foam boundary in the graph. The steps involved in this process are as follows:

First, the graph strength was calculated to increase the clarity of foam boundary. The kernel of Lap 3 is shown in Figure 10.

Second, the Canny edge detection algorithm was carried out. A suitable filter was applied to smooth the graph and remove the noise line. The intensity gradients of the graph were found. Nonmaximum suppression was applied to get rid of spurious response to edge detection. Then, double threshold was applied to determine potential edges. Edge by hysteresis was tracked and the detection of edges was finalized by suppressing the other edges that were weak and not connected to strong edges.

Third, a cycle of erosion, scanning, segmentation, and filling was carried out. Erosion could remove the interference of point or line to eliminate the graph noise. Scanning could identify the completed foam boundary. Segmentation could separate the completed foam from other parts. Then, filling could fill and number the completed foam. The other uncompleted foam could be recognized on the next cycle. This cycle would be ended when all the foams were recognized. The strategy of the modified particle analyzer is shown in Figure 11.

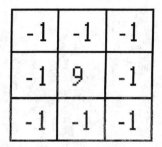

Figure 10. Kernel of Lap 3.

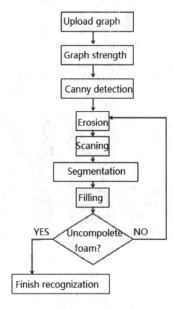

Figure 11. Strategy of the modified particle analyzer.

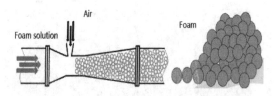

Figure 12. The generation progress of foam.

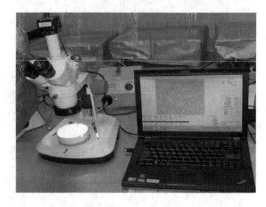

Figure 13. Modified particle analyzer and stereomicroscope with aided light.

4 APPLICATION OF MODIFIED PARTICLE ANALYZER

4.1 Test progress

Fire-fighting foam was liquid-based foam for fire suppression, such as Class A foam, AFFF (aqueous film-forming foam), P (Protein), and FP (fluorine protein) (Yilixiati S, 2015). Fire-fighting foam could be used to cool the fire and coat the fuel, preventing its contact with oxygen (Fu X, 2012). These foams were produced by charging air into foam solution, as shown in Figure 12. The foam was a nonequilibrium system, and its structure evolution could be analyzed by modified particle analyzer (Zhao H, 2016).

The produced foam should be collected by a foam receiver. Then, the receiver was fixed under the stereomicroscope's objective lenses. The aided light helped to capture graph with higher brightness, as shown in Figures 13 and 17.

4.2 Analysis progress

The foam graph captured by a stereomicroscope with aided light at 120 s is shown in Figure 14. The analyzed foam graph with bold boundaries is shown in Figure 15. Most of the completed foams were recognized. The particle size distribution of these foams is shown in Figure 16.

The foam parameters would change with time. Then the foam graph was captured by a stereomicroscope with aided light at 360 s, as shown in Figure 17. The analyzed foam graph with bold boundaries is shown in Figure 18. Most of the completed foams were recognized. The particle size distribution of these foams is shown in Figure 19.

The foam particle distributions of foam graph at 120 and 360 s were different. Other parameters could also be studied using modified particle analyzer, including average diameter, aspect ratio, and foam number per unit area. The diameter of foam was any straight line segment that passed through the center of the circle foam whose endpoints lie on the circle. To ellipse fire-fighting foam, the diameter was an average diameter of equivalent circle. Aspect ratio was the ratio of its sizes in different dimensions. For fire-fighting foam, the aspect ratio was the ratio of short axis to long axis, which was no less than 1. Foam number per unit area was the number of foam bubbles per unit area. This number changes with the breakup of foam, as shown in Table 1.

The average diameter increased with time (Hu Z, 2013). This foam coarsening was driven by air

Figure 14. Class A foam graph at 120 s.

Figure 17. Class A foam graph at 360 s.

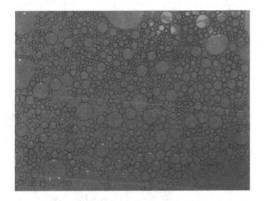

Figure 15. Recognized Class A foam graph at 120 s.

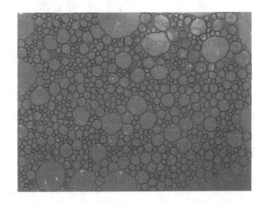

Figure 18. Recognized Class A foam graph at 360 s.

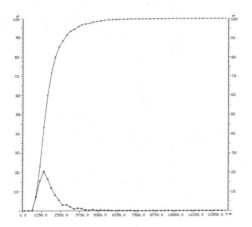

Figure 16. Foam particle distribution of Class A foam graph at 120 s (X-axis: average diameter; Y-axis: distribution percentage and cumulative percentage).

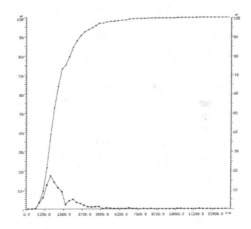

Figure 19. Foam particle distribution of Class A foam graph at 360 s (X-axis: average diameter; Y-axis: distribution percentage and cumulative percentage).

pressure. The foam bubble's inner air pressure was higher than normal atmospheric pressure outside foam bubble. Then, the average diameter would increase. At the same time, some bubble's film might break up and two neighboring bubbles might turn into one bubble, which also prolonged the bubble's average diameter. The aspect ratio decreased with time. This was due to the liquid's drainage in the bubble film. The film's liquid between foam bubbles flowed downward due to gravity. Then, the

Table 1. Parameters of foam graphs in 120 and 360 s.

Parameters	120 s	360 s
Average diameter/μm	161.32	224.51
Aspect ratio	0.99	0.98
Foam number per unit area	25.7	14.8

Figure 20. Study of foam's particle bubble in the vertical direction.

circular bubble's film became thinner and thinner. Then, the shape of the bubble might change because of the different air pressure in neighboring bubbles. More circular bubbles became polygon bubbles (Le Merrer M, 2012). Foam numbers per unit area decreased with time. The reason was foams would break up by larger difference in air pressure and thinner foam film as drainage. Therefore, series quantitative data of foam bubble could be obtained using modified particle analyzer (Gai M, 2016; Uehara T, 2014).

The future study should focus on the application of modified particle analyzer, including the change between foams produced by different generating devices, the foam particle distribution curve's fitting by function like Rosin–Rammler equation (Tambun R, 2016) and the particle bubble change of the foam in the vertical direction, as shown in Figure 20.

5 CONCLUSION

In summary, the modified particle analyzer of fire-fighting foam was designed by introducing Canny operator and applied to study fire-fighting foam. The foam graphs could be obtained by stereomicroscope with aided light. Then, the foam boundary could be recognized by modified particle analyzer. Finally, the particle size distribution could be obtained with other parameters, including average diameter, aspect ratio, and foam number per unit area. This analyzer would be applied to characterize fire-fighting foam in the future.

ACKNOWLEDGMENTS

This research work was supported by 2014 Tianjin Postdoctoral Science Foundation and National Science and Technology Support Program No. 2014BAK17B01.

REFERENCES

Foams: physics, chemistry and structure[M]. Springer Science & Business Media, 2013.

Fu X, Bao Z, Chen T, et al. Application of compressed air foam system in extinguishing oil tank fire and middle layer effect[J]. Procedia Engineering, 2012, 45: 669–673.

Gai M, Frueh J, Hu N, et al. Self-propelled two dimensional polymer multilayer plate micromotors[J]. Physical Chemistry Chemical Physics, 2016, 18(5): 3397–3401.

Hu Z, Luo H, Du Y, et al. Fluorescent stereo microscopy for 3D surface profilometry and deformation mapping[J]. Optics express, 2013, 21(10): 11808–11818.

Le Merrer M, Cohen-Addad S, Höhler R. Bubble rearrangement duration in foams near the jamming point[J]. Physical review letters, 2012, 108(18): 188301.

Sawamura S, Ichitani M, Ikeda H, et al. Foaming property and foam diameter of matcha varies with particle size[J]. Journal of the Japanese Society for Food Science and Technology-Nippon Shokuhin Kagaku Kogaku Kaishi, 2012, 59(3): 109–114.

Tambun R, Furukawa K, Hirayama M, et al. Measurement and Estimation of the Particle Size Distribution by the Buoyancy Weighing–Bar Method and the Rosin–Rammler Equation[J]. Journal of Chemical Engineering of Japan, 2016, 49(2): 229–233.

Uehara T. Numerical Simulation of Foam Structure Formation and Destruction Process Using Phase-Field Model[C]//Advanced Materials Research. Trans Tech Publications, 2014, 1042: 65–69.

Xia J J, Fu X C, Zhang X Z, et al. Design and Application of Endoscope with Aided Lens for Observation of Fire-Fighting Foam[C]//Applied Mechanics and Materials. Trans Tech Publications, 2013, 271: 823–828.

Yang F, Zhang G L, Yang F, et al. Pedogenetic interpretations of particle-size distribution curves for an alpine environment[J]. Geoderma, 2016, 282: 9–15.

Yilixiati S, Wojcik E, Zhang Y, et al. Patterns, Instabilities, Colors, and Flows in Vertical Foam Films[C]//APS March Meeting Abstracts. 2015, 1: 1142P.

Zhao H, Liu J. The Feasibility Study of Extinguishing Oil Tank Fire by Using Compressed Air Foam System[J]. Procedia Engineering, 2016, 135: 61–66.

Advances in Energy Science and Equipment Engineering II – Zhou, Patty & Chen (Eds)
© *2017 Taylor & Francis Group, London, ISBN 978-1-138-71798-5*

Leak-before-break analysis of a primary pipe

Ran Liu, Mei Huang, Kai Li Sun & Jun Liu
North China Electric Power University, Beijing, China

ABSTRACT: The Leak Before Break (LBB) assessment is widely used as an integrity evaluation method to avoid large rupture accident. In this paper, axial cracks were inserted into the model to assess the LBB behavior of the pressurized pipe after static stress analysis. The crack length when the leakage can be detected is computed through two-phase critical flow model and the elliptical shape assumption. The critical crack length is analyzed by the *J*-integral instability assessment diagram. The results of the analysis revealed that the value of crack driving force *J*-integral augmented significantly along with the half crack length c increasing. There was a nonlinear increase in the Crack Opening Displacement (COD) versus the increments of half crack length *c*. The detectable leakage crack length $2c_{det}$ was two times smaller than the critical crack length $2c_{cri}$. This work demonstrates that the primary pipe confirms to the LBB criteria.

1 INTRODUCTION

With the development of the economy and the promotion of the public awareness of environmental protection, the utilization of nuclear power and other clean energy has become the main direction of energy development. Nuclear power has a greater advantage over other ways: low operating costs and modular construction. The safe operation of related equipment and structure has always been the concerns of the design and construction of nuclear power plant (Liang et al. 2015, Zhong 2012).

The primary pipe and the reactor pressure vessel are important parts of the primary coolant boundary to be the second barrier to guarantee the safety of nuclear power (Lin 2008). Especially in case of the severe accident, it is very important to ensure the integrity of primary coolant circuit to contain the core radioactive fission products. At the same time, The primary pipe and the reactor pressure vessel have been running under harsh environment of high temperature, high pressure, water hammer and so on. So the service induced defects is inevitable during the design life. The study shows that there are some cases of the service induced defects of the primary coolant circuit around the world (Zhang 1996). For example, in 1970 the experts of the United States found reheated cracks in the heat affected zone under the stainless steel clad layer of parent metal, its length is 0.2–20 mm, and the depth is 1–4 mm. In 1979 pressure vessel nozzle of the surfacing layer of base metal occurs many cracks in a French nuclear power plant, general depth size is about 2–3 mm, the deepest is 7 mm, the length is 12–15 mm.

Since the 1980s, the nuclear experts found that the probability of occurrence of double ended shear fracture accidents in nuclear power plants is very small, and treating it as a design basis accident will result in many problems outweigh the benefits. The Leak Before Break (LBB) assessment is widely used in nuclear power plants as an integrity evaluation method to avoid large rupture accident (Li et al. 2011).

There is an assumption that a through-wall crack exits in a pressurized pipe or vessel, then the liquid leakage will occur. And the crack will remain stable before the monitoring systems find it. That is the key principle of the LBB behavior (Lv et al. 2015).

Lv et al. (2014) assessed the effect of thermal aging on the structural integrity of nuclear pipe. Gong et al. (2013) applied a LBB approach to a dissimilar metal welded joint of an AP1000 reactor pressure vessel. In this paper, some works about LBB analysis addressed in preliminary pipe lines are carried out.

2 LBB ASSESSMENT

2.1 *Material properties and geometry model*

The material of the primary pipe investigated is SA508-3 mild carbon steel with a mean density of $\rho = 7850$ kg/m³. The elastic modulus and the average thermal expansion coefficient of SA508-3 mild carbon steel at different temperature are shown respectively in Table 1.

The geometry selected is the preliminary pipe containing a axial partial-through crack with crack

Table 1. Mechanical performance parameters of SA508-3 mild carbon steel at different temperature.

T°C	E × 10¹¹ Pa	α × 10⁻⁶/°C
25	2.04	11.22
50	2.03	11.45
100	2.00	11.79
150	1.97	12.14
200	1.93	12.47
250	1.90	12.78
300	1.86	13.08
350	1.83	13.40

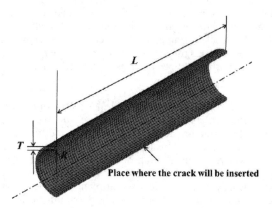

Figure 1. FE meshes for uncracked pipe.

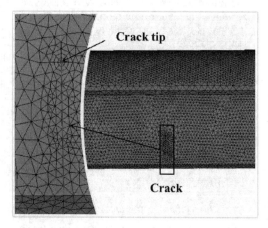

Figure 2. FE meshes for cracked pipe.

depth a and crack length c. The inner radius and thickness of the pressure vessel shell are $R_i = 380$ mm and $T = 40$ mm respectively. In the critical case of $a/T \to 1$, it is a axial through-wall crack.

2.2 FE analysis

The 3-d FE model was built for the pressurized pipe. Figure 1 and 2 depict typical FE meshes for uncracked and cracked pressurized pipe. Taking advantage of symmetry conditions, only half of the pipe was modeled using twenty-node isoparametric brick elements with uniformly reduced integration (solid186 in ANSYS). The element number for uncracked model is 4350 and for cracked models ranged from 102221 to 457458.

For the loading conditions, the water pressure inside the pressure pipe of 17 MPa, that is the service pressure, was imposed as a distribute load on the inner surface in the FE model. And the bending moment under normal operation and upset condition are 3068 and 3616 kNm respectively. That was applied to the pipe end the operating temperature and environment reference temperature is 343°C and 20°C. The average convection coefficient between the outer face and environment is 8.4 W/(m²·°C) (Luan 2008). The movement of the pressure pipe end was restricted in axial and circumferential directions, and the symmetry boundary condition was applied on the symmetry planes.

2.3 LBB procedure

LBB is defined by the European commission as "the failure model of a cracked piping is a leaking through-wall crack which may by timely and safely detected by the available systems and which does not challenge the pipe's capacity to withstand any design loading" (Bourga et al. 2015). The LBB is applicable to the ductile materials and fracture mechanics principle are used to demonstrate that a flaw will develop through-wall allowing sufficient and stable leakage that can be detected before catastrophic rupture of the structure occurs.

The following analysis should be done in LBB analysis (Dundulis et al. 2013):

1. Determination of most loaded part of system.
2. Determination of through-wall crack critical length.
3. Determination of crack opening function.
4. Determination of leak rate function.
5. Analysis of postulated crack growth.
6. Compliance to the LBB requirements.

There are two important criteria for demonstrating LBB: 1) the length of initial through-wall crack (developed from a sub-critical crack) is smaller than the length of critical through-thickness crack determined at most critical loading condition; 2) the time for detection of leak is shorter than the time needed for an initial through-wall crack grows to a critical length (Ren et al. 2015).

2.4 Determination of critical crack length

The critical crack length is determined by the crack growth stability analysis on the base of elastic-plastic fracture mechanics theory. The crack driving force, J-integral of J (load, $2c$), is a function of the bending moment on the pipe end as well as total through-wall crack length $2c$ (Li et al. 2006). At the same time, J-resistance to ductile fracture can be obtained from testing and expressed in the term of $J_R(\Delta c)$ as a function of ductile crack extension Δc. In this paper, the $J_R(\Delta c)$ curve is cited from RCC-MR A16. In the critical situation that rapid ultimate fracture will just occur, it should confirm to the following two equations (Li 2012),

$$J(load, 2c) = J_R(\Delta c) \qquad (1)$$

$$\frac{E}{\sigma_o^2}\left(\frac{dJ(load, 2c)}{dc}\right)_{\Delta T} = \frac{E}{\sigma_o^2}\frac{dJ_R(\Delta c)}{dc} \qquad (2)$$

where Δc and σ_o donates the crack extension in the axial direction and reference stress respectively.

2.5 Determination of detectable leakage crack length

The detectable leakage crack length calculation has been widely analyzed, and two-phase flow using the following equation (Gong 2013, Xu 1995):

$$m = G \times A = A \times 0.61\sqrt{2P\rho(1-\eta)} \qquad (3)$$

where m is leakage rate and G is the mass flux per unit area, A is COA and P is the inner pressure of vessel, ρ denotes the fluid density, and η is the critical pressure ratio. According to the USNRC methodology of the LBB analysis, m is 0.5 kg/s and ρ is 992.34 kg/m³ for the water in preliminary pipe lines. P is 17 MPa for the internal pressure in preliminary pipe lines and the saturated pressure of pure water is 15.19 MPa under 340°C, so η is 0.89 (Xu 1995). For the crack in the pipe, the FE analysis is used to estimate the Crack Opening Displacement (COD) to calculate the corresponding COA and the detectable leakage crack length for the given loading condition.

3 RESULTS AND DISCUSSION

3.1 Critical crack length calculation

The calculation of the critical crack length is an important of LBB assessment. The critical crack length was determined by the J-integral stability assessment diagram approach and is explained in section 2.3. The crack driving force J(load, $2c$) was directly extracted from the finite element results with different through-wall crack lengths. The material resistance $J_R(\Delta c)$ curves can be obtained from RCC-MR A16. Figure 3 shows the determination of the critical crack length of the pressurized pipe.

Figure 3 shows the critical crack length for the pressurized pipe is 180.2 mm.

3.2 Detectable leakage crack length calculation

There is an assumption that the crack opening profile is oval shape in the calculation of COA. Then the COA calculating confirms to the following equation:

$$A = \pi c \delta \qquad (4)$$

where A and δ denote the COA value and COD value, respectively, c is the half crack length.

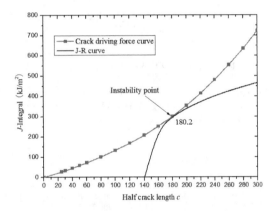

Figure 3. Determination of critical crack length.

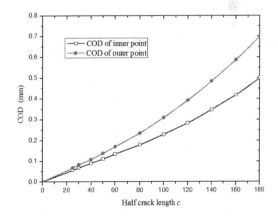

Figure 4. Variations of crack opening displacement.

Figure 4 shows the variation of COD values obtained from the FE analysis with through-wall crack length $2c$. The detectable leakage crack length calculated by Eqs. (3) and (4) is 44.93 mm.

4 CONCLUSIONS

The 3-d FE model was built for the pressurized pipe. The circumferential through-wall cracks were postulated at the pressurized pipe, and accurate J-integral calculations were carried out, and then the LBB curve was constructed. The main results obtained as follows:

1. The value of crack driving force J-integral augmented significantly with the increased half crack length c.
2. There was a linear increase in the Crack Opening Displacement (COD) versus the increments of half crack length c.
3. The detectable leakage crack length $2c_{det}$ was two times smaller than the critical crack length $2c_{cri}$.
4. This work demonstrates that the pressurized pipe confirms to the LBB criteria.

AUTHOR'S INFORMATION

Ran Liu, North China Electric Power University, Fracture mechanics, +8618811587699, huangmei_team@163.com

REFERENCES

Bourga, R. et al. 2015. Leak-before-break: Global perspectives and procedures. International Journal of Pressure Vessels and Piping, 2015. 129–130: p. 43–49.

Dundulis, G., Janulionis, R. & Karalevius, R. 2013. The application of leak before break concept to W7-X target module. Fusion Engineering and Design. 2013.88(11): p. 3007–3013.

Gong, N. et al. 2013 Leak-before-break analysis of a dissimilar metal welded joint for connecting pipe-nozzle in nuclear power plants. Nuclear Engineering and Design, 255: p. 1–8.

Li, Q., Cen, P. & Zhen, H.D. 2011. Introduction of the Application of the LBB Technology in Nuclear Power Plant High Power Lines. Nuclear Power Engineering, 2011(S1): p. 189–191.

Li, Z. 2012. Applied fracture mechanics[M]. Beijing: Beijing Aeronautics and Astronautics University Press.

Li, Z.A., Zhang, J.W., & Wu, J.H. 2006. Fracture theory and defect evaluation of process equipment[M]. Chemical Industry Press.

Liang, W.B., Liu T. & Liu M.S. 2014. Three-dimensional finite element analysis and LBB assessment of nuclear pipe with crack. pressure vessel, 2014(11): p. 56–60.

Lin, C.G. 2008. The advanced active safety AP1000 nuclear power plants. Beijing: Atomic Energy Press.

Luan, C.Y. 2008. ANSYS analysis and strength calculation of pressure vessel, Beijing: China Water Power Press.

Lv, X.M. et al. 2014. Effect of thermal aging on the leak-before-break analysis of nuclear primary pipes. Nuclear Engineering and Design. 280: p. 493–500.

Lv, X.M. et al. 2015. Leak-before-break analysis of thermally aged nuclear pipe under different bending moments. Nuclear Engineering and Technology, 47(6): p. 712–718.

Ren, X.B. et al. 2015. Leak-before-break analysis of a pipe containing circumferential defects. Engineering Failure Analysis. 58: p. 369–379.

Xu, J. 1995. Two-Phase Critical Discharge of Initially Saturated or Subcooled Water Flowing in Sharp-Edged Tubes at High Pressure[J]. Journal of Thermal Science,1995.03:193–199.

Zhang, Z.G. 1996. The probability fracture mechanics applied in the pressure vessel[M]. Beijing: China petrochemical Press.

Zhong, L.X., 2012. Leak before break (LBB) analysis of the pressure vessel with irradiation damage. NCEPU. p. 46.

Drop in multilayered porous media

Guangyu Li, Xiaoqian Chen & Yiyong Huang
National University of Defense Technology, Changsha, Hunan, P.R. China

ABSTRACT: Spontaneous imbibition is a very important aspect of nature phenomena. Its mechanism is widely used in scientific research and industry. In this paper, theoretical model for the transport of a droplet through the layered porous media was presented. The transport time of a droplet through the layered porous media was calculated. On the basis of the theoretical model, a contrast of transport time between a 0.004 m long droplet and a 0.002 m long droplet transport through the same layered media was made. The conclusion is useful for the design of layered porous media.

1   INTRODUCTION

Capillary imbibition in porous media is a very important aspect of nature phenomena. Its mechanism is widely used in many scientific areas and industry, for example, pharmacology, composite materials, and textile industry. In practice, several porous materials used in industry are designed in a layered configuration, and the flow through layered porous media is a general and important one.

A considerable amount of research has been done on liquid flow through porous media in the last decade. Experiment measurements were carried out by Sarkar to observe the transport property in 20 layers of Fabric A on the top of 20 layers of Fabric B. The experimental result agrees well with the theoretical prediction. Parker established the Lattice–Boltzmann model of multiphase transport in porous media to predict the real-world fluid physics phenomena. Peng investigated the interface rise for the flow in a capillary with a nonuniform cross section distribution along a straight center axis. Prommas deducted a combined mass and thermal model for a convective drying of multilayered packed beds.

In this paper, zero-dimensional capillary rise theoretical model was constructed in layered porous media and used to predict transport time of one droplet flow through such materials. The result is essential to the design of porous materials for applications.

2   THEORETICAL FRAMEWORK

Some assumption must be made for the theoretical model construction. First, the droplet is incompressible. Second, the flow is Newtonian and continuous. A droplet flows upward through a multilayered porous media with external pressure P, as shown in Figure 1. The multiporous media include three porous layers of different properties, such as particle size, porosity, and permeability. $h_i(i = 1,2,3)$ presents the thickness of the *ith* layer of the porous media, h_a and h_r represent the distances of the advances and receding interfaces of the liquid droplet, and P represents the external pressure.

According to the momentum conservation law, the modified Darcy equation can be related as:

$$\frac{\mu \dot{h}_a (h_a - h_r)}{K_a} = P + P_a - P_r - \rho g(h_a - h_r) \tag{1}$$

where μ is the liquid viscidity, K_a denotes the permeability of the porous media, P_a, P_r is the capillary

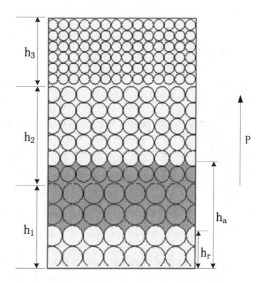

Figure 1.   Sketch of multilayered porous media.

pressure of the advances and receding interfaces, respectively, ρ is the density of the liquid, and g is the acceleration due to gravity.

According to the mass conservation law, the velocities of the two interfaces can be related as:

$$A_f \phi_r \dot{h}_r = A_f \phi_a \dot{h}_a \qquad (2)$$

where A_f stands for cross-sectional areas of the porous media and ϕ is the porosity.

The capillary rise model for rigid beads proposed by Comiti and Renuad was applied to analyze liquid transport properties of layered porous media. This model assumes the packed beads to consist of a bundle of identical cylindrical tortuous pore, and this concept has been adopted by various researchers to characterize the process of capillary imbibition in porous media. The capillary pressure is defined by:

$$P_c = \frac{2\gamma_{lv}\cos\theta}{r_{eff}} \qquad (3)$$

where γ_{lv} is the liquid–vapor surface tension, θ is the contact angle of the liquid against the solid, which is considered constant in this paper, r_{eff} is the effective capillary radius, which is defined as:

$$r_{eff} = \frac{2\phi}{A(1-\phi)} \qquad (4)$$

where A is the specific surface area, a ratio of wetted surface area to pore surface area to volume of solid, which is related to glass bead diameter, d_s, as:

$$A = \frac{6}{d_s} \qquad (5)$$

The permeability K_a can be calculated as:

$$K_a = \frac{\phi^3 d_b^2}{150(1-\phi)^2} \qquad (6)$$

The capillary pressure P_a, P_r in equation (1) can be obtained by substituting Equations (4) and (5) into Equation (3). Consequently, the governing equation for the advancing capillary flow in the multilayered porous media with both advancing and receding interfaces can be obtained by combining Equation 1 with Equation 2.

3 NUMERICAL ANALYSIS

A material having three layers (Figure 1) is used to analyze the transport properties, with each layer

being a fractal porous medium, where a droplet is placed between the first layer and the second layer, at time $t = 0$. The capillary flow of droplet due to absorption takes place first in the first layer, then in the second layer, and so on. And at time t, it moves in a straight line, covering a distance of $(h_a - h_r)$, under the action of P. In this paper, three types of particles were chosen for each layer, whose sizes are listed in Table 1.

The first layer is filled with particles A in Table 1, the second layer is B, and C is put into the top. The respective thicknesses are $h_1 = 0.01\,m$, $h_2 = 0.03\,m$, and $h_3 = 0.05\,m$. Integrating functions (1) and (2) with the initial condition, $h_r = 0.001\,m$, $h_a = 0.005\,m$, the distance traveled by the droplet can be obtained, and the external pressure P is 5000 Pa. The liquid used in this numerical analysis is water, and $\rho = 1000\,kg/m^3$, $\mu = 0.001\,Pa\cdot s$, $\gamma_{lv} = 0.072\,J/m^2$, and the liquid–solid contact angle is 75°.

In Figure 2, droplets rise from the bottom to the top. At the beginning of the transport, the droplet moves slowly because of the zero initial velocity, then the velocity begins to slow down; for the capillary pressure difference between the first layer and the second layer (Figure 3), a great capillary difference is shown. When the droplet passes

Table 1. Particle size.

Symbol	Size (μm)	d_s (μm)	A (μm⁻¹)	ϕ
A	38~45	40.5	0.148	0.351
B	90~106	99.6	0.060	0.366
C	150~180	169.6	0.035	0.371

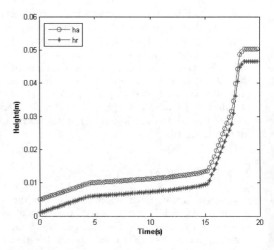

Figure 2. Height vs. time.

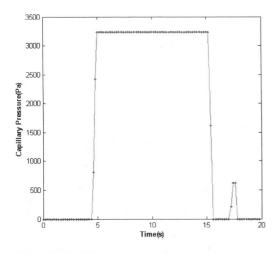

Figure 3.   Capillary pressure vs. time.

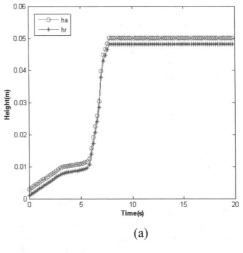

(a)

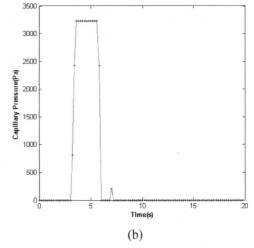

(b)

Figure 4.   Volume-reduced droplet transport through the layered porous media.

through the span between the layers, it moves fast and reaches the top.

Both Figures 2 and 3 reveal that when the droplet passes through the area between layers, most of the external pressure is used to overcome the capillary pressure, and the droplet moves slowly. Consequently, the droplet can move faster through the porous media by reducing the volume of the droplet. Integrating functions (1) and (2) with the initial condition, $h_r = 0.001\,m$, $h_a = 0.003\,m$, and the other condition will be the same. The result is shown in Figure 4 and Table 2.

A smaller droplet through the layered porous media is shown in Figure 4 (a). The velocity of the small droplet increases rapidly for the effect of the capillary pressure as shown in Figure 4 (b). Only 7.2 s was taken to reach the equilibrium height.

In Table 2, n denotes the number of droplets through the layered porous media in 1 min. Figure 4 shows that the droplet through the layered porous media moves faster than the bigger one. The time of capillary pressure kept is shorter than that for the bigger droplet to pass through the porous media. Table 2 shows that, it takes 18 s to reach the top for a 0.004 m long droplet, and only 7.2 s for a 0.002 m long droplet. In 1 min, three 0.004 m long droplets were pushed through the layered porous media; however, eight 0.002 m long droplets were transported through the layered porous media. That means, the whole volume transport through the porous media of the 0.002 m length droplet is larger than that of the 0.004 m long droplet. To conclude, compared with the small-volume droplet, the larger-volume droplet was pushed through the porous media in 1 min.

Table 2.   Transport time contrast.

Droplet length (m)	Equilibrium time (s)	n
0.004	18	3
0.002	7.2	8

4   CONCLUSION

In this paper, a theoretical model for the transport of a droplet through the layered porous media was presented. The transport time was calculated on the basis of the theoretical model. The result shows that the larger the droplet is, the more time it took to pass through the layered porous media. Finally, a contrast of transport time was made between a

0.004 m long droplet and a 0.002 m long droplet transport through the same layered porous media. The result shows that the whole volume for short droplets through the media is larger than that of long droplets in the same time. This result is useful for designing layered porous media.

REFERENCES

Cai & Hua, An analytical model for spontaneous imbibition in fractal porous media including gravity, Colloids and Surfaces A: Physicochem. Eng. Aspects 414 (2012) 228–233.

Cai & Yu, Fractal Characterization of Spontaneous Co-current Imbibition in Porous Media, Energy Fuels 2010, 24, 1860–1867.

Comiti & Renaud, 1989. A new model for determining mean structure parameters of fixed beds from pressure drop measurements: application to beds packed with parallelepipedal particles. Chem. Eng. Sci. 44, 1539–1545.

Parker & Duval, Lattice-Boltzmann Model Development For Capillary Pressure Gradient Gas/Liquid Phase Separation in Microgravity, 41st International Conference on Environmental Systems 17–21 July 2011, Portland, Oregon.

Peng & Parker, Analytical investigation of free surface flow in multi-layer porous media, Colloids and Surfaces A: Physicochem. Eng. Aspects 380 (2011) 213–221.

Prommas, Theoretical and experimental study of heat and mass transfer mechanism during convective drying of multi-layered porous packed bed, International Communications in Heat and Mass Transfer 38 (2011) 900–905.

Sarkar & Qian, Transplanar water transport tester for fabrics, Meas. Sci. Technol. 18 (2007) 1465–1471.

Stevens & Sedev, The uniform capillary model for packed beds and particle wettability, Journal of Colloid and Interface Science 337 (2009) 162–169.

Yu, Analysis of flow in fractal porous media, Appl. Mech. Rev. 61 (2008). 050801–1.

Advances in Energy Science and Equipment Engineering II – Zhou, Patty & Chen (Eds)
© 2017 Taylor & Francis Group, London, ISBN 978-1-138-71798-5

Interfacial crack between linear FGM and homogeneous material of cylindrical shell with Reissner's effect

Xiao Chong
Changping School of Noncommissioned Officer, PLA Academy of Equipment, Beijing, China

Yao Dai, Jing-wen Pan & Wei-hai Sun
Academy of Armored Force Engineering, Beijing, China

ABSTRACT: The material properties of functionally graded materials (FGMs) are assumed to be linear functions of y with their gradient direction perpendicular to the interface. The crack is oriented in circumferential direction. Base on Reissner theory, the governing equations are derived for weak-discontinuous problems of cylindrical shells. The higher order crack-tip fields for the interfacial crack between FGM and homogeneous material cylindrical shells are investigated. By using the eigen-expansion method, the high order displacement fields and the crack tip fields are obtained.

1 INTRODUCTION

FGMs are widely used in energy, chemical industry, aviation, aerospace and other departments due to their excellent performance. In the engineering, FGMs mainly appear in the coating and interface layer, forming lots of interface structures. Due to the feature of production technology, there are a number of interface defects in structures. Furthermore, due to differences in material performance across interfaces (i.e. the differentials of material parameters at interfaces are different, which is defined the physical weak-discontinuous), the interfacial fracture becomes the main form of structure failure. Consequently, the study on weak-discontinuous fracture of the FGM cylindrical shell is of great significance.

The transverse shear deformation must be considered in Reissner's theory to analysis of fracture mechanics, which eliminates the inherent defects in the application of classical plate and shell theories. The higher order crack-tip fields of Reissner's FGMs plates obtained by Liu (Liu Chuntu, 1983). Zhang (Zhang Lei, 2012) studied the crack problem of Reissner's functionally graded spherical and cylindrical shells and the higher order crack-tip fields are obtained. In this paper, the physical weak-discontinuous problem of an interfacial crack is considered and the high order crack tip fields are obtained.

2 THE BASIC EQUATIONS

The elastic modulus function form of FGMs is assumed to be

$$E^{(\mathrm{II})} = E^{(\mathrm{I})}(1+\beta y) = E^{(\mathrm{I})}(1+\beta r \sin\theta) \qquad (1)$$

where, $E^{(\mathrm{I})}$ is the elastic modulus of homogeneous material, and β are the non-homogeneity parameter. The variation of Poisson's ratios has very insignificant effect on the stress intensity factor of non-homogeneous materials (Delale F, 1988). So, Poisson's ratio is assumed to be the constant μ.

Assumed the transverse loading of the shell is to be zero, the governing equations for shells with Reissner's effect are

$$
\begin{cases}
D^{(k)}\left[\dfrac{\partial^2 \varphi_x^{(k)}}{\partial x^2} + \dfrac{1-\mu}{2}\dfrac{\partial^2 \varphi_x^{(k)}}{\partial y^2} + \dfrac{1+\mu}{2}\dfrac{\partial^2 \varphi_y^{(k)}}{\partial x \partial y}\right] \\
\quad + C^{(k)}\left(\dfrac{\partial w^{(k)}}{\partial x} - \varphi_x^{(k)}\right) + \dfrac{h^3}{24(1+\mu)}\dfrac{\partial E^{(k)}}{\partial y} \\
\quad \times \left(\dfrac{\partial \varphi_x^{(k)}}{\partial y} + \dfrac{\partial \varphi_y^{(k)}}{\partial x}\right) = 0 \\[4pt]
D^{(k)}\left[\dfrac{\partial^2 \varphi_y^{(k)}}{\partial y^2} + \dfrac{1-\mu}{2}\dfrac{\partial^2 \varphi_y^{(k)}}{\partial x^2} + X\dfrac{1-\mu}{2}\dfrac{\partial^2 \varphi_x^{(k)}}{\partial x \partial y}\right) + \\
C^{(k)}\left(\dfrac{\partial w^{(k)}}{\partial y} - \varphi_y^{(k)}\right) + \dfrac{h^3}{24(1-\mu^2)}\dfrac{\partial E^{(k)}}{\partial y}(\dfrac{\partial \varphi_y^{(k)}}{\partial y} \\
\quad + X\mu\dfrac{\partial \varphi_x^{(k)}}{\partial x}\Bigg] = 0 \\[4pt]
C^{(k)}\left(\nabla^2 w^{(k)} - \dfrac{\partial \varphi_x^{(k)}}{\partial x} - \dfrac{\partial \varphi_y^{(k)}}{\partial y}\right) + \kappa \dfrac{\partial^2 \psi^{(k)}}{\partial y^2} \\
\quad + \dfrac{5h}{12(1+\mu)}\dfrac{\partial E^{(k)}}{\partial y}\left(\dfrac{\partial w^{(k)}}{\partial y} - \varphi_y^{(k)}\right) = 0
\end{cases}
\qquad (2)
$$

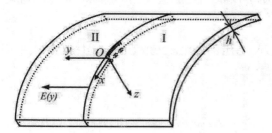

Figure 1. The interfacial crackThe boundary condition of the crack surface is.

$$\nabla^2\nabla^2\psi^{(k)} + \kappa B^{(k)}\frac{\partial^2 w^{(k)}}{\partial y^2}$$

$$-2\frac{dE^{(k)}}{E^{(k)}dy}\left(\frac{\partial^3\psi^{(k)}}{\partial y^3} + \frac{\partial^3\psi^{(k)}}{\partial x^2\partial y}\right)$$

$$+\frac{d^2E^{(k)}}{E^{(k)}dy^2}\cdot\left(\mu\frac{\partial^2\psi^{(k)}}{\partial x^2} - \frac{\partial^2\psi^{(k)}}{\partial y^2}\right) \qquad (3)$$

$$+2\left(\frac{dE^{(k)}}{E^{(k)}dy}\right)^2\left(\frac{\partial^2\psi^{(k)}}{\partial y^2} - \mu\frac{\partial^2\psi^{(k)}}{\partial x^2}\right) = 0$$

where $D^{(k)} = \frac{E^{(k)}h^3}{12(1-\mu^2)}$, $C^{(k)} = \frac{5E^{(k)}h}{12(1+\mu)}$, $k = 1, 2, 3$ and h is thickness of the shell (Figure 1).

$$\begin{cases} \dfrac{\partial\phi_y^{(k)}}{\partial y} + \mu\dfrac{\partial\phi_x^{(k)}}{\partial x} = 0, \quad \dfrac{\partial\phi_x^{(k)}}{\partial y} + \dfrac{\partial\phi_y^{(k)}}{\partial x} = 0, \\ \dfrac{\partial w^{(k)}}{\partial y} - \phi_y^{(k)} = 0 \qquad\qquad\qquad y = 0 \\ \dfrac{\partial^2\psi^{(k)}}{\partial x^2} = 0, \quad \dfrac{\partial^2\psi^{(k)}}{\partial x\partial y} = 0 \end{cases} \qquad (4)$$

3 HIGHER ORDER CRACK TIP FIELDS

The crack tip stress field would be equipped with the same square root singularity as that of homogeneous materials when the material prosperities of different composite materials at the interfaces are continuous (Jin Z. H, 1994). Therefore, the generalized displacements $\varphi_r, \varphi_\theta, w$ and $\psi^{(k)}$ can be expressed as follows

$$\phi_x^{(k)} = \sum_{n=1}^{\infty} f_n^{(k)}(\theta)r^{\frac{n}{2}}, \quad \phi_y^{(k)} = \sum_{n=1}^{\infty} g_n^{(k)}(\theta)r^{\frac{n}{2}},$$

$$w^{(k)} = \sum_{n=1}^{\infty} j_n^{(k)}(\theta)r^{\frac{n}{2}}, \quad \psi^{(k)} = \sum_{n=1}^{\infty} \psi_n^{(k)}(\theta)r^{\frac{n+2}{2}} \qquad (5)$$

where, $f_n^{(k)}(\theta), g_n^{(k)}(\theta), j_n^{(k)}(\theta)$ are eigen-functions and $k = 1, 2, 3$.

Substituting Eq. (5) into Eq. (2) and Eq. (3), the coefficients of $r^{-3/2}, r^{-1}, \ldots, r^{n/2-2}$ are linear independent, the each coefficient term must be zero. The obtained equations are solved and the eigenfunction are derived as follows

$$\begin{cases} f_1^{(k)} = A_{12}^{(k)}\sin\dfrac{3\theta}{2} + A_{11}^{(k)}\cos\dfrac{3\theta}{2} + \dfrac{\mu+9}{\mu+1}A_{12}^{(k)}\sin\dfrac{\theta}{2} + \\ \dfrac{3\mu-5}{\mu+1}A_{11}^{(k)}\cos\dfrac{\theta}{2} \\ g_1^{(k)} = A_{11}^{(k)}\sin\dfrac{3\theta}{2} - A_{12}^{(k)}\cos\dfrac{3\theta}{2} + \dfrac{\mu-7}{\mu+1}A_{11}^{(k)}\sin\dfrac{\theta}{2} + \\ \dfrac{\mu-7}{\mu+1}A_{12}^{(k)}\cos\dfrac{\theta}{2} \\ j_1^{(k)} = A_{13}^{(k)}\sin\dfrac{\theta}{2} \\ \psi_1^{(k)} = A_{15}^{(k)}\sin\dfrac{3\theta}{2} + A_{14}^{(k)}\cos\dfrac{3\theta}{2} + A_{15}^{(k)}\sin\dfrac{\theta}{2} + 3A_{14}^{(k)}\cos\dfrac{\theta}{2} \end{cases} \qquad (6)$$

$$\begin{cases} f_2^{(k)} = A_{21}^{(k)}\cos\theta + A_{22}^{(k)}\sin\theta \\ g_2^{(k)} = -A_{22}^{(k)}\cos\theta - \mu A_{21}^{(k)}\sin\theta \\ j_2^{(k)} = A_{23}^{(k)}\cos\theta \\ \psi_2^{(k)} = -A_{24}^{(k)} + A_{24}^{(k)}\cos2\theta \end{cases} \qquad (7)$$

$$\begin{cases} f_3^{(k)} = A_{31}^{(k)}\cos\dfrac{3\theta}{2} + A_{32}^{(k)}\sin\dfrac{3\theta}{2} + \dfrac{3(\mu+1)}{\mu-7}A_{31}^{(k)}\cos\dfrac{\theta}{2} \\ \quad +\dfrac{3(\mu+1)}{3\mu+11}A_{32}^{(k)}\sin\dfrac{\theta}{2} - \dfrac{4N_mA_{13}^{(k)}}{3\mu^2+8\mu-11}\cdot\sin\dfrac{\theta}{2} \\ \quad -\delta_{2k}\beta\left[\dfrac{1}{4}A_{11}^{(k)}\sin\dfrac{5\theta}{2} - \dfrac{1}{4}A_{12}^{(k)}\cos\dfrac{5\theta}{2}\right. \\ \quad \left. -\dfrac{3\mu-13}{4(3\mu+11)}A_{11}^{(k)}\sin\dfrac{\theta}{2} + \dfrac{5\mu-43}{4(\mu-7)}A_{12}^{(k)}\cos\dfrac{\theta}{2}\right] \\ g_3^{(k)} = \dfrac{7\mu-1}{3\mu+11}A_{32}^{(k)}\cos\dfrac{3\theta}{2} + \dfrac{3\mu-5}{\mu-7}A_{31}^{(k)}\sin\dfrac{3\theta}{2} \\ \quad -\dfrac{3(\mu+1)}{3\mu+11}A_{32}^{(k)}\cos\dfrac{\theta}{2} + \dfrac{3(\mu+1)}{\mu-7}A_{31}^{(k)}\sin\dfrac{\theta}{2} \\ \quad +\dfrac{4(3\mu+1)N_mA_{13}^{(k)}}{3(8\mu+3\mu^2-11)}\cos\dfrac{3\theta}{2} + \dfrac{4N_mA_{13}^{(k)}}{8\mu+3\mu^2-11} \\ \quad \cdot\cos\dfrac{\theta}{2} + \left[\dfrac{A_{12}^{(k)}}{4}\sin\dfrac{5\theta}{2} + \dfrac{3\mu-17}{3(\mu-7)}A_{12}^{(k)}\sin\dfrac{3\theta}{2}\right. \\ \quad +\dfrac{3\mu-13}{4(\mu-7)}A_{12}^{(k)}\sin\dfrac{\theta}{2} + \dfrac{A_{11}^{(k)}}{4}\cos\dfrac{5\theta}{2} + \dfrac{9\mu+13}{3\mu+11} \\ \quad \left.\cdot A_{11}^{(k)}\cos\dfrac{3\theta}{2} + \dfrac{21\mu+101}{4(3\mu+11)}A_{11}^{(k)}\cos\dfrac{\theta}{2}\right]\delta_{2k}\beta \end{cases} \qquad (8)$$

916

$$\left\{\begin{aligned}
j_3^{(k)} &= \frac{2(\mu-1)A_{11}^{(k)}}{\mu+1}\cos\frac{\theta}{2} + \frac{2(\mu+7)A_{11}^{(k)}}{3(\mu+1)}\cos\frac{3\theta}{2} \\
&+ \frac{2(\mu-1)A_{12}^{(k)}}{\mu+1}\sin\frac{\theta}{2} + A_{33}^{(k)}\sin\frac{3\theta}{2} - \frac{A_{13}^{(k)}}{12}\left(\cos\frac{3\theta}{2}\right. \\
&\left.+ 3\cos\frac{\theta}{2}\right)\delta_{2k}\beta + \frac{\kappa}{C^{(1)}}\left(\frac{A_{15}^{(k)}}{16}\sin\frac{5\theta}{2} - \frac{7A_{15}^{(k)}}{8}\sin\frac{\theta}{2}\right. \\
&\left.- \frac{A_{14}^{(k)}}{16}\cos\frac{3\theta}{2} + \frac{3A_{14}^{(k)}}{16}\cos\frac{5\theta}{2} - \frac{9A_{14}^{(k)}}{8}\cos\frac{\theta}{2}\right) \\
\psi_3^{(k)} &= A_{35}^{(k)}\sin\frac{5\theta}{2} + A_{34}^{(k)}\cos\frac{5\theta}{2} - A_{35}^{(k)}\sin\frac{\theta}{2} - 5A_{34}^{(k)} \\
&\cdot \cos\frac{\theta}{2} - \frac{\kappa B^{(1)}}{32}A_{13}^{(k)}\left(\sin\frac{\theta}{2} + \sin\frac{3\theta}{2}\right) \\
&+ \frac{\delta_{2k}\beta}{4}\left(3A_{14}^{(k)}\sin\frac{3\theta}{2} + 3A_{14}^{(k)}\sin\frac{\theta}{2} - A_{15}^{(k)}\cos\frac{3\theta}{2}\right. \\
&\left.- 3A_{15}^{(k)}\cos\frac{\theta}{2}\right)
\end{aligned}\right.$$

$$\left\{\begin{aligned}
f_4^{(k)} &= A_{41}^{(k)}\cos 2\theta + A_{42}^{(k)}\sin 2\theta - \frac{\mu+1}{\mu-3}A_{41}^{(k)} \\
&+ \frac{N_m A_{23}^{(k)}}{\mu^2+2\mu-3} \\
g_4^{(k)} &= A_{42}^{(k)}\cos 2\theta - \frac{2\mu A_{41}^{(k)}}{\mu+3}\sin 2\theta - \frac{\mu+1}{2}A_{42}^{(k)} \\
&- \frac{\mu A_{23}^{(k)} N_m}{\mu^2+2\mu-3}\sin 2\theta \\
j_4^{(k)} &= A_{43}^{(k)}\cos 2\theta - \frac{A_{22}^{(k)}}{2}\sin 2\theta - \frac{\mu-1}{4}A_{21}^{(k)} + \frac{\kappa}{C^{(1)}}A_{24}^{(k)} \\
\psi_4^{(k)} &= A_{44}^{(k)}\cos 3\theta + A_{45}^{(k)}\sin 3\theta - 3A_{45}^{(k)}\sin\theta - A_{44}^{(k)}\cos\theta
\end{aligned}\right.$$

$$(9)$$

$$\left\{\begin{aligned}
f_5^{(k)} &= A_{51}^{(k)}\cos\frac{5\theta}{2} + A_{52}^{(k)}\sin\frac{5\theta}{2} - \frac{5(\mu+1)}{5\mu+13}A_{52}^{(k)}\sin\frac{\theta}{2} \\
&- \frac{5(\mu+1)}{\mu+9}\cdot A_{51}^{(k)}\cos\frac{\theta}{2} \\
&+ N_m\left\{\left[\frac{35\mu^2+44\mu+233}{96(5\mu+13)(\mu-1)}\frac{\kappa}{C^{(1)}}\cdot A_{15}^{(k)}\right.\right. \\
&\left.+ \frac{12(\mu+1)A_{33}^{(k)} - (39\mu+71)A_{12}^{(k)}}{3(5\mu+13)(\mu-1)(\mu+1)}\right]\sin\frac{\theta}{2} \\
&+ \left[\frac{\kappa}{C^{(1)}}\frac{\mu+53}{32(9+\mu)}A_{14}^{(k)} - \frac{9\mu-71}{3(\mu-1)(\mu+1)(9+\mu)}A_{11}^{(k)}\right] \\
&\cdot \cos\frac{\theta}{2} + \left[\frac{5}{64}\frac{\kappa}{C^{(1)}}A_{15}^{(k)} - \frac{A_{12}^{(k)}}{(\mu-1)(\mu+1)}\right]\sin\frac{3\theta}{2} \\
&+ \left[\frac{3}{64}\frac{\kappa}{C^{(1)}}A_{14}^{(k)} - \frac{A_{11}^{(k)}}{(\mu-1)(\mu+1)}\right]\cos\frac{3\theta}{2} \\
&\left.- \frac{1}{192}\frac{\kappa}{C^{(1)}}\cdot\left(A_{15}^{(k)}\sin\frac{7\theta}{2} + 3A_{14}^{(k)}\cos\frac{7\theta}{2}\right)\right\}
\end{aligned}\right.$$

$$(10)$$

$$\begin{aligned}
&+ \delta_{2k}\gamma\left\{-A_{31}^{(k)}\cdot\frac{(15\mu+31)(\mu+1)}{4(5\mu+13)(\mu-7)}\sin\frac{\theta}{2}\right. \\
&+ \left[\frac{27\mu^3+270\mu^2+99\mu-272}{48(3\mu+11)(\mu-1)(\mu+9)}\right. \\
&\left.\cdot N_m A_{13}^{(k)} + \frac{(\mu+1)(9\mu-71)}{4(\mu+9)(3\mu+11)}A_{32}^{(k)}\right]\cos\frac{\theta}{2} \\
&- \frac{3(\mu+1)}{4(\mu-7)}A_{31}^{(k)}\cdot\sin\frac{3\theta}{2} + \\
&\left[\frac{3\mu^2+11\mu-16}{16(\mu-1)(3\mu+11)}N_m A_{13}^{(k)} + A_{32}^{(k)}\cdot\frac{3(\mu+1)}{4(3\mu+11)}\right] \\
&\cdot\cos\frac{3\theta}{2} + \delta_{2k}\gamma^2\left[\frac{3\mu^2-13\mu-64}{16(\mu-7)}A_{12}^{(k)}\sin\frac{3\theta}{2}\right. \\
&+ \frac{50\mu^3-165\mu^2-1152\mu-2089}{48(5\mu+13)(\mu-7)}\cdot A_{12}^{(k)}\sin\frac{\theta}{2} \\
&+ \frac{21\mu^2+77\mu-24}{16(3\mu+11)}\cdot A_{11}^{(k)}\cos\frac{3\theta}{2} \\
&+ \frac{210\mu^3+1715\mu^2+2832\mu+175}{48(\mu+9)(3\mu+11)}A_{11}^{(k)}\cos\frac{\theta}{2} \\
&\left.+ \frac{\mu-5}{48}\left(A_{12}^{(k)}\sin\frac{7\theta}{2} + A_{11}^{(k)}\cos\frac{7\theta}{2}\right)\right] \\
&\cdots\cdots
\end{aligned}$$

where $\delta_{2k} = \begin{cases} 0, & k=1 \\ 1, & k=2 \end{cases}$, and $B_{ij}^{(k)}$ are undetermined coefficients $(i=1,\dots,n;\ j=1,2,3)$ [5].

4 CONTINUOUS CONDITIONS

Substituting Eqs. (6)–(10) into Eq. (5), the generalized displacement fields in homogenous material and FGMs region are obtained, and the stress fields will be obtained based on the relationship between the generalized displacement and stress. The stress equations are subject to the following continuous conditions of stress

$$\sigma_y^{(1)} = \sigma_y^{(2)},\ \tau_{xy}^{(1)} = \tau_{xy}^{(2)},\ \tau_{zy}^{(1)} = \tau_{zy}^{(2)}, \quad \theta=0 \tag{11}$$

The relations between undetermined coefficients can be obtained as

$$A_{11}^{(1)} = A_{11}^{(2)}, \quad A_{12}^{(1)} = A_{12}^{(2)}, \quad A_{13}^{(1)} = A_{13}^{(2)} \tag{12}$$

$$A_{14}^{(1)} = A_{14}^{(2)}, \quad A_{15}^{(1)} = A_{15}^{(2)}, \quad A_{31}^{(1)} = A_{31}^{(2)} - \frac{3\mu-29}{12(\mu+1)}\beta A_{12}^{(2)} \tag{13}$$

$$A_{32}^{(1)} = A_{32}^{(2)} + \frac{21\mu+53}{12(\mu+1)}\beta A_{11}^{(2)}, \quad A_{33}^{(1)} = A_{33}^{(2)} \tag{14}$$

$$A_{34}^{(1)} = A_{34}^{(2)} + \frac{\beta}{4}A_{15}^{(2)}, \quad A_{35}^{(1)} = A_{35}^{(2)} + \frac{3\beta}{4}A_{14}^{(2)} \tag{15}$$

$$A_{51}^{(1)} = A_{51}^{(2)} - \frac{15\mu - 17}{20(3\mu + 11)}\gamma A_{32}^{(2)}$$
$$- \frac{(15\mu + 7)(3\mu^2 + 35\mu + 64)}{240(\mu + 1)(\mu - 1)(3\mu + 11)}\gamma N_m A_{13}^{(2)} \quad (16)$$
$$- \frac{345\mu^2 + 3323\mu + 6898}{240(3\mu + 11)}\gamma^2 A_{11}^{(2)}$$

$$A_{52}^{(1)} = A_{52}^{(2)} + \frac{25\mu + 57}{20(\mu - 7)}\gamma A_{31}^{(2)}$$
$$+ \frac{55\mu^2 - 147\mu - 1434}{240(\mu - 7)}\gamma^2 A_{12}^{(2)} \quad (17)$$

$$A_{53}^{(1)} = A_{53}^{(2)} - \frac{28\mu + 57}{30(\mu + 1)}\gamma A_{11}^{(2)}$$
$$- \frac{137}{320}\frac{\kappa}{C^{(1)}}\gamma A_{14}^{(2)} + \frac{11}{160}\gamma^2 A_{13}^{(2)} \quad (18)$$

$$A_{54}^{(1)} = A_{54}^{(2)} + \frac{1}{4}\gamma A_{35}^{(2)} - \frac{1}{96}\gamma\kappa B^{(1)} A_{13}^{(2)} - \frac{7\mu + 8}{16}\gamma^2 A_{14}^{(2)} \quad (19)$$

$$A_{55}^{(1)} = A_{55}^{(2)} - \frac{5}{4}\gamma A_{34}^{(2)} + \frac{3\mu - 8}{48}\gamma^2 A_{15}^{(2)} \quad (20)$$

Substituting Eqs. (12)–(20) and Eqs. (6)–(9) into Eq. (5), the generalized displacement fields in homogenous material and FGMs regions are obtained finally (Chong Xiao, 2014).

5 CONCLUSIONS

The governing equations are derived for interfacial crack between linear FGM and homogeneous Material of cylindrical shells with Reissner's effect. The higher order crack tip fields are obtained by assuming that the material parameter of FGM changes in linear function. The non-homogeneous material parameter β first appeared in the third order field. The curvature only occurs in the third order and higher order items, and both have evident effect on the higher order items. All the singular items of the weak-discontinuity crack tip fields are continuous in the interface, but not in the higher order items. The crack-tip fields obtained also have the same property of eigen-function as Williams' solution. They can be applied to the fracture parameter calculation and crack-tip stress analysis in different materials, loads and structures, which is of extensive applicability.

ACKNOWLEDGEMENTS

The research is supported by the National Natural Science Foundation of China (No.11172332)

REFERENCES

Chong Xiao, 2014. The higher order crack-tip fields for physical weak-discontinuous problems of functionally graded plates, shells and piezoelectric materials, Academy of Armored Force Engineering (Ph D. thesis), Beijing.

Delale F., Erdogan F. 1988. Interface crack in a non-homogeneous elastic medium, *International Journal of Engineering Science*. 6: 559–568.

Jin Z. H., Noda N., 1994. Crack-tip singular fields in nonhomogeneous materials, *Journal of Applied Mechanics*. 3: 738–740.

Liu Chuntu, 1983. Stresses and deformations near the crack tip for bending plate, *ACTA Mechanica Solid Sinica*. 3: 441–448.

Zhang Lei, 2012. Analysis of the higher order crack-tip asymptotic fields for the functionally graded spherical shell and cylindrical shell with Reissner's effect, Academy of Armored Force Engineering (Ph D. thesis), Beijing.

In-situ cross-linking of horseradish peroxidase on ZnO nanowires for decolorization of azo dye

Huaiyan Sun, Xinyu Jin, Nengbing Long & Ruifeng Zhang
Faculty of Materials Science and Chemical Engineering, Ningbo University, Ningbo, Zhejiang, P.R. China

ABSTRACT: A ZnO nanowires/ macroporous SiO$_2$ composite was used as support to immobilize Horseradish Peroxidase (HRP) simply by in-situ cross-linking method. As long-chained cross-linker Diethylene glycol Diglycidyl Ether (DDE) was adsorbed on the surface of ZnO nanowires, the cross-linked HRP was quite different from the traditional Cross-Linking Enzyme Aggregates (CLEAs) on both structure and catalytic performance. The immobilized HRP showed high activity in the decolorization of azo dyes. The effect of various conditions such as the loading amount of HRP, solution pH, temperature, contact time and concentration of dye were optimized on the decolorization. The decolorization percentage of Acid Blue 113 and Acid black 10 BX reached as high as 95.4% and 90.3%, respectively. The immobilized HRP exhibited much better resistance to temperature and pH inactivation than free HRP. The storage stability and reusability were greatly improved through the immobilization.

1 INTRODUCTION

The textile industry generates large amounts of liquid effluent pollutants every year (Robinson et al. 2001, Savin & Butnaru 2008). Azo dyes constitute a major class of dyes used in textile processing, contributing to about 70% of total dyes consumption (Farabegoli et al, 2010; Kuberan et al., 2011). These dyes have detrimental effect on human health as well as aquatic life, and can be hardly degraded in the environment because of their resistance to the oxidizing agents, light and water due to their chemical structure (Torres et al, 2003; Boucherit et al., 2013). It is necessary to treat industrial wastewaters containing azo dyes before their final discharge into the environment.

As compared with the physico-chemical methods, decolorization by enzymatic degradation has attracted more attention because of mild reaction condition and environment-friendness (Ulson de Souza et al, 2007), their technical feasibility and advantages on cost-effectiveness have been confirmed in many studies (Li et al, 2013). The oxidative degradations of azo dyes have been catalyzed by many kinds of peroxidases (Husain, 2009; Solís et al, 2012), one of them is Horseradish peroxidase (HRP). HRP is known to degrade a wide spectrum of aromatic compounds such as phenols, anilines as well as dyes in the presence of H$_2$O$_2$ (Karam, Nicell 1997; Preethi et al, 2013). Up to now numerous analytical applications of HRP have been developed because of its high activity and broad substrate specificity (Monier et al, 2010). However,

the HRP enzyme with a large range of application in the field of bioremediation and biotechnology suffers from inadequate stability. A rise in temperature and changes of pH beyond the critical level play an important role in the disruption of enzymatic activity and denaturation of HRP (Kalaiarasan, Palvannan, 2015).

Advantages such as activity beyond the optimum temperature and pH, long storage and high reuse stability could be realized through the immobilization of HRP (Magnin et al, 2003; Hassani et al, 2006; Mohamed et al, 2013). The immobilizations have been carried out by covalently attachment on poly (ethylene terephthalate) fibers (Arslan, 2010) and polysulfone (Celebi et al, 2013), or by entrapment in polyacryamide gel and alginate acid (Mohan et al, 2005). The azo dyes are usually nagative-charged, so a positive-charged support would be more suitable than a neutral one for the immobilized HRP to catalyst the dye degradation. For example, on a β-cyclodextrine/chitosan complex HRP was adsorbed and cross-linked with glutaraldehyde, the immobilized HRP showed much higher activity and stability than the free HRP in the decolorization of azo dyes (Karim, 2012). Cross-linked HRP was also prepared on a polycationic films (Bayramoğlu et al, 2012), it showed higher decolorization efficiency for dye degradation as compared with free HRP, whereas the HRP in the form of cross-linking enzyme aggregates (CLEAs) without support showed lower activity than the free HRP (Jiang et al, 2014). Obviously effect of static interaction on the biodegradation of azo dye is not negligible. From the point of stability and tolerance towards hydrogen peroxide an inorganic support such as mesoporous material seems to be much better than polymeric ones. A lot of references have been reported about the immobilization of HRP on mesoporous silica (Li, Takahashi 2000; Ikemoto et al, 2010; Yasutaka et al, 2011; Wan et al, 2012), however, probably due to the diffusional limitation of reactants and products within the narrow pores, the immobilized HRP on mesoporous silica has not been widely used for the biodegradation of dye.

The combination of macroporous silica with ZnO nanowire might be a reasonable strategy to design an inorganic support that could be used to immobilize HRP for the degradation of azo dye. The macroporous silica with much larger pore size than mesoporous silica can greatly reduce the diffusional limitation, while the ZnO nanowires possess high surface area and ample surface positive charges, endowing the nano-composite with strong adsorbing ability. In our previous work the nano-composite of ZnO nanowires and macroporous silica was prepared via an in-situ growth of ZnO nanowires in macro-pores of three-dimensional SiO$_2$ (Zhang et al, 2009; Shang et al, 2014; Shang et al, 2015a), the specific surface area of the composite is 233 m^2/g, the support exhibited strong adsorbing ability towards lipase with an average loading amount of 196.8 mg/g. In this work ZnO nanowires were capable of adsorbing bi-epoxy compound that can react with amino groups on HRP under proper conditions, an in-situ cross-linking between HRP molecules can therefore be realized on the surface of ZnO nanowires (see Fig. 1). Quite different from the traditional CLEA preparations in which the cross-linker was directly added to the reaction systems, in the current operation the cross-linker was adsorbed on the support before the reaction so that its amount and dispersion could be well-controlled. This method provided an approach to avoid excessive cross-linking of the immobilized enzyme and

make it possible to get higher catalytic efficiency than above-mentioned CLEAs. Furthermore, the bi-epoxy cross-linker has longer chain than glutaraldehyde, it can help to remove the influence of the carrier on the activity of the enzyme.

In the present study, the HRP was immobilized on the nano-composite support through a controlled cross-linking method and applied to decolorize two azo dyes (their structures are shown in Fig. 1). The influence of conditional parameters such as loading amount of enzyme, pH, temperature, and dye concentration on the process of decolorization were investigated. The kinetic parameters (K_m and V_{max}) of both free and immobilized HRP were calculated according to the Lineweaver-Burk plot. Finally, storage stability and reusability of the immobilized HRP were also tested to show the advantage of immobilized HRP.

2 MATERIALS AND METHODS

2.1 Materials

Horseradish peroxidase (HRP, EC.1.11.1.7, pI: 5.3, 150 U/mg) was purchased from Source leaves Biotechnology Co. (Shanghai, China). Bisphenol A epoxy resin (the epoxy value is 0.44), PEG600, PEG1000, PEG2000, Ammonia solution, Zinc acetate dehydrate, Zinc nitrate hexahydrate, buffer salts, hydrogen peroxide, pyrogallol, Diethylene glycol diglycidyl ether (DDE) was purchased from Aladdin Industrial Co. (Shanghai, China). Acid blue 113 and Acid black 10 BX were purchased from Boyle Chemical Co. Ltd (Shanghai, China) and Dynasty Chemical Co. Ltd (Ningbo, China), respectively. All other reagents were used of analytical grade and without further purification.

2.2 Preparation of the support

Nano-composite support was synthesized according to our previous report (Zhang et al, 2009; Shang et al, 2014; Shang et al, 2015a; Shang et al, 2015b). The three-dimensional (3D) SiO$_2$ was prepared by coating silica on the surface of a template polymer and a sequent calcination of the silica/polymer composite. The growth of ZnO nanowires in the macroporous SiO$_2$ was carried out through two steps: ZnO crystal seed pre-treatment and in situ hydrothermal synthetic process under a properly controlled condition.

2.3 Immobilization of HRP on the composite supports

The supports were cut into cubes with a size of $2 \times 2 \times 2$ (mm). In the first step, supports were soaked in ethanol solution of cross-linker

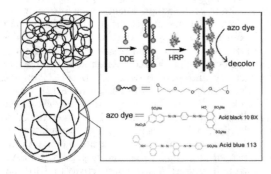

Figure 1. Schematic illustration of HRP cross-linked on nano-composite support for decolorization of two azo dyes.

DDE. The concentration of solution was about 20–30 mg/mL. 10 minutes later samples were taken out of the solutions and completely dried at 70 °C for 3 h. Then the dried samples were washed with cyclohexane for three times and dired again for next experiments. In the second step, HRP solutions were prepared by adding different amounts of HRP powder to phosphate buffer (20 mL). The mixture was shaken for 20 min, and then collected supernatant by centrifugation for 10 min at 25 °C. The supports with cross-linker were soaked and shaken in 10 mL of HRP solution, the adsorption proceeded for 2 h at pH 7, then the samples were taken out from the solution and sealed in a bottle. The following cross-linking was carried out at 15 °C for 15 h. After that the samples were washed with deionised water and buffer solution, finally stored at 4°C in a refrigerator.

2.4 Loading assay

The amount of immobilized HRP (P_0) was calculated by the difference of the concentration of the HRP before and after adsorption in the buffer solution. The protein content of HRP was determined by measuring the protein concentration in the solution using a Bio-Rad protein assay kit (Sigma–Aldrich) with bovine serum albumin as a standard, the absorbance of the reaction mixture was recorded at 595 nm (Malania et al. 2013). According to the Bradford method (Bradford 1976), the P_0 is calculated from the formula:

$$P_0 = \frac{(C_1 - C_2)V_1 - C_3V_2}{m} \quad (1)$$

$$\text{Loading efficiency (\%)} = \frac{P_0}{C_1V_1} \times 100\% \quad (2)$$

C_1 and C_2 (mol/mL) are the concentration of the HRP initial and final in the reaction medium respectively; C_3 (mol/mL) is the concentration of the HRP in the buffer solution for washing the sample; V_1 is the volume of the reaction medium (mL); V_2 is the volume of the buffer solution for washing the sample (mL) respectively; m is the weight of the supports (g).

2.5 Enzyme assays for free and immobilized HRP

HRP activity was measured at 30 °C, using pyrogallol and H_2O_2 as substrates. Typical tests were carried out by adding 500 µL of H_2O_2 (0.05 mol/L) to 2.5 ml of phosphate buffer 0.1 mol L^{-1}, pH 6.0 containing 0.013 mol/L of pyrogallol and 1.0 µg of free HRP. The progress of the reaction was monitored by the formation of purpurogallin, the oxidized product of pyrogallol at 420 nm (λ_{max} of purpurogallin). In triplicate experiments, the activity values

were found to be varied within ± 2–5%. Tests for the immobilized enzyme were performed in the same conditions used for free enzyme, except that the reaction were maintained with stirring, and interrupted by separation of the HRP from the reaction mixture by filtration in a Buchner funnel before the spectrophotometric readings. One unit of HRP activity (U) is defined as the amount of HRP capable of producing 1 µmol of purpurogallin in 1 min under the specified reaction conditions (Malania et al., 2013).

$$\text{Immobilization efficiency (\%)} = \frac{E_f P_0 m}{E_1 C_1 V_1} \times 100\% \quad (3)$$

where E_1 is the initial HRP activity (U/mg), E_f is the final immobilized HRP activity (U/mg).

2.6 Decolorization efficiency

The experiments were conducted in Erlenmeyer flask which containing the 10 mL dye solution, 8 U enzyme and 10 mL of 8 mM H_2O_2. The percentage of dye decolorization efficiency was determined by measuring the decrease of the color at the wavelength of the maximum absorbance of Acid blue 113 (566 nm) (Preethi et al. 2013) and Acid black 10 BX (617 nm) (Mohan et al. 2005) respectively:

$$\text{Decolorization (\%)} = [(C_0 - C_f) / C_0] \times 100\% \quad (4)$$

where C_0 and C_f are the concentration of the azo dye in the initial and final solution for a certain period of time (mg/L). To obtain maximum dye decolorization, a series of experiments was carried out by varying the process parameters such as loading amount of enzyme, solution pH, reaction time, temperature and dye concentration on the dye decolorization. All sets of experiments were reproduced at least three times at the same operating conditions, and the discrepancy was below 5%.

2.7 Kinetic properties

The Michaeli's constant value (K_m) and maximum rate (V_{max}) was determined by measuring the initial reaction rates of free and immobilized HRP using different concentrations of Acid blue 113 as substrate in phosphate buffer (0.05 M, pH 7.0) at 25°C. The variation in the substrate-removal rate as a function of the substrate concentration was described by the Michaelis–Menten model. The model has the form (Kim et al. 2012):

$$v = \frac{v_{max} \times C_s}{k_m + C_s} \quad (5)$$

where v is the rate of substrate utilization, V_{max} is the maximum rate of substrate utilization, C_S is the initial substrate concentration, and K_m is the half-substrate constant indicating enzyme-substrate affinity. The Michaelis–Menten model parameters were determined graphically by constructing the Lineweaver–Bulk plot.

2.8 Storage stability

For testing the storage stability of enzymes, free and immobilized HRP in phosphate buffer (pH 7.0) were stored in a refrigerator at 4 °C for several days. Then the enzyme samples were taken out at each time point and the decolorization of Acid blue 113 and Acid black 10 BX by free and immobilized HRP were measured under optimum conditions.

2.9 Reusability of immobilized HRP

The reusability of the immobilized HRP was assayed up to twelve sequential cycles for the decolorization of dyes. At the end of each reaction cycle, the immobilized HRP was recovered by filtration and washed three times with phosphate buffer (pH 7.0), then repeated with a fresh aliquot of substrate.

3 RESULTS AND DISCUSSION

3.1 Cross-linking of HRP on the nano-composite support

The characterization of the composite support was published in the previous work (Zhang et al, 2009; Shang et al, 2014; Shang et al, 2015a; Shang et al, 2015b), a schematic diagram is given in Fig. 1 to describe the structure of the composite support. The large-sized macroporous SiO_2 acts as a framework, it has good mechanical stability and three-dimensional (3D) pass-through pores in size of 0.5~1 μm. The ZnO nanowires with diameter of 15–20 nm and morphology in random coil are dispersed in the SiO_2 pores. In the adsorption experiment the adsorbed cross-linker should be completely washed with cyclohexane to ensure that the non-adsorbed molecules are removed, the remained cross-linker can be detected by Raman spectra on the composite support but not on the macroporous SiO_2. Figure 2 shows Raman spectra of composite support and sample adsorbed with epoxy cross-linker. The spectrum a has three absorption bands at 582, 1156 and 1715 cm^{-1}, respectively, assigned to the characteristic peaks of ZnO nanowires, the macroporous SiO_2 has no Raman signal. The spectrum b shows the characteristic peak at 1259 cm^{-1}, it should be assigned to

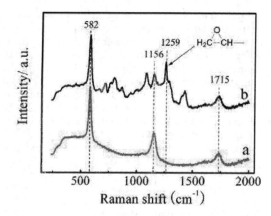

Figure 2. Raman spectra of composite support before (a) and after adsorption with cross-linker (b).

the stretching vibration of epoxy group, indicating that the epoxy cross-linker can be adsorbed on the composite support. The amount of adsorbing cross-linker was determined to be 8~10 wt.% by weight loss experiments at 650 °C for 1 hour.

The effect of HRP concentration on loading amount and loading efficiency was investigated. A higher concentration of enzyme naturally leads to a higher enzyme loading amount, but the loading efficiency decreased. Since the amount of DDE adsorbed on the composite supports changed in a very narrow range (8~10 wt.%), the amount of HRP that could be cross-linked was also limited, as the concentration of HRP increased, more and more free HRP would be left in the filtrate and washing buffer solutions, resulting in a drop of the loading efficiency. This experiment showed that the optimized concentration was 4 mg/mL. The loading amount was 161.3 mg of HRP/1 g support and efficiency was 80.65%, the corresponding immobilization efficiency was 75.3%.

3.2 Loading amount

The optimization of the loading amount of enzyme was carried out aiming at high efficiency of decolorization with low enzyme loading amount. Figure 3 shows the effect of loading amount on the decolorization. The support itself can partly adsorb azo dye, when using the pure support to treat the dye solution of Acid blue 113 and Acid black 10 BX, the decolorization were 11.4% and 16.2%, respectively. The major role in decolorization of the dye is enzymatic catalysis. Increasing the loading amount of HRP resulted in a gradual increase in the dye decolorization rates, as the loading amount was 107.5 mg/g-support, the decolorization of Acid blue 113 and Acid black 10 BX was up to 95.8% and 84.5%, respectively. Subsequent

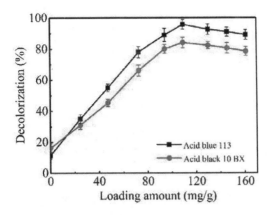

Figure 3. Effect of loading amount on the decolorization efficiency, Conditions: buffer pH 7.0, 30 °C, 35 min,
dye concentration (50 mg/L).

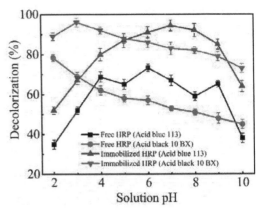

Figure 4. Effect of pH on the decolorization of dye by
free HRP [30 °C; 45 min; dye concentration (30 mg/L)]
and immobilized HRP [30 °C; 35 min; dye concentration
(50 mg/L)].

increase in loading amount imposed a negative
effect on the dye decolorization. The HRP molecule was negative-charged as long as the solution
pH (7) above the isoelectric point (5.3), there would
be a repelling force between HRP and dyes bearing
SO_3^- groups, the positive-charged ZnO nanowires
exhibited attracting force towards both HRP and
dyes and speeded up the decolorization reaction.
As more and more HRP was immobilized on the
support the positive effect of ZnO nanowires on
the reaction would be counteracted, leading to a
slight decrease of decolorization rate. The number
of negative charge on dye molecule can also effect
the decolorization percentage. The decolorization
of Acid blue 113 with two SO_3^- groups was always
higher than that of Acid black 10 BX bearing four
SO_3^- groups in cases of both the free and immobilized HRP catalysis. Similar observations were
reported in earlier studies (Celebi et al, 2013).

3.3 Solution pH

The performance of enzyme activities strongly
depended on the solution pH value. Figure 4 shows
effect of the solution pH on the decolorization catalyzed by free and immobilized HRP, respectively.
It can be seen that the decolorizations of both dyes
under catalysis of immobilized HRP was less sensitive to pH compared to its free counterpart, and the
free HRP reached a maximum dye decolorization at
pH 2.0 and 6.0 for Acid blue 113 and Acid black 10
BX, whereas the optimal pH for immobilized HRP
rose to 3.0 and 7.0, respectively. The slight shift of
the optimal pH after HRP immobilization was also
observed in a number of prior studies (Bayramoğlu
et al, 2012; Rai et al, 2012; Jiang et al, 2014). This
is due to the change in the microenvironment
around the immobilized enzyme towards a cationic

environment and the separation of the enzyme
from the bulk solution. The unequal H^+ displacement in the microenvironment around the immobilized enzyme with bulk solution can lead to a
more-alkaline condition for the optimal enzymatic
reaction. Free HRP got a highest decolorization of
78.4% and 73.3% for Acid blue 113 and Acid black
10 BX, respectively. The immobilized HRP provided a highest decolorization of 96% and 92.4%
for Acid blue 113 and Acid black 10 BX, respectively. The immobilized HRP showed catalysis less
sensitive to pH compared to its free counterpart,
which may be attributed to the protection effect
provided by the stable binding with the composite support. Strong interactions, including cova-
lent cross-linking, have been suggested to provide
intramolecular forces that prevent conformational
changes in enzymes (Sanjay & Sugunan, 2006). The
enzyme protection reinforced by enzyme immobilization has been reported by a number of other
investigators (Bayramoğlu & Arıca 2008, Kim et
al, 2012) who demonstrated that enzyme immobilization restricted denaturation or unfolding of
enzymes in response to sudden pH changes.

3.4 Temperature

Reaction temperature is another important
parameter affecting the activity of the enzyme,
consequently, the percentage of decolorization.
The decolorization process was carried out under
different temperatures varying from 10°C to
60°C, the results are illustrated in Figure 5. Both
free and immobilized HRP kept high activities
when the temperature maintained at 20~30°C. As
the temperature above 30°C, the decrease in the

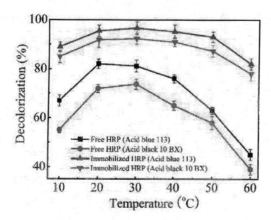

Figure 5. Effect of temperature on the decolorization of dye by free HRP [buffer pH 6.0 for Acid blue 113 and 2.0 for Acid black 10 BX; 45 min; dye concentration (30 mg/L)] and immobilized HRP [buffer pH 7.0 for Acid blue 113 and 3.0 for Acid black 10 BX; 35 min; dye concentration (50 mg/L)].

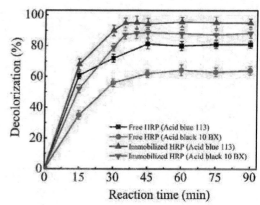

Figure 6. Effect of reaction time on the decolorization of dye by free HRP [buffer pH 2.0 for Acid blue 113 and 6.0 for Acid black 10 BX, 30 °C, dye concentration (30 mg/L)] and immobilized HRP [buffer pH 3.0 for Acid blue 113 and 7.0 for Acid black 10 BX, 30 °C, dye concentration (50 mg/L)].

decolorization rates of both dyes with free HRP was more pronounced compared to the immobilized form, after 45 min at 60°C the rates were found to be 45% and 39% for Acid blue 113 and Acid black 10 BX, respectively. The immobilized HRP kept catalytic activities much higher than free HRP at all the range of temperature, after only 35 min reaction more than 78% of decolorization was found at 60°C. This improvement mainly resulted from the difference in the microenvironment between immobilized and free HRP, and the multipoint attachment between the cross-linked HRP could improve the rigidification of the protein and protect it from denaturation.

3.5 Time

The investigation of reaction time is always necessary because it will be helpful to determine the shortest time necessary for obtaining the highest decolorization and so enhancing cost-effectiveness. Figure 6 shows the dye decolorization against time with two azo dyes under the catalysis of free and immobilized HRP. Under the catalysis of free HRP dye decolorization stopped increasing as the reaction time exceeded 45 min, at that time only 81.3% of Acid blue 113 and 64.1% of Acid black 10 BX were degraded. The immobilized HRP made the dyes decolorized more effectively than did the free HRP. Only after 35 min of reaction the decolorization reached the maximum of 95.4% and 90.3% for Acid blue 113 and Acid black 10 BX, respectively. When HRP was immobilized on the calcium alginate beads, significantly high reaction time (240 min) is required to degrade the

Acid blue 113 (Preethi et al, 2013). The improvement of decolorization efficiency might be owing to two important reasons: (1) the substrate can be enriched through adsorption on support, making the catalytic reaction easier; (2) the macroporous structure of support is benefit for the improvement of the mass-transfer rate of substrate to the active site of the enzymes, which is beneficial to exert the optimal catalysis of enzyme and shortening the reaction time.

3.6 Concentration of azo dye

Concentration of the substrate present in the aqueous phase has significant influence on any enzyme-mediated reaction. If the amount of enzyme is kept constant the degradation rate of the dye will increase with the substrate concentration before it reaches the maximum. Once reaching the equilibrium state any further addition of the substrate will not change the dye removal. The dependence of dye removal on the dye concentration were investigated using both the free and the immobilized HRP, the results were presented in Figure 7. The free HRP showed the equilibrium point at 30 mg/L dye concentration, the corresponding dye removal were 23.5 mg/L and 19.6 mg/L for Acid blue 113 and Acid black 10 BX, respectively. The immobilized HRP showed the equilibrium point at 50 mg/L dye concentration, the dye removal were 46.3 and 41.6 mg/L for Acid blue 113 and Acid black 10 BX, respectively. Obviously the decolorization rates by the immobilized HRP were higher than those by free HRP. These observations could be attributed to the nano-composite support's adsorbing ability and

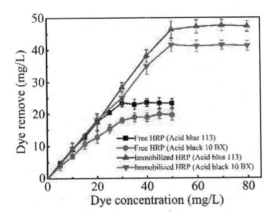

Figure 7. Effect of dye concentration on the decolorization of dye by free HRP (buffer pH 2.0 for Acid blue 113 and 6.0 for Acid black 10 BX; 30 °C; 45 min) and immobilized HRP (buffer pH 3.0 for Acid blue 113 and 7.0 for Acid black 10 BX; 30 °C; 35 min).

Table 1. Kinetic parameters of free and immobilized HRP under optimum conditions.

	V_{max} (mM/min)	K_m (mM)
Free HRP	0.069	0.073
Immobilized HRP	0.557	0.052

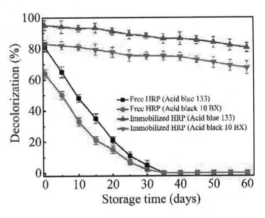

Figure 8. Storage stabilities of free and immobilized HRP for degradation Acid blue 113 and Acid black 10 BX under optimum conditions.

the improvement of immobilized HRP's catalytic capacity. Those may be presumed to be the cut-off concentration of the both dyes for the optimum removal at the specified experimental conditions.

3.7 Kinetic parameters of free and immobilized HRP

The kinetics of the free and immobilized HRP were investigated follows the Michaelis–Menten equation. Lineweaver-Burk plots (double reciprocal plot) were made for both free and immobilized HRP, the results were summarized in Table 1. The decrease in K_m value and the increase in V_{max} value of immobilized HRP confirmed that the greater stability of the immobilized HRP to the dye compared to the free one. HRP immobilized in nano-composite support showed lower K_m values and higher V_{max} value than that of calcium alginate beads (Preethi et al, 2013), which demonstrated that the diffusional limitation of reactants and products can be evidently reduced within this support. The large interconnected pores of composite supports can enhance the accessibility of substrates to the enzyme active sites and the short diffusion path of the pore walls can accelerate the material exchange.

3.8 Storage stability

Figure 8 shows the long-term stability of free HRP and immobilized HRP at a common storage temperature of 4 °C, at pH 7.0. The decolorization of free HRP dropped much faster than the immobilized HRP which almost lost all of its catalytic

activity of Acid blue 113 and Acid black 10 BX over 35 days, whereas immobilized enzymes were stable, showing higher decolorization values for long storage. A slight drop of decolorization was observed during 60 days storage of the immobilized HRP, 80.4% and 67.5% of initial activity retained for decolorization of Acid blue 113 and Acid black 10 BX, respectively. A leaching test was carried out and pyrogallol was selected as substrate to detect HRP in storage solution at different timings. Almost no increase of absorbance at 420 nm (λ_{max} of purpurogallin) was observed during 60 days of testing, indicating that leaching of HRP had been inhibited by in-situ cross-linking. Therefore, gradual deactivation of the immobilized HRP might be the reason for the drop of decolorization during storage. Generally speaking, immobilized enzymes had very long storage lifetime compared to free HRP which is considered to be a favorable feature for usage in applications.

3.9 Reusability

The reusability is one of the most important factors for reducing overall cost of industrially applied enzymes. Unlike free enzyme, immobilized enzyme could be easily separated from reaction solution and reused. Figure 9 shows the decolorization of both dyes treated by immobilized HRP during 12 cycles.

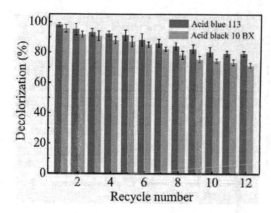

Figure 9. Reusability of immobilized HRP for degradation Acid blue 113 and Acid black 10 BX under optimum conditions.

The immobilized HRP retained 79.4% (Acid blue 113) and 71.1% (Acid black 10 BX) decolorization after 12 cycles. The gradual decline of the decolorization in the subsequent cycles could be correlated with enzyme inactivation during the catalytic process. It has been reported that the HRP immobilized on phospholipid-templated titania for oxidizing (Direct Black-38) retained 64% of initial decolorization efficiency after 5 cycles (Jiang et al., 2014), and the immobilized HRP on activated PET fibers for oxidizing (methyl orange) retained 69.6% of initial decolorization efficiency cycle after 5 cycles (Arslan, 2010). From the comparison it could be concluded that the in-situ cross-linking of HRP on the nano-composite support was an effective method to improve the reusability and catalytic performance of the immobilized HRP.

4 CONCLUSIONS

1. HRP was successfully immobilized on the nano-composite support through in-situ cross-linking method using epoxy cross-linker with long chain. The immobilized HRP was proved to be an effective biocatalyst in decolorization of azo dyes.
2. The immobilized HRP decolorized 95.4% of Acid blue 113 and 90.3% of Acid black 10 BX within 35 min at 30 °C, and exhibited much better resistance to temperature and pH inactivation than free HRP. The presence of ZnO nanowires accelerated the decolorizing reaction by attracting HRP and dyes together. The enzymatic kinetic constants showed the improved affinity and selectivity of the immobilized enzyme toward the substrate. The storage stability and reusability could

be greatly improved compare with free form and other types of immobilized HRP.

ACKNOWLEDGEMENTS

This research was financially supported by Zhejiang Provincial Natural Science Foundation (No. LY15B010002) and Wang Kuan-Chen Foundation.

REFERENCES

Arslan, M. (2010). Immobilization horseradish peroxidase on amine-functionalized glycidyl methacrylate-g-poly (ethylene terephthalate) fibers for use in azo dye decolorization. Polym. Bull. 66, 865–879.

Bayramoğlu, G., B. Altintas, & M.Y. Arica (2012). Cross-linking of horseradish peroxidase adsorbed on poly-cationic films: utilization for direct dye degradation. Bioproc. Biosyst. Eng. 35, 1355–1365.

Bayramoğlu, G., & M.Y. Arıca (2008). Enzymatic removal of phenol and p-chlorophenol in enzyme reactor: Horseradish peroxidase immobilized on magnetic beads. J. Hazard. Mater. 156, 148–155.

Bayramoğlu, G., I. Gursel, M. Yilmaz, & M.Y. Arica (2012). Immobilization of laccase on itaconic acid grafted and Cu(II) ion chelated chitosan membrane for bioremediation of hazardous materials. J. Chem. Technol. Biot. 87, 530–539.

Bradford, M.M. (1976). A rapid and sensitive method for the quantitation of microgram quantities of protein utilizing the principle of protein-dye binding. Anal. Biochem. 72, 248–254.

Boucherit, N., M. Abouseoud, & L. Adour (2013). Degradation of direct azo dye by Cucurbita pepo free and immobilized peroxidase. J. Environ. Sci. 25, 1235–1244.

Celebi, M., M.A. Kaya, M. Altikatoglu, & H. Yildirim (2013). Enzymatic decolorization of anthraquinone and diazo dyes using horseradish peroxidase enzyme immobilized onto various polysulfone supports. Appl. Biochem. Biotechnol. 171, 716–730.

Farabegoli, G., A. Chiavola, E. Rolle, & M. Naso (2010). Decolourization of Reactive Red 195 by a mixed culture in an alternating anaerobic–aerobic sequencing batch reactor. Biochem. Eng. J. 52, 220–226.

Hassani, L., B. Ranjbar, K. Khajeh, H.N. Manesh, M.N. Manesh, & M. Sadeghi (2006). Horseradish peroxidase thermostabilization: the combinatorial effects of the surface modification and the polyols. Enzyme. Microb. Technol. 38, 118–125.

Husain, Q. (2009). Peroxidase mediated decolorization and remediation of wastewater containing industrial dyes: a review. Rev. Environ. Sci. Bio. 9, 117–140.

Ikemoto, H., Q.J. Chi, & J. Ulstrup (2010). Stability and catalytic kinetics of horseradish peroxidase confined in nanoporous SBA-15. J. Phys. Chem. C 114, 16174–16180.

Jiang, Y.J., C.C. Cui, L.Y. Zhou, Y. He, & J. Gao (2014). Preparation and characterization of porous horseradish peroxidase microspheres for the removal of phenolic compound and dye. Ind. Eng. Chem. Res. 53, 7591–7597.

Jiang, Y.J., W. Tang, J. Gao, L.Y. Zhou, & Y. He (2014). Immobilization of horseradish peroxidase in phospholipid-templated titania and its applications in phenolic compounds and dye removal. Enzyme Microb. Technol. 55, 1–6.

Kalaiarasan, E., & T. Palvannan (2015). Efficiency of carbohydrate additives on the stability of horseradish peroxidase (HRP): HRP-Catalyzed removal of phenol and malachite green decolorization from wastewater. Clean-Soil Air Water 43, 846–856.

Karam, J., & J.A. Nicell (1997). Potential applications of enzymes in waste treatment, J. Chem. Technol. Biotechnol. 69, 141–153.

Karim, Z., R. Adnan, & Q. Husain (2012). A β-cyclodextrin–chitosan complex as the immobilization matrix for horseradish peroxidase and its application for the removal of azo dyes from textile effluent. Int. Biodeter. Biodegr. 72, 10–17.

Kim, H.J., Y. Suma, S.H. Lee, J.A. Kim, & H.S. Kim (2012). Immobilization of horseradish peroxidase onto clay minerals using soil organic matter for phenol removal. J. Mol. Catal. B: Enzym. 83, 8–15.

Kuberan, T., J. Anburaj, C. Sundaravadivelan, & P. Kumar (2011). Biodegradation of azo dye by Listeria sp. Int. J. Environ. Sci. 1, 1760–1770.

Li, B., & H. Takahashi (2000). New immobilization method for enzyme stabilization involving a mesoporous material and an organic/inorganic hybrid gel. Biotechnol. Lett. 22, 1953–1958.

Li, X., J. Zhang, Y. Jiang, M. Hu, S.N. Li, & Q.G. Zhai (2013). Highly efficient biodecolorization/degradation of Congo Red and Alizarin Yellow R by Chloroperoxidase from Caldariomyces fumago: catalytic mechanism and degradation pathway. Ind. Eng. Chem. Res. 52, 13572–135729.

Magnin, D., S. Dumitriu, & E. Chornet (2003). Immobilisation of enzymes into a polyionic hydrogel: ChitoXan. J. Bioact. Compat. Pol. 18, 355–373.

Malania, R.S., S. Khanna, & V.S. Moholkar (2013). Sonoenzymatic decolourization of an azo dye employing immobilized horseradish peroxidase (HRP): A mechanistic study. J. Hazard. Mater. 256–257, 90–97.

Mohamed, S.A., A.L. Al-Malki, T.A. Kumosani, & R.M. El-Shishtawy (2013). Horseradish peroxidase and chitosan: Activation, immobilization and comparative results. Int. J. Biol. Macromo. 60, 295–300.

Mohan, S.V., K.K. Prasad, N.C. Rao, & P.N. Sarma (2005). Acid azo dye degradation by free and immobilized horseradish peroxidase (HRP) catalyzed process. Chemosphere 58, 1097–1105.

Monier, M., D.M. Ayad, Y. Wei, & A.A. Sarhan (2010). Immobilization of horseradish peroxidase on modified chitosan beads. Int. J. Biol. Macromo. 46, 324–330.

Preethi, S., A. Anumary, M. Ashokkumar, & P. Thanikaivelan (2013). Probing horseradish peroxidase

catalyzed degradation of azo dye from tannery wastewater. SpringerPlus 2, 341–347.

Rai, A., A. Prabhune, & C.C. Perry (2012). Entrapment of commercially important invertase in silica particles at physiological pH and the effect of pH and temperature on enzyme activity. Mat. Sci. Eng. C: Mater. 32, 785–789.

Robinson, T., G. McMullan, R. Marchant, & P. Nigam (2001). Remediation of dyes in textile effluent: a critical review on current treatment technologies with a proposed alternative. Bioresour.Technol. 77, 247–255.

Sanjay, G., & S. Sugunan (2006). Enhanced pH and thermal stabilities of invertase immobilized on montmorillonite K-10. Food Chem. 94, 573–579.

Savin, I.I., & R. Butnaru (2008). Wastewater characteristics in textile finishing mills. J. Environ. Eng. Manage. J. 7, 859–864.

Shang, C.Y., W.X. Li, & R.F. Zhang (2014). Immobilized Candida antarctica lipase B on ZnO nanowires/macroporous silica composites for catalyzing chiral resolution of (R, S)-2-octanol. Enzyme Microb. Tech. 61–62, 28–34.

Shang, C.Y., W.X. Li, F. Jiang, & R.F. Zhang (2015a). Improved enzymatic properties of Candida rugosa lipase immobilized on ZnO nanowires/macroporous SiO_2 microwave absorbing supports. J. Mol. Catal. B: Enzym. 113, 9–13.

Shang, C.Y., W.X. Li, & R.F. Zhang (2015b). Immobilization of Candida rugosa lipase on ZnO nanowires/macroporous silica composites for biocatalytic synthesis of phytosterol esters. Mater. Res. Bull. 68, 336–342.

Solís, M., A. Solís, H.I. Pérez, N. Manjarrez, & M. Flores (2012). Microbial decolouration of azo dyes: A review. Process Biochem. 47, 1723–1748.

Torres, E., I. Bustos-Jaimes, & S. Le Borgne (2003). Potential use of oxidative enzymes for the detoxification of organic pollutants. Appl. Catal. B: Environ. 46, 1–15.

Ulson de Souza, S.M., E. Forgiarini, & A.A. Ulson de Souza (2007). Toxicity of textile dyes and their degradation by the enzyme horseradish peroxidase (HRP). J. Hazard. Mater. 147, 1073–1078.

Wan, M.M., W.G. Lin, L. Gao, H.C. Gu, & J.H. Zhu (2012). Promoting immobilization and catalytic activity of horseradish peroxidase on mesoporous silica through template micelles. J. Colloid Interf. Sci. 377, 497–503.

Yasutaka, K., Y. Takato, K. Takashi, M. Kohsuke, & Y. Hiromi (2011). Enhancement in adsorption and catalytic activity of enzymes immobilized on phosphorus- and calcium-modified MCM-41. J. Phys. Chem. B 115, 10335–10345.

Zhang, R.F., N.B. Long, & L.L. Zhang (2009). Preparation of 3-dimensional SiO_2 structures via a templating method. Thin Solid Films 517, 6677–6680.

Advances in Energy Science and Equipment Engineering II – Zhou, Patty & Chen (Eds)
© 2017 Taylor & Francis Group, London, ISBN 978-1-138-71798-5

Extraction and determination of α-solanine in tomato samples with UPLC/Q-TOF–MS

J.Y. Liang, Z.L. Chen, Z.M. Li, X.X. Yuan, Y.T. Wang & Z. Dong
Institute of Agricultural Quality Standards and Testing Technology Research,
Shandong Academy of Agricultural Sciences, Jinan, China
Key Laboratory of Testing Technology on Food Quality and Safety of Shandong Province, Jinan, China

Y. Zhao
Tai'an Animal Husbandry Bureau, Taian, China

ABSTRACT: α-Solanine is a glycoalkaloid in the species of the nightshade family. In plants, it serves as natural defenses against phytopathogens. However, at high concentration, α-solanine is considered toxic to both animals and humans. Thus, there is a need of developing a new method for the rapid extraction and accurate analysis of α-solanine. In this study, we first found that immature tomato fruit contained a small amount of α-solanine and then developed a method for the extraction and analysis of α-solanine in tomato. The tomato samples were extracted with supersonic extraction by methanol solution made from a mixture of 1mmol ammonium acetate plus 1% formic acid. This solution greatly improved the recovery as compared to that of other extraction solutions. Thereafter, the mixture was purified by Oasis HLB solid-phase extraction and isolated by UPLC BEH C18 chromatographic column (100 mm × 2.1 mm × 1.8 μm) to rake Q-TOF analysis of α-solanine using positive-ion mode of electrospray ionization (ESI). The analysis duration was only 6 min. Linearity range was 0.01–10 mg/kg, with linear coefficient of 0.9999; the Limit of Detection (LOD) (S/N = 3) was 0.0005 mg/kg; and the Limit of Quantification (LOQ) (S/N = 10) was 0.001 mg/kg. The sensitivity of this method was higher than that achieved by other methods. When being spiked at the levels of 0.01, 0.05, and 0.1 (mg/kg), the recoveries of α-solanine were in the range of 70.2–85.4%, and the Relative Standard Deviations (RSDs) were in the range of 2.6–4.2%. Six different varieties of mature and immature tomatoes were tested by this method. The results indicated that that all the immature tomatoes contained a small amount (about 9 μg/kg) of α-solanine, but α-solanine was not detected in mature tomatoes. At the same time, the TIC was analyzed and α-chaconine was detected in immature samples, but no α-chaconine was found in the immature tomatoes.

1 INTRODUCTION

Glycoalkaloids are the secondary metabolites of plants, especially the members of the family Solanaceae, such as pepper, eggplant, and tomato. Glycoalkaloids in food have attracted substantial research attention (Manrique-Carpintero et al. 2014, Petersson et al. 2013, Sawai et al. 2014, Tata et al. 2014). These plants synthesize a variety of compounds, which serve as natural defenses against phytopathogens including fungi, viruses, bacteria, insects, and worms (Meziani et al. 2015). For example, glycoalkaloids α-solanine and α-chaconine are produced in varieties of potato and potato leaves. However, at high concentration, they are considered toxic to both animals and humans (Buyukguzel et al. 2013). The highest safe level of total potato glycoalkaloids for human consumption was estimated to be about 1 mg/kg Body Weight

(BW) while the acute toxic dose was estimated to be 1.75 (mg/kg BW) and a lethal dose may be 3–6 (mg/kg BW) (Friedman, Henika and Mackey 2003). According to Machado et al. (Machado, Toledo and Garcia 2007), these toxins have been found in all parts of the potato plant. Symptoms of glycoalkaloid poisoning include colic pain in the abdomen and stomach, gastroenteritis, diarrhea, vomiting, fever, rapid pulse, low blood pressure, and neurological disorders. However, in addition to their toxic effects, studies in the last decade have also demonstrated that these glycoalkaloid compounds may possess beneficial properties such as anticancer and anti-inflammatory effects, depending on the dose and conditions of use (Friedman 2015, Kenny et al. 2013, Lv et al. 2014, Mohsenikia et al. 2013, Sun et al. 2014). It was reported that both potato and eggplant contained solanine while only tomato contained tomatidine

(Friedman 2004). However, there have been no detailed experimental data to verify that the tomato indeed contains α-solanine.

Methods used for the extraction of glycoalkaloids from the secondary metabolites of plants can be classified into ultrasound-assisted extraction (Hossain et al. 2014a, Hossain et al. 2014b) and solvent extraction method (Cabral et al. 2013). Both the methods have their own advantages and disadvantages. For instance, ultrasound-assisted extraction, a commonly used method, has the characteristics of high speed, little solvent usage, and easy operation as compared with other methods. The quantitative methods include Thin-Layer Chromatography (TLC), spectrophotometric method, electrochemical determination (Wang et al. 2013), High-Performance Liquid Chromatography (HPLC) (Maurya et al. 2013), Ultraperformance Liquid Chromatography (UPLC) tandem mass spectrometry (UPLC-MS/MS) (Carlier et al. 2015, Zhang, Cai and Zhang 2014), ultraperformance liquid chromatography coupled to time-of-flight mass spectrometry (UPLC-TOF/MS) (Tai et al. 2014), and enzyme-linked immunosorbent assay (ELISA) (Espinoza et al. 2014). The most commonly used method is HPLC-UV, but its earlier stage of processing is complicated and requires solid-phase extraction. Liquid Chromatography–Mass Spectrometry (LC-MS) is considered as the major method for solanine inspection because of its high sensitivity and perfect selectivity. UPLC is featured with good LC separation effect and high analysis speed so that it is suitable for rapid inspection of multicomponents. The time-of-flight mass spectrometry (TOF-MS) can accurately measure the molecular weight and is suitable for accurate qualitative analysis of components of complex stroma, so UPLC-TOF-MS has become a valid analysis method for multicomponents of complex stroma. In this study, we used UPLC-TOF-MS method to analyze various tomatoes and found that this method could improve extraction and isolation efficiency, simplify steps, and greatly improve the Limit of Detection (LOD). This method was applied to quantitatively measure the contents of α-solanine in immature and mature tomato fruits.

2 EXPERIMENTAL

2.1 Test materials and reagents

Tomatoes were purchased from vegetables base in Liangzhuang Town in Tai'an City of Shandong Province, China. The representative tomato fruit samples were obtained by combining fruits of six different tomato cultivars that produce large tomatoes, green tomatoes (still green after mature), yellow large tomatoes, cherry red tomatoes, cherry green tomatoes, and cherry yellow tomatoes at different stages of ripening. These samples were cleaned and then dissolved into liquid for subsequent analysis.

The α-solanine standard product (purity >99%) was provided by Sigma (Saint Louis, MS, USA), methanol (MS grade) was provided by Merck (Germany), acetonitrile (MS grade) was purchased from Sigma Aldrich (Germany), and formic acid (HPLC grade) was obtained from Sigma. The water for the experiments was purified using a Milli-Q System (Millipore Corp, Billerica, MA, USA). Ammonium acetate (HPLC-grade) was purchased from Sigma and Oasis®HLB cartridge (3 mL, 60 mg) from Waters (Milford, MA, USA).

2.2 Instruments

ACQUITY Ultra Performance LC ultra-high-performance liquid chromatography, ACQUITY UPLC BEH C18 (50 mm × 2.1 mm, i.d. 1.7 μm); Waters Xevo Q-Tof quadrupole time-of-flight mass spectrometry were purchased from Waters; T25BS2, high-speed dispersion homogenates were obtained from IKA (Germany). 3K30 high-speed centrifuge was purchased from Sigma; Laborota 4000 rotary evaporator was provided by Heidolph (Germany); and AE-240 electronic scale was obtained from Mettler-Toledo (Switzerland).

2.3 Pretreatment of samples

2.3.1 Extraction of α-solanine

Fresh tomato liquid (10g) was added to 10 mL of 0.1% formic acid–methanol (30:70 V/V) containing 1 mmol/L of ammonium acetate solution. The solution was mixed, processed with ultrasound (55Hz, 30min), and the whole procedure was conducted twice. The conditioning fluids were combined and subsequently evaporated to dryness under a stream of air at 45°C and the dried residue was dissolved in 1mL of 50% methanol-extracting solution. Finally, the extracting solution containing α-solanine was obtained.

2.3.2 Purification

Clean Oasis HLB solid phase (3 mL, 60 mg) with 2mL of methanol and water, respectively, were treated with extracting solution to drain under vacuum at the rate of 1 mL/min, carried with 1mL of chloroform liquid containing 5% methanol, and then eluted with 2 mL of methanol. The eluted solution was evaporated with nitrogen gas, and the residuals were dissolved with 1 mL of 0.1% formic acid–methanol solution (30:70, V/V) containing 1mmol/L of ammonium acetate and then filtered using a 0.22-μm nylon membrane filter.

Table 1. Time table of Ultraperformance Liquid Chromatographic (UPLC) gradient.

Time (min)	Flow rate (mL/mi)	Mobile phase A(%)	Mobile phase B(%)	Gradient curve
0	0.30	98	2	–
1	0.30	98	2	6
4	0.30	2	98	6
4.5	0.30	2	98	6
5	0.30	98	2	6
6	0.30	98	2	1

2.4 Detection of α-solanine

2.4.1 Calibration curves

Six calibration points of α-solanine were taken from the blank samples spiked with different standard samples to prepare a series of standard solutions of 1, 10, 25, 50, 75, and 100 μg/L for testing. Mass spectrogram of level-1 mass spectrometry was taken as the quantitative basis to extract the parent ion to quantitatively calculate the quantitative ion and then obtain the equation of linear regression.

2.4.2 UPLC and mass spectrometry conditions

Waters UPLC BEH C18 column (1.7 μm, 50 mm × 2.1 mm) was used as the separation column; mobile phase A was 0.1% methanoic acid plus 1 mmol/L ammonium acetate solution; mobile phase B was acetonitrile solution; the elution gradient is presented in Table 1. Column temperature was set at 35°C; room temperature was 25.2°C; and injection volume was 5 μL.

Mass spectrometry (MS) conditions were set as follows: electrospray ionization mode, positive-ion mode; TOF running mode; V mode; resolution ratio, 10000–12000; scanned area, m/z 100–1000; voltage of capillary tube (capillary), 3.0 kV; ion source temperature, 120°C; sample cone voltage (sample cone), 40 V; extraction cone voltage (extraction cone), 4.0 V; cone gas flow (Cone), 50 L/h; desolvation temperature, 350°C; desolvation gas flow (desolvation), 800 L/h; tuning solution; and leucine enkephalin (m/z 556.2771), 50 pg/μL.

3 RESULTS AND DISCUSSION

3.1 Optimization of ionization conditions and mass spectrometry conditions

A-Solanine contains three types of saccharides with different hydroxide radicals, which are easy for ionization. Therefore, in this experiment, the positive-ion mode was adopted. In order to obtain optimal ionization effect of α-solanine molecular ion, we surveyed the ionization effects of the solvents with four different preparation ratios of methanol, acetonitrile, 0.1% formic acid, and 0.1% formic acid-methanol (30:70, V/V) containing ammonium acetate at 1 mmol/L. Among these solvents, methanol and acetonitrile did not display chromatographic peak, while acetonitrile and 0.1% formic acid displayed the asymmetry peak. Thus, we decided to use the 0.1% formic acid–methanol (30:70, V/V) containing ammonium acetate at 1 mmol/L as the mobile phase of α-solanine.

All the mass spectrum parameters were adjusted and optimized so that the ion signal strength could be in the range of 200–400 count/s with the least disturbance. Q-TOF MS can extract the quality of quasi-molecular ion as quantification, so it can eliminate the disturbance of ion peak with similar mass number. Therefore, it can obtain good chromatographic peak by level-1 mass spectrum scanning, and the quantitative method can be simplified greatly. Molecular ion peak of α-solanine was determined as m/z 868.5058.

3.2 Selection of chromatographic conditions

Four different chromatographic columns, including Waters XTERRA chromatographic column (1.8 μm), UPLC BEH (1.8 μm), and HILIC (1.7 μm) chromatographic column, were initially investigated. The effect of isolating α-solanine showed that the isolation effect of UPLC BEH C18 chromatographic column (1.8 μm) was the optimal one. α-Solanine is a type of glycoside with a large steroid structure containing three saccharides. Among these columns, BEH C18 and Waters XTERRA chromatographic columns are of inversed-phase chromatography, whereas HILIC is of reversed-phase chromatography. Thus, HILIC chromatography is also named hydrophilic interaction chromatography. In this experiment, the retention time of α-solanine was only 0.1 min because its conservation is poor with HILIC column. Waters XTERRA chromatographic column is a kind of hybrid particle chromatographic column with good conservation, but its sensitivity is poorer than that of the BEH C18 column.

The experiment compared the isolation effects of three groups of mobile phase (Figures 1 and 2). Group A was 0.1% formic acid–acetonitrile, group B was 0.1% formic acid–methanol, and group C was 0.1% formic acid–methanol containing ammonium acetate at 0.1 mmol/L. The experimental results showed that the sensitivity of the group without acetonitrile was higher. When ammonium acetate was added, the obtained chromatogram displayed better resolution and higher sensitivity. Therefore, we adopted 0.1% formic acid–methanol containing ammonium acetate at 0.1 mmol/L as the

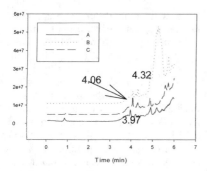

Figure 1. Total ion peaks of 100ppb standard of different mobile phase, group A is 0.1% formic acid–acetonitrile, group B is 0.1% formic acid–methanol, and group C is 0.1% formic acid–methanol containing 0.1 mmol /L ammonium acetate.

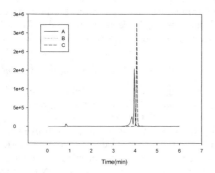

Figure 2. Extracting ion peaks of α-solanine with 100ppb standard of different mobile phase, A, B, and C are the same as those in the caption to Figure 1.

mobile phase. Under these conditions, α-solanine could be isolated well and its retention time was 4.06 min.

3.3 Optimization of the methods for pretreatment of samples

The final method for the extraction of α-solanine from tomato samples was determined based on the modification and consideration of the varied procedures as applied to potato samples (Jensen et al. 2007, Moco et al. 2006). With extraction method for potato, such as that used by Jaromir Lachman (Lachman et al. 2013) and others, recovery of α-solanine was only about 52%, because the matrix of potato is different from that of tomato. The tomato samples contain saccharides, protein, fat, malic acid, citric acid, vitamins, and other substances, and the acidic materials are the major ones. Thus, we treated the samples by mixing solution extraction and solid-phase extraction simultaneously and compared the extraction effects of methanol, methanol containing formic acid, and formic acid–methanol

containing ammonium acetate. The results showed that the extraction effect of 1% formic acid–methanol-containing ammonium acetate at 1mmol/L was the optimal one. Recovery of α-solanine was about 75%. After the effects of different solid-phase extraction columns were examined, the disturbance background of Oasis HLB solid-phase column was the weakest one and its recovery was the highest one. With more hydroxyl saccharides, the polarity of α-solanine is higher; thus, methanol was chosen for solvent extraction. At the same time, as α-solanine is alkaline, addition of formic acid can improve its solubility in methanol. Because the matrix in tomatoes is more complex and contains acid, ammonium acetate was added to the methanol solution of formic acid to form a buffer solution to further increase the solubility of α-solanine and to improve the extraction yield. Therefore, 1% formic acid of methanol containing 1mmol ammonium acetate was used as the extraction solvent and Oasis HLB solid-phase column as decontaminating column were adopted to process the samples.

3.4 Other alkaloids in tomatoes

Adopting time-of-flight mass spectrometry can obtain accurate molecular weight of the drugs, and it can also obtain accurate qualitative result although there is no standard substance. The most substances contained in tomato, especially in immature tomato, are tomatidine (Friedman 2013, Caprioli, Cahill and James 2013a), which has been reported elsewhere. However, if the immature sample contains α-chaconine, the difference between α-solanine and α-chaconine is that there are different amounts of three saccharides in substituent group and the connecting positions are different. After being calculated by accurate mass number of $[M + H]^+$, molecular number of $C_{45}H_{73}NO_{14}$ was 852.5109. A higher peak was obtained by extracting the total ion peak, as shown in Figure 3, which shows that immature tomato contained both α-solanine and α-chaconine.

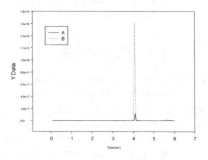

Figure 3. Extracting ion peaks of α-solanine (A) and α-chaconine (B) in immature tomato.

3.5 Linearity range and detection limit

The mass spectrograms of α-solanine standard products with different concentrations were measured, the concentrations of α-solanine standard product were taken as abscissa, and peak area was taken as ordinate to a draw standard curve of α-solanine standard product. Regression equation of standard solution was: Area = 1148.97 × Amt+34.1348; r = 0.9999. The linear relation was good when α-solanine concentrations were in the range of about 0.01–10 mg/kg, and the correlation coefficient was 0.9999, which successfully met the method requirements. When the lowest concentration of signal-to-noise ratio (SNR) for three times of quantitative ion chromatography peak was taken as the limit of detection (LOD), the limit of quantitation (LOQ) was 1μg/L (S/N≥3). When the lowest concentration of signal-to-noise ratio (SNR) for 10 times of quantitative ion chromatography peak was taken as LOQ, the LOQ was 10 μg /L (S /N≥10).

The analytical characteristics of the developed methods, including linearity, LODs, LOQs, accuracy precision, and recoveries, were also investigated. LOD and LOQ, defined as the peak giving a response equal to a blank signal plus 3 and 10 times the standard deviation of the noise, respectively, were calculated. The linear range of this method was 0.01–10mg/kg with the linear coefficient of 0.9999. The LOQ for this method was 0.0005 (mg/kg) and the LOD was 0.001 (mg/kg) for α-solanine. These results highlighted the good sensitivity that can be achieved with the full-scan method for the determination of α-solanine in tomatoes mentioned above. Caprioli et al. (Caprioli et al. 2013b, Caprioli et al. 2014, Caprioli et al. 2013a) used liquid chromatography LTQ Orbitrap mass spectrometry method to measure α-chaconine and α-solanine in soil or potato samples. They reported that with this method, the linear range was 0.025–1 mg/kg with linear coefficient (R = 0.997), the best percentage of recovery was about 90%, and the LOD and LOQ were 0.01 and 0.025 (mg/kg), respectively. Compared with those reported by Caprioli et al. mentioned above, the linear range, the linear correlation coefficient, percentage of recovery, LOD, and LOQ obtained with this newly developed high-performance liquid chromatography/tandem mass spectrometry were better with higher sensitivity.

3.6 Precision and accuracy

We found that the mature tomatoes did not contain α-solanine, so those purchased from market were taken as blank samples and spiked with the standard α-solanine solutions at different concentrations. The samples were extracted and puri-fied in accordance with the methods described in Section 2.3 after the samples and the standard solution were mixed completely. Each concentration was measured six times. The recovery and standard deviation were calculated and presented in Table 2. The results showed that recoveries of spiking α-solanine standard were in the range of 70.2–85.4% and the Relative Standard Deviations (RSDs) were in the range of 2.6–4.2%. When the spiking level of α-solanine was 0.01mg/kg, its recovery was lower, possibly due to its low ion strength. The main reason is that the extraction and purification processes are not perfect, and matrix inhibitory effect may occur during electrospray ionization. However, its accuracy can meet requirements of the method (RSD<15%).

3.7 Detection of the content of α-solanine in real tomato samples

The developed method was used to test and analyze the content of α-solanine in different immature tomatoes collected from Tai'an City of Shandong Province, China, and the mature tomatoes purchased from local farmer market. The identification information of chromatography and level-1 mass spectrum fingerprints of different tomatoes are shown in Figure 4, and the quantitative determination results are presented in Table 3. Six cultivars of tomatoes included cherry green tomato (still green after becoming mature), yellow cherry large tomato, cherry red tomato (become red after maturing), nor-

Table 2. Date of accuracy and precision of tomato sample.

Spiking level	Average recovery (%)	Relative standard deviations (mg/kg) (%)
0.01	70.2	4.2
0.05	85.4	3.5
0.1	84.5	2.6

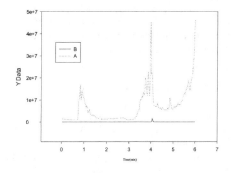

Figure 4. Total ion peak (A) and extracting ion peaks (B) of α-solanine in tomato.

Table 3. α-Solanine content of field-grown tomatoes at different stages of maturity (in ug/kg of fresh weight).

Tomato variety	Cherry	Standard
Ripe		
Green	ND	ND
Yellow	ND	ND
Red (standard)	ND	ND
Unripe		
Green	8.6	7.4
Yellow	9.8	8.2
Red	8.5	7.6

mal green tomato (still green after maturing), yellow tomato (become yellow after maturing), and normal red tomato (green when immature but become red after maturing). These six cultivars of tomatoes were also subdivided into matured tomatoes and immature tomatoes, and thus, there were 12 types of tomatoes. As shown in Table 3, the mature tomatoes were presented in the first four lines while the immature tomatoes were presented in the later four lines. It can be seen from Table 3 that α-solanine was not detected in all the mature tomatoes, no matter the tomato was cherry tomato or normal tomato and no matter the tomato was green, yellow, or red. A small amount of α-solanine was detected in the immature tomatoes. For instance, the immature fruit of green cherry tomato, green large tomato, yellow cherry tomato, and yellow large tomato contained α-solanine at levels of 8.6, 7.4, 9.8, and 8.2 ug/kg, respectively. The normal immature fruit of cherry tomato contained α-solanine at 8.5 ug/kg while the normal immature large tomato contained α-solanine at 7.6 ug/kg. They contained α-solanine at the mean level of 10ug/kg. They all contained α-solanine regardless of different cultivars. These results indicated that no matter what the cultivars of the tomatoes are, their immature fruit contained small amount of α-solanine, whereas their matured fruits did not contain α-solanine. Thus, although the contents of α-solanine in immature fruits of these tomatoes were relatively low, caution must be taken when one eats immature fruit of tomatoes. One should not eat immature tomatoes or eat them as little as possible.

4 CONCLUSIONS

In this study, we first discovered that the immature fruit of six tomato cultivars contained small amount of α-solanine, and then developed a method for the extraction and determination of α-solanine in tomatoes. It targets six different tomato cultivars with 1% formic acid in methanol containing 1mmol ammonium acetate extraction solution, HLB col- umn purification, and ultra-high-performance liquid chromatography–electrospray ionization–quadrupole time-of-flight mass spectrometry for detecting α-solanine. The precision and accuracy for various solutions, purification columns, mobile phases, and chromatographic columns were optimized. The sensitivity of this method was higher than that obtained with other methods. As a result, all the immature tomatoes contained small amount of α-solanine, but matured tomatoes contained no α-solanine, and the same is true for α-chaconine.

REFERENCES

Buyukguzel, E., K. Buyukguzel, M. Erdem, Z. Adamski, P. Marciniak, K. Ziemnicki, E. Ventrella, L. Scrano & S. A. Bufo (2013) The influence of dietary alpha-solanine on the waxmoth Galleria mellonella L. *Arch Insect Biochem Physiol. 83*, 15–24.

Cabral, E. C., M. F. Mirabelli, C. J. Perez & D. R. Ifa (2013) Blotting assisted by heating and solvent extraction for DESI-MS imaging. *J Am Soc Mass Spectrom. 24*, 956–65.

Caprioli, G., M. G. Cahill & K. J. James (2013a) Mass Fragmentation Studies of α-Tomatine and Validation of a Liquid Chromatography LTQ Orbitrap Mass Spectrometry Method for Its Quantification in Tomatoes. *Food Analytical Methods. 7*, 1565–1571.

Caprioli, G., M. G. Cahill, K. J. James & S. Vittori (2014) Liquid Chromatography-Orbitrap Mass Spectrometry Method for the Determination of Toxic Glycoalkaloids and their Aglycons in Potato Upper Soil. *Journal of the Brazilian Chemical Society*.

Caprioli, G., M. G. Cahill, S. Vittori & K. J. James (2013b) Liquid Chromatography–Hybrid Linear Ion Trap–High-Resolution Mass Spectrometry (LTQ-Orbitrap) Method for the Determination of Glycoalkaloids and Their Aglycons in Potato Samples. *Food Analytical Methods. 7*, 1367–1372.

Carlier, J., J. Guitton, L. Romeuf, F. Bévalot, B. Boyer, L. Fanton & Y. Gaillard (2015) Screening approach by ultra-high performance liquid chromatography-tandem mass spectrometry for the blood quantification of thirty-four toxic principles of plant origin. Application to forensic toxicology. *Journal of Chromatography B Analytical Technologies in the Biomedical & Life Sciences. 975*, 65–76.

Espinoza, M. A., G. Istamboulie, A. Chira, T. Noguer, M. Stoytcheva & J. L. Marty (2014) Detection of glycoalkaloids using disposable biosensors based on genetically modified enzymes. *Analytical Biochemistry. 457*, 85–90.

Friedman, M. (2004) Analysis of biologically active compounds in potatoes (Solanum tuberosum), tomatoes (Lycopersicon esculentum), and jimson weed (Datura stramonium) seeds. *Journal of Chromatography A. 1054*, 143–155.

Friedman, M. (2013) Anticarcinogenic, cardioprotective, and other health benefits of tomato compounds lycopene, alpha-tomatine, and tomatidine in pure form and in fresh and processed tomatoes. *J Agric Food Chem. 61*, 9534–50.

Friedman, M. (2015) Chemistry and Anticarcinogenic Mechanisms of Glycoalkaloids Produced by Eggplants, Potatoes, and Tomatoes. *Journal of Agricultural & Food Chemistry. 63.*

Friedman, M., P. R. Henika & B. E. Mackey (2003) Effect of feeding solanidine, solasodine and tomatidine to non-pregnant and pregnant mice. *Food Chem Toxicol. 41*, 61–71.

Hossain, M. B., B. K. Tiwari, N. Gangopadhyay, C. P. O'Donnell, N. P. Brunton & D. K. Rai (2014a) Ultrasonic extraction of steroidal alkaloids from potato peel waste. *Ultrasonics Sonochemistry. 21*, 1470–1476.

Hossain, M. B., B. K. Tiwari, N. Gangopadhyay, C. P. O'Donnell, N. P. Brunton & D. K. Rai (2014b) Ultrasonic extraction of steroidal alkaloids from potato peel waste. *Ultrason Sonochem. 21*, 1470–6.

Jensen, P. H., B. J. Harder, B. W. Strobel, B. Svensmark & H. C. B. Hansen (2007) Extraction and determination of the potato glycoalkaloid α -solanine in soil. *International Journal of Environmental Analytical Chemistry. 87*, 813–824.

Kenny, O. M., C. M. Mccarthy, N. P. Brunton, M. B. Hossain, D. K. Rai, S. G. Collins, P. W. Jones, A. R. Maguire & N. M. O'Brien (2013) Anti-inflammatory properties of potato glycoalkaloids in stimulated Jurkat and Raw 264.7 mouse macrophages. *Life Sciences. 92*, 775–782.

Lachman, J., K. Hamouz, J. Musilova, K. Hejtmankova, Z. Kotikova, K. Pazderu, J. Domkarova, V. Pivec & J. Cimr (2013) Effect of peeling and three cooking methods on the content of selected phytochemicals in potato tubers with various colour of flesh. *Food Chem. 138*, 1189–97.

Lv, C., H. Kong, G. Dong, L. Liu, K. Tong, H. Sun, B. Chen, C. Zhang & M. Zhou (2014) Antitumor Efficacy of α-Solanine against Pancreatic Cancer In Vitro and In Vivo. *Plos One. 9*, e87868.

Machado, R. M. D., M. C. F. Toledo & L. C. Garcia (2007) Effect of light and temperature on the formation of glycoalkaloids in potato tubers. *Food Control. 18*, 503–508.

Manrique-Carpintero, N. C., J. G. Tokuhisa, I. Ginzberg & R. E. Veilleux (2014) Allelic variation in genes contributing to glycoalkaloid biosynthesis in a diploid interspecific population of potato. *Theoretical & Applied Genetics. 127*, 391–405.

Maurya, A., N. Manika, R. K. Verma, S. C. Singh & S. K. Srivastava (2013) Simple and reliable methods for the determination of three steroidal glycosides in the eight species of Solanum by reversed-phase HPLC coupled with diode array detection. *Phytochemical Analysis. 24*, 87–92.

Meziani, S., B. D. Oomah, F. Zaidi, A. Simon-Levert, C. Bertrand & R. Zaidi-Yahiaoui (2015) Antibacterial activity of carob (Ceratonia siliqua L.) extracts against phytopathogenic bacteria Pectobacterium atrosepticum. *Microbial Pathogenesis. 78*, 95–102.

Moco, S., R. J. Bino, O. Vorst, H. A. Verhoeven, G. J. De, T. A. van Beek, J. Vervoort & C. H. de Vos (2006) A liquid chromatography-mass spectrometry-based metabolome database for tomato. *Plant Physiology. 141*, 1205–1218.

Mohsenikia, M., A. M. Alizadeh, S. Khodayari, H. Khodayari, S. A. Kouhpayeh, A. Karimi, M. Zamani, S. Azizian & M. A. Mohagheghi (2013) The protective and therapeutic effects of alpha-solanine on mice breast cancer. *European Journal of Pharmacology. 718*, 1–9.

Petersson, E. V., N. Nahar, P. Dahlin, A. Broberg, R. Tröger, P. C. Dutta, L. Jonsson & F. Sitbon (2013) Conversion of Exogenous Cholesterol into Glycoalkaloids in Potato Shoots, Using Two Methods for Sterol Solubilisation. *Plos One. 8*, e82955-e82955.

Sawai, S., K. Ohyama, S. Yasumoto, H. Seki, T. Sakuma, T. Yamamoto, Y. Takebayashi, M. Kojima, H. Sakakibara & T. Aoki (2014) Sterol side chain reductase 2 is a key enzyme in the biosynthesis of cholesterol, the common precursor of toxic steroidal glycoalkaloids in potato. *Plant Cell. 26*, 3763–3774.

Sun, H., C. Lv, L. Yang, Y. Wang, Q. Zhang, S. Yu, H. Kong, M. Wang, J. Xie & C. Zhang (2014) Solanine induces mitochondria-mediated apoptosis in human pancreatic cancer cells. *Biomed Research International. 2014*, 85–117.

Tai, H. H., K. Worrall, Y. Pelletier, K. D. De & L. A. Calhoun (2014) Comparative metabolite profiling of Solanum tuberosum against six wild Solanum species with Colorado potato beetle resistance. *Journal of Agricultural & Food Chemistry. 62*, 9043–9055.

Tata, A., C. J. Perez, T. S. Hamid, M. A. Bayfield & D. R. Ifa (2014) Analysis of Metabolic Changes in Plant Pathosystems by Imprint Imaging DESI-MS. *Journal of the American Society for Mass Spectrometry. 26*, 641–648.

Wang, H., M. Liu, X. Hu, M. Li & X. Xiong (2013) Electrochemical Determination of Glycoalkaloids Using a Carbon Nanotubes-Phenylboronic Acid Modified Glassy Carbon Electrode. *Sensors. 13*, 16234–16244.

Zhang, X., X. Cai & X. Zhang (2014) [Determination of alpha-solanine, alpha-chaconine and solanidine in plasma and urine by ultra-performance liquid chromatography-triple quadrupole mass spectrometry]. *Se Pu. 32*, 586–590.

935

Determination of residues of fluroxypyr-meptyl and its metabolite in maize, maize straw and soil by high-performance liquid chromatography

J.Y. Liang, Z.L. Chen, Z. Dong & Z.M. Li
Institute of Agricultural Quality Standards and Testing Technology Research, Shandong Academy of Agricultural Sciences, Jinan, China
Key Laboratory of Testing Technology on Food Quality and Safety of Shandong Province, Jinan, China

Y. Zhao
Tai'an Animal Husbandry Bureau, Taian, China

ABSTRACT: In this study, a novel method was developed for the simultaneous determination of fluroxypyr-meptyl and fluroxypyr in maize by High-Performance Liquid Chromatography (HPLC) with an Ultraviolet (UV) detector. Fluroxypyr-meptyl was extracted with acetonitrile-water mixture, followed by salting out, and cleaned-up by Graphitized Carbon Black (GCB). Fluroxypyr was extracted with acidic acetonitrile-water mixture, and cleaned-up by ionic columns (Waters MAX). The elutants were evaporated to appropriately dry under a nitrogen stream, and dissolved in 2 mL of acetonitrile-0.1% acetic acid mixture (70:30, V/V). The residues were analyzed by a HPLC method with an UV detector. The average recoveries of fluroxypyr-meptyl and fluroxypyr were in the range of 81.1–103.3% at spiked levels of 0.01, 0.05 and 0.5 (mg/kg), respectively, with the Relative Standard Deviation (RSD) between 0.3% and 2.3%. The half-life of fluroxypyr in maize straw in 2015 was 4.0 d in testing site located Anhui province and 2.6 d in antoher testing site located in Shandong province. The half-life of fluroxypyr in soil in 2015 was 4.6 days in testing site in Anhui and 4.9 days in another testing site in Shandong. The final residue results indicated that the maximum residue of fluroxypyr in maize was lower than the limited value of 0.05 mg/kg (maximum residue limit value) as well as 0.01 mg/kg (LOQs). This study developed a good method for the simultaneous analysis of fluroxypyr-meptyl, fluroxypyr and, perhaps, other similar compounds.

1 INTRODUCTION

Pesticides (or herbicides) are a group of agro-chemicals used extensively in farmland to enhance the selective chemical control of weeds, insects, and fungi. Although application of pesticides to soils has greatly improved food production, it has also brought about substantial contamination of pesticides to plant, soil and aquatic eco-systems. Unlike other hydrophobic contaminants such as polycyclic aromatic hydrocarbons, most of the commercial pesticides can easily enter the tissues of living organisms and give rise to bioaccumulation there. Over the last decade, increasing amounts of pesticides have been detected in both plants and soils (Omari and Frempong, 2016) and consequently, there is a need to evaluate residual dynamics and the final residues of pesticide residues and to develop more pesticide-accumulating plants and precise methods for monitoring and remediating the contaminated plants and soils or surfer water (Ulen et al, 2014).

Fluroxypyr-meptyl and its metabolite, fluroxypyr (4-amino-3,5-dichloro-6-fluoro-2-pyridyloxyacetic

acid), are the selective post-emergent systemic herbicide registered for use on wheat, barley, oats, maize and fallow crop land. They were developed by the Dow Chemical Company (now Dow Agro Sciences) and released commercially in 1985 (Hu et al, 2011). Fluroxypyr-metyl is usually commercialized as an ester (Starane) that is rapidly degraded to fluroxypyr acid (Pang, Wang and Hu, 2016). As a herbicide, fluroxypyr is more toxic to human cells than its declared active principles are (Mesnage et al, 2014). China, the European Union, and other countries have formulated strict standards for limiting its residue level. For instance, in the European Union, the maximum residue limits of the fluroxypyr-metyl and fluroxypyr in maize are set at 0.05 mg/kg (European Commission 2013); and in China, the maximum residue limits in maize are also set at 0.05 mg/ kg (GB 2763-2014).

Fluroxypyr belongs to the pyridine family with two pKa values, i.e. pK_1 at 3.46 resulted from deprotonation of the carbonyl group and pK_2 at 10.89 resulted from deprotonation of the amino group. It is usually commercialized as an ester of fluroxypyr-meptyl, which is rapidly degraded to its active acidic form, fluroxypyr. Both fluroxypyr-meptyl and fluroxypyr were found to be absorbed by plants soon after being applied as the hormone herbicides, causing harmful effects on plants inducing plant deformity and twist (Lazartigues et al, 2013; Yang et al, 2016).

The choice of the appropriate method for extraction of pesticides residues depends on the type of matrix. Finding a simpler and easier method to extract the compounds with very different physicochemical properties remains an analytical challenge. Furthermore, the complex sample matrix may contain abundant quantities of chlorophyll, lipids, sterols, and other components that can interfere with good sample analysis. Recently, the quick, easy cleanup, effective, rugged, and safe (QuEChERS) pretreatment, which was introduced in 2003, has been widely used as pesticide multi-residue methods by many governments, organizations, and laboratories, especially in vegetables, fruits, and many other matrices(David et al. 2015; Zhang et al, 2016b; Zhang et al, 2016a; Wu et al, 2016) etc.

To date, numerous analytical chromatography methods for detection of fluroxypyr residues have been published. For example, fluroxypyr was analyzed by Gas Chromatography (GC) (Halimah, Tan and Ismail 2004), Gas Chromatography-Mass Spectrometry (GC-MS) (Raeppel et al, 2015, Tao and Yang 2011), Ultraviolet-Matrix-Assisted Laser Desorption/Ionization (UV-MALDI) mass spectrometry(Ivanova 2015), High Performance Liquid Chromatography (HPLC) (Peng, Pang and Deng 2011), liquid chromatography and mass spectrometry(LC-MS/MS) (Zhao et al, 2015, Santilio et al., 2011). However, there have been limited reports about the methods used to analyze fluroxypyr-meptyl residues in different matrices. Only a few reports are available about the analysis of these residues in rice, soil and water by GC-MS (Wang et al, 2011), enantiomeric separation by HPLC (Wang et al. 2008, Wang et al. 2006), residue analysis of fluroxypyr-meptyl in wheat and soil by GC–ECD (Hu et al, 2011), and multi-class pesticide residues determined by LC-ESI-MS/MS (Kujawski and Namieśnik, 2011). In 2016, Pang et al. described the simultaneous determination of fluroxypyr-meptyl and fluroxypyr in maize by LC-MS (Pang et al, 2016). Gas chromatography is most frequently converted to more volatile derivatives. Although the use of LC–MS/MS for residue determination notably increases the selectivity of the method, it may overestimate the residue. In some occasions, the complexity of the sample can exceed the selectivity of MS detection, resulting in the false positive findings(Niu et al, 2011). For reducing this risk, liquid chromatography can be a better choice.

To date, there have been no reports about the method for simultaneous determination of fluroxypyr-meptyl and its metabolite residue. Thus, this study aimed to develop a simpler and more versatile HPLC-UV method with modified QuEChERS sample preparation for simultaneous determination of fluroxypyr-meptyl and its metabolite residue. The influences of experiment conditions on the determination were also systematically investigated. This method was further validated in different matrices including soil, maize grain, and maize straw. The fluroxypyr-meptyl in three matrices was quantitatively analyzed by monitoring the quantity of GCB in d-SPE, and fluroxypyr was quantitatively analyzed by monitoring the quantity of ionic columns (Waters MAX).

2 EXPERIMENT

2.1 Instruments and equipment

Shimadzu 20 AT high performance liquid chromatograph with a variable wavelength detector set at 212 nm was obtained from Shimadzu Corporation (Japan). Heidolph rotary evaporator was obtained from the Heidolph Co., Ltd (Schwabach, Germany). Rotating centrifuge was obtained from Sigma (Saint Louis, MO, USA). Vortex spin oscillator was obtained from IKA Industrial Equipment Co., Ltd (German). Ultrasonic extractor was obtained from the Kunshan Ultrasonic Instrument Co., Ltd (Kunshan, Jiangsu, China). OA-SYS pressure blowing concentrator was obtained from the U.S. Organomation Co., Ltd (Berlin, MA, USA). Organic membrane (0.22 μm) was obtained from the Pall Co., Ltd. (Port Washington, NY, USA) and single-channel pipettes were obtained from the Eppendorf Co., Ltd (Holstein, German). Agela Venusil MP C18 150A 4.6 × 250mm was purchased from Aela Agela Technologies (Wilmington, DE, USA)

2.2 Reagents

The fluroxypyr-meptyl (≥ 97.8% purity) was provided by the Shanghai Chemical Reagent Factory (Shanghai, China). Fluroxypyr (97.0% purity) was purchased from Agro-Environmental Protection Institute of Ministry of Agriculture (Tianjin, China). The mixed mode anion exchange solid phase extraction column MAX (60 mg/3mL) was obtained from Waters Co., Ltd (Vallejo, CA, USA). The acetonitrile and acetic acid were obtained from Fisher Scientific Co. (Waltham, MA, USA). Methanol was obtained from Merk Co., Ltd. (German). Anhydrous magnesium sulfate ($MgSO_4$), sodium chloride (NaCl) and acetic acid (analytical grade) were purchased from Jinan Chemical Reagents Company (Jinan, Shandong, China). Graphitized Carbon Black (GCB) (120–400 mesh) were purchased from Agela Technologies (Tianjin, China).

A series of standard solution of fluroxypyr-meptyl was prepared with a stock standard solution at 100 ug mL^{-1} by diluting it with methanol into 10.0, 5.0, 1.0, 0.50, 0.20, 0.10, and 0.01 (ug/mL), respectively. A series of standard solution of fluroxypyr were prepared by dissolving 10 mg fluroxypyr standards in methanol into a 10 mL-volumetric flask as the stock solution, which was mixed well and diluted with methanol to obtain 10.0, 5.0, 1.0, 0.50, 0.20, 0.10, and 0.01 (ug/mL) for spiking experiment and calibration. All the standard solutions were freshly prepared, filtered through nylon membrane filters (0.22 μm), and stored in the dark at 4°C.

2.3 Designs of field experiments

The field trials included a dissipation study and a final residue study. The testing sites were two areas located in Anhui and Shandong provinces, respectively, with test was conducted in 2015. Every test was applied to a land of 30m^2 with triplicates.

2.3.1 Experiments of dissipation dynamics and final residue disposition of Fluroxypyr-meptyl and fluroxypyr in maize plants and soil

2.3.1.1 Experiments of dissipation dynamics
The maize field that had not been applied with fluroxypyr-meptyl was selected for conducting the experiments. The area of each plot was 30 m^2 and triplicates were set for each plot. Protection rows were set between plots. A blank plot was also set as the control. The dispersible oily suspension of 12% fluroxypyr-meptyl was diluted 50 times with water and directionally sprayed one time on the maize plants at the early stage (with plant height of about 15 cm). Among which, the amount of fluroxypyr-meptyl applied (the active ingredient) on maize plants was 29.4 g active ingredient per hectare, which was 1.5 times that of the recommended high dose and was equivalent to the formulated dose of 105 g per hectare. The samples of maize straw and soil were collected at 2h, 1, 3, 5, 7, 10, 14, 21, 28 and 35 d after application of fluroxypyr-meptyl. For maize straw samples, the whole maize plants were collected, cut into small pieces and ground into powder. The samples were either directly analyzed or stored at -20°C freezer for subsequent analysis.

Another blank plot with the area of 10 m^2 was also selected. The oily suspension of fluroxypyr-meptyl was directly sprayed on the soil surface when maize plants grew to the stage of 3–5 leaves. The applied amount of fluroxypyr-meptyl (the active ingredient) for soil was 196 g active ingredient per hectare, which was 10 times that of the recommended high dose and was equivalent to the formulated dose of 700g per hectare. The soil samples were collected at 2h, 1, 3, 5, 7, 14, 21, 30, 45 and 60 d after application. The application of same volumes of clean water was set as the negative control. Each treatment was repeated three times. For collection of soil samples, 10 spots were randomly selected and 1–2 kg soil was collected from the 0–10 cm layer with an auger boring. The debrides such as small pieces of rocks, weeds and plant roots and stems were removed. After being mixed well, soil samples (500 g) were collected with quartering method and put into the sample container, labeled correctly and stored at -20°C freezer for subsequent analysis.

2.3.1.2 Experiments of the final residues of Fluroxypyr-meptyl and fluroxypyr
The experimental designs and plot areas for these experiments were the same as those described in section 1.3.1 but two doses, high dose and low dose, were set for both maize plants and their soil samples, respectively. The active integrant of the high dose was 29.4 g active ingredient per hectare, which was the recommended dose and was equivalent to the formulated dose of 105 g/hectare. The active integrant of the low dose was 19.6 g active ingredient per hectare, which was the recommended dose equivalent to the formulated dose of 70 g/hectare. Fluroxypyr-meptyl was applied one time at the early stage of maize seedling. Each treatment was repeated three times. The samples of soil, maize straw and maize grains in the control group and treated groups were collected at the milk stage and mature stage, respectively. The samples collected at both stage were mixed evenly and then divided equally. These samples were either directly analyzed or stored at 20°C freezer for subsequent analysis.

2.4 Sample extraction

Homogenized soil (10 g), maize grain (5 g), and maize straw (5 g) were weighed and put into a 50 mL-polypropylene centrifuge tube, separately. For extraction of soil samples, 10 mL of acetonitrile, 500μL of acetic acid, and 5 mL of water were added to the soil sample and extracted for 15 min in an ultrasonic bath. Anhydrous NaCl (2 g) and anhydrous MgSO$_4$ (3 g) were added, vortexed vigorously for 10 min, and centrifuged at 4,000 r/min for 5 min. The supernatant was transferred to a fresh tube, and then 10 mL of extracting solution was added. The ultrasonic extraction was repeated. The supernatants were combined and purified twice. Finally, the acetonitrile layer was evaporated to near dryness with a vacuum rotary evaporator at 40°C. Two (2) mL of methanol was added to dissolve the dried sample. The solution was then filtered through 0.22μm nylon membrane filter, the filtered solution was used for subsequent injection. For extraction of maize straw samples, the volume of water added was changed from 5 mL to 2 mL for straw samples in the extraction step because water content in straw samples is higher than that of the soil samples. The other procedures were the same as those for extraction of maize grain samples.

2.5   Sample purification

Soil samples were not purified with dispersive solid-phase extraction (d-SPE).

For the detection of fluroxypyr-meptyl, maize grain, and maize straw were purified with GCB cartridge and conditioned with 5, 10, and 10 mL of methylene dichloride–methanol (95/5, v/v), respectively. Fifteen (15) mL of eluent was evaporated to near dryness with a vacuum rotary evaporator at 40°C. Two (2) mL of mobile phase was added to dissolve the sample and solution was filtered through 0.22μm nylon membrane filter and used for injection.

For detection of the fluroxypyr, maize grain, and maize straw were purified with the MAX cartridge. Five (5) mL of the methyl alcohol and 5 mL of water were added successively to activate the MAX columella. When the solvent liquid level flowed into the adsorption filler surface, the sample extracting solution was added, and then the cartridge was rinsed with 5 mL of 2% ammonia water. The eluted solutions were discarded. Finally, fluroxypyr was eluted with 12 mL of 2% ammonium formic acid–methyl alcohol solution. All the eluents were collected, and blown by nitrogen gas in a water bath at 40°C to nearly dry at a constant volume of 2 mL by the mobile phase. After being filtered with 0.22 μm water system filter membrane, fluroxypyr was measured with High-Performance Liquid Chromatography (HPLC).

2.6   Recovery assay

Samples of the untreated soil, maize grain and maize straw were fortified with fluroxypyr-meptyl and fluroxypyr standard solutions to reach the concentrations of 0.01, 0.05, and 0.5 (mg/kg), respectively. They were then processed according to extraction procedures described above. Five replicates for each concentration were analyzed.

2.7   Assay determination

A reverse-phase C_{18} chromatographic column (250 × 4.6 mm, i.d. 5 μm) was used as the separation column and maintained at 30°C. Mobile phases A and B were 0.1% acetic acid water and acetonitrile:methanol (1:3, v/v), respectively. Herbicides were separated by the following gradient program: gradient was firstly increased from 30% B to 70% B for 15 min and then increased from 70% B to 90% B for 17 min. Solvents were maintained at 90% B for 10 min and then, returned to initial conditions with an equilibration for 30 min. The flow rate was 1.00 mL min^{-1}. The injection volume was 20 μL and the detection wavelength was 212 nm.

3   RESULTS AND DISCUSSIONS

3.1   Research on appropriate extraction methods

The experimental conditions were systematically optimized by investigating the effects of the main parameters on the modified QuEChERS sample preparation as discussed below. Modified QuEChERS involved extraction solutions with acidic acetonitrile. Although fluroxypyr is the metabolite of fluroxypyr-meptyl, it differs largely from fluroxypyr-meptyl in chemical property. The fluroxypyr-meptyl is an ester with weak alkaline and weak polarity while the fluroxypyr is an amphoteric compound belonging to the pyridine family with two pKa values, i.e. pK_1 at 3.46 resulted from deprotonation of the carbonylgroup and pK_2 at 10.89 resulted from deprotonation of the amino group. Thus, the fluroxypyr is more soluble under acidic conditions with strong polarity. For fluroxypyr-meptyl, the recoveries were higher with both the acetonitrile (93.5%) and acidic acetonitrile (90.4%). However, for fluroxypyr, its recovery was much lower with acetonitrile (16.4%) but was higher with acidic acetonitrile (92.6%). Therefore, acidic acetonitrile could be used to simultaneously extract fluroxypyr-meptyl and fluroxypyr. Thus, we selected it as the extracting solution.

3.2   Research of purification methods

For soil samples, the matrices were simpler and did not need to be purified with the solid phase extraction column. For maize grains and maize straws, the solid phase extraction column was selected for purifying them. The results showed that with the solid phase extraction column, little amounts of both fluroxypyr-meptyl and fluroxypyr were retained in C_{18} filler, especially for fluroxypyr, which was only about 36%; and thus, the effects of gathering and purification could not be achieved. Moreover, fluroxypyr-meptyl and fluoxypyr molecules contained an amidogen, and showed weak alkaline. Therefore, a PCX cation exchange column was selected and used for gathering and purification. In this experiment, the recovery of fluroxypyr-meptyl was about 80% whereas the recovery of fluroxypyr was only about 45%. This difference in recovery rate may be due to the reason that while both fluroxypyr-meptyl and fluroxypyr are alkaline, they are much weaker alkalines. Due to the p, π conjugate effect of pyridine amino, alkaline amino is much weaker than normal. Thus, recovery of fluroxypyr is much poorer. Moreover, cleannet PestiCarb/NH$_2$ column was a mixed column with graphitized carbon black and amidogen, which could effectively remove pigments, organic acids and fatty acids. Therefore, when cleannet Pesti-Carb/NH$_2$ column was used, the recovery was about 95% for fluroxypyr-meptyl and 63% for fluroxypyr. This may be due to the reason that fluroxypyr is an acid absorbed in column. As an acid, MAX anion exchange column was selected for purifying fluroxypyr. MAX columns contain ammonium cation, which exchanged with the acidic substances. Thus, fluroxypyr could be well separated from other impurities and its recovery was much higher (about 95%). This purification method is more convenient and quicker. Thus, MAX anion exchange column was selected as the purification method for fluroxypyr while cleannet PestiCarb/NH$_2$ column was selected as the purification method for fluroxypyr-meptyl.

3.3 Selection of liquid phase conditions

According to PDA, the maximum absorption wavelength of fluroxypyr was 212 nm, as shown in Figure 1(A) while that of fluroxypyr-meptyl was 211 nm as shown in Figure 1(B). Thus 212 nm was selected as the maximum absorption wavelength for fluroxypyr-meptyl and fluroxypyr.

The text should fit exactly into the type area of 187 × 272 mm (7.36" × 10.71"). For correct settings of margins in the Page Setup dialog box (File menu) see Table 1.

Fluroxypyr-meptyl possesses relatively weak polarity. When 90% acetonitrile was used as the mobile phase, the retention characteristics of fluroxypyr-meptyl were manifested obviously. However, fluroxypyr possesses strong polarity with acidic property. When the water-acetonitrile mixture was used as the mobile phase, no fluroxypyr was obtained. However, when 0.1% of acetic acid was added to the water, it showed a favorable retention performance and a higher separation degree in the C18 column. This improvement may be due to the reason that as a strong carboxylic acid ion-pairing reagent, acid can shield

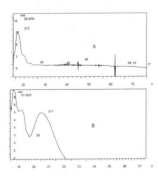

Figure 1. A: Maximum absorption wavelength for fluroxypyr and B: Maximum absorption wavelength for fluroxypyr-meptyl.

Table 1. Data of accuracies and precisions of corn and soil samples.

| Spiking level (mg/kg) | Sample | Fluroxypyr-meptyl | | Fluroxypyr | |
		Average recovery (%)	Relative standard (%)	Average recovery (%)	Relative standard (%)
maize grain	0.01	97.6	1.6	95.8	0.7
	0.05	89.6	2.3	95.4	0.6
	0.5	88.0	1.7	81.1	0.8
maize straw	0.01	96.5	0.7	96.5	0.7
	0.05	99.7	0.9	99.7	0.9
	0.5	97.8	1.4	97.8	1.4
soil	0.01	94.4	0.3	93.4	1.4
	0.05	103.3	0.7	84.0	1.6
	0.5	86.7	0.8	91.5	1.1

the non-polar group of C18 stationary phase by activating a dewatering force that makes the carboxylic acid ion of the chromatographic column "exposed" on the surface, and this allows the ion exchange effect with the fluroxypyr to achieve good chromatographic separation, with the advantages of a sharp peak, non-tailing, and good reproducibility. More importantly, fluroxypyr-meptyl could be separated well from fluroxypyr. The mobile phase ratio was optimized repeatedly by the experiment. With the isocratic mobile phase, there was always the appearance of a big interference peak about maize straw and maize to fluroxypyr as shown in Figure 2. No matter how the proportions of mobile phase were changed, the large peak always produced interference. However, with the gradient mobile phase, this problem was resolved completely, as shown in Figure 3. Gradient mobile phase was selected with obvious retention and poor retention characteristics. This experiment employed different mobile phases and the results indicated that order of different acidic mobile phases that made the peak of the target material sharper and more beautiful was as follows: 0.1% acetic acid > 0.1% formic acid > water. Thus, 0.1% acetic acid was selected as the mobile phase. The standard HPLC spectrum of the fluroxypyr-meptyl and fluroxypyr were shown in Figure 4. The change of mobile phase was an innovative HPLC method for detection of fluroxypyr-meptyl and fluroxypyr, which offers a good reference for the simultaneous analysis of fluroxypyr-meptyl, fluroxypyr and similar compounds.

3.4 Method validation

The standard solutions of fluroxypyr-meptyl and fluroxypyr with three files of concentration were added into the blank maize grain, maize straw and soil samples, and each file was repeated five times.

The recovery rate was tested by the analysis method described above. The results were shown in Table 1. The average recovery rates were in the range of 86.7–103.3% with a Relative Standard Deviation (RSD) of 0.3–2.3% for fluroxypyr-meptyl. For fluroxypyr, the average recovery rates were in the range of 81.1–99.7% with RSDs of 0.7–1.6%. The recovery rates for HPLC experiments in soil were shown in Figure 5.

According to the recovery rate experiment, the minimum detectable amount of the instrument was 0.1 ng under the chromatographic conditions described above. The minimum detectable concen-

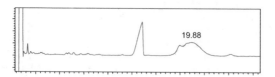

Figure 2. The high-performance liquid chromatography of the fluroxypyr in maize straw sample with the isocratic mobile phase.

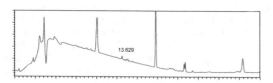

Figure 3. The high-performance liquid chromatography of the fluroxypyr in maize straw sample with the gradient mobile phase.

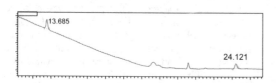

Figure 4. The standard high-performance liquid chromatography of the fluroxypy and fluroxypy-meptyl (peak 13.685 is for fluroxypy and peak 24.121 is for fluroxypy-meptyl).

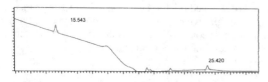

Figure 5. The high-performance liquid chromatography of the fluroxypy and fluroxypy-meptyl in soil sample (peak 15.543 is for fluroxypy and peak 25.420 is for fluroxypy-meptyl).

trations of fluroxypyr-meptyl and fluroxypyr in the maize grains and the soil were 0.01 mg/kg and 0.01 mg/ kg, respectively.

3.5 Dissipation of two herbicides in different ecosystems under field conditions

In this study, we also investigated the dissipation patterns of the dispersable and oily suspensions of fluroxypyr-meptyl and fluroxypyr in maize straw and soil. The samples of maize straw and their corresponding soil were collected at different intervals after the application of fluroxypyr-meptyl. These two herbicides were pretreated with different pretreatment methods but both them were detected with the same HPLC method. The dynamic equation for fluroxypyr, the half-life (d) and the dissipation patterns of these two herbicides were shown in Figures 6 and 7 and Table 2.

The dissipation kinetics of two herbicides were determined by plotting the concentration (mg/kg) against time (d). The dissipation kinetics followed

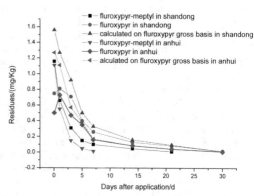

Figure 6. Degradation curve of fluroxypy, fluroxypy-meptyl and alculated on fluroxypyr gross in soil.

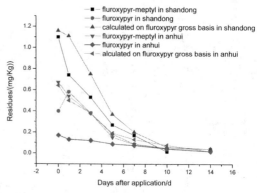

Figure 7. Degradation curve of fluroxypy, fluroxypy-meptyl and alculated on fluroxypyr gross in maize straw.

Table 2. Dynamic degradation equations of fluroxypyr in maize straw and soil.

Experiment site	Matrix	Degradation equation[1]	Determination coefficient (R)	Half-life (days)
Shandong	Soil	$C_t = 1.2862e^{-0.142t}$	−0.9873	4.9
	Maize straw	$C_t = 1.3438e^{-0.268t}$	−0.9845	2.6
Anhui	Soil	$C_t = 0.9074e^{-0.15t}$	−0.9798	4.6
	Maize straw	$C_t = 0.5341e^{-0.172t}$	−0.9860	4.0

1) C_t: terminal concentration (mg/kg); t: time (d).

Table 3. Residues of fluroxypyr-meptyl and fluroxypyr in maize grain and soil in 2015.

Sample type	Dosage gactive ingredient per hectare	Days after spraying		Residue (mg kg⁻¹)	
		Shandong	Anhui	Shandong	Anhui
Green maize grain	19.6	48	40	0.01	0.01
	29.4	48	40	0.01	0.01
Maize grain	19.6	68	60	0.01	0.01
	29.4	68	60	0.01	0.01
Soil	19.6	68	60	0.01	0.01
	29.4	68	60	0.01	0.01

the first order kinetic equation $C_t = C_0 \exp(-kt)$, where C_t represents the concentration of the herbicide (mg/kg) at time (d), C_0 represents the initial concentration, and k is the rate constant. The half-life was equal to the ratio of ln 2 and k.

It can be seen from the data shown in Figures 6 and 7 that the amounts of fluroxypyr-meptyl residue in soil and maize straw were increased with prolonging the time after being applied while the amounts of fluroxypyr residue in soil and maize straw were initially increased and generally reached the maximal values one day after being applied but decreased with prolonging time thereafter. However, for the fluroxypyr residue in maize straw in Anhui province, its level was decreased with prolonging time. This observation is different from that made by Pang et al. [7], whose study indicated that the amount of fluroxypyr residue was decreased with prolonging the time after being applied. Our analysis indicated that the metabolism rate of fluroxypyr-meptyl into fluroxypyr was not so rapid and it reached the maximal value one day after being applied. According to *"Pesticide Maximum Residue Limits (MRLs) in China 2763–2015"*, the amount of fluroxypyr-meptyl is calculated based on fluroxypyr. The amount of fluroxypyr-meptyl was converted to that of fluroxypyr and then the sum of fluroxypyr-meptyl was calculated to determine the dynamic dissipation rate of the residues. The amounts of residues were decreased with prolonging the time. The dissipation rate of fluroxypyr was more rapid in maize straw than that in soil. At 10 days after application of herbicide, the dissipation rate of fluroxypyr in maize straw could reach more than 90% while at 14 days after application of herbicide, the dissipation rate of fluroxypyr in soil could reach more than 90%. The dissipation curves fitted the first-order dynamic equation. Its half-lives in maize straw were 2.6 and 4.0 d while the half-lives of fluroxypyr in soil were 4.9 d and 4.6 d, respectively.

3.6 Terminal residues of two herbicides in milk stage, mature stage maize grain and soil

The final residue results of fluroxypyr-meptyl and fluroxypyr in milk stage, mature stage, maize grain and soil in two sites located in Shandong and Anhui provinces in 2015 with 12% fluroxypyr-meptyl were shown in Tables 3, which indicated that there were no detectable residues of fluroxypyr-meptyl and fluroxypyr in milk stage, mature stage, maize grain and soil, and the residue levels were lower than the limited value of 0.0.05 mg/kg (maximum residue limit value) as well as 0.01 mg/kg (LOQs). Obviously, it would be accepted to spray the formulation at the recommended dosage due to their terminal residues in soil, maize grain, and maize straw.

4 CONCLUSION

This study developed a method to detect the residues of fluroxypyr-meptyl and fluroxypyr in maize straw and soil. Its pretreatment method was simpler and more efficient. Acetic acid was used as the mobile phase. With gradient mobile phase which could be washed to separate completely from interference peak and possessed the advantages of a sharp peak, non-tailing, and good reproducibility. More importantly, fluroxypyr-meptyl could be simultaneously separated well from fluroxypyr. The detection method was applied to study the metabolic patterns of maize grain, straws and soil.

The results in this experiment showed that the metabolism rates of fluroxypyr in maize straw and soil were fast, with a half-life of 2.6–4.9 days, which is relatively close to that (0.6–3.6d) reported from a study by Pang et al(Pang et al. 2016). Those short half-lives indicated their appropriate stability and safety.

The terminal residues of two herbicides in edible maize grain and non-edible matrices of soil were all lower than LOQs and MRLs. 12% fluroxypyr-meptyl with recommended dosages were proved to be safe for controlling weeds. The dissipation study and terminal residue assay were useful not only for environmental protection but also for food safety.

REFERENCES

David, A., C. Botias, A. Abdul-Sada, D. Goulson & E. M. Hill (2015) Sensitive determination of mixtures of neonicotinoid and fungicide residues in pollen and single bumblebees using a scaled down QuEChERS method for exposure assessment. *Anal Bioanal Chem. 407*, 8151–62.

Halimah, M., Y. A. Tan & B. S. Ismail (2004) Method development for determination of fluroxypyr in soil. *J Environ Sci Health B., 39*, 765–77.

Hu, J.-Y., Y.-Q. Hu, Z.-H. Zhen & Z.-B. Deng (2011) Residue Analysis of Fluroxypyr-meptyl in Wheat and Soil by GC–ECD. *Chromatographia. 74*, 291–296.

Ivanova, B. (2015) Erratum to: Solid-state UV-MALDI mass spectrometric quantitation of fluroxypyr and triclopyr in soil. *Environ Geochem Health*.

Kujawski, M. W. & J. Namieśnik (2011) Levels of 13 multi-class pesticide residues in Polish honeys determined by LC-ESI-MS/MS. *Food Control. 22*, 914–919.

Lazartigues, A., M. Thomas, C. Cren-Olive, J. Brun-Bellut, Y. Le Roux, D. Banas & C. Feidt (2013) Pesticide pressure and fish farming in barrage pond in Northeastern France. Part II: residues of 13 pesticides in water, sediments, edible fish and their relationships. *Environ Sci Pollut Res Int. 20*, 117–25.

Mesnage, R., N. Defarge, J. Spiroux de Vendomois & G. E. Seralini (2014) Major pesticides are more toxic to human cells than their declared active principles. *Biomed Res Int. 2014*, 179691.

Niu, Y., J. Zhang, Y. Wu & B. Shao (2011) Simultaneous determination of bisphenol A and alkylphenol in plant oil by gel permeation chromatography and isotopic dilution liquid chromatography-tandem mass spectrometry. *J Chromatogr A. 1218*, 5248–53.

Omari, R. & G. Frempong (2016) Food safety concerns of fast food consumers in urban Ghana. *Appetite. 98*, 49–54.

Pang, N., T. Wang & J. Hu (2016) Method validation and dissipation kinetics of four herbicides in maize and soil using QuEChERS sample preparation and liquid chromatography tandem mass spectrometry. *Food Chem. 190*, 793–800.

Peng, X., J. Pang & A. Deng (2011) [Determination of seven phenoxyacid herbicides in environmental water by high performance liquid chromatography coupled with three phase hollow fiber liquid phase microextraction]. *Se Pu. 29*, 1199–204.

Raeppel, C., M. Fabritius, M. Nief, B. M. Appenzeller, O. Briand, L. Tuduri & M. Millet (2015) Analysis of airborne pesticides from different chemical classes adsorbed on Radiello(R) Tenax(R) passive tubes by thermal-desorption-GC/MS. *Environ Sci Pollut Res Int. 22*, 2726–34.

Santilio, A., P. Stefanelli, S. Girolimetti & R. Dommarco (2011) Determination of acidic herbicides in cereals by QuEChERS extraction and LC/MS/MS. *J Environ Sci Health B. 46*, 535–43.

Tao, L. & H. Yang (2011) Fluroxypyr biodegradation in soils by multiple factors. *Environ Monit Assess. 175*, 227–38.

Ulen, B. M., M. Larsbo, J. K. Kreuger & A. Svanback (2014) Spatial variation in herbicide leaching from a marine clay soil via subsurface drains. *Pest Manag Sci. 70*, 405–14.

Wang, L., J. Xu, P. Zhao & C. Pan (2011) Dissipation and residues of fluroxypyr-meptyl in rice and environment. *Bull Environ Contam Toxicol. 86*, 449–53.

Wang, P., D. Liu, S. Jiang, Y. Xu, G. Gu & Z. Zhou (2008) The chiral resolution of pesticides on amylose-tris(3,5-dimethylphenylcarbamate) CSP by HPLC and the enantiomeric identification by circular dichroism. *Chirality. 20*, 40–6.

Wang, P., S. Jiang, D. Liu, H. Zhang & Z. Zhou (2006) Enantiomeric resolution of chiral pesticides by high-performance liquid chromatography. *J Agric Food Chem. 54*, 1577–83.

Wu, Y. L., R. X. Chen, L. Zhu, Y. Lv, Y. Zhu & J. Zhao (2016) Determination of ribavirin in chicken muscle by quick, easy, cheap, effective, rugged and safe method and liquid chromatography-tandem mass spectrometry. *J Chromatogr B Analyt Technol Biomed Life Sci. 1012–1013*, 55–60.

Yang, F., T. Mipam, L. Sun, S. Yu & X. Cai (2016) Comparative testis proteome dataset between cattleyak and yak. *Data Brief. 8*, 420–5.

Zhang, Q., B. Gao, M. Tian, H. Shi, X. Hua & M. Wang (2016a) Enantioseparation and determination of triticonazole enantiomers in fruits, vegetables, and soil using efficient extraction and clean-up methods. *J Chromatogr B Analyt Technol Biomed Life Sci. 1009–1010*, 130–7.

Zhang, Y., D. Hu, S. Zeng, P. Lu, K. Zhang, L. Chen & B. Song (2016b) Multiresidue determination of pyrethroid pesticide residues in pepper through a modified QuEChERS method and gas chromatography with electron capture detection. *Biomed Chromatogr. 30*, 142–8.

Zhao, H., J. Xu, F. Dong, X. Liu, Y. Wu, X. Wu & Y. Zheng (2015) Simultaneous determination of three herbicides in wheat, wheat straw, and soil using a quick, easy, cheap, effective, rugged, and safe method with ultra high performance liquid chromatography and tandem mass spectrometry. *J Sep Sci. 38*, 1164–71.

Influence of fly ash on the pore structure and performance of foam concrete

Zhongwei Liu

Department of Materials Science and Engineering, Xi'an University of Technology, Xi'an, China
Sichuan College of Architectural Technology, Deyang, China

Kang Zhao

Department of Materials Science and Engineering, Xi'an University of Technology, Xi'an, China

Chi Hu

Sichuan College of Architectural Technology, Deyang, China

Yufei Tang

Department of Materials Science and Engineering, Xi'an University of Technology, Xi'an, China

ABSTRACT: Foam concrete with varying dry densities (nominally 400, 500, 600, 700, and 800 kg/m3) were prepared using ordinary Portland cement (42.5R), fly ash, and vegetable protein foaming agent as raw materials, by adjusting the fly ash content through a physical foaming method. The performance of cement paste as well as the structure and distribution of air pores were characterized by a rheometry, SEM, vacuum water saturation instrument, and image analysis software. Effects of the fly ash content on the relative viscosity of the cement paste, as well as pore structure and strength of the hardened foam concrete sample were also explored. Results showed that the fly ash content can crucial influence the size, distribution, and connectivity of pores in foam concrete. When the fly ash content of the foam concrete increased from 0% to 50%, the compressive strength of foam concrete increased first and then decreased. The same changes were observed in anti-frost property of foam concrete.

1 INTRODUCTION

Foam concrete is light weight, thermally insulated, sound insulated, environment friendly, and quickly manufactured. Thus, this material is widely applied in heat insulation, pipeline backfill, and subgrade treatment (Farzadnia, N., 2015). The current research focuses mainly on the influence of various admixtures on the strength, heat-conducting property, anti-frost property, and sound-absorbing performance of foam concrete (Chen, 2013; Ali A. Sayadi, 2016; Liu, 2013; Alengaram, 2013; Zhang, 2015; Yang, 2014; Hu, 2016).

Admixtures are widely available, cheap, environment friendly, and energy efficient. When added to concrete, admixtures can adjust the pore structure of foam concrete and influence its performance. Chindaprasirt et al. (2011) mentioned that replacing a part of cement or sand with high-calcium fly ash can reduce the shrinkage of foam concrete and improve its compressive strength. They also observed the pore features. Chen et al. (2014) studied the preparation of foam concrete with circulating fluidized bed fly ash. They found that

calcium oxide and aluminate cement can accelerate coagulation and harden the foam concrete mixed with fly ash, and the optimal contents of the two materials in foam concrete are 8% and 2%. Yang et al. (2013) applied fly ash in alkali-activated slag foam concrete and concluded that fly ash can improve the rheological property of alkali-activated slag foam concrete. The compressive strength of alkali-activated slag foam concrete can be improved when the added amount of fly ash is less than 15%; otherwise, the compressive strength decreases. Bentz et al. (2011) studied the preparation of concrete using the optimized particle size method, selected the proper particle size distribution of cement and fly ash, and replaced 35% of cement with fly ash to maintain the strength at 1 and 28 days. Jitchaiyaphum et al. (2011) suggested that replacing cement with fly ash can decrease pore size but increase compressive strength. Hilal et al. (2015) studied the influence of silica fume, fly ash, and plasticizer on the distribution of pore size and further examined the influence of pore size distribution on the compressive strength of foam concrete. All of the above-mentioned studies

analyzed the influence of fly ash on the performance of foam concrete. However, research on the
influence of fly ash on the pore structure of foam
concrete and the adjusting mechanism of its pore
structure is limited. Thus, the current study examines the influence of fly ash on the rheological
property of cement paste and the pore structure
of foam concrete. This study provides reference for
the preparation of foam concrete with characteristics of light weight, high performance, and strong
frost resistance.

Table 1. The physical properties of cement.

Material	P.O42.5R
Blaine fineness (m²/kg)	343
Initial setting time (min)	91
Final setting time (min)	210
Soundness	qualified
3d compressive strength (MPa)	28.7
28d compressive strength (MPa)	48.9
Density (g/cm³)	3.06

Table 2. The main chemical composition
of cement (wt%).

Compositions (%)	Content
SiO_2	21.6
Al_2O_3	4.9
Fe_2O_3	2.5
CaO	63.4
MgO	1.8
SO_3	2.14
Na_2O	0.14
K_2O	0.37
Loss of ignition	3.15

2 TEST

2.1 Raw materials

Ordinary Portland cement used was P.O.42.5R
cement provided by the Deyang Lisen Cement
Plant. Tables 1 and 2 and Figure 1(a) show the
physical properties, chemical components, grain
size, and surface morphology of the cement. Level
I fly ash was acquired from the Jiangyou Thermal
Power Plant. Tables 3 and 4 and Figure 1(b) show
the physical properties, chemical components, grain
size, and surface morphology of the fly ash. Meanwhile, the plant protein foaming agent was provided
by Sichuan Xinhan Corrosion Protection Engineering Co., Ltd.

2.2 Preparation

The cement content (g), fly ash content (g),
water content (g), foam content (mL), and
water–binder ratio of various experimental
samples were configured according to Table 5.
According to Table 5, cement, fly ash, and water
were placed into a 15 dm³ horizontal type mixer
(GH-15, Beijing Guanggui Jingyan Foamed
Concrete Science & Technology Co., Ltd.) and
mixed under the mixing speed of 40 r/min for
2 min at 25°C to generate paste. The relative

Table 3. The physical properties of fly ash.

material	fly ash
Particle size (45 um sieve rate) (%)	2.9
Ratio of water demand (%)	91
water content (%)	0.20%
Activity index (%)	66
Density (g/cm³)	2.6

(a) Cement×1000

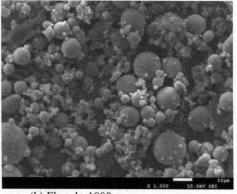

(b) Fly ash×1000

Figure 1. SEM images of raw materials.

viscosity of each paste was measured imme-diately after mixing. Meanwhile, the foaming agent was diluted with water at 1:15 ratio. Next, protein foam produced by a foam generator (ZK-FP-20, Beijing Zhongke Zhucheng Building Materials Co., Ltd.) was injected into a foam concrete mixer and stirred for

2 min. Next, The slurry was placed in a 100 mm × 100 mm × 100 mm mold and then maintained in a static at $20 \pm 2°C$ with relative humidity (RH) of 60% for 24 h. After demold-ing, The specimens were stored in a fog room $(20 \pm 2°C; RH > 95\%)$ for curing for 28 d.

2.3 Test method

Relative viscosity was tested by using a rotary vis-cometer (NXS-11 A, Chengdu Instrument Fac-tory, China). Slices (8 mm × 5 mm × 5 mm) of the test specimens were obtained from six directions. The reaction was terminated through hydration with absolute ethyl alcohol and dried in an oven at 60°C until a constant mass was obtained. The microstructures of the specimens were determined by using a scanning electron microscopy (SEM; Hitachi JSM-7500F). After binarization process-ing of the SEM images was conducted. Finally, Image-Pro Plus 6.0 image analysis software was used to analyze and extract data on the pore char-acteristics of the specimens.

Table 4. The main chemical composition of fly ash (wt%).

Compositions (%)	Content
SiO_2	57.8
Al_2O_3	24.84
Fe_2O_3	4.6
CaO	4.74
MgO	2.45
SO_3	2.6
Na_2O	1.2
TiO_2	0.5
Loss of ignition	0.9

Table 5. Mix proportions of raw materials.

Mixes designation	Design density (kg/m³)	Cement (g)	Fly ash (g)	Water (g)	w/b	Fly ash content (%)	Foam (mL)
F400–0	400	2909	0	1745	0.60	0	5301
F400–10	400	2637	293	1758	0.60	10	5264
F400–20	400	2362	590	1771	0.60	20	5227
F400–30	400	2082	892	1784	0.60	30	5190
F400–40	400	1798	1199	1798	0.60	40	5152
F400–50	400	1509	1509	1811	0.60	50	5113
F500–0	500	3636	0	2073	0.57	0	4735
F500–10	500	3297	366	2088	0.57	10	4690
F500–20	500	2952	738	2103	0.57	20	4645
F500–30	500	2602	1115	2119	0.57	30	4599
F500–40	500	2247	1498	2135	0.57	40	4552
F500–50	500	1887	1887	2151	0.57	50	4505
F600–0	600	4364	0	2400	0.55	0	4169
F600–10	600	3956	440	2418	0.55	10	4116
F600–20	600	3542	886	2435	0.55	20	4063
F600–30	600	3123	1338	2454	0.55	30	4008
F600–40	600	2697	1798	2472	0.55	40	3953
F600–50	600	2264	2264	2491	0.55	50	3896
F700–0	700	5091	0	2698	0.53	0	3633
F700–10	700	4615	513	2718	0.53	10	3572
F700–20	700	4133	1033	2738	0.53	20	3510
F700–30	700	3643	1561	2758	0.53	30	3447
F700–40	700	3146	2097	2779	0.53	40	3383
F700–50	700	2642	2642	2800	0.53	50	3318
F800–0	800	5818	0	2909	0.50	0	3183
F800–10	800	5275	586	2930	0.50	10	3115
F800–20	800	4723	1181	2952	0.50	20	3045
F800–30	800	4164	1784	2974	0.50	30	2975
F800–40	800	3596	2397	2996	0.50	40	2903
F800–50	800	3019	3019	3019	0.50	50	2830

Volume density and compressive strength tests of the foam concrete complied with the Chinese *Foamed Concrete* standard (JG/T 266–2011). The compressive strength of the specimens was measured by a fully automatic, constant stress testing machine (JYE-300 A, Beijing Jiwei Testing Instrument Co., Ltd., China) under a loading rate of 200 N /s. The true density (ρ) of the specimens was tested in accordance with the *Cement Density Measurement Method* standard (GB/T 208–2014), The volume density of the specimens was denoted by ρ_1. Then, the porosity of specimens was determined using the following formula:

$$P = \frac{\rho - \rho_1}{\rho} \times 100\% \qquad (1)$$

where P is the porosity of the specimens (%), ρ is the true density of the specimens (kg/m³), and ρ_1 is the volume density of the specimens (kg/m³).

The material mass under the water—saturated state of the test specimens was measured using an intelligent concrete vacuum water saturation instrument (SW-6, Beijing Shengshi Weiye Science & Technology Co., Ltd). The open porosity of the foam concrete was assessed. Open porosity (p_K) and closed porosity (p_B) were calculated using Equations (2), and (3).

The p_K of the specimens was calculated as

$$P_K = \frac{m_2 - m_1}{V} \times \frac{1}{\rho_W} \times 100\% \qquad (2)$$

where m_1 is the dry material mass (kg), m_2 is the material mass under the water—saturated state (kg), ρ_W is the water density (kg/m³), and V is the natural volume of the material.

Meanwhile, the p_B of the specimens were determined as follows:

$$P_B = P - P_K \qquad (3)$$

Following *GB/T50082-2009: Testing methods for long-term and long-lasting performance of ordinary concrete*, a 100 mm × 100 mm × 100 mm cubic test piece was placed in a DW-40 cryogenic box (produced by Tianjin Huanan Experimental Instrument Co., Ltd.). The slow-freezing method was employed to test the anti-frost property of foam concrete. The results were tested after 50 freezing and thawing cycles.

3 RESULT ANALYSIS

3.1 Influence of fly ash content on the rheological property of cement paste

Cement paste is a kind of non-Newtonian fluid, and The slope of the fitting curve represents the relative viscosity of the cement paste (experiment parameters: laboratory temperature is 25°C, A system):

$$\eta = \frac{\tau}{D_S} \qquad (4)$$

where η is the relative viscosity, τ is the shear stress and D_S is the shearing rate.

Figure 2(a) shows the influence of fly ash content on the rheological property of the cement paste of foam concrete with 500 kg/m³ density (w/b = 0.57). Equation (4) shows that the relative viscosity of cement paste decreased from 0.018 Pa·s to 0.014 Pa·s when fly ash content increased from 0% to 50%. Figure 2(b) shows the influence of fly ash content on the rheological property of the cement paste of foam concrete with 800 kg/m³ density (w/b = 0.50). The relative viscosity of cement paste decreased from 0.035 Pa·s to 0.020 Pa·s when fly ash content increased from 0% to 50%. This result is attributed to the "morphological effect" of fly ash. Fly ash features a spherical microsphere, whose surface. Figure 1(b) shows that fly ash could be used as an effective lubricant to reduce cohesion in paste and change the rheological property of mixtures. When fly ash content increased, the relative viscosity of cement paste decreased. The relative viscosity of cement paste with a water–binder ratio of 0.50 decreased in a significantly faster rate than that of cement paste with a water–binder ratio of 0.57. This finding is attributed to the high cohesion and internal friction between pastes with small water–binder ratio and the improved consistency. Fly ash exhibited improved lubrication effect in pastes with high consistency.

3.2 Influence of fly ash content on the pore structure of foam concrete

Figures 3 and 4 show the SEM images of the foam concrete samples with densities of 500 and 800 kg/m³ and different fly ash contents. The figures indicate that 1) connected pores were few when fly ash content was high and 2) connected pores were few when dry density was high. The influence of fly ash content on the porosity of foam concrete was also tested (Figure 5). Figure 5(a) shows that the open porosity of samples F500–0, F500–10, F500–20, F500–30, F500–40, and F500–50 decreased gradually from 43.06% to 35.9%, whereas the close porosity increased gradually from 32.5% to 38.93%. Figure 5(b) indicates that the open porosity of samples F800–0, F800–10, F800–20, F800–30, F800–40, and F800–50 decreased gradually from 39.1% to 36.55%, whereas the close porosity increased steadily from 22.62% to 24.22%. The results suggest that 1) when fly ash content increased, connected pores were few, open porosity was low, and close porosity was high. This phenomenon is

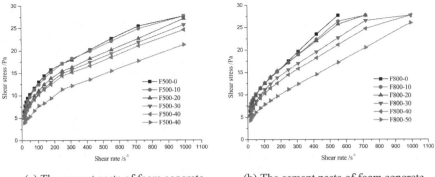

(a) The cement paste of foam concrete with 500 kg/m³ density

(b) The cement paste of foam concrete with 800 kg/m³ density

Figure 2. Effects of fly ash content on the rheological property of cement paste.

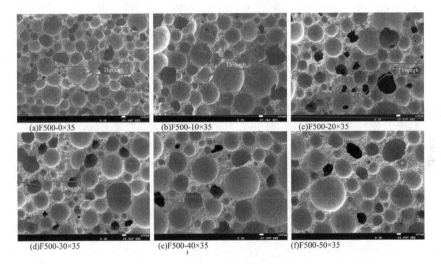

(a)F500-0×35   (b)F500-10×35   (c)F500-20×35

(d)F500-30×35   (e)F500-40×35   (f)F500-50×35

Figure 3. SEM micrographs of the 500 kg/m³ foamed concretes with different fly ash content.

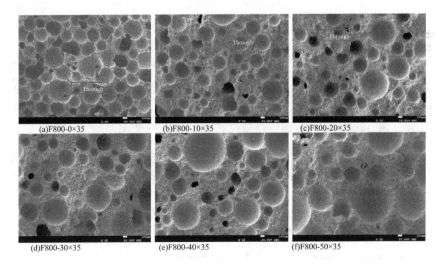

(a)F800-0×35   (b)F800-10×35   (c)F800-20×35

(d)F800-30×35   (e)F800-40×35   (f)F800-50×35

Figure 4. SEM micrographs of the 800 kg/m³ foamed concretes with different fly ash content.

attributed to two main reasons. On the one hand, the increase in fly ash content decreased relative viscosity. The combination of additional foams decreased the total surface area of foams but increased the cement paste wrapped on the surface of foams. Accordingly, the pore wall thickened, the number of connected pores decreased, open porosity decreased, and close porosity increased. On the other hand, given that the particle surface of fly ash was smooth and its grading was favorable, migration and attachment to the surface of foams in the paste easily occurred. Thus, pore wall thickness increased [15–16], thereby decreasing connected pores and open porosity and increasing closed porosity. The results also suggest that, 2) when dry density increased, the connected pores of foam concrete and open porosity decreased and closed porosity elevated. This result is due to that foam concrete with a low dry density exhibited a thin pore wall and additional open pores. The pore wall thickened when dry density increased. Subsequently, connected pores and open porosity decreased and close porosity increased.

Figure 6 shows the influence of fly ash on the pore diameter distribution of foam concrete. Figure 6(a) shows that the small pore proportions (<100 um) of samples F500–0, F500–10, F500–20, F500–30, F500–40, and F500–50 were 13.91%, 11.84%, 10.48%, 8.66%, 8.09%, and 4.32%, respectively. The medium pore proportions (100–400 um) were 72.39%, 72.58%, 73.03%, 73.62%, 73.09%, and 76.63%. The large pore proportions (>400 um) were 13.70%, 15.58%, 16.48%, 17.72%, 18.82%, and 19.05%. Figure 6(b) indicates that the small pore proportions (<100 um) of samples F800–0, F800–10, F800–20, F800–30, F800–40, and F800–50 were 18.15%, 14.03%, 10.97%, 10.67%, 8.70%, and 7.00%, respectively. The medium pore proportions (100–400 um) were 78.21%, 81.49%, 83.58%, 83.00%, 82.96%, and 83.67%. The large pore proportions (>400 um) were 3.64%, 4.47%, 5.45%, 6.33%, 8.35%, and 9.33%. The results suggest that the pore diameter of foam concrete was mainly distributed within the range of 100–400 um. When fly ash content increased, the small pore proportion (<100 um) decreased, the medium pore proportion

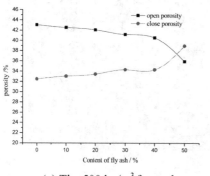

(a) The 500 kg/m³ foamed concretes

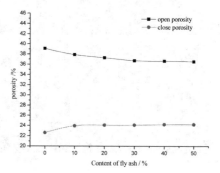

(b) The 800 kg/m³ foamed concretes

Figure 5.   Effect of fly ash on the porosity.

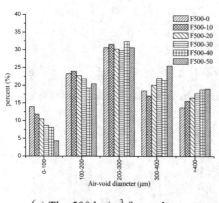

(a) The 500 kg/m³ foamed concretes

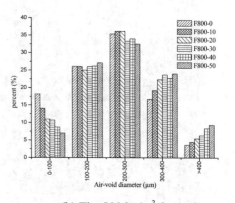

(b) The 800 kg/m³ foamed concretes

Figure 6.   Effects of fly ash content on air voids diameter distribution.

(100–400 um) slightly changed, and the large pore proportion (>400 um) increased. These results are attributed to two reasons. On the one hand, when fly ash content increased, the hydration reaction of the paste decelerated the coagulation, hardening duration lengthened, the foam combined, and the growth duration lengthened. On the other hand, when fly ash content increased, the relative viscosity of the paste decreased gradually and the bubble-maintaining capacity of the paste weakened. Thus, small bubbles easily merged and grew during stirring [14].

3.3 Influence of fly ash content on the compressive strength of concrete

Table 6 shows the measured value of dry density and the compressive strength of foam concrete at 28 days. On the basis of Table 6, the relationship between the two is shown in Figure 7. The figure indicates that a power relationship existed between relative dry density and the compressive strength of foam concrete at 28 days. Varying fly ash content resulted in different relational expressions.

The relationship between the compressive strength of foam concrete at 28 days with different designed densities and fly ash contents (Figure 7) is fitted in Figure 8. The figure shows that that the optimal fly ash contents for foam concrete with densities of 400, 500, 600, 700, and 800 kg/m³ were 36.4%, 29.9%, 27.9%, 24.8%, and 23.9%, respectively. The power relationship between dry density and the optimal fly ash content was $y = 1326.9x^{-0.6045}$ and $R^2 = 0.9737$. The findings indicate that 1) the compressive strength of foam concrete increased first and then decreased when fly ash content increased. The optimal fly ash con-

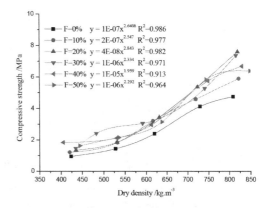

Figure 7. Effect of the dry density on the compressive strength at different fly ash content.

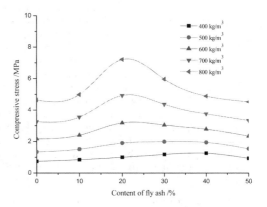

Figure 8. Effect of fly ash content on the compressive strength at different density.

Table 6. Dry apparent density and compressive strength of foamed concrete.

Percentage of fly ash (%)	Dry apparent density (kg/m³)	Compressive strength (MPa)	Percentage of fly ash (%)	Dry apparent density (kg/m³)	Compressive strength (MPa)
0	423.5	0.95	30	434.6	1.46
	528.0	1.44		482.5	2.41
	619.7	2.40		591.5	3.06
	729.3	4.13		740.7	5.27
	808.8	4.75		815.8	7.36
10	420.2	1.22	40	405.3	1.84
	531.8	1.83		530.5	2.12
	616.0	3.21		612.7	2.96
	719.2	4.60		748.7	5.76
	821.8	5.90		829.8	6.68
20	435.2	1.35	50	444.2	1.63
	531.5	1.85		535.7	2.15
	631.2	3.44		637.7	3.15
	723.2	5.37		743.8	5.84
	819.7	7.60		848.5	6.39

tent for foam concrete with different dry densities varied. Such difference is due to two reasons. On the one hand, when fly ash content increased from zero to the optimal amount, the small pore proportion decreased, the pore wall thickened, and the number of connected pores and pores of irregular shapes decreased (Figures 3 and 4). Given that stress concentration decreased under stress, the strength of foam concrete increased. On the other hand, when fly ash content increased from the optimal amount to 50%, cement content decreased and the amount of $Ca(OH)_2$ from the hydration reaction of cement paste was insufficient. Part of fly ash could not generate a "pozzolanic activity" because of the lack of adequate $Ca(OH)_2$. The hydration products with bubble-maintaining capacity generated by the hydration reaction of paste decreased. The initial setting time duration lengthened. During the initial maintenance period, foams within the paste could easily merge, thereby decreasing the number of pores of foam concrete. When the pore diameter increased and the pore distribution became uneven, the stress concentration decreased the strength of foam concrete. The findings also indicate that (2) the optimal fly ash content decreased when the dry density of foam concrete increased. The amount of foam added to prepare foam concrete with a large dry density was low. Thus, the total superficial area of the foam decreased, thereby decreasing the total fly ash content attached to the surface of the foam.

3.4 Influence of fly ash content on the anti-frost property of foam concrete

Table 7 shows the influence of fly ash on the anti-frost property of foam concrete. The results suggest that (1) the anti-frost property of foam concrete increased first and then decreased with fly ash content. The same trend was observed in the compressive strength of foam concrete. When the fly ash content of the foam concrete with 500 kg/m³ density increased from 0% to 50%, the mass loss rate of the foam concrete decreased from 2.51% to 2.14% first and then increased to 2.35%. Its strength loss rate decreased from 20.86% to 16.59% first and then increased to 18.13%. When the fly ash content of the foam concrete with 800 kg/m³ density increased from 0% to 50%, the mass loss rate of the foam concrete decreased from 2.26% to 1.81% first and then increased to 2.30%. Its strength loss rate decreased from 10.23% to 7.69% first and then increased to 10.99%. The results also suggest that (2) the anti-frost property of foam concrete improved when the density of foam concrete was high. Comparisons of the best anti-frost property of the foam concrete samples with densities of 500 and 800 kg/m³ indicate that the mass loss rate of F500–30 was 2.14% and its strength loss rate was 16.59%. The mass loss rate and strength loss rate of F800–20 were 1.81% and 7.69%, respectively. These results are due to two reasons. On the one hand, the compressive strength of foam concrete was an important factor for foam concrete to resist internal stress when water freezes. On the other hand, the compressive strength and anti-frost property of foam concrete were positively related to the pore structure of foam concrete. The combined data analysis results in Section 3.2 for the influence of fly ash on the pore structure of foam concrete show that the foam concrete with 800 kg/m³ density exhibited a smaller total porosity and open porosity than that with 500 kg/m³ density. Thus, the water within open pores froze and internal stress

Table 7. Effect of fly ash on the frost resistance of foamed concrete.

Mixes designation	Dry density before the frozen (kg/m³)	Dry density after the frozen (kg/m³)	Loss rate of weight (%)	Compressive strength before the frozen (MPa)	Compressive strength after the frozen (MPa)	Loss rate of compressive strength (%)
F500-0	521.0	504.0	2.51	1.44	1.19	20.86
F500-10	526.8	497.5	2.41	1.83	1.54	18.83
F500-20	528.5	484.0	2.32	1.85	1.57	17.55
F500-30	491.0	504.0	2.14	2.41	2.07	16.59
F500-40	521.5	490.0	2.29	2.12	1.81	17.44
F500-50	529.7	507.5	2.35	2.15	1.82	18.13
F800-0	808.8	790.5	2.26	4.75	4.31	10.23
F800-10	828.7	795.0	2.07	5.90	5.41	9.09
F800-20	812.6	793.0	1.81	7.60	7.06	7.69
F800-30	808.8	777.0	1.99	7.36	6.76	8.95
F800-40	821.8	803.5	2.23	6.68	6.10	9.56
F800-50	843.5	785.0	2.30	6.39	5.76	10.99

caused by volume expansion decreased. Internal, close, and small pores could alleviate the pressure caused by the freezing of water. Thus, the alternative acting force caused by freezing and thawing decreased and the anti-frost property of foam concrete improved.

4 CONCLUSIONS

1. The addition of fly ash could decrease the relative viscosity of cement paste, increase the rheological property of foam concrete, and help improve the performance of foam concrete;
2. Fly ash content could influence the diameter, shape, distribution, and pore connection of foam concrete. When fly ash content increased, the compressive strength of foam concrete first increased and then decreased.
3. The optimal fly ash content for foam concrete with different densities varied. The optimal fly ash contents to prepare foam concrete with densities of 400, 500, 600, 700, and 800 kg/m^3 were 36.4%, 29.9%, 27.9%, 24.8%, and 23.9%, respectively. The power relationship between the dry density and the optimal fly ash content was $y = 1326.9x^{-0.6045}$ and $R^2 = 0.9737$.
4. The anti-frost property of foam concrete improved first and then weakened when fly ash content increased. The same changes were observed in the compressive strength of foam concrete. The anti-frost property of foam concrete improved when the density of foam concrete was high.

ACKNOWLEDGMENTS

The authors would like to acknowledge the support from the National Natural Science Foundation of China (no. 51372199).

REFERENCES

Alengaram, U.J., Muhit, B.A.A., Jumaat, M.Z.B., & Jing, M.L.Y. (2013). A comparison of the thermal conductivity of oil palm shell foamed concrete with conventional materials. Materials & Design, 51(5), 522–529.

Ali A. Sayadi, Juan V. Tapia, Thomas R. Neitzert, & G. Charles Clifton. (2016). Effects of expanded polystyrene (eps) particles on fire resistance, thermal conductivity and compressive strength of foamed concrete. Construction and Building Materials, 112, 716–724.

Bentz, D.P., Hansen, A.S., & Guynn, J.M. (2011). Optimization of cement and fly ash particle sizes to produce sustainable concretes. Cement & Concrete Composites, 33(8), 824–831.

Chen, W., Tian, H., Yuan, J., & Tan, X. (2013). Degradation characteristics of foamed concrete with lightweight aggregate and polypropylene fibre under freeze–thaw cycles. Magazine of Concrete Research, 65(12), 720–730.

Chi Hu, Hui Li, Zhongwei Liu, & Qingyuan Wang. (2016). Research on properties of foamed concrete reinforced with small sized glazed hollow beads.

Chindaprasirt, P., & Rattanasak, U. (2011). Shrinkage behavior of structural foam lightweight concrete containing glycol compounds and fly ash. Materials & Design, 32(2), 723–727.

Farzadnia, N. (2015). Properties and applications of foamed concrete; a review. Construction & Building Materials, 101(Part 1), 990–1005.

Hilal, A.A., Thom, N.H., & Dawson, A.R. (2015). On void structure and strength of foamed concrete made without/with additives. Construction & Building Materials, 85, 157–164.

Jitchaiyaphum, K., Sinsiri, T., & Chindaprasirt, P. (2011). Cellular lightweight concrete containing pozzolan materials. Procedia Engineering, 14(11), 1157–1164.

Liu, M.Y.J., Alengaram, U.J., Jumaat, M.Z., & Mo, K.H. (2013). Evaluation of thermal conductivity, mechanical and transport properties of lightweight aggregate foamed geopolymer concrete. Energy & Buildings, 72(2014), 238–245.

Rashed, A.I., & Williamson, R.B. (1991). Microstructure of entrained air voids in concrete, part I. Journal of Materials Research, 6(9), 2004–2012.

Rashed, A.I., & Williamson, R.B. (1991). Microstructure of entrained air voids in concrete, part II. Journal of Materials Research, 6(11), 2474–2483.

Yang, K.H., & Lee, K.H. (2013). Tests on alkali-activated slag foamed concrete with various water-binder ratios and substitution levels of fly ash. Journal of Building Construction & Planning Research, 01(1), 8–14.

Yang, Y. (2014). Utilization of circulating fluidized bed fly ash for the preparation of foam concrete. Construction & Building Materials, 54(3), 137–146.

Yang, Y. (2014). Utilization of circulating fluidized bed fly ash for the preparation of foam concrete. Construction & Building Materials, 54(3), 137–146.

Zhang, Z., Provis, J.L., Reid, A., & Wang, H. (2015). Mechanical, thermal insulation, thermal resistance and acoustic absorption properties of geopolymer foam concrete. Cement & Concrete Composites, 62, 97–105.

Sol-gel hybrid layer encapsulated aluminum pigments using organic silane acrylate resin/TEOS as precursors

A.H. Gao & Y.W. Huang
School of Chemistry and Chemical Engineering, Zhao Qing University, China

ABSTRACT: In order to enhance the chemical resistance and resin consistence, aluminum pigments were encapsulated by inorganic-organic hybrid layer using organic silane acrylate resin PMBV and Tetraethyl Orthosilicate (TEOS) as precursors through sol-gel process. Lightness value test showed the organic-inorganic hybrid film kept the metallic appearance of aluminum pigments well. Peeling test demonstrated that the compatibility of modified aluminium pigments with resin was improved obviously. And acid-dipping test showed the inorganic-organic hybrid encapsulated aluminum pigments had good anti-corrosion.

1 INTRODUCTION

Flaky aluminum pigments are one popular effective pigments in the fields of paints, inks and plastics industries due to their special metallic appearance and floating property. There are two important limitations in the use of aluminum pigments. One is the poor anti-corrosion in corrosion media caused by the chemical activity of aluminum and aluminum oxide. The other is their poor compatibility with most polymers, so they can be peeled off easily when used in coatings or inks.

In the last two decades, researches on the field of anti-corrosion of aluminum pigments showed that SiO_2 encapsulation was extraordinarily efficient to improve anti-corrosion (Anthony, 2000; Emregül, K. C. 2003). However, it did little to enhance the compatibility of aluminum pigments with polymers, for the surface properties of aluminum pigments or SiO_2 were distinctly different from that of ploymer. So, the compatibility should rely on encapsulating aluminum pigments with polymers, and lots of researches have confirmed it (Kiehl A., 2002; Yang X. F., 2001; Pi P. H., 2012).

To combine SiO_2 encapsulation and polymers encapsulation on aluminum pigments could be a workable technique to solve these two problems. In this work, aluminum pigments were surface modified by a sol-gel hybrid layer using organic silane acrylate resin poly (methyl methacrylate-n-butyl acrylate-vinyl trieth oxysilane) (PMBV) and Tetraethoxyl Silane (TEOS) as co-precursors through sol-gel process. The compatibility of modified aluminum pigments with polymers and their corrosion resistance properties were discussed. The structure and morphology of aluminum pigments were characterized by means of SEM.

2 EXPERIMENT

2.1 Materials

The resin PMBA is made by us through in-situ copolymerization of MMA, BA and TEOS. The silane content of PMBA is 3.89% and the molecular weight is 3700. TEOS, absolute alcohol and ammonia water are all analytical grade. The aluminum pigments are flaky particles with median size of 30 μm (Tianlong Trade Co., Ltd., China).

2.2 Preparation of $PMBV$-SiO_2/Al pigments

Firstly 1.5 g aluminum pigments and 50 mL ethanol were put into a four neck-round bottom flask, which was connected to a condenser, a thermometer, and a nitrogen gas inlet/outlet, respectively. The solution was stirred at room temperature for 1 hr and then pre-heated to 55°C. A mixture of PMBV (dissolved in 10 mL dimethylbenzene), TEOS and 30 mL ethanol and another mixture of ammonia, distilled water and 30 mL ethanol were added drop-by-drop over 1 hr to the pigments dispersion simultaneously. The mixing solution was further stirred for 6 hrs before vacuum filtered. The filter cake was washed with ethanol and acetone for several times, and then dried under vacuum at 50°C for 24 hrs.

2.3 Preparation of coating containing aluminum pigments

Aluminum pigments, acrylic resin and thinner in the proportion of 3:63:34 (mass ratio) were first mixed and then sprayed onto a black plastic board. The plastic board was subsequently be dried in thermostatic drier at 60°C to constant weight.

2.4 Characterization

SEM (XL-30, Philips Co.) was used to observe the morphology of aluminum pigments and coating. Peeling test was carried out as reference (Gao A. H., 2012) to evaluated the adhension of Al pigments in coating made in 2.3. Color-eye (Model 3100, Macbeth Co.) was used to determined L value of coating in order to reflect the effect of encapsulation or corrosion on the lightness of pigments (Mardalen J., 2008). The chemical stability of aluminum pigments was tested by dipping the board with painting partially in pH = 1 HCl solution at 25°C for 24 hrs. L-values of coating before and after dipping test were measured and their difference (ΔL_a) was calculated.

3 RESULTS AND DISCUSSIONS

3.1 Preparation mechanism

Organic silane acrylate resin PMBV containing–Si-O-CH$_3$ groups can participate in sol-gel process as the same as TEOS. So silica based organic-inorganic hybrid material can be formed through sol-gel process using PMBV and TEOS as precursors and alkali as catalyst in the ethanol/water median. There are large numbers of –OH groups on the surface of aluminum pigments. They can also participate in condensation reaction with –Si-OH groups in the hydrolysis product of PMBV and TEOS to form a chemical linkage, Si-O-Al, between aluminum and resin. This chemical bond linkage may produce strong interaction of silica based organic-inorganic hybrid material with aluminum surface (Gao A. H. 2010).

The hybrid film formed by PMBV and TEOS containing acrylate resin promotes the compatibility of modified pigments and resin used in coating

according to similar compatible principle. Moreover, the film can also hinder the diffusion of H$_2$O between the aluminum surface and corrosion solution, then effectively reduce the corrosion rate of aluminum pigments.

3.2 Optimization of reaction parameters

3.2.1 Effects of ammonia

As a catalyst in the sol-gel process, ammonia concentration has great influence on the rates of hydrolysis and condensation of silane. When the concentration of ammonia increases, the rate of hydrolysis and condensation of silane also increases. It is favorable for encapsulation of aluminum. However, the result of SEM in Figure 1 shows that the surface of sample encapsulated with 3 mL ammonia (Figure 1a) is smoother than that with 4 mL (Figure 1b) or 5 mL ammonia (Figure 1c). It is probably because of the hydrolysis-condensation was speed up as the increased concentration of ammonia solution. Then SiO$_2$ particles generated on the surface of aluminum particles, leading to uniform encapsulation.

3.2.2 Effects of water

When the amount of precusors and ammonia is kept constant, increasing the amount of water will greatly promote the rate of hrdrolysis. However, the concentration of silane in the system will be reduced significantly when the amount of water is increased, which reduces the hydrosis and condensation reactions. What's more, the hydrophilic groups in PMBV may cause the compatibility of the hydrolysis products of precursors decrease, resulting in self-condensation rather than co-condensation. From Figure 2, we can find there are granules on the surface of samples in Figure 2b and Figure 2c.

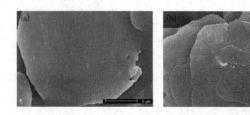

Figure 1.   Effects of ammonia (a) 3 mL; (b) 4 mL; (c) 5 mL.

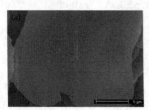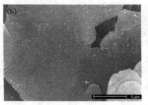

Figure 2.   Effects of water (a) 3 mL; (b) 4 mL; (c) 5 mL.

Those granules may be the self-condensation product of precursors. The surface of PMBV-SiO₂/Al with 3 mL water is smooth and uniform.

3.2.3 Effects of precursors

L-value is usually used to illustrate the lightness of a color. More brightly a color, higher its L-value, and vice versa. Aluminum has a silvery lustre. In general, L-value of aluminum will decrease when aluminum is corroded. So, the difference of L-value before and after corroded can be an indicator of corrosive degree of metal surfaces. The chemical stability of different samples in acid aqueous media of pH = 1 at 25°C is shown in Figure 3. Comparsion studies of different samples dipped in acid for 24 hrs show lightness difference of raw aluminum pigments is much larger than that of encapsulated samples. Sample encapsulated with the ratio of PMBV/TEOS 1.1 and 0.45 displays the best stability in acid media.

3.3 Properties of encapsulated aluminum pigments

3.3.1 Lightness difference analysis

L-values of aluminum samples are listed in Table 1. L-value of raw aluminum pigments is 93.92, and that of modified aluminum pigments is 93.36. As the encapsulated layer is not completely transparent, lightness of aluminum pigments will decrease after encapsulated. However, the drop rate of lightness caused by encapsulated is just 0.53%. The hybrid layer keeps lightness of aluminum pigments well.

Anti-corrosion effectiveness of encapsulated layer is also characterized by L-value. L-value of coating containing raw aluminum sample after acid dipping is 43.97. ΔL_a of raw aluminum sample is 49.95, demonstrating poor anti-corrosion of aluminum sample. However, ΔL_a value of PMBV-SiO₂/Al sample is 0.8, indicating excellent anti-corrosion property of encapsulated layer. It may be attributable to shielding effect of SiO₂ and its acid resistance. In a word, the PMBV-SiO₂ encapsulated layer on aluminum pigments not only holds metallic appearance of aluminum pigments well, but also enhances the anti-corrosion of aluminum pigments. In additional, it improves the compatibility of aluminum pigments to resins.

3.3.2 Adhesion property

Results of peeling test are in Table 1. The adhesion of encapsulated samples in coating is stronger than that of bare aluminum flakes. It may be attributable to the polyacrylate grafted on the surface of aluminum pigments, which shows especially similar characteristic to resin used in coating and improves the compatibility of aluminum pigments with coating.

4 CONCLUSIONS

1. Organic-inorganic hybrid layer coated aluminum pigments were prepared using organic silane acrylate resin PMBV and TEOS as precursors and ammonia as catalyst through sol-gel process.
2. PMBV-SiO₂/Al pigments showed good anti-corrosion.
3. Lightness value test showed that the organic-inorganic hybrid film maintained the metallic appearance of aluminum pigments well. Peeling test demonstrated that that the compatibility of PMBV-SiO₂/Al with resin had been improved obviously.

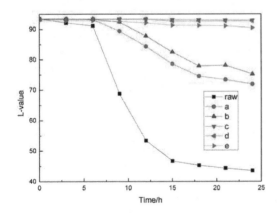

Figure 3. Effects of ratio of PMBV/TEOS on lightness difference of aluminum pigments (a) 4.8; (b) 1.5; (c) 1.1; (d) 0.45; (e) 0.17.

Table 1. Lightness and W_p of raw and modified aluminum pigments.

Sample	L-value	$W_p/g \cdot m^{-2}$
raw Al	93.92	6.80
PMBV-SiO₂/Al	93.36	0.66
raw Al after acid dipping	43.97	–
PMBV-SiO₂/Al after acid dipping	93.12	–

REFERENCES

Anthony, 2000. Paint pretreatment for aluminum. *Metal Finishing* 98(6): 74–79.

Emregül, K. C., 2003. The effect of sodium molybdate on the pitting corrosion of aluminum. *Corrosion Science* 45(11): 2415–2433.

Gao A. H., 2010. Modification of aluminium pigments via encapsulated with polyacrylate and silicon dioxide. *Journal of South China University of Technology (Natural Section Edition)* 38(2): p. 1–5.

Gao A. H., 2012. Preparation and characterisation of aluminium pigments encapsulated by composite layer containing organic silane acrylate resin and SiO$_2$. *Pigment and rsein technology* 41(3): 49–53.

Kiehl A., 2002. Corrosion inhibited metal pigments. *Macromolecular Symposia* 187(1): 109–120.

Mardalen J., 2008. Time and Cost Effective Methods for Testing Chemical Resistance of Aluminum Metallic Pigmented Powder Coatings. *Progress in Organic Coating* 63(1): 49–54.

Pi P. H., 2012. Effects of acid treatment on adhesive performance of encapsulated aluminium pigments on plastic sheets. *Canada journal of chemical and engineering* 90: 1224–1230.

Yang X. F., 2001. Use of a sol-gel conversion coating for aluminum corrosion protection. *Surface and Coatings Technology* 140(1): 44–50.

Research status and progress of self-compensation lubricating composite materials at high temperatures

Chao Li & Yanjun Wang
School of Mechanical Engineering, University of Jinan, Jinan, China
Shandong Machinery Industry Association, Jinan, China

Ke Qin
Shandong Machinery Industry Association, Jinan, China

Shu Long Li
School of Mechanical Engineering, University of Jinan, Jinan, China

ABSTRACT: With the rapid development of science and technology, it has been found that lubricating oil and grease cannot be used to lubricate parts and components when working at high temperature, high vacuum, high speed, and heavy load conditions. These challenges have promoted the development of advanced high-temperature solid lubricant materials. In this paper, the research status and progress of self-lubricating composite materials at high temperature are introduced and their application and lubrication mechanism are narrated. Furthermore, several methods to prepare self-lubricating materials are introduced, and the author prepared the M3/2/TiC metal ceramic materials. However, some problems in this field are proposed.

1 INTRODUCTION

As relative motion is inevitable in any machine or mechanism, friction and wear occur. Lubrication is one of the most effective measures to reduce friction. In general, lubricating oil is used to reduce friction and wear; however, with the rapid development of science and technology, it has been found that lubricating oil and grease cannot be used to lubricate parts and components when working at high temperature, high vacuum, high speed, and heavy load conditions. It is necessary to use new type of solid lubricant to prevent wear and contamination of the system, so the unique advantages of self-lubricating materials are shown (M.Q. Xue, 2013). This type of material has broad application prospects in special conditions. In this paper, the development and progress of self-compensation lubricating composite materials at high temperature are narrated from different angles.

2 THE RESEARCH STATUS AND TREND

Self-lubricating materials at high temperature can be divided into metal matrix self-lubricating composite material, self-lubricating alloy, and ceramic self-lubricating composite (F.L. Han, 1984). The preparation of solid self-lubricating composite material is in need of adding solid and additional component to the matrix. Meanwhile, it is prepared by certain prepara-tion technology; besides, it has certain strength and lubricating performance (X.M. Feng, 2007).

2.1 Metal matrix self-lubricating composite material

Metal matrix self-lubricating composite material has the properties of toughness and plasticity; besides, it has certain processing and deformation properties. Its strength and hardness can meet the bearing capacity and wear resistance of lubrication film. Furthermore, it could adapt to the different chemical environment, atmospheric environment, high temperature, high vacuum environment, and so on (M.S. Shi, 2000). In the 1970s, the cor-responding self-lubricating composite materials were prepared by adding solid lubricant to tung-sten, molybdenum, chromium, niobium, and other refractory metals possessing good wear resistance, so as to reduce friction. In order to improve the mechanical properties of the material, we can add WC, NbC, VC, TiC, ZrC, Al_2O_3, and other hard phases (FEDORCHENKO H M., 1985). However, the refractory metal has obvious disadvantages that they are expensive, complex to form, and difficult to sinter; thus, it was replaced by copper base, iron base, and so on.

Common iron-based self-lubricating composite material has Fe/C, Fe-Cu/C, Fe-Mo/C, and so on (Y.D. Peng, 2008; W. F. Wang, 2005). However, it generally has higher hardness, and it is easy to cause damage to the spouse member by reacting with graphite disperses hard points during use. Common silver-based self-lubricating composite material has Ag/MoS_2, Ag/MoS_2-C, Ag/WS_2, $Ag/NbSe_2$, and so on. The study showed that those silver-based composites could keep the friction factor less than 0.2 below 300°C. Nickel and cobalt are commonly used as base material as they are used in self-lubrication at high temperatures. Nickel base alloy exhibits excellent mechanical properties above 500°C, it could work at high temperature and high stress (G.L.Chen, 1988), and its surface is susceptible to form NiO layer that has good plasticity and adhesion by oxidation. It is good for reducing wear. In addition, nickel base alloy is a kind of high-temperature solid lubricant and its study results are attractive. Lanzhou Institute of Chemical Physics, Chinese Academy of Sciences (J. Zh. Liu, 1993), conducted a systematic study and developed several kinds of sulfur and sulfur-free Ni base alloy by high-temperature self-lubrication material by PM process and intermediate-frequency excitation induction heating high-temperature hot-pressing and achieved satisfactory results. Therefore, it is expected to become the materials that could make friction parts of advanced engine.

Silver base self-lubricating materials have low coefficient of friction, small contact resistance, and high conductivity. Silver–graphite composite materials are widely used in electric contact parts, brush, bearings, slider, bushing, and contact bushing of chemical solutions (J.L. Johnson, 1978). The Ag-MoS_2 self-lubricating composite bearing material could be used under ultrahigh vacuum and strong radiation conditions. The Ag-Ta-MoS_2-graphite self-lubricating brush material can be used for space electricity transmission and signal transmission mechanism (Y.B. Liu, 1991).

In the preparation of self-lubricating composite materials, we often add various combinations of additives to reduce the friction properties of materials. When researching and developing powder metallurgy high-temperature metal matrix solid self-lubricating materials, we must choose material matrix and lubricating additives reasonably and determine the best content of them according to the specific working conditions. However, the understanding about this kind of composite material's mechanism of high-temperature tribological characteristics is still very insufficient. For the specific application, we still need detailed engineering data to solve the problem of production and design.

2.2 Nonmetallic base self-lubricating composite materials

Nonmetallic materials mainly refer to some polymer materials such as teflon (PTFE), nylon (PA),

polyformaldehyde (POM), polyethylene (PE), and other engineering plastics. In general, these materials are lightweight and have small friction coefficient, but their mechanical properties, heat resistance, and heat transfer performance are not ideal. In order to achieve particular purposes, we often add some wear-resistant materials, such as carbon fiber, glass fiber, disulfide platinum and some organic compounds to increase the strength and abrasion resistance of solid self-lubricating composite materials, so we could obtain the comprehensive performance with excellent physical, chemical, and mechanical aspects. Zhang Zh Zh et al. (2005) developed a kind of self-lubricating composite material containing metal oxide PTFE base. They found that adding Pb_3O_4 or Cu_2O can reduce the friction coefficient of PTFE, and among them, the antifriction effect of Pb_3O_4 is the best. Its friction coefficient decreases with the increase of load and its wear resistance is far better than that of the pure PTFE when the load is less than 300 N.

There is a wide range of applications in sliding bearings, gears, slider, and so on, because the nonmetallic base self-lubricating composite materials could save much metal and lubricating grease, consume less energy, and realize oil-less self-lubrication.

2.3 Ceramic self-lubricating composite

Self-lubricating ceramics can be classified into two categories: one is metal ceramic composite materials, which have good mechanical and self-lubricating properties at good wettability; the other is forming self-lubricating ceramic composite materials by adding solid lubrication components to ceramics and its composite material. Chinese scientists have conducted extensive research in this topic and achieved certain results. For example, Wang J B et al. (1997) prepared Ni-WC-PbO department of self-lubrication metal ceramic materials; Jiang Y et al. (1999) prepared graphite/ZTA ceramic base self-lubricating composite materials with graphite as solid lubrication components; and Wang F prepared Al_2O_3/hBN base self-lubricating composite materials with hBN as solid lubrication components. From this, we could realize ceramic materials' self-lubricating ability by adding solid lubricant.

With the development of science and technology, the self-repairing solid lubricating materials having no environmental pollution, wear resistance, long life, and low friction have a wide range of applications in electronics, biology, communications, aerospace, aviation, and other high-tech fields. Owing to the most promising development direction in the field of lubrication solid self-lubricating materials, they have attracted increasing attention. The future research focuses on the study of the theory of the self-lubricating material, new solid self-lubricating materials, the study of the new way of self-lubricating, structural design of new materials, and so on.

2.4 *The lubrication mechanism*

Sweat contributes to regulating body temperature, keeping water and salt balance, and wetting the body skin, which are the main functions of sweat glands in the human body. Figure 1 is the sweat pore structure diagram, showing sweat glands, ducts, and sweat pore (T.T.Han, 2016). When environment temperature is greater than 30°C, the body starts to sweat and the sweat glands secrete sweat and excrete it into the surface of the skin by the catheter. The self-compensation lubricating composite materials at high temperature is prepared according to the structure of human body sweat and its sweat principle.

All kinds of solid self-lubricating materials have some common features. Most importantly, they generate continuous, stable, molecular orientation transfer film on the friction surface and convert the friction between friction surface into the friction between solid lubricant, so as to reduce the friction and wear. When metal base solid self-lubricating materials rub with dual pieces, firstly, friction factor becomes higher. After a period, the solid lubricant is transferred to the grinding of the surface of the metal, and gradually formed a layer of continuous, stable solid lubricating film. At the same time, the grain size of solid lubricants become directional arrangement under the shear stress, so it is called molecular orientation transfer membrane (D.Zh. Xing, 2009). Friction coefficient and wear resistance of solid self-lubricating materials depend on the physical and mechanical properties of solid lubricant and grinding material. Besides, it is closely related to the continuous and stable degree of the transfer film generated. In high-temperature and high-pressure working environment, friction heat is produced due to the effect of friction. It makes the precipitation solid lubricant separate out from the surface of friction along the micropore to form lubricant film as shown in Figure 2 (Y. Han, 2015). Therefore, the friction coefficient of the contact surface is reduced, wear and tear is decreased, and the service life is improved.

Solid lubricant could form film under the action of shear force in friction process. Solid lubricating film can be fully opened to cover the friction surface as much as possible, depending not only on the nature of the solid lubricant but also on the interaction of solid lubrication film and metal substrate and the transfer of solid lubricant on the metal substrate. Over the years, various research on transition of lubricating film have been made, but there is no unified and accurate conclusion about the interaction between metal and lubricating film in the mechanical action, free energy effect,

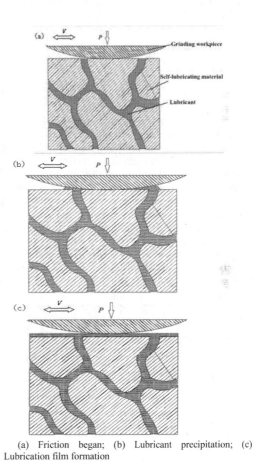

(a) Friction began; (b) Lubricant precipitation; (c) Lubrication film formation

Figure 2. Self-lubricating composite material of lubrication film formation mechanism.

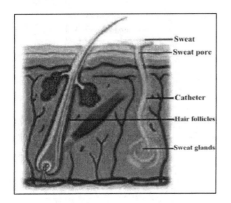

Figure 1. Sweat pore structure diagram.

electrostatic adsorption, chemical effect, and the roles of polarity (H.D. Wang, 2009).

2.5 *Precipitation mechanism of solid lubricants*

Under the action of thermal stress (or friction temperature), the material's organization structure will produce phase volume changes. Each phase of the material will expand and form stress between each group of yuan when heated. Solid components (Figure 3) will suffer from matrix grain's compressive stress σ_M. On the contrary, the matrix grain also suffers from the solid lubrication group yuan grain's positive pressure σ_i. When $\sigma_M = \sigma_i$, stress will cancel and cause the whole volume's expansion. When $\sigma_M - \sigma_i > \tau$ (solid lubricant shear stress), solid lubricating film will separate into interoperable friction surface to a certain extent. In the process of sliding, the friction deformation is mainly conducted on membrane because of solid lubricant film's adhesion, and the shear strength is lower than that of pure metals (Y.Y. Dong, 1987). This is the root cause of the solid lubricant antifriction effect.

2.6 *Some existing problems in this area*

At present, there is no complete and systematic theory of friction and wear for metal matrix solid self-lubricating composite materials. There are a variety of factors like metal matrix, lubrication, the lubrication film, friction pair, atmosphere, environment, and so on, in the friction system of this kind of materials. The current theoretical research mainly involves the explanation of this phenomenon, the related theoretical research is not enough in-depth, so it is difficult to provide reference and guidance effect for design and development of material (Ch.Ch. Wang, 2012). In addition, fiber-reinforced composite material has good tribological properties, but the research about its performance is still rare at high temperature. Then, because of the limited range of solid lubricants, it is difficult to meet its require-

ments in new period, so it is urgent to develop and research new type of high-temperature solid lubricant. Furthermore, it is difficult to unify the lubrication performance and mechanical performance of materials. Finally yet importantly, tribological properties of nanomaterials have been paid attention to in recent years, so it is worth to focus on their high-temperature tribological performance.

3 PREPARATION OF SELF-LUBRICATING MATERIALS

The main preparation method of the self-lubrication material at high temperature basically involves powder metallurgy method, smelting method, and in situ synthesis method.

3.1 *Powder metallurgy method*

The powder metallurgy method puts a certain components ratio of metal and ceramic phase substrate material into the vacuum hot-pressing sintering furnace and prepares the self-lubricating composite material at high temperature according to the certain sintering process curve. The preparation method requires that metallic phase and ceramic could wet with each other, and it also needs to strictly control the heating speed. The author prepared M3/2/TiC metal ceramic materials by the powder metallurgy method, and the specific preparation process is shown in Figure 4 and the sintering process curve is shown in Figure 5 (Y.J. Zhang, 2005). With the sintering process, the particles will bond each other and gradually form the sintering neck. Furthermore, the parameters such as intensity and density of sintered body will be improved.

The author prepared the metal ceramic substrate by the above technology. The sample matrix is shown in Figure 6-a and the polished sample is shown in Figure 6-b. The author used Scanning

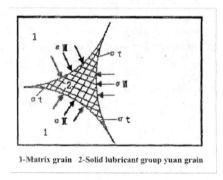

1-Matrix grain   2-Solid lubricant group yuan grain

Figure 3.   Self-lubricating materials group of yuan thermal stress effect diagram.

Figure 4.   Process diagram of preparation techniques.

Electron Microscope (SEM) to observe the pore structure of the matrix, and the fracture pore morphology is shown in Figure 7. Figure 7(a) and

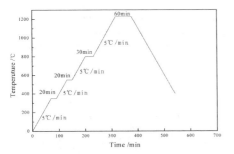

Figure 5. Diagram of the sintering process curve.

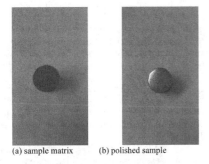

(a) sample matrix   (b) polished sample

Figure 6. Diagram of sintering process curve.

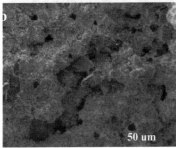

Figure 7. 1250°C sintering matrix pore morphology SEM images.

(b) represent the horizontal and vertical fracture pore morphology of SEM images. Figure 6 shows that the pore diameter of matrix transverse fracture is similar, the shape of the pore circle, and the longitudinal fracture pore approaches tubular morphology. Furthermore, because of the higher connectivity of matrix pore, this structure is easy to infiltrate and separate out the solid lubricant.

3.2 *Smelting method*

Compared with the powder metallurgy method, smelting method's process is relatively simple. It could prepare a complex–shaped component and material with good mechanical properties. Wang Z S et al. (2009) prepared NiAl-28Cr–5.6Mo-0.25Ho-0.15Hf (atomic fraction.%) eutectic alloy by the method, and it has good self-lubricating wear-resistant performance at 700–900°C.

3.3 *In situ synthesis method*

The basic principle of in situ synthesis is generating one or several strengthening phase particles by the reaction of different elements in certain conditions. It could achieve the goal of improving composite material properties (L.F.Cai, 2005). Zhang H et al. (2002) prepared the Ti-53 Al-xB alloy composite material and studied its microstructure with the method. However, this method will lead to uneven material composition because of the higher temperature that it is difficult to control in the process of preparation.

4 CONCLUSIONS

Improvement of lubricating technology not only can obtain enormous economic benefits, but also it is favorable to environmental protection and reasonable utilization of resources. The tribology basic of high-temperature solid self-lubricating materials and applied research have gained widespread attention and made substantial progress. In conclusion, the future solid lubricating materials are developing toward directions of resisting high temperature, high voltage, high speed, and other harsh conditions. Besides, it realizes the development in microscale by combining with nanotechnology. The overall trend is that high polymer material replaces the traditional metal materials in low temperature, and metal self-lubrication material competes with ceramic in the field of high temperature and develops to higher temperature to realize effective lubrication of higher temperature (W.F. Yang, 2007). The self-lubrication at high temperature has attracted more and more attention because of its unique application environment.

Therefore, it also becomes the focus of research about how to prepare self-lubrication, which has both good self-lubricating properties and certain strength at high temperature.

ACKNOWLEDGMENT

The authors would like to thank the National Natural Science Foundation of China for the financial support (ID: 51275208).

REFERENCES

Cai L.F. & Y. Zh. Zhang. (2005). Applications and development of in-situ synthesis for material preparation. Metal Heat Treatment. 30(10), 1–6.

Chen G.L. (1988). High temperature alloy, Beijing university of iron and steel technology.

Dong Y.Y. (1987). The tribological properties of composite materials and its development trend. Solid Lubrication. 7(3), 293–300.

Fedorchenko H.M. (1985). Friction composite of powder metallurgy. K.X. Li, transl. Beijing: Powder Metallurgy Institute of Beijing, 168–215, 190–193.

Feng X.M. & Ch.C. Zhang. (2007). Composites. Chongqing: Chongqing University Press, 1–3.

Han F.L. (1984). Powder metallurgy machine parts. China Machine Press.

Han T.T. (2016). Preparation and two-dimensional phase field simulation of microporous metal ceramic matrix with interconnected structure. Jinan: School of mechanical engineering, University of Jinan. 11–15.

Han Y. (2015). Design and simulation analysis of microporous structure in infiltration-type self-lubrication composites matrix. Jinan: School of mechanical engineering, University of Jinan. 11–14.

Jiang Y.& Y.H. Xia. (1999). Study on friction and wear properties of Graphite-ZAL self-lubrication ceramic composites. Powder Metallurgy Technology. 17(4), 273–276.

Johnson J.L. & L.E. Morberly. (1978). High-current brushes, Pactc II: Effect of brush and ring materials, IEEE Trans. Compo. Hybr. Manuf. Tech. Vol. CHMT-1 36–41.

Liu Y.B. & S.C. Lim. (1991). Friction and wear of self-lubricating metal-matrix composites, Journal of the Institutiion of Engineers, Singapore. 31(5), 57–62.

Peng Y.D. & J.H. Yi, Y. (2008). L. Properties of Fe-Cu-C alloys microwave sintered at different temperatures. Journal of Central South University: Science and Technology. 39(4), 723–728.

Shi M.S. (2000). Solid lubricating materials. Beijing: China Petrochemical Press, 14.

The Chinese Mechanical Engineering Society and Tribology. (1986). Lubrication Engineering [M]. Beijing: China Machine Press.

Wang Ch.Ch. & R.Ch. Wang. (2012). Research progress of metallic solid self-lubricating composites. The Chinese Journal of Nonferrous Metals. 22(7).

Wang H.D. & B.Sh. Xu. (2009). Technology and application of solid lubrication friction film. Beijing: National defence industry press. 6–74, 273–275.

Wang J.B. & J.J. Lv. (1997). Study on the tribological properties of SiC-Ni-Co-Mo-PbO high-temperature self-lubricating cermet material. Tribology. 17(1), 25–31.

Wang W.F. (2005). Effect of alloying elements and processing factors on the microstructure and hardness of sintered and induction-hardened Fe-C-Cu alloys. Materials Science and Engineering. 402(1/2), 92–97.

Wang Zh.Sh. & J.T. Guo. (2009). High temperature wear behavior of NiAl-Cr(Mo)-Ho-Hf eutectic alloy. Acta Metallurgica Sinica. 45(3), 297–301.

Xing, D. Zh. (2009). The study of heat-driven process of sweating simulation in high temperature based on pore structure characteristics. Wuhan: School of mechanical and electrical engineering, Wuhan University of Technology. 7–11.

Xue M.Q. & D.Sh. Xiong. (2003). Tribological study on solid lubricating materials at high temperatures. Ordnance Material Science and Engineering. 26(6), 58–62.

Yang W.F. (2007). Solid self-lubricating materials and their research trends. Lubrication Engineering. 32(12).

Zh J. Liu & M.A. Ou Yang. (1993). The effects of alloying elements on the tribological behaviors of the high temperature, self-lubricating nickel-base alloys containing sulphur. Tribology. 13(3), 193–199.

Zhang H. & W.L. Gao. (2002). TiAlB alloy in the morphology of primary TiB_2 and its formation mechanism. Jounrnal of Beijing University of Aeronautics and Astronautics. 28(5), 540–542.

Zhang Y.J. & Z.M. Liu. (2005). Design of diffusng self-lubricating and wear-resisting ceramic sinter. Materials for Mechanical Engineering. 29(9),12–15.

Zhang Zh. Zh. & P.X. Cao. (2005). Study on tribological and mechanical properties of PTFE composites. Polymer Materials Science and Engineering. 21(2), 189–192.

Advances in Energy Science and Equipment Engineering II – Zhou, Patty & Chen (Eds)
© 2017 Taylor & Francis Group, London, ISBN 978-1-138-71798-5

Fe/Mo-supported mesoporous silica as efficient heterogeneous catalyst for aerobic Baeyer–Villiger oxidation

Chunming Zheng, Chuanwu Yang, Dongying Lian, Shubin Chang & Jun Xia
State Key Laboratory of Hollow-fiber Membrane Materials and Membrane Processes,
School of Environmental and Chemical Engineering, Tianjin Polytechnic University, Tianjin, China

Xiaohong Sun
Key Laboratory of Advanced Ceramics and Machining Technology, Ministry of Education,
School of Materials Science and Engineering, Tianjin University, Tianjin, China

ABSTRACT: Highly ordered mesoporous silica with bimetallic Fe-Mo were successfully synthesized by magnesiothermic reduction method and evaluated in the aerobic Baeyer–Villiger oxidation using benzaldehyde as sacrificing agent and air as oxidant. The obtained results suggested that Fe-Mo@MPSi exhibited higher catalytic performance, high conversion (99.5%) of cyclohexanone, and high selectivity (99.9%) of ε-caprolactone. The good catalytic activity was ascribed to Fe and Mo on the wall of ordered mesoporous silica, which could accelerate the mass transfer of cyclohexanone and ε-caprolactone. In addition, the cooperative role of Fe and Mo toward the oxidation of cyclohexanone was discussed.

1 INTRODUCTION

In recent years, aerobic Baeyer–Villiger oxidation using heterogeneous catalysts is quite attractive and challenging, which could directly convert ketones to corresponding lactones (ten Brink, Arends et al. 2004). As lactone is an important monomer in the preparation of various fine chemicals, monomers for polymerization, and pharmaceuticals, numerous protocols have been made to study the B-V oxidation. Among these protocols, Mukaiyama method is an important protocol, which uses aldehydes as sacrificing agent and air as oxidant. This system can avoid using peracetic acid and corrosive hydroperoxides as the oxidizing agent, which required expensive production processes. However, most of the studies on the B-V oxidation of cyclohexanone require large amounts of aldehydes as sacrificing agents over the stoichiometric ratio (e.g., 3 equiv.), resulting in large amounts of benzoic acid. If the efficiency of the sacrificing agent becomes higher, this system will become more environment friendly. Recently, B-V oxidation of cyclohexanone has been catalyzed by different heterogeneous catalysts. Nevertheless, these systems suffer from the limitation of complicated synthesis of catalysts, relatively low activity, longer reaction time, or much consumption of aldehyde.

The molybdenum-based catalysts also have been used in a number of reactions, including esterification, hydrogenation reactions, and decomposition of ammonia catalyst. Recently, we have found that MoO_3/SiO_2 nanocomposites are effective for the

acetalization of benzaldehyde and exhibit enhanced catalytic activities and good reusability. Meanwhile, in consideration of the broad distribution and low cost of iron, previous studies showed that Fe^{3+} as catalytic active site into mesoporous silica material such as MCM-41 and MCM-48 was responsible for the catalytic activities in the aerobic Baeyer–Villiger oxidation.

Herein, we report the facile synthesis of Fe-promoted molybdenum supporting mesoporous silica (which is denoted as Fe-Mo@MPSi) by the method of magnesiothermic reduction and impregnation. It is found that Fe-Mo/MPSi for Baeyer–Villiger oxidation of cyclohexanone using benzaldehyde as sacrificing agent and air as oxidant shows remarkable catalytic performance and recycling performance.

2 EXPERIMENT

2.1 *Chemicals*

Pluronics (P123, $PEO_{20}PPO_{70}PEO_{20}$) was purchased from Sigma Aldrich. Tetraethoxysilane (TEOS, 99.5%), hydrochloric acid (36 wt%), magnesium powder, and hydrofluoric acid (10 wt%) were purchased from Tianjin Chemical Corp. Organic reagents used in the Baeyer–Villiger catalytic reaction were purchased from Sigma Aldrich. All other chemicals were used without further purification.

2.2 Synthesis of catalysts

Fe-Mo@MPSi was prepared by low-temperature magnesium reduction and high-temperature method. First, SBA-15 mesoporous templates were prepared through the method reported by Zhao et al. (Zhao, Feng et al. 1998). Then, 1 g of the as-prepared SBA-15 and magnesium powder (0.44 g) were placed in an agate mortar and mixed uniformly, and then flatted on the magnetic boat, followed by calcination in a following argon–hydrogen mixture in a tubular furnace at 500°C for 6 h. After cooled to the ambient temperature, hydrochloric acid (2 mol/L) and hydrofluoric acid (10 wt%) were used to etch, followed by filtration and washed by deionized water and ethanol each, and dried at 60°C in vacuum. Finally, 0.5 g of the as-prepared MPSi was added into the solution (3 mL), which consists of ferric nitrate (0.3269 g) and ammonium molybdate (0.0869 g) in the beaker (50 mL), and then magnetically stirred at ambient temperature in order to completely dry the mixture, followed by drying in vacuum at 60°C. Finally, the resultant solid was calcined in a tubular furnace at 200°C for 2 h in argon gas and the desired products were obtained, which are denoted as Fe-Mo@MPSi.

2.3 Characterization

Powder X-ray diffraction patterns of the samples were determined with Cu Kα radiation on Rigaku/mac. Transmission electron microscope (Hitachi H-7650) observations were used to confirm metallic dispersion of the samples.

2.4 Baeyer–Villiger oxidation of cyclohexanone

Baeyer–Villiger oxidation of cyclohexanone to ε-carprolactone was conducted in the liquid reaction. Typically, the reaction mixture consisting of catalyst (50 mg), cyclohexanone (2 mmol), benzaldehyde (4 mmol), and 1,2-dechlore (10 mL) was stirred vigorously, heated to 50°C for 4 h, and air (20 mL/min) was purged into the mixture in a three-neck round bottle flask equipped with a reflux condenser. The obtained products were transferred to a HP 9790 gas chromatograph equipped with a polycapillary column and hydrogen flame ionizaton detector. After the reaction, the catalyst was recovered and washed by ethanol, dried at 60°C, and used further for recycling test.

3 RESULTS AND DISCUSSION

3.1 Features of material

In order to investigate the internal pore structure of these catalysts, TEM images are shown in Fig. 1.

Figure. 1(A) shows the typical well-defined hexagonal symmetry mesoporous structure along the channel direction. Fig. 1(B) shows the TEM image of Fe-Mo@MPSi, where hexagonal arrangement of pores are quiet clear and the bimetallic metals of Fe and Mo are uniformly dispersed on the wall and channel of pores. The above study indicates that the texture of mesoporous silica material (MPSi) is still maintained by the incorporation of Fe-Mo bimetal, which could ensure the large specific surface area and increase the contact between the substrate of Baeyer–Villiger oxidation and the metal active sites. Furthermore, it could accelerate the mass transfer of reactants.

Figure 2 shows large-angle XRD pattern of MPSi, Fe@MPSi and 4%Fe-4%Mo@MPSi. It is evident from the figure that the diffractions at 47°, 61°, 69°, and 76° match with the values reported in the literature (Chen, Bao et al. 2012, Xing, Zhang et al. 2013), suggesting that mesoporous silica material was obtained after etching with appropriate amount of HCl and HF. When Fe and Mo were incorporated on MPSi, these diffractions still existed, but their intensity decreased and peak width broadened. This suggests that the structure

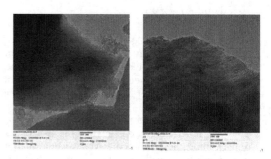

Figure 1. TEM images of as-prepared MPSi (A), 4%Fe-4%Mo/MPSi nanoparticles (B).

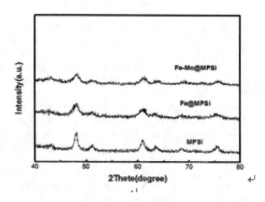

Figure 2. Large-angle XRD pattern of MPSi, Fe@MPSi, and 4%Fe-4%Mo@MPSi.

of mesoporous silica still kept but uniformity of pore channel changed owing to a small amount of the metal nanoparticles blocking the pore channel.

3.2 Catalyst performance and recyclability

The catalytic behaviors of various mesoporous silica materials for Baeyer–Villiger oxidation of cyclohexanone in the presence of benzaldehyde are presented in Table 1. For all cases, ε-caprolactone selectivity reached 100% on the basis of the internal standard method. In the presence of siliceous MPSi, it resulted in poor catalytic performance, cyclohexanone conversion (13.2%), and ε-caprolactone yield (13.2%), similar to the result without catalyst (Table 1, entries 1 and 6). When Fe and Mo are introduced into the MPSi, Fe@MPSi showed higher conversion (85.8%), whereas Mo@MPSi gave lower catalytic activity for cyclohexanone conversion (16.3%). It was noticed that, Fe and Mo supported on MPSi resulted in higher cyclohexanone conversion (99.5%) and benzaldehyde conversion (92.6%), which may be due to the cooperative role of Fe and Mo, increasing the efficiencies of the mass transfer. Meanwhile, it was also noted that, higher loading of Mo decreased the conversion of cyclohexanone and yield of ε-caprolactone, probably attributed to the fact that loading of Mo led to the formation of MoO_3 agglomerates, limiting the diffusion of reagent and products.

The recyclability of the catalyst is an important parameter, which determines their practical application. Figure 3 shows the catalytic performance of 4%Fe-4%Mo@MPSi after each regeneration cycle. There was a slight loss in conversion of cyclohexanone from 99.4% to 93.8%, which may be due to the loss of catalytic activity, that is, Fe and Mo on the surface of mesoporous silica may be washed away during filtration, washing, and drying

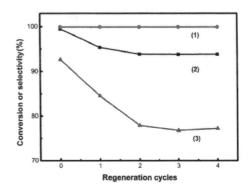

Figure 3. Reusability of 4%Fe-4%Mo@MPSi. (1) cyclohexanone conversion, (2) ε-caprolactone selectivity, (3) benzaldehyde conversion. For reaction conditions, see Table 1.

(Shukla, Wang et al. 2010). This suggests that 4%Fe-4%Mo@MPSi is a stable, efficient, and recyclable catalyst under the present reaction condition.

4 CONCLUSIONS

In conclusion, bimetallic Fe and Mo on mesoporous silica disperse homogenously in the pores and surface of the mesoporous silica, which create a rigid and dispersed structure. The effects of temperature, reaction time, and reusability on the aerobic Baeyer–Villiger oxidation of cyclohexanone to ε-caprolactone were also studied. This catalyst exhibited higher catalytic performance. Furthermore, it could be easily separated, show stable catalytic activity, and be reused at least up to four times without any significant loss.

ACKNOWLEDGMENTS

This work was supported by the National Natural Science Foundation of China (51202159, 51208357, 51472179 and 51572192), Doctoral Program of Higher Education, Ministry of Education (20120032120017), and General Program of Municipal Natural Science Foundation of Tianjin (13 JCYBJC16900, 13 JCQNJC08200).

Table 1. Catalytic results of aerobic Baeyer–Villiger oxidation of cyclohexanone to ε-caprolactone using different catalysts.

Entry	Catalysts	Conv. a/%	Yield b/%	Conv. c%
1	MPSi	13.2	13.2	10.8
2	8%Mo-MPSi	16.3	16.3	20.8
3	8%Fe-MPSi	85.8	85.8	66.4
4	4%Fe-4%Mo-MPSi	99.5	99.5	92.6
5	4%Fe-8%Mo-MPSi	78.0	78.0	61.0
6	No catalyst	11.4	11.4	9.8

Reaction conditions: catalysts (50 mg), DCE (10 mL), cyclohexanone (2 mmol), benzaldehyde (4 mmol), air (20 mL/min), 50°C, 6 h. [a]Conversion of cyclohexanone based on GC analysis. [b]Yield of ε-caprolactone. [c] Conversion of benzaldehyde; Dodecane used as an internal standard.

REFERENCES

Chen, K., Z. Bao, J. Shen, G. Wu, B. Zhou & K.H. Sandhage 2012. Freestanding monolithic silicon aerogels. *Journal of Materials Chemistry* **22**(22): 16196–16200.

Shukla, P., S. Wang, H. Sun, H.M. Ang & M. Tadé 2010. Adsorption and heterogeneous advanced oxidation of phenolic contaminants using Fe loaded mesoporous

SBA-15 and H_2O_2. *Chemical Engineering Journal* **164**(1): 255–260.

Ten Brink, G.J., I.W. Arends & R.A. Sheldon 2004. The Baeyer-Villiger reaction: new developments toward greener procedures. *Chem Rev* **104**(9): 4105–4124.

Xing, A., Zhang, J., Bao, Z., Mei, Y., Gordin, A.S. & Sandhage, K.H., 2013. A magnesiothermic reaction process for the scalable production of mesoporous silicon for rechargeable lithium batteries. *Chemical Communications* **49**(60): 6743–6745.

Zhao, D.Y., Feng, J.L., Huo, Q.S., N. Melosh, G.H. Fredrickson, B.F. Chmelka & G.D. Stucky 1998. Triblock copolymer syntheses of mesoporous silica with periodic 50 to 300 angstrom pores. *Science* **279**(5350): 548–552.

Advances in Energy Science and Equipment Engineering II – Zhou, Patty & Chen (Eds)
© *2017 Taylor & Francis Group, London, ISBN 978-1-138-71798-5*

Preparation and microstructures of porous materials with a strengthened pore wall

K.W. Tian, M.M. Zou, J. Zhang, X.B. Liu & S.Q. Yuan
Ningbo Branch of North Material Science and Engineering Institute, Ningbo, China

W.X. Zhang
Xi'an Technological University, Xi'an, China

ABSTRACT: Al-Mg porous materials with strengthened pore wall were prepared by self-melting reaction at low temperatures. Porous materials with different pore size were prepared at different temperatures, and their microstructures were observed by a metallurgical microscope. Microhardness test and XRD analysis were also carried out. The results show that the porosity of the prepared porous materials increases with the increase of mass fraction of magnesium powder; however, the increasing range is becoming smaller. The pore size is mainly related to the sintering temperature and sintering time. The porous materials with larger pore structure and its pore wall strengthened by $Al_3 Mg_2$ intermetallic composite phase can be prepared at temperatures above 462°C

1 INTRODUCTION

With the increase of application scope, the metal porous material is developing toward the integration of structure and function (He, 2006; Lu, 2006; Gong, 2010). Compared with the metal material, the metal porous material possesses certain strength and toughness, low density, large specific surface area, and high damping properties, which makes it one of the research hotspots (Liu, 2001; Liu, 2004). However, there still exists a series of problems in the preparation process as follows:

1. The additives added in the preparation process of porous materials, such as foaming agents, volatiles, and ceramic particles, are difficult to remove (Mi, 2007).
2. Pore diameter is large and the strength of pore wall is lower (Chawla, 2005).
3. High preparation temperature and pore-wall defects lead to poor performance and poor performance reproduction.

Al-Mg porous materials with strengthened pore wall were prepared by self-melting reaction at low temperatures and without additives of foaming agents, volatiles, and ceramic particles. The prin-

ciple and methods for preparing metal porous material described in this paper can be applied to the preparation of other porous materials, which provides a new way to prepare high-performance porous materials.

2 EXPERIMENT

In this paper, the Al-Mg alloy was used as research object, and the raw material powder supplier and specification are listed in Table 1.

The preparation process flow of metal porous material is as follows: pretreatment of powder, proportioning, mixing, cold press forming, and sintering (self-melting). Then, the metal porous material with pore-wall strengthening was prepared.

The Al powder and Mg powder were respectively dried at temperatures of 200°C for 2 h and 120°C for 1 h. The Mg powder of mass fractions of 10%, 20%, and 30% and the rest of Al powder were weighed. The proportioned powders were then put into SYH-5 mixer for a 40 min mixing. The mixed powders were cold-pressed into a φ30 mm blank at a pressure of 400 MPa, then sintered at 400–470°C for several minutes.

Table 1. Raw material supplier and specification.

Powders	Specification	Impurity	Suppliers
Mg powder	CP-150M	–	Northeast Magnesium Powder Factory
Al powder	Industrial pure 150M	$Fe + Si \leq 0.6\ O_2 + H_2O \leq 0.3$	Northwest Aluminum Plant

The microstructures were observed with MeF-3 metallurgical microscope and JSM-840 SEM.

3 RESULTS AND DISCUSSIONS

3.1 *Effect of sintering temperature on pore characteristics*

The 80%Al–20%Mg was prepared. The Al-Mg binary diagram is shown in Fig. 1. The sintering process can be divided into four stages according to the sintering temperature ranges and the binary diagram: T < 437°C < 437°C ≤ T < 451°C, 451°C ≤ T < 462°C, T ≥ 462°C. The liquid phases were formed in different temperature ranges, except for the first temperature range. The pore formation rules were observed and obtained for the above four sintering temperature ranges.

Experimental results:

When the sintering temperature is below the eutectic temperature at Mg-rich end, pores will not be formed, regardless of the sintering time. Fig. 2 shows the microstructure of samples sintered at

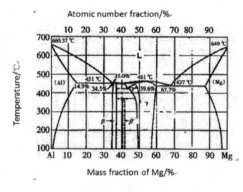

Figure 1.   Al-Mg binary diagram.

400°C for 4 h, and there is little change in the shape of Mg particle.

When the sintering temperature is at 440°C, the shape of Mg particle is still unchanged, but there is a small amount of reaction phases in the grain boundary region (see Fig. 2b).

When the sintering temperature is 455°C, the small pores will be formed in the boundary area of Al and Mg particles. The size of pores increases with the increase of sintering temperature and sintering time. Fig. 2c and Fig. 2d shows sintering at 10 min and 2 h, respectively.

When the sintering temperature is more than 462°C, for example, at 470°C, the microstructures of samples sintered for 3 min, 8 min, and 10 min are shown in Fig. 2e, Fig. 2f, and Fig. 2 g, respectively.

It can be seen that the pore is initially formed in the grain boundary of Al and Mg particles, and most of the Mg particles do not melt in samples sintered for 3 min, but Mg particles almost disappear (completely melt) in samples sintered for 8 min. When the sintering time reaches 10 min, the round pores with 30–50 um diameter are formed in the center of particle, and the pore-wall material is homogeneous.

In solid-state sintering (T < 437°C), because the melting point of Al and Mg is similar and there is little difference between their diffusion rates, the porous material cannot be prepared.

When the sintering temperature is between 437°C and 462°C, the liquid phases form slowly, and firstly produce at regions of higher local energy or in the interface between Mg particles and Al particles, so more than one liquid-phase region will probably be produced for each Mg particle. Because liquid phase is generally considered incompressible, so during the sintering, the sample can be expanded, but in the subsequent cooling process, the shrinkage degree of liquid is larger than that of solid, the

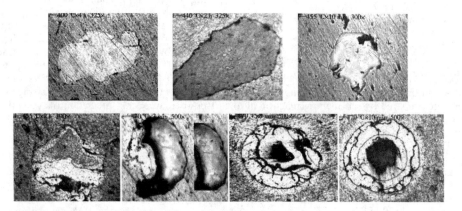

Figure 2.   Pore characteristics of Al-20%Mg sintered at different temperatures for different time.

pores are generated. In this temperature range, the amount of liquid phases is smaller, and the porosity of sintered materials is relatively low, which are not suitable for porous materials.

When the sintering temperature is more than 462°C, the Al-Mg alloy system enters the rapid melting stage. When the sintering time is relatively short, the magnesium particles are not completely melted, the morphology of pores is similar to that of the pores sintered at 437–462°C. With prolonged sintering time, the magnesium particles will gradually be melted and the liquid phase increases. When the magnesium particles are just completely melted and cooled, the liquid in the side of the aluminum particles and magnesium particles will unevenly nucleate and grow, resulting in the location of the shrinkage pore not in the center of the liquid phase, but close to the solid–liquid boundary (see Fig. 2f). With continuous extension of the sintering time, the magnesium particles will be completely melted; at this time, the heterogeneous nucleation of liquid phase is easier on the aluminum side, the liquid will directionally solidify, and the final shrinkage will tend to be in the center of the liquid-phase region, see Fig. 2g.

3.2 Porosity of Al-Mg porous material

The porosities of Al-Mg porous material prepared under sintering temperature 470°C for 10 min are listed in Table 2. It can be seen that the porosity of the porous material increases with the increase in the mass fraction of magnesium, but the increasing range is becoming smaller. The main reason is that, during the preparation period of low temperature self-melting reaction, the blank composed of the mixed Al powder and Mg powder will maintain at the temperature of the liquid–solid two-phase region for some time, atomic interdiffusion between magnesium and aluminum element occurs, and mixed zone of liquid and solid forms. In the subsequent cooling process, the liquid solidifies and shrinks, and pores form in the blank. According to the Al-Mg binary diagram, at the temperature of 470°C, the liquid in the two-phase region increases,

and the degree of shrinkage increases, resulting in the increase of porosity. The reason for the smaller increasing range of porosity is that the volume of liquid phase formed by the interdiffusion of elements in the sintering process increases and overlaps with each other.

3.3 XRD analysis and microhardness of strengthened pore wall

The 80%Al-20%Mg was sintered at 470°C for 10 min. The microhardness of the sintered sample was measured (see Fig. 3). The microhardnesses (HV0.02/10) in positions A, B, C, and D are 41, 55, 302, and 31, respectively.

X-ray diffraction analysis was also carried out on the strengthened pore wall (see Fig. 4). It can be seen that the material is mainly composed of pure aluminum, aluminum solid solution containing magnesium α-Al and $\beta(Al_3 Mg_2)$ intermetallic compound phase.

Therefore, it can be seen that the shrinkage pore produced by the low-temperature self-melting reaction of powder metallurgy and the directional solid-

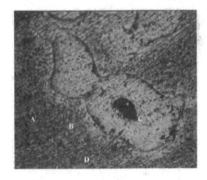

Figure 3. Sampling position for microhardness measurement.

Table 2. Composition of Al-Mg porous material and porosity.

Properties	Mass fraction of Mg (%)		
	10%	20%	30%
Mass in gas	33.0043	37.0124	28.0065
Mass in water	19.00161	19.6225	13.5447
Measured density	2.3594	2.1284	1.9366
Theoretical density	2.5604	2.4330	2.3176
Relative density	0.9215	0.8748	0.8356
Porosity	7.85	12.52	16.44

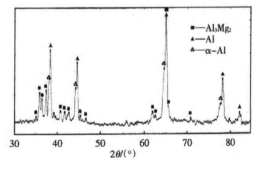

Figure 4. XRD analysis result.

ification of liquid phase can be used as a new type of pore structure. In the process of solidification, a strengthened layer is formed at the edge of the pore, so the composite pore structure is obtained.

Combined with the XRD analysis result, because zones A and B are mainly composed of pure aluminum or α-Al phase (solid solution with a small amount of magnesium), the microhardness is relatively lower. Zone C is composed of α-Al phase strengthened by $\beta(Al_3 Mg_2)$ intermetallic compound phase, the microhardness is obviously higher than that of zones A and B because of the high microhardness of $\beta(Al_3 Mg_2)$ phase. Although the phase composition of zone D is the same as that of zone C, the microhardness is very low, because the area is too close to the cavity.

4 CONCLUSIONS

Al-Mg porous materials with strengthened pore wall can be prepared by low-temperature self-melting reaction, and the pore is near-spherical. When the sintering temperature is in the range of 437–462°C, the sintered porous material with irregular shape, small volume, and low porosity is prepared. When the sintering temperature is more than 462°C, the sintered porous material with high porosity and strengthened pore wall can be obtained.

The quantity of pore is mainly related with the added amount of Mg, and the size of pore is determined by the sintering temperature, sintering time, and Mg powder size.

REFERENCES

Chawla N. & Deng X (2005). Microstructure and mechanical behavior of porous sintered steels. *Materials Science and Engineering A.*390, 98.

Gong, L.Y., Y.Y. Gao & H.A. Zhang (2010). Mechanical model and viscoelastic properties of porous materials. *Journal of Hunan University of Science and Technology*: Natural Science Edition.25, 34.

He, D.P., S.Y. He. & J.T. Shang (2006). Progress and Physics of ultra light porous metals. *Advances in Physics.*26, 346.

Liu, P.S. (2004). *Introduction to Porous Material*, Tsinghua University Press. Beijing, 25

Liu, P.S., T.F. Li. & C Fu (2001). Application of porous materials. *Functional Material.*32, 12.

Lu, T.J., D.P. He. & C.Q. Chen (2006). Multi functional properties of ultra light porous metal and its application, *Advances in Mechanics.*36, 517.

Mi, G.F., H.Y. Li & X.Y. Liu (2007). Preparation and properties of spherical porous high temperature alloy materials. *Mechanical Engineering Materials.*31,13.

Tang J.L. & D.B. Zeng (2007). *Cast Nonferrous Alloy and Its Melting*, China Water Power Press. Beijing, 30.

Advances in Energy Science and Equipment Engineering II – Zhou, Patty & Chen (Eds)
© *2017 Taylor & Francis Group, London, ISBN 978-1-138-71798-5*

Flow stress behavior of spray-formed Al-9Mg-0.5Mn-0.1Ti alloy during hot compression process

C.H. Fan, Z.Y. Hu, L. Ou, J.J. Yang & X.C. Liu
School of Metallurgical Engineering, Hunan University of Technology, Zhuzhou, P.R. China

X.H. Chen
CRRC Zhuzhou Electric Locomotive Co. Ltd., Zhuzhou, P.R. China

ABSTRACT: The flow stress behavior of spray-formed Al-9Mg-0.5Mn-0.1Ti alloy was studied using thermal simulation tests on a Gleeble-3500 machine over deformation temperature range of 300–450°C and strain rate of 0.01–10 s^{-1}. The results show that the flow stress behavior was sensitive to deformation parameters. The peak stress level and steady flow stress of the alloy increased with decreasing deformation temperature and increasing strain rate. The flow stress behavior can be represented by the Zener-Hollomon parameter Z in the hyperbolic sine equation with the hot deformation activation energy of 184.2538 kJ/mol. The constitutive equation of spray-forming Al-9Mg-0.5Mn-0.1Ti alloy was established.

1 INTRODUCTION

The Al-Mg alloy has excellent properties such as corrosion resistance, weldability, moderate strength and low cost, which makes it an attractive material in automotive, shipping and aerospace industry (Williams 2003 & Tolga 2014). However, it is difficult to obtain high-strength and high-toughness by conventional casting method (Gang 2011) due to the coarse grain and dendrite segregation caused by the low cooling rate. In addition, eutectic usually forms at $W(Mg) > 5\%$ and the alloy matrix exhibits lower solid solubility of Mg. Spray forming technology is recognized as an ideal method to prepare high Mg content Al-Mg alloy with high cooling rate, high level of solute Mg, fine grains as well as no macro segregation of alloy composition, which can compensate the deficiencies of the casting method (Dai 1998). Compared with the as-cast alloy ingot, however, the hot-compression deformation of the spray forming alloy is more difficult due to the high solute Mg content in the Al-Mg alloy (Chen 2009). It is very significant to establish the reasonable processing and basic parameters before plastic processing (e.g. the flow stress) (Hogg 2007). In recent years, many authors have reported the flow stress behavior of as-cast Al-Mg alloys during hot compression process. Jobba et al.(2015) investigated the flow stress and word-hardening behavior of Al-Mg alloys and reported that the flow stress was a significant influence factor for plastic deformation capacity of cast-state Al-Mg alloy. Sheikh et al. (2008) investigated the

effect of flow stress on the plastic deformation behavior of cast-state AA5083 alloy by establishing model. Wang et al. (2006) researched the flow stress behavior of spray-formed 5 A06 aluminum alloy during hot compression process. However, all of these reports are focused on flow stress behavior of low Mg content of Al-Mg alloy, and little reports are on high Mg content, especially high Mg content of spray formed Al-Mg alloy.

In this study we will carry out an isothermal hot compression test on Gleeble-3500 machine to investigate the flow stress behavior of spray-formed Al-9Mg-0.5Mn-0.1Ti alloy in a wide range of strain, deformation temperature and strain rate. The quantitative relationship between flow stress and the deformation parameters will be explored to establish the constitutive equation of the alloy.

2 EXPERIMENTAL METHOD

The spray forming preform of Al-9Mg-0.5Mn-0.1Ti alloy was produced by a self-developed spray forming machine (SD380). The chemical composition (wt.%) of the deposit preform is as follows: 9.0 Mg, 0.5 Mn, 0.1Ti and 91.4 Al. The plate with cross section size of 12 mm × 100 mm used in this investigation were extruded (using 1250 t extruder) from the preform at extrusion temperature of 450°C and extrusion ratio of 15:1.

Isothermal and constant-strain rate compression tests were carried out on a Gleeble-3500 testing system at temperatures of 300°C, 350°C,

400°C and 450°C and strain rates of 0.01 s⁻¹, 0.1 s⁻¹, 1 s⁻¹, 5 s⁻¹ and 10 s⁻¹. Before hot compression tests, the specimens were heated to the deformation temperature at the heating rate of 5°C/s and held for 3 min at isothermal conditions so as to obtain the uniform deformation temperature, and then hot compressed to a true strain of 0.8. Both ends of the cylinder sample were daubed lubricant with the chemical component of 75% graphite, 20% engine oil and 5% nitric acid from mesitylene fat.

3 RESULTS AND DISCUSSION

3.1 Flow stress behavior

Fig. 1 shows a series of true-stress true-strain curves of the as-extruded plate of the spray-forming Al-9 Mg-0.5 Mn-0.1Ti alloy compressed at 300–450°C and various strain rates. As the stress rate increasing, the obvious stress peak can be only observed at higher deformation temperature (see Fig. 1a), the peak can be observed until the temperature is up to 450°C at the stress rate of 5 s⁻¹ (see Fig. 1b). Further increasing the strain rate

to 10 s⁻¹, as shown in Fig. 1b, the curves exhibit multiple peaks. At same strain rate, the flow stress decrease significantly with the deformation temperature increasing, as shown in Fig. 1a. However, when the deformation temperature keeps constant, the flow stress increased with the stress rate increasing. As shown in Fig. 1b, at the lowest stress rate, all of the true stress-true strain curves at deformation temperature from 350°C to 450°C have an obvious stress peak, which exhibit initial flow softening on reaching a single peak stress.

The effect of deformation temperatures and strain rates on peak flow stress of the spray-forming Al-9Mg-0.5Mn-0.1Ti alloy is shown in Fig. 2. It can be obviously found that the peak flow stress is sensitive to deformation temperature and strain rate. The peak flow stress decreases as deformation temperature increases or stain rate decreases. At low deformation temperature and high strain rate, the dynamic recovery or recrystallization is restrained. This is because the mobility of dislocation is put off, there is no sufficient time for energy accumulation and dislocation annihilation. But at high deformation temperature and low strain rate, the average kinetic energy of metal atoms accumulates, which can improve the dislocation mobility. Therefor, the dynamic recovery or dynamic recrystallization may occur and the peak flow stress will decrease.

3.2 Constitutive equation

The Arrhenius equation is widely used to describe the relationship among flow stress, strain rate and temperature at hot deformation conditions (Sellars 1966 & Spigarelli 2003). There are three forms of the flow stress function as follows:

$$\dot{\varepsilon} = A_1 \, \sigma^{n_1} \exp\left(\frac{-Q}{RT}\right), (\alpha\sigma < 0.8) \tag{1}$$

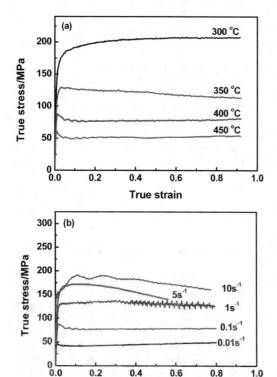

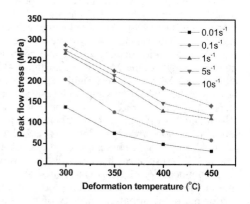

Figure 1. True stress-true strain curves of the spray-formed Al-9Mg-0.5Mn-0.1Ti alloy during hot compression deformation.(a) $\dot{\varepsilon} = 0.1$ s⁻¹, (b) $T = 400$°C.

Figure 2. Peak flow stress in the hot compression of the experimental alloy at different deformation conditions.

$$\dot{\varepsilon} = A_2 \exp(\beta\sigma)\exp(\frac{-Q}{RT}), (\alpha\sigma > 1.2) \qquad (2)$$

$$\dot{\varepsilon} = A[\sinh(\alpha\sigma)]^n \exp(\frac{-Q}{RT}) \qquad (3)$$

where a, n, P, A, A_1 and A_2 are constants: $a(MPa^{-1})$ ($a = P/n$) is the parameter of the stress level; n is the stress exponent; A, A_1 and A_2 (s^{-1}) are structure factors; $\sigma(MPa)$ is the flow stress; $\dot{\varepsilon}$ (s^{-1}) is the strain rate; T (K) is the deformation temperature; and Q (J/m) is an activation energy for deformation.

In the present work, the relation ships between peak flow stress and strain rate are obtained, the n_1 and β are calculated from the slopes of the plots of $\ln\dot{\varepsilon} - \ln\sigma$ and $\ln\dot{\varepsilon} - \sigma$, as shown in Fig. 3(a) and (b).

Zenner and Sellars also proposed that the flow stress behavior can be represented by the Zener-Hollomon parameter Z in the hyperbolic sine equation (Sellars 1966). The equation is given as follows:

$$Z = \dot{\varepsilon}\exp(\frac{Q}{RT}) \qquad (4)$$

Then, the formula (1) into equation (4) is available:

$$Z = \dot{\varepsilon}\exp(\frac{Q}{RT}) = A[\sinh(\alpha\sigma)]^n \qquad (5)$$

Taking the nature logarithm of both sides of Eq. (1) and Eq. (3) and derivation, we can obtain:

$$\ln\dot{\varepsilon} = \ln A - Q/(RT) + n\ln[\sinh(\alpha\sigma)] \qquad (6)$$

And then the deformation activation energy Q can be expressed by:

$$Q = nR\frac{\partial\ln[\sinh(\alpha\sigma)]}{\partial(1/T)}\bigg|_{\dot{\varepsilon}} = R\frac{\partial\ln\dot{\varepsilon}}{\partial\ln[\sinh(\alpha\sigma)]}\bigg|_{T}$$
$$\cdot\frac{\partial\ln[\sinh(\alpha\sigma)]}{\partial(1/T)}\bigg|_{\dot{\varepsilon}} \qquad (7)$$

$$K = \frac{\partial\ln[\sinh(\alpha\sigma)]}{\partial(1/T)} = \frac{Q}{nR} \qquad (8)$$

where n is the mean slope of the $\ln\dot{\varepsilon} - \ln[\sinh(\alpha\sigma)]$ plots at different temperatures and K is the mean slope of $\ln[\sinh(\alpha\sigma)] - 1/T$ plots at various strain rates, as shown in Fig. 3(c) and (d).

Taking the natural logarithm of both sides of Eq. (5) and derivation, we can obtain:

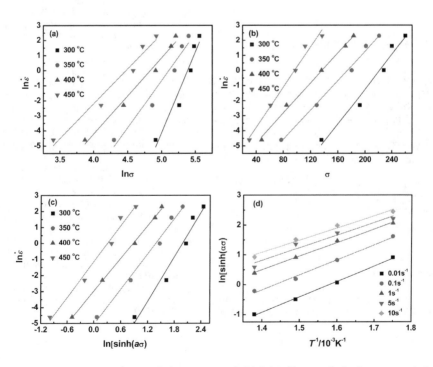

Figure 3. Relationship between $\dot{\varepsilon}$, σ, $\ln\dot{\varepsilon}$, $1/T$ and $\ln[\sinh(\alpha\sigma)]$: (a) $\ln\dot{\varepsilon} - \ln\sigma$,, (b) $\ln\dot{\varepsilon} - \sigma$, (c) $\ln\dot{\varepsilon} - \ln[\sinh(\alpha\sigma)]$, (d) $\ln[\sinh(\alpha\sigma)] - 1/T$.

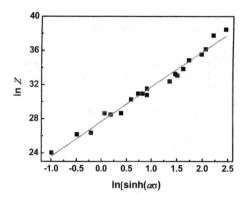

Figure 4.   Relationship between $\ln Z$ and $\ln[\sinh(\alpha\sigma)]$.

Table 1.   Material constants of the experimental alloy.

Calculation times	$A(\text{s}^{-1})$	n	$\alpha(\text{MPa}^{-1})$	$Q(\text{kJ/mol})$
1	1.3067×10^{10}	4.8900	0.0121	210.6522
2	1.2543×10^{10}	4.3516	0.0137	186.7743
3	1.0479×10^{10}	3.9367	0.0145	155.3349
Average value	1.2029×10^{10}	4.3928	0.0134	184.2538

$$\ln Z = \ln \dot{\varepsilon} + \frac{Q}{RT} = \ln A + n\ln[\sinh(\alpha\sigma)] \qquad (9)$$

Fig. 4 shows the relationship $\ln Z - \ln[\sinh(\alpha\sigma)]$ of the experimental alloy. The material constants of the alloy obtained from the experimental data are listed in Table 1. By substitution of the above values into Eq. (3), the constitutive equation of spray-formed Al-9Mg-0.5Mn-0.1Ti alloy is obtained and shown as follows:

$$\dot{\varepsilon} = 1.2029 \times 10^{10} [\sinh(0.0134\sigma)]^{4.3928}$$
$$\exp\left(\frac{-184253.8}{RT}\right) \qquad (10)$$

4   CONCLUSIONS

The hot compressive behaviors of spray-forming Al-9Mg-0.5Mn-0.1Ti alloy were studied at the tem-perature from 300°C to 450°C and the strain rate range from 0.01 s^{-1} to 10 s^{-1}. The results show that the flow stress of the spray-forming Al-9 Mg-0. 5Mn-0.1Ti alloy are sensitive to deformation parameters, and the spray-forming Al-9 Mg-0. 5Mn-0.1Ti alloy has high work hardening ability. The peak stress level and steady flow stress increase with decreasing deformation temperature and increasing strain rate, which can be represented by the Zener-Hollomon parameter Z with the hyper-bolic sine equation. The constitutive relationship of experiment alloy has been developed to describe deformation temperature and strain rate depend-ence of the flow stress. The deformation activation energy Q is calculated to be 184.2538 kJ/mol.

REFERENCES

Chen Z.H., C.H. Fan, Z.G. Chen & W. Li, (2009). Den-sification of large-size spray-deposited Al-Mg alloy square preforms via a novel wedge pressing technol-ogy. Mater. Sci. Eng. A. 506, 152–156.

Dai S.L., J.P. Delplanque & E.J. Lavernia (1998). Micro-structural characteristics of 5083 Al alloys processed by reactive spray deposition for net-shape manufac-turing. Metall. Mater. Tran. A. 29, 2597–2611.

Gang D.U., W. Yang, D.Shan & L.J. Rong (2011). Hard-ening behavior of the as-cast Al-Mg-Sc-Zr alloy. Acta Metall. Sin. 47, 311–316.

Hogg S.C., I.G. Palmer, L.G. Thomas & P.S. Grant (2007). Processing, microstructure and property aspects of a spraycast Al-Mg-Li-Zr alloy. Acta Mater. 55, 1885–1894.

Jobba M., R.K. Mishra & M. Niewczas (2015). Flow stress and work-hardening behaviour of Al-Mg binary alloys. Inter. Jour. Plast. 65, 43–60.

Sellars C.M. & W.J. Mctegart (1966). On the mechanism of hot deformation. Acta Meter. 14, 1136–1138.

Sheikh H. & S.Serajzadeh (2008). Estimation of flow stress behavior of AA5083 using artificial neural net-works with regard to dynamic strain ageing effect. J. Mater. Pro. Tech. 196, 115–119.

Spigarelli S., E. Evangelista & H. Mcqueen (2003). Study of hot workability of a heat treated AA6082 alumi-num alloy. Scripta Mater. 49, 179–183.

Tolga D. & S. Costas (2014). Recent developments in advanced aircraft aluminium alloys. Mater. Design. 862–871.

Wang Z.F., H. Zhang & Z.H. Chen (2006). Flow stress behaviors of spray deposited 5 A06 aluminum alloy under hot compression deformation. J. Nonferrous Metals 11, 1938–1944.

Williams J.C. & Starke E.A. (2003). Progress in struc-tural materials for aerospace systems. Acta Mater. 51, 5775–5799.

Advances in Energy Science and Equipment Engineering II – Zhou, Patty & Chen (Eds)
© 2017 Taylor & Francis Group, London, ISBN 978-1-138-71798-5

Preparation, characterization and release behavior of 5-Fu loaded PCL/CoFe$_2$O$_4$ magnetic composite microspheres

Tianxiang Zhang, Xin Tong, Lian Liu, Guangshuo Wang & Pei Wang
Department of Materials Science and Engineering, Dalian Maritime University, DaLian, P.R. China

ABSTRACT: In recent years, magnetic drug-loaded microspheres as a new functional material have received more and more attention. 5-Fluorouracil (5-Fu) as anti-cancer drug model was loaded on the poly(ε-caprolactone) (PCL) magnetic composite microspheres. The products were characterized by the means of FTIR, XRD, VSM, SEM, UV-vis. The saturation magnetization of the drug-loaded magnetic composite microspheres is 29.91 emu/g with superparamagnetic propertie. With the increase of the amount of 5-Fu, the drug-loading quantity increased first and then decreased, and the encapsulation efficiency is gradually reduced. When the amount of 5-Fu was 0.3 g, the drug-loading and encapsulation efficiency reached the maximum, the encapsulation efficiency was 27.5% and the drug-loading rate was 29.6%, respectively. Finally, the released properties of the drug-loaded composite microspheres was investigated, and found that the release of 5-FU showed a sustained and controlled behavior.

1 INTRODUCTION

Magnetic drug-loaded microspheres can be used as a suitable method to embed magnetic particles and drugs in the polymer materials, which have good performance of targeting, biocompatibility and biodegradability (Yin, 2010; Gao, 2011; Ajay Kumar Gupta, 2003). When it is injected into the body as a drug carrier, the effect of external magnetic field can be used to reach and gather in the lesion site, then release the drug to achieve purposes of targeted, reduce the toxic and side effect (Pinna, 2009; Duan, 2006; Wang, 2009).

Although there are more powerful drugs to be developed, people pay more attention to the way the drug can be delivered to the target cells and the subsequent release of the drug. In various drug delivery systems, polymers and magnetic carriers are the most representative (Zhu, 2009; Jang, 2012). The polymer carrier can maintain the drug concentration in the body by regulating the drug release rate. The rate of which applied to the polymer drug carrier of external stimulation of physical or chemical reaction control or in vivo biological environment control.The magnetic controlled release technology is mainly to put the drug carrier in the alternating magnetic field. Drug delivery by manipulating the magnetic carrier of a drug carrier can be quickly and continuously in a predetermined time, and continuously maintained in the target area. Therefore, a substantial increase in the local area of the drug concentration, but also the negative impact on non target cells may also be reduced (Cai, 2012).

In this experiment, the polymer skeleton material poly(ε-caprolactone) (PCL) and magnetic nanomaterials CoFe$_2$O$_4$ by emulsion solvent evaporation method to prepare magnetic microspheres (Roberto L. Sastre, 2004). PCL has good biocompatibility and biodegradability, it is a kind of linear polyester compound, and the molecular chain is arranged in a stacking and folding way. The structure is made up of five (-CH2-) and a polar ester group, which makes it easy to be biodegradable and processed (Yavuz CT, 2009). CoFe$_2$O$_4$ is a kind of spinel structure of magnetic metal oxide material with spinel is a class of ionic crystals, belongs to equiaxed crystal (H. Shen, 2010). The magnetic properties of CoFe$_2$O$_4$ are derived from the magnetic moment of the 3d electron layer in the structure, which has a very high resistivity and good magnetic properties (Lebourg M, 2010). The model drug 5-Fu, it is a widely used in cancer treatment of anti drug metabolism through the inhibition of thymidylate synthase and inhibition of DNA synthesis (Hiep NT, 2010). However, because of its toxicity, its clinical application is limited. Magnetic field effect of CoFe$_2$O$_4$ nanoparticles can kill tumor cells. Drug and suitable magnetic materials can be assembled into the magnetic nanoparticles in the blood vessel. Therefore, it has been the focus of research and development.

In this paper, PCL/CoFe$_2$O$_4$/5-Fu magnetic targeted drug delivery microspheres were prepared by solvent evaporation method, and their magnetic properties, drug loading, encapsulation efficiency and drug release properties were investigated, and then seek the best drug dosage.

2 MATERIALS AND METHODS

2.1 Materials

$CoFe_2O_4$ was prepared in laboratory, $PCL(M_w = 4,000)$ was purchased from Solvay (USA), PVA and 5-Fu was purchased from Guangzhou chemical reagents wholesale Department (China), CH_2Cl_2 was purchased from Tianjin Kermel Chemical Reagent Company (China), Absolute Ethyl Alcohol and EDTA(elhylene diamine tetraacetic acid) was purchased from Liaoning Emerging Reagent Co., Ltd. (China), HCL was purchased from Beijing chemical liquid Co., Ltd, Dialysis bags was purchased from Sinopharm Group Chemical Reagent Co., Ltd. (China).

2.2 Preparation of $CoFe_2O_4$ magnetic particles

Co-precipitation method is the most commonly used method to prepare magnetic nano materials. Compared with other methods, the method is simple and easy to operate. It is in aqueous solution containing Co^{2+} and Fe^{3+}, mixed with stirring reaction in alkaline conditions and then generates precipitation, then separated the product and dried in a vacuum for 24 hours to obtain magnetic nano particles. The ionic reaction equation is: $Co^{2+} + 2Fe^{3+} + 8OH^- = CoFe_2O_4 + 4H_2O$

2.3 Preparation of 5-Fu/$CoFe_2O_4$/PCL micropheres

Preparation of magnetic drug loaded microspheres by emulsion solvent evaporation method (Si, 2016). Taking quantitative $CoFe_2O_4$ and PCL dissolved in CH_2Cl_2, making magnetic and skeleton materials uniformly dispersed in the oil phase, and then the solution of a certain quality of 5-Fu powder of PVA water solution dropwise adding stirring at a certain speed of oil phase, getting the colostrum after emulsion for 10 minutes. Then the resulting colostrum was slowly added to the PVA solution, and obtain the double emulsion after emulsion, Continue stirring 24 hours at room temperature to volatile solvent. The resulting product is washed, filtered and dried to a constant weight.The experimental procedure is shown as Figure 1.

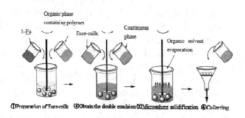

①Preparation of Fore-milk ②Obtain the double emulsion ③Microsphere solidification ④Collerting

Figure 1. The preparation of drug carrier magnetic microspheres.

2.4 Determination of the maximum absorption wavelength and standard curve of 5-Fu

The Certain quality 5-Fu was dissolved in 0.1 mol/L HCL solution, and the maximum wavelength was determined by UV spectrophotometer in the wavelength of 200–500 nm.

Accurately weighing 0.02 g 5-Fu dissolved in 0.1 mol/L HCL solution, and set the volume at 100 ml. And drawing the solution 1 ml, 2 ml, 3 ml, 4 ml, 5 ml, 6 ml in the 50 ml capacity of the bottle and fix capacity, using 0.1 mol/L HCL solution as a blank control, using UV spectrophotometer to determine the absorbance value of A 265 nm and linear regression, getting 5-Fu in 0.1 mol/L dilute HCL solution in the standard curve square.

2.5 Determination of the properties of drug-loaded microspheres

Taking appropriate magnetic microspheres with CH_2Cl_2 stirring and dissolving, with 0.1 mol/L dilute HCL extraction 3 times after completely dissolved. Collecting the upper 5-Fu dilute HCL solution to 50 ml, with UV spectrophotometer to determine the absorbance of A 265 nm, and then making it into the standard curve equation of 5-Fu in 0.1 mol/L dilute HCL, then getting the 5-Fu content.

The Entrapment Efficiency (EE) and drug-load quantity (DL) of magnetic drug-loaded microspheres is calculated as follow formula:

$$EE\,(\%) = \frac{W_0 - W_C}{W_0} \times 100\% \qquad (1)$$

$$DL\,(\%) = \frac{W_0 - W_C}{W_m} \times 100\% \qquad (2)$$

In the Formula: W_0 represents the total amount of 5-Fu which added in microspheres, W_C represents the amount of 5-Fu in the supernatant fluid, W_m represents the quality of magnetic drug-loaded microspheres.

Dynamic dialysis method is adopted to accurately take 20 mg magnetic drug-loaded microspheres under the treated dialysis bag, and add 10 ml phosphate buffer solution, then dialysis bag is sealed and placed in a conical flasks with 40 ml phosphate buffer. Put the conical flask in constant temperature water bath at 37°C, Draw 10 ml solution at regular time,at the same time, add 10 ml phosphate buffer in a conical flask. The absorption value was determined by UV spectrophotometer at 265 nm wavelength, and the amount of drug release was determined by standard curve equation. The cumulative release percentage of 5-Fu was calculated according to the following formula.

Cumulative Drug Release Rate$\% = \sum_{t=0}^{t=\infty} Mt/Mo \times 100\%$

$$(3)$$

In the Form, Mt is the release amount of 5-Fu in buffer solution at time t, and M_0 is the amount of 5-Fu microsphere in experimental determination.

Accurately weighing magnetic microspheres for four groups, with each group of 20 mg, and other conditions are the same as above.

3 RESULTS AND DISCUSSIONS

3.1 X-ray diffraction analysis

As seen in Figure 2, 5-Fu diffraction peak at $2\theta = 29°$ have a strong characteristic peaks, and PCL/$CoFe_2O_4$ have the roughly same obvious diffraction peaks position with 5-Fu/PCL/$CoFe_2O_4$, at $2\theta = 22°$, other peaks are not much difference. The characteristic absorption peaks of 5-Fu and $CoFe_2O_4$ are not shown in 5-Fu/PCL/$CoFe_2O_4$, this maybe the interaction between PCL are stronger than $CoFe_2O_4$ and 5-Fu, so the crystallization behavior of other substances became weakened, so the crystallization peak in the figure are disappear, showing only the diffraction peak location of PCL.

3.2 FTIR analysis

As seen in Figure 3, the broad peak at $3134\,cm^{-1}$ is the N-H in(-NH-CO-) stretching vibration peak;where the $1667\,cm^{-1}$ is the coincidence of the $C = O$ and $C = C$ stretching vibration absorption;the bond stretch bending vibration at $1430\,cm^{-1}$ is C-H of (-CF-CH-); the peak at $1250\,cm^{-1}$ is stretching vibration peak of C-N. It can be seen from the infrared spectrum of PCL/$CoFe_2O_4$, where $1730\,cm^{-1}$ is the characteristic absorption peak of ester bond ($C = O$) stretching vibration, the strong peak at $2960\,cm^{-1}$ is C-H stretching vibration peak, and the width peak

of $3440\,cm^{-1}$ is the expansion vibration absorption peak of -OH. Compare the infrared spectrum of 5-Fu/$CoFe_2O_4$/PCL and PCL/$CoFe_2O_4$, the characteristic peaks of both have no big change in the whole, probably because the strength of the PCL is stronger than that of the PCL, magnetic substances and drugs, the characteristic peak of magnetic material and the medicine are weakened.

3.3 SEM observation

As seen in Figure 4, magnetic drug-loaded microspheres showed as regular spherical, and the surface is relatively smooth. At a certain agitation speed, the average particle size of magnetic drug-loaded microspheres was larger than that of polymer microspheres, which may be due to the magnetic drug-loaded microspheres with magnetic $CoFe_2O_4$ and 5-Fu.The local release of microspheres of figure(d)

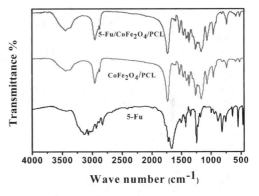

Figure 3. FTIR spectrum of(a)5-Fu drug, PCL/$CoFe_2O_4$ magnetic microspheres and 5-Fu/$CoFe_2O_4$/PCL drug carried magnetic microspheres.

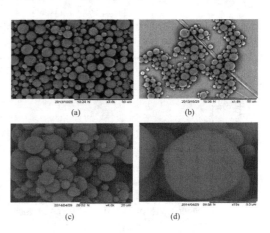

Figure 4. SEM images of 5-Fu/$CoFe_2O_4$/PCL drug-loaded magnetic microspheres at different magnification: (a) 3.0 k, (b) 1.8 k, (c) 4.0 k, (d) 15 k.

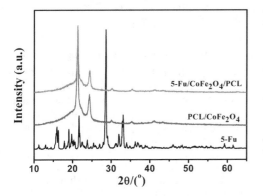

Figure 2. XRD patterns of 5-Fu/$CoFe_2O_4$/PCL, PCL/$CoFe_2O_4$ and 5-Fu magnetic microspheres.

shows that there are holes on the surface of the microspheres, which may be caused by continuous phase evaporation. This also provides a channel for the release of the drug in the microspheres, which can prolong the drug release time and play a role in the delayed release of the drug. From figure (a), (b), (c), we can also see some of the microspheres adhere some other smaller particles at the surface, this may be due to a part of the 5-Fu or magnetic material adhesion in the microspheres which are not coated inside of the microspheres.

3.4 VSM analysis

It can be seen clearly from Figure 5 that the specific saturation magnetization of $CoFe_2O_4$ was 43.438emu/g, and the specific saturation magnetization of 5-Fu/$CoFe_2O_4$/PCL was 29.91 emu/g, which compared with the added equal amount magnetic materials of the magnetic microspheres were reduced, it is probably because of the magnetic microspheres added to the drug, the ratio of the magnetic material after the coating has been reduced, and thus the specific saturation magnetization were reduced. It can be seen from the figure that the coercivity of the hysteresis loops is 0, which indicates that the magnetic microspheres have super paramagnetic properties.

3.5 Determination of 5-Fu standard curve

As seen in Figure 6, 5-Fu maximum absorption peak at 265 nm.

As seen in Figure 7, in the concentration range of 0.004–0.024 mg/ml, the concentration and absorbance of 5-Fu was at a good linear relationship. Linear regression equations (linear regression with c-A):

$$A = 58.84C + 0.019, r2 = 0.999 \qquad (4)$$

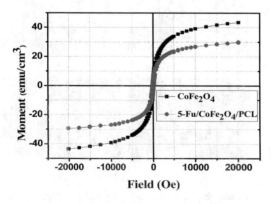

Figure 5. Magnetic hysteresis loop of 5-Fu/$CoFe_2O_4$/PCL and $CoFe_2O_4$ drug carried magnetic microspheres.

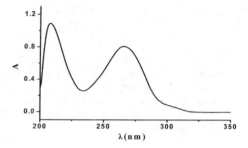

Figure 6. Ultraviolet spectrophotometer of 5-Fu.

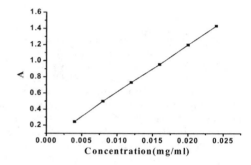

Figure 7. The standard curve of 5-Fu in 0.1 mol/L HCL solution.

In the form: The linear range is 4~24 mg/ml, A represents the absorbance, C represents the drug concentration (mg/ml).

As seen in Figure 8, in the concentration range of 0.004–0.024 mg/ml, the concentration and absorbance of 5-Fu was at a good linear relationship. 5-Fu at the wavelength of 265 nm spectral absorption of the linear regression equation (with c-A for linear regression):

$$A = 49.17C + 0.025, r2 = 0.999 \qquad (5)$$

In the form: The linear range is 4~24 mg/ml, A represents the absorbance, C represents the drug concentration (mg/ml).

3.6 Drug loading and release properties in vitro

Put 0.1,0.2,0.3,0.4 g 5-Fu into 10 ml 0.1 mol/L PVA solution, ultrasonic 60 min. Then, stirring the PVA solution at the same time, slowly dropping into CH_2Cl_2 solution which has dissolved quantitative PCL and $CoFe_2O_4$, then the oil water emulsion is formed, and put it into the PVA water solution, at the same time accompanied by stirring, emulsifying 30 min, adding ionized water, evaporating solvent overnight, filtration, washing and drying to get the product.

As can be seen from the Table 1, with the increase of the amount of 5-Fu, the drug load-

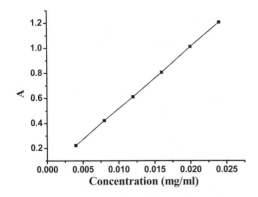

Figure 8. The standard curve of 5-Fu in PBS solution.

Table 1. The effect of 5-Fu to the envelopment efficiency and drug-loading rate.

Quantity of 5-Fu(g)	EE (%)	DL (%)
0.1	14.6%	10.7%
0.2	21.7%	18.5%
0.3	27.5%	29.6%
0.4	10.3%	8.8%

ing increases first and then decreases, and the encapsulation efficiency is gradually reduced. The increase of 5-Fu content in the ratio of raw materials will increase the drug loading, but not necessarily improve the encapsulation efficiency of the drug. When the amount of 5-Fu was 0.4, the encapsulation efficiency and drug loading were the least, which may be due to excessive drug dosage, and the solubility of the drug in the solution was too small. While 5-Fu was at 0.3 g, the drug loading and entrapment efficiency of magnetic drug-loaded microspheres were the largest, comparing with the magnetic chitosan microspheres prepared as usual, the 5-Fu drug loading rate and encapsulation efficiency all have been improved, which also showed that the combination of poly.

This also shows that the PCL as the substrate material can be a good drug combination. For magnetic drug-loaded microspheres, lower drug loading may be due to magnetic drug loaded microspheres not only contains drugs, but also contains magnetic material, so that the amount of 5-Fu in the microspheres is relatively small, then it leads to low drug loading rate.

4 CONCLUSIONS

In this paper, the preparation of magnetic drug loaded microspheres was prepared by the method of double emulsion solvent evaporation to prepare magnetic drug-loaded microspheres, the $CoFe_2O_4$ as magnetic core, the PCL as the matrix material, and 5-Fu as drug model.

The results of the experiment were evaluated with magnetic properties, encapsulation efficiency, drug loading rate and drug release rate. Results shows that the specific saturation magnetization of $5\text{-}Fu/CoFe_2O_4/PCL$ was 29.911 emu/g, which compared with the added equal amount magnetic materials of the magnetic microspheres were reduced. The coercivity of the hysteresis loops is 0, which have superparamagnetism. In magnetic drug loaded microspheres, when 5-Fu was added to 0.3 g, the drug loading and encapsulation efficiency of magnetic drug-loaded microspheres were the largest. Magnetic drug-loaded microspheres showed good in vitro release effect. Overall consideration about the results of encapsulation efficiency and drug loading, select 5-Fu amount of 0.3 g for the best drug concentration.

ACKNOWLEDGMENTS

The support of this work was provided by the National Natural Science Foundation of China, (No. 30870633).

REFERENCES

Ajay Kumar Gupta, Mona Gupta. Synthesis and surface engineering of iron oxide nanoparticles for biological applications [J].Biomaterials,2005,26: 3995–4021.B.W. Bestbury, R-matrices and the magic square, J. Phys. A 36, 1947 (2003).

Cai, Q., J.Z. Bei, S.G. Wang. In Vitro Study on the Drug Release Behavior from Polylactide-based Blend Matrices [J], Advanced Technologies Polymers, 13(2002):534–540.

Duan, J.F., J. Du, B. Zheng. Preparation and Drug-Release Behavior of 5-Fluorouracil-Loaded Poly(lactic acid-4-hydr oxyproline-polyethylene glycol) Amphipathic Copolymer Nanoparticles [J].Applied Polymer Science,2006,10.1002: 2654–2659.

Gao, X., B. Kan, M.L. Gou, et al. Preparation of Anti-CD40 Antibody Modified Magnetic PCL-PEG-PCL Microsphere [J].Biomedical Nanotechnology, 2011, 7:285–291.

Hiep NT, Lee BT. Electro-spinning of PLGA/PCL blends for tissue engineering and their biocompatibility [J], Mater Sci Mater Med 2010; 21:1969–1978.

Lebourg M, Anton JS, Ribelles JLG. Hybrid structure in PCL-HAp scaffold resulting from biomimetic apatite growth. [J] Mater Sci Mater Med 2010; 21:33–44.

Pinna N, Grancharov S, Beato P, Bonville P, Antonietti M, Niederberger M Magnetite nanocrystals non-aqueous synthesis, characterization, and solubility[J]. Chem Mater, 2009, 17:3044–3049.

Roberto L. Sastre, M. Dolores Blanco, Rosa Olmo, et al. Preparation and Characterization of 5-Fluorouracil-loaded Poly(ε-Caprolactone) Microspherefor Drug Administration [J], Drug Development Research 63(2004):41–53.

Shen, H. X.X. Hu, F. Yang, J.Z. Bei, S.G. Wang. An injectable scaffold:rhBMP-2-loaded poly(lactide-co-glycolide)/ hydroxyapatite composite microspheres. Acta Biomater 2010;6:455–465.

Si, J.H., Z.X. Cui, Q.T. Wang, et al. Biomimetic composite scaffolds based on mineralization of hydroxyapatite on electrospun poly(ε-caprolactone)/nanocellulose fibers [J], Carbohydrate Polumers, 143(2016):270–278.

Tae-Sik Jang, Eun-Jung Lee, Hyoun-Ee Kim. Hollow porous poly(ε-caprolactone) microspheres by emulsion solvent extraction [J], Materials Letters 72(2012):157–159.

Wang, D.S., J.G. Li, H.P. Li. Preparation and drug releasing property of magnetic chitosan-5-fluorouracil nano-particles [J]. Transactions of Nonferrous Metals Society of China, 19(2009):1232–1236.

Yavuz CT, Prakash A, Mayo JT, Colvin VL. Magnetic separations from steel plants to biotechnology[J]. Chem Eng Sci, 2009, 64:2510–2521.

Yin, H., S. Yu, Philip S.Synthesis and properties of poly(D,L-lactide) drug carrier with maghemite nanoparticles[J]. Materials Science and Engineering C, 2010, 30:618–623.

Zhu, L.Z., J.W. Ma, N.Q. Jia. Chitosan-coated magnetic nanoparticles as carriers of 5-fluorouracil: Preparation, characterization and cytotoxicity studies [J]. Colloids and Surfaces B: Biointerfaces, 2009, 68(1): 1–6.

Improving the thermal behavior of silver-sintered joint by mixing silver nanoparticles of a different size

Yong Xiao
Yik Shing Tat Industrial Co. Ltd., Shenzhen, P.R. China

Shuai Wang
Shanghai Radio Equipment Research Institute, Shanghai, China

Ming Yang
Micro-Joining Center, Korea Institute of Industrial Technology (KITECH), Incheon, Republic of Korea

Jianxin Wu
Yik Shing Tat Industrial Co. Ltd., Shenzhen, P.R. China

Zhihao Zhang
Department of Materials Science and Engineering, College of Materials, Xiamen University, Xiamen, P.R. China

Mingyu Li
State Key Laboratory of Advanced Welding Production Technology, Shenzhen Graduate School, Harbin Institute of Technology, Shenzhen, China

ABSTRACT: In this paper, a modified reduction reaction in an aqueous solution was developed to prepare Ag nanoparticles with diameter ~80 nm (Ag NP-80). To improve the compactness of the sintering structure, a certain amount of Ag nanoparticles with diameter ~10 nm (Ag NP-10) was added into the Ag NP-80 paste to fill the space between the large particles. During sintering, the smaller particles act as the filling material to reduce the space between larger particles and require sintering energy. Consequently, the thermal conductivity and the thermal expansion performance of the sintered joints greatly improved. The result indicate that the composite Ag NP is a promising bonding material in high-power electronic applications.

1 INTRODUCTION

Recently, the improvement of the global environmental awareness poses serious challenges to electronic packaging industry as many EU directives, such as ROHS, ELV, and WEEE, stressing that lead was damaging to human life as well as to the global environment (Suganuma. K, 2009; Y. Yamadaa, 2007; Chidambaram. V, 2011). Many studies focused on the development of lead-free materials with high signal transfer rate, thermal conductivity, and reliability have been carried out. Among all the alternatives, Ag nanoparticle (Ag NP) paste has been regarded as the most promising candidate because of its low jointing temperature and supreme properties (Suganuma. K, 2012; Mei. Y, 2012). Different from most of joining materials, such as solders and conductive adhesives, Ag NP paste can be used for jointing at a temperature much lower than the Ag melting point; meanwhile, the joints can work at a temperature much higher

than the joining temperature. The reason is the large surface energy of Ag particles with small size, which can provide high sintering driving force (Siow. K, 2012). In general, Ag NP paste contains organic carriers, surface active agents, solvents, and Ag NP. Until now, various methods have been proposed to prepare Ag nanopar-ticles with different sizes and improve the reliability of the sintered joint (Yan. J., 2012; Hu. P, 2010; Ide. E, 2005; Alarifi. H, 2011). However, some problems still need to be clarified to further improve the property of the Ag NP and the sintered joint.

This paper aims to achieve low-temperature sintering of Ag NP paste at atmospheric pressure. To improve the density and thermal conductivity of the sintered joint, a novel idea combining the Ag NP with different sizes was proposed in this paper. The smaller particles could act as the filling material to reduce the space between larger particles and required sintering energy, shown as Fig. 1. As a result, the thermal performance was

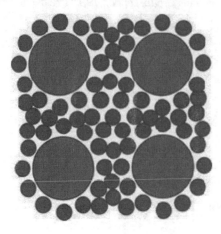

Figure 1. Illustration of the filling mechanism of the blended two types of nanoparticles.

greatly improved and other properties were investigated to test the sintered paste.

2 EXPERIMENT

2.1 Preparation of large Ag NP

The large Ag NP used in this study was prepared by a modified reduction reaction. A solution of 0.68 g AgNO$_3$ in 500 ml of deionized water was heated to 80°C, and 40 mL aqueous solution of 1% sodium citrate dehydrate was added dropwise to the silver nitrate solution with vigorous stirring. Then, the solution was heated to 90°C and kept for 30 min with stirring. Finally, the large Ag NP can be obtained after centrifugation and flocculation. The particle size is about 80 nm, and the morphology of the Ag NP was polyhedron rather than sphere, as shown in Fig. 2.

2.2 Preparation of small Ag NP and mixed-size Ag NP paste

Small Ag NPs were produced by the flocculation of the Carey Lea's colloidal (Frens G, 1969; Wang, S, 2012). The size of the small nanoparticles was around 10 nm observed by transmission electron microscopy (TEM, Tecnai G2 F20, FEI), as shown in Fig. 3 (a). The large Ag NP and small Ag NP were mixed at a weight ratio of 1:9 under the assistance of ultrasound. The distribution of the two types of Ag NP was recorded by TEM observation, as shown in Fig. 3 (b). As previously discussed, large Ag NPs were surrounded by the smaller ones due to the different surface energy between the two types of nanoparticles. Because the small nanoparticles have larger surface area

Figure 2. SEM picture of large Ag NP.

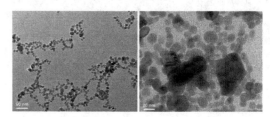

Figure 3. TEM images of (a) small Ag NP and (b) mixed-size Ag NP.

and higher surface energy, they can attach to the surface of the large nanoparticles easily.

2.3 Sintering process of the mixed-size Ag NP paste

The as-prepared mixed Ag NPs were heated in an oven (Binder FD23, Germany) from room temperature to the objective sintered temperature at a heating rate of 5°C/min in air atmospheres, followed by an isothermal hold for 30 min.

3 RESULTS AND DISCUSSION

3.1 Sintering of the mixed-size Ag NP paste

Fig. 4 shows the DSC and TGA profiles of the composite Ag NP paste, reflecting the thermal characteristics of the Ag metallo-organic nanoparticles. The exothermic peak at around 220°C should be caused by the surface sintering reaction of the Ag NP or the recrystallization of strained nano-Ag particles (Moon, K.S., 2005; Yu. H, 2012). This phenomenon should be related to the nanoparticle size rather than the organic shell, because the DSC profile of the organic component cannot demonstrate this exothermal peak. The fluctuation between 150 and 220°C should be caused by the accompanied endothermic peaks resulting from the decomposition of organic shell, which would dehydrate and decompose if the sintering temperature exceeded 150°C. The weight loss of the paste was only 1.52%, indicating that the Ag content in the composite Ag NP paste was higher than most of the other Ag NP pastes.

The mixed-size Ag nanoparticles were heated at 150, 200, and 250°C for 30 min to investigate the organization of sintered paste. Fig. 6 shows the TEM images of these sintered joints. When the paste was sintered at 150°C, the Ag NPs were connected through their surface and sintered without external pressure. Because the organic shell (i.e., sodium citrate dehydrate) covered on the Ag NP started to decompose at 150°C, half an hour of sintering was insufficient to remove most of the organic dispersant. Therefore, it can be expected that the atom diffusion between particles was obstructed, resulting in the low growth rate of the Ag particles. As shown in Fig. 5 (a), the diameters of the grain size were just several tens of nanometers. With the increase in the sintering temperature, the Ag particles became larger, about 85 nm, and the organic component seemed to be reduced significantly, as shown in Fig. 5 (b). In addition,

when the sintering temperature reached 250°C, the particles further coalesced and some bridging paths were formed. Fig. 5 (c) indicates that the organic shell was completely decomposed, and a 3D connection network was generated at a higher sintering temperature.

3.2 Characteristics of sintered Ag NP paste

The XRD profile of mixed-size nano Ag paste sintered at different temperatures was obtained by using D/max-2500/PC, Rigaku, Japan. As shown in Fig. 6 (a), the XRD pattern of sintered product agreed with that of silver-3C, syn. Through calculation, the grain size of Ag NP paste sintered at different temperatures was determined from 20.4 nm to 30.1 nm with the increase of temperature from 150 to 280°C. This change is consistent with the sintering principle and the solid-state diffusion mechanism that high temperature provides great driving force to decompose the dispersant and facilitate the growth of Ag grains.

3.3 Thermal performance of the sintered Ag NP paste

The thermal performance of sintered mixed-size Ag NP pastes was determined by a laser flash apparatus (LFA477, Netzsch, Germany), and the thermal conductivity data as a function of sintered temperature is presented in Fig. 7. The profile of thermal conductivity value can be divided into three parts. First,

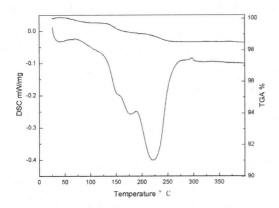

Figure 4. DSC and TGA results of the mixed-size Ag NP paste.

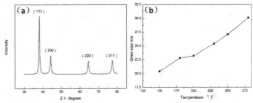

Figure 6. (a) XRD pattern of Ag NP paste sintered at 150°C, (b) grain size measured by XRD of pastes sintered at different temperature.

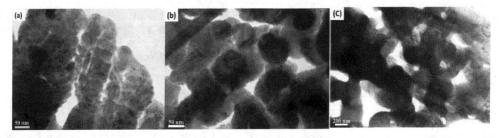

Figure 5.   TEM images of mixed-size Ag NP paste sintered at (a) 150°C, (b) 200°C, and (c) 250°C.

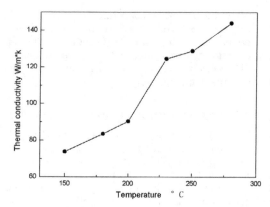

Figure 7. Thermal conductivity of Ag NP paste as a function of sintering temperature. All the samples were held at the sintered temperature for 30 min.

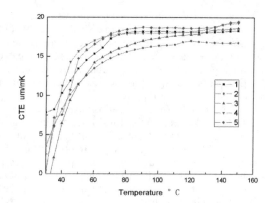

Figure 8. Coefficient of thermal expansion of the Ag NP paste sintered at 150°C. The sample was tested five times from room temperature to 150°C at a rate of 5°C/min.

from 150 to 200°C, the thermal property increased slowly, which should be caused by the insufficient removal of the organics capped on the Ag particles. It should be noted that the thermal diffusivity of these organic residues is very low, which greatly reduced the heat transfer of the entire joint. Besides, the existence of the numerous grain boundaries and particle surfaces at the low temperature further contributed to enhancing phonon reflection and dropping the thermal properties. Then, the thermal conductivity value increased suddenly from 90 to 125 W/m K at the temperature range of 200–230°C, indicating that the recrystallization of the Ag NP occurred around this temperature. In this stage, the surface melting and atom diffusion promoted necking and coalescence. The numbers of surface and boundaries were reduced and many heat dissipation paths were generated. When the paste was sintered at a temperature higher than 220°C, the Ag grains continued to grow and the sintering process advanced to form a densified organization. Compared to the sintered paste composed of only small Ag nanoparticles (Wang. S, 2012), this mixed-size Ag NP paste exhibited improved thermal performance. This phenomenon should be due to the fact that the addition of large particles reduced grain boundaries and surfaces, which weakened the phonon reflection. On the contrary, adequate amounts of small particles could still provide enough sintering driving force to ensure the formation of densified structure.

To evaluate the thermal properties of the composite Ag NP paste, coefficient of thermal expansion (CTE) test was performed. A material with lower CTE means lower change in volume under different thermal conditions, which is very important in the application of thermal interfacial materials. In this study, five tests were performed on each sample, as shown in Fig. 8. During the first two

tests, CTE of the sintered nano-Ag paste slightly dropped, whereas it greatly increased to a constant value after the third test. This phenomenon should be due to the decomposition of the organic layer on Ag NP during CTE test. During the first two tests, the heat in the test tube can promote the organic decomposition and complete sintering process, resulting in shrinkage of sample and therefore lower values. After three tests, the decomposition reaction and sintering process had been completed. Therefore, the properties of the sample maintained at a constant value. Compared with Sn63Pb37 and Sn3.5 Ag solders, the CTE of the mixed-size Ag NP paste is much lower.

4 CONCLUSIONS

In summary, Ag NP with diameter about 80 nm were prepared through a reduction reaction and mixed with small Ag NP particles. With the help of smaller Ag NP, which can fill the space between large particles, the Ag paste sintered joint was greatly improved in density as well as thermal properties. After sintering at 220°C, the thermal conductivity of the sintered joint reaches 115 W/m K. Besides, because of the low requirement of sintering energy for smaller Ag NP particles, the sintering temperature can be down to as low as 150°C. This study shows that the mixed Ag NP paste is a promising bonding material in high-power electronics industry.

ACKNOWLEDGMENT

This work was supported by Shanghai Science and Technology Committee Yangfan Plan Project under Grant No. 15YF1411700.

REFERENCES

Alarifi. H, Hu. A, Yavuz. M, Zhou. YN. 2011. Silver Nanoparticle Paste for Low-Temperature Bonding of Copper. Journal of Electronic Materials 40: 1394–1402.

Chidambaram. V & Hattel. J & Hald. J. 2011. High-temperature lead-free solder alternatives. Microelectron. Engineering 88: 981–989.

Frens G, Overbeek JTG. 1969. Carey Lea's colloidal silver. Colloid Polym Sci 233: 922–9.

Hu. P & O'Neil. W & Hu. Q. 2010. Synthesis of 10 nm Ag nanoparticle polymer composite pastes for low temperature production of high conductivity films. Applied Surface Science 257: 680–685

Ide. E & Angata. S & Hirose. A & Kobayashi. K.F. 2005. Metal–metal bonding process using Ag metallo-organic nanoparticles. Acta Materialia 53: 2385–2393.

Mei. Y & Chen. G & Lu. G & Chen. X. 2012. Effect of joint sizes of low-temperature sintered nano-silver on thermal residual curvature of sandwiched assembly. International Journal of Adhesion & Adhesives 35: 88–93.

Moon, KS & Dong, H. & Maric, R & Pothukuchi. S. & Hunt. A. & Li. Y. & Wong. C.P. 2005. Thermal behavior of silver nanoparticles for low-temperature interconnect applications. Electronic Materials 34:168–175.

Siow. K 2012. Mechanical properties of nano-silver joints as die attach materials. Journal of Alloys and Compounds 514: 6–19.

Suganuma. K & Kim. S.J. & Kim. K.S. 2009. High-temperature lead-free solders: properties and possibilities. JOM Journal of the Minerals 61(1): 64–71.

Suganuma. K & Sakamoto.S & Kagami. N & Wakuda. D & Kim K.S. & Nogi. M. 2012. Low-temperature low-pressure die attach with hybrid silver particle paste. Microelectronics Reliability 52: 375–380.

Wang. S & Ji. H & Li. M & Wang. C. 2012. Fabrication of interconnects using pressureless low temperature sintered Ag nanoparticles. Materials Letters 85: 61–63.

Wang. S & Ji. H & Li. M & Wang. Q. 2012. Fabrication of interconnects using pressureless low temperature sintered Ag nanoparticles. Materials Letters 85: 61–63.

Yamadaa. Y. & Y. Takakub & Y. Yagia & I. Nakagawac & T. Atsumic & M. Shiraic & I. Ohnuma & K. Ishida. 2007. Reliability of wire-bonding and solder joint for high temperature operation of power semiconductor device. Microelectronics Reliability 47: 2147–2157.

Yan. J & Zou. G. & Wu. A & Ren. J & Yan. J & Hu. A & Zhou. Y. 2012. Pressureless bonding process using Ag nanoparticle paste for flexible electronics packaging. Scripta Materialia 66: 582–585.

Yu. H & Li. L & Zhang. Y. 2012. Silver nanoparticle-based thermal interface materials with ultra-low thermal resistance for power electronics application. Scripta Materialia 66: 931–934.

Protection-type extrusion die structure for aluminum profiles

Rurong Deng & Xuemei Huang

Guangzhou Vocational College of Science and Technology, Guangzhou, China

ABSTRACT: In this paper, methods for the determination of semi-hollow profiles were introduced. A new protective die structure was analyzed, and the selection of design parameters was introduced. The structure has a good effect on improving the strength and life of the die, and the quality of the products was shown. It is a structure worthy of promotion.

1 INTRODUCTION

With the rapid development of the national economy, as well as the continuous development of modern processing technology, the application of aluminum profile is more and more extensive and the field is becoming increasingly wider. Therefore, the emergence of varieties and specifications are also high. In many varieties and specifications, there is a type known as the "semi-hollow profiles." This type of material is not closed in the hollow part, but cannot use the traditional flat die structure for extrusion. This is because, in the extrusion process, the cantilever surface of the die must bear a lot of positive pressure. During elastic deformation of the die, the cantilever will produce the axial and lateral bending and lose stability, thereby leading to thin wall and scraping. Plastic deformation will lead to the failure of the cantilever, so that the die strength of this type of material is very difficult to guarantee. In order to reduce the positive pressure on the surface of the cantilever, we improve the pressure bearing capacity of the cantilever, produce the qualified product, so as to improve the die life. In recent years, extrusion workers, experts, scholars, and engineering technicians worldwide have carried out a lot of research, and some new die structures are obtained, such as the hollow die with a penetrating plane and the hollow die with a mosaic structure. Through a practical example, a protection-type hollow extrusion die was introduced, which is used for reference.

2 DETERMINATION OF SEMI-HOLLOW PROFILES

As shown in Figure 1, the ratio of A to R in the form of W^2 and the ratio of the profile of the open width to the profile is called a semi-hollow profile, which is referred to as the value of the tongue is larger than that presented in Table 1.

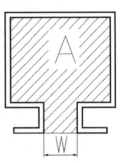

Figure 1. The signal of tongue ratio.

Table 1. Open width (W) of profile and the maximum of tongue ratio (R_{max}).

W/mm	1.0–1.5	1.6–3.1	3.2–6.3	6.4–12.6	12.7–
R_{max}	2	3	4	5	6

In practice, although some of the nonhollow profiles have a opening width of more than 12.7, the tongue ratio is less than 6 and the cantilever area is very large. In the design of the die, the strength calculation formula of the solid die is used to calculate the intensity. When the safety factor is less than 2, it can be determined as a half-hollow profile.

3 THE STRUCTURE DESIGN OF A PROTECTION-TYPE HOLLOW DIE

In Figure 2, the tongue ratio of the profile, which is calculated, is 1.32, according to the size and characteristics of the profile. Select the 18 MN extrusion machine, its cylinder diameter is Φ185 mm, selection of die size is 165 mm × 240 mm (diameter × thickness). In conventional solid die structure of extrusion, after calculation, the die safety coefficient is 1.3 and die strength is insufficient to guarantee, and hence cannot select conventional solid die structure.

Assuming that the distance profile cantilever edge 2–5 mm as a virtual core size, to determine the size of the core, the distance from cantilever root to core edge should be 5–10 mm, as shown in Figure 3.

The male die of porthole die is designed with the core, and the cantilever as a female die hole was arranged in the welding chamber of hollow die. The cantilever section is protected by a central portion of the male die. When the male die and female die assemble, there is a certain gap between the core end of the male die and the welding chamber plane of female. The practical experience value is 0.5–1.2 mm. The larger the size of the cantilever, the larger the gap value, but not more than 50% of the wall thickness. This gap is also known as the stress gap, which is used to eliminate the pressure of the cantilever caused by the male die center during the compression or the elastic deflection. Hence, in the extrusion process, the core may be

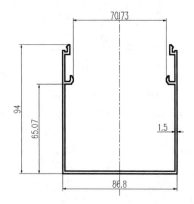

Figure 2.   Profile section.

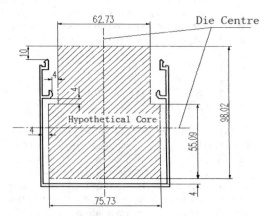

Figure 3.   Signal of hypothetical core.

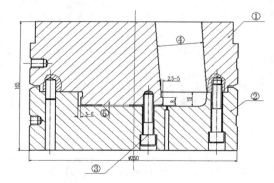

Figure 4.   Die structure.

downward flexing, but will not touch the surface of the cantilever plane welding chamber. Therefore, it is not the force in the cantilever.

Thus, as long as the strength and stiffness of the male die is ensured, the positive pressure of the cantilever surface is reduced or eliminated. In the extrusion process, the majority of the positive pressure is supported by the male die, thus the deviation of the wall thickness of the end of the cantilever is stable. It can ensure the quality of the profile, more importantly ensure the strength of the die and improve the life of the die.

Illustration for Fig. 4:

①Male die ②Female die ③The screw of H13 high-strength steel ④The verge of parallel feeder hole ⑤The stress gap of from 0.5 to 1.2 mm

Although the male die under normal pressure protects most of the cantilever, there are still some exposure to the core of the external cantilever. At the same time, because of the role of friction in the process of extrusion, it will also exert a certain downward force on the cantilever. The elastic deformation of the cantilever results, which leads to a certain thinning of the end of the cantilever. In order to eliminate this phenomenon, we can set a high-strength bolt, combine the die to produce a certain pre-tightening force. At the end of the cantilever direction from the center of a position, the cantilever will be "hanging" in the male die. The elastic deflection of the cantilever can be overcome, thus ensuring the uniformity of the wall thickness, as shown in Figure 4.

4   DIE DESIGN KEY POINTS

4.1   Layout of the porthole

As shown in Figure 5, when the porthole is arranged, it is to be noted that the porthole is not arranged at one end of the cantilever supporting edge (the cantilever root). The area difference of each porthole is controlled at about 10%, and the feeder ratio is 30–40%, which can reduce the extrusion pressure and increase the strength of the die. In the corner of profiles and the root position of the cantilever, the porthole will be extended by 5 degrees. The purpose is to reduce the pressure and increase the amount of metal in these parts, so that the metal supply of these forming is more severe than the low flow rate. Thereby, increasing the metal flow velocity in these parts results in the consistent velocity of the flow.

4.2 Structure of the core and the type of feeder

The bearing length of core is 5–8 mm and the die-relief under the neck of the core is 3–5 mm. The longer the bearing length, the lower the value of the die relief. In order to reduce the compression area of the center portion of the die, the parallel oblique way or chamfering on the feeder hole in the porthole can be used, as shown in Figure 5.

4.3 Wall thickness compensation

Although protected mode on the cantilever surface is provided with a hanging screw, in practice, the end wall thickness of the cantilever still needs to be left with the compensation of 0.15–0.1 mm. After all, in the process of extrusion, the friction force is also very large and the extrusion process is also influenced by the process parameters.

4.4 Welding chamber and bearing belt

The bridge piers should be split under mode corresponding to the welding chamber, to reduce the span of bridge and improve the strength of the mold. Welding chamber height should not be too large, in order to ensure the core rigidity and stability. Welding chamber height is greater; the greater the welding force, the higher the cantilever force. It will reduce the strength of the cantilever.

Welding chamber height can choose a conventional depth as small as 2–3 mm, as shown in Figure 6.

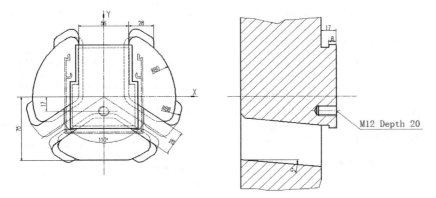

Figure 5.   Signal of male die.

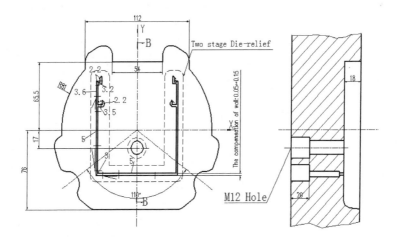

Figure 6.   Chamber and bearing of female die.

5 CONCLUSIONS

The protection-type hollow extrusion die described in this paper is simple and easy to manufacture. The practical application shows that:

1. The die structure has the semi-hollow profile, more target, more wide range, and is suitable for all kinds of semi-hollow sections, especially for the semi hollow profile with large section.
2. In the extrusion process, the new die structure can reduce the formation of the hollow parts of the cantilever of the force area, a decline of more than 80%, so as to greatly reduce the force of the cantilever during metal extrusion and greatly improve the strength of the mold.
3. The new structure can change the stress state of the cantilever, so that the cantilever structure of the cantilever beam changes into simply supported beam, reduces the stress of the mold, and improves the mold strength.
4. Compared with the traditional mold life, the die life of new structure can be greatly improved. The die life can reach 10 tons or more; at the same time, the quality of the profile can be better to ensure that it can avoid the wall thickness and surface roughness profiles.

REFERENCES

Deng Rurong, Huang Xuemei. Analysis of the die design for aluminum hollow profiles with large section[J]. Light Alloy Fabrication Technology, 2015, 43(3): 41–46.

Deng Rurong, Li Changquan. Design of spread shunt of the extrusion die large section aluminum profile[J]. Light Alloy Fabrication Technology, 2014, 42(1): 39–43.

Deng Rurong, Zeng Lei. Design of flow-guided die for aluminum solid profiles with the large section[J]. Light Alloy Fabrication Technology, 2014, 42(11): 33–37.

Guang Wang, Jian Liu. Technology study of coarse grain ring of 2024 aluminum alloy extrusion bar[J], Light alloy fabrication technology, 2013(04): 36~40.

Guang-Lei Lin. Study on the production process of 7A04-T6 aluminum alloy rod [J], Light alloy fabrication technology, 2007(11): 33~36.

Huang Xuemei, Zeng Lei, et al. Analysis of the die design of aluminum alloy profiles for radiators with high-density fins[J]. Light Alloy Fabrication Technology, 2015, 43(5): 36–39.

Jian-xin Xie, Jing-an Liu. Metal extrusion theory and technology[M]. Metallurgical Industry Press, Beijing, 2001: 28–31, 68~69.

Jing-an Liu. Aluminum alloy extrusion die design, use and maintenance. [M], Metallurgical Industry Press, Beijing, 1999: 159~167.

Ke-Wei Zhang, Lei Zhang. Effect of pre-stretching on microstructure and mechanical properties of 2A12 aluminum alloy bar[J], Light alloy fabrication technology, 2013(11): 37~40.

Ru-rong Deng, Lei Zeng, The design of flow-guided die in large solid section [J], Light alloy fabrication technology, 2014 (11): 29~31.

Ru-rong Deng. To determine the key parameters of aluminum extrusion die design by shunt[J], Light alloy fabrication technology, 2002, 30(2): 23~24.

Yu Mingtao, LI Fuguo. Simulation extrusion process of the sketch hollow aluminum profile based on infinite volume method[J]. Die and Technology, 2008(4): 40–43.

Zu-tang Wang, Xin-quan Zhang. Research on the design technology of the flow guided die for aluminum profile extrusion die[J], Light alloy fabrication technology,1992(01): 38~42 In Chinese.

Research on the stability of interface microstructure of thermal barrier coatings under cyclic thermal loading

Jian Sun, Yingqiang Xu, Wanzhong Li & Kai Lv
Northwestern Polytechnical University, Xi'an, China

ABSTRACT: Thermal barrier coating system (TBCs) can effectively improve the thermal efficiency of thermal power equipment. However, little attention has been devoted to study fatigue characteristics of TBCs under cyclic thermal loading. In this paper, according to the stress-strain accumulation theory and dissipation energy under cyclic thermal loading, the stability criteria of interface microstructure of TBCs based on static shakedown theorem and kinematic shakedown theorem were proposed respectively. Then, the influence of different interface microstructure on the stability of TBCs was analyzed through finite element methods. Finally, the simulation results were proved by thermal fatigue tests of TBCs. The results of this study demonstrate that the interface microstructure with higher curvature will lead to a lower stability and the criteria based on shakedown theorem can effectively determine the stability of TBCs under multiple cyclic thermal loading.

1 INTRODUCTION

Thermal barrier coatings (TBCs) are widely used in turbines for propulsion and power generation. And they act as a thermal insulator and an oxidation-inhibitor, which used to increase the heat resistance of the blades. However, instability of TBCs occurred under multiple cyclic thermal loading. With the development of initial undulate interface and enhancement of material mismatch effect, the thermal stress of TBCs will be more complicated under cyclic thermal loading which causes the instability and ultimate failure of TBCs (Evans, 2001; Ranjbar-Far, 2010; Karlsson, 2002).

Investigations have been undertaken to broaden the basic understanding of the stabilities of TBCs under thermal cycling. Tolpygo and Clarke (Tolpygo, 2000; Tolpygo, 2000) measured the residual compressive stress in a-Al$_2$O$_3$ oxide films and studied the buckling and spalling of thermally-grown alumina films. Evans, Karlsson and He etc (Karlsson, 2001; He, 2002; Balint, 2005) studied the influence of thermal mismatch on TBCs. The influence of interface geometry is critical to the state of TBCs. However, most former researches mainly focus on the stress-strain state and deformation under single thermal cycle, little attention has been devoted to study stability of TBCs with different interface microstructures under multiple cyclic thermal loading.

The primary purpose of this research is to study the influence of different interface microstructures on the stability of TBCs under multiple cyclic thermal loading. Furthermore, the behavior of stress-strain accumulation in micro-zones of TBCs is studied by static shakedown theorem, while the stability of the whole system is analyzed through kinematic shakedown theorem.

2 STABILITY CRITERIA

Two criteria were utilized to judge the stability of interface microstructure of thermal barrier coatings. The first one was shown in Equation (1), which is based on static shakedown theorem.

$$f\left(s^1 - \bar{\alpha}^0\right) = \frac{1}{2}\left(s^1 - \bar{\alpha}^0\right)\left(s^1 - \bar{\alpha}^0\right) - k_i(T)^2 = 0 \qquad (1)$$

where S is deviator stress tensor, $i = 1, 2, 3, 4$ denotes the different materials of TBCs, α is the back stress, k is relating to the yield strength of the material.

The second criterion described in Equation (2) is based on the kinematic shakedown theorem which is proposed for stability analysis of elastic perfectly plastic materials subject to cyclic or repeated loading.

$$\int_V (\sigma^E : \varepsilon^p + T\eta^p)dV \leq \int_V \Gamma(\varepsilon^p, \xi, \eta^p)\,dV \qquad (2)$$

where, $\sigma^E : \varepsilon^p + T\eta^p$ denotes the thermo-mechanical loading, and $\sigma^E : \varepsilon^p$ for the mechanical loading, $T\eta^p$ for the thermal loading, $\Gamma(\varepsilon^p, \xi, \eta^p)$

denotes a function for the rate of dissipation power in terms which can be obtained from first and second law of thermodynamics and Clausius-Duhem Inequality, as follows:

$$\dot{\Gamma} = \boldsymbol{\sigma} : \dot{\boldsymbol{\varepsilon}}^p - \boldsymbol{\chi} \cdot \dot{\boldsymbol{\xi}} + T\dot{\boldsymbol{\eta}}^p \qquad (3)$$

where χ denotes the dual internal variable (G. Borino, 2000), where ξ means internal variable, $\dot{\eta}^p$ is plastic entropy ratio.

Before the TBCs reach shakedown state, residual stress-strain of the interface microstructure and dissipation energy of TBCs are altering with the cyclic thermal loading. The stability of TBCs is analyzed by the relationship described in (2).

3 NUMERICAL MODEL

3.1 Geometry model

The model consists of four layers: superalloy Substrate (SUB), bond coat (BC), Thermally Grown oxide (TGO), and top coat (TC, ceramic layer). The interface around TGO is very rough due to the manufacturing process (Tolpygo, 2004). To accurately describe the interface geometry of TGO, sinusoidal model were applied to simulate the interface geometry. Different wavelength (2λ) and different amplitude (m) of sinusoid were utilized to simulate various TGO undulate interface microstructures. The width of the model is 30 μm (λ), the thickness of the TC is 200 μm (h1), the thickness of the BC is 100 μm (h2), and the thickness of the SUB is 100 μm (h3). The thickness of the TGO is 1 μm, as shown in Fig. 1.

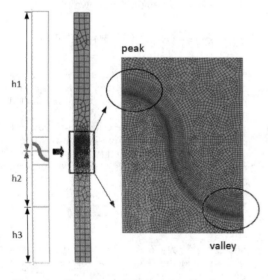

Figure 1.   Geometry and FEA model of TBCs.

3.2 Material properties

The TBCs have different mechanical properties under cyclic thermal loading, each layer of TBCs was considered to be isotropic. The TGO, TC and SUB layers were treated as elastic materials and the BC was elastoplastic. The mechanical properties and thermal properties were temperature dependent from 25°C to 1000°C, and the material properties just reference to literature (Aktaa, 2005; Hu, 2010; Busso, 2006; Zhou, 2007; Kang, 2009; Mao, 2006; Yang, 2002).

3.3 Boundary condition and loading

The Multi Point Constraint (MPC) was applied on the right edge of model. This condition allows the affected mesh nodes to move freely but simultaneously in the horizontal direction. and their displacement in the vertical direction is possible. The horizontal displacement at the left side of the model was cancelled by applying an axial symmetry condition. Nodes at the bottom side were constrained in the Y direction. And to avoid the rigid displacement of the overall system, the lower left corner node in the model was chosen to fix.

Furthermore, the research mainly focuses on the influence of interface microstructure on the stability of TBCs, hence, creep effect was not considered. Cyclic thermal loading imposed a homogeneous temperature on the whole system following the thermal cycle. For the case of single cycle, the load consists of two steps, heating from 25°C to 1000°C in 600 s, followed by a cooling step from 1000°C to 25°C in 600 s. For the multi-cycles case, the thermal loading spectrum is the accumulation of single thermal cycle process.

4 RESULTS AND DISCUSSIONS

4.1 Residual stress in TBCs

The residual stress of TBCs with different interface structure in single cyclic thermal loading were shown in Fig. 2 without considering the TGO, the maximum stress exists in BC, whereas less stress exists in TC and SUB layers. Furthermore, compared with the material properties of each layer, BC is apt to failure. The stress in Y direction plays a main role in BC delamination failure. Therefore, the residual stress in Y direction in BC layer is taken as the key factor in evaluating the effect of the various interface microstructures on the stability of TBCs and the results obtained from simulation are presented in Table 1.

According to the results, σ_y in valley zones was compressive and maximum was 1142.88 MPa, which was higher than the maximum tensile σ_y in peak zones. The phenomenon that tensile stress existed in peak zones while compressive stress

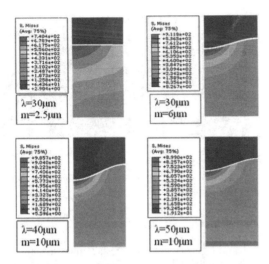

Figure 2. The residual stress of TBCs with different interface structure.

Table 1. Residual stress σ_y varying with interface in BC.

Number	Wavelength $2\lambda/\mu m$	Amplitude $m/\mu m$	Peak/ MPa	Valley/ MPa
1	30	2.5	394.4	−741.356
2	30	6	400.875	−910.556
3	30	25	683.881	−1142.88
4	40	10	373.262	−982.492
5	50	10	368.789	−895.228
6	60	10	350.838	−834.577

existed in valley zones was in accord with the results acquired from multilayer-concentric-circle model (Hsueh, 2000). Furthermore, the reason for the phenomenon lies in the different material properties between BC and TC.

With constant wavelength, maximum stress in valley and peak zones increased with increasing amplitude. Moreover, with constant amplitude, maximum stress in valley and peak zones decreased with increasing wavelength. The reason for the alteration is that different wavelength and amplitude will induce the change of curvature of interface structure. As curvature increases, the stress concentration in BC will increase, which explains why the maximum stress changes.

4.2 The stability of TBCs

The first criterion was applied to judge the stability of TBCs. Stress-strain evolution behaviors of BC layer for $\lambda = 30$ μm and m = 6 μm under thermal cycles were shown in Fig. 3. The evolution behaviors in X and Y directions around valley zones

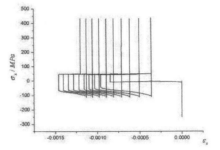

(a) Evolution behaviors of ε_x and σ_x in valley zone

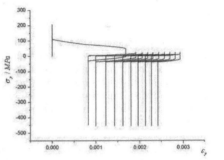

(b) Evolution behaviors of ε_y and σ_y in valley zone

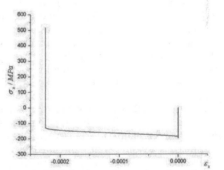

(c) Evolution behaviors of ε_x and σ_x in peak zone

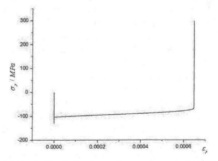

(d) Evolution behaviors of ε_y and σ_y in peak zone

Figure 3. Stress-strain evolution behaviors of maximum stress node in BC.

995

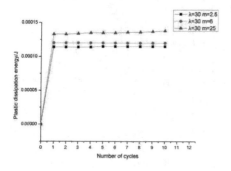

Figure 4. Evolution behaviors of plastic dissipation energy.

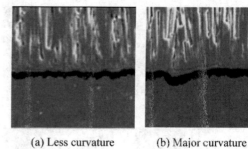

(a) Less curvature   (b) Major curvature

Figure 5. Microstructure of the TGO interface under cyclic thermal loading.

were shown in (a) and (b). Where ε_x was free at the initial state in the first cycle, which is different from the rest cycles. Then, the ε_x increases at the latter stage with the increasing temperature and reached its maximum with the maximum temperature. When TBCs enters decreasing temperature stage, ε_x decreases and σ_x changes from compressive to tensile and increases till the end of the first cycle. In the rest cycles, as TBCs serves in cyclic thermal loading, ε_x accumulates but the increment of decreases in each cycle, which indicates the TBCs has a tendency to be stable. And similar evolution ε_x behaviors appears between σ_y and ε_y as illustrated in Fig. 1(b). Additionally, in the peak zones, the evolution behaviors of stress-strain are similar within each cycle shown in Fig. 1(c) and (d), which means no strain energy accumulation occurs in peak zones, therefore the TBCs is also stable.

The second criteria was utilized to analyze the stability of TBCs through plastic dissipation energy, furthermore its evolution behaviors in TBCs with different interface microstructure under cyclic thermal loading were illustrated in Fig. 4. The dissipation energy increases at the first, and it remains steady in the latter cycles, which indicates the TBCs becomes stable in following stages. Therefore, the TBCs tends to be stable through the analysis of dissipation energy.

Moreover, the results acquired by numerical simulation were verified by experiments. Cyclic thermal fatigue (TMF) tests of TBCs specimens, about 20 mm in diameter and 3 mm thick SUB, 200 µm thick Y_2O_3–ZrO_2 TC layer, 150 µm CoNiCrAlY BC were carried out in ambient air using standard furnace. The polished sections were then imaged by Scanning Electron Microscopy (SEM) for multiple cycles in Fig. 5. The thickness of TGO in Fig. 5 (a) is thinner than the TGO layer in Fig. 5 (b) due to oxidation. Cracks have already existed in Fig. 5 (b) which means the larger curvature is inclined to instability. Therefore, the experimental results were accord to simulation results.

4 CONCLUSIONS

The influence of different interface microstructure on the stability of TBCs under single cyclic loading was analyzed through numerical simulation. The results which obtained according to static shakedown theorem demonstrate that the interface microstructure with major curvature will lead to a lower stability. Therefore, TBCs will serve longer if the interface is smoother and with less undulate geometry. The stability criterion based on kinematic shakedown theorem of TBCs was proposed, and the results show that the stability criterion can effectively determine the stability of different interface model under cyclic thermal loading, which indicates that the method can be used for TBCs failure judging.

ACKNOWLEDGMENT

This work were financially supported by the National Natural Science Foundation of China (Grant No. 11072196, 5875214, 10672134) and the Natural Science Basic Research Plan in Shaanxi Province of China (Program No. 2015 JM1009).

REFERENCES

Aktaa, J., K. Sfar, D. Munz. Assessment of TBC systems failure mechanisms using a fracture mechanics approach. Acta Mater 53: 4399–4413(2005).

Balint, D.S., J.W. Hutchinson. An analytical model of rumpling in thermal barrier coatings. J Mech Phys Solids 53: 949–973(2005).

Borino, G. Consistent shakedown theorems for materials with temperature dependent yield functions. Int J Solids Struct 37: 3121–3147(2000).

Busso, E.P., Z.Q. Qian. A mechanistic study of microcracking in transversely isotropic ceramic–metal systems. Acta Mater 54: 325–338(2006).

Evans, A.G., R.D. Mumm, J.W. Hutchinson. Mechanisms controlling the durability of thermal barrier coatings. Prog Mater Sci, 46: 505–553 (2001).

He, M.Y., J.W. Hutchinson, A.G. Evans. Large deformation simulations of cyclic displacement instabilities in thermal barrier systems. Acta Mater 50: 1063–1073(2002).

Hsueh, C.H., E.R. Fuller. Analytical modeling of oxide thickness effects on residual stresses in thermal barrier coating. Scripta Mater 42: 781–787 (2000).

Hu, H.J., J.Y. Zhang, X.G. Yang. Numerical study of failure mechanisms on plasma sprayed thermal barrier coatings. Journal of Aerospace Power 25: 1085–1091(2010).

Kang, G.Z., Q.H. Kan, L.M. Qian, Y.J. Liu,. Ratchetting deformation of superelastic and shape memory NiTi alloys. Mech. Mater 41: 139–153(2009).

Karlsson, A.M., A.G. Evans. A Numerical Model for the Cyclic Instability of Thermally Grown Oxides in Thermal Barrier Systems. Acta. Mater 49: 1793–1804(2001).

Karlsson, A.M., J.W. Hutchinson, A.G. Evans. A fundamental model of cyclic instabilities in thermal barrier systems. J Mech Phys Solids. 50:1565–1589(2002).

Mao, W.G., Y.C. Zhou. Modeling of residual stresses variation with thermal cycling in thermal barrier coatings, Mech. Mater 38: 1118–1127(2006).

Ranjbar-Far, M., J. Absi, G. Mariaux, F. Dubois. Simulation of the effect of material properties and interface roughness on the stress distribution in thermal barrier coatings using finite element method. Mater Des 31: 772–781(2010).

Tolpygo, V.K., D.R. Clarke, K.S. Murphy. Evaluation of interface degradation during cyclic oxidation of EB-PVD thermal barrier coatings and correlation with TGO luminescence. Surf Coat Technol 188–189: 62–70(2004).

Tolpygo, V.K., D.R. Clarke. Spalling failure of α-alumina films grown by oxidation: I. Dependence on cooling rate and metal thickness. Mater Sci Eng. 278: 142–150(2000).

Tolpygo, V.K., D.R. Clarke. Spalling failure of α-alumina films grown by oxidation: II. Decohesion nucleation and growth. Mater Sci Eng. 278: 151–161(2000).

Yang, X.G., R. Geng. The Analysis of 2D Temperature and Thermal Stress of TBC-Coated Turbine Vane. Journal of Aerospace Power 17: 432–436(2002).

Zhou, C.G., N. Wang, H.B. Xu. Comparison of thermal cycling behavior of plasma-sprayed nanostructured and traditional thermal barrier coatings. Mater Sci Eng. 24, 452–453: 569–574(2007).

Advances in Energy Science and Equipment Engineering II – Zhou, Patty & Chen (Eds)
© 2017 Taylor & Francis Group, London, ISBN 978-1-138-71798-5

Seismic assessment of existing RC frames based on the story damage model

Ning Ning
School of Civil Engineering, Qingdao University of Technology, Qingdao, China

Wenjun Qu
Department of Building Engineering, College of Civil Engineering, Tongji University, Shanghai, China

Z. John Ma
Department of Civil and Environmental Engineering, University of Tennessee, Knoxville, Tennessee, USA

ABSTRACT: This paper deals with the investigation of seismic assessment method based on the story damage model. Low-frequency cyclic loading tests of two spatial Reinforced Concrete (RC) frames are carried out. The damage, failure patterns, and the skeleton curves of RC frames are experimentally studied. Finite-element analyses are carried out to investigate the effects of influencing factors on the behavior of RC frames. The factors considered are axial compression ratio, concrete strength, reinforcement ratio of slabs, thickness of the slabs, and the stiffness of the transverse beams. On the basis of tests and FEM data, a story damage model is established. The proposed model considers the failure pattern of the RC frames.

1 INTRODUCTION

In Wenchuan earthquake (May 12, 2008, China) most Reinforced Concrete (RC) frames suffered huge damages. Those RC frames were collapsed due to "strong beam weak column." Studies on the damage assessment of RC frames should be carried out to investigate the damage behavior during strong earthquakes. Many research efforts have been made in this field (Chung, 1989; Rodriguez, 2009; Teran, 2005).

The earliest and simplest damage models were either ductility-based or with introducing the parameter of the inter-story deformation (Newmark, 1963; Sameh, 2001). These models did not consider the cumulative damage, which failed to reflect the real performance of structures in earthquake. Recently, researchers (Williams, 1995; Kunnath, 2006) have found that repeat load and maximum displacement play an important role in characterizing the actual damage. On the basis of testing data from beams and columns, Ang et al. (Park, 1985) proposed a damage model, as shown in Eq. (1), for components. The model was widely accepted and used in damage assessment of RC members:

$$D = \frac{x_m}{x_{cu}} + \beta \frac{\int dE}{F_y x_{cu}} \qquad (1)$$

where x_m = maximum deformation under earthquake; x_{cu} = ultimate deformation under monotonic loading; F_y = yeild strength; and dE = incremental absorbed hysteretic energy. Parameter β can be represented in terms of Eq.(2):

$$\beta = (-0.447 + 0.073\lambda + 0.24n_0 + 0.314\rho_t) \times 0.7^{\rho_w} \qquad (2)$$

where λ = shear span ratio; n_0 = axial compression ratio; ρ_t = longitudinal reinforcement ratio; and ρ_w = confinement ratio.

In order to reflect the damage of structure during earthquakes, the damage models for whole structure were proposed (Park, 1985; Gu, 1997; Ghobarah, 1999). Park et al. (Park, 1985) later proposed the damage model of structures by integrating the damage of components:

$$D = \sum_i \lambda_i D_i \qquad (3)$$

$$\lambda_i = \frac{E_i}{\sum E_i} \qquad (4)$$

where λ_i = energy dissipation proportion of the member, which can be calculated by Eq. (4).

These models were based on the local damage at the component level. The global damage was evaluated by combining component damages using a weighted-average method. However, there exist very limited testing data to verify the relationship. Damages of earthquakes such as Wenchuan earthquake indicate story damage, and failure pattern should be considered in a damage model. The objective of this paper is to assess the damage of the RC frames during strong earthquakes, using the proposed damage model on the basis of story damage.

2 EXPERIMENTAL PROGRAM AND FEM ANALYSIS

2.1 Experimental program

Two 1:2.5 scaled spatial RC frames were tested under low-frequency cyclic loading: RC frame without cast in situ slabs (RC-1) and RC frame with cast in situ slabs (RC-2). It should be noted that the material properties, member cross-sectional dimensions, and the reinforcement details of RC-1 were the same as those of RC–2. All specimens were designed according to the Chinese concrete structure code (GB50010–2008) (GB50011–2008,2008). The slab was 50 mm deep. Details of specimens are shown in Figure 1(a)–(c).

The average compressive strength of the measured concrete prism (150 mm × 150 mm × 300 mm) was fc = 28.4 MPa, and the elastic modulus was

2.85 × 104 MPa for the first floor of RC frames. The fc and the elastic modulus of the second floor were 25.1 MPa and 2.64 × 104 MPa, respectively. According to the Chinese Code GB50010–2008, steel bars of HPB235 and HRB335 were adopted as transverse and the longitudinal steel bars in this paper. The details of steel bars are classified in Ning (2014).

2.2 Tests result

For the RC-1, the flexural cracks formed at the ends of the first floor beams and columns when the lateral displacement reached 0.2% story drift. The concrete cracks developed with the increase of lateral displacement. No new crack appeared after 1.5–2.0% story drift, and the original cracks increased with the increase of cycles. When the specimen was close to failure, the flexural hinges formed at all beams ends while there was no hinge formed at columns except the first story column feet. The failure pattern of RC-1 was typical "strong column-weak beam."

For the RC-2, similar phenomena were observed during the tests, but the numbers of cracks of columns were more than those of beams compared to RC-1. Diagonal cracks at an inclination of 45° appeared at the corner of the slabs. When the lateral displacement reached the 1% story drift, torsional cracks were observed on the surface of the transverse beams, and the joint diagonal shear cracks appeared. After 2.0% story drift, new cracks were not observed while the original cracks developed with an increase of lateral displacement. The failure pattern of RC-2 turned to "strong beam-weak column."

The skeleton curve is shown in Figure 2. The test results are shown in Ning (2014).

2.3 FEM analysis

Three-dimensional spatial frame models of RC-1 and RC-2 have been developed using nonlinear structural analysis software ABAQUS. The comparison of skeleton curve is shown in Figure 3. The analysis result shows that the FEM analysis data fit the test results well.

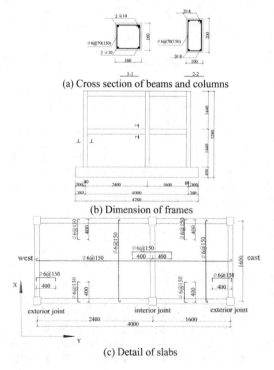

(a) Cross section of beams and columns

(b) Dimension of frames

(c) Detail of slabs

Figure 1. Design of specimens.

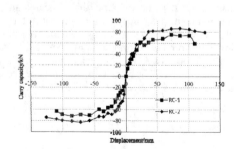

Figure 2. Skeleton curve of frames.

The compression ratio, concrete strength, reinforcement ratio of slabs, thickness of the slabs, and the stiffness of the transverse beams are regarded as influence factors in FEM analysis. Results of variation of these factors are shown in Tables 1–5.

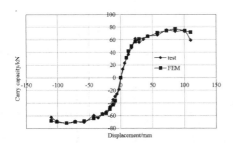

(a) RC-1

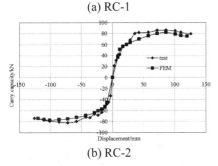

(b) RC-2

Figure 3. Comparison of skeleton curve of frames.

Table 1. Axial compression ratio.

Models	East column	Middle column	West column
RC-3	0.10	0.18	0.13
RC-4	0.20	0.36	0.26
RC-5	0.30	0.54	0.39
RC-6	0.40	0.72	0.52
RC-7	0.50	0.90	0.65

Table 2. Concrete strength.

Models	Concrete strength/MPa
RC-8	20
RC-9	25

Table 3. Reinforcement ratio of slabs.

Models	Reinforcement	Reinforcement ratio (%)
RC-10	Φ6@200	0.283
RC-11	Φ6@100	0.556
RC-12	Φ8@150	0.670
RC-13	Φ8@100	1.005

Table 4. Thickness of the slabs.

Models	Thickness (mm)
RC-14	30
RC-15	40
RC-16	60

Table 5. Stiffness of the transverse beams.

Models	Dimension of transverse beam
RC-17	$100 \times 100(4\Phi8)$
RC-18	$100 \times 300(4\Phi8)$

3 DAMAGE MODEL

The damage model of RC structures is proposed in Eq. (5). The model considers story damage of each floor, which reflects the story damage effect on the whole structure:

$$D = \sum_{i=1}^{n} w_i D_i \qquad (5)$$

$$D_i = \frac{x_{mi}}{x_{cui}} + \beta_i \frac{E_{hi}}{F_{yi} x_{cui}} \qquad (6)$$

where w_i = weighted-average, $w_i = \frac{(n+1-i)E_{hi}}{\sum_{i=1}^{n}(n+1-i)E_{hi}}$;

n = layers of frames; D_i = story damage index of ith floor, given by Eq. (6); x_{cui} = ultimate displacement of ith story under monotonic loading; x_{mi} and E_{hi} = story maximum deformation and cumulative hysteretic energy in earthquake, respectively; F_{yi} = yield shear force under monotonic loading; and β_i = coefficients of ith story. Test and FEM analysis results indicate that axial compression ratio and column-to-beam strength ratio play an important role in failure mode and damage in earthquake. Thus, β_i is calculated by Eq. (7):

$$\beta_i = (A + B\overline{\lambda}_i + C\overline{n}_{0i} + D\overline{\eta}_i)0.7^{\rho_{wi}} \qquad (7)$$

where $\overline{\lambda}_i$ = average column shear span ratio of ith story; $\overline{n}_{0i}$ = average axial compression ratio of ith story; $\overline{\eta}_i$ = average value of column-to-beam strength ratio of ith floor; ρ_{wi} = confinement ratio of column; and A, B, C, and D are coefficients to be fit. β is derived by D = 1 and D = 0.

The results of parameters of each model are shown in Tables 6 and 7.

So, β_i is expressed in Eq.(8) by fitting analysis. The damage model for RC frames is expressed by Eq. (9):

Table 6. Parameters of the first story.

Models	$\overline{\lambda_1}$	$\overline{n_{01}}$	$\overline{\eta_1}$	$\rho_{w1}(\%)$	x_{cu1}	x_{m1}	E_{h1}	F_{y1}
RC-1	4.28	0.012	2.23	0.504	150	57.5	111149	72.2
RC-2	4.28	0.012	0.81	0.504	154	60	166982.9	77.16
RC-3	4.28	0.120	1.19	0.504	173.39	60	224412.3	95.82
RC-4	4.28	0.237	1.56	0.504	174.445	60	255292.9	98.6
RC-5	4.28	0.411	1.67	0.504	181.25	52.8	208642.1	111.8
RC-6	4.28	0.548	1.31	0.504	139.935	39.5	149395.2	106
RC-8	4.28	0.012	0.93	0.504	116.665	25.3	52875.33	75.02
RC-9	4.28	0.012	0.87	0.504	119.75	39.5	84652.44	76.64
RC-10	4.28	0.012	0.76	0.504	156.25	60	135080	72.78
RC-11	4.28	0.012	0.95	0.504	156.93	60	143413.3	73.16
RC-12	4.28	0.012	0.85	0.504	157	60	140836.6	73.16
RC-13	4.28	0.012	0.93	0.504	156.25	60	149001.8	72.16
RC-14	4.28	0.012	0.84	0.504	164.335	60	137673.6	72.7
RC-15	4.28	0.012	0.83	0.504	159.11	60	133818.9	72.94
RC-16	4.28	0.012	0.80	0.504	151.975	60	136925.1	73.88
RC-17	4.28	0.012	0.95	0.504	155.555	60	147392.7	74.78
RC-18	4.28	0.012	0.78	0.504	151.665	60	147509	75.02

Table 7. Parameters of the second story.

Models	$\overline{\lambda_2}$	$\overline{n_{02}}$	$\overline{\eta_2}$	$\rho_{w2}(\%)$	x_{cu2}	x_{m2}	E_{h2}	F_{y2}
RC-1	4.28	0.006	1.12	0.504	150	57.5	77239.14	50.54
RC-2	4.28	0.006	0.41	0.504	154	60	116039	54
RC-3	4.28	0.114	0.60	0.504	173.39	60	155947.5	67.07
RC-4	4.28	0.231	0.78	0.504	174.445	60	177406.9	69.02
RC-5	4.28	0.404	0.84	0.504	181.25	52.8	144988.6	78.26
RC-6	4.28	0.542	0.67	0.504	139.935	39.5	103817	74.2
RC-8	4.28	0.006	0.47	0.504	116.665	25.3	36743.87	52.51
RC-9	4.28	0.006	0.44	0.504	119.75	39.5	58826.27	53.65
RC-10	4.28	0.006	0.38	0.504	156.25	60	93869.15	50.94
RC-11	4.28	0.006	0.48	0.504	156.93	60	99660.09	51.21
RC-12	4.28	0.006	0.43	0.504	157	60	97869.5	51.21
RC-13	4.28	0.006	0.47	0.504	156.25	60	103543.6	50.51
RC-14	4.28	0.006	0.42	0.504	164.335	60	95671.48	50.89
RC-15	4.28	0.006	0.42	0.504	159.11	60	92992.79	51.05
RC-16	4.28	0.006	0.40	0.504	151.975	60	95151.34	51.71
RC-17	4.28	0.006	0.48	0.504	155.555	60	102425.4	52.34
RC-18	4.28	0.006	0.39	0.504	151.665	60	102506.3	52.51

$$\beta_i = (-0.23 + 0.073\overline{\lambda_i} + 0.027\overline{n_{0i}} + 0.003\overline{\eta_i})0.7^{100\rho_{wi}} \tag{8}$$

$$D = \sum_{i=1}^{n} \frac{(N+1-i)E_i}{\sum_{i=1}^{n}(N+1-i)E_i}$$

$$\left[\frac{x_{mi}}{x_{cui}} + (-0.23 + 0.073\overline{\lambda_i} + 0.027\overline{n_{0i}} + 0.003\overline{\eta_i})\right.$$

$$\left. 0.7^{100\rho_{wi}}\frac{E_{hi}}{F_{yi}x_{cui}}\right] \tag{9}$$

4 CONCLUSION

In this paper, we proposed a new framework for seismic assessment of existing RC frames. On the basis of the results of the experimental investigation and FEM analysis, the following conclusions are drawn:

1. Low-frequency cyclic loading test of two RC spatial frames was carried out to study the damage, failure pattern, and the skeleton curves of the frames.

2. The seismic damage model for RC frame structures is proposed on the basis of test and analysis data. The model considered the story damage. The failure pattern was also considered by introducing the column-to-beam strength ratio.

ACKNOWLEDGMENTS

The authors wish to acknowledge the financial support by the National Science Foundation of China for Youth (51508289). Visit to University of Tennessee Knoxville by the first author and visit to Tongji University, China, by the third author were made possible by Department of Civil and Environmental of University of Tennessee Knoxville, USA.

REFERENCES

Chung, Y.S., C. Meyer, M. Shinozuka. Modeling of concrete damage. ACI Struct J. 186, 259(1989).

GB50011-2008. *Chinese Code for seismic design of buildings*. Beijing, 2008.

Ghobarah, A., A. Abou-Elfath. Response-based damage assessment of structure. Earthquake Engineering and Struct. Dyn., 28,79(1999).

Gu, X., Shen Z. Damage analysis reinforced concrete structures under earthquake series. *Proceedings pf ICCBE-VII,* Seoul, Korea, 1997.

Kunnath, S.K. Cumulative seismic damage of reinforced concrete bridge piers. *University at Buffalo, State University of New York, NCEER Report:* 97–2006

Newmark, N.M., C.P. Siess, M.A. Sozen, Moment-Rotation Characteristics of Reinforced Concrete and Ductility Requirement for Earthquake Resistance. *Proceedings, 30th Annual convention of SEAOC*, 1963.

Ning, N., W. Qu, and P. Zhu. Role of Cast-in-Situ Slabs in RC Frames under low frequency cyclic load. Eng Struct, 59, 28(2014).

Park, Y.J., A.H.-S. Ang, Y.K. Wen. Seismic damage analysis of reinforced concrete buildings. J Struct Eng., 111, 740(1985).

Park, Y.J., A.H.-S. Ang. Mechanistic seismic damage model for reinforced concrete. J Struct Eng., ASCE, 111, 722(1985).

Rodriguez, M.E., D. Padilla. A damage index for the seismic analysis of reinforced concrete members. J. Earthq Eng. 13, 36(2009).

Sameh, S.F., Mehanny, G. Gregory. Deierlein. Seismic damage and collapse assessment of composite moment frames. J. ASCE, 127, 1045(2001).

Teran, G.A., J.O. Jirsa. A damage model for practical seismic design that accounts for low cycle fatigue. Earthq Spectra. 21, 803(2005).

Williams, M.S., R.G. Sexsmith. Seismic Damage Indices for Concrete Structures: A State-of-Art Review. Earthq Spectra, 11, 320(1995).

Advances in Energy Science and Equipment Engineering II – Zhou, Patty & Chen (Eds)
© 2017 Taylor & Francis Group, London, ISBN 978-1-138-71798-5

Study on creep characteristics of backfill under step loading

Shuguo Zhao

College of Mining Engineering, North China University of Science and Technology, Tangshan, China

Dongliang Su

HeBei Province Key Laboratory of Mining Development and Safety Technique, Tangshan, China

ABSTRACT: The time effectiveness of the backfill often affects the long-term stability of the mining engineering. In order to analyze the creep characteristics of backfill, uniaxial creep test of backfill under step loading were conducted by using the computer-controlled shear creep test machine BLW-3000. The result shows that the backfill has a significant creep deformation and a tendency of strain hardening. The greater the stress, the faster is the decay of the elastic modulus of backfill with time. The strain rate of backfill shows an exponential decay, and the deformation modulus decreases with time, which is described by Logistic function. MATLAB software is used for fitting the creep curve by using the custom equation fitting function, and the data of fitting results and tests were analyzed, indicating larger values of attenuation creep. In general, the model can be well described under multistage loading backfill creep characteristics.

1 INTRODUCTION

The mining industry has brought great economic benefits to several countries of the world (Benzaazoua, 2008); however, it also produces a series of problems. The wastes from mining enterprises, such as waste rock, tailings, and others, will occupy a lot of farms and pollute the water and soils. Failure of tailings dam will lead to huge geological disasters and economic problems (Fall, 2009; Nasir, 2010). Because of the cost savings, less requirement of cultivated land, and efficient and safe mining resources, cemented backfill technology, especially the tailing-cemented backfill, has great development.

A large number of scholars have studied tailings backfill materials and their mechanical properties. HAN Bin (Bin, 2012) used orthogonal experimental design ratio of aggregates and tailings methods. Optimization of backfill mix in terms of slurry density, cement dosage, mass ratio of aggregates and tailings, slurry fluidity indices, and backfill strength indices were investigated. CHEN Qiusong (Qiusong, 2015) explored the effect of magnetized water on flocculating sedimentation. The changing laws of flocculating sedimentation velocity and mass fraction of the underflow of unclassified tailings were studied under different magnetization conditions. In order to improve the stability of the filling body, a series of unconfined compressive strength (UCS) tests were carried out on both nonreinforced and fiber-reinforced cemented tailings by X. W. and Yi, G. (Yi, 2015).

Most of these studies are focused on the optimization of material ratio and strength design and few of them on the creep characteristics of backfill. Creep is the phenomenon that the material changes with time under the action of constant stress, and it began with metal materials. The creep tests of limestone, shale, and sandstone were carried out by Griggs D T in 1939, and the results show that when the load reaches 12.5–80% of the uniaxial compressive strength of the rock, it can produce significant creep deformation (Griggs, 1939). After that, more and more scholars studied the creep characteristics of rock materials (Jun, 2014; Hua, 2012; Hong, 2015; Liang, 2015; Sone, 2014).

This paper is based on the existing research results of rock creep, and a uniaxial creep test under step loading of backfill was carried out. The deformation characteristics of backfill under different stresses were analyzed based on the test data. The creep curves were fitted by Burgers model, and the fitting precision and error are analyzed.

2 CREEP TESTS

2.1 Test equipment

The strength tests of backfill were conducted in WHY-300 Press. The creep tests of backfill under step loading were conducted using the computer-controlled shear creep test machine BLW-3000. The maximum load of this instrument can reach 3000 kN; the whole operation can be carried out using computer, and the real-time test data are collected.

2.2 Test methods

In this paper, we studied the creep of backfill by same ratios (cement:sand ratio is 1:6; concentration 70% by weight). The test specimen size is 100 mm × 100 mm × 100 mm. The standard curing period was 28 days.

Before backfill creep tests, the uniaxial compression tests were carried out, and the single axial compressive strength of backfill was used as the basis of the creep test. The literature study on the creep of rock is generally considered. The long-term strength of the rock is 75–85% of the uniaxial compressive strength (UCS). In this paper, 75% UCS is considered as the last stage in the creep test. The test processes and calculation were performed according to Chinese National Standard GB/T50081–2002 <Standard for test method of mechanical properties on ordinary concrete>. The uniaxial compressive strength of backfill is 2.48 MPa. The creep tests of backfill were conducted in RLW-3000 microcomputer-controlled shear creep machines. Because of the rapid deformation of the early stage of the creep test, 3h data acquisition interval is 30 s, and the data are collected after every 3 min. Loading rate is 30 N/s. When the deformation rate is less than 0.001 mm/h (or loading time to achieve 24h), the next level of loading is carried out.

3   CREEP TEST ANALYSIS

Fig. 1 shows the experimental data. This test adopts the step loading method; the test data cannot be analyzed directly and hence should be processed.

At present, the Boltzmann superposition principle and CHEN's method are used to deal with the data of the step loading creep test data.

A comparative study of two methods of data processing in Z. Xianwei's paper shows that Chen's method can obtain better creep curve of single loading by step loading test (Xian, 2010). The experimental data were used in CHEN's method. CHEN's method and its mathematical basis were introduced in detail by Z. Xiaowen (Xian, 2010).

Tailing cemented backfill has obvious creep behavior (Fig. 2). At each stage of stress, the deformation of backfill shows instantaneous deformation, decay creep stage, and constant creep stage. Because of the limitation of test time and conditions, accelerated creep of backfill did not occur.

After instantaneous deformation, the strain rate can be considered infinite. In the first stage of stress (0.62 MPa), backfill creep rate is gradually reduced to 4.96×10^{-7}/h (Fig. 3), which is considered close to zero. Creep deformation is stable at 7.96×10^{-3}/h, and the creep curves of the backfill only show the decay creep stage. In the third stage of stress (1.86 MPa), the final strain rate is 7.17×10^{-5}/h, which is two orders of magnitude higher than the strain rate of 0.62 MPa, the deformation of backfill shows the decay creep and the constant creep stage.

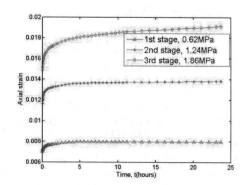

Figure 2.   Strain–time diagram of Chen's method.

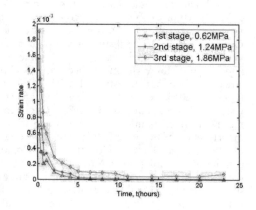

Figure 3.   Strain rate and fitting curve.

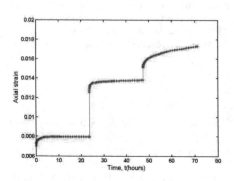

Figure 1.   Creep test curve of backfill.

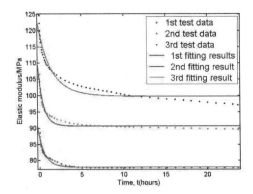

Figure 4. Test curve and fitting curve of elastic modulus.

Table 1. Logistic function parameters.

Stress/MPa	a/MPa	b/10^{-2}	c/t^{-1}	R^2
0.62	78	9.85	1.194	0.9645
1.24	90	12.18	1.431	0.9416
1.86	99.83	16.13	0.5418	0.9401

where t is the time (h), E (t) is the elastic modulus at any moment, and a, b, and c are constant, $0 \le b \le 1$.

When t = 0, $E(0) = a/(1 - b)$ can describe the instantaneous elastic deformation of backfill. When t → ∞, $E(\infty) = a$, long-term elastic modulus of backfill can be expressed. The parameters Logistic function and determination coefficient R^2 are shown in Table 1.

The fitting results show that Logistic function is better for describing the change trend of elastic modulus under low stress; under high stress, the fitting effect is poor; and the error increases with time. Finally, the fitting results completely deviate from the experimental data.

Variation law of elastic modulus of backfill with time (Fig. 4), the instantaneous elastic modulus, and the long-term elastic modulus of backfill increase with time. The instantaneous modulus and elastic modulus of backfill under the first-stage load are 88.6 MPa and 88 MPa, respectively. Under the third-stage load, they are 124.7 MPa and 99.83 MPa, with growth rates of 39.95% and 27.99%, respectively. It shows that backfill has obvious strain hardening tendency under the step loading. The specimens showed a trend of strain hardening.

At the beginning of the tests at each level, the elastic modulus is larger than that at other times. The elastic modulus of backfill decreases with the increase of time, but its change rate decreases gradually and tends to become constant. Under the first-stage loading, the elastic modulus of backfill gradually decreases and tends to become stable at 78 MPa. At the other two levels, the elastic modulus of backfill is kept a constant velocity reduction in a certain rate change, and the greater the stress, the faster the change rate. When the stress in 0.62 MPa, the internal pore and fracture of backfill decreased gradually and new fractures are not produced, so the elastic modulus of backfill is maintained at 78 MPa. As the stress continues to increase, the specimens continue to be compacted, and the new fracture begins to germinate and expand. Deformation of backfill is subjected to the combined action of hardening and damage, both of them gradually reaching a steady state. On the macroscopic level, a second creep stage appears and the elastic modulus of constant velocity decreases. It will lead to the creep damage of the specimen of backfill when the time is long enough.

The variation law of elastic modulus of backfill can be described by Logistic function:

$$E(t) = \frac{a}{1 - b \times \exp(-ct)} \qquad (1)$$

4 CREEP MODEL AND PARAMETER IDENTIFICATION

Creep behavior is mainly to study the strain and time of the material in the process of rheology. This paper focuses on the creep characteristics of backfill. Empirical formula and creep models are generally used to study the creep behavior of materials. The empirical method is used to carry out a series of experiments on the material, draw the regression curve, and fit the rheological formula. The advantage of this method is that the fitting effect is good, the shortcoming is only suitable for the test results, promotion to other tests, and poor adaptability. The model method is used to describe the creep behavior of the material under different loads by the different creep models, which are composed of three elements: elastic body, plastic body, and viscous body. These elements are formed by series connection and parallel connection. The creep model method has the advantages of clear physical significant, simple form, and so on, and is widely applied in the study of creep properties of materials. By the analysis of the test results, the creep deformation of backfill has an obvious decay creep and constant creep stage. And the Burgers model has the characteristics of instantaneous elastic deformation, decay creep, and constant creep, which can be used to describe the creep behavior of backfill. The Burgers model is composed of Maxwell model and Kelvin model in series (Fig. 5).

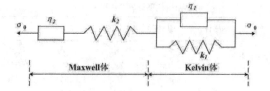

Figure 5. Burgers mechanics model.

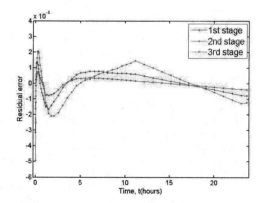

Table 2. Burgers model parameters.

Stress/ MPa	k_1/ MPa	k_2/ MPa	η_1/ MPa	η_2/ GPa	R^2
0.62	846	86.59	614.3	13.39	0.970
1.24	412.2	51.59	195	31.88	0.972
1.86	163	40.23	217.4	9.79	0.985

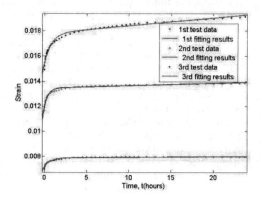

Figure 6. Test data and fitting curve.

Figure 7. Residual error of fitting.

Burgers model creep equation:

$$\varepsilon = \frac{\sigma_0}{k_2} + \frac{\sigma_0}{\eta_2}t + \frac{\sigma_0}{k_1}\left(1 - e^{-\frac{k_1}{\eta_1}t}\right)$$

(2)

In this paper, the MATLAB software is used for fitting the creep curve by using the custom equation fitting function, the fitting curve, and test curve (Fig. 6). The determination coefficient of fitting results at each level of Burgers model all reached 0.95 (Table 2). It can be seen that the fitting curves are in good agreement with the experimental curves, and the Burgers model can describe the creep characteristics of backfill under step loading.

Fitting results at each level show that the large residual error produced in creep deformation is less than 4 h (Fig. 7). This stage corresponds to the decay creep stage. The largest error at $t = 0$ h is the instantaneous deformation of backfill. Thus, Burgers model cannot describe the instantaneous deformation of backfill very well, and leads to a large residual error of fitting at a short period of time.

When $t\rightarrow0+$, the backfill strain rate is infinite, and Burgers model in the vicinity of the zero moment fitting effect is poor. In general, the fitting accuracy of the Burgers model at each level is quite high. Over time, each level of error will increase slightly, and this difference increased

significantly under high stress. The reason is that the creep equation of the Burgers model with $\varepsilon = \sigma_0/k_2 + \sigma_0 \times t/\eta_2$ to the asymptote, the slope of the creep curve has been small in theory. The test curve of step loading gradually decreases to a stable value or slightly increased under the long-term load, and this difference leads to the difference between the fitting curve and the test curve, which increases with time.

5 CONCLUSION

The filling material has significant creep characteristics, under the condition of step loading. In the first stage of stress (0.62 MPa), creep strain accounts for 12.06% of the total strain, when the stress is increased to 1.86 MPa, the proportion is 22.10%. The creep deformation of backfill is obviously influenced by the stress.

The elastic modulus of backfill increases with time. And the elastic modulus of backfill under the first-stage load is 88.6 MPa and that under the third-stage load is 124.7 MPa, with a growth of 39.95%. It shows that backfill has obvious trend of strain hardening under step loading; the greater the stress, the faster the elastic modulus of backfill decays with time. The interaction between the pressure of the primary fracture of backfill and the germination and expansion of the fracture is the cause of these changes.

The Burgers model was used for fitting the test curve based on the analysis of the creep characteristics of backfill. The model can describe the creep characteristics of backfill under step loading very well, and its physical significance is clear. The difference between the fitting results and the test data was analyzed. The residual error is mainly concentrated in the decay creep stage. Enhancing the fitting precision of the curve of the decay creep stage through the improvement of the creep model theory is the next research direction.

ACKNOWLEDGMENTS

The authors are grateful for the financial support from Hebei Province (Grant No: E2015209172).

REFERENCES

Benzaazoua, M., B. Bussière, I. Demers, M. Aubertin, É. Fried, A. Blier, Integrated mine tailings management by combining environmental desulphurization and cemented paste backfill: Application to mine Doyon, Quebec, Canada. J. Miner. Eng. **21**, 330-340.(2008)

Bin, H., W, AIxiang W, Yiming, Optimization and application of cemented hydraulic fill(CHF) with low strength aggregate and extra fine grain full tailings[J]. J. Cent. South U. (Science and Technology), **43**, 2357-2362(2012).

Fall, M., D. Adrien, J.C. Célestin, M. Pokharel, M. Touré, Saturated hydraulic conductivity of cemented paste backfill. J. Miner. Eng. **22**, 1307-1317(2009).

Griggs, D., Creep of rock. J. J. Geo. **47**, 225-251(1939).

Hongwei, Y., X. Jiang, N. Wen, 2015. Experimental study on creep of rocks under step loading of seepage pressure. J. Chinese Journal of Geotechnical Eng. **37**, 1613-1619(2015).

Huabin, Z., W. Zhiyin, Z. Yanjie, The whole process experiment of salt rocks creep and identification of model parameters. J. Acta Petrol. Sin. **33**, 904-908(2012).

Junbao, W., L. Xinrong, G. Jianqiang, Creep properties of salt rock and its nonlinear constitutive model. J. J. China Coal Soc. **39**, 445-451(2014).

Liang, C., L. Jianfeng, W. Chunping, Creeping behavior of beishan granite under different temperatures and stress conditions. J. Chin. J. Geo. Eng. **34**, 1228-1235(2015).

Nasir, O., M. Fall, Coupling binder hydration, temperature and compressive strength development of underground cemented paste backfill at early ages. J. Tunn. Undergr. Sp. Tech. **25**, 9-20(2010).

Qiusong, C., Z, Qinli, W, Xinming, Experimental study on effect of magnetized water on flocculating sedimentation of unclassified tailings. J. J. Cent. South U. (Science and Technology), 4256-4261(2015).

Sone, H., M. D. Zoback, Time-dependent deformation of shale gas reservoir rocks and its long-term effect on the in situ state of stress. J. Int. J. Rock Mech. Min. (69):120-132(2014).

Xianwei, Z., W. Changming, Z. Shuhua, Comparative analysis of soft clay creep data processing method. J. J. JiLin U. (Earth Science Edition), **40**, 1401-1408(2010).

Yi, X. W., G. W. Ma, A. Fourie, Compressive behavior of fibre-reinforced cemented paste backfill. J. Geotext. Geomembranes.**43**, 207-215(2015).

Advances in Energy Science and Equipment Engineering II – Zhou, Patty & Chen (Eds)
© *2017 Taylor & Francis Group, London, ISBN 978-1-138-71798-5*

Influence of the filling material and structure of railway subgrade on its dynamic responses under simulated train loading

Junhua Xiao

Key Laboratory of Road and Traffic Engineering of the Ministry of Education, Tongji University, Shanghai, China

ABSTRACT: In order to explore the influences of the filling material and structure of railway subgrade on its dynamic responses (i.e., dynamic elastic deformation, accumulative plastic deformation, and dynamic stress attenuation) under cyclic train loading, field tests of dynamic responses of railway subgrade under simulated train loading were conducted in four subgrade sections with different filling materials and structures. The total thickness of each subgrade bed was constant, that is, 3.0 m. The thicknesses of the surface layers of subgrade bed, which were filled with well-graded gravel, were 0.7, 0.5, 0.65 (plus a 0.05 m layer of bituminous concrete on the top), and 0.9 m, respectively, for sections A, B, C, and D; and the bottom layers of subgrade bed for A and B were filled with weathered granite, but for C and D, Xiashu clay were used. Test results indicate that, for each section, subgrade elastic deformation and ultimate accumulative deformation were smaller than 1 and 5 mm, respectively, representing the high quality of railway subgrade. Furthermore, section A achieving the smallest elastic deformation and the largest dynamic stiffness was recommended for railway subgrade structure. In addition, a 0.05 m layer of bituminous concrete on the subgrade top was also recommended.

1 INTRODUCTION

A stable subgrade provides a solid foundation for railway track (Selig, 1994; Li, 2015). In 1960, because of the problems in subgrade design, filling materials, and construction methods, the subgrade of Shinkansen, Japan, encountered excess settlement, which led to an interruption of railway operation, and it took more than 2 years to repair the subgrade distress (WJRC, 2002). Countries that independently construct high-speed railway, such as Japan, France, Germany, and China, have spent a long time in the research and development of railway subgrade. However, significant differences in the design of subgrade and the use of filling material exist among countries (WJRC, 2002; National Railway Administration of PRC, 2014; German Railway Standard, 2008; UIC, 1994; Hu, 2010). Burrow et al. (2007) found that the thickness of calculated trackbed layers varied significantly by using the design approaches of railway track foundation in the United States, the United Kingdom, Europe, and Japan (Burrow, 2007). Therefore, they considered that it was necessary to have a thorough knowledge of the methodologies employed together with their inherent assumptions.

The upper part of railway subgrade is the most susceptible part to cyclic train loading and environmental variations, and its design is the key for track foundation design. Designations for this part vary among countries. For simplicity, this paper addresses it as SUBGRADE BED in accordance to Chinese Code for Design of High Speed Railway (TB10621-2014) (National Railway Administration of PRC, 2014). In general, the subgrade bed should be strengthened in railway subgrade design, to provide a solid foundation for track with enough strength, high stiffness, and long-term stability. However, the strengthening methods in countries are different in structures and filling materials (National Railway Administration of PR, 2014). Therefore, it is meaningful to study the behaviors of different subgrade beds under train loading and to make an optimized design in the future.

This paper introduced a group of full-scale field tests of the dynamic responses of subgrade with different thickness of the surface layers of subgrade bed, as well as different filling materials of the bottom layers of subgrade bed under simulated cyclic train loading. Consequently, the effects of the thickness and stiffness (comparing different filling materials) of both the surface layer and the bottom layer of subgrade bed on dynamic responses of railway subgrade were investigated, and the appropriate subgrade structure and filling material were suggested.

2 SUBGRADE DESIGN FOR CYCLIC LOADING TESTS

Four different subgrade sections marked as A, B, C, and D were built for cyclic loading tests. For each subgrade, the cross section was trapezoidal, as shown in Figure 1. The width of the subgrade surface was 13.8 m, the slope rate was 1:1.5, and the height was 4.5 m, of which the thickness of the subgrade bed (including the surface layer and the bottom layer) was 3.0 m and the rest 1.5 m was embankment fill.

The surface layer of the subgrade bed was made of well-graded gravels (National Railway Administration of PRC, 2014; Ministry of Railways of PRC, 2005), with thicknesses of 0.7, 0.5, 0.65, and 0.9 m for sections A, B, C, and D, respectively, and the top of the surface layer of C was covered by a 0.05 m layer of bituminous concrete (WJRC, 2002; National Railway Administration of PRC, 2014; Ministry of Railways of PRC, 2005). The bottom layers of the subgrade bed of A and B were filled with weathered granite, but for C and D, Xiashu clay was used. Below the subgrade bed, the embankments were filled with Xiashu clay. The relative compactions of the filling materials for the bottom layer of the subgrade bed and embankment were 95% and 90%, respectively (National Railway Administration of PRC, 2014). The natural subsoil below the subgrade was improved by vacuum preloading with prefabricated vertical drain method. The cyclic loading test was carried out after subgrade settlement caused by self-weight finished. The specified design of the tested subgrade is shown in Table 1.

In regard to the properties of the two filling materials of weathered granite and Xiashu clay, the former had the main mineral compositions of quartz and feldspar, with a coefficient of curvature of 1.08, a coefficient of nonuniformity of 14.71, a dry density of 2.15 kg/m³, and an optimum moisture content of 7.2%. For the latter, all particles passed No. 200 (0.075 mm) US standard sieve. The soil had a liquid limit of approximately 38.4, a plasticity index of approximately 21.2, a dry density of 1.80 kg/m³, and an optimum moisture content of 15.9%. In addition, the average unconfined compressive strength and compression modulus of undisturbed Xiashu clay sample were 776 kPa and 47.28 MPa, respectively.

3 TEST PROCESSES

The dynamic stress pulse on subgrade surface induced by moving train can be simulated by haversine pulse, and two adjacent axles under the same bogie generate approximately a single stress pulse (Liu, 2010). In this testing, dynamic stress on the subgrade induced by a passenger train with the axle load of 16 t running at 200 km/h was simulated. The stress amplitude was calculated as 67 kPa using the formula: $\sigma = 2.6P (1 + \alpha v)$ [10], where P is the static axle load, α is a velocity factor (0.003), and v is the train speed. In addition, loading frequency of the stress pulse ranged from 3 to 14 Hz, determined by the distance of two bogies, vehicle length, and train speed (Liu, 2010).

As shown in Figure 2, the in situ excitation system was a vibrator seated on a concrete foundation. The concrete foundation, which was 11.5 t in weight and acted as a counterweight in tests, was 2.6 m long, 1.2 m wide, and 1.5 m high. The vibrator generated vibrations by rotating a centrifugal mass block. The self-weight of the vibrator was 4 t and the excitation forces could be set from 0 to 300 kN, with a maximum frequency of 30 Hz. In this testing, the frequency of cyclic loading was selected as 10–13 Hz.

Figure 1.   Subgrade cross section and sensors layout.

Table 1.   Design of the tested subgrade.

Section	Range	Length (m)	Thickness of the surface layer of subgrade bed (m)	Thickness of the bottom layer of subgrade bed (m)	Filling material of the bottom layer of subgrade bed	Notes
A	k0 + 40 ~ k0 + 45	5	0.7	2.3	Weathered granite	
B	k0 + 45 ~ k0 + 50	5	0.5	2.5	Weathered granite	
C	k0 + 50 ~ k0 + 55	5	0.65	2.3	Xiashu clay	With additional 0.05 m layer of bituminous concrete on the top
D	k0 + 55 ~ k0 + 60	5	0.9	2.1	Xiashu clay	

Figure 2. In situ excitation system on subgrade surface.

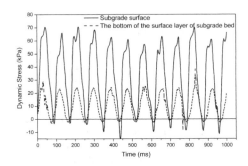

Figure 3. Typical dynamic stress waveform in different subgrade depths (section A).

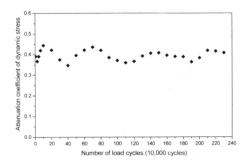

Figure 4. Typical attenuation coefficient of dynamic stress for the surface layer of the subgrade bed (section A).

Table 2. Average attenuation coefficient of dynamic stress.

A	B	C	D	Notes
0.393	0.492	0.352	0.341	The attenuation coefficient of dynamic stress for a 0.05 m layer of bituminous concrete was 0.85

The sensors layout in the subgrade is depicted in Figure 1. The dynamic elastic deformation sensors were directly set on the top of concrete foundation as the concrete foundation always kept close contact with subgrade surface during tests. The data of subgrade dynamic stress, dynamic elastic deformation, and accumulative plastic deformation were collected every 100,000 times of load cycles.

4 TESTS RESULTS AND ANALYSIS

4.1 Attenuation of dynamic stress in subgrade

The dynamic stress applied on subgrade by moving train is one of the major factors affecting subgrade stability and permanent deformation. To ensure long-term stability of subgrade, subgrade dynamic stress should not exceed the dynamic strength of filling materials, and the surface layer of the subgrade bed should significantly attenuate the dynamic stress on the subgrade surface in order to reduce the dynamic stress transmitted to underlying weak subgrade soil (National Railway Administration of PRC, 2014). Figure 3 shows the typical dynamic stress waveform in different subgrade depths for section A under cyclic loading; Figure 4 shows the attenuation coefficient of dynamic stress for the surface layer of subgrade bed (i.e., the ratio of dynamic stress on the bottom of the surface layer of subgrade bed and on subgrade surface). It can be seen that the peak dynamic stress on the subgrade surface during tests was approximately equal to the set value of 67 kPa, and the average attenuation coefficient of dynamic stress was 0.393.

Table 2 shows the average attenuation coefficients of dynamic stress for the surface layer of subgrade bed for four different sections.

It can be seen from Table 2 that the average attenuation coefficient of dynamic stress for the surface layer of subgrade bed ranged from 0.341 to 0.492. The attenuation coefficient was influenced by both the thickness and stiffness of the surface layer of the subgrade bed. Comparing section A with section B, the surface layer of the subgrade bed of section A was 0.2 m thicker than section B, and consequently section A attenuated an additional 10% of dynamic stress more than section B. This represents that a thicker surface layer of subgrade bed (filled with well-graded gravel) increased the attenuation effect of dynamic stress. Similarly, section C, with a 0.05 m layer of bituminous concrete on the top of subgrade and a thinner surface layer of subgrade bed than section D, nearly matched section D in dynamic stress attenuation. This indicates that the attenuation effect also increased with increasing subgrade bed stiffness (i.e., bituminous concrete), so the 0.05 m layer of bituminous concrete played a conspicuous role in dynamic stress attenuation in tests.

1013

4.2 Subgrade elastic deformation and dynamic stiffness under cyclic loading

Subgrade elastic deformation is triggered by instantaneous dynamic load of moving train. The value of subgrade elastic deformation directly affects the stabilities of subgrade and track bed, and subgrade elastic deformation reflects in track deformation, which affects safe running and comfortably of high-speed train. Figure 5 gives an example of the variation of subgrade elastic deformation with load cycles in section A, indicating the average value of 0.32 mm, which was much smaller than the generally accepted upper limit of subgrade elastic deformation for high-speed railway, that is, 1 mm (National Railway Administration of PRC, 2014).

Figure 6 shows the typical variation of subgrade dynamic stiffness with load cycles. When subgrade reached a steady forced vibration state under cyclic loading, its dynamic stiffness can be directly obtained from the dynamic stress and elastic deformation, if the influence of damping is neglected. In Figure 6, the average subgrade dynamic stiffness of Section A was 201.4 MPa/m.

Table 3 shows the elastic deformation on subgrade surface and subgrade dynamic stiffness for different sections in tests.

Table 3. Subgrade elastic deformation and dynamic stiffness.

	A	B	C	D
Elastic deformation on subgrade surface (mm)	0.320	0.412	0.370	0.430
Subgrade dynamic stiffness (MPa/m)	201.4	151.4	170.1	148.5

It can be seen from Table 3 that, for all sections in tests, the elastic deformation on subgrade surface ranged from 0.32 to 0.43 mm, that is, all smaller than 1 mm; and the dynamic stiffness of subgrade ranged from 148.5 to 201.4 MPa/m. Subgrade elastic deformation showed a negative trend with the thickness and stiffness (i.e., different filling materials) of the surface layer of subgrade bed as well as the stiffness of the bottom layer of subgrade bed. In this testing, section A achieved the smallest elastic deformation and the largest dynamic stiffness. Although the thickness or stiffness of the surface layers of subgrade bed of sections C and D were larger than that of section A, the dynamic stiffness of the two sections were smaller than that of A, because the bottom layers of subgrade bed of sections C and D were filled by weaker material (i.e., Xiashu Clay). Consequently, subgrade dynamic stiffness of sections C and D were approximately the same as that of section B, while section B had the thinnest surface layer of subgrade bed but had a stiffer bottom layer of subgrade bed (i.e., weathered granite). These results indicate that an optimized subgrade dynamic stiffness requires comprehensive matching of the thickness and stiffness (i.e., different filling materials) of the layer structure of subgrade. In addition, a higher dynamic stiffness of section C than section D shows that the 0.05 m layer of bituminous concrete covered on subgrade surface significantly increased the subgrade dynamic stiffness.

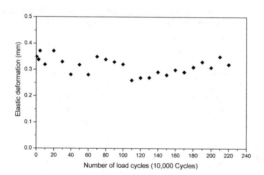

Figure 5. Typical subgrade elastic deformation during cyclic loading tests (section A).

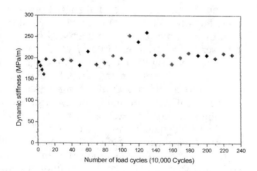

Figure 6. Typical subgrade dynamic stiffness during cyclic loading tests (section A).

4.3 Subgrade accumulative plastic deformation

The development of accumulative plastic deformation (average values of the left and right sensors) on subgrade surface with load cycles is shown in Figure 7.

It can be seen from Figure 7 that the accumulative plastic deformation developed fast in the initial stage of load cycles and became stable gradually. The accumulative plastic deformation became stable after approximately 1.8, 1.8, 1.0, and 2.5 million load cycles for sections A, B, C, and D, respectively. Table 4 lists the accumulative deformations on subgrade surface observed for four sections.

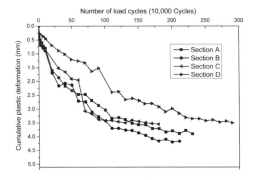

Figure 7. Subgrade accumulative plastic deformation during cyclic loading tests.

Overall, the ultimate accumulative deformations were smaller than 5 mm, meeting the requirement that subgrade permanent deformation due to train traffic load should not exceed 5 mm (German Railway Standard, 2008). In addition, the accumulative deformations of different subgrade bed structures differed slightly, representing the appropriate abilities of resistance to permanent deformation for both weathered granite and Xiashu clay as the filling materials for the bottom layer of subgrade bed. A comparison of Tables 2 and 4 indicates that the accumulative plastic deformation is related to the attenuation effect of dynamic stress in the subgrade. The higher the attenuation of dynamic stress in the subgrade, the smaller the accumulative deformation of the subgrade.

5 CONCLUSIONS

In situ cyclic train loading simulation tests on railway subgrade were conducted in four subgrade sections to explore the effects of the filling material and structure of railway subgrade on its dynamic responses. For each section, subgrade elastic and accumulative deformation were smaller than 1 and 5 mm, respectively, representing the high quality of railway subgrade. However, section A achieved the smallest subgrade elastic deformation and the largest dynamic stiffness, so this type of structure and filling material for subgrade bed is recommended. In addition, the following conclusions were drawn:

1. Both weathered granite and Xiashu clay had appropriate abilities of resistance to permanent deformation when filling the bottom layer of subgrade bed, but subgrade dynamic stiffness and elastic deformation were significantly affected by the stiffness (i.e., comparing weathered granite with Xiashu clay) of the bottom layer of subgrade bed.
2. A 0.05 m layer of bituminous concrete on subgrade top significantly increased the subgrade dynamic stiffness and promoted the attenuation of dynamic stress in subgrade. Therefore, it is also recommended in future.
3. The increase in the thickness or stiffness of the surface layer of subgrade bed promoted the attenuation of dynamic stress in the subgrade. The higher the attenuation of dynamic stress in the subgrade, the smaller the accumulative deformation of subgrade.

However, the effects of the variations in environmental conditions (e.g., variations of water content, drying, and watering cycle) on dynamic responses of railway subgrade were not considered in this paper. This part could be concerned as contents for future studies.

REFERENCES

Burrow, M.P.N., D. Bowness, G.S. Ghatatora, Proc. IMechE, J. Rail and Rapid Transit **221**, 1 (2007).
German Railway Standard, *Erdbauwerke planen, bauen und instand halten* (2008).
Hu, Y.F., N.F. Li, *Theory of ballastless track-subgrade for High speed railway* (2010).
Li, D.Q., J. Hyslip, T. Sussmann, etc., *Railway Geotechnics* (2015).
Liu, J.K., J.H. Xiao, Journal of Geotechnical and Geoenvironmental Engineering 136, 833 (2010).
Ministry of Railways of PRC. *Interim design provisions for newly-built passenger railway line at speed of 200 to 250km/h* (2005).
National Railway Administration of PRC, *Code for Design of High speed Railway (English Edition)* (TB 10621–2014) (2014).
Selig, E.T., J.W. Waters, Track Geotechnology and Substructure Management (1994).
UIC, *Earthworks and Track-Bed Layers for Railway Lines* (UIC Code 719 R) (1994).
WJRC, *Construction and maintenance standards for Shinkansen track* (2002).

Advances in Energy Science and Equipment Engineering II – Zhou, Patty & Chen (Eds)
© *2017 Taylor & Francis Group, London, ISBN 978-1-138-71798-5*

The experimental verification showing principal stress is not the extreme stress in elastic theory

Jianqing Zhou, Xiaoming Li, Shuanghua Huang, Wenba Han & Bingqing Cai
Panzhihua University, Panzhihua, China

ABSTRACT: This study discovered a phenomenon that the mass point in unit cell cannot maintain balance in current elastic theory. Under different stress states, the absolute values of all equilibrium stress on the mass point are greater than the absolute values of principal stress. Thus, based on the new concept of point stress balance, this research study introduced a new formula of stretch-shear combined deformation. The new formula can explain the issue that in the state of stretch-shear, the constructions are destroyed much easier than the state of compress-shear. Besides that based on the new concept of point stress balance, this study also establishes a new theory of strength that is much more accurate than the third and fourth strength theories, which has also been validated in the Damage Mechanics National Key Laboratory of Tsinghua University. On comparing experimental data, the errors that are calculated from the new theory are only 1%, while calculated based on the third and fourth strength theories, they are 14.2% and 18.2%, respectively. According to balanced biaxial tension of Q235, the errors of the new theory are only 2.3% while those of the first, second and third strength theories are all 31%. Thus, this study confirms the validity of the new theory.

1 THE CONTRADICTION THAT A PARTICLE ON THE STRETCHING OBLIQUE SECTION OF A SINGLE-PULLING POLE CANNOT KEEP BALANCE

1.1 *Any point on the stretching oblique section of a single-pulling pole is unbalanced*

A single-pulling pole is shown in Figure 1(a) (Liu, 2000). As we all know, any particle in a pole is subject to equal but opposite tensile stress and in equilibrium. However, the current theory of elasticity cannot justify this assumption.

Considering that the axial tension is F and the cross-sectional area is A, we can derive the normal stress σ_α and the shearing stress τ_α on the slope k-k (an angle α with the vertical direction) with the current theory of elasticity as follows (Zhao, 2002).

$$\sigma_0 = \frac{F}{A} \quad (1)$$

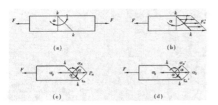

Figure 1.

Defining the area of the cross-section of the slope k-k as A_α, we get

$$A_\alpha = \frac{A}{\cos\alpha} \quad (2)$$

If F_α represents the horizontal internal forces of the slope k-k, to keep the pole in balance, we assume

$$F = F_\alpha$$

As the internal forces are distributed evenly, the stress P_α of the slope k-k is

$$P_\alpha = \frac{F_\alpha}{A_\alpha} = \frac{F}{A_\alpha}$$

Substituting Eq. (2) in the above equation, we get[2.3]

$$P_\alpha = \frac{F}{A}\cos\alpha = \sigma_0 \cos\alpha \quad (3)$$

P_α can be resolved into two parts: the normal stress σ_α, which is perpendicular to the slope, and the shear stress τ_α, which is tangential to the slope (see Figure 1(c)). Then

$$\sigma_\alpha = P_\alpha \cos\alpha = \sigma_0 \cos^2\alpha \quad (4)$$

$$\tau_a = P_a \sin\alpha = \sigma_0 \cos\alpha\sin\alpha = \frac{\sigma_0}{2}\sin 2\alpha \qquad (5)$$

Note: Formulas (4) and (5) are to maintain a balance between the normal stress and the shear stress of the left side of the pole. Nevertheless, it may not be appropriate for either particles of the slope k-k. This can be easily proved. Take point a in Figure 1(c), which is subjected to three forces, σ_0, σ_a and τ_a. The composition stress of σ_a and τ_a is P_a.

$$P_a = \sqrt{\sigma_a^2 + \tau_a^2} = \sqrt{(\sigma_0\cos^2\alpha)^2 + (\sigma_0\cos\alpha\sin\alpha)^2}$$
$$= \sigma_0\cos\alpha$$

It is obvious that the stress of two sides is not in equilibrium, that is

$$\sigma_0 \neq \sigma_0\cos\alpha$$

The above formulas indicate that particle a is in an unbalanced state, and as we all know, it is not realistic.

1.2 The normal stress and shear stress of the particles on the slope to maintain balance

To ensure that any particle on the slope k-k is in balance, $\sigma_{0,left}$ must be equal to $\sigma_{0,right}$, as shown in Figure 1 (d). It is obvious that any particle in a pole is subject to equal but opposite tensile stress and in equilibrium. Particle a should always be in balance and this conclusion has no relationship with the distribution of the slope k-k. To keep the particles on the slope k-k in balance, we can use the force-analytic method[4,5]. That is,

$$\sigma'_a = \sigma_0\cos\alpha \qquad (6)$$

$$\tau'_a = \sigma_0\sin\alpha \qquad (7)$$

Only formulas (6) and (7) can meet the condition that any particle in the pole is in balance. Because,

$$\sigma_{0,right} = \sqrt{(\sigma'_a)^2 + (\tau'_a)^2}$$
$$= \sqrt{(\sigma_0\cos\alpha)^2 + (\sigma_0\sin\alpha)^2} = \sigma_{0,left}$$

And particle a is in equal and opposite stress and in balance. Thus, formulas (6) and (7) are the ones to be used.

1.3 The conclusion of the unit body balance and particle balance

① There are fundamental differences between the balance of the body unit and the particle. When analyzing the balance of the body unit, the area of stress is inevitably taken into consideration, no matter how tiny the body unit is. According to the definition of a particle, that is a particle is of no size, there is no need to consider about the area.

② When using micro-body balance to get the stress on the oblique section, even if the micro-body tends to be infinitesimal, its stress is not equal to the particle balance stress of its neighbourhood.

③ In the elastic theory, micro-regular hexahedron is used to calculate stress. And, the stress is calculated under the unit body balance instead of the particle balance.

④ The stress of each unit body surface is in the non-concurrent force system. As the unit body approaches to the particle, the stress tends to be in the concurrent force system, and on this occasion, the analytical force method can be used.

2 THE PARTICLE BALANCE STRESS IN THE STATE OF TWO-ORIENTED TENSILE AND SHEAR STRESS

Figure 2(a) shows the state of two-oriented tensile and shear stress. The stress of particle a is shown in Figure 2(b), and its resultant stress is

$$\sum x = \sigma_x + \tau$$
$$\sum y = \sigma_y + \tau$$

Then, the particle balance stress is given by

$$\sigma'_a = \sqrt{\left(\sum x\right)^2 + \left(\sum y\right)^2} = \sqrt{(\sigma_x + \tau)^2 + (\sigma_y + \tau)^2} \qquad (8)$$

Formula (8) is the particle balance stress in the state of two-oriented stress.
When $\sigma_x = \sigma_y = \sigma$

$$\sigma'_a = \sqrt{2}(\sigma + \tau) \qquad (9)$$

And when $\sigma_y = 0$, from formula (8),

$$\sigma'_a = \sqrt{(\sigma_x + \tau)^2 + \tau^2} = \sqrt{\sigma_x^2 + 2\sigma_x\tau + 2\tau^2} \qquad (10)$$

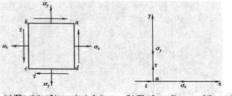

(a) The state of two-oriented stress (b) The force diagram of the particle a

Figure 2.

1018

Formula (10) is the balance stress under combination of tensile and shear stress.

The condition is established by using particle balance stress.

$$\sqrt{\sigma_x^2 + 2\sigma_x\tau + 2\tau^2} \leq [\sigma] \tag{11}$$

The new intensity formula (11) is different from the third and fourth strength theories for the unit body balance[6,7].

$$\sqrt{\sigma^2 + 4\tau^2} \leq [\sigma] \tag{12}$$

$$\sqrt{\sigma^2 + 3\tau^2} \leq [\sigma] \tag{13}$$

In the formula, $[\sigma]$ represents material allowable tensile stress.

3 THE PARTICLE BALANCE STRESS IN THE STATE OF THREE-ORIENTED STRESS

Figure 3 shows an equilateral triangle cone cut from a micro-cube. When the equilateral triangle cone is in balance, the stress component on the inclined section (ABC) in x, y, and z axes can be deduced. Set the outward normal direction of inclined section to be n, and the direction cosine is

$$\cos(n, x) = l$$

$$\cos(n, y) = m$$

$$\cos(n, z) = n$$

The projection of the total stress σ_α on the inclined section (ABC) in the x, y, and z axes is

$$X_n = -\left(\tau_{yx}m + \tau_{zx}n + \sigma_x l\right) \tag{14a}$$

$$Y_n = -\left(\tau_{xy}l + \tau_{zy}n + \sigma_y m\right) \tag{14b}$$

$$Z_n = -\left(\tau_{zx}l + \tau_{zyz}m + \sigma_z n\right) \tag{14c}$$

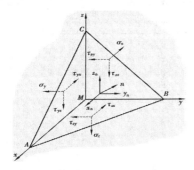

Figure 3.   The state of three-oriented stress.

Then, the total stress σ_α on the inclined section (ABC) is

$$\sigma_\alpha = \sqrt{X_n^2 + Y_n^2 + Z_n^2}$$
$$= \sqrt{\begin{array}{c}(\tau_{yx}m + \tau_{zx}n + \sigma_x l)^2 + (\tau_{xy}l + \tau_{zy}n + \sigma_y m)^2 \\ + (\tau_{zx}l + \tau_{zyz}m + \sigma_z n)^2\end{array}}$$

$$\tag{15}$$

From the new theory, σ_α is the stress on the slope to maintain the balance of the triangular pyramid instead of the particles on the slope (ABC). When this micro-triangular pyramid tends to be infinitely small, the slope (ABC) tends to point M. In this occasion, the stress of the M point is the balance stress for every point.

Obviously, the projection of the total stress for the point M in the x, y, and z axes is

$$X_M = -\left(\tau_{yx} + \tau_{zx} + \sigma_x\right) \tag{16a}$$

$$X_M = -\left(\tau_{yx} + \tau_{zx} + \sigma_x\right) \tag{16b}$$

$$X_M = -\left(\tau_{yx} + \tau_{zx} + \sigma_x\right) \tag{16c}$$

The total stress σ^* for point M is

$$\sigma^* = \sqrt{X_M^2 + Y_M^2 + Z_M^2}$$
$$= \sqrt{\begin{array}{c}(\tau_{yx} + \tau_{zx} + \sigma_x)^2 + (\tau_{xy} + \tau_{zy} + \sigma_y)^2 \\ + (\tau_{zx} + \tau_{zyz} + \sigma_z)^2\end{array}} \tag{17}$$

By comparing formula (15) with formula (17), the stress on the the the inclined section to keep the balance of the truncated body is related to the direction cosine of the inclined section. The particle balance stress is only associated with the size of stress, and has nothing to do with the direction cosine. Since the direction cosine is less than or equal to 1, the stress on the inclined section is always less than the particle balance stress.

If the shear stress is zero, which means $\tau_{yx} = \tau_{zx} = \tau_{xy} = \tau_{zy} = \tau_{yz} = \tau_{xz} = 0$, formula (16) can be changed into

$$X_M = -\sigma_x$$

$$Y_M = -\sigma_y$$

$$Z_M = -\sigma_z$$

The total stress becomes

$$\sigma^* = \sqrt{\sigma_x^2 + \sigma_y^2 + \sigma_z^2} \tag{18}$$

Formula (18) is the particle balance stress only with normal stress in the state of three-oriented stress.

4 THE EXPERIMENTAL VERIFICATION OF PRINCIPAL STRESS BASED ON THE THIRD AND THE FOURTH STRENGTH THEORY IS LESS THAN THE BALANCE STRESS OF THE PARTICLE

We can see the force direction that acts on the body in Figure 4(a). Figure 4(b) shows the force state of the shear surface. Experimental aircraft ensures the position of the cutter at the center. Using the centerline to divide the area into two parts, the force state of the inner point is shown as Fig. 4(c). Strict sense, there exists smaller normal stress in the y direction, that is

$$\sigma_y = \frac{Q_y}{A} = \frac{4 \times 10^3}{340 \times 10^{-3} \times 9 \times 10^{-3}} = 1.3 \text{ (MPa)}$$

We can see that the value is too small, and hence it can be ignored.

At the same time, the stress concentration is restrained due to the dull cutting blade (because the stress concentration has no influence on the comparing result of the third strength theory) experimental data and the corresponding stress.

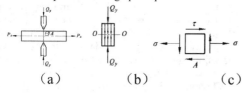

Experiment design principle

(a)　　　　(b)　　　　(c)

Figure 4.　Uniaxial tensile and double shear.

Table 1.　The test specimen chart of uniaxial tension and shear.

Item Number	Pulling force Px/kN	Tension area A/m²	Tensile stress $\sigma_x = \frac{P_x}{A}$ /MPa	Shear force Qy/kN	Shear force area $A_r = \frac{A}{2}$ m²	Shear stress τ_y/MPa $\tau_y = \frac{Q_y}{A_r}$	Tensile strength limit σ_b/MPa	Note
1	14.1	531 × 10⁻⁶	26.6	2	265.5	7.5	43	Data of Pulling force Px and shear force Qy: Data of number 1 and number 2 were from testing and automatic record. Data of number 3 and number 4 were handwritten test record. (Because of fault recorder.)
2	15.2	531 × 10⁻⁶	28.6	3.2	265.5	12	43	
3	17.2	531 × 10⁻⁶	32.4	2.5	265.5	9.4	43	
4	16.0	531 × 10⁻⁶	30.1	4	265.5	15	43	

Table 2.　The comparison of the fracture table of the maximum principal stress, the third strength theory and fourth strength theory, and the calculation of the fracture stress under condition of a particle balance stress.

Item Number	Test tensile stress σ_x/MPa	Test shear stress τ_y/MPa	The maximum principal stress $\sigma_{max} = \frac{\sigma}{2} + \sqrt{\left(\frac{\sigma}{2}\right)^2 + \tau^2}$ /MPa	The third strength theory $\sigma_{r3} = \sqrt{\sigma^2 + 4\tau^2}$ /MPa	The fourth strength theory $\sigma_{r4} = \sqrt{\sigma^2 + 3\tau^2}$ /MPa	Balance of a particle stress strength theory $\sigma'_{\sigma r} = \sqrt{\sigma^2 + 2\sigma\tau + 2\tau^2}$ /MPa	The material tensile strength limit σ_b /MPa
1	26.6	7.5	33.55	30.6	29.6	40.2	43
2	28.6	12	32.97	37	35	42.1	43
3	32.4	9.4	34.92	37.5	36.3	42.6	43
4	30.1	15	36.3	42.5	39.8	45.1	43
Average value	–	–	34.44	36.9	35.2	42.5	43
Percentage error of the fracture stress and ultimate strength	–	–	19%	14.2%	18.2%	1%	–

In order to verify the correctness of the new point balance theory, a series of tensile–shear experiments were conducted in the Damage Mechanics State Key Laboratory of Tsinghua University. The ultimate strength of PVC is $\sigma_b = 43$ MPa.

The experimental results are shown as Table 1.

The results based on the third and fourth strength theories and the new point balance theory are shown as Table 2.

5 CONCLUSION

The error between new stretch-shear theory and experiments was only 1%, while the errors between third and forth strength theories were 14.2% and 18.2%, respectively. The discrepancy value between maximum principal stress σ_{max} and σ_b is 19%, which declares that it is dangerous to use the classical strength theory to design engineering. Besides that, it also validates the correctness and accuracy of the new stretch-shear theory as well as the point balance theory.

REFERENCES

Huang yan. *Engineering Elasticity*. Beijing: Tsinghua University Press, 1982.

James M. Gere *Mechanics of Materials*. Beijing: China Machine Press, 2002.

Liu hongwen. *Mechanics of Materials*. Beijing: Higher Education Press, 2000.

Pobertl. Mott. *Applied Strength of Materials*. Chongqing: Chongqing University Press, 2005.

Qian weichang. Ye kaiyuan. *Elastic Mechanics*. Beijing: Science Press, 1956.

Shan huizu. *Mechanics of Materials*. Beijing: Higher Education Press, 2005.

Zhao jiujiang. Zhang shaoshi. Wang chunxiang. *Mechanics of Materials*. Harbin: Harbin Institute of Technology Press, 2002.

Advances in Energy Science and Equipment Engineering II – Zhou, Patty & Chen (Eds)
© *2017 Taylor & Francis Group, London, ISBN 978-1-138-71798-5*

Review on the bearing capacity and seismic analysis of pre-stressed high-strength concrete piles

Hang Chen
Guangzhou Institute of Building Science, Guangzhou, China
School of Civil Engineering, Guangzhou University, Guangzhou, China

Hesong Hu
Guangzhou Institute of Building Science, Guangzhou, China

Tong Qiu
Department of Civil and Environmental Engineering, Pennsylvania State University, University Park, PA, USA

ABSTRACT: Pre-stressed High-strength Concrete (PHC) piles have been used in a wide variety of buildings due to their economic advantages, controllable quality, and relatively low-carbon footprint. In this paper, the main methods to calculate their bearing capacity are reviewed. It was found that almost all of these methods are based on pile strength or empirical relationships. The commonly used seismic analysis approaches (quasi-static analysis and dynamical characterization methods) are also reviewed. These methods mainly focus on obtaining the lateral soil-pile interaction and dynamical properties of the soil-pile system. In our opinion, the diverse failure modes of the entire soil-pile system should be considered and characterized to establish the fundamental theory to guide the design optimization for taking the full advantages of PHC piles. The idea to modify the quasi-static analysis by comparing the quasi-static and seismic results through experiments is also proposed. This idea can reveal the frequency-dependent response of piles under seismic loading, which is the key factor to establish a desirable quantitative assessment system for seismic analysis of pile foundations.

1 INTRODUCTION

Eighty percent of buildings that are 10 stories or higher are supported by pile foundations in China. The rapid urbanization of China has increased the demand for bearing capacity and geological adaptability of pile foundations, and hence promoted significant research for their design and analysis (Nie, 2001; Kishida, 1998; Liu, 2005; Xu, 2013). Pre-stressed High-strength Concrete (PHC) piles, which overcome the disadvantages of high cost and corrosion susceptibility of steel piles, and large consumption of raw materials and lack of quality control of cast-in-place piles, have been used in a wide variety of buildings due to the demand for low-carbon footprint and sustainability from the architectural industry (Long, 1992; Svetinskii, 1995; Yu, 2008; Xia, 2013; Liu, 2012; Choi, 2015; Tung, 2014). In the last five years, the average annual growth of the demand for PHC piles in China is about 11.9%, which translates into a market of 600 million meters (i.e. 10 billion dollars) by 2016 (Wang, 2009).

According to the historical experience of Japan and the latest version of Chinese national standard GB 13476 (The national standard of the People's Republic of China: Pretensioned Spun Concrete Piles, 2009), the trend of PHC piles is towards larger diameter and greater depth. The continuous improvement of construction techniques has also substantially enhanced their performance (Nie, 2001; Kim, 2009; Wang, 2011). Therefore, appropriate theories for assessing the bearing capacity and seismic performance of PHC piles are urgently needed to guide their engineering applications, taking full advantages of these piles (e.g., economics, controllable quality, environmentally friendly and high strength). However, according to a survey by Liu et al. (Liu, 2013) as shown in Figure 1, which includes 100 pipe piles in Tianjin (China) between 2003~2008, the average ratio of the design bearing capacity to the ultimate value based on material strength is just around 0.43, which demonstrates an inadequate utilization of the material strength.

On the other hand, under seismic load, a pile sustains an inertial force and bending moment from the superstructure at the pile head and distributed multi-directional pile-soil interaction force along the pile. Due to the different properties between soft and stiff soil stratums, shear concentration exists at the stratum interface and generates

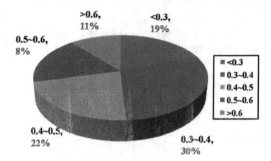

Figure 1. Distribution of ratio of design bearing capacity to ultimate value calculated from material strength for piles surveyed by Liu et al. (Liu, 2013).

additional bending moment in the pile. Complex seismic loading conditions lead to diverse failure modes of the piles (Liu, 1999; Prakash, 2014), further increasing the difficulties to establish an evaluation method for their seismic performance.

Here, the main methods to calculate the bearing capacity of PHC piles are first reviewed, followed by reviews of soil-pile interaction modes and dynamical analysis and shaking table tests to evaluate the seismic performance of PHC piles. Last, the future direction of PHC piles is discussed.

2 BEARING CAPACITY

As a basic property of pile foundations, the bearing capacity is calculated systematically and differently in each country. In China, the vertical, bending and shear bearing capacities are calculated individually according to loading forms.

2.1 Vertical bearing capacity

The Chinese national standard GB50007 (The national standard of the People's Republic of China: Code for Design of Building Foundation, 2011) and industry standards JGJ94 (The industry standard of the People's Republic of China: Technical Code for Building Pile Foundations, 2008) and JGJ106 (The industry standard of the People's Republic of China: Technical Code for testing of Building Foundation Piles, 2014) suggest two methods based on material strength and empirical coefficients, respectively.

2.1.1 Based on material strength
The fundamental formula is as follows

$$R \leq \omega (f_c - \sigma_{pc}) A \tag{1}$$

where R is the allowable vertical bearing capacity; f_c is the ultimate compressive strength of the

pile concrete; σ_{pc} is the effective pre-stress; A is the effective area of the pile cross-section; and ω is the ratio of the allowable bearing capacity to ultimate bearing capacity, i.e. a safety factor considering strength loss during manufactering and piling processes, which is generally taken as 0.25~0.3.

Based on Eq. (1), many scholars have proposed different simplified or modified methods (The national drawing collection of building standard design of the People's Republic of China, 2010; Wang, 2011; Tang, 2014; Tang, 2015). The National Building Standard 10G409 (The national drawing collection of building standard design of the People's Republic of China, 2010) considers the potential buckling of a pile, while simplifying the effect of pre-stress as

$$R \leq \varphi \Psi_c f_c A \tag{2}$$

where Ψ_c is a reduction factor considering pile construction and pre-stress; and ϕ is a stability factor related to the aspect ratio when the pile passes through liquefying or soft soil. Wang et al. (Wang, 2011) investigated the relationship between the residual pre-stress and reinforcement ratio. They found that a larger reinforcement ratio gives lower effective residual pre-stress and a larger allowable vertical bearing capacity according to Eq. (1). Tang and Kuang (Tang, 2014) discussed the relationship between pre-stress σ_{pc} and reduction factor Ψ_c in Eq. (2) or safety factor ω in Eq. (1). They proposed an important trade-off relation that, as the pre-stress increases, the bearing capacity decreases according to Eq. (1), while better manufacturing process gives larger reduction or security factor and hence greater bearing capacity. Tang et al. (Tang, 2015) also analyzed the complex effects of residual pre-stress, concrete-filled core and reinforced ratio on the bearing capacity. A comparison between theoretical analysis and experimental results shows that the concrete-filled core (especially reinforced) can improve the bearing capacity, and the improvement is more significant for piles with larger diameters.

2.1.2 Based on empirical coefficients
In addition to in-situ test, a method based on empirical coefficients can also be used to obtain the vertical bearing capacity of a pile foundation, if the geological parameters are available, through a simple form as

$$R = Q_{uk}/K = (q_{pa}A + u_p\Sigma q_{sia}l_i)/K \tag{3}$$

where Q_{uk} is the ultimate bearing capacity for a single pile; $K=2.0$ is the safety factor; u_p is the pile perimeter; l_i is the depth of ith soil layer; and q_{pa} and q_{sia} are the empirical coefficients corresponding to end-bearing and frictional piles, respectively, which are decided by stratum properties and piling techniques.

Regardless of the construction technique, the plugging effect, squeezing effect and pore water pressure of soil need to be assessed, which has led many scholars to continuously modify load transfer models of pile-soil system (Lehane, 2002; Yang, 2015; Zhou, 2015; Li, 2008; Li, 1992; Poulos, 1980; Seed, 1957). Lehane et al. (Lehane, 2002) demonstrated that the bearing capacity is sensitive to the properties of the soil under the pile ring, instead of those under the pipe core, and the plugging effect and sealing pile end for small-diameter piles can lead to a significant increase of actual bearing capacity. Considering the plugging effect in sand, Yang et al (Yang, 2015) investigated the weakening of bearing capacity induced by a thin and soft stratum and the aging effect that the bearing capacity will increase over time. From the design point of view, Xia and Dong (Xia, 2004) demonstrated the enhancement of lateral restraint due to spiral stirrup can improve the axial compressive strength of pile concrete by experiments and analytical calculations.

2.2 Bending bearing capacity

Based on the assumption of plane cross-section and small deformation, the National Building Standard 10G409 gives the cracking moment M_{cr} and ultimate moment M_u for pipe piles without concrete-filled core as

$$M_{cr} = (f_{tk} + \gamma\sigma_{pc})W_0 \qquad (4)$$

$$M_u = \alpha_1 f_c A(r_1 + r_2)\sin(\pi\alpha)/2\pi + r_p f_{pyc} A_p \sin(\pi\alpha)/\pi \\ + (f_{py} - \sigma_{p0})r_p A_p \sin(\pi\alpha_t)/\pi \qquad (5)$$

where f_{tk} is the tensile strength of pile concrete; γ ($= 1.9\sim2.0$) is a modification factor accounting for plasticity and casting technique of the pile; W_0 is the effective elasitc bending modulus of the pile section; r_1 and r_2 are the inner and outer diameter of the pipe, respectively; A_p and r_p are the area and distributed diameter of pre-stressed steels, respectively; f_{py} and f_{pyc} are the tensile and compressive strengths of the pre-stressed steels, respectively; σ_{p0} is the stress of the pre-stressed steels when the axial stress of the concrete vanishes at the acting point of the resultant of all steels; α_1 is the ratio of compressive stress of the effective concrete rectangle stress diagram to f_c; α_t is the yielding of the pre-stressed steel in concrete rectangle stress diagram; and α is the ratio of compressive area of the effective concrete rectangle stress diagram to A.

Song et al. (Song, 2007) derived similar expressions as Eqs. (4) and (5) for pipe piles with

concrete-filled core, which are verified by experiments. Tang et al. (Tang, 2013) demonstrated that the enhancement of bending bearing capacity by concrete-filled core is more pronounced for piles with larger diameters. The experiments conducted by Liu et al. (Liu, 2007) also showed that concrete-filled core can restrain crack propagation during cycling load, and hence improve the ductility of pile section. From the point of pile design, a simplified calculation method based on tables and first order differentiation is proposed by Ji et al. (Ji, 1996) through an optimization method by searching for the extreme values.

2.3 Shear bearing capacity

For a beam with symmetric soil section, the upper and lower edges suffer from the maximum normal stress under bending moment, while the maximum shear stress occurs at the neutral axis when sustaining a shear force. Therefore, the weakening effects of the ring section on the shear capacity are more significant than that on the bending capacity. However, almost none of the Chinese national or industry standards include corresponding clauses to guide the design for shear bearing capacity, except the latest edition of the National Building Standard 10G409 (Xia, 2004) as

$$V \le (tI/S_0)\,[(2\varphi_t f_t + \sigma_{pc})^2 - (\sigma_{pc})^2]^{1/2} \qquad (6)$$

where V is the allowable shear force; I is the moment of inertia of the pile section; t is the wall thickness of the pipe pile (i.e. $t = r_2 - r_2$); S_0 is the static moment for half of the pile section; and φ_t is a modification factor considering the concrete tensile strength.

Based on experiments, Zhang et al. (Zhang, 2008) studied the enhancement of concrete-filled core on the shear bearing capacity and the ductility of PHC piles, and proposed modification parameters about the stirrup and concrete-filled core. Zheng et al. (Zheng, 2014) analyzed the shear failure modes (i.e. shear compression and diagonal compression failures) of PHC piles with extra longitudinal reinforcement and proposed formula considering the modification of shear span ratio. It is also demonstrated by Finite Element Analysis (i.e. FEA) and experiments conducted by Chen et al. (Chen, 2013) that a larger pile diameter and wall thickness benefits the shear bearing capacity.

2.4 Complex bearing capacity

Almost all pile foundations bear complex loads including axial force, shear force, bending moment and even torsion. Therefore, Fan and Meng (Fan, 2015) established an FEA model, where a stubby

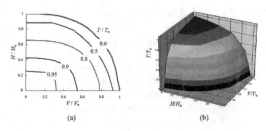

(a)   (b)

Figure 2. Failure envelopes on different loading planes (i.e. V-H, V-T, T-H and V-H-T) for a stubby pile (Fan, 2015).

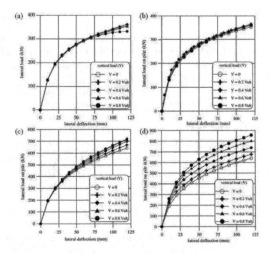

Figure 3. Lateral load-deflection curves for: (a) simultaneously applied vertical and lateral loads in loose sand; (b) vertical load prior to lateral load in loose sand; (c) simultaneously applied vertical and lateral loads in dense sand; and (d) vertical load prior to lateral load in dense sand (Karthigeyan, 2006).

pile is under complex loading, to investigate the bearing capacity and failure mechanism of piles under the combined vertical load V, lateral load T and torsion H. The failure envelopes on different loading planes (i.e. V-H, V-T, T-H and V-H-T) are shown in Figure 2.

Karthigeyan et al. (Karthigeyan, 2006) also employed FEM to study a pile in sand subjected to combined axial and lateral loads, which is verified by the experiments by Karasev et al. (Karasev, 1997) and Comodromos et al. (Comodromos, 2003). The results indicate that the loading sequence plays a significant role only for dense sand, where the increase of lateral stiffness for the pile is more pronounced for larger vertical pre-load (up to 40% when the pre-load reaches 80% of the ultimate vertical bearing capacity as shown in Figure 3). It was found that a relaxed pile head

restraint, larger shear strength, friction angle and dilatation angle of the soil and smaller aspect ratio of the pile also benefit the pile's bearing capacity under complex loads.

3 SEISMIC PERFORMANCES

Most of the analysis of bearing capacity is concerned on the pile itself. However, complex soil-pile interactions play a key role in the internal stresses developed along the pile. In seismic analysis, soil also plays an indispensable role in the load transfer and relieve from the bedrock. Therefore, a large part of the existing studies on seismic performance of pile foundations focuses on soil-pile interactions.

Considering the complex and diverse failure mechanisms in a soil-pile system, two approaches are commonly employed in the seismic analysis: 1) quasi-static method (i.e. the maximum peak values of dynamic variables are simplified as quasi-static ones to analyze soil-pile interactions and pile responses) and 2) dynamical characterization (i.e. characterizing the seismic properties of soil-pile system by dynamical soil-pile interactions, hysteresis curves etc.).

3.1 Quasi-static method

Abghar and Chai (Abghari, 2014) proposed a quasi-static nonlinear method to analyze soil-pile interactions based on the Winkler foundation model. The proposed method involves two steps as shown in Figure 4. Under seismic loading, the deformation field of soils in the free field is solved first. The inertial forces, which are calculated as the product of the maximum acceleration and the mass of upper structure, are then applied onto the pile head to initiate a static analysis, where the lateral reaction forces are characterized by a series of 'p-y' curves for different soil types and depths based on the Winkler model.

After considering pile group effect, Castelli and Maugeri (Castelli, 2009; Castelli, 2013) used a hyperbolic 'p-y' relationship to solve the pile deflection as

$$p(z) = \frac{\left[y_p(z) - y_s(z)\right]}{\dfrac{1}{\zeta_m E_{si}(z)} + \dfrac{\left|y_p(z) - y_s(z)\right|}{f_m p_{lim}(z)}} \tag{7}$$

where z is the depth; $y_p(z)$ and $y_s(z)$ are the lateral deflection of pile and soil, respectively; $p(z)$ is the distributed force on pile with the converse direction as $y_p(z)-y_s(z)$; E_{si} is the initial slope of 'p-y' curve; p_{lim} is the ultimate value of reaction force;

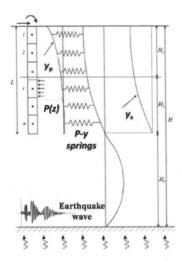

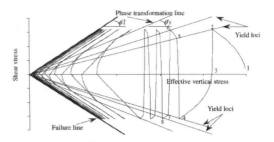

Figure 5. Model of soil stress path model used by Liyanapathirana et al. (Chatterjee, 2015).

k_h is a lateral load coefficient proportional to the depth. A pile is discretized as elements with two degrees of freedom (i.e. deflection and rotation) by Chatterjee et al. (Liyanapathirana, 2014). Instead of solving the displacement field of soil in the free field, the effective soil pressure (the product of P, the maximum horizontal acceleration, and the lateral load coefficient) is exerted along the pile to obtain the deformation and stresses distributions along the pile.

Xu et al. (Xu, 2013) and Norris et al. (Norris, 1986) employed the strain wedge model to obtain the reaction force factor of the soil-pile interface, in order to carry out the nonlinear analysis of laterally loaded single piles in sand, as shown in Figure 6. Their results demonstrate the important role of pile diameter in soil-pile interactions. Obtaining the soil pressure due to surcharge by simplified analytical method or experiments first, Wang and Yang (Wang, 2009) established the formula to calculate the ultimate reaction force by a plastic analysis of the strain wedge, which provides an effective theoretical basis to manage the surcharge in construction sites.

In addition, a variational method developed by Shen and Teh (Shen, 2014) provides a new approach to model the soil with stiffness increasing with depth. The settlement calculation is also modified by considering the effects of initial frictional resistance by Zhao et al. (Zhao, 2013) in quasi-static analysis.

3.2 Dynamical characterization

The 'p-y' curves to characterize soil-pile interactions and the soil properties are not constant during dynamical loading. The energy dissipation and buffering effects of soil are fundamental factors in protecting a pile from failures induced by earthquakes. Therefore, dynamic 'p-y' curves (i.e. dynamic soil-pile interactions) and the characterization of dynamic responses of the entire soil-pile system during seismic loading are the two main

Figure 4. Mechanical model of quasi-static nonlinear method proposed by Abghar and Chai (Abghari, 2014) and Castelli and Maugeri (Castelli, 2009; Castelli, 2013).

and f_m is the parameter for pile group effect. Their numerical results are verified by experiments and consistent with other numerical methods.

Based on the isotropic elasticity and nonslippage and delamination at the soil-pile interface, the incremental differential equations that combine the finite difference method with the Mindlin solution (concentrated force acted on semi-infinite foundation) are given by Tabesh and Poulos (Tabesh, 2001) and Elahi et al. (Elahi, 2014; Elahi, 2010). They simplified the upper structure as a mass with a single degree of freedom, which is denoted as m_0 with a stiffness $K_x = m_0(2\pi f)^2$, where f is the fundamental frequency. The loads at pile head are determined by the dynamical movement of the ground surface and the fundamental frequency of the upper structure. The elastoplastic behavior and damping of soil are also considered by setting an upper limit for the reaction force during iterative calculations and a damping factor in kinetic equations, respectively.

The numerical method proposed by Liyanapathirana and Poulos (Chatterjee, 2015) involves a reduction of the stiffness and strength of soil, as shown in Figure 5, due to the nonlinearity and pore water pressure, which is consistent with centrifuge experiments.

The governing equation for piles subjected to the combined axial and lateral loads is

$$EI(d^4y_p/dz^4) + P(d^2y_p/dz^2) + k_hy_p = 0 \qquad (8)$$

where EI is the bending stiffness of pile; P is the axial force parabolically distributed along pile; and

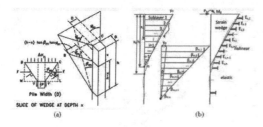

Figure 6. Strain wedge in the passive pressure zone analyzed by Xu et al. (Xu, 2013) and Norris et al. (Norris, 1986) (Norris, 1986): (a) single soil layer and (b) multiple soil layers.

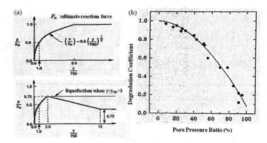

Figure 7. (a) Cyclic 'p-y' curve constructed from the static 'p-y' under low-speed cycling load by Matlock (Matlock, 1998) and (b) degradation coefficient of the reaction force versus pore pressure ratio in the experiments by Liu and Dobry (Liu, 1999).

dynamic methods in analyzing the seismic performances of piles.

3.2.1 Dynamic 'p-y' curves

For soft clay, Matlock (Matlock, 1998) established the cyclic 'p-y' curve from the static 'p-y' curve under low-speed cycling load (to provide sufficient time for the dissipation of pore water pressure), as shown in Figure 7(a), so that the model can capture the effects of soil-pile gap and modulus degradation, and hence can be used in a quasi-static analysis combined with peak load.

Experiments are conducted to exert cycling lateral load on a pile in loose sand immediately after centrifugal testing in order to study the effects of pore water pressure by Liu and Dobry (Liu, 1999). The degradation coefficient of the reaction force decreases with the pore pressure ratio as shown in Figure 7(b), which is due to the natural vibration of soil, inertia of the pile and upper structures, the formation of soil-pile gap and the cyclic degradation of soil.

Boulanger et al. (Boulanger, 2014) decomposed the dynamical 'p-y' curve into three components, i.e. elastic (p-y_e), plastic (p-y_p), and gap (p-y_g) parts. Kinematic hardening model is employed in the plastic part, and the gap part consists of a nonlinear closure spring (p^c-y_g) in parallel with a nonlinear drag spring (p^d - y_g), as shown in Figure 8(a). Their analytical results are verified by centrifugal testing and provide an effective guidance to capture the dynamical characteristics of soil-pile interactions, as show in Figure 8(b). However, more experiments are needed to obtain the parameters and check the applicability of the proposed model.

Wang et al. (Wang, 2009) employed the Novak layer method to calculate the dynamic impedance of soil in order to analyze the horizontal dynamic soil-pile interactions during the passage of Rayleigh waves. It was found that the Poisson's ratio, the loading frequency, and the pile/soil stiffness ratio make significant impacts on the dynamic response of the pile.

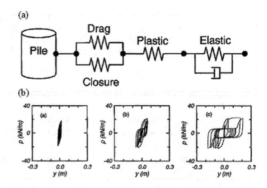

Figure 8. (a) Mechanical model for dynamical 'p-y' curve and (b) calculated 'p-y' behaviour at 2.1 m depth in soft clay under different dynamical loadings by Boulanger et al. (Boulanger, 2014).

3.2.2 Characterization during cyclic loading

Based on the traditional static analysis, Gazetas and Dobry (Gazetas, 1984) established a solution for piles under lateral dynamic steady-state harmonic load $P = P_0 \exp(i\omega t)$

$$K + i\omega C = P_0/y_d \qquad (9)$$

where P_0 is the amplitude of the harmonic load; ω is the angular frequency of the load; y_d is the complex displacement of the pile; and $K = K(\omega)$ and $C = C(\omega)$ are the stiffness and damping (including material and structural damping) coefficients, respectively. Through an FEA, the dynamical complex stiffness K is an approximate to the static solution $K_0 = P_0/y_d(f = 0, z = 0)$ in most cases, except a necessary correction for the resonant dip in a stiff, rock-like formation. The damping factor can be calculated by the frequency ω and the static normalized deflection profile $\chi_s(z) = y_d(f = 0)/ y_d(f = 0, z = 0)$.

Irshad and Akhtar (Irshad, 2014) solved the dynamic differential equation for the shear wave propagated from the bedrock:

$$\mathrm{d}^4 \mathbf{y}_p(z)/\mathrm{d}t^4 - \lambda^4 \mathbf{y}_p(z) = \alpha \mathbf{y}_s(z) \qquad (10)$$

where

$$\mathbf{y}_p(z) = [e^{-\lambda z}\ e^{\lambda z}\ e^{-i\lambda z}\ e^{i\lambda z}]\ [D_1\ D_2\ D_3\ D_4]^{\mathrm{T}}$$

$$\mathbf{y}_s(z) = [e^{-\lambda z}\ e^{\lambda z}\ e^{-i\lambda z}\ e^{i\lambda z}]\ [C_1\ C_2\ C_3\ C_4]^{\mathrm{T}}$$

$$\lambda^4 = [m_p\omega^2 - (K+i\omega C)]/EI$$

$$\alpha = (K+i\omega C)/EI$$

$\mathbf{y}_p(z)$ and $\mathbf{y}_s(z)$ are the displacement distribution of the pile and soil without pile, respectively; m_p is the pile mass. The stiffnesses K is obtained by matching the maximum moment in Beam on the Dynamic Winkler foundation (BDWF) model to that calculated by 3D FEA, and the damping factor C is adopted from Gazetas and Dobry (Gazetas, 1984). They employed the above solution to analyze an end-bearing pile embedded in two soil layers with highly contrasting stiffness, and demonstrated the existence of the predicted intensive moment located near the interface between two soil layers, which should be adequately considered in pile design.

Combining the boundary integration method, FEA and boundary element method, Fan et al. (Fan, 2014) also analyzed the pile response with cap during harmonic lateral load at the bedrock in detail. They found that the frequency plays an important role in controlling the ground displacement for any geological conditions. They also provided the typical curve of the dimensionless displacement versus frequency (as shown in Figure 9) with two characteristic values (i.e. α_{o1} and α_{o2}), which are sensitive to soil properties, pile to soil stiffness ratio, pile head restraint, and pile group effects.

Shaking table test is also an important tool for studying the seismic behaviour of pile foundations. Su and Li (Su, 2008) employed bi-directional centrifugal shakers to study the seismic performance of pile group (shown in Figure 10(a)). The symmetric response of the pile group, found during unidirectional vibration as shown in Figure 10(b), no longer existed during bi-directional shaking. However, due to more rapid increase of pore water pressure, the ground vibration waves were greatly reduced for bi-directional cases.

Kim et al. (Yang, 2010) studied the dynamical pile group effects, which were found to be dependent on the pile length, earthquake frequency and intensity. It was proved by Suzuki et al. (Suzuki, 2014) that when the natural period of the super-

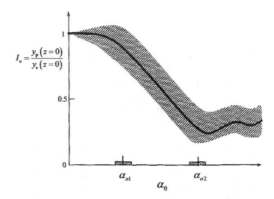

Figure 9. Typical curve of the dimensionless displacement versus frequency (Fan, 2014).

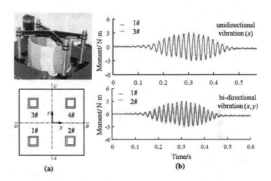

Figure 10. (a) Shaking table test conducted by Su and Li and (b) moment in piles during unidirectional and bi-directional vibrations (Su, 2008).

structure is shorter than or close to that of the ground, the inertia force of superstructure primarily determines the stresses in piles. On the contrary, if the natural period of the superstructure is much longer than that of the ground, the dominant factor to control the stresses in piles is the ground displacement.

Lou et al. (Lou, 2001) also demonstrated that soil-pile interaction and an increase of the damping factor decrease the fundamental frequency of the upper structure both in the vertical and horizontal directions. The experiments by Wu et al. (Wu, 2011) showed that, during the upward propagation of seismic waves, the low-frequency excitations will be amplified, while the high-frequency excitations will be attenuated significantly. Rong et al. (Rong, 2012) (Rong, 2012) noticed the improvement of ductility and stiffness degradation of a PHC pile by using non-prestressed reinforcement in shaking table test.

4 DISCUSSIONS

Pile foundations support the gravity of upper structures and objects in the building during the entire life cycle, as well as the horizontal and bending loads due to wind, asymmetric load etc. During earthquakes, compressive and shear waves from the bedrock also induce more complex stresses distributions in piles.

Although the bearing capacity is systematically calculated in each country or territory (The national standard of the People's Republic of China, 2011; The industry standard of the People's Republic of China, 2008; The industry standard of the People's Republic of China, 2014; The national drawing collection of building standard design of the People's Republic of China, 2010), almost all of them are based on the pile strength or empirical methods without clear mechanisms. Few progresses have been made to explain the similarities and differences among various construction techniques.

Therefore, to reveal the failure mechanisms of pile foundations and then establish reliable method for bearing capacity assessment during complex loading, complex soil-pile interactions and diverse failure modes should be considered. The failure envelopes of different loading planes should be combined together to characterize the bearing capacity of the pile foundations, which is the key to design optimization for a more predictable and controllable failure mode.

On the other hand, only the reduction and amplification factors, which are based on the probabilistic characteristics and liquefaction susceptibility, are proposed for the seismic calculation in the existing standards of China (The national standard of the People's Republic of China, 2010). Recent studies have mainly focused on the lateral soil-pile interaction (Abghari, 2014; Castelli, 2009; Castelli, 2013; Chatterjee, 2015; Xu, 2013; Norris, 1986; Shen, 2014) and dynamical properties of the soil-pile system (Matlock, 1998; Liu, 1999; Boulanger, 2014; Wang, 2009; Gazetas, 1984; Irshad, 2014; Fan, 2014; Su, 2008; Yang, 2010; Suzuki, 2014; Lou, 2001; Wu, 2011; Rong, 2012), but few of them investigated the effect of construction techniques.

But almost all soil-pile interactions involve coupled horizontal reaction force around the pile side, friction resistance along the pile length and end-bearing resistance at the pile end. According to the Newton friction relation, the presence of horizontal reaction force will alter the distribution of friction resistance due to the slippage between soil and pile. Next, the change of modulus due to shear deformation of soil will influence the deformation compatibility between pile and soil. At the same time, settlement is also affected by the re-distribution of frictional and end-bearing resistance. Therefore, seismic analysis with bi-directional coupling should be carried out and related to the failure modes in bearing capacity calculations.

Almost no efforts, except that by Matlock (Matlock, 1998), are devoted to the combination of quasi-static and dynamic analysis methods. Seismic waves have a wide frequency bandwidth, which is very important to the ground response and structural responses (Tabesh, 2001; Elahi, 2014; Elahi, 2010; Gazetas, 1984; Irshad, 2014; Fan, 2014; Yang, 2010; Suzuki, 2014; Lou, 2001; Wu, 2011). Therefore, various frequency-dependent phenomena cannot be revealed by quasi-static analysis methods. On the other hand, components of real earthquake are complex, random and uncertain so that it is impossible to verify the safety of a particular structure by testing it against countless simulated earthquakes. Therefore, it is desirable to combine the quasi-static method with the dynamic characterization so that the variables used in quasi-static analysis can describe the effects of frequency-rich seismic shaking.

In summary, the main methods to calculate the bearing capacity against vertical, bending, shear loading and their combinations are reviewed. Almost all of these methods are based on pile strength or empirical method without clear mechanism. In our opinion, the failure of a soil-pile system (including the failure of soil-pile interface, shear failure of the soil, damage to concrete, and excessive settlement etc.) should be considered and characterized by phase diagrams of failures to establish the fundamental theory to guide the design optimization for a more predictable and controllable failure mode. The commonly-used seismic analysis approaches (i.e. quasi-static analysis and dynamical characterization methods) are also reviewed. These methods mainly focus on obtaining lateral soil-pile interaction and dynamical properties of the soil-pile system. The idea to compare the quasi-static analysis method and the real seismic results through in-situ or numerical experiments to obtain the dynamic modification to the quasi-static analysis is proposed to reveal the frequency-dependent response of piles under seismic loading. This idea may lead to a desirable quantitative assessment system for seismic analysis of pile foundations.

ACKNOWLEDGEMENTS

The authors acknowledge the supports from Guangdong Provincial Science and Technology Foundation (Grant No. 2015B020238014 and 2014TQ01Z014) and China Postdoctoral Science Foundation (Grant No. 2016M592471).

REFERENCES

Abghari, A., J. Chai, J. In Performance of Deep Foundations Under Seismic Loading ASCE, 45 (2014).

Boulanger, R.W., C.J. Curras, B. L Kutter, D.W. Wilson, A. Abghari. J. Geotech. Geoenviron. Eng. 125, 750 (2014).

Castelli, F., M. Maugeri, J. Geotech. Geoenviron. Eng. 135, 1440 (2009).

Castelli, F., M. Maugeri, J. Geotech. Geoenviron. Eng. 139, 1882 (2013).

Chatterjee, K., D. Choudhury, H.G. Poulos. Comput. Geotech. 67, 172 (2015).

Chen, K., *The reseach on shear performance of prestressed concrete pipe pile* (mater degree thesis of Hefei University of Technology, 2013).

Choi, Y., D.C. Kim, T.H. Kim, Mar. Georesour. Geotec. 34, 474 (2015).

Comodromos, E.M., Geotech. Eng. 34, 123 (2003).

Elahi, H., H.G. Poulos, M. Moradi, A. Ghalandarzadeh, In GeoCongress 2012@State of the Art and Practice in Geotechnical Engineering ASCE, 2273 (2014).

Elahi, H., M. Moradi, H.G. Poulos, A. Ghalandarzadeh, Comput. Geotech. 37, 25 (2010).

Fan, K., S. Hayashi. J. Geotech. Engrg. 117, 1860 (2014).

Fan, Q., X. Meng, In Advances in Pile Foundations, Geosynthetics, Geoinvestigations, and Foundation Failure Analysis and Repairs ASCE, 109 (2015).

Gazetas, G., R. Dobry, J. Geotech. Engrg. 110, 20 (1984).

Irshad, A., N.K. Akhtar, In GeoCongress 2006@sGeotechnical Engineering in the Information Technology Age ASCE, 1, 2014.

Ji, G., Chin. Rural Water Hydro. 7, 29 (1996).

Karasev, O.V., G.P. Talanov, S.F. Benda, Soil Mech. Found. Eng. 14, 173 (1977).

Karthigeyan, S., V.V.G.S.T. Ramakrishna, K. Rajagopal, Comput. Geotech. 33, 121(2006).

Kim, Y.S., J.B. Lee, S.K. Kim, J.H. Lee. Automat. Constr. 18, 737 (2009).

Kishida, S., J. Struct. Const. Eng. 510, 123 (1998).

Lehane, B.M., M.F. Randolph, J. Geotech. Geoenviron. Eng. 128, 198 (2002).

Li, J., J. Zhou, Rock Soil Mech. 29, 449 (2008).

Li, X., J. Liu, Chin. J. Geotech. Eng, 14, 9 (1992).

Liu, B., J. Li, X. Zhang, H. Sheng, Ind. Constr. 37, 46 (2007).

Liu, C., Z. Zhang, H. Mu, seismic performance and new progress of presressed concrete pipe pile (Chian Communications Press, 2013).

Liu, H., Earthq. Eng. 1, 37 (1999).

Liu, H.X., J. Oper. Res. Soc. 446–449, 1649 (2012).

Liu, J., H. Liu, J. Huangshi I. Tech. 21, 24 (2005).

Liu, L.R., Dobry, In Workshop on New Approaches to Liquefaction Analysis, No. FHWA-RD-99–165 (1999).

Liyanapathirana, D.S., H.G. Poulos. In GeoCongress 2006@sGeotechnical Engineering in the Information Technology Age. ASCE, 1, (2014).

Long, R.P., Corrosion of steel piles in some waste fills (Transportation Research Board, 1992)

Lou, M., W. Wang, H. Ma, T. Zhu, J. Tongji Univ. (Nat. Sci.) 29, 763 (2001).

Matlock, H., In Proceedings of offshore technoligy conference 124, 1998.

Nie, R., W. Leng, Q. Yang, Y. Frank Chen, P.I. Civil Eng-Geotec. (to be published)

Norris, G., In Proceedings of the 3rd international conference on numerical methods in offshore piling, 361, 1986.

Poulos, H.G., E.D. Davis, *Pile Foundation Analysis and Design* (John Wiley, 1980).

Prakash, S., V.K. Puri, Geotechnical Earthquake Engineering and Soil Dynamics IV ASCE, 1 (2014).

Rong, X., H. Di, Y. Li, Chin. Concrete Cement Prod. 9, 32 (2012).

Seed, H.B., L.C. Reese, T. Am. Soc. Civ. Eng. 122, 731 (1957).

Shen, W.Y., C.I. Teh, J. Geotech. Geoenviron. Eng. 130, 878 (2014).

Song, Y., B. Liu, J. Li, H. Sheng, X. Zhang, J. Hefei Univ. Tech. (Nat. Sci.) 35, 607 (2007).

Su, D., X. Li, Rock Soil Mech. 29, 603 (2008).

Suzuki, H., K. Tokimatsu, Soil Found. 54, 699 (2014).

Svetinskii, E.V., M.S. Gaidai, Soil Mech. Found. Eng. 32, 63 (1995)

Tabesh, A., H.G. Poulos, J. Geotech. Geoenviron. Eng. 127, 757 (2001).

Tang, G., H. Kuang, Build. Struct. 44, 6 (2014).

Tang, M., Y. Qi, Z. Zhou, H. Hu, Build. Struct. 45, 67 (2015).

Tang, M., Y. Qi, Z. Zhou, H. Hu, Chin. J. Geotech. Eng, 35, 1075 (2013).

The industry standard of the People's Republic of China: *Technical Code for Building Pile Foundations*, JGJ94 (2008).

The industry standard of the People's Republic of China: *Technical Code for testing of Building Foundation Piles*, JGJ106 (2014).

The national drawing collection of building standard design of the People's Republic of China: *Prestressed Concrete Pipe Piles*, 10G409 (2010).

The national standard of the People's Republic of China: *Code for Design of Building Foundation*, GB 50007 (2011).

The national standard of the People's Republic of China: *Code for Seismic Design of Buildings*, GB 50011 (2010).

The national standard of the People's Republic of China: *Pretensioned Spun Concrete Piles*, GB 13476 (2009).

Tung, N.K., W. Li, D.T. Nam, D.M. Ngoc, N.M. Thanh, Electron. J. Geotech. Eng., 19, 6047 (2014).

Wang, G., M. Yang, Chin. J. Undergr. Space Eng. 5, 1530 (2009).

Wang, H., S. Shang, Z. Zhou, F. Zhou, J. Huan Univ. (Nat. Sci.) 36, 1 (2009).

Wang, L.Y., J.T. Wang, J. Rail. Eng. Soc. 6, 6 (2011).

Wang, Q., L. Chen, H. Kuang, B. Yu, Build. Struct. 41, 113 (2011).

Wang, Y., Shanxi Architect. 35, 95 (2009).

Wu, X., X. Jiang, Eng. Mech. 28, 201 (2011).

Xia, C., T. Dong, Soil Eng. Found. 18, 46 (2004)

Xia, X., H. Xu, H.D. Xu, R.J. Gu, Adv. Mater. Res. 772, 193 (2013).

Xu, L., F. Cai, G. Wang, K. Ugai, Comput. Geotech. 51, 60 (2013).

Xu, W., Y. Hou, Appl. Mech. Mater. 353–356, 533 (2013).

Yang, E.K., J.I. Choi, J.T. Han, M.M. Kim, J. Korean Geotech Soc. 26, 77 (2010)

Yang, Z., W. Guo, F. Zha, R.J. Jardine, C. Xu, Y. Cai, J. Geotech. Geoenviron. Eng. 141, 04015020 (2015).

Yu, H.K., J. Korean. Sol. Energy Soc. 28, 56 (2008).

Zhang, X., B. Liu, Y. Lin, H. Sheng, Build. Struct. 38, 11 (2008).

Zhao, M., S. Liu, P. Yin, M. Liu, J. Cent. South Univ. (Nat. Sci.) 44, 3625 (2013)

Zheng, G., T. Zhang, Q. Li, Chin. Civil Eng. J. 47, 97 (2014).

Zhou, X., G. Fang, In *MATEC Web of Conferences*, 22, 04024 (2015).

Advances in Energy Science and Equipment Engineering II – Zhou, Patty & Chen (Eds)
© *2017 Taylor & Francis Group, London, ISBN 978-1-138-71798-5*

Ultimate bearing capacity of concrete-filled-steel-tubular circular stub columns under axial compression

Xiaowei Li

School of Civil Engineering, Panzhihua University, Panzhihua, China
Department of Tunnel and Underground Engineering, Southwest Jiaotong University, Chengdu, China

Wei Chen, Xuewei Li, Yukun Quan & Hongfen Nian

School of Civil Engineering, Panzhihua University, Panzhihua, China

ABSTRACT: This study aims to develop a versatile method to obtain the ultimate bearing capacity of Concrete-Filled-Steel-Tubular (CFST) circular columns. Based on the united strength theory, the formula for calculating the ultimate bearing capacity of CFST circular columns is presented. Including the intermediate principal stress and Tension-Compression-Ratio (TCR), the formula is applicable to columns with core concrete confined by materials such as metal materials, non-metal materials, elastic-perfectly-plastic materials and hardening materials. Besides, the formula presented is validated by the data from the literature.

1 INTRODUCTION

As a kind of composite construction, the CFST columns are formed by filling concrete in thin-wall steel tubes; the steel tubes only need be filled by plain concrete, and the concrete in the steel tube does not have to be reinforced by steel bars. Confined by steel tubes, the core concrete is subjected to three-dimensional compressive stress, and the physical and mechanical properties of the core concrete are changed qualitatively, such as increasing the compressive strength, improving the deformation capacity and changing the property from a brittle material into a ductile material. Furthermore, supported by the inner core concrete, the external thin steel tube improves its ability to resist local buckling. Thus, the inner core concrete and the external steel tube can mitigate each other's weakness, play to their strengths, and maximize the effectiveness of both materials (Cai, 2003; Zhong, 2003; Han, 2000).

There are various methods which can be used to obtain the ultimate capacity of CFST columns. For example, the regressive method based on experimental data, the superposition method according to mechanical equilibrium conditions and deformation compatibility conditions and the composite-design-index method based on the analysing results of the Finite Element Method (FEM). Based on the united strength theory, the formula for calculating the ultimate bearing capacity of CFST circular columns was presented. Including the intermediate principal stress and TCR, the formula is applicable to columns with core concrete confined by metal and non-metal materials.

2 UNIFIED STRENGTH THEORY

Based on the twin shear strength theory, M.H. Yu (Yu, 2004) developed a unified strength theory. The theory includes the contribution of σ_2 to material strength, and uses a uniform mechanical model to descript the plastic features of the material. The mathematical expression for the unified strength theory is simple and given as follows.

$$F = \sigma_1 - \alpha(b\sigma_2 + \sigma_3)/(1+b)$$
$$= \sigma_t (\sigma_2 \le (\sigma_1 + \alpha\sigma_3)/(1+\alpha)) \tag{1}$$

$$F = (\sigma_1 + b\sigma_2)/(1+b) - \alpha\sigma_3$$
$$= \sigma_t (\sigma_2 \ge (\sigma_1 + \alpha\sigma_3)/(1+\alpha)) \tag{2}$$

$$b = ((1+\alpha)\tau_o - \sigma_t)/(\sigma_t - \tau_o)$$
$$= ((1+\alpha) - B)/(B-1)$$
$$\alpha = \sigma_t / \sigma_c$$
$$B = \sigma_t / \tau_o \tag{3}$$

where, σ_t is the tensile ultimate strength of the material, σ_c is the compressive ultimate strength, τ_o is the shear ultimate strength, α is the TCR, varying from 0.77 to 1.0 for ductile metal materials, from 0.33 to 0.77 for brittle metal materials, and generally less than 0.5 for geotechnical materials, B is the tension shear ratio and b is the participant factor of intermediate principal shear stress.

3 THEORETICAL DERIVATIONS

In the elastic stage, the Poisson's ratio of steel varies in a small scope, which can be treated as a constant of 0.283. At a low stress stage, the Poisson's ratio of concrete can be taken as 0.167, while, at a high stress stage, the Poisson's ratio of concrete will increase with the development of internal micro-cracks and be more than 0.5. Once the Poisson's ratio of concrete is more than the Poisson's ratio of steel, the external circular steel tube will uniformly confine the internal core concrete along the perimeter of the core concrete. At this moment, the core concrete is compressed in three directions, its strength will increase and its ductility will improve. Furthermore, supported by the inner core concrete, the external thin steel tube improves its ability to resist local buckling (Han, 2000). The calculation model of CFST circular columns is shown in Figures 1 and 2.

Using the unified strength theory as the yield criterion of the tube (Zhao, 2000; Wang, 1982), the plastic ultimate internal pressure of the tube is calculated as follows.

$$
\begin{aligned}
p &= \frac{\sigma_s}{1-\alpha}\left(\left(\frac{r_i}{r_i+t}\right)^{\frac{2(1+b)(\alpha-1)}{2+2b-b\alpha}}-1\right) \\
&= \frac{\sigma_s}{1-\alpha}\left[(1+\mu_t/2)^{\frac{2(1+b)(1-\alpha)}{2+2b-b\alpha}}-1\right]
\end{aligned}
\tag{4}
$$

where, p is the confining pressure applied to the interface between the steel tube and the core concrete, r_i is the radius of the internal wall of the steel tube, and t is the thickness of the steel tubular wall.

The ultimate strength of the CFST stub columns, i.e., the average strength of core concrete $f_{e,c}$, is given as

$$
f_{e,c} = \frac{N_u}{A_c} = \frac{1}{A_c}\left[\sigma_{cp}A_c + \sigma_{zp}A_s\right]
\tag{5}
$$

Figure 1.   Internal pressure of steel tube.

Figure 2.   Confining force applied to concrete.

where, A_c is the cross-sectional area of the core concrete, A_s is the area of the steel tube, σ_{cp} is the compressive stress of the core concrete along the axis of the tube, and σ_{zp} is the compressive stress of the steel tube along the axis of the tube.

$$
A_c = \pi r_i^2
\tag{6}
$$

$$
A_s = ((2r_i+2t)^2-r_i^2)/4 = \mu_t A_c
\tag{7}
$$

where, μ_t is the steel ratio.

$$
\mu_t = A_s/A_c \approx 2t/r_i
\tag{8}
$$

The nonlinear equation of the yielding criterion of the core concrete (Cai, 2003) can be written as

$$
\sigma_{cp} = f_c[1+c(p)]
\tag{9}
$$

$$
c(p) = 1.5\sqrt{p/f_c}+2p/f_c
\tag{10}
$$

According to the thick cylinder theory in elastic mechanics (Wang, 1982; Zhao, 2000), σ_{zp} is given as

$$
\sigma_{zp} = \frac{pr_i^2}{(r_i+t)^2-r_i^2} = \frac{4p}{4\mu_t+\mu_t^2}
\tag{11}
$$

Substituting Eqn. 10 into Eqn. 9, and combining Eqn. 5, Eqn. 9 and Eqn. 11, the average strength of core concrete $f_{c,c}$ can be calculated as follows.

$$
f_{e,c} = f_c+1.5\sqrt{pf_c}+2p+\frac{4p}{4+\mu_t}
\tag{12}
$$

Substituting Eqn. 4 into Eqn. 12, we get

$$
\begin{aligned}
f_{e,c} &= f_c+1.5\sqrt{\frac{f_c\sigma_s}{1-\alpha}\left[(1+\mu_t/2)^{\frac{2(1+b)(1-\alpha)}{2+2b+b\alpha}}-1\right]} \\
&\quad +\left(2+\frac{4}{4+\mu_t}\right)\left(\frac{\sigma_s}{1-\alpha}\right)\left[(1+\mu_t/2)^{\frac{2(1+b)(1-\alpha)}{2+2b-b\alpha}}-1\right]
\end{aligned}
\tag{13}
$$

The ultimate strength of the CFST circular column is

$$N_u = A_c f_{c,c} \tag{14}$$

The confinement index of CFST columns is given as

$$\theta = \mu_t \frac{\sigma_s}{f_c} \tag{15}$$

4 DISCUSSIONS

4.1 Influence of α and b on the ultimate strength of CFST circular columns

Setting the ultimate compressive strength of concrete f_{cu} as 30 Mpa and the tensile strength of the steel tube σ_s as 310 Mpa, Figure 3 shows that the ultimate compressive strength ratio $f_{c,c}/f_{cu}$ of CFST columns decreases with the increase of α, and Figure 4 shows that the ultimate compressive strength ratio $f_{c,c}/f_{cu}$ of CFST columns increases with the increase of the participant factor of intermediate principal shear stress of b.

4.2 Influence of θ on the compressive stress of the steel tube along the axis of the tube σ_{zp}

Setting α as 0.1, b as 0, σ_s as 310 Mpa, and f_{cu} as 30 Mpa, Figure 5 shows that σ_{zp}/σ_s decreases with the increase of θ. It indicates that the compressive stress of the steel tube along the axis of the tube contributes less to the compressive strength of CFST columns with the increase of the confinement index of CFST columns, and that the confinement effect of the steel tube contributes more to the compressive strength of CFST columns after yielding the steel tube. The compressive stress of the steel tube shall satisfy the requirement of $0 \leq$

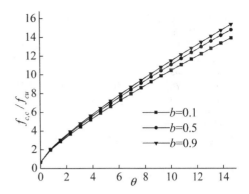

Figure 4. Relationship between b and $f_{c,c}/f_{cu}$.

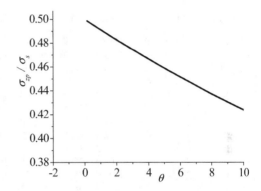

Figure 5. Relationship between σ_{zp}/σ_s and θ.

$\sigma_{zp}/\sigma_s \leq 1$. As the confinement index θ is less than 0.281, the above-mentioned requirement cannot be satisfied, therefore, the confinement index θ shall be more than 0.281 for practical purposes.

5 VERIFICATION BASED ON TESTED RESULTS

Generally, the TCR of structural steel is equal to 1. As α is equal to 1, the unified strength theory is changed to the unified yielding criterion. As factor b is set as constant, the unified yielding criterion is changed to a yielding criterion which is defined or not defined until now. Using $\alpha = 1$, and finding the limit of Eqn. 4, the following equation can be obtained.

$$p = \sigma_s \frac{1+b}{2+b} \ln\left(1 + \frac{\mu_t}{2}\right) \tag{16}$$

Setting the shear yielding strength τ_s equal to the half of the tensile yielding strength σ_s, i.e., B = 2,

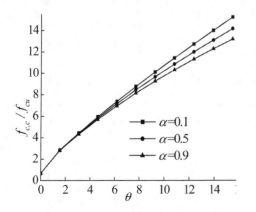

Figure 3. Relationship between α and $f_{c,c}/f_{cu}$.

and $b = 0$, the unified yielding criterion is changed to the Tresca criterion, which is the lower limit of the ductile metal criterion. Using the ratio of the shear yielding strength τ_s to the tensile yielding strength σ_s as 0.577, i.e., $B = 1.733$ and $b = 0.364$, the unified yielding criterion is changed to the linear approximation of the Mises criterion.

Based on the existing benchmark data, S.H. Cai presented a formula to predict the compressive strength of CFST columns (Cai, 2003), which is given as

$$N_0 = A_c f_c (1 + \sqrt{\theta} + 1.1\theta) \qquad (17)$$

The literature (Cai, 2003; Han, 2000; Cai, 1984; Tang, 1982) had presented the tested results of the compressive strength of CFST columns, which is taken as a benchmark data to verify the validity of the formulae presented in this paper. As the related references did not give the tested data of τ_s and σ_s of steel tubes, we suppose that the exterior steel tube is a material which conforms to the Tresca criterion, i.e., $b = 0$, and a material which conforms to the linear approximation equation of the Mises criterion, i.e., $b = 0.364$, respectively. Table 1 gives the results predicted by the formula presented by S.H. Cai and the results predicted

Table 1. Results obtained from test and prediction.

Number	D_t_l /mm	σ_s /Mpa	f_{cu} /Mpa	θ	N_t /kN	N_M /kN	ρ_M	N_T /kN	ρ_T	N_C /kN	ρ_C
G_34	204_2_880	235.40	12.80	1.12	1069.00	1100.00	0.03	1010.00	−0.06	881.00	−0.18
G_35	204_2_880	235.40	12.80	1.12	1040.00	1100.00	0.05	1010.00	−0.03	881.00	−0.15
G_50	204_2_880	235.40	48.40	0.29	1638.00	2153.00	0.24	2042.00	0.25	1898.00	0.16
G_21	273_8_1100	307.00	34.70	1.69	5580.00	6359.00	0.12	5784.00	0.04	5023.00	−0.10
G_32	273_8_1100	307.00	12.80	4.60	4040.00	4906.00	0.18	4380.00	0.08	3646.00	−0.10
G_33	273_8_1100	307.00	12.80	4.60	3844.00	4906.00	0.22	4380.00	0.14	3646.00	−0.05
G_57	273_8_1100	307.00	49.00	1.20	5296.00	7187.00	0.26	6590.00	0.24	5815.00	0.10
G_38	96_5_410	410.00	12.80	11.87	912.00	1177.00	0.23	1038.00	0.14	865.00	−0.05
G_39	96_5_454	410.00	12.80	11.87	843.00	1177.00	0.28	1038.00	0.23	865.00	0.03
G_44	96_5_450	410.00	35.00	4.30	1044.00	1386.00	0.25	1239.00	0.19	1062.00	0.02
G_45	96_5_450	410.00	35.00	4.30	1167.00	1386.00	0.16	1239.00	0.06	1062.00	−0.09
G_48	96_5_400	410.00	48.40	3.11	1177.00	1493.00	0.21	1343.00	0.14	1165.00	−0.01
G_49	96_5_400	410.00	48.40	3.11	1172.00	1493.00	0.22	1343.00	0.15	1165.00	−0.01
G_58	96_5_400	410.00	48.40	3.11	1074.00	1493.00	0.28	1343.00	0.25	1165.00	0.08
G_59	96_5_400	410.00	48.40	3.11	1123.00	1493.00	0.25	1343.00	0.20	1165.00	0.04
G_1	166_5_660	274.60	33.00	1.65	1744.00	2180.00	0.20	1984.00	0.14	1727.00	−0.01
G_2	166_5_660	274.60	33.00	1.65	1695.00	2180.00	0.22	1984.00	0.17	1727.00	0.02
G_3	166_5_660	274.60	33.00	1.65	1705.00	2180.00	0.22	1984.00	0.16	1727.00	0.01
G_4	166_5_660	274.60	33.00	1.65	1735.00	2180.00	0.20	1984.00	0.14	1727.00	0.00
G_12	166_5_660	274.60	36.50	1.49	1863.00	2256.00	0.17	2059.00	0.11	1800.00	−0.03
G_15	166_5_660	274.60	36.50	1.49	1873.00	2256.00	0.17	2059.00	0.10	1800.00	−0.04
G_16	166_5_660	274.60	36.50	1.49	1697.00	2256.00	0.25	2059.00	0.21	1800.00	0.06
G_22	166_5_660	274.60	36.50	1.49	1736.00	2256.00	0.23	2059.00	0.19	1800.00	0.04
G_23	166_5_660	274.60	36.50	1.49	2030.00	2256.00	0.10	2059.00	0.01	1800.00	−0.11
G_29	166_5_660	274.60	36.50	1.49	2109.00	2256.00	0.07	2059.00	−0.02	1800.00	−0.15
G_21	273_8_1100	313.00	35.40	1.69	5690.00	6485.00	0.12	5899.00	0.04	5132.00	−0.10
G_32	273_8_1100	313.00	12.40	4.80	4120.00	4953.00	0.17	4418.00	0.07	3670.00	−0.11
G_57	273_8_1100	313.00	47.80	1.25	5400.00	7206.00	0.25	6601.00	0.22	5812.00	0.08
G_36	121_12_500	300.00	12.40	20.00	2465.00	2161.00	−0.14	1899.00	−0.23	1691.00	−0.31
G_42	121_12_500	300.00	12.40	20.00	2550.00	2462.00	−0.04	2187.00	−0.14	1980.00	−0.22
Z_69_84	100_2.5_300	442.00	39.20	1.82	845.00	1033.00	0.18	938.00	0.11	809.00	−0.04
Z_70_102	100_2.5_300	249.00	43.40	0.93	684.00	747.00	0.08	689.00	0.01	614.00	−0.10
Z_70_106	100_2_300	241.00	43.40	0.71	548.00	659.00	0.17	612.00	0.12	550.00	0.00
Z_70_107	100_1.5_300	237.00	43.40	0.51	515.00	574.00	0.10	538.00	0.05	489.00	−0.05
Sccs1_1	131_2.3_396	323.30	53.40	0.67	1250.00	1370.00	0.09	1274.00	0.02	1146.00	−0.08
Sccs2_1	111_2_339	353.60	53.40	0.75	894.00	1045.00	0.14	968.00	0.08	867.00	−0.03
Sccs3_1	114_3.2_337	353.60	53.40	1.21	1140.00	1381.00	0.17	1266.00	0.11	1116.00	−0.02
Sccs4_1	133_3.5_397	323.30	53.40	1.03	1440.00	1722.00	0.16	1584.00	0.10	1405.00	−0.02

by combining Eqns. 12, 13, 16 with Eqn. 14. The error between the predicted result and the tested result can be defined by the coefficient of ρ, which is given as

$$\rho = \frac{N - N_t}{N_t} \tag{18}$$

In Table 1, notation N_t is the tested result, N_M is the result obtained from the linear approximation equation of the Mises criterion, i.e., $b = 0.364$, N_T is the result obtained from the equation of the Tresca criterion, i.e., $b = 0$, N_C is the result obtained from the equation presented by S.H. Cai, ρ_M is the error between the predicted result obtained from the linear approximation equation of the Mises criterion and the tested result, ρ_T is the error between the predicted result obtained from the equation of the Tresca criterion and the tested result and ρ_C is the error between the predicted result obtained from the equation presented by S.H. Cai and the tested result. Notation D is the exterior diameter of the circular steel tube, t is the thickness of the circular steel tube and l is the length of the circular steel tube. The data of the samples beginning with capital letter G is referred to references 1 and 8; the data of the samples beginning with capital letter Z is referred to reference 9 and the data of the samples beginning with letters Sccs is referred to reference 3.

Figure 6 shows the relationship between ρ_M and N_t, the relationship between ρ_T and N_t and the relationship between ρ_C and N_t. Comparing the results obtained from the Mises criterion, Tresca criterion, and the equation presented by S.H. Cai with the tested results, it is found that the error between the predicted results obtained from aforementioned three equations and the tested results is all acceptable. The predicted results obtained from N_M are larger than those obtained from N_T and N_C.

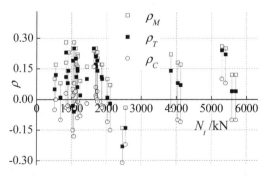

Figure 6. Relationships between ρ_M and N_t, ρ_T and N_t, ρ_C and N_t.

The Tresca criterion corresponds to the lower limit of the yielding surfaces of the metal material, and the results predicted by N_T are closer to the tested results, and can be used as a design standard.

As Eqn. 4 takes into consideration the TCR α and the participant factor of intermediate principal shear stress b, the formulae presented in this paper are applicable to concrete columns confined by metal materials or by non-metal materials. Suppose the concrete columns are confined by non-metal materials, the parameters α and b of the unified strength theory should be certified by the tested results of clear small samples.

For practical purposes, the formula presented should take into consideration the effects of the initial imperfection, initial eccentricity and slenderness ration of the samples on the ultimate strength of CFST columns. This can be settled by introducing a combined influence factor γ, which can be calculated by using the following equation.

$$f_{c,c} = f_c \gamma \left(1 + 1.5\sqrt{\frac{p}{f_c}} + 2\frac{p}{f_c}\right) + \frac{4p}{4 + \mu_t} \tag{19}$$

6 CONCLUSIONS

Based on the united strength theory, the formulae for predicting the ultimate bearing capacity of CFST circular columns are presented. Including the intermediate principal stress and the TCR, the formula is applicable to columns with core concrete confined by materials such as metal materials, non-metal materials, elastic-perfectly-plastic materials and hardening materials. The applicability of formulae presented is verified by the tested results from the literature. The results predicted by N_T are closer to the tested results, and can be used as a design standard. If the parameters of the exterior confining material such as the tensile strength, the compressive strength and the shear strength are obtained from clear small samples, the formulae presented should be amended by a combined influence factor γ, which can be determined by Eqn. 19.

ACKNOWLEDGEMENTS

This study was financially supported by the Fund Program of Sichuan Province Key Lab for Comprehensive Development and Utilization of Industrial Solid Waste (SC-FQWLY201305, SC-FQWLY201403), Sichuan Education Department Fund Program (16ZB0475) and Panzhihua University Doctor Fund Program (bkqj2015004).

REFERENCES

Cai, S.H. *Modern steel tube confined concrete structures* (CCP, Beijing, 2003).

Cai, S.H., Z.Q. Jiao, J. Build. Struc. **6**, 13–29 (1984).

Han, L.H. *Concrete filled steel tubular structures—Theory and Applications* (SP, Beijing, 2000).

Tang, G.Z., B.Q. Zhao, H.X. Zhu, J. Build. Struc. **1**, 13–31 (1982).

Wang, R., Z.H. Xiong, W.B. Huang, *Plastic Mechanics Basis* (SP, Beijing, 1982).

Yu, M.H. *Unified Strength Theory and Its Applications* (Springer, Berlin., 2004).

Zhao, J.H. *Strength Theory and Its Applications* (SP, Beijing, 2000).

Zhao, J.H., Y.Q. Zhang, H.J. Liao, Z.N. Yin, Chin. J. Appl. Mech. **17**, 157–161 (2000).

Zhong, S.T. *The concrete filled steel tubular structures* (TUP, Beijing, 2003).

Advances in Energy Science and Equipment Engineering II – Zhou, Patty & Chen (Eds)
© *2017 Taylor & Francis Group, London, ISBN 978-1-138-71798-5*

Study on the alert value of construction safety monitoring for a formwork support system

Jian Sua & Ren-jin Chen
Guangzhou Institute of Building Science Co. Ltd., Guangzhou, China

ABSTRACT: Formwork support systems are widely used in construction in China. Formwork support systems consist of wooden templates, main beams, minor beams, U-head jacks, steel tubular columns and base jacks. Construction safety monitoring is important during concreting to prevent accidents. However, the empirical monitoring alert value is generally incorrect. Damage or failure of any part of the formwork support system may cause collapse. A new method, namely the maximum load method, to determine the alert value, based on the capacity of templates, beams and vertical columns, is proposed in this paper. A flowchart for this method is also proposed. Finally, in-situ tests have been performed, the results of which demonstrate the applicability of the suggested method to determine the monitoring alert value method.

1 INTRODUCTION

Formwork is a temporary construction structure designed to guide the solidification of fresh concrete to a desired size and shape. In China, High Formwork Support Systems (HFSS) are always with height exceeding 8 m, span exceeding 18 m, total construction loads over 15 kN/m², and concentrated line load over 20 kN/m. Many efforts have been made to study their performcance. Octavian George Ilinoiu (Ilinoiu, 2006), Rose and Sujeethra (Rose, 2014), Zhang and Rasmussen (Zhang, 2010), and Samer Barakat (Baraka, 2011) independently proposed different design theoretical methods for formworks. Zhang et al. (Zhang, 2011) and Liu et al. (Liu, 2010) presented their experimental research studies, respectively. The studies of Huang et al. (Huang, 2000), Zhang et al. (Zhang, 2015) and Xue et al. (Xue, 2012) are devoted to the construction safety monitoring of formworks.

HFSS collapse accidents would cause heavy casualties and huge economic losses, resulting in adverse social impact (Yuan, 2006; Du, 2004; Xie, 2002). Many researchers have analyzed the mechanism of catastrophic collapse during construction. According to the surveys (Yuan, 2006; Du, 2004; Xie, 2002), the failures of structures during construction were commonly induced by ineffective real-time safety monitoring. HFSS consists of wooden templates, which directly constitute the mold for the solidification of concrete, main beams, minor beams, U-head jacks, steel tubular columns and base jacks (Wei, 2010). Damage or failure of any part of HFSS may cause collapse. It is important to monitor the working condition of HFSS

during concreting, as well as reasonable design and construction in accordance with corresponding technical codes, for the purpose of minimizing safety accidents. There are two main methods to monitor the safety of HFSS: the manual method by optical instruments and the automatic real-time method using sensors.

The arrangement of measuring points (or observation points) and the alert values are two key factors to the effectiveness of HFSS construction safety monitoring. Only the outer periphery of HFSS can be monitored using the manual method due to the limitation of observing conditions, while, taking advantages of various sensors, the entire HFSS can be monitored by the automatic real-time method, including the deformation, axial force and inclination of support columns. The alert values are generally calculated from the formulas for design of templates, or engineering experience. The low alert value may cause frequent false alarm, while the high alert value will make the monitoring invalid. It is critical to set up an appropriate alert value before the monitoring of HFSS. The traditional method to determine the alert value is introduced in this paper, and then we propose a new method, namely the maximum load method, which gives an appropriate construction safety monitoring alert value.

2 CURRENT METHODS TO DETERMINE ALERT VALUES

Formworks and support systems should be sized to support all the weight produced by fresh concrete construction besides the live load itself: materials (G_k), equipment and personnel (Q_{1k}), vibration (Q_{2k}), and wind load (ω_k). There are two ways to determine the alert value:

1. Based on the design checked formulas. First substituting combination load into Eqs. (1)–(3) respectively. If the deflections of the template, minor beam and main beam meet the allowable value, substitute the combination load into Eq. (4) and (5), and check whether the vertical support column has enough strength and stability capacity. Then, the monitoring alert value is given by the deflection of the template. The deflection of the template is

$$w_1 = \frac{5q_s l_s^4}{384 E_s I_s} \tag{1}$$

The deflection of the minor beam is

$$w_2 = \frac{K_w q_{mi} l_{mi}^4}{100 E_{mi} I_{mi}} \tag{2}$$

The deflection of the main beam is

$$w_3 = \frac{K_w F_{ma} l_{ma}^3}{100 E_{ma} I_{ma}} \tag{3}$$

The compression stress of the vertical support column is

$$\sigma = \frac{N_{st}}{A_{st}} \le f_{st} \tag{4}$$

The stability of the vertical support column is

$$\frac{N_{st}}{\varphi A_{st}} \le f_{st} \tag{5}$$

$$N_{st} = q l_a l_b \tag{6}$$

where, K_w is the coefficient of deflection, which can be found in appendix C in (JGJ 162-2008, 2008), q_s is the line load of the template, $q_s = q b_s$, q is the area load of the template, b_s is the effective width of the template, q_{mi} is the line load of the minor beam, $q_{mi} = q b_{mi}$, b_{mi} is the space of the minor beam, and F_{ma} is the concentrated load applied to the main beam, $F_{ma} = 1.1 q b_{mi} l_{mi}$. l_s, l_{mi} and l_{ma} are the effective span of the template, minor beam and main beam, respectively. E_s, E_{mi} and E_{ma} are elastic modulus of the template, minor beam and main beam, respectively. I_s, I_{mi} and I_{ma} are the second moment of the cross-section of the template, minor beam and main beam, respectively. A_{st} is the area of the

vertical support column, f_{st} is the yield strength of the vertical support column, φ is the stability coefficient of the vertical support column, N_{st} is the axial force of the vertical support column, l_a is the transverse distance of the column, and l_b is the longitudinal distance of the column.

The alert value given by the deflection of the template under design combination load without considering the temporary overload caused by the accumulation of the fresh concrete is too low. Therefore, false alarms occur.

2. Based on the construction experiments. An alert value of 10 mm is usually adopted in different projects during HFSS construction safety monitoring. It is obviously unreasonable for the different HFSS to get the same alert value through the engineering experience.

3 PROPOSED METHOD

3.1 Maximum load

The templates, beams and columns are the main bearing parts of HFSS. The concrete will leak if the templates and beams are broken or deformed to a large extent, and the support system will collapse if the columns yield or are unstable. To determine the monitoring alert value, the capacity of these components should be considered.

Considering the failure of the formwork templates, the maximum load of the supporting system is

$$q_1 = \frac{f_s W_s}{0.125 l_s^2 b_s} \tag{7}$$

Considering the failure of the minor beam, the maximum load of the supporting system is

$$q_2 = \frac{f_{mi} W_{mi}}{0.1 l_{mi}^2 b_{mi}} \tag{8}$$

Considering the failure of the main beam, the maximum load of the supporting system is

$$q_3 = \frac{f_{ma} W_{ma}}{1.1 \times 0.175 l_{mi} b_{mi} l_{ma}} \tag{9}$$

Considering the stability of the vertical column, the maximum load of the supporting system is

$$q_4 = \frac{f_{st} \varphi A_{st}}{l_a l_b} \tag{10}$$

Considering the failure of the vertical column, the maximum load of the supporting system is

$$q_5 = \frac{f_{st} A_{st}}{l_a l_b} \tag{11}$$

Considering the failure of the base jack and U-head jack, the maximum load of the supporting system is given by

$$q_6 = \frac{40000}{1.15 \times 1.1 \times b_{mi} \times l_{mi}} \tag{12}$$

where, f_s, f_{mi}, and f_{ma} are design strength values of templates, minor beams and main beams, respectively. W_s, W_{mi} and W_{ma} are the section modulus of templates, minor beams and main beams, respectively.

The maximum allowable load of the HFSS is given as

$$q_u = \min\{q_1, q_2, q_3, q_4, q_5, q_6\} \tag{13}$$

3.2 Determination of the monitoring alert value

HFSS remains working well while the load on the template is less than q_u. In other words, the HFSS maximum load is q_u, not design combination load. The maximum axial force of the column is given by substituting q_u into Eq. (6), as

$$N = q_u l_a l_b \tag{14}$$

Substituting q_u into Eq. (1) ~ (3), total displacement Δ is

$$\Delta = w_1 + w_2 + w_3 + \delta \tag{15}$$

$$\delta = \frac{N_{st} H}{E_{st} A_{st}} \tag{16}$$

where, δ is the axial deformation of the vertical column, given by Eq. (16), E_{st} is the elastic modulus of the vertical column, and H is the effective height of the vertical column.

The displacement alert value of HFSS is given by Eq. (15) and the force alert value of HFSS is given by Eq. (14). The flow chart for determining the alert value of monitoring HFSS is shown in Figure 1.

4 EXPERIMENTAL MEASUREMENTS

4.1 HFSS description

In-situ tests are conducted for a HFSS supporting concrete box girder with a height of 4.8 m, as shown in Figure 2. The span and width of the girder are 23.4 m and 9.9 m, respectively. Eight water boxes are distributed on the templates of HFSS to simulate the 20 kN/m² total construction loads. The columns of the HFSS are Φ48 × 3.2 mm Q234 steel. The templates are 18 mm composite wooden fiber plates. The minor beams are 100 mm × 100 mm square timbers. The main beams are 100 mm × 46 mm × 4.5 mm Q234 channel steels. The parameters of the HFSS are shown in Table 1.

4.2 Monitoring alert value determination

The value of $q_1 = 34$ kN/m² when substituting the parameters of the template into Eq. (7). Substituting the parameters into Eqs. (8) and (12) gives $q_2 = 66$ kN/m² and $q_6 = 78$ kN/m². Substituting the parameters of main beams into Eq. (9) gives $q_3 = 102$ kN/m². $q_4 = 73$ kN/m² and $q_5 = 113$ kN/m² are calculated from Eq. (9) and Eq. (10). Substituting $q_1 \sim q_6$ into Eq. (13) gives the maximum load of the HFSS allowable $q_u = 34$ kN/m².

Substituting q_u into Eq. (14), Eq. (16), Eq. (1), Eq. (2) and Eq. (3), respectively, the alert value of the vertical column is 28 kN, the axial elastic deformation of the vertical column is 1.45 mm,

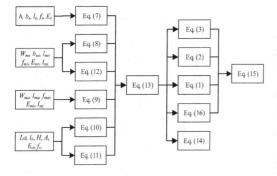

Figure 1. Flow chart for calculating the monitoring alert value.

Figure 2. High formwork support system.

Table 1. Parameters of the HFSS.

	Template	Minor beam	Main beam	Column
Thickness (mm)	18	/	/	/
Effective width (mm)	1000	450	/	/
Effective span (mm)	450	900	900	/
Transverse distance of the column (mm)	/	/	/	900
Longitudinal distance of the column (mm)	/	/	/	900
Strength (N/m²)	16	15	205	205
Elastic modulus (N/m²)	6000	9500	2.06×10^5	2.06×10^5
Second moment of cross section (mm⁴)	4.86×10^5	8.33×10^6	1.74×10^6	1.14×10^5
Section modulus (mm³)	5.40×10^4	1.67×10^5	3.48×10^4	4.73×10^3

Table 2. Monitoring alert value.

Sensor	Monitoring parameter	Warning value	Alert value
F (kN)	Column axial force	22.4	28.0
D (mm)	Template displacement	8.8	11.0

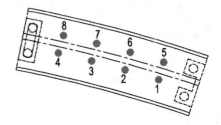

Figure 3. Layout of the monitoring points.

the deflection of the template is 6.23 mm, the deflection of the minor beam is 0.86 mm, and the deflection of the main beam is 0.56 mm. The deformation alert value of the HFSS is given by Eq. (9) to be 9.1 mm. Finally, the alert value is calculated to be 11 mm, considering the contact gap and sliding of the components. The warning value is 80% of the alert value. The relevant value is given in Table 2.

4.3 Monitoring data and results analysis

Eight monitoring points are selected in the HFSS as shown in Figure 3. A displacement senor and a force senor are set up in each monitoring point. A wireless formwork support system safety intelligent monitor is used. It takes 12 h to fill the boxes with water to simulate the concreting loads on the templates up to 20 kN/m². The monitored maximum force of each vertical column and displacement of the each measuring point of the templates are shown in Figures 4 and 5. Measuring point No. 8 encountered the maximum axial force and measuring point No. 4 the maximum displacement.

The axial force time history of point No. 8 is shown in Figure 6 and the displacement time history of point No. 4 is shown in Figure 7. The monitoring value is less than the alert value which represents the ultimate capacity of the HFSS. The test results show that the HFSS is safe and the maximum load method can provide reasonable safety assessment of HFSS and make the monitoring effectively.

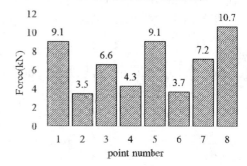

Figure 4. Maximum axial force of columns.

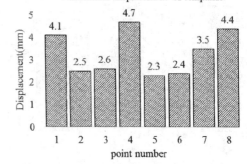

Figure 5. Maximum displacement of templates.

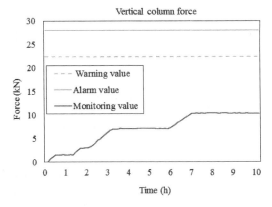

Figure 6. Axial force time history of monitoring point 8.

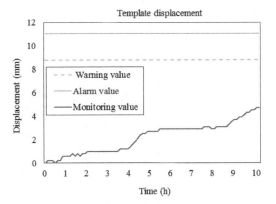

Figure 7. Displacement time history of monitoring point 4.

5 CONCLUSION

In this paper, a new method to determine the alert value of HFSS construction safety monitoring is introduced. The capacity of the critical parts of the HFSS, including templates, main beams, minor beams and vertical columns, is highlighted and taken into account in the proposed method, the flowchart of which is also given and explained in detail. Finally, in-situ tests have been performed, the results of which demonstrate the applicability of the suggested method to determine the monitoring alert value method.

ACKNOWLEDGEMENT

The authors acknowledge the support from Guangdong Provincial Science and Technology Foundation (Grant No. 2016 A040403067). J. Su acknowledges the support from national key research and development plan (Grant No. 2016YFC0802500).

REFERENCES

Baraka, S. Jord. J. Civil. Eng. **5** 1 (2011).
Du, R. *Formwork support of concrete project*, 422 (2004). (**in Chinese**).
Huang, Y., W. Chen, H. Chen, T. Yen, Y. Kao, C. Lin, Com. & Struc. **78** 681 (2000) (**in Chinese**).
Ilinoiu, O. Civil. Eng. Dim. **8** 47 (2006).
JGJ 162-2008, Prof. std of PRC (2008) (**in Chinese**).
Liu, H., Q. Zhao, X. Wang, T. Zhou, D. Wang, J. Liu, Z. Chen, Eng. Struc. **32** 1003 (2010) (**in Chinese**).
Rose, A., M. Sujeethra, Int. J. Lat. Technol. Eng. Manag. & Sci. **3** 49 (2014).
Wei, C, Ph.D. Thesis, Hei. Un. Tech. (2010) (**in Chinese**).
Xie, J., B. Xiao, Buil. Struc. **4** 71 (2002) (**in Chinese**).
Xue, X., N. Shi, X. Chen, C. Wang, Q. Zhao, Y. Luo, J. Con. In. Tec. **7** 140 (2012) (**in Chinese**).
Yuan, X., W. Jin, Z. Lu, X. Liu, T. Chen, China Civil Eng. J. **5** 43 (2006) (**in Chinese**).
Zhang, H., K. Rasmussen, Struc. Saf. **32** 393 (2010).
Zhang, J., L. Cai, J. Miao, X. Wang, H. Xiao, Y. Zhou, J. Shen. Jian. Un. (Nat. Scien.) **27** 685 (2011) (**in Chinese**).
Zhang, L., C. Wang, G. Song, Shoc. & Vib. **2015** 1 (2015) (**in Chinese**).

Advances in Energy Science and Equipment Engineering II – Zhou, Patty & Chen (Eds)
© 2017 Taylor & Francis Group, London, ISBN 978-1-138-71798-5

Performance-based seismic performance analysis of steel frame structures

Hengchao Chen
Zunyi Vocational and Technical College, Zunyi Guizhou, China

ABSTRACT: In order to overcome the limitations of the seismic design of current structures according to limit bearing capacity including strength and ductility, the Performance-Based Seismic Design (PBSD) on the structural has become an important issue of widespread concern. Through the analysis of the Performance-Based Seismic Design (PBSD), a five story steel frame model was established by software SAP2000, a nonlinear static pushover analysis was carried out, and the performance of the steel frame under the earthquake structure was evaluated. The results show that the displacement based design method is the key point of the current research and application. The future of seismic design will be the performance-based seismic design, but before accepting the performance-based design and using it widely, there are a lot of research studies to be carried out.

1 INTRODUCTION

Performance-based seismic design theory has received wide attention by scholars from various countries since it was proposed in 1990s. Unlike the traditional philosophy of seismic design which is only to ensure life safety, this theory is committed to achieve multiple performance objectives of building structures and requires the design of a structure that has predictability under the earthquake that may occur in the future. A large number of earthquake disasters show that the structure is under the action of non-elastic large deformation, member failure, etc. Energy dissipation capacity and plastic deformation capacity are the main reasons for the collapse of a structure. Steel structures as the main structure of our architecture, studying their performance of various stages, form initial flexibility, and gradually degrade until the overall instability collapse has an important theoretical significance and practical value. Therefore, based on the nonlinear seismic response analysis of the structure, the performance based seismic performance evaluation of a steel frame structure is carried out. Comprehensive use of theoretical analysis, numerical simulation study and some valuable results are obtained.

2 THEORETICAL RESEARCH OF PERFORMANCE—BASED SEISMIC PERFORMANCE ANALYSIS OF STEEL FRAME STRUCTURES

The Performance-Based Seismic Design (PBSD) is a kind of seismic design system which has been

paid more attention in the world in recent years. For the performance of the structure, that is, before designing to achieve the kind of structural performance, the knowledge on how the design can meet the requirements has not been mastered by structural engineers.

The description of the performance-based seismic design released by SEAOC is "performance design should choose a certain design standards, proper structure, proper planning and structure of the scale, ensure the detail design of building structure and non-structure members and control the quality and long-term maintenance level of construction, guarantee the buildings under seismic action in a certain level and structural damage does not exceed a specific limit states" (SEAOC Vision2000, 1995). The performance based seismic design of ATC-40 is described as "The design criteria of the structure are represented by a series of structural performance objectives that can be achieved, mainly aimed at the concrete structure and uses the ability-based design principle (Federal Emergence Management Ageney (FEMA), 1997)." The performance based seismic design of FEMA is described as "based on the different strength of the seismic effect, the different performance objectives are achieved. In the analysis, the design of the static elasticity and the elastic plastic time history analysis are used to obtain a series of performance levels, and the top displacement of the structure is used to define the level of performance of the structure (Applied Technology Council (ATC), 1996)."

Although the description of different agencies of the performance-based seismic design is not identical, the main idea of the performance-based seismic design is common. The performance design requires specified performance goals (the level of seismic, structural performance level and structure of performance objectives determined) at first before the design, then choose the appropriate method according to the different performance objectives and propose different seismic fortification criteria, so that the structure has clear performance levels during its structure design period under different levels of earthquake action, and it makes the structure of the entire life cycle cost effective (Bayat M R, 2010).

The performance objectives refer to the performance level of the structure under different earthquake levels, including three contents: the seismic level, structural performance level and structural performance objectives.

The earthquake levels are typically expressed as a probability over a period of time. Different countries have different seismic norms, and there are three levels of seismic norms in our country, namely "Big earthquake does not fall", "minor earthquake can repair", and "small earthquake is not bad."

The performance level refers to the maximum extent of damage of buildings to withstand, including the damage to the structures and non-structural components of the structures. Considering the safety of the structure and economic and social effects, countries have different provisions on performance levels. The performance level is based on the definition of the extent of the structural damage, and the extent of damage is determined by the structure of the reaction parameters, which can be embodied as stress, deformation, energy dissipation index, acceleration and other parameters that reflect the specified limits.

3 DISPLACEMENT BASED SEISMIC ANALYSIS METHOD

The performance-based seismic design method mainly includes the displacement based design method, comprehensive design method, reliability based analysis method, design method based on damage performance, etc (Wang, 2014). The structure under the design ground motion deformation value can well reflect the performance of the building structure. Therefore, the displacement based design method can well describe the structural damage situation in the earthquake process or after the earthquake for a period of time, analyze structure quantitative, control structure displacement limits under different levels of earthquake, and intuitively get the structural performance levels.

3.1 The method of capacity spectrum

The capacity spectrum method is adopted by the American ATC-40. The basic idea is to set up two lines of the same benchmark: one is the static pushover analysis (Pushover) structure capacity curve (i.e., the curve of bottom shear V and top displacement D) convert to the curve of acceleration spectrum and displacement spectrum, referred to as the capacity spectrum (ADRS, acceleration and displacement response spectrum); the other one is the acceleration response spectrum obtained from input earthquake, and converse the standard spectrum to the demand spectrum of the acceleration-displacement relationship according to the requirements of seismic ductility demand. Then, put the capacity spectrum and demand spectrum in the same coordinate system in comparison, and the intersection of the two curves as the target displacement. By comparing the displacement of the performance point to the allowable value of the displacement, determine whether the structure meets the seismic requirements.

3.2 Direct displacement based design method

The direct displacement based design method is based on the relationship between the deformation and stiffness, the displacement and the strength of the structure under the action of a horizontal earthquake (i.e., non-linear response spectrum), so as to solve the structural stiffness and strength directly and hope the design of structures could reach the target displacement when the actual earthquake happens. For a multi-degree of freedom system, we need to assume a reasonable lateral mode first, and the multi-degree of freedom system is converted to an equivalent single degree of freedom system to determine its equivalent coefficient (Huo, 2012).

4 PUSHOVER ANALYSIS OF STEEL FRAME STRUCTURES

In this paper, the direct displacement based seismic design method of steel frame structures is studied, and the performance index of the structure is determined from the displacement angle between the layers when the displacement index is determined (Shao, 2014).

The core problem of displacement based on the seismic design method is the analysis of the non-linear state of the structure. Therefore, the static pushover analysis (Pushover), as a relatively simple elastic-plastic analysis method, is an analysis method based on the method of displacement-based seismic design at home and abroad widely, also known as nonlinear static method analysis.

The pushover analysis method is applied to the unchanged vertical load on the structure, while applying a distribution horizontal load. With the increase of the monotonic loading, the component gradually yields, which clearly reflects the various aspects of the structure elastic plastic properties under strong earthquake, and could well reflect the structural deformation and localized plastic deformation mechanism, as well as the member of beam and column section failure and failure sequence, so as to find out the weak parts of the structure.

Before conducting displacement based seismic design, we need to understand the structure of the elastic-plastic performance and accurately predict the elastic-plastic response of the structure under normal earthquake and rare encounter earthquake. Steel frames in the action of rare earthquake will enter into the plastic stage, and the internal force distribution is very different. So, it is of high necessity to analyze the elastic and plastic behavior of steel frames (Shao, 2014).

After pushover analysis, the structure of the performance points is obtained, according to the corresponding structural deformation of the performance points. The value of the seismic performance of the structure is evaluated by the following 3 aspects:

1. Lateral displacement: whether the limit value of the elastic-plastic limit is satisfied by the seismic code.
2. Displacement angle between the layers: whether the displacement angle limit of the elastic-plastic layer can meet the requirements of the seismic code.
3. Local deformation of the member: to test whether the plastic hinge of the beam and column is more than one performance level.

5 CASE STUDY

The project is a five storey steel frame structure with 3 meters high, and the steel selection adopts Q345. The H-shaped beams of H-300 × 200 × 8 × 12 (section depth × flange width × wed thickness × flange thickness) and the H-shaped columns of H-500 × 200 × 11 × 19 (section depth × flange width × wed thickness × flange thickness) were selected for numerical analysis. Figure 1 shows the finite element model.

5.1 Definition of plastic hinges

Two kinds of definitions of the plastic hinge method are provided by SAP2000: one is user-defined plastic hinge characteristics and another is the program in accordance with the norms of

Figure 1. Structural model.

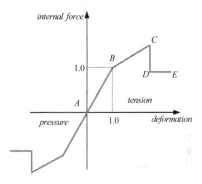

Figure 2. Constitutive relation of plastic hinge.

ATC40 and FEMA356 in America (Fig. 2). In this paper, the default plastic hinge is selected in SAP2000 and the plastic hinge should be set at the maximize internal force of the elastic stage, where the position of the structure is the first to reach the yield. For the beam column element, the general situation of the maximum internal force is at both ends, so set the hinge moment (M-M) at the beam end, and set the bending shear hinge (V) and the bending hinge (P-M-M) in column ends.

5.2 Lateral loading mode

Three lateral loading modes are used in this example: ① gravity and vibrating type 1, equivalent to the laterally inverted triangular lateral load patterns; ② gravity and X-direction acceleration, equivalent to the longitudinal lateral load patterns; ③ gravity and Y-direction acceleration, equivalent to the transverse lateral load patterns.

In this paper, the lateral displacement and the displacement angle between the layers in the case of multiple earthquake are shown in Table 1. The period of the model was calculated using SAP2000 as shown in Table 2.

Table 1. Lateral displacement and inter storey drift angle.

Number of layers	X-direction /mm	Y-direction /mm	X-direction storey drift angle	Y-direction storey drift angle
5	28.1	34.2	1/818	1/766
4	25	25	1/621	1/878
3	21	20	1/643	1/621
2	17	17	1/974	1/527
1	10	8	1/857	1/571

Table 2. Period and frequency of modal.

Vibration mode number	Period /s	Frequency /Hz
1	0.7062	1.4159
2	0.6464	3.23940
3	0.6463	3.68534
4	0.2114	4.7303
5	0.1939	5.5166
6	0.1926	501923

5.3 Analysis of the results of the steel frame structure in the elastic stage

According to the requirement of the design of the response spectrum, the seismic fortification intensity is 8 degree, the basic design of the earthquake acceleration is 0.2 g, and the characteristic period of the site is 0.4 s. As a result of calculation, the maximum vertex displacement is 23.128 mm and the maximum base shear is 180.32 kN.

According to the frequent earthquake pushover analysis method, the vertex displacement is 22.335 mm and the maximum base shear is 173.71 kN. Compared with the results of the response spectrum method, the vertex displacement deviation is 3.4%, the base shear deviation is 3.7%, and the deviation is not more than 5%.

According to the seldomly occurred earthquake pushover analysis method, the vertex displacement is 136.8 mm, the maximum base shear is 634.127 kN, and the structure in the rare encounter earthquake occurred under the yield, but the plastic deformation is in the allowable range. Under the action of rare earthquake, the structure occurred in rare lower yield, but the plastic deformation is within the allowable range.

5.4 Plastic hinge distribution

The distribution of the plastic hinge is shown in Figure 3. The plastic hinge first appeared at the

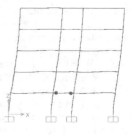

a Loading pattern 1

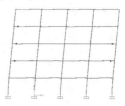

b Loading pattern 2

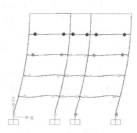

c Loading pattern3

Figure 3. The plastic hinge distribution of the structure.

edge of the cross-beam in the first layer of the transverse frame under the action of loading mode 1 and loading mode 3, and the plastic hinge first appeared at the edge of the cross beam in the first layer of the longitudinal frame under loading mode 2 (Fig. 3). As can be seen from Figure 3, plastic hinges of the framework appear on the beam, mostly conforming to the design requirements of "strong column and weak beam".

Based on the above evaluation, the project meets the requirements of the application. If a local component does not meet the plastic limit, it needs to be strengthened locally, but not to change the performance of the whole structure.

6 CONCLUSION

In this paper, the performance-based seismic performance analysis of steel frame structures was studied and the basic principle of the performance-based seismic design is described. The conclusions of this studied are as follows:

1. Based on the seismic code of our country, the pushover analysis is carried out on a five storey frame structures by software SAP2000, and the performance of the structure is evaluated by using a kind of reaction parameters. The results show that the displacement based design method is the key point of the current research and application.
2. The displacement based seismic design method is better than the traditional seismic design based on bearing capacity, and more beneficial to ensure the structural performance under strong earthquake action. The method can be designed and calculated at different performance levels, and can effectively control the performance of the structure under different earthquake levels.

REFERENCES

Applied Technology Council (ATC). Seismic Evaluation and Reotrift of Existing Conerete buildings [C]. ATC40, 1996.

Bayat M R. Performance-based plastic design of earthquake resistant steel structures: concentrically braced frames, tall moment frames, plate shear wall frames [D]. Arlington, Texas: The University of Texas at Arlington, USA, 2010.

Federal Emergence Management Ageney (FEMA). NEHRP Guidelines for the Seismic Rehalilitaton of Buildings [C]. FEMA273, 1997.

Jian-Hua Shao, Bai-Jie Tang. Seismic Performance Evaluation of Steel Frame-Steel Plate Shear Wall Using Pushover and IDA [J]. Applied Mechanics and Materials 2014 (578).

Jing-Si Huo, Cong-Ling Hu, Jin-Qing Zhang, Yu-Rong Guo. Analysis of dynamic behavior and ductility of steel moment frame connections [J]. Journal of Civil, Architectural & Environmental Engineering, 2012, 34(Suppl 1):149–154.

SEAOC Vision 2000. Pelformance-based Seismic Engineering [C]. Sacarmento. CA, 1995.

Yan Wang, Li-Ting Dong. Experimental study and finite element analysis of seismic performance on enhanced joints in steel frame [J]. Journal of Building Structures, 2014, 35(Suppl 1):94–100.

Advances in Energy Science and Equipment Engineering II – Zhou, Patty & Chen (Eds)
© *2017 Taylor & Francis Group, London, ISBN 978-1-138-71798-5*

FE analysis on impact performance of composite steel-concrete beam-to-column joints

Hengchao Chen

Zunyi Vocational and Technical College, Zunyi Guizhou, China

ABSTRACT: This paper addresses the behavior of steel beam-column connections that consist of steel beams and columns, as well as composite beam-column connections that consist of steel columns, beams, and concrete slabs. The two kinds of connections under study were simulated with a tridimensional numerical model using nonlinear FEA software ABAQUS, which is based on the Finite Element Method (FEM). The dynamic response and internal force development of joints were obtained by moment-rotation and axial force-rotation curves of joints. Analysis of the results led to the conclusion that the composite joints have credible rotation and moment capacities in elastic and plastic stages enter the catenary stage later and thus improve the capacity of resisting progressive collapse.

1 INTRODUCTION

Progressive collapse is a situation where local failure of a primary structural component leads to the collapse of adjoining members which, in turn, leads to additional collapse. And, the total damage is disproportionate to the original cause (GSA, 2009). Along the development of research on progressive collapse of structures, the major focus has put emphasis on performance studies of joints and components rather than the analysis and design of the whole structure (S Jeyarajan, 2015). When a primary support column is removed by exploding or impacting, whether the remaining structure parts will be capable of effectively redistributing loads is very important. The performance of beam-to-column joints is essential for structural tie action.

In this study, finite element program ABAQUS is used to develop the numerical simulation of the traditional beam-to-column joints that welded flange-bolted web connection and beam-to-column composite joints. The deformation modes, bearing capacity, the displacement of midspan, the internal force development and the dynamic response of beam-to-column substructures under impact loads were studied.

2 BRIEF DESCRIPTION OF FE MODELING TECHNIQUE

The building dimensions of portal frames mainly include column spacing, rigid frame span, roof slope and eaves height.

Within the current research project, the FE model was designed in order to investigate the performance of steel and steel-concrete composite joints under impact loads. The steelwork part of the connection uses an end-plane welded to the beam and bolted to the column flange. For the purpose of comparing, the bare steel model was also employed with the same condition as the composite steel-concrete joints (Marcela N. Kataoka, 2014).

All the steel members (beams, column and endplates) were made of Q345. The model was made of H-section beams (H250 x 125 x 6 x 9) and H-shaped steel columns (H200 x 200 x 16 x 16). The bolt connectors were M16 of class 10.9. The sheeting of composite flooring was ignored so as to facilitate the modeling process. The composite action between the steel beam and concrete was provided by the headed studs (SDΦ16 x 60) that

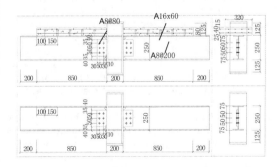

Figure 1. Model construction details (dimensions in mm).

were friction welded to the upper of the beam. The concrete class was of C30. The slab reinforcing bars were made of steel grade HRB300 (Abdolreza Ataei, 2016). The model construction details of steel and steel-concrete composite joints are presented in Figure 1.

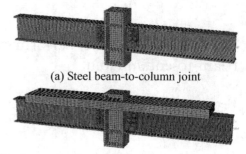

(a) Steel beam-to-column joint

3 DESCRIPTION OF FINITE ELEMENT MODELING TECHNIQUE

FE models developed and numerical simulations performed for the purpose of this study are based on the FE package ABAQUS. The 3D FE modeling technique is used to study numerically the behavior of steel and composite steel-concrete beam-to-column joints. A general 3D FE view of modeled steel and steel-concrete composite joints is depicted in Figure 2. The beam-to-column joint is modeled with two half-span on the left and right side of the column. The half-span is hinged at the reference points for simulating the point of inflections. The top of the column is connected with a hammer that is emulated to discrete rigid. Define a reference point on the discrete rigid to exert mass and velocity for simulating the impact loading. In the calculation, the model is chosen to consider the interaction between the interfaces that adopt the surface-surface contacting element. The tangential friction model is replaced by coulomb friction; the normal contact is hard contact and allows interface separation.

3.1 Modeling of materials and element types

In addition to geometric nonlinearity, the materials employed for joint components are also a major source of nonlinearity. In order to simulate the material performance under impact loading, the strain rate effect should take into consideration by the Cowper-Symonds model (Symonds P S, 1967) for steel components and the model in CEB (Comit Euro-international du Beton, 1990, Wiltshire Redwood Books, 1993) for concrete. In FE analysis, the plastic-damage model available in ABAQUS is used to model the concrete material. The idealized concrete stress-strain relationship is diagrammatically shown in Figure 3. The material properties for steel structural components of steel beams, columns and end-plates are given in Figure 4.

Steel structural components, concrete slabs and bolts are modeled using eight-node linear brick elements with reduced integration C3D8R. Beam element type B32, based on Timoshenko beam theory allowing for shear strain deformations, is used to model the stud. Reinforcements are modeled with T3D2 (Feng, 2007; Beatriz Gil, 2013).

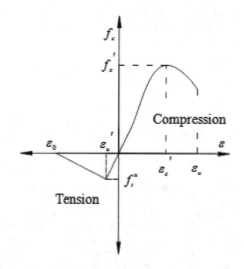

(b) Composite steel-concrete beam-to-column joint

Figure 2.   3D FE model of steel and composite steel-concrete beam-to-column joints.

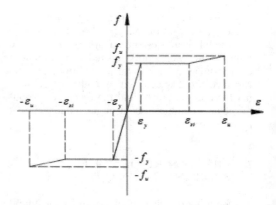

Figure 3.   Idealized uniaxial stress-strain.

Figure 4.   Idealized trilinear stress-strain relationship for concrete slab relationship for steel structural components.

3.2 Results of finite element modeling and validation of FE technique used

The main aim of this study is to trace numerically the behavior of beam-to-column composite joints with concrete slab.

1. The deformation of joints

Deformation in the mid-span of steel and composite joints is given in Figure 5. As shown in Figure 5, the results show that both of the two kinds of joints generate large deformation, crippling appears clearly on the top flange, and the lower flange and web produce large strain. Compared with the steel joint, the composite joints' strain distribution is smaller, which can cause increased deformation in the connection of beams, being unfavorable to stress of joints.

2. The mechanical performance of joints

Moment-rotation and the axial force curve for steel and composite joints are shown in Figure 6. The internal force-time history of plastic hinge in steel and composite joints is shown in Figure 6. The variation tendencies of the bending moment and axial force by two joints are same. For moment-time history, the moment would keep growing in a gradual, linear fashion at the beginning phase, then enter into platform segments of rising slowly, and the moment plummets at the end of the curve.

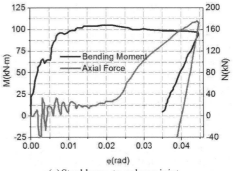

(a) Steel beam-to-column joint

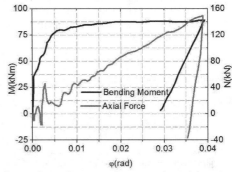

(b) Composite steel-concrete beam-to-column joint

Figure 6. Moment-rotation and axial force-rotation curves for steel and composite steel-concrete beam-to-column joints.

The difference is that the moments of steel joints reach the maximum in platform segments. For axial force-time history, the values of axial force are negative and undulate at the beginning phase, then enter into the linear-rising phase, and finally the linear decrease stage. The model formed a plastic hinge at the beam of steel connections, but not at composite connections.

Considering the concrete slab of the composite joints, the change in the moment of the plastic hinge at the steel beam section is little, although the axial force is rather changeable. The composite joint has the stronger capability of the whole deformation than the steel joint. And the composite joint has smaller deformation under the action of the same load. According to the catenary model in the literature (Ding, 1998; Yang, 2001), these two types of joints in the plastic-hinge stage lack catenary action. Based on the above, one can draw a conclusion that the composite joint has larger rotation compared to the steel joint in the elastic stage and plastic stage, enters the catenary stage later and thus improves the capacity of resisting progressive collapse.

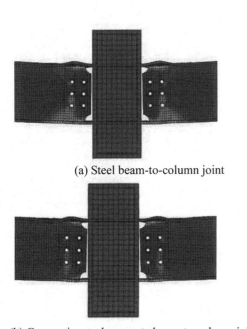

(a) Steel beam-to-column joint

(b) Composite steel-concrete beam-to-column joint

Figure 5. Deformation in the mid-span of joints.

4 CONCLUSIONS

This study investigates the dynamic collapse behaviour of steel and composite connections. The impact analysis of steel and composite connections shows that the steel connection with the concrete slab model has great influence on the behaviour. The conclusions of this study are as follows:

1. The composite connections' strain distribution is smaller than that of steel connections, because of the concrete slabs, which cause increased deformation in the connection of beams.
2. The composite connections have greater anti-collapse capacity than the steel connections, because the composite connections have credible rotation and moment capacities and enter the catenary stage later.

REFERENCES

Abdolreza Ataei, Mark A. Bradford, Xinpei Liu. Experimental study of flush end plate beam-to-column composite joints with precast slabs and deconstructable bolted shear connectors [J]. Structures. 2016(7), 43–58.

Beatriz Gil, Rufino Goni, Eduardo Bayo. Experimental and Numerical Validation of a New Design for Three-dimesional Semi-rigid Composite Joints [J]. Engineering Structures, 2013(48).

Comit Euro-international du Beton. CEB-FIP MODEL CODE [S], 1990. Wiltshire Redwood Books, 1993.

Feng Fu, Dennis Lam, Jianqiao Ye. Parametric Study of Semi-rigid Composite Connections with 3D Finite Element Approach [J]. Engineering Structures, 2007 (29).

GSA, Progressive Collapse Analysis and Design Guidelines for New Federal Office Building and Major Modernization Project [S]. The US General Services Administration, 2009.

Jeyarajan S, J Y Richard Liew, CG Koh. Enhancing the Robustness of Steel-Concrete Composite Buildings under Column Loss Scenarios [J]. International Journal of Protective Structures, 6(3)(2015), 529–550.

Marcela N. Kataoka, Ana Lucia H.C.El Debs. Parametric study of composite beam-column connections using 3D finite element modelling [J]. Journal of Constructional Steel Research, 2014(102), 136–149.

Symonds P S. Survey of Methods of Analysis for Plastic Deformation of Structures under Dynamic Loading [J]. Brown University, Division of Engineering Report, BU/NSRDC, 1967, 1–67.

Yang B, Tan K H. Behaviour of Steel Beam-column joints Subject to Catenary Action under A Column-removal Scenario [J]. Journal of Structural Engineering, ASCE. Submitted for Publication.

Yun-Sun Ding. Load Requirements for Steel Structure of Light-weight Buildings [J]. Building Structure, 1998 (08).

Advances in Energy Science and Equipment Engineering II – Zhou, Patty & Chen (Eds)
© 2017 Taylor & Francis Group, London, ISBN 978-1-138-71798-5

Study on rock mass discontinuities of a rock slope based on stereographic projection

Ji-qing Yang
Department of Civil Engineering, Kunming University of Science and Technology, Kunming, China
Department of Civil Engineering, Yunnan Agricultural University, Kunming, China

Rui Zhang & Jing Li
Department of Civil Engineering, Yunnan Agricultural University, Kunming, China

Xiu-shuo Zhang
Department of Civil Engineering, Kunming University of Science and Technology, Kunming, China

ABSTRACT: This paper presents the results of geotechnical investigation of the rock mass discontinuities of a copper ore district in China. Firstly, geometrical parameters of rock mass discontinuities were analyzed to obtain the distribution pattern of joint fissure and histogram of roughness. Secondly, with the help of software, the density contour of polar stereographic projection was plotted, and the attitude of the preferred structural plane, related to rock mass stability, was determined. Finally, a comparison of statistical results and physical measurements was made which has shown good consistency. The results of the discontinuity data were interpreted with the application of stereographic projection techniques for quantifying the stability of rock mass. Finally, by combining with analysis of the preferred structural plane, the strength and stability of the engineering rock mass could be identified.

1 INTRODUCTION

The characteristics of a rock mass, such as anisotropy, heterogeneity and deformation irregularity, are related to its internal discontinuities and joint fissure. These are the most important characters differing from other rock blocks in which sometimes a small size of fracture could change the mechanical properties of a rock mass dramatically. The analysis of structural stability of a rock mass is on the basis of a comprehensive collection of geometrical data and the statistical methods. Stereographic projection is considered as an ideal technique to interpret results. Therefore, the study of stability and mechanical properties of a rock mass depends on the distribution pattern of its discontinuity (Zhao, 1998; Zhang, 2002; Du, 1999).

Stereographic projection defines the relationship between the spatial orientations of geometrical parameters (points, lines and planes) and their angular shapes. It involves not only the size of the area, but also shows the line length and distance between points. Firstly, stereographic projection makes geometrical parameters parallel translated to the center of the sphere, which then projects points, lines and planes to the sphere, and therefore represents these geometrical parameters with point coordinates. Secondly, the sphere is projected onto a plane from the south (or north) pole. Finally, a particular bijective mapping is obtained (Du, 1999; Gao, 1987).

Rock mass discontinuities in a specific district are highly diverse. The attitude, spacing, continuity and roughness of rock mass discontinuities are also varied. Thus, it is difficult to show every single discontinuity for its great complexity, and hard to make statistical analysis as well (Shi, 2011). Therefore, stereographic projection is used as an alternative to represent the attitude of discontinuities with pole projection.

In this study, original data were collected based on physical measurement of the rock mass discontinuities of a copper ore district in China (Tang, 2008; Xu, 2008; Chen, 2011). And then, geometrical parameters of rock mass discontinuities were analyzed by software which was combined with stereographic projection to simulate the density contour of polar stereographic projection, and to determine the rock mass stability. Finally, with analysis of the preferred structural plane, the state of the engineering rock mass could be identified (Yang, 2011).

2 INVESTIGATION OF DISCONTINUITIES IN ROCK MASS

The scan-line method was used to investigate the joint fissure, and a layout of physical arrangement for survey lines is shown in Figure 1. The measurement was set along with the outcrop of the rock slope at a height of 1 m, and it helped to determine the position of each structural factor. The basic point was set at the start of the investigation point, and then making the tape tighten horizontally. The structural factors were measured and analyzed from the basic point along the line direction. All the measurements were carried out within the range of 1 m around the survey line.

The investigated rock slope was washed with high pressure water jet before measurement, and records were made according to the requirements of the survey as mentioned above.

The main elements of the survey included attitude (strike angle, dip direction angle, and dip angle), spacing, trace length, roughness and continuity of rock mass discontinuities. Based on the physical situation and regulatory requirement, 235° B68 transverse drift, 300° north ore drift, 135° top of the rock slope supporting, 122° B38 top edge supporting, 96° B54 southeast ore drift, 40° B34 transverse drift, and other 6 locations with the slope of a horizontal line at 600 were chosen as investigation sites. The total length of the survey was 280.59 m, with 1224 joints, 18 weak intercalated layers and 1 fault plane also been investigated. The main categories of the rock mass included marble, biotite schist, tuff, dolomitic marble and biotite garnet films.

Investigation results showed that the joint of the rock slope is relatively developed, which were divided into five groups according to the attitude shown as follows: (1) the slope of a horizontal line at 600, 235° northwest ore drift, dip angle 280°~340°, an average of 31°, strike N34E, angle 65°~90°, an average of 80°; (2) 300° north ore drift, dip direction 108°~134°, an average of 12°, strike N20E, angle 63°~81°, an average of 70°; (3) 135° top of the rock slope supporting, dip direction 279°~304°, an average of 29°, strike N22E, angle

44°~58°, an average of 50°; (4) 96° B54 southeast ore drift, dip direction 18°~72°, an average of 45°, strike N40S, angle 35°~88°, an average of 62°; (5) 270° northwest ore drift, dip direction 133°~167°, an average of 15°, strike N58E, angle 38°~96°, an average of 75°.

3 GEOMETRICAL PARAMETERS OF ROCK MASS DISCONTINUITIES

Based on the statistical results of rock mass discontinuities and the analysis of joint fissure by software, the distribution pattern of joint fissure, stereographic projection, the attitude of the preferred structural plane, and the density contour of polar stereographic projection were plotted. In this study, marble was used as the main subject for statistical analysis, and this method could be applied to other similar joint fissures.

3.1 Spacing of discontinuities

Spacing of discontinuities is the most important parameter to represent the intensive level of discontinuities, which can be shown with the density of discontinuities, and it includes three different types. The first type is the density of lines, also known as frequency, which means the number of discontinuities per unit length. The second type is the density of planes, which means the number of discontinuities per unit area, and it could be obtained by sampling. The third type is the density of volume, which means the number of discontinuities per unit volume, also known as volumetric joints.

The average spacing of discontinuities in the linear density can be determined by Eq. (1).

$$d_p = L/N \qquad (1)$$

where d_p is the average spacing of joints, L the length of measured lines, and N is the number of joint fissure intersecting with measured lines.

The number of joints in volume can be determined by Eq. (2).

$$j_v = S_1 + S_2 + S_3 + ... + S_n + S_k \qquad (2)$$

where j_v is the number of joints in volume, S_n is the number of joints above measured line at the height of 1 m and n is the normal vector, and S_k is the number of joints in each cubic meter of rock mass.

The measurement was set along with the typical outcrop of the rock slope, with the dimension not less than 2 m × 5 m. The extended length of the joint which is greater than 1 m should be taken

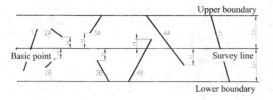

Figure 1. The layout of physical arrangement for survey lines.

into account. In rock engineering standards, the quality of the rock mass was determined by two parameters, which were the compressive strength R_c and integrity of the coefficients K_v. K_v could also be obtained by using Table 1 when it is difficult to make measurement.

In this way, an appropriate indicator could be applied to determine the degree of development of the rock mass discontinuities. Spacing of discontinuities is an important indicator representing the integrity of the rock mass. With smaller spacing, the rock mass is divided more seriously, and the integrity of the rock mass is getting worse. The statistical results have shown that the average spacing of joint fissure of rock mass discontinuities was 0.3 m as shown in Figure 2. The percentage of spacing within 10 cm~20 cm was more than 75%, which indicated intensive joints and highly developed discontinuities.

3.2 Trace length

The trace length is the length of joint fissure in the outcrop of the rock mass. With longer trace lengths, the rock mass is more affected. As the rock mass was deeply buried with only a small exposed part available for survey, it was not possible to measure the extended length. Therefore, the estimation of the total trace length based on measurement was hard to achieve.

The trace length of discontinuities has been studied for a long time. Priest S.D. and Hundson J.A.

Table 1. The parameter list of j_v and K_v.

j_v (joints/m³)	3	3~10	10~20	20~35	>35
K_v	>0.75	0.75~0.55	0.55~0.35	0.35~0.15	>0.15

made a hypothesis that the distribution pattern of the trace length of discontinuities was negatively exponential, and therefore obtained the formula as shown in Eq. (3) by the scan-line method.

$$L = C/2(1 - \sqrt{(n-r)/n}) \qquad (3)$$

where L is the average trace length of discontinuities, C is the truncated semi-trace length, n is intercutting of the truncated semi-trace length in population, and r is the number of semi-trace length which is less than C in sampling. This formula is only applied in the situation of sample sizes large enough.

The trace length of discontinuities could also be determined by sampling methods. Firstly, a rectangular area with a length of a and a width of b located at the outcrop of the rock slope was taken as the sampling area. Secondly, the relationships between the sampling size, the number of discontinuities, and the trace length were determined. Thirdly, the formula for calculating the average of trace lengths was conducted based on the interaction between discontinuities and sampling size.

3.3 Captions/numbering

The fluctuation and roughness of discontinuities have great impacts on the shear properties of the structural plane. With greater roughness, the shear resistance of discontinuities is higher, and the quality of the rock mass is better, despite the direction. The roughness of the rock mass in this study was divided into step type, wavy and straight-type as shown in Fig. 3.

Most discontinuities of the rock mass in the sampling area were straight-type, with only a few samples were considered as wavy type. And the step type was rare which would destabilize the structure of the rock mass.

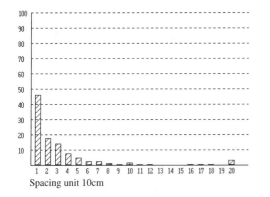

Figure 2. The distribution pattern of joint fissure.

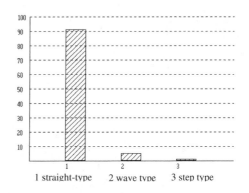

Figure 3. The roughness of joint fissure.

3.4 *Dip angle*

The dip angle is one of the important parameters influencing the spatial distribution of the discontinuities. Most of the discontinuities of joint fissure were steeply inclined, with dip angles ranging between 70° and 90° as shown in Figure 4. The stability of the rock mass was good in normal circumstances due to the steep angle. But, if the outcrop of the rock slope was fluctuated, when the excavated rock was cut into isolated blocks, it would be easy to fall off, causing safety problems. Therefore, close attention should be paid to the unstable rock mass on the top of the rock slope in engineering practice.

3.5 *Dip direction*

As can be seen from the dip direction histogram (Fig. 5), mainly for the dip direction between 40° and 60°, the stability of the rock slope could be preliminarily determined by comparing the preferred orientation of the dip joint with the dip direction.

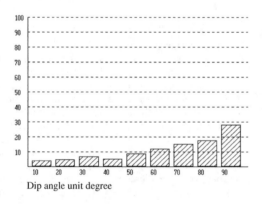

Dip angle unit degree

Figure 4. The dip angle of joint fissure.

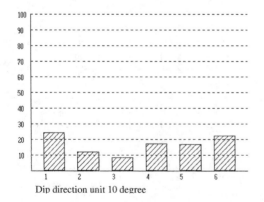

Dip direction unit 10 degree

Figure 5. The dip direction of joint fissure.

4 THE DENSITY CONTOUR OF POLAR STEREOGRAPHIC PROJECTION AND ANALYSIS OF THE PREFERRED STRUCTURAL PLANE

The presence of differences in the discontinuities is an important indicator that distinguishes the rock mass from each other, and therefore determines the engineering characteristics of the rock mass. More discontinuities indicate that the rock mass is divided into smaller blocks, so that the complexity of geometry would increase with degree of crushing. The structure of rock mass discontinuities is often generated by a certain combination with the regular distribution. On the one hand, they usually develop in groups with a certain degree of dispersion, but on the other hand, the development of discontinuities in different groups, such as the number and density, is often extremely uneven. Regular and complex distribution patterns would be presented after being subjected to repeated tectonic movements. In order to determine the number of rock slope and development of discontinuities in groups, as well as their spatial distribution, the stereographic projection method and pole density map method were applied to compare the development of discontinuities in groups (She, 2011).

In order to evaluate the geological conditions of the rock slope and the stability, the entire slope was zoned according to engineering geological conditions. Finally, the density contour of polar stereographic projection was obtained by equal area net (Schmidt net) (Fig. 6.).

In the rock mass structural analysis, the integral stability could be determined by stereographic projection, which was plotted based on the attitude of the preferred structural plane and its interaction with the rock slope. The preferred joint fissure of the rock mass and the location of the rock slope

Figure 6. The density contour of polar stereographic projection.

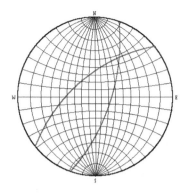

Figure 7. The stability analysis of preferred structural plane.

were projected on the stereographic projection map, and the results shown that the first and second groups of preferred structural planes created unstable blocks (Fig. 7).

As the impact is high on the stability of the rock mass, further engineering measures should focus on the safety problems, such as cleaning up the roadway pumice or reinforcement by anchors (Hou, 2011).

5 CONCLUSIONS

The main conclusions drawn from investigation and statistical results are presented as follows. (1) The distribution pattern of joints, roughness, dip angle, dip direction, and the density contour of the polar stereographic projection have shown good consistency with physical measurements. (2) The spacing of discontinuities was obtained by employing the scan-line method, and the optimal survey length of different rock mass continuities could be determined at certain confidence levels. (3) Stereographic projection is an ideal

fractal-geometry method to analyze the planes and linear space by calculating the projection ratio, and therefore, the stability of the rock slope influenced by rock mass discontinuities could be quantified and expressed by diagrams. (4) The properties of the rock mass depend on its internal structural characteristics, and thus the strength and stability of the engineering rock mass could be identified by analysis of the preferred structural plane.

REFERENCES

Dong-mei She, Sheng-xie Xiao. The western traffic science and technology, Application of stereographic projection method in slope stability analysis, (2011).

Fei Zhang et al. Non ferrous metals, Study on the fractal of the pole figure of the rock structure plane in the plane of the red flat pole. 54(5):21–23 (2002).

Ji-qing Yang, Yi Li. Theory of urban construction, Research on the contribution of artificial neural network to slope engineering, 10:252–255 (2011).

Jun-zhi Chen, Ji-qing Yang et al. Yunnan metallurgy, Study on the 3D modeling of rock mass, 10:3–6 (2011).

Ke-peng Hou, Shi-ye Yang, Hui-ming Yang. Non ferrous metals, Stability analysis of rock slope based on structural plane, 63 (2011).

Lei Gao. Rock mechanics of mine. Beijing: Machinery Industry Press, 70 (1987).

Li-qin Chen, Cheng Chen et al. Engineering technology, Based on stereographic projection of structural plane analysis, 28:108–112 (2014).

Peng Tang, Ze-min Xu. Mining research and development, Mechanism analysis of rock structure weathering front expansion, 28(4): 9–11 (2008).

Shi-gui Du. The engineering properties of rock mass structural plane. Beijing: Earthquake Press, 60 (1999).

Ze-min Xu. Earth Science Frontier, Slope unsaturated zone with low permeability rock structure weathering front expansion process, 15(4):258–268 (2008).

Zhao, W., C. Tang. China mining industry, Measurement theory of structural plane spacing and trace length. 7(3):36–37 (1998).

Zhen-ming Shi, Xian-li Kong. Engineering geology. Beijing: China Architectural Industry Press, 53 (2011).

The effect of vertically increasing the number of outrigger systems with a belt in concrete high-rise buildings under wind loading

Abla Krouma & Osama Mohamed
Department of Civil Engineering, Abu Dhabi University, Abu Dhabi, UAE

ABSTRACT: A variety of lateral force resisting systems exist to meet the demand for taller buildings typically at locations where land is at a premium. The design of tall buildings is typically governed by the stiffness and drift rather than strength, and one effective system for control of drift is the outrigger-belt system. Outrigger-belt systems engage a stiff core and perimeter columns in order to improve the structure stiffness and reduce the lateral displacements. This study assesses the effect of the number of outriggers and their locations in reducing the lateral drift. The study showed that the optimum location of one outrigger system for maximum drift reduction is the mid-height of the building height. If two-outrigger systems are used, the optimum location for reducing the lateral drift is to place one system at one third of the building height and the second system and three quarters of the building height. The lowest drift was obtained for a system of three outriggers when outriggers are placed at one quarter, one third, and three quarters of the building height. Adding forth outrigger at the top of the building, known as the hat, reduces displacements only by 6%. The hat outrigger offers minimal contribution to the control of lateral drift.

1 INTRODUCTION

The need of high-rise buildings has increased in many cities over the last few decades to maximize the utilization of land. The first tall building in Chicago was constructed after the 1871 great fire, followed by significant increase in the number of tall buildings. The use of tall structures extended from office to residential buildings in 21st century (Sandelin, 2013).

The number of high-rise buildings is increasing every year, and based on the latest research conducted by the Council on Tall Buildings and Urban Habitat (CTBUH) in 2015, Asia has 81 new tall buildings which represent 76% of the total new tall buildings in 2015 followed by Middle East with 8.5% which equals to 9 new buildings. Currently, the tallest building in the world is Burj Khalifa in Dubai which was completed in 2010 with 828-meter height. The tower has outrigger systems represented by shear walls every 30 stories (Gabe, 2015; Emaar).

Nowadays, high-rise buildings are used as architecture and aesthetic phenomena in capitals, and in order to achieve the desired height, structural systems have evolved. Structural systems progressed from rigid frames to tube and outrigger systems in 1960s. Once the building height exceeds the normal limit, the design concept changes. Khan (1969) concluded that lateral displacement under lateral load controls the design when the building height exceeds 10 stories with lateral stiffness dominating over strength. One of the effective methods to control the drift is the outrigger systems, which provides lateral stiffness by engaging both a stiffness central core with perimeter columns (Ali, 2007).

This paper presents a review on outrigger systems and their role in reducing lateral displacements and increasing building stiffness. A case study tall building was developed to model and investigate the optimum location and number of outrigger systems vertically across the building height. The analysis was done using the finite element program ETABS, produced by Computers and Structures. Emphasis is on gravity and wind load only, in order to demonstrate the concept. The contribution of outriggers at the top of the building, also known as hat, is also examined.

2 OUTRIGGER SYSTEMS' PURPOSE AND TYPES

The main purpose of the outrigger is to place horizontally system structural elements that connect the building core to perimeter columns in order to increase the structure stiffness. As shown in Figure 1, the outrigger system acts as an arm between the core and exterior columns to reduce the drift and increase the overall lateral stiffness.

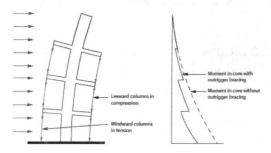

Figure 1. The core and outrigger system behaviour (Choi, 2012).

In addition to reduced lateral deformation, flexural resistance is also enhanced.

The addition of the belts at the level of the outriggers increases the efficiency of the lateral load resisting system. The reduction of the displacements and moments depends on the stiffness of outriggers, core and columns. It also depends on floor plan dimensions and the vertical location of the outriggers (Choi, 2012; Taranath, 2011).

Systems with outriggers exhibit higher flexural resistance and drift control. The core forces at the base of the core will be reduced as well. The core and outrigger system types are classified based on the material type and number of stories (Choi, 2012; Ho, 2016).

These systems consist of vertical elements, such as shear walls, deep beams, and bracing. The outriggers in the floor level are of two types: direct outriggers, which are located on the short load path between the core walls and the column, and indirect outriggers, which are located on the building perimeter and known as belts. These two types could be used separately or together based on the building floor plan and conditions (Choi, 2012).

The outrigger system floor is often used for mechanical equipment and services. The number and location of mechanical floors in each building differs based on the country design code, total area, dimensions, type and most importantly, the architecture design of the building. Typically, a minimum of one mechanical floor exists in a tall building (Khajehpour, 2001). In addition, the location of the outrigger systems will differ depending on the building height and the drift value which needs to be controlled (Choi, 2012).

3 CASE STUDY OVERVIEW

The case study examined in this paper is a symmetric 60-story reinforced concrete building with flat slab floor systems and exterior 1000×500 mm deep beams, along with a central concrete core.

The building is modelled as an office building with a 250 mm thick flat slab floor system. The total height of the building is 210.5 meters, including a 4-meter high lower level and 3.5-meter high typical floors. The symmetric floors consist of 6 and 5 bays in x- and y-direction, respectively. Spans are 7-meters long and are equal in each direction. The floor plan with and without the outrigger system is shown in Figures 2 and 3.

The outriggers consist of shear walls with thickness similar to the core thickness, and deep belt at the perimeter of each floor to engage the exterior columns. The concrete compressive strength of floor and beams is 50 MPa, while that of columns and shear walls is 70 MPa. The cross-section dimensions of the structural elements are shown in Table 1.

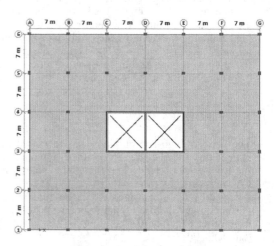

Figure 2. Floor plan without outrigger system.

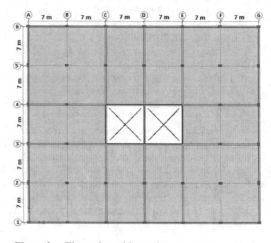

Figure 3. Floor plan with outrigger system.

1062

Table 1. Structural elements sections in mm.

Floors number	Columns sections (mm)	Shear wall sections (mm)
Base to 20	1200×1000	500
21 to 40	1000×800	350
41 to 60	800×600	250

The applied loads on floors are as follows: the live load equals to 3.88 KN/m^2 and the superimposed dead load was considered to be 2.25 KN/m^2. The wall load on the exterior beams is equal to 2.6 KN/m^2. The load combination used for comparison is DL+0.5 LL+0.7 W as per ASCE 7–10, appendix C, under serviceability conditions (American Soicety of Civil Engineers (ASCE), 2013).

The case study modelling was done using ETABS v15.2, and the design was based on ACI 318M-08 requirements (American Concrete Institute (ACI), 2008). The analysis results included displacement comparison using different numbers of outriggers in the building at different vertical locations. A total of seventeen cases were used in conducting the study and evaluating the effect of the number and location of outriggers on reduction of lateral deformation.

4 CASE STUDY ANALYSIS RESULTS AND DISCUSSION

4.1 One outrigger system in different vertical locations

In this section, the effect of the location of one outrigger in controlling lateral drift is examined. The maximum lateral deformation obtained by using one outrigger system at five different vertical locations across the height of the building is shown in Figures 4 and 5. The comparison showed that using one outrigger at 1/2 of the building height provides the least drift in x and y directions which is compatible with previous studies (Kian, 2001; Nanduri, 2013). It is worthy to note that outrigger systems offer not only reduction in lateral deformation, but also contributes to mitigation of the potential for progressive collapse due to loss of a critical vertical load-carrying element (Mohamed, 2011). When using outrigger systems at the optimum location, the displacement decreased by 42% in the x-direction and 35% in the y-direction compared to the same building without outriggers. Using an outrigger system at the top of the building reduced the drift by 17% and 10% in x- and y-directions, respectively. Therefore, a top floor outrigger contributes the least to decreasing lateral deformations.

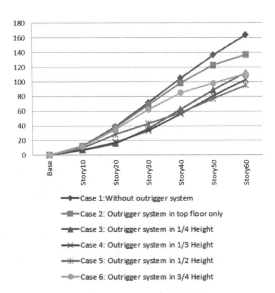

Figure 4. Comparison of the displacements in different vertical locations for one outrigger system in the x-direction.

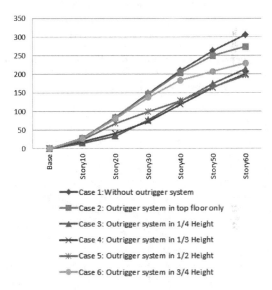

Figure 5. Comparison of the displacements in different vertical locations for one outrigger system in the y-direction.

4.2 Two outrigger systems in different vertical locations

In order to determine the optimum location of two-outrigger systems, six positions were examined. Determination of lateral deformation showed that using two outriggers at 1/3 and 3/4 of the building height provides the maximum reduction in lateral deformation compared to all other positions, as shown in Figures 6 and 7. The reduction percentages of the displacements were 57% and 52% in x- and y-directions, respectively, compared to the

case without outriggers. In addition, using a roof-level outrigger system offered minimal contribution to controlling lateral deformations.

4.3 Three and four outrigger systems in different vertical locations

Five locations of three and four outrigger systems were investigated to determine optimal locations for control of lateral deformations and the results are shown in Figures 8 and 9. The comparison showed that using three outriggers at 1/4, 1/2 and 3/4 of the building height provides the maximum reduction in lateral deformation.

The percentage reduction in lateral displacements in x- and y-directions was 66% and 63%, respectively, compared to the case without outriggers. The use of one of the three or four outrigger systems in the top floor was least effective in reducing lateral deformation. Nonetheless, when three and four outriggers are used, the loss in effectiveness due to the roof-level is smaller than when fewer total outriggers are used, as demonstrated in the previous section.

4.4 Comparison between different numbers of outrigger systems in different vertical locations

The comparison included the best cases which provide the least lateral displacements among the

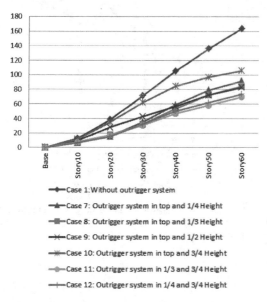

Figure 6. Comparison of the displacements in different vertical locations for two outrigger systems in the x direction.

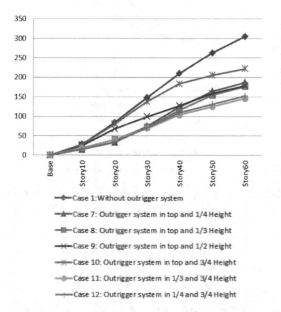

Figure 7. Comparison of the displacements in different vertical locations for two outrigger systems in the y direction.

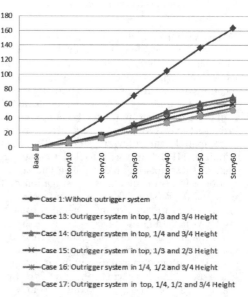

Figure 8. Comparison of the displacements in different vertical locations for three and four outrigger systems in the x-direction.

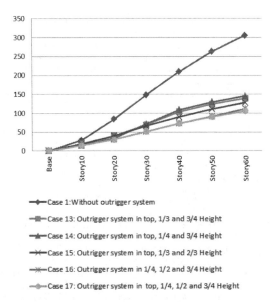

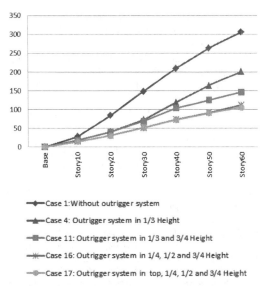

Figure 9. Comparison of the displacements in different vertical locations for three and four outrigger systems in the y-direction.

Figure 11. Comparison of the displacements in best locations for one, two, three and four outrigger systems in the y-direction.

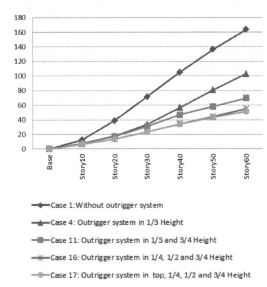

Figure 10. Comparison of the displacements in best locations for one, two, three and four outrigger systems in the x-direction.

seventeen cases for one, two, three and four outrigger systems as shown in Figures 10 and 11.

Using two outriggers at 1/3 and 3/4 of the building height compared to using one outrigger at 1/2 of the building height reduced displacements by 27% in the x-direction and 26% in the y-direction.

The use of three outriggers at 1/4, 1/2 and 3/4 of the building height compared to using two outriggers at 1/3 and 3/4 of the building height reduced displacements by 21% in the x-direction and 23% in the y-direction. Adding forth outrigger at the top to the case where three outriggers are used at 1/4, 1/2 and 3/4 of the building height reduced displacements only by 6% in the x-direction and y-direction.

5 CONCLUSION

The outrigger systems are efficient and economical in controlling the lateral drifts for high-rise buildings. In case of using one outrigger system in the vertical direction, the system located at half of the building height reduces the displacement by approximately 39% compared to the case without outrigger systems. The use of two outrigger systems at 1/3 and 3/4 of the building height reduces lateral displacement by approximately 55% compared to the case without outrigger systems. Using three outriggers at 1/4, 1/3 and 3/4 of the building height reduces lateral displacements by 65%, while adding fourth outrigger at the top level increases the lateral displacement reduction to 67%.

However, the improvement in the displacements due to increasing the number of outriggers in the vertical direction decreases as the number of outriggers is increased. For example, using two outriggers at 1/3 and 3/4 of the building height compared to one outrigger at half of the building height reduces displacements by 26%. The lateral displacement reduction is only 22% when using three outriggers at 1/4, 1/2 and 3/4 of the building height. The addition of the forth outrigger at the top compared to the three optimal outriggers at 1/4, 1/2 and 3/4 of the building height adds an additional lateral reduction of 6%, which may not always motivate the structural designer to consider adding this hat outrigger.

Generally, a roof-level outrigger contributes minimally to control of lateral deformation.

REFERENCES

Ali, M., Moon, K. *Structural Developments in Tall Buildings: Current Trends and Future Prospects*, Architectural Science Review, 50(3), 205–223 (2007).

American Concrete Institute (ACI), *Building Code Requirements for Structural Concrete (ACI 318M-08) and Commentary* (2008).

American Soicety of Civil Engineers (ASCE), *Minimum Design Loads for Buildings and Other Structures* (ASCE/SEI 7-10) (2013).

Choi, H., Ho, G., Joseph, L., Mathias, N. *Outrigger Design for High-Rise Buildings: An output of the CTBUH Outrigger Working Group*, The Council on Tall Buildings and Urban Habitat (CTBUH), Chicago (2012).

Choi, H., Tomasetti, T. Joseph, L. *Outrigger System Design Considerations*, International Journal of High Rise Buildings, 1(3), 237–246 (2012).

Emaar Properties PJSC. Burj Khalifa. [Online]. http://www.burjkhalifa.ae/en/the-tower/structures.aspx

Gabe, J., M., Gerometta, M. *CTBUH Year in Review: Tall Trends of 2015 and Forecasts for 2016*, The Council on Tall Buildings and Urban Habitat (CTBUH), Review report (2015).

Ho, G. *The Evolution of Outrigger Systems in Tall Building*, International Journal of High Rise Buildings, 5(1), 21–30 (2016).

Khajehpour, S. *Optimal Conceptual Design of High-Rise Office Buildings*, University of Waterloo, Waterloo, Ontario, Doctor of Philosophy Thesis (2001).

Kian, P., Siahaan, F. *The Use of Outrigger and Belt Truss System for High-Rise Concrete Buildings*, Dimensi Teknik Sipil, 3(2) (2001).

Mohamed, O.A. *Modelling of Steel Structures for Progressive Collapse Mitigation: Contribution of Lateral Force Resisting System*, The 13th. International Conference on Civil, Structural and Environmental Engineering Computing—CC2011, Civil-Comp Proceedings, ISSN 1759–3433, (2011).

Nanduri, P.M.B., Suresh, B., Hussain, MD. *Optimum Position of Outrigger System for High-Rise Reinforced Concrete Buildings Under Wind And Earthquake Loadings*, American Journal of Engineering Research (AJER), 2(8), 76–89 (2013).

Sandelin, C., Budajev, E. *The Stabilization of High-rise Buildings: An Evaluation of the Tubed Mega Frame Concept*, Department of Engineering Science, Applied Mechanics, Civil Engineering, Uppsala University, Uppsala (2013).

Taranath, B. *Structural analysis and design of tall buildings; steel and composite construction*, New York: CRC Press (2011).

Advances in Energy Science and Equipment Engineering II – Zhou, Patty & Chen (Eds)
© 2017 Taylor & Francis Group, London, ISBN 978-1-138-71798-5

Development of the relation between maximum moment and maximum displacement of the anchor cable frame beam based on design theory

YanLei Zhang, WanKui Ni, HuanHuan Li, QingQing Zhu & Bo Jing
College of Geological Engineering and Geomatics, Chang'an University, Xi'an, China

ABSTRACT: In order to early warn and forecast the slopes or landslides treated by the prestressed anchor cable frame beam along an expressway in Shaanxi province, a typical section of landslide treatment along the expressway was studied. Based on the Winkler model and the design theory, the relationship between the maximum moment and maximum displacement of the beam was divided into two situations of the same or different anchoring forces to study. First, under the condition of the same anchoring forces, an expression for the critical length of the cantilever end was determined by means of data fitting when the position of maximum moment changed. Then, the accurate position of the maximum moment was ascertained. Second, according to the exact position of the maximum moment on the beam, the relationship between the maximum moment and maximum displacement was obtained through a formula derivation when the anchoring forces were same. Third, in the case of different anchoring forces, the relationship between the two was directly determined through data fitting of First Optimization. Finally, the formulas were proved reasonable by an engineering example. In this way, as long as the related parameters of the landslide are known, the relationship between the maximum moment and maximum displacement will be obtained based on the design scheme of the prestressed anchor cable frame beam. Relying on the monitoring data of displacement, we can apply the research results to early warning and forecasting of the landslides.

1 INTRODUCTION

As a new type of retaining structure, the prestressed anchor cable frame beam has been widely used in the reinforcements of slopes (Xia, 2002; Shi, 2004). At present, there have been few studies on the anchor cable lattice of foreign scholars, and the foreign literature on this aspect can rarely be seen. However, domestic scholars have done lots of studies on the calculation methods of internal force (Li, 2000; Yang, 2002; Xiao, 2002; Song, 2004), design methods (Liu, 2004; Liang, 2006), numerical analysis methods [9] and experiments [10] of the prestressed anchor cable frame beam, while the relation between maximum moment and maximum displacement of the prestressed anchor cable frame beam has been less studied.

In recent years, expressways have developed rapidly in Shaanxi province, but the landslide and other disasters have also increased. According to incomplete statistics, in the period of the construction for highways in the mountainous area of Shaanxi province, as many as 200 landslides with obvious instability signs had been found. The stability of the landslide section has become the key factor that affects the safe operation of expressways. Because the prestressed anchor cable frame beam is one of the main measures for the landslides

along the expressway in Shaanxi province, the control effect and the early warning and forecasting of the landslide after the treatment have become especially important.

2 THEORETICAL BASIS

For the moment, the calculation methods of internal force for the prestressed anchor cable frame beam, which are commonly used, mainly have two kinds: one kind is to assume that the distribution of subgrade reaction is linear, and then use the method of inverted beam or continuous beam to calculate; this method does not strictly calculate internal force according to actual stress characteristics of the prestressed anchor cable frame beam. The other kind is to use the method of elastic foundation beam, which is mainly divided into the Winkler elastic foundation model, the double parameter elastic foundation model and the elastic half-space foundation model. By comparison, the Winkler model is the simplest model and it is the most widely used in engineering, but the calculation processes of two other kinds of models are both complex.

Generally, in practical engineering, the prestressed anchor cable frame beam is often regarded as a finite length beam to calculate, and the longitudinal and transverse beams are considered separately. This study uses the Winkler elastic foundation model, and considering the boundary conditions of the frame beam, $M_0 = 0$ and $Q_0 = 0$, referring to the literature (Ninth Design, 1983), the expression of the moment and displacement of the prestressed anchor cable frame beam can be shown as follows:

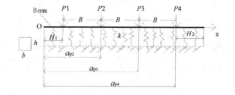

Figure 1. Schematic diagram of the Winkler model.

$$
\left.
\begin{aligned}
M(x) &= -y_0 \frac{kb}{\lambda^2} F_3(\lambda x) + \theta_0 \frac{kb}{\lambda^3} F_4(\lambda x) + \frac{1}{\lambda} \sum_{i=0}^{j} P_i F_2(\lambda(x - a_{p_i})) \\
y(x) &= y_0 F_1(\lambda x) - \theta_0 F_2(\lambda x)/\lambda + \frac{kb}{4\lambda} \sum_{i=0}^{j} P_i F_4(\lambda(x - a_{p_i}))
\end{aligned}
\right\}
\tag{1}
$$

In the above formula, k is the coefficient of foundation resistance, b is the width of the cross-section, λ is the characteristic coefficient, and the formula for λ is shown as follow: $\lambda = (kb/(4EI))^{0.25}$, among them, E is the elastic modulus of concrete and I is the moment of inertia, the formula of which is shown as follows: $I = bh^3/12$, where h is the height of the cross-section; P_i is the anchoring force of the i cable which is distributed to longitudinal and transverse beams in the vertical direction, and the distributive formula of anchoring force refers to literature (Zhou, 2007); j is the number of concentrated load on the left side of the calculation section, and when the value of j is zero, the value of P_0 will be zero, too; a_{p_i} is the distance between the i cable and the beam's end; if we assumed H_1 as the origin of coordinates, a_{p_i} can be shown as follows: $a_{p_i} = H_1 + (i-1)B$, $i = 1, 2, 3, ..., n$, in the formula, H_1 and H_2 are the lengths of the cantilever end, B is anchor spacing, and n is the number of anchor cable; $F_1(\lambda x)$, $F_2(\lambda x)$, $F_3(\lambda x)$ and $F_4(\lambda x)$ are Krylov functions, whose expressions are expressed as formula (2); y_0 and θ_0 are the initial parameters, which can be obtained by substituting $M_0 = 0$ and $Q_0 = 0$ into the expressions of moment and shear force according to the literature (Ninth Design, 1983), and the expressions are shown in formula (3). In the expression, L is the length of the beam. A schematic diagram of the Winkler model is shown in Figure 1.

$$
\left.
\begin{aligned}
F_1(\lambda x) &= \cosh(\lambda x)\cos(\lambda x) \\
F_2(\lambda x) &= 0.5[\cosh(\lambda x)\sin(\lambda x) + \sinh(\lambda x)\cos(\lambda x)] \\
F_3(\lambda x) &= 0.5\sinh(\lambda x)\sin(\lambda x) \\
F_4(\lambda x) &= 0.25[\cosh(\lambda x)\sin(\lambda x) - \sinh(\lambda x)\cos(\lambda x)]
\end{aligned}
\right\}
\tag{2}
$$

3 DERIVATION OF THE RELATIONSHIP

In the design of the prestressed anchor cable frame beam, the design values of anchoring forces may be same or different. Considering the applied prestress and the degree of the loss of prestress, the anchoring forces of the anchor cable frame beam may also be same or different. Therefore, the relationship between the maximum moment and maximum displacement can be divided into same or different anchoring forces to study.

3.1 Same anchoring forces

For the slopes or landslides reinforced by the prestressed anchor cable frame beam along the highways in Shaanxi province, the design values of anchoring forces of the same beam are generally same in design. If the same prestress is applied with no attenuation or attenuation to the same degree, the anchoring forces of the same beam will be the same.

When the anchoring forces of a beam are same, that is Pi = P, the expression of the moment and displacement of the beam can be simplified, and then the relationship between them is obtained, which is shown in formula (4). In the formula, y1 and θ1 can be obtained according to formula (3), and they are shown as formula (5).

$$
\left.
\begin{aligned}
y_0 &= \frac{\lambda}{kb} \cdot \frac{F_3(\lambda L) \cdot \sum_{i=0}^{n} P_i F_2(\lambda(L - a_{p_i})) - F_4(\lambda L) \cdot \sum_{i=0}^{n} P_i F_1(\lambda(L - a_{p_i}))}{[F_3(\lambda L)]^2 - F_2(\lambda L) \cdot F_4(\lambda L)} \\
\theta_0 &= \frac{\lambda^2}{kb} \cdot \frac{F_3(\lambda L) \cdot \sum_{i=0}^{n} P_i F_1(\lambda(L - a_{p_i})) - F_2(\lambda L) \cdot \sum_{i=0}^{n} P_i F_2(\lambda(L - a_{p_i}))}{[F_3(\lambda L)]^2 - F_2(\lambda L) \cdot F_4(\lambda L)}
\end{aligned}
\right\}
\tag{3}
$$

$$y = M \cdot \frac{\lambda^2}{kb} \cdot \frac{y_1 \cdot kb\lambda F_1(\lambda x) + \theta_1 \cdot kbF_2(\lambda x) + 4\lambda^2 \sum_{i=0}^{j} F_4(\lambda(x - a_{p_i}))}{y_1 \cdot kb\lambda F_3(\lambda x) + \theta_1 \cdot kbF_4(\lambda x) - \lambda^2 \sum_{i=0}^{j} F_2(\lambda(x - a_{p_i}))} = M \cdot \frac{\lambda^2}{kb} \cdot \frac{T_1(x)}{T_2(x)} \tag{4}$$

$$y_1 = y_0/P, \quad \theta_1 = \theta_0/P \tag{5}$$

By analyzing formula (4), we can find that only when the locations of the maximum moment and maximum displacement on the beam are the same, their relationship can be directly obtained by formula (4). Assuming they are all located in $x = xmax$, their relationship can be shown as follow:

$$y_{max} = M_{max} \frac{\lambda^2}{kb} \cdot \frac{T_1(x_{max})}{T_2(x_{max})} \tag{6}$$

However, when the positions of the maximum moment and maximum displacement are not consistent, the relationship between them cannot be derived from formula (4).

Through the study of a large number of calculations and materials of the slopes or landslides reinforced by the prestressed anchor cable frame beam along the highways in Shaanxi province, we found that the maximum displacement is basically located in the middle of the beam or joints at a smaller distance from the middle of the beam, but the position of the maximum moment is not fixed.

Through the research, it is found that in the case of the same anchoring forces, the maximum moment is always located at the node of anchor cable at the end of the beam or in the middle of the beam, but the accurate position is related to the length of the cantilever end of the beam. There exists a critical length of the cantilever end named Hcr when the position of the maximum moment changes. And Hcr is mainly related to the coefficient of foundation resistance (x1), moment of inertia (x2), and anchor spacing (x3). The different values of x1, x2 and x3 are substituted into the expression of moment, and then the value of moment is obtained. Through constant trial and observation, Hcr is finally obtained corresponding to x1, x2 and x3. Then, we used the software named First Optimization to fit the data and get the expression of Hcr, which is shown in formula (7).

$$H_{cr} = \frac{p_1 + |AC| + |ADA^T| + p_{11} \ln(x_1)x_2x_3}{1 + |AE| + |AFA^T| + p_{21} \ln(x_1)x_2x_3} \tag{7}$$

The correlation coefficient of formula (7) is 0.997, and the fitting coefficients can be seen in Table 1. In formula (7), A is the parameter matrix and C, D, E and F are fitting coefficient matrixes, and they are all shown below:

$$A = [\ln(x_1), x_2, x_3]; \quad C = [p_2, p_3, p_4]^T; \quad E = [p_{12}, p_{13}, p_{14}]^T;$$

$$D = \begin{bmatrix} p_5 & p_8 & p_9 \\ 0 & p_6 & p_{10} \\ 0 & 0 & p_7 \end{bmatrix}; \quad F = \begin{bmatrix} p_{15} & p_{18} & p_{19} \\ 0 & p_{16} & p_{20} \\ 0 & 0 & p_{17} \end{bmatrix}.$$

Therefore, when the lengths of the cantilever end are all less than Hcr, the maximum moment is located in the middle of the beam or joints at a smaller distance from the middle of the beam; however, when H1 or H2 is greater than Hcr, the maximum moment is located at the node of anchor cable at the end of the beam.

In summary, only when the cantilever lengths are all less than Hcr, we can think that the positions of the maximum moment and maximum displacement are consistent. In such a situation, assuming that H1 was less than H2, the relationship between the maximum moment and maximum displacement can be obtained by formula (6), which is shown as (8) and (9). In the formula, ycr is the maximum displacement on the beam and Mmax is the maximum moment on the beam.

$$n \text{ is an odd number, } y_{cr} = M_{max} \cdot \frac{\lambda^2}{kb} \cdot \frac{T_1|_{x=H_1 + B(n-1)/2}}{T_2|_{x=H_1 + B(n-1)/2}} \tag{8}$$

$$n \text{ is an even number, } y_{cr} = M_{max} \cdot \frac{\lambda^2}{kb} \cdot \frac{T_1|_{x=H_1 + B \cdot n/2}}{T_2|_{x=H_1 + B \cdot n/2}} \tag{9}$$

3.2 Different anchoring forces

For the two kinds of situations, the same anchoring forces but the different positions of the maximum moment and maximum displacement, the different anchoring force, I, use the software named First Optimization to fit the data in order to set up the relationship between the maximum moment and maximum displacement of the beam.

The fitting process is shown below: first, the values of all parameters are substituted into the expressions of the moment and displacement of the beam, and then the values of moment and displacement are obtained; second, select the values of the maximum

Table 1. The values of the coefficients of H_{cr}.

Coefficient	Value	Coefficient	Value	Coefficient	Value	Coefficient	Value
P_1	−449.72	P_7	5.45326	P_{13}	556546.9	P_{19}	−2.51487
P_2	87.43	P_8	−6768.86	P_{14}	−78.977	P_{20}	−35963.76
P_3	55339.37	P_9	0.00152	P_{15}	1.21042	P_{21}	3942.25
P_4	−74.961	P_{10}	122277.4	P_{16}	−1114236.7		
P_5	−2.56545	P_{11}	−7662.81	P_{17}	10.7784		
P_6	−2176327.5	P_{12}	9.03299	P_{18}	−42886.9		

moment and maximum displacement, and then a lot of values of the two can be obtained by changing the parameters; third, use First Optimization to fit the data, and then the relationship between the maximum moment and maximum displacement is obtained, which is shown in formula (10).

$$y_{cr} = \frac{P_1 + |QR| + |QSQ^T| + p_{29}\ln(x_1)\ln(x_2)x_3x_4x_5x_6}{1 + |QT| + |QUQ^T| + p_{57}\ln(x_1)\ln(x_2)x_3x_4x_5x_6}$$

(10)

In formula (10), x1 is the coefficient of foundation resistance, x2 is the elastic modulus of concrete, x3 is the moment of inertia, x4 is the anchor spacing, x5 is the length of a cantilever end, x6 is the maximum moment, ycr is the maximum displacement, Q is a parameter matrix, and R, S, T and U are fitting coefficient matrixes, and they are all shown below:

The fitting coefficients are from p1 to p57, which are shown in Table 2. The fitting correlation coefficient is 0.998.

The formula is mainly used in the frame beam whose concrete grade is C25. Anchor spacing varies from 3 m to 5 m and size combinations are 0.3 m × 0.3 m, 0.3 m × 0.4 m, 0.4 m × 0.4 m, 0.4 m × 0.5 m and 0.5 m × 0.5 m.

For the early warning and forecasting of the prestressed anchor cable frame beam, taking the safety reserve into account, a criterion can be given as follows: $y > y_{cr}/1.25$. In this criterion, y is the measured displacement of the beam. In order to calculate conveniently, assuming H_1 is less than H_2, the calculation flow chart of the relationship between the maximum moment and maximum displacement of the prestressed anchor cable frame beam is shown in Figure 2.

$$Q = [\ln(x_1), \ln(x_2), x_3, x_4, x_5, x_6]; \quad R = [p_2, p_3, p_4, p_5, p_6, p_7]^T; \quad T = [p_{30}, p_{31}, p_{32}, p_{33}, p_{34}, p_{35}]^T$$

$$S = \begin{bmatrix} P_8 & P_{14} & P_{15} & P_{16} & P_{17} & P_{18} \\ 0 & P_9 & P_{19} & P_{20} & P_{21} & P_{22} \\ 0 & 0 & P_{10} & P_{23} & P_{24} & P_{25} \\ 0 & 0 & 0 & P_{11} & P_{26} & P_{27} \\ 0 & 0 & 0 & 0 & P_{12} & P_{28} \\ 0 & 0 & 0 & 0 & 0 & P_{13} \end{bmatrix}; \quad U = \begin{bmatrix} P_{36} & P_{42} & P_{43} & P_{44} & P_{45} & P_{46} \\ 0 & P_{37} & P_{47} & P_{48} & P_{49} & P_{50} \\ 0 & 0 & P_{38} & P_{51} & P_{52} & P_{53} \\ 0 & 0 & 0 & P_{39} & P_{54} & P_{55} \\ 0 & 0 & 0 & 0 & P_{40} & P_{56} \\ 0 & 0 & 0 & 0 & 0 & P_{41} \end{bmatrix}$$

Table 2. The values of the coefficients of y_{cr}.

Coeffi-cient	Value (10^{-5})	Coeffi-cient	Value (10^{-5})	Coeffi-cient	Value (10^{-5})	Coeffi-cient	Value (10^{-5})	Coeffi-cient	Value (10^{-5})
P_1	43.167	P_{13}	0.00232	P_{25}	−105.82	P_{37}	−466.411	P_{49}	351.533
P_2	−10.876	P_{14}	−31.15	P_{26}	25.124	P_{38}	$−1.68 \times 10^{-7}$	P_{50}	−37.220
P_3	21.615	P_{15}	26.461	P_{27}	−0.0311	P_{39}	85.549	P_{51}	0.151
P_4	25.338	P_{16}	5.871	P_{28}	0.0826	P_{40}	37.087	P_{52}	0.656
P_5	−40.979	P_{17}	27.357	P_{29}	−0.0987	P_{41}	0.009	P_{53}	21.386
P_6	63.965	P_{18}	−0.718	P_{30}	163.137	P_{42}	−8.933	P_{54}	58.593
P_7	1.767	P_{19}	17.678	P_{31}	−212.472	P_{43}	0.380	P_{55}	2.501
P_8	20.875	P_{20}	−8.527	P_{32}	$−2.91 \times 10^{-5}$	P_{44}	204.153	P_{56}	7.370
P_9	11.868	P_{21}	−39.47	P_{33}	24.873	P_{45}	222.278	P_{57}	7.382
P_{10}	−0.015	P_{22}	0.494	P_{34}	17.215	P_{46}	47.062		
P_{11}	11.617	P_{23}	5.222	P_{35}	137.276	P_{47}	0.706		
P_{12}	65.282	P_{24}	−1.5	P_{36}	300.204	P_{48}	194.390		

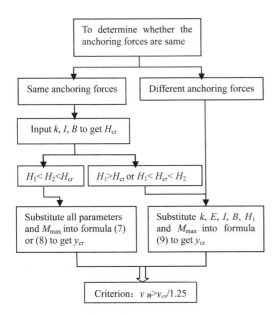

Figure 2.   Calculation flow chart of the relationship.

4   ENGINEERING EXAMPLE

We will take Shunxichang landslide treatment in the literature (Zhu, 2009) as an example. According to the formation lithology and engineering geological profile, the value of k is 100000 kN/m³. Carry on the analysis to the cross beam named H_2. Because the design values of anchoring forces are all 500 kN, we can consider the example as the same anchoring forces. The values of all basic parameters are shown below: $k = 100000$ kN/m³, $E = 2.8 \times 10^7$ kPa, $b = 0.4$ m, $h = 0.5$ m, $I = 0.4 \times 0.5^3/12 = 0.004167$ m⁴, $B = 4$ m, $H_1 = H_2 = 2$ m, and $L = 12$ m.

Substituting k, I and B into fitting formula (7), we can obtain $H_{cr} = 1.379$ m. Substituting all basic parameters into the expression of moment, by changing the cantilever length, through computation and observation, we get the theoretical value of H_{cr}' to be 1.391 m. The error is 0.012 m and the relative error is 0.86%, which prove that formula (7) is reasonable.

Because H_1 and H_2 are both greater than H_{cr}, the maximum moment should be located at the node of anchor cable at the end of the beam. As a matter of fact, by substituting all basic parameters into the expression of moment, we can also find that the maximum moment is located at the node of anchor cable at the end of the beam. At the same time, we can obtain the values of M_{max} and y_{cr}', which are shown as follows: $M_{max} = 196.96$ kN · m and $y_{cr}' = 3.648$ mm. Substituting k, E, I, B, H_1 and M_{max} into fitting formula (10), we can obtain y_{cr} to be 3.808 mm. The error is 0.16 mm and relative error is 4.4%, which prove that formula (10) is reasonable.

5   CONCLUSION

This study used the Winkler elastic foundation model. Based on the conventional design data of the prestressed anchor cable frame beam along the expressways in Shaanxi province, the relationship between the maximum moment and maximum displacement on the beam was divided into the same or different anchoring forces to research. Through a series of formula derivations and data fitting of software, the relationship between the maximum moment and maximum displacement of the prestressed anchor cable frame was eventually established, and the formulas were verified reasonable by an engineering example. For such a landslide with structural deformation monitoring, as long as related parameters of the landslide and prestressed anchor cable frame beam are known, we can get the maximum displacement of the frame beam, and then we can determine whether the structure is damaged. The research result has an important practical significance for early warning and forecasting of the landslides.

ACKNOWLEDGMENTS

This research was supported by the special funds of scientific and technological innovation for the provincial state-owned capital operating budget (2013gykc-018) and the transportation research projects of traffic and transportation department of Shaanxi province in the year of 2013 (13–28 K).

REFERENCES

Li D. F., Y. L. Zhang, Rock. Soil Mech, **21**, 170(2000).
Liang Y., D. P. Zhou, G. Zhao, Rock. Mech. Eng, **25**, 318(2006).
Liu X. L., Z. M. Zhang, J. H. Deng, Rock. Soil Mech, **25**, 1617(2004).
Ninth Design. Research Institute. China Shipbuilding Industry Corporation, *Calculation of elastic foundation beam and rectangular plate* (National Defence Industry Press, 1983).
Shi Y. W., Highway, **02**, 89(2004).
Song C. J., D. P. Zhou, Highway, **07**, 76(2004).
Wu L. Z., R. Q. Huang, Rock. Soil Mech, **27**, 605(2006).
Xia X., D. P. Zhou, Rock. Soil Mech, **23**, 242(2002).
Xiao S. G., D. P. Zhou, Geotech Eng, **24**, 479(2002).
Yang M., H. T. Hu, C. J. Lu, et al, Rock. Mech. Eng, **21**, 1383(2002).
Zhou J. X., G. X. Li, S. M. Yu, *Foundation engineering* (Tsinghua University Press, 2007).
Zhu B. L., M. Yang, H. T. Hu, et al, Rock. Mech. Eng, **24**, 697(2005).
Zhu D. P., E. C. Yan, K. Song, Rock. Mech. Eng, **28**, 2947(2009).

Advances in Energy Science and Equipment Engineering II – Zhou, Patty & Chen (Eds)
© 2017 Taylor & Francis Group, London, ISBN 978-1-138-71798-5

Hypothesis and verification of single-degree-of-freedom deformation in vertical progressive collapse of RC frames

Jian Hou & Shuli Sun
Department of Civil Engineering, Xi'an Jiaotong University, Xi'an, P.R. China

ABSTRACT: Column loss scenarios can be used to study the progressive collapse process and resistance of multistory Reinforced Concrete (RC) buildings. Under gravity loading following sudden column removal, multistory RC frames tend to progressive collapse in the vertical direction. In this study, a single-degree-of-freedom deformation hypothesis was proposed based on different floors. Therefore, based on the superposition principle, the total progressive collapse resistance of multistory RC frames can be obtained through the resistance of each floor. The single-degree-of-freedom deformation hypothesis was verified by the experiment (a single-story one-third scale RC frame) and numerical calculation (a single-story and a two-story one-third scale RC frames). During the progressive collapse of the two-story frame, the vertical displacements of the top of the removal column in the first floor and the second floor were almost the same. Due to the same size and reinforcement of beams and slabs in the first floor and the second floor, the vertical displacements and the reinforcement stresses of the other corresponding locations in two floors were also almost the same. The total progressive collapse resistance was equal to two times the resistance of the first or the second floor.

1   INTRODUCTION

Since the collapse of Ronan Point in England in 1968, caused by gas explosion, increased attention to the subject of progressive collapse has continued through the ensuing decades. Progressive collapse is defined as the spread of initial local failure from element to element, eventually resulting in the collapse of an entire structure or a disproportionately large part of it (ASCE, 2010). These progressive collapse accidents resulted in a number of casualties and significant property loss. Many countries have initiated investigations on progressive collapse resistance and they have published design codes, specifications and guidelines (ASCE, 2010; GSA, 2013; DoD, 2009). However, the location and strength of extreme accidental events that cause initial local failure are unforeseen, and the construction cost estimate of almost all buildings is not unlimited. Therefore, the current design codes and guidelines are not considered to completely satisfy the progressive collapse design requirements.

Further research is necessary to aid better understanding of the mechanisms of progressive collapse resistance of structures and ultimately to seek the establishment of rational methods for the assessment of structural robustness under extreme accidental events. Some researchers have examined structural progressive collapse experimentally (Yi, 2008; Pham, 2013; Sadek, 2011). Other researchers have investigated structural performance using nonlinear static or dynamic procedures (Alashker, 2011; Marjanishvili, 2006). However, duo to the limitations of the experimental conditions and costs, or the numerical analysis accuracy, Reinforced Concrete (RC) beam-column substructures or single-story frame structures are often considered as the object of study in the literature. For progressive collapse of multistory RC frame structures, the coupling effects of different floors are in need of additional experimental or numerical study.

In this paper, based on the experiment and numerical calculation, the process and resistance of progressive collapse of a single-story and a two-story RC frame model were investigated. The hypothesis of vertical single-degree-of-freedom deformation was proposed based on different floors. By comparing the deformation and reinforcement stress of the first and the second floor, the single-degree-of-freedom deformation hypothesis in the vertical direction was verified. Based on the superposition principle, the total progressive collapse resistance of multistory RC frames can be obtained through the resistance of each floor.

2   EXPERIMENTAL PROGRAM AND FINITE ELEMENT MODELING

2.1   *Experimental program*

A four-span, eight-story and four-bay RC frame structure was designed considering the concrete design code and seismic design code of China. For the purpose of the collapse experiment, a third-scale model of a segment of the original frame ground story was constructed for the collapse experiment. The floor height of the model frame is 1100 mm. The floor plan details of the reinforcement and cross-sectional dimensions used in the model frame are listed in Figure 1. The middle column was removed when the model was built in the laboratory, but the corresponding frame joint was intact. In fact, the support of the removed side column was considered, namely, the prototype frame was intact when it was designed according to design codes.

Figure 2 depicts the details of the instrumentation layout and test setup. As illustrated in Figure 2, the load was placed on the top of the center column that was removed by a MTS servo actuator. The load was placed with displacement control at 3 mm per minute. Downward displacements of the top of the removed center column were imposed until failure. Frame collapse was defined in this study as the rupture of tension steel bars in the floor beams; in fact, the progressive collapse-resistance capacity reached the peak in the same time.

2.2 Finite element modeling

Due to the limitations of experimental conditions and cost, the structural response information

(a) Photograph

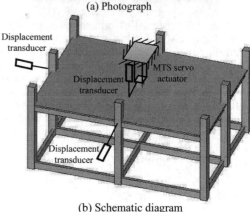

(b) Schematic diagram

Figure 2. Test setup and instrumentation layout.

obtained in the experiment was not enough. A computational study of the response of progressive collapse of a single-story and a two-story RC frame model was carried out using explicit time integration in LS-DYNA (Hallquist, 2007). For the numerical frame models, the cross-sectional dimensions and reinforcement were the same as those of the experiment. For the two-story frame model, the cross-sectional dimensions and reinforcement of the second floor was the same as those of the first floor.

The analysis accounted for both material and geometrical nonlinearities, which included fractures from element erosion. Concrete was represented by solid elements that were finely meshed in the model and beam elements were used as reinforcing bars in the model. Steel properties of reinforcing bars were modeled with a piecewise-linear plasticity model (Material 24 in LS-DYNA) using stress-strain curves derived from tensile test data. A continuous surface cap model (Material 159 in LS-DNA) was used to model the concrete material. The exterior column bottoms were assumed to be fixed. In the computational study, the loading mode was the same as the experiment. Due to the symmetry, only 1/4 of the whole model was

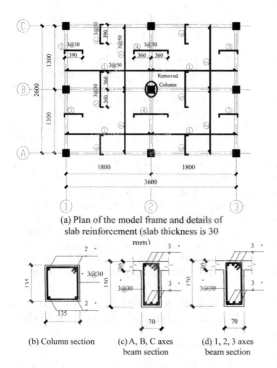

(a) Plan of the model frame and details of slab reinforcement (slab thickness is 30 mm)

(b) Column section   (c) A, B, C axes beam section   (d) 1, 2, 3 axes beam section

Figure 1. Details of the model frame (Unit: mm).

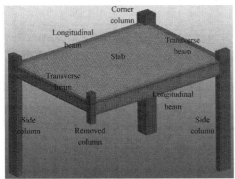

(a) Single-story frame

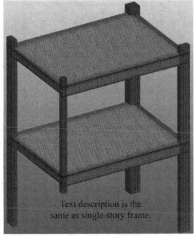

(b) Two-story frame

Figure 3. Finite element quarter-model of single-story and two-story model frames.

established in the computation. The overview of the model is illustrated in Figure 3.

Defining a one-dimensional contact interface (Contact_1d in LS-DYNA) was used to model bond-slip behavior between the beam elements representing reinforcing bars and the solid elements representing concrete. Bond slip was not considered for the column longitudinal and transverse reinforcement. The solid elements using the CONSTRAINED_LAGRANGE_IN_SOLID card constrained the beam elements representing the reinforcing bars.

3 SINGLE DEGREE OF FREEDOM DEFORMATION HYPOTHESIS

Based on the experimental process and result of progressive collapse of the single-story RC frame model, it can be observed that the progressive collapse spread only in the vertical direction.

A detailed experimental result was described in the literature (Hou, 2016). Therefore, the hypothesis of vertical single degree of freedom deformation was proposed in this study. As shown in Figure 4, for a multistory RC frame structure, the progressive collapse occurs only in the region adjacent to the initial removal column and spread only in the vertical direction. And during the process of progressive collapse, the other columns on the top of the removal column do almost the translation motion of the rigid body in the vertical direction, and the vertical displacement of these columns is almost the same. Based on the hypothesis of vertical single-degree-of-freedom deformation, the progressive collapse resistance of a multistory RC frame can be obtained by each single-story RC frame substructures using the superposition principle, as shown in Figures 4c and 4d.

Figure 5 shows the experimental and numerical results of the progressive collapse resistance of single-story and two-story frame models. As shown

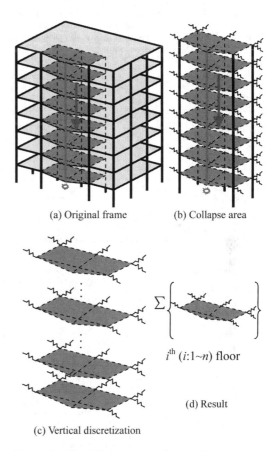

(a) Original frame   (b) Collapse area

$\sum \left\{ \quad \right\}$

i^{th} (i:1~n) floor

(d) Result

(c) Vertical discretization

Figure 4. Simplified process of vertical progressive collapse of multi story frame structures (the shadow part is the collapse area).

in Figure 5, the progressive collapse resistances of the single-story model are 135.7 KN (experimental) and 143.9 KN (numerical), respectively, and the progressive collapse resistance of the two-story model is 303.2 KN. And, the vertical displacements

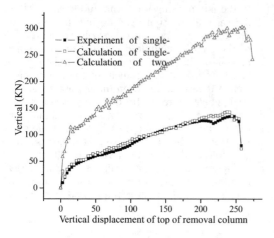

Figure 5.   Vertical load versus downward displacement of the top of the removal column.

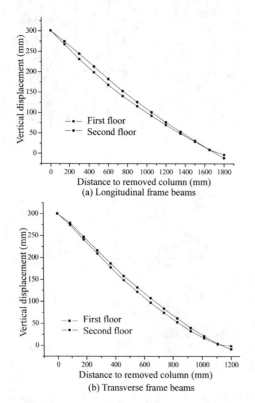

Figure 6.   Vertical displacement of different positions on frame beams of the two-story frame at collapse limit state.

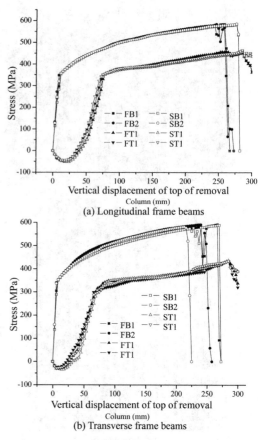

Figure 7.   Longitudinal reinforcement stress at beam ends near removal column (FB/T: bottom/top reinforcement of first floor; SB/T: bottom/top reinforcement of second floor).

of the top of the removal column corresponding to the resistances of single- and two-story models are 245 mm and 255 mm, respectively. It can be found that the progressive collapse resistance of the two-story model is almost two times that of the single-story model, and the corresponding vertical displacements of the top of the removal column of two models are almost the same. Figure 6 shows the vertical displacement of different positions on the frame beams of the two-story model at collapse limit state. It can be observed from Figure 6 that the vertical deformations of the first floor and the second floor are almost the same at collapse limit state. Figure 7 shows the longitudinal reinforcement stresses at beam ends near the removal column. As shown in Figure 7, the corresponding longitudinal reinforcement stresses in the first floor and the second floor are also almost the same during the process of progressive collapse. Based on the above analysis, the hypothesis of vertical single-degree-of-freedom deformation was proved to be accurate.

4 CONCLUSIONS

In this study, the processes and resistances of progressive collapse of a single-story and a two-story RC frame models were investigated based on the experiment and numerical calculation. The hypothesis of vertical single degree of freedom deformation was proposed, and a simplified calculation method of progressive collapse resistance of multistory RC frame structures was established based on the proposed hypothesis and superposition principle. By comparing the deformations and the reinforcement stresses of the first and the second floor in the two-story model, and the progressive collapse resistances of single-story and two-story models, the hypothesis of vertical single-degree-of-freedom deformation was proved to be reliable and accurate.

ACKNOWLEDGMENTS

This work was supported by the National Natural Science Foundation of China (Grant No. 51208421).

REFERENCES

ASCE. Minimum design loads for buildings and other structures. SEI/ASCE 7–10, Reston, VA: American Society of Civil Engineers (2010).

Alashker, Y., H.H. Li, S. El-Tawil. Approximations in progressive collapse modeling. Journal of Structural Engineering 137(9): 914–924 (2011).

DoD. Unified facilities criteria, design of buildings to resist progressive collapse. Washington, DC: Department of Defense (2009).

GB50010-2010. National standard of the People's Republic of China, code for design of concrete structures. Beijing: Ministry of Housing and Urban-Rural Development (2010).

GB50011-2010. National standard of the People's Republic of China, code for seismic design of buildings. Beijing: Ministry of Housing and Urban-Rural Development (2010).

GSA. Alternate path analysis and design guidelines for progressive collapse resistance. Washington, DC: General Services Administration (2013).

Hallquist, J.O. LS-DYNA keyword user's manual, version 971. Livermore, CA: Livermore Software Technology Corporation (2007).

Hou, J., L. Song, H.H. Liu. Progressive collapse of RC frame structures after a center column loss. Magazine of Concrete Research 68(8): 423–432 (2016).

Marjanishvili, S., E. Agnew. Comparison of various procedures for progressive collapse analysis. Journal of Performance of Constructed Facilities 20(4): 365–374 (2006).

Pham, X.D., K.H. Tan. Experimental study of beam-slab substructures subjected to a penultimate-internal column loss. Engineering Structures 55: 2–15 (2013).

Sadek, F., J.A. Main, H.S. Lew, Y.H. Bao. Testing and analysis of steel and concrete beam-column assemblies under a column removal scenario. Journal of Structural Engineering 137: 881–892 (2011).

Yi, W.J., Q.F. He, Y. Xiao, S.K. Kunnath. Experimental study on progressive collapse-resistant behavior of reinforced concrete frame structures. ACI Structural Journal 105: 433–439 (2008).

Advances in Energy Science and Equipment Engineering II – Zhou, Patty & Chen (Eds)
© 2017 Taylor & Francis Group, London, ISBN 978-1-138-71798-5

Dynamic responses of the planar linkage mechanism with impact in three clearance joints

Zheng Feng Bai, Xin Shi & Ji Jun Zhao
Department of Mechanical Engineering, Harbin Institute of Technology, Weihai, Shandong, P.R. China

ABSTRACT: The effect of impact in multiple clearance joints on dynamic responses of the mechanism is studied. The normal force in the revolute clearance joints is modeled using a continuous contact force model. The friction effect is considered using a modified Coulomb friction model. The dynamics model of a planar four-bar mechanism with three clearance joints is established. The dynamic responses of the mechanism with multiple clearances are investigated with four different cases. The output responses of mechanical systems are obviously vibration and presents higher peaks of its values. The level of the contact peaks in each clearance joint increases with the increase of the number of clearance joints. Therefore, when more joints are considered as clearance joints, the reaction force of each clearance joints of mechanism is changed more obviously, which is represented by vibration with higher peaks. The effects of clearance on dynamic responses of mechanical systems are more obvious when considering more clearance joints.

1   INTRODUCTION

In general dynamic analysis of multibody mechanical systems, it is assumed that the kinematic joints are ideal or perfect, that is, clearance effects are neglected. However, in a real mechanical kinematical joint, a clearance is always existence due to assemblage, manufacturing errors and wear (Varedi, 2015; Bai, 2013; Bai, 2013; Flores, 2009; Erkaya, 2009). The movement of the real mechanism is deflected from the ideal mechanism and the motion accuracy decreases due to the joint clearances. These clearances in joints also cause impact loads, modify the dynamic response of the system, justify the deviations between the numerical predictions and the experimental measurements and eventually lead to important deviations between the projected behavior of the mechanism and their real outcome (Zhao, 2011; Bai, 2012; Erkaya, 2010; Zhao, 2011; Flores, 2010).

Over the last few decades, effects of clearance on dynamic responses of mechanisms using theoretical and experimental approaches have been studied by many researchers. All the studies indicate that contact and impact in clearance joints will lead to important effects on dynamics responses of the mechanism (Bai, 2014; Koshy, 2013; Bai, 2012). However, some of these studies were limited to the single clearance joint. These research studies focus less on the interaction between multiple clearances. Thus, this paper studies the dynamic responses of the planar linkage mechanism with multiple revolute clearance joints. A four-bar mechanism with three clearances is used as the numerical example to implement the investigation. The dynamics equations of

the four-bar mechanism with three clearance joints are established and the dynamics simulation is made to analyze the dynamic responses of the four-bar mechanism with impact in clearance joints.

2   DYNAMICS MODELLING OF FOUR-BAR MECHANISM WITH CLEARANCE JOINTS

2.1   Model of the mechanism with clearance joints

The model of the four-bar mechanism with three clearances is shown in Figure 1. The four-bar mechanism consists of four rigid bodies that

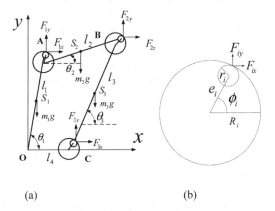

(a)   (b)

Figure 1.   The model of the mechanism with clearance. ((a) Force analysis of the mechanism with clearance; (b) the sketch of a clearance joint).

represent the crank, coupler, follower, ground, and one ideal revolute joint connecting the ground to the crank. Three revolute joints with clearance exist between the crank and the coupler (Joint A), between the couple and follower (Joint B), and between the ground and the follower (Joint C). The clearance in joint is usually very small, so it is zoomed-in as shown in Figure 1(b) in order to express clearly.

2.2 Dynamics and kinetics equations of mechanism of with clearance joints

The crank is the driving link and rotates at a constant angular velocity. Then, the dynamics and kinetics equations of the four-bar mechanism with clearances are established as equation (1).

$$\begin{cases} r_1(F_{1x}\sin\phi_1 - F_{1y}\cos\phi_1) + l_1(F_{1x}\sin\theta_1) - m_1gl_1\cos\theta_1/2 = 0 \\ m_2\ddot{x}_2 = F_{1x} - F_{2x} \\ m_2\ddot{y}_2 = F_{1y} - m_2g - F_{2y} \\ J_2\ddot{\theta}_2 = r_2(F_{2x}\sin\phi_2 - F_{2y}\cos\phi_2) + l_2(F_{2x}\sin\theta_2 + F_{1x}\sin\theta_2 - F_{2y}\cos\theta_2 - F_{1y}\cos\theta_2)/2 \\ \qquad + (r_1 + e_1)(F_{1y}\cos\phi_1 - F_{1x}\sin\phi_1) \\ m_3\ddot{x}_3 = F_{2x} - F_{3x} \\ m_3\ddot{y}_3 = F_{2y} - m_3g - F_{3y} \\ J_3\ddot{\theta}_3 = l_3\cos\theta_3(F_{2y} + F_{3y})/2 - l_3\sin\theta_3(F_{2x} + F_{3x})/2 + (r_2 + e_2)(F_{2y}\cos\phi_2 - F_{2x}\sin\phi_2) \\ \qquad + r_3(F_{3x}\sin\phi_3 - F_{3y}\cos\phi_3) \end{cases} \quad (1)$$

where, l_1, l_2, l_3 is the length of each link, separately. The mass of each link is m_1, m_2, m_3, separately. l_4 is the length of the frame. r_i and R_i are the radius of the journal and bearing in each clearance joint, where $i = 1, 2, 3$ corresponding to clearance joint A, B and C. $c_i = R_i - r_i$ is the radial clearance. θ_j is the angle of each link to the frame, where $j = 1, 2, 3$. F_{ix} is the x-direction force in each clearance joint and F_{iy} is the y-direction force in each clearance joint. e_i is the clearance vector of each clearance joint.

2.3 Contact condition of clearance joints

Based on Figure 1(b), the normal deformation of contact point can be obtained through clearance vector and the radial clearance. The normal deformation is calculated by

$$\delta_i = e_i - c_i, i = 1, 2, 3 \quad (2)$$

Furthermore, the contact condition is expressed as

$\delta_i < 0$ Not contact;
$\delta_i = 0$ Begin approaching or separating;
$\delta_i > 0$ The deformation of the contact.

Define ϕ_i as the angle between x axis and the connecting line of the centers of bearing and journal for each clearance joint. e_{ix} and e_{iy} are the projection

of the centre distance on x and y axes, from Figure 1(b), which can be obtained as follows.
For clearance joint A:

$$e_{1x} = x_{s2} - \frac{1}{2}l_2\cos\theta_2 - l_1\cos\theta_1$$
$$e_{1y} = y_{s2} - \frac{1}{2}l_2\sin\theta_2 - l_1\sin\theta_1 \quad (3)$$

For clearance joint B:

$$e_{2x} = x_{s2} + \frac{1}{2}l_2\cos\theta_2 - \frac{1}{2}l_3\cos\theta_3 - x_{s3}$$
$$e_{2y} = y_{s2} + \frac{1}{2}l_2\sin\theta_2 - \frac{1}{2}l_3\sin\theta_1 - x_{y3} \quad (4)$$

For clearance joint C:

$$e_{3x} = x_{s3} - \frac{1}{2}l_3\cos\theta_3 - l_4$$
$$e_{3y} = y_{s3} - \frac{1}{2}l_3\sin\theta_3 \quad (5)$$

Further, $e_i = \sqrt{e_{ix}^2 + e_{iy}^2}$, $\phi_i = \tan^{-1}(e_{iy}/e_{ix})$.

3 CONTACT FORCE MODEL IN CLEARANCE JOINT

3.1 Normal contact force model

In this study, for revolute joint with clearance, the dry contact between the journal and bearing is modeled using the continuous contact force model proposed by Lankarani and Nikravesh and the expression of the continuous contact force model is expressed as follows (Lankarani, 1990):

$$F_n = K\delta^n\left[1 + \frac{3(1 - c_e^2)\dot{\delta}}{4\dot{\delta}^{(-)}}\right] \quad (6)$$

where K is the contact stiffness coefficient of the impact body, δ is the deformation, $\dot{\delta}$ is the relative deformation velocity, c_e is the coefficient of restitution and $\dot{\delta}^{(-)}$ is the initial relative velocity of the impact point.

3.2 Tangential friction force model

The tangential contact characteristic of clearance joint is represented using the tangential friction force model. In this paper, the friction model in the clearance joint is considered as dry contact and there is no lubrication. The most famous one of the friction model is the Coulomb friction model, which is used to represent the friction response in impact and contact processes. In this paper, a modified Coulomb friction model is used to represent the friction response between the journal and bearing (Bai, 2013). The friction coefficient, which is not a constant, is introduced in the modified Coulomb friction model. The friction coefficient is a function of tangential sliding velocity, which can represent the friction response in the impact and contact process as well as the viscous and micro-slip phenomenon in relative low-velocity case more accurately. And also, the modified Coulomb friction model can avoid the case of abrupt change of friction in numerical calculation due to the change of velocity direction.

The expression tangential friction force is shown as

$$F_t = -\mu(v_t)F_n \frac{v_t}{|v_t|} \qquad (7)$$

where friction coefficient $\mu(v_t)$ is a function of tangential sliding velocity.

Furthermore, the step function is introduced in equation (1), which is expressed as

$$u(\delta_i) = \begin{cases} 0 & \delta_i < 0 \\ 1 & \delta_i \geq 0 \end{cases} \qquad (8)$$

Therefore, the forces in the clearance joint can be calculated as

$$F_{ix} = u(\delta_i)(F_{in}\cos\phi_i + F_{it}\sin\phi_i) \\ F_{iy} = u(\delta_i)(F_{in}\sin\phi_i - F_{it}\cos\phi_i) \qquad (9)$$

where F_{in} is the normal contact force in the clearance joint and calculated using Eq. (6). F_{it} is the tangential friction force in the clearance joint and calculated using Eq. (7).

4 DYNAMICS SIMULATION AND RESULTS

4.1 Parameters of the four-bar mechanism

In the dynamic simulation, the crank is the driving link and rotates at a constant angular velocity of 300 r/min. The initial configuration corresponding to the crank and ground is vertical and the initial angular velocity is zero. Initially, the journal and bearing centers are coincident. The clearance size in each joint is 0.2 mm. The structural parameters for the planar four-bar mechanism are presented in Table 1. Table 2 provides the parameters which were used in the simulation of the four-bar mechanism with multiple revolute clearance joints.

Four different cases are studied to investigate the dynamic response of the four-bar mechanism. That is, the four-bar mechanism is simulated without a clearance joint and with one, two and three clearance joints, respectively. The four cases are presented in Table 3.

4.2 Results and discussion

The dynamic responses of the four-bar mechanism with different number of clearance joints are presented in Figure 2. Figure 2 shows that the angular acceleration of the follower presents significant differences between the dynamic responses of the mechanism when modelled with one, two and three clearance joints. When more joints are modelled as clearance joints, the dynamic responses of the mechanism are changed more obviously, which is represented by vibration with higher peaks.

Furthermore, the contact forces at clearance joint B are presented in Figure 3. As shown

Table 1. Structural parameters for the four-bar mechanism.

Parameter	Values
Crank length l_1/m	0.400
Couple length l_2/m	0.300
Follower length l_3/m	0.530
Framework length l_4/m	0.135
Mass of Crank m_1/kg	1.2964
Mass of Couple m_2/kg	0.9869
Mass of Follower m_3/kg	1.7027
Inertia of crank J_1/kg m²	0.018696
Inertia of couple J_2/kg m²	0.008265
Inertia of follower J_3/kg m²	0.042135

Table 2. Material and dynamic simulation parameters for the four-bar mechanism.

Parameter	Values
Young's modulus E/Gpa	207
Coefficient of restitution	0.9
Dynamic coefficient of friction	0.01
Crank speed V/r min⁻¹	300
Initial angle $\theta/°$	90

Table 3. Case description for dynamics simulation.

	Case description
Case 1	Ideal joint without clearance, $c_1 = c_2 = c_3 = 0$
Case 2	Considering one clearance in joint B, $c_1 = c_3 = 0$, $c_2 = 0.2$ mm
Case 3	Considering two clearances in joint B and C, $c_2 = c_3 = 0.2$ mm, $c_1 = 0$
Case 4	Considering three clearances in joint A, B and C, $c_1 = c_2 = c_3 = 0.2$ mm

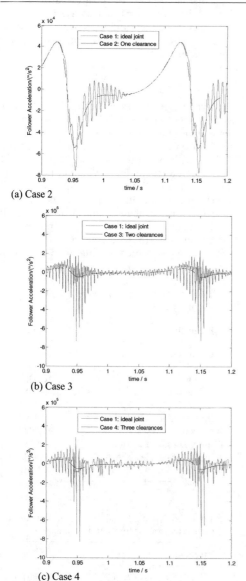

(a) Case 2

(b) Case 3

(c) Case 4

Figure 2. Angular acceleration of the four-bar mechanism ((a) case 2: with one clearance; (b) case 3: with two clearances; (c) case 4: with three clearances).

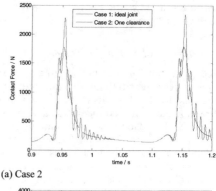

(a) Case 2

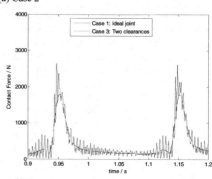

(b) Case 3

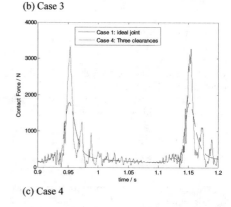

(c) Case 4

Figure 3. Contact forces in clearance joint B ((a) case 2: with one clearance; (b) case 3: with two clearances; (c) case 4: with three clearances).

in Figure 3, the contact forces of joint B are significantly different when the mechanism was modelled with one, two and three clearance joints. It shows that when joint B is modelled as the clearance joint and other joints are modelled as the ideal joint, the contact force is vibration with high peaks. However, when Joint A and Joint C are also modelled as clearance joints, the reaction force of joint B is obvious vibration and presents the higher peaks of the contact forces. It indicates that when more joints are modelled as clearance joints, the contact force at each clearance joint is more obvious vibration with higher peaks, which explains the acceleration characteristics in Figure 2.

4 CONCLUSIONS

In the presented work, the effects of contact in multiple clearance joints on dynamic responses of planar linkage mechanisms are investigated. The constraint of the clearance joint is considered as force constraint. The normal force in revolute clearance joints is modeled using a nonlinear continuous contact force model. The friction effect is considered using a modified Coulomb friction force model. The dynamics equations of a planar four-bar mechanism with three clearance joints are established. Then, four different case studies are presented and discussed.

The output responses of the four-bar mechanism are obviously vibration and presents higher peaks of values. It indicates that the angular acceleration of the mechanism is significantly different when the mechanism was modeled with one, two and three clearance joints. The results also show that the level of the contact peaks in clearance joints increases with the increase in the number of clearance joints. Therefore, when more joints are considered as clearance joints, the reaction force of each clearance joint of the four-bar mechanism is changed more obviously, which is represented by vibration with higher peaks. It indicates that multiple clearances interact and couple with one another, which increases the contact forces in clearance joints and decreases the performance of the mechanism.

ACKNOWLEDGMENTS

This work was supported by the National Natural Science Foundation of China (Grant No.: 51305093) and the Natural Science Foundation of Shandong Province (Grant No.: ZR2013EEQ004). The project supported by the "China Postdoctoral Science Foundation funded project (Grant Nos. 2012M520723; 2014T70317)" is also gratefully acknowledged.

REFERENCES

Bai, Z.F., J. Chen, Y. Sun, IJST—T. Mech. Eng. **38**, 2, pp. 375–388 (2014).
Bai, Z.F., Y. Zhao, Int. J. Mech. Sci. **54**, pp. 190–205 (2012).
Bai, Z.F., Y. Zhao, Int. J. Non-Linear Mech. **48**, pp. 15–36 (2013).
Bai, Z.F., Y. Zhao, J. Chen, Tribol. Int. **64**, 85–95 (2013).
Bai, Z.F., Y. Zhao, Precis. Eng. **36**, pp. 554–567 (2012).
Bai, Z.F., Y. Zhao, X.G. Wang, Sci. China Phys. Mech. **56**, 8, pp. 1581–1590 (2013).
Erkaya, S., I. Uzmay, Multibody Syst. Dyn. **24**, pp. 81–102 (2010).
Erkaya, S., I. Uzmay, Nonlinear Dyn. **58**, pp. 179–198 (2009).
Flores, P., Mech. Mach. Theory. **44**, pp. 1211–1222 (2009).
Flores, P., Nonlinear Dyn **61**, pp. 633–653 (2010).
Koshy, C.S., P. Flores, H.M. Lankarani, Nonlinear Dyna. **73**, pp. 325–338 (2013).
Lankarani, H.M., P.E. Nikravesh, ASME—J. Mech. Design **112**, pp. 369–376 (1990).
Varedi, S.M., H.M. Daniali, M. Dardel, Nonlinear Dyn. **79**, pp. 1578–1600 (2015).
Zhao, Y., Z.F. Bai, Acta Astronaut. **68**, pp. 1147–1155 (2011).
Zhao, Y., Z.F. Bai, T. Japan Soc. Aeronaut. S. **53**, 182, pp. 291–295 (2011).

Advances in Energy Science and Equipment Engineering II – Zhou, Patty & Chen (Eds)
© 2017 Taylor & Francis Group, London, ISBN 978-1-138-71798-5

Study on a novel and lightweight five-layer architecture and its application

Bi Liang

School of Computer Science, Sichuan University of Arts and Science, Dazhou, China

ABSTRACT: Developing Web application systems with the classical three-layer architecture achieved by S2SH (Struts2, Spring and Hibernate) has many defects such as the difficulty in integration and the long development cycle. Moreover, Web systems developed by such method require large space and long start-up time, thus this paper aims at coping with these problems mentioned above and presents a novel and lightweight five-layer architecture based on Spring framework and the effective combination of the three-layer architecture & MVC (Model, View and Controller) pattern. The architecture is achieved by using Spring IoC (Inversion of Control), Spring MVC, Spring JDBC (Java Data Base Connectivity) and etc. And that architecture is successfully applied to the Ancient Building Heritage Exhibition System (ABHES), which proves that the lightweight five-layer architecture is practical and feasible. What's more, compared with the classical three-layer architecture, using this five-layer architecture to develop the ABHES under the same environment not only has shorten the development cycle of the system by 26.3%, but also reduced the start-up time of the system by 4.9 seconds on average.

1 INTRODUCTION

At present, the classical three-layer architecture achieved by S2SH has been widely applied to Web application systems, such as e-commence, online bank, online office and so on (Yang, 2012). It divides an entire Web application system into three layers, namely, the presentation layer, the business layer and the persistence layer. And different layers undertake different tasks, in which the task of the presentation layer is accomplished by Struts2 technology, the task of the business layer is completed by Spring technology, and the task of the persistence layer is realized by Hibernate technology. What's more, the whole functions of the system are completed by the cooperation among all layers.

However, since the classical three-layer architecture achieved by S2SH mainly adopts three different technologies from different communities to develop the Web application system, incompatible problems often occur in the latter stage while the system is integrated, for instance it may be hard to integrate the jar files, the XML (Extensible Markup Language) files, the java codes and some other files. These problems, on the one hand, lead to the difficulty in integrating the system, and on the other hand, they not only prolong the development cycle of the system, but also increase the cost of developing a system. Moreover, the Web system developed by such an approach requires large space and long start-up time. In order to solve these problems, in this paper a new and lightweight five-layer

architecture, which is based on Spring framework, the three-layer architecture and MVC pattern, is proposed and then applied to the ABHES, and as a result, satisfying results are obtained.

2 CORE THEORIES

2.1 *Spring framework*

Spring framework is a popular solution for the lightweight turnkey enterprise application program at present. It consists of seven modules: Spring Core, Spring AOP (Aspect Oriented Programming), Spring ORM (Object Relational Mapping), Spring DAO (Data Access Object), Spring Web, Spring Context and Spring Web MVC (Li, 2008). Among them, the Spring Core module is the most basic and important part for the whole Spring framework, which provides the function of Dependency Injection (DI) and the management function of IoC container. The Spring DAO module has implemented a lightweight encapsulation for JDBC to provide support for JDBC operation. The Spring Web MVC module provides a complete solution for MVC, and to develop the Web application system through using the module is able to combine with IoC container well.

2.2 *Three-layer architecture*

The three-layer architecture is a current mainstream of the multi-layer architecture model, from the top to the bottom they are: the presentation layer, the business layer and the persistence layer (Yang, 2015). The presentation layer is used to display data, receive data from the user's input and provide an interactive interface for the user. The business layer is the core part of the system, mainly concentrating on the development of business rules, implementation of business processes and etc. The persistence layer is mainly responsible for the access to database, which can access to the database system, binary files, text documents, XML documents and etc. Moreover, different layers can be achieved by different technologies in the practical application, and S2SH is a current mainstream technology.

2.3 MVC pattern

MVC is currently a popular software design pattern, which is composed of three components (that is the Model, the View and the Controller). Among them, the Model processes data logic, the View handles data presentation, the Controller deals with the interaction of users (Zhang, 2011). The purpose of using MVC is to achieve the code separation from the M to the V, so that the same program can use different forms of representation. The C is to ensure the synchronization of the V and the M, namely, once the M is changed, the V will be updated accordingly. The core idea of MVC is to separately implement the view function, model function and control function of the application in different parts (also called layers), which can increase the reusability of the code and reduce the coupling degree between data description and application operation.

3 KEY TECHNOLOGIES

3.1 Spring IoC

IoC is the core technology of Spring framework. It realizes the important idea of "decoupling" in Spring framework, and through DI, a class can be connected to another one by configuring XML instead of by coding, which belongs to the technology of the Spring Core module. Moreover, Spring is able to automatically creates all objects that will be used in the application system in the future via IoC container. And then properties of the object will be set, meanwhile the connection of the object will also be configured. When these objects are called, IoC container will actively respond. Through this way of evaluation, the life cycle management of the object, the dependence relationship of the object and other issues are transferred to IoC container, which relieves the programming burden of developers.

3.2 Spring MVC

Spring MVC, one of the seven modules in Spring framework, is a MVC framework technology based on DispatcherServlet. DispatcherServlet is not only the controlling and forwarding center of Spring MVC but also the center of other designs. It is responsible for dispatching each request to the corresponding Handler. After Handler processes, the corresponding views and models or only models or only views will be returned, and even sometimes there is nothing to return. Moreover, Spring MVC technology makes full use of the advantage of non-invasive programming in Spring framework, which means the developer only needs to configure XML files simply without the need to implement any related Spring interface.

3.3 Spring JDBC

Spring JDBC is an ORM framework technology, which has lightly encapsulated JDBC. It is composed of four different packages, namely, the Core, the Datasource, the Object, and the Support. The most critical package is the Core, in which the most important class is the JdbcTemplate. The JdbcTemplate completes CRUD (Create, Retrieve, Update and Delete) operation to data by providing appropriate templates and auxiliary classes, and then solves the problems of the tedious operation and the code duplication of traditional JDBC towards database, so Spring JDBC belongs to the technology of the Spring DAO module. Moreover, due to the fabulous isolation design of Spring framework, Spring JDBC package can be used independently from the Spring Context library.

4 STUDY OF THE FIVE-LAYER ARCHITECTURE

4.1 The design of the five-layer architecture

Based on Spring framework and the effective combination of the three-layer architecture and MVC pattern, this paper presents a new and lightweight five-layer architecture (namely, the presentation layer, the control layer, the business layer, the persistence layer and the database layer), which is constructed through the Spring Core module, the Spring Web MVC module and the Spring DAO module in Spring framework and achieved through Spring related technologies mainly including Spring IoC, Spring MVC, Spring JDBC and etc. The detailed design of the five-layer architecture is shown in Figure 1 below (Liang, 2014).

From Figure 1, the design of the five-layer architecture can reduce the coupling degree among each part of the Web application system and allow

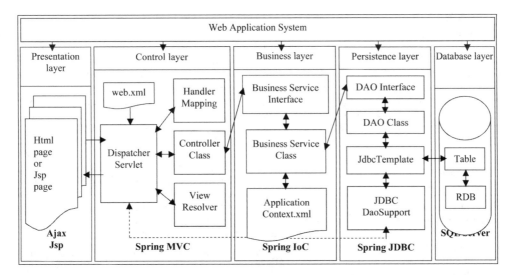

Figure 1.   The five-layer architecture.

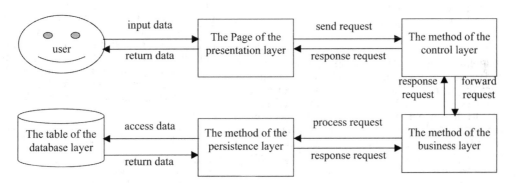

Figure 2.   The working process of the five-layer architecture.

the developer to only focus on some part of the whole system, which is beneficial for the division of labor of developers and enhances the maintainability and expansibility of the system. Besides, the five-layer architecture mainly uses Spring related technologies to realize the Web application system. Compared with the classical three-layer architecture achieved by S2SH, its technology is much simpler and can effectively avoid the incompatible problems of the jar files, the XML files and other files in the latter stage of the system integration, so it is relatively much easier to integrate both codes and XML configuration files.

4.2   Tasks of each layer in the five-layer architecture

The presentation layer is the UI (User Interface) part of the Web application system, taking the responsibility for data input validation, data output presentation and page state management in the process of interaction between the user and the system, which is mainly realized through Ajax or Jsp technology. The task of the control layer is to receive the user request from the presentation layer, and then control and dispatch the request to other layers for processing. That process is mainly completed by Spring MVC. The business layer, which lies between the control layer and the persistence layer, plays a connecting role in the process of data exchange. It is specifically designed to complete the business functions of the system, and the business is encapsulated into bean component to be managed by IoC container.

The persistence layer is the interface for database connection. It is a set of classes or components that is responsible for extracting or modifying the data from one or more than one databases, and used to define, select, update and delete data, even manage and meet the requirements of Web applications for data. All these tasks are completed by Spring JDBC. The database layer displays the persistence performance of the system state and provides data services for the entire Web application system. It is completed by the relational database tool–SQL (Structured Query Language) Server, mainly including the relational pattern, the storage procedure and so on.

4.3 The working process of the five-layer architecture

The five-layer architecture is applicable for the development of the Web application system and its basic working principle keeps consistent with the running process of the Web application system. The collaborative working process of five layers is shown in Figure 2.

5 APPLICATION OF THE FIVE-LAYER ARCHITECTURE

In this paper, the proposed five-layer architecture is applied to the ABHES, and through the example, the specific usage of such architecture and its advantages are well elaborated. At first, according to the presented five-layer architecture, we explicitly divide the ABHES into five units, they are the presentation layer unit, the control layer unit, the business layer unit, the persistence layer unit and the database layer unit, which are designed as shown in Figure 1 above. Then according to related technologies of the five-layer architecture, we set MyEclipse 8.5 as the development platform, Tomcat 7.0 as the server and employ Spring IoC, Spring MVC, Spring JDBC, Jsp and other technologies to realize the ABHES. The specific software and hardware environment is shown in Table 1.

The presentation layer of the ABHES is realized mainly by Ajax and Jsp technologies, and

Table 1. Basic development environment.

Category	Basic configuration
Hardware	Asus 15.6" Laptop, Intel Core i5-5200U 2.2 GHz, 4 GB RAM, 500 GB HDD, 100 Mb/s Ethernet, etc.
Software	Microsoft Windows 7, SQL Server 2008, JKD 1.7, MyEclipse 8.5, Tomcat 7.0, Spring 3.2, Spring MVC 3.2, Struts 2.3, Hibernate 3.5, jQuery 2.1, Dreamweaver 8, Flash 10, Panorama Studio Pro V2, etc.

user's experience is enhanced by the combination of Web3D, jQuery and other technologies (Xu, 2015). There are many presentation pages, such as the index.html, the ancientBuildPicture.htm, the ancientBuildWeb3D.html, the ancientBuild-Web3D_dengxiaopingguju.html, the register.jsp, the login.jsp, the video.jsp and the news.jsp, etc. For the database layer of the system, the classical relational database–SQL Server 2008 is applied to store and manage data. The ancientBuildDB, the name of the database of the system, contains many data tables. They are the ancientBuildTable, the imageTable, the videoTable, the newsTable, the adminTable and the userTable, etc. The methods for specific implementation of the key layers including the control layer, the business layer and the persistence layer are as follows.

1. The implementation of the control layer: The control layer of the ABHES is to receive, parse and distribute the request from users and its purpose is to achieve the correct jump of the request. These functions are completed by the Spring MVC technology. The main control files of the system are the AncientBuildController. java, the UserController.java, the AncientBuild-Web3DController.java, the ImageController. java, the VideoController.java and the News-Controller.java, etc, among which the core code of the AncientBuildWeb3DController.java is as follows.

```
......@Controller@RequestMapping("/ancientBuildWeb
3D")
    public class AncientBuildWeb3DController{
    @RequestMapping("/viewWeb3D_dazhouhanjue")
    public String viewWeb3D_dazhouhanjue (){
    return "viewWeb3D_dazhouhanjue";}
    @RequestMapping("/viewWeb3D_dengxiaoping
guju")
    public String viewWeb3D_dengxiaopingguju (){
    return "viewWeb3D_dengxiaopingguju";}.....}
```

2. The implementation of the business layer: The business layer is to achieve related business functions of the ABHES. They are mainly completed by Spring IoC and DI, which can simplify the coding and management of business beans. There are some important business beans in the ABHES, including the Ancient-BuildService.java, the UseService.java, the ImageService.java, the VideoService.java and the NewsService.java, etc. Besides, the most critical task of this layer is to configure the ApplicatonContext.xml file, which is the core of the whole system. Through it, the upper layer and the lower layer are organically integrated and the key configuration information is as follows.

```
<beans xmlns="http://www.springframework.org/
schema/beans"......>
    <context:component-scan base-package="com.
ancientBuild"/>
    <bean id="dataSource" class="org.apache.
commons.dbcp.BasicDataSource"....../>
    <bean id="jdbcTemplate" class="org.spring
framework.jdbc.core.JdbcTemplate" p:dataSource-ref=
"dataSource"/>
    <bean id="transactionManager" class="org.
springframework.jdbc.datasource.DataSource
TransactionManager"p:dataSource-ref="dataSource"/>
    <mvc:annotation-driven/>
    <bean class="org.springframework.web.servlet.
mvc.annotation.AnnotationMethodHandlerAdapter"/>
    <mvc:resources location="/image/" apping=
"/image/**"/>......</beans>
```

3. The implementation of the persistence layer:
   The persistence layer of the ABHES is responsible for the CRUD operation of the data in
   the ancientBuildDB database, which is realized through the Spring JDBC technology. The
   related data access classes of the system mainly
   include the AncientBuildDAO.java, the User-
   DAO.java, the ImageDAO.java, the VideoDAO.
   java and the NewsDAO.java, etc. Among these,
   the main code from the class for deleting the
   information of ancient building heritage in the
   AncientBuildDAO is as follows.

```
public boolean delAncientBuild ( AncientBuild ancientBuild )
{
    final String sql="delete from ancientBuildTable where
build_id=?";
    Object[] params=new Object[] {ancientBuild.
getBuild_id()};
    int i=JdbcTemplate.update(sql, params);
    if (i > 0) {return true;} else {return false;} }
```

6 ANALYSIS OF THE EXPERIMENT RESULTS

Under the same development environment, we
adopt the five-layer architecture presented in this
paper to design and implement the ABHES and

make a comparison between the lightweight five-
layer architecture and the classical three-layer
architecture in respect of development cycle,
number of code lines and software size, etc. The
specific data obtained are shown in Table 2 below.

As can be seen from Table 2, compared with the
classical three-layer architecture, using the lightweight five-layer architecture to develop the case
under the same environment not only has shorten
the development cycle by 26.3%, but also reduced
the start-up time by 4.9 seconds on average. Therefore, using the novel five-layer architecture to
develop the Web application system has the following characteristics in a certain degree.

1. Clearer design: The five-layer architecture is
   based on Spring framework and effectively
   combines with the classical three-layer architecture and MVC pattern. It properly and reasonably divides a complete Web application system
   into five layers according to different functions
   so that the shortcoming of unclear layering
   of the classical three-layer architecture can be
   avoided.
2. Shorter development cycle: The clear layered
   design of the five-layer architecture makes
   the division of labor of the system clearer.
   Although different developers are responsible
   for a different layer in the five layers, the corresponding modules of the five layers can be
   developed at the same time with no interference,
   which effectively shorten the development time
   of the system.
3. Shorter start-up time: It can be shown from the
   actual case that through the full use of Spring's
   new services (such as annotation, DI, scanning,
   etc.) and Spring container's powerful management function, the amount of code and XML
   configuration file can be greatly reduced, which
   makes the size of the Web application system
   smaller. So it takes less time to start the Web
   application system.
4. Lower cost: The Web application system developed by the constructed five-layer architecture
   not only has shortened the development cycle
   but also made the follow-up maintenance and

Table 2. Comparison on applications developed with two architectures.

Type of architecture	Number of developers (person)	Number of software layers (layer)	Development cycle (day)	Number of code lines (KLOC)	Software size (GB)	Software average start-up time (second)
Classical three-layer architecture	5	3	156	11.29	2.38	9.07
Lightweight five-layer architecture	5	5	115	6.38	1.27	4.21

expansion easier, so to some extent these advantages effectively help to reduce the cost in developing and maintaining the system.

7 CONCLUSIONS

Presently, the multi-layer architecture is a kind of design pattern commonly used in developing the Web application system. In this paper, we not only construct and realize a new and lightweight five-layer architecture mainly based on Spring framework, but also appropriately apply it to the ABHES. The example in this paper indicates that the five-layer architecture is practical and effective. To some extent, the five-layer architecture is a progress of software development ideas, because it further solves some problems in the development of software, such as the difficulty in integration, the long development cycle and the high cost, etc, and provides the basic idea for the elegant system that has little configuration, clear code, good maintainability, low cost and other good characteristics.

However, the multi-layer architecture reduces the performance of the system in a certain degree. In other words, the multi-layer architecture reduces the performance of the system in exchange for the maintainability of the system. Therefore, in actual development, we cannot over-divide or under-divide the layers of the Web application system. Instead, there should be an appropriate limit, just like five layers.

ACKNOWLEDGEMENTS

This work is supported by the Science and Technology Project of Archives Bureau in 2014 (2014-X-65), and the General Project of Education Department of Sichuan Province in 2016 (16ZB0362).

REFERENCES

Li, X.P., Y.F. Xiao, Y. Su, Study of Web-based framework based on J2EE multi-tier architecture, Chinese Journal of Application Research of Computers, **25**, 1429–1431 (2008).

Liang, B., D.J. Liu, L.L. Xiao, Research on the construction of immovable historical relics digital platform based on the five-layer architecture, Chinese Journal of Computing Technology and Automation, **4**, 115–118 (2014).

Xu, J.H., B. Song, R. Ding, Research of spring MVC+Hibernate+jQuery development framework application, Chinese Journal of Computer Application, **34**, 42–46 (2015).

Yang, L.T. Management system of teaching resource based on SSH framework, National Conference on Information Technology and Computer Science (CITCS 2012), 803–806 (2012).

Yang, Z.G., Design and implementation of enterprise office automation system based on three layer architecture, Chengdu: University of Electronic Science and Technology of China, 6–10 (2015).

Zhang, L., Application of paging component based on Model-View-Controller pattern, Chinese Journal of Computer Engineering, **37**, 255–257 (2011).

Advances in Energy Science and Equipment Engineering II – Zhou, Patty & Chen (Eds)
© 2017 Taylor & Francis Group, London, ISBN 978-1-138-71798-5

Programming of optimization design for a green building envelope

Jianmin Zhou, Zheng Cheng & Shiyu Pu
College of Civil Engineering, Tongj University, Shanghai, China

ABSTRACT: Based on the concept of green building, the energy saving design for building an envelope was carried out. Several common methods of unsteady-state heat conduction in building an envelope were compared. It was concluded that the calculations by the finite difference method and finite element method are more accurate, while the computation procedure is inefficient and time consuming. The response factor method is more suitable for the calculation of unsteady-state heat conduction in engineering. Based on the basic principle of the response factor method, a computing program for heat transfer calculation was developed by the *Matlab* software. In addition, a program of energy saving optimization design for building an envelope was developed, which can provide optimization design suggestions in consideration of the economic factor.

1 INTRODUCTION

The sustainable development is the common subject of human being. It is necessary for the construction industry to switch to the efficient green development mode from the traditional high-consumption development mode. Due to the current inevitable development trend of building, the green building is the only way to realize this transition. In design of green building, much effort should be put into the energy conservation, which can be well reflected by the proportion of it in various green building evaluation methods around the world. In the LEED, the USA, the weight of "the energy saving of building" is 23.0% (Wang, 2004); in the BREEAM, the UK, the weight of "energy resources" is 19% (Ma, 2013). Each of them is the maximum value in all the evaluation items. In the <Assessment standard for green building> (GBT 50378–2014), China, the weight of "energy saving and energy utilization" is 0.24 in the design stage and 0.19 in the operation stage for the residential building; the value is 0.28 in the design stage and 0.23 in the operation stage for the public building. The weight is always bigger than that of other items. Thus, to conduct the energy saving design is the primary way to achieve the concept of green building.

As the member separating the building from the outdoor thermal environment, the building envelope plays an important role in energy conservation. The thermal performance of the building envelope has large influence on the cold load and heat load of the building, affecting the heating and cooling energy consumption at last. For the thermal performance design of the building envelope, the index control method and the trade-off calculation

are adopted in the Chinese code. The code limits the value of various indexes, such as the heat transfer coefficient, thermal inertia, etc. The trade-off calculation should be conducted if the index limitation is not satisfied. When it comes to the trade-off calculation, the heat transfer problem of the building envelope is involved. The calculation methods of heat transfer are numerous and complex, which are inconvenient for engineers to design.

Furthermore, in the practical energy saving design process, the designer used to check whether the indexes meet the requirements of the code, rather than calculate the practical heat transfer of the building envelope, which leads to the conservative and uneconomical design. It is necessary to develop a program to help the designers to evaluate the thermal performance of the building envelope, and can also optimize the thickness and cost further according to the required thermal performance.

2 ENERGY SAVING DESIGN METHOD

Building energy consumption is mainly made up of the heat transfer of the building envelope and the energy consumption of air infiltration. The index control method and the trade-off calculation are adopted in the practical energy saving design in China.

2.1 Index control method

The index control method is established by selecting the specified shape factor and window-wall ratio, calculating the heat consumption that satisfies the energy saving requirements on the reference building, and decomposing the heat transfer coefficients of each part of the envelope. The expression of the energy saving design is therefore

$$L_i \leq L_{i,reg} \tag{1}$$

where L_i—shape factor, window-wall ratio, heat transfer coefficient of building envelope, shading coefficient, thermal resistance, thermal inertia, etc.;

$L_{i,reg}$—the limit value of L_i, which is irrelevant to the HDD (CDD).

2.2 Trade-off calculation

The trade-off calculation is actually the calculation of the practical heat consumption. The expression of the energy saving design is therefore

$$q_H \leq q_{H,reg} \tag{2}$$

where:

Q_H—the index of the heat consumption of the building, W/m²;

$q_{H,reg}$—the limit value of heat consumption, W/m², which is determined by the NDD, CDD and the number of building storey.

When it comes to the trade-off calculation, the heat transfer problem of the building envelope is involved. Different calculation methods are adopted in various design regions of building thermal. In severe cold and cold regions, it could be analysed by the theory of steady-state heat transfer. For heat insulation in summer, however, it should be analysed by the theory of unsteady-state heat transfer.

3 ANALYSIS AND COMPARISON OF CALCULATION METHODS FOR UNSTEADY-STATE HEAT CONDUCTION IN THE BUILDING ENVELOPE

When the building envelope is subjected to the changing environment, the analysis must be modified to take into account the change in internal energy of the envelope with time. This problem is referred to as the unsteady-state heat conduction.

Nowadays, there are many methods for solving the transient heat conduction problem: the Finite Volume Method (FVM), the Spectrum Element Method (SEM), the Boundary Element Method (BEM), the Finite Element Method (FEM), the Finite Difference Method (FDM), etc. However, these methods are based on the solution of large partial differential equations, which is time consuming and inefficient. The widely-used response factor method, compared to the mentioned methods above, can directly express the thermal performance by the response factors.

The solution of the response factor is the most important and basic part in the procedure of the response factor method. There are mainly two methods: the conventional method (Laplace transform) and the state space method.

The number of response factors is infinite in theory. Nevertheless, the response factors also decay quickly with time. Considering the large quantity and fast decay of response factors, Ceylan and Myers (1980) proposed the Conduction Transfer Function (CTF).

Since the exterior walls and roofs are made of multi-layer materials, Elisabeth Kossecka and Jan Kosny (1997) put forward the concept of equivalent wall to solve the complex thermal structure.

In summary, there are many methods to solve the transient heat conduction problem of the building envelope, which can be mainly divided into the numerical method and response factor method. It is necessary to conduct some comparisons to find out the suitable analysis method for the unsteady-state heat transfer problem.

In the following sections, to make the comparison among the finite element method, the finite difference method and the various response factor methods, the response factors and heat transfer of three composite walls were calculated through these methods, as shown in Tables 1–3.

3.1 Basic parameters of the calculated composite walls

In order to compare the various heat conduction calculation methods of composite walls, three walls of different weight were chosen for comparison. The basic parameters of composite walls are listed above.

3.2 Calculation and comparison of response factors of composite walls

In order to save the calculation time, a computing program for response factors was developed by the *Matlab* software. Two examples of reference (Pedersen, 1997) were selected to verify the correctness of the calculation of the response factor. Due to the length limitation, it would not be described here. In addition, a computing program for the equivalent wall was also developed. After determining the thermal parameters of the equivalent wall, the program will use the response factor program to calculate the response factor.

The calculation results of response factors are plotted below (Figure 1, Figure 2 and Figure 3).

Table 1. Parameters of the lightweight composite wall (W-light).

Layer	Outer layer			Middle layer			Inner layer		
Thickness	0.005			0.05			0.005		
Average density	Density	Specific heat	Thermal conductivity	Density	Specific heat	Thermal conductivity	Density	Specific heat	Thermal conductivity
466.7	1300	1.05	0.26	300	1.31	0.06	1300	1.05	0.26

Table 2. Parameters of the medium-weight composite wall (W-medium).

Layer	Outer layer			Middle layer			Inner layer		
Thickness	0.005			0.145			0.05		
Average density	Density	Specific heat	Thermal conductivity	Density	Specific heat	Thermal conductivity	Density	Specific heat	Thermal conductivity
875	1300	1.05	0.26	300	1.31	0.06	2500	0.92	1.74

Table 3. Parameters of the heavy-weight composite wall (W-heavy).

Layer	Outer layer			Structural layer		
Thickness	0.05			0.2		
Average density	Density	Specific heat	Thermal conductivity	Density	Specific heat	Thermal conductivity
2060	30	1.38	0.042	2500	1.74	0.92

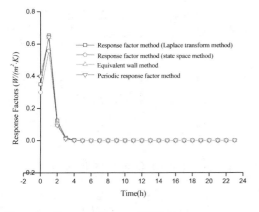

Figure 1. Response factors of W-light.

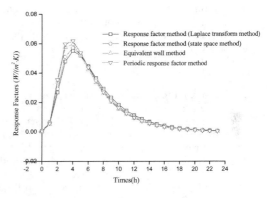

Figure 2. Response factors of W-medium.

From Figure 1, Figure 2 and Figure 3, it can be seen that

1. For a lightweight composite wall, the results of the response factor calculated by the Laplace transform method, state space method, equivalent wall method and periodic response factor method are similar. The difference among them turns great gradually as the weight of the thermal material increases.

2. The results of the periodic response factor method are significantly larger than those of other methods. This is because the convergence of response factors becomes slower with the increase of the thermal material. As a result, the superposition of the response factors increases obviously.

3. As the weight of the thermal material increases, the difference between the equivalent wall

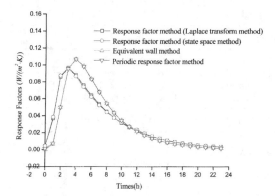

Figure 3.   Response factors of W-heavy.

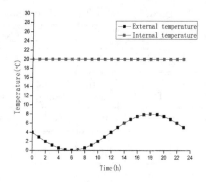

Figure 4.   Internal and external temperature.

method and Laplace transform method turns larger. This phenomenon may result from adopting linear interpolation in the equivalent wall method.

In general, for simple multi-layer walls without thermal bridge, it is more reliable to adopt the state space method to calculate the response factors.

3.3   Calculation and comparison of heat transfer of composite walls

In order to compare the heat transfer calculated by various methods, it is assumed that the external temperature varies according to the sine curve and the internal temperature is constantly 20 (Figure 4). The calculation results of response factors are plotted below (Figure 5, Figure 6, and Figure 7).

From Figure 5, Figure 6 and Figure 7, it can be seen that

1. The maximum difference of transient heat flow between the response factor method and finite difference method is 6.10 W/m² (W-light), 2.01 W/m² (W-medium) and 4.46 W/m² (W-heavy), respectively. The average difference is 2.83 W/m² (W-light), 0.90 W/m² (W-medium) and 2.57 W/m² (W-heavy), respectively.
2. The maximum difference of transient heat flow between the response factor method and finite element method is 6.05 W/m² (W-light), 1.57 W/m² (W-medium) and 3.28 W/m² (W-heavy), respectively. The average difference is 2.66 W/m² (W-light), 0.78 W/m² (W-medium) and 1.63 W/m² (W-heavy), respectively.

From the results, it can be seen that the calculation difference between the numerical methods and response factor methods is acceptable. Although the calculation of the finite difference method and finite element method is more accurate, they consume much more time during the calculation than

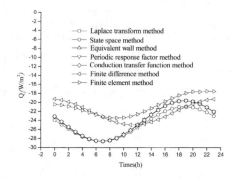

Figure 5.   Heat transfer of W-light.

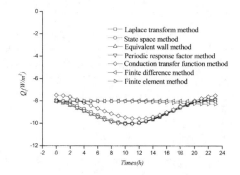

Figure 6.   Heat transfer of W-medium.

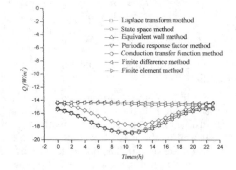

Figure 7.   Heat transfer of W-heavy.

the response factor methods. In conclusion, it is suggested to adopt the response factor method to solve the transient heat conduction problem.

4 PROGRAMMING OF ENERGY SAVING OPTIMIZATION DESIGN FOR THE BUILDING ENVELOPE

In order to evaluate the thermal performance of building conveniently and accurately, and to optimize the thickness and cost of the envelope according to the thermal performance, a program for optimizing the thermal performance of the building envelope is developed by using *Matlab* software based on the existing response factor program and equivalent wall program. The relation of the modules is shown in Figure 8.

4.1 *Functions and features of the program*

The main function of the program is to give the thermal resistance, the cost per square meter and heat transfer coefficient of the building envelope, and to determine whether the heat transfer coefficient meets the requirements in various regions, after considering the thermal performance and cost. According to the requirements of the heat transfer coefficient in various regions, taking cost minimization as the optimization objective function, the program can give the optimal thickness of the wall. The program interface is shown in Figure 9 and Figure 10.

This program has the following features:

1. It supports both data input in the program interface and batch data input from the text;
2. The users can consider whether to optimize a layer thickness by setting the layer whether it is a structural layer;
3. The users can set the minimum thickness and thickness increment of each layer in the envelope during the optimization;
4. It can display the analysis results with bar chart intuitively.

4.2 *Calculation example for the program*

In order to verify the effectiveness and correctness of the energy saving optimization design program, six common exterior wall systems were analysed by the program. Due to the length limitation of the paper, only the partial optimization results of the No.3 wall (Figure 11) are listed below.

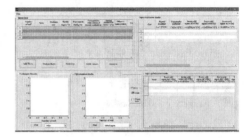

Figure 9.   Initial interface of the program.

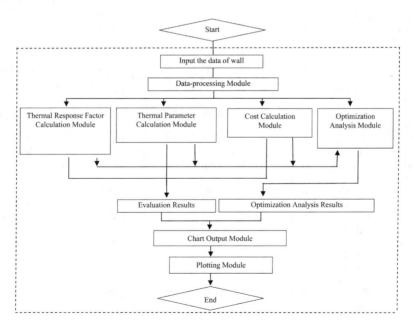

Figure 8.   Flow chart of the program of energy saving optimization design for the building envelope.

1095

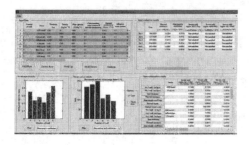

Figure 10.　Result interface of the program.

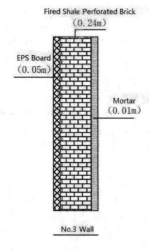

Fired Shale Perforated Brick
(0. 24m)

EPS Board
(0. 05m)

Mortar
(0. 01m)

No.3 Wall

Figure 11.　Diagram of the No.3 wall system.

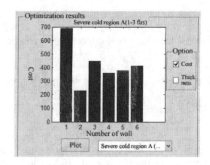

Figure 12.　Cost optimization result in severe cold region A.

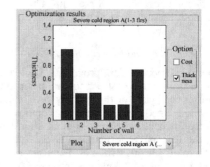

Figure 13.　Thickness optimization result in severe cold region A.

Table 4.　Optimization results of No.3 wall.

Type of region		Layer 1	Layer 2	Layer 3	Total thickness	Thermal resistance	Heat transfer coefficient	Thermal inertia index	cost per m²
Severe Cold Region A	≤3 storeys (0.25)	0.15	0.055	0.01	0.215	4.022	0.249	9.165	358.6
	4~8 storeys (0.40)	0.125	0.055	0.01	0.19	3.427	0.291	6.437	308.2
	≥9 storeys (0.50)	0.09	0.055	0.01	0.155	2.593	0.386	3.451	237.6
Severe Cold Region B	≤3 storeys (0.30)	0.075	0.055	0.01	0.14	2.236	0.447	2.469	207.4
	4~8 storeys (0.45)	0.07	0.055	0.01	0.135	2.117	0.472	2.181	197.3
	≥9 storeys (0.55)	0.06	0.055	0.01	0.125	1.879	0.532	1.665	177.2
Hot-summer and Cold-winter Region	Shape coefficient ≤0.40 (0.8/1.0)	0.025	0.055	0.01	0.09	1.046	0.956	0.485	106.6
	Shape coefficient >0.40 (1.0/1.5)	0.035	0.055	0.01	0.1	1.284	0.779	0.723	126.8

From Table 4 it can be seen that the optimization program is of strong practicability. The program can not only calculate the thermal performance of the original wall, but also provide reference advice for various climatic regions. In summary, the energy saving optimization design program can help engineers to evaluate the thermal performance of the building envelope and provide suggestions about cost and thickness optimization for design.

5 CONCLUSION AND PROSPECT

In this paper, several common methods of unsteady-state heat conduction in a building envelope were compared. For the calculation of heat transfer, although the calculations by the finite difference method and finite element method are more accurate, they consume much more time than the conduction transfer function method. Therefore, it is suggested to adopt the response factor method to calculate the unsteady-state heat conduction in engineering. Furthermore, according to the basic principle of the response factor method, a computing program for the response factor was developed by the *Matlab* software. The program is convenient to calculate the response factor of various kinds of wall.

Subsequently, a program for optimizing the thermal performance of the building envelope was developed based on the existing response factor program and equivalent wall program. The results of the calculation example showed that the program can not only calculate the thermal performance of the building envelope, but also optimize the thickness and cost. In other words, it can help engineers by providing suggestions about cost and thickness optimization for a design. However, this program can be only applied in calculation of the building envelope. It is necessary to study the energy consumption simulation of the whole building, and to develop the building energy consumption simulation software in the future.

REFERENCES

Ceylan, H.T., G.E. Myers, ASME Journal of Heat Transfer, 102(1), 115–120 (1980).
J.S. Ma, Energy Saving. Nonferrous Metall. J. **29.3**, 42–45(2013).
Kossecka, E., J. Kosny,, J. Buil Phys, **20.3**, 249–268 (1997).
Ministry of housing and urban rural development of the people's Republic of China, *Assessment standard for green building*, GBT 50378–2014.
Ministry of housing and urban rural development of the people's Republic of China, *Design Standard for Energy Efficiency of Public Buildings*, GB50189–2015.
Pedersen, C.O., D.E Fisher, R.J. Liesen, ASHRAE Transactions **103**, 459–468 (1997).
Wang, Q.Q., SHENZHEN Civ. Struct. **1**, 18–24 (2004).

Advances in Energy Science and Equipment Engineering II – Zhou, Patty & Chen (Eds)
© 2017 Taylor & Francis Group, London, ISBN 978-1-138-71798-5

Optimization design of propellant grain's shape parameters based on structural integrity analysis

Hao Xu, Futing Bao & Ze Lin

Science and Technology on Combustion Internal Flow and Thermal-structure Laboratory, Northwestern Polytechnical University, Xi'an, Shaanxi, China

ABSTRACT: An optimization process based on structural integrity analysis was carried out to get better propellant grain's shape with higher safety factor. Script flows were introduced to help grain's parametric modeling and automatic structural integrity analysis. Combined with suitable genetic algorithm, the optimization got the optimal results in solidification and ignition process, which were close to the theoretical calculation result. The results showed the grain has the optimal shape with lower maximum strain and higher safety factor. The analysis data also showed that radius of the transition circular arc has the highest sensitivity to the optimal target. The whole process was easy and convenient to realize, which would improve the design efficiency obviously.

1 INTRODUCTION

Solid Rocket Motor (SRM) is a kind of chemical rocket engine, which gains kinetic energy by igniting the propellant grain in the combustion chamber. Grain design greatly affects the performance of the SRM, its geometry determines the structural integrity level. Researchers can get optimized design results by several modern methods. Universal CAE analysis instruments provide many kinds of custom interfaces for designers to achieve the parametric modeling on the grain. Combined with advanced optimization algorithms (Nisar, 2008), such as Genetic Algorithm (GA) and sequential Quadratic Programming Algorithm (SQP), grain designers can obtain reasonable results with better structural integrity level, larger grain filling fraction, larger impulse or smaller quality (Kamran, 2010; Kamran, 2011; Lei, 2011; Lei, 2011; Lei, 2011). For those grains with complex surfaces, such as umbrella plate structure shape and wheel shape, the optimization procedure includes too much parameters, and takes a long time.

Aiming at improving the efficiency and precision of the optimization procedure, a star grain analysis is carried out. The results show that reasonable parameter choices and optimization procedure design can improve the SRM's design level.

2 STAR GRAIN MODELING METHOD

2.1 Simplified model

Star grain has complex surface. It has been simplified with plane strain model, as shown in Figure 1.

Two assumptions are as followed: the whole surface of the grain was smooth without any defects of cracks or sharp points; the grain was evenly distributed in axial ways, and strictly met circular symmetry structure.

Main design parameters include outside Diameter (D), grain thickness (e1), radius of the transition circular arc (r), radius of star angle arc (r1), angle coefficient (ε), star edge angle (θ), length of grain (L) and number of the star angle (n).

2.2 Script flow

With the help of scripting language, such as Python (Zhong, 2006; Zhang, 2011), researchers can build the model automatically in CAE software. Main works in the modeling scripts include setting the self-defined parameters, coordinates and constraints, opening the dimensioning and referencing external file contents. Some script command flow examples are as below:

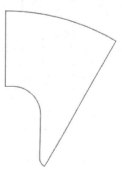

Figure 1. Star grain section.

Arc By Center Ends (center = (,), point1 = (,), point 2 = (,), direction = CLOCKWISE)...

s. *Arc By Center Ends (center = (,), point1 = (,), point2 = (,), direction = CLOCKWISE)...*

s. *Coincident Constraint (entity1 = v[ ], entity2 = g [ ], addUndoState = False)...*

s. *Oblique Dimension (vertex1 = v[ ], vertex2 = v[ ])...*

file = open('D:/temp/putin.txt','r')...

3 AUTOMATIC STRUCTURAL INTEGRITY ANALYSIS

3.1 *Viscoelastic mechanics finite element method*

Suppose that the grain is an isotropy viscoelastic material. In the range of small-strain theory, the constitutive equation is established as below.

$$\sigma = \int_0^t 2G(t-\tau)\frac{de}{d\tau}d\tau + I\int_0^t K(t-\tau)\frac{d\Delta}{d\tau}d\tau \quad (1)$$

where, σ is the Cauchy stress, $G(t)$ is the Shear relaxation kernel function, $K(t)$ is the Relaxation kernel function, e is the Part of strain deviator (shear deformation), Δ is the Part of volume deviator (volume deformation), t is the current time, τ is the past time, I is the unit tensor.

Prony series is introduced to describe shear relaxation modulus and volume relaxation modulus as below.

$$G = G_\infty + \sum_{i=1}^{n_G} G_i \exp(-\frac{t}{\tau_i^G})$$
$$K = K_\infty + \sum_{i=1}^{n_K} K_i \exp(-\frac{t}{\tau_i^K}) \quad (2)$$

Or

$$G = G_\infty + G_0 \left[\sum_{i=1}^{n_G} g_i \exp\left(-\frac{t}{\tau_i}\right) \right]$$
$$K = K_\infty + K_0 \left[\sum_{i=1}^{n_K} k_i \exp\left(-\frac{t}{\tau_i}\right) \right] \quad (3)$$

where, G_i is the weight of the shear modulus, G_∞ is the shear modulus when t tends to infinity. K_i is the weight of the volume modulus, K_∞ is the volume modulus when t tends to infinity. τ_i is the relaxation time of each weight of each Prony series, G_0 and K_0 is the shear modulus and volume modulus under the instantaneous load.

Lastly, the relative modulus is introduced.

$$g_i = G_i / G_0$$
$$g_i = K_i / K_0 \quad (4)$$

Where, $G_0 = G_\infty + \sum_{i=1}^{n_G} G_i$, $K_0 = K_\infty + \sum_{i=1}^{n_K} K_i$.

3.2 *Analysis procedure*

Figure 2 shows a typical structural integrity analysis procedure.

For SRM's grain, the aim of its structural integrity analysis is to get the maximum equivalent strain value in the working process, especially the solidification and ignition process. The finite element analysis is based on reasonable assumptions as below.

1. The propellant grain is isotropic viscous-elastic material. Although the grain is composed of polymeric particles with different diameters actually, we suppose that it's homogeneous in every direction.
2. The Poisson's ratio remains unchanged.
3. The temperature field of the combustion chamber is uniform.

The whole procedure can be defined with Python, the language interface is available and convenient.

3.3 *Meshing*

The meshing of the model mainly uses quadrilateral elements, and the element type chooses plane strain element. After distributing the grid seeds on the edges, the whole grain model selects global seeds whose distribution density is 0.5. Figure 3 shows the meshing result.

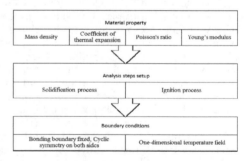

Figure 2. Structural integrity analysis procedure.

Figure 3. Meshing result.

4 OPTIMIZATION OF THE GRAIN SHAPE

The script flows established above can automatically complete the grain's parametric modeling and the structural integrity analysis. This is the basis of facilitating the optimization.

4.1 Optimization model

To reduce the calculation work, the design variables are chosen as follows: star edge angle (θ), angle coefficient (ε), radius of the transition circular arc (r) and radius of star angle arc (r1).

The constraint condition is keeping the grain's initial burning area (A_b) unchanged.

$$A_b = 2ns' L \tag{5}$$

Where,

$$s' = \frac{l\sin\varepsilon\frac{\pi}{n}}{\sin\frac{\theta}{2}} + l(1-\varepsilon)\frac{\pi}{n}$$
$$+(e+r)\left(\frac{\pi}{2}+\frac{\pi}{n}-\frac{\theta}{2}-ctg\frac{\theta}{2}\right)$$

The objective function is the grain's maximum strain. The optimization program should find the optimal design variables to get the minimum of the grain's maximum strain.

$$\begin{cases} \min f(X), X = \{x_1, x_2, x_3 \ldots x_n\} \\ g(X) = C \end{cases} \tag{6}$$

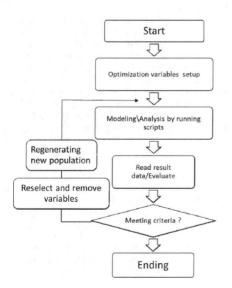

Figure 4. Procedure of MIGA.

where, $f(X)$ is the grain's maximum strain, $g(X)$ is the initial burning area.

4.2 Optimization algorithm

Multi-Island Genetic Algorithm (MIGA) is suitable for the problems which has no explicit correlation function, like the grain's structural integrity analysis. Figure 4 shows the procedure of MIGA.

5 EXAMPLE AND RESULT ANALYSIS

5.1 Initial setup

Define the grain's initial sizes as: D = 60 mm, e1 = 10 mm, r = 2 mm, r1 = 1.5 mm, ε = 0.8, $\theta/2$ = 33.50°. The parameters to be optimized are r, r1, $\theta/2$, ε. The material property sets are as Table 1 shows.

The temperature of the grain change at a speed of 2.25°C/h from 68°C to –40°C, which simulates the solidification process. The igniting process is defined as: the pressure load increases from 0 to 12 MPa in 1 second.

5.2 Optimization results

The final results are listed by Table 2.

Figure 5 shows the initial strain contours and optimal strain contours of the solidification process. The maximum strain is optimized from 5.022e-02 to 4.223e-02. Figure 6 shows the same contents of the igniting process. The maximum strain is optimized from 1.128e-01 to 7.910e-02. The safety factor is significantly improved after optimization as Table 3 shows.

Table 1. Initial gap size.

Mass density (Ton/mm³)	Coefficient of thermal expansion (1/°C)	Poisson's ratio	Young's modulus (GPa)
1.77e-009	9.5e-005	0.498	0.2199

Table 2. Optimizaion results of the variables.

Variables	Floor limit	Upper limit	Initial value	Optimal value
r (mm)	0	5.00	2.00	3.80
r1 (mm)	0	5.00	1.50	1.34
$\theta/2$(°)	0	90.00	33.50	29.14
ε	0.4	0.9	0.8	0.87

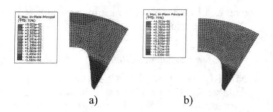

a)   b)

Figure 5. Initial and optimal strain contours in solidification.

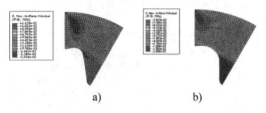

a)   b)

Figure 6. Initial and optimal strain contours in Ignition.

Table 3. Safety Factor.

Process	Solidification		Ignition	
	Initial	Optimal	Initial	Optimal
Ultimate strain (ε_m)	0.2		0.4	
Max strain (ε)	5.022 e-02	4.223 e-02	1.128 e-01	7.910 e-02
safety factor (FS)	3.98	4.74	3.55	5.06

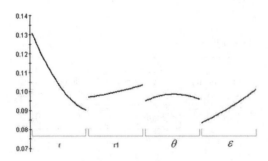

Figure 7. Variables' influence to maximum strain.

5.3 Results analysis

The initial strain can be calculated by formula (1)-(4) with a result of $\varepsilon_{max}^* = 5.194\text{e-}2$, which is very close to the CAE result. It proves that the script runs basically correct.

The variables' influence to the objective function value is arranged, as Figure 7 shows, they are, in order, r, r1, θ and ε.Obviously, r and ε have greater influences to the results, and r has the highest sensitivity in the whole process.

6 CONCLUSION

An optimization of propellant grain's shape parameters based on structural integrity analysis is carried out, main conclusions are as follows.

1. Script languages can simplify the parametric modeling and the automatic structural analysis, which is quick and easy for researcher to learn and use.
2. The whole optimization process aims at minimizing the maximum strain in different working condition, and keeps the initial burning area unchanged. The results show the optimization is effective and runs stably.
3. The results show that variables have different influence to the optimal target, those who have high sensitivity should get more attentions in grain design, such as radius of the transition circular arc.

REFERENCES

Kamran A. Design and Optimization of 3D Radial Slot Grain Configuration. Chinese Journal of Aeronautics. **4**(2010).

Kamran A. Design and optimization of solid rocket motor Finocyl grain using simulated annealing. Journal of Aerospace Power. **4**(2011).

Lei L, G.J. Tang, J.B. Duan. Shape Optimization of Wheel Shape Grain Based on Parameterized Modelin., Journal of National University of Defense Technology. **1**, 33(2011).

Lei L, J.B. Duan, Z.B. Shen, Shape optimization of grain umbrella slot based on parameterized modeling. Journal of Solid Rocket Technology. **5**, 34(2011).

Lei L, Z.B. Shen, G.J. Tang. Shape optimization of star shape grain considering structure integrity and loading fraction. Journal of propulsion technoloty. **2**, 32(2011).

Nisar K, Zeeshan Q. A Hybrid Optimization Approach for SRM FINOCYL Grain Design. Chinese Journal of Aeronautics. **6** (2008).

Zhang, Q., Y. Ma, S.C. Li. Method and Application of Second developed ABAQUS Based on Python. Ship Electronic Engineering. **2**(2011).

Zhong, T.S., F. Wei. Second Development for Fore Treatment of ABAQUS Using Python Language. Journal of Zheng zhou University. **1**(2006).

Advances in Energy Science and Equipment Engineering II – Zhou, Patty & Chen (Eds)
© 2017 Taylor & Francis Group, London, ISBN 978-1-138-71798-5

Micropore dispersing synthesis of CaCO₃ in the presence of hexylic acid

Ziling Yan, Yao He, Chen Zhang, Xiaoshi Tu & Zhezhe Deng
Hubei Provincial Key Laboratory of Green Materials for Light Industry, Hubei University of Technology, Wuhan, China

Jiuxin Jiang
School of Materials and Chemical Engineering, Hubei University of Technology, Wuhan, China

ABSTRACT: As the research continues, the control on the polymorph and morphology of calcium carbonate ($CaCO_3$) becomes a hot topic because its applications are limited by these parameters. In the present work, $CaCO_3$ powder was synthesized by dispersing CO_2 gas, via a micropore plate, into calcium hydroxide (Ca(OH)2) slurry in the presence of hexylic acid as the template. The polymorph, morphology of $CaCO_3$ particles and the interaction between $CaCO_3$ particles and hexylic acid were investigated by X-Ray Diffraction (XRD), Transmission Electron Microscopy (TEM) and Fourier Transform Infrared (FTIR) spectroscopy, respectively. XRD patterns indicate that the addition of hexylic acid cannot change the polymorph of $CaCO_3$ particles. However, the remarkable difference in morphology for different amounts of hexylic acid shows that hexylic acid can alter the morphology of $CaCO_3$ particles. FTIR spectra verify the interaction between $CaCO_3$ particles and hexylic acid.

1 INTRODUCTION

Based on micro-dispersion, one of micro-mixing technology which has tremendous mass transfer because the path of molecular diffusion greatly shortens and homogenous mixing will take place in extremely short time (millisecond to microsecond grade) only depending on molecular diffusion. Wu *et al.* (2010) have developed a new reactor to intensify gas-liquid mass transfer, in which a micro-pore glass plate with average pore size about 10 μm was adopted as the gas disperser and this method was defined as micropore dispersion.

There are some literature reports on the application of fatty acid in the preparation of $CaCO_3$ and its surface modification (Wang, 2010; García-Carmona, 2003; Trana, 2010). Using lauric acid as an organic substrate, Wang *et al.* have synthesized $CaCO_3$ powder which consists of most rod-like particles with a diameter of 200–400 nm and length of 2–4 μm, and a few ellipse-like particles with a diameter of about 100 nm. García-Carmona *et al* (2003) have investigated the different impact of sodium stearate on the morphology of $CaCO_3$ particles at different concentrations of dissolved Ca^{2+} (García-Carmona, 2003). Trana *et al.* have successfully modified nano-$CaCO_3$ particles with sodium stearate *via* adding at the end of carbonation stage, and well dispersed nano-$CaCO_3$ particles and rod-like or spindle-like aggregates of primary particles are obtained (Trana, 2010).

We have reported the synthesis of $CaCO_3$ particles *via* micropore dispersion in the presence of oleic acid (Jiang, 2011), lauric acid, palmitic and stearic acid (Jiang, 2014; Jiang, 2015) in our previous research studies. All the results indicate that the addition of these fatty acids has no obvious influence on the polymorphs of $CaCO_3$ particles and has a remarkable impact on their morphologies. However, as far as we know, the effect of hexylic acid on the polymorph and morphology of $CaCO_3$ particles has rarely involved. Therefore, the effect of different amounts of hexylic acid on the polymorphs and morphologies of $CaCO_3$ particles is focused in the present work.

2 EXPERIMENT

2.1 *Materials and reaction equipment*

Ca(OH)₂ and absolute ethyl alcohol are of analytical grade, hexylic acid is of chemical grade, and the above reagents are purchased from Sinopharm Chemical Reagent Company. Water is distilled water and the purity of CO_2 exceeds 99.5%. The schematic of reaction equipment is shown in Figure 1 and the micropore disperser is made

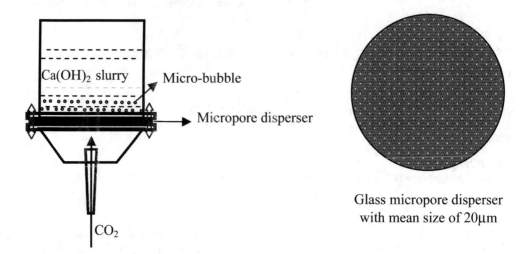

Glass micropore disperser
with mean size of 20μm

Figure 1. Schematic of micropore dispersion reaction equipment.

of glass with an average pore diameter of about 10–20 μm.

2.2 Preparation of CaCO₃ powder

5 g of pre-ground $Ca(OH)_2$ was added in distilled water to prepare 100 mL $Ca(OH)_2$ slurry. Hexylic acid was dissolved in absolute ethyl alcohol to form a solution with the concentration of 0.1 mol/L and a certain amount of this solution was added to $Ca(OH)_2$ slurry. The weight ratio of hexylic acid to $CaCO_3$ was selected as 0, 0.5, 1.0,

1.5, 2.0, 2.5, 3.0, 3.5 and 4.0%, and the corresponding sample number was marked as CH0, CH0.5, CH1.0, CH1.5, CH2.0, CH2.5, CH3.0, CH3.5 and CH4.0, respectively. CO_2 gas was introduced into the mixture and the reaction was not stopped until the pH value dropped to about 7. The final suspension was centrifuged at 8000 rpm for 10 min and the supernatant solution was discharged, and then the solid was re-dispersed in absolute ethyl alcohol. This process was repeated and sediments were dried in the oven at 110°C for 4 hr.

2.3 Characterization of CaCO₃ particles

The polymorphs of $CaCO_3$ were characterized by X-ray diffraction (D/MAX-RB, RIGAKU). The morphology was observed using a transmission electron microscope (Tecnai, G20). For TEM observation, the powder was dispersed in absolute alcohol, dropped onto carbon-covered copper grids placed on filter paper, and dried at room temperature. In order to investigate surface characteristics, Fourier Transform Infrared (FTIR) spectroscopy was performed on a NEXUS with the KBr pellet method.

3 RESULTS AND DISCUSSION

3.1 Effect of hexylic acid on the polymorphs of CaCO₃

To investigate the effect of hexylic acid and its amount on the polymorphs of $CaCO_3$, the powder was characterized by XRD and the results are shown in Figure 2. From XRD patterns, it can be found that calcite is the only crystalline phase in all samples with different amounts of hexylic acid.

Rhombic calcite, needle-like aragonite and spherical vaterite are three typical polymorphs of $CaCO_3$. Different crystalline phases can be obtained from different preparation methods. Generally, calcite is the only crystalline phase during the carbonation of $Ca(OH)_2$ aqueous slurry (Jiang, 2011; Jiang, 2014; Jiang, 2015), the vaterite phase emerges in the reaction system of $Ca^{2+}–CO_3^{2-}$ (Chen, 2010), $Ca^{2+}–CO_2$ (Watanabe, 2009) or mineralization for a long period (Guo, 2006), and the aragonite phase generates at high reaction temperatures (Nan, 2008).

According to our previous research, calcite is the only phase in $CaCO_3$ prepared from the same method without any additives (Jiang, 2011). The present result indicates that the addition of hexylic acid cannot alter the crystal type of $CaCO_3$, which is in accordance with our previous conclusion that the addition of fatty acid cannot change the polymorphs of $CaCO_3$ during the carbonation of $Ca(OH)_2$ slurry (Jiang, 2014; Jiang, 2015).

3.2 Effect of hexylic acid on the morphologies of CaCO₃

To study the effect of the addition of hexylic acid on the morphologies of $CaCO_3$ particles, the morphologies of sample CH0, CH0.5, CH1.5, CH2.5 and CH3.5 were observed on a transmission electron microscope and the images are illustrated in Figure 3, from which it can be found that

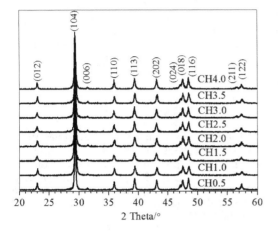

Figure 2. XRD patterns of CaCO3 with different addition of hexylic acid.

there is obvious differences in morphology for different amounts of hexylic acid.

Although some particles with sizes less than 100nm exist in the sample without addition of hexylic acid, lots of large particles and serious aggregation can be found in the sample in the absence of hexylic acid, as shown in Figures 3a and 3b. In the sample with addition of 0.5 wt% of hexylic acid, although the aggregates are not scattered yet, most aggregates become regular spindles and the average particle size slightly decreases, as shown in Figures 3c and 3d. The similar morphology is found in the sample when the amount of hexylic acid is increased to 1.5 wt%, as shown in Figure 3e. When the amount of hexylic acid is increased to 2.5 wt%, more broken spindle particles are found than integrated spindles, as shown in Figures 3f and 3g. In addition, the average particle size further decreases. Rod-like particles are found in the sample with the amount of hexylic acid reaching 3.5 wt%, as shown in Figures 3h and 2i. These results indicate that the addition of hexylic acid has remarkable influence on the morphology of $CaCO_3$ particles.

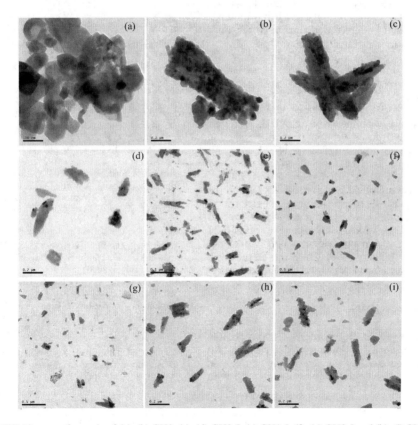

Figure 3. TEM images of sample of (a)–(b) CH0, (c)–(d) CH0.5, (e) CH1.5, (f)–(g) CH2.5 and (h)–(i) CH3.5.

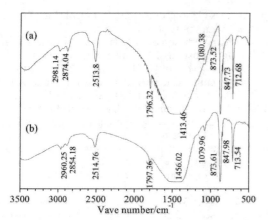

Figure 4. IR spectra of sample (a) CH0 and (b) CH3.5.

3.3 *Interaction between hexylic acid and CaCO₃ particles*

Generally, the fatty acid can interact with inorganic particles when they co-exist in the water medium. To investigate the interaction between hexylic acid and CaCO₃ particles, FTIR spectra of the as-prepared CaCO₃ powder in the absence and presence of hexylic acid were recorded and the corresponding spectra are illustrated in Figure 4.

For both cases, according to Holmgreen *et al*'s research (Holmgreen, 1994), the characteristic absorption peaks of calcite emerge at the wave number of 873.52 cm⁻¹, 712.68 cm⁻¹ and at 873.61 cm⁻¹, 713.54 cm⁻¹, respectively, which is in accordance with the result of XRD. However, the characteristic absorption peaks of aragonite also appear in both samples, at 847.73 cm⁻¹ and 1080.38 cm⁻¹ for CH0 and at 847.98 cm⁻¹ and 1079.98 cm⁻¹ for CH3.5. Despite this, their diffraction peaks cannot be observed in XRD patterns maybe because of the trace amount. The absorption occurs at around 2924.81 cm⁻¹ and 2854.09 cm⁻¹ in sample CH3.5, which is characteristic of H-C-H asymmetric and symmetric stretching vibrations, respectively (Gilbert, 2000), which verifies the interaction between hexylic acid and CaCO₃ particles.

4 CONCLUSIONS

In summary, CaCO₃ powder was successfully synthesized *via* micropore dispersing carbonation in the presence of hexylic acid as the template in this study. Hexylic acid molecules interact with CaCO₃ particles during the carbonation process. The addition of hexylic acid has obvious influence on the morphologies of CaCO₃ particles, although it cannot change the polymorphs of CaCO₃. Compared with the case in the absence of hexylic acid, when different amounts of hexylic acid is added in the reaction system, the aggregates become spindle or rod-like and their size decreases to some extent, although the morphology and the average size of a single particle have no evident difference.

ACKNOWLEDGEMENT

This work was funded by the Hubei Provincial Department of Education Research Program (No. D20151405), the Open Foundation of Hubei Provincial Key Laboratory of Green Materials for Light Industry (No. 2013-2-8) and the Hubei Natural Science Foundation (No. 2009CDA030).

REFERENCES

Chen, Y.X., X.B. Ji, X.B. Wang, Mater. Lett. **64**, 2184 (2010).

García-Carmona, J., J. Gómez-Morales, R. Rodríguez-Clemente, J. Colloid Interf. Sci. **261**, 434 (2003).

Gilbert, M., I. Sutherland, A. Guest, J. Mater. Sci. **35**, 391 (2000).

Guo, X.H., S.H. Yu, G.B. Cai, Angew. Chem. Int. Ed. **45**, 3977 (2006).

Holmgreen, A., L.M. Wu, W. Spectrochim Acta **50**, 1857 (1994).

Jiang, J.X., D.D. Xu, Y. Zhang, S.P. Zhu, X.P. Gan, J.N. Liu, Powder Technol. **270**, 387 (2015).

Jiang, J.X., J. Liu, C. Liu, G.W. Zhang, X.H. Gong, J,N, Liu, Appl. Surf. Sci. **257**, 7047 (2011).

Jiang, J.X., Y. Zhang, X. Yang, X.Y. He, X.X. Tang, J.N. Liu, Adv. Powder Technol. **25**, 615 (2014).

Liu, Q., Q. Wang, L. Xiang, Appl. Surf. Sci. **254**, 7104 (2008).

Nan, Z.D., X.N. Chen, Q.Q. Yang, X.Z. Wang, Z.Y. Shi, W.G. Hou, J.Colloid Interf. Sci. **325**, 331 (2008).

Trana, H.V., L.D. Tranb, H.D. Vua, H. T, Colloids Surface A **366**, 95 (2010).

Wang, C.Y., C. Piao, X.L. Zhai, N. Hickman, J. Li, Powder Technol. **200**, 84 (2010).

Watanabe, H., Y. Mizuno, T. Endo, X.W. Wang, M. Fuji, M. Takahashi, Adv. Powder Technol. **20**, 89 (2009).

Wu, G.H., Y.J. Wang, S.L. Zhu, J.D. Wang, Powder Technol. **172**, 82 (2007).

Zhao, Z.X., L. Zhang, H.X. Dai, Y.C. Du, X. Meng, R.Z. Zhang, Y.X. Liu, J.G. Deng, Micropor. Mesopor. Mat. **138**, 191 (2011).

Advances in Energy Science and Equipment Engineering II – Zhou, Patty & Chen (Eds)
© *2017 Taylor & Francis Group, London, ISBN 978-1-138-71798-5*

The tribological mechanism of the Tris (phosphino) borato silver(I) complex lubricant additive on a textured steel surface

Ningning Hu, Songquan Wang, Jinhe Wu & Na Wu
School of Mechanical and Electrical Engineering, Jiangsu Normal University, Xuzhou, China
School of Mechanical and Electrical Engineering, China University of Mining and Technology, Xuzhou, China

ABSTRACT: Micro-textured surfaces with grooves on 52100 disks are fabricated by picosecond laser RAPID, and then the silver precursor $[PhB(CH_2PPh_2)_3AgPEt_3]$ is used as the lubricant additive to study the tribological properties under high temperatures. In this study, we will discuss the lubricating mechanism of the coupling effect including the friction pair surface and the lubricant additive. The results showed that it is not conductive to form a dynamic pressure oil film when the sliding speed is low. A part of the contact surface is formed by monolayer or multilayer lubricating films, and the rest is through the interaction of the solid micro-convex body; at this moment, the surface texture is the decisive factor to the friction coefficient. The performance of the lubricating oil and additives is decisive to the friction coefficient with the increase of velocity; the best mass percentage of the additive in lubricant at high temperatures is 2%.

1 INTRODUCTION

With the advancement of science and technology, development of modern large-scale mechanical equipment towards extreme environment such as high speed and heavy load is in progress. Because of the unpredictability of the friction pair lubrication state (for example, the temperature and type, the size and quantity of particles when serious wear occurs due to particulate pollution formation), the lubrication system mutation problem is difficult to predict and control. Therefore, when the lubrication failure or working temperature exceeds the upper limit of design, the lubrication system should release the lubricating materials at different rates and quality according to the different temperatures of actual conditions as high temperature lubrication additives which have the effect as backup lubrication (Ali, 2015; Braun, 2014). M.N. McCain (HAN, 2013) synthesized a series of phosphine boric acid silver precursor complexes, and these precursors generated a silver thin film in glass and steel surfaces through the Aerosol-Assisted Chemical Vapour Deposition (AACVD) method, and then discussed its tribological properties as the silver film. X.W. Qi (Hao, 2015; Jatti, 2015) prepared nanoscale serpentine powder as a high temperature lubricating oil additive using the high energy ball milling method, and experiment proved that the serpentine powder can form a layer of self-repairing film on the friction

contact surface at 400°C, reducing the friction and wear of the mating surfaces. In recent years, surface micro-texture technology has become one of the most important research directions in the field of tribology. A micro textured surface can not only effectively improve the lubrication performance between two surfaces, but also reduce surface wear (MNMCCAIN, 2008; Nan, 2005; Ning-ning, 2012; Ning-ning, 2014; S.H, 2015; XWQI, 2011). However, different surface textures and even different directions of surface texture have great influence on friction and wear. Therefore, when measuring the tribological properties of a high temperature lubricating additive, it is also necessary to consider its performance in association with a textured surface.

In this paper, we prepare a rectangular steel surface texture at first; then, we study the high temperature tribological behavior using Tris (phosphino) borato silver(I) complexes as lubricant additives onto the laser-textured surface, thus discussing the coupled mechanism of the friction pair surface and the lubricant additives.

2 PREPARATION OF THE STEEL SURFACE TEXTURE

Plane steel samples of AISI52100 were cut from flat plates as laser processing materials. Before surface texture, the plane samples were ground and polished to a mirror finish with sandpapers and diamond paste; then, the samples were put into absolute ethanol for ultrasonic cleaning and blow dry. Through roughness measurement, it is found that surface roughness Ra is always less than 0.2 µm.

A groove texture was processed by using pico-second laser RAPID (the product name is Lumera RAPID picosecond laser) on plane steel samples. In order to process regular groove surface texture, we chose a pulse frequency of 500 kHz, 0.10 W, the beam diameter is 8–10 µm and the depth of the groove produced is 0.375 µm after each single laser passing. The processed groove texture has a width of 30 µm, the groove depth is 1.7 µm, and the interval between the grooves is 40 µm. A diamond polishing liquid sample with a diameter of 0.1 µm was used to polish the processed surface again. The topographies of the textured samples were assessed with an optical interferometer (Zygo New Image 200). Figure 1 shows the surface topography of the samples.

3 HIGH TEMPERATURE FRICTION TEST OF THE TEXTURED SURFACE

According to the literature (X, 2011), it is clear that the tris (phosphino)borato silver(I) complexes $[PhB(CH_2PPh_2)_3AgPEt_3]$ as high temperature lubricant additives have excellent abrasion wear properties, and their optimal content is 2%. On this basis, we firstly add volume 2% concentration silver complexes in based lubricating oil additives, and then test the tribology performance of lubricant additives on the textured surface under high temperatures.

The tribological behaviors of the samples were evaluated on a CETR UMT-2 multifunctional Tribometer with a ball-on-disk geometry at temperature 300°C (Xing, 2014) under rotation drive with its load $5N$. The samples were divided into 6 groups with experimental groups as shown in Table 1. Samples were all AISI 52100 steel plates with a grooved surface texture, the running direction and texture direction of the test ball (the material of the ball is 45 steel) are 90°, 0° and 45°, respectively. The lubricating oil is composed of special engine lubricating oil SAE J2359 and additives containing 2% ($[PhB(CH_2PPh_2)_3AgPEt_3]$). Test speeds are respectively 4 rpm, 9 rpm, 22 rpm, 43 rpm, 94 rpm, 130 rpm, 174 rpm, 217 rpm, 261 rpm, 304 rpm, 347 rpm, 391 rpm and 434 rpm. The corresponding linear velocities are respectively 0.01 m/s, 0.02 m/s, 0.05 m/s, 0.1 m/s, 0.2 m/s, 0.3 m/s, 0.4 m/s, 0.5 m/s, 0.6 m/s, 0.7 m/s, 0.8 m/s, 0.9 m/s, and 1 m/s.

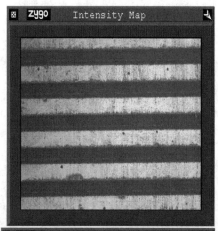

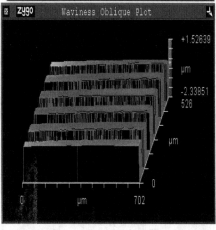

Figure 1. Texture groove three-dimensional surface topography.

Table 1. Test groups.

Samples	Siding orientation	Orientation angles	Lubricant
1	┤┤┤┤►	90°	Base oil
2	──►	0°	Base oil
3	╱╱╱╱►	45°	Base oil
4	┤┤┤┤►	90°	Base oil +2%Ag additive
5	──►	0°	Base oil +2%Ag additive
6	╱╱╱╱►	45°	Base oil +2%Ag additive

4 TEST RESULTS

4.1 Results analysis

Figure 2 shows the friction coefficient produced during lubricated sliding between the textured steel samples and the counter-body, with and without the Ag-based additive, at a temperature of 300°C. We can get the following conclusions:

1. The friction coefficients with lubricant additives are significantly lower than the friction coefficients with base oil. This is determined by the chemical characteristics of additive [PhB $(CH_2PPh_2)_3AgPEt_3$]. According to the thermogravimetric analysis of silver complexes reported in the literature (QI, 2011), the silver element in the complexes is fully released at a high temperature of 300°C, while silver has good lubricating and anti-friction performance, the friction performance of the lubricating oil with additives will be improved significantly.

2. In the lubricating environment with additives, when the velocity is under 0.6m/s, the friction coefficient is the highest along with the steel ball movement direction being parallel to the groove, the friction coefficient at an angle of 45° is lower, and in the vertical direction is the lowest. The friction coefficient develops at the same trend. This gap is decreased with the increasing velocity. When the velocity increases to 1 m/s, the friction coefficient is almost the same, and the friction coefficient is about 0.046. It means that the faster the velocity, the smaller the effect of the surface texture on the friction coefficient. This is because that with the increasing sliding velocity, the lubricating oil can form a stable dynamic pressurized film in the meatus, and the oil film thickness is much larger than the surface roughness, so there is no direct contact between solid interfaces, so that the surface texture cannot have a significant effect on the friction coefficient. At this time, the friction coefficient is mainly determined by the performance of lubricating oil and additives.

3. The friction coefficient produced by the lubricating oil containing additives is larger when the speed is less than 0.6 m/s. The friction coefficient when the steel ball movement direction is parallel to the groove texture is obviously higher than the friction coefficient when the movement and the groove are perpendicular. When the speed is smaller, the friction coefficient is increased rapidly when the motion is parallel to the groove direction, and it is higher than that produced when the sliding direction is perpendicular to the groove orientation. When the speed reduces to 0.01 *m/s*, the friction coefficient in the vertical motion direction is 0.061, the friction coefficient in the parallel motion direction increase to 0.088, and approaches the friction coefficient of the oil without additives at high temperature. This is because the low sliding velocity does not allow the onset of a hydrodynamic lubrication regime and the solid surface gap is very small (boundary or mixed lubrication regime). A part of the contact surface is supported through a single molecular layer or multi-molecular layer of the lubricant, and the rest is supported by solid micro-convex body interactions. Therefore, the surface texture direction is the decisive factor for the friction coefficient.

4.2 The tribological mechanism analysis of surface texture

According to the literature (Xing, 2014), it has been proved that the surface texture and motion direction have a decisive effect on the friction property under certain conditions. So, we mainly analyze the tribological mechanism of the textured surface in this section.

In the mixed lubrication regime, the relationship between the running direction of the material and the direction of texture has great influence on the fluid dynamics, and the thickness of the oil film and the bearing capacity of the oil film are also affected. In certain movement direction and movement speed conditions, "micro-wedge" will be formed between the texture surface and the spouse surface. If the lubricating oil is a newtonian fluid and is a layered fluid (there is no sliding between the lubricating oil and the solid boundary), and the specimen has a grooved surface, the dual steel ball has a smooth surface, while the lubrication is

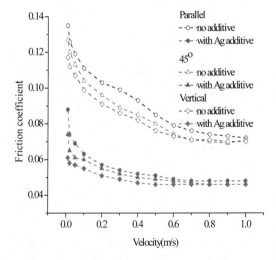

Figure 2.   The friction coefficients.

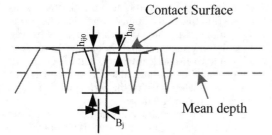

Figure 3.   Micro-wedge model.

stable as shown in Figure 3. According to the literature (YIN, 2012), the bearing capacity of a single micro-wedge can be obtained:

$$w_{ij} = \frac{6U\,\eta LB_i^2}{h_{ij1} - h_{ij0}} \left[\ln \frac{h_{ij1}}{h_{ij0}} - \frac{2(h_{ij1} - h_{ij0})}{h_{ij1} + h_{ij0}} \right] \qquad (1)$$

where, U is the sliding speed, L is the width of the groove texture, Bj is half the width of the groove, that is the wedge length, h_{ij0} is the minimum gap of two surfaces, and h_{ij1} is the maximum gap.

Because each rough peak in the counter-body surface will have a dynamic pressure on the textured surface, the micro-wedge bearing capacity of a groove in the dual is

$$W_{ja} = \frac{1}{m} \sum_{i=1}^{mj} w_{ij} \qquad (2)$$

In which, m is the number of dual parts roughness peak.

Assuming that the number of grooves in the static specimen is n, therefore

$$W_h = \sum_{j=1}^{n} W_{ja} = \sum_{j=1}^{n} \frac{1}{m_j} \sum_{i=1}^{mj} w_{ij} \qquad (3)$$

w_h is the total capacity of the groove surface dynamic membrane. In this chapter, the surface of the dual steel ball is considered as a smooth surface, so $m = 1$. Then,

$$W_h = \sum_{j=1}^{n} W_{ja} = \sum_{j=1}^{n} w_{ij} \qquad (4)$$

According to the micro-wedge model, under the condition of sufficient lubrication, when the dual steel ball moves perpendicular to the groove texture, every time it passes through a trench, the micro-wedge model on the surface has great influence on the fluid dynamics. The surface will continuously produce a convergence-dispersed micro-wedge

flow channel, and the bearing capacity of the oil film will be greatly enhanced, the surface friction coefficient will be reduced and the wear and tear will also be reduced. However, when the sliding direction of the dual ball is parallel with surface groove texture, it is difficult to form a micro-wedge runner, so the surface cannot change the bearing capacity of the oil film.

The contact width of the friction pair is small because of the ball-on-disk geometry. When sliding is parallel to the groove texture, it can only cover a handful of grooves, and with the increase of running distance, the lubricating oil stored in the groove will run down the texture direction. Therefore, when the friction pair runs parallel to the groove texture, it can not only produce continuous micro-wedge channels, but also lead to the occurrence of starved lubrication phenomena as the sliding time increases. From equation 1, it can be seen that the bearing capacity of the fluid in the micro-wedge of the surface groove is proportional to the film viscosity. The viscosity of base oil is lower at 300°C, so the bearing capacity is also reduced, and the friction coefficient of the groove tribo-couple is greatly increased. At this time, the influence of texture direction on the oil film bearing capacity is decreased. In other words, when there is no additives under high temperature, regardless of the friction pair operation direction, perpendicular or parallel to the texture, the difference of the friction coefficient is very small. While the high temperature additives produced a large number of small silver particles suspended in the lubricating oil, increasing the lubricating oil viscosity, and silver particles increased the bearing capacity of the lubricating oil film, the friction coefficient of the lubricant containing additives is obviously lower than that of the base oil.

5   CONCLUSIONS

Firstly, the steel specimen with micro-rectangular texture surfaces is prepared, on the basis of which, the research is mainly focused on the tribological effect of the silver complex additives [PhB $(CH_2PPh_2)_3AgPEt_3$] and micro-rectangular surfaces under the high temperature environment, and also the related mechanism is discussed and analyzed. The main conclusions are as follows:

1. It is difficult to form a micro-wedge runner when the matching parts slide in a direction parallel to that of the grooved surface under the condition of base oil. The bearing capacity of the oil film is therefore not enhanced by the grooved texture and will lose most of the lubricating oil on this occasion. Therefore, the friction coefficient is the

largest at this point when parallel to the groove texture. Secondly, the lubrication performance is the best when the steel ball slides in a direction perpendicular to that of the grooved surface.

2. After adding 2% of lubricant additives, the lubricating silver particles in the additives are released at high temperature, thus improving the lubrication performance in general. Among them, the low-speed condition is the most adverse one for the formation of a continuous lubricating film and the relation orientation of surface texture has the most important effect on the friction coefficient. At high speed, a stable dynamic pressure oil film can be formed between the two surfaces and the film thickness is much larger than the surface roughness; therefore, the lubricating oil containing 2% silver complexes shows excellent high temperature lubricating performance.

ACKNOWLEDGMENT

This work was financially supported by the National Natural Science Foundation of China (51305177), the Natural Science Foundation of Jiangsu Province of China (BK20130229) and the National Natural Science Foundation of China (51505201).

REFERENCES

Ali F, Kaneta M & Krupka I (2015). Experimental and numerical investigation on the behavior of transverse limited microgrooves in EHL point contacts [J.] Tribology International, 84:81–89.

Braun D, Greiner C & Schneider J (2014). Efficiency of laser surface texturing in the reduction of friction under mixed lubrication [J], Tribology International, 77:142–147.

Han Yun-yan, Zhang Lin & Feng Da-peng (2013). TriboLogical behaviors of Zinc Dialkyl Phosphate as a lubricantadditive [J]. Tribology, 33(2):118–122.

Hao, XQ, Pei, SY, Wang L, etc (2015). Microtexture fabrication on cylindrical metallic surfaces and its application to a rotorbearing System [J]. International Journal of Advanced Manufacturing Technology, 78:1021–1029.

Jatti VS, Sing TP (2015). Copper oxide nano-particles as friction-reduction and anti-wear additives in lubricating oil [J], Journal of Mechanical Science and Technology, 29(2):793–798.

MNMCCAIN & SSCHNEIDER (2008). Tris (phosphino) borato silver(I) Complexes as precursors for metallic silver aerosol-assisted chemical vapor deposition [J]. Inorganic Chemistry, 47(7):2534–2542.

Nan F, Xu Y, Xu, BS. Tribological Performance of Attapulgite Nanofiber/Spherical Nano-Ni as Lubricant Additive [J]. Tribology Letters, 56(3):531–541.

Ning-ning Hu, Ji Guang Han & and Bo Hu (2012). Orientation effects on tribological behavior of laser textured surface [J]. Journal of Computational and Theoretical Nanoscience. 9(12):2113–2115.

Ning-ning Hu, Ji Guang Han & Bo Hu (2014). Effect of silverorganic precursor on friction and wear. Industrial Lubrication and Tribology [J]. 2014, 66(3):468–472.

Qi X W, Z N Jia & Y L Yang (2011). Characterization and autorestoration mechanism of nanoscale serpentine powderas lubricating oil additive under high temperature [J]. Tribology International, 44(7–8):805–810.

Qi X W, Z N Jia & Y L Yang (2011). Influences of nanoscale serpentine content on self-repairing performance of metalpairs under high temperature [J]. Advanced Science Letters, 4(3):844–850.

S H, C H Chen (2015). Effects of micro-wedges formed between parallel surfaces on mixed lubrication—Part II: modeling [J]. Tribology Letters, 19(2):83–91.

Xing Y Q, Deng J X & Zhou Y H (2014). Fabrication and Tribologic alproperties of Al_2O_3/TiC ceramic with nano-textures and WS2/Zr soft-coatings [J]. Surface & Coatings Technology, 258:699–710.

Yin Bifeng, Lu Zhentao, Liu Shengji, etc (2012). Theoretical and experimental research on lubrication performance of laser surface texturing cylinder liner[J]. J Mechanical Engineering, 48(21):91–96.

Advances in Energy Science and Equipment Engineering II – Zhou, Patty & Chen (Eds)
© 2017 Taylor & Francis Group, London, ISBN 978-1-138-71798-5

Optical properties of Li, Na, and K doped ZnO investigated by first principles

Yanfang Zhao, Ping Yang, Bing Yang, Daojun Zang, Shuai Xu & Yan Wang
Laboratory of Advanced Design, Manufacturing and Reliability for MEMS/NEMS/OEDS,
Jiangsu University, Zhenjiang, China

ABSTRACT: To understand the optical behaviors of pure ZnO, Li-doped ZnO, Na-doped ZnO and K-doped ZnO, the band structure and optical properties of pure ZnO and Li, Na, and K-doped ZnO are investigated by first principles based on DFT. The changing trend of volume is in agreement with the size of ionic radius. The DOS shows that the doping ZnO is the p-type ZnO. The imaginary part has three response peaks, which are 1.76 eV, 6.42 eV, and 10.16 eV, respectively. With the doping of Li, Na, and K, the response peaks move to the higher energy direction (blue shift). The absorption spectra of Li, Na, and K doped ZnO are always larger than that of pure ZnO in the UV region, and the reflection spectra of Li, Na, and K doped ZnO are always larger than that of pure ZnO in the visible and infrared regions, which may be an important application value for some UV, visible and infrared light optical elements. Meanwhile, the change of energy loss spectroscopy indicates that Li, Na, and K-doped ZnO have a blue shift when the critical point turns from the metal characteristics to the dielectric properties.

1 INTRODUCTION

Because of its wide band gap (Eg = 3.37 eV) at room temperature and high exciton binding energy (up to 60 meV), wurtzite ZnO is a new promising direct bandgap photonics materials in the aspects of ultraviolet (Tang, 1998) and blue light-emitting diodes (Yang, 2007), and has great application potential to solar cells (Bar, 2005), liquid crystal displays (Sasaki, 2005), gas sensors (Wang, 2003), ultraviolet semiconductor lasers (Ma, 2007) and transparent conductive films (Zhou, 2006). Currently, n-type doped ZnO has been successful in doping, such as III group elements B (Lokhande, 2001), Al (Song, 2002; Kim, 2000), Ga (Christopher, 2010), In (Luna-Arredondo, 2005); however, it is more difficult for p-type conversion. As a suitable dopant as acceptor, the V group element N (Guo, 2001) has got widespread concerns. However, the disadvantages of its poor activity, difficult to bond with Zn, and strong repulsive interactions between N and N make it difficult for doping. In order to overcome these shortcomings of N doping, co-doping with IIIA elements has often been used (Yang, 2012; Sumiyaa, 2004; Chen, 2008) to further improve the solubility of the N element. The co-doping method is difficult to produce and complex in the experimental procedure, researchers turn their eyes to the group I elements by replacing Zn atoms to achieve acceptor doping. At present, some experimental and theoretical research results about Li (Zhu, 2011), Na (Wu, 2010), and K

(Shanmuganathan, 2013) -doped ZnO have been reported. For example, in the experimental research aspect, Linhua Xu et al.(2011) studied the 1at.% K-doped ZnO thin films by the sol-gel method on Si substrates with different temperatures and the results showed that all the samples showed strong ultraviolet emission and weak blue emission, and the sample annealed at 500°C showed the best crystalline quality and strongest ultraviolet emission. S.S. Shinde et al. (2013) investigated the photoelectron chemical properties of highly mobilized Li-doped ZnO thin films prepared by the spray pyrolysis technique, and found that the Li-doped ZnO films prepared for 1 at% doping possessed the highest electron mobility of 102 cm^2/Vs and a carrier concentration of 3.62×10^{19} cm^{-3}. Z. Zheng et al. (Zheng, 2013) investigated and found that the conductivity conversion from n-type to p-type with targeted Na doping content increased up to more than 1% by the pulsed laser deposition method. In the theoretical research aspect, Hu Xiao-Ying et al. (Hu, 2012) studied the electronic structures of pure and Li-N, Li-2N codoped wurtzite ZnO systems by the density-functional theory based on first principles, and found that it was p-type by co-doping Li and N, and the carrier concentration was enhanced in the Li-2N codoped system. However, the studies of comparison of Li, Na, and K doped ZnO on the optical properties by theory are not many. This project studied the optical properties of Li, Na, and K doped ZnO by using first principles, and made a certain application significance for the ZnO.

2 MODELS AND COMPUTING METHOD

The calculations were performed using the Cambridge Sequential Total Energy Package (CASTEP) (Watanabe, 1999), which used the plane wave pseudopotential method based on the DFT. The ionic potential was dealt with by the Vanderbilt ultra-soft pseudopotential; the valence electron configuration of Zn, O, Li, Na, and K atoms were chosen as Zn-3d^{10}4s^2, O-2s^{2}2p^4, Li-2s^1, Na-3s^1, and K-4s^1, respectively. The Perdew-Burke-Ernzerhof (PBE) functional of GGA was processed by the exchange-correlation between the electrons. The experimental lattice constants of wurtize ZnO are $a = b = 0.325$ nm, $c = 0.521$ nm, $c/a = 1.602$ and $\alpha = \beta = 90°$, $\gamma = 120°$. The bond length of Zn-O along the c-axis is 0.1992 nm and the other directions is 0.1973 nm. The model is built as a $2 \times 2 \times 2$ supercell which includes 32 atoms. One Zn atom is replaced by a Li (or Na, K) atom and the doping content is 6.25%. The cutoff energy is 400 eV and the k-point of brillouin zone is $3 \times 3 \times 2$. The self-consistent convergence precision is 5×10^{-6} eV per atom.

3 RESULTS AND DISCUSSIONS

3.1 Electronic structure

Before analyzing the optical properties of Li, Na, and K doped ZnO, the band gap of pure ZnO, Li, Na, and K doped ZnO were investigated firstly. Figures 1(a)–(d) show the band structure of pure ZnO, Li-doped ZnO, Na-doped ZnO and K-doped ZnO. The short dot line in the 0eV is the Fermi level. From Figure 1(a), it can be seen that the Conduction-Band Minimum (CBM) and the Valence-Band Maximum (VBM) are at the same k-point (G), which indicates that ZnO is a direct gap semiconductor. The band gap of pure ZnO is 0.73 eV, which is in good agreement with 0.74 eV

calculated by Fang-wei Xie et al. (2012) and 0.75 eV calculated by Osuch et al. (2006) by first-principles methods. However, it is much smaller than the experimental data (3.37 eV), which is a common problem of LDA or GGA and it can be corrected by scissors approximation on optical properties. Figures 1(b)–(d) show the band structure of Zn0.9375Li0.0625O, Zn0.9375Na0.0625O and Zn0.9375K0.0625O. It can be seen that the Fermi levels are all in the valence band, which indicates that the doping of Li, Na, and K leads to the p-type ZnO. Table 1 and Table 2 show the comparison of ionic radius and the volume. Compared with Table 1 and Table 2, the changing trend of volume is in agreement with the ionic radius.

3.2 Optical properties

The optical properties of the medium can be described by the complex dielectric response function $\varepsilon(E) = \varepsilon_1(E) + i\varepsilon_2(E)$ in the linear response range, in which $\varepsilon_1 = n^2 - k^2$ and $\varepsilon_2 = 2nk$. As we all know, the imaginary part $\varepsilon_2(E)$ is the overview of optical properties for any material. The imaginary part $\varepsilon_2(E)$ and real part $\varepsilon_1(E)$ can be calculated by the dispersion relation of Kramers-Kronig. Meantime, other optical parameters such as absorption coefficient $\alpha(E)$, reflectance $R(E)$, and energy loss spectroscopy $L(E)$ can be obtained from $\varepsilon_2(E)$ and $\varepsilon_1(E)$. The computational formula is described as follows:

$$\alpha(E) = 2\sqrt{2}\pi[\sqrt{\varepsilon_1(E)^2 + \varepsilon_2(E)^2} - \varepsilon_1(E)]^{1/2}/\lambda \quad (1)$$

$$R(E) = \left|\sqrt{\varepsilon(E)} - 1\right|^2 / \left|\sqrt{\varepsilon(E)} + 1\right|^2 \quad (2)$$

$$L(E) = \varepsilon_2(E)/[\varepsilon_1^2(E) + \varepsilon_2^2(E)] \quad (3)$$

where, λ is the wavelength in vacuum.

Table 1. The values of ionic radius.

Element	Ionic radius/Å
Zn^{2+}	0.60
Li$^+$	0.59
Na$^+$	0.97
K$^+$	1.33

Table 2. the volumes of different doping condition.

Supercell	Ionic radius/Å
ZnO	398.35
Zn$_{0.9375}$Li$_{0.0625}$O	394.76
Zn$_{0.9375}$Na$_{0.0625}$O	403.97
Zn$_{0.9375}$K$_{0.0625}$O	412.5164

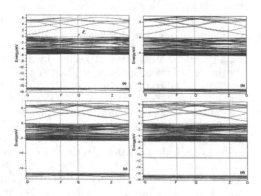

Figure 1. The band structure of (a) pure ZnO; (b) Zn$_{0.9375}$Li$_{0.0625}$O; (c) Zn$_{0.9375}$Na$_{0.0625}$O; (d) Zn$_{0.9375}$K$_{0.0625}$O.

Figure 2 shows the imaginary part of pure ZnO and different doping conditions. The imaginary part of the dielectric function determines the nature of the absorption spectrum. From Figure 3, it can be seen that in the case of pure ZnO, the first threshold energy point is around 1.76 eV. When the energy is greater than 1.76 eV, the value of the imaginary part rapidly increases and the peak of the optical spectrum corresponds to the electron transitions between valence band and conduction band. There are three major peaks in the imaginary part of the dielectric function, which are located at 1.76 eV, 6.42 eV, and 10.16 eV. Wherein, the first peak (near 1.76 eV) is mainly decided by the optical transitions from the upper valence band O2p state to the Zn4s state, the second peak (near 6.42 eV) is mainly decided by the optical transitions from the lower valence band Zn3d state to the O2s state, and the third peak (10.16 eV) mainly comes from optical transitions of the below valence band Zn3d states to the conduction band O2p states. After doping of Li, Na, and K, the three peaks of the

imaginary part of the dielectric function shifted to the higher energy directions (blue shift), which is considered as the cause of Li, Na, and K doping.

Figure 3 shows the real part spectra of pure ZnO and different doping ZnO. The real part spectra of the dielectric function indicate the reflective properties of the material. It can be seen from the figure that in the case of Li, Na, and K doping, respectively, the real part peaks all move to the high energy direction. In the photon energy of 0 ~ 5 eV, the real part peak position of the dielectric function increases during doping, which indicates that the incorporation of Li, Na, and K in the low energy region causes higher reflectivity.

Figure 4 shows the real absorption spectra of pure ZnO and different doping ZnO. It can be seen from the figure that in the UV region of 100–400 nm, the absorption range of K, Na, and Li significantly increased sequentially compared with pure ZnO, which is consistent with the trend of the imaginary part of the dielectric function. It may supply a certain potential application prospect for some UV light optical element.

Figure 5 shows the relationship between reflection spectra of pure ZnO, Li-doped, Na-doped and K-doped ZnO with wavelength. It can be found that the reflectivity of Li, Na, and K doped ZnO increases significantly compared with pure ZnO. It is obvious that the reflection spectra of Li, Na, and K doped ZnO are always larger than that of pure ZnO in the visible and infrared regions, which may be an important application value for some visible and infrared light optical elements.

Figure 6 shows the relationship between energy loss spectra of pure and Li, Na, and K doped ZnO with energy. The energy loss spectroscopy is used to describe the energy loss when a fast electron passes through the material; its sharp maxima often relates to the existence of the plasma oscillation. From Figure 7, it can be seen that the energy

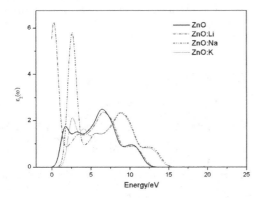

Figure 2. The imaginary part $\varepsilon_2(E)$ of pure ZnO and different doping ZnO.

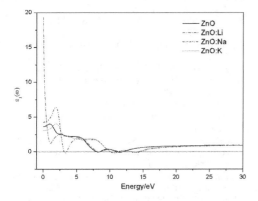

Figure 3. The real part $\varepsilon_1(E)$ of pure ZnO and different doping ZnO.

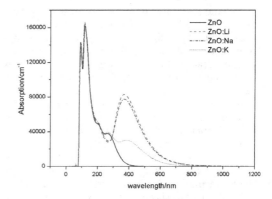

Figure 4. The real absorption of pure ZnO and different doping ZnO.

1115

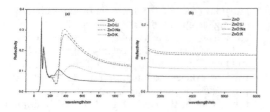

Figure 5. The reflectivity of pure ZnO and different doping ZnO in the wavelength of (a) 0–1200 nm; (b) 1200–8000 nm.

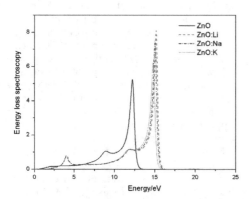

Figure 6. The relationship of energy loss spectroscopy of pure ZnO and different doping ZnO with energy.

loss spectra of Li, Na, and K-doped ZnO moves to the higher energy direction compared with pure ZnO successively, which indicates that Li, Na, and K-doped ZnO have a blue shift when the critical point turns from the metal characteristics to the dielectric properties.

4 CONCLUSIONS

The band structure and optical properties of pure ZnO and Li, Na, and K-doped ZnO are investigated by first principles based on DFT. The changing trend of volume is in agreement with the size of the ionic radius. From the DOS, it can be seen that all the Fermi levels of TDOS of doping ZnO move to the valence band, which indicates that the doping ZnO is the p-type ZnO. The optical parameters such as the dielectric response function and absorption coefficient $\alpha(E)$ have been investigated. The imaginary part has three response peaks, which are 1.76 eV, 6.42 eV, and 10.16 eV, respectively. With the doping of Li, Na, and K, the response peaks move to the higher energy direction (blue shift). The absorption spectra of Li, Na, and K doped ZnO are always larger than that of pure ZnO in the UV region, and the reflection spectra of Li, Na,

and K doped ZnO are always larger than that of pure ZnO in the visible and infrared regions, which may be an important application value for some UV, visible and infrared light optical elements. Meanwhile, the energy loss spectra of Li, Na, and K-doped ZnO move to the higher energy direction compared with pure ZnO successively, which indicates that Li, Na, and K-doped ZnO has a blue shift when the critical point turns from the metal characteristics to the dielectric properties.

ACKNOWLEDGEMENT

The authors would like to acknowledge the support of Jiangsu Province Science Foundation for Youths (BK20130537), the Innovative Foundation for Doctoral Candidate of Jiangsu Province, China (KYLX_1018 and CXZZ13_0655), the support of the National Natural Science Foundation of China (51165004), the support of the Natural Science Foundation of Guangxi Province (No. 2012GXNSFDA053029) and the Special Natural Science Foundation for Innovative Group of Jiangsu University during the course of t0his work.

REFERENCES

Bar, M., J. Reichardt, A. Grimm, I. Kotschau, I. Lauermann, K. Rahne, S. Sokoll, M. C. Lux-Steiner, C. H. Fischer, L. Weinhardt, E. Umbach, C. Heske, C. Jung, T. P. Niesen, S. Visbeck, *J. Appl. Phys.* **98**, 5 (2005).

Chen, K., G. H. Fan, Y. Zhang, S. F. Ding, *Acta.Phys. Sin.* **57**, 5 (2008).

Christopher, W. G., K. S. Ajaya, J. B. Joseph, J. R. Brandon, F. H. Maikel, H. H. Paul, S. G. David, D. P. John, *Thin Solid Films* **519** (2010).

Guo, X. L., H. Tabata, T. Kawai, *Journal of Crystal Growth* **223**, 1 (2001).

Hu, X. Y., H. W. Tian, L. J. Song, P. W. Zhu, L. Qiao, *Acta Phys. Sin.* **61**, 4 (2012) (in Chinese).

Kim, H. K., S. H. Han, T. Y. Seong, *Appl. Phys. Lett.* **77** (2000).

Lokhande, B. J., P. S. Patil, M. D. Uplane, *Phys. B,* **302–303** (2001).

Luna-Arredondo, E. J., A. Maldonado, R. Asomoza, D. R. Acosta, M. A. Melendez-Lira, M. L. Olvera, *Thin Solid Films,* **490**, 2 (2005).

Ma, X., P. Chen, D. Li, *Appl. Phys. Lett.* **91**, 25 (2007).

Osuch, K., E. B. Lombardi, W. Gebicki, *Phys. Rev. B* **73**, 7 (2006).

Sasaki, A., W. Hara, A. Matsuda, N. Tateda, S. Otaka, S. Akiba, K. Saito, T. Yodo, M. Yoshimoto, *Appl. Phys. Lett.* **86**, 23 (2005).

Shanmuganathan, G., I. B. Shameem Banu, S. Krishnan, B. Ranganathan, *J. Alloy. Compd.* **562** (2013).

Shinde, S. S., C. H. Bhosale, K. Y. Rajpure, Photoelectrochemical properties of highly mobilized Li-doped ZnO thin films. *J. Photoch. Photobiolo. B* **120** (2013).

Song, D. Y., A. G. Aberle, J. Xia, *Appl. Surf. Sci.* **195** (2002).

Sumiyaa, M., A. Tsukazaki, S. Fuke, A. Ohtomo, H. Koinuma, M. Kawasaki, *Appl. Surf. Sci.* **223**, 1(2004).

Tang, Z. K., G. K. L. Wong, P. Yu, M. Kawasaki, A. Ohtomo, H. Koinuma, Y. Segawa, *Appl. Phys. Lett,* **72**, 25 (1998).

Wang, Z. L., *Adv. Mater.* **15,** 5 (2003).

Watanabe, K., M. Sakairi, H. Takahashi, *J. Electroanal. Chem.* 473, 1 (1999).

Wu, C. L., Q. G. *J. Lumin.* **130** (2010).

Xie, F., P. Yang, P. Li, L. Q. Zhang, *Opt. Commun.* **285**, 10 (2012): 2660-2664.

Xu, L. H., F. Gu, J. Su, Y. L. Chen, X. Y. Li, X. X. Wang, *J.Alloy. Compd.* **509**, 6 (2011).

Yang, J. J., Q. Q. Fang, B. M. Wang, C. P. Wang, J. Zhou, Y. Li, Y. M. Liu, Q. R. Lv, *Acta Phys. Sin.* **56**, 2 (2007) (in Chinese).

Yang, T. H., J. M. Wu, *Acta Materialia* **60**, 8 (2012).

Zheng, Z., Y. F. Lu, Z. Z. Ye, H. P. He, B. H. Zhao, *Mat. Sci. Semicon. Proc.* **16** (2013).

Zhou, X., S. Q. Wang, G. J. Lian, G. C. Xiong, *Chin. Phys.* **15,** 1 (2006).

Zhu, J. Y., S. H. Wei, *Solid State Communications* **151** (2011).

Advances in Energy Science and Equipment Engineering II – Zhou, Patty & Chen (Eds)
© 2017 Taylor & Francis Group, London, ISBN 978-1-138-71798-5

Effects of zinc ions on structural and Mössbauer properties of nanoferrites $Li_{0.5-0.5x}Zn_xFe_{2.5-0.5x}O_4$

Yun He, Fengwu Du, Hui Li, Zhiqing Luo, Zhimin Ji, Jinpei Lin, Zeping Guo & Jie Song
College of Physics and Technology, Guangxi Normal University, Guilin, China

Qing Lin
College of Physics and Technology, Guangxi Normal University, Guilin, China
College of Medical Informatics, Hainan Medical University, Haikou, China

ABSTRACT: Polycrystalline samples $Li_{0.5-0.5x}Zn_xFe_{2.5-0.5x}O_4$ (x = 0, 0.1, 0.3, 0.5, 0.7, 0.9) were prepared by the sol-gel auto-combustion method. TG and DSC analyses showed that the best calcining temperature in this series of samples is 550°C. Zn^{2+} doping affects the ordered distribution of the lithium, iron ion in the B site. The average grain size of the sample becomes smaller along with the replacement of zinc ions, while the lattice constants increase with the increase of the Zn^{2+} content. The particle size distribution is almost uniform, and the crystallinity of the samples is relatively well. The change of Mössbauer spectra absorption areas could confirm the change in ion content. On addition of Zn^{2+}, the absorption area of sextet in the A site decreases, and that in the B site increases gradually. Mössbauer spectra analysis shows that with Zn^{2+} ion concentration increasing, the ferromagnetic phase gradually transforms into a paramagnetic phase.

1 INTRODUCTION

Lithium ferrite is an important class of spinel. For excellent properties like high Curie temperature, high electrical resistivity, high saturation magnetization, high magnetocrystalline anisotropy, low magnetostriction, and low cost, lithium ferrite has been intensively applied in memory cores, microwave devices, and electromagnetic absorbing materials (Kuroda, 1969; Argentina, 1974). In the spinel ferrite, non-magnetic zinc ions (Zn^{2+}) have strong tendency to occupy the tetrahedral site (A) (Sun, 2007). The molecular magnetic moment of ferrite could be adjusted via changing the doping content of zinc ions. Thus, the static magnetic properties and the microstructures could be controlled effectively, and the ferrite could meet the needs of different applications (Gharagozlou, 2015; Zhang, 2014). In this paper, polycrystalline samples $Li_{0.5-0.5x}Zn_xFe_{2.5-0.5x}O_4$ were prepared by the sol-gel auto-combustion method. The main purpose is to investigate the effect of substitution of zinc on microstructural and magnetic properties of lithium ferrite.

2 EXPERIMENT

2.1 Sample preparation

Samples were prepared by the sol-gel auto-combustion method. The specific operation is as follows: the analytical grade $LiNO_3$,

$Zn(NO_3)_2 \cdot 6H_2O$, $Fe(NO_3)_3 \cdot 9H_2O$ samples were accurately weighed according to the stoichiometric ratio $Li_{0.5-0.5x}Zn_xFe_{2.5-0.5x}O_4$ (x = 0, 0.1, 0.3, 0.5, 0.7, 0.9), and dissolved thoroughly in deionized water. The analytical grade citric acid ($C_6H_8O_7 \cdot H_2O$) was added into the mixed solution. The molar ratio of citric acid to total metal nitrates (n(CA):n(M)) was taken as 1:1. Then, ammonia is added to change the pH value to 7. The beakers which contained the sol are placed into thermostat water baths and heated at 80°C. In the course of heating, the sol was constantly stirred until it is transformed into gel. After 12 h of aging time, the wet gel was put into a electrothermal blowing dry box and dried at 120°C for 6 h. A part of the resultant dry gel is employed to a thermogravimetric test. A little absolute ethyl alcohol is dropped on the rest of the gel. Being ignited in aria at indoor temperature, the dry gel gives rise to a self-propagating reaction, which formed a fluffy flocculent substance. After fully ground in a agate mortar, fine powders were obtained. The powders were annealed in a muffle furnace at a temperature of 800°C for 2 h. Samples of target were obtained after natural cooling.

2.2 Characterization

The thermal decomposition behavior of the gel was analyzed using a thermal analysis instrument (SDT Q600, America). The crystalline structure of the sample was obtained using X-ray diffraction

(D8 Advance, Germany) with Cu K_α radiation (λ=0.15405 nm). The microstructure of particles was observed by scanning electron microscopy (SU8020, Japan). The Mössbauer spectra were observed using a conventional Mössbauer spectrometer (Fast Com Tec PC-moss II, Germany) at room temperature, in constant acceleration modeat. γ-ray was provided by a ^{57}Co source in a rhodium matrix.

3 RESULTS AND DISCUSSION

3.1 TG-DSC analysis

TG, DTG and DSC curves for the dry gel of $Li_{0.5}Fe_{2.5}O_4$ are presented in Figure 1. As shown in the figure, the system loses about 8.5% of the weight, because of the evaporation of the remaining molecular water and crystal water (Inbanathan, 2014). From about 202°C to 214.6°C, the weight loss of the system is nearly 76.3%, with a sharp exothermic peak locating in the vicinity of about 219.4°C. This phenomenon corresponds to the decomposition reaction of citrates and nitrate in the gel. The reaction is a kind of self oxidation-reduction reaction which releases great heat (Al-Hilli, 2012; West, 1967).

Thereafter, the system loses about 2.6% of the weight, with a subdued exothermic peak locating in the vicinity of about 219.4°C, corresponding to the crystallization process of the system (Mohammed, 2012). After this, the system maintains almost invariant weight, and forms a relatively stable phase after 550°C. With temperature increasing, the weight of the system increases slowly, and the corresponding grain will continue to grow. In consideration of the time-delay effect caused by the heating rate of 10°C/min,

the crystalline temperature should be slightly lower than 550°C (Upadhyay, 2004; Bayoumy, 2014). So, 550°C is chosen as the sintering temperature.

3.2 XRD analysis

Figure 2 shows the XRD patterns of $Li_{0.5-0.5x}Zn_x Fe_{2.5-0.5x}O_4$ ferrite powders calcined at 550°C for 2h. Compared with the standard patterns given in International Centre for Diffraction Data (ICDD) files, the XRD patterns all exhibit a cubic spinel structure. The sample of x = 0 matches the single phase of $Li_{0.5}Fe_{2.5}O_4$ (file no: PDF#49-0266, space group P4$_1$32). The sample of x = 0.3 and x = 0.5 matches the single phase of $Li_{0.435}Zn_{0.195}Fe_{2.37}O_4$ (file no: PDF#37-1471, space group Fd3m). The sample of x = 0.9 matches mixed phases of $ZnFe_2O_4$ (file no: PDF#82-1049, space group Fd3m) and ZnO (file no: PDF#36-1451, space group P63mc), and the former is the major phase whilst the latter is the minor phase. The sample of x = 0.1 matches mixed phases $Li_{0.5}Fe_{2.5}O_4$ and $Li_{0.435}Zn_{0.195}Fe_{2.37}O_4$, and the sample of x = 0.7 is in a transition state between $Li_{0.435}Zn_{0.195}Fe_{2.37}O_4$ and $ZnFe_2O_4$. In the range of 20~70°, the main diffraction peaks of the XRD patterns are (220), (311), (222), (400), (422), (511) and (440). Furthermore, the sample of x = 0 has the characteristic peaks of space group P4$_1$32: (210), (211), (310), (320), (321), (421), which means that this kind of lithium ferrite is in an ordered state (Schieber, 1964; Navrotsky, 1968).

According to Figure 3, with an increasing amount of zinc content, the diffraction peaks get broader and broader, and the position of the peaks

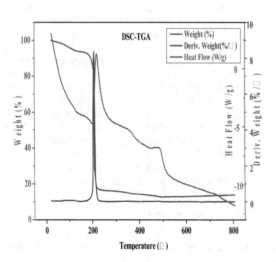

Figure 1. TG-DSC curves for the dry gel of $Li_{0.5}Fe_{2.5}O_4$.

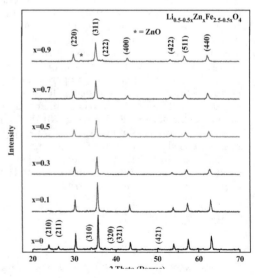

Figure 2. XRD of $Li_{0.5-0.5x}Zn_x Fe_{2.5-0.5x}O_4$ ferrite.

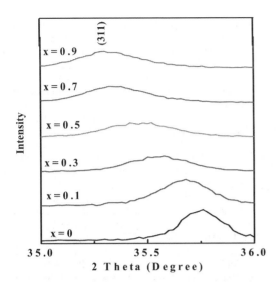

Figure 3. Drifting of the main diffraction peaks.

Table 1. The XRD datum of $Li_{0.5-0.5x}Zn_xFe_{2.5-0.5x}O_4$ samples.

Sample (x)	Average crystallite size (Å)	Lattice parameter (Å)	Density (g·cm−3)
0	548	8.32817	4.7624
0.1	375	8.34328	4.8144
0.3	297	8.37041	4.9217
0.5	274	8.39574	5.0299
0.7	291	8.42325	5.1319
0.9	275	8.43156	5.2674

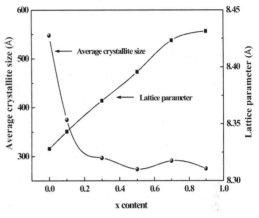

Figure 4. The average crystallite size and lattice parameter of $Li_{0.5-0.5x}Zn_xFe_{2.5-0.5x}O_4$.

have an obvious trend to the small angle. According to Scherrer Equation: $D = \frac{K\lambda}{\beta cos\theta}$ (Verma, 2102; Ridgley, 1970), increasing the Full Width At Half Maximum (FWHM) of the diffraction peak and decreasing of diffraction angle stand for decreasing of average crystallite size. In other words, the doping of Zn^{2+} causes the decrease of the grain size, which is consistent with the result of cell refinement.

Table 1 shows the average crystallite size, lattice parameters, and X-ray density date of $Li_{0.5-0.5x}Zn_x$ $Fe_{2.5-0.5x}O_4$ after cell refinement. Figure 4 visualizes the relationship between the XRD parameters and zinc doping content. According to Shannon et al (Shannon, 1970), the lattice parameters of composite metal compound $M_pM'_qX_s$ are related to the effective radius of ions (Sawant, 2015). With the increase of Zn^{2+} doping, the average crystallite size of the sample presents a tendency to decrease.

3.3 SEM analysis

The SEM micrographs of $Li_{0.5-0.5x}Zn_xFe_{2.5-0.5x}O_4$ (x = 0.1, 0.5, 0.9) ferrites calcined at 550°C for 2h and the particle size distribution histogram are shown in Figure 5. It can be observed from the SEM micrographs that the particle size distribution is almost uniform, and the crystallinity of the samples is relatively ideal. The magnetic interaction force between particles is difficult to avoid, making samples to agglomerate (Navrotsky, 1968). Agglomeration in the samples of high Zn^{2+} content is relatively weak, which may be attributed to the magnetic interaction being relatively weak in these samples. The results can also be explained by the views about ion placeholder and temperature effects

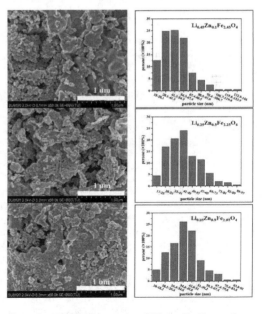

Figure 5. SEM micrographs of $Li_{0.5-0.5x}Zn_xFe_{2.5-0.5x}O_4$.

1121

on the surface of the crystal. The average particle sizes are larger than the corresponding XRD grain sizes, because the sample of each particle is composed of a number of grains (Verma, 2012).

3.4 *Mössbauer spectrum analysis*

Figure 6 shows the room temperature Mössbauer spectra of $Li_{0.5-0.5x}Zn_xFe_{2.5-0.5x}O_4$. Isomer shift in Table 2 is in the range of 0.1~0.5 mm/s, according to the literature (Ridgley, 1970). Iron ions in this series of samples are all Fe^{3+}. For $Li_{0.5-0.5x}Zn_xFe_{2.5-0.5x}O_4$ with x = 0, the Mössbauer spectra exhibit two normal Zeeman-split sextets, corresponding to Fe^{3+} at tetrahedral and octahedral sites, respectively.

The sextet with the larger isomer shift is assigned to the Fe^{3+} ions in the B site and the other is assigned to the Fe^{3+} ions in the A site. Isomer shift mainly reflects the overlap degree of the s shell electron cloud of the $Fe^{3+}-O^{2-}$ system (Verma, 2012; Shannon, 1970), so the isomer shift in the B site is larger than that in the A site. The changing of Mössbauer spectra absorption areas could confirm the change in ion content. On addition of Zn^{2+}, the absorption area of sextet in the A site decreases, and that in the B site increases gradually.

As the Zn^{2+} content continues to increase, the probability of the nearest neighbor A sites of Fe^{3+} in the B site occupied by Zn^{2+} continues to increase at the same time. When Zn^{2+} is doped, the nearest neighbor A sites of Fe^{3+} in the B site will have a probability that is dominated by Zn^{2+}, which leads to the change in the hyperfine field of the B site. According to the reports about the magnetism of this series of samples, excessive Zn^{2+} will weaken the magnetic property, and the sample will transform to a paramagnetic state. The spectra of the sample of x = 0.5 consist of a set of paramagnetic bimodal spectra and a set of relaxation sextet. The proportion of paramagnetic bimodal spectra in the sample of x = 0.7 significantly enhances. Mössbauer spectra of the samples of x = 0.9 almost collapse to a paramagnetic double line spectrum (Upadhyay, 2004).

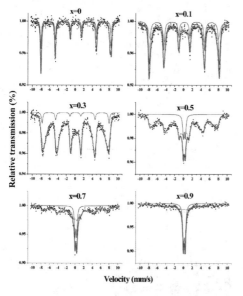

Figure 6. Mössbauer spectra of $Li_{0.5-0.5x}Zn_xFe_{2.5-0.5x}O_4$.

Table 2. The Mössbauer spectra of $Li_{0.5-0.5x}Zn_x$ $Fe_{2.5-0.5x}O_4$.

Sample (x)	Component	IS (mm/s)	QS (mm/s)	Γ (mm/s)	A0 (%)
0	Sextet(A)	0.144	−0.035	0.358	56.2
	Sextet (B)	0.239	−0.015	0.276	43.8
0.1	Sextet(A)	0.184	−0.010	0.442	40.8
	Sextet (B)	0.192	−0.056	0.482	59.2
0.3	Sextet(A)	0.138	0.038	0.380	6.50
	Sextet (B)	0.198	−0.043	0.411	93.5
0.5	Sextet (B)	0.143	−0.070	0.426	89.8
	Double	0.245	0.450	0.367	10.2
0.7	Sextet (B)	0.211	−0.117	0.015	77.9
	Double	0.230	0.395	0.328	22.1
0.9	Sextet (B)	0.197	−0.034	0.665	56.9
	Double	0.236	0.406	0.370	43.1

4 CONCLUSIONS

In this study, we compose a series of Zn^{2+} doped lithium ferrite $Li_{0.5-0.5x}Zn_xFe_{2.5-0.5x}O_4$ (x = 0, 0.1, 0.3, 0.5, 0.7, 0.9) nano-particles via the sol-gel auto-combustion method. TG and DSC analyses showed that the best calcining temperature in this series of samples is 550°C. The analysis of XRD patterns shows that all the samples exhibit a single-phase cubic spinel structure except the sample of x = 0.9 with a second phase of ZnO. The Zn^{2+} doping affects the ordered distribution of lithium, iron ions in the B site. The average grain size of the sample becomes smaller along with the replacement of zinc ions, while the lattice constants increase with the increase of the Zn^{2+} content. The SEM analysis shows that the particle size decreases and the aggregates are improved. The Mössbauer spectra analysis shows that with Zn^{2+} ion concentration increasing, the ferromagnetic phase gradually transforms into a paramagnetic phase.

ACKNOWLEDGMENTS

This work was supported by the National Natural Science Foundation of China (NO.11364004,

11164002) and the Graduate student excellent thesis cultivation plan of Guangxi Normal University (NO.A410213000001).

REFERENCES

Al-Hilli, M.F., S. Li, K.S. Kassim. Journal of Magnetism and Magnetic Materials, 2012, 324(5), 873–879.

Argentina, G.M., P.D. Baba. IEEE Transactions on, 1974, 22(6), 652–658.

Bayoumy, W.A.A. Journal of Molecular Structure, 2014, 1056, 285–291.

Gharagozlou, M., R. Bayati. Superlattices and Microstructures, 2015, 78, 190–200.

Inbanathan, S.S.R., V. Vaithyanathan, J.A. Chelvane, G. Markandeyulu, K.K. Bharathi. Journal of Magnetism and Magnetic Materials, 2014, 353, 41–46.

Kuroda, C., T. Kawashima. IEEE Transactions on, 1969, 5(3), 192–196.

Mohammed, K.A., A.D. Al-Rawas, A.M. Gismelseed, A. Sellai, H.M. Widatallah, A. Yousif, M.E. Elzain, M. Physica B, Condensed Matter, 2012, 407(4), 795–804.

Navrotsky, A., O.J. Kleppa. Journal of Inorganic and Nuclear Chemistry, 1968, 30(2), 479–498.

Ridgley, D.H., H. Lessoff, J.D. Childress. Journal of the American Ceramic Society, 1970, 53(6), 304–311.

Sawant, V.S., K.Y. Rajpure. Journal of Magnetism and Magnetic Materials, 2015, 382, 152–157.

Schieber. M., Journal of Inorganic and Nuclear Chemistry, 1964, 26(8), 1363–1367.

Shannon, R.D., C.T. Prewitt. Journal of Inorganic and Nuclear Chemistry, 1970, 32(5), 1427–1441.

Sun, C., K. Sun. Solid state communications, 2007, 141(5): 258–261.

Upadhyay, C., H.C. Verma, S. Anand. Journal of applied physics, 2004, 95(10), 5746–5751.

Verma, K., A. Kumar, D. Varshney. Journal of Alloys and Compounds, 2012, 526: 91–97.

West, R.G., A.C. Blankenship. Journal of the American Ceramic Society, 1967, 50(7), 343–349.

Zhang, J., Y.M. Zhang, C.Y. Hu, Z.Q. Zhu, Q.J. Liu. Eighth China National Conference on Functional Materials & Applications, 2014, 873, 304–310.

Advances in Energy Science and Equipment Engineering II – Zhou, Patty & Chen (Eds)
© 2017 Taylor & Francis Group, London, ISBN 978-1-138-71798-5

A transfer-matrix method model for an inhomogeneous metal/SiC contact

Lingqin Huang & Xiaogang Gu
School of Electrical Engineering and Automation, Jiangsu Normal University, Xuzhou, China

ABSTRACT: A theoretical model based on the Gaussian distribution of inhomogeneous barrier heights is proposed to analyze the electrical characteristics of metal/SiC contacts. The transmission coefficient of the electron movement across the Schottky barrier is estimated by using the transfer-matrix method. The comparison of theoretical simulated forward electrical characteristics with the experimental results concludes that the proposed model could correctly predict the characteristics of metal/SiC contacts under forward bias.

1 INTRODUCTION

In order to develop high-performance SiC-based devices, one of the main technological problems is the fabrication of operational Schottky contacts to SiC. Therefore, it is imperative to understand the knowledge of barrier formation and the conduction process involved at the interface of metal contacts to SiC on a fundamental basis (Wang, 2016). As we all know, standard Thermionic Emission (TE) theory should predict the barrier height and ideality factor independent of measurement temperature (*T*) (Sze, 1981). However, we often observe an increase in barrier height but a decrease in ideality factor with increasing temperature in metal/SiC contacts. It is common to ascribe the experimental abnormal temperature-dependent electrical characteristics to the barrier inhomogeneities at the interface, which could be fitted to a Gaussian distribution function. In this study, to get better understanding of the barrier properties and the transport mechanism of metal contacts to SiC, a model for the electrical characteristics of metal/SiC contacts with Gaussian distribution of inhomogeneous barrier heights is proposed to study the forward characteristics of metal/SiC theoretically and compared to experimental results of our previous studies.

2 THE INHOMOGENEOUS BARRIER HEIGHT MODEL

It is known that the current transport at the interface of a meta-semiconductor system could be roughly categorized into TE over the Schottky barrier, Field Emission (FE) near the surface Fermi level (E_{FS}) or Thermionic-Field Emission (TFE) at an energy between these two mechnisms as shown in Fig. 1 (a).

The current density of metal contacts to semiconductor as a function of *V* could be given by (Sze, 1981; Gehring, 2004; Grover, 2012)

$$
\begin{aligned}
J(V) &= J_{s \to m} - J_{m \to s} \\
&= \frac{A^* T}{k} \int_{E\min}^{\infty} T(E_x) dE_x \int_0^{\infty} [f_s(E) - f_m(E + qV)] dE_\rho
\end{aligned}
\tag{1}
$$

The total energy *E* is the sum of E_x, associated with the velocity component v_x transversal to the barrier plane. E_ρ is the velocity component parallel to the Schottky interface. E_{min} is the minimum energy value for the tunneling process to occur. $T(E_x)$ is the transmission probability, f_s and f_m are the quasi-Fermi functions of the semiconductor and metal, respectively, as

$$
f_s(E) = \frac{1}{1 + \exp(E_{F,s} + qV - E_x)/kT},
\tag{2}
$$

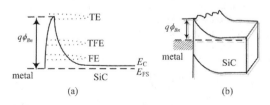

Figure 1. (a) Electron potential energy diagram of a Schottky barrier and (b) a two-dimensional energy band diagram of a Gaussian distributed barrier.

$$f_m(E) = \frac{1}{1 + \exp(E_{F,m} - E_x)/kT}, \quad (3)$$

where $E_{F,m}$ and $E_{F,s}$ are the Fermi energies in the metal and semiconductor, respectively. In the thermal equilibrium condition, the Fermi levels (E_F) on both sides will line up. Assuming that $T(E_x)$ only depends on the transversal energy component. And under the assumption of isotropic Fermi-Dirac distribution, Eq. (1) can be evaluated as follows (Sze, 1981).

$$J(V) = (A^*T/k)$$
$$\times \int_{E_{min}}^{\infty} T(E_x) \ln\left[\frac{1 + \exp(E_F - E_x + qV)/kT}{1 + \exp(E_F - E_x)/kT}\right]dE_x \quad (4)$$

This expression includes both tunneling current $(E_x > q\phi_{Bn})$ and thermionic emission current $(E_{min} < E_x < q\phi_{Bn})$. In the view of thermionic emission, the transmission probability $T(E_x)$ is simply assumed to be unity for $E_x > q\phi_{Bn} + E_F$, and zero for $E_x < q\phi_{Bn} + E_F$. Therefore, in this simple case, the current density is given by (Werner, 1991)

$$J(V) = (A^*T/k)$$
$$\times \int_{q\phi_{Bn}+E_F}^{\infty} \ln\left[\frac{1 + \exp(E_F - E_x + qV)/kT}{1 + \exp(E_F - E_x)/kT}\right]dE_x, \quad (5)$$

resulting into the standard TE model (Sze, 1981)

$$J(V) = A^*T^2 \exp(-\frac{q\phi_{Bn}}{kT})\left[\exp(\frac{qV}{kT}) - 1\right], \quad (6)$$

Considering that the metal/SiC interface is electrically inhomogeneous and barrier heights could be Gaussianly distributed as shown in Fig. 1(b)

$$P(q\phi_{Bn}) = \frac{1}{\sigma_0\sqrt{2\pi}} \exp\left[-\frac{(q\phi_{Bn} - \overline{q\phi_{Bn}})^2}{2\sigma_0^2}\right], \quad (7)$$

Where $\overline{q\phi_{Bn}}$ is the mean barrier height with standard deviation σ_0. Therefore, in the presence of such a Gaussian distribution of barrier inhomogeneities, the overall current density could be expressed as

$$J_I(V) = \int J(q\phi_{Bn}, V)P(q\phi_{Bn})d(q\phi_{Bn}). \quad (8)$$

where $J(q\phi_{Bn}, V)$ is the current density. Considering that the Gaussian function decays dramatically from the mean value, the integration limits was set to be $\overline{q\phi_{Bn}} \pm 8\sigma_0$ in the numerical implementation.

3 TRANSFER-MATRIX METHOD (TMM)

It is known that the calculation of $T(E_x)$ requires the plane-wave solution of the one dimensional Schröinger equation. The most common approach is the Wentzel-Kramers-Brillouin approximation. However, this approximation has limited accuracy for rapidly varying potential profiles. By comparison, the Transfer-Matrix Method (TMM) is a powerful and relatively simple calculation approach. Therefore, in this study, TMM is employed to calculate the values of $T(E_x)$ (Miller, 2008).

According to the TMM, an arbitrarily shaped potential barrier is divided into equipotential layers. Within each layer, the wave function in each layer can be written as the sum of an incident and a reflected wave.

$$\Psi_m(x)$$
$$= A_m \exp(ik_m x) + B_m \exp(-ik_m x) \, m = 1, 2, 3, \ldots, n \quad (9)$$

where A_m and B_m are the forward and backward wave amplitudes, respectively. The wave number is

$$k_m = \sqrt{(2m_m^*/\hbar^2)(E - E_F)}, \quad (10)$$

where m_m^* is the mass of the particle in a given layer of the structure. A_m, B_m, k_m and m_m^* are assumed constants for each layer m. Inserting the boundary conditions for the wave function into (9), the transmitted wave of a layer relates to the incident wave by a complex transfer matrix (Ando, 1987)

$$\binom{A_m}{B_m} = M_m\binom{A_{m-1}}{B_{m-1}} 1 \leq m \leq n \quad (11)$$

where

$$M_m = \frac{1}{2}\begin{bmatrix}(1+S_m)1+S_m \\ 1-S_m \quad 1-S_m\end{bmatrix}\begin{bmatrix}\exp(-ik_m d_m)0 \\ 0 \qquad \exp(ik_m d_m)\end{bmatrix} \quad (12)$$

where

$$S_m = (k_{m+1}/K_m)(m_m^*/m_{m+1}^*) \quad (13)$$

Therefore, the transmitted wave in region N could be estimated from the incident wave by subsequent multiplication of transfer matrices (Jonsson, 1990)

$$\binom{A_N}{B_N} = \prod_{m=1,2,3\ldots N-1} M_m\binom{A_1}{B_1} \quad (14)$$

Assuming that there is no reflected wave in Region N ($B_N = 0$) and $A_1 = 1$, Eq. (14) could be simplified to

$$\begin{pmatrix} A_N \\ 0 \end{pmatrix} = \prod_{m=1,2,3....N-1} M_m \begin{pmatrix} 1 \\ B_1 \end{pmatrix} = \begin{bmatrix} M_{11} & M_{12} \\ M_{21} & M_{22} \end{bmatrix} \begin{pmatrix} 1 \\ B_1 \end{pmatrix} \quad (15)$$

The transmission probability could be calculated by

$$T(E) = (k_N / k_1)(m_1^* / m_N^*)|A_N| \quad (16)$$

4 RESULTS

The model is validated against the experimental results. The contacts under examination in this study are the Pt/4H-SiC Schottky barriers from our previous studies (Huang, 2013; Huang, 2015). We investigated the barrier properties of Pt contacts to lightly doped ($N_d \sim 1 \times 10^{16}$ cm^{-3}) and highly doped ($N_d \sim 1 \times 1018$ cm^{-3}) 4H-SiC. We found that the origin of the barrier inhomogeneities could be related to the partially pinning of the surface Fermi level and a resultant spatial variation of the surface state density on 4H-SiC. The variations in barrier height which is temperature dependent could be fitted well to a single Gaussian distribution for the lightly doped sample. The degree of barrier inhomogeneities could be controlled by surface pretreatment (Huang, 2013). For the highly doped sample (Huang, 2015), the anomalous temperature dependence of the barrier height could be explained in terms of a double Gaussianly distributed barrier heights. However, the average barrier height obtained from Gaussian distribution at the low temperature range is lower than the actual value, which may be due to the tunneling current at low temperatures.

The simulations were done based on MATLAB with the TMM using the same parameters as on the experimental data. Figures 2 (a) and (b) show the numerical forward characteristics of Pt contacts to lightly and highly doped 4H-SiC with a partial pinning Fermi level (with the hydrogen plasma pretreatment for 2 min) at 300 K in (Huang, 2013) and (Huang, 2015), respectively, which are calculated by using the experimental obtained values of σ_0 and $q\phi_{Bn}$. The results of the standard TE model with homogeneous barriers are also shown for comparison. It can be seen that both simulated and experimental characteristics are deviated from the TE model with homogeneous barriers, indicating that barrier inhomogeneities exist at the contact area of both lightly and highly doped samples. For the lightly doped sample, a good fit is achieved

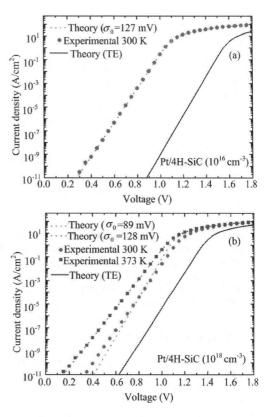

Figure 2. Calculation results for Pt/4H-SiC contact under forward characteristics by using the experimental parameters, and comparison with experimental data.

for the experimental σ_0 (127 mV) (see Fig. 2 (a)). However, for the highly doped sample as shown in Fig. 2 (b), using $\sigma_0 = 89 meV$, there is a big discrepancy between the simulated characteristic and the experimental data. When T is higher than 323 K, the data calculated from the $\sigma_0 = 128 meV$ value are in good agreement with the experimental characteristics, similar to that for the lightly doped sample (the data for $T = 375$ K are also shown in Fig. 2 (b)). It is confirmed that there is likely a tunneling current component by the TFE mechanism at the contact interface at low temperatures. TE is not suitable for interpreting the carrier transport as analyzed in our previous work.

We have also investigated the dependence of $q\phi_{Bn}$ on σ_0 using the TMM implementation. When $\sigma_0 = 0mV$, $q\phi_{Bn} = 1.95$ eV is obtained. This value is consistent with the theoretical Schottky–Mott value, close to the experimental $q\phi_{Bn}$ (1.93 eV) for the sample with a uniform electronic interface achieved by the hydrogen plasma pretreatment for 3 min and a subsequent 400°C annealing process. When σ_0 increases, the deviation from the theory value (1.95 eV) increases. The value of $\overline{q\phi_{Bn}}$ is correspondingly decreased. The experimental values $(\sigma_0, q\phi_{Bn})$ for the samples with a partially pinned fermi level in our previous work are consistent with the theoretical values obtained from the proposed model. However, it is noted that although a homogeneous barrier could be observed for a highly doped sample in the low temperature range and the contact when the SiC surface Fermi level is totally pinned (Huang, 2013), the $(\sigma_0, q\phi_{Bn})$ values are not consistent with the theoretically obtained values, which requires further investigation.

5 CONCLUSION

The electrical characteristics of a metal/SiC contact was theoretically studied. A Gaussian distribution of the inhomogeneious barrier height model based on the transfer-matrix method was proposed and verified by the the experimental Pt/4H-SiC results from our previous studies.

ACKNOWLEDGEMENTS

This work was supported by the National Natural Science Foundation of China (No.11547136), the Natural Science Foundation of Jiangsu province, China (Grant No. SBK2015040637) and the Natural Science Foundation of the Jiangsu Higher Education Institutions of China (Grant No.15KJB510010).

REFERENCES

Ando, Y. and T. Itoh, J. Appl. Phys. 61 (1987) 1497.

Gehring, A., S. Selberherr, IEEE Trans. Device Mater. Reliability. 4 (2004) 306–319.

Grover, S., G. Moddel, Solid-State Electron. 67 (2012) 94–99.

Huang, L., D. Wang. J. Appl. Phys. 117 (2015) 204503.

Huang, L., F. Qin, S. Li, D. Wang, Appl. Phys. Lett. 103 (2013) 033520.

Jonsson, B., S. T. Eng, IEEE J. Quantum Electron. 26 (1990) 2025–2035.

Miller, D. A. B. Quantum Mechanics for Scientists and Engineers, Cambridge Univ., New York, 2008.

Sze, S. M. Physics of Semiconductor Devices, Wiley, New York, 1981.

Wang, Z., W. Liu, C. Wang, J. Electron. Mater. 45 (2016) 267–284.

Werner, J. H., H. H. Güttler, J. Appl. Phys. 69 (1991) 1522.

Effects of Al ions on some properties of $Zn_{0.4}Mn_{0.6}Al_xFe_{2-x}O_4$ ferrites by the sol-gel combustion method

Jinpei Lin, Yunlong Wang, Hu Yang, Hangyu Xu & Yanbing He
College of Physics and Technology, Guangxi Normal University, Guilin, China

Ruijun Wang
College of Medical Informatics, Hainan Medical University, Haikou, China

Qing Lin
College of Physics and Technology, Guangxi Normal University, Guilin, China
College of Medical Informatics, Hainan Medical University, Haikou, China

Yun He
College of Physics and Technology, Guangxi Normal University, Guilin, China

ABSTRACT: $Zn_{0.4}Mn_{0.6}Al_xFe_{2-x}O_4$(x = 0, 0.1, 0.2, 0.3, 0.4, 0.5) ferrites were synthesized by the sol-gel combustion method with citric acid as the complexing agent, and the mole ratio of metal ions and citric acid is 1:1. All samples were sintered at 550°C. The structural and magnetic properties of MnZn ferrites were studied. The XRD analysis shows that the structure of the samples is cubic spinel ferrite. The SEM analysis shows that the grain size of the samples is homogeneous, and has good crystallinity. The Mössbauer spectrum analysis shows that both Fe ions are in the Fe3+ state. The quadrupole splitting QS of all samples is small, which indicates that the charge distribution of the nucleus has no little change, and maintains symmetry.

1 INTRODUCTION

Mn-Zn ferrites are one of the important functional materials because of their soft magnetic properties. They are widely used in electromagnetic equipment that records heat, high frequency transformers, sensors, switches, and memory devices due to their characteristics such as high permeability, high saturation magnetization, high stability, and low coercive force, and low power loss (Angermann, 2010; Rath, 1999; Zheng, 2008). In the spinel lattice structure of MnZn ferrite, the change in the magnetic properties and microstructure of MnZn ferrites depends on the solubility of cations in the ferrite lattice and the occupying positions of tetrahedral or octahedral sites (Hu, 2011; Gabal, 2013). Y.Y. Meng et al (Meng, 2012) studied the structure and magnetic properties of $Mn(Zn)Fe_{2-x}RE_xO_4$ ferrite synthesized by the co-precipitation and refluxing method, whose research results showed that the grain size was approximately 10 nm to 20 nm. The saturation magnetization strongly depends on the RE content. M.A. Gabal et al (Gabal, 2013) studied the influence of Al-substitution on structural, electrical and magnetic properties of MnZn ferrites, and research results showed that the grain

size was approximately 5 nm to 38 nm and the saturation magnetization decreased with the increase of Al^{3+} ion concentration. Because the substitution of Fe^{3+} ions by nonmagnetic Al^{3+} ions, which have stronger preference for occupying the octahedral sites, decreases the magnetic moment and consequently the net magnetization through weakening A-B interactions and disturbing of the spin ordering. In this paper, $Zn_{0.4}Mn_{0.6}Al_xFe_{2-x}O_4$ (x = 0, 0.1, 0.2, 0.3, 0.4, 0.5) ferrites were synthesized by the sol-gel combustion method using citric acid as the complexing agent, with a pH of 7 and a CA/MN molar ratio of 1. All samples were sintered at 550°C. The literature mainly investigated the microstructure and the magnetic properties of the samples. The aim of this study is to investigate the structural and magnetic properties of ferrite powders by doping small amounts.

2 EXPERIMENT

2.1 Sample preparation

$Zn_{0.4}Mn_{0.6}Al_xFe_{2-x}O_4$(x = 0, 0.1, 0.2, 0.3, 0.4, 0.5) ferrites were synthesized by the sol-gel combustion method. The desired proportions of high-purity

(AR grade) iron, zinc, manganese nitrates and citric acid were weighed and dissolved in distilled water of 80 ml. The mole ratio of metal ions and citric acid is 1:1. Ammonia was dropped to control the pH = 7 of the mixed solution. The solution was stirred continuously with motor stirrer to form a homogeneous solution at 80°C until the water evaporated to obtain wet gel. The resultant wet gel was dried at 120°C in a drying oven for 12 hours to form dry gel, which is ignited in air at room temperature. The dried gel burnt in a self-propagating combustion way to form loose powders. All samples were sintered at 550°C to obtain nanosize particles.

2.2 Characterization

The thermal decomposition behavior of gel was analyzed using a thermal analysis instrument (SDT Q600, America). The crystalline structure of the sample was observed using X-ray diffraction (D8 Advance, Germany) with Cu Kα radiation (λ = 0.15405 nm). The microstructure of particles was observed by scanning electron microscopy (SU8020, Japan). The Mössbauer spectra were observed using a conventional Mössbauer spectrometer (Fast Com Tec PC-moss II) at room temperature, in constant acceleration modeat. The γ-ray was provided by a ^{57}Co source in a rhodium matrix.

3 RESULTS AND DISCUSSION

3.1 XRD analysis

At room temperature (300k), all data of the samples were fitted by JADE 5. As shown in Figure 1, the structure of the samples is cubic spinel ferrites, and main diffraction peaks of (220), (311), (400), (422), (511), and (440) were observed. The samples were single cubic spinel ferrite, and no impurity phase was detected.

However, for x = 0, the samples' impurity phase was detected, and impurities were phases of Fe_2O_3 (Gimenes, 2012), because sintering in aerobic environment increased the number of phases of Fe_2O_3 continuously with the continuous increase of Mn^{2+} ion concentration. Also, it is possible that the decomposition reaction of ferrites happened (Hu, 2010; Dhiman, 2008) as follow: $MnFe_2O_4 + [O] \rightarrow Fe_2O_3 + Mn_2O_3$. The measured ferrite nanoparticle sizes calculated by the Debye-Scherrer (Dhiman, 2008) formula from the (311) peak of XRD patterns. The formula used is $D = (0.9\lambda)/(\beta\cos\theta)$.

Where, D is the average grain size, λ is the wavelength of copper (0.15405 nm), β is the Full Width At Half Maximum (FWHM) of the main (311) peak, and θ is the Bragg diffraction angle. All the

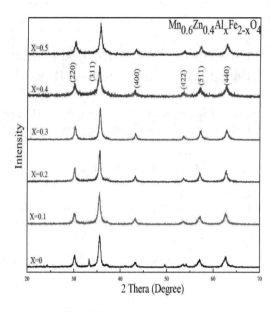

Figure 1. The XRD patterns of samples.

XRD data are listed in Table 1. With the continuous increase of Mn^{2+} ion concentration, the grain size is about 15.7 nm to 20.4 nm. The lattice constant is obtained by the following formula (Hassan, 2012): $a = d_{hkl}\sqrt{h^2 + k^2 + l^2}$.

Where, (hkl) is the Miller index and d is the crystal plane spacing. With the continuous increase of Al^{3+} ion concentration, the lattice constant gradually decreased, since the ionic radius of Fe^{3+} (0.67Å) is larger than the ionic radius of Al^{3+} (0.51Å) (Haralkal, 2013). Figure 2 shows an obvious displacement of the peak position to higher angles observed for the (311) reflection peak, indicating a decrease in the lattice parameter.

The density of X-Ray decreased with the continuous increase of Al^{3+} ion concentration. The density of X-Ray depends on the lattice constants and the molar mass of the sample (Asif, 2013). It can be determined by the following formula: $\rho_x = \frac{8M}{Na^3}$. Where, M and N are the molar mass of the sample and Avogadro constant, respectively, and a is the lattice constant. This may be because M/a^3 gradually decreased with the continuous increase of Al^{3+} ion concentration.

3.2 SEM analysis

At room temperature (300k), the SEM images of $Mn_{0.6}Zn_{0.4}Al_{0.1}Fe_{1.9}O_4$ and $Mn_{0.6}Zn_{0.4}Al_{0.3}Fe_{1.7}O_4$ ferrites are captured, which are shown in Figure 3. It can be seen that the particles were square, and the grain size of samples is homogeneous, and has good crystallinity. Figure 4 shows the grain size

Table 1. The XRD datum of $Zn_{0.4}Mn_{0.6}Al_xFe_{2-x}O_4$ samples.

Sample (x)	Average crystallite size (Å)	Lattice parameter (Å)	Density (g·cm⁻³)
0	193	8.389	5.283
0.1	173	8.387	5.220
0.2	204	8.370	5.194
0.3	193	8.365	5.122
0.4	157	8.362	5.083
0.5	177	8.360	5.004

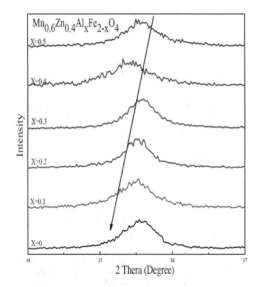

Figure 2. The (311) peak of sample enlarge figure.

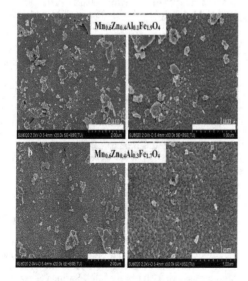

Figure 3. The SEM pictures of $Mn_{0.6}Zn_{0.4}Al_{0.1}Fe_{1.9}O_4$ and $Mn_{0.6}Zn_{0.4}Al_{0.3}Fe_{1.7}O_4$ ferrites.

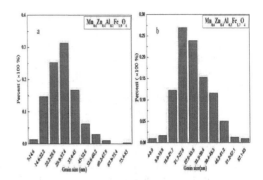

Figure 4. The grain size distribution of $Mn_{0.6}Zn_{0.4}Al_{0.1}Fe_{1.9}O_4$ and $Mn_{0.6}Zn_{0.4}Al_{0.3}Fe_{1.7}O_4$ ferrites.

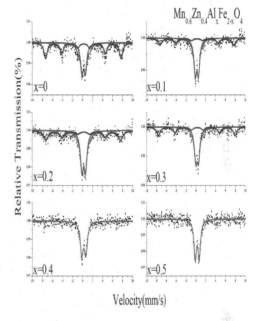

Figure 5. Mössbauer Spectrometer of $Zn_{0.4}Mn_{0.6}Al_xFe_{2-x}O_4$ ferrites.

Table 2. The Mössbauer spectra of $Zn_{0.4}Mn_{0.6}Al_xFe_{2-x}O_4$ ferrites.

Sample (x)	IS (mm/s)	QS (mm/s)	Γ (mm/s)	A₀ (%)
0	0.193	0.690	0.693	48.1
	0.196	0.038	0.802	51.9
0.1	0.209	0.709	0.720	53.4
	0.165	−0.097	1.184	46.4
0.2	0.227	0.582	0.696	58.3
	0.213	−0.017	0.838	41.7
0.3	0.196	0.660	0.665	61.1
	0.221	−0.088	0.987	38.9
0.4	0.214	0.806	0.711	100
0.5	0.229	0.826	0.826	100

1131

distribution of $Mn_{0.6}Zn_{0.4}Al_{0.1}Fe_{1.9}O_4$ and $Mn_{0.6}Zn_{0.4}$ $Al_{0.3}Fe_{1.7}O_4$ ferrite powders. The average grain size of $Mn_{0.6}Zn_{0.4}Al_{0.1}Fe_{1.9}O_4$ and $Mn_{0.6}Zn_{0.4}Al_{0.3}Fe_{1.7}O_4$ ferrites obtained by a statistical method is about 30.9 nm and 32.4 nm, respectively. The statistical average grain size as seen in the SEM picture is larger than crystalline size of XRD. This indicated that the samples are polycrystalline ferrites.

3.3 *Mössbauer spectrum analysis*

At room temperature (300 k), Figure 5 shows the Mössbauerspectra of $Zn_{0.4}Mn_{0.6}$ $Al_xFe_{2-x}O_4$ ferrites fitted by Mosswinn 3.0, and all the data are listed in Table 2. For x = 0, 0.1, 0.2 and 0.3, the spectra of the samples were fitted by a set of Zeeman six line and a set of paramagnetic double line. For x = 0.4 and 0.5, the spectra of the samples were fitted by a set of paramagnetic double line. The paramagnetic phase exists because the surrounding of the part of iron ions has the nearest disordered spin magnetic moment, and the existence of the Zeeman six line magnetic pattern is due to the superexchange interaction between the magnetic ions at A- and B-sublattices. The ferromagnetic phase completely transformed into the paramagnetic phase with the increase of Al^{3+} ion concentration.

As we all known that the change of isomer shift is connected with the s-electron charge density of Fe^{3+} ions, it depends on the change of the electronic density of the shielding effect of the s, p and d orbital electrons. As is known from the literature (Shalendra, 2008), the isomer shift is 0.1~0.5 mm/s relative to Fe metal, which indicates that both Fe ions are in the Fe^{3+} state. The isomer shift is 0.6~1.7 mm/s relative to Fe metal, which indicates that both Fe ions are in the Fe^{2+} state. As can be seen in Table 2, the isomer shift distributed in the 0.19 mm/s to 0.22 mm/s region, which indicates that both Fe ions are in the Fe^{3+} state. For two sets of Zeeman six line-spectrum, the two sextets correspond to Fe^{3+} on the A-site and B-site, respectively, in the spinel lattice. Generally speaking, it is known that the isomer shift value for octahedral sites is more than that for tetrahedral sites, because the bond separation Fe^{3+}- O^{2-} is larger for the B-site than for the A-site as the overlapping of the orbital of Fe^{3+} ions is large at the A-site in cubic spinel ferrites (Li, 2007; Inbanathan, 2014). The isomer shift has significant change with the increase of Al^{3+} ion concentration, which shows that substitution of Al^{3+} ions change the electron density distribution of the s-shell of the iron ion. The quadrupole splitting QS indicates the symmetry of irons' surrounding (Mathew, 2006; Haining, 2013). The quadrupole splitting QS of all samples is small, which indicates that the charge distribution of the nucleus has no little change, and maintains symmetry.

4 CONCLUSIONS

The XRD analysis shows that the structure of the samples is cubic spinel ferrite. The SEM analysis shows that the grain size of the samples is homogeneous, and has good crystallinity. The Mössbauer spectrum analysis shows that the isomer shift is distributed in the 0.19 mm/s to 0.22 mm/s region, and both Fe ions are in the Fe^{3+} state. The quadrupole splitting QS indicates the symmetry of irons' surrounding (Mathew, 2006; Haining, 2013). The quadrupole splitting QS of all samples is small, which indicates that the charge distribution of the nucleus has no little change, and maintains symmetry.

ACKNOWLEDGMENTS

This work was supported by the National Natural Science Foundation of China (NO.11364004, 11164002) and the Graduate student excellent thesis cultivation plan of Guangxi Normal University (NO.A410213000001).

REFERENCES

Angermann, A., J. Töpfer, K.L. Da Silva, K.D. Becker. Journal of Alloys and Compounds, 2010, 508, 433–439.

Asif Iqbal, M., Misbah-ul-Islam, Irshad Ali, Hasan M. Khan, Ghulam Mustafa, Ihsan Ali. Ceramics International, 2013, 39, 1539–1545.

Dhiman, R.L., S.P. Taneja, and V.R. Reddy. Advances in Condensed Matter Physics, 2008, 1155, 1–7.

Gabal, M.A., A.M. Abdel-Daiem, Y.M. Al Angari, I.M. Ismail. Polyheedron, 2013, 57, 105–111.

Gimenes, R., M.R. Baldissera, M.R.A. da Silva, C.A. da Silveira, D.A.W. Soares, L.A. Perazolli, M.R. da Silva, M.A. Zaghete. Ceramics International, 2012, 38, 741–746.

Haining Ji, Zhongwen Lan, Zhiyong Xu, Huaiwu Zhang, and Gregory J. Salamo. IEEE Transactions Magnetics, 2013, 49(7), 4277–4280.

Haralkal, S.J., R.H. Kadam, S.S. More, Sagar E. Shirsath, M.L. Mane, Swati Patil, D.R. Mane. Materials Research Bulletin, 2013, 48, 1189–1196.

Hassan Hejase, Saleh S. Hayek, Shahnaz Qadri, Yousef Haik. Journal of Magnetism and Magnetic Materials, 2012, 324, 3620–3628.

Inbanathan, S.S.R., V. Vaithyanathan, J. Arout Chelvane, G. Markandeyulu, K. Kamala Bharathi. Journal of Magnetism and Magnetic Materials, 2014, 353, 41–44.

Lijun Zhao, Zhaoyang Han, Hua Yang, Lianxiang Yu, Yuming Gui, Weiqun Jin, Shouhua Feng. Journal of Magnetism and Magnetic Materials, 2007, 309, 11–14.

Mathew George, Asha Mary John, Swapna S. Nair, P.A. Joy, M.R. Anantharaman. Journal of Magnetism and Magnetic Materials, 2006, 302, 190–195.

Meng, Y.Y., Z.W. Liu, H.C. Dai, H.Y. Yu, D.C. Zeng, S. Shukla, R.V. Ramanujan. Powder Technology, 2012, 229, 270–275.

Ping Hu, De-an Pan, Xin-feng Wang, Jian-jun Tian, Jian Wang, Shen-gen Zhang, Alex A. Volinsky. Journal of Magnetism and Magnetic Materials, 2011, 323, 569–573.

Ping Hu, Hai-bo Yang, De-an Pan, Hua Wang, Jian jun Tian, Shen gen Zhang. Journal of Magnetism and Magnetic Materials, 2010, 322, 173–177.

Rath, C., K.K. Sahu, S. Anand, S.K. Date, N.C. Mishra, R.P. Das. Journal of Magnetism and Magnetic Materials, 1999, 202, 77–84.

Shalendra Kumar, A.M.M. Farea, Khalid Mujasam Batoo, Chan GyuLee, B.H. Koo, Ali Yousef, Alimuddi. Physica B, 2008, 403, 3604–3607.

Zheng, Z.G., X.C. Zhong, Y.H. Zhang, H.Y. Yu, D.C.Zeng.Journal of Alloys and Compounds, 2008, 466, 377–382.

Advances in Energy Science and Equipment Engineering II – Zhou, Patty & Chen (Eds)
© 2017 Taylor & Francis Group, London, ISBN 978-1-138-71798-5

Fatigue resistance and corrosion resistance of electroplated Ni-Fe-W alloy coating on coiled tubing

Wenwen Song
CNPC Tarim Oilfield Company Oil and Gas Engineering Research Institute, Kuerle, Sinkiang, China

Tao Guo & Fei Zhou
CNPC Liaohe Oilfield Branch Company Drilling and Production Technology Research Institute, Panjin, Liaoning, China

ABSTRACT: At present, there is certain blindness in the selection of Ni-Fe-W alloy coating for coiled tubing used in different corrosion environments. In terms of QT-900 coiled tubing electroplated with Ni-Fe-W alloy, the corrosion environment of the oil field was simulated to study the impact of Ni-Fe-W coating on the fatigue resistance and corrosion resistance against CO_2 of QT-900 coiled tubing by means of a fatigue test, the corrosion resistance test under high temperature and pressure, metallographic microscope, SEM, and EDS. Research findings suggested that Ni-Fe-W alloy coating showed favorable corrosion resistance against CO_2 with lower uniform corrosion rate, and it changed little with the increase of CO_2 partial pressure. Ni-Fe-W alloy coating significantly improved the fatigue life of coiled tubing, with all the cracks of coiled tubing with Ni-Fe-W coating caused by fatigue failure originated from the surface, penetrating the coating to the substrate, and the cracks close to the leakage location were dense while those 350 mm away from the leakage location were sparse.

1 INTRODUCTION

As the development conditions of oil and gas fields are turning increasingly strict, Chinese oil field equipment have been faced with severe challenges. Ever since the 1970s, Enhanced Oil Recovery (EOR) by means of CO_2 injection has been widely used in various oil fields followed by serious corrosion when CO_2 permeates into oil drilling, production, gathering, and transportation systems. To improve the corrosion resistance of pipes, various measures have been taken, such as corrosion inhibition, anti-corrosive alloy steels, electrochemical protection, and coating protection. Moreover, there is still another effective approach that has broad application prospects—electric-deposition of a compact layer of Ni-Fe-W alloy coating on the surface of metal materials. This approach boasts of high strength, high density, good electrical and thermal conductivity, a little coefficient of thermal expansion, strong oxidation and corrosion resistance, and excellent abrasive resistance. Currently, with the increasing development of Ni-Fe-W alloy electroplating technology, abrasion-proof pipes and pillars plated with Ni-Fe-W alloy coating have gone into service in various oil plants, such as Shengli oil field, with the preferable effect achieved (Ren, 2011). However, due to lack of systematic

knowledge on the performance of Ni-Fe-W alloy coating, there is certain blindness in selecting Ni-Fe-W alloy coating materials for coiled tubing used in different corrosion environments. This study focuses on the mechanical fatigue and corrosion resistance against CO_2 of coiled tubing plated with Ni-Fe-W alloy coating, in hope of providing a theoretical foundation for the fatigue life prediction of Ni-Fe-W alloy coating and providing some references for material selection of oil fields.

2 TEST

2.1 Substrate pre-treatment

The QT-900 coiled tubing was selected as the substrate with the following chemical composition (mass fraction, %): 0.140 C, 0.350 Si, 0.770 Mn, 0.013 P, 0.002 S, 0.580 Cr, 0.180 Ni, 0.270 Cu, 0.100 Mo, 0.020 Ti, and allowance of Fe. The pipe dimension was φ 38.0 mm × 1.4 m and the thickness of the pipe wall was 5.0 mm. Pre-treatment is as follows: performing degreasing treatment with the 50 g/L PA30-IA degreasing agent for 10 min at normal temperature; preparing a solution by mixing the rust remover (main components: organic acid, sodium alkyl sulfate, hexamethylene tetramine, and polyethylene glycol) and 8% hydrochloric acid at the mass rate of 1:3 to remove rust for 1~2 min at room temperature; carrying out plating activation treatment using 25% ammonia for etching for 20 min; lastly, cleaning and drying.

2.2 Ni-Fe-W alloy electroplating

A direct current electrodeposition method was employed by using Type XJL-7232 stable DC power supply, and parameters employed are as follows: current density 6 A/dm^2, temperature 40°C, pH 7~8, and time 10 min. A citric acid system was selected to be the plating solution with the following composition: 8 g/L NiSO$_4 \cdot$ 7H$_2$O (measured by Ni), 30 g/L Na$_2$ WO$_4 \cdot$ 2H$_2$O (measured by W), 12 g/L NH$_4$OH, 1.6 g/L benzene sulfonic amide and 72 g/L citric acid (C$_6$H$_8$O$_7$). The depth of coating was 50–100 μm.

2.3 Property measurement and characterization of coating

1. Composition and morphology. An Olympus Type GX-51 metallographic microscope was used to observe the cross-section morphology of the coating and measure its depth; a Type JSM-6390 scanning electron microscope was used to scan the cross-section of coiled tubing plated with Ni-Fe-W alloy coating and analyze chemical composition changes of the coating and substrate.
2. Adhesion force. The bending test was carried out by cutting out a section of Ni-Fe-W–alloy-coated coiled tubing of dimension 200 mm × 10 mm × 5 mm and placing it on the bending tester. To measure the binding effect of coating and the substrate, the up-down reverse bending test was performed with an angle of bending of 60° according to the standards specified in GB/T 5270-2005 until the pipe was fractured. The bending test was aimed to bend or fracture the sample by exerting external force. With different stress levels, component force would occur between the coating and substrate. When the component force exceeded adhesive strength, the coating would be peeled off from the substrate. The stripping or fragmentation of coating could be deemed as the manifestation of unsound adhesion force between the coating and substrate.
3. Fatigue resistance. The mechanical fatigue test of Ni-Fe-W–alloy-coated coiled tubing was made on a Type PLW-100 electro-hydraulic servo-controlled fatigue testing machine: the bottom of the sample was fixed while the upper end was loaded with a bending angle of 30° at normal temperature. The sample was straightened and bended once for each cycle; the setting pressure internal the pipe was 34.5 MPa with a bending radius of 1219 mm and a welded joint placed at the most strict position—at the very opposite close to the bending mold; the tension and bending was repeated until the coating was fractured or cracked. The DPT-5 solvent-removal dyeing penetration flaw detecting agent was applied to failed cracks at the coating surface for penetrant flaw detection for 15 min to analyze the morphology and extension mechanism of coating cracks.

4. Corrosion resistance. 320-grit, 600-grit, 800-grit, and 1,200-grit abrasive papers were used to polish the 50 mm × 10 mm × 3 mm standard coupon processed by Ni-Fe-W-alloy-coated coiled tubing. It was fixed on the sample fixture after cleaning, drying and weighing. There were three parallel samples fixed on each of the four groups of fixtures; the fixtures and samples were put into a Type TFCZ-25/250 HPHT magnetic-drive reactor for the dynamic corrosion test with experimental environment simulating the operating condition of an oil well in the Tarim oilfield: 406.23 mg/L NaHCO$_3$, 161.23 mg/L Na$_2$SO$_4$, 24 420.00 mg/L CaCl$_2$, 6 419.88 mg/L MgCl$_2 \cdot$ 6H$_2$O, 126 954.59 mg/L NaCl, and 616.57 mg/L KCl; highly pure N$_2$ was pumped to remove oxygen for 4 h, and CO$_2$ pressure was set to be 2, 3, 4, and 6 MPa, respectively; the fluid velocity was 1 m/s with temperature at 40°C and the corrosion time was 168 h. One sample was selected from the four groups each when the experiment was finished. The samples were scanned using a scanning electron microscope to analyze the coating surface morphology and element composition of the corrosion product after they were cleaned and dried without damaging the corrosion products at the surface. The other two samples in each group were dipped into the hydrochloric acid corrosive liquid to remove the corrosion products. Afterwards, they were weighed after cleaning and drying. The weight-loss method was employed to calculate the corrosion rate; a metallographic microscope was used to measure the corrosion pitting depth.

3 RESULTS AND DISCUSSION

3.1 Morphology and composition of Ni-Fe-W alloy coating

The surface of Ni-Fe-W alloy coating was smooth and uniform without bubbles, peeling, or stains. Figure 1 shows the metallographic morphology of Ni-Fe-W-alloy-coated coiled tubing: the white

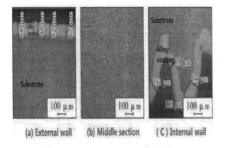

(a) External wall   (b) Middle section   (C) Internal wall

Figure 1.   Metallographic structure of the cross-section of coiled tubing coated with Ni-Fe-W alloy.

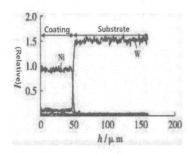

Figure 2.   The line-scanning energy spectrum of the cross-section of Ni-Fe-W-alloy-coated coiled tubing.

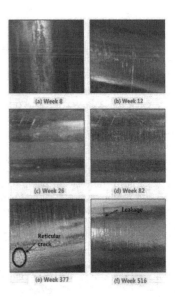

Figure 3.   The macro-morphology of the Ni-Fe-W alloy coating surface with different numbers of cycles (10 x).

bright layers at the internal and external walls of coiled tubing are Ni-Fe-W alloy coatings, of which the average depth of coating at the external wall is 54.0 μm and that at the internal wall is 98.5 μm. When the bending test was finished, the cracked or fractured part was placed under the metallographic microscope for observation with findings that no stripping phenomenon was observed even though there was edge wrapping of the coating, evidencing that the coating was stably adhered to the substrate with high quality.

See Figure 2 for the line-scanning energy spectrum of the cross-section of Ni-Fe-W-alloy coated coiled tubing. As shown in Figure 2, the contents of elements Ni and W in the coating were relatively high while the content of Fe was low; the contents of elements C and Mn in the substrate and coating were low and basically did not change.

3.2 *Fatigue resistance of Ni-Fe-W alloy coating*

3.2.1 *Fatigue resistance*

Figure 3 shows the results of the alloy coating fatigue test.

The following are the results that can be seen from Figure 3: (1) Hoop micro-cracks occurred at the coating surface (Figure 3a) when the fatigue test was in its 8th week, and the micro-cracks became denser as the number of cycles increased (Figures 3b and 3c); (2) In the 82nd week of the fatigue test, longitudinal micro-cracks occurred (Figure 3d) at the coating surface; (3) both longitudinal and hoop micro-cracks tend to be denser and formed reticular cracks with the increase of the number of cycles. The morphology of micro-cracks formed at the coating surface in the 377th week can be seen in Figure 3e; (4) when the test came to the 516th week, as a result of fatigue, the cracks expanded and caused leakage that led to coiled tubing failure. It can be seen in the figure that the surface close to the leakage turned from smooth to rough (Figure 3f). In terms of common QT-900 coiled tubing, its fatigue life in the same experimental conditions is 245 times (Zhu, 2004). Thus, it may be inferred that Ni-Fe-W alloy coating has evidently enhanced the fatigue life of coiled tubing.

Direct current electrodeposition of Ni-Fe-W alloy coating is a process during which new crystal faces are formed while previous ones disappear continuously. Due to the effect of surface tension, there are more vacancies on the new crystal face than the inside. An excessive amount of vacancies are frozen at high speed of deposition, making deposition of hole lines perpendicular to the coating surface difficult, thus resulting in the edge dislocation perpendicular to the coating surface. While this leads to the tensile stress within the coating owing to lattice distortion because the atoms surrounding the dislocation deviate equilibrium positions (Zhao, 1992). The structure of Ni-Fe-W alloy coating is a substitutional type solid solution where Ni is the solvent atom and W is the solute atom. The atomic size of the substrate is similar to that of the Ni atom. The larger atomic size of the W atom results in a larger lattice constant, however, the increasing interatomic force has made the grain size smaller, which reduces the fatigue crack threshold value. Element W in the coating can improve the stack fault energy at the material surface in a significant level (Ling, 1992), making the dislocation in the coating liable to alternative slide and climbing (Yu, 2007). For this reason, the alloy coating failure could be observed when the test cycle came to week 516.

3.2.2 Surface penetrant flaw detection

The penetrant flaw detection of the failed Ni-Fe-W alloy coating surface suggested that there were dense hoop and longitudinal cracks close to the leakage; clear hoop cracks could be observed at the position about 150 mm away from leakage, resulting in the morphology of longitudinal cracks; sparse longitudinal cracks could be observed at sites which were 350 mm away from leakage. Alternating stress was concentrated at the leakage during the bending fatigue test. It was hoop micro-cracks that firstly emerged at the stress concentration point, and longitudinal micro-cracks appeared when the hoop micro-cracks expanded to a certain extent and a part of the energy was distributed in the axial direction because of axial tension stress. Both the hoop and longitudinal micro-cracks grew increasingly denser and finally formed reticular cracks. As the cracks expanded, the energy was consumed gradually, and further away it was observed from the stress concentration point, the less visible the morphology of cracks turned.

3.2.3 State of surface crack

The metallographic morphology of the cross-section at the leakage of failed Ni-Fe-W alloy coating and at sites which were 350 mm away from leakage is shown in Figure 4. It can be inferred from Figure 4 that cracks stem from surface coating instead of the interface between the coating and substrate, evidencing that the coating and substrate were connected with each other tightly with good quality; the stress concentrates around the leakage with dense cracks penetrating the coating to the substrate; the intermittent extension of cracks is stagnated with an oversized morphology (Figure 4b) in the presence of alternating stress; cracks at sites which were 350 mm away from the leakage are sparse and all penetrate the coating. A majority of the cracks have extended to the

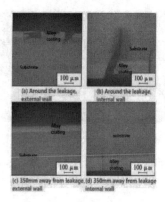

Figure 4. Metallographic morphology of the cross-section at the leakage of failed Ni-Fe-W alloy coating and at sites which were 350 mm away from leakage.

substrate and a minority of them to the interface between the coating and substrate.

3.3 Corrosion resistance of Ni-Fe-W alloy coating

Results of the coupon corrosion test suggested that the average corrosion rate of Ni-Fe-W alloy coating was 0.022 0, 0.029 8, 0.018 7, and 0.0 35 0 mm/a when the partial pressure of CO_2 was 1, 3, 4, and 6 MPa, respectively. According to the standards specified in NACE RP-0775-2005, slight corrosion of Ni-Fe-W alloy coating could be observed when the partial pressure of CO_2 was 2.4 MPa; moderate corrosion could be observed when the partial pressure of CO_2 was 3.6 MPa; the corrosion rate changed a little with the increase of the partial pressure of CO_2, suggesting that Ni-Fe-W alloy coating showed favorable corrosion resistance against CO_2. The macro-corrosion morphology at the Ni-Fe-W alloy coating surface suggested that there was a thin layer of corrosion products at the alloy coating surface without obvious local corrosion at the aforementioned partial pressure of CO_2.

Figure 5 shows the SEM morphology of corrosion at the Ni-Fe-W alloy coating surface with different partial pressures of CO_2.

As shown in Figure 5, there is basically no local corrosion at the Ni-Fe-W alloy coating surface where few corrosion products can be observed.

Table 1 shows the EDS composition at the corrosion surface of Ni-Fe-W alloy coating when the partial pressure of CO_2 is 3 MPa. Since there were few corrosion products, elements Ni and W mostly originate from the coating in which a small amount of harmful elements P and S were contained.

Figure 5. SEM morphology of corrosion at the Ni-Fe-W alloy coating surface with different partial pressures of CO_2.

Table 1. EDS composition at sample surface when the partial pressure of CO_2 is 3 MPa.

Element	C	O	P	S	Ni	W
w/%	4.98	7.20	2.48	2.05	61.38	21.91
A/%	19.09	20.70	3.68	2.94	48.11	5.48

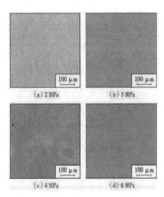

Figure 6. SEM morphology of Ni-Fe-W alloy coating with corrosion product removed after corrosion under different partial pressures of CO_2.

The presence of element W strengthened interatomic force, enhanced thermal stability, reduced porosity, and improved coating density. Moreover, the fact that the content of element W exceeded 21% suggested that the alloy was amorphous alloy that generated a compact and stable passive film to restrain dissolution activity of the alloy in the process of corrosion. The presence of element Ni would form a Ni_2S_3 corrosion product film at the coating surface and covered the sample surface,

playing a buffer and corrosion inhibition role. As a result, Ni-Fe-W alloy coating boasts of satisfactory corrosion resistance.

The macro-morphology of Ni-Fe-W alloy coating with corrosion product removed after corrosion under different partial pressures of CO_2 suggested that there was still metallic luster at the coating surface where no visible pit morphology could be observed. Figure 6 shows the SEM morphology of Ni-Fe-W alloy coating with corrosion product removed after corrosion under different partial pressures of CO_2. It can be seen in Figure 6 that there was no visible corrosive morphology at the coating surface at different partial pressures of CO_2, evidencing that Ni-Fe-W alloy coating had excellent corrosion resistance, and that the variation of partial pressure of CO_2 impacted little on its corrosion resistance.

4 CONCLUSION

1. Ni-Fe-W alloy coating has favorable corrosion resistance against CO_2 with low uniform corrosion rate that changes little with the increase of partial pressure of CO_2. The average thickness of Ni-Fe-W alloy coating is 54.0 μm and 98.5 μm at the external wall and internal wall of coiled tubing, respectively. Compared with QT-900 coiled tubing that serves as the substrate, Ni-Fe-W alloy coating mainly contains Ni and W, and a little of element Fe.
2. In the 8th week of the fatigue test, hoop microcracks occur at the Ni-Fe-W alloy coating surface; cracks grow denser with the increase of the number of cycles of the test. Longitudinal micro-cracks also occur in week 82; reticular micro-cracks can be observed in week 377; and the coiled tubing plated with Ni-Fe-W alloy coating fails because of leakage in week 516. Ni-Fe-W alloy coating has evidently improved coiled tubing's fatigue life.
3. Cracks in the failed Ni-Fe-W alloy coating caused by fatigue all stem from the coating, turn denser around the leakage, and penetrate the coating to the substrate. Cracks at sites which were 350 mm away from the leakage appear sparse; all of these have also penetrated the coating; a majority of them have extended to the substrate, and a minority of them to the interface between the coating and substrate.

REFERENCES

Ling Chao, Li Guobin and Meng Xianling. Study on the relationship between fatigue crack growth threshold and grain size application of dislocation theory [J].

Journal of Hebei University of Technology, 1992, 21(3): 67–72.

Ren Chengqiang, Cao Ranwei, Zheng Yunping, et al. Study on CO_2 Corrosion Behavior of Pipeline Steel [J]. Natural Gas and Oil, 2011, 29(2): 58–59.

Yu Xingfu, Tian Sugui and Du Hongqiang et al. Influences of elements W and Co on fault energy of Ni-Al alloy stacking fault energy [J]. Rare Metal Materials and Engineering, 2007, 36(12): 49–50.

Zhao Zuxin. Stress within and Dislocation in the Nickel Coating [J]. Journal of Northeast University of Technology, 1992, 21(5): 205–207.

Zhu Xiaoping. Analysis of the Fatigue Life of Coiled Tubing in the Presence of Bending and Internal Pressure [J]. Drilling & Production Technology, 2004, 27(4):73–75.

Advances in Energy Science and Equipment Engineering II – Zhou, Patty & Chen (Eds)
© 2017 Taylor & Francis Group, London, ISBN 978-1-138-71798-5

Applicability study of a Ni-P clad layer in an RNH₂-CO₂-H₂S-Cl-H₂O corrosion environment

Wen-wen Song
CNPC Tarim Oilfield Company Oil and Gas Engineering Research Institute, Kuerle, Sinkiang, China

Tao Guo
CNPC Liaohe Oilfield Branch Company Drilling and Production Technology Research Institute, Panjin, Liaoning, China

Ze-liang Chang
CNPC Tarim Oilfield Company Oil and Gas Engineering Research Institute, Kuerle, Sinkiang, China

Fei Zhou
CNPC Liaohe Oilfield Branch Company Drilling and Production Technology Research Institute, Panjin, Liaoning, China

ABSTRACT: By stimulating corrosion environment of a reboiler at the bottom of a regeneration tower, 304 stainless steel and a Ni-P clad layer were evaluated. The results showed that the uniform corrosion rate of both 304 stainless steel and the Ni-P clad layer at 105 °C was higher than 90°C, 304 stainless steel corrosion was relatively less, and the Ni-P clad layer appeared cracking and obscission in two stimulating corrosion environments.

1 INTRODUCTION

The electroless plating Ni-P is produced by the method of autocatalytic reduction using a reducing agent on the surface of the active materials without current. The introduction of the technology was during the 1980s in China. The technology has formed a certain scale of production capacity in East China, North China, northwest and other regions at present. Its output value has reached millions of dollars. The output value of the Baoji petroleum machinery factory of electroless plating nickel reached 200 million yuan. The value of the consumption of nickel plating solution in Iaoyang petrochemical fiber company reached 70 million yuan in 1991. A large number of electroless plating nickel have also been used in Daqing Oilfield, Zhongyuan Oilfield, Jinling Petrochemical Corporation, Guangzhou Petrochemical Company and so on. The electroless plating Ni-P has been chosen as the anticorrosion coating of oil field production equipment in Daqing, Shengli, Jianghan, Zhongyuan, and Xinjiang Oilfields.

But the corrosion performance research studies of Ni-P coating in different media are very less, especially in some acidic corrosion environments, or multi agent corrosive environments which are very

common in Tarim Oilfield. Therefore, the research is highly necessary. The weight loss method is always used in the corrosion resistance of the Ni-P coating. And the corrosion performance is judged from the corrosion rate. In this paper, the corrosion of Ni-P coating in RNH₂-CO₂-H₂S-Cl-H₂O corrosion environment is studied in detail. And its applicability in this environment is analyzed in detail too, so as to solve the practical problems in the field.

Natural gas exploited from gas wells contains acidic components, such as H_2S and CO_2, which not only significantly reduce its caloric value but also lead to severe corrosion of metal equipment in the downstream working sections. For this reason, de-sulfurization of natural gas is needed before it is transported to the users or further processing. Corrosion of de-sulfurization systems generally contains electrochemical corrosion, chemical corrosion, stress corrosion caused by carbonates or sulfides, and hydrogen bubbling with corrosion medium including CO_2-H_2S-H_2O, RNH_2-CO_2-H_2S-H_2O, and other corrosion contaminants. Methyldiethanolamine solution doesn't corrode metal from the perspective of working medium of the regeneration tower and reboiler. Once the solution has undergone the regenerative process, most H_2S and

Table 1. Conditions of corrosion simulation test.

Test temperature (°C)	90°C (liquid phase)	105 °C (gas-liquid)
Solution	27% MDEA barren solution	
Desalting and deoxidizing water in the tower	K^+: 10.01; Na^+: 335; Ca^{2+}: 42.44; Mg^{2+}: 49.74;: 10.22;: 426.83; Cl^-: 422.44	
H_2S content in the raw gas in the tower (%)	0.84	
CO_2 content in the raw gas in the tower (%)	4.91	
Heat-stable salt (mg/L)	Chloride (Cl^- 1100 mg/L), sulfate (16500 mg/L)	
Material	304 stainless steel, Ni-P coating	

CO_2 have become acidic gas, however, there is still a small amount of them in the solution and they remain a principal factor of corrosion in the presence of water (Luo, 2006).

Chemical plating, also known as auto-catalytic plating, is a process of metal ion deposition by means of metal autocatalysis by adding a reducing agent into the solution without current. Ni-P eletroless plating is a technology developed based on chemical plating by forming a clad layer at the equipment surface to provide the surface with effective protection and keep it away from corrosion of corrosive medium (Zhu, 2013).

This experiment focuses on the applicability of Ni-P electroless plating in the corrosive environment of a reboiler at the bottom of a regeneration tower in the processing plant of an oil field, in hope of providing reference for tube selection of the reboiler in the oil field.

2 EXPERIMENTAL MATERIAL AND METHOD

The current material of tubes used in the reboiler at the bottom of the regeneration tower in the processing plant of the oil filed is 304 stainless steel. An evaluation test that simulated the corrosion environment of 304 stainless steel and Ni-P plated samples was performed.

The substrate material for the Ni-P plated sample was P110 that was processed into a $50 \times 10 \times 3$ mm coupon. The chemical plated samples underwent cleaning, weighing, and dimensional measurement. The samples made up of 304 stainless steel were polished by 320-grit, 600-grit, 800-grit, and 1200-grit abrasive papers and then received cleaning and degreasing, as well as weighing and dimensional measurement after drying by cool air; the samples were mounted on the sample holder and mutually insulated with each other; then the samples and the sample holder were put into a Type TFCZ-25/250 HPHT magnetic-drive reactor as a whole for the corrosion test. Distilled water was used to wash the

corrosive medium at the sample surface followed by water removal with absolute alcohol and drying when the test was completed. Once the corrosion product film was removed, the samples were weighed by a Type FR-300MKO electronic balance to calculate the weightlessness and average corrosion rate of samples; a Type JSM-5800 scanning electron microscope was used to observe the corrosion morphology of the sample surface. The test condition is given in Table 1.

3 RESULTS AND DISCUSSION

The average corrosion rates of 304 stainless steel and Ni-P plated materials at 90°C were 0.0010 mm/a and 0.0510 mm/a, respectively. Figure 1 shows the micromorphology of samples before cleaning. It can be inferred from the figure that slight corrosion and machining marks could be observed at the surface of the 304 stainless steel sample; while the corrosion product layer at the surface of the Ni-P plated sample was thick and there were cracks at the surface of the Ni-P plated sample.

Figure 2 shows the micromorphology of samples after cleaning. As shown in the figure, no obvious local corrosion could be observed on the 304 stainless steel sample, while there were damages in various degree and obscission at the surface of the Ni-P plated sample.

In the co-existing condition of gas and liquid phases at 105 °C, the average corrosion rates of the 304 stainless steel sample and Ni-P plated sample were 0.0034 mm/a and 0.1126 mm/a, respectively. Figure 3 presents the micromorphology of samples after cleaning from three parts, i.e. gas phase, gas-liquid interface, and liquid phase. It can be seen in the figure that there are more corrosion products at the gas-liquid interface, and machining marks could be observed at the surface of the 304 stainless steel sample in both gas and liquid phases while there were thick corrosion products covering the surface of the Ni-P plated sample with severe corrosion, and that there were also cracks on the corrosion products of Ni-P coating.

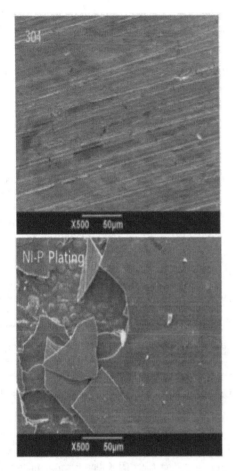

Figure 1. Micromorphology of samples before cleaning.

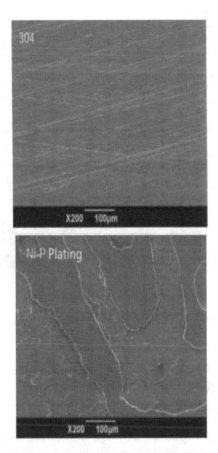

Figure 2. Micromorphology of samples after cleaning.

The micromorphology of samples before cleaning is shown in Figure 4. As is shown, no visible local corrosion on the 304 stainless steel sample was observed while there was local corrosion at the surface of the Ni-P plated sample.

Ni-P amorphous alloy coating prepared by chemical plating is troubled by various problems, such as high porosity and low deposition efficiency (Gao, 2004). A high salt content and Cl⁻ concentration in RNH_2-CO_2-H_2S-Cl--H_2O medium environment will cause serious corrosion to the coating. Due to the fact that Cl⁻ is strong in penetrability because of its small radius and permeates existing pores or defects in the coating more easily than other ions in the presence diffusion or electric field, the coating will react with the substrate, thus leading to corrosion. In the meantime, the potential difference between the coating and substrate makes it vulnerable to electrochemical corrosion that undermines the protective effect of coating.

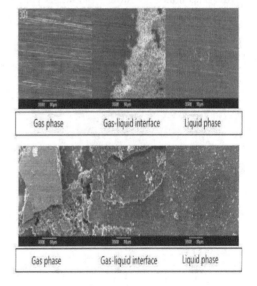

Figure 3. Micromorphology of samples before cleaning.

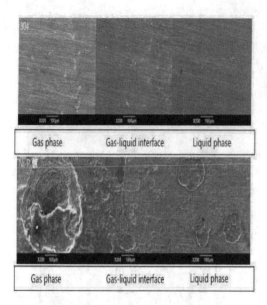

Figure 4.   Micromorphology of samples after cleaning.

When H$_2$S contacts water molecules, the following reactions will take place at the coating surface:

$$H_2S \rightarrow HS^- + H^+$$

Ions released from the reaction are hydrogen depolarization agents that are easy to absorb electrons to promote anodic reaction, thus resulting in hydrogen sulfide corrosion. HS- and S2- generated by H$_2$S in water-bearing medium are able to maintain cathodic reaction that separates hydrogen atoms which combine into hydrogen molecules; instead, the hydrogen atoms are absorbed at the coating surface and expand and gather to coating

defects. In the hydrogen accumulation area, tensile stress is generated within the coating whose stress also changes causing cracks at the coating surface followed by obscission because of corrosion (Du, 2006). Local obscission of coating will lead to small cathode and large anode, aggravating local corrosion (Zong, 2009).

4   CONCLUSION

1. The uniform corrosion rates of both 304 stainless steel and Ni-P plated materials in the co-existing condition of gas and liquid phases at 105°C are all higher than those at 90°C.
2. Cracks and obscission of Ni-P plating occur in both corrosion simulation conditions.

REFERENCES

Du Yanbin, Zhu Liqun, Xue Zhen et al. Corrosion Resistance of Ni-P Amorphous Composite Coating in the Environment Containing H$_2$S/CO$_2$ [J]. Acta Materiae Compositae Sinica: 2006, 23 (6): 92~97.

Gao Y, Zheng J Z, Zhu M, et al. Corrosion Resistance of Electrolessly Deposited Ni-P and Ni-W-P Alloys with Carious Structures[J].Materials Science and Engineering A, 2004, 381: 98~103.

Luo Guomin, He Fengru, Xu Shili. Corrosion Reason Analysis and Countermeasures for De-sulfur Regenerator and Reboiler [J]. Chemical Fertilizer Design: 2006, 44 (4): 40~51.

Zhu Xinliang. Analysis on the Application of Ni-P Eelectroless Plating in Tubing Corrosion Prevention of Heat Exchanger [J]. China Petroleum and Chemcial Standard and Quality: 2013 (11):32.

Zong Yun, Guo Rongxin and He Bingqing etc. Influences of different constituents of plating solution on the structure and performance of Ni-W-P alloy layer of chemical deposition [J]. Surface Technology, 2009, 38 (6): 9~12.

Study on the blending coal replacement and combustion rate for the BF PCI process

Zhengqing Zhao
School of Metallurgy and Ecological Engineering, University of Science and Technology Beijing, Beijing, China

Lingling Zhang
School of Civil and Environmental Engineering, University of Science and Technology Beijing, Beijing, China

Daqiang Cang
School of Metallurgy and Ecological Engineering, University of Science and Technology Beijing, Beijing, China

Yu Li
State Key Laboratory of Advanced Metallurgy, University of Science and Technology Beijing, Beijing, China

Dejian Pei
School of Metallurgy and Ecological Engineering, University of Science and Technology Beijing, Beijing, China

ABSTRACT: In order to find a new way to calculate the pulverized coal replacement rate and its combustion rate in BF tuyere for Pulverized Coal (PC) Injection (PCI), a mathematical model was developed based on both the theoretical replacement rate and PCI pyrolysis heat calculation formula in this paper. The model is more accurate than the conventional test for determining the different properties of PC and the different mixing ratios of the blending coal for PCI in Anyang Steel Co., and 60% of higher volatile bituminous coal in the blending coal is found to be the best ratio. Practical results of the model in the BF PCL of Anyang Steel Co. showed that the coke rate was reduced and the PC injection rate was increased.

1 INTRODUCTION

Coal injection is the key technology of the blast furnace to cut the costs and increase the efficiency. Evaluation indices of coal injection are as follows: the composition of coal, grindability, explosiveness, the ash fusion point, the pneumatic conveying properties of pulverized coal and coal combustion performance, the replacement ratio of pulverized coal. The sampling in the laboratory or field, and the measurement and calculation of the burning rate reflect that the combustibility of coal is not easy. There are many limitations in calculation results; and there are also so many supposed factors in calculation of the coal replacement ratio which reflects the effect of the application. Therefore, the comparability is poor. We use the simple mathematical model to calculate the combustion rate of pulverized coal. Additionally, the results for the traditional calculation of the coal replacement ratio are corrected by calculating the coal decomposition heat. In conjunction with other indicators we measured, we achieve the evaluation of using all kinds of coal in collocation.

Simultaneously, based on the actual situation of Ansteel, we put forward reasonable coal collocation and the effect of application in the blast furnace, which is 2200 m³.

2 THE BASIC PROPERTIES OF PULVERIZED COAL

Prior to 2008, coal samples of the blast furnace are substances with low volatility. From 2008, we have included the Shenmu bituminous coal to increase the proportion of coal and cut costs. The initial ratio of Shenmu coal is 1/3 and the volatility is between 16–18%. Chemical properties of conventional coal in Ansteel are shown in Table 1 and Table 2, and granular control is listed in Table 3.

2.1 *Proximate analysis and elemental analysis of pulverized coal*

Types and composition of pulverized coal commonly used in Ansteel are shown in Tables 1 and 2.

Table 1.   Proximate analysis of injecting coal (%).

Coal type	Moisture (M)	Ash content (A)		Volatile component (V)		Fixed carbon (FC)	
	ad	ad	d	ad	d	ad	d
Shenmu	3.64	8.00	8.28	30.07	31.23	58.29	60.49
Lu'an	0.83	10.14	10.11	10.14	10.22	78.92	79.56
Hebi	0.86	9.44	9.52	12.55	12.66	77.15	77.82
Fengfeng	1.10	10.92	11.05	10.64	10.76	77.30	78.19
Xinlong	0.92	9.84	9.93	8.84	9.00	80.40	81.16

Table 2.   Elemental analysis of injecting coal (%).

Coal type	Carbon (C)		Hydrogen (H)		Sulfur (S)		Nitrogen (N)		Oxygen (O)	
	ad	d	ad	d	ad	d	ad	d	ad	d
Shenmu	72.65	75.39	4.37	4.54	0.42	0.44	0.92	0.95	10.02	10.40
Lu'an	81.91	82.57	3.76	3.79	0.30	0.30	1.22	1.23	1.87	1.88
Hebi	81.43	82.14	3.85	3.89	0.40	0.40	1.15	1.16	2.87	2.89
Fengfeng	80.26	81.19	3.55	3.59	0.40	0.40	0.98	0.99	2.75	2.78
Xinlong	81.73	82.47	3.52	3.55	0.44	0.44	1.10	1.11	2.47	2.49

Table 3.   Result of grindability, explosiveness, ash fusion point and liquidity.

Coal type	Returned fire Length/mm	HGI	ST	HT	Flowability Indices	Floodability Indices
Shenmu	>400	55	1295	1295	61	77.5
Lu'an	No data	93	>1450		51	67
Hebi	No data	104	>1450		54	67
Fengfeng	Spark in fire source	80	>1450		59	69
Xinlong	No data	85.7	>1450		60	71.5
Shenmu:Lu'an 1:2	No data		>1473		51	68.5
Shenmu:Lu'an 1:1	30		1458	1463	58	78
Shenmu:Lu'an 2:1	115		1374	1340	52.5	74
Shenmu: Hebi 1:2	No data		>1473		49	70.5
Shenmu: Hebi 1:1	40		1433	1448	55	78.5
Shenmu: Hebi I 2:1	55		1380	1388	61	81.5
Shenmu: Fengfeng 1:2	No data		>1473		58	75
Shenmu: Fengfeng 1:1	No data		1458	1465	60	75.5
Shenmu: Fengfeng 2:1	55		1379	1386	59	75.5
Shenmu: Xinlong 1:2	No data		>1473		59	69
Shenmu: Xinlong 1:1	No data		1461	1466	62	76
Shenmu: Xinlong 2:1	No data		1382	1388	62	77.5

Among five kinds of coal as described above, Shenmu bituminous coal possesses high volatility, and Xinlong coal belongs to anthracite. The rest of the three coal varieties are lean coal, and their ingredients are similar. In terms of the sulfur content, all kinds of coal meet the requirements of coal injection in BF. The reason is that the sulfur content is lower in coal, where in the difference between the highest and lowest is only 0.14%.

2.2 *Measurement of grindability, explosiveness, ash fusion point and liquidity indices*

The results of measured indices are shown in Table 3.

As shown in Table 3, Shenmu bituminous coal showed the worst grindability, which is 55, and others have good grindability. Shenmu bituminous coal also has explosiveness; conversely, the four remaining coal varieties and blended coal of Shenmu and the others have no explosiveness. Therefore, the safety of blended coal injection can be assured without taking special explosion-proof measures. Shenmu bituminous coal is the worst in ash melting property, while the four remaining coal and blended coal have a high ash fusion point. All kinds of coal meet the requirements of coal injection in BF. The distribution interval of the flowability index is 56–61 for single coal, and the distribution interval of the flowability index is 49–62 for binary mixed coal. It can thus be seen that the flowability of pulverized coal we measured is below average. Most floodability indices are distributed from 60 to 80. As we know, the floodability of all coal varieties is strong. Therefore, we must take effective measures to prevent spillage in bunker unloading and the uneven phenomenon of coal injection (Zheng, 2014).

3 CALCULATION OF THE PULVERIZED COAL COMBUSTION RATE

Predecessors have conducted a lot of research on pulverized coal combustion in the tuyere area, and established a complex mathematical model (Zhao, 1999; Chen, 2014; Kong, 2011; Li, 2014). They also reported the combustion mechanism of pulverized coal and the distribution of the product in the tuyere area. While we focus our attention on the overall effect of combustion or the average effect in tuyere and raceway, our calculation is based on the model of the combustion rate which is established in the literature (Tang, 1996; Ding, 1999). The assumptions of the model are as follows: (1) the gas flow in the raceway is free jet; (2) carbon-oxygen reaction at high temperature: $C + 1/2O_2 = CO$; (3) in the raceway, we do not take the interaction between pulverized coal or pulverized coal and coke into consideration; and the interaction between pulverized coal and iron slag isn't considered.

3.1 Calculation of volume in tuyere and raceway

According to the assumption of the model, the raceway formed before the tuyere area is shown in Fig. 1.

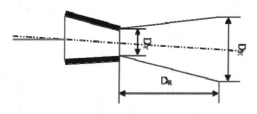

Figure 1. The structure diagram of tuyere.

The depth of the raceway:

$$\frac{D_R}{D_T} = 0.521 \times \left(\sqrt{\frac{\rho_{gb}}{\rho_c}} \times \frac{U_{ob}}{\sqrt{g \times D_c}} \right)^{0.8} \tag{1}$$

The volume of the raceway:

$$V_f = \pi/12 \, D_R \left(D_T^2 + D_I D_E + D_E^2 \right) \tag{2}$$

3.2 Calculation about the time of pulverized coal in the raceway zone

The time of pulverized coal in the raceway zone:

$$\tau = V_f / Vg \tag{3}$$

$$Vg = (V_{g0} + V_{g0} \times (Yo_2)_0 \times X \times B) \times T/T_0 \tag{4}$$

$$X = \eta \times 1/[22.4 \times (E_c/12 - E_o/32)] \tag{5}$$

$$B = 22.4 \times (E_O/32 + E_C/12 + E_H/2 + E_N/28) \tag{6}$$

3.3 Chemical reaction constant of pulverized coal combustion

Pulverized coal combustion includes three processes: (1) chemical reaction; (2) diffusion of gas in the gas film on the surface of particles; (3) turbulence diffusion of pulverized coal and oxygen.

Rate constant K can be expressed as:

$$\frac{1}{K} = \frac{1}{K_c} + \frac{1}{K_f} + \frac{1}{K_t} \tag{7}$$

$$K_c = 6.5 \times 10^5 \times T^{0.5} \times \exp(-22000/T) \tag{8}$$

$$K_f = 3.7 \times 10^{-5} \times T^{0.5}/(D_P \times P) \times \left(T/273 \right)^{1.75} \tag{9}$$

$$K_t = F_k \times (3.6 \times 10^{-2}) \times (\rho_P \times D_P) \times \frac{U_f}{D_f} \\ \times \frac{1 + B \times (Y_{O2})_0 \times X}{(Y_{O2})_0 \times X} \times \frac{T}{273} \times \frac{1}{P} \tag{10}$$

where $U_f = \dfrac{D_R}{\tau}$ $D_f = \left[4 \times V_f/(D_R \times \pi) \right]^{0.5}$

3.4 Combustion rate of pulverized coal in the tuyere region

The following formula can be deduced to calculate the combustion rate of pulverized coal in the tuyere region:

$$\eta = \frac{24}{E_c \times R_g} \times \frac{6}{\rho_P \times D_P} \times \frac{1}{1 + B \times (Y_{O_2})_0 \times X} \\ \times (Y_{O_2})_0 \times \frac{P}{T} \times K \times \tau \tag{11}$$

where, D_R is the depth of the raceway (m); D_T is the diameter of tuyere (m); D_C is the coke particle size (m); ρ_{gb} is the density of gas (kg/m³); ρ_c is the density of coke (kg/m³); U_{ob} is the air speed in the nose of tuyere (m/s); g is the acceleration of gravity (9.81 m/s²); D_E is the diameter of the end of the tuyere raceway, $D_E = D_T + 2tga \cdot D_R$; a is the semi-divergent angle 9–13°; V_{g0} is the blast volume of each tuyere (m³/s); X is the coal-to-air ratio (kg coal/Nm³); Ec, E_h, En, Eo, and Es (percentage) are the components of pulverized coal (C, H, N, O, S), $(Yo)_0$ is the mole fraction of oxygen in the blast; η is the combustion rate of pulverized coal; ρ_P is the density of pulverized coal (kg/m³); D_P is the pulverized coal particle size (m); P is the pressure of the tuyere raceway (atm); T is the tuyere raceway temperature (K); F_k is the correction factor which reflects the mixed state of coal and oxygen. $F_k = 1.0$; R_g is the gas constant, where $R_g = 0.082054$ m³·atm/k•mol.

Property parameters and process parameters of pulverized coal are as follows: T0 = 1373(K),

Table 4. The calculations of pulverized coal's calorific value, decomposition heat, replacement ratio and burning rate.

No.	Coal and proportion	Calorific value kJ/kg	Decomposition heat of coal kJ/kg	Replacement ratio 1	Replacement ratio 2	Burning rate
1	Shenmu	30155.93	1098.26	0.724	0.700	0.7000
2	Luan	30198.27	2569.61	0.744	0.749	0.6500
3	Hebi	30197.36	2562.75	0.736	0.747	0.6550
4	Fengfeng	30193.08	1880.40	0.751	0.743	0.6540
5	Xinlong	30196.74	2271.47	0.749	0.747	0.6530
6	Shenmu10% and Lu'an	30240.54	2375.78	0.745	0.786	0.6543
7	Shenmu20% and Lu'an	30231.14	2233.80	0.743	0.777	0.6612
8	Shenmu30% and Lu'an	30221.74	2091.83	0.740	0.767	0.6639
9	Shenmu40% and Lu'an	30212.34	1949.85	0.738	0.758	0.6658
10	Shenmu50% and Lu'an	30202.94	1807.87	0.735	0.748	0.6696
11	Shenmu60% and Lu'an	30193.54	1665.90	0.733	0.738	0.6744
12	Shenmu70% and Lu'an	30184.13	1523.92	0.731	0.729	0.6801
13	Shenmu80% and Lu'an	30174.73	1381.94	0.728	0.719	0.6834
14	Shenmu90% and Lu'an	30165.33	1239.96	0.726	0.710	0.6868
15	Shenmu10% and Hebi	30239.69	2369.83	0.738	0.786	0.6577
16	Shenmu20% and Hebi	30230.38	2228.54	0.736	0.776	0.6621
17	Shenmu30% and Hebi	30221.08	2087.26	0.735	0.767	0.6649
18	Shenmu40% and Hebi	30211.77	1945.97	0.733	0.757	0.6677
19	Shenmu50% and Hebi	30202.46	1804.69	0.732	0.748	0.6742
20	Shenmu60% and Hebi	30193.16	1663.40	0.730	0.738	0.6781
21	Shenmu70% and Hebi	30183.85	1522.12	0.728	0.729	0.6803
22	Shenmu80% and Hebi	30174.54	1380.83	0.727	0.719	0.6853
23	Shenmu90% and Hebi	30165.24	1239.55	0.725	0.710	0.6878
24	Shenmu10% and Fengfeng	30212.77	1778.78	0.750	0.761	0.6589
25	Shenmu20% and Fengfeng	30206.45	1703.17	0.747	0.754	0.6654
26	Shenmu30% and Fengfeng	30200.14	1627.55	0.744	0.747	0.6676
27	Shenmu40% and Fengfeng	30193.82	1551.94	0.741	0.740	0.6691
28	Shenmu50% and Fengfeng	30187.51	1476.33	0.738	0.734	0.6723
29	Shenmu60% and Fengfeng	30181.19	1400.71	0.735	0.727	0.6757
30	Shenmu70% and Fengfeng	30174.88	1325.10	0.732	0.720	0.6818
31	Shenmu80% and Fengfeng	30168.56	1249.49	0.729	0.714	0.6846
32	Shenmu90% and Fengfeng	30162.25	1173.88	0.726	0.707	0.6874
33	Shenmu10% and Xinlong	30230.94	2115.87	0.749	0.778	0.6577
34	Shenmu20% and Xinlong	30222.60	2002.81	0.746	0.769	0.6599
35	Shenmu30% and Xinlong	30214.27	1889.74	0.743	0.761	0.6666
36	Shenmu40% and Xinlong	30205.94	1776.67	0.740	0.752	0.6682
37	Shenmu50% and Xinlong	30197.60	1663.60	0.738	0.743	0.6716
38	Shenmu60% and Xinlong	30189.27	1550.53	0.735	0.735	0.6760
39	Shenmu70% and Xinlong	30180.93	1437.47	0.732	0.726	0.6796
40	Shenmu80% and Xinlong	30172.60	1324.40	0.729	0.717	0.6842
41	Shenmu90% and Xinlong	30164.27	1211.33	0.726	0.709	0.6872

T = 2200 (K), DT = 0.12 (m), P = 3.5 (atm), DP = 0.000074 (m), FK = 1, $(Y_{O2})_0$ = 0.24, ρ_p = 560 (kg/m³), α = 13°, ρgb = 0.94 (kg/m³), Vgo = 2.43 (m³/s), ρc = 450(kg/m³), Dc = 0.015 (m), and Uob = 200 (m/s).

The calculation results are shown in Table 4. The combustion efficiency of the pulverized coal in the tuyere raceway is 65–70%, and the combustion performance is good. Among them, the burning rate of Shenmu is up to 70%, and the combustion rate of other coal varieties is approximate. The results show that the combustion rate of pulverized coal is 40–75% (Zhao, 2000), and the calculated results are in agreement with the results of tuyere.

4 CALCULATION AND ANALYSIS OF REPLACEMENT RATIO OF COAL

The replacement ratio of coal is the heat ratio of carbon into which infection fuel and coke are converted under relative stable smelting conditions, which is the basis of high temperature. With the increase of coal rejection rate and development of mixing injection technology, the traditional calculation method, the heat absorption is the empirical value 1005 kJ/kg, has a big error because of the existence of the heat absorption effect during the pulverized coal burning. Therefore, this paper uses Gass's law to calculate the decomposition heat of coal by the statistic formula to get a more accurate replacement ratio of pulverized coal.

The replacement ratio formula is as follows (Zhang, 2001; Wu, 2008; Cheng, 1996):

$$R = \frac{\left[C^C + C^H - 0.66(s) - (0.11 + 0.16b) \times (A) - C^X\right]_{inj} \times B - 0.11(1 - B)}{\left[C^C + C^H - 0.66(s) - (0.11 + 0.16b) \times (A) - C^X\right]_{coke} + 0.145} \tag{12}$$

where, B = $(SiO_2 * R - (CaO))/(CaO)_{effective}$; B is the utilization factor of coal in the furnace;

$$C_i = \frac{q_{co} \times (C_i + Q_{Bi})}{Q_C} \tag{13}$$

$$C_i^H = \frac{121020 \times (H_2)_i \times \eta_H \times b_H + (13450 \eta_H \times b_H - 16035)(H_2O)_i}{Q_c} \tag{14}$$

$$b_r = \frac{(SiO_2)_i \cdot R - (CaO)_i}{(CaO)_{effective}} \tag{15}$$

$$C^X = \frac{Q_d}{Q_C} \tag{16}$$

$$Q_c = \frac{22.4 \cdot \omega_B^t}{2 \times 12(O_2)_B} \tag{17}$$

where, Q_{Bi} is the heat into the hearth by blast of unit fuel, K_J/Kg; i is the type of coke or fuel injection; R is binary-component basicity of slag; ω_B^t is the enthalpy of blast at t, $K_J/m^3 \cdot °C$;

Where Q_d is the decomposition heat of coal, Q_d = (the heat produced by coal burned into CO_2, H_2O(vapour) and N_2) Qtotal − (the heat measured) Q_f

$$Q_{total} = 340.67w_{cd} + 177.13w_{od} + 1210.0w_{Hd} \tag{18}$$

where w_{Cd} stands for fixed carbon in coal by dry ingredients, (%); w_{Od} stands for oxygen fixation of coal by dry ingredients, (%); w_{Hd} stands for hydrogen of coal by dry ingredients, (%);

Considering the water that the coal contains, the result in the above formula need correction.

$$Q'_d = (1 - w_{H2O}) Q_d - 124.5 * 10^3 w_{H2O}/18 (kJ/kg) \tag{19}$$

Q_d is the decomposition heat of dry coal (kJ/kg); Q'_d is the position decomposition of coal in consideration of water (kJ/kg).

Using the formula for calculating the heat of coal according to the literature (Liu, 1989).

$$Q_{D,w}^f = 8633.5 + 254.4C^f + 859.7H^f - 154.8O^f + 54.1S_Q^f - 99.3A^f - 145.1W^f \ (kJ/kg) \tag{20}$$

The results are shown in Table 4.

From the data in Table 4, it can be seen that, for single coal, the replacement ratio is the lowest in Shenmu coal containing high volatility. The replacement ratio is the highest in Fengfeng coal, followed by Xinlong coal, and Luan and Hebi coal are in the middle. Taking into account the coal varieties, except Shenmu coal, the properties of other coal are similar. Therefore, when we calculate the combination of coal varieties, we only calculate the replacement ratio of the coal mixed by Shenmu coal and the other coal from 10% to 90%.

The result calculated by decomposition heat is replacement ratio 1, and replacement ratio 2 is the result of the fixed decomposition heat of 1005 KJ / Kg. The comparison of the two calculation results is shown in Figure 2. As shown in Table 4 and Fig. 2, with the increasing ratio of Shenmu coal containing high volatility, the replacement ratio of mixed coal reduces. The order of the replacement ratio of the coal calculated by fixed decomposition heat is Shenmu, Fengfeng, Xinlong, Hebi, and Lu'an. When the ratio of Shenmu coal is low, the replacement ratio is large and larger than the largest single coal. The order of the replacement ratio of the coal calculated by decomposition heat is Shenmu, Hebi, Lu'an, Xinlong, and Fengfeng. The replacement ratio of mixed coal is between the two single coals' repalcement ratio. Obviously, the replacement ratio calculated by decomposition heat is more reasonable.

To analyze the reasonable blending relationship conveniently by the combustion rate and replacement ratio, the combustion rate and replacement ratio of pulverized coal in Table 4 are presented in Figure 3.

It can be seen from Figure 3 that with the increase of proportion of Shenmu coal, the coal combustion rate increases. The combustion rate of Shenmu and Lu'an with different ratios is low. Other collocation burning rates are similar. The change of the burning rate can be divided into three different regions. The ratio is under 20%; with the increase of bituminous coal, the burning rate increased rapidly from 20% to 40%, and the increasing rate goes up to a limited degree, from 80% to 90%, slowly and the combustion rate of the mixture is equal.

With the proportion of Shenmu coal increasing, the replacement ratio almost decreases linearly; the Shenmu and Hebi coal collocation replacement ratio is the lowest. The ratio is beyond 80%, The replacement ratio of a variety of coal blending is similar. Considering the coal combustion rate and replacement ratio, the ratio of Shenmu coal of about 50% should be more reasonable. In actual

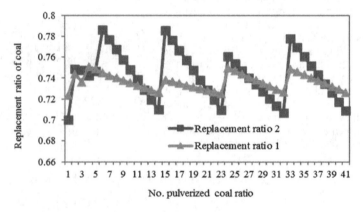

Figure 2.   The comparison of the replacement ratio calculated by different ways.

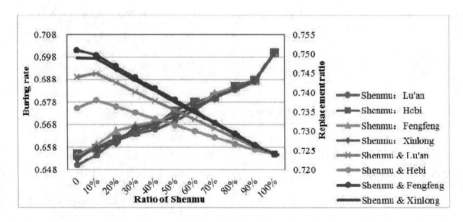

Figure 3.   Combustion rate of mixed coal and theoretical calculation of the displacement ratio.

1150

Table 5. Main technical and economic index of Ansteel 2200 m³ blast furnace.

Time	TFe%	Utilization coefficient t/m³ • d	Coke ratio kg/t	Coal ratio kg/t	Fuel ratio kg/t	Oxygen enrichment %	Air volume Ten thousand M³/h	Wind pressure KPa	Wind temperature °C	Top pressure KPa	[Si] %	[S] %
2008	58.50	2.01	414	106	504	1.57	22.77	331	1106	184	0.52	0.023
2009	58.17	2.13	387	120	503	1.84	24.17	331	1140	185	0.46	0.018
2010	57.19	2.30	353	154	508	3.32	24.65	360	1169	223	0.39	0.018
2011	56.73	2.30	374	155	529	3.38	25.44	361	1150	219	0.37	0.017
2012	56.66	2.24	385	140	548	2.83	25.23	354	1139	212	0.40	0.020

production, due to the increase in the wind temperature or oxygen enrichment, the replacement ratio of coal is improved by the decomposition of coal. Although the burning rate also increases, in comparison, the replacement ratio will be improved more. Therefore, the reasonable ratio of Shenmu coal in practical production should be 50–60%.

5 RESULTS IN INDUSTRIAL PRACTICE

Ansteel 2200 m³ blast furnace is put into operation in October 2005. Since the coal injection system builds lingeringly, in March 2006, coal injection officially began. Fengfeng and Xinlong coal is the main blowing coal in the initial phase, Thereafter, Lu'an coal was introduced. In April 2007, we tried a small amount of Shenmu of high volatility with 5000 tons (volatile 33–38%). Then, we developed a research plan to achieve stable quantity of coal, systematically increase the coal rate and decrease the coke ratio with the improvement of a coal injection system. In 2008, we officially began blowing bituminous coal. The ratio of coal is controlled in 30%. In 2010, the ratio of coal is improved to 60%, and we achieved the effect of greatly reducing the coke rate. Table 5 shows the main technical and economic indexes of the 2200 m³ blast furnace. As shown in the table, the coal ratio reached 154 kg/t in 2010. The coal ratio of mixed injection increased to 160 kg/t in March 2010, and the coke rate reduced to 330 kg/t. The result of the test is satisfying when we consider the grade of the furnace, which falls about 1% compared to 2008s' 58.50 and factors of the increasing slag.

6 CONCLUSIONS

1. The calculated maximum combustion rate and replacement ratio of blending coal and the best industrial practice results are the foundation for coal selection and optimization of the mixing rate of blending coal.
2. The theoretical replacement ratio of coal is more reasonable by the calculation of decomposition heat of coal.
3. For blowing Shenmu bituminous coal in the Aanyang Steel company, the reasonable ratio was about 60% when the air temperature was over 1100°C and oxygen enrichment over 3%. In actual production, with 60% of the Shenmu bituminous coal, the increase of the coal rate would significantly reduce the coke rate.

REFERENCES

Chuan Chen, Shusen Cheng, JUSTB, 36 (2): 266–234 (2014).
Dewen Kong, Jianliang Zhang, et al, JISR, 23 (11): 4 (2011).
Guanglin Tang, Shuren Na, JBUIST, 15 (3): 37–40 (1996).
Jianliang Zhang, Tianjun Yang, JUSTB, 23 (4): 308–310 (2001).
Lanbo Cheng. Beijing Metallurgical Industry Press, 325–326 (1996).
Ligang Ding, Baowei Li, Chushao Xu, JBUIST, 18 (3): 178–182 (1999).
Shengli Wu, Xiaobo Yu, JUSTB, 30 (12): 1432–1438 (2008).
Tianxin Liu, Shijie Han, CT, 4,7–10 (1989).
Yongqing Li, Xiaohui Zhang, Jiayuan Zhang, J CSU(ST), 45 (7): 2432–2439 (2014).
Zhengqing Zhao, Daqiang Cang, EMI, 33 (5): 39–42. (2014).
Zhengqing Zhao, Yongqiang Guo, RIS, 200 (2): 1 (2000).
Zuqiao Zou, Xuexin Sun, JHUST, 27 (5): 97–99 (1999).

Advances in Energy Science and Equipment Engineering II – Zhou, Patty & Chen (Eds)
© 2017 Taylor & Francis Group, London, ISBN 978-1-138-71798-5

Effect of electrode material on electro-discharge machining of nickel-based superalloy

Yan Li, Jinjuan Fu & Jianmei Guo
Beijing Institute of Electro-Machining, Haidian, Beijing, China
Beijing Key Laboratory of Electrical Discharge Machining Technology, Haidian, Beijing, China

ABSTRACT: The electrode material plays an important role in electro-discharge machining. It affects the electro-discharge efficiency and the machining quality. Two kinds of electrode materials including copper and EDM-C3 were compared. The similarities and differences of the relevant properties were analysed. Nine experiments were designed for each kind of electrode to get a comprehensive evaluation. In order to evaluate the performance of different materials, the same set of electrical parameters such as peak current, pulse on time, duty cycle and servo voltage were selected. Meanwhile, material removal rate and tool wear rate were selected as the quality targets. Experiments on the two electrodes were carried out. The experimental results show that EDM-C3 electrode has better performance in the case of Material Removal Rate (MRR) and Tool Wear Rate (TWR). The outcome of the present research work will be a considerable aid to industries for efficiency improvement in electro-discharge machining of nickel-based superalloy.

1 INTRODUCTION

Developments of aerospace industries are in great demand for superalloy. Nickel-based superalloy has excellent inherent characteristics. The alloy is often applied to manufacture engines of liquid-fueled rockets and aircraft parts. It is also used in manufacturing modern advanced turbines of compressor. The alloy has the advantages of high-strength, corrosion resistance and superior mechanical properties. It has good welding characteristics with outstanding resistance to the post-weld cracking. Moreover, it has good tensile, fatigue and creep characteristics (Yilbas, 2016; Oezkaya, 2016). However, properties such as low thermal conductivity, rapid work hardening, ability to react with tool materials under atmospheric conditions, formation of built-up edge and presence of abrasive carbides in their microstructure cause poor machinability and restrict the wider usage (Senthilkumaar, 2012).

In order to overcome the poor machinability in conventional processes, nonconventional processes are increasingly used in superalloy. Electro-Discharge Machining (EDM) offers considerable advantages over the conventional techniques. EDM is one of the most widely used non-conventional material removal processes (Abbas, 2007). EDM process is based on removing material from a part by means of a series of repeated electrical discharges between tool called the electrode and the workpiece in some medium (Rajurkar, 2013). EDM is based on thermoelectric energy. There is no direct contact between the electrode and the workpiece, so EDM can eliminate mechanical stresses, chatter and vibration problems under conventional machining. There were many valuable researches on machining nickel-based superalloy by EDM. In literature (Klocke, 2014), advanced wire-EDM capabilities for the manufacture of fir tree slots in Inconel 718 were evaluated. The precision, surface roughness, minimization of white layer formation and contamination were measured. In literature (Li, 2016), thinning process of EDM to reduce the recast layer thickness of film cooling holes on turbine blade was investigated. Experimental investigation to determine the machining parameters in machining Inconel 738 was conducted. In literature (Atzeni, 2015), the surface quality of Inconel 718 machined by wire electrical discharge machining was studied.

Literature review reveals, though a number of attempts have been made until now to enhance the accuracy, productivity and utility of the process, effect of electrode material on electro-discharge machining of Inconel 625 has not been attempted yet. To address this issue, the present research work proposes an experimental investigation on machinability of Inconel 625 alloy by using copper and EDM-C3 electrodes. The performances such as Material Removal Rate (MRR) and Tool Wear Rate (TWR) were studied.

2 EXPERIMENTAL DETAILS

The experimental study was carried out on an EDM machine tool manufactured by Beijing Institute of Electro-Machining, the process parameters can be set as needed. Figure 1 shows a photograph of the experimental setup. Moreover, a circular shaped tool was taken to perform the experiment. A rectangular shaped workpiece made of Inconel 625 alloy was placed in the parallel-jaw vice. The parallel-jaw vice was fixed on the machining table. The workpiece and the electrode were measured before and after the machining process. Initially, the electrode tool and the workpiece were fully immersed in the dielectric kerosene fluid. Then the power supply was applied to the EDM machine. The power supply produces electrical discharge sparks between the workpiece and the tool.

Inconel 625 belongs to a kind of nickel-based superalloy. The chemical components of Inconel 625 are shown in Table 1. Some physical and mechanical properties are listed in Table 2.

Inconel 625 has advantages of high yield strength, outstanding fatigue resistance, good welding characteristics and excellent corrosion resistance in severe conditions. But everything has two sides. Inconel 625 is one of the alloys that have relatively poor machinability in the conventional machining processes, due to its work-hardening nature and low thermal conductivity. For Inconel 625, EDM is a preferred machining method due to its advantages like reduced machining stresses.

EDM is a complex machining process, which includes mechanical, electrical and fluidic interaction. Many factors such as electrode parameters can influence the processing effects. Electrode material is an important factor. Many kinds of materials can be used for electrode, such as brass, copper, graphite, copper-tungsten alloy and so on. Given the workpiece properties, copper electrodes (Figure 2) and EDM-C3 electrodes (Figure 3) were taken to perform the experiments.

Copper and EDM-C3 have been applied to the mould and die industries. Pure copper belongs to the group of common metals. Some properties of copper and EDM-C3 are listed in Table 3.

EDM-C3 electrode is a graphite electrode infiltrated with copper. EDM-C3 has the quality of

Figure 1. Experimental setup.

Table 1. Main metal compositions of Inconel 625.

Element	Ni	Nb	Mo	Fe	Mn
Wt.%	Bal.	3.90	8.80	3.00	0.03
Element	Cr	Ti	C	Si	Al
Wt.%	21.2	0.25	0.02	0.16	0.15

Table 2. Physical and mechanical properties of Inconel 625.

Properties	Value
Density (g/cm³)	8.44
Melting range (°C)	1290–1350
Thermal conductivity (W/m·K)	12.1
Modulus of elasticity (GPa)	205
Tensile strength (MPa)	830
Yield strength(0.2% offset) (MPa)	410

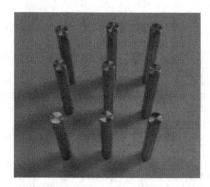

Figure 2. Copper electrodes.

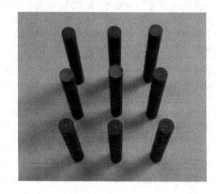

Figure 3. EDM-C3 electrodes.

Table 3. Properties of electrode materials.

Electrode materials	Copper	EDM-C3
Density (g/cm³)	8.9	3.05
Melting range (°C)	1065–1083	1100
Electrical resistivity/($\Omega \cdot$ m)	1.7×10^{-8}	3.2×10^{-6}
Thermal conductivity (W/m $\cdot$ K)	401	135

Table 4. Experimental parameters.

No.	Material	IP (A)	Ton (µs)	DC	SV (V)
1	Copper	24	40	3	45
2	Copper	24	50	4	60
3	Copper	24	60	5	70
4	Copper	36	40	4	70
5	Copper	36	50	5	45
6	Copper	36	60	3	60
7	Copper	48	40	5	60
8	Copper	48	50	3	70
9	Copper	48	60	4	45
10	EDM-C3	24	40	3	45
11	EDM-C3	24	50	4	60
12	EDM-C3	24	60	5	70
13	EDM-C3	36	40	4	70
14	EDM-C3	36	50	5	45
15	EDM-C3	36	60	3	60
16	EDM-C3	48	40	5	60
17	EDM-C3	48	50	3	70
18	EDM-C3	48	60	4	45

two materials. Compared with the copper material, the EDM-C3 material has lower density and better machinability. The material of copper in the open porosity increases thermal conductivity and lowers the electrical resistivity of graphite. Although EDM-C3 contains copper, the main component is graphite. Therefore EDM-C3 is classified as a kind of graphite. Graphite is widely used as a kind of electrode material for EDM. Graphite grades are often divided by the particle size. The average particle size of EDM-C3 is less than 5 µ.

In EDM, many electrical and non-electrical parameters can potentially affect the energy efficiency. Electrical parameters are about power supply unit. Non-electrical parameters include but not limited to polarity, work fluid, processing depth, processing area and nozzle flushing. Among the process parameters, electric parameters are important factors. Various electric parameters i.e. peak current (IP), pulse on time (Ton), Duty Cycle (DC) and Servo Voltage (SV) were selected from the range of acceptable values. A set of parameters were selected to evaluate the performance of different materials. In order to get a comprehensive evaluation, nine experiments were designed for each kind of electrode. Experimental parameters are listed in Table 4.

3 RESULTS AND DISCUSSION

EDM performance is evaluated by the primary performance parameters, i.e., MRR and TWR. In the present investigation, the influence of input parameters on response variables like MRR and TWR was examined. MRR is an important criterion determining the effectiveness of the machining operation. MRR is measured by reduction in volume to the machining time. Once the machining is done, the time is automatically recorded. MRR can be expressed as:

$$MRR = \frac{V}{t} \tag{1}$$

where,
MRR – Material removal rate (mm³/min),
V – Volume of material removed from workpiece (mm³)
T – Constant time (min).

TWR not only cares about the electrode wear, but also cares about the machining volume. TWR can be defined as "the ratio of volume of electrode wear to the volume of workpiece wear". TWR can be expressed as:

$$TWR = \frac{EWR}{MRR} \times 100\% \tag{2}$$

where,
TWR – Tool wear rate (%),
EWR – Volume of electrode wear (mm³/min).

The experimental results of copper and EDM-C3 are shown in Table 5. Although the same electrical parameters were used, i.e., the same discharge energy was released, the experimental results of the two kinds of electrodes are different.

Compared with copper electrodes, the MRR of graphite electrodes can be changed by

$$\lambda = \frac{V_{Gr} - V_{Cu}}{V_{Cu}} = \frac{\sum_{i=1}^{9} V_{Gri} - \sum_{i=10}^{18} V_{Cui}}{\sum_{i=10}^{18} V_{Cui}} \approx 18.0\% \tag{3}$$

where,
λ – Percentage change of MRR between EDM-C3 and copper,
V_{Gr} – Sum of MRR for EDM-C3,
V_{Cu} – Sum of MRR for copper,
V_{Gri} – MRR of the ith experiment for EDM-C3,
V_{Cui} – MRR of the ith experiment for copper.

It can be seen that, by using EDM-C3 electrodes, the material removal rate of maching Inconel 625 was increased by 18.0%.

Table 5. Experimental results.

No.	Material	MRR (mm³/min)	TWR (%)
1	Copper	5.2294	1.8528
2	Copper	15.7788	0.7496
3	Copper	18.6557	0.5789
4	Copper	25.0462	4.7585
5	Copper	17.0887	2.4768
6	Copper	14.2884	2.0950
7	Copper	30.9211	4.3361
8	Copper	24.0628	10.6402
9	Copper	14.1635	11.5493
10	EDM-C3	10.5816	5.2534
11	EDM-C3	18.5511	2.9066
12	EDM-C3	18.8019	1.4319
13	EDM-C3	23.5431	5.5781
14	EDM-C3	21.9795	2.0107
15	EDM-C3	22.8457	3.7248
16	EDM-C3	26.0276	4.3746
17	EDM-C3	32.0438	4.1713
18	EDM-C3	20.5919	2.9886

Compared with copper electrodes, the TWR of EDM-C3 electrodes can be changed by

$$\eta = \frac{\theta_{Gr} - \theta_{Cu}}{\theta_{Cu}} = \frac{\sum_{i=1}^{9} \theta_{Gri} - \sum_{i=10}^{18} \theta_{Cui}}{\sum_{i=10}^{18} \theta_{Cui}} \approx -16.2\% \quad (4)$$

where,

η – Percentage change of TWR between EDM-C3 and copper
θ_{Gr} – Sum of TWR for EDM-C3
θ_{Cu} – Sum of TWR for copper
θ_{Gri} – TWR of the ith experiment for EDM-C3
θ_{Cui} – TWR of the ith experiment for copper

It can be seen that, by using EDM-C3 electrodes, the tool wear rate of machining Inconel 625 was decreased by 16.2%.

From the experimental results, it can be found that EDM-C3 electrode has better performance in the case of MRR and TWR. Although EDM-C3 electrode has higher price, from the economic point of view, EDM-C3 is a better choice especially for complex structures. The increased productivity of EDM-C3 is an important factor for the overall cost reduction. Dozens of electrodes are usually needed to complete machining of a complex structure. Additional electrodes mean additional material cost. In addition, compared with graphite, copper has poor machinability. Due to the ductile characteristic of copper, it is often gummy. Fees and speeds of machining copper must be appropriate. It means that copper electrode needs additional machining cost. Moreover, replacing additional electrodes means more labour cost and more factory cost. So EDM-C3 is the preferred electrode material.

4 CONCLUSION

In EDM process, many factors can influence the processing effects. Electrode material is an important factor especially for superalloy. Two different materials including copper and EDM-C3 were selected. Different properties of the two materials were analysed. Experiments based on different electrical parameters were carried out. The results show that, by using EDM-C3 electrodes, the material removal rate was increased by 18.0% and the tool wear rate was decreased by 16.2%. It can be seen that EDM-C3 is the better material for electro-discharge machining of nickel-based superalloy.

ACKNOWLEDGMENTS

This work was financially supported by Beijing Natural Science Foundation (No. 3154033). The authors would also like to thank the anonymous reviewers whose comments greatly helped in making this paper better organized and more presentable.

REFERENCES

Abbas N.M., D.G. Solomon, M.F. Bahari, A review on current research trends in Electrical Discharge Machining (EDM), International Journal of Machine Tools and Manufacture. 47 (2007) 1214–1228.

Atzeni E., E. Bassoli, A. Gatto, et al., Surface and Sub Surface Evaluation in Coated-Wire Electrical Discharge Machining (WEDM) of INCONEL Alloy 718, Procedia CIRP. 33 (2015) 388–393.

Klocke F., D. Welling, A. Klink, et al., Evaluation of advanced wire-EDM capabilities for the manufacture of fir tree slots in Inconel 718, Procedia CIRP. 14 (2014) 430–435.

Li C.J., Y. Li, H. Tong, et al., Thinning process of recast layer in hole drilling and trimming by EDM, Procedia CIRP. 42 (2016) 575–579.

Oezkaya E., N. Beer, D. Biermann, Experimental studies and CFD simulation of the internal cooling conditions when drilling Inconel 718, International Journal of Machine Tools and Manufacture. 108 (2016) 52–65.

Rajurkar K.P., M.M. Sundaram, A.P. Malshe, Review of electrochemical and electrodischarge machining, Procedia CIRP. 6 (2013) 13–26.

Senthilkumaar J.S., P. Selvarani, R.M, Arunachalam, Intelligent optimization and selection of machining parameters in finish turning and facing of Inconel 718, International Journal of Advanced manufacturing and technology. 58 (2012) 885–894.

Yilbas B.S., H. Ali, N. Al-Aqeeli, et al., Laser treatment of Inconel 718 alloy and surface characteristics, Optics & Laser Technology. 78 (2016) 153–158.

Advances in Energy Science and Equipment Engineering II – Zhou, Patty & Chen (Eds)
© 2017 Taylor & Francis Group, London, ISBN 978-1-138-71798-5

Numerical analysis of rainfall equipment in the geotechnical centrifuge

Jiandong Li, Yuting Zhang & Wenbin Pei
Tianjin Research Institute for Water Transport Engineering, M.O.T, Tianjin, China
National Engineering Laboratory for Port Hydraulic Construction Technology, Tianjin, China

ABSTRACT: A CFD analysis software, Fluent, was used for numerical simulations to study the working performance of the rainfall equipment with different hole-distance rainmaking boards, which was designed to work in the geotechnical centrifuge. The results show that raindrops are well distributed around the centre line under the rainmaking holes when the hole-distance is 5 mm. The no-rain region between two rainmaking holes is much smaller when the hole-distance is 5 mm compared to when it is 10 mm. The results also show that the existence of the adjacent rainmaking holes has no appreciable effect on the working performance of the rainfall equipment when it is subjected to a high centrifugal force. This rainfall equipment also provides a feasible way to simulate uniform rainfall in a geotechnical model test.

1 INTRODUCTION

Soil slope is most likely to slide or collapse under the impact of strong and continuous rainfall (Ochiai, 2004; Moriwaki, 2004; Wu, 2004), and this phenomenon becomes more obvious as the amount of rainfall increases (Liu, 2006). There are two main research methods employed for studying slope failure: (i) post-disaster analysis and (ii) model tests. The reasons for damage and damage mechanisms can be gathered by field research after the slope failure through a post-disaster analysis, but this method is not replicable. But, model tests can be used to reproduce the failure process by adjusting different parameters of human-made slopes to understand the damage mechanism of slope failure. Compared to post-disaster analysis method, high-quality experimental data can be obtained with model tests by strictly controlling the experimental conditions (Zhou, 2000). So, the model tests are more advanced in studying the process and mechanism of slope failure.

Geotechnical centrifuge model tests have proved to be particularly valuable in revealing mechanisms of deformation and collapse and are widely applied in geotechnical engineering, especially in the study on deformations and failures of soil slope, because they can reproduce the weight stress field and the deformation process related to self-weight (Myoung, 1982; William, 1988; You, 2000). It is an ideal test to study slope stability under the conditions of rainfall in the geotechnical centrifuge by building a reduced scale version of the prototype according to the scaling laws. The stability of the slope correlates closely with rainfall, in order to understand the influence of rainfall on slopes. The rainfall equipment in the geotechnical centrifuge becomes indispensable in geotechnical centrifuge model tests. While simulating rainfall in a high centrifugal force field, the working performance of rainfall equipment will be affected by high gravity in the direction of centrifugal acceleration. So studying the working performance of the rainfall equipment in high gravity field gains significance.

In this paper, the distribution of rainfall using different rainmaking boards was studied with VOF method. The results can provide necessary basis for theories of analysis and computational method for developing and designing an artificial rainfall equipment that employs the geotechnical centrifuge.

2 SCALING LAW OF THE RAINFALL MODEL

The scaling law must ensure stress similarity between the model and the corresponding prototype. When studying the landslide induced by rainfall, rainfall infiltration is assumed to be the water-flow boundary condition, which is enforced on the slope surface. Therefore, model scale effects of rainfall should be consistent with rainfall infiltration (Zhang, 2007).

According to Darcy's law,

$$v = ki \tag{1}$$

where v is the superficial seepage velocity, k the permeability coefficient and i the hydraulic gradient.

In hydrodynamics, intrinsic permeability K is denoted by an equation taking into account shape, dimension and filler particle:

$$K = \frac{vk}{\rho g} \qquad (2)$$

where v is the fluid dynamic viscosity and ρ is the fluid density.

Permeability coefficient in Darcy's law is an equation concerning gravity acceleration when pore fluid in model is the same as in the prototype,

$$k_m = Nk_p. \qquad (3)$$

Hydraulic gradient is an unchanged parameter with different gravity accelerations:

$$i_m = i_p. \qquad (4)$$

Following this,

$$v_m = i_m k_m = i_p N k_p = N v_p. \qquad (5)$$

The seepage velocity in the model is N times that of the prototype. Consequently, the scale of rain intensity q is the same as seepage velocity:

$$q_m = N q_p. \qquad (6)$$

3 RAINFALL EQUIPMENT

A rainfall equipment has been designed to study its performance in a high gravity field produced by the geotechnical centrifuge. The exploded view of the rainfall equipment is shown in Figure 1.

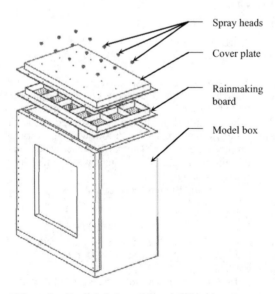

Spray heads

Cover plate

Rainmaking board

Model box

Figure 1. Exploded view of the rainfall equipment.

The model box of the rainfall equipment working in the geotechnical centrifuge is situated at the bottom of the basket. The rainmaking board is a multi-hole board installed on the model box. A cover plate is used to cover the rainmaking board. Raining cells are formed between the rainmaking board and the cover plate. Spray heads, which connect water supply system through water inlet pipes, are installed on the top of raining cells. There is an electromagnetism flow valve on each water inlet pipe. Electromagnetism flow valves can be used to control the water flow in every inlet pipe, so the water volume in every raining cell will not be influenced by the installation positions and lengths of the inlet pipes. This arrangement can overcome non-uniform distribution of water in a high centrifugal field. Water is well distributed in the raining cells by water atomization where water flows through the spray heads. Water drops fall into the model box through the multi-hole board to achieve a uniform rainfall effect. Different rain intensities can also be achieved by regulating the water inlet valves.

The rain intensity standards (from the light rain to the torrential rain) adopted by Chinese meteorological department are shown in Table 1.

In this paper, the working performance of the rainfall equipment in the geotechnical centrifuge with different hole-distances on the rainmaking boards was studied. Torrential rain (140 mm/12 h) was selected as the rainfall condition. The hole-diameter on the rainmaking board was 0.5 mm. To ensure uniform rainfall in the whole of the model box, the effective area of the rainmaking board was of the same size as the model box (1 m × 0.6 m) and the centrifugal acceleration was 100g. The parameters of two different rainmaking boards in the rainfall equipment are shown in Table 2.

Table 1. Rain intensity standards.

Types	Rainfall in 12 h (mm)	Rainfall in 12 h (mm)
Light rain	<5.0	<10.0
Moderate rain	5.0–14.9	10.0–24.9
Heavy rain	15.0–29.9	25.0–49.9
Rainstorm	30.0–69.9	50.0–99.9
Heavy rainstorm	70.0–139.9	100.0–249.9
Torrential rain	>140.0	>250.0

Table 2. Parameters of rainmaking boards.

Hole distance	5 mm	10 mm
Hole numbers	120 × 200	60 × 100
Flow velocity of the holes	2.471 m/min	9.885 m/min

1158

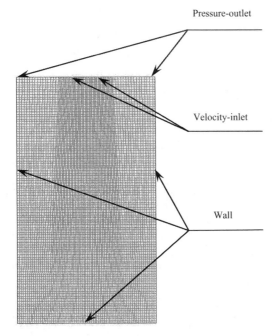

Pressure-outlet

Velocity-inlet

Wall

Figure 2.   Finite element model of the rainfall equipment.

4   FINITE ELEMENT MODEL

Numerical analysis of the rainfall equipment in the geotechnical centrifuge involving two kinds of dissolved flow, involving air and water, show that it is a gas–liquid two-phase flow. The Volume of Fluid (VOF) method was adopted for calculation in this study. To study the working performance of the geotechnical centrifuge rainfall equipment in high centrifugal force field, a rectangular region was selected as the calculation region. Boundary conditions of the finite element model are shown in Figure 2. The left, right and bottom face of the calculation region are considered as walls, and velocity-inlet on the top face was taken to simulate rainfall through rainmaking holes. The flow velocity of water inlet was set to the corresponding rain intensity. In order to balance the air pressure between the inside and outside of the rainfall equipment, pressure outlets were set at the two ends of the top face. Considering the computation accuracy and efficiency of the numerical simulation, grid distribution has equal intervals and local mesh refinement technique is applied under the rainmaking holes in the rainfall calculation region.

5   RESULTS AND DISCUSSIONS

To illustrate the working performance of the rainfall equipment in a high centrifugal force field,

two and three rainmaking holes in the calculation region were taken as examples. In this paper, by noting the water flowing into the calculation region through the rainmaking holes, the distribution of raindrops can be obtained.

5.1   Water drops down from rain making board

As shown in Figure 3, water flow through the spray heads is atomized in the raining cells and well distributed on the rainmaking board. Water drops falling from the small holes on the rainmaking board are converted into the small droplets. The droplets are accelerated to high speed in a very short time by application of a high centrifugal force field produced by the geotechnical centrifuge. In this process, droplets break up into smaller droplets and finally fall down on the reduced scale slope in the model box.

As shown in Figure 4, the droplet deforms because of the combined action of the self-weight, air resistance and the surface tension of the small droplet. The existence of velocity difference and the air-driven pressure changes the bottom of the round droplet to a flat one. The bottom of the droplet expands and the top of droplet becomes sharp. Finally, the droplet breaks up into smaller droplets in the edge parts when air-driven pressure breaks through the surface tension of the droplet.

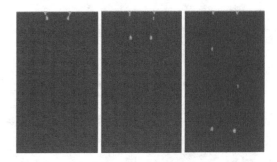

Figure 3.   Contour of rainfall in the rainfall equipment.

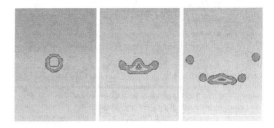

Figure 4.   Contour of droplet deformation and breakup.

5.2 Two rainmaking holes in the calculation region

As shown in Figures 5 and Figure 6, the black dots indicate the volume fraction of the phase 2 (water), which can be considered as the distribution of raindrops. Water drops fall into the model box through the rainmaking board and then break up into small raindrops. Raindrops have been well distributed around the centre line of the rainmaking holes, compared with the working performance of the rainfall equipment with different hole-distances of 5 and 10 mm. Raindrops distribute under the rainmaking holes. There is an apparent no-rain region between the two rainmaking holes when the hole-distance is 10 mm. When the hole-distance was reduced to 5 mm, the no-rain region between the two rainmaking holes becomes much smaller. From this arrangement, a well-distributed rainfall can be modelled.

5.3 Three rainmaking holes in the calculation region

In order to study the working performance of three rainmaking holes in the calculation region,

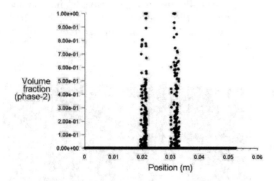

Figure 5. Volume fraction of water (2 rainmaking holes, hole-distance is 10 mm).

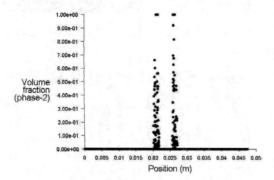

Figure 6. Volume fraction of water (2 rainmaking holes, hole-distance is 5 mm).

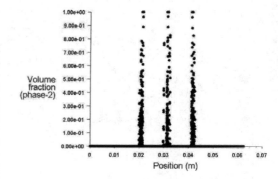

Figure 7. Volume fraction of water (3 rainmaking holes, hole-distance is 10 mm).

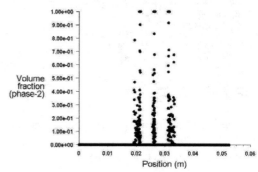

Figure 8. Volume fraction of water (3 rainmaking holes, hole-distance is 5 mm).

as shown in Figures 7 and 8, the working performance is the same like the two rainmaking holes in the calculation region. Raindrops are well distributed around the centre line of the rainmaking holes when the hole distance is 5 mm. The no-rain region between the two adjacent rainmaking holes is much smaller compared to the simulation when the hole-distance is 10 mm. It can therefore be concluded that the existence of adjacent rainmaking holes has little appreciable effect on the working performance of the rainfall equipment under the condition of high centrifugal force.

6 CONCLUSIONS

In this paper, numerical simulation was carried out to study the working performance of the rainfall equipment operating with the geotechnical centrifuge with different rainmaking boards. The conclusions are as follows.

Water drops fall into the model box through the multi-hole board and break up into smaller raindrops. The raindrops are well distributed around the centre line of the rainmaking holes.

This arrangement can overcome non-uniform distribution of water in a high centrifugal field. It is a feasible method for developing rainfall equipment in the geotechnical centrifuge.

The no-rain region under rainmaking holes will be much smaller when the hole-distance is smaller on the rainmaking board. It is a more beneficial way to simulate uniform rainfall.

Under the condition of high centrifugal force, the existence of the adjacent rainmaking holes has no appreciable effect on the working performance of the rainfall equipment.

REFERENCES

Liu Cuirong, Wang Huaguang, Influnce of the rainfall infiltration in the stability of soil slope [J], Railway engineering, 2006(2), 66–68. (In Chinese).

Moriwaki H, Inokuchi T, Hattanji T, et al. Failure process in a full-scale landslide experiment using a rainfall simulator [J]. Landslides, 2004, 1(4): 277–288.

Myoung Mo Kim, Hon Yim Ko. Centrifugal testing of soil slope models [J]. Tran. Res. Rec., 1982, 872: 7–14.

Ochiai H, Okada Y, Furuya G, et al. A fluidized landslide on a natural slope by artificial rainfall [J]. Landslides, 2004,1(3): 211–219.

William H. Craig. Centrifuge modeling for site-specific prototypes [C]// Centrifuge'88. Rotterdam: Balkema, 1988: 485–494.

Wu Jun-jie, Wang Cheng-hua, Li Guang-xin. Influence of matric suction in unsaturated soils on slope stability [J]. Rock and Soil Mechanics, 2004, 25(5): 732–736. (In Chinese).

You Xin-hua, Li Xiao. Current status and prospect of application of centrifugal model test to slope engineering [J]. J. Eng. Geology, 2000, 8(4): 442–445. (In Chinese).

Zhang Min, Wu Hongwei. Rainfall simulation techniques in centrifuge modeling of slope [J]. Rock and Soil Mechanics, 2007, 28 supp.: 53–57. (In Chinese).

Zhou Pei-hua, Zhang Xue-dong, Tang Ke-li. Rainfall installation of simulated soil erosion experiment hall of the state key laboratory of soil erosion and dryland farming on loess plateau [J]. Bulletin of Soil and Water Conservation, 2000, 20(4):27–30 (In Chinese).

Advances in Energy Science and Equipment Engineering II – Zhou, Patty & Chen (Eds)
© 2017 Taylor & Francis Group, London, ISBN 978-1-138-71798-5

Utilization of calcium carbide residue in developing unburned bricks

Qi Zhao, Jing Li & Hongtao Peng
College of Water Conservancy and Civil Engineering, China Agricultural University, Beijing, China

Bayaertu
Ordos Elion Desert Biomass Energy LLC, Elion Desert Economy Business Group, Inner Mongolia, China

Shuai Zhang, Xinping Zhang, Zhiyuan Yao, Qiongyu Liu & Jiahui Zhang
College of Water Conservancy and Civil Engineering, China Agricultural University, Beijing, China

ABSTRACT: Calcium Carbide Residue (CCR) is a by-product of the hydrolysis of calcium carbide and commonly regarded as a solid waste. Reuse of waste material for various applications is attracting great interest. The properties of CCR from Dalate County branch of Yili Energy Company Limited located at the Inner Mongolia Province in northern China have been studied in this paper. The mineral composition of CCR was determined by rotating anode X-ray polycrystalline diffractometry. The major minerals present in CCR are portlandite (95%), hydrocalumite (3%) and calcite (2%). In the presence of water and some additives (such as sodium bicarbonate, silicate sodium hydrogen), the reaction of SiO_2 and Al_2O_3, other raw materials (such as coal gangue, coal gasification furnace slag) and Ca^{2+} and OH^- in CCR mainly leads to the formation of hydrated calcium silicate and hydrated calcium aluminate in an unburned brick. In this reaction, CCR plays the role of an alkaline activator. A method to determine the CCR content for producing unburned bricks is recommended. A positive correlation was shown between compressive strength of the unburned brick and the content of CCR (20%–40%).

1 INTRODUCTION

As the PVC market continues to grow sharply, oil resources become very valuable. The cost of producing PVC by the petroleum ethylene process has increased dramatically. Therefore, carbide acetylene method may be preferred for the production of PVC. This method is recognized as a mature technology making use of abundant raw materials. The cost of production is also competitive. Therefore, the carbide acetylene method is a sustainable way to produce PVC in China (Gao, 2009). Other industrial processes, such as metal cutting, polyvinyl alcohol, lime nitrogen and dicyandiamide will also produce a large amount of CCR. As the output of domestic PVC through the carbide acetylene method and the demand for carbide further increase, it is predicted that more CCR will be produced in the next few years. As CCR stockpiling not only needs storage space, but also pollutes environment, some enterprises have enhanced the comprehensive utilization of CCR (Yan, 2007).

2 CCR USED IN THIS STUDY

The sample of CCR was collected from Dalate Branch of Inner Mongolia Elion Clean Energy Company Limited and is shown in Fig. 1.

The mineral composition of CCR and the X-ray quantitative analysis were conducted at Beijing North Yanyuan Micro Structure Test Center. The X-ray diffraction pattern is shown in Fig. 2. Mineral compositions of the CCR are shown in Table 1.

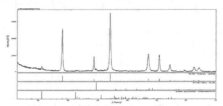

Figure 1. The sample of CCR.

Figure 2. X-ray diffraction result.

3 METHOD FOR PRODUCING UNBURNED BRICK

The method for producing unburned bricks is shown in Fig. 3. Brick is produced by mixing raw materials (such as fly ash, coal gangue), cement and CCR in a certain proportion. Water is added to the mixture and it is stirred well. Then the mixture is subjected to pressing molding and left for natural conservation or water steam curing (Hu, 2011).

Materials with high content of SiO_2 and Al_2O_3 are selected as the raw materials to produce unburned bricks. The analyzed chemical composition and their content are shown in Tables 2 and 3 based on the quantitative analysis of the materials.

By comparing the two materials, it is found that the content of SiO_2 is two or three times that of Al_2O_3 and they together constitute about 60 percent of coal gangue or coal gasification furnace slag, which meets the standards for making unburned bricks. After mixing and stirring the materials, the activity index of some materials (coal residue, coal gasification

furnace slag) and chemical reaction rate were found to be low. The additives such as sodium bicarbonate and silicate sodium hydrogen can help overcome this problem. Hydrated reaction requires a sufficient quantity of water. After compaction, the specimens were stored in a room to cure at room temperature ($20 \pm 2°C$) and relative humidity (95%) for 28 d. The additives (sodium bicarbonate, silicate sodium hydrogen) react with water, providing an alkaline environment for producing unburned bricks. SiO_2 and Al_2O_3 in the materials (coal residue, coal gasification furnace slag) react with Ca^{2+} and OH^- in CCR. This leads to the formation of SiO_4^{2-} and AlO_2^- as products. SiO_2 and Al_2O_3 in additive materials and $Ca(OH)_2$ in CCR undergo hydration reactions:

$$mCa(OH)_2 + SiO_2 + xH_2O = mCaO \cdot SiO_2 \cdot xH_2O$$
$$mCa(OH)_2 + Al_2O_3 + xH_2O = mCaO \cdot Al_2O_3 \cdot xH_2O$$

These reactions will produce hydrated calcium silicate and hydrated calcium aluminate. The hydration reaction takes place in two steps (dissolution

Table 1. Mineral composition proportion of the CCR.

Mineral	Quartz	Plaster	Calcium silite	Calcite	Dicalcium silicate	Portlandite	Hydrocalumite
Proportion (%)	/	/	/	2	/	95	3

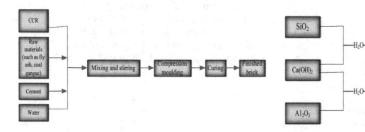

Figure 3. Method for producing unburned bricks.

Figure 4. The processes of reaction between different raw materials.

Table 2. Chemical compositions of coal gangue.

Chemical compositions	SiO_2	Al_2O_3	$FeCO_3$	$CaCO_3$	Fe_2O_3	CaO	MgO	K_2O
Content (%)	48.2	15.8	11.0	13.0	4.9	2.0	2.4	1.6

(The sample of coal gangue used for experiments is from Wulanmulun Site, Inner Mongolia, China.)

Table 3. Chemical compositions of coal gasification furnace slag.

Chemical compositions	SiO_2	Al_2O_3	Fe_2O_3	CaO	MgO	TiO_2	K_2O	Na_2O	P_2O_5	MnO
Content (%)	41.1	17.2	8.57	6.48	1.23	0.79	0.75	0.40	0.38	0.10

(The sample of coal gasification furnace slag used for experiments is from Inner Mongolia, China.)

of SiO_2 and Al_2O_3 and hydration leading to crystallization) (Wang, 2005). The content of $Ca(OH)_2$ in CCR is high. So CCR will provide alkaline excitation to some extent. The reaction processes of raw materials are shown in Fig. 4.

4   ANALYSIS OF CCR CONTENT

In the production of unburned bricks, the amount of CCR to be added should be determined. $Ca(OH)_2$ is produced by the crystallization of CCR. A part of $Ca(OH)_2$ and CO_2 will react to form $CaCO_3$. In addition, $Ca(OH)_2$ is slightly soluble in water and produces a small amount of OH^- radicals. The Si–O and Al–O bonds in the raw material crack and form some free unsaturated active bonds that can react with $Ca(OH)_2$ to produce a small amount of calcium silicate hydrate and calcium aluminate hydrate. CCR and cement rapidly react with water and produce a variety of hydration products. Also, CCR effectively stimulates the activity of some other raw materials. Therefore, the amount of CCR should not be too small. However, if the amount of CCR is too high, there will be a residual amount of CCR left without reaction. As a result, it may affect the strength of the brick. Thus, the content of CCR should not be too high (Wang, 2003). In order to ensure the right amount of CCR, the following reactions should be given attention:

$$SiO_2 + xCa(OH)_2 + mH_2O = xCaO \cdot SiO_2 \cdot nH_2O$$
$$Al_2O_3 + yCa(OH)_2 + mH_2O = yCaO \cdot Al_2O_3 \cdot nH_2O$$

When the $x = y = 1$, the reaction formulation shows that 1 kg SiO_2 and 1 kg Al_2O_3 matches 1.23 kg and 0.72 kg $Ca(OH)_2$ (Meng, 2006). In the production of unburned bricks, if we ignore the content of SiO_2 and Al_2O_3 in the CCR and cement, and assume that the content of SiO_2 is a, the degree of activity is m, content of Al_2O_3 is b, the degree of activity is n in the raw materials (for example, coal

gangue, coal gasification furnace slag), the content of $Ca(OH)_2$ in the CCR is c and cement hydration produces $Ca(OH)_2$ whose content is t. Theoretically, the amount of CCR should be S times the raw material (such as coal gangue, coal gasification furnace slag). S is determined from the following formula:

$$S = (1.23am + 0.72bn - t)/c \qquad (1)$$

5   RESULTS OF THE ANALYSIS

At present, there has been a plenty of research done in China on CCR unfired bricks. Analysis of these research results shows that the main raw materials for producing unfired bricks are fly ash and CCR. Slag, sand and cement are used as an admixture. The research results of unfired bricks are shown in Table 4. The table shows that the compressive strength of unfired brick can fully meet the quality requirements of the highest grade (MU10-grade) of fly ash brick in the Chinese standard (JC239-91), and the compressive strength even reached a value of 20.1 MPa.

By analyzing the above data, it is possible to draw a relationship between the compressive strength of the unburned bricks and CCR unburned bricks, as shown in Fig. 5. A positive correlation was shown

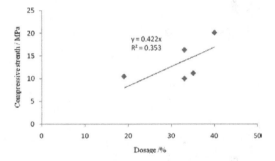

Figure 5.   The relationship between the compressive strength and the dosage of CCR.

Table 4.   The research results contrast of unfired bricks.

		(Hu et al. 2011)[7]	(W. 2003) [8]	(M. 2011) [9]	(W et al. 2010)[10]	(F. 2011) [11]
Raw material	CCR (Wt.%)	33	19	33	40	30–40
	Fly ash (Wt.%)	40	/	34	30	40–45
	Coal cinder (Wt.%)	/	71	/	/	20–30
	Cement (Wt.%)	25	/	/	/	6–8
	Sand (Wt.%)	/	/	33	/	/
	Admixtures (Wt.%)	2	2	/	10	/
Process	Curing condition	Nature	Nature	180°C	Nature	Steam
Performance of unfired brick	Compressive strength (MPa)	16.3	10.5	≥10	20.1	11.2
	Bending strength (MPa)	1.36	3	/	6.1	/

1165

between compressive strength of the unburned brick and the content of CCR (20%–40%).

6 CONCLUSION

1. After assessing the chemical composition of the raw materials used in the production of unburned bricks, formula (1) can be used appropriately to control the amount of added CCR.
2. The analysis of mineral composition of CCR from Dalate County branch of Yili Energy Company Limited located at Inner Mongolia Province in northern China indicated that the CCR can be used for producing unburned bricks.
3. A positive correlation was found between the compressive strength of the unburned brick and the content of CCR (20%–40%).

ACKNOWLEDGMENTS

This work was supported by the National Undergraduate Training Programs for Innovation and Entrepreneurship in China (No. 15056150).

REFERENCES

Feng, Z.G. Improved semi dry process for carbide slag brick. Science & Technology Information, 73(2011).

Gao, M. & Wen, B.M. Study on raw material properties of calcium carbide slag cement. Cement Guide for New Epoch, 1–2 (2009).

Hu, W.X. & Zhang, T.G. Composition and mechanical properties of carbide slag and fly ash brick of new environmental protection. New Chemical Materials, 85 (2011).

Meng, C.C. Discussion on the feasibility of using fly ash and carbide slag. Journal of Nantong Textile Vocational Technology College, 8–9(2011).

Wang, M. Brickmaking with carbide slag and cinder. New Wall Materials, 22(2003).

Wang, W.B. Experimental study on the activity of fly ash and high content of fly ashsilica. Northwestern Polytechnical University, 21(2005).

Wang, Y.B. & Zhu, S.M. Study on Preparation and properties of unburned carbide slag and fly ash brick. New Building Materials, 19(2010).

Yan, X.H. & Li, S.Y. Comprehensive utilization of carbide slag to produce block. Polyvinyl Chloride, 44 (2007).

Advances in Energy Science and Equipment Engineering II – Zhou, Patty & Chen (Eds)
© 2017 Taylor & Francis Group, London, ISBN 978-1-138-71798-5

Experimental study on a concrete-filled round-ended steel tube under eccentric compression

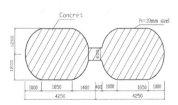

Dan Zhou
Changjiang Survey, Planning, Design and Research Limited Company, Wuhan, China

Erlei Wang
Design and Research Institute of Wuhan University of Technology, Wuhan, China

ABSTRACT: Using the round-ended Concrete-Filled Steel Tubular (CFST) tower of Houhu bridge in Wuhan as the reference, large-sized specimens were designed and round-ended CFST columns were subjected to eccentric loads for carrying out experiments. The failure mechanisms, load–displacement relations, load–longitudinal strain relations and load–circumferential strain relations were analysed. Experimental results indicate that round-ended CFST has a high loading capacity under eccentric loads, and eccentricity is the main factor affecting the mechanical performance of round-ended CFST. As eccentricity increases, bearing capacity is reduced. Section deformation coupling constitutive models of concrete show that with increasing eccentricity, the axis position continues to move to the load side. The compression zone of the steel pipe enters a plastic state, but tension zone is still in the elastic state in the failure stage.

1 INTRODUCTION

At the present stage, research on a special-shaped concrete-filled steel tube mainly focuses on T-shaped or L-shaped sections, and there is no report on a concrete-filled round-ended steel tube as a compression member prior to the construction of Houhu Bridge, which employed a concrete-filled round-ended steel tube for its pylon. Using the engineering practice of Houhu Bridge as the reference, the research team conducted stress measurement, theoretical analysis of finite element model and related research on concrete-filled round-ended steel tube pylon (Meng, 2008; Li, 2008; Feng, 2008; Xie, 2011). Although various studies provide good guidance for engineering practices, the research on the concrete-filled steel tube members under eccentric compression is not deep enough (Mohammod, 1997; N, 2001; Zhong, 1997), especially on the round-ended section concrete-filled steel tubes. In this paper, an experimental study was carried out on a concrete-filled round-ended steel tube under eccentric compression.

2 PROJECT PROFILE

Houhu Bridge is located in the Houhu section beyond the extension line of Xiaotian Road, Chuanlong Main Road in Wuhan, with a bridge span arrangement as follows: 12 × 20 m (pre-stressed hollow concrete slab) + (34 + 56 + 128 m) (prestressed concrete cable-stayed bridge with single pylon and cable plane) + 17 × 20 m (prestressed hollow concrete slab) = 798 m. The main bridge is a pre-stressed concrete winged, box-girder cable-stayed bridge with a single pylon and one cable plane, employing the fixed pier–pylon–beam system, in which a round-ended section concrete-filled steel tube is used for its pylon (Meng, 2008; Li, 2008). The pylon section of the bridge is shown in Fig. 1. The pylon has a vertical spacing of 150 cm and is connected by eight connecting rods of a concrete-filled rectangular-section steel tube. The connecting rod has a length of 80 cm along the bridge, a length of 60 cm across the bridge and a height of 50 cm. The steel tubes were welded on-site using Q235 steel plates of a thickness of 20 mm,

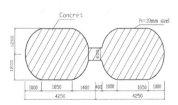

Figure 1. Cross-section of round-ended CFST power (mm).

and for the concrete core, C50 self-compacting micro-expansive concrete was used.

3 EXPERIMENTAL CONTENT

3.1 Member design

In order to avoid too many defects in the fabricating process of the members due to smaller size as well as accidental factors, which might influence the accuracy of experimental data, the specimens were fabricated in accordance with the engineering practice ratio of 1:10. As the connecting rod of the concrete-filled steel tube requires a complicated fabricating procedure, which may easily cause initial defects, making it difficult to reflect the genuine stress mechanism of the members, single-limb concrete-filled steel tubes were chosen for fabrication in accordance with the symmetry principle.

In accordance with the *Technical Specification for Concrete-Filled Steel Tubular Structures* (CECS28:90), for concrete-filled steel tube members, the degree of strength of concrete core should not be lower than C30 or the tube wall thickness should not be less than 4 mm. Restricted by the tonnage of loading-test machines, C30 concrete was used and 4 mm steel plate was chosen to weld into the round-ended steel tube in the members in this work. A set of specimens was fabricated by employing eccentric compression; the specimen section is presented in Fig. 2 and the specific parameters are given in Table 1.

3.2 Mechanical properties of the materials

Tensile tests were conducted on steel plate standard specimens, and the method employed follows the *Metallic Materials-Tensile Testing at Ambient*

Table 1. Specimen data.

Specimen	fcu,k	Steel ration $\rho\%$	L/D	$\dfrac{f_s A_s}{f_c A_c}$	Eccentricity	Eccentricity ratio
P-1	C30	4	4.8	0.66	9	0.46
P-2	C30	4	4.8	0.66	13	0.68
P-3	C30	4	4.8	0.66	17	0.88

Temperature (GB/T228-2002) standard. The steel plate thickness was 3.8 mm, yield strength calculated as $f_y = 317 \ N/mm^2$ and elastic modulus was $E_a = 2.07 \times 10^5 \ N/mm^2$.

In accordance with the requirements in the *Standard for Test Method of Mechanical Properties on Ordinary Concrete* (GB/T 50081-2002), the C30 concrete ratio of cement: sand: aggregate: water was set at 1: 1.12: 1.89: 0.34. The sand used was in the particle size range of 5–25 mm, and 32.5 ordinary Portland cement was used. By use of a test coupon formed under the same conditions for 28 days of curing, the compressive strength of the concrete cube was measured as 34.8 N/mm^2.

3.3 Specimen fabrication

Concrete-filled round-ended steel tubes were fabricated by welding steel plate section bars. The steel plates were first cut into pieces according to the length specified in the design and joined together by welding to guarantee quality. Square steel plates were welded at the bottom of the specimen as roof and floor, which were centred to the geometric centre of the steel tubes. The edges of the roof and floor exceeded the outer diameter of the specimens by 20 mm and a round hole with a diameter of 100 mm was made in the centre of the roof to provide an opening for pouring concrete. While pouring concrete, a vibrating tube with a diameter of 5 cm was inserted for jolt ramming. The thickness of the concrete layer filled each time was about 30–40 cm and the specimens were left for natural curing outdoors.

3.4 Measuring point arrangement and measuring method

The experiment was conducted on the 500-ton universal compression testing machine in the Engineering Structure Test Center at Wuhan University of Technology, by applying different load eccentricities and transmission of load through blade hinges set on both ends of the specimens. Considering the actual load-bearing capacity of the engineering structure and the feasibility of the experiment, eccentricity was controlled within

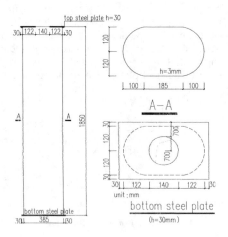

Figure 2. Cross-section of specimen.

the section. Specific data and eccentricity of the specimens are given in Table 1. The loading device used for eccentric compression column experiment is as shown in Fig. 3 and the strain measuring points are shown in Fig. 4. The measuring point displacement and strain were determined using the DH3815 static strain measurement system. An electronic dial gauge was used to measure the lateral deflection at the middle of the member along its length, and the resistor disc was used to measure the longitudinal strain and circumferential strain on surface of the steel tube.

The monotonic loading classification was adopted in the experiment. At the initial stage, 1/12 to 1/15 of the ultimate load was added for 4 min at each level. When the load exceeded 50% of the ultimate load, it was reduced to be 1/20 to 1/25 of the ultimate load at each level. After the load reached 0.8 times of the ultimate load, more intricate loading classification was applied. When the load approximated the ultimate load, slow continuous loading was employed until the specimen deformed too much to read the data and then the experiment was stopped. Experiment time for each specimen was approximately 1–1.5 h.

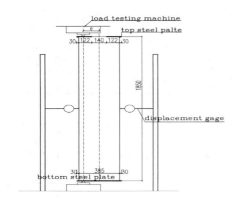

Figure 3. Test setup.

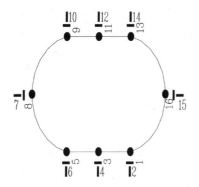

Figure 4. Plain views of strain gauge displacement meter layout.

4 RESULTS AND ANALYSIS OF THE ECCENTRIC COMPRESSION EXPERIMENT

4.1 *Specimen failure modes*

During the loading process, as the load increased, bulking occurred on the steel tube wall at the midspan of the section; the steel tube weld further cracked at the round shape and at the rectangular interface, presenting the characteristics of strength failure. When failure occurred to specimen P-1, concrete crushed out, indicating that the member material had reached its ultimate strength under the condition of low eccentric compression. When failure occurred to specimens P-2 and P-3, steel tube weld cracked slightly, the mid-span deflection was relatively large and the steel tube bulking was not significant. With an increase in eccentricity, the mid-span deflection exhibited a non-linear growth trend. When the rate of pumping of loading-test machine failed to follow the rate of strain, the strain gauge was unable to record the readings as the resistance wire was ruptured.

4.2 *Experimental results*

Fig. 5 shows the load–transverse displacement curve under the load effects of different eccentricities. The load–transverse displacement curve experienced three stages of linear ascending, curvilinear ascending and descending, the proportion of which in the loading process changed according to the differences in eccentricity. With an increase in eccentricity, the bearing capacity of the specimens was reduced. Under the working condition of a relatively low eccentricity, the elastic working stage was the longest (approximately 75% of the ultimate load), with a correspondingly steeper load–transverse displacement curve, weaker deformation ability, higher bearing capacity of the members,

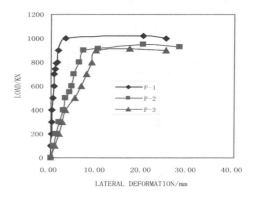

Figure 5. Load versus lateral deformation curves of specimens.

a longer course of non-linear development, but a relatively short horizontal course of development. Under the working condition of a relatively high eccentricity, the elastic working stage was relatively short (approximately 50% of the ultimate load).

Fig. 6 shows the longitudinal strain curve of different loading stages at the middle section of the specimens (x-axis coincides with the long axis of the round-ended section, with the coordinate centre at the section centroid). The x-axis is the horizontal axis and the vertical axis represents longitudinal strain value. The analysis showed that the middle section remained as a plane throughout different loading stages, while the position of neutral axis moved towards the direction where the load was applied with an increase in load.

Fig. 7 displays the relationship curve between the measured load N at different measuring points with different eccentricities on the members under eccentric compression and the steel tube longitudinal strain ε_L in the middle section. The numerical value of ε_L was negative in the compressive zone and positive in the tensile zone. It is observed that the longitudinal strain presented a gradual increase in the compressive zone before the yielding of steel; after the yielding, there was an increase in strain rate of steel. When the experiment was stopped, the side of the steel tube at the middle section

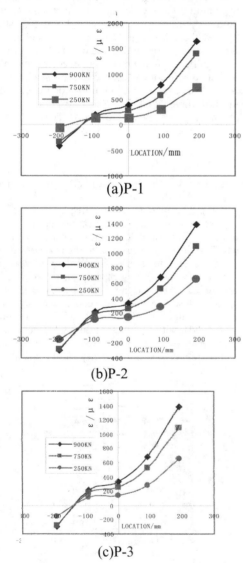

(a)P-1

(b)P-2

(c)P-3

Figure 6. Longitudinal strain of specimens in center section.

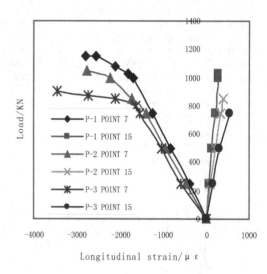

Figure 7. Relation of load versus longitudinal strain.

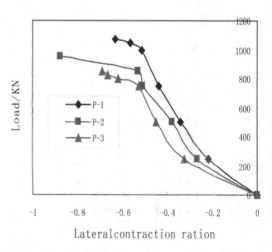

Figure 8. Relation of load versus coefficient of lateral deformation.

closer to the eccentricity yielded under compression, while the side of the steel tube farther to the eccentricity was within the elastic stage of tension and therefore was yet to yield.

Fig. 8 is the relationship curve of load–Poisson ratio of concrete-filled round-ended steel tube members under eccentric compression. The analysis of the figure leads to the conclusion that the Poisson ratio before the steel yielding was negative and remained basically unchanged on the side under larger compression in the compressive zone; after the steel tube yielded, the absolute value of Poisson ratio increased rapidly over 0.5, indicating that the steel tube presented significant confinement effect on the concrete.

5 CONCLUSIONS

The experiment indicates that concrete-filled round-ended steel tubular eccentric compression column has a relatively good elastic–plastic working capability and ductile capability. The deformation of the middle section conforms to the hypothesis of plane section, and the position of neutral axis continuously moves to the side with load effect with the increase of eccentricity. Eccentricity is a major factor affecting the mechanical behaviour of round-ended steel tubular eccentric compression column: with an increase in eccentricity, the bearing capacity of the specimens continuously decreased. Under the working condition of a relatively low eccentricity, the elastic working stage was the longest with relatively weaker deformation ability; under the working condition of a relatively high eccentricity, the elastic working stage of the load–transverse

displacement curve was relatively shorter, with the bearing capacity of the specimen reduced drastically. When the load reached the ultimate stage, the edges of the compressive zone of steel tubes with weld cracked at the interface of the circular arc segment and all straight-line segments reached the yielding stage, while the other side of the steel tube was still within the elastic stage of tension.

REFERENCES

Bin Li. Study on the Interaction between Steel Tube and Concrete of Round-ended Concrete-Filled Steel Tube Tower Column [D]. Wuhan: Master's thesis of Wuhan university of technology, 2008.

Jianxiong Xie. Key Techniques Study on Design and Construction for Large Size Coupled Round-ended Steel Tube-Filled Concrete Cable Stayed Bridge [D]. Wuhan: Doctor's dissertation of Wuhan university of technology, 2011.

Meng Chen. Analysis on Mechanical Behaviors of concrete-filled Steel Tubular Tower of Houhu Cable—stayed Bridge [D]. Wuhan: Master's thesis of Wuhan university of technology, 2008.

Mohammod Shames abd M Ala Saadeghvaziri. State of the art of concrete-filled steel tubular columns [J]. ACI Structural Journal, 1997, 94–S51:558–571.

Shanmugan N E, Lakshmi B. State of the art report on steel-concrete composite columns [J]. Journal of Constructional Steel Research, 2001, 57:1041–1080.

Shantong Zhong. Concrete Filled Steel Tube (revised edition) [M]. Harbin: Heilongjiang science & technology Press, 1994.

Taohua Feng. Study on stress of node of concrete Filled Steel Tubular of Houhu Cable-stayed Bridge [D]. Wuhan: Master's thesis of Wuhan university of technology, 2008.

Advances in Energy Science and Equipment Engineering II – Zhou, Patty & Chen (Eds)
© 2017 Taylor & Francis Group, London, ISBN 978-1-138-71798-5

Adsorption kinetics of an aliphatic-based superplasticizer in CFBC ash–Portland cement paste

Jingxiang Liu & Litao Ma
Anhui Singular Environmental Protection Co. Ltd., Suzhou, Anhui, China

Jingyong Yan & Yuyan Wang
School of Environmental and Material Engineering, Yantai University, Yantai, China

Xin Wang
Construction Engineering Quality Inspection Center of Suining County, Xuzhou, Jiangsu, China

ABSTRACT: Because of their high pozzolanic activity, Circulating Fluidized Bed Combustion (CFBC) ashes can be potentially used as supplementary cementitious materials for concrete production. This paper investigates the adsorption kinetics of an aliphatic-based superplasticizer when it is mixed with CFBC ash–cement pastes. UV-visible absorption spectroscopy was used to evaluate the adsorption kinetics. The results show that CFBC ash–cement pastes possess higher adsorption capacity than the Pulverized Coal Combustion (PCC) fly ash–cement pastes. In addition, CFBC ash–cement pastes exhibit lower workability with higher slump loss. It is concluded that the high adsorption of the superplasticizer diminishes water-reducing efficiency.

1 INTRODUCTION

Circulating Fluidized Bed Combustion (CFBC) ashes are produced by the combustion of coal through an injection of limestone or dolomite for sulfur capture. As the annual production of CFBC ashes in China is 20 million tons, it is urgent to identify potential reutilization of an otherwise landfill material. Because of the rich content of active SiO_2 and Al_2O_3, CFBC ashes have been reported to exhibit good pozzolanic activity (Behr-Andres, 1994; Zheng, 2009; Chindaprasirt, 2010; Xu, 2011; Li, 2012) and recognized as a potential supplementary cementitious material to partially replace cement in the production of concrete.

CFBC ashes are produced at a much lower temperature (850–900°C) than Pulverized Coal Combustion (PCC) fly ashes (1200–1400°C), a widely used supplementary cementitious material in concrete. The physical properties as well as the chemical composition of CFBC ashes are therefore distinct from those of PCC fly ashes. For example, the content of unburnt carbon in CFBC ashes is higher than that in PCC fly ashes. Further, the shape of CFBC particles is irregular, which gives it a loose and porous surface structure (Zheng, 2009; Qian, 2008).

The adsorption effect of cement or PCC fly ashes on water-reducing agents has been studied extensively (Singh, 1992; Yoshioka, 2002; Termkhajornkit, 2004). To the best of our knowledge; however, no study has been done on the efficiency of reduction of water in CFBC ash–cement system. There is a need to study the adsorption of water-reducing admixtures in the CFBC ash–cement paste. This paper investigates the effects of CFBC ashes on the adsorption of an aliphatic-based superplasticizer in the ash–cement pastes.

2 EXPERIMENTAL PROGRAM

2.1 *Materials*

The chemical composition of CFBC fly and bottom ashes, PCC fly ashes and ordinary Portland (PO42.5) cement clinker were evaluated by XRF and are summarized in Table 1. The fly ashes were tested as-received, and the bottom ashes were milled so that 97% of the bottom ash particles pass through the 80 μm sieve. PO42.5 cement clinker was ground by a small laboratory grinding machine for 2 h to ensure 97.5% of the particles pass through the 80 μm sieve. Then, 5% of gypsum was added to the cement to control the setting. The initial and

Table 1. Chemical composition (by mass) of coal ashes and Portland cement clinker.

Sample	SiO_2	Fe_2O_3	Al_2O_3	CaO	MgO	Na_2O	K_2O	SO_3	Free lime	LOI	Sum
CFBC fly ashes	37.54	5.84	23.10	10.52	1.29	1.17	0.55	4.80	3.03	13.24	98.05
CFBC bottom ashes	56.08	4.91	24.28	3.62	1.11	1.97	0.79	1.48	0.93	4.87	99.05
PCC fly ashes	53.91	4.12	28.81	4.83	2.68	1.20	0.44	0.98	0.95	2.03	99.00
Cement clinker	20.70	3.56	4.94	62.49	3.38	0.11	0.82	0.48	0.82	1.41	97.89

Table 2. Physical characteristics of coal ashes.

Sample	Water demand (%)	Specific surface area (m^2/kg)	Average particle size (µm)
CFBC fly ashes	183	391.4	21.40
CFBC bottom ashes	172	297.6	27.59
PCC fly ashes	95	402.5	20.33

final settings of the resulting cement were measured at 135 and 252 minutes, respectively. The aliphatic-based superplasticizer with a solid content of 27.39% was used in this study.

As shown in Table 1, Loss on Ignition (LOI) of CFBC ashes is higher than that of PCC fly ashes and cement clinker. The higher LOI is mainly a result of a high amount of unburnt carbon and some residual calcite left after the sulfur capture process (Anthony, 2001) in CFBC ashes.

Table 2 summarizes the physical properties of the CFBC ashes and the PCC fly ashes. As can be seen, the water demand of CFBC ashes is nearly twice that of PCC fly ashes. This may be attributed to the high unburnt carbon content in the CFBC ashes and its porous surface structure. The CFBC bottom ashes have a larger particle size with lower specific surface area as compared to the CFBC and the PCC fly ashes.

2.2 Tests

2.2.1 Adsorption kinetics of the aliphatic superplasticizer in CFBC ash–cement pastes

To understand the adsorption kinetics of the aliphatic superplasticizer in CFBC ash–cement pastes, three ash–cement pastes incorporating different ashes were prepared. The pastes were mixed in a proportion of 70: 30: 70: 1.2 (cement: ash: water: superplasticizer) and the ashes investigated were

PCC fly ash, CFBC fly ash, and CFBC bottom ash. PCC fly ashes were used as the control in this study.

A 5 liter bench mixer was used to prepare the mix. Cement and ash were first dry-mixed for 5 minutes. Water and superplasticizer were then added to the powder and mixed continuously for another 150 minutes. Samplings of the paste were done at 20, 45, 90, 120 and 150 minutes. Each paste sample was then processed by the TD5A-WS low speed tabletop centrifuge to separate free unadsorbed superplasticizer solids suspended in the paste. The supernatant was collected and the residual concentration of the superplasticizer was measured by the UV-visible spectrophotometer.

The adsorption of the superplasticizer in ash–cement pastes can be calculated based on the following equation.

$$\Gamma = (C_0 - C) \cdot v/1000 \, W \tag{1}$$

where Γ is the adsorption amount, mg/g; C_0 is the initial concentration of the superplasticizer, mg /L; C is the residual concentration of the superplasticizer, mg /L; v is the volume of the superplasticizer solution, mL; and W is the quality of the mix of coal ashes and cement, g.

2.2.2 Workability of CFBC ash–cement pastes with aliphatic superplasticizer

To evaluate the viability of the CFBC ash–cement paste, three ash–cement pastes incorporating different ashes were prepared. The pastes were mixed in the proportion 70: 30: 35: 1.2 (cement: ash: water: superplasticizer) and the ashes investigated were PCC fly ash (control), CFBC fly ash, and CFBC bottom ash.

The cement and ash powder were first mixed for 5 minutes followed by the addition of water and superplasticizer, which was then mixed for another 3 minutes. Mini slump flow test was performed at different time intervals up to 90 minutes to evaluate the loss of slump flow of each mix. The test follows the Chinese standard GB/T8077-2000 on methods for testing uniformity of concrete admixture.

3 RESULTS AND DISCUSSION

3.1 *Adsorption kinetics of aliphatic superplasticizer in CFBC ash-cement pastes*

Figure 1 shows the adsorption amount of the aliphatic superplasticizer in coal ash–cement pastes as a function of time t. It can be seen that the adsorption of the aliphatic superplasticizer in the CFBC ash–cement paste is greater than that in the PCC fly ash–cement paste. The general trend shows that the amount of adsorption increases with time. The adsorption rate (i.e. the slope of the curve), however, reduces with time. No significant change was observed for 120 to 150 minutes. This may be attributed to the initial setting of the pastes, which reduces the mobility of the superplasticizer. It is concluded the adsorption of the aliphatic superplasticizer in coal ash–cement pastes can be reasonably captured and described within the first 150 minutes.

3.2 *Workability of CFBC ash–cement paste with aliphatic superplasticizer*

The measured slump flow of the three ash–cement pastes at the pre-determined time interval up to 90 minutes was plotted against time as shown in Fig. 2. As can be seen, the slump flow of PCC fly ash–cement paste is higher than that of the CFBC ash–cement paste. The loss of slump flow is greater in the CFBC ash–cement as compared to that in the PCC fly ash–cement paste.

It has been reported that the slump flow of cement paste is proportional to the adsorption amount of water-reducing agents on the surface of cement (Li, 2011). In the present study, CFBC ashes show higher adsorption ability than the PCC fly ashes. The measured slump flow of CFBC ash–cement paste, however, is lower than that of the PCC fly ash–cement paste. This may be attributed

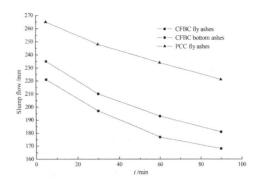

Figure 2. Loss of slump flow in ash–cement pastes.

to the high water demand of CFBC ashes as shown in Table 2, which reduces the amount of available free water in the paste, resulting in lower slump flow. In addition, a high amount of unburnt carbon in CFBC ashes may adsorb and trap the aliphatic superplasticizer, which reduces the adsorption of the superplasticizer on ash and cement particles. In addition, the porous and loose surface of CFBC ashes may adsorb some amount of superplasticizer on the internal surface, which diminishes the efficiency of water reducer.

4 CONCLUSIONS

This paper investigated the effect of CFBC ashes on the adsorption kinetics of the aliphatic-based superplasticizer in the ash–cement paste.

CFBC ash–cement pastes possess higher adsorption ability than PCC fly ash–cement pastes. CFBC ash–cement pastes show lower slump flow and higher loss of slump flow as compared to the PCC fly ash–cement paste.

PCC fly ash is generally understood as a supplementary cementitious material that improves concrete workability. When CFBC ashes are used as a supplementary cementitious material, however, potential loss on workability and water-reducing efficiency need to be considered in the mix design.

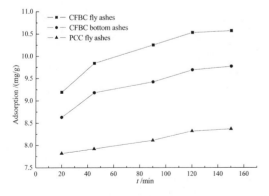

Figure 1. Adsorption of the aliphatic superplasticizer versus time in coal ash–cement pastes.

REFERENCES

Anthony EJ, Granatstein DL. Sulfation phenomena in fluidized bed combustion systems. Prog Energ Combust Sci 2001; 27(2): 215–236.

Behr-Andres CB, Hutzler NJ. Characterization and use of fluidized-bed-combustion coal ash. Journal of Environmental Engineering 1994; 120(6):1488–1506.

Chindaprasirt P, Rattanasak U. Utilization of blended fluidized bed combustion (FBC) ash and pulverized coal combustion (PCC) fly ash in geopolymer. Waste Manage 2010; 30(4): 667–672.

Li Shun, Yu Qijun, Wei Jiangxiong, et al. Effects of molecular mass and its distribution on adsorption behavior of polycarboxylate water reducers. J Chin Ceram Soc 2011; 39(1): 80–86.

Li Xiangguo, Chen Quanbin, Huang Kuaizhong, et al. Cementitious properties and hydration mechanism of circulating fluidized bed combustion (CFBC) desulfurization ashes. Constr Build Mater 2012; 36: 182–187.

Qian Jueshi, Zheng Hongwei, Song Yuanming, et al. Special properties of fly ash and slag of fluidized bed coal combustion. J Chin Ceram Soc 2008; 36(10): 1396–1400.

Singh NB, Dwivedi MP, Singh NP. Effect of superplasticizer on the hydration of a mixture of white portland cement and fly ash. Cem Concr Res 1992; 22(1): 121–128.

Termkhajornkit P, Nawa T. The fluidity of fly ash–cement paste containing naphthalene sulfonate superplasticizer. Cem Concr Res 2004; 34(6): 1017–1024.

Xu Huizhong, Song Yuanming, Liu Jingxiang, et al. Method for Determining the Hydration Reaction Kinetics of Coal Ashes. J build mater 2011;14(4): 564–568.

Yoshioka K, Tazawa E, Kawai K, et al. Adsorption characteristics of superplasticizers on cement component minerals. Cem Concr Res 2002; 32(10): 1507–1513.

Zheng Hongwei, Wang Zhijuan, Qian Jueshi, et al. Pozzolanic reaction kinetics of coal ashes. J Wuhan Univ Technol 2009; 24(3): 488–493.

Advances in Energy Science and Equipment Engineering II – Zhou, Patty & Chen (Eds)
© 2017 Taylor & Francis Group, London, ISBN 978-1-138-71798-5

Double-mixture concrete mix design in one hydropower station and the performance test

Hongfen Nian, Qian Zhang & Lixia He
Department of Civil Engineering and Architecture, Panzhihua University, Panzhihua, China

ABSTRACT: There is no fly ash factory near the hydropower station, making the transport cost of fly ash high due to distance. In order to supply reliable admixture sources with the desired performance characteristics and economy, the hydropower project made full use of the local raw material sources and also allowed intake of double admixtures as the admixture of concrete, which has been successfully applied to various types of concrete. All check indexes of this type of concrete meet the requirements of design and it has good adaptability so that it reduces the project cost and offers good technical, economical and social benefits.

1 INTRODUCTION

The hydropower station is located in the lower reaches of Lancang River in Yunnan in Xishuangbanna Dai Autonomous Prefecture. It is a multipurpose project for power generation, shipping and tourism with an installed capacity of 1750 MW. The hub consists of the remaining water, drainage, power generation, sand flushing and navigation structures, and the maximum height of roller-compacted concrete gravity dam is 114 m, and the total amount of concrete is about $393 \times 3,672.73$ ft³.

Due to the lack of fly ash factory near the Jinhong Hydropower Station, in addition to the transport cost of fly ash being high, the amount of supplied of fly ash is insufficient to meet the requirement of construction intensity during the peak period. Therefore, the admixture material becomes one of the key factors to restrict the construction of the concrete dam. In order to provide reliable sources, ideal performance and economically viable admixture for engineering, the aggregate admixture in Jinghong and the nearby areas was investigated and its quality was assessed. The available source material was screened for selection, combination and optimization. Finally, an admixture and concrete mix was proposed, which met the technical requirements of the project.

2 SELECTION AND TEST OF RAW MATERIAL

2.1 Cement

Through the experimental study, the physical and mechanical properties of 42.5-class ordinary Portland cement (rotary kiln) produced by Pu'er Cement.

Factory in Simao prefecture are as shown in Table 1.

Table 1. Physical mechanical performance of the cement.

Cement variety	Fineness (%)	Density (g/cm³)	Normal consistency (%)	Setting time (h:min)		Rupture strength (MPa)		
				Initial set	Final set	3d	7d	28d
Pu'er 42.5-class Ordinary Portland (1st batch)	2.6	3.15	25.4	2:54	4:31	4.6	7.48	8.89

Cement variety	Compressive strength (MPa)			Heat of hydration (kJ/kg)			
	3d	7d	28	1d	3d	5d	7d
Pu'er 42.5-class Portland cement	24.5	49.9	65.4	179.65	254.01	278.37	294.12

It can be seen from Table 1 that the technical indexes of the 42.5-class ordinary Portland cement produced by Pu'er Cement Factory in Simao prefecture meets the current national standard.

2.2 Admixture

There are two kinds of admixtures. One mixes the water-granulated manganese and iron slag of Jinghong Manganeisen Factory and the limestone powder of Xiaomengyang Daheishan Quarry (JMH, for short). The other one mixes the water-granulated manganese and iron slag of Menghai Manganeisen Factory and limestone powder of Xiaomengyang Daheishan Quarry (MMH, for short).

2.2.1 Physical properties of admixture

Jinghong water-granulated manganese and iron slag has the following properties: density, 3.07 g/cm³, fineness: 2.1%, water content: 0.08%.

Limestone powder of Xiaomengyang Daheishan Quarry has the following properties: density, 2.70 g/cm³, fineness: 32.9%, water content: 0.09%.

Menghai Dianxin water-granulated slag has the following properties: density, 2.96 g/cm³, fineness, 5.5%, water content: 0.09%.

Menghai Xiongtai water-granulated slag has the following properties: density, 3.01 g/cm³, fineness, 5.9%, water content, 0.17%.

2.3 Aggregate

The aggregate consists of the the natural gravel material in downstream river island stock ground; diorite and artificial coarse aggregate and the natural sand in the dam foundation; and diorite and artificial aggregate in the dam foundation. Fine aggregate mainly is made up of the fine sand with an average particle size of 0.46–0.53 mm, fineness modulus of 2.39–2.70, stable quality and less weak particles. The coarse aggregate is the natural gravel, with an apparent unit weight of 2.69 g/cm³ and water absorption of 0.69–1.14. The three-grade normal concrete is determined by vibrating stacking weight test: its aggregate gradation ratio is small stone: medium stone: large stone = 30:30:40; for two-grade normal concrete, the aggregate gradation ratio is the small stone: medium stone = 45:55.

2.4 Additive

The additive is used to improve and enhance the performance of concrete, whose quality will affect the concrete quality, progress and project cost. Good-quality and high-water reducing admixture is one of the important characteristics to reduce the unit water use of concrete and temperature control.

The normal concrete of the dam uses the ZB-1A water reducer and ZB-1G air entraining agent produced by Zhejiang Longyouwuqiang Concrete Additive Co. Ltd. RCC uses the Glenium26 macromolecule water reducer (GM26 for short) produced by Shanghai Maisite Building Materials Co., Ltd., and ZB-1G air entraining agent produced by Zhejiang Longyouwuqiang Concrete Additive Co., Ltd.

3 DAM CONCRETE PARTITION AND KEY TECHNICAL INDEXES

Tl.e dam of the hydroelectric power station consists of the roller-compacted and normal mass concrete. The concrete strength of roller-compacted mass concrete is mainly C9015, while that of the normal mass concrete is mainly C9020 and C2820, and concrete with other strength grades are present in low amounts.

According to the working conditions, strength, impermeability, resistance to erosion and crack of the dam, the dam concrete is divided into four parts as shown in Table 2.

Table 2. Dam concrete partition table.

Partition	Part	Key technical indexes	Comment
①	Inside the dam	C15 W6 F50	Roller compacted concrete
②	Foundation pad, dam crest, base plate of stilling pool, partial side wall	C20 W8 F100	Normal concrete
②	Around the steel tube, workshop, ship lift	C20 W8 F100	Normal concrete
③	Upper overflow surface, gate pier, around the diversion bottom tunnel	C25 W8 F100	Normal concrete
③	Lower overflow surface, upper base plate of stilling pool, partial side wall, around the sand flushing bottom hole	C30 W8 F100	Normal concrete
④	Upper and lower impervious barrier	C20 W8 F100	Roller compacted concrete (two grade aggregate)

Table 3.   Normal concrete mix table.

Strength grade	Grading	Sand ratio (%)	Double-mixture Name	Dosage (%)	Water cement ratio	Water reducer ZB-1A Dosage (%)	Air entraining agent ZB-1G Dosage (1/10000)	Material amount per cubic metre (kg/m³) Water	Cement	Additive	Sand	Stone	Slumps (cm)	Air content (%)
Pu'er 42.5-class ordinary Portland cement and natural aggregate														
$C_{90}15$	Third	28	JMH55	30	0.60	0.5	0.1	98	114.3	49	594.8	1568.5	5.5	4.6
$C_{90}20$	Third	28		30	0.55	0.5	0.1	98	124.7	53.5	591.3	1559.2	4.6	4.6
$C_{90}20$	Second	34		30	0.55	0.6	0	115	146.4	62.7	694.1	1382.7	6.7	4.6
$C_{28}25$	Second	33		30	0.45	0.5	0.1	112	174.2	76.5	665.1	1385.8	5.2	4.8
C2825	Third	27		30	0.45	0.5	0	94	146.2	62.7	566.0	1569.1	5.8	4.7
C2820	Second	34		30	0.50	0.5	0.1	113	158.2	67.8	691.0	1376.5	5.8	4.6
C2820	Third	27		30	0.50	0.5	0.1	95	133.0	57.0	569.6	1579.2	5.9	4.0
C2815	Third	28	JMH37	30	0.55	0.5	0.1	98	124.7	53.5	591.3	1559.2	4.6	4.6
C9015	Third	28		30	0.60	0.5	0.1	98	114.3	49	594.5	1567.6	5.0	4.4
(Pu'er 42.5-class ordinary Portland cement, natural sand + artificial gravels)														
$C_{90}15$	Third	32	JMH55	30	0.60	0.7	0	108	126.0	54.0	667.0	1545.1	4.2	3.7
$C_{90}20$	Third	32		30	0.55	0.7	0	110	140.0	60.0	659.9	1528.7	5.2	4.5
$C_{90}20$	Second	39		30	0.55	0.7	0.1	125	159.1	68.2	780.1	1327.3	5.0	5.2
$C_{28}25$	Second	38		30	0.45	0.7	0.1	122	189.8	81.3	748.9	1329.2	5.2	5.2
C2825	Third	31		30	0.45	0.7	0	106	164.9	70.7	633.2	1536.4	4.3	4.7
$C_{28}20$	Second	39		30	0.50	0.7	0.1	125	175.0	75.0	772.5	1314.5	6.5	5.4
$C_{28}20$	Third	31		30	0.50	0.7	0	106	148.4	63.6	639.3	1551.4	4.8	4.6
$C_{90}15$	Third	32	JMH37	30	0.60	0.7	0	113	131.8	56.5	660.2	1529.3	4.6	4.6
$C_{90}15$	Third	32		30	0.60	0.7	0	108	126.0	54.0	667.0	1545.1	4.2	3.7

4 CONCRETE MIX DESIGN

Binding material: The concrete used in the test is the 42.5-class ordinary Portland cement produced by Pu'er Cement Factory in Simao prefecture and 42.5-class Portland cement produced by Pu'er Cement Factory in Simao prefecture for review. The admixture uses the JMH double-mixture material, in a 5:5 mix proportion for the eight mix ratio normal concrete tests.

The relationship between mixing strength R_p and design strength R_s is as follows:

$$R_p = \frac{R_S}{(1 - t.C_v)} \tag{1}$$

where R_p—concrete mixing strength;
R_s—concrete design strength;
C_v—concrete strength deviation coefficient;
t—probability parameter of assurance rate.

5 DAM NORMAL CONCRETE MIX DESIGN

In the normal concrete mix design, the cement uses the Pu'er 42.5-class ordinary Portland, and the aggregate uses the natural aggregate in the river island (the natural aggregate for short), natural sand in river island + diorite and artificial gravel in the dam foundation (the natural sand + artificial gravel for short), and the admixture uses JMH55 double-mixture material.

Preparation test for the two kinds of aggregate: the normal concrete admixture dosage is 30%. The unit water use of natural second-class concrete is 112–115 kg/m³, and the sand ratio is 33–34%; the unit water use of third-class concrete is 94–98 kg/m³, and the sand ratio is 27–28%; when using the natural sand + artificial gravel aggregate, the unit water use of second-class concrete is 122–125 kg/m³, and the sand ratio is 38–39%; the unit water use of third-class concrete is 108–113 kg/m³, and the

Table 4. Normal concrete performance test results table.

Strength grade	Aggregate type	Double-mixture material	Water cement ratio	Compressive strength (MPa)		Ultimate tensile ($\times 10^{-6}$)		Elastic modulus ($\times 10^4$ MPa)	
				28d	90d	28d	28d	90d	28d
$C_{90}15$	Natural aggregate	JMH55	0.60	24.1	26.7	1.31	79.9	79.5	2.888
$C_{90}15$			0.60	23.8	28.3	1.72	74.9	80.1	3.139
$C_{90}20$			0.55	28.0	31.4	1.89	77.2	88.2	3.205
$C_{90}20$			0.55	26.3	29.9	2.10	70.3	90.6	3.314
$C_{28}25$			0.45	32.9	–	2.37	97.8		3.473
$C_{28}25$			0.45	34.4	–	2.52	90.7		3.631
$C_{28}20$			0.50	29.3	–	2.07	86.2		3.340
$C_{28}20$			0.50	31.1	–	2.44	93.5		3.333
$C_{28}15$			0.55	28.0	–	1.89	77.2		3.205
$C_{90}15$	Natural sand + artificial gravels	JMH55	0.60	24.6	28.5	1.31	66.7	78.7	2.608
$C_{90}15$			0.60	22.8	27.4	1.46	71.7	76.6	2.515
$C_{90}20$			0.55	28.3	32.5	1.73	83.8	85.2	2.863
$C_{90}20$			0.55	22.4	30.4	1.39	75.8	80.3	2.653
$C_{28}25$			0.45	39.0	–	2.56	92.7	–	3.309
$C_{28}25$			0.45	40.3	–	2.54	98.7	–	3.363
$C_{28}20$			0.50	34.6	–	2.31	85.3	–	3.059
$C_{28}20$			0.50	35.8	–	2.49	84.1	–	3.149
$C_{28}15$			0.55	28.3	–	1.73	83.8	–	2.863

Strength grade	Concrete dry shrinkage rate ($-\varepsilon_t \times 10^{-6}$)					Impermeability grade >W6	Frost resistant grade >F100
	3d	7d	28d	60d	90d		
$C_{90}15$	11.96	112.42	387.49	494.44	521.43	>W6	>F100
$C_{90}15$	13.31	102.40	403.11	529.07	558.76	>W8	>F100
$C_{90}20$	21.55	134.42	397.77	506.53	552.02	>W8	>F100
$C_{90}20$	31.03	143.21	394.17	515.90	567.05	>W8	>F100
$C_{28}20$	40.25	157.91	393.60	504.79	554.58	>W8	>F100

sand ratio is 31–32%. The concrete mix is as shown in Table 3.

6 PERFORMANCE TEST OF NORMAL CONCRETE

The results of the determination of physical and mechanical properties of the normal concrete in each part of the dam is as shown in Table 4.

7 SUMMARY

According to test results, it is proved that the concrete mix proportion meets the engineering requirements fully. The tests showed that the main project of the power station is made of two kinds of cement, two kinds of aggregates and two proportions of the mixtures (JMH55, JMH37). Its strength, durability, frost resistance, anti-permeability and other indicators are designed to meet the requirements of the design.

REFERENCES

Concrete structural engineering construction and acceptance [S]. GB 50204–92.
Hydraulic concrete construction specifications[S]. SDJ 207–82.
Li Lian. Research on the Double-mixing Technology of Concrete. Construction Economy, pp. 70–73, 2007.
Shengjiang Lin. Mix proportion design of high performance concrete with double mixing and its engineering application, Shanxi Architecture. pp. 104–105, 2012.
Wei YY. The application of the double-mixing technology of slag powder and fly ash on concrete. Shanxi Architecture. pp. 54–56, 2010.

Advances in Energy Science and Equipment Engineering II – Zhou, Patty & Chen (Eds)
© 2017 Taylor & Francis Group, London, ISBN 978-1-138-71798-5

Analysis of asphalt layer shear stress based on the contact condition between layers

Min Luo
China Aiport Construction Group Corporation of CAAC, R&D Center, Beijing, China
Beijing Super-Creative Technology Co. Ltd., Beijing, China

Yiqiong Xie
CCCC First Highway Consultants Co. Ltd., Xi'an, China

Bing Hui
School of Highway, Chang'an University, Xi'an, China

ABSTRACT: In order to study the influence of interlayer contact state on the asphalt shear strength, we analysed different contact states of asphalt layer shear stress distribution laws using the Bisar3.0 software. It mainly considers three different contact states between pavement layers (layer–layer), layer–base under standard design state (state 1) and heavy-load high-temperature state (state 2). The results show that shear stress of the layers is minimum in the continuous contact state, and maximum in the smooth contact state, and the maximum shear stress appears on each side position of the bottom asphalt layer. The stress increases sharply in the heavy-load and high-temperature state. Shear stress of pavement layers (layer–layer) is significantly larger than that of layer-base; therefore, the focus should be on strengthening the construction quality of the asphalt layers and a continuous contact between the layers must be ensured.

1 INTRODUCTION

In China, asphalt pavement with semi-rigid base is applied widely, but with the increase in the number of heavy vehicles, numerous slipping diseases of different degrees are found on on pavement. Slip diseases generally appear in the sections that bear big horizontal force caused by vehicle acceleration, deceleration, turning, etc. When the bond strength between layers is weak, pavement slip failure appears easily. Pavement deformation and cracking affect the smoothness of the road, and with further damage, there will also appear pits and mud pumps, which seriously affect the pavement performance, reduce social benefits and economic benefits of the road.

In current pavement design specification, the pavement structure is simplified as an elastic layered system structure with layers in a completely continuous state, but in reality, the state between different layers of pavement is not completely continuous. It is a kind of a bonding state, because the structure of pavement layers is made up of different materials or a different construction sequence. The influence of interlayer bonding material selection is not the same in construction. The management of materials and other factors indicate that design specification does not conform to reality. Therefore, simulation of a reasonable interlayer contact state in the pavement structure is used in the research the pavement structure's stress state. It is important to improve the pavement design method.

In this paper, by using Bisar3.0 program, asphalt layer shear stress distribution rules are analysed under different layer contact conditions in standard design states and heavy-load high-temperature states.

2 PAVEMENT STRUCTURE MODEL AND THE SELECTION OF PARAMETERS

Pavement structure, design parameters, and the loading mode are shown in the following tables. In these experiments, 20°C elastic modulus is set as the recommended value in "design specifications for highway bituminous pavement", 65°C modulus is determined by the formula for "change rule of asphalt pavement strength and maintenance":

$$\frac{E_t}{E_c} = 1.0 - 1.5 \lg \frac{T}{C}$$

Type in the: Et—elastic modulus of T;
Ec—elastic modulus of Standard temperature (C), generally C is 20°C, as shown in Table 1.

The load schema uses the formula applied in the Belgian experience showing relationship between grounding area and axle load (A = 0.008P+152 where A is the grounding area, cm²; P is the load, N). The schema represents the actual situation. The wheel pressure and ground area increase with an increase in the axle load and the center distance remains 31.95 cm. Because vehicles that bear more heavy cargo transport tend to have higher strength and higher-pressure tires in statistics, the horizontal load coefficient is 0.2, as shown in Table 2.

3 SIMULATED EXPERIMENTAL RESULTS

3.1 Asphalt layer shear stress distribution under completely continuous status

The contact conditions are set to be completely continuous between asphalt layers and asphalt-base layers. The asphalt layer shear stresses are calculated respectively under modes 1 and 2 by Bisar3.0 and the program calculation is based on the theory of elastic layered system. The results are shown in Figure 1.

We can see from the chart that under the two conditions (modes 1 and 2), the maximum shear stress of asphalt layer appears on the road surface, and it increases from 0.3999 MPa in mode 1 to 0.5849 MPa in mode 2, with an increase of 46.3%. The interlayer shear stress between surface layer and the middle layer increases from 0.2278 MPa in mode 1 to 0.3428Mpa in mode 2, with an increase of 50.5%. The interlayer shear stress between middle layer and the bottom layer increases from 0.2097 MPa in mode 1 to 0.3768 MPa in mode 2, with an increase of 79.7%. The calculation results show that asphalt layer shear stress under the high-temperature and heavy-load condition is significantly high.

Table 1. Pavement structure and calculation parameters.

Structure	Material	Thickness (m)	Elastic modulus (Mpa) 20°C	65°C	Poisson's ratio
Surface layer	AC-13	0.04	1400	325	0.25
Middle layer	AC-20	0.06	1200	279	0.25
Bottom layer	AC-25	0.08	1000	232	0.25
Base	Cement stabilized gravel	0.18	1600	1600	0.25
Subbase	Cement stabilized gravel	0.36	1400	1400	0.25
Subgrade			35	35	0.35

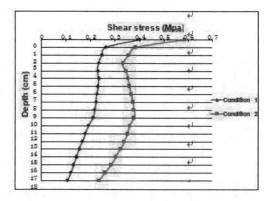

Figure 1. Figure of shear stress distribution under completely continuous status.

Table 2. Loading mode.

Ground pressure (Mpa)	Tire pressure (KN)	Radius (cm)	The grounding area (cm²)	Temperature (°C)	Horizontal load (KN)	Remarks
0.7	25	10.65	356	20	5	Standard design status
1.0	76	15.55	760	65	15.2	Heavy load and high temperature status

3.2 *Asphalt layer shear stress distribution under completely continuous condition between asphalt layer, and a different contact status between asphalt layer and the base layer*

The contact conditions are set to be completely continuous between asphalt layers. The contact condition between asphalt layer and the base layer is semi-continuous semi-sliding ($\alpha = 0.5$) and completely smooth ($\alpha = 0.99$).

The asphalt layer shear stresses are calculated respectively in modes 1 and 2 by Bisar3.0 and the program calculation is based on the theory of elastic layered system. The results are shown in Figure 2.

We can see from the chart that in these two conditions (modes 1 and 2), asphalt layer shear stress distribution under the condition of semi-continuous semi-sliding between asphalt layer and the base layer is the same as completely continuous status basically. The maximum shear stress increases from 0.4009 MPa in mode 1 to 0.5886 MPa in mode 2, with an increase of 46.8%. The interlayer shear stress between surface layer and the middle layer increases from 0.2285 MPa in mode 1 to 0.3523 MPa in mode 2, with an increase of 54.2%. The interlayer shear stress between the middle layer and the bottom layer increases from 0.2152 MPa in mode 1 to 0.4075 MPa in mode 2, with an increase

of 89.4%. In modes 1 and 2, under completely smooth status between asphalt layer and the base layer, shear stress in each layer increases with the increasing depth of the asphalt layer. The maximum shear stress increases from 0.4055 MPa under standard status to 0.6081 MPa under heavy-load high-temperature status, with an increase of 50%.

Thus, under heavy-load and high-temperature condition, the contact status between the asphalt layer and base layer deteriorates, which will lead to the sharply increased shear stress of the asphalt layer, and the possibility of disease emergence increases significantly.

3.3 *Asphalt layer shear stress distribution under completely continuous between asphalt layer and base layer, different contact status between asphalt layers*

The contact conditions are set to be completely continuous between the asphalt layer and the base layer. The contact condition between asphalt layers is semi-continuous semi-sliding ($\alpha = 0.5$) and completely smooth ($\alpha = 0.99$). The asphalt layer shear stresses are calculated respectively in modes 1 and 2 by Bisar3.0 and that the program calculation is based on the theory of elastic layered system. The results are shown in Figure 3.

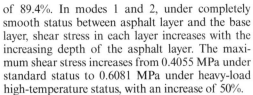

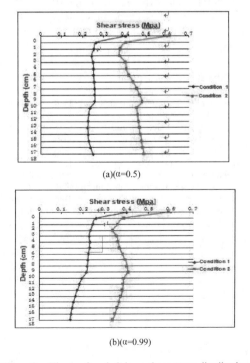

(a)(α=0.5)

(b)(α=0.99)

Figure 2. Figure of asphalt layer shear stress distribution under completely continuous between asphalt layer, different contact status between asphalt layer and base layer.

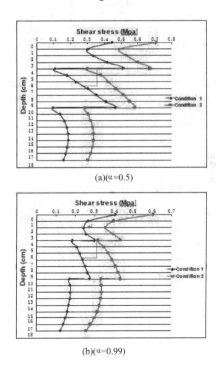

(a)(α=0.5)

(b)(α=0.99)

Figure 3. Figure of asphalt layer shear stress distribution under completely continuous between asphalt layer and base layer, different contact status between asphalt layers.

We can see from the figure that under the two conditions of modes 1 and 2, the asphalt layer shear stress distributions under semi-continuous semi-sliding status are as the same as completely smooth status basically. Three extreme values appear in the asphalt layer of road surface, road surface under 4 cm and the road surface under 10 cm respectively. In the semi-continuous semi-sliding condition, numerical value increases from 0.4068 MPa to 0.6041 MPa, with an increase of 48.5%; under completely smooth conditions, maximum shear stress appears on the road surface in modes 1 and 2, and the numerical value increases from 0.5076 MPa to 0.7114 MPa, with an increase of 40.1%.

3.4 *Asphalt layer shear stress distribution under completely smooth status*

The contact conditions are set to be completely smooth between asphalt layers, and also between asphalt layer and the base layer. The asphalt layer shear stresses are calculated respectively under modes 1 and 2 by Bisar3.0 and the program calculation program is based on the theory of elastic layered system. The results are shown in Figure 4.

Figure 4. Asphalt layer shear stress distribution under completely smooth status.

We can see from the chart that the asphalt layer shear stress distribution is consistent between modes 1 and 2 under a completely smooth status. Three extreme values appear in asphalt layer: bottom surface layer, middle layer and bottom layer, respectively. The numerical value of shear stress increases from 0.5271 MPa under standard design state to 0.7128 MPa under high-temperature heavy-load state, with an increase of 35.2%.

4 CONCLUSION

1. With different contact states between asphalt layers, shear stress distribution of asphalt layers varies, but from the results, we can see that the position of maximum shear stress lies in the bottom of each layer. Thus the most unfavorable position is the bottom of each layer. With completely smooth states, the maximum shear stress will grow exponentially, which is adverse for asphalt layer as it increases its stress.
2. The shear stress of the asphalt layers greatly increases under deterioration in the condition of the contact between the asphalt layers than between asphalt layer and the base layer.
3. With different contact status between asphalt layers, the shear stress increases rapidly under heavy-load and high-temperature state compared with the standard design state, which pushes the possibility of disease increase greatly, which results in asphalt pavement rutting, shoving etc.
4. We should control the strength and construction quality between asphalt layers and the bonding layer, and between asphalt layer and the base layer, to establish a strict quality control system. It should be ensure that the contact between layers is in a continuous status.

ACKNOWLEDGEMENTS

The paper was sponsored by the Shaanxi Provincial Transportation Research Project (No. 12–15 K), Special Fund for Basic Scientific Research of Central Colleges, Chang'an University (No. 310821141004, 310821153104 and 310821151006), Inner Mongolia Transportation Research Project (No. NJ-2015-31). Their support and assistance are gratefully acknowledged.

REFERENCES

Chen Chao. Research on shear strength between the layers of the asphalt pavement [D]. Changsha: Changsha University of Science and Technology, 2008.

Sha Qing-lin. phenomenon and prevention of early damage of Expressway Asphalt Pavement [M]. Beijing: China Communications Press, 2001.

Shen Jin-an. Asphalt and asphalt mixture road performance [M]. Beijing: China Communications Press, 2001.

Shen Jin-an. Design method of confluence of foreign asphalt pavement [M]. Beijing: China Communications Press, 2004:176–194.

Zhang Deng-liang. Asphalt and asphalt mixture [M]. Beijing: China Communications Press, 1993.

Advances in Energy Science and Equipment Engineering II – Zhou, Patty & Chen (Eds)
© *2017 Taylor & Francis Group, London, ISBN 978-1-138-71798-5*

Erosion damage mechanism analysis of tempered glass in a wind-blown sand environment

Yun-Hong Hao

Inner Mongolia Key Laboratory of Civil Engineering Structures and Mechanics, Civil Engineering Department, Inner Mongolia University of Technology, Hohhot, Inner Mongolia, China
Civil Engineering Department, Inner Mongolia University of Technology, Hohhot, Inner Mongolia, China

Ru-Han Ya, Yong-Li Liu & Hui Li

Civil Engineering Department, Inner Mongolia University of Technology, Hohhot, Inner Mongolia, China

ABSTRACT: To understand the environmental characteristics of sandstorm frequency in Midwestern Inner Mongolia, sediment–air injection method was used in an experimental study to determine erosion on tempered glass, and the influence of the different factors (impact angle, impact velocity, erosion time and abrasive feed rate) were discussed. The surface damage mechanism micrograph of the tempered glass after erosion was observed by Laser Scanning Confocal Microscope (LSCM). The results obtained show that the erosion rate of tempered glass increases with an increase in the impact angle and impact velocity, and the erosion rate shows an exponential relationship with the impact velocity. The maximum erosion rate is reached at an impact angle of 90° and an impact velocity of 35 m/s. At low impact angles, material mass loss is mainly caused by cutting, while crack propagation is the main factor at high impact angles. Erosion rate increases at first, then decreases with the increase of erosion time and finally tends to be stable. The erosion rate increases in the initial stages and decreases in the secondary stage with an increase in the abrasive feed rate. As seen in LSCM images, erosion damage increases rapidly with an increase in the impact angle and impact velocity. The erosion area is becoming more regular as the impact angle increases. In addition, roughness of eroded surface increases with an increase in the impact angle and impact velocity.

1 INTRODUCTION

The Badain Jaran, Tengger, Ulan Buh, and Hobq deserts are the deserts in China and Maowusu and Hunshandake are sandy lands. These deserts and sandy lands are widely distributed in western and north-western Inner Mongolia (Chen, 2011; Wang, 2011). Sandstorm occurs frequently in these regions, as the weather of Inner Mongolia is influenced by adverse weather conditions, such as strong winds, sandy soil, etc. The glass surface of the buildings is struck by sand particles blown by wind. Particles attack the glass maintenance structure and desquamates the materials, and threatens the safety of tenements, which is harmful for the structure's durability. Tempered glass is extensively used in the maintenance structure of the buildings due to its high transparency, good corrosion and impact resistance (Hu, 2002; Arjula, 2008).

Erosion wear damages materials by strike material surface with a flowing particle. There are two kinds of wear: liquid–solid wear and gas–solid wear. Liquid–solid wear, which is caused by a fluid carrying solid particles, impacts the material's surface, damaging it. The related research is mature (Bousser, 2013; Liu, 2009); Gas–solid wear is defined as solid particles carried by gas impacting the material surface, leading to material surface damage (Dong, 2004; G. P. TILLY, 1973; G. P. TILLY, 1969). Liquid–solid wear is a fluid carrying solid particles impacting material surface, damaging it.

The brittle material damage model of erosion based on crack propagation and chipping was developed by Finnie and Sheldom, and wear morphology characteristic research is significant in the study of wear processes occurring in erosion analysis. In 1965, the first Scanning Electron Microscopic (SEM) images of particles was obtained from healthy synovial joints, and wear topography analysis touched another level, but the topography of erosion with a gray image was not directly observed by SEM. A smaller field limited its view, even though SEM is routinely used in wear particle research. With the use of Laser Scanning Confocal Microscopy (LSCM) in the biomedical field for the analysis of human tissue in 1980s, topography analysis has developed to a three-dimensional analysis. This paper describes a method based on the acquisition of a three-dimensional surface morphology of erosion damage (López-Cepero, 2005; Jia, 2012; Muller, 2013; Y. Tian, 2012).

2 SAMPLE PREPARATION AND LABORATORY EQUIPMENT

2.1 Sample preparation

A 5-mm-thick tempered glass was chosen for analysis in this test. The size of sample is 70 mm × 70 mm × 5 mm, and it was cleaned with alcohol before weight measurement every time. The average weight of the sample is 58.4532 g. The sediment–air injection method was used in erosive wear test. The sand used in the experiment was obtained from Hobq desert, and the shape of sand particle was irregular and angular, as shown in Fig. 1.

As shown in Table 1, the Hobq desert mainly has five kinds of sand particle sizes: Coarse sand content (>0.5 mm) less than 1%, very fine sand (<0.05 mm) less than 10%, medium sand and fine sand (between 0.25 and 0.05 mm) content of 88%. So the sand particle size between 0.25 and 0.05 mm was adopted in the experiments.

In this experiment, the factors such as impact angle, impact velocity and erosion time were

considered. Impact velocity of 20, 25, 30 and 35 m/s was chosen. The impact angle was fixed at 15°, 30°, 45°, 60°, 75° and 90°. Erosion time was set to 3, 6, 9, 12 and 15 min. An abrasive feed rate of 60, 90, 110, 180, 240 and 260 g/min was selected. The erosion morphology in this investigation is based on these variables.

2.2 Laboratory equipment

The laboratory equipment included wind-sand laboratory experiment erosion system, measuring equipment and morphology analysis instrument. The measuring equipment included precision electronic analytical balance (with an accuracy of 0.0001 g) and anemometer (with a maximum range of 50 m/s). Morphology analysis was done through a Laser Confocal Microscope (LSCM).

3 RESULTS AND DISCUSSION

3.1 Impact of impact angle on tempered glass erosion rate

The typical ductile and brittle material erosion model curve of erosion rate with the impact angle is shown in Fig. 2(a). In contrast with brittle material,

Figure 1. The shape of sand.

Table 1. Desert sand particle size distribution in the Hobq desert, Midwestern Inner Mongolia.

Particle size/mm	>0.5	0.5~ 0.25	0.25~ 0.1	0.1~ 0.05	<0.5
Content Variation (%) range	0.60~ 0.72	1.64~ 1.84	49.88~ 52.44	36.44~ 38.20	8.16~ 9.7
Average	0.66	1.74	50.82	37.53	8.96

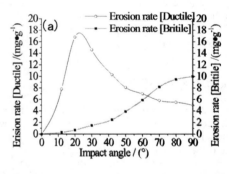

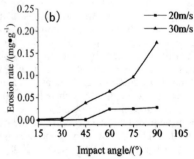

Figure 2. The erosion rate vs impact angle of model and test (a) The erosion rate vs impact angle of ductile and brittle material (b) Erosion rate vs impact angle of tempered glass.

ductile material erosion requires two-stage mechanisms to explain the erosion rate due to impact angle. Erosion damage in the initial stage is dominated by the incidence angle and in the secondary stage, it is dominated by material characteristics. The mass loss of ductile materials is due to cutting. So the erosion rate increases with the impact angle in the initial stage but decreases in the secondary stage. The erosion rate at 20°–30° is larger than at 90°. Erosion of the brittle material erosion depends upon the incidence angle. The maximum erosion rate appears at the maximum impact angle.

Fig. 2(b) shows the relationship between erosion rate and impact angle at an impact velocity of 20 and 30 m/s, respectively, and an abrasive feed rate of 110 g/min. It can be observed that the erosion rate increases with the impact angle, and the maximum erosion rate corresponds to the maximum impact angle. Erosion of the brittle material such as tempered glass in this investigation is dependent on impact angle as shown in Fig. 2(a) and (b). Cutting action and crack propagation appears when irregular sand particles impact the material surface, but cutting action is not the main reason for mass loss because of the poor toughness of the brittle materials. Two kinds of crack are seen in the crack propagation action: radial crack and lateral crack. Criss-crossing crack produce mass loss, which increases with impact angle.

3.2 *Effect of impact velocity on tempered glass erosion rate*

The effect of impact velocity on erosion rate of tempered glass is shown in Fig. 3 at an impact angle of 60° and 90° and an abrasive feed rate of 110 g/min. The erosion rate increases with impact velocity. An imaginary line represents the fitting curve of each erosion graph. The erosion rate and impact velocity have an exponential relationship, which was obtained by the fitted curve function $\alpha = a \times (1 - b^x)$. (a and b are constant; x is the impact velocity).

This experimental result can be explained as follows: at low impact velocity, the kinetic energy of the sand particle imparted to the surface of the tempered glass is lower than at high impact velocity. The higher impact velocity leads to a higher kinetic energy and impact strength. So, impact strength increases with the impact velocity, causing erosion surface damage more seriously.

3.3 *Impact of erosion time on tempered glass erosion rate*

The effect of erosion time on the erosion rate of tempered glass was shown in Fig. 4, at a 90° impact angle and abrasive feed rate of 110 g/min. Three stages (increasing stage, reducing stage and steady stage) are included in material mass loss.

The material mass loss increases within the first 7 min and reaches a maximum at the erosion time of 7 min. From 7 to 12 min, the surface of tempered glass surface appears with erosion pits on the eroded areas, and these pits limit the reverse direction of the sand particle at the same incident condition. Changing the direction of rebound particles prevents the onset of second erosion, which reduces the material mass loss effectively, and the material enters the steady-state erosion state finally.

3.4 *Impact of abrasive feed rate on tempered glass erosion rate*

Fig. 5 shows the relationship between erosion rate and abrasive feed rate at an impact angle of 90° and impact velocity of 30 m/s. The erosion rate does not increase with the abrasive feed rate. Another phenomenon occurs. The erosion rate increases first and then decreases with abrasive feed rate. It can be assumed that each sand particle has experiences the same impact velocity and same impact angle.

The first stage can be explained as follows: as the abrasive feed rate is low, the kinetic energy of

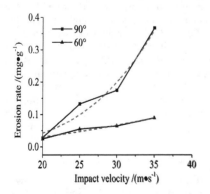

Figure 3.   Erosion rate vs impact velocity fitting curve.

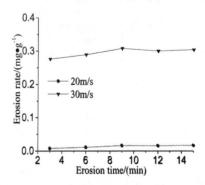

Figure 4.   Erosion rate vs erosion time of tempered glass.

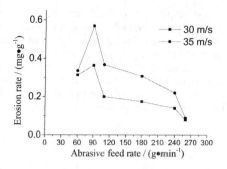

Figure 5. Erosion rate vs abrasive feed rate of tempered glass.

each sand particle is lower, and impact strength is lower than the threshold for fracture of tempered glass. As material surface is impacted continually by sand particles, newly generated cracks together with criss-crossing cracks lead to material mass loss.

4 EROSION DAMAGE MECHANISM ANALYSIS

4.1 *Erosion damage mechanism at impact angle of 15°*

In LSCM images, different colours represent different erosion depths. The unit of depth is μm. The magnification is ten times based on the initial three-dimensional image as shown in Fig. 6.

Fig. 6 shows the three-dimensional topography image at an impact angle of 15°, impact velocity of 30 m/s, and an abrasive feed rate of 110 g/min.

Glass erosion has two kinds of mechanisms to explain the erosion phenomenon. Mass loss caused by cutting and ploughing is probably due to crack or crack propagation of the erosion system. Relative to the high impact angle, it has a single reverse direction at low impact angle, and reverse direction is dependent on the original impact site, while erosion topography is dependent on the proportions of impact angle. In terms of above-mentioned, sand particle contact area is smaller on the glass surface at a low impact angle, causing shallow erosion damage.

4.2 *Erosion damage mechanism at impact angle of 90°*

Fig. 7 shows the three-dimensional erosion area of test sample surface at impact angle of 90°, impact velocity of 30 m/s and abrasive feed rate of 110 g/min.

It is seen that erosion damage zone increases with the impact angle, which is larger and more regularly around the original impact site compared

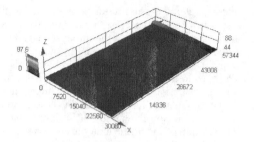

Figure 6. Three-dimensional erosion topography image at 15° impact angle and 30 m/s impact velocity.

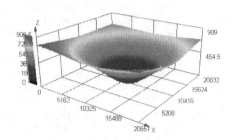

Figure 7. The three-dimensional erosion topography image at 90° impact angle and 30 m/s impact velocity.

with 15° impact angle. The mass loss is because of crack propagation. The erosion area resembles concentric circles from inside to outside. This result indicates that erosion direction is dependent on the impact angle, and the extent of erosion damage is dependent on the vertical force distribution. Serious damage occurs due to the formation of cracks and crack propagation by vertical force distribution increases with an increase of the impact angle.

4.3 *Erosion damage mechanism at an impact angle of 90° and impact velocity of 20 m/s*

Fig. 8 shows a three-dimensional image of test sample surface at impact angle of 90°, impact velocity of 20 m/s, and abrasive feed rate of 110 g/min.

The erosion area is similar to that at 90° impact angle, but damage is does not vary much from that at 30 m/s impact velocity. Compared with images of erosion damage at 20 and 30 m/s, the shape of the erosion area is related to the impact angle, while impact velocity makes little contribution. Erosion depth is influenced by the increasing impact energy, which increases with impact velocity. Therefore, erosion damage at 30 m/s was relatively serious.

4.4 *Surface roughness analysis*

Surface roughness is defined as surface irregularities. It was not only used to analyse surface topography, but to evaluate many fundamental

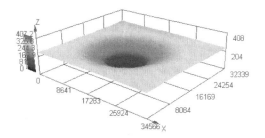

Figure 8. The three-dimensional erosion topography image at 90° impact angle and 20 m/s impact velocity.

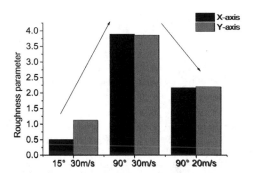

Figure 9. Surface roughness of tempered glass depending on impact angle and impact velocity.

problems such as friction, contact deformation, etc. The formula for calculation is as follows:

$$R_a = \frac{1}{l}\int_0^l |Z(x)| dx \qquad (1)$$

R_a—is the arithmetic average height parameter; l—sampling length.

Fig. 9 shows the roughness values at different erosion conditions, which includes impact angle and impact velocity. Ra increases with impact angle and impact velocity. The curve in Fig. 9 expressed as red and black pillar denote X-axis and Y-axis surface roughness values of the three-dimensional erosion images. It is seen that the surface damage is serious at a high impact angle compared with 15° and 90°, and roughness distribution is in agreement with this result.

The damage zone is dependent on the impact angle, and it has one reverse direction. This, in turn, depends on the original impact size at a low impact angle and is uniformly distributed around the original impact size at a high impact angle. The roughness distribution shown in X-axis smaller than that the low impact angle on the Y-axis. This looks similar to the X-axis and Y-axis roughness values at a high impact angle.

5 CONCLUSIONS

In conclusion, it is found that the erosion of brittle material such as tempered glass is dependent on impact angle, and erosion rate increases with impact angle and impact velocity. The relationship between the erosion rate and impact velocity is approximately an exponential relationship. The erosion mass loss increases at first and then decreases with an increase in erosion time, finally entering the steady erosion state. The erosion rate increases with an increase of abrasive feed rate at first, and then decreases with an increase in the abrasive feed rate. LSCM images showed the damaged area is dependent on the impact angle, and its shape tends to be circular. The erosion width increases with an increase in the impact angle. Erosion damage becomes more serious with an increase of impact velocity and impact angle. Roughness increases with the increase of impact angle and impact velocity.

REFERENCES

Arjula, S. Harsha, A.P. and Ghosh, M.K. Journal of Materials Science 3, 1757 (2008).
Bousser, Martinu, L. Klemberg-Sapieha, J.E. Journal of Materials Science 48(16), 5543 (2013).
Chen Yi, Shang Ke-zheng, and Wang Shi-gong. Journal of Desert Research 25, 132 (2011)(China).
Dong Gang, Bai Wan-jin, Zhang Jiu-yuan and Ren Bu-fan. Journal of Materials Science & Engineering 22, 909(2004) (China).
Hu Chun-yuan, Yang Mao, Yang Cun-liang, Liu Yong-mao, Jiang You-ze and Zhang Run-huan. Journal Of Arid Land Resources and Environment 16, 71(2002) (China).
Jia Peng and M. Zhou. Chinese Journal of Mechanical Engineering 25, 1224(2012) (China).
Liu Bao-lin, Gao De-li, Yang Jing-zhou, Fang Ming-hao, Wu Xiao-xian. METALMINE 396, 132(2009) (China).
López-Cepero and J.M. Key Engineering Materials 290, 280(2005).
Muller and K.H. Anal Bioanal Chem 405, 7117(2013).
Tian Y. and Wang J. Wear, 282–283, 59(2012).
Tilly G.P. Wear 14, 63(1969).
Tilly G.P. Wear 23, 87(1973).
Wang Wen-biao and Ma Jun-jie. Journal of Inner Mongolia Forestry Science & Technology 37, 27(2011) (China).

Advances in Energy Science and Equipment Engineering II – Zhou, Patty & Chen (Eds)
© 2017 Taylor & Francis Group, London, ISBN 978-1-138-71798-5

Bitumen modification using polyethylene terepthalate

Gatot Rusbintardjo

Department of Civil Engineering, Faculty of Engineering, Sultan Agung Islamic University, Jawa Tengah, Indonesia

ABSTRACT: This study was conducted to investigate the suitability of using plastic bottle as a bitumen modifier. The plastic bottle fulfilled all of bitumen modification requirements and it was chosen to take advantage of plastic bottle waste, which could help to reduce environmental pollution. Plastic bottle was burned, crushed and sieved, resulting in fine particles passing the sieve # 200. The fine particle of the plastic bottle was called Polyethylene Terepthalate (PTE). PTE was mixed into bitumen pen of penetration grade 80/100 using four different amounts of PTE. This mixing process was conducted at a temperature 1400°C, a mixing time of 40 min, and a stirring mixing speed of 500 rpm. Penetration, softening point, and specific gravity tests were performed. The results of penetration and softening point tests then were used to determine Penetration Index (PI) of the PTE-MB. PI is used to measure the susceptibility of bitumen to the change in temperature. From the PI value, it was shown that bitumen mixed with 15% of PTE has the highest PI, which was used as a binder of Stone Mastic Asphalt (SMA) mixtures. The Marshall test procedure was conducted to determine the strength of the SMA. The results show that when 15%-PTE-MB is used as a binder, SMA has a higher strength compared to when unmodified binder is used. It can be concluded that PTE is suitable to be used as a bitumen modifier.

1 INTRODUCTION

Pavements can be broadly classified into two types: flexible and rigid. Flexible pavement is the most used in the world at the moment. Figure 1 shows basic flexible pavement structure.

In most asphalt pavements, stiffness in each layer or lift is greater than that in the layer below and and less than that in the layer above (Gatot, 2011). This could be understood from the load distribution (Figure 2) where the stress at the surface layer is higher than that of the layer below. The surface layer has to withstand the maximum stress and also the changing conditions of the environment. Therefore, this surface layer usually consists of the 'best' and most costly materials. Also, this layer is always 'bound', that is, mixed with a 'binder', in this case asphalt cement or bitumen binder, to prevent the raveling of materials due to traffic, as well as to provide a dense surface to prevent ingress of water, unless it is an open-graded friction course. Therefore, the surface layer has two major components: bitumen binder and aggregates.

Bitumen is exposed to a wide range of load and weather conditions; however, it does not have good engineering properties, because it is soft in a hot environment and brittle in a cold weather. To prevent the occurrence of pavement distress, it is important to use reinforced bitumen to improve its mechanical properties. Modifying

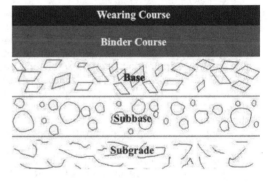

Figure 1. Basic flexible pavement structure [1].

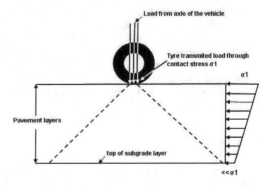

Figure 2. Load distribution on flexible pavement [1].

bitumen with additives to strengthen its mechanical properties has been the practice in many forms for over 150 years, but there is a renewed interest. This resurgence in interest can primarily be attributed to the following factors (Robert, 1996; King, 1999):

1. The increase demand on HMA pavements. Traffic volume, and traffic loads, as well as tyre pressures have increased significantly in recent years, causing premature rutting of HMA pavements.
2. The new binder specifications recommended by Strategic Highway Research Program (SHRP) in March 1993 requires the bitumen binder to meet the stiffness requirements at high as well as low pavement surface temperatures. Most base bitumen does not meet these requirements in the regions with extreme climatic conditions and, therefore, modification is needed.
3. There is an environmental and economic pressure to dispose of some waste materials and industrial by-products as additive in HMA.
4. Public agencies willingness to pay a higher first cost for pavements with a longer service life or which will reduce the risk of premature distress (failure).

This paper reports the results using PTE-modified bitumen. In this research, 5, 10, 15, and 20% (four different amounts) of very fine of PTE (having a grain size diameter of 0.075 mm or passing sieve #200) was added into the bitumen penetration grade 80/100. Penetration at 25°C and softening point test were conducted on the seven mixtures of PTE–Bitumen and the base bitumen to determine the Penetration Index (PI), the parameter that represents the temperature susceptibility of the binder.

Besides determining PI, penetration and softening point tests were also conducted to know the change in the properties of the base bitumen in case of its hardness. The influence of adding PTE to the bitumen was also examined by using PTE–Bitumen mixtures as a binder of SMA mixtures. The results show that 15% PTE content in the binder has the higher PI value, higher Marshall stability as well as the stiffness of the SMA mixtures. Details of the test results are given in the Results and Discussion section, and from the test results, it can be concluded that PTE was can be used as additional substance to the bitumen binder, especially the mixture containing 15% PTE.

2 MATERIALS

Two basic materials used in this research were PTE and crude oil base bitumen penetration grade 80/100, and the other material was aggregate.

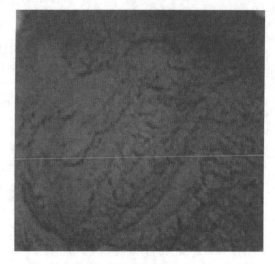

Figure 3. Powder of PTE 0.075 mm particles size.

2.1 Polyethylene Terepthalate (PTE)

PTE as shown in the Figure 3 is the very fine brown powder passing sieve #200 or has a particle size 0.075 mm, resulting from burning and crushing of plastic bottles.

2.2 Bitumen

Bitumen used in this research was of the penetration grade 80/100 produced by Pertamina Co. Ltd., the Indonesian state-owned oil company. Penetration at 25°C and softening point test conducted by the author show that the bitumen has a penetration of 83, softening point at 39°C, and specific gravity of 1.02.

2.3 Aggregate

Aggregates used for this research were taken from the quarry in Bodri River, located around 80 km from West Semarang, the city where the research was performed. The type of boulders of raw materials is basalt stone, which has resistance to abrasion of small-size coarse aggregates, which when measured by the Los Angeles Machine was 90%.

3 RESEARCH METHODOLOGY

The research methodology was divided into three main activities. The first activity was mixing PTE with bitumen. An amount of 5, 10, 15, and 20% of PTE was mixed to the base bitumen. The percentage of the base bitumen different from the amount of PTE, and the mean amount of PTE was 15% and the amount of bitumen was 85%. PTE was

Table 1. Gradation limit for combined aggregate of SMA mixtures.

Sieve size (mm)	% Passing	Sieve size (mm)	% Passing
19.0	100	2.36	16–24
12.5	85–100	0.600	12–16
9.5	65–75	0.300	12–15
4.75	20–28	0.075	8–10

Table 2. Penetration, softening point and specific gravity test results.

% PTE in the binder	Penetration at 25° (dmm)	Temperature of Softening Point (°C)	Specific Gravity
0	98.70	41.25	1.036
5	81.50	41.00	1.212
10	77.42	43.25	1.411
15	64.88	44.50	1.367
20	57.33	46.00	1.391

mixed with base bitumen using propeller mixer at a mixing temperature of 140°C, 500 revolutions per minute (rpm) stirring mixing speed, and 30 minutes of mixing time.

After thorough mixing, the second activity was penetration at 25°C, the softening point. Specific gravity test was conducted to determine value of Penetration Index (PI). Penetration test was carried out on two samples and each sample was taken five reading or tests, while softening point test was conducted on four samples.

The third research activity was Marshall Mix Design and Test. This activity started with aggregate characterization. In this research, Bina Marga's (Directorate General of Highways, Ministry of Public Works Republic of Indonesia) (Directorate, 1995) mixed design for Stone Mastic Asphalt (SMA) mixtures was used, and the gradation of the aggregate is shown in Table 1.

Marshall stability is performed to examine the strength of SMA mixtures. The optimum binder content was 6.8% taken from the range values of the specification design requirement of 5–7%. Marshall Stability test was taken according to ASTM D1559-92.

4 RESULTS AND DISCUSSION

This section was organized based on the outline of the research methodology. The sequences of information of the research are similar to the ones presented in that section. The laboratory works are divided into four groups and the test results of each group is discussed and analyzed. Statistical analysis using student t distribution and regression model are used to analyze the results of all the tests. The details of statistical analysis for penetration and softening point test as well as PI value are given in the following sub sections.

4.1 Penetration test

The value of the penetration test at 25°C of the bitumen will decrease with the increasing of PTE content. This hypothesis is tested by using student

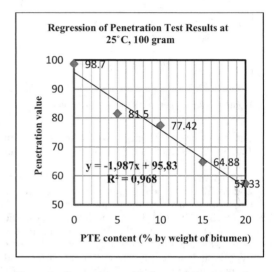

Figure 4. Regression model of penetration test results.

t distribution and simple regression model. The results of penetration test are shown in column 2 of Table 2. The regression model developed based on all results of the penetration test by using statistical analysis is given in Fig. 4. It shows that test data have a confidence interval of 95%, and this means that all penetration values are true and fit the hypothesis. Meanwhile, coefficient of correlation R = 0.98 as well as coefficient of determination R2 (R-square) = 0.968 show that penetration test results are significant.

4.2 Softening point

There was an inverse relationship between the penetrations and the temperature of the softening point of the binder. The lower the penetration value of the binder the harder the binder is and higher the temperature of softening point. The test results of temperature of the softening point are shown in column 3 of Table 2. The regression model of softening point resulted from statistical analysis using student t-test indicate that all softening point test

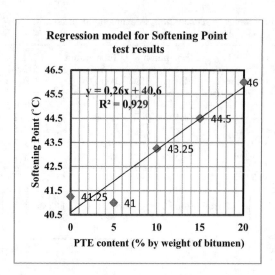

Figure 5. Regression model for softening point test results.

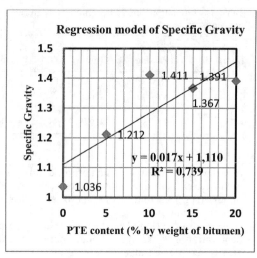

Figure 6. Regression model of specific gravity.

data have a confidence interval of 95%, as shown in Fig. 5. This shows that all softening point test values are true and fit thehypothesis. The regression model gives a coefficient of correlation R of 96% and coefficient of determination R2 (R-square) of 0.929, which are very significant test results.

Specific Gravity. The hypothesis of specific gravity is that higher the PTE content in the binder, the higher of specific gravity will be. This hypothesis is tested by using Student's t-test and simple regression model. The values of specific gravity are shown in column 4 of Table 2, and simple regression model resulting from statistical analysis is shown in Fig. 6. The values of specific gravity indicate that all specific gravity values have a confidence interval of 95%, and this shows that all specific gravity values are true and fit the hypothesis. The coefficient of correlation R was 86% and coefficient of determination R2 (R-square) was also 0.739, which indicate the very significant figure of the test results. These results could be understood since PTE mixed with the bitumen was in form of fine particles, so that the higher the Buton-NRA content, the higher the unit weight of the binder.

4.3 Penetration index (temperature susceptibility)

Table 3 gives the results of Penetration Index (PI) calculation. The PI value shows the temperature susceptibility of the binder. Pfeiffer and Van Doormaal, as quoted by Robert et al. (Robert, 1996), explain that the lower the PI values of the binder (bitumen), the higher its temperature susceptibility. According to Read, J. and Whiteoak, D.

Table 3. Penetration index resulted by Pen/SP value.

% PTE content in the bitumen	Penetration Index (PI)
0	−2
5	−1,6
10	−1,59
15	−1,0
20	−1,5

(Read, 2003). The value of PI ranges from about −3 for highly temperature susceptible bitumen to about +7 for low-temperature susceptible bitumen. The results of PI show that the binder with 15% of PTE content has the highest PI value or less temperature susceptibility.

The simple regression model for PI values using statistical analysis of Student's t-test indicate that all PI data have confidence interval of 95% as shown in Fig. 7. The coefficient of correlation R was only 0.71 and coefficient of determination R2 (R-square) was also only 0.501, which show that the test results were not significant.

4.4 Storage stability

The storage stability test was conducted to evaluate the possible separation of PTE from the bitumen under storage. The test procedure was conducted in accordance with ASTM D5892 standards. The test procedure was as follows: immediately after mixing was finished, PTE-MB was poured into 25.4 mm by 139.7 mm aluminium tube and was heated to 165°C for 1 and 3 days in the oven. The selection of

1196

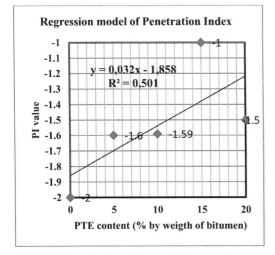

Regression model of Penetration Index

$y = 0,032x - 1,858$
$R^2 = 0,501$

PI value

PTE content (% by weigth of bitumen)

Figure 7. Regression model for Penetration Index.

Table 4. Softening point test results for storage stability.

PTE-BM	Temp. SP at top	Temp. SP at bottom	Different Temp. SP	Status
5% PTE	43,40°C	44,25°C	0,85°C	Stable
10% PTE	44,50°C	45,12°C	0,62°C	Stable
15% PTE	46,75°C	47,63°C	0,88°C	Stable
20% PTE	48,25°C	48,50°C	0,25°C	Stable

storage days was based on the estimation of road construction delay. At the end of the test period, samples were placed in the freezer at −10°C for 4 h to solidify the PTE-MB. Upon removing the tube from the freezer, samples were cut into three equal length portions. Softening point (TR&B) test was performed to the top and bottom parts of the samples. The difference of TR&B between top and bottom portions was used to evaluate PTE-MB's stability. The difference of TR&B should be controlled within 2°C so that the PTE-MB can be properly stored (Gatot, 2013). The results of storage ability test are given in Table 4.

From Table 4, it is clear that PTE-modified bitumen is stable and can be properly stored.

4.5 Marshall test results

As mentioned above, based on the PI value, the PTE–bitumen mixture with 15% PTE has a higher PI value, namely −1. This mean that 15% PTE-BM has a higher temperature susceptibility to the change of temperature. Therefore, 15% PTE-BM was used as a binder of Stone Mastic Asphalt (SMA) mixtures. As a comparison, original bitumen starbit

Table 5. Marshall test results.

No.	Parameter	Original Bitumen	15% PTE-BM
1	Bulk density	2.22	2.23
2	VIM (%)	7.34	6.95
3	Stability (kg)	551.25	676.15
4	Flow (mm)	0.94	1.05
5	Stiffness	586.44 kg/mm	643.85 kg/mm

(Asphalt Starbit) was also used as a binder of SMA mixtures.

From the experiment, the optimum asphalt content for SMA was found to be 6.8% by weight of aggregate. Marshall test results are given in Table 5.

The results of Marshall test given in Table 5 show that SMA mixtures using 15% PTE-BM as the binder have good properties compared to original bitumen as the binder.

5 CONCLUSIONS

From the results obtained in this study, the following conclusions can be drawn:

1. PTE will increase the resistance of binder so as to increase the temperature susceptibility.
2. Binder containing PTE has a high Marshall stability. It means that it has a good strength.
3. Binder containing PTE also has a high Marshall flow, which means that it has a good flexibility.

The three conclusions show that PTE can significantly be used to modified bitumen. However, as mentioned in the introduction, this research is the first part to gather information on using PTE as an additive of the bitumen, rheological testing as well as other test in addition to Marshall test for HMA mixtures like indirect tensile resilient modulus test, wheel tracking test, need to be performed.

REFERENCES

American Society for Testing and Materials (ASTM) (1992). ASTM D1559–92: Standard Test Method for Resistance to Plastic Flow of Bituminous Mixtures Using Marshall Apparatus. Philadelphia U.S.: ASTM International.

American Society for Testing and Materials (ASTM). ASTM D5892 standard test method for storage stability determination of bitumen modifier. Philadelphia US: ASTM International.

Directorate General of Bina Marga (Highway), Ministry of Public Works, Republic of Indonesia. International Competitive Bidding, Second Highway Sector Investment Project. Volume Three—General Specification (1995). pp. 6.55–6.63.

Gatot Rusbintardjo (2011), *Oil Palm Fruit Ash (OPFA) Modified Bitumen—New Binder for Hot-Mix Asphalt (HMA) Pavement Mixtures.* LAP LAMBERT Academic Publishing GmbH & Co. KG, Heinrich-Böcking-Str. 6–8, 66121 Saarbrücken, Gemany (2011), pp. 2,3.

Gatot Rusbintardjo, Mohd Rosli Hainin, Nur Izzi Md. Yusoff (2013), *Fundamental and rheology properties of oil palm fruit ash modified bitumen,* Journal Construction and Building Materials Vol. 49 (2013) pp. 703–711. Journal homepage: www.elsevier.com/locate/conbuildmat.

King, G., King, H., Pavlovich, R.D., Epps, A.L., and Kandhal, P.S. (1999). Additives in Asphalt. *Journal of Association of Asphalt Paving Technology.* Vol. 68, pp. 32–69.

Read John and David Whiteoak, The Shell Bitumen Hand Book. Fifth edition, Thomas Telford Publishing Thomas Telford Ltd., 1 Heron Quay, London E14 4 JD, 2003, pp. 136–139.

Robert, F.L., Kandhal, P.S., Brown, E.R., Dah, Y.L., and Kennedy, T.W. (1996). *Hot Mix Asphalt—Materials, Mixture Design and Construction.* 2nd edition. NAPA Education Foundation, Lanham, Maryland. pp 448–463.

Robert, F.L., Kandhal, P.S., Brown, E.R., Dah, Y.L., and Kennedy, T.W. (1996). Hot Mix Asphalt—Materials, Mixture Design and Construction. 2nd edition. NAPA Education Foundation, Lanham, Maryland. pp. 50–.

Advances in Energy Science and Equipment Engineering II – Zhou, Patty & Chen (Eds)
© 2017 Taylor & Francis Group, London, ISBN 978-1-138-71798-5

Experimental study on the loading capacity of high titanium slag reinforced concrete slim beams

Shuanghua Huang & Xiaowei Li
Civil and Architectural Engineering College, Panzhihua University, Panzhihua, China

Long Xie, Hanfeng Duan & Song Zhong
Architecture and Civil Engineering College, Xihua University, Chengdu, China

ABSTRACT: In this experiment, six test beams were designed and tested, which included four high titanium slag reinforced concrete slim beams and two ordinary reinforced concrete slim beams. The cross-sectional size and length of the test beam is $150 \times 250 \times 1500$ mm. Experiments on normal section bending performance of high titanium blast furnace slag-reinforced concrete slim beams and ordinary reinforced concrete slim beam of different reinforcement ratios and intensity levels are carried out. The research shows that (1) under the conditions of same concrete strength and reinforcement ratio, the cracking moment, the yield moment and ultimate load of high titanium slag concrete beams are higher than those of ordinary concrete beams and the safety factor of flexural carrying capacity is also higher. (2) Mean strain along the height of the cross-section of high titanium slag concrete beam distributes into two upper and lower triangles and fits with horizontal section assumption. (3) In the concrete design of same strength and same reinforcement ratio, the deflection of C30 and C40 high titanium slag-reinforced concrete beams under the same load is smaller and stiffness is stronger than ordinary beams. The above conclusions provide technical support for high titanium blast furnace slag concrete slim beams in structural engineering applications.

1 INTRODUCTION

In order to make full use of industrial waste, in Panzhihua region, gravel and sand consisting of high titanium slag aggregates are used to produce coarse aggregate and fine aggregate to prepare high titanium slag concrete. Then this concrete is widely used in engineering constructions. However, the feasibility of the use of this particular structure of high titanium slag concrete slim beam has not been proved practically. In this paper, flexural properties of high titanium slag reinforced concrete slim beam are studied.

2 EXPERIMENTAL PROGRAM

2.1 Design of test pieces

Combined with the existing equipment in Panzhihua University Structural Engineering Laboratory and the current code for structure design, in this experiment, six test beams were designed and tested, including four high titanium slag reinforced concrete slim beams (number recorded as LA1, LA2, LB1, LB2), and two ordinary reinforced concrete

beams (number denoted by LA, LB). The cross-sectional size of the test pieces is 150×250 mm and length is 1500 mm. While producing test pieces, samples used for testing material performance were kept separately. Every time six poured concrete test pieces are left along with three steel test pieces of each diameter, and the mechanical properties of concrete and steel materials are shown in Tables 1 and 2. Table 3 shows the details of the reinforcement of beams.

2.2 Experimental loading and survey

The ends of a simply supported beam are pivotally mounted on a steel bending girder platform.

Table 1. Mechanical properties of concrete and steel materials.

Types	Levels (N/mm²)	Yield strength (N/mm²)	Ultimate strength (N/mm²)	Modulus of elasticity (Gpa)
Φ16	HRB335	388	620	200
Φ8	HPB235	312	503	210

Table 2. Designed mix ratios of reinforced concrete slim beam.

Beam numbers	Designed strength	Water-cement ratio	Materials amount (kg/m³)		High titanium heavy slag sand (ordinary sand)	Blast furnace slag gravel (ordinary gravel)	Fly	Cube compressive strength
			Cement	Water				
LA1 LA2	C30	0.6	327	230	726	1192	87	38.5
LA	C30	0.45	440	198	(372)	(1400)	0	32.7
LB1 LB2	C40	0.5	3961	230	627	1229	104	48.9
LB	C40	0.44	452	199	(352)	(1407)	0	42.3

Table 3. Reinforcements of each beam.

Beam numbers	Longitudinal reinforcement	Reinforcement rates (%)	Stirrup
LA1	3Φ16	1.61	Φ8@60
LA2	4Φ16	2.14	Φ8@60
LA	3Φ16	1.61	Φ8@60
LB1	3Φ16	1.61	Φ8@60
LB2	4Φ16	2.14	Φ8@60
LB	3Φ16	1.61	Φ8@60

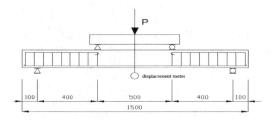

Figure 1. Schematic diagram of the test load manner.

In order to eliminate the effect of shear on the section curve, through centralized loading, two symmetrical points are formed under the action of simply assigned supported beam of the test beam. The method of test loading is shown in Figure 1. In the case of ignoring the self-weight, and taking only bearing bending without shearing into account, in order to prevent the concrete from local compression damage and to ensure the evenness of the contact stress, the loaded point features a 200 × 80 × 12 mm steel plate, and 1000 kN hydraulic pressure testing machine is used for loading. Computer data acquisition system is used for real-time data acquisition.

2.3 Experimental method

2.3.1 LA1 beam
When loading member LA1 to 53 kN, the first vertical crack occurs at about 50 mm away from mid-span, the width of which is less than 0.05 mm. When loaded to 67 kN, new cracks appear more rapidly and develop to three, slowly growing in length to about 30–50 mm. When loaded to 96 kN, the crack width increases to 0.18 mm. When loaded to 136 kN, both fracture length and width rapidly increase. The maximum crack width touches 0.4 mm and appears in the lower load points of the hinge bearing. The crack width of mid-span also increases to 0.35 mm. Loaded to 190 kN, the local concrete displays the crisp phenomenon. Around the original main cracks, a lot of horizontal fractures can be seen, which develop like a tree. Below the load point, the maximum fracture width reaches 0.7 mm. Loaded to 210 kN, the samples enter the "ductile" stage. The load increased at a significantly slower rate, and the cracks occur sharply. There is a surge in the deflection of the beam. When the load increases to 226 kN, the concrete load decreases slightly after being crushed.

2.3.2 LA2 beam
When loading the member LA2 to 59 kN, the first vertical cracks appeared at the mid-span. The width is less than 0.05 mm, and length is about 30 mm. Loaded to 110 kN, the crack length rapidly widens. The crack length is about 70–80 mm, and the width is about 0.0–0.15 mm. The number of fractures also increase. Loaded to 181 kN, both the crack length and width are rapidly increasing. The crack width is 0.05–0.2 mm, and the maximum crack width reaches 0.32 mm. After loading to 250 kN, loading speed significantly slows, and cracks expand dramatically. There is a surge in the

Table 4. Loads of various beams.

Number	Measured cracking load P_{cr}/kN	Measured yield load P_{y1}/kN	Theoretical yield load P_y/kN	Measured ultimate load P_u/kN
LA1	53	136	133	226
LA2	59	181	177	280
LA	54	135	133	212
LB1	65	126	118.5	214
LB2	62	163	150.5	263
LB	64	120	118.5	209

deflection of the beam. When loaded to 280 kN, beam reaches its ultimate strength. The load is decreased after the concrete is pressed crisply.

2.3.3 Others beams

The damage mode of each beam is substantially similar. On reaching a certain load, the beam begins to crack. When it reaches the theoretical yield load, the development of cracks accelerates and deflection begins to occur. After reaching the ultimate load, the upper part of the concrete is pressed crisply and the beam is completely destroyed. The measured cracking load P_{cr}, measured yield load P_{y1}, theoretical yield load P_y and measured ultimate load P_u are shown in Table 4:

3 ANALYSIS OF EXPERIMENTAL RESULTS

3.1 Horizontal section assumption analysis

In the test, from top to bottom in the middle span of the beam, three strain gauges are successively pasted on the top, middle and lower parts to, respectively, measure concrete strains at the height of 0, 75 and 150 mm. Then the distribution of mean strains of concrete along the height of each concrete beam under certain loads are obtained, as shown in Fig. 2.

As can be seen from Figure 2, as the load increases, the moment constantly increases, and the mean strain of the beam subsequently increases, always forming two basic triangles, that is, mean strain is in agreement with the assumption of the horizontal section.

3.2 Load–deflection at middle span analysis

During the experiment, displacement meter at middle span of the beam always records displacement values in the middle span position. Displacement of each beam at middle span with load variation is shown in Figure 3.

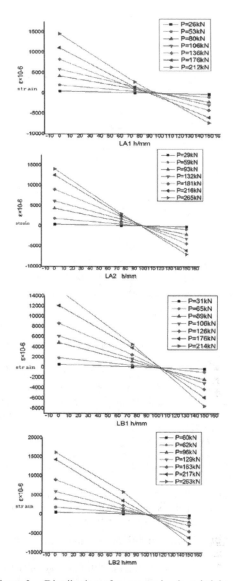

Figure 2. Distribution of mean strain along height of high titanium blast furnace slag slim beam.

According to the analysis of Figure 3, C30 and C40 concrete slim beams have similar changing modes, that is, in the concrete design of same strength and same reinforcement ratio, the deflection of high titanium blast furnace slag concrete beams under the same load is slightly smaller than that of the ordinary beams. In other words, the stiffness of high titanium blast furnace slag concrete slim beams is marginally stronger than ordinary concrete slim beams; by the comparison of slim beams of LA1-LA2 and LB1-LB2, it can be found that when the concrete strength

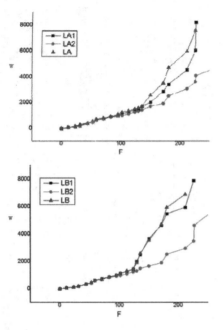

Figure 3. Variation of displacement at middle span of beam with load.

is the same, the higher the reinforcement ratio is, the smaller the deflection of the slim beam at middle span will be, which is in line with theoretical predictions.

4 CONCLUSIONS

In this study, by experimental research and theoretical analysis of six beam samples, the following basic conclusions can be drawn:

1. Under the conditions of same concrete strength and reinforcement ratio, the cracking moment, the yield moment and ultimate load of high titanium slag concrete beams are higher than those of ordinary concrete beams and safety factor of flexural carrying capacity is also higher.
2. The mean strain along height of the cross-section of high titanium slag concrete beam distributes into two upper and lower triangles and fits with the horizontal section assumption.

3. In the concrete design of same strength and same reinforcement ratio, the deflection of C30 and C40 high titanium slag reinforced concrete beams under the same load is smaller and stiffness is stronger than ordinary beams.

REFERENCES

Chen Hui-fa. Constitutive equation of civil engineering materials. [M]. Huazhong Science and Technology Press, 2001.

Chen Quan, Shi Yong-jiu, Wang Yuan-qing, Chen Hong, Zhang Yong. Loading Capacity of Steel-Concrete Composite Slim Beam [A]. Seventh International Symposium on Structural Engineering for Young Experts [C], Tianjin, China, August 28–31 2002.

Chen Quan, Shi Yong-jiu, Wang Yuan-qing, Chen Hong, Zhang Yong. System analysis of multilayer light steel frame structure with a combination of slim beams [J]. Building Structure, 2002.

Chen Quan. Loading capacity analysis of composite slim beam [D]. Beijing: Tsinghua University, 2002.

Chen Wei, Huang Shuang-hua, Sun Jun-kun, Jiao Tao. Experimental research on normal section intensity of high titanium blast furnace slag reinforced concrete beam [J]. Sichuan Building Science, 2009(4): 51–53.

GB 50010—2010. Code for Design of Concrete Structures [S]. AQSIQ, 2010.

GB/T 50152—2012. Standards of Test Methods for Concrete Structures [S]. 2012.

Jiang Hai-min, Mou Ting-min, Ding Qing-jun. Research on working performance of high titanium heavy slag concrete [J]. Concrete, 2011(5): 125–127.

Li Xiao-wei, Chen Wei, Li Xue-wei. Experimental study on seismic behaviours of high titanium heavy slag aggregates high-strength concrete column [J]. Building Structure, 2013(9): 96–100.

Sun Jin-kun, Chen Wei, Huang Shuang-hua, LI Ying-min. Mechanics performance of complex high titanium heavy slag reinforcement concrete beam [J]. Advanced Materials Research, 2011, 1068(168): 2013–2020.

Sun Jun-kun, Huang Shuang-hua, Nian Hong-fen, Cheng Min, Li Bing. Experimental study on optimization design of mix ratio of complex high titanium heavy slag pavement concrete. [J]. Concrete, 2011(8): 135–137.

Sun Jun-kun, Chen Wei, Li Ying-min, Zhou Wen-feng. Experimental research on bonding performance between complex high titanium heavy slag concrete and steel bars [J]. Sichuan Building Science, 2010(4): 216–219.

Advances in Energy Science and Equipment Engineering II – Zhou, Patty & Chen (Eds)
© 2017 Taylor & Francis Group, London, ISBN 978-1-138-71798-5

Strain dependent dynamic properties of clay–gravel mixtures

Kang Fei, Jinxin Xu, Jian Qian & Wei Hong
Institute of Geotechnical Engineering, Yangzhou University, Yangzhou, China

ABSTRACT: Clay–gravel mixtures are widely distributed in nature and are frequently used in civil engineering projects. To improve the understanding of the effects of coarse particles on the strain dependent dynamic properties, a series of cyclic triaxial tests was conducted on clay-gravel mixtures with various coarse contents and confining stresses. It was observed that the normalized shear modulus decreased and damping ratio increased with the increase in the coarse content. However, the effect of coarse particles was negligible at high shear stain levels. The influence of the confining stress on the strain dependent dynamic properties was found to be small. Based on the test results, empirical relationships were proposed to estimate the normalized shear modulus and the damping ratios at different strain levels. In addition, a numerical analysis of an ideal rockfill dam with clay-gravel core was performed to demonstrate the effects of the coarse content on the dynamic response.

1 INTRODUCTION

In the practice of geotechnical engineering, coarse particles are often added to the clay to improve the strength and the stiffness. For example, to reduce the post-construction settlement and the arching effect of earth-rockfill dams, a certain amount of gravels are usually added to the clay core. In road engineering, gravels are also often added to the embankment fill to obtain a better performance. These kinds of soils are usually called as clay-gravel or clay-aggregate mixtures (Fei, 2016). A complete understanding of the normalized shear moduli and damping characteristics of such composite materials is very important to analyze the dynamic response of the corresponding soil structures.

In the past decades, numerous studies have been performed to investigate the dynamic behaviors of clays. Hardin and Black (Hardin, 1968) studied the effects of void ratio and effective confining stress on the shear modulus by torsional vibrations tests. Kim and Novak (Kim, 1981) investigated the effects of confining stress on the strain dependencies of shear modulus and damping ratio. A series of cyclic triaxial tests was conducted by Kokusho et al. (Korkusho, 1982) to study the effects of consolidation histories. The extensive laboratory tests made by Seed and Idriss (Seed, 1970), Hardin and Drnevich (Hardin, 1972) showed consistently that the plasticity index is an important factor. A various of empirical relationships has been proposed to evaluate the strain-dependent shear moduli and damping characteristics (Mayora, 2016; Brennan, 2005). However, the applicability of these empirical formulas is limited to pure clays.

The presence of gravels in the mixture is expected to change its dynamic behaviors. Based on the results of dynamic hollow cylinder tests of sand-clay mixtures, empirical relationships to estimate the strain dependent shear modulus and damping ratio were proposed by Yamada et al. (Yamada, 2008; Yamada, 2008). In their studies, the effects of sand content were represented by equivalent plasticity index. But for clay-gravel mixtures, the plasticity index cannot be obtained by conventional laboratory tests due to the existence of coarse particles, their formulas are not applicable. Yamada et al. (Yamada, 2008; Yamada, 2008) also concluded that both the initial shear modulus and the damping ratio increased with the increase in sand content, while the opposite trend was observed for the normalized shear modulus. Similar observations have been made by Shafieeh and Ghate (Shafiee, 2008). However, a different pattern in dynamic properties with the coarse gravel content was observed by Meidani et al. (Meidani, 2008). Based on the data obtained from 54.3%, 64.1% and 73.5% gravel content samples, they concluded that the samples with higher gravel content have higher normalized shear modulus and lower damping ratio. One possible reason is that for the tested range of gravel content, the gravels were in contact with one another, and the formed coarse particle structure dominated the dynamic behaviors.

As reviewed, the strain dependent dynamic properties of clay-gravel mixtures have not been comprehensively studied yet. The published experimental data are very limited, and some findings are conflicting. To obtain a full understanding of the effects of the coarse content on the dynamic behaviors of clay-gravel mixtures, a series of stress controlled cyclic triaxial tests were performed in this study. In addition, a numerical analysis of an ideal rockfill dam with clay-gravel core was performed to demonstrate the effects of the coarse content on the dynamic response.

2 EXPERIMENTAL PROGRAM

2.1 Test materials

The clay used in the investigation was kaolinite clay. Its liquid limit and plastic limit were 43 and 21, respectively. The specific gravity of the clay was 2.62. For all the specimens, the dry density of the clay was kept constant at 1.55 g/cm³. The sampling water content water content of the clay was 21%.

Glass beads were used as the coarse fraction added into the clay. The specific gravity of the glass beads was 2.54, which is similar to that of natural gravels. The diameter of the glass beads was 1.4 cm. The main purpose of using the uniform-sized glass beads was to eliminate the effects of shape and graduation of coarse particles. The volume of the glass beads relative to the total volume of the mixture P was selected as 0, 8%, 16%, 24%, and 32%, respectively. These coarse aggregate contents cover the normal range used in the composite clay core of dams.

2.2 Specimen preparation and test procedure

The cyclic triaxial tests were performed by GDS dynamic triaxial system. The specimens were 10 cm in diameter and 20 cm in height. All the test specimens were prepared using a moist tamping technique. First, the glass beads were mixed with the dry clay based on the designated coarse content. Then, the required amount of water was added to the soil mixture, and the soil was mixed thoroughly to ensure homogeneous distribution of coarse particles. Subsequently, the mixed soil was kept inside a plastic bag for 3 days to moisture homogenization. Afterward, the soil was compacted using a 2.5-kg rammer to the desired dry density of the clay. The number of rammer blows was determined by trial and error. After the compaction, the specimens were saturated by vacuum-pumping for 24 h.

During the cyclic triaxial tests, the specimens were first isotropically consolidated to a mean confining pressures σ'_m of 100, 200, and 400 kPa. Then, the specimens were subjected to cyclic sinusoidal loads with a loading frequency of 1 Hz in the undrained condition. To study the strain dependent dynamic characteristics of the soils, 10 loading steps with different amplitudes were employed for each specimen. In each loading step, the cyclic load was applied 10 cycles. At the end of each loading step, the drainage valve was opened to dissipate the developed excess pore-water pressure.

3 TEST RESULTS

3.1 Equivalent shear modulus

The variations of the equivalent shear modulus G_{eq} versus the single amplitude shear strain γ_a at various

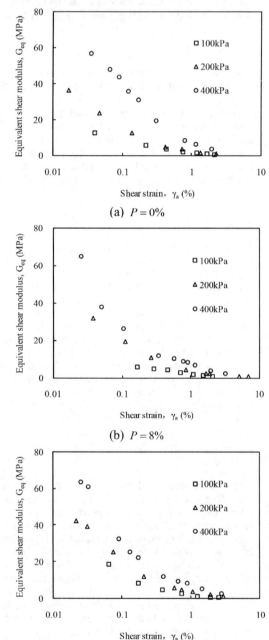

Figure 1. (Continued)

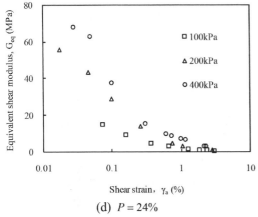

(d) $P = 24\%$

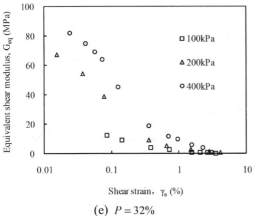

(e) $P = 32\%$

Figure 1. Equivalent shear modulus versus shear strain.

coarse contents and initial confining stresses are presented in Figure 1. In the figure, the equivalent shear modulus was determined from the slope of the secant line connecting the two extreme points of the hysteresis loop at the 8th cycle. By fitting the data of each specimen, the initial shear modulus G_0 can be obtained.

Figure 2 shows the relationship between initial shear modulus and confining stress for all the coarse contents considered in the present study. It is observed that both the effective confining stresses and the coarse content have remarkable effects on the initial shear modulus. The initial shear modulus at a given coarse content tends to increase with the increase in the confining stress, and the relation can be described approximately by a power rule.

$$G_0 = k_0 P_a \left(\sigma'_m / P_a \right)^n \tag{1}$$

where P_a is the atmospheric pressure; k_0 and n are test parameters, the determined values are shown in Table 1.

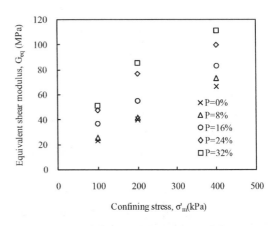

Figure 2. Initial shear modulus versus confining stress.

Table 1. Summary of k_0 and n.

$P(\%)$	0	8	16	24	32
k_0		272	280	374	500	565
n		0.60	0.65	0.60	0.52	0.53

For the specimens with the same confining stress, the initial shear modulus increased with the increase in the coarse content and the rate of increase was faster when the coarse content was larger than 16%. This phenomenon can be explained from the point of the view of force chains. For the specimens with relatively low coarse contents, the coarse particles were far away from each other, the dynamic behavior was controlled by the clay. When more coarse particles were added into the host clay, due to the heterogeneous matrix structure, the clay fractions trapped between coarse particles were consolidated at higher confining stress and were stiffer than the other part of the clay. These relative stiffer clay fractions formed the force chain together with coarse particles. Therefore, the larger coarse content leaded to a stronger force chain and a higher initial shear modulus.

As observed in Figure 1, the effect of coarse particles in increasing the equivalent shear modulus is also obvious. However, at large shear strains beyond about 1%, the effect became negligible and each specimen showed the same stiffness, irrespective of the variation in the coarse content. This result occurred because the clay-gravel skeleton has already collapsed at such large strains, and the clay fraction dominated the equivalent shear modulus.

3.2 Normalized shear modulus

It is customary to investigate the modulus reduction by normalizing the equivalent shear modulus with the initial shear modulus. Figure 3 shows the strain-

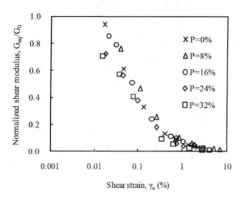

Figure 3. Relationship between normalized shear modulus and shear strain for different coarse content ($\sigma'_m = 200\ kPa$).

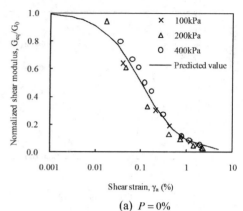

(a) $P = 0\%$

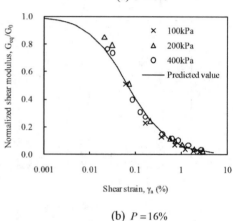

(b) $P = 16\%$

Figure 4. Relationship between normalized shear modulus and shear strain for different confining stress.

dependent normalized shear modulus G_{eq}/G_0 with different coarse contents at a confining stress of 200 kPa. It is evident that samples with high coarse content have lower normalized shear modulus with the increase in strain level. The similar conclusion has also been made by Yamada et al. (Yamada, 2008). One reason for this finding is that the deformation induced by the cyclic loading occurred mainly in the clay since the stiffness of individual coarse particles was much greater than that of the clay. The actual shear strain developed in the clay matrix increased with the coarse content. In consequence, the shear modulus reduced faster.

It is general recognized that for sands, the rate of reduction in shear modulus with strain becomes greater as the confining stress decreases, while there is practically no effect of the confining stress on the strain dependency of clays (Ishihara, 1996). To investigate the influence of the confining stress on the normalized shear modulus, the $G_{eq}/G_0 \sim \gamma_a$ of samples with two coarse contents of $P = 0\%$ and 16% at three different confining stresses of 100, 200, 300 kPa are shown in Figure 4(a) and 4(b) respectively. As observed, each clay-gravel mixture sample exhibited an almost unique shear modulus reduction curve irrespective of the confining stress. It is clear that within the adopted coarse contents, the confining stress has no influence on the strain dependent normalized shear modulus.

The hyperbolic model has been used widely to describe the nonlinear stress strain relations of pure clays. Based on this model, the relationship for the shear modulus in the cyclic loading can be expressed as

$$\frac{G_{eq}}{G_0} = \frac{1}{1 + k_1 \gamma_a} \tag{2}$$

where k_1 is the test constant. For pure clay specimens, k_1 was obtained as 980. The computed

$G_{eq}/G_0 \sim \gamma_a$ by Equation (2) is shown in Figure 4 as the solid line. The computed values are found to agree well with the test values of pure clays. As discussed earlier, G_{eq}/G_0 decreases with the increase in the coarse content, which implies a larger value of k_1 should be used in Equation (2). Based on test results, a relationship of hyperbolic type can be assumed between k_1 and the coarse content. Hence, the relationship between the normalized shear modulus and the shear strain of clay-gravel mixtures can be expressed as follows:

$$\frac{G_{eq}}{G_0} = \frac{1}{1 + k_1 \left(1 + \dfrac{P}{a + bP}\right) \gamma_a} \tag{3}$$

where a, and b are test parameters and are estimated to be 0.24 and 0.37 respectively; P is the volume content of gravels in decimal representation. The predicted $G_{eq}/G_0 \sim \gamma_a$ relationship for the specimen with 16% coarse content is also presented in Figure 4(b). There is a good agreement between the predicted and the test values.

3.3 Damping ratio

The relationships between the damping ratio D and the shear strain γ_a with different coarse contents at a confining stress of 200 kPa are shown in Figure 5. It is observed that both the shear strain level and the coarse content have remarkable effects on the damping ratio. Similar to that of pure clays, the damping ratios of clay-gravel mixtures increased with increasing shear strain to a stable value D_{max} when the shear strain approached a extremely large value. It may also be seen that the damping ratio has a tendency to increase as the coarse content increases. This may be explained in terms of the slippage between the clay and coarse particles, which resulting in a larger energy loss than that of the pure clay. Because the number of contact points between gravels and the clay increased with the increase in the coarse content, samples with larger coarse content showed larger damping ratios. However, when the shear strain was greater than 1%, the clay has been fully sheared and the effect of coarse particles was not significant, the values of D_{max} of clay-gravel mixtures are found to be approximately 0.23, which is close to that of the pure clay. It should be noted that mixtures with extremely high coarse content may have a different trend, since the coarse particles will control the dynamic behavior in those cases.

Figure 6 shows the $D \sim \gamma_a$ of samples with two coarse contents of $P = 0\%$, 16% at three different confining stresses of 100, 200, 300 kPa. As observed, the test data under the three confining stress are concentrated in a narrow band area, the effect of confining stress on the strain dependent damping is negligibly small. Similar trends are also observed at other coarse contents.

According to the test results, there is a good correlation between the damping ratio D and the normalized shear modulus G_{eq}/G_0. The following empirical relationship was proposed

$$D = D_{max}\left(1 - G_{eq}/G_0\right)^{\beta} \qquad (4)$$

where D_{max} is the maximum damping ratio, which can determined by the asymptotic value of the damping ratio curve; β is the test parameter; G_{eq}/G_0 is the normalized shear modulus computed by Equation (3). Because the maximum damping ratio did not change significantly with the coarse content, and the shape of damping ratio curve at different coarse content can be represented by the normalized shear modulus, the average value of test data of the pure clay can be used to estimate the value of D_{max} and β. In this study, D_{max} and β were obtained as 0.23 and 1.5 respectively. The fitted curves are compared with the test data in Figure 6, a good agreement is observed.

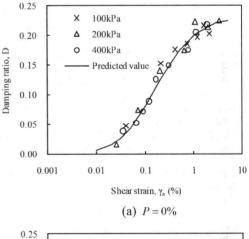

(a) $P = 0\%$

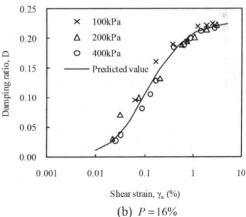

(b) $P = 16\%$

Figure 6. Relationship between damping ratio and shear strain for different confining stress.

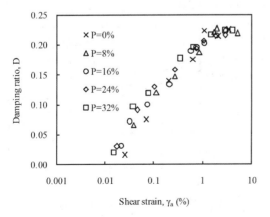

Figure 5. Relationship between damping ratio and shear strain for different coarse content $\left(\sigma_m' = 200 kPa\right)$.

4 NUMERICAL CASE STUDY

4.1 Finite element model and input motion

To demonstrate the effects of the coarse content on the dynamic response, a 2D seismic analysis was carried out for an ideal rockfill dam with clay-gravel core. The cross section of the dam is shown in Figure 7. The crest has a width of 15 m and a height of 100 m. The upstream and downstream slopes of the dam are 1V:2.0H, and those of the core zone are 1V:0.2H.

For the purpose of this example, the dam–foundation interactions were neglected by assuming that the foundation is rigid. The horizontal ground accelerations recorded during the Koyna earthquake was used as the input motion. In the analysis, the input Peak Ground Acceleration (PGA) was scaled to 0.1 g, 0.3 g, and 0.5 g respectively. The vertical acceleration component was not included in the analysis.

4.2 Material models and parameters

Three kinds of core material were accounted in the analysis: (i) pure clay, (ii) clay mixed with 16% gravels, and (iii) clay mixed with 16% gravels. The dynamic behaviors of the rockfill and the core were both modeled using equivalent linear visco-elastic model, which described through G_0, $G_{eq}/G_0 \sim \gamma_a$, and $D \sim \gamma_a$. Equation (1) was used to determine the values of G_0, and the $G_{eq}/G_0 \sim \gamma_a$ and $D \sim \gamma_a$ curves were obtained by Equation (3) and (4) respectively. The parameters of core material were determined according to the test data in this study, and those of the rockfill were estimated from typical values in the literature. All the parameters used in the analysis are summarized in Table 2.

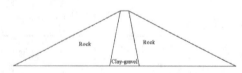

Figure 7. Cross-section of the dam.

Table 2. Dynamic parameters of dam materials.

| | Core | | | |
	$P = 0\%$	$P = 16\%$	$P = 32\%$	Rockfill
$\rho(g/cm^3)$	1.84	1.94	2.04	2.20
k_0	272	374	565	2150
n	0.6	0.6	0.53	0.4
k_1	980	980	980	2240
a	0.24	0.24	0.24	–
b	0.37	0.37	0.37	–
D_{max}	0.23	0.23	0.23	0.2
β	1.5	1.5	1.5	1

4.3 Dam response results

4.3.1 Peak absolute acceleration

Figure 8 shows the distribution of computed peak absolute acceleration a_{max} along the dam centerline for the case of $P = 0\%$. In the figure, z is the distance measured from dam base and H is the dam height. For the input PGA of 0.1 g, 0.3 g, and 0.5 g, the maximum peak absolute accelerations were 0.44 g, 1.22 g, and 1.40 g respectively. It is observed that the computed accelerations were almost equal to or little smaller than the input PGA over about 80% of the dam height and exhibited large amplifications in the crest region. The similar distributions of a_{max} were also observed for other cases of coarse contents.

To investigate the effect of the coarse content on the peak acceleration, the relationships between the acceleration amplification coefficient α at the dame crest and the coarse content are presented in Figure 9.

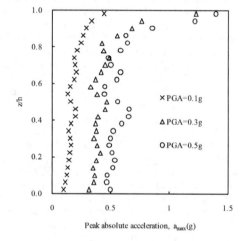

Figure 8. Distribution of peak acceleration at the centerline $(P = 0\%)$.

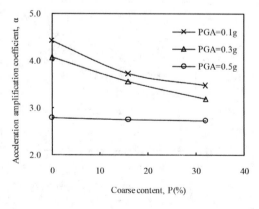

Figure 9. Acceleration amplification coefficient versus coarse content.

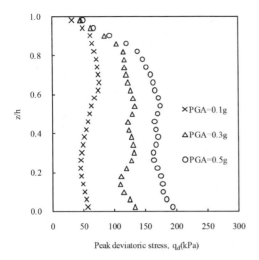

Figure 10. Distribution of peak deviatoric stress at the centerline ($P = 0\%$).

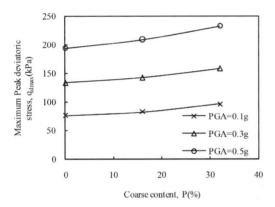

Figure 11. Peak deviatoric stress versus coarse content.

It is evident that the peak acceleration decreased with the increase in the coarse content. For example, the acceleration amplification coefficient reduced from 4.43 to 3.48 as P increased from 0 to 32% at an input PGA of 0.1 g. The decrease in the acceleration response is believed to be related to the larger shear modulus at higher coarse contents. However, because the shear modulus drops faster for the case with higher coarse content, the difference of the computed accelerations was very small at strong earthquakes. This is corresponding to the observation from the test that the effect of coarse content on the equivalent shear modulus was negligible at large shear strain levels.

4.3.2 *Peak deviatoric stress*
Figure 10 shows the peak deviatoric stress q_d profiles at different input intensity levels for the case of $P = 0\%$. For the PGA of 0.1 g, 0.3 g, and 0.5 g,

the maximum peak deviatoric stresses were 76 kPa, 133 kPa, and 193 kPa respectively.

The relationships between the maximum peak deviatoric stress q_{dmax} and the coarse content are given in Figure 11. It is noted that the maximum peak deviatoric stress increased remarkably with the coase content even at large input intensity levels. This is because a large density of the clay-gravel mixture was used in the analysis for a high coarse content. Therefore, in order to guarantee the safety of the dam, the effects of the coarse content on the dynamic shear strength should be studied.

5 CONCLUSIONS

The normalized shear modulus decreased and damping ratio of clay-gravel mixtures increased with the increase in the coarse content, and the influence of the confining stress was small. However, the effect of coarse particles was negligible at high shear stain levels. The proposed empirical relationships provided reasonable estimation to the strain dependent dynamic properties. The numerical results of an ideal rockfill dam shows that the peak acceleration at the dam crest decreased with the increase in the coarse content at low input intensity level. The maximum peak deviatoric stress increased remarkably with the coase content.

ACKNOWLEDGEMENTS

The authors acknowledge support from the Natural Science Foundation of Jiangsu Province (No. BK20141279) and the Qing Lan Project (No. 20160512) by the Jiangsu Province Government.

REFERENCES

Brennan, A.J., N.I. Thusyanthan, S.P. Madabhushi, Evaluation of shear modulus and damping in dynamic centrifuge tests, J. Geotech. Geoenviron. Eng., **131**(12):1488–1497, (2005).
Fei, K. Experimental study of the mechanical behavior of clay–aggregate mixtures, Eng. Geol., **210**, 1–9, (2016).
Hardin, B.O., V.P. Drnevich, Shear modulus and damping in soils: measurement and parameter effects, ASCE J. Soil Mech. Found., **98**(6), 603–624, (1972).
Hardin, B.O., W.L. Black, Vibration modulus of normally consolidated clay, ASCE J. Soil Mech. Found., **94**(2), 353–369, (1968).
Ishihara, K. *Soil behaviour in earthquake geotechnics*, Clarendon Press, (1996).
Kim, T.C., M. Novak, Dynamic properties of some cohesive soils of Ontario, Can. Geotech. J., **18**(3), 371–389, (1981).

Korkusho, T., Y. Yoshida, Y. Esashi, Dynamic properties of soft clay for wide strain range, Soils Found., **22**(4),1–18, (1982).

Mayoral, J.M., E. Castañon, L. Alcantara, T. Simon, Seismic response characterization of high plasticity clays, Soil Dyn. Earthqu. Eng., **84**: 174–189, (2016).

Meidani, M., A. Shafiee, G. Habibagahi, M.K. Jafari, Y. Mohri, A. Ghahramani, C.S. Chang, Granule shape effect on the shear modulus and damping ratio of mixed gravel and clay, Iran. J. Sci. Technol., **32**(B5), 501–518, (2008).

Seed, H.B., I.M. Idriss, Soil moduli and damping factors for dynamic response analyses, *Report No. EERC 77–10*, University of California, Berkeley, (1970).

Shafiee, A., R. Ghate, Shear modulus and damping ratio in aggregate-clay mixtures: an experimental study versus ANNs prediction, J. Appl. Sci., **8**(18), 3068–3082, (2008).

Yamada, S., M. Hyodo, R.P. Orense, S.V. Dinesh, T. Hyodo, Strain-dependent dynamic properties of remolded sand-clay mixtures, J. Geotech. Geoenviron. Eng., **134**(7), 972–981, (2008).

Yamada, S., M. Hyodo, R.P. Orense, S.V. Dinesh, Initial shear modulus of remolded sand-clay mixtures, J. Geotech. Geoenviron. Eng., **134**(7), 960–971, (2008).

Advances in Energy Science and Equipment Engineering II – Zhou, Patty & Chen (Eds)
© 2017 Taylor & Francis Group, London, ISBN 978-1-138-71798-5

Effect of artificial pozzolan on the strength and resistance to chloride ion penetration of marine concrete

Baoshi Jiang, Hu Juna & Haiyang Zhu

College of Civil Engineering and Architecture, Hainan University, Haikou, China

ABSTRACT: An alternative equivalent method was used to investigate the impact of the artificial pozzolan content on the strength and resistance to corrosion of sulfur aluminate cement marine concrete to reduce the application costs. First, 42.5 MPa sulfate aluminate cement mixed with artificial pozzolan was used to prepare a fast early-strength concrete. An orthogonal experimental method was used to arrive at the optimal concrete design for the C30 concrete. Then, the effect of artificial ash on marine concrete strength and resistance to chloride ion penetration were studied. Studies show that the at the combination of 400 kg/m³ of cement plus 20% ash content, or 310 kg/m³ of cement plus 20% to 35% ash content, the concrete strength and resistance to chloride ion permeability are high, reaching the Q-IV level of Chinese specification.

1 INTRODUCTION

The cost of sulfate aluminate cement is higher than that of Portland cement. Adding ash and other admixtures to the concrete has important practical significance in reducing their application costs. Many Chinese scholars have carried out research on the sulfur aluminate cement concrete. X. Wang et al. (Wang, 2011) studied the effect of the addition of the admixtures [m(slag):m(ash) = 2:1] to the sulfoaluminate cement concrete's compressive strength and impact on its impermeability. The results showed that admixtures that provided the early and late strength to concrete significantly reduced permeability, and the higher the content, compressive strength, the more it is easy to reduce impermeability. X. Jiang et al (Jiang, 2011) studied the chloride ion penetration of sulfur cement concrete with mixed material content (the mixed material is 1:1 slag and fly ash). Studies show that with the increase in admixtures, anti-chlorine ion permeability decreases. Zhao et al (Zhao, 2011) studied the mechanism of resistance to chloride ion corrosion of the sulfur cement concrete, and the results showed that sulfur cement concrete has good resistance to chloride corrosion performance and at a lower water–cement ratio, this performance will be enhanced. S. Yao et al (Yao, 2010) summarizes the impact of different admixtures (slag, zeolite powder, fly ash) on the late strength of sulfoaluminate, which leads to different degrees of improvements for retraction, too short setting time and other defects, indicating that with an increase in incorporation of admixtures, the early

hydration slowed down and early strength reduced, but this has little effect on the late strength. D. Zhang et al (Zhang, 2010) studied the concrete rheology and microstructure of the sulfur cement and observed that the polycarboxylate superplasticizer and amino and naphthalene superplasticizer for cement have good slurry rheological properties. S. Huang et al (Huang, 2011) studied the effect of setting time, hydration process, hydration product types by adding concrete retarder and hardening accelerator to the fast hard sulfur cement.

Based on the results from research, it is clear that commonly used cement admixtures are volcanic ash, slag, fly ash, silica fumes, etc. They can be used alone or in combination in order to improve concrete properties of the cement. Currently admixture research on ordinary Portland cement has been more comprehensive and mature, which basically concerns their chemistry and microstructure of hydration products.

But for sulfoaluminate cement, admixtures in the current study were mainly slag, fly ash and their mixture. Our research shows how the volcanic ash, slag, fly ash and their admixtures affecting the hydration properties of this cement is extremely rare. The reason is the territorial distribution of volcanic ash, as volcanic areas may distribute lava or ash, and the high cost of its acquisition is detrimental to their large-scale application. The artificial pozzolan formed by ash fly ash, slag, basalt and others is an excellent choice. Its chemical composition is shown in Table 1, and its chemical composition is similar to natural pozzolan.

Table 1. Artificial chemical composition of volcanic ash.

Ingredient name	Al_2O_3	CaO	Fe_2O_3	K_2O	MgO	MnO	Na_2O	SiO_2	SO_3	TiO_2
content (%)	14.95	9.1	10.74	1.51	4.05	0.145	4.43	52.21	0.0759	2.08

In this paper, 42.5 MPa by mixing artificial pozzolan with sulfate aluminate cement, fast early strength concrete was prepared to reduce the application costs as an equivalent of cement. Artificial pozzolan content, sand ratio and water–cement ratio were investigated by the orthogonal experiment method to ascertain the strength requirements of C30 concrete. Under the optimal concrete design from the orthogonal experiment, the paper also studies the relationship between artificial pozzolan on the strength of the cement and its resistance to chloride ion permeability.

2 THE ORTHOGONAL DESIGN

In order to investigate the impact of artificial pozzolan on concrete strength, water–cement ratio, artificial pozzolan content and sand rate were as the three factors considered in the study. Each factor had four levels, and an orthogonal experiment was designed to find the best values of each factor. Increasing the strength of concrete is the main goal of the orthogonal experiment. In order to ensure the work performance of concrete, an appropriate amount (0.8% of cement and pozzolan) of water reducer was added.

2.1 The orthogonal design parameters

2.1.1 Compressive strength and water–cement ratio
The strength of concrete has an inverse relationship with water–cement ratio; water–cement ratio is the most direct factor that determines the strength of concrete. High-strength concrete mostly incorporates silica fume, ground slag, fly ash and other mineral admixtures and super plasticizer. So the relationship formula of water–cement ratio and strength is not the same for different concrete. The design strength of ordinary concrete is determined in accordance with its 28-day strength. The relationship between sulfate aluminate cement's concrete strength and water–cement ratio refers to the design strength of the Portland cement for 28 days. As for the early-strength concrete, a smaller water–cement ratio leads to a higher strength. Taking into account the convenience of construction design, the concrete design mainly follows the "Ordinary concrete design procedures (JGJ 55–2011)". Thus, the initial water–cement ratio of early strength concrete is about 0.4.

2.1.2 Water and sand rate
In the selected conditions for water–cement ratio, the amount of water represents the number in the cementation material. The amount of water in the ordinary concrete depends mainly on the concrete slump and aggregate properties. China's current concrete design code is based on the value of water consumption of concrete as determined by concrete slump and maximum particle size of the stone. Due to the incorporation of early-strength concrete superplasticizer and a high influence of water in artificial pozzolan modified concrete, slump is largely determined by the size of the water reducer and artificial pozzolan content. Thus water and slump degree have no significant correlation.

The amplitude of aggregate properties is small; the impact on water is not high. To minimize the amount of cement, the volume of water chosen was around 170 kg/m³. Sand ratio is the ratio of the amount of the fines and aggregate, since the impact of artificial volcanic ash is not very significant, taking into account that the cohesiveness and retention are relatively good after artificial volcanic ash is added. Therefore, a smaller value (30%) can be added, according to the sand ratio of ordinary concrete.

2.2 Exploratory experiments

In order to reduce blindness of the tests, before the formal experiment, in accordance with the above principles and selected parameters, exploratory experiments were conducted. using the assumed apparent density method and considering the right amount of water reducer for the water-reducing efficiency of artificial volcanic ash. The amount of water was 168 kg/m³ and water–cement ratio was 0.4. A 30% sand and concrete slump of 10–30 mm, and the superplasticizer dosage of 0.8% a apparent density of 2460 kg/m³ were the initial conditions. Orthogonal test was designed on this basis to find the optimal mixture ratio.

2.3 The content of the orthogonal test program

On the basis of a tentative trial design and its results, water–cement ratio, ash content and sand ratio and artificial pozzolan are the three factors considered and each factor takes four levels in the orthogonal design. The concrete strength was the assessment indicator to find the optimum mix of early-strength concrete. The orthogonal design factors and levels are shown in Table 2, and the orthogonal table L_{16} (4^5) was selected. The test program, range analysis process and the 3D concrete compressive strength test results are shown in Table 3.

It can be seen that the range B is greater than range A, which is higher than range C, as shown in Table 3. The effect of B is high, followed by A but the effect of C is minimal.

Table 2. Factors and levels of the concrete in the orthogonal experimental design.

Factors	A—artificial pozzolan (%)	B—water–cement ratio	C—sand ratio (%)
Level 1	10	0.4	30
Level 2	20	0.45	35
Level 3	30	0.5	40
Level 4	40	0.55	45

Table 3. Test program and results.

Column	1	2	3	3d strength (MPa)
Factor	A	B	C	
Test1	1	1	1	47.8
Test 2	1	2	2	38.1
Test 3	1	3	3	33.0
Test 4	1	4	4	30.5
Test 5	2	1	2	39.6
Test 6	2	2	1	36.3
Test 7	2	3	4	29.4
Test 8	2	4	3	28.4
Test 9	3	1	3	38.4
Test 10	3	2	4	29.9
Test 11	3	3	1	29.3
Test 12	3	4	2	23.7
Test 13	4	1	4	29.7
Test 14	4	2	3	25.9
Test 15	4	3	2	23.0
Test 16	4	4	1	18.2
mean value 1	37.4	38.9	32.9	
mean value 2	33.4	32.6	31.1	
mean value 3	30.3	28.7	31.4	
mean value 4	24.2	25.2	29.9	
range	13.1	13.7	3.00	

First, the direct analysis of single indicator gave an insight on the primary and secondary factors and different combinations (results shown in Table 2). The results of using various factors and their impact were also ascertained (see Fig. 1).

Factor A—Artificial pozzolan: Artificial pozzolan content and 3D strength have a linear relationship. For every 10% increase in dosage, the strength decreased by 10%. When the content exceeds 20%, the strength decreased further. So the optimum dosage of pozzolan content evaluated from 3D strength is 10%. But the strength with 20% dosage remains the standard to meet the specification requirements. Considering the economy, the best dosage of artificial pozzolan is 20%.

Factor B—water–cement ratio: There is a linear relationship between water–cement ratio and strength. The larger the water–cement ratio, the lower the strength. Taking the 3D compressive strength as the assessment indicator, the lower the water–cement ratio, the higher the concrete strength, so the water–cement ratio of 0.4 was found to be appropriate.

Factor C—sand ratio: It has little impact on the strength overall. Its role is to ensure enough lubrication between the coarse aggregate filled with cement mortar, greater friction between the coarse aggregate particles, mixing was less mobile and low slump. Considering all these, a sand ratio of 0.35 was selected.

From the above analysis, the best combination is A2B1C1, namely the optimum water–cement ratio is 0.4, artificial pozzolan content is 20%, the sand rate is 0.3.

2.4 Artificial pozzolan effects on the strength and resistance to chloride ion penetration

Keeping the water cement ratio as 0.4, sand rate as 0.35, water reducing agent at 1.2%, cementing material consumption at 400 kg/m³, concrete strength and resistance to chloride ion penetration were studied by varying the ash content between 0% and 40%. Concrete strength was measured by a pressure test using 150 mm × 150 mm × 150 mm standard cube specimen. The electric flux measurement method in the section 7.2 of "Ordinary

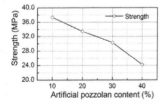 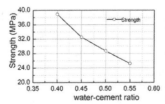 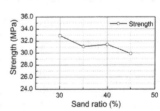

Figure 1. The relationship between artificial pozzolan, water–cement ratio, sand ratio and strength.

concrete and long-term durability test method standards GB-T50082" was used to assess the concrete's resistance to chloride ion penetration. The results are shown in Table 4.

The results shown in Table 1 indicate that when ash content of more than 30%, 3-day strength is reduced by 24% than the non-doped cement. The electric flux was at a same level, but at 56 days, the strength was significantly reduced. The overall electric flux improved slightly, with a content reduction of 20% from the initial strength, which is within an acceptable range, while the electric flux is minimized. The reason for this phenomenon is the amount of cement is larger than amount of cement in the previous scheme. At a cement consumption of 400 kg/m³, although using ash replaced 40% of the amount of cement, sulfur cement concrete has a higher strength to meet the C30 concrete strength rating requirements. According to Table 5, the concrete has a high resistance to chloride ion penetration in the Q-IV and Q-V grade.

3 ARTIFICIAL POZZOLAN EFFECTS AT A MINIMUM DOSAGE ON STRENGTH AND RESISTANCE TO CHLORIDE ION PENETRATION

Seen from the above test content, the 3D strength of the concrete in ash mixed with sulfur aluminate cement-base is determined by its cement. At a certain amount of water, a small water–cement ratio means more amount of cement, and so the strength will be improved. At this time, more artificial volcanic ash can be incorporated to reduce cement consumption. Four schemes are presented here to compare the strength of its 3-day and 28-day strength as well as its resistance to chloride ion penetration. The amount of cement in accordance with Section 5.2.5 of "Corrosion Prevention Technical Specifications for Concrete Structures of Marine Harbor Engineering (JTJ275–2000)" is not less than 300 kg/m³. Concrete resistance to chloride ion permeability using the flux measurement method is provided in Section 7.2 of the "Ordinary Concrete Long-term and Durable Performance Test Method Standards (GB-T50082) ".

As can be seen from Table 6, the 28-day strength of these four programs with the ratio to 3-day strength increased by about 10% to 25%, and they meet the strength requirements of C30 concrete. From Tables 2 and 7, it is shown that the electric flux in the four mixed program is Q-IV within the target range, and they have good resistance to chloride ion permeability. From the table, it can be seen from BJ1 and BJ2 that the chloride ion permeability coefficient increased with an increase in the admixtures. This conclusion is consistent with the literature (Zhao, 2011). The reason is that the increase of blending quantity reduces the cement content in cementation materials, and so with increasing ash content, the chloride ion permeability decreases.

Table 4. Artificial pozzolan effects on the strength and resistance to chloride ion penetration.

Test No.	Ash content %	3-day strength MPa	56-day strength MPa	56-day electric flux C
HS0	0	48.0	55.9	644.7
HS1	10	45.4	42.3	512.7
HS2	20	40.5	49.0	445.4
HS3	30	34.3	47.0	674.0
HS4	40	36.4	47.5	612.64

Table 5. Concrete resistance to chloride ion penetration grading (electric flux method).

Grade	Q-I	Q-II	Q-III	Q-IV	Q-V
Electric flux Q_s(C)	$Q_s \geq 4000$	$2000 \leq Q_s < 4000$	$1000 \leq Q_s < 2000$	$500 \leq Q_s < 1000$	$Q_s < 500$

Note: Quoted from "Concrete Quality Control Standards GB50164–2011".

Table 6. Economic programs and their strength and electric flux test results.

Test No.	Artificial pozzolan (%)	Water–cement ratio	Sand ratio (%)	Superplasticizer (%)	Cement content (kg/m³)	3d strength (MPa)	28d strength (MPa)	Electric flux (C)
BJ1	20	0.4	35	1.2	336.0	40.8	50.1	690.6
BJ2	25	0.4	35	1.2	315.0	38.9	48.3	759.7
BJ3	30	0.38	35	1.2	309.5	38.1	42.9	762.6
BJ4	35	0.36	35	1.2	303.3	37.4	44.4	946.4

Correspondingly, the electric flux and the amount of cement in Table 7 also show this characteristic.

4 CONCLUSIONS

In this paper, the orthogonal experiment was used to ascertain the best mixture ratio for early strength concrete to reduce its application cost. Studies have shown that when the sulfur aluminate cement content is 400 kg/m^3, ash content is 20%, the concrete strength and resistance to chloride ion permeability is optimum. When the cement content is about 310 kg/m^3, and artificial pozzolan content varies from 20% to 35%, concrete strength and resistance to chloride ion permeability still are good, reaching the Q-IV level of specification for the strength C30.

ACKNOWLEDGEMENTS

The authors wish to express their gratitude for the financial support to the Scientific Research Starting Foundation of Hainan University (NO.kyqd1402), the Hainan Natural Science Foundation (20155214, 20155211), the Academic Innovation Program of Hainan Science Association for the Youth scientific and technological excellence (201505), the National Science Foundation of China (51368017), the Key Research & Development Science and Technology Cooperation Program of Hainan Province (ZDYF2016226), Science and Technology Program of Hainan Province (ZDXM2015117), Post-doc Research Fund of China (2015M580559), the Scientific Research Project of Education Department of Hainan Province (Hnky2016ZD-7, Hnky2015–10), the SRF for ROCS, MOHRSS (MOHRSS [2014] 240) and the Research Project of Ministry of Housing and Urban Rural Development (2016-k5–060).

REFERENCES

Huang, S., C. Wu, R. Yang. The impacts of concrete admixtures on sulfur aluminate cement hydration process. China concrete and cement products, 2011, 1:7–12. (In Chinese)

Jiang, X.H., Z.X. Guo, Q.Z. Liu, K.F. Sun. Effect of mixed material content on chloride ion penetration of sulfur cement concrete. 21st century building materials, 2(2): 23–27, 2011. (In Chinese)

Wang, X., Y. Zhang, J. Luan, J. Li. Influence on the compressive strength and the impermeability of concrete of sulphoaluminate cement with composite admixture. Shandong chemical industry, 40:57–60, 2011. (In Chinese)

Yao, S., G. Hu, M. Wang, Y. Ding. Study on effect of sulfur aluminate cement performance of different admixtures. Guangdong Building Materials, 9:14–16, 2010. (In Chinese)

Zhang, D., M. Wang, Z. Ding, X. Bai, M. Zhang. Rheology and micro-structure analysis of sulphoaluminate cement concrete. China Powder Industry, 4(13):13–18, 2010. (In Chinese)

Zhao, J., G. Cai, D. Gao. Analysis of mechanism of Resistance to Chloride ion erosion of Sulphoaluminate cements concrete. Journal of building materials, 14(3):357–361, 2011. (In Chinese)

Advances in Energy Science and Equipment Engineering II – Zhou, Patty & Chen (Eds)
© 2017 Taylor & Francis Group, London, ISBN 978-1-138-71798-5

Experimental research and finite element analysis on fire resistance of joints of end-plate connections in steel structure

Yu Wang, ZhaoBo Zhang & Hong Hai
School of Civil Engineering, Shenyang Jianzhu University, Shenyang, China

ABSTRACT: In order to provide a theoretical basis and a reasonable structure form for fire resistance design, the fire-resistance performance of end-plate joints in steel structure was studied. The key parts and the failure load of end-plate connection joints were studied through the full scale fire test, and beam-column joints were simulated under different working conditions using the finite element analysis software ANSYS. Simulation results were compared with the experimental results. The load-displancement curves under different temperatures were obtained. It is founded that in order to ensure the ductility of structure under elevated temperatures, the premature failure of fillet welds and bolts in end-plate connections should be avoided. When the temperature exceeds 550°C, the tensile and shear failures in the contact parts of bolts and end-plate and beam web are dangerous zones. To guarantee the stability of structure under fire, the stiffening ribs should be used in engineering to prevent structural collapse.

1 INTRODUCTION

Steel building construction offers many benefits to structural engineers, such as lower cost, lighter weight, low maintenance and fast construction. As a consequence, portal frames, steel structures and steel-concrete composite structures are widely used for high-rise and office buildings in many places around the world. However, there are two problematic issues closely associated with a steel-framed structure: corrosion and fire. As far as fire is concerned, steel is a non-combustible material, but the mechanical properties of structural steel are affected by temperatures. One of the most important factors to describe the fire performance of a structure is fire resistance, which is defined as the time during which structural elements can withstand the standard provisions of a fire test. Based on this, the author intends to use the full-scale fire test and the finite element analysis to research on the mechanical properties and heat-response characteristic under high temperature of joints of end-plate connections. The appropriate constructional measure is given and the theoretical basis of the fire-resistant design is provided.

2 TEST SURVEY

2.1 Specimen of test

According to the structure and stress characteristics of joints of end-plate connections, the beam, column and end-plate were used rolled H-section steel. The section size of column was H250 × 250 × 9 × 11, and the section size of beam was H294 × 200 × 8 × 12. The thickness of end-plate was 9 mm, and the bolt was used M20 high strength bolts (grade 8.8), and diameter was 22 mm. Test component parameters were shown in Figure 1. The load heating mode was used constant temperature heating, and temperature rasing curve was used natural fire curve.

2.2 Experimental device

The experiment was used the fire test furnace of building elements whose effective size of furnace body was 3.0 m × 1.5 m × 1.5 m, and the working temperature was 0~1200°C. The fuel was used liquefied petroleum gas, and the data acquisition of the furnace temperature was used industrial

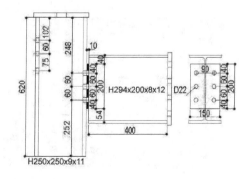

Figure 1. Design details of experimental component.

computer and independent collection with real-time curve. The loading system was used hydraulic loading system, as shown in Fig. 2.

2.3 *Loading mode*

This inclined tensile force was applied by using an assembled loading system. The steel connection specimen was connected to the support beam via a rigid connection, and the top end of the column was fixed on the reaction frame through two support Macalloy bars. The angle (α) between the furnace bar and the beam axis determined the ratio of shear and tensile components applied to the connection. This load ratio was determined before the testing by introducing the initial load angles of 45° or 55°. During the testing, movements of the furnace bar and rotation of the steel beam changed this load ratio leading to failure of a specimen at a different shear/tension ratio. The magnitude and angle of this inclined force were monitored and recorded throughout the experiments. A tension meter was mounted at the jack rod and the link rod in order to measure the specific value of the applied force during the experiments, as shown in Fig. 3.

Figure 2. Test unit.

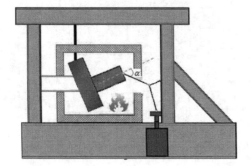

Figure 3. Test unit diagram.

3 EXPERIMENTAL PHENOMENA AND RESULTS ANALYSIS

3.1 *Experimental phenomena*

With the temperature rising, the end plate firstly occurs shear failure, and then beam web occurs shear failure, as shown in Fig. 4. The main reasons are as follows: ① Most of the columns and beams are covered with ceramic fibers, so the elevated temperature of columns and beams is far below the end plate near the site; ② The connection of the end plate and beam would occur greater stress, so it is easy to become weak parts; ③ The loading angle of the test component make tension and shear forces to work together in near the end plate parts, thus the ultimate bearing capacity and yield stress of the part would be decreased.

3.2 *The results of the test analysis*

From experimental results of end-plate connections component (Table 1) can be seen, the angular displacement of the corner area of the junction and maximum failure load would be decreased with the temperature rising, so the region belongs to the weak areas, the damage is easy to be occured in the part, it does not comply with the seismic design requirements about "strong pole piece and weak lever". In order to improve the structural strength design value, the stiffening rib must be set to prevent structural collapse.

Figure 5 shows the load—displacement curve of different applied loading angles. As can be seen from the figure, the curve clearly shows the nonlinear response of joints of end-plate connections. With the temperature rising, the carrying capacity of joints of end-plate connections would gradually be decreased. It turned out that the temperature should have a significant impact on the carrying capacity. But when the joints of end-plate connections withstanded high temperature, its ability of rotating the angle have not been changed. It is

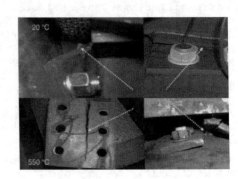

Figure 4. Destruction form of the end plate.

Table 1. Experimental results of end-plat connections component e.

Test No.	Angular displacement connection area	The maximum failure load (kN)
End-plate 35-20	8.7	193.00
End-plate 35-450	4.3	91.36
End-plate 35-550	3.8	69.51
End-plate 35-650	3.5	33.55
End-plate 45-20	8.9	150.30
End-plate 45-450	3.7	65.50
End-plate 45-550	3.3	45.10
End-plate 45-650	3.6	28.95

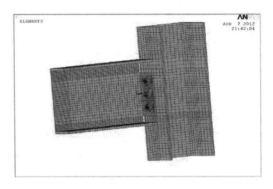

Figure 6. Model of ANSYS.

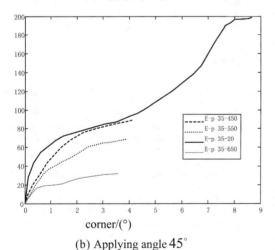

(a) Applying angle $35°$

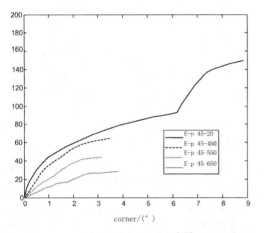

corner/(°)

(b) Applying angle $45°$

Figure 5. Load and displacement curves.

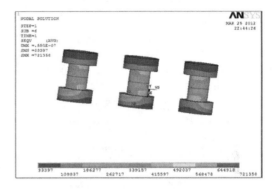

Figure 7. Bolt force diagram.

4 FINITE ELEMENT CALCULATION

A three dimensional numerical model was created for a flexible end-plate connection, using the ANSYS finite element code, in order to investigate its resistance and ductility at ambient and elevated temperatures. This model started with the creation of individual components such as bolts, end-plates, beams and columns, and then assembled these components into a connection, as shown in Fig. 6 and Fig. 7. When the mesh were divided, the thickness of the grid is required to make a transition, and the density of the grid is set to a moderate degree, which will be beneficial to the simulation of higher precision. Then the SOLID70 element is used to simulate the thermal state. Because of the symmetry of the structure, the 1/2 structure is used to model.

5 COMPARISON AND ANALYSIS OF TEST RESULTS AND FINITE ELEMENT RESULTS

Load—displacement curve were compared in Figure 8. As can be seen from the figure, the two curves almost coincide with the increase of the load in the

worth mentioning that joints of end-plate connections rotation capacity at 650°C were higher than 450°C and 550°C.

1219

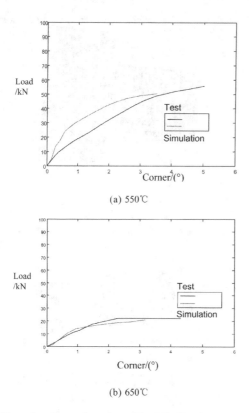

(a) 550℃

(b) 650℃

Figure 8. Comparison chart of load displacement curves.

elastic stage and the elastoplastic stage. It can be founded that the ultimate bearing capacity of the structure is different by comparing of two load—displacement curves when the material reaches the yield. In general, the ultimate bearing capacity of the finite element simulation is slightly higher than the experimental value.

The reasons for the differences were analyzed:

1. Due to human factors, steel components in the process will be produced a part of the welding residual strain and the component stiffness was reduced.
2. The actual test environment is disturbed by outside interference of various factors, while the ideal finite element model was existed with pure and no interference in virtual space environment, and different environment will greatly affect the mechanical properties of components. Therefore, the results of the difference in test and simulation would be obtained.
3. In the production process of the test member, it is difficult to avoid some error between the actual size of the model and the exact dimensions of the test component; it is also an important reason that the size deviation result in some error of the test and simulation.

6 CONCLUSIONS

1. Numerical simulation calculation results and experimental results show that the numerical simulation is an effective research tools, and simulation method can give experimental research and engineering application to bring great convenience.
2. Comparisons between the finite element simulation and the component-based modelling were presented, which demonstrated the simplified connection model to be capable of representing the complex behaviour of steel joints. This new simplified method will be used to analyze the performance of a Rugby-post sub-frame under fire conditions.
3. When the temperature exceeds 550°C, the portion of contacting the bolt and end-plate is easy to be occured tension and shear failure and cause the structure to collapse. In order to ensure the stability of beam-to-column connections under fire, the stiffening rib should be set to prevent structural collapse.

ACKNOWLEDGMENTS

The authors would like to acknowledge the financial support of Shenyang Science and Technology Planning Project (F14-186-1-00), Major production safety accident prevention and control of key technologies of science and technology projects (2013-10), Major production safety accident prevention and control of key technologies of science and technology projects (2013-12), General fund project of Shenyang Jianzhu University (2014-077).

REFERENCES

American Society of Civil Engineering (ASCE) [S]. New York: Manual and Reports on Engineering pratice NO.78 Structural Fire Protection, ASCE, 2008.
ECCS (European Convention for Constructional Steelwork) European Recommendations for the Fire Safety of Steel Structures, 2005.
Ershad Ullah Chowdhury. Behaviour of fibre reinforced polymer confined reinforced concrete columns under fire condition [D]. Queen's University, 2009.
Harmathy T Z. Deflection and failure of steel supported floors and beams in a fire [S]. ASTM STP422, Symposium on Fire Methods-Restraint and smoke, Philadelphia, ASTM Publications, 2009.
Theodorou Y. Mechanical Properties of Grade 8.8 Bolts at Elevated Temperatures [D]. University of Sheffield, 2006.
Wang Y Z, Behavior and design of composite beam in fire with considering global structural effect [D]. Shanghai: Tongji University, 2008.
Zehfuss, J., D. Hosser. A parametric natural fire model for the structural fire design of multi - storey buildings [J]. Fire Safety Journal, 2006(42):115–12.

Advances in Energy Science and Equipment Engineering II – Zhou, Patty & Chen (Eds)
© 2017 Taylor & Francis Group, London, ISBN 978-1-138-71798-5

Experimental study on acoustic emission characteristics of the coal failure process under different confining pressures

Huiming Yang

State Key Laboratory of Gas Disaster Detecting, Preventing and Emergency Controlling, Chongqing, China
Chongqing Research Institute Co. Ltd., China Coal Technology and Engineering Group, Chongqing, China

ABSTRACT: In order to study the effect of confining pressure on coal failure characteristics and Acoustic Emission (AE) evolution law, the coal failure experiments of coal under uniaxial and conventional triaxial stresses are carried out by using rock mechanics testing machine and AE monitoring instrument. The results show that mechanism of coal failure is changed gradually from brittle tensile mode to ductile shear failure under the action of confining pressure, and the energy and distribution characteristics of AE events also undergo a change. Under low confining pressure, the AE event energy is concentrated in the break moment around the peak stress area. On the contrary, the energy distribution of AE events is relatively dispersed with higher confining pressure. In elastic deformation stage of coal, the quantity of AE event under triaxial stress is significantly lower than that under the uniaxial load. In the post-peak stage, under the effect of confining pressure, the coal failure process shows a progressive damage characteristics, and AE activities gradually reduce, and the "relative decline period" lasts for a long time.

1 INTRODUCTION

With the progressive increase of mining depth, there is an obvious change of mechanical properties of coal or rock and in-situ stress in deep mines compared to shallow. The complex "high stress, high temperature, high gas pressure and strong mining disturbance" environment (three high one strong) was encountered in deep mines. The interaction effect of coal bump and gas outburst is more significant. The form of disaster is gradually transformed from a single disaster into a coupling one, and the index threshold of the coupling disaster is lower than that of the single disasters. For example, the stress-dominated outburst still occurred during excavation in some deep mines after gas drainage to meet the requirement of outburst prevention regulation. Therefore, coal mine production safety faces greater technical challenges, and a study of pre-warning and prevention methods of coal or rock dynamic disasters in deep mines is urgently needed. Using Acoustic Emission (AE) technology to realize the monitoring and pre-warning of stress-dominant dynamic disasters is an important development in the direction of deep mine disaster prevention. AE is the elastic stress wave generated by the energy release of coal rock material destruction under load (Hardy, 2003; Jiang, 2000; Zou, 2004). So the AE method is a monitoring and pre-warning technology reflecting changes of stress and stability of coal or rock mass

(Yang, 2015; Zuo, 2011). Meanwhile, studies show that acoustic emission method is applicable to the monitoring and early warning of stress-dominant dynamic disaster in deep mines (Yang, 2015; Yang, 2013).

AE characteristics of coal or rock failure process is the theoretical basis of AE technology in the monitoring and early warning of mine dynamic disasters, and a number of achievements in this aspect have been published. Mogi (1962) conducted the rock failure test and found that AE activity occurs in four patterns by turn with the loading process: weak, sporadic or small, gradually increasing with the load increase, a sharp increase near rock failure (Kiyoo, 1962). Zhang Ru (2006), Cao Shugang (2007) found the "relatively quiet period" phenomenon that the rate of AE events declined before the failure of coal (rock), and viewed it as the precursory characteristic of rock failure. Li Yushou (2011) conducted AE characteristic study of coal samples failure under uniaxial compression, triaxial compression and pore water pressure. Su Chengdong (2013) conducted scale effect study on AE activities of coal failure under uniaxial load with different sizes of coal samples, and found that failure of different size samples show magnitude of differences in AE activity. Li Hongyan (2014) carried out the precursor information study on acoustic emission of coal with different bursting proneness under stress loading, and found that, as the level of coal's outburst proneness increases, the counts of coal vibration and energy of acoustic emission were concentrated in the high stress region. The time proportion of the elastic stage increases while time proportion of plastic stage decreases. Cao Anye (2015) found that the releasing form of AE energy changes from solitary earthquake type to swarm type along with the increasing loading rate, and the peak AE energy increases greatly under a higher loading rate.

The research results are of great significance for the AE precursor characteristics of coal or rock instability and disaster pre-warning. Deep coal stress state is different from the shallow stress, and stress state has a larger impact on the characteristics of deformation and failure and the AE characteristics. Therefore, this paper analyses the AE evolution characteristics of coal failure process under different conditions of confining stress by means of uniaxial and triaxial compression tests, discusses the stress state effect on the AE characteristics of coal. It is meaningful to the improve the AE pre-warning technology of underground structural instability and dynamic disasters in deep mines.

2 MATERIALS AND METHODS

2.1 Materials

Test coal samples were taken from domestic weak bursting liability coal seam. According to the requirements for samples, the coal was cut into Φ50 mm × 100 mm cylindrical samples in the vertical direction of the bedding, where the non-parallelism of coal sample less than 0.05 mm. All of coal samples are shown in Figure.1.

2.2 Experiment equipment and method

The loading equipment of raw coal failure testing used TAW2000 computer controlled tri-axial testing machine, The device can offer 2000 KN maximum axial pressure, which can be used to conduct uniaxial and tri-axial loading tests. Acoustic emission monitoring is done using two-channel acoustic emission monitoring equipment, produced by

Figure 1. Photo of coal samples.

Table 1. Time parameter settings of AE acquisition.

Parameters	Time settings[μs]
Peak definition time (PDT)	150
Hit definition time (HDT)	300
Hit lockout time (HLT)	500

Beijing Shenghua technology company. The type of sensor is SR150 N and its' frequency ranges from 22 to 220 kHz. According to the waveform characteristics of AE in coal failure process, the test optimized the time parameters of acquisition settings, which are shown in Table 1.

In uniaxial compression test, AE monitoring system employs two sensors for data acquisition, which are fixed symmetrically to sides of the coal sample with tape, and contact faces are coupled with vaseline. In triaxial compression test, the sensors are fixed to the triaxial cylinder wall with an adhesive tape. During the triaxial test, first, confining pressure is exerted on the sample to the specified value (5 MPa, 7.5 MPa, 10 MPa). Second, load axial stress remains stable until the sample is completely damaged. In both tests, axial loading is controlled by displacement, and loading velocity is 0.005 mm/min. AE monitoring is done simultaneously with the loading in experiment.

3 EXPERIMENTAL RESULTS

In geological mechanics engineering, the monitoring and pre-warning of rock structure stability and dynamical disasters are usually based on AE activity frequency and energy characteristics analysis. Therefore, the event rate and energy parameters are mainly used to analyse the AE evolution characteristics of coal failure under different stress conditions to explore the influence effect of stress environment on the AE in this paper.

3.1 Experimental results of uniaxial compression failure

The test results of uniaxial compression failure is shown in Figure 2. The AE characteristics of coal failure from the beginning to the failure stage can be divided into several stages as follows:

1. Micro-crack closure stage: Only a small number of AE events are generated by contact and friction of crack surface in the closure or dislocation process of original micro-crack in the coal sample. Due to fewer new cracks, crack extension is not obvious, and the AE event rate and energy are low in this stage.

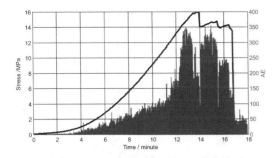

(a) AE event rate and stress curve

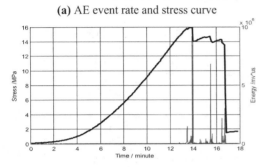

(b) AE energy and stress curve

(c) Photos of coal after failure

Figure 2.   Test results of a coal sample of weak bursting liability.

2. Elastic deformation stage: With the stress increasing, micro-cracks in coal are initiated and propagate, which causes AE activity to become active gradually, and event rate gradually increases.

3. Plastic deformation stage: AE activity increases sharply with the increase in stress, and gradually achieves the maximum in failure process of coal in this stage. And the crack activity gets into an unstable development phase. The internal crack propagates quickly, and gradually concentrates to become a macro-crack or macro-fracture. The coal begins to present localization damage in this stage.

4. Residual strength stage: The stress of weak bursting liability drops sharply after the peak strength of coal, and the AE activities also subsequently fall rapidly. In this stage, macro-crack

can be seen on the coal surface, the elastic energy declines sharply, and the energy release is lower.

3.2 Experimental results of triaxial compression failure

The test results of triaxial compression failure of coal under 5 MPa, 7.5 MPa and 10 MPa confining pressure are shown in Figs. 3 to 5. The AE characteristics of coal under triaxial stress drawn from the results are as follows:

1. Compared with uniaxial loading, crack closure stage in triaxial compression is very short, and the concave of early stress curve is not obvious. The results indicate that the most internal cracks of coal are already closed under the

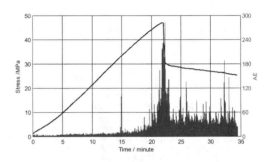

(a) AE event rate and stress curve

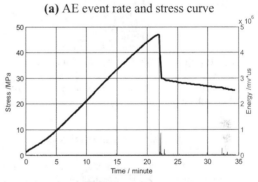

(b) AE energy and stress curve

(c) Photos of coal after failure

Figure 3.   Test result of a coal sample under 5 MPa triaxial compression.

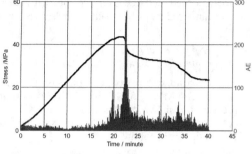

(a) AE event rate and stress curve

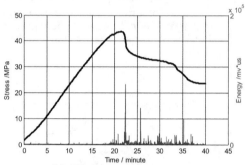

(b) AE energy and stress curve

(c) Photos of coal after failure (7.5MPa)

Figure 4. Test result of a coal sample under triaxial compression (7.5 MPa).

(7.5MPa).

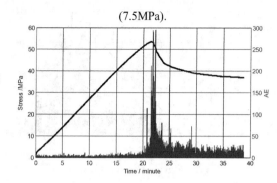

(a) AE event rate and stress curve

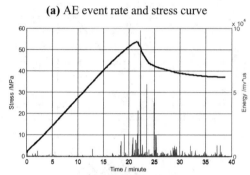

(b) AE energy and stress curve

(c) Photos of coal after failure (10MPa)

Figure 5. Test results of a coal sample under triaxial compression (10 MPa).

effect of confining pressure in this stage, and so the AE activity is not seen.

2. In the elastic deformation stage, the stress–strain curve is linear. Due to the effect of confining pressure, the whole strength of the coal is enhanced and axial load is not strong enough to break the coal. The internal micro-cracks slowly develop, and the AE activity in the phase is still not active.

3. In coal yield stage, the stress–strain curve gradually flattens and deviates from the straight line. The development of internal micro-cracks is gradually accelerated, and AE event rate and energy increase gradually. Near the peak stress, AE events rate and energy reach a maximum, which indicates that the macro fracture surface is forming.

4. In post-peak stage, axial bearing strength of coal is gradually reduced. Due to the formation of the fracture surface, crack extension activity is weakened and AE event rate is gradually reduced.

5. Compared with the results of uniaxial compression, the corresponding strains of coal samples under peak stress in triaxial compression are increased, and post-peak deformations are also increased, AE activity "drop" duration is longer than the one under uniaxial stress. Meanwhile, with the increase of confining pressure, the stress drop velocity in post-peak stage gradually slows. All these indicate that the coal gradually tends to ductile fracture with an increase in confining pressure.

4 ANALYSIS AND DISCUSSION

By comparing the AE results of coal under uniaxial and triaxial compression tests, it can be found that the confining pressure has an important influence on the failure characteristics and AE of coal. The effects of confining pressure on the coal mechanical failure characteristics are shown through two aspects. On the one hand, the compressive strength of coal is significantly increased under the influence of confining pressure. The strain amplitude of coal is increased obviously under the triaxial compression condition. The failure mode of coal is changed from brittle failure to ductile failure. On the other hand, under uniaxial compression of the macroscopic crack structure of weak bursting liability coal shows characteristics of multiple cracks with a small angle to the direction of load. It indicates that coal failure is controlled mainly by multiple cracks, and the crack extension mechanism is tensile failure. The macroscopic crack under the triaxial compression is single and present at a large angle. Moreover, with the increase of confining pressure, the angle of the failure fracture to the loading direction also increases, which indicates that the main failure mechanism is the shear failure.

The effect of confining pressure on AE evolution characteristics of coal are mainly manifested as follows: (1) By comparing the AE event energy feature under different confining pressures, it can be found that the distribution of AE event energy is relatively dispersed under the high confining pressure, while the energy is mainly concentrated on the stress peak area of the failure time under low confining pressure. With the increase of confining pressure, the energy of main crack event is gradually reduced and dispersed to nearly the peak stress, which indicates that the failure characteristics of coal is changed from drastic failure to progressive failure with the increase of confining pressure. (2) Compared with uniaxial compression, the AE signal of the coal under triaxial compression in the elastic deformation stage is significantly reduced, and with the increase of confining pressure, there is a decreasing trend of AE activity in this stage. (3) In the post-peak stage, with the effect of confining pressure, the coal failure presents a progressive damage of ductile failure characteristics, AE activity decreases gradually after a peak stress. Compared to the brittle failure under uniaxial compression, the AE "relative drop" period of coal is more obvious.

In the deep mine, coal and rock are in a high stress environment, and the deformation mechanism of coal changes from the brittle failure in shallow to ductile failure. But under the influence of mining, the stress path of coal failure is more complex, especially the coal damage is more drastic when the horizontal stress is unloaded. The experimental study[14] also indicated that in the damage process of rock under triaxial compression, with the increase of confining pressure, the brittleness of coal is weakened, and its ductility enhanced. Under the same confining pressure, unloading confining pressure before the peak stress can lead to the worst crushing and brittleness of coal failure. Therefore, confining pressure has an important influence on the failure mechanism and AE characteristics of coal or rock. The study in this paper is focused on the effect of confining pressure only by the loading stress path. Studies are still needed to be carried out to study the failure characteristics and AE characteristics of coal and rock under real stress path to clarify the AE evolution characteristics and pre-warning information of dynamic disaster in deep mine.

5 CONCLUSIONS

1. Confining pressure has an important influence on the coal failure mechanism. The failure mechanism of weak bursting liability coal under uniaxial stress is transformed gradually from brittle tensile failure to ductile shear failure under the effect of confining pressure.
2. The energy distribution characteristics of AE events also change under the confining pressure effect. Under low confining pressure, the AE event energy is concentrated in the break moment around the peak stress area, while the energy distribution of AE events is relatively dispersed in a higher confining pressure, which indicates that the failure characteristics of coal is changed from drastic failure to progressive failure with the increase of confining pressure.
3. AE activity distribution of coal failure process is also changed under the effect of confining pressure. In elastic deformation stage, the quantity of AE events under triaxial stress is significantly fewer than those under uniaxial load. With the increase of confining pressure, there is a decrease trend of AE activity in this stage. In the post-peak stage, under the effect of confining pressure, coal failure process shows the progressive damage characteristics, and AE activities gradually reduce. Here the "relative decline period" lasts for a long time, while it is short in case of a uniaxial compression.

ACKNOWLEDGEMENTS

This work was financially supported by the National Natural Science Foundation of China (51604298, 51574280), the State Key Research Development Program of China (2016YFC0801402), the Science and Technology Innovation Foundation Program of China Coal Research Institute (2015ZYJ002), and the General Program of China Coal Technology and Engineering Group Chongqing Institute (2015YBXM39).

REFERENCES

Cao Anye, Jing Guangcheng, Dou Linming, et al. Damage evolution law based on acoustic emission of sandy mudstone under different uniaxial loading rate. Journal of Mining & Safety Engineering, 32(06): 923–928+935, (2015).

Cao Shugang, Liu Yanbao, Zhang Liqiang. Study on characteristics of acoustic emission in outburst coal. Chinese Journal of Rock Mechanics and Engineering, 26(s1):2794–2799, (2007).

Hardy, H.R. Acoustic Emission/Microseismic activity—Volume1 Principles, Techniques and Geotechnical Applications. A.A. Balkerma Publishers, (2003).

Jiang Hai-kun Zhang Liu Zhou Yong-sheng. Behavior of acoustic emission time sequence of granite in deformation and failure process under different confining pressures. Chinese Journal of Geophysics, 43(6): 812–823, (2000).

Kiyoo Mogi. Study of Elastic Shocks Caused by the Fracture of Heterogeneous Materials and its Relations to Earthquake Phenomena. Bull. Earthq. Res. Inst., 40(1): 125–173, (1962).

Li Hongyan, Kang Lijun, Xu Zijie, et al. Precursor information analysis on acoustic emission of coal with different outburst proneness. Journal of China Coal Society, 39(2): 384–388, (2014).

Li Yu-shou, Yang Yong-jie, Yang Sheng-qi et al. Deformation and acoustic emission behaviors of coal under triaxial compression and pore water pressure. Journal of University of Science and Technology Beijing, 33 (06): 658–663, (2011).

Su Cheng-dong, Guo Bao-hua, Tang Xu. Research on acoustic emission characteristics of Zhangcun coal samples in two sizes subject to uniaxial compression. Journal of China Coal Society, 38(A01): 12–18, (2013).

Yang Huiming. Current status and development trend of AE identification method in pre-warning of coal and rock dynamic disaster. Mining Safety and Environment Protection, 42(04): 91–94, (2015).

Yang Huiming. Study and application of acoustic emission identification model of coal-gas dynamic disaster. Shandong, Qingdao: Shandong university of science and technology, (2013).

Yu Jin, Li Hong, Chen Xu et al. Experimental study of permeability and acoustic emission characteristics of sandstone during processes of unloading confining pressure and deformation. Chinese Journal of Rock Mechanics and Engineering, 33(01): 69–79, (2014).

Zhang Ru, Xie Heping, Liu Jianfeng etc. Experimental study on acoustic emission characteristics of rock failure under uniaxial multilevel loadings. Chinese Journal of Rock Mechanics and Engineering, 25(12):2584–2588, (2006).

Zou Yinhui. Preliminary study on coal and rock acoustic emission mechanism and relevant experiments. Mining Safety and Environment Protection, 31(01): 31–33, (2004).

Zuo Jianping, Pei Jianliang, Liu Jianfeng, et al. Investigation on acoustic emission behavior and its time-space evolution mechanism in failure process of coal-rock combined body. Chinese Journal of Rock Mechanics and Engineering, 30(8): 1564–1570, (2011).

Advances in Energy Science and Equipment Engineering II – Zhou, Patty & Chen (Eds)
© *2017 Taylor & Francis Group, London, ISBN 978-1-138-71798-5*

Experimental study on the strength parameter of unsaturated loess

Xuelian Wang & Yunfeng Zhao
Xi'an Institute of Geological Survey, China Geological Survey, Shanxi, China
Chang'an University, Shanxi, China

ABSTRACT: The problem of unsaturated soil arises in actual engineering. As an important research objective of unsaturated soils, the strength of loess problem is of widely concern. Based on the theory of unsaturated soil, the value of the friction angle of the suction parameters can be determined through direct shear test and determination of matrix suction. The test shows that with the increase of moisture content, the cohesive force decreases and the internal friction angle remains unchanged. Through mapping the soil–water characteristic curve, the relationship between moisture content and matrix suction can be obtained, which is an exponential relationship. With the increase of moisture content, matrix suction reduces gradually. When the suction is lower than a certain value, the friction angle of suction is constant.

1 INTRODUCTION

Loess is the typical unsaturated soil, and its strength reduces sharply with water, which is the main cause of a large number of engineering construction deformation damages and it also triggers landslide and collapse, which are geological hazards in the loess area. Air and water exist in the unsaturated loess at the same time. How the pore air pressure and pore water pressure affect the strength of the unsaturated loess, and various geological disasters and slope stability issues triggered by water have been the concern of researchers now. Finding the strength characteristics of unsaturated loess, selecting the parameters of unsaturated loess strength formula and predicting accurately the stability of unsaturated loess slope have great significance to the engineering construction and the safe operation of projects in the loess area.

For loess in the Huang Ling of Shanxi province as the research object,, the loess strength parameter values under different moisture contents are studied using the simple system sampling and laboratory test and the measured curves of soil water characteristic curve combine, which is analyzed and then the related parameters of unsaturated loess strength formula are obtained. The parameters not only have important theoretical significance for the loess mechanics study, but also provide a scientific basis on the loess landslide disaster prediction and prevention in Huang Ling, Shanxi province.

2 THE LOESS STRENGTH PARAMETER TEST UNDER DIFFERENT MOISTURE CONTENTS

2.1 *The basic physical indicators of the soil sample*

In this paper, the soil sample is taken from artificial loess (Q_2) slope in the Yang Gou village, Huang Ling, and the basic physical indicators are shown in Table 1.

2.2 *Consolidation slow shear test*

After drying the soil sample, the gravity water content of undisturbed soil samples is respectively configured at 0%, 5%, 10%, 15%, 20%, 25%, 30%, which forms a total of seven groups of soil sample. Slow shear tests with the seven samples with the same moisture content but varying vertical pressure as 50, 100, 200, 300, 400, 500, and 600 kPa, and keeping the shear rate at 0.02 mm/min are done, and the results are recorded.

According to the Coulomb Law $\tau = \sigma \tan \phi + c$. The data obtained from the above experiments and its results provide a linear fit (as shown in Figure 1). The R^2 of the linear fit are all greater than 0.97. The cohesive force and internal friction angle of the soil sample under different moisture contents can be obtained by fitting a linear straight line.

As shown in Figure 1, the intersection point of the extension line of fitting line and ordinate gives the cohesive force, and the straight angle is taken as

Table 1. The basic physical indicators of the testing loess.

soil	dry density (g/cm³)	natural moisture content (%)	specific gravity G_S	plastic limit $\omega_p/\%$	liquid limit $\omega_L/\%$	plasticity index I_p	fraction and content/(%)		
							sand >0.05 mm	silt 0.00~0.05 mm	clay <0.005 mm
loess	1.54	20.5	2.71	23.99	29.82	5.83	7.5	65.4	27.1

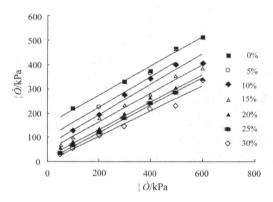

Figure 1. Relationship between the shear strength and stress for the different water contents.

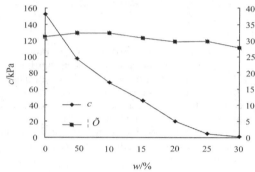

Figure 2. The influence of moisture content on cohesive force and internal friction angle.

the internal friction Angle. It can be seen that the internal friction angle is nearly a constant and cohesive force decreases with an increase of moisture content. The cohesive force and friction angle under different moisture content values concluded from Figure 1 can be shown in Table 2 and Figure 2.

As can be seen from Figure 2, the cohesive force obviously decreases with the increase of water content. At the same time, the angle of internal friction does not change with the change of moisture content, which remains close to 30°. Many studies (Jinqian, 1997) have reported that the angle of internal friction decreases with the increase of moisture content. Because this experiment is a drainage experiment and the influence of matrix suction is ruled out, the above value of φ is a valid value and remains unchanged.

2.3 The matrix suction test

In indoors, the matrix suction can be obtained through the pressure plate test, penetration drying test, controlling the humidity test and centrifugal dehydration experiment, by drawing the soil water characteristic curve of the unsaturated soil. This experiment adopts a special soil tension meter for testing, which is designed for measuring soil matrix suction at a small scale.

The tension meter used in the test is TEN tension meter, and the specific procedure is: make a hole, deep about 8 cm, in the center of the soil sample,

Table 2. The cohesive force and the friction angle for the different water contents.

w/%	0	5	10	15	20	25	30
c/kPa	152.1	97.7	67.9	45.5	19.4	4.9	0.4
$\varphi/°$	31	32	32.1	30.5	29.4	29.6	27.5

which is square (about 30 cm length) taken from the sampling point. Insert the exhaust and water-injection tension meter in the hole vertically, while the gap fills with the same moisture sand as the soil. Filling the soil samples with the water is continued until the soil samples reach near saturation, a tape is used to seal the sample. Then the sample moisture content is changed by means of natural air drying, and the data is read by showing by tension meter per injection level, which is the matrix suction under the moisture content. The suction under different moisture contents can be obtained in turn. The test results are shown in Figure 3.

Many scholars put forward different soil water characteristic curve equations. However, soil water characteristic curve of different soils is not the same. There are many test parameters that are difficult to determine. The fitting curve put forward by the related literature is mostly index curve. Therefore, based on experimental data in this paper, the soil water characteristic curve equation of unsaturated loess (as shown in Figure 3) is concluded, which can be used as a simple and practical equation.

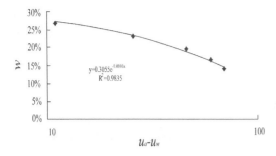

Figure 3. The soil-water characteristic curve.

3 TO DETERMINE THE PARAMETERS OF THE UNSATURATED LOESS STRENGTH FORMULA

3.1 Unsaturated loess strength characteristics

The strength of the unsaturated soil consists of the following four parts: effective cohesive force, friction between grains, matrix suction and the solute suction. As the ion concentration in pore solution of the common unsaturated soil is very small, the solute suction can often be ignored. Therefore, the research of the unsaturated soil strength characteristics mainly focuses on the contribution to the strength of the matrix suction.

Shear strength formula of unsaturated loess can be written as:

$$\tau_f = (\sigma - u_a)\tan\phi' + c' + (u_a - u_w)\tan\phi^b \quad (1)$$

Cohesive force of unsaturated loess is made up of two parts: the cohesive force in saturated state, which can be obtained by routine test, and the shear strength caused by matrix suction. ϕ' is the effective internal friction angle and ϕ obtained from the consolidation test in our study show the valid values of ϕ'. Known as the friction angle of suction, the parameter ϕ^b is the angle of the curve, which shows increases of the shear strength caused by an increase in the matrix suction, while we want to get the unsaturated soil parameter values in this study.

In the formula (1), $c' + (u_a - u_w)\tan\phi^b$, the unsaturated soil cohesive force c is present. The same kind of soil has the same physical characteristics, and its effective cohesive force c' is constant. So the curve slopes with the suction $(u_a - u_w)$ as the abscissa are the same, which are respectively with $(u_a - u_w)\tan\phi^b$ as the ordinate and c as the ordinate. Therefore, the paper focuses on the friction angle of suction, which is the straight line angle obtained by fitting a straight line between the cohesive force c and the matrix suction $(u_a - u_w)$. After calculating parameters ϕ^b, the shear strength of unsaturated soil can be ascertained.

3.2 The selection of parameter value of unsaturated loess strength

According to soil water characteristic curve obtained by the matrix suction test, matrix suction can be calculated keeping the moisture at 5%, 10%, 15%, 20%, 25%, 30%. Based on direct shear test and the above matrix suction values along with the results of the analysis, the relationship between the cohesive force and the matrix suction can be obtained for different values of moisture content, which are respectively 5%, 10%, 15%, 20%, 25%, 30%, which are shown in Table 3.

According to the data in Table 3, by linear fitting, the linear angle (Figure 4) obtained is the required parameter ϕ^b.

Most of the literature (Yi, 2002; Qianling, 2001) shows that ϕ^b changes generally along with the change of matrix suction. But Gan and Escario (Gan, 1988) prove that when the soil suction is below a certain value (such as inlet value of soil), $\phi^b = \phi$, and the experimental results are the same. When the matrix suction is over a certain range of suction, ϕ^b is not significant. Figure 4 shows when the range of the suction is 1.7–172.4 kPa, cohesive force and the suction has a linear relationship. ϕ^b is a certain value in a limited range of suction. By fitting a straight line in the figure, we can get that the straight angle ϕ^b as 30.1°, which is close to 30°, the effective internal friction angle obtained by the direct shear test before. Essentially, the friction angle of the suction ϕ^b reflects comprehensively the interaction between water and soil of the unsaturated soil. The angle of suction has a great

Table 3. The cohesive force and the matrix suction for different water contents.

w/%	5	10	15	20	25	30
c/kPa	97.7	67.9	45.5	19.4	4.9	0.4
u_a-u_w/kPa	172.4	106.4	67.7	40.3	19.1	1.7

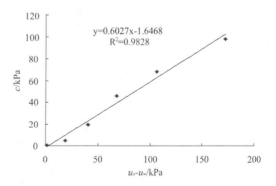

Figure 4. The relationship between cohesive force and matrix suction.

1229

influence on the research of the shear strength. So it has a great significance in engineering.

4 CONCLUSIONS

Based on the loess in Yang Gou village, Huang Ling of Shanxi province as the research object, the strength parameters of unsaturated loess is determined by a simple and easy operating test. The paper shows the relationship between moisture content and cohesive force by the direct shear test under different values of moisture content. Drawing the soil water characteristic curve by the matrix suction test, the relationship between the moisture content and suction can be obtained. By analyzing the data, the strength parameter values ϕ^b of the unsaturated loess is also obtained. The conclusion of this paper is as follows:

1. Moisture content has an effect on the shear strength of loess. It has bigger influence on the cohesive force, while its impact on internal friction angle is low. With the increase of moisture content, cohesive force reduces gradually and the internal friction angle remains unchanged.
2. By drawing the soil water characteristic curve, the relationship between moisture content and matrix suction can be obtained. The relationship between matrix suction and moisture content is exponential, and matrix suction is gradually reduced with an increase in moisture content.
3. Based on loess strength parameters and soil water characteristic curve under different moisture contents by means of the analysis and

calculation, it is concluded that the friction angle of suction ϕ^b is a constant, about 30.1°, in certain range of matrix suction.

REFERENCES

Bishop, A.W. et al. *Factors controlling the strength of partly saturated cohesive soils.* Research Conf. of Shear Strength of Cohesive Soils, University of Colorado (1960).

Fredlund, D.G., H. Rahardjo. *Soil mechanics for unsaturated soils.* New York: John Wiley & Sons, inc, (1993).

Gan, J.K., D.G. Fredlund, H. Rahardjo. *Determination of shear parameters of an unsaturated soil using the direct shear test.* Can. Geotech. J. **25**(8): 500–510 (1988).

Jinqian, D., L. Jing. *Strength characteristics of unsaturated loess.* Chin. J. Geotech. Eng., **19**(2):56–61 (1997).

Kun, H., W. Junwei, C. Gang, Z. Yang. *Testing study of relationship between water content and shear strength of unsaturated soils.* Rock Soil Mech., **33**(9): 2600–2604 (2012).

Ning, L., J.L. Willianm. *Unsaturate Soil Mechanics.* John wiley & Sons, inc, (2004).

Qianling, M., Y. Hailin, Q. Lunfeng. *The contribution of matric suction to shear strength of unsaturated soils.* Rock Soil Mech., **22**(6): 423–426 (2001).

Shengxia, H., Z. Yundong, C. Zhenghan. *Test study on strength character of unsaturated and undisturbed loess.* Rock Soil Mech., **26**(4): 0660–04 (2005).

Shuo, W., L. Zhanhui. *Test Study on Influence Factor of Shear Strength of Unsaturated Remolded Loess.* J. Shijiazhuang Tiedao Univ., **23**(3): 0086–05 (2010).

Yi, H., Z. Yinke. *The theory of soil-water characteristic curve and structure strength for unsaturated soil.* Rock Soil Mech., **23**(3): 268–277 (2002).

Advances in Energy Science and Equipment Engineering II – Zhou, Patty & Chen (Eds)
© 2017 Taylor & Francis Group, London, ISBN 978-1-138-71798-5

Study on the influence-law of various factors on granite aerated concrete

Shuangshuang Hou

School of Materials Science and Engineering, University of Jinan, Jinan, Shangdong, China
GCBM Technology Center, Shenzhen Gangchuang Building Material Co. Ltd., Shenzhen,
Guangdong, P.R. China

Wenhong Tao

School of Materials Science and Engineering, University of Jinan, Jinan, Shangdong, China

ABSTRACT: Because of the depletion of natural resources and the need for continuous improvement of the ecological awareness through environmental protection, how to make use of granite powder resources has become a new field of research by experts and scholars. This paper aims to study the influence of different fineness values of granite powder on the properties of aerated concrete under the condition of different water cement ratio, aluminum powder content and cement content, and this study also focuses on the influence of water cement ratio, aluminum powder content, cement content and the fineness of granite powder. The microstructure and hydration products are studied by Scanning Electron Microscopy (SEM). The comprehensive utilization of granite powder and the pollution of environment were improved.

1 INTRODUCTION

Autoclaved Aerated Concrete (AAC) usually is made from quartz-rich sand, lime, cement, and calcium sulfate with traces of aluminum powder as the pore-forming agent (Narayanan, 2000). These components are mixed with high amounts of water and molded to produce a cellular green body by H_2-gas generation at atmospheric pressure, and then autoclaved at 200°C under saturated steam pressure for several hours(André, 1999). Calcium Silicate Hydrate (CSH) phases are formed by reactions between silicates and CaO of the bonding agent.

The traditional raw material for the preparation of aerated concrete is fly ash. But with the change in industrial policy and gradual reduction in thermal power generation, fly ash is in short supply. And the gap between supply and demand is becoming increasingly acute.

Granite is rich in resources, but due to various reasons, its degree of development is low. Granite powder is an industrial waste residue, having a large yield and can cause serious pollution. Dry granite powder forms dust. The comprehensive application of granite powder is important in the building material industry. Granite is used in autoclaved aerated concrete block in the production of raw materials, which is one of its important applications.

This study uses granite powder as the main material and the influence of water cement ratio, aluminum powder, cement, and grinding time on the performance of aerated concrete is studied by experiments.

2 RAW MATERIALS

2.1 *Granite powder*

The granite powder with different grinding times are used in this experiment. The chemical compositions are shown in Table 1, XRD analysis as shown in Figure 1. Granite of three different fineness are ground out by ball mill and the grinding time is controlled for 15, 20 and 25 minutes. Then the performance test was carried out after drying at 105°C. The density was tested by Lee pycnometer. The measurement of specific surface area was tested by Blaine permeability apparatus; The measurement of fineness was done by negative pressure sieve method. The relationship between grinding time and powder properties are shown in Table 2.

It can be seen from Table 1 that the chemical composition of granite is mainly made up of SiO_2, Al_2O_3, Na_2O, K_2O and Fe_2O_3. From Figure 1. we can see that the crystallization properties of granite is good. The main mineral components are quartz, feldspar and a small amount of mica. Studies show that SiO_2 in quartz state is involved in the reaction of hydration products with higher compressive strength, The more the presence of quartz in the granite the better. It can be seen from Table 2 that the specific surface area increases and sieve residue percentage decreases with the increase of grinding time, which indicates that with the increase of grinding time, the granite powder becomes thin.

The particle size distribution is one of the most important physical indexes of cement mixed materials. The particle size distribution of raw materials has a great influence on the performance of concrete. The three different varities of granite powder fineness were detected by laser particle size analyzer. The results of the particle size distribution are shown in Figures 2 to 4. The average particle

Table 1. Chemical composition of granite powder (wt %).

SiO$_2$	Al$_2$O$_3$	CaO	Fe$_2$O$_3$	Na$_2$O	K$_2$O
72.05	13.47	2.05	2.09	3.04	5.43

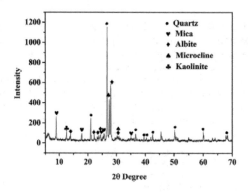

Figure 1. XRD analysis of granite.

Table 2. The relationship between grinding time and properties of granite powder.

Types of materials	Density (g/cm^3)	Specific surface area (m^2/Kg)	Sieve residue percentage (%)
15 min	2.615	380.9	5.21
20 min	2.615	440.7	4.81
25 min	2.615	523.0	3.20

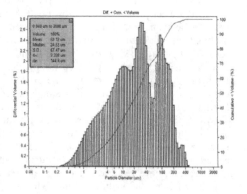

Figure 2. Laser particle size distribution of grinding 15 min granite powder.

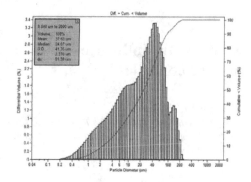

Figure 3. Laser particle size distribution of grinding 20 min granite powder.

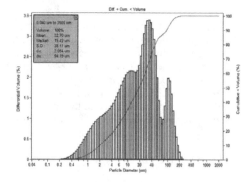

Figure 4. Laser particle size distribution of grinding 25 min granite powder.

size (mean) and median diameter (median) of each figure are gradually smaller with the increase of grinding time, which indicates that the particle size of granite powder has become smaller and smaller. In addition, from the particle distribution plot of the fine granite powder, it can be seen that independent distribution peaks exists in the coarse particles. The peak can be regarded to be formed with some difficult ground minerals, i.e. mica.

2.2 Other materials

The cementation materials used in the experiment are P.O42.5 ordinary Portland cement, and the main performance indicators are shown in Table 3. The effective CaO content of the lime used in this experiment is not less than 90%, and its specific compositions can be seen in Table 4. Chemical composition analysis of desulfurization gypsum is shown in Table 5. The solid part of the aluminum paste is higher than 65%, active aluminum content is greater than 85%, fineness of 0.075 sieve is ≤3.0%, gas generation rate is in the range 40%–60%,16 min ≥ 90%, 30 min ≥ 99%. The main component of the air lead agent used is the rosin thermal polymer.

Table 3. Performance indexes of cement.

Condensation time (min)		Compressive strength (MPa)		Bending strength (MPa)	
Initial condensation	final condensation	3d	28d	3d	28d
150	280	29.5	48	5.3	7.2

Table 4. Chemical composition of raw lime (wt%).

CaO	MgO	SiO$_2$	Al$_2$O$_3$	Fe$_2$O$_3$	SO$_3$	Loss
95.17	2.36	1.02	0.63	0.44	0.21	0.17

Table 5. Chemical composition of desulfurization gypsum (wt%).

SiO$_2$	Al$_2$O$_3$	CaO	MgO	Fe$_2$O$_3$	SO$_3$	NaO	K$_2$O
2.69	0.74	31.68	0.65	0.56	42.53	0.04	0.14

3 TEST SCHEME AND RESULTS

The following parameters are determined according to the preliminary experimental results: lime 15 wt%, gypsum 5wt%, air entrained agent 0.2%, curing temperature of 60°C, curing time of 4h, autoclave pressure of 1.0 MPa and an autoclave time of 8h. Test results are shown in Table 6.

Because the content of cement and granite powder is 80%, the content of the granite powder changes with the change of the cement content.

3.1 Effect of various factors on dry density

Figure 5 shows the effects of the factors on dry density. It can be seen from the figure that compared with cement content and grinding time that aluminum powder content and water cement ratio have a great influence on dry density. This is because in a certain range, with the increasing of water cement ratio, the fluidity of the slurry becomes better and thickening speed slows down. The resistance in the process of aluminum powder paste gas production is reduced, and so the gas is relatively smooth. The slurry maintains proper shear stress, the block gets a good pore structure (Yang, 2012). So the dry density is greatly reduced. In a certain range, with the increase of aluminum powder, the amount of gas increases, which leads to a great reduction in dry density. With the increase of water cement ratio and aluminum powder content, dry density decreases obviously. With the increase of cement content, dry density increases first and then decreases. With the increase of grinding time, dry density greatly decreases first and then increases slowly. It is well known that a variation in cement content and grinding time of granite powder has a certain influence

on dry density. When the water cement ratio is 0.60, the aluminum powder content is 0.2%, the cement content is 10%, and the grinding time is 20 minutes, the water absorption of the block is low.

3.2 Effect of various factors on water absorption radio

The effects of various factors on water absorption radio are shown in Figure 6. It can be seen from the figure that compared with other factors, aluminum powder content has a great influence on water absorption ratio. With the increase of aluminum powder content and water cement ratio, water absorption ratio obviously increases. With the increase of cement content, dry density increases first and then decreases. And with the increase of grinding time, water absorption ratio increases first and then decreases. When the water cement ratio is 0.55, the aluminum powder content is 0.1%, the cement content is 18%, the grinding time is 15 minutes, the water absorption of the block is low.

3.3 Effect of various factors on compressive strength

The effects of various factors on compressive strength are shown in Figure 7. The compressive strength is an important index to measure the performance of aerated concrete. With the increase of water cement ratio and aluminum powder content, compressive strength obviously decreases. This is because in a certain range, with the increasing of water cement ratio, the fluidity of the slurry becomes better and thickening speed slows down. The resistance in the process of aluminum powder paste gas is reduced, so the gas is relatively smooth. The slurry maintains proper shear stress, and the block gets good pore structure. At this point, the porosity of the block is large, but the pore structure is good. Therefore, the compressive strength decreased slowly. And with the increase of aluminum powder content, the block gas increases. So the compressive strength decreases. Within a certain range, with the increase of cement content, the compressive strength increases. Cement can be hydrolyzed to form Ca(OH)$_2$ and C-S-H gel in the process of static and stop of slurry, and the latter makes a good contribution to the early strength of the material. When the grinding time is 25 minutes, the water cement ratio is 0.55, the aluminum powder content is 0.1% and the cement content is 18%, the compressive strength of block is large.

3.4 Effect of various factors on compressive strength

The species, structure and morphology of aerated concrete hydration products were studied by SEM. The results are shown in Figure 8.

Table 6. The test results.

Number	Water cement ratio	Aluminum powder (wt%)	Cement (wt%)	Grinding time (min)	Dry density (Kg/m³)	Water absorption rate (%)	Compressive strength (Mpa)
A_1	0.60	0.10	10	15	622	59.5	4.3
A_2	0.60	0.15	14	20	557	63.3	3.0
A_3	0.60	0.20	18	25	592	64.1	3.1
A_4	0.57	0.10	14	25	643	57.5	4.6
A_5	0.57	0.15	18	15	583	59.6	3.0
A_6	0.57	0.20	10	20	584	65.3	2.6
A_7	0.55	0.10	18	20	669	55.7	4.9
A_8	0.55	0.15	10	25	605	59.2	3.7
A_9	0.55	0.20	14	15	692	60.8	3.1

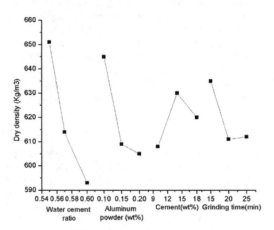

Figure 5. Effect of various factors on dry density.

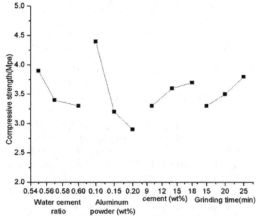

Figure 7. Effect of various factors on compressive strength.

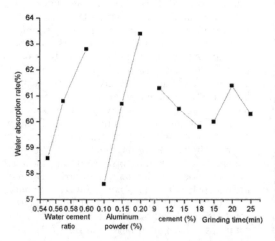

Figure 6. Effect of various factors on water absorption ratio.

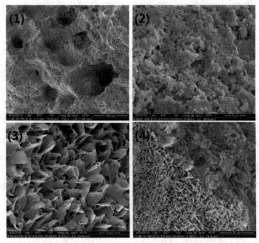

Figure 8. Microstructure of aerated concrete.

In the Figure, (1) shows the pore structure of aerated concrete under low magnification. (2) is a diagram representing the micro morphology before the autoclave pressure of aerated concrete. It can be seen that there are a large number of non-hydration products before autoclaving. Before autoclaving, the temperature is low. The internal structure of block is loose (Narayanan, 1999). At this time, the hydration reaction is slow, and so no large area hydration takes place. The main hydration products are C-S-H gel and rod-shaped ettringite, and tobermorite crystals are not produced. In the Figure, (3) and the left of (4) are hydration products in pores after autoclaving. After autoclaing, the hydration products obviously increased and the structure was compact. The hydration products are a large number of tobermorite and C-S-H gel. At the right of (4) is the hydration product of cement, which has a lot of gel.

4 CONCLUSIONS

In this study, the feasibility of using granite powder for the production of autoclaved aerated concrete has been demonstrated. The above experimental results can be obtained when the other conditions are clear. The water–cement ratio, aluminum powder content, cement content and grinding time on the performance of the block has a certain degree of influence. The influence of various factors are considered comprehensively. Water cement ratio and aluminum powder content have a great influence on dry density, compressive strength and water absorption ratio. Cement content and grinding time have a certain degree of influence. The main hydration product is tobermorite, C-S-H gel and a small amount of water of garnet.

REFERENCES

André Hauser, Urs Eggenberger, Thomas Mumenthaler, Fly ash from cellulose industry as secondary raw material in autoclaved aerated concrete. *Cement and Concrete Research,* 29 (1999) 297–302.

Jie Yang, The influence of several main factors on the performance of the steam aerated concrete. *New building materials*, 2012(4):70–73. (In Chinese)

Narayanan N, Ramamurthy K. Structure and properties of aerated concrete: a review. *Cement and Concrete Composites*, 22, 321–329 (2000).

Narayanan, N. Influence of composition on the structure and properties of aerated concrete, MS Thesis, IIT Madras, June 1999.

Advances in Energy Science and Equipment Engineering II – Zhou, Patty & Chen (Eds)
© 2017 Taylor & Francis Group, London, ISBN 978-1-138-71798-5

Workability and mechanical property of concrete modified by limestone and slag blended cement

Zhaodong Wang, Shuangshuang Hou & Xinghua Fu
School of Materials Science and Engineering, University of Jinan, Shandong, China
GCBM Technology Center, Shenzhen Gangchuang Building Material Co. Ltd., Shenzhen,
Guangdong, P.R. China

ABSTRACT: In the process of mining and processing limestone, a large amount of limestone dust is produced. It not only pollutes the environment but also causes the waste of resources. Many studies focus on the use of limestone powder to produce concrete. In this study, three different particle size distributions of limestone powder blended with slag powder is added into a concrete mixture as the replacement of cement. The water-binder ratio, the ratio of the limestone powder and slag powder (L/S ratio) and the dosage of the water reducing agent are changed to explore their effect on the workability and properties of the Limestone Composite Powder (LCP) concrete by orthogonal experiment, and obtain the optimal ratio. The mineral admixtures of the control concrete are fly ash and slag, and the mixing proportion of the control concrete is same as the S5 of the LCP concrete proportion. Experimental results show that to obtain the optimal ratio of the LCP concrete, the water-binder ratio, fineness of limestone, L/S ratio and dosage of the water reducing agent are 0.40, 799 m²/kg, 4/5 and 2.5%, respectively.

1 INTRODUCTION

Limestone is one of the most widely distributed mineral resources in China. The cutting and processing of the limestone produces a lot of limestone waste. The accumulation of limestone waste is an additional cost to industry and pollutes the environment. So many researches focus on the blending of limestone with concrete.

Skaropoulou A et al researched the effect of the limestone cement on sulphate attack (Skaropoulou, 2013; Sotiriadis, 2013). Celik K et al blended the volcanic pozzolan and the limestone as the replacement for cement to study its effect on the self-compacting and sustainable concrete (Celik, 2014; Ghrici, 2007).

2 RAW MATERIALS

In this study, the Portland cement supplied by the Shandong cement supplier was used. And the cement type is classified in China standards as P.O 42.5. The supplied fly ash is of Grade II class F. Slag powder supplied is of S75 grade. Limestone was ground for (a) 25 min, (b) 30 min, and (c) 35 min in the ball mill to obtain three kinds of fineness. For the fine aggregate, the fineness modulus of river sand was 2.9. For coarse aggregate, in the continuous gradation gravel, the particle diameter of gravel was 5–25 mm. Polycarboxylate superplasticizer, supplied by Jinan Changqing concrete batching plant, is used as the water-reducing

agent. Subsequently, the physical performance of raw materials was tested.

2.1 Specific gravity test

In Table 1, the density of the limestone powder, slag powder, fly ash and cement was tested respectively according to GB/T 208–2014.

2.2 Laser particle size analysis

Limestone was ground to three different fineness values of limestone powder. In Figure 1, the corresponding curve of cumulative volume and particle diameter of the limestone powder are shown. The detailed data of the limestone powder, slag powder and fly ash were tested via laser particle size analyzer (Beckman Coulter LS 13320), as shown in the Table 2.

2.3 Particle morphology analysis

The particle morphology picture of limestone, slag, fly ash and cement was obtained via polarizing microscope (AXIO Scope.A1), as shown in Figure 2. The particle morphology of the mineral

Table 1. Specific gravity of raw materials [g·cm⁻³].

	Limestone	Slag	Fly ash	Cement
Specific gravity	2.62	2.88	2.17	3.14

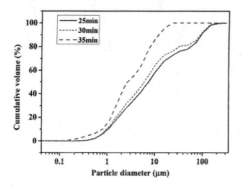

Figure 1. Laser particle size analysis of the limestone powder.

Table 2. Laser particle size analysis data.

	Lime- (a)	Lime- (b)	Lime- (c)	Slag	Fly ash
Mean particle diameter [μm]	26.90	23.01	4.766	29.95	34.66
Median particle diameter [μm]	7.321	5.980	2.527	23.66	25.11
Unit volume surface area [cm²/ml]	20927	22778	39988	12007	11306
Specific surface area [m²/Kg]	799	869	1526	417	520
d10 [μm]	1.031	0.997	0.792	1.217	3.414
d90 [μm]	93.72	89.33	11.87	53.55	74.72

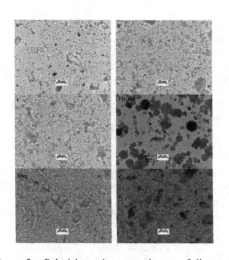

Figure 2. Polarizing microscope image of limestone powder (a, b, c), slag powder, fly ash, cement.

Table 3. Chemical analysis of limestone, slag, fly ash and cement.

	Lime- (%)	Slag (%)	Fly ash (%)	Cement (%)
SiO_2	8.41	30.34	46.44	22.66
Al_2O_3	2.49	10.47	33.75	4.49
Fe_2O_3	1.21	0.55	8.50	2.29
CaO	84.58	45.06	2.67	65.54
MgO	1.25	9.37	0.79	1.04
Na_2O	--	0.44	--	--
K_2O	0.35	0.57	--	--
SO_3	--	--	0.05	2.25
Loss	0.66	3.2	7.80	0.89

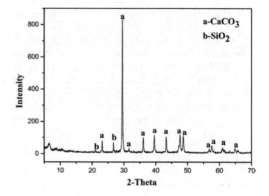

Figure 3. The XRD pattern of limestone powder.

Table 4. The design table of orthogonal experiment.

Factors	Level 1	Level 2	Level 3
A	0.40	0.41	0.42
B (m²/Kg)	799	869	1526
C	1/2	4/5	5/4
D (%)	2.5	2.6	2.7

Note: A stands for water-binder ratio; B stands for fineness of the limestone powder; C stands for ratio of the tuff powder and slag powder (L/S ratio); D stands for addition amount of water reducing agent. When the fineness of the tuff powder reaches the required level, it is feasible for tuff powder to replace fly ash.

admixtures has an important effect on the workability of concrete.

2.4 Chemical component analysis

The chemical components of binding materials were tested by x-Ray Fluorescence spectrometer (XRF). In Table 3, the binding materials are listed. Slag and fly ash are of amorphous structure,

Table 5. The elaborate mixture proportions of concrete.

Mix no.	W-b ratio	Water reducing agent (%)	Water (Kg)	Sand (Kg)	Coarse aggrgate (Kg)	Cement (Kg/%)	TCP (Kg/%) A	B	C	Slag (Kg/%)
S1	0.40	2.50	177	776	988	243/55	66/15			133/30
S2	0.40	2.60	177	776	988	243/55		88/20		111/25
S3	0.40	2.70	177	776	988	243/55			111/ 25	88/ 20
S4	0.41	2.70	181	774	986	243/55	88/20			111/30
S5	0.41	2.50	181	774	986	243/55		111/ 25		88/ 20
S6	0.41	2.60	181	776	986	243/55			66/15	133/30
S7	0.42	2.60	186	772	983	243/55	111/ 25			88/ 20
S8	0.42	2.70	186	772	983	243/55		66/15		133/30
S9	0.42	2.50	186	772	983	243/55			88/20	111/25
F	0.41	2.50	181	774	986	243/55	Fly ash:111/25			88/ 20

so only the tuff powder was tested by x-ray dif-fractometer (Bruker D8, Germany), as shown in Figure 3.

3 EXPERIMENTAL PROCEDURE

The mineral admixtures replaced 45% cement in this study, consisting of limestone powder (or fly ash) and slag powder. The water binder ratio, the fineness of the limestone powder, the ratio of the limestone powder with slag powder (L/S ratio) and the amount of water reducing agent were the four variable factors. And the factor levels design table is the one shown in Table 4.

To compare the workability and mechanical prop-erties of the LCP concrete and the fly ash concrete, the ratio of the fly ash and the slag powder was kept at 5/4. The mixture proportion of the fly ash con-crete is same as S5 of the LCP concrete. The detailed mixture ratio of the concrete was listed in Table 5. The slump of the concrete was tested by slump tester according to GB 50080-2002. And the mould of the concrete is 100 mm × 100 mm × 100 mm. The temperature and humidity were maintained at 20°C and 95%, respectively. When the concrete samples were maintained for 7 days and 28 days, the compression strength of concrete samples was tested according to GB 50081-2002.

4 RESULTS AND DISCUSSION

The workability and mechanical properties of the concrete are listed in Table 6.

4.1 Workability of concrete

According to the results listed in Table 6, the range analysis of the slump of the concrete was shown in Table 7.

Table 6. The workability and mechanical property results of concrete.

Mix no.	Factors A	B (m²/Kg)	C	D (%)	Compressive strength (MPa) 7d	28d	Slump value (mm)
S1	0.40	799	1/2	2.5	41.45	58.52	212
S2	0.40	869	4/5	2.6	44.18	58.52	196
S3	0.40	1526	5/4	2.7	38.43	51.55	190
S4	0.41	799	4/5	2.7	38.51	52.63	210
S5	0.41	869	5/4	2.5	37.21	52.16	204
S6	0.41	1526	1/2	2.6	40.09	54.59	198
S7	0.42	799	5/4	2.6	37.40	52.06	208
S8	0.42	869	1/2	2.7	36.48	50.98	197
S9	0.42	1526	4/5	2.5	39.52	56.27	215
F	0.41	520	5/4	2.5	37.94	52.04	225

Table 7. Range analysis of slump of concrete [mm].

	A	B	C	D
k1	199	210	202	210
k2	204	199	207	201
k3	207	201	201	199
Range	8	11	6	11
Optimal solution	A3	B1	C2	D1

Note: ki is the average value of sum of each factor level.

According to Table 7, the major factors that affect on the slump of the LCP concrete are the fine-ness of the limestone powder (B) and the amount of water reducing agent (D) followed by the water binder ratio (A) and the L/S ratio (C). the optimal solution is A3, B1, C2 and D1. The fineness of the limestone powder is finer, and the requirement of

1239

water to coat the surface of particles is high. So the level 1 of B factor is more beneficial for the workability of concrete. Meanwhile, the polycarboxylate superplasticizer is sensitive to the amount of tiny particles, and so taking into consideration various factors, the level 1 becomes the optimal level in the D factor (Felekoğlu, 2008). However, the slump of the fly ash concrete is 225 mm, the reason of which is that the particle morphology of the fly ash is spherical and the limestone powder is an irregular polygon according to the Figure 2. From Figure 2, the particle morphology of the cement can also be understood to be a polygon. Thus the fly ash has a better influence on the workability of concrete. But the slump of the LCP concrete can satisfy the requirement for pump concrete.

4.2 Mechanical property of concrete

In Table 6, the compressive strength results of the concrete have been listed. The 7d compressive strength of the N5 concrete is lower with the fly ash concrete, but the 28d compressive strength of the N5 concrete is higher than the fly ash concrete. Thus the limestone powder can replace fly ash in cement when it is blended with the slag powder. And the 28d compressive strength results of all LCP concrete samples reach the compressive strength design standard of C40 according to JGJ 55–2011. The compressive strength design standard of C40 is 48.23 MPa.

The range analysis of the compressive strength of the LCP concrete is shown in Table 8. The water binder ratio has a major effect on the 7d compressive strength of the LCP concrete. Because the hydration speed of the slag powder is slower than the cement, and the limestone powder only has a filling effect. The amount of the water reducing agent has a major influence on the compressive of the LCP concrete followed by the L/S ratio. When concrete

age reaches 28d, the slag powder is activated. The major reason for the amount of the water reducing agent having an effect on the strength of concrete is the amount of the pores in the concrete. The 7d compressive strength of the TCP concrete has an optimal solution is A1, B3, C2, D2, and the optimal solution of the 28d compressive strength is A1, B1, C2 and D1. The relationship curves reflecting the compressive strength and the level of various factors are detailed in Figure 4.

In Figure 4, it is clear that the lower water binder has a higher compressive strength. However, the fineness of the limestone powder makes a slight distinction to the compressive strength. The limestone powder just has a filling effect, and almost does not participate in the hydration reaction of the cement. The reason is that limestone has a good crystallinity according to Figure 3. Meanwhile, limestone is mainly composed of calcium carbonate and quartz according to Table 3. The ratio of the limestone powder and the slag powder has an optimal ratio, as shown in Figure 4. When the L/S ratio is 4/5, the compressive strength of the LCP concrete shows the highest improvement. In this case, the limestone powder could

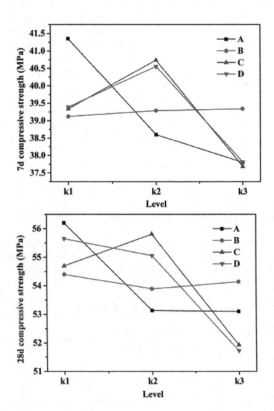

Figure 4. The compressive strength of the LCP concrete.

Table 8. Range analysis of compressive strength of the LCP concrete [MPa].

Age		A	B	C	D
7d	k1	41.35	39.12	39.34	39.39
	k2	38.60	39.29	40.74	40.56
	k3	37.80	39.34	37.68	37.81
	Range	3.55	0.22	3.06	2.75
	Optimal solution	A1	B3	C2	D2
28d	k1	56.20	54.40	54.70	55.65
	k2	53.13	53.89	55.81	55.06
	k3	53.10	54.14	51.92	51.72
	Range	3.10	0.51	3.89	3.93
	Optimal solution	A1	B1	C2	D1

form a good particle size distribution with the slag powder. The amount of the water reducing agent has a major effect on the slump of the concrete. When the workability of the concrete is very low, the pores in the concrete are exported with difficulty. In Figure 4, the lower amount of the water reducing agent has a higher strength. The reason is that tiny particles have a negative effect on the polycarboxylate superplasticizer so that the lower amount of the water reducing agent has a higher slump shown in Table 7. So the optimal amount of the water reducing agent is 2.5%.

5 CONCLUSION

The LCP concrete has a lower slump than the fly ash concrete, but the slump of the LCP concrete is sufficient as the pump concrete. The 28d compressive strength results of all LCP concrete samples meet the design compressive strength of C40 concrete. The limestone is a feasible replacement for cement. The workability and mechanical properties of the LCP concrete with regard to four factors of the LCP concrete are the water binder ratio of 0.4, fineness of the limestone at 799 m^2/kg, the L/S ratio at 4/5, and 2.5% water reducing agent.

REFERENCES

Celik K, Jackson M D, Mancio M, et al. High-volume natural volcanic pozzolan and limestone powder as partial replacements for portland cement in self-compacting and sustainable concrete[J]. CCC, 2014, **45**: 136–147.

Felekoğlu B, Sarıkahya H. Effect of chemical structure of polycarboxylate-based superplasticizers on workability retention of self-compacting concrete [J]. CBM, 2008, **22**(9): 1972–1980.

Ghrici M, Kenai S, Said-Mansour M. Mechanical properties and durability of mortar and concrete containing natural pozzolana and limestone blended cements[J]. CCC, 2007, **29**(7): 542–549.

Skaropoulou A, Sotiriadis K, Kakali G, et al. Use of mineral admixtures to improve the resistance of limestone cement concrete against thaumasite form of sulfate attack[J]. CCC, 2013, **37**: 267–275.

Sotiriadis K, Nikolopoulou E, Tsivilis S, et al. The effect of chlorides on the thaumasite form of sulfate attack of limestone cement concrete containing mineral admixtures at low temperature [J]. CBM, 2013, **43**: 156–164.

Workability and mechanical property of tuff composite powder concrete

Zhaodong Wang
School of Materials Science and Engineering, University of Jinan, Shandong, China
GCBM Technology Center, Shenzhen Gangchuang Building Material Co. Ltd., Shenzhen, Guangdong, P.R. China

Wenhong Tao
School of Materials Science and Engineering, University of Jinan, Shandong, China

Xinghua Fu
School of Materials Science and Engineering, University of Jinan, Shandong, China
GCBM Technology Center, Shenzhen Gangchuang Building Material Co. Ltd., Shenzhen, Guangdong, P.R. China

ABSTRACT: Because of the uncertainty in supply and demand of fly ash in parts of southern China, the feasibility of adding Tuff Composite Powder (TCP) to concrete as the replacement for cement was explored in this study. The tuff composite powder concrete is consisted of tuff powder and slag powder. Meanwhile, the fineness of tuff powder, the ratio of tuff powder and slag powder (T/S ratio), the water–binder ratio and the addition of water reducing agent are the variable factors used to design the orthogonal experiments to study the workability and mechanical properties of TCP concrete. The experimental results are as follows: (1) Tuff composite powder can be used as a partial replacement for cement. Compared with the fly ash concrete, the 28 days compressive strength results of the TCP concrete were higher. (2) The best slumps of the TCP concrete are 185 mm, which reaches the requirement of pumping concrete. (3) The optimistic factors of TCP concrete are as follows: the specific surface area of tuff powder is 561 m^2/kg; the ratio of tuff powder and slag powder (T/S) is 1/2; the water-binder ratio is 0.42; the addition of the water reducing agent is 3.1%.

1 INTRODUCTION

At present, the common mineral admixtures of concrete are fly ash, slag powder, silica fume and so on, especially the use of fly ash and slag is popular (Li, 2010; Li, 2016). The fly ash is in short supply in parts of southern China; however, the tuff resources have not been fully taken advantage (Mu, 2000). Thus many researchers focus on the study of new replacement for cement.

Topçu İ B et al mixed tuff into hardened self-consolidating concrete as the aggregate (Topçu, 2010). Yu L et al researched the pozzolanic activity of the tuff powder in the cement-based composite, and they found that the pozzolanic activity of the tuff improves continuously with the extension of the curing time (Yu, 2015). Yılmaz B et al prepared the mortar via blending the zeolitic tuff into cement, and found that the clinoptilolite is beneficial for the strength of mortar (Yılmaz, 2007). But the effect of fineness of the tuff powder on its hydration reaction is necessary.

2 RAW MATERIALS

The following raw materials were used: cement, P.O 42.5, produced by Shandong Cement Factory; Tuff powder: tuff was ground at (a) 25 min (b) 30 min and (c) 35 min. Fly ash: II grade fly ash, supplied by Jinan Huangtai Power plant; Ground slag: S75 grade ground slag; Water reducing agent: polycarboxylate superplasticizer, from Jinan Changqing concrete batching plant; Fine aggregate: the fineness modulus of river sand was 2.9; Coarse aggregate: continuous gradation gravel, the particle diameters of gravel was 5–25mm.

2.1 Specific gravity test

The specific gravity of the tuff powder, slag powder, fly ash and cement was tested respectively according to the standard GB/T 208–2014. The results were as shown in Table 1.

Table 1. Specific gravity of raw materials [g·cm⁻³].

	Tuff	Slag	Fly ash	Cement
Specific gravity	2.65	2.88	2.17	3.14

Table 2. Laser particle size analysis data.

	Tuff (a)	Tuff (b)	Tuff (c)	Slag	Fly ash
Mean particle diameter [μm]	42.42	34.47	15.77	29.95	34.66
Median particle diameter [μm]	14.06	14.10	7.370	23.66	25.11
Unit volume surface area [cm²/ml]	13642	14833	18900	12007	11306
Specific surface area [m²/Kg]	516	561	714	417	520

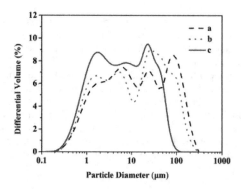

Figure 1. Laser particle size analysis of the tuff powder (a: 25 min, b: 30 min, c: 35 min).

2.2 Laser particle size analysis

Laser particle size analysis of the tuff powder, slag powder and fly ash were tested via laser particle size analyzer (Beckman Coulter LS 13320), as shown in the Figure 1 and Table 2.

2.3 Particle morphology analysis

The particle morphology of mineral admixtures had an important influence on the working performance. The particle morphology picture of raw materials was obtained via polarizing microscope (AXIO Scope.A1) in Figure 2.

2.4 Chemical component analysis

The chemical components of raw materials were tested by x-Ray Fluorescence spectrometer (XRF), and the results of raw materials were listed in

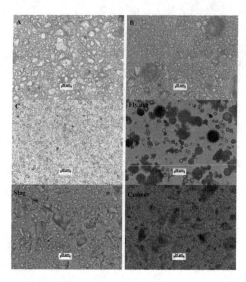

Figure 2. Polarizing microscope image of tuff powder (A, B, C), slag powder, fly ash, cement.

Table 3. Laser particle size analysis data.

	Tuff (%)	Slag (%)	Fly ash (%)	Cement (%)
SiO₂	74.57	30.34	46.44	22.66
Al₂O₃	15.32	10.47	33.75	4.49
Fe₂O₃	2.22	0.55	8.50	2.29
CaO	0.28	45.06	2.67	65.54
MgO	0.42	9.37	0.79	1.04
Na₂O	0.30	0.44	--	--
K₂O	6.49	0.57	--	--
SO₃	--	--	0.05	2.25
Loss	0.4	3.2	7.80	0.89

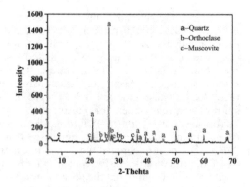

Figure 3. The XRD pattern of tuff powder.

Table 3. Slag and fly ash consisted of an amorphous structure, so only the tuff powder was tested by x-ray diffractometer (Bruker D8, Germany), as shown in Figure 3.

3 EXPERIMENTAL PROGRAM

The TCP concrete is the mixture of coarse aggregate, tuff composite powder, cement, natural sand, water and water reducer. The mineral admixtures of control concrete consist of fly ash composite powder (the ratio of tuff powder and slag powder was 5/4). The mixing amount of cementitious materials was 442 kg/m³. The mixing amount of the tuff composite powder (or fly ash composite powder) was 45% of the cementitious materials. The density of concrete was 2384 kg/m³.

The fineness of the tuff powder, the ratio of the tuff powder and slag powder (T/S), water-binder ratio and the addition of water reducing agent were variable factors, as shown in Table 4. The elaborate mixture proportions of concrete were listed in Table 5. The slump of the fresh concrete was tested by the slump meter according to GB 50080–2002. The size of concrete mould is 100 mm × 100 mm × 100 mm. The concrete samples were casted and maintained according to GB 50081–2002. When the concrete samples were maintained 7 days and 28 days, the compressive of concrete samples was tested.

4 RESULTS AND DISCUSSION

The results of workability and mechanical properties of concrete were shown in Table 6. The mechanical properties of concrete were tested according to GB 50081-2002.

4.1 Workability of concrete

From Table 6 can be seen, when the ratio of tuff powder to slag powder is 1:2, the TCP concrete has a better workability than 4/5 T/S ratio and 5/4 T/S ratio. But the workability of the fly ash concrete is 240 mm, because the particle morphology of fly ash is sphercial and the particle morphology of tuff powder is an irregular polygon. Meanwhile, the specific surface area of fly ash is similar to the tuff powder ground for 25 min, which is even lower than tuff powder ground for 30 and 35 min. Thus the water requirement of normal consistency of tuff powder is so high than fly ash that the

Table 6. The workability and mechanical properties of concrete [mm].

Mix no.	Factors				Compressive strength (MPa)		Slump value (mm)
	A	B (m²/Kg)	C	D (%)	7d	28d	
N1	0.40	516	1/2	2.9	40.47	58.39	140
N2	0.40	561	4/5	3.0	39.65	56.62	55
N3	0.40	714	5/4	3.1	37.30	56.43	40
N4	0.41	516	4/5	3.1	39.11	58.33	71
N5	0.41	561	5/4	2.9	34.99	52.50	40
N6	0.41	714	1/2	3.0	39.36	55.39	145
N7	0.42	516	5/4	3.0	35.78	52.19	40
N8	0.42	561	1/2	3.1	38.13	55.96	185
N9	0.42	714	4/5	2.9	37.56	52.50	45
F	0.41	520	5/4	2.9	37.97	51.21	240

Table 4. The factors and level design of the orthogonal experiment.

Factors	Level	Level 1	Level 2	Level 3
A		0.40	0.41	0.42
B (m²/kg)		516	561	714
C		1/2	4/5	5/4
D (%)		2.9	3.0	3.1

Note: A stands for water-binder ratio; B stands for fineness of tuff powder; C stands for ratio of tuff powder and slag powder (T/S ratio); D stands for addition amount of water-reducing agent.

Table 5. The elaborate mixture proportions of concrete.

Mix no.	W-b ratio	Water reducing agent (%)	Water (kg)	Sand (kg)	Coarse aggrgate (kg)	Cement (kg/%)	TCP (kg/%)			Slag (kg/%)
							A	B	C	
N1	0.40	2.90	177	776	988	243/55	66/15			133/30
N2	0.40	3.00	177	776	988	243/55		88/20		111/25
N3	0.40	3.10	177	776	988	243/55			111/25	88/20
N4	0.41	3.10	181	774	986	243/55	88/20			111/30
N5	0.41	2.90	181	774	986	243/55		111/25		88/20
N6	0.41	3.00	181	776	986	243/55			66/15	133/30
N7	0.42	3.00	186	772	983	243/55	111/25			88/20
N8	0.42	3.10	186	772	983	243/55		66/15		133/30
N9	0.42	2.90	186	772	983	243/55			88/20	111/25
F	0.41	2.90	181	774	986	243/55	Fly ash:111/25			88/20

workability of the TCP concrete becomes almost unviable.

The range analysis of orthogonal test of every factor on the workability of the TCP concrete is shown in Table 7.

It can be seen from Table 7 that the influence of the four factors on the slump of the TCP concrete from large to small is C, D, B, and A.

4.2 Mechanical properties of concrete

In Table 6, the compressive strength results of concrete are shown. The compressive strength of 28d TCP concretes is higher than the fly ash concrete. The water-binder ratio of 0.42 for the TCP concrete is higher for the 28d compressive strength TCP concrete than the fly ash concrete. The filling effect of the tuff powder is better than fly ash. Thus when the fineness of the tuff powder is fine enough, it is feasible that the tuff powder replaces the fly ash as a substitute of cement. The range analysis of orthogonal test on mechanical properties of TCP concrete is displayed in Table 8.

According to the results in Table 8, the effect of every factory on the 7d compressive strength of the TCP concrete from large to small is C, A, B and

D, and the 28d compressive strength of the TCP concrete from large to small is A, C, D and B. The corresponding relationship between the compressive strength of the TCP concrete and the level of various factors is described in Figure 4.

Figure 4 shows that the lower water-binder ratio (A) could improve the compressive strength of the TCP concrete. And the water-binder ratio is the major factor influencing the 28d compressive strength of the TCP concrete. But the major factor of the 7d compressive strength of the TCP concrete is the ratio of tuff powder and slag powder (C), because cement does not undergo enough degree of hydration reaction. Thus the filling effect of tuff composite powder is a major factor to enhance the strength of the TCP concrete. However, the slag powder has higher pozzolanic activity than tuff powder, because slag has an amorphous structure. According Figure 3, tuff consists of quartz, orthoclase and muscovite. So tuff is inert. When the content of slag is higher in the tuff composite powder, the tuff composite makes a bigger contribution to the compressive strength of the TCP concrete. Meanwhile, the additional amount

Table 7. Range analysis of orthogonal test on workability of TCP concrete.

	A	B	C	D
k1	78	84	157	75
k2	85	93	57	80
k3	90	77	40	99
Range	12	16	117	24
Optimal solution	A3	B2	C1	D3

Note: ki is the average value of sum of each factor level.

Table 8. Range analysis of orthogonal test on mechanical property of TCP concrete [MPa].

Age		A	B	C	D
7d	k1	39.14	38.45	39.32	37.67
	k2	37.82	37.59	38.77	38.26
	k3	37.16	38.07	36.02	38.18
	Range	1.98	0.86	3.3	0.59
	Optimal solution	A1	B1	C1	D2
28d	k1	57.14	56.30	56.58	54.46
	k2	55.41	55.03	55.82	54.73
	k3	53.55	54.77	53.71	56.91
	Range	3.59	1.53	2.87	2.18
	Optimal solution	A1	B1	C1	D3

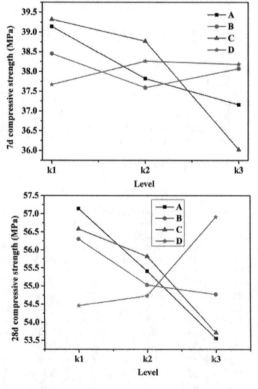

Figure 4. Effect of the level of various factors on compressive strength of TCP concrete.

of water reducing agent (D) could improve the mechanical properties of the TCP concrete. The reason for this is that water reducing agent is beneficial for the slump of concrete so that the bubble could be exhausted easily from the concrete. Tuff powder (B, level 1) at 516 m^2/kg has more coarse particles so that it forms a better grain composition with sand. However, according to the effect of the 561 m^2/kg and 714 m^2/kg of the tuff powder on the compressive strength of the TCP concrete, when the tuff powder becomes finer, it has the higher pozzolanic activity.

5 CONCLUSION

Comparing with the fly ash concrete, the 28 days compressive strength results of the N5 and N7 concrete were higher than fly ash concrete. And the fineness of the tuff powder of N7 is similar to that of fly ash. However, the water-binder ratio of the N7 is higher than the fly ash concrete. Above all, the tuff composite powder can be used as the replacement for cement. But the slumps of all kinds of the TCP concrete are lower than the fly ash concrete. In consideration of the workability and mechanical properties of the TCP concrete, the optimistic factors of TCP concrete are as follows: the water-binder ratio is 0.42; the specific surface area of tuff powder is 561m^2/kg; the ratio of tuff powder and slag powder (T/S) is 1/2; the addition of the water reducing agent is 3.1%.

REFERENCES

Li, W.T. Application of slag in concrete [J]. WSTE,2016, **22**(1):103–105 (In Chinese)
Li, Z.Y. Application of fly ash in concrete [J]. HPWC-SWDP, 2010(**8**):165–167 (In Chinese)
Mu, C. S. Physical and chemical properties and its development and utilization of tuff [J]. CMI,2000, **19**(3):17–19 (In Chinese)
Topçu İ B, Uygunoğlu T. Effect of aggregate type on properties of hardened self-consolidating lightweight concrete (SCLC) [J]. CBM, 2010, **24**(7): 1286–1295
Yu L, Zhou S, Deng W. Properties and pozzolanic reaction degree of tuff in cement-based composite [J]. ACC, 2015, **3**(1): 71–90
Yılmaz B, Uçar A, Öteyaka B, et al. Properties of zeolitic tuff (clinoptilolite) blended Portland cement [J]. BE, 2007, **42**(11): 3808–3815

Advances in Energy Science and Equipment Engineering II – Zhou, Patty & Chen (Eds)
© 2017 Taylor & Francis Group, London, ISBN 978-1-138-71798-5

Low-cyclic loading tests on concrete joints with non-demolition templates

Hongyang Xie
Nanchang Construction Engineering Group Limited Company, Jiangxi Province, China

Wentao Xu, Kang Ni & Depu Kong
College of Civil Engineering and Architecture, Nanchang Hangkong University, Jiangxi Province, China

ABSTRACT: Non-demolition template is introduced to replace the concrete cover of ordinary concrete beams and columns, and Low-cyclic loading tests on reinforced concrete frame joints with non-demolition templates are carried out. In the load control stage and the beginning of displacement control stage, the non-demolition templates are firmly bonded to the main concrete, and the two work together to bear the load. In the late displacement control stage, a few of non-demolition templates are observed to be peeled off the main concrete, and the bond failure and the strength failure of the main concrete occurred approximately simultaneously. The failure modes, load-displacement hysteresis curve, energy dissipation capacity and displacement ductility of all the tested joints are systematically studied. The results show that concrete frame joints with non-demolition templates have similar energy dissipation capacity to, and better displacement ductility than ordinary concrete frame joints.

1 INTRODUCTION

Non-demolition template can be introduced to replace the concrete cover of ordinary concrete beams and columns, bonded to the main concrete and bearing the load together with the main concrete. When imposed by static load, the co-working performance of the non-demolition template and the main concrete is reliable (Shan, 2014; Xue, 2014; Zhao, 2005). However, the co-working performance under seismic action has not been studied.

In this paper, low-cyclic loading tests for reinforced concrete joints with non-demolition templates are carried out by controlling load and displacement. The corresponding failure patterns, hysteretic curves, energy dissipation and ductility are analyzed.

2 NON-DEMOLITION TEMPLATE

Non-demolition template is made up of cement, sand, stone, glass fibre mesh cloth, concrete admixture and water according to a certain process. Its shape is determined in accordance with the dimensions of the column or beam which it attaches to, and its thickness ranges from 30 to 40 millimetres. There are some pits of certain shape on the inner

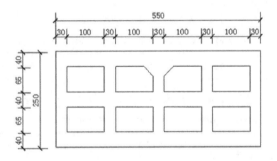

Figure 1. Typical shape of a non-demolition template.

surface of the template, designed to strengthen the bond between the template and the concrete. Figure 1 shows a typical shape of a piece of non-demolition template.

3 CONCRETE FRAME JOINT SAMPLES

Three reinforced concrete column-beam joint samples covered with non-demolition templates are made. The concrete grade is C30, longitudinal reinforcements use HRB335 bars, and stirrups adopt HPB235 bars. Figure 2 shows the section

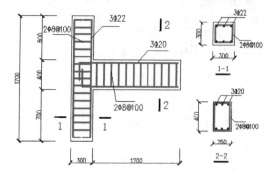

Figure 2. Section dimensions and reinforcement distribution of the joint sample.

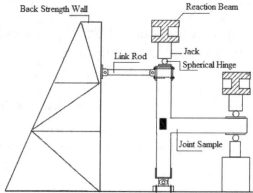

Figure 3. Test device.

dimensions and the reinforcement distribution of the joints.

4 LOADING TEST

Figure 3 shows the test device where the joint sample is subjected by a constant vertical load on the top of the column and a cyclic load at the end of the beam. Figure 4 is a picture of the test. A total of three joint samples are tested.

4.1 Vertical load on the top of the column

A fixed vertical load of 50 kN is applied by a jack on the top of the concrete column. The load is kept constant during the whole test.

4.2 Low frequency cyclic load at the end of beam

A low frequency cyclic load is applied at the end of beam. The loading procedure is divided into two stages as load control stage and displacement control stage.

4.2.1 Load control stage

The expected yield load P_y is supposed to be 80 kN. The load control stage is divided into 4 steps in which a cyclic load of 24 kN, 40 kN, 60 kN and 80 kN is respectively applied. For each step, the load cycles only once. In the final step, the applied load is kept increasing until the tensile longitudinal reinforcements in the beam yield, thus determining the actual yield load.

4.2.2 Displacement control stage

When the beam begins to yield, the test is conducted by controlling the vertical displacement at the end of the beam. At first the yield displacement, Δ_y, is cycled twice. Then the controlled displacement, which all cycled twice, is gradually increased by the fixed increment of Δ_y until the

Figure 4. A picture of the test.

bearing capacity decrease to less than 70% percent of the yield load.

4.3 Data collection

During the experiment, the following data are observed and collected: the constant vertical load on the top of the concrete column, the cyclic load at the end of the beam, and the cyclic displacement at the end of the beam. Besides, the cracks are also recorded.

5 FAILURE PATTERN OF THE JOINT

The crack development and joint failure process are observed. When the load is increased to about

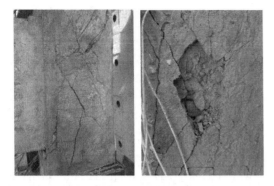

Figure 5.   Cracks in the templates when failure occurs.

Table 1.   Special loads at the end of the beam.

Sample number	1	2	3
Cracking load (kN)	38	40	58
Yield load (kN)	79	78	78
Peak load (kN)	128	127	126
Failure load (kN)	76	79	77

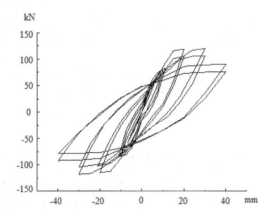

Figure 6.   Hysteretic curve.

40 kN, cracks begin to appear at the tensile region of the non-demolition template. When the load rises to about 60 kN, cracks appear in the core area of the joint. When the load continues up to about 80 kN, the beam begins to yield, and the corresponding vertical displacement at the end of the beam is about 10 mm. During the load control stage, the non-demolition templates stick firmly to the main concrete.

Thereafter the displacement control stage begins. At first, the yield displacement, which is 10 mm, cycles twice. Then the displacement

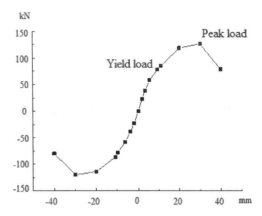

Figure 7.   Skeleton curve.

is increased step by step with an increment of 10 mm. After the controlled displacement of 20 mm cycled twice, crossing cracks appear in the core area of the joint. When the cyclic displacement rises to 30 mm, in the core area of the joint, the edge of one non-demolition template is observed to warp off the main concrete. When the cyclic displacement increases to 40 mm, the main concrete and the templates are crushed together, and some plates are peeled off. Figure 5 shows the cracks in the templates when failure occurs. Some special loads at the end of the beam are listed in Table 1. The hysteretic curve is shown in Figure 6. The load-displacement skeleton curve, shown in Figure 7, is acquired based on hysteretic curve.

6   ENERGY DISSIPATION CAPACITY

The plumpness of the hysteresis loop reflects the energy dissipation capacity. The hysteresis loop corresponding to the maximum cycling load is selected to calculate the equivalent viscous damping coefficient ξ based on the following definition.

$$\xi = \frac{1}{2\pi} \frac{S_{ABCD}}{S_{OBE} + S_{OFD}} \tag{1}$$

where S_{ABCD} is the shaded area enclosed by the closed curve $ABCD$ in the selected hysteresis loop as shown in Figure 8. S_{OBE} and S_{OFD} are the area of the two corresponding triangles. The energy dissipations and equivalent viscous damping coefficients of all the three tested joint samples are listed in Table 2.

Y.L. Wang (Wang, 2010) suggests that the equivalent viscous damping coefficient of outer concrete joint is about 0.17, which is close to the results listed in Table 2.

1251

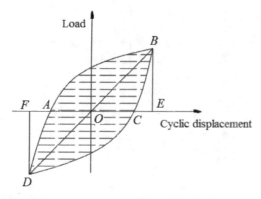

Figure 8. Selected hysteresis loop.

Table 2. Energy dissipations and equivalent viscous damping coefficients.

Sample number	Energy dissipation (kN·mm)	Equivalent viscous damping coefficient
1	3761	0.193
2	3697	0.185
3	3828	0.195

7 DUCTILITY

Ductility is defined as the deformation capacity of the yielded joint without remarkable decline in strength and rigidity. Here displacement ductility factor μ is adopted to measure the ductility of the joint.

$$\mu = \frac{\Delta_u}{\Delta_y} \quad (2)$$

where Δ_u is the ultimate vertical displacement, and Δ_y is the yield displacement at the end of the beam. The displacement ductility factors of all the three joint samples are listed in Table 3.

J. Cai (Cai, 2005) has conducted a statistical analysis to determine the displacement ductility factor of concrete frame joint based on a collection of 150 joint samples, and he suggests that the mean value of the displacement ductility factor is 3.801. Results in Table 3 indicate that concrete joints with non-demolition templates, whose displacement ductility factors all exceed 4.0, have a better ductility than ordinary concrete frame joints.

Table 3. Displacement ductility factor.

Sample number	Δ_y (mm)	Δ_u (mm)	μ
1	10	40	4.00
2	10	40	4.00
3	9	40	4.44

8 CONCLUSIONS

Low-cyclic loading tests on concrete joint samples with non-demolition templates are conducted, and the failure modes, load-displacement hysteresis curve, skeleton curve and energy dissipation capacity of all the tested joints are systematically studied. On basis of the test results, we can reach the following conclusions.

1. In the load control stage and the beginning of displacement control stage, the non-demolition templates are firmly bonded to the main concrete, and they cooperate quite well to bear the load.
2. In the late displacement control stage, the bond failure between the template and the main concrete occurs almost simultaneously with the strength failure of the main concrete.
3. The Energy dissipation capacity of concrete joint with non-demolition templates is close to that of ordinary concrete frame joints.
4. The displacement ductility of concrete joint with non-demolition templates is better than that of ordinary concrete frame joints.

REFERENCES

Cai, J., Q. Zhou, X.D. Fang, Earthquake Resistant Engineering and Retrofitting. 27(3), 2 (2005)

Shan, W.T. Experimental study on composite shear wall of cement formwork and concrete under cyclic loading (Master's thesis, North China University of Technology, 2014)

Wang, Y.L. Experimental study on the behaviors of exterior joints of RC frame under low cyclic reversed loading (Master's thesis, Harbin Institute of Technology, 2010)

Xue, J.Y., Y.Z. Bao, R. Ren, G. Wang, CHN. Civil Eng. J. 47(10), 1 (2014)

Zhao, J.M. The Experimental study and non-linear FEM analysis of thin shear wall without demoulding of reinforced concrete (Master's thesis, HeBei University of Technology, 2005)

Analysis of the seismic behavior of FRP-reinforced damaged steel fiber-reinforced high-strength concrete frame joints

Tingyan Wang

Research Center of New Style Building Material and Structure, Zhengzhou University, Zhengzhou, China
School of Civil Engineering and Communication, North China University of Water
Resources and Electric Power, Zhengzhou, China

Daoying Gao

Research Center of New Style Building Material and Structure, Zhengzhou University, Zhengzhou, China

Junwei Zhang

Research Center of New Style Building Material and Structure, Zhengzhou University, Zhengzhou, China
The Construction Department, Henan Agricultural University, Zhengzhou, China

ABSTRACT: Based on the experimental results of two FRP-reinforced damaged steel fiber-reinforced high-strength concrete frame exterior joints under low-cycle loading, the numerical simulation with nonlinear finite element method was adopted to analyze the behavior and to explore the influence of FRP reinforcement style on the seismic behavior of these damaged steel fiber-reinforced high-strength concrete frame joints. The result shows that the reinforcing type of sticking carbon fiber sheet at 45° in the joint core area is better than that of the carbon fiber sheet at 0°, and the analysis results and the experimental results quite agree with each other.

1 INTRODUCTION

The frame joints is the weak link in the steel fiber-reinforced high-strength concrete frame structure and can lead to severe disasters including the collapse of the building if damaged. The old constructions are faced with the problem of the damage of concrete members in the earthquake. There are often some danger hidden in the frame joints, which need to use the FRP reinforcement technology to reinforce the structure so that the building is guaranteed to function well. But the study on strengthening damaged steel fiber reinforced concrete frame joints with FRP has been given only sparse attention Therefore, this research analyzes the effect of FRP on the seismic behavior of damaged steel fiber reinforced high-strength concrete frame joints based on the experimental study on the two models of damaged steel fiber reinforced high-strength concrete frame joints under low cyclic loading.

2 EXPERIMENTAL CONDITIONS

2.1 Design of the model

The plane combination between pillars at the end of the middle layer of the bearing frame and the inflection point of beams is selected as the model. The size and reinforcement drawing, designed according to the *Code for Design of Concrete Structures* (GB50010-2002), are shown in Figure 1. The steel fiber used for experiment is ingot milling fiber (AMI04-32-600) with the volume fraction of 1.0%, draw ratio of 35–40, equivalent diameter of 0.94 mm, tensile strength ≥700 MPa, adulteration

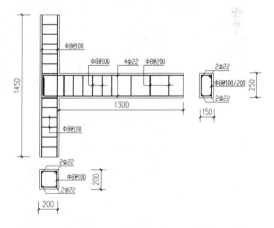

Figure 1. The sizes and reinforcements of the specimen (unit:mm).

range of 125 mm and 50 mm respectively in beam end and two sides of the pillar end at core area. The cement used is 42.5# high-strength Portland cement, in which the largest particle size is 20 mm, the fineness modulus of medium-coarse sand is 2.91 and the aggregate gradation is fair. The concrete strength is C60. The water reducer is JKH-1 highly efficient powder water reducer (FDN) with reducing rate of 18–25% (Zhang, 2011; Gao, 2012; Zhan, 2013).

2.2 Strengthening scheme

First, low cyclic loading is applied to simulate the earthquake load that damages the model of steel fiber-reinforced high-strength concrete frame joints. Next, FRP sheet is used to reinforce the joint model. The scheme is shown in Table 1.

Locate the cracks on the surface of the concrete, measure the width and repair them. The small independent chinks with the width of ≤0.2 mm should be repaired with an adhesive of good permeability and then sealed. For large independent and penetrating cracks, the periphery should be sealed first. Then the high-strength adhesive can be applied. For dormant crack of width≥0.5 mm, the crack has to be filled up. First, chisel a U-shape groove along the crack with concrete cutting disc. Then clean the crack with acetone, paste the grout nipple, and fill up the crack with a sealant. Under the sustained compression made by the air compressor, the structural repair adhesive is injected into cracks. The nooks of the beams are beveled by the structural adhesive. Burnish the surface with the sander when the adhesive is solidified.

Next, reinforce the damaged joints by FRP. Polish the concrete surface until the tectorium is removed and the smooth structure interface appears. Intensively mixed fat material is then spread on the concrete surface evenly with brush, and its thickness should not be too thin. Paste the processed fiber sheet to the due position and roll it with the special roller along the stress direction for several times till the connection resin is extruded from the fiber. By this method, the bubble is squeezed out so that the fiber sheet is pressed closely to the concrete surface. Then apply the impregnating resin to the

Table 1. Strengthening scheme of the joints.

Model	Strengthening scheme of the damaged model
1	Treat the crack; paste one piece of FRP at the angle of 45° across another piece in the core area; paste two pieces of FRP at the beam.
2	Treat the crack; paste two pieces of FRP vertically at the angle of 0° in the core area; paste two pieces of FRP at the beam.

surface of fiber sheet. Repeat the process above to paste more pieces of fiber sheet. The FRP sheet, made by hand and by impregnating the substrate with fiber, is home-made FRP.

2.3 Loading scheme

The experiment applies multichannel electro-hydraulic servo dynamic fatigue test system to conduct low-cyclic loading. The pillar end will endure axial load from 2000 kN oil jack fixed on the heavy frame. First, the pillar top is put under the axial load from the oil jack. When the compression ratio reaches the predesigned value of 0.3, keep it unchanged during the experiment. Then apply low cyclic loading by the electro-hydraulic servo actuator. The first two cycles load with a control force, of which the first cycle loads to 75% of the yield load calculated value so as to simulate the stress condition on normal occasions. Since the key part of the study lies in the phase of plastic deformation, in the second cycle, the model is loaded directly to the yield state. Displacement control is then used for loading. Staged-loading is adopted according to the multiples of displacement of the pillar end in the yielded state. Recycle twice at each displacement level until the maximum load value of the nth cycle is lower than 85% of the highest load value when the model is damaged.

3 FINITE ELEMENT ANALYSIS MODEL

3.1 The constitutive structure relation of the material

The axial compressive stress–strain curve of high-strength concrete uses the code for design of concrete structures (GB 50010-2002). The tensile stress–strain relationship of high-strength concrete is described as follows: the upper period denotes the line and lower period the index. The elastic modulus before crack is the same as that of the initial elastic modulus, and after the crack, the concrete is considered as non-brittle material. That is to say, the concrete tensile strength of the crack did not decline to zero immediately, but it drops with the widening of the crack. For high-strength concrete, the parameter of the calculation formula is from reference (Guo, 1997).

At the yielding period, the reinforcement stress has a small change, but strain increases dramatically, and the stress–strain curve is a 0.01E slope line. After the formation of the plastic hinge, the ultimate deformation of the plastic area concrete will seldom exceed 0.006, so the deformation of reinforcement undergoes a small change at the hardening period, and the stress–strain curve is a 0.01E slope line, taking E′ = 0.01E (Liu, 2007). As metal material, steel fiber takes the ideal plastic stress–strain relationship as the reinforcement. The damage of steel fiber concrete mainly occurs because of the pulling out of steel fiber from the concrete, and not because of the broken of the steel fiber. Therefore, taking the tensile stress when steel fiber is pulling out as the tensile strength of steel fiber, it is as assumed to be 250 MPa (Zhao, 1999) in this paper.

3.2 Concrete failure criterion and the crack treatment

Considering the concrete failure criterion taking the William–Warnke model with five parameters, the formula is $F/fc − S \geq 0$. In this formula, F is the function of the main stress σ_1, σ_2, σ_3, S is the destroying section of the main stress σ_1, σ_2, σ_3 and the five parameters f_t, f_c, f_{cb}, f_1, f_2, and f_t, f_c, f_{cb}, f_1, f_2 stand for the single-axe tensile strength, the axes' compression strength, the biaxial compression strength's constant pressure, the biaxial compression with ambient pressure and the multi-axial compression strength with ambient pressure. The the ANSYS method is used. When the ambient pressure is small, the failure surface can be made certain by two parameters f_t and f_c, and the other three are: $f_{cb} = 1.2_{fc}$, $f_1 = 1.45f_c$, $f_2 = 1.725f_c$ (Moaveni, 2008).

In the analysis, the dispersion crack model is used to treat frame joint cracking process. After the cracks emerge, subsequent loading will produce sliding or shearing on the crack surface. In analysis, introducing split plane shear transfer coefficient (β_t) when crack opens and split plane transfer coefficient (β_c) to simulate the loss of the concrete shearing capacity is done. β_t has a great influence on the calculation results when the crack emerges, and generally it is between 0.3 and 0.5, and in this paper $\beta_t = 0.3$. β_c has little influence on the result of monotonic loading of the load-carrying specimen, and generally it is between 0.9 and 1.0, and in this paper $\beta_c = 0.9$. Without considering the concrete crush, it is hard to simulate the concrete failure process, and the normal failure surface will cause premature crush invalidation. Therefore, this paper uses the expanded crush plane to consider the restriction effect of the steel fiber to the concrete, and the parameter of the axis compressive strength of the failure criterion is $1.2f_c$ to $2f_c$.

3.3 Choosing elements

The reinforcement uses Link8 unit and concrete uses Solid 65 unit. FRP sheet uses the shell41 film unit. The steel fibers are large in number and also they are distributed in chaos. So it is impossible to model the role of each single steel fiber.

Therefore, we take the steel fibers as distributing micro reinforcements. They are distributed into the concrete Solid65 evenly according to volume ratio, and together with concrete, they compose integral model, that is steel fiber distributes along with the axes of the coordinates, and in the volume ratio is distributed in each direction according to unit side length ratio. The the influence of the steel fiber distribution on the effective coefficient and orientation coefficient is considered. Moreover, because the steel fiber's tensile strength is the tensile strength when steel fiber is pulling out, there is no need to consider the bond-slip between the steel fiber and the concrete. For the concrete and reinforcement, the discrete model is used for modeling. The bond-slip between the two uses Combin39 for simulation, and the bond-slip constitutive construction relation uses the four line segment model recommended by the CEB-FIPMC90 standard (Si, 2009).

For the concrete and FRP sheet, the discrete model is used for modeling. The bond-slip between the two uses Combin39 for simulation. Due to the plane perpendicular to the fiber direction of slip, loading process does not occur, so in the analysis, it can correspond to nodes of the direction of coupled degrees of freedom. Only the vertical slip to the FRP needs to be considered and the slip along the fiber direction. For the transversal U-shaped wrapped FRP sheet, it is assumed that the relative slip concrete is mainly along the beam height direction with respect to the FRP sheet, and the joint of concrete and FRP sheet node displacement coupling in the other two directions.

3.4 Unit birth and death

Because the reinforcement of the frame joints have been carried out on the basis of stress on the joints, although the cracks have been repaired before the reinforcement, there are still micro cracks in the concrete body. Strengthening the finite element analysis without taking into account the impact of micro cracks will cause the structure to have just passed the test, and the actual results do not agree with the test. Therefore, in the finite element analysis, the method of element birth and death is used to consider the influence of micro cracks. In the finite element analysis, the FRP sheet units were killed in the load step before cracking, and the nodes were treated as the unreinforced fame. On the basis of this, the FRP sheet unit is activated in the load step of the crack, so that it is involved in the later work.

3.5 Mesh partition

When we do analysis by taking the whole model of joints, the size and the shape of the element mesh

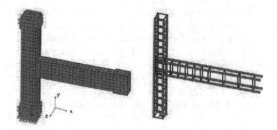

Figure 2. Finite element model of frame exterior joints.

will both influence the results of the analysis. If the meshes are divided too meticulously, it will cause the instability of numerical calculation, and if the meshes are divided too coarsely, it will influence precision. Only the right division of meshes can result in convergence. By comparison, this paper does mesh partition by 5 cm, and the corresponding mesh partition needs to satisfy certain geometry topology rule. It is necessary to consider the thickness of the protective layer of the beam and the column in modeling, so in this paper, besides 5.0 cm, the measurements 3.5 cm and 4.0 cm are also considered and unit generation is shown in Figure 2.

3.6 Boundary conditions and load application

In order to simulate the form of joint destruction under the low cycle loading, we supply X direction with the two lateral surfaces of the upper and lower end of the column translation displacement constraint, and supply X, Y, Z three directions with the translation displacement constraint at the lower end of the column. The Y axial force is placed at the column's top end, and Y reversed force or displacement is placed at the beam ends. In the nonlinear analysis, by defining load step, the load sub-step divides the load into a series of load increments. To solve by finite element method, we make a series of linear approaching in the load increments to achieve balance. Considering the nonlinear influence of steel fiber concrete, we adopt Newton–Raphson method in the nonlinear iteration. Displacement convergence criterion is used and the convergence precision is 1%. In order to improve the convergence of nonlinear analysis, adaptive descend gene, linear searching, forecasting and dichotomy will be used at the same time.

4 RESULTS AND DISCUSSION

Table 2 shows the comparative result of the damaged frame joints reinforced by FRP. According to the ratio of the experimental results and the analysis results, the average value of the yield load is

Table 2. Result of the experiment and analysis.

Model		1			
YS	Load (kN)	36.00	36.52	34.04	38.04
	Displacement (mm)	8.98	14.03	10.37	13.86
US	Load (kN)	47.39	53.07	40.27	56.10
	Displacement (mm)	17.89	41.91	19.92	41.67
	Model	2			
YS	Load (kN)	30.06	35.98	30.06	37.96
	Displacement (mm)	7.40	16.14	7.69	16.36
US	Load (kN)	47.44	56.66	42.98	58.88
	Displacement (mm)	25.14	35.05	24.98	40.72
Condition		BR	AR	BR	AR
Means		Experiment		Analysis	

YS—Yield State, US—Ultimate State, BR—Before reinforcement, AR—After reinforcement.

1.011. The standard deviation of the yield load is 0.043, and the coefficient of variation of the yield load is 0.042. The average value of the yield displacement is 1.049, the standard deviation of the yield displacement is 0.064, and the coefficient of variation of the yield displacement is 0.061. The average value of the ultimate load is 0.963, the standard deviation of the ultimate load is 0.087, and the coefficient of variation of the ultimate load is 0.091. The average value of the ultimate displacement is 1.066, the standard deviation of the ultimate displacement is 0.074, and the coefficient of variation of the ultimate displacement is 0.069. Thus the results of the analysis and the experimental results quite agree with each other.

According to the results of the analysis, the yield load of reinforced model 1 is 1.12 times of the previous one. The yield displacement of reinforced model 1 is 1.34 times of the previous one. The ultimate load of reinforced model 1 is 1.39 times of the previous one. The ultimate displacement of reinforced model1 is 2.09 times of the previous one. Model 2 also presents a similar result. The yield load of reinforced model 2 is 1.26 times of the previous one. The yield displacement of reinforced model 2 is 2.13 times of the previous one. The ultimate load of reinforced model 2 is 1.37 times of the previous one. The ultimate displacement of reinforced model 2 is 1.63 times of the previous one. Thus improvements in the yield load, yield displacement, ultimate load and ultimate displacement all reach the original condition, which can meet the standard of seismic appraisal.

The major problem of the reinforced Model 1 is the crazing of the concrete along the margin of fiber sheet, while the problem of Model 2 after strengthening is the rip of fiber sheet in the core area. Although the ultimate load of the two models is similar, the ultimate displacement of reinforced Model1 is larger than Model 2 mainly because of the mechanical behavior and fiber direction of fiber sheet. Model 2 with fiber sheet pasted at 0° vertically in the core area only provides vertical constraint while Model 1 with cross fiber sheet enhances both vertical and horizontal deformation capability of the core area concrete.

4.1 Hysteretic curve and skeleton curve

The hysteretic and skeleton curves of the damaged frame joints model reinforced by FRP are shown in Figures 3–10. The result shows that compared with the original model, the hysteretic curve of the reinforced damaged joint model is full. Meanwhile, the ultimate load, ultimate displacement and times of circulation all increase to some degree. The coverage of the skeleton curve enlarges, which meets the standard of the seismic appraisal. For example, compared with the original Model 1, the reinforced Model 1 has a fuller hysteretic curve and more circulation times, whose ultimate load and ultimate displacement is 1.39 and 2.09 times respectively of the previous one. For the reinforced Model 2,

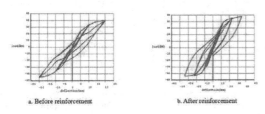

a. Before reinforcement   b. After reinforcement

Figure 3. Comparison of the hysteretic curves of Model 1 (experimental results).

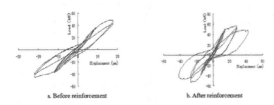

a. Before reinforcement   b. After reinforcement

Figure 4. Comparison of the hysteretic curves of Model 1 (analysis results).

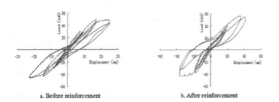

a. Before reinforcement   b. After reinforcement

Figure 5. Comparison of the hysteretic curves of Model 2 (experimental results).

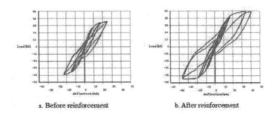

a. Before reinforcement   b. After reinforcement

Figure 6. Comparison of the hysteretic curves of Model 2 (analysis results).

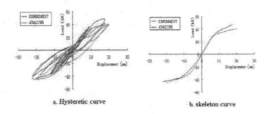

a. Hysteretic curve   b. skeleton curve

Figure 7. Comparison of the hysteretic curves and skeleton curve of Model 1 (before).

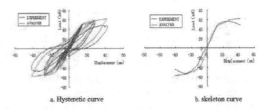

a. Hysteretic curve   b. skeleton curve

Figure 8. Comparison of the hysteretic curves and skeleton curve of Model 1 (after).

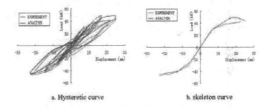

a. Hysteretic curve   b. skeleton curve

Figure 9, Comparison of the hysteretic curves and skeleton curve of Model 2 (before).

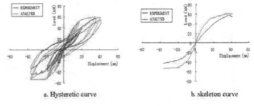

a. Hysteretic curve   b. skeleton curve

Figure 10. Comparison of the hysteretic curves and skeleton curve of Model 2 (after).

1257

the hysteretic curve is also fuller. The ultimate load and ultimate displacement is 1.37 and 1.63 times respectively of that before while the circulation times remain the same.

Compared with the reinforced Model 2, the reinforced Model 1 has a fuller hysteretic curve, larger coverage of the skeleton curve and more circulation times. The ultimate load of Model 1 is 0.95 times of Model 2, and the ultimate displacement is 1.02 times of Model 2. The result indicates that the cross fiber sheet at the angle of 45° exerts a stronger constraint on core area concrete than the vertical fiber sheet. The shear capacity and the deformation capacity of the joint core area are enhanced, thus improving the energy dissipation capacity.

4.2 Energy dissipation capacity

The comparison of the energy dissipation capacity of damaged frame joints model reinforced by FRP is shown in Table 3. The result shows that after reinforcement the energy dissipation capacity of the model reaches the original level, which meets the standard of seismic appraisal. For example, dissipation and strain energy of the reinforced Model 1 are 4.58 times and 2.98 times of the previous values, respectively. The ratio of dissipation energy and strain energy is 1.53 times compared to the earlier one. The dissipation energy and strain energy of the reinforced Model 2 are 2.13 times and 1.65 times of that before respectively. The ratio of dissipation energy and strain energy is 1.30 times of the previous model.

The dissipation energy and strain energy of reinforced Model 2 are 0.86 times of Model 1. The ratio of dissipation energy and strain energy of Model 2 and Model 1 are the same. The result indicates that the cross-fiber sheet at the angle of 45° exerts a stronger constraint on core area concrete than the vertical fiber sheet, thus enhancing the energy dissipation capacity.

4.3 Displacement ductility

The comparison of the displacement ductility of damaged frame joints reinforced by FRP is shown in Table 4. The yield displacement and ultimate displacement of Model 1 are 1.34 times and 2.09 times of the previous one, while the displacement ductility is 1.57 times of that before. The yield displacement and ultimate displacement of Model 2 are 2.13 times and 1.63 times of the earlier model, while the displacement ductility is 0.77 times of the earlier value. The result shows that the displacement ductility of the model with cross fiber sheet at the angle of 45° meets the standard of seismic appraisal, but the model with fiber sheet at the angle of 0° provides a bad displacement ductility.

4.4 Bearing capacity degeneration

The bearing capacity reduction coefficient of the damaged frame joints reinforced by FRP is shown

Table 4. Displacement ductility of joints.

Model number	1			
YD (mm)	8.98	14.03	10.37	13.86
UD (mm)	17.89	41.91	19.92	41.67
DD	1.99	2.99	1.92	3.01
Model number	2			
YD (mm)	7.40	16.14	7.69	16.36
UD (mm)	25.14	35.05	24.98	40.72
DD	3.40	2.17	3.25	2.49
Condition	BR	AR	BR	AR
Means	Experiment		Analysis	

YD—Yield Displacement; UD—Ultimate Displacement; DD—Displacement Ductility.

Table 5. Comparison of bearing capacity reduction coefficient of joints.

Model number	1			
YL(kN)	36.00	36.52	34.04	38.04
UL(kN)	47.39	53.07	40.27	56.10
Δy	0.969	0.937	0.996	0.967
$2\Delta y$	0.901	0.930	0.969	0.946
Model number	2			
YL(kN)	30.06	35.98	30.06	37.96
UL(kN)	47.44	56.66	42.98	58.88
Δy	0.949	0.963	0.993	0.994
$2\Delta y$	0.950	0.957	0.977	0.956
$3\Delta y$	0.909	-	0.982	-
Condition	BR	AR	BR	AR
Means	Experiment		Analysis	

YL—Yield Load; UL—Ultimate Load.

Table 3. The energy dissipation capacity of joints.

Model number	1			
DE	1344.1	5194.7	1184.3	5424.0
SE	3394.2	8870.9	3504.0	10438.2
DE/SE	0.4	0.59	0.34	0.52
Model number	2			
DE	2160.7	4161.1	2175.9	4643.0
SE	5314.9	7559.1	5424.0	8948.9
DE/SE	0.41	0.55	0.4	0.52
Condition	BR	AR	BR	AR
Means	Experiment		Analysis	

DE—Dissipation energy; SE—Strain energy.

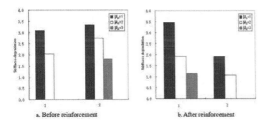

a. Before reinforcement

b. After reinforcement

Figure 11. Stiffness comparison of joints (Experimental results).

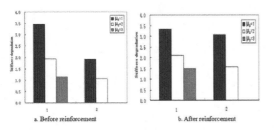

a. Before reinforcement

b. After reinforcement

Figure 12. Stiffness comparison of joints (Analysis results).

in Table 5. The result shows that different methods of reinforcement barely lead to any differences in the reduction of the bearing capacity. For example, when displacement ductility is Δy, the bearing capacity reduction coefficient of Model1 is 0.97 times of the previoius value, while the its value through Model 2 is 1.00. When displacement ductility is 2Δy, the bearing capacity reduction coefficient of Model 1 is 0.98 times of that before while Model 2 is 0.98. When displacement ductility is 3Δy, the Model 2 is damaged, which is reinforced. From this we see that the bearing capacity reduction of the models after the reinforcement is worse than before.

4.5 Stiffness degradation

The loop and linear stiffness of the damaged frame joint model reinforced by FRP is shown in Figures 11 and 12. At every ductility factor, the stiffness of Model 2 is lower than Model 1, and the circulation times of Model 2 are fewer than Model 1. The result indicates that the stiffness of reinforced model with cross fiber sheet at the angle of 45° meets the standard of seismic appraisal, but the model with fiber sheet at the angle of 0° degenerates fast.

5 CONCLUSIONS

From the experiment of reinforcing the damaged steel fiber reinforced concrete frame joints by two different FRP reinforcement methods under low cyclic load, we can draw the following conclusions:

- To paste carbon fiber sheet at the angle of 45° in core area after crack treatment can lead to batter reinforcement performance than at the angle of 0°.
- Although these two methods exert a little influence on the degradation of bearing capacity, the pasted fiber sheet increases the constraint of the core area concrete so that the bearing capacity as well as the energy dissipation capacity of the model is enhanced.
- The strengthening method of pasting carbon fiber sheet at 45° in the joint core area prevents the rigidity degeneration and improves the displacement ductility of the model. On the contrary, the method of pasting carbon fiber sheet at 0° speeds up the rigidity degeneration and reduces the displacement ductility.
- The finite element analysis result proves the rationality of the element type, material constitutive relation and failure criterion used in the finite element analysis.

REFERENCES

Gao Danying, Zhang Junwei, "Experimental investigation on seismic behavior of steel fiber reinforced high-strength concrete frame exterior joints", Industrial construction, 42.5., 84–89., 2012.

Guo, Z.H., "Strength and deformation of concrete," China,1997.

Liu, Z.P., Nonlinear numerical simulation of pre-stressed concrete and reinforced concrete beam based on ANSYS. Xinjiang: Xinjiang agricultural university. 2007.

Moaveni, S., "Finite element analysis theory and application with ANSYS," China, 2008.

Si, B.J., Z.G. Sun, Q.H. Ai, and D.S. Wang, "Sensitive analysis and model modification for finite element analysis of R/C bridge piers under cyclic loading," Engineering mechanics, China, vol. 26,2009, pp. 174–180.

Zhang Junwei, Tian Xiangyang, Wang Tingyan, Gao Danying, "Influence of FRP reinforcement style on seismic behavior of FRP reinforced damaged steel fiber reinforced high-strength concrete frame joints", Concrete, 286.8., 24–28., 2013.

Zhang Junwei, Wang Tingyan, "Experimental investigation on seismic behavior of high-strength concrete frame exterior joints partially reinforced by steel fiber", Concrete, 261.7., 13–16., 2011.

Zhao, G.F., S.M. Peng, C.K. Huang, "Steel fiber reinforced concrete structures," China, 1999.

Advances in Energy Science and Equipment Engineering II – Zhou, Patty & Chen (Eds)
© 2017 Taylor & Francis Group, London, ISBN 978-1-138-71798-5

Characteristics analysis of cavitation flow in a centrifugal pump

Yuliang Cao
College of Power Engineering, Naval University of Engineering, Wuhan, Hubei, China

Guo He
Department of Management Science, Naval University of Engineering, Wuhan, Hubei, China

Yongsheng Su & Tingfeng Ming
College of Power Engineering, Naval University of Engineering, Wuhan, Hubei, China

ABSTRACT: In order to study the characteristics of cavitation flow in a centrifugal pump, by Zwart cavitation model and RNG turbulent model with modified turbulent viscosity, the cavitation flow in centrifugal pump was numerically simulated by computational fluid dynamics. First, cavitation performance of centrifugal pump was calculated steadily, and the predicted curve of head-NPSHa was compared with the experimental results. Then, two cavitation situations with NPSHa of 2.25 m and 1.02 m were calculated transiently. Time–frequency characteristics of pressure fluctuation at volute tongue and pump outlet were analyzed, and characteristics of cavity vapor shape on impeller blades were illustrated. The study found that the main frequency of pressure fluctuation in the centrifugal pump is rotation frequency and its multiplier. When NPSHa decreases from 2.25 m to 1.02 m, volume of cavity vapor on impeller suction side increases greatly, and many small vapor bubbles grow and break up at the end of vacuum, leading to the increasing of noise and spectral peaks in high-frequency band of pressure fluctuation.

1 INTRODUCTION

Cavitation is a common phenomenon in fluid machinery, which occurs when the local pressure falls below the liquid's saturation vapor pressure. The cavitation flow, including quasi-turbulence, multi-phase flow and phase change, is much complex and unsteady.

Numerical simulation has played an important role in the study of cavitation. By numerical methods of Computation Fluid Dynamics (CFD), the cavitation flow of different hydraulic machines, such as centrifugal pumps (Wang, 2016; Zhu, 2016), axial-flow pumps (Wang, 2016; Zhu, 2016), hydrofoils (Wang, 2016; Zhu, 2016) and marine propellers (Wang, 2016) has been studied.

In most cases, cavitation is undesirable, because it leads to significant decrease in performance and causes vibrations, noise, and mechanical damage of the centrifugal pumps [1]. Many researches have studied the cavitation of centrifugal pumps. Liu Houlin et al visualized the cavitation flow in a centrifugal pump by high-speed camera and numerical simulation (Zhu, 2016). Zhu Rongsheng et al numerically studied the cavitation performance of first stage impeller of centrifugal charging pump in nuclear power stations (Zhu, 2015). Based on spectral methods, Kristoffer et al proposed a sensitivity parameter to analyse the cavitation of centrifugal pumps (Zhu, 2016; Zhang, 2015).

However, the characteristics and mechanism of cavitation flow in centrifugal pump have not been fully understood yet (Zhu, 2016). In order to detect and prevent cavitation, the characteristics and mechanism of cavitation flow in centrifugal pump should be understood first. By Zwart cavitation model and RNG turbulent model with modified turbulent viscosity, the cavitation flow in centrifugal pump was numerically calculated steadily and transiently. The predicted results were compared with the experimental results. The characteristics of pressure fluctuation were analyzed by time–frequency methods, and the characteristics of cavity vapor shape were illustrated.

2 NUMERICAL SIMULATION

2.1 *Geometries, mesh and boundary conditions*

Geometries of centrifugal pump are modeled by UG software. Flow domains of centrifugal pump include four parts: inlet pipe, impeller, wear-ring clearances and volute, as are shown in Figure 1. The main parameters of the centrifugal pump are shown in Table 1.

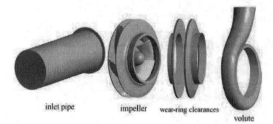

inlet pipe   impeller   wear-ring clearances   volute

Figure 1.   Geometries of centrifugal pump.

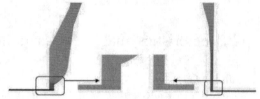

Figure 2.   Mesh of wear-ring clearances.

Table 1.   Main parameters of centrifugal pump.

Physical quantity	Unit	Value
Specific speed	/	130
Pump inlet diameter	mm	126
Pump outlet diameter	mm	100
Number of blades	/	6
Rotation speed	r/min	1480
Designed flow rate	m³/h	101
Wear-ring clearances	mm	0.5
Impeller outlet width	mm	26.5

Figure 3.   Mesh of impeller and volute.

All flow domain parts are assumed to be meshed hex grids by ICEM software. In order to reduce the effects of grids number on numerical results, a grid dependence study at the designed flow rate when *NPSHa* is 2.25 m is carried out. The results demonstrate that the pump head increases with mesh grid number at first, then changes less than 1% when the grid number increases from 3901034 to 4902103. So, the mesh grid number 4301034 is taken. The mesh of wear-ring clearance, rotor and volute is shown in Figures 2 and 3. During the numerical simulation, pump inlet is set to be total pressure; pump outlet is set to be the velocity; all surfaces are set to be no slip wall. In order to capture the cavitation flow near the surface, Yplus of walls is kept less than 10.

2.2 Cavitation model and turbulent model

Zwart cavitation model, reported by Zwart Gerber Belamri (Zhu, 2016), has been widely used in the numerical simulation of the cavitation flow. The form of Zwart-Gerber-Belamri cavitation model is as follows:

$$\dot{m} = F_{vap} \frac{3\alpha_{nuc}(1-\alpha_v)\rho_v}{R_b} \sqrt{\frac{2}{3}\frac{P_v - P}{\rho_l}}, P < P_v$$
$$\dot{m} = F_{cond} \frac{3\alpha_v\rho_v}{R_b} \sqrt{\frac{2}{3}\frac{P - P_v}{\rho_l}}, P > P_v \quad (1)$$

where, F_{vap} is cavitation coefficient, default number is 50. F_{cond} is condensation coefficient, default number is 0.01. α is nucleation site volume fraction, default number is 5×10^{-4}. R_b is bubble radius, default number is 10^{-6} m. α_v is vapor volume fraction, ρ_v is the vapor density, P_v is saturation pressure, P is pressure of mixed flow, ρ_l is water density.

In order to account the effects of turbulent on cavitation flow, we adopted the approach reported by Singhal (Zhu, 2016). It works as follows:

$$P_v' = P_v + \frac{P_{turb}}{2} = P_v + 0.195\rho_m k \quad (2)$$

where k is turbulent kinetic energy and ρ_m is the density of the mixed-flow. The cavitation flow is a mixture of liquid and vapor, and the turbulent viscosity of the cavitation flow should be less than single flow. So, we adopted the method provided by Delgosha to reduce the mixture turbulent viscosity of the standard RNG model (Zhang, 2015). The method has been used in many studies (Zhu, 2016; Zhang, 2015). It is defined as

$$\mu_t = f(\rho_m)C_\mu \cdot k^2 / \varepsilon$$
$$f(\rho_m) = \rho_v + (\rho_v - \rho_m / \rho_v - \rho_l)^n(\rho_l - \rho_v) \quad (3)$$

The numerical simulations are performed by using the commercial CFD code ANSYS-CFX 15.

The modified saturation pressure and turbulent model were added to ANSYS-CFX 15 by CEL expression.

2.3 Cavitation performance

The available net positive suction head (*NPSHa*) of centrifugal pump is calculated as

$$NPSH_a = \frac{p_s}{\rho g} + \frac{v_s^2}{2g} - \frac{P_v}{\rho g} \qquad (4)$$

where, p_s is absolute pressure at pump inlet, v_s is average inlet velocity. In the numerical simulation, cavitation is achieved by gradually decreasing the inlet total pressure.

Figure 4 shows the experimental and numerical predicted results at designed flow rate. The predicted results are in good agreement with the experimental results. Pump head remains steady when *NPSHa* decreases from 6 m to 2.25 m. When *NPSHa* continues to decrease from 2.25 m to 1.02 m, the cavitation intensity increases rapidly and the cavity vapor volume on impeller blades increases greatly, leading to the sudden decrease of pump head. The cavity vapor on impeller blades is shown in Figure 5.

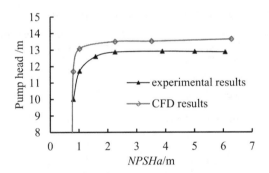

Figure 4.   Relationship between head and *NPSHa*.

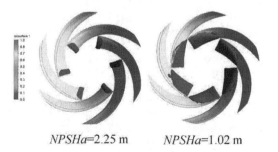

NPSHa=2.25 m   NPSHa=1.02 m

Figure 5.   Cavity vapor on impeller blades.

Figure 6.   Position of monitor points.

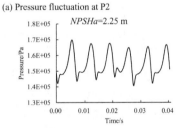

(a) Pressure fluctuation at P2

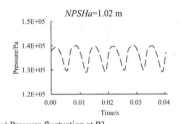

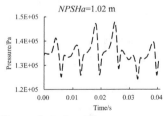

(b) Pressure fluctuation at P1

Figure 7.   Pressure fluctuation at P2 and P1 in a rotation period.

$NPSH_3$, referred to as the critical net positive suction head, is the net positive suction head of pump when pump head decreases by 3%. The experimental $NPSH_3$ is 1.4 m. The numerical predicted $NPSH_3$ is 1.22 m, and the absolute error is 0.18 m.

3 ANALYSIS OF CHARACTERISTICS

3.1 Transient simulation

In order to analyze the characteristic of cavitation flow in the centrifugal pump, a transient simulation is carried out. Two cavitation situations with $NPSHa$ of 2.25 m and 1.02 m were simulated transiently. The time step is set to be 1.126e-4 s, in which the impeller would rotate 1°. The total time is $7T_0$, and T_0 is the time of rotation period. The convergence criterion is 1e-4. The monitor points are shown in Figure 6. P1 is at volute tongue and P2 is at the center of the pump outlet.

3.2 Characteristics of pressure fluctuation

3.2.1 Time analysis
In a rotation period, pressure fluctuation at monitor points is shown in Figure 7. The pressure fluctuation at P2 presents has peaks and troughs, when $NPSHa$ is 1.02 m and 2.25 m. Because the flow at P2 is smooth and pressured, it is not sufficiently influenced by the cavity vapor of impeller blades. The pressure fluctuation at P2 is mainly influenced by the impeller rotation.

The pressure fluctuation at P1 is different with that at P2. It presents several sub-peaks as well as 6 main peaks, which become more obvious when $NPSHa$ decreases from 2.25 to 1.02 m. Because monitor point P1 is closer to the cavity vapor of impeller blades, the pressure fluctuation is seriously influenced by cavitation.

3.2.2 Frequency analysis
The frequency of pressure fluctuation at monitor points is shown in Figure 8. The gray spectrum is $NPSHa = 2.25$ m, and blue spectrum is $NPSHa = 1.02$ m. In order to keep the same frequency resolution, the length of data is set to be equal.

The pressure fluctuation at P1 is shown in the upper part. In the frequency band of 0–2000 Hz, it presents several spectral peaks, such as at 148, 296, and 444 Hz, which are rotation frequency and its multiplier. The energy of spectral peaks decreased when $NPSHa$ decreased from 2.25 m to 1.02 m. In the frequency band of 2000–4000 Hz, noise greatly increased and spectral peaks also increased. The pressure fluctuation at P2 is shown in the lower part. It shows the similar characteristic with P1. In the frequency band of 0–2000 Hz, the main spectral peaks are rotation frequency and its multiplier. In the frequency band of 2000–4000 Hz, noise and spectral peaks increased.

3.3 Characteristics of cavity vapor shape

In order to analyze the reasons for the increasing of noise and spectral peaks, the shape of cavity

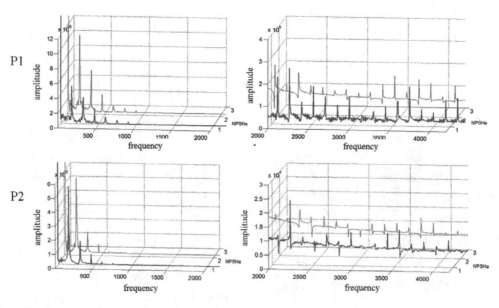

Figure 8. Frequency of pressure fluctuation at point P1 and point P2.

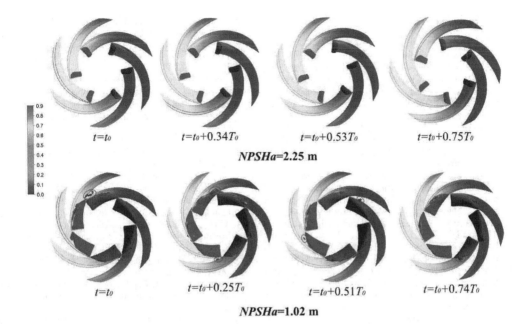

Figure 9.   Cavitation vapor shape on impeller in a rotation period.

vapor on impeller blades is analyzed, as shown in Figure 9. The blade with red edge is a reference, which rotates with time.

When *NPSHa* is 2.25 m, the cavity vapor mainly is generated on the suction side along the leading edge. In a rotation period, the shape of cavity vapor almost remains the same.

When *NPSHa* is 1.02 m, the volume of cavity vapor on the impeller blades suction side greatly increased. The vapor extends from leading edge to trailing edge. In addition, many little vapor bubbles grow up at the end of vacuum, and the shape and position of the little vapor bubbles change with time. It shows that the little vapor bubbles grow and break continually. The growth and break of vapor bubbles lead to the increasing of noise and spectral peaks.

4   CONCLUSIONS

The cavitation flow in centrifugal pump was numerically simulated steadily and transiently by Zwart cavitation model and RNG turbulent model with modified turbulent viscosity. The numerically predicted head-*NPSHa* curve is in good agreement with the experimental curve. The absolute error of predicted $NPSH_3$ is 0.18 m. In a rotation period, the pressure fluctuation at volute tongue presents six main peaks and several sub-peaks, which becomes more obvious with the decrease of

NPSHa. The main frequency of cavitation flow in centrifugal pump is rotation frequency and its multiplier. When *NPSHa* decreases from 2.25 m to 1.02 m, many little vapor bubbles grow up at the end of the vacuum. The growth and breaking up of little vapor bubbles lead to the increasing of noise and spectral peaks in high-frequency band.

ACKNOWLEDGMENTS

This work was financially supported by the National Natural Science Foundation of China (51306205) and the Natural Science Foundation of Hubei Province (2015CFB700).

REFERENCES

Cao Yuliang, He Guo, Ming Tingfeng, et al. Transient cavitation analysis of a mix-flow pump by modified turbulent viscosity. Journal of Harbin Engineering University, 2016, **37** (5): 678–683. (in Chinese)

Delgosha C O, Reboud J L, Delannoy Y. Numerical simulation of the unsteady behaviour of cavitating flows. Int. J. Numer. Methods Fluids, 2003, **42**, 527–548.

Huang Biao, Zhao Yu, Wang Guoyu. Large Eddy Simulation of Turbulent Vortex-Cavitation Interactions in Transient Sheet/Cloud Cavitating Flows. Computers & Fluids, 2014, **92**: 113–124.

Ji Bin, Luo Xianwu, Peng Xiaoxing, et al. Numerical analysis of cavitation evolution and excited pressure fluctuation around a propeller in non-uniform wake. International Journal of Multiphase Flow, 2012, **43**: 13–21.

Ji Bin, Luo Xianwu. Three-dimensional Large Eddy Simulation and Vorticity Analysis of Unsteady Cavitating Flow Around a Twisted Hydrofoil. Journal of Hydrodynamics, 2013, **25** (4): 510–519.

Johann F. G. Centrifugal Pumps (Third Edition). Springer, 2014: 287–292.

K K Mckee, G L Forbes, I Mazhar, et al. A vibration cavitation sensitivity parameter based on spectral and statistical methods. Expert systems with Applications, 2015, **42**: 67–78.

Ki-Han Kim Georges Chahine, Jean-Pierre Franc Ayat Karimi. Advanced Experimental and Numerical Techniques for Cavitation Erosion Prediction. Springer Dordrecht Heidelberg New York London, 2013.

Liu Houlin, Liu Dong-xi, Wang Yong, et al. Experimental Investigation and Numerical Analysis of Unsteady Attached Sheetcavitating Flows in a Centrifugal Pump. Journal of Hydraulics, 2013, **25** (3): 370–378.

Mckee K K, Forbe G, Mazhar I, et al. Cavitation Sensitivity Parameter Analysis for Centrifugal Pumps Based on Spectral Methods. In World congress on engineering asset management, Daejeon, South Korea, 2012.

Singhal, A K, Vaidya N, Leonard A D. Multi-Dimensional Simulation of Cavitating Flows Using a PDF Model for Phase Change. ASME FED Meeting, Vancouver, Canada, 1997.

Wang Yong, Liu Houlin, Liu Dongxi, et al. Application of the Two-Phase Three-Component Computational Model to Predict Cavitating Flow in a Centrifugal Pump and its Validation. Computers and Fluids, 2016. (to be published)

Zhang Bo, Wang Guoyu, Zhang Shuli, et al. Evaluation of a modified RNG κ-ε model for computations of cloud cavitating flows. Transactions of Beijing Institute of Technology. 2008, **28** (12): 1065–1069. (in Chinese)

Zhang Desheng, Shi Weidong, Pan Dazhi, et al. Numerical and Experimental Investigation of Tip Leakage Vortex Cavitation Patterns and Mechanisms in an Axial Flow Pump. Journal of Fluids Engineering, 2015, **137**/121103:1–14.

Zhang Desheng, Shi Weidong, Wu Suqing, et al. Numerical and Experimental Investigation of Tip Leakage Vortex Trajectory in an Axial Flow Pump. ASME 2013 Fluids Engineering Division Summer Meeting, Incline Village, Nevada, USA, 2013, **1B**, FEDSM2013–16058: 1–14.

Zhu Rongsheng, Fu Qiang, Liu Yong, et al. The research and test of the cavitation performance of first stage impeller of centrifugal charging pump in nuclear power stations. Nuclear Engineering and Design, 2016, **300**: 74–84.

Zwart P J, Gerber A G, Belamri T. A Two-Phase Model for Predicting Cavitation Dynamics. Fifth International Conference on Multiphase Flow. Yokohama, Japan, 2004.

Advances in Energy Science and Equipment Engineering II – Zhou, Patty & Chen (Eds)
© 2017 Taylor & Francis Group, London, ISBN 978-1-138-71798-5

Air entrainment and energy dissipation by flow types over stepped weir structure

Jin-Hong Kim

Department of Civil and Environmental Engineering, Chung-Ang University, Seoul, Korea

ABSTRACT: Air entrainment and energy dissipation by flow types over the stepped weir structure were performed through hydraulic experiments. Nappe flow occurs at low flow rates and for relatively large step height. Its dominant features are air pocket, nappe impact and subsequent hydraulic jump. At larger flow rates, skimming flow occurs with formation of recirculating vortices. Air entrainment occurs through free-falling nappe impact and subsequent hydraulic jump in the nappe flow, and occurs from the step edges in the skimming flow. Energy dissipation occurs through the jet impact and the subsequent hydraulic jump in the nappe flow and occurs through maintaining the recirculation vortices between step edges in the skimming flow regimes. Average values of the oxygen transfer are 0.45 in the nappe flow and 0.28 in the skimming flow, and Efficiencies of energy dissipation in the nappe flow and in the skimming flow are 70~95(%) and 60~90(%), respectively. From these results, the stepped weir structure is found to be efficient for oxygen transfer and for energy dissipation.

1 INTRODUCTION

Weir, the typical crossing structure, is installed across the river for water intake or navigation. It is installed against the flow direction, thus it makes deterioration of water quality by reducing flow velocity and extending retention time of the river flow. For that reason, it is necessary to improve the water quality by oxygen transfer through air entrainment when the water flows over the weir.

Weir structures are useful for air entrainment by the stepped type of the overflow section. The flows over the stepped weir are characterized by the large amount of self-entrained air. Air entrainment by macro-roughness is efficient in water treatment because of the strong turbulent mixing. The macro roughness of the steps leads to an increase in the thickness of the turbulent boundary layer. Where the boundary layer reaches the free surface, air is entrained at the so-called inception point of air entrainment (Henry, 1985). Thus, the stepped weirs have been used for a long time for the purpose of aeration, and they will be built along polluted and eutrophic streams to control the water quality through reoxygenation, denitrification and Volatile Organic Component (VOC) removals (Henry, 1985).

Weir structures are also useful for energy dissipation by the stepped type of the flow section. They are characterized by significant flow resistance and associated energy dissipation taking place on the steps (Chanson, 2000). Thus, they are installed to protect the stream bed against scour since they are assuming a role for energy dissipation and for size reduction of retention basin or bottom protection, and for cutting costs. Figure 1 shows the typical case of stepped weir in Korea.

This study presents the oxygen transfer through the air entrainment and the energy dissipation

Figure 1.   Stepped type of weir structure.

by the flow types over the stepped weir structure. Hydraulic analysis on the oxygen transfer and the energy dissipation by the nappe flow and the skimming flow were performed through the hydraulic experiments.

2 FLOW TYPES

Flow over stepped weir structures are characterized by the two types: nappe flow and skimming flow shown in Figure 2 (Chanson, 1993). At low flow rates and for relatively large step height, nappe flow occurs. The water bounces from one step onto the next one. Dominant flow features include an enclosed air cavity, a free-falling nappe, a nappe impact and a subsequent hydraulic jump.

At larger flow rates and for relatively small step height, skimming flow occurs. The flow skims over the step edges. The water flows down in a coherent stream where external edges determine a pseudo-bottom defined by the straight line that connects the edges of each step (Chanson, 1993). In a skimming flow, the free surface on the upper steps is clear and transparent. A turbulent boundary layer develops along the step edges. When the outer edge of the boundary layer reaches the free surface, free surface aeration takes place.

For intermediate flow rates, the flow exhibits strong splashing and droplet ejections at any position downstream of the inception point of free surface aeration: i.e. the transition flow regime. The transition flow has a chaotic appearance with numerous drop ejections that are seen to reach heights of up to 3 to 8 times the step height (Chanson, 2004). The transition between nappe and skimming flow is related to the flow rate, step slope, and local flow properties. However this distinction does not seem to create well defined limits as for each geometric configuration (Fratino, 2000).

3 OXYGEN TRANSFER AND ENERGY DISSIPATION

Oxygen transfer through air entrainment occurs mainly from behind the trailing edge of the stepped weir structures (Kim, 2003). Air bubbles form and proceed to downward direction becoming larger in volume, and are finally broken during proceeding upward. Dissolved oxygen is stored with breaking of the air bubbles, and this would give the good habitat condition downstream of the stepped weir. Hydraulic jump makes the air entrainment more active.

The efficiency of the oxygen transfer E is used for representing occurrence of air entrainment (Avery, 1978);

$$E = (C_d - C_u) / (C_s - C_u) \qquad (1)$$

where C_d and C_u are contents of dissolved oxygen measured at downstream and upstream point, respectively, and C_s is the saturated contents of the dissolved oxygen. Since the oxygen transfer is affected by the water temperature, E is substituted by E_{20} (Gulliver, 1990);

$$\frac{\ln(1 - E_T)}{\ln(1 - E_{20})} = 1.0 + \alpha(T - 20) + \beta(T - 20)^2 \qquad (2)$$

where E_T and E_{20} are efficiencies of the oxygen transfer at temperature $T°C$ and the reference temperature $20°C$, respectively. α and β are constants as $\alpha = 0.02103°C^{-1}$, $\beta = 8.621 \times 10^{-5} °C^{-2}$.

The dissipation efficiency and the mechanisms that determine its effectiveness are defined using different evaluation processes for nappe and skimming flow regimes (Fratino, 2000). In the first case, energy dissipation is due to nappe impact on the underlying water cushion and subsequent hydraulic jump. In contrast, most of the energy is dissipated in maintaining the recirculation vortices beneath the pseudo-bottom formed by the edges of the steps in the skimming flow regimes.

Dissipation equation for the nappe flow regions is represented by (Fratino, 2000),

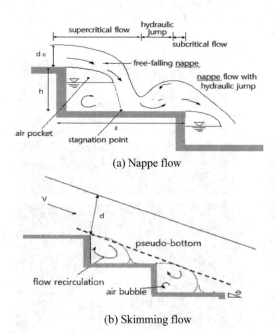

(a) Nappe flow

(b) Skimming flow

Figure 2. Sketch of nappe and skimming flow (Chanson, 1993).

$$\frac{\Delta H}{H_{\max}} = 1 - \frac{H_{res}}{H_{\max}} = 1 - \frac{\lambda + 0.5\lambda^2}{H_{weir}/k + 0.5}$$

$$\lambda = \frac{\sqrt{2}}{1.06 + \sqrt{h/k + 1.5}} \qquad (3)$$

where ΔH is the head difference between upstream and downstream of the stepped weirs, $H_{\max}$ is the maximum total head upstream of the structure, H_{res} is the residual head downstream of the structure, H_{weir} is the structure height, k is the overflow depth at the crest of the structure, and h is the step height.

Dissipation equation for the skimming flow regions is represented by (Fratino, 2000),

$$\frac{\Delta H}{H_{\max}} = 1 - \frac{\left[\dfrac{f}{8\sin\alpha}\right]^{1/3}\cos\alpha + 0.5\left[\dfrac{f}{8\sin\alpha}\right]^{2/3}}{\dfrac{H_{weir}}{k} + 1.5}$$

$$f = \frac{8g\sin\alpha d^2}{q^2}\frac{R}{4} \qquad (4)$$

where α is the slope of the stepped drop structure, d is the uniform flow depth upstream of the structure, R is the hydraulic radius, q is the discharge per unit width, and f is friction factor representing aeration properties at the overflow sections.

4 LABORATORY EXPERIMENTS

Figure 3 shows the experimental arrangements. The typical model of the stepped weir was made of waterproof plywood. It was installed in a recirculatory tilting flume of 0.6 m wide, 0.4 m deep and

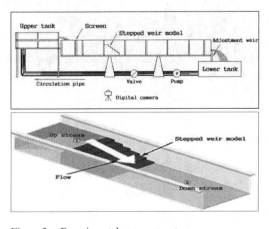

Figure 3. Experimental arrangement.

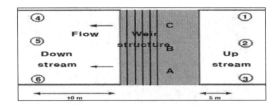

Figure 4. Measuring points at weir structure.

15 m long. The sidewall of the flume was made of glass and a transparent scale was attached to the side wall to see the flow features well. A damper was laid at the upstream section of the flume to reduce the turbulence and to assure the hydraulic feed having negligible kinetic components.

Water level was regulated by the down-stream adjustment weir. The discharge which was controlled by a valve in a feed-back loop could be measured with a v-notch at the upper tank.

The stepped weir model was 0.6 m wide and 0.31 m high. Five different slopes of the model were selected (1:2.0, 1:1.7, 1:1.5, 1:1.2 and 1:0.7). Hence, in case of the slope 1:2.0, the model was 0.6 m wide, 0.62 m long, 0.31 m high and on a slope of 30°. The number of steps was 6, and each step was 0.4 m wide, 0.10 m long and 0.05 m high.

Flow velocity was measured by using an electromagnetic current meter (model; MI-ECM4). To check the flow pattern, dye injection and a digital camera (model; Olympus c-5050z) with a strong light were used.

Efficiencies of the oxygen transfer and the energy dissipation are estimated by measuring contents of the dissolved oxygen and the total head at the upstream and downstream point of the structure, respectively. Figure 4 shows the measuring points at the weir structure.

Here, point A and C is the right and left side of the streams, respectively. Point B is the midpart of the stream. All the data are measured 5 m upstream and 10 m downstream of the structure for considering data consistency.

5 RESULTS AND DISCUSSIONS

Flow over stepped weir are characterized by three types: nappe flow, skimming flow and transition flow. They are shown in Figure 5.

Nappe flow occurs at low flow rates and for relatively large step height. The flow is accelerated in the downstream direction until a deflected nappe took place. Dominant features of an air pocket, nappe impact and subsequent hydraulic jump occur apparently. The inception of free surface aeration took place at the first deflected nappe

1269

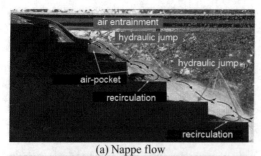

(a) Nappe flow

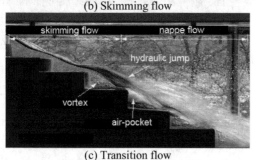

(b) Skimming flow

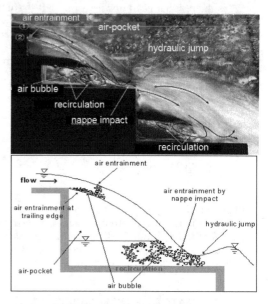

(c) Transition flow

Figure 5.   Nappe flow and skimming flow.

although some bubbles were trapped in cavities immediately upstream of the nappe take-off.

At larger flow rates, skimming flow occurs with formation of recirculating vortices between the main flow and the step corners. The flow direction of air-water mixture was almost parallel to the pseudo-bottom formed by the step edges although shapes of the recirculating vortices beneath the main flow alternate from step to step.

Transition between the nappe and skimming flow occurs. In this case, the skimming flow and the nappe flow occurs at upper steps and at lower steps, respectively. It does not have the quasi-smooth free surface appearance of skimming flow, nor the distinctive succession of free falling nappes observed in nappe flow. In transition flows down the step slope, the upstream flow is non-aerated. The free surface exhibited however an undular profile in phase with almost the same wave length as the stepped invert profile.

Air entrainment in the nappe flow and in the skimming flow is shown in Figures 6 and 7, respectively.

In the nappe flow, free surface aeration was observed at both the upper and the lower nappes with additional air entrainment at the impact followed by nappe breakup. Air entrainment occurs from the step edge, but most air is entrained through a free-falling nappe impact and subsequent

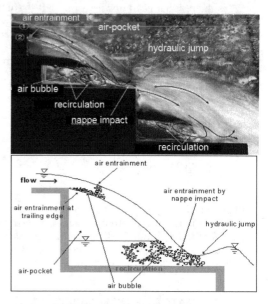

Figure 6.   Air entrainment in nappe flow.

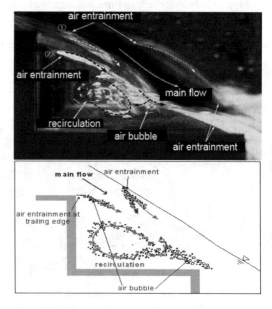

Figure 7.   Air entrainment in skimming flow.

hydraulic jump. Air pocket also has an important role to the air entrainment. In the skimming flow, air entrainment occurs from the step edges. Downstream of the inception point, the flow is highly aerated at each and every step with very significant splashing.

Air entrainment was more active in the nappe flow than in the skimming flow, main reason of which is due to the nappe impact and the subsequent hydraulic jump, whereas they don't occur in the skimming flow.

Figure 8 shows the relationship between the oxygen transfer and the flow parameters by flow types. Flow condition changes from a nappe flow to a skimming flow as the flow velocity and Froude number increase. The transition between nappe and skimming flow was shown to occur at region of $v = 0.56$–0.79(m/s) and Fr = 1.32–1.51. Undular profile of the free surface, acceleration above filled cavities and deceleration at nappe impact at that region as was shown (Chanson, 2004).

Oxygen transfer becomes smaller and reaches to minimum at the region of a transition flow, but recovers in the region of skimming flow. It occurs more actively in the region of nappe flow than in the skimming flow. This is because the air entrainment is made mainly through a free-falling nappe impact, a hydraulic jump and an air pocket in the region of nappe flow, while they don't occurr in

the region of the skimming flow. Oxygen transfer in the region of transition flow is not so high compared with nappe flow and skimming flow, since the flow in this region is not aerated but exhibits just an undular profile.

The average values of the oxygen transfer in the region of the nappe flow and in the region of the skimming flow are about 0.45 and 0.28, respectively. It reveals that average value of the oxygen transfer at the riparian riffles is about 0.085 (Kim, 2015), thus the stepped weir structure is found to be efficient for oxygen transfer and for treatment of water quality associated with substantial air entrainment.

Figure 9 shows the relationships between the energy dissipation and the ratio of overflow depth and step height with values of step slopes in the region of nappe flow. Here, h is the step height and L is the step length. Hence, h/L is the step slope. All the other parameters are already explained in equation (3). The straight line is the theoretical results of equation (3).

Energy dissipation is proportional to the step height and is inversely proportional to the overflow depth, but it is not proportional to the step slope, which means that it takes place mainly through the nappe impact on the underlying water cushion and the subsequent hydraulic jump.

Experimental values except for those of $h/L = 0.50$ showed the similar results to those of (Fratino, 2000) and the theoretical results, which is due to the longer interval of occurrence of the hydraulic jump.

Figure 10 shows the relationships between the energy dissipation and the ratio of structure height and overflow depth with parameters of slope angles in the region of skimming flow. Two curves are the theoretical results of equation (4) with f equal to one and α equal to $60°$ and $6°$, respectively (Fratino, 2000).

Energy dissipation is related to the structure height and is not proportional to the step slope,

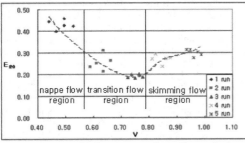

(a) Oxygen transfer to flow velocity

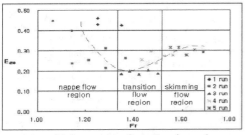

(b) Oxygen transfer to Froude number

Figure 8.   Relationship between oxygen transfer and flow parameters.

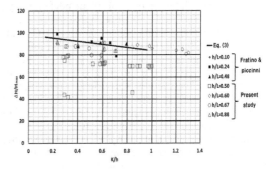

Figure 9.   Energy dissipation in the nappe flow.

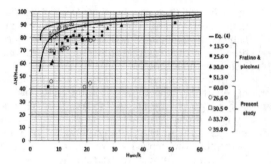

Figure 10.   Energy dissipation in the skimming flow.

dissipation occurs through the nappe impact on the underlying water cushion and the subsequent hydraulic jump in the nappe flow and occurs through maintaining the recirculation vortices between step edges in the skimming flow regimes. The average values of the oxygen transfer are 0.45 in the nappe flow and 0.28 in the skimming flow, which is higher than value at the riparian riffles, and the efficiencies of energy dissipation in the nappe flow and in the skimming flow are about 70~95(%) and 60~90(%), respectively. From these results, the stepped weir structure is found to be efficient for oxygen transfer and for energy dissipation.

which is the similar case in the nappe flow, since it occurs through maintaining the recirculation vortices between the step edges.

Theoretical results overestimate the experimental data although they show smaller differences when the step slope decreases. This seems to arise from the assumption of the constant value of friction factor f, which depends on slope angle, overflow depth, flow discharge and hydraulic radius as is shown in equation (4).

In Figures 9 and 10, efficiencies of energy dissipation in the nappe flow and in the skimming flow are about 70~95(%) and 60~90(%), respectively. Energy dissipation in the nappe flow was still more active than in the skimming flow due to the nappe impact on the underlying water cushion and the subsequent hydraulic jump. From these results, the stepped weir structure is also to be efficient for energy dissipation.

6   CONCLUSIONS

In this study, oxygen transfer through the air entrainment and the energy dissipation by the flow types at the stepped weir structure were performed through the hydraulic experiments. Nappe flow occurs at low flow rates and for relatively large step height. Dominant features of an air pocket, nappe impact and subsequent hydraulic jump occurs. At larger flow rates, skimming flow occurs with formation of recirculating vortices. At a transition flow regime, skimming flow at upper steps and nappe flow at lower steps occur simultaneously. It does not have quasi-smooth free surface nor distinctive succession of free falling nappes. Air entrainment occurs from the step edge in the nappe flow, but most air is entrained through a free-falling nappe impact and hydraulic jump. In the skimming flow, air entrainment occurs from the step edges. Energy

ACKNOWLEDGEMENTS

This research was supported by a grant (12-TI-C02) from Advanced Water Management Research Program funded by Ministry of Land, Infrastructure and Transport of Korean Government.

REFERENCES

Avery S.T. and P. Novak, "Oxygen transfer at hydraulic structures." Journal of the Hydraulics Division, ASCE, Vol. 104, No. 11, pp. 1521–1540 (1978).

Chanson H. "Self-aerated flows on chute and spillways." Journal of the Hydraulics Division, ASCE, Vol. 119, No. 2, pp. 220–243 (1993).

Chanson H. and L. Toombes, "Hydraulics of stepped chutes: The transition flow." Journal of the Hydraulic Research, IAHR, Vol. 42, No. 1, pp. 43–54 (2004).

Chanson H., Y. Yasuda and I. Ohtsu, "Flow resistance in skimming flow: A critical review." Proceedings of the International Workshop on Hydraulics of Stepped Spillways, Zurich, Switzerland, Vol. 42, No. 1, pp. 43–54 (2000).

Fratino U. and A.E. Piccinni, "Dissipation efficiency of stepped spillways" Proceedings of the 9th International workshop on hydraulics of stepped spillway, Zurich, Switzerland, pp. 103–110 (2000).

Gulliver J.S., J.R. Thene, and A.J. Rindels, "Indexing gas transfer in self-aerated flows." Journal of the Environmental Engineering, ASCE, Vol. 116, No. 3, pp. 503–523 (1990).

Henry T. Air-water flow in hydraulic structures. A Water Resources Technical Publication, Engineering Monograph No. 41, pp. 251–285 (1985).

Kim J.H. "Oxygen Transfer through Air Entrainment in Riparian Riffles: A Case Study of Seomjin River, Korea." Jurnal Teknilogi, Vol. 74, No. 3, pp. 21–26 (2015).

Kim J.H. "Water quality management by stepped overflow weir as a method of instream flow solution." Proceedings of the First International Conference on Solutions of Water Shortage and Instream Flow Problems in Asia. Incheon, Korea, pp. 24–36 (2003).

Advances in Energy Science and Equipment Engineering II – Zhou, Patty & Chen (Eds)
© 2017 Taylor & Francis Group, London, ISBN 978-1-138-71798-5

Classified analysis on energy consumption of machine tools

Shuang Liu & Jie Meng
Chongqing University of Science and Technology, Huxi University Town, Shapingba District, Chongqing, China

Jin Chen
Chongqing College of Electronic Engineering, Huxi University Town, Shapingba District, Chongqing, China

ABSTRACT: Despite China's rapid growth of machine tools throughput, there's strong and emergent need to design energy-efficient machine tools. This paper uses a typical machine process to look into the energy consumption of machine tools. This process is classified into three categories according to energy consumption feature. The internal rule of energy consumption of these processes are discussed in detail, especially that of start-up process. The case study shows that those manufacturing processes which choose low spindle speed may be more reasonable in view of energy consumption.

1 INTRODUCTION

It is reported that China's yield of machine tools is booming in recent years. Despite this speedy growth, the import of machine tools is still in great demand (Bohao, 2006). This may be partly ascribed to our weakness in energy consumption, which is an important evaluation indicator of machine tools apart from cost and efficiency (Hongying, 2011). At the same time, the energy shortage and environmental contamination are becoming the top two serious problems (Weidou, 2008). This lead to the emergent need to convert the formal resource-sacrificed manufacturing into the brand new environmental-oriented green manufacturing. Consequently, it is important to probe into the energy consumption of machine tools, which plays significant role in a green manufacturing system. this paper aims at a basic analysis on the energy consumption of machine tools, so as to pave the way to design energy-efficient machine tools.

2 CHARACTERISTIC OF ENERGY CONSUMPTION OF MACHINE TOOLS

2.1 Monitoring the energy consumption of machine tools

To analyze the characteristics of the energy consumption of machine tools, a workpiece is chosen for monitoring the power expenditure during its manufacturing process.

This whole manufacturing process includes the following steps: Firstly start up the machine tools to the target spindle speed. Then the cutter traverse rapidly to cutting point, where the cutting of the cylinder begin. After that, the external turning of the stepped shaft is carried out, following by corresponding round off. Afterwards, the external turning of the smallest cylinder and corresponding smoothing is performed. At last, the fillet on the right end face is machined out. This manufacturing process is realized by a CNC program displayed in Table 1.

Table 1. Program code used for manufacturing the chosen workpiece.

Number	Program Code
1	N0000 %25
2	N0010 G00X80 Z80
3	N0020 T0101
4	N0030 M03 S500
5	N0040 G00 X38 Z2
6	N0050 G01 Z-38 F80
7	N0060 G00 X40 Z2
8	N0070 X34
9	N0080 G01 Z-25
10	N0090 G02 X38 Z-27 R2 F20
11	N0100 G00 X38 Z2
12	N0110 X30
13	N0120 G01 Z-10 F80
14	N0130 G02 X34 Z-12 R2 F20
15	N0140 G00 X80 Z80 F80
16	N0150 T03
17	N0160 G00 X38 Z-36.5
18	N0170 G01 X-1 F20
19	N0180 G00 X80
20	N0190 Z80
21	N0200 M30

During the above process, the power expenditure of the machine tools is monitored by a power meter. The power curve along time is drawn in Figure 1.

2.2 Sort feature of energy consumption of machine tools

Through the power curve in Figure 1, we see different feature of energy consumption in different operation type. The most common type of operation is idling, when the spindle keeps revolving without cutting operation. Idling includes processes like rap-id traverse, feeding, retracting, etc. As the spindle speed is a constant during each of these processes, the idling power fluctuates in a relatively short range just like the saw tooth of the curve indicates. Thus, we can use the average value of these processes to represent the basic energy requirement to run a ma-chine tool. We call this average value as idling power. Subgraph c in Figure 2 gives out several idling processes which

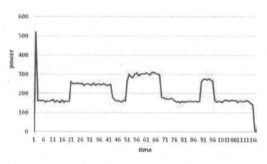

Figure 1. Power consumption of a typical manufacturing process on machine tools.

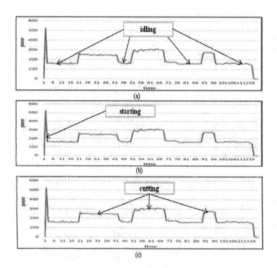

Figure 2. Classified power consumption of machine tools.

occurs at different time in a complete manufacturing process.

The second type of operation happens when the ma-chine tool starts up. The power curve increases from zero to a very high point in an extremely short time, and then declines rapidly to a point where it equals to the idling power. The accelerated speed is very high so as to save the time and energy in this operation. Hence, the highest curve point in this staring process usually is also the peak of the whole curve which includes the following cutting process. This first-climb-then-drop trend can be seen in the power consumption curve of every start up operation.

Besides idling and starting, the remaining type of operation is cutting, which is also the most important one as it's responsible for shaping the workblank to final product. Hence, during this phase, the power consumption is always higher than that of idling. The excessive power consumption is mainly used for remove material from the workpiece while a proportion of it is expended as increased friction loss, heat dissipation, etc. One point to notice is that the power value during different cutting phase is also different. This is due to different material remove rate or diverse process parameters in these cutting phase. This phenomenon can also be explained by the theory that the additional load loss coefficient is correlated to spindle speed.

3 CASE STUDY AND CLASSIFIED ANALYSIS ON ENERGY CONSUMPTION OF MACHINE TOOLS

3.1 Energy consumption during start up

The aforementioned sort feature of energy consumption of machine tools during start up shows how power changes in the course time. However, the relationship of energy consumption among different start up process is still unknown. Hence, several start up processes are monitored to look into the internal connection.

One of these processes is introduced in Table 2 which indicates the change of voltage along with time.

From Table 2, we see voltage stepped up along with time, no matter forward or reverse. Figure 3 gives the visual exhibition of this law.

The corresponding change of current during this same process is also given out in Table 3.

From Table 3, we see current increases unsteadily at first, nevertheless it drops quickly after a certain peak. This rule is the same either the spindle revolving is forward or reverse, which is displayed in Figure 4.

Table 2. Change of voltage along with time.

time	0	0.05	0.1	0.15	0.2	0.25	0.3
forward	0	48	48	48	48	48	114
reverse	0	3	3	3	76	71	71
time	0.35	0.4	0.45	0.5	0.55	0.6	0.65
forward	114	155	155	194	194	210	210
reverse	71.4	128	128	128	167	220	220
time	0.7	0.75	0.8	0.85	0.9	0.95	□
forward	231	231	247	247	246	246	□
reverse	217	238	246	246	247	247	□

Table 3. Change of current along with time.

time	0	0.05	0.1	0.15	0.2	0.25	0.3
forward	0	4.19	4.19	4.19	4.19	4.19	10.6
reverse	0	3.66	0.35	0.35	0.35	12.5	12.5
time	0.35	0.4	0.45	0.5	0.55	0.6	0.65
forward	10.6	14.1	14.1	12.6	12.6	8.02	8.02
reverse	12.5	13.5	13.5	17.6	17.6	7.25	7.85
time	0.7	0.75	0.8	0.85	0.9	0.95	□
forward	6.98	6.98	8.18	8.18	7.63	7.61	□
reverse	8.69	8.69	7.63	7.63	7.84	7.84	□

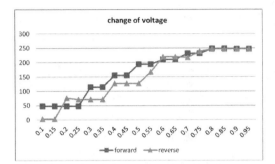

Figure 3. Variation of voltage.

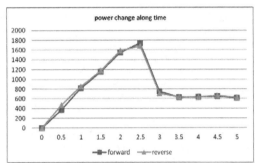

Figure 5. Variation of power.

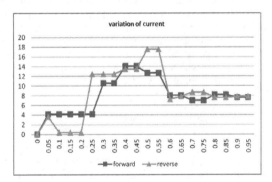

Figure 4. Variation of current.

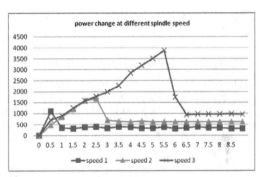

Figure 6. Comparing the power change at different spindle speed.

If we multiple voltage in Table 2 and current in Table 3, then the product is power. Thereby, we get a new table of start-up power, and its figure which resembles greatly to Figure 5.

It can be easily drawn from Figure 5 that the rule of power variation is regard less of the direction of revolving. Hence, only one direction is chosen to while comparing the energy consumption among different spindle speed. Thus, we get Figure 6, which presents different power variation at the above mentioned different start up processes.

As shown in Figure 6, different spindle speed lead to different power change. The most obvious differences are power peak, duration of start-up process, and the time used to reach power peak. So the conclusion can be drawn that these three parameters in a start-up process are strongly related to the spindle speed of machine tools. The first two relationship are provided in Figure 7 and Figure 8.

The power peak in Figure 7 goes up in accordance with spindle speed. The identical relation between duration of start-up process and spindle speed can be seen in Figure 8.

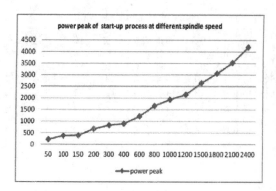

Figure 7.   Power peak at different spindle speed.

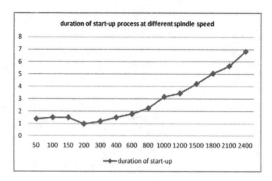

Figure 8.   Duration of start-up at different spindle speed.

3.2 *Energy consumption during idling*

The law of energy consumption during idling is very similar to that of start-up process, which means the power consumption has a strong correlation to spindle speed. This relation is clearly demonstrated in Figure 9, and can be represented by appropriate curve fitting.

3.3 *Energy consumption during cutting*

The energy consumption during cutting process is composed of the energy of idling at the same spindle speed, the energy needed to remove the material from the workpiece, energy consumed in heat dissipation, energy wasted in mechanical friction, etc. Thereinto, the first two occupies the vast majority of the total energy consumption. It's already known from the above analysis that power consumption of idling strongly relates to spindle speed. Meanwhile, the energy required for workpiece removal is also related to spindle speed, which is the most important one of the three cutting parameters in

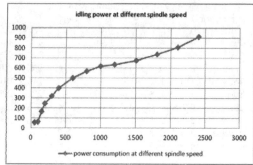

Figure 9.   Comparing idling power at different spindle speed.

a machining process. It thus appears that a large proportion of energy consumption during cutting is similarly related to the spindle speed.

4   CONCLUSION

A manufacturing process is classified into three main categories according to the power expenditure during this process. The feature of each category is detailedly discussed, particularly that of the start-up process. It is revealed that the power peak during start-up, the duration of start-up, idling power, and cutting power are all strongly related to spindle speed. From this, those manufacturing processes which choose low spindle speed may be more reasonable in view of energy consumption.

ACKNOWLEDGMENT

The authors kindly acknowledges the funding provided by the National Natural Science Foundation (Grant No. 51305474), National High Technology Research and Development Program (Grant No. 2014AA041506), and Science and technology research project of Chongqing Municipal Education Committee (Grant No. KJ1503006).

REFERENCES

Association for manufacturing excellence, Green manufacturing, Post and Telecom Press, (2010).
Bohao S., Chin. Ind. Newsp. **2**, 17–21 (2006).
Hongying Y. , Chin. Ind. Newsp. Net, http://www.cinn.cn/wzgk/wsm/250455.shtml (2011).
Weidou N., C. Zhen, L. Zheng, Chn. Energy **12**, 5–9 (2008).

Advances in Energy Science and Equipment Engineering II – Zhou, Patty & Chen (Eds)
© 2017 Taylor & Francis Group, London, ISBN 978-1-138-71798-5

The measurement for rotary axes of 5-axis machine tools

Fangyu Pan, Yuewei Bai, Li Nie, Xiaogang Wang, Kai Liu & Xiaoyan Wu
Shanghai Second Polytechnic University, Industrial Engineering, Shanghai, China

ABSTRACT: In the paper, the errors of 5-axis machine tools are introduced, especially geometric errors. According to the structure of the 5-axis machine tool, the error modelling is discussed. Based on the modelling, the measurement methods for rotary axes are introduced. They are laser interferometer method and laser tracker method. The principles of the two methods are also analyzed. By experiments, the errors of rotary axes are given, which prove the feasibility of the two methods.

1 INTRODUCTION

Equipment manufacturing industry is athe most significant basic one, which related to national economy and the people's livelihood. The 5-axis machine tools play a key role in equipment manufacturing and are supposed to be the most important mechanical processing method. Compared with traditional 3-axis machine tools, the 5-axis ones have more extra rotary axes, which makes them more flexible. Due to longer transport chain, 5-axis machine tools' accumulative errors are larger, which make their precision worse. As accuracy is the key character for machine tools, improving it is a hot topic of research. Generally speaking, there are two methods to achieve above goal. One is traditional way, error avoiding, which means careful design, lean manufacturing and perfect installation. Thus, the cost is high, especially when the machines' precision has a reached a certain level; the other one is error compensation, which achieves high accuracy by bringing in additional new errors to offset the existing errors. This method is more economical and convenient, and so it is welcomed by researchers.

There are three steps to compensate for the machine tools' errors. The first step is error modeling to establish the relationship between all the errors, The second is the error measurement to find out each error' magnitude. The third is error compensation, which is carried out based on above two steps. Therefore, the paper focuses on the first two steps: error modeling and measurement.

2 ERROR MODELING

In order to model for machine tools, it is necessary to find out the source of errors that influence the accuracy of the machine tools.

2.1 The source of errors

Generally speaking, there are three aspects of errors involved, shown as Table 1, which are caused by machine tools, machining process and detection, respectively (Abdul, 2010).

According Table 1, errors caused by machine tools account for 50% of all errors, which are geometric errors and thermal errors. Geometric errors are the machine tools' inherent errors, due to the defects of machine tools' manufacture and assembly. Thermal errors, just as its name implies, are related to temperature, whose variation may lead to machine tool deformation. In this paper, we mainly focus on the geometric errors.

2.2 The geometric errors of machine tools

An object without any restriction will have six degrees of freedom. For each axis of machine tools, five degrees of freedom should be constrained,

Table 1. The errors deteriorating precision of machine tools.

Machine tool errors	Geometric errors	22%	50%
	Thermal errors	28%	
Machining process errors	Cutting tool errors	13.5%	35%
	Champ errors	7.5%	
	Thermal errors and elastic deformation errors of workpiece	6.5%	
	Operation errors	7.5%	
Detection errors	Uncertain errors	10%	15%
	Installation errors	5%	

which means the axis only can move along one direction in ideal conditions. However, in fact, due to imperfect manufacture, the axis can have micro movements in every direction, so there are six errors for each axis. Take x-axis for example; the six errors are three translational errors and three angular errors. The former errors are positioning error $\delta_x(x)$, horizontal straightness error $\delta_y(x)$ and vertical straightness error $\delta_z(x)$, respectively; the later ones are roll error $\varepsilon_x(x)$, pitch error $\varepsilon_y(x)$ and yaw error $\varepsilon_z(x)$. Therefore, the other two translational axes of machine tools have similar errors, given as follows.

Y-axis: $\delta_y(y), \delta_x(y), \delta_z(y), \varepsilon_y(y), \varepsilon_x(y), \varepsilon_z(y)$

Z-axis: $\delta_z(z), \delta_x(z), \delta_y(z), \varepsilon_z(z), \varepsilon_x(z), \varepsilon_y(z)$

Rotary axes are similar with translational ones, they also have 6 errors for each rotary axis, shown as follows.

C-axis: $\delta_x(C), \delta_y(C), \delta_z(C), \varepsilon_y(C), \varepsilon_x(C), \varepsilon_z(C)$

A-axis: $\delta_x(A), \delta_y(A), \delta_z(A), \varepsilon_y(A), \varepsilon_x(A), \varepsilon_z(A)$

Besides above errors, errors between axes also exist, which usually are called perpendicularity. They are $\varepsilon_{xy}, \varepsilon_{yz}, \varepsilon_{xz}, \varepsilon_{C-xy}, \varepsilon_{CA}$.

2.3 The structure of 5-axis machine tools

A 5-axis machine tool usually consists of three translational axes and two rotary axes. According to location of rotary axes, 5-axis machine tools can be divided into three categories: RRTTT, RTTTR and TTTRR (Wang, 2013).

RRTTT: Two rotary axes are linked together, and one of them connects with the machine table, as shown in Fig. 1. Generally speaking, c-axis can rotate from 0 to 360 degrees, but A-axis only from

30 degree to –120 degree. This kind of structure limits the load of the workpiece, especially when A-axis is located at a large angle (Yang, 1998).

TTTRR: Two rotary axes link together, and one of them connects with spindle and cutting tool, as shown as Fig. 2. The C-axis can rotate from 0 to 360 degrees, but A-axis cannot rotate full circle. It is worth mentioning that the axis of A is not fixed, but instead, it is mobile, which makes the cutting tool more flexible. Besides, the machine table doesn't rotate with any axis.

RTTTR: Two rotary axes are separated. One connects with machine table and the other connects with cutting tool, as shown in Fig. 3. The character of the structure is between that of RRTTT and TTTRR.

In the paper, the TTTRR structure machine tool will be taken, for example, to introduce the process of error modeling.

2.4 Error modeling

As shown in Fig. 4, it represents the grand; 1 is workpiece; 2 is gantry, moving along Y-axis; 3 is glide board, running along the X-axis; 4 is the column,

Figure 2. TTTRR.

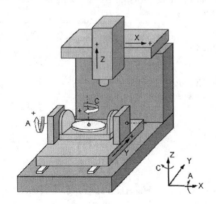

Figure 1. RRTTT.

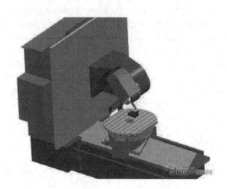

Figure 3. RTTTR.

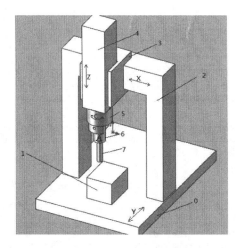

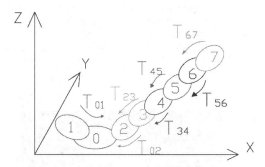

Figure 4.   The diagram of a TTTRR machine tool.

Figure 5.   The kinematic chain of 5-axis machine tool.

gliding along Z-axis; 5 is swivel head, rotating around C-axis; 6 is rotary head, revolving around A-axis; 7 represents the cutting tool. The TTTRR machine tools can be divided into two branches. One is workpiece branch, and the other is cutting tool branch. In an ideal situation, the end of workpiece branch should coincide with the end of cutting tool branch when machining. However, due to errors, they can't be same with each other and the difference is the machine error, which we hunt for. In order to describe the position and the gesture of each moving part in its own coordinate system, Denavit-Hartenberg matrixes are used. Meanwhile, homogeneous transformations are utilized to transform the position and gesture between different coordinate systems.

The difference of position and gesture between cutting tool tip and the end of workpiece can be calculated in the same coordinate system. The two branches have the same base that are grand. Therefore, they should be transformed into the grand coordinate system. Tool tip can be expressed in the

cutting tool coordinate system 7 and transformed to the coordinate system 6 and 6 to 5. Then it can be transformed to the machine coordinate system 0 step by step as shown Fig. 5 and Eq. 1; the workpiece branch is similar, described in Eq. 2. Hence, the final error we want is given in Eq. 3

$$^0T_t = T_{02}T_{23}T_{34}T_{45}T_{56}T_{67}{}^7T_t \tag{1}$$

$$^0T_w = T_{01}{}^1T_w \tag{2}$$

$$E = {}^0T_t - {}^0T_w \tag{3}$$

3   THE PRINCIPLE OF THE MEASUREMENT

The measurement of three translational axes is relatively mature, so the paper focuses on the two rotary axes' measurements. Two methods will be introduced to collect the errors of rotary axes. One is the laser interferometer method and the other is the laser tracker method.

3.1   Laser interferometer method

In order to explain the principle of measurement of rotary axes, the basic principle of laser interferometer should be introduced first. It uses interference phenomenon to do the measurements. As shown in Fig. 6, laser beam from the laser head's launch hole go to the spliter, which separates it into two beams. One beam goes up to the fixed reflector, and the other beam runs to the mobile reflector. Two beams are both returned back by the reflectors and meet with other at spliter where the interference phenomenon happens. According to the times of interference, the distance L (the mobile reflector moves) can be calculated. It is the principle of length measurement. Turning table measurement also uses in theory, shown as Fig. 7. When measuring, two distances are measured at the same time. Specifically speaking, when working, supposed that the rotary axis rotates α degree in the positive direction, but in fact, due

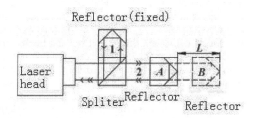

Figure 6.   The basic principle of laser interferometer.

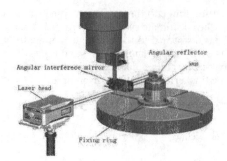

Figure 7.   The measurement of rotary axes.

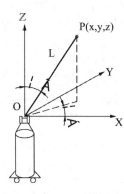

Figure 8.   The principle of laser tracker.

to the error, it may rotate $\alpha + \Delta$ degree. Then the angular reflector will revolve at α in negative direction by built-in high precision turning table. Thus, the error Δ can be calculated by the change in two beam's distances (Renishaw 2005).

3.2   Laser tracker method

In the industry, the laser tracker usually uses spherical coordinates to gather information. It consists of a laser tracker head, a tracker body, a reflector and a tracker controller. Its core components are a laser tracker head, two angle encodes and electric machine. The laser tracker head integrates a laser interferometer, which can achieve distance measurement, while the angle encodes can measure the angle. When working, the original point O sets in the rotary center of the laser tracker's head, and the reflector is fixed on the target (the point P) whose coordinate is supposed as (x, y, z), as shown in Fig. 8. The laser interferometer in the laser tracker measures the distance L and two encodes collect the horizontal angle α and vertical angle β, so the point P's coordinate can be calculated in Eq. 4 (Bridges, 2001).

$$\begin{cases} x = L*\sin\beta*\cos\alpha \\ y = L*\sin\beta*\sin\alpha \\ z = L*\cos\beta \end{cases} \qquad (4)$$

4   EXPERIMENTS

Due to the limit of condition, two experiments are done in different 5-axis machine tools, which are both TTTRR structures.

4.1   Experiment by laser interferometer method

The experiment is done in small testing 5-axis machine tool. Its range of Y-axis is from 0 to -1000 mm and that of X-axis is from 0 to -1200 mm, while the Z-axis is from 0 to -500. The rotary axis C is from 0 to 360 degree and A-axis is from -90 to 90 degree, shown as Fig. 9.

When measuring C-axis, the equipment is fixed with spindle and the layout can be used to measure all the ranges, as shown in Fig. 10.

However, due to the backlash, it is necessary to arrange overtravel, which makes A-axis measurement

Figure 9.   The testing 5-axis machine tool.

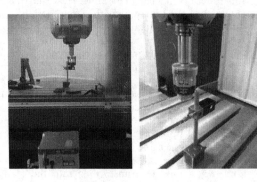

Figure 10.   The measurement of the C-axis.

range be shorter, only from –85 to 85 degree, as shown in Fig. 11.

The result of C-axis is given in Fig. 12. There are two crews in forward direction (FW) and backward direction (BW), respectively, which occur at 80 degree.

The largest value is 4.517199 arc-seconds at 280 degree, as shown as Fig. 12.

The result of A-axis is shown in Fig. 13. It also has a crew in each direction, which occurs at –45 degree. In particular, the error's direction is opposite between forward direction and backward direction, whose value is 5.197873 and –3.114599 arc-seconds, respectively.

Figure 11.   The measurement of the A-axis.

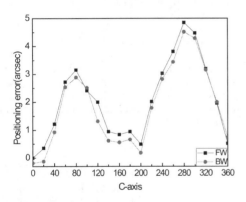

Figure 12.   The result of the C-axis.

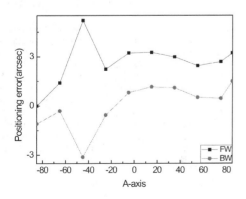

Figure 13.   The result of the A-axis.

4.2   *Experiment by laser tracker method*

The experiment is done in large gantry machine tool, whose X-axis can move from 0 to 6000 mm,

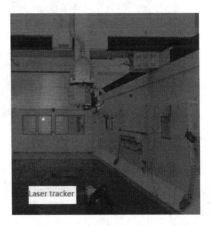

Figure 14a.   The measurement of the C-axis (laser tracker).

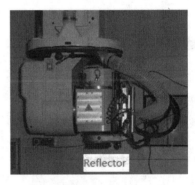

Figure 14b.   The measurement of the C-axis (reflector).

Figure 15a.   The measurement of the A-axis (refelector).

Figure 15b. The measurement of the A-axis.

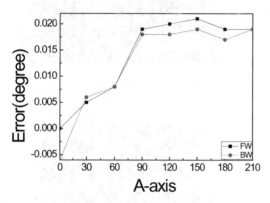

A-axis

Figure 17. The result of the A-axis.

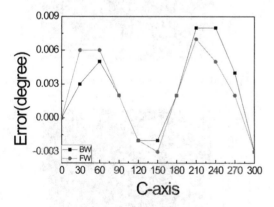

C-axis

Figure 16. The result of the C-axis.

Y-axis can run from 0 to 2000 mm and Z-axis' range is from 0~2000 mm. C-axis can revolve full circle and A-axis can rotate from 0 to 210 degree.

Compared with the laser interferometer method, the setup of laser tracker method is much easier. It only needs to fix the reflector on the target. The C-axis' setup is given in Fig. 14 and the A-axis is shown as Fig. 15.

The results of the C-axis and A-axis are given in Fig. 16 and Fig. 17, respectively.

5 CONCLUSION

In the paper, the error sources of 5-axis machine tools are introduced, especially the geometric errors. Meanwhile, 5-axis machine tools' structures and their classification are elaborated. According to the structure, the error modelling is discussed. Based on the modelling, the measurement methods for rotary axes are introduced, which are laser interferometer method and laser tracker method. The principles of the two methods are also analyzed. By experiments, the errors of rotary axes are given, which prove the feasibility of the two methods. However, the two methods only give the positioning errors of the rotary axes and the rest errors are not mentioned. Therefore, the rest errors of rotary axes need further research.

ACKNOWLEDGEMENT

The Key Academic Discipline Projects of Mechanic Engineering of Shanghai Second Polytechnic University under Grant (XXKZD1603); the School Fund of Shanghai Second Polytechnic University (EGD16XQD05).

REFERENCES

Abdul Wahid Khan. Calibration of 5-axis machine tools[D]. Beijing: Beihang University. 48–49(2010).

Bridges, B., and K. Hagan, Laser tracker maps three-dimensional features, The industrial physicist, 8, 1(2001).

Renishaw Co., The benefits of laser systems that use remote interferometer optics for linear, angular and straightness measurements.

Shih-Ming Wang, Han-Jen Yu and Hung-Wei Liao. Study of a new and low-cost measurement method of volumetric errors for CNC five-axis machine tools. Transactions of the Canadian Society for Mechanical Engineering, 37, 3(2013).

Yang Jianguo, Pan Zhihong, Xue Bingyuan. Kinematics modeling for geometric errors and thermal errors of NC machine tools [J]. Machinery Design & Manufactury. 5, 31(1998).

Advances in Energy Science and Equipment Engineering II – Zhou, Patty & Chen (Eds)
© *2017 Taylor & Francis Group, London, ISBN 978-1-138-71798-5*

An experimental study on the biomass central heating intelligent tobacco-curing system

Yali Guo
Guizhou Provincial Tobacco Company Qianxinan Branch, Xingyi, China

Zuguo Liu
College of Mechanical Engineering, Guizhou University, Guiyang, Guizhou, China
Guizhou Mechanical and Electrical Equipment Engineering Technology Research Center Co. Ltd., Guiyang, Guizhou, China

Zhaoxian Hong
Guizhou Provincial Tobacco Company Qianxinan Branch, Xingyi, China

Dabin Zhang & Yang Cao
College of Mechanical Engineering, Guizhou University, Guiyang, Guizhou, China
Guizhou Mechanical and Electrical Equipment Engineering Technology Research Center Co. Ltd., Guiyang, Guizhou, China

Feng Wang
Guizhou Provincial Tobacco Company Qianxinan Branch, Xingyi, China

ABSTRACT: To advance tobacco curing technology, biomass briquette fuel is used as a substitute coal for in tobacco curing, and also to realize automation control mode of one-to-n bulk curing barns instead of traditional family-style tobacco curing method. According to the theory, biomass briquetting and briquette fuel and automation control and compound control of the whole system and each single curing barn can be realized by means of PLC automation control. On this basis, the biomass central heating intelligent tobacco-curing system was proposed, for which biomass briquette fuel is used as a substitute for coal. The experimental study shows that 140,000 T biomass energy of tobacco stems and cabo can be used as a fuel for tobacco curing up to 500,000 mu (= 0.0667 hectares), equivalent to the consumption of 93,750 T of coal.

1 INTRODUCTION

Curing technology plays a vital role in assuring tobacco quality. The biomass central heating intelligent tobacco-curing system is a new type of efficient tobacco curing equipment system, which is widely used in the tobacco curing system (Changrong, 2005; Chaopeng, 2009; Zenghan, 2000; Roslee, 2010). Presently, junked tobacco stems and powder, and waste stalks are mainly disposed of by burning or burning, which causes environmental pollution and also entails spending money. Thus, the biomass central heating intelligent tobacco-curing system is recommended to be employed, and briquette tobacco stems and cabo are used as biomass fuel for tobacco curing instead of coal to realize the objective of "turning waste into wealth, energy-saving and cost reduction, and increasing benefits". Thus, the pressure of environmental protection be relieved, fuel can be largely saved, and the quality of tobacco can be improved because tobacco will be free from soot; besides, the recycling of tobacco stems helps in reducing tobacco virus disease and plant diseases and insect pests; most importantly, tobacco growers' income will increase and burden will decrease. They will become more enthusiastic about tobacco growing, and the fundamental policy of the state with respect to saving energy, reducing emission and recycling economy is implemented. In total, it is a measure that benefits the nation and the people.

2 TECHNICAL PRINCIPLES AND EXPERIMENTAL METHODOLOGY

2.1 System design and operating principle

A one-to-twenty biomass fuel hot water heating intelligent tobacco-curing system consists of biomass briquetting fuel at atmospheric pressure in a hot water boiler. Heat exchanger installed in curing barns, and heat supply pipeline is regulated by PLC automation control system, which is employed for system control.

Biomass briquetting fuel is different from coal. The amount of volatile matter in coal is 10–20%, while that (H_2, O_2, CH_4, CO, C_nH_n, etc.) of biomass briquetting fuel is 60–70%. The volatile matter consists of a gas fuel with a high heating value. When the combustion temperature reaches above 300°C, the volatile substances rapidly precipitate out and thermal cracking occurs, generating hydrocarbon of low molecular mass (such as H2, O2, CH4, CO, etc.) and granular black carbon, which sufficiently burns with replenished oxygen to eliminate black smoke and tar, reduce soot and dust, and increase the burn-off rate of fuel and conversion efficiency of thermal energy. The main reactions are:

$$2H2O + C \ (biomass) = CO + 2 \ H2$$

$$C \ (biomass) + 2 \ H2 = CH4$$

$$C + O2 \rightarrow C \ O2 \uparrow CH4 + 2 \ O2 = C \ O2 + 2 \ H2O$$

Hot water temperature is around 95°C heat the biomass briquetting fuel. The atmospheric pressure of hot water boiler is provided by a hot water circulating pump to supply the heat required by curing barns in the form of reverse return in parallel via the heat exchanger, and the return water in the system enters the boiler again, to circulate continuously (Shuang, 2007; Lei, 2004; Xin, 2009; Y. Shaopeng, 2009; Z. Wenbin, 2009). The PLC control device can assure to the realize separate control of each curing barn. The curing temperature is controlled in stages according to the three-step tobacco-curing process. When the required temperature is reached, the electric ball valve automatically turns off. When the upper limit of humidity is reached, the moisture removal window is automatically opened (Rongzu, 2001).

2.2 Main technical pathway

The study on model selection of straw chopping and briquetting machine involves design and development of biomass semi-gasification "one-to-twenty" hot water boiler, design and development of heat supply pipeline and control system of "one-to-twenty" central heating curing barn, verification and analysis of curing test, technological improvement and application demonstration.

The design of boiler and system pipeline and material selection were performed considering the energy consumption for curing unit weight of tobacco, the thermal load of full-load "one-to-twenty" curing barn, boiler thermal efficiency, system pipeline thermal efficiency, and radiator thermal efficiency. The automatic temperature and humidity control system was designed with PLC technology for each curing barn based on the technical principle of tobacco curing (S. Chaopeng, 2008; Z. Yanzhe, 2003; R. Qiangqiang, 2008; Z. Jun, 1999) to realize automatic temperature and humidity control of each curing barn at different curing stages.

2.3 Research contents

A chopping test of tobacco stems of different water contents was conducted (Z. Jun, 1998; Y. Juan, 2002; lee, 2010). Test results show that the tobacco stems of high water content (>45%) are unable to be chopped or are chopped to realize the product even at low quantity; generally, the production of chopping tobacco stems of water content lower than 20% is high; the production in unit time of chopping tobacco stems of the same water content varies from machine to machine. Based on the comprehensive evaluation, 9FX-80 chopping machine was selected, which is superior in respect of quality and production (see Table 1).

Table 1. Analysis of test data of two biomass straw chopping machines and conclusion.

Model	9FX-60			9FX-80		
	Power (Kw)	Revolving speed (rPm)	Sieve diameter (mm)	Power (Kw)	Revolving speed (rPm)	Sieve diameter (mm)
Item	18.5	2100	12	30	2000	12
	Tobacco stems water content ≤20%			Tobacco stems water content≤20%		
Production (Kg/h)	300 (average)			800 (average)		
Evaluation	Bad			Good		
Conclusion			9FX-80 was selected.			

Table 2. Analysis of test data of two briquette forming equipment and conclusion.

Model	9SYS32-1000			9SYS36-800II		
Item	Power (Kw)	Revolving speed (rPm)	Discharge hole diameter (mm)	Power (Kw)	Revolving speed (rPm)	Discharge hole dimension (mm)
	22	160	φ30	37	140	32 × 32
Water content of tobacco stems to be briquetted (%)	18–35			18–28		
Density (g/cm³)	0.8–1.1			0.8–1.1		
Production (Kg/h)	320			800–2000		
Evaluation	Bad			Good		
Conclusion			9SYS36-800II was selected.			

A test for the influence of pretreatment process, humidity, and grain diameter differences on briquette forming effect of chopped tobacco stems or cabo was conducted, and 9SYS36-800II circular mould briquette forming machine was selected based on the test results. Please see Table 2 for the test data.

3 RESEARCH ON INTEGRATION TECHNOLOGY OF THE BIOMASS CENTRAL HEATING CURING SYSTEM

3.1 Determination of main parameters of the biomass hot water boiler

Total heat load: N = 52,000 kcal/h. barn ÷ 0.85 (Σ heat Exchanger) ÷ 0.95 (Σ delivery pipe) × 20 barns = 64,400 kcal/h. barn × 20 barns = 128.8 × 104 kcal/h. Considering the synchrony of tobacco feeding, a chain type atmospheric pressure hot water boiler of heat load of 120 × 104 kcal/h was selected. Biomass fuel consumption per unit time: CT = 120 × 104 kcal/h÷0.75 (Σ boiler) ÷3800 kcal/kg = 421 kg/h. Boiler heating capacity: 1.4 Mw, boiler pressure: 0 MPa (see Fig. 1, Process Flow Diagram of Heat Supplying of Biomass Hot Water Boiler).

Since secondary and tertiary oxygenating devices were employed, the combustible gas of high calorific value (H2, CH4, CO, O2) is produced and the granular black carbon precipitating out in the primary combustion were thoroughly burned, so that the primary burn-off rate of fuel and the conversion efficiency of thermal energy were largely increased.

Smoke free and dust free: Since secondary oxygenating device was employed, no black smoke was emitted, and the Ringelman emittance was ≤1 grade.

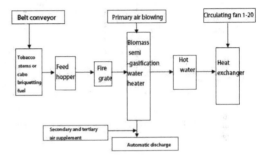

Figure 1. Process flow diagram of heat supplying of biomass hot water boiler.

3.2 Research on piping design of the biomass hot water boiler system

To guarantee the thermal load balance of each curing barn, a reverse return piping layout was adopted (see Fig. 2).

As shown in Fig. 3, when temperature rise in curing the barn is required, PLC control automatically opens valve 7. Hot water enters heat exchanger 4 and dissipates heat to the barn. Circulating fan 2 blows downward to heat exchanger 4 and blows hot air into the tobacco-curing area via the bottom outlet of the partition wall, and inlet air of circulating fan is drawn from the upper outlet of the partition wall to finish the continuous circulating temperature rise process of tobacco-curing area.

When the required temperature is reached, PLC automatically closes control valve 7 to stop heating the barn, but circulating fan 2 still works to keep the air flow in tobacco-curing area without heating.

When it is required to dehumidify the curing barns, PLC control automatically closes air door 1,

1285

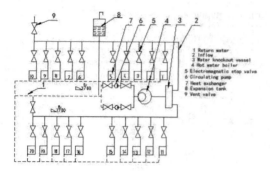

Figure 2. One-to-twenty biomass fuel hot water heating intelligent tobacco-curing system diagram.

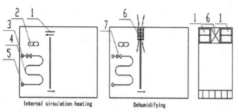

Figure 3. Schematic diagram of hot water boiler central heating curing barn.

when positive pressure is formed in tobacco-curing area under the action of circulating fan 2 to automatically open the dehumidifying port of waste heat recoverer 6 so that wet gas is discharged and fresh air is supplied. The function of waste heat recoverer 6 is to recover the heat of wet hot air discharged from curing barns, and heat the incoming fresh air with the recovered heat, so that heat loss due to dehumidifying is reduced. The temperature of fresh air entering curing barns is increased, realizing comprehensive benefits.

When the required humidity is reached, PLC automatically opens valve 7 again, and the waste heat recoverer is closed.

3.3   Automatic control system of biomass hot water boiler central heating and curing

PLC (microcomputer) was adopted to control the whole heating and temperature rise process, which has characteristics of high automation degree and low labor intensity. ① The heat exchanger installed in each barn is provided with electric ball valve. The supply of hot water is controlled via the electric ball valve to control the temperature of barn, so as to realize thermostatic control. ② Automatic dehumidifying windows are provided. When the set humidity is exceeded, the dehumidifying windows are automatically opened and vice versa. ③ Each curing barn is provided with a control cabinet, and control cabinets are connected with the central control center via data cable, to realize separate and central control.

4   PERFORMANCE TEST AND CURING TEST RESULTS OF BIOMASS HOT WATER BOILER CENTRAL HEATING AND CURING SYSTEM

4.1   No-load performance test result analysis

No-load heating up test was performed over the one-to-twenty hot water boiler curing barns using tobacco stems and cabo briquette as the fuel, and the temperature, heating up time, the initial reading of electric meter, and the use level of briquette fuel of each bulk-curing barn were recorded. Figs. 4 to 7 show the test results.

The no-load test results show that using a no-load one-to-twenty biomass semi-gasification hot water boiler and high-temperature flue gas boiler in the central heating intelligent tobacco-curing

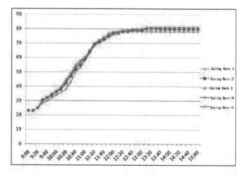

Figure 4. No-load test result of No. 1 to No. 5 hot water boiler curing barns.

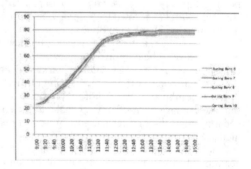

Figure 5. No-load test result of No. 6 to No. 10 hot water boiler curing barns.

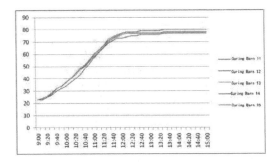

Figure 6. No-load test result of No. 11 to No. 15 hot water boiler curing barns.

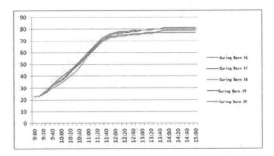

Figure 7. No-load test result of No. 16 to No. 20 hot water boiler curing barns.

system, it took 6 hours for the temperature of a curing barn to increase from 23°C to 81°C at an average speed of 9.67°C per hour, meeting the no-load performance requirements of bulk curing barn. For the no-load test, the total energy consumption was 3.54T of tobacco stem briquette fuel on average. The energy consumption per unit time was 590 kg tobacco stems briquette fuel/hour, and the power consumption was 262 KWH

4.2 Tobacco curing test of the biomass semi-gasification hot water boiler system

① One-to-three curing test: No. 1 to 3 barns were filled with tobacco for curing with the three-step curing process (under the control of intelligent tobacco-curing control cabinet), and the temperature, humidity, time, the initial reading of electricity meter and the consumption of briquette fuel were automatically recorded. Figs. 8 to 10 show the temperature and humidity change and time.

The three barns were filled with 10,656 kg fresh tobacco in total (wherein: barn 1 with 3,640 kg, barn 2 with 3,568 kg and barn 3 with 3,448 kg green tobacco), and 1332 kg dry tobacco was the output (wherein barn 1 output 455 kg, barn 2 output

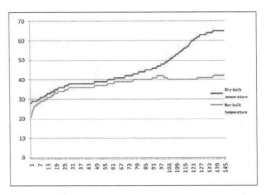

Figure 8. Temperature, humidity and time change of curing test of barn 1.

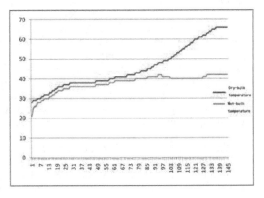

Figure 9. Temperature, humidity and time change of curing test of barn 2.

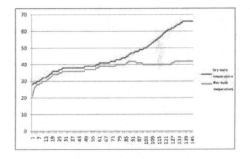

Figure 10. Temperature, humidity and time change of curing test of barn 3.

446 kg, and barn 3 output 431 kg). The heat value of the consumed tobacco stems briquette fuel was 2,800 kcal/kg, 4.15 T tobacco stems briquette fuel was consumed for test in total, and 1,090 KWH electricity was consumed. Consumption per unit dry tobacco: 3.12 kg tobacco stems briquette fuel

per 1 kg dry tobacco, 0.82 KWH per 1 kg dry tobacco.

The test results show that when three barns are filled with tobacco, the dehumidifying effect of curing barns was not satisfactory, heating was slow, and sooting happened locally. The main reason is that the resistance of hot-water steel–aluminium compound fin type heat exchanger is higher than that of coal heating bulk curing barn heat exchanger, and the wind pressure of the circulating fan (7# fan/1.5 Kw heated air circulating fan used for coal curing barn whose penetrating power is insufficient) is insufficient to open the dehumidifying windows. Besides, the consumption of briquette fuel and electricity for test was large, mainly becauselittle thermal load was consumed, and a high-power equipment drove a low-power one.

② One-to-twenty curing test: Five curing barns were taken as a unit. Specifically, No. 1 to 5 barns were filled with tobacco for curing by loading and unloading five times a day with the three-step curing process, and the temperature, humidity, time, the initial reading of electricity meter and the consumption of briquette fuel were automatically recorded. Figs. 6–20 to 6–23 for the automatically recorded temperature, humidity and time data of curing barns with different distances from the boiler.

The curing test lasted from September 11, 2010 to September 27, 2010, and 40 times of curing were carried out in total with 2 bakes a barn; 140.856 T fresh tobacco was filled, 17.718 T dry tobacco was output in total (442.95 kg dry tobacco per bake), and some 145 hours was taken (see Tables 6–3 and 6–4). For the test, the heat value of the tobacco stems briquette fuel was 2800 kcal/kg, 46.82 T tobacco stems briquette fuel and 13,960 KWH electricity were consumed. Consumption per unit dry tobacco: 2.64 kg tobacco stems briquette fuel per 1 kg dry tobacco, 0.79 KWH per 1 kg dry tobacco.

Curing test results show that the temperature of the curing barn far away from the biomass hot water boiler rose slowly. After improvement by replacing the 1.5 Kw blade and dynamo of 7# heated air circulating fan with a 2.2 Kw one, the one-to-twenty biomass hot water central heating intelligent tobacco-curing system roughly can meet the requirements of temperature and humidity of each curing process step.

4.3 Test comparison with ordinary single coal-fired curing barn

For purpose of test, two treatment processes were adopted and were not repeated. Treatment process I: tobacco stems and cabo briquette fuel, and one-to-twenty biomass hot water boiler central heating curing barn; treatment process II (control group): coal, and ordinary single bulk curing barn. The same part of tobacco of the same breed and quality was used for curing test with three-step tobacco-curing technology of bulk curing barn, and the tobacco quality, energy consumption and time consumption were recorded, as shown in Fig. 11.

Table 3. Record chart of measured data of no-load performance test of different curing barns.

Treatment process	Heating rate (°C/hour)	Temperature difference among five points at one floor (°C)	Maximum temperature (°C)	Remarks
Treatment process I	9.67	1.2	81	6 h from 23°C to 81°C
Treatment process II (control)	26.56	1.8	67.7	110 minutes from 19°C to 67°C

Item	Treatment Process I	Treatment Process II	Difference Value between Treatment Process I and Control Group
fuel consumption per 1kg dry tobacco (Kg/)	2.64	1.45	1.19
coal fuel cost per 1kg dry tobacco (yuan/)	0.79	1.45	-0.66
Electricity consumption per 1kg dry tobacco (Kwh)	0.79	0.69	0.10
Electricity cost per 1kg dry tobacco (yuan)	0.36	0.31	0.05
Labor cost per 1kg dry tobacco (yuan)	0.10	0.48	-0.38
Total cost per 1kg dry tobacco (yuan)	1.25	2.24	-1.00

Figure 11. Comparison of labor and energy consumption of different curing barns.

According to the table, the cost of biomass curing barn is lower than that of ordinary coal heating curing barn by 1.00 yuan/kg dry tobacco, or 44.64%. On this basis, a biomass curing barn can save 3,482 yuan a year

5 CONCLUSION

Adopting 3PCC to control the process of heating up and dehumidifying of one-to-twenty biomass hot water boiler curing barn according to three-step tobacco-curing technology can meet the requirements for temperature and humidity of the curing barn; equipping each curing barn with one intelligent control cabinet, which is communicated with the master control cabinet, can realize separate and central control. The hot water:central heating intelligent tobacco-curing system solved the problem of optimum energy match. The one-to-twenty curing barn was used as an option to realize the mode of intensive factory-based tobacco-curing. On the premise of guaranteeing flue-cured tobacco quality and energy saving, PLC automatic control can establish compound control over the whole system and each single curing barn, namely the functions of central control and separate control.

ACKNOWLEDGEMENT

This paper was funded by the Guizhou Provincial Tobacco Company Qianxinan Branch S & T project "The Popularization and Application of Intelligent Biomass Barn and Briquette Fuels".

REFERENCES

Changrong, G., P. Jianbin, S. Chaopeng. Tobacco Science & Technology. J. 11(2005): 34–36.

Chaopeng, S., C. Jianghua, X. Zicheng, et al. Acta Tabacaria Sinica. J. 15(2009): 83–85.

Chaopeng, S., Journal of Hebei Agricultural Sciences. J. 12(2008): 58–60.

Juan, Y., Z. Mingchuan, S. Yi, et al. Journal of Shanghai Jiaotong University. J., 36(2002): 1475–1480.

Jun, Z., S. Changdong, H. Chunli, et al. Industrial Boiler. J. 3(1999): 22–24.

Jun, Z., Y. Jianwei, X. Yiqian. Journal of Combustion Science and Technology. J. 4(1998): 63–68.

Kyong·hwan lee, sea chenon oh. Korean J Chem Eng. J. 27(2010): 139–143.

Lei, L., F. Min, G. Baoxing. Sichuan Chemical Industry. J. 7 (2004): 9–12.

Qiangqiang, R., Z. Changsui. Journal of Fuel Chemistry and Technology. J. 36(2008): 232–235.

Rongzu, H., S. Qizhen. Thermal Analysis Kinetics. M. Beijing: Science Press, 2001. 48–49.

Roslee Othman, M., Young—Hun Park, T. An Ngo. Korean J Chem Eng. J. 27(2010): 163–167.

Shaopeng, Y., X. Yong. Energy Conservation. J. 9 (2009): 6–9.

Shuang, W., W. Ning, Y. Lijun, et al. Journal of Chinese Electrical Engineering Science. J., 27(2007): 102–106.

Wenbin, Z., Z. Longquan. Chinese Agricultural Mechanization. J. 6 (2009): 90–93.

Xin, C., C. Jun. Contemporary Farm Machinery. J. 12(2009): 16–17.

Yanzhe, Z., L. Yi, L. Jiping. Journal of Cellulose Science and Technology. J. 11 (2003): 57–61.

Zenghan, X., W. Nengru, C. Yan, et al. Journal of Anhui Agricultural Sciences, J. 28(2000): 795–798.

Advances in Energy Science and Equipment Engineering II – Zhou, Patty & Chen (Eds)
© 2017 Taylor & Francis Group, London, ISBN 978-1-138-71798-5

Design and experimental research on a tobacco curing system with heat supply by biomass gasification

Xiangdan Hu
Guizhou Provincial Tobacco Company Qianxinan Branch, Xingyi, Guizhou, China

Chaojing Yu
School of Mechanical Engineering, Guizhou University, Guiyang, Guizhou, China
Guizhou Mechanical and Electrical Equipment Engineering Technology Research Center Co. Ltd., Guiyang, Guizhou, China

Yali Guo & Weilin Chen
Guizhou Provincial Tobacco Company Qianxinan Branch, Xingyi, Guizhou, China

Dabin Zhang & Yang Cao
School of Mechanical Engineering, Guizhou University, Guiyang, Guizhou, China
Guizhou Mechanical and Electrical Equipment Engineering Technology Research Center Co. Ltd., Guiyang, Guizhou, China

Feng Wang
Guizhou Provincial Tobacco Company Qianxinan Branch, Xingyi, Guizhou, China

ABSTRACT: To promote the development of recycling of ecological tobacco agriculture, this paper analyzes components of biomass materials, studies the characteristics of pyrolysis and gasification and looks at gas and heat production characteristics of biomass, mainly including tobacco stem. A study is conducted on tobacco leaf flue-curing system with biomass gasification in place of coal, according to pyrolysis and gasification characteristics of biomass materials and tobacco leaf flue-curing process. The key devices of tobacco leaf flue-curing system with heat supply by biomass gasification are designed with the objective of improving the performance of the heat supply system. No-load and flue-curing tests on the system are conducted. The results show that the system meets tobacco leaf flue-curing process, reduces average comprehensive energy consumption cost by 51.7% compared to curing barn fueled by coal and has significant economic, ecological and social benefits.

1 INTRODUCTION

The development and utilization of biomass energy is one of the effective measures in relieving the pressure on energy and environment in China and establishing an energy system for sustainable development (X. Min, 2008). Biomass pyrolysis and gasification technology is adapted to the current technological and economic development level and is a key technology for biomass energy conversion and utilization (Xiong, 2012). Biomass fuel has high volatile content and carbon activity and lower sulphur content and ash content than coal. They have low emission of SO_2 and NO_x in the utilization process and cause significantly less air pollution and acid rain. Therefore, the study on the application of tobacco leaf flue-curing technology

with biomass gasification in place of coal not only meets national development policies and direction of low-carbon economy and recycling agriculture and the general trend of development of modern tobacco agriculture, but also mainly uses waste biomass resources such as crop straw, including tobacco stalk and stem and are good for resolving environmental pollution caused by straw combustion in rural areas, in reducing energy consumption cost, saving coal resources and realizing energy saving and emission reduction.

In recent years, people understanding the development and utilization of biomass energy as environment friendly. The study on application of gasification technology raises the attention of people again (Z. Hailong, 2014). The Chinese tobacco industry has made some explorations of heat supply by bioenergy gasification, but has not made a substantial breakthrough in the improvement of curing barn equipment, biomass gasification preparation and heat exchanger. As Chinese tobacco leaf flue-curing mainly involves fuel coal and has low flue-curing efficiency, high labor intensity, high cost and environmental pollution, this paper focuses on tobacco leaf flue-curing system with heat supply by biomass gasification by its emphasis on the improvement of heat supply system according to pyrolysis and gasification characteristics of biomass and tobacco leaf flue-curing process.

2 ANALYSIS ON BIOMASS CHARACTERISTICS

Though biomass materials have various shapes, their internal components and pyrolysis and gasification characteristics are similar. The analysis of physical and chemical properties and pyrolysis and gasification characteristics of typical biomass materials help us further understand the properties of biomass and design pyrolysis and gasification processes.

2.1 Physical and chemical properties of biomass

Biomass refers to the carbohydrate produced by photosynthesis. Its main components include cellulose, semicellulose, lignin and inert ash etc. The elementary composition mainly includes C, H, O, N and S (Q. Ling, 2012). Chemical components and elements of some biomass straws are analyzed. The results show that the components of different biomasses are not completely the same. The content of cellulose, semicellulose, lignin, C, H ad O are respectively 30%–45%, 15%–31%, 13%–22%, 35%–45%, 4.5%–5.5% and 34%–46%. S content is very low.

2.2 Pyrolysis and gasification characteristics of biomass

Biomass pyrolysis and gasification mean that incomplete combustion of biomass materials with oxygen or oxygenic substances in air the as gasification agent forces organic hydrocarbons with high molecular weight to undergo cleavage, combustion and reduction reactions and their conversion into combustible gas such as CO, H2 and CH4 with low molecular weight. It mainly includes drying, pyrolysis, combustion and reduction reaction processes. Pyrolysis is the thermal chemical reaction of decomposing cellulose, semicellulose and lignin in biomass into solid, liquid and gaseous products and is the crucial process in the whole reaction (X. Min, 2008).

2.2.1 Biomass pyrolysis process

To understand the influence of parameters such as reaction temperature and time in the thermal reaction of biomass on product conversion rate, this paper establishes a pyrolysis kinetic model and obtains kinetic parameters such as reaction activation energy and frequency factor, which can provide theoretical basis for the reactor design. The main thermal analysis methods include thermogravimetry (TG), Differential Thermal Analysis (DTA) and Differential Scanning Calorimetry (DSC). Based on thermogravimetric analysis, kinetic model of pyrolysis reaction can be obtained as (Basu, 2009; lee, 2010):

$$\frac{dx}{dT} = \frac{A}{\beta}\exp\left(-\frac{E}{RT}\right)(1-x)^x \tag{1}$$

where, A, β, E, T, R, x and n respectively refer to frequency factor, temperature rise rate, activation energy, reaction temperature, gas constant, reactant conversion rate and reaction order.

The following is obtained after the logarithm is taken for the above formula:

$$\ln\left(\frac{dx}{dT}\right) - n\ln(1-x) = \ln\left(\frac{A}{\beta}\right) - \frac{E}{RT} \tag{2}$$

Enable

$$Y = \ln\left(\frac{dx}{dT}\right) - n\ln(1-x),\ x = \frac{1}{T},\ k = -\frac{E}{R},\ b = \ln\frac{A}{\beta},$$

then

$$Y = kX + b \tag{3}$$

The formula above is a straight line, k is the slope and b is the intercept. For the same biomass material, its activation energy E and frequency factor A are constant (X. Min, 2008). Reaction order n can be obtained through test data fitting.

The pyrolysis experiment of tobacco stem is carried out with the temperature rising rate of 20°C/min. The thermogravimetric curve (TG) and differential thermogravimetric curve (DTG) of the tobacco stem obtained are shown in Fig. 1.

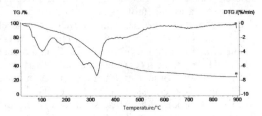

Figure 1. TG and DTG curves of pyrolysis of tobacco stem.

Table 1. Kinetic characteristics of tobacco stem.

Heating rate (°C/min)	n	Temperature range (°C)	Fitting formula	E (KJ/mol)	A	R
20	1.5	150–600	Y = −3193.4X+5.455	28.58	1.53×10^7	0.9945

Table 1 shows its kinetic characteristics obtained by pyrolysis kinetic analysis on tobacco stem. The results show that pyrolysis of tobacco stem includes dehydration, severe weight loss and slow weight loss. Main pyrolysis occurs within the area of 200°C–560°C (G. De-hong, 2011). When the heating rate is 20°C/min, activation energy of pyrolysis of tobacco stem is 28.58 KJ/mol and frequency factor is 1.53×10^7.

3 OVERALL STRUCTURE AND KEY DEVICES DESIGN

The tobacco leaf flue-curing system with biomass gasification in place of coal mainly includes gasifier, heat exchanger, curing barn facilities and control equipment. With existing bulk curing barn of flue-cured tobacco as the research object, this paper focuses on the improvement of performance of heat supply system, integrates technologies of flue-curing system with heat supply by biomass gasification according to system integration technology and forms complete equipment and technology of the system. It realizes automatic control over temperature rise by heating and humidity elimination of curing barn according to tobacco curing process with 3PCC and humidity modules. To improve the gas production efficiency of gasifier and heat exchange efficiency of heat exchanger, the key devices such as gasifier and heat exchanger are studied and designed.

3.1 Design of biomass gasifier

On the basis of establishing the replacement of coal with biomass gasification in tobacco leaf flue-curing, this study improves gasification efficiency of straw gasifier and fuel adaptability and improves the design of the existing gasifier in order to meet the demand of heat supply ability of tobacco leaf flue-curing. According to requirements of tobacco leaf flue-curing system and gasifier pattern and features, up-draft gasifier is designed (Surjosatyo, 2014). Fig. 2 shows the structural representation of up-draft biomass gasifier (Fang, 2012):

3.1.1 Body design of up-draft gasifier

As gasification efficiency of biomass per unit mass under standard state is $\eta = Q_{QT}G_V/Q_{WL}$, hearth volume can be obtained as (Peng, 2008):

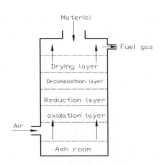

Figure 2. Structural representation of up-draft biomass gasifier.

$$V_{LT} = \frac{V_{QT}Q_{QT}}{\eta Q_{WL}\rho} \tag{4}$$

where V_{QT}, Q_{QT}, G_V, Q_{WL} and ρ are respectively gas production, heat value of cold gas, cold gas yield, heat value of material and density of biomass after compaction.

Flue-curing test is conducted with tobacco stem as a test material. Considering the total straw fuel quantity required for each curing of dry tobacco (600 kg dry tobacco), gas production rate can be obtained, i.e. $V_{QT} = 2520$ m³. According to the test, gas heat value of tobacco stem $Q_{QT} = 5020$ KJ/m³, heat value of tobacco stem $Q_{WL} = 10040$ kJ/kg, compaction density of tobacco stem is 80 kg/m³ and gasification efficiency $\eta = 70\%$. Therefore, hearth volume of a gasifier is calculated as $V_{LT} = 22.5$ m3. To guarantee that each reaction area of up-draft gasifier has an appropriate height so that gasifier can operate in a stable condition (Bin, 2011), height–diameter ratio is taken as 2.5 and it is calculated that gasifier height $H = 2.26$ m and diameter $D = 1.13$ m.

3.2 Heat exchanger design

The study shows that the heat exchange effect of heat exchanger is directly proportional to heat radiating area and heat flow path length. Through comparative analysis with heat exchanger of coal-fired furnace and heat flow path length, biomass gasifier heat exchanger is designed in accordance with the heat radiating area and flow path length (L. Xuefang, 2011), as shown in Figure 3.

Keeping in mind the inadequate efficiency of heat exchanger of coal-fired furnace, this study improves biomass gasifier heat exchanger, designs loop and straight-type heat exchange boxes, increases effective heat radiating area of heat exchanger and heat flow path length and reduces heat energy loss. Figures 4 and 5 show a comparison of heat flow path of coal-fired furnace and biomass gasifier. Table 2 shows the result of their effective heat radiating area and heat flow

Table 2. Statistical table of heat radiating area and heat flow path length of coal-fired furnace and biomass gasifier.

Name	Overall effective radiating area (m²)	Heat flow path length (mm)
Biomass gasifier	14.3	3760
Coal-fired furnace	9	2790
Ratio (biomass gasifier/coal-fired furnace)	1.59	1.35

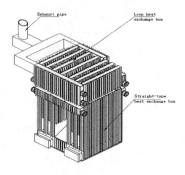

Figure 3. Biomass gasifier heat exchanger.

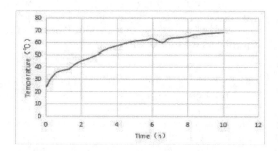

Figure 6. Temperature rise in no-load test.

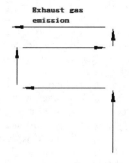

Figure 4. Heat flow path of coal-fired furnace.

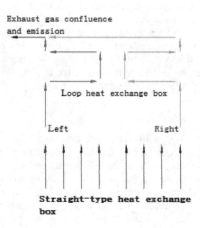

Figure 5. Heat flow path of biomass gasifier.

path length. The overall effective heat radiating area increases by 59% and overall heat flow path increases by 35%. Heat exchange rate increases and heat energy loss decreases.

4 FLUE-CURING TEST

The stability of performance, flue-curing effect and energy consumption and flue-curing costs such as manpower of the system are tested through no-load and flue-curing tests according to principles of tobacco leaf flue-curing process. The project team conducted a test in Songlin Tobacco Science Park in Bijie City.

4.1 No-load test

No-load heating test is conducted according to principles of tobacco leaf flue-curing process, mainly involving tobacco stalk biomass material, so as to test the temperature increase rate of curing barn. Fig. 6 shows its test result. The test result shows that the temperature of curing barn increases from 23.6°C to 68.0°C within 10 h and rate of increase in temperature is 4.44°C/h when gasifier is used for heat supply for no-load curing barn. Therefore, heat supply with gasifier can meet the requirement of curing barn for heat during flue-curing and guarantee

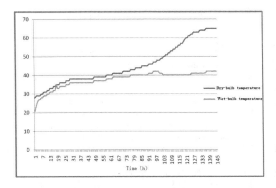

Figure 7. Temperature and humidity change in flue-curing test.

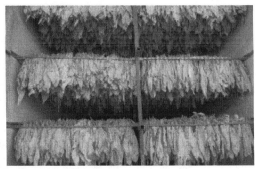

Figure 8. Appearance quality of dry tobacco in flue-curing test.

Table 3. Energy consumption comparison between barn of biomass gasifier and barn fueled by coal.

Barn type	Dry tobacco mass (kg)	Fuel consumption (kg)	Fuel cost (Yuan)	Power consumption (kilwatt hour)	Power cost (Yuan)	Total cost (Yuan)	Unit energy consumption cost (Kg/Yuan)
Barn fueled by coal	427	631	631	231	103.95	735	1.72
Test barn	442	1288	258	248	111.6	367	0.83

Note: coal price 1000 Yuan/T, biomass fuel price 200 Yuan/T and electricity price 0.45 Yuan/kilowatt hour.

normal operation of tobacco leaf flue-curing (J. Duzhong, 2010).

4.2 Tobacco flue-curing test

To obtain flue-curing effect, energy consumption and cost of tobacco leaf flue-curing system with heat supply by biomass gasification, flue-curing test is conducted based on tobacco leaf flue-curing process, which mainly involves tobacco stalk biomass material. Dry-bulb temperature and wet-bulb temperature in the flue-curing process are obtained as shown in Fig. 7. Fig. 8 shows the appearance quality of tobacco leaves. Table 3 shows its energy consumption compared to traditional curing barn fueled by coal. The results of flue-curing test show that the process of tobacco leaf flue-curing with heat supply by biomass gasification meets the requirements of the tobacco flue-curing process. Tobacco leaves obtained by flue-curing have good leaf structure, oil content, identity and chrominance. The quality of tobacco leaf is equivalent to that obtained by curing barn fueled by coal. Meanwhile, in tobacco leaf flue-curing with heat supply by biomass gasification, dry tobacco per kilogram can save fuel cost by 0.89 Yuan and average comprehensive energy consumption by 51.7% and reduce the emission of SO_2, CO_2 and NO_X. It is good for protecting the environment and reducing the emission of room temperature gas.

5 CONCLUSION

1. This paper conducted an analysis on components of typical biomass materials and a study on elementary composition and pyrolysis and gasification characteristics, tested pyrolysis and gasification characteristics of tobacco stem based on thermogravimetry, obtains its thermogravimetric curve (TG) and differential thermogravimetric curve (DTG) and provides a theoretical basis for the design of the reactor.

2. It completes the study on design of heat supply system with biomass gasification appropriate to tobacco leaf flue-curing, which has stable operation, replaces coal and meets the demand of tobacco leaf flue-curing and provides technical support for promoting the development of cycling ecological tobacco agriculture.

3. The results of no-load and flue-curing tests on tobacco leaf flue-curing system with heat suppl biomass gasification show that dry tobacco per kilogram consumes a biomass fuel of 2.91 kg and electricity of 0.56 kilowatt hours. Compared to ordinary curing barn fueled by coal, average comprehensive energy consumption cost is reduced by 51.7% and it is obviously energy efficient. Flue-curing quality is equivalent to ordinary bulk curing barn. The system has no influence on tobacco leaf flue-curing quality and meets tobacco leaf flue-curing process.

ACKNOWLEDGEMENT

This paper was funded by the Guizhou Provincial Tobacco Company Qianxinan Branch S & T project "The Popularization and Application of Intelligent Biomass Barn and Briquette Fuels".

REFERENCES

Basu P, Kaushal P. Chemical Product and Process Modeling. J. 4 (2009).

Bin, L., C. Hanping, Y. Haiping, W. Xianhua, Z. Shihong. Transactions of the CSAE. J. 7: 270–273 (2011).

De-hong, G., X. Cheng, G. Hong-yan. Journal of Guizhou University (Natural Sciences). J. 4: 33–36 (2011).

Duzhong, J., T. Shen, S. Jiangbo, C. Qingsong, Yuanfang. Chinese Agricultural Science Bulletin. J. 14 (2010): 392–395.

Fang Jin. The Design and Experiment of Suction Style on Biomass Gasifier (Anhui Polytechnic University, 2012).

Hailong, Z. *Research on New Energy Development in China* (Jilin University, 2014).

Kyong-hwan lee,chenon oh. Korean J Chem Eng. J. 27: 139–143 (2010).

Ling, Q. *Study on Biomass Pyrolysis Kinetics and Mechanisms* (Tsinghua University, 2012).

Min, X. *Mechanism and Experimental Study on Biomass Gasification and Pyrolysis* (Tianjin University, 2008).

Peng, L., W. Jie, W. Wei-xin. Journal of Agricultural Mechanization Research. J. 5: 76–78 (2008).

Surjosatyo A, Vidian F, Nugroho Y S. Journal of Combustion. J. 2014 (2014).

Xiao Xiong, Z., G. Yi Chen, Y. Wang. Advanced Materials Research. Trans Tech Publications. C. 512: 552–557 (2012).

Xuefang, L., G. Jiangfeng, X. Mingtian, C. Lin. Chinese Science Bulletin. J. 11: 869–873 (2011).

Advances in Energy Science and Equipment Engineering II – Zhou, Patty & Chen (Eds)
© 2017 Taylor & Francis Group, London, ISBN 978-1-138-71798-5

Electronic structure and half-metallicity of Heusler alloy Co$_2$ZrSn

Zhi Ren, Jian Jiao, Yang Liu, Jinjian Song & Xiaohong Zhang
School of Mathematics and Physics, North China Electric Power University, Baoding, P.R. China

Heyan Liu
School of Material Science and Engineering, Hebei University of Technology, Tianjin, P.R. China

Songtao Li
School of Mathematics and Physics, North China Electric Power University, Baoding, P.R. China

ABSTRACT: The site preference, electronic structure and magnetic properties of Co$_2$ZrSn have been studied by first-principles calculations and the stability of Cu$_2$MnAl type and Hg$_2$CuTi type structures has been tested in this paper. The Cu$_2$MnAl type is more favorable than the Hg$_2$CuTi type structure for Co$_2$ZrSn compound and the equilibrium lattice parameter is 6.30Å. It is found that Co$_2$ZrSn alloy has an energy gap in the minority spin direction at the Fermi level (EF) whereas the other spin band is strongly metallic. As a result, Co$_2$ZrSn alloy is predicted to be a half-metal with 100% spin polarization of the conduction electrons at the EF. The calculated total magnetic moment is 2.00 µB per unit cell, which is in line with the Slater-Pauling curve of Mt = Zt-24. Co atom-projected spin moment is 1.14 µB, which mainly determines the total moment. Such an alloy may be a promising material for future spintronics devices.

1 INTRODUCTION

At present Heusler alloys have attracted much attention for their potential applications as half-metallic materials (Groot, 1983; Dieny, 1991; Julliere, 1975). Usually, a half-metal is semiconductor-like in the minority-spin band at the Fermi level (E_F), whereas the majority-spin band is strongly metallic, which results in a complete (100%) spin polarization of the electrons at E_F. Many magneto electronic devices depend on an imbalance in the number of majority-spin and minority-spin carriers at E_F. Heusler alloys family has become a hotspot since de Groot *et al.* reported the conception of the half-metal licity arises from electronic structure calculations for the Heusler NiMnSb (Groot, 1983). Particularly, Heusler alloys containing Co atoms possess strong ferromagnetic with high Curie temperatures, and most of the predicted half-metallic Heusler alloys belong to the Co$_2$YZ families (Kandpal, 2006; Huang, 2011). To improve spin injection efficiency in magnetic multilayer structures, we need materials with high spin polarization (Schmidt, 2000; Felser, 2007; Mahdi, 2006). The theoretical predictions of 100% spin polarization in FM and ferrimagnetic Heusler alloys have stimulated research on half-metallic materials as well as the development of devices using them. Therefore, they are the ideal choice for spin injection electrodes. All these make research-

ers take great efforts to investigate the properties of them and to explore new Heusler alloys.

Many exciting properties have been found in the Co$_2$-based Heusler alloys, for example Co$_2$MnSi compound is the promising material for spintronics applications due to a several ideal physical properties like high Curie temperature (T_c = 985 K) (Brown, 2000) and widely selected as the ferromagnetic layer to achieve a giant tunneling magnetic resistance (Fu, 2016). There is some information nowadays regarding the Heusler alloys containing 4d and 5d elements. M. Benkabou *et al.* (Benkaboua, 2015) predicted Heusler alloys CoRhMnZ (Z = Al, Ga, Ge, Si) by using first principle calculations. B. Zhao *et al.* (Zhao, 2015) studied Pt doping on the Co based Heusler alloy. Moreover, the Co$_2$ZrZ (Z = Al, Si, Sn) (Kanomata, 2005; Zhang, 2006; Jiu, 2007) alloys have been studied by theoretical methods and experiments.

In this paper, we studied the site preference in Co$_2$ZrSn by first-principles calculations. It is found that Co$_2$ZrSn has a prospective potential for applications as half-metal and the total spin moment is 2.00 µ$_B$ per unit cell.

2 COMPUTATIONAL METHOD

The electronic structure was calculated by means of CASETP code based on pseudopotential method with a plane-wave basis set (Payne, 1992; Segall, 2002). The interactions between the atomic core and the valence electrons were described by the ultrasoft pseudopotential (Vanderbilt, 1990). The electronic exchange correlation energy was treated under the Generalized Gradient Approximation (GGA) (Perdew, 1996). For all cases, a plane-wave basis set cut-off of 500 eV was used. A mesh of $25 \times 25 \times 25$ k-points was employed for Brillouin zone integrations. These parameters ensured good convergences for total energy. The convergence tolerance for the calculations was selected as a difference on total energy within 1×10^{-6} eV/atom. These parameters ensured good convergences for total energy.

3 RESULTS AND DISCUSSION

3.1 Site preference and lattice parameter

Heusler alloy has a stoichiometric composition of X_2YZ, where X and Y atoms are transition metal elements, and Z atom is a main group element, which has four interpenetrating face-center-cubic (fcc) lattice. Generally, there are two possible structures in Heusler alloy: Hg_2CuTi-type and Cu_2MnAl-type (Fig. 1). In Hg_2CuTi type structure, the X atoms occupy the A (0 0 0) and B (1/4 1/4 1/4) site, and the Y atom enters the C(1/2 1/2 1/2) site and Z atom occupies the D(3/4 3/4 3/4) site in the Wyckoff coordinates. While in Cu_2MnAl type structure, the X atoms occupy the A and the C site, Y atom enters the B site and Z atom occupies the D site. For 3d transition metal elements, this preference of transition metal elements is determined by the number of their valence electrons, the elements with less electrons prefer to occupy the B site, while others prefer to enter the (A, C) sites (Kandpal, 2007).

In order to investigate the site preference of Co and Zr atoms, we undertook structural optimization for both Hg_2CuTi type and Cu_2MnAl type

structures. Fig. 2 shows the calculated total energy for Co_2ZrSn compound as functions of lattice parameters for both the Hg_2CuTi type and Cu_2MnAl type structures. The zero point has been chosen as the energy of the equilibrium lattice parameter. It is obvious that the energy of Cu_2MnAl type is lower compared with that of Hg_2CuTi-type. So, in Co_2ZrSn, the Co atoms prefer occupying the A and C sites and form Cu_2MnAl type structure. The corresponding equilibrium lattice parameter is 6.30Å. The following calculations are mainly based on the Cu_2MnAl type structure at the theoretical equilibrium lattice parameter.

3.2 Electronic structure

The electronic structure of the Co_2ZrSn alloy has been studied by first-principles calculations. The calculated total and partial Density of States (DOS) for the Co_2ZrSn Heusler alloy are shown in Fig. 3.

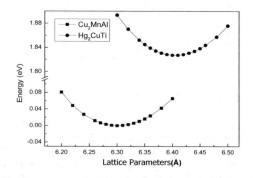

Figure 2. Calculated total energy for Co_2ZrSn compound as functions of lattice parameters for both the Hg_2CuTi type and Cu_2MnAl type structures, respectively. The zero of the energy has been chosen as the energy of the equilibrium lattice parameter.

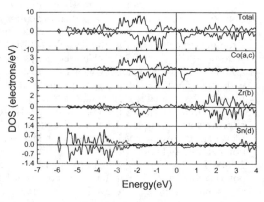

Figure 3. The calculated total and partial density of states for the Cu_2MnAl type Co_2ZrSn Heusler alloy under the equilibrium lattice parameter 6.30Å.

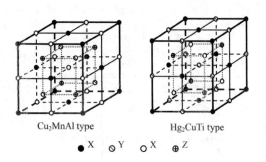

Figure 1. Schematic diagram of the Cu_2MnAl and Hg_2CuTi type.

In this figure, we choose the majority-spin states as positive value and the minority-spin states as negative value. It can be seen that the Fermi level just falls in a energy gap in the minority-spin states, while in the majority spin, there is still DOS exist, which result in a 100% spin polarization ratio P at E_F. Here the spin polarization ratio is calculated as the value of $(N_\uparrow - N_\downarrow)/(N_\uparrow + N_\downarrow)$, where $N_\uparrow$ and $N_\downarrow$ are the majority-spin and minority-spin DOS at E_F, respectively. All these make Co_2ZrSn a possible half-metal and worth further experimental investigations.

In the minority-spin and majority-spin directions, the bonding peaks between -3.0 eV and -0.5 eV are mainly from Co(A, C) atoms. The antibonding peak in the minority-spin also arises from d states of Co (A, C) atoms. The states between 0 eV and 1.0 eV come from d states of Zr atoms and form antibonding band in the minority. It is known that the bonding hybrids are localized mainly at the higher-valence transition metal atom sites and the unoccupied antibonding states mainly the lower-valence transition metal atom site (Galanakis, 2002). The Ge s electrons mainly contribute to the total DOS below -7 eV and are not presented in Fig.3. In addition, in the cubic crystal field the Co d states are split into a doublet with e_u symmetry and a triplet with t_{1u} symmetry, respectively. As discussed by Galanakis $et\ al.$ (Galanakis, 2006), the energy gap of the minority spin is determined by the energy separation between the e_u and t_{1u} the states produced by the Co-Co hybridization. Similar results have also been observed in Co_2ZrSi (Jiu, 2007) alloys.

Majority-spin and minority-spin band structure for the Cu_2MnAl type Co_2ZrSn Heusler alloy under the equilibrium lattice parameter 6.30Å is shown in Fig.4. In majority-spin, the top of valence band overlaps with E_F, exhibiting metallic characters at

E_F. However, the minority-spin exhibits an indirect G-X gap, E_F is exactly located in the band gap of 0.41 eV. In summary, Co_2ZrSn alloy has an ideal half-metallic band structure.

3.3 Magnetic properties

The total and atom-projected spin moment of the Co_2ZrSn alloy have been studied by first-principles calculations. The calculated total moment is 2.00 μ_B, is mainly determined by the parallel aligned Co (A) and Co (C) spin moment. Co (A, C) atom moment is 1.14 μ_B, Zr possesses a small spin magnetic moment of -0.36 μ_B, and Sn atom moment is 0.08 μ_B. It is known that in half-metallic Heusler alloys, the total moments M_t have integral values and follow the Slater-Pauling rule of $M_t = Z_t - 24$, where Z_t is the number of valence electrons(Galanakis, 2002). Co_2ZrSn has 26 valence electrons and the total magnetic moment should be 2.00 μ_B, which agrees with our calculations quite well.

The calculation is under their theoretical equilibrium lattice constant and Co_2ZrSn alloy exhibits half-metallic characters. However, a small change in the lattice parameter may shift the E_F with respect to the minority gap, which obviously affects the half-metallic character (Block, 2004). In experimental studies, such as melt spinning or ball milling, the strain may be quite large and will make the lattice constant deviate from the ideal one. Furthermore, the growth of thin film materials becomes possible by modern techniques. The lattice constant of the thin films is strongly influenced by the relationship between the half-metallic character and the lattice constant for a given material from both theoretical and technical aspect (Luo, 2008). Therefore, it is meaningful to study the dependence of the magnetic properties on the lattice parameters for Co_2ZrSn.

4 CONCLUSIONS

We discussed the site preference, electronic structure and magnetic properties of Heusler alloy Co_2ZrSn using first-principles calculations. It is shown that Co_2ZrSn alloy is a half-metal. We found the compound prefers to crystallize in Cu_2MnAl type structure rather than Hg_2CuTi type. The equilibrium lattice parameter is 6.30Å. In the DOS of Co_2ZrSn, the Majority-spin states display metallic character and minority-spin states exhibit semiconductor nature at E_F. Thus, Co_2ZrSn alloy has a 100% spin polarization and presents half-metallic character. It is found that the total magnetic moment is 2.00 μ_B per unit cell, which is in line with the S-P rule of $M_t = Z_t-24$. This total

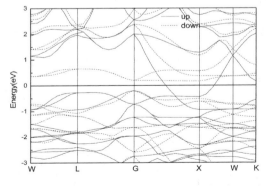

Figure 4. Majority-spin and minority-spin band structure for Cu_2MnAl type Co_2ZrSn Heusler alloy under the equilibrium lattice parameter 6.30Å.

moment is mainly determined by the antiparallel aligned Mn spin moments. As a result, Co_2ZrSn alloy may be a promising material for future spintronics devices and have advantages in some technical applications.

ACKNOWLEDGEMENT

This work has been supported by the National Science Foundation of China (No. 11004055), the Fundamental Research Funds for the Central Universities (2016YQ11, 2016MS134, 2014MS162) and the Program for New Century Excellent Talents in University (NCET-12–0844).

REFERENCES

Benkaboua, M., H. Racheda, b, A. Abdellaouia, D. Racheda, R. Khenatac, M.H. Elahmara, B. Abidria, N. Benkhettoua, S. Bin-Omrand, J ALLOY COMPD. **647**, 276 (2015).

Block, T., M.J. Carey, B.A. Gurney, O. Jepsen, Phys. Rev. B **70**, 205114 (2004).

Brown, P.J., K.U. Neumann, P.J. Webster, K R A Ziebeck, J. Phys. Condens. Matter **12**, 1827 (2000).

Dieny, B., V.S. Speriosu, S.S.P. Parkin, B.A. Gurney, D.R. Wilhoit, D. Mauri, Phys. Rev. B **43**, 1297 (1991).

Felser, C., G.H. Fecher, B. Balke, Angew. Chem. Int. Edn. **46**, 668 (2007).

Galanakis, I. Ph. Mavropoulos, P.H. Dederichs, J. Phys. D: Appl. Phys. **39**, 765 (2006).

Galanakis, I. and P.H. Dederichs, Phys. Rev. B **66**, 174429 (2002).

Galanakis, I., P.H. Dederichs, N. Papanikolaou, Phys. Rev. B **66**, 134428 (2002).

Huang, H.M., S.J. Luo, K.L. Yao, Physica B **406**, 1368 (2011).

Huarui Fu, Caiyin You, Yunlong Li, Ke Wang, Na Tian, J. Phys. D: Appl. Phys. **49** 195001 (2016).

Jiu, J.Y. and J.I. Lee, J. Korean Phys. Soc. 51, 155 (2007).

Julliere, M. Phys. Lett. A **54**, 225 (1975).

Kandpal, H.C., G.H. Fecher, C. Felser, J. Phys.: D Appl. Phys. **40**, 1507 (2007).

Kandpal, H.C., G.H. Fecher, C. Felser, Phys. Rev. B **73**, 094422 (2006)

Kanomata, T., T. Sasaki, H. Nishihara, H. Yoshida, T. Kaneko, S. Hane, T. Goto, N. Takeishi, S. Ishida, J. Alloys Compd. **393**, 26 (2005).

Luo, H.Z., H.W. Zhang, Z.Y. Zhu, L. Ma, S.F. Xu, G.H. Wu, X.X. Zhu, C.B. Jiang, H.B. Xu, J. Appl. Phys. **103**, 083908 (2008).

Mahdi Sargolzaei, Manuel Richter, Klaus Koepernik, Ingo Opahle, Helmut Eschrig, Igor Chaplygin, Phys. Rev. B **74**, 224410 (2006).

Payne, M.C., M.P. Teter, D.C. Allan, J.D. Joannopoulos, Rev. Mod. Phys. **64**, 1065 (1992).

Perdew, J.P., K. Burke, M. Ernzerhof, Phys. Rev. Lett. **77**, 3865 (1996).

Schmidt, G., D. Ferrand, L.W. Molenkamp, A.T. Filip, B.J. van Wees, Phys. Rev. B R **62**, 4790 (2000).

Segall, M.D., P.L.D. Lindan, M.J. Probert, C J Pickard, P J Hasnip, S J Clark, M C Payne, J. Phys.: Cond. Matt. **14**, 2717 (2002).

Vanderbilt, D. Phys. Rev. B **41**, 7892 (1990).

Wei Zhang, Zhengnan Qian, Yu Sui, Yuqiang Liu, Wenhui Su, Ming Zhang, Zhuhong Liu, Guodong Liu, Guangheng Wu, J. Magn. Magn. Mater. **299**, 255 (2006).

Zhao, B, X Xu, IEEE Transactions on Magnetics. **51**, 2600704 (2015).

de Groot, R.A., F.M. Mueller, P.G. van Engen, K.H.J. Buschow, Phys. Rev. Lett. **50**, 2024 (1983).

Advances in Energy Science and Equipment Engineering II – Zhou, Patty & Chen (Eds)
© *2017 Taylor & Francis Group, London, ISBN 978-1-138-71798-5*

Analysis of erosion and wear characteristics of chrome plating on the gun barrel

Shuo Zhao, Guohui Wang, Xiangrong Li, Shanshan Dong & Pengfei Fan
Department of Arms Engineering, Academy of Armored Force Engineering, Beijing, China

ABSTRACT: The gun barrel erosion wear problem seriously affects the operational performance of the gun, and the gun barrel is needed to withstand the harsh environment of gunpowder burning. In order to reduce the wear and tear of the inner bore surface of the gun, the anti-erosion wear resistance needs to be improved. The technology of electroplating chromium on gun barrel in foreign countries is studied. From the point of view of applicability, the technology of electroplating chromium plating is analyzed, and the characteristics of ablation and wear of the inner surface of the gun and chrome-plated gun pipe on the performance of the gun are discussed in two cases. In view of the complexity of the erosion and wear of chrome plated tube at present, the wear condition of the inner bore surface is observed by the peep bore instrument. Combined with an example, the wear characteristics of chrome plated pipe are analyzed.

1 INTRODUCTION

In the process of gun firing, the barrel bore continues to bear the function of propellant gas at high temperature and high pressure. The erosion of the bore surface and destruction of the body tube is extremely serious. At present, the research on the erosion and wear of gun barrel is becoming more and more complex, which is due to the use of a series of measures to reduce the wear and tear of the bore in the body, including the use of granular propellant, adding corrosion inhibitor, liquid propellant medicine and optimizing the internal chamber structure (LIU, 2013), plasma spraying, nitride and refractory metal liner. The body tube bore surface chromium plating technology is widely used at home and abroad, and the technology can effectively ease the metal wear of the body tube bore surface base, but the erosion and wear of inner bore surface is more complicated. Therefore, the paper analyses the erosion and wear characteristics of the gun barrel plating chromium plating process.

2 TECHNOLOGY OF ELECTROPLATING CHROMIUM IN THE GUN BARREL

At present, anti-erosion and wear of various types of caliber gun tubes are addressed using chrome plating technology. This technology was in research and development for a long time, and can be used for the mass production of the gun barrel. The slow erosion of barrel of the gun's bore surface and wear is an obvious effect, and gun tube's service life is greatly improved. So it becomes the main technology of the gun barrel making.

Electroplating chromium technology is one of the more mature methods to address gun barrel's erosion wear resistance. The performance of the coating is good, but there are some problems, such as coating micro crack stress and so on. The United States Army evaluated a series of methods as an alternative for chromium electroplating. The results show that the advantages of chrome plating technology is obviously higher than that of the other technologies. The coating evaluation table as follows (Gert, 2002):

According to the requirements of the performance indicators listed in Table 1, chromium plating technology almost meets all the targets, but it causes pollution to the environment. The technology of magnetron sputtering is the most close to the performance of electroplating chromium technology, but it is in the phase of the adhesion and therefore not suitable for the processing and use of the gun barrel (Chiang, 2006). So there is no alternative technology that can be used in the production of artillery tube to be widely used. According to the expert research, the requirement of the ideal gun barrel's anti-erosion and wear coating process should be satisfied. The coating is uniform, and applicable to 30~50 times of the large bore pipe and rifled tube. Coating has good adhesion, and it is best to form a metallurgical combination. The coating has no crack, no micro hole, displays good integrity, low stress and requires no precision machining. The temperature should be lower

Table 1. Evaluation of the application of the United States Army on various coating processes.

Objective function	Process technology					
	Electrodeposition in molten salts	Plasma spraying	Magnetron sputtering	Chromium plating	Ion substrate	Explosion welding
Large diameter pipe autofrettaged stress			√	√	√	
No precision machining requirements	√		√	√	√	
Excellent adhesion	√		To be determined	√		√
Suitable for large diameter (>50 times)			√	√		
Suitable for line gun	√			√		
Appropriate processing depth/width ratio			√	√		
Acceptable deposition rate	√	√	√	√		√
Drying process		√	√		√	√
Eliminate air and water pollution			√		√	To be determined
Removal of hazardous materials		√	√		√	To be determined

than the self-tightening of the body tube after the deposition process, that is less than 357°C, in order to achieve the protection of autofrettaged stress. High deposition rate, low processing cost and environmental protection are the other advantages (Michael, 2006).

At present, the United States Army is still using chromate plating technology in the body tube for artillery equipment in active service and in the research to prevent erosion and wear. For example, in U.S. Army's large caliber gun, the active M1 A2 main battle tank has a 120 mm gun tube chamber and bore made of of chrome handle. The M109 A6 self-propelled howitzer 155 mm gun barrel and breech chamber components are chrome plated; In research of MCS system for chromium plating, parts contains the gun tube chamber and bore, showing that degree of chrome plating is high in the United States. The Walt Flit Arsenal in large caliber gun barrel bore are hard chrome plated that its surface layer thickness can reach 127 μm and coating thickness of the barrel bore length difference value shall not be more than 50 μm. Otherwise coating requires reprocessing (Wang, 2012). In addition, the chromium electroplating technology of the United States Army uses the standard dip coating technique, for the U.S. Army 120 mm gun tube manufacturing process.

Chromium plating technology has long been used in various types to prevent gun barrel's erosion and wear, which exposes a series of process defects. The more severe is the erosion of the bore surface of the body. Once the chromium plating layer is peeled off, the erosion and wear of the surface of the body tube is more serious, which increases the wear of the metal matrix. Therefore, all the countries in the promotion of chromium plating technology is further improved, and the development of alternative technology is sought to solve the body tube's bore ablation wear problem.

3 ANALYSIS OF ABLATION WEAR CHARACTERISTICS OF CHROME PLATED BODY TUBE

3.1 Erosion and wear characteristics of the inner chamber of the traditional gun barrel

For the erosion of traditional gun barrel, bore wear law has been widely recognized. The ablation wear condition of the body along the pipe's axial direction experiences varying degrees of wear, including the severe wear section, uniform wear section and muzzle wear section (Yu, 2010). Severe wear section occurs due to the serious destruction of the barrel's bore diameter. This wear is easily measured, for certain types of artillery, for the tests in severe wear regions, a fixed point is selected, and the inner diameter of the point is used as one of the evaluation conditions of the life of the body. Therefore, the research results of the serious wear section of the gun barrel are relatively mature (Jin, 2014). The erosion and wear of the gun barrel is known to start from the crack, and gradually

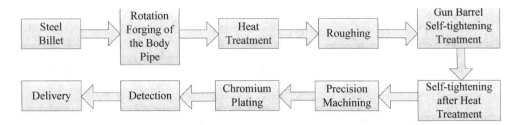

Figure 1. The U.S. Army 120 mm gun tube manufacturing process.

form a network. Conroy observed that the trend and range of crack network will influence the erosion and wear. General powder gas along the axial crack occurs, but in case of radial cracks, propellant gas backflow phenomenon is seen. The action of propellant gas and bore action time are prolonged, increasing the heat, raising the temperature, resulting in a serious erosion and wear (Conroy, 2006).

The identification tests of a certain cannon specialized to tube diameter cannonballs shot at different stages of measurement were done. The detailed measurement of this type of gun barrel rifling's wear shows that the starting position of the areas are the worst. Respectively on shooting test, strength test and battle fire rate test of body tube, the diameter of candle was measured, and the measurement results were analyzed. The measurement results are as shown in Table 2.

From Table 2, the data analysis of barrel muzzle end along the tail end direction shows that wear data increases gradually. This is because the initial part of the rifling is near the area of serious wear and tear, and along the direction of the muzzle, the wear is gradually weakened; With an increase in the number of bullets fired, the barrel's inner bore wear is aggravated gradually. The table shooting test, strength test and battle fire rate test over three stages of the barrel bore with a cross-section wear increased significantly, and especially after the strength test showed that the tube line wear increased sharply. The wear trend of the bore line section is shown in Figure 2.

3.2 Erosion and wear characteristics of the inner chamber of the chrome plated body

After a gun factory field research found that the current chrome plating process has been generally applied to the body tube manufacturing, and parts of the gun are given a full bore chrome plating process. The method can be used in batch processing and manufacturing, and is widely used in China and abroad. The chrome layer has high melting point and high hardness, which can largely reduce the erosion and wear of metal on the inner surface

Table 2. Wear measurement data of a gun barrel (Unit: mm).

From the Muzzle end	Line section wear		
	Shooting test	Strength test	Battle fire rate test
3700	105.07	105.32	105.48
3800	105.07	105.37	105.59
3900	105.07	105.42	105.67
4000	105.07	105.47	105.74
4100	105.07	105.55	105.80
4200	105.07	105.66	105.91
4300	105.07	105.74	105.99
4400	105.08	105.80	106.14
4500	105.09		
4600	105.05		
4700	105.13		

Note: Table space for rifling wear caused by measurement cannot accurately position and not measure the results.

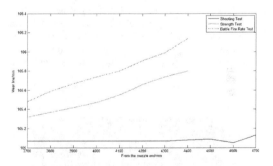

Figure 2. The gun bore line section wear.

of the inner bore. However, the research shows that the chrome plated body tube is the final scrap, mainly due to the inner surface of the chrome layer peeling off instead of wear. After removal of hydrogen treatment, the remaining hydrogen and other impurities in the release process, the chromium plating layer volume shrinks due to the formation of surface micro cracks. With the increase in the number of bullets fired, these cracks began to gradually extend, the general chrome layer thickness is about 0.08–0.12 mm. It's easy to extend to the chrome layer below, and the metal–substrate reaction between the powder gas and body tube in the chamber leads to the formation of gas phase compounds. They occur through the crack overflow, and the reaction takes place for a long time, caused by chromium plating layer because of metal substrate consumption and the formation of crater. When the bullet continues shooting, the high temperature and high pressure of the harsh environment lead to the chrome layer peeling off, and the metal substrate directly exposed to the powder gas. So the body tube undergoes erosion and wear (Zhang, 2001). The surface temperature increased in the chrome layer in the barrel boredue to expansion, forming an island shaped crack. The edge of the uplift is the high-speed movement of the projectile erosion, repeated after chrome plating layer peeling. The heat is directly transmitted to the metal, accelerating the erosion damage. As shown in Figure 3 peep bore instrument is used for a certain type of gun barrel bore surface inspection. From the figure, it can be seen that there is surface irregular ablation crater in the body tube bore. Melting phenomenon occurs around the ablation crater, and the strip shape of the shadow part is the projectile belt residual on the inner bore surface. The red circle in figure b clearly shows the visible chrome layer cracks, and extends the formation of cross shaped cracks. The phenomenon of peeling off the chrome layer is seen in the blue circle.

Chrome plated gun in the early stage of the use of the body wear is not obvious, and the oxidation resistance and carburizing resistance of chromium layer are better than that of gun steel. After long-term use. the coating barrel wear regularity is not obvious, and each bore wear state shows variations. There is no regular peeling off of the chromium layer, which leads to the great fluctuation of the interior ballistic performance of the gun. The known gun chrome tube type testing of firing projectiles' number with muzzle velocity and chamber pressure relationship were recorded in detail, such as shown in Figures 4 and 5.

Figure 4 shows firing projectiles number before firing is 300, although there exists velocity fluctuation, but the fluctuation range is small, and varies between 1740 m/s and 1750 m/s. It meets the requirements of gun muzzle velocity. When the shooting projectile number reaches 337, the velocity increases significantly, reaching 1751 m/s. However, after 441 rounds of firing, the muzzle velocity decreased slightly larger than before the wave velocity fluctuation, and interior ballistic performance is not stable. After 520 rounds of firing, projectile muzzle velocity begins to fall rapidly, and the curve appeared 600 trough when the speed reaches the minimum value of 1715 m/s. At this point, the chamber pressure is at minimum of 4531 kg/cm^2, and the erosion and wear of the inner tube is serious. When the projectile is fired, which

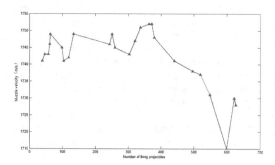

Figure 4. Number of firing projectiles and muzzle velocity variation trend.

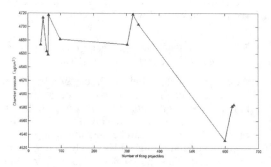

Figure 5. Number of firing projectiles and chamber pressure variation trend.

a

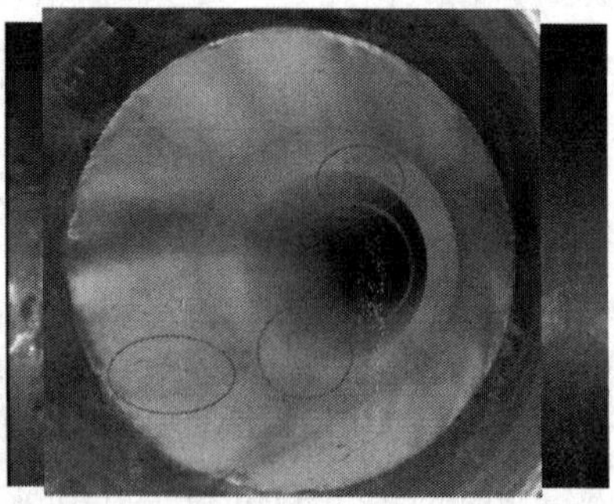

b

Figure 3. Gun barrel erosion and wear of the inner bore.

is closed to the bore of the barrel, and the interior ballistic performance is influenced by the powder gas leakage.

Figure 5 is divided into three parts. The first part is the number of firing projectiles is less than 100. The chamber pressure changes between 4650 and 4720 kg/cm^2, which is the normal phenomenon in the early use of the new gun tube; When the number of firing projectiles reaches about 300, the data of the chamber pressure is increased, but the change is not obvious; When the number of firing projectiles reaches about 600, the pressure of the chamber is obviously decreased, and the minimum value is reached. The analysis of the above two figures show that the number of firing projectiles is about 0–300, the barrel bore shows almost no wear, and the interior ballistic performance of the gun is not obvious. When the firing is more than 300, the chrome layer begins to wear off and continues to launch and in another 100 firings, the chromium layer is almost worn out. During this period, there is no law to follow the wear state of the bore in the barrel, and the change of the interior ballistic performance is difficult to grasp. When the firing number reaches about 400, the metal of the barrel body begin to wear, and the interior ballistic performance of the gun begins to decline, which is consistent with the change of the wear law of the traditional gun barrel.

REFERENCES

Chiang, Kuang-Tsan Kenneth. Method for Magnetron Sputter Deposition. United States Patent Application, 2006, 25:17–19.

Conroy P J, Leveritt C S, Hirvonen J K, et al. The role of nitrogen in gun tube wear and erosion [R]. Maryland: Army Research Lab Aberdeen Proving Ground MD Weapons and Materials Research Directorate, 2006.

Gert Schlenkert. Method of Providing a Weapon Barrel with an Internal Hard Chromium Layer [J]. US6467213B1, 2002, 10:22.

Jin Wenqi, Feng Sanren, Xu Da. Extrapolation Technology and Engineering Practice for Gun Barrel Life in the Approval Test [M]. National Defense Industry Press, 2014.

Liu Zhiwei, Wang Qilin, Yu Pengming. Damage mechanism of gun bore [J]. Journal of Sichuan Ordnance, 2013, 34(3):47–49.

Michael J.Audino, Use of Electroplated Chromium in Gun barrels [J].US Army RDECOM-ARDEC-Bcnct Laboratorics, DoD Metal Finishing Workshop, Washington, 2006, May, DC22–23.

Wang Baosheng. Research on anti-ablation materials and technology of artillery [R]. China North Group Corporations: The 210 Research Institute of China North Group Corporations, 2012.

Yu Wei, Tian Qingtao, Yu Xudong, et al. A review of research on the erosion and wear of gun bore [J]. Journal of Sichuan Ordnance, 2010, 31(2): 97–99.

Zhang Xifa, Lu Xinghua. Gun interior ballistics of the ablation [M]. National Defense Industry Press, 2001.

Advances in Energy Science and Equipment Engineering II – Zhou, Patty & Chen (Eds)
© *2017 Taylor & Francis Group, London, ISBN 978-1-138-71798-5*

Optimal parameters design and analysis of multi-stage forging of a micro/meso copper fastener

Gow-Yi Tzou & Un-Chin Chia
Department of Mechanical and Automation Engineering, Chung-Chou University of Science and Technology, Yuanlin, Changhua, Taiwan

Shih-Hsien Lin & Dyi-Cheng Chen
Department of Industrial Education and Technology, National Changhua University of Education, Changhua, Taiwan

ABSTRACT: This study uses the SolidWorks drawing software to construct realistic mold and dies of 2D fasteners, and import 2D mold-and-dies and wire drawings to commercial software. Assuming constant shear friction between the dies and workpiece to perform multi-stage cold forging forming simulation analysis, Taguchi method with the finite element simulation has been used for mold-and-dies parameters design optimization simulation. The forging force optimization is used to explore the effects on effective stress, effective strain, velocity field, die stress, and shape of product. The influence of the forging process of a micro/meso copper fastener can be determined, and the optimal parameters assembly can be obtained in this study. It is noted that the punch design innovation can reduce the forging force and die stress.

1 INTRODUCTION

Shah et al. (Shah, 1974) used Finite Element Method (FEM) to perform cold forging simulation analysis of thread head, besides comparing simulation results with the experiment to verify the acceptance of this model. Vickers et al. (Vickers, 1975) used plasticine, aluminum, 6061-T6 aluminum alloy as the simulated materials, using the experiment to analyze cold forging of thread head. MacCormack et al. (MacCormack, 2001) used FEM simulation to analyze the die of hexagonal bolt in the multi-stage cold forging, to increase die's life and successfully reduce die stress by 17.7%. Asnafi (Asnafi, 1999) used FEM to analyze the dies of cold forging, and the reasons for failure of dies can be found through the experimental method. Sun et al. (Sun, 2013) developed a stainless automotive battery fastener to perform the processing animation simulation and FEM analysis of multi-stage cold forging. They proposed the pass schedule plan to carry out the experiments and compared the results with FEM simulation. The results were in the good agreement. Shih et al. (Shih, 2014) proposed FEM simulation and experimental verification on multi-stage forming of flange sleeve. They used Deform 3D commercial software to design the pass schedule and the dies dimensions, and realistic experiments were performed to verify the acceptance of FEM simulation. Engel et al. (Engel, 2002) first used basic research to realize what is microforming. Wang et al. (Wang, 2007) explored the size effects of the cavity dimension on the microforming ability during the coining process. Gau et al. (Gau, 2007) utilized an experimental study to investigate into the size effects on flow stress and formability of aluminum and brass for micro-forming; the size effect reduces the flow stress. Engel et al. (Engel, 2006) continued to explore tribology in micro-forming; the tribology concept for micro-forming is different from traditional tribology. Chen (Chen, 2015) proposed a robust design on equal channel angle extrusion of Ti-6 Al-4V using Taguchi method; they combined FEM simulation with Taguchi method to obtain the optimal parameter assembly to arrive at a robust design.

This study changes the traditional punch design of a micro/meso copper fastener forging to conduct the forging force optimization to explore the effects of effective stress, effective strain, velocity field, die stress, and shape of the product. The optimal parameter assembly and the influence of the control factors can be obtained.

2 FEM SIMULATION

2.1 *Material properties and product specifications*

Figure 1 shows a micro/meso copper fastener for each pass. The copper fastener in the second stage is the specification of final product. The workpiece is a copper, C2600; the flow stress is $s = 553.547\varepsilon^{0.26}$, which is an elastic-plastic material. This study uses Deform-2D to perform FEM simulation; the simulation conditions

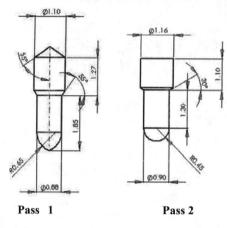

Pass 1	Pass 2

Figure 1. Diagram of a micro/meso copper fastener for each pass.

Table 1. FEM simulation conditions.

Workpiece	C2600
Young's modulus (E)	110GPa
Poisson's ratio (v)	0.28
Power law	$\sigma = 553.547\varepsilon^{0.26}$
Object type	Elasto-plastic
Number of elements	8800
Front die	Rigid body
Velocity (V_o)	0.1 mm/s
Shear friction (m)	0.2
Rear die	Rigid body
Velocity (V_o)	0 mm/s
Shear friction (m)	0.2

are summarized in Table 1. The friction is assumed to be a constant shear friction. Figure 2 shows schematic diagram before and after forming the first pass. Figure 3 shows a schematic diagram of punch and die for the first pass. In Figure 3, the major four parameters are shown as punch angle (α), die angle (β), die bottom height (H), and die filet (R). Taguchi method can be used to carry out the force optimization with these four control factors. The orthogonal table, $L_9(3^4)$, can be used to do experimental plan, including four control factors and three levels. The signal noise ratio can be used to obtain the optimal parameter assembly of reducing the forging force.

3 OPTIMIZATION AND RESULTS

Figure 4 shows control factors and levels for forging force optimization. According to the layout in Figure 4, $L_9(3^4)$ can be used to obtain S/N of forging force. The results of the analysis are

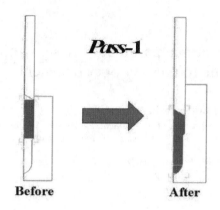

Figure 2. Schematic diagram before and after forming for the first pass.

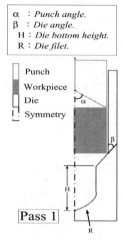

Figure 3. Schematic diagram of punch and die for the first pass.

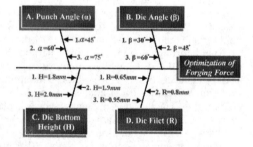

Figure 4. Control factors and levels for forging force optimization.

summarized in Table 2. According to data in the table, the fourth simulation experiment (IV) has the maximum forging force and the minimum S/N; the ninth simulation experiment (IX) has the

Table 2. $L_9(3^4)$ simulation results for forging force.

EXP	A (α)	B (β)	C (H)	D (R)	Force (kN)	S/N
I	1	1	1	1	2.05	−6.235
II	1	2	2	2	1.90	−5.575
III	1	3	3	3	1.90	−5.575
IV	**2**	**1**	**2**	**3**	**2.54**	**−8.097**
V	2	2	3	1	2.20	−6.848
VI	2	3	1	2	2.29	−7.197
VII	3	1	3	2	2.32	−7.310
VIII	3	2	1	3	1.88	−5.483
IX	3	3	2	1	1.72	−4.711

Table 3. S/N response characteristics.

S/N	A (α)	B (β)	C (H)	D (R)
Level 1	**−5.7951**	−7.2138	−6.3050	**−5.9314**
Level 2	−7.3806	−5.9689	**−6.1274**	−6.6938
Level 3	−5.8345	**−5.8275**	−6.5778	−6.3850
Effect	1.5855	1.3864	0.4503	0.7625
Rank	1	2	4	3

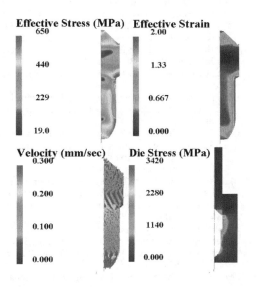

Figure 5. The effective stress, the effective strain, the velocity field, the die stress for pass 1.

minimum forging force and the maximum S/N, where the optimal parameters assembly is $A_3B_3C_2D_1$ (i.e. $\alpha = 75°$, $\beta = 60°$, H = 1.9 mm, R = 0.65 mm). Table 3 shows S/N response characteristics, form this table, the punch angle (α) influences to forging force very much, then next is die angle (β). The optimal parameters assembly is $A_1B_3C_2D_1$ (i.e. $\alpha = 45°$, $\beta = 60°$, H = 1.9 mm, R = 0.65 mm). Using this assembly to run Deform-2D, the forging force

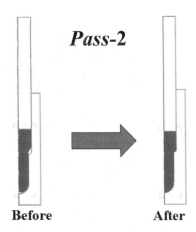

Figure 6. Schematic diagram of before and after forming for the second pass.

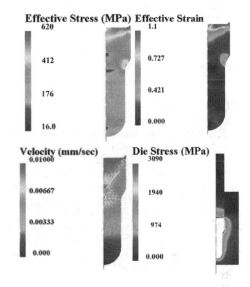

Figure 7. The effective stress, the effective strain, the velocity field, the die stress for pass 2.

obtained is 1.53 kN. The optimal force (1.53 kN) is lower than that obtained from the orthogonal table (1.72 kN). Let this optimal force be compared to the worst force in Table L_9 (3^4). The forging force is improved by 47.67% in that case. If this optimal force is compared to the best force in Table 2, L_9 (3^4), the forging force is improved by 12.4%.

The effective stress, the effective strain, the velocity field, the die stress can be simulated under the optimal parameters assembly shown in Figure 5. The maximum effective stress occurring in die angle location is 650 MPa, and the maximum strain occurring in die angle location is 2.0 mm/mm. The maximum velocity occurring in the head of the fastener is 0.3 mm/sec, and the die stress for

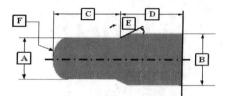

Pass_2			
	Specification (S)	Deform-2D (F)	Error(%)
A	Φ0.9±0.01mm	Φ0.9	0
B	Φ1.16±0.01mm	Φ1.16	0
C	1.75±0.02mm	1.746	0.22%
D	1.1±0.02mm	1.085	1.3%
E	30°±2°	30°	0
F	0.45R±0.1R	0.45R	0

Figure 8. Dimension comparisons of simulations and specifications.

WC mold is 3420 MPa; this die stree still is under the allowable stress of WC.

The forging parts in the first pass can be transferred to the second pass to form the final product, as shown in Figure 6. The effective stress, the effective strain, the velocity field, the die stress can be simulated under the optimal parameters assembly shown in Figure 7. The maximum effective stress occurring in die angle location is 620 MPa. The maximum strain occurring in die angle location is 1.1 mm/mm, the maximum velocity occurring in the head of fastener is 0.01 mm/sec. The die stress for WC mold is 3690 MPa.

4 CONCLUSIONS

This study proposes the forging force optimization combining FEM simulations and Taguchi method to obtain the optimal parameters assembly, the effective stress, the effective strain, the velocity field, and the die stress. Moreover the influence rank to the forging process can be determined in this study. The major results can be summarized as follows:

1. The optimal parameters assembly is $A_1B_3C_2D_1$ (i.e. $\alpha = 45°$, $\beta = 60°$, $H = 1.9$ mm, $R = 0.65$ mm). The forging force can be improved 47.6% compared to the worst force occurred in Table 2, L_9 (3^4). Besides the forging force is also improved 12.4% compared to the best force in Table 2, L_9 (3^4).
2. The influence rank to the forging force is punch angle >die angle>die filet >die bottom height.

3. At the die angle location, the maximum effective stress, the maximum effective strain, and the die stress are occurred. Especially for the die stress, the die stress is still under the allowable stress of WC mold.
4. The dimensions simulated can compare with the specifications, the error is under 2% shown in Figure 8. It indicates the FEM simulations can be accepted.

ACKNOWLEDGMENTS

The authors thank the project's support by the Ministry of Economic Affairs (MOEA) and Ministry of Science and Technology (MOST) to complete this research and the financial support for the trips from Chung-Chou University and MOST to attend the international conference of RPMSE 2016.

REFERENCES

Asnafi, N. On tool stresses in cold heading of fasteners, Engineering Failure Analysis. 6, 321–335 (1999).

Chen, D.C., G.Y. Tzou, Y.J. Li, Robust design on equal channel angle extrusion of Ti-6 Al-4V using Taguchi method, International Journal of Material Science, 5, 1 (2015), doi:10.12783/ijmsci.2015.0501.

Engel, U. Tribology in microforming, Wear, 260, 265–273 (2006).

Engel, U. and R. Eckstein, Microforming—from basic research to its realization, J. of Materials Processing Technology, 125–126, 35–44 (2002).

Gau, J.T., C. Ptincipe, J. Wang, An experimental study on size effects on flow stress and formability of aluminum and brass for microforming, Journal of Materials Processing Technology, 184, 42–46 (2007).

MacCormack, C., J. Monaghan, A finite element analysis of cold-forging dies using two-and three-dimensional models, J. of Materials Processing Technology, 118, 286–292 (2001).

Shah, S.N., S. Kobayashi, Rigid-plastic analysis in cold heading by the matrix method, Proc. 15th International Conference on Machine Tool Design Research, 603–608 (1974).

Shih, C.W., G.Y. Tzou, K.H. Chang, FEM simulation and experimental verification on multi-stage forming of flange sleeve, Journal of Chinese Society Mechanical Engineers, 35, 243–250 (2014).

Sun, M.C., G.Y Tzou., L.A. Zheng, Processing animation simulation and FEM analysis of multi-stage cold forging of stainless automotive battery fastener, Indian Journal of Engineering & Materials Sciences, 20, 219–224 (2013).

Vickers, G.W., A. Plumtree, R. Sowerby, J.L. Duncan, Simulation of the heading process, J. Engineering Materials and Technology, 126–135 (1975)

Wang, C.J., D.B. Shan, J. Zhou, B. Guo, L.N. Sun, Size effects of the cavity dimension on the microforming ability during coining process, J. of Materials Processing Technology, 187–188, 256–259 (2007).

Advances in Energy Science and Equipment Engineering II – Zhou, Patty & Chen (Eds)
© 2017 Taylor & Francis Group, London, ISBN 978-1-138-71798-5

Study on the conductive property of DL05 cable steel

Qi-chun Peng, Ming-ming Deng & Li Deng
Key Laboratory of Ferrous Metallurgy and Resources Utilization, Ministry of Education, Wuhan University of Science and Technology, Wuhan, China

Jing-bo Xu, Tian-hui Xi & Wei Li
Wuhan Iron and Steel (Group) Company, Wuhan, China

ABSTRACT: With rapid economic development, copper clad steel wire is widely applied to various fields. This paper undertook a study on the conducting properties of the internal DL05 steel core. Through theoretical analysis, we conclude that cable steel doping leads to a decrease in conductivity because the solute elements destroy the periodic lattice structure of the metal element. This increases the probability of electron wave scattering and leads to an increase in resistance. The DL05 steel belongs to the category of clean steel. When the solute concentration is a, the lower the content of solute elements is, the higher the conductivity of DL05 steel will be. So by controlling the cleanliness and reducing the degree of impurity elements, we can effectively improve its conductivity. Through plant data acquisition and analysis, we find that total element content and manganese content have significant effect on the conductivity of DL05 steel. Therefore, for enhancing toughness and elongation, minimizing the content of manganese can improve the conductivity of the cable steel.

1 INTRODUCTION

Because of high cost and scarcity, the traditional single metal copper wire has become increasingly unable to meet the needs of the market. Bi-metallic composite materials are being accepted by many enterprises based on its high strength, good conductivity, light weight, low cost and many other advantages (Peng, 2000; He, 2007; Fei, 2001). Copper clad steel is such a material. The copper on the surface is a good conductor of high-frequency signals, and the internal steel core has a good strength and toughness (Wang, 2002; Wu, 2007). The conductivity of the internal steel core required is higher than 16% at 20°C. However, many plants cannot meet this requirement. Therefore, this study was done to understand the conductive mechanism of cable steel, with the aim of improving its conductivity.

2 CONDUCTIVE MECHANISM ANALYSIS

As we all know, different materials have different conductivities, and according to their conductivity values, we can classify them into different categories, as shown in Table 1.

Conducting materials contain pure metals and alloys. The ρ value of pure metals is 10^{-8} to 10^{-7} and that for the the alloys is 10^{-7} to 10^{-5}. Based on

Table 1. ρ value of different materials $/\Omega \cdot m$.

material	conductor	semiconductor	insulator
ρ	$<10^{-5}$	$10^{-5}\sim10^{9}$	$>10^{9}$

the values, we find that pure metal has a stronger conductivity than alloys.

Figure 1 shows resistance value of different metal elements. Where Mn content is less than Fe in DL05 steel, it has a very high resistance, and the high-resistance anomalies of Mn are related to its internal crystal structure.

At absolute temperature (0 K), metal crystal is an ideal space lattice structure. So the electron wave will not be hindered and scattered when passing through it. When the temperature rises, impurity elements, alloying elements and crystal defects will cause abnormal crystal structure distortion. These damaged lattice structures will lead to additional scattering (incoherent scattering) of the electron wave, similar to suspended particles affecting light propagation, and the scattering of delayed electron wave propagation in the metal conductor, so that the resistance is generated in these distortions, reducing the conductivity of the metal conductor. This is the essence of the resistor.

According to the Matthissen and Vogt's earlier studies, since the concentration of solute atoms in a solid solution is small, the interaction between them

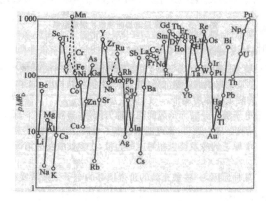

Figure 1.　Resistance value of metal elements in the periodic table (Shun, 2013).

can be ignored. The resistance of the solid solution can be expressed by basic resistance $\rho(T)$ and heterogeneous atoms resistor ρ'. In fact, as a first approximation, the contribution of different scattering mechanisms to the total resistance can be summed by the Matheson's law, which is expressed as:

$$\rho = \sum_i \rho_i = \rho' + \rho(T) \tag{1}$$

where $\rho(T)$ is only affected by the temperature; ρ' is affected by the foreign atom concentration, dislocations, vacancies and so on, and independent of temperature.

Iron vacancies and dislocations affect its resistance, as shown in Table 2.

When a solid solution is formed between different metals, conductivity decreases. When the heterogeneous atoms dissolve into the original lattice, the lattice structure of the original organization will be twisted and deformed. This will destroy the periodical change in the potential field of the lattice. The probability of scatter increases when the electron wave goes through a conductor, and its resistance increases. Even when a good conductivity metal dissolved in a less conductive metal, it will destroy the integrity of the original metal lattice structure, resulting in decreased conductivity, and therefore, $\sigma_{pure\ mental} > \sigma_{alloys}$ (σ represents conductivity). Lattice distortion affects the resistance of a solid solution, and the chemical interaction between each component of the solid solution also affects its conductivity.

In general, the maximum resistance of binary alloys appear in the fraction of solute atoms at 50%, and it may be much higher than the single metal. Iron and solute elements and their relationship with conductivity is shown in Figure 2. When the solute concentration is low, the lower content of solute component, namely the alloy composi-

Table 2.　Effect of dislocation, vacancy of metallic iron resistance.

Metal	$\Delta\rho_{dislocation} / \Delta N_{dislocation}$ /$10^{-19}\ \Omega \cdot cm^3$	$\Delta\rho_{vacancy} / C_{vacancy}$ /$10^6\ \Omega \cdot cm \cdot (atom\%)^{-1}$
Fe	/	2.0

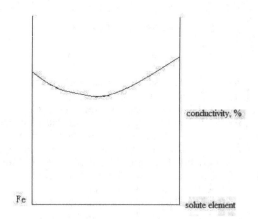

Figure 2.　Iron and solute elements in the relationship with electrical conductivity.

tion is close to pure iron, and the conductivity of iron is higher.

3　PLANT DATA ACQUISITION AND ANALYSIS

In DL05 steel, except the basic element Fe, the content of other elements (C, Si, P, S, N, Ni, Cr, Mo, Cu, Mn, Als) are very low, and the content of manganese is high. Their magnitude is close, and all of them have an influence on the conductivity and other properties of DL05 steel. We explored the relationship between manganese and conductivity in DL05 steel, through data acquisition and analysis of a steel plant from January to November 2014. The relationship between $W_{\Sigma total}$, $W_{\Sigma total(excluding Mn)}$, W_{Mn} and conductivity in DL05 steel are shown as Figures 3 to 5.

Figure 3 shows that $W_{\Sigma total}$ in DL05 steel varies from 0.093% to 0.135% and has a negative correlation with electrical conductivity. The higher its value, the lower the conductivity. So improving the cleanliness of the steel and reducing the total content of the steel elements can help to increase conductivity.

Figure 4 shows that total element content (excluding Mn) in DL05 steel is mainly concentrated in

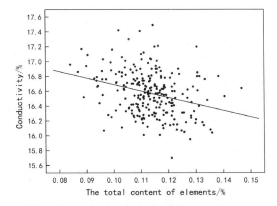

Figure 3. The relationship between the total element content and the conductivity of DL05 steel.

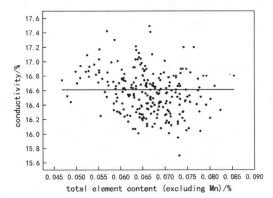

Figure 4. The relationship between the total element content (excluding Mn) and the conductivity of DL05 steel.

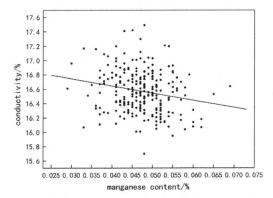

Figure 5. The relationship between w[Mn] and the conductivity of DL05 steel.

the 0.046–0.073%, which is <0.10%. The correlation between element content and the conductivity is not significant.

Figure 5 shows that in DL05 steel, the manganese concentration varies between 0.035% and 0.058%, with an average of 0.045%. When manganese content increases, cable steel conductivity decreases and therefore the steel of manganese and its conductivity was negatively correlated.

On comparing Figures 4 and 5, the total element content (excluding Mn) in DL05 steel has an insignificant relationship with its conductivity. After the addition of manganese, the relationship between conductivity and total element content is obvious. Therefore, manganese content controls the conductivity of the DL05 steel and plays a vital role.

4 MEASURES TO IMPROVE THE CONDUCTIVITY

Through the above study, we can understand the significant impact of the cleanliness of liquid steel and manganese content on steel conductivity. In order to improve the conductivity of the cable steel, we look at the liquid steel smelting process in the following:

1. Increasing the circulation of liquid steel and argon gas flow rate at the time of the RH treatment leads to a stable operation and prevents the reaction of raw materials with carbon.
2. With an appropriate increase in the CaO content of the slag, basicity of the slag increases, lowering the temperature and increasing the FeO content. Increasing the amount of slag improve thermodynamics and kinetic conditions of liquid steel dephosphorization. Estimating the activity of CaO ash according the thickness of slag reduces the "back P" to control the phosphorus content in steel/
3. For the optimization of desulfurization agent, the use of low-sulfur synthetic slag should be chosen with the right w [Mn]/w [S] to control the content of manganese and sulfurin the steel.
4. In sorting process optimization, reasonable ore, combined with boiling decarbonization furnace are vacuum degassed to control the content of residual elements in steel.

Before and after taking these measures, the conductivity of steel is shown in Figures 6 and 7.

Figures 6 and 7 show that the average total element content in DL05 steel before taking the abovementioned measures is 0.14%. The average total element content in DL05 steel after taking the measures is 0.114%. Reducing the total content

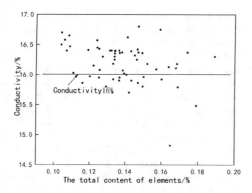

Figure 6.   Steel conductivity before taking measures.

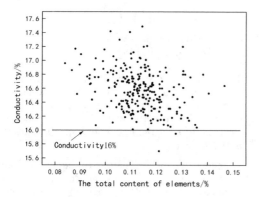

Figure 7.   Steel conductivity after taking measures.

Table 3.   Steel conductivity before and after taking measures/%.

Conductivity of DL05 steel	Minimum	Maximum	(≥ 16)
Before taking measures	14.8	17.08	63.25%
After taking measures	15.7	17.19	93.75%

of elements will improve the conductivity by over 16%, as shown in Table 3. Meanwhile, points in Fig. 7 are concentrated on the average content of 0.114 percent and those in Fig. 6 are scattered. It shows the importance of these measures to increase the conductivity.

5   SUMMARY

1. When passing through the conductor, the electron wave will be hampered by crystal defects, which leads to scattering. Attenuation is the essence of conductor resistance.
2. The conductivity of the pure metal is stronger than that of the alloy metal. When solute concentration in at a low level, the lower is content of solute components, and the higher the conductivity of iron. Since DL05 steel belongs to the category of clean steel, cleanliness control and reducing impurity elements are important to improve its conductivity.
3. The content of Mn has a significant effect on the conductivity of DL05 steel. Therefore, for enhancing toughness and elongation, minimizing the content of manganese can improve the conductivity of the cable steel.
4. Reducing the total content of doping elements in steel has a distinct effect in improving the conductivity.

REFERENCES

Fei Baojun, Fei Mingbing, Chen Zhenghong. High electric conduction property of composite copper-clad steel wire. J. IEEE, 15, 15–20(2001).

He Fei, Liu Limei, Zhong Yun etc. The production process of Copper-clad steel wire. J. Surface Technology, 36,78–80, 90(2007).

Peng Dashu, Liu Langfei, Zhu Xuxia etc. Metal layered composite material research status and prospects. J. Materials Review, 14, 23–24(2000).

Shun Jinyu, Tian Xiaohui, Yuan Yizhong. Materials Physics. M. Shanghai: East China University of Technology Press, 65(2013).

Wang Qingjuan, Du Zhongzhe, Wang Heibo etc. Characteristics and status of copper-clad steel wire production process. J. Wire and Cable, 4,15–17, 27(2002).

Wu Yunzhong. Research on drawing method of clad bimetallic wire of copper clad aluminum, copper clad steel. D. Dalian Maritime University, 2007.

Advances in Energy Science and Equipment Engineering II – Zhou, Patty & Chen (Eds)
© 2017 Taylor & Francis Group, London, ISBN 978-1-138-71798-5

Characterization of acoustic signals during a direct metal laser sintering process

Dongsen Ye
Department of Mechanical Engineering, National University of Singapore, Singapore
Department of Automation, University of Science and Technology of China, Hefei, China
Institute of Advanced Manufacturing Technology, Chinese Academy of Science, Changzhou, China

Yingjie Zhang
Department of Mechanical Engineering, National University of Singapore, Singapore

Kunpeng Zhu
Institute of Advanced Manufacturing Technology, Chinese Academy of Science, Changzhou, China

Geok Soon Hong & Jerry Fuh Ying Hsi
Department of Mechanical Engineering, National University of Singapore, Singapore

ABSTRACT: The quality of parts during Direct Metal Laser Sintering (DMLS) processing has been studied for improving the process and repeatability. Acoustic signal processing is one of the most important approaches to monitoring the workpiece's built-quality during a powder-based additive manufacturing process. This paper reports the relationship between acoustic signals, laser power as well as its laser scanning speed. The variety of acoustic signal Power Spectrum Density (PSD) is presented and then the mechanism of acoustic signal formation is elaborated. A good mapping between acoustic signals and laser parameters has been found during the DMLS process. This lays a good foundation for monitoring the process and quality by acoustic signal and will enhance the part quality during the powder-based laser sintering and melting processes in the future.

1 INTRODUCTION

Metal-based Additive Manufacturing (AM) technology shows excellent prospects for application due to short processing lead time, low tooling investment and environmental sustainability (Gao, 2015). Direct Metal Laser Sintering (DMLS), which is based on Powder Bed Fusion (PBF), is one of the most important metal-based AM processes with immense potential (Gibson, 2015). Complex metal components that cannot be manufactured by a traditional process can be produced by DMLS with shorter lead time and geometrical freedoms of creation. This AM technology fabricates the metal prototypes and tools directly from the Computer-Aided Design (CAD) data with flexibility in freedom and innovative bonding techniques (Guo, 2013). Unfortunately, with all benefits DMLS can offer, delamination between base plates and inaccuracy among various orientations are its shortcomings. The quality and repeatability of parts still hamper their widespread use largely because sintered parts are still relatively soft, rough and porous (Tapia, 2014). This is particularly true in industrial applications with stringent requirements on part quality such as the aeronautical and medical applications. Thus, improving part quality and repeatability becomes a vitally important issue in DMLS process (Mahesh, 2004). To improve the part accuracy and quality, optimization of DMLS process parameters is crucial. The research about particle size distribution for determining part density, surface quality and mechanical properties has been reported by Spierings (Spierings, 2011). Kruth used bridge curvature method to assess and compare laser scan patterns, laser parameter settings and more fundamental factors on residual stresses (Kruth, 2012).

To enhance part quality and repeatability, one approach aims to overcome this challenge recently been receiving much attention for on-line monitoring and process control. Process control can limit this variability generally, but it is impeded by lack of adequate process measurement methods (Mani, 2015).

Measurement science barriers and challenges prevent the widespread use of DMLS technology. It is necessary to develop in-process measurement method, which currently has certain drawbacks to enable monitoring of the DMLS process.

For in-process monitoring, sampling accurate signals is seen to be extremely significant. Signals used for monitoring the stabilization and quality in metal AM processes mainly include optical, thermal, ultrasound and acoustic signals as illustrated in Figure 1. For optical signals, there are two experimental setups designed for data processing. One involves monitoring by off-axial (external) sensors that can receive various optical signals, but the signals are easily affected by the clamping position, angle and distance. The other setup is the one involving a coaxial sensor design. This method is not affected by any obstacles. However, it will affect the stability of the whole optical system, depending on lens characteristics. In addition, it will cause failure easily because it is highly sensitive to dynamic plasma and vapor. Electrical signals in DMLS processing are not easy to obtain because experimental setup for charge sensors is troublesome. Signals from ultrasonic wave have high frequency and these signals are too weak and complex to be obtained. Besides, it is not suitable for sintering components with a cavity structure. Acoustic signals are easy to be disturbed by environmental noises. However, if noise reduction works well or features are extracted accurately, acoustic signals can be used for on-line monitoring during DLMS processing. Acoustic signals have been successfully used in laser welding or cladding process with in situ monitoring scheme (Khosroshahi, 2010).

In this paper, acoustic signal was recorded for monitoring the characteristics during DLMS process. The Power Spectrum Density (PSD) of acoustic signal changing with various values of laser power and laser scanning speed was analysed and the forming mechanism was discussed through a theoretical analysis in detail.

2 MECHANISMS OF ACOUSTIC SIGNAL FROM DMLS PROCESS

Acoustic signals are formed by the vibration of the substance and then spread through the medium. In DMLS process, vibration phenomena occur on target surface including melting, vaporization, condensation, plasma formation and laser absorption waves. If the surface temperature of powder is below the vaporization point (2750°C for steel) and above the melting point (1535°C for steel), acoustic signals are mainly created by the vibration from the friction of flow medium with liquid or solid matter, and flow motion. During laser sintering or melting, a melt pool forms among the metal powder, which is surrounded by metal vapour and plasma. A mixed substance of vapour and plasma will rush up from the gap between melt pool and metal powder. Friction forms from the flow medium while it is in contact with the melt pool and metal powder, respectively. Afterwards, melt pool and metal powder vibrate due to friction with the mixed substance in moving and vibrating. Vibration energy of the matter is transmitted into the medium in the form of a wave, resulting in an acoustic signal. In addition, when the metal vapour and plasma fill up the gap between the melt pool and metal powder, pressure inside will be much larger than that of outside. Vapour and plasma rush out along gas pressure gradients around the melt pool, which enhances the wave vibration speed. Besides, vapour and plasma bubbles formed by the trapped air inside the melt pool are trying to escape. The local gas volume inside will be expanded by high pressure to generate acoustic signals.

If surface temperature of powder rises above the vaporization point of metal powder, a vast amount of metal vapour is generated over the melted area, forming an internal vapour cavity on the metal powder where heat absorption increases and multiple reflections occur subsequently.

3 EXPERIMENT SETUP FOR MONITORING SINTERING CHARACTERISTIC BY MICROPHONE

All the experiments were done using the High-Temperature Laser Manufacturing System II (HTLMS-II) with Cyberlaser software. A Q-switched YLR-200-SM-AC laser with 1064 nm wavelenth and 20 nm pulse duration were used for was used for sintering the metal powder. During the experiment, the metal powder of 304 Stainless Steel was applied. The acoustic signals were sampled by an electret condenser microphone with response frequency of 20–16,000 Hz and a sample frequency of 44,100 Hz. All sampled signals were

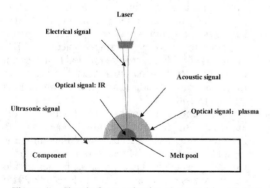

Figure 1. Signals for monitoring the metal-based AM process.

processed with MATLAB 2015b (8.6.0.267245) and executed on a High Performance Computing (HPC) cloud service with 500 GB workplace, running on the CentOS operating system.

The experiment setup is illustrated in Figure 2.

As shown in Table 1, in laser power difference experiment at Test Nos. 1–5, the laser power changed from 40 W to 200 W at a scanning speed of 100 mm/s and laser frequency of 1,000 Hz. In scanning speed difference experiment at Test Nos. 6–13, the scanning speed varied from 60 mm/s to 270 mm/s at a laser power of 120 W and laser frequency of 10,000 Hz. As illustrated in Test Nos.14–17, acoustical signals at laser frequencies 6,000 Hz, 8,000 Hz, 10,000 Hz and 12,000 Hz with different scanning speeds between 180 mm/s and 240 mm/s were recorded. During all the experiments, the hatch density and layer thickness were fixed at 0.1 mm and 0.2 mm respectively. The experimental

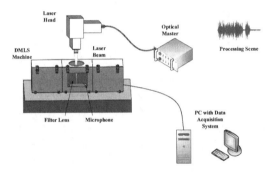

Figure 2.   Experimental setup.

Table 1.   Parameter settings for experiments.

Test no.	Laser frequency [Hz]	Laser power [W]	Scanning speed [mm/s]
1	10000	40	100
2	10000	80	100
3	10000	120	100
4	10000	160	100
5	10000	200	100
6	10000	120	60
7	10000	120	90
8	10000	120	120
9	10000	120	150
10	10000	120	180
11	10000	120	210
12	10000	120	240
13	10000	120	270
14	6000	120	180~240
15	8000	120	180~240
16	10000	120	180~240
17	12000	120	180~240

environment was protected in argon gas. Every process was tested and recorded four times and lasted 1.5 s. Contrast experiments with and without ventilation were processed respectively.

4   RESULTS AND DISCUSSION

Figure 3 illustrates the laser scanning pattern (Figure 3(a)) and acoustic signal (Figure 3(b)) of the DLMS process against time when performing laser sintering on one layer and along the boundary. As can be seen in Figure 3(b), acoustic signal varies considerably during the sintering process. In terms of the signature of acoustic signals, two phases can be roughly separated, i.e. internal laser sintering (0–10.1 s) and boundary laser sintering (10.1–11 s). Acoustic wave is generated by metal powder vibration and fluid pulsation. During DMLS, the pulsation of the laser, melt pool and metal powder are sources of the acoustic signal. As internal pattern will be sintered first at each layer, there is much fluctuation of acoustic amplitude during the internal pattern. Laser for the scanned boundary will reheat the sintered metal powder in the internal pattern again and thus the signal will be smooth and steady as shown in Figure 3(b).

The acoustical signal at laser power 80 W, laser frequency 10,000 Hz without ventilation was processed by Fast Fourier Transform (FFT) analysis as shown in Figure 4. The laser sintering acoustic signal mainly distributed among 20–2,000 Hz and 9,500–10,500 Hz as illustrated in Figure 4(a).

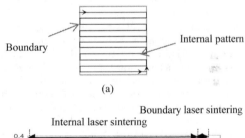

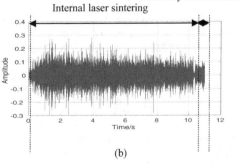

Figure 3.   The acoustic signal variation over time (a) laser sintering pattern; and (b) acoustic signal.

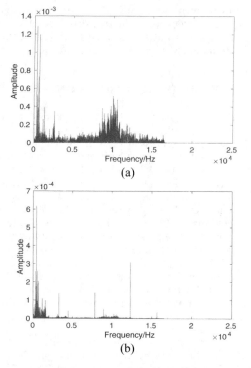

(a)

(b)

Figure 4. Acoustic signal analysis in frequency domain (a) without laser processing, and (b) with laser processing.

For analysis on the frequency distribution peaks, comparable experiments and acoustic signal analyses by FFT without laser processing were conducted as shown in Figure 4(b).

When there is no laser sintering, the first peak in the acoustic signal frequency is mainly not more than 2,000 Hz and amplitude of the frequency band is very low as shown in Figure 4(b). This acoustic signal mainly comes from machine operation, fan rotating in the host computer of laser machine, environment noise and their harmonic response. The second peak appears near 10,000 Hz as shown in Figure 4(a), when conducting the experiment at different laser frequencies, the second peak will change its location as the laser frequency changes (Figure 5). The second frequency peak corresponds to laser processing frequency. When substance and flow vibration frequency are equal to the laser frequency, the resonant interaction occurs. As shown in Figure 5, PSD peak of vibration frequency matches with the laser frequency. The vibration of medium molecules with other frequencies has relatively weak PSD. For analysis on characteristics of acoustic signal forming during DMLS, PSD analyses at different laser powers and laser scanning speeds were done for further study.

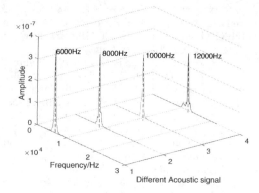

Figure 5. PSD array of laser with different frequency.

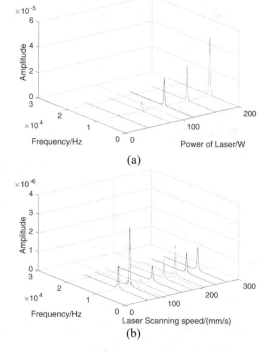

(a)

(b)

Figure 6. PSD array (a) different laser powers, and (b) different scanning speeds.

Figure 6 shows the PSD array of different laser powers and laser scanning speed in laser frequency at 10,000 Hz after band-pass filtering. The frequency higher than 10,500 Hz and lower than 9,500 Hz were not used in the experiment. Zero at different power value stands for no processing state. The PSD peak value increases with the added laser power in Figure 6(a). In Figure 6(b), two maximum values appear in the PSD array from a laser scanning speed of 60 mm/s to 270 mm/s. There is an abrupt augmentation at a scanning speed 90 mm/s and the

PSD peaks are relatively weak at the scanning speed of 60 mm/s and 120 mm/s. PSD peak goes up initially from the scanning speed of 150 mm/s to 210 mm/s and decreases later from 210 mm/s to 270 mm/s. The whole trend of PSD array with the laser scanning speed variation is shown in Figure 6(b). More accurate parameters can be selected if more elaborated parameter intervals are separated.

During the scanning speed changing process, the laser energy absorbed at unit time by the metal powder varies so that the mechanical vibration and vapour perform at different degrees as indicated by the changing acoustic signals. If the metal powder surface temperature generated by the laser energy overcomes the vaporization point, evaporation will start, and it not only removes heat but also particles. A proper choice of laser power and scanning speed is essential to avoid local overheating which can lead to dislocations, or surface corrugation.

The absorbed laser energy can be evaluated using the laser irradiance $I(t)$—energy per unit area and unit time. With the reflectivity R_o and the absorption coefficient α, absorbed laser energy of metal powder is

$$I_{abs} = I(t)\alpha(1-R_0)exp(-\alpha z) \tag{1}$$

where z the irradiated depth is. The laser beam temporal shape is based on the Gaussian profile:

$$I(t) = \left(\frac{\Phi}{\sigma\sqrt{2\pi}}\right)exp\left[\frac{-(t-t_0)^2}{2\sigma^2}\right] \tag{2}$$

where Φ is the laser fluence (Schaaf, 2010). Laser irradiance is assumed to be fully absorbed at the discontinuity and a part appears as an additional enthalpy of vapour and plasma. The wave velocity in terms of the irradiance is

$$u = \left[\frac{2(\gamma^2-1)I_{abs}}{\rho_0}\right]^{\frac{1}{3}} \tag{3}$$

where γ is the adiabatic index of the vapour, and ρ_0 is the mixed substance density. The vapour and plasma internal energy can be estimated from

$$W_1 = Mu^2\gamma\left[(\gamma^2-1)(\gamma+1)\right] \tag{4}$$

where M is vapour and plasma molecular mass. The width δ_m of the melt layer is

$$\delta_m = \left(\frac{k}{v}\right)\left(\frac{T_{lv}-T_{sl}}{T_{lv}-\dfrac{\Delta H_{lv}}{c_P}}\right) \tag{5}$$

where k is the heat diffusivity, v is the scanning speed, T_{lv} and T_{sl} are liquid–vapour and solid–liquid equilibrium temperature respectively, ΔH_{lv} is the latent heats, and c_P is the heat capacity ratio (Martin, 1986).

According to the equations (1)–(3) above, the vapour and internal energy of the plasma become proportional to the square of the wave velocity, which increases with the laser irradiance I_{abs}. As acoustic signals are mainly aroused by vapour and plasma motion, acoustic signal energy is in proportion to the wave energy (4). Melt layer width will decrease with the increasing scanning speed. The wave velocity is near invariant before the vaporization point. The acoustic signal energy increases with the mass of vapour and plasma molecular heated by laser irradiance. If the temperature overcomes the vaporization point, evaporation will start, and the evaporation not only removes heat but also particles, which is in terms of cavity in the component. The acoustic signal energy will rise sharply. This is consistent with the experiment phenomenon shown in Figure 6.

Furthermore, from the melt equation (5), very higher scanning speed leads to a shorter duration of laser illumination over the metal powder. It is due to lack of enough laser power energy to generate intense vibration for melt pool boiling. Narrow δ_m of the melt layer will cause the metal powder to not bond with each other perfectly when the scanning speed is too high or laser power is too small. This will be performed in the range of the sound energy. It is corresponding to the phenomenon as depicted at laser power 40 W in Figure 6(a) and scanning speed 270 mm/s in Figure 6(b).

Generally, when both laser power and scanning speed are higher, the sintering result will be satisfactory. However, this is also limited by the laser power equipment and scanning pattern parameters, which cannot be precisely set manually. However, the whole process performed will not be relevant if improper laser parameter settings are used. Only the appropriate process parameters used can lead to better and stronger printed metal structures.

5 CONCLUSIONS

In this paper, acoustic signals generated during the DMLS process were sampled and utilized for on-line monitoring. Experiment results showed that there was a good correlation between the laser frequency, laser power, as well as laser scanning speed and acoustic signals. Through the investigation of the acoustic signal, information on the laser scanning characteristics can be extracted. The second frequency peak is more promising for detecting the laser scanning attributes. It is shown from the study that there is a good mapping between the acoustic signals and laser scanning status as well as the resulting laser sintering quality. It provides a good reference for the research on the acoustic signal correlating the different laser parameters during the DMLS process. The results will lead to future monitoring techniques in DLMS and lay a good foundation for real-time control of metal printing process.

Future studies will be carried out on part qualities such as surface roughness, porosity, density and composition of the powder mixture interpreted via acoustic signals. Defects can be predicted automatically for quality monitoring and feedback control.

ACKNOWLEDGEMENT

The authors would like to acknowledge the support from the China Scholarship Council and National University of Singapore through research attachment for this work.

REFERENCES

Gao W., Y. Zhang, D. Ramanujan, K. Ramani, Y. Chen, C.B. Williams, P.D. Zavattieri, Comput Aided Design. 69, 67 (2015).

Gibson I., D.W. Rosen and B. Stucker, *Additive manufacturing technologies*. Springer(2015).

Guo N., M.C. Leu, Front Mech Eng. 8, 220 (2013).

Khosroshahi M.E., M. Hadavi and M. Mahmoodi, Appl Surf Sci 256, 7427 (2010).

Kruth J.P., J. Deckers, E. Yasa, R. Wauthlé, P I Mech Eng B-J Eng. 226, 989 (2012).

Mahesh M., Y.S. Wong, J.Y.H. Fuh, H.T. Loh, Rapid Prototyping J. 10, 126 (2004).

Mani M., B. Lane, A. Donmez, S. Feng, S. Moylan, R. Fesperman, *Measurement science needs for real-time control of additive manufacturing powder bed fusion processes*. NISTIR, Gaithersburg, 8036 (2015).

Martin A., *Laser-beam interactions with materials-physical principles and applications*. Springer (1986).

NIST, *Measurement science roadmap for metal-based additive manufacturing*. US Department of Commerce, NIST, Energetics Incorporated (2013).

Schaaf P., *Laser processing of materials: Fundamentals, applications and developments*. Springer (2010).

Spierings A.B., N. Herres and G. Levy, Rapid Prototyping J. 17, 201 (2011).

Tapia G., A. Elwany, J Manuf Sci E-T Asme. 136, 060801-7 (2014).

Advances in Energy Science and Equipment Engineering II – Zhou, Patty & Chen (Eds)
© 2017 Taylor & Francis Group, London, ISBN 978-1-138-71798-5

The bow defect and spring back in the roll forming process with pre-heating

Ya Zhang, Dong-Hong Kim & Dong-Won Jung
Department of Mechanical Engineering, Jeju National University, Jeju-si, Jeju, Republic of Korea

ABSTRACT: With the development of the automotive industry, the sheet forming process has acquired a new appeal of safety, environmental protection and energy saving. The emergence of high-strength steel meets the requirement of those high-impact factors and not only can achieve the same strength but also effectively reduce the material thickness. High-strength steel forming technique has been widely recognized in the market, so the insiders pay close attention to this technology. However, with the increase of strength, the elongation and plastic property of the sheet are reduced, the forming performance decreases and it is difficult to achieve deformation at room temperature. At the same time, because of the work hardening, the product is easy to crack and has a large spring back angle. To solve this problem, some researchers proposed a method of applying pre-heating to the sheet before the forming process. With the pre-heating process, the microstructure of the steel undergoes a change, thereby affecting the plastic property and the elongation of the sheet, reducing the deformation resistance and making the sheet easier to deform. In this paper, we have set up a roll forming model with the high-strength steel using the ABAQUS software. The sheet in these models have been measured for the contour bow defect of displacement and the stress is used for analysis of the deformation of the sheet. The spring back also has been developed in the simulations. At last, experiments were carried out to verify the validity of the proposed models.

1 INTRODUCTION

The sheet after roll forming has defects such as buckling, spring back, wave, bow and so on. Those defects arise due to many forming parameters and sometimes they depend upon the combination of several factors (Panthi, 2010; Cho, 2003; Shirani, 2006; Abvabi, 2015). Compared to other forming technologies, roll forming has the advantage of high production effectiveness, low cost and ease of getting a complicated cross-sectional profile. Usually roll forming happens at room temperature, the roll forming also called cold roll forming.

A kinematical approach is proposed in G. Nefussi and P. Gilormini's paper for predicting the optimal shape and the deformed length of a metal sheet during cold-roll forming. Tao Zhou and Zhao Yang (Zhou, 2015) has analysed the influence of final rolling speeds varying from 0.1 m/s to 0.4 m/s on the microstructure, textures, mechanical properties and stretch formability of AZ31 alloy sheet rolled at 550°C. Vitalii Vorkov and Richard Aerens (Vitalii, 2014) studied the spring back of different types of high-strength steels using finite element modelling. The shell element and solid element had been compared and the deflection of the

sheet during and after bending had been measured according to the images record by a camera.

In this paper, we mainly focus on the analysis of cold roll forming defects under different conditions. We analyze the bow defect and spring back with the pre-heating process. The simulation and experiment have been set up to study the influence of the temperature on spring back and bow defect.

2 THE MATERIAL PROPERTIES

There are three kinds of materials used in the current work. The mechanical properties of the materials are shown in Table 1.

Table 1.   The mechanical properties of the materials.

	Yield strength (MPa)	Ultimate tensile strength (MPa)	Young's Modulus of elasticity (GPa)	Poisson's ratio
SPFH 590	448	671	250	0.29
Stainless steel	251	340	250	0.30
Mild steel	205	520	250	0.30

3 THE INFLUENCE OF PRE-HEATING IN ROLL FORMING PROCESS WITH SIMULATION METHOD

Four simulations were done with the ABAQUS software. The explicit/dynamic simulation process has been used in these simulations. The material used in the simulation is SPFH590. The temperature in the simulations has been set as room temperature, 50°C, 150°C and 250°C. The spring back and bow defect have been compared at difference temperatures to reveal the relationship between the temperature and forming parameters.

In the simulations, we assumed that the temperature is constant and that there are no other forms of heat transfer or heat radiation. The sheet element type will be S4R and the step time in the simulation will be 1 second. Only half of the sheet has been considered in the simulation because of symmetry. The friction between the roll and sheet are considered to be frictionless. The interaction of the rolls and sheet will be surface-to-surface contact. The rolls are a rigid body and have no other form of deformation all around the forming process. The models of the roll forming in simulations are shown in Fig. 1. The width and length of the sheet is 60 mm and 746 mm. The distance between the rolls is 220 mm. The roll forming direction has been shown on the figure.

The simulation conditions for pre-heating are the same as for the simulation under room temperature. The sheet was modelled with shell element, in which the element type is S4R. The interaction between the rolls and sheet is surface-to-surface interaction. The von Mises yield criterion has been used in the simulation. The hardening assumed in the simulations is isotropic. Although there are many kinds of hardening modules in the ABAQUS, isotropic hardening is the most suitable in this situation considering the accuracy, time and cost. The boundary conditions have been set by the displacement condition and the sheet is fixed, rolls pass through the sheet and deformation occurrs.

In the roll forming process, the spring back is one of the most important defects of the product. The sheet after forming to a certain shape will continue to deform due to the residual stress and this deformation is called spring back. The spring back results have are shown in Fig. 2 as per the simulation results. We calculated nine spring back angles along the longitudinal direction of the sheet. From the simulation results, we can know that the spring back angles decrease with the increasing of sheet temperature. This means that a better accurate sheet deformed at elevated temperature. The spring back angle below a temperature of 250°C is 1.6% smaller than the spring back angle at room temperature. Due to the sudden force change at the start and the end of the forming process, the spring back angle of the first point and the end point are considered to be inaccurate.

Bow defect was measured with the displacement of the element in the centre line of the sheet. The results are shown in Fig. 3. The maximum bow

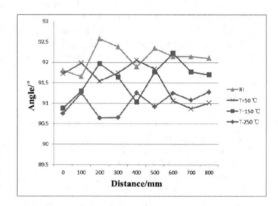

Figure 2. The spring back angles with different temperatures.

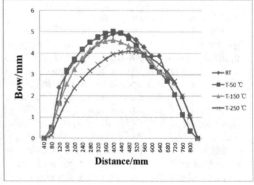

Figure 3. The bow displacement under different temperatures.

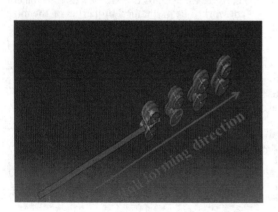

Figure 1. The model of roll forming in simulation.

displacement in room temperature is 4.94 mm. When the temperature increases to 250°C, the bow displacement in the same location is 4.10 mm. This means that pre-heating of the sheet can reduce the extent of the bow defect to a certain amount.

The kinetic energy is shown in Fig. 4. We can know from the simulation that kinetic energy does not significantly change with temperature. The deviation under 1% is acceptable in the simulations. This is because the velocity in the simulation has not changed with the temperature. The internal energy, which has been shown in Fig. 5, decreases with the increase of temperature.

According to the energy conservation laws, the work due to forming force is reduced with an increase in temperature, which leads to a decrease of the working energy of the roll. So the internal energy decreases with the increase of temperature. Besides, we defined the temperature in the simulations. Changing the material properties has no influence on the internal energy. Along the forming process, the kinetic energy is smaller than 5% of the internal energy, which means the simulations are reliable.

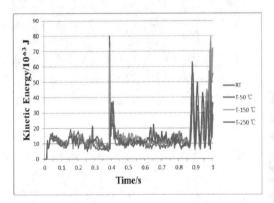

Figure 4. The kinetic energy under different temperatures.

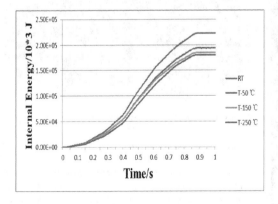

Figure 5. The internal energy under different temperatures.

4 THE ROLL FORMING EXPERIMENT WITH PRE-HEATING

Depending on the requirements, heating can be completed before, during and after roll forming. The experiment has been established with the butane heating. There will be seven experiments with three kinds of steels. Fig. 6 shows the heating location using butane torch in these experiments.

We understand from Table 2 that the experiments have been done based on the different heating locations. Other experiment factors are the same in those experiments. The velocity of the roll, the inner distance of the roll and sheet dimension are the same. The butane torch has been fixed in a frame at a height of 800 mm. There are totally four torches used in the forming process and the sheet passes through the rolls from right to the left. By controlling the heating location and torch quantity, we can control the experimental conditions easily. The velocity of the sheet in the experiments will be 10.3 mm/s. The inner distance of the rolls is 220 mm. The thicknesses of the sheets are 0.82 mm and the gap between upper roll and lower roll is set to be 1 mm. The high strength sheet has a length of 746 mm and the length of the mild steel is 750 mm. The length of the stainless steel will be 810 mm. The widths of all the sheets are 60 mm.

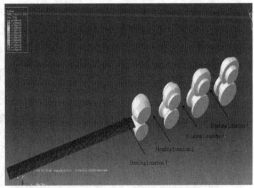

Figure 6. The heating locations in the experiments.

Table 2. The experimental conditions.

Experiment no.	Heating location
1	1
2	2
3	3
4	4
5	1, 2
6	1, 2, 3
7	1, 2, 3, 4

The temperature has been observed with IR camera. As shown in Fig. 7(a), the temperature on the sheet with butane torch heating increases up to 229°C at the heating point. Fig. 7(b) shows the temperature of the flame, and the maximum temperature is 120.80°C.

Schematically illustrated in Figs. 8 to 10 are the experimental results of three materials with different heating methods. The experimental results are acceptable and no big defects occurred. The black part seen on the sheet was formed due to oxidation. Besides we can see the head part of the sheet has not been oxidized as same with the other part of the sheet. This means that the result of this part is not reliable and should be ignored. The spring back angles in the experiments are measured by a protractor. Also, we have calculated the bow displacements of the sheet. The bow displacements have been defined as the maximum displacement of the sheet along the vertical direction of the sheet.

Figs. 11 to 13 shows spring back angle of the three materials with different heating locations. We observe from the figures that the spring back angles in the room temperature of three materials are the widest. The spring back angle decreases with an increase in temperature. The spring back in some

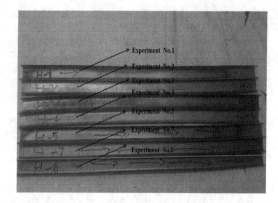

Figure 8. The experiment result of high strength steel.

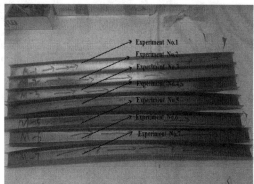

Figure 9. The experiment results of stainless steel.

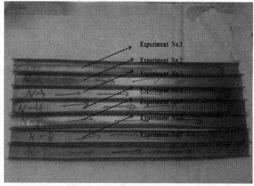

Figure 10. The experiment results of mild steel.

experiments with a high temperature has a bigger spring back angle. This may be due to the deviation of the measurement of the spring back angle. In the all three materials, the spring back angle shares the same tendency, as the spring back angle decreases with heating. As heating continues and the temperature increases, the spring back angle decreases.

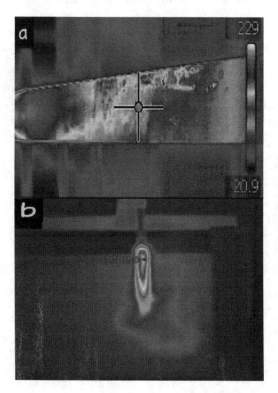

Figure 7. The temperature measured in the experiment.

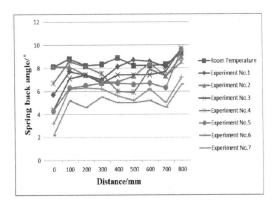

Figure 11.   The spring back angles of high strength steel.

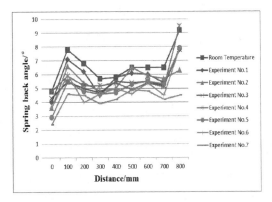

Figure 12.   The spring back angles of mild steel.

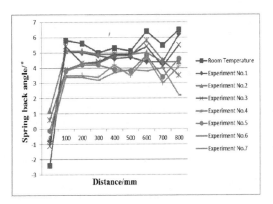

Figure 13.   The spring back angles of stainless steel.

Also, the heating positions of the sheet have no significant influence on the spring back angle. The spring back angles of different heating locations are almost the same. The high strength steel has the maximum spring back angles in the three kinds of materials.

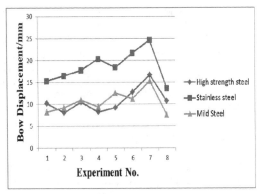

Figure 14.   The bow displacements of three different materials.

Fig. 14 shows the bow displacement of the sheet at different heating locations of different materials. Bow displacement is affected by many factors and one of the most important factors is the tension along the edge of cross-section. This tension has a tendency to make the cross-section longer in the longitudinal direction. On the other side, the tension is insufficient for the whole cross-section of the sheet and makes the sheet curved. We know that stainless steel has a bigger bow displacement compared with other two kinds of materials. For all three kinds of materials, the bow displacement becomes bigger with the increase of temperature. The bow displacement of stainless steel is bigger than the other two kinds of materials because stainless steel used in the experiment is a little longer than the other two materials.

5   CONCLUSIONS

The influence of pre-heating in bow defect and spring back in roll forming process has been studied using simulation and experimental methods. The simulation results show that the pre-heating has a positive influence on the forming defects of spring back but have a negative influence on bow displacement. The forming force of the rolls decreases with the increase of temperature. The internal energy increases with the temperature.

The experiment results of the roll forming with pre-heating have the same tendency with simulation results. The spring back at room temperature is bigger than the other sheets in the pre-heating process. The bow displacement in the simulations are much smaller than that in the experiment, which is because in the experiments, flatteners have not been used. For the simulations, because of the boundary conditions, the sheet can move forward in a more stable manner. The mild steel has the maximum bow displacement of the three kinds of materials, and it increases with temperature. The simulation and experiment results can help us design and optimize the roll forming process. The results gives guidance on the roll design, optimizes the forming parameters and also helps to predict the metal flow.

ACKNOWLEDGEMENT

This research was supported by the Basic Science Research Program through the National Research Foundation of Korea (NRF) funded by the Ministry of Education (2014009199).

REFERENCES

Abvabi, A., B. Rolfe, P.D. Hodgson, M. Weiss, Int. Jou. of Mec. Sci. **10**, 124 (2015).

Cho, J.R., S.J. Moon, Y.H. Moon, S.S. Kang, Jou. of Mat. Pro. Tech. **141**, 109 (2003).

Nefussi, G., P. Gilormini, Int. Jou. of Mec. Sci. **35**, 867 (1993).

Panthi, S.K., N. Ramakrishnan, Meraj Ahmed, Shambhavi S. Singh, M.D. Goel, Mat. and Des. **31**, 657 (2010).

Shirani Bidabadi, B., H. Moslemi Naeini, M. Salmani Tehrani, H. Barghikar, Jou. of Con. Ste. Res. **118**, 243 (2006).

Tao Zhou, Zhao Yang, Dong Hu, Jou. of All. and Com. **652**, 434 (2015).

Vitalii Vorkov, Richard Aerens, Pro. Eng. **81**, 1005 (2014).

Polymer light-emitting diode with resistivity optimized p-type Si anode

Qiaoli Niu & Hengsheng Wu
Key Laboratory for Organic Electronics and Information Displays and Institute of Advanced Materials
National Jiangsu Syngerstic Innovation Center for Advanced Materials (SICAM), Nanjing University
of Posts and Telecommunications, Nanjing, P.R. China

Yongtao Gu & Yanzhao Li
State Key Laboratory for Mesoscopic Physics, School of Physics, Peking University, Beijing, P.R. China

Wenjin Zeng
Key Laboratory for Organic Electronics and Information Displays & Institute of Advanced Materials
National Jiangsu Syngerstic Innovation Center for Advanced Materials (SICAM), Nanjing University
of Posts and Telecommunications, Nanjing, P.R. China

Yong Zhang
Institute of Optoelectronic Material and Technology, South China Normal University, Guangzhou, P.R. China

ABSTRACT: P type Silicon was used as the anode of Polymer Light-Emitting Diode (Si-PLED) instead of traditional Indium Tin Oxide (ITO) anode. The Si-PLED has a device structure of p-Si/SiOx (2 nm)/poly (2-methoxy-5-(2'-ethylhexyl-oxy)-1,4-phenylenevinylene) (MEH-PPV) (70 nm)/Sm (10 nm)/ Au (20 nm). Optoelectronic performances of Si-PLEDs as a function of resistivity of p-Si anode were investigated. As the increasing of resistivity of p-Si, the luminance of Si-PLEDs enhanced first and then decreased. That may be caused by the different hole concentration of p-Si anodes and electric charge density of dipole layer formed in the p-Si/MEH-PPV interface. In consideration of the luminance efficiency of Si-PLED, an optimal resistivity of p-Si was obtained, which is $0.1\Omega \cdot cm$.

1 INTRODUCTION

The current trend towards optoelectronic integration on Silicon, leads to the hot researches on light sources fabricated directly onto silicon integrated circuits. Unfortunately, silicon light-emitting devices suffer from low emission efficiencies due to the indirect-band gap characteristic (Xu, 2015; Mastronardi, 2012). One replaceable strategy is to grow a III-V group semiconductor with a direct-band gap on a Si substrate (Leong, 1997; Zhang, 2001). However, dislocations and stress in lattice mismatch of Si/III-V composite materials are great obstacles for fabricating efficient lighting devices. On the other side, Organic Light-Emitting Diodes (OLEDs) have attracted considerable attentions due to their potential applications in flat-panel displays and solid-state light sources since the report of efficient OLEDs by Tang and Van Slyke in 1987 (Tang, 1987). The combination of Si technology and organic light emission will open up a promising route to the integration of lighting devices on Si, including active matrix displays. Furthermore, organic materials are amorphous, rather than

crystalline, so there is no lattice mismatch between organic materials and Si. Recently, Si-based organic light-emitting diodes (Si-OLEDs) have been paid attention to (Parker, 1994; Heinrich, 1997; Zhou, 1999; Li, 2013; Zhang, 2012; Zhang, 2011; Liu, 2011). However, those reports are mostly focused on the Si-OLEDs with small molecular organic materials. Polymer based light-emitting diodes are particularly attractive because they can be processed by solution technology, e. g., inkjet printing, for low-cost production (Burroughes, 1990). Park et al. demonstrated for the first time the fabrication and characterization of Si-based Polymer Light-Emitting Diodes (Si-PLEDs) using both 10^{21} cm^{-3} doped n-and p-Si as anodes and an external quantum efficiency of 0.07% was obtained (Parker, 1994). In this manuscript, p type Silicon was used as the anode of Polymer Light-Emitting Diode (Si-PLED) instead of traditional Indium Tin Oxide (ITO) anode. The Si-PLED has a device structure of p-Si/SiOx (2 nm)/poly (2-methoxy-5-(2'-ethyl-hexyl-oxy)-1,4-phenylenevinylene) (MEH-PPV) (70 nm)/Sm (10 nm)/Au (20 nm). Optoelectronic performances of Si-PLEDs as a function of resistivity of p-Si anode were investigated. As the increasing of resistivity of p-Si, the luminance of Si-PLEDs enhanced first and then decreased. That may be caused by the different hole concentration of p-Si anodes and electric charge density of dipole layer formed in the p-Si/MEH-PPV interface. In consideration of the luminance efficiency of Si-PLED, an optimal resistivity of p-Si was obtained, which is 0.1 Ω • cm.

2 EXPERIMENT

All the p-Si wafers used here were (100) oriented, whose resistivities around 30, 9, 1, 0.1, 0.02, and 0.002 Ω • cm. Before using, the p-Si wafers were routinely cleaned in a sequence of ultrasonic rinses in acetone, ethanol, and de-ionized water for 5 min each, and then etched for 1 min in a 4% HF solution to remove the native oxide. After being blown dry, p-Si wafers experienced a 400°C 40 mins oxidation for growing an about 2 nm SiO_x layer on the surface. After that, Al was deposited on the backside for forming ohm contract. Then, p-Si wafers were immediately transferred into a N_2 atmosphere dry box with oxygen and water concentration less than 1ppm. A 70 nm-thick Poly (2-methoxy-5-(2'-ethylhexyloxy)-1,4-phenylenevinylene) (MEH-PPV) was spin-coated on the top of thin thermal SiO_x modified p-Si surface from chlorobenzene solution as the emissive layer. It was then heated at 80°C for 0.5 h to remove the residual solvent. Finally, a 10 nm-thick Sm and 20 nm-thick Au layers were evaporated sequentially on top of MEH-PPV as semi-transparent composite cathode. The current-voltage and luminance-voltage characteristics of all devices were measured in air at room temperature by using HP 34410 A and a Si photodiode sensor calibrated by a PR705 spectrometer.

3 RESULTS AND DISCUSSIONS

The current-voltage characteristics of the Si-PLEDs with resistivities of p-Si vary from 30, 9, 1, 0.1, 0.02, to 0.002 Ω • cm are depicted in Figure 1. It can be seen that the current density of Si-PLEDs increase monotonically with the increasing of resistivity of p-Si. With the same bias voltage, the higher the resistivity of p-Si anode is, the larger the current density of Si-PLED will be. For all devices here, the only difference was the resistivity of p-Si anode. Therefore, the current density of Si-PLED increased because of the enhancement of hole-injection current. This indicates that hole-injection ability of Si-PLED can be tuned by altering the resistivity of p-Si anode. This phenomenon of tuneable hole-injection ability has been

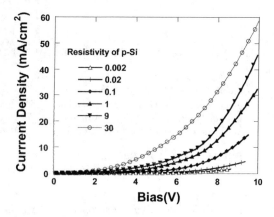

Figure 1. Current density-voltage characteristics of Si-PLEDs with the resistivities of p-Si vary from 0.002, 0.02, 0.1, 1, 9 to 30 Ω • cm.

reported in p-Si anode OLED, the lower resistivity of p-Si anode corresponds to the higher hole concentration and hole injection current of Si-OLED (Zhang, 2012; Xu, 2006). However, it is opposite to that in the p-Si anode PLED reported here.

To understand the different phenomena between Si-PLED and Si-OLED, we believe that an interfacial dipole layer is formed at the p-Si/MEH-PPV interface due to interactions between the inorganic semiconductor and polymer. As a result, an abrupt shift of vacuum level occurred at the interface, as shown in Figure 2. Similar results were observed for the metal/organic materials interface by Hill et al (Hill, 1998) and the semiconductor/organic materials interface by Chen et al (Chen, 2009).

When MEH-PPV is fabricated on the surface of p-Si anode, hole diffuses from p-Si to MEH-PPV because of the hole concentration difference between them, and accumulates in MEH-PPV near the p-Si/MEH-PPV interface. Therefore, interfacial dipole layer formed at the p-Si/MEH-PPV interface between the positive charged layer in MEH-PPV and electrons left over in p-Si after the diffusion of holes. The electric field direction of dipole layer points from MEH-PPV to p-Si, as shown in Figure 3, which was taken as the positive direction.

Therefore, the electric potential energy of MEH-PPV was lower than that of p-Si, we use ΔE_{di} representing the absolute value of the difference. Due to this positive interface dipole moment, a significant shift of the vacuum level for MEH-PPV and a boosting of the barrier height for hole injection occurs at the p-Si anode/MEH-PPV interface. The diagrams of energy levels of p-Si and MEH-PPV before and after contact were shown in Figure 2. The hole injection barrier Φ_{Bh} composed of two parts: the difference between the conduction band

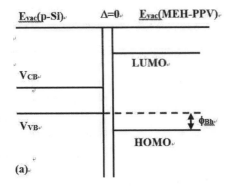

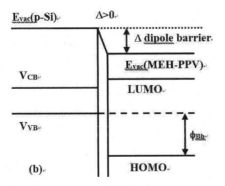

Figure 2. Schematic of energy diagrams of p-Si/MEH-PPV interface without (a) and with (b) an interface dipole Δ. Φ_{Bh} is hole injection barrier, E_{vac}(p-Si) and E_{vac}(MEH-PPV) are vacuum levels of p-Si and MEH-PPV, respectively.

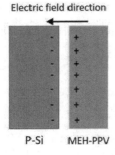

Figure 3. The diagram of interface dipole layer in the p-Si/MEH-PPV interface.

of p-Si anode and the Highest Occupied Molecule Orbit (HOMO) of MEH-PPV (marked as ΔE_b), and ΔE_{di}. That is to say, $\Phi_{Bh} = \Delta E_b + \Delta E_{di}$. And ΔE_{di} was affected by the resistivity of p-Si anode. The lower the resistivity of p-Si is, the higher the hole concentration of anode will be. That will cause

the higher density of dipole layer formed by the diffusion of holes. And therefore, the absolute value of ΔE_{di} will become larger. For p-Si anode with the lower resistivity, the corresponding Si-PLED had lower current density at the same driving voltage due to the decrease of hole injection. Hence, the current density of Si-PLED decreased as the decreasing of resistivity of p-Si, as shown in Figure 1.

Figure 4 shows luminance-voltage characteristics of Si-PLEDs with different resistivities of p-Si anodes. With the same bias voltage, the luminance of Si-PLEDs enhances first as the increasing of resistivity of p-Si and then decreases. The maximum luminance has been observed with a resistivity of 0.1 Ω • cm of p-Si. The values of maximum luminous efficiency and maximum power efficiency both show the same trend as the luminance; increas-

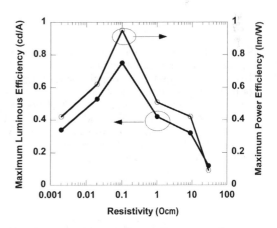

Figure 4. Luminance-voltage characteristics of the Si-PLEDs with different resistivities of p-Si.

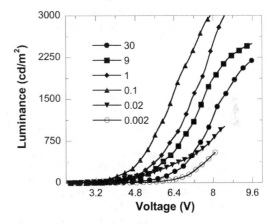

Figure 5. The maximum luminous efficiency and maximum power efficiency of Si-PLEDs as a function of resistivity of p-Si.

ing first with the resistivity of p-Si and reaching a maximum of 0.75cd/A and 0.95lm/w for $0.1\Omega \cdot$ cm, respectively. The maximum luminous efficiency and maximum power efficiency of Si-PLEDs as a function of p-Si resistivity were shown in Figure 5. It indicates that there is an optimum p-Si resistivity in consideration of Si-PLED's performances. That could be because of the hole-injection tuneable ability of p-Si anode by adjusting its resistivity. The higher the electrical resistivity of the p-Si anode is, the smaller the injection barrier of hole is. Best performance of Si-PLED can be obtained when the electron and hole current matched best at a certain value of resistivity, which is $0.1\Omega \cdot$ cm in this case. When the resistivity of p-Si is larger than $0.1\Omega \cdot$ cm, there is more hole than electron in Si-PLED, the balance between hole and electron currents was destroyed, which led to the decrease of luminous and power efficiency. This work is our first time attempting on p-Si anode PLEDs. We believe the luminance efficiency will be improved after optimizing the device structure.

4 CONCLUSIONS

In summary, we fabricated p-Si anode PLEDs with different resistivities of p-Si. The experimental results show that the performance of Si-PLED is strongly dependent on the resistivity of p-Si anode. The hole-injection ability enhances with the increasing of resistivity due to the reducing of dipole layer intensity between p-Si anode and MEH-PPV. With a resistivity of $0.1\Omega \cdot$ cm, electron and hole current of Si-PLED matched best, and the optimal maximum luminous efficiency and maximum power efficiency were achieved, respectivel.

ACKNOWLEDGEMENTS

This work was supported by the Key Laboratory of Mesoscopic Physics (Peking University, Grant No. 4172022004Q1), the Natural Science Foundation of Nanjing University of Posts and Telecommunications (NUPTSF Grants NY213044), the National Natural Science Foundation of China (Grant No. 61504066), Education Institutions of China (15 KJB430024), and Natural Science Foundation of Jiangsu Province, China (BK20150838)

REFERENCES

Burroughes, J.H., D.D.C. Bradley, A.R. Brown, R.N. Marks, K. Mackay, R.H. Friend, P.L. Burns, and A.B. Holmes, Light-emitting diodes based on conjugated polymers, Nature 347, 539(1990).

Chen, W.J., B.K. Li, H.T. He, J.N. Wang, H.L. Tam, K.W. Cheah, X.C. Cao, Y.Q. Wang, G.J. Lian, G.C. Xiong, Interface dipole formation between GaMnAs and organic material, J. Phys.: Conf. Ser. 193, 012105(2009).

Heinrich, L.M.H., J. MuLler, U. Hilleringmann, K.F. Goser, A. Holmes, D.H. Hwang, R. Stern, CMOS-compatible organic light-emitting diodes, IEEE Trans. Electron Devices 4, 1249(1997).

Hill, I.G., A. Rajagopal, A. Kahn, Y. Hu, Molecular level alignment at organic semiconductor-metal interfaces, Appl. Phys. Lett. 73, 662(1998).

Leong, D., M. Harry, K.J. Reeson, K.P. Homewood, A silicon/iron-disllicide light-emitting diode operating at a wavelength of 1.5 μm, Nature 387, 686(1997).

Li, Y.Z., Z.L. Wang, Y.Z. Wang, H. Luo, W.J. Xu, G.Z. Ran, G.G. Qin, Electron-irradiated n+-Si as hole injection tunable anode of organic light-emitting diode Appl. Phys. B 110, 95–99(2013).

Liu, N., M.M. Shi, Y.Z. Li, Y.W. Shi, G.Z. Ran, G.G. Qin, M. Wang, H.Z. Chen, The enhanced electron injection by fluorinated tris-(8-hydroxy-quinolinato) aluminum derivatives in high efficient Si-anode OLEDs, J. Lumin. 131, 199–205(2011).

Mastronardi, M.L., E.J. Henderson, D.P. Puzzo, Y. Chang, Z.B. Wang, M.G. Helander, J. Jeong, N.P. Kherani, Z. Lu, G.A. Ozin, Silicon Nanocrystal OLEDs: Effect of Organic Capping Group on Performance, Small 8, 3647–3654(2012).

Parker, I.D., H.H. Kim, Fabrication of polymer light-emitting diodes using doped silicon electrodes, Appl. Phys. Lett. 64, 1774(1994).

Tang, C.W., S.A. VanSlyke, Organic electroluminescent diodes, Appl. Phys. Lett. 51, 913(1987).

Xu, A.G., G.Z. Ran, Z.L. Wu, G.L. Ma, Y.P. Qiao, Y.H. Xu, B.R. Yang, B.R. Zhang, G.G. Qin, Effects of resistivity of a p-Si chip on the light-emitting efficiency of a top-emission organic light-emitting diode with the p-Si chip as the anode, phys. stat. sol. (a) 203, 428–434(2006).

Xu, K., Q. Yu, G. Li, Increased Efficiency of Silicon Light-Emitting Device in Standard Si-CMOS Technology, IEEE J. Quantum Elect. 51, 1(2015).

Zhang, P.H., V.H. Crespi, E. Chang, S.G. Louie, M.L. Cohen, Computational design of direct-bandgap semiconductors that lattice-match silicon, Nature 409, 69(2001).

Zhang, X.W., H.P. Lin, J. Li, F. Zhou, B. Wei, X. jiang, Z.L. Zhang, Elucidations of weak microcavity effect and improved pixel contrast ratio in Si-based top-emitting organic light-emitting diode, Curr. Appl. Phys. 12, 1297–1301(2012).

Zhang, X.W., X. Jiang, W.Q. Zhu, Z.L. Zhang, X.Y. Liu, H. Wang, H.R. Xu, Efficient fluorescence from 9,10-bis(m-tolylphenylamino) anthracene doped into a blue matrix in Si-based top-emitting organic light-emitting diode, Thin Solid Films, 519, 6595–6597(2011).

Zhou, X., J. He, L.S. Liao, M. Lu, Z.H. Xiong, X.M. Ding, X.Y. Hou, F.G. Tao, C.E. Zhou, S.T. Lee, Enhanced hole injection in a bilayer vacuum-deposited organic light-emitting device using a p-type doped silicon anode, Appl. Phys. Lett. 74, 609(1999).

Advances in Energy Science and Equipment Engineering II – Zhou, Patty & Chen (Eds)
© 2017 Taylor & Francis Group, London, ISBN 978-1-138-71798-5

Influence of carbon additive on the selective laser sintering of silica

Shuai Chang
Department of Mechanical Engineering, National University of Singapore, Singapore
State Key Laboratory of Advanced Welding and Joining, Harbin Institute of Technology, Harbin, China

Liqun Li
State Key Laboratory of Advanced Welding and Joining, Harbin Institute of Technology, Harbin, China

Li Lu & Jerry Ying Hsi Fuh
Department of Mechanical Engineering, National University of Singapore, Singapore

ABSTRACT: Selective Laser Sintering (SLS) is one of the additive manufacturing techniques that can be used to synthesize three-dimensional objects for different applications. Presently, notable progress has been achieved in laser-based processing of metallic materials in this field. However, it is yet to be fully applied to ceramic material applications without the addition of a polymer binder. Besides the limitation of instinct brittleness and low thermal shock resistance of the ceramic material, the main laser sintering hurdles lie in the poor absorptivity of oxide ceramics to near-infrared laser, which is widely used in commercial Selective Laser Sintering (SLS) machines. This study explores an alternative solution to overcome the poor absorptivity. The influence of carbon additive on the improved absorption of silica powder to fiber laser is investigated in this paper. The improvement of absorptivity with doping carbon was also discussed. The characteristics of the absorptivity on the laser sintering process were also investigated. This approach makes direct selective laser sintering of oxide ceramics possible. Silica parts with low volume shrinkage of less than 1% have been achieved via direct selective laser sintering.

1 INTRODUCTION

Oxide ceramics exhibits unique properties such as good mechanical strength, combined with superior wear and thermal resistance, which could meet the requirements for a growing demand of high-end precision engineering applications. As traditional manufacturing processes, e.g., dry pressing, slipcasting, tape casting and injection molding requirements. The technique of fabricating ceramics is difficult or even impossible for producing complex shapes, (Tang, 2011; Wu, 2009). Notable progress in Additive Manufacturing (AM) technologies allows for a rapid freeform fabrication of complex geometries that cannot be achieved easy by the conventional techniques. Among the different AM processes, Selective Laser Sintering (SLS) has been developed introduced in the late 1980s is the most promising technique to process ceramics components because of its ability to create Three-Dimensional (3D) net-shaped parts in one operation (Kruth, 1998). It has become more attractive for manufacturing of complex polymeric or metallic parts directly. However, the Selective Laser Sintering (SLS) of ceramics without using polymer binder or post-processing have not been successful due to the brittleness and low thermal shock resistance material properties (Ber-

trand, 2007; Shishkovsky, 2007). Besides, the main hurdle for laser sintering is its very poor absorptivity of oxide ceramics to near-infrared laser, which is widely used in commercial Selective Laser Sintering (SLS) machines (Tolochko, 2000). Several researches attempted to attain direct SLS of ceramics through use of lower-melting point ceramic coatings. But it was still not successful because the absorbance is even weaker after coating with transparent ceramic-like silica although the coating can reduce cooling stress and improve homogeneity slightly (Kruth, 2007).

The study on laser powder absorptivity is useful to the understanding of the Selective Laser Sintering (SLS) process because it allows one to obtain prior information that is crucial to determine a more uniform and suitable processing window for laser sintering (Duley, 2012). Continuous-wave CO_2 and Nd:YAG fiber lasers are typical laser sources for most of the SLS machines. In comparison with Nd:YAG fiber laser with wavelength of 1070 nm, the continuous wave CO_2 laser emitting at 10.6 μm is more easier to be absorbed by most of ceramic materials, but the spot size is much larger in operation than that of the fiber laser (Gahler, 2006; Qian, 2013). A larger spot size will decrease the dimensional accuracy and power intensity. However, silica is almost non-absorbent in Nd:YAG fiber laser emitting at 1070 nm (Qian, 2013;

Juste, 2014). Therefore, the objective of this paper is to develop an approach to overcome poor absorptivity. The results reported here aim to study the effect of laser absorptivity on the direct Selective Laser Sintering (SLS) process.

2 EXPERIMENT

2.1 Principle of SLS of materials

The Selective Laser Sintering (SLS) technique as depicted in Fig. 1 is a powder-based AM technique in which using a Nd:YAG fiber laser beam to fuse powder materials to develop a 3D near-net-shape object layer-by-layer. Based on the digital design, a CAD/CAM model is sliced into thin powder layers of less than 100 μm (Ko, 2007; Kruth, 2005). Fabrication of the final parts with a SLS method comprises the following steps:

1. Thin layer of free-packed powder is deposited onto the build platform.
2. Selectively laser scans over and fuses the new layer via sintering according to the sliced data.
3. The building platform is lowered by a height equal to the layer thickness.
4. Steps 1 to 3 are repeated until the completion of laser sintering of all layers.

In the process of interaction between the ceramic powder granules and the laser beam, only a small portion of laser energy is absorbed by ceramic powder and the rest is reflected as shown in Fig. 1. The amount of the absorbed radiation is a key factor that determines whether the laser processing could be successful or not. The degree of absorption is normally represented by "absorptance". The Absorptance (A) is defined as the ratio of energy absorbed by materials to the incident laser energy. In our experiments, Reflectance (R) will be measured. Reflectance is the ratio of reflected laser energy to the total laser energy. Then A is calculated as

$$A = 1 - R$$

2.2 Materials and sample preparation

The silica powder used in the present work is of the code SS1206 supplied by Industrial Powder,

USA. This powder (99.5%) with d50 30 ± 2 μm size is spherical in shape. The spherical shape of silica powder can be seen in Fig. 2. As shown in Fig. 3, the particle size distribution ranges from less than 1 μm to 60 μm (measured by laser diffraction on LS 100 Q, Coulter International Corporation, USA). Then, 0.2 wt.% of fine active carbon was doped into silica powder to improve its laser absorptivity. The Ball Mill (from Planetary Mono Mill, FRITSCH GmbH, Germany) at a low rotating rate of ~150 rpm was used to mix powder sufficiently and evenly.

Ceramic parts were fabricated using a self-developed SLS system. It is equipped with a galvano scanner and one fiber laser with a maximum power of 200 W with a wavelength of 1070 nm. The beam focusing diameter is approximately 100 μm. It provides a $100 \times 100 \times 150$ mm³ build envelope and the whole process is carried out under a construable air atmosphere.

The main parameters for the SLS processing are illustrated in Fig. 4, including (1) laser Power (P); (2) laser spot size (fz); (3) layer thickness (D); (4) Hatch Space (HS), defined as the distance between the centerline of the two adjacent parallel laser tracks; (5) scan speed (v) and (6) scanning strategy. The optimized parameters for the SLS of silica doped with

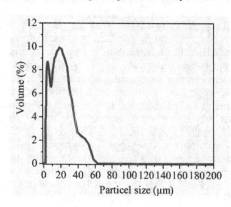

Figure 2. SEM image of spherical silica particles.

Figure 3. Particle size distribution of the spherical silica powder.

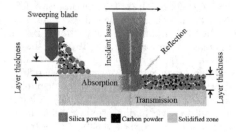

Figure 1. Schematic of SLS.

carbon were listed in Table 1. Experiments were conducted for the laser sintering a matrix of 8×8 mm^2 squares by varying the main parameters mentioned above. This optimized process was subsequently used to fabricate 3D ceramic parts for testing.

The furnace sintering process was carried out in a high temperature chamber furnace (Carbolite HTF 17/5/3216P1, UK). The heating rate was set as 2°C/min for an increase from room temperature to 1,200°C, and then kept for 5 h.

Finally, it was cooled down naturally. The whole furnace sintering process is much more cost-effective and less time consuming due to freedom of de-binding process needed.

2.3 *Sample characterization*

The morphology and microstructure of the SLS silica was investigated under Zeiss SUPRA™ 40VP Field Emission Scanning Electron equipped with Thermo Noran System SIX EDS system. Absorptance was measured with ultraviolet-visible light-near infrared spectrometer (UV-1800UV-VIS spectrophotometer, Shimadzu, Japan). Crystal phase identification was obtained on X-ray diffraction patterns (Bruker AXS D8 Advance XRD instrument, running with a voltage of 40 kV and a current of 40 mA, with Cu Kα radiation, $\lambda = 0.15418$ nm).

3   RESULTS AND DISCUSSIONS

The SEM micrograph and chemical map of silica powder with carbon addition are shown in Fig. 5. The carbon element is evenly distributed on powder bed. The ball mill at low rotating speed has achieved uniform mixing without crushing spherical silica granules. As introduced previously, fine carbon was doped into the silica granules to enhance the absorp-

tivity to laser beam. Adding carbon exhibits a high absorptivity at 1070 nm. On the other hand, carbon would generate additional heat when reacting with oxygen in the air and it was proved that no reaction with silica took place, as shown in the XRD pattern later. Hence, there is no risk to induce undesirable new phase, which may have effects on the final properties. Fig. 6 shows the clear improvement of absorptance by adding only 0.2 wt.% of carbon. The absorptance to fiber laser ($\lambda = 1070$ nm) reaches ~8%, which is up to four times higher than pure silica. The benefit of carbon additive was directly demonstrated in the comparison between SLS of silica with and without carbon addition (Fig. 7). The powders with carbon were successfully joined to form a solid plate by laser beam. However, it did not achieve good melting in the SLS of pure silica powders due to limited absorptivity.

Fabrication of 3D silica part has been performed with optimized parameters according to Table 1.

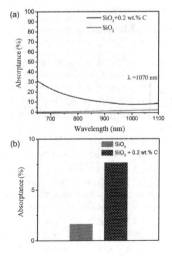

Figure 5.   (a) SEM of the silica–carbon composite particles with 0.2 wt.% of carbon additive, chemical map of, (b) carbon, (c) silicon and (d) oxygen, using EDX for the ceramic composite.

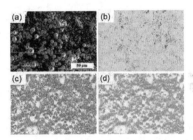

Figure 6.   (a) Absorptance analysis with UV light-near infrared spectrometer and (b) Comparison of the absorptance to 1070 nm wavelength between pure silica and silica–carbon composite.

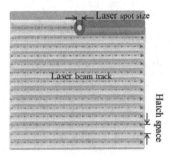

Figure 4.   Key parameters of the SLS process.

Table 1.   Optimized laser sintering parameters.

Laser power	Hatch space	Scan speed	Spot size	Layer thickness
80 W	140 μm	60 mm/s	120 μm	100 μm

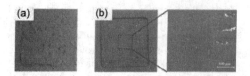

Figure 7. (a) Appearance of SLS samples using pure silica and (b) silica–carbon composite (*P*: 80 W, *V*: 60 mm/s).

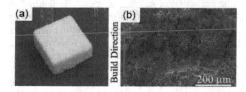

Figure 8. (a) Silica cubic sample of 30 layers and (b) cross section of the multi-layered sample.

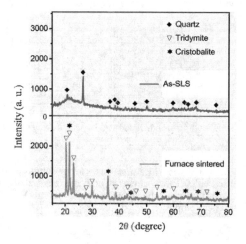

Figure 9. X-ray diffraction analysis.

Thirty layers of silica cubic sample (Fig. 8(a)) were successfully produced by SLS and subsequent furnace sintering. Non-sintered or green particles at the layer interfaces were not observed (Fig. 8 (b)) in the cross-section of the sample. The macro pores and cracks were also not detected. The linear and volume shrinkage: $\Delta L/L$, $\Delta W/W$, $\Delta H/H$ and $\Delta V/V$, were measured and the results were 0.54%, 0.47%, 1.7% and 0.43%, respectively in length, width, height and volume. The relative higher shrinkage in height is due to the loser layer packing in the build direction. A good bonding and extremely low shrinkage was obtained due to direct SLS without use of polymer binder.

XRD analysis of SLS parts (Fig. 9) shows that the structures are well crystallized into quartz phase after laser sintering. The transformation of quartz into cristobalite and tridymide after a furnace sintering post-treatment have also been investigated using X-Ray Diffraction (XRD). Moreover, there is no carbon phase or any composite phase detected as a consequence of carbon addition.

4 CONCLUSIONS

This work highlights the influence of carbon additive on the SLS of silica. The greatest benefit from carbon additive is to improve absorptivity of fiber laser to silica material. Moreover, no undesirable phase was introduced during the laser and post-furnace sintering. The multi-layered well-bonded 3D part with very low shrinkage was successfully fabricated without visible cracks. Moreover, our approach is also suitable for other SLS processing of reflective oxide ceramics. This approach makes direct selective laser sintering of oxide ceramics possible. The detailed study of mechanical properties on the laser-sintered ceramic composts remains to be done in future.

ACKNOWLEDGMENT

This work was financially supported by the overseas exchange of NUS-HIT, NUS Strategic Research Programme fund and Morgan Advanced Materials Pt Ltd, Singapore.

REFERENCES

Bertrand, P., Bayle, F., Combe, C., Gœuriot, P. and Smurov, I., Appl Surf Sci. 254, 989–992 (2007).
Duley, W., Springer Science & Business Media. (2012).
Gahler, A., Heinrich, J.G. and Guenster, J., J. Am. Ceram. Soc. 89, 3076–3080 (2006).
Juste, E., Petit, F., Lardot, V. and Cambier, F., J Mater Res. 29, 2086–2094 (2014).
Ko, S.H., Pan, H., Grigoropoulos, C.P., Luscombe, C.K., Fréchet, J.M. and Poulikakos, D., Nanotechnology. 18, 345 (2007).
Kruth, J.P., Leu, M.C. and Nakagawa, T., CIRP Ann. Manuf. Technol. 47, 525–540 (1998).
Kruth, J.P., Levy, G., Klocke, F. and Childs, T.H.C., CIRP Ann. Manuf. Technol 56, 730–759 (2007).
Kruth, J.P., Vandenbroucke, B., Vaerenbergh, V.J. and Naert, I., Proc. AFPR. (2005).
Qian, B. and Shen, Z., J. Asian Ceram. Soc. 1, 315–321 (2013).
Shishkovsky, I., Yadroitsev, I., Bertrand, P. and Smurov, I., Appl Surf Sci. 254, 966–970 (2007).
Tang, H.H., Chiu, M.L. and Yen, H.C., J. Eur. Ceram. Soc. 31, 1383–1388 (2011).
Tolochko, N.K., Khlopkov, Y.V., Mozzharov, S.E., Ignatiev, M.B., Laoui, T. and Titov, V.I., Rapid Proto J. 6, 155–161 (2000).
Wu, H., Li, D., Tang, Y., Sun, B. and Xu, D., J. Mater. Process. Technol. 209, 5886–5891 (2009).

Advances in Energy Science and Equipment Engineering II – Zhou, Patty & Chen (Eds)
© *2017 Taylor & Francis Group, London, ISBN 978-1-138-71798-5*

Structural and electrical properties of the novel liquid PMMA oligomer electrolyte

Famiza Abdul Latif, Norashima Kamaluddin, Ruhani Ibrahim,
Sharil Fadli Mohamad Zamri & Nabilah Akemal Muhd Zailani
Faculty of Applied Sciences, Universiti Teknologi MARA, Shah Alam, Selangor, Malaysia
Core Frontier Materials and Industry Application, Research Management Institute,
Universiti Teknologi MARA, Shah Alam, Selangor, Malaysia

ABSTRACT: In order to indulge the advantages of polymer as electrolyte, a new liquid-type polymer electrolyte was synthesized and investigated using liquid PMMA oligomer as the host and 1-ethyl-3-methylimidazolium bis(trifluoromethylsulfonyl)imide ([EMIM]Tf$_2$N) ionic liquid as the doping agent. This new doped liquid poly(methyl methacrylate) (PMMA) oligomer exhibited the highest ionic conductivity of 5.679×10^{-3} Scm^{-1} at room temperature when 30 wt.% of [EMIM]Tf$_2$N was added into the system. The complexation that occurred between the polymer and the doping material were further investigated using Fourier Transform Infrared Spectroscopy (FTIR).

1 INTRODUCTION

The need of having smaller batteries with high energy and power density are becoming crucial due to the advancement in the communications and information technology such as smart phones and Tablet PCs. Polymeric materials have became the most extensively studied electrolyte host due to its flexible characteristic that make them possible to be fabricated into thin films and exhibit wide range of electrical properties. However, film type polymer electrolytes exhibit poor adhesion properties that cause additional resistance layer, i.e. the air gap, for the ionic conduction. On the other hand, gel type polymer electrolytes are mechanically un-stable.

Therefore, in this study, a new type of liquid polymer electrolyte base on PMMA oligomer has been synthesized and investigated. To the best of our knowledge, this liquid type polymer electrolyte has not been extensively explored. PMMA is chosen because of its unique properties that showed stability towards electrode (Appetecchi, 1995). In addition, it is a non-explosive and nonflammable material and when it leaks it will not harm the users like the common aqueous based electrolytes do. Since this liquid PMMA is not commercially available, it needs to be synthesized. Therefore, this liquid PMMA electrolyte was prepared by free radical polymerization technique because it is the easiest method that requires no stringent conditions.

In recent years, the addition of ionic liquid to polymer electrolytes has been adopted as a means

to improve the conductivity of a polymer electrolyte system (Hu, 2007). Room Temperature Ionic Liquids (RTILs) that consist of bulky cations, are considered to be suitable electrolyte salts for polymer-in-salt system due to their excellent properties such as good thermal and electrochemical stability, high ionic conductivity, negligible vapor pressure and non-flammability (Tokuda, 2004; Noda, 2001). Ionic liquids also meet the requirements of plasticizing salts and offer improved thermal and mechanical properties to flexible polymers. To date many different forms of ionic liquid incorporated polymer electrolyte have been reported (Saroj, 2013; Cheng, 2007; Mohammad, 2013; Khairul, 2010). Therefore, in this work, this liquid PMMA oligomer was doped with 1-ethyl-3-methylimidazolium bis(trifluoromethylsufonyl)imide [EMIM]Tf$_2$N ionic liquid. The complexation between the polymer and this IL and its effect on the electrical properties of this new liquid-type polymer electrolyte were investigated and presented in this paper.

2 EXPERIMENT

2.1 Synthesis of liquid PMMA oligomer electrolyte

5 mL of MMA monomer (ACROS) (Mw = 100.12 g/ mole) was added with 10% of benzoyl peroxide (MERCK). The test tubes were then soaked in a hot water bath for a few minutes to accelerate the polymerization process. The polymerization of PMMA was then allowed to continue at room

temperature for another 24 hours. After 24 hours, various weight percent of [EMIM]Tf$_2$N (SIGMA) IL were added as doping material.

2.2 Fourier transform infrared spectroscopy analysis

The polymer-ion complexation was determined by PERKIN ELMER Fourier transform infra red spectrophotometer Spectrum One in the frequency range of 4000–600 cm^{-1} at a resolution of 16 cm^{-1}.

2.3 Impedance spectroscopy

To measure the impedance of this liquid PMMA oligomer electrolyte, a Hioki 3532–50 LCR HiTester was used to perform the impedance (Z) measurement for each sample over the frequency range of 100 Hz–1 MHz. From the Cole-Cole plots obtained, the bulk resistance, R_b of each sample was determined and hence the conductivity (σ) of the samples were then calculated using Eq. (1) (Nik, 2005).

$$\sigma = l/R_b A \tag{1}$$

L is the distance of the electrodes (cm), A is the effective contact area of the electrodes and the electrolyte (cm^2) and R$_b$ is the bulk resistance (Ω) of the samples.

3 RESULTS AND DISCUSSION

3.1 Characterization of liquid PMMA oligomer electrolyte

3.1.1 FT-IR analysis

Figure 1 exhibits the FTIR spectra for liquid-based PMMA oligomer doped with [EMIM]Tf$_2$N ionic liquid. It was observed that carbonyl (C = O) group of PMMA at ~1700 cm^{-1} has been shifted to higher wavenumber with reduced intensity indicating that interaction has occurred between the C = O group of PMMA and the imidazolium cation [EMIM]$^+$ of the IL. The presence of this [EMIM]$^+$ peak were detected between 900–700 cm^{-1}. Peaks at ~1353 cm^{-1} and ~1060 cm^{-1} were assigned to the assymetric S-N-S and C-SO$_2$-N stretching modes of IL and become more pronounced upon increasing the concentration of IL indicating an increased in the number on [Tf$_2$N]$^-$ anions as the amount of IL increased. The appearance of these characteristic peaks of IL confirmed the infusion of IL in the liquid PMMA oligomer.

3.1.2 Electrical studies

Figure 2 shows the complex impedance plot for the highest ionic conducting liquid PMMA

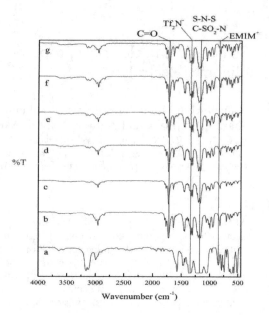

Figure 1. FTIR spectra for (a) pure [EMIM]Tf$_2$N ionic liquid and liquid PMMA oligomer doped with (b) 5 wt.%, (c) 10 wt.%, (d) 15 wt.%, (e) 20 wt.%, (f) 25 wt.% and (g) 30 wt.% of [EMIM]Tf$_2$N respectively at room temperature.

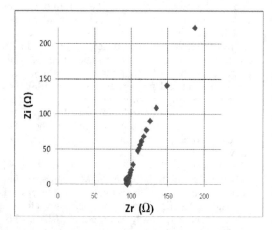

Figure 2. Complex impedance plot for the highest conducting liquid-based PMMA oligomer/30 wt.% [EMIM] Tf$_2$N polymer electrolyte at room temperature.

oligomer containing 30 wt.% of [EMIM]Tf$_2$N at room temperature. It consists of a spike which corresponds to the interfacial impedance of the electrolyte.

With the addition of [EMIM]Tf$_2$N ionic liquid into the liquid PMMA oligomer, high ionic conductivity of ~10^{-3} Scm^{-1} was achieved. It was found that the ionic conductivity increased as the weight percentage of [EMIM]Tf$_2$N ionic liquid increased (Table 1). This was due to an increased in the number of charge present in the liquid oligomer system which has been confirmed from the FTIR analyses. This new liquid type polymer electrolyte has shown comparable or better ionic conductivity when compared to the film or gel type PMMA-based electrolyte systems that have been previously studied (Rajendran, 2004; Xie, 2012).

This 30% [EMIM]Tf$_2$N doped system also exhibited the highest value of dielectric constant (Fig. 3) and dielectric loss (Fig. 4) indicating that it contain the highest number of charge as compared to other liquid electrolyte system. The sharp rose of the plots at low-frequency region demonstrating the occurrence of space charge polarization at the electrode-electrolyte interface (Ramesh, 2011; Govindara, 1995) hence confirming non-Debye dependence.

Table 1.   Average ionic conductivity of doped [EMIM] Tf$_2$N liquid PMMA oligomer at room temperature.

wt.% [EMIM]Tf$_2$N	Conductivity ($\times 10^{-3}$) (Scm^{-1})
5	0.511
10	1.46
15	2.323
20	2.69
25	4.543
30	5.679

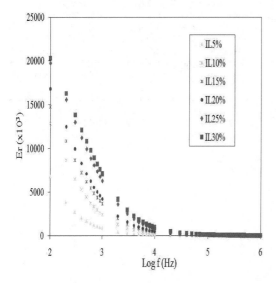

Figure 3.   Dielectric constant vs. frequency for all doped [EMIM]Tf$_2$ N liquid PMMA oligomers at room temperature.

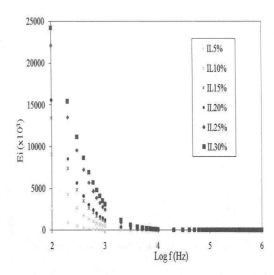

Figure 4.   Dielectric loss vs. frequency for all doped [EMIM]Tf$_2$N liquid PMMA oligomers at room temperature.

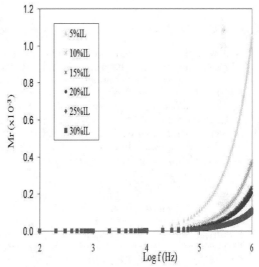

Figure 5.   Real part of electric modulus vs. frequency for all doped [EMIM]Tf$_2$ N liquid PMMA oligomers at room temperature.

The variations of electric modulus for all doped [EMIM]Tf$_2$N liquid PMMA oligomers at room temperature are shown in Figures 5 and 6. The plots exhibit a long tail at lower frequencies that is associated with long-range ionic motion (Majid, 2007), in which ions can successfully hop from one site to a neighbouring site. This finding further confirms the non-Debye behavior in the samples.

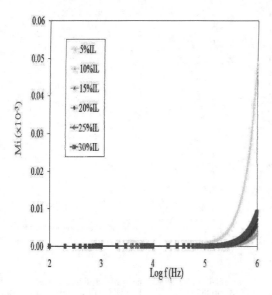

Figure 6. Imaginary part of electric modulus vs. frequency for all doped [EMIM]Tf₂N liquid PMMA oligomers at room temperature.

4 CONCLUSION

A new liquid type polymer electrolyte based on PMMA oligomer has been successfully synthesized. The highest ionic conductivity achieved was 5.679×10^{-3} Scm^{-1} when only 30 wt.% of [EMIM] Tf₂N was added which is comparable with other PMMA-type electrolytes that have been previously studied. Therefore, it can be concluded that this liquid polymer electrolyte is a potential material in electrochemical applications and can be further modified and explored.

ACKNOWLEDGEMENT

Financial support and technical support from Malaysia Toray Science Foundation (MTSF) (Ref: CRO-CA/13.206 [File Ref: 13/G191]), The Ministry of Higher Education (FRGS grant) and UiTM are highly acknowledged.

REFERENCES

Appetecchi, G.B., F. Croce, B. Scrosati, Kinetics and stability of the Lithium Electrode in Poly (methyl metacrylate)- Based Gel Electrolytes, *Electrochimia Acta*, **40(8)**, 991–997 (1995).

Cheng, H., C. Zhu, B. Huang, M. Lu, Y. Yang, *Electrochimica Acta*, **52**, 5789–5794 (2007).

Govindaraj, G., N. Baskaran, K. Shahi, P. Monoravi, *Solid State Ion*, **76**, 47–55 (1995).

Hu, C., Z. Changbao, H. Bin, L. Mi, Y. Yong, Synthesis and electrochemical characterization of PEO-based polymer electrolytes with room temperature ionic liquids, *Electrochimica Acta*, **52**, 5789–5794 (2007).

Jaipal Reddy M, Sreekanth T, Subba Rao UV, *Solid State Ion*, **126**, 55–63 (1999)

Khairul Anwar, N., R.H.Y. Subban, N.S. Mohamed, *Materials*, **5**, 2609–2620 (2010).

Majid, S.R. & A.K. Arof, *Phys B Condens Matter*, **390**, 209–215 (2007).

Mohammad, S.F., N. Zainal, S. Ibrahim, N.S. Mohamed, *Int. J. of Electrochem. Sci.*, **8**, 6145–6153 (2013).

Nik Aziz, N.A., N.K. Idris, M.I.N. Isa, Proton conducting polymer electrolytes of methylcellulose doped ammonium fluoride: Conductivity and ionic transport studies. *Journal of the Physical Sciences*, **5(6)**, 748–753 (2010).

Noda, A., K. Hayamizu, M. Watanabe, Pulsed-gradient spin-echo 1H and 19F NMR ionic diffusion coefficient, viscosity, and ionic conductivity of non-chloroaluminate room-temperature ionic liquids. *J. Phys. Chem. B*, **105(20)**, 4603–4610 (2001).

Rajendran, S., M. Sivakumar, R. Subadevi, Investigations on the effect of various plasticizers in PVA-PMMA solid polymer blend electrolytes. *Materials Letters*, **58**, 641–649 (2004).

Ramesh, S., C.W. Liew, K. Ramesh. *Journal of Non-Crystalline Solids*, **357(10)**, 2132–2138 (2011).

Saroj, A.L., R.K. Singh, S. Chandra, *Materials Science and Engineering*, **B 178**, 231–238 (2013).

Tokuda, H., K. Hayamizu, K. Ishii, H.S. Md Abu, M. Watanabe, Physicochemical properties and structures of room temperature ionic liquids. 1. variation of anionic species. *J. Phys. Chem., B* **108(42)**, 16593–16600 (2004).

Xie, Z.L., H.B. Xu, A. Gebner, M.U. Kumke, M. Priebe, K.M. Fromm, A. Taubert. *Journals of Materials Chemistry*, **22(16)**, 8110–8116 (2012).

Advances in Energy Science and Equipment Engineering II – Zhou, Patty & Chen (Eds)
© 2017 Taylor & Francis Group, London, ISBN 978-1-138-71798-5

Synthesis, x-ray crystal structure and photoluminescence properties of niobium complex: Sol-gel preparation nanoparticles of Nb_2O_5

Mahsa Armaghan & Mohsen Amini
Department of Chemistry, Science and Research Branch, Islamic Azad University, Tehran, Iran

Mostafa M. Amini
Department of Chemistry, Shahid Beheshti University, G.C., Tehran, Iran

ABSTRACT: Stoichiometric reaction of benzo[h]quinolin-10-ol with Niobium(V) ethoxide in toluene at room temperature resulted in formation of [Nb4-μ-(OC2H5)2-μ-(O)4(C13H9NO)2(OC2H5)6] (1). Complex 1 was characterized by IR, 1H, 13CNMR and fluorescence spectroscopes. The molecular structure of 1 was determined by single-crystal X-ray diffraction. The coordination geometries around the metal ions are distorted octahedral, which share an edge through bridged alkoxy and oxygen groups. Each benzo[h]quinolin-10-ol ligand is also chelated to one metal ion through its pyridine nitrogen and phenolate oxygen atoms. Complex 1 show fluorescence with an emission centered at 481 nm. Nano-crystalline Nb2O5 was synthesized via sol–gel processing of [Nb4-μ-(OC2H5)2-μ-(O)4(C13H9NO)2(OC2H5)6]. TGA-DTA and X-ray powder diffraction confirmed the formation of Nano—crystalline Nb2O5 which was compared with nanosize of Nb2O5 from hydrolysis of Niobium(V) ethoxide.

1 INTRODUCTION

The world market for nanotechnologies and nano-materials is rapidly growing. Advances in nanotechnology are expected to lead the vast growth of demand for the novel materials. Nowadays Nano-crystalline metal oxide materials have been studied because possessing special electrical, electro-optic, electromechanical, magnetic, and other properties are of key importance in various applications (Surnev, 2013; Drobot, 2007).

New homometallic polyoxoalkoxides with auxiliary ligands have generated considerable interest as precursors for preparation of nano-crystalline metal oxides by sol-gel processing. Because of the growing importance of nano—crystalline niobium oxide materials (physical properties and acidic catalysts), research on niobium(V) polyoxoalkoxides with chelating ligands that could be used as single-source molecular precursors for the preparation of Nb oxide is therefore welcome (Nowak, 1999).

In this paper, we report the synthesis and characterization of novel moisture sensitive homotetrametallic complexe of Nb^V with simple chelating ligand, benzo[h]quinolin-10-ol. The formation of Nb_2O_5 nanoparticles from sol-gel processing (hydrolysis,-condensation) and thermal decomposition of the title complex and niobium(V) ethoxide is also investigated (Mohammadnezhad, 2010; Bayot, 2005; Amini, 2008).

2 EXPERIMENTAL SECTION

2.1 *Synthesis of the precursors*

All manipulations were carried out under nitrogen using standard Schlenk techniques. Solvents were dried and distilled under nitrogen before use. Niobium(V) chloride, ethanol, benzo[h]quinolin-10-ol were purchased from Merck and used without further purification. Niobium(V) ethoxide was prepared according to reported procedures (Mirzaee, 2005). Complex **1** was prepared by mixing benzo[h]quinolin-10-ol (0.19 g, 1.0 mmol) with $Nb(OCH_2CH_3)_5$ (0.318 g, 1.0 mmol) in toluene (5 mL). The mixture was stirred for a day and the solvent was removed under reduced pressure to furnish a yellow solid. The solid was recrystallized from a mixture of 1,2-dichloromethane and *n*-hexane. Suitable crystals of the complex for single-crystal structure determination were isolated from the solution after some days at −5°C, mp $300 \geq$ °C. Red solid (0.37 g, 52%), Anal. Calcd for $C_{48}H_{70}Cl_4N_2Nb_4O_{16}$: C, 39.91; H, 4.88; N, 1.94; found: C, 39.80; H, 4.41; N, 1.88%. IR (KBr) (v_{max}/cm⁻¹): 3458, 1634, 1228, 1094, 806, 612. ¹H NMR (300.13 MHz, $CDCl_3$) δH: 0.75 (t, 4H, −CH₃), 0.88 (t, 18H, −CH₃), 1.50 (t, 4H, −CH₃), 3.72 (q, 12H, −OCH₂-), 4.42 (q, 4H, −OCH₂-), 4.70 (q, 4H, −OCH₂-), ligand protons 7.20–8.32 (m, 8H, H-Ar).¹³C {¹H} NMR (75.47 MHz, $CDCl_3$) δC:

11.4, 14.1, 17.7, 18.1, 20.6, 22.6, 29.0, 31.5, 34+.5, 34.6, 53.4, 58.4, 67.1, 69.3, 117.0, 117.8, 119.3, 120.3, 120.8, 124.5, 127.5, 129.6, 129.8, 129.9, 130.4, 135.9, 138.2, 145.0, 148.8, 159.4, 161.3.

2.2 Hydrolysis of the precursors

Stable sols and gels were obtained by stoichiometric hydrolysis of the prepared complex and niobium(V) ethoxide precursors. Hydrolysis is performed by adding water diluted in ethanol and toluene under nitrogen atmosphere and vigorous stirring. Precipitation occurs readily when the hydrolysis ratio $h = H_2O/Nb$ is 5 (h = 5). The gels were dried under reduced pressure till powders were obtained. In order to obtain pure phase crystallization temperature, TGA-DTA analysis of powders was studied. Then thermal decomposition of powders at appropirate tempreture was performed to prepare nano-crystalline Nb_2O_5 Nano-crystalline Nb_2O_5 was characterized by X-ray powder diffraction.

3 RESULT AND DISCUSSION

Compound **1** was prepared by stoichiometric reaction of benzo[h]quinolin-10-ol with niobium(V) ethoxide in toluene under inert atmosphere using standard Schlenk techniques. After stirring the reaction mixture for a day, all volatiles were removed under reduced pressure to furnish yellowish solids. The solid was crystallized from a mixture of 1,2-dichloroethane and *n*-hexane and crystals were obtained after some days at −5°C.

3.1. X-ray crystal structures

3.1.1 $[Nb_4\text{-}\mu\text{-}(OC_2H_5)_2\text{-}\mu\text{-}(O)_4(C_{13}H_9NO)_2 (OC_2H_5)_6]$

Crystallographic data shows that **1** is centro-symmetric complex. The molecular structure of complex **1** is shown in Figure 1 and crystallographic data is listed in Table 1. It can be seen that the geometries at the niobium atoms in complex is

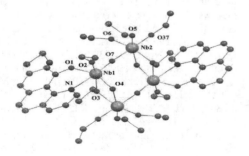

Figure 1.   The molecular structure of **1**.

Table 1.   Crystal data and refinement details for **1**.

Complex	1
Empirical formula	$C_{48} H_{70} Cl_4 N_2 Nb_4 O_{16}$
Formula weight	1444.50
Crystal size (mm)	$0.37 \times 0.25 \times 0.22$
Crystal color	Colorless/Block
Temperature (K)	120(2)
Wavelength (Å)	0.71073
Crystal system	Monoclinic
Space group	*P21/n*
a (Å)	11.2137(4)
b (Å)	16.4161(4)
c (Å)	15.9688(5)
α (°)	90.00
β (°)	93.774(3)
γ (°)	90.00
v (Å³)	2933.25(16)
Z	2
Absorption coefficient (mm⁻¹)	1.008
Calculated density (mg/m³)	1.635
F(000)	1464
θ Range for data	2.15–29.15
Reflections collected	21200
Unique reflections [R(int)]	0.0476
Data/restrains/parameters	7866 /0/339
Goodness–of–fit on F2	1.055
Final R indices [I > 2σ(I)]	R1 = 0.0451
	wR2 = 0.0882

distorted octahedral. Four distorted octahedral niobium atoms share on four edge through bridging ethoxy and oxygen groups. The two distorted octahedral Nb connect to one chelating ligand, a terminal ethoxy, and bridge oxygens and ethoxy groups and other two distorted octahedral Nb connected to one terminal and bridging ethoxy group and bridging oxygen groups. In addition, each benzo[h] quinolin-10-ol ligand chelated to the one Nb atom through its pyridine nitrogen and phenolate oxygen atoms. In this structure two bridging oxygen groups are *trans* to the pyridine nitrogen, and the other two ones are *trans* to the bridged alkoxy oxygen atoms.

3.2 Spectroscopic studies

3.2.1 Fluorescence properties of prepared complex

Fluorescence spectra of the benzo[h]quinolin-10 -ol ligand and the complex **1** in a metanol solution are given in Figure 2. The prepared complex have broader fluorescence emission at 782 nm compared to the ligand that centered at 481 nm. There is a 301 nm blue shift after coordination of ligand to the niobium atom and formation of complex. The blue shift of **1** with respect to ligand can be

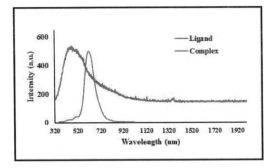

Figure 2. Fluorescence spectra of ligand and complex 1 in a methanol solution at room temperature.

attributed to the presence of electron withdrawing alkoxy groups. Notably, spectrofluorometric results showed the fluorescence of the Nb-complex is due to the chelate complex formation. The spectra of the ligand and complex have similar shapes therefore, the emission of complex originated from $\pi^* \rightarrow \pi$ transition (Frevel, 1995).

3.3 Preparation of Niobim (V) oxide from

3.3.1 Complex 1 and niobium ethoxide thermal behavior of niobium pentoxide xerogels

Thermal analysis curve of gel obtained from sol-gel process of complex **1** is shown in Figure 3. The gel obtained after drying, contained organic groups chemically bonded to the oxide backbone. More precisely, as shown from IR measurement, chelating organic ligand and ethoxide groups are present in the gel (IR (KBr) (v_{max}/cm^{-1}): 3362, 2959, 1629, 1587, 1466, 1435, 1354, 1330, 1260, 1096, 1032, 802, 711, 806, 600). No significant weight loss is observed below 200°C, whereas two weight loss (26% and 11%) can be seen in the TGA curve between 200 and 700°C correspond to the exothermic peak in DTA curve, suggesting that organic species (organic ligand and ethoxide groups) are removed and burnt. Therefore two precursors, niobium(V) ethoxide and complex **1**, were hydrolyzed and then calcined at 700°C for 3 hours to remove all organic residue. The white powder was prepared at 700°C from niobium(V) ethoxide and complex **1** Nb$_2$O$_5$ 1, Nb$_2$O$_5$ 2 respectively which were stored for X-ray powder diffraction analysis.

3.3.2 X-Ray powder Diffraction (XRD)

X-ray diffraction experiments was performed on xerogels after calcination at 700°C for 3 hours under a flow of air. According to the X-ray diffraction patterns shown in Figure 4, well crystallized pure phase Nb$_2$O$_5$ was developed (Nb$_2$O$_5$ 1 and Nb$_2$O$_5$ 2 from Niobium(V) ethoxide and complex **1** respectively) in hexagonal crystal system (Nowak,

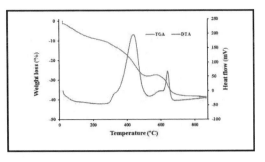

Figure 3. TGA-DTA of niobium pentoxide xerogels (ligand/Nb = 0.5, H$_2$O/Nb = 5).

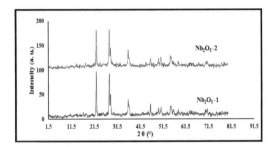

Figure 4. XRD pattern of prepared Nb$_2$O$_5$ from two precursors.

1999). The size of the Nb$_2$O$_5$ nanoparticles calculated from the XRD line broadening technique is 32 nm (Frevel, 1995).

4 CONCLUSION

A new niobium complex **1** was synthesized and characterized by spectroscopic methods. Single-crystal X-ray diffraction analysis showed that the compound has a centrosymmetric structure with four niobium-atom having a coordination number six with chelating organic ligand and terminal and bridging ethoxies groups. Complex **1** exhibited fluorescence emission at room temperature in solution. Nb$_2$O$_5$ nanoparticles were synthesized by a simple calcination of gel obtained from sol-gel hydrolysis of niobium (V) ethoxide and **1**. The size of the Nb$_2$O$_5$ nanoparticles were determined. Thus [Nb$_4$-μ-(OC$_2$H$_5$)$_2$-μ-(O)$_4$(C$_{13}$H$_9$ NO)$_2$(OC$_2$H$_5$)$_6$] may be a suitable precursor for a simple preparation of nanoscale Nb$_2$O$_5$ materials.

ACKNOWLEDGEMENTS

The authors thank professor Mostafa M. Amini (Shahid Beheshti University) to provide some chemicals and his useful comments.

REFERENCES

Amini, M.M., M. Mirzaee, F. Yaganeh, H. R. Khavasi, P. Mirzaei, and S. W. Ng, *Transition Met. Chem.* **33**,79, (2008).

Bayot, D., B. Tinant, and M. Devillers, *Inorg. Chem.*, **44**, 1554, (2005).

Drobot, D.V., P. A. Shcheglov, E. E. Nikishina, and E. N. Lebedeva, *Inorg. Mater.*, **43**, 493, (2007).

Frevel, L. K., and H. W. Rinn, *Anal. Chem.*, **27**, 1329, (1995).

Mirzaee, M., M. M. Amini, M. Sadeghi, F. Yeganeh Mousavi, and M. Sharbatdaran, *Ceramics—Silikáty*, **49**, 40, (2005).

Mohammadnezhad, G., M. M. Amini and H. R. Khavasi, *Dalton Trans.*, **39**, 10830, (2010).

Nowak, I., and M. Ziolek, *Chem. Rev.*, **99**, 3603, (1999).

Surnev, S., A. Fortunelli, and F. P. Netzer, *Chem. Rev.*, **113**, 4314, (2013).

Advances in Energy Science and Equipment Engineering II – Zhou, Patty & Chen (Eds)
© 2017 Taylor & Francis Group, London, ISBN 978-1-138-71798-5

Combined effects of fluid inertia and curved shapes in convex circular squeeze films

Jaw-Ren Lin

Department of Mechanical Engineering, Nanya Institute of Technology, Taoyuan City, Taiwan

ABSTRACT: The combined effects of fluid inertia forces and curved shapes in convex circular squeeze film characteristics are investigated in this paper. Applying the momentum integral method, a pressure gradient equation is derived. Comparing with the non-inertia case, the film pressure and the film force of the squeeze film including the influences of the fluid inertia forces and the curved shapes are presented and discussed through the variation of the density parameter and the convex shaped parameter.

1 INTRODUCTION

Squeeze film mechanisms play an important role in many areas of engineering applications, for example, the bio-lubricated joints, hydraulic dampers, cam and followers, approaching gears, automotive components, etc. Many contributors have investigated the squeeze film behaviors in different systems, for example, the finite rectangular plates by Archibald (Archibald, 1956), Wu (Wu, 1972) and Hamrocks (Hamrock, 1994), the circular disks by Wu (Wu, 1970), Lin (Lin, 1996) and Khonsari and Booser (Khonsari, 2001), the curved plates by Murti (Murti, 1975) and Gupta and Vora (Gupta, 1980), the spherical bearings by Gould (Gould, 1971) and Lin (Lin, 1996), and the cylinder-plate mechanisms by Lin and Lu (Lin, 2014). In these contributions, the Reynolds equation is derived from the momentum equation together with the continuity equation. The squeeze film pressure is solved from the Reynolds equation. The load capacity and the squeeze film time are then evaluated. However, these studies neglect the effects of fluid inertia forces on the squeeze film performances. With the development of modern engineering, the influences of fluid inertia forces may become more and more significant with increasing speeds. In a study of rectangular parallel plates, a pressure gradient equation including the effects of fluid inertia forces is derived by Pinkus and Sternlicht (Pinkus, 1961), and the squeeze film characteristics are then investigated. According the results, the influences of fluid inertia forces yield a significant effect on the load capacity and the squeeze film time in rectangular squeezing plates. However, influences of curved shapes in squeeze films are not included.

In the present study, the combined effects of fluid inertia forces and curved shapes in convex circular squeeze films are investigated. A pressure gradient equation is derived by using momentum integral method across the film height. Analytical expressions for the film pressure and the load capacity are derived. Comparing with the non-inertia case, the squeeze film characteristics taking into account the fluid inertia forces and the curved shapes are presented in terms of the variation of the density parameter and the convex shaped parameter.

2 ANALYSIS

Figure 1 presents the squeeze film configuration of convex circular surfaces with a separated film height h. The convex disk of radius r_a is approaching the lower surfaces with a squeezing velocity: $v_s = -\partial h/\partial t$.

Assume that the lubricant in the film region is an incompressible fluid and that the thin film hydrodynamic lubrication theory of Pinkus and Sternlicht (Pinkus, 1961) is applicable. Then, the momentum equations including the convective inertia terms, the continuity equations and the equation of the squeezing motion expressed in axial symmetrical coordinates (r, θ, y) can be written as

$$\rho\left(u\frac{\partial u}{\partial r} + v\frac{\partial u}{\partial y}\right) = -\frac{\partial p}{\partial r} + \eta\frac{\partial^2 u}{\partial y^2} \quad (1)$$

$$\frac{\partial p}{\partial y} = 0 \quad (2)$$

$$\frac{1}{r}\frac{\partial(ru)}{\partial r} + \frac{\partial v}{\partial y} = 0 \quad (3)$$

$$v_s = -\partial h / \partial t$$

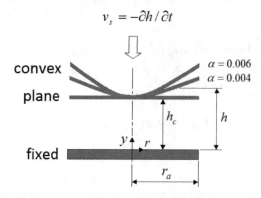

Figure 1. Squeeze film configuration of convex circular surfaces.

$$2\pi \int_{y=0}^{h} rudy = \pi r^2 v_s \qquad (4)$$

where u and v are the velocity components in the r and y directions, ρ is the fluid density, and η is the fluid viscosity. The boundary conditions for velocity components are

$$u = 0, v = 0, \text{ at } y = 0 \qquad (5)$$

$$u = 0, v = -v_s, \text{ at } y = h \qquad (6)$$

In this study, we propose a convex surface with a local film height h expressed as

$$h = h_c + \alpha \cdot \frac{r^2}{r_a} \qquad (7)$$

where h_c is the central minimum film height and α denotes the convex shaped parameter. Since the thin film theory is applicable, fluid inertia forces can be treated as constant over the film thickness. Applying the momentum integral method across the film Pinkus and Sternlicht (Pinkus, 1961), one can rewrite the momentum equation.

$$\frac{\rho}{h} \int_{y=0}^{h} \left(u \frac{\partial u}{\partial r} + v \frac{\partial u}{\partial y} \right) dy + \frac{\partial p}{\partial r} = \eta \frac{\partial^2 u}{\partial y^2} \qquad (8)$$

Utilizing the continuity equation (3) and the velocity boundary conditions (5) and (6), one can achieve the momentum integral equation.

$$\eta \frac{\partial^2 u}{\partial y^2} = \frac{\partial p}{\partial r} + \frac{\rho}{h} \left[\left(\frac{1}{r} + \frac{\partial}{\partial r} \right) \int_{y=0}^{h} u^2 dy \right] \qquad (9)$$

Define a modified pressure gradient m_p as follows.

$$m_p = \frac{\partial p}{\partial r} + \frac{\rho}{h} \left[\left(\frac{1}{r} + \frac{\partial}{\partial r} \right) \int_{y=0}^{h} u^2 dy \right] \qquad (10)$$

Then, momentum integral equation can be rewritten as

$$\frac{\partial^2 u}{\partial y^2} = \frac{1}{\eta} m_p \qquad (11)$$

Solving this equation subject to the velocity boundary conditions, one can obtain

$$u = \frac{y^2 - hy}{2\eta} \cdot m_p \qquad (12)$$

Substituting this expression into the equation for the squeezing motion and performing the integration, one can derive obtain the modified pressure gradient.

$$m_p = -\frac{6\eta r}{h^3} v_s \qquad (13)$$

Substituting the expressions of u and m_p into the definition of the modified pressure gradient, one can derive the pressure gradient equation for the convex circular squeeze films including the combined effects of fluid inertia forces and curved shapes.

$$\frac{\partial p}{\partial r} = -\frac{6\eta v_s r}{h^3} \left\{ 1 + \frac{3\rho v_s}{20\eta} \left[h - \frac{2\alpha}{3r_a} \cdot r^2 \right] \right\} \qquad (14)$$

The non-dimensional quantities are introduced as follows.

$$r^* = \frac{r}{r_a}, h_c^* = \frac{h_c}{h_{c0}}, h^* = \frac{h}{h_{c0}} \qquad (15)$$

$$p^* = \frac{p}{p_a}, \beta = \frac{h_{c0}}{r_a} \qquad (16)$$

$$v_s^* = \frac{6\eta r_a^2 v_s}{p_a h_{c0}^3}, \delta = \frac{\rho h_{c0}^4 p_a}{6\eta^2 r_a^2} \qquad (17)$$

where h_{c0} represents the initial central minimum film height, β is the initial height-to-radius ratio, p_a denotes the ambient pressure, and δ is the density parameter. One can obtain the non-dimensional pressure gradient equation.

$$\frac{\partial p^*}{\partial r^*} = -\frac{r^* v_s^*}{h^{*3}} \cdot \left\{ 1 + \frac{3\delta v_s^*}{20} \left[h^* - \frac{2\alpha}{3\beta} \cdot r^{*2} \right] \right\} \qquad (18)$$

Integrating the non-dimensional pressure gradient equation with the boundary conditions: $p^* = 1$ at $r^* = 1$, one can obtain the film pressure.

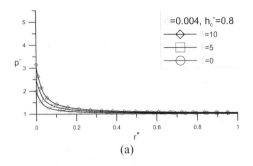

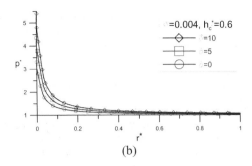

(a)   (b)

Figure 2.   Film pressure $p*$ versus the radial coordinate $r*$ for different h_c^* and δ under the shape parameter $\alpha = 0.004$.

$$p^* = 1 + v_s^* \int_{r^*}^1 \frac{r^* dr^*}{h^{*3}} + \frac{3\delta v_s^{*2}}{2\beta} \int_{r^*}^1 \frac{r^*(3\beta h^* - 2\alpha r^{*2}) dr^*}{30 h^{*3}}$$

(19)

Integrating the film pressure over the film region, one can obtain the film force.

$$f = 2\pi \int_{r=0}^{r_a} (p - p_a) r dr \qquad (20)$$

Using the non-dimensional expression, the non-dimensional film force is obtained after performing the integration.

$$f^* = \frac{32 f}{p_a \pi r_a^2} = f_0^* v_s^* + f_1^* \delta v_s^{*2} \qquad (21)$$

$$f_0^* = 64 \cdot \int_{r^*=0}^1 \left(\int_{r^*}^1 \frac{r^*}{h^{*3}} dr^* \right) \cdot r^* dr^* \qquad (22)$$

$$f_1^* = \frac{16}{5\beta} \cdot \int_{r^*=0}^1 \left[\int_{r^*}^1 \frac{r^*(3\beta h^* - 2\alpha r^{*2}) dr^*}{h^{*3}} \right] r^* dr^* \qquad (23)$$

Taking the initial height-to-radius ratio $\beta = 0.005$ and the reference value $v_s^* = 1$ as Lin (Lin, 2012), the non-dimensional film force can be evaluated.

3   RESULTS AND DISCUSSION

According to the above analysis, the combined effects of fluid inertia forces and curved shapes in convex circular squeeze film characteristics are dominated by two parameters: the density parameter δ characterizing the inertia force effects, and the shaped parameter α characterizing the convex shapes. When the value of the density parameter $\delta = 0$, the result is the case of the squeeze film neglecting the fluid inertia forces.

Figure 2 shows the film pressure $p*$ versus the radial coordinate $r*$ for different h_c^* and δ under

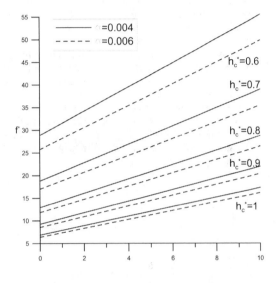

Figure 3.   Film force $f*$ versus the density number δ for different h_c^* under the shape parameters $\alpha = 0.006$ and $\alpha = 0.004$.

the shape parameter $\alpha = 0.004$. Comparing with the non-inertia case ($\delta = 0$) at the film height $h_c^* = 0.8$, the influences of fluid inertia forces ($\delta = 5, 10$) result in a higher film pressure. Decreasing the film height down to a lower value $h_c^* = 0.6$, the more the influences of density parameter are observed to increase the central film pressure.

Figure 3 describes the film force $f*$ versus the density parameter δ for different h_c^* under the shape parameters $\alpha = 0.006$ and $\alpha = 0.004$. Since the film force is evaluated by integrating the film pressure over the film region, similar tendencies are obtained from the combined influences of the fluid inertia forces and the shapes of convex surfaces. For the convex surface under the shape parameter $\alpha = 0.006$ at the film height $h_c^* = 1$, the film force $f*$ is observed to increase with increasing values of the density parameter δ. Decreasing the values of the film height ($h_c^* = 0.9, 0.8, 0.7, 0.6$), further effects of the fluid inertia forces on the film force are obtained. It is also observed that decreasing the value of the shape parameter ($\alpha = 0.004$) the convex circular squeeze film results in a higher value of the film force. On the whole, the combined effects of fluid inertia forces and convex curved surfaces on the film force are more pronounced for the squeeze film operating at a smaller shape parameter, a larger density parameter, and a lower film height.

4 CONCLUSIONS

According to the momentum integral method, a pressure gradient equation has been derived for the study of fluid inertia force effects on the convex curved circular squeeze films.

It is found that both of the fluid inertia forces and the shapes of convex surfaces show significant influence on the circular squeeze film characteristics. On the whole, the combined effects of fluid inertia forces and convex curved surfaces on the increased film force are more pronounced for the squeeze film operating at a smaller shape parameter, a larger density parameter, and a lower film height.

ACKNOWLEDGEMENTS

This study is supported from the extension of the support by the Ministry of Science and Technology of Republic of China: MOST 105-2221-E-253-001.

REFERENCES

Archibald, F.R. Trans. ASME **78**, A231 (1956).
Gould, P., ASME J. Lubr. Techno. **93**, 207 (1971).
Gupta, J.L., K.H. Vora, ASME J. Lubr. Techno. **102**, 48 (1980).
Hamrock, B.J. *Fundamentals of fluid film lubrication* (McGraw-Hill, New York, 1994).
Khonsari, M.M., E.R. Booser, *Applied tribology—bearing design and lubrication* (John Wiley & Sons, Inc., New York, 2001).
Lin, J.R., Int. J. Mech. Sci. **39**, 373 (1996).
Lin, J.R., R.F. Lu, Indu. Lubr. Tribol. **66**, 505 (2014).
Lin, J.R., STLE Tribol. Trans. **39**, 769 (1996).
Lin, J.R., Tribol. Int. **49**, 110 (2012).
Murti, P.R.K. ASME J. Lubr. Techno., **97**, 650 (1975).
Pinkus, O., B. Sternlicht, *Theory of hydrodynamic lubrication* (McGraw-Hill, New York, 1961).
Wu, H. ASME J. Lubr. Techno. **92**, 593 (1970).
Wu, H. ASME J. Lubr. Techno. **94**, 64 (1972).

Processing and tribological properties of aluminum/graphite self-lubricating materials

Xicong Ye, Binbin Wu, Yang Li & Haihua Wu
Hubei Key Laboratory of Hydroelectric Machinery Design and Maintenance, China Three Gorges University, Yichang, China

ABSTRACT: The framework of graphite was prepared by Selective Laser Sintering (SLS). Aluminum/ graphite self-lubricating materials were prepared by the casting process. To develop the wettability of the composites, the impacts of different technological parameters were investigated. The results show that when the wetting time is 40 s and the preheating temperature is 200°C, the composites have the best wettability. The composites were fabricated under these operating parameters. The tribological properties of the composites were investigated on dry wear test conditions. The Coefficient of Friction (COF) of aluminum/graphite materials ranged from 0.42 to 0.50, which is about 28% lower than the base material, thus greatly improving the self-lubricating property of the matrix material.

1 INTRODUCTION

As a commonly used solid self-lubricating material, aluminum/graphite material combines the advantages of both the metal and graphite. Its salient properties, such as self-lubricating property, high heat resistance, favorable electrical and thermal conductivity, and the specific modulus, specific strength and heat resistance exceed those of the base metal. At present, self-lubricating materials have been successfully prepared by introducing base metal, such as aluminum matrix, and Ti-based, Mg-based and Cu-based materials (Wang, 2012). Aluminum is extensively used due to its light weight, higher heat treatability, cheap price, non-polluting property and flexible preparation process (Fan, 2012). There are many methods to fabricate the aluminum/graphite self-lubricating materials. Casting process (Akhlaghi, 2009) and powder metallurgy method (Liu, 2008) are more mature. The casting method is the most suitable method for large-scale industrial production of self-lubricating materials. In this paper, the framework of graphite was prepared by Selective Laser Sintering (SLS), using casting process to fabricate the aluminum/graphite self-lubricating materials. To develop the wettability of the composites, the author investigated the impact of different operating parameters. The friction and wear behaviors of the composites were analyzed.

2 PROCESSING AND TREATMENT OF THE FRAMEWORK OF GRAPHITE

Different from the conventional processes of aluminum/graphite materials, this paper used Selective Laser Sintering (SLS) to prepare the spatial ordered structure of graphite, using Al-Mg alloy as the matrix alloy (as shown in Table 1). By investigating the wetting time and the preheating temperature of this experiment, the optimum technological parameters were acquired. In addition, the tribological properties of the composites were investigated.

2.1 Processing of the spatial ordered structure of graphite

Figure 1(a) shows the design of the spatial ordered structure of graphite. The design of the computer

Table 1. Contents of alloying elements.

Composition	Content (wt%)
Al	Balance
Mg	2.50
Si	0.25
Cu	0.10
Zn	0.10
Mn	0.10
Cr	0.20
Fe	0.40

solid model of graphite is based on Pro/E. The data format of three-dimensional structure modeling is converted to STL and data is sent to the SLS 3D printer. The type of SLS 3D printer is HK S500. The molding material was prepared by mixing scaly graphite powder and phenolic resin powder. The process parameters were as follows: filled distance of 0.1 mm, filled power of 13 W, and operating temperature 20–40°C. The forming process was as follows: the date of the spatial ordered structure of graphite was sliced to provide data layer by layer to the laser and optical system. When the forming process began, a thin layer of the molding material powder was deposited on the top surface of a container by a powder leveling drum. The optical system can transform the laser beam with circular shape into a long and thin beam. The laser coupled to the computer-controlled optical system was then scanned selectively over the power surface. Where the part cross-section in the database was solid, the laser was switched on and the powder under the beam was sintered to itself and to the previously sintered layer. Where the part cross-section was nonexistent, the laser was switched off. The powder was not fused and remains loose to be removed and recycled, once the part was completely formed. The length of the laser beam was altered according to the geometry of the portions to be produced under the control of computer. On accomplishing the first layer, the base is lowered and a second layer of loose powder is spread over the previous layer. The process is repeated by altering the shape of each scanning layer. A complete three-dimensional object was created, as shown in Figure 1 (b), which can be easily removed from the loose, unsintered powder surrounding the object.

2.2 Treatment of the spatial ordered structure of graphite

Because the scaly graphite powder is lipophilic and hydrophobic, its wettability is poor. The spatial ordered structure of graphite adsorbs a high content of impurity and water. Furthermore, the intensity of the spatial ordered structure of graphite is low. In order to improve the wettability and intensity, the spatial ordered structure of graphite should be preheated. According to the literature (Dong, 2008), the preheating temperature is 300°C and the preheating time is 40–50 min.

2.3 Experimental method

In this paper, the liquid alloy was prepared by the process of the high-frequency inductive heating method, and the aluminum/graphite self-lubricating materials were prepared by liquid casting. The operating parameters were: the wetting time (10 s, 20 s, 30 s, 40 s, 50 s) and the preheating temperature (100°C, 150°C, 200°C, 250°C, 300°C). The wettability of the composites was studied by varying these operating parameters. By studying the wettability of the aluminum/graphite materials in different operating parameters, the optimized parameters were obtained. The Aluminum/graphite materials are prepared in the optimized parameters. The Coefficient Of Friction (COF) of the aluminum/graphite materials was tested by JP150-W-II friction and wear tester. The test was carried out under dry wear test conditions. The operating parameters were: sliding speed 450 r/min, normal load 200 N, and the sample size 25 mm × 25 mm (as shown in Figure 2).

Figure 1. (a) Computer solid model of graphite and (b) complete three-dimensional object.

Figure 2. Prepared sample.

3 RESULTS AND DISCUSSION

3.1 Influence of wetting time

According to the experimental method, the influences on the wettability, different wetting times and preheating temperature were investigated by single factor experiments. Fixed preheating temperature (250°C) to study the influence on the wettability and different wetting times (10 s, 20 s, 30 s, 40 s, 50 s) were investigated. As Figure 3 shows, at a fixed preheating temperature, the wettability decreases with an increase in wetting time. When the wetting time was 40 s, the aluminum/graphite materials had the best wettability. During the processing of aluminum/graphite materials, an interfacial reaction occurs, and the interfacial reaction product is Al_4C_3. With the increase in wetting time, the interface reaction will be faster. An appropriate level of interfacial reaction is good for wettability. So with the increase in wetting time (from 10 s to 40 s), the wettability of the composites became better. However, as Al_4C_3 is a brittle phase, when the wetting time exceeded 40 s, it produces a high level of Al_4C_3. The interface strength decreases. The wettability becomes poor.

3.2 Influence of preheating temperature

According to the results of Section 3.1, the optimum wetting time is 40 s. By fixing the wetting time (40 s), preheating temperature was varied (100°C, 150°C, 200°C, 250°C, 300°C) to study its influence on the wettability. The wetting angle of graphite and aluminum alloys decreased as the temperature increases. Figure 4 shows that with the increase in preheating temperature, the wettability became better. When the temperature exceeded 200°C, the change of wettability of the composites was not obvious. The results showed that the optimum preheating temperature is 200°C.

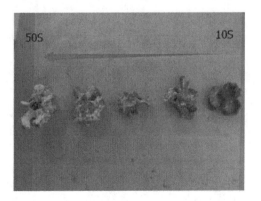

Figure 3. Interface combination status of composites at different wetting time.

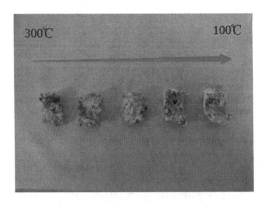

Figure 4. Interface combination status of composites at different preheating temperature.

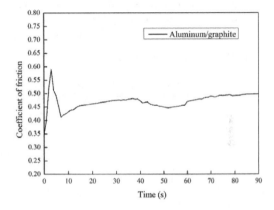

Figure 5. The friction coefficient of Aluminum/graphite materials.

3.3 Coefficient of friction

Figure 5 shows the variation of the Coefficient of Friction (COF) with time for aluminum/graphite materials at constant sliding speed of 450r/min. The results indicated that at the initial stage of friction and wear, the friction coefficient of aluminum/graphite materials increased. The Coefficient of Friction (COF) of aluminum/graphite materials was unstable and fluctuated. This is called the running-in stage. At stable wear stage, the friction coefficient of the composites remained stable. In this case, the friction coefficient of aluminum/graphite materials ranged from 0.42 to 0.50. The friction coefficient of aluminum/graphite materials decreased by 28% compared to aluminum alloys. This is because during the process of friction, the aluminum/graphite composites formed a layer of graphite, which reduced the direct contact of aluminum alloys and the friction disc. The layer of graphite reduced adhesive wear and played a good role in reducing the friction coefficient.

4 CONCLUSIONS

1. In this investigation, the design of the computer modelling of graphite was based on Pro/E. The data format of three-dimensional structure modeling was converted to STL and the data was sent to SLS 3D printer. The framework of graphite was prepared by Selective Laser Sintering (SLS).
2. The best process parameters were optimized with single factor experiments. When the wetting time was 40 s and the preheating temperature was 200°C, the aluminum/graphite composites displayed the best wettability.
3. The aluminum/graphite materials were prepared by the casting method. The aluminum/graphite materials had obvious self-lubricating property under dry wear test conditions. The coefficient of friction ranged from 0.42 to 0.50, which is about 28% lower than the base material.

ACKNOWLEDGMENTS

The authors would like to acknowledge the support of the foundation of Hubei Key Laboratory of Hydroelectric Machinery Design & Maintenance, China Three Gorges University, and the technical assistance of China Three Gorges University of Materials and Chemical Engineering.

REFERENCES

Akhlaghi F, Zare-Bidaki A. Influence of graphite content on the dry sliding and oil impregnated sliding wear behavior of Al 2024–graphite composites produced by in situ powder metallurgy method [J]. Wear, **266(1–2)**:37–45 (2009).

Dong Z.G. Study on the preparation and performance of graphite/Al composite material [D]. Northeastern University (2008).

Fan J. Development and Application of Particulate Reinforced Aluminum Matrix Composites [J]. Aerospace Materials & Technology (2012).

Liu Z.G, Tian Z M, Yao G C, et al. Development of Graphite Particles Reinforced Aluminum Matrix Self-Lubricating Composites Prepared by Casting [J]. Foundry (2008).

Wang C.C, Wang R.C, Peng C.Q, et al. Research progress of metallic solid self-lubricating composites [J]. Zhongguo Youse Jinshu Xuebao/Chinese Journal of Nonferrous Metals, **22(7)**:1945–1955 (2012).

Advances in Energy Science and Equipment Engineering II – Zhou, Patty & Chen (Eds)
© 2017 Taylor & Francis Group, London, ISBN 978-1-138-71798-5

Separation of *n*-propanol–toluene azeotropic mixture by extractive distillation

Ke Tang, Jiao Liu & Jinjuan Xing

School of Chemistry and Chemical Engineering, Bohai University, Liaoning Jinzhou, China

ABSTRACT: *n*-Propanol and toluene form a minimum azeotrope, and extractive distillation is a suitable method to separate the mixture. In this work, a systematic study of the separation of *n*-propanol–toluene mixture using an entrainer by applying the method of extractive distillation was performed. In order to break the original binary azeotrope, *o*-xylene was chosen as the entrainer. Vapor–Liquid Equilibrium (VLE) of *n*-propanol–toluene–*o*-xylene system was determined experimentally, and UNIFAC model was the best one to predict this system. The feasibility analysis of this process was validated through simulation by a chemical process simulation software and the optimal operating conditions for the extractive distillation process were found to be valuable for industrial applications. In the simulation, *n*-propanol and toluene, whose purity is more than 98.73% and 97.92%, respectively, was obtained.

1 INTRODUCTION

Binary mixtures, which are commonly used in the material, pharmaceutical and fine chemical industries, form azeotropes or close-boiling mixtures, and they are characterized as behaving like a pure component when subjected to a distillation process. It is impossible to achieve the separation of such azeotropic mixtures through conventional methods. Therefore, extractive distillation is a common method employed to separate these kinds of mixtures by using a third component, which may be a salt solution or a solvent, called the entrainer (Kotai, 2007). The function of the entrainer is to increase the relative volatility of the mixture; in this way, the components will boil separately, which allows the collection of heavier components (solvent and heavy key) at the bottom of the distillation column and the light key component at the top (Batista, 1997; Langston, 2005).

n-Propanol–toluene mixtures are commonly encountered in the new material preparation field, especially in the separation of boron isotope by chemical reaction. The *n*-propanol–toluene mixture forms a minimum boiling point azeotrope at a mole composition of 0.6083 of *n*-propanol and at a constant pressure of 101.3 kPa (Lin, 2008; Hwang, 2009).

The aim of this work is to study the Vapor–Liquid Equilibrium (VLE) of *n*-propanol–toluene–*o*-xylene system, and furthermore to establish industrial operating conditions for the extractive distillation process of *n*-propanol–toluene by using *o*-xylene as the entrainer.

2 VAPOR–LIQUID EQUILIBRIUM (VLE) OF *N*-PROPANOL–TOLUENE–*o*-XYLENE SYSTEM

The ternary vapor–liquid equilibrium (VLE) of *n*-propanol–toluene–*o*-xylene system was determined experimentally. Briefly, mixtures of *n*-propanol–toluene with *o*-xylene at 101.3 kPa were prepared at a vapor–liquid equilibrium. After the VLE was achieved (constant temperature throughout the system), samples of condensed vapor and liquid were taken for GC analysis. Each assay was made in duplicate. Figure 1 shows the

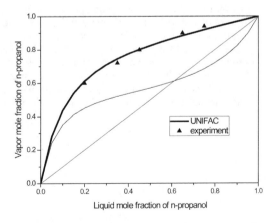

Figure 1. Pseudo-binary equilibrium curve for *n*-propanol–toluene with o-xylene at 101.3 kPa. Thick and thin lines predicted by UNIFAC model for the pseudo-binary and binary systems, respectively. ▲ experimental data.

pseudo-binary diagram obtained when a solvent's feed molar ratio (E/F) using three thermodynamic models (Wilson, NRTL, and UNIFAC) were analyzed to determine which was the best model to predict the system performance. The obtained results show that the UNIFAC model presented the highest accuracy.

3 SIMULATION OF THE PROCESS

The separation of *n*-propanol–toluene mixture was carried out with the help of an entrainer, *o*-xylene. The simulation of the process was performed with a chemical process simulation software. The UNIFAC model was used to predict the activity coefficients and ideal gas model to predict the fugacity coefficient.

The flowsheet of the extractive distillation process is presented in Figure 2. The process has two columns. One column is for extractive distillation and the other is for solvent recuperation. The azeotropic mixture and the solvent are fed into the first column, in which the components boil separately, which allows the less volatile components to be collected at the bottom and the light key component at the top. The bottom product is fed to the second column, in which the solvent is recovered and recycled to the extractive distillation column.

3.1 Sensitivity analysis

To establish the operating conditions for the extractive distillation process, a sensitivity analysis was conducted.

3.1.1 Effect of azeotropic mixture feed stage
In Figure 3, the effect of azeotrope feed stage on the molar composition of distillate in the extractive column is presented. It can be observed that as the azeotropic feed stage was closer to the reboiler, the mole fraction of *n*-propanol in distillate increased. The purity reached a maximum when feed stage was 38. This meant that the biggest separation took place in the rectifying section. In order to obtain

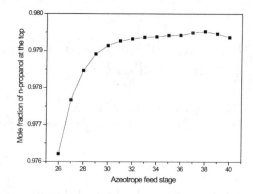

Figure 3. Effect of azeotrope feed stage in the extractive distillation column with o-xylene on distillate composition.

the *n*-propanol composition required in the distillate and decrease the reboiler energy consumption, the stage near the reboiler should be chosen as the suitable azeotrope feed stage.

3.1.2 Effect of reflux ratio
Figure 4 shows that the change of molar reflux ratio exerted a great influence on the distillate composition and reboiler energy consumption. The results showed that the purity reached a maximum with the increasing molar reflux, because as the reflux rate increased, the solvent was diluted in the column. The results also showed that the reboiler energy consumption increased sharply as the molar reflux ratio increased. The liquid in the reflux should be vaporized; therefore, the molar reflux ratio should be as low as possible to bring down the reboiler energy consumption.

3.1.3 Effect of solvent-to-feed ratio (S/F)
The solvent to feed ratio (S/F) exerts a direct effect on the distillate purity. In Figure 5, before S/F molar ratio touched a value of 2, the content of *n*-propanol kept increasing, but after reaching E/F = 2, it remained stable. This occurred because more solvent may enlarge the relative volatility of the *n*-propanol–toluene mixture; however, the purity of *n*-propanol was so high that it is not necessary to increase S/F sequentially. The increasing value of S/F increased the liquid flow in the reboiler. Although the liquid brought by the solvent remained in liquid phase in the column, it may absorb heat there. Taking into consideration the situation that high S/F ratios increased the energy consumption in reboiler, S/F should not be too high.

3.1.4 Effect of entrainer feed stage
The results shown in Figure 6 allow establishing a maximum in the distillate molar fraction for all the tested conditions when the solvent is fed on stage 27. As the solvent feed stage approached the condenser, *n*-propanol content touched its maximum

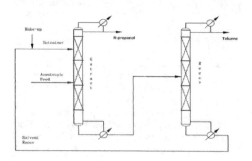

Figure 2. Extractive distillation flowsheet.

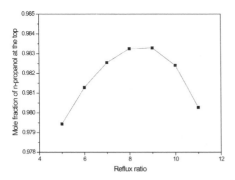

Figure 4. Effect of azeotrope feed stage in the extractive distillation column with o-xylene on distillate composition.

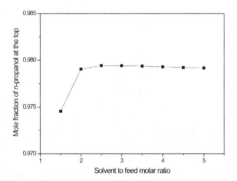

Figure 5. Effect of S/F molar ratio in the extractive distillation column with o-xylene on distillate composition.

value, after which point it decreased. The decrease occurred because the vaporization of *o*-xylene entered the column, which became part of the vapor that flowed up to the condenser and that was withdrawn as distillate.

Based on the sensitivity analysis above, the best configuration and operating conditions of the columns were established and presented in Table 1.

3.2 Simulation results

The simulation results were presented in Table 2. It was shown that the obtained *n*-propanol mole composition at the top of the extractive column was 0.9873, and the toluene mole composition at the recovery column top was 0.9792. Meanwhile, *o*-xylene, whose mole composition was 1, was obtained at the bottom of the recovery column.

In Figure 7, the temperature profile in the extractive distillation column showed a temperature change at stage 27 due to the entrainer inlet, whereas at stage 38, the temperature had a minimal decrease due to the azeotropic mixture feeding.

Figure 8 presents the liquid and vapor molar flow rate profiles in the extractive column. The liquid molar flow exposed two significant changes due

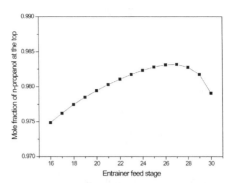

Figure 6. Effect of solvent feed stage of the distillation column with o-xylene on distillate composition.

Table 1. Optimal data for the distillation columns.

Parameter	Extractive column	Recovery column
Azeotropic feed flowrate, kmol/h	1	–
Number of stages	50	40
Solvent feed stage	27	–
Azeotrope feed stage	38	18
Solvent feed temperature, °C	25	–
Azeotrope feed temperature, °C	25	–
Mole fraction of toluene in azeotropic feed	0.6083	–
Solvent feed flowrate, kmol/h	4	–
Mole fraction of DMF in solvent feed	1	–
Reflux ratio	8	8

Table 2. Simulation results for the process flowsheet.

	n-Propanol	Toluene	SOLVREC
T/°C	97.13	106.45	150.83
mole flow kmol/h	0.6083	0.3914	5
mole fraction			
n-Propanol	0.9873	0.0205	0
toluene	0	0.9792	0
o-xylene	0.0127	0.0003	1.000

to the solvent and the azeotropic mixture feed at stages 27 and 38, respectively. Meanwhile, the vapor molar flow rate remained constant along the column, except at stage 27 where the solvent was fed. This change was caused by the liquid-phase vaporization due to the *o*-xylene inlet temperature.

In Figure 9, it was observed that the concentration of *n*-propanol kept decreasing, from the top of the column to the bottom. The *o*-xylene composition was approaching zero, from the top of the column to the bottom. The main product at the top of the column was *n*-propanol and at the bottom,

1353

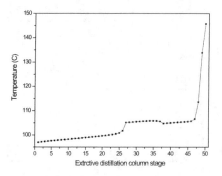

Figure 7. Extractive distillation column temperature profile.

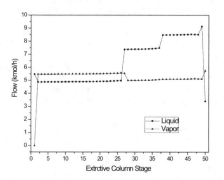

Figure 8. Liquid and vapor molar flowrate profiles in the extractive distillation column.

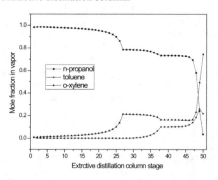

Figure 9. Vapor-phase concentration profiles of n-propanol, toluene and o-xylene in the extractive distillation column.

the main components were toluene and o-xylene. Meanwhile, in Figure 10, for the liquid molar composition profiles, n-propanol and o-xylene had an important change at stage 27, in which the n-propanol composition decreased rapidly and o-xylene increased due to the entrainer inlet. The feed of azeotropic mixture at stage 38 caused an increase in n-propanol composition, whereas a decrease in o-xylene composition was observed. The n-propanol composition decreased along the column and at the bottom, only o-xylene and toluene were present.

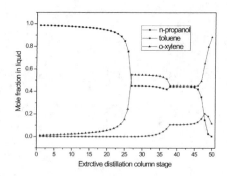

Figure 10. Liquid-phase concentration profiles of n-propanol, toluene and o-xylene in the extractive distillation column.

Toluene exposed only one significant change at stage 38 where the binary mixture was fed.

4 CONCLUSION

O-Xylene was chosen as the entrainer to achieve the separation of the azeotropic mixture of n-propanol–toluene by extractive distillation. UNIFAC model was the best model to predict Vapor–Liquid Equilibrium (VLE) of n-propanol–toluene–o-xylene system. The feasibility of the process was confirmed via simulation and experiments. Sensitivity analysis was performed to obtain the best conditions and configuration for extractive and recovery columns. The temperature, composition, and molar flow profiles obtained for the extractive distillation column were consistent. In the simulation, highly purified n-propanol and toluene were obtained.

ACKNOWLEDGMENTS

The authors gratefully acknowledge the financial support of the Doctoral Scientific Research Foundation of Liaoning Province (grant number 201501198), the Scientific Research Foundation of Liaoning Educational Committee (grant number L2015011), the Doctoral Scientific Research Foun-dation of Bohai University (grant number 0515bs036-1).

REFERENCES

Batista E. , A. Meirelles, J. Chem. Eng. Jpn. 3 (1), 45–51 (1997).

Hwang I., S. Park, J. Lee, J. Chem. Eng. Data. 54 (3), 1041–1045 (2009).

Kotai B., P. Lang, G. Modla, Chem. Eng. Sci. 62, 6816–6826 (2007).

Langston P., N. Hilal, S. Shingfield, Chem. Eng. Proc. 44, 345–351 (2005)

Lin C.N., Solvent Handbook (2008).

Advances in Energy Science and Equipment Engineering II – Zhou, Patty & Chen (Eds)
© 2017 Taylor & Francis Group, London, ISBN 978-1-138-71798-5

Synthesis and properties of the photochromic nano/micrometer copolymer brushes with spiropyran and methyl methacrylate segments

Tianxiang Zhao
Tianjin Municipal Key Lab of Fiber Modification and Functional Fiber, Tianjin Polytechnic University, Tianjin, P.R. China

Peng Xi
Tianjin Municipal Key Lab of Fiber Modification and Functional Fiber, Tianjin Polytechnic University, Tianjin, P.R. China
State Key laboratory of Polymer Physics and Chemistry, Institute of Chemistry, Chinese Academy of Sciences, Beijing, P.R. China

Dengkun Shu, Mengjiao Ma & Bowen Cheng
Tianjin Municipal Key Lab of Fiber Modification and Functional Fiber, Tianjin Polytechnic University, Tianjin, P.R. China

ABSTRACT: Photochromic nano/micrometer copolymer brushes with spiropyran and methyl methacrylate segments (SiO_2-PSPMA-*co*-PMMA) were prepared by atom transfer radical polymerization (ATRP) technology. The structure and morphology of the photochromic nano/micrometer copolymer brushes were analyzed through FT-IR, TGA, SEM and TEM. The results show that the copolymer brush contains a multilayer structure. The nano-silica particle is in the nucleus. The second part contains the spiropyran derivative (SPMA) units, which had been grafted to the surface of the nano-silica particle. The last part is made up of methyl methacrylate (MMA) units, which can improve the compatibility among photochromic nano/micrometer copolymer brush and other materials. The diameter of SiO_2-PSPMA-*co*-PMMA is 354 nm. Furthermore, the photochromic performances of the photochromic nano/micrometer copolymer brushes were analyzed by fluorescence spectrophotometer and UV-visible spectrophotometer. The results show that the as-prepared samples have good photochromic properties.

1 INTRODUCTION

Recently, intelligent polymer materials (Sun, 2012; Yerushalmi, 2005) have attracted much attention because of their quick stimuli–responsive performances in different external environments, such as temperature (Alem, 2008), pH value (Hensarling, 2009), light (Brunsen, 2012), ionic strength (Kitano H, 2006) and other specific chemicals (Yu, 2010). Especially, the photochromic intelligent polymer material, which plays an important role in many areas, such as bio-medical applications (Xu, 2016), the field of security (Chen, 2013), military applications (Mu, 2015) was the topic of interest.

Spiropyran and spiropyran derivatives are important photochromic intelligent materials. When spiropyran and spiropyran derivatives were exposed in the UV light, their fluorescence properties were detected via the isomerization reactions

of molecular structure (Figure 1). After the UV light was removed, the photochromic properties of spiropyran and spiropyran derivatives can prolong for 20min. And then, the color of samples become weak gradually until they disappear

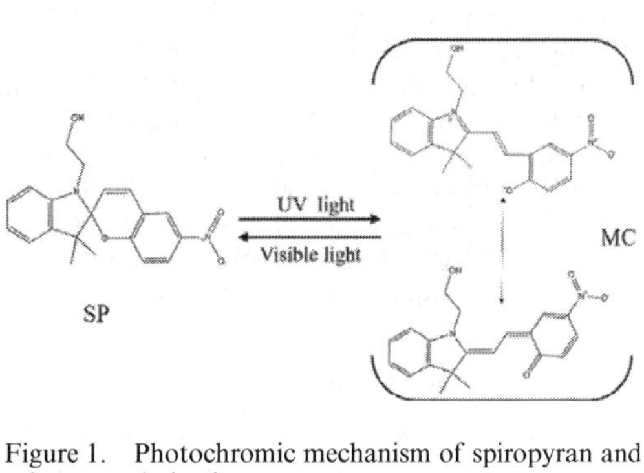

Figure 1. Photochromic mechanism of spiropyran and spiropyran derivatives.

entirely. The process can be repeated for an unlimited amount of time. Based on the photochromic properties of spiropyran and spiropyran derivatives, some photochromic intelligent materials with spiropyran functional segments have been prepared by blend (Mele, 2006), doping (Yamano, 2009) and graft polymerization (Jin, 2010) technologies. Although as-prepared materials all show the photochromic properties of spiropyran, the precipitation or low grafting rate limits their applications. Thus the design and synthesis of long-living and high-performance photochromic intelligent material is still an important challenge.

In this research, considering the advantages of nano-silica particles (such as-ordered uniform pore network, large pore volume, high specific surface, and the surface silanol groups can be modified (Liu, 2004)), we prepared a type of photochromic nano/micrometer copolymer brushes with spiropyran and methyl methacrylate segments through atom transfer radical polymerization. As they possess uniform pore network and high specific surface of nano-silica particles, the spiropyran units were fully introduced into as-prepared photochromic intelligent materials. The multi-layer structure design affords a good compatibility between as-prepared photochromic intelligent material and other polymers. The photochromic nano/micrometer PET fibers were also prepared via electrospun technology. The results indicated that as-prepared nano/micrometer copolymer brushes and PET fibers have all good photochromic properties.

2 EXPERIMENT

2.1 Materials and reagents

Ethylsilicate, CuBr, toluene, ethanol pentamethyldieth-ylen (PMDETA), trifluoroacetic acid, methylmetacrylate (MMA) were purchased from Tianjin Fengchuan Chemical Reagent Co., Ltd. Polyethylene terephthalate, 3-Aminop-ropyl-triethoxysilane (KH-550), triethylamine were purchased from Tianjin heowns Biochemical Technology Co., Ltd. Ethyl 2-bromoisobutyrate (EBIB) and 2-bromoisobutyryl bromide (BIBB, 98%) were purchased from Aladdin Bio-Chemical Technology Co., Ltd. 10-(2-Methacryloxyethyl)-3',3'-dimethyl-6-nitrospiro-(2H-1-benzopyran-2,20-indoline) (SPMA) was prepared according to the literature (Huang, 2011). All of the solvents and chemicals were used as supplied without further purification.

2.2 Analytical methods

The structure of the samples was analyzed by Fourier transform infrared spectroscopy (FT-IR, Gangdong FTIR-650). The morphologies of

samples were characterized by field emission scanning electron microscopy (SEM, Hitachi S-4800) and transmission electron microscopy (TEM, Hitachi H7650). Thermal gravity analyzer (TGA, NetzschSTA 449F3) was used to characterize the thermal properties of the samples. The samples were tested in the temperature range 35°C–800°C; heating rate was 10°C/min. Photochromic properties of the samples were analyzed by UV-visible spectrophotometer (Perkin Elmer Lambda 35) and fluorescence spectroscopy (Gangdong F-380).

2.3 The preparation of the photochromic nano/micrometer copolymer brushes

The synthetic route to photochromic nano/micrometer copolymer brushes via atom transfer radical polymerization was shown in Figure 2. The detailed process mainly includes the following four aspects.

2.3.1 Preparation of SiO_2-NH_2

The nano-silica particles were prepared by Stober method (Bogush, 1988). First, 22 ml aqueous ammonia (28%) and 200 ml absolute ethanol were added in a 500 ml single-neck flask, The mixture was incubated at 40°C for 2 h. Then 21 ml TEOS and 2 ml KH-550 were added quickly. After moderate agitation for 6 h, the temperature was increased to 85°C by heating, and kept in reflux condenser for 2 h. Finally, the product SiO_2–NH_2 were obtained by high-speed centrifugation, and washed with anhydrous ethanol three times, and dried under 60°C vacuum in vacuum oven.

2.3.2 Preparation of SiO_2-Br

In a sequence, 2.0 g SiO_2-NH_2, 20 ml toluene and 4 ml trimethylamine were added into three-neck flask under N_2 atmosphere. The mixture was uniformly dispersed with ultrasonic centrifugation. And then, the solution of 4ml BIBB and 10ml toluene were added dropwise in an ice-water bath; the drop time was controlled within 30min. At last,

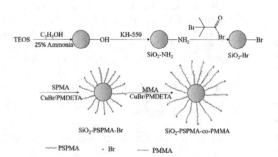

Figure 2. Synthetic route to photochromic nano/micrometer copolymer brushes.

the mixture was stirred under 25°C for 24 hours. The product (SiO$_2$-Br) was obtained by high-speed centrifugation, washed with deionized water and toluene alternately, and dried under 60°C vacuum in vacuum oven.

2.3.3 Preparation of SiO$_2$-PSPMA-Br

The atom-transfer radical polymerization was realized in a Schlenk flask. Prior to the joining reagent degassing the bottle, the roasted bottle was vacuum-filled with nitrogen three times. Then, 1.1 gSPMA,1.422 g SiO$_2$-Br, 10 ml toluene and 10 μl PMDETA were added into the Schlenk flask. After further degassing was done by frozen-pump-thaw cycle three times, 0.018 g CuBr was added rapidly. The reaction was allowed to take place at 80°C for 8 hours under N$_2$ atmosphere. Then the mixture was poured into methanol. As-prepared precipitatewas dissolved in THF. And then, the solution was poured into methanol to precipitate. The process was repeated several times. The product (SiO$_2$-PSPMA-Br) was obtained by high-speed centrifugation and dried in vacuum at 60°C.

2.3.4 Preparation of SiO$_2$-PSPMA-co-PMMA

A mixture of 0.5 g SiO$_2$-PSPMA-Br, 10 ml MMA, 5 μl EBIB, 25 μl PMDETA and 10 ml toluene were added sequentially. And then, after further degassing with freeze-pump-thaw cycle for three times, 0.033 g CuBr was added rapidly. The reaction was took place at 80°C for 24 hours under N$_2$ atmosphere. The reacted mixture was exposed to air. The subsequent processing was done according to the above mentioned method.

2.4 The preparation of photochromic nano/micrometer PET fibers

The spinning solution consisted of 0.40 g PET, 0.05 g trifluoroacetic acid and 0.55 g SiO$_2$-PSPMA-co-PMMA. The electrospinning parameters were 15 kV voltage, 20 cm receiving distance, 60 μL/min perfusion, 20°C and 59% humidity.

3 RESULTS AND DISCUSSION

3.1 FTIR analysis of SiO$_2$-PSPMA-b-PMMA

Figure 3 shows the FT-IR spectra of SiO$_2$-NH$_2$, SiO$_2$-Br, SiO$_2$-PSPMA and SiO$_2$-PSPMA-co-PMMA. As shown in Figure 1a, the characteristic absorption band of SiO$_2$ appeared at 1100 cm^{-1}. The broad band centered around 3452 cm^{-1} was assigned to the hydroxyl group, which was attributed to residual water on the surface of SiO$_2$-NH$_2$ nanoparticles. The peaks at 802 and 953 cm^{-1} were assigned to the Si-O-Si symmetrical stretch vibration and Si-OH asymmetric bend vibration peaks.

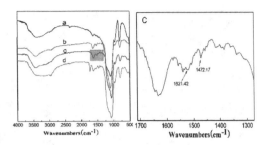

Figure 3. FT-IR spectra of a SiO$_2$-NH$_2$, b SiO$_2$-Br, c SiO$_2$-PSPMA-Br, d SiO$_2$-PSPMA-co-PMMA.

In Figure 3b, the band at 1649 cm^{-1} was attributed to the O = C-NH bending vibration peak. The results show that the immobilization of initiator on the surface of nano-silica particle was complete. In Figure 3c, the new peaks appearing at 1521 cm^{-1} and 1472 cm^{-1} were the C = C characteristic peaks of benzene ring skeleton vibration in the photochromic monomer. In Figure 3d, a new band at 1737 cm^{-1} appears, which is a characteristic peak of the C = O present in PMMA of the compound SiO$_2$–PSPMA-co-PMMA. These results suggested that PMMA has been grafted from the surface of SiO$_2$–PSPMA-Br nanoparticles.

3.2 The morphologies of the of SiO$_2$-PSPMA-b-PMMA

In order to observe the morphology of SiO$_2$-PSPMA-co-PMMA, TEM and SEM were employed. As shown in Figure 4a, the size of the nano-silica particles is uniform; the shape of the particles is regular spherical; the diameter of the particles is at 290–295 nm; the surface of the particles is smooth. After grafting the polymer brush, it can be seen that there is a layer of polymer on the surface of the nanospheres (Figure 4b and c). The diameter of SiO$_2$-PSPMA-co-PMMA touched 354 nm. Figure 4d-f are the SEM images of SiO$_2$-PSPMA-co-PMMA. It can be found that SiO$_2$-PSPMA-co-PMMA has a multi-layer structure.

3.3 The TGA of SiO$_2$-PSPMA-co-PMMA

The TGA was used to analyze the thermal performance of the obtained SiO$_2$-PSPMA-co-PMMA. In the Figure 5, it can been found that the sample has a good thermal stability below the 253°C. When the temperature increases beyond 253°C, the sample begins to decompose, and the mass loss arises from the graft polymer. When the temperature reaches 390°C, the sample is completely decomposed. Good thermal properties are a prerequisite for a material to be used as a filling material.

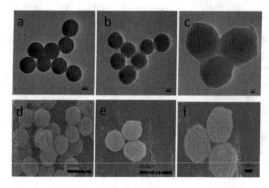

Figure 4. TEMphotos: a nano-SiO$_2$; b and c SiO$_2$-PSP-MA-co-PMMA;SEM photos: d, e and f SiO$_2$-PSPMA-co-PMMA.

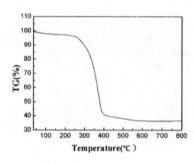

Figure 5. The TG curve of SiO$_2$-PSPMA-co-PMMA.

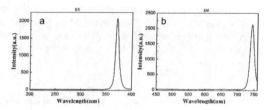

Figure 6. Fluorescence excitation and emission spectra of SiO$_2$-PSPMA-co-PMMA.

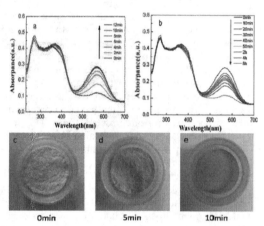

Figure 7. UV–vis absorption spectra of SiO$_2$-PSPMA-co-PMMA:a under UV (365 nm) irradiation; b under visible light (570 nm) irradiation.

3.4 The fluorescence properties of SiO$_2$-PSPMA-co-PMMA

The fluorescence properties were characterized by the fluorescence spectroscopy. Figure 6a is the excitation spectrum. It can be found that there is a significant excitation peak at 365 nm. Correspondingly, Figure 6b shows a strong emission peak appearing at 735 nm. It is because the spiropyran segments of polymer brushes occur in the isomerization reaction under the UV excitation; the closed form (spiropyran, SP) was transformed into the open form (merocyanine, MC). At the same time, the chromophore was generated; after absorption of high-energy ultraviolet light, the excess energy is released in the fluorescence mode.

3.5 UV–vis absorption spectra of SiO$_2$-PSPMA-co-PMMA

The UV-light absorption property of SiO$_2$-PSP-MA-co-PMMA was also determined by UV-vis absorption spectra. In Figure 7a, a strong absorption band appeared at 570 nm, and the colorless nano particles became purple (Figure 7e). Figure 7c shows the pictures of SiO$_2$-PSPMA-co-PMMA at

different UV irradiation times; it is intuitive to see the change in color of the samples. Figure 7b gives UV absorption curves of complete isomerization sample at different time periods under visible light. In Figure 7b, it is obvious that the fade rate was decreased gradually; the photochromic rate from colorless to purple is higher than the photochromic rate from purple to colorless. The analysis results indicate that thephotochromic performance of SiO$_2$-PSPMA-co-PMMA is good. As a photochromic material, it is widely applied in many fields.

3.6 The morphology and photochromic prosperity of photochromic nano/micrometer PET fiberswithSiO$_2$-PSPMA-co-PMMA

The morphology of the electrospun fibers with SiO$_2$-PSPMA-co-PMMA were analyzed by SEM. In the Figure 8a, it can be found that the surface of as-prepared fiber is smooth, and the SiO$_2$-PSPMA-co-PMMA is well dispersed in the nano fibers. The result indicated that the SiO$_2$-PSPMA-co-PMMA had good compatibility with other fiber-forming polymers. Under the UV light, the as-prepared

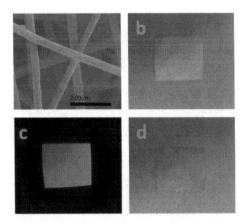

Figure 8. aSEM photos of the electrospun fibers with-SiO₂–PSPMA-co-PMMA; b and c digital photos under visible and UV lights.

fiber samples show a bright red color. When the UV light was removed, the sample show a purple color. The results confirm that as-prepared fibers have good photochromic property.

4 CONCLUSION

The novel photochromic nano/micrometer copolymer brushes were prepared by the living radical polymerization (ATRP). The spiropyran photochromic polymer brushes were grafted on the surface of nano-silica particles. FT-IR was used to characterize the changes of the groups in the molecular structure. The multi-layer structures of the obtained samples were observed by SEM and TEM. And the photochromic functions was analyzed by UV-visible spectrophotometer and fluorescence spectroscopy. Under the 365 nm UV irradiation, the molecular structure of polymer brushes was changed into MC form from the SP form. Through the isomerization reaction of the spiropyran segments, SiO₂–PSPMA-co-PMMA shows different colors under UV and visible light. The SEM images of the electrospun fibers show that the as-prepared SiO₂–PSPMA-co-PMMA had good compatibility with other polymers. Based on analyzed properties of SiO₂–PSPMA-co-PMMA, it can play a significant role in the field of security, and can serve as a good photochromic filler.

REFERENCES

Alem H, Duwez A S, Lussis P, et al. J. Membr. Sci., 2008, 308(1): 75–86.
Bogush G H, Tracy M A, Zukoski C F. J. Non-Cryst. Solids, 1988, 104(1): 95–106.
Brunsen A, Cui J, Ceolín M, et al. Chem. Commun., 2012, 48(10): 1422–1424.
Chen Q, Chen F, Yan Y. BioResources, 2013, 8(1).
Hensarling R M, Doughty V A, Chan J W, et al. J. Am. Chem. Soc., 2009, 131(41): 14673–14675.
Huang C Q, Wang Y, Hong C Y, et al. Macromol. Rapid Commun., 2011, 32(15): 1174–1179.
Jin Q, Liu G, Ji J.J. Polym. Sci. Part A: Polym. Chem., 2010, 48(13): 2855–2861.
KitanoH, Anraku Y, Shinohara H. Biomacromolecules, 2006, 7(4): 1065–1071.
Liu X, Ma Z, Xing J, et al. J. Magn. Magn. Mater.2004, 270(1): 1–6.
Mele E, Pisignano D, Varda M, et al. Appl. Phys. Lett. 2006, 88(20): 203124.
Mu L, Fang L. ActaAgrScand Sect B-Soil Pl, 2015, 65(8): 735–746.
Sun L, Huang W M, Ding Z, et al. Materials & Design, 2012, 33: 577–640.
Xu B, Feng C, Hu J, et al. ACS Appl. Mat. Interfaces, 2016, 8(10): 6685–6692.
Yamano A, Kozuka H. J. Phys. Chem. B, 2009, 113(17): 5769–5776.
Yerushalmi R, Scherz A, van der Boom M E, et al. J. Mater. Chem., 2005, 15(42): 4480–4487.
Yu K, Kizhakkedathu J N. Biomacromolecules, 2010, 11(11): 3073–3085.

Advances in Energy Science and Equipment Engineering II – Zhou, Patty & Chen (Eds)
© 2017 Taylor & Francis Group, London, ISBN 978-1-138-71798-5

Microscopic phase-field model and simulation on the γ′ rafting behavior in Ni-base alloy under external stress

Lijuan Lei, Songyu Wang & Jian Wen

College of Information and Management Science, Henan Agricultural University, Zhengzhou, China

ABSTRACT: The binary microscopic phase field kinetic model coupled with microscopic elastic theory is developed under external stress field, which describes the progress of the atom clustering and ordering by atom occupation probability. The model is used to study the γ′ coarsening behavior for Ni-18.4 at.% Al binary alloys at 1270 K. The simulation results show that the precipitates morphological evolution is in good agreement with experiment results in the γ′ coarsening process. Microscopic phase field method can distinctly simulate the phase structure and atomic configuration, as well as the anti-phase boundary of ordered phase in atomic scale. The hypothesis for regular solution, the Vagard hypothesis of elastic constant and lattice constant as well as the isotropy constant hypothesis of diffusion coefficient have an important influence on the simulation results. It is found that stress magnitude, direction and applied time have significant influence on the rafting. The rafting of γ′ particles can be obviously more coarsened with the increasing of external stress and applied time.

1 INTRODUCTION

The lattice mismatch between precipitate phase and matrix will result in the increasing of elastic strain energy in solid state transformation. An applied stress can introduce an additional lattice mismatch strain in binary or complex alloys; it also induces the coupling elastic strain energy (Li, 2011). On the other words, the lattice mismatched precipitates will coarsen preferentially in a parallel (P-type) or perpendicular (N-type) direction to the applied stress, which is known as rafting (Gururajan, 2007). Khachaturian described the influence of lattice misfit strain energy on the behavior of precipitates in the process of phase transformation early (Khachaturyan, 1983). Zhou investigated the directional coarsening of γ′ precipitates in single crystal Ni–Al alloys under external load by an integrated phase field model (Zhou, 2007). At present, there have been a number of attempts to determine the type of rafting under the elastic stress. For example, the energy-based approaches examine the elastic energy of isolated precipitates under external stress to deduce energetically favorable particle shapes and orientations (Durga, 2013). Thermodynamic approaches, on the other hand, examine the instantaneous chemical potential contours around mismatch inhomogeneous particles under external stress (Cotturaa, 2012). The kinetic studies of rafting are based on embedded-atom method (Svoboda, 2012) or continuum

phase method (Li, 1999). The above approaches are effective on the rafting morphology of particles in mesoscopic scale, but it cannot reveal the atomic configuration and phase structure of the precipitates and matrix, especially the anti-phase domain boundary and the point defects in ordered structure. Microscopic phase-field method is an important dynamical method to describe the evolution of field variable (density, concentration, occupation probability) with time (Yang, 2012; Lu, 2008).

In this paper, the microscopic phase field kinetic model will be developed by coupling with microscopic elastic theory proposed by Khachaturian (Khachaturyan, 1983). Then, the model is used to simulate the rafting behavior of L12-structure γ′ phase under external stress in binary Ni-base alloys.

2 MICROSCOPIC PHASE-FIELD MODEL

Based on the microscopy diffusion theory of R. Poduri (Poduri, 1998), the kinetic evolution model of binary alloys for microscopic phase-field is established under external stress. Atomic configurations and the phase morphologies are described by atomic occupation probability function Φ (r, t), which represents the probability that a given lattice site r is occupied by an atom at time t. The binary system equilibrium state is homogeneous disordered state at high temperature. If the aging temperature of homogeneous disordered system is lowered, the system will be destabilized due to the atomic ordering and coalescence. The unstable-stable state evolution is a highly nonlinear complex process under the aging temperature, which can be described by the Onsager microscopic diffusion equation proposed by Khachaturian (Khachaturyan, 1983). The kinetic equation of solute atomic occupation probability function $\Phi(r, t)$ in the site r can be written as:

$$\frac{\partial \Phi(r,t)}{\partial t} = \frac{c_0(1-c_0)}{k_B T} \sum_{r'} L(r-r') \frac{\delta F}{\delta \Phi(r',t)} + \xi(r,t) \quad (1)$$

where k_B is the Boltzmann constant, T is the temperature, $L(\vec{r}-\vec{r}')$ is a constant related to exchange probabilities of a pair of atoms at lattice sites r and r' per unit time, F is the total Helmholtz free energy of the system. $\xi(r, t)$ denotes the thermal noise item and obey the Gaussian distribution. It meets the requirements of the fluctuation dissipation theory (Lifshitz, 1980).

$$\langle \xi(r,t) \rangle = 0 \quad (2)$$

$$\langle \xi(r,t)\xi(r',t) \rangle = -2k_B TL(r-r') \, \delta(t-t') \, \delta(r-r') \quad (3)$$

In the mean-field approximation, the total free energy of the binary dispersion lattice system is given by:

$$F = \frac{1}{2}\sum_{r}\sum_{r'} V(r-r')\Phi(r)\Phi(r') + k_B T$$
$$\sum_{r}\left\{\Phi(r)\ln\Phi(r) + [1-\Phi(r)]\ln[(1-\Phi(r)]\right\} \quad (4)$$

with

$$V(r-r') = V^{ch}(r-r') + V^{el}(r-r') \quad (5)$$

where $V^{ch}(r-r')$ is the chemical interactive energy between a pair of atoms at site r and r' in the distortionless lattice; $V^{el}(r-r')$ is the elastic interaction energy in distortion lattice. The chemical interaction energy is given as:

$$V_{\alpha\beta}(r-r') = W_{\alpha\alpha}(r-r') + W_{\beta\beta}(r-r') - 2W_{\alpha\beta}(r-r') \quad (6)$$

where $W_{\alpha\beta}(r-r')$ is the pairwise interaction energy at orthoscopic lattice site r and r'. Based on the microscopic elastic theory proposed by Khachaturian (Khachaturyan, 1983), the embedding influence of precipitate particles on the total elastic energy of binary system can be described as:

$$E^{el} = \frac{1}{2}\sum_{r}\sum_{r'} V^{el}(r-r')\Phi(r)\Phi(r')$$
$$= \frac{1}{2}\sum_{r} C^0_{ijkl}\varepsilon^\alpha_{ij}\varepsilon^\alpha_{kl} - \sum_{r} \Delta C_{ijkl}\varepsilon^\alpha_{ij}\varepsilon^0_{kl}\overline{\Delta\Phi^2(r)}$$
$$+ \frac{1}{2}\sum_{r} C^0_{ijkl}\varepsilon^0_{ij}\varepsilon^0_{kl}\overline{\Delta\Phi^2(r)}$$
$$- \frac{1}{2}\sum_{g} \frac{1}{(2\pi)^3} n_i\sigma^*_{ij}\Omega_{jk}\sigma^*_{kl}n_l |\Delta\Phi(g)|^2 \quad (7)$$

where C^0_{ijkl} is the systemic average elastic constant, $\Delta C_{ijkl} = C^p_{ikl} - C^m_{ijkl}$ is the difference of elastic constant between precipitation phase and matrix, ε^α_{ij} is the external strain, ε^0_{ij} is the dilatation coefficient of lattice constant for the concentration, $\sigma^*_{ij} = C^0_{ijkl}\varepsilon^0_{kl} - \Delta C_{ijkl}\varepsilon^\alpha_{kl}$ is the systemic effective stress, $n = \{n_i, n_j, n_k\} = g/g$, which g is the reciprocal lattice vector, g is the absolute value of g and n is the unit normal vector of precipitates interface. The variation derivative of the elastic energy E^{el} related to the $\Phi(r)$ can be written as

$$\sum_{r'} V^{el}(r-r')\Phi(r') = -2\Delta C_{ijkl}\varepsilon^\alpha_{ij}\varepsilon^0_{kl}\Delta\Phi(r) +$$
$$C^0_{ijkl}\varepsilon^0_{ij}\varepsilon^0_{kl}\Delta\Phi(r) - \left\{n_i\sigma^*_{ij}\Omega_{jk}\sigma^*_{kl}n_l\Delta\Phi(g)\right\}_r \quad (8)$$

The mass of all kinds of atoms meets the conservation equation:

$$\sum_{r} \frac{\partial \Phi(r)}{\partial t} = 0 \quad (9)$$

Substituting Eq. (4) and (9) into Eq. (1), and performing Fourier transformation of equations, then the microscopic phase field kinetic equation in reciprocal space can be written as:

$$\frac{\partial \Phi(g,\tau)}{\partial \tau} = L(g)\left[V(g)\Phi(g,\tau) + k_B T\left\{\ln\frac{\Phi(r,\tau)}{1-\Phi(r,\tau)}\right\}_g\right]$$
$$+ \xi(g,\tau) \quad (10)$$

where $\tau = c_0(1-c_0)t/(k_B T)$ is the dimensionless time, $\left\{\ln[\Phi(r,\tau)/(1-\Phi(r,\tau))]\right\}_g$ and $V(g)$ are the corresponding Fourier transformations in the real space. Using equation (5), $V(g)$ is given by

$$V(g) = V^{ch}(g) + V^{el}(g) \quad (11)$$

Using the Eq. (8), the Fourier transformation of the interatomic elastic interaction can be written by:

$$V^{el}(g) = -2\Delta C_{ijkl}\varepsilon^\alpha_{ij}\varepsilon^0_{kl} + C^0_{ijkl}\varepsilon^0_{ij}\varepsilon^0_{kl} - n_i\sigma^*_{ij}\Omega_{jk}\sigma^*_{kl}n_l \quad (12)$$

The $V^{ch}(g)$ can be written as the sum of interatomic interactive energy for the neighbor lattices

$$V^{ch}(g) = \sum_r W(r)e^{igr} \qquad (13)$$

By assuming atomic jumps between nearest neighbor sites only, and using the condition that the total number of atom in the system are conserved, for a f. c. c lattice, the following equation can be written

$$L(g) = -4L^0[3 - \cos\pi h \cdot \cos\pi k - \cos\pi k \cdot \cos\pi l - \cos\pi l \cdot \cos\pi h] \qquad (14)$$

where L^0 is proportional to the jump probability between the nearest neighbor sites at a time unit (Poduri, 1998).

3 NUMERICAL SIMULATION RESULTS AND DISCUSSION

Compared with the continuum phase-field, microscopic phase-field has fewer phenomenological parameters. If inputting the interatomic interactive energy obtained from computation or experiment to parameter equations, then we can obtain the morphological evolution of the phase structure and organization in the process of phase transformation.

By the equilibrium of the γ and γ′ phase for 12.5at.%Al at 1270K in Ni-Al binary phase-diagram (Wang, 1998), the Ni-18.4at.% Al alloy will precipitate the precipitates that the volume fraction is about 50% at same temperature, and the interatomic interaction energy in alloys are W^1 = 122.3, W^2 = −6.0, W^3 = 16.58, W^4 = −6.82 (Poduri, 1998). The elastic constant of the γ and γ′ phase and the other physical parameters are presented in the Table 1, from which we can find that γ matrix for f. c. c structure and the γ′ phase of Ll2 structure have the same square structure on the (001) plane.

The simulation is performed in a square lattice consisting of 2048 × 2048 unit cells; and the space

Table 1. The crystal and physical parameters of γ, γ′ phase and other physical parameters.

Parameters	Units	γ	γ′	Ni	Al
C_{11}	G Pa	206.8	229		
C_{12}	G Pa	148.5	161		
C_{44}	G Pa	94.4	125		
Crystal structure		f. c.c	L12	f. c.c	f.c. c
Lattice constants	nm	0.3528	0.3572	0.3520	0.4050
Atomic mass				58.6934	26.982
Density	g /cm³			8.902	2.6989
Mol volume	cm³/mol			6.6	10.0

increasing step Δx = a/2. We initialize the computed area is 0.363 μm × 0.363 μm by the lattice constant of γ phase.

Figure 1 (a) shows the γ′ phase morphology of Ni-18.4 at.% Al alloys for 5.18 hours at 1270 K under sustained tension stress **σ** = 200 MPain the [100] direction. The particles are lengthened under external tension stress, meanwhile all the particles are directional distribution along the [110] direction. The coarsening orientation of particles is

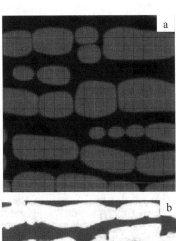

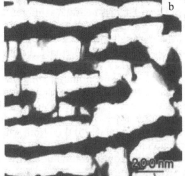

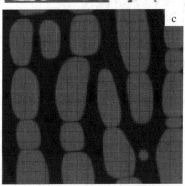

Figure 1. The comparison of γ′ phase rafting simulation: (a) microscopic phase-field; (b) the TEM image for Ni-13.3 at.%Al-8.8 at.% Mo alloys (Paris, 1977); (c) the stress is $\sigma_{0000}^{\alpha} = -200$ Mpa.

related to elastic constant and elastic aeolotropism of precipitates and matrix (Khachaturyan, 1995). The effective strain energy E^{ema} related to the precipitate morphology can be given by:

$$E^{ema} = \frac{1}{16\pi^3} \sum_g n_i \sigma_{ij}^* \Omega_{jk} \sigma_{kl}^* n_l \left| \Phi_\alpha(g) \right|^2 \qquad (15)$$

where σ_{ij}^* is the effective stress, which can be used to the computation of effective strain ε_{ij}^*.

$$\varepsilon_{ij}^* = S_{ijkl}^0 \sigma_{kl}^* = \varepsilon_{ij}^0 - \Delta\varepsilon_{ij}^\alpha \qquad (16)$$

where $S_{ijkl}^0 = \left(C^0\right)_{ijkl}^{-1}$ is average compliance coefficient.

$$\Delta\varepsilon_{ij}^\alpha = S_{ijkl}^0 \Delta C_{klmn} S_{mnop}^0 \sigma_{op}^\alpha \qquad (17)$$

$\Delta\varepsilon_{ij}^\alpha$ is the effective external strain, which is the first approximation of systemic strain for the external stress (Khachaturyan 1995). Precipitates will grow along a direction with respect to the minimum strain energy.

The effective stress can be used to study the effect of the external stress field on the particle morphology. Because There is nearly no difference between the elastic constant of precipitate and matrix of Ni-18.4 at.% Al alloy in the aging process, the elastic constant of matrix can replace the average elastic constant, thus $C_{11}^0 - C_{12}^0 - 2C_{44}^0 < 0$. According to the Eq. (15), the precipitates will grow along the (100) plane if E^{el} can obtain the minimum value (Khachaturyan 1995). We can acquire the strain ε_{ij}^α on the interface of precipitate under the tension stress σ_{ij}^α in the (100) direction, which causes the precipitates elongation in the [100] direction and the Poisson compression in the [010] and [001] direction, then $\varepsilon_{xxx}^\alpha > 0$ and $\varepsilon_{yy}^\alpha = \varepsilon_{zz}^\alpha < 0$.

The Table 1 shows the difference of elastic constant between γ' phase and γ matrix is $\Delta C_{ijkl} > 0$ for Ni-18.4 at.% Al. According to the Eq. (16), we can obtain the $\varepsilon_{xxx}^* < \varepsilon_{yy}^* < \varepsilon_{zz}^*$, thus the problem is simplified as the one that how to optimize the shape of rectangular precipitate. If $C_{11}^0 - C_{12}^0 - 2C_{44}^0 < 0$ and $0 < \varepsilon_{xxx}^* / \varepsilon_{yy}^* < 1$, the minimum strain energy will cause the formation of plate precipitates on the (010) and (001) plane. Whereas, the applied compression stress in the [100] direction will cause the formation of plate precipitates on the (100) plane based on the minimum strain energy. Fig. 1(b) shows the rafting morphology of precipitates for L12 structure using experimental methods; Fig. 1(c) describes the simulation of microscopic phase-field under sustained tension stress $\sigma_{xxx}^\alpha = -200 Mpa$. Moreover, the rafting simulation of γ' phase using continuum phase-field is shown in Fig. 4 (a) and

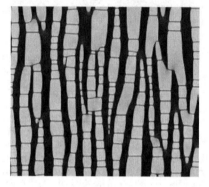

Figure 2. The γ' phase rafting morphology simulation using continuum phase-field [16]; (a) $\sigma_{xxx}^\alpha = 200$ MPa; (b) $\sigma_{xxx}^\alpha = -200$ MPa.

(b). Compared with the experimental results, it is not difficult to find that the rafting morphology simulation of γ' particles using phase-field methods is accorded with experiment and theory analytical results.

The morphological evolution of γ'-particles is the directional growth along the stress direction due to the effect of external stress, diffusion energy and lattice distortion energy. Besides above three factors, inter-particles anti-phase boundary also has marked influence on the formation of bamboo shape of multi-particles. The rafting morphology of γ' particles for microscopic phase field (Fig. 1 (a), (c)) is similar to the simulation of continuum phase field (Fig. 2 (a), (b)), and the width of particles is about 100 nm, which is in good agreement with the experimental data (Fig. 1 (b)). From the simulation of microscopic phase-field (Fig. 1 (a), (c) and Fig. 3 (a), (c)), we find that the applied stress is a determinant factor on morphology of particles. With the external stress increasing, we can obtain the elongate particles morphology in a shorter time. With the aging time lengthening, the rafting morphology of precipitates is more obvious under the same external stress.

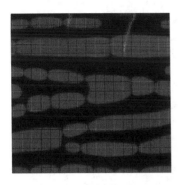

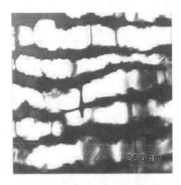

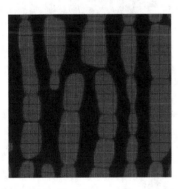

Figure 3. The rafting morphology after the increase of the external stress σ_{ij}^a: (a) microscopic phase-field; (b) the TEM image (Paris, 1977); (c) the stress is $\sigma_{xxxx}^\alpha = -300$ MPa.

4 CONCLUSIONS

In this paper, the binary microscopic phase field kinetic model coupled with microscopic elastic theory is developed under external stress field, which solves the equation on reciprocal space by adopting 2-D projections for a 3-D system and gives the visible graphic expression of atomic occupation probabilities. The morphological evolution

of precipitates for Ni-18.4 at.% Al alloy at 1270 K is accorded with the experimental results and continuum phase-field simulation in the coarsening process. In the microscopy phase field model, the hypothesis for regular solution, the Vagard hypothesis of elastic constant and lattice constant as well as the isotropy constant hypothesis of diffusion coefficient have an important influence on the simulation results. The model parameters can be optimized through experimental results.

From the comparison of microscopic phase-field and continuum phase-field with experimental results, we find that microscopic phase-field can distinctly simulate the phase structure, atomic configuration and the anti-phase domain boundaries of the ordered phase particles in atomic scale without external stress.

By the simulation results for rafting behavior of γ' phase using microscopic phase-field and continuum phase-field, and comparing with experimental results under the sustained stress, it is found that the magnitude, direction and applied time of external stress have marked influence on the rafting behavior of γ' phase. With the increasing of external stress and applied time, the rafting behavior is more obvious in the direction of applied stress.

REFERENCES

Cotturaa, M., Y. Le Bouar, A. Finel, et al. J. Mech. Phys. Solids[J], 2012,60:1243.

Durga, A., P. Wollants, N. Moelans. Model. Simul. Mater. Sc. [J], 2013,21: 055018.

Gururajan, M.P., T.A. Abinandanan. Acta Mater. [J], 2007, 55: 5015.

Khachaturyan, A.G. Theory of Structural Transformations in Solids[M]. Wiley, New York, 1983.

Khachaturyan, A.G., S. Semenovskaya, T. Tsakalakos. Phys. rev. B [J], 1995, 52: 15909.

Li, D.Y., L. Q. Chen. Acta Mater. [J], 1999, 47: 247.

Li, D.Y., L. Q. Chen. Scripta Mater. [J], 1997, 37: 1271.

Li, Y.S., Y.Z. Yua, X.L. Cheng, et al. Mat. Sci. Eng. A[J], 2011,28:8628.

Lifshitz, E.M., L. P. Pitaevskii. Statstical Physic, Pergamon Press [M], Oxford, 1980.

Lu, Y.L., Z. Chen, Y.X. Wang. Mater. Lett. [J], 2008, 62:1385–1388.

Paris, O., M. Fahrmann, E. Fahrmann, et al. Acta mater. [J], 1977, 45:1085.

Poduri, R., L. Q. Chen. Acta Mater. [J], 1998, 46:1719.

Svoboda, J., F.D. Fischer, D.L. McDowell. Acta Mater. [J], 2012, 60: 396.

Wang, Y., D. Banerjee, C. C. Su, et al. Acta Mater. [J], 1998, 46:2983.

Yang, K., M.Y. Zhang, Z. Chen, et al. Comp. Mater. Sci. [J], 2012, 62: 160.

Zhou, N., C. Shen, M.J. Mills, et al. Acta Mater. [J], 55 (2007) 5369–5381.

Advances in Energy Science and Equipment Engineering II – Zhou, Patty & Chen (Eds)
© 2017 Taylor & Francis Group, London, ISBN 978-1-138-71798-5

Size-dependent optical properties of ZnO nanorods

Chunping Li & Peilin Wang
Aviation University of Air force, Changchun, China

Xin Chen
Changchun University of Science and Technology, Changchun, China

ABSTRACT: Nanometer-sized ZnO crystals were prepared successfully by a simple wet-chemical synthesis method. The rods with diameters from 30 to 200 nm were put through morphology control experiments. X-ray Diffraction (XRD) spectra show that there are no differences among the products, but the laser power dependent photoluminescence (PL) is closely related to the size of the nanorods. PL intensity and the full width at the half maximum (FWHM) and the time resolved PL are compared and discussed. Barely observed visible emission band indicates the good optical quality of all the synthesized ZnO nanorods.

1 INTRODUCTION

Zinc oxide (ZnO) is a remarkable semiconductor due to its wide range of applications including luminescent displays, optical wave-guides, ultraviolet laser sources, ultraviolet detectors, electrochemistry and photochemistry materials(Wang, 2006; Mohammed, 2011; Huang, 2001; Ü. Özgür, 2005; Lin, 2014). It is attractive because of its direct wide band gap (Eg~3.3 eV at room temperature), large exciton binding energy, its low cost and environmental friendly nature (Huang, 2001). The large exciton energy ensures an efficient exciton emission at room temperature under low excitation energy. Low-dimensional ZnO nanostructures are fundamental elements of nanodevices. Significantly, physical properties are dependent on the structures of nanomaterials (Li, 2006; Han, 2004; Shi, 2005; Li, 2013).

The use simple growth route to grow desired low-dimensional ZnO and study the structure-related properties are the objective of our work. In this paper, we will mainly report the size-dependence of the optical properties of ZnO nanorods synthesized by simple wet chemical methods. It has been proved that the preparation of ZnO via wet chemical routes provides a promising option for the large-scale production of nanomaterials, which is a simpler crystal-growth technology, resulting in a potentially lower cost for ZnO-based nanodevices (Li, 2007). Although high-yield syntheses of ZnO one-dimensional nanocrystals have been achieved, simultaneous control of their morphology, surface structure, and monodispersed distribution is still a significant challenge. Different-sized ZnO nanorods were prepared by controlling the growth condition. The size-dependent optical properties of the exciton luminescence are investigated and compared.

2 CHARACTERIZATION TECHNIQUES

The Scanning Electron Microscope (SEM) measurements were used to determine the morphology of ZnO nanoproducts. The structural characterization of the synthesized ZnO nanorods was performed by XRD on D/max-2500 copper rotating-anode X-ray diffractometer with monochromic Cu Ka radiation (40kV, 100 mA).

Low laser power exciting PL measurements were performed by a micro-Raman spectrometer (Lab-Ram HR800, JY, France). The laser excitation source is the 325 nm line of a He–Cd laser. The laser power-dependent PL excitation with relatively high laser power was explored by using a pulsed YAG laser operating at 266 nm.

Time-resolved PL was done by the femtosecond (fs) time-resolved spectra mode-locked Ti:sapphire laser (266 nm, 150 fs pulse width, 1000 Hz repetition rate). The excitation laser pulse is focused on the sample by a lens and the excitation intensity is 40 μJ/cm^2. A photon counting streak camera (Hamamatsu C2909) was used for the measurements. The temporal and spectral resolution of the instrument response function is less than 10 ps and 1 nm. All the experiments were carried out at room temperature.

3 RESULTS AND DISCUSSION

3.1 Morphology and structural characterization

Figure 1 shows the morphology of the three kinds products. High-yield ZnO nanorods are well-defined crystals with diameters of around 30 nm (Fig.1 (a), sample a), 60 nm (Fig.1 (b), sample b) and 200 nm (Fig.1 (c), sample c), respectively, and their lengths are around 1.6, 1.9, and 2.4 μm, respectively, which proves the large aspect ratios of the nanorods. Details of the ZnO nanorod fabrication have been described elsewhere (Li, 2013).

Figure 2 shows the XRD spectra of the three kinds nano products. All diffraction peaks can be indexed to wurtzite hexagonal ZnO, which indicates that the ZnO nanorods have a hexagonal

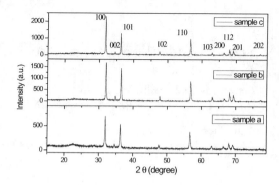

Figure 2. XRD spectra of the three kinds of ZnO samples.

wurtzite structure. No diffraction peaks from other phases or other elements are detected. The results reveal that preferential orientations are all along the (100) direction.

3.2 Optical properties

The effect of the laser power dependence of PL is discussed. First, the spectra shown in Fig. 3 is the result excitation at low laser power density (4 kW/cm^2) and collected by micro-Raman spectrometer. The PL spectra show peaks located at around 388.4, 388.1, 386.5 nm for sample c, b, a, respectively, and the full width at the half maximum is about 28.3 nm, 31.4 nm, 34.4 nm, respectively. Obviously, PL intensity increases with the increasing diameters of the nanorods. The single strong ultraviolet (UV) emission band with narrow FWHM identifies the free excitonic emission from its peak energy (Huang, 2001; Li, 2006). Visible emission bands associated with Zn interstice or oxygen vacancy or surface state emission are barely observed, which indicates the good optical quality of all the synthesized ZnO nanorods, but the PL intensity decreases with the decrease in diameters. The multi-peaks fitting results show that there is not much difference for the three kinds of samples (not shown in the figure), which indicates that phonon-assisted effects can be neglected for these different sized rods.

Photoluminescence properties of the prepared ZnO nanorods are also examined by power-dependent measurements, which were carried out by using a pulsed YAG laser operating at 266 nm. Power-dependent PL spectra of the three kinds of ZnO nanorods have the similar evolutional tendency. The power-dependent PL spectra of sample b, as a representative, are given in Fig. 4. The visible emission bands are still very weak when the excitation intensity increased from 0.2 mw to 7.5 mw. The FWHM is increasing from 18 to 31.2 nm, and the

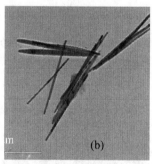

Figure 1. SEM of the three kinds of ZnO nanorods.

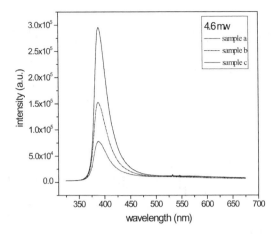

Figure 3. PL spectra of the three kinds of samples excited by He-Cd laser.

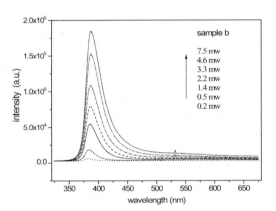

Figure 4. Power dependent PL spectra of sample b excited by YAG laser.

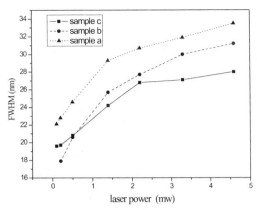

Figure 5. Power-dependent FWHM of PL spectra.

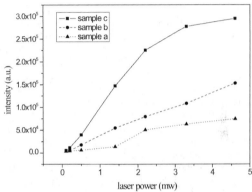

Figure 6. Power-dependent peak intensity of PL spectra.

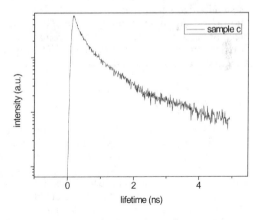

Figure 7. Time-resolved PL of sample c.

peak of UV emission shifts from 381.5 to 387.6 nm with the increase in laser power.

The compared FWHM and peak intensity are shown in Figs. 5 and 6, respectively. The results show that the FWHM and peak intensity of the three kinds products will increase with the increase in laser power, but the speed of increase slows down and tends to a saturated value. Relatively, the thick rods have narrow FWHM and high PL intensity.

Picosecond time-resolved measurements were employed to investigate the PL dynamics under femtosecond laser pulses at 266 nm excitation. Fig. 7 shows the normalized time-resolved spectra. The exciton lifetime is an important parameter related to material quality.

Generally, the PL decay rate of a semiconductor can be analyzed by double exponential fitting (Li, 2006): $A_1 \exp(-t/\tau_f) + A_2 \exp(-t/\tau_s)$. Here τ_f and τ_s are the fast decay time and the slow decay time, respectively. A1 and A2 are the decay con-

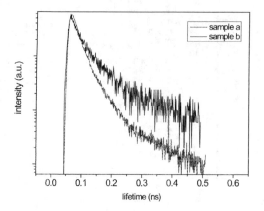

Figure 8. Time-resolved PL of sample a and b.

stants. Our time-resolved PL results show that the decay fits well to a biexponential function with a fast component 25 ps and a slower component 141 ps for sample a, a fast component 75 ps and a slower component 310 ps for sample b, and a fast component 531 ps and a slower component 2.4 ns for sample c, respectively.

Normally, non-radiative recombinations give rise to a fast relaxation of the excited carriers, while the slow decay term is usually due to the radiative recombination of free or localized excitons. The dominant fast decay term in relatively fine samples is induced by the effective non-radiative recombination. The size-dependence of time-resolved PL spectra of ZnO nanorods show that with decreasing size of microcrystals, the surface-to-volume ratio becomes larger, and this is why the rod with a smaller diameter has larger nonradiative relaxation rates. Additional sources of the non-radiative recombination may be caused by the trapping of carriers by deep centers at defects and/or impurities and multiphonon scattering.

4 CONCLUSION

In summary, we have successfully prepared single-crystalline ZnO nanorods with different diameters and with large aspect ratios using the improved microemulsion method. PL and time-resolved PL spectra of zinc oxide were recorded as a function of the size of the products. The relatively larger FWHM and weak PL intensity for narrow-sized ZnO rods are mainly due to nonradiative recombination channels and this process are also activated by thermalization. The difference of phonon-assisted effects for the different rods can be neglected at low laser excitation intensity. PL measurements also confirm the good optical quality of the obtained ZnO nanorods. These high-quality single crystalline ZnO nanorods represent good candidates for further potential applications.

ACKNOWLEDGMENT

Chunping Li thanks the financial support of the National Natural Science Foundation of China under Grant No. 51272285.

REFERENCES

Chunping Li, Li Zhang, Changjie Liu, etc., Exciton localization and stimulated emission of ZnO nanorods, Key Engineering Materials, 538 (2013) 161–164.

Han Shalish, Henryk Temkin, and Venkatesh Narayanamurti, Size-dependent surface luminescence in ZnO nanowires, Phy. Rev. B, 69 (2004) 245401.

Huang, M.H., S Mao, H. Feick, H.Q. Yan, et al., Room temperature ultraviolet nanowire nanolasers, Science, 292 (2001) 1897–1899.

Li, C.P., L. Guo, Z.Y. Wu, etc., Phtoluminescence and time-resolved photoluminescence of star-shaped ZnO nanostructures, Solid State Communications, 139 (2006) 355–359.

Li, C.P., Y.Z. Lv, L Guo, H.B. Xu, X.C. Ai, J.P. Zhang, Raman and excitonic photoluminescence characterizations of ZnO star-shaped nanocrystals, Journal of Luminescence 122–123 (2007) 415–417.

Mohammed Riaz, Jinhui Song, Omer Nur, Zhong Lin Wang, Magnus Willander, Study of the Piezoelectric Power Generation of ZnO Nanowire Arrays Grown by Different Methods, Adv. Funct. Mater., 21(2011) 628–633.

Özgür, Ü., Ya. I. Alivov, C. Liu, A. Teke, et al., A comprehensive review of ZnO materials and devices, 98 (2005) 041301.

Pei Lin, Xiang Chen, Xiaoqin Yan, etc., Enhanced photoresponse of Cu2O/ZnO heterojunction with piezo-modulated interface engineering, Nano Research, 7, 6 (2014) 860–868.

Shi, W.S., B. Cheng, L. Zhang, E.T. Samulski, Influence of excitation density on photoluminescence of zinc oxide with different morphologies and dimensions, J. App. Phy. 98 (2005) 083502.

Wang Z.L., and J.H. Song, Piezoelectric Nanogenerators Based on Zinc Oxide Nanowire Arrays, Science, 14 (2006) 242–246.

Xiuyan Li, Jian Wang, Jinghai Yang, etc., Size-controlled fabrication of ZnO micro/nanorod arrays and their photocatalytic performance, Materials Chemistry & Physics, 2013, 141(2–3) 929–935

Advances in Energy Science and Equipment Engineering II – Zhou, Patty & Chen (Eds)
© *2017 Taylor & Francis Group, London, ISBN 978-1-138-71798-5*

Relationship between non-uniformity and its distribution inside the specimen

Binbin Xu

Tianjin Port Engineering Institute Ltd. of CCCC, Tianjin, China
Key Laboratory of Geotechnical Engineering of Tianjin, Tianjin, China
Key Laboratory of Geotechnical Engineering, Ministry of Communication, Tianjin, China

Wei Si

Tianjin Port Engineering Institute Ltd. of CCCC, Tianjin, China

ABSTRACT: Using the proposed evaluating method, the variation of non-uniformity of the mean effective stress and the stress ratio and their distribution along the radius direction are discussed and explained in detail. It is found that there are several peak for the non-uniformity of mean effective stress but only one peak for the stress ratio. As the apparent shear strain increases, the distribution of mean effective stress and stress ratio along the radius also varies and the outer is always faster than the inner.

1 INTRODUCTION

In order to clarify the soil properties and verify the validity of proposed constitutive model, various in-door experiments have been carried out to check the mechanical behavior of soil under different stress path, such as plane strain test, oedometer test and triaxial test, in which the triaxial compression test is the most popular test. However, the stress state in the triaxial test is not three dimensional and it cannot be called a true triaxial test. The hollow torsion test is usually regarded to be the most similar to the element-wise sample to research the mechanical response of certain soil (Asaoka & Noda 1995, Jin et al. 2010). However, during the torsion test there would be non-uniform deformation which would influence the apparent behavior of the specimen (Hight et al. 1983, Sayao & Vaid 1996). (Xu et al. 2016) proposed an evaluating method to judge the non-uniformity quantitatively.

In this paper, the non-uniformity of mean effective stress and stress ratio are shown and their variations are explained in detail.

2 VARIATION OF MEAN EFFECTIVE STRESS

Figure 1 is the non-uniformity of mean effective stress by marking the maximum and minimum deviation with (a), (b) and (c) which corresponds to the apparent shear strain 1.83%, 3.54% and

11.0%. Figure 2 shows the relationship between apparent shear strain and mean effective stress p' calculated by taking the specimen as a whole mass, where (a), (b) and (c) represent the same apparent shear strain. The distribution of p' at each node along the radius at different apparent shear strains is presented in Figure 3, where it can be seen that the development of p' at the inner radius always follows the tendency of p' at the outer radius. The details about the variation of deviation of p' will be explained as follows:

1. Initially p' along the radius is same and equal to the cell pressure and therefore the deviation of p' is zero as shown in Figure 3, then according

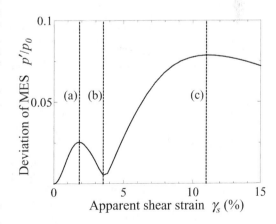

Figure 1. Deviation of mean effective stress p' versus apparent shear strain.

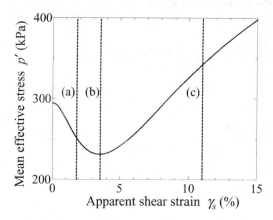

Figure 2. Mean effective stress p' when taking the specimen as a whole mass versus apparent shear strain.

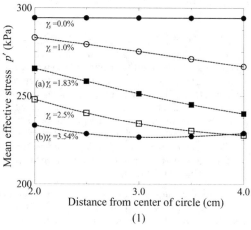

(1)

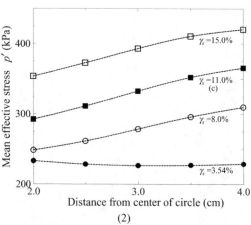

(2)

Figure 3. Mean effective stress p' of each node along the radius.

to the progress of effective stress path as shown in Figure 2 p' begins to decrease. However, due to the rapid decrease at outer radius, the non-uniformity will begin to increase until the first peak around (a), that is $\gamma_s = 1.83\%$;

2. After p' at outer radius reaches the minimum (b) in Figure 2, that is the positive dilatancy occurs, p' at outer radius tends to increase whereas p' at inner radius is still on the way to the minimum. At around $\gamma_s = 3.54\%$ the non-uniformity decreases to almost zero;

3. After p' at both outer radius and inner radius passes over the minimum, the non-uniformity increases again due to the rapid increasing at outer radius;

4. Finally p' at outer firstly reaches the critical state and keeps constant, correspondingly the non-uniformity begins to decrease after the second peak around (c), that is $\gamma_s = 11\%$;

According to the above analysis, the deviation of p' firstly reaches the maximum and then follows minimum when the soil-water couple effect (two-phase) is considered, which is different from the one-phase results the deviation of p' increases monotonically.

3 VARIATION OF STRESS RATIO

Similarly, the non-uniformity of stress ratio η is shown in Figure 4 as marked by (a) at the maximum value when the apparent shear strain is 2.45%. As the apparent shear strain increases, η firstly increases to the maximum and then decreases again as shown in Figure 5. Therefore, when both of η at outer and inner radius increase, the deviation of η will also increase. But when η at outer radius decreases, the difference between η at outer and inner radius also becomes decreasing. Finally, when η reaches the critical state, that is η is equal to M, the deviation of η turns to be almost zero.

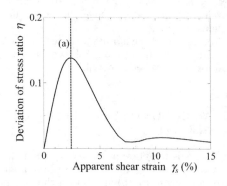

Figure 4. Deviation of stress ratio η versus apparent shear strain.

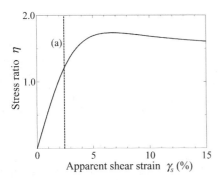

Figure 5. Stress ratio η when taking the specimen as a whole mass versus apparent shear strain.

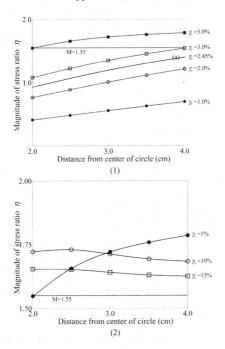

Figure 6. Stress ratio η of each node along the radius.

The corresponding distributions of η from inner radius to outer radius are shown in Figure 6. In addition, the relationship among stress ratio η, threshold between hardening and softening M_s and threshold between plastic compression and plastic expansion M are shown in Figure 7 at various apparent shear strain. It can be seen that i) all the elements along the radius turns from hardening and plastic compression to hardening and plastic expansion; ii) the element locating at the outer radius always develops faster than that at the inner radius.

Correspondingly, the deviation of shear strain ε_s, the magnitude of shear strain ε_s and distribution of ε_s along the radius versus the apparent shear strain γ_s are drawn in Figures 8–10 respectively.

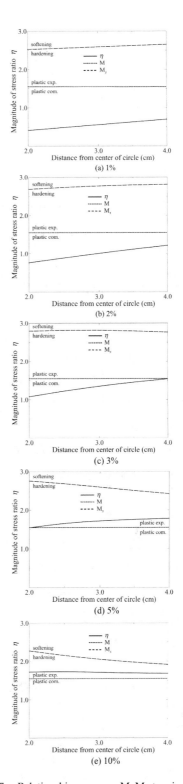

Figure 7. Relationship among η, M_s M at various shear strain.

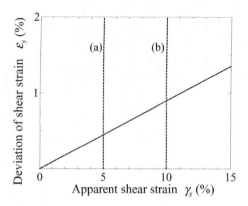

Figure 8. Deviation of shear strain ε_s versus apparent shear strain.

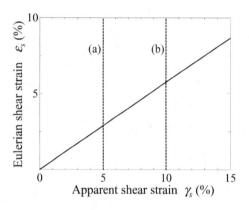

Figure 9. Shear strain ε_s when taking the specimen as a whole mass versus apparent shear strain.

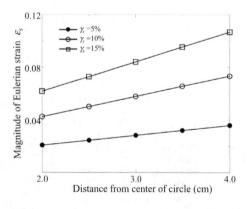

Figure 10. Shear strain ε_s of each element along the radius.

4 CONCLUSIONS

In this paper, the variations of non-uniformity of mean effective stress and stress ratio along with the apparent behavior are discussed in detail. Moreover, the magnitude and distribution of above variables along the radius directions at different stages are also explained. The conclusions are as follows:

1. The non-uniformity of mean effective stress firstly reaches the maximum and then follows minimum when the soil-water couple effect (two-phase) is considered, which is different from the one-phase results the non-uniformity of mean effective stress increases monotonically.

2. As the apparent shear strain increases, the non-uniformity of stress ratio firstly increases to the maximum and then decreases again. When the stress ratio at the outer radius decreases, the difference between the stress ratio at outer and inner radius also becomes decreasing. When the stress ratio reaches the critical state, the non-uniformity of stress ratio turns to be almost zero.

REFERENCES

Asaoka, A. & Noda, T. 1995. Imperfection-sensitive bifurcation of Cam-clay under plane strain compression with undrained boundaries, Soils and Foundations, 35(1): 83–100.

Binbin Xu. 2016. A Numerical Evaluation Method of Non-uniformity in Hollow Torsional Test. Submitted to IFCAT2016.

Hight, D.W., Gen, A. and Symes, M.J. 1983. The development of a new hollow cylinder apparatus for investigating the effects of principal stress rotation in soils, Géotechnique, 33(4): 355–383.

Jin, Y., Ye, B. and Zhang, F. 2010. Numerical simulation of sand subjected to cyclic load under undrained conventional triaxial test, Soils and Foundations, 50(2): 177–194.

Sayao, A. and Vaid, Y.P. 1996. Effect of intermediate principal stress on the deformation response of sand, Can. Geotech. J., 33: 822–828.

Advances in Energy Science and Equipment Engineering II – Zhou, Patty & Chen (Eds)
© 2017 Taylor & Francis Group, London, ISBN 978-1-138-71798-5

Effect of multi-point input on seismic response of large-span isolated structures with different supporting forms

Z.Y. Gu, W.Q. Liu & S.G. Wang

School of Civil Engineering, Nanjing University of Technology, Jiangsu, Nanjing, China

ABSTRACT: The non-stationary multi-point ground motion was generated by traditional trigonometric series method. Isolated and non-isolated structural models of 120 m × 60 mwith multi-point support or boundary support were established in the finite element software. Uniform input and multi-point input were successively applied to all the structures. The analysis results of dynamic seismic response showed that the multi-point effect factor of lower chords and web members for models with different supporting forms differs greatly. It can be drawn that it is very important to take spatial variation of both horizontal and vertical ground motion into consideration when we study the large-span structures, so is the select of supporting forms.

1 INTRODUCTION

With the rapid development of the design of the building structure and the construction technology, a great amount of large-span structures which are more than one hundred meters has emerged at home and abroad. The large distances between the supports of large-span structures can lead to inconsistent ground motion inputs there (multi-point input effect), which will cause distinct seismic response compared with that under uniform input (Qu T.J, Wang Q.X., 1993). Under the multi-point input, if the supporting forms of the structure are different, the distribution of internal force and the pattern of response will differ greatly.

Based on the above research background and practical engineering, two rectangular large-span plate truss structures were selected as research objects, which had boundary and multi-point supports and respectively. Firstly, the non-stationary multi-point ground motion was artificially generated. Secondly, the non-isolated and isolated large-span models with two kinds of supporting forms were established by the finite element analysis software. Then the calculation of the maximum axial force of the upper members was carried out under horizontal multi-point input, vertical multi-point input horizontal uniform input and vertical uniform input. At last the difference of the seismic response between the non-isolated and the isolated structures was analyzed. All the above can provide some theoretical basis for the seismic analysis and design of large-span truss structures.

2 MOTION EQUATION

The dynamic equilibrium equation of structures under multi-point input can be described as:

$$\begin{bmatrix} M_{ss} & M_{sb} \\ M_{bs} & M_{bb} \end{bmatrix} \begin{Bmatrix} \ddot{u}_s \\ \ddot{u}_b \end{Bmatrix} + \begin{bmatrix} C_{ss} & C_{sb} \\ C_{bs} & C_{bb} \end{bmatrix} \begin{Bmatrix} \dot{u}_s \\ \dot{u}_b \end{Bmatrix} + \begin{bmatrix} K_{ss} & K_{sb} \\ K_{bs} & K_{bb} \end{bmatrix} \begin{Bmatrix} u_s \\ u_b \end{Bmatrix} = \begin{Bmatrix} R_s \\ R_b \end{Bmatrix} \tag{1}$$

where, M is the mass matrix; C is the damping matrix; K is the stiffness matrix; ss, bb and sb represent degrees of freedom of superstructure, the bearing and their coupling term; u_s is the absolute vector of displacement for the superstructure; u_b is the absolute vector of displacement for the bearing; R_s, R_b represent the vector of bearing reaction; M_{sb} is 0 for lumped mass model.

Ignore the items whose value is very small, then the equilibrium equation of superstructure can be obtained:

$$M_{ss}\ddot{u}_s + C_{ss}\dot{u}_s + K_{ss}u_s = -K_{sb}u_b \tag{2}$$

When the known quantity is ground displacement, Equation (2) is called absolute displacement input model for solving structural response.

3 GENERATION OF THE NON-STATIONARY MULTI-POINT GROUND MOTION

The non-stationary multi-point ground motion is generated by traditional trigonometric series method based on spectral representation. The generated acceleration time history can take both the spatial non-stability of the ground motion and correlation into consideration at the same time. In order to avoid the complicated operation of complex matrix, a simplified generating method is adapted, which can consider traveling wave effect and partial coherence effect separately. For the large-span structures, the coherence function of the ground motion can be written as (Harichandran R.S. & Vanmarcke E.H., 1986; Kanai K., 1957):

$$\gamma_{ij}(i\omega) = |\gamma_{ij}(i\omega)|\exp\left(-i\theta_{ij}(\omega)\right) \tag{3}$$

where $|\gamma_{ij}(i\omega)|$ is hysteresis coherent function, which can measure correlation of seismic motion between any two points; $\theta_{ij}(\omega)$ is phase angle between two points that can reflect the influence of traveling wave effect, which can be represented with the distance between two points, $d_{ij} = d_i - d_j$, and the apparent wave velocity, V_{app}:

$$\theta_{ij}(\omega) = \frac{\omega d_{ij}}{V_{app}} = \frac{\omega}{V_{app}}\left(d_i - d_j\right) \tag{4}$$

Assume that the auto-power spectrum density function of each point on the ground is the same usually, that is $S_0(\omega)$, therefore, the cross power spectral density function between any two points is:

$$S_{ij}(i\omega) = |\gamma_{ij}(i\omega)|\exp\left(-i\omega\frac{(d_i - d_j)}{V_{app}}\right)S_0(\omega) \tag{5}$$

The cross power spectral density matrix of each point on the ground is:

$$S(i\omega) = \begin{bmatrix} S_{11}(i\omega) & S_{12}(i\omega) & \dots & S_{1M}(i\omega) \\ S_{21}(i\omega) & S_{22}(i\omega) & \dots & S_{2M}(i\omega) \\ \dots & \dots & \dots & \dots \\ S_{M1}(i\omega) & S_{M2}(i\omega) & \dots & S_{MM}(i\omega) \end{bmatrix} \tag{6}$$

Substitute Equation (5) into Equation (6), and the power spectrum matrix of multi-point ground motion can be written as:

$$S(i\omega) = BRB^H S_0(\omega) \tag{7}$$

where

$$B = diag\left[\exp\left(-i\omega\frac{d_1}{V_{app}}\right), \exp\left(-i\omega\frac{d_2}{V_{app}}\right), \right.$$
$$\left. \dots \exp\left(-i\omega\frac{d_M}{V_{app}}\right)\right] \tag{8}$$

$$R = \begin{bmatrix} 1 & |\gamma_{12}| & \dots & |\gamma_{1M}| \\ |\gamma_{21}| & 1 & \dots & |\gamma_{2M}| \\ \dots & \dots & \dots & \dots \\ |\gamma_{M1}| & |\gamma_{M2}| & \dots & 1 \end{bmatrix} \tag{9}$$

R is non-negative definite real symmetric matrix, which can become the product of the lower triangular matrix, Q, and its transpose, Q^H:

$$R(\omega) = Q(\omega)Q^H(\omega) \tag{10}$$

Equation (7) becomes:

$$S(i\omega) = BQQ^H B^H S_0(\omega)$$
$$= \left[\sqrt{S_0(\omega)}BQ\right]\left[\sqrt{S_0(\omega)}BQ\right]^H \tag{11}$$

Make

$$L = \sqrt{S_0(\omega)}BQ \tag{12}$$

Then

$$A_{im}(\omega) = \sqrt{4S_0(\omega)\Delta\omega}|Q_{im}(\omega)| \tag{13}$$

$$\beta_{im}(\omega) = \tan^{-1}\left(\frac{\mathrm{Im}\,e^{-i\omega\frac{d_i}{V_{app}}}}{\mathrm{Re}\,e^{-i\omega\frac{d_i}{V_{app}}}}\right) = -\omega\frac{d_i}{V_{app}} \tag{14}$$

Therefore, the multi-point stationary ground motion can be obtained:

$$a_i(t) = \sum_{m=1}^{i}\sum_{n=1}^{N}A_{im}(\omega_n)\cos\left[\omega_n\left(t - \frac{d_i}{V_{app}}\right) + \phi_{mn}(\omega_n)\right] \tag{15}$$

The three-segment intensity envelope function was put forward by Amin and Ang (Amin M., & Ang A.H.S., 1968):

$$f(t) = \begin{cases} (t/t_1)^2 & 0 \le t \le t_1 \\ 1 & t_1 \le t \le t_2 \\ \exp\left[-c(t - t_2)\right] & t_2 \le t \end{cases} \tag{16}$$

where c is attenuation coefficient; t_1 and t_2 are start and end time of strong motion duration.

Multiply Equation (15) by Equation (16), the non-stationary multi-point ground motion will be obtained.

In the paper, we assume site fortification intensity is 8 degree; site classification is II; damping ratio is 0.05; classification of design earthquake is the second one; $V_{app} = 200$ m/s; the structures are under frequent seismic action.

The auto-power spectrum uses Kanai-Tajimi spectra (Tajimi H., 1960):

$$S(\omega) = \frac{\omega_g^4 + 4\zeta_g^2 \omega_g^2 \omega^2}{(\omega_g^2 - \omega^2)^2 + 4\zeta_g^2 \omega_g^2 \omega^2} S_0 \quad (17)$$

where $\omega_g = 23.3$ rad/s², $\zeta_g = 0.42$, $S_0 = 36.8$ cm²/s³; hysteresis coherent function uses Harichandran and Vanmarcke model:

$$\gamma_{ij}(i\omega) = |\gamma_{ij}(i\omega)| \exp\left(-i\frac{\omega d_{ij}}{V_{app}}\right) \quad (18)$$

$$|\gamma(\omega, d_{ij})| = A \exp\left[-\frac{2d_{ij}}{\alpha\theta(\omega)}(1 - A + \alpha A)\right]$$
$$+ (1 - A)\exp\left[-\frac{2d_{ij}}{\theta(\omega)}(1 - A + \alpha A)\right] \quad (19)$$

$$\theta(\omega) = K\left[1 + \left(\omega/\omega_0\right)^b\right]^{-0.5} \quad (20)$$

The value of parameters are determined as $A = 0.736$; $\alpha = 0.147$; $K = 5210$; $\omega_0 = 6.85$; $b = 2.78$.

Due to site classification is II, some value of parameters can be determined: $t_1 = 0.8$ s; $t_2 = 7$ s; $c = 0.35$.

Program in MATLAB according to simplified method of generating ground motion and the above parameters, so order to generate all the acceleration time history required in the following sections. Transform the acceleration time history into displacement time history, and then we can use absolute displacement model mentioned in the last section as input of the two structures.

4 STRUCTURE MODELS

The model 1 and model 2 are multi-point supporting and boundary supporting rectangular structures with the same plane size 120 m × 60 m. Both of the models are consist of the upper plate truss and the lower frame. The grid form of the superstructure is orthogonal spatial square pyramid. The height of truss is 3.45 m. The cross section of members is steel tube. Elastic modulus of steel is 206Gpa. The uniformly distributed loads, including dead load 0.55 kN/m² and live load 0.50 kN/m² are applied to the upper chords. The uniformly distributed loads, including dead load 0.30 kN/m² and live load 0.30 kN/m² are applied to the lower chords. The material of the lower frame beams and the supporting columns are C40. There are two stories and the storey-height is10 m. The finite

Figure 1. FEM model of model 1.

Figure 2. FEM model of model 2.

Table 1. Natural period of two models.

Model 1			Model 2		
Modal order	Isolated	Non-isolated	Modal order	Isolated	Non-isolated
1	0.957	2.570	1	0.610	2.364
2	0.946	2.559	2	0.528	2.350
3	0.918	2.407	3	0.520	2.249
4	0.595	0.816	4	0.496	1.149
5	0.582	0.806	5	0.490	1.117
6	0.561	0.776	6	0.398	0.876
7	0.483	0.595	7	0.378	0.780
8	0.464	0.583	8	0.330	0.741
9	0.442	0.562	9	0.307	0.638
10	0.435	0.483	10	0.284	0.613

element models of the two structures are shown in Figure 1 and Figure 2.

Set isolation layer at the bottom of supporting columns by using lead rubber bearings and natural rubber bearings. Then the isolation layer will be equipped with the capacity of vertical bearing, variable horizontal stiffness, horizontal elastic recovery and large energy dissipation.

The modal analysis of the two models is carried out. The first 10 order natural vibration periods of the two models before and after isolation are listed in Table 1. The frequencies of the two non-isolated structures distribute closely which can reflect complexity of structural dynamic characteristics. The period of structures after isolation becomes obviously longer.

5 SEISMIC ANALYSIS RESPONSE

The eighth degree frequent earthquakes are applied to model 1 and model 2 at the bottom of all columns, in the X direction, Y direction and Z direction. Under multi-point input and uniform input, response analysis of all structures is carried out by using the finite element software SAP2000 (Liang J.Q, Ye J.H, 2003; Shen, S.G.et al., 2008; Pan, D.G.et al., 2001; Bai, F.L. & Li, H.N., 2010; Yang, Q.S. et al., 2008). In order to research the influence of multi-point effect on the structural seismic response, define a parameter as follows:

$$\gamma_m = \frac{the\ internal\ force\ of\ members\ under\ multi-point\ input}{the\ internal\ force\ of\ members\ under\ uniform\ input}$$

The control members of the upper truss are the lower chords and web members which connect with supporting columns of the lower frame. The dotted lines in Figure 3 and Figure 4 denote the to the control members for the two models.

5.1 Response under X-directional earthquake action

Under X-directional earthquake action, all the multi-point effect factors of axial forces are shown in Figure 5 and Figure 6.

It can be seen that the multi-point effect factors of lower chords and web members for non-isolated model1 and model 2 are almost much greater than 1 under X-directional earthquake action. It can illustrate that it is very important

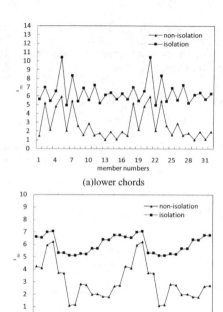

(a)lower chords

(b) web members

Figure 5.　Multi-point effect factors of axial forces for model 1, under X-directional earthquake action.

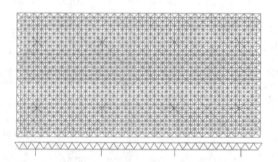

Figure 3.　Position of control members for model 1.

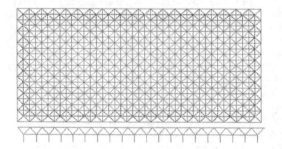

Figure 4.　Position of control members for model 2.

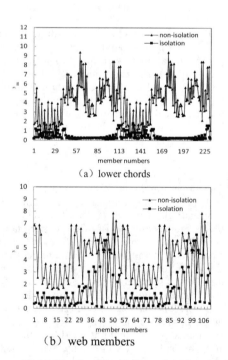

（a）lower chords

（b）web members

Figure 6.　Multi-point effect factors of axial forces for model 2, under X-directional earthquake action.

to consider spatial variation of ground motion when we study the seismic response of large-span structures. The multi-point effect factors of isolated model 1 are still greater than 1, and the effect grows compared to non-isolation. However, nearly half of the multi-point effect factors for isolated model 2 are less than 1, and the effect fades compared to non-isolation. All the results show the importance of supporting forms for large-span structures.

5.2 Response under Y-directional earthquake action

Figure 7 and Figure 8 show the multi-point effect factors of axial forces for model 1 and model 2, respectively, under Y-directional earthquake action. It can be seen that most of multi-point effect factors for non-isolated model 1 are greater than 1, while all those of model 2 are bigger than 1. Whether lower chords or web members, the multi-point effect of axial forces for model 1 has

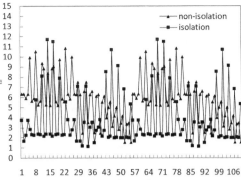

(a) lower chords

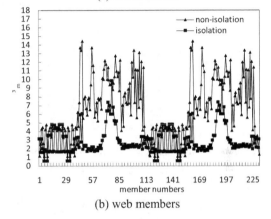

(b) web members

Figure 8.　Multi-point effect factors of axial forces for model 2, under Y-directional earthquake action.

an obvious grow after isolation. However, the multi-point effect of axial forces for isolated model 2 decline mostly.

5.3 Response under Z-directional earthquake action

Figure 9 and Figure 10 show the multi-point effect factors of axial forces for model 1 and model 2, respectively, under Z-directional earthquake action.

Form the Fig. 9 and Fig. 10, we can see the multi-point effect factors of model 1 and model 2 have just a little difference, and the regularities are similar, under vertical earthquake action. The axial forces of lower chords or web members for two models under multi-point excitation are obviously bigger than those under uniform excitation. It can be drawn that it is very important to take spatial variation of vertical ground motion into consideration when we study the large-span structures.

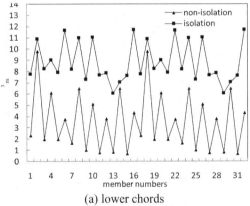

(a) lower chords

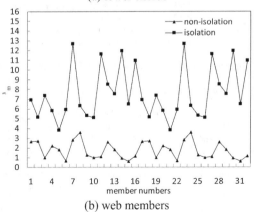

(b) web members

Figure 7.　Multi-point effect factors of axial forces for model 1, under Y-directional earthquake action.

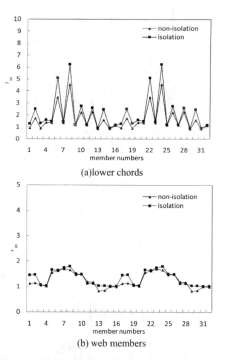

Figure 9. Multi-point effect factors of axial forces for model 1 under Z-directional earthquake action.

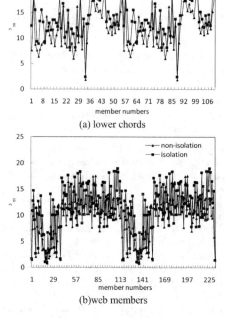

Figure 10. Multi-point effect factors of axial forces for model 2 under Z-directional earthquake action.

6 CONCLUSION

Based on the simplified generating method of non-stationary multi-point ground motion and the establishment of rectangular large-span models with boundary or multi-point supports by the finite element analysis software, the difference of the seismic response between the non-isolated and the isolated structures with two supporting forms was analyzed through the calculation of the maximum axial force of the upper members under horizontal multi-point input, vertical multi-point input horizontal uniform input and vertical uniform input. It can be drawn that it is very important to take spatial variation of both horizontal and vertical ground motion into consideration when we study the large-span structures, so is the select of supporting forms.

REFERENCES

Amin. M. & Ang, H.S. (1968). Non-stationary Stochastic Model of Earthquake Motion. Engineering Mechanics Division, 94(EM2): 559–583.

Bai, F.L. & Li, H.N. (2010). Seismic response analysis of large-span spatial truss structures under multi-point input. Engineering mechanics, 27(7):67–73.

Harichandran R.S. & Vanmarcke E.H.(1986). Stochastic variation of earthquake ground motion in space and time. Journal of Engineering Mechanics, 112(2):154–174.

Kanai K (1957). Semi-empirical formula for the seismic characteristics of the ground. Bulletin of earthquake research institute, 35:70–76.

Liang, J.Q. & Ye, J.H (2003).Seismic response analysis of large-span spatial grid structures under multi-point input. Journal of Southeast University, 33(5):625–630.

Pan, D.G., Lou, M.L. & Fan, L.C. (2001). Status of seismic response analysis of large-span structures under multiple support excitations. Journal of Tongji University,29(10):1213–1219.

Qu, T.J. & Wang, Q.X. A (1993). progress in research on seismic response analysis under multi-support excitation, 9(3):30–36.

Shen, S.G., Zhang, W.G. & Zhu, D. (2008). Seismic response of large-span hangar structure under multi-point input. China Civil Engineering Journal, 41(2):17–21.

Tajimi H. (1960). A statistical method of determining the maximum response of a building structure during an earthquake. Proceeding of Second World Conference on Earthquake Engineering, Tokyo and Kyoto, Japan, 2: 1467–1482.

Yang, Q.S., Liu, W.H. & Tian, Y.J. (2008). Response analysis of national stadium under multi-point input. China Civil Engineering Journal, 41(2):35–41.

Advances in Energy Science and Equipment Engineering II – Zhou, Patty & Chen (Eds)
© 2017 Taylor & Francis Group, London, ISBN 978-1-138-71798-5

Experimental measurement of cloud-point and bubble-point for the {poly(vinyl acetate) + CO₂ + ethanol} systems

Teng Zhu, Houjian Gong & Yajun Li
School of Petroleum Engineering, China University of Petroleum (East China), Qingdao, China

Mingzhe Dong
Department of Chemical and Petroleum Engineering, University of Calgary, Calgary, Canada

ABSTRACT: Phase equilibrium data were reported for the ternary system at temperatures ranging from (30 to 65)°C and pressures ranging from (0 to 40) MPa. The cloud-point and bubble-point data of this ternary mixtures were measured using the supercritical CO_2 phase equilibrium apparatus. The phase behavior for these mixtures were measured with CO_2 mass fraction of 0.04, 0.12, 0.20, 0.28, 0.36, 0.44, 0.52, 0.60, 0.68, 0.76, 0.80, 0.84 and 0.88 with a fixed mass ratio of PVAc and ethanol (1:7) in this study. The p-x phase equilibrium diagram for the ternary system shows that the cloud-point or bubble-point pressure increase as the addition of CO_2. The cloud-point curves take on the typical appearance of a lower critical solution temperature behavior.

1   INTRODUCTION

CO_2 has been used as environmentally friendly solvent in polymerization processes, polymeric materials production, and polymer purification; CO_2 has also applied in petroleum engineering as CO_2-based drilling fluid and CO_2-based fracturing fluid (Jang, Yong, & Byun, 2015). However, the application of CO_2 in petroleum engineering is limited because of the low viscosity and low density of the supercritical CO_2. It is necessary to increase the viscosity and density of CO_2 by adding polymer and cosolvents under moderate temperatures (30 to 70°C) and pressures (0 to 40 MPa) for oil and gas development process (Gu, Zhang, & She, 2013).

As a weak solvent which forms weak complexes with basic functional groups in polymers (Yuan & Teja, 2010), CO_2 can not dissolve in most of polymers under moderate conditions (<100°C, <50 MPa). Poly (vinyl acetate) (PVAc) has been shown to exhibit reasonably high solubility in CO_2 because of the characteristics of amorphous structure and low melting point (Shen et al., 2003). In spite of this, PVAc can be dissolved above 30 MPa with relatively low molecular weights (Sevgi Kilic et al., 2007). Therefore, some cosolvent need to be added to shift the cloud-point curve of (polymer + CO_2) system at relatively low temperatures and pressures (Arce & Aznar, 2005).

A cosolvent can greatly decreased the cloud-point pressure of (polymer + CO_2) system because it can reduce the free volume difference between

polymer and CO_2 (And & Lee, 2005). Previous work has shown many experimental data for phase equilibrium in (polymer + CO_2 + cosolvent) systems (Hossain & Teja, 2014). Polar cosolvents such as ethanol, acetate acid, and ethyl acetate et al. are often added to increase the solubility of polymer in CO_2 (Matsuyama & Mishima, 2006). As the CO_2-philic homopolymer, experimental data for the binary system (PVAc + CO_2) has been reported many times (Tan, Bray, & Cooper, 2009). However, there are few reports about the influence of cosolvent on the solubility of PVAc in CO_2. In this work, the liquid-liquid or vapor-liquid equilibrium of (PVAc + CO_2 + ethanol) ternary systems were determined at constant pressure and temperature. Cloud-point and bubble-point curves were obtained at pressures up to 40 MPa and temperatures up to 65°C in order to determine the effect of the mass fraction of CO_2 on the phase behavior of (PVAc + CO_2 + ethanol) mixtures.

2   EXPERIMENTAL METHODS

2.1   *Materials*

CO_2 was obtained from Qingdao Tianyuan industrial Gas Co. Ltd. with mass fraction of more than 99.99%. Poly (vinyl acetate) (PVAc) with narrow molecular weight distribution ($M_w = 50$ kg·mol⁻¹, PDI = 1.13; polydispersity index (PDI) = M_w/M_n) was supplied by Sigma-Aldrich Co. Ltd., Ethanol with purity of more than 99.9% was obtained from

Guangdong Xilong Chemical Co. Ltd., All reagents were used without further purification.

2.2 *Apparatus and procedure*

The cloud-point and bubble-point pressures of the ternary systems were measured using the supercritical CO_2 phase equilibrium apparatus. A schematic diagram of the apparatus is shown in Figure 1. The apparatus mainly includes a view cell components, a supercharging system and a cloud-point measurement device. The view cell contains two sapphire windows on both side of it in order to allow light source through it. The volume of the view cell can be varied between 150 mL and 350 mL utilizing a servo motor to control the piston movement.

In a typical experiment, a known amount of PVAc and ethanol were loaded into the view cell, then the view cell was purged with nitrogen several times using a vacuum pump. After that, CO_2 from a cylinder was pressurized into a storage tank using the supercharging system. After the pressure in the storage tank was up to 40 MPa, the globe valve was opened so the high pressure CO_2 could flowed into the view cell. Then, the supercharging system was removed and the cloud-point measurement device was installed. A parallel light source was fixed on one side of the sapphire window, while a photoresistance which connected to a multimeter was fixed on the other side. Then, the pressure was decreased with a rate of about $\Delta p = 0.5$ MPa min^{-1} until the polymer-rich phase or polymer-poor phase was separated from the mixture. In this process, the resistance displayed on the multimeter increased rapidly until out of range. Then, the pressure increased with the same rate until the mixture became transparent homogeneous system again, and the resistance back to the previous values. The phase transition point was taken as the derivative extremum of the resistance-pressure curves. The liquid-liquid or vapor-liquid phase equilibrium can be distinguished significantly by visual observation. The experiments were repeated at least three times (Görnert & Sadowski, 2008).

3 RESULT AND DISCUSSION

The cloud-point or bubble point phase behavior of the (PVAc + CO_2 + ethanol) ternary mixtures were measured. The maximum deviation was $\Delta w = \pm 1.0\%$ for mass fraction, $\Delta T = \pm 0.1$ K for temperature and $\Delta p = \pm 0.1$ MPa for pressure.

The p-x phase diagram and experimental data for the {PVAc(1) + CO_2(2) + ethanol(3)} ternary systems are shown in Figure 2. and Table 1. The mass ratio of PVAc and ethanol is fixed to 1:7 in all experiment. The p-T phase diagram of the {PVAc(1) + CO_2(2) + ethanol(3)} ternary systems show a positive slope in the temperature range of 308.2–338.2 K and pressure up to 40 MPa. The phase behavior curve of this system was the LCST type (Baek & Byun, 2016). The p-x phase diagram shows that the cloud-point and bubble-point pressure increase as the addition of CO_2. The bubble-point pressures were measured to determine the vapor-liquid phase equilibrium with $w_2 = 0.04$–0.52 CO_2, and the cloud-point pressures were measured to determine the liquid-liquid phase equilibrium with $w_2 = 0.68$–0.88 CO_2 from 308.2 to 338.2 K.

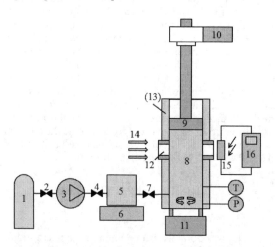

Figure 1. The schematic diagram of supercritical CO_2 phase equilibrium apparatus. (1) CO_2 cylinder; (2), (4), (7) globe valve; (3) system; (5) storage tank; (6) balance; (8) high-pressure view cell; (9) piston; (10) servo motor; (11) magnetic-coupled drive; (12) sapphire window; (13) oil bath heating jacket; (14) light source; (15) photoresistance; (16) multimeter.

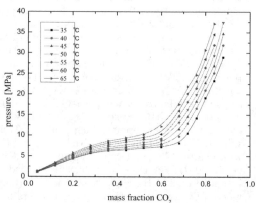

Figure 2. p-x phase equilibrium diagram for the {PVAc(1) + CO_2(2) + ethanol(3)} ternary systems with different CO_2 content measured in this study, a fixed mass ratio of PVAc and ethanol is 1:7 in all experiments.

Table 1. Experimental data of the {PVAc(1) + CO$_2$(2) + ethanol(3)} ternary systems with different CO$_2$ content measured in this study, a fixed mass ratio of PVAc and ethanol is 1:7 in all experiments.

T/°C	P/MPa	Transition
$w_1 = 0.12, w_3 = 0.84$		
35.0	1.07	BP
40.0	1.12	BP
45.0	1.17	BP
50.0	1.20	BP
55.0	1.24	BP
60.0	1.29	BP
65.0	1.34	BP
$w_1 = 0.11, w_3 = 0.77$		
35.0	2.67	BP
40.0	2.75	BP
45.0	2.89	BP
50.0	3.01	BP
55.0	3.12	BP
60.0	3.22	BP
65.0	3.33	BP
$w_1 = 0.10, w_3 = 0.70$		
35.0	4.37	BP
40.0	4.46	BP
45.0	4.65	BP
50.0	4.88	BP
55.0	5.16	BP
60.0	5.46	BP
65.0	5.76	BP
$w_1 = 0.09, w_3 = 0.63$		
35.0	5.70	BP
40.0	5.86	BP
45.0	6.12	BP
50.0	6.45	BP
55.0	6.83	BP
60.0	7.15	BP
65.0	7.52	BP
$w_1 = 0.08, w_3 = 0.56$		
35.0	6.34	BP
40.0	6.52	BP
45.0	6.85	BP
50.0	7.23	BP
55.0	7.66	BP
60.0	8.14	BP
65.0	8.63	BP
$w_1 = 0.07, w_3 = 0.49$		
35.0	6.45	BP
40.0	6.73	BP
45.0	7.21	BP
50.0	7.67	BP
55.0	8.39	BP
60.0	8.88	BP
65.0	9.38	BP
$w_1 = 0.06, w_3 = 0.42$		
35.0	7.01	BP
40.0	7.35	BP

(*Continued*)

Table 1. (*Continued*).

T/°C	P/MPa	Transition
45.0	7.88	BP
50.0	8.32	BP
55.0	9.03	BP
60.0	9.52	BP
65.0	10.20	BP
$w_1 = 0.05, w_3 = 0.35$		
35.0	7.20	BP
40.0	7.62	BP
45.0	8.01	BP
50.0	8.77	BP
55.0	9.17	BP
60.0	10.30	CP
65.0	12.15	CP
$w_1 = 0.04, w_3 = 0.28$		
35.0	8.10	CP
40.0	9.73	CP
45.0	11.51	CP
50.0	13.15	CP
55.0	14.70	CP
60.0	15.94	CP
65.0	17.53	CP
$w_1 = 0.035, w_3 = 0.245$		
35.0	10.47	CP
40.0	12.89	CP
45.0	14.55	CP
50.0	16.25	CP
55.0	18.17	CP
60.0	20.14	CP
65.0	21.85	CP
$w_1 = 0.03, w_3 = 0.21$		
35.0	14.14	CP
40.0	16.54	CP
45.0	18.25	CP
50.0	19.88	CP
55.0	21.43	CP
60.0	23.10	CP
65.0	24.67	CP
$w_1 = 0.025, w_3 = 0.175$		
35.0	19.10	CP
40.0	20.96	CP
45.0	22.74	CP
50.0	24.21	CP
55.0	26.43	CP
60.0	28.51	CP
65.0	30.66	CP
$w_1 = 0.02, w_3 = 0.14$		
35.0	23.23	CP
40.0	25.08	CP
45.0	26.84	CP
50.0	29.34	CP
55.0	31.89	CP
60.0	34.51	CP
65.0	37.08	CP

(*Continued*)

Table 1. (*Continued*).

T/°C	P/MPa	Transition
$w_1 = 0.015$, $w_3 = 0.105$		
35.0	28.93	CP
40.0	31.84	CP
45.0	34.66	CP
50.0	37.40	CP

CP-cloud point, BP-bubble point.

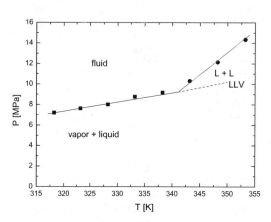

Figure 3. Effect of $w_2 = 0.60$ CO_2 on the phase behavior of the {PVAc(1) + CO_2(2) + ethanol(3)} mixtures. -•-: fluid—(liquid + liquid) transition; -■-: fluid—(vapor + liquid) transition; dash line: suggested extension of the LLV.

Figure 3 shows a p-T phase diagram for the {PVAc(1) + CO_2(2) + ethanol(3)} ternary systems with $w_1 = 0.05$, $w_3 = 0.35$. The cloud-point pressure curve takes on the typical appearance of a lower critical solution temperature (LCST) boundary. The phase diagram shows a vapor-liquid curve at 308.2–328.2 K or a liquid-liquid curve at 333.2–338.2 K. The phase behavior curve intersects a fluid---(vapor + liquid) (VL) curve at ca. 342 K and ca. 9.0 MPa. A liquid and vapor phase coexists at pressure below this curve, and the VL curve switches to a {liquid(1) + liquid(2) + vapor}(LLV) curve at temperature higher than about 342 K. The slope of this {PVAc(1) + CO_2(2) + ethanol(3)} LCST curve at (vapor + liquid) region and (liquid + liquid) region were ca. 0.10 MPa/K and 0.20 MPa/K, respectively.

4 CONCLUSIONS

A novel high-pressure variable volume and visual apparatus for phase equilibrium measurements of (PVAc + CO_2 + ethanol) ternary systems was developed utilizing a photoresistance to reduce the experiment deviation.

The experimental results for the ternary systems reveal quantitatively the effect of pressure,

temperature, as well as the composition of CO_2 on the phase behavior of the ternary systems. The cloud-point or bubble-point pressure increase as the addition of CO_2. When the {PVAc(1) + CO_2(2) + ethanol(3)} ternary system with $w_1 = 0.05$, $w_3 = 0.35$, the cloud-point pressure curve takes on the typical appearance of the LCST boundary.

REFERENCES

And, H.S.B., & Lee, D.H. (2005). Phase Behavior of Binary and Ternary Mixtures of Poly (decyl acrylate)–Supercritical Solvents–Decyl Acrylate and Poly(decyl methacrylate)–CO_2–Decyl Methacrylate Systems. *Industrial & Engineering Chemistry Research, 45*(10), 3373–3380.

Arce, P.F., & Aznar, M. (2005). Phase behavior of polypropylene + n-pentane and polypropylene + n-pentane + carbon dioxide: modeling with cubic and non-cubic equations of state. *Journal of Supercritical Fluids the, 34*(2), 177–182.

Baek, S.H., & Byun, H.S. (2016). Bubble-point measurement for the binary mixture of propargyl acrylate and propargyl methacrylate in supercritical carbon dioxide. *Journal of Chemical Thermodynamics, 92*, 191–197.

Chen, Z.H., Cao, K., Yao, Z., & Huang, Z.M. (2009). Modeling solubilities of subcritical and supercritical fluids in polymers with cubic and non-cubic equations of state. *Journal of Supercritical Fluids, 49*(2), 143–153.

Görnert, M., & Sadowski, G. (2008). Phase-equilibrium measurement and modeling of the PMMA/MMA/carbon dioxide ternary system. *Journal of Supercritical Fluids the, 46*(3), 218–225.

Gu, Y., Zhang, S., & She, Y. (2013). Effects of polymers as direct CO_2 thickeners on the mutual interactions between a light crude oil and CO_2. *Journal of Polymer Research, 20*(2), 1–13.

Hossain, M.Z., & Teja, A.S. (2014). Modeling phase equilibria in CO 2 +polymer systems. *Journal of Supercritical Fluids the, 96*, 313–323.

Jang, Y.S., Yong, S.C., & Byun, H.S. (2015). Phase behavior for the poly(2-methoxyethyl acrylate)+supercritical solvent+cosolvent mixture and CO_2+2-methoxyethyl acrylate system at high pressure. *Korean Journal of Chemical Engineering, 32*(5), 958–966.

Matsuyama, K., & Mishima, K. (2006). Phase behavior of CO_2 +polyethylene glycol+ethanol at pressures up to 20 MPa. *Fluid Phase Equilibria, 249*(1–2), 173–178.

Sevgi Kilic, Stephen Michalik, Wang, Y., ‡, J.K.J., †,, And, R.M.E., †, ‡, & †, E.J.B. (2007). Phase Behavior of Oxygen-Containing Polymers in CO_2. *Macromolecules, 40*(4), 1332–1341.

Shen, Z., Mchugh, M.A., Xu, J., Belardi, J., Kilic, S., Mesiano, A.,. .. Enick, R. (2003). CO_2-solubility of oligomers and polymers that contain the carbonyl group. *Polymer, 44*(5), 1491–1498.

Tan, B., Bray, C.L., & Cooper, A.I. (2009). Fractionation of Poly (vinyl acetate) and the Phase Behavior of End-Group Modified Oligo(vinyl acetate)s in CO_2. *Macromolecules, 42*(20), 7945–7952.

Yuan, Y., & Teja, A.S. (2010). Reprint of: Extension of a compressible lattice model to CO_2 + cosolvent + polymer systems. *Journal of Supercritical Fluids the, 55*(1), 358–362.

Study of the behavior of 3d-shell electrons in ZnO-based varistor ceramics with Nb_2O_5 additions

Tie-Zhu Yang
Department of Basic Courses, Zhengzhou College of Science and Technology, Zhengzhou,
Henan Province, P.R. China

Qi-Tao Zhu
Department of Basic Courses, College of Information and Business, Zhongyuan University
of Technology, Zhengzhou, Henan Province, P.R. China

Shu-Sheng Meng & Zhi-Qiang Zhang
Department of Basic Courses, Zhengzhou College of Science and Technology, Zhengzhou,
Henan Province, P.R. China

Yan-Qiong Lü & Wen Deng
Department of Physics, Guangxi University, Nanning, Guangxi Province, P.R. China

ABSTRACT: The behavior of 3d electrons in ZnO based varistor ceramics doped with different Nb_2O_5 content from 0.05 to 0.5 mol% has been studied by using positron coincidence Doppler broadening techniques. It has been found that the 3d electron signal in sample a03 is relatively high as compared to other samples. As the content of Nb^{5+} increases, which gives rise to the decrease in the probability of positron annihilation with 3d electrons. The d-d interactions are weakened and the positron annihilation probability with 3d electron decreases with the addition of Nb_2O_5 up to 0.2 mol%. When the content of Nb_2O_5 is more than 0.2 mol%, the above mentioned changes will be reversed. The nonlinear coefficient α and the leakage current I_L of the ZnO ceramics decrease with the d-d interaction, moreover, the threshold field E_B increases with the d-d interaction.

1 INTRODUCTION

ZnO varistor ceramics are useful functional ceramic materials and are extensively used in electronic circuits to sense and limit transient voltage surges and to do so repeatedly without being destroyed. They show a strong nonlinear current–voltage characteristic. The highly non-linear I-V behavior is caused by the formation of back-to-back Schottky barriers at the ZnO grain boundaries (Xu, 2014; Hofstätter, 2013; Daneu, 2013; Pillai, 2013). The electrical properties of ZnO varistor ceramics are directly related to the additions and microstructure(Xu, 2015; Izoulet, 2014). The microstructure and electrical properties of ternary $ZnO - V_2O_5 - Mn_3O_4$ varistor ceramics with different amounts of Nb_2O_5 (from 0.05 mol% to 0.25 mol%) were investigated by Choon-W. Nahm (Nahm, C.-W., 2012). His results show that, for Nb_2O_5-doped ceramics, SEM (scanning electron microscope) data analysis revealed the average grain size for Nb_2O_5-doped ceramics was larger than Nb_2O_5-free ceramics. The breakdown field increased from 947 to 4521V/cm with an increase in the amount of Nb_2O_5, whereas

a further addition caused it to decrease down to 4374 V/cm at 0.25 mol%.

The addition which influences the defects and electronic structure is known to be the crucial factors to control the electrical properties of ZnO varistor ceramics. Positron annihilation techniques are well suited to investigate defects and electron density in materials, and they are well established to detect open volume and negatively charged centers in solids (West, R. N., 1973; Xue, 2014; Deng, 1999). The positron lifetime presents information about the electron density of the medium in which it is annihilated. The higher the electron density, the shorter the positron lifetime, and vice versa. The positron annihilation rate in any state reflects the electron density "seen" by the positron in that state (West, R. N., 1973; Brusa, R., 2001). Coincidence Doppler broadening spectrum provides information about the one-dimensional momentum distribution of the annihilating positron–electron pair. The 3d electron signal of materials and ceramics can be extracted from the coincidence Doppler broadening spectra (Deng, 2006; Alatalo, 1995; Szpala, 1996; Kirkegaard, 1981).

In this work, positron lifetime and coincidence Doppler broadening spectra of pure Zn, Nb samples and $ZnO - Bi_2O_3 - TiO_2$ (ZBT) system with different Nb_2O_5 content from 0.05 to 0.5 mol% (a01, a02, a03, a04, successively) have been measured, and obtained information of 3d electron and defect in the samples.

2 EXPERIMENT

2.1 Sample preparation

The samples, based on the ZBT system, were produced following the solid state reaction technique of preparing ceramics. The raw materials used in this work were analytical grades of ZnO (99.9%), Bi_2O_3 (99.5%), T_iO_2 (99.9%) and Nb_2O_5 (99.5%). The compositions used was $(100 - 1.70-x)\%$ ZnO + 0.70% Bi_2O_3 + 1.00% TiO_2 + $x\%Nb_2O_5$ with $x = 0.05, 0.10, 0.20$ and 0.50 and (see Table 1). The oxides were mixed and ball-milled in agate pot and balls of agate with distilled water for 12 h. The aqueous slurry obtained is dried at 300°C for 5 h. After drying, the powders were granulated and pressed into pellet shapes with a diameter of 15 mm and thickness of 2 mm at a pressure of 100 Mpa.

The samples were sintered at 1250°C for 2 h in air and furnace cooled to room temperature. They were heated at a rate of 4°C/min to the sintering temperature. In the process of sintering, compacts were surrounded with powders of matching components and covered with crucibles to reduce evaporation of low-melting point components.

2.2 Parameters of coincidence Doppler broadening spectra and positron lifetime spectra of the varistors

Positron lifetime spectra of pure metals Zn and Nb were measured by the conventional fast-fast coincidence ORTEC system with a time resolution of 240 ps (FWHM, full width at half maximum) under our experimental conditions. ^{22}Na source of approximately 3.7×10^5 Bq was encapsulated in kapton foil, and the source was sandwiched between two identical sample pieces. The measurements of positron lifetime were performed at room temperature (25°C). Coincidence Doppler

Table 1. Chemical composition of the samples (mol%).

Samples	ZnO	Bi_2O_3	TiO_2	Nb_2O_5
a01	93.00	0.70	1.00	0.05
a02	93.00	0.70	1.00	0.10
a03	93.00	0.70	1.00	0.20
a04	93.00	0.70	1.00	0.50

broadening spectra were measured with a two-detector coincidence system (2 high purity Ge detectors and multi parameter analyzer). About 10^5 counts were accumulated for each spectrum.

2.3 Electric performance

The pellets were embedded within a dielectric glass layer. Then two opposite surfaces of the coated samples were polished and an aluminum electrode deposited by sputtering on each side. The current-voltage of the specimen was measured using a $V–I$ source/ measure unit (CJP CJ1001, China). The breakdown field (E_B) was measured at a current density of 1.0 mA/cm²and the leakage current density (I_L) was measured at 0.80 E_B. In addition, the coefficient of nonlinearity (α), for all of the samples studied, was determined by $\propto= 1/\log(E_2 - E_1)$ where E_1 and E_2 are the electric fields corresponding to $I_1 = 1.0$ mA/cm² and $I_2 = 10$ mA/cm², respectively. The V–I characteristics were measured at room temperature.

3 RESULTS AND DISCUSSION

3.1 3d electrons in ZnO based varistor ceramics

In data analysis, POSITRONFIT EXTENDED program (Kirkegaard, 1981) was used. The background of the Doppler broadening spectrum is reduced remarkably (with a peak to background ratio of more than 10^5) by using of the coincidence technique. The 511 keV line is Doppler broadened $(511 + \Delta E)$ due to the longitudinal momentum p_L component of the annihilating positron-electron pair.p_L is correlated to the Doppler shift ΔE by $p_L = 2\Delta E / c$, where c is the light velocity.

Ratio curves are well—established to distinguish the difference between different Doppler broadening spectra, dividing every spectrum by the spectrum of a reference specimen (Cz-Si) (Deng, 2006; Zhu, 2009). All of the spectra have been normalized to a total area of 10^6 from 511 to 530 keV (p_L from 0 to $74.3 \times 10^{-3}m_0c$) and a smoothing routine on nine points was applied before the ratio is taken. Fig. 1(a) shows the ratio curves for pure Zn and Nb. The ratio curves for a01, a02, a03, a04 are shown in Fig. 1(b) and Fig. 1(c). The abscissa is the longitudinal momentum p_L component of the annihilating positron-electron pair in 10^3m_0c.

It can be seen in Fig. 1(a) that the ratio curve for Zn shows a very high peak at about $17.3 \times 10^{-3}m_0c$, which is due to positron annihilation with 3d electron of Zn atom, and the ratio curve for Nb show a low peak at about $10.1 \times 10^{-3}m_0c$, which is mainly due to positron annihilation with 4d electrons of Nb atom. That the ratio curve for Zn is higher

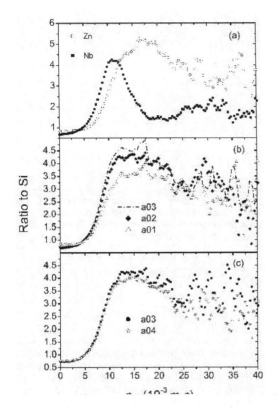

Figure 1. The ratio curves for pure Zn, Nb and ZBT system with different Nb_2O_5 content.

indicates the probability of positron annihilation in Nb atom is lower than in Zn atom.

Our results (see Fig. 1(b)) indicate that the ratio curve of the samples increases with Nb_2O_5 content up to 0.2 mol% and further addition of Nb_2O_5 decreases the ratio curve (see Fig. 1(c)). This can be interpreted as follows: the height of the peak of the ratio curve of the ZnO varistor ceramics depends on the probability of positron annihilation with d electrons and the defect concentration in the ZnO varistor ceramics, that is the height of the peak of the ratio curve increases with the probability of positron annihilation with d electrons, while it will decrease with the increase of the defect concentration (Dupasquier, 2004). This means that in samples a01, a02 and a03, the d-d interactions are enhanced and the probability of positron annihilation with d electrons will increase with the increasing of Nb_2O_5 content, therefore the height of the peak in the ratio curve of the ZnO ceramics will increase. On the other hand, the vacancy compensation in Nb_2O_5 doped ZnO varistor ceramics will be built due to the product of substitutional solid solution, which leads to the decrease of Zn^{2+}

and the increase of Nb^{5+} in ZnO grains, generated by a small amount of Nb^{5+} will replace the Zn^{2+}. Furthermore, Nb_2O_5 doping promoted the growth of ZnO grains, which leads to the decrease of the grain boundaries in unit volume, and therefore the height of the peak of the ratio curve increased. The reason for the decrease in the doppler broadening ratio curves with Nb_2O_5 content greater than 0.2 mol% may be due to the formation of spinel phases of $Zn_3Nb_2O_8$, $ZnNb_2O_6$ and Zn_2TiO_4 with high resistance[20-24] in the grain boundary, which hindered the growth of ZnO grains, so that the grains become smaller and grain boundary defects will increase.

From the above discussion, it has been found that the 3d electron signal in sample a03 is relatively high as compared to others. In Nb_2O_5 doped ZnO varistor ceramics, the height of the peak in the ratio curve of the sample a03 is higher than others, which illustrates probability of positron annihilation with 3d electrons is greatest.

3.2 Parameters of positron lifetime spectra of the ZnO varistor ceramics

The defects in the ZBT system doped with Nb_2O_5 can be further characterized by positron lifetime measurements. The spectra of the ceramics were fitted by using three decay lifetime components after source corrections $(\tau_s = 375 ps, I_s = 8.7\%)$ and background subtractions. Values of the lifetimes (τ_1, τ_2, τ_3) and the corresponding intensities (I_1', I_2', I_3') are different with the different samples. τ_1 is the positron annihilate in the free state of the bulk in the samples; τ_2 is the positron annihilate at the defects; τ_3 is the result of positron annihilate at the surfaces of the samples or the surface of large voids. The long-life component $\tau_3 (\approx 1200\ ps)$ in each sample with a small intensity I_3' ($<1\%$) was considered to be the result of positron annihilated at the surfaces of the samples (Sui, 1991; Lynn, 1977; Wang, 1996; Chen, 2004; Chen, 2005). It is consequently disregarded in our discussion. The first two intensities I_1' and I_2' are re-normalized and marked as I_1 and I_2. The in-between life component τ_2 was due to the lifetime of positron annihilated at the defects. The mean positron lifetime was determined by $\tau_m = I_1\tau_1 + I_2\tau_2$. The parameters of the positron lifetime spectra in the bulk of tested ceramics are shown in Table 2.

It can be seen in Table 2 that the values of the mean positron lifetime τ_m decrease with the addition of Nb_2O_5 up to 0.2 mol% and then increase beyond this composition. In sample a03, the mean positron lifetime τ_m is shortest, which means that the defects in sample a03 are the least and the electron density is the largest. In sample a04, the mean positron lifetime τ_m prolongs and I_2 is bigger, this

Table 2.   Parameters of the positron lifetime spectrum of samples.

Samples	τ_1/ps	τ_2/ps	I_1/%	I_2/%	τm/ps
a01	175 ± 4	326 ± 11	65.4	34.6	227.246
a02	178 ± 2	367 ± 19	84.3	15.7	207.673
a03	169 ± 4	310 ± 17	73.9	26.1	205.801
a04	164 ± 5	393 ± 14	66.1	33.9	241.631

interprets that the defects in ZnO varistor ceramics increase and the electron density decrease with further increase in the content of Nb_2O_5. From what has been discussed above, we find that the lifetime data agree well with the results of Doppler broadening spectra.

3.3   Electrical properties of Nb_2O_5 doped zinc oxide varistor ceramics

The behavior of breakdown field (E_B), nonlinear coefficient(α) and leakage current I_L as a function of Nb_2O_5 amount were indicated graphically in Fig. 2.:

The I_L value increases with the addition of Nb_2O_5 up to 0.10 mol% and further addition of Nb_2O_5 decreases the I_L value slightly. When the Nb_2O_5 content is 0.05 mol%, I_L value exhibited a minimum value ($80\mu A$) (see Fig. 2(a)). The breakdown field (E_B) value decreases from 50 to 15 V/mm with Nb_2O_5 content up to 0.2 mol% and then increases beyond this composition. As indicated in Fig. 2(b). This can be explained by the grain size and the breakdown voltage per grain boundaries. The breakdown voltage increases and leakage current I_L decreases with the number of grain boundaries per unit volume. The value of α is maximum when the amount of Nb_2O_5 is 0.05 mol%, However, further addition in the amount of Nb_2O_5 (0.1 mol%) caused nonlinear coefficient to decrease to minimum (see Fig. 2(c)). This can be interpreted as follow: The potential barrier depending on the electronic states at the grain boundaries in accordance with amount of Nb_2O_5 are closely related to the behavior of α. As a matter of fact, the nonlinear coefficient increases with the potential barrier (Nahm, 2012). The electrical conduction in the obtained ceramic materials is controlled by the grain boundaries (Glot, 2015). In the Nb_2O_5-doped ceramics, the average grain size has a fluctuation. The Nb_2O_5 may vary the density of interface states with the transport of the defect ions toward the grain boundaries. Therefore, the increase or decrease of α with an increase in the amount of Nb_2O_5 is attributed to the increase or decrease of potential barrier height at the grain boundaries[7,32].

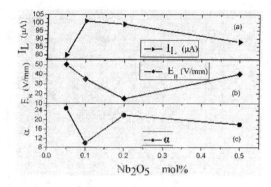

Figure 2.   The curves for leakage current (at the top of the graph, part (a)), breakdown field (In the middle of the graph, part (b)) and nonlinear coefficient (at the bottom of the graph, part (c)) as a function of Nb_2O_5.

4   CONCLUSIONS

1. The ratio curve for Zn shows a very high peak at about 17.3×10^{-3} m_0c, which is due to positron annihilation with 3d electron of Zn atom.
2. The 3d electron signal in ZBT system doped with 0.2 mol% Nb_2O_5 is relatively high as compared to other ZnO varistor ceramics. In Nb_2O_5 doped ZnO varistor ceramics, as the content of Nb_2O_5 is 0.2 mol%, the height of the peak in the ratio curve of the samples is higher than others, which illustrates the greatest probability of positron annihilation with 3d electrons, and with the increase of Nb_2O_5 content more than 0.2 mol%, the d-d interactions are weakened, the probability of positron annihilation with 3d electrons will decrease.
3. The breakdown field (E_B) decreases and the leakage current I_L increases with the d-d interaction.

ACKNOWLEDGEMENTS

This work was supported by the key scientific research projects in universities of Henan Province of China (Grant No. 15B430012), and College of information and business, Zhongyuan University of Technology in 2016 for the scientific research project (Grant No. ky1616).

REFERENCES

Alatalo, M.; Kauppinen, H.; Saarinen, K.; Puska, M.; Mäkinen, J.; Hautojärvi, P.; Nieminen, R. Phys. Rev. B. 1995, 51, 4176.
Brusa, R.; Deng, W.; Karwasz, G.; Zecca, A.; Pliszka, D. Appl. Phys. Lett. 2001, 79, 1492.

Chen, Z.; Kawasuso, A.; Xu, Y.; Naramoto, H.; Yuan, X.; Sekiguchi, T.; Suzuki, R.; Ohdaira, T. Phys. Rev. B. 2005, 71, 115213.

Chen, Z.; Maekawa, M.; Yamamoto, S.; Kawasuso, A.; Yuan, X.; Sekiguchi, T.; Suzuki, R.; Ohdaira, T. Phys. Rev. B. 2004, 69, 035210.

Daneu, N.; Gramc, N. N.; Rečnik, A.; Kržmanc, M. M.; Bernik, S. J. Eur. Ceram. Soc. 2013, 33, 335.

Deng, W.; Huang, Y.; Brusa, R.; Karwasz, G.; Zecca, A. J. Alloys. Compd. 2006, 421, 228.

Deng, W.; Huang, Y.; Brusa, R.; Karwasz, G.; Zecca, A. J. Alloys. compd 2005, 386, 103.

Deng, W.; Zhong, X.; Huang, Y.; Xiong, L.; Wang, S.; Guo, J.; Long, Q. Sci.China Ser. A: Maths. 1999, 42, 87.

Deng, W.; Zhu, Y.; Zhou, Y.; Huang, Y.; Cao, M.; Xiong, L. Rare Metal Mater. Eng. 2006, 35, 348.

Dupasquier, A.; Kögel, G.; Somoza, A. Acta Mater. 2004, 52, 4707.

Glot, A.; Bulpett, R.; Ivon, A.; Gallegos-Acevedo, P. Physica B: Condens. Matter 2015, 457, 108.

Hofstätter, M.; Nevosad, A.; Teichert, C.; Supancic, P.; Danzer, R. J. Eur. Ceram. Soc. 2013, 33, 3473.

Izoulet, A.; Guillemet-Fritsch, S.; Estournès, C.; Morel, J. J. Eur. Ceram.Soc. 2014, 34, 3707.

Kim, D.-W.; Kim, J.-H.; Kim, J.-R.; Hong, K.-S. Jpn. J. Appl. Phys. 2001, 40, 5994.

Kirkegaard, P.; Eldrup, M.; Mogensen, O. E.; Pedersen, N. J. Comput. Phys. Commun. 1981, 23, 307.

Liou, Y.-C.; Chen, H.-M.; Tsai, W.-C. Ceram. Int. 2009, 35, 2135.

Lynn, K.; MacDonald, J.; Boie, R.; Feldman, L.; Gabbe, J.; Robbins, M.; Bonderup, E.; Golovchenko, J. Phys. Rev. Lett. 1977, 38, 241.

Nahm, C. W. J. Am. Ceram. Soc. 2012, 95, 2093.

Nahm, C.-W. Ceram. Int. 2012, 38, 5281.

Nahm, C.-W. Ceram. Int. 2013, 39, 3417.

Nenasheva, E.; Kartenko, N. J. Eur. Ceram. Soc. 2006, 26, 1929.

Nie, Z.; Lin, Y.; Wang, F. J. Eng. Mater. Technol. 2015, 137, 031010.

Pillai, S. C.; Kelly, J. M.; Rameshc, R.; McCormackad, D. E. J. Mater. Chem 2013, 100, 3268.

Shihua, D.; Xi, Y.; Li, Y. J. Electroceram. 2008, 21, 435.

Sui, M.; Lu, K.; Deng, W.; Xiong, L.; Patu, S.; He, Y. Phys. Rev.B 1991, 44, 6466.

Szpala, S.; Asoka-Kumar, P.; Nielsen, B.; Peng, J.; Hayakawa, S.; Lynn, K.; Gossmann, H.-J. Phys. Rev. B. 1996, 54, 4722.

Wang, B.; Wu, C.; Deng, W.; Tang, S.; Jin, X.; Chuang, Y.; Li, J. J. Appl. Phys. 1996, 79, 2587.

West, R. N. J.Adv. Phys. 1973, 22, 263.

Xu, D.; He, K.; Yu, R.; Tong, Y.; Qi, J.; Sun, X.; Yang, Y.; Xu, H.; Yuan, H.; Ma, J. Mater. Technol.: Adv. Funct. Mater. 2015, 30, A24.

Xu, D.; Jiang, B.; Cui, F.-d.; Yang, Y.-t.; Xu, H.-x.; Song, Q.; Yu, R.-h. J. Cent. South Univ. 2014, 21, 9.

Xue, R.; Chen, Z.; Xue, Y.; Dai, H.; Li, T.; Chen, J. J. Supercond. Nov. Magn. 2014, 27, 1201.

Zhu, Q.-T.; Yang, T.-Z.; Cheng, X.-X.; Zhang, W.; Huang, Y.-Y.; Deng, W. Nucl. Instrum. Methods Phys. Res., Sect.B. 2009, 267, 3159.

Analysis of the bin packing problem

Qianpan Jiang
Columbia University, NY, USA

ABSTRACT: In this paper, we study the classical bin packing problem and analyze the concept of problem definition and possible variants of the basic problem. Focusing on classical bin packing problem, we analyse several commonly used heuristic algorithms, clarify the algorithms and discuss the boundedness of those algorithms. We also explore some related or derived problems for bin packing problem.

1 INTRODUCTION

As a significant problem in combinatorial optimization, bin packing problem is studied since early 1970s. In the classical version of the bin packing problem, one is given an infinite supply of bins with a capacity C and a list of items with sizes no larger than C, and the problem is to pack the items into a minimum number of bins so that the sum of the sizes in each bin is not greater than C. In simpler terms, a set of items is to be put into a minimum number of bins with same sum constraint.

Since bin packing is NP-hard, the approximation algorithm for bin packing is also well studied. In standard bin packing, the input is a sequence of items, which are to be packed into unit capacity bins, without exceeding the capacity. The goal is to minimize the number of bins used. Thus, compared with optimal results, the ratio is used as a measurement for the quality of an algorithm. There are also several standard judgements as worst case bound and average case bound.

Besides the basic version above, bin packing problem has a lot of variants that are well-studied. Bin packing is also a dual problem for other NP-hard problems. This problem can also be converted into problems in other fields. In bin packing, we further consider a maximization variant to maximize the number of bins with a total size of items of at least one. In the paper, we propose a bin that has more than one item assigned to the bin, and consider two variants where all items have the same size, so that all bins can contain the same number of items.

In section 2, we give a detailed definition of bin packing problem and discuss the possible variants of this problem. We also provide several measurements for related algorithm. In section 3, we perform analysis on classical heuristic algorithms by discussing the performance and bound of them.

2 DEFINITIONS AND CLARIFICATION

This section introduces a formal definition of bin packing problem by using mathematical notation to describe and specify the problem. Also this section lists general versions and variants of bin packing problem and clarifies the variations that will be studied in this article.

2.1 Problem

In the classical bin packing problem, an instance of the problem consists of an infinite supply of bins with capacity C, and a set of items (or elements) $a_1, ..., a_n$ each item has an associated size s_1 hat $\forall i = 1, ..., n$, $s_i \leq C$. The goal is to pack these items into bins so as to minimize the number of bins used.

The problem is also equivalent to partition a set of items into m parts, and for all $j = 1, ..., m$ arts, the sum of size of the items in the j-th part is no more than C (the capacity of bins), and m is minimized.

Since in classical version of bin packing problem the bin capacity C is the same for all bins, we can regard it as a scale factor. Then after scaling, the problem becomes as follows: Given a set of items with rational sizes between 0 and 1, we want to pack these items into the smallest possible number of unit size bins.

2.2 Problem variations

For one-dimensional bin packing problem, there are several variants that are widely studied by people. Firstly, the bin size can be varied, i.e. more than one bin size for the bins and unlimited supply for each size. In that case, the number of different item sizes can be finite or infinite, and for both situations, we could have limitation on sizes like $s_i \leq a$

or $s_i \geq a$. We can also set restriction on the number of items packing in one bins as a simplification.

Second, for the items, traditional bin packing means just partitioning items to different bins, but there is another widely studied version that allows segmentation of items. Then the items can be split into fragments and fractionally assigned to different bins. With the difference of the size before and after segmenting items, there are two variants as BPP-SIF and BPP-SPF (Marco, 2014). We will study the latter variant in section 4.2. Also constraints of cardinality can be set in that case, namely only at most k items in each bin.

A more complex version named as generalized bin packing problem (GBPP) (Mauro,2012). Divide items into two groups: compulsory and non-compulsory. Each item is equipped with size and profit, and each bin is characterized by volume and cost.

Then the goal becomes packing all compulsory items and load a subset of profitable non-compulsory items into certain bins as to minimize the difference between total cost of bins used and total profit of items selected.

As for the algorithms to approximate bin packing problem, there are three types: on-line, semi-on-line and off-line. The online algorithm only see items one by one and the off-line algorithm can get all information about items and bins. And the semi-online algorithm is the algorithm that can repack items, or get information of the later element before handling the current one, or assume some pre-ordering of elements.

There are also several derived problems from bin packing. In max pack problem, the number of bins is to be maximized under the restriction that a new bin cannot be opened for item s_i if s_i an be put into any current bins. In the mincap problem, the n items must be packed into given m bins, and the target is to find a smallest bin capacity C such that there exists a packing of all the items in m bins of capacity C.

2.3 Problem analysis

2.3.1 Worst case
The worst case bound for an algorithm refers to the ratio between the worst possible performance of the algorithm and the performance of optimized algorithm; in some contexts, it is also called performance guarantee and competitive analysis. Worst case results provide a guarantee that result for the algorithm never worse than optimal result by a certain percentage. The worst case bound is also divided into two kinds. Here we denote R_A as the absolute worst-case ratio and R_A^∞ the asymptotic worst case ratio. Then denote the optimal result as OPT and $A(L)$ is the number of bins used

with input item as list Land use algorithm A. These two bounds are defined as:

$$R_A \equiv \inf\left\{ r \geq 1 \mid \frac{A(L)}{OPT} \leq r, \forall L \right\} \tag{1}$$

$$R_A^\infty \equiv \inf\left\{ r \geq 1 \mid \exists N > 0, \frac{A(L)}{OPT} \leq r, \forall L, OPT \geq N \right\} \tag{2}$$

2.3.2 Average case
The average case bound for an algorithm is preferred, especially for probabilistic analysis, where the size of items are generated independently and randomly from continuous uniform distribution. The measurement is called expected waste, calculated by difference between number of bins used times the capacity and sum of size of items packed.

2.4 Clarification

In this article, we only consider one-dimensional version of the bin packing problem. The main focus of the study is on the classical version of bin packing as stated in section 2.1, so this paper will not include generalized bin packing and other variants like bins with different sizes. And thus we can assume the problem is scaled and capacity for all bins is 1. But the fragmentation bin packing will be discussed since this is a close related problem. For performance, we only consider worst case bound.

3 HEURISTIC METHODS

For bin packing problems, there are four classical heuristic algorithms: First Fit, Best Fit, First Fit Decreasing and Best Fit Decreasing. Those algorithms are based on natural heuristic and greedy concepts, but work efficiently in practice. Multiple analysis is based on these algorithms including bound analysis and time analysis. Also several algorithms are derived from these algorithms with improved performance.

3.1 First fit

The main idea of First Fit algorithm is to pack each item to the first bin fits; if the item does not fit into any currently open bin, open a new one. Formal algorithm is:

1. Index bins as $B_1, ..., B_m$. Initially all bins are empty.
2. Pack items $a_1, ..., a_n$ one by one. To pack a_i, find the least j such that $\sum_{k, a_k \ in \ B_j} s_{ak} + s_{ai} \leq C$, put a_i in B_j.

Here we denote the number of bins used by First Fit algorithm as *FF*. For the absolute ratio, the benchmark is Simchi-Levi (David, 1994), which shows that the absolute performance ratio for First Fit is no more than 1.75. More specifically, use c to represent the number of items with size larger than 1/2. Then for all lists, we have conclusions: If c is even, $FF \leq 1.75 * OPT$; and if c is odd, $FF \leq 1.75 * OPT - 1/4$.

An improved result given by Xia and Tan (Ullman, 2010) shows that the absolute performance ratio of First Fit algorithm is not more than 12/7. The latest result is from Dosa (György, 2013), which gives a proof that the absolute approximation ratio is exactly 1.7. This result means that for any instance of bin packing, $FF \leq [1.7 * OPT]$. As for asymptotic ratio, the worst case bound of First Fit algorithm is first given. Ullman proved the asymptotic approximation ratio is 1.7. He also proved that $FF \leq 1.7 * OPT + 3$. And in the later research of Ullman and Johnson (David, 1974), the additive term was improved to 2. Garey proved that $FF \leq [1.7 * OPT]$ in Michael, 1976. A significant update in Xia's article [3] proved a more exactly additive coefficient for the bound as $FF \leq 1.7 * OPT + 0.7$ for all lists.

3.2 Best fit

Best Fit algorithm packs each item into the non-empty bin with smallest residual capacity that can contain it, namely the most full bin where it fits, possibly opening a new empty bin if no such bin exists. Detailed algorithm is as below:

1. Index bins as $B_1, ..., B_m$ initially all bins are empty.
2. Pack items $a_1, ..., a_n$ one by one. To pack a_i, find the least j value such that $\sum_{k, a_k \text{ in } B_j} s_{ak} + s_{ai} \leq C$, where $\sum_{k, a_k \text{ in } B_j} a_k$ as large as possible, and put a_i in B_j.

Here we denote the number of bins used by Best Fit algorithm as *BF*. Though the *FF* and *BF* algorithms are different, they shared a lot of similar conclusion in worst case bound.

For the absolute ratio, the benchmark is Simchi-Levi (David,1994), which shows that the absolute performance ratio for Best Fit is not more than 1.75. More specifically, use c to represent the number of items with size larger than 1/2, then for all lists we have conclusions: If c is even, $BF \leq 1.75 * OPT$; and if c is odd, $BF \leq 1.75 * OPT - 1/4$.

The proofs of absolute ratio bound for First Fit and Best Fit are the same and both based on the lemma 3.1.

Lemma 3.1. For the j-th bin in the algorithm, if j $ then the item packed in j-th bin before the bin is more than half full cannot be put into bin opened before j.

Proof. For First Fit algorithm, this is obvious according to the definition of algorithm. And for Best Fit algorithm, suppose there exists an item i which is packed into bin j before j is more than half full and item i can fit bin k which is opened before j. So according to our algorithm, denote the current size of bin k and j as S_{kS} and S_j since we put i in bin j, $S_k < S_j < 1/2$ must hold. Here i cannot be the first item in j since if so, bin j will not be opened and i will be packed in k. Then assume the first item in j is item 1, then $S_l < S_j < 1/2$. Then since $S_l + S_k < 1$, the l should be put in bin k, which contradicts the fact that l is in j.

The latest result is from Dosa and Sgall (Ullman, 2010) where gives a proof that the absolute approximation ratio is exactly 1.7. This result means for any instance of bin packing, $BF \leq [1.7 * OPT]$.

As for asymptotic ratio, the worst case bound of Best Fit algorithm is firstly given in (György, 2014). Ullman proved the asymptotic approximation ratio is 1.7. And it also proved that $BF \leq 1.7 * OPT + 3$ And the later research of Ullman and Johnson (David, 1985), the additive term is improved to 2. An update in [5] proved that $BF \leq [1.7 * OPT]$ and this result leads to new additive coefficient for the bound as $BF \leq 1.7*OPT + 0.9$ for all lists.

3.3 First fit decreasing

The First Fit Decreasing algorithm is based on First Fit algorithm but items are sorted by size in decreasing order before the placement.

Due to the pre-sorting phase in the algorithm, the running time for the first part must take O(nd + n log n) time. The packing procedure processes items sequentially and thus have a complexity of O(n). So the total running time of the algorithm is bounded by O(n log n).

Here we denote the number of bins used by First Fit algorithm as FFD. The first result for absolute performance bound of First Fit Decreasing algorithm is provided by Simchi-Levi (David, 1994), where gives ratio as 1.5, namely $FFD < 1.5 \, OPT$, which is better than the result known for First Fit as 1.7. The proof of absolute worst case bound is also based on lemma 3.1.

The latest result in (György, 2013) implies that the worst case absolute ratio is 4/3 and the worst case is reached at $OPT = 6$ and $FFD = 8$.

And for the asymptotic worst case performance, First Fit also shows better quality than the First Fit. Johnson gives the result that $R_{FFD}^{\infty} = 11/9$ in Johnson's paper (David, 1974). Specifically the result is $FFD \leq 11/9 * OPT + 4$ for all possible list of item. And for different values of the largest item size $\alpha \in (0,1]$, Johnson provided result as:

$$R_{FFD}^{\infty}(\alpha) = \begin{cases} \dfrac{11}{9} & \dfrac{1}{2} < \alpha < 1 \\[2mm] \dfrac{71}{60} & \dfrac{8}{29} < \alpha < \dfrac{1}{2} \\[2mm] \dfrac{7}{6} & \dfrac{1}{4} < \alpha < \dfrac{8}{29} \\[2mm] \dfrac{23}{20} & \dfrac{1}{5} < \alpha < \dfrac{1}{4} \end{cases} \qquad (3)$$

An improved result in the additive term is presented by Baker (Brenda, 1985) with a much shorter proof and he concluded that $FFD \le 11/9 * OPT + 3$. The simplest proof is given by Yue (Yue, 1991) and the additive coefficient is reduced to 1, and later Yue published (Li, 2000) with Li, where they proved that the bound could be tightened to 7/9 and Not conjectured the tight term as 5/9.

The state-of-art result is raised by Dosa (György, 2007) as $FFD \le 11/9 * OPT + 6/9$. This result also proved to be tight in this paper as sketch and Dosa articulated the final proof in (György, 2013). The proof is pages long and the author believes that there is no simple way to give the proof. And since the proof of tightness is given, the least additive term for First Fit Decreasing algorithm should be 2/3.

3.4 Best fit decreasing

The First Fit Decreasing algorithm is based on Best Fit algorithm but items are sorted by size in decreasing order before the placement.

Here we denote the number of bins used by Best Fit algorithm as BFD. The result for absolute worst case bound is provided by Simchi-Levi (David, 1994) as $BFD < 1.5\ OPT$, which is better than the result known for Best Fit as 1.7. The proof of absolute worst case bound is also based on lemma 3.1. As for the asymptotic worst case bound, Johnson proposed Theorems 3.2 and 3.3 in (David, 1974).

Theorem 3.2. For all possible lists of item, $FFD \le 11/9 * OPT + 4$, and $BFD \le 11/9 * OPT + 4$. And thus $R_{FFD}^{\infty} = R_{BFD}^{\infty} = 11/9$.

Theorem 3.3. For each $k \ge 2$ exists a list L with $OPT = k$ and $FFD(L) = BFD(L) \ge 1/9 * OPT - 2$.

3.5 Colour illustrations

Besides those heuristic algorithms above, there are some algorithms for bin packing problem that are based on First Fit and Best Fit algorithms, and also based on other heuristics.

3.5.1 Refined First Fit Decreasing. (RFFD)
This algorithm is proposed by Yao [13] as the first improved offline algorithm based on FFD.

The algorithm is rather complex and is based on separating items into several pieces with each piece including items for certain size range, and put the item by first fit into bins by first fit follows certain rules related to the piece the item belongs to. This algorithm reaches the ratio as $R_{FFD}^{\infty} \le 11/9 - 10^{-7}$ with time complexity as $O(n^{10} log\ n)$.

3.5.2 Modified First Fit Decreasing. (MFFD)
Johnson and Garey proposed MFFD (David, 1985). The algorithm also divides items into several groups by size, and the main idea is to try to pack items as pairs with size (1/6, 1/3) into bins that are more than half full, and when packing those pairs, pack the largest such items first and then put the smallest items fit. Except the processing for such item pairs, other items are packed as FFD. This algorithm has asymptotic worst case ratio as 71/60 that is better than FFD and the running time of this algorithm is $O(n\ log\ n)$.

3.5.3 Next Fit algorithm (NF)
This algorithm packs items in the following method: Packed as the first item in first bin, then for all j-th items that $j \ge 2$, pack it into the bin that contains $(j - 1)$-th item, if not fit, close that bin and put j-th item into a new bin. In this algorithm, only one bin is open at any time. This algorithm has asymptotic ratio = 2.

3.5.4 Worst Fit algorithm (WF)
The idea for Worst Fit algorithm is to pack items in the bin with smallest current size, and if no such bin exists, put item into a new bin. The asymptotic ratio of this algorithm is the same as next fit algorithm.

3.5.5 Good Fit algorithm (GF)
Sgall introduced this algorithm in (Jiri, 2012). The main idea is to require item to be packed to the fullest bin it fits or any bin that has size > 1/2 and also allows closing of bins. The exact rules are as follows:

Packing rule. Pack the new item into an arbitrary more than half-full bin if fits, else put it into an arbitrary bin if fits, otherwise open a new bin.

Closing rule. If more than three bins are opened, one bin except the newly opened one can be closed if its size is $\ge 5/6$ or $\ge 2/3$ or $\ge 1/2$.

The asymptotic ratio of this algorithm is ≤ 1.7.

4 CONCLUSIONS

In this article, we give an exact definition of classical bin packing problem, conclude the possible variants from it, and specify the commonly used measurement of related algorithms as worst case bound and average case bound. We also analyse four commonly used heuristic algorithms for the classical bin packing problem: First Fit, Best Fit, First Fit Decreasing and Best Fit Decreasing algorithms. We declare the algorithms and give important results for the boundedness of those algorithms. And we also explore some derived algorithm on those heuristic algorithms.

REFERENCES

Andrew Chi-Chih Yao. 1980. New algorithms for bin packing. Journal of the ACM (JACM) 27, 2 (1980), 207–227.

Brenda S Baker. 1985. A new proof for the first-fit decreasing bin-packing algorithm. Journal of Algorithms 6, 1 (1985), 49–70.

David S Johnson and Michael R Garey. 1985. A 7160 theorem for bin packing. Journal of Complexity 1, 1(1985), 65–106.

David S. Johnson, Alan Demers, Jeffrey D. Ullman, Michael R Garey, and Ronald L. Graham. 1974. Worst case performance bounds for simple one-dimensional packing algorithms. SIAM J. Comput. 3, 4 (1974), 299–325.

David Simchi-Levi. 1994. New worst-case results for the bin-packing problem. Naval Research Logistics 41, 4 (1994), 579.

György Dósa and Jiri Sgall. 2013. First Fit bin packing: A tight analysis. In STACS. Citeseer, 538–549.

György Dósa and Jiri Sgall. 2014. Optimal analysis of Best Fit bin packing. In Automata, Languages, and Programming. Springer, 429–441.

György Dósa. 2007. The tight bound of first fit decreasing bin-packing algorithm is FFD (I)11/9OPT (I)+6/9. In Combinatorics, Algorithms, Probabilistic and Experimental Methodologies. Springer, 1–11.

Jiri Sgall. 2012. A new analysis of Best Fit bin packing. In Fun with Algorithms. Springer, 315–321.

Li Rongheng and Yue Minyi. 2000. A tighter bound for FFd algorithm. Acta Mathematicae Applicatae Sinica16, 4 (2000), 337–347.

Marco Casazza and Alberto Ceselli. 2014. Mathematical programming algorithms for bin packing problems with item fragmentation. Computers & Operations Research 46 (2014), 1–11.

Mauro Maria Baldi, Teodor Gabriel Crainic, Guido Perboli, and Roberto Tadei. 2012. The generalized bin packing problem. Transportation Research Part E: Logistics and Transportation Review 48, 6 (2012), 1205–1220.

Michael R Garey, Ronald L Graham, David S Johnson, and Andrew Chi-Chih Yao. 1976. Resource constrained scheduling as generalized bin packing. Journal of Combinatorial Theory, Series A 21, 3 (1976), 257–298.

Minyi Yue. 1991. A simple proof of the inequality FFD(L) leq 11/9 OPT (L)+ 1, ∀ for the FFD bin-packing algorithm. Acta mathematicae applicatae sinica 7, 4 (1991), 321–331.

Ullman, J.D. 1971. The performance of a memory allocation algorithm. Princeton University. Binzhou Xia and Zhiyi Tan. 2010. Tighter bounds of the First Fit algorithm for the bin-packing problem. Discrete Applied Mathematics 158, 15 (2010), 1668–1675.

Computer simulation & computer and electrical engineering

Advances in Energy Science and Equipment Engineering II – Zhou, Patty & Chen (Eds)
© 2017 Taylor & Francis Group, London, ISBN 978-1-138-71798-5

Noise reducing effect of aluminum foam in 66 kV and 220 kV electrical substations

Qianqian Yang, Wenxiao Qian, Guoli Zhang, Haipeng Shi & Jingyuan Liu
State Grid East Inner Mongolia Electric Power Research Institute, China

Ming Chen
Liaoning Rontec Advanced Material Technology Co. Ltd., China

ABSTRACT: With the present attention given to the electrical substation noise problem, the application of soundproof and noise reducing material gains significance. This work evaluated the noise reducing effect of aluminum foam in two substations: a 66 kV Kundu substation and a 220 kV Chengyuan substation. The results were compared with the environmental data and the effect of 304 L stainless steel. It shows that the soundproof and noise reducing effect of the aluminum foam in the substation area is significantly superior to that of the 304 L stainless steel. The shielding effect of the aluminum foam box is more compelling with higher environmentally audible noises. The maximum shielding effectiveness can reach 20 dB. Viscous absorption and Helmholtz resonance sound absorption are the mechanisms mainly involved in the aluminum foam.

1 INTRODUCTION

In recent years, with the economic development and the urbanization process, shortage of land resources are noticeable. Not only the buildings and the population density are rapidly expanded around the electrical substations, but also more and more substations have been constructed in city centers or densely populated area to meet the increasing electric power consumption per capita (Li,2015). In the meantime, the noise pollution caused by substations is drawing continuous attention as one of the world's four major forms of pollutions.

According to "Emission standard for industrial enterprises noise at boundary" promulgated by Chinese government at the end of 2008, noise emissions from many substations in service are close to or above the standard limits. Citizens are voicing their protest by invoking their right-protection awareness, and the petitions and mass incidents are on the rise owing to the noise environment of substation. To solve this problem and create a quiet neighborhood, necessary actions need to be taken by the electricity bureaus. In "Macro Strategic Study on China's Environment" introduced by the state council, noise pollution prevention and control by the electrical substations is listed as a key issue (Zhu,2012). In addition, the Twelfth Five Plan makes it clearly that "the urban noise pollution needs to be effectively controlled". This is the first time that this topic of noise pollution due to substation is being included in the National Five-year Plan.

Noises in the substations are caused by equipment such as transformers, reactors, filters and fan cooling devices (Wang,2008; Wu,2010). Low-frequency electromagnetic and mechanical noises produced by the transformers are the main sources of noise (Chen,2006). At present, two kinds of measures have been taken to reduce the noise exposure from the substations. One is to reduce the noises at their sources. This involves device modification within the substations and has disadvantages such as more time consumption, high cost and poor feasibility. Furthermore, not more than 15 dB noises can be reduced (Geng,2013). The other measure is to separate the living environment from the sound source. Sound waves are blocked and weakened on their propagation pathway using sound absorption materials and soundproof configurations, such as large sound barriers along the substation boundary and "box-in" packaging technique towards the main devices (Chen,2006; Zhou,2009; Chen,2008; Ou,2005). The latter has gained increasing attention due to its noticeable cost-effectiveness, smallscale construction and uninterrupted installation.

Aluminum foam that first appeared in the 1980s is a fast developing advanced material with excellent physical and mechanical properties. This eco-friendly material has been widely employed in military defense, aerospace, automotive and railway construction field due to its light weight, high specific strength/stiffness and energy/sound absorption qualities (Banhart,2001; Lefebvre, 2008). It also can be used as a noise-reduction material in the substationarea (Du,2015).

In this work, the enclosure made of aluminum foam has been utilized to evaluate its noise reducing effect both in the area inside and outside the enclosing wall of the substation of two substations: a 66 kV Kundu substation and a 220 kV Chengyuan substation in Tongliao, Inner Mongolia Autonomous region. The results were compared with the environmental data and the effect of 304 L stainless steel.

2 EXPERIMENT

2.1 *Material*

304 L stainless steel was selected as a control group material to inspect the sound absorption effect of the aluminum foam. The aluminum foam was supplied by Liaoning Rontec Advanced Material Technology Co., Ltd. As shown in fig. 1, the porosity of aluminum foam panels is about 50–70%, and the cell size is about 5–10 mm. Two 1 m × 1 m × 2 m enclosures were made by aluminum foam panels (1#) and 304 L stainless steel (2#) respectively to run the noise reduction test. Fig. 2 shows the aluminum foam enclosure. Except the shell structure, the enclosure also consists of frame, shielded door, shielded honeycomb, optical light, ground connection and universal wheels. The thickness of the adopted aluminum foam is 10 mm and the stainless steel's thickness is 1.5 mm.

2.2 *Methodology*

In this work, HS6298 Multifunction Noise Analyzer was used, with the sound frequency ranging from 20 to 10 KHz and the noises from 35 to 135 dB. Based on the requirement of Emission Standard for Industrial Enterprises Noise at Boundary (GB 12348-2008) and Environmental Quality Standard for Noise (GB 3096-2008), this measurement was conducted in the surrounding area with a distance of 1–8 m from the power transformer inside the

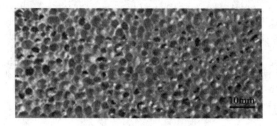

Figure 1.   Morphology of the aluminum foam surface.

Figure 2.   Aluminum foam enclosure.

Table 1.   Weather condition on methodology.

Substation Name	Weather	Temp	Humidity	Wind speed
66 KV Kundu substation & 220 KV chengyuan substation	Cloudy	−1 to −11° C	35%	4.5 m/s

substation and the area with a distance of 1–15 m outside the enclosing wall of the substation respectively. Both the environmental audible noises and the ones within the shielding boxes during the daytime (6:00-22:00) were evaluated. Equivalent continuous A-weighted sound level was measured at the height above 1.2 m for an overall time of 1 min.

The weather conditions on the day of measurement are shown in Table 1.

3 RESULTS AND DISCUSSION

3.1 *Noise reducing effect*

The audible noises measured from the 66 kV Kundu substation and the 220 kV Chengyuan substation are shown in fig. 3 and fig. 4 respectively. The results suggest that for the 66 kV Kundu substation, the maximum environmental value of audible noises in the daytime next to the transformer and outside the substation was 55 dB, and not surprisingly, the latter ones were generally lower.

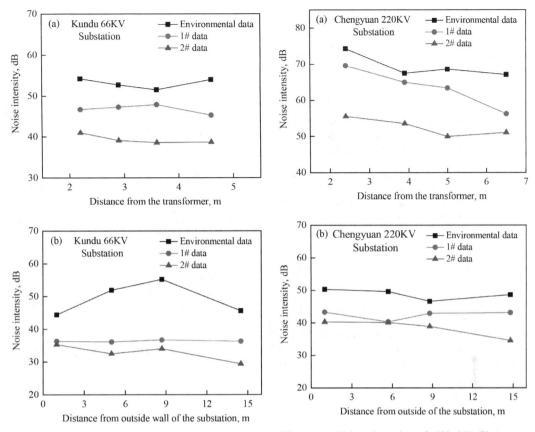

Figure 3. Noises intensity of 66 kV Kundu substation.

Figure 4. Noises intensity of 220 kV Chengyuan substation.

While for the 220 kV Chengyuan substation, the maximum daytime environmental noises reached as high as 75 dB, which were far beyond the legal limit (65 dB). The audible noises were reduced by increasing the measuring distance from the transformer and enclosing the wall for both substations.

After shielding with a stainless steel box (1#) and an aluminum foam box (2#), the environmentally audible noises next to the transformer and outside the substation were significantly reduced. Furthermore, the noise reduction effect of aluminum foam was greater than that of stainless steel, with a noise value as low as 35 dB. As shown in fig. 4(a), the higher the environmental audible noise, the better the shielding effect of the aluminum foam box can reach. The maximum shielding effectiveness was as high as 20 dB.

3.2 Sound absorption mechanisms of aluminum foam

Every material processes sound absorption effect to some degree. The main mechanisms of aluminum foam corresponding to this are viscous absorption and thermal conduction absorption. When sound waves enter a media from another, the latter would absorb the acoustic energy by viscous resistance between the interfaces. Meanwhile, the speed difference between adjacent mass points would cause friction, leading to the transfer of acoustic energy into thermal energy (Sun,2014). Thermal energy would then be transferred into internal energy by thermal conduction.

Another sound absorption mechanism is Helmholtz resonance sound absorption (Liang,2008). There are numerous "small holes" inside the aluminum foam, acting as a Helmholtz resonator. Moreover, the connections between the "small holes" together with the micro-cracks inside the aluminum foam form sound wave propagation channels, resulting in abounding abrupt changed cross sections and bypass type of Helmholtz resonance chambers. This makes the aluminum foam a brilliant sound absorption material.

4 CONCLUSION

The daytime noises of the transformer in 220 kV Chengyuan substation exceed the legally permissible limit, causing certain health threats to the staff on duty.

Noise reducing effect is more significant for aluminum foam compared with stainless steel. The higher the environmental audible noises, the better the shielding effect of the aluminum foam box can reach. The maximum shielding effectiveness was as high as 20 dB.

Therefore, aluminum foam is an ideal sound reducing material, which can be used in sound insulating casings of electric transformers, sound barriers close to the sound sources and barriers on the enclosing walls.

ACKNOWLEDGEMENT

The authors would like to thank State Grid East Inner Mongolia Electric Power Research Institute General Public Fund financial support (Project code: 526604150001), Liaoning Rontec Advanced Material Technology Co., Ltd., and Tsinghua University for the valuable discussions and technical support.

REFERENCES

Banhart, J., Pro. Mater Sci. **46**, 553(2001).

Chen, Q., Z. H. Li, Elec. Power Environ. Prot. **22**, 49(2006).

Chen, Z. P., Q. Li, Elec. Power Environ. Prot. **24**, 61(2008).

Du, H. Y., X. Su, X. Chen, J. K. Nie, X. F. Kong, Q. S. Xu, B. Wang, Y. H. Tang, Int. Conf. Adv. Energy Environ. Sci., 1400(2015).

Geng, M. X., J. Wu, C. C. An, P. H. Lv, X. C. Bai, Shanxi Elec. Power **41**, 80(2013).

Lefebvre, L. P., J. Banhart, D. C. Dunand, Adv. Eng. Mater **10**, 775(2008).

Li, N., D. M. Tian, D. P. Shan, C. Y. Zang, C. Zheng, Y. J. Wei, High Volt. App. **51**, 0139(2015).

Liang, L. S., Study on Acoustics Property and Application of Closed-cell Aluminum Foam [D]. Northeastern University, 13(2008).

Ou, Y., J. Y. Gao, Noise and Vib. Control, 20(2005).

Sun, M. Q., B. Wei, H. L. Zhang, Hebei J. Ind. Sci. Technol. **31**, 65(2014).

Wang, Z. H., J. G. Zhou, L. Su, J. H. Jiang, East China Elec. Power **36**, 1336(2008).

Wu, X. X., Y. P. Yang, J. Liu, W. Han, East China Elec. Power **38**, 0887(2010).

Zhou, J. G., L. H. Li, Y. Du, F. Chi, Elec. Power **42**, 75(2009).

Zhu, Z. X., Y. Han, J. K. Nie, F. Y. Yang, W. M. Xiao, X. Chen, Elec. Power **45**, 57(2012).

Advances in Energy Science and Equipment Engineering II – Zhou, Patty & Chen (Eds)
© 2017 Taylor & Francis Group, London, ISBN 978-1-138-71798-5

Water simulation of swirl-type lance in vanadium extraction process

Yong-hui Han, Da-qiang Cang & Rong Zhu

School of Metallurgical and Ecological Engineering, University of Science and Technology Beijing, Beijing, China

ABSTRACT: The stirring effects of conventional oxygen lance and swirl-type oxygen lances (swirl angles of 5°, 10°, 13°) were studied by using water model. It is found that 10° swirl-type oxygen lance has the best stirring effect with mixing time of 43.5 s. Industrial experiments shows that in 100t BOF smelting effect of swirl-type oxygen lance is better than conventional oxygen lance, vanadium content of semi-steel is decreased by 0.007%, and carbon content is increased by 0.17%.

1 INTRODUCTION

Vanadium extraction is the selective oxidation process of vanadium and carbon in hot metal, during which the temperature should be controlled at a lower level, and the force generated from chemical reaction of carbon and oxygen is comparatively less (Yang, 2010; Shin, 2007; Taylor, 2006). The physical force of the oxygen lance is the main source of the stirring power, and dynamic condition in hot metal is poor (Bai, 2012; Huang, 2000).

The impulse of conventional oxygen lance can be decomposed into radial and axial forces. Besides radial and axial forces, swirl-type oxygen lance can produce swirl tangential component which can enlarge the impact area of pool surface and reduce the mixing time of hot metal. In this article, the stirring effects of conventional oxygen lance and swirl-type oxygen lance by water model were studied, and industrial tests were carried in 100t BOF.

2 ESTABLISHING THE MODEL

Geometry determined. Based on the similarity principle, physical model established in this experiment model is the model of 100t converter. The geometric model and the prototype are remained similar and the geometric similarity ratio is kept at 6:1. Oxygen lance parameters of prototype and model are shown in Table 1.

Bath depth and lance height of prototype and model are shown in Table 2.

Steam parameters determined. Model modified Froude number (Frm') is equal to prototype (Frp') to ensure dynamic similarly:

Table 1. Oxygen lance parameters of prototype and model.

		Prototype	Model
Mach		1.98	1.98
Hole number		5	5
Swirl angle/(°)		0/5/10/13	0/5/10/13
Center hole /mm	Throat	27.6	4.6
	Exit	35.6	5.9
Ring hole /mm	Throat	32.8	5.5
	Exit	42.3	7.1

Note: 0° on behalf of the conventional oxygen lance.

$$Frm' = Frp' \tag{1}$$

$$\frac{u_m^2}{gL_m} \cdot \frac{\rho_{gm}}{\rho_{lm} - \rho_{gm}} = \frac{u_p^2}{gL_p} \cdot \frac{\rho_{gp}}{\rho_{lp} - \rho_{gp}} \tag{2}$$

After the conversion the function becomes:

$$\frac{Q_m}{Q_p} = \left(\frac{L_m}{L_p} \right)^{\frac{5}{2}} \cdot \left(\frac{\rho_{gp}}{\rho_{gm}} \right)^{\frac{1}{2}} \cdot \left(\frac{\rho_{lm} - \rho_{gm}}{\rho_{lp} - \rho_{gp}} \right)^{\frac{1}{2}} \tag{3}$$

In the formula: u_m, u_p represent characteristic velocity, m/s; ρ_{gm}, ρ_{gp} represent gas density, kg/m³; ρ_{lm}, ρ_{lp} represent liquid density, kg/m³; L_m, L_p represent the size of feature, m; g is acceleration of gravity, m/s²; Q_m, Q_p are gas flow, Nm³/h. Subscripts l and g respectively represent liquid and gas; subscripts m and p respectively represent model and prototype.

Gas flow of prototype and model are shown in Table 3.

Table 2. Bath depth and lance height of prototype and model.

Parameter	Bath depth/mm	Lance height/mm
Prototype	1428	1400/1500/1600/1700/1800
Model	246	233/250/267/283/300

Table 3. Gas flow of prototype and model.

Parameter	Prototype	Model
Top gas flow/(Nm³·h⁻¹)	17000	76.7
	18000	81.2
	19000	85.7
Bottom gas flow/(Nm³·h⁻¹)	120	0.51

3 EXPERIMENT RESULTS AND ANALYSIS

3.1 Effects of oxygen lance height and gas flow on mixing time

In Figure 1, the mixing time is relatively shorter with 10° swirl-type oxygen lance at each lance height. The average mixing time is only 40 s at 233 mm lance height.

In Figure 2, the mixing time with swirl-type oxygen lance is shorter than conventional oxygen lance. The stirring effect of 10° swirl-type oxygen lance is the best, which can effectively improve dynamic conditions of BOF.

3.2 Effects of lance height and gas flow on impact depth

In Figure 3, with the increasing lance height, the overall impact depth is generally decreasing. The impact depth is the minimum with the lance height of 300 mm.

In Figure 4, the impact depth becomes greater when gas flow is greater. The impact depth of 10° swirl-type oxygen lance is the minimum, while the impact depth of 5° swirl-type oxygen lance is the maximum.

3.3 Effects of lance height and gas flow on impact diameter

In Figure 5, although the lance height is improving, there is no big variation between the impact diameters of different lances.

Figure 6 shows that, with the increasing gas flow, the impact diameters of 10°, 13° swirl-type oxygen lance and conventional lance slightly increases with no big change, while the impact diameter of 5° swirl-type decreases.

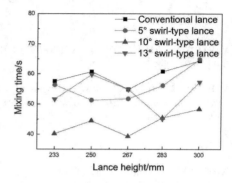

Figure 1. Mixing time of different lance height.

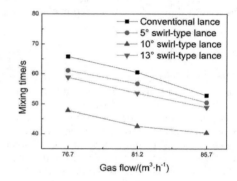

Figure 2. Mixing time of different gas flow.

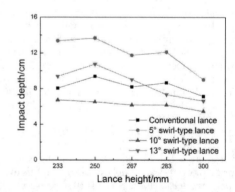

Figure 3. Impact depth of different lance height.

The foregoing analysis demonstrates that effects of lance height and gas flow on impact diameter are not large. The impact diameter of 10° swirl-type oxygen lance is the biggest, and the impact diameter of 5° swirl-type oxygen lance is the smallest.

As all said above, during the application of swirl-type oxygen lance, when swirl angle increases, tangential force also increases, so does the impact diameter. When the swirl angle is too large,

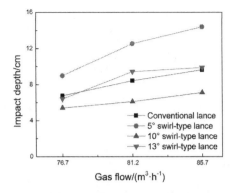

Figure 4. Impact depth of different gas flow.

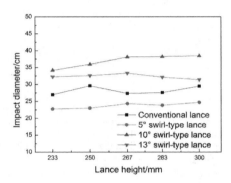

Figure 5. Impact diameter of different lance height.

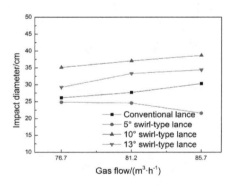

Figure 6. Impact diameter of different gas flow.

resistance is also increased, so the bath stirring ability decreases instead.

3.4 *Effects analysis of swirl-type oxygen lance and conventional oxygen lance*

Figure 7 shows that the shortest mixing time is 43.5 s with 10° swirl-type oxygen lance. If swirl angle is too large or too small, the bath stirring ability cannot be improved. Best lance heights are

233 mm and 267 mm, corresponding to the field lance heights of 1400 mm and 1600 mm.

4 INDUSTRIAL TESTS

The industrial tests of 10° swirl-type oxygen lance and conventional oxygen lance were carried in 100t BOF. Comparison of vanadium and carbon content of semi-steel was analyzed by 23 heats for each test.

From Figure 8, when using swirl-type lance, vanadium content of semi-steel is between 0.030% and 0.040%, with an average of 0.039%, and the average vanadium oxidation rate is 84.1%. When using conventional lance, vanadium content of semi-steel is between 0.035% and 0.055%, with an average of 0.046%, and the average vanadium oxidation rate is 80.5%. Vanadium content of semi-steel is decreased by 0.007%. when using swirl-type lance, and vanadium oxidation rate is increased by 3.6%. The use of swirl-type oxygen lance can improve dynamic conditions in bath.

In Figure 9, when using conventional lance, carbon content of semi-steel is between 3.12% and 3.79%, with an average of 3.53%; when using

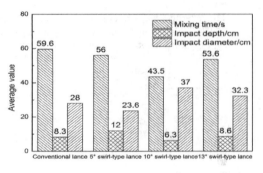

Figure 7. Comparison effects of different oxygen lances.

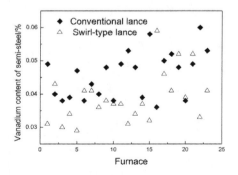

Figure 8. Distribution of vanadium content.

1405

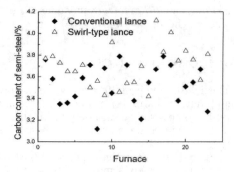

Figure 9. Distribution of carbon content.

swirl-type lance, carbon content of semi-steel is between 3.42% and 4.12%, with an average of 3.70%. Carbon content of semi-steel when using swirl-type is increased by 0.17%.

5 CONCLUSIONS

1. Swirl-type oxygen lance can improve bath stirring ability. The stirring effect of 10° swirl-type oxygen lance is the best, and the mixing time is only 43.5 s
2. The average carbon content of semi-steel is 3.70%, which is increased by 0.17% compared to that of conventional oxygen lance. The average

vanadium content is 0.039%, which is decreased by 0.007% compared to that of conventional oxygen lance. The vanadium oxidation rate is 84.1%, which is increased by 3.6%.
3. Industrial tests shows that the effect of 10° swirl-type oxygen lance is better than conventional oxygen lance, which is beneficial to improve BOF dynamic conditions and promote the oxidation of vanadium.

REFERENCES

Bai Ruiguo. Research and development of vanadium extraction and utilization technology of chengde steel [J]. Hebei metallurgy,2012194 (2): 5–9.

Huang Daoxin. Vanadium extraction and steelmaking [M]. Beijing: Metallurgical Industry Press, 2000: 21–22.

Shin D Y, Wee C H Kim M S. Distribution behavior of vanadium and phosphorus between slag and molten steel metals and materials international [J]. 2007, 13(2): 171–176.

Taylor R R, Shuey S A, Vidal E E, etc. Extractive metallurgy of vanadium containing titaniferous magnetite ores: A review [J]. Minerals and Metallurgical Processing, 2006, 23 (2): 80–86.

Yang Shouzhi. Vanadium metallurgy [M]. Beijing: Metallurgical Industry Press, 2010.

Advances in Energy Science and Equipment Engineering II – Zhou, Patty & Chen (Eds)
© 2017 Taylor & Francis Group, London, ISBN 978-1-138-71798-5

Numerical simulation of the shear band process on the initial defect

Shi-wei Hou, Shao-po Guo, Xin Zhang & Xiao Liu
School of Civil Engineering, Shenyang Jianzhu University, Shenyang, China

ABSTRACT: The geotechnical material undergoes significant deterioration. The holistic failure process usually is associated with the formation of macroscopic strain localization band. A numerical analysis study on the progressive failure of the soil based on the initial defect is conducted. The progressive failure of the sample under plane strain and triaxial conditions is simulated. Based on the initial defect research on clearly appointed position location, a further analysis with random distribution of defect nodes or elements is done to understand the process of shear band. The random number from another program is introduced into the model file to realize the random distribution defect in the numerical model. The results show that the different strain localization bands caused by initial defect present the competition process until the peak value of stress–strain curve. The whole stress–strain relationship is slightly softened after the leading control deformation band is determined. The deformation located at the band during the following process, while non-localization band response change to a small extent and main axial of stress change. The stress–strain curves of the characteristic elements in the localization and non-localization areas present significant hardening and softening character. The percent and setting method of initial defect both contribute to the progressive process of the shear band.

1 INTRODUCTION

Soil is an uneven frictional material, and the internal stress field in the soil under the load is uneven. Shear failure occurs first on the shear strength of the particle shear, and the excess stress is passed on to neighboring particles. The process shows progressive failure characteristics. With the development of plastic deformation, the deformation mode is developed from the gradual homogeneous deformation to the one confined within the finite width strip of the inhomogeneous deformation. This strain concentration is formed in a limited area. The formation of the shear strain localization area is called the shear band. There exist different scales of shear bands in nature, from the micro shear bands in rock and soil to tens of meters, even hundreds of kilometers long with giant shear. The progressive failure of the soil is a basic subject in the field of geotechnical engineering (Zhao, 2003). The generation of shear band is usually early signs of the overall instability and failure of soil, and the study on the failure mechanism of localization mechanism should have an important guiding significance to understand the soil and stability analysis of practical engineering in the progressive failure process.

2 PROGRESSIVE RESEARCH METHODS

The research on progressive failure of soil mass can be used by three kinds of research methods, which is experimental research, theoretical research and numerical calculation. Oda (1998) studied the change of porosity obtained by testing the light elastic granular material, and the porosity of the shear band significantly increased. The development of the micro-light elastic particles of the tour test not only made the particle mechanics to develop some basic theories, but it also directly gave birth to the particle discrete element program (Rechenmacher, 2006). Given the difficulty in explaining the start of the strain localization phenomenon and the change of physical properties after the formation of shear band in the traditional theory, scholars have put forward a series of theories and methods such as the bifurcation theory (Ming, 2004), complex theory (Pietruszczak, 1993), the nonlocal strain theory (Bazant, 1988), the cosserat theory (Borst, 1991) and the gradient plasticity theory (Larsy, 1988). Xue-bin wang (2001) has done a lot of work in the initial defects. The stress and strain characteristics of soils under different test conditions was obtained by experimental observation and measurement. Then based on the existing theory, the theory to explain the experimental phenomena would be proposed. The theoretical research results were applied to the numerical calculation to predict and reproduce the experiment phenomenon, so as to realize the purpose of fully understanding the progressive failure process. The research focus is to find and track the starting and developing process of the shear band of the sample. Theoretical analysis solved the reasons of localized shear band and the changes of soil properties during the progressive failure process, and it provided the basis for the establishment of the reasonable constitutive model. The difficulty of a numerical calculation is to solve the softening calculation technology of the finite element and to describe the problem of the mesh distortion and dependency in the generation of shear band. The initial inhomogeneity of a soil and the trigger of the shear band are the basis of numerical simulation.

This article looks, from the perspective of numerical simulation, the formation of macroscopic shear band in the progressive failure process, and mainly analyses the effects of the initial inhomogeneity on shear band formation. The shear band formation process of the triaxial specimen and soil slopes is simulated by using the general numerical calculation software. Based on the initial defect research on clearly appointed position location, further analysis with random distribute defect nodes or elements is done to present the process of shear band. The numerical simulation results of triaxial specimen analyzed the formation mechanism of the shear band. The soil slopes example can be combined with the strength reduction method. It can be used for the analysis of safety factor as well as the random initial defects. Shear strength reduction factor is defined as the ratio of the maximum shear strength and the shear stress generated by the external load in the slope under the condition that the load remains constant. The formation process of progressive failure of soil slope is studied by simulating the shear band trigger of the slope. The model of the shear band formation and the interaction between different shear bands are also studied.

3 ESTABLISHMENT OF THE ABAQUS MODEL

3.1 Plane strain test specimen

The rock and soil are uneven friction material. The initial inhomogeneity has an influence on the deformation process and strength of the soil. It is necessary to simulate the initial heterogeneity of soil to reproduce the true progressive failure process of the soil. In practice, the initial defect location is not certain, which may be on the surface of the sample, or in the interior of the sample, based on the simulation results (Hou, 2010) of explicitly specifying the defect position, taking into consideration the fact that the defect position is not explicitly specified, a part of defective elements or nodes are generated randomly according to the number of elements and nodes, and the generation and development of shear band with initial random distribution of defects is analyzed.

Modified Cambridge model is used in the finite element method, The simulation parameters of the normal consolidated soil parameters are taken from the Helwany book (Helwany,2007), in which the isotropic compression coefficient is $\lambda = 0.174$, the coefficient of resilience is $\kappa = 0.026$, the initial void ratio is $e_0 = 0.889$, the Poisson's ratio is $\mu = 0.28$, the critical state stress ratio is $M = 1$, the tensile and compressive stress ratio parameters is $K = 1$, the initial yield strength is $a_0 = 50$ kPa, and the permeability coefficient is $k = 25$ mm/s. The plane strain element with pore pressure CPE8RP is used.

In order to consider the constrained conditions, the sample size is taken as 50 mm × 100 mm, The simulation process takes place in two steps. The cell pressure 100 kPa is put in first step, strain controlled loading is put in second step, and the vertical loading rate is 0.008 mm/s, far less than the coefficient of permeability. Pore water is considered in full flow, and pore pressure of the sample is uniform. The top and bottom of the specimen are the displacement boundary conditions; the vertical strain of each point is always equal during the loading process; the left and right sides of the specimen are the boundary conditions of force, the bottom center of the specimen is fixed.

3.1.1 Drainage condition

The shear band formation process is studied considering the drainage condititon. The initial yield strength of the initial defect element is 15 kPa, and vertical compression occurs up to 30% of the axial strain. The location of initial element and corresponding vertical strain and shear strain are shown in figure 1. The result shows that under the drainage condition, the initial defect element or area can introduce the strain concentration, but the finite mesh is still uniform.

For the same calculation model, now the undrainage condition is analyzed. Figures 2–4 show the total vertical strain for the same location condition of the defect element, such as 1 corner defect element, 2 top defect elements and 4 corner defect elements. The result shows that under the undrainage condition, the initial defect element or area can not only introduce the strain concentration, but also cause the concentration of the finite mesh. This means that continuous macro shear band is formed.

The vertical compression displacement is 10 mm, 20 mm, 30 mm respectively in figures 2–4. The strain localization area effect by the initial defect element is limited in the same length compared with the sample width. This can be seen in all the 10 mm condition results. The shear band is developed along 45° angle with the sample top for both corner and top defect elements conditions.

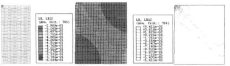

(a) vertical strain and shear strain of 1 corner defect element

(b) vertical strain and shear strain of 2 top defect elements

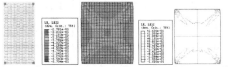

(c) vertical strain and shear strain of 4 corner defect elements

Figure 1.   Computation of initial defect with different locations.

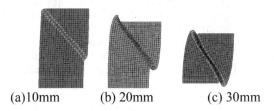

(a)10mm   (b) 20mm   (c) 30mm

Figure 2.   Pore pressure and different strain of intitial defect at left corner point.

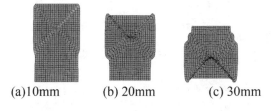

(a)10mm   (b) 20mm   (c) 30mm

Figure 3.   Different strain of intitial defect at left and right corner point.

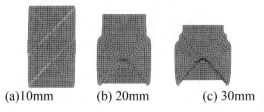

(a)10mm   (b) 20mm   (c) 30mm

Figure 4.   Different strains of initial defect at four corner points.

Now we can conclude that the affected region has the same length for both top and bottom defect elements. So they meet at the sample's middle part at the beginning of the deformation. The conclusion was verified for 10 mm and four corner elements condition in figure 4.

The setting of initial defect element can induce the macro shear band, especially for undrainage condition. And the symmetric defect boundary condition induces the symmetric strain localization area. The unsymmetric defect boundary condition induces the unsymmetric strain localization area, as the whole strain mode shown in figures 1–4.

3.1.2   Defect node

Geometric model is divided into $20 \times 40 = 800$ elements, a total of 2521 nodes, and 49 defect nodes are randomly generated, which account for about 2%, the distribution of the defect nodes, as shown in figure 5. The compression coefficient of defect node is 0.08, and the other is 0.174. The distribution of the defect nodes is shown in figure 6.

The formation process of shear zone is analyzed in this case. Representative analysis steps 1, 14, 15, 32, 33, 44, 53, 122 are used to analyze the effects of random defect elements, and the corresponding vertical strain contour map is shown in figure 6. After consolidation the defect points of the initial random distribution form, the strain concentration region is shown in figure 6 (a). The strain is concentrated at the defect nodes. To continue loading, the deformation of each defect point is gradually connected as shown in figure 6 (b). A continuous competition between the strain concentration regions connected to a network is shown in steps 15–33, Dominant shear band appears in step 44. After the position of the dominant shear band is determined, deformation is all concentrated in this region; In the other strain concentrated region, the strain has the process of resilience in steps 44–53. The sample on the right side of strain localization region expands, and shear band has a wider trend as the last strain contour in step 122. The final failure mode of the specimen is the formation of an asymmetric cross shear band. The peak shear stress curve is step 33. There is no obvious shear band in figure 6 (e). The different shear bands are in the process of competition, which make them reach the peak value. The curves of shear stress and pore water pressure with the loading process are shown in figure 7. After that, the peak stress strain curve is decreased obviously. The stress exhibits the softening properties after the dominant shear band form.

3.2   Axisymmetric sample

The symmetric model geometry of three axis is 50 mm × 100 mm, and the sample model contains

Figure 5.   Location of initial defect nodes.

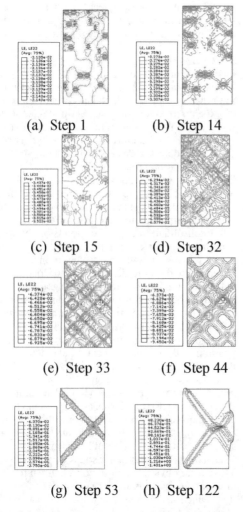

(a) Step 1   (b) Step 14

(c) Step 15   (d) Step 32

(e) Step 33   (f) Step 44

(g) Step 53   (h) Step 122

Figure 6.   Progressive process of vertical log strain.

2373 nodes and 1920 elements. A defect part contains six nodes in a element, that is, a defect element named 1644 is located at the top of the sample, and the location of the selected feature element is shown in figure 8 (a). The top of sample doesn't drain. The pore pressure unit is used, and the normal unit strength is 1.4. The initial yield strength

value is 50 kPa, The stress ratio of the defect unit named 1644 is 1. The initial yield strength value is 15 kPa. The logarithmic shear strain nephogram is shown in figure 8 (b).

A defect unit causes the sample to tilt from the results of the nephogram, and the deformation concentrated area is formed under the condition of non-drainage. The initial defect unit 1644 generates the unit 1406, 1633, 1664 of the concentration area and the unit 1164, 1885 of the non concentration area. Three elements 1406, 1633, 1885 are selected to extract three stress–time history curves as shown in Figure 9.

The relationship of the stress variation with time at different location elements shows the process of the initial consistent growth, and it displays an obvious difference after the strain concentration bands appear. The stress increases, and reaches the stages of gradual reduction and quick reducing. The strength and the initial yield strength of the defect element is reduced, the stress increases in the late stage, in the initial stage, and the level of stress increases significantly. From non-shear band region, the representative element named 1885 of stress–strain curves, the stress vertical force decreases sharply, until it is lower than the horizontal stress. It shows that the principal stress axis rotates. The vertical stress is transformed from the large principal stress to the minor principal stress. When the vertical stress reaches the peak value, the

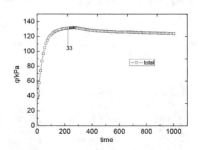

Figure 7.   Curves of shear stress.

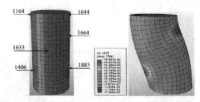

(a) representative elements (b) Logarithmic shear strain nephogram

Figure 8.   Location of initial defect element and repre- sentive elements.

horizontal decreases at the same time. Strain in the diagonal region of the non-shear band shows the first compression, after stretching. The element 1633 of the central surface exhibits softening characteristics. The horizontal stress in the 2 direction of the shear band located on the 23 surface reduces, and the horizontal stress of the 1 direction is basically unchanged. The vertical and horizontal stresses of the up and bottom of the shear band are both increased.

The calculation of the different defect nodes content is carried out. The critical state stress ratio is 1.4. The initial yield strength of the defect node is 30 kPa, The end of the boundary condition is constrained in the first step, The midpoint of the bottom keeps the horizontal constraint, the vertical displacement of the bottom keeps constraints, and the top of sample drains; The second step, based on the above conditions, the top of the sample is compressed without drainage. Defect nodes account for about 2% of the whole nodes, the vertical compression is 0.03 m, the sample has the bulging deformation, but there is no formation of macroscopic shear bands, the distribution of defects and the nephogrampores of pore ratio are shown in figure 10.

3.3 Soil slope sample

The geometric model of the slope is shown in Figure 11. The soil bulk density is 20 kN/m³, the

cohesion is 12.38 kPa, the friction angle is 20°, the elastic modulus is 100 MPa, and the Poisson's ratio is 0.35. The sample model contains 390 nodes and 349 elements. The four-node plane strain element CPE4 is used.

In the slope model, 49 defect nodes are randomly generated, which accounts for about 5%. The distribution of the defect nodes is shown in figure 12 (a).

The formation process of shear zone is analyzed in this case. Representative analysis steps 3, 4, 100 are used to analyze the effects of random defect element. The corresponding progress of the plastic strain nephogram is shown in figure 12. The plastic zone appears at the bottom of the slope as shown in figure 12 (b), Plastic zone of the slope bottom is developing upward in step 4, and due to the influence of the random defect unit, the plastic zone of the slope appears as shown in figure 12 (c). The two parts of the plastic zone appear in the competition state. After the competition, the plastic zone of starting from the slope bottom dominates. And the dominant plastic zone determines the development of the strain concentration region. A shear band through the whole slope is formed, and the stress–strain relationship exhibits the softening properties. The plastic strain in the non-dominant

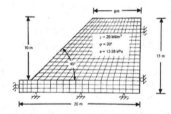

Figure 11. Sketch map of slope model.

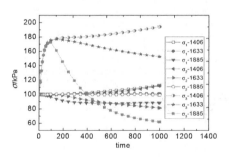

Figure 9. Stresses of representative elements.

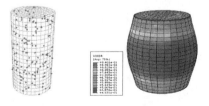

(a) 10% position of defect node (b) nephogram of pore ratio

Figure 10. Location of initial defect nodes and nephogram results.

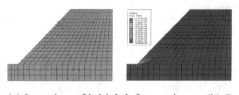

(a) Location of initial defect nodes   (b) Step 3

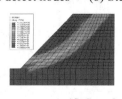

(c) Step 3   (d) Step 3

Figure 12. Initial defect nodes and strain plot of slope model.

region has the process of resilience as shown in the nephogram of the plastic strain.

4 CONCLUSION

In this paper, the author analyzes the generation and development of shear bands with random distribution of the initial defect. The following conclusions are drawn:

The setting of initial defect element can induce the macro shear band, especially for undrainage condition. And the symmetric defect boundary condition induces the symmetric strain localization area, and the unsymmetric defect boundary condition induces the unsymmetric strain localization area.

There is a competition between the strain localization regions caused by the initial defect, until the peak value of stress and strain occurs. The whole stress–strain relationship shows slight softening characteristics.

After the dominant deformation zone is determined, the stress–strain relationship exhibits the softening characteristics, the concentrated deformation develops, the deformation of the non-localized region has a slight rebound, and the rotation of principal stress axes is generated, and the vertical stress is transformed from the large principal stress to the minor principal stress.

The percent and setting method of initial defect both contribute to the progressive process of shear band. The initial yield strength, the critical state stress ratio and compression parameters as the defect condition are calculated as the starting of the excitation shear band.

The stress–strain curves of characteristic elements in localization and non-localization areas present significant hardening and softening character.

ACKNOWLEDGEMENTS

This work was financially supported by the National Natural Science Foundation of China (51308355), Project supported LiaoNing Province, colleges and universities outstanding young scholar growth plan (LJQ2014058), Ministry of Housing and Urban-Rural Development of the People's Republic of China Project (2015-K3-025) and Discipline Content Education Project of Shenyang Jianzhu University(XKHY2-03).

REFERENCES

Bazant Z P, Lin F-B. Non-local yield limit degradation[J]. Int J Numer Methods Engng, 1988,26:1805–1823.

de Borst R. Simulation of strain localization:a reappraisal of the Cosserat continuum[J]. Engineering Computation,1991,8:317–332.

Helwany S. Applied soil mechanics with ABAQUS applications[M]. USA: John Wiley & Sons, Inc, 2007.

Hou Shi-wei, LU De-chun, DU Xiu-li. Simulation and fo-rmation mechanism analysis of shear band in soil. Journal of disaster prevention and mitigation engineering[J]. 2010,30(5):544–549.

Larsy D, Belytschko T. Localization limiters in transient problems[J]. Int J Solids Structures, 1988,24:581–597.

Ming Li. Continuing equilibrium assumption over-restricts bifurcation condition in the classical localization theory[J]. Inernational Journal of Plasticity, 2004,20:2047–2061.

Oda M, Kazama H, Konishi J. Effects of induced anisotropy on the development of shear bands in granular materials[J]. Mechanics of Materials, 1998, 28(1):103–111.

Pietruszczak S, Niu X. On the description of localized defo-rmation[J]. Int J Numer Analyt Meth Geomech,1993,17:791–805.

Rechenmacher A L. Grain-scale processes governing shear band initiation and evolution in sands[J]. Journal of the Mechanics and Physics of Solids, 2006,54(1):22–45.

Wang Xue-bin, PAN Yi-shan, et al. Simulation of triaxial compression and localization of deformation[J]. Rock and soil mechanics, 2001,22(3):323–326.

Zhao Xihong, Zhang Qihui. Experiment and numerical analysis of shear band in soil [M]. Beijing: China Machine Press, 2003.

Advances in Energy Science and Equipment Engineering II – Zhou, Patty & Chen (Eds)
© 2017 Taylor & Francis Group, London, ISBN 978-1-138-71798-5

DEM simulation of granular materials under direct shearing considering rolling resistance

Jin-kun He
China Petroleum Pipeline Research Institute, Langfang Hebei, China

ABSTRACT: This paper investigated the effect of rolling resistance on the shear behavior of granular materials under the direct shear condition. A 3D Discrete Element Method (DEM) model of the direct shear test was developed and a contact model considering rolling resistance was proposed in this study. Numerical simulations were conducted to study the effect of the rolling friction and the rolling stiffness on the stress–strain and volumetric behavior of granular materials. The results from the particle scale such as the evolutions of force chain and coordinate number are also presented in this study. Finally, the results from the angle of energy dissipation are also presented. It is shown that rolling resistance parameters, rolling friction coefficient and rolling stiffness coefficient can affect the energy dissipation, however, in different ways.

1 INTRODUCTION

Granular materials are a kind of special inhomogeneous, multi-phase geo-material and widely encountered in many complex geotechnical engineering problems, for example, building foundations, high embankments, rockfill dams, slope stability problems. In the past few decades, many researchers have paid great attention to study the stress–strain behavior and shear strength of granular materials (i.e. Bolton,1984; Cavarretta, 2010; Mahmud, 2012; Guo, 2013).

The direct shear test is one of the most widely used methods to measure the strength and stiffness of granular materials. Although the test has some disadvantages such as deformation and stress fields are strongly non-uniform within the box, the interface area diminishes during shearing, the principal stresses are not known and the shear strength is larger than the one from triaxial tests or simple shear tests, It has still widely used due to its simplicity and its ability to determine important properties such as drained strength envelope parameters, flow function, angle of internal friction, dilatancy angle, wall friction angle and cohesion (Kozicki, 2013).

In recent years, researchers have made many attempts to study the meso-structure and inhomogeneity of granular materials under direct shear test using DEM. Cui and O'Sullivan (Cui, 2006) were the first to perform a comparison of physical test data and direct shear test DEM model. Zhao et al. (Zhao, 2015) investigated the effect of the particle angularity in light of its importance

in angular particle assemblies by introducing a general contact force law for arbitrarily shaped particles, using the Discrete Element Method (DEM). Suhr and Six (Suhr,2016) introduced a model for stress-dependent inter-particle friction in DEM simulations, and conducted a series of numerical direct shear tests to validate experimental results on single and paired glass beads. Salazar et al. (Salazar, 2015) developed a direct shear test model considering rolling friction and compared numerical results against experimental data.

In this paper, a commercial software of discrete element method in 3D condition, Particle Flow Code in Three Dimensions (PFC3D), is used to simulate mechanical behavior of granular material under direct shear. In order to consider shape effect of particles, a rolling resistance model, which has been validated as a simple way of modeling the effects of non-sphere and irregular particle shapes (Iwashita,1998), is implemented into PFC3D as the contact model of granular materials. The simulating results suggest that this model considers the discrete nature of granular materials and provides a new insight. Constitutive modeling is used to investigate the shear behavior of at the scale of a particle form.

2 DEM SIMULATION

In conventional DEM, such as PFC3D, rolling resistance of particles is not involved in inter-particle contact behaviors. Particles in contact are assumed to rotate freely and offer no resistance to rotation or to be fixed and can't rotate, and sliding rather than rolling dominates the shear strength and dilatancy of granular materials. However, recent experimental observations have shown that interparticle rolling plays a key role in controlling the stress–stain and deformation behavior of granular materials (Belheine, 2009). Based on the geolaboratory results, many researchers have developed several new contact models taking rolling resistance into account, and have conclusively proven that rolling resistance is a significant parameter influencing the shear and deformation behavior of granular materials (Jiang, 2005; Estrada, 2011; Ai, 2011). In order to take rolling resistance into account, a user-defined constitutive model is implemented into PFC3D. User-defined contact constitutive models can be added to PFC3D by writing the model in C++ language, compiling it as a Dynamic Link Library (DLL) file and loading the DLL into PFC3D whenever needed(Itasca Consulting Group,2005).

2.1 Implementation of the rolling resistance model

In this paper, a user-defined constitutive model considering rolling resistance is installed into PFC3D. This model is composed of normal, tangential, and rolling contact components, as shown in Fig. 1.

The normal and tangential behavior of this model is just the same of definition of PFC3D built-in linear contact model (Itasca Consulting

Group,2005).The normal stiffness is a secant modulus and relates the total normal force to the total normal displacement. The shear stiffness is a tangent modulus and relates the increment of shear force to the increment of shear displacement. The normal and shear forces are given by:

$$F_i^n = K^n \cdot u_i^n \tag{1}$$

$$\Delta F_i^s = K^n \cdot \Delta u_i^s \tag{2}$$

where K^n and K^s are the stiffness of normal and tangential model, respectively. The contact stiffnesses are computed assuming that the stiffnesses of the two contacting spheres acting in series, so that K^n and K^s are defined by following equations:

$$K^n = \frac{k_n^{[A]} \times k_n^{[B]}}{k_n^{[A]} + k_n^{[B]}} \tag{3}$$

$$K^s = \frac{k_s^{[A]} \times k_s^{[B]}}{k_s^{[A]} + k_s^{[B]}} \tag{4}$$

where the superscripts [A] and [B] denote the two spheres in contact.

The slip model describes the constitutive behavior tangential to the particle contact between two particles. It is defined by the friction coefficient at the contact μ, where μ is taken to be the friction coefficient of the two contacting entities. The slip model starts working at the contact i, when the following condition is satisfied:

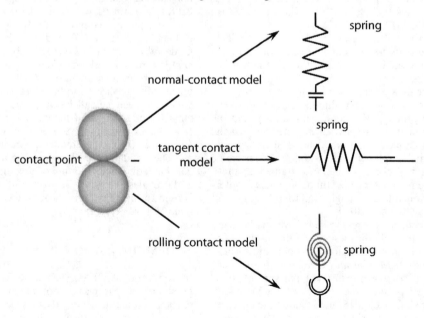

Figure 1. Rolling resistance model.

$$|F_i^s| \geq \mu \cdot |F_i^n| \qquad (5)$$

The rolling contact model includes an elastic spring and a roller representing the contact rolling resistance. M_i, which is balanced by the rolling resistance provided by the elastic spring, represents the rotational moment, and is written as:

$$M_i = k_r \cdot \theta_r \qquad (6)$$

where θ_r is the relative rolling vector between two particles, and k_r is the rolling stiffness and its value is obtained as(Oda, 2000):

$$k_r = 2 \cdot r \cdot F_i^n \cdot J_n \qquad (7)$$

The roller for rolling starts working when M_i exceeds a value of ηN_i:

$$|M_i| \geq \eta N_i \qquad (8)$$

where η is the coefficient of rolling resistance, with a dimension of length.

2.2 DEM simulation of direct shear tests

A shear box with its scale of 50 cm × 50 cm × 40 cm is established. The shear box is separated horizontally into two equal boxes by an imaginary shear plane. The shear box is enclosed by ten rigid walls including two horizontal walls at the top and bottom, four vertical walls at the front and back, and four vertical walls at the left and right of the specimen, shown in Fig. 2.

The radius expanding method is applied in the simulations reported in this paper to generate the numerical specimen. The balls were first randomly generated inside the shear box with an initial ball diameter smaller than their intended final diameter. The diameter of the balls was then increased to the final diameter for the analysis and contacts are formed. The Particle Size Distribution (PSD) of final specimen is shown in Fig. 3. The particles assembly formation is illustrated in Figure 4 (a).

After generation, the specimen was consolidated by moving the top wall using a servo algorithm. The consolidation procedure stopped when the confining stress of the specimen reached a given value and a complete equilibrium condition was achieved. The given confining stress is 1200 kPa in this study. The time step was set to around 10^{-6} seconds for each computation cycle. The numerical processing time depends on the number of balls in the model and on the computation power available. The Measurement Sphere (MS) routine in PFC3D is used to calculate local stresses and porosities. The MS routine computes the stress tensor from forces at contacts averaged over the volume of the selected MS. Stresses and void ratios reported in the following sections are calculated at the specimen core using a central measurement sphere with a radius of 15 cm.

Then the specimen was sheared by moving the lower box in a constant velocity. In this study, the shearing velocity is set as 1.00 mm/s. The shearing processer was continued until the shear strain in Y-direction reached 16.0%. The final specimen after shearing is illustrated in Fig. 4(b). The force chain evolution is shown in Fig. 5.

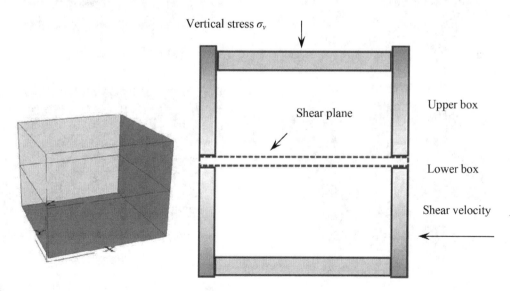

Figure 2.   Schematic diagram of the direct shear box.

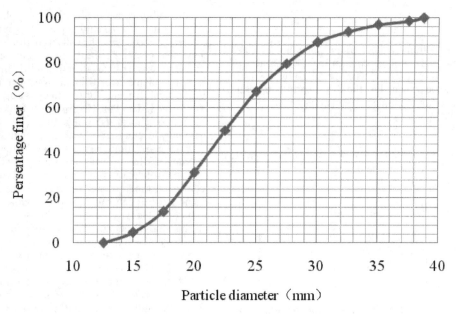

Figure 3.   Particle size distribution.

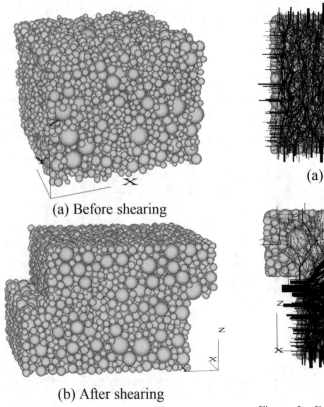

(a) Before shearing

(b) After shearing

Figure 4.   Numerical specimen.

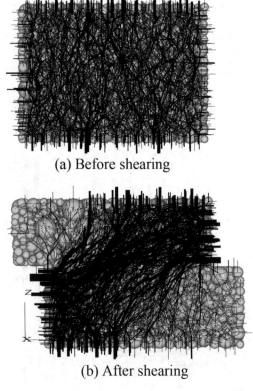

(a) Before shearing

(b) After shearing

Figure 5.   Force chain before shearing and after shearing.

3 SIMULATION RESULTS

3.1 *The effect of rolling resistance on the overall mechanical behavior*

The effects of rolling stiffness on the shear stress σ_y and volumetric strain are shown in Fig. 6. It is shown in Fig. 6(a) that the shear stress depends on rolling stiffness coefficient during all the shearing procedure. And the shear stress increases with increasing rolling stiffness coefficient at the same shear strain. The stress–strain behaviors of the granular materials exhibit a strain hardening characteristic for all rolling friction coefficients.

Fig. 6(b) illustrates that evolutions of volumetric strain versus shear strain for different rolling stiffness coefficients. The curves of volumetric strain show that volumetric strain is also dependent on rolling stiffness coefficient. And a higher rolling stiffness coefficient leads to a high dilatancy of granular materials.

The effects of rolling friction on the shear stress σ_y and volumetric strain are shown in Fig. 7. The results presented in Fig. 7(a) illustrate that the shear stress is initially dependent on the rolling friction coefficient until the shear strain is about 4%. After that point, the shear stress displays a strong dependency on rolling friction coefficient.

The shear stress increases with an increasing rolling friction coefficient at first, and when the rolling friction coefficient reaches nearly 0.3, the increment of shear stress becomes very small.

The relationship of rolling friction coefficient and volumetric strain is presented in Fig. 7(b). The volumetric strain is initially dependent on rolling friction coefficient just like shear stress, and after shear strain reaches nearly 2%, the curve of rolling friction coefficient equaling 0.1 exhibits a stronger compressibility.

The shear strength is one of the most important and useful properties that can be achieved in direct shear test. Fig. 8 presents the shear strengths for different rolling stiffness coefficients and rolling friction coefficients. The results illustrate that there is a linear relationship between shear strengths and natural logarithm of rolling stiffness coefficients, while the relationship of shear strengths and rolling friction coefficients follows a hyperbolic model.

3.2 *The effect of rolling resistance on the micro-mechanical behavior*

Fig. 9 presents the evolutions of coordinate number with different rolling resistance coefficients. It is shown in all curves that the coordinate

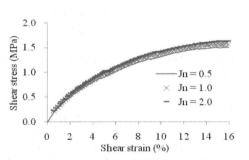

(a) Shear stress vs. Shear strain

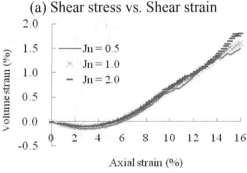

(b) Volumetric strain vs. Shear strain

Figure 6. The effect of rolling stiffness on stress–strain and volumetric behavior.

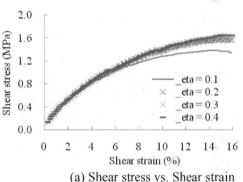

(a) Shear stress vs. Shear strain

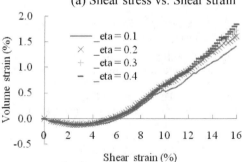

(b) Volumetric strain vs. Shear strain

Figure 7. The effect of rolling friction on stress–strain and volumetric behavior.

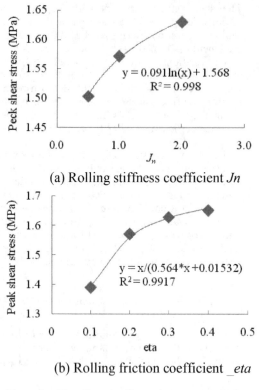

(a) Rolling stiffness coefficient J_n

(b) Rolling friction coefficient eta

Figure 8. The effect of rolling resistance on peck shear stress.

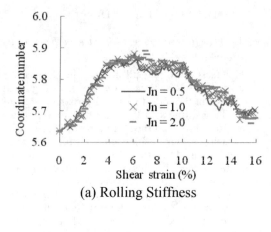

(a) Rolling Stiffness

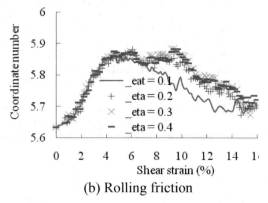

(b) Rolling friction

Figure 9. The effects of rolling resistance on coordinate numbers.

number first increases with shear strain until shear strain reaching nearly 5.0%. Then the coordinate number stays nearly constant. When shear strain reaches 9.0%, the coordinate number decreases with increasing shear strain until the end of shearing. It is also found that coordinate number has more dependency on rolling friction than rolling stiffness.

3.3 The effect of rolling resistance on the energy behavior

Fig. 10 presents the evolutions of input energy, friction energy and the ratio of the friction energy and input energy with different rolling stiffness coefficients. The curves in Fig. 10(a) shows that input energy increases with increasing shear strain, and a higher rolling stiffness coefficient leads to a higher input energy at the same shear strain. Fig. 10(b) illustrates that friction energy exhibits an independency on rolling stiffness coefficient. The evolutions of ratio of the friction energy and input energy are shown in Fig. 10(c). The energy ratio depends on rolling stiffness during the whole shearing procedure. The energy ratio decreases

with increasing rolling stiffness at the initial stage. Then after the shear strain reaches nearly 10%, the tendency overturns.

Fig. 11 presents the evolutions of input energy, friction energy and the ratio of the friction energy and input energy with different rolling friction coefficients. It is shown in Fig. 11(a) and Fig. 11(b) that both input energy and friction energy are independent on rolling friction coefficient until a certain shear strain value is reached. Then both input energy and friction energy increase with increasing shear strain, and a higher rolling friction coefficient leads to a higher input energy and friction energy at the same shear strain. The evolutions of ratio of the friction energy and input energy are shown in Fig. 10(c). It is clear that rolling friction influences the energy ratio in a totally different way than rolling stiffness. The energy ratio exhibits an independency on rolling friction until shear strain reaches nearly 4.0%, then the energy ratio increases with increasing rolling stiffness until end of tests.

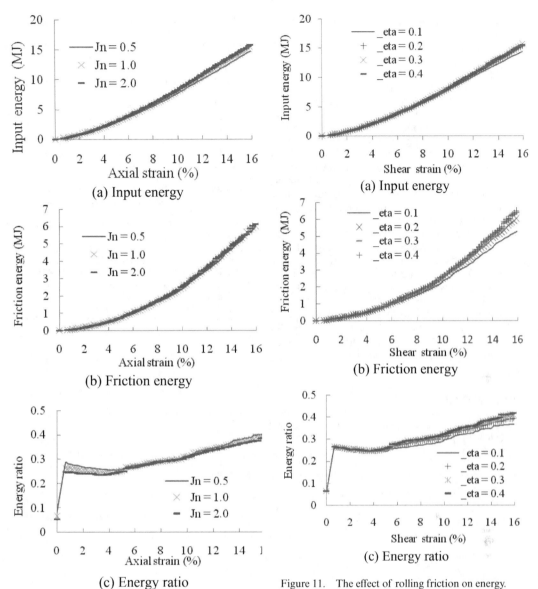

(a) Input energy

(b) Friction energy

(c) Energy ratio

Figure 10.   The effect of rolling stiffness on energy.

(a) Input energy

(b) Friction energy

(c) Energy ratio

Figure 11.   The effect of rolling friction on energy.

4   CONCLUSIONS

In this paper, a numerical direct shear test is established using 3D discrete element method. To consider the effect of particle shape, a user-defined contact model, rolling resistance model, is implemented into the numerical model. This model has two coefficients. One is the rolling stiffness coefficient and the other is the rolling friction coefficient. A series of numerical direct shear tests are conducted to discover how these two coefficients

influence the behavior of granular materials under the direct shear condition. The effects of rolling stiffness and rolling friction on overall behavior, microscopic behavior and energy behavior are investigated. The conclusions gained in this paper are shown as follows:

a. Rolling stiffness and rolling friction influence the overall shear behavior of granular materials significantly and differently in direct shear test. Both shear stress and volumetric strain depend on rolling stiffness during shearing procedure, while they are independent on rolling friction in the initial stage of shearing. The difference is also shown in the results of shear strength. There is a linear relationship between shear strengths and natural logarithm of rolling stiffness coefficients, while the relationship of shear strengths and rolling friction coefficients follows a hyperbolic model.

b. The evolutions of coordinate number with different rolling stiffness coefficients and rolling friction coefficients illustrate that there are three stages during the tests. In the first stage, coordinate number increases with shear strain; next coordinate number maintains nearly a constant; in the final stage, coordinate number decreases with shear strain. Coordinate number has more dependency on rolling friction than rolling stiffness.

c. The effects of rolling stiffness and rolling friction on energy behavior are also very different. The results show that rolling stiffness has little influence on friction energy, while friction energy shows a strong dependency on rolling friction. The ratio of friction energy and input energy initially decreases with rolling stiffness and then the tendency overturns. Unlikely, the ratio of friction energy and input energy is initially independent on rolling friction.

REFERENCES

Ai, J., J. F. Chen, J. M. Rotter, et al., Powder Technol 206,3(2011).

Belheine, N., J. P. Plassiard, F. V. Donzé, et al, Comput Geotech 36,1(2009).

Bolton, M. D. *The strength and dilatancy of sands* (Cambridge University Engineering Department, 1984).

Cavarretta, I., M. Coop, C. O'sullivan, Géotechnique 60, 6(2010).

Cui L, O'sullivan C. Geotechnique 56(2006).

Estrada, N., E. Azéma, F. Radjai, et al, Phys Rev E 84, 1(2011).

Guo, N., J. Zhao, Comput Geotech 47(2013).

Itasca Consulting Group, Inc. 2005a. PFC3D (Particle Flow Code in Three Dimensions), Version 3.1. Minneapolis, USA.

Iwashita, K., M. Oda, J Eng Mech 124,3(1998).

Jiang, M. J., H. S. Yu, D. Harris, Comput Geotech 32,5(2005).

Kozicki, J., M. Niedostatkiewicz, J. Tejchman, et al, Granul Matter 15,5(2013).

Mahmud Sazzad, M., K. Suzuki, A. Modaressi-Farahmand-Razavi, Int J Geomech 12,3(2012).

Oda, M., K. Iwashita, Int J Eng Sci 38,15(2000).

Salazar, A., E. Sáez, G. Pardo, Comput Geotech 67(2015).

Suhr, B., K. Six, Powder Technol 294(2016).

Zhao, S., X. Zhou, W. Liu, Granul Matter 17,6(2015).

Finite element modelling of concrete confined by hoops based on ABAQUS

Xiang Zeng
College of Civil Engineering and Architecture, Hainan University, Haikou, China
Hainan Institute of Development on International Tourist Destination, Haikou, China

ABSTRACT: Concrete confined by hoops is common in Reinforced Concrete (RC) structures and the behaviour of hoop-confined concrete is a classic topic. Relatively few studies have been conducted using the advanced finite element model based on the widely used general-purpose simulation tool such as ABAQUS and ANSYS. In fact, it is more accessible for advanced finite element modelling using the general-purpose simulation tool than to compile the corresponding program code. Moreover, advanced finite element analysis is a powerful means to investigate the mechanism of the confinement effect of hoops. This paper presents a new uniaxial compression stress–strain relationship of concrete confined by square hoops and a nonlinear finite element analysis model for analysis of the behaviour of a square hoop-confined RC column under concentric compression based on ABAQUS. In the finite element model, the concrete damaged plastic model was used to describe the behaviour of confined concrete in conjunction with the proposed uniaxial compression stress–strain relationship of confined concrete. The results indicated that the finite element model can simulate the realistic behaviour of confined RC columns. Thus, this shows that the proposed uniaxial compression constitutive model in conjunction with the concrete damaged plasticity material model in ABAQUS is suitable for simulating the behaviour of the hoop-confined concrete.

1 INTRODUCTION

Concrete confined by hoops exists widely in RC structures due to its increase in ductility and strength, which is dependent on the confinement stress caused by hoops. Many various empirical or semi-empirical uniaxial compression models have been developed for describing the compression stress versus strain relationship of hoop-confined concrete by using the regression analysis based on experimental data or simplified theoretical studies (Park,1982; Sheikh, 1982; Mander,1988; Saatcioglu,1992; Le´geron,2003; Bousalem,2007; Samani,2012). Most of the models are very practical.

However, advanced numerical models are needed to study the mechanism of the confinement effect of hoops, and to investigate the details of behaviour of hoop-confined concrete members such as the distribution of strain and stress in concrete and steel, and confinement pressure in core concrete. With the advancement of the computer power, an increasing number of modelling studies using finite element methods have emerged in recent years (Liu,2000; Kwon,2002; Zeng,2014; Song,2011; Faria,2004; Foster,1998). One of the main interests of many such studies has been focused on the development of the constitutive

model (Liu,2000; Kwon,2002; Zeng,2014), and some of them have investigated the effectiveness of certain confinement arrangements in special structural member designs (Zeng,2014; Song,2011; Faria,2004; Foster,1998).

In fact, it is more accessible for advanced finite element modelling to use the general-purpose simulation tools than to compile the corresponding program code because the general-purpose simulation tools usually provide the convenient visual modelling interface and the main work for the user is to choose the rational parameter setup and material models in order to efficiently predict the behaviour of confined concrete structure. This paper describes a new finite element modelling of concrete confined by hoops using a general-purpose simulation tool such as ABAQUS. A new uniaxial compression stress–strain relationship of concrete confined by square hoops was proposed to simulate the behaviour of hoop-confined concrete using the concrete damaged plasticity model in ABAQUS. For the purpose of verifying the finite element model, axially loaded RC columns confined by hoops were simulated, for which experimental data are widely available. By comparison of the numerical results and experimental data, the effectiveness of the proposed uniaxial compression stress–strain relationship and finite element modelling of confined concrete was examined.

2 BRIEF DESCRIPTION OF THE EXPERIMENT

The experiment used to verify the finite element model was conducted by Sheikh and Uzumeri (Sheikh,1980). Five specimens with a volumetric ratio between 0.8% and 2.39% from the experiment were simulated. The length of the test region, in which the hoops were placed at specified spacing, was about 610 mm (Figure 1). The scheme of the hoops in the four specimens and the details of the test specimens are shown in Figure 1. To ensure that the failure would occur in the test region of the column, the tapered ends of the column were further confined with the help of welded boxes. All the specimens were applied on a concentric load. For more details about the experiment, see the paper by Sheikh and Uzumeri (Sheikh,1980).

3 MATERIAL MODELLING OF CONCRETE

In ABAQUS, the concrete damaged plasticity model is widely used to describe the behaviour of concrete, which is composed of plasticity model and linearly damaged model. If no damage parameter is defined, then the model is equal to the plasticity model. Here, only the plasticity model is used because it can well describe the behaviour of concrete in a RC column under an axial load. The plastic model use the yield function proposed by Lubliner et al. (Lee,1998) and incorporates the modifications proposed by Lee and Fenves (Lubliner,1989) to account for different evolutions of strength under tension and compression, and a non-associated potential flow is adopted in the model. The model has pressure dependence of strength.

The modulus of elasticity is assumed to be constant and is equal to $4730(f_c)^{0.5}$ that is determined by ACI-318, where f_c (N/mm^2) is the cylinder strength of concrete. The Poisson ratio of concrete is assumed to be constant and is equal to 0.2.

The parameters for the plastic model of unconfined and confined concrete are dilation angle, eccentricity, ratio of the biaxial compression strength to uniaxial compression strength of concrete, the ratio of the second stress invariant on the tensile meridian to that on the compressive meridian and viscosity parameter, whose corresponding values are 30°, 0.1, 1.16, 0.667, and 0.0001. Moreover, the uniaxial behaviour of concrete in compression and tension should be provided by the user of ABAQUS to the concrete damaged plastic model.

3.1 A new uniaxial compressive stress–strain relationship of hoop-confined concrete

As shown in Figure 2, it seems that it is difficult to reasonably predict the behaviour of confined RC columns under concentric compression using the stress–strain relationship of unconfined concrete in conjunction with the concrete damaged plastic model in ABAQUS. Here, the stress–strain relationship of unconfined concrete described in the *CEB-FIP Model Code 1990* is used. A solution for the problem is the development of a new strain hardening/softening function for confined concrete, which is dependent on the confining pressure (Tao,2013).

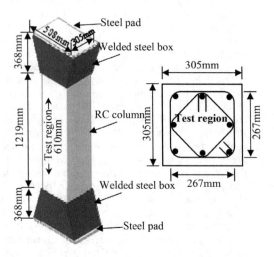

Figure 1. Scheme of hoops and details of the test specimens.

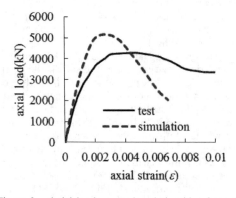

Figure 2. Axial load vs. strain relationships from the test and simulation of specimen 4 A3–7 tested by Sheikh and Uzumeri (Sheikh,1980).

A new uniaxial compressive stress–strain relationship for confined concrete considering the confining effect of hoops was developed in this study, which is similar to the model proposed by Tao et al. (Tao,2013), as shown in Figure 3. The ascending branch of the stress–strain relationship of unconfined concrete is appropriate to be used to represent the curve OA of confined concrete until the peak strength f_{cc} is reached. After that, a plateau (from Point A to Point B) is included to represent the increased peak strain of concrete from confinement. During this stage, any increase in concrete strength from confinement will be captured in the simulation through the interaction between the hoops and concrete. Beyond Point B, a softening portion with increased ductility resulting from confinement is defined.

A model proposed by Wang et al. (Wang,1978) is used to describe the ascending curve OA:

$$Y = \frac{AX + BX^2}{1 + CX + DX^2} \quad 0 \le \varepsilon \le \varepsilon_c \quad (1)$$

where $X = \varepsilon/\varepsilon_c$ and $Y = \sigma/f_{cc}$; $A = 1.323$, $B = -0.809$, $C = -0.676$, and $D = 0.191$. f_{cc} is the peak strength of confined concrete and is equal to $0.85f_c$ by considering a strength-reduction factor related to column shape, size and the difference between the strength of *in situ* concrete and the strength determined from standard cylinder tests. ε_c is the axial strain corresponding to concrete cylinder strength and the value is determined according to the *Fib Model Code for Concrete Structures 2010*.

The strain at Point B (ε_{cc}) for the concrete model is determined by the following equation proposed by Samani and Attard (Samani,2012):

$$\frac{\varepsilon_{cc}}{\varepsilon_c} = e^k, \ k = (2.9224 - 0.00367f_{cc})\left(\frac{f_B}{f_{cc}}\right)^{0.3124+0.002f_{cc}} \quad (2)$$

where f_B is the confining stress provided to the concrete at Point B. f_B is proposed by Le´geron and Paultre (Le´geron,2003):

$$f_B = \rho_{se}f_h, \quad \rho_{se} = k_e\frac{A_{sh}}{sc} \quad (3)$$

where f_h is the stress in confinement reinforcement at the peak strength of confined concrete. ρ_{se} is the effective sectional ratio of confinement reinforcement in one direction perpendicular to the side of columns. s and c are the spacing of hoops and the diameter of the core measured centre-to-centre of hoops, respectively. A_{sh} is the section area of hoops within the spacing s and perpendicular to the side of columns. k_e is the geometrical effectiveness

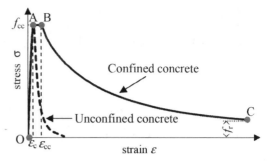

Figure 3. Stress–strain model proposed for confined concrete.

coefficient of confinement, which is proposed by Mander et al. (Mander,1988):

$$k_e = \frac{\left(1 - \dfrac{\sum w_i^2}{6c_xc_y}\right)\left(1 - \dfrac{s'}{2c_x}\right)\left(1 - \dfrac{s'}{2c_y}\right)}{1 - \rho_c} \quad (4)$$

where w_i is the ith clear distance between adjacent longitudinal bars; s' is the clear spacing of transverse reinforcement; ρ_c is the ratio of the area of longitudinal steel to the area of core section; c_x and c_y are the core dimensions to the centrelines of perimeter hoop in two directions along the two sides of a RC column, respectively, which are equal to c for the square RC column.

In equation (3), f_h is determined by the following equation proposed by Le´geron and Paultre (Le´geron,2003):

$$f_h = \begin{cases} f_h = f_{hy} & \kappa \le 10 \\ f_h = \min(f_{hy}, \dfrac{0.25f_{cc}}{\rho_{sc}(\kappa - 10)}) & \kappa > 10 \end{cases} \quad (5)$$

$$\kappa = f_{cc}/(\rho_{se}E_s\varepsilon_c) \quad (6)$$

where f_{hy} is the yield strength of hoops; κ is a parameter used to determine whether yielding of transverse reinforcement occurs at the peak strength of confined concrete. E_s is the steel modulus of elasticity.

For the descending branch of the concrete model (BC), shown in Figure 3, an exponential function proposed by Binici (Binici,2005) was used, which is defined by:

$$\sigma = f_r + (f_{cc} - f_r)\exp\left[-\left(\frac{\varepsilon - \varepsilon_{cc}}{\alpha}\right)^\beta\right] \quad \varepsilon \ge \varepsilon_{cc} \quad (7)$$

where f_r is the residual stress, as shown in Figure 7; α and β are parameters determining the shape of the softening branch. f_r is equal to $0.1f_{cc}$ as proposed by Tao et al. (Tao,2013) and β is set to 0.8 based on regression analysis. The parameter α is determined as:

$$\alpha = 0.002 + 0.39I_{e50} \tag{8}$$

where I_{e50} is the effective confinement index at 50% of the maximum stress in the descending branch of the stress–strain curve, which is defined as [5]:

$$I_{e50} = \rho_{se}f_{hy}/f_{cc} \tag{9}$$

Hence, different trial values α and β were used until best-fit values were obtained to ensure the predicted $N - \varepsilon$ (axial load vs. axial strain) curves match with the measured curves. It was found that α can be expressed as a function of I_{e50}. Equation (8) was then developed on the basis of regression analysis.

3.2 Other uniaxial compressive stress–strain relationships for concrete

The uniaxial compressive stress–strain relationship proposed by Wang et al. (Wang,1978) was used for the cover concrete of all specimens.

The stress–strain model for concrete confined by the steel box presented by Tao et al. (Tao,2013), which considers the confinement effect of the steel tube on the plastic behaviour of concrete, was used to describe the compressive behaviour of confined concrete at the end of the specimens.

3.3 Uniaxial tensile model

The tensile behaviour of concrete is assumed to be linear elastic until the tensile strength is reached. The post-failure behaviour is specified by applying a fracture energy cracking criterion. The fracture energy is specified directly as a material property in the model and a linear loss of strength after cracking is assumed. The value of fracture energy G_F in N/m is provided by the *Fib Model Code for Concrete Structures 2010*:

$$G_F = 73f_c^{0.18} \tag{10}$$

where f_c is the compressive strength in MPa.

4 MATERIAL MODELLING OF REBAR AND STEEL PLATE

Isotropic elastic–plastic model was used to describe the behaviour of the rebar. The stress–strain relationship for the steel rebar consists of two linear stages (i.e. elastic and hardening), and the hardening modulus was taken as $0.01E_s$. The steel plate at the ends of the columns (Figure 1) was elastic, with an elastic modulus of 2.06×10^5 MPa.

5 FINITE ELEMENT MODEL OF CONFINED RC COLUMN

The steel rebar was modelled using 2-node linear 3D truss element (T3D2) in the Explicit element library. Both the steel plate and the concrete were modelled as 8-node brick elements (C3D8R), with three translation degrees of freedom at each node. The approximate global mesh size of 50 mm for the concrete body and the approximate global mesh size of 25 mm for the steel cage can provide precise simulation results. The FEM mesh is illustrated in Figure 4.

Embedded region constraint was used in the model to embed the steel reinforcement cage (embedded elements) into the concrete block (host elements). The translational degrees of freedom of the embedded node are constrained to the interpolated values of the corresponding degrees of freedom of the host element. However, these rotations are not constrained by the embedding.

General contact combining penalty friction formulation for the tangential behaviour and a contact pressure model in the normal direction in explicit module was used to simulate the interaction between the contact surfaces of steel boxes and the ends of concrete columns. A tie constrain was used to define the interaction between the steel pad at the end of the column and the corresponding end surface of concrete, which fuse together two regions even though the meshes created on the surfaces of the regions may be dissimilar.

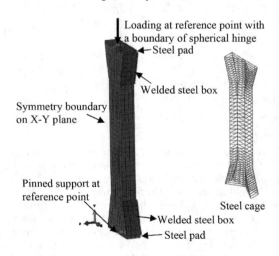

Figure 4.   Boundary condition and FEM meshes.

As shown in Figure 4, one-half model with symmetry boundary on the X–Y plane was used to reduce the computation cost. The kinematic coupling constraint was used to constrain the motion of the end surface of the specimen to the motion of a reference point. The axial load was applied to the top reference point, with translational degrees of freedom in the directions Y and Z and rotational degrees of freedom like a spherical hinge. Pinned support was applied on the bottom reference point.

Displacement loading mode was used in the FEM model. The amplitude curve with a smooth step was used for the static loading application in ABAQUS /Explicit and enough step time is configured to minimize the kinematic and inertia effect.

6 VERIFICATION OF THE FINITE ELEMENT MODEL

Figure 5 shows the comparison of $N - \varepsilon$ (axial load vs. axial strain) curves from the experimental and predicted results. Here, the axial strain is an average value of the test region in the specimen. It can

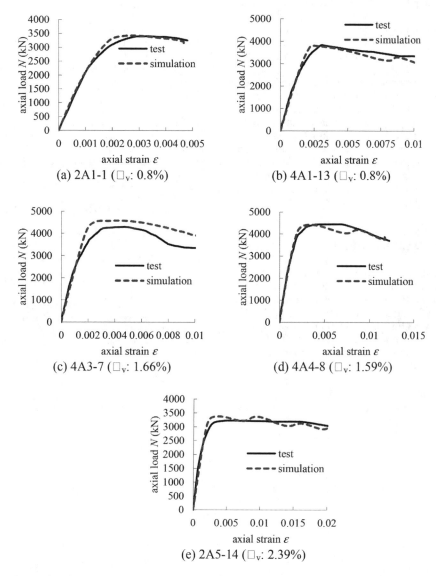

Figure 5. Comparison between the predicted and measured $N - \varepsilon$ curves.

be seen that the developed finite element model predicts the $N - \varepsilon$ curves well.

7 CONCLUSION

In this paper, experimental data were collected and used to develop a new finite element model to simulate the behaviour of confined RC columns under axial compression. The following conclusions can be drawn based on the results of this study:

1. A new uniaxial compressive stress–strain relationship for confined concrete was developed to be used in modelling hoop-confined concrete in a three-dimensional finite element model.
2. An explicit finite element model was developed for square RC columns with consideration of the confinement effect of hoops.
3. The developed finite element model with the proposed constitutive law predicts the behaviour of the laterally confined RC columns well. Thus, the proposed uniaxial compressive stress–strain relationship in conjunction with the concrete damaged plasticity model has the ability to describe the behaviour of hoop-confined concrete and can be used to further investigate the behaviour of hoop-confined RC members under different types of load.

ACKNOWLEDGEMENT

This research was supported by the Project of Natural Science Foundation of Hainan Province (No. 20165208), the Scientific Research Starting Foundation of Hainan University (No. kyqd1534) and the Tianjin University – Hainan University Collaborative Innovation Fund (No. HDTDU201603). All these sources of financial support are highly appreciated.

REFERENCES

ACI Committee 318, *Building code requirements for structural concrete and commentary* (American Concrete Institute, Farmington Hills, 2008).

Binici, B., Eng. Struct. **27**, 1040–1051 (2005).

Bousalem, B., N. Chikh, Mater. struct. **40**, 605–613 (2007).

Comité Euro-International du Béton, *CEB-FIP Model Code 1990* (Thomas Telford House, London, 1993).

Dassault Systèmes, *Abaqus Theory Guide 4.5.2 Damaged plasticity model for concrete and other quasi-brittle materials* (http://129.97.46.200:2080/v6.13/books/stm/default.htm?startat=ch01s05ath08.html).

Faria, R., N.V. Pouca, R. Delgado, J. Earthq. Eng. **8**, 725–748 (2004).

Fib, *Fib Model Code for Concrete Structures 2010* (Ernst & Sohn, Berlin, 2013).

Foster, S. J., J. Liu, S.A. Sheikh, J. Struct. Eng. **124**, 1431–1437 (1998).

Kwon, M., E. Spacone, Comput. Struct. **80**, 199–212 (2002).

Lee, J., G.L. Fenves, J. Eng. Mech. **124**, 892–900 (1998).

Le´geron, F., P. Paultre, J. Struct. Eng. **129**, 241–252 (2003).

Liu, J., S.J. Foster, Comput. Struct. **77**, 441–451 (2000).

Lubliner, J., J. Oliver, S. Oller, E. Oñate, Int. J. Solids Struct. **25**, 299–329 (1989).

Mander, J.B., M.J.N. Priestley, R. Park, J. Struct. Eng. **114**, 1804–1826 (1988).

Park, R., M.J.N. Priestlye, W.D. Gill, J. Struct. Eng. **108**, 929–950 (1982).

Saatcioglu, M., S.R. Razvi, J. Struct. Eng. **118**, 1590–1607 (1992).

Samani, A.K., M.M. Attard, Eng. Struct. **41**, 335–349 (2012).

Sheikh, S. A., S.M. Uzumeri, J. Struct. Div. **106**, 1079–1102 (1980).

Sheikh, S.A., S.M. Uzumeri, J. Struct. Div. **108**, 2703–2722 (1982).

Song, Z. H., Y. Lu, Comput. Concrete **8**, 23–41 (2011).

Tao, Z., Z.B. Wang, Q. Yu, J. Constr. Steel Res. **89**, 121–131 (2013).

Wang, P. T., S.P. Shah, A.E. Naaman, J. Struct. Div. **104**, 1761–1773 (1978).

Zeng, X., B. Xu, Eng. Mech. **31**, 190–197 (2014).

Advances in Energy Science and Equipment Engineering II – Zhou, Patty & Chen (Eds)
© 2017 Taylor & Francis Group, London, ISBN 978-1-138-71798-5

The finite element simulation of shape memory alloy pipe joints

Wei Wang & Bo Wang
School of Civil Engineering, Shenyang Jianzhu University, Shenyang, China

ABSTRACT: The stress distribution of shape memory alloy pipe joints is analyzed by the software ABAQUS. The stress distribution during the process of heating up is simulated by this finite element software. The relationship between the radial compressive stress and various influencing factors is discussed systematically. The simulation demonstrates that the compressive stress between the connected pipe and SMA joint is influenced by many factors, such as the wall thickness of SMA joint, the temperature and the expanding rate. In order to satisfy the needed compressive stress, we consider comprehensively to select the appropriate SMA pipe wall thickness, hole enlargement ratio and connecting pipe diameter.

1 INTRODUCTION

Serious corrosion problems have been existed in oil and gas pipelines, which lead to huge economic losses (Brinson, 1993; Lang, 2006; Ogawa,1993; Graesser, 1991). The main reason for the large amount of corrosion existing in the joints of pipeline is the original pipe connected by welding, on the one hand, needing high technical requirements, and on the other hand the joint itself having a lot of disadvantages. NiTi shape memory alloys (NiTi SMA) have excellent shape memory properties. So they were applied earliest in military aircraft hydraulic pipeline connections. Now they have been widely used in aerospace and other fields of pipe connection. Iron base shape memory alloys (Fe SMA) have many advantages, such as high strength, high restored deformation and good property of corrosion resistance as well as their phase transition temperature Ms near the room temperature; in addition to the above advantages, the alloy of Fe SMA is easy to process and its price is cheaper than that of NiTi SMA. Due to these advantages, this alloy is widely used in pipeline connection in the field of petroleum, chemical industry and so on.

The Fe SMA pipeline joints are easy to keep at room temperature, due to their inverse phase transition temperatures above room temperature. Because the inner diameter of pipe joint is smaller than the outer diameter of connected pipe, the joint diameter can be expanded at room temperature. Then the joint can be set on the outside of the connected pipe completely. And then because of the increasing temperature, the heated pipe joint can also make the SMA joint induce the shape memory effects. The radius of the joint become narrower, at last the joint and pipelines connect together tightly (Qi, 2009; Tan, 1991; Zhao, 2005; Tan, 2000).

In this paper, the ABAQUS software and FORTRAN were used to analyze the fastening force of shape memory alloys in many aspects. An executable program was formulated using FORTRAN language, which used a simplified constitutive relation model of SMA. A FORTRAN program was run in ABAQUS to simulate the whole process of producing deformation recovery and restoring force of SMA pipe joints after heating. This paper systematically discusses the relationship between the radial compressive stress of SMA pipe joints and various influencing factors.

2 THE SIMPLIFIED CONSTITUTIVE RELATION MODEL OF SMA

At room temperature conditions, the radius of SMA pipe joints were expanded first, and then the pipe joint was heated up. Because the connected iron pipe hindered the recovery of SMA pipe joint, the recovery stress is generated in the SMA pipe joint, that is the radial compressive stress in the SMA pipe joint, which is meant to achieve the goal of fastening connected pipes.

The recovery stress of SMA basically followed the Clausius–Clapeyron equation. It's a function of strain, temperature, the rate of stressing and other variable (Tan, 2008)

$$\frac{d\sigma}{dT} = \frac{\Delta H^{m-p}}{T_0 \varepsilon} \approx C_A \qquad (1)$$

Under constant constraints, the recovery stress and temperature approximately follow a linear relationship (Zhao, 2002):

$$\sigma_h = \sigma_h(\varepsilon, \ T) = C_A\big(T - T_0(\varepsilon_C)\big) \qquad (2)$$

According to the experimental curve of stress and temperature, the result basically meets the formula (2), and the C_A measured by experiments approximately equal to 13.8 MPa/°C.

Liu xiao Peng and Jin Wei, etc. conducted a series of research for SMA under the constraining conditions. In order to facilitate calculations, the recovery stress and the temperature can be approximately as a linear relationship (Liu, 2004). That is,

$$\varepsilon(T)=\begin{cases} \alpha_M(T-T_0), & T \leq A_S \\ \alpha_M(A_S-T_0)-\beta(T-A_S), & A_S \leq T \leq A_f \\ \alpha_M(A_S-T_0)-\varepsilon_r^m+\alpha_A(T-A_f), & T \geq A_f \end{cases}$$
(3)

where α_A and α_M are the thermal expansion coefficient of austenite and martensite respectively.

$$\beta=\frac{\varepsilon_r^m}{A_f-A_s},$$

$$\alpha_A=\frac{\pi}{A_f-A_s},$$

$$\alpha_M=\frac{\pi}{M_f-M_s},$$

ε_r^m is the largest recoverable strain.

ABAQUS software not only can perform the standard finite element analysis. At the same time, the software also has the very good open characteristics. The software provides the subroutine interface for users to design nonstandard analysis program to meet the needs. So, in the absence of material constitutive relation, boundary conditions, contact conditions, and loading conditions, the user can through the subroutine of ABAQUS as an extension to the main program.

The user-defined material behaviors (UMAT) is a Fortran program interface, which is provided to the users of ABAQUS to define the material properties. The user subroutines can be used to define some material constitutive relation not contained in the database of ABAQUS.

Through the UMAT interface definition of supplementary material constitutive relation, an arbitrary number of material parameters can be read, as difference between the main program and UMAT exists with respect to data transfer and shared variables. So you must follow the general structure of writing format of UMAT.

According to equations 1 to 3, a material subroutine using Fortran language in Abaqus software is programmed. Different sizes of pipe fittings system and analysis of various factors which influence the fastening force are simulated.

3 ESTABLISHING THE MODEL

The SMA pipe joints were heated in ABAQUS software, and then the shape memory alloy pipe coupling will produce the martensite phase transformation. The stress distribution of the pipe joints can be analysed through the simulation results. The model diagram of shape memory alloy pipe joints as shown in figure 1.

The inner radius of the connected steel pipe is 34 mm, and the outer radius is 42 mm; the inner radius of SMA pipe joint is 33 mm, and the outer radius is 38 mm. The expanding rate is 3%.

The parameter of SMA as follows: modulus of elasticity of 100% austenite Da = 67,000 MPa, martensite modulus elasticity Dm = 26300 MPa, transformation temperature $M_f = -9$°C, Ms = 0°C, As = 80°C, A_f = 250°C, the critical stress $\sigma_s^{cr} = 100$ MPa, $\sigma_f^{cr} = 170$ MPa; transformation constants $C_M = 8$ MPa/°C, $C_A = 13.8$ MPa/°C, The size of connected pipe are Ra = 38 mm, R_b = 42 mm respectively.

Due to the assembled model, the model of SMA pipe joint and connected pipe overlapped, so don't set any constraints between them. Uniformly distributed load was added along the pipe wall, which just expanded the radius of SMA pipe joint, as shown in figure 2.

In the step of thermo-mechanical analysis, SMA pipe joint and connected pipe are set temperature constraints. Due to the shape memory alloy martensitic phase, the transformation effect is significantly higher than the thermal expansion effect. Therefore thermal expansion effect is neglected in the simulation process. The assembled component is meshed and the grid size is 0.005. The selected thermal coupling is a hexahedral grid cell. The meshing diagram is shown in figure 3.

An UMAT of constitutive relations subroutine was imported in ABAQUS. The model was heated

Figure 1. Model diagram of shape memory alloy pipe connecting piece.

for 16 minutes in the temperature field. The stress distribution is shown in figure 4, and it can be seen from the results that the maximum radial stress is 46 MPa.

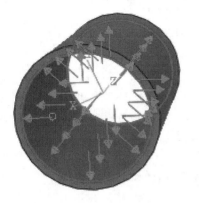

Figure 2. Expanding pipe joint diameter.

Figure 3. Meshing diagram.

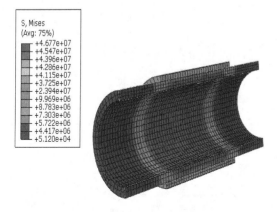

Figure 4. Stress distribution after heating in the temperature field.

4 ANALYSIS OF SIMULATION RESULTS

The size and other parameters are changed in the process of the simulation, and all the parameters that influence the fastening force of the shape memory alloy pipe joints were compared. The conclusions are as follows:

When the expanding rate of SMA joint from 0.03 to 0.13, the thickness of SMA joint is 12 mm, and the outer radius of connected pipe is 42 mm. The radial pressure decreases from 83.2 MPa to 32.3 MPa. This is mainly because the deformation of SMA joint increases with the increase of hole enlargement ratio, which lowered the recovery stress of SMA pipe joint. So, in practical application, while increasing the rate of hole enlargement can be convenient for the construction, but it will also reduce the recovery stress of SMA pipe joint, and affect the values of fastening force.

When the wall thickness of SMA joint varies from 2 mm to 20 mm, the expanding rate of SMA joint is 3%, and the outer radius of connected pipe is 42 mm. The radial pressure increases from 17.1 MPa to 82.1 MPa. But in the actual project if we blindly increase the thickness of the SMA pipe joint, it can cause uneven changes of the stress and strain in the process of heating. This will make the martensitic SMA to produce non-uniform phase transformation. This reduces the ability to recover the shape and strength of SMA pipe joint.

When the outer radius of connected pipe is increased from 42 mm to 62 mm, the expanding rate is 3%, and wall thickness of SMA joint is 12 mm. The radial pressure decreases from 94.0 MPa to 66.9 MPa. During the simulation process of ABAQUS, the temperature gradient was taken into account, so the radial compressive stress reduced obviously with the increase of outer radius of connected pipe. Therefore the size of the connected pipe should be considered in the practical application.

5 CONCLUSIONS

As can be seen from the simulation results, the radial compressive stress increases with the increase of wall thickness of SMA pipe joint, and the radial compressive stress decreases with the increase of hole enlargement ratio of SMA pipe joint and outer radius of connected pipe. Therefore, in engineering practice, selecting suitable thickness of SMA pipe wall, hole enlargement ratio and outer radius of connected pipe are needed to achieve optimal results.

ACKNOWLEDGEMENT

This work was financially supported by the following funders: the Youth Foundation of National

Natural Science (51308357), the Science and College and Universities Outstanding Young Scholar Growth Plan in Liaoning province (LJQ2015091), and Subject Cultivation Plan of Shenyang Jianzhu University (XKHY2–11).

REFERENCES

Brinson, L.C, *One-dimensional constitutive behavior of shape memory alloy: thermo-mechanical derivation with non-constant material functions and redefined martensite internal variable* [J]. Journal of Intelligent Material Systems and Structures, 4(2), 229–242(1993).

Graesser E.J., Cozzarelli F A. *Shape memory alloys as new materials for aseismic isolation* [J]. J. of Eng. Mech, 117 (11), 2590–2608(1991).

Lang, X.Q., Z.Y. Zhao, H. Gong, Q.Z. Liu, *Accident statistical analysis and safety operation measures of the oil-gas pipeline* [J]. Journal of Security and Technology, 6(10), 15–17(2006).

Liu, X.P., W. Jin, M.Z. Chao, R. Yang. *Effect of constraint transformation on recovery strain of $Ti_{44}Ni_{47}Nb_{b9}$ Alloy* [J], Acta Metallurgica Sinica, 2004, 40(4):363.

Ogawa K, Kajiwara S. *Hrem Study of Stress-Induced TransFormation Structures in an Fe-Mn-Si-Ni-Cr Shape Memory Alloy* [J]. Mat Trans JIM, 34(12),1169–1176(1993).

Qi, A.H. *The development current situation and problem analysis of oil-gas pipeline in our country* [J]. International Petroleum Economics. 12 (1), 57–59(2009).

Tan S.M., Lao J.H., Yang S.W. *Influence of grain size on shape memory effect of poly crystalline Fe-Mn-Sialloys* [J]. Scripta Metal Lurgicaet Materialia, 25 (11): 2813–2615(1991).

Tan Z. *Engineering alloy hot property* [M], Beijing: metallurgical industry press, 2008.

Tan, S.M., J.H. Lao, S.W. Yang, *Influence of grain size on shape memory effect of poly crystalline Fe-Mn-Sialloys* [J]. Scripta Metal Lurgicaet Materialia, 25(11), 2813–2615 (1991).

Zhao C.X., Liang G.Y. *Titanium and nitro guenon shape memory effect of Fe-Mn-Si-Cr-Ni alloys* [J]. Scripta Metallurgica et Materialia 38 (7): 1163–1168.

Zhao, L.CH., W Cai, Y.F. Zheng. *Shape memory effect and superelasticity of Alloys* [M]. Beijing: national defence industry press, 2002.

Advances in Energy Science and Equipment Engineering II – Zhou, Patty & Chen (Eds)
© 2017 Taylor & Francis Group, London, ISBN 978-1-138-71798-5

Study on the effect of random cable length error on the structural behaviour of a ridge bar dome

Ailin Zhang

College of Architecture and Civil Engineering, Beijing University of Technology, Beijing, China
Beijing Engineering Research Centre of High-rise and Large-span Prestressed Steel Structure, Beijing University of Technology, Beijing, China

Chao Sun

College of Architecture and Civil Engineering, Beijing University of Technology, Beijing, China

Xuechun Liu

College of Architecture and Civil Engineering, Beijing University of Technology, Beijing, China
Beijing Engineering Research Centre of High-rise and Large-span Prestressed Steel Structure, Beijing University of Technology, Beijing, China

ABSTRACT: A new roof structure of a ridge bar dome is proposed in this paper, and the effect of random error of cable length on the structural behaviour is studied. First, the mechanism of behaviour and failure modes of the ridge bar dome is analysed and compared with cable dome. Then, through the Monte Carlo method and finite element analysis, the effect of random cable length error on the component stress, displacement and ultimate bearing capacity have been researched and compared for the ridge bar dome and the cable dome. The results show that the ridge bar dome overcomes the early relaxation problem of the cable dome and has a better bearing capacity. The effect of cable length error on the structural behaviour of the ridge bar dome is much less than the cable dome, with no cable relaxation and no strut buckling under the design load. For the cable dome, the manufacture error in of ridge cables should be controlled within 5 mm to avoid early failure of cable relaxation and serious deformation.

1 INTRODUCTION

Cable dome is the first civil structure inspired by the tensegrity principle (Geiger, 1986). With the advantages of lightweight and architectural impact, it became popular as roofs for arenas and stadiums in recent years. However, in practical engineering, the internal force of the cables on upper layer, namely the ridge cables, would decrease and cause relaxation under excessive external load, thus resulting in the weakening of structural stiffness (Zhang, 2012). In addition, it's difficult to control the manufacturing errors of flexible cables in the construction process.

To solve the deficiencies of cable domes, a new structure of a ridge bar dome is proposed by replacing the flexible ridge cables with steel bars. The steel bars can bear compression force so that they won't relax and cause local failure. Moreover, steel bars can reduce the material costs and improve the precision of manufacture and positioning, making the ridge bar dome more suitable to engineering application.

As tensegrity structures, the prestress of the cable dome and the ridge bar dome both depend on the change of cable length during the tensioning process. And to control the prestress distribution conveniently, the construction method of "fixed length cable" is always adopted as shown in Figure 1, which makes the anchor end of cable

Figure 1. Unadjustable anchor end of the cable.

unadjustable so that we can tension the cable from its original length to the connection directly through jack equipment (Deng, 2012; Tian, 2011). The construction method of "fixed length cable" is economical and efficient, avoiding time and labour consumption to adjust cable force repeatedly. But if there is a big error in the original cable length, the prestress distribution and geometry shape will seriously deviate from design value after completing construction, which will impact the subsequent mechanical properties under the external load. Therefore, it's necessary to study the influence of cable length error on the structural behaviour of the ridge bar dome.

At present, the research on such tensegrity structure mainly focus on form finding (Juan, 2008; Tibert, 2011) and structural behaviour (Fu, 2004; Ge, 2012), but the researches on the effect of cable length error are only a few. Quirant et al. (Quirant, 2003) studied the effect of cable length error to tensegrity gird through Monte-Carlo method. Gao (Zhang, 2011) and Zhang (Gao, 2004) adopted the orthogonal test to study the sensitivity of manufacturing error for the suspen-dome and cable dome. (Tian, 2011) studied construction tolerance of spoke structure through reliability theory and experiment test. However, studies on the effects of cable length error on the performance of tensegrity dome under load conditions remain blank.

In this paper, a new roof structure of a ridge bar dome is proposed as an improvement plan for the cable dome, and the effects of random cable length error on the structural behaviour of the ridge bar dome is investigated. First, the mechanical performance and failure modes of the ridge bar dome is analysed and compared with cable dome without consideration of cable length error. Then through the combination of finite element analysis and Monte Carlo method, the effects of random cable length error on the component stress, displacement and ultimate bearing capacity are investigated and compared for the two structures.

2 THE MECHANICAL BEHAVIOUR OF RIDGE BAR DOME WITHOUT CABLE LENGTH ERROR

2.1 *Computing model and load cases*

In this paper, the computing model is transformed on the basis of a Geiger cable dome, the roof structure of Erdos gymnasium in China (Zhang, 2012), with a span of 71.2 m and a height of 5.5 m. By replacing the flexible ridge cables with steel bars, the original cable dome is converted to ridge bar

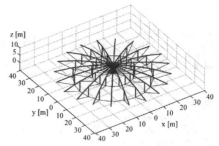

(a) Perspective view of ridge bar dome

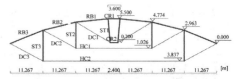

(b) Profile view and component arrangement of ridge bar dome

Figure 2.   Ridge bar dome.
Note: The thin lines represent cables, and bold lines represent steel tubes (struts and ridge bars).

dome, as shown in Figure 2. The ridge bars are three circular steel tubes ($\Phi 299 \times 8$) with hinge joints at both ends, which bears tension force at the initial prestress state. Under external load, the internal force of ridge bars declines gradually and turns into compression. The maximum calculation length of ridge bars is 11.2 m, and slenderness ratio is 108, which is designed to guarantee the ridge bars from bulking failure under the applied load.

Table 1 shows the section parameters, initial prestress and material properties of different components in ridge bar dom. Using FEM software Ansys14.5 for structural analysis, diagonal cables (DC1, DC2 and DC3) and hoop cables (HC1 and HC2) are simulated by link 180, with the elasticity modulus of 1.95×10^5 MPa and the real constant set as tension only. Struts (ST1, ST2 and ST3) and ridge bars (RB1, RB2 and RB3) are also simulated by link 180, with the elasticity modulus of 2.06×10^5 MPa and the real constant set as bi-direction behaviour. The centre rings (CR1 and CR2) are simulated by Beam 188. The boundary conditions of the computing model are set as hinge constraints, and prestress is introduced by initial strain.

The design dead load is 0.8 kN/m², and live load is 0.5 kN/m². For the structure after tension construction without applied load, i.e., the prestressed state, the load combination is: self-weight + prestress. For the structure under design load, i.e., the design load state, the load combination is:

Table 1. Design of the components.

Components	Section	Prestress/MPa	Material strength/MPa
CR1	Square tube	107.9	—
CR2	$300 \times 300 \times 20$	35.6	—
ST1	$\Phi 32$	-18.8	$\Psi f = -207$
ST2	$\Phi 38$	-37.0	-184
ST3	$\Phi 65$	-48.1	-170
RB1	$\Phi 299 \times 8$	73.0	$\psi f = -118$
RB2	$\Phi 299 \times 8$	99.0	-118
RB3	$\Phi 299 \times 8$	147.9	-118
DC1	$\Phi 194 \times 8$	339.7	$f_y = 1336$
DC2	$\Phi 194 \times 8$	442.1	$f_u = 1670$
DC3	$\Phi 219 \times 12$	347.5	
HC1	$3\Phi 40$	402.1	$f_y = 1336$
HC2	$3\Phi 65$	351.8	$f_u = 1670$

Note: For cables, f_y denotes the yield strength and f_u denotes the ultimate strength. For struts and ridge bars, f denotes the designed compression strength and ψ denotes the stability factor in Chinese code GB20017-2003.

self-weight + prestress + 1.2 × dead load + 1.4 × live load.

2.2 Analysis of the loading process

The whole loading process of the structure is analysed through Newton–Raphson method, with material nonlinearity and geometric nonlinearity considered. The maximum horizontal and vertical displacement occurs on the upper node of ST2, and the corresponding load–displacement curves are shown in Figure 3, where the load factor λ is defined as the ratio of external load to the design load (1.66 kN/m²). The stress curves of some main components are represented in Figure 4.

It can be seen that the load–displacement curves rise linearly at early stage. When $\lambda = 1$, i.e., under the design load, the inner ridge bar remains in tension state. When $\lambda = 1.57$, the stress of RB1 turns into compression state. When $\lambda = 2.41$, inflection points are found both in the load–displacement curves and stress curves, indicating the structural stiffness decreases obviously due to the relaxation of inter-diagonal cables (DC1). However, the whole structure can continue to bear load.

When $\lambda = 3.04$, the maximum vertical displacement reaches 0.29 m, that is 1/250 of the span. So the structure fail to satisfy the deflection requirements specified in Chinese code JGJ257-2012. When $\lambda = 5.21$, the outer struts (ST3) buckle and result in the failure of whole structure. At that moment, the minimum stress of inner ridge bar is −65 MPa, still below its stability capacity, and the maximum stress in hoop cables is 1105 MPa, not exceeding its yield strength.

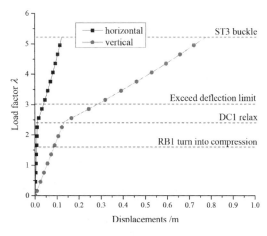

Figure 3. Load–displacement curves of the ridge bar dome.

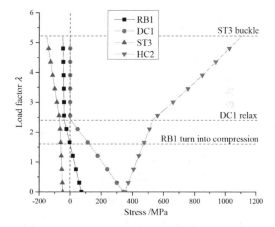

Figure 4. Stress curves of components in the ridge bar dome.

2.3 Comparison with cable dome

To further demonstrate differences of mechanical properties between the ridge bar dome and the original cable dome, the same loading process analysis is carried out for the cable dome, and its load–displacement curves and components stress curves are shown in Figure 5 and Figure 6 respectively. Except for the material difference of ridge components, the computing models for such two structures share the same conditions (including materials of other components, cross-section, prestress and load conditions).

It can be seen that when $\lambda = 1.56$, the stress of inner ridge cable (RC1) of the cable dome decreases to 0, which results in local relaxation failure and weakens the structural stiffness significantly. When

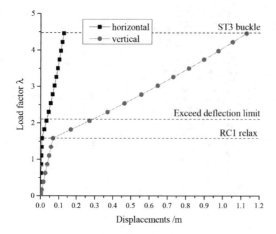

Figure 5. Load–displacement curves of the cable dome.

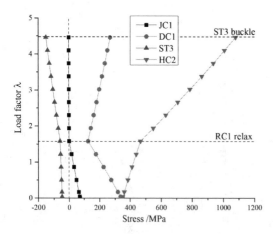

Figure 6. Stress curves of components in the cable dome.

$\lambda = 4.46$, the outer strut buckles (ST3) and results in the failure of whole structure, and there is no more buckling or strength failure occurs in other struts and cables.

Through comparison of the two structures, it can be found that in the ridge bar dome, the relaxation of DC1 happens at a higher load factor, about 1.55 times that of RC1's relaxation in cable dome. Thus the weakening of structural stiffness is delayed, and the structural elastic limit and ultimate bearing capacity is improved.

3 THE SAMPLES OF RANDOM CABLE LENGTH ERROR

In the case of ridge bar dome, the use of steel bars as upper components not only promotes the bearing capacity, but also improves the manufacturing precision of components. Therefore, the structural performance of ridge bar dome is further investigated considering the effects of manufacturing precision.

Table 2 represents the maximum allowable error for cable length specified in JGJ257-2012. Since all cable lengths in the calculation example are less than 50 m, so 15 mm is adopted as the maximum allowable error. Assuming the processing of cables follows 3σ quality management standard, then the random samples of cable length error δ based on normal distribution are generated by Monte Carlo method, with mean 0 and variance $\sigma = 5$ mm, namely $\delta \sim N(0,5^2)$. And Latin hypercube sampling is adopted (Olssin, 2003) to improve the sampling efficiency. Such random cable length error is introduced into the computing model for analysis.

All cables are fixed-length cables with initial length error, because in practical engineering, the hoop cable is connected by four cables and each part passes through five nodes below the struts. So the length error of each part will be only 1/5th of the sample value. For the rigid components such as struts, ridge bars and centre rings, according to Chinese code GB50205-2001, their allowable manufacturing error is 1 mm, which is far smaller than that of cables and can be ignored in the following analysis.

The above random samples of cable length error are generated by MATLAB, and then incorporated to finite element model through temperature load. By use of ANSYS parametric design language

Table 2. Allowable manufacturing error of cable length.

Cable Length L (m)	Allowable Error (mm)
≤50	±15
50<L≤100	±20
>100	L/5000

Table 3. Influence of the sample capacity.

Structural responses	Sample capacity			
	600	800	1000	1200
u_{HC2}/MPa	351.31	351.28	351.24	351.26
σ_{HC2}/ × 10^6	4.60	4.54	4.56	4.55
u_{N2} / × 10^{-5}m	8.28	−5.80	−2.35	−2.11
σ_{N2} / × 10^{-9}	7.27	7.21	7.18	7.19

Note: u_{HC2} and σ_{HC2} denotes the mean value and standard deviation of the stress of HC2 respectively. u_{N2} and σ_{N2} denotes the mean value and standard deviation of the displacement the upper node of ST2 respectively.

(APDL), the structural stress and displacement responses in different load cases are obtained.

To determine the sample capacity, the correlation between sample capacity and the statistics of structural responses of ridge bar dome in prestress state is shown in Table 3. Thus it can be seen that the mean value of the examined variables converges at 1000, which demonstrates the sample capacity is sufficient. The sample capacity is determined as 1200 finally.

4 THE EFFECTS OF RANDOM CABLE LENGTH ERROR ON THE STRUCTURAL BEHAVIOUR

4.1 *Deviation of component stress*

Figure 7 presents the stress statistics of ridge bar dome with random cable length error considered,

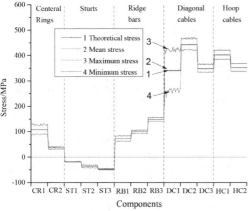

(a) Prestress state

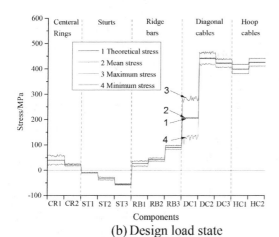

(b) Design load state

Figure 7. Stress statistics of the ridge bar dome.

under prestress state and design load state. The mean stress of components in prestress state are all identical to the design values, while the maximum and minimum stress deviate from the design values to varying extent. If the stress deviation rate is defined as the ratio of stress deviation from the design value, the maximum stress deviation rate of DC1 is 27.5%, exceeding the allowable limit 10%, which has been specified in JGJ257-2012. For other cables, the prestress deviation rates are all below 10% and satisfy the requirements. For the rigid components, although their manufacturing error are not considered, the stress deviation is inevitably generated in order to achieve a new balance state resulting from the cable length error.

Figure 7(b) indicates that as the effects of random cable length are considered, under the design load, the maximum and minimum stress of all components in ridge bar dome still remain within the safety range, with no failure of cable relaxation, cable breaking, or strut buckling.

By contrast, Figure 8 presents the stress statistics of cable dome with random cable length error considered. It can be seen that the minimum stress of RC1 is 0, so RC1 would relax in prestress state, leading cable dome to a premature local failure. This is because the excessive length errors of ridge cables which only exists in cable dome change the structural behaviour significantly. Besides the RC1's relaxation, the maximum stress deviation rates of DC1 and CR1 is 84.8% and 247% respectively, much higher than that of ridge bar dome. These two components are both directly connected to ridge cables and very sensitive to these length errors.

4.2 *Deviation of initial shape and displacement*

The displacement statistics of the two structures are investigated under prestress state and design load. Under prestress state, for ridge bar dome, the maximum displacement caused by random cable length error is 28.4 mm, and that of cable dome is 50.0 mm. Therefore, the effect of cable length error on the initial design shape of the two structures are both not obvious.

Figures 9 and 10 present the node displacement statistics of the two types of structures under design load state. It can be seen that there is an obvious increment between the design value and the maximum value, indicating the cable length error can significantly increase the displacement of the two structures under external load. The increment of horizontal displacements is much bigger than in the vertical direction. This is because the Geiger type tensegrity dome has few components connected in the lateral direction and the structural response, especially the horizontal displacement is very sensitive to the asymmetric distribution of structural stiffness and external load (Fu, 2004; Ge, 2012). Since the random cable length errors

can greatly disturb the symmetry of structural stiffness, as a result, the node displacements of the two structures increase obviously under load state.

As shown in Figure 9, for ridge bar dome the maximum vertical and horizontal displacement are 54.5 mm and 9.4 mm respectively, still in an acceptable range. As shown in Figure 10, for cable dome, the maximum vertical and horizontal displacements are 115.9 mm and 688.2 mm respectively, which are far greater than those of ridge bar dome. The maximum deformation in cable dome is shown in Figure 11. It can be seen that due to the random cable length error, the ST2 of cable dome would incline seriously under design load and fail to meet the deformation requirement. This is because the cable length error can lead to RC1's relaxation, thus resulting in the lack of balancing force on the upper node of ST2 in radial direction. Then the node has to move horizontally to transmit the increasing load and get a new balance state.

4.3 Bearing capacity of relaxation and buckling

The cable length error can change the stress and shape distribution of the structures, and may result in the structures reaching limit state before the expected load. Hence, the load factor correlated with two limit states, cable relaxation and strut buckling, is defined as the structural bearing factor λ_r and λ_b. After introducing the random cable length error satisfying $\delta{\sim}N(0,5^2)$, $\delta{\sim}N(0,3.33^2)$ and $\delta{\sim}N(0,1.67^2)$, namely the allowable cable length errors $[\delta] = 15$ mm, 10 mm and 5 mm respectively, the two bearing factors are obtained through the loading analysis.

Figure 12 presents the average bearing factors of the two structures adopting different $[\delta]$

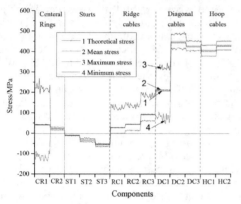

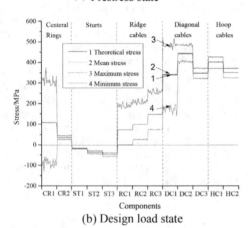

Figure 8. Stress statistics of the cable dome.

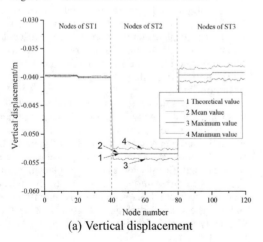

(a) Vertical displacement

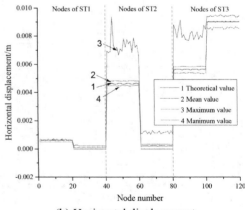

(b) Horizontal displacement

Figure 9. Displacement statistics of the ridge bar dome.

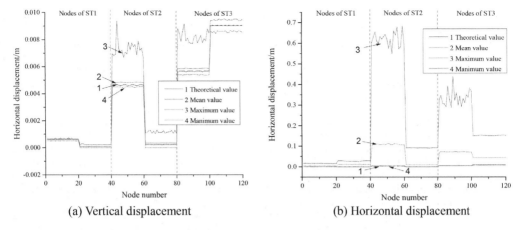

(a) Vertical displacement   (b) Horizontal displacement

Figure 10. Displacement statistics of the cable dome.
Note: The upper and lower nodes of ST1 are numbered from 0 to 20 and from 20 to 40 respectively. The upper and lower nodes of ST2 from 40 to 60 and from 60 to 80. And the upper and lower nodes of outer ST3 from 80 to 100 and from 100 to 200.

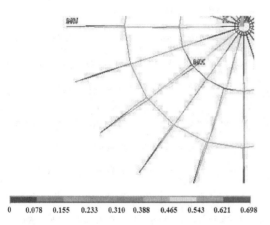

Figure 11. The maximum displacement distribution of the cable dome with cable length error.

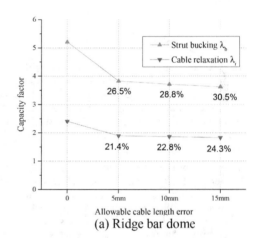

(a) Ridge bar dome

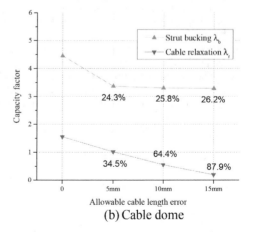

(b) Cable dome

Figure 12. Bearing capacity factors under different allowable length errors.

values, with numbers on the curve representing the decrease of such bearing factors compared to structures with no cable length error. For the two structures, their bearing factors both decrease considerably with the increase of [δ]. This is because the random cable length errors disturb the symmetry of the two Geiger type tensegrity dome and weaken their structural stiffness significantly. For ridge bar and cable domes, their buckling bearing factor λ_b declines 28.6% and 25.4% respectively. So in the design of the two structures, it's necessary to guarantee sufficient safety margin to avoid the early buckling failure caused by cable length errors.

Comparing the two structures, it can be seen that the bearing factors of ridge bar dome are higher than cable dome. This is because the DC1' relaxation in ridge bar dome occurs later than the RC1's relaxation in cable dome, so the structural stiffness of ridge bar dome declines later under a bigger load level. The λ_b of ridge bar dome is 3.72, 13% higher than that of cable dome. The λ_r of ridge bar dome is greater than 1, indicating that no cable relaxes and no strut buckles under the design load.

However, in the case of cable dome, when $[\delta]$ exceeds 5 mm, the λ_r is close to 0, which demonstrates that excessive cable length error can result in relaxation failure at quite low load level. From the deformation results as shown in Figure 11, it can be deduced that the struts would incline seriously and fail to meet the normal serviceability. Therefore, the length error of ridge cables in cable dome should be controlled within 5 mm. Meanwhile, the prestress level of the whole structure should be increased appropriately, so as to prevent the early relaxation of ridge cables caused by cable length error.

5 CONCLUSION

In this paper, a new roof structure of a ridge bar dome is proposed and analysed with respect to mechanical behaviour and failure mode. The effect of random cable length error on the structural performance of the ridge bar dome is investigated and compared with cable dome, the following conclusions have been drawn.

1. In the whole loading process, the structure experiences the following states, such as the relaxation of inter-diagonal cables, the displacement exceeding allowable limit and the buckling of outer struts. By using steel tubes which can bear compression force as the upper component, ridge bar dome overcomes the early relaxation problem of the cable dome and has a better bearing capacity.
2. For Geiger type tensegrity structures, random cable length error can distort the symmetry of structural stiffness, thus resulting in the buckling bearing capacity of the ridge bar dome and cable dome decreasing by 28.6% and 25.4% respectively. It is necessary to guarantee

sufficient safety margin to avoid the early buckling failure in the design work.
3. In the case of cable dome, the manufacturing error in of ridge cables should be controlled within 5 mm. Otherwise, the excessive length error would cause cable relaxation and serious incline of struts, leading the structure failure before the expected load.
4. Thanks to the high manufacture accuracy of upper components, the effect of cable length error on the ridge bar dome is much smaller than the cable dome in terms of stress and displacement. With the random cable length error considered, the buckling bearing capacity of the ridge bar dome is 13% higher than the cable dome, and no cable relaxes under the design load. So the ridge bar dome has a better reliability and more suitable for engineering applications.

REFERENCES

Deng, H., R.M. Song. J. Build. Struc. 33, 71–78 (2012).
Fu, F., J. Constr. Steel Res. **61**, 23–35 (2004).
Gao, B.Q., E.H. Weng. J. Zhejiang Univ. Sc. **5**, 1045–1052 (2004).
Ge, J.Q., A.L. Zhang, X.G. Liu. J. Build. Struc. 33, 1–11 (2012).
Geiger, D.H., A. Stefaniuk, D. Chen. *Shells, Membranes and Space Frame, Proceedings IASS Symposium.* (IASS, Madrid, 1986).
Juan, S.H., J.M. Tur. Mech. Mach. Theory **43**, 859–881 (2008).
Olssin, A., G. Sandberg, O. Dahlblom. Struct. Saf. **25**, 47–68 (2003).
Quirant, J., M.N. Kazi-Aoual, R. Motro. Eng. Struct. **25**, 1121–1130 (2003).
Tian, G.Y., Y.L. Guo, B.H. Zhang. J. Build. Struc. **32**, 11–18 (2011).
Tibert, A.G., S. Pellegrino. Int. J. Space Struc. **26**, 241–255 (2011).
Zhang, A.L., X.C. Liu, J. Li. J. Build. Struc. 33, 54–59 (2012).
Zhang, G.J., J.Q. Ge, S. Wang. J. Build. Struc. 33, 12–22 (2012).
Zhang, J.H., Z.Q. Wang, Y.G. Zhang. Adv. Mater. Res. **163**, 75–81 (2011).

Numerical simulation of rheological behavior of structure plane in rock mass

Zhengqi Lei & Guoxin Zhang
State Key Laboratory of Simulation and Regulation of Water Cycle in River Basin, China Institute of Water Resources and Hydropower Research, Beijing, China

ABSTRACT: Sorts of structure planes with different genetic types and various sizes exist in the rock mass. The rheology of structural planes often plays an important role in the aging deformation and long-term strength of the rock mass. In the long-term practice, scholars have proposed several constitutive models to describe the rheological behavior of rock mass. In this paper, we establish the recurrence formula of creep strain for each step of calculation by introducing the generalized Kelvin rheological model to the discontinuous deformation analysis method. The DDA equilibrium equation, which takes shear creep deformation of structural plane into account, has been deduced based on the minimum potential principle. On this basis, the extension development of the currently available DDA computing program has been done. The creep curves of a laboratory test of structural planes are used to validate the extended DDA program. The result shows that the approach taken in this paper is effective.

1 INTRODUCTION

Rheology is the phenomenon of gradual change of force and deformation over time under the constant external condition (Zhou, 1990). The rheological behavior of rock mass contains two aspects. Under the condition of constant stress, the phenomenon that the deformation increases with time is creep. Under the condition of constant strain, the phenomenon that the stress decreases with time is stress relaxation. Rheology is one of the important reasons for the large deformation and instability of the underground engineering, the foundation of the structure, and the rock slope (Ding, 2005). For instance, Kuyu diversion tunnel is a 9.8 km long tunnel in Xinjiang Province of China. During the 100 days after the excavation of the tunnel, the surrounding rock had produced a continuous deformation and the cumulative deformation had increased by 50% (Zhang, 2011). A variety of supporting measures were adopted during the construction of a highway tunnel in Japan. However, the surrounding rock had still deformed for 300 days, and the maximum deformation had reached 93 cm (Yu, 1998). In 2009, the slope of a large hydropower station in China began to creep into the river after the hydropower station impounded. The maximum deformation rate reached 20–30 mm/day. The total deformation has reached more than 40 m up to now. The study of rheological behavior of rock mass is of great significance for the interpretation of the time-dependent deformation in engineering.

Rock mass is a kind of complex geological body. There are a large number of 2D geological interfaces in the rock mass, such as joints, fissures, and faults. These geological interfaces are collectively called structural planes. The rock mass is cut into blocks with different shapes and sizes according to different combinations of structural planes. Both the rock blocks and the structure planes, which compose of the rock mass together, exhibit rheological behavior (Gong, 2011). The factors that affect the rheology of the rock block and the structure plane mainly include stress level, water content, and temperature. Furthermore, the rheology of the structural plane is also affected by the occurrence, filling, and weathering degree. At present, research works conducted worldwide are mainly confined to the rheological properties of the intact rock, while the understanding of the rheological law of the structural plane is still not enough.

Discontinuous Deformation Analysis (DDA) is a discrete numerical method (Shi, 1985). The method is mainly used to study the block system, which is cut by the natural joint and fissure. In the DDA method, the displacement and deformation of the blocks are unknown. On the basis of an efficient contact search algorithm, all the possible contact forms between blocks could be obtained. Compared with the continuous method, the DDA method could effectively simulate the contact transformation and the large deformation of the block system (Hatzor, 2004; Portales, 2009; Zhang, 2003). In this paper, we establish the recurrence formula of creep strain for each step of calculation by introducing the generalized Kelvin rheological model to the discontinuous deformation analysis method. The DDA equilibrium equation, which takes shear creep deformation of structural plane into account, has been deduced based on the minimum potential principle. On this basis, the extension development of the currently available DDA computing program has been done. The creep curves of a laboratory test of rock mass structural planes are used to validate the extended DDA program.

2 RHEOLOGICAL MODEL

The creep constitutive model of rock mass is a mathematical physical model, which is used to describe the stress–strain–time relationship of rock mass. In the long-term research, scholars have put forward several creep constitutive models of the rock mass materials. From the form, these constitutive models could be roughly divided into empirical formula, combined model, and integral model. At present, the combined model is often used in numerical simulation. According to the elastic, plastic, and viscous properties of rock, the combined model sets up some basic components, which could be combined to describe creep properties of different rocks.

The generalized Kelvin model is made up of elastic element, which reflects the viscoelastic creep behavior of rock mass. In this paper, the model is used to describe the shear creep of rock mass structural plane. The components and creep curve of the generalized Kelvin model are shown in Figures 1 and 2.

The creep equation of the generalized Kelvin model is:

$$\varepsilon(t) = \varepsilon^e + \varepsilon^c = \frac{\sigma_0}{E_1} + \frac{\sigma_0}{E_2}\left(1 - e^{-\frac{E_2}{\eta}t}\right) = \sigma_0 J(t) \quad (1)$$

where ε^e is the instantaneous elastic strain; ε^c is the strain of Kelvin; E_1 and E_2 are elasticity moduli; η is the coefficient of viscosity; t is the creep time; and $J(t)$ is the creep compliance.

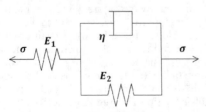

Figure 1. Components of the generalized Kelvin model.

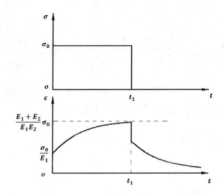

Figure 2. Creep curve of the generalized Kelvin model.

3 DDA EQUATION

The tangential contact deformation in DDA method is shown in Figure 3. The point P_0, which belongs to edge P_2P_3 is assumed to be the lock point of P_1. The tangential contact force is zero when P_1 is in the position of P_0. It is assumed that the vertex P_1 moves to P_4 along the direction of P_2P_3. Then, a shear spring should be arranged between P_0 and P_1 to calculate the tangential contact force. The coordinates of P_0, P_1, P_2, P_3, and P_4 are (x_i, y_i), $i = 0, 1, 2, 3,$ and 4, respectively. The displacement of P_0, P_1, P_2, P_3, and P_4 are (u_i, v_i), $i = 0, 1, 2, 3,$ and 4, respectively.

In the calculation process of DDA, time t is divided into a series of time periods. The equilibrium equation is established in each time period to obtain the stress increment. It can be seen from formula 1 that the creep deformation of the structure plane is influenced by the whole history of the stress. Therefore, the recurrence relation of creep strain in each time step should be established. As shown in Figure 4, time t is divided into a series of moments. Taking the moments t_{n-1} and t_n as the research object, the corresponding stress increments of the moments are $\Delta\sigma_{n-1}$ and $\Delta\sigma_n$. The creep deformation for the time period is:

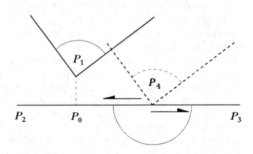

Figure 3. Tangential contact deformation in DDA method.

1440

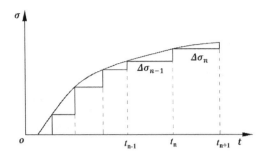

Figure 4. Stress increment of each step.

$$\Delta\varepsilon^c_{t_{n+1}} = \varepsilon^c(t_{n+1}) - \varepsilon^c(t_n)$$
$$= \Delta\sigma_0\left[J(t_{n+1}-t_0) - J(t_n-t_0)\right]$$
$$+ \Delta\sigma_1\left[J(t_{n+1}-t_1) - J(t_n-t_1)\right] \qquad (2)$$
$$+ \cdots$$
$$+ \Delta\sigma_{n-1}\left[J(t_{n+1}-t_{n-1}) - J(t_n-t_{n-1})\right]$$
$$+ \Delta\sigma_n J(t_{n+1}-t_n)$$

The creep compliance $J(t)$ is:

$$J(t) = \frac{1}{E_1} + \frac{1}{E_2}\left(1 - e^{-\frac{E_2}{\eta}t}\right) \qquad (3)$$

Substituting formula 3 into formula 2:

$$\Delta\varepsilon^c_{n+1} = \left(1 - e^{-\frac{E_2}{\eta}\Delta t_{n+1}}\right)\omega_{n+1} + \Delta\sigma_n J(t_{n+1}-t_n) \qquad (4)$$

$$\omega_{n+1} = \frac{\Delta\sigma_0}{E_2}e^{-\frac{E_2}{\eta}(t_n-t_0)} + \frac{\Delta\sigma_1}{E_2}e^{-\frac{E_2}{\eta}(t_n-t_1)} + \cdots$$
$$+ \frac{\Delta\sigma_{n-1}}{E_2}e^{-\frac{E_2}{\eta}(t_n-t_{n-1})} \qquad (5)$$

In the same way, we obtain:

$$\Delta\varepsilon^c_n = \left(1 - e^{-\frac{E_2}{\eta}\Delta t_n}\right)\omega_n + \Delta\sigma_{n-1} J(t_n-t_{n-1}) \qquad (6)$$

$$\omega_n = \frac{\Delta\sigma_0}{E_2}e^{-\frac{E_2}{\eta}(t_{n-1}-t_0)} + \frac{\Delta\sigma_1}{E_2}e^{-\frac{E_2}{\eta}(t_{n-1}-t_1)} + \cdots$$
$$+ \frac{\Delta\sigma_{n-2}}{E_2}e^{-\frac{E_2}{\eta}(t_{n-1}-t_{n-2})} \qquad (7)$$

Comparing formula 5 and formula 7, we obtain:

$$\begin{cases} \omega_{n+1} = \omega_n e^{-\frac{E_2}{\eta}\Delta t_n} + \frac{\Delta\sigma_{n-1}}{E_2}e^{-\frac{E_2}{\eta}\Delta t_n} \\[2mm] \omega_1 = \frac{\Delta\sigma_0}{E_2} \end{cases} \qquad (8)$$

Formulas 6, 7, and 8 constitute a set of recursive formulas, which can be used to calculate the tangential creep deformation at any time.

The potential energy of the shear spring caused by the tangential creep deformation is:

$$\Pi_s = \frac{K_s}{2P_s}d^2$$
$$= \frac{K_s}{2(1-q_n K_s)}\left(\{H\}^T\{D_{(i)}\} + \{G_{(j)}\}^T\{D\} + \Delta\varepsilon^c_n\right)^2$$
$$= \frac{K_s}{2(1-q_n K_s)}[\ \{D_{(i)}\}^T\{H\}\{H\}^T\{D_{(i)}\}$$
$$+ \{D_{(j)}\}^T\{G\}\{G\}^T\{D_{(j)}\} + 2\{D_{(i)}\}^T\{H\}\{G\}^T\{D_{(j)}\}$$
$$+ 2\Delta\varepsilon^c_n\{D_{(i)}\}^T\{H\} + 2\Delta\varepsilon^c_n\{D_{(j)}\}^T\{G\} + \{\Delta\varepsilon^c_n\}^2 \qquad (9)$$

where matrices $\{D_{(i)}\}$ and $\{D_{(j)}\}$ are:

$$\{D_{(i)}\} = \{D_{i(1)}, D_{i(2)}, D_{i(3)}, D_{i(4)}, D_{i(5)}, D_{i(6)}\} \qquad (10)$$

$$\{D_{(j)}\} = \{D_{j(1)}, D_{j(2)}, D_{j(3)}, D_{j(4)}, D_{j(5)}, D_{j(6)}\} \qquad (11)$$

Matrices $\{H\}$ and $\{G\}$ are:

$$\{H\} = \frac{1}{l}\left[T_{(i)}(x_1,y_1)\right]^T\begin{Bmatrix} x_3 - x_2 \\ y_3 - y_2 \end{Bmatrix} \qquad (12)$$

$$\{G\} = \frac{1}{l}\left[T_{(j)}(x_0,y_0)\right]^T\begin{Bmatrix} x_2 - x_3 \\ y_2 - y_3 \end{Bmatrix} \qquad (13)$$

where $T_{(i)}(x_1,y_1)$ is the displacement function of point P_1 and $T_{(j)}(x_1,y_1)$ is the displacement function of point P_0.

Therefore, shear spring matrix $[\mathbf{K}]$ is:

$$\begin{cases} \dfrac{K_s}{1-q_n K_s}\{H\}\{H\}^T \\[3mm] \dfrac{K_s}{1-q_n K_s}\{H\}\{G\}^T \\[3mm] \dfrac{K_s}{1-q_n K_s}\{G\}\{H\}^T \\[3mm] \dfrac{K_s}{1-q_n K_s}\{G\}\{G\}^T \end{cases} \qquad (14)$$

Load matrix $[\mathbf{F}]$ caused by shear spring is:

$$\begin{cases} \dfrac{-K_s}{1-q_n K_s}\Delta\varepsilon^c_n\{H\} \\[3mm] \dfrac{-K_s}{1-q_n K_s}\Delta\varepsilon^c_n\{G\} \end{cases} \qquad (15)$$

1441

In addition, elastic submatrix, inertia submatrix, load matrix, and initial stress submatrix of the block system are integrated into the overall balance equation:

$$\begin{bmatrix} K_{11} & K_{12} & \cdots & K_{1n} \\ K_{21} & K_{22} & \cdots & K_{2n} \\ \cdots & \cdots & \cdots & \cdots \\ K_{n1} & K_{n2} & \cdots & K_{nn} \end{bmatrix} \begin{Bmatrix} D_1 \\ D_2 \\ \cdots \\ D_n \end{Bmatrix} = \begin{Bmatrix} F_1 \\ F_2 \\ \cdots \\ F_n \end{Bmatrix} \qquad (16)$$

4 VERIFICATION

Aiming at the development characteristics of the high slope of a hydropower station under the condition of saturated water, the shear rheological test device of the weak structure plane has been developed in the literature (Cao, 2008). In the literature, the rheological properties of the weak structural planes in coal measure strata are studied, and the shear strain time curves under different stress states are obtained. In this paper, we simulated the test, as described in Cao (2008), with the extended DDA program. As shown in Table 1, the rheological parameters of the generalized Kelvin model are determined by inverse analysis. The validity of this program is verified by comparing the results of DDA numerical simulation and the experimental results in the literature (Cao, 2008).

The comparison between the DDA results and the test results are shown in Figures 5, 6, 7, and 8.

In the figures above, the simulation results are in agreement with the experimental results. The comparison of the numerical results with the shear creep test results shows that the extended DDA method can accurately simulate the mechanical

Table 1. Rheological parameters of structural plane.

f	c (MPa)	E_1 (MPa)	E_2 (MPa)	η (MPa*h)
0.325	0.071	56.80	14.71	$1.1*10^4$

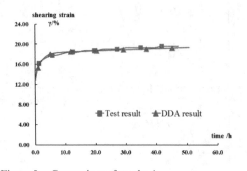

Figure 5. Comparison of results A.

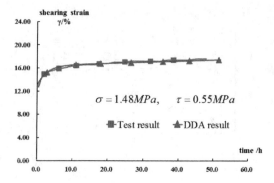

Figure 6. Comparison of results B.

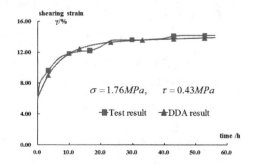

Figure 7. Comparison of results C.

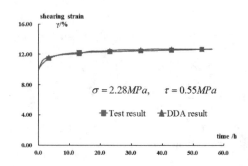

Figure 8. Comparison of results D.

behavior of the structure plane under various stress states by choosing appropriate rheological parameters.

5 CONCLUSION

On the basis of the research work in this paper, the following understanding can be obtained: (1) rheology is an inherent mechanical property of rock mass. Under the long-term load, the stress and strain state and the deformation and failure characteristics of rock mass change with time. The time effect is significant. Therefore, it is necessary to consider the rheological characteristics for the evaluation of long-term stability of rock mass engineering. (2) In this paper, the generalized Kelvin model and discrete numerical method are combined to simulate the shear creep deformation of rock mass structure plane. Comparison of experimental results with laboratory test results shows that the method proposed in this paper is effective. (3) For the soft rock, the rheological behavior of rock block is also very significant. In addition to considering the creep deformation of the structural plane, the creep deformation of the rock block should also be considered.

ACKNOWLEDGMENT

This work was sponsored by the National Natural Science Foundation of China (GN: 51579252).

REFERENCES

Cao, Y.J., R.Q. Huang, H.M. Tang, Chinese Journal of Rock Mechanics and Engineering, **27**, 3732–3739, (2008).

Ding, X.L., Institute of Rock and Soil Mechanics, Chinese Academy of Science, (2005).

Gong, C., Chengdu University of Technology, (2011).

Hatzor, Y.H., A.A. Arzi, Y. Zaslavsky, Int. J. Mech. Min, **41**, 813–832, (2004).

Portales, H., N. Pinna, M.P. Pileni, J. Phys. Chem. A, **113**, 4094–4099, (2009).

Shi, G.H., R E Goodman, Int. J. Numer. Anal. Met, **9**, 541–556, (1985).

Yu, Y., Modern Tunnelling Technology, **20**, 46–51, (1998).

Zhang, G.X., X.F. Wu, Chinese Journal of Rock Mechanics and Engineering, **22**, 1269–1275, (2003).

Zhang, Q.L., N. Li, X. Qu, Chinese Journal of Rock Mechanics and Engineering, **30**, 2196–2202, (2011).

Zhou, W.Y., *Advanced Rock Mechanics* (China Water & Power Press, Beijing, 1990).

Advances in Energy Science and Equipment Engineering II – Zhou, Patty & Chen (Eds)
© 2017 Taylor & Francis Group, London, ISBN 978-1-138-71798-5

Calculation of the pile foundation considering the influence of the karst

Zhangjun Dai, Shanxiong Chen, Xichang Xu & Shanglin Qin

State Key Laboratory of Geomechanics and Geotechnical Engineering, Institute of Rock and Soil Mechanics,
Chinese Academy of Sciences, Wuhan, China

ABSTRACT: Simulation of the influence of the karst on the pile foundation has become a widespread concern, which is of great significance to the foundation design and construction of the karst area. In this paper, modulus stiffness reduction method is used for the soft-filling material inside the cave, to simulate the secondary consolidation deformation of the filling due to the interference of environment. The results show that the effect of upper load on the pile settlement deformation is the most significant, and the existence of karst has a certain effect on the increase of pile settlement. Under the influence of the cave consolidation settlement, the karst cave area appears to have negative side friction. Furthermore, the consolidation settlement of the cave filler significantly increases the axial force of the pile.

1 INTRODUCTION

Karst is widely distributed in China, in all types of soluble rocks, especially carbonate rocks. The total area occupied by karst is about 344.3 km², accounting for more than 1/3 of the total land area. The adverse effect of karst on buildings has attracted increasing attention (Jiang, 2014; Sun, 2013; Zhao, 2004; Zheng, 2011). The development of karst destroys the integrity of the rocks, and also greatly reduces their stability and strength. Under other factors such as the action of the vibration and additional load, they may cause roof collapse, sudden sinking of ground, and structural damage, resulting in significant losses (Yi, 2013; Poulos, 2013). Therefore, how to simulate the influence of karst on the pile foundation has become a widespread concern, which is of great significance to the foundation design and construction in the karst area (Hua, 2014; Lodigina, 2014).

2 SIMULATION METHOD CONSIDERING THE INFLUENCE OF THE KARST

Pile construction is a rebalancing process for beaded cave and its filler. Because in the actual project, a lot of soft plastic or plastic stuffing exist in the beaded cave, after millions of years of consolidation settlement and geological evolution, the soft filling material has been in a relatively balanced steady state. When the initial stress state of the surrounding rock and the filling are not damaged, the soft filler will not have additional deformation. However, in the pile foundation construction, because of the structure stiffness, the pile body has

large stiffness difference with the soft filler. When the pile passes through the beaded cave, strong disturbance will damage the balance state of the soft filler inside the caves, which will stimulate stress release effect of the cave filling. Then, soft filler is bound to consolidate and deform, to resist the work done by the stress release, until the energy environment in the karst cave is in a new equilibrium state.

In this paper, modulus stiffness reduction method is used for the soft-filling material inside the cave, to simulate the secondary consolidation deformation of the filling due to the interference of environment because under the same stress condition, the reduction of modulus stiffness will produce new deformation. Therefore, by using the modulus stiffness reduction method, it can reflect not only the stress release effect of the filler in the cave but also the consolidation deformation of the filler.

3 COMPUTATIONAL MODEL

According to the geometrical position of the pile and the karst cave, the relationship between the pile foundation and the stratum is close to the axial symmetry, as shown in Figure 1. Therefore, using the axisymmetric model, the numerical simulation of pile foundation and karst cave is studied, and the geometric model is established in Figure 2.

The PLANE 82 element and triangle mesh are used in the finite-element model, Drucker–Prager is chosen as the constitutive model, on both sides of the model, the lateral displacement is restricted, the upper part of the model is free surface, and the vertical displacement is limited in the bottom of the model. Contact elements are set between the pile foundation and formation. The material parameters are shown in Table 1.

The friction material is set on the contact pair, the friction coefficient is 0.3, and the initial bite force is 9 kPa, By taking the load of pile foundation and the upper load of pile foundation of 624 kPa into consideration.

4 ANALYSIS OF THE INFLUENCE OF KARST CAVE ON THE DEFORMATION OF PILE FOUNDATION

4.1 *Foundation settlement deformation*

The settlement deformation of the foundation in different load stages of pile construction, upper load, and cave effect is shown in Figure 3.

Figure 1.   Axisymmetric model.

Figure 2.   Geometric model.

Under the pile body weight and the upper load, the stratum settlement deformation is a typical settlement basin, and the pile settlement is greater than the settlement of the foundation in the same layer. Under the effect of the consolidation settlement of the filling in the cave, there is a new settlement deformation in the foundation around

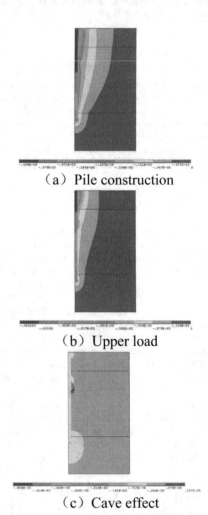

（a）Pile construction

（b）Upper load

（c）Cave effect

Figure 3.   Settlement deformation of the foundation.

Table 1.   Calculation parameters.

Material	Lithology	Density (kN/m³)	Elastic modulus (MPa)	Poisson ratio	c (kPa)	φ (°)
Foundation	Pebble	21	50	0.27	10	40
	Medium weathered rock	22	180	0.26	80	35
	Slightly weathered rock	23	1000	0.23	150	40
Pile	C40 reinforced concrete	26	32500	0.20	–	–
Karst cave	Soft filler	13	10	0.4	5	8–10

the pile, but the settlement and the influence range are all small.

4.2 *Pile settlement deformation*

The settlement deformation of the pile in different load stages of pile construction, upper load, and cave effect is shown in Figure 4.

Under the pile body weight and upper load, the pile top settlement is maximum and the settlement at the pile bottom is minimum. Under the consolidation settlement of the cave, the settlement of the pile near the cave area increases, which similar to the settlement of the pile top. The farther the lower part of the pile from the cave, the smaller is the settlement.

Figure 5 shows the settlement curve of the pile under the influence of the pile foundation

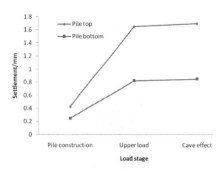

Figure 5. Settlement curve of the pile under different load stages.

construction, the upper load, and the karst cave. It can be seen, after the completion of the pile foundation construction, due to the weight of the pile, the settlements at the pile top and pile bottom are 0.43 and 0.25 mm, respectively. The pile has small amounts of compression deformation. Under the upper loads, the pile settlement increases significantly, the settlement at the pile top increases to 1.65 mm, and the settlement at the pile bottom increases to 0.82 mm. The settlement difference between the pile top and bottom increases significantly, which shows that the upper load further causes compressive deformation of the pile. Under the influence of the cave filling consolidation, the pile settlement deformation increment is very small, the settlement at the pile top is 1.70 mm, and the settlement at the pile bottom is 0.85 mm. The differential settlement of the pile top and pile bottom has a slight increase of 0.02 mm, which shows that the influence of the karst on the settlement deformation of the pile top is larger than that of the pile bottom.

The settlement on pile top caused by the pile weight load accounts for about 25%, that caused by the upper load accounts for about 72%, and that caused by karst accounts for about 3%. Therefore, the effect of the upper load on the pile settlement deformation is the most significant, and the existence of karst has a certain influence on the increase of pile settlement.

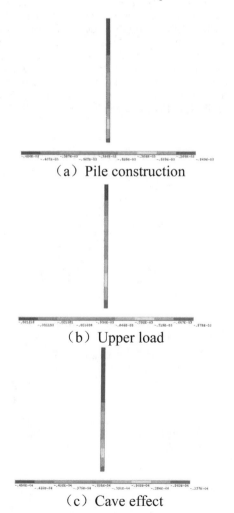

（a）Pile construction

（b）Upper load

（c）Cave effect

Figure 4. Settlement deformation of the pile.

5 ANALYSIS OF THE INFLUENCE OF KARST CAVE ON THE STRESS OF PILE FOUNDATION

5.1 *Pile side friction*

The side friction of the pile in different load stages of pile construction, upper load, and cave effect is shown in Figure 6.

1447

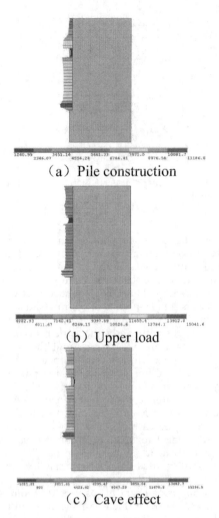

(a) Pile construction

(b) Upper load

(c) Cave effect

Figure 6. Side friction of the pile.

Under the weight of the pile, the side friction increases with the increase of the depth, and the value is stable after reaching the peak value. In the vicinity of the karst cave, the side friction is low, and through the cave area, the side friction is restored to the peak state.

Under the upper load, the side friction increases, in addition to the cave area, and the value is the same as that of the cave area, but still smaller than that of its upper and lower sides.

Through reduction of the elastic modulus of the soft filler inside the cave, under the influence of the cave consolidation settlement, the karst cave area appears negative side friction, and in the central region of the cave appears as the maximum negative side friction (−1.01 kPa), which slowly decreases on both sides until the positive side friction is reached.

5.2 Axial force of the pile

Figure 7 shows the variation of the axial force of the pile under the upper load and the karst cave effect. Under the upper load of the pile, the axial force at the pile top is maximum and the axis force decreases with depth and is minimum at the pile bottom. The reduction of the elastic modulus of the soft filling inside the cave and the consolidation settlement of the cave filler lead to the sudden reversal increase of pile axial force. When this reversal increase of axial force reaches a certain value, the pile axial force decreases with the depth. Comparing the axial forces in the lower part of the pile at the karst area under these two load stages, it can be known the consolidation settlement of the cave filler significantly increases the axial force of the pile.

Figure 8 shows the additional axial force of the pile caused by the cave effect. From the pile top to the top of the cave with the depth of 0–9 m, the axial force of pile is almost 0 kN. When the cave distribution area is reached with the depth of 9~12.6 m, the axial force of pile linear increases from 0 kN to a maximum value of 110 kN, which then declines to a steady value of 95 kN and remains unchanged until the bottom of the pile.

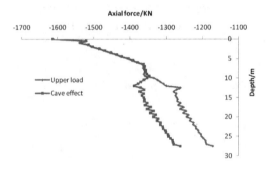

Figure 7. Variation of the axial force of the pile under different load stages.

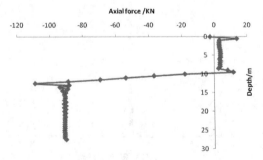

Figure 8. Axial force of the pile caused by the cave effect.

1448

6 CONCLUSION

Modulus stiffness reduction method is used for the soft-filling material inside the cave to simulate the secondary consolidation deformation of the filling due to the interference of environment. It can reflect not only the stress release effect of the filler in the cave but also the consolidation deformation of the filler. The settlement on pile top caused by the pile weight load accounts for about 25%, that caused by the upper load accounts for about 72%, and that caused by karst accounts for about 3%. The effect of upper load on the pile settlement deformation is the most significant, and the existence of karst has a certain effect on the increase of pile settlement. Under the influence of the cave consolidation settlement, the karst cave area achieves negative side friction. And the consolidation settlement of the cave filler significantly increases the axial force of the pile.

REFERENCES

Hua, S., & Engineering, D.O. Numerical simulation research on the stability of transmission tower pile foundations in a karst area of guangdong province. Carsologica Sinica, 33(1), 44–50. (2014).

Jiang, C., Liu, L., & Wu, J.P. A new method determining safe thickness of karst cave roof under pile tip. Journal of Central South University, 21(3), 1190–1196. (2014).

Lodigina, N., & Sharapov, R. Calculation of pile foundations at the karst areas. International Conference on Mechanical Engineering, Automation and Control Systems. IEEE. (2014).

Poulos, H.G., Small, J.C., & Chow, H. Foundation design for high-rise tower in karstic ground. Geotechnical Special Publication (229), 720–731. (2013).

Sun, Y., Zhang, Z., & Zhang, H. Analysis on parameter sensitivity of pile foundations stability in karst areas based on the theory of grey relation. Chinese Journal of Underground Space & Engineering, 9(2), 297–303. (2013).

Yi, J., He, G.J., Liu, S.S., Li, Z.Y., & Zheng, Z.E. Construction method and numerical analysis on the bearing capacity of the large diameter and abyssal pile located in complex karst area. Advanced Materials Research, 639–640, 688–693. (2013).

Zhao, M.H., Cao, W.G., Peng-Xiang, H.E., & Yang, M.H. Study on safe thickness of rock mass at end of bridge foundation's pile in karst and worked-out mine area. Rock & Soil Mechanics. (2004).

Zheng, W.G., Xie, Y.C., & Xue, X.B. Selection of pile foundations in karst areas. Yantu Gongcheng Xuebao/chinese Journal of Geotechnical Engineering, 33, 404–407. (2011).

Advances in Energy Science and Equipment Engineering II – Zhou, Patty & Chen (Eds)
© 2017 Taylor & Francis Group, London, ISBN 978-1-138-71798-5

Analysis of seismic response of bedding slope based on displacement characteristics

Qian Zhang
Structure Health Monitoring and Control Institute, Shijiazhuang Tiedao University, Shijiazhuang, China

Haixia Zhang
Beijing General Municipal Engineering Design and Research Institute Co. Ltd., Beijing, China

Weigang Zhao & Jingchun Wang
Structure Health Monitoring and Control Institute, Shijiazhuang Tiedao University, Shijiazhuang, China

ABSTRACT: In order to discuss the influence rules of slide mass thickness and seismic wave parameters (amplitude, frequency and duration) on deformation characteristics of bedding rock slope, the finite element analysis software ABAQUS and the protective effect of anti-slide piles were studied by the response simulation of slope subjected to seismic load under three working conditions of supporting without piles, supporting with single row of anti-slide piles and supporting with two rows of anti-slide piles providing reference for similar projects. The results showed: the maximum cumulative displacement of bedding slope increased with the increase of seismic load amplitude and vibration time, while the increasing rate was also increasing; furthermore, the maximum cumulative displacement of bedding slope decreased with the increase of slide mass thickness under the same conditions. Simultaneously, the influence of high-frequency seismic waves on bedding slope decreased with the increase of slide mass thickness. Finally, according to analysis of the displacement of slope midline and the acceleration-time curve of measuring key point, obvious anti-seismic effect of anti-slide pile in the reinforcement of bedding rock slope was verified providing reference to similar projects.

1 INTRODUCTION

The failure mechanism and reinforcement of the high-steep slope under seismic load are important problems in the slope protection research. The shear force and thrust of thick-type bedding slope are relatively large under normal conditions with sudden and short process of failure, so the slope instability can cause much more serious damage and the bolt supporting is generally difficult to achieve the desired effect (Xiang, 1955). As a kind of more effective supporting structure, anti-slide pile is widely used in the reinforcement engineering of bedding slope having broad application prospect (Liu, 2006).

However, because of the sudden and unpredictable occurrence of earthquakes, it is difficult to observe the dynamic failure process of rock slope on site and verify the reinforcement effect. Therefore, numerical simulation method was adopted to predict the stability of the ground motion, the dynamic damage range and the failure mode of the slope. Through detailed investigation of the engineering geological conditions of the dangerous

rock mass in high risk areas, the physical model or numerical model of the rock mass was established to study the geometrical, physical and mechanical characteristics of the dangerous rock mass (Cai, 2000) eously, the protective effect of anti slide piles could also be verified by numerical simulation.

J.P. Stewart (Stewart, 2003) presented a probability sieve analysis method through calculation of seismic dynamic response and sliding displacement based on the nonlinear viscoelastic model. Li Haibo (Xiao, 2007) used UDEC software to study slope stability under the influence of parameters: slope height, seismic wave (such as amplitude and frequency) and gave a series of meaningful conclusions. Pradel (Daniel, 2010) btained the safety factor and critical failure mode of the pile by using the finite difference software. S.L. Kramer (Kramer, 1997; Robert, 1999) et al presented a nonlinear seismic displacement model based on the Newmark method combining with the dynamic response and the potential risk of the sliding body. On the basis of the Newmark method, E.M. Rethje (Rathje, 1999) developed a more complex seismic displacement model according to the interaction between plastic deformation and seismic response of the slide mass. However, the research on the seismic gradual failure instability of the bedding rock slope is still in its initial stage.

2 RESPONSE CHARACTERISTICS OF BEDDING ROCK UNDER SEISMIC LOAD

Currently, ABAQUS is one of the relatively advanced and widely used large universal nonlinear finite element analysis softwares in dealing with geotechnical engineering, which includes abundant model libraries of material and defines the material characteristics controlled by tools such as user subroutine.

In this study, the displacement response characteristics of bedding rock under seismic load was studied by the finite element analysis software ABAQUS to mainly analyze the influence of sliding mass thickness and seismic wave parameters (amplitude, frequency and duration) on deformation characteristics of bedding rock slope.

2.1 Calculation model and boundary conditions

In order to research the influence rules of sliding mass thickness and seismic wave parameters (amplitude, frequency and duration) on deformation characteristics of bedding rock slope, the simplified computing model was established shown in Figure 1. The lengths of X, Y, Z directions of model slope mass were 200 m, 120 m, 100 m respectively. The lengths of top and bottom platforms were both 60 m, and the height of bottom platform was 30 m. The thicknesses of slide mass on the surface of the slope were 10 m, 20 m, 30 m respectively.

In the simulation process of slope response under seismic load using ABAQUS, the D–P material yield criterion was adopted in the slide mass, bedrock and anti-slide pile for the better description of strength characteristics of rock material; the viscous boundary condition was applied on the bottom of model to eliminate the reflection effect of wave, and the free field boundary conditions were applied on the other boundaries of model. For the influence of damping, the kinetic energy attenuation of slide mass was realized by Rayleigh damping to factually simulate the seismic load. The simulated seismic wave was the cosine shear stress wave applied on the bottom of the model with the acceleration time history of $a = \lambda \cos (2\pi f t)$, and the influence of the thickness, amplitude, frequency in slide mass and the seismic wave duration on the slope stability was studied by controlling above variables.

（a）Thickness of 10m

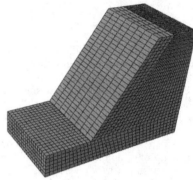

（b）Thickness of 20m

（c）Thickness of 30m

Figure 1.　Building of model.

2.2 Influence analysis of amplitude on bedding slope displacement

The change rules of maximum cumulative displacement of slope with the amplitude of input seismic waves were shown in Figure 2 concerning different slide thicknesses of 10 m, 20 m and 30 m. The frequency and duration were taken as 0.1 Hz and 10 s, and the amplitude was taken as 0.25 m/s², 0.5 m/s², 0.75 m/s² and 1 m/s². It showed that the maximum cumulative displacement of bedding slope increased with the increase of seismic load amplitude, and its increasing rate was also increasing; the permanent displacement of bedding slope decreased with the increase of slide mass thickness under the same conditions.

2.3 Influence analysis of vibration time on bedding slope displacement

The change rules of maximum cumulative displacement of slope with the time of vibration were shown in Figure 3 concerning different slide thicknesses. The frequency and amplitude were taken as 0.1 Hz and 1 m/s2, and the duration was taken as 1 s, 2 s, 4 s and 10 s. It showed that the maximum cumulative displacement of bedding slope increased with the increase of vibration time and decreased with the increase of slide mass thickness under the same conditions.

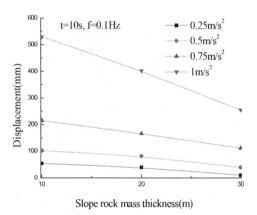

Figure 2. Curves of bedding slope displacement affected by amplitude.

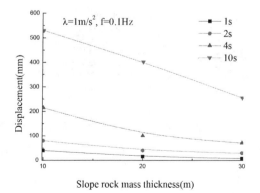

Figure 3. Curves of bedding slope displacement affected by vibration time.

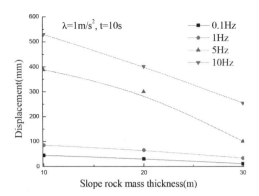

Figure 4. Curves of bedding slope displacement affected by frequency.

2.4 Influence analysis of frequency on bedding slope displacement

The change rules of maximum cumulative displacement of slope with the frequency of seismic load were shown in Figure 4. The vibration time and amplitude were taken as 10 s and 1.25 m/s², and the frequency was taken as 0.1 Hz, 1 Hz, 5 Hz and 10 Hz. It showed that the maximum cumulative displacement of bedding slope decreased with the increase of frequency, and the maximum cumulative displacement of slope caused by low-frequency (0.1 Hz, 1 Hz) seismic waves was significantly higher than that case caused by high-frequency seismic waves; the maximum cumulative displacement of bedding slope decreased with the increase of slide mass thickness under the same conditions. At the same time, the influence of high-frequency seismic waves on bedding slope decreased with the increase of slide mass thickness.

3 RESPONSE CHARACTERISTICS OF BEDDING ROCK UNDER SEISMIC LOAD

Because the support project was in a seismically active area, the stress and strain situation of support project under seismic load should be analyzed to ensure the safety of the construction and operation of slope project. The simulated seimic wave was the cosine shear stress wave with the acceleration time history of $a = \lambda \cos(2\pi f t)$, and the amplitude and frequency of the acceleration remained the same taking as $\lambda = 0.31$ m/s², f = 0.1 Hz. The time-history curve of seismic acceleration with change of vibration duration was shown in Figure 5.

The displacement curves of the midline in slide mass under three working conditions of supporting without piles, supporting with single row of anti-slide piles and supporting with two rows of

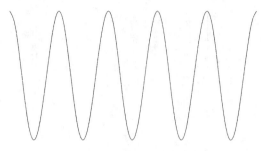

Figure 5.　Acceleration time history curve.

anti-slide piles concerning vibration times of 1 s, 2 s, 4 s and 10 s through the modeling of ABAQUS were obtained (shown in Figure 6).

It can be seen from above curves that the cumulative displacement of slope is 529.9 mm when the vibration lasts for 10 s under the condition of no piles, and the deformation of slope is very obvious. The cumulative displacement of slope narrows to 185.9 mm under the condition of single row of piles

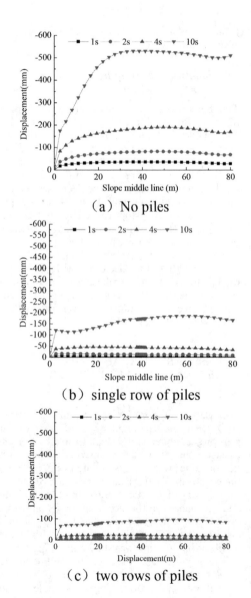

（a）No piles

（b）single row of piles

（c）two rows of piles

Figure 6.　Displacement curve of middle line.

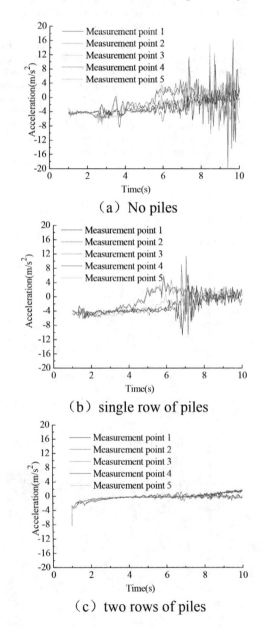

（a）No piles

（b）single row of piles

（c）two rows of piles

Figure 7.　The acceleration-time curve of measuring point.

and is less than 100 mm under the condition of two rows of piles; the displacement values at different times are significantly smaller, and the displacement change of slope along the length direction is more harmonious.

The acceleration-time curves (shown in Figure 7) of five representative points in the midline of slope under three working conditions are picked up for comparison showing that the rangeability and asynchronous condition of acceleration of measuring points are "no piles" > "single row of piles"> "two rows of piles" exposing the obvious anti-seismic action of anti-slide pile in the reinforcement of bedding rock slope.

4 CONCLUSIONS

In order to research the influence rules of sliding mass thickness and seismic wave parameters (amplitude, frequency and duration) on deformation characteristics of bedding rock slope, the finite element analysis software ABAQUS, and the protective effect of anti-slide piles was researched by the response simulation of slope subjected to seismic load under three working conditions of supporting without piles, supporting with single row of anti-slide piles and supporting with two rows of anti-slide piles providing reference for similar projects. Some meaningful conclusions are obtained as follows:

1. The maximum cumulative displacement of bedding slope increased with the increase of seismic load amplitude and vibration time, while the increasing rate was also increasing;
2. The maximum cumulative displacement of bedding slope decreased with the increase of slide mass thickness under the same conditions. Simultaneously, the influence of high-frequency seismic waves on bedding slope decreased with the increase of slide mass thickness.

3. According to analysis of the displacement of slope middle line and the acceleration-time curve of measuring key point, obvious anti-seismic effect of anti-slide pile in the reinforcement of bedding rock slope was verified, providing reference to similar project.

REFERENCES

Cai F, Ugai K. Numerical analysis of the stability of a slope reinforced with piles. Soils and Foundations, **40**(1): 73–84 (2000).

Daniel Pradel, Jason Garner, Annie. Design of drilled shafts to enhance slope stability. 2010 Earth Retention Conference, 920–927 (2010).

Kramer S L, Smith M W. Modified Newmark model for seismic displacements of compliant slopes. Journal of Geotechnical and Geoenvironmental Engineering, **123**(7): 635–644 (1997).

Liu Xinrong, Liang Ninghui. Research progress and application of anti-slide piles in slope engineering. Chinese Journal of geological hazard and control, **17**(1): 56–62 (2006).

Rathje E M, Bray J D. An examination of simplified earthquake-induced displacement procedures for earth structures. Canadian Geotechnical Journal, **36**(1): 72–87 (1999).

Robert W D. Modified Newmark model for seismic displacements of compliant slopes. Journal of Geotechnical and Geoenvironmental Engineering, **125**(1): 86–90 (1999).

Stewart J P, Blake T F, Hollingsworth R A. A screen analysis procedure for seismic slope stability. Earthquake Spectra, **19**(3): 697–712 (2003).

Xiang Hai Liu. Studies on mechanism of the landslide and anti-shear strength of sliding layer surface. Sixth set of "Landslide Anthology", China Railway Publishing House, 125–142 (1955).

Xiao Keqiang, Li Haibo, Liu Yaqun. Analysis of safety factor of bedding rock slope under seismic loads. Chinese Journal of rock mechanics and engineering, **26**(12): 2385–2394 (2007).

Advances in Energy Science and Equipment Engineering II – Zhou, Patty & Chen (Eds)
© 2017 Taylor & Francis Group, London, ISBN 978-1-138-71798-5

Design and research of indoor environmental monitoring experiment system based on WSN technology

Jianjun Wang, Su-lan Hao & Yihao Wang
Anhui Technical College of Mechanical and Electrical Engineering, Wuhu, China

ABSTRACT: Wireless sensor networks can be applied in many fields, with the people's demand of smart home environment of design scheme based on WSN indoor environment monitoring system, including system requirements analysis and system structure design. The system design principle could solve the problems related to the comfort and safety of the living environment.

1 INTRODUCTION

Intelligent indoor environment monitoring system mainly deploys a certain number of wireless sensor nodes in the interior, which are used to measure various parameters of the complex indoor environment, such as temperature, humidity, illumination, and combustible gas parameters, monitor the abnormal situation in the environment and the wireless sensor network for timely discovery, and remind people to take timely measures. Thus, the wireless sensor network intelligent indoor environment monitoring system meets the requirements of monitoring function, environment temperature and humidity, illumination and cocoa-burning gas parameters monitoring, personnel position determination, and alarm mechanism (Zhang, 2008).

2 MONITORING SYSTEM ARCHITECTURE

The system logical structure of wireless sensor network technology is shown in Figure 1. The system can be divided into two parts: monitoring network and terminal monitoring system. Wireless sensor monitoring network consists of three nodes: the base station node, routing node, and terminal node (Wang, 2009; Zhang, 2011). A number of different types of sensor nodes will be placed in the monitored region, all nodes through wireless self-organizing established platform, real-time acquisition, processing environment parameters, and the multihop transmission to the terminal monitoring system, so as to realize the monitoring of the indoor environmental condition.

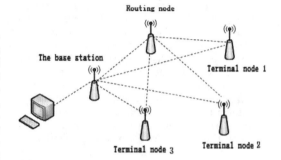

Figure 1. System logical structure.

The monitoring center includes a computer and monitoring software to realize real-time monitoring, storage, analysis, and display of indoor environmental factors, so as to achieve a good man–machine interaction interface. As shown in Figure 1, the routing node analyzes the received data, which are then uploaded to the base station, whose node processing power and storage capacity are relatively high, as it not only receives data of terminal nodes, but also analyzes the process. The base station node transmits the data received from the sensor network to the monitoring center of the host computer and links the sensor network with the external network to realize the conversion between the two communication protocols. The terminal node is a common sensor node, which is generally used for battery power supply. In this system, the terminal nodes are deployed in the laboratory, which can not only be used as a beacon node, but also can collect data of environmental parameters such as temperature, humidity, and illumination.

3 SYSTEM PLATFORM DESIGN

The hardware equipment used in this experiment system is a complete set of equipment, which is produced by the California University, including the model for the IRIS-XM2110 node, the MIB520 programming interface board, and the MTS310 sensor board (Yuan, 2008).

The Iris node hardware, as shown in Figure 2, supports multiple sensor plates or data acquisition board. The sensor board comprises a sensing module: temperature, sound, light, humidity, pressure, and sensor board have unified 51 pin expansion interface used to connect the node. In this experimental system, the data acquisition information and the node location information are the focus of attention.

The hardware programming interface board used in the experiment system is MIB520, as shown in Figure 3. IRIS nodes and Mica series of node programming and communication can be used as MIB520. Any one node and MIB520 programming board can be used as a base station; when MIB520 is connected to the computer, two independent virtual ports (n and n + 1) are provided, which can be used in the device manager. A smaller port number n is used for online Mote node to download the program, whereas the larger the port number (n + 1) is used for data transmission.

In this experiment, we collect the temperature information using MTS310 sensor. The MTS310 sensor board includes light, temperature,

Figure 2.   IRIS sensor node.

Figure 3.   Interface board.

Figure 4.   Sensor board.

humidity, sound sensor, dual axis accelerometer, a buzzer, and a biaxial magnetometer, and the sensing module and node of mica series are compatible. Data acquisition board has a flexible external interface, if we need to gather additional information on the environment through the external interface to extend the sensor type, increase the node function, and improve the MTS310 sensor board, as shown in Figure 4.

The software platform of this experimental system uses MoteWork, which supports a variety of wireless sensors and OEM devices, especially suitable for low-power operation of the network. The operating system used by the sensor network node is TinyOS operating system, which is a new type of sensor network operating system, which was originally written in assembly language and C language. However, in order to improve the efficiency of the development of the wireless sensor network applications and for more convenient use of the operating system, the NESC language, NESC language support component-based programming, component/module of thought, and the TinyOS execution model based on the event were used.

4   ENVIRONMENTAL MONITORING SYSTEM DESIGN

In the system, wireless sensor network environment monitoring system consists of three nodes: the terminal node, the routing node, and the base station node (Hu, 2010). The terminal node is one whose location is known, and the parameters of the surrounding environment can be collected in real time. In the network, the coordinates of the nodes and the collected environment parameters can be transmitted to other nodes through the transmission of the data packets. Routing node from the terminal nodes receive data packet information and get the terminal nodes' location coordinates and the RSSI value, and through calculation and processing, its position coordinates are calculated and the terminal node temperature value is transmitted to the base station node by XSniffer.

The main task of the base station node is to establish a wireless ad hoc network, which is responsible for the initialization of various devices, channel selection, address allocation, and so on. The base station node program flowchart as shown in Figure 5. The base station node networking is in accordance with the data structure in the program header file defined to scan the network channel selection. When the network is successfully established, there will be an indicator light on the node.

In the intelligent monitoring system of indoor environment, the routing node is a very important network equipment, which plays a very important role, which is mainly responsible for the processing and transmission of the data of the terminal node. These data include terminal node to monitor the temperature information, node coordinates information, the RSSI value, and so on. Routing node execution task of forwarding information, receiving the terminal node join request, coming from the terminal nodes of data by one-hop or multihop transmission to the base station node, through the corresponding destination address or a communication channel to complete the data forwarding; docking received data on the other hand. A pair of receiving packets receive the ambient temperature parameter processing data in Celsius; the second is the receiving terminal node location information and intensity indicating RSSI value, and then through signal transmission model formula for processing, finally calculate the location information of the routing nodes (we usually say mobile nodes), and then upload it to the base station node. The routing node can act as a mobile node, can be installed in the personnel, can be placed in certain equipment, and can also be placed in the laboratory. The routing node program flowchart is shown in Figure 6.

The indoor environment intelligent monitoring system has two functions: (1) to locate the terminal node, that is, the beacon node, its location information, and node number is already known; (2) to collect the temperature parameters of the surrounding environment. These two functions are implemented by the same node. The following is the program flow diagram of the terminal node.

When the routing node receives the data frame, according to the received environmental parameters for and the conversion formula, the received terminal voltage is transformed into a Celsius temperature value. In addition, on the basis of the strength of the received signal, we can calculate the distance between them and the terminal node, then the positioning algorithm based on the measured location information of the nodes, routing nodes to calculate the position coordinate of the nodes. Finally, node degrees Celsius temperature value is sent to the monitoring host computer that displays the terminal node power for application to join the network, gather the environment parameters, and acquire periodic broadcasting data and location

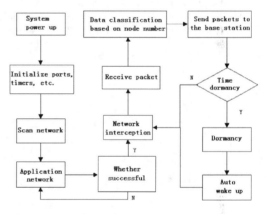

Figure 6.   Routing node.

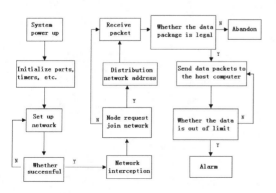

Figure 5.   Base station node.

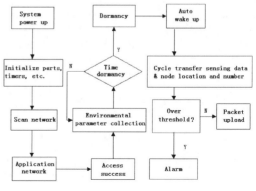

Figure 7.   Terminal node.

information of the nodes in certain distance range. The implementation of the localization algorithm will be described in the following chapters.

Assuming that the data sent by each node conflict, by adding a timer on each node, and each node set timer intervals cannot be multiple relation exists, to avoid signal sending conflict. Usually, each node has its own definition. If it is not properly defined, it will not receive the data sent from other nodes (Lin, 2007). The transmission data structure of the terminal node must correspond to the receiving data structure of the routing node, and the message sending and receiving interface must be consistent.

5 CONCLUSION

On the basis of the research of wireless sensor network and indoor environment, a whole design scheme of indoor environment intelligent monitoring system based on wireless sensor network technology is proposed. In addition to the base station node, routing node and terminal node are used for the preparation of the program. Through the base station node and the sensor network for data transmission, we realize the indoor environment information, data acquisition, data processing, data transmission, temperature data, the unknown node location, and historical data query. Thus, we could achieve the desired function of the system.

REFERENCES

Chunyi Hu, Changgeng Li. Design and implementation of a temperature and humidity acquisition system for wireless sensor networks. Computer Measurement and Control. 2010, 18(5): 1199–1201.

Hao Zhang, Qianchuan Zhao. Indoor positioning system for Bluetooth mobile phone. Computer applications 2011, 31(11): 3152–3156.

Jichun Wang. Research on localization of nodes in wireless sensor networks. University of Science and Technology of China Press, 2009.

Junxia Zhang, Yang Wang, Shanliang Li. Based on the wireless sensor network positioning system design. Computer Engineering and Application. 2008, 44(17): 67–71.

Yaping Lin, Songtao Ye. Design and implementation of data acquisition prototype system for wireless sensor networks in the TinyOS environment. Computer Science, 2007, 34(9): 409–411.

Yuan Lu, Jun Li. Research on the positioning technology of indoor environment. Electronic Test. 2008, 4: 19–23.

Advances in Energy Science and Equipment Engineering II – Zhou, Patty & Chen (Eds)
© 2017 Taylor & Francis Group, London, ISBN 978-1-138-71798-5

Structure of the integrated IP-bearing network routing optimization analysis and design

Hui-Kui Zhou
Nanchang Institute of Science and Technology, Nanchang City, Jiangxi Province, China

Mu-Dan Gu
Jiangxi Modern Polytechnic College, China

ABSTRACT: Because the IP network cannot meet the needs of many Internet applications at present, the existing IP network through the network structure, protocol optimization, and route optimization configuration after modification, compensate for IP itself trying to transfer to a certain extent of birth defects. Taking a telecom CDMA network IP metropolitan area network status as the main object of study, the development of integrated telecom CDMA network to IP-bearing network made an exploratory research, planning, and design. By analyzing the experimental results, comprehensive IP-bearing network has good maintainability and expansibility, which can provide higher-bandwidth resources and meet the demand of future data explosion.

1 INTRODUCTION

The flexibility of IP technology is an important reason for the current 3G core network and the application of IP technology, as well as the future 3G bearer network and even to transition to the whole packet domain. 3G IP carrying network is a concept, shown from the viewpoint of mobile operators (Hu, 2015; Zhao, 2014), that contains different levels of network element, including across provincial backbone network, metropolitan area formed within the various 3G core, and an access node of the network, including various signaling and application platform of the network connection. For both mobile operators independent networking for 3G service and a full-service operator, a variety of closed-type businesses of bearing platform will also complete end-to-end QOS guarantee, and 3G of IP carrying network importance is self-evident. Lightly loaded network can not only greatly reduce the equipment processing delay, but also provide enough basic guarantee of the bandwidth after the failure.

2 ROUTING AND FORWARDING STRATEGY AS A WHOLE

IP-integrated carrier network using IP packets to carry voice, data, and multimedia information can not only fully inherit the existing IP metropolitan area network to provide Internet Service Provider (ISP) (Bartlett, 2014), but also be able to provide

voice, IPTV, cloud services such as rich and unique new business. With many advantages of the technology and cost, it is the development trend of the network bearing of the next generation. According to the nature of transfer of IP network, it is unable to meet the current Internet application demand, to the existing IP network through the adjustment of network structure, protocol optimization, routing optimization means to transform to in a certain extent make up for its IP tried to transfer the birth defects.

Network Customer Edge (CE) to the city, as an independent unit network, CDMA network access, according to each city's CDMA network construction scale of the physical topology, can be divided into three cases, corresponding to the routing group as shown in Fig1.

From the network business development and maintenance management and various aspects of comprehensive consideration, an independent private AS number is assigned to each of the CE network as the resources allocated by the group company.

Customer Edge (CE) network domain IGP agreement with ISIS routing protocol, routing design with flat structure, and all ISIS domain CE routers together Level2 store area, ISIS routing protocol only bearing CE equipment and internal Loopback address Internet address. CE network inside all relay link Internet ports enables MPLS LDP label distribution protocol (SHI, 2013).

Customer Edge (CE) as MPLS VPN network, deployment of MP-BGP, the MP-distributed BGP

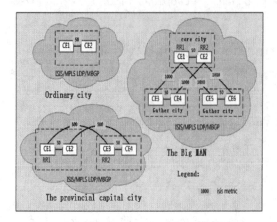

Figure 1. Overall routing and forwarding strategy map.

Table 1. Routing priority.

Route type	Route preference/AD
Direct attached	0*
Static routes	1 or 5
EBGP	10
IS-IS level 2 internal	21
IS-IS level 1 internal	20
IS-IS level 2 external	26
IS-IS level 1 external	25
OSPF internal	30
OSPF external	35
RIP	50
IBGP & Local BGP routes	170
The manual summary routing	100

protocol VPNv4 route, given the MP easy deployment and maintenance of Border Gateway Protocol (BGP), each CE network to select two sets of CE as VRR, two sets of VRR belong to the same reflection cluster.

1. For the ordinary city network only two CE do not need to configure VRR.
2. For the situation of network capital city, each bureau site chooses a CE as VRR.
3. For large metropolitan area network, select CE as the core in Taiwan VRR.

Customer Edge (CE) between two sets of MP-IBGP neighbor relationship, other CE as the client and two sets of VRR to establish the MP-IBGP neighbor relationship.

3 ROUTING DESIGN AND FORWARDING CONFIGURATION

3.1 *Modify routing and priority configuration requirements*

The routing protocol priority needs of unified planning in view of the Customer Edge (CE) networks of the future expansion of the default value of different values in Table 1, router routing protocol priority/unified management distance-modified values as presented in Table 1.

3.2 *Modify the routing priority configuration*

Cisco router, open the Border Gateway Protocol (BGP) (Martin, 2013) deterministic-med (disabled by default; but do not open the BGP always-compare-med—the default off). Avoid potential routing loops and unified for failure to carry the Med attribute routing as MED = 0. The possible introduction of using the following means is a simplified static route to Border Gateway Protocol (BGP):

```
#
router isis
    level 2
        external-preference 26    #level 2 Exterior routing priority 26
        preference 21             #level 2 Internal routing priority 21
        wide-Metrics-only
    exit
exit
#
BGP Internal routing priority 170
#
configure router static-route 1.0.0.0/8 black-hole   preference 100
# Summary routing routing priority 100
```

3.3 *ISIS protocol configuration*

In IS-IS system, Intermediate System (IS) is equivalent to the router in the TCP/IP system, which is the basic unit of routing and routing information in the IS-IS protocol. ES is equivalent to the host system in TCP/IP.

```
#ISIS Global configuration
router isis
    level-capability level-2        # The router is set to the LEVEL 2      exit
    area-id 86.4809.0020.1151.6812.8004.00
    overload-on-boot timeout 300    # Set the ISIS OVERLOAD BIT, waiting for the
300s
    spf-wait 1 50 100               # SPF calculation time
    lsp-wait 1 0 1                  # Time distribution of LSP
    level 2
        external-preference 26
        preference 21
        wide-Metrics-only
    exit
    interface "system"             # The system interface passive way
        level-capability level-2
        passive
    exit
    interface "ge-6/1/3"
        interface-type point-to-point    # Configuration interface to p2p mode
        lsp-pacing-interval 33
        level 2
            hello-interval 3
        exit
    exit
```

1462

3.4 *MP-BGP protocol configuration*

Traditional Border Gateway Protocol (BGP-4) (Kan, 2012) can only manage IPv4 unicast routing information, for the use of other network layer protocols (such as IPv6), the spread of cross-autonomous systems will be subject to certain restrictions. In order to provide support for a variety of network layer protocols, IETF is extended to BGP-4, the MP-BGP is formed, and the current MP-BGP standard is RFC 4760 (Extensions for BGP-4 Multiprotocol, BGP-4 multi-protocol extension). MP-BGP uses the address family (Family Address) to distinguish different network layer protocols, some values on the address family can refer to Assigned 1700 (Numbers RFC). At present, the system implements a variety of MP-BGP applications, including the expansion of the IPv6 and the expansion of the VPN9.

VPN PE exchange visits logic is described as follows:

```
ip vrf CDMA-MGNT
    rd RD-CDMA-MGNT
    route-tagart export CDMA-MGNT_RT
    route-tagart import CDMA-MGNT_SPOKE_
    RT
ANHui _RT
    route-tagart import CDMA-MGNT_ ANHui
    RT
    route-tagart import CDMA-MGNT_RT
ip vrf CDMA-PI-1
    rd RD-PI-1
    route-tagart export PI-1_RT
    route-tagart export CDMA-MGNT_SPOKE_
    RT
    route-tagart import PI-1_RT
    route-tagart import CDMA-MGNT_RT
ip vrf CDMA-PI-2
    descrption for VPDN
    rd RD-PI-2
    route-tagart export PI-2_RT
    route-tagart export CDMA-MGNT_SPOKE_
    RT
    route-tagart import PI-2_RT
    route-tagart import CDMA-MGNT_RT
ip vrf CDMA-RP
    rd RD-ANHUI_RP
    route-tagart export RT-ANHUI-RP-Hub
    route-tagart export CDMA-MGNT_SPOKE_
    RT
    route-tagart import RT—ANHUI-RP-Spoke
    route-tagart import CDMA-MGNT_RT
ip vrf  CDMA-IT
    rd RD-IT
    route-tagart export IT_RT
    route-tagart import IT_RT
ip vrf CDMA-NGN
    rd RD-CDMA-NGN
    route-tagart export RT-CDMA-NGN
    route-tagart import RT-CDMA-NGN
ip vrf CDMA-NGN-MGNT
    rd RD-CDMA-NGN-MGNT
    route-tagart export RT-CDMA-NGN-MGNT
    route-tagart import RT-CDMA-NGN-MGNT
```

In order to avoid the CE-PE due to the error caused by the extended send-community topology confusion and ensure the safety of AAA, AAA routing only injected PI-0/PI-1/PI-2/RP device routing, and all users are not injected into the AAA VRF, as shown in Table 2.

Bearer network should have enough flexibility to adapt to the development of 3G network architecture and business. In the early stage of the construction of IPv6 network, the network size and the amount of traffic are small, which is often used in the connection mode. Carrying the proposed operators of 3GIP early IPv6 experiment network construction according to the actual situation of the tunnel, the utility model has the following advantages: it does not need several IPv6 Router equipment and dedicated links can significantly reduce investment. When the IPv6 core network reaches a certain size, and when the amount of data is large enough, we can consider the deployment of MPLS backbone IPv6.

3.5 *Packet domain network device address planning*

The address of the packet domain network device follows the following planning principles:

1. To facilitate centralized management, the same element should allocate contiguous address space;
2. The public interaction interface and equipment to maximize the use of the public address, for with no connection to the public network equipment as far as possible the use of private address;
3. To fully consider the future of new technology and new business address needs, we should eliminate the future communication model and business model of the impact of unfavorable factors and ensure that the future expansion of the equipment can be carried out smoothly.

Table 2. VPN parameter planning.

VPN	RD1 (PE1)	RD1 (PE2)	export1	export2	import1	import2
CDMA-MGNT	4134:999	4134:999	4134:99900		4134:99901	4134:99900
CDMA-PI1	4134:1001	4134:1101	4134:100100	4134:99901	4134:100100	4134:99900
CDMA-PI2	4134:1002	4134:1102	4134:100200	4134:99901	4134:100200	4134:99900
CDMA-IT	4134:1015	4134:1115	4134:101500		4134:101500	
CDMA-NGN	4134:1020	4134:1120	4134:102000		4134:102000	
CDMA-NGN-MGNT	4134:1021		4134:102100		4134:102100	

Table 3. VPN parameter planning.

VPN	RD1 (CE1)	RD1(CE2)	Export1	export2	import1	import2
CDMA-PIO	4134:3000	4134:3100	4134:300000	4134:301199	4134:300000	4134:301100
CDMA-PI1	4134:3001	4134:3101	4134:300100	4134:301199	4134:300100	4134:301100
CDMA-PI2	4134:3002	4134:3102	4134:300200	4134:301199	4134:300200	4134:301100
CDMA-AAA	4134:3011	4134:3111	4134:301100		4134:301100	4134:301199
CDMA-IT	4134:3015	4134:3115	4134:301500			4134:301500
CDMA-NGN	4134:3020	4134:3120	4134:302000		4134:302000	
CDMA-NGN-MGNT	4134:3021		4134:302100		4134:302100	

At present, devices such as PDSN and AAA need public Internet and do not need IP addresses. R-P network using IP network organization use private address. The original Unicom R-P network planning and allocation of 172.16.0.0/12 private address segment, in order to save address and the cutover project Telecom R-P network by Unicom R-P network yet using private address space.

CE device address and Internet address mainly refer to the loopback0 address and the physical interface between PE and CE. Its planning principle is to maintain the continuity of the address segment, which is easy to aggregate and manage.

1. CE loopback( ) address by the group of unified planning, the address at the same time for the address, only in the CN2 effective, not on the radio;
2. CE on the PE interface and the CE interface address by the group assigned to the provinces, the provincial companies in accordance with the ordinary city of 64 IP, the provincial capital city of 128 IP further distribution;
3. CE and 163 backbone Internet addresses provided by the backbone network; CE and the address of the metropolitan area network are manually provided.

4 CONCLUSIONS

The flexibility of IP technology is an important reason for the current 3G core network and the application of IP technology as well as the future 3G bearer network to transition to the whole packet domain and IP network cannot meet the current Internet application requirements. In this paper, the existing IP network through adjusting the network structure, protocol optimization, routing optimization means transformation, in a certain extent make up for its IP tried to transfer the birth defects. Experimental results show that the IP integrated bearer network has good maintainability and scalability and can provide more bandwidth resources to meet the needs of the explosive growth of the future data. Rich three functions provide higher transmission efficiency and support integrated service bearing. With the aid of the general characteristics of IP network, on the one hand, we can reduce the technical threshold of operation and maintenance personnel and reduce maintenance costs, and on the other hand, can result in rapid fusion of Internet and Intranet.

ACKNOWLEDGMENT

This work was financially supported by the educational reform subject (JXJG-13-27-8).

REFERENCES

Bai Zegang, IPRAN rapid end-to-end business configuration of complex scenes, the optical communication research, 2013.

Bao-Qiang Kan, Jian-Huan Fan. Interference activity aware multi-path routing protocol [J]. EURASIP Journal on Wireless Communications and Networking, 2012.

Chunhui Zhao. Phase analysis and statistical modeling with limited batches for multimode and multiphase process monitoring [J]. Journal of Process Control, 2014.

Chunhui Zhao. Phase analysis and statistical modeling with limited batches for multimode and multiphase process monitoring [J]. Journal of Process Control, 2014.

guo-ping Chen, with polish zhao, IPRAN carrying scheme based on network protocol control tunnel research, "television technology", 2013.

Kelly Bartlett, Junho Lee, Shabbir Ahmed et al. Congestion-aware dynamic routing in automated material handling systems [J]. Computers & Industrial Engineering, 2014(70).

Małgorzata Kossowska, Gabriela Czarnek, Eligiusz Wronka et al. Individual differences in epistemic motivation and brain conflict monitoring activity [J]. Neuroscience Letters, 2014.

Ning Hu, BaoSheng Wang. Cooperative monitoring BGP among autonomous systems. Xin Liu Security Comm. Networks [J]. 2015.

Shelly Salim, Sangman Moh. On-demand routing protocols for cognitive radio ad hoc networks [J]. EURASIP Journal on Wireless Communications and Networking, 2013.

SHI Suixiang. Research on cloud computing and services framework of marine environmental information management. Acta Oceanologica Sinica. 2013(10):57–66.

Sotirios Tsiachris, Georgios Koltsidas, Fotini-Niovi Pavlidou. Junction-Based Geographic Routing Algorithm for Vehicular Ad hoc Networks [J]. Wireless Personal Communications, 2013.

Vasilis Sourlas, Paris Flegkas, Leandros Tassiulas. A novel cache aware routing scheme for Information-Centric Networks [J]. Computer Networks, 2014(59).

Wooyoung Kim, Martin Diko, Keith Rawson. Network Motif Detection: Algorithms, Parallel and Cloud Computing, and Related Tools, Tsinghua Science and Technology. 2013(5): 469–489.

Advances in Energy Science and Equipment Engineering II – Zhou, Patty & Chen (Eds)
© 2017 Taylor & Francis Group, London, ISBN 978-1-138-71798-5

A new neural network-based signal classification method for MPSK

Liu Wang & Yubai Li
School of Communication and Information Engineering, University of Electronic Science and Technology of China, Chengdu, China

ABSTRACT: Noncooperative communication, such as signal interception for satellite communication, surveillance, and threat analysis, has little prior knowledge, which brings a great challenge for the automatic modulation recognition. Therefore, the automatic modulation recognition is very attractive and it became a hot topic in wireless communication system. Because MPSK is widely used in various noncooperative communication systems, in this paper, we propose an automatic modulation recognition algorithm for MPSK signals. The algorithm takes high-order cumulants as characteristic parameters of neural network to recognize the MPSK signal. The simulation results show that our algorithm can recognize 2PSK, 4PSK, and 8PSK signals correctly with probability higher than 91% when SNR ≥ 0 dB. Moreover, when SNR ≥ 3 dB, more than 90% 16-PSK signals can be identified correctly. Compared with state-of-the-art techniques, which only work efficiently for MPSK signals with SNR ≥ 5 dB, our algorithm has an additional advantage on low signal-to-noise recognition.

1 INTRODUCTION

Wireless communication plays an important role in modern communication. Because of the increasing use of digital signals in novel technologies, such as software radio, recent research studies have focused on identifying these signal types. With the rapid development of modern wireless communication technology, different types of modulation method have become more and more complex, which promotes us to study the automatic modulation recognition, which can be applied to many applications and thus plays an important role in various wireless communication systems, such as, reconnaissance, communication confrontation, and electronic warfare. Given a received communication signal, the goal of modulation recognition is to identify the modulation format and estimate the modulation parameters without any prior knowledge about the signal. As many communication systems adopt MPSK modulation, it is particularly important to find an efficient automatic recognition algorithm for MPSK signals.

This paper is organized as follows: the related work is presented in Section II and the recognition algorithm based on neural network is discussed in Section III. In Section IV, we present the simulation results. Finally, we conclude the paper in Section V.

2 RELATED WORKS

Some relevant works have been addressed in other papers, in which various classification methods have

been proposed and many attractive results are shown. For example, A.K. Nandi and E.E. Azzouz extracted some parameters on time domain and frequency domain and used pattern recognition method to classify parameters (Nandi, 1998). The method is the basis for many later algorithms. However, with the development of low SNR wireless communication, the method cannot be applied to signal modulation recognition in low SNR, as it needs SNR ≥ 15 dB. Consequently, Liu Ning and Guo Shuxia abstracted some additional parameters, which were used to distinguish modulation types with SNR ≥ 3 dB (Liu, 2010). This algorithm can only be used to identify ASK, FSK and PSK, but not determine the modulation decimal.

Besides, many likelihood-based methods were addressed for the purpose of modulation signal identification. For example, Huan et al. used likelihood-based method in decision theory approach for MPSK modulation classification (Chung, 1995; Hameed, 2009), which can recognize MPSK signals at low SNR. However, the recognition rate of this method is relatively low while the computation complexity of this method is relatively large. Zhao et al. also proposed a classification method for MPSK signals based on the maximum likelihood criterion (Zhao, 2004), which can identify all types of MPSK signal when SNR ≥ 10 dB but powerless for signals in low SNR.

M Pedzisz et al. posed a method to combine constellation with fourth-order cumulants (Pedzisz, 2005). Because of the limitation of algorithm, it effectively identifies 2PSK, 4PSK, and 8PSK; however, it cannot identify 16PSK signal.

Table 1. Description of notations.

Notations	Description
$r(t)$	Received signals
$s(t)$	Sent signals
$n(t)$	The additive Gaussian white noise in the channels
$\Phi_x(\omega)$	The first characteristic function of a random variable x
$\Psi_x(\omega)$	The second characteristic function of a random variable x
m_k	K-moments of a random variable
c_k	K order cumulant of a random variable
$M_{p,q}$	The k moments of complex stationary signals
$C_{p,q}$	The k order cumulant of complex stationary signals

Mahmoud M.Shakra and Ehab M.Shaheen used higher-order cumulants too (Shakra, 2015), but owing to the relationship between higher-order cumulants and the energy of signals, their algorithm does not make energy normalization for higher-order cumulants. It is not practical.

3 NEURAL NETWORK-BASED RECOGNITION ALGORITHM

To make the paper more readable, we define some notations as follows:

3.1 Signal models

Signals interfered by noise in the channel could be expressed as follows:

$$r(t) = s(t) + n(t) \qquad (1)$$

where $n(t)$ is the additive white Gaussian noise and $x(t)$ is the modulated signal, which depends on modulation type.

MPSK signal models can be expressed as follows:

$$s(t) = \sum_{n=-\infty}^{\infty} a_n \sqrt{E} * p(t - nT_s - t_0) * \exp\left[j\left(2\pi f_c t + \varphi_0\right)\right] \quad (2)$$

where $a_n \in [\exp(j2\pi(m-1)/M), m = 1, 2, \dots M]$, M is the modulation decimal, E is the energy of per symbol, $p(t)$ is the energy-normalized baseband pulse, T_s is the symbol duration, t_0 is the timing deviation, f_c is the carrier frequency, and φ_0 is the initial phase of carrier.

3.2 Higher-order cumulants

Let the probability density function of the random variable x be $f(x)$ and distribution function

be $F(x)$. Then, the first characteristic function of x can be expressed as:

$$\Phi_x(\omega) = E[e^{j\omega x}] = \int_{-\infty}^{+\infty} e^{j\omega x} dF(x) = \int_{-\infty}^{+\infty} e^{j\omega x} f(x) dx \quad (3)$$

The second characteristic function of x can be expressed as:

$$\Psi_x(\omega) = \ln \Phi_x(\omega) \qquad (4)$$

The K-order moment of random variable x is defined as:

$$m_k = E[x^k] = \int_{-\infty}^{+\infty} x^k f(x) dx \qquad (5)$$

$$m_k = (-j)^k \Phi_x^{\ k}(0) \qquad (6)$$

If the first characteristic function is expanded by a Taylor series, we can get the expression represented by higher moments:

$$\Phi(\omega) = 1 + \sum_{k=1}^{n} \frac{m_k}{k!} (j\omega)^k + O(\omega^n) \qquad (7)$$

The higher-order cumulants of random variable x is defined as:

$$c_k = (-j)^k \Psi_x^{\ k}(0) \qquad (8)$$

Similarly, if the second characteristic function is expanded by a Taylor series, we can get the expression represented by higher-order cumulants:

$$\Psi(\omega) = \ln \Phi(\omega) = \sum_{k=1}^{n} \frac{c_k}{k!} (j\omega)^k + O(\omega^n) \qquad (9)$$

According to formulas (7) and (9), we can get the following expression:

$$\begin{aligned}
\Phi(\omega) &= 1 + \sum_{k=1}^{\infty} \frac{m_k}{k!} (j\omega)^k = \exp\left[\sum_{k=1}^{\infty} \frac{c_k}{k!} (j\omega)^k \right] \\
&= 1 + \sum_{k=1}^{\infty} \frac{c_k}{k!} (j\omega)^k + \frac{1}{2!} \left[\sum_{k=1}^{\infty} \frac{c_k}{k!} (j\omega)^k \right]^2 + \cdots \\
&\quad + \frac{1}{n!} \left[\sum_{k=1}^{\infty} \frac{c_k}{k!} (j\omega)^k \right]^n + \cdots
\end{aligned} \quad (10)$$

According to formula (10), by comparing the coefficients in both sides of the equation, we can get relationship between higher moments and higher-order cumulants:

$$c_1 = m_1 \qquad (11)$$
$$c_2 = m_2 - m_1^2 \qquad (12)$$
$$c_3 = m_3 - 3m_1 m_2 + 2m_1^2 \qquad (13)$$
$$c_4 = m_4 - 3m_2^2 - 4m_1 m_3 + 12m_1^2 m_2 - 6m_1^4 \qquad (14)$$

3.3 Algorithm procedure

The steps of the algorithm are shown in Figure 1.

1. The receiver receives signals and makes Hilbert transform, followed by down-conversion. The purpose of low-pass filter is to stop out-band noise.
2. Calculate higher-order cumulants of the baseband signal and then calculate the parameters F1, F2, and F3.
3. A large amount of sample data needed to train the neural network with supervision (Hagan, 2002). By studying large numbers of samples, the neural network can be used for modulation classification.

Suppose the receiver receives a signal r ($r = s+n$), where s and n are independent. According to the characteristics of higher-order cumulants, we could get:

$$Cum(r) = Cum(s) + Cum(n) \tag{15}$$

According to modern signal processing theory, it is easy to get the first and second characteristic functions of a Gaussian signal (assuming that mean is μ, variance is δ^2):

$$\Phi(\omega) = \exp\left(j\omega\mu - \frac{1}{2}\delta^2\omega^2 \right) \tag{16}$$

$$\Psi(\omega) = j\omega\mu - \frac{1}{2}\delta^2\omega^2 \tag{17}$$

According to formula (8), we can get the higher-order cumulants of Gaussian signal:

$$c_1 = \mu \tag{18}$$

$$c_2 = \delta^2 \tag{19}$$

$$c_n = 0 \quad n = 3,4,5,\cdots \tag{20}$$

Therefore, higher-order cumulants of Gaussian white noise signals is zero. Thus, the higher-order cumulants of the received signal is equal to those of the useful signals. In addition, different types of modulation signals have different high-order

cumulants (Wang, 2009; Zhao, 2015; Li, 2006), and we use higher-order cumulants as the characteristic parameter to identify the type of modulation, which can effectively suppress noise. Therefore, this method is very effective for modulation recognition signals interfered by Gaussian white noise.

The high moment of complex stationary signals X(t) is expressed as follows:

$$M_{p+q,q} = E[X^p(X^*)^q] \tag{21}$$

The higher-order cumulants of complex stationary signals X(t) is expressed as follows:

$$C_{p+q,q} = cum[X,\ldots,X,X^*,\ldots X^*] \tag{22}$$

Several common higher-order cumulants are calculated by high moments expressed as follows:

$$C_{20} = M_{20} \tag{23}$$

$$C_{21} = M_{21} \tag{24}$$

$$C_{40} = M_{40} - 3M_{20}^2 \tag{25}$$

$$C_{41} = M_{41} - 3M_{21}M_{20} \tag{26}$$

$$C_{42} = M_{42} - |M_{20}|^2 - 2M_{21}^2 \tag{27}$$

$$C_{60} = M_{60} - 15M_{40}M_{20} + 30M_{20}^3 \tag{28}$$

$$C_{63} = M_{63} - 6M_{41}M_{20} - 9M_{42}M_{21} + 18(M_{20})^2M_{21} + 12M_{21}^3 \tag{29}$$

$$C_{80} = M_{80} - 28M_{60}C_{20} - 35M_{40}^2 + 420M_{40}M_{20}^2 - 630C_{20}^4 \tag{30}$$

The theoretical values of the high-order cumulants of MPSK are shown in Table 2:

According to the value of higher-order cumulants of MPSK signal, we can construct three parameters:

$$F1 = \frac{|C41|}{|C42|} \quad F2 = \frac{|C40|}{|C42|} \quad F3 = \frac{|C80|}{|C42|^2} \tag{31}$$

Given the theoretical value of three parameters (F1, F2, and F3), many of the existing methods may use the pattern recognition method. Each higher-order cumulant is used to distinguish a

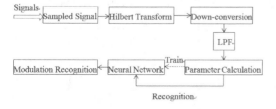

Figure 1.   Steps of the algorithm.

Table 2.   Higher-order cumulants of MPSK signal.

Higher-order cumulants	$|C_{40}|$	$|C_{41}|$	$|C_{42}|$	$|C_{80}|$
BPSK	$2E^2$	$2E^2$	$2E^2$	$272E^4$
QPSK	E^2	0	E^2	$34E^4$
8PSK	0	0	E^2	E^4
16PSK	0	0	E^2	0

class of signal. Recognition rules are shown in Figure 2.

If we use pattern recognition method, as shown in Figure 2, for modulation-type classification, we choose t(F1), t(F2), and t(F3) as the thresholds of F1, F2, and F3, respectively. An appropriate value of t(F1) could more effectively recognize BPSK signal. If F1 < t(F1) and F2 > t(F2), the signal will be regarded as QPSK. After these judgments, if the signal, neither BPSK nor QPSK, analyzes further calculation result, the signal F3 > t(F3) will be sentenced as 8PSK, otherwise, be sentenced as 16PSK.

In this way, we can identify the modulation type in a certain SNR; however, in lower SNR, it becomes helpless, mainly due to the method mentioned above using a fixed threshold. In low SNR, there are different degrees of deviation about F1, F2, and F3, fixed threshold method at low SNR cannot distinguish signals.

In this paper, we proposed a neural network-based classification algorithm to identify the type of modulation signal. Let F1, F2, and F3 be parameters consisting of feature vectors to train the network. Neural network method for modulation recognition at low SNR will correctly identify more signals.

4 SIMULATIONS

Simulation process and result of the algorithm proposed in this paper are expressed as follows:

First, parameter simulation of each modulation scheme and the simulation conditions were assumed: Simulate BPSK, QPSK, 8PSK, 16PSK signals, the carrier frequency fc = 15 MHz, the sampling rate fs = 60 MHz, symbol rate is 3 MHz, the length of each signal contain 512 symbols, SNR in the range [0 dB, 10 dB], for a certain SNR,

after 100 times independent simulation, calculating its average for each SNR, and the simulation results are shown in Figures 3–6.

According to the simulation results, the simulation results and theoretical values of the three parameters of MPSK are basic anastomosis, which proves that the theory of higher-order cumulants is suitable for the identification of MPSK signals.

The use of neural network-based approach to the classification and simulation is presented as follows:

First, we determine the specific structure of the neural network. In this paper, the neural network classifier uses BP neural network structure, and the feedback signal is only in the training stage of the neural network. The specific structure of the neural network is as follows: the first layer is input layer, which contains three input node, the second

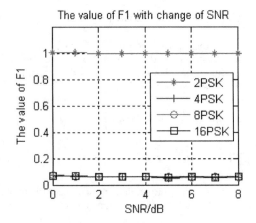

Figure 3. F1 change with SNR.

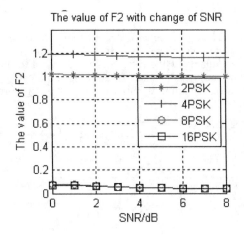

Figure 4. F2 change with SNR.

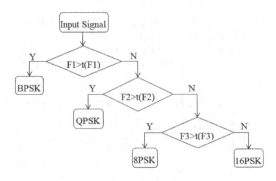

Figure 2. Decision method for MPSK modulation identification.

1468

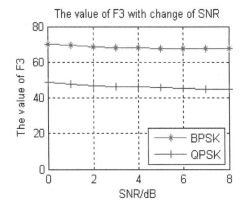

Figure 5. F1 change with SNR.

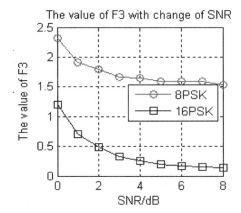

Figure 6. F2 change with SNR.

layer is hidden layer containing 10 neurons, the last output layer that contains four output neurons. In the training process of the network uses gradient descent method to adjust the network parameters.

Then, it generates the simulation data to train the network. We also set that the simulation uses BPSK, QPSK, 8PSK, and 16PSK signals, the carrier frequency fc = 15 MHz, the sampling rate fs = 60 MHz, symbol rate is 3 MHz, the length of each signal contain 512 symbols, SNR in the range [0 dB, 10 dB], for a certain SNR, 100 times independent simulation, and then store F1, F2, and F3.

Finally, after training of the neural network, the correct recognition rate is shown in Table 3.

5 CONCLUSION

In this paper, we propose a signal classification method, which identifies BPSK, QPSK, 8PSK,

Table 3. The identified correct rate of neural network.

SNR Types	0 dB	3 dB	4 dB	6 dB	8 dB	10 dB
2PSK	100%	100%	100%	100%	100%	100%
4PSK	100%	100%	100%	100%	100%	100%
8PSK	91%	95%	96%	96%	98%	100%
16PSK	37%	90%	93%	99%	100%	100%

and 16PSK signals by taking higher-order cumulants as the feature extraction parameters of neural network. In our algorithm, the receiver does not need any prior knowledge; thus, it is very suitable for identifying the noncooperative communication signal. Simulation result shows that the proposed methods can effectively identify the MPSK modulated signals, even in the case of low SNR. In details, the correct identification rate of our method is higher than 91% for BPSK, QPSK, and 8PSK signals with SNR ≥ 0 dB and larger than 90% for 16PSK with SNR ≥ 3 dB.

ACKNOWLEDGMENT

The authors would like to express their gratitude to all those who helped during the writing of this paper. They gratefully acknowledge the help of thier supervisor Professor Li Yubai, who has offered them valuable suggestions in the academic studies. Without his consistent and illuminating instruction, this thesis could not have reached its present form.

REFERENCES

Chung-Yu Huan and A. Polydoros, Likelihood Methods for MPSK Modulation Classification, *IEEE Transactions on communications*, 1493–1504 (1995).

Hagan, M.T., H.B. Demuth, M. Beale, Neural network design, China Machine Press, (2002).

Hameed, F., O.A. Dobre, D.C. Popescu, On the Likelihood-Based Approach to Modulation Classification, *IEEE Transactions on wireless Communication*, 5884–5892 (2009).

Haykin, S., Neural Network: A Comprehensive Foundation, 2nd Edition, China Machine Press, ISBN 7-111-12759-5.

Li, P., F.P. Wang, Algorithm for Modulation Based on High-order Cumulants and Subspace Decomposition, *8th international Conference on Signal Processing*, 16–20 (2006).

Liu, N., S.X. Guo, B. Liu, Investigation On Signal Modulation Recognition in the Low SNR, *2010 International Conference on Measuring Technology and Mechatronics Automation*, 528–531 (2010).

Nandi, A.K., E.E. Azzouz, Algorithm for Automatic Modulation Recognition of Communication Signals, *IEEE Transactions on Communications*, **46**, 431–436 (1998).

Pedzisz, M., A. Mansour, Automatic modulation recognition of MPSK signals using constellation rotation and its 4th order cumulant, *Digital Signal Processing*, 295–304 (2005).

Shakra, M.M., E.M. Shaheen, Automatic Digital Modulation Recognition of Satellite Communication Signals, *Radio Science Conference (NRSC) 2015 32nd National*, 118–126 (2015).

Wang, L.X., Y.J. Ren, Recognition of Digital Modulation Signals Based on High Order Cumulants and Support Vector Machines, *2009 ISECS International Colloquium on Computing, Communication*, 271–274 (2009).

Zhao, Y., H. Jiang, Recognition of Digital Modulation Signals Based on High-order Cumulants, *2015 International Conference on Wireless Communications & Signal Processing (WCSP)*, 15–17 (2015).

Zhao, Z.J., T. Lang, A Classification Method for MPSK Signals Based on the Maximum Likelihood Criterion, *In Proceedings of the IEEE 6th Circuits and Systems Symposium on Emerging Technologies: Frontiers of Mobile and Wireless Communication*, 385–388 (2004).

Advances in Energy Science and Equipment Engineering II – Zhou, Patty & Chen (Eds)
© 2017 Taylor & Francis Group, London, ISBN 978-1-138-71798-5

A novel algorithm of pedestrian dead reckoning based on multiazimuth movement pattern

Yu Liu, Fan Zhou, Yong-Le Lu, Yun-Mei Li, Xin Zhang & Yang Cao
Chongqing Municipal Level Key Laboratory of Photoelectronic Information Sensing and Transmitting Technology, Chongqing University of Posts and Telecommunications, Chongqing, China

ABSTRACT: Pedestrian navigation based on Micro-Electro-Mechanical System (MEMS) sensor usually achieves real-time pedestrian positioning by using Pedestrian Dead Reckoning (PDR) algorithm to calculate the coordinate of pedestrian's position. Traditional PDR only considers forward movement pattern, which is not applicable to the actual multiazimuth human movement pattern. Hence, a novel algorithm of PDR based on multiazimuth movement pattern is proposed. We use the data of three-axis accelerometer to step detection and stride length estimation. On the basis of the analysis of the accelerometer data, the pattern recognition method for the four types of movement pattern is discussed and the dead reckoning equation, which is applied to pedestrian multiazimuth movement pattern, is derived. Experimental results based on smartphone platform show that the positioning error is less than 3.2% by using the proposed algorithm for the actual multiazimuth human movement pattern, and the accuracy of navigation is therefore improved.

1 INTRODUCTION

In recent years, the research of pedestrian navigation based on MEMS inertial sensor components has been becoming a hot topic. Inertial system is an autonomous navigation system, which does not depend on the external information, not radiate energy outward, and not need external reference system either. Hence, having little electromagnetic interference from outside, inertial system can work in the air, earth's surface, and underwater, and it can provide speed, heading orientation, and attitude angle information continuously (Donovan, 2012). The generated navigation information can not only have good stability and continuity, but also have high data update rate.

Many studies aimed at improving pedestrian navigation accuracy focus on the sensor calibration and optimization algorithm. X. Niu proposed a method of inertial sensor calibration about temperature drift (Niu, 2013). A partial decentralized system based on stepwise inertial navigation and stepwise dead reckoning was presented by Nilsson, which adopted two inertial measurement units strapped to feet. The complexity of the algorithm was reduced, which greatly improved the cost of hardware (Nilsson, 2013). The work done by Schindhelm in Reference 4 explored the capabilities of using inertial sensors for the detection of the pedestrian movement pattern, but it only studied the forward movement pattern.

The actual pedestrian movement includes four directional movement patterns such as walking forward, walking backward, walking left transverse, and walking right transverse. If those movement patterns are not distinguished, it will lead to navigation path being opposite to or perpendicular to the actual movement and cause a great error. To solve this problem, a novel algorithm of PDR based on multiazimuth movement pattern is proposed in this paper.

2 SYSTEM FRAMEWORK

The system architecture of PDR algorithm based on multiazimuth movement pattern is shown in Fig. 1. During data processing, the data of three-axis gyroscope, three-axis accelerometer, and three-axis magnetometer are preprocessed first, the heading orientation of each pedestrian step is then given through the data fusion, and the pedestrian walking state and stride length are achieved by the accelerometer data processing. We use peak detection method and phase inversion method to distinguish movement patterns among walking forward, walking backward, walking left transverse, and walking right transverse. Finally, the algorithm of PDR based on multiazimuth movement pattern is performed to calculate the user's locations in a real-time manner.

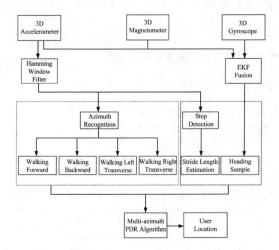

Figure 1.   System architecture.

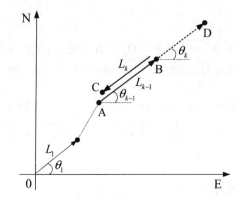

Figure 2.   Model for PDR algorithm.

3   DEAD RECKONING

Traditional PDR algorithm is shown in Eq. (1), which uses the heading orientation and step length to calculate user's exact coordinates (Zhang, 2015; Levi, 1996). The disadvantage of the algorithm is that heading orientation is the same among walking forward, walking backward, walking left transverse, and walking right transverse, which will therefore make the solved positioning path appear walking forward always. Hence, a large error is generated, thereby affecting its practical application:

$$\begin{cases} E(k) = E(k-1) + L_k \cos\theta_k \\ N(k) = N(k-1) + L_k \sin\theta_k \end{cases} \quad (1)$$

where θ_k and L_k stand for the heading angle and the stride length of step k, respectively. Notations $E(k)$ and $N(k)$ denote the east and north directions, respectively.

3.1   Dead reckoning for walking longitudinal

As shown in Fig. 2, a pedestrian walks from point A to B, and then walks back to point C from B. This process has two valid strides with no change in the heading. If we use the traditional algorithm of dead reckoning, the pedestrian's position will be at point D, which is opposite to the actual path. In this case, the traditional algorithm of dead reckoning is not applicable.

$$\begin{cases} E(k) = E(k-1) + S_k L_k \cos\theta_k \\ N(k) = N(k-1) + S_k L_k \sin\theta_k \end{cases} \quad (2)$$

The PDR algorithm is shown in eq. (2), which is suitable for walking forward and backward

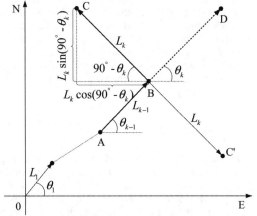

Figure 3.   Model for PDR algorithm.

movement patterns, where θ_k and L_k stand for the heading angle and the stride length of the step k, respectively, and S_k is the flag of the movement pattern. When the pedestrians walk forward, $S_k = 1$; when the pedestrian walks backward, $S_k = -1$; and for other walking patterns, $S_k = 0$.

3.2   Dead reckoning for walking transverse

As shown in Fig. 3, when a pedestrian walks from point A to B, then walks left transverse from point B to C, the heading orientation has not been changed. $E(k)$ is reduced by $L_k \cos(90° - \theta_k)$ and $N(k)$ is increased by $L_k \sin(90° - \theta_k)$. However, when pedestrian walks from point A to B, and then walks right transverse from point B to C', there is no change in the heading orientation. $E(k)$ is increased by $L_k \cos(90° - \theta_k)$ and reduced by $L_k \sin(90° - \theta_k)$. If we use traditional PDR, the pedestrian's position will be at point D, which is perpendicular to the actual walking path.

Therefore, the traditional algorithm of dead reckoning is not applicable.

If we only consider walking movement, transverse and dead reckoning algorithm can be expressed as:

$$\begin{cases} E(k) = E(k-1) + U_k L_k \cos(90° - \theta_k) \\ N(k) = N(k-1) - U_k L_k \sin(90° - \theta_k) \end{cases} \quad (3)$$

Eq. (3) is only suitable for the movement pattern, such as walking left transverse and walking right transverse, where θ_k and L_k stand for the heading angle and the stride length of step k, respectively, and U_k is the flag of the movement pattern. When the pedestrian walks left transverse, $U_k = -1$; when the pedestrian walks right transverse, $U_k = 1$; and for other walking patterns, $U_k = 0$.

3.3 Pedestrian dead reckoning for multiazimuth movement pattern

The pedestrian movement is a multiazimuth hybrid movement pattern, which includes walking forward, walking backward, walking left transverse, and walking right transverse. In this paper, the hybrid movement pattern dead reckoning algorithm can be given as follows:

$$\begin{cases} E(k) = E(k-1) + (1-|U_k|)S_k L_k \cos\theta_k \\ \qquad + (1-|S_k|)U_k L_k \cos(90° - \theta_k) \\ N(k) = N(k-1) + (1-|U_k|)S_k L_k \sin\theta_k \\ \qquad - (1-|S_k|)U_k L_k \sin(90° - \theta_k) \end{cases} \quad (4)$$

where θ_k and L_k are the heading angle and the stride length of step k, respectively, S_k is the movement pattern flag for walking forward and backward, and U_k is the flag of the movement pattern when the pedestrian walks left and right transverse. When the pedestrian walks forward, $S_k = 1$, $U_k = 0$; when he/she walks backward, $S_k = -1$, $U_k = 0$; when he/she walks left transverse, $S_k = 0$, $U_k = -1$; when he/she walks right transverse, $S_k = 0$, $U_k = 1$; and for other walking patterns, $S_k = 0$, $U_k = 0$. Therefore, S_k and U_k can be determined by the azimuth recognition.

4 WALKING-STATE DETECTION AND AZIMUTH RECOGNITION

4.1 Step detection and stride length estimation

The output waveform of the accelerometer original data has a phenomenon with multipeaks, which causes the error of the gait. Hence, the original data of the accelerometer value need to be preprocessed, all kinds of noise and burr should be eliminated as much as possible, and the features of the original data should be kept as much as possible. By using filter processing technology, the accelerometer curve is smoother and more suitable for feature extract and classification. Therefore, the Hamming window filter technique is used to process original data, and the window length of the filter is 5 (Liu, 2015). After the filtering, the accelerometer data show some regular changes and each valid stride corresponds to a peak, a zero, and a valley value, which is similar to a sine wave. Therefore, peak detection method is used for the steps of valid strides of the four movement patterns (Kang, 2015). Fig. 4 shows the filtered accelerometer waveform, revealing every peak of the pedestrian step. Therefore, the effectiveness of the gait detection algorithm is verified.

According to the empirical model (Pei, 2012), we have:

$$step\_length = C\sqrt[4]{A_{max} - A_{min}} \quad (5)$$

where step_length is the step length, A_{max} and A_{min} stand for the maximum and minimum of three-axis accelerometer modulus value, respectively, which are obtained from the step detection, and C can be calculated as follows:

$$C = d_{real}/d_{estimated} \quad (6)$$

where d_{real} is the real distance and $d_{estimated}$ is estimated reference trajectory distance.

4.2 Azimuth recognition

Through comparison and analysis of the experimental data, the phase difference of the Y-axis acceleration waveform is 180° between pedestrian walking forward and backward and the phase difference of the X-axis acceleration waveform is also

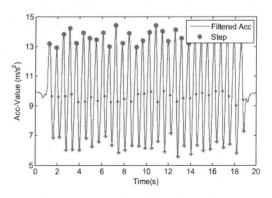

Figure 4.   Results of step detection.

180° between pedestrian walking left transverse and right transverse. That is to say, the acceleration value always increases first and then decreases along the direction of walking.

Fig. 5(a) and (b) shows the filtered waveforms in Y-axis when a pedestrian walks forward and backward, respectively. It can be seen that each stride is marked out, and the phase difference of Y-axis is 180°. The smartphone is handheld with the X-axis pointing to the right direction, Y-axis to the forward direction, and Z-axis to the downward direction during the walking. The waveform with Y-axis accelerometer appeared to increase first and then decreased, which means the pattern of walking forward. If the waveform with Y-axis accelerometer appeared to decrease first and then increased, then the pattern of walking backward.

5 TESTING RESULTS

Choosing the football field in CQUPT as the experimental site, the positioning accuracy of the proposed algorithm is tested. As shown in Fig. 6(a), the football field is 110 m long and 75 m wide.

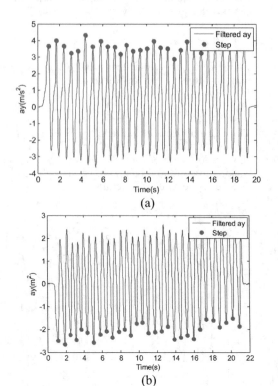

Figure 5. Waveforms of (a) walking forward and (b) walking backward.

A pedestrian holding a phone walks around the football field in a loop (i.e., the starting position and ending position coincide) for our testing. That is to say, the pedestrian first walks forward from point A to B, and then walks right transverse from point B to point C, walks backward from point C to point D, and finally walks left transverse from point D to A. The distance between the starting position and the ending position on the tracked trace (or called tracking error) is selected to evaluate the accuracy of positioning. The smaller the distance between the starting position and ending position is, the higher the accuracy of positioning

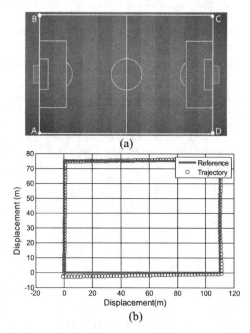

Figure 6. Simulation results of (a) experimental site and (b) walking simulation trajectory of pedestrians.

Table 1. Tracking results.

Testing No.	Reference distance (m)	Trajectory distance (m)	Tracking error (m)
1	370	358.24	11.76
2	370	358.97	11.03
3	370	359.23	10.77
4	370	358.71	11.29
5	370	359.24	10.76
6	370	359.75	10.25
7	370	360.02	9.98
8	370	358.84	11.16
9	370	358.94	11.06
10	370	359.09	10.91

is proved to be. When the algorithm of PDR based on multiazimuth movement pattern is used to deal with the sensor data, the measured walking trajectory is shown in Fig. 6(b).

Table 1 shows the tracking results by using pedestrian navigation algorithm, for which, 10 groups of data from 10 different people are used to evaluate the accuracy of positioning. For the trace with the total length of 370 m, the tracking error is only within 11.8 m and the distance error between the starting point and the end point is within 3.2%.

6 CONCLUSION

In this paper, we propose a novel algorithm of PDR based on multiazimuth movement pattern by using MEMS inertial sensor components. It can not only improve the accuracy of step detection under hybrid movement pattern, but also distinguish four types of movement patterns. Furthermore, the algorithm significantly improves the positioning accuracy, which can decrease positioning error to less than 3.2%. In conclusion, pedestrian multiazimuth movement pattern dead reckoning algorithm proposed in this paper can be effectively applied to the real lives of people, which has a great value in engineering applications.

REFERENCES

Donovan, G.T., IEEE J. Ocean. Eng. 37, 431–445 (2012).
Kang, W., Y. Han, IEEE Sens. J. 15, 2906–2916 (2015).
Levi, R.W., T. Judd, U.S. Patent No. 5583776 (1996).
Liu, Y., Y. Chen, L. Shi, Z. Tian, M. Zhou, L. Li. J, Commun. 10, 520–525 (2015).
Nilsson, J.O., D. Zachariah, I. Skog, P. Händel, Eurasip, J. Adv. Sig. Pr. 2013, 1–17 (2013).
Niu, X., Y. Li, H. Zhang, Q. Wang, Y. Ban, Sensors 13, 12192–12217 (2013).
Pei, L., J. Liu, R. Guinness, Y. Chen, H. Kuusniemi, R. Chen, Sensors 12, 6155–6162 (2012).
Schindhelm, C.K., Personal Indoor and Mobile Radio Communications (PIMRC), 2012 IEEE 23rd International Symposium on. IEEE, 2454–2459 (2012).
Zhang, H., J. Zhang, D. Zhou, W. Wang, J. Li, F. Ran, Y. Ji, J. Sens. 2015, 1–13 (2015).

Advances in Energy Science and Equipment Engineering II – Zhou, Patty & Chen (Eds)
© 2017 Taylor & Francis Group, London, ISBN 978-1-138-71798-5

Implementation of an automatic positioning system for light sheet fluorescence microscopy

Cong Niu
Department of Design Manufacturing and Engineering Management, University of Strathclyde, Glasgow, Scotland, UK

Wanfang Che & Yun Bai
Key Laboratory of Complex Aviation System Simulation, Beijing, China

ABSTRACT: Light Sheet Fluorescence Microscopy (LSFM) is a fluorescence microscopy technique which is widely used in cell biology, especially embryonal development observations. In this paper, we introduced a microscope automatic positioning system implementation for LSFM. Based on existing stepper motor and micro-stepping technique, we conducted hardware mechanical design and auxiliary software implementation. Systematic evaluations showed the system achieved good speed accuracy and position accuracy. We believe this research would serve as a guide for LSFM related automatic positioning system design.

1 INTRODUCTION

Light Sheet Fluorescence Microscopy (LSFM) uses a plane of light to optically section and view tissues or whole organisms that have been labeled with a fluorophore (Santi, 2011). In contrast to other microscopy techniques such as confocal microscopy, only a thin slice of the specimen is illuminated perpendicularly to the direction of observation in LSFM. LSFM offers higher resolution, better optical sectioning capability, faster imaging speed compared with other techniques (Greger, 2007). What's more, it is able to image thicker tissues (>1 cm) with reduced photobleaching and phototoxicity. This non-destructive method produces well-registered optical sections that are suitable for three-dimensional reconstruction and is widely used in developmental biology, such as longtime observations of embryonal development in different model organisms (Verveer, 2007).

The first implementation of LSFM is Orthogonal Plane Fluorescence Optical Sectioning microscopy or tomography (OPFOS), which was developed by Voie et al. (Voie, 1993) in 1993. A number of LSFM devices have been developed now, including TSLM (Fuchs, 2002), mSPIM (Huisken, 2007), HROP-FOS (Buytaert, 2007), OPM (Dunsby, 2008), OCPI (Holekamp, 2008), DSLIM (Keller, 2008; Keller, 2010), TSLIM (Santi, 2009), HiLo (Mertz, 2010). A typical LSFM device consists of following parts: illuminator, specimen chamber, detection

apparatus and computer-controlled motorized micropositioners for specimen movement.

LSFM has separate optical paths for illumination and detection. For three-dimensional reconstruction of large specimen larger than the distance of the confocal parameter of the light sheet, the specimen is scanned in the x-axis and z-axis, which requires a fine-tuned automatic positioning system for the specimen chamber. To ensure image quality, this system must be accurate and stable and the rotation speed should be interchangeable. What's more, a user-friendly computer program is needed to control this position system.

To remove unwanted vibration in image capturing and provide flexible control, a stepper motor with micro-stepping driving method is a very good choice. This micro-stepping driving way can meet different object and resolution requirements. It is a stepper motor driving strategy to achieve lower vibration, lower noise, better positioning and better angles resolution.

In this study, we aim to implement an automatic positioning system for the light sheet fluorescence microscopy with existing stepper motor and micro-stepping technique.

2 MICRO-STEPPING STEPPER MOTOR

dsPICDEM MCSM Development tool kit is used for stepper motor development. The dsPICDEM MCSM Development Board dsPIC33FJ32MC204

Plug-In-Module and a stepper motor are included in this development kit. The dsPIC33FJ32MC204 is a 16-bit microcontroller that is specifically designed for digital signal control and motor control. With dsPIC33FJ32MC204, 1/128 micro-stepping is achieved. What's more, the rotating speed of the stepper motor is adjustable under any micro-stepping level with this tool kit.

The weight of original stepper motor (42H03) in the tool kit is 340 g, which is too heavy for a micro-scope. So we use stepper motor SY20STH30 in our design. This stepper motor weighs 60 g. What's more, SY20STH30 can provide comparative holding torque (180 g/cm compared to 200 g/cm of 42H03) and has much smaller size (20 by 20 by 30 millimeter dimensions). All these considerations make SY20STH30 a suitable solution for positioning system design.

3 MECHANICAL HARDWARE DESIGN

Mechanical hardware design should consider following functions. First, the stepper motor should be capable of being installed on the hardware. Second, transmission parts should be able to transmit the torque and speed from the motor axis to the object holder. Third, the object holder should be able to hold different objects including microtube (10 millimeter diameters) and film holder (4.7 millimeter diameter), etc.

A section of aluminum angle bar serves as the mounting base for stepper motor and object holder. Stepper motor is screwed on the aluminum bar with 2 fasteners. The object holder is a 12 millimeter outer diameter aluminum tube. The bearing is screwed onto the aluminum angle with 2 fasteners, which have a 12 millimeter inner diameters and are used to hold the object holder. The transmission part utilizes 2 timing pulleys (each has 28 teeth). And a 250 millimeter long timing belt is used to connect the pulleys. A mounting seat is provided on the aluminum angle bar which is appended to the mounting bracket of the microscope. This mounting seat has 2 M6 fasteners. One fastener is horizontal. The other one is vertical to the bracket of the microscope. The 2 M6 fasteners touch each other and make the installation more firm for the mounting seat and microscope bracket. What's more, a spacer is added for the mounting seat, so the bracket of the microscope can be segregated from the stepper motor. This spacer is a 3 millimeter aluminum screwed onto the side wall with 4 M2 countersunk bolts and 4 M2 nuts. Fig. 1 is the blue print of mechanical hardware.

Mechanical hardware need fit with bracket of the microscope, stepper motor and the bearing. So we carefully designed the dimensions of the

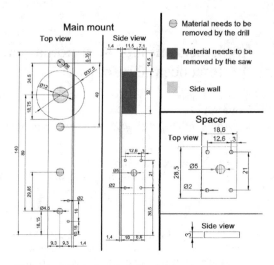

Figure 1. Blueprint of mechanical hardware.

Figure 2. Blueprint of the object holder.

hardware to ensure compatibility with existing materials. The 20*20 mm aluminum angle bar was manufactured as shown in Fig. 1. Note the 5 mm holes on the side of the mount and top of the spacer are aligned and tapped with M6 screw thread. Object holder has a 12 mm outer diameter aluminum tube with 8 M3 thumb screws threaded into the front section of the tube, as shown in Fig. 2.

As shown in Fig. 2, the shaft is drilled in eight 2.5 mm holes and taped in M3. So the M3 thumb screw can be used as the holder for different object. The timing pulley was reamed from 4 mm to 12 mm and installed onto the aluminum tube. The distance of two pulleys is calculated based on the length of timing belt. In Fig. 1, total length needed is about 248 mm, a 250 mm long timing belt can be used with lower tension.

4 SOFTWARE DESIGN

The software should be able to control the stepper motor to receive and execute instructions such as changing the rotation angle and speed from the computer. This can be accomplished by an embedded software and a PC interface software.

The embedded software works on the development board. It contains multiple function modules to control and schedule the development board. A series of functions can be implemented with this software, including receiving data from PC through the Universal Serial Bus (USB), processing the data and generating instructions for position and speed control, sending instructions to the corresponding module. We implemented our embedded software based on the demonstration software provided in the dsPICDEM MCSM development tool kit. After comprehensive analysis of the demonstration system, we figured out the workflow of the entire process and modified the software to achieve our requirements.

PC interface software takes the parameters of speed and rotational position from the operator, and then these parameters are encoded and sent to the development board. Entire software was developed with C# language, an object-oriented programming language for Graphical User Interface (GUI) design.

5 POSITIONING SYSTEM OVERVIEW

We first manufactured aluminum angle bar, spacer and object holder respectively based on the parameters in the blue print. After completing these three parts, the mechanical hardware was assembled as in Fig. 3. In the magnified red box, there is a set screw. Black, green, red and blue wires of the stepper motor were connected to M1, M2, M3 and M4 in the development board correspondingly.

The final version of the embedded software is a fully functional microscope position control system. We removed unnecessary functions and modified the code to achieve 1/128 micro-stepping.

Figure 3.   Completed mechanical hardware.

Adjustable stepper motor driving speed and precise angular position control was also achieved in this embedded system.

First the initialization function is called, which initializes the I/O pins, AD converter, PWM module, debugging function, speed and position parameters. In the infinite loop, the software gets the speed and position data from the serial port buffer and performs Longitudinal Redundancy Check (LRC) to validate the data. Then speed and position execution function constructs the PWM driving wave. This wave is utilized to drive the stepper motor.

After fundamental initialization such as serial port initialization, the PC interface software waits for the user's operation. First it needs to connect to the appropriate Communication (COM) port, which is provided by the computer when the development board driver is installed. Then speed and position of the stepper motor can be adjusted with this interface.

Since this automatic positioning system is part of the microscope and will be integrated with other systems such as automatic imaging system, we implemented it as a dynamic link library. This library can provide all functions to control the automatic positioning system.

6 MULTI-PARAMETER EVALUATION

After installing the automatic positioning system onto the mounting bracket of the microscope, we systematically evaluated the performance of this system concerning thermal feature, speed accuracy and position accuracy.

The thermal feature of the system is shown in Fig. 4. It shows the system is able to work continuously under 40 degrees centigrade.

We set the speed of the stepper motor at 500 degrees per second and measured the actual speed to evaluate the speed accuracy. Since the speed of the stepper motor is determined by driving frequency, we measured the output voltage and current in the windings. Data showed the

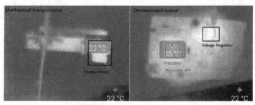

Figure 4.   Thermal features of microscope automatic positioning system. It is captured with a smartphone thermal camera after the stepper motor had worked for 5 minutes.

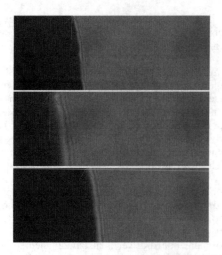

Figure 5. Camera captured video frames of a black film. Three frames (from top to bottom) were captured from the original position, +1 degree position and –1 degree position. It was captured using the longest object holder with stepper motor speed at 50 degrees per second. Camera is basler aca1920-25uc and pixel size is 2.2 by 2.2 μm.

stepper motor rotates 1step (1.8 degrees) in about 3.575 ms, so the rotation speed is 503 degrees per second. Speed accuracy is 500/503 = 99.4%, which is excellent for this system.

Position accuracy is a changeable parameter. It depends on several variables such as the length and centering of the object holder, weight and eccentricity of the object and the stepper motor speed. Fig. 5 is three frames from a video captured by Basler acA1920–25uc camera when the stepper motor rotated +1 and -1 degree (clockwise direction is positive).

In Fig. 5, the left side is the edge of the black film. Three reference pixels (red arrow in Fig. 5) are the seams of black film edge and focused diffraction pattern. Three reference pixels all have 0 vertical axis. Horizontal axis are 571, 361 and 714 respectively. The position error is 571–361 = 210 pixels. Each pixel is 2.2 μm. The error on the sensor is 462 μm. The objective lens is 20 times. So the actual error is 23.1 μm. This error is considerable but avoidable. This error is mainly produced from the backlash because the timing belt is loosely tensioned to prevent vibration. The backlash error can be avoided by tensioning the timing belt beforehand, which can be done by controlling the stepper motor rotation on the desired direction first. Additional factors may also cause this error, such as the bearing precision, centering of the object and machining accuracy.

7 CONCLUSIONS

Accounting for the various advantages of LSFMs, they are being developed and improved in a number of laboratories at a rapid pace. In this paper, we implement the automatic positioning system for the LSFMs. This system has controllable speed and position and can be integrated with other systems. To overcome the vibration problem for high precision system driven by the stepper motor, high level micro-stepping is utilized. The overall accuracy of this system is feasible but can be improved by better manufacture technique, better hardware design and better mechanical components.

To reduce the position error, two precision stepper motors can be equipped for X and Y axis control on the pedestal of the microscope mounting bracket. With these two stepper motors it is possible to counteract the position error with auxiliary image processing algorithm. This algorithm will detect any undesired position change and calculate the correction value for two stepper motors in real time. So the centering process can be carried out by two stepper motors. This is scheduled in future work.

REFERENCES

Buytaert J.A., and J.J. Dirckx, J Biomed Opt, 12, 014039 (2007).

Dunsby, C., Opt Express, 16, 20306–16 (2008).

Fuchs, E., J. Jaffe, R. Long, and F. Azam, Opt Express, 10, 145–54 (2002).

Greger, K., J. Swoger, and E.H. Stelzer, Rev Sci Instrum, 78, 023705 (2007).

Holekamp, T.F., D. Turaga, and T.E. Holy, Neuron, 57, 661–72 (2008).

Huisken J., and D.Y. Stainier, Opt Lett, 32, 2608–10 (2007).

Keller, P.J., A.D. Schmidt, A. Santella, K. Khairy, Z. Bao, J. Wittbrodt, and E.H. Stelzer, Nat Methods, 7, 637–42 (2010).

Keller, P.J., A.D. Schmidt, J. Wittbrodt, and E.H. Stelzer, Science, 322, 1065–9 (2008).

Mertz J., J. Kim, J Biomed Opt, 15, 016027 (2010).

Microchip. *dsPICDEM MCSM Development Board.* 2016; Available from: http://www.microchip.com/Developmenttools/ProductDetails.aspx?PartNO=DM330022.

Santi, P.A., and J Histochem Cytochem, 59, 129–38 (2011).

Santi, P.A., S.B. Johnson, M. Hillenbrand, P.Z. Grand-Pre, T.J. Glass, and J.R. Leger, Biotechniques, 46, 287–94 (2009).

Verveer, P.J., J. Swoger, F. Pampaloni, K. Greger, M. Marcello, and E.H. Stelzer, Nat Methods, 4, 311–3 (2007).

Voie, A.H., D.H. Burns, and F.A. Spelman, J Microsc, 170, 229–36 (1993).

Intrusion detection using boosting combined with data sampling techniques

Shi Chen, Xiaojun Guo, Zhiping Huang & Zhen Zhuo
College of Mechatronics and Automation, National University of Defense Technology, Changsha, P.R. China

ABSTRACT: For meeting the challenge brought by the rapid growth of network security threat, a number of researchers studied the Intrusion Detection Systems (IDSs) throughout the network security field. IDSs are usually subjected to multiclass imbalance, which leads to lower detection rate of minority classes. In this paper, we present an ensemble method, combining boosting with sampling techniques to address the problem of multiclass imbalance for IDSs. During each iteration of boosting, the size of the temporary balanced training data set is maintained to be the same with the original data set by under-sampling the large classes at random weight and over-sampling the small classes using SMOTE. In addition, information gain-based feature selection method is adopted to rank the features and the top 24 features are selected for building models. The experiments are conducted on NSL-KDD data set. The results show that the proposed method is successful in improving the classification performance of minority classes in terms of Recall, Precision, and F-Measure.

1 INTRODUCTION

Machine learning methods have been widely applied to improve the accuracy of network attack identification. Nevertheless, researchers have found that the class distribution in IDSs is highly skewed (Qazi, 2012). Learning from class-imbalanced data sets tends to generate biased classifiers that tend to focus on the majority class and neglect the minority examples. Therefore, it is a challenge to build an effective classification model if the data set is class imbalanced. Most recent research works on class-imbalanced data sets for intrusion detection dedicated to improve classification performance by using data sampling techniques (Qazi, 2012; Tesfahun, 2013), boosting algorithms (Natesan, 2012; Sornsuwit, 2015), and combining them (Yueai, 2009). On the one hand, data sampling methods balance the class distribution in the training data set by either randomly removing examples of majority class or adding novel events for minority class. Random resampling is the simplest sampling method that includes Random Over-Sampling (ROS) and Random Under-Sampling (RUS). Another famous "intelligent" sampling technique is synthetic minority over-sampling technique (SMOTE) (Chawla, 2002), which generates novel examples for the rare class by incorporating with k-nearest neighbors method. On the other hand, boosting is another effective algorithm that can be employed to handle the skew class distribution, and AdaBoost (Freund, 1996) is the most common

boosting algorithm. Moreover, some improved boosting algorithms combined with other methods have been put forward to solve the class imbalance problem, such as AdaCost (Fan, 1999), RareBoost (Mahesh, 2001), SMOTEBoost (Chawla, 2003), and RUSBoost (Seiffert, 2010; Weiss, 2004).

The work of this paper aims to study the impact of sampling techniques and boosting algorithms on the multiple class-imbalanced NSL-KDD data set, which is widely used as training and testing data set for IDSs. For dealing with the multiclass imbalance problem, we propose an improved boosting algorithm combining RUS and SMOTE to increase the detection rate of minority attack classes.

2 RELATED WORK

Sampling techniques have attracted great interest in research on class imbalance. ROS and RUS are two simple sampling methods that randomly duplicate rare class instances and remove majority class instances, respectively. Several improved sampling algorithms have been proposed. One of the famous over-sampling methods is SMOTE, which searches k-nearest neighbors for generating novel synthetic examples for rare class (Chawla, 2002). Compared with ROS, SMOTE is more complex and time consuming, but will not cause over-fitting. Yen and Lee (Yen, 2006) proposed several under-sampling methods based on clustering for improving the classification performance of K clusters, some representative majority class examples are singled out from each cluster based on designed distance formulas.

Boosting algorithm is another technique that is used to improve the classification accuracy for class-imbalanced data set. Schapire (Schapire, 1990) uses the constructor method to demonstrate that multiple weak learners can be promoted to a strong learner, which is the rudiment of boosting algorithms. Freud and Schapire (Freund, 1996) proposed AdaBoost algorithm, which improves boosting algorithm and is more feasible. Combining the boosting algorithms with other techniques can effectively improve the performance of classifier, which becomes an importance research aspect in the area of machine learning. Chawla (Chawla, 2003) proposed SMOTEBoost algorithm based on the combination of SMOTE and Ada-Boost. Seiffert (Seiffert, 2010) proposed a hybrid algorithm combining RUS and AdaBoost (named RUSBoost) for solving the class imbalance problem. By comparing with SMOTEBoost, RUSboost is more efficient and often obtains comparable performance in terms of four performance metrics.

Data sampling and boosting algorithms have been widely applied in IDSs to improve performance while learning from imbalanced training data. Qazi and Raza (Qazi, 2012) explored the effect of combination of feature selection, SMOTE, and under-sampling technique on the KDDCup99 data set. They found that the under-sampling techniques are more efficient than SMOTE when the original data set is class-imbalanced. Natesan (Natesan, 2012) used AdaBoost algorithm to improve the attack detection rate in IDSs, and it selected several classifiers, such as Bayes Net, Decision tree, and Naive Bayes, as weak learners. They concluded that the Decision Tree and Naive Bayes classifiers as weak learners in AdaBoost could achieve better performance than other several classifiers. Sornsuwit and Jaiyen (Sornsuwit, 2015) concentrated on the detection of minority attack examples, such as U2R and R2 L. In their work, correlation-based algorithm is adopted for removing redundant features, and AdaBoost is applied to construct the ensemble of weak learners including Naïve Bayes, Decision Tree, MLP, k-NN, and SVM. The trial results reveal that the AdaBoost-based ensemble of weak classifier outperforms single classifier.

3 BOOSTING ALGORITHM COMBINED WITH RUS AND SMOTE

The boosting algorithm proposed in this paper is based on the multiclass AdaBoost algorithm proposed by Zhu. The multiclass AdaBoost is an ensemble-learning method that can improve the multiclass classification performance of weak classifiers by constructing a linear form of all base classifiers. In the process of boosting iteratively, the weights of training examples are adaptively adjusted according to the prediction results. That is, if the examples are predicated falsely during the current iteration, their weights will be assigned a higher value during next iteration. After all iterations are finished, all testing examples are predicated as a class based on a weighted vote of all weak learners.

In our proposed method, RUS and SMOTE sampling techniques are adopted to select examples at weighted random for creating a temporary balanced training data set. Because the class distribution of the given data set is highly skewed in this paper (see Table 1), if we only use SMOTE to generate new examples for minority classes, the classifiers training time will be greatly increased because of a larger training data set obtained. RUS is also used to maintain the number of examples in the training data set by removing some examples from the majority classes. However, the use of RUS will cause the loss of information that contained within the deleted examples, which will reduce the accuracy of majority classes. Combining RUS with AdaBoost algorithm can effectively solve this problem. While the information of some examples

Table 1. Selected features.

Rank	Features	Information gain
1	src_bytes	1.031066
2	service	0.861024
3	diff_srv_rate	0.725548
4	flag	0.705471
5	dst_bytes	0.661823
6	same_srv_rate	0.657095
7	dst_host_diff_srv_rate	0.653126
8	count	0.604614
9	dst_host_srv_count	0.592836
10	dst_host_same_srv_rate	0.576413
11	dst_host_serror_rate	0.565873
12	serror_rate	0.548323
13	dst_host_srv_serror_rate	0.538136
14	srv_serror_rate	0.514508
15	logged_in	0.43935
16	dst_host_srv_diff_host_rate	0.377375
17	dst_host_same_src_port_rate	0.343376
18	dst_host_count	0.300444
19	srv_count	0.235109
20	srv_diff_host_rate	0.213594
21	dst_host_rerror_rate	0.135349
22	protocol_type	0.120556
23	dst_host_srv_rerror_rate	0.119999
24	rerror_rate	0.109055

may be lost during a given iteration of boosting, it will likely be contained during other iterations (Seiffert, 2010). As a result, during each iteration of boosting, we are keeping the size of each temporary training data set to be equal to that of original data set by random under-sampling of the majority classes and generating novel synthetic examples for minority classes using SMOTE rather than simply making a copy of original examples.

Algorithm 1 Boosting algorithm combined with RUS and SMOTE

Input: Data set S of examples $(x_1, y_1), \cdots, (x_N, y_N)$ with $y_i \in \mathbf{Y}, |\mathbf{Y}| = m$; number of iterators, T

Output: The ensemble classifier:

$$C^*(x) = \arg\max_y \sum_{t=1}^{T} \alpha_t I(C_t(x) = y)$$

where $I(A)$ is an indicator function that if A is true, $I(A)=1$ or 0 otherwise.

Begin:

1 Initialize $D_1(i) = 1/N$ for all $i = 1, \cdots, N$

2 do **for** $t = 1, 2, \cdots, T$

(1) Create temporary balanced training data set S'_t with distribution D_t and replacement by using RUS and SMOTE.

    a. Random under-sampling of each majority class with weight distribution D_t and replacement at percentage $(1/m) \cdot 100\%$.

    b. Generate $(N/m) - N_j$ new examples for minority classes by using SMOTE.

(2) Call weak learner, train the classifier C_t on data set S'_t.

(3) Calculate the error rate of the classifier $C_t : \mathbf{X} \to \mathbf{Y}$.

$$\varepsilon_t = \sum_{i=1}^{N} D_t(i) I(C_t(x_i) \neq y_i)$$

If $\varepsilon_t > 0.5$, then set $t = t - 1$ and go to (1).

(4) Calculate the weight factor of the classifier C_t:

$$\alpha_t = \log\frac{1-\varepsilon_t}{\varepsilon_t} + \log(m-1)$$

(5) Update weight distribution D_t:

$$D_{t+1}(i) = D_t(i) \cdot \exp(\alpha_t I(C_t(x_i) \neq y_i))$$

$$D_{t+1}(i) = \frac{D_{t+1}(i)}{Z_t}$$

where Z_t is a normalization constant and $Z_t = \sum_i D_{t+1}(i)$.

End.

Algorithm 1 shows the proposed boosting algorithm combined with RUS and SMOTE. Each of the N examples is represented by (x_i, y_i) where x_i is a feature vector in $\mathbf{X}$ and y_i is the ith class label in $\mathbf{Y}$ that has m classes. Let C_t be the base classi-

fier based on weak learner on iteration t trained by the temporary training data set and $C_t(x_i)$ be the prediction result of ith example on iteration t. Let α_t denote the weight factor of C_t and $D_t(i)$ denote the ith example's weight. In step 1 of algorithm 1, the weight of each example is initialized to $1/N$. In step 2, T base classifiers are iteratively trained in five substeps (1–5). We create a temporary balanced training data set S'_t by using RUS and SMOTE in step (1). For each majority class whose size is over N/m, random under-sampling N/m examples with weight distribution D_t and replacement is carried out. On the contrary, we use SMOTE to create $(N/m) - N_j$ novel temporary synthetic examples for the jth rare class whose size is under N/m. As a result, the size of the temporary training data set is equal to that of the original data set, and the class distribution is balanced. Then, we call weak learner and train the base classifier C_t in step (2). The error rate ε_t of the classifier C_t is calculated based on the weight distribution D_t in step (3). Step (4) calculates the weight factor α_t of the classifier C_t, and α_t is used to update the weight distribution D_t in step (5). The final ensemble classifier $C^*(x)$ is outputted based on the weighted vote of T base classifiers after T iterations of step 2. In this paper, the iterations T of boosting algorithms is set to 10. The C4.5 classifier (Salzberg, 1993) is applied in this paper as the weak classifier of the boosting algorithms.

4 EXPERIMENTS

4.1 *NSL-KDD data set and preprocessing*

The NSL-KDD data set used in our work is proposed by Tavallaee (Tavallaee, 2009) to alleviate some of the problems in KDDCup99 data set as described by (McHugh, 2000). In NSL-KDD data set, each attack example can be classified as one of the following four classes: Denial of Service (DOS) attack, Probe Attack, User to Root (U2R) Attack, and Root to Local (R2 L) Attack. The IDS in this paper is considered as a five-class classification problem. We use the KDDTrain+_20Percent (25,192 examples) of NSL-KDD in our experiments.

Data preprocessing in this paper includes feature selection, nominal features encoding, and normalization. For removing the redundant and irrelevant features and reducing the time of calculations, a feature selection method based on information gain is performed on the original data set to rank the 41 features of NSL-KDD. Then, the top 24 features, whose values of Information Gain are over 0.1, are selected for experiments.

Considering the nominal features are not suitable for SMOTE, it is necessary to encode them to numeric features. We just transform the values of nominal features into integer numbers. In addition, data normalization is an effective procedure for preventing the very large features dominating the very small ones. In our work, numeric features are normalized by Eq. (1):

$$\hat{x} = \frac{x - \bar{x}}{\sigma} \tag{1}$$

where $\bar{x}$ and σ are the mean and standard deviation of feature x, respectively.

4.2 Performance metrics

When the class distribution is skewed, it is more reasonable to use Precision, Recall, F-Measure, and other similar metrics as metrics rather than overall accuracy and error rate. These performance metrics for our experiments are described as follows:

$$Recall = \frac{TP}{TP + FN} \tag{2}$$

$$Precision = \frac{TP}{TP + FP} \tag{3}$$

$$F - Measure = \frac{(1 + \beta^2) \cdot Recall \cdot Precision}{\beta^2 \cdot Recall + Precision} \tag{4}$$

where TP, FN, and FP denote true positive, false negative, and false positive, respectively. In formula (7), β denotes the relative significance of Recall and Precision. Typically, $\beta = 1$.

4.3 Results and analysis

This section presents the results of our experiments and compares our proposed method with its components (Adaboost, RUS/SMOTE, and base classifier C4.5) in terms of Recall, Precision, and F-Measure performances. The experiments are conducted on a 64 bit Windows operating system, which has 2.90GHz Intel Pentium CPU and 8 GB of RAM. The feature selection algorithm, called InfoGainAttributeEval, in WEKA 3.7.13, is used to rank the features of NSL-KDD. The code of the proposed boosting algorithm is written in R programming language. Ten-fold cross-validation is used to perform the experiments. The selected features and corresponding Information Gain values are ranked in Table 1.

Table 2 and Figures 1–3 show the comparisons of the proposed method with AdaBoost, RUS/SMOTE+C4.5 and single C4.5 classifier for each class. From Figures 1 and 2, we can see that, although the Recall of RUS/SMOTE+C4.5 outperforms that of C4.5 for Probe, U2R, and R2 L, there is an obvious reduction of Precision because of the loss of information that contained within the removed examples. It results in the reduction of F-Measure for all classes. According to Table 2 and Figure 3, by combining boosting with RUS/SMOTE, the proposed method shows the best

Table 2. Comparative results of the proposed method and its components for each class.

Class	Methods	Recall	Precision	F-Measure
Normal	Proposed method	0.998	**0.997**	0.997
	AdaBoost	**0.999**	0.997	**0.998**
	RUS/SMOTE +C4.5	0.973	0.996	0.984
	C4.5	0.996	0.993	0.994
DOS	Proposed method	**0.999**	0.999	0.999
	AdaBoost	0.999	**1.000**	**1.000**
	RUS/SMOTE +C4.5	0.995	0.994	0.994
	C4.5	0.999	0.999	0.999
Probe	Proposed method	**0.995**	0.994	**0.995**
	AdaBoost	0.992	**0.998**	0.995
	RUS/SMOTE +C4.5	0.988	0.934	0.960
	C4.5	0.978	0.983	0.980
R2 L	Proposed method	**0.928**	0.933	0.930
	AdaBoost	0.919	**0.990**	**0.953**
	RUS/SMOTE +C4.5	0.909	0.560	0.693
	C4.5	0.847	0.932	0.887
U2R	Proposed method	0.455	**0.625**	**0.526**
	AdaBoost	0.364	0.571	0.444
	RUS/SMOTE +C4.5	**0.545**	0.130	0.211
	C4.5	0.182	0.333	0.235

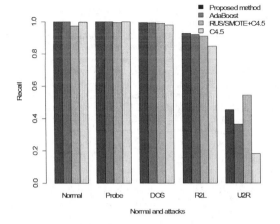

Figure 1. Recall of the four methods for each class.

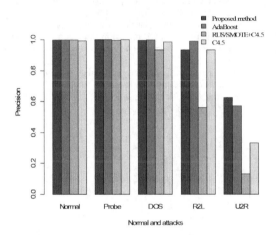

Figure 2. Precision of the four methods for each class.

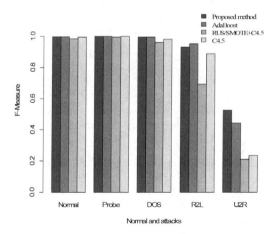

Figure 3. F-Measure of the four methods for each class.

performance for the minority class Probe and U2R among the four methods with an F-Measure of 0.995 and 0.526, respectively. Although the proposed method does not achieve the best performance for class Normal, DOS, and R2 L, the Recall, Precision, and F-Measure of the proposed method are very similar to that of AdaBoost and higher than that of RUS/SMOTE+C4.5 and C4.5. Overall, our proposed method can effectively improve the detection performance of the minority attacks for IDSs.

5 CONCLUSION

In order to settle the matter of class imbalance in intrusion detection, we propose a hybrid method combining boosting algorithm with data sampling techniques for improving the detection performance of minority attacks in this paper. Our proposed method is evaluated on the NSL-KDD data set based on 10-fold cross-validation. For reducing the complexity of training model and balancing the class distribution, we maintain the number of examples in training data set by under-sampling the majority classes at random and generating novel synthetic examples for minority classes using SMOTE during the iteration of boosting. The trial results show that the proposed method can significantly improve the detection performance for minority attacks (including class Probe, R2 L, and U2R) and maintain satisfactory detection performance for majority classes (include class Normal and DOS).

ACKNOWLEDGMENTS

This work was supported by the National High Technology Research and Development Program of China (Grant No. 2015 AA7115089) and the National Natural Science Foundation of China (Grant No. 61374008).

REFERENCES

Chawla, N.V., A. Lazarevic, L.O. Hall, K.W. Bowyer, Smoteboost: Improving prediction of the minority class in boosting. *7th European Conference on Principles and Practice of Knowledge Discovery in Databases*, Croatia: Springer Verlag. 107–119 (2003).

Chawla, N.V., K.W. Bowyer, L.O. Hall, W.P. Kegelmeyer, Smote: Synthetic minority over—sampling technique, Journal of Artificial Intelligence Research, 16:321–357 (2002).

Fan, W., S.J. Stolfo, J. Zhang, C.P. K., Adacost: Misclassification cost-sensitive boosting. *16th Int Conf Mach Learn.* 97–105 (1999).

Freund, Y., R.E. Schapire, Experiments with a new boosting algorithm. *ICML*. 148–156 (1996).

Mahesh, V.J., K. Vipin, C.A. Ramesh, Evaluating boosting algorithms to classify rare cases: Comparison and improvements. *Data Mining*. (2001).

McHugh, J., Testing intrusion detection systems: A critique of the 1998 and 1999 darpa intrusion detection system evaluations as performed by lincoln laboratory, ACM Transactions on Information and System Security, 3, 4:262–294 (2000).

Natesan, P., P. Balasubramanie, G. Gowrison, Improving the attack detection rate in network intrusion detection using adaboost algorithm, Journal of Computer Science, 8, 7:1041–1048 (2012).

Qazi, N., K. Raza, Effect of feature selection, smote and under sampling on class imbalance classification. *14th International Conference on Modelling and Simulation*, IEEE. 145–150 (2012).

Salzberg, S.L. Book review: C4.5: Programs for machine learning by j. Ross quinlan. Morgan kaufmann publishers, inc., 1993. Springer; 1994. p. 235.

Schapire, R.E., The strength of weak learn ability, Machine learning, 5, 2:197–227 (1990).

Seiffert, C., T.M. Khoshgoftaar, J. Van Hulse, A. Napolitano, Rusboost: A hybrid approach to alleviating class imbalance, IEEE Transactions on Systems, Man, and Cybernetics—Part A: Systems and Humans, 40, 1:185–197 (2010).

Sornsuwit, P., S. Jaiyen, Intrusion detection model based on ensemble learning for u2r and r2l attacks. *7th International Conference on Information Technology and Electrical Engineering (ICITEE)*, IEEE. 354–359 (2015).

Tavallaee, M., E. Bagheri, W. Lu, A. Ghorbani, A detailed analysis of the kdd cup 99 data set. *Second IEEE Symposium on Computational Intelligence for Security and Defense Applications (CISDA)*(2009).

Tesfahun, A., D.L. Bhaskari, Intrusion detection using random forests classifier with smote and feature reduction. *International Conference on Cloud & Ubiquitous Computing & Emerging Technologies*, IEEE. 127–132 (2013).

The NSL-KDD dataset: http://nsl.cs.unb.ca/nsl-kdd/.

Weiss, G., Mining with rarity: A unifying framework, ACM SIGKDD Explorations Newsletter, 6, 1:7–19 (2004).

Yen, S.-J., Y.-S. Lee, Under-sampling approaches for improving prediction of the minority class in an imbalanced dataset.344, Berlin, Heidelberg: Springer Berlin Heidelberg. 731–740 (2006).

Yueai, Z., C. Junjie, Application of unbalanced data approach to network intrusion detection. *First International Workshop on Database Technology and Applications*, 140–143 (2009).

Zhu, J., H. Zou, S. Rosset, S. Rosset, Multi-class adaboost, Statistics and Its Interface, 2:349–360.

Advances in Energy Science and Equipment Engineering II – Zhou, Patty & Chen (Eds)
© 2017 Taylor & Francis Group, London, ISBN 978-1-138-71798-5

An improved algorithm for detecting SIFT features in video sequences

Bin Chen, Jianjun Zhao & Yi Wang
Weapon Science and Technology Department, NAAU, Yantai, China

ABSTRACT: Although SIFT algorithm has great advantage of extracting the invariant features, it is a challenge to use the algorithm in real time because of time complexity. This paper improves the SIFT algorithm by limiting the region of extrema detection. The radius of the region is updating with the global motion vectors between the continuous frames of a video. With the improved algorithm, we obtained the areas in which the extrema is detected. The algorithm keeps the amount of SIFT features and is proved to be more efficient by decreasing the time cost during extrema detection.

1 INTRODUCTION

Scale Invariant Feature Transform (SIFT) is a computer vision algorithm to detect and describe the local features in the images. It constructs scale space, finds extreme points in the space, and calculates their positions and scale and rotation invariant (CHANG, 2006). SIFT features, which are independent of the image size and rotation, achieve good results in the condition with noise and light changing. In terms of invariant feature extraction, SIFT feature has a great advantage (BATTIAT, 2007). However, the algorithm needs to build scale space and Gaussian pyramid, which is time consuming. In order to detect the SIFT features in the images, it is necessary to do extrema detection, with every point meeting the detecting condition. Therefore, the task of real-time video stabilization and object tracking using SIFT features still has great challenges.

This paper presents an improved SIFT feature detection method, which is divided into three stages:

- Acquiring the SIFT features of the previous frame;
- Updating the radius of neighborhood;
- Acquiring the SIFT features of posterior image.

The radius of neighborhood area adjusts dynamically according to the global motion vector between successive frames (CHEN, 2011). The method reduces the times of extrema detection and improves the efficiency of the algorithm.

2 SIFT FEATURE

2.1 Scale space

The basic idea of scale space theory is to introduce a parameter of scale in the model of image processing. By varying the introduced parameter, the visual information at different scales can be acquired for getting the essential characteristics of the images.

Scale space meets the requirement of visual invariance. In other words, the scale space of a image is not affected by the variation of the gray level and contrast ratio of the images. At the same time, scale space operator should meet translation invariance, scale invariance, Euclidean invariance, and affine invariance.

A scale space of an image can be expressed as $L(x, y, \sigma)$, which is defined as a convolution of a Gaussian function $G(x, y, \sigma)$` and the original image $I(x, y)$ shown in (1):

$$L(x, y, \sigma) = G(x, y, \sigma) * I(x, y) \tag{1}$$

$G(x, y, \sigma)$ is shown in (2):

$$G(x, y, \sigma) = \frac{1}{2\pi\sigma^2} e^{-\frac{(x-m/2)^2+(y-n/2)^2}{2\sigma^2}} \tag{2}$$

where (x, y) represents the position of the pixel and σ is the scale space factor proportional to the smooth level and scale of images. Large scale corresponds to the appearance of the images, and small scale describes the details of the images.

2.2 The process of SIFT

SIFT algorithm (David, 1999; David, 2004) operates in four stages.

2.2.1 Extrema detection of scale space
The first step of constructing scale space is to build Gaussian pyramid. Making the assumption that the size of the original image is M1 × N1 and the

size of the top-tier is Mn × Nn, the number of the pyramid's layer satisfies (3):

$$n = \log_2(\min(N_1, M_1)) - \log_2(\min(N_n, M_n)) \qquad (3)$$

Gaussian differential operator, which is expressed as (4), can produce stable image features. Thus, generating Gaussian differential pyramid instead of the Gaussian pyramid can achieve key points, which are more stable. Gaussian difference images can be generated by subtracting two adjacent images in the Gaussian pyramid. Each image group in Gaussian pyramid contains P subpictures, and Gaussian differential pyramid only holds P-1 subpictures in each group:

$$\begin{aligned} &D(x, y, \sigma) \\ &= (G(x, y, k\sigma) - G(x, y, \sigma)) * I(x, y) \qquad (4) \\ &= L(x, y, k\sigma) - L(x, y, \sigma) \end{aligned}$$

Flowchart of constructing the Gaussian pyramid is shown in Figure 1.

After obtaining the Gaussian differential pyramid, extremum judgment is applied to every pixel meeting the condition of key point detection in every image. As extrema detection requires comparing the pixel with 26 adjacent points in the scale space, only middle layers of Gaussian differential pyramid can perform the extrema detection. Therefore, in order to ensure K extreme points in each set of images, the Gaussian pyramid with K + 2 subpictures in each group is necessary. As the Gaussian differential pyramid is generated from the Gaussian pyramid, each image set in Gaussian pyramid needs K + 3 subpictures. In order to ensure the highest spatial sampling rate, the common solution is to double the scale of the image; in other words, double the size of the image, using bilinear interpolation.

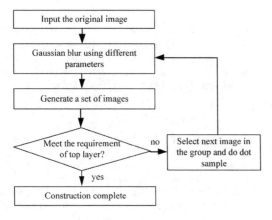

Figure 1. Process of constructing the Gaussian pyramid.

2.2.2 Key points locating

As the detected extreme points are distributed in discrete space in the first stage, it is necessary to remove key points with low contrast and unstable edge points to enhance the stability of the match and improve noise immunity. The location and scale of the key points are determined by function fitting.

2.2.3 Determine the direction

In the Gaussian pyramid, the algorithm computes the gradient and the directional distribution of every key point in each neighbor window with radius equals 3σ. Gradient histogram divides 360° into 36 regions averagely. The direction with peak value in the gradient histogram is considered as the principal direction. At the same time, the directions with values higher than 80% of the peak value are regarded as subdirections. In conclusion, SIFT feature points can be represented as key points containing position, scale, and orientation.

2.2.4 Description of key point

Finally, the algorithm creates a descriptor for each key point and describes it with a set of vectors to ensure that all key points are not affected by light, viewing angle changes, and so on. SIFT descriptor catches 4 × 4 windows around each key point and computes gradient of eight directions in each subwindow. Finally, it generates a 128-dimensional vector. This vector is actually a unique abstract of the region information.

3 IMPROVED DETECTION METHOD

The extreme points detected in scale space composes the initial SIFT key points set. By removing some unstable edge points, the SIFT key points set is established. Extrema detection stage is the most important one in SIFT algorithm.

3.1 Time complexity of extrema detection

Suppose image sequence contains S frames with resolution of M × N. We build n-tier pyramid according to equation (3) and obtain images with K scale by Gaussian blur in every tier. The resolution of image in ith layer is $\left\lfloor \dfrac{M}{2^{i-1}} \right\rfloor \times \left\lfloor \dfrac{N}{2^{i-1}} \right\rfloor$. As the extreme points on the edges of the images cannot be detected, the number of pixels to be detected in the ith layer equals Pi, as shown in (5):

$$P_i = \left\lfloor \frac{M}{2^{i-1}} - 1 \right\rfloor \times \left\lfloor \frac{N}{2^{i-1}} - 1 \right\rfloor \qquad (5)$$

The total number of pixels to perform extrema detection meets (6):

$$P = 26 \sum_{i=1}^{n} P_i (S-1)(K-3) \qquad (6)$$

where S is determined by the length and the frame rate of the video sequence and K determines the number of scales that can be detected in a set of images with default value equals to 5. Therefore, the effective reduction of Pi is a feasible way to reduce the amount of detecting pixels and improve the speed of SIFT algorithm.

3.2 Improved algorithm

Existing video sequence acquired by digital sensors generally has a high frame rate. The camera motion among continuous frames is tiny in the obtained video sequence when the sensor moves. Thus, SIFT feature points in the posterior image are mostly located in the neighborhood of corresponding points in previous frame. As shown in Figure 2, we limited the radius of the neighborhood to get a collection, which contains every neighborhood of all feature points in the previous frame, reducing the amount of pixels involved in extrema detection.

The radius of neighborhood is limited by (7):

$$r = \begin{cases} k \left| \overrightarrow{mj} \right| & j = 1 \\ k \dfrac{\left| \overrightarrow{mj} \right| + \left| \overrightarrow{m_{j-1}} \right|}{2} & 2 \le j \le S-1 \end{cases} \qquad (7)$$

where S is the number of frames in a video sequence, k is a controlled parameter to avoid failure of detection caused by mutation of the sensor's speed.

4 EXPERIMENTS AND RESULTS

By comparing the original and improved algorithm, Experiment 1 measures the effectiveness of the improved one, including the number of matching points and time cost. The regional images to detect the extrema are also acquired.

In Experiment 2, we chose the first three frames in the test set to run the improved algorithm.

4.1 Test set of Experiment 1

The test video set shown in Figure 3 is shot by an UAV, selecting nine sequential frames (frames 71–79), testing the effectiveness of the algorithm.

Resolution of images is 240×320. The parameters in equation (6) are set as below: S = 9, K = 5, n = 4, k = 2.

4.2 Results of experiment 1

The number of feature points in every frame acquired by SIFT feature detection algorithm is shown in Table 1.

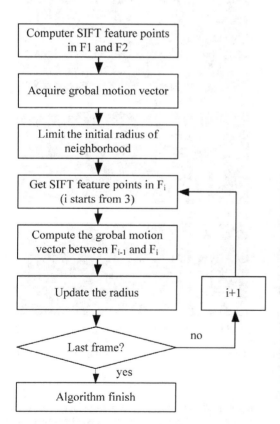

(a) Frame 71   (b) Frame 72   (c) Frame 73

(d) Frame 74   (e) Frame 75   (f) Frame 76

(g) Frame 77   (h) Frame 78   (i) Frame 79

Figure 3.   Test set of experiment.

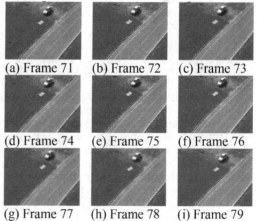

Figure 2.   Process of the improved algorithm.

Table 1. Number of SIFT feature points.

Frame	71	72	73	74	75	76	77	78	79
Number	45	47	38	44	39	41	29	35	56

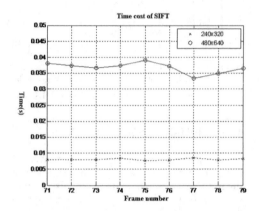

Figure 4. Cost of extrema detection using SIFT algorithm.

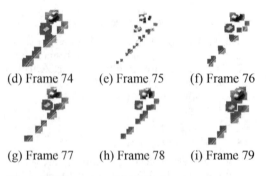

(d) Frame 74 (e) Frame 75 (f) Frame 76

(g) Frame 77 (h) Frame 78 (i) Frame 79

Figure 5. The region of extrema detection.

Table 2. Number of improved SIFT feature points.

Frame	71	72	73	74	75	76	77	78	79
Number	45	47	38	42	33	37	26	39	54

Figure 4 shows that, when increasing the image resolution to 480×640, pixels meeting the condition of extreme detection increases greatly. As a result, the time cost of detection increases dramatically.

The radius of the neighborhood is initialed by the global motion vector obtained from the first and second frames. With the radius, we generate a regional image, in which the pixels detect the extrema. The whole set of regional images is generated repeatedly, as shown in Figure 5.

The number of SIFT feature points obtained by the improved algorithm are shown in Table 2.

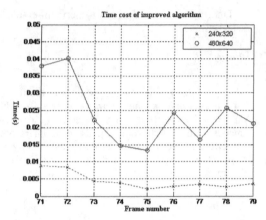

Figure 6. Cost of extrema detection using improved algorithm.

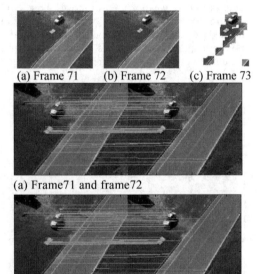

(a) Frame 71 (b) Frame 72 (c) Frame 73

(a) Frame71 and frame72

(b) Frame72 and frame73

(c) Frame73 and frame74

Figure 7. Matching result of the improved algorithm.

By improving the algorithm, the efficiency of the extrema detection process is greatly increasing under various resolutions. Figure 6 shows the time cost of the improving algorithm under two resolutions.

Experimental results show that the improved algorithm decreased the time cost during extrema detection.

4.3 *Result of experiment 2*

The improved algorithm is applied to the first three images in Experiment 1. The features are detected and matched, as shown in Figure 7. As the result shows, the SIFT features detected by the improved method are well performed in the matching process.

5 CONCLUSIONS

This paper improves the SIFT algorithm aiming at the issue that the extrema detection of SIFT algorithm is time consuming. By computing global motion vector between adjacent images, the improved algorithm defines the region to be detected in the posterior frame. The radius of neighborhood updates by computing new global motion vectors in subsequent frame pairs. The improved algorithm reduces the times of extrema detection and improves the efficiency of SIFT algorithm.

Experiments show that the improved algorithm limited the radius of neighborhood dynamically and ensured the amount of SIFT feature points. At the same time, we obtained the region to be detected in every image, reduced the times of extrema detection effectively, and decreased the time cost significantly.

REFERENCES

Battiat O S, Gallo G, Puglisi G. Sift feature tracking for video stabilization [C]. International conference on Image Analysis and Processing, Modena 2007.
Chang H C, Lai S H, Lu K R. A Robust Real-time Video Stabilization Algorithm [J]. Journal of Visual Communication and Image Representation, 2006, 17(3):659–673.
Chen Qi-li, Song Li, Yu Song-yu. A Overview of Video Stabilization [J]. Video Engineering, 2011,35(7): 15–17.
David G. Lowe. Distinctive Image Features from Scale-invariant Keypoints [C]. International Conference on Computer Vision, Springer US, 2004: 91–110.
David G. Lowe. Object Recognition from Local Scale-invariant Features [C].
International Conference on Computer Vision, Corfu, Greece: Springer US, 1999: 1150–1157.

Advances in Energy Science and Equipment Engineering II – Zhou, Patty & Chen (Eds)
© 2017 Taylor & Francis Group, London, ISBN 978-1-138-71798-5

Research on polarization dehazing algorithm based on atmospheric scattering model

Yanhai Wu, Jing Zhang & Kang Chen
College of Communication and Information Engineering, Xi'an University of Science and Technology, China

ABSTRACT: A polarization dehazing algorithm based on atmospheric scattering model is presented to solve the low contrast and fuzzy problems of images acquired by machine vision system under the condition of fog. First, the algorithm uses polarization imaging detection system to obtain the best and worst conditions of two polarization images, estimates atmospheric light intensity with the quadtree segmentation method, determines the degree of polarization and transmittance by vertical and horizontal atmospheric light intensity, and eventually realizes the image dehazing. Experimental results show that the algorithm has adaptable scene, and its rehabilitation effect is fresh and natural, proving that this algorithm has certain visual advantages.

1 INTRODUCTION

In foggy weather, the reflection luminous flux of the objects are partly reflected or absorbed, caused by the scattering effect of atmospheric particles, which leads to the attenuation of reaching incident light to the observer, furthermore causing worse natural scene image color and contrast and harming the application of the features of image and extraction of image. Thus, the research on improving the reliability of the visual system and robustness of the image dehazing of the visual system is of utmost importance.

The existing dehazing technologies include single image and multiple image dehazing. At present, scholars worldwide are mainly focusing on single optical image dehazing, and have achieved certain results, while resulting in the limitation of geometric scene oneness, access to information, and range of application. The He-method (He, 2010) needs images to satisfy the a priori assumption for dark gray, while this assumption is false in the region of the large sky area. The restoration algorithms for multiple fog-degraded images are of two main types: one algorithm dehazes by multiple fog-degraded images from different weather conditions, such as dichromatic atmospheric scattering model proposed by Narasimhan (Narasimhan, 2003); the other one is to use light polarization characteristics of the implementation of many images to haze removal. Compared with the traditional optical imaging technology, the polarization imaging detection technology has many advantages in target detection and recognition, and hence it is widely applied by scholars worldwide,

especially in military applications. In terms of polarization image dehazing, Schechner and some others (Schechner, 2000; Schechner, 2003) obtained the depth information of the scene based on the polarization characteristics of atmospheres and eventually restored the fog-degraded images. The algorithm has advantages like low computing complexity and instant dehazing ability, but needs to select area manually to estimate the parameters of the image. Peng Wenzhu et al. (Peng, 2011) segmented the sky using edge detection and best normal distribution searching algorithm, thus estimating the atmospheric light intensity. The dehazing effect of this algorithm is remarkable in removal of the slight mist, but the overall recovery effect appears darker and lower tonality. Zhou Pucheng et al. (Zhou, 2011) defined a polarization image with dark channel to get the atmospheric light intensity and degree of polarization and proposed an adaptive algorithm for dehazing images. Miyazki et al. (Miyazaki, 2013) tracked two specific goals on traffic image with the different distances and obtained the depth information of the scene by estimating the distances, thus calculating the atmospheric light intensity and polarization degree to dehaze.

In the dehazing method for multiple fog-degraded images above, the restoration effect of dehazing near scene is satisfying, but the recognition and recovery of vague or invisible scene in far distance is not ideal. In this paper, we propose a dehazing algorithm based on the atmospheric scattering physical model. First, the algorithm uses polarization imaging detection system to obtain two polarization images under the best and worst conditions, estimates atmospheric light intensity with the quadtree segmentation method, determines the degree of polarization and transmittance by vertical and horizontal atmospheric light intensity, and eventually realizes the image dehazing.

2 OPTICAL MODEL

Because of the scattering of atmosphere particles, the incident light into optical imaging path is combined by the atmospheric scattering light and reflection light from targets. Moreover, the atmospheric scattering light, the reflection light from targets, and the synthesis of light are partially polarized light:

$$I(x,y) = D(x,y) + A(x,y) \tag{1}$$

where $I(x, y)$ represents light intensity of the point (x, y) of the image; $D(x, y)$ represents directly transmitted light describing the attenuation process of reflection light from target; and $A(x, y)$ represents atmospheric light, showing effects to imaging from other light in the atmosphere, and the offset of the image colors and brightness. $D(x, y)$ and $A(x, y)$ can be expressed as:

$$D(x,y) = J(x,y)t(x,y) \tag{2}$$
$$A(x,y) = A_\infty(1 - t(x,y)) \tag{3}$$

where J is the light intensity of scene; A_∞ is the atmospheric light intensity from infinity; and t can be expressed as:

$$t(x) = e^{-\beta(\lambda)d} \tag{4}$$

where d is the distance between scene point and observation point; λ is the wavelength of light; and β represents atmospheric scattering coefficient, assuming that the atmosphere is a homogeneous medium, so β is global.

After installing a polarizing lens in the front of the camera, we could get two images with the polarization angle as $\partial = \theta_{//}$ and $\partial = \theta_{\perp}$, respectively, $I^{//}$ and $I^{\perp}$. Defining $\Delta I = I^{\perp} - I^{//}$ as polarization differential images and defining $I = I^{\perp} + I^{//}$ as polarization summing images, polarization can be defined as:

$$p = \frac{I^{\perp} - I^{//}}{I^{\perp} + I^{//}} = \frac{\Delta I}{I} \tag{5}$$

Assuming that particle scattering sources of any scattering particles are from the same spot at the same time, an incident plane would be formed by the light incident on scattering particles and the connection line from the imaging system to the scattering particles. Atmospheric light could be decomposed into two polarized components, parallel and perpendicular to the incident plane, denoted by $A^{\perp}$ and $A^{//}$, respectively. The polarization degree of $A(x, y)$ for atmospheric light is:

$$P_A = \frac{A^{\perp} - A^{//}}{A^{\perp} + A^{//}} = \frac{\Delta A}{A} \tag{6}$$

Assuming that positive transmission light is natural light, the energy of positive transmission light through any direction is 1/2 times its original value. Therefore, two images from orthogonal polarization direction generated by imaging system can be respectively represented as:

$$I^{\perp} = D/2 + A^{\perp} \tag{7}$$
$$I^{//} = D/2 + A^{//} \tag{8}$$

3 DEHAZING ALGORITHM

3.1 Restoration of fog-degraded image

From formulas (3), (5), (6), (7), and (8), we can calculate the transmittance t as:

$$t(x,y) = 1 - \frac{\Delta I(x,y)}{A_\infty P_A} \tag{9}$$

Combining formulas (1), (2), (3), and (9), the scene radiation brightness J can be obtained as follows:

$$J(x,y) = \frac{\Delta I(x,y) - P_A I(x,y)}{\Delta I(x,y)/A_\infty - P_A} \tag{10}$$

where ΔI is the polarization differential image and polarization and image I are the known variables. As long as we could estimate the atmospheric light intensity as A_∞ from infinite distance and the atmospheric light polarization degree as P_A, we can restore the scene brightness J.

3.2 Estimation of atmospheric light

Schechner et al. (Schechner, 2000; Schechner, 2003) manually chose the pixel with the highest fog concentration value in fog-degraded image as the intensity of the light atmosphere, A_∞. However, in the actual scene, the works are often more complex, and the error of relevant parameters of estimation is larger by selecting image area manually, which cannot achieve the goal of adaptive dehazing. Peng et al. (Peng, 2011) used the best normal distribution searching algorithm to separate the polarization image area and took the maximum brightness value as the atmospheric optical brightness in the sky. This algorithm can approximately estimate the atmospheric light value, and it will not be affected by other targets on the images. However, for images with bright scene or many white buildings, the algorithm also produces error and processes slowly. In order to estimate the atmospheric light value rapidly and accurately, this paper uses the quadtree segmentation method [8] to improve the above algorithm. First, we calculate the channel minimum of the image to avoid the estimation error resulting from the existence of maximum in partial image. Then, we separate the channel minimum image into four regions, calculate the average brightness of all regions, and repeat the separation on the maximum average region, until the area is smaller than the given threshold. The general threshold value is 20×20. Finally, we calculate the average $\overline{R}, \overline{G}, \overline{B}$ of this block's original R, G, B channel, getting the atmospheric light A_∞. The method has high convergence rate, high positioning accuracy, and reduces time complexity.

3.3 Estimation of polarization degree P_A and transmittance of atmospheric scattering t

The polarization degree of atmospheric light $A(x, y)$ is shown in formula (6), where d tends to infinity, $t(x) \to 0$, so the formula can be expressed as $P_A \to \dfrac{A_\infty^\perp - A_\infty^{//}}{A_\infty^\perp + A_\infty^{//}}$, and P_A can be obtained. At the same time, the t can be calculated by the equation $t(x, y) = 1 - \dfrac{\Delta I(x, y)}{A_\infty P_A}$. All parameters in this paper have been solved so far, and the image could be restored.

4 EXPERIMENTAL RESULT

In order to verify the validity of the algorithm, the polarization camera is used to obtain polarization image in fog weather conditions. The following simulation experiment is conducted in Windows 7 operating system, and processor with dominant frequency of 3.1GHz, PC with system memory of 4GB, and MATLAB 2013software platform are used.

4.1 Comparison of subjective visual effects

Figure 1 shows the four groups of images. the first two pictures of each group are polarization images with the input orthogonal angle, called optimal polarization image and the worst polarization image. The experimental results prove that the overall image restored by Peng algorithm (Peng,

(a)   (b)   (c)   (d)

Figure 1.   Comparison of simulation results. (a) Optimal polarization image, (b) Worst polarization image, (c) Peng algorithm, (d) Proposed algorithm.

2011) is slightly dark and weak in dehazing effect, layering, and brightness. On the contrary, the image restored by the algorithm presented in this paper shows rich details, clear structure, and especially superiority in avoiding the halo and color in the sky area, and meanwhile, the restoring color is similar to the actual scene, performing certain visual advantages.

4.2 Objective analysis and comparison

The present popular objective evaluation system is presented by Hautiere et al., which is named visible edge gradient method (Hautière, 2008). The letter e is the number ratio of new increment visible edge set. $e = (n_r - n_0)/n_0$; n_0 and n_r represent the number of visible edge of the original image and the dehazing image, respectively; and r is the average gradient ratio, $r = g_r/g_0$, where g_r represents the average gradient of dehazing image and g_0 represents the average gradient of the original image. The time required for the algorithm is t. The greater the

Table 1. Objective evaluation of the simulation results.

Image	Defog algorithm	e	r	t/s
First group	Peng algorithm	0.0721	1.2914	1.1448
	Proposed algorithm	1.5597	3.3531	1.0204
Second group	Peng algorithm	2.7326	3.0563	0.9534
	Proposed algorithm	3.815	4.9898	0.863
Third group	Peng algorithm	2.9759	8.0956	0.8697
	Proposed algorithm	7.4024	10.6923	0.9225
Fourth group	Peng algorithm	1.892	7.1912	0.9642
	Proposed algorithm	3.7931	7.3765	0.9655

values of e and r, the better the image restoration is. The smaller the value of t, the better is the real-time performance. Table 1 shows the four groups of test results of the image in Figure 1.

As we can see from Table 1, the indices e and r of the proposed algorithm are superior to that of Peng algorithm. The dehazing time costs of two types of algorithms are similar, but on the whole, the algorithm proposed in this paper is better than Peng algorithm.

5 CONCLUSION

Because of the scattering effect of the particles in atmosphere, fog-degraded image has the characteristics of low contrast and fuzzy details, which brings inconvenience to the application. Therefore, the restoration of the fog-degraded image and clear imaging are of great significance. To solve the problems of the method based on polarization filtering, a polarization dehazing algorithm based on atmospheric scattering model is presented in this paper. In the first place, the algorithm adopts polarization imaging detection system to get two polarization images of the best and worst conditions and estimates atmospheric light intensity with the quadtree segmentation method, automatically optimizing the atmospheric light intensity and other parameters, estimating the degree of polarization and transmittance by vertical and horizontal atmospheric light intensities, and implementing the image dehazing finally. Experimental results show that this algorithm has high adaptability, higher contrast ratio, clearer image details, and rich tonality, showing certain visual advantages.

REFERENCES

Hautière, N., J.P. Tarel, D. Aubert, et al. Blind Contrast Enhancement Assessment by Gradient Ratioing at Visible Edges [J]. Image Analysis & Stereology, Vol. 27(2): 87–95 (2008).

He, K., J. Sun, X. Tang. Single Image Haze Removal Using Dark Channel Prior. [C]// 2013 IEEE Conference on Computer Vision and Pattern Recognition. IEEE, 2010: 2341–2353.

Kim, J.H., W.D. Jang, J.Y. Sim, et al. Optimized contrast enhancement for real-time image and video dehazing[J]. Journal of Visual Communication & Image Representation, Vol. 24(3): 410–425 (2013).

Miyazaki, D., D. Akiyama, M. Baba, et al. Polarization-Based Dehazing Using Two Reference Objects [C]// IEEE International Conference on Computer Vision Workshops. 2013: 852–859.

Narasimhan, S.G., S.K. Nayar. Contrast restoration of weather degraded images [J]. Pattern Analysis & Machine Intelligence IEEE Transactions on, Vol. 25(6): 713–724 (2003).

Peng, W.Z. Polarization image defog algorithm based on atmospheric scattering model [J]. electronic technology measurement, Vol. 34(7): 43–45 (2011).

Schechner, Y.Y., S.G. Narasimhan, S.K. Nayar. Instant Dehazing of Images Using Polarization[C]// IEEE Computer Society Conference on Computer Vision & Pattern Recognition. IEEE, 2001: 325.

Schechner, Y.Y., S.G. Narasimhan, S.K. Nayar. Polarization-based vision through haze.[J]. Applied Optics, Vol. 42(3): 511–525 (2003).

Zhou, P.C., M.G. Xue, H.K. Zhang. Using polarization filtering automatic image to fog [J]. Chinese Journal of image and graphics newspaper, Vol. 16(7): 1178–1183 (2011).

Advances in Energy Science and Equipment Engineering II – Zhou, Patty & Chen (Eds)
© 2017 Taylor & Francis Group, London, ISBN 978-1-138-71798-5

An improved polarization image defogging algorithm

Yanhai Wu, Kang Chen, Jing Zhang & Xingyu Hou
College of Communication and Information Engineering, Xi'an University of Science and Technology, Xi'an, China

ABSTRACT: In foggy conditions, because of the scattering and adsorption of the suspended particles in the atmosphere, the sensor receives the low-contrast image and produces blurred image. In this paper, an improved polarization image defogging algorithm is proposed. First, a front-end Polaroid camera acquires the best and worst atmospheric polarization image, calculates the mean number of the sky region as the light intensity, combines vertical and horizontal atmospheric light intensity to estimate the image's polarization ratio and the transmittance of the atmosphere, and ultimately realizes the defogging. The experimental results show that this algorithm can improve not only the definition of image but also the contrast of the image.

1 INTRODUCTION

Fog is a very common natural phenomenon. In recent years, with the development of science and technology, fog and haze occur frequently, especially the northern regions of China suffer from haze invasion in winter. When the suspended particles increased in the atmosphere, it has a combined function of light scattering and absorption for haze formation. Haze is not only harmful to people's health, but causes fuzziness in the image received by the sensor. Therefore, it is significant to study how to carry out the image defog.

The current image defogging algorithm is divided into the image enhancement and the restoration method, which is based on atmospheric scattering physical model in (Fan, 2010; Jing, 2011). The image enhancement method can be used in a wide range, can effectively improve the contrast of the image, highlight the detailed information of the image, and improve the effect of image details, but the prominent parts of the information of image will cause a certain degree of loss. The method of image restoration evolves from the reverse degradation process, compensates for the distortion caused by the degradation process in order to obtain a clear image without the interference or the optimal value of the defogging image, and improves the quality of the fog image. Image enhancement includes histogram equalization (Kim, 2002) and homomorphism filtering (Yafei, 2013). Histogram equalization through the frequency histogram for each gray level realizes the overall image contrast enhancement and displays some new information in Kim (2002) and Stark (1996). Homomorphism filtering is a method combining frequency domain filter with gray change, through the fog image transformation to frequency domain, then smoothly by homomorphism filter, strengthening the high-frequency information and reducing the low-frequency information of the image to improve the quality of image processing technique in the literature (Yafei, 2013). Image restoration, which includes dark channel theory in Kaiming (2011) and polarization image defogging algorithm in Wenzhu (2011) and Yong (2009), the dark channel theory was proposed by Do He Kaiming in 2009 by counting 5000 pieces clear image. It is found that in a local area of the vast majority of clear image, there is always at least one color channel in the pixel with a relatively low value. When the fog image has a large number of white areas, the dark channel theory will be distorted. Polarization image defog method is obtained by two images of the best and worst states in the polarization state, and the defogging image is received by calculating the light intensity and transmittance. The use of edge detection and optimal normal distribution search algorithm was proposed in ref. [8] to segment the sky area, compute the atmospheric light intensity, and complete the image defog. This algorithm chooses the maximum value within the region of the optimal normal distribution as the sky brightness, which increases time consumption.

Aiming at the shortage of the above algorithms, an improved polarization image defogging algorithm is proposed, which utilizes the segmentation regions of the sky proposed in Yue (2012) to isolate sky region in polarization image and calculate the mean value of the sky area as atmospheric light intensity, A_∞. The vertical $A_\infty^\perp$ and horizontal $A_\infty^\parallel$ components of atmospheric light intensity estimate the degree of polarization, use the polarization degree p and $A_\infty^\perp$, A_∞'' to calculate the transmittance $t(x)$, and finally obtain a clear image.

2 ATMOSPHERIC PHYSICAL MODEL

The atmospheric scattering physical model is formed by a variety of effects on the scattering and absorption of sunlight by suspended particles in the atmosphere. The model is expressed as:

$$E(d,\lambda) = D_0(\lambda)e^{-\beta(\lambda)d} + A_\infty(\lambda)(1-e^{-\beta(\lambda)d}) \qquad (1)$$

where d is the distance from the photographed object to the sensor (scene depth), λ is the wavelength of visible light, $D_0(\lambda)$ is the object surface reflection of total light intensity, A_∞ is the infinity atmospheric light intensity, and β is the scattering coefficient of the air.

3 PRINCIPLE OF POLARIZATION IMAGE

By analyzing the physical model of atmospheric scatter, we can further simplify the model as:

$$I(x) = J(x)t(x) + A_\infty(1-t(x)) \qquad (2)$$

where $I(x)$ is the observed image intensity, $J(x)$ is a clear image, $t(x)$ is the transmittance of the atmosphere, and A_∞ is the atmospheric light intensity. First, a polarizing plate is added to the front end of the camera to obtain the best and worst polarization state images (as shown in Fig. 1). If we calculate the transmittance and the light intensity, A_∞, the image $I(x)$ can be approximated as a combination of the best and worst state polarization images and $J(x)$ can be recovered by means of formula (3).

$$J(x) = \frac{I'' + I^\perp - A_\infty(1-t(x))}{t(x)} \qquad (3)$$

In Eq. (3), the range of the transmittance is $[0,1]$;

(a) Best-state   (b) Worst-state
   polarization   polarization

Figure 1.   Different polarization states of the same scene.

4 PARAMETERS ESTIMATION

4.1 Estimation of the atmospheric light intensity

First, we find the sky region, and in this paper, we use the segment method of the sky area, which is proposed in the literature [10], then by dilation and erosion to further refine the sky region and ultimately calculate the mean value of the isolated sky area as atmospheric light intensity. The specific steps are as follows:

1. From left to right, the gray histogram of the polarization image is scanned and the corresponding maximum gray value μ is found.
2. Set up the initial value σ (in general $\sigma = 1$), the gray level histogram through the range of $[\mu - 2\sigma, \mu + 2\sigma]$ is divided into three sections, the mean gray value of three ranges are $m1, m2, m3$.
3. Calculating the mean value of $m1, m2, m3$ is s, t and $s = (m1 + m2)/2$, $t = (m2 + m3)/2$ the condition for updating iteration is $\sigma = (t - s)/4$.
4. Repeat steps (2) and (3) until the condition $\sigma < 1$ is satisfied. The final value σ is the best value.
5. If the image is gray, the sky is segmented according to the segmented region $[\mu - 2\sigma, \mu + 2\sigma]$. If it is a color image, the segmented range is $[h_1, h_2]$ (among $h_1 = \min\{u_R - 2\sigma_R, \mu_G - 2\sigma_G, \mu_B - 2\sigma_B\}$, $h_2 = \max\{\mu_R - 2\sigma_R, \mu_G - 2\sigma_G, \mu_B - 2\sigma_B\}$) and the result of the segment is shown in Fig. 2 (a). In this paper, the segmentation of sky region uses the expansion and erosion to further improve the results. The segmentation result is shown in Fig. 2 (b).
6. Marked all black position in Fig. 2, the average of the corresponding position of the fog image is obtained as the value of the atmospheric light intensity.

4.2 Estimation of the degree of polarization

In formula (2), $A_\infty(1-t(x)) = A$. According to the definition of polarization degree, we obtain formula (4):

$$p = \frac{A^\perp - A''}{A'' + A^\perp} \qquad (4)$$

In Eq. (4), $A^\perp$ and A'' are respectively the vertical and horizontal components of A. When the scene

(a) Segmentation in ref.[8] (b) Segmentation in this paper

Figure 2.   Result of the sky region segmentation.

depth tends to infinity and the transmittance tends to zero, Eq. (4) can be expressed as:

$$p \rightarrow \frac{A_\infty^\perp - A_\infty''}{A_\infty^\perp + A_\infty''} \tag{5}$$

4.3 Estimation of transmittance

According to Eqs. (9) and (10) in ref. [9], formula 6 can be expresses as:

$$I^\perp - I'' = A^\perp - A'' \tag{6}$$

According to Eqs. 4 and 5, Eq.7 can be shown as:

$$t(x) = 1 - \frac{I^\perp - I''}{A_\infty p} \tag{7}$$

5 ALGORITHM FLOWCHART

First, according to the camera to obtain the best and worst states of polarization image, we reuse the method proposed in ref. [10] to segment the sky region of the polarization image, use expansion and corrosion to deal with the segmentation sky region, and calculate the mean value of the corresponding position of polarization image as the atmospheric light intensity. In accordance with the vertical and horizontal components of atmospheric light intensity to calculate the polarization degree and transmittance, we finally get the defogging image. The algorithm flowchart is shown in Fig. 3.

6 EXPERIMENTAL RESULTS

In this paper, the experiment uses a computer with Inter(R) Core(TM) i3–2330 CPU@2.2 GHz, 4G memory, Windows 7 operating system, and MATLAB R2013b experimental platform. The

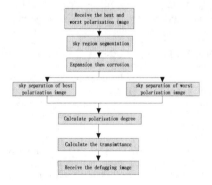

Figure 3. Algorithm flowchart.

proposed algorithm will be compared to the results of the literature (Wenzhu, 2011), and the algorithm defogging effects are shown in Fig. 4.

6.1 Subjective evaluations

Best polarization and worst polarization original algorithm results

There are four groups of images in Fig. 4; from top to bottom is the first group to the fourth group, it can be seen that the original polarization image algorithm was improved in the fog image in different degrees and there are a large number of fog in the results. In the second group, the outline of the near window is blurred; on the left-hand side of the building and on the right-hand side of the tip-shaped building were shrouded in mist in the third group. The algorithm proposed in this paper, except for removing the fog more thoroughly, the outline of far building in the second group is more clearly visible. On the left- and right-hand sides, tip-shaped buildings were clearly identifiable, and in the fourth group, both the pedestrian contour and leaves are clearly visible.

6.2 Objective evaluation

Because the subjective evaluation has a great impact on the individual, therefore, in this paper, we choose the number ratio of the visible edge e, the average gradient r, and the standard deviation δ as the objective parameters to evaluate the results. The number of the visible edge indicates

Figure 4. Results of the literature algorithm defogging effects.

Table 1. Defogging objective image quality evaluation of Fig. 4.

Image group	Algorithm name	e	r	δ
First	Alg.[8]	0.1088	1.1395	0.1456
	Paper Alg.	0.5366	1.7461	0.3384
Second	Alg.[8]	0.1455	1.9232	0.0999
	Paper Alg.	1.5499	3.9037	0.2600
Third*	Alg.[8]	0.3034	3.9037	0.0790
	Paper Alg.	4.6762	5.4633	0.9765
Fourth	Alg.[8]	0.2665	1.8961	0.1882
	Paper Alg.	0.9499	2.7992	0.2929

the ratio of the number of new visible edge to the original image edge, which represents the algorithm ability to recover the image. The ratio of mean gradient indicates the mean gradient of the restored to the original, which indicates the quality of the restored. The standard deviation expresses the discrete degree; the higher the discrete process, the greater the image contrast. The greater e, r, δ indicates the better of the quality of the image. The calculation formula is as follows:

$$e = \frac{n_r - n_o}{n_o} \tag{8}$$

$$\bar{r} = \frac{\bar{g}_r}{\bar{g}_o} \tag{9}$$

$$\delta = \frac{\sum_{x=1}^{M}\sum_{y=1}^{N}[f(x,y) - \bar{f}(x,y)]^2}{M*N-1} \tag{10}$$

In Eq. 7, n_r indicates the number of visible edge of the restored, n_o indicates the number of visible edge of the fog image, $\bar{g}_r$ indicates the mean gradient of the restored, $\bar{g}_o$ indicates the mean gradient of the fog image, $f(x, y)$ indicates the pixel values of the restored, $\bar{f}(x,y)$ indicates the mean value of pixels, and M, N are the size of the image.

From the objective parameters presented in Table 1, the number of visible edge of the algorithm is about five times higher than that of the polarization image method; the ratio of average gradient is about 1.5 times higher than the polarization image method, and standard deviation increased slightly.

7 CONCLUSIONS

A lot of fog is still left in the result of the fog removing algorithm for polarization image, and an improved polarization image defog algorithm

is proposed in this paper. The experimental results show that this algorithm solves the shortcoming that a lot of fog still remain in the restored of the polarization image algorithm, the results of this algorithm is not only fog removal more thoroughly but also satisfied people's visual requirements subjectively. However, this algorithm is further improved, such as: the results of the algorithm in this paper still have some white areas in the sky, which will cause a certain degree of visual illusion.

ACKNOWLEDGMENTS

This work was supported by the National Natural Science Foundation of Shaanxi Province (Grant No. 2015 JQ6221) and Scientific and Technological Research Project of Science and Technology Department of Shaanxi Province.

REFERENCES

Fan G, Zi-xing C, Bin X, ET on. Review and prospect of image dehazingtechniques [J]. Journal of Computer Applications, Vol. 30(9): 2417–2421(2010); (In China).

Jing Y, Dongbin X, Qingmin L. Image defogging: a survey [J]. Journal of Image and Graphics, Vol. 16(9): 1561–1576(2011); (In China).

Kaiming H, Jian S, Xiaosu T. Single image haze removal using dark channel prior [J]. IEEE.Transactions on Pattern Analysis and Machine Intelligence. Vol. 33(12): 2341–2353 (2011).

Kim J.Y, Kim LS, Hwang SH. An advanced contrast enhancement using partially overlapped sub-block histogram equalization [J]. Circuits and Systems for Video Technology. Vol. 11(4): 475–484(2002).

Stark J.A., WJ Fitzqerald. An Alternative Algorithm for Adaptive Histogram Equalization [J].Graphical Models and Image Processing. Vol. 58(2): 180–185(1996).

Wenzhu P. Polarization dehazing algorithm based on atmosphere scattering model [J]. Electronic Measurment Technology. Vol. 34 (7):43–46(2011); (In China).

Yafei Z, Minghong X. Block—DCT based homomorphic filtering algorithm for color image enhancement [J]. Computer Engineering and Design, Vol. 34(5): 303–306(2013); (In China).

Yafei Z, Minghong X. Colour image enhancement algorithm based on his and local homomorphic filtering [J]. Computer Applications and Software. Vol. 30 (12): 303–306(2013); (In China).

Yong W, Mugen X, Qin-chao H. Polarization DehazingAlogorithm Based on Atmosphere Background Suppression [J]. Computer Engineering, Vol. 35(4): 271–275(2009); (In China).

Yue D, Yan-jie W, Jingyu L. ET on. Improvement of enhancement algorithm for aerial image [J]. Laser & Infrared. Vol. 42(9): 1080–1085(2012); (In China).

Advances in Energy Science and Equipment Engineering II – Zhou, Patty & Chen (Eds)
© 2017 Taylor & Francis Group, London, ISBN 978-1-138-71798-5

Design of forest fire collection system based on OMAP-L138

Zhijian Yin & Zhaopan Wu

School of Communications and Electronics, Jiangxi Science and Technology Normal University, Nanchang, China

ABSTRACT: With the continuous development in the national economy of China, people pay more and more attention to the monitoring and early warning of forest fires, which is one of the major disasters endangering the forest. With the increase in the severity of forest fire identification problems, forest fire identification based on image has become the main research direction. Through processing and analysis of images, which are captured by the image sensors that are placed in the forest, we determine whether forest fire disaster or hidden dangers are happening presently. This paper takes the OMAP-L138 company's TI as the main processor and uses the special CMOS camera sensor 0V2640 to complete the system's image acquisition function. Besides, it also performs the system's various software and hardware tests. The function of forest image acquisition is achieved by DSP hardware platform.

1 INTRODUCTION

The forest is the main terrestrial ecosystem and also the treasure-trove of human resources. It can not only adjust the natural climate, but also produce the oxygen for human survival. Forest fires occur frequently in China. Many countries, including China, are aware of the serious harm of forest fires and have tried to take various measures to minimize the loss of forest fires and protect forests and living species from damage. Monitoring and early warning of forest fires is one of the effective measures. In particular, if the early stage of forest fire is complete, protection of forest trees and the circumjacent people's life and property is significance. In this paper, a real-time acquisition system of forest image for the early warning of forest fires is studied.

2 SYSTEM FRAMEWORK

The overall system design mainly includes image information acquisition, image signal processing, and real-time display of image. The image acquisition and processing system based on DSP can meet the requirement of portability, simple operation, fast processing, and low power consumption in the system design. The design of reasonable system architecture allows the chip to show a greater performance. Therefore, on the basis of meeting system design requirements, the system can reduce the complexity and improve the stability. The system design mainly includes three parts: fire image acquisition, fire image storage and processing,

and back-end image display. Its basic work flow is shown in Figure 1.

Digital image acquisition part is mainly composed of the CMOS image sensor and the high-speed AD converter. The sensor lens is used to gather the light reflected by objects on the image sensor, and the high-speed AD converter is used to convert the analog signal of the image signal into digital signal. Image processing is mainly performed by the high-performance DSP processor of TI. The collected digital signal is preprocessed (gray-level transformation) and stored in the external Flash memory. The image display part adopts VGA monitor to display the collected image signal and the transformed image signal. In this paper, the hardware design scheme of forest fire image acquisition, processing, and display system is based on OMAP-L138 processor design, as shown in Figure 2.

2.1 *Image acquisition part of the circuit design*

In the design of this system, OV2640 image acquisition chip is integrated together by the image sensor and the driving chip, so the module power supply,

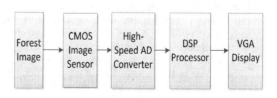

Figure 1. Basic work flow of the system.

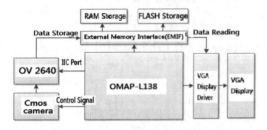

Figure 2. Hardware structure of the system.

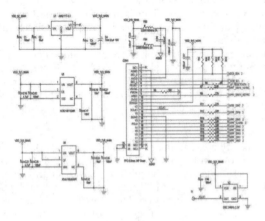

Figure 3. Driving circuit of OV2640 module.

the crystal part, and the signal output part are just designed in the design of the circuit. The specific hardware circuit diagram is shown in Figure 3.

Because the supply voltages in the kernel, I/O port, and the analog part were different, LM1117-3.3, XC6219B132MR, XC6219B282MR, three kinds of power management chip are used to produce the voltage of 3.3, 1.3, and 2.8V for the power supply module in the design. R1, D1, as the 3.3V power indication, are used to indicate whether the power supply module was normal. A π-type filter constituted by C33, C36, FB3, and FB4 is used to filter out the high-frequency interference signal of power supply part. The crystal part, with 24 MHz active crystal, provides a reference clock signal for the module.

OV2640 module uses the Serial Camera Control Bus (SCCB). The chip uses a three-wire structure bus mode to achieve the function control of the vast majority of the MiniVision series image chip. It is composed of SCCB_E, SIO_C, and SIO_D, and its bus structure is shown in Figure 4.

SCCB_E is the signal-enabled chip. When the signal level is low, it is valid. The start of transmission data is indicated in the falling edge of the pulse. The end of the data transmission is

indicated in the rising edge, and the enable signal is maintained at a low level in the process of data transmission. When it is high, the bus is idle. When SIO_C signal is high, it is valid, and when the bus is idle, it is pulled high. When the transmission is started, the SIO_C that is pulled low indicates the beginning of the data transmission. A high level in the transmission process indicates that the data are being transmitted. Therefore, SIO_D data changes can only occur when the SIO_C is low, a transmission time is defined as t_{CYC} and the minimum 10us. Its timing diagram is shown in Figure 5.

2.2 System reset circuit design

Reset circuit is a circuit device used to restore the circuit to the starting state. In practice, it is generally used to rerun the program to facilitate debugging and the observation of the running results. The design requirements of DSP reset circuit are: after the system power is on, it needs to provide a 0.2s reset pulse and there are strict requirements for the reset pulse width and setting time when the system is working. This makes the need to use a special reset chip to achieve the reset function of the system in the designing circuit. In this system design, the chip TPS3705 of TI is used to achieve the reset function of the whole system, mainly for the DSP processor system reset. The basic working principle of TPS3705 chip is as follows:

The power supply is switched on. When the power supply voltage VDD is higher than 1.1V, the RESET pins are in a state of disconnection. Thereafter, we monitor the power supply voltage VDD. As long as it is below the threshold voltage, we can make the RESET to reset (active low). Output delay of internal timer is returned to the control reset time to ensure proper system reset. The

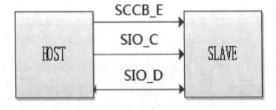

Figure 4. Bus structure of SCCB.

Figure 5. Control timing diagram of SCCB.

RESET output is high when the VDD is higher than the threshold voltage. When the power supply voltage is lower than the threshold voltage, the RESET once again enables the output to be low. Delay time Tdt = 200 mS, and it does not require internal capacitance. TPS3705 does not require external components for use. This reduces the failure rate of the entire system. The basic principle of the chip diagram is shown in Figure 6. It can work normally in the 2.5, 3.3, and 5V, and other kinds of different power supply voltage. The maximum operating current is only 50uA, and the operating temperature range is from −40 to 85°C. TPS3705 is widely used in microcontroller or microprocessor, industrial equipment, programmable control, automotive systems, intelligent instruments, wireless communication systems, and other fields.

C4 is the input decoupling capacitor, which is used to filter the high-frequency noise in the power supply and prevent noise of the power supply into other parts of the circuit through the power lines.

2.3 *VGA display driver design*

In the design of this system, the VGA display is driven by the special drive chip CS7123. CS7123 is an integrated chip, which has three high-speed AD converter. The chip contains three high-speed AD converter, 10-bit DA converter for video output, standard TTL input, and complementary output of high-impedance analog current source. It has three independent 10-bit AD input ports. It can work under not only a single power supply 5V, but also the single power 3.3V. It is mainly used in digital video system (1600 × 1200@100HZ), high-resolution color images, digital radiofrequency modulation, image processing, instrumentation, and video signal reproduction.

The basic work flow of the chip is: clock signal provided by the external, the analog RGB signal is converted by the three-way RGB digital signal, and then output to the display screen. Its basic operation sequence is shown in Figure 7. In practice, the chip can also be used as a three-way independent DA.

In the design of the system, the driving circuit diagram of CS7123 is shown in Figure 8. In this circuit, the data signal and the clock control signal are provided by the processor OMAPL-138.

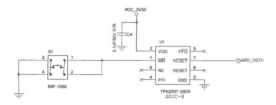

Figure 6. Reset circuit of TPS3705 chip.

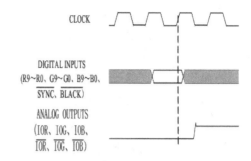

Figure 7. Basic work sequence of CS7123.

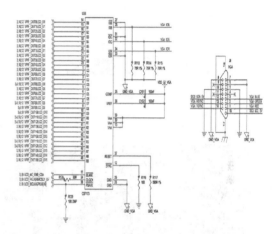

Figure 8. Drive schematic diagram of CS7123.

The development board, which provides VGA bus interface, is convenient for developers to use.

3 SYSTEM SOFTWARE IMPLEMENTATION

In the design, this paper takes low-power floating point digital signal processor OMAP-L138 of TI as the core. The running mechanism of parallel pipeline is used, the processor's running program is written in C language, and various performance indices of the system are realized, including system initialization, image acquisition, video encoding and decoding, image storage, image processing, and image output display. The flowchart of the whole system is shown in Figure 9.

4 SYSTEM TEST

In the comprehensive test of the system running result, we mainly test the real-time performance

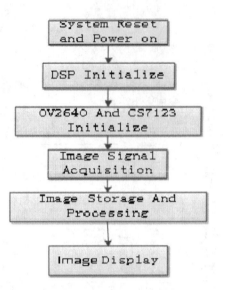

Figure 9. Software running flowchart.

Figure 10. VGA display test.

of the output image of the test system. Therefore, we can take photos in the laboratory as the real-time output and display the images captured by the camera in the VGA display. Test results are shown in Figure 10.

Through the test results, we know that the VGA display can show the digital images, which are captured by the camera in real time, and the clarity of the system output images meet the design requirements.

5 CONCLUSION

In this paper, we study an image acquisition system for forest fire based on OMAP-L138. The system can meet the real-time requirements of forest fire image acquisition and processing, and has important significance for the prevention of forest fires.

REFERENCES

Along Ren. Research on the underground image acquisition system based on DSP [D]. Xi'an Petroleum University. 2011.28–30.

Guoqing Chou. Research on flame recognition and detection technology based on image features [J]. Automation and Instrumentation. 2012[3].6–7.

Hongdi Li and Feiniu Yuan. Image smoke detection using Pyramid texture and edge features [J]. Chinese Journal of Image and Graphics. 2015[6].2–4.

Rule Huang. Study on forest fire smoke recognition method based on video image [D]. Beijing Forestry University. 2011.26–30.

Yuting Wang. Research and design of forest fire recognition system based on image features [D]. Northeast Forestry University. 2015.29–31.

Advances in Energy Science and Equipment Engineering II – Zhou, Patty & Chen (Eds)
© 2017 Taylor & Francis Group, London, ISBN 978-1-138-71798-5

Fuzzy c-means clustering algorithm based on density

Laiquan Liu
Department of Digital Design, Hainan College of Software Technology, Qionghai, Hainan, P.R. China

Yanrui Lei
Department of Network Engineering, Hainan College of Software Technology, Qionghai, Hainan, P.R. China

Wentian Ji
Department of Software Engineering, Hainan College of Software Technology, Qionghai, Hainan, P.R. China

Yan Chen
Department of Digital Design, Hainan College of Software Technology, Qionghai, Hainan, P.R. China

ABSTRACT: Along with the application of data mining, the clustering algorithms are gradually applied to the specific dataset analysis. The clustering algorithms are basically divided into fuzzy clustering algorithm, neural network clustering algorithm, clustering algorithm based on objective function optimization, hierarchical clustering algorithm, etc. Actually, clustering is an important research content in the fields of data analysis, knowledge discovery, intelligent decision, etc. As one of the most widely applied fuzzy clustering algorithms, Fuzzy C-means (FCM) clustering algorithm is different from such hard clustering algorithms as K-means clustering algorithm and hierarchical clustering algorithm. Specifically, the concepts of membership degree and fuzziness are introduced to broaden the application scope thereof, but the clustering effect is significantly influenced by the random selection of the initial cluster center point. In allusion to the problem regarding the random selection of the initial cluster center point, an FCM cluster center point selection method based on density is proposed in this paper to initialize the initial center point of the FCM clustering algorithm and improve the disadvantage of the original algorithm. The proposed algorithm is experimentally proved to be feasible.

1 INTRODUCTION

Along with the continuous development of the information technology, massive data are required to be stored, network information sources are exponentially expanded, and the volume and the variety of the information stored in databases are also increased. In fact, the increase of data diversity and complexity makes people have to find stronger data analysis method in order to better provide assistance to scientific researchers and decision makers. In order to solve above problems and extract the useful information hidden in massive data, more and more scholars devote to data mining research, thus promoting the continuous development of data mining theory and technology (Cabena, 1997).

Data mining, an important step of knowledge discovery, refers to the process of mining and extracting knowledge from massive data. As an emerging multidisciplinary application field, knowledge discovery and data mining are playing an important role in the decision support activities

of various industries. Data mining involves multi-disciplinary knowledge including database, statistics, neural network, machine learning, pattern recognition, etc. Potential knowledge can be extracted from databases through data mining, and the corresponding results can be used for information management, decision support, etc. Therefore, data mining is regarded as one of the most important academic forelands for database system and has great prospect in the information industry (Kaile, 2014). Main contents of data mining are as shown in Fig. 1.

In the data mining field, the pattern recognition is usually divided into supervisory pattern recognition and non-supervisory pattern recognition. In the supervisory pattern recognition, the specific cluster information of each cluster has been known before classification and certain algorithm is subsequently adopted for the classified dataset for machine learning in order to obtain the cluster characteristics of each cluster and finally obtain the correct classifier for the sample sets; then, the mentioned classifier is adopted to correctly classify the unknown samples.

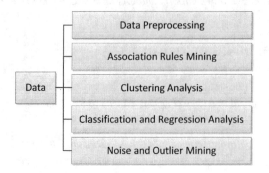

Figure 1. Main contents of data mining.

Obviously, the supervisory pattern recognition needs the prior knowledge of the classified datasets. However, the non-supervisory pattern recognition can form specific classification information during classification according to the sample element similarity or distance rather than need any prior knowledge of the sample sets.

Clustering analysis, also called as non-supervisory classification (Jain, 2010), is a pattern recognition method without any prior knowledge and a non-supervisory learning process. Namely, without the guidance of any obvious information, the potential similar patterns are discovered according to the sample object similarity in order to maximize the similarity in the same cluster and the difference in different clusters. Such clustering process usually includes the following three steps (Handl, 2005):

1. Data conversion: this step mainly includes variable selection, data standardization, distance function selection, etc.,
2. Clustering algorithm selection: this step aims at determining relevant parameters of the algorithm and clustering the given data;
3. Evaluation of clustering results: this step usually needs various clustering effectiveness verification methods.

In traditional clustering algorithms, the sample data are strictly classified into a certain cluster (Rui, 2005) 0 and described as "either one or the other" in a hard and clear way. However, not all objects can be clearly classified in actual environment, and most objects usually have fuzzy cluster characteristics, namely "double-sided" properties. For example, when describing an object, people usually adopt some fuzzy equivocal languages, e.g.: "the car has large noise" and "this person is very tall". Obviously, the hard classification method is inapplicable to the classification of these objects. Although it is very easy for us to understand such fuzziness, yet the computers cannot handle the fuzzy objects according to the previous clustering algorithms. The reason is that the previous computer technology application is based on classical mathematics for solving and handling the objects with accurate information rather the objects with strong fuzzy information. The generation of massive fuzzy information in daily life promotes experts to start researching the method for solving such problems.

In the fuzzy clustering algorithm, membership degree is adopted to represent the subordination of each cluster. Different from traditional hard classification, the membership degree value range is expanded to the whole continuous closed interval [0, 1] from traditional discrete space in the interval of {0, 1}, and the cluster membership degree sum of the samples is equal to 1. The fuzzy clustering results enable people to understand the relationship between the samples and each cluster and obtain the sample membership degree for each cluster. Such membership degree can not only reflect characteristic fuzziness, but also more objectively reflect the actual world. Therefore, the fuzzy clustering algorithm has become the popular method for clustering analysis and research (Spragins, 2005).

The FCM clustering algorithm aims at converting the clustering problem into objective function optimization to realize the intensive data clustering. The iterative strategy is adopted to solve the non-supervisory classification problem. Due to simple method, rapid calculation speed and easy promotion, the proposed algorithm is popular among various industries. However, relevant research shows that the constrained nonlinear programming function constructed by the mean square error approximation method in the FCM clustering algorithm is a non-convex function. If there are many local peaks in the sample data, the algorithm can be easily trapped in the local peaks and fail to obtain the global optimal solution (Sheng, 2008).

2 FCM CLUSTERING ALGORITHM

A dataset $X = (x_1, x_2, ..., x_n)$ including n samples is divided into c clusters (C_1, C_2, ..., C_c) during the clustering process to obtain the classification matrix U(X). U(X) is expressed as $U(X) = [\mu_{ij}]_{c \times n}$ (i = 1,...,c, j = 1,...,n), where u_{ij} is the membership degree of sample x_j corresponding to cluster C_i. In general conditions, the clustering results should meet the following requirements:

$$\begin{cases} U_{i=1}^c C_i = X \\ C_i \cap C_j = \varnothing & i, j = 1,...,c; i \neq j \\ C_i \neq \varnothing & i = 1,...,c \end{cases} \quad (1)$$

According to relevant mathematical theory, cluster C_i is determined by following formula:

$$\begin{cases} C_i = \left\{ X_j \left\| X_j - V_i \right\| \le \left\| X_j - V_p \right\|, X_j \in X \right\}, \\ p \ne i, p = 1, \dots, c \\ V_j = \frac{\sum_{x_j \in c_i} x_j}{|c_i|}, i = 1, \dots, c \end{cases} \quad (2)$$

Where $\| \ \|$ represents sample distance measurement, v_i is the cluster center of cluster C_i, and $|C_i|$ represents the number of the samples in cluster C_i. The sum of squared errors, usually adopted as the clustering criterion in the clustering process, is represented as follows:

$$\text{SSE} = \sum_{i=1}^{c} \sum_{x_j \in c_i} \left\| x_j - v_i \right\|^2 \quad (3)$$

For each sample point in the given dataset, the error refers to the distance to the nearest cluster center. Generally speaking, under the given cluster number, the purpose of the clustering process is to obtain the clustering classification with the minimum sum of squared errors.

For the classification matrix U(X) in the hard clustering classification, each sample only belongs to one cluster (Sledge, 2010). c hard classifications HCM of sample X can be represented as follows:

$$M_{\text{HCM}} = \begin{cases} U \mid \mu_{ij} \in \{0,1\} \forall i, j; \\ 0 < \sum_{j=1}^{n} \mu_{ij} < n \, \forall i; \\ \sum_{i=1}^{c} \mu_{ij} = 1 \forall j \end{cases} \quad (4)$$

However, it is sometimes unsuitable to include one sample only into one cluster. For example, a certain sample object may belong to multiple clusters and even to each cluster in c clusters. Through loosening the membership degree constraints in hard classification HCM, c probability classifications PCM and fuzzy classifications FCM can be represented as follows:

$$M_{\text{HCM}} = \begin{cases} U \mid \mu_{ij} \in \{0,1\} \forall i, j; 0 < \sum_{j=1}^{n} \\ \mu_{ij} < n \, \forall i; \max_{1 \le i \le c} \mu_{ij} > 0 \, \forall j \end{cases} \quad (5)$$

$$M_{\text{HCM}} = \begin{cases} U \mid \mu_{ij} \in \{0,1\} \forall i, j; 0 < \sum_{j=1}^{n} \\ \mu_{ij} < n \, \forall i; \max_{1 \le i \le c} \mu_{ij} > 0 \, \forall j \end{cases} \quad (6)$$

The relationship among hard classification, fuzzy classification and probability classification is as follows: the hard classification is a special fuzzy classification, and the fuzzy classification is a special probability classification.

Professor Zadeh proposed the fuzzy set concept for the first time in 1965 (Zadeh, 1965) 0, and

Ruspini subsequently proposed the fuzzy clustering model in 1969 (Ruspini, 1969). Based on the contents of the fuzzy set and the fuzzy clustering model, relevant scholars have proposed many different fuzzy clustering algorithms (Liu, 2007). Dunn gave the definition of the fuzzy C-means model in 1973 (Dunn, 1973), and Bezdek promoted the fuzzy C-means model and proposed the fuzzy C-means clustering in 1974 (Bezdek, 1974).

The fuzzy C-means algorithm is a data clustering method (Wei, 2008) based on objective function optimization. The cluster results can reflect the membership degree of each sample datum to the cluster center, wherein the membership degree is valued in the interval of [0,1]. The mean square error approximation theory is adopted therein to construct the constrained nonlinear programming function. In this way, the clustering problem is solved through the objective function, and the sum of the intra-cluster squared errors is a universal form of the clustering objective function.

The fuzzy C-means algorithm can be adopted for the automatic classification of the sample data. Specifically, standard function J is optimized to classify the samples into the corresponding clusters; then, the optimized standard function is adopted to determine the membership degree of the sample data relatively to the center point of the affiliated cluster. J represents the sum of the squared errors between the sample data and the center points of the affiliated cluster thereof.

As a classification algorithm, the final objective of the algorithm is to minimize the quadratic sum of the distances between the samples in various clusters and the cluster center.

The values of membership degree u_{ij} and cluster center V_i are expressed as follows:

$$u_{ij} = \frac{1}{\sum_{k=1}^{c} \left(\frac{d_{ij}}{d_{kj}} \right)^{\frac{2}{m-1}}} \quad (7)$$

$$v_i = \frac{\sum_{j=1}^{n} (u_{ij})^m x_j}{\sum_{j=1}^{n} (u_{ij})^m} \quad (8)$$

The fuzzy C-means clustering algorithm is a simple-iteration and dynamic-clustering process, and the specific flow chart is as shown in Fig. 2:

Each iteration in the execution process of the fuzzy C-means clustering algorithm is implemented towards the objective function reduction function. If the sample space includes multiple minimum points and the initial cluster center is randomly selected around a local minimum point, then the algorithm may converge to the local mini-

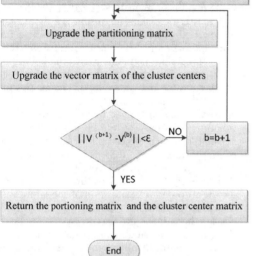

Figure 2. Flow charts of the fuzzy C-means clustering algorithm.

Table 1. Information of some pharmaceutical stocks in shanghai stock market.

No.	a	b	c	d
1	0.21	6.98	0.40	16.37
2	0.27	14.67	0.23	24.00
3	0.26	7.55	0.19	14.97
4	0.20	6.35	0.28	12.32
5	0.38	10.92	0.05	15.90
6	0.25	6.25	0.03	15.95
7	0.37	7.99	0.07	15.61
8	0.17	4.45	0.43	14.60
9	0.18	6.73	0.17	10.93
10	0.74	10.13	0.93	22.45

mum (Zhang, 2009). Moreover, the fuzzy C-means algorithm is sensitive to the initial data and can be easily trapped in the local minimum, so the global optimal solution cannot be obtained.

In order to solve the problem regarding the random selection of the initial cluster center in the fuzzy C-means clustering algorithm, a fuzzy C-means clustering algorithm based on density is proposed in this paper. Specifically, the correlation density of the sample data is calculated to optimize

the initial cluster center; then, the correlation density of the clustering data is regarded as the basis for determining the initial cluster center so as to improve the algorithm efficiency.

3 IMPROVEMENT OF FUZZY C-MEANS CLUSTERING ALGORITHM BASED ON DENSITY

The fuzzy C-means clustering algorithm based on density is abbreviated as NFCM, the NFCM algorithm aims at setting the distance for the selected initial cluster center and purposefully selecting the initial cluster center according to certain rules. Firstly, the density of each sample point is calculated and the cluster number is set as x; then, the first x sample points with large density are obtained and the distance among x sample points must be a certain preset distance; then, x points can be taken as the initial cluster center points, thus not only ensuring the data compactness in the same cluster, but also ensuring the separation degree of the initial cluster centers.

Step 1: set clustering number as c ($2 \leq c \leq n$), fuzzy weighted index number m as 2, threshold value ε for iteration termination as e, initial iteration value as b = 0, distance among the sample data as dis and maximum iteration as b_{max};

Step 2: calcualte the distance between each sample points and each other sample point in the clustering data to generate the distance matrix;

Step 3: take the sample point with maximum density as the nth (n = 1) center point of the initial cluster center;

Step 4: take the sample point with the second largest density as the candidate sample point;

Step 5: If the distance between the candidate sample point and the nth center point can meet the preset distance requirement, then take the candidate sample point as the n+1th ($n \leq c$) center point; or else, repeat Step 4 till finding c cluster center points.

Step 6: continue the iterative dynamic clustering process of the fuzzy C-means clustering algorithm to output the classification matrix and the cluster center matrix.

4 EXPERIMENT RESULTS

In this section, the numerical experiment is adopted to verify the algorithm effect. Tab. 1 shows the information of some pharmaceutical stocks in Shanghai stock market, sourced from literature [17]. Four numerical attributes are listed therein, namely: a: earnings per share, b: net assets income rate, c: cash flow per share, d: share price.

The author adopts NFCM to cluster the data in literature (Sun, 2003), and compares the clustering result with that in literature (Sun, 2003). The cluster center only has an insignificant deviation, as shown in Table 2.

50, 75, 100, 125 and 150 data are respectively randomly selected for testing, as shown in Figs. 3–6. The calculation data of the fuzzy C-means clustering algorithm in literature (Sun, 2003) is adopted for the clustering analysis efficiency comparison with the proposed NFCM algorithm. Fig. 7 is the operation time comparison between the two algorithms. According to Fig. 7, the fuzzy C-means clustering algorithm has certain operation time advantage under small data volume. However, along with the increase of the data volume, such advantage becomes less and less obvious, and the NFCM algorithm gradually presents the advantages thereof.

Table 2. Cluster center of each attribute calculated by NFCM.

Attribute	Cluster center		
	1	2	3
a	0.21	0.38	0.74
b	6.61	10.53	14.67
c	0.05	0.28	0.83
d	11.63	15.57	23.23

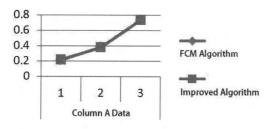

Figure 3. Comparison of column a cluster centers.

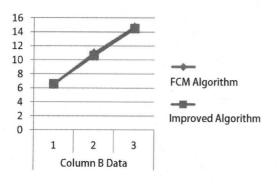

Figure 4. Comparison of Column B cluster centers.

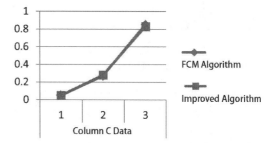

Figure 5. Comparison of Column C cluster centers.

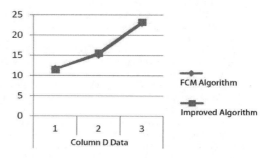

Figure 6. Comparison of Column D cluster centers.

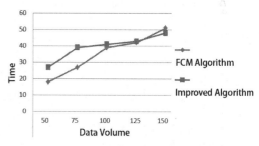

Figure 7. Comparison of execution time change along with data volume.

5 CONCLUSIONS

For optimizing the initial cluster center point of the fuzzy C-means clustering algorithm, the fuzzy C-means clustering algorithm based on density (NFCM) is proposed in this paper. NFCM algorithm is firstly adopted to generate the distance matrix of the clustering data, and then the density of the sample points and the preset distance are combined to select the initial cluster center point.

The efficiency of the NFCM algorithm is improved more or less on the basis of the fuzzy C-means clustering algorithm. In future, the fuzzy C-means clustering algorithm will be continuously researched and NFCM algorithm will be further improved to provide theoretical basis and further improve algorithm efficiency and application scope.

ACKNOWLEDGEMENT

This work is supported by Natural Science Foundation of Hainan Province of China (20156232).

REFERENCES

Bezdek J.C. Fuzzy Mathematics in pattern classification [D]. Ithaca, NY: Cornell University, 1974.

Cabena P. Discovering Data Mining From Concept to Implementation [J]. New Jersey: Prentice Hall Inc. 1997: 1–47.

Dunn J.C. A fuzzy relative of th ISODATA process and its use in detecting compact well-separated clusters [J]. Journal of Cybemetics, 1973, 3(3): 32–57.

Handl J. Knowlege j, Kell D.B. Computational cluster validation in post-genomic data analysis [J]. Bioinformatics, 2005, 21(15): 3201–3212.

Jain A.K, Data clustering: 50 years beyond k-means [J]. Pattern Recognition Letters, 2010, 31(8): 651–666.

Kaile Zhou. Theoretical and applied Research on Fuzzy C-means Clustering and Its Cluster Validation [D]. Hefei: Hefei University of Technology, 2014.

Leiming Studies on New Fuzzy Clustering Algorithms [D]. Tianjing: Tianjing University, 2006.

Liu X, Pedrycz W. The development of fuzzy decision trees in the framework of axiomatic fuzzy set logic [J]. Applied Soft Computing, 2007, 7(1): 325–342.

Rui Xu, Donald Wunsch. Survey of clustering algorithms [J]. IEEE Transaction on Neural Networks, 2005, 16(3): 645–678.

Ruspini E.H. A new approach to clustering [J]. Information and Control, 1969, 15(1): 22–32.

Sheng Li, Zhou Kaiqi, Deng Guannan. AnInitializaion Method for Fuzzy C-means Clustering Algorithm Based on Crid and Density[J].Computer Application and Software. 2008,3: 22–23.

Sledge I.J, Bezdek J.C, Havens T.C, Keller J.M. Relational generalizations of cluster validity indices [J]. IEEE Transactions on Fuzzy Systems, 2010, 18(4): 771–786.

Spragins J. Learning without a teacher [J]. IEEE Transaction of Information Theory, 2005, 23(6): 223–230.

Sun jianxun, Chen Mianyun, Zhang Shuhong. Mining Quantitative Association Rules with Fuzzy Method [J]. Computer Engineering and Applications, 2003(18): 190–192.

Wei Wang, Chun-heng Wang, Xia Cui, Ai Wang. Fussy C-Means Text Clustering with Supervised Feature Selection. International Conference on Fuzzy Systems and Knowledge Discovery. 2008.

Zadeh L.A. Fuzzy sets [J]. Information and Control, 1965, 8(3): 338–353.

Zhang Hui-zhe, Wang jian. Improved Fuzzy C means-Clustering Algorithm Based on Selecting Initial Clustering Centers [J].Computer Science, 2009, 36(6).

Sympathetic detonation analysis of the short-range double-explosion sequence in the fuze

Peng Liu, He Zhang & Shaojie Ma

School of Mechanical Engineering Nanjing University of Science and Technology, Nanjing, Jiangsu, China

ABSTRACT: With the rapid changes in the battlefield environment and situation in modern war, the requirements of some conventional ammunition have become increasingly higher; the demand for multi-explosion sequence is more prominent than that of single-explosion sequence. Detonation of the closed-explosion sequence is a very common problem because of the small ammunition space. The theory of the dual detonators at short range in the fuze and the attenuation of shock wave in the flameproof medium are analyzed. In this paper, asynchronous anisotropic detonation of short-range double detonators is performed using AUTODYN. The experimental result shows that plexiglass is a kind of good explosion-proof medium and the minimum edge distance without being sympathetic detonation is 8 mm.

1 INTRODUCTION

Fuze safety is one of the most important requirements of the fuze design. It requires that fuze is safe in several areas, such as production, service processing, filling, launch until the delay arming, and is armed under specified conditions. The design of fuze safety system depends on the special requirements of the weapon system, which is one of the core technologies in fuze design. The value of ammunition becomes higher and higher, so it requires the fuze safety system to improve the reliability of arming.

The output of single-path explosion sequence has been unable to meet the requirements of some special bombs. Therefore, the demand for multi-channel explosion is increasing. Besides, in the limitation of conventional ammunition volume, a fuze safety and arming mechanism is required to achieve the safety control of multichannel explosion sequence. In this paper, we studied in short-range two-way explosion sequence and asynchronous anisotropic output safety control technology of fuze to ensure that this explosion sequence is safe in peacetime and armed in the specified time and to ensure that the two explosive sequences work without interference. Jiao Qingjie from Beijing Institute of Technology conducted sympathetic detonation experiments for detonators in cardboard of different thicknesses (Zhao, 2006). Zhang Hua from Liaoning Technical University carried out a research on antiexplosive coupling in the wood of the detonators (Zhang, 2004), where the gap distance is large and not suitable for use in small structure fuze. Therefore, this paper

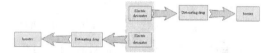

Figure 1. Dual anisotropic asynchronous initiation explosion sequence.

aims at the research on sympathetic detonation of short-range detonator, as shown in Figure 1.

2 SYMPATHETIC EXPLOSION STUDY ON TWO-WAY EXPLOSION SEQUENCE IN THE FUZE

To ensure that two-way detonating explosive trains do not interfere and two detonators have normal detonation reliability, the explosion-proof technology of two detonators is studied at short distance. This section focuses on the research of the sympathetic detonation of detonator and attenuation theory of the shock waves in the explosion-proof medium, analyzes the antiknock properties of several dense media, and designs the explosion-proof medium and dimension parameters of the detonator.

3 THE DISTANCE OF THE SYMPATHETIC EXPLOSION BETWEEN TWO DETONATORS

The structure of the dual detonators in the fuze is shown in Fig. 2. The active detonator explodes first, then the passive does. The passive detonator

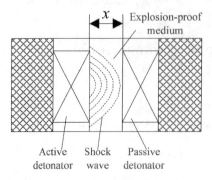

Figure 2. Dual-detonator structure model.

Table 1. Transmission distance of several detonator detonation.

Sample	Detonating drug	Quantity (mg)	Reinforcing cap	Height (mm)	Distance (mm)
A	NHN	300	Iron	21.7–22.2	40
B	LA	150	Aluminum	13.3–14.0	75
C	LA	300	Aluminum	15.0–15.4	70
D	KD	150	Aluminum	14.0–15.5	65
E	KD	300	Aluminum	16.3–17.5	55

will be detonated as long as the active detonator shockwave energy reaches a certain value.

The distance of several side sympathetic explosions, as shown in Table 1, is found in the literature (Zhu, 2004).

The transmission distance of several detonator detonations is too large and is not suitable for the fuze, so a new flameproof technology is sought.

4 NUMERICAL CALCULATION OF THE SHOCK WAVE IN THE FLAMEPROOF MEDIUM

The shock impedance of the detonator explosive is larger than the explosion shock impedance of the medium. The shockwave pressure formula of the interface between detonator and flameproof medium is (Zhang, 2000):

$$u_x = \frac{D}{\gamma+1}\left\{1+\frac{2\gamma}{\gamma-1}\left[1-\left(\frac{p_x}{p_H}\right)^{\frac{\gamma-1}{2\gamma}}\right]\right\} \quad (1)$$

where u_x is the wave particle velocity, γ is the polytrophic index, p_x is the shockwave pressure in the

Table 2. Values of a and b of several explosion-proof media.

Explosion-proof medium	a	b
Ly-12	5.48	1.30
45 steel	3.57	1.92
Plexiglass	2.87	1.88

interface, D is the detonation velocity, and p_H is the detonator detonation pressure. There is a linear relationship between the shockwave velocity in the medium D_m and u_x:

$$D_m = a + bu_x \quad (2)$$

where the values of a and b are associated with explosion-proof medium characteristics, as shown in Table 2 (Chen, 1991):

The initial shockwave pressure in the explosion-proof medium is:

$$p_x = \rho_m D_m u_x = \rho_m(a + bu_x)u_x \quad (3)$$

The shock wave in the medium conforms to the exponential attenuation (Wang, 1996; Zhao, 2011):

$$p = p_x e^{-\alpha x} \quad (4)$$

where α is the shockwave attenuation coefficient in the flameproof medium.

5 CRITICAL JUDGMENT OF SHOCK INITIATION

There are two criteria about detonation of critical shock: pressure criterion and energy criterion. The pressure criterion of detonator initiation is that the critical impact pressure, p, is greater than the critical pressure, p_c (Khasainor, 1997). The main sign of the explosion is a sudden sharp rise in pressure in the surrounding explosion medium. Sudden rise in pressure is a direct factor of damaging. The detonator explosive reaction time is within the order of magnitude for 10-6s -10-7s. Therefore, the object in contact with or close to it will be deformed, damaged, or moved. The detonation shock initiation of detonators has issues under short pressure pulses. In this paper, the critical energy criterion is adopted (Walker, 1969). The shockwave pressure of detonator is denoted by p:

$$p^2 t \geq E_c \quad (5)$$

where E_c is the critical initiation energy. When the main detonators shockwave energy is higher than the

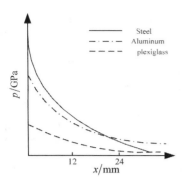

Figure 3. Relation curves of pressure and propagation distance in the medium.

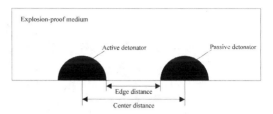

Figure 4. The simulation model.

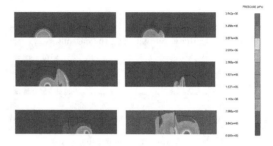

Figure 5. Pressure flow distribution of the passive detonator sympathetic detonation.

critical initiation energy, the launched detonators will be detonated. Figure 3 shows the experimental setup in ref. [8]: the decay curves from several media explosion shockwave pressure with propagation distance.

As Figure 3 shows, the same type of explosive detonators are at the same environment after detonation, the initial pressure of the shock wave formed in the steel material is the highest, but also the decay is the fastest, and the initial pressure in the plexiglass is minimum, with the slowest decay. And all aspects of the aluminum material are in the middle.

6 SIMULATION OF THE DUAL-DETONATOR SYMPATHETIC DETONATION

In this section, we use dynamics simulation software AUTODYN to carry out simulation on the detonation of closed dual detonator, which established a two-dimensional symmetry finite-element model and further explored the transmission attenuation of the main detonator shockwave and detonation condition of sent detonator. The detonators used aluminum cap and were simplified when modeling.

The model plane is shown in Fig. 4, using Lagrange algorithm. The grid and material coupled together and the grid varied with the movement of the material. The boundary movement and deformation of the model structure can be observed clearly.

Detonator charge was made up of primary explosive and main charge. First, the primary explosive was equivalent to main charge, and then it was replaced by reduced columnar charge TNT equivalent, regardless of the factor of primary explosive converting main charge to 0.6, the TNT equivalent coefficient of main charge was 1.3, the diameter of detonator was 5 mm, and the height of TNT grain was 10 mm. The JWL state equation of TNT is shown below (Li, 2009):

$$P = A(1 - \frac{\omega}{R_1 V})e^{-R_1 V} + B(1 - \frac{\omega}{R_2 V})e^{-R_2 V} + \frac{\omega}{V}E_0 \quad (6)$$

where P is the pressure of detonation products; V is the hematocrit of unit volume; E_0 is the initial internal energy; R_1, R_2, and ω are constants: $R_1 = 4.15$, $R_2 = 0.95$, $\omega = 0.3$; A and B are material parameters, whose values are 3.712×10^{11} and 3.23×10^9 Pa, respectively, in the model; and the initial internal energy $E_0 = 4.29$MJ/kg.

In Figure 5, the center of double detonator was 12 mm away from another, the diameter of detonator was 5 mm, the edge distance was 7 mm, and the pressure flow field distribution used steel as proof media. Main detonators used central initiation, in which the whole process of the shockwave delivery of main detonators and the detonation of sent detonators can be seen, and the detonation way was eccentric detonation. Therefore, using steel as explosion media required maximum proof distance.

While organic glass is used as explosion-proof medium, detonation pressure flow distribution for 12 mm edge distance of primary detonator is shown in Figure 6. In the figure, the moment of explosion is 3.1s.

It can be seen that in the attenuation process of shock wave, the detonated detonator is under very small impact pressure, not to be sympathetic detonated. When explosion-proof medium is replaced by aluminum or steel, the detonator exhibited sympathetic detonation, as shown in Figure 3. The

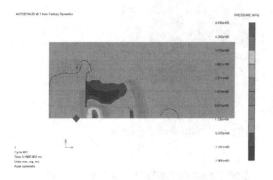

Figure 6. Pressure flow distribution of the passive detonator sympathetic detonation in the medium.

Table 3. Result of close-dual-detonator-induced detonation experiences.

No.	Edge distance	Medium	Results	Remarks
1	5 mm	Aluminum	Passive detonator is detonated	——
2	5 mm	Plexiglass	Passive detonator is detonated	——
3	8 mm	Aluminum	Passive detonator is detonated	——
4	8 mm	Plexiglass	Passive detonator is intact	OK
5	10 mm	Aluminum	Passive detonator is intact	OK
6	10 mm	Plexiglass	Passive detonator is intact	OK

authors of ref. [6] adopted pressure as the critical initiation criterion and judged whether detonators in dense medium are affected by the impact wave into sympathetic detonated. The experimental results show that when the peak pressure of shock wave of detonator is more than 129GPa, sympathetic detonation occurs, whereas the simulation results show that when the peak pressure is above 1.5GPa, the detonator shows sympathetic detonation, whose simulation reliability is validated.

7 EXPERIMENTS OF THE DUAL-DETONATOR SYMPATHETIC DETONATION

Dual-channel explosion sequence sympathetic detonation experiments with several medium were conducted; the purpose of these experiments is to test the flameproof reliability of dual-channel explosion sequence under the flameproof state. Because of the high cost of microsecond detonator and limited funds for the experiment, only six experiments were performed. Considering the poor flameproof

performance of steel, the experiment was conducted only on aluminum and organic glass. Experimental select edge distances were 5, 8, and 10 mm. The experiment results are shown in Table 3.

When the edge distance is larger than 8 mm, plexiglass can prevent the passive detonator being detonated, but the reliability of the flameproof medium needs more experiments to be validated. According to the reliability requirement, we can determine the induced detonation probability (such as less than 0.01%), so as to determine the safe distance X0.0001 of the dual detonators.

8 SUMMARY

The main reason for making the passive detonator sympathetic detonation is the shock wave, and it is very effective to regard the attenuation material as the explosion-proof medium. The parameter cannot be the same as the fact, and the model is simplified. Thus, the simulation and experimental results are not completely consistent. The plexiglass is a fine explosion-proof medium through the simulation and experiments; it can set the minimum edge distance to 8 mm, without sympathetic detonation.

REFERENCES

Chen Xirong, Wang Ke. Characteristcs of Attenuation of Shock Waves in Barriers of Different Materials [J]. Acta Armamentarii, 1991,(2): 75–80.

Khasainor, B.A., et al. On the Effect of Grain Size on Shock Sensitivity of Heterogeneous High Explosives. Shock wave, 1997(7): 89–105.

Li Shunbo, Dong Zhaoxing. Numerical Simulation for Spread Decay of Blasting Shock Wave in Different media [J]. Journal of Vibration and Shock, 2009, 28(7): 115–119.

Walker F. E, Wasley R. J. Critical Energy for the Shock Initiation of Heterogeneous Explosives [J]. Explosive Stoffe, 1969, 17(1): 9–13.

Wang Haifu, Feng Shunshan. An Approximate Theoretical Model for the Attenuation of Shock Pressure in Solid Materials [J]. Acta Armamentarii, 1996, 17(1): 79–81.

Zhang Hua, Guo Biaoyi. Research on Anti-explosive Coupling Performance of Detonator in Wood [J]. Express Information of Mining Industry, 2004, (12): 26–27.

Zhang Junxiu. Explosion and its Application Technology [M]. Beijing: Beijing National Defence Industry Press, 2000.

Zhao Haixia, Xu Xinchun. Attenuation Model of Shock Wave in Different Materials Gap [J]. Chinese Journal of Explosives and Propellants, 2011, 34(6): 84–87.

Zhao Yaohui, Jiao Qingjie. Characteristics of Detonator Sympathetic Explosion in Densified Medium [J]. Chinese Journal of Energetic Materials, 2006, 14(3): 224–2.

Zhu Shunguan, Mu Jingyan. Studies on Characteristics of Side Sympathetic Explosion of Detonators [J]. Initiators and Pyrotechnics, 2004, (2): 33–35.

Advances in Energy Science and Equipment Engineering II – Zhou, Patty & Chen (Eds)
© 2017 Taylor & Francis Group, London, ISBN 978-1-138-71798-5

A new real-time calculating method for TDOA and FDOA passive location

Yiming Zhou
Research Institute of CETC, Jiaxing Zhejiang, China

Jiawei Zhu
Science and Technology on Communication Information Security Control Laboratory, Jiaxing Zhejiang, China

ABSTRACT: By researching on the Differential Evolution (DE) algorithm and passive localization model of two-station Time Difference of Arrival (TDOA) and Frequency Difference of Arrival (FDOA), a new calculating method based on DE algorithm is proposed for real-time locating. High precision position estimating employing time cumulate TDOA and FDOA values is also solved by this method. The cases using both Instantaneous and multiple measurements are simulated, which show the more excellent performance on real-time effect and location precision comparing with Unscented Kalman Filter (UKF) algorithm. The location precision almost achieves the Cramer-Rao Low Bound (CRLB). Due to the simple principle and easy realization, this method is suitable for project application.

1 INTRODUCTION

The problem of passive source location has been of considerable interest for many years. It has found wide applications in many areas including radar, sonar, and wireless communications (Li, 2003; Xie, 2008). For a stationary emitter, one common technique is to measure the TDOAs of the source signal to a number of spatially separated receivers. Each TDOA defines a hyperbola in which the emitter must lie. The intersection of the hyperbolae gives the source location estimate. When the velocity between source and receivers exists, FDOA measurements can be used in addition to TDOAs to decrease the number of receivers. Therefore, two moving stations with TDOA and FDOA measurement systems are enough to locate a fixed target. This method has been proved its properties of high precision, real-timing and acceptable system complexity (Schmidt, 1980; Lu, 2006; Zhao, 2012).

Many research and references focusing on such systems were found for the moment, but only several people made mention of the calculating method. Determining the source position from TDOA and FDOA measurements obtained at a single time instant is not a trivial task. This is because the source location is nonlinearly related to TDOA and FDOA measurements. Lu only discussed the location precision in his paper (Schmidt, 1980). An analytic method (Lu, 2006) and a closed-form solution (Zhao, 2012) were proposed to solve the problem of locating the fixed emitter using TDOA and FDOA information, which are too complicated to be applied.

The unscented Kalman filter is a viable technique source location. The UKF requires TDOA and FDOA measurements at different times, and an initial transient period is needed before it produces a good source location estimate (Zhan, 2007). The UKF has no optimality properties, and its performance depends on the accuracy of the linearization in the measurement equation. Furthermore, a good initial guess close to the true solution is often required to start the estimation of a target.

We will present in this paper a new attractive solution to the stationary source position using TDOA and FDOA measurements from two moving receivers. The proposed solution does not require initial solution guess and attains the CRLB when the TDOA and FDOA estimating precision are moderate. Higher precision position estimating employing time cumulate TDOA and FDOA values is also solved by this method. The simulation examples using once and multiple TDOA and FDOA measurements show the more excellent performance on real-time effect and location precision comparing with UKF algorithm.

2 LOCATION SCENARIO

We shall consider the general case of stationary source localization using TDOA and FDOA measurements by two moving receivers.

Real-world collection systems often employ a pair of separate collectors. The signals received by the individual collectors are from the same transmitter, but shifts in time and frequency are inherent due to the different paths travelled by the two signals. In these configurations, the two received signals can be processed to determine the TDOA and FDOA between the two collectors. With exact knowledge of the collectors' positions, successive TDOA and FDOA measurements can be plotted to determine the location of the associated emitter.

Cartesian coordinates simplify the model by assuming that the emitter is fixed and collectors are moving on or around a flat earth. Obviously, real-world systems are three-dimensional, but Figure 1 is in two dimensions solely for ease of depiction.

In Figure 1, $A1(x_1, y_1)$, $A2(x_2, y_2)$, and $T(x, y)$ represent the two collectors and the emitter, all located at the coordinate positions shown. The symbols r_1 and r_2 are the relative position vectors between each of the collectors and the emitter, while v_1 and v_2 are the respective velocity vectors. It is important to note that Figure 1 represents instantaneous "snapshot" of a generic system's geometry. Since the collectors are moving at their respective velocities, the geometry changes with each passing instant of time. This is precisely why the TDOAs and FDOAs in a system are time-varying in nature.

The TDOA between two signals is simply the difference in time that it takes two signals to travel down their respective paths from the emitter to the associated receivers. For the geometry shown in Figure 1, the TDOA between A1 and A2 is therefore:

$$\tau = (r_1 - r_2) / c \tag{1}$$

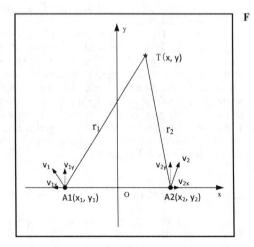

Figure 1. 2-D Emitter-Collector Geometry.

where τ represents the TDOA, c is the speed of light. The vectors r_1 and r_2 are determined simply by calculating the difference between their x and y coordinates and those of the emitter,

$$r_i = \sqrt{(x_i - x)^2 + (y_i - y)^2}, \, i = 1, 2.$$

The FDOA between two signals is simply the difference between their two Doppler shifts. From Figure 1, the Doppler shift between one of the collectors and the emitter can be defined as

$$f_d = \frac{1}{\lambda} \cdot \left(\frac{(x_1 - x) \cdot v_{1x} + (y_1 - y) \cdot v_{1y}}{r_1} - \frac{(x_2 - x) \cdot v_{2x} + (y_2 - y) \cdot v_{2y}}{r_2} \right) \tag{2}$$

where f_d represents the FDOA, f_0 is the signal's constant wavelength.

Based on equation (1) and (2), the emitter's position x and y can be acquired theoretically. Because of the measurement error, the result will be an estimated value. Next chapter will give the new method to estimate the source position.

3 PROPOSED SOLUTION

3.1 Differential evolution algorithm

DE is a global optimization technique, whose kernel idea is to produce the trial vectors according to the manipulation of the target and difference vectors (Price, 1996; Storn, 1997). If the trail vector yields a better fitness than a predetermined population member, the new trail vector will be adopted into the vector base. DE now has been applied in many areas (Liu, 2014; Secmen, 2013). Next, discuss the commonly used operators and parameters in DE.

The classic differential evolution search begins with a randomly initiated population of n_p N-dimensional parameter vector. Each variable $x_{i,gn} = (x_1, x_2, ..., x_{np})$ is a solution to the optimization problem, where $i = 1, 2, ..., n_p$ is the index of the variable in the population and $gn = 0, 1, ..., g_{max}$ is the subsequent generations. For each individual $x_{i,gn}$ in the population, a mutant vector is $y_{i,gn}$ produced according to the following formula:

$$y_{i,gn}^j = \begin{cases} x_{a,gn}^j + q(x_{b,gn}^j - x_{c,gn}^j) & rand(j) \leq cr \text{ or } j = randn(i) \\ x_{i,gn}^j & otherwise \end{cases} \tag{3}$$

where a, b, c are randomly chosen from $[1, n_p]$, $1 \leq i \neq a \neq b \neq c \leq n_p$, and $rand(j)$ is a uniformly distributed random number, $randn(i) \in [1, 2, ..., g_{max}]$. The element of mutant vector $y_{i,gn}$ is generated by the differential mutation, whenever a randomly generated number between [0, 1] is less than or equal to the crossover rate ($cr \in [0, 1]$) value. Otherwise, it is equal to the corresponding element of the individual $x_{i,gn}$. q is the mutation intensity for the vector differences and usually a positive constant factor $\in [0, 2]$. For relatively small values of q (e.g. 0.5) mutant vectors become less deviated from their generators (i.e. last generation).

For a minimization optimization, the selections are described as

$$x_{i,gn+1} = \begin{cases} y_{i,gn} & f\left(y_{i,gn}\right) < f\left(x_{i,gn}\right) \\ x_{i,gn} & otherwise \end{cases} \quad (4)$$

where $f(\bullet)$ is the objective function to be minimized. It is to say that if the new vector $y_{i,gn}$ produced by differential mutation and crossover operations yields a lower value of the objective function, it would replace the corresponding individual $x_{i,gn}$ in the next generation.

From above descriptions, if the vectors in the population are similar, the vector difference tends to be zero, which may lead to the poor population diversity and the premature convergence. This case may occur at the last generations.

3.2 Algorithm application

The kernel idea of our solution is to search for a target based on DE method ensuring that the FDOA and TDOA of two signals between the two collectors and the emitter is most close to the measurements.

The objective function to be optimized can be defined as follow.

$$f = \tilde{f}_d + \tilde{\tau} \quad (5)$$

where $\tilde{f}_d$ and $\tilde{\tau}$ are the normalization values of FDOA and TDOA estimation error.
$\hat{f}_d = mean\left[\left(\hat{f}_{d1} - f_{d1}\right)^2, \left(\hat{f}_{d2} - f_{d2}\right)^2, \cdots, \left(\hat{f}_{dn} - f_{dn}\right)^2\right] / \sigma_{f_d}^2$,
$\tilde{\tau} = mean\left[\left(\hat{\tau}_1 - \tau_1\right)^2, \left(\hat{\tau}_2 - \tau_2\right)^2, \cdots, \left(\hat{\tau}_n - \tau_n\right)^2\right] / \sigma_\tau^2$. $\hat{f}_{di}$ and $\hat{\tau}_i$ are FDOA and TDOA estimations calculated by an individual from DE population between the two receivers at different positions, while the f_{di} and τ_i represent the measurements from real world respectively. σ_{f_d} and σ_τ are the estimation error of FDOA and TDOA. n is the cumulative numbers until the current location operation. For example, n can be the latest 5 or 10 times, but at least $n = 1$.

The location algorithm based on DE can be summarized as following operations.
• Initialization

The population amount m, generation number g, q and cr values, should be set. An expected area that the target may exist also needs to be limited. As an example of figure 1, coordinate X and Y of this area can belong to $[-300,300]$ km and $[5,300]$ km separately. Then population of the initial targets is

$$X(t) = \left[X_1(t), X_2(t), \cdots, X_M(t)\right] \quad (6)$$

where, $X_i(t) = [x_{ti}, y_{ti}]$, represents the ith individual coordinate in tth generation. It's a 2-dimensional vector.
• Differential mutation and crossover

For each initial target in the population, The mutant coordinate vectors are generated first in this step.

$$v_i(t+1) = X_{r1}(t) + q \times [X_{r2}(t) - X_{r3}(t)], \ i \neq r1 \neq r2 \neq r3 \quad (7)$$

where $r1$, $r2$ and $r3$ are integers $\in (1,2,...,m)$ selected randomly and mutually different and also different from i.

For each x and y in $v_i(t+1)$, a crossover operator is applied to give birth to offsprings. The coordinate x and y are calculated by following formula separately.

$$u_{ij}(t+1) = \begin{cases} v_{ij}(t+1) & if \ rand(0,1) \leq cr \ or \ j = j_{rand} \\ x_{ij}(t) & otherwise \end{cases} \quad (8)$$

where $u_{ij}(t+1)$, $v_{ij}(t+1)$, $x_{ij}(t)$ are the x and y values in crossover vector, mutant vector and initial target vector respectively. Here, $j = 1,2 \in j_{rand}$ is integer $\in (1,2)$. Such operation make sure that at least one value in $u_{ij}(t+1)$ comes from $v_i(t+1)$.
• Selection

In the selection stage offspring compete with their primary parents; better parameter vectors for the next generation are selected. If an offspring vector gives a lower cost function, then it is selected in place of its primary parent for the next generation, otherwise the primary parent remains in the next generation.

$$X_i(t+1) = \begin{cases} U_i(t+1) & if \ f\left(U_i(t+1)\right) \leq f\left(X_i(t)\right) \\ X_i(t) & otherwise \end{cases} \quad (9)$$

where, $U_i(t+1) = [u_{i1}(t+1) \ u_{i2}(t+1)]$, f is defined in formula (5).

Repeat the above three steps till the last generation, the final estimation of the target can be chosen in the population who gives a least cost according to objective function.

4 SIMULATIONS

We shall consider the general case of stationary source localization using TDOA and FDOA measurements.

The two-dimensional (2-D) location geometry is shown in Fig. 1, where the two planes at $[-25000, 0]$ and $[250000, 0]$ meter both with velocity $[200, 0]$ m/s are used to locate the emitting source at position [50000, 200000] through TDOA and FDOA measurements. The receiver positions and velocities are perturbed as $\sigma_p = 10$m, $\sigma_v = 5$m/s. The receiver location vector and velocity vector are Gaussian distributed, and they are independent of each other. the noise standard deviations relative to the true values of the TDOA σ_τ^2 and FDOA and σ_{fd}^2 measurements are $(30 \times 10^{-9})^2$ and 1 respectively. The frequence of the emitter is 300MHz. The algorithmic parameters listed in 3.2 are set as follow: $m = 100$, $q = 0.4$, $cr = 0.8$, $g = 50$.

Simulations on instantaneous location and cumulate time location are both evaluated as below. The simulations are done using Matlab program.

4.1 Instantaneous location

The proposed algorithm can be easily applied for real time location estimation using the current measurements. The following simulations in this section just focus on the beginning of location. Figure 2 depicts the population difference in one of these Monte Carlo simulations. The populations at the last generation are close to the real target, and their average value can be calculated to denote the estimation. Figure 3 shows these averages according to the scene above by 100 times of Monte Carlo simulations. The CEP (circular error probable, 50%) estimation precision of single time instant of location is almost 6% of distance.

Given the different target under the same scene, we can obtain the following result as Table 1.

4.2 Cumulate time location

The proposed method can increase the precision by serial observation on stationary target. For the scenario in chapter 4.1, if the time varying TDOA and FDOA measurements are acquired per second

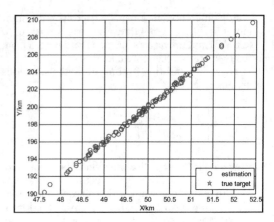

Figure 3. Results of 100-time Monte Carlo simulations.

Table 1. Simulating results for different targets.

Target (km)	Estimation	Error (R%)
[100,200]	[91.2,182.2]	8.9
[150,150]	[156.9,157.1]	4.7
[−80,180]	[−85.1,186.8]	4.3
[0,220]	[5.2,217.6]	2.6

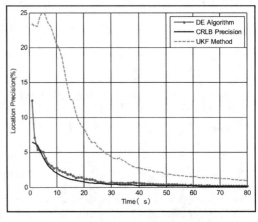

Figure 4. Comparison of UKF method and CRLB.

and each locating adopts all the measurements cumulated from beginning to the current time. Each simulation run simulates 80 seconds and repeats over 100 runs.

The location accuracy comparisons between DE algorithm, UKF method and CRLB precision are given in Figure 4. Comparing with the UKF, the DE converges faster and shows better precision. It is evident that reduced location errors have been obtained by the DE method, which shows a better

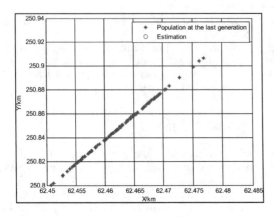

Figure 2. Population at the last generation.

ability to quickly converge to the optimal solution. Reaching the CRLB also corroborates the perfect performance on Real-time TDOA and FDOA Passive Localization.

5 CONCLUSION

This paper presents a stationary emitter geo-location method based on DE algorithm using the time varying TDOA and FDOA of a signal received at two moving receivers. It does not have the initialization and local convergence problem as in the conventional linear iterative method. The synthesis results show that the DE converges faster and the estimated accuracy of the source position is shown to achieve the CRLB. Simulations are included to examine the algorithm's performance and compare it with the UKF method.

REFERENCES

Li, Z.H., D.W. Feng, Z.K. Sun, et al, Passive Location and Tracking Algorithm for Air-emitters and Its Observability Analysis. Journal of Astronautics 24(5), 473–477. (2003).

Liu, B., H. Aliakbarian, Z. Ma, G.A.E. Vandenbosch. An Efficient Method for Antenna Design Optimization Based on Evolutionary Computation and Machine Learning Techniques. IEEE Transactions on Antennas and Propagation, 62(1), 7–18. (2014).

Lu, A.N. Two Aircraft TDOA/DD Passive Localization. Journal of UEST of China 35(1), 17–20. (2006).

Price, K. Differential Evolution A Fast and Simple Numerical Optimizer. Biennial Conference of the North American Fuzzy Information Processing Society 8, 524–527. (1996).

Schmidt, R.O. An Algorithm for Two-receiver TDOA/ FDOA Emitter Location. Tech. Memo. 15, 1229. (1980).

Secmen, M., M.F. Tasgetiren. Ensemble of differential evolution algorithms for electromagnetic target recognition problemIET Radar Sonar Navig., 7(7), 78–788. (2013).

Storn, R., K. Price, Differential Evolution - A Simple and Efficient Heuristic for Global Optimization over Continuous Spaces. Journal of Global Optimization, 11, 341–359. (1997).

Xie, K., Y.G. Chen, X.H. Li, Y. Shen, Passive Location for Netted Radar on Distributed Jammers. Acta Electronica sinica 36(6), 1164–1168. (2008).

Zhan, R., J. Wan, Iterated Unsecented Kalman Filter for Passive Target Tracking. IEEE Trans. Aerosp. Electron. Syst., 43(3), 1155–1163. (2007).

Zhao, K., D.N. Qi, A Tracking TDOA/FDOA Joint Location Algorithm Based on UKF. Computer Technology and Development 22(5), 127–129. (2012).

Advances in Energy Science and Equipment Engineering II – Zhou, Patty & Chen (Eds)
© 2017 Taylor & Francis Group, London, ISBN 978-1-138-71798-5

Video target tracking in clutter environment based on relevant particle filter

De-jing Jiang & Tao Sun
Xuzhou Institute of Technology, Xuzhou, China

ABSTRACT: In order to realize robust video object tracking in the clutter environment, particle filter method, in cases of correlated system noise and observation noise, was studied. First, video target movement behavior in the clutter environment was modeled by introducing Cauchy–Gaussian mixture noise models on the basis of establishing the common nonlinear system model. On the basis of the above established motion models, the proposal distribution function nature of noise correlation was analyzed in detail. Then, the mathematical expression of optimal proposal distribution function involved in non-Gaussian noise was deduced by the standards of minimum variance weighting conditions. Finally, the detailed implementation steps of the above algorithm were shown, and this algorithm was applied to the video target tracking system in clutter environment. The experimental results indicated that the proposed nonlinear filtering algorithm was effective, and the algorithm had a good stability, especially for extreme cases of large-area occlusion and sudden night light change. In addition, compared with standard particle filter algorithm, the overall tracking accuracy of proposed algorithm was improved by nearly 40%.

1 INTRODUCTION

Video target tracking technology is a fundamental problem in the fields of artificial intelligence, video surveillance, intelligent transportation, and so on. Video signals is so seriously interfered that the aliasing of surveillance target is easily caused by some cluttered environments such as multitarget occlusion, disappearance and reappearance of target, and sudden night light change. It brings great challenge for traditional methods of tracking and detection with the arising of the above situation (Morris, 2010; Li, 2012).

In recent years, with the development of computer and nonlinear filter technology, many researchers have adopted nonlinear filtering technology to study video target tracking, and gained much valuable results. Sun et al. (Sun, 2007) proposed a similar Kalman filter method. Because linear degree of the system was artificially set, practical application range of the method was restricted. Wu et al. (Wu, 2014) realized robust tracking of video object with occlusion by presenting a Relative Discriminative Histogram of Oriented Gradients (RDHOG)-based particle filter approach. The prerequisite condition of the approach been used in many fields assumed that the process noise and observation noise of the system were independent of each other, so it was suitable only for some slight occlusion circumstances.

In reality, large occlusion of fast-moving vehicle exists. To solve this problem, in this paper, we explored the PF method of noise-related cases and applied this method to several different video target tracking problems under noise environment based on the complex background of video target tracking problem.

2 TRADITIONAL PARTICLE FILTER ALGORITHM

2.1 Common model of the system

The common model of nonlinear particle filter is as follows (Wang, 2010):

$$\begin{cases} x_{k+1} = f_k(x_k, u_k) + \Gamma_k w_k \\ y_k = h_k(x_k) + v_k \end{cases}. \tag{1}$$

where x_k and y_k are the status value and observed value of system in k moment, respectively; $f(\cdot)$ and $h(\cdot)$ are dynamic function of system model and observation model, respectively, which together with x_0 decide dynamic model of the whole filter system; u_k is used to control input vector; w_k and v_k are process noise and measured noise of system, respectively; and Γ_k is the input matrix of process noise.

In this paper, the filter algorithm of nonlinear system with correlated noise is studied.

2.2 Conventional particle filter algorithm

Conventional Particle Filter (CPF) algorithm provides an approximation approach for random models. The main idea of CPF algorithm is to estimate posterior probability density of a system by a set of particle collection strategies with given weight (Trivedi, 2010; Li, 2012; Sun, 2007; Wu, 2014).

As described in formula 1, CPF can be summarized as two steps: forecast and update. $\left\{x_{0:k}^i, \omega_k^i\right\}_{i=1}^N$ is expressed as a set of sample particle collection of posterior probability density of system $p(x_{0:k} | Y_k)$. Among them, $\left\{x_{0:k}^i\right\}_{i=1}^N$ is a set of sample particle collection with given corresponding weight, and $\left\{\omega_k^i\right\}_{i=1}^N$ is the corresponding weight that satisfies the all-state set of system with $x_{0:k} = \{x_0, x_1, ..., x_k\}$ from 0 to k moment for $\sum_{i=1}^N \omega_k^i = 1$. On the basis of measuring messages of $Y_k = \{y_1, y_2, ..., y_k\}$ and the CPF's idea, formula 2 can be derived as:

$$p(x_{0:k} | Y_k) \approx \sum_{i=1}^N \omega_k^i \delta(x_{0:k} - x_{0:k}^i) \tag{2}$$

where

$$\omega_k^i \propto \omega_{k-1}^i \frac{p(y_k | x_k^i) p(x_k^i | x_{k-1}^i)}{q(x_k^i | x_{k-1}^i, z_k)} \tag{3}$$

$q(x_k | \cdot)$ is the proposal distribution function, also known as the important probability density function. Generally, formula 4 can be used as prior distribution of proposal distribution function:

$$q(x_k^i | x_{k-1}^i, z_k) = p(x_k^i | x_{k-1}^i) \tag{4}$$

Substituting formula 4 into formula 3, formula 5 can be obtained:

$$\omega_k^i \propto \omega_{k-1}^i p(y_k | x_k^i) \tag{5}$$

where the case is assumed that process noise v_k and measured noise e_k are independent of one another for white Gauss noise, and the condition is satisfied, that is, $q_k = r_k = 0$.

3 NEW ALGORITHM OF ASSOCIATED NOISE PARTICLE FILTER

3.1 Analysis of relevant information for noise

Measured noise is regarded as additive noise, and formula 1 can be redefined as formula 6:

$$\begin{cases} x_{k+1} = f(x_k, v_k) \\ y_l = h(x_l) + e_l \end{cases} \tag{6}$$

where joint posterior density function of measured value and process state can be regarded as:

$$p(X_k | Y_k) \propto p(y_k | X_k, Y_{k-1}) \\ \times p(x_k | X_{k-1}, Y_{k-1}) p(X_{k-1} | Y_{k-1}) \tag{7}$$

Correlation of noise can be expressed as Figure 1, based on ref. [6].

As shown in Figure 1, the correlations mainly focus on forward–backward correlation of time. The main purpose of considering the correlation is to seek right decomposition form for joint posterior density function of noise $p(v_i, e_j)$. For noise independence, the following formulas can be given:

$$p(v_i, e_j) = p(v_i) p(e_j) \tag{8}$$

$$p(x_k | X_{k-1}, Y_{k-1}) = p(x_k | x_{k-1}) \tag{9}$$

$$p(y_k | X_k, Y_{k-1}) = p(y_k | x_k) \tag{10}$$

Noise vector sequence of a moment $(v_{k-1}, e_{k-1})^T$ can be assumed as independent of each other, and the following formulas can be given based on Figure 1:

$$p(x_k | X_{k-1}, Y_{k-1}) = p(x_k | x_{k-1}, y_{k-1}) \tag{11}$$

$$p(y_k | X_k, Y_{k-1}) = p(y_k | x_k) \tag{12}$$

Therefore, joint density function of measured value and process state can be decomposed as shown in formula 13:

$$p(v_{k-1}, e_{k-1}) = p(v_{k-1} | e_{k-1}) p(e_{k-1}) \tag{13}$$

3.2 Modeling of non-Gaussian noise

To solve the problems of large-area occlusion, sudden night light change, and so on, a two-parameter Cauchy–Gaussian mixture noise model is introduced as follows:

$$f_\alpha(x) \approx \varepsilon \frac{\gamma}{\pi(x^2 + \gamma^2)} + (1 - \varepsilon) \frac{1}{2\sqrt{\pi}\gamma} e^{-\frac{x^2}{4\gamma^2}} \tag{14}$$

where the mixing rate of model is given as:

$$\varepsilon = \frac{4 - \alpha^2}{3\alpha^2} \tag{15}$$

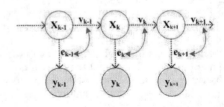

Figure 1. Correlation diagram of process and measured noises.

where steady-state distribution parameter is $0<\alpha<2$, γ is rate of α. Noise model of formula 5 can be express as:

$$f(x) \approx \varepsilon \frac{\gamma}{\pi((x-v)^2 + \gamma^2)} + (1-\varepsilon)\frac{1}{2\sqrt{\pi}\sigma}e^{\frac{(x-\mu)^2}{4\sigma^2}} \quad (16)$$

where v is the position parameter of Cauchy distribution peak that is expressed as pixel values of position for central point of current image, γ is the width corresponding half-maximum of Cauchy distribution, and μ and σ are, respectively, the mean and variance of Gaussian distribution.

3.3 Deriving the optimal proposal distribution function

Key of traditional particle filter algorithm is how to choose proposal distribution function, so optimal proposal distribution function will be derived in the case of correlated non-Gaussian noise based on weighted criteria of minimum variance. To differentiate noise correlation from noise independence, two assumptions will be given on the basis of Gaussian noise (Wang, 2010).

Assumption 1: Process noise w_k and measured noise v_k of system are Gaussian noise, which have the following statistics features:

$$\begin{cases} E(w_k) = q_k, Cov(w_k, w_j^T) = Q_k \delta_{kj} \\ E(v_k) = r_k, Cov(v_k, v_j^T) = R_k \delta_{kj} \\ Cov(w_k, v_j^T) = S_k \delta_{kj} \end{cases} \quad (17)$$

where Q_k is the covariance value of process noise that is symmetric nonnegative definite matrix, R_k is the covariance value of measured noise that is symmetric positive definite matrix, and δ_{kj} is Kronecker function, which is satisfied as the following feature:

$$\delta_{ij} = \begin{cases} 1, i = j \\ 0, i \neq j \end{cases} \quad (18)$$

Assumption 2: The initial state x_0, process noise w_k, and measured noise v_k have complementary relationship. x_0 obey Gauss normal distribution, and the mean and covariance of x_0 can be expressed as:

$$\begin{cases} \hat{x}_0 = E(x_0) \\ P_0 = Cov(x_0) = |E[(x_0 - \hat{x}_0)(x_0 - \hat{x}_0)^T]| \end{cases} \quad (19)$$

On the basis of the above two assumptions, optimal proposal distribution function of formula 1 is deduced, and the state space model of formula 1 can be modified as shown in formula 20:

$$\begin{cases} x_{k+1} = f_k(x_k) + G_k v_k \\ y_k = h_k(x_k) + e_k \end{cases} \quad (20)$$

where process noise v_k and measured noise e_k are related, dependency relationship of noise can be expressed as $p(y_k, x_{k+1} | x_k)$ in Figure 1, as follows:

$$p(y_k, x_{k+1} | x_k) \neq p(x_{k+1} | x_k)p(y_k | x_k) \quad (21)$$

where y_k and x_{k+1} are not independent in the case of given x_k. In standard-particle filter algorithm, the form of proposal distribution function is $q(x_k | X_{k-1}, Y_k)$. On the basis of Markov property, independent Y_{k-1} and x_k can express as:

$$q(x_k | X_{k-1}, Y_k) = q(x_k | x_{k-1}, y_k) \quad (22)$$

According to the interdependency of y_{k-1} and x_k, formula 22 can be express as:

$$q(x_k | X_{k-1}, Y_k) = q(x_k | x_{k-1}, y_k, y_{k-1}) \quad (23)$$

Proposal distribution function of posterior distribution can expressed as:

$$\begin{aligned} q(x_k | x_{k-1}, y_k, y_{k-1}) &= p(x_k | x_{k-1}, y_k, y_{k-1}) \\ &= \frac{p(x_k, y_k | y_{k-1}, x_{k-1})}{p(y_k | y_{k-1}, x_{k-1})} \\ &= \frac{p(y_k | x_k)p(x_k | x_{k-1}, y_{k-1})}{p(y_k | y_{k-1}, x_{k-1})} \end{aligned} \quad (24)$$

Therefore, the optimal proposal distribution function with noise-related cases can be expressed as:

$$q(x_k | x_{k-1}, y_k, y_{k-1}) = \frac{p(y_k | x_k)p(x_k | x_{k-1}, y_{k-1})}{p(y_k | y_{k-1}, x_{k-1})} \quad (25)$$

Under the background of non-Gaussian noise, initial results are as follows:

$$x_0 \sim f(\varepsilon, \gamma, \hat{x}_{1|0}, P_{1|0}) \quad (26)$$

$$\begin{pmatrix} v_k \\ e_k \end{pmatrix} \in f\left(\varepsilon, \gamma, 0, \begin{bmatrix} Q_k & S_k = 0 \\ S_k^T & R_k \end{bmatrix}\right) \quad (27)$$

In a more general sense, non-Gaussian noise model can also be express as:

$$\begin{aligned} &p\left(\begin{pmatrix} x_{k+1} \\ y_k \end{pmatrix} | x_k\right) \\ &= f\left(\varepsilon, \gamma, \begin{pmatrix} f(x_k) \\ h(x_k) \end{pmatrix}, \begin{bmatrix} G_k Q_k G_k^T & G_k S_k \\ S_k^T G_k^T & R_k \end{bmatrix}\right) \end{aligned} \quad (28)$$

From formula 28, we can see that independent noise case was widely adopted at present in the case of $s_k = 0$, and the case of $s_k \neq 0$ will be studied in this paper.

3.4 Implementation steps of a new relevant noise particle filter algorithm

According to the above analyses, the specific implementation steps of the algorithm are as follows.

Step 1: Initialization. Sampling from a given initial distribution is $x_0^i \sim p(x_0)$, set of initial points $\{x_0^i\}_{i=1}^N$ are gained, and even distributional corresponding weight is as follows:

$$\omega_0^i = \frac{1}{N} \tag{29}$$

Step 2: Weight calculation. The case of $k = k+1$ is set. According to formula 25, sampling results are as follows:

$$x_k^i \sim q(x_k \mid x_{k-1}, y_k, y_{k-1}), i = 1, 2, ..., N \tag{30}$$

And, corresponding weight is as follows:

$$\omega_k^i = \omega_{k-1}^i p(y_k \mid x_k^i)$$
$$= \omega_{k-1}^i \frac{p(y_k \mid x_k^i) p(x_k^i \mid x_{k-1}^i)}{q(x_k^i \mid x_{k-1}^i, z_k)}, i = 1, 2, ..., N \tag{31}$$

Combining with the weight-normalized method, the following formula can be obtained:

$$\omega_k^i = \omega_k^i \bigg/ \sum_{i=1}^N \omega_k^i \tag{32}$$

Step 3: Resampling. If the criterion of $N_{eff} = 1 \bigg/ \sum_{i=1}^N (\omega_k^i)^2 < N_{threshold}$ is satisfied, relevant resampling will be operated so that original weighted sample $\{x_{0:k}^i, \omega_k^i\}_{i=1}^N$ is mapped to equal-weight sample $\{x_{0:k}^i, 1/N\}_{i=1}^N$.

Step 4: Outputting results. First, state estimation is:

$$\hat{x}_k = \sum_{i=1}^N \omega_k^i x_k^i \tag{33}$$

Second, variance estimation is:

$$P_k = \sum_{i=1}^N \omega_k^i (x_k^i - \hat{x}_k)(x_k^i - \hat{x}_k)^T \tag{34}$$

4 EXPERIMENT AND ANALYSIS

4.1 Simulation experiment of the model

The following strong nonlinear Gaussian system model is adopted to verify the effectiveness of the algorithm proposed in this paper (Wang, 2010):

$$\begin{cases} \mathbf{x}_{k+1} = \begin{bmatrix} x_{1,k+1} \\ x_{2,k+1} \\ x_{3,k+1} \end{bmatrix} = \begin{bmatrix} 3\sin(x_{2,k}) \\ x_{1,k} + x_{3,k} \\ 0.2E^{-0.5x_{1,k}}(x_{2,k} + x_{3,k}) \end{bmatrix} + \begin{bmatrix} 1 \\ 1 \\ 1 \end{bmatrix} \omega_k \\ z_k = x_{1,k} + E^{-3x_{2,k}x_{3,k}} + v_k \end{cases} \tag{35}$$

Unlike in refs. [5] and [7], ω_k and v_k are Cauchy–Gaussian mixture noise. Formula 14 is adopted to calculate density function of noise, where $\varepsilon = 0.1$, $\gamma = .05$, and other parameter settings are as follows:

$$q_k = 0.2, Q_k = 0.04, r_k = 0.3, R_k = 0.09 \tag{36}$$

$$x_0 = [-0.7, 1, 1]^T \tag{37}$$

$$\hat{x}_0 = [-0.7, 1, 1]^T, P_0 = I \tag{38}$$

Standard Particle Filter (SPF) algorithm, Unscented Kalman Filter (UKF) algorithm of ref. [5], and the algorithm proposed in this paper are adopted to simulate different model of system. And, state estimation results and curve of correlative tracking root mean squared error are given as follows.

4.1.1 Simulation of irrelevant noise situation

As we can see in formula 17, S_k is equal to 0, if ω_k and v_k are irrelevant. The state of x_1 is tracked, as shown in Figures 2 and 3.

As we can see in Figures 2 and 3, because of the use of non-Gaussian noise, tracking error is greater when system noise and measured noise are irrelevant ($S_k = 0$). SPF algorithm and the proposed algorithm can accurately track x_1 state, and their errors are moreover the same.

Simulations of independent noise situation show that the proposed algorithm can be adopted to effectively dispose filter tracking in normal situation.

4.1.2 Simulation of relevant noise situation

As we can see in formula 17, S_k is not equal to 0, if ω_k and v_k are relevant, and S_k is set at 0.1. The state of x_3 is tracked, as shown in Figures 4 and 5.

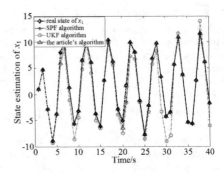

Figure 2.   Estimation curve of state x_1.

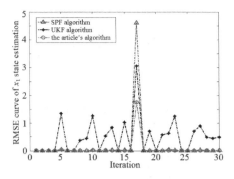

Figure 3. Root Mean Squared Error (RMSE) curve of x_1 state estimation.

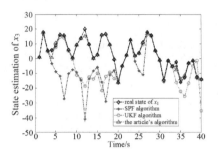

Figure 4. Estimation curve of state x_3.

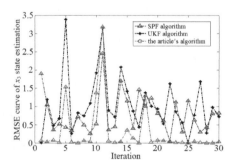

Figure 5. RMSE curve of x_3 state estimation.

Figure 6. Tracking results in the case of occlusion.

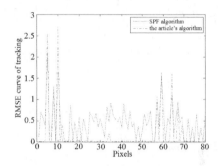

Figure 7. Tracking RMSE cure in the case of occlusion.

Figure 8. Tracking results in the case of sudden night light change.

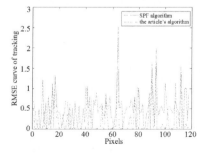

Figure 9. Tracking RMSE cure in the case of light mutation.

As we can see in Figures 4 and 5, because of the use of relevant noise, the system model gradually mismatches so that the tracking error of state x_3 gradually increases by using SPF algorithm when the system noise and measured noise are relevant ($S_k = 0$). UKF algorithm of ref. [5] has large error so that it cannot satisfy the need of tracking accuracy, and the proposed algorithm has high accuracy. As shown in tracking RMSE curve of Figure 5, with increasing time, RMSE of SPF and UKF algorithm are cumulative and growing large, and the proposed algorithm maintain g better tracking result so that the RMSE curve gradually converge to 0. The results show that the algorithm proposed in this paper has high accuracy and stability.

4.2 Tracking experiment of reality video

To investigate the effectiveness of the proposed algorithm, real-time tracking of moving cars is studied in situations of large-area occlusion and sudden night light change.

The algorithm proposed in this paper is compared with SPF algorithm, and the results are shown in Figures 6–9. In experiments, the red box indicates tracking results of the proposed algorithm and blue box indicates tracking results of SPF algorithm. The results show that the proposed algorithm has better tracking stability than that of the SPF algorithm. Figures 7 and 9 show that tracking accuracy of the proposed algorithm is 40% higher than that of the SPF algorithm.

5 CONCLUSIONS

In this paper, system state model of relevant noise situation is established and proposal distribution function of relevant noise situation is analyzed.

The proposal distribution function expression of non-Gaussian relevant noise situation is studied based on Cauchy–Gaussian mixture noise model. And, optimal proposal distribution function is deduced. Finally, the video target tracking algorithm in clutter environment based on relevant particle filter is given.

The experiments show that the tracking accuracy of the proposed algorithm is 40% higher than that of the SPF algorithm.

REFERENCES

Li W, Cao J, Wu D. Journal of Computers, 12, 7 (2012).
Morris B, Trivedi M M. IEEE Intelligent Systerms, 25, 3 (2010).
Saha S, Gustafsson F. IEEE Transactions on Signal Processing, 60, 9 (2012).
Sun Z, Sun J, Song J, Qiao S. Opt. Precision Eng., 15, 2 (2007).
Wang X X, Zhao L, Pan Q, Xia Q X, Hong W. Control and Decision, 25, 9 (2010).
Wang X X, Zhao L, Xia Q X, Cao W, Li L. Control Theory & Applications, 27, 10 (2010).
Wu B F, Kao C C, Jen C L, Li Y F. IEEE Transactions on Industrial Electronics, 61, 8 (2014).

Advances in Energy Science and Equipment Engineering II – Zhou, Patty & Chen (Eds)
© 2017 Taylor & Francis Group, London, ISBN 978-1-138-71798-5

Design of the spiral cruising orbit for noncooperative targets at the geosynchronous orbit

Haijun Hou, Zhi Li, Yasheng Zhang & Yanli Xu
Equipment Academy, Beijing, China

ABSTRACT: Many high-value spacecraft, with reconnaissance and communication payload, are often placed in Geosynchronous Orbit (GEO), a regime in which a service spacecraft is difficult to deploy. On-Orbit-Service (OOS) by a microplatform has been an effective manner in space surveillance and service application. A new orbit type is put forward for rendezvous, proximity, and surveillance operations with the noncooperative targets in GEO, namely the spiral cruising orbit. The primary mission of the spiral cruising orbit is carried out in short-distance, high-precision, and omni-directional detection of the targets. The parameters description and design method of the orbit configuration are developed. And the design index is indicated with constrains of the maneuver ability and the payload capability of the platform. The result of orbit design is validated using STK.

1 INTRODUCTION

Many high-value spacecraft (e.g., Earth observation, communication, and reconnaissance) are placed in GEO belt. Observation and imaging in a short distance can easily obtain the orbit catalog, shape, and optical or electromagnetism information of an interested target by a space microplatform. The United States launched the Mitex and the GSSAP in pairs successively in 2006 and 2014, respectively, conducting the surveillance and approaching operations to the GEO targets (Meng, 2014; David, 2012). The mission reveals a potential way to implement OOS by placing a space robot in a quasi-geosynchronous orbit and inspires the interest to explore a feasible, practical, and cost-effective method for surveillance of the noncooperative GEO targets. Technologies that enable the sort of OOS by a microplatform could be directly applied to these missions (Barbee, 2011). The orbit design is a key technology.

In this paper, we explore the design method of the spiral cruising orbit, a new type of orbit for short-distance surveillance and approaching operations with noncooperative targets in GEO. Spacecraft in spiral cruising orbit can cruise around the target orbit in a spiral way. Therefore, detection of multitargets can be realized only with a single microplatform.

2 CONCEPT AND DESCRIPTIONS

The spiral cruising orbit is a relative orbit with reference to the whole or a certain arc section of a

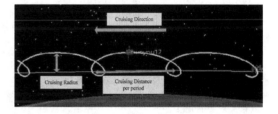

Figure 1. Configuration of spiral cruising orbit.

specific target orbit. Through designing the proper orbital elements (e.g., eccentricity, inclination, semi-major, RAAN, true anomaly, and argument of Perigee), the cruising platform moves relative to the target orbit in a way similar to spiral precession, forming a spiral flying-around relative orbit configuration. Figure 1 shows a typical configuration of a spiral cruising orbit.

The configuration parameters and description of spiral cruising orbit are shown in Table 1, in which the five entries are the key design parameters of a spiral cruising orbit.

The cruising direction determines the precession direction of the cruising orbit relative to the target's orbit. The direction may be positive, reverse, or round-trip, up to the relative motion between the cruising orbit and the target's orbit.

The cruising velocity largely determines the precession velocity of the spiral cruising orbit relative to the target orbit, calculating in cruised radian per unit time, which is directly related to the difference in semi-major axis between the cruising spacecraft and the target.

Table 1. Description parameters of spiral cruising orbit.

	Parameter	Descriptions
1	Cruising direction	Precession direction relative to the target spacecraft
2	Cruising velocity	Cruised radian per unit time
3	Cruising radius	Maximum distance between the cruising orbit's projection in the target orbit plane and the target orbit
4	Cruising inclination	Inclination of the cruising orbit relative to the target orbit plane
5	Initial phase	Initial position of the cruising spacecraft in the target orbit coordinate system

The cruising radius is the maximum distance between the cruising orbit's projection in the target orbit plane and the target orbit, which mainly influences how much details of the target can be obtained by the cruising spacecraft.

The cruising inclination is the angle between the cruising orbit and the target orbit plane, which mainly determines the direction that the servicer will detect the target spacecraft, ensuring the information monitored in the mission (such as the position of key components and radio signals in a particular direction).

The initial phase is the initial position of the cruising spacecraft in the target orbit coordinate system, in other words, the observation angles at the initial time. The initial phase is mainly determined by the observation condition.

3 DESIGN METHOD

As described above, a spiral cruising orbit has a relative motion configuration. Compared with the design of the traditional relative motions such as satellite formation (Kapila, 2000), however, the traversal cruising orbit is designed with reference to a space orbit, rather than a spacecraft. Therefore, it is necessary to find a new or improved orbit design method.

The design method of the cruising orbit is proposed based on Hill Equation, which is expressed as follows (Fehse, 2003):

$$
\begin{cases}
\ddot{x} - 2n\dot{y} - 3n^2 x = 0 \\
\ddot{y} + 2n\dot{x} = 0 \\
\ddot{z} + n^2 z = 0
\end{cases}
\tag{1}
$$

When the initial state is obtained, the relative motion is determinate. As it is known, the state in the target's plane is coupled. After mathematical transformation, we get the relative motion trajectory of the cruising spacecraft in the target plane, described as:

$$
\frac{(x - x_{c0})^2}{b^2} + \frac{(y - y_{c0} + 1.5 x_{c0} n t)^2}{(2b)^2} = 1
\tag{2}
$$

Here:

$$
\begin{cases}
x_{c0} = \dfrac{2\dot{y}_0 + 4n x_0}{n} \\
y_{c0} = \dfrac{n y_0 - 2\dot{x}_0}{n} \\
b = \sqrt{(x_0 - x_{c0})^2 + (\dfrac{\dot{x}_0}{n})^2}
\end{cases}
\tag{3}
$$

$(x_0, y_0, z_0, \dot{x}_0, \dot{y}_0, \dot{z}_0)$ is the initial state of the cruising spacecraft in the orbital coordinate system of the target and n is the angular rate of the target orbit.

When $x_{c0} \neq 0$, the relative motion trajectory in the plane is an ellipse whose center drifts in the along-track direction of the target orbit. The center's drift velocity is proportional to x_{c0}, and the drift distance in each period is:

$$
L = \left| 3\pi x_{c0} \right|
\tag{4}
$$

The cruising velocity is:

$$
V = 1.5 x_{c0} n
\tag{5}
$$

The cruising direction is determined by the sign of x_{c0}, that is:

$$
\begin{cases}
x_{c0} < 0 & \text{Positive direction cruising} \\
x_{c0} > 0 & \text{Reverse direction cruising}
\end{cases}
$$

The cruising radius, R, can be determined by b and x_{c0}, as:

$$
R = b + \left| x_{c0} \right| = \sqrt{\left(\frac{2\dot{y}_0}{n} + 3x_0 \right)^2 + \left(\frac{\dot{x}_0}{n} \right)^2} + \left| \left(4x_0 + 2\frac{\dot{y}_0}{n} \right) \right|
\tag{6}
$$

The time required for the cruising orbit to complete a spiral cruising is directly related to the cruising velocity. The higher the cruising velocity is, the shorter the traversal period will be. The traversal period can be approximately estimated by the following formula:

$$T \approx 2\pi a_T / V \qquad (7)$$

where a_T is the semi-major of the target orbit.

Figure 2 can be seen that when the cruising velocity is fixed, the longer the semi-major axis of the target orbit is, the longer the traversal period will be; when the length of the semi-major axis of the target orbit is fixed, the higher the cruising velocity is, the shorter the traversal period will be. With regard to the same target orbit, there is an approximate inverse relation between the traversal period, T, and the cruising velocity, V.

As the trajectory center is constantly drifting with the change of time, a certain target satellite is usually taken as the reference point in orbit design. We can suppose that the relative motion state meets the following condition at the initial moment:

$$y_{c0} = y_0 - \frac{2}{n}\dot{x}_0 = 0 \qquad (8)$$

The initial position of the cruising spacecraft in the radical direction is:

$$x_0 = R' \cos\theta \qquad (9)$$

where R' is the projection of the cruising radius R in the target orbit plane; θ is the phase angle based on the x-axis, with counterclockwise direction as the positive direction.

Additionally, in the target orbital coordinate system, the initial relative motion state of the spiral cruising spacecraft in the normal direction satisfies:

$$\begin{cases} z_0 = x_0 tg\varphi \\ \dot{z}_0 = \dot{x}_0 tg\varphi \end{cases} \qquad (10)$$

where φ is the angle between the projection on the $x-z$ plane and the x-axis in the target orbital coordinate system.

If the target orbital element is known, when the cruising velocity, V, cruising radius R, phase angle

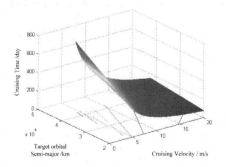

Figure 2.   Relationship of $a_T - V - T$.

θ, and normal angle φ are given, we can combine Eqs. (4)–(10), then the initial relative motion state of the cruising spacecraft in the target orbital coordinate system can be obtained. Then, according to the coordinate transformation relation, the position vector and velocity vector of the cruising spacecraft can be obtained in the inertial coordinate system.

4   CONSTRAINTS IN ORBIT DESIGN

The spiral cruising orbit is mainly used for the reconnaissance and surveillance mission. The service platform conducts the mission by approaching operation with a visible or infrared payload. The imaging quality is closely related to the capability of the camera fixed in the satellite. Also, it is susceptible in the range of relative velocity and observation angle between the surveillance camera and the target. As it is difficult to ensure a proper observation angle for every target during the whole cruising period, only the constraints of the cruising radius and the cruising velocity are focused on in this paper.

The effective range of a CCD camera is described as (Liu, 2009):

$$R_1 = \frac{R_d D}{2.44 \lambda Q} \qquad (11)$$

where D is the size of the optical entrance aperture; λ is the wavelength; Q is the imaging quality factor; and R_d is the minimum spatial resolution needed for imaging. If the optical entrance aperture is $D = 0.3\,m$, the wavelength is $\lambda = 0.5 um$ the quality factor $Q = 1.1$, $R_d = 0.5m$, then the maximum working distance of the service platform in the spiral cruising is about 111.8 km.

Both the cruising radius and cruising velocity are functions of the relative position and velocity; and there exists a coupling relation between the two. When the cruising radius R is given, the cruising velocity can be calculated from Eqs. (5) and (6):

$$R = b + \left| \frac{2}{3n} V \right| \qquad (12)$$

Extremely, when $b = 0$, the cruising velocity reaches its maximum value. In this case, the spiral cruising orbit degrades into a linear cruising orbit. For maintaining a spiral cruising configuration, the cruising velocity of a spiral traversal cruising orbit should meet the following constraint:

$$|V| \le \frac{3}{2} nR \qquad (13)$$

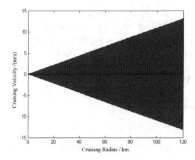

Figure 3. Value range of cruising velocity changing with cruising radius.

As it is shown in Fig. 3, a higher cruising velocity produces a larger cruising radius. The fact that higher-quality imaging requires a shorter range between the two spacecraft means that the responsive ability improvement of the cruising spacecraft can be achieved at the cost of some imaging quality. Therefore, in designing a spiral cruising orbit, indices of the two sides must be balanced so as to achieve an optimal overall effect.

5 SIMULATION AND RESULTS

The orbit elements of the target are: $e = 0$, $i = 0°$, $\Omega = 0°$, $w = 0°$, $M = 0°$, and $a = 42164 km$. The spiral cruising radius is $R = 100 km$, the cruising velocity $V = 4 m/s$, the initial phase angle in the target orbit plane $\theta = -90°$, and the initial angle in the normal direction $\varphi = 0°$. It can be calculated that the position vector and the velocity vector of the cruising spacecraft in the inertial coordinate system are:

$$\vec{r} = \begin{bmatrix} 42164.1 \\ 103.269 \\ 0 \end{bmatrix} km \quad \vec{v} = \begin{bmatrix} -0.00375497 \\ 3.07319 \\ 0 \end{bmatrix} km/s$$

The STK is used to simulate the spiral cruising orbit, as shown in Figure 4.

As it is shown in Fig. 4, the orbit in the middle is the GEO, that is, the target orbit; parallel to the GEO, two lines on the top and at the bottom are the boundaries, 100 km higher and lower than the GEO, respectively; the red trajectory in the middle is the spiral cruising orbit.

Fig. 5 shows the changes of the relative distance between the service platform and the target orbit in 60 days. The result shows that the cruising radius is within 100 km. In this case, the service platform can stay in the vicinity of the target orbit for a long time, and can conduct close reconnaissance and surveillance mission of multiple spacecraft in the target orbit.

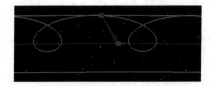

Figure 4. Spiral cruising orbit shown by STK.

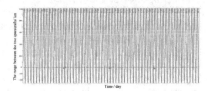

Figure 5. Relative distance between the service platform and target orbit (in 60 days).

6 SUMMARY

In this paper, we focus on the design method of the spiral cruising orbit, a relative orbit for noncooperative targets surveillance at the geosynchronous orbit. The relative orbit configuration is described. The relation between the parameters of the spiral cruising orbit and the relative state is studied. Constrains of the capability of the payload are considered in the orbit design. Simulation is conducted by STK. A more precise dynamic model and the control strategy will be focused on in the next work.

REFERENCES

Barbee B. W., Carpenter J. R., Heatwole S., et al. A guidance and navigation for rendezvous and proximity operations with a non-cooperative spacecraft at geosynchronous orbit [J]. Journal of the Astronautical Science, 2011, 58(3):389–408.

David A. B. DARPA's Phoenix Project [EB/OL]. Tactical Technology Office: Briefing prepared for the on orbit satellite servicing workshop GSFC, 2012.

Fehse W. Automated rendezvous and docking of spacecraft [M]. Cambridge university press, 2003.

Kapila V, Sparks A G, Buffington J M, et al. Spacecraft formation flying: dynamics and control [J]. Journal of Guidance, Control, and Dynamics, 2000, 23(3): 561–564.

Liu Yunmeng, Chai Jinguang, Wang Xuhui, et al. Visible characteristic and calculation of detecting distance for space target [J]. Infrared Technology. 2009. 31(1):23–26.

Meng Bo, Huang Jianbin, Li Zhi, et al. Introduction to American approaching operation satellite Mitex in geoststionary orbit and its inspiration [J]. Spacecraft Engineering, 2014, 23(3): 112–118.

U.S. Air Force Fact Sheet. Geosynchronous Space Situational Awareness Program (GSSAP). 2014.

Advances in Energy Science and Equipment Engineering II – Zhou, Patty & Chen (Eds)
© *2017 Taylor & Francis Group, London, ISBN 978-1-138-71798-5*

Study on a management information system of ship materials

Zhangxuan Yin, Yajun Wang & Huijie Yan
Dalian Polytechnic University, Dalian, China

ABSTRACT: In order to realize the information and data of the ship materials management, this paper analyzes the current problems in the management of ship materials, with the use of advanced computer technology and modern communications technology, to build the ship materials management information system, which is proposed to increase the efficiency of the management of ship materials. Combined with practical application requirements, the complex characteristics of ship materials management are analyzed and studied. By designing and constructing the platform of ship materials management information system, it analyzes the basic functions of the system and data processing applications. By designing and building the overall framework and operation mode of ship materials management information system, the current problems in the management of ship materials are solved effectively through analysis, research, and design to influence the efficiency of the enterprise.

1 INTRODUCTION

Materials management plays an important role in ship enterprise and involves a wide range of application, including materials procurement plan, technical support, financial control, and other aspects (Xu, 2000). There are different kinds of machinery and electronic equipment, data services, equipment inside the ship, so it is necessary to adopt the modern information management system to meet the needs of ship life (Feng, 2010). To a certain extent, the management of ship materials determines the endurance of the ship; therefore, a good job in the management of ship materials will ensure the safety of ship navigation (Wang, 2012).

China is fully capable for the research and design of a set of high-efficiency, low-cost, suitable ship materials management system. Lin Hua (2012) introduced COSCO modern materials management mode in the study of materials management based on COSCO shipping (Lin, 2012). Qian Yun (2006) stressed that strengthening the control of procurement cost by using the ERP principle (Qian, 2006); Ji Juan (2010) designed a set of materials accounting control management system (Ji, 2010), enabling BOM design materials, picking control, warehouse control, materials management, accounting-fed approval process management, and other functions, by solving the disadvantages of traditional manual work. Lv Guoqing (2007) applied materials classification theory in the management of materials procurement and procurement of ships (Lv, 2007) and realized the control of procurement cost. Bi Li (2011) proposed ship ERP system on the use of materials classification and coding by

allowing perfect materials into or out of the library management approach (Bi, 2011). Wang Ping (2010) discussed the theory and method of ship integrated manufacturing cost management, which provided guidance for the establishment of the cost management system of ship integrated materials (Wang, 2010). In materials inventory management, all materials should not use the same approach of management (He, 2006). If the appropriate high-efficiency procurement plan is used, it is necessary to classify materials required in a reasonable way, strengthen the selection and management of suppliers, and the level of materials management of enterprises will be greatly improved (Cui, 2001).

This paper will from the perspective of the needs of a company, combined with the specific materials distribution of the company; the platform of ship materials management information system is designed, the basic functions of the system and data processing applications are analyzed, and the overall framework and operation mode of ship materials management information system is built.

2 MODELING AND ANALYSIS OF MATERIALS INFORMATION SYSTEM PLATFORM

Ships in the process of long-term voyage will have a shortage of living materials and equipment spare parts. At this point, we need to apply shore-based total service platform for deployment supplies. In this process, we will produce a lot of materials data forms that are used to record the dynamic information distribution. However these materials data sheets are distributed in the hands of departments and relevant staff of different functions, making it difficult to query the whole process and the specific implementation of the whole process. If we are on the track of the entire process of observation, we will need to keep tracking of key data throughout the execution of the business form formed. At this point, we should make use of information management system to achieve the computer's inventory management and monitor the whole process of materials management (Zhao, 2011). We can grasp the dynamic information materials by tracking the overall implementation process, so the collation and integration of the key data is the core of the management of materials distribution. Only to establish a relationship between interlocking materials distribution in the process of implementation of the data form and the key data of these forms are collected in real time during the execution of the actual business to realize the allocation of materials dynamic information processing. On the basis of the whole process of the allocation of the materials, the key data forms are generated and extracted to establish a model form of critical data on materials management, as shown in Figure 1.

Model $L = \{L_1, L_2\}$ represents the type of materials, $P = \{P_1, P_2, \ldots P_{31}\}$ represents the order type of materials in the warehouse, $S = \{S_1, S_2, \ldots S_{41}\}$ expresses the order details, $M = \{M_1, M_2 \ldots M_{41}\}$ expresses materials-type information, G_N represents shelf information, W represents the type of supplies, N denotes the supply company information, E represents staff information, and F represents shore-based information. As can be seen from the model, it will produce a series of form data in the implementation process of the entire

Table 1. Materials management information table.

Data list	Key data forms
L_1	Living materials
L_2	Equipment spare parts
P_1, P_2, P_3	Storage order type
P_{11}, P_{21}, P_{31}	Storage order number and order date
S_1, S_2, S_3, S_4	Storage order details
$S_{11}, S_{21}, S_{31}, S_{41}$	Materials composition, materials number
M_1, M_2, M_3, M_4	Materials information
$M_{11}, M_{21},$ M_{31}, M_{41}	Materials name, number, unit price, quantity
G_1, G_2, G_3	Shelf information
W	Type of materials supply
N	Supply company name, honors level
E	Staffs information
F	Total shore-based services platform.

business. We can obtain a list of orders and data inventory from the model, as shown in Table 1, thereby tracking the business execution in the process of materials, as long as tracking and tracing service execution recorded during these critical data and forms through the data form and the various departments to determine the process and information of the corresponding task. Tracking the implementation of dynamic information realizes the information management of materials allocation through the correlation of data forms.

Tracking the key data node in the model can achieve the business execution schedule and safety of materials of tracking and monitoring the implementation of business to enable enterprises to adjust procurement strategy and transport strategy timely and ensure the timely delivery of purchase orders and materials. After the materials procurement planning, the need for following timely schedule when supplies not provided in accordance with the plan, it should reflect various departments and adjust the timely deployment of goods in order to meet the needs of the ship's supplies (Zhang, 2006).

3 BASIC FUNCTIONS OF INFORMATION SYSTEM OF SHIP MATERIALS

Data form is the core content of information management system, extracting the execution of business critical data forms that reflect the progress of the implementation of business information and materials security in front of the process analysis and modeling. Analyzing the data form and the relationship between them in the process of business execution to establish a data chain based on key data form a materials management system,

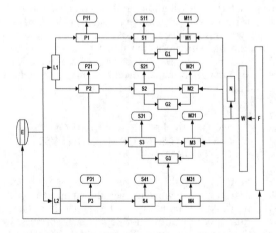

Figure 1. A model form of critical data on materials management.

with the support of the coherence of data. The data link of the materials data information system is shown in Figure 2.

On the basis of the data link in Figure 2, the relationship among the various data in the execution process is established. By making the business execution in the process of information as the chain even together through this interlocking data forms information, there will not be any information loss and disruption.

The main functions of the system are designed as follows (Yang, 2010):

1. Supplier Information Management: It is divided into three modules: ① Order invoice information, ② Materials category information, and ③ Materials-related certificate. The suppliers who meet the requirements of the company should be classified, and the management materials provided by suppliers should be managed in different ways as the basic data in the database to provide to the staff. In this way, all the information records and data operation management of suppliers in all procurement plans are unified.

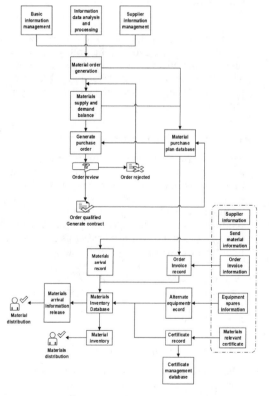

Figure 2. Data link of the materials data information system.

2. Materials order generation: It is based on the materials demand, order inventory, procurement agreement, and other procurement information to generate orders. Materials management information system is the expansion of materials procurement management applications that is based on the standard code information and the upstream design department of the information interface, the expansion of materials procurement management applications.

3. Materials supply and demand balance: Staff check the details of the goods, input the details of the goods provided by suppliers that is based on the qualified supplier information management in the basic information of the system, and make a comprehensive assessment of supply and demand balance.

4. Generate purchase order: By selecting suppliers, the system calculates the purchase price on the basis of quotations from suppliers information automatically submitted to the relevant departments for approval. The system will establish a unique order number for each purchase plan, the order number and the number of each materials are bounded to become unique identify as a follow-up to the operation of the database while establishing the bill of procurement plans to facilitate follow-up query statistics (Zheng, 2012).

5. Order review: After the order is completed, the system will submit to the relevant departments automatically according to the actual demand and order audit successful system automatically submitted to the next program that generates orders for the contract. If purchase orders plan does not pass, it will be returned to the preparation of the procurement staff to modify the program automatically, after the completion of the modification to repeat the above operation until successful.

6. Generate purchase order: If procurement plan details pass, the system assists the user to generate purchase order automatically, which can be printed, and other management operations and order contract administration maintain the execution status of the contract in real time.

7. Materials arrival record: Warehouse clerk statistics physical arrival of goods, purchasing department statistics order invoice. When confirming statistics, we only need to select the corresponding materials procurement plan coded information, and the system will automatically complete the information related to the information recorded to change the inventory and modify the procurement plans and so on. And the materials to the account information are automatically sent to the various departments of the company in order to facilitate the various relevant departments and staff to inquiry (Molnar, 2004).

8. Materials certificate management: The certificate provided by the suppliers of materials is stored in the graphics file server, establishing the materials certificate information database to manage and query the certificate.

9. Equipment information management: The supplier will provide the relevant information of the equipment to the company in the form of electronic form, and the procurement department staff will match the corresponding equipment information according to the procurement plan form. The system saves data files automatically and adds the sequence number for each data file to "procurement plan form number + device information sequence table" as the primary key for each data file management as a postmanagement strategy and inquiry.

10. Materials arrival information release: The system uses Web mode to publish the implementation of materials procurement plan to the whole company, and each department can screen and use information according to their needs.

11. Materials distribution management: System database edit out materials requisitioned forms and the library form, print related data forms, improve the system of materials management data, and update the status of materials data information automatically. Warehouse management personnel check the department of materials application procedures of the application materials, and the company will issue materials delivery information.

4 OVERALL FRAMEWORK AND APPLICATION OF SHIP MATERIALS MANAGEMENT INFORMATION SYSTEM

The overall structure of the system is mainly composed of three parts: the company client management system (shore), ship client management system (ship), and communication systems. The overall structure of ship materials management information system is shown in Figure 3.

4.1 Company management system

The main functions of the company's management system are the management of the integrated information, summarizing, analyzing, and dealing with the dynamic information of all materials. Using JAVA development language, the system will be released on Server Web and Server Application through the three-layer structure of the development model. As shore-based systems need to han-

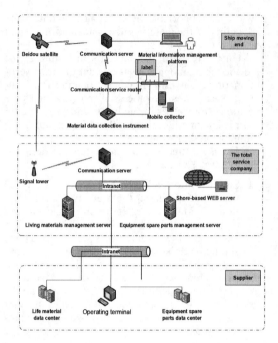

Figure 3. Overall structure of ship materials management information system.

dle large amounts of data, the Oracle database system is the best choice. The ability of the new distributed and extensible database can be used for remote control and management, which is convenient for the daily maintenance of the system administrator.

4.2 Ship terminal management system

The main function of the ship's side is the management of the data information receiving and materials application and receiving and allocation for the company. Also using JAVA development language and using two-layer structure, each platform installs SQL Server2008 database, which has the following features: easy installation and maintenance, low cost, simple management and high efficiency, suitable for ship version of the system to use, and installation of C/S client program on the ship, which can connect single ship and shore service platform for information and data management.

4.3 Communication systems

The communication system adopts two types of connection modes:

1. Shore-based communications and communication server is connected directly, a large amount

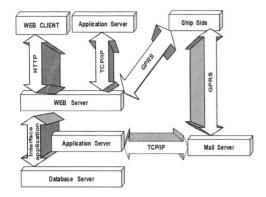

Figure 4. Overall structure model of the system.

of data is stored in the Oracle database, and the system sends and receives data through the mail server.

2. Ship terminal uses GPRS equipment to communicate, the data are stored in the server SQL2008 database, and mails are exchanged via mobile Internet directly with shore-based server.

All communication systems are based on XML standard electronic data exchange. The system can change business data or other data into the same XML document to support data exchange with other application (Sarma, 2002).

The overall structure model of the system is shown in Figure 4.

The system is mainly composed of database services, business logic processing application service, interface application system, Web services (based on B/S and C/S mixed mode), the client application (using JAVA development, B/S browser), XML message exchange system, wherein the Web Server, Application Server, Database Server are scalable architecture to improve the integrity and reliability of the system (Finkenzeller, 2003).

5 CONCLUSION

In this paper, we studied and analyzed the operation process of ship materials management, designed the ship materials management information system, constructed ship materials management information system platform, and realized the information recording and tracking of key data form. The company can track the whole process of the implementation of the business according to the key data form, to form a strong all-round protection for the materials. After using the materials management system, we can fully intuitively query the status of the transfer of materials, inventory, and other information to enhance the efficiency of enterprises.

REFERENCES

Bi Li. Research on the ERP system of Shipbuilding Enterprises—the management of materials in the warehouse. J. *Science and technology information*, (2011).

Cui Sheng. Shipbuilding enterprises traditional materials management and product oriented materials management. J. *Ship supplies and market*, 2(2001).

Feng Liang. RFID-based supply chain and warehouse management study design. D. Shan Dong University, (2010).

Finkenzeller, K. RFID Handbook Fundamentals and Applications in Contactless Smart Cards and Identification. M. New York, USA: John Wiley and Sons, Inc, (2003).

He Jing. Warehousing management and inventory control. J. *Intellectual Property Publishing house*, 8(2006).

Ji Juan, Xia Fang chen, Tu Haining, et al. Research and development of material accounting control management system. J. *Automation and instrument*, (2010).

Lin Hua. Study on COSCO shipbuilding modernization material management mode of. J. *Ship supplies and markets*, (2012).

Lv Guoqing. Research on procurement management strategy of ship manufacturing enterprises based on material classification. D. Harbin: *Harbin Institute of Technology*, (2007).

Molnar D, Wagner D. Privacy and security in library RFID: Issues, practices, and architectures. C. Proc of Computer and Communications Security. USA, (2004).

Qian Yun. Shipbuilding enterprise procurement control method. J. *Ship supplies and markets*, (2006).

Sarma S.E, Weis S.A, Engels D.W. RFID systems and security and privacy implications. C. Cryptographic Hardware and Embedded Systems-CHES 2002. Lecture Notes in Computer Science 2523, Redwood Shores, CA, USA, Springer-Verlag, (2002).

Wang Ping. Theory and method of ship integrated manufacturing management. M. Beijing: *Science press*, (2010).

Wang Yaqi, Zhao Hong, Ren Wenyuan. Intermediate product-oriented material management mode research. J. *Ship materials and markets*, (2012).

Xu Boli. Materials to be effective management. J. *Market and ship supplies*, (2000).

Yang Guihai, Lu Mingyu, Hu Hu. Design and implementation of material management system for ship enterprise. J. Dalian Maritime University, 8(2010).

Zhang Guangming, Wang Yaping, Wang Hui, Xie Pingshun. Cooperative progress of shipbuilding supply chain collaboration. J. *Journal of Jiangsu University of Science and Technology*, 6(2006).

Zhao Bao, often Guijiao. The application of informatization construction of. J. *Coal economic management in material management*, 4–337(2011).

Zheng Shijun, Yue Yueshen, Hou Xiaoming. Dredging ship equipment management information platform solution. J. *China navigation*, (2012).

Advances in Energy Science and Equipment Engineering II – Zhou, Patty & Chen (Eds)
© 2017 Taylor & Francis Group, London, ISBN 978-1-138-71798-5

Simulation analysis of welding precision on friction stir welding robot

Haitao Luo, Jia Fu, Min Yu & Changshuai Yu
Shenyang Institute of Automation, Chinese Academy of Sciences, Shenyang, China

Tie Liu
Steelmaking Central Factory, Ansteel Corporation Co. Ltd., Anshan, Liaoning, China

Xiaofang Du
Dalian CIMC Logistics Equipment Co., Ltd., Dalian, Liaoning, China

Guangming Liu
Northeastern University, Shenyang, Liaoning, China

ABSTRACT: In this paper, a new type of Friction Stir Welding (FSW) robot is introduced for the welding of large complex thin-walled cured surface. The welding mechanism of the FSW robot is different from the ordinary mechanical processing, which has extremely bad mechanical load at the mixing head. Insufficient stiffness of the welding equipment may have serious influence on the geometry precision of welding. Through the analysis of multibody system simulation, the ram structure, which is most sensitively affected by welding load, is equivalent to flexible body. The stiffness of joint is taken into account to set up the rigid-flexible coupling dynamic model of the whole robot, and the displacement error of spindle end face is simulated under the worst working condition (melon-disk welding condition). A kind of effective assessment method is provided for the stiffness and welding precision of the FSW robot.

1 INTRODUCTION

The FSW technology is an advanced solid-state welding process, gradually popular over the past 20 years. Compared with the traditional fusion welding, it has good mechanical properties at welded seams, simple techniques, no subsequent processing, green and environmental protection, and many other advantages. Therefore, it has been applied in various fields (Dong, 2007; Luan, 2002). The FSW robot is used to implement this kind of welding process. Under the control of welding mechanism, its design needs to meet the welding load in different working conditions. The load is different from conventional cutting and drilling using the blade to remove material. Instead, it is through the interaction between stirring head and welded parts so that the magnitude of the load is far beyond that the ordinary mechanical processing makes (Xia, 2002). Therefore, the stiffness and dynamic properties of the FSW robot plays a decisive role on the welding precision of terminal seams.

In general, large complex equipment is made up of many parts, and they are connected through specific joints. For example, CNC machine consists of different types of large structures through bearing, screw, and guide rail with sliding blocks

(Ebrahimi, 2006). Therefore, we call it multi-rigid-body system, and various components and their integrations will not happen to be out of shape. In some special occasions, if the deformation of parts or joints has impact on the precision of the end of the tool, the large equipment is called flexible multibody system (Ghaisas, 2004).

Because there is no deformation among each workpiece, the manufacture error mainly comes from equipment's system error. And in overloading or high-speed conditions, the structure deformation of their own or insufficient stiffness of joints will have an important effect on the end of tools, which can increase the system error. For analysis of multi-rigid-body system, completely rigid hypothesis cannot reflect the real effect of structural parts and joints (Wu, 2004). Adopting the idea of flexible body can link the deformation of parts to their movement and consider the coupling relationship between them, which makes the final analysis results become more accurate.

With the analysis method of multi-rigid-body system becoming increasingly mature, more people start an in-depth study of flexible multibody system. At present, various scientific research institutions worldwide have conducted several studies on this kind of analysis method, and the related research areas include aerospace, railway vehicles, NC-machine, and robots (Rubio, 2007; Hagiu, 1997). Because the products developed by researching facility or manufacturing firm are large heavy-duty structures, the deformation of all parts in the equipment will produce certain effect on other structures. Ignoring these factors may make the equipment's global error bigger. Therefore, we need rigid-flexible coupling dynamics method or fully flexible body dynamics method to carry out relevant research work.

In this paper, by using large-scale finite-element analysis software ANSYS, we generate the flexible ram body of the FSW robot and improve quality, center of mass, moment of inertia, and other important information into the flexible body. In addition, the flexible body also contains modal frequency and relevant modal information about vibration mode, and we call it the modal neutral file. In *ADAMS* software, flexible body is used to replace the ram, to eventually create the rigid-flexible coupling dynamic model of the FSW robot. On the basis of the coupling model, considering the flexible of ram into the whole dynamic model, it makes the simulation results more accurate.

2 COMPOSITION OF THE FRICTION STIR WELDING ROBOTS

2.1 *Simulation model introduction*

The FSW robot simulation model mainly consists of the components shown in Fig 1. They are X-axis component, Y-axis component, Z-axis component, AB-axis component, and mixing head component.

Here, the first three joints of X-Y-Z-axis are ball screw and linear guide rail drive, and AB-axis is rotational joint. The mixing head is made up of an insertion shaft formed by a rotating joint and agitator shaft formed by a screw pair. Wherein, the lead of X-axis screw pair is 12 mm; the lead of Y- and Z-axes spiral deputy is 20 mm; and the lead of welding insertion shaft screw pair is 1 mm.

In order to carry out the simulation analysis smoothly, we need to take necessary simplification and consolidation for the model. The principle of simplified model is: do not destroy the original structure as much as possible, and some equivalent approximate method is used to simulate the robot's real machining conditions. These include:

1. In order to avoid excessive constraints and ensure that the simulation results are correct, single-screw drive simulation of the FSW robot Y-axis is taken.
2. For the connection of rail and slider, the factors of structural strength and balance and oth-

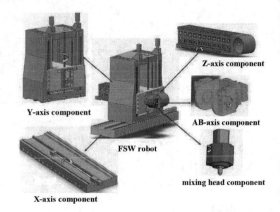

Figure 1. Composition of the simulation model.

ers must be taken into consideration, multiple sets of slide and rail is used to ensure structural support and slide. In the process of multibody system simulation, all rail slider can be merged into a translation pair that can simulate both real effects between the two parts combined into one, without considering the binding portion of the case under load, and the species does not simplify the impact of the end of the welding accuracy.
3. For the gravity compensation mechanism, there are two vertical upward forces, which are fixed, with their size and weight equal to counterweight at the position of connection between constant saddle and ropes.

For the column top pressure on the pulley, we can apply a vertical downward constant force on the top of column corresponding to the pulley-installed direction to simulate the column positive pressure on the wire rope.
4. The parts without relative motion have been simplified and merged.

2.2 *Five types of typical working conditions*

Depending on the type of welding tasks, the FSW robot welders condition can be divided into five categories: cylindrical girth weld, cylinder longitudinal seam welding, melon bottom girth weld, melon top girth weld, and melon valve welding, as shown in Fig 2. These categories are defined as follows:

1. Cylindrical girth weld: The main task is completing the welding of rocket and missile barrel along the circumferential direction. In the process of welding, XAB axis is fixed and the posture of welding tool keeps level, and Y- and Z-axes provide offset compensation function. Thus, the robot needs to provide linkage func-

tion of YZ-axis and the axis of turntable in the cylindrical ring seam welding condition.

2. Cylinder longitudinal seam welding: The main task is completing the welding of rocket and missile barrel along the cylinder generatrix direction. In the process of welding, AB-axis and turntable are fixed, the mixing head moves along the vertical welding seam from the bottom to the top (Y-axis positive), and X- and Z-axes provide offset compensation. Thus, the robot needs to provide linkage function of XYZ-axis in the cylinder longitudinal seam welding condition.

3. Melon bottom girth welding: The main task is completing the welding between the rocket or the melon of missile and the cylinder transition section. In the process of welding, XB-axis is fixed, and Y- and Z-axes provide offset compensation, and the A-axis provides attitude deviation compensation. Therefore, the robot needs to provide linkage function of YZ-axis and the axis of turntable in melon bottom girth welding condition.

4. Melon top girth welding: The main task is completing the welding between the rocket or the melon of missile and the top cover. The robot needs to provide linkage function of YZ-axis and the axis of turntable, which is similar to melon bottom girth welding.

5. Melon-disk welding: The main task is completing the welding between the melons of sphere or ellipsoid parts.

The welding trajectory is from the bottom, and the axis direction of the stirring head is the law of the surface welders. The interpolation motion of YZ-axis and A-axis makes up mixing head melon flap trajectory. In practice, the seam has some error in the X-direction, so the X-axis offset compensation function is required. Thus, the robot needs to provide linkage function of XYZA-axis in the melon-disk welding condition.

The solid-phase welding of rocket body, aircraft, and missile cover structure can be achieved by the five typical welding conditions described

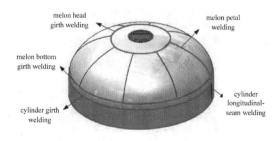

Figure 2. Five kinds of typical working conditions.

Table 1. Working mode and linkage situation of motion axes under every typical working condition.

Conditions	Fixed shaft	Movement of the shaft	Compensation axis
Cylindrical girth welding	XAB	C	YZ
Cylinder longitudinal seam welding	ABC	Y	XZ
Melon bottom girth welding	XB	C	YZA
Melon top girth welding	XB	C	YZA
Melon-disk welding	BC	YZA	X

above and the five typical conditions of the FSW robot shown in Fig 2. In each typical welding condition, the mode of each robot axis motion and linkage condition is shown in Table 1. For irregular welding seam, because the entire robot has seven degrees of freedom, it can achieve any welding of space complex surfaces, such as skin space shuttle, spacecraft "humanoid" mast, and the cabin space station spacewalk hole.

3 ESTABLISHING A RIGID-FLEXIBLE COUPLING DYNAMIC MODEL

A variety of mechanics principle can be used in multibody systems dynamic model, including law of the conservation of energy, Lagrange's equations, Hamiltonian principle, and Kane's equations. The establishment of Lagrange equation is based on the point of energy, and the advantages are that it can be easily programmed, can easily establish the model of forward dynamic or inverse dynamics, realized recursive forms of model, and convenient to add control feedback. Adopting the method of Lagrange equation model is relatively mature, the Lagrange equation model method is used in a lot of multibody system analysis software and so does the multibody system dynamics software *ADAMS* of this paper (Yu, 2003).

3.1 Basic theory of flexible body

Flexible body can be seen as a node set of finite-element model, and its deformation can be treated as linear superposition of modal vibration mode (Lambert, 2006). As shown in Figure 3, point P is a node on flexible body, point P' is the position after the deformation in the process of movement, B is the flexible body coordinate system, and G is the fundamental coordinate system.

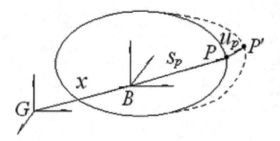

Figure 3. Moving schematic of flexible-body on node.

The position of B object coordinate system in the fundamental coordinate G is $\vec{x} = (x \quad y \quad z)$, the direction shown by Euler angles is $\vec{\psi} = (\varphi \quad \theta \quad \psi)$, and the modal coordinate is $q = \{q_1, q_2, ..., q_m\}^T$ (M is the number of modal coordinates). Therefore, the generalized coordinates of the flexible body is as follows:

$$\vec{\zeta} = \{x \; y \; z \; \varphi \; \theta \; \psi \; q_i \; (i=1,...,M)\}^T = \{\bar{x} \; \bar{\psi} \; \bar{q}\}^T \quad (1)$$

The position vector of node P on the flexible body can be shown as follows:

$$\vec{r}_p = x + {}_B^G A \left(s_p + u_p \right) \quad (2)$$

where ${}_B^G A$ —The transformation matrix of coordinate system B relative to coordinate system G;

s_p —the position of point P in the coordinate system B before the body deformation;

u_p —the direction vector of point P' relative to point P. $u_p = \Phi_p q$, where Φ_p is the modal sub-block matrix of point P' moving degree of freedom.

The speed of point P is:

$$\vec{v}_p = \left[I - {}_B^G A \left(\tilde{s}_p + \tilde{u}_p \right) B \, {}_B^G A \Phi_p \right] \dot{\xi} \quad (3)$$

where the wave symbol is the position vector of the nonsymmetric matrix. Matrix B is defined as transformation matrix that takes the derivative of Euler angles with respect to time. Thus, kinetic energy and potential energy can be obtained and shown as:

$$T = \frac{1}{2} \int_v \rho v^T v dV = \frac{1}{2} \dot{\xi}^T M(\xi) \dot{\xi} \quad (4)$$

$$V = V_g(\xi) + \frac{1}{2} \xi^T K \xi \quad (5)$$

The motion differential equations of flexible body are established by the Lagrange multiplier method:

$$M\ddot{\xi} + \dot{M}\dot{\xi} - \frac{1}{2} \left[\frac{\partial M}{\partial \xi} \dot{\xi} \right]^T$$

$$+ K\xi + f_g + D\dot{\xi} + \left[\frac{\partial \Psi}{\partial \xi} \right]^T \lambda = Q \quad (6)$$

where K is the modal stiffness matrix; D is the modal damping matrix; f_g is the general gravity, Q is the general external force; λ is the Lagrange multiplier of constraint equations; ξ and $\dot{\xi}$ are the general coordinates of flexible body and its time derivative, respectively; M and $\dot{M}$ are the mass matrix of flexible body and its time derivative, respectively; $\partial M / \partial \xi$ is the mass matrix of flexible body partial derivative of generalized coordinates that is a $(M+6) \times (M+6) \times (M+6)$ tensors; and M is the modal.

3.2 Establishing the coupled model

Each part of the mass distribution, the moment of inertia, and geometry information of Friction Stir Welding (FSW) robot dynamic model is very important. In general, the kinetic parameters need to be identified by certain methods. Because of its complex structure and changing position-shape configuration, it is difficult to directly obtain the result of the theoretical distinction. In this section, we directly obtain the geometric modeling of robot, material parameters, and assembly relationship information through *ADAMS* to get the relative kinetic parameters. On this basis, constraints, force, and motion relations are appropriately added on the model to establish the digital virtual prototype model of Friction Stir Welding (FSW) robot.

Friction Stir Welding (FSW) robot is composed of XYZ-axis, AB-axis, and stir-welding head, and the end welding operation is executed by each joint motor driving. A three-dimensional entity model of the robot is established by *SolidWork*, then the model is imported into *ADAMS* specified material properties for each component, and motion pair and drive is also added on the component. In order to simulate the effect of the integration of flexible joint, a six-dimensional stiffness damping element is used to simulate the stiffness and damping of the junction that combines with the guide rail, slide block, screws and screw nuts, and the integration of a bearing in the connection position. In order to consider the precision error of welding, seam is generated by cantilever of the ram, which is regarded as a flexible body. Because the coupled dynamics simulation is extremely time-consuming and consumes resources, other parts are temporarily considered as rigid body.

Here, fixed joint is added between the base of the robot and the earth, and the integration of guide and screw is mobile and rotation joint, and rotation connection is used at the end of the mixing head. Its welding load is equal to the load at the end of the mixing head in the static analysis. Because of the influence of the barycenter compensation mechanism, the model exerts a vertical upward force 30000 N on both sides on the top of the saddle and a vertical downward reaction force 60000 N on the mounting surface of pulley block, that is, on both sides of the pillar top.

For the movement of each joint motor, motion planning is made in advance, and the corresponding expressions or spline interpolation data points are given in order to achieve the expected effect of planning. The coupled dynamics simulation model of FSW robot is shown in Figure 4.

In the finite-element analysis software ANSYS, the slippery pillow of the FSW robot is made flexible. The material properties, unit type, and unit properties are given, and then mesh grid parameters are selected. According to constraint relations of the big arm with other components in the *ADAMS* environment, the outreach is defined well in the *ANSYS*, which can make the flexible body and the model of other components to establish the right connections.

Figure 4.   Rigid-flexible model of the FSW robot.

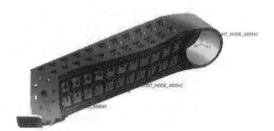

Figure 5.   Flexible-body model of ram.

After the definition of outreach point, *ANSYS* command needs to be prepared. Bearing area can be rigid by using Beam 4 unit, and interface of *ADAMS* and *ANSYS* is used to generate the modal neutral file (.mnf file). The file generated in ANSYS is read by *ADAMS/View*, and the ram in the multi-rigid-body dynamics model is replaced by the flexible body. The ram flexible body model generated by ANSYS is shown in Figure 5.

4   PRECISION SIMULATION OF MELON-DISK WELDING WORKING CONDITION

The configuration of the robot will be changed in order to meet the welding spot position when the FSW robot is welding. In the process, the configuration of the robot is continually varying, and the magnitude and direction of the inertia load on the component are also continually varying. The inertia loading includes angular velocity, angular acceleration, and acceleration at the vehicle mass center. They are the principal causes of inertial force and inertial moment. Because of the stiffness and the different mass distributions on different configuration of robots, the inertia load on each component is also different. In the process of the dynamic rigid-flexible coupling simulation, we know from static analysis that the working condition of the melon-disk welding has worst welding configurations. Therefore, the melon-disk welding working condition is selected to research the precision of robot welding seam.

Because the joint part and the ram part of the FSW robot are flexible, the end of mixing head under the action of external force will produce deformation, and the deformation amount influences the precision of robot welding seam. It is worth saying that the stiffness of different types of integration changes with the external load at different times. We should introduce the different stiffness of the integration to simulation calculation in the process of simulation by the calculation of separate manual or query sample handbook. We can get the displacement curve and the trajectory curve along the XYZ-axis of the end of the FSW robot welding head in the workpiece coordinate system through the simulation and analysis of rigid model, the rigid-flexible coupling with no-load and the rigid-flexible coupling coupled with a load model.

As shown in Figure 6, there are still some errors in the three-simulation curve, no matter changing with the XYZ-axis of workpiece coordinate system or the overall track changes of the end of mixing head. The error is caused by the flexible body ram and the dynamic stiffness of the joint part, and the variation amount is different in different configurations and different working conditions.

In order to verify the welding precision of the FSW robot under melon-disk welding condition, in the process of the above three simulations, we will respectively extract the displacement along the welding normal direction of mixing needle point on the shaft shoulder end face, and calculate the difference to obtain the error. Finally, in the process of the three simulations, the displacement error curve along the weld normal direction of mixing needle point on the shaft shoulder end face will be obtained when the FSW robot is under melon-disk welding condition, as shown in Figure 7.

Through the above simulation, we can get the following conclusions:

1. Considering the condition of the integration stiffness and the ram flexibility, in this process, the FSW robot moves from the starting point to the end of the weld without load, the whole simulation only considers the effect of gravity in addition to the gravity compensation mechanism of the external force. Compared with the simulation results of the rigid model, the maximum displacement error is 0.28 mm when the mixing needle point on the shaft shoulder end face of the FSW robot along the weld normal direction.

2. In the same way, considering the condition of the dynamic stiffness and the flexibility of ram, the robot completes the whole welding process with load. Compared with the simulation results of the rigid model, the maximum displacement error is about 0.22 mm, when the mixing needle point on the shaft shoulder end face of the FSW robot is along the weld normal direction. The error reduces 0.06 mm compared with the previous one. It shows that the deformation of the robot is recovered with load.

3. When the difference between the two simulation results is obtained, the welding precision of SFW robot will be got under melon-disk welding condition. From the figure, we can see the error precision is 0.06 mm. It eliminates the effect caused by static deformation of ram due to gravity.

According to the design indices of the SFW robot overall stiffness, the maximum deformation should be less than 0.1 mm when the mixing head is on the shaft shoulder end face of the FSW robot along the weld normal direction under the effect of the largest inserting resistance feeding resistance and rotating torque. The final design of the FSW robot can satisfy the given precision.

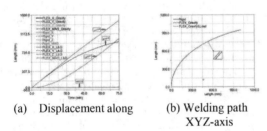

(a)   Displacement along (b) Welding path
                          XYZ-axis

Figure 6.   Displacement curve under mixing head workpiece coordinates.

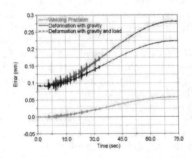

Figure 7.   Three types of weld simulation errors under melon condition.

5   CONCLUSION

In this paper, we introduced a kind of the FSW robot used to weld large, thin-walled curved surface, and the linkage motion condition of different axis under different welding conditions has been given. In order to verify the effect of robot system stiffness for welding precision, a dynamics model is built based on the virtual prototype of multibody system according to the actual welding condition. The error of the whole robot in the most serious condition (melon-disk welding condition), in the end time configuration (the maximum elongation of ram), is obtained through the rigid-flexible coupling dynamic simulation of the FSW robot.

By considering the dynamic stiffness of junction and the structure flexibility of the weakest ram, the worst working condition of melon-disk welding condition is simulated, and the welding precision of the whole robot is 0.06 mm, which can satisfy the given design index. By the simulation analysis, the judgment can be passed on the welding performance of the robot more realistically. The simulation result obtained is an important reference and guidance for subsequent structure improvement and validation.

ACKNOWLEDGMENTS

This work was supported by the National Science Foundation of China (Grant No. 51505470) and Dr. Start-up Fund in Liaoning province (No. 20141152).

REFERENCES

Chunlin Dong, Guohong Luan. Friction stir welding has been set sail in China. Advanced Welding Technology, 154–157 (2007).

Deshun Xia. The application of friction stir welding in the rocket. Missiles and Space Vehicles, **4**: 27–32 (2002).

Ebrahimi S., and Eberhard P. Rigid-Elastic Modeling of Gear Wheels in Multi-body Systems. Multi-body System Dynamics, **16**: 55–71 (2006).

Ghaisas N., Wassgren C.R., and Sadeghi F. Cage instability in cylindrical roller bearings. J.of Trib, **12**: 681–689 (2004).

Guohong Luan, Delun Guo, Tiancang Zhang. The application of friction stir welding in aircraft manufacturing industry. Aeronautical Manufacturing Technology, **11**: 20–24(2002).

Hagiu G.D., and Gafitanu M.D. Dynamics characteristics of high Speed angular contact ball bearings. Wear, **211**: 22–29 (1997).

Lambert R.J., Pollard A., and Stone B.J. Some characteristics of rolling-element bearings under oscillating conditions. Part 2: experimental results for interference-fitted taper-roller bearings. Proceedings of the Institution of Mechanical Engineers-Part K-Journal of Multi-body Dynamics, **20**: 171–179 (2006).

Miao Yu, Ji Zhao. Just a 5-dof virtual axis machine to rigid-flexible coupling dynamics model. Journal of Changchun University, **13**: 7–10 (2003).

Rubio H., Gareiaprada J.C., and Castejon C. Dynamic analysis of rolling bearing system using Lagrangian model Vs. FEM code. 12th IF To MM World Congress, Besancon. France, **13**: 18–21 (2007).

Wu J.S., and Chiang L.K. Dynamic analysis of an arch due to a moving load. Journal of Sound and Vibration, **26**: 511–534 (2004).

Advances in Energy Science and Equipment Engineering II – Zhou, Patty & Chen (Eds)
© 2017 Taylor & Francis Group, London, ISBN 978-1-138-71798-5

Optimal fixture locating layout design for sheet metal based on BA and FEA

Bo Yang, Zhongqi Wang, Yuan Yang & Yonggang Kang

The Ministry of Education Key Laboratory of Contemporary Design and Integrated Manufacturing Technology, Northwestern Polytechnical University, Xi'an, China

ABSTRACT: Fixture layout directly affects the dimensional quality of a deformable workpiece during the manufacturing process. One of the main tasks of Computer Aided Fixture Design (CAFD) is to determine the optimal positions of locators for reducing excessive deformation of a sheet metal workpiece. To this end, an approach based on the direct coupling of the Bat Algorithm (BA) with Finite Element Analysis (FEA) is proposed in this paper, to design the optimal fixture locating layout to reduce the deformation of sheet metal and improve the manufacturing precision. The FEA is used to calculate the deformation of sheet metal and evaluate the performances of different fixture locating layouts. The BA is applied to find the optimal fixture locating layout. Finally, a case study of sheet metal is conducted based on the "4-2-1" locating scheme to verify the application of the proposed method in the optimization of fixture locating for sheet metal parts.

1 INTRODUCTION

Sheet metal parts are widely used in aviation and automotive industries (Saadat, 2009; Yu, 2008). However, they are subject to deformation during the manufacturing process because of their thin wall and low rigidity. In order to reduce excessive deformation of sheet metal parts that are subject to loading, Cai et al. proposed the "N-2-1" (N ≥ 3) locating principle and indicated that this principle was more suitable for locating sheet metal parts than the "3-2-1" principle (Cai, 1996). One of the main tasks of Computer Aided Fixture Design (CAFD) is to determine the optimal number and positions of locators on a primary datum plane based on the "N-2-1" locating principle, in order to reduce excessive deformation and improve the dimensional quality of sheet metal during the manufacturing process. To this end, many scholars and technicians have conducted extensive research on fixture layout analysis and optimization over the past few years.

Initially, finite element analysis (FEA) was used to compute the deformation of sheet metal in order to analyze and evaluate the performances of different fixture locating layouts (Amaral, 2005; Siebenaler, 2006). Later on, Krishnakumar and Melkote (Krishnakumar, 2000) used the GA and FEA to find the optimal fixture layout to reduce the deformation of steel metal parts. Liao (Liao, 2003) proposed a GA-based method integrated with FEA to optimize the number of locators as well as their

positions in sheet metal assembly fixtures. Lai et al. (Lai, 2004) combined the GA with FEA to obtain the final fixture layout that can effectively control the influence of the fixture layout on the flexible sheet metal assembly deformation. Prabhaharan et al. (Prabhaharan, 2007) presented the fixture layout optimization methods using the GA and the Ant Colony Algorithm (ACA) separately integrated with FEA to reduce the dimensional and form errors of the workpiece. Based on the FEA and GA, Chen et al. (Chen, 2008) built a multi-objective model for fixture layout and clamp force optimization not only to reduce the excessive deformation of the workpiece but also to increase the uniform distribution of deformation. Padmanaban et al. (Padmanaban, 2009) applied the ACA-based optimization method coupled with FEA to optimize the fixture locating layout so that the elastic deformation of the workpiece was reduced. Kumar and Paulraj (Kumar, 2012) proposed a method to analyze and optimize the fixture layout using the GA with FEA for reducing the geometric dimensional errors of the workpiece. Abedini et al. (Abedini, 2014) developed a GA-based optimization method to obtain the optimal layout of locators for minimum machining error. By integrating FEA to compute the workpiece deformation for a given fixture layout, Dou et al. (Dou, 2012) described the applications and comparisons of the GA, improved GA, PSO, and improved PSO for the fixture layout locating optimization to reduce the elastic deformation of the workpiece. Xing et al. (Xing, 2015) proposed a new method to optimize fixture layout by using a Non-domination Sorting Social Radiation Algorithm (NSSRA) combined with FEA to satisfy the requirements of assembly tolerance. In summary, the direct coupling of evolutionary algorithms with FEA has already become the main stream approaches for fixture layout optimization to control the deformation of the workpiece.

Metaheuristic algorithms such as harmony search, firefly algorithm and cuckoo search algorithm have now become powerful methods for solving complex engineering optimization problems. Recently, a new metaheuristic algorithm called the Bat Algorithm (BA), which was inspired by the echolocation behavior of microbats, was proposed by Yang (Yang, 2010). Furthermore, Yang indicated that the BA is superior to the GA and PSO in terms of accuracy and efficiency.

In this paper, a new approach to integrate the BA directly with FEA is proposed to obtain the optimal fixture locating layout in order to reduce the deformation of sheet metal and improve its manufacturing precision. This paper is organized as follows. Section 2 presents the constructions of the optimization model of the sheet metal fixture layout. Section 3 introduces the basic principle of the BA. Section 4 describes the flowchart of the proposed method for the fixture locating layout optimization for sheet metal. Section 5 presents a case study based on the "N-2-1" (N ≥ 3) locating principle subject to its dead weight to verify the application of the proposed method. Finally, Section 6 summarizes the conclusions of the study.

2 OPTIMIZATION MODEL

In order to provide more support to reduce excessive deformation under the loading condition, the sheet metal part is always located based on the so-called "N-2-1" (N ≥ 3) locating principle during the manufacturing process. According to this principle, there are "N" locators on the primary/

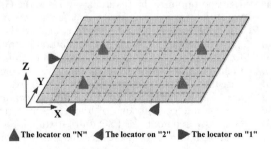

Figure 1. "N-2-1" locating principle.

first datum plane of the sheet metal part, and "2" and "1" on the second and third datum planes, respectively. Figure 1 illustrates the typical "4-2-1" principle. The locators' number "N" is determined by the dimensional specification of the sheet metal part. In fact, one of the main tasks of CAFD is the optimization of the positions of "N" locators to reduce the deformation of the sheet metal part.

The optimization model for the fixture locating layout of sheet metal can be defined as:

$$
\begin{cases}
\text{Find: } \mathbf{X} = \left[\mathbf{x}_1, \mathbf{x}_2, \cdots, \mathbf{x}_i, \cdots, \mathbf{x}_j, \cdots, \mathbf{x}_N \right], \\
\text{Min: } F(\mathbf{X}) = \sqrt{\dfrac{\sum_{i=1}^{K} \varepsilon_i^2(\mathbf{X})}{K}}, \\
\text{s.t. } \mathbf{x}_i \in \Omega \text{ and } \mathbf{x}_j \in \Omega, \\
\qquad \mathbf{x}_i \neq \mathbf{x}_j.
\end{cases}
\tag{1}
$$

where $\mathbf{X}$ is the design variable vector representing the fixture locating layout; $\mathbf{x}_i$ and $\mathbf{x}_j$ are the position vectors of the ith and jth locating points, respectively, with $i = 1, \cdots, N$ and $j = 1, \cdots, N$; N is the number of locating points; K is the number of all the finite element nodes of the sheet metal part; ε_i is the normal displacement of the ith finite element node; and Ω denotes the domain of all the node sets of the finite element model for the sheet metal part.

3 BAT ALGORITHM (BA)

Bats are fascinating mammals and microbats use a type of sonar called echolocation to detect prey, avoid obstacles, and locate their roosting crevices in the dark. By idealizing the echolocation behavior of bats, the bat algorithm can be formulated. In the BA, the basic steps of the algorithm can be summarized as the pseudo code (Yang, 2010; Gandomi, 2012), as shown in Figure 2.

For each bat (i), new solutions $\mathbf{X}_i^t$ and velocities $\mathbf{V}_i^t$ in a d-dimensional search space at the iteration t can be calculated as:

$$f_i = f_{\min} + \left(f_{\max} - f_{\min} \right) \beta \tag{2}$$

$$\mathbf{V}_i^t = \mathbf{V}_i^{t-1} + \left(\mathbf{X}_i^t - \mathbf{X}_* \right) f_i \tag{3}$$

$$\mathbf{X}_i^t = \mathbf{X}_i^{t-1} + \mathbf{V}_i^t \tag{4}$$

Here $\beta \in [0,1]$ is a random vector drawn from a uniform distribution; $\mathbf{X}_*$ is the current global best solution, which is located after comparing all the solutions among all the n bats; $f_{\min}$ and $f_{\max}$ are the

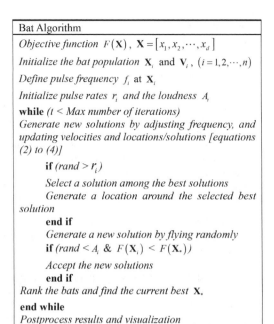

Bat Algorithm

Objective function $F(\mathbf{X})$, $\mathbf{X} = [x_1, x_2, \cdots, x_d]$

Initialize the bat population $\mathbf{X}_i$ *and* $\mathbf{V}_i$, $(i = 1, 2, \cdots, n)$

Define pulse frequency f_i *at* $\mathbf{X}_i$

Initialize pulse rates r_i *and the loudness* A_i

while *(t < Max number of iterations)*
Generate new solutions by adjusting frequency, and updating velocities and locations/solutions [equations (2) to (4)]

    if *(rand > r_i)*

    Select a solution among the best solutions
    Generate a location around the selected best solution

    end if
    Generate a new solution by flying randomly
    if *(rand < A_i & $F(\mathbf{X}_i) < F(\mathbf{X}_*)$)*

    Accept the new solutions
    end if
Rank the bats and find the current best $\mathbf{X}_*$

end while
Postprocess results and visualization

Figure 2.   Pseudo code of the BA.

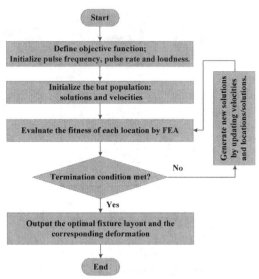

Figure 3.   The flowchart of the optimal fixture layout design for sheet metal based on the BA and FEA.

minimum and maximum pulse frequencies, respectively. In the local search part, once a solution $\mathbf{X}_{old}$ is selected among the current best solutions, a new solution $\mathbf{X}_{new}$ for each bat is generated locally by a local random walk:

$$\mathbf{X}_{new} = \mathbf{X}_{old} + \varepsilon A^t \qquad (5)$$

where $\varepsilon \in [-1,1]$ is a random number and $A^t = < A_i^t >$ is the average loudness of all the bats at the iteration t.

4   METHOD

In this paper, a new approach to optimize the fixture locating layout of sheet metal is proposed based on the "N-2-1" locating principle by the direct coupling of the BA with FEA. The fixture locating layout is considered as the design variable, while the overall sheet metal deformation under its dead weight is considered as the objective. The finite element model for the fixture locating layout of sheet metal is established to compute the objective fitness function value for each generation in the iterative optimization procedure, and the BA is applied to find the optimal design variable for the minimum fitness. The flowchart shown in Figure 3 explains the procedure of the proposed algorithm.

5   CASE STUDY

In this section, the fixture locating layout optimization for sheet metal by the direct coupling of the BA with FEA is illustrated by using an aluminum alloy sheet metal part. Its fixture locating scheme given by "N = 4" is shown in Figure 4. The sheet metal has dimensions of 400 mm × 400 mm × 1 mm and the physical properties of the material are listed in Table 1. The "4" locating points on the primary datum plane are L1, L2, L3, and L4, and the "2" locating points on the second datum plane are L5 and L6, while the "1" locating point on the third datum plane is L7. The coordinates of the fixed locating points L5, L6, and L7 are set as (133, 0), (267, 0), and (0, 200). The locating points to be optimized are L1, L2, L3, and L4 and their corresponding coordinates are denoted by (x1, y1), (x2, y2), (x3, y3), and (x4, y4), respectively. In the finite element model, the grid size is selected as 1 mm × 1 mm and the mesh type is S4R.

In the developed optimization model, the fixture locating layout is considered as the design variable, while the overall sheet metal deformation under its dead weight is considered as the objective. The commercial finite element software ABAQUS® (Simulia, 2012) is used to compute the objective fitness function value for a given fixture locating layout. The BA is applied to find the optimal design variable $\mathbf{X}$ for the minimum $f(\mathbf{X})$ on the platform of MATLAB®, respectively.

After several times of trials, the population size is fixed as 20 and the maximum iteration number is

Figure 4.   The initial fixture locating layout of the aluminum alloy sheet metal part.

Table 1.   The physical properties of the material.

	Value
Mass density	2.8×10^{-3} g/mm^3
Young's modulus	7.12×10^4 MPa
Poisson's ratio	0.33

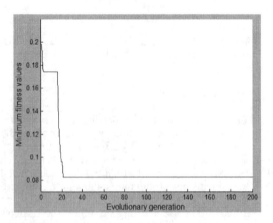

Figure 5.   The minimum fitness values obtained using the BA.

Table 2.   The best solution of the BA.

Optimal layout of L1, L2, L3, and L4	Minimum fitness value
(32,150), (254,75), (87,343), (321,288)	0.0825 mm

set as 200. Both the initial loudness and pulse rate in the BA are 0.5.

The optimization procedure of the minimum fitness by the BA for the fixture locating layout optimization is shown in Figure 5. The results for the minimum fitness and the corresponding fixture locating layout of the best solution are given in Table 2. From Figure 5 and Table 2, it can be inferred that the BA steadily reaches the minimum value after a few iterations. Following the optimization, the deformation of sheet metal can be reduced by 58%.

6   CONCLUSION

In this paper, a new method to integrate BA directly with FEA is proposed to deal with the problem of fixture locating layout optimization for sheet metal. The optimization model for the fixture locating layout is developed using the BA to reduce the overall sheet metal deformation through the iterative FEA method. The conclusions are drawn as follows:

1. The BA-based optimization method for the fixture locating layout design for sheet metal is studied and verified using an aluminum alloy sheet metal part. The results indicate that the BA coupled with FEA is capable of obtaining the optimal fixture locating layout for the sheet metal part.
2. The proposed approach can be further extended for fixture layout design and optimization at complex machining, assembly, and measuring stages.
3. In the near future, the BA will be compared with the GA and PSO in terms of search accuracy and efficiency in the optimization of the fixture locating layout for sheet metal.

ACKNOWLEDGMENTS

This work was supported by the National Natural Science Foundation of China (Grant No. 51375396).

REFERENCES

Abedini, V., M. Shakeri, M.H. Siahmargouei, Analysis of the influence of machining fixture layout on the workpiece's dimensional accuracy using genetic algorithm. Proc. IMechE Part B: J. Eng. Manuf., **228**, 11:1409–1418 (2014).

Amaral, N., J.J. Rencis, Y. Rong, Development of a finite element analysis tool for fixture design integrity verification and optimisation. Int. J. Adv. Manuf. Technol., **25**, 5:409–419 (2005).

Cai, W., S.J. Hu, J.X. Yuan, Deformable Sheet Metal Fixturing: Principles, Algorithms, and Simulations. ASME J. Manuf. Sci. Eng., **118**, 3:318–324 (1996).

Chen, W.F., L. Ni, J. Xue, Deformation control through fixture layout design and clamping force optimization. Int. J. Adv. Manuf. Technol., **38**, 9:860–867 (2008).

Dou, J.P., X.S. Wang, L. Wang, Machining fixture layout optimisation under dynamic conditions based on evolutionary techniques. Int. J. Prod. Res., **50**, 15:4294–4315 (2012).

Krishnakumar, K., S.N. Melkote, Machining fixture layout optimization using the genetic algorithm. Int. J. Mach. Tools Manuf., **40**, 4:579–598 (2000).

Kumar, K.S., G. Paulraj, Geometric error control of workpiece during drilling through optimisation of fixture parameter using a genetic algorithm. Int. J. Prod. Res., **50**, 12:3450–3469 (2012).

Lai, X.M., L.J. Luo, Z.Q. Lin, Flexible assembly fixturing layout modeling and optimization based on genetic algorithm, Chin. J. Mech. Eng., **17**, 1:89–92 (2004).

Liao, Y.G. A genetic algorithm-based fixture locating positions and clamping schemes optimization. Proc. IMechE. Part B: J. Eng. Manuf., **217**, 8:1075–1083 (2003).

Matlab 8.0, http://www.mathworks.com/, (2012).

Padmanaban, K.P., K.P. Arulshri, G. Prabhakaran, Machining fixture layout design using ant colony algorithm based continuous optimization method. Int. J. Adv. Manuf. Technol., **45**, 9:922–934 (2009).

Prabhaharan, G., K.P. Padmanaban, R. Krishnakumar, Machining fixture layout optimization using FEM and evolutionary techniques. Int. J. Adv. Manuf. Technol., **32**, 11–12:1090–1103 (2007).

Saadat, M., L. Cretin, R. Sim, F. Najafi. Deformation analysis of large aerospace components during assembly. Int. J. Adv. Manuf. Technol., **41**, 1:145–155 (2009).

Siebenaler, S.P., S.N. Melkote, Prediction of workpiece deformation in a fixture system using the finite element method. Int. J. Mach. Tools Manuf., **46**, 1:51–58 (2006).

Simulia, D.S. Abaqus 6.12 documentation, Providence, Rhode Island, US, (2012).

Xing, Y.F., M. Hu, H. Zeng, Y.S. Wang, Fixture layout optimisation based on a non-domination sorting social radiation algorithm for auto-body parts. Int. J. Prod. Res., **53**, 11:3475–3490 (2015).

Yang, X.S. *Nature-Inspired Metaheuristic Algorithms.* 2nd Ed., Luniver Press, Frome, UK, (2010).

Yang, X.S., A.H. Gandomi, Bat algorithm: a novel approach for global engineering optimization. Eng. Comput., **29**, 5:464–483 (2012).

Yang, X.S. A new metaheuristic bat-inspired algorithm, In: *Nature Inspired Cooperative Strategies for Optimization* (NISCO 2010) (Eds. J.R. Gonzalez et al.), Studies in Computational Intelligence, Springer Berlin, **284**, Springer, 65–74 (2010).

Yu, K.G., S. Jin, X.M. Lai, Y.F.X. Modeling and analysis of compliant sheet metal assembly variation. Assem. Autom., **28**, 3:225–234 (2008).

Advances in Energy Science and Equipment Engineering II – Zhou, Patty & Chen (Eds)
© *2017 Taylor & Francis Group, London, ISBN 978-1-138-71798-5*

Superpixel-based graph ranking for robust object tracking

Ruitao Lu, Wanying Xu, Yongbin Zheng & Xinsheng Huang
College of Mechatronic Engineering and Automation, National University of Defense Technology, Changsha, China

ABSTRACT: Online object tracking is a challenging issue because the appearance of an object tends to change due to intrinsic or extrinsic factors. In this study, we propose a novel object tracking algorithm based on superpixel-based graph ranking, which takes the intrinsic relationships of mid-level visual cues into account. First, the superpixel-base graph which consists of three types of subgraphs is constructed to encode local affinity information. Second, the query set selection mechanism is proposed based on the clustering method to integrate both foreground and background information. The confidence map is obtained by performing the ranking function on the constructed graph. Finally, object tracking is formulated as a transductive learning issue, and the optimal target location is determined by maximum a posterior estimation on the confidence map. During the procedure, a dynamic updating scheme is proposed to address appearance variations and alleviate tracking drift. A series of experiments and evaluations on various challenging image sequences are performed, and the results show that the proposed algorithm performs favorably against other state-of-the-art methods.

1 INTRODUCTION

Object tracking is one of the most important issues in computer vision, which has been widely applied in surveillance, activity analysis, classification and recognition from motion (Yilmaz, 2006; Li, 2006; Comaniciu, 2003). Although significant progress has been made over the past few decades, object tracking still remains challenging due to significant appearance changes caused by shape deformation, pose variation, motion blur, occlusion, camera motion and illumination.

With rich information of structure and great flexibility, superpixels have been one of the most effective representations widely applied in computer vision (Achanta, 2010; Levinshtein, 2012). X. Ren et al (Ren, 2007) presented a novel tracking method based on superpixels, and the paper considered the tracking task as a figure-ground segmentation. To tackle the problem of computational complexity and adaptability, S. Wang et al (Wang, 2011) formulated the tracking task by computing a confidence map which is obtained by superpixels grouping, and the best candidate is obtained by maximum a posterior estimation. Most of the papers take superpixels grouping into account for object detecting and tracking, this kind of relationships among features could be called extrinsic relationships. However, intrinsic relationships which contain local affinity information are also important for visual tracking.

In recent years, significant research has been performed regarding discriminative tracking methods (Ren, 2007; Wang, 2011; Babenko, 2009; Kalal, 2010; Wang, 2015). These methods consider visual object tracking as a binary classification problem by identifying targets from backgrounds. As one kind of the discriminative learning method, Graph-based transductive learning method (Zhang, 2014; Zhou, 2004; Yang, 2013) study the intrinsic geometric structure of both labeled and unlabeled samples and can thus explore the affinity relationships among vertices. Zhang et al. (Zhang, 2014) proposed a graph-embedding-based learning method in which a graph structure is designed to reflect the properties of the sample distributions. Yang et al. (Yang, 2013) exploited manifold ranking to detect salient regions in images, in which the image was presented by a close-loop graph that incorporated local grouping cues and boundary priors. The saliency of the image elements was obtained based on their relevancies to the labeled queries. However, the graphs constructed in these methods cannot represent the manifold effectively especially among unlabeled data. This limited work may fall down the performance of the algorithms.

In this study, we propose a novel object tracking algorithm based on superpixel-based graph ranking, which takes the intrinsic relationships of mid-level visual cues into account. Three subgraphs, namely, the full connection graph, the k-regular graph, the k-nearest neighbor graph, are constructed to encode high-order affinity relationships. The query set selection mechanism is proposed based on the clustering method to integrate both foreground and background information. The confidence map is obtained by performing the ranking function on the constructed graph. Then, object tracking is then formulated as a transductive learning problem with superpixel-based ranking analysis. The optimal target location is obtained by maximum a posterior estimation on the confidence map. In addition, a dynamic updating scheme is proposed to address appearance variations and alleviate tracking drift. A series of experiments and evaluations on various challenging sequences are performed, and the results show that the proposed algorithm performs favorably against other state-of-the-art methods.

2 BAYESIAN INFERENCE FRAMEWORK

Object tracking problem can be considered as a Bayesian inference framework in a Markov model (Wang, 2011), (Ross, 2008; Mei, 2009) with hidden state variables. Given a set of observed images $Y_t = \{y_1, y_2, \dots y_{t-1}, y_t\}$ at the t-th frame, the target state x_t can be computed by maximum a posterior estimation over the samples:

$$\hat{x}_t = \arg\max_{x_t^i} p(y_t^i \mid x_t^i) p(x_t^i \mid x_{t-1}), i = 1, 2, 3, \dots, N \quad (1)$$

where N is the number of samples, x_t^i is the i-th sample of the state x_t, and y_t^i is the image path predicted by x_t^i. The posterior probability $p(x_t \mid Y_t)$ can be inferred using the Bayesian theorem recursively:

$$p(x_t \mid Y_t) \propto p(y_t \mid x_t) \int p(x_t \mid x_{t-1}) p(x_{t-1} \mid Y_{t-1}) dx_{t-1} \quad (2)$$

where $p(y_t \mid x_t)$ denotes the observation model and $p(x_t \mid x_{t-1})$ denotes the dynamic model.

The dynamic model $p(x_t \mid x_{t-1})$ represents the correlation of states between two consecutive frames. The state transition distribution $p(x_t \mid x_{t-1})$ can be modeled by a Gaussian distribution around its counterpart in x_{t-1}: $p(x_t \mid x_{t-1}) = N(x_t; x_{t-1}, \Psi)$, where Ψ is a diagonal covariance matrix.

The observation model $p(y_t \mid x_t)$ denotes the likelihood of the observation y_t at state x_t. In this study, $p(y_t \mid x_t)$ is related to the confidence value, which is computed by performing the ranking function on the constructed superpixel-based graph.

3 SUPERPIXEL-BASED GRAPH RANKING

3.1 Ranking function

The graph $G = (A, E, w)$ is constructed with a vertex A, an edge set E, and a positive edge weight W. The edges between vertices $(\mathbf{a}_i, \mathbf{a}_j)$ represent their similarity in feature space. Using L_1 distance and Laplace kernel to measure the edge weights in W:

$$w_{ij} = k_L(\mathbf{a}_i, \mathbf{a}_j) = \exp(-|\mathbf{a}_i - \mathbf{a}_j| / \sigma_e^2) \quad (3)$$

where σ_e is a positive parameter that controls the strength of the weight. The sum of the i-th row of W is denoted by d_{ij} which is the diagonal element in the diagonal matrix D.

For a graph partition problem, the vertices linked by an edge are likely to have the similar label; vertices lying on a densely linked graph are likely to have the similar label. The cost function is defined as (Zhou, 2004):

$$\Phi(f) = \alpha \sum_{i,j=1}^{n} w_{i,j} \left| \frac{f_i}{\sqrt{d_{ii}}} - \frac{f_j}{\sqrt{d_{jj}}} \right|^2 + (1-\alpha) \sum_{i=1}^{n} |f_i - y|^2 \quad (4)$$

where $\alpha \in [0,1)$ is regularization parameter, y is the indication vector.

The learning task is to minimize the cost function with respect to f, which is described by:

$$f^* = \arg\min_f \Phi(f) \quad (5)$$

The result of the function can be computed by:

$$f^* = (1-\alpha)(I - \alpha S)^{-1} y \quad (6)$$

where $S = D^{-1/2} W D^{-1/2}$.

3.2 Superpixel-based graph construction

In this study, the vertices are represented by the segmented superpixels in surrounding area of the target location. To effectively capture the underlying high-order affinity relationships among vertices, three types of subgraphs are constructed: the full connection graph, the k-regular graph and the k-nearest neighbor graph. The weighted matrix in this study can be represented as follows:

$$W = \begin{pmatrix} W^{ll} & W^{lu} \\ W^{ul} & W^{uu} \end{pmatrix} \quad (7)$$

where W^{lu} and W^{ul} are symmetrical.

A. The full connection subgraph W^{ll}. The subgraph W^{ll} represents the distributions among labelled queries, which is constructed with a full connection graph. The weight of labelled queries can be denoted as:

$$w_{ij}^{ll} = 1/n_l \qquad (8)$$

where n_l is the number of labelled queries.

B. The k-regular subgraph W_{lu}. The subgraph W_{lu} represents the distributions between labelled queries and unlabelled queries. To effective utilize the spatial relationship, the graph is constructed with k-regular graph, whose nodes are connected to both its neighbourhoods and the nodes which have common boundaries with the neighborhood ($k = 2$). This can be denoted as:

$$w_{ij}^{lu} = \begin{cases} \exp(-|\mathbf{a}_i - \mathbf{a}_j|/2\sigma_e^2) & if \ \mathbf{a}_j \in krg(\mathbf{a}_i) \\ 0 & others \end{cases} \qquad (9)$$

where $\mathbf{a}_i$ is labeled query and $\mathbf{a}_j$ is unlabeled query. We set $k = 2$ and $\sigma_e^2 = 0.1$ in the weights W_{lu} in the experiments.

C. The k-nearest neighbors subgraph W_{ll}. The subgraph W_{ll} represents the distributions among unlabelled queries. To utilize the relationship in feature space, k-nearest neighbor graph is constructed whose nodes are connected to its neighbourhoods in feature space.

$$w_{ij}^{uu} = \begin{cases} \exp(-|\mathbf{a}_i - \mathbf{a}_j|/2\sigma_e^2) & if \ \mathbf{a}_j \in knn(\mathbf{a}_i) \\ 0 & others \end{cases} \qquad (10)$$

where $\mathbf{a}_i$ and $\mathbf{a}_j$ are unlabeled queries. We set $k = 7$ and $\sigma_e^2 = 0.1$ in the weights W_{lu} in the experiments.

3.3 Query set selection mechanism

Two query sets are selected for ranking: positive query set Q^+ and negative query set Q^-. The classifier based clustering method is constructed to generate the query sets.

Firstly, M frames with the ground truth of the target are used for training. The surrounding area of the target is segmented into N_t superpixels, which can be defined as $\mathbf{a}_{t,r}(t = 1,...,M, r = 1,...,N_t)$. The feature pool $F = \{\mathbf{a}_{t,r} | t = 1,...,M, r = 1,...,N_t\}$ is cluster into n different clusters $Cluter_i(i = 1,...,n)$ by Mean Shift clustering [15]. The positive cluster $Cluter^+_j(j = 1,...,N_+)$ is defined as $S_j^+/S_j^- > \lambda(\lambda > 1)$, where S^+ denotes that $Cluter_j$ contains the local areas of superpixels located in target area, and S^- represents the local areas outside the target area.

With a new test frame, the positive query set Q^+ is obtained by finding k-nearest neighbourhood of the center of the $Cluter^+_j(j = 1,...,N_+)$:

$$Q^+ = \{Q_j^+ | Q_j^+ \\ = \underset{r=1,...,N_t}{\arg\min}(dis(\mathbf{a}_{M_t,r}, Cluter_j^+)), j = 1,...,N_+\} \quad (11)$$

Analogously, the negative cluster is defined $Cluter_j^-(j = 1,...,N_-)$ if $S_j^+/S_j^- < \lambda$. Based on the visual attention theories, the negative query set Q^- contains two parts:

$$Q^- = Q_1^- \cup Q_2^- \qquad (12)$$

where Q_1^- is obtained from negative cluster, and Q_2^- is obtained from the boundary of surrounding area:

$$\begin{cases} Q_1^- = \{Q_{1,j}^- | Q_{1,j}^- = \underset{r=1,...,N_t}{\arg\min}(dis(\mathbf{a}_{M_t,r}, Cluter^-)), j = 1,...,N_-\} \\ Q_2^- = \{\mathbf{a}_{M_t,r} | \mathbf{a}_{M_t,r} \in boundary, r = 1,...,N_t\} \end{cases} \quad (13)$$

3.4 Observation model based on the confidence map

Based on the positive query set, indication vector $y^+ = [y_1, y_2,..., y_n]$ is defined where $y_i = 1$ if $\mathbf{a}_i \in Q^+$ and $y_i = 0$ for other vertices. The graph weight W^+ is computed based on the Q^+ and other unlabeled samples using Eq(7). The confidence map C_m^+ belonging to the target is obtained:

$$C_m^+ = f^{+*} = A^+ b^+ \qquad (14)$$

where the optimal affinity matrix $A^+ = (1 - \alpha)$ $(I - \alpha S^+)^{-1}$.

Analogously, based on the graph weight W^- and indication vector y^-, the confidence map belonging to the background C_m^- can be written as:

$$C_m^- = f^{-*} = A^- b^- \qquad (15)$$

where $S^- = (D^-)^{-1/2} W^- (D^-)^{-1/2}$.

The final confidence map combining the two confidence maps can be written as

$$C_m = C_m^+ \times (1 - C_m^-) \qquad (16)$$

Figure 1 shows the construction of the final confidence map. It is obvious that a good target candidate not only has a higher confidence value, but also covers more part of foreground regions. The observation model is defined as follows:

$$p(\mathbf{y}_t^i | \mathbf{x}_t^i) = C_t^i = v_t^i \exp(-|\mathbf{s}_t^i - \hat{\mathbf{s}}_t^i|_2))/|\mathbf{s}_t^i| \qquad (17)$$

1553

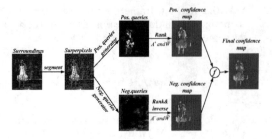

Figure 1. An illustration of the primary steps of the final confidence map construction.

where $v_t^i = \sum f_{i,j}, i = 1,...,N$ denotes the sum of confidence values of candidate target areas. $s_t^i = [h_t^i, w_t^i]$ is an area vector of the i-th sample of the state $\mathbf{x}_t$, h_t^i is the height of the area, and w_t^i is the width. And $\hat{\mathbf{s}}_t$ denotes the estimation of the target area at the t-th frame using linear estimation. The candidate target with the highest confidence value is considered to be the object state estimation in the current frame.

3.5 Occlusion handling and updating scheme

Because the tracker is based on confidence map, it can overcome the partial occlusions or very brief occlusions automatically. We compute the ratio $\tau = |1 - \max(C_t^i)/\bar{C}_t|$, where C_t^i denotes the confidence rate of the i-th sample at the t-th frame. Two threshold θ_0 and θ_1 are defined to describe the degree of occlusion (e.g., $\theta_0 = 0.2$ and $\theta_1 = 0.8$ in this paper). If $\tau > \theta_0$, it means the target undergoes a heavy occlusion, the x, y translation is estimated by Kalman filtering with a constant acceleration for tracking, the other parameters is the same as $t-1$. And we discard this sample without update. If $\tau < \theta_1$, we put the samples into the training dataset and delete the oldest one. And the appearance model constructed by the training dataset is updated for every W frames.

4 EXPERIMENTAL RESULTS

The proposed algorithm was written in MATLAB 2011 and ran at 1 frame per second on a Core 2.0GHz Dual Core PC with 2 GB of RAM. The number of frames for training was set to 5. The area of the surrounding region was set 1.5 times of the size of target area. The number of superpixels was set to 300. The observation of the target candidate was normalized to 16×16 pixels, and the number of samples was set to 300 in all experiments. The threshold of query generator λ was set to the range of 2 and 4. The positive edge weight σ_e was defined as $\sigma_e^2 = 0.1$, and the negative edge

Figure 2. Qualitative evaluation of the 6 challenging image sequences, being Lemming, Carscale, Woman, Caviar1 Basketball and Singer1.

weight was defined as $\sigma_e'^2 = 0.8\sigma_e^2$. In the mean shift clustering, the bandwidth was set to the range between 0.15 and 0.25. The update frequency W was set to 8. The balance weight α is 0.99.

To evaluate the performance of the proposed algorithm, we use a set of 6 challenging image sequences (Wu, 2013). We compare the proposed tracking methods with 7 state-of-the-art visual trackers including visual tracking by fragment tracker (FRAG) (Adam, 2006), incremental visual tracker (IVT) (Ross, 2008), L1 tracker (Mei, 2009), MIL tracker (Babenko, 2009), Mean Shift (MS) (Comaniciu, 2000), decomposition (VTD) algorithm (Kwon, 2010) and PN tracker (Kalal, 2010). Both qualitative and quantitative evaluations are presented in this section.

4.1 Qualitative evaluation

We test our method with 6 challenging image sequence. The image sequence include pose variation (Lemming, Carscale), occlusion (Woman, Caviar1), illumination change and cluttered background (Basketball, Singer1). As shown in Fig 2, the proposed tracker achieves favourable results with most of the sequences compared with the other state-of-the-art methods tested in this paper.

The reason for these results can be attributed to the constructed superpixel-based graph can capture the underlying high-order affinity relationships among the background and foreground samples. The dynamic updating scheme allows the proposed tracker to manage significant pose variations well throughout these sequences.

Table 1. The center location error (CLE) (in pixels). The Bold fonts indicate the best performance while the italic fonts indicate the second best performance.

Image sequence	FRAG	IVT	L1T	MIL	VTD	P-N	MS	Ours
Lemming	149	93.4	184	25.6	86.9	*23.2*	108	**10.1**
Singer1	22	8.5	*4.6*	15.2	**4.1**	32.7	33.1	5.6
Carscale	28.8	61.6	37.2	*31.6*	39.7	84.8	40.7	**6.3**
Caviar1	5.7	45.2	120	48.5	**3.9**	5.6	9.5	*5.2*
Woman	101	182	132	125	119	*22.2*	131	**9.9**
Basketball	*11.7*	115	125	106	177	130	77.1	**8.9**

Table 2. The overlap rate (ORE). The Bold fonts indicate the best performance while the italic fonts indicate the second best performance.

Image sequence	FRAG	IVT	L1T	MIL	VTD	P-N	MS	Ours
Lemming	0.13	0.18	0.13	*0.53*	0.35	0.49	0.40	**0.71**
Singer1	0.34	0.66	0.70	0.34	**0.79**	0.41	0.36	*0.77*
Carscale	0.37	*0.46*	*0.46*	0.42	0.37	0.35	0.24	**0.78**
Caviar1	0.68	0.28	0.28	0.25	**0.83**	0.70	0.58	*0.77*
Woman	0.15	0.13	0.15	0.17	0.15	*0.51*	0.13	**0.68**
Basketball	*0.65*	0.07	0.17	0.23	0.03	0.07	0.40	**0.70**

4.2 Quantitative evaluation

For thorough evaluations, two evaluation criteria are used to assess the performance of tracking algorithms: the center location error and the overlap rate (Everingham, 2010). The average errors are presented in Table 1. The proposed tracker performs favorably against the other methods tested on these sequences.

A frame is counted when the center location error is within a certain threshold. The overlap rate is defined as $score = area(ROI_T \cap ROI_G / ROI_T \cup ROI_G)$, where ROI_T is the tracked bounding box and ROI_G is the ground truth bounding box. Table 2 presents the average overlap rates. The proposed tracker outperforms other tracking algorithms. The success of the proposed tracker can be attributed to the effective discriminative appearance model with mid-level visual representation.

5 CONCLUSION

In this study, we propose a new tracking algorithm based on a superpixel-based graph, which is constructed by encoding the local affinity information using three types of subgraphs to capture high-order affinity relationships. Object tracking is formulated as a transductive learning issue, and the optimal target location is obtained by maximum a posterior estimation based on the confidence map. Both quantitative and qualitative evaluations on challenging image sequences against several state-of-the-art algorithms demonstrate the accuracy and robustness of the proposed tracker.

ACKNOWLEDGMENTS

This study was supported in part by the National Science Foundation of China under Grants NO. 61403412.

REFERENCES

Achanta, R., K. Smith, A. Lucchi, P. Fua, S. Susstrunk. Slic superpixels. Technical report, EPFL, Tech. Rep. (2010).

Adam, A., E. Rivlin, I. Shimshoni. Robust fragments-based tracking using the integral histogram. *In Proceedings of CVPR*, New York, NY, USA, 798–805 June (2006).

Babenko, B., M.H. Yang, S. Belongie. Visual tracking with online multiple instance learning. *In Proceedings of CVPR*, Miami Beah, Florida, USA, 983–990 June (2009).

Comaniciu, D., V. Ramesh, P. Meer. Kernel-based object tracking. IEEE Trans. Pattern Anal. Mach. Intell. **25**, 564–577 (2003).

Comaniciu, D., V. Ramesh, P. Meer. Real-Time tracking of Non-Rifgid Objects using MeanShift. *In Proceedings of CVPR*, Hilton Head Island, SC, USA, 142–149 June (2000).

Everingham, M., L.V. Gool, C.K. Williams, J. Winn. Zisserman A. The pascal Visual Object Classes (VOC) challenge. Int. J. Comput. Vis. **88**, 303–338 (2010).

Kalal, Z., J. Matas, K. Mikolajczyk. P-n learning: Bootstrapping binary classifiers by structural constraints," *in Proceedings of CVPR*, San Francisco, CA, USA, 49–56 June (2010).

Kwon, J., K.M. Lee. Visual tracking decomposition. *In Proceedings of CVPR*, San Francisco, CA, USA, 1269–1276 June (2010).

Levinshtein A., Sminchisescu C., Dickinson S. Optimal Image and Video Closure by Superpixel Grouping. Title of Presentation. Int. J. Comput. Vis. **100**, 99–119 (2012).

Li, X., W. Hu, C.H. Shen. Zhang Z. Asurvey of appearance models in visual object tracking. ACM Transactions on Intell. Syst.Technol. **4**, 58 (2006).

Mei, X., H. Ling. Robust visual tracking using l1 minimization. *In Proceedings of CVPR*, Miami Beah, Florida, USA, 1436–1443, June (2009).

Ren, X.F., J. Malik. Tracking as repeated figure/ground segmentation. *In Proceedings of CVPR*, Minneapolis, Minnesota, USA, 1–8 June (2007).

Ross, D.A., J. Lim, R.S. Lin, M.H. Yang. Incremental learning for robust visual tracking. Int. J. Comput. Vis. **77**, 125–141 (2008).

Wang, B., L. Tang, J. Yang, B. Zhao, S. Wang. Visual Tracking Based on Extreme Learning Machine and Sparse Representation. Sensors. **15**, 26877–26905 (2015).

Wang, S., H.C. Lu, F. Yang, M.H. Yang. Superpixel tracking. *In Proceedings of ICCV*, Barcelona, Spain, 1323–1330 November (2011).

Wu, Y., J. Lim, M.H. Yang. Online Object Tracking: A Benchmark. *In Proceedings CVPR*, Portland, OR, USA, 2411–2418 June (2013).

Yang, C., L.H. Zhang, H.C. Lu, X. Ruan, M. Yang. Saliency Detection via Graph-Based Manifold Ranking. *In Proceedings CVPR*, Portland, OR, USA, 3166–3173 June (2013).

Yilmaz, A., O. Javed, M. Shah. Object tracking: A survey. ACM Comput. Surv. **38**, 1–45 (2006).

Zhang, X.Q., S.Y. Chen. Graph-Embedding-Based Learning for Robust Object Tracking. IEEE Trans. Ind. Electron. **61**, 1072–1084 (2014).

Zhou, D., J. Weston, A. Gretton. Ranking on data manifold. *In Proceedings of NIPS*, Vancouver, Canada, 169–176 December (2004).

Advances in Energy Science and Equipment Engineering II – Zhou, Patty & Chen (Eds)
© 2017 Taylor & Francis Group, London, ISBN 978-1-138-71798-5

The Chebyshev spectral collocation method of nonlinear Burgers equation

Lizheng Cheng & Hongping Li
Hunan International Economics University, Changsha, China

ABSTRACT: The nonlinear Burgers equation is a hot issue in the field of computational fluid dynamics, which has nonlinear convection term and diffusion term. In this paper, we give a numerical solution of the nonlinear Burgers equation by a combination of finite difference and Chebyshev spectral collocation. The nonlinear Burgers equation is solved by use of the Crank-Nicolson Leap-frog scheme in time, and the Chebyshev spectral collocation method in space. Finally, numerical experiments are carried out to illustrate the efficiency of the proposed method.

1 INTRODUCTION

Burgers equation can be used as a mathematical model to describe many physical phenomena, in particular, it can be used as a mathematical model of fluid flow phenomena. It is a very important and basic partial differential equation in fluid mechanics, which is widely used in aerodynamics, turbulent flow, heat transfer, transportation flow, groundwater pollution and other fields. It has some properties of the Navier-Stokes equation, hence, it can be regarded as a simple Navier-Stokes equation. Burgers equation can be used to study the properties of the Navier-Stokes equation by numerical simulation. However, It usually produces shock waves in this equation, therefore, it brings some difficulties to the numerical solution. The further research on it will help us solve other non linear problems.

At the same time, the exact smooth solution of the Burgers equation can be constructed, therefore, many nonlinear Burgers equation can be solved numerically by means of model which can be numerically tested. As a result, it is important to study the efficient numerical method for the Burgers equation. The numerical approaches of the Burgers equation have been developed as follows: the finite element method, finite difference method and Chebyshev spectral method and spectral collocation method (Fletcher, 1983; Shen, 2007; Luo, 2008; Jin, 1994). In 1915, Bateman-Lrstly got the stable solution of one dimensional viscous Burgers equation(Bateman, 1915). In some special initial boundary value condition, the exact solution of the Burgers equation can be arrived, but these exact solutions are usually in the form of infinite series. Therefore, We must calculate much

more items to get the high accuracy, and the calculation process is also very complex.

The paper is organized as follows. Section 1 mainly introduces the physical background of the Burgers equation and the main numerical methods. Section 2 mainly illustrates the basic principle of the Chebyshev-Galerkin method of the nonlinear Burgers equation. In section 3, we design the numerical scheme and algorithm of the nonlinear Burgers equation. The numerical experiments of the nonlinear Burgers equation are presented in section 4 to verify the efficiency of the method.

2 THE CHEBYSHEV SPECTRAL COLLOCATION METHOD

Let $\Omega=(a, b)$ is an open set,we consider the following nonlinear Burgers equation with initial and bound conditions:

$$\begin{cases} \dfrac{\partial u}{\partial t} = \varepsilon \dfrac{\partial^2 u}{\partial x^2} - u \dfrac{\partial u}{\partial x}, & a < x < b, 0 < t \le T, & (1) \\[2mm] u(0,t) = u(1,t) = 0, & t > 0, & (2) \\[2mm] u(x,0) = u_0(x), & a < x < b. & (3) \end{cases}$$

where u is the speed, ε is the viscosity constant and $u_0(x)$ is the initial function.

The collocation method, or more specifically the collocation method in the strong form, is fundamentally different from the Galerkin method, in the sense that it is not based on a weak formulation. Instead, it looks for an approximate solution which enforces the boundary conditions in (2) and collocates (1) at a set of interior collocation points.

We describe below the collocation method for the two-point boundary value problem (1) with the general boundary conditions (2) and initial conditions (3). Notice that the nonhomogeneous boundary conditions can be treated directly in a collocation method so there is no need to "homogenize" the boundary conditions as the Galerkin methods. Now we consider Chebyshev spectral collocation method in this paper.

Let $\left\{x_j\right\}_{j=0}^{N}$ be the Chebyshev-Gauss-Lobatto points on the $\Omega = (-1, 1)$, $\omega = (1-x^2)^{-1/2}$ be the corresponding weight function, and let h_j be the Chebyshev basic polynomials associated with $\left\{x_j\right\}_{j=0}^{N}$. Denoting

$$D=(d_{kj}=h_j'(x_k))_{k,j=0,1,\cdots,N}, l_j = u_N(x_j),$$

We have

$$\int_a^b f(x)dx = \sum_{k=0}^{N} x_k \omega(x_k),$$

$$u_N = \sum_{j=0}^{N} l_j h_j(x), \quad u_N(x_k) = \sum_{j=0}^{N} l_j h_j(x_k) = l_k,$$

$$u_N'(x_k) = \sum_{j=0}^{N} l_j h_j'(x_k) = \sum_{j=0}^{N} d_{kj} l_j$$

$$= \sum_{j=1}^{N-1} d_{kj} l_j + d_{k0} l_0 + d_{kN} l_N,$$

$$u_N'(x_k) = \sum_{j=0}^{N} l_j h_j'(x_k) = \sum_{j=0}^{N} (D^2)_{kj} l_j$$

$$= \sum_{j=1}^{N-1} (D^2)_{kj} l_j + (D^2)_{k0} l_0 + (D^2)_{kN} l_N.$$

From the paper (Shang, 1996): if the initial function such that $u_0(x) \in H^m(\Omega), m \geq 2$, then equations (1)–(3) have a unique solution u and $(\frac{\partial}{\partial t})^j u \in C(0,T; H^{m-j}(\Omega)), \quad j \geq 0, \ m - j \geq 0$ and j is integer. In addition, there exists a constant $C(\|u_0\|_m)$ such that

$$\left\|(\frac{\partial}{\partial t})^j u\right\|_{m-j} \leq C\left(\|u_0\|_m\right).$$

3 NUMERICAL SCHEME

In the equations (1)–(3), We use the Crank-Nicolson Leap-frog scheme in time, and the Chebyshev spectral collocation method in space. Let the time step be dt, $u^k(*)$ is the approximate solution of the exact solution $u(*,kdt)$, where u^k is the (N+1)-dimension vector, $u^k(j) = u^k(x_j)$ n = 1, 2, ..., [T/dt]([ ] is rounding), therefore, the equatons (1)–(3) can discrete as following:

$$\frac{u^{n+1} - u^{n-1}}{2dt} = \frac{\varepsilon}{2} D^2(u^{n+1} + u^{n-1}) - u^n Du^n \ .$$

Let

$$u^{n+1} = \begin{pmatrix} u(x_0,(n+1)dt) \\ u(x_1,(n+1)dt) \\ \vdots \\ u(x_N,(n+1)dt) \end{pmatrix}, u^n = \begin{pmatrix} u(x_0,ndt) \\ u(x_1,ndt) \\ \vdots \\ u(x_N,ndt) \end{pmatrix},$$

$$u^{n-1} = \begin{pmatrix} u(x_0,(n-1)dt) \\ u(x_1,(n-1)dt) \\ \vdots \\ u(x_N,(n-1)dt) \end{pmatrix},$$

Then $A = I - \varepsilon dt D^2, B = I + \varepsilon dt D^2$, Where $D = D_{kj}$, $k, j = 1,2,\cdots,N-1$, I is (n-1)-dimension unit vector, which leads to the matrix equation as following:

$$\begin{cases} Au^{n+1} = Bu^{n-1} - u^n Du^n, & 0 \leq n \leq [T/dt], & (4) \\ u(x_0,t) = u(x_N,t) = 0, & t > 0, & (5) \\ u(x_j,0) = u_0(x_j), & j = 0,1,\cdots,N. & (6) \end{cases}$$

4 NUMERICAL EXAMPLE

In this section, we present below numerical experiment using the Chebyshev spectral Collocation

Table 1. The discrete maximum errors.

$dt = 10^{-k}$	n = 32	n = 64	n = 128
K = 2	0.00256548563519249	5.33391163407559e-05	0.0532202205148147
K = 3	0.00255379863620286	2.38707334714317e-05	3.22153911880285e-07
K = 4	0.00255384305444051	2.38982668693266e-05	3.80712372738401e-09

method designed in the previous section. All computations are performed with matlab on a personal computer. We consider the following problem:

$$\frac{\partial u}{\partial t} = \varepsilon \frac{\partial^2 u}{\partial x^2} - u \frac{\partial u}{\partial x}, \quad \varepsilon > 0.$$

The exact solution is

$$u(x,t) = \kappa \left[1 - \tanh\left(\frac{\kappa(x - \kappa t - x_c)}{2\varepsilon} \right) \right],$$

Where the parameter $\kappa > 0$ and the center $x_c \in \mathbb{R}$.

Let $\varepsilon = 0.1, \kappa = 0.5, x_c = -3, x \in (-5,5)$, $T = 12$, and impose the initial value $u(x,0)$ and boundary conditions $u(\pm 5, t)$ by using the exact solution. Use the Crank-Nicolson leap-frog scheme to in time (see (2)–(3)), and the Chebyshev collocation method in space to solve the equation. The time step size are $dt = 10^{-k}$, $k = 2, 3, 4$, the number of Chebyshev spectral collocation points are 32, 64, 128 at $T = 12$.

In Figure 1 and table 1, n is the number of Chebyshev spectral collocation points, dt is the time step. From the figure 1, we can observe the approximate solutions converge exponentially to the exact solution. In addition, from the table 1, when the time step value is smaller, the Chebyshev spectral collocation method can get the higher

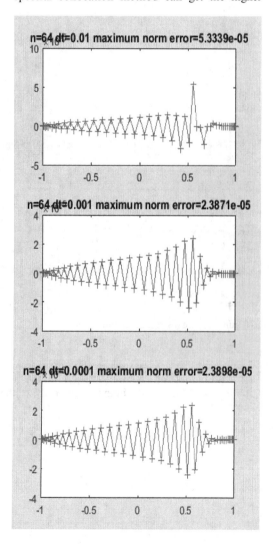

Figure 1. (*Continued*)

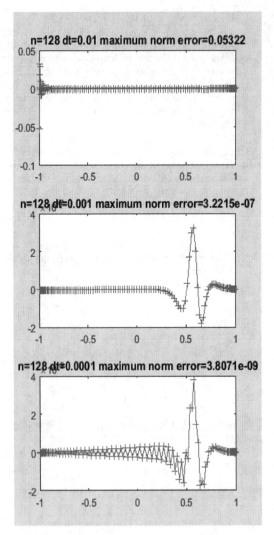

Figure 1. The discrete maximum errors.

accuracy. Table 1 shows: when the time step $dt = 10^{-4}$, the number of Chebyshev spectral collocation points are 128, the maximum errors can get the accuracy of 10^{-9}.

5 CONCLUDING REMARKS

We developed an efficient Chebyshev spectral collocation method for nonlinear Burgers equation. There are some advantages of using Chebyshev spectral collocation method, e.g., its implementation is simple and the method seems convenient to handle nonlinear or more complicated problems. In addition, our program avoid a large number of circulation structure, and our method can quickly and effectively calculate the results of high accuracy. This is the other advantage in dealing with nonlinear problems. How to efficiently solve the more complication problems and how to analysis the error will be the further work.

REFERENCES

Fletcher, C.A. A comparison of finite element and finite difference solution of the one-and two-dimensional Burgers equations [J]. J Comp Phys, 1983(51): 159–188.

Shen, J., T. Tang. High Order Numerical Methods and Algorithms [M]. Beijing: Science press, 2007.

Luo zhendong, Liu ruxun. Mixed finite element analysis and the numerical simulation of Burgers equation [J]. Computational mathematics, 2008, 21(3): 257–268.

Jin xiaokang, Liu ruxun, Jiang bocheng. Computational Fluid Dynamics [M]. Changsha: National University of Defense Technology, 1994.

Bateman. Some tecent researches on the motion of fluids [J]. Monthly Weather Rev, 1915(43): 163–170.

Benton, E., G.W. Platzman. A table of solutions of the one dimensional Burgers equation [J]. Quart Appl Math, 1972(30): 195–212.

Shang, Y.D. Initial boundary value problem for a class of generalized Burgers type equations [J]. Mathematica Applicata, 1996(9): 166–171.

Shen, J. Effcient spectral-Galerkin method II. Direct solvers for second- and fourth-order equations by using Chebyshev polynomials [J]. SIAM J Sci Comp, 1995(16): 74–87.

Battles, Z., L.N. Trefethen. An extension of Matlab to continuous functions and Operators [J]. SIAM J Sci Comp, 2004(25): 1743–1770.

Canuto, C., M.Y. Hussaini, A. Quarteroni, T.A. Zang. Spectral Methodsin Fluid Dynamics [M]. New York: Springer Verlag, 1988.78.

The study and design of using hybrid development framework WeX5 to develop "takeaway online" apps

Gongjian Zhou

School of Management, Tan Kah Kee College, Xiamen University, Zhangzhou, Fujian, China

ABSTRACT: This paper focuses on and spreads out from the development of Apps. After comparing the development modes of the current mainstream mobile Apps, the paper introduces the famous open source, free, Hybrid-type WeX5 development framework and uses it to develop the "takeaway online" apps whose functions are outstanding and applications are wide. This paper comprehensively analyzes and explains all the steps to design the App, including the analysis and design of the system's function requirements, the process of data interaction, the design of the database and the implementation, debugging, packaging and releasing of the system. The design methods, technologies, and ideas mentioned in the paper will have a strong reference and commercial application value for App developers.

1 INTRODUCTION

In recent years, with the rapid development of mobile Internet technology and the rapid adoption of smart phones, tablet PCs and other mobile devices, bringing out explosive growth of mobile apps application, the traditional PC Web application has gradually been replaced by mobile Apps of Android, Apple or other mobile Apps. Besides, more and more developers start to pay attention to and plunge into the development of mobile Apps. However, because different system platforms are not compatible, data processing technologies are not unified, the development of mobile Apps often needs to be designed and developed separately for different systems, which leads to high cost and the complexity of maintenance and update. (Xu, 2014; Hui, 2016) Therefore, choosing a good App development framework and development mode has become the key point to reduce development costs and improve the development efficiency.

2 MOBILE APP DEVELOPMENT MODEL

The mobile App development technology involves a wide-ranging scope, including different terminal systems (Android, iOS, WP, etc.), UI standards and technologies (html5, css3, js etc.), server-side technology (Java, Php, C#, etc.). In terms of the development mode, it is divided into three categories: native App (native App development mode) and web App (web App development mode), as well as Hybrid App (hybrid App development mode). (Xu, 2014; Hui, 2016) Which kind of development modes you are going to use will have an effect

on the development cycle, operating results and application scope of Apps.

1. **Native App**: This mode consists of two parts: the "cloud + client-side Apps". All the UI elements, data content, logical framework of the Apps are installed on a mobile terminal. After the code is compiled, the OS Device API is needed, so as to phones with different operating systems, such as IOS, Android or other systems, different languages and frameworks should be used for development. To develop Android Apps, developers need to master java language and understand the Eclipse. (Xu, 2014) To develop iOS Apps, developers need to master objective-c language and learn Xcode. This mode has the advantages of good security and interaction, but the cost of development and maintenance is high and it is also difficult to develop. (Hui, 2016).

2. **web App**: The pattern consists of two parts: "HTML5 cloud website + client-side Apps". The client-side App only needs to install its framework and the code will be run in a browser. The browser will invoke the Device API, and the application data can be extracted each time you open the App from the cloud server. (Xu, 2014; Hui, 2016) To develop web Apps, developers need to master the HTML5, CSS3, JavaScript and other technologies. This mode has the advantages of good cross-platform support, low development cost and short cycle, but its security is not good and it lacks interactive experience.

3. **Hybrid App**: This kind of App is developed from the previous two Apps and it has the

advantages of the two Apps, which are Native App's good user interaction and Web App's low cost of cross-platform Web App development. The development of Hybrid Apps is dominated by the mutual calling between JS and Native and it can achieve the goal of "a development, multiple runs" from the development level. The mainstream development platforms include PhoneGap, AppCan, AppMobi, Titanium, etc., which are all based on the open source webkit core, use HTML5 standard, fit various mobile terminal modes, allow developers to customize the widget, and they can also be well used in the development of application in various industries. With the Hybrid app development modes becoming more and more mature, nowadays many enterprise-scale customers are considering Hybrid app solutions. (Xu, 2014; Hui, 2016).

3 WEX5 DEVELOPMENT FRAMEWORK

WeX5 is a free, cross-end and fast App development framework which is developed by Beijing Justep Software Co. Ltd through using Apache v2.0 open source license. Based on phonegap (cordova)'s hybrid App development mode and the MVC design pattern, data and view can be separated; Meanwhile, page description and code logic can also be separated. (Pan, 2014) The back-end server is fully open, and it can be connected to a variety of back-end middle wares or cloud services via http, Web socket protocols (java, node, php, .net, etc.). UI front-end system is based on the w3c's html5 + css3 + js, jquery and bootstrap are introduced to optimize its page and this system is of high efficiency.

The IDE of WeX5 is based on eclipse, and it provides a fully visual, component-based, drag and drop, guide-based, template-based development environment that integrates hundreds of components. Developers can also customize the components, and easily use the equipment of mobile phones such as cameras, maps, LBS positioning, and contacts, etc. (Pan, 2014) WeX5 supports various debugging

modes, such as browser debugging, device debugging, and native debugging, and it provides Intelligent code prompt function, and unlimited and multi-mode packaging and releasing services of apps which can be encrypted. This provides maximum convenience for application developers and the total amount of code can be reduced by 80%, which helps them easily respond to all kinds of complex data applications. Once developed, it can be released and run on a variety of front-end platforms. Currently, WeX5 has supported iOS ipa, android apk, wechat service number/Enterprise applications, web app and other light application development, as shown in Figure 1.

4 THE APPLICATION AND DEVELOPMENT OF "TAKEAWAY ONLINE" APPS

"Takeaway online" Apps use WeX5 framework for development, belonging to a single store business mode. Through multiple ways of packaging and releasing provided by WeX5, application stores can turn those Apps into the Android App, iOS App, wechat App and web App and then release them, which will help to improve the stores' brand image, increase sales, and stay close to their customers. Currently, the apps are mainly be applied to all independent businesses which can provide local online services, such as fast-food restaurants, snack bars, fruit shops, supermarkets, snack shops, dessert shops, bakeries, milk tea shops and so on.

4.1 The analysis and design of the system's function requirements

The App system is divided into two parts: mobile client-side and server-side. The client-side is installed on the mobile terminal device, which can provide the following functions: displaying dishes information, adding goods or services to the shopping cart, placing orders and paying, viewing the order history, managing users' information, and other functions. The server-side provides a background

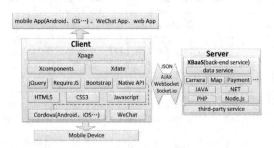

Figure 1.   Wex5 overall technical architecture.

Figure 2.   The system function module structure diagram.

management functions for stores, mainly including add management of dishes, delete management of dishes, modify management of dishes, query management of dishes, category management of dishes, order processing and registered user information management, as shown in Figure 2.

4.2 The data interaction process and the database design

The process of the system's data interaction can be divided into three parts: front-end, back-end and database, the front end can use the back-end services through making a request to AJAX, the back-end services then access the database, save or extract the data, and ultimately return the data to the front page in the form of response. The front page is composed of different components which are divided into two categories: display component (Panel, Contents, List, Input, Image, Button, etc.) and data component (Baas Data, Data, etc.). The back-end services can provide QueryFood, QueryOrder, QueryUser, Save and other actions, which are responsible for the interaction of the three database tables (Takeout_food, Takeout_order, Takeout_user) to finally realize the interaction of the whole system's data, as shown in Figure 3.

According to the previous analysis of the function requirements and data interaction process, we have designed the three data in the database table: Takeout_food information table for storing dishes, Takeout_order table which is used to store the order information, and Takeout_user table which is used to store user information.

4.3 System design and implementation

The system uses the integrated studio development platform provided by WeX5, and Php + Mysql development kit to design Apps. It mainly includes the design and implementation of back-end server-side services, business logic functions and the display modules of mobile terminals, as shown in Figure 4. We use some model resources provided by Studio application development platform, including UI2, Native and Baas as design aids. UI2 is the front page model. HTML5 + CSS3 + JS technology can be used to layout and interact with the

integrated display components and data components of UI2 so as to display the page. The pages we mainly develop include: dishes list page, shopping cart page, the user information page and order page. Native model is mainly used to debug the source code and then package and generate Apps. Baas is a back-end service model. On one hand it provides back-end services to the front page, and the services take action as an interface and the action used in the front page is defined in the service and called by the API function. On the other hand it provides convenient access to the program database for procedures. Through using Baas Data component associated back-end service's query action and save action, we can realize the functions of data adding, deleting, changing and checking, as shown in Figure 5.

4.4 System debugging and publishing

In studio which is WeX5 development's tool, you should start Tomcat and switch to debug mode, then you can dynamically compile the client-side page (.w). through UIServer. When the page (.w) makes a request for arrival in UIServer, UIServer

Figure 4.   The data table model diagram.

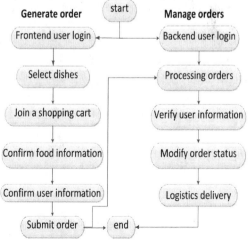

Figure 5.   Generate and manage order flow chart.

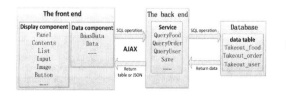

Figure 3.   The system data exchange process.

Figure 6.   The mobile client page display map.

will first check whether there is the cache of compiling conversion result (standard html page) in this page. (Pan, 2014) If there is the cache, then you can use the cache response to return; if there is no cache, it will first compile the page (.w) into a standard html page, and then cache response and return. After checking that all compilation is correct, you can start the integrated digital signature technology in studio and package the procedures through starting the packaging mode of "Start smart update UI resources". It will produce Android apk package and iOS ipa package, and then you can download them to the corresponding mobile terminals to debug and release them.

5   CONCLUSION

The development of Apps is a complex project. In order to develop Apps with good performance and stable systems, it requires that developers should have an in-depth understanding of the working principles of the operating systems of mobile terminals and proficiently master the relevant key technologies. To choose your own development mode and development framework is the key to helping you get twice the result with half the effort during the process of developing Apps. After two months' trial operation, the "takeaway online" App mentioned in this paper shows good stability, scalability and portability. Moreover, it also shows that its procedures are easy to install and maintain, its functions are multiple, its applications are wide and its commercial value is very high, as shown in Figure 6. The current function of the system is only for single-store system, you can add multiple shops business mode to the later upgraded version of the system, making it become a mobile application side "takeaway online" platform. Inviting more shops to join in the system will improve its value in promotion and drainage.

ACKNOWLEDGEMENT

Fund Project: Education Research Projects for Young and Middle-aged Teachers in Fujian Province (JB14212).

REFERENCES

Hui Ye. Comparison of three mobile APPs (application) development methods [EB/OL]. http://mt.sohu.com/20160516/n449620466.shtml, 2016–05–16/2016–06–27.

http://docs.wex5.com, 2015–10–20/2016–06–27.

Pan Chunhua, Li Junjie. Cross platform application of smart phone based on PhoneGap [J]. Computer Systems & Applications, 23(7):106–109 (2014).

Wex5 Documentation Center [EB/OL].

Xu Wei. Comparative analysis and case development of cross platform mobile development framework [D]. Changchun: Jilin University, 2014.

Advances in Energy Science and Equipment Engineering II – Zhou, Patty & Chen (Eds)
© 2017 Taylor & Francis Group, London, ISBN 978-1-138-71798-5

Computer simulation of the MTF of the infrared imaging system based on the OGRE compositor

Min Li, Min Yang, Yaxing Yi, Yibin Yang & Yanan Wang
Xi'an High-Tech Research Institute, China

ABSTRACT: The simulation of the physical effects of the infrared imaging system is important for accurate infrared imaging. The Modulation Transfer Function (MTF) is the most widely used modeling method. An MTF model of the infrared imaging system is developed in this paper. Based on the OGRE compositor and GPU programming, each frame of the original image is obtained with Fourier transform. The total MTF of the infrared imaging system is calculated according to the MTF of each subsystem. After multiplying the total MTF, the frequency domain image is inversely Fourier transformed. After processing each frame of the original image using the aforementioned operations, the corresponding material is obtained, showing the real-time simulation of the physical effects of the infrared imaging system. The experimental results indicate that the proposed method can achieve a strong sense of reality and meet the real-time requirements.

1 INTRODUCTION

At present, the infrared thermal imaging technology, with its strong military requirements and application background, has become the research focus at home and abroad. The simulation of the physical effects of the infrared imaging system is important for the accurate processing of infrared imaging. Because of the influence of the infrared detector, the image quality is degraded, and a certain gap remains between the actual imaging and the ideal imaging. In order to make the IR imaging simulation more realistic, the infrared imaging system needs to take the typical physical effect into consideration.

The simulation of the physical effects of an infrared imaging system is critical to establish a reasonable mathematical model for the imaging mechanism of each imaging unit in the system. Ideal images act as input and are processed to observe the imaging effect. Currently, there are four modeling methods: modulation transfer function method, the method of ray tracing point, diffusion function method, and image pixel processing-based method (Wang, 2012). The MFT method can naturally combine the effects of the optical system, the detector system, and the signal processing system, and each unit of the imaging effect can be independently analyzed. At the same time, the model includes most of the physical phenomena to describe the imaging effect accurately.

2 MTF MODEL OF THE INFRARED IMAGING SYSTEM

Assuming that the infrared imaging system is a space invariant linear system, according to the linear filtering and signal processing theory, the Point Spread Function (PSF) is a function describing the analytical ability of point source in the optical imaging system. The output of the imaging system can be obtained by convoluting the input image and the point spread function. First, the MTF is the Fourier transform of the PSF of the photoelectric imaging system in the frequency domain. Then, the input image is multiplied by the MTF and output though the inverse Fourier transform (Wang, 2012). Thus, the convolution cooperation on airspace can be converted into multiplication in the frequency domain, reducing the amount of calculation. Denoting $i(x,y)$ as the input image, $o(x,y)$ as the output image, and $MTF_{total}(f_x, f_y)$ as the total MTF of the imaging system, the process can be described as follows:

$$o(x,y) = IFFT\left[FFT\left[i(x,y) \right] \times MTF_{total}(f_x, f_y) \right] \tag{1}$$

The MFT describes a transfer ability of the contrast of input images at various frequencies for a system, and represents the frequency throughput of the system. It is a function of the spatial frequency (Wang, 2012). With the increase in spatial frequency,

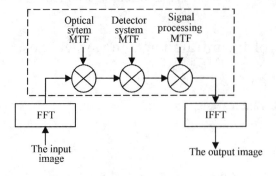

Figure 1. MTF model of the imaging system.

the MTF becomes smaller and the corresponding image is extremely blurred, which shows that the resolution of the general imaging system is not high. According to the modular principle, the infrared imaging system consists of an optical system, a detector system, a signal processing system, and by multiplying its MTF, the total MTF of the whole system can be obtained, as shown in Figure 1.

2.1 MTF of the optical system

The optical system will focus the infrared radiation on the focal plane of the detector. The influence factors of the MTF are the aperture diffraction effect and geometrical aberration, and the corresponding transfer functions are MTF_{dif} and MTF_{abe}.

For a diffraction-limited system, the MTF expression of a circular aperture optical system can be represented as follows:

$$MTF_{dif}(f) = \frac{2}{\pi} \left\{ arccos\left(\frac{f}{f_{oc}}\right) - \left(\frac{f}{f_{oc}}\right) \left[1 - \left(\frac{f}{f_{oc}}\right)^2 \right]^{1/2} \right\} (f \leq f_{oc}) \quad (2)$$

where $f = \sqrt{f_x^2 + f_y^2}$ is the spatial frequency, $f_{oc} = D/\lambda$ is the spatial-cut-off frequency, D is the entrance pupil diameter, and λ is the central wavelength, which can be set as the intermediate value of the working wavelength. When the spatial frequency is greater than the cut-off frequency, the MTF is zero.

For a non-diffraction-limited optical system, the system-specific aberration has an impact on the MTF. Because non-paraxial ray tracing through the optical system is not consistent with the result of paraxial ray tracing, there is a deviation between the actual and ideal imaging, which is called aberration. Robert R. Shannon proposed an empirical formula for the aberration effect,

which is described approximately with a square wave aberration W:

$$MTF_{abe}(f) = 1 - \left(\frac{W}{A}\right)^2 \left[1 - 4\left(f/f_{oc} - 0.5\right)^2 \right] \quad (3)$$

The MTF of the optical system is the product of the MTF of aperture diffraction effects and the MTF of geometrical aberration. That is:

$$MTF_{opt}(f) = MTF_{dif}(f) \cdot MTF_{abe}(f) \quad (4)$$

2.2 MTF of the detector system

Detector is the key component of the infrared imaging system. It is placed in the focal plane of the optical system, completing the infrared radiation sampling in the field, at the same time converting the infrared radiation signal to the voltage signal by the photovoltaic effect. From the perspective of the linear filter, filtering characteristics of the detector can be decomposed into spatial filter and time filter (Wang, 2012).

The modulation transfer function of the spatial filtering effect is MTF_{ds}. Because the pixel of the general CCD detector is even square, its airspace PSF is a rectangular function. Assuming that the area of a rectangular detector is $a \times b$, the space angle is α and β, the response function is the rectangular box function $Rect(x/\alpha) \times Rect(y/\beta)$, Fourier transforming to the product of sinc functions disporting from the spatial frequency (f_x, f_y). Thus, the MTF of the spatial filtering effect can be expressed as:

$$MTF_{ds}(f_x, f_y) = \left|\frac{sin(\pi\alpha f)_x}{\pi\alpha f_x}\right| \left|\frac{sin(\pi\beta f_y)}{\pi\beta f_y}\right|$$
$$= \left|sinc(\pi\alpha f_x)\right| \left|sinc(\pi\beta f_y)\right| \quad (5)$$

The time filtering effect is equivalent to a low-pass filter, and its modulation transfer function is given by:

$$MTF_{dt}(f) = \frac{1}{\sqrt{1 + (f/f_0)^2}} \quad (6)$$

where f_0 is the spatial frequency when the time response is 3dB, $f_0 = 1/2\pi\tau_d$, and τ_d is the time constant of the detector.

MTF of the detector system is as follows:

$$MTF_{det}(f) = MTF_{ds}(f) \cdot MTF_{dt}(f) \quad (7)$$

2.3 MTF of the signal processing system

The signal processing circuit of the infrared imaging system can be regarded as a combination of different filters, including the low-pass, high-pass filter and the enhanced circuit filter. The corresponding main physical effects in the imaging process are: low-pass filter, high-pass filtering, high-frequency lifting, and CCD signal transfer effect.

A low-pass filter and high-pass filter in the spatial frequency transfer function are given by:

$$MTF_{e1}(f) = \sqrt{\frac{1}{1+(f/f_{0l})^2}} \quad (8)$$

$$MTF_{e2}(f) = \frac{f/f_0}{\sqrt{\left[1+\left(f/f_{0h}\right)^2\right]}} \quad (9)$$

where f_{0l} and f_{0h} are the spatial frequency of the low-pass circuit and the high-pass circuit, respectively, at 3dB, which can be calculated by the time frequency, and the corresponding relationship between the time frequency and the spatial frequency is $f = f_t/\omega$, with ω being the scan angular frequency and $f_t = 1/2\pi RC$, and R and C are resistance and capacitance, respectively.

High-frequency lifting circuit is an enhancing circuit, which can smooth the frequency response of the system when the proportion of noise is big. Setting the highest lifting point as f_{max} and the lifting order as $K(\geq 1)$, the MTF is given by:

$$MTF_{e3}(f) = 1 + \frac{K-1}{2}\left[1-\cos\left(\frac{\pi f}{f_{max}}\right)\right] \quad (10)$$

When the signal charge is transferred through CCD, the signal attenuates to some extent. Setting the transfer efficiency as η, the digit of CCD as N, and the clock frequency as f_{tc}, the MTF of CCD transfer is given by:

$$MTF_{CCD}(f) = exp-\left\{-N(1-\eta)\left[1-\cos\left(2\pi\frac{f}{f_{tc}}\right)\right]\right\} \quad (11)$$

MTF of signal processing system is:

$$MTF_{sig}(f) = MTF_{e1}(f)\cdot MTF_{e2}(f) \\ \cdot MTF_{e1}(f)\cdot MTF_{CCD}(f) \quad (12)$$

So the total MTF of the infrared imaging system is the product of the MTF of the optical system, the detector system and the signal processing system, which is given by:

$$MTF_{total} = MTF_{opt}\cdot MTF_{det}\cdot MTF_{sig} \quad (13)$$

3 OGRE COMPOSITOR

After a frame of the scene image is output, there is a need for adding the effect of the infrared imaging system. In the OGRE rendering engine, the subsystem that is used to add special after-effects is called the late compositor. Compositor chain combines multiple compositors in accordance with the order, where the output of the last compositor is the input of the next compositor, to realize the superposition of the effects of multiple compositors. The way in which the compositor communicates with each other is called "texture transfer". In this paper, three materials are created and combined in a certain order.

First, a new material file is created. Then, according to the three mathematical models of the infrared imaging effect, using the CG language to program the corresponding fragment shader, they are added to the material file. The three materials are called Ogre/Compositor/optical, Ogre/Compositor/detection, and Ogre/Compositor/signal, which is the core factor to realize the MTF model. In fact, the content of the material file is calculated for each frame image. By using GPU programming, each frame of the original image can be obtained in real time, which is then Fourier transformed, MTF processed, and inverse Fourier transformed.

Second, a compositor file is created and a scene texture is defined. The API inputs the current rendering scene that has no effect, to pad the scene texture in the video memory. Furthermore, it uses the scene texture and the optical system effect material to cover and fill the scene texture, that is, the scene texture is both input and output. In the same way, the scene texture is updated with the detector system and signal processing effect materials. The pseudocode of the compositor is as follows:

```
compositor infrared
{
texture scene //define a scene texture
target scene
{
input previous// Pad the scene texture with the
current rendering scene that has no effect
}
target scene
{
material Ogre/Compositor/optical //Use the
optical system effect material
input 0 scene
}
target scene
{
```

material Ogre/Compositor/detection // Use the detector system effect material
```
    input 0 scene
    }
    target_output
    {
    material Ogre/Compositor/signal // Use the signal processing system effect material
    input 0 scene
    }
}
```

The process builds a compositor named infrared, and only one scene texture "scene" is used. It is iterated several times, which saves the memory space.

Finally, in the application, the compositor is added and parsed by using the compositor manager in the OGRE resource manager. Each compositor can be associated with only one viewport through compositor instance, and the rendering result is output to a certain viewport.

4 RESULTS OF THE SIMULATION

Based on the MTF model of the infrared detector, the parameters of the detector are listed in Table 1.

The results of the simulation using Matlab are shown in Figure 2. The MTF curve shows that the MTF decreases with the increase in spatial frequency.

In order to add the effects of the detector to the infrared scene, based on the OGRE compositor, with C++ programming under the platform of VS2008, the computer simulation of three kinds of the MTF effect is carried out. The final imaging effects are as follows.

From the results of the simulation, we can see, in general, that the output image becomes blurred,

Table 1. The parameters of each subsystem.

Optical system	Entrance pupil diameter: 168 mm Mean square wave aberration: 10
Detector system	Time constant: 20 μs Space angle: 150°

Signal processing
system
	Frequency in low-pass 3dB:
	    400000 Hz
	Frequency in high-pass 3dB: 0 Hz
	Highest lifting frequency:
	    400000 Hz
	Lifting order: K = 1
	Efficiency of transfer: $\eta = 0.99$
	Digit of CCD:N = 2
	Clock frequency:
	    $f_{tc} = 2000000HZ$

and the resolution and contrast are reduced. Meanwhile, the average rate of the simulation reduces from 120 to only 112 frames/s, but still meeting the real-time requirements, as shown in Figure 3.

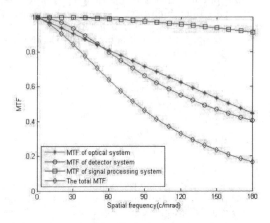

Figure 2. The MTF graph.

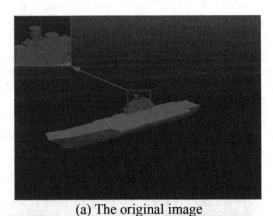

(a) The original image

(b) Added effect of the optical system

Figure 3. (*Continued*)

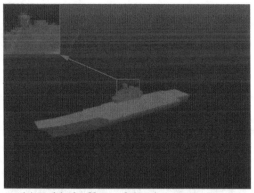

(c) Added effect of the detector system

(d) Added effect of the signal processing
system

(e) The final image

Figure 3.   The results of MTF processing.

5   CONCLUSION

This paper builds an MTF model of the infrared imaging system, using the OGRE rendering engine, based on the OGRE compositor and GPU programming to realize the physical imaging effects of each subsystem and the total MTF effects through their superposition. The experimental results are reasonable and indicate that proposed model has a good real-time performance.

REFERENCES

Holly Wenaas, Computer-generated correlated noise images for various statistical distributions. SPIE Proceedings on Stochastic and Neural Methods in Signal Processing, Imaging Processing and Computer Vision, 410–421, (1991).

Wang, SX.W., K. Li, S.L. Wang, Research on modeling and simulation methods of infrared imaging, Computer & Digital Engineering, 40 (2012).

Zhang, F.Q., Y.F. Zhang, Q. Deng, Model construction and simulation of electro-optical imaging system based on MTF, Laser & Infrared, 45 (2015).

Zhao, R.B., S.H. Zhao, X.L. Hu, A CPU-GPU collaboration based computing parallel algorithm for MTF degradation of remote sensing simulation in images, Computer Engineering & Science, 37 (2015).

Advances in Energy Science and Equipment Engineering II – Zhou, Patty & Chen (Eds)
© 2017 Taylor & Francis Group, London, ISBN 978-1-138-71798-5

Study on an algorithm for calculating the turning attitude of agricultural wheeled mobile robots in a complex terrain

Bin Liu & Yan Ren

Henan Forestry Vocational College, Luoyang, Henan, China

ABSTRACT: The study of agricultural wheeled mobile robots is a particular area of focus in the 2025 Chinese intelligent manufacturing plan, and the stability of turning attitude control of agricultural wheeled mobile robots during walking has been the emphasis of wheeled robot research. This is because traditional technology that applies only Kalman filtering to calculate the angular velocity output by the gyro cannot meet the actual demand for turning attitude control of agricultural wheeled mobile robots. As MEMS gyros may show greater integral floating and the accelerometer is sensitive to non-gravitational acceleration, in this paper, the robot speed measured by GPS is introduced into the calculation of attitude, and this data is used to correct the angular velocity of Kalman filtering. Based on the comparison between the corrected results and the angular velocity by using only Kalman filtering, the actual vehicle test results indicate that the proposed algorithm has higher accuracy in calculation and has a certain reference value.

1 INTRODUCTION

The roll and pitch angle of agricultural wheeled mobile robots is an important attitude information about robots walking in a complex terrain. Inertial components such as gyros and accelerometers are usually used in the attitude stabilization system of wheeled robots. With the development of MEMS technology, the performance of the MEMS gyro and the MEMS accelerometer has been increasingly improved and widely used. On the one hand, if the gyro is used alone for calculating attitude, the results will show integral floating over time. On the other hand, if the accelerometer is used alone for that x_b purpose, no accurate dynamic roll angle can be obtained by calculating gravitational acceleration in three coordinate components. This is because wheeled robots may generate centripetal acceleration and other non-gravitational influences during walking, especially when the robot is turning, and the accelerometer is very sensitive to acceleration due to non-gravitational forces. Therefore, the attitude cannot be accurately obtained by calculating the gravitational acceleration in the three axes.

To solve the aforementioned problems, scholars at home and abroad have done a lot of research work. In a study (e.g., Skog, 2006), a dual quaternion algebra was used to reinterpret the basic principles of the SINS (Serial Inertial Navigation System) and a fast coordinate conversion algorithm was presented based on iterating Newton–Raphson and a half-Gaussian filter. In another study (e.g., Demoz,1998), a dominant

complementary filter with a gain adjustment mechanism was presented, which was suitable for low-cost inertial measurement units for estimating the turning attitude of wheeled mobile robots. Furthermore, in a paper (e.g., Sergio, 2005), the static initial alignment of robot attitude was realized based on the static initial alignment algorithm by using the data of processed magnetic field strength and acceleration. In another paper (e.g., Lance, 2003), a traditional AHRS was presented to improve the reliability of the measurement system by increasing redundancy. In a study, (e.g., Peter, 2003), the AHRS was placed on a rotating platform fixedly connected to the 1024-line grating encoder to test the measurement precision and tracking performance of the attitude angle. In another study (e.g., Kevin, 2012), the Jacobean matrix was used to make the state transition function linear and update the state error covariance. In order to obtain unbiased estimates of the gravitational acceleration in three coordinate axes, in this paper, the robot speed measured by GPS will be corrected and the corrected result will be introduced into the calculation of the angular velocity of Kalman filtering.

2 ATTITUDE MEASUREMENT SYSTEM DESIGN

In order to meet the requirements of the attitude measurement system, STM32 is used as the main control chip with processor resources such as

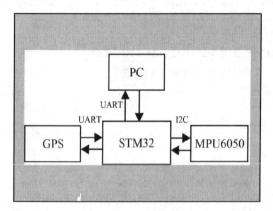

Figure 1. The system block diagram.

timer, interrupter, and serial communication, and the microprocessor is enabled to communicate with MPU6050 by simulating an I2C bus by using a software. The system block diagram is shown in Figure 1.

MPU6050 is a motion sensing chip produced by InvenSense as an integration of the three-axis accelerometer and the three-axis gyroscope. Speed sensor is a high-precision GPS sensor with a data update frequency of 20 Hz transducers. The speed measurement accuracy is 0.03 m/s. The communication with the microprocessor is based on a serial port. The calculated result is uploaded to the computer through the serial port, and the data is displayed and recorded in real time using the afore-mentioned software written in LabVIEW.

3 ATTITUDE CALCULATION ALGORITHMS

We first define the robot body coordinate system $Ox_b y_b z_b$ with in the forward direction along the longitudinal axis, y_b in the right direction along the horizontal axis, and z_b in the upward direction along the vertical axis.

3.1 Angular velocity integral

Gyro is used to obtain the angular velocity of robots in turning motion. The gyro is not likely to be affected by robot vibrations, and there is little noise detected in angular velocity signals (e.g., Rogerio, 2012). The roll angle can be obtained by integrating the angular velocity around the axis x_b based on the following calculation formula (e.g., Adrian, 2006):

$$\theta_k = \left(\omega_k - \omega_{bias\_k} \right) dt + \theta_{k-1} \tag{1}$$

where θ_k is the roll angle between the axis y_b of the gyro and the horizontal plane at the present moment; θ_{k-1} is the roll angle at the previous moment; ω_k is the angular velocity around the axis x_b at the present moment; ω_{bias_k} is offset of the angular velocity around the axis x_b at the present moment; dt is the integral time, that is, the sampling period of the angular acceleration.

If only one gyro is used for attitude measurement without assistance from other sensors, errors related to the zero offset stability will grow rapidly with the increase in the integration period, thus leading to obvious rotation of the equipment, even if the equipment is still at the moment.

3.2 Speed differential measured by GPS

The relationship between the forward acceleration and the robot speed measured by GPS is given by:

$$Acc_f = \frac{v(t_k) - v(t_k - \Delta t)}{\Delta t} \tag{2}$$

where Acc_f is the robot forward acceleration along the longitudinal axis of the robot body, $v(t_k)$ is the ground speed measured by GPS at the present moment, $v(t_k - \Delta t)$ is the ground speed measured by GPS at the previous moment, and Δt is the sampling period of GPS.

The component a_x of the corrected gravitational acceleration on the axis x is given by:

$$a_x = A_{x,out} - Acc_f \tag{3}$$

3.3 Attitude calculation of the corrected accelerometer by GPS

The accelerometer uses the gravity vector and its projection on the coordinate axis to determine the angle of inclination (e.g., Christopher, 2010). Basic trigonometric identities can be used to calculate the pitch angle, which is given by:

$$\theta_{Acc} = \tan^{-1}\left(\frac{A_{y,out}}{\sqrt{a_x^2 + A_{z,out}^2}} \right) \tag{4}$$

where θ_{Acc} is the roll angle between the axis y of the accelerometer and the horizontal plane at the present moment; a_x is the component of the gravitational acceleration on the axis x after correction by GPS; $A_{y,out}$ is the component of the gravitational acceleration on the axis y; and $A_{z,out}$ is the component of the gravitational acceleration on the axis z.

3.4 Kalman filtering

Kalman filtering is an optimal recursive state estimation algorithm for the linear system. Suppose that the system state equation and measurement equation can be expressed as (e.g., Geiger, 2008):

$$\begin{cases} x_k = Ax_{k-1} + BU_k + \omega_k \\ \quad y_k = Hx_k + v_k \end{cases} \tag{5}$$

where x_k is a state variable; y_k is an observed variable; A is the state-transition matrix from time $k-1$ to time k; U_k is the system input control vector; B is the gain matrix of the input control vector; H is the gain matrix from the state variable to the observed variable; ω_k is the input noise; and v_k is the observed noise. Suppose that both the input noise and the observed noise obey the normal distribution, and let the covariance of the input noise be Q and that of the observed noise be R, then the recursive Kalman filter is as follows:

Pre-estimation of the state variable:

$$\hat{X}_{k|k-1} = A\hat{X}_{k-1} + BU_k \tag{6}$$

Pre-estimation of the error covariance:

$$P_{k|k-1} = AP_{k-1}A^T + Q \tag{7}$$

Filtering gain update:

$$K_k = P_{k|k-1}H^T \left[HP_{k|k-1}H^T + R \right]^{-1} \tag{8}$$

Status update:

$$\hat{X}_k = \hat{X}_{k|k-1} + K_k \left(y_k - H\hat{X}_{k|k-1} \right) \tag{9}$$

Error covariance update:

$$P_k = \left(I - K_k H \right) P_{k|k-1} \tag{10}$$

where $\hat{X}_k$ is the estimation of the real value X_k of the current state; $\hat{X}_{k|k-1}$ is the pre-estimation based on $k-1$ times to k times; A is the state-transition matrix from $k-1$ times to k times; B is the gain matrix of the input control vector; H is the gain matrix from the state variable to the observed variable; y_k is the observed variable of k times; K_k is the modified weighting for observed deviation $\left(y_k - H\hat{X}_{k|k-1} \right)$; P_k is the error covariance matrix of the estimated value of the present state; $P_{k|k-1}$ is the pre-estimated error covariance matrix; and I is the identity matrix.

In the program, parameters are set as:

$$A = \begin{bmatrix} 1 & -0.05 \\ 0 & 1 \end{bmatrix}, B = \begin{bmatrix} 0.05 \\ 0 \end{bmatrix}, H = \begin{bmatrix} 1 & 0 \end{bmatrix},$$
$$\hat{X}_0 = \begin{bmatrix} -1.62 \\ 0 \end{bmatrix}, P_0 = \begin{bmatrix} 1 & 0 \\ 0 & 1 \end{bmatrix}, Q_0 = \begin{bmatrix} 0.2 & 0 \\ 0 & 0.1 \end{bmatrix},$$
$$R = 2.$$

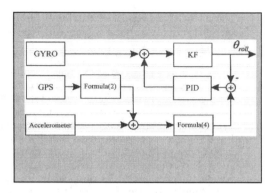

Figure 2. Block diagram of the algorithm.

3.5 Fusion algorithm of GPS and inertial element data

Integral floating of the gyro is corrected by the PID control algorithm, as shown in Figure 2:

$$u(k) = K_P e(k) + K_I \sum_{j=0}^{k} e(j) + K_D \left[e(k) - e(k-1) \right] \tag{11}$$

$$e(k) = \theta_{ag}(k) - [u(k-1) + \omega_{gyro\_x}(k)dt + \theta_{roll}(k-1)] \tag{12}$$

where $u(k)$ is the corrected controlled variable of the angular velocity at the present moment; K_P is the proportional gain; K_I is the integration constant; K_D is the differential constant; $e(k)$ is the error calculated at the present moment; $\theta_{ag}(k)$ is the pitch angle calculated by the fusion algorithm of the accelerometer and GPS data at the present moment; $\omega_{gyro\_x}(k)$ is the angular velocity output on the axis x of the gyro at the present moment; dt is the integral time; and $\theta_{roll}(k-1)$ is the roll angle calculated at the previous moment.

In the program, PID parameters are set as:

$$K_P = 1, K_I = 0.1, K_D = 0.01.$$

4 ACTUAL VEHICLE TEST AND DATA ANALYSIS

The experimental site selected for conducting the test on the agricultural wheeled mobile robot was an open agricultural land, where there were few buildings blocking GPS signals and the site is wide and flat. After calibrating the sensor before the test, measurement is conducted in four cases, that is, the robot walks at a speed of 5, 10, and 18 m/h, respectively, when turning left and at a speed of 18 km/h when turning right.

1573

The attitude is first calculated jointly by the accelerometer and the acceleration obtained by robot speed differential, and then the integral floating generated by the correction of the gyro is removed. In this paper, a commercial operation stability test system is used to test the accuracy of the algorithm. For the data of the robot when turning clockwise at a speed of 10 km/h taken as an example, the proposed algorithm is compared with the angular velocity calculated only by using Kalman filtering, the results of which are shown in Figure 3. It is evident from the figure that there is noticeable integral floating as time passes if only Kalman filtering is used to calculate the angular velocity, and it is unable to meet the need of the turning stability calculation of wheeled mobile robots. The algorithm used in this paper effectively suppresses the integral floating of the gyro.

According to the statistical analysis of errors, the use of only Kalman filtering to calculate the angular velocity leads to an angle error variance ranging from 11.8870 to 12.7017, which is unable to meet the turning stability test requirements of wheeled robots. The angle error variance calculated by the fusion algorithm of GPS/INS data is 0.2339, which is smaller than the former. The experimental results indicate that the proposed algorithm can generate better results than the acceleration calculation generated by only using Kalman filtering, as shown in Figure 4.

5 CONCLUSIONS

A new data fusion algorithm was presented in this paper for the turning attitude control of a wheeled mobile robot. The wheeled robot speed measured by GPS is used to correct the accelerometer data output and the corrected value after the introduction of attitude calculation into Kalman filtering, in order to fuse with the data of the angular velocity sensor for attitude measurement.

According to the field test and simulation analysis, this algorithm is valuable for popularized applications in the turning stability test system of agricultural wheeled mobile robots.

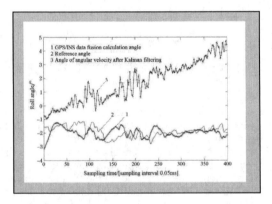

Figure 3.   Comparison of the calculation results of the roll angle.

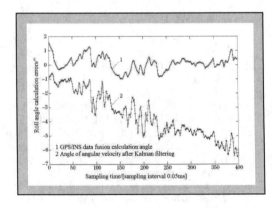

Figure 4.   Calculation errors of the roll angle.

REFERENCES

Adrian, Schumacher. Integration of a GPS aided Strapdown Inertial Navigation System for Land Vehicles [D]. Stockholm: Royal Institute of Technology (2006).
Christopher J. Fisher. Using an Accelerometer for Inclination Sensing [Z/OL].
Demoz G, Roger C, J D Poewell. A low-cost GPS/inertial Attitude Heading Reference System (AHRS) for general aviation applications [C] (1998).
Geiger W., J. Bartholomeyczik, U. Breng, et al. MEMS IM-U for AHRS applications. Position, Location and Navigation Symposium, IEEE/ION, Monterey, CA (2008).
Kevin Murrant, Attitude and Heading Reference System for Small Unmanned Aircraft Collision Avoidance Maneuver-s. Memorial University of Newfoundland (2012).
Lance Sherry, Chris Brown, and Ben Montazed et al. Performance of automotive-grade MEMS sensors in low cost AH-RS for general aviation. Digital Avionics Systems Conference, Indianapolis, USA (2003).
Peter S, Integrating GPS and INS with a tightly coupled Kalman filter—using proximity information for compensating signal loss [D] (2003).
Rogerio R. L, Leonardo A.B. T, Performance Evaluation of Attitude Estimation Algorithms in the Design of an AHRS for Fixed Wing UAVs [C] (2012).
Sergio D, Javier A, Low cost navigation system for UAV's [J] (2005).
Skog I, Handel P, Calibration of a mems inertial measurementunit [C] (2006).

Advances in Energy Science and Equipment Engineering II – Zhou, Patty & Chen (Eds)
© 2017 Taylor & Francis Group, London, ISBN 978-1-138-71798-5

Analysis of the obstacle-climbing performance of two wheel-rod hybrid robots

Jie Huang, Yujun Wang, Ruikun Quan & Kaiqiang Zhou
School of Computer and Information Science, Southwest University, Chongqing, China

ABSTRACT: Considering that traditional mobile robots have low walking efficiency, poor climbing ability, insufficient velocity, or complex structural design, a new type of two wheel-rod hybrid robots is presented in this paper. The robot can not only effectively walk on a flat ground, but also has a better crossing ability on an irregular surface, especially the continuous stairs. First, the structure of the wheel with the connecting rod and the design of the whole robot are introduced in a simple way. The length of links and the maximum height of surmountable obstacles are given at the same time. Then, the obstacle-climbing performance of two wheel-rod hybrid robots is mainly determined by kinetic and kinematic analysis, including the locomotive efficiency on a flat ground, the overstepping ability on stairs, or any other obstacles. Locomotive efficiency is determined by four ways of actions, and the obstacle-surmounting ability of the robot is maintained by the mutual switching of the wheels and the connecting rods. The validity and feasibility of this wheel-rod mechanism is confirmed by conducting experiments using a real machine. The results of the analysis proves that two wheel-rod hybrid robots have a high walking efficiency and sufficient velocity, can climb continuous stairs easily, and can adapt to any other non-structural terrains with high efficiency.

1 INTRODUCTION

In recent years, all kinds of mobile robots have been presented and designed, and a lot of research achievements have been made. They have been developed in various application fields, such as building inspection and security, space and undersea exploration, military, reconnaissance, and warehouse services (Muir, 1987). Mobile robots are designed with locomotive mechanisms according to the environment of the application field. The locomotive mechanisms of mobile robots can be divided into three types: wheeled robots, legged robots, and tracked robots (Woo, 2007).

It is well known that each locomotive type of robot has its inherent advantages and disadvantages. Wheeled mobile robots have a simple structural design and high walking efficiency, but have poor adaptability to irregular terrains (Tan, 2011). Legged mobile robots can easily adapt to unstructured environments, but have poor mobility on a flat ground (Takahashi, 2006). Tracked mobile robots have a certain obstacle-climbing ability, but they are bulky and require high-power driving at the same time.

In view of these disadvantages of legged and tracked robots, this paper proposed a connecting-rod mechanism that can help the robot cross obstacles and cannot influence it in walking on

any terrains. Therefore, a new type of locomotive mechanism is presented. This mechanism combines with the advantages of connecting rods and wheels to have high walking efficiency and obstacle-surmounting ability.

2 STRUCTURAL DESIGN OF A TWO WHEEL-ROD HYBRID ROBOT

For the purpose of developing a mobile robot with factors such as high walking efficiency, strong climbing ability, sufficient velocity, simple structural design, and lightweight, this paper elaborately analyzed the features of three main types of locomotive mechanism: wheeled, legged, and tracked (Woo, 2007). Then, it proposes a new type of two wheel-rod hybrid robots, which combines the walking efficiency of the wheeled mechanism with the overstepping ability of the connecting-rod structure. In this section, the wheels with the connecting rods and robots with this special structure will be elucidated.

2.1 Wheel with the connecting rod

The traditional wheel mechanism can yield high walking efficiency, which was proved thousands of years ago. However, in general, the common round wheels are center of the wheel in the center position, which may result in poor obstacle-climbing ability.

However, connecting rods can maintain strong climbing performance by using structural features itself, which can rely on the static friction force to overcome obstacles easily when the robot encounters any type of terrains. Therefore, this paper proposes a model of wheels with connecting rods, which can not only sustain the locomotive efficiency of the wheel mechanism, but also yield an excellent obstacle-climbing performance for the connecting rod structure. On the other hand, the wheel with the connecting rod has a simple structure and control.

The structure consists of a round wheel and a connecting rod. The link can rotate around the wheel by the bearing or any other mechanical device. The closer the installing position of the bearing to the edge of the wheel is, the better it will be. The structure of the wheel with a connecting rod is shown in Fig. 1.

In this mechanical structure, the round wheel can be seen as a driving wheel and the link moves with the wheel by a physical connection. It is assumed that the junction of the round wheel and the link is point P, the power shaft pass point is O, and eccentricity of the wheel is OP. Points A and B are two arbitrary points on the edge of the wheel. As shown in Fig. 2, when the link encounters an obstacle, the wheel can rotate around the connecting rod. At this point, the maximum height of the crossing obstacle H is given by:

$$H = R + 2e \tag{1}$$

Where R is the radius of the driving wheel and e is the eccentricity, which is the distance between the junction P and the power shaft O.

From the above discussion, we can safely draw the conclusion that the larger the eccentricity e is, the better the obstacle-climbing performance of the robot is. In fact, the value of the eccentricity e is in the range of $(0, R)$. The supporting point P should be as close as possible to the edge of the wheel; at this point, the mechanism can achieve the maximum height of the crossing obstacle, which is $3R$. When the height of obstacle-climbing reaches

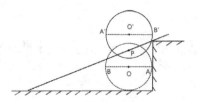

Figure 2. The principle of the robot's obstacle-climbing.

$3R$, there is one condition as follows: when the supporting point P reaches the maximum value in a cycle, the connecting rod will likely cross the obstacle.

2.2 Structure of the robot

The two wheel-rod hybrid robot uses a symmetric structure design. It includes two wheels with a connecting rod, which is already described above, and a machine body that is used to include some control components, power sources, and so on. There are two same types of electric motor in the machine body, which can drive each wheel independently (Li, 2012).

Fig. 3 shows a model of a two wheel-rod hybrid robot. It has two wheels with high obstacle-climbing performance, walking efficiency, and a simple mechanical design. It can cross obstacles easily and by simply switching between the wheels and the connecting rods. Furthermore, it can keep sufficient velocity through two round wheels.

The front parts of the connecting rods must make to be jagged or skidproof to have crossing obstacles. However, the front parts of the connecting rods should have an original angle, which is defined as the angle ψ, such that the end of the connecting rod is relative to the horizontal plane.

The maximum height of the robot's obstacle-climbing can reach $3R$, which is described in section 2.1. In addition, when it comes to $3R$, the angle between the link and the ground is θ and ψ is equal to θ (see Fig. 4).

$$\psi = \theta = \arctan\left(\frac{2R - e}{R}\right) \tag{2}$$

Therefore, considering the extreme situation, the length of the front and rear parts for the connecting rod are, respectively, L_1 and L_2:

$$L_1 = \frac{R}{\cos 2\theta} = \frac{R}{\cos 2\psi} \tag{3}$$

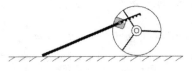

Figure 1. The structure of the wheel with a connecting rod.

$$L_2 = \frac{R + e}{\sin \theta} \tag{4}$$

1576

Figure 3. The model of the robot.

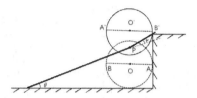

Figure 4. The principle of the robot obstacle-climbing with initial angle.

3 OBSTACLE-CLIMBING PERFORMANCE OF A TWO WHEEL-ROD ROBOT

In this section, the property and efficiency of two wheel-rod hybrid robots will be analyzed from three aspects: locomotion on a flat ground, climbing continuous stairs, and other obstacles.

3.1 *Locomotion on a flat ground*

It is well known that wheeled mobile robots can have sufficient velocity and walking efficiency on plane surfaces. In this paper, although a connecting rod is added to each wheel, it cannot influence the performance of the mobile robot. When the robot moves on a flat ground, the links rotate passively around the round wheels without influencing the movement of the wheels, which can maintain the center of gravity in the same position at all time to ensure the static stability of the robot.

In general, there are four ways of action, which include moving forwards, moving backwards, turning left, and turning right, to verify the locomotion performance of the mobile robot on a flat ground. The wheeled mechanism is adopted by the robot; therefore, forward and backward movements can be switched with sufficient velocity and the phenomenon of ups and downs cannot exist. For the locomotive mechanism for turning operations, this paper uses a method or principle that is defined by the wheels of the device on the left and right sides, respectively, with different velocities or rotating in a reverse direction. The advantage of this method is that each wheel drives independently.

3.2 *Climbing continuous stairs*

As described in the previous section, the proposed linked mechanism offers extensive adaptability to various rough terrains, especially continuous stairs. When the robot encounters obstacles, it can automatically switch to the obstacle-climbing mode, in which the connecting rod and the wheel can switch to each other, and, therefore, the robot has a very good obstacle avoidance ability.

Fig. 5 shows the process of climbing continuous stairs by two wheel-rod hybrid robots.

The process of climbing a stair by the robot can be divided into four steps (Li, 2013).

Step 1: initial state. The connecting rods first encounter the stair, as shown in Fig. 5(a). In this state, the robot cannot cross the first stair, and it must continuously drive the wheels rotating and find a suitable opportunity to switch to the next state.

Step 2: another initial state. In this state, first, two wheels come across the stair. Then, there is always a chance of the connecting rods to reach the obstacle. Additionally, step 1 can always jump to step 2, as shown in Fig. 5(b).

Step 3: obstacle surmounting by means of an eccentric pattern. In this state, the links and wheels switch to each other. First, the connecting rods meet the stair. Then, the wheels rotate continuously to support the connecting rod, the links of which are equal to an axis of rotation. The links can lift the whole body of the robot, as shown in Fig. 5(c).

Step 4: accomplishing obstacle surmounting. Through the eccentric locomotion of the device, the robot can always accomplish climbing one stair and continuously prepare for climbing the next stair (see Fig. 5(d)). The initial state of climbing the next stair is likely to involve step 1 or step 2, or and both are possible.

As mentioned above in the process of surmounting obstacles, it is assumed that the angular velocity of the motor is ω, the torque capacity of the motor is T, the distance of the driving raft and the link is e, and the radius of the driving wheel is R.

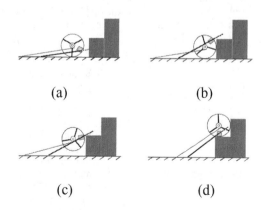

(a)　　　　　　　　(b)

(c)　　　　　　　　(d)

Figure 5. The process of climbing continuous stairs.

Therefore, the rotating force of the wheel around the joint in step 3 is given by:

$$F_{ROT} = T/_e \qquad (5)$$

When the robot moves on a flat ground, the driving force exerted by the motor is given by:

$$F_{DRI} = T/_R \qquad (6)$$

From the above discussion, we can easily conclude that R is obviously larger than e in the design of the robot. More powering force is needed in the process of climbing continuous stairs, and this method can make the robot cross obstacles relatively smoothly (Li, 2013).

3.3 Other obstacles and terrains

The robot proposed in this paper can not only climb stairs easily, but also cross many other obstacles. Through the kinematic and kinetic mechanism, the robot can easily cross other types of crossing obstacle, which relies on the connecting rods and the wheels. The other three types of crossing obstacle are small barriers (e.g. square tube or stone, where the width of the obstacle is smaller than the length of the robot's body), platforms (multiple small obstacles spliced together), and mating surface groove (the case in which some gaps are present in the obstacles).

Indeed, the robot can cross many other terrains efficiently. For example, it can cross a sandy terrain by the wheels with sufficient velocity. The robot can even travel easily and effectively across mountains.

4 EXPERIMENTAL RESULTS

Based on the above description and analysis, a real prototype of the proposed mobile robot is fabricated to verify the moving performance and the obstacle-climbing ability. It is mainly composed of three parts: the driving round wheel assembly, the proposed connecting-rod mechanism, and the robot body. There are two same motors in the robot's body, which can drive each wheel independently to guarantee the turning of the robot on any terrains easily.

Some physical and necessary experiments are performed, including moving forwards and backwards, turning left and right, jumping small barriers, crossing platforms, climbing continuous stairs, and walking on any other irregular ground. In this section, the robot's ability to climb continuous stairs will be mainly described and analyzed, in order to verify the obstacle-crossing ability. There are two difficulties encountered when climbing continuous stairs, compared with a single step or other simple obstacles (Siciliano, 2013). On the one hand, a continuous step has a certain slope gradient, necessitating the robots to increase the driving torque. On the other hand, there is a great irregularity compared with the tilted plain, necessitating that the structure of the robot be designed as simple as possible. In this paper, the designed and selected size of stairs is determined according to the standard size of the stairs in the building, and other reasonable and representative sizes have

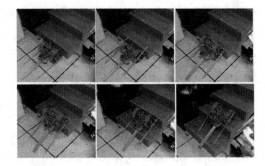

Figure 6. The process of climbing continuous stairs by a two wheel-rod hybrid robot.

Table 1. Functional testing experiments of the mobile robot.

Terrains	Functional testing experiments	Successes	Failures	Success rate
Flat ground	Moving forwards and backwards	100	0	100%
	Turning left and right	100	0	100%
	Stopping	100	0	100%
Continuous stairs	Crossing obstacles	98	2	98%
Other irregular terrains	Crossing small barriers	100	0	100%
	Crossing platforms	99	1	99%
	Crossing mating surface groove	91	9	91%

also been selected (Yu, 2010). The width, height, and slope gradient of the staircase are 28.5 cm, 15.5 cm, and 28.54 degrees, respectively (Qiao, 2015).

Fig. 6 shows the process of climbing continuous stairs by a two wheel-rod hybrid robot. This process of climbing obstacles is mainly divided into six steps, which describes that the robot climbs one step and the action of crossing the next step is the same as the first one.

Table 1 presents the experimental and testing results obtained by performing a series of experiments to test the stability of the robot. Each functional test was repeated 100 times.

5 SUMMARY

This paper proposed a new type of two wheel-rod hybrid robot, which could not only move optionally on any terrains and maintain sufficient velocity on a flat ground with the locomotion method based on wheel locomotion, but also shows obstacle-crossing performance with high efficiency based on the connecting-rod mechanism. the proposed method was verified experimentally using a real machine.

The proposed connecting rod mechanism improved the obstacle-surmounting ability of the robot, which was verified by the modeling and the kinematic and kinetic analysis. In addition, the design of two wheels and independent driving allowed the robot to control easily and to move freely with excellent locomotive efficiency.

ACKNOWLEDGMENT

This work was financially supported by the "Innovative Research Project for Postgraduate Students of Chongqing, China" (CYS2015050) and the "Fundamental Research Funds for the Central Universities" (XDJK2016D022, XDJK2016E071).

REFERENCES

Ge G., Y. Wang. On Obstacle Crossing Research for Quadruped Eccentric Wheeled-Legged Robot. J. of Southwest China Normal University (Natural Science Edition), 39(10), 95–100, (2014).

Li J. Designing and Analyzing for an Wheel-Ski-Style obstacle-surmounting Robot. Master Degree, Southwest University, China. 24–29, (2013).

Li J., Y. Wang, et al. Static Stability Analysis on Three-Foot Gait of Eccentric-Type Legged Hexapod Robot. J. of Henan Institute of Education (Natural Science Edition). 21(1), 40–43, (2012).

Muir P.F., C.P. Neuman. J. Robot, Syst, 4(2), 28–340, (1987).

Qiao G., G. Song, et al. A Wheel-legged Robot with Active Waist Joint: Design, Analysis, and Experimental Results. Journal of Intelligent & Robotic Systems, 1–18, (2015).

Siciliano B., O.Khatib. Handbook of Robotics. Peking: China Machine Press, 641–642, (2013).

Takahashi M., K. Yoneda, S. Hirose. Rough Terrain Locomotion of a Leg-Wheel Hybrid Quadruped Robot. IEEE International Conference on Robotics & Automation, 1090–1095, (2006).

Tan X., Y. Wang, X. He. The Gait of a Hexapod Robot and Its Obstacle-surmounting Capability. World Congress on Intelligent Control & Automation, 303–308, (2011).

Woo C., H. Choi, et al. Optimal Design of a New Wheeled Mobile Robot Based on a Kinetic Analysis of the Stair Climbing States, Journal of Intelligent & Robotic Systems, 49(4), 325–354, (2007).

Yu Y., K. Yuan, et al. Dynamic Model of All-Wheel-Drive Mobile Robot Climbing over Obstacles and Analysis on Its Influential Factors. Robot, 30(1), 1–6.

Advances in Energy Science and Equipment Engineering II – Zhou, Patty & Chen (Eds)
© 2017 Taylor & Francis Group, London, ISBN 978-1-138-71798-5

Research on the evaluation system of technology progress in China's express industry

Chun-hui Yuan, Chen Wang & Xiao-long Li
School of Economics and Management, Beijing University of Posts and Telecommunications, Beijing, China

ABSTRACT: Based on the policy environment of "Internet plus", express companies have been trying to save costs and improve service quality by using new technical methods. Thus, the evaluation system of technology progress has a certain role to play in guiding the development direction of express companies. Based on the calculation system set by the Ministry of Science and Technology, this paper first takes the "informatization" as the main clue to formulate the index system in China's express industry, and then uses the combination of Delphi and Analytic Hierarchy Process (AHP) techniques to fix the weight of each indicator. Finally, this paper provides the basis and reference for the future study for upgrading and improving the index system, and puts forward the policy suggestions of technology progress in China's express industry.

1 INTRODUCTION

With the rapid development of information technology, a huge productivity has been created, which brings a "new atmosphere" for the service industry. As an important part of the modern service industry, business requirements in the express industry have shown rapid growth driven by e-commerce. According to the State Post Bureau statistics, in the past five years, domestic online shopping transactions supported by the express industry have exceeded 3 trillion Yuan annually in China, and the business volume and operating income have increased by 7.8 times and 3.8 times, respectively. However, the express industry in China is still in its early developmental stage, and there are still a number of bottlenecks in technology and network management that leads to the predicament of decelerated growth and high cost with low profit. Before 2005, the profit margin of China's express industry was about 35%, and now it has reduced to 5% or less in the vast majority of express companies. Therefore, high technologies have been the only way for sustainable development in the express industry. With the rise of big data, cloud computing, networking, mobile Internet, and other information technology applications, the transformation and upgrading of technology development in service and business mode have been imminent. Regarding the opening, sharing, and integration of Internet thinking as the clue, the express industry should change itself from a labor-driven industry into a technology-driven industry to bring out a profound impact on the society.

Thus, it can be seen that an academic theory study of technology progress in China's express industry is necessary. It will be beneficial to understand the current situation by establishing the evaluation system and calculating index weights. In addition, clear and specific indicators will lead to deeper studies, which in turn will provide ideas and directions for the actual breakthrough in technology improvement.

2 THE CONSTRUCTION OF THE EVALUATION SYSTEM

There are a wide variety of methods for choosing indicators in technology progress, including the authoritative technology progress assessments released by the Ministry of Science and Technology in China, which is based on the input–output system that is mainly related to personnel, technology equipment, R & D, and other aspects. As the result of the corresponding measurement in different research fields, currently, academia uses an innovative theory to establish a comprehensive index system for services and management in technology progress, which includes three dimensions such as input system, output system, and innovative organization, and innovative organization that can be defined by the management mechanism and other three aspects (Shi Shude, 2012). Based on national conditions in China, this paper proposes a mature indicator framework of the express industry, which contains four dimensions: technology investment, technology output, service innovation, and management innovation. (The basic content follows the secondary indicators of technology progress assessments released by the Ministry of Science and Technology.)

In order to facilitate the calculation process, this paper will label three-level indicators in sequence. The three-level indicators are, respectively, labeled as F, S, and T.

2.1 Technology investment (F1)

Investment in science and technology shows the level of emphasis on technology progress and the degree of reserve supporting in every industry. Science and technology investment includes advanced equipment investment and senior personnel investment. The express companies take advantages of a series of advanced equipment of handheld terminals, automated sorting machines, intelligent storage centers, and call centers to network the entire express delivery process. Based on the Internet of Things (IoT), the use of Radio Frequency Identification (RFID), global positioning technology systems, and remote data exchange, the process of customer services has reduced production costs and optimized the customer experience (see Figure 1). In addition, from the perspective of education and career, senior personnel investment includes bachelor degree personnel and technical personnel investment.

2.1.1 Technology equipment investment (S1)

Table 1. Index of technology equipment investment.

Third-level index	Index explanation
Investment intensity of advanced equipment purchasing (T1)	Ratio of the acquisition cost of automation equipment (including mobile devices, sorting, handling, transportation equipment, and handheld terminals) to the total revenue
Investment intensity of information system construction (T2)	Ratio of development and acquisition costs of information systems to the total revenue

2.1.2 Senior personnel investment (S2)

Table 2. Index of senior personnel investment.

Third-level index	Index explanation
Ratio of scientific and technical personnel (T3)	Ratio of employees working in scientific and technical jobs to the total number
Ratio of employees of bachelor degree or above (T4)	Ratio of employees with bachelor degree or above to the total number

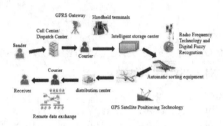

Figure 1. Technology application of the operation process in the express industry.

2.2 Technology output (F2)

Technology output directly explains the investment results and efficiency. There are two second-level indicators to indicate the technology output: service benefit and technology achievements. Service benefit considers the improvement of service level and efficiency mainly from the aspects of express delivery speed, information processing, and customer feedback. Technology achievements measure the impact on output based on four areas: industry technical standards, patents, scientific papers, and technical prizes.

2.2.1 Service benefit (S3)

Table 3. Index of service benefit.

Third-level index	Index explanation
Express delivery speed (T5)	Average number of days during the express service
Efficiency and quality of express information processing (T6)	Update speed and query convenience for the express information
Overall satisfaction of the express service (T7)	Customer satisfaction in every aspect of the express service

2.2.2 Technology achievement (S4)

Table 4. Index of technology achievement.

Third-level index	Index explanation
Publishing number of industry standards (T8)	Number of standards released by express industry authorities
Number of patent licensing (T9)	Number of invention and utility model patents
Publishing number of science and technology papers (T10)	Number of scientific and technological papers published in professional journals
Number of awards for technology activities (T11)	Number of awards in the technical and related professional field

2.3 Service innovation (F3)

Input and output can be regarded as the premise and results of technology progress, and service innovation is the process and detailed description in technology progress. Today, the driving force of e-commerce cannot be underestimated, and business tentacles of express companies have extended into respective service areas. For example, special value-added services (e.g., payment collection, frozen distribution, logistics solutions) rely on the technical development of express companies. At the same time, thanks to the growth of mobile Internet technology, mobile APPs, and social platforms which continue to promote service and marketing channels gradually shifting from offline to online, and accelerate the integration of offline and online channels.

2.3.1 Business extension (S5)

Table 5. Index of business extension.

Third-level index	Index explanation
Express delivery volume of e-commerce business (T12)	Number of express packages provided for e-commerce
Proportion of the value-added service revenue (T13)	Ratio of the value-added service revenue to the total revenue

2.3.2 Channel expansion (S6)

Table 6. Index of channel expansion.

Third-level index	Index explanation
Conditions of existing online channels (T14)	Improvement of online services, including websites, APPs, WeChat public number, call center, etc.
Integration of online and offline channels (T15)	Situation of the channel integration of online and offline services

2.4 Management innovation (F4)

Management innovation concerns whether to establish standards of information systems at the technical management level. The general idea of technology construction in express companies is shown in Figure 2, which can measure integrity, innovation, and technology iteration cycle of the technology architecture.

2.4.1 Technology management system (S7)

Table 7. Index of the technology management system.

Third-level index	Index explanation
Technical architecture construction (T16)	Improvement and innovation of the technical architecture of companies, which strictly comply with the technical management process
Technical management operation (T17)	Operating period of the technology management system of companies, and situation of an update on optimization

3 MEASUREMENT OF THE EVALUATION SYSTEM

At present, the calculation method of the evaluation system is still in the process of exploration among academia, especially in the application of various statistical indicators. However, the qualitative or quantitative analysis method is relatively one-sided and cannot be achieved according to scientific and reasonable principles. Therefore, the main focus is on the combination of qualitative and quantitative research methods, and the calculation of synthetic weight by using the Delphi and AHP techniques (Ma, 2009; Vidal, 2011) (Ma Rui & Zhang Wentao, 2009, Vidal L A & Marle, 2011). A comprehensive measurement is generally applied to the research field of technology progress (Skibniewski, 1992) (Skibniewski M J & Chao L C, 1992), and several new technology products need to be studied in depth to determine business opportunities (Cho, 2013) (CHO J & Lee J., 2013).

3.1 Rating method

There are two main types of scoring methods in the Delphi method: first, by setting different expert levels, the importance of experts is combined with their scores for compositing the final weight (Gao Yanhong, Yang Jianhua, etc., 2013). Second, experts mark in accordance with their own authority and familiarity, and propose weights depending on the importance of each sub-indicator to develop an expert weight questionnaire after reviewing three times (Zhao Liguo & Xiao Hui Luan, 2008). The calculation process collects a total of eight copies

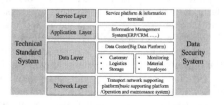

Figure 2. Technology architecture design in express companies.

of the expert scoring table, including doctoral students and teachers at the Beijing University of Posts and Telecommunications, as well as the express company staff. As the degree of familiarity among experts is different in the express industry, this paper will first use the grading method. The scoring criteria are as follows:

Table 8. Scoring criteria in the Delphi method.

Scale	Meaning
1	Compared with two elements, the former and the latter are of the same importance
3	Compared with two elements, the former is a little more important than the latter
5	Compared with two elements, the former is more important than the latter
7	Compared with two elements, the former is strongly more important than the latter
9	Compared with two elements, the former is extremely more important than the latter.
2,4,6,8	The intermediate values of the above adjacent judgment
Reciprocal	If the importance of the elements i and j are in the proportion of a_{ij}, the importance of the elements j and i are in the proportion of $a_{ij} = 1/a_{ji}$

3.2 Index weight calculation

In the 1970s, the American scholar T. L. Saaty initiated the Analytic Hierarchy Process (AHP), and evaluated indicators comprehensively in different dimensions (Saaty, 1977). In view of the scheme evaluation and performance comparison, the AHP is advantageous to the weight calculation of each indicator and each level of indicators (Luke Georghiou & David Roessner, 2000). Following the process and innovation by most scholars, recently, the AHP has been applied to the evaluation system in relevant research fields.

Based on the judgment matrix constructed by expert rating, the weights of third-level indicators can be obtained with MATLAB software. In accordance with the bottom-up level, the overall index weights can be achieved. The calculation method is as follows:

$$\text{Weight of upper level} = \frac{\Sigma(\text{weight of lower level})}{\max(\Sigma(\text{weight of lower level}))} \quad (1)$$

The comprehensive weight of the index system is provided in the following table.

The results obtained from the consistency test are meaningful and not complicated. The consistency checking process needs to focus on whether the value of the random Consistency Ratio (CR) is less than 0.1. If the CR is greater than 0.1, it needs

Table 9. Index weight of the evaluation system.

First-level index	W	Second-level index	W	Third-level index	W
F1	0.251	S1	0.135	T1	0.0677
				T2	0.0677
		S2	0.116	T3	0.0677
				T4	0.0483
F2	0.389	S3	0.155	T5	0.0580
				T6	0.0483
				T7	0.0483
		S4	0.234	T8	0.0677
				T9	0.0591
				T10	0.0591
				T11	0.0483
F3	0.224	S5	0.107	T12	0.0584
				T13	0.0483
		S6	0.118	T14	0.0584
				T15	0.0591
F4	0.135	S7	0.135	T16	0.0677
				T17	0.0677

to adjust the judgment matrix constantly to satisfy the consistency test. Based on the calculation, λ_{max} (maximum characteristic root) = 17.0713, CI (consistency index) = 0.0045, and CR = 0.0028 < 0.1. Then, the satisfactory consistency of results between the theoretical analysis and the empirical analysis is obtained. Therefore, the weight calculation process can be used as a reference basis for the evaluation of technology progress in the express industry.

4 CONCLUSION

The results from the study indicate that there is little difference between each indicator, and no extreme values exist. To a certain extent, it is proved that the construction of the evaluation system is basically in line with the requirements and has a certain reference value. In terms of the scores of three-level indicators, the values of technical equipment investment, technical personnel investment, industry standard released, and technology management system are slightly larger, which demonstrate the importance of three fundamental parts, namely information technology, personnel, and mechanism, and provide views and guidance for all-round development of technology progress in express companies. For second-level indicators, the outputs of technology achievements and service benefits are prominent, indicating that companies still need to pay attention to the results and objectives in business management, especially the scientific research results at the technical level. However, the weight of business extension and channel expansion is relatively low, and there is little impact on technology progress in China's express industry.

From the perspective of the express industry, promoting technology progress should accelerate the construction of information and technology industry standards, and create a scientific research platform to encourage technological innovation and knowledge sharing. For express companies, the requirements are more specific: in order to attach importance to invest in science and technology and develop sophisticated personnel, it is crucial to carry out high-quality and efficient research projects, as well as establish a good relationship with key talents in universities.

ACKNOWLEDGMENTS

This work was supported by the Situation Evaluation and Policy Research of Scientific and Technological Progress in China Postal Industry (Grant No. 2013GXS4B095) and Intellisense Third-party Express Logistics Cloud Service Platform's Research and Application (Grant No. 2014BAH23F07).

REFERENCES

Cho J., J. Lee, Expert Syst. Appl. **40**, 17 (2013).

Gao Y.H., J.H. Yang, F. Yang, Sci. & Technol. Prog. Pol. **30**, 5 (2013).

Georghiou L., D. Roessner, RP, **29**, 22 (2000).

Ma R., W.T. Zhang, S.M. Sun, CS, **36**, 4 (2009).

Saaty T.L., J. Math. Psycho. **15**, 48 (1977).

Shi S.D., STI, **12**, 4 (2012).

Skibniewski M.J., L.C. Chao, J. Constr. Eng. M. **118**, 17 (1992).

Vidal L.A., F. Marle, J.C. Bocquet, Expert Syst. Appl. **38**, 18 (2011).

Zhao L.G., X.H. Luan, X.Y. Zhu, Coast. Ent. Sci. & Technol. **11**, 2 (2008).

Robust patch-based visual tracking with adaptive templates constraint

Ruitao Lu, Wanying Xu, Yongbin Zheng & Xinsheng Huang
College of Mechatronic Engineering and Automation, National University of Defense Technology, Changsha, China

ABSTRACT: A novel patch-based algorithm for robust object tracking is proposed in this study. To effectively capture the information of the target, the object appearance model is constructed by the adaptive selection mechanism. The discriminative patches of the appearance model are more robust to match. The vote maps that are obtained by matching the target patches independently are fused for determining the new location of the object. At last, a dynamic updating scheme is proposed to address appearance variations and alleviate tracking drift. Experiments and evaluations on various challenging image sequences are performed, and the results show that the proposed algorithm performs favorably against other state-of-the-art methods.

1 INTRODUCTION

Object tracking is one of t Lie most important issues in computer vision, which has been widely applied in surveillance, activity analysis, classification and recognition from motion (Yilmaz, 2006). Although significant progress has been made over the past few decades, object tracking still remains challenging due to significant appearance changes caused by shape deformation, pose variation, motion blur, camera motion, occlusion and illumination.

Typically, a visual tracking system consists of four modules (Li, 2006): object initialization, appearance model, motion model, and object localization. Object initialization can be manual or automatic; the appearance model estimates the likelihood of a state belonging to the object class; the motion model describes the correlation of the different states of an object over time; object localization is performed by a search mechanism or maximum a posterior estimation based on the motion model.

In recent years, significant research has been performed regarding patch-based tracking methods (Hager, 2004; Adam, 2006; Zhu, 2008; Jose, 2013; Liu, 2011; Zhong, 2014). These methods perform well in pose changes and occlusion. Hager et al. (Hager, 2004) proposed a new iterative scheme for matching kernel-modulated histograms, and the paper introduced objective functions optimizing more elaborate parametric motion models based on multiple spatially distributed kernels. Adam et al. (Adam, 2006) developed a fragment-based appearance model that considered the target as a collection of image patches. Their approach presents good tracking results in substantial partial occlusions, but fails as the number of occluded patches increases. Zhu et al. (Zhu, 2008) dealt with occlusions by dividing the object in smaller blocks, and explored the similarity in local color and texture features for each of the blocks. Jose Bins et al. (Jose, 2013) presented a patch-based tracking method; each patch was tracked individually, and the individual displacement vectors were combined in a robust manner to obtain the accurate tracking results. Liu et al. (Liu, 2011) use a local sparse representation to model the appearance of target patches, and a sparse coding histogram is used to represent the basis distribution of the target. Zhong et al. (Zhong, 2014) proposed a robust visual tracking method via patch-based appearance model driven by context-awareness and attentional selection. The local background estimation was used to update the appearance model which provided the robust tracking results.

Although the use of multiple patches tends to present more robust tracking results, the presence of occlusions (particularly heavy occlusions) and pose variations are still a challenge. In this study, we proposed a novel patch-based tracking method with adaptive templates constraint. To effectively capture the information of the target, the appearance model of the object is constructed by the adaptive selection mechanism. The vote maps that are obtained by matching the target patches independently are fused for determining the new location of the object. At last, we propose a dynamic updating scheme to address appearance variations and alleviate tracking drift. A series of experiments and evaluations on various challenging sequences are performed, and the results show that the proposed algorithm performs favorably against other state-of-the-art methods.

2 OUR METHOD

We present the details of the proposed tracking algorithm in this section. Specifically, our method works as follows:

A novel adaptive selection mechanism is presented to select the discriminative patches for matching; a set of confidence maps are calculated by matching the target patches represented by histogram; then, we fuse the confidence maps by an effective scheme and obtain the new location of the target in the frame; At last, we propose a dynamic updating scheme to address appearance variations and alleviate tracking drift.

2.1 Appearance model based on adaptive templates constraint

The principles of the selection mechanism of the patches are:

1. Representative: the patches need to cover the target as far as possible.
2. Discriminative: the patches in templates set have more discriminative information, and reduce the redundant information as far as possible.

Based on the above mentioned, we present a novel selection mechanism of the patches with clustering method. If the size of the target is $m \times n$, we select N_h horizontal patches for matching. The size of horizontal patch is $(m/6) \times (n/2)$, and the size of the vertical patch is $(n/2) \times (m/6)$. Each patch is represented by the raw histogram, and the distance of the patches is measured by Euclidean distance. We take horizontal patch for example, and the steps of the selection mechanism are presented as Algorithm 1.

Algorithm 1 Adaptive selection mechanism.

Step 1: Typically, the center of the template is more reliable, so the patch in the center of the target is selected as the first patch F_1.

Step 2: The candidate patches is collected by the sliding windows from the top-left of the target.

Step 3: We compute the distance between every candidate patches and the first patch F_1; we choose the patch as F_2 which has the largest distance.

Step 4: For each candidate patch, its distance to F_1 is d_1 and to F_2 is d_2, we reserve the smaller distance $x = \min(d_1, d_2)$; we select the patch as template patch F_3 if the candidate patch with the largest distance of x.

Step 5: Repeat the Step 4, until the number of the template patches is N_h.

Analogously, we obtain the vertical patches by the adaptive selection mechanism. Fig. 1 shows which patches are collected by the selection mechanism.

2.2 Matching and decision fusion

In order to utilize ratios of difference and sum of the corresponding bins of two histograms, we use the Chi-square distance to measure similarity of two histograms, which is describe by

$$D(\mathbf{x},\mathbf{y}) = \exp\left(-\alpha \sum \frac{(\mathbf{x}_i - \mathbf{y}_i)^2}{\mathbf{x}_i + \mathbf{y}_i} \right) \qquad (1)$$

where x and y represent two vectors, and x_i and y_i are the i-th corresponding bins, and α control parameter.

For a new testing frame t, we obtain a vote map $V_i^P(\cdot,\cdot)$ for every template patch F_i based on the Chi-square distance. To handle outliers resulting from occluded patches, we obtain the final recognition result by combing a set of vote maps:

$$V^P = \left\{ V_i^P(\cdot,\cdot) \,|\, i = 1, 2, ..., N_p \right\} \qquad (2)$$

where $N_p = N_h + N_v$. More specifically, at each location (x, y), we order the corresponding vote values $V^P(x,y) = \{V_i^P(x,y) \,|\, i = 1,2,...,N_p\}$, and choose the Q-th smallest values, to construct the fusion confidence map (as shown in Fig. 2):

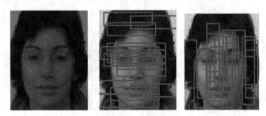

Figure 1. The appearance model of the target.

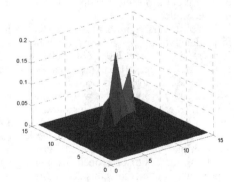

Figure 2. The fusion confidence map.

1588

$$C(x,y) = \sum_{r=1}^{Q} V^{p}(x,y)_{r} \qquad (3)$$

Typically, Q is set as 25%. The robust estimator is designed as follows:

$$(x^{*}, y^{*}) = \arg\max_{(x,y)} C(x,y) \qquad (4)$$

2.3 Updating scheme

Because the appearances of the target and background may change, updating the template set is necessary (Babenko, 2009; Kalal, 2010). However, minor errors may be introduced and accumulate if the template set is updated too frequently, resulting in a drifting problem. In this paper, we address this problem by updating the template set dynamically.

The patch P_i of the new object location is considered as the foreground patch if $D(P_i, F_i) > \theta_1$, where F_i is the corresponding patch in the appearance model. And the foreground patch is updated by component-wise convex combination of P_i and F_i:

$$F_i = \beta F_i + (1 - \beta) P_i \qquad (5)$$

which β is a forgetting factor.

The patch P_i is deemed to occluded patch if $D(P_i, F_i) < \theta_2$. For occluded patch, we do not update the template for alleviating tracking drift.

3 EXPERIMENTAL RESULTS

Our approach was implemented in MATLAB 2011 on a Core 2.0 GHz Dual Core PC with 2 GB memory. We did not use any motion model to predict the target position.

The number of horizontal patches is defined as $N_h = 18$, and the number of vertical patches is set as $N_v = 18$. The control parameter α is set to 20. In updating scheme, the forgetting factor β is set to 0.85. The thresholds θ_1 and θ_2 are set to 0.1 and 0.04.

For comparison, we evaluate our tracker against two state-of-art tracking methods (Frag (Adam, 2006) and MS (Comaniciu, 2000)) on four challenging sequences including the occlusions and pose variations. Both qualitative and quantitative evaluations are presented in this section.

3.1 Qualitative evaluation

Fig. 3 shows the qualitative evaluation of the four sequences. In the *FaceOcc2* sequence, the target undergoes heavy occlusion and pose change.

MS is drift away from the target when significant occlusion occurs. Frag and the proposed method perform well in this sequence.

In the *Girl* sequence, the target undergoes the significant pose variations. MS and Frag drift away from the ground truth, because the lake of the effective updating scheme. The proposed method achieve the favorably result.

In the *FaceOcc1* sequence, a face is heavily occluded by a magazine. Although all tracker can track the target, MS has the large drifting errors than the others tracking algorithms.

In the *Subway* sequence, the target is clustered by the other person, and the object is occluded in some frames. As shown in Fig. 3, MS and Frag drift away from the ground truth when the occlusion occurs in the clustered background. The proposed method achieve the best tracking result can be attributed to the robust appearance model with adaptive templates constraint. The dynamic updating scheme of patches provides the proposed tracker with robustness against appearance variations.

3.2 Quantitative evaluation

For quantitative evaluation, two evaluation criteria are used to assess the performance of tracking algorithms. The center location error is the distance between the central location of the tracked target and the ground truth. The average errors are presented in Table 1. The proposed tracker achieves the lower drifting errors than the others tracking algorithms in all consequences. Our tracker

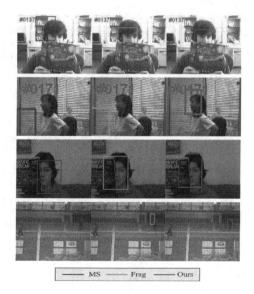

Figure 3. Qualitative evaluation of the four sequences, including *FaceOcc2*, *Girl*, *FaceOcc1* and *Subway*.

Table 1. The Overlap rate (ORE). The Bold fonts indicate the best performance.

Sequence	Mean shift	Fragment	Ours
FaceOcc2	11.56%	50.17%	**54.78%**
Girl	28.42%	41.39%	**49.22%**
FaceOcc1	61.62%	70.53%	**74.13%**
Subway	2.77%	47.42%	**68.86%**

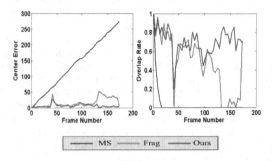

Figure 4. The center location errors and the overlap rates of each algorithm for the *Subway* sequences.

Table 2. The Center Location Error (CLE) (in pixels). The Bold fonts indicate the best performance.

Sequence	Mean shift	Fragment	Ours
FaceOcc2	78.55	38.70	**18.10**
Girl	30.28	23.51	**9.72**
FaceOcc1	25.57	17.19	**15.44**
Subway	140.42	16.17	**6.81**

outperforms other trackers as the proposed patch-based appearance model is robust to pose variation and occlusions. The updating scheme improves the tracking stability and accuracy.

In addition, the overlap rate (Everingham, 2010) is used to evaluate the stability of each algorithm. Given the tracking result ROI_T and the ground truth bounding box ROI_G, the overlap rate is defined as $score = \dfrac{area(ROI_T \cap ROI_G)}{area(ROI_T \cup ROI_G)}$. Fig. 4 shows the center location errors and the overlap rates of each algorithm for the *Subway* sequences. Table 2 presents the average overlap rates. The success of our tracker can be attributed to the effective patch-based appearance model with spatial representation and discriminative information.

4 CONCLUSIONS

In this study, we propose a novel patch-based tracking method with adaptive templates constraint. To effectively capture the information of the target, the appearance model of the object is constructed by the adaptive selection mechanism. The vote maps that are obtained by matching the target patches independently are fused for determining the new location of the object. A dynamic updating scheme is presented to account for appearance variations and to reduce the tracking drift. Both quantitative and qualitative evaluations on challenging image sequences against several state-of-the-art algorithms demonstrate the accuracy and robustness of the proposed tracker.

ACKNOWLEDGMENT

This study was supported in part by the National Science Foundation of China under Grants NO. 61403412.

REFERENCES

Adam A, E. Rivlin, I. Shimshoni. Robust fragments-based tracking using the integral histogram. *In Proceedings of CVPR*, New York, NY, USA, 798–805 June (2006).

Babenko B, M. H. Yang, S. Belongie. Visual tracking with online multiple instance learning. *In Proceedings of CVPR*, Miami Beah, Florida, USA, 983–990 June (2009).

Comaniciu D, V. Ramesh, P. Meer. Real-Time tracking of Non-Rifgid Objects using MeanShift. *In Proceedings of CVPR*, Hilton Head Island, SC, USA, 142–149 June (2000).

Everingham M, L.V. Gool, C.K. Williams, J. Winn. Zisserman A. The pascal Visual Object Classes (VOC) challenge. Int. J. Comput. Vis. 88, 303–338 (2010).

Hager G. D, M. Dewan, C. V. Stewart. Multiple kernel tracking with SSD. *In Proceedings of ICCV*, 790–797 June (2004).

Jose B, L. Leandro, R. Claudio. Target Tracking Using Multiple Patches and Weighted Vector Median Filters. J Math Imaging Vis. 45, 293–307 (2013).

Kalal Z, J. Matas, K. Mikolajczyk. P-n learning: Bootstrapping binary classifiers by structural constraints," in *Proceedings of CVPR*, San Francisco, CA, USA, 49–56 June (2010).

Li X, W. Hu, C.H. Shen. Zhang Z. Asurvey of appearance models in visual object tracking. ACM Transactions on Intell. Syst. Technol. 4, 58 (2006).

Liu B, J. Huang, C. Kulikowski, L. Yang. Robust tracking using local sparse appearance model and K-selection. *In Proceedings of CVPR*, 1–8 June (2011).

Yilmaz A, O. Javed, M. Shah. Object tracking: A survey. ACM Comput. Surv. 38, 1–45 (2006).

Zhong B, Y. Chen, Y Shen. Robust tracking via patch-based appearance model and local background estimation. Neurocomputing. 123, 344–353 (2014).

Zhu L, J. Zhou, J. Song. Tracking multiple objects through occlusion with online sampling and position estimation. Pattern Recognition. 41(8), 2447–2460 (2008).

Advances in Energy Science and Equipment Engineering II – Zhou, Patty & Chen (Eds)
© 2017 Taylor & Francis Group, London, ISBN 978-1-138-71798-5

A multi-sensor-based information fusion optimization algorithm for the detection of ceramic shuttle kiln temperature

Yonghong Zhu, Wei Wang & Junxiang Wang
School of Mechanical and Electronic Engineering, Jingdezhen Ceramic Institute, Jingdezhen, China

ABSTRACT: In view of the present single detection method and low detection accuracy of ceramic shuttle kiln temperature, a new kind of multi-sensor-based information fusion optimization algorithm for the detection of burning zone temperature is proposed in this paper according to the fundamental characteristics of the burning zone of ceramic shuttle kiln. First, different temperatures corresponding to the three primary colours (RGB) of flame images are obtained by an expert system. Second, real-time three primary colour values of the kiln flame images taken by a CCD camera are obtained by using the K-mean clustering segmentation algorithm. Third, the corresponding temperature value of the image in the expert system is obtained by the least-squares method. Finally, the temperature values obtained from the multi-sensor are fused, and the burning zone temperature of ceramic shuttle kiln is obtained by using the Kalman filter to smooth the fusion temperature. In conclusion, the simulation results indicate that the proposed algorithm has a good effect.

1 INTRODUCTION

China is a big ceramic producer, and its ceramic product export occupies a large share in foreign trade. However, compared with other developed countries, there are still some shortcomings in terms of ceramic production, such as the structure of kiln type, the use of heat energy, thermotechnical testing, environmental protection and the control system. Therefore, the improvement in the thermal efficiency of ceramic kiln is regarded as one of the key technologies of research and development by the relevant state departments in China. Ceramic shuttle kiln is one of the most widely used modern intermittent kilns in ceramic industry (Saidur, 2011; He, 204; Dinh, 2007). Its detection and control methods directly influence the quality and energy efficiency of ceramic products.

Ceramic shuttle kiln is a kind of complex object with large time-delay, strong nonlinearity, multivariable coupling and variable parameters. The quality of ceramic products depends to a large extent on the temperature of the sintering process. In order to achieve the effective control of the burning zone temperature of ceramic shuttle kiln, it is necessary to study a set of effective temperature recognition algorithms for the burning zone. At present, the burning zone temperature of ceramic shuttle kiln is mainly detected through thermocouples in industrial production. With the ageing of the thermocouple, its detection precision becomes much lower. It is also difficult to obtain

accurate temperature results by only extracting the characteristics of the flame image (Tian, 2011; Fan, 2014; Zhang, 2014; Han, 2009). In addition, the Kalman filtering algorithm (Peng, 2009; Xia, 2013) has a strong ability to control errors in a small range. Based on the above situations, a new kind of temperature information fusion optimization algorithm for ceramic shuttle kiln is proposed in this paper. In this method, the temperature point detection method and the flame image recognition method are first combined to measure the burning zone temperature, and then the temperature data obtained is smoothed to obtain the actual burning zone temperature value by using the Kalman filtering algorithm. The experimental results indicate that the proposed algorithm is feasible and effective.

2 TEMPERATURE DETECTION BASED ON MULTI-SENSORS

The temperature of ceramic shuttle kiln can be detected and recognized by multi-sensors. These sensors can be divided into two categories. The first category is the non-contact temperature identification sensor such as CCD camera which recognizes the temperature by shooting flame images in real time. The second one is the contact temperature sensor such as thermocouple which detects the firing zone temperature by point detection. The two kinds of method are discussed below.

2.1 Flame image recognition based on CCD

An industrial camera is placed in front of each fire hole and shoots kiln flame images in real time. The flame images obtained are pre-processed and segmented in order to extract flame parts. The K-means clustering algorithm is used to segment the flame image. This algorithm has advantages such as good stability, low computational complexity and fast arithmetic speed. The specific process is shown in Fig. 1.

Combined with the flame image characteristics of ceramic shuttle kiln, according to the experience and through the experimental verification, the K-mean clustering segmentation effect of the flame image on ceramic shuttle kiln is best if K is equal to 3. The segmentation results are shown in Fig. 2.

2.2 Temperature detection by thermocouples

Nowadays, the temperature of ceramic shuttle kiln is detected mostly by thermocouples. A small number of thermocouple sensors are installed on the inner wall of the kiln, and the kiln temperatures are detected by thermocouple sensors. The principle of this temperature detection method is as follows. The temperature sensed by the thermocouple sensor is transformed into the corresponding electrical signal, and then the electrical signal is amplified and transformed into the digital signal by the A/D converter. The digital signal is sent into a computer and processed. After obtaining the consistency temperature measurement values, the information fusion is carried out to obtain the accurate measurement results.

Thermocouples are widely used in industrial temperature measurements. It has many advantages such as high measurement precision, wide measurement range, small heat inertia, rapid measurement, simple structure and easy to use.

The basic principle of a thermocouple is that two different materials of conductors A and B are connected to a closed loop in which there is electric current if the temperatures T and T0 of the two ends are different. At this point of time, there exists an electromotive force between A and B, namely thermo-electromotive force $E_{AB}(T,T_0)$, as shown in Fig. 3.

The thermoelectric potential of a thermocouple is approximately given by:

$$E_{AB}(T,T_0) = \frac{K}{e}(T-T_0)\ln\frac{n_A}{n_B} = E(T) - E(T_0) \quad (1)$$

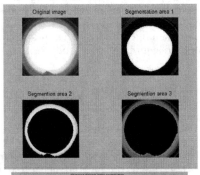

Figure 2. Clustering segmentation images.

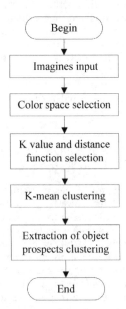

Figure 1. Flowchart of the K-mean clustering algorithm.

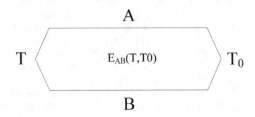

Figure 3. Thermoelectric effect of a thermocouple.

where $E(T) = E(T_0) + E_{AB}(T, T_0)$. Here, both A and B conductors are thermodes, T node is referred to as the work side, T_0 node is referred to as the reference side usually at a constant temperature.

3 INFORMATION FUSION OPTIMIZATION ALGORITHM FOR THE DETECTION OF CERAMIC SHUTTLE KILN TEMPERATURE

On the basis of the above method, a new kind of information fusion optimization algorithm for the detection of ceramic shuttle kiln temperature is proposed as follows:

1. Obtain the different temperatures corresponding to R, G, B by the expert system;
2. Obtain the R, G, B values after image segmentation;
3. Calculate the temperature value corresponding to the image by the least-squares method;
4. Fuse the temperature data obtained by the multi-sensor and smooth them by using the Kalman filtering method.

3.1 *The acquisition of different temperatures corresponding to R, G, B by the expert system*

Expert system is a kind of intelligent computer program that includes knowledge and reasoning. The knowledge and experience of experts in some fields are put in it, and special problems can be solved by knowledge and experience. A complete expert system mainly includes knowledge base, inference engine, working database, explanation machine, knowledge acquisition machine and user machine interface. Their relationship is shown in Fig. 4.

The knowledge base includes fact knowledge base and expert knowledge base in which the specialized knowledge and data required in a certain field are stored, and it is the composition base of the expert system. There are two kinds of form for the acquisition of the knowledge base. One form is that the knowledge base is obtained by the definition, facts and theory of related academic works and textbooks. The another form is that the knowledge

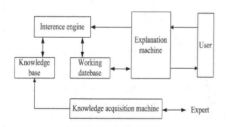

Figure 4. Expert system.

base is the practical experience obtained by experts in the long-term work process. These theoretical knowledge and practical experience constitute knowledge base in the expert system. The working database is used to store the initial data of the field and the data generated in the reasoning process. The inference engine is used for the rules memory in order to make the expert system work harmonically. The explanation machine is used to explain the behaviour of the expert system for users.

In the expert system, the colours corresponding to the standard R, G, B are red [255, 0, 0], orange [255, 125, 0], yellow [255, 255, 0] and white [255, 255, 255], and the corresponding temperatures are 800°C, 925°C, 1050°C and 1311°C, respectively. The temperature values from 0°C to 1311°C are stored in the knowledge base of the expert system.

3.2 *The acquisition of R, G, B values after image segmentation*

Each pixel of the flame image is composed of tri-phosphor R, G, B components, that is, $f(:, :, (R, G, B))$. Therefore, the image tri-phosphor extracted constitutes a new three greyscale, that is:

$$f_R(:, :) = f(:, :, 1), f_G(:, :) = f(:, :, 2), f_B(:, :)$$
$$= f(:, :, 3) \qquad (2)$$

where $f(:, :, 1)$ is the R component corresponding to the image $f(:, :, (R, G, B))$, $f(:, :, 2)$ is the G component corresponding to the image $f(:, :, (R, G, B))$ and $f(:, :, 3)$ is the B component corresponding to the image $f(:, :, (R, G, B))$.

3.3 *The acquisition of the temperature values corresponding to the image by the least-squares method*

For a set of observational data $(x_i, y_i)(i = 0, 1, \cdots, n)$, the function $y = p(x, a_0, a_1, \cdots, a_m)$ between the dependent variable y and the independent variable x is $\sum_{i=0}^{n} \delta_i^2$, which is required to make the quadratic sum of the error $\delta_i = f(x_i) - y_i$ at the fixed point x_i minimum, where, $a_k(k = 0, 1, 2, \cdots, m)$ is the unknown parameter. Thus, the following polynomial can be solved:

$$p(x) = a_0 \varphi_0(x) + a_1 \varphi_1(x) + \cdots + a_m \varphi_m(x), m < n \qquad (3)$$

such that

$$I(a_0, a_1, \cdots, a_m) = \sum_{i=0}^{n} \rho_i [p(x_i) - y_i]^2$$
$$= \sum_{i=0}^{n} \rho_i [a_0 \varphi_0(x_i) + a_1 \varphi_1(x_i) + \cdots + a_m \varphi_m(x_i) - y_i]^2 \qquad (4)$$

is minimum. According to the necessary condition of the extreme value, we have:

$$\frac{\partial I}{\partial a_k} = 2\sum_{i=0}^{n} \rho_i[a_0\varphi_0(x_i) + a_1\varphi_1(x_i)$$
$$+\cdots+a_m\varphi_m(x_i) - y_i]\,\varphi_k(x_i) = 0 \qquad (5)$$
$$k = 0,1,\cdots,m$$

where x is R, G, B values obtained by image segmentation and y is R, G, B values in the expert system. By applying the least-squares method to R, G and B values obtained by partial segmentation images, we can obtain the accurate temperature, as given in Table 1.

3.4 Information fusion optimization algorithm

As the measurement results of temperature sensors with equal precision indicate normal distribution characteristics, the temperature data of ceramic shuttle kiln can be fused by the arithmetic mean value algorithm and the partial estimation method, which is as follows. The temperature values obtained by the multi-sensor are divided into two groups. The temperature values of the first group are the temperatures obtained from the expert system after image processing, and the temperature values of the second group are the temperatures obtained directly from thermocouple sensors. The consistent data of each group is obtained from the two groups of temperature data. Then, the mean value of the two groups of measurement data is estimated to obtain the fusion value $\hat{T}^+$ closer to the true temperature so that the accurate temperature measurement result of ceramic shuttle kiln can be obtained to eliminate the uncertainty in the measurement process.

The specific algorithm is as follows.

Assume that the two groups of data are T_{11}, $T_{12}, \ldots, T_{1m}$ and $T_{21}\ T_{22}, \ldots, T_{2n}\ (m < n)$, respectively. Their arithmetic mean values are respectively:

$$\bar{T}_{(1)} = \frac{1}{m}\sum_{p=1}^{m} T_{1p},\ \bar{T}_{(2)} = \frac{1}{n}\sum_{p=1}^{n} T_{2q} \qquad (6)$$

Table 1. Temperature values obtained by the least-squares method.

R	G	B	Temperature (°C)
241	192	10	992
249	250	21	1076
252	250	73	1128
255	251	110	1165
245	237	117	1172
252	253	148	1203
254	254	183	1238
255	255	218	1273

Their corresponding standard errors are respectively:

$$\hat{\sigma}_{(1)} = \sqrt{\frac{1}{m-1}\sum_{p=1}^{m}(T_{1p} - \bar{T}_{(1)})^2},$$
$$\hat{\sigma}_{(2)} = \sqrt{\frac{1}{n-1}\sum_{q=1}^{n}(T_{2q} - \bar{T}_{(2)})^2} \qquad (7)$$

Considering the two groups of measurement results at the same time, assume that there is no statistic about temperature measurements in the previous measurement results, namely the variance of the previous measurement results $\hat{\sigma}^- = \infty, \text{so}(\hat{\sigma}^-)^{-1} = 0$.

According to the estimation theory, we can obtain the variance of temperature fusion values as:

$$\hat{\sigma}^+ = [(\hat{\sigma}^-)^{-1} + H^\tau R^{-1}H]^{-1}$$
$$= \left\{ [1\ \ 1]\begin{bmatrix} 1/\hat{\sigma}_{(1)}^2 & 0 \\ 0 & 1/\hat{\sigma}_{(2)}^2 \end{bmatrix}\begin{bmatrix} 1 \\ 1 \end{bmatrix} \right\}^{-1}$$
$$= \frac{\hat{\sigma}_{(1)}^2\hat{\sigma}_{(2)}^2}{\hat{\sigma}_{(1)}^2 + \hat{\sigma}_{(2)}^2} \qquad (8)$$

where $H = [1\ \ 1]^T$ is the coefficient matrix of the measurement equation and the covariance of measurement noise is R.

$$R = E[vv^\tau] = \begin{bmatrix} E[v_{(1)}{}^2] & E[v_{(2)}v_{(1)}] \\ E[v_{(2)}v_{(1)}] & E[v_{(2)}{}^2] \end{bmatrix}$$
$$= \begin{bmatrix} \hat{\sigma}_{(1)}{}^2 & 0 \\ 0 & \hat{\sigma}_{(2)}{}^2 \end{bmatrix} \qquad (9)$$

According to the batch estimation method, we can derive $\hat{T}^+$ as follows:

$$\hat{T}^+ = [\hat{\sigma}^+(\hat{\sigma}^-)^{-1}]\hat{T}^- + [\hat{\sigma}^+ + H^\tau R^{-1}]T$$
$$= [\hat{\sigma}^+ + H^\tau R^{-1}]T \qquad (10)$$

Substituting eq. (8) and eq. (9) into eq. (10), we can obtain:

$$\hat{T}^+ = \frac{\hat{\sigma}_{(1)}{}^2\hat{\sigma}_{(2)}{}^2}{\hat{\sigma}_{(1)}{}^2 + \hat{\sigma}_{(2)}{}^2}[1\ \ 1]\begin{bmatrix} 1/\hat{\sigma}_{(1)}^2 & 0 \\ 0 & 1/\hat{\sigma}_{(2)}^2 \end{bmatrix}\begin{bmatrix} \bar{T}_{(1)} \\ \bar{T}_{(2)} \end{bmatrix}$$
$$= \frac{\hat{\sigma}_{(2)}{}^2}{\hat{\sigma}_{(1)}{}^2 + \hat{\sigma}_{(2)}{}^2}\bar{T}_{(1)} + \frac{\hat{\sigma}_{(1)}{}^2}{\hat{\sigma}_{(1)}{}^2 + \hat{\sigma}_{(2)}{}^2}\bar{T}_{(2)} \qquad (11)$$

$\hat{T}^+$ is the temperature value of data fusion by multi-sensor parameter estimation.

Detecting the temperatures of ceramic shuttle kiln by using sensors at the same time t, the results obtained are summarized in Table 2.

Table 2. Multi-sensor-based temperature measurement results.

Multi-sensor temperature measurement results			
Sensor	S_1	S_2	S_3	S_4
Measurement value (°C)	1301	1298	1303	1299
Sensor	S_5	S_6	S_7	S_8
Measurement value (°C)	1297.6	1300.8	1303.5	1299.3

Its mean value is $\overline{T}_{(8)} = \frac{1}{8}\sum_{i=1}^{8} T_i = 1300.275°C$. The first four temperatures are the temperatures obtained in the expert system after image processing, and the last four temperatures are the temperatures obtained directly by the temperature sensor. The mean value and variance of the temperature data detected by two groups of sensors are respectively:

$$\overline{T}_{(1)} = 1300.25°C \quad \hat{\sigma}_{(1)} = 2.24$$
$$\overline{T}_{(2)} = 1300.05°C \quad \hat{\sigma}_{(2)} = 2.5?$$

Calculating the fusion temperature value of eight measurement data by using the above temperature fusion formula, we obtain $\hat{T}^+ = 1300.16° C$.

It is obvious that the data fusion result is better than the measurement result of a single sensor.

Kalman filtering algorithm

Without considering the control signal, the mathematical model of the Kalman filter can be expressed as:

$$x(k) = A(k)x(k-1) + B(k)w(k)$$
$$y_v(k) = C(k)x(k) + v(k) \tag{12}$$

where x is n-dimensional estimation state variables, y is an m-dimensional output vector, A is the state transition matrix, B is the noise input matrix, C is the output matrix, $w(k)$ is the process noise signal and $v(k)$ is the measurement noise signal.

In the context of minimum variance, the optimal estimate value based on the Kalman filter theory is given by the following recursive formulas:

Estimate covariance:

$$p(k) = A(k)P(k-1)A^T(k) + B(k)Q(k-1)B^T(k) \tag{13}$$

Filter gain:

$$M_n(k) = P(k)C^T(k)\left[C(k)p(k)C^T(k) + R(k)\right]^{-1} \tag{14}$$

Mean square error matrix:

$$P(k) = [I - M_n(k)C(k)]p(k) \tag{15}$$

Optimal estimation value of the state variable:

$$\hat{x}(k) = A(k)\hat{x}(k-1) + M_n(k) \left[y_v(k) - C(k)A(k)\hat{x}(k-1)\right] \tag{16}$$

Error covariance:

$$err\,cov(k) = CP(k)C^T \tag{17}$$

where Q(k) is the observation noise covariance, R(k) is the measurement noise covariance, and I is the unit matrix.

The Kalman filtering algorithm is as follows.

1. **Calculating initial state** $x(0)$

Initial state can be obtained through the system. Assuming that the estimate value is the expected value, that is, $\hat{x}(0) = E[x(0)]$. From

$$p(0) = E\{[x(0) - \hat{x}(0)][x(0) - \hat{x}(0)]^T\},$$

we can have $p(0) = \text{var}[x(0)]$.

2. **Calculating** $p(1)$ **and** $M_n(1)$

$$p(1) = A(1)P(0)A^T(1) + B(1)Q(0)B^T(1) \tag{18}$$

$$M_n(1) = P(1)C^T(1)\left[C(1)p(1)C^T(1) + R(1)\right]^{-1} \tag{19}$$

3. **Calculating the optimal estimate**

$$\hat{x}(1) = A(1)\hat{x}(0) + M_n(1) \left[y_v(1) - C(1)A(1)\hat{x}(0)\right] \tag{20}$$

4. **Calculating the mean square error matrix** $P(1)$

$$P(1) = [I - M_n(1)C(1)]p(1) \tag{21}$$

5. **Repeating steps 2, 3 and 4, we obtain the best estimate** $\hat{x}(k)$ **at any time.**
6. **Finally, calculating the output** $y_v(k) = C(k)\hat{x}(k)$

4 SIMULATION RESEARCH

In order to verify the effectiveness of the temperature information fusion algorithm for ceramic shuttle kiln, an experiment was conducted to detect and recognize the temperature of ceramic shuttle kiln by the point detection method and the flame image recognition method. According to steps

(1), (2) and (3) of the temperature information fusion optimization algorithm for ceramic shuttle kiln, the flame image data is calculated to obtain a group of corresponding temperature values. At the same time, another group of temperature data is obtained by thermocouples. Each point temperature can be obtained by fusing a group of flame image recognition temperature data and a group point detection temperature data. Then, according to step (4) of the proposed algorithm for ceramic shuttle kiln but without using the Kalman filtering algorithm, the temperature simulation curve obtained is shown in Fig. 5. Here, it can be noted that the temperature curve has measurement noise. Hence, the temperature shown in Fig. 5 is the sampling temperature with measurement noise.

Finally, according to step (4) of the proposed algorithm for ceramic shuttle kiln and using the Kalman filtering algorithm, the temperature shown in Fig. 5 is smoothed to obtain the temperature curve, as shown in Fig. 6.

As can be seen from Fig. 6, the proposed algorithm for ceramic shuttle kiln is effective. In the burning zone temperature detection process of ceramic shuttle kiln, the corresponding measurement point temperature data can be automatically obtained by the proposed method so that the labour intensity of workers is reduced. Moreover, it is conducive to the intelligent control of ceramic shuttle kiln temperature and to the improvement in the production efficiency of ceramic shuttle kiln.

5 CONCLUSIONS

The digital images taken by a CCD camera are segmented by using the K-mean clustering algorithm to obtain the three primary colour values corresponding to the images. Then, the temperature corresponding to the expert system is obtained by using the least-squares method, and the kiln temperature with measurement noise is obtained by the temperature information fusion method. Finally, the sampling temperature is smoothed by using the Kalman filter so as to achieve the expected temperature detection and recognition effect. The simulation results indicate that the proposed algorithm is effective. This method will provide a theoretical basis for the effective detection of ceramic shuttle kiln temperature and has an important engineering application value. Meanwhile, it also provides a new way and a new idea for the intelligent detection of ceramic shuttle kiln temperature.

ACKNOWLEDGEMENT

This work was financially supported by the National Natural Science Foundation of China (No. 61563022 and No. 61164014) and the Jiangxi Province Natural Science Foundation of China (No. 20152 ACB20009).

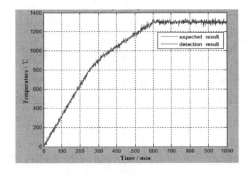

Figure 5. Temperature curve with measurement noise.

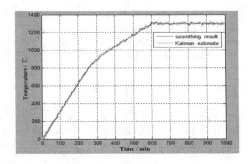

Figure 6. Temperature curve obtained by using the Kalman filtering algorithm.

REFERENCES

Dinh N.Q., N.V. Afzulpurkar, Simul. Model. Pract. & Th., 15, 1239–1258 (2007).
Evensen G., Ocean Dyn., 53, 343–367 (2003).
Fan C.D., Y.J. Zhang, H.L. Ouyang, L.Y. Xiao, Acta Auto. Sinica (In Chinese), 40, 2480–2489 (2014).
Han P., X. Zhang, B. Wang, W.H. Pan, Proc. CSEE (In Chinese), 18, 22–26 (2008)
He F., P.Y. Song, M.F. Jin, J. Wen, J. Wuhan Univ. Tech. (In Chinese), 36, 135–139 (2014).
Koh J.Y., J.C. Ming Teo, W.C. Wong, Proc. the 2014 IEEE ICCS,14, 263–267 (2014).
Jiaocheng M., L. Jun, Y. Qiang, Mater. Trans., 55, 1319–1323 (2014).
Peng D.C., SW.G. (In Chinese), 8, 32–34 (2009).
Saidur R., M.S. Hossain, M.R. Islam, H. Fayaz, H.A. Mohammed, Renew. Sust. Energ. Rev., 15, 2487–2500 (2011).
Tian Q., Y. Yong, L. Gang, Instrum. & Meas. Tech. Conf. (I2MTC), IEEE, 1–4 (2011).
Xia N., T.S. Qiu, J.C. Li, S.F. Li, Acta Electronica Sinica (In Chinese), 41, 148–152 (2013).
Zhang H., Z. Zou, J. Li, J. Cent. S Univ. Tech. (In Chinese), 15, 39–43 (2008).

Failure cause analysis of a solid rocket motor nozzle insert

Lin Sun & Weihua Hui
Science and Technology on Combustion Internal Flow and Thermal-Structure Laboratory, Northwestern Polytechnical University, Xi'an, Shaanxi, China

Lian Hou
The 210th Institute of the Sixth Academy of CASIC, Xi'an, Shaanxi, China

ABSTRACT: A thermo-structural analysis method was developed to investigate the crack cause of a solid rocket motor nozzle insert. The experimental result indicated that there is an occurrence of radial cracking. To explore the cause of this occurrence, a quasi-1D flow was first calculated as the boundary condition for thermo-structural simulation. Then, temperature and stress conditions were obtained. The results of the analysis showed that although the compressive stress did not exceed the compressive strength, the extrusion stress was larger than its strength. This was found to be the main cause of the crack. To improve this status, both improvement in the material property and optimization of the nozzle design should be considered. The simulation result agreed well with the experimental result.

1 INTRODUCTION

Modern Solid Rocket Motors (SRMs) show better performance at higher temperatures and pressures (Tang, 2013; Bao, 2015). This proposes a higher demand for materials used in nozzles, especially for nozzle inserts. Modern nozzles usually use composite materials such as carbon/carbon as inserts. Researchers have made great efforts in studying their thermo-structural responses acting as the ablation material (Morozov, 2009; Keswani, 1985; Bian, 2012; Zheng, 2011; Liu, 2012).

However, for cost reduction, in some solid rocket motors, graphite is used as the nozzle insert material. Graphite material has a good ablation resistance performance and its density is much lower than that of other ablation materials. However, its thermo-physical characteristic is so poor that severe cracking could occur, which will lead to SRM failure (Chen, 1991). Research into graphite has mainly focused on its manufacturing (Gu, 1999; Gu, 2001; Xie, 2015), property test (Wang, 2002; Su, 2015), and new technology in improving graphite property (Su, 2003; Chen, 2004; Su, 2004). Graphite insert failure often occurs during operation (Carrier, 2007; Crane, 2007), and at the initial time of firing, graphite may also fail (Xiong, 2016).

Based on a graphite insert failure that occurred during a nozzle test, a thermo-structural analysis is carried out to discuss the cause. It is found that the simulation result agree reasonably well with the test result.

2 NOZZLE OUTLINE AND TEST PROBLEM

Figure 1 shows the 3D geometry structure of the nozzle. It includes inserts made of T705 graphite, an ablation layer made up of carbon–phenolic resin, a heat insulator made up of silica–phenolic resin, and a support layer made up of structural steel. This nozzle has the characteristic of a 2D axisymmetric model, which will be discussion later.

A study (Chen, 1991; Xiong, 2016) has pointed out that graphite insert may crack when subjected to high temperatures and pressures. During the test of this nozzle, the insert is broken at the end of it which contacts with the ablation layer of the divergent region. Figure 2 shows the crack of throat insert after the experiment.

Figure 1.   Nozzle geometry outline.

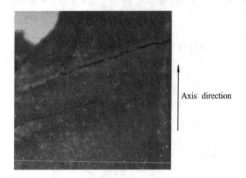

Figure 2.   Crack of throat insert after the experiment.

3   ANALYSIS MODEL AND SCHEME

To explore the cause of this phenomenon and provide a basis for structure optimization and product quality control, this paper uses the thermal–mechanical coupled analysis method. By referring to the stress status of the insert, especially the extrusion stress, a detailed failure cause analysis is carried out.

3.1   Material property

Material property is an important input for finite element method analysis. The analysis takes the variation property of different temperatures of graphite T705 into consideration. The extension strength of graphite T705 used in this nozzle is 11.02 MPa, and the compressive strength is 43.54 MPa.

3.2   Boundary conditions and assumptions

According to test conditions and several impact factors during analysis, the specific boundary conditions are as follows:

1. The inner flow of the nozzle is calculated by the quasi-1D model, and the forced convection between the flow and the nozzle wall is calculated by Bartz formulation. The recovery temperature of the gas and the forced convection coefficient along the nozzle axis are shown in Figure 3.
2. The inner flow pressure is imposed onto the wall of the nozzle, which is shown in Figure 4. In addition, the radial direction outlined at the head of the nozzle is constrained to move along the radial direction but not along the axial direction.
3. The nozzle insert and the surrounding parts are treated as the contact problem. The initial gap sizes are defined in Figure 5 and Table 1.

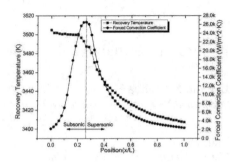

Figure 3.   Heat environment of the nozzle.

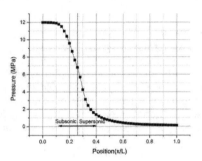

Figure 4.   Pressure load of the inner wall.

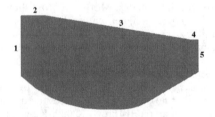

Figure 5.   Contact region definition.

Table 1.   Initial gap size.

Contact pairs	1	2	3	4	5
Gap size (mm)	0	0	0.1	0.1	0.16

Some other assumptions that are required to be added are as follows:

1. The complicated phenomena of erosion and pyrolysis behavior of erosion and heat insulation materials are neglected;
2. The connecting part of the nozzle and the chamber has no heat exchange, and other parts experience free convection;
3. The contact thermal resistance is assumed to be a constant of $1.5e^{-3}$ m$^2 \cdot$ K/W (Zheng, 2011), which ignores the variation of thermal resistance caused by environmental changes such as contact status and temperature;

4. The friction coefficient between the nozzle parts is 0.2.

3.3 *Analysis procedure*

The analysis contains only one simulation step with all the necessary freedom elements, and the coupled analysis is carried out on all element matrices or load vectors of physical variables. The analysis progress lasts for 5s.

4 RESULTS AND ANALYSIS

4.1 *Temperature and stress*

Figure 6 shows the temperature distribution graphs of different moments during the experiment. It is obvious that after the nozzle works for about 0.2s, the maximum temperature reaches up to about 3000K and above. When the nozzle stops working, the maximum temperature reaches up to about 3462.6K.

Figure 7 shows the von Mises stress distribution graphs of different moments during the experiment. It can be observed that the maximum von Mises stress locates at the support layer of the nozzle and the stress level of the other parts is acceptable. It can be inferred that the reliability of the support layer made of structural steel should be reconsidered.

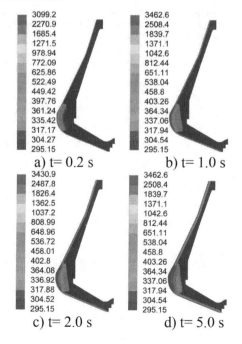

Figure 6. Temperature distribution graphs of different moments (K).

4.2 *Cause exploration of insert failure*

Figure 2 shows the occurrence of radial cracking during the experiment. To explore the cause of cracking, Figure 8 shows the axial stress distribution graphs of different moments during the experiment.

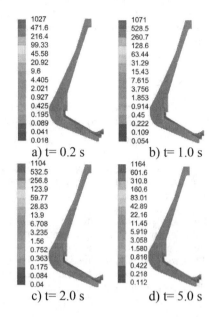

Figure 7. von Mises stress distribution graphs of different moments (MPa).

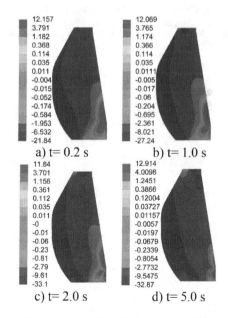

Figure 8. Axial stress distribution graphs of different moments (MPa).

It is clear that from the initial time of firing, the extrusion stress exceeds the extension strength, especially the small region contacted by the expansion region and the back surface. However, for the compressive stress of the insert, it is close to the compressive strength.

From Figure 2 and Figure 8, it can be inferred that the cause of the crack results from the big extrusion stress, which occurs at the initial time of firing.

5 CONCLUSION

A thermo-structural analysis is carried out to study the failure cause of the graphite nozzle insert.

1. During firing of the rocket, the firing impact will lead to large thermal stress. If the strength of the material is not enough, crack will occur;
2. The main cause of the crack is that the extrusion stress exceeds the extension strength. To improve the nozzle response at high temperatures and pressures, both enhancement of the material property and optimization of the nozzle design should be considered;
3. The simulation result agreed well with the experimental result. It provides the future design of this nozzle with great support.

REFERENCES

Bao F.T. *Design of Solid Rocket Motor* (China Astronautic Publishing House, Beijing, 2015).

Bianchi D, Nasuti F, Martelli E.J. Space. & Rocket. **46**, 492 (2012).

Carrier D. AIAA SPACE 2007 Conference & Exposition, 65 (2007).

Chen L.Q., S.X. Wang, S.Y. Zhang, X. Hou. J. Solid Rocket Technol. **27**, 58 (2004).

Chen R.X. *Solid Rocket Motor Design and Study* (China Astronautic Publishing House, Beijing, 1991).

Crane D J. JANNAF Interagency Propulsion Committee, 16th Nondestructive Evaluation/25th Rocket Nozzle Technology/38th Structures and Mechanical Behavior Joint Subcommittee Meeting, (2007).

Gu J.L, Y. Gao, F.Y. Kang, W.C. Shen. New. Carbon. Mater. **16**, 53 (2001).

Gu J.L, Y. Leng, Y. Gao, F.Y. Kang, W.C. Shen. New. Carbon. Mater. **14**, 22 (1999).

Keswani S T. Combust. Sci. Technol. **42**, 145 (1985).

Liu Q, Sheng Z F, Shi H B, et al. J. Propuls. Technol. **33**, 468 (2012).

Morozov E V, J.F.P. Pitot de la Beaujardiere. Compos. Struct. **91**, 412 (2009).

Su J.M., L.Q. Chen, S.X. Wang, X. Hou, G.L. Li, Z.M. Hao, W. Cheng. J. Solid Rocket Technol. **26**, 58 (2003).

Su J.M., L.Q. Chen, S.X. Wang, X. Hou, G.L. Li. J. Solid Rocket Technol. **27**, 69 (2004).

Su J.M., Q. Xie, J. Feng, Y. Zhu, M. Zhang, Z.C. Xiao. J. Solid Rocket Technol. **38**, 554 (2015).

Tang J.L., P.J. Liu. *Principle of Solid Rocket Motor* (National Defense Industry Press, Beijing, 2013).

Wang J.G., Q.G. Guo, L. Liu, J.R. Song. New. Carbon. Mater. **17**, 13 (2002).

Xie H.Z., H. Cui, R. Zh. Li, S.Y. Qin, J.T. Sun, M.W. Zheng. Mater. Rev. **29**, 53 (2015).

Xiong B., Y.J. Bai, M. Tang. J. Solid Rocket Technol. **39**, 179 (2016).

Zheng Q. Thermo-structural Analysis and Failure Behavior of C/C Composite Throats. Master's Thesis. Harbin Institute of Technology, Harbin, China, 2011.

Advances in Energy Science and Equipment Engineering II – Zhou, Patty & Chen (Eds)
© *2017 Taylor & Francis Group, London, ISBN 978-1-138-71798-5*

Design of the pintle controlled solid rocket motor's thrust control system

Chen Cheng, Fu-ting Bao & Hao Xu
College of Astronautics, Northwestern Polytechnical University, Xi'an, China

ABSTRACT: According to pintle controlled solid rocket motor's characteristic, the thrust modulation model was built. Two kinds of the thrust control systems were designed which were the location-based control system and the pressure-based control system, and the effects of the control systems were compared through the simulation. The simulation indicated that both the two kinds of systems can control the thrust effectively. The location-based control system is easy to establish due to its simple structure, and its response speed is fast. But its accuracy and the environmental adaptability is not good enough. The pressure-based control system's structure is more complex. Although its response speed is slower, it has higher accuracy, better environmental adaptability, and the compensation ability when the grain' burning surface is unsteady or the nozzle erosion occurs.

1 INTRODUCTION

With the advantages of simple structure, small volume, convenient maintainability, long storage period, and high reliability, the Solid Rocket Motor (SRM) has been widely used in various rockets, missiles, attitude and orbit control thrusters, ejection escape systems, and weather or anti-hail rockets. And owing to the high maneuverability, quick response and strong ability to survive, it has become the standard means of propulsion. The SRM takes advantage in the productivity and the operation performance, but the difficult to gain the changeable thrust reduces the flexibility and limits its application range. The application of the pintle technology in the SRM can not only keep its own advantages in production, storage, operation and the response speed, but also give the SRM the ability of the active control or trust modulation like the liquid rocket engine. So the pintle controlled solid rocket motor leads to a revolutionary breakthrough in the SRM area (Ostrander M J, 2000; Jin J, 2014; Bergmans J L, 2003 & Li J, 2007).

The U.S. Army Aviation and Missile Command (AMCOM) is developing pintle technology for controllable thrust propulsion. The technology program is investigating several technical areas: modeling and simulation, pintle motor design and performance prediction, pintle and nozzle design, materials testing, actuation and control mechanisms, and ammonium nitrate propellants. These various technology areas are being focused for future generation U.S. Army tactical missiles (Burroughs S L, 2001). The pintle technology attracts widely attentions in the world.

One of the key technologies of the pintle controlled SRM is how to modulate the thrust accurately, steady, and fast. To achieve this goal, this paper built the thrust modulation model of the pintle controlled SRM, designed two kinds of the thrust control systems according to its characteristics, and then simulated the effects of the control systems.

2 THRUST MODULATION MODEL

The pintle controlled SRM modulate its thrust through changing the nozzle throat area by a moving pintle in the nozzle. The servomachanism drives the pintle moving along the center line to change the nozzle throat area which causes the pressure of the combustion changing and leads to the thrust changing in the end. When the pintle moves back, the nozzle throat area increases, which leads to the decrease of the combustion pressure and the thrust. When the pressure cannot keep the propellant burning anymore, the motor will flame out. On the contrary, when the pintle moves forward, the nozzle throat area decreases to increase the combustion pressure and the thrust.

The pressure of the combustion and the nozzle throat area obey the equation below:

$$\frac{V_c}{RT_f}\frac{dP}{dt} = \rho_p A_b a P^n - \frac{PA_t}{c^*} \qquad (1)$$

Where V_c is the free volume of the combustion, R is the gas constant, T_f is the burning temperature of the propellant, ρ_p is the density of the propellant, A_b is the burning area of the propellant, a is coefficient of the burning rate, n is the pressure exponent, and c^* is the characteristic velocity of the propellant.

According to the thrust equation:

$$F = C_F P A_t \tag{2}$$

$$C_F = \Gamma \sqrt{\frac{2k}{k-1}\left[1-\left(\frac{P_e}{P}\right)^{\frac{k-1}{k}}\right]} + \frac{A_e}{A_t}\left(\frac{P_e}{P}-\frac{P_a}{P}\right) \tag{3}$$

$$\Gamma = \sqrt{k}\left(\frac{2}{k+1}\right)^{\frac{k+1}{2(k-1)}} \tag{4}$$

Where C_F is the coefficient of thrust, k is specific heat ratio, P_e is the pressure on the nozzle outlet, P_a is the environmental pressure, and A_e is the nozzle outlet area.

$$\frac{A_e}{A_t} = \frac{\Gamma}{\left(\frac{P_e}{P}\right)^{\frac{1}{k}}\sqrt{\frac{2k}{k-1}\left[1-\left(\frac{P_e}{P}\right)^{\frac{k-1}{k}}\right]}} \tag{5}$$

$\frac{P}{P_e}$ and $\frac{A_e}{A_t}$ obey an approximate relation of $\frac{P}{P_e} = a_k \frac{A_e}{A_t}+b_k$. Where, a_k and b_k is decided by k. In conclusion, the thrust and the nozzle throat area obey the equation below:

$$F = PA_t\left\{\Gamma\sqrt{\frac{2k}{k-1}\left[1-\left(a_k\frac{A_e}{A_t}+b_k\right)^{\frac{1-k}{k}}\right]}\right.$$
$$+\frac{A_e}{A_t}\left(\frac{1}{a_k\frac{A_e}{A_t}+b_k}-\frac{P_a}{P}\right)\right\}$$
$$=\left(\rho_p c^* a\frac{A_b}{A_t}\right)^{\frac{1}{1-n}}$$
$$A_t\left\{\Gamma\sqrt{\frac{2k}{k-1}\left[1-\left(a_k\frac{A_e}{A_t}+b_k\right)^{\frac{1-k}{k}}\right]}\right.$$
$$\left.+\frac{A_e}{A_t}\left[\frac{1}{a_k\frac{A_e}{A_t}+b_k}-\frac{P_a}{\left(\rho_p c^* a\frac{A_b}{A_t}\right)^{\frac{1}{1-n}}}\right]\right\} \tag{6}$$

Except the environmental pressure P_a, the other parameters in equation (6) are all motor design parameters. So the thrust is not only controlled by the nozzle throat area which is decided by the location of the pintle in the nozzle, but also related to the environmental pressure.

3 DESIGN OF THE THRUST CONTROL SYSTEM

As mentioned above, the environmental pressure has an impact on the thrust. If the control system doesn't include the thrust feedback, the influence of the environmental pressure can't be considered, so the thrust can't be controlled accurately. In the missile flight condition, the thrust is hard to measure and form an available feedback to build a thrust closed-loop control. In this case, two kinds of the thrust control systems are designed below, which are the location-based control system and the pressure-based control system.

3.1 The location-based control system

The location-based control system is designed based on the accurate control of the pintle's location. The structure is showed in figure 1. The controller calculates the pintle's location according to the command signal of the thrust, and sends it to the location server, then the location server drives the pintle to the calculated location to change the nozzle throat area and export the required thrust.

According to the equation (6), the location can be calculated after calculating the nozzle throat area. The method to calculate the nozzle throat area is stated in the reference 2. The calculation needs various parameters, most of which are hard to measure. Therefore, the theoretical equation is not suitable to design the controller directly. Through designing the pintle's contour, the pintle's location can be an approximate liner related to the thrust under the same environmental pressure. The controller can be designed as $s = aF + b$. Where, a and b are calculated by fitting the pintle's location and the thrust under the same environmental pressure, which can be measured through experiments.

The location-based control system is easy to establish due to its simple structure. Its accuracy is decided by the degree of linear between the pintle's location and the thrust, and its error mainly comes

Figure 1. Structure of the location-based control system.

from the difference between the actual environmental pressure and the environmental pressure when the control parameters are measured.

3.2 The pressure–based control system

The pressure-based control system is designed based on accurate control of the combustion pressure to control the thrust. The structure is showed in figure 2. The thrust controller calculates the combustion pressure according to the command signal of the thrust and the actual environmental pressure, and sends it to the pressure control loop to export the required combustion pressure. In the pressure control loop, the electromotor adjusts its rotate speed according to import pressure signal from the thrust controller to drive the pintle moving along the nozzle, and then change the nozzle throat area to export the required combustion pressure and thrust.

3.2.1 Design of the thrust controller

According to the pressure balance equation $P_{eq} = \left(\rho_p c^* a \dfrac{A_b}{A_t} \right)^{\frac{1}{1-n}}$ and equation (6), the combustion pressure can be calculated. But for the same reason of the location-based system, the equation can be hardly used to design the thrust controller directly. Researches have shown that the relationship between the combustion pressure and the thrust is approximately linear in most cases. So the thrust controller also can be designed to be linear. The relation between the combustion pressure and the thrust can be fitting to in a certain environmental pressure by experiment. According to the thrust equation $F = \dot{m}_d u_e + (P_e - P_a)A_e$, the final thrust controller is corrected to be like this:

$$P = aF + b + aA_e(P_a - P_{a0}) \qquad (7)$$

Where Pa_0 is the environmental pressure when the parameters a and b is measured, P_a is the actual environmental pressure. The correction can reduce the error from the environmental pressure.

3.2.2 Design of the pressure control loop

The response's accuracy and speed of the pressure control loop largely decide the effect of the con-

trol system. The structure of the pressure control loop is showed in figure 3. This kind of control method doesn't care about the accurate location of the pintle, it adjust the combustion pressure by controlling the forward and backward rotate speed of the electromotor to drive the pintle moving. The controller in the loop exports a rotate speed signal to the electromotor.

The control loop is designed based on the signal neuron adaptive PSD control. This controller is adaptive by combining the PSD adaptive algorithm and the single neuron PID controller. The PSD algorithm is presented by Marsik and Strejc which needs no identification. Because the pintle controller SRM is disposable, if using the traditional PID controller, the control parameters tuning needs a large number of simulations and experiments, which cost much. Although a large number of experiments are done, the performance is hardly to be guaranteed when the operating condition changes. The single neuron PSD controller can solve the problem well through combining the PSD algorithm and the single neuron PID controller's advantages.

The structure of the single neuron PID controller is shown in figure 4. According to the single neuron's structure, the connecting weights w_1, w_2, w_3 respectively correspond to proportional, integral, and differential coefficients of the PID, and the three state variables x_1, x_2, x_3 of the neuron respectively correspond to the proportional, integral, and differential part, which respectively reflect the error's change, accumulation, and first order difference. K is the proportional coefficient of the neuron. The control signal $u(k)$ is produced by self-learning of the neuron through modulating the connecting weights w_1, w_2, w_3. The process of

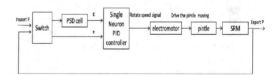

Figure 3. Structure of the pressure control loop.

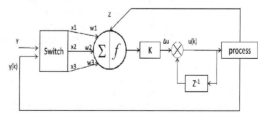

Figure 4. Structure of the single neuron PID controller.

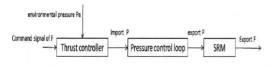

Figure 2. Structure of the pressure-based control system.

self-learning equals to the tuning process of the PID parameters. The performance of the signal neuron PID controller is decided by the parameter K, so the number of the parameters which need to be tuned reduces from three to only one, and the times of the tuning can be reduced a lot[6–8].

The PSD adaptive algorithm (Hunt K J, 1992) establishes the performance indicator according to the error of the process. The adaptive control law is produced only according to the desired export and the actual export, without identifying parameters of the process. The problem of the parameters tuning of PID controller can be solved by adjusting the parameter K automatically through combining the PSD adaptive algorithm with the single neuron PID controller.

The PSD adaptive algorithm is as follow:

$$K(k) = \begin{cases} K(k-1) + CK(k-1)/T_v(k-1), \\ sign[e(k)] = sign[e(k-1)] \\ 0.75K(k-1), \\ sign[e(k)] \neq sign[e(k-1)] \end{cases} \quad (8)$$

$$T_v(k) = T_v(k-1) + Lsign[|e(k)-e(k-1)| \\ -T_v(k-1)|e(k)-2e(k-1)+e(k-2)|] \quad (9)$$

Where, $0.025 \leq C \leq 0.05, 0.05 \leq L \leq 1$.

4 SIMULATIONS

4.1 Effects comparison of the control systems

The simulations for the two kinds of systems of thrust control are designed under different environmental pressure conditions. When the controllers are designed, the controller parameters are fitted under the same designed environmental pressure $(P_{a0} = 101325 Pa)$. For the location-based system, as shown in figure 5, the thrust responds quickly, but the accuracy is not high enough. And the error will grow bigger as the difference between the actual environmental pressure and the designed environmental pressure increases. So the location-based system shows the poor adaptability to the environment. For the pressure-based system, as shown in figure 6, its control accuracy is higher, and has low data sensitivity to the environmental pressure. As shown in figure 7, the import combustion pressure produced by the thrust controller can adjust itself according to the change of the environmental pressure, which gives the controller better adaptability to the environment. Figure 8 shows the combustion pressure response under designed environmental pressure. The pressure response needs a period of time which leads to thrust responds slower than the location-based system does. And overshoot exists because of the pressure control loop.

4.2 Simulations of unsteady burning surface and nozzle erosion

The controllers are designed according to the standard burning surface and the nozzle throat area before the motor starts to work. But when the motor is working, the actual burning surface can hardly match the standard designed state, and usually appears to oscillate. The nozzle throat area will change when the nozzle erosion occurs as well. The two factors have strong impacts on the combustion pressure which decides the thrust control.

A sine signal with amplitude of 5% is added to the original burning surface to simulate the situation of unsteady burning surface occurring. And a slope signal is added to the original nozzle throat area to simulate the situation of nozzle erosion occurring here.

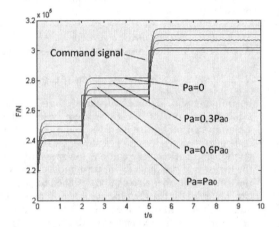

Figure 5. The thrust response under different environmental pressure for the location-based control system.

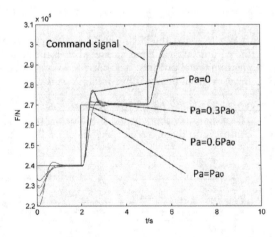

Figure 6. Thrust response under different environmental pressure for the pressure-based control system.

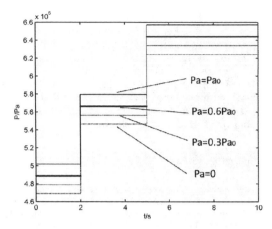

Figure 7. The import combustion pressure produced by the thrust controller of the pressure-based control system.

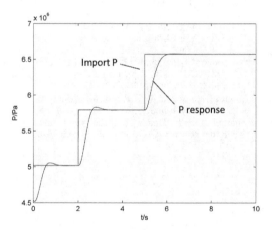

Figure 8. The combustion pressure response of the pressure-based control system under Pa = Pa₀.

As shown in figure 9 and 10, for the location-based system, the oscillation of burning surface and change of nozzle throat area will both cause the change of the combustion pressure because the location-based system just control the pintle's location. Either of these two situations occurs, the export thrust will deviate from the expected number, and the thrust control will produce error. The location-based system doesn't have the ability to compensate the situation of unsteady burning surface and nozzle erosion.

As shown in figure 11 and 12, for the pressure-based system, the combustion pressure can adjust itself to follow the expected value well to export the right thrust because the pressure is feedback. So the pressure-based system has the ability to compensate the situation of unsteady burning surface and nozzle erosion.

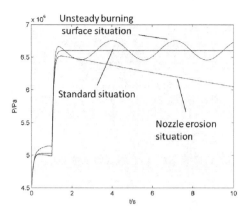

Figure 9. The combustion pressure in the situation of unsteady burning surface and nozzle erosion for the location-based system.

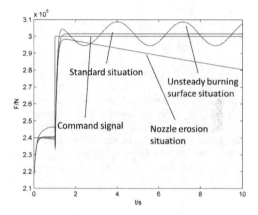

Figure 10. The thrust response in the situation of unsteady burning surface and nozzle erosion for the location-based.

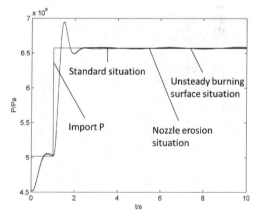

Figure 11. The combustion pressure in the situation of unsteady burning surface and nozzle erosion for the pressure-based system.

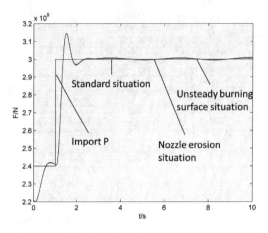

Figure 12. The thrust response in the situation of unsteady burning surface and nozzle erosion for the pressure-based system.

5 CONCLUSIONS

Two kinds of thrust control system were designed and simulated for the pintle controlled SRM. One system is location-based, the other one is pressure–based. The location-based control system is easy to establish due to its simple structure and fast response speed, but its accuracy and environmental adaptability is far from satisfaction. It doesn't have the ability to compensate the situation of unsteady burning surface and nozzle erosion. The pressure-based control system's structure is more complex. Although its response speed is slower because of the pressure control loop, its accuracy is higher, and it can adjust itself to adapt the environment well. What's more, the pressure-based control system applies the pintle controlled SRM for the ability to compensate the situation of unsteady burning surface and nozzle erosion.

REFERENCES

Bergmans J L, Di Salvo R. AIAA, 4968: 1–7 (2003).
Burroughs S L. AIAA/ASME/SAE/ASEE Joint Propulsion Conference and Exhibit, 37 th, Salt Lake City, UT. (2001).
Ding J, Xu Y M. Control Engineering of China, 11, 27–30 (2004).
Hunt K J, Sbarbaro D, Żbikowski R, et al. Automatica, 28, 1083–1112 (1992).
Jin J, Ha D S, Oh S. Journal of the Korean Society of Propulsion Engineers, 18, 19–28 (2014).
Li J, Wang Y, et al. Journal of Solid Rocket Technology, 30, 505, (2007).
Ostrander M J, Bergmans J L, Thomas M E, et al. AIAA/ASME/SAE/ASEE Joint Propulsion Conference and Exhibit, 36 th, Huntsville, AL. (2000).
Yabuta T, Yamada T. IEEE transactions on systems, man, and cybernetics, 22,170–177 (1992).
Zhang M, Xia C, Tian Y, et al. 2007 IEEE International Conference on Control and Automation. IEEE, 617–620 (2007).

Advances in Energy Science and Equipment Engineering II – Zhou, Patty & Chen (Eds)
© 2017 Taylor & Francis Group, London, ISBN 978-1-138-71798-5

Information processing, organizational learning, knowledge sharing, and technology ability effects on firm performance

Kuo-En Huang

Institute of Marine Environment and Engineering, National Sun Yat-sen University, Kaohsiung, Taiwan

William-S Chao

Institute of Information Management, National Sun Yat-sen University, Kaohsiung, Taiwan

ABSTRACT: The goal of this study is to understand how Information processing, organizational learning, knowledge sharing, and technology ability effects on firm performance. The use of comparative methods in this research includes Multiple Regression Analysis (MRA) and Fuzzy-set Qualitative Comparative Analysis (FsQCA). These analyses demonstrate that an FsQCA can successfully identify conditions that are adequate for successful firm performance outcomes. The results indicate that an FsQCA outperforms an MRA and successfully models both types of data with causal complexities. This study's findings provide useful insights into how firms' members should reinforce their collaborative behaviors and activities to enhance their competitive advantages.

1 INTRODUCTION

Research has found that enterprise are investing ever more resources into information technology based systems to manage their knowledge resources, with the premise that this will lead to strategic competitive advantage, thus impacting firm performance (Sambamurthy, Bharadwaj, & Grover, 2003).

Enterprise is increasingly facing rapidly changing business environments. To thrive in such uncertain environments, enterprise is under constant pressure to leverage their organizational learning capabilities.

Information processing, with regard to firm performance, could potentially provide a competitive advantage for firms. For an enterprise to achieve a competitive advantage, all functions must be interrelated. The operational function adds value by using an enterprise's resources effectively and by producing goods and services that satisfy the needs and requirements of customers (Singh, 2008).

Despite significant progress in answering is the question of how information technology contributes to firm performance. Nevertheless, with some notable exceptions,

First, few studies have empirically examined the link between information processing and firm performance.

Second, the role and articulation of the underlying mechanisms through which information technology ability improve firm performance remain unclear (Bharadwaj, 2000).

Finally, from an empirical perspective, many of the prior studies linking information technology

and related ability with firm performance do not fully address the issues related to reactive measures and unobserved enterprise heterogeneity.

The study adopts the relational view of the Resources-Based Theory (RBT) and the Resource-Based View theory (RBV) of the enterprise to explain the information processing, organizational learning, knowledge sharing, and technology ability effects on firm performance. Resource-based perspectives, we argue that organizational capabilities are rent-generating assets, and they enable enterprise to earn above-normal returns.

2 LITERATURE REVIEW AND HYPOTHESES

2.1 Information processing

Information processing is the capability to provide information and data to users with appropriate levels of security, timeliness, connectivity, reliability, confidentiality, access, and accuracy as well as the ability to tailor these in response to changing business needs and directions. Technology supported information management enables higher-order business capabilities, which in turn influence firm performance (Sambamurthy, Bharadwaj, & Grover, 2003).

2.2 Organizational learning

Organizational learning plays a vital role when a customer is building a relationship with a company and is a relatively stable and conscious tendency of the relationship a customer creates with retailers of a particular product category.

Organizational learning and firm performance have increasingly become key determinants of enterprise' competitive advantages. Organizational learning is a key component of knowledge management, as it helps in codifying the repository of available knowledge in firm performance ability, and increases over time (Liebowitz, 1999).

2.3 Knowledge sharing

Knowledge sharing is an important component of knowledge management, as it helps in codifying the repository of available knowledge in an organization and increasing it over time (Liebowitz, 1999). Failure to acquire new knowledge sharing can cause a firm to be unable to keep up with technological progress in an industry and to anticipate shifts in customer requirements since it may be incapable of producing technology standards. Knowledge sharing has been well emphasized as important factors to successful entrepreneurship over the last decades.

2.4 Technology ability

A team's technology ability enables a firm to turn knowledge into new product quality management, services, or processes to support innovation. Technology ability can reduce the costs involved in supply chains. Prior studies show evidence that technology ability significantly impacts firm performance (Rothaermel & Alexandre, 2009).

2.5 Firm performance

Firm performance management ability influences various measures of firm performance by allowing business leaders to review and take corrective actions on any potential or actual slippages proactively and in a timely manner. Firm performance is the only sustainable source of advantage, so managers must link their core competence to different types of strategies across time. We control for enterprise size and industry sector in our firm performance models to account for any difference in firm performance attributable to enterprise resources or inter-industry differences (Capon, Farley, & Hoenig, 1990).

2.6 Hypotheses

From a review of the marketing and supply chain literature, preliminary in-depth interviews with 21 managers, and Exploratory Factor Analysis (EFA), four groups of concepts emerged as influencers of firm performance: information processing, organizational learning, knowledge sharing, and technology ability. The following hypotheses capture the influence of these constructs; Fig. 1 shows the relationships. Based on the previous research mentioned above, we test the following hypotheses:

H1. Information processing has a positive effect on firm performance.

H2. Organizational learning has a positive effect on firm performance.

H3. Knowledge sharing has a positive effect on firm performance.

H4. Technology ability has a positive effect on firm performance.

3 RESEARCH METHODS

3.1 Data collection procedure

In the current method, 26 items capture information processing, organizational learning, knowledge sharing, technology ability and firm performance. Lee and Choi (2003); Petter, Straub, and Rai (2007) employ an eight-item scale to measure information processing. The six-item scale in Lee and Choi (2003); Gold, Malhotra, and Segars (2001) study offers a good tool to measure organizational learning. The present study uses a four-item instrument from (Pavlou & El Sawy, 2006) to measure knowledge sharing. (Roberts, Galluch, Dinger, & Grover, 2012) present a three-item instrument to determine technology ability. Janz, Colquitt, and Noe (1997); Choi, Lee, and Yoo (2010) five-item scale provides a suitable tool to measure firm performance. These same scales report that they have strong psychometric properties with acceptable reliability and validity.

Questionnaires were distributed to 21 businesses, after a revision and consulting with 5 reference scholar and 21 managers. These firms are listed in the MARBO investment Weekly as the top 500 manufacturing firms of 2015, obtained 250 effective responses. The total response rate was 50%. Additionally, 180 respondents (72% of the 250 effective responses) were function managers or other managers on the senior management team. A pre-test on the questionnaire comprising 26 items was carried out with the help of academic researchers to improve the content and appearance of the instrument, as well as to conduct factors analysis.

SPSS statistics 21, FsQCA software was used to carry out Exploratory Factor Analysis (EFA) and path analysis of all constructs. More than 250 valid questionnaires were collected and Cronbach's α value reached 0.7, which was suitable for factor analysis. Their responses suggest that only minor cosmetic changes were required and that all statements could be retained.

This study sets the alpha coefficient at 0.6. Information processing is Cronbach Coefficient Alpha

Standardized 0.89. Organizational learning is 0.85. Knowledge sharing is 0.74. Technology ability is 0.61. Kaiser's Measure of Sampling Adequacy: Overall MSA = 0.92 > 0.5; all of the individual MSA values 0.88–0.95 > 0.5, which are suitable for factor analysis. The eigenvalue is greater than or equal to 1 and can be divided into four factors. The eigenvalue value is 8.8, 1.49, 1.25, and 1.15. Test of appropriate of Correlation matrix (50% R value > 0.3).

All measures of the survey instrument are developed from the literature. Where appropriate, the manner in which the items were expressed is adjusted to the context of supply chains. The items measure the subjects' responses on a five-point Likert scale ranging from strongly disagree (1) to strongly agree (5).

To check for potential bias of a single informant, the consistency between the data collected from managers and senior managers was verified. A Chi-square analysis of the industry distribution of the respondents shows no difference from the industry distribution of all the firms used in the survey.

3.2 Regression analysis criteria

The regression coefficient evaluates the relationship between individual independent and dependent variables. This study uses (p < 0.05) as its significance interval. R^2 ranges between 0 and 1, indicating the magnitude of the independent variables' effects on changes as exhibited by the dependent variable. The coefficient of determination (R^2) expresses the goodness of fit between a model and its data (Neter & Wasserman, 1974). Values closer to 1 indicate that more of the estimate variance is determined by the XY influence (Draper & Smith, 1981), but do not prove a causal relationship. See Table 1.

3.3 FsQCA calibration

Ragin (2008) shows that calibration is possible using three breakpoints for qualitative norms, for example, full membership (= 1.0), entirely non-members (= 0.0), and the maximum ambiguity cross point (= 0.5). To analyze whether combinations of variables are necessary and/or sufficient to influence outcomes, the FsQCA analyzes all the antecedents and uses a membership function to express the relationships between combinations and outcomes (Woodside & Zhang, 2012).

3.4 FsQCA consistency and coverage

The index assesses the extent to which cause or causal combination accounts for an outcome (Ragin, 2008). Values exceeding 0.75 are representative of the observed phenomena. The coverage index is analogous to R^2. A consistency index is analogous to a correlation, and represents a sub-

Table 1. The multiple regression model.

Model summary[b]

Model	R	R square	Adjusted R square	Std. error of the estimate
1	.489[a]	.239	.226	.57189

		Change statistics			
R square change		F change	df1	df2	Sig. F change
.239		19.210	4	245	.000

Anova[b]

Model		Sum of squares	df	Mean square	F	Sig.
1	Regression	25.131	4	6.283	19.210	.000[b]
	Residual	80.128	245	.327		
	Total	105.259	249			

Coefficients[a]

Model		Unstandardized coefficients		Standardized coefficients		
		B	Std. error	Beta	t	Sig.
1	(constant)	1.916	.281		6.823	.000
	IM	.270	.083	.291	3.241	.001
	IL	.338	.084	.321	4.005	.000
	KA	.050	.096	.042	.521	.603
	AA	−.227	.062	−.255	−3.644	.000

Unstandardized coefficients		Correlations			Collinearity statistics	
Lower Bound	Upper Bound	Zero-order	Partial	Part	Tolerance	VIF
1.363	2.470					
.106	.434	.073	.083	.075	.584	1.721
.172	.504	.412	.354	.342	.705	1.419
−.139	.238	.096	−.103	−.093	.642	1.559
−.349	−.104	.158	.003	.002	.614	1.628

set of the relation of antecedent(s) to outcome(s) (Ragin, 2008). The FsQCA uses a consistency and coverage index to evaluate antecedents and their combinations (Ragin, 2008).

4 MRA RESULTS

The data model analysis. The Model Summary shows the explanatory power of independent variables. The $R^2 = 0.23$ indicates that they explain the

Table 2. FsQCA results.

Subset/superset analysis

Outcome: EP

	Consistency	Coverage	Combined
IP * OL * KS * TA	0.819987	0.653594	0.740958
IP * OL * KS	0.824013	0.783241	0.815938
IP * OL * TA	0.814224	0.672238	0.746991
IP * KS * TA	0.803217	0.662990	0.728280
IL * KS * TA	0.805386	0.671977	0.737768
IP * OL	0.809730	0.849877	0.834805
IP * KS	0.800760	0.806525	0.803256
IP * TA	0.793226	0.681678	0.729184
OL * KS	0.804695	0.822610	0.816281
OL * TA	0.791667	0.698529	0.733394
KS * TA	0.785539	0.687908	0.723056
IP	0.780746	0.882046	0.813348
OL	0.756949	0.912173	0.787577
KS	0.757105	0.865196	0.767029
TA	0.758531	0.717320	0.698411

23% of observed variation in the dependent variable. In a post-hoc test for the effectiveness of independent variables, the prediction of results of the four input variables and the three variables are significant. One independent variable demonstrates no significant effect on the dependent variable.

Multiple regression analysis results reach a significant level, but only 23% of the predictive power of explanatory power, which represents the regression model; this is insufficient to predict information processing, organizational learning, knowledge sharing, and technology ability effects on firm performance. Therefore, this study employs FsQCA to verify the consistency coverage of the model. See Table 2.

4.1 FsQCA results

This study uses conventional MRA and fuzzy set qualitative comparative analysis FsQCA to analyze the data. FsQCA is an analytical tool that uses fuzzy set theory and differentiates itself from the conventional statistical methods (Ragin, 2008). Ragin (2008) stress the importance of aiming for high consistency over high coverage. Consistency and coverage test results for FsQCA yield the relationships. The results show that (FsQCA) captures relationships better and has better predictive capabilities than MRA does.

This technique differs from conventional statistical methods (Ragin, 2008). Woodside and Zhang (2012) provide more details on how to perform calibrations. This study explores the same variables. To demonstrate predictive validities, this study presents a prediction analysis.

The results for the analysis of data using FsQCA the results reveal coverage of 0.65, a combination of 0.74 and consistency of 0.82. The high consistency score is strong evidence that the antecedent combination is sufficient to produce the outcome (firm performance).

5 CONCLUSIONS AND IMPLICATIONS

Multiple regression analysis results reach a significant level, but only 23% of the predictive power of explanatory power, which represents the regression model; this is insufficient to predict firm performance. Therefore, this study presents the FsQCA to verify the consistency and coverage of the model. The results reveal coverage of 0.65 with a combination of 0.74 and consistency of 0.82. The antecedent combination includes all four independent variables. These analyses demonstrate that the FsQCA successfully identifies conditions sufficient for the firm performance outcome.

This study has five key managerial implications:

A. The whole process, including applying EFA extraction, exploring the factors impacting firm performance and building the main construct, confirms the four vital factors regarding firm performance in the sophisticated relationships among the factors in this study.
B. Integrated information systems have various functions, such as the application of information technology and firm performance. Firms can then strengthen the capability of information processing and the variety of collaboration among team members.
C. Organizational learning has a positive impact on business firm performance. Proving new technology for the improvement of existing products and services is a necessary factor for enterprises.
D. Knowledge sharing is a necessary condition. The results of this study show that businesses with high capability will handle operational problems efficiently. Knowledge sharing achievement often shapes the composition of internal and external resources.
E. Technology ability can enhance firm performance as well as the output performance of enterprise; intense competition in the supply chain requires assessments of customer views and attention to quality management processes. Enterprise should integrate technology to underpin technology ability assurance and enhancement.

Future research could examine, and even compare, different types of enterprise managements, and in different geographical areas.

REFERENCES

Bharadwaj A., A resource-based perspective on information technology capability and firm performance: An empirical investigation, MIS Quarterly 24 (1), 2000, 169–196.

Choi S.Y., Lee H., and Yoo., The impact of information technology and transactive memory systems on knowledge sharing, application, and team performance: a field study, MIS Quarterly 34 (4), 2010, 855–870.

Draper N.R., and Smith, H., Applied regression analysis, 1981, John Wiley & Sons; New York.

Gold A.H., Malhotra A., and Segars A.H., Knowledge management: An organizational capabilities perspective, Journal of Management Information Systems 18 (1), 2001, 185–214.

Hendricks K.B., and Singhal V.R., Firm Characteristics, total quality management, and fnancial performance, Journal of Operations Management 19, 2001, 269–285.

Janz B.D., Colquitt J.A., and Noe R.A., Knowledge worker team effectiveness: The role of autonomy, interdependence, team development, and contextual support variables, Personnel Psychology 50 (4), 1997, 877–904.

Kaplan R.S., and Norton D.P., The balanced scorecard: Measures that drive performance, Harvard Business Review, 70 (1), 1992, 71–79.

Lee H., and Choi B., Knowledge management enablers, processes, and organizational performance: An integrative view and empirical examination, Journal of Management Information Systems 20 (1), 2003, 179–228.

Liebowitz J., Knowledge management handbook, 1999, CRC Press; Boca Raton, FL.

Neter J., and Wasserman W., Applied linear statistical models: Regression, analysis of variance, and experimental designs, 1974, Published R.D. Irwin; Vancouver, CA.

Pavlou P.A., and El Sawy O.A., From IT leveraging competence to competitive advantage in turbulent environments: The case of new product development, Information Systems Research 17 (3), 2006,198–227.

Petter S., Straub D., and Rai A., Specifying Formative Constructs in Information Systems Research, MIS Quarterly 31 (4), 2007, 623–656.

Ragin C., Redesigning social inquiry: Fuzzy sets and beyond, 2008, Chicago University Press; Chicago.

Roberts N., Galluch P.S., Dinger M., and Grover V., Absorptive capacity and information systems research: Review, synthesis, and directions for future research, MIS Quarterly 36 (2), 2012, 625–648.

Rothaermel F.T., and Alexandre M.T., Ambidexterity in technology sourcing: The moderating role of absorptive capacity, Organization Science 20 (4), 2009, 759–780.

Sambamurthy V., Bharadwaj A., and Grover V., Shaping agility through digital options: Reconceptualizing the role of information technology in contemporary firms, MIS Quarterly 27 (2), 2003, 237–263.

Singh J., Distributed R&D, cross-regional knowledge integration and quality of innovative output, Research Policy 37, 2008, 77–96.

Woodside A.G., and Zhang M., Identifying x-consumers using causal recipes: Whales and jumbo shrimps casino gamblers, Journal of Gamble Study 28 (1), 2012, 13–26.

Advances in Energy Science and Equipment Engineering II – Zhou, Patty & Chen (Eds)
© *2017 Taylor & Francis Group, London, ISBN 978-1-138-71798-5*

A grinding robot with rapid constraint surface modelling and impedance model-based intelligent force control

Fei Wang, Zhong Luo & Hongyi Liu
School of Mechanical Engineering and Automation, Northeastern University, Shenyang, China

ABSTRACT: As the most important parts of a robotic grinding process, constraint surface modelling and force/position control algorithm of the robot is discussed in this paper. To accomplish rapid automatic detecting and modelling of workpieces' inner surface, a rapid translational detection strategy is developed. A series of key points of the axial section contour, which reflects the contour's big curvature changes, are recorded to roughly express the inner surface's model. Considering the imprecision of the constraint surface's model, a real-time adjusting fuzzy logic of reference trajectories in the impedance model is proposed. The adjustment depends on the real-time force and position sensor output. It is feasible in situations where there is no accurate model of the constraint surface or when geometric errors exist. A simple velocity PID control model is introduced to effectively reduce the impact when the robot's movement changes from free motion to constraint motion. Moreover, a PD method based on adaptive fuzzy estimation is used to compensate the joint friction that exists in the robot. The experimental results indicate that the proposed integrated control algorithm of the grinding robot is valid.

1 INTRODUCTION

Grinding is an important and expensive manufacturing process. Since industrial robots have been widely used in manufacturing, it is a tendency to use these robots to accomplish the grinding process. A lot of research on grinding robots has been published.

A. Robertsson established a robot platform to implement force-controlled grinding, finishing and deburring experiments (Robertsson, 2006). During experiments, off-line programming is needed. The main purpose of off-line programming is path generation, which is based on workpieces' CAD model and collision avoidance functionality. Force control-based robotic machining production was developed by ABB Co. in 2008 using a PI-based force control model with off-line programming (Tan, 2008). The pre-programming in this system concentrates on the lead-through teaching and path learning. The curved surface machining robot developed by CNYD can be used to grind curved surfaces of large—scale vanes with consistently high quality.

In these robot systems, path generation in the off-line programming and force control algorithm in the grinding process are the two most important parts. Moreover, the accurate information of workpieces, i.e. their location, contour, stiffness and other characteristics, are the main vital effects that influence the grinding quality and system stability. Even a small difference in the location or contour of workpieces will cause a huge change in the contact force between the robot manipulator's end-effector and the workpiece. However, the off-line programming is usually time consuming.

The purpose of this paper is to develop a robot grinding strategy with less time consumption and comparatively high quality. It includes a rapid detection strategy of workpieces' inner contour and an impedance model-based intelligent force control method. The joint friction compensation and the principle of velocity determination are also discussed in this paper.

2 STRUCTURE OF THE GRINDING ROBOT SYSTEM

The grinding robot, which can grind the inner surface of rotary cavities with a constant material removal thickness, is shown in Figure 1. The robot has 5-DOF, including a translational joint, two rotational joints, a tool and a shell's rotational movement (Wang, 2013).

The sketch map of the robot's control system is shown in Figure 2. The main parts of the system are IPC, motion controller, AC servo motors and several sensors. The main controller functionally includes status monitor, collision detection, processing control and knowledge base. During the contour detection or grinding process, the decision of processing control is based on the feedback of encoders, laser sensor and force sensor. The control flowchart is shown in Figure 3.

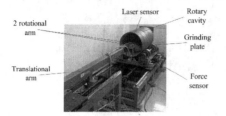

Figure 1. Robot grinding the inner surface of the rotary cavity.

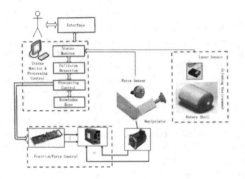

Figure 2. Sketch map of the grinding robot's control system.

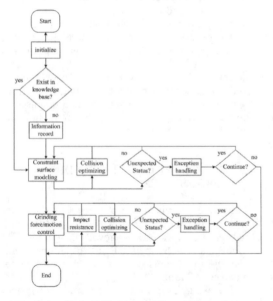

Figure 3. Control flowchart of the grinding robot.

3 MODELLING OF THE ROTARY CAVITY'S INNER SURFACE

In the cavity's crafting process, the contour of the cavity always has a large difference with respect to its CAD model, as shown in Figure 4. With the addition of location error that occurs when a workpiece is placed in the jig, as shown in Figure 5, the CAD model cannot be used directly in the robot's grinding process. Therefore, the inner surface's model needs to be established to guarantee the accurate grinding process.

3.1 Rapid translational detection strategy

When detecting and modelling the cavity's inner surface, the CAD model could be used as a reference. The detection of the inner surface's longitudinal section curve along the axial direction can be conducted with a laser sensor, as shown in Figure 6.

3.2 Model of the cavity's inner surface

Following the rapid translational detection strategy shown in Figure 6, the key point series of the jth longitudinal section curve could be recorded, as shown in Figure 7 (with $j = 1, 2, \ldots n$, assuming that the amount of the detection curve is n).

The model of the cavity's inner surface can be roughly expressed by the discrete point set of the whole n series of the longitudinal section recorded point.

However, the model shows imprecision. It only includes several key points that can reflect big curvature changes of the contour. Therefore, an intelligent force control algorithm is introduced to adapt to the imprecision.

Figure 4. Shape error of an inner surface's contour.

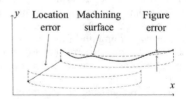

Figure 5. Location and shape error of an inner surface's contour.

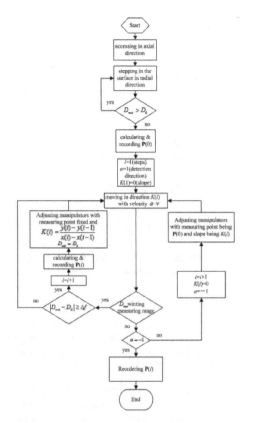

Figure 6. Flowchart of the rapid translational detection strategy.

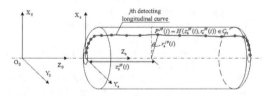

Figure 7.   jth recorded point series of the cavity's inner surface.

4 GRINDING FORCE/POSITION CONTROL ALGORITHM

Since there exists imprecision in both the contour model and the location of the inner surface, the force control algorithm should be adaptive to this error. A fuzzy force control strategy based on the impedance control model is proposed.

4.1 Dynamic model of the grinding robot

The dynamic model of the grinding robot is defined by:

$$M_x(x)\ddot{x} + C_x(x,\dot{x})\dot{x} + G_x(x) = J^{-T}\tau + F_c \qquad (1)$$

Here $M_x(B) = J^{-1}MJ^{-1}, C_x(x,\dot{x}) = J^{-T}(C - MJ^{-1}\dot{J})J^{-1}$ and $G_x(x) = J^{-T}G$. M, C, G, τ and J are, respectively, inertia, centripetal and Coriolis, gravitational terms, input torque and Jacobian matrix in the joint coordinate. $x, \dot{x}$ and $\ddot{x}$ are the actual position, velocity and acceleration of the robot manipulator, respectively.

4.2 Impedance model-Based intelligent force control

4.2.1 Impedance control model

The impedance mathematic model can be expressed as follows (Wang, 2010):

$$M_d(\ddot{x}_d - \ddot{x}) + D_d(\dot{x}_d - \dot{x}) + K_d(x_d - x) = F_d - F_c \qquad (2)$$

where M_d, D_d and K_d are respectively the desired inertia, damping and stiffness matrices of the manipulator-environment system. F_d is the desired contact force between manipulator's end-effector and the constraint surface. $x_d, \dot{x}_d$ and $\ddot{x}_d$ is respectively the desired trajectory, velocity and acceleration.

4.2.2 Fuzzy adjusting strategy of the reference trajectory

To promote adaptability of the control algorithm, an intelligent prediction and adjusting of the reference trajectory is introduced in the impedance model (Wang, 2010).

The reference trajectory intelligent adjusting law can be established in discrete form as follows:

$$x_r(k+1) = x_r(k) + \Delta x_{ec}(k) \qquad (3)$$

Where k is defined as the current moment. The term $\Delta x_{ec}(k)$ is an adjustment based on the predicted location and geometric error:

$$\Delta x_{ec}(k) = x_{ec}(k+1) - x_{ec}(k) + k_c \cdot e_{xp}(k) \qquad (4)$$

Where $e_{xp}(k) = \Delta x_{ep}(k) - \Delta x_{ep}(k-1)$ represents the tendency of the adjusting error. $x_{ep}(k) = x(k) - \dfrac{F_c(k)}{K_{ec}}$ represents the approximately actual position of the constraint surface. k_c is the adapting scale factor determined by a fuzzy controller.

The input of the fuzzy controller is $\left|e_{xp}(k)\right|$. The linguistic variables assigned to e_{xp} and k_c are EX and KC. The fuzzy rules are given in Table 1.

Where VB, B, M, S and VS, respectively, denote very big, big, middle, small and very small. KB,

BK, KM, SK and KS, respectively, denote very big, big, middle, small and very small.

4.2.3 *Control law of the robot manipulator*

Replacing the desired trajectory with reference, $u \equiv \ddot{x}$ is defined as the control rule. It could be deduced as follows:

$$u = \ddot{x}_d + M_d^{-1}(D_d \dot{e}_r + K_d e_r - F_d + F_e) \qquad (5)$$

Where $e_r = x_r - x, \dot{e}_r = \dot{x}_r - \dot{x}$.

The join torque control law of the force control model with online intelligent adjusting reference trajectory could be as follows.

$$\tau = J^T[M_x u + C_x \dot{x} + G_x - F_e] \qquad (6)$$

4.3 *Impact restrainability*

The grinding process can be divided into free motion and constraint motion, depending on whether or not the end-effector of the robot contacts the rocket inner shell.

Compared with control law given in formula (6), a simple PID control rule is obviously more suitable for the control of the manipulator's free motion:

$$\tau = K_P \cdot e + K_I \int e dt + K_D \cdot \dot{e} \qquad (7)$$

where K_P, K_I, K_D, respectively, represent the proportion, integral and differential coefficient. The term e is the error of the joint position, and $\dot{e}$ is correspondingly the error changing rate.

Due to the fact that the initial velocity at the collision moment between the grinding wheel and

Table 1. Fuzzy control rules for the output KC.

	EXP				
	VB	B	M	S	VS
KC	KB	BK	KM	SK	KS

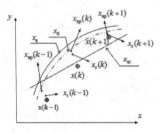

Figure 8. Reference trajectory adjusting.

the grinded surface is not equal to zero, it causes a serious impact on the contact force and even loses control of the robot manipulator.

In order to solve the problem, the device that can reduce the stiffness of the grinding wheel is designed. In addition, the error of the joint position e and the error rate of the position $\dot{e}$ are replaced by the error of the velocity $\dot{e}$ and the error rate of the velocity $\ddot{e}$ in the free motion controller formula (7). The corresponding control law is represented in formula (8):

$$\tau = K_P \cdot \dot{e} + K_I \int \dot{e} dt + K_D \cdot \ddot{e} \qquad (8)$$

Considering the fact that the robot manipulator moves with a small initial velocity in free motion, the impact force of the grinding wheel will be effectively reduced when the manipulator converts from free motion to constraint motion.

4.4 *Joint friction*

In the robot system, joint friction is one of the major nonlinear factors that affect robot performance. In particular, static friction significantly affects the performance of the system.

Joint friction compensation can be divided into model-based compensation and non-model compensation (Wang, 2006; Jatta, 2006). Since static friction is unpredictable and its exact model is unable to be established, researchers usually use a PD controller with a high gain stage to compensate the static friction in machines. Such a compensation can normally reduce the friction and impact to a certain extent, but can also reduce the system's stability and robustness.

Taking advantage of the adaptive fuzzy compensation algorithm proposed by Wang Yongfu (Wang, 2006), when the manipulator is bound state, by adding friction compensation τ_f to the output joint torque, the system control rules represented by formula (6) can be changed to:

$$\tau = J^T[M_x u + C_x \dot{x} + G_x - F_e] + \tau_f \qquad (9)$$

where τ_f is described in detail in the literature (Wang, 2006).

In conclusion, the control law of the manipulator's free motion presented in formula (10) changes to:

$$\tau = \tau_f + K_P \cdot \dot{e} + K_I \int \dot{e} dt + K_D \cdot \ddot{e} \qquad (10)$$

4.5 *Force/position control model*

The integrated control rule of the robot system is established by considering the joint friction

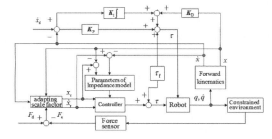

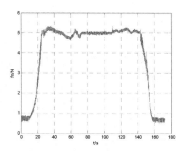

Figure 9. Scheme of the impedance model-based force control.

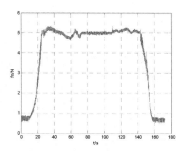

Figure 10. Normal force response with expected 5N.

and impact in a grinding process, as shown in Figure 9.

5 DETECTING AND GRINDING EXPERIMENT

By detecting the inner surface of the shell with a laser sensor and then executing the grinding process based on the proposed fuzzy force control algorithm, with an expected contact force 5N, the force response can be obtained, as shown in Figure 10.

Figure 10 shows that the contact normal force between the manipulator's end-effector and the constrained inner surface is controlled to be around its expected value. In the very beginning of the grinding process, the manipulator's movement changes from free motion to constraint motion,

with only a small fluctuation in the force response. The big fluctuation of the force response is caused by drastically changing the curvature of the constrained surface. The error of the normal force is within the range of [-0.3, 0.2] N.

6 CONCLUSIONS

A grinding robot system is introduced in this paper. Taking advantage of the proposed rapid translational detection strategy and the impedance model-based intelligent force control algorithm, a cavity's inner surface can be rapidly and automatically detected and grinded. The rapid translational detection strategy provides a way to record key points of the inner surface's contour, which reflects the contour's big curvature changes. On the basis of the discrete set of recorded points, an impedance model-based fuzzy force control algorithm, with real-time adjusting reference trajectories, is proposed to counteract the imprecision of the constraint surface. In addition, taking the impact and joint friction into consideration, the force control algorithm is proved to be feasible. The grinding force and material removal thickness can be controlled to be approximately consistent.

REFERENCES

Jatta F., G. Legnani, A. Visioli., Industrial Electronics, IEEE Transactions, 53, 604 (2006).

Robertsson A., T. Olsson, R. Johansson, etc.. Proceedings of the 2006 IEEE/RSJ International Conference on Intelligent Robots and Systems, 2743 (2006).

Tan F. S., J. G. Ge. Journal of Shanghai Electric Technology, 1(2), 35 (2008).

Wang F., H. Y. Liu, Z. Luo. Optics and Precision Engineering, 21(6),1479 (2013).

Wang F., Z. Luo, H. Y. Liu, L. Wang. 2010ROBIO, 1555 (2010).

Wang Y. F., T. Y. Chai. Chinese Journal of Scientific Instrument, 27(2), 186 (2006).

Advances in Energy Science and Equipment Engineering II – Zhou, Patty & Chen (Eds)
© *2017 Taylor & Francis Group, London, ISBN 978-1-138-71798-5*

An automatic CFD program for the rocket nozzle

YanJie Ma, FuTing Bao & WeiHua Hui
School of Astronautics, Northwestern Polytechnical University, Xi'an, China

ABSTRACT: In this paper, a Java program is developed using the secondary development of ProE5.0 and Ansys15.0. With the input of a few parameters with the GUI, the program can automatically solve CFD problems for the rocket nozzle. An example case has been carried out and the results indicate that the program can regenerate the geometrical model properly, obtain fine mesh, set up the physical conditions and boundary conditions, and make the iteration correctly. The program can be used to accelerate the design and evaluation of the rocket nozzle.

1 INTRODUCTION

As an effective and reliable way to study fluid dynamics, CFD has been widely used in the design and evaluation of rocket nozzles (Ebrahimi, 1993; Koutsavdis, 2002; Gross, 2004; Igra, 2016). Currently, there are many kinds of software, such as open source software or commercial software, that can be used for CFD, but each kind will take substantial amounts of time to set lots of parameters to solve a regular CFD problem. Fortunately, for rocket nozzles, the same topology can often be shared, and similar physical and mathematical conditions are often present, which make the CFD process possible to automate by a program.

A CFD problem generally involves three steps: geometrical modeling, meshing, and boundary conditions setup and iteration. Using the secondary development of ProE5.0 and Ansys15.0, a Java program is developed in this paper, by which all the three steps can be realized automatically for the rocket nozzle. Figure 1 shows the workflow of the program.

2 PARAMETRIC GEOMETRICAL MODELING

A specific geometrical model of a rocket nozzle can be identified by a group of parameters, as shown in Figure 2, and Table 1 provides the meaning of each parameter.

A parametric model of a double-arc nozzle is built in ProE5.0 as the template model in the program, with the parameters mentioned above as the parameters for the model file. A GUI is designed, which can be used to input the parameters for generating the geometrical model they need. Using the ProE-Java interface J-Link (He, 2008), the program can regenerate a new nozzle immediately, with the set of parameters input from the GUI.

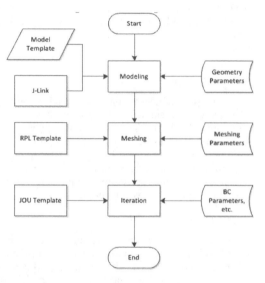

Figure 1. Program workflow.

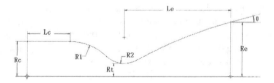

Figure 2. Identification of a nozzle by parameters.

3 AUTOMATIC MESHING

In a regular CFD solution procedure, meshing often takes the most part of manual time. However, the automation of meshing with ICEM in Ansys15.0 can be realized by running a TCL language script. First, a template is created for the TCL script (suffixed with "rpl") to be suitable for

Table 1.   The meaning of geometrical parameters.

Parameter	Meaning
Rc	Radius of the combustion chamber
Lc	Length of the combustion chamber
R1	Radius of the beginning of the convergent section
R2	Radius of the throat arc
Rt	Radius of the throat
Le	Length of the expanded section
Re	Radius of the exit
θ	Angle of the exit

Table 2.   The meaning of meshing parameters.

Parameter	Meaning
TP_inlet	Total pressure of the inlet
SP_inlet	Static pressure of the inlet
TT_inlet	Total temperature of the inlet
SP_outlet	Static pressure of the outlet
TT_outlet	Total pressure of the outlet
k	Specific heat ratio of the working gas
R_m	Gas constant of the working gas
μ	Kinetic viscosity
y^+	Expected value of y^+

the model discussed in the last section. Users only need to input a few necessary parameters from the GUI, and the program will complete the rest of the mission: modifying the TCL script with parameters, and running it automatically to obtain a mesh file to be used at iteration by Fluent later on. The parameters that need to be input are listed in Table 2.

3.1   Block split

When using ICEM to mesh an axial symmetry model, O-Grid can often be used to obtain a fine meshing. With the TCL script, the block built based on the model is split into five sub-blocks along the axial. Then, O-Grid split is carried out to split the sub-blocks. Since different models have an identical template, the script is suitable for all the models regenerated by this program.

3.2   Cell height near the wall

To ensure the iteration convergence and obtain a credible result, many factors must be considered. When it comes to the rocket nozzle, the most important factors are the turbulence model and the near-wall treatment. These factors require the corresponding cell height near the wall. In the flow field of rocket nozzles, the velocity of the fluid often has a wide distribution, ranging from subsonic to supersonic. Other parameters, such as temperature and pressure, can change rapidly along the flow field. The realizable k-ε model is recommended (Raheem, 2004), as well as the enhanced wall treatment. With the combination of these two options, the simulation can obtain a fine result, both in the near wall region where the Reynolds number is low, and in the main flow region where the Reynolds number is often high.

To determine the first cell height near the wall, a variable y^+ is introduced:

$$y^+ = \frac{\rho U_r y}{\mu} \qquad (1)$$

where y is the first cell size, ρ is the density of the gas, and μ is the viscosity of the gas. U_r can be determined by:

$$U_r = \sqrt{\frac{\frac{1}{2} C_f \rho U_\infty^2}{\rho}} \qquad (2)$$

where C_f is the skin friction coefficient and for internal flows:

$$C_f = 0.079 \, \mathrm{Re}^{-0.25}. \qquad (3)$$

In some problems, the flow near the wall can be important; on these conditions, $y^+ \approx 1$ is needed to obtain more number of small cells near the wall. On other conditions, y^+ should be between 30 and 300. The GUI allows users to input the expected value of y^+, and using the equations mentioned above, the first cell size can be calculated. The calculation will be carried out automatically while the TCL script is running.

After the necessary calculations, the modified TCL script will drive the ICEM CFD to produce a mesh file for the specific nozzle model that is generated before. The mesh file will also be used in iteration.

It should be noted that for the rocket nozzle, the value of y^+ near the wall at the throat point will be always the largest, thus both the expected value of y^+ input from the GUI and calculated cell height near the wall are at the throat point.

4   AUTOMATIC ITERATION

Fluent module in Ansys15.0 can be run in a batch mode using a journal file that is written in the Scheme language. Like automatic meshing, a journal file template is first created and can be modified properly by the Java program, with the input parameters from the GUI.

As the program is developed specifically for the rocket nozzle, some physical models and algorithms are chosen automatically, rather than by users. The program uses a steady density-based solver. Furthermore, the realizable k-ε model is used as the turbulence model, with enhanced wall treatment, as mentioned previously. The density scheme for the working gas is ideal gas, and the specific heat is calculated using the following equation:

$$C_p = \frac{k}{k-1} R_m.$$ (4)

The boundary condition for the inlet is set to be the pressure inlet, and the outlet is the pressure outlet. The parameters needed are obtained from the GUI of the meshing module. The initialization will be carried out and the variables of the whole flow field will be set using the values of the inlet.

After the basic setup by running the journal file, the iteration will begin automatically, with the number of iteration steps input by users. Moreover, the case and data will be saved after the iteration.

5 AN EXAMPLE CASE AND RESULTS

As an example case, the flow field simulation of a rocket nozzle is carried out. The geometrical parameters of the nozzle are listed in Table 3. After inputting the parameters from the GUI, the Java program regenerates a new model from the template file, which will be used for further analysis.

Another set of parameters for meshing the geometrical model are listed in Table 4. Automatic meshing is carried out by the program and the mesh is obtained, as shown in Figure 3.

Table 3. Geometrical parameters.

Parameter	Value	Parameter	Value
Rc	16 mm	Lc	20 mm
R1	25 mm	R2	12 mm
Rt	6 mm	Le	50 mm
Re	25 mm	θ	15°

Table 4. Meshing parameters.

Parameter	Value	Parameter	Value
TP_inlet	4000000 Pa	SP_inlet	3999990Pa
TT_inlet	3000 K	SP_outlet	101325Pa
TT_outlet	300 K	k	1.2
R_m	287.41 J/kg K	μ	1.79e-5 kg/ms
y^+	200		

Lastly, the number of iteration steps is input. The program modifies the journal file template and runs it automatically to set up the physical options and boundary conditions, begin iteration, and save the case and data afterwards. Figure 4 shows the contour of the temperature of the flow field. The static temperature decreases along the flow from

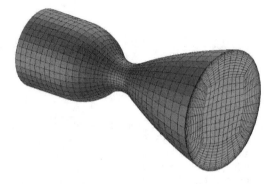

Figure 3. Mesh for the model.

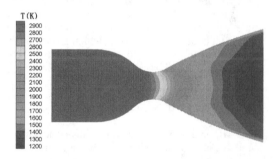

Figure 4. Temperature contour.

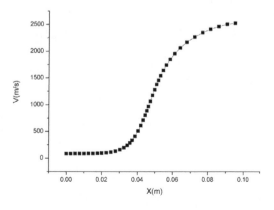

Figure 5. Velocity distribution along the axis.

1621

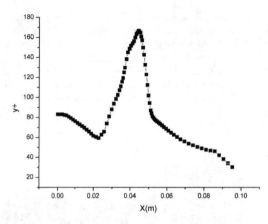

Figure 6. Y⁺ distribution along the wall.

3000 K to about 1200 K, due to the acceleration of the gas. Figure 5 shows the velocity distribution along the axis. The velocity increases from 0 on the inlet to about 2500 m/s on the outlet. These results indicate that the automatic CFD program is valid.

Figure 6 shows the y^+ distribution along the wall. The maximum value of y^+ is about 170, meeting the requirement of the expected value of 200. This indicates that the program can generate qualified mesh as users' need. Meanwhile, y^+ hits the maximum value at the throat point, which ascertains the validity of the method to determine the first layer height of the wall based on the throat point.

6 CONCLUSION

An automatic CFD program is developed for the rocket nozzle. With the program, users can generate nozzle models, obtain a fine mesh, and then implement iteration procedure by just inputting a few necessary parameters. The use of this program can decrease the manual operating time distinctly, and make the CFD for the rocket nozzle easy to accomplish. The validity of the result has been verified by a case.

The program can only treat the models properly with the specific topology, and its applicability to other models is still need to be developed.

REFERENCES

Ebrahimi H. B., CFD validation for scramjet combustor and nozzle flows, part 1, AIAA-**93**-1840 (1993).

Gross A., C. Weiland, Numerical simulation of separated cold gas nozzle flows, Journal of Propulsion and Power, Vol. **20**, No. 3 (2004).

Igra D., Numerical simulation of nozzle starting flow, Journal of Spcecraft and Rockets, Vol. **53**, No. 1 (2016).

Koutsavdis E. K., G. Stuckert, A numerical investigation of the flow characteristics of plug nozzles using fluent, AIAA-**2002**-0511 (2002).

Pei-ying HE, LI Yue-qin, Developing the computer aided fix design system based on ProE/J-Link, Journal of Engineering Graphics, No. **4** (2008).

Raheem S. A., V. Babu, Numerical simulations of unsteady flows in solid rocket motors, 10th AIAA/CEAS Aeroacoustics Conference, 2004, **10**.2514 (2004).

Advances in Energy Science and Equipment Engineering II – Zhou, Patty & Chen (Eds)
© 2017 Taylor & Francis Group, London, ISBN 978-1-138-71798-5

The fire command and dispatching system based on precise positioning

Dejing Cui, Shutao Guo & Likun Xiong
Hengde Digital Choreography Technology Co. Ltd., Qingdao Economic and Technological Development Zone, China

ABSTRACT: Considering the issues of the existing fire application platform in high-rise buildings such as lack of indoor location and indoor map support, this paper proposes a fire command and dispatching system based on precise positioning. By means of the TC-OFDM indoor and outdoor positioning technology, which is integrated with the base station positioning system, based on the SIP network communication technology protocol by developing a series of service protocols interacting within the system, we develop a multi-functional fire command and dispatching platform. The platform solves the key problems of the indoor and outdoor map service, the tracking and display of indoor and outdoor location information, and the information sharing and synchronization between the fireman's mobile terminal, command center command system and mobile command system. The experimental results indicate that the platform is stable, accurate, and reliable.

1 INTRODUCTION

Fire is one of the frequent disasters that causes a serious threat to human survival and development, leading to serious damage at high frequency, time, and space span. With the development of social science and technology, the information system of fire alarm and command dispatching has been used widely during firefighting and emergency rescue operations.

Both at domestic and foreign levels, based on the Geographic Information System (GIS), 3G communication, virtual reality, global positioning technology, video surveillance, and other technologies are required to build a large number of fire command and dispatching systems (CHEN, 2007; LV, 2008; Sheng, 2005). However, due to the limitation of the positioning time, positioning accuracy and complex indoor environment conditions, the GPS and the Beidou navigation positioning system are difficult to be located in the interior area. This is because transmitted microwaves of the positioning system are too weak and the frequency is too high to spread along a straight line to penetrate through walls; thus, the signal will not reach the interior area.

Although several types of fire emergency rescue command system have been developed, it is limited by the development of indoor positioning technology and indoor map, so that the fire emergency rescue command system cannot make up for fire command and dispatch to provide positioning and indoor map information. In other words, the existing fire application platform lack support

for indoor positioning and indoor map, the commander at the scene of a fire relies mainly on his experience and intuition to command troops and implement the emergency rescue activities. This will inevitably lead to the command system that has great blindness and lack of scientific basis, thereby greatly reducing the success rate and efficiency of fire rescue operations. In order to reduce casualties and losses due to fire, the fire brigade is required to have an efficient command system and a scientific decision-making system.

TCOFDM (Li, 2015; YU, 2015) is the new signal system that integrates both positioning and communication signals. Based on this technology, which saves the cost of indoor positioning, the current mobile communication network becomes both a communication network and a high-precision positioning network. The indoor positioning accuracy is improved several times, which breaks the bottleneck of the indoor high-precision positioning technology and realizes the high-precision indoor positioning and indoor and outdoor seamless positioning.

In this paper, by means of the TC-OFDM indoor and outdoor positioning technology, we develop a fire control system based on the precision positioning technology. The system aims to establish a set of positioning networks in one of the fire command and control systems, which can provide more detailed and accurate space location information for the fire emergency rescue command. The fire command center can accurately display the rescue scene personnel and facility distribution, deployment of troops, and other location information. This will greatly improve the accuracy and efficiency of fire rescue operations, so that the decision-making of fire emergency rescue activities can be carried out more efficiently and with scientific basis. It will greatly enhance the efficiency of China's firefighting and rescue activities during emergency rescue, public safety, networking, smart city, and other wide area indoor and outdoor location services to lay an important foundation.

2 TC-OFDM

At present, a lot of research has been done on the positioning technology based on the ground mobile communication network, which can be used to locate indoor users because of the good coverage of ground network signals. The basic positioning principle adopted by all types of mobile communication network-based positioning system is generally similar. They use detection characteristic parameters of the communication signal mobile station and the multiple fixed position transceiver machine to estimate the geometric position of the target mobile station, such as radio wave field strength, propagation time, time difference, and incidence angle.

The wide area indoor signal coverage is achieved by mobile BSs with high-precision time synchronization. Terminals demodulate the TC-OFDM signals and obtain the navigation message for positioning. The TCOFDM system can provide seamless outdoor and indoor positioning in a wide area with meter accuracy. The TC-OFDM positioning system uses the integration of positioning and communication signals to achieve a high-precision wide area indoor and outdoor seamless positioning. First, the communication signal and the timing signal are uplinked to the satellite by the uplink station. Second, the communication signal and the timing signal are transmitted to the terrestrial broadcasting station by the satellite. Third, the communication base station receives the signal transmitted by the satellite for high-precision time synchronization. Finally, the communication base station will combine high-precision positioning signals with the OFDM signal to form positioning signals based on TC-OFDM with high synchronization accuracy. The system architecture is shown in Figure 1.

TC-OFDM signal is transmitted through the mobile base station network; its transmission process is shown in Figure 2.

The TC-OFDM signal is transmitted by the base station with high-precision synchronization, and the system can be carried out to obtain the wide area indoor signal coverage. The TC-OFDM signal can be obtained by the terminal and the navigation

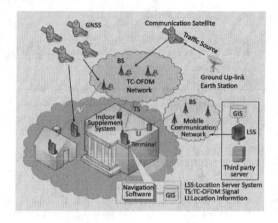

Figure 1. System architecture of TC-OFDM.

Figure 2. The transmission of the TC-OFDM signal.

message. In addition to the indoor environment, the TC-OFDM signal can also help the GNSS system obtain higher outdoor positioning accuracy, robust localization, and improved time to first fix.

3 SIP

SIP (SI, 2005) (Session Initiation Protocol, RFC 3261) is an Internet conferencing and telephony signaling protocol developed by the Internet Standards Organization (IETF). SIP is an application layer protocol that can establish, modify, or suspend multimedia sessions or calls. It is an ASCII-based protocol, which provides "dating" service on the Internet.

Because SIP is an application layer signaling protocol, it is not a complete communication system solution. It requires a combination of other solutions or protocols to achieve the entire system. For example, the Real-time Transport Protocol (RTP) (RFC1889) is used to transmit audio and video and other real-time streaming data.

Compared with the traditional H.323 protocol, SIP has obvious advantages:

Excellent scalability greatly improves the processing capability of the system;

Closely integrated with the Internet to make communication easier and convenient;

Remarkable openness can not only provide good support for mobile phones, PDA, and other mobile devices, but also provide online instant

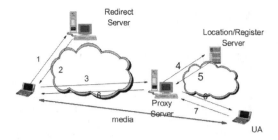

Figure 3. The principle of SIP.

communication, voice and video data transmission, and multimedia applications, as shown in Figure 3.

The SIP and HTTP (client–server protocol) have a similar structure. The client sends a request to the server. Thus, the server processes the request and sends a response to the client. The request and response form a transaction.

4 THE FIRE COMMAND AND DISPATCHING SYSTEM BASED ON PRECISE POSITIONING

4.1 System design

Command and communication system is the basic neural network of firefighting, rescue organization, and command, and timely, effective, and reliable command and communication is the key to ensuring the smooth running of the firefighting and rescue operations. Based on the precision positioning technology of the fire command and dispatching platform to provide the building's internal floor information, indoor positioning information, and real-time communication, the support fire rescue is commanded. The platform is composed of a server, a common user terminal, a fireman's handheld terminal, a command center command system, and a mobile command system.

The core problem of fire command and dispatch is the information communication between different firefighting equipment and systems. Initially, each information isolated island will be combined based on the precise positioning technology of the fire command and control platform, which will in fact realize interconnection and converged communications of a multi-system, to provide a powerful tool for on-site command.

The main functions are as follows:

1. The integration of indoor and outdoor maps to realize the relationship between outdoor maps and indoor maps. The indoor map of the building can be entered by clicking on the outdoor building, and thus the building's indoor map

organization and the map browsing of the building unit can be realized.

2. The analysis of the indoor rescue path. According to the floor information, the indoor evacuation passageway of each floor in the same building is automatically connected to generate the floor evacuation network. On this basis, the function of the indoor rescue path analysis can be realized based on the path analysis algorithm.

3. The sharing and display of indoor and outdoor positioning information. The system can receive and analyze the data of the mobile terminal of firefighters and the mobile terminal of common users, and sends the received information to the command center server in real time. The system can dynamically track and display the received mobile terminal information of mobile terminal information in the indoor map.

4.2 System function introduction

This section focuses on the function of the command platform server, the command center command system, and the mobile command system.

4.2.1 The server
Server is the core of the whole system. It uses the international standard SIP protocol to design the signaling protocol, and combines with VOIP soft switch technology to achieve registration authentication for each subsystem of the system. It is logically divided into the following types: registration server, proxy server, media processing server, and map server. It provides communication services, indoor and outdoor map service, fire control information sharing, and synchronization service for the command system.

4.2.2 Command center command system
Command center command system is used in the fire command hall on the PC. Fire control commands personnel through the system to track the location of firefighters in real time, to carry out the plan query management, command and dispatch, and evacuation of victims. Its core functions include indoor and outdoor map display, real-time tracking of the location information of firefighters, and real-time communication between the terminals.

4.2.3 Mobile command system
The function of the mobile command system is similar to the command center command system. The difference is that it is used on the mobile terminal, so that the personnel can use the platform to carry out the command and dispatch in time on the fire command vehicle.

4.2.4 *Mobile terminal*

The mobile terminal is divided into ordinary user handheld terminal and fireman's handheld terminal. Common user terminal can transmit the location information of the trapped mass to the server. The fireman's handheld terminal will transmit the location information and firefighters' real-time information to the server.

4.3 *Principle of the system*

The system is designed based on the IP soft switch architecture, and uses SIP as the signaling protocol, through RTP protocol transmission media, which is combined with the TC-OFDM positioning technology. Mobile terminal position information and fire and real-time information will be transmitted to the server. The server receives and maintains the information from the mobile terminal, and realizes the fire information sharing and synchronization, meanwhile providing indoor and outdoor location service and map service for the fire command system.

5 EXPERIMENTAL RESULTS

In order to verify the correctness of the method, we use C# as the experimental platform to realize the system. Through several experiments, a set of positioning networks can be established in one of the fire command platforms, as shown in Figure 4.

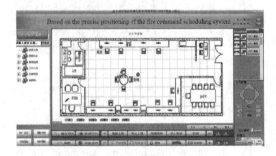

Figure 4. Interface of the system.

6 CONCLUSIONS

In this paper, a firefighting command and dispatching platform is developed, which is common and practical. It uses the indoor and outdoor integrated geographic information system technology and the indoor positioning technology. The system provides accurate positioning information of fire facilities, firefighters, and trapped people in high-rise buildings and personnel intensive areas. It solves the key problems such as indoor and outdoor map service, tracking and display of indoor and outdoor positioning information, information communication, and sharing and synchronization between different subsystems of fire control and the dispatching platform.

REFERENCES

Chen Tao, Weng Wenguo, SUN Zhanhui. Architecture of a fire emergency response system based on fire model. Journal of Tsinghua University (Science and Technology), 2007, 47(6): 863–866.

Li Xiaoyang, Deng Zhongliang, Ma Wenxu. Design and Implementation of LBS Platform based on Indoor & Outdoor Positioning Technology TC-OFDM. 2015, (z1).

LV Xin-chi, JIN Jing-tao, Zhang Hao. Research on technique system of communications for fire fighting emergency response command and control. Fire Science and Technology, 2008, 27(5): 350–355.

Sheng Jialun Liu Zaitao. Key technique and general design of the earthquake emergency command system. Seismological and Geomagnetic Observation and Research, 2005, 26(4): 107–116.

Si Duan-Feng, Han Xin-Hui, Long Qin. A Survey on the Core Technique and Research Development in SIP Standard. Journal of Software, vol. 16, No.2, 2005 pp. 239–250.

Yu Rui, Lu Nan, Zhang Haohao. Design and Implementation of Campus LBS System Based on Indoor-Outdoor Positioning. Chinese Journal of Electron Devices, 38(2), 2015: 464–468.

Advances in Energy Science and Equipment Engineering II – Zhou, Patty & Chen (Eds)
© *2017 Taylor & Francis Group, London, ISBN 978-1-138-71798-5*

Construction of a smart meter life-cycle management system in industry 4.0 mode based on the cloud platform and RFID

Xiangqun Chen, Rui Huang, Hao Chen, Liman Shen, Longmin Bu & Dezhi Xiong
Hunan Electric Power Metrology Center, Changsha, Hunan Province, China

Renheng Xu
Harbin Institute of Electrical Instrument, Harbin, Heilongjiang Province, China

ABSTRACT: In order to meet the need of the State Grid Corporation for the life-cycle management of a massive smart meter, this paper uses the cloud platform based on cloud computing technology and Radio Frequency Identification Devices (RFID) to design a smart meter based on RFID communication, a meter detection system based on the RFID smart meter, a smart storage management system based on both the RFID smart meter and the cloud platform, and the life-cycle management system of the smart meter based on the cloud platform. The entire system implemented functions such as fast parameter setting, rapid detection, automatic storage, and fast copy reading for the smart meter, as well as shortened time and improved efficiency by using RFID communication for its speedy communication and strong group-reading ability. At the same time, the whole system realized a large amount of data collection, data management, data filtering and extraction, and data mining and analysis to help smart production and service in industry 4.0 mode by building the cloud platform management system.

1 INTRODUCTION

It has been estimated that the proportion of smart meter users served by the State Grid Corporation System will reach up to 300 million. In the long term, taking into consideration China's 1.3 billion people, the total number of installed smart meters in China would be 500 million. Since 2010, State Grid Corporation has begun to do centralized bidding for smart meters. Consequently, every provincial and municipal power system faces the management problem of how to manage a huge number of smart meters rapidly and efficiently.

As an advanced identification technology that is developing rapidly and has broad application prospects, RFID has become much robust in reading number, distance, accuracy, and many other aspects, as well as has broad applications. RFID is a form of technology that replaces bar code with a wide range of applications, and it will become the next mainstream recognition technology. The RFID-adopted smart meter management system can overcome the defects of ordinary electronic energy meters, such as single function, poor anti-theft effect, backward meter reading, and easily damaged and polluted IC card. In order to adapt to the trend of meter intelligent management, the RFID technology is applied for the transmission of electric power information, to better reflect

contact-free, information security, and other advantages of RFID.

In order to meet the requirements of the State Grid Corporation and domestic customers, and at the same time meet the foreign users' requirements, this study aims to establish an efficient cloud service platform for a smart meter life cycle management system in industry 4.0 mode based on the cloud computing and RFID technology. In addition, it also performs calibration testing, test verification, and data analysis for smart meter products systematically to develop a smart meter life cycle management system. Moreover, it can not only design, guide, test, and rotate on-site products for manufacturers through this system and reach the bidding requirements of the State Grid Corporation, but also improve our electrical instrumentation industry's integrated management capability of digital network products. Through this research, it will eventually implement meter production in intelligence, meter management in intelligence, and service in intelligence based on industry 4.0 mode.

2 RFID SMART METER BASED ON INDUSTRY 4.0 MODE

To construct a smart meter life-cycle management system based on the RFID technology, first, it is necessary to transform the RFID technology on smart meters, and add the RFID communication module to the original structure of the smart meter. The RFID smart meter based on industry 4.0 mode will support the functions that are required by production automation in industry 4.0 mode to be implanted into the smart meter, and provide all the necessary data that production automation requires, such as table number, manufacturer, parameters, date, and other data. It can also realize to connect and measure the network based on an advanced two-way interactive communication network and implement its smart architecture (Philips, 2002).

The RFID smart meter combines mainly the sampling measurement unit, the power supply unit, the charge control unit, the communication unit, the key display unit, the function controlling unit, and the carrier unit with RFID chips, thus increasing the RFID communication function of the smart meter. The structure of the RFID smart meter is shown in Figure 1.

3 RFID SMART METER DETECTION SYSTEM BASED ON INDUSTRY 4.0 MODE

Based on the integrated and intelligent principle of industry 4.0 mode, the RFID smart meter detection system can interconnect all products and equipment as well as networks. Therefore, the smart meter detection system based on industry 4.0 mode connects smart meter calibration sets and the full life-cycle cloud platform management system, and it is connected via the RFID technology. After the construction of each smart meter, the interior will have a RFID chip, and each chip has a globally unique identifier, so that each smart meter will have a unique identification code for registering and entering the smart meter-wide life-cycle cloud platform (Samuel, 2013).

The ARM processor and the DSP processor will be integrated into an embedded system that can implement an automatic smart meter calibration device. The automatic calibration device has a variety of communication ways, such as RS485, RFID, and TCP/IP networks, which can connect with a calibration table, program-controlled power sources, a standard table, and a cloud platform. Moreover, this device has an adaptive RFID frequency hopping function, which can not only automatically connect each band RFID device, but also avoid interfering with the signals of other RFID devices. When the automatic calibration software is programmed into an ARM embedded system, the data obtained from the checking platform can be classified, and the checklist data can be obtained by DSP calculations, and then they are passed into the test table for verification. The block diagram of the composition of the smart meter detection system is shown in Figure 2. The whole system consists of a computer with background control, micro-controller of communication control, programmable digital signal source, standard table, and error display module.

The smart meter detection system uses the ARM series processor and the DSP digital signal processor. S3C6420 uses the ARM11 core, which contains instruction data cache of 16 KB and instruction data TCM of 16 KB. When the ARM core voltage is 1.1 V, they can run up to 553 MHz and further up to 667 MHz, and they can be equipped with an open, scalable 32-bit embedded operating system when the ARM core voltage is 1.2 V. The TMS320F28335 digital signal processor is a floating-point DSP controller of TMS320C28X series of TI company. It has many features, such as high accuracy, low cost, low power consumption, high performance, highly integrated peripherals, data and program storage capacity, and more precise and rapid A/D conversion.

Calibration table and smart meter communications abandon the traditional RS485 and infrared

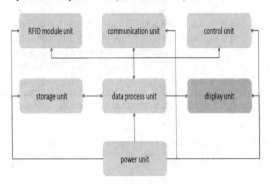

Figure 1. Principle block diagram of the RFID smart meter.

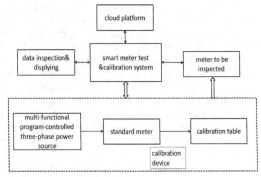

Figure 2. The structure diagram of the smart meter detection system.

communication smart meter, and use the RFID technology. Through an automatic checklist, it can read every piece of information on the smart meter from the cloud platform, and pass the calibrating and checking data into the smart meter via the RFID. Finally, it can pass the final status of the smart meter onto the cloud platform for a record.

4 SMART METER WAREHOUSE MANAGEMENT SYSTEM RESEARCH AND EQUIPMENT DEVELOPMENT BASED ON THE CLOUD PLATFORM

Work efficiency can be greatly improved by combining the cloud platform with the warehouse management system, and the complete warehouse management of the smart meter, as well as the development of the surplus high-frequency reader, operate the electronic tag and avoid manual errors.

4.1 Reader modular design method

Warehouse management system mainly uses the RDI reader and electronic tag communication of the smart meter to complete reading or erasing tag data and information, and the reader can be used for transmitting and receiving RF signals via an external antenna. The reader can also have functions such as reading and writing, displaying, and processing data, and can connect with a computer or other systems to complete the operations related to the RF tag. Therefore, the hardware modular design is a crucial step for developing the smart meter warehouse management system based on the cloud platform. The reader developed in this article uses the modular design method, including the RFID RF module, the core control module, the power module, and the communication module. The core control module uses the ARM Cortex-M3 processor, and the RF module uses the indyR2000 chip as the core. Figure 3 shows the schematic of the structure of the reader (Samuel, 2013).

4.2 Realization of the smart meter warehouse management system

This system is mainly used to manage internal information data; its use is limited to enterprise LAN. Also, because of the interaction of hardware and data acquisition, it should be capable of responding quickly and passing large amounts of data. In view of the above characteristics, we will use the developing mode based on C/S, and give full play to the PC processing and the fast response ability of the client. The system will develop a multi-tier application architecture based

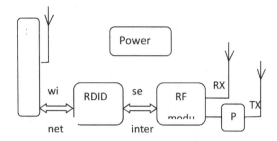

Figure 3. Principle block diagram of the reader structure.

on the net platform. The overall system architecture is divided into three levels, namely the client, the application server, and the database server of the three-tier logical structure, which is called the presentation layer, the business physical layer, and the data layer in the system architecture. Because of this three-tier structure, interface presentation, logic processing, and data management are divided into three logically independent but functionally interdependent parts. The logic layer depends on the interface layer to transfer and display data, and it also depends on data management to store and provide data. Data management depends on logic processing to transfer data. Therefore, the three layers are logically independent, functionally interdependent, and distinctively unique in level, and they form an overall close relationship.

Smart meter warehouse management is the core of the entire smart meter management system, according to the actual needs and specification of warehouse, the RFID-based warehouse cargo management system can be divided into the following functional blocks: a single table to check operation, in-storage and out-storage management systems, and information inquiry.

5 SMART METER FULL LIFE-CYCLE MANAGEMENT PLATFORM BASED ON INDUSTRY 4.0 MODE

We can establish a complete smart meter full life-cycle management platform based on industry 4.0 mode, and through RFID in the networking technology and the combination of the smart meter and the cloud platform system, it can implement all-round monitoring and management of the smart meter, including smart meter intelligent design, calibration factory testing, warehouse management, field operations, maintenance records, and scrap disposal. It is particularly critical to establish a complete and rational cloud platform management system for a smart meter full life-cycle management, which can implement tracing back to the source of smart meter full life-cycle management and improve industrial efficiency, and promote technological progress.

5.1 Study on the smart meter system based on UHF radio frequency technology

The system uses the structure combined with structures B/S and C/S, and uses WEB to show the reporting modules and functions that each subsystem required remote operations, which is easy for user access, data registration, portability, and good user experience.

5.1.1 Smart meter tracking system

It is shown in Figure 4 and uses the combination of RFID and smart meters to implement the entire tracking of the smart meter from design to the factory test calibration, maintenance of scrapped, and form trace ability of core parts.

The database adopted is a relational database that can handle large amounts of data while maintaining the integrity of the data, and the database offers many advanced data management and distribution capabilities. Such a database has sufficient data integrity protection capabilities, including complex transaction support, with a good performance-price ratio.

It collects the actual operating data of the RFID smart meter, and for the massive multi-format content data and status information, via a variety of clients after the acquisition, together with tens of thousands of visits and requests for action, it will increase pressure on the system server in the form of high concurrent. Moreover, the article also studies the sharing pattern of data and efficient distributed storage methods, and uses grid computing technology to reasonably correspond data block and server, to identify bad data and effective data mining, and do the data aggregation when data has a high-degree similarity in characteristics configuration.

The system provides a variety of ways of information checking, and is convenient to use. It provides a variety of statistical methods such as pie charts, histograms, and other graphs; and it also provides a function of complete report printing; the entire system complete each operation in graphical interface, which is easy to use.

5.1.2 Run evaluation system

RFID meter collects voltage, current, power, frequency, harmonics, and other real-time data in working, combined with error data obtained in the field cycle test, and carries out a comprehensive assessment of the impact of complex and real power grid operating environments on the meter accuracy and reliability. The evaluation of the meter state is divided into two parts: component evaluation and overall

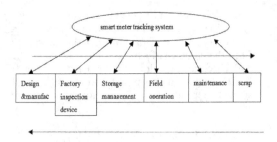

Figure 4. System information tracking and trace smart meter detection system.

evaluation. Component evaluation includes four parts: meter, PT, CT, and secondary circuit. Component evaluation should also consider the mark deduction both of the individual component state and the components in total. When all the components are evaluated as a normal state, the overall evaluation is a normal state; when any one component state is attention state, abnormal state, or critical state, the overall evaluation is the most serious state, which establishes appropriate data model of the evaluation system according to the massive large amount of data filtering analysis of the mobile client.

We will develop the public cloud platform on WeChat or mobile client APP, and set different permissions for different customer groups. The function of the smart meter traceability system is its terminal applications.

5.2 Study of a big data cloud platform

Figure 5 shows the data collection of a big data cloud platform, which makes hierarchical extracting, clearness, filtering, and other data pre-processing from the correlation between the results of a large amount of data. Moreover, it stores data safely, quickly, and efficiently through distributed storage. For massive multi-format data analysis, exploration and mining, exploring patterns and characteristics of the data, complex mathematical statistics and analyzing models are developed, and changes in information and value behind data are investigated. Through combining the unified big data analysis platform based on cloud computing, database storage, and Map Reduce framework, an efficiently processing big data analysis platform is developed, which is structured, semi-structured, and even unstructured for enterprises. Based on this platform, clients can achieve the transition of data assets from a cost center to a profit center, and driving business by data (Anonymous, 2008).

By connecting the master host and segment host of multiple nodes with the database via the Internet, applications access data by the master

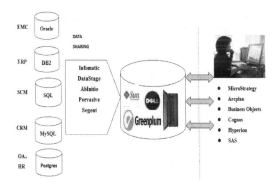

Figure 5. The framework of a big data cloud platform system.

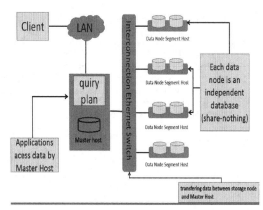

Figure 6. Diagram showing the structure of the big data analysis platform.

host, and each storage node in the Internet is an independent database, i.e. they do not share with each other. Moreover, they transfer data between the multiple storage node and the master host. Segment servers connect with each other via the Internet, and finish the same task; from users' point of view, it is a server system. Its essential feature is the server segment (each segment server works as nodes) connected by the Internet, each node only accesses its own local resources including memory, storage, etc., which is a completely non-shared structure (share-nothing). Therefore, it has the best scalability and can extend without any limit theoretically; 512 nodes and thousands of CPU can be interconnected with the current technology. Each node runs its own database and operating system, but each node cannot visit the storage of other nodes. The information exchange between each node is implemented via the Internet, and this process is known as data redistribution.

The cloud system shown in Figure 6 can collect data from each collection point throughout the country, super-efficiently, credibly, and independently, and preprocess the data of various types through the constituted configuration and strategy at the same time. Through the data layer of the business system, the service layer achieves functional abstraction, and uses a Service-Oriented Architecture (SOA) to connect different functional units of the system applications with the contract via well-defined interfaces between services, clean technology difference between different applications, and allow different applications coordinate with each other, which enables communication and integration between different services It also provides a unified and transparent data access interface to the users of data, applications, and services, and efficiently stores and accesses huge amounts of semi-structured data and structured data but without a strong correlation; meanwhile, it supports

distributed real-time and off-line calculation and analysis of such data (Anonymous, 2010).

6 CONCLUSION

To construct a smart meter life-cycle management system in industry 4.0 mode based on the cloud platform and RFID, all the crucial data of the life–cycle management system can be recognized by RFID including the design of the smart meter, factory inspection settings, warehouse management, field operations, maintenance, and the scrap. Moreover, it uses the anti-collision mechanism of RFID to implement multichannel, multi-tag, and simultaneous reading and writing; it also implements multi-parameter settings, rapid detection, on-line fault diagnosis, data storage capacity, strong real-time processing capability and remote data transmitting and receiving ability. The construction of the cloud platform can help clients know smart meter life cycle management quickly and easily, and it achieves data collection, data filtering and extraction, data statistic analysis, inquiry and reporting of large amounts of data of the smart meter. This system shortens the time it takes for the circulation process, accelerates turnaround time, reduces cycle time, and reduces costs in disguised form. Based on the support of RFID identification technology, asset management, inventory management, the data collection process becomes very simple. This system reduces the workload of staff, and reduces the work intensity of staff by using the RFID smart meter management system.

REFERENCES

Anonymous. Microsoft Releases SQL Server 2008 [J]. Worldwide Databases, 2008, V01. 20(9).

Anonymous. Professional SQL Server 2008 Internals andTroubleshooting [J]. M2Presswire, 2010.

Philips Semiconductors. Mifare MF RC500 highly integrated ISO14443 Areader IC data sheet [M]. The Netherlands: Philips Corporation, 2002.

Samuel Fosso Wamba, Abhijim Anand. A literature review of RFID—enabled Healthcare applications and issues [J]. International Journal of Information-Management, 2013, V01. 33(5).

Samuel Fosso Wamba. A literature review of RFID—enabled healthcare applications and issues [J]. International Journal of Information Management, 2013, 33(5).

Advances in Energy Science and Equipment Engineering II – Zhou, Patty & Chen (Eds)
© *2017 Taylor & Francis Group, London, ISBN 978-1-138-71798-5*

Smart meter design based on RFID technology

Xiangqun Chen, Rui Huang, Hao Chen, Liman Shen, Dingying Guo & Dezhi Xiong
Hunan Electric Power Metrology Center, Changsha, Hunan Province, China

Renheng Xu
Harbin Institute of Electrical Instrument, Harbin, Heilongjiang Province, China

ABSTRACT: In this paper, the basic principle of single-phase Watt-hour meter based on RFID (Radio Frequency Identification, RFID) is introduced, its design proposal of applying RFID into energy meter communication is put forward, and its circuit design, hardware principle diagram and work flowchart, and the process of achieving RFID communication are minutely focused.

1 INTRODUCTION

RFID technology has many advantages such as anticollision reading information, antifouling label, easy modification and reading of information in the label, and label information storage. Furthermore, it is easy to meet many application needs of the power industry and realize many management functions of power industry during working, such as preventing power stealing, remote meter reading, and asset management. To meet the need of smart meter life cycle management, it can greatly improve circulation efficiency, shorten the cycle, and save manpower to design a type of smart meter using RFID to realize communication, which can use RFID to recognize and communicate all the data from the smart meter manufacturing, factory testing data, warehouse management, field operations, maintenance and scrapping, using anticollision mechanism of RFID to realize multichannel, multitag reading and writing at the same time.

2 PRINCIPLE OF RFID

RFID technology uses radio frequency to get non-contact bidirectional communication and automatically identify the target object and access to relevant data. RFID system basically consists of three major parts: electronic tag (Tag), reader (Reader), and data exchange and management system (Processor). RFID tag (or RF card, transponder, etc.) is composed of coupling components and chips, which contain encryption logic, serial electrically erasable, programmable read-only memory and EEPROM, microprocessor CPU, RF transceiver, and related circuit. Tag has the functions of smart reading and encrypted communication;

it exchanges data via radio waves and reader, and working energy is provided by radio frequency pulses emitted by the reader. Reader sometimes is called Finder, reader, or readout device, primarily composed of wireless transceiver modules, antennas, control module, and interface circuits. The reader can transfer the reading–writing commands of the host or encrypted data sent from the host to the tag, and the data returned from tag after the decryption can also be sent to the host. The main functions of data exchange and management system are to complete the storage and the management of data information and the reading–writing control for cards.

The RFID system works as follows: the reader encodes the information to be sent, loads it at the carrier signal of certain frequency, and sends it outside via antenna, and the tag going into the working area receives a pulse signal, and the related circuit in the chip will demodulate, decode, and decrypt the signal, and then judge the command requests, password, and permission. After the judgment, if

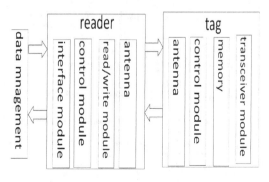

Figure 1. Principle of RFID.

their corresponding passwords are not compatible with the permissions, it returns an error message. In the case that everything is consistent: if it is a reading command, the control logic circuit will read the related information from the memory, after encryption, encoding, modulation, and then send to the reader via the antenna, Reader demodulates, encodes, and encrypts the received signal, and then send to the central information system for data processing; if it is a reading–writing command to modify information, the internal charge pump caused by controlling logic enhances the operating voltage, and provides the contents of flash EEPROM for modification. RFID principle is shown in Figure 1.

3 HARDWARE DESIGN OF RFID SMART METER

The hardware circuit of single-phase energy meter based on RFID consists of several parts, including measurement unit, radio frequency interface unit, display unit, clock unit, keyboard processing unit, memory unit, relay control unit, communication unit, CPU monitoring unit, and power supply circuit unit. Prepaid single-phase energy meter principle block diagram is shown in Figure 2.

Microcontroller adopts MSP430F147 single-chip, launched by TI company, with low power consumption and high performance. The measurement unit adopts special ADE7755 energy metering chip for ADI's high reliability of the United States, storage adopts nonvolatile ferroelectric memory FM24CL04 and EEPROM memory 24C64, real-time clock adopts low-power real-time clock chip RTX8025, lithium-ion battery is used in

the power meter to ensure the normal operation, off-electric monitoring unit monitors the work situation of linear power network in real time, RF reading–writing interface chip adopts the dedicated chip U2270B of recognizing IC (integrated circuit) reading–writing base station, RF card selects low-cost reading–writing E5550 card of TEMIC series, LCD drive uses low-power driver chip, PCF8576DT of PHILIPS company, energy meter and the information interaction of upper computer uses near-infrared photoelectric communications way, high-frequency output pulse of ADE7755 chip calculated by single-chip computer outputs low-frequency pulse, and visually reflects the user's electricity consumption displayed by LED, load control uses magnetic latching relay to control the load switching, for the convenience of users and staff to obtain information of electric energy meter, which uses single buttons for metering information queries.

4 READING–WRITING CHIP OF RF-INTERFACE

The dedicated chip U2270B of recognizing IC reading–writing, as shown in Figure 3 base station, is the communication interface between the microprocessor and an RF card. The chip consists of an antenna driver, supply circuit, oscillator, frequency tuning circuit, low-pass filter, amplifier output control circuit, and so on (TEMIC, 2002).

On-chip power, oscillator circuit, and antenna driver constitute the powered energy conversion circuit, and the internal oscillation frequency can be fixed at the desired frequency by adjusting the size of the resistor RF pin of the U2270B, then

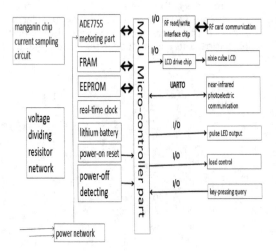

Figure 2.   Principle of RFID smart meter.

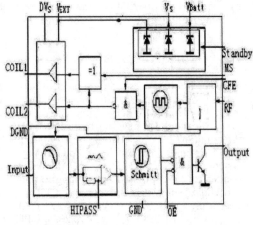

Figure 3.   Dedicated chip U2270B of recognizing IC reading–writing.

the RF field of the frequency is formed near the antenna by the amplification of antenna driver. By controlling the CFE-end of the chip to control the starting and stopping of the oscillator and the RF field produced by the antenna with a narrow interval, the different width represents data "0" and "1," thereby completing the process of writing the data sent from the base station into RF chip cards. Chip antenna creates weak load modulation according to the modulation of the RF chip cards, signal processing circuit, composed of chip internal low-pass filter, amplifier, Smith trigger, and output control circuit, will transfer this weak input signal into a signal required by the microcontroller, so as to achieve the objective of RF card read.

U2270B RF base station chip is mainly used for data modulation, RF transmitting and receiving, and data demodulation of tasks. U2270B is a low-cost, perfect-performance, low-frequency RF base station chip, whose main features are as follows (Jia, 2004; Jia, 2003):

1. The carrier oscillator can generate the oscillation frequency of 100–150 kHz, and can be fine-tuned with an external resistor, whose typical application frequency is 125 kHz;
2. Typical data rate is 5 kbps (125 kHz);
3. Supports two modes of power supply: (1) +5V DC power supply and (2) +12V batteries for cars using;
4. With a microprocessor interface that can be directly connected to the microcontroller;
5. Applies to Manchester and double-phase encoding;
6. Has a microprocessor-compatible interface;
7. With a function of voltage output, which can supply power to microprocessors and other peripheral circuits and keep the function of voltage output at the same time, and supply power to microprocessor or other peripheral circuits;
8. With low power consumption in standby mode, it can greatly reduce base station power consumption;
9. Apply to e5530/e5550/e5560 of TEMIC RF card reading and writing operations.

5 RF CARD CHIP

In this study, E5550 chip is used to design RF chip. It is a low-cost reading–writing RF IC chip produced by TEMIC and ATMEL company (USA). It uses low-power and low-voltage CMOS structure through noncontact inductance coupling to get power. It contains 264 bit nonvolatile memory, 224 bit of which can be provided for users' free using. It has functions of protecting storage block and protecting password. Bit transfer rate can be

selected. According to the needs, it can select RF frequency of 8, 16, 32, 40, 50, 64, 100, and 128 data to transfer frequency rate. It provides many methods such as a binary code, amplitude shift keying, frequency shift keying, Manchester coding, and bi-phase coded modulation. It has a variety of working methods to choose from (Hwan, 2004).

5.1 Structure of RF chips

E5550 RF chip is a complete system by itself, and the chip internal structure is shown in Figure 4. The interior of the chip includes an antenna, analog front end, modulator, writing decoder, bit rate generator, mode register, controller, testing logic, reset circuit, memory, high-voltage generator, and other modules, whose parts are described below (Luo, 2004):

1. Analog front end.
Analog signal processing front-end generally connects RF chip antenna directly and provides power for the chip, through RF chip antenna and inductance coupling of power meter interface chip U2270B antenna and transfers two-way data through a magnetic field. It mainly consists of a rectifier circuit, which can rectify AC voltage responding to antenna, clock signal pickup circuit, shut-off load, which is used for transferring data to a base station, and magnetic gap detection, which is used for receiving data from the base station.
2. Modulator
Modulator is used for modulating the data transferred from a base station, and the modulation method can be selected by users between Manchester code, binary code, and bi-phase encoding. Considering the convenience of base station decoding, Manchester encoding is selected in general.

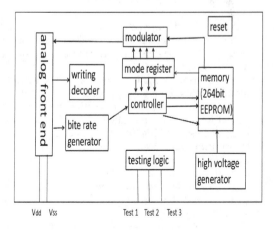

Figure 4. E5550 RF chip internal structure.

1635

3. Writing decoder.

Writing operation to E5550 is accomplished by strictly controlling the time of magnetic gap, the data "0" and "1" correspond to a different field width. The loops are mainly used for detecting real data stream from the magnetic field with a clearance.

4. Bit rate generator.

E5550 has multiple data transferring rate available for users to choose; therefore, it can generate a corresponding clock signal according to the data transferring rate set by users through the bit rate.

5. Mode register.

Mode register is a 32-bit RAM, mainly used for storing the user-mode set data read from EEPROM "0" data blocks, and it performs operations with a single call automatically after every power-on reset or program.

6. Host controller.

According to the working mode read from EEPROM "0" data blocks when it is power on, the host controller circuit controls reading–writing access to register, transfer data, process wrong information, decode the opcode of transferring stream data, and validate the password in the password protection mode.

7. Memory.

E5550 card inside has 264-bit EEPROM memory area, which is divided into eight blocks (BLOCK), each block with a 33-bit, 4-byte length, and No. 0 for block lock (LOCK). It is important to note that once the lock bit is set to "1", data of the corresponding blocks will not be able to make any changes, and the LOCK is not recoverable, and it cannot be launched with the others to the base station.

6 DESIGN OF RF INTERFACE

E5550 card and low-cost RF base station chip card U2270B form the complete RF card application system with high ratio of performance and price, widely used in many fields.

RF interface circuit, using U2270B short-distance and close coupling applications circuit, completes two-way data communication between the E5550 RF cards of the same TEMIC series, and microprocessor MSP430F147 takes charge of data transmitting and decoding of the received data. The MSP430F147 controls CFE feet to transmit data, the returned data of TEMIC series RF card uses Manchester encoding, and U2270B transmits data to the MSP430F147 for encoding through the OUTPUT PIN [5].

E5550 RF card system of single-phase energy meter consists of an RF card read–write interface chip U2270B, antennas, and passive RF card E5550. It uses wireless RF to exchange data between U2270B and E5550, which receives the electromagnetic wave sent from U2270B and generates the power required by the signal conditioning card. Through adjusting the external resistor parameter, U2270B fixes the internal oscillation frequency at 125 Hz, and then the RF field of the frequency is formed near the antenna by the amplification of antenna driver. Reading–writing channel module of chip is a signal processing circuit composed of low-pass filter circuits, amplifiers, Smith trigger, and output control circuit. By controlling the chip to achieve the starting and stopping of the oscillator, and RF field is produced by the antenna with a narrow interval, the different width separately represents data "0" and "1", so that completing the process of writing the data sent from U2270B into RF chip cards. Chip antenna creates weak load modulation according to the modulation of the RF chip cards, and read–write channel achieves the goal of reading E550 card through transferring the weak input signal to the signal microcontroller required (TEMIC, 2002).

Chip U2270B of RF reader interface writes data to the card by way of short-gap interrupting the RF field, and different width of RF field between two gaps represents separately 0 and 1. Writing card program encodes data through controlling on the starting and stopping of antenna loading and achieves all kinds of controlling functions of IC card through sending various combinations of data flow to IC card. For cards by U2270B while reading the data, RF card in Manchester RF modulation sends the digital signal into RF form. Card-reading process uses software to simulate signal timing and automatically check whether the synchronization signals are synchronized or not. Furthermore, it decodes the software according to the encoding chosen and finally stores the data stream from the decoded in designated storage areas in a reasonable order.

7 CONCLUSION

The main innovation of this paper is to present a plan of designing an RF interface composed of TEMIC E5550 card and base station chip U2270B. This design gives full play to the advantages of RF card without power supply, noncontact, and the interface is stable and reliable. RF technology with a high degree of information integration and security has been integrated into the mainstream of today's information technology, RFID energy meter is not vulnerable to outside attack, combined with appropriate communication protocols, and it can use RFID Group read properties to achieve rapid detection, fast meter-copying, and rapid logistics. RFID single-phase energy meter not only has charging function, but also can connect with related smart meter data management system through the reader, realize all data communication through RFID, greatly improve the efficiency, and conform to the international demand of meter.

REFERENCES

Hwan Lei, Zhang Baoping. The development of TEMIC RF card reader series, Electronic Technology Application, 2004 (4): 22–24.

Jia Zhenguo, Xu Lin, Gu Chunmiao. Principles and applications of RF base station chip U2270B. International Electronic Elements, 2004(1): 66–68.

Jia Zhenguo, Xu Lin. Read / write RFID chip E5550 and its applications. International Electronic Elements, 2003(12): 41–43 [20]. Luo Fan, Wang.

Luo Fan, Hwan Lei, Zhang Baoping. The development of TEMIC RF card reader series. Electronic Technology Application, 2004 (4): 22–24.

TEMIC Semiconductors. Read/Write Base Station IC U2270B Data Sheet (2002).

TEMIC Semiconductors. Read/Write Base Station IC U2270B Data Sheet, 2002.

Zhang Guoyun, Jin Weixiang, Peng Shiyu. The development of New RF IC card reading and writing control system. Journal of ChangSha University of Electric Power (Nature Science version).

Advances in Energy Science and Equipment Engineering II – Zhou, Patty & Chen (Eds)
© 2017 Taylor & Francis Group, London, ISBN 978-1-138-71798-5

Confirming warehouse financing process modeling and analysis

Zhijian Wang, Jun Zhang, Hua Yin & Pei Shen
Information Science School, Guangdong University of Finance and Economics, Guangzhou, China

ABSTRACT: Design and innovation of supply chain financial products is a complex problem. Conventional methods are not only inefficient and time-consuming, but also unreliable. To solve this problem, a Petri net-based modular approach is adapted in this paper. Confirming warehouse financing, a common and basic service in supply chain financial products, is selected as an example to explain this method. Business process of this service is introduced first, then basic structures of Petri net supporting modular modeling are investigated, after that we build a hierarchical business model to describe the business process with the idea of modularization using Petri net. The accuracy and performance of the model are analyzed. The product design can be optimized according to the result.

1 INTRODUCTION

The nature of supply chain finance is to provide comprehensive financial services for a single upstream/downstream enterprise or several enterprises in a supply chain. The implementation of supply chain finance should be based on the characteristics of that supply chain's structure and details of corresponding transaction. It relies on the credit of the core business, the degree of self-liquidating for individual transaction, or the value of goods in circulation (Wil, 2013).

Confirming warehouse financing is a process with the participation of warehouse regulators (usually a logistics company). Financing enterprise, the core enterprise (seller), warehousing, and banking regulators are the four signs for confirming warehouse business cooperation agreement. Warehousing regulators provide credit guarantees; the seller provides buyback guarantee; finally, the bank provides its bankers' acceptances for financing companies.

With confirming warehouse financing, suppliers can achieve bulk sales, increase its operating profit, reduce bank financing, lower capital costs, guarantee receivables, and improve capital efficiency. On the contrary, buyers can get the convenience of financing and solve financial difficulties for full purchase, so as to achieve bulk purchases and reduce costs.

However, traditional integrated supply chain finance product development approaches expose a lot of deficiencies with the development of e-commerce. They are not easy to meet different requirements from customer quickly. Banks now face the problems of rapid and low-cost supply chain financial products innovation.

2 PROCESS OF CONFIRMING WAREHOUSE FINANCING

General processes of confirming warehouse financing are as follows: First, the buyer obtains line of credit for warehouse receipts pledge from a bank, then the bank will review the supplier's credit status and his/her ability to repurchase, next the bank and the supplier will sign a repurchase agreement and quality assurance agreement, and the bank signs a storage regulatory agreement with the warehousing regulator. After receiving a notification from the bank, which declares that it agrees to finance the buyer, the supplier delivers to the designated warehouse and gets a receipt. The seller is required to deposit an acceptance margin in the bank, which is 30% of the total turnover. The supplier pledged the warrant to the bank; the bank then opens an acceptance bill

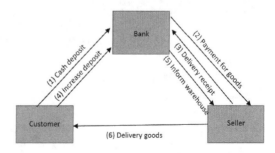

Figure 1. Process of confirming warehouse financing.

to the supplier, where the purchaser is the drawer and the supplier is the recipient. As the buyer gradually pay the guaranty money, the bank proportionately release the right to deliver goods to the buyer until the margin account balance is equal to the amount of the bill. When the bill expires, if the margin account balance is insufficient, providers will repurchase the pledges on warehouse receipt on due date.

Such advance payment guarantee delivery applies to some occasions that customers want to purchase large quantities in order to obtain preferential treatment. Its business processes [2] are illustrated in Figure 1.

3 PETRI NET-BASED MODELLING TECHNOLOGY

A Petri net is composed of places and transitions. In general, places are represented by circles,

rectangles indicate transitions, and places and transitions are connected by arcs. Places contain tokens represented by a black dot. Places are used to indicate conditions or media; transitions are used to indicate operation or transmission. In general, a token indicates a specific thing or abstract information. It is assigned to a place in the initial state. If certain conditions are met, corresponding transitions are fired on behalf of a process completion.

There are four basic Petri net structures:

1. If the Petri net modeling business processes are executed sequentially, such a structure is called a sequential structure, which is shown in Figure 2.
2. If more than one task can be performed in any order or performed simultaneously, such a structure is called a parallel structure, which is shown in Figure 3.

Figure 2.   Sequential structure.

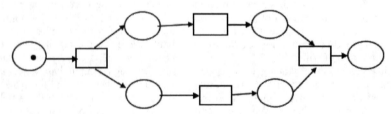

Figure 3.   Parallel structure.

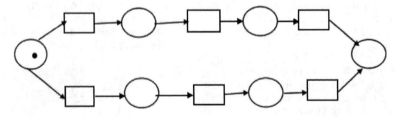

Figure 4.   Selection structure.

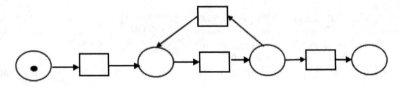

Figure 5.   Looping structure.

3. When there is a choice between two or more tasks, such a structure is called selection structure, which is shown in Figure 4.
4. In the ideal case, a task runs no more than once for each case, but sometimes we need to perform a task repetitively. Such structure is a looping structure, which is shown in Figure 5.

In practice, Petri nets would normally be extended in order to achieve better business process description.

4 PROCESS MODEL

The confirming warehouse financing process model in Petri nets is shown in Figure 6.

Places and transitions of the Petri nets model in Figure 6 correspond to the processes and entities in Figure 1; the details are explained in Table 1.

The submodel of the section is shown in Figure 7, where B1–B3 refer to different departments in a bank and B4 refers to the customer. D1 is the auditing process, D2 is approving, and D3 is contract signing.

5 MODEL RUNNING AND ANALYSIS

The process of supply chain finance product business is in chronological order and is performed according to certain procedures. In order to make the model run correctly, it is necessary to extend

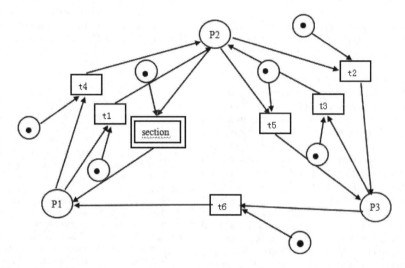

Figure 6.   Confirming warehouse financing process model.

Table 1.   Places and transitions in Figure 6.

Place	Meaning	Transitions	Meaning
P1	Customer	t1	Apply for financing
P2	Bank	Section	Review, approve, and sign a contract
P3	Seller	t2	Payment
P4–P10	Review mechanism and supervision mechanism for submitted information on each node in the supply chain	t3	Delivery receipt
		t4	Increase deposit
		t5	Inform warehouse
		t6	Delivery goods

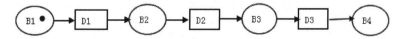

Figure 7.   Submodel of section.

Table 2. Time for business activities.

Activities	Time	Activities	Time
Apply for financing	3	Increase deposit	2
Section	10	Inform warehouse	1
Payment for goods	1	Delivery goods	1
Delivery receipt	1		

Table 3. Time stamp for transitions.

Transition	Time stamp	Transition	Time stamp
t1	0	t4	3
Section	1	t5	4
t2	2	t6	5
t3	2		

this Petri nets model with time factors. Time spent for various business activities and time stamps for different transitions are supposed in Tables 2 and 3.

Next, we should analyze the correctness of the model. Normal operation of a model is required. In general, no deadlock or conflict is allowed in such model, and these are some basic requirements. The model in Figure 6 uses P1 as a starting point and return to P1 eventually; it can complete the entire operation correctly. Different timestamps are set for every transition in advance; this ensures that the model is run sequentially in chronological order in accordance with those business processes.

This model in Figure 6 is relatively simple. According to the results obtained after several tests, the average model response time is about 0.2 s and the completion time of the entire process is 19 days. According to actual needs, customers can also calculate the efficiency of the receipt; payback period and business receipts can be calculated by the bank; and the seller can also calculate his/her income. These results can be used as a reference by all parties for the analysis to determine whether he/she should be involved in such a business.

6 CONCLUSIONS

Nowadays, modular approach is widely used. It is involved in many aspects of technological innovation. Modular approach can accelerate product development and reduce risk; it helps enterprises to deal with uncertainty about the future. Because the modular approach has its particular advantages, it is being valued by industry and academia increasingly. Petri net-based modular modeling method plays an important role in supply chain finance product design and test.

ACKNOWLEDGMENTS

This work was supported by Foundation for Technology Innovation in Higher Education of Guangdong Province, China (No. 2013 KJCX0085), Degrees and Graduate Education Reform Project of Guangdong University of Finance and Economics (2014ZD01), Guangdong Natural Science Foundation (2014 A030313609), and Philosophy and Social Sciences planning project of Guangdong Province, China (No. GD15CGL11).

REFERENCES

Fan, G, H Yu, L Chen, D Liu, Petri net based techniques for constructing reliable service composition, *Journal of Systems & Software*, 86(4), pp. 1089–1106 (2013).

Fu Xinhua, Luo Hu, Xiao Mingqing, Method of Resource Optimization for Test System Based on Timed Coloured Petri-Net. *Systems Engineering-Theory & Practice*, 30(9), pp. 1672–1678 (2010).

Jianxin (Roger) Jiao, Timothy W.Simpson, Zahed Siddique, Product family design and platform-based Product development: A state-of-the-art review. *Journal of Intelligent Manufacturing*, 18, pp. 5–29 (2007).

Richard N. Langlois, Modularity in technology and organization. *Journal of Economic Behaviour & Organization*, 49(1), pp. 19–37 (2002).

Sun Hong, Huang Qing, Modularized Management for Financial Product Innovation in Commercial Banks. *Commercial Research*, 9, pp. 1–4 (2005).

Supply Chain Finance group, Shenzhen Development Bank - China Europe International Business School, *Supply Chain Finance*. Shanghai Far East Publishers, pp. 73–75 (2009).

Wil M.P. van der Aalst, Arthur H.M. ter Hofstede, Mathias Weske, Business Process Management: A Survey. *Lecture Notes in Computer Science*. 2678, pp. 1–12 (2003).

Advances in Energy Science and Equipment Engineering II – Zhou, Patty & Chen (Eds)
© *2017 Taylor & Francis Group, London, ISBN 978-1-138-71798-5*

Analysis of the axial trapping force of the laser based on T-matrix

Yingjie Zhang
North China Electric Power University, Baoding, Hebei, China

ABSTRACT: T-matrix is an efficient method to calculate the trapping force of optical tweezers. This method depends only on the composition, size, shape, and orientation of particles, and is independent of the incident field. In this paper, the basic principles and calculation involved in the T-matrix method are reviewed. The calculation steps of trapping force are also summarized.

1 INTRODUCTION

Optical tweezers were first developed by Ashkin (White, 2000). Optical tweezers technology aims at trapping and manipulating microscopic particles by a focused laser beam. It is associated with zero injury and no mechanical contact with particles, so it is widely used in the metal particles, biologically active cells, and organelles. A simple optical trap is shown in Figure 1. The particles are typically suspended in water or an aqueous solution. If they are sufficiently transparent, they will be attracted to be the focus of the beam and be stably trapped there. If the particles are too strongly absorbing or too reflective, the radiation pressure due to absorption or reflection will push them out of the trap.

The optical forces are usually divided into gradient force, which acts toward areas of higher irradiance, and scattering forces, including absorption and reflection forces. The optical forces and torques result from the transfer of momentum and angular momentum from the trapping beam to the particle. Many approximate methods are often used to calculate the optical forces, such as geometric optics or Rayleigh approximations or electromagnetic scattering theory (Daniel, 2000), which is based on the relationship of the particle size and the wavelength of incident light. When the particle size is much smaller than the laser wavelength, in general a$<<$ λ/20, we can use the Rayleigh approximation. When the particle size is much larger than the laser wavelength, we can use geometry approximation. When the particle size is between the two, namely the particle size is close to the wavelength of incident laser, we can use electromagnetic scattering theory.

2 T-MATRIX

2.1 T-matrix method

The T-matrix method relates the incident wave and a scatter. We can write the incident field as a superposition of a discrete basis set of functions $\psi_n^{(inc)}$, which is a solution of the Helmholtz equation (Nieminen, 2001):

$$U_{inc} = \sum_n^\infty a_n \psi_n^{(inc)} \qquad (1)$$

where a_n is the expansion coefficient of the incident wave.

Similarly, the scattered wave can be written as:

$$U_{scat} = \sum_k^\infty p_k \Psi_k^{(scat)} \qquad (2)$$

where p_k is the expansion coefficient of the scattered wave.

We can write, $p = Ta$ \qquad (3)

where a and p are the column vectors, which express the scattering properties of the object, and T is the T-matrix or transition matrix. The T-matrix method can be used for scalar waves or vector waves in a variety of geometries.

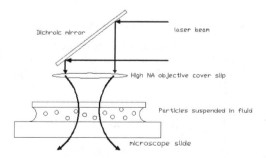

Figure 1. Optical tweezers.

Table 1. Comparison of calculation methods.

Methods	Calculation steps	Advantage	Disadvantage
FEM	1. The laser beam generating an optical trap is equivalent to electromagnetic field source changing with time. 2. The space is divided into discrete lattice, then the field change of each lattice is calculated by taking a certain time step, as shown in the figure	1. The theory is simple and can be used to calculate arbitrary shape and random component particles. 2. The calculation is very accurate.	1. They only can calculate relative small particles, if the volume increases, the amount of computation will grow very fast. 2. If we need more accurate results, we need to finely divide space, and the amount of computation will grow very fast. 3. This division can be performed only on a finite volume of space, so the selection of boundary condition is very important. If the far field cannot be ignored, it is to be included in the near field through appropriate transformation.
FDTD	1. The space is discretized into a fine mesh. 2. Iterations of the field between the grid spacing of the mesh and the time step.		
DDA	1. The particles trapped in the optical tweezers are discretized into a cubic array of polarizable points with each point representing a dipole. 2. The scattering field of each dipole is calculated.	1. The method needs not be considered the boundary condition of electromagnetic fields. 2. It can be used to deal with any arbitrary shape and refractive index distribution of the particles. 3. It can be used to separate two or more particles trapped in the optical tweezers.	1. The amount of calculation of the method is relatively large. 2. When the composition and orientation of the particles are changed in the optical tweezers, the entire process is recalculated.
T-matrix	1. Calculate the elements of T-matrix. 2. Expand the incident and scattering fields in terms of regular vector spherical wave functions. 3. Calculate incident expansion coefficients. 4. Calculate the scattering expansion coefficients.	1. It depends only on composition, size, and shape of the particle, orientation and the wavelength of the incident field. 2. The T-matrix only needs to be calculated once for any particular particles.	No

The elements of T-matrix are numbers, rather than functions of angles. The elements of the T-matrix are independent of direction, polarization, and other spatial properties of the incident light. It depends only on the particle, its composition, size, shape, and orientation. Therefore, the T-matrix remains the same for different directions of illumination. Therefore, it only needs to be calculated once for any particle and need not repeat calculations. This is a significant advantage over many other methods, such as FEM, FDTD, and DDA (Zemanek, 1998).

2.2 Trapping as a scattering problem

In the T-matrix method, the incident trapping field is expressed as a sum of regular Vector Spherical Wave Functions (VSWFs) (Doic, 1997; Waterman, 1971):

$$E_{inc}(r) = \sum_{n=1}^{\infty} \sum_{m=-n}^{n} [a_{mn} RgM_{mn}(kr) + b_{mn} RgN_{mn}(kr)]$$

(4)

where

$$RgM_{mn}(kr) = (-1)^m d_n \exp(im\phi) \times j_n(kr) C_{mn}(\theta) \quad (5)$$

$$RgN_{mn}(kr) = (-1)^m d_n \exp(im\phi)$$
$$\times \left\{ \begin{array}{l} \dfrac{n(n+1)}{kr} j_n(kr) \\ P_{mn}(\theta) + [j_{n-1}(kr) \\ -\dfrac{n}{kr} j_n(kr)] B_{mn}(\theta) \end{array} \right\}, \quad (6)$$

$$B_{mn}(\theta) = \hat{\theta} \frac{d}{d\theta} d_{0m}^n(\theta) + \hat{\phi} \frac{im}{\sin\theta} d_{0m}^n(\theta) \quad (7)$$

$$C_{mn}(\theta) = \hat{\theta} \frac{im}{\sin\theta} d_{0m}^n(\theta) - \hat{\phi} \frac{d}{d\theta} d_{0m}^n(\theta) \quad (8)$$

$$P_{mn}(\theta) = \hat{r} d_{0m}^n(\theta), \quad (9)$$

$$d_n = \left(\frac{2n+1}{4\pi n(n+1)} \right)^{\frac{1}{2}} \quad (10)$$

where $j_n(kr)$ are spherical Bessel functions and $d_{0m}^n(\theta)$ are Wigner d functions.

Similarly, the scattered fields are expressed as:

$$E_{scat}(r) = \sum_{n=1}^{\infty} \sum_{m=-n}^{n} [p_{mn} M_{mn(kr)} + q_{mn} N_{mn}(kr)] \quad (11)$$

where M_{mn} (kr) and N_{mn} (kr) are the same as RgM_{mn} (kr) and RgN_{mn} (kr), respectively, and the spherical Bessel functions are replaced by spherical Hankel functions of the first kind, $h_n^{(1)}(kr)$.

In practice, the field expansions and T-matrix are terminated at some $n = N_{max}$. Assuming a scatter is contained within a radius r_0, $N_{max} \approx kr_0$ is usually adequate. If we can choose $N_{max} = kr_0 + 3\sqrt[3]{3kr_0}$, good convergence results are acquired.

2.3 Calculation of the T-matrix

When introducing the T-matrix, Waterman used the Extended Boundary Condition Method (EBCM) to calculate the T-matrix. The EBCM remains widely used, and can be considered the "standard" method for the calculation of the T-matrix. In the EBCM, the internal field within the particle is expanded in terms of regular VSWFS. Therefore, the main disadvantage of the EBCM is to restrict to isotropic and homogeneous particles. Additionally, if the particles have aspect ratios different from 1, the computation will be ill-conditioning.

In the ECBM, rather than considering the coupling of the incident and scattered fields directly, the coupling between the incident and internal (the RgQ matrix), and scattered and internal fields (the Q matrix) is calculated, and T-matrix is found from these (T = −RgQQ⁻¹) (Wiscombe, 1986).

We consider some general principles of calculating T-matrices and applied the point-matching method to calculate the T-matrix for particles devoid of symmetry. This method avoids the time-consuming surface integrals required by the magnetic fields similarly to those for the electric fields (Nieminen, 1991):

$$H_{inc}(r) = \frac{1}{\kappa_{medium}} \sum_{n=1}^{N_{max}} \sum_{m=-n}^{n} a_{mn} N_{mn}(kr) + b_{mn} M_{mn}(kr) \quad (12)$$

$$H_{int}(r) = \frac{1}{\kappa_{particle}} \sum_{n=1}^{N_{max}} \sum_{m=-n}^{n} c_{mn} RgN_{mn}(kr) + d_{mn} RgM_{mn}(kr) \quad (13)$$

$$H_{scat}(r) = \frac{1}{\kappa_{medium}} \sum_{n=1}^{N_{max}} \sum_{m=-n}^{n} p_{mn} N_{mn}(kr) + q_{mn} M_{mn}(kr) \quad (14)$$

where κ_{meduim} is the wavenumber of the field in the surrounding medium and $\kappa_{particle}$ is the wavenumber inside the particle.

We considered a single scatterer, centered on the origin, contained entirely within a radius r_0, and with a surface specified by a function of angle:

$$r = r(\theta, \phi), \quad (15)$$

The boundary conditions which match the tangential fields on the surface of the scatterer are:

$$\hat{n} \times (E_{inc}(r) + E_{scat}(r)) = \hat{n} \times E_{int}(r) \quad (16)$$

$$\hat{n} \times (H_{inc}(r) + H_{scat}(r)) = \hat{n} \times H_{int}(r) \quad (17)$$

where $\hat{n}$ is a unit vector normal to the surface of the particle.

The expansion coefficients are c_{nm}, d_{nm}, p_{nm}, and q_{nm}, so each point gives four independent equations. We can generate a grid of 2 Nmax (Nmax+2) points with equal angular spacings in each of θ and ϕ. In order to ensure a dependent result, it is very important to use an over-determined linear system by least-squares solution. The values of the VSWFs at these points on the particle surface are calculated, and we can use the values of the T-matrix obtained by column-by-column calculation.

The intensity distributions of scattering field of various particles are shown in Figure 2.

3 CALCULATION OF THE OPTICAL TWEEZERS

The general procedure for the calculation of optical forces and torques using T-matrix method is straightforward:

1645

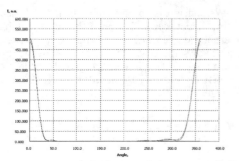

a. Intensity distribution of the scattered field of rotating ellipsoids.

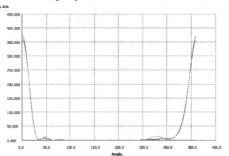

b. Intensity distribution of the scattering field of cylinders, whose length-to-diameter ratio is 2.

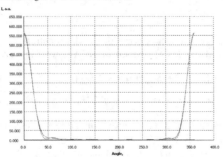

c. Intensity distribution of the scattering field of general Chebyshev particles.

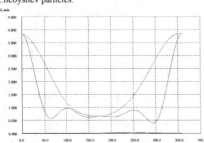

d. Intensity distribution of the scattering field of Chebyshev particles, whose deformation degree is 1 and the deformation parameters is 0.1.

Figure 2.   Intensity distribution of scattered field.

1. Calculate the T-matrix;
2. Calculate the incident beam coefficients a and b;
3. Calculate the scattered field coefficients assuming $p = Ta$;
4. Calculate the force and torque.

Optical tweezers force can be expressed as:

$$F = QnP / c, \qquad (18)$$

where n is the refractive index of the medium in which the particle is immersed, P is the beam power, c is the speed of light in free space, and Q is known as Q efficiency factor or Q factor, which is a dimensionless number and expresses the size of optical tweezers. The axial trapping efficiency, Q_z, is:

$$Q_z = \frac{2}{P} \sum_{n=1}^{n_{max}} \sum_{m=-n}^{n} \frac{m}{n(n+1)} \mathrm{Re}(a_{nm}^* b_{nm} - p_{nm}^* q_{nm}) - \frac{1}{n+1}$$

$$\left[\frac{n(n+2)(n-m+1)(n+m+1)}{(2n+1)(2n+3)} \right]^{\frac{1}{2}}$$

$$\times \mathrm{Re}(a_{nm}a_{n+1,m}^* + b_{nm}b_{n+1,m}^* - p_{nm}p_{n+1}^* - q_{nm}q_{n+1,m}^*) \qquad (19)$$

The torque efficiency about the z-axis is:

$$\tau_z = \sum_{n=1}^{n_{max}} \sum_{m=-n}^{n} m(|a_{nm}|^2 + |b_{nm}|^2 - |p_{nm}|^2 - |q_{nm}|^2) / P \qquad (20)$$

where

$$P = \sum_{n=1}^{n_{max}} \sum_{m=-n}^{n} |a_{nm}|^2 + |b_{nm}|^2, \qquad (21)$$

This torque includes contributions from both spin and orbital components, both of which can be calculated by similar formulas.

4   CONCLUSION

T-matrix method is a general, accurate, and effective method that is widely used to analyze electromagnetic scattering properties of single particle, two particles, and random discrete medium surface in the optical tweezers. It also calculates force and moment of the optical tweezers. It is used especially in the determination of particle size and the incident light wavelength of similar cases. In this paper, the principle and calculation method of T-matrix and the optical tweezers are summed up. With the growing maturity of the application of

T-matrix method in optical tweezers and optical tweezers structure tending to diversification, T-matrix method will be used to trap multiparticle fiber optical tweezers and holographic optical tweezers numerical calculation and application of optical tweezers in the future. It provides theoretical guidance for molecular biology, colloid chemistry, and experimental atomic physics.

REFERENCES

Daniel, A. White. Vector finite element modeling of optical tweezers Comp. Phy. Comm Vol. 128 (2000).

Doic, A., T. Wriedt. Plane wave spectrum of electromagnetic beams Optics Comm. Vol. 136 (1997) 114–124.

Nieminen, T.A., etc, Calculation of the T-matrix: general considerations and application of the point-matching method J. Opt. Soc. Am. Vol. A 8 (1991), 871–882.

Nieminenm, T.A. Numerical modelling of optical trapping. Comp. Phy. Comm Vol. 142 (2001). 468.

Waterman, P.C. Symmetry, unitarity, and geometry in electromagnetic scatter, Phys. Rev. D3, (1971) 825–839.

White, D.A. Numerical modeling of optical gradient traps using the vector finite element method. J Comput Phys Vol. 159 (2000): 13–37.

Wiscombe, W.J., and A. Mugnai Single scattering from nonspherical Chebyshev particles: a compendium of calculation, NASA Ref. Pub (1986) 1.1157.

Zemanek, P. Optical trapping of Rayleigh particles using a Gaussian standing wave, Optics Comm. Vol. 151 (1998) 273–285.

Advances in Energy Science and Equipment Engineering II – Zhou, Patty & Chen (Eds)
© 2017 Taylor & Francis Group, London, ISBN 978-1-138-71798-5

A review of three methods on the study of traffic problems

Jianfeng Chi & Weihao Li
Beijing University of Posts and Telecommunication, Beijing, China

ABSTRACT: Traffic problems affect the daily life of people in urban area greatly. In this paper, we review three major methods, graph-and-complex-network method, human mobility-based method, and traffic flow-based method, to study the traffic problem. We present some typical research works on these methods and then summarize the characteristics and limits of these studies using these methods. Finally, we present some other relevant methods and give our perspective on future studies.

1 INTRODUCTION

With the development of urbanization, the scale of cities has expanded and roads have increased rapidly. As a result, the complexity of roads and highway are consequently increased. More problems thus arise.

Traffic has been one of the most concerned problems faced by people all over the world. Traffic congestion causes a tremendous loss of time, and even lead to serious air pollution problem. Nearly all major provincial capital cities in China suffer serious traffic congestion problem, and many drivers in these cities lose nearly twice the time on the roads due to traffic congestion.

Safety problems are also the focus of the public. Serious traffic accidents, such as car crashes, lead to loss of millions of dollars every year in many countries. Understanding the pattern of traffic accident—how these accidents happen, in what way people can prevent them, how to ease the loss caused by accidents after they happened—are of great practical value to policy maker and the authority.

Most of these problems happened because of the bad design of roadways network. Some places, such as central business districts and densely populated area, still suffer congestion problems despite good design of roads. Therefore, many researchers are trying to study the road network to solve traffic problem like these. Also, the rapid development of the research of big data enables the authority to obtain related information like people's work places, their homes, and their commuting time. In addition, many studies have focused on traffic flow to study the interaction between vehicles and infrastructures.

In this paper, we review three different methods to study the road traffic problems: graph and complex network, human mobility, and traffic flow and traffic state. These methods provide different views and aspects to help us understand road traffic problem, enabling different focuses and some kinds of relations and connections in the study in this field.

2 GRAPH AND COMPLEX NETWORK

The fundamental infrastructure and lifeline of a city—the road network—affects the whole city in many aspects, for it is the basis of socioeconomic activities as well as whole city transportation of people and cargo. Some researchers try to understand operational efficiency and some other characteristics from the perspective of graph theory and complex network theory.

When using this method to explore the inner characteristics of road network, there are two ways of representing road segment in a graph. One is segment-based primal graph, in which a road segment represents an edge, while the junctions between road segments are regarded as nodes (Curtin, 2007). The other is segment-based dual graph, in which road segments are taken as nodes while the junctions between roads are mapped to edges. The former representation is easier to understand and capture the spatial data through digitalizing and the latter one is more helpful in real-world analysis, for our focus of interest is road itself, as shown in Figure 1.

In the real world, we are often blocked by traffic congestion or other events that lead to road useless—we cannot choose that road when we traffic. Yingying Duan and Feng Lu describe road network robustness as the ability of road networks under a wide range of attack (Duan, 2013). They analyzed the road network robustness in six cities by first reconstructing the road into a set of communities, in which the roads are more closely related, using the theory of complex network.

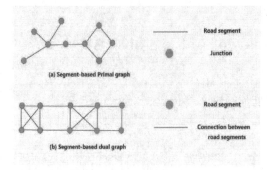

Figure 1. Two ways of representing road segment in a graph.

Then, they further simulated that a series of attack randomly or intentionally removes the nodes and the corresponding edges in the road networks to see the changes in related metrics, such as clustering coefficient and average path length. By doing so, the process of the road network collapse can be detected and the robustness can be well understood. The results reveal that the structure is robust under random failure but fragile under intentional attack.

In a graph, centrality measures the relative importance of a node. Different types of centrality measures the nodes in different aspects (Asgari, 2013): What types of centrality measurement we choose to explore the network depending on the way we attach importance to the node. In road network, the centrality investigated most in road network is betweenness centrality. It characterizes the number of shortest path passes on a node. Aisan Kazerani and Stephan Winter studied the betweenness centrality for its potential to characterize traffic flow in road networks (Kazerani, 2009). They found that traditional definition of betweenness centrality cannot be applied to road network for there are dynamic and more complicated road networks. Thus, they modify the traditional definition of betweenness centrality to apply to road network-characterizing nonuniform distribution of travel demand and time dependency.

Graph, as a simple way to represent road network, is often favored by researchers to combine with other data to study traffic problem together. Road network is often transformed into graph as the first step to set up traffic models. Gonzalez et al. (Gonzalez, 2007) proposed a novel road network partition approach on the basis of the road hierarchy with the model of graph. The road hierarchy was used to partition the network into area, and Gonzalez et al. used the data of drivers' driving pattern and speed pattern fitting in the graph model to provide adaptive fastest routes close to optimal to the driver.

3 HUMAN MOBILITY

Thanks to the development of big data resources of all kinds—call detail records and GPS and social media from cell phone—we are able to extract useful information from the abundant amounts of data. At present, travelers usually carry mobile phones, which enables us to get the trajectory of the traveler movement. Indeed, recent studies focusing on human mobility have attracted great attention (Brockmann, 2006; González, 2008; Song, 2010; Qu, 2010), and studies need to explore how human mobility affects traffic condition using big data sources, as shown in Figure 2.

Serdar Çolak, Lauren Alexander and their team in MIT found that cell phone CDRs can help to figure out people's home and work place. They explored the use of Call Detail Records (CDR) to generate Origin-Destination (OD) trips of different purposes (home, work, and other) and present a replicable procedure: data filtering, stay extraction, activity inference, and trip estimation, to process CDRs (Alexander, 2015; Colak, 2015; Toole, 2015) in order to extract information relevant to trip generation. Their works can help estimate the traffic demand in different times of a day and thus help the policy maker to better understand how people move in the city.

Furthermore, Serdar Çolak, Antonio Lima, and Marta C González used CDRs again to gain deep understanding of congested traffic in urban area (Colak, 2016). By detecting the home and work location of millions of users and their location shifts in the morning peak hours and parsing OpenStreetMap road network data, they modeled travel demand and supply and presented the relationship between time lost in congestion and the ratio of road supply to travel demand in five cities. They further analyzed how different levels of

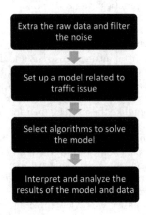

Figure 2. Steps of dealing with traffic issues with human mobility.

awareness of social goods (or selfish routing by drivers) cause the overall traffic system gains and traffic relieves.

Real-time data can even help detect traffic anomalies like accidents protests and sports events. Bei Pan (Pan, 2013) used two types of data to do so: GPS trajectories of vehicles and social media data. They identified the anomaly according to the drivers' suddenly changed routing behavior on an urban road network and describes the detected anomaly by mining representative term from the social media that people posted when anomaly happened.

Predicting how human will move is very helpful for policy makers to deal with the issue of traffic congestion. Now it seems possible to achieve this goal. Song C (Song, 2010) explored the limits of predictability in human dynamics by studying the mobility patterns of anonymized mobile phone users. By measuring the entropy of each individual's trajectory, they found a 93% potential predictability in user mobility across the whole user base even though there are significant differences in the travel patterns.

The analysis of traffic problem with human mobility can be summarized in the following steps: first, extra the raw data and filter the noise based on our needs; second, set up a model to solve our problem we concerned; then, select algorithms to solve our model; finally, interpret and analyze the results of the model and data.

4 TRAFFIC FLOW

Traditionally, researchers focus on the analysis of the traffic flow and traffic state, and the systematic theories and models are well established (Treiber, 2013). Traffic flow is often characterized by speed, density, traffic state, and other parameters. Methods generally use differential and difference equations, probability statistics, and physics-based method in micro-, macro-, and meso-scales, as shown in Figure 3.

Although there are a large amount of traffic-flow-based analysis, researchers still further investigated in this field. Ehsan Azimirad presented a novel fuzzy model and a fuzzy logic controller for an isolated signalized intersection (Azimirad, 2010). They applied the state-space equation to formulate the averaging time of vehicles in an isolated signalized intersection and further proposed a fuzzy model and fuzzy traffic controller that can optimally control traffic flows under normal and exceptional traffic conditions. Simulation results showed an average of 83% improvement of waiting time in one single intersection.

Traffic state, with respect to the traffic flow, is a way to characterize the traffic condition. How

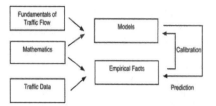

Figure 3. Procedures of dealing with traffic problem using traffic flow method.

traffic states change and evolve arouse the interest of many researchers. Wang L (Wang, 2014) studied the transition characteristics and distinct regularity of traffic-state pattern during the traffic evolution process from the viewpoint of transition network model. He used the data flow of multiple road sections from Network of Shenzhen's Nanshan District, China, to demonstrate that the methodology permits extraction of many useful traffic transition characteristics, including stability, preference, activity, and attractiveness and the relationships between these characteristics, helping to understand the complex behavior of the temporal evolution features of traffic patterns.

5 CONCLUSION AND PERSPECTIVES

In this paper, we reviewed three different methods to study the road traffic problems: graph and complex network, human mobility, and traffic flow and traffic state. With different views and aspects, these methods help us understand road traffic problem from different angles. Also, these studies have some kinds of relations and connections.

The three methods mentioned above have their own characteristics and limits: Graph theory and complex network theory have the great potential to understand the urban road network from its topology—the basis of network, which may enable us to view the problem from a fundamentally new aspect. However, studies on this field are still in rudimentary phase, focuses of these studies are dispersive, and the models are not well established. With the data of human mobility from a variety of sources, data statistics and analysis provide opportunities and angles to fulfill the new aspect for the traffic problem. Research has shown that data can even aid a taxi driver to find the most profitable area in a city. How to interpret traffic data and convey its intrinsic meaning will be the mainstream in the future study. Algorithm in the field Machine Learning may help us to do so. Studies on traffic flow are very thorough and comprehensive while their future development is very limited.

In addition to the aforementioned three methods, several other methods can be adopted to study the traffic problem. Image processing is one the promising way to do so. Images and video that provide us a source of the real-time traffic data can be further processed to gain more insight to the traffic behavior and traffic anomaly. Satellite central control on cars with computers leads to an efficient car management. All these methods lead us to the development of intelligent transportation (Wang, 2012).

REFERENCES

Alexander, L., Jiang, S., Murga, M. and Gonzalez, M.C. Origin–destination trips by purpose and time of day inferred from mobile phone data. *Transportation Research Part C-emerging Technologies* (2015).

Asgari, F., Gauthier, V. and Becker, M. A survey on Human Mobility and its applications. *Computer Science* (2013).

Azimirad, E., Pariz, N. and Sistani, M.B.N. A Novel Fuzzy Model and Control of Single Intersection at Urban Traffic Network. *IEEE Systems Journal*, 4, 1 (2010), 107–111.

Brockmann, D., Hufnagel, L. and Geisel, T. The scaling laws of human travel. *Nature*, 439, 7075 (2006), 462–465.

China's major urban traffic analysis report in first quarter of 2016. 2016.

Colak, S., Alexander, L.P., Alvim, B.G., Mehndiratta, S. and Gonzalez, M.C. Analyzing Cell Phone Location Data for Urban Travel: Current Methods, Limitations, and Opportunities. *Transportation Research Record* (2015).

Colak, S., Lima, A. and Gonzalez, M.C. Understanding congested travel in urban areas. *Nature Communications*, (72016).

Curtin, K.M. Network Analysis in Geographic Information Science: Review, Assessment, and Projections. *Cartography & Geographic Information Science*, 34, 2 (2007), 103–111.

Duan, Y. and Lu, F. Structural robustness of city road networks based on community. *Computers Environment & Urban Systems*, 41, 9 (2013), 75–87.

Gonzalez, H., Han, J., Li, X., Myslinska, M. and Sondag, J.P. *Adaptive fastest path computation on a road network: a traffic mining approach*. City, 2007.

González, M.C., Hidalgo, C.A. and Barabási, A.L. Understanding individual human mobility patterns. *Nature*, 453, 7196 (2008), 779–782.

Kazerani, A. and Winter, S. *Can Betweenness Centrality Explain Traffic Flow*. City, 2009.

Pan, B., Zheng, Y., Wilkie, D. and Shahabi, C. *Crowd sensing of traffic anomalies based on human mobility and social media*. City, 2013.

Song, C., Koren, T., Wang, P. and Barabási, A.L. Modelling the scaling properties of human mobility. *Nature Physics*, 6, 10 (2010), 818–823.

Song, C., Qu, Z., Blumm, N. and Barabasi, A. Limits of Predictability in Human Mobility. *Science*, 327, 5968 (2010), 1018–1021.

Toole, J.L., Colak, S., Sturt, B., Alexander, L.P., Evsukoff, A. and Gonzalez, M.C. The path most traveled: Travel demand estimation using big data resources. *Transportation Research Part C-emerging Technologies* (2015).

Treiber, M., Kesting, A. and Thiemann, C. *Traffic flow dynamics: data, models and simulation*. Springer, 2013.

Wang, G.F., Song, P.F. and Zhang, Y.L. Review on Development Status and Future of Intelligent Transportation System. *Highway* (2012).

Wang, L., Chen, H. and Li, Y. Transition characteristic analysis of traffic evolution process for urban traffic network. *The Scientific World Journal*, 2014 (2014).

Advances in Energy Science and Equipment Engineering II – Zhou, Patty & Chen (Eds)
© 2017 Taylor & Francis Group, London, ISBN 978-1-138-71798-5

Detection of malware based on apps' behavior

Shaoming Chen & Yiyang Wang
Guangdong Branch of National Computer Network Emergency Response Technical Team/Coordination Center of China, China

Yajun Du
School of Mathematics and Computer Engineering, Xihua University, China

ABSTRACT: With the increasing number of Android devices, there is simultaneous increase in malware apps that perform various activities behind the scene, such as misusing sensitive information of the users and signing up victims to subscription services. Hence, malware identification is a critical issue. To solve this problem, a new, hybrid approach is proposed in this paper. The method analyzes the potential relevance of the apps' requesting permission history and used history. Thus, we can identify malwares and inform users about the risk of apps before installation. An experiment illustrates that our method can effectively identify malicious apps and protect the user's information security.

1 INTRODUCTION

With the rapid development of communication technology, mobile phone has become a fundamental necessity, and is closely related to people's lives. About 84% of all phones, worldwide, are Android devices, the majority of which are unprotected. Given Android's prominence and general phone vulnerability, it is not surprising that the majority malicious mobile attacks are designed for the Android mobile operating system (Paul, 2016). Android is an open-source project, open to developers with too many interfaces, which also provides a chance for some bad developers. At the same time, the user can install the application through a variety of channels, resulting in a large number of applications with malicious code flooding in the network, causing damage to the user. Malware can be detected statically or dynamically. If a malware is detected without actually running or executing, it is called static analysis. If a malware is detected by executing and understanding its behaviors, the technique is called dynamic analysis. In this paper, we present a method called BIA for batch analysis of application's initial permissions and the applying permissions when used.

2 RELATED WORK

We discuss the detection of malware techniques shortly, which will be used for comparison, and indicate their advantages and disadvantages.

Suleiman Y. et al. have presented a novel approach to alleviate the problem of detecting malware based on Bayesian classification models obtained from static code analysis. The models are built from a collection of code and app characteristics that provide indicators of potential malicious activities. The detection strategy leverages the applications' reliance on the platform APIs and their structured packaging to extract certain properties. These properties then form the basis for Bayesian classifier, which is used to determine whether a given Android app is harmless or suspicious. In order to obtain the feature sets for building the Bayesian model, a Java-based Android package profiling tool for automated reverse engineering of the APK files is implemented (Shaikh, 2015). In practice, it is difficult to reverse app and is time-consuming too.

Bayer et al. presented a method where binary is run in open-source PC emulator Qemu to monitor its security-relevant activities by analyzing Windows native call or API call. It did not change binary to prevent detection by malware and used hooks and breakpoints implanted in relevant API and native libraries (Bayer, 2006). The method destroys the integrity of the operating system or malicious program, and it can be used for testing the integrity of software or malicious programs.

Hamandi K. et al. demonstrated the working of SMS malware and proposed their solution. They studied and identified various solutions to prevent this kind of malware. They suggested the practical solutions such as the user must always be notified of the receipt of the message as certain SMS malwares hide notifications. Apart from this, application must also be prevented from receiving certain SMSs by setting its own priorities, and the user must grant explicit permissions for every SMS, that is, permissions must be modified during runtime (Hamandi, 2013).

Willems et al. implemented CWSandbox using hook function for API and system calls (Willems, 2007). It captures behavior with respect to file system, registry manipulation, network communications, and OS interaction. It implements hook by replacing the first five bytes of API with noncondition jump to monitor function and also captures passed parameters for analysis (Choudhary, 2015).

Takayuki et al. built a system to assess and present the risk level of an Android app to the user at the time of installing the application (Takayuki, 2012).

FlowDroid is a static taint analysis system that is fully based on Android context and objects. FlowDroid uses configuration files and decompiles an Android app in order to construct an inter-procedural control-flow graph, which is used to find potential privacy leaks that are either caused by carelessness or malicious intention (Arzt, 2014).

This paper is organized as follows. In Section 3 we describe the proposed method of detection of malware. In Section 4, the experimental results by the proposed method are presented. Section 5 contains conclusions and further directions for research.

3 DETECTION OF MALWARE BASED ON APPS' BEHAVIOR (BIA)

Android operating system is designed on the basis of Linux kernel and is developed by Google. Android has a layered architecture, including the Linux kernel layer, middle layer, and application layer, which can provide consistent services for the upper layer, masking the differences of the current layer and lower layer. Android security model is similar to Linux, as it is designed on Linux kernel. The main part of Android security model is permission mechanism, which limits applications to access user's private data (i.e., telephone numbers and contacts), resources (i.e., log files), and system interface (i.e., Internet and GPS). In permission mechanism, the phone's resources are organized by different categories, and each category corresponds to one kind of accessed resource (Pooja, 2016).

Our method is inspired by Takayuki (2012) and illustrated as follows:

1. In a class of applications, we regard the permissions set as permission set P.
2. Given a class of applications in the safe app store, at the time of installing the applications, we consider the frequency of each permission in the permission set P the original degree O:

$$O(p_i) = c/n \qquad (1)$$

where c is the frequency of permission p_i and n is the total number of applications.

3. Given a class of applications, when we consider the frequency of each permission in the permission set P, the support degree S is:

$$S(p_i) = c/n \qquad (2)$$

where c is the frequency of permission p_i and n is the total number of applications.

4. The total weight of permissions is the sum of the original degree and support degree. However, the two degrees are of unequal importance. For example, the significance of the support degree is greater than the original degree. Different weights must be assigned to each degree. We defined the weights as below: $W_1 = a$, $W_2 = a(1-a)$, where a accords with $0 < a < 1$. W_1 is assigned to the original degree. W_2 is assigned to the support degree. The total weight of permission is indicated as follows:

$$W(p_i) = O(p_i)*w_2 + S(p_i)*w_1 \qquad (3)$$

5. The permissions are ranked by their total weight, and the top-rank permissions are selected to be compared with the permissions, which a new application required before installation. Thus, we can detect whether the application is malicious and let the user decide whether to install or not.

Figure 1 shows the structure of the method proposed in this paper.

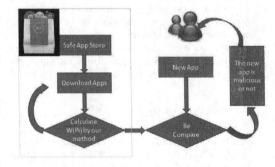

Figure 1. Structure of the method.

1654

4 VALIDATION

For this experiment, the details of the procedures followed in this experiment can be summarized in the following steps:

1. We choose 10 dangerous application permissions, which can be easily used by malicious programs. They are permissions such as CHANGE_NETWORK_STATE, DELETE_CACHE_FILE, INTERNET, READ_CONTACTS, SEND_SMS, READ_SMS, WRITE_EXTERNAL_STORAGE, ACCESS_FINE_LOCATION, CALL_PHONE, READ_OWNER_DATA.

2. We choose 10 apps of music from a secure and reliable application store. These apps are Kugou music, QQ music, Netease music, Xiami music, Baidu music, Kuwo music, Duomi music, Migu music, Love music, and Love 4G. We use "P_1," "P_2," "P_3," "P_4," "P_5," "P_6," "P_7," "P_8," "P_9," and "P_{10}" represent CHANGE_NETWORK_STATE, DELETE_CACHE_FILE, INTERNET, READ_CONTACTS, SEND_SMS, READ_SMS, WRITE_EXTERNAL_STORAGE, ACCESS_FINE_LOCATION, CALL_PHONE, READ_OWNER_DATA, respectively, and "*" represents the application to apply for the permission. We get the permissions when they are installed and are shown in Table 1.

   We calculate original degree $O(p_i)$: $O(p_1) = 10/10 = 1$, $O(p_2) = 1/10 = 0.1$, $O(p_3) = 10/10 = 1$, $O(p_4) = 7/10 = 0.7$, $O(p_5) = 6/10 = 0.6$, $O(p_6) = 4/10 = 0.4$, $O(p_7) = 2/10 = 0.2$, $O(p_8) = 0/10 = 0$, $O(p_9) = 1/10 = 0.1$, $O(p_{10}) = 3/10 = 0.3$.

3. We obtain the required permissions of 10 applications when they are running during 1 week and are shown in Table 2.

   We calculate original degrees (p_i): $S(p_1) = 10/10 = 1$, $S(p_2) = 1/10 = 0.1$, $S(p_3) = 10/10 = 1$, $S(p_4) = 5/10 = 0.5$, $S(p_5) = 5/10 = 0.5$, $S(p_6) = 4/10 = 0.4$, $S(p_7) = 2/10 = 0.2$, $S(p_8) = 0/10 = 0$, $S(p_9) = 1/10 = 0.1$, $S(p_{10}) = 2/10 = 0.2$.

4. We calculate the total weight of permissions based on $W(p_i)$. We set a as 0.7. They are $W(p_1) = O(p_1)*a(1-a) + S(p_1)*a = 0.910$, $W(p_2) = 0.091$, $W(p_3) = 0.910$, $W(p_4) = 0.497$, $W(p_5) = 0.476$, $W(p_6) = 0.364$, $W(p_7) = 0.182$, $W(p_8) = 0$, $W(p_9) = 0.091$, $W(p_{10}) = 0.182$.

5. The permissions are ranked by their total weight, and the top-rank permissions are selected to be compared with the permissions, which a new application required before installation. If a new application's required permissions are included in the ranked permissions set and is in the top n position (we set n as 8), we think this new application is safe and reliable. Otherwise, we believe that the application is risky, that is,

a malicious program. We choose 10 categories of applications including music, 10 applications for each category. Comparing the methods of Bayer (Bayer, 2006), Willems (Willems, 2007), and our approach, the experiment proved that the method can be effective to detect malicious programs. The result is shown in Tables 3, 4, and Figure 2.

Table 1. Permissions of music apps during installation.

App	P_1	P_2	P_3	P_4	P_5	P_6	P_7	P_8	P_9	P_{10}
Kugou	*		*	*						
QQ	*		*	*		*				
Netease	*		*	*						
Xiami	*		*	*						*
Baidu	*		*		*				*	*
Kuwo	*		*	*	*		*			
Duomi	*	*	*	*	*		*			
Migu	*		*	*	*	*				*
Love	*		*		*	*				
Love 4G	*		*		*	*				
Total	10	1	10	7	6	4	2	0	1	3

Table 2. Permissions of music apps when be used.

App	P_1	P_2	P_3	P_4	P_5	P_6	P_7	P_8	P_9	P_{10}
Kugou	*		*	*						
QQ	*		*			*				
Netease	*		*	*						
Xiami	*		*							*
Baidu	*		*		*				*	*
Kuwo	*		*	*	*		*			
Duomi	*	*	*	*	*		*			
Migu	*		*	*		*				
Love	*		*		*	*				
Love 4G	*		*		*	*				
Total	10	1	10	5	5	4	2	0	1	2

Table 3. Detection of malware from 10 apps of music.

New app	Our method	In fact
Poweramp	Safe	Safe
Kugou	Safe	Safe
QQ	Malicious	Safe
Spotify	Safe	Safe
Guitar	Safe	Safe
Violin	Safe	Safe
Tianlai	Safe	Safe
DJ	Safe	Safe
Yoyo	Malicious	Malicious
Joox	Safe	Safe
Accuracy rate	90%	

Table 4.   Detection of malware from 10 classes of apps.

Categories	By Bayer	By Willems	By ours
Music	80%	80%	90%
Social	100%	100%	100%
Office	70%	60%	80%
News	90%	90%	90%
Shooting	80%	90%	90%
Financial	90%	80%	100%
Medical	80%	80%	80%
Game	60%	70%	70%
Sports	70%	80%	90%
Shopping	80%	80%	80%

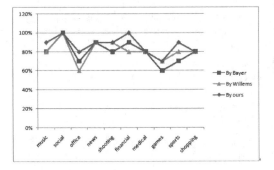

Figure 2.   Experimental result.

5   CONCLUSIONS

In this paper, a method called BIA for batch analysis of apps' initial permission and the applying permission when used was presented. In the method, we showed that it is possible to benefit from the analysis of permission. Experimental result is encouraging. This paper contributes to the improvement of detection of malware in several ways. First, we selected the apps from safe store and extracted the permissions, which the app required and used. These permissions can accurately confirm the security program. Second, it demonstrated that the usage records of a normal user can greatly improve detection of malware.

Further research will concentrate on the optimization of the parameters a, n, and the automation of our approach.

ACKNOWLEDGMENTS

This work was one of the projects supported by the National Nature Science Foundation (Grant Nos. 61271413, 61472329, and 61532009).

REFERENCES

Arzt, S., S. Rasthofer, C. Fritz et al., Flowdroid: Precise context, flow, field, object-sensitive and lifecycle-aware taint analysis for android apps, SIGPLAN Not. 49 (6) (2014) 259–269.

Bayer, U., A. Moser, C. Krugel and E. Kirda, Dynamic Analysis of Malicious Code, Journal in Computer Virology, vol. 2(1), pp. 67–77, (2006).

Choudhary, S.P., and Miss Deepti Vidyarthi, A Simple Method for Detection of Metamorphic Malware using Dynamic Analysis and Text Mining, Procedia Computer Science 54 (2015) 265–270.

Google, Android Home Page, 2009. (http://www. android. com).

Google, Android Security and Permissions, 2013. (http://d.android.com/guide/topics /security.html).

Hamandi, K. et al., Android SMS Malware: Vulnerability and Mitigation, Advanced Information Networking and Applications Workshops (WAINA), 2013 27th International Conference, vol., no., pp.1004, 1009, 25–28 March 2013.

Paul McNeil et al., SCREDENT: Scalable Real-time Anomalies Detection and Notification of Targeted Malware in Mobile Devices, Procedia Computer Science 83(2016) 1219–1225.

Pooja Singh, Pankaj et al., Analysis of Malicious Behavior of Android Apps, Procedia Computer Science 79 (2016) 215–220.

Shaikh Bushra Almin and Madhumita Chatterjee, A Novel Approach to Detect Android Malware, Procedia Computer Science 45(2015) 407–417.

Takayuki Matsudo et al., A Proposal of Security Advisory System at the Time of the installation of Applications on Android OS, 15th IEEE International Conference on Network-Based Information System (NBiS), 2012.

Willems, C., T. Holz and F. Freiling, Toward Automated Dynamic Malware Analysis using CWSandbox, IEEE Security and Privacy, vol. 5(2), pp. 32–39, (2007).

Advances in Energy Science and Equipment Engineering II – Zhou, Patty & Chen (Eds)
© 2017 Taylor & Francis Group, London, ISBN 978-1-138-71798-5

Analysis of the queuing mechanism of the VANET communication nodes

Yatie Xiao & Luqun Li
Shanghai Normal University, Shanghai, Xuhui District, China

ABSTRACT: The VANET is a huge dynamic communication network composed of vehicles and road-side infrastructures with communication function. As parts of the VANET, the communication node message queuing quality directly affects the performance of the entire VANET. On the basis of the analysis of the VANET communication node characteristics, this paper introduces the inverse number of communication nodes message queue and quantifies the quality of message queues. According to the quantified parameters and results, it could analyze the communication message queues of the VANET contrapuntally and provide reliable advices for optimizing the communication performance of the VANET.

1 INTRODUCTION

VANET performance analysis has become a hot spot of ad hoc research in recent years. The communication queuing mechanisms of the VANET, which is good or not directly related to the network transmission delay, will have a significant impact on the performance of the VANET communication. Because the VANET is based on ad hoc network, and the ad hoc connection model is not unique, the analysis of the VANET communication nodes message queuing models is very significant, which should take varieties of the message queuing model and its scope of application into account, for example,

I. M/M/1 queuing model. In this queue, there is only one service node and M queuing nodes. Those customer nodes should obey the queuing rules provided by the service node.

II. M/M/C queuing model. There are C service nodes in this queue model, M queuing customer nodes could select different service nodes to transfer data (queuing capacity K is not considered here), actually selecting different service nodes to transfer message means message transmission mechanism is different.

III. M/M/∞ queuing model. In this queue model, the number of service nodes is unlimited. This situation does not exist in the VANET in nature, which will not be discussed in this paper.

Typically, there are about three types of message queuing states in the VANET, which are shown in Figure 1.

The first queuing state indicates a single service node queuing model. In this model, there is only one service node, and all the customer nodes must

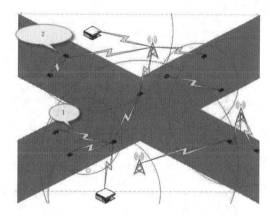

Figure 1. Different types of message queuing states of communication nodes in the VANET.

obey the queuing mechanism. The service node provides message transmission service once a time; the second state indicates the multiservice queuing model of the VANET. Customer communication nodes select different service nodes to communicate with by using some appropriate mechanism in this model. There are no effects between service nodes when providing services. Besides, there is a simple state that a node does not make any data transmission in the VANET, which will not be discussed in this paper.

Therefore, during the analysis message queuing process of the communication node, how to express the VANET node message queuing model and how to calculate the inverse number of communication nodes in different message queuing model, all of these reasonably require careful consideration. These factors not only affect analysis of the VANET nodes message queuing mechanism, but also have an important impact on the message queuing optimization.

2 DEFINITION

The VANET consists of vehicles and roadside infrastructures with communication functions in a region, based on the analysis of node message queuing model. This paper introduces parameters P and λ, which are message queuing parameters needed in the VANET node message queuing analysis process. P is expressed as a node message queuing priority in the communication queuing process; the smaller the value of P, the higher the priority level it has. The parameter λ is defined as the quality of the entire queuing process at a certain time; according to its definition, the larger the value of λ, the poorer the quality of the queuing model.

In general, the VANET is defined in a region of M*M (km²), assuming the connection number of communication nodes (vehicles and infrastructures, which can transfer data with others) is N in the region at a certain time, each communication node maximum open signal transmission radius is R_i (usually R_i is set to about 250 m), the connected state between communication nodes V_i and V_j is defined as C_{ij}, and the value of C_{ij} is defined as 1 or 0 (1 shows connected, which means nodes can transfer data with each other; 0 indicates disconnected state). The communication node connected states in the entire VANET is shown in Figure 2.

Communication nodes in the red Box have more than two adjacent communication nodes, which are all connected. According to the queuing theory, there are many types of queuing situation in the VANET.

It should be noted that the prerequisite of communication nodes in the VANET queuing situation established is the communication distance D_{ij} between nodes being less than the minimum open-signal transmission radius R of the connected nodes, and the service order is First-In-First-Out (FIFO). On the basis of the above content, it can obtain the relevant queuing formula and the prerequisite conditions for the establishment of the definition above.

3 ANALYSIS

It is easy to know, as the communication node of the VANET can communicate freely with other nodes, that mean nodes signal transmission radius R can meet the requirements of communication freely. In this paper, message queuing model is simplified as the connection state model, and the service order is FIFO, so a connection state of a node V_i may exist in the following situations: I at a certain time; a communication node V_i of the VANET does not participate in the data transmission process with any node, which means that at that time, this node is not in a communication queue.

Therefore, it can be seen that the connection state of the node C = 0 and the inverse number of queue is $\lambda = 0$. This model of queuing is shown in Figure 3.

II at a time: The arrival of nodes is in accordance with Poisson flow arrival. If node V_i only performs data transmission with anther node V_j, and node V_j connected to node V_i also connects with other communication nodes, there are data transmissions between these nodes, so the message queuing behavior of the situation above appears. According to the queuing theory, the message queuing model in this situation conforms to the M/M/1 model.

The queuing process of communication nodes with a single service node is shown in Figure 4, which displays communication nodes waiting for data transmission at a time in a message queue.

Suppose node V_i connects with another node C, and node C makes connection with other M nodes,

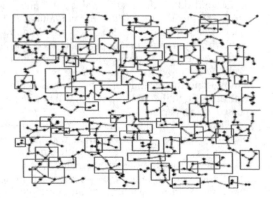

Figure 2. Communication nodes' queuing situation in a certain moment in the region.

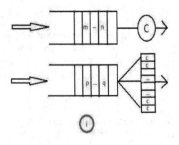

Figure 3. Node V_i does not join communication queues, meaning that it does not connect with any node, although there are many service nodes that could provide communication service.

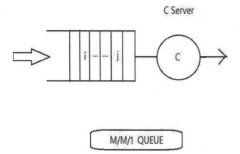

Figure 4. A node entering a communication queue with only one server. The message queue has more than one node waiting for data transmission service.

according to Poisson input stream and single service queuing model, the connection probability of node V_i with node C, p_c, and the data transmission probability p_t between two nodes are 100%. There are about $N-1$ nodes in the front of the node V_i in this message queue, and those $N-1$ nodes need different service times $T_k (1 \le k \le N-1)$, which is shown in Figure 5.

When node V_i joins the queue, before it receive message service, it should wait for the total time T, that is, $T = T_1 + T_2 + \cdots + T_k + \cdots + T_{N-1} = \sum_{k=1}^{N-1} T_k (1 \le k \le N-1)$, because node V_i joins the queue with probability 100%, so the weighted service time $T_w = 100\% * T = T = \sum_{k=1}^{N-1} T_k$. Based on the above content, the weighted inverse number λ of the communication queue established is

$$\lambda = \sum_{i=N}^{1} \sum_{j=i-1}^{1} a_{ij} \quad \begin{pmatrix} a_{ij}=0 & P_i - P_j \ge 0 \\ a_{ij}=1 & P_i - P_j < 0 \end{pmatrix} \quad (1)$$

In equation (1), a_{ij} is the inverse state of node V_i and node V_j queuing priority. A large λ indicates that the message queue of communication between nodes is not reasonable, which may cause a big communication time delay and influence data transmission services. Therefore, it should adjust nodes' message transmission order relatively by some mechanism.

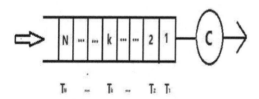

Figure 5. N nodes waiting for packet service in a queue, with each node requiring different time T_k.

III at a moment, the communication node V_i transfer data with one of plurality of communication service nodes, and there are several queuing behaviors between nodes. This situation mentioned above is in line with the multiline servers M/M/C model. Multiserver queuing process model is shown in Figure 6.

In the multiservice node queues, suppose that there are m communication service nodes. If node V_i joins a communication queue k ($k \le m$) with probability p_k, and in the queue k, there are N_k nodes waiting for packet service, then it can get the expression $\sum_{k=1}^{m} p_k = 1$. Therefore, in all queues, each queue needs service time $T_k (1 \le k \le m)$, which is shown in Figure 7.

According to the total service time in a queue, equation (2) is obtained as:

$$T_k = T_{1_k} + T_{2_k} + \cdots + T_{i_k} + \cdots + T_{N_k} = \sum_{i=1}^{N_k} T_{i_k} (1 \le i \le N_k) \quad (2)$$

where T_k is the total time of service needed in queue k. It is clear that higher of the value of T_k, the more time node has to wait.

Besides, an inverse number λ_k exists in the queue k, according to equation (1), which can be obtained as:

$$\lambda_k = \sum_{i=N_k}^{1} \sum_{j=i-1}^{1} a_{ij} \quad \begin{pmatrix} a_{ij}=0 & P_i - P_j \ge 0 \\ a_{ij}=1 & P_i - P_j < 0 \end{pmatrix} \quad (3)$$

In equation (3), N_k is the nodes number of waiting for service in the queue k, when node V_i selects queue k and joins the queue k, the inverse number of the queue k will be rebuild.

Suppose in this process, for a node, the probability of selecting queues p is different, there are m

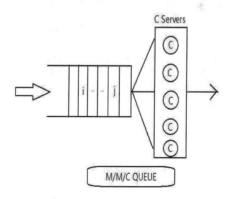

Figure 6. Communication nodes with a certain probability p waiting for data transmission with C servers, located in different communication queues.

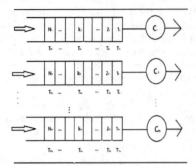

Figure 7. Different queues with difference in waiting time T.

queues. If the probability of selecting queue k $(k \le m)$ is p_k, then it can get a weighted inverse number λ_{kw} and a weighted service time T_{kw} of queue k in equation set (4).

$$\lambda_{kw} = p_k \sum_{i=N_k}^{1} \sum_{j=i-1}^{1} a_{ij} \begin{pmatrix} a_{ij}=0 & P_i - P_j \ge 0 \\ a_{ij}=1 & P_i - P_j < 0 \end{pmatrix}$$

$$T_{kw} = p_k \sum_{i=1}^{N_k} T_{i_k} (1 \le i \le N_k) \tag{4}$$

According to the definition given above, node V_i should select a queue with the smallest weighted inverse number λ_{min} and T_{min}, which is shown in equation set (5):

$$\lambda_{min} = min \left\{ \lambda_{kw} = p_k * \sum_{i=N_k}^{1} \sum_{j=i-1}^{1} a_{ij} \right.$$
$$\left. \begin{pmatrix} a_{ij}=0 & P_i - P_j \ge 0 \\ a_{ij}=1 & P_i - P_j < 0 \end{pmatrix} k = 1,2,3...m \right\}$$

$$T_{min} = min \left\{ T_{kw} = p_k * \sum_{i=1}^{N_k} T_{i_k} (1 \le i \le N_k) \right.$$
$$\left. k = 1,2,3...m \right\} \tag{5}$$

where λ_{min} is the smallest weighted inverse number of queues in the VANET. It depends on a_{ij} and p_i, T_{min} is the smallest weighted service time of queues, which depends on nodes service time T, equation (5) gives reliable specific research directions for optimization of the message queues in the VANET.

4 CONCLUSION

On the basis of the analysis of the VANET node queuing model, this paper introduces the inverse number of queuing to quantify the quality of nodes queuing. By analyzing the specific quantitative value of the inverse number in communication queues, it can provide a basis for optimizing the VANET communication.

Next, based on the weighted inverse number of VANET nodes queue, we use the appropriate sorting algorithm to adjust the order of queuing nodes in the queue, at the same time taking the service time into adjusting the queue order. It reduces the inverse number overall, so it can optimize the VANET communication queues and improve entire nodes packet transmission efficiency of the VANET.

REFERENCES

Borodin A, Kleinberg J, Raghavan P, et al. Adversarial queuing theory [J]. Journal of the Acm, 2001, 48(1):13–38.
Dantzig G.B, Harvey R.P, Mcknight R.D. Updating the Product Form of the Inverse for the Revised Simplex Method. [J]. Journal of the Acm, 1965, 12(4):603.
Dousse O, Thiran P, Hasler M. Connectivity in ad-hoc and hybrid networks [J]. IEEE Infocom, 2002, 2:1079–1088 vol.2.
Kafsi M, Papadimitratos P, Dousse O, et al. VANET Connectivity Analysis [J]. Computer Science, 2009, abs/0912.5527.
Stewart W.J. Probability, Markov chains, queues, and simulation. The mathematical basis of performance modeling [J]. Princeton Univers Press, 2009: xviii+758.

Advances in Energy Science and Equipment Engineering II – Zhou, Patty & Chen (Eds)
© 2017 Taylor & Francis Group, London, ISBN 978-1-138-71798-5

Development and application of antiterrorism command and control system

Peng Sun, Huihui Zhang & Fei Liu
Hengde Digital Choreography Technology Co. Ltd., Qingdao, China

ABSTRACT: Rapid development of technology, while fear of violent doctrine sources, has become more complex, as violent terrorists attack in diverse ways. In recent years, international terrorism and violent extremist act frequently, causing not only serious casualties and property loss but also long-term psychological impact of terror to the people. Antiterrorism command and control system, by means of a wide range of network system, receives a precise positioning of riot police with the use of wearable terminal and sends real-time audio and video data to the command platform, so that commanders can be effective and rationally allocate police officers in order to achieve fast and accurate effects against violence elements and minimize its harmful consequences.

1 INTRODUCTION

International terrorist violence is a prominent factor affecting the activities of world security and stability (Chen, 2015). In recent years, extreme violent elements are highly active in many countries and regions, including China, which have suffered varying degrees of violence threats and terrorism harm.

With the advent of transportation, communication, and other developments of technology and the coming Internet community, personnel mobility and rapid dissemination of information have been significantly improved. At the same time, violent terrorist attacks plan and contact more stealthily and violent extremists flow easier, greatly increasing the difficulty of counterterrorism riot. Increased focus on counterterrorism riot work in technology and the establishment of global interdepartmental antiterrorist riot scientific work are imperative (Cao, 2014). Several countries and regions in the world use a variety of information and data involving transportation, shopping, talk, video, chat, Web browsing, e-mail address, and other Internet technologies. The use of big data analysis can be in forms of effective intelligence gathering, investigation and early warning, and detection to improve antiterrorism riot capacity and reduce the cost of the same. Antiterrorism command and control system should be established.

2 SYSTEM OVERVIEW

Antiterrorism command and control system is the integration of hardware and software systems of network-based precise positioning information and real-time audio and video data communications technology. Antiterrorism antiriot personnel collect live audio and video information using wearable terminal. While dispatching officers can use antiterrorism antiriot personnel's wearable terminal to determine the actual positions of the police officers at the scene, combining with real-time return of live audio and video information makes a reasonable and accurate command and control, to quickly and effectively combat violence fear molecules, greatly enhance the ability of on-site processing, and reduce casualties and property losses.

The system is mainly composed of a network base station, audio and video servers, location servers, network switches, integrated dispatching platform, and wearable terminals. The main network base station constructs a local area network as information data transmission vector, which may be existing Wi-Fi network, 3G/4G communication network, and a temporary base station onboard. Audio and video server is a device used for real-time data signal analysis, processing, compression, and transmission, from which we can achieve grouping cluster intercom of one-to-one, one-to-many, many-to-one, or many-to-many and quick return of live video. The location server can obtain accurate real-time location information of each officer, with the network transmitting it to dispatching integrated platform, simultaneously integrated dispatching platform sends live audio, video, and voice commands to any police officer or real wearable terminal. Every antiterrorism & riot police is equipped with a unique ID wearable terminal, recording the details of police officers and equipment.

3 NETWORK POSITIONING

Global Positioning System (GPS) is the most common way to get location information outdoor. In recent years, with the rapid development of wireless networks and mobile communications technology, Assisted Global Positioning System (A-GPS), combining GPS and cellular networks, has been gradually applied in various emergency and Location-Based Services (LBS) (Schulzrinne, 2003). However, because of the susceptibility of satellite signal to various obstructions, GPS/APGS and other satellite positioning technologies are not suitable for indoor occasions. The emergence and development of network technology makes indoor positioning a strong complement to the GPS.

3.1 Positioning architecture

The positioning system architecture is shown in Figure 1. Wearable terminal with restrictions like computing power, storage capacity, and battery capacity simply complete the signal acquisition work and submit signal characteristics to the server side, so the positioning calculation must be performed by the positioning server. On the basis of load-balancing considerations, the network server in response to a location request and the location server operating position calculation are separated, which are connected in the same local area network, responsible for locating and positioning calculation in response to a request from wearable terminal. Local area network can be an existing public Wi-Fi, 3G/4G mobile networks, as well as a vehicle network base station providing temporary network.

3.2 Position calculation

Wearable terminal in the system, by means of the network, submits signal strength feature vector to the server. The network server conveys the received data to the location server in the same way, which

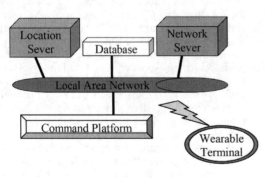

Figure 1. Positioning architecture.

then queries database based on existing reference point and related positioning calculation to obtain location information of the wearable terminal.

Whether 3G/4G mobile communications network or Wi-Fi network, both the signal indoors have transmission loss. Signal strength sent by antiterrorism & riot personnel wearable terminal to the Web server is given by the following equation:

$$Pr(d) = Pr(d_0) - 10nlg(d/d_0) + x \qquad (1)$$

where d is the distance between the transmitting end and the receiving end (m), d_0 is the reference distance, which is generally 1 m, $Pr(d)$ is the signal power received by network server, $Pr(d_0)$ is the signal power received from the reference point d_0, X is a Gaussian random variable with an average of 0, reflecting changes of received signal power at a certain distance, and n is the path loss exponent, which is a value related to the environment, combined with practical experience to take four (Fan, 2013). The network server sends the received data to the location server. On the basis of the obtained database and signal strength feature vectors, the location server determines the coordinates of the three known nodes as (x_1, y_1), (x_2, y_2), and (x_3, y_3), the distance between them and the desire node are d_1, d_2, and d_3. Assuming the desire position node coordinates (x, y), the distance formula is:

$$\begin{cases} (x-x_1)^2 + (y-y_1)^2 = d_1^2 \\ (x-x_2)^2 + (y-y_2)^2 = d_2^2 \\ (x-x_3)^2 + (y-y_3)^2 = d_3^2 \end{cases} \qquad (2)$$

According to the least-squares method, the desired node coordinate vector is:

$$X = (A^T A)^{-1} A^T \qquad (3)$$

where

$$A = \begin{bmatrix} 2(x_1 - x_3) & 2(y_1 - y_3) \\ 2(x_2 - x_3) & 2(y_2 - y_3) \end{bmatrix} \qquad (4)$$

The physical distance between the desire node and other three known nodes is measured. Circles are drawn with the three known nodes as center and the measured three physical distances as radius. In theory, these three circles and the desire node should intersect, that is, the intersection of the three circles is the position of the desire node. In fact, in the indoor environment, the signal undergoes a sharp decline after encountering an obstacle, coupled with errors caused by measurement tools and other reasons. Therefore, the general received signal strength values become smaller.

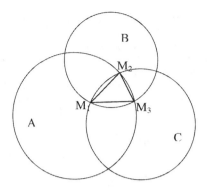

Figure 2. Circumferential triangle centroid diagram.

Reflected on estimated distance between the desire node and known node, the results are shown in Figure 2. The three circles (A, B, and C) intersect in three points (M_1, M_2, and M_3) to form a triangle. The desire node position should be in the common intersection triangle area, while centroid of the triangle is usually taken as the required estimate of node location.

Figure 2 does not reflect the size of influence from the known nodes to the desire node, which affects the positioning accuracy, so by the following weighting factor to compensate the influence level:

$$X = (x_1/(d_1 + d_2) + x_2/(d_2 + d_3) + x_3/(d_3 + d_1))/ (1/(d_1 + d_2) + 1/(d_2 + d_3) + 1/(d_3 + d_1)) \quad (5)$$

$$Y = (y_1/(d_1 + d_2) + y_2/(d_2 + d_3) + y_3/(d_3 + d_1))/ (1/(d_1 + d_2) + 1/(d_2 + d_3) + 1/(d_3 + d_1)) \quad (6)$$

The positioning accuracy can be improved plus the weights based on the above, whereby the desire node position coordinates are:

$$Q(x,y) = \frac{\sum_{i=1}^{3}(w_i, p_i(x,y))}{\sum_{i=1}^{3} w_i} \quad (7)$$

where $P_i(x, y)$ represents the coordinates of known nodes, $w_i = d_{max}/(d_i)^g$, g is a parameter, which can be adjusted according to the actual environment.

4 REAL-TIME DATA TRANSMISSION

Dispatching in terrorism riot scene focuses on the effectiveness of audio and video information, but the security and confidentiality of the information has no expectations. Therefore, the Real-time Transport Protocol (RTP) can meet the communication requirements of the system. RTP is an application layer transport protocol designed by the IETF (Internet engineering task force) for real-time audio and video transmission, which is located above the User Datagram Protocol (UDP), functionally independent of the underlying transport and network layers (Jiang, 2009). However it cannot exist as a separate layer, generally used the low-layer UDP for real-time audio and video multicast or unicast, enabling multipoint or single-point transmissions of audio and video data.

RTP, built on the connection-oriented underlying protocol, can be created for the connectionless underlying protocol as well. It is independent with respect to the transport layer, and generally consists of two components: Real-time Transport Protocol (RTP) and Real-time Transport Control Protocol (RTCP) (Li, 2005). RTP packets also consist of two parts: a header and a payload; the header format is shown in Table 1.

V: RTP version number, accounting for two, the current protocol version 2.

P: filling mark, accounting for one, if P = 1, then the end of the packet will be filled with one or more additional octets, they are not part of the payload.

X: extension flag, accounting for one, if X = 1, then the RTP header is followed by an extension header.

CC: CSRC counter, accounting for four, indicating the number of CSRC identifiers.

M: mark, accountings for one, different payloads have different meanings, for video, mark the end of one; for audio, mark the beginning of the session.

PT: payload type, accounting for seven, for indicating the type of RTP packet payload, such as GSM audio and JPEM images.

SN: serial number, accounting for seven, to identify RTP packet serial number by the sender, with each sending a message, the serial number increased. The receiver detects packet loss by the serial number, then reorder packets to recover the data.

TS: time stamp, accounts for 32, the time stamp reflects the time of the first sample of the octet RTP packets. The receiver uses TS to calculate the delay and jitter and run synchronization control.

Synchronization Source (SSRC) Identifier: accounting for 32, used to identify the synchronization signal source. The identifier is randomly

Table 1. RTP packet format.

Bit No.						
0–1	2	3	4–7	8	9–15	16–31
V	P	X	CC	M	PT	SN
TS						
Synchronization Source Identifier						
Contributing Source Identifier						

selected, synchronizing signal, two participate in the same video conference cannot have the same SSRC.

Contributing Source (CSRC) Identifier: each CSRC identifier representing 32, from 0 to 15. Each special CSRC identifies all sources included in the RTP packet payload (Jenkac, 2006).

From Table 1, it is known that RTP packet does not contain a "length" field. RTP data are segmented by the underlying UDP and then organized into a number of UDP packets for transmission. RTP is a protocol for user to carry data with real-time characteristics, with the RTCP as supporting, which is the message sent by the receiver, responsible for monitoring the quality of network service, communication bandwidth, and information transmission over the Internet, and send this information to the sending end. The protocol structure is shown in Table 2.

RC/SC: receiver/sender reports count. The number of report blocks contained in the packet, the effective value is zero.

Length: the size of the RTCP packet (32 words minus 1), including the header and any gap (introducing offset such that 0 become valid values, and avoiding complex RTCP packet scanning process infinite loop phenomenon, and the use of 32-bit word count method is to avoid the check on the validity of a multiple of 4).

RTCP packet has five types: Send Report (SR), Receive Report (RR), source description item (SDES), BYE, and application (APP), as shown in Table 3.

The continuity of video and audio streams on the timeline requires real-time transmission of high-bandwidth networks, while allowing certain error rate of data transmission and data loss. As video and audio on the timeline are not very rel-evant, the real-time data have higher reliability. With the use of UDP over RTP/RTCP for audio and video data encapsulation, packaging, and synchronization, digital network audio and video signal transmission can make delay to a minimum. Receiver and sender of the system is in the same local area network, so there is sufficient bandwidth, the real-time audio and video data transmission and transmission quality can have very good results by RTP/RTCP.

5 EXPERIMENT AND APPLICATION

Experiments in a business hotel, whose rooms are distributed in six floors (see the floor plan in Figure 3).

Five rooms (609, 623, 630, 643, and 648) are randomly selected as the location experimental points. Testers are free to choose their action trajectory in the hotel, and occur within a predetermined test room time to time. The recorder observes and records the number of testers accurately appearing in a preset room according to the system integrated platform, repeatedly tests 100 times. Scan cycle of the system is 1 s and the number of scans is 120. Data collection took about 1 h. To verify the accuracy of positioning, use the mean of multiple scanning as positioning signal feature, and the scan cycle is 2 s. The system displays the screen shown in Figure 4.

The statistical results of each room are shown in Figure 5. The positioning accuracy of each random test points is high.

After testing in the physical environment, the accuracy of positioning experiments is within 5 m. Field information, the venue environment around testers, can be easily obtained by audio communication and live video exchange. Therefore, the system can be easily and efficiently applied in shopping malls, hospitals, leisure hotels, railway

Table 2.   RTCP packet format.

Bit No.				
0–1	2	3–7	8–15	16–31
V	P	RC/SC	PT	length
TS				
Synchronization Source Identifier				
Contributing Source Identifier				

Table 3.   RTCP packet type.

Type	Meaning
200	SR (sender report)
201	RR (receiver report)
202	SDES (source descriptions items)
203	BYE
204	APP (application)

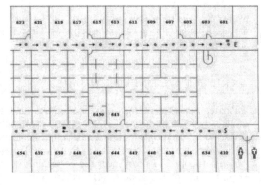

Figure 3.   Business hotel floor plan.

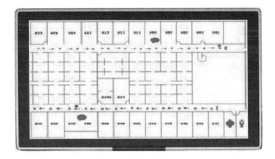

Figure 4. Comprehensive platform display.

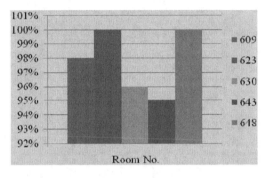

Figure 5. Each test position point positioning result.

stations, airports, and other public areas of counterterrorism and antiriot scene venues integrated scheduling.

REFERENCES

Chen, B., Y.Q. Jiang, *Defense technology industry Scientific Development Forum*, 35(2015).

Cao, K., C.Y. Li, B. Wang, Machine China, **2**, 46(2014).

Schulzrinne, H., S. Casner, *A Transport Protocol for Real-Time Applications*, 50(2003).

Fan, C.H., Z.J. Ji, Ordnance Industry Automation, **6**, 301(2013).

Jiang, J.G., Z.P. Su, Y. Li, ACTA ELECTRONICA SINICA, **9**, 67(2009).

Li, Y.L., R.F. Ma, L. Zuo, MICROELECTRONICS & COMPUTER, **8**, 22(2005).

Jenkac, H., T. Stockhammer, *The Official Publication of the European Association for Signal Processing* (EURASIP), 86(2006).

Advances in Energy Science and Equipment Engineering II – Zhou, Patty & Chen (Eds)
© 2017 Taylor & Francis Group, London, ISBN 978-1-138-71798-5

Fast-SVM applied to large scale wireless sensor networks localization

Fang Zhu

School of Computer and Communication Engineering, Northeastern University at Qinhuangdao, Hebei, Qinhuangdao, China

Junfang Wei

School of Resource and Material, Northeastern University at Qinhuangdao, Hebei, Qinhuangdao, China

ABSTRACT: Sensor node localization is one of research hotspots in the applications of Wireless Sensor Networks (WSNs) field. In recent years, many scholars proposed some localization algorithms based on machine learning, especially Support Vector Machine (SVM). Localization algorithms based on SVM have good performance without pairwise distance measurements and special assisting devices. But if detection area is too wide and the scale of wireless sensor network is too large, the each sensor node needs to be classified many times to locate by SVMs, and the location time is too long. It is not suitable for the places of high real-time requirements. To solve this problem, a localization algorithm based on Fast-SVM for large scale WSNs is proposed in this paper. The proposed Fast-SVM constructs the minimum spanning by introducing the similarity measure and divided the support vectors into groups according to the maximum similarity in feature space. Each group support vectors is replaced by linear combination of "determinant factor" and "adjusting factor". The "determinant factor" and the "adjusting factor" are decided by similarity. Because the support vectors are simplified by the Fast-SVM, the speed of classification is evidently improved. Through the simulations, the performance of localization based on Fast-SVM is evaluated. The results prove that the localization time is reduce about 48% than existing localization algorithm based on SVM, and loss of the localization precision is very small.

1 INTRODUCTION

Wireless sensor network is a multi hop self-organizing wireless communication network system by deploying a large number of cheap micro sensor nodes in the monitoring region. Its purpose is to perceive, gather and process the information detected by the tiny sensors in the coverage area, and send the useful information to the observers and control centers (Liu, 2009). Localization is the key to many WSN applications, such as network management, environmental monitoring, target tracking and routing. WSN node localization technology has gained wide attention and become one research hotpot of WSN.

There is a lot of work done in the field of localization in WSNs. WSN node localization can be divided into two major categories. One is the localization scheme in the centralized manner, such as Semidefinite Programming (YS, 2015) and Multidimensional Scaling (Morral, 2015). The other category is the localization scheme in the distributed manner, such as DV-Hop (GJ, 2016), MDS-MAP (CY, 2015) and MDL (YKim, 2009). The localization algorithm in distributed manner has lower computation and communication costs, so it is suitable for large-scale wireless sensor networks.

Support Vector Machine (SVM) is a machine learning methods, evolving from the statistical learning theory. SVM has high generalization ability and good classification precision. Moreover, SVM can solve the overfitting problem effectively. So SVM has been applied to a number of fields, such as classification and regression fields. There has been some proposed localization algorithm based on SVM in (Pan, 2006; Dai, 2014; Kim, 2014; Safa, 2014; Keji, 2014). But if detection area is too wide and the scale of wireless sensor network is too large, the each sensor node needs to be classified many times by a large number of SVM to locate itself, so the location time is too long. This problem will lead to localization algorithm impractical in the fields of high real-time requirements. In this paper, we will try to improve classification speed of SVM to meet the requirement of real-time location.

2 LOCALIZATION BASED ON SVM

It is assumed that N wireless sensor network nodes {S1,S2,S3,…,SN} are deployed in a 2D geographic area $[0,D]2(D > 0)$, and the communication range R of each node is the same. Two nodes can communicate with each other if no signal blocking entity exists between them and their geographic distance is less than their communication range. Two nodes are said to be "reachable" from each other if there exists a one-hop or multi-hop path between them. There are k < N nodes{Si}(i = 1→k) with known their own location that are called beacon nodes in the wireless sensor network. These beacon nodes are reachable from each other. The remaining nodes{Sj}(j = k + 1→N) can reach to the beacon nodes.

The 2D area is divided into k cells. Each dimension is divided into M parts. Therefore, the area covered by the network is divided into M × M square cells for 2D area. Each cell constitutes the label for training data. Intuitively, x-coordinate has $M = 2^m$ classes cx_i, and the y-coordinate has $M = 2^m$ classes cy_i. The model assumes that each node exists in class [cx_i, cy_i] in 2D area. In other words, respectively each node exists in [(i-1) D/M, iD/M] × [(j-1)D/M, jD/M] unit. Simply, we assign the center of each unit with (x_i, y_i) and we assign the center of estimated unit as the predicted position. If the above prediction is indeed correct, the location error for each node is at most $D/(M\sqrt{2})$. The connectivity information would be gathered by beacons from the network and would be sent to the head beacon where the SVM algorithm is running. The SVM method will build a model by training. This built model would be broadcasted to all the nodes even they are beacons or not. Then each non beacon node uses this model to compute its cell in which it resides.

Let $(x(S_i), y(S_i))$ denote the true (to be found) coordinates of node S_i's location, and $h(S_i, S_j)$ the hop-count length of the shortest path between nodes S_i and S_j. Each node S_i is represented by a vector $S_i = <h(S_i, S_1), h(S_i, S_2), h(S_i, S_k)>$. The training data for SVM is the set of beacons{S_i} (i = 1→k). We define the kernel function as a Radial Basis Function because of its empirical effectiveness.

$$K(S_i, S_j) = e^{-\gamma\|s_I - s_J\|_2^2} \qquad (1)$$

Obviously, localization algorithm based on SVM described above is a multi-class problem, but SVM can only solve tow-class problem. So, how to solve multi-class problem is the key to localization algorithm. In this paper, decision tree is used to solve the multi-class problem.

Take X-dimension for example, X-dimension is divided into M parts, so x-coordinate has M–1 classes cx_i {cx_1, cx_2, ..., cx_{M-1}}. Each class cx_i contains nodes with the x-coordinate greater or equal to iD/M. According to the decision-tree strategy, M–1 binary SVM need to be trained, and localization of each node needs $\log_2 M$ binary

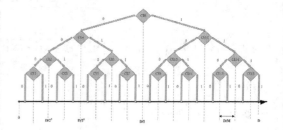

Figure 1. Decision tree used for X-dimension classification.

classifications. For the Y-dimension, the processing method is the same as the X-dimension. The Figure 1 shows the situation that X-dimension is divided into 16 parts.

3 FAST CLASSIFICATION METHOD FOR SVM (FAST-SVM)

The basic theory of SVM shows that the decision time is proportional to the number of support vectors. The basic principle of fast classification method is searching for a reduction vector set which contains less support vector and replaces the original support vectors. The method can be expressed as the following:

$$f(\mathbf{x}) = \text{sgn}(\sum_{i=1}^{n} \alpha_i^* y_i K(\mathbf{x}_i, \mathbf{x}) + b^*)$$
$$= \text{sgn}(\sum_{i=1}^{m} \alpha_i^* y_i K(\mathbf{x}_i, \mathbf{x}) + b^*) \quad m \ll n \qquad (2)$$

According to above principle, this paper proposed a fast classification method based on similarity analysis in feature space. The basic idea is that support vectors are divided into several groups by analyzing the similarity of support vectors in feature space. Support vectors have largest similarity in each group. Then the "determinant factor" and the "adjusting factor" are found in each group by some rules. Finally, the linear combination of "determinant factor" and "adjusting factor" is used to fit the weighted sums of support vectors in feature space.

$$\sum_{i=1}^{n} \alpha_i^* y_i \varphi(\mathbf{x}_i) = \sum_{i=1}^{m} \sum_{j=1}^{v_m} \alpha_{ij}^* y_{ij} \varphi(\mathbf{x}_{ij})$$
$$\approx \sum_{i=1}^{m} (\beta_{i1}\varphi(\mathbf{x}_{i\max}) + \beta_{i2}\varphi(\mathbf{x}_{i\min})) \qquad (3)$$

Where n is the number of support vectors, m is the number of groups, v_i is the number of support

1668

vectors in group i, and $\mathbf{x}_{i\max}$ is the determinant factor, $\mathbf{x}_{i\min}$ is the adjusting factor.

Equation (2) is taken into equation (3, and the decision function of support vector machine can be expressed as the following:

$$f(\mathbf{x}) = \text{sgn}(\sum_{i=1}^{m}(\beta_{i1}K(\mathbf{x}_{i\max},\mathbf{x}) + \beta_{i2}K(\mathbf{x}_{i\min},\mathbf{x})) + b^*)$$

(4)

3.1 The similarity coefficient

The cosine between vectors has rotation-scale invariance, so it can well reflect the characterizations of geometric distributions of the support vectors in feature space. This paper selects the cosine value of angle between support vectors in high-dimensional space as the standard of similarity measure:

$$C_{ij} = \cos(\varphi(x_i),\varphi(x_j))$$
$$= \frac{\varphi(x_i)\cdot\varphi(x_j)}{\sqrt{(\varphi(x_i)\cdot\varphi(x_i))(\varphi(x_j)\cdot\varphi(x_j))}}$$
$$= \frac{K(x_i,x_j)}{\sqrt{K(x_i,x_i)K(x_j,x_j)}}$$

(5)

When $C_{ij} = 1$, the angle is $0°$, it shows that two vectors is very similar to each other; when $C_{ij} = 0$, the angle is $90°$, it shows that two vector are orthogonal and no correlation. The equation (5) shows that the original space is certainly mapped into a finite-dimension feature space where the inner product can be expressed by the kernel function to avoid the restrictions and speculation of the unknown mapping function $\varphi(\mathbf{x})$. The similarity coefficient constructs the similarity matrix C that is the data base of the proposed method. Where C is a symmetric matrix.

$$\mathbf{C} = \begin{bmatrix} C_{11}^2 & C_{12}^2 & \cdots & C_{1n}^2 \\ C_{21}^2 & C_{22}^2 & \cdots & C_{2n}^2 \\ \vdots & \vdots & & \vdots \\ C_{ij}^2 & C_{n2}^2 & \cdots & C_{nn}^2 \end{bmatrix} C_{ij}^2$$

(6)

3.2 Grouping support vectors based on the minimum spanning tree

First, a weighted acyclic connected graph G(V, E) is generated in the feature space. The vertices of the graph G are images of support vectors in the feature space, $V = \{\varphi(x_i), \varphi(x_j)|i \neq j\}$. 1- C_{ij}^2 is the weight of edge, $E = \{1-c_{ij}^2|i \neq j\}$. Second, the support vectors are grouped according to similarity. The similarity of support vectors is the largest in each group. And the similarity of support vectors

is the least between groups. For this purpose, minimum spanning tree is introduced. A minimum spanning tree Tr is generated from the connected graph G. In the Tr, each edge is constructed by two vertexs which similarity is largest. After the minimum spanning tree Tr is obtained, vertex and edge is grouped by the "long edge" of Tr.

The "long edge" is defined as following:

Definition 1: XY is an edge of the spanning tree. If the length of XY is obviously greater than the average length of edges which starting point is the X and Y, XY is defined as long edge.

According to the definition of the long edge, the group rule is that subtrees at both ends of the "long edge" are grouped into different groups. In the figure 2, the support vectors A, B, C, D are grouped into group1, and Y, E, G, F are grouped into group2. Under this group rule, the scale of subtrees is the smallest and the similarity is the largest between the connected nodes in subtrees.

3.3 The particular factors and the related coefficient

In order to reduce the support vectors, determinant factor and adjusting factor need to be decided in each group. The selection of these particular factors in groups is the key to the algorithm, so the selection rules are described as following.

1. Make sure that each determinant factor has the largest similarity among the support vectors in its own group.
2. Make sure that each adjusting factor has the greater adjustable range.

Based on the above rules, the support vector with the approximately largest similarity in each group is selected as the determinant factor, and the support vector with the approximately smallest similarity in each group is selected as the adjusting factor.

The related coefficients of the particular factors need to be obtained to fit the weighted sums of support vectors in each group. The weighting coefficients β_{i1} and β_{i2} of group i can be got by solving the Programming Problems as following:

$$\min f(\beta_{i1},\beta_{i2})$$
$$= \min\left\|\sum_{j=1}^{vi}\alpha_{ij}^* y_{ij}\varphi(x_{ij}) - (\beta_{i1}\varphi(x_{i\max} + \beta_{i2}\varphi(x_{i\min}))\right\|^2$$
$$= \min(\beta_{i1}^2 k(x_{i\max},x_{i\max}) + \beta_{i2}^2 k(x_{i\min},x_{i\min})$$
$$+ 2\beta_{i1}\beta_{i2}k(x_{i\max},x_{i\min})$$
$$- 2\beta_{i1}\sum_{j=1}^{vi}k(x_j,x_{i\max}) - 2\beta_{i2}\sum_{j=1}^{vi}k(x_j,x_{i\min})$$
$$+ \sum_{j=1}^{vi}\sum_{k=1}^{vi}k(x_j,x_k))$$

(7)

4 SIMULATION

We conducted the simulations on a network of 1000 sensor nodes randomly distributed in the 100 m × 100 m 2D area. In the network, the beacon nodes are selected at random. Three different beacon populations were considered: 20 percent of the WSN size (n = 200 beacon nodes), 25 percent (n = 250 beacon nodes), 30 percent (n = 300 beacon nodes). The communication range is 10 m. The x-dimension and y-dimension is divided into 128 parts (M = 128). All simulation experiment is done on a PC machine with CPU-Intel Core2 Duo E7500, 2 G memory. The simulation platform is matlab 7.0 and libsvm software is used for SVM classification. The kernel of SVM is RBF. Fast classification method proposed in section 3 is used to preprocess the support vector set before localization.

4.1 Fast-SVM versus SVM with different beacon population

We compared localization algorithm based on Fast-SVM proposed in the section 3 and based on SVM at different beacon population of 20%, 25% and 30%. The nodes distribution is shown as figure 2, figure 3, and figure 4.

The result is shown in table 1. The results suggest that localization accuracy by Fast-SVM is lower than SVM, but not significant. Since the support vectors of each SVM are reduced by Fast-SVM, accuracy of classification is decreased. However, Fast-SVM method possibly reduces the impact of interference to classificatio.

From the table 1, we can see that the average localization Error by the Fast-SVM is higher about 0.1 m than by the SVM. The localization error E is described as follow:

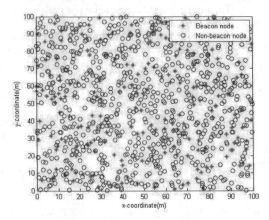

Figure 2. Nodes distribution at beacon population of 20%.

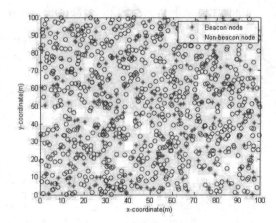

Figure 3. Nodes distribution at beacon population of 25%.

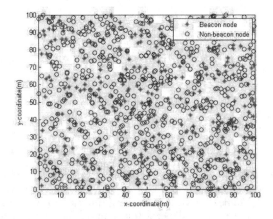

Figure 4. Nodes distribution at beacon population of 30%.

Table 1. Average localization error comparison of Fast-SVM versus SVM.

Beacon Population	Average localization error of SVM (m)	Average localization error of Fast-SVM (m)
20%	3.21	3.33
25%	2.0	2.11
30%	1.22	1.29

Table 2. Average localization time comparison of Fast-SVM versus SVM.

Beacon Population	Average localization time of SVM (s)	Average localization time of Fast-SVM (s)
20%	1.22	0.62
25%	1.28	0.67
30%	1.35	0.73

$$E = \frac{\sum_{i=1}^{k} \sqrt{(\hat{x}_i - x_i)^2 + (\hat{y}_i - y_i)^2}}{k} \qquad (8)$$

Where $(\hat{x}_i, \hat{y}_i)$ is the estimated coordinate, (x_i, y_i) is the true coordinate, k is the number of unknown nodes.

Table 2 shows the simulation result of Fast-SVM versus SVM localization time consumption. When the beacon population is 20%, 25% and 30%, the Fast-SVM localization time is about 0.62 s, 0.67 s and 0.73 s, the SVM localization time is about 1.22 s, 1.28 s and 1.35 s. The simulation results show that the total localization time is reduced about 48%.

ACKNOWLEDGMENTS

This work is supported by the National Natural Science Foundation of China under Grant No. 61401083 and the Natural Science Foundation for Young Scholars of Hebei Province of China under Grant No. F2014501139.

REFERENCES

CY, Z., X. T, D. HR. Distributed Locating Algorithm MDS-MAP (LF) Based on Low-Frequency Signal. Computer Science and Information Systems, **12**,1289 (2015).

GJ, H., Z. CY, L. TQ, S. L. A multi-anchor nodes collaborative localization algorithm for underwater acoustic sensor networks. Wireless Communications and Mobile Computing, **16**,682 (2016).

H, L., L. YKim, et al. Grouping Multi-Duolateration Localization Using PartialSpace Information for Indoor Wireless Sensor Networks. IEEE Trans. on Consumer Electronics, **55**, 1950 (2009).

J, X., Q.H, Dai H, et al. Wireless Sensor Network Localization Based On A Mobile Beacon And Tsvm. Cybernetics & Information Technologies, **14**, 98 (2014).

Keji, M., F. Congling, Y. Fei, W. Peng, Ch. Qingzhang. Node localization algorithm in wireless sensor networks based on SVM. Computer Research and Development, **51**, 2427–2436 (2014).

Kim W, Park J, Kim HJ, et al. A multi-class classification approach for target localization in wireless sensor networks. Journal of Mechanical Science & Technology, **28**, 323 (2014).

Liu, H., S. Xiong, Q. Chen. Localization in wireless sensor network based on multi-class support vector machines. 5th International Conference on Wireless Communications, Networking and Mobile Computing 2009, 4 (2009).

Morral, G, Bianchi, P. Distributed on-line multidimensional scaling for self-localization in wireless sensor networks. Signal Processing, **120**, 88 (2015).

Pan J.J, Kwok J. T, Chen Y. Multidimensional Vector Regression for Accurate and Low-Cost Location Estimation in Pervasive Computing. IEEE Trans. on Knowledge and Data Engineering, **18**, 1181 (2006).

Safa H. A novel localization algorithm for large scale wireless sensor networks. Computer Communications, **45**, 32 (2014).

YS, Y., W. HY, S. XH, H. K, Z. XH. TDOA-Based Source Collaborative Localization via Semidefinite Relaxation in Sensor Networks. International Journal of Distributed Sensor Networks, 1 (2015).

Advances in Energy Science and Equipment Engineering II – Zhou, Patty & Chen (Eds)
© 2017 Taylor & Francis Group, London, ISBN 978-1-138-71798-5

Research status and prospect of computer rooms in China

Wei Liu
The College of Post and Telecommunication of Wuhan Institute of Technology, Wuhan, China

Liang He
Beijing New Building Materials Public Limited Company, Beijing, China

ABSTRACT: With the development of computer technology and application in different fields, the green construction and management promotion have become a hot research topic. In this paper, we overviewed the research of the computer room in construction, management, and environment and predicted the future research direction of the computer room. Cloud computer room, cloud desktop technology, student participation in the management, and human comfort promotion will be the research focus.

1 INTRODUCTION

With the development of social information, computer applications have spread to all sectors of society as a general tool. In order to meet the requirements of teaching, examination, and computing, the scale and level of computer room escalated in schools, scientific research institutes, and enterprises. The construction and management of computer room experience in accordance with the relevant standards of construction and management with adequate professional teachers is on demand with the construction and management of part-time teachers, trending to cloud computer room. In the construction, the lack of systematic design concept, availability, scalability, and equipment redundancy are the problems. In the management, complexity of user, lack of management personnel, and establishment of imperfect system result in management problems. In the actual environment, more people consider using a variety of monitoring tools to ensure room stability and comfort promotion of room, such as electromagnetic radiation protection, VOC release, and others.

2 CONSTRUCTION OF A COMPUTER ROOM

With the advent of the information age, the computer room of school provides an important learning environment for future information talent, whose construction and application level will be reflected in the school of modern teaching environment and research strength. Detailed studies have been carried out on the room of the building by scholars in China, particularly on the green reconstruction of

computer room and cloud computer room because of the lack of systematic design concept, availability, scalability, and equipment redundancy of the room, as well as the increasing demand of computer room in elementary, middle, and high schools.

Zhanzhen Wei (Wei, 2012) introduced standardized network center room of the building from hard and soft conditions, respectively, including the room location, size, floor, walls, ceiling, windows, dust, noise, static electricity, wiring and health, temperature, humidity, lighting, air-conditioning, fire, and other environmental factors. Xuefei Xu (Xu, 2014) introduced the concerns on the planning stage from location, indoor environmental balance, reasonable cabling, and security measures. Jiling Zheng (Zheng, 2013) pointed out the need to consider the decoration material, equipment, mine design, and power supply during the construction based on actual experience of their computer room. Also, the integrated wiring, network access points, and heat dissipation need to lay the foundation for future expansion.

Ji Zhang (Zhang, 2011) overviewed the Chongqing Liang jiang New Area Soil and Water international cloud-based Pacific Telecommunications cloud computer room project and pointed out the need for careful planning system requirements, design objectives and constraints, the main building, preliminary design, and construction design at research phase after determining the design standards. The power and HVAC systems are the issues that need to be resolved, while the server will be powered off at any time with no backup power supply system, and the CPU of high-density cabinet will increase by 10°C/min with no reliable HVAC system support.

In order to solve the problem of energy consumption, Mao Lin (Li, 2013) conducted a detailed analysis

Table 1. Effect of water-cooling and air-cooling.

Item	Water-cooling	Air-cooling
Safety	High (requires a variety of media)	Low (only needs to add freon)
Adaptability in high cold area	Need to adjust the ratio of antifreeze	No special maintenance
Operating cost	High	Low
Cost of equipment	High (needs to add cold water piping and control panel)	Low (no additional device)
Cooling effect	Good	Good
Maturity of technique	Large range of options	Small range of options

Figure 1. Effect of equipment layout to air circulation.

Table 2. Energy consumption before and after reconstruction.

	Before reconstruction	After reconstruction
Total power of IT equipment	180 kW	150 kW
Total power of air-conditioning	182 kW	68 kW
Energy consumption of UPS	38.4 kW	33 kW
Energy consumption of illumination	12.59 kW	12.59 kW
Energy consumption of power system	14.8 kW	14.8 kW
Total power	427.79 kW	278.39 kW
Calculate ability	4.3 trillion calculations/s	11.5 trillion calculations/s
PUE	2.37	1.85
EER of IT equipment	238.89 billion/degree	766.67 billion/degree

on water-cooling and air-cooling and the proposed green reconstruction and considered the actual needs of the room (Table 1 and Figure 1), implementation of on-demand cloud room, and rational and efficient layout equipment. The PUE value of domestic room reduced from 2.37 to 1.85, close to that of the international advanced room through the reconstruction, compared with the energy consumption before and after reconstruction in Table 2.

3 MANAGEMENT MODEL

Computer room as a school-based public teaching base, bears basic courses, such as the school computer literacy, C language programs, and many comprehensive works, such as computer network technology, applied technology, student exams, online enrollment, teacher curriculum, school information, social workers, level examinations, with characteristics of complex user, mobility, wide range coverage, lack of systematic design, and so on. Therefore, room management is of great significance as an important part of computer teaching management. At present, China's computer room management is difficult, with less professional management and maintenance. Therefore, computer teachers were exploring the joint management model of teacher and student, developing automatic room management systems and cloud-based management technology.

For the computer room management, scholars (Ma, 2014; Huang, 2012; Fang, 2013; Liu, 2011; Ding, 2011; Yang, 2013) proposed the views of management and maintenance in accordance with the study of their own school's computer room and formed consensus on management system by creating hardware and software maintenance, sanitizing, and reducing the workload of teachers. Yingjie Lan (Lan, 2014) conducted a detailed analysis of traditional room management whose shortcomings include trivial work, heavy workload, and lack of cooperation between full-time and part-time teachers. The management and maintenance need adopt the joint mode of teacher and student so as to create a good learning environment, while improving students' ability to maintain computer and competitiveness, and

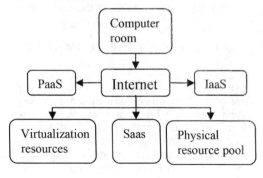

Figure 2. Management structure of computer room-based cloud computing.

to liberate time of full-time teachers for teaching reform and research work. Huaijin Liu also believed that room management had personnel shortage and management problems. There was a need to encourage students who had preferably computer skills room management. Songjiang Wang (Wang, 2012) achieved automatic room management system after an in-depth study of the system startup and exit, network communications and control, process hiding and protection, data encryption and protection technologies using Microsoft Visual Basic 6.0 as a programming tool and MS SQL Server 2000 as the database management system. Yuanyuan Wu (Wu, 2015) analyzed the current problems in the management of university computer room and designed the cloud desktop room management system. It proved the advantages of improving teaching security, stability, software installation, and quicker update through the application. Jian Huang (Huang, 2012) also used cloud technology in management, and the application results proved advantages of security, convenience, simplicity, and economical (Figure 2).

4 PROMOTION OF ENVIRONMENT

The adjustment of room environment is achieved by various monitoring systems, including temperature and humidity monitoring, water leakage monitoring, dynamic environment monitoring, screen monitoring, and smoke monitoring. The monitoring system of domestic room was mainly used to reduce the frequency of failure, troubleshoot staff time, reduce energy consumption, and ensure safe operation of the equipment room. Xu Zhou (Zhou, 2013) used "decentralized collection, centralized management" mode to monitor the temperature and humidity, power quality, switches, UPS, leaking, and precision air-conditioning systems. The results showed that it could effectively adjust the room temperature and humidity, give rapid response alarm, fully guarantee the life of the equipment, and improve management efficiency (Figure 3).

The unattended remote monitoring system designed by Dacong Jiang (Jiang, 2014) that monitors the physical environment and operating parameters of power, air-conditioning, servers, switches, and high- and low-voltage power equipment could reduce the probability of equipment failure and maintenance costs. The simulation results of central and top of the cabinet by Luping Zhu (Zhu, 2014) showed that hot spots exist in the far corner inside cabinet. It needs to increase the number of outlet or reduce local server within the apparatus to control the temperature (Figures 4–7).

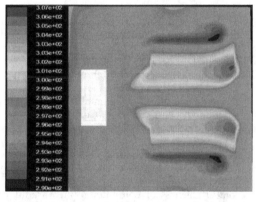

Figure 4. Temperature distribution of the cabinet center.

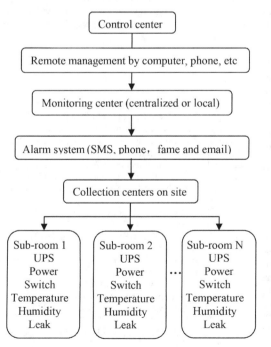

Figure 3. Topology of monitoring system.

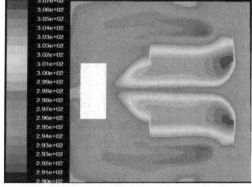

Figure 5. Temperature distribution of the cabinet top.

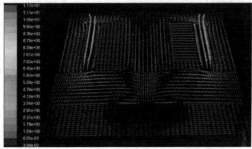

Figure 6.   Airflow vector of floor.

Figure 7.   Airflow vector of cabinet top.

Table 3.   Score of evaluation.

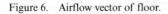

Evaluation	Score (point)						
Parameters	Grade	5	4	3	2	1	0
Temperature (°C)	F_T	24~26	20~24	16~20	12~16	8~12	≤ 8
			26~28	28~30	30~32	32~34	≥ 34
Relative humidity (%)	F_{RH}	50~60	40~50	30~40	20~30	10~20	≤ 10
			60~65	65~70	70~75	75~80	≥ 80
Oxygen (%)	F_{O2}	≥ 20	≥ 19	≥ 18	≥ 15	≥ 13	< 13
Carbon dioxide (%)	F_{CO2}	≤ 0.07	≤ 0.1	≤ 0.15	≤ 0.2	≤ 0.4	> 0.4

The monitoring system needs to detect environmental comfort parameters for people, thereby ensuring the safe operation of equipment. Yun Ou (Ou, 2013) used SCM technology and sensor technology to monitor temperature, humidity, CO_2, O_2, and other core parameters inside the room and evaluated the environmental comfort by a five-point mode, and the score can seen in Table 3. With the development of information education, elementary-, middle-, and high-school students start using computer room to learn, while the environment has a significant impact on growth stage. Therefore, it is urgent to carry out research on the comfortableness of computer room. In addition, the accumulation of electromagnetic radiation from a large number of computers and VOC emission from decoration materials within the room will have some impact on students. Therefore, these hot issues related to the indoor environment also need to be included in the room management.

5   CONCLUSIONS

The room requirements are bound to become more sophisticated with the development of society and computer deepening to all walks of life, and the complexity of user, lack of systematic design, and part-time teacher management issues will become more prominent. Therefore, building scalable, easy-to-update computer room and establishing management and maintenance system are urgent

to solve this problem. With the rapid increase of elementary-, middle-, and high-school computer rooms and increasing attention on the indoor environment, the electromagnetic radiation pollution and VOC emission will also be included in the room management. Therefore, cloud computer room, new management tools and system, and environmental comfort will be the future research directions.

REFERENCES

Ding, H.R., Pioneer. sci. & technol. mon., 13 (2011).
Fang, S.L., S.P Chen. Changsha univ. J. 27, 5 (2013).
Huang, A., Pop. sci. & techonol. 14, 5 (2012).
Huang, J., Comput. Program. skill & maint., 14 (2012).
Jiang, D.C., Netw. Secur. techonol. & appl., 10 (2014).
Lan, Y.J., Netw. Secur. techonol. & appl., 5 (2014).
Li, M., J. Zou, S.J Jin. Inf. techonol., 2 (2013).
Liu, H.J., C.B Zhou, W.T Zhen. Bull. sport sci. technol. 19, 5 (2011).
Ma, S.H., Netw. Secur. techonol. & appl., 10 (2014).
Ou, Y., Lab. sci. 16, 3 (2013).
Wang, S.J., Softw. eng. (UESTC, ChenDu, 2012).
Wei, Z.Z., Value eng., 22 (2012).
Wu, Y.Y., Jilin teach. inst. eng. technol. J. 31, 10 (2015).
Xu, X.F., Liaoning high. vocat. J. 16, 2 (2014).
Yang, J., Guide sci. & educ., 9 (2013).
Zhang, J., Chongqing archit. 10, 98 (2011).
Zheng, J.L., Comput. knowl. techonol. 9, 33 (2013).
Zhou, X., Internet th. technol., 11 (2013).
Zhu, L.P., C.G Mao, J.D Zhang. Equip. geophys. Prospect. 24, 5 (2014).

Advances in Energy Science and Equipment Engineering II – Zhou, Patty & Chen (Eds)
© 2017 Taylor & Francis Group, London, ISBN 978-1-138-71798-5

A software tool for the generation of the typical meteorological year in different climates of China

Haixiang Zang, Miaomiao Wang, Zhinong Wei & Guoqiang Sun
College of Energy and Electrical Engineering, Hohai University, Nanjing, China

ABSTRACT: The correct selecting of Typical Meteorological Year (TMY) is an important factor for the field of solar energy and many other areas. This study presents the Finkelstein-Schafer (FS) statistical method for the generation of TMY. Also, a software tool is developed, relying on Matlab/GUI development tools, to generate TMYs for 35 cities covering six climatic types in China based on the latest and accurate recorded weather data (air dry-bulb temperature, relative humidity, wind velocity and global solar radiation). The data used are obtained from the China meteorological station in the period of 1994–2009.

1 INTRODUCTION

Solar radiation data are very important for the application of solar energy systems and always vary from year to year (Zang, 2012). So, there is a need to form a customized solar radiation database, which can represent the long-term averaged solar radiation over a year (Bulut, 2009). A representative file for year duration like Typical Meteorological Year (TMY) is usually applied for scientific studies and engineering applications in photovoltaic related fields (Jiang, 2010).

In the past, several methods have been developed to generate TMYs. Hall et al. (1978), Layi Fagbenle (1995), Petrakis et al. (1998), Gazela and Mathioulakis (2001), Kalogirou (2003), Bulut (2004), Sawaqed et al. (2005), Chan et al. (2006), Lhendup and Lhundup (2007), Rahman and Dewsbury (2007), Janjai and Deeyai (2009), Jiang (2010), Ebrahimpour (2010), Zang et al. (2012), Wong et al. (2012), Ohunakin et al. (2013) and Pusat et al. (2015) made efforts to generate TMYs for different locations around the world.

However, studies carried out in China are few, and the software tool for TMY generation in different climates of China is not developed. In the present study, the software tool is developed for obtaining the TMYs based on Matlab/GUI tool. The modified TMY method, proposed in our previous research (Zang, 2012), is recommended in this study.

2 DATA USED

China is a large country whose climate can be divided into six major types by applying the temperature

strip method (Zang, 2012), including Tropical Zone (>8000°C), Subtropical Zone (4500°C–8000°C), Warm Temperate Zone (3400°C–4500°C), Mid Temperate Zone (1600°C–3400°C), Cold Temperate Zone (<1600°C) and Tibetan Plateau Zone. In this study, measured weather data of 35 meteorological stations covering all over the climatic types in the period of 1994–2009 are selected to form TMY.

3 METHOD USED

In this paper, the modified TMY method used in our previous researches (Zang, 2012), based on the Finkelstein-Schafer (FS) statistical method, is suggested to form the TMY. A complete year (TMY) can be determined by combining 12 Typical Meteorological Months (TMMs) using the FS statistical method based on eight daily meteorological indices: maximum, minimum and mean air dry-bulb temperature (Max DBT, Min DBT, Mean DBT), minimum and mean relative humidity (Min RH, Mean RH), maximum and mean wind velocity (Max WV, Mean WV), and Daily Global Solar Radiation (DGSR).

Based on the variation between the long-term and yearly CDF in the month concerned, five candidate months that have CDF closest to the respective long-term distributions are selected from each month of the different years. In addition, the following function is firstly used to measure the variation by calculating the empirical CDF for each meteorological index:

$$S_n(x) = \begin{cases} 0 & for \quad x < x_1 \\ (i-0.5)/n & for\ x_i \le x < x_{i+1} \\ 1 & for \quad x \ge x_n \end{cases} \quad (1)$$

where $S_n(x)$ is the value of the CDF of index x. n is the total number of meteorological indices. Then, the values of FS statistics for each of the meteorological indices being considered are available by:

$$FS_x(y,m) = \frac{1}{N} \sum_{i=1}^{N} \left| CDF_m(x_i) - CDF_{y,m}(x_i) \right| \qquad (2)$$

where $FS_x(y,m)$ is FS statistic for the year y, month m. N is the number of daily readings of the month (e.g., $N = 31$ for March).

Next, a Weighted Sum (WS) of the FS statistics is derived by applying the weighting factors, WF_x, to the FS statistics values, corresponding to each specific month in the selected period:

$$WS(y,m) = \frac{1}{M} \sum_{x=1}^{M} WF_x * FS_x(y,m) \qquad (3)$$

where WF_x is the novel weighting factor (shown in Table 1). M is the number of meteorological indices considered (8 in this study). The five months with lowest WS values are chosen to be the candidate months.

A selection process is adopted for selecting the TMM from the five candidate months, starting with calculating the Root Mean Square Difference (RMSD) of global solar radiation (Jiang, 2010; Hall, 1978; Fagbenle, 1995; Petrakis, 1998; Gazela, 2001; Kalogirou, 2003; Bulut, 2004; Sawaqed, 2005; Chan, 2006; Lhendup, 2007; Rahman, 2007; Janjai, 2009; Ebrahimpour, 2010; Wong, 2012; Ohunakin, 2013; Pusat, 2015; Zang, 2012; Pissimanis, 1988):

$$RMSD = \left[\frac{\sum_{k=1}^{N}(H_{y,m,k} - H_{ma})^2}{N} \right]^{1/2} \qquad (4)$$

where $RMSD$ is the root mean square difference of global solar radiation. $H_{y,m,k}$ is the value of daily global solar radiation for the year y, month m and day k. H_{ma} is the long-term mean value of global solar radiation for the month m.

Finally, the month with the minimum RMSD is selected as the TMM, and a typical meteorological year is the combination of the 12 TMMs.

Table 1. Weighting factors for TMY method.

Max DBT	Min DBT	Mean DBT	Min RH	Mean RH	Max WV	Mean WV	DGSR
1/24	1/24	3/24	1/24	2/24	2/24	2/24	12/24

4 THE SOFTWARE TOOL FOR TMY GENERATION

The software tool generates TMY based on the measured data of relative humidity, air dry-bulb temperature, wind velocity, and global solar radiation in 35 cities. A series of menus provide a user with the appropriate function selections as well as a graphical result outlay, even the user is lack of profound computational knowledge.

The initial section of the program is an interface with 3 options, "About TMY", "Next" and "Exit". With the "About TMY" option, a user can read the definition and application of TMY. Moreover, this option allows the user to have a good understanding of the selection process of TMY presented above. The "Next" and "Exit" selections make it possible for one to decide the operation of the program.

The second section focuses on meteorological data collection and processing for inclusion in the TMY Generation. One climatic type among the 6 varied types and a city (part of the selected climatic type) can be determined by the user. With the "Display" option, the interface shows the data of eight daily meteorological indices of each individual month over the available years. A typical meteorological year is generated for Lhasa (which belongs to the Tibetan Plateau Zone) using this software tool, as a test run. See Lhasa data in Figure 1.

The TMM Generation is the third and fourth section, which are accomplished by employing the methodology presented in Part 2. With the "Next" option on the "Data collection and data processing" screen, the user employs the TMM selection procedure in two stages. First, filter months for each month in the hope that 5 candidate months can be selected, and then determine the TMM from the candidate months.

Within the frame of the five candidate month selection, the TMY generation routine

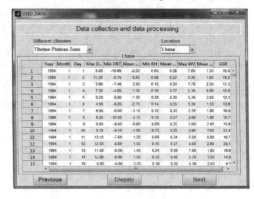

Figure 1. Interface of data collection and data processing.

incorporates the application of the FS method on the meteorological data. In this method, the FS statistic can be obtained by comparing the CDF for a meteorological index (e.g., Max DBT, January, 2009), month with the respective CDF for the long-term composite of all the years involved. The mean difference of the short-term and long-term CDF is the value of the FS statistic. Figure 2 shows the values of eight meteorological indices of January in 2009.

Next is the application of weighting factors, which are selected according to the existing experience on the influence of the meteorological indices utilized on the simulated application. See Table 1 for the actual distribution of the weighting factors in this study. The importance of the effect of the particular meteorological index, to which each one of them is assigned, on the behavior of a solar energy conversion system or building is indicated by the values of the weighting factors. It can be seen that the GSR gets the highest value of the weighting factor for its most far-reaching influence.

Accordingly, a weighed sum of the FS statistics is created, and this value is assigned to the respective month. The smaller the value is, the better the approximation to a TMM will be. All individual months are ranked in ascending order of WS values, and the selected candidate months are those 5 months which have the lowest values. With the "Display" option in Figure 3, the WS values of each month in the period of 1994–2009 will appear. What's more, the "Next" option makes it possible that a user can see the five candidate months intuitively shown in Figure 4. In addition, Figure 4 provides the information about the RMSD of the mean distribution of the Daily Global Solar Radiation (DGSR) for each year of each month, with respect to the mean long-term daily distribution. The month with minimum RMSD of DGSR among the 5 highest ranked months in ascending order of WS values is then selected for the TMM. A complete typical meteorological year is finally constructed in Figure 5 consisted of the 12 TMMs.

The final section of the program contributes to the graphical analysis of the results. The user

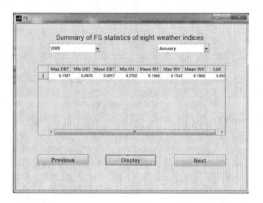

Figure 2. FS statistics of eight weather indices of each individual month in Lhasa.

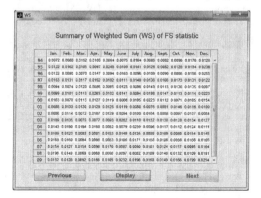

Figure 3. WS values of each month in the available years.

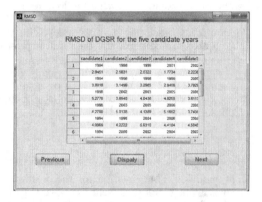

Figure 4. RMSD of the mean daily values of global solar radiation of the candidate months.

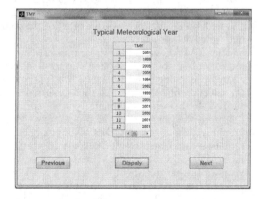

Figure 5. TMY constructed from selected months.

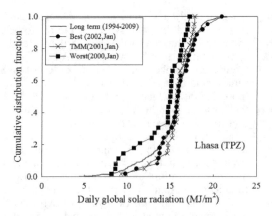

Figure 6. Comparison of long-term CDF with short-term CDF of daily global solar radiation in Lhasa.

can check whether the selected months of the TMY is adequate to represent the long-term sets of weather data using the "Next" option on the "Typical Meteorological Year" interface screen. For the concerned city, the comparison of the daily global solar radiation between the long-term and an individual month (January) is created with this option. Lhasa is selected as an example to illustrate the comparison patterns, and the result is plotted in Figure 6.

From Figure 6, it can be seen that daily global radiation CDF for January 2002 is most similar to the long-term CDF, while daily global radiation CDF for January 2000 is least similar.

5 CONCLUSIONS

The modified TMY method for the creation of a typical meteorological year is presented. The TMYs for 35 representative cities in China based on the latest measured data is generated via a developed software tool using Matlab/GUI. This tool allows the user to perform a task with ease, and without intimate knowledge of all of the computational details.

The "TMY Generation" program is a self-contained and flexible tool. It allows a user to choose a kind of climatic type and a city, which is part of the selected climatic type, and then determines the 12 TMMs to form a complete typical year. Furthermore, the final part of the program provides a graphical output of the CDFs of the global solar radiation for the entirety of at least 10 years, for the selected TMY, and for the best year and worst year, for further treatment.

From the application of the program for the generation of a TMY for the city-Lhasa, in the Tibetan Plateau Zone, it can be seen that the methodology implemented leads to acceptable results. What's more, the user interface is made up of a series of windows in which the user can make a decision to enter the relevant input data or review the results.

ACKNOWLEDGMENTS

The research is supported by National Natural Science Foundation of China the NSFC Program (Program No. 51507052), the China Postdoctoral Science Foundation (Program No. 2015M571653), and the Fundamental Research Funds for the Central Universities (Program No. 2015B02714). The authors also thank the China Meteorological Administration.

REFERENCES

Bulut, H. International Journal of Green Energy 6, 173 (2009).

Bulut, H. Renew. Energ. **29**, 1477 (2004).

Chan, A.L.S., T.T. Chow, S.K.F. Fong, J.Z. Lin, Energ. Convers. Manage. **47**, 87 (2006).

Ebrahimpour, M. Maerefat, Energ. Convers. Manage. **51**, 410 (2010).

Gazela, M., E. Mathioulakis, Sol. Energy. **70**, 339 (2001).

Hall, I.J., R.R. Prairie, H.E. Anderson, E.C. Boes, *the 1978 annual meeting of the American Society of the international solar energy society* (1978).

Janjai, S., P. Deeyai, Appl. Energ. **86**, 528 (2009).

Jiang, Y.N. Energy. **35**, 1946 (2010).

Kalogirou, S.A. Renew. Energ. **28**, 2317 (2003).

Layi R. Fagbenle, Energ. Convers. Manage. **36**, 61 (1995).

Lhendup, T., S. Lhundup, Energy. Sustain. Dev. **11**, 5 (2007).

Ohunakin, M.S. Adaramola, O.M. Oyewola, R.O. Fagbenle, Appl. Energ. **112**, 152 (2013).

Petrakis, M., H.D. Kambezidis, S. Lykoudis, A.D. Adamopoulos, P. Kassomenos, I.M. Michaelides, S.A. Kalogirou, G. Roditis, I. Chrysis, A. Hadjigianni, Renew. Energ. **13**, 381 (1998).

Pissimanis, D., G. Karras, V. Notaridou, K. Gavra, Sol. Energy. **40**, 405 (1988).

Pusat, S., İ. Ekmekçi, M.T. Akkoyunlu, Renew. Energ. **75**, 144 (2015).

Rahman, I.A., J. Dewsbury, Build. Environ. **42**, 3636 (2007).

Sawaqed, N.M., Y.H. Zurigat, H. Al-Hinai, Int. J. Energ. Res. **29**, 723 (2005).

Wong, S.L., K.K.W. Wan, D.H.W. Li, J.C. Lam, Build. Environ. **56**, 321 (2012).

Zang, H., Q. Xu, H. Bian, Energy. **38**, 236 (2012).

Zang, H., Q. Xu, P. Du, K. Ichiyanagi, Int. J. Photo-energy. 2012, 1 (2012).

Range-extended electrical vehicle performance simulation research based on power demand prediction

Limian Wang
China Agricultural University, China
Beijing Automotive Industry Advanced Technical School, China

Du Chen, Shumao Wang & Zhenghe Song
China Agricultural University, China

Dihua Yi & Yueyuan Wei
Beijing Electric Vehicle Co. Ltd., China

Zhenyu Zhang
Beijing Automotive Industry Advanced Technical School, China

Jinlong Zhou
Beijing Electric Vehicle Co. Ltd., China

ABSTRACT: In this paper, distribution control system of prototype vehicle is studied on the basis of one Range-Extended Electrical Vehicle (RE-EV) prototype vehicle. This system includes a Vehicle Control Unit (VCU) level and an Auxiliary Power Control Unit (APU). VCU calculates the vehicle power distribution according to driver's demand, and APU coordinates the engine and generator to work. With reference to the control strategy, the operating point of range extender is taken as the optimization target. The sequential index smoothing method will predict the power of the drive motor and optimize the power distribution between range extender and battery, which is emphatically studied in this paper. The simulation method is used to optimize the vehicle control strategy, and the result shows that the sequential index smoothing method can improve the work range of the range extender and thus optimize the power distribution of the whole vehicle.

1 INTRODUCTION

The Electric Vehicle (EV) is recognized as an effective way to relieve the energy crisis and environmental deterioration, but there are some factors which limit the marketization and industrialization development of electric vehicle at the present phase, such as the battery power capacity, lifetime, and cost of EV. The range-extended electrical vehicle is seen as a good way to extend the driving range, and it has technology advantages of both hybrid electric vehicles and pure electric vehicles.

Foreign companies such as AVL, MAHLE, and FEV, which are engaged in many years of research and practice on the range-extended EV, mainly focus on the research field of the design and the control effect of the extended range engine, which have formed some mature application such as using the small displacement, miller cycle engine,

and so on. However, studies on the process control strategy of vehicle equipped with RE are different because of different vehicle performance targets. In this paper, we take a prototype range-extended EV as the research object and set different powertrain parameters. We design the control strategy using simulation software such as MATLAB and Simulink and built the prediction model of drive motor to make some integrated simulation and control strategy verification of the whole vehicle.

2 RANGE-EXTENDED EV STRUCTURE AND CONTROL STRATEGIES

The structure characteristic of RE-EV is that, one engine and generator is added to EV as shown in Figure 1, so the RE-EV has the characteristics of EV (Mi, 2014). Compared with the traditional hybrid electric vehicle, the engine power of the

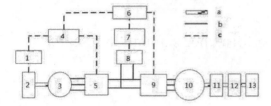

Figure 1. System structure of RE-EV.
Notes:
1. Engine control unit
2. Engine
3. Generator
4. Range extender control unit
5. Generator control unit
6. Vehicle control unit
7. Battery management system
8. Battery
9. Drive motor control unit
10. Drive motor
11. Transmission box
12. Transmission shaft
13. Vehicle wheel
a. Mechanical transmission
b. Electrical power supply
c. Control signal connection

Table 1. Basic parameters of powertrain.

System	Parameter type and unit	Value
Engine	Peak power/kw	84/6000
	Peak torque/Nm	142/4800
Generator	Rated power/kw	35
	Peak torque/Nm	180
	Max speed/rpm	6000
Drive motor	Rated power/kw	50
	Peak power/kw	100
	Max speed/rpm	6000
Battery	Capacity at 20°C/kwh	9.07
	Rated voltage/V	350.4
	Continuous discharge power/kw	33

RE-EV is small and the motor and battery power of the RE-EV are relatively high.

The vehicle powertrain parameter is shown in Table 1. Drive motor is the main power source. The range extender control unit is composed of an engine and a generator, the main function of which is to increase the driving range of electric vehicles. The control of the range extender is related to mutual coupling of many variables, which is the key to optimize the performance of electric vehicle. Currently, there are two control strategies: thermostat and power following. The thermostat needs high-performance battery, which will affect the battery lifetime. The power following can improve the battery running state, but needs high-performance engine, and the running range is wide, so the fuel consumption is high (Bott, 2005; Qu, 2013).

The power following control strategy is adopted in this paper. The engine power will follow the change in the drive motor demand power. Significant influence factors of system operation include battery SOC, range-extended working state (working or stopping), ranger-extended instant power changing, and vehicle speed (Qu, 2013; Botti, 2005). In order to set the operating point of the engine in the relatively high-efficiency range, and meanwhile meet the requirements that the engine must follow the rapid changes of the motor power, the motor power prediction method is introduced to obtain the average steady-state component of motor demand power in this paper. The final operating point is determined simultaneously by considering the high-efficiency area of the engine, and thus both the battery and the drive motor performance are combined to a certain extent.

3 ESTABLISHMENT OF MOTOR POWER PREDICTION MODEL

In the power following control strategy, the drive motor power will change fast, following the vehicle state, and it is difficult to use mathematic model to describe this change. In this paper, we use the time series prediction and build the sequential index smoothing model to predict the drive motor power on the basis of historical data of motor power. The principle is that, using the weighted average value to predict future observation value and giving more weighted coefficient to the observation value, which are more close to the prediction value, and the weighted value is distributed by the exponential law, while the weighted factor changes exponentially according to sequential prediction data, and the sequential index is an actual geometrical progression (Wang, 2009).

We set P_{t1} as the actual drive motor power and S_{t1} as the smooth predictive value at t1 moment, so the motor power is predicted by the sequential index smoothing model as follows:

$$S_{t1+1} = \alpha P_{t1} + (1 - \alpha)S_{t1}$$

where α is a model factor, whose value is between 0 and 1. In the significant change sequential, prediction value has small dependency to the nearby sequential data, so we can use small α, otherwise, we use the big α. The first smoothing data are the first actual data in the model as follows:

$$S_1 = P_1$$
$$S_{t1+1} = \alpha P_{t1} + \alpha(1 - \alpha)P_{t1-1} + \alpha(1 - \alpha)^2 P_{t1-2} + \alpha(1 - \alpha)^3 P_{t1-3} + \alpha(1 - \alpha)^4 P_{t1-4} + \dots$$

In the vehicle-controlling process, the prediction motor power will be input to the range extender as the steady-state power demand and APU optimizes the operating points according to the feedback power and current engine working state. The transient state power demand is met by the battery. In this model, α is 0.5, so $\alpha(1-\alpha)^4$ is 0.03125. The following component is much smaller to be ignored. The sequential index is 5.

4 SIMULATION RESULT ANALYSIS

In this paper, RE-EV performance is simulated based on the NEDC cycle. Figure 2 shows the motor prediction model value and no prediction model value, and it also shows that the original motor power changes fast and the power following will make the range extender operating points change fast. After using the motor sequential index smoothing model, the motor demand power changes more smoothly. In low-demand power changing conditions, the prediction value follows faster than the original demand power and prediction value follows more slowly with bigger demand power. This strategy can make the acceleration faster in the low vehicle speed conditions. In high vehicle speed conditions, the lost battery power will fill up by the range extender to keep the acceleration performance. At the same time, the predictive power shows hysteresis under the condition of a large drop in the motor demand power, but it is good for stable operation and smooth unloading of range extender engine.

Figure 3 shows the process of output power and battery SOC value trends. It is seen from the figure that in NEDC cycle, battery SOC is set at 30% initially and reduces to 28% at end of the cycle. From the whole process of view, SOC decreases slowly

and the battery energy is mainly used for lower power demand in pure electric operating mode in the initial phase. When range extender engine operates, the batteries do not need continuous discharge. Coupled with energy recovery and timely charging of RE, it causes little change of SOC value. In the late NEDC cycle, with the increase in vehicle driving power demand, RE mainly used to meet the output power to drive the vehicle. At the same time, according to the power demand prediction, the VCU coordinates energy distribution and tries to make RE work in the economic region with battery power supplement, which drives the vehicle together. Thus, SOC decreases rapidly in the high vehicle speed region at the later stage of the cycle.

Figure 4 shows the change curve of the output power of the battery. Visibly, the battery power frequently changes. However, in addition to the late segment of high power demand, the battery maximum output power is relatively stable, and it is always within the limits of continuous output power that the battery allows to output.

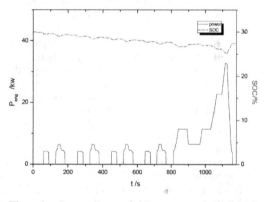

Figure 3. Power change of the range extender and SOC change.

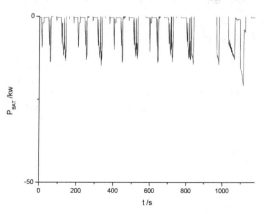

Figure 4. Change of output power of the battery.

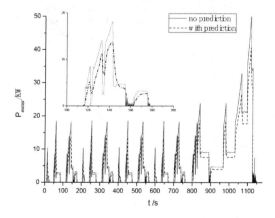

Figure 2. Influence of the sequential index smoothing model on motor demand power.

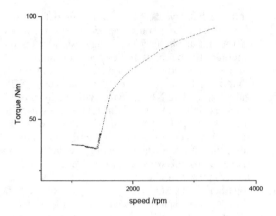

Figure 5.   Operating conditions of RE.

After the introduction of the prediction model, the batteries are mainly in response to transient changes in power demand under the coordination of vehicle control unit, which has played an electrically driven responsive advantage. At the same time, it meets the limits of the battery output performance, and to some extent, it also protects the battery.

Figure 5 shows the extended-range engine operating points during the NEDC cycle. It is evident from the figure that the engine operating points are approximately centered around a curve, which covers a wide area of the engine speed, so there must be some points falling outside the economical region. However, overall, RE operating region is different with the traditional surface area conditions. In the role of prediction models and control strategies, the work area is concentrated at the smoothing curve from 1800 to 3800rpm. The condition of alternating mutation with external input disturbance has been effectively controlled, which is conducive to approach the optimization point of the fuel consumption under these conditions.

5   CONCLUSIONS

A motor power prediction model is established based on mathematical method. In low-demand power changing conditions, the prediction value follows fast, whereas it follows more slowly with higher power demand. The predictive power shows hysteresis under the condition of a large drop in the motor demand power.

The batteries are mainly in response to transient changes in power demand under the coordination of vehicle control unit after the motor power prediction control strategy brought in RE-EV control system. Especially in the conditions of lower vehicle power demand, it showed an electrically driven responsive advantage. In high power demand conditions, although the battery discharge power increases, it still can be maintained within the range of the permitted battery power.

ACKNOWLEDGMENTS

This work was funded by National 13th Five-Year Science and Technology Plan Project (Grant No. 2016YFD070010), Beijing Municipal Science and Technology Commission Project (Grant No. Z151100001615017), and National Science and Technology Support Project (Grant No. 2015BAG05B00).

REFERENCES

Haitao Mi, Dongjin Ye & Yuanbin Yu, Optimization of the Control Strategy for Range Extended Electric Vehicle, Automotive Engineering, 2014, 36(8):899–903.

Jean J Bott, M James Grieve, John A MacBain. Electric Vehicle Range Extension Using an SOFC APU[C]. SAE Paper 2005-011172.

Jean J Botti, M James Grieve, John A MacBain. Electric Vehicle Range Extension Using an SOFC APU[C]. SAE Paper 2005-011172.

Lijie Wang, Xiaozhong Liao, Yang Gao, & Shuang Gao,Summarization of modeling and prediction of wind power generation, Power System Protection and Control, 2009, 37(13):118–121.

Xiaodong Qu, Qingnian Wang & Yuanbin Yu, A Study on the Control Strategy for APU in a Extended-range EV, Automotive Engineering, 2013, 35(9):763–768.

Advances in Energy Science and Equipment Engineering II – Zhou, Patty & Chen (Eds)
© 2017 Taylor & Francis Group, London, ISBN 978-1-138-71798-5

Research on an intelligent warning system of highway geological hazards

Jinbo Song
Jiangxi Vocational and Technical College of Communication, Nanchang, China

Maozong Zeng
Jiangxi Communication Design and AMP, Research Institute Co. Ltd., Nanchang, China

ABSTRACT: Frequent geological disasters to the mass production safety in recent years has brought a huge risk, so addressing disaster monitoring and timely detection of disaster types and location are particularly important. The monitoring system discussed in this paper can be used to analyze the disaster information, according to the early warning model to judge, and provide a variety of warning methods such as sound, graphics flicker, and text information. The system is practical and the geological hazard analysis is correct, which is effective for the prediction and early warning of geological hazards.

1 INTRODUCTION

China has a vast territory, complicated natural conditions, complex geological structure, and large areas. The high rainfall concentration in the East Asian monsoon region, coupled with China's large population, uneven distribution, and severe economic activity of human society has led to frequent occurrence of mountain flood disasters and obvious regional differences. Relevant experts in the field of research on the causes of flood disasters in China pointed out that topographic and geological conditions are the basis of flood disaster formation, rainfall is the motivating factor, and the dominant factors of flood formation and irrational human economic and social activities accelerated and exacerbated the occurrence and harm of the degree of mountain flood disasters. The cause of flash flood disaster is attributed to three factors: geological landform, rainfall, and human activity. A case study of Southwest China is taken as an example, which is characterized by flood-induced debris flow and landslide hazards. The debris flow events and outbreaks accounted for 47.77% and 39.45%, respectively, thereby accounting for 24.89% of the area of flood disaster prevention and control area in the eastern monsoon region. The area of mountain flood disaster prevention and control in this area is only 16.97% of the total area of China, but the number of debris flow gully and the number of outbreaks accounted for 36.99% and 31.06%, respectively. Landslide disaster in the eastern monsoon region and the national landslide in the proportion of the total number reached 48.04% and 42.23%, respectively. Therefore, the establishment of geological disasters early warning system based on the county in the southwest region of the comprehensive prevention and control of geological disasters is an important guarantee for the sustainable development of our country, which is of great significance for China's economic development and social stability.

Geological disaster is an active dynamic environment change process, whose prediction and prevention requires huge system engineering, because it involves huge data. In accordance with the traditional manual input information, although the file management mode is difficult to manage such large geological disasters information that consumes huge manpower and time, effectiveness and efficiency are very high. The introduction of GIS technology can not only effectively manage the important database of geological hazards, such as spatial information, but also effectively manage the various human factors and factors that affect the occurrence of disasters. Internet of things, cloud computing, smart sensors, as well as IPV6 and ZigBee, and other technologies also provide early warnings of disaster, and the transmission of information has been largely protected.

2 RESEARCH STATUS AND EXISTING PROBLEMS

The current state of the country is mainly manifested in the lack of network and hardware, and the software of R & D system has been limited. In recent years, a part of the unit began to operate flood disaster early warning system to implement the design and development, but most of them are made based on a single point, and information cannot be shared. A few provinces have established a centralized system of the whole province, and they do not reflect the rapid response of the administrative level of town, township, county, and a single way of warning, warning result just passable.

Study on early warning mode is based on the prediction model and all kinds of warning and a lot of studies, but the effect is not optimistic. Therefore, in the hardware under the current conditions, warning value detection is still based on the critical level, and real-time performance is not high. This kind of early warning system in the current still has problems such as rigid model, and generally used early-warning model of fixed, nonflexible use of various settings to meet the changeable climate in terrain areas. There are a number of GISs based on the design of flash flood disaster detection, and early warning system in the early warning effect is simple, not intuitive, the final decision is not expected to support the auxiliary role of prayer. With the rapid development of Internet-related technologies, the development of traditional GIS is more extensive, which changes the GIS data and the application of transmission and access methods, making GIS a widely used tool.

3 KEY TECHNOLOGIES OF MONITORING AND THE EARLY WARNING SYSTEM

3.1 Internet of things technology

Things (Internet of things) concept was proposed in 1999, with a very simple definition: all items through the radio frequency identification and other information sensing devices connected to the Internet achieve intelligent identification and management. Internet of things through intelligent sensing, recognition technology and pervasive computing, ubiquitous network integration applications, known as the Internet, is the world's third wave of information industry development. Internet of things is regarded as the application of Internet development, application innovation is the core of the development of the Internet of things, and the user experience as the core of innovation 2 is the soul of the development of the Internet of things.

Early warning system in the design of communication acquisition equipment ensures the timely information submitted in extreme weather, usually by the standby channel (GPRS/GSM and other forms of VHF radio communication design). This design can only solve the blockage of communication channel (mobile signal instability or VHF due to weather disruptions caused by the communication failure). However, in practice, especially

in major natural disasters such as Wenchuan earthquake and Zhouqu landslides, to bring the main equipment failure deformation and collapse from the mountain and the buildings, the original design is completely incapable of action.

Constructed on the basis of the ZigBee and IPv6 automatic networking technology, multirouting, and multichannel communication redundancy backup scheme, the relay site through the ZigBee communication module is less costly and has appropriate redundant layout to achieve in one or several points of communication breakdown, which can choose other equipment by the communication range of the router conveying the information to the information center, to make the information collection ability of the whole warning system more perfect in the face of disaster.

3.2 WebGIS subsystem

WebGIS is the combination of Web technology platform and GIS technology, which is a kind of analysis, processing, storage, compatibility, display, and application of geographic information system in the Internet network environment. WebGIS is produced with the development of database technology, network technology, and computer technology, which will develop with the development of these technologies.

Compared with the traditional WebGIS, GIS has a wider range of access and platform independence, which can greatly reduce the system cost, increase scalability, realize balanced and efficient calculation of load, and so on. On the basis of the Internet standard and Intranet, it can be shown that WebGIS can connect all the technical data with access to the Internet, and the part affected in the LAN or separated from the database will not provide users WebGIS access to the Internet, as long as the system can access the WebGIS system. As the user is operating GIS data in the Internet network environment, to get appropriate permissions, we can follow the requirements of the authority to perform a part of the GIS functions, including query, annotation, and other related information. At the same time, WebGIS can also model updating data, when the system administrator of the system in the data update and upload, all users of WebGIS systems can carry out activities related to the latest data after login, solve the problem of synchronization in very important data, and improve the efficiency of information acquisition and analysis.

WebGIS is distributed GIS. WebGIS architecture has three modes: distributed model, based on the end of the model, and based on the server side of the model. The traditional mode of GIS is based on the good side; in this mode, all operations are concentrated in the end; however, this model requires the terminal with high performance, and the cross platform support is not high, the user experience and interactive RIA WebGIS are just passable, using a distributed model based on the number of simple GIS operations at the end of execution. Complex operations are executed on the server, which can realize the load-balancing system and improve efficiency. At the same time, the WebGIS-based Internet has the characteristics of strong interaction, rich experience, and platform independence of RIA application.

3.3 IPv6 and ZigBee technology

ZigBee, translated as "purple bee," is similar to bluetooth. Sensor is a new short-range wireless communication technology, which is used for sensing and control applications (and Control). ZigBee has been widely used in the networking industry chain in M2M industry, such as smart grid, intelligent transportation, intelligent mobile terminal, home furnishing finance, POS, supply chain automation, industrial automation, intelligent building, fire protection, public security, environmental protection, meteorology, remote sensing, digital medical, agriculture, forestry, water, coal, petrochemical, and other fields.

IPv6 is the abbreviation of Protocol Version Internet 6, in which Protocol Internet translates to "Internet Protocol." IPv6 is IETF (Internet Engineering Task Force, Engineering Task Force Internet) designed to replace the current version of the IP protocol (IPv4) of the next-generation IP protocol. The current version number of the IP protocol is 4 (referred to as IPv4), and its next version is IPv6.

Ad hoc network and IPv6 based on ZigBee technology is in a particular situation of each terminal having a Zigbee network module in the geographical range within a certain geographical scope and the no communication beyond the set of network communication system. Then, each terminal can automatically find each other, forming a small interconnection network, and the terminal can move freely within the communication range, and can automatically add and delete, for network real-time refresh. This is the self-organizing network.

4 SYSTEM DESIGN

4.1 Function module design

The water and rainfall information obtained from automatic monitoring station data transmission and channel transmission, through the

data receiving software is stored in the database. Artificial observation of water and rainfall information through the voice telephone reporting mode automatically stored in the database, or by other artificial forecast collected by artificial entry way into the database. The information of the simple monitoring station can be stored in the database after completion. Higher authorities forward the relevant information after processing, in accordance with the unified data format into the database. Reserved weather, land, and other departments of information interface, through information collection and early warning platform; and meteorological, land, and other departments for information exchange, after processing into the database.

Geological disaster-based data management system function structure shown in Figure 4-4 includes the basic geographic data management, basic geological data management, data management, geological disaster site rainfall feature vector data management, image data management, and geological hazard database management subsystem function structure. Title, author, and affiliation frame: Place the cursor on the T of Title at the top of your newly named file and type the title of the paper in lower case (no caps except for proper names). The title should not contain more than 75 characters). Delete the word Title (do not delete the paragraph end). Place the cursor on the A of A.B. Author(s) and type the name of the first author (first the initials and then the last name). If any of the coauthor has the same affiliation as the first author, add his/her name after an ampersand (&) (or a comma if more names follow). Delete the words A.B. Author, and so on, and place the cursor on the A of Affiliation. Type the correct affiliation (Name of the institute, city, state/province, country). Now delete the word Affiliation. If there are authors linked to other institutes, place the cursor at the end of the affiliation line just typed and give a return. Now type the name(s) of the author(s) and after a return the affiliation. Repeat this procedure until all affiliations have been typed.

All these texts fit in a frame, which should not be changed (width: exactly 187 mm (7.36"); height: exactly 73 mm (2.87") from top margin; Lock anchor).

4.2 Thematic data management of geological hazards

Geological disaster thematic data management aims at keeping data related to collapse, landslide, debris flow, ground subsidence, ground crack, ground subsidence, dangerous slopes, seven types of disaster, data conversion, data update (input, update, delete), storage and integration of geological disaster prevention data, distribution of thematic data in damage-prone areas. The system provides query statistics, space making analysis, output, and all kinds of disasters in a professional thematic map.

The vector data, graphic data, and attribute data are combined, so that the user can operate the data at the same time with the familiar data. Disaster characteristics are composed of the logical name of disaster, graph information, attribute information in the database, the process of development, and change of the disaster. Disaster management system provides disaster management tool featuring simple and intuitive control and operation of disaster characteristics. The user only needs to focus on the task at hand, and need not care about the mechanisms by which data entry and operation of the data entry form become simple and convenient, can use the check box drop-down box, select the form, reduce the text entry time, improve work efficiency, and ensure and improve the consistency and reliability of data.

4.3 Functional structure design for forecasting and early warning analysis

Forecast and early warning of geological disaster analysis subsystem of geological disaster prediction and early warning using the configured parameters, support automatic, manual intervention and timing, and other means of forecasting and early warning analysis, rainfall isoline map, rainfall forecast map, contour map, including early warning results, analysis results; support parameter analysis of forecasting and early warning management for the prediction of different personnel, time-generated forecast results are analyzed. Geological disaster prediction and early warning indicators are in line with the Department of Land and Resources and the relevant provisions of the land and resources bureau.

4.4 Graphics publishing

PDA, the network of geological disasters information release system, is an effective means to the timely release of disaster warning information and link information to the people's community and the disaster area, as well as the disaster management network extension and service management application of the grassroots land. Published results include not only the use of television programs to the public release, but also the use of the site to the public release. The former focuses on a wide range of regional disaster forecasting and early warning, which focuses on the detailed forecast and early warning of the disaster point, as well as includes

information on major disasters in the early warning area. Geo-environmental information includes geological and environmental management of the relevant industry information, standards and norms, and disaster prevention guidelines.

4.5 Integrated management and emergency command sequence diagram design

Comprehensive management of geological disaster and emergency command is managed by professional users. Users first enter the main interface to the system, click to enter into the comprehensive management of geological disaster and emergency command interface, then load with various disaster point data, and finally analyze the disaster point data. The analysis result guides emergency rescue command scheme.

5 CONCLUSION

This paper mainly discusses the analysis of the project of highway engineering geological disaster intelligent early warning system and the actual demand. The system of each link and the technical characteristics required for the implementation of comprehensive utilization are in accordance with the design requirements of the system and the components, respectively, for each subfunction of the design and implementation of the detailed design and implementation and for the potential risks encountered in the development process. It realizes risk identification, estimation, evaluation, and a series of process specification, principles, and methods. Risk monitoring means a detailed explanation and guidance.

ACKNOWLEDGMENT

This work was supported by Science and Technology Project of Jiangxi Provincial Department of transportation (2015T0060 Gjj151423).

Any opinions, findings, and conclusions or recommendations expressed in this paper are those of the authors and do not necessarily reflect those of the sponsor.

REFERENCES

Gao Gao, Peng Liming. The design of the robot body for the detection of the cable in the Hangzhou Bay sea crossing bridge [J]. Shanxi architecture, 2012, 38 (22): 174–175.
Li Liwei, Shi Rong, Dong Zhenhua, et al. Structural design and dynamic simulation of cable crawling

robot [J]. Mechanical and electrical engineering technology, 2014 (4): 25–29.

Lv Tiansheng, Luo Jun. The development of the climbing mechanism of the Cable Painting Robot (CPR) [J]. Mechanical design and manufacturing engineering, 1999 (4): 32–34.

Peng Liming, Chen Lei, Gao Zhiyong. A design and research of the robot for the detection of the cable with adjustable cable force, 2012, 28 (3): 21–23.

Peng Liming, Gao Zhiyong, Chen Lei, et al. Design and analysis of a multi roller frame type climbing robot [J]. Mechanical science and technology, 2014 (4).

Qccounting fine, Zhou Yu, Ye Qingwei, et al. The detection method of cable force of cable-stayed bridge based on ERA algorithm [J]. Journal of electronic measurement and instrument, 2010, 24 (8): 743–747.

Yu Chaoyang. Research on [D]. Cable testing robot South China University of Technology, 2009 Layout of text.

Zhang Ni. Germany Elbe bridge detection system [J]. World bridge, 2015 (5): 94–95.

Advances in Energy Science and Equipment Engineering II – Zhou, Patty & Chen (Eds)
© 2017 Taylor & Francis Group, London, ISBN 978-1-138-71798-5

Rheology modeling of multiphase flow

Zhihong Han, Shuyang Liu & Yonghong Zhu
Jingdezhen Ceramic Institute, Jingdezhen, China

ABSTRACT: Most multiphase flow models use viscosity as an important parameter, however, the viscosity of mixed fluid sometimes is unavailable or difficult to measure especially in small-scale flowing condition. In order to solve this problem, a two-phase wedge-sliding model is developed in this paper and this model is non-viscosity dependent. Two variables of Drift-inhibition angle and expasion-inhibition angle were defined and their expressions were deduced from the model, which can well index the phase-drift trend of mixed fluid. The study also found that there is an optimal ratio in the ingredients design of mixed fluid and mixed fluid can only be stable when the volume ratio of heavier phase is smaller than this optimal value.

1 INTRODUCTION

The phase separation and sediment of mixed fluid is a multi-scale and multi-factors coupling process, especially as micro-particle-added lubricant used in the porous material, rheological behaviors present a very complex phenomenon.

Some multiphase flow models have been successfully proposed by researchers to describe the rheological behaviors in the certain application before, and nearly all these models regard viscosity as an important even key parameter. Among them, there are five typical models, such as the Power law model, Bingham model, Herschel-Bulkley model, Casson model and the Ree-Eying model etc., are widely used in industrial field in their respective adaptable conditions. Early in 1962, Higgins and Leighton (R.V. Higgins, 1962 & 1962) presented a fast method to calculate thoroughly the performance of two-phase flow in reservoir rock with complex geometry. By using Stokesian Dynamics, Nott and Brady (R. Nott, 1994) conducted dynamic simulations of the pressure-driven flow in a channel of a non-Brownian suspension at zero Reynolds number. Ali and Mohamed (A. El Afif, 2009) derived a nonlinear 3D model that investigates the flow–diffusion–structure interaction occurring in mixtures. This is the result of the application of current computer technology and advanced research method in the research field of multi-phase fluids.

But there is no evidence shown that viscosity of mixed fluid is related with the phase separation and sediment at a long-term or stable state; The ingredients separation and sediment of mixed fluid is a long-term accumulated process under stress; In addition, the viscosity of mixed fluid is varying with the temperature, stress, and time, and is difficult to measure even unavailable.

Therefore, when present conventional models are inappropriate to describe the rheological behaviors of mixed fluid accurately, it is necessary to develop a new model with non-viscosity dependence.

2 CHECKING OF THE MODEL 1

A number of balls with same diameter but two different weights are random tiled and formed as a close-packed hexagon-shape which was exactly circumscribed by the inner surface of the barrel without stacking.

According to the theory of mechanics, rotating balls around center had lower kinetic energy, which explained the final distribution of brown (heavier) balls. This result also meets the principle of minimum entropy in the circulation system.

Supposing that all these balls were the structure micro-cell of the mixed fluid, accordingly the heavier phase will have a trend to drift along the stress gradient to lower kinetic energy field. The researches of Cohen (Y. Cohen, 1985) on the polymer-phase flow problem, Hatzikiakos (1991 & 1992) and Kalika (1987) on the flow of different density polyethylene, and Yoshimura (1985) on the suspension of clay and paraffin oil-water emulsion etc., all obtain the similar conclusion that larger molecules phase has trend to drift along the way of stress reduction in the non-uniform stress field, and cause the significantly viscosity decreasing the area of higher stress.

3 CHECKING OF THE MODEL 2

Assuming that the mixed fluid were a uniform mixture of heavier micro-cell (added phase) A and basic micro-cell (basic phase) B, this well-mixed fluid could be categorized as a typical rheological stability question of two-phase flowing fluid.

Extracting a unit cell from the mixed fluid and assuming the faying surface of two phases in a cell was a wedge, the basic two-phase wedge-sliding model would be generated as shown in Fig. 1.

The action line of the forces is perpendicular to the wedge surface.

Where, δ_y is the yield stress of cell (N/mm^2), which is related to the environmental temperature and pressure and assumed to be a constant here; λ_δ is the prior coefficient of power exponential and defined as,

$$\lambda_\delta = \frac{e^{(\rho_A \cdot \varphi_A + \rho_B \cdot \varphi_B)/2}}{e^{\rho_A \cdot \varphi_A} + e^{\rho_B \cdot \varphi_B}} \quad (1)$$

Clearly, $0 < \lambda_\delta < 1$ and which is no unit.

The λ_k is the elastic-plastic anti-force coefficient of the cell.

Obviously, F_f is the maximum static friction of two phases when flow reached its steady state, so F_f can be defined as:

$$F_f = \lambda_f \cdot (F_{kA} + F_{kB}) \quad (2)$$

The acting direction of friction is parallel to the wedge surface but opposite with the trend of movement.

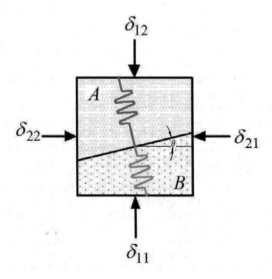

Figure 1. The unit cell of the two-phase wedge-sliding model.

The λ_f is the maximum static friction coefficient between two phases, which is related to the environmental temperature and pressure and assumed to be a constant here.

Defining $\varphi_B = \lambda$, $\lambda \in (0,1)$, then $\varphi_A = 1 - \lambda$. Substituting φ_A and φ_B by λ, the equation above can be rewritten as:

$$F_f = \lambda_f \cdot \frac{\rho_A^2 \cdot (1-\lambda)^2 + \rho_B^2 \cdot \lambda^2}{\rho_A \cdot (1-\lambda) + \rho_B \cdot \lambda} \cdot \frac{e^{\lambda_\delta \cdot \delta_y} + e^{-\lambda_\delta \cdot \delta_y}}{e^{\lambda_\delta \cdot \delta_y} - e^{-\lambda_\delta \cdot \delta_y}} \quad (3)$$

Defining:

$$C_\rho = \frac{\rho_A^2 \cdot (1-\lambda)^2 + \rho_B^2 \cdot \lambda^2}{\rho_A \cdot (1-\lambda) + \rho_B \cdot \lambda} \quad (4)$$

4 THE STRESSES EQUILIBRIUM ANALYSIS

Assuming the mixed fluid steadily flows caused by the uniform pressure difference $\Delta P = P_1 - P_2$ between two ends in a slot.

At the parallel direction of wedge surface,

$$(\delta_{12} - \delta_{11}) \cdot \sin\theta + (\delta_{22} - \delta_{21}) \cdot \cos\theta - F_f = 0 \quad (5)$$

At the perpendicular direction of wedge surface,

$$(\delta_{12} - \delta_{11}) \cdot \cos\theta + (\delta_{22} - \delta_{21}) \cdot \sin\theta - (F_{kA} + F_{kB}) = 0 \quad (6)$$

The longitudinal pressure differential and the transversal expansion-inhibition force are:

$$\delta = \frac{\Delta p}{A_p} = \delta_{12} - \delta_{11} \quad (7)$$

$$C_h = \delta_{21} - \delta_{22} \quad (8)$$

Where, A_p is the transversal surface area of the fluid in the slot.

Defining $\theta_c = \arctan\left(\frac{C_h}{\delta}\right)$, the equation above can be rewritten as:

$$\sqrt{\delta^2 + C_h^2} \cdot \sin(\theta - \theta_c) = \lambda_f \cdot \lambda_k \cdot C_\rho \cdot \frac{e^{\lambda_\delta \cdot \delta_y} - e^{-\lambda_\delta \cdot \delta_y}}{e^{\lambda_\delta \cdot \delta_y} + e^{-\lambda_\delta \cdot \delta_y}} \quad (9)$$

Then the wedge angle can be obtained by solving the Eq. (9):

$$\theta = \arcsin[\frac{\lambda_f \cdot \lambda_k \cdot C_\rho}{\sqrt{\delta^2 + C_h^2}} \cdot \frac{e^{\lambda_\delta \cdot \delta_y} - e^{-\lambda_\delta \cdot \delta_y}}{e^{\lambda_\delta \cdot \delta_y} + e^{-\lambda_\delta \cdot \delta_y}}] + \theta_c$$

$$\theta \in [-\frac{\pi}{2}, +\frac{\pi}{2}] \quad (10)$$

It's clearly that θ and θ_c respectively indicate the moving trend of phase B in the longitudinal and transversal directions within cells, so θ is named as drift-inhibition angle, and θ_c as expansion-inhibition angle here.

Substituting $\lambda_0 = \Phi_B$, then the initial joint stress modulus is gotten:

$$C_{\rho 0} = \frac{\rho_A^2 \cdot (1 - \lambda_0)^2 + \rho_B^2 \cdot \lambda_0^2}{\rho_A \cdot (1 - \lambda_0) + \rho_B \cdot \lambda_0} \qquad (11)$$

Accordingly, the initial drift-inhibition angle is:

$$\theta_0 = \arcsin[\frac{\lambda_f \cdot \lambda_k \cdot C_{\rho 0}}{\sqrt{\delta^2 + C_h^2}} \cdot \frac{e^{\lambda_\delta \cdot \delta_y} - e^{-\lambda_\delta \cdot \delta_y}}{e^{\lambda_\delta \cdot \delta_y} + e^{-\lambda_\delta \cdot \delta_y}}] + \theta_c \qquad (12)$$

5 CONCLUSION

1. The study shows that these two variables can well index the phase-drift trend of mixed fluid.
2. An optimal volume ratio of heavier gradient was found and its calculation formula was deduced.

ACKNOWLEDGMENTS

This work was financially supported by the National Natural Science Foundation of China (No. 61563022 and No. 61164014) and Jiangxi Province Natural Science Foundation of China (No. 20152 ACB20009).

REFERENCES

Cohen, Y. and A.B. Metzner. "Apparent slip flow of polymer solutions", J. Rheol. vol. 29, no. 1, pp. 67–102, 1985.

El Afif, A. and M. El Omari. "Flow and mass transport in blends of immiscible viscoelastic polymers", Rheol. Acta. vol. 48, no. 3, pp. 285–299, 2009.

Hatzikiakos, S.G. and J.M. Dealy, "Wall slip of molten high density polyethylene. II: Capillary rheometer studies", J. Rheol. vol. 36, no. 4, pp. 703–741, 1992.

Hatzikiakos, S.G., J.M. Dealy, "Wall slip of molten high density polyethylenes. I: Sliding plate rheometer studies", J. Rheol. vol. 35, no. 4, pp. 497–523, 1991.

Higgins, R.V. and A.J. Leighton, "A Computer Method to Calculate Two-Phase Flow in Any Irregularly Bounded Porous Medium", J. Petrol. Technol. vol. 14, no. 6, pp. 679–683, 1962.

Higgins, R.V. and A.J. Leighton, "Computer Prediction of Water Drive of Oil and Gas Mixtures Through Irregularly Bounded Porous Media Three-Phase Flow", J. Petrol. Technol. vol. 14, no. 9, pp. 1048–1054, 1962.

Kalika, D.S., M.M. Dean, "Wall slip and extrudate distortion in linear low-density polyethylene", J. Rheol. vol. 31, no. 8, pp. 815–834, 1987.

Nott, R., Prabhu and F. Brady, John. "Pressure-driven flow of suspensions: simulation and theory". J. Fluid Mech. vol. 275, pp. 157–199, 1994.

Yoshimura, A.S. and R.K. Prudhomme. "Wall slip corrections for couette and parallel disk viscometers", J. Rheol. vol. 32, no. 1, pp. 53–67, 1985.

Advances in Energy Science and Equipment Engineering II – Zhou, Patty & Chen (Eds)
© *2017 Taylor & Francis Group, London, ISBN 978-1-138-71798-5*

Analysis of the effect of thermal environment on the natural frequency of fiber-optic microphone membrane

Zheng-Shi Zheng & Xi-Chui Liu
School of Energy and Environment, Southeast University, Nanjing, China

ABSTRACT: Natural vibration characteristics of fiber-optic microphone membrane are affected by the change of temperature, which leads to the performance degradation of the microphone in acoustic measurement in thermal environment. In this paper, we study the influence rule and calculation method of natural vibration characteristic of fiber-optic microphone membrane structure in the thermal environment by the finite-element analysis method. The research results show that the natural frequency of the membrane structure is influenced by the change of material property parameters and thermal stress in the thermal environment. With the increase in temperature, the natural frequency affected by elastic modulus decreases approximately linearly, while the natural frequency affected by the thermal stress is nonlinear. The effect of thermal stress on the natural frequency is very high. The effect of thermal stress on the natural frequency is related to the form of thermal stress. When the temperature varies in a narrow range, the natural frequency decreases with the increase of the compressive stress caused by the thermal expansion. And as the temperature keeps increasing, the critical temperature of thermal buckling is reached and the mode shape changes significantly, and the tensile stress caused by severe bending deformation leads to a significant increase in the natural frequency.

1 INTRODUCTION

1.1 *Background*

With the rapid development of optical fiber technology, it has been widely used for the measurement of temperature, pressure, sound, and other physical parameters. Fiber-optic microphone is a new type of sound sensor based on optical fiber technology, which collects the sound signal by measuring the changes in light intensity, phase, and polarization state, which are caused by the changes in sound pressures (Yang, 2002). Currently, fiber-optic microphone has a wide range of application in many fields such as biomedical, bridge inspection, and oil exploration. Because of the advantages of high temperature resistance, corrosion resistance, and high sensitivity (Grattan, 2000), fiber-optic microphone has better performance than that of the conventional microphone in acoustic measurements in high-temperature environment, for example, furnace temperature distribution measurement by acoustic technique (Shen, 2007), leak detection of furnace tube (An, 2007), temperature detection of aircraft engine (Yu, 2015), and the detection of combustion vibration (Konle, 2011).

1.2 *Problem statement and research method*

At present, the research on the effect of thermal environment on fiber-optic microphone is insufficient. The effect of temperature changes on sensor needs to be considered in high-temperature environment. With the changes in temperature, physical parameters of fiber-optic microphone membrane material as well as the thermal stress of membrane structure change, which leads to the changes of the structural stiffness and natural frequency of membrane (Huang, 2009; Wang, 2015). Natural frequency is an important characteristic to measure the performance of fiber-optic microphone, which will directly affect the accuracy of acoustic measurements. Therefore, the analysis of the natural vibration characteristics of membrane plays an important role in the performance design of fiber-optic microphone.

In this paper, we take the titanium alloy membrane structure of fiber-optic microphone as the research object, using finite-element analysis software ANSYS to establish the analysis model. This paper investigates the influence rules and calculation method of natural vibration characteristic of fiber-optic microphone membrane structure in the thermal environment by the finite-element analysis method, which provides theoretical support for the structure design of fiber-optic microphone in thermo-acoustic measurements in the future.

2 THEORETICAL ANALYSIS

2.1 Basic principle of fiber-optic microphone

The basic principle (Cook, 1979) of fiber optic microphone is shown in Figure 1. The sound wave signal causes the change of membrane displacement, which leads to the change of the reflected light intensity. The displacement of the membrane can be obtained by measuring the change of reflected light, and then the acoustic wave signal is obtained. The frequency characteristics and sensitivity characteristics, which indicate the natural frequency and the displacement of the center point, are the most important performance parameters over the entire conversion process in thermo-acoustic measurements. Microphone sensitivity depends on whether the acoustic signal frequency is close to the natural frequency, which means the resonant response. Therefore, the study on the effect of temperature changes on the natural frequency is meaningful for the structure design of fiber-optic microphone.

2.2 Thermal stress distribution

The influence of temperature change on the natural vibration characteristics of the membrane structure can be obtained by the thermal mode analysis based on the finite-element method. The basic ways of thermal modal finite-element analysis are as follows.

For the plane stress, the element total strain $\{\varepsilon\}$ can be calculated according to the linear thermal stress theory proposed by Duamelle Neumann (Kong, 1999):

$$\{\varepsilon\} = \{\varepsilon\}_E + \{\varepsilon\}_T \tag{1}$$

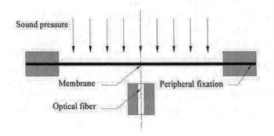

Figure 1.   Basic principle of fiber-optic microphone.

where $\{\varepsilon\}_E$ is the elastic strain and $\{\varepsilon\}_T$ is strain.

The thermal strain, $\{\varepsilon\}_T$, of the isotropic materials is:

$$\{\varepsilon\}_T = \{\alpha\Delta T \quad \alpha\Delta T \quad 0\}^T \tag{2}$$

where α is the coefficient of linear expansion of isotropic material and ΔT is the changes of temperature.

For elastic problems, the element stress $\{\sigma\}$ is:

$$\{\sigma\} = [D]\{\varepsilon\}_E \Rightarrow \{\sigma\} = [D](\{\varepsilon\} - \{\varepsilon\}_T) \tag{3}$$

where $[D]$ is the plane stress elasticity matrix, which is related to the elastic modulus and Poisson's ratio.

The elastic strain energy of the element is only related to the elastic strain; therefore, the elastic strain energy of the element is (Wei, 2004):

$$
\begin{aligned}
U^e &= \frac{1}{2}\int_{V_e} \{\varepsilon\}_E^T [D]\{\varepsilon\}_E \, dV \\
&= \frac{1}{2}\int_{V_e} \{\varepsilon\}^T [D]\{\varepsilon\} \, dV + \frac{1}{2}\int_{V_e} \{\varepsilon\}_T^T [D]\{\varepsilon\}_T \, dV \\
&\quad - \int_{V_e} (\{\varepsilon\}^T [D]\{\varepsilon\}_T) \, dV
\end{aligned} \tag{4}
$$

The relationship between element strain and nodal displacement is:

$$\{\varepsilon\} = [B]\{\delta\}^e \tag{5}$$

where $[B]$ is the plane stress geometry matrix. Therefore, equation (4) is transformed as follows:

$$U^e = \frac{1}{2}\{\delta^e\}^T [K]^e \{\delta^e\} - \{\delta^e\}^T \{Q\}_T^e + C \tag{6}$$

where $[K]^e$ is the element stiffness matrix, $\{Q\}^e$ is the element heat load, which is an equivalent load caused by thermal expansion of the structure, and C is related to the changes of temperature and independent of the nodal displacement.

According to the principle of potential energy minimization:

$$\frac{\partial U^e}{\partial \{\delta\}^e} = 0 \Rightarrow [K]^e \{\delta\}^e = \{Q\}_T^e \tag{7}$$

The unknown element nodal displacement is solved by the combination of the above formula with the known boundary condition. The element strain is calculated by formula (5), and then the thermal stress distribution is obtained by formula (2).

2.3 Structural stiffness matrix

The strain caused by the membrane deformation is composed of three parts: strain caused by the displacement u v, strain caused by the bending of the plate, and the additional strain caused by the deflection w. The bending strain energy of isotropic plate is:

$$U_b^e = \frac{D}{2} \int_A \{ (\frac{\partial^2 w}{\partial x^2})^2 + (\frac{\partial^2 w}{\partial y^2})^2 + 2\mu \frac{\partial^2 w}{\partial x^2} \frac{\partial^2 w}{\partial y^2}$$
$$+ 2(1-\mu)(\frac{\partial^2 w}{\partial x \partial y})^2 \} dA + \frac{1}{2} \int_A \{ \sigma_x t (\frac{\partial w}{\partial x})^2 \quad (8)$$
$$+ \sigma_y t (\frac{\partial w}{\partial y})^2 + 2\tau_{xy} t \frac{\partial w}{\partial x} \frac{\partial w}{\partial y} \} dA$$

where $D = \dfrac{Et^3}{12(1-\mu^2)}$ and t is the thickness of plate.

The second item in equation (8) represents the strain energy caused by the initial stress in the plane, which can be described by matrix as:

$$\frac{1}{2} \int_A \begin{Bmatrix} \frac{\partial w}{\partial x} \\ \frac{\partial w}{\partial y} \end{Bmatrix}^T \begin{bmatrix} \sigma_x & \tau_{xy} \\ \tau_{xy} & \sigma_y \end{bmatrix} \begin{Bmatrix} \frac{\partial w}{\partial x} \\ \frac{\partial w}{\partial y} \end{Bmatrix} t dA \quad (9)$$

The deflection of the plate is expressed by shape function [N] and the bending displacement $\{\delta_b^e\}$ of the plate element as follows:

$$w = [N]\{\delta_b^e\} \quad (10)$$

Then, the bending strain energy of plate element was obtained as:

$$U_b^e = \frac{1}{2} \{\delta_b^e\}^T [K_b]\{\delta_b^e\} + \frac{1}{2} \{\delta_b^e\}^T [K_T]\{\delta_b^e\} \quad (11)$$

where

$$[K_b] = \int_A [B]^T [D][B] dA, \quad (12)$$

$$[K_T] = \int_A [g]^T \begin{bmatrix} \sigma_x & \tau_{xy} \\ \tau_{xy} & \sigma_y \end{bmatrix} [g] t dA \quad (13)$$

$[K_b]$ is the thin-plate linear small deformation stiffness matrix and $[K_T]$ is the initial stress stiffness matrix related to the thermal stress.

Then, the initial stress of stiffness matrix can be obtained by substituting the thermal stress distribution into equation (11). Therefore, the stiffness matrix of the thin plate in the thermal environment can be expressed as the linear superposition of the initial stress stiffness matrix and the linear small deformation stiffness matrix, the structural stiffness matrix is expressed as follows:

$$[K] = [K_b] + [K_T] \quad (14)$$

2.4 Thermal modal analysis

Mode is the intrinsic property of the structure; in the case of ignoring the damping effect, the modal equation of the vibrating membrane is derived as:

$$([K] - \omega^2[M])\{\varphi\} = 0 \quad (15)$$

where [K] is the stiffness matrix of the structure, [M] is the mass matrix of the structure; ω is the natural circle frequency; and φ is the modal shape vector.

It can be considered that the influence of temperature change on the mass matrix of structure [M] is negligible. The changes of the mode and modal shapes of the structure are mainly caused by the change of the stiffness matrix of the structure (Wang, 2003). It can be known from the upper section that the influence of temperature change on the structural stiffness matrix is mainly reflected in two aspects:

1. The effect on linear small deformation stiffness matrix $[K_b]$. The formula of structure stiffness matrix including linear elastic matrix [D], which contained elastic modulus and Poisson ratio, which changes with the change of temperature.
2. The effect on the initial stress stiffness matrix $[K_T]$. The thermal deformation is restricted by the surrounding of the membrane when the temperature changes, and then the membrane shows tensile or compressive stress, leading to the change of the initial stress stiffness matrix. Further, the initial stress stiffness matrix is related to the form of thermal stress: $[K_T]$ is negative if the thermal stress is compressive, which leads to the natural frequency of the structure decreases. By contrast, the matrix is positive if the thermal stress is tensile, which leads to the natural frequency of the structure increases.

Therefore, it is needed to synthetically consider the effect of temperature change on material property parameter and thermal stress. The natural frequencies and natural modes of the membrane structures can be obtained by solving formula (15).

3 NUMERICAL SIMULATION EXAMPLE

3.1 Finite-element modeling

Fiber-optic microphone membrane is clamped by the means of peripheral fixation. A peripheral fixed-circle membrane is subjected to uniformly distributed sound pressure, which can be described as the small deflection bending of circular plates. The thickness of the circular membrane is set to 1 mm, and the radius of the membrane is set to 100 mm. In this paper, Finite-element model was established with finite-element software ANSYS. In order to observe the change of the vibration mode of the membrane better, the axisymmetric two-dimensional simplified model is established to simulate the circular membrane by means of the axisymmetric SHELL element. As shown in Figure 2, the coordinate origin is located in the center of the circular membrane. The X-axis displacement degree of freedom constraint and Z-axis rotation degree of freedom constraint are applied to the center point, and all degrees of freedom are imposed on the boundary. Applying pressure loads on all surfaces, set the initial reference temperature to 20°C. The material of membrane is titanium alloy TA15 with density 4450 kg/m³ and Poisson ratio 0.39. Other physical parameters are shown in Table 1.

3.2 Finite-element analysis

3.2.1 Comparative analysis

The temperature range in typical high-temperature thermal environment is considered in this paper. The modal analysis of the membrane structure is carried out under uniform temperature field at 20–600°C. First, the trend of the change of the preceding six-order natural frequency of the membrane is analyzed by only considering the change of the elastic modulus.

As shown in Figure 3, with the increase of temperature, the natural frequency affected by elastic modulus is decreased approximately linearly. This is because the elastic modulus of the membrane material decreases with the increase in temperature, which leads to the decrease of the linear small

deformation stiffness matrix $[K_b]$. Moreover, the influence degree of temperature becomes more obvious with the increase of the modal order.

Then, the thermal stress distribution is taken as an additional load on the structure. The trend of the change of the preceding six-order natural frequency of the membrane is analyzed by only considering the change of the thermal stress.

Figure 4 shows that the natural frequency fluctuates dramatically with the change in thermal stress,

Table 1. Physical property parameters of the titanium alloy TA15[14].

$T/$ °C	$E/$ GPa	$\alpha/$ 10^{-6}°C^{-1}	$\lambda/$ W·m^{-1}.°C^{-1}	$C/$ J·Kg^{-1}.°C^{-1}
20	118	7.85	6.0	509
100	108.7	8.55	8.8	545
200	101.2	9.10	10.2	587
300	95.8	9.55	10.9	628
400	89.7	10.00	12.2	670
500	80.0	10.30	13.8	712
600	64.1	10.45	15.1	755

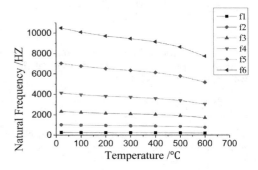

Figure 3. Trend of the change of the preceding six-order natural frequency of the membrane analyzed by only considering the change of the elastic modulus.

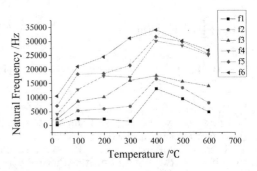

Figure 4. Trend of the change of the preceding six-order natural frequency of the membrane analyzed by only considering the change of the thermal stress.

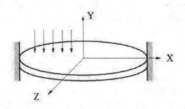

Figure 2. A peripheral fixed circle membrane subjected to uniformly distributed sound pressure.

and the relationship between the natural frequency and thermal stress is nonlinear and irregular. It can be concluded that, compared with the influence of the elastic modulus, the thermal stress changes have different effects on the natural frequency of the membrane.

From the statistical perspective, this paper further analyzes and compares the effects of the two factors by calculating the standard deviation of each order of natural frequency.

The standard deviation, σ, can reflect the fluctuation characteristics of the natural frequency with the change of temperature, and the calculation formula is as follows:

$$\sigma = \sqrt{\frac{1}{N}\sum_{i=1}^{N}(x_i - \mu)^2} \qquad (16)$$

According to Table 2, it can be seen that the standard deviation of the natural frequency in the influence of thermal stress is much larger than that in the influence of elastic modulus. This indicates that, although the change of natural frequency is affected by elastic modulus and thermal stress change, the nonlinear change of the natural frequency is mainly caused by the change of the thermal stress.

In addition, the standard deviation of natural frequency considering two types of factors is slightly smaller than the standard deviation under the influence of thermal stress. This is because the elastic modulus under the influence of the natural frequency is approximately linear, which will reduce the fluctuation degree of the overall natural frequency variation, decreasing the nonlinearity of overall natural frequency variation.

3.2.2 Analysis of the effect of thermal stress

In the previous section, we showed that the natural frequency fluctuates dramatically with the change of thermal stress, and the relationship between the natural frequency and thermal stress is nonlinear and irregular, which is necessary to analyze further. On the basis of theoretical analysis, the increase or decrease of the initial stress stiffness matrix is

related to the form of thermal stress. In order to analyze the influence of thermal stress on the natural frequency further, only the change of first-order mode shape under the change of thermal stress is analyzed, where the temperature range reduced to 20–100°C. The change trend of the first-order natural frequency under the influence of thermal stress is shown in Figure 5.

Next, we analyze the change trend in combination with the first-mode shape at different temperatures.

As shown in Figure 6 (a) (b), the first-mode shape has not changed significantly in the temperature range of 20–30°C. And the thermal stress caused by thermal expansion is mainly manifested as the compressive stress. With the increase of the compressive stress, the thermal stress stiffness matrix decreases, which leads to the decrease of natural frequency.

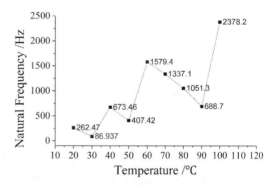

Figure 5. First-order natural frequency change trend under the influence of thermal stress change.

Table 2. Standard deviation of each order of natural frequency in thermal environment.

	f1	f2	f3	f4	f5	f6
Factor	Hz	Hz	Hz	Hz	Hz	Hz
Both elastic modulus and thermal stress	4405	4862	5014	8571	7726	7339
Thermal stress	5212	5872	6296	11146	10225	9610
Elastic modulus	21	84	190	337	574	857

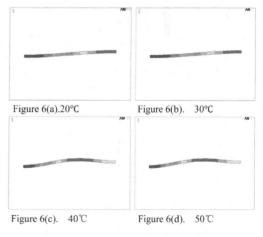

Figure 6(a). 20°C   Figure 6(b). 30°C

Figure 6(c). 40°C   Figure 6(d). 50°C

Figure 6. First-order modal shape in the temperature range of 20–50°C.

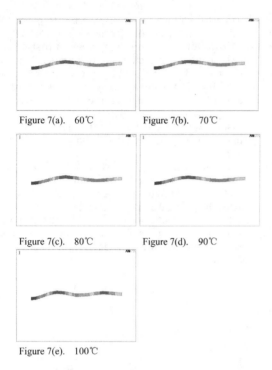

Figure 7(a).   60℃   Figure 7(b).   70℃

Figure 7(c).   80℃   Figure 7(d).   90℃

Figure 7(e).   100℃

Figure 7.   First-order modal shape in the temperature range of 20–50°C.

When the temperature rises to 40°C (as shown in Figure 6(c)), the structure began to appear larger, with bending and deformation, and this state of the expansion and extension changed the mode shape greatly. At this time, the thermal stress is mainly manifested as tensile stress, and the thermal stress stiffness matrix increases, which leads to the increase of the natural frequency.

The similar change rules were found in the temperature range of 60–100°C. There is no significant change in the first-order mode shape in the temperature range of 60–100°C (as shown in Figure 7 (a)(b)(c)(d)), and the natural frequency decreases with the increase in the compressive stress caused by the thermal expansion. When the temperature increases to 100°C (as shown in Figure 7(e)), the mode shape changes significantly, and the tensile stress caused by severe bending deformation leads to a significant increase in the natural frequency.

To summarize, considering the temperature variation within narrow range, the natural frequency decreases with the increase of the compressive stress caused by the thermal expansion.

And as the temperature keeps rising, the critical temperature of thermal buckling is reached, and the mode shape changes significantly, and the tensile stress caused by severe bending deforma-

tion leads to a significant increase in the natural frequency. Such changes revolving in the whole process of the temperature increase.

4   CONCLUSION

In this paper, we investigated the influence rules and calculation method of natural vibration characteristic of fiber-optic microphone membrane structure in the thermal environment by finite-element analysis method, and the research results are as follows:

1. The effect of temperature on the natural frequency of the fiber-optic microphone membrane should be considered in two aspects: one is the change of material properties caused by temperature, mainly reflected in the influence on the elastic modulus. Elastic modulus of membrane structure decreased with the increase of temperature, which leads to the approximately linear downward trend of natural frequency. Second, because of thermal expansion caused by the change of membrane structure thermal stress, the natural frequency of membrane changes.
2. The increase or decrease of the initial stress stiffness matrix is related to the form of thermal stress. When the temperature varies within narrow range, the natural frequency decreases with the increase of the compressive stress caused by the thermal expansion. And as the temperature keeps rising, the critical temperature of thermal buckling is reached and the mode shape changes significantly, and the tensile stress caused by severe bending deformation leads to a significant increase in the natural frequency. Such changes revolving in the whole process of the temperature increase.

REFERENCES

An, Lian-Suo. & Ma, Hui. & Zhang, Jian. (2007). Experimental Study on Leakage Acoustic Wave in Boiler Superheater. *Electric Power Science and Engineering*: 23(4), 35–37.

Cook, R.O. & Hamm, C.W. (1979). Fiber optic lever displacement transducer. *Applied Optics*: 18(19), 3230–3241.

Grattan, K.T.V. & Sun, T. (2000). Fiber optic sensor technology: an overview. *Sensors and Actuators A*: Physical, 82(1), 40–61.

Huang, Shi-Yong. & Wang, Zhi-Yong. (2009). The Structure Modal Analysis with Thermal Environment. *Missile and Space Vehicle*: 5, 50–52.

Kong, Xiang-Qiang. (1999). *Thermal stress finite element analysis*. Shanghai: Shanghai Jiao Tong University press.

Konle, H.J. & Paschereit, C.O. & Röhle, I. (2011). Application of Fiber-Optical Microphone for Thermo-Acoustic Measurements. *Journal of Engineering for Gas Turbines and Power: 133*(1), 011602.

Liu, Qing. (2004). *The Aerothermoelastic Analysis of Missile Structure.* Xi'an: Northwestern Polytechnical University.

Shen, Guo-Qing. (2007). *Study of Real-time Monitoring on Furnace Temperature Field Based on Acoustic Theory.* Baoding: North China Electric Power University.

Wang, Hong-Hong. & Chen, Huai-Hai. & Cui, Xu-Li, et al. Thermal Effect on Dynamic Characteristics of a Missile Wing. *Journal of Vibration, Measurement & Diagnosis:* 3, 015.

Wang, Xu-Cheng. (2003). *Finite element method.* Beijing: Tsinghua University press.

Wei, Te. & Bao, Hai. & Zhong, Bo. (2004). *Theoretical analysis and application of thermal stress.* Beijing: China Power Press.

Yan, Ming-Gao. (2002). *China Aeronautical Materials Handbook. Beijing:* Standards Press of China.

Yang, Hua-Yong. (2002). *Research on the mathematical model and key technology of the Reflective Intensity Modulated Fiber-Optic sensor.* Changsha: Institute of Materials Engineering & Automation, National University of Defense Technology.

Yu, Yuan-Yuan. (2015). *Research on Temperature Measurement System of High-Temperature Gas Based on Acoustics.* ShenYang: ShenYang Aerospace University.

Advances in Energy Science and Equipment Engineering II – Zhou, Patty & Chen (Eds)
© 2017 Taylor & Francis Group, London, ISBN 978-1-138-71798-5

Calculation and comparative analysis of extension length for shallow extended reach horizontal well under multi-limited conditions

Qimin Liang, Lingdong Li & Xinyun Liu
Research Institute of Petroleum Exploration and Development, Beijing, China

Yi Zhang
School of Petroleum Engineering, China University of Petroleum (East China), Qingdao, China

Jiali Zang
Shanghai Petroleum and Natural Gas Co. Ltd., Shanghai, China

ABSTRACT: Extended reach well technology represents the most advanced drilling technology in the world, and it has become the most important measure for developing the offshore oil and gas resources. Down-hole drag, torque and the rated pump pressure are three key factors affecting its mechanical reach ability. The commonly used down-hole string drag and torque calculation formula was explained in this paper. Aiming at the shallow extended reach horizontal well with single curve profile and two dimensions, the extension lengths of "single curve" trajectory under the certain vertical depth are calculated and comparatively analyzed if screw drilling tools and rotary steering drilling system are used. The primary extension limitation conditions of the screw drilling tools are no sine bulking and no compression at the surface when sliding, and torsional strength can meet the requirements of rotary drilling and flow rate can reach the cuttings transportation when drilled to this depth. The extension limitations under three operation conditions of the rotary drilling system are calculated, and the primary extension limitation conditions are no sine bulking and no compression at the surface when tripping in. Both of setting the kick-off point upper and decreasing the build-up rate are better choice for the extension of shallow extended reach horizontal wells with single curve profile. For the certain trajectory with the certain build-up rate under the same other conditions, rotary drilling system can obtain better extension than screw drilling tool, and more obvious advantage is achieved with greater build-up rate.

1 INTRODUCTION

Extended reach well technology represents the most advanced drilling technology in the world, and it has become the most important measure for developing the offshore oil and gas resources (Wu et al., 2002; Jiang Shiquan,1999; Su and Dou, 2003; Song et al., 1998; Shen and Tan 2000). The longer horizontal section of the extended reach well, the larger the contact area is, which could greatly reduce the number of required wells and rigs and improve the oil and gas recovery in the end. In the meantime, questions such as "which is the primary condition of constraining extended reach well's extension into the reservoirs", "how far could the extension reach", and "what is the relationship between extension and trajectory variations" still remain. It is known that down-hole drag and torque are two key factors affecting the mechanical reach ability of extended reach wells, and lots of attention has been payed to the down-hole string drag and torque calculation during the drilling and case run-

ning process (Gao et al., 2003; Wang et al., 2005; Adewuya O peyem iA, et al., 2012; Adit, 2006; Yan et al, 2010). Thus, based on the extended reach well model with certain vertical depth and single curve profile and the commonly used down-hole string drag and torque calculation formula, the largest extension length of horizontal section under different limitations (friction drag, torque and hydraulic condition) is computed and comparatively analyzed through simplifying the computation conditions. Then the primary condition of constraining its extension and the variation laws of the largest extension length and build-up rate are illustrated, which serves to the better development of extended reach well drilling technology.

2 GETTING STARTED AND TORQUE CALCULATION FORMULA

Friction factor is an integrated coefficient that is consists of friction coefficient and equivalent friction coefficient caused by all additional resistance, and it is larger than the friction coefficient. The friction factor is decided by multiple factors such as drilling pipes, casing pipes, reservoir rock properties, the lubricating properties of drilling fluids and borehole cleanliness, which is generally achieved through inverse computation with field data. When the drag and torque are calculated theoretically, all the above factors should be idealized in order to get calculation results with simplified models. The adopted empirical value of friction factor is "the empirical conference value of friction factor" as shown in Table 1.

If the influences of elements like joints' connections and centralizers, as well as the pipe dynamic effect are ignored, the simplified soft string model of friction resistance and torque is obtained at the assumption of it bringing into correspondence with rigid borehole axis. As shown in Figure 1, the force analysis of each infinitesimal arc segment with length of dl at ideal well trajectory, and the corresponding calculating formulas of axial force, drag and torque are as follows (Yan et al.,2011):

$$T_{i+1}=T_i+\left(W_g\mathrm{d}l\cos\alpha\pm\mu N_i\right) \quad (1)$$

$$M_i+1=M_i\pm\mu N_i r \quad (2)$$

Table 1. Empirical conference value of friction factor.

Drilling fluid type	Friction factor	
	Casing	Open hole
Synthetic based	0.1	0.13
Mineral base	0.17	0.21
Polymer water-based	0.22	0.25
Salt polymer water based	0.26	0.29
KCl salt water	0.41	0.44
Sea water	0.5	0.5

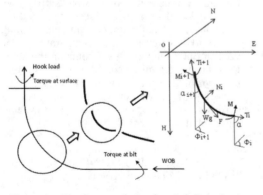

Figure 1. The force analysis of each infinitesimal arc segment in the drill string.

$$N_i=\sqrt{\left(T_i\Delta\phi\sin\alpha\right)^2+\left(T_i\Delta\alpha+W_g\mathrm{d}l\sin\alpha\right)^2} \quad (3)$$

$$F=\pm\,\mu N_i \quad (4)$$

Where, T_{i+1}, T_i are axial stress at the top and down side of the No.i infinitesimal arc; M_{i+1}, M_i are torque at the top and down side of the No.i infinitesimal arc; N_i is contact normal stress of the No.i infinitesimal arc with well bore; Wg is buoyant weight of each infinitesimal arc; μ is friction factor of sliding; r is radius of drill string; F is drag; α, $\Delta\alpha$, $\Delta\phi$ are mean inclination, inclination increment and azimuth increment.

Both axial and lateral forces distribution from bottom to top can be obtained for the whole drill string. The above formulas have become the classic type for calculating drag and torque due to their simplification, practicality and high accuracy, which are also applied in Wellplan module of the Landmark software.

3 CALCULATION AND COMPARATIVE ANALYSIS OF THE EXTENSION LENGTH FOR SHALLOW EXTENDED REACH HORIZONTAL WELL WITH SINGLE CURVE UNDER DIFFERENT BOTTOM HOLE ASSEMBLY

When reservoir depth and build-up rate have certain values, the single curve type profile can be used to describe the trajectory of the horizontal well (Chen and Guan, 2005). If the build-up rate of single curve trajectory is 4°/30 m, then the required minimum vertical depth is 430 m. It is assumed that the reservoir depth is 400m, single curve trajectories with build-up rates of 4.5°/30 m, 8°/30 m,12°/30 m respectively are studied for calculation convenience. The 9-5/8 inch casing above the landing point and the borehole of the horizontal section is 8-1/2 inch open hole are also assumed. Bottom hole assembly with PDM and rotary steering system are used during the drilling process of horizontal wells, and the whole drill string is like this: two joints of 5" heavy weight drill pipe combined with one joint of 7" drill collar are taken as one stand in the section with the inclination no more than 30°, following which is 5" heavy weight drill pipe with length of 160m, and the bottom part are 5" drill pipe and bottom hole assembly. Influence of different limitation conditions on extension length as well as the influence of different build-up rates on the largest extension length under the same limitation condition are analyzed for these two types of drill assemblies respectively. Drilling fluid is assumed have the density of 1.138 g/cm³, plastic viscosity of 14 mPa*s and yield point value of 14.843 Pa; friction coefficient is set as 0.32 for open hole and 0.3 for casing. In our calculation, 10 tons weight on bit and 2kN*m torque are applied for the rotary drilling mode, while 5 tons weight on bit and 1kN*m torque are applied for sliding drilling mode.

3.1 Bottom hole assembly with PDM

When the PDM drilling assembly is applied to drill horizontal section, the trajectory must be adjusted through continuous sliding drilling mode (Su, 1999; Di, 2000; Liang et al., 2015; Liang et al., 2012; Liang, 2013; Liang et al., 2013). Thus, there is no sinusoidal buckling along the drill string and drill strings on wellhead without compression, under these harshest conditions of sliding drilling mode the largest extension lengths are calculated. By using the above formulas, the extension lengths of trajectories with 3 kinds of build-up rates are calculated respectively, the result is as shown in Figure 2. Where, no buckling occurs when the hook load of single curve trajectory with 4.5°/30 m build-up rate is 0; there is sinusoidal buckling at 170 m of single curve trajectory with 8°/30 m build-up rate when hook load is 18.1 tons; sinusoidal buckling also occurs at 225m (where the inclination is 10°) of single curve trajectory with 12°/30 m build-up rate when hook load is 25 tons.

As shown in Figure 2, both the largest length and largest extension length decrease with the increasing of build-up rate when all other conditions stay the same.

The extension lengths when sliding with PDM drill assembly mode under single curve trajectory with 3 different build-up rates are calculated, and the relationship of torque and these build-up rates under rotary drilling mode is shown in Figure 3.

It can be concluded from Figure 3 that the wellhead torque when rotary drilling at the largest extension length of sliding drilling decreases with the increasing of build-up rate of single curve

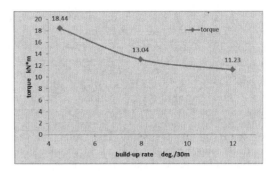

Figure 3. The relationship of torque at the extension length in rotation and build-up rate.

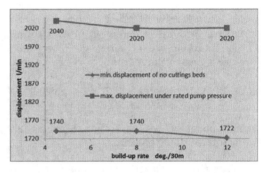

Figure 4. The relationship of displacement at the extension length in rotation and build-up rate.

trajectory, and it never exceed the rated torsional strength.

With the assumption of 25MPa rated pump pressure and 1300 horsepower rated pump power, when sliding using PDM drill assembly in the extension lengths which are determined under the condition of single curve trajectories with 3 different build-up rates, the responding maximum displacement and the required minimum displacement for hole cleaning is shown in Figure 4.

As shown in Figure 4, under the condition of rated pump pressure, when the build-up rate of single curve trajectory increases, the displacement at the largest extension of sliding drilling varies in a limited range, all those displacements could meets the requirement of cuttings transportation.

Based on the above calculation results, it can be found that the largest extension length of horizontal well with single curve profile primarily depends on no sinusoidal buckling occurs along the whole drill string when sliding and no compression in the drill string at surface. Furthermore, the largest extension length could be reached by setting kick-off point upper to get decreased build-up rate actually.

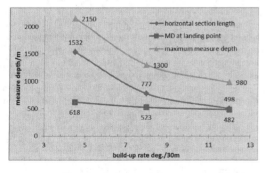

Figure 2. The relationship of extension length when sliding with PDM drill assembly and build-up rate.

As shown in Figure 4, under the condition of rated pump pressure, when the build-up rate of single curve trajectory increases, the displacement at the largest extension of sliding drilling varies in a limited range, all those displacements could meets the requirement of cuttings transportation.

Based on the above calculation results, it can be found that the largest extension length of horizontal well with single curve profile primarily depends on no sinusoidal buckling occurs along the whole drill string when sliding and no compression in the drill string at surface. Furthermore, the largest extension length could be reached by setting kick-off point upper to get decreased build-up rate actually.

3.2 Bottom hole assembly with RSS

The trajectory can be adjusted freely when using rotary steering system during the rotary drilling process, and a smoother trajectory can be created than that of PDM drilling. Taking consideration of the same conditions of PDM drill assembly case, the largest extension length is calculated respectively for single curve horizontal well with different build-up rates (4.5°/30m, 8°/30m, 12°/30m) during the rotary drilling process.

3.2.1 Extension length when no sinusoidal buckling occurs along the whole drill string during tripping in and no compression in the drill string at surface

When using RSS there is only rotary drilling but no sliding mode, so tripping in is taken as the condition. In this calculation, all conditions are the same with that of the PDM case except for the tripping in velocity is 25m/min. In the tripping in process, the depth where buckling starts to occur for the whole drill string and no compression on the drill string at surface can be taken as the limited extension depth. The calculated largest extension length result is shown in Figure 5.

As seen from Figure 5, both the largest extension length and largest horizontal section length decrease gradually with the increasing of build-up rate based on the conditions we talked about above.

3.2.2 Extension length under the condition of torsional strength of drill string

In the condition of using rotary steering assembly to create single curve trajectory with 3 build-up rates, the depth where the torque value extremely approaches the rated torsional strength can be taken as the limited extension depth. The results of calculated largest extension length are shown in Figure 6.

As seen from Figure 6, both the largest extension length and the largest horizontal section length increase gradually with the increasing of build-up rate in the rotary drilling process less than the limitation of torsional strength.

3.2.3 Extension length under the condition of hydraulics

Supposing that rated pump pressure is 25MPa and the rated pump power is 1300 horsepower. In the condition of using rotary steering assembly to create single curve trajectory with 3 build-up rates, the depth where the offered maximum displacement under the rated pump pressure just satisfies the minimum displacement of cuttings transportation can be regarded as the Max. hydraulic extension depth, and relevant calculation results are shown in Figure 7.

It can be concluded from Figure 7 that both the largest extension length and the largest horizontal section length decrease gradually with the increasing of build-up rate during the rotary drilling process with limitation of rated pump pressure.

3.3 Abstract extension length of two kinds of drill assembly comparison under the multi-conditions

Drilling assembly with PDM and rotary steering system are used for creating horizontal well with

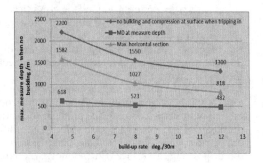

Figure 5. The relationship of the extension length under the following condition: RSS drill assembly no bulking and no compression when tripping in and build-up rate.

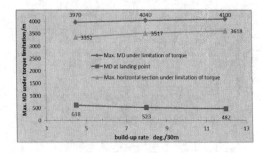

Figure 6. The relationship of the extension length under the following condition: torsion strength of RSS drill assembly in rotation and build-up rate.

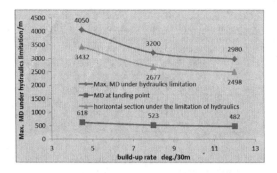

Figure 7. The relationship of the extension length under the following condition: hydraulics of RSS drill assembly in rotation and build-up rate.

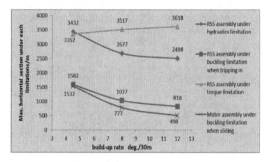

Figure 8. The relationship of the extension length under the multi condition of RSS and PDM drill assemblies and build-up rate.

single curve profile of 3 different build-up rates, and the responding limited extension length is shown in Figure 8.

Based on calculation results, rotary steering drilling assembly could reach longer extension for the shallow horizontal well with single curve profile, and this advantage becomes more obvious when the build-up rate increases. Thus, setting kick-off point upper to reduce build-up rate could help extend horizontal section for this kind of two-dimensional horizontal well with single curve profile.

4 CONCLUSIONS

The extension capacity of two-dimensional shallow extended reach horizontal well with single curve under various limitations are calculated and comparatively analyzed in this study. The results can't be taken as reference value in the field application because the calculation model and conditions have been simplified, but the relevant disciplines

is meaningful for the design and construction of extended reach wells.

Aim to the drill assembly with PDM, the largest extension length of horizontal well with single curve profile primarily depends on no sinusoidal buckling occurs along the whole drill string when sliding and no compression in the drill string at surface. When rotary drilling at this calculated largest extension length, torque at surface less than the limitation of torsional strength, and the displacements could meets the requirement of cuttings transportation. Both the largest length and largest extension length decrease with the increasing of build-up rate when all other conditions stay the same.

Aim to the drill assembly with rotary steering system, the extension length are calculated under 3 conditions: no buckling for the whole drill string and no compression on the drill string at surface during the tripping in process, torque at surface less than the limitation of torsional strength during the rotary drilling process, and no cutting beds occur when pump pressure no more than the rated one. Where, no buckling for the whole drill string and no compression on the drill string at surface during the tripping in process can be taken as the primary condition of constraining its extension. Under the same build-up rate and other conditions, rotary steering drilling assembly could reach longer extension for the shallow horizontal well with single curve profile, and this advantage becomes more obvious when the build-up rate increases.

Setting kick-off point upper to reduce build-up rate could help extend horizontal section for shallow horizontal extended reach well with single curve profile.

REFERENCES

Adewuya, O peyem iA, etc, 2012. A Rubust Torque And Drag Analysis Approach for well Planning and Drillstring Design. In Proc. *IADC/SPE* 39321.

Adit, G., 2006. Planning and Identifying the Best Technologies for Extended Reach Wells. In Proc. SPE 106346.

Chen, T.G., Guan Z.C., 2005. Theory & techniques of drilling engineering [M]. Dongying, China University Of Petroleum Press, P210.

Di, Q.F., 2000. Guide force calculation and analysis of sliding assembly when compound drilling. J. *Oil Drilling and Production Technology*. 22, 14–16.

Gao, D.L.,Tan C.J., Li, W.Y., 2003. Research on numerical analysis of drag and torque for xijiang extended reach wells in south-china sea. J. *Oil Drilling and Production Technology*, 25,7–12.

Jiang, S.Q., 1999. Developments of Extended-Reach Drilling and Its Application. J. *China Offshore Oil And Gas.* 11,1–8.

Liang, Q.M., Liu X.Y., Shi L.B., Tao, Y., Li, L.D., Hu, G., 2015. Application of combined drilling mode of pilot hole & reaming in the control of directional well trajectory. J. *Oil Drilling and Production Technology*. 37, 9–11.

Liang, Q.M., 2013. Discuss on Safe Sector of Deflecting Tool Face in Cluster Directional Well Drilling. J. *Petroleum Machinery*. 9, 41–43.

Liang, Q.M., Gao, D.L., Wang, S.W., 2012. Study and Practice of Downhole Tool Face Control Method in Directional Drilling. J. *Petrochemical Industry Application*. 31, 10–14.

Liang, Q.M., Zhu, Y., Xie, Y., Li, C., 2013. Difficulties and Strategies of Directional Well Trajectory Control in Xinjiang XinKen Region. In Proc. the *26th ECOS*, Guilin, China, Section29, G007.

Offshore Drilling Mannual [M]. Beijing: Petroleum Industry Press, 2009:587.

Shen, W., Tan S.R., 2000. Key technologies in the large extended reach drilling. J. *Oil Drilling and Production Technology*, 22, 21–26.

Song, Y.L., Dong, L.J., Li, Z.W., 1998. Oversea Developments of Extended-Reach Driling. J. *Petroleum Drilling Techniques*, 4, 4–8,12.

Su, Y.N., Dou, X.R., 2003. General condition and technical difficulties of extended reach drilling and its requirements on tools and instrument. J. *Oil Drilling and Production Technology*. 25, 6–10.

Su, Y.N., 1999. Selection of the steerable drilling tool and overall design principles and methods. J. *Oil Drilling and Production Technology*. 03, 86–90.

Wang, X.T., Wang, H.G., Chen, Z.X., Tang, X.P., 2005. Analysis on friction drag and torque from extended reach wells and the influence on well drilling depth. J. *China Petroleum Machinery*, 33,6–9.

Wu, S., Li, J., Zhang, Y., 2002. Current Situation of Extended-Reach Drilling Techniques. J. *Petroleum Drilling Techniques*. 30, 17–19.

Yan, T., Li, Q.M., Wang, G.Y., Li, J.H., Bi, X.L., 2011. Segmental calculation model for torque and drag of drillstring in horizontal wells. J. *Journal of Daqing Petroleum Institute*. 35, 69–72.

Yan, T., Zhang, F.M., Liu, W.K., Zou, Y., Bi X.L., 2010. Mechanical analysis on the limit extended capacity for an extended reach well. J. *Drilling and Producing Technology*. 33, 4–7.

Advances in Energy Science and Equipment Engineering II – Zhou, Patty & Chen (Eds)
© 2017 Taylor & Francis Group, London, ISBN 978-1-138-71798-5

Modeling of temperature and pressure distribution of underground gas storage during gas injection and production

Hong Zhang
Research Institute of Petroleum Exploration and Development, PetroChina and Drilling Research Institute, CNPC, Beijing, China

Ruichen Shen, Wentao Dong & Xiugang Liu
Drilling Research Institute, CNPC, Beijing, China

ABSTRACT: Accurate prediction of Underground Gas Storage (UGS) temperature and pressure distribution is necessary for improvement of wellbore integrity management and tubular design. Coupling differential equations describing temperature, pressure, velocity and density distribution along UGS wellbore under actual working conditions are presented based on fluid mass, momentum and energy conservation equations and the overall process of heat transfer between the surrounding rocks and the internal flowing gas. The equations can be solved by the fourth-order Range-Kutta method numerically. An application example is used to verify the model. The calculated temperature and pressure are compared with the results of WellCat software and thus prove the high precision of the model. Also factors affecting wellbore temperature and pressure distribution such as inlet temperature and pressure during gas injection, gas flow rate during injection and production and operation duration time are analyzed by this model. As a result, the injection pressure at wellhead has the largest impact on the bottom-hole pressure but has little effect on the bottom-hole temperature. The temperature and pressure vary greatly over time in the early period of injection and production.

1 INTRODUCTION

The casing and tubing strings of Underground Gas Storage (UGS) are in a state of varied complex stress for a long term due to cyclic change of temperature and pressure resulted from alternative injection and production of UGS. Predicting the temperature and pressure (*T, P*) along wellbore precisely is critical for the stress calculation and design of tubular string.

The *T, P* distribution has been the focus for decades in the petroleum research domain. As early as 1962, Ramey presented an approximated method of calculating temperature along wellbore for single-phase incompressible fluid or idea gas flowing through injection and production wells based on the principle of steady heat transfer. Shiu& Beggs (1980) simplified Ramey's method by optimizing the correlated parameters and introducing the relaxation length parameter. Hagoort (2005 & 2013) evaluated the precision of Ramey's model and concluded that Ramey's method ignored the early transient heat transfer period, leaving the temperature overestimated obviously. He also presented a simple analytical model for temperature prediction in gas wells. The mechanism models are

often used to predict pressure distribution based upon basic theory of fluid mechanics and multiphase flow, which are built by local instantaneous conservation equations, such as Hagedorn & Brown correlations describing the pressure drop of gas-liquid two-phase steady flow in vertical wells and Beggs & Brill (1973) model for pressure drop prediction of gas-water two-phase unsteady flow in deviated wells. All the above models addressed temperature and pressure separately and were solved by section iterative method. The physical parameters were averaged by segments.

According to the thoughts of different methods, models predicting temperature and pressure distribution along wellbore can be grouped into three categories: (1) only the temperature is concerned and the pressure can be obtained separately, such as Ramey's method; (2) by using the average temperature along the overall wellbore or a certain segment, the iterative formulae for the pressure can be derived, such as the famous Cullender & Smith method; (3) the flow in wellbore is considered as steady flow and the heat transfer within the surrounding rocks is seen as unsteady. Based on the exact analytical solution for stable heat source, the equations on heat loss vs time are obtained.

In combination with energy conservation equation, the coupling model of T, P can be built. Actually the T, P distribution and the flow velocity, the physical parameters of the fluid influence each other. Therefore, the coupling temperature and pressure should be worked out together instead of separately. In this paper, a coupling model concerning the temperature, pressure, velocity and density along wellbore during cyclic injection and production of UGS is proposed and solved numerically through the fourth-order Range-Kutta method, which provides basic data for the design and operation of UGS.

2 MODEL ESTABLISHMENT

2.1 Basic assumptions

1. Gas flow in the tubing is one-dimensional steady flow. Parameters of heat transfer, physical properties and flow rate are only related to the location of cross section;
2. One-dimensional steady heat transfer from the flowing gas to the interface between cement and formation occurs, while the heat transfer from the surrounding rocks to the interface between cement and formation is seen as transient;
3. Heat exchange only occurs in the radial direction and the exchange in the flowing direction is neglected;
4. The geothermal gradient is constant so the formation temperature varies with depth linearly;
5. The surrounding rocks are homogeneous and extend infinitely;
6. Gas starts to flow under a certain temperature and pressure and in the form of turbulent.

With the wellhead as the origin, the coordinate system is established as Figure 1. The length element of a deviated pipe dz is illustrated as the control volume with an inclination angle of θ. The gas

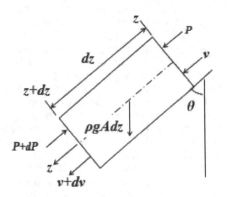

Figure 1. Length element analysis of pipe flow during injection.

flow through the pipe at a mass flow rate of w and the density, velocity, pressure and temperature at z point are (T, P, ρ, v), respectively.

Under the initial conditions, the tubing is full of the working gas which has reached thermal equilibrium. During injection and production, gas flows at constant mass flow rate in the tubing and the heat exchange and loss due to friction along the flowing direction can be neglected, which enables the use of a single comprehensive heat transfer coefficient to describe the series thermal resistance of tubing wall, fluid in annulus between casing and tubing, casing wall and cement sheath.

2.2 Basic differential equations

Under the condition of steady flow, the mass conservation equation for the length element dz is:

$$\frac{dv}{dz} = -\frac{v}{\rho}\frac{d\rho}{dz} \tag{1}$$

The unit mass of gas moves downward or upward under the impact of gravity, pressure and friction. In the light of principles of fluid mechanics, the momentum balanced equation is:

$$\frac{dP}{dz} = -\rho g \cos\theta - \frac{\rho v dv}{dz} - \frac{f\rho v|v|}{2d} \tag{2}$$

v is positive during injection and vice versa.

The energy flowing into and out of the control volume includes internal energy, kinetic energy, potential energy and pressure energy. In addition, the sum of internal energy and pressure energy forms enthalpy of the gas in control volume, i.e. $H = E + PV$.

The heat transferred radially from gas to the surrounding rocks is dQ. dQ is positive when gas is produced. The energy conservation equation is:

$$H_z + \frac{1}{2}mv_z^2 - mgz\cos\theta = $$
$$H_{z+dz} + \frac{1}{2}mv_{z+dz}^2 - mg(z+dz)\cos\theta + dQ$$

i.e.

$$-\frac{dq}{dz} = \frac{dh}{dz} + v\frac{dv}{dz} - g\cos\theta \tag{3}$$

where q is the heat loss of unit mass of gas and h is the specific enthalpy of the gas.

dq can be obtained by analyzing the radial heat transfer through tubing, annulus, casing and cement sheath, as shown in Figure 2.

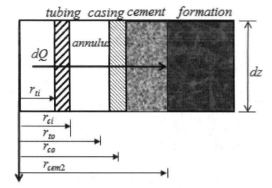

tubing casing cement formation

Figure 2. Schematic of radial heat transfer of the system.

The heat transferred from the flowing gas to the interface between cement and formation equals to that from the interface to surrounding rocks, i.e.

$$dq = \frac{2\pi r_{ti} U_{ti}}{w}\left(T - T_{cem2}\right)dz = \frac{2\pi k_e}{w f\left(t_D\right)}\left(T_{cem2} - T_e\right)dz$$

i.e.

$$\frac{dq}{dz} = \frac{2\pi r_{ti} U_{ti} k_e}{w\left(k_e + r_{ti} U_{ti} f\left(t_D\right)\right)}\left(T - T_e\right) \quad (4)$$

According to the fundamental principles of thermodynamics, the differential specific enthalpy dh/dz is:

$$\frac{dh}{dz} = \left(\frac{\partial h}{\partial P}\right)_T \frac{dP}{dz} + \left(\frac{\partial h}{\partial T}\right)_P \frac{dT}{dz}$$

Considering the definition of isobaric heat capacity and Joule–Thomson coefficient, the differential specific enthalpy can be simplified as follows:

$$\frac{dh}{dz} = -C_J C_P \frac{dP}{dz} + C_P \frac{dT}{dz} \quad (5)$$

Combining equations (3), (4) and (5), we can reach equation (6):

$$\frac{2\pi r_{ti} U_{ti} k_e}{w\left(k_e + r_{ti} U_{ti} f\left(t_D\right)\right)}\left(T - T_e\right) - C_J C_P \frac{dP}{dz} \\ + C_P \frac{dT}{dz} + v\frac{dv}{dz} - g\cos\theta = 0 \quad (6)$$

Based on the gas state equation $T\rho = MP/ZR$, the following relation holds:

$$\frac{d\rho}{dz} = \frac{M}{RZT}\left(\frac{dP}{dz} - \frac{P}{T}\frac{dT}{dz}\right) \quad (7)$$

The flow rate and density at z point are:

$$v = \frac{Q_{sc}}{A}\frac{P_{sc}}{T_{sc}}\frac{T}{P}Z = 4.95\times10^{-3}\frac{Q_{sc}}{d^2}\frac{T}{P}Z \quad (8)$$

Integrating equations (1), (2), (6), (7) and (8) we can obtain the coupling differential equations concerning wellbore temperature, pressure, density and velocity:

$$\begin{cases} \dfrac{dT}{dz} = C_J \dfrac{dP}{dz} + \dfrac{\dfrac{v^2}{\rho}\dfrac{d\rho}{dz} + g\cos\theta - C\left(T - T_e\right)}{C_P} \\[2ex] \dfrac{dP}{dz} = v^2\dfrac{d\rho}{dz} - \rho g\cos\theta - \dfrac{f\rho v|v|}{2d} \\[2ex] \dfrac{d\rho}{dz} = \dfrac{M}{RZT}\left(\dfrac{dP}{dz} - \dfrac{P}{T}\dfrac{dT}{dz}\right) \\[2ex] v = 4.95\times10^{-3}\dfrac{Q_{sc}}{d^2}\dfrac{T}{P}Z \end{cases} \quad (9)$$

where

$$C = \frac{2\pi r_{ti} U_{ti} k_e}{w\left(k_e + r_{ti} U_{ti} f\left(t_D\right)\right)}$$

Gas is assumed to exist above the working fluid level in the annulus and the modes of heat transfer in the annulus include radiation and natural convection. Thus the comprehensive thermal resistance of the system is:

$$\frac{1}{U_{ti}} = \frac{1}{h_t} + \frac{r_{ti}\ln\left(r_{to}/r_{ti}\right)}{k_{tub}} + \frac{r_{ti}}{r_{to}\left(h_c + h_r\right)} \\ + \frac{r_{ti}\ln\left(r_{ci}/r_{to}\right)}{k_{cas}} + \frac{r_{ti}\ln\left(r_{cem2}/r_{co}\right)}{k_{cem}} \quad (10)$$

In practical, the forced convection between gas and the tubing wall and the heat conduction of tubing wall and casing wall have little effect on the total thermal resistance.

The initial conditions are as follows:

$$P(z_0) = P_0, T(z_0) = T_0, \rho(z_0) = MP_0/RZ, \\ v(z_0) = w/A\rho(z_0)$$

z_0 is the point at wellhead for the case of injection and the point at bottom-hole during production.

3 NUMERICAL SOLUTION

3.1 Calculation of physical parameters

Z-factor of natural gas can be obtained by Dranchuk method (P M Dranchuk, et al, 1975) or AGA8-92DC model (Danhua Li, et al, 2011):

$$Z = 1 + \left(A_1 + A_2/T_r + A_3/T_r^3\right)\rho_r + \left(A_4 + A_5/T_r\right)\rho_r^2 + A_5A_6\rho_r^5/T_r + \left(A_7\rho_r^2/T_r^3\right)\left(1 + A_8\rho_r^2\right)e^{-A_8\rho_r^2} \quad (11)$$

where $\rho_r = 0.27P_r/(ZT_r)$, the values of constants A_i ($i = 1,2,3...,8$) are:

$A_1 = 0.315062$, $A_2 = -1.04671$, $A_3 = -0.578327$, $A_4 = 0.535308$
$A_5 = -0.612329$, $A_6 = -0.104888$, $A_7 = 0.681570$, $A_8 = 0.684465$

The isobaric heat capacity of natural gas from regression analysis is:

$$C_P = 1697.5107 P^{0.0661} T^{0.0776} \quad (12)$$

Joule–Thomson coefficient can be obtained as follows:

$$C_J = \frac{R}{C_P} \frac{\left(2r_A - r_B T - 2r_B BT\right)Z - \left(2r_A B + r_B AT\right)}{\left(3Z^2 - 3Z + A - B - B^2\right)T} \quad (13)$$

where

$$A = \frac{r_A P}{T}, B = \frac{r_B P}{T}, r_A = \frac{0.42747\beta T_c^2}{P_c}, r_B = \frac{0.08664 C_p T_c}{P_c},$$

$$\beta = \left[1 + m\left(1 - T_r^{0.5}\right)\right]^2, m = 0.48 + 1.574\omega - 0.176\omega^2$$

According to Jain's formula, the friction coefficient for gas flowing in pipe is:

$$\frac{1}{\sqrt{f}} = 1.14 - 2\lg\left(\frac{\varepsilon}{d_{tub}} + \frac{21.25}{Re^{0.9}}\right) \quad (14)$$

The function of transient heat transfer dimensionless time in the surrounding rocks is (Hasan, et al, 1991):

$$f\left(t_D\right) = \begin{cases} \left(0.5\ln t_D + 0.4063\right)\left(1 + 0.6/t_D\right) & (t_D > 1.5) \\ 1.1281\sqrt{t_D}\left(1 - 0.3\sqrt{t_D}\right) & (10^{-10} \leq t_D \leq 1.5) \end{cases} \quad (15)$$

where $t_D = \alpha t/r_{wb}^2$, α is the thermal diffusion coefficient of formation.

3.2 Numerical solution to the model

The whole length of wellbore is divided into n segments with the same length as the calculation segment on the basis of wellbore diameter, types of inner and outer fluid and formation properties. The equations can be solved segment by segment successively. In each segment, the inclination angle and physical parameters are seen as constant.

The differential equations can be solved by using the fourth-order Ranger-Kutta method. From equation (9) and the above physical parameters, the right terms of the four differential equations are the functions of P, T, ρ, v. Denote the functions as F_i ($i = 1, 2, 3, 4$), i.e.

$$\frac{dT}{dz} = F_1\left(P,T,\rho,v\right), \frac{dP}{dz} = F_2\left(P,T,\rho,v\right),$$
$$\frac{d\rho}{dz} = F_3\left(P,T,\rho,v\right), \frac{dv}{dz} = F_4\left(P,T,\rho,v\right)$$

Take the length of calculation segment as h and the following parameters are built based on the function rules and initial conditions (z_0, P_0, T_0, ρ_0, v_0).

$$\begin{cases} A_i = F_i\left(z,T,P,\rho,v\right) \\ B_i = F_i\left(z + \frac{h}{2},T + \frac{h}{2}A_1, P + \frac{h}{2}A_2, \rho + \frac{h}{2}A_3, v + \frac{h}{2}A_4\right) \\ C_i = F_i\left(z + \frac{h}{2},T + \frac{h}{2}B_1, P + \frac{h}{2}B_2, \rho + \frac{h}{2}B_3, v + \frac{h}{2}B_4\right) \\ D_i = F_i\left(z + \frac{h}{2},T + hC_1, P + hC_2, \rho + hC_3, v + hC_4\right) \end{cases}$$

Then the temperature iterative formula for segment j and $j+1$ can be written as:

$$T^{(j+1)} = T^{(j)} + \frac{h}{6}\left(A_1 + 2B_1 + 2C_1 + D_1\right)$$

where $i = 1, 2, 3, 4; j = 1, 2,..., n$;

Similarly we can obtain the iterative formulae for pressure, density and velocity.

Repeat the same procedure above to calculate the other concerned parameters of each segment successively until the required depth is reached.

4 EXAMPLE ANALYSIS

4.1 Model verification

A UGS vertical well is taken as an example to verify the model established and the results are compared with those calculated from WellCat software to validate the reliability of the model. The UGS is reconstructed in a depleted gas reservoir and the operation pressure is 11.7~28.0 MPa. The demand of peak shaving requires the injection or production gas volume of each well no less than $40 \times 10^4 m^3/d$. The undisturbed reservoir temperature is 90°C and the temperature at surface is 25°C.

The other basic parameters of the well are shown as Table 1.

A corresponding MATLAB program was designed based on the above model to solve the equations numerically in combination with the data in Table 1. The calculation segment is 1m and the results are shown as Figure 3 and Figure 4.

From Figures 3 and 4, we can see the prediction results of the model built are consistent with those of WellCat software and the relative error between the two methods is very small. Wellhead and bottom-hole are the positions where the maximum relative errors occur during production and injection, respectively.

4.2 Influencing factors analysis

4.2.1 Inlet temperature of gas

The inlet temperature is changed and the other related parameters in Table 1 remain the same. The results of T, P distribution during gas injection is shown as Figure 5 and Figure 6.

As Figure 5 shows, the inlet temperature has a significant impact on the temperature profile along wellbore, especially on the temperature near wellhead and has less impact on the temperature profile at the lower section.

In Figure 6, the pressure at bottom-hole decreases with the increase of inlet temperature, mainly because the pressure drop due to acceleration and friction rises as the velocity of flow

Table 1. Basic parameters for T, P, ρ, v prediction.

Parameter type	Value
Well depth/m	2500
Cement outer radius /mm	320
Casing outer radius /mm	273.0
Casing inner radius /mm	220.5
Tubing outer radius /mm	114.3
Tubing inner radius /mm	99.57
Wellhead pressure during injection/MPa	24.80
Injection rate/m^3/d	40×10^4
Production rate/m^3/d	80×10^4
Bottom-hole pressure during production/MPa	28
Injection time duration/d	50
Production time duration/d	20
formation thermal diffusion coefficient/m^2/s	1.03×10^{-6}
Tubular heat conductivity/$W/(m.K)$	45.35
Cement conductivity/$W/(m.K)$	0.983
Formation conductivity/$W/(m.K)$	1.592
Wellhead temperature during injection/$^\circ C$	35
Friction coefficient	0.015
Gas relative density	0.554

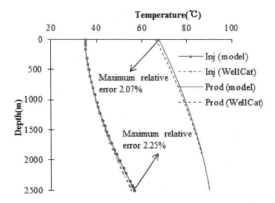

Figure 3. Calculation results of temperature distribution during gas injection and production.

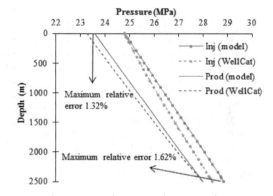

Figure 4. Calculation results of pressure distribution during gas injection and production.

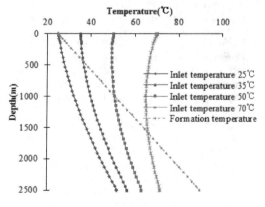

Figure 5. The effect of inlet temperature on temperature distribution during gas injection.

increases which is caused by temperature rise. Overall, inlet temperature does not have much impact on the distribution of pressure.

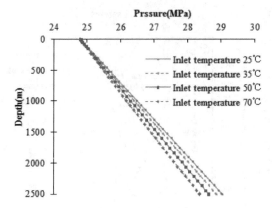

Figure 6. The effect of inlet temperature on pressure distribution during gas injection.

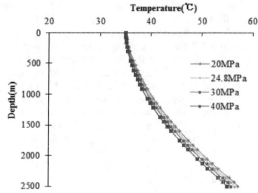

Figure 7. The effect of inlet pressure on temperature distribution during gas injection.

4.2.2 Inlet pressure of gas

Similarly, we analyzed the effect of inlet pressure with the model as shown in Figure 7 and Figure 8.

In Figure 7, change of inlet pressure of gas has little impact on the temperature profile. The temperature along the lower section of wellbore shows a slight decrease as inlet pressure increases.

In Figure 8, as a whole, the pressure increases significantly as inlet pressure rises and the pressure increment at bottom is slightly greater than that at wellhead. So adjusting the pressure at wellhead is the most effective method to influence the bottom-hole pressure during injection.

4.2.3 Gas injection and production rate

Also we calculated the corresponding temperature and pressure with the injection and production rate of $40 \times 10^4\ m^3/d$, $80 \times 10^4\ m^3/d$ and $120 \times 10^4\ m^3/d$, respectively to analyze the effect of flow rate. The results are shown in Figure 9 and Figure 10.

From Figure 9, we can see that during injection the temperature at the lower section of wellbore decreases as flow rate rises and during production the temperature at the upper section of wellbore increases as gas flow rate rises. Because the heat transfer between gas and formation reduces as flow velocity of gas increases. As a result, both the increment of gas temperature at bottom–hole during injection and the decrement of gas temperature at wellhead during production decrease.

From Figure 10, both the bottom-hole pressure during injection and the wellhead pressure during production decrease as gas flow velocity increases due to the increase of frictional pressure drop and acceleration pressure drop.

4.2.4 Job duration

The variations of bottom-hole pressure and temperature with different injection time are calculated as Figure 11.

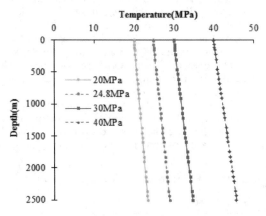

Figure 8. The effect of inlet pressure on temperature distribution during gas injection.

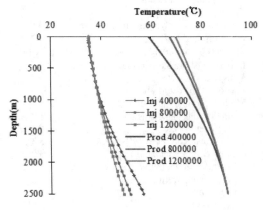

Figure 9. The effect of injection and production rate on temperature distribution.

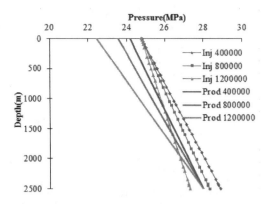

Figure 10. The effect of injection and production rate on pressure distribution.

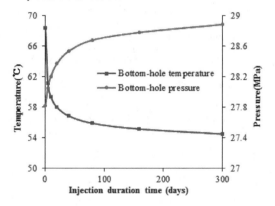

Figure 11. The effect of injection time on the bottom-hole pressure and temperature.

From Figure 11, we can see the bottom-hole temperature drops dramatically in the initial periods of gas injection. As injection time goes on, the temperature decreases slowly. After a certain period of time, the effect of duration on temperature can be neglected and the heat transfer is considered to be steady at this moment. The trend of bottom-hole pressure over time is opposite to that of temperature.

5 CONCLUSIONS

A set of coupling differential equations concerning temperature, pressure, density and velocity distributions along UGS wellbore during gas injection and production has been presented based on mechanism model and the equations can be solved numerically through the fourth-order Ranger-Kutta algorithm. A practical example is used to verify the model by comparing the T, P results from the model built and WellCat software, which shows little difference between the results from the

two methods. Also we analyzed factors affecting T, P distribution such as inlet temperature, inlet pressure, gas injection and production rate and operation duration using the proposed model. As a result, during injection, bottom-hole temperature increases with the increase of inlet temperature, decreases with the increase of injection rate and almost unaffected by inlet pressure. Also bottom-hole pressure decreases with the increase of inlet temperature and injection rate, and increases significantly with inlet pressure increasing. During production, as gas production rate increases, well-head temperature increases and wellhead pressure decreases. In the early periods, duration of operation impacts T, P distribution greatly.

NOMENCLATURE

A—open area of tubing, m^2;
C_J—Joule Thomson coefficient, K/Pa;
C_P—isobaric heat capacity of gas, $J/(kg.K)$;
d_{tub}—tubing diameter, mm;
f—friction coefficient, dimensionless;
$f(t_D)$—dimensionless time function of transient heat transfer;
H—enthalpy, J;
h—specific enthalpy of gas, J/kg;
h_c—natural convection heat transfer coefficient of annulus fluid, $W/(m^2.K)$;
h_r—radiation heat transfer coefficient of annulus walls, $W/(m^2.K)$;
h_t—forced convection heat transfer coefficient of gas and tubing inner wall, $W/(m^2.K)$;
k_{cas}—thermal conductivity of casing, $W/(m.K)$;
k_{cem}—thermal conductivity of cement, $W/(m.K)$;
k_e—thermal conductivity of formation, $W/(m.K)$;
k_{tub}—thermal conductivity of tubing, $W/(m.K)$;
M—The average molecular weight gas, kg/mol;
P—gas pressure distribution along wellbore, Pa;
P_c—critical pressure of gas, Pa;
P_r—reduced pressure of gas, dimensionless;
P_{sc}—pressure at standard conditions, Pa;
Q_{sc}—gas flow rate at standard conditions, m^3/d;
q—heat loss of unit mass of fluid, J/kg;
R—gas constant, $R = 8.314 \, J/(mol.K)$;
R_e—Reynolds number, dimensionless;
r_{ti}—inner radius of tubing, m;
r_{to}—outer radius of tubing, m;
r_{ci}—inner radius of casing, m;
r_{co}—outer radius of casing, m;
r_{cem2}—radius of cement-formation interface, m;
r_{wb}—radius of wellbore, m;
t—injection and production duration, s;
t_D—dimensionless production time;
T_e—undisturbed temperature of surrounding rocks, K;

T_{cem2}—temperature at cement-formation interface, K;

T_c—gas critical temperature, K;

T_r—reduced temperature of gas, dimensionless;

T_{sc}—temperature at standard conditions, K;

U_{ti}—comprehensive heat transfer coefficient, $W/(m.K)$;

v—gas flow rate, m/s;

w—gas mass flow rate, kg/s;

Z—compressibility factor of gas, dimensional;

z—position coordinates in the flow direction, m;

α—Formation thermal diffusion coefficient, m^2/s;

ρ—gas density, kg/m^3;

θ—inclination angle,°;

ε—absolute roughness of tubing wall, m;

ω—gas eccentric factor, dimensionless

REFERENCES

Beggs D.H, Brill J.P. 1973. A Study of Two-Phase Flow in Inclined Pipes [J]. Journal of Petroleum Technology, 25(5):607–617.

Durrant, A.J, Thambynayagam, R.K.M, Durrant, A.J. 1986. Wellbore Heat Transmission and Pressure Drop for Steam/Water Injection and Geothermal Production: A Simple Solution Technique [J]. Spe Reservoir Engineering, 1(2):148–162.

Gao Y, Liu J, Zhang J. 2013. Wellbore temperature and pressure coupling analysis in directional well [J]. Reservoir Evaluation & Development.

Hagoort, J. 2005. Prediction of wellbore temperatures in gas production wells [J]. Journal of Petroleum Science & Engineering, 49(1):22–36.

Hagoort, J. 2013. Ramey's Wellbore Heat Transmission Revisited [J]. Spe Journal, 9(4):465–474.

Hasan A.R, Kabir C.S. 1991. Heat Transfer during Two-Phase Flow in Wellbores; Part I–Formation Temperature[C]// Society of Petroleum Engineers.

Liao X, Liu L. 2003. New Idea for Borehole Pressure and Temperature Analysis of Gas Wells [J]. Natural Gas Industry, 23(6):86–87.

Ramey, H.J. 1962, Wellbore Heat Transmission [J]. Journal of Petroleum Technology, 14(4):427–435.

Shiu, K.C, Beggs H.D. 1980. Predicting Temperatures in Flowing Oil Wells[J]. Journal of Energy Resources Technology, 102(1):97–97.

Su Z, Liu Y, Zhang J. 2010. Study on AGA8–92DC Method of Gas Deviation Factor Calculation [J]. Well Testing, 19(6):29–36.

Zeng X L, Liu Y H, Yu-Jun L I. 2003. Mechanism Model for Predicting the Distributions of Wellbore Pressure and Temperature [J]. Journal of Xian Petroleum Institute, 18(2): 40–44.

Zeng Zhijun, Hu Weidong, Liu Jingcheng. 2010. A mechanical analysis of the gas testing string used in HTHP deep wells [J]. Natural Gas Industry, 2010, 30(2):85–87.

Advances in Energy Science and Equipment Engineering II – Zhou, Patty & Chen (Eds)
© *2017 Taylor & Francis Group, London, ISBN 978-1-138-71798-5*

An analysis of flexibility resources under the energy Internet

D. Liu, L. Ji & H. Hu
School of Economics and Management, North China Electric Power University, Beijing, China

ABSTRACT: With the development of economy, the demand of energy supply and flexibility resource is increasing under the energy Internet. First, this paper carries the definition of flexible resources under energy Internet. Next, the demand and mechanism of flexibility resources are discussed and the existing forms of flexible resources in the existing energy supply and demand system are summarized. And then, various aspects of flexibility characteristics are expounded and the existing problems of flexibility resources are discussed. Finally, in this paper, some suggestions are put forward for the development of flexibility resources.

1   INTRODUCTION

By the end of 2015, the total installed capacity of renewable energy in China reached 480 million kilowatts, accounting for 32% of all power generation sources installed, including 140 million kilowatts of wind power installation and 43 million kilowatts of photovoltaic installation. In 2015, China's renewable energy generating capacity reached 1 trillion and 300 billion kWh, accounting for 23% of the total generating capacity, of which the wind power generation capacity is 185 billion 100 million kWh. However, in 2012–2014, the national average abandoned wind rate was as high as 17.12%, 11%, and 8% respectively; the abandoned light rate in some photovoltaic regions reached up to 22%. Renewable energy resources were wasted seriously (Hassan 2016). The advancements of green revolution and low carbon energy conservation and environmental protection promote the development and utilization of renewable energy sources.

In sharp contrast to the problem, with the development of economy and the adjustment of the industrial structure, the total demand of energy is increasing and the type of energy demand is more diversified. Take the supply of electric energy as an example: according to an analysis of the 95598 telephone traffic of the National Power Grid Corp Southern Customer Service Center, telephone traffic is increasing in recent years, of which 90% business is consulting (Yang 2014), which means customers' demand for the use of electricity is diversified.

Energy Internet is a new form of energy industry development, which means Internet integrates deeply with energy production, transmission, storage, consumption, and energy market (Liu 2015). It has many features, such as intelligent equipment, multi-functional coordination, information symmetry, supply and demand dispersion, flat system, open system transaction, and so on (Chen 2015). Thus, under the energy Internet, the energy supply and demand system will be very complex. But the demand of various energy sources, renewable energy consumption and multi-flexible interaction will increase the complexity of the energy supply and demand system and aggravate the instability, which requires more flexibility resources to maintain the balance and stability of the system.

Carrying out flexibility resource analysis with the energy Internet provides reference for energy Internet's planning, design, and construction, and also provides technical analysis for the smooth operation and multiple possibilities of energy trade. In this paper, the definition of flexibility resources is firstly clarified. Secondly, it will analyze the mechanism and role in energy Internet construction. Thirdly, it elaborates its form and existing problems. Finally, some development suggestions are put forward.

2   DEFINITION OF FLEXIBLE RESOURCES

In the United States, flexibility is defined as the capacity to cope with load (demand) and power generation (supply). And they pointed out that, due to the change of power demand and the uncertainty of schedulability, regardless whether the uncertainty of high penetration of consumption is significant for renewable energy, flexibility is very important.

A high proportion of renewable energy consumption in the United States shows that a high proportion of renewable energy consumption will not damage the reliability of the electric power system, and the cost is not high, because flexibility makes the importance of a high proportion of renewable energy sources accepted gradually.

Flexibility resources refer to the resources that can increase the flexibility and elasticity for the energy supply and demand system, and serve for the dynamic balance of the energy supply and demand system (Yao 2015). In the system, flexibility resources mainly include severe peak load regulation units of thermal power generation, large hydropower, multi-level scheduling mechanism and this means, high responsive load, pumped storage, and other storage mechanisms, and each part's flexibility is detailed in Section 5.

The system can be regulated and control quickly and accurately by using flexibility resources, thereby realizing balance between the supply and demand, and satisfy the diversified energy supply and demand meanwhile.

3 FUNCTION MECHANISM OF FLEXIBILITY RESOURCES

Take the power grid as an example, when wind power and photovoltaic grid are connected, as shown in Figure 1, a graph of the wind power daily output in an area shows that the maximum wind power output changes frequently and violently in the whole year. Overall, wind power resources are rich in winter and spring, but short in summer and autumn.

It can be seen that, due to the uncertainty of wind power output, network uncertainty increases. In the same way, if any side's uncertainty in the energy supply and demand system increases, it will affect the stability of the whole system.

With the energy Internet, multiple energy sources interact flexibly, the energy demand is diversified, the proportion of renewable energy consumption is increasingly high, thereby resulting in instability and the difficulty in balancing increases (Huang 2015). In order to meet the diversified needs and achieve a high proportion of renewable energy consumption, the amount of flexibility resources increase, which will make the system more flexible and elastic (Yao 2015).

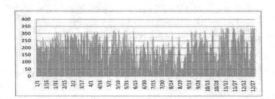

Figure 1. The diagram of the daily output of wind power in one part.

4 THE ROLE OF FLEXIBLE RESOURCES IN THE CONSTRUCTION OF THE ENERGY INTERNET

Flexibility, elasticity, and economy of flexibility resources will play an important role in the construction of energy Internet. The specific content is as follows.

Flexibility resources can reduce the system's complexity of the energy Internet (trüker 2011).

Flexibility resources can reduce the design difficulty of the energy Internet.

Flexible resources can enhance inclusive of energy Internet (Wang 2014).

Flexibility resources can reduce the cost of energy Internet infrastructure construction, which is the physical prerequisite for the realization of energy interconnection.

5 THE EXISTENCE FORM OF FLEXIBILITY RESOURCES

The energy supply and demand system will become a synthesis process which contains "Source - Net - charge - storage" under the energy Internet. Flexibility resources exist in different forms in the four aspects. The following will resume from the above-mentioned four aspects:

5.1 "Source" side

In the energy supply and demand system, "source", which means energy supply, is the original motive power of the entire energy supply and demand system. In the power system, typical flexibility resources mainly include the following:

5.1.1 Severe peak load regulation units of thermal power generation

The capacity of peak load regulation not only depends on the ratio of the minimum load and the maximum load, but also depends on the boiler's ability to adapt to high and low loads. A domestic boiler whose peak depth is 70% can be called as severe peak load regulation units of thermal power generation of flexibility resources theoretically, but the actual peak depth is 40%–50% (Yang 2014).

5.1.2 Large hydropower with capacity of long-term regulation

Long-term regulation refers to save excess water over the years when the reservoir capacity is large enough, and then use abundant water to fill the reservoir. The annual regulation, which is distributed in many dry years called as multi-year regulation, is also called long-term regulation. The large hydropower station or hydropower units, which with long-term regulation capacity or multi-year regulation capacity, can be controlled flexibly (Wang 2013).

5.1.3 Large scale thermal power unit of the sectional control coal mill

It is similar to severe peak load regulation units of thermal power generation; it can adjust the thermal power unit's output through coal grinding efficiency, so as to realize control flexibly.

5.2 Net side

In the energy supply and demand system, "net" is the energy's transport channel, and it is a bridge which links energy supply and demand.

5.2.1 Multi-level bidirectional coordinated scheduling

Multi-level bidirectional coordinated scheduling refers to each levels connecting interactively, scheduling coordinately, thereby achieving interconnection and interworking with multi-stage energy network in the energy scheduling, such as national, regional, provincial, municipal, and other multi-level scheduling.

5.2.2 Coordinated scheduling with multi-time scales

Coordinated scheduling with multi-time scale, that spans multiple time dimensions (including short, medium and long-terms, time, days and months, seasons and years multiple time dimensions) (Manfrida 2014) is used to schedule and achieve coordinated optimization and flexible regulation.

5.2.3 Scheduling transaction of multi-energy complementary

Scheduling transactions of many kinds of energy complementaries refer to achieving flexible regulation and control by mutual coordination, scheduling transactions, thereby optimizing the energy supply.

5.2.4 Micro balance scheduling of centralized optimization and decentralized optimization

Micro-balance scheduling of centralized optimization and decentralized optimization refers to centralized optimization and decentralized optimization for the energy supply and demand problem, in order to achieve micro-balance scheduling and flexible regulation, and therefore, the scheduling system can be simplified.

5.2.5 Network of multiple energy sources converge and interconnect

A network of multiple energy sources converge and interconnect, which refers to the mutual conversion, in order to achieve energy allocation and mutual coordination of energy resources. It can increase the flexibility of the energy supply and demand system.

5.3 Charge side

In the energy supply and demand system, the "charge" side is the energy demand side, and it is the end of the entire energy supply and demand, but also is the core of the energy supply and demand system.

Available typical flexibility resources mainly include demand side resources that can be regulated and excitation response, which is main electric vehicle.

5.4 Storage side

In the whole energy supply and demand system, the application of energy storage can increase the system's flexibility and reduce the system's operation risk. Available typical flexibility resources include the following:

5.4.1 Various types of small energy storage measures and equipment

Storage is flexible. Therefore, various types of energy storage measures and energy storage equipment all belong to flexibility resources.

5.4.2 Pumped storage (Jayasekara 2014)

A pumped storage power station uses electric energy pumped water to the reservoir at low load, and then discharge water to the hydropower station at peak load. In the power system, redundant electric power is converted into high value electrical energy at a peak period, and it also fit for frequency adjustment, phase, stabilized cycle, and voltage, and it is a standby for accidents. At the same time, it can improve the efficiency of the thermal power station and nuclear power station and achieve flexible control.

5.4.3 Compressed air energy storage

Compressed air energy storage referring to electrical energy is used to compress air at low load, makes air seal in high pressure in the abandoned mines, submarine gas storage tank, caves, expired oil and gas wells or new gas wells, to promote steam turbine power generation by releasing compressed air at peak load. The latest compressed air energy storage efficiency reached up to 50% and the energy efficiency of the cold and heat comprehensive system reached up to 70% or so (Wang 2014).

6 FLEXIBILITY RESOURCES PROBLEMS AT PRESENT

Under the energy Internet, there are still many problems related to the flexibility resources at present, and the following are discussed from three aspects:

6.1 Technical problems

At present, vacancies exist in flexible conversion technology of various energy forms under the energy Internet. The technology of energy routing, energy allocation, and other aspects is still immature.

6.2 Economic problem

According to the current situation of the development of market and technology, both from the supply side and from the demand side, the flexibility cost which can improve the consumptive level of renewable energy, is very high and is much higher than the gain under flexible resource control, so that it cannot be accepted.

At the same time, the lack of free and flexible market operation mechanism of the energy power industry and the effective return channel of flexibility results in the limited development. And because the demand type is different, it is difficult to form scale economy.

6.3 Policy and market environment problem

Take the power grid as an example:

There is no ancillary services market. There are many problems in ancillary services of flexibility resources.

There is no auxiliary service market. There is no commercialization of the flexibility resources, without reducing the flexibility resources value. There is no pricing mechanism and evaluation mechanism, and so it is difficult to be reasonably approved.

7 ENLIGHTENMENT

Under the energy Internet, the demand of flexibility resources has been growing. At the same time, the existing problem is still serious. Therefore, the development of flexible resources should be promoted from the following aspects:

To promote the transformation of severe peak load regulation of thermal power units and build the infrastructure of the source side.

To build a flexible direct current transmission network and infrastructure of the net side.

To push forward the demand side response vigorously and promote the user side participation in operation management of the power network.

To promote the construction of energy storage facilities and increase the flexibility of the power system.

To build the ancillary services market and rationally quantify the value of the ancillary service.

To build the demand side of the electricity market and reduce the cost of electricity.

To promote new techniques, research and development, and application of new materials in order to enhance the flexibility of the power system.

To provide policy support and promote development and consumption of renewable energy.

To improve the forecast and planning level and promote the coordinated development of "Source network charge storage".

8 CONCLUSION

In this paper, the definition, demand and mechanism, and existing forms of flexibility resources under energy Internet are expounded. Also, the existing problem is analyzed and the enlightenment of the development of flexible resources is put forward. Next, how to solve these problems should be studied.

REFERENCES

Chen, Q & Liu, D (2015). Business models and market mechanisms of E-Net (1). *Power System Technology, 2015, v. 39; No. 38411:3050–3056.*

Hassan, S & Khosravi, A (2016). A systematic design of interval type-2 fuzzy logic system using extreme learning machine for electricity load demand forecasting. *International Journal of Electrical Power & Energy Systems, 2016, 82:1–10.*

Huang, R & Pu T (2015). Design of hierarchy and functions of regional energy internet and its demonstration applications. *Automation of Electric Power Systems, 2015, v. 39; No. 55909:26–33+40.*

Jayasekara, S & Halgamuge, S (2014). Optimum sizing and tracking of combined cooling heating and power systems for bulk energy consumers. *Applied Energy, 2014, 118(1):124–134.*

Liu, D & Zeng, M (2015). Business models and market mechanisms of E-Net (2). *Power System Technology, 2015, v. 39; No. 38411:3057–3063.*

Manfrida, G & Secchi, R (2014). Performance prediction of a Small-Size Adiabatic Compressed-Air Energy Storage system. *International Journal of Thermodynamics, 2014, 18(2).*

Strüker, J & Weppner, H (2011). Intermediaries for the internet of energy—Exchanging smart meter data as a business model. *European Conference on Information Systems, Ecis 2011, Helsinki, Finland, June 2011.*

Wang, G & Ma, X (2014). Research on Combined Peak Load Regulation with Hydropower, Thermal Power and Nuclear Power Plants. *Advanced Materials Research, 2014, 986–987:465–469.*

Wang, X & Zhang, L (2013). Evaluate Pumped-Storage Unit in Power System with Multi-Type Units. *Applied Mechanics & Materials, 2013, 392:586–592.*

Yang, Q & Jiang, J (2014). Design and implementation of power marketing customer service decision support system. *Electronic Design Engineering.*

Yang, S & Liu, J (2014). Model and Strategy for Multi-time Scale Coordinated Flexible Load Interactive Scheduling. *Proceedings of the Csee, 2014, 34(22): 3664–3673.*

Yao, J & Gao, Z (2015). Understanding and prospects of Energy Internet. *Automation of Electric Power Systems, 2015, v. 39; No. 57323:9–14.*

Advances in Energy Science and Equipment Engineering II – Zhou, Patty & Chen (Eds)
© 2017 Taylor & Francis Group, London, ISBN 978-1-138-71798-5

Development of an optimized neural network model to predict the gas content of coal beds

Yan-sheng Guo

College of Geoscience and Surveying Engineering, China University of Mining and Technology (Beijing),
Beijing, China
School of Fundamental Education, Beijing Polytechnic College, Beijing, China

Zi-xuan Cui

College of Petroleum Engineering, China University of Petroleum (Beijing), China

ABSTRACT: The coal gas content is the foundation of resource evaluation. Theoretically, the logging response can reflect the hydrocarbon content of the coal seam, but the value of the correlation coefficient between the coal gas content and well logging parameters is very small when their relationship is quantitatively analyzed. To accurately predict the coal bed's gas content, on one hand, the concept of logging parameters attribute is introduced based on the theory of statistics and the lumped parameter method is adopted, and on the other hand, the BP neural network is optimized by using the genetic algorithm. A mathematical model of well logging parameters and coal gas content is built and the coal gas content is predicted. According to the results, the application of composite parameters and optimized algorithm can improve the prediction effect.

1 INTRODUCTION

The gas content of the coal seam is not only the basis of resource evaluation but also the important index of making a development plan. The logging curve is applied to predict the gas content of the coal seam because of its less cost, wider data coverage, rich and high resolution information, all of which can make up for the drawback of coring, testing, and coal core analysis (Guo Yan-sheng, 2014). With an increase in the gas content of the coal seam, the logging response of the natural gamma, density, and natural potential should be low and compensated neutron, resistivity, acoustic time, and bore diameter should be high theoretically (Hou Jun-sheng, 2000). However, due to the influence of various geologic factors on the coal bed gas content, there is a complex non-linear relationship between them, and some of them are even random and fuzzy; therefore, it is difficult to express their intrinsic relationship by using traditional methods (Meng Zhao-ping, 2014). The artificial neural network method has strong non-linear approximation ability. The BP neural network is the most widely used adaptive learning algorithm currently. Based on the gradient descent method to search for the optimal solution, BP is easy to fall in the local minimum point; and thus, the prediction accuracy is affected. A genetic algorithm has the ability of global search; therefore, in this paper, a genetic algorithm is used to optimize the neu-

ral network weights, and an optimization model is established to predict the Coal Bed Methane (CBM) content.

2 OPTIMIZATION OF LOGGING PARAMETER ATTRIBUTE

The correlation between the coal bed gas content and mean logging parameter is found weak in the south of Qinshui Basin. The logging parameter attribute is introduced to analyze the logging curve of the CBM (Coal Bed Methane) reservoir in order to improve the correlation between the coal bed gas content and logging parameters. In addition, the logging parameters are optimized by using complex parameters and mathematical methods.

2.1 *The principle of preference*

The correlativity and cross-correlation between the measured coal bed gas content and the attributes of the reservoir's logging parameters are analyzed according to the following equation (Guo Yan-sheng, 2004):

$$r = \frac{\sum_i (x_i - \bar{x})(y_i - \bar{y})}{\sqrt{\sum_i (x_i - \bar{x})^2}\sqrt{\sum_i (y_i - \bar{y})^2}} \quad (1)$$

Thus, the logging parameters' attributes with high correlation and low cross-correlation coefficient are selected for coal bed methane content prediction.

2.2 Extraction of logging parameter attributes

Equations (2)–(4) were used to extract the mean attribute, amplitude variation property, and Root Mean Square (RMS) of the well logging curves. These logging curve attributes were normalized and then correlated with the gas content and cross-correlated with themselves to be preferred.

The equation for calculating the mean amplitude attribute is as follows:

$$\bar{x} = \frac{\sum_{i=1}^{N} x_i}{N} \qquad (2)$$

The equation for calculating the variation in the amplitude attribute is as follows:

$$Var = \frac{1}{N} \sum (x_i - \bar{x})^2 \qquad (3)$$

The RMS (Root-Mean-Square) amplitude attribute equation is as follows:

$$RMS = \sqrt{\frac{1}{N} \sum_{i=1}^{N} x_i^2} \qquad (4)$$

where, x_i is the amplitude of the logging curve; N is the number of sampling points in the CBM reservoir; $\bar{x}$ is the mean attribute; Var is the variation in the amplitude; and RMS is the Root-Mean-Square attribute.

2.3 Compound parameters and mathematical methods to improve the correlation

The logarithm and multi-parameter attribute compound method were used to improve the correlation between the gas content and the attributes of the logging parameters.

3 GENETIC ALGORITHM (GA) TO OPTIMIZE THE BP NEURAL NETWORK (BPNN) TO PREDICT THE COAL SEAM GAS CONTENT

The relationship between the gas content of the coal seam and the logging parameters not only is fuzzy and uncertain, but also has obvious multiple solutions and non-linear characteristics. The BP neural network has good non-linear approximation ability for learning samples, but the algorithm has the following three weaknesses: slow convergence rates, local optimization, and generalization ability. The genetic algorithm introduced the principle of biological evolution into the coding string group formed by using optimization parameters. According to the principles based on which the fitness value of the individual is retained, poor fitness is eliminated, and the new group inherited the previous generation of information and even better than the previous generation, the network can be optimized. The genetic

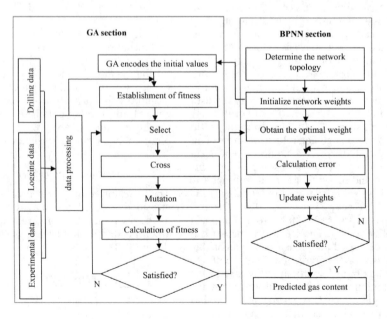

Figure 1. The technology route of the genetic algorithm to optimize the neural network model.

algorithm was used to optimize the initial weights and thresholds of the neural network, and finally the optimal coalbed methane content prediction model was obtained (Shi Feng, 2010). The technical route is shown in Figure 1.

4 APPLICATION EXAMPLE

4.1 *Logging parameter optimization*

Taking 3# coal seam of the Shanxi Formation of Lower Permian in the south in Qinshui Basin as the research object, through the coalbed gas content desorption experiment and coal bed methane logging data, the relationship between the coalbed gas content and logging parameter attributes was analyzed. The mean of Spontaneous Potential (SP), logarithm of the effective depth (H), the mean logarithm of the Microspherical Focusing Resistivity (R_{MSF}), the logarithmic ratios of the Deep lateral Resistivity (RD) and the Shallow lateral Resistivity (RS), the amplitude variations product of the natural gamma (GA) and the Compensated Neutron Logging (CNL), the ratio of the mean density (DEN) and acoustic time (AC) of the square root amplitude attribute were used as the basic feature of the forecast model. Table 1 shows the effect of compounding or mathematical calculation.

4.2 *BP neural network model optimized by genetic algorithm*

A three-layer BP neural network was established. The BP neural network consists of six input nodes,

Table 1. Correlation coefficient comparison between before and after compound.

| Logging parameters | Correlation coefficient between logging parameters and gas contents | | | Compound equation | Correlation coefficient after compound |
	Mean	Variation in amplitude	Root mean square		
SP	0.40	−0.21	0.39		0.40
R_{MSF}	−0.16	0.10	−0.07	$LN(R_{MSF})$	−0.44
RD	−0.11	−0.12	−0.13	$\dfrac{LN(R_D)}{LN(R_S)}$	0.32
RS	−0.09	−0.17	−0.11		
GA	−0.07	0.10	−0.08	$GA \times CNL$	0.18
CNL	−0.03	0.07	−0.03		
DEN	−0.12	−0.06	−0.10	$\dfrac{DEN}{AC}$	0.15
AC	0.14	0.09	0.14		
H			0.77	$LN(H)$	0.79

Table 2. The weights and thresholds of the GABP model.

| Weight | Input layer to the hidden layer weights (wij) | | | | | | Threshold |
	1	2	3	4	5	6	
1	3.387	−0.134	0.288	−0.451	−0.197	0.049	0.867
2	0.216	−0.845	0.071	0.598	0.398	0.006	0.472
3	0.216	0.853	1.754	0.554	1.117	0.297	1.53
4	−0.528	−1.825	0.580	0.255	0.219	0.553	0.533
5	0.077	0.867	0.113	0.563	0.558	−0.133	−0.789
6	0.354	0.469	−1.800	0.761	0.053	1.037	0.798
7	0.680	0.498	1.455	0.712	−2.377	−0.258	1.318
8	0.724	0.214	0.084	−0.540	0.275	0.704	2.017
9	0.871	0.331	1.407	−1.009	0.216	0.209	0.253
10	0.896	1.874	0.618	−0.699	1.331	−0.273	0.318
11	0.817	−0.060	1.103	−0.042	0.728	0.117	0.811

| Weight | Hidden layer to the output layer weights (Ti) | | | | | | | | | | | Threshold |
	1	2	3	4	5	6	7	8	9	10	11	
	0.235	0.039	−0.575	0.343	0.389	0.209	0.267	−0.338	0.109	0.356	0.541	0.239

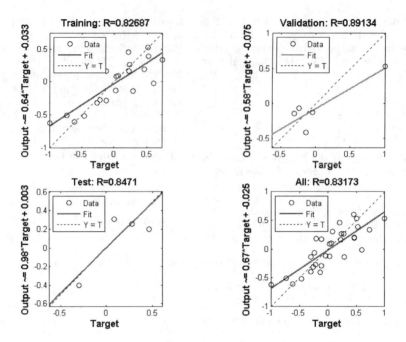

Figure 2. Graphs showing the error analysis of the predictive and actual values.

11 hidden nodes, and one output node. The six logging parameters are the input layer and the coalbed methane content is the output layer. The optimized prediction model (GABP) weights and threshold are shown in Table 2.

4.3 Results and error analysis

The samples were divided into training samples, validity samples, and test samples. Error analyses of the gas content between prediction and laboratory are shown in Figure 2. It can be seen that the BP neural network prediction model optimized by genetic algorithm optimization has a good effect.

5 CONCLUSIONS
AND RECOMMENDATIONS

1. The correlation between the coalbed methane content and logging parameters had found that statistical attributes were not significantly improved when compared to the mean attribute. Since the mean attribute calculation is simple, and it is recommended to build the model with mean attributes.
2. It is suggested that the correlation between the gas content and the logging parameters can be improved by using the compound and the mathematical method.
3. The BP neural network weights and thresholds can be optimized by using the genetic algorithm,

which can improve the accuracy of the prediction model. It is also suggested that the method can be used to optimize the network structure.

ACKNOWLEDGMENTS

This work was financially supported by the China National Natural Science Foundation (41172145) and Scientific Research Project of Beijing Polytechnic College (bgzyky201530 bgzykyz201407).

REFERENCES

Guo Yan-sheng & Meng Zhao-ping. 2014. Study on Prediction Method of Coalbed Methane Content Based on Regression Model. *Coal Engineering* 46(9):112–115.
Guo Yan-sheng, Meng Zhao-ping, Yang Rui-zhao et al. 2004. Seismic Attributions Analysis And it's Application in Predicting Thickness of Coal. *Journal of China University of Mining and Technology* 33(5):557–562.
Hou Jun-sheng, 2000. *Logging evaluation method of coalbed gas reservoir and its application*. Beijing: Metallurgical Industry Press.
Meng Zhao-ping, Guo Yan-sheng & Zhang Ji-xing 2014. Application and Prediction Model of Coalbed Methane Content Based on Logging Parameters. *Coal Science and Technology* 42(6):25–30.
Shi Feng, Wang Xiaochuan, YU Lei et al. 2010. *Matlab neural network analysis of 30 cases*. Beijing: Beihang University Press.

Dynamic modelling of boiler combustion system based on improved online support vector machine

Qi Shen, Xiao Wu, Yiguo Li & Jiong Shen
Key Laboratory of Energy Thermal Conversion and Control of Ministry of Education, Southeast University, Nanjing, China

Xia Zhou
School of Mechanical and Electrical Engineering, Jinling Institute of Technology, Nanjing, China

ABSTRACT: Boiler combustion optimization is an important means to achieve energy conservation and emission reduction in power plant. Nowadays, most combustion optimization methods are based on the steady-state model of the boiler combustion system, making it difficult to achieve dynamic optimization under variable load conditions. Therefore, this paper proposes an improved online adaptive least squares support vector machine dynamic modeling algorithm. The offline support vector screening is completed firstly to reduce the modeling sample size and ensure the sparsity of support vector. Then three support vector online updating strategies including replacement, addition and deletion are employed to accommodate the algorithm better adapt to the variation of object characteristics. The proposed algorithm is used to develop a dynamic model of a 600 MW unit boiler combustion system. Simulation results have demonstrated the developed model can reflect the dynamic characteristic of the boiler efficiency and NOx emission under different load condition accurately. Compared with the original steady-state model obtained by online adaptive least squares support vector machine algorithm, the newly developed model has not only higher precision and forecasting ability but also simplified model structure and can greatly reduce online computation burden. Moreover, the proposed model will be the base for future research on the dynamic optimization control strategy of boiler combustion system.

1 INTRODUCTION

Nowadays, boiler combustion optimization mainly adopts nonlinear modeling based on data combined with intelligent optimization algorithm. The boiler combustion characteristic model is developed at first, then operating parameters involved are optimized by maximizing boiler efficiency while minimizing boiler emission (Gu, 2010; ZHOU, 2010; WANG, 2004; Wang, 2007; Chang, 2004). However, similar methods are all essentially based on the steady-state model, thus the boiler combustion system cannot be optimized under variable load condition. Moreover, the influence of modeling error on optimization cannot be removed when large model variation occurs due to its lack of model parameters online correction function. Therefore, developing a dynamic model with online adaptive function is vitally important for the dynamic optimization of boiler combustion.

To address the problems mentioned above, this paper proposes an improved online adaptive least squares support vector machine (FVS-ALS) algorithm based on the combination of online sparse support vector machine regression algorithm and online Adaptive Least Squares Support Vector machine (ALS-SVM) algorithm (Gu, 2011). Firstly, the offline support vector screening is completed to reduce the required modeling sample size and ensure the sparsity of support vectors, which will accelerate the speed of model online running and updating. Then three support vector online updating strategies including replacement, addition and deletion are employed to enable the algorithm better adapt to the variation of object characteristics.

The above algorithm is further used to establish the dynamic model of a 600 MW unit boiler combustion system. The simulation results show that the proposed model has higher prediction accuracy than traditional steady-state model. Moreover, it can adapt to combustion system characteristic transformation caused by the variation of load, coal quality and equipment and so on.

2 THE IMPROVED ONLINE ADAPTIVE LEAST SQUARES SUPPORT VECTOR MACHINE ALGORITHM

2.1 *Brief introduction of ALS-SVM(Gu, 2011)*

Suppose that the training sample is set as T = {(x$_1$, y$_l$), ..., (x$_l$, y$_l$)}, in which $\mathbf{x}_i \in R^n$, $y_i \in R$, $i = 1, 2, ... l$.

The decision function is $f(\mathbf{x}) = \mathbf{w}^T \Phi(\mathbf{x}_i) + b$, and the sample data is used to solve the minimization of structural risk objective function, as is shown in formula (1)

$$\min_{\mathbf{w},b} \frac{1}{2} \|\mathbf{w}\|^2 + c \sum_{i=1}^{l} \xi_i^2 \qquad (1)$$

s.t.

$$y_i = \mathbf{w} \cdot \Phi(\mathbf{x}) + b + \xi_i, i = 1,2,\ldots l \qquad (2)$$

In which, $\Phi(\mathbf{x}_i)$ is the high-dimensional space nonlinear mapping of $\mathbf{x}_i$, $\mathbf{w}$ and b are decision function parameters, c is the reguarization plarameter, ξ_i is the prediction error of training sample. The problem mentioned above can be transformed into formula (3) by using related Lagrange function.

$$\begin{bmatrix} 0 & 1 & 1 & \cdots & 1 \\ 1 & K(\mathbf{x}_1,\mathbf{x}_1)+\dfrac{1}{2c} & K(\mathbf{x}_1,\mathbf{x}_2) & \cdots & K(\mathbf{x}_1,\mathbf{x}_l) \\ 1 & K(\mathbf{x}_2,\mathbf{x}_1) & K(\mathbf{x}_2,\mathbf{x}_2)+\dfrac{1}{2c} & \cdots & K(\mathbf{x}_2,\mathbf{x}_l) \\ \vdots & \vdots & \vdots & \ddots & \vdots \\ 1 & K(\mathbf{x}_l,\mathbf{x}_1) & K(\mathbf{x}_l,\mathbf{x}_2) & \cdots & K(\mathbf{x}_l,\mathbf{x}_l)+\dfrac{1}{2c} \end{bmatrix}$$
$$\begin{bmatrix} b \\ \alpha_1 \\ \alpha_2 \\ \vdots \\ \alpha_l \end{bmatrix} = \begin{bmatrix} 0 \\ y_1 \\ y_2 \\ \vdots \\ y_l \end{bmatrix} \qquad (3)$$

Define

$$\mathbf{H} = \begin{bmatrix} K(\mathbf{x}_1,\mathbf{x}_1)+\dfrac{1}{2c} & K(\mathbf{x}_1,\mathbf{x}_2) & \cdots & K(\mathbf{x}_1,\mathbf{x}_l) \\ K(\mathbf{x}_2,\mathbf{x}_1) & K(\mathbf{x}_2,\mathbf{x}_2)+\dfrac{1}{2c} & \cdots & K(\mathbf{x}_2,\mathbf{x}_l) \\ \vdots & \vdots & \ddots & \vdots \\ K(\mathbf{x}_l,\mathbf{x}_1) & K(\mathbf{x}_l,\mathbf{x}_2) & \cdots & K(\mathbf{x}_l,\mathbf{x}_l)+\dfrac{1}{2c} \end{bmatrix}$$

$\mathbf{y} = [y_1, y_2, \cdots, y_l]^T$; $\mathbf{e} = [1,\cdots,1]_{1\times l}^T$; $\boldsymbol{\alpha} = [\alpha_1,\alpha_2,\cdots,\alpha_l]^T$;

then we can get

$$\begin{cases} \alpha = H^{-1}y - H^{-1}e \cdot \dfrac{e^T H^{-1} y}{e^T H^{-1} e} \\ b = \dfrac{e^T H^{-1} y}{e^T H^{-1} e} \end{cases} \qquad (4)$$

In which $K(\mathbf{x}_i,\mathbf{x}_j) = \langle \Phi(\mathbf{x}_i), \Phi(\mathbf{x}_j) \rangle$ is kernel function.

Substituting (4) into (5), the decision function can be expressed as

$$f(\mathbf{x}) = \sum_{j=1}^{l} \alpha_j K(\mathbf{x},\mathbf{x}_j) + b \qquad (5)$$

When large deviations occur in the model prediction, the algorithm updates online in the following procedure.

Firstly, a sample point which is the closest to the current new sample point $\mathbf{x}_{new}$ is found in the original sample, as is shown in formula (6)

$$i = \arg\left(\min_{k=1,\cdots,l} \|\mathbf{x}_{new} - \mathbf{x}_k\|\right) \qquad (6)$$

Secondly, exchange the last row and the last column of the i-th sample in matrix H with the corresponding row and column, then we have $\mathbf{H}_1$, as is shown in formula (7)

$$\mathbf{H}_1 = \mathbf{I}_{Ri \leftrightarrow Rl} \mathbf{H} \mathbf{I}_{Li \leftrightarrow Ll} \qquad (7)$$

$$\mathbf{H}_1 = \begin{bmatrix} K(\mathbf{x}_l,\mathbf{x}_l)+\dfrac{1}{2c} & \cdots & K(\mathbf{x}_l,\mathbf{x}_i) & \cdots & K(\mathbf{x}_l,\mathbf{x}_l) \\ \vdots & \ddots & \vdots & \ddots & \vdots \\ K(\mathbf{x}_i,\mathbf{x}_l) & \cdots & K(\mathbf{x}_i,\mathbf{x}_i)+\dfrac{1}{2c} & \cdots & K(\mathbf{x}_i,\mathbf{x}_l) \\ \vdots & \ddots & \vdots & \ddots & \vdots \\ K(\mathbf{x}_l,\mathbf{x}_l) & \cdots & K(\mathbf{x}_l,\mathbf{x}_i) & \cdots & K(\mathbf{x}_l,\mathbf{x}_l)+\dfrac{1}{2c} \end{bmatrix}$$

In which, $\mathbf{I}_{Li \leftrightarrow Ll}$ represents exchanging the i-th row and the l-th row in an identity matrix; $\mathbf{I}_{Ri \leftrightarrow Rl}$ exchanging the i-th column and the l-th column in an identity matrix.

$$\mathbf{I}_{Ri \leftrightarrow Rl} = \mathbf{I}_{Li \leftrightarrow Ll} = \mathbf{I}_{Ri \leftrightarrow Rl}^{-1} = \mathbf{I}_{Li \leftrightarrow Ll}^{-1}.$$

Define $\mathbf{H}_1 = \begin{bmatrix} G & \mathbf{g}_i \\ \mathbf{g}_i^T & K(\mathbf{x}_i,\mathbf{x}_i)+\frac{1}{2c} \end{bmatrix}$

$$\mathbf{H}_1^{-1} = \begin{bmatrix} \mathbf{h}_{11} & \mathbf{h}_{12} \\ \mathbf{h}_{21} & \mathbf{h}_{22} \end{bmatrix} = \mathbf{I}_{Ri \leftrightarrow Rl} \mathbf{H}^{-1} \mathbf{I}_{Li \leftrightarrow Ll} \qquad (8)$$

The related parameters of the new sample can be calculated according to the following formulas:

$$\mathbf{g}_{new} = \begin{bmatrix} K(\mathbf{x}_1,\mathbf{x}_{new}),\cdots,K(\mathbf{x}_{i-1},\mathbf{x}_{new}), \\ K(\mathbf{x}_l,\mathbf{x}_{new}),K(\mathbf{x}_{i+1},\mathbf{x}_{new}),\cdots,K(\mathbf{x}_{l-1},\mathbf{x}_{new}) \end{bmatrix}_{l \times 1}^T \qquad (9)$$

$$k_{new} = K\left(\mathbf{x}_{new}, \mathbf{x}_{new}\right) + \frac{1}{2c} \tag{10}$$

$$r_{new} = k_{new} - \mathbf{g_{new}^T G^{-1} g_{new}} \tag{11}$$

$$\mathbf{G}^{-1} = \mathbf{h}_{11} - \mathbf{h}_{12}\mathbf{h}_{22}^{-1}\mathbf{h}_{21} \tag{12}$$

$$\mathbf{H}_2^{-1} = \begin{bmatrix} \mathbf{G}^{-1} + \mathbf{G}^{-1}\mathbf{g_{new}}r_{new}^{-1}\mathbf{g_{new}^T G^{-1}} & -\mathbf{G}^{-1}\mathbf{g_{new}}r_{new}^{-1} \\ -r_{new}^{-1}\mathbf{g_{new}^T G^{-1}} & r_{new}^{-1} \end{bmatrix} \tag{13}$$

Finally, the coefficients of the decision function after updating can be obtained as is shown in formula (14).

$$\begin{cases} \boldsymbol{\alpha}^* = \mathbf{H}_2^{-1}\mathbf{y}^* - \mathbf{H}_2^{-1}\mathbf{e} \cdot \dfrac{\mathbf{e^T H_2^{-1} y^*}}{\mathbf{e^T H_2^{-1} e}} \\[2mm] b^* = \dfrac{\mathbf{e^T H_2^{-1} y^*}}{\mathbf{e^T H_2^{-1} e}} \end{cases} \tag{14}$$

In which, $\mathbf{y}^* = [y_1, \cdots, y_{i-1}, y_l, y_{i+1}, \cdots y_{l-1}, y_{new}]^T$.

2.2 Algorithm improvement

2.2.1 Offline support vector screening
Since Least Square Support Vector Machine (LSSVM) does not have sparsity, it is necessary to carry out offline support vector screening to remove redundant support vectors, which may also help simplify model structure and decrease online computation.

A nonlinear mapping $\Phi(\mathbf{x})$ is designed to map l samples $\left(\mathbf{x}_1, \mathbf{x}_2, \cdots, \mathbf{x}_l\right)$ to high dimensional feature space, and the corresponding mapping vector is $\left(\Phi(\mathbf{x}_1), \Phi(\mathbf{x}_2), \cdots \Phi(\mathbf{x}_l)\right)$, from which a subset can be chosen as the representative of the whole training set[8].

Suppose that any vector in $\boldsymbol{\Phi}_S\left(\Phi(\mathbf{x}_{s1}), \Phi(\mathbf{x}_{s2}) \cdots \Phi(\mathbf{x}_{sm})\right)$ has linear representation by $\left(\Phi(\mathbf{x}_1), \Phi(\mathbf{x}_2), \cdots \Phi(\mathbf{x}_l)\right)$, which means that there exists $A = [a_1, a_2, \cdots a_m]^T$ that satisfies $\Phi(\mathbf{x}_i) = \boldsymbol{\Phi}_S \cdot A$. If there exists corresponding high dimensional feature vector $\Phi\left(\mathbf{x}_{new}\right)$ of any new training sample $\mathbf{x}_{new}$, the closet feature vector $\tilde{\boldsymbol{\Phi}} = \boldsymbol{\Phi}_S \cdot A$ can be found in the original feature space.

$$\delta = \frac{\left\| \Phi\left(\mathbf{x}_{new}\right) - \tilde{\Phi} \right\|^2}{\left\| \Phi\left(\mathbf{x}_{new}\right) \right\|^2}$$
$$= \frac{\left(\Phi\left(\mathbf{x}_{new}\right) - \boldsymbol{\Phi}_S \cdot A\right)^T \left(\Phi\left(\mathbf{x}_{new}\right) - \boldsymbol{\Phi}_S \cdot A\right)}{\left\| \Phi\left(\mathbf{x}_{new}\right) \right\|^2} \tag{15}$$

Suppose there exists $A = [a_1, a_2, \cdots a_m]^T$ which can make the European distance δ reach minimum value, then we have

$$\frac{\partial \delta}{\partial \mathbf{A}} = \frac{2\left(\Phi_S^T \Phi_S\right)A - 2\Phi_S^T \Phi\left(\mathbf{x}_{new}\right)}{\Phi\left(\mathbf{x}_{new}\right)^T \Phi\left(\mathbf{x}_{new}\right)} \tag{16}$$

Set

$$\frac{\partial \delta}{\partial \mathbf{A}} = 0 \Rightarrow \mathbf{A} = \left(\Phi_S^T \Phi_S\right)^{-1} \Phi_S^T \Phi\left(\mathbf{x}_{new}\right) \tag{17}$$

The matrix $\left(\boldsymbol{\Phi}_S^T \boldsymbol{\Phi}_S\right)$ composed of feature vectors is reversible according to the principle of linear algebra.

Substitute (17) into (15), and we can get:

$$\min_A \delta = 1 - \frac{1}{\left\| \Phi\left(\mathbf{x}_{new}\right) \right\|^2} \left(\Phi\left(\mathbf{x}_{new}\right)^T \Phi_S \left(\Phi_S^T \Phi_S\right)^{-1} \Phi_S^T \Phi\left(\mathbf{x}_{new}\right)\right) \tag{18}$$

Which can be rewritten as:

$$\min_A \delta = 1 - J_{new} \tag{19}$$

$$J_{new} = \frac{\left(\Phi\left(\mathbf{x}_{new}\right)^T \Phi_S \left(\Phi_S^T \Phi_S\right)^{-1} \Phi_S^T \Phi\left(\mathbf{x}_{new}\right)\right)}{\left\| \Phi\left(\mathbf{x}_{new}\right) \right\|^2}$$
$$= \frac{K_{Snew}^T K_{SS}^{-1} K_{Snew}}{K\left(\mathbf{x}_{new}, \mathbf{x}_{new}\right)} \tag{20}$$

In which,

$$\mathbf{K}_{SS} = \begin{bmatrix} K\left(\mathbf{x}_{s1}, \mathbf{x}_{s1}\right) & \cdots & K\left(\mathbf{x}_{sm}, \mathbf{x}_{s1}\right) \\ \vdots & \ddots & \vdots \\ K\left(\mathbf{x}_{sm}, \mathbf{x}_{s1}\right) & \cdots & K\left(\mathbf{x}_{sm}, \mathbf{x}_{sm}\right) \end{bmatrix};$$

$$K_{Snew} = \left[K\left(\mathbf{x}_{s1}, \mathbf{x}_{new}\right) \quad \cdots \quad K\left(\mathbf{x}_{sm}, \mathbf{x}_{snew}\right) \right]^T.$$

Meanwhile, δ can be expressed as $\delta = \sin^2(\theta)$, in which θ is the angle between $\Phi\left(\mathbf{x}_{new}\right)$ and $\tilde{\Phi}$ ($0 \le \delta \le 1$). When the two vectors get closer, $\theta \to 0$, $\delta \to 0$, therefore, whether the new training sample is the new support vector can be judged according to the value of δ. Under given threshold value ρ, if $|1 - J_{new}| \ge \rho$, add the sample into the training set, otherwise remove it due to the existence of the linear relation of the sample in the training set. The smaller ρ is, the stricter the sample screening is. Thus, a simplified support vector set can be obtained by using this method.

2.2.2 Online updating algorithm improvement
The online updating of original online least squares support vector machine algorithm (Gu, 2011) is carried out only through the replacement of support vectors. To better adapt to combustion

system characteristic transformation caused by the variation of load, coal quality and equipment and so on, the improved algorithm proposed in this paper adds the addition and deletion strategy of support vectors.

2.2.2.1 The addition strategy of support vectors

If no sample can be found in original sample set to be similar to the new sample, which can be expressed as

$$\min_{k=1,\cdots,l}\|\mathbf{x}_{new},\mathbf{x}_k\| > \max D \tag{21}$$

In which, maxD is the threshold used to determine whether the two vectors are similar, then the new sample $\{\mathbf{x}_{new}, y_{new}\}$ can be directly added into the sample set to calculate new $\mathbf{H}_3^{-1}$, as is shown in formula (22), and new decision function coefficients $\boldsymbol{\alpha}^* = [\alpha_1^*, \alpha_2^* \ldots, \alpha_l^*, \alpha_{new}^*]_{1\times(l+1)}^T$ and b* can be calculated through formula (23).

$$\mathbf{H}_3^{-1} = \begin{bmatrix} \mathbf{H}^{-1} + \mathbf{H}^{-1}\mathbf{g}_{new}r_{new}^{-1}\mathbf{g}_{new}^T\mathbf{H}^{-1} & -\mathbf{H}^{-1}\mathbf{g}_{new}r_{new}^{-1} \\ -r_{new}^{-1}\mathbf{g}_{new}^T\mathbf{H}^{-1} & r_{new}^{-1} \end{bmatrix} \tag{22}$$

$$\begin{cases} \alpha^* = \mathbf{H}_3^{-1}\mathbf{y}^* - \mathbf{H}_3^{-1}\mathbf{e} \cdot \dfrac{\mathbf{e}^T\mathbf{H}_3^{-1}\mathbf{y}^*}{\mathbf{e}^T\mathbf{H}_3^{-1}\mathbf{e}} \\ b^* = \dfrac{\mathbf{e}^T\mathbf{H}_3^{-1}\mathbf{y}^*}{\mathbf{e}^T\mathbf{H}_3^{-1}\mathbf{e}} \end{cases} \tag{23}$$

In which $\mathbf{y}^* = [y_1, \cdots, y_l, y_{new}]^T$.

2.2.2.2 The deletion strategy of support vectors

As with the increase of support vectors, the developed model will be more complicated and require larger storage space. Moreover, the online updating speed will be also affected. So when the sample size reaches the maximum value, the deletion strategy of support vectors will be used on condition that large model error occurs.

Firstly new sample is added into the training set, and relevant parameters are calculated as formula (22). Then use formula (24) to find the most similar two samples in historical data, and remove the old sample in time order.

$$\{k,m\} = \arg \min_{k,m=1,\cdots N, k\neq m}\|\mathbf{x}_k - \mathbf{x}_m\| \tag{24}$$

In which N is the maximum number of samples set in advance.

Suppose the k-th sample needs to be removed, exchange the last row and the last column of the k-th sample in matrix $\mathbf{H}_3^{-1}$ with corresponding row and column, as is shown in formula (25).

$$\mathbf{H}_4^{-1} = \mathbf{I}_{Ri\leftrightarrow R(N+1)}\mathbf{H}_3^{-1}\mathbf{I}_{Li\leftrightarrow L(N+1)} = \begin{bmatrix} \mathbf{h}_{11}^* & \mathbf{h}^{*12} \\ \mathbf{h}_{21}^* & h_{22}^* \end{bmatrix} \tag{25}$$

Define

$$G^{*-1} = \mathbf{h}_{11}^* - \mathbf{h}_{12}^* h_{22}^{*-1} \mathbf{h}_{21}^* \tag{26}$$

Then, new decision function coefficients are calculated according to formula (27).

$$\begin{cases} \alpha^* = \mathbf{G}^{*-1}\mathbf{y}^* - \mathbf{G}^{*-1}\mathbf{e} \cdot \dfrac{\mathbf{e}^T\mathbf{G}^{*-1}\mathbf{y}^*}{\mathbf{e}^T\mathbf{G}^{*-1}\mathbf{e}} \\ b^* = \dfrac{\mathbf{e}^T\mathbf{G}^{*-1}\mathbf{y}^*}{\mathbf{e}^T\mathbf{G}^{*-1}\mathbf{e}} \end{cases} \tag{27}$$

2.3 FVS-ALS computation procedure

FVS-ALS algorithm includes two parts: offline computing and online computing:

1. Offline calculation steps:
   - S1: Organize sample data according to the input output structure of dynamic model.
   - S2: Screen the support vectors according to formula (19).
   - S3: Calculate off-line parameters of boiler combustion dynamic model by using filtered samples, as is shown in formula (4).

2. Online calculation steps:
   - S1: Calculate the error err between model prediction and actual value.
   - S2: Judge whether the error exceeds the maximum allowable error ERR, If $|err| < ERR$, the dynamic model does not need to update, the model parameters remain unchanged, and jump to step 1; if $|err| > ERR$, correct the dynamic model, and jump to step 3.
   - S3: If $\min_{k=1,\cdots,l}\|\mathbf{x}_{new} - \mathbf{x}_k\| \leq \max D$, replace the corresponding sample, search for the corresponding sample in formula (6) and jump to step 4; Otherwise, consider the new sample as new support vector according to $\min_{k=1,\cdots,l}\|\mathbf{x}_{new} - \mathbf{x}_k\| > \max D$, and add it to the training set. Then jump to step 5.
   - S4: Update the model parameters through sample replacement, as is shown in formula (14), and jump to step 1.
   - S5: Compare the sample number n with the given maximum sample number N. If n<N, then jump to step 6, otherwise, jump to step 7.
   - S6: Update model parameters according to the addition strategy of support vectors in formula (23), and jump to step 1. S7: Call the deletion strategy of support vectors to ensure the total amount of sample less than N, meanwhile re-adjust the model parameters according to formula (27), and jump to step 1.

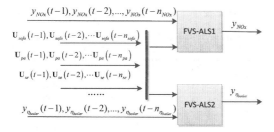

Figure 1. The input and output structure diagram of boiler combustion dynamic modeling.

3 DESIGN OF THE BOILER COMBUSTION DYNAMIC MODEL

The boiler combustion system of a 600 MW subcritical unit is taken as the modeling object in this paper. The SG-2028/17.5-YM908 boiler produced by Shanghai Boiler Works Ltd is a tangentially firing, once reheat, single furnace balanced draft, solid state slag, closed seal, all-steal π-type drum boiler. The furnace has six-layer primary air, seven-layer secondary air and seven-layer overfire air after later furnace reformation.

The boiler combustion dynamic model built in the paper considers the total amount of coal and air injected into the furnace, the auxiliary air, and oxygen and so on as the main factors that influence the emission and efficiency of the boiler. In which, the seven damperopenings of secondary air $\mathbf{U}_{se}$ represent the influence of seven-layer secondary air; the six damper openings of primary air $\mathbf{U}_{pa}$ represent the influence of six-layer primary air; the seven damper openings of overfire air $\mathbf{U}_{sofa}$ represent the influence of seven-layer overfire air. The model outputs are the NOx emission $\mathbf{y}_{NOx}$ and the boiler efficiency $\mathbf{y}_{\eta boiler}$. In comparison with steady-state modeling, the dynamic modeling considers orders of the input and the output variables. The structure of the developed dynamic model of the boiler combustion system is shown in Fig. 1.

In which, n_{pa}, n_{se} and n_{sofa} represent orders of the input variables of the model respectively, and n_{NOx} and $n_{\eta boiler}$ represent orders of the output variables of the model.

4 THE ESTABLISHMENT AND VALIDATION OF BOILER COMBUSTION DYNAMIC MODEL

4.1 Sample preparation

The dynamic model of boiler combustion system is developed based on routine operating data. The sample time is 20 seconds. The sample data is organized according to the model structure shown

in Fig. 1, in which 1000 groups of data are chosen as the training sample and the other 15000 groups of data are chosen as the test sample. To improve the model performance, the training data needs to cover the operation range of test condition. Besides, sample data normalization should be performed using (28) before model training.

$$x' = (x - \bar{x}) / \sigma_x \qquad (28)$$

In which, x, x' are the sample value before and after normalization respectively; $\bar{x}$, σ_x are the mean value and standard deviation of the sample data respectively.

4.2 Model parameter

Radial basis function is employed as the kernel function of FVS-ALS, ALS-SVM and normal LSSVM respectively due to its preferable function, as shown in formula (29).

$$K\left(\mathbf{x}_i, \mathbf{x}_j\right) = \exp\left(-\left\|\mathbf{x}_i - \mathbf{x}_j\right\|^2 / 2\sigma^2\right) \qquad (29)$$

In which, σ is a width parameter of the kernel function.

Model parameters settings of different algorithms in following simulations are shown in Table 1.

4.3 Simulation comparison

The boiler efficiency model and NOx emission model are developed by FVS-ALS and ALS-SVM, and the performance comparisons are shown in Table 2; the predicted value and actual value are shown in Fig. 2. The simulation results indicate that modeling by FVS-AL and ALS-SVM has good dynamic characteristics which can clearly show how the NOx emission and the boiler efficiency change with the load. Moreover, both the precision and predictive ability are satisfied. Compared with ALS-SVM model, the update time is restricted to the millisecond level due to the reduced original sample and simplified model configuration of the FVS-ALS model, thus providing enough time for the dynamic optimization algorithm in the same control period. The whole run time of FVS-ALS model is less than one fifth that of ALS-SVM model, therefore higher operating efficiency can be obtained.

The comparison results of steady-state model and dynamic model based on FVS-ALS algorithm are shown in Table 3, and the predicted value and actual value are shown in Fig. 3. The variation trend of boiler NOx emission and boiler efficiency obtained by the steady-state model is basically the

Table 1. The comparisons among different algorithm model parameters.

Algorithm	Model	SVM filter setpoint ρ	Maximum permissible error ERR	nuclear parameter σ	regularization parameter c
FVS-ALS	NOx model	0.005	$5/mg \cdot m^{-3}$	500	1000
	Efficiency model	0.005	0.2/%	400	800
ALS-SVM	NOx model	/	$5/mg \cdot m^{-3}$	500	1000
	Efficiency model	/	0.2/%	400	800
LSSVM	NOx model	/	/	500	1000
	Efficiency model	/	/	400	800

Table 2. The performance comparisons between FVS-ALS and ALS-SVM algorithms.

Algorithm	Model	Modeling sample group number	Model update number	Error standard deviation	Update date/s	Whole run time/s
FVS-ALS	NOx model	288	285	$1.54/mg \cdot m^{-3}$	0.043	23.9
	Efficiency model	17	26	0.0207/%	0.0013	
ALS-SVM	NOx model	700	410	$1.5/mg \cdot m^{-3}$	0.35	137
	Efficiency model	500	27	0.0299/%	0.2	

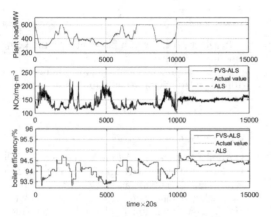

Figure 2. The dynamic modeling comparison between FVS-ALS and ALS.

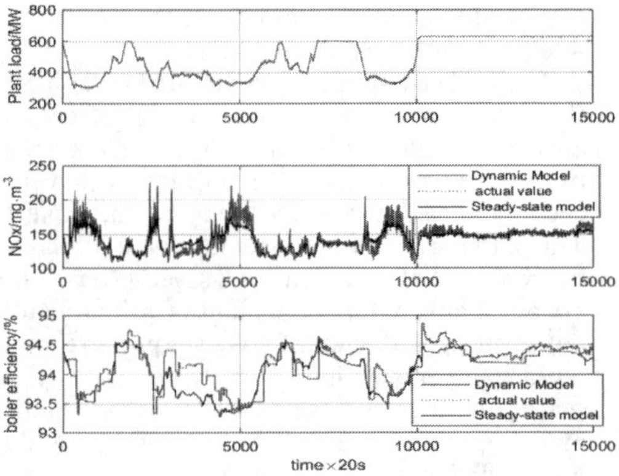

Figure 3. Comparison of dynamic modeling and steady state modeling.

Table 3. The comparison between steady-state and dynamic modeling.

	Model	Modeling sample group number	Model update number	Error standard deviation
Dynamic model	NOx model	288	285	$1.54/mg \cdot m^{-3}$
	Efficiency model	17	26	0.0207/%
Steady-state model	NOx model	501	3350	$8.5/mg \cdot m^{-3}$
	Efficiency model	223	4235	0.23/%

same as the actual value, however the modeling precision is too low to meet the later requirements of dynamic optimization.

The adaptivity of FVS-ALS model and LSSVM model is compared in Fig. 4. Compared with LSSVM model, FVS-ALS model has better sparsity since support vector screening has been executed. Moreover, when large deviation appears, it can update the model parameters online according to the new sample and thus reduce model forecasting error.

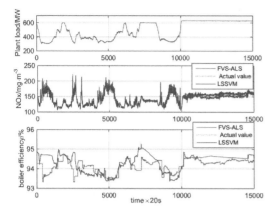

Figure 4. Online Adaptability verification.

5 CONCLUSIONS

This paper proposes an improved online adaptive least squares support vector machine dynamic modeling algorithm, by which a dynamic model of the boiler combustion system is developed based on the operating data of a 600 MW unit. Simulation results have indicated that the developed model can reflect the dynamic characteristic of the boiler efficiency and NOx emission changing with different load condition accurately. Compared with the original steady-state model obtained by online adaptive least squares support vector machine algorithm, the newly model has not only higher precision and forecasting ability but also simplified model structure and reduced online calculation. Moreover, the proposed model will be the base for future reserch on the dynamic optimization control strategy of boiler combustion system.

REFERENCES

Chang X, Jian-hong L.U, Yuan Z, et al. A boiler combustion global optimization on efficiency and low NO (x) emissions object[J]. Proceedings of the CSEE, 2006, 26(4): 46–50.

Chang X.U, L Jian-hong, Zhen Yuan. Minimal Resource Allocation Networks and Application for A Power Station Boiler [J]. Proceedings of the Csee, 2004, 11.

Gu Y, Zhao W, Wu Z. Online adaptive least squares support vector machine and its application in utility boiler combustion optimization systems[J]. Journal of Process Control, 2011, 21(7): 1040–1048.

Gu Y, Zhao W, Wu Z. Online adaptive least squares support vector machine and its application in utility boiler combustion optimization systems[J]. Journal of Process Control, 2011, 21(7): 1040–1048.

Gu Y.P, Zhao W.J, Wu Z.S. Combustion optimization for utility boiler based on least square-support vector machine[C]. Chinese Society for Electrical Engineering, 2010, 30(17): 91–97.

Guo Z, Song Z, Mao J. An improved online least squares support vector machines regression algorithm [J]. Control and Decision, 2009, 1: 031.

Keerthi S.S, Lin C.J. Asymptotic behaviors of support vector machines with Gaussian kernel[J]. Neural computation, 2003, 15(7): 1667–1689.

Leung F.H.F, Lam H.K, Ling S.H, et al. Tuning of the structure and parameters of a neural network using an improved genetic algorithm[J]. IEEE Transactions on Neural networks, 2003, 14(1): 79–88.

Liu J, Zio E. An adaptive online learning approach for Support Vector Regression: Online-SVR-FID[J]. Mechanical Systems and Signal Processing, 2016, 76: 796–809.

Wang C.L, Zhou H, Li G.N, et al. Support vector machine and genetic algorithms to optimize combustion for low NO (x) emission [C]. Proceedings of the Chinese Society of Electrical Engineering. 2007, 27(11): 40–44.

Wang D, Jiang B. Online sparse least square support vector machines regression[J]. Control and decision, 2007, 22(2): 132.

Wang P, Li L, Chen Q, et al. Research on Applications of artificial Intelligence to Combustion Optimization in a Coal-Fired Boiler [J]. Proceedings of the Csee, 2004, 4.

Xinran Z, Zhaosheng T, Xingjun J. Predictive control using sparse online non-bias LSSVM[J]. Journal of Electronic Measurement and Instrument, 2011, 25(4): 331–337.

Zhou H, Zhao J.P, Zheng L.G, et al. Modeling NO x emissions from coal-fired utility boilers using support vector regression with ant colony optimization[J]. Engineering Applications of Artificial Intelligence, 2012, 25(1): 147–158.

Zhou X, Teng Z. An Online Sparse LSSVM and Its Application in System Modeling [J]. Journal of Hunan University (Natural Sciences), 2010, 4: 009.

Advances in Energy Science and Equipment Engineering II – Zhou, Patty & Chen (Eds)
© *2017 Taylor & Francis Group, London, ISBN 978-1-138-71798-5*

A compositional analysis for the flashed-off oils of the liquid–liquid equilibrium of (deasphalted-oil + *i*-butane) systems

Chuanbo Yu & Jianmei Deng
Panzhihua University, Panzhihua, China

Chong Zhang & Suoqi Zhao
China University of Petroleum, Beijing, China

ABSTRACT: For consuming the DeAsphalted-Oil (DAO) in high-efficiency, the DAO was split into the Light DeAsphalted-Oil (LDAO) and the Heavy DeAsphalted-Oil (HDAO) by using a novel de-asphalting process called "SELEX-Asp". This process was tested at the plant scale in 2010; however, the lack of data and model for the phase equilibrium of the (DAO + solvent) system hindered the optimization of the dea-sphalting process. Therefore, in this work, the liquid–liquid equilibrium of (LDAO/ HDAO + *i*-butane) systems was investigated. The results showed that the extraction yield for the LDAO (28.95–70.43 wt%) is greater than that of the HDAO (10.03–20.76 wt%). The results of the composition analysis for DAOs and their flashed-off oils showed that, (1) the carbon number distributions for components appear as left-skewed curves and (2) the dominant components are C30–90. The results revealed that the quantity of DAO has a significant influence on the extraction yield and the quality of extracted oils.

1 INTRODUCTION

With the depletion of light crude oils, the Solvent DeAsphalting process (SDA) is widely used in the refinery plants for exacting the light fraction from the vacuum residue. The common SDA might obtain one DeAsphalted Oil (DAO) and one De-Oil Asphalt (DOA). For consuming DAO in high-efficiency, we created a selective extraction of the de-asphalting process "SELEX-Asp" (Zhao et al. 2009) to split DAO into the two DAOs: the Light DAO (LDAO) and the Heavy DAO (HDAO). The merit is that the light components are enriched in LDAO and the resins are enriched in HDAO. This process was tested at the plant scale in 2010 (Zhao et al. 2010); however, the lack of data and model for the phase equilibrium of the (DAO + solvent) system hindered the optimization of the SELEX-Asp process.

In our previous studies, the liquid–liquid equilibrium of the (DAO + *i*-butane) system by using SRK-EOS (Yu et al. 2015) is simulated. The model might predict both the extraction yield and the phase behaviors; nevertheless, it failed to predict well the compositions distribution at the same time.

The simulated distillation can quickly simulate the boiling point distillation and obtain the composition distribution (Neer & Deo 1995, Wen et al. 2007, Zhao 2009). In a recent paper, the traditional equilibrium constant, k-value, was used for describing the composition distribution

of the liquid–liquid equilibrium of (Athabasca bitumen + ethane) systems (Kariznovi et al. 2013). This work focused on the liquid–liquid equilibrium of (Athabasca bitumen DAO + i-butane). The phase experiment was performed with accurate PVT equipment under the appropriate conditions for the splitting liquid–liquid phase. The DAOs and the flashed-off oils were subjected to the high-temperature simulated distillation for obtaining the carbon number distributions from C22 to C120.

2 EXPERIMENTAL SECTION

2.1 *Procedure*

Figure 1 shows the procedure of the liquid–liquid equilibrium. Firstly, the LDAO and HDAO are obtained from the first-stage and the second-stage separator of the SELEX-Asp process. The solvent is pentane. The representative conditions are as follows: the pressure is 5 Mpa, temperature is the 160°C, and the ratio of the solvent to the oil sample (S/O) is the 4.0. The detailed experimental methods have been described elsewhere (Fan et al. 2011).

Secondly, the liquid–liquid equilibrium was performed with accurate PVT equipment, RUSKA 2370–601 A. The "DAO + Solvent" systems were split into two conjugated phases: the light phase L_1 and the heavy phase L_2. Considering compre-

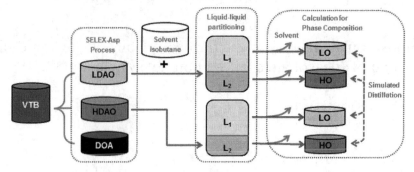

Figure 1. Schematic of the procedure.

hensively the operation parameters for the plant scale, the rheology of DAO, and the splitting of liquid–liquid, the experimental conditions were set as the pressure is 4–7 Mpa, temperature is 80–120°C, and the ratio of the solvent to DAO (S/O) is 1.4–3.4. The detailed operational methods have been described elsewhere (Zhao et al. 2006, Yu et al. 2015).

Thirdly, the phase L_1 and phase L_2 were flashed at ambient temperature and atmosphere, respectively. The flashed-off Light Oil (LO) and the Heavy Oil (HO) were subjected to the analysis of oil components.

2.2 Materials

LDAO and HDAO were extracted from the Athabasca bitumen vacuum tower bottoms (VTB, > 520°C) by using the SELEX-Asp process (Zhao et al. 2010). The properties of the VTB, LDAO, and HDAO are shown in Table 1.

2.3 Analytical method

2.3.1 Phase compositions analysis

The LP or HP was flashed at the ambient temperature and atmosphere. The samples of evolved oils were collected for the weight analysis and the compositions analysis.

2.3.2 High-Temperature Gas Chromatography analysis

The DAOs and the flashed-off oils (LO and HO) were subjected to the High-Temperature Gas Chromatography (HT-GC) simulated distillation for obtaining the carbon number distributions. The HT-GC analysis was performed by using the Agilent 6890 N with a HT-750 capillary column (Zhou et al. 2014) (similar to ASTM D7500–14). By using this method, the boiling point of eluted components was up to 750°C. Therefore, it is often applied in the compositional analyses for the heavy vacuum residues.

Table 1. List of properties of deasphalted oil.

Properties		VTB	LDAO	HDAO
Specific gravity, 15.5°C		1.0548	1.0042	1.0336
100 °C viscosity (mPaS)		26480	454.0	5473
Molecular weight (MW)		1147	648	856
Molecular formula			$C_{45}H_{69}N_{0.28}$	$C_{59}H_{85}N_{0.48}$
Average, T_b (K)			859.8	889.7
Residue carbon, wt%		24.9	10.5	20.0
SARA	Saturate, wt.%	7.8	14.2	10.3
	Aromatic, wt.%	41.5	55.8	47.0
	Resin, wt.%	32.6	29.9	38.2
	Asphalt, wt.%	18.1	0.1	4.5

3 RESULTS AND DISCUSSION

3.1 Solvent choice

Because the DAO has no low boiling point component and no asphaltene, its distribution of components is narrower than the bitumen or vacuum residue. Therefore, the equilibrium conditions for the LLE are more rigorous than the usual heavy oil. We tried propane, n-butane, i-butane, and n-pentane as solvents. Experiments showed that propane and n-butane are not suitable for the DAO in this study. The reason is that the solvability for n-propane is too small, while that of n-butane or n-pentane is too large.

Figure 2 showed the P-T phase diagram of the LDAO + isobutene system. From point A to D demonstrates a representative process of phase transition at the temperature of 90°C: (1) when the pressure is beyond 7.5 Mpa, point A, the phase system is a mono liquid phase, and it looks dark; (2) when the pressure decreases down to 7.5 Mpa, point B, the phase system is split into two liquid phases, and the light phase appears dark red in color; (3) when the pressure decreases to 2.6 Mpa, point C, the first bubble appears, and the phase system becomes three phases; (4) when the pressure decreases to point D, one of the liquid phases completely vaporizes leaving only two phases (L-V). The results of the pre-experiment show i-butane and provide the enough operation flexibility for splitting and sampling the light phase and the heavy phase.

3.2 *Liquid–liquid equilibrium*

Table 2 lists the weight fractions of each liquid phase, the weight fraction of oil in each liquid phase, and the extraction yield of the light oil under equilibrium conditions. The extraction yield is defined as the mass of oil in the light phase divided by the mass of oil in the feed.

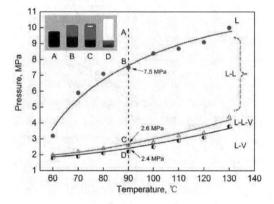

Figure 2. Graph showing the phase behavior of LDAO + *i*-butane (S/O = 2.3 w/w).

Table 2 shows that there is significantly more mass of the light liquid phase, L_1, than the heavy liquid phase, L_2. For an LDAO + i-butane system, the weight fraction of phase L1 is about 78%; the weight fraction of phase L_2 is about 22%. Moreover, phase L_1 has a much higher solvent concentration than phase L_2. Phase L_1 is solvent-enriched and phase L_2 is oil-enriched. For the LDAO + *i*-butane system, the solvent concentration is about 80% in phase L_1; the oil concentration is about 67% in phase L_2. It indicates that most amount of the solvent existed in phase L_1 during the process of liquid–liquid separation. Therefore, it suggests that the recovery of the solvent should carefully be evaluated.

The results show the extraction yield for the LDAO (28.95–70.43, wt%) is greater than that for the HDAO (10.03–20.76, wt%). The reason is that the content of the light component for LDAO is greater than that of HDAO. It indicated the liquid–liquid separation is more effective for LDAO than HDAO.

As temperature increases from 80°C to 120°C, the extraction yield of the light component for the LDAO decreases from 70.43% to 39.13%. The extraction yield of the light component for the HDAO decreases from 17.50% to 10.03%. It indicated that the extraction yield decreases with an increase in the temperature.

As the pressure increases from 4.0 MPa to 7.0 MPa, the extraction yield of the light component for the LDAO increases from 42.87% up to 61.67%. It indicated that the extraction yield increases with an increase in the pressure.

As the S/O increases from 1.4 to 3.4, the extraction yield of the light component for the LDAO increases from 28.95% up to 61.63%. It indicated that the extraction yield increases with an increase in the S/O.

Table 2. Properties for the liquid–liquid equilibrium of the (LDAO/HDAO + i-butane) system.

Conditions				wt%		wt%		wt%
Materials	p/MPa	T/°C	S/O	L_1	L_2	LO	HO	Extraction yield
LDAO	5.0	80	2.3	86.51	13.49	19.32	8.11	70.43
LDAO	5.0	90	2.3	81.49	18.51	18.13	11.36	61.49
LDAO	5.0	120	2.3	73.43	26.57	11.68	18.17	39.13
LDAO	4.0	90	2.3	75.63	24.37	12.59	16.78	42.87
LDAO	7.0	90	2.3	81.09	18.91	18.55	11.53	61.67
LDAO	5.0	90	1.4	65.10	34.90	11.72	28.78	28.95
LDAO	5.0	90	3.4	85.26	14.74	15.89	9.38	61.63
HDAO	5.0	80	2.3	73.62	26.38	5.12	24.17	17.50
HDAO	5.0	90	2.3	71.86	28.14	5.43	24.25	18.31
HDAO	5.0	120	2.3	71.44	28.56	2.91	26.13	10.03
HDAO	4.0	90	2.3	72.36	27.64	3.51	25.37	12.16
HDAO	7.0	90	2.3	71.31	28.69	4.08	25.88	13.62
HDAO	5.0	90	1.4	62.88	37.12	4.72	33.73	12.27
HDAO	5.0	90	3.4	81.79	18.21	5.72	15.43	20.76

In general, these results indicated that the increase of the pressure and S/O will enhance the extraction of the light oil; however, the temperature has a negative influence.

3.3 Compositional analyses of extracts and residues

Figure 3 shows the boiling point curves for DAOs and their flashed-off oils. At the final point of 750°C, the recovered rates for DAOs and their flashed-off oils cannot reach up to 100%. For example, the recovered rates of LDAO, LDAO–LO, and LDAO–HO are 94.4%, 95.8%, and 86.8%, respectively. It indicated that those oil samples cannot completely be distilled. For LDAO, LDAO–LO, and HDAO–LO, about 3% of components are heavier than C120+. For HDAO, LDAO–HO, and HDAO–HO, about 15% of components are heavier than C120+.

When comparing LDAO with LDAO–LO and LDAO–HO, the recovered rate of the LDAO–LO is greater than the LDAO at the same boiling point, and that the recovered rate of the LDAO is greater than that of LDAO–HO. It causes that the boiling point curves of the LDAO–LO lie on the right side of the LDAO; and that the boiling point curves of the LDAO–HO lies on the left side of the LDAO. However, three boiling point curves are closed with each other. It indicated that those oil samples have similar distributions of components. The reason is that the distribution of components for the LDAO of HDAO is very narrow, and which is narrower than the bitumen or vacuum residue.

Figure 4 shows the component distribution for DAOs and their flashed-off oils. According to the boiling point of eluted components, the oil sample might be sliced into a series of pseudo-components with various carbon numbers (Kariznovi et al. 2013). The mass weight of components is equal to

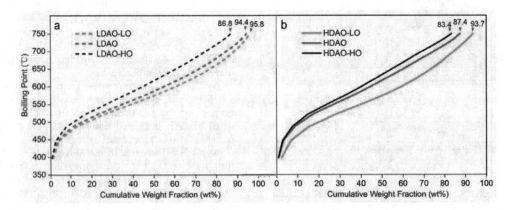

Figure 3. Boiling point curves for DAOs (LDAO and HDAO) and their flashed-off oils (representative separation conditions: 90°C, 5 MPa and S/O = 2.3 w/w).

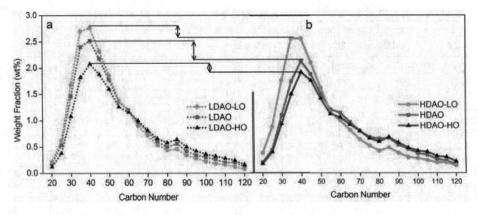

Figure 4. Compositional analysis for raw DAOs (LDAO and HDAO) and their flashed-off oils (representative separation conditions: 90°C, 5 MPa, and S/O = 2.3 w/w).

the eluted mass according to the boiling point of n-paraffin. For visualizing wells, the weight fractions were displayed in the form of the five-point average value in order to eliminate the data noise.

Figure 4a shows the component distribution for the LDAO and its flashed-off equilibrium phases. The range of the component distribution is from C22 to C120. The reason is that: (1) the DAO has no low boiling point component which is lighter than C22, and (2) the heavy component C120°C cannot be distillated at the final point 750°C (Ferreira & Aquino Neto 2005). Three oil samples are composed of components C22 to C120 and mainly contain components from C30 to C90. In the range of C30–C55, the LDAO–LO is evidently higher than raw LDAO, and LDAO–HO is lower than raw LDAO; while the component is beyond C55+, the order for the component abundance is opposite. It means that the light components are trendily extracted into the LDAO–LO and the heavy components are enriched into the LDAO–HO. The big difference also reveals that the light components C30–C55 have the high selectivity of dissolution by using the *i*-butane solvent.

Figure 4b shows that the component distributions for the HDAO are similar to the LDAO. Nevertheless, in the range of C30–C55, the component distribution for the HDAO–LO is evidently higher than that of the raw HDAO. It indicated that the quality of light phase oils extracted from HDAO is better than that of raw HDAO.

The results of compositional analysis showed the following: (1) the carbon number for DAOs and their flashed-off oils appear as left-skewed distributions; (2) the range of dominant components is C30–90; and (3) the big difference for the carbon number distributions mainly appears in the range of C30–C55.

4 CONCLUSIONS

This work investigated the phase equilibrium experiment and the component analysis during the liquid–liquid separation of the (LDAO/HDAO + i-butane) system. The results of phase experiments showed the extraction yield for the LDAO (28.95–70.43, wt%) is greater than that of the HDAO (10.03–20.76, wt%) by using i-butane as the solvent. The results of composition analysis for DAOs and their flashed-off oils showed that: (1) the carbon number for components appears as left-skewed distributions; (2) the range of dominant components is C30–90; and (3) the quality of extracted oils from the LDAO is better than that of HDAO.

The results revealed that the quantity of the DAO has a significant influence on the extraction yield and the quality of extracted oils.

ACKNOWLEDGMENTS

This work was supported by the National Natural Science Foundation of China (Nos NSFCU1162204 and 21176254), the Science Foundation of Panzhihua University, the Doctoral Foundation of Panzhihua University, and the Key Fund Project of Education Department of Sichuan Province.

REFERENCES

ASTM D7500–14. *Determination of Boiling Range Distribution of Distillates and Lubricating Base Oils—in Boiling Range from 100 to 735°C by Gas Chromatography*.

Fan, M., X. W. Sun, Z. M. Xu, S. Q. Zhao, C. M. Xu and K. H. Chung (2011). Softening Point: An Indicator of Asphalt Granulation Behavior in the Selective Asphaltene Extraction (SELEX-Asp) Process. *Energy & Fuels*, 25(7): 3060–3067.

Kariznovi, M., H. Nourozieh and J. Abedi (2013). Experimental Determination of k-Values and Compositional Analysis of Liquid Phases in the Liquid–Liquid Equilibrium Study of (Athabasca Bitumen + Ethane) Systems. *Journal of Chemical & Engineering Data*, 58(6): 1772–1780.

Neer, L. A. and M. D. Deo (1995). Simulated distillation of oils with a wide carbon number-distribution. *Journal of Chromatographic Science*, 33(3): 133–138.

Wen, L.-x., J.-h. Liang and M.-z. Cai (2007). Determination of Distribution of Residual Oil Fraction by Chromatographic Simulated Distillation Method. *Journal of Instrumental Analysis*, 26(2): 270–273.

Yu, C. B., C. Zhang, X. Y. Guo, Z. M. Xu, X. W. Sun and S. Q. Zhao (2015). Liquid-liquid equilibrium of iC4+DAO of VTB: Experiment and modeling. International Conference on Energy Equipment Science and Engineering, ICEESE 2015, May 30, 2015 - May 31, 2015, Guangzhou, China, CRC Press/Balkema.

Zhao, H.-j. (2009). Wide distribution of boiling point of simulated distillation for crude oil. *Modern Scientific Instruments*, (3): 73–76.

Zhao, S., C. Xu, X. W. Sun, K. H. Chung and Y. Xiang (2010). China Refinery Tests Asphaltenes Extraction Process. *Oil & Gas Journal*, 108(12): 52–58.

Zhao, S., C. Xu, R. A. Wang, Z. Xu, X. Sun and K. H. Chung (2009). A deep separation method and processing system for the separation of heavy oil through granulation of coupled post-extraction asphalt residue. U.S. Patent 7,597,794,.

Zhao, S. Q., R. A. Wang and S. X. Lin (2006). High-Pressure Phase Behavior and Equilibria for Chinese Petroleum Residua and Light Hydrocarbon Systems. Part I. *Petroleum Science & Technology*, 24(3/4): 285–295.

Zhou, X., Y. Zhang, S. Zhao, K. H. Chung, C. Xu and Q. Shi (2014). Characterization of Saturated Hydrocarbons in Vacuum Petroleum Residua: Redox Derivatization Followed by Negative-Ion Electrospray Ionization Fourier Transform Ion Cyclotron Resonance Mass Spectrometry. *Energy & Fuels*, 28(1): 417–422.

Advances in Energy Science and Equipment Engineering II – Zhou, Patty & Chen (Eds)
© 2017 Taylor & Francis Group, London, ISBN 978-1-138-71798-5

Numeric modeling of the electric conductor's ice-shedding phenomenon with ice adhesive and cohesive failures

K.P. Ji, B. Liu, X.Z. Fei & X.P. Zhan
China Electric Power Research Institute, Beijing, China

C. Deng & L. Liu
Jibei Electric Power Research Institute, Beijing, China

ABSTRACT: Ice shedding may cause a significant vertical vibration of electrical conductors and ground wires and further lead to mechanical and electrical accidents. However, the induced ice-shedding effect was barely considered in previous studies. By analyzing the forces (gravity, inertia force, adhesive force, and cohesive force) acting on the ice accretion, a judging condition was proposed to simulate the induced ice-shedding effect in FE (Finite Element) analysis. After modeling a full-scale single span transmission line with ice-shedding test, it is found that the results obtained by the proposed FE model in both static and dynamic analysis agree well with those measured in the test. Besides, it indicates that the previous numerical method may underestimate the adverse influence of ice shedding on transmission lines. Thus, it is recommended that the proposed method should be used in the future to ensure the safety of overhead transmission lines in cold regions.

1 INTRODUCTION

Atmospheric icing is one of the common disaster threatening the security of overhead electric transmission lines in cold region countries, such as Russia, Canada, China, America, Japan, United Kingdom (UK), Finland, Norway, Sweden, and Iceland (Farzaneh, 2008). It may lead to several serious electric accidents, such as flashover, short circuit and outage, and mechanical accidents, such as rupture of the insulator and fitting, wires or cable breakage, cross-arm deformation, and even tower failure (McClure et al., 2002).

During January and February 2008, a series of heavy ice storms hit south China. For the State Grid Company of China alone, the storms result in a total of 170 464 towers' failure and 11 965 towers' damage of the 10 to 110 kV transmission lines, and cause 1 327 towers to collapse and 381 towers were damaged of the 220 kV and above transmission lines, and 707 sub-transformer stations were shut down (Hu, 2008). The total direct economic loss is approximately 17 billion US dollars and other 62 billion US dollars for the electric network repairing and rebuilding. In addition, for the South China Grid Company, the storms caused a total of 1 673 towers' collapse, 905 towers' damage, and 2 728 conductors and ground wires breakage for lines of 110 kV and above (Cao et al., 2009). Among the collapsed and damaged towers, 338 towers are from the 500 kV lines and 915 towers are from 220 kV lines, taking an account of 26.9% and 73.1%, respectively. Besides, there are 106 telecommunication optical cables breakage, among which 54 cables are OPGWs (optical fiber composite overhead ground wires) and 52 ADSS (All-Dielectric Self-Supporting optical fiber cable), with an average optical cable rupture rate of 18%.

Ice shedding is a phenomenon that ice deposited on electric cables suddenly detach and cause the cable jumping dramatically due to change of ambient conditions (wind load and temperature), thermal de-icing, and mechanical de-icing (Jamaleddine et al., 1993). A significant transient dynamic response of the overhead towers–lines system following sudden ice shedding may lead to the vibration of conductors and ground wires, and swing of the insulator string, as may further result in electric accidents such as flashover and short circuit for insufficient clearance among conductor–ground, conductor–ground wires, and conductor–tower, and mechanical accidents, such as damages of hardware, fittings, cables, insulators, and towers.

The study on ice shedding of transmission lines can date back to early 1940s (Morgan and Swift, 1964). Morgan and Swift conducted a series of ice-shedding tests on a five-span full-scale 132 kV line section to validate the maximum jump height calculation equation proposed by the author with analytical method and to study the influence of different insulators' configuration on the jump height (Morgan and Swift, 1964). Jamaleddine et al. carried out several test scenarios on a two-span reduced scale laboratory physical model, each span with the length of 3.322 m (Jamaleddine et al., 1993). Because of the difficulty in forming real ice on test lines, in the above-mentioned two studies, the ice was modeled by hanging lumped masses along the span and ice shedding was simulated by sudden release of the lumped masses, which was called as the "lumped mass" or "lumped force" method. Also, Jamaleddine et al. were the first to introduce the nonlinear FE (Finite Element) method into dynamic response analysis of transmission lines following ice shedding, and their numerical model was successfully validated by carrying out physical tests (Jamaleddine et al., 1993). And then, Roshan Fekr and McClure used a new method to simulate ice load and ice shedding, which is called as the "changing density" method (Roshan Fekr and McClure, 1998). In this method, the ice load was considered by changing the equivalent of the electric conductor between static analysis and dynamic analysis in the FE model. The influence of ice thickness, span length, suspension point elevation difference, and uneven icing were systematically studied, which gives a deep understanding of the ice-shedding phenomenon. Kollar et al. studied the difference of three types of ice-shedding patterns, i.e. sudden ice shedding from the whole span, ice shedding with certain velocity, and large chunks of ice shedding (Kollar et al., 2011). The influence of the spacer was also studied and the numerical model was validated by carrying out ice-shedding tests. Kollar et al. also studied the dynamic response of the bundled conductor after ice shedding (Kollar and Farzaneh, 2013), and Wu et al. proposed an analytical model to calculate the ice-shedding height of the conductor bundle (Wu et al., 2016). Li et al. used a "composite element" method in ice-shedding analysis (Li et al., 2008), which was first proposed by Kalman et al. for mechanical de-icing of transmission lines (Kalman et al., 2007).

Although there are plenty of studies on ice shedding, most of them only focus on the dynamic response of transmission lines following initial ice shedding and take it for granted that the mass of the dynamic system kept constant during the whole process. In fact, after initial ice shedding from a certain location of the span, more ice accretion at other locations may be shed off due to the dramatic vibration of the conductors, as may cause more serious transient loads to the towers–lines system. This effect was first proposed by Ji et al. as the "induced ice shedding effect" (Ji et al., 2015). In this paper, the proposed numerical method is partially validated by comparing with a full-scale ice-shedding test and previous ice numerical model.

2 NUMERICAL MODEL

2.1 Concept of induced ice shedding

In previous studies, ice shedding was simulated by using the lumped mass method, changing density method, or composite element method, but only the initial ice shedding was considered, where other ice accretion is assumed staying along the span and cannot be shed off during the jump of the conductor following initial ice shedding. However, in the real world, the remaining ice (besides the initial ice shedding) may detach from the cable due to insufficient adhesive force between the ice-cable interface and cohesive force within the ice. That is, if the resultant of the inertial force of the ice chunk and ice weight exceeds the summation of the adhesive force and cohesive force, the ice chunk will certainly be shed off during the vibration of the conductor. The judging condition used to describe this criterion can be written as Equation 1, as shown below (Ji et al., 2015):

$$F_{inertia} \pm G \geq F_{cohesive} + F_{adhesive} \tag{1}$$

where, G is the gravity, $F_{inertia}$ is the inertia force, $F_{cohesive}$ is the cohesive force, and $F_{adhesive}$ is the adhesive force.

Due to the fact that the adhesive force is almost zero as a result of the thin liquid film appearing at the ice–cable interface under natural ice-shedding conditions, the adhesive term in Equation 1 is ignored in the following part of this paper. In commercial FE software, the acceleration term of the element is easier to be extracted and used in defining the ice failure criterion, and the judging condition in Equation 1 can be rewritten, in terms of acceleration, as Equation 2, which shows that the critical acceleration needed to shed off the ice deposit (Ji et al., 2015).

$$a \geq \frac{8(D - D_{cable})\tau_{cohesive}}{\pi\rho_{ice}(D^2 - D_{cable}^2)} \pm g \tag{2}$$

where a is the vertical acceleration acting on the ice, D_{cable} is the diameter of the cable, D is the outer diameter of the ice profile, $\tau_{cohesive}$ is the cohesive strength, ρ_{ice} is the ice density; "+" is for the upper part of the ice deposit, and "–" is for its lower part. It should be noted that the acceleration is independent of length, for the length term is eliminated during the derivation.

2.2 FE model of the iced conductor

Non-linear FE analysis software ADINA is employed for numerical simulation in this study. The electrical cable is modeled by two-node iso-parametric truss element, with two translational DOFs (Degrees of Freedom) at each node, and the tension-only non-linear elastic material is used, where the specific value of each parameter should be adjusted according to the conductor types. Ice is modeled by using a two-node iso-beam element, with six DOFs at each node, and the bilinear material model is used, the specific value of which is set the same as these in (Kalman et al., 2007).

Due to the induced ice shedding, the system stiffness and mass matrix will change, and the element of ice in the matrices should be updated at each time step where there is induced ice shedding. The governing equations of motion can be expressed as Equation 3 in the incremental FE form (Bathe, 2006).

$$^{t+\Delta t}M\,^{t+\Delta t}\ddot{U}^{(k)} + \,^{t+\Delta t}C\,^{t+\Delta t}\dot{U}^{(k)} + (\,^{t+\Delta t}K_L + \,^{t+\Delta t}K_{NL})\Delta U^{(k)} = \,^{t+\Delta t}R - \,^{t+\Delta t}F^{(k-1)} \quad (3)$$

where, $^{t+\Delta t}\ddot{U}^{(k)}$ is the vector of accelerations obtained in iteration (k), $^{t+\Delta t}\dot{U}^{(k)}$ is the vector of velocities obtained in iteration (k), $\Delta U^{(k)}$ is the vector of incremental displacements obtained in iteration (k), $^{t+\Delta t}M$ is the mass matrix, $^{t+\Delta t}C$ is the damping matrix, $^{t+\Delta t}K_L$ is the linear strain incremental stiffness matrix, $^{t+\Delta t}K_{NL}$ is the non-linear strain (geometric or initial stress) incremental stiffness matrix, $^{t+\Delta t}R$ is the vector of external applied nodal point loads, $^{t+\Delta t}F^{(k-1)}$ is the vector of nodal point forces corresponding to the displacements $^{t+\Delta t}U^{(k-1)}$, and $^{t+\Delta t}U^{(k)} = \,^{t+\Delta t}U^{(k-1)} + \Delta U^{(k)}$ = vector of displacements in iteration (k) (Bathe, 2006).

3 VALIDATION OF FE SIMULATION

3.1 Ice-shedding experiment

To validate the numerical model proposed in Section 2, the ice-shedding test on a full scale physical test line is used, which is conducted by Meng et al. (Meng et al., 2011). A total of 12 ice-shedding scenarios were carried out in the test and the scenario numbered B3 (single span, 100% ice shedding) is taken as example in this work. The ice load was simulated by using a lumped mass method, with ten sand bags hanging on the cable, as shown in Figure 1. Ice shedding was modeled by using a remote control device to cut the rope. The span length is 235 m and the conductor's parameters are listed in Table 1.

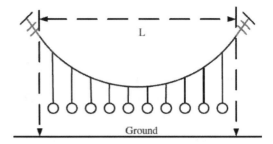

Figure 1. Schematic of the ice-shedding test (Meng et al., 2011).

Table 1. List of conductor parameters (LGJ-630/45).

Parameter	Unit	Value
Cross-section	mm²	666.55
Diameter	mm	33.6
Rated strength	kN	148.7
Mass per unit length	kg/m	2.06
Young's modulus	GPa	63
Thermal expansion coefficient	1/°C	2.09e–05

3.2 Numerical simulation and validation

In the experiment, the static maximum sag and static tension of the conductor were measured before ice shedding, and the time histories of the midpoint displacement and horizontal tension were also recorded for comparison.

Following the modeling method in Section 2.2, the numerical model of the test can be established with ADINA. After static analysis, it is found that the numerical results of the static sag and tension are 6.13 m and 45.61 kN, when compared with the measured values of 6.70 m and 48.10 kN, where the maximum error is about 8.66%. The difference mainly results from the missing of the insulator at the two ends of the conductor in the FE model, for no specific parameters of the insulator is provided in the reference. Also, the centenary chain shape of the conductor in the FE model is different from that in the test, where the lumped mass affects the configuration. Thus, the difference of the results from the FE model and test is acceptable.

Furthermore, the time histories of the midpoint and horizontal tension obtained from numerical simulation and experiment are compared (Figure 2).

From Figure 2, the time histories of the concerned parameters are exactly predicted. The maximum values and time history curves of the numerical simulation agree well with those measured in the test. The difference between tensions is less than that between displacements, as is in

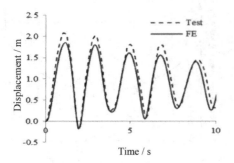

(a) Midpoint displacement

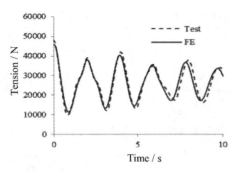

(b) Cable horizontal tension

Figure 2. Comparison of numerical and measured results.

consistent with the static analysis. The reason is that the insulator in the test may aggravate the vertical vibration.

After static and dynamic analysis of the FE model and by comparison with the experimental results, it concludes that the FE model proposed is partially validated and that it is capable of modeling the ice-shedding phenomenon, since there is no test considering the induced ice-shedding effect available in open literature.

3.3 Induced ice-shedding modeling

To further demonstrate the ability of the proposed modeling method, a case with partial initial ice shedding is employed in this work. The concerned line section is 100 m in length, of which 88 m is covered by ice symmetrically from the midpoint. The initial ice shedding is 44 m at the middle of the span (Ji et al., 2015).

The case is modeled both by using the previous changing density method without considering the induced ice-shedding effect and the method proposed in this study considering the induced ice-shedding method, respectively.

It shows that the maximum jump height obtained by using the present model is 75.9% greater than that by using the previous model, and the minimum residual cable tension is 11.6% less than that obtained by using the previous model. The reason for the difference is that there is another 50% ice is shed off in the present numerical simulation. This means that, the previous method may underestimate the adverse impacts of ice shedding, which may result in potential threats for the mechanical security of overhead transmission lines.

4 CONCLUSIONS

In this study, the induced ice-shedding phenomenon is taken into consideration during the transient dynamic analysis of overhead transmission lines in cold regions following natural ice shedding. The phenomenon always occurs in reality, but it is not being considered in most of previous studies. By checking the forces balance of gravity, inertia force, and adhesive force at the ice–cable interface and cohesive force within the ice, a judging condition to model the induced ice-shedding effect in FE simulation is obtained. And then, the FE model is used to calculate the static and dynamic characteristics of a full-scale line section following ice shedding. A comparison of experimental and simulation results shows that the proposed numerical model is capable of predicting both the static and dynamic response of the test line under ice shedding, and the FE model is partially validated. After comparing with previous numerical method without considering the induced ice-shedding effect, it shows that the previous method may underestimate the adverse influence of ice shedding on transmission lines. Thus, it is recommended that the proposed method should be used in the future during the design and safety evaluation of overhead transmission lines in cold regions.

ACKNOWLEDGMENT

This study was financed by the State of Grid of China's scientific project—the Applicability and Anti-Disaster Ability Improvement of Compact Transmission lines (No. SGTYHT/14-JS-188).

REFERENCES

Bathe, K. J. 2006. *Finite Element Procedures,* Upper Saddle River, New Jersey, USA, Prentice Hall.
Cao, M., Luo, X., Shi, S., Zhang, Z. & Gao, S. 2009. Research and Discussion on Dynamic Warning Program of On-line Icing Monitoring System Based on

the Growth Rate of Icing. *Southern Power System Technology,* 3, 186–188.

Farzaneh, M. 2008. *Atmospheric Icing of Power Networks,* Springer Netherlands.

Hu, Y. 2008. Analysis and Countermeasures Discussion for Large Area Icing Accident on Power Grid. *High Voltage Engineering,* 34, 215–219.

Jamaleddine, A., Mcclure, G., Rousselet, J. & Beauchemin, R. 1993. Simulation of ice-shedding on electrical transmission lines using ADINA. *Computers and Structures,* 47, 523–536.

Ji, K., Rui, X., Li, L., Zhou, C., Liu, C. & Mcclure, G. 2015. The Time-Varying Characte-ristics of Overhead Electric Transmission Lines Considering the Induced-Ice-Shedding Effect. *Shock and Vibration,* 2015, 1–8.

Kalman, T., Farzaneh, M. & Mcclure, G. 2007. Numerical analysis of the dynamic effects of shock-load-induced ice shedding on overhead ground wires. *Computers and Structures,* 85, 375–84.

Kollar, L. E. & Farzaneh, M. 2013. Modeling sudden ice shedding from conductor bundles. *IEEE Transactions on Power Delivery,* 28, 604–611.

Kollar, L. E., Farzaneh, M. & Van Dyke, P. 2011. Modeling of cable vibration following ice shedding propagation.

Li, L., Xia, Z., Fu, G. & Liang, Z. 2008. Ice-Shedding Induced Vibration of A Long-Span Electric Transmission Tower-Line System. *Journal of Vibra-tion and Shock,* 27, 32–34+50+180.

Mcclure, G., Johns, K. C., Knoll, F. & Pichette, G. 2002. Lessons From The Ice Storm Of 1998 Improving The Structural Features Of Hydro-Québec's Power Grid. *the 10th International Workshop on Atmospheric Icing of Structures.* Brno, Czech Republic.

Meng, X., Wang, L., Hou, L., Fu, G., Sun, B., Macalpine, M., Hu, W. & Chen, Y. 2011. Dynamic characteristic of ice-shedding on UHV overhead transmission lines. *Cold Regions Science and Technology,* 66, 44–52.

Morgan, V. T. & Swift, D. A. 1964. Jump height of over-head-line conductors after the sudden release of ice loads. *Proceedings of the Institution of Electrical Engi-neers,* 111, 1736–1746.

Roshan Fekr, M. & Mcclure, G. 1998. Numerical mod-elling of the dynamic response of ice-shedding on electrical transmission lines. *Atmospheric Research,* 46, 1–11.

Wu, C., Yan, B., Zhang, L., Zhang, B. & Li, Q. 2016. A method to calculate jump height of iced transmis-sion lines after ice-shedding. *Cold Regions Science and Technology,* 125, 40–47.

Advances in Energy Science and Equipment Engineering II – Zhou, Patty & Chen (Eds)
© 2017 Taylor & Francis Group, London, ISBN 978-1-138-71798-5

Research on the agricultural materials e-commerce and information service cloud platform

Rongmei Zhang, Wenling Hu & Guangquan Fan
College of Information Technology, Hebei University of Economics and Business, Shijiazhuang, China

ABSTRACT: Internet+ brings a new opportunity for the agriculture development, and the agricultural electronic commerce is the main form of "Internet+ Agriculture". Firstly, the functions of the e-commerce of agricultural materials are analyzed in the paper. Then, the framework of the agricultural e-commerce and information service cloud platform is designed based on cloud computing. Next, the key technologies are discussed to build up the platform. These will provide the technical support to accomplish the agriculture e-commerce platform. The development of the platform will enhance to improve the level of agricultural informatization, and to promote the development of the modern agriculture.

1 INTRODUCTION

In China 2015 "Internet+" strategies, the "Internet + agriculture" has become one of the focuses of attention. The "internet+ agriculture" is a cross-border integration of agriculture with new generation of information technologies such as mobile Internet, cloud computing, big data and Internet of things and so on. The "internet+ agriculture" is a new mode of modern agriculture based on the internet platform, and it is also an important means to promote the development of modern agriculture. The rapid development of rural informatization has laid down the foundation for "Internet+ agriculture". According to statistics, by the end of 2015, the internet users of China's agricultural population reached 1.95 million, and the online shopping users accounted for 22.4 percent of the total rural internet users.

A large number of agricultural e-commerce websites emerged over the past ten years in China, but most of these websites are mainly used as the show windows of the enterprises or the platforms to publish information, and just a few were used as a trading place (Tan, 2011). Besides, the most farmers nowadays are still used to purchasing agricultural materials offline, because they are not familiar with online shopping. Fortunately, the electronic business enterprises, such as Alibaba, Sunning and JD, have set up e-commerce service stations at rural areas in the last two years. The agricultural e-commerce is moving into the stage of exploration and practicing. So far, e-commerce is still the major form of "internet+ agriculture". So, the information service platform based o2o, which provides the agricultural materials trading and other information service for farmers, is being received more attention than ever.

Based on function requirements of the electronic commerce system of agricultural materials, we design the framework of cloud platform of the agricultural e-commerce and information service using cloud computing technology, and discuss the key technologies to build up the platform. The development of the platform will promote the application and development of agricultural e-commerce in China.

2 THE FUNCTION ANALYSIS OF THE AGRICULTURE E-COMMERCE INFORMATION SERVICE PLATFORM

The e-commerce information service platform of agricultural materials is a comprehensive information system based on e-commerce. The platform should provide electronic trading service and other information services. The system has two types of users, one is agricultural producers who are farmers, ranchers, agricultural cooperatives, etc.; another is the suppliers of agricultural materials including the retailers and producers of agricultural materials. The former can buy agricultural materials, browse the related information, study agricultural technology, publish information about demand and supply on the platform, and the latter should sell online their agricultural materials on the platform, which can broaden the market sales channels. The system has five subsystems, these are as follows:

2.1 *Membership subsystem*

Membership is mainly refers to the farmers, ranchers, farm cooperatives and agricultural enterprises, etc. They complete the registration on the platform, manage their own information, trade online, check the order status, evaluate the goods, write online massage, release the information of demand and supply, etc.

2.2 Administer subsystem

The administers are the managers of the platform, they manage the data of the platform including trading information. The functions of this subsystem consists of (a) management information of goods as well as demand and supply, (b) setting up the knowledge base, (c) release and management of agricultural knowledge and policy, (d) auditing the suppliers' information, (e) processing orders, (f) replying the massage.

2.3 Supplier subsystem

The suppliers, who are usually the agricultural enterprises and retailers of agricultural materials, can complete the supplier registration, manage their own enterprise information, process orders, release information of demand and supply.

2.4 Information service subsystem

This is an information service platform, including goods information service, agricultural knowledge information service, policy and law information service, demand and supply information service and so on. The farmers can learn agricultural technology and policy anytime and anywhere without going out of home on the platform.

2.5 Goods and service recommending subsystem

Whether to provide the users with accurate recommendations in goods and services or not is the key of an e-commerce platform to operate. Recommending customers the personalized goods and service will improve the user experience and customer stickiness. The recommendation strategies of goods and service in the system includes:

- New goods recommendation: recommending new goods online.
- Hot goods recommendation: recommending top 5 goods based on order sales.
- Agricultural materials recommendation based on expert knowledge.

The knowledge base is build by gathering agricultural expert experience and knowledge in the system. The knowledge-base is composed of recommended rules. A rule is shown as follows:
if wheat then urea
When the users buy wheat seed, the system will give a list of fertilizers like urea to be selected.

- Agricultural material recommendation based on the users' region:
Since agricultural production activities have strongly regional characteristics, the crops are different in different regions. When the users buy seed, the platform will recommend most fitting crop seeds according to the user's region.
- Agricultural material recommendation based on association rules:
The platform will analyze offline the data of shopping basket with MapReduce Apriori algorithm to mine the association rules among the agricultural materials. If the users choose an item, the platform will push a list of associated goods according to the association rules.
- Agricultural materials recommendation based on collaborative filtering algorithm:
The Hybrid recommendation model of the collaborative filtering based on items and users is build according to users' ratings record about items. Platform will recommend the most related agricultural materials to users based on association rules.

3 THE ARCHITECTURE OF THE AGRICULTURAL MATERIALS E-BUSINESS INFORMATION SERVICE CLOUD PLATFORM

Because of large-scale, virtualization, high reliable, scalable, universal, cheap features, cloud computing has been utilized in many areas, such as science applications (Keahey, 2008), data mining (Yu, 2012), database application (Cheng, 2012), parallel computing (Li, 2011), and so on. Cloud computing is a collection of resource services that are made up of a large number of hardware and software resources linked on internet in certain form. The main cloud services include IaaS, PaaS, SaaS, etc (Lucky, 2009; Luo, 2011; Fang, 2012). In cloud environment, a enterprise may apply to buy the cloud services with the pay-as-you-go model without any infrastructure or software investments. They consume resources as a service and pay only for resources that they use. Hence, the enterprise can easily deploy and operate the corresponding computation and storage resources they use. For the small and medium e-commerce enterprises of agriculture, an agricultural e-commerce cloud platform will provide new technical support and service model for their development. Meanwhile, the enterprises will get more broad developing opportunities by using the platform.

Based on the three types of cloud services, the framework of cloud platform of agricultural electronic commerce information service is designed as three layers, as shown in Figure 1. The three layer are, from bottom to top, the cloud computing

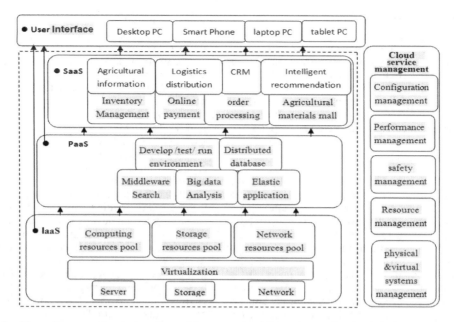

Figure 1. The architecture of agricultural materials e-commerce information service cloud platform.

infrastructure resources pool layer, the supporting platform in the middle layer, and application layer of e-commerce software services.

3.1 The infrastructure layer

Infrastructure layer provides a large number of devices like servers, storage devices, network devices. Through virtualization of web server, database server, storage equipment, network infrastructure, the platform forms computing resources pool, storage resources pool, and network resources pool, and provides scalable computation resources and infrastructure for agricultural e-commerce information platform in order to complete the application computation, data storage, network transmission. Nowadays, many enterprises in China, such as Tencent, Baidu, Alibaba, Huawei and Jingdong, provide cloud services like cloud storages, cloud servers and cloud backup services.

3.2 The platform support layer

The platform support layer provides the users with a set of software and middleware services to develop, run and operate an agricultural e-commerce system, including operating system (Window Server, Linux, Hadoop), development tools (Eclipse, JDK, Android, ASP.NET etc.), search engine, database management system (such as Sql Server, Oracle, MySql), big data storage and query management

(such as DFS, HBase, etc.), algorithm library (Mahout) and so on.

3.3 The application service layer

The application service layer is the terminal services, including basic services and business services. The basic services consist of the unified security and permissions authentication, and other services. The users of the agricultural e-business system are divided into four types: normal users, members users, the supplier users, the system administrator. The users can get the corresponding services of information and application according to the roles and their own privileges. The business services include the information services for the normal users, such as agricultural technology, policies and regulations, the supply and demand information; application services for the member users such as purchasing agricultural materials and services recommendation, while business services consist of selling agricultural materials for agricultural enterprises. The application service layer also provides various forms of internet access interface for users, such as mobile phone, tablet PC, laptop PC and desktop PC, in order to improve the user experience. In the application layer, the user certification, membership information management, product information management, order information management, intelligent recommendation module and other function modules are encapsulated into Web services to complete SaaS.

4 KEY TECHNOLOGIES

The implementation of agricultural e-commerce information service cloud platform involves many technologies, from construction of the cloud platform to the realization of the application system, mainly including virtualization technologies, web service technologies, data processing technologies and intelligent recommendation technologies.

4.1 Virtualization technologies

The virtualization technologies implement the "infrastructure as a service" cloud service. The infrastructure mainly includes the hardware of server, storage and network. Through virtualization technologies, the platform combines multiple physical servers, storage, network and other resources into one to build as an unified infrastructure cloud, and then create many virtual machines to provide users in IaaS service model. The virtual machines can be dynamically adjusted and migrated as users need, which makes the network resources to achieve the dynamic, flexibility, high availability and performance. Virtualization technologies consist of hard partition (such as Sun Domain) and soft partition (such as VMware vCloud, Citrix Xen, RedHat KVM) [8].

Virtualization technologies implement "platform as a service" cloud services. An unified public platform is built to supply the users in PaaS by software virtualization for the operating system, database system deployed on the hardware. Then, the users can deploy their applications on the platform without building as well as maintaining the operating environment, which can achieve the PaaS.

4.2 Big data storage and processing technologies

With wide application of e-business, there are large amounts of data sets generated by the users trading. Hence, it is necessary to store, analyze, process the massive data in the cloud system. Apache Hadoop is a open source project to support large data analysis. Hadoop is a distributed cluster-based computing framework, using MapReduce programming model [9–10]. Hadoop can process large data sets and execute distributed computations on a large cluster of commodity PCs (called as calculation nodes). Meanwhile, Hadoop also provides a sets of stable and reliable interfaces for applications. A cluster can expend itself from a single server to hundreds or even thousands of calculation nodes, each node provides the parallel computing and distributed storage.

4.2.1 The clustered computing architecture

A cluster consists of a large amounts of inexpensive commodity hardware (i.e. computing nodes). These nodes are placed on racks, each rack can contain 8 to 64 nodes linked into a network. A cluster consists of a single master and multiple chunk servers and is accessed by multiple clients. Each of these is typically a commodity Linux machine running a user-level server process. The file system based clustered computing is called as the distributed file system, i.e. DFS (Dean, 2005; Huang, 2015). Files are divided into fixed-size chunks. Each chunk is identified by an immutable and globally unique 64 bit chunk handle assigned by the master at the time of chunk creation. Chunk servers store chunks on local disks as Linux files and read or write chunk data specified by a chunk handle and byte range. Each chunk is replicated on multiple chunk servers. By default, we store three replicas. The master maintains all file system metadata. This includes the namespace, access control information, the mapping from files to chunks, and the current locations of chunks. It also controls system-wide activities such as chunk lease management, garbage collection of orphaned chunks, and chunk migration between chunk servers. The master periodically communicates with each chunk server in HeartBeat messages to give it instructions and collect its state. The chunk server is called as DataNodes which are responsible for node data storage, handling file read/write requests of clients, as well as creating, deleting and copying data blocks.

4.2.2 MapReduce programming model

MapReduce is a programming model and an associated implementation for processing and generating large data sets. MapReduce provides parallelized processing model and programming interfaces. Two phases of data processing are Map and Reduce. In the stage of Map, the MapReduce library in the user program first splits the input files into m pieces of typically 16 megabytes to 64 megabytes (MB) per piece, and then each piece of data is further decomposed into a number of key/value pairs (K1; V1) and passed to the map function. The map function will takes a set of input key/value pairs, and produces a set of intermediate output key/value pairs (K2; V2). In Reduce stage, the reduce function merges together these values V2 in the same key K2 to form a possibly smaller set of values, then calculate and form the output results (K3; V3). Finally, all the results of reduces will be summarized and appended to a final output file (Dean, 2005; Huang, 2015)

4.2.3 Distributed database system Hbase

Hbase is a component of the Hadoop Apache project, it is NOSQL. The Hbase implements column storage data by the version. Big data sheet can be read and written in real time in Hbase system (Huang, 2015).

4.2.4 Data warehouse system hive

Hive is also a part of Apache Hadoop project. It provides the collection, query and analysis of a large data sets. Hive also can realize special query by HiveSQL like SQL. Meanwhile, it can maintain the traditional graphics operations when HiveSQL cannot adequately express these complex logics (Huang, 2015).

4.3 Intelligent recommendation technologies

The personalized and accurate recommend has become new direction of e-commerce in the future. Customer structure, click through rate, the buying cycle and interest will produce large amounts of data on the e-commerce platform. Through the collection, integration, processing, analysis and using data, e-commerce system can accurately recognize consumer tastes and willingness, and initiatively provide personalized and accurate products and services for users, so as to enhance the user experience and increase sales. Therefore, for agricultural e-commerce, the intelligent recommendation engine is one of the core functions

The traditional data mining tools, such as Weka, SPSS, Matlab etc., have been not completed fast processing for big data. Mahout and Spark which are Apache Hadoop open source project have achieved a variety of data mining algorithms in clustered computing environment, including clustering, classification, collaborative filtering and frequent item sets mining and so on (Anand, 2012; Haralambos, 2009; Dietmar, 2011). The users can be divided into different groups by using classification or clustering algorithm. Then recommendation algorithm such as frequent item sets mining algorithm and collaborative filtering algorithm will recommend goods that the users are interested in. The big data analysis technologies will bring huge business value for the e-business companies, achieve high value-added services, further enhance their economic benefits by accurate position marketing to increase customers.

4.4 Web services

The goal of agricultural electronic commerce information service platform is to provide quality information services as much as possible for individual and enterprise users including farmers, agricultural cooperatives, agricultural enterprises and so on. However, the networking equipments are different, such as desktop computers, notebook computers, mobile phones and panel computers Thus, web service technologies are key technologies to implement the application layer.

Web service is a technology based on SOA, and is a new platform to build interoperable distributed applications. There are three types of job roles and three operations in the architecture of Web service. These roles are the service provider (server), service requester (client), a service registry, and these operations include publishing, finding and binding services (Yue, 2004). Web service is a functional program with web interface. The functional program can be easily access through the standard Internet Protocol (HTTP). Each web service specifically defines only interface, which expose an API access outside by soap through HTTP. It completely shielding the difference of between software platforms, or development languages, Internet access devices, and realize the communication between the different platforms and systems different programming language.

Web Services uses XML to encode and decode data, and uses SOAP to transfer data, uses WSDL to describe service, uses UDDI to discovery and access the metadata service.

In the agricultural e-business system, Using the web service technologies packages the database operation modules into web services, including data operation layer, business logic layer and representation layer, and then release the web services on Internet, such as the information service of agricultural goods, order information, supplier and agricultural knowledge, etc.. This makes interoperability between the different platforms such as the web and mobile phone client platform, development languages implemented.

5 CONCLUSIONS

With the advance of the national "Internet+" action plan, the influence of Internet has entered a new stage for agriculture. Thus, it is necessary and feasible to build an agricultural e-commerce information service platform. Cloud computing is very suitable for agricultural e-commerce because of its advantages of shared infrastructure, saving investment, ubiquitous access and its reliable and secure data storage centres. Appling cloud computing into the agricultural e-commerce to build the cloud service platform will break the bottleneck of terminal, technique, cost and management, and reduce the threshold for farmer to access to the agricultural e-commerce. As long as have the farmers a smart phone, they can trade online so as to cultivate the farmers' habits of shopping online. When farmers are accustomed to the Internet, transaction and consumption patterns of farmers will be dramatically changed, so to promote the popularization and development of agricultural e-commerce.

ACKNOWLEDGEMENT

This work was supported by the social science foundation project of Hebei province (No. HB15GL092)

and the scientific and technical project of Hebei province (No. 15214706D)

REFERENCES

Anand Rajaraman, Jeffrey David Ullman. *Big data: Mining of Massive Datasets*. Cambridge University Press (2012).

Chenghua Li, Xinfang Zhang, Hai Jin, et al. MapReduce: a New Programming Model for Distributed Parallel Computing. Computer Engineering & Science, **33**(3):129–135 (2011).

Dean J, Ghemawat S. MapReduce: Simplified Data Processing on Large Clusters. Communications of the ACM, **51**(1):107–113 (2005).

Dietmar Jannach, Markus Zamker, Alexander Felferning, et al. *Recommender system: An introduction*. Cambridge University Press (2011).

Haixia Tan, Min Zhang. Hindrances for the Development of Farm Produce E-commerce in China and the Countermeasures. Logistics Technology, **3**(3):121–124 (2011).

Haralambos Marmanis, Dmitry Babenko. *Algorithms of the Intelligent Web*. Manning Publications (2009)

Junzhou Luo, Jiahui Jin, Aibo Song, et al., Cloud computing: architecture and key technologies. Journal on Communications, **32**(7):3–21 (2011).

Keahey K, Figueiredo R, Fortes J, et al., Science clouds: Early experiences in cloud computing for scientific applications. *In Proc of Workshop on Cloud Computing and its Applications* (2008).

Kun Yue, Xiaoling Wang, Aoying Zhou. Underlying Techniques for Web Services: A Survey. Journal of Software, **15**(3):428–440 (2004).

Le Yu, Jian Zheng, Bin Wu, et al., BC-PDM: Data Mining, Social Network Analysis and Text Mining System Based on Cloud Computing. *The 18th ACM SIGKDD international conference on knowledge discovery and data mining* (2012).

Lucky, Robert W. Cloud Computing. IEEE Spectrum, **46**(5):27 (2009).

Wei Fang, Xuezhi Wen, Wubin Pan, et al. Cloud computing: Conceptions, key technologies and application. Journal of Nanjing University of Information Science and Technology: Natural Science Edition, **4**(4): 351–361 (2012).

Yihua Huang, Kaixiang Miao. *Understanding big data: Big data processing and programming*. China Machine Press (2015).

Ying Cheng, Yunyong Zhang, Lei Xu, et al., Research on Large-Scale Data Processing Based on Hadoop and Relational Database. Telecommunication Science, (11):47–50 (2010).

Advances in Energy Science and Equipment Engineering II – Zhou, Patty & Chen (Eds)
© 2017 Taylor & Francis Group, London, ISBN 978-1-138-71798-5

The design of remote display systems of refrigerated vehicles

Yuan Liu & Yanan Zeng

College of Engineering and Technology, Tianjin Agricultural University, Tianjin, China

Shijie Yan

College of Food Science and Biological Engineering, Tianjin Agricultural University, Tianjin, China

ABSTRACT: The remote display system of refrigerated vehicles can display positions of vehicles in real time by utilizing GPS/GPRS technology. This system is mainly controlled by using the Micro-Controller Unit (MCU), thereby acquiring longitude, latitude, and time data. Temperature, humidity, and concentration of CO_2 data of refrigerated vehicles are transported by using the GPRS module to the database of the host computer, the management platform of which is developed based on the B/S framework. Users can remotely monitor the vehicles through the displaying of the positions and working status via the remote display system.

1 INTRODUCTION

Refrigerated vehicles are the most critical infrastructures of cold-chain logistics in transport and storage. During transport, the goods in the vehicles would be damaged by the environmental changes, thereby resulting in considerable economic losses. Since the refrigerated vehicles are utilized widely, especially in hostile environments, the safety of transport should be taken into account. By using GPS/GPRS technology, the data of multiple vehicles can be transported to the control platform simultaneously. Users can manage the refrigerated vehicles by acquiring information through remote monitoring via login access to the software platform, thereby improving the efficiency of management (Huang, 2013).

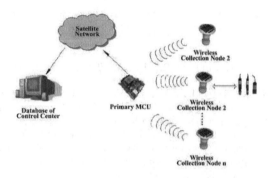

Figure 1. Schematic of a network structure of a refrigerated vehicle.

2 MONITORING SYSTEM OF REFRIGERATED VEHICLES

The network structure of a refrigerated vehicle is presented in Figure 1. As the main controlling chip, MCU exchanges data and instructions with end-nodes through wireless communication. Parameters and thresholds are also set by using the MCU. To retain the parameters in the set range, the system is adjusted automatically when the parameters of end-nodes are detected out of the set range. Simultaneously, the MCU exchanges information with the monitoring centre and instructions and alarms through the GSM/GPRS module.

The data of temperature, humidity, and concentration of CO_2 are collected by using end-nodes. By renting of a common end-node, users can monitor and control the parameters in refrigerated vehicles through a software platform.

3 GPS POSITIONING

The GPS in the system is aUblox-M6–0 module with the signal output of standard NEMA-0183 protocols. The standard information of this format outputs per second when GPS receives signals. The $GPGGA statement is utilized to parse the longitude, latitude, and time data required by the vehicle terminal. The $GPGGA statements are as follows:

$GPRMC, Greenwich Mean Time (hhmmss), positioning is valid or not <A(valid) or V(not valid)>, latituded dmm.mmmm (degree, point), latitude N (northern hemisphere) or S (southern hemisphere), longtituded dmm.mmmm (degree, point), longtitude E (east longitude) or W (west longitude), ground speed (knot), ground direction (degree), UTC date (ddmmyy), mode*checksum, <CR>, <LF> (Shen, 2013).

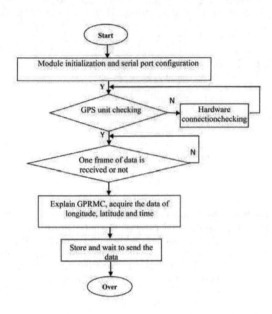

Figure 2. Flowchart showing the acquirement of positioning information from GPS.

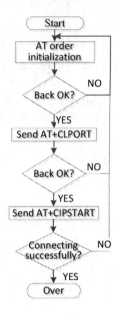

Figure 3. Flowchart showing the program chart of network connection and data transport of GPRS.

Example, \$GPRMC, 062559.00, A, 3957.40922, N, 11618.7561, E, 0.006, 65.52, 091202, A*57

MSP430 is used as the control centre of the vehicle. The low-power MSP430 performs well in portable devices with low electricity supplements. The data transported from the GPS module are received through a serial port in MSP430. The process of data transporting of GPS is shown in Figure 2 (Guo, 2011).

4 GPRS TRANSPORTATION

GPRS is transported with TCP/IP protocol in the system. To guarantee the steady transport of the signal, the only IP address of the destination port is identified by using a handshake response four times.

The IP address and serial port number of the host computer with connection to the outside net are determined by software and required in the communication with host computer. To maintain stability of data transport, the data are stored in arrays in a coded form (Yuan, 2016). The data are received when the st signal is detected by the host computer and written in an array. The responding signal is sent when the ed signal is detected, which means the signal is received. Figure 3 shows the program chart of network connection and data transport of GPRS. Figure 4 shows the flowchart of data transport. The data of an end-node comprise temperature, humidity, and concentration of

Figure 4. Flowchart of data transport.

CO_2, each of which consumes 2 bytes, which is 6 bytes in total. The checksum is added before data are sent.

5 DESIGN OF A MANAGEMENT AND DISPLAY PLATFORM

A software management system is composed by vehicles management, position display, login page, and user-management. This system is designed based on the B/S structure. Users can login to the browser-based interface with user names and passwords. The page of position display reveals the position and track of the vehicles. User-management sidebar is set to manage users' permissions. An administrator is in the highest permission level, who can change any settings in the user-management platform. A first-level account and second-level account are created with different permission levels to manage vehicles. The vehicles are positioned in the map after the data of a vehicle are written into the management system. The written data include name and telephone number of the owner and fleet customers and the type and license plate number of vehicles. The data table can be edited based on the actual situation.

5.1 Build the development environment

The system is built in a Windows environment by utilizing JAVA language, Eclipse development environment tools, SQLServer database and a Model-View-Controller (MVC).

5.2 Design of the software interface

5.2.1 Design of the login interface
Users enter into the management system after entering the correct and legal account and password through the login interface with the prompting of login information. The login interface is blurred to ensure that users would be focused on the login information. A command prompt window would appear when the input account or password is incorrect.

The login panel is a custom control, thereby inheriting the JPanel. By drawing a background picture, the login panel possesses the background effect to prevent other controls.

5.2.2 Design of data communication
A TCP program is written by using the Socket program. In a GPRS module, the information of GPS is sent to an IP address and a serial port number by using an AT instruction. After that, data are obtained out from serial ports by using the Socket method. Though these two applied programs are both communicated with TCP protocol, they also have priorities. In this system, the GPRS module sends information to the program of the host computer, thereby acting as a server program, while the host computer program acts as a client program. The management system can be updated out of the platform and program by the technology of Web Service.

5.2.3 Design of vehicles management
The design of vehicles management is based on the support of the database. The vehicle information mainly includes the following:

1. Management of vehicle information
This part can be completed on all vehicle information browsing and on functions such as add, modify, delete, and other functions by using the garage attendant.
2. Management of vehicle storage and library
This part is completed on the registration of vehicle storage and library.
3. Management of vehicle cargo
This part is completed on the registration of the type, weight, and ownership of the vehicle cargo.

6 ANALYSIS OF DATA TEST

6.1 Analysis of GPS data

The GPS module should be powered under test. The MCU is accessed to reach the GPS, thereby receiving GPS data from serial ports. The information of longitude and latitude are extracted from the GPS. The lights keep shining when the GPS is just powered on, which means that the GPS module is searching for the satellite's signal. If the satellite's signal is searched, the lights flash per second, which implies that the signal is received.

6.2 GPRS test

A GPRS module is manipulated by AT instructions (Jiang, 2006). SIM900 A is powered at 5V and about 800 mA by using the USB of a computer. The card slot of SIM900 A modules is

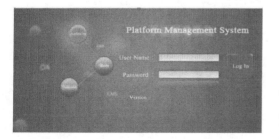

Figure 5. The login interface of the platform.

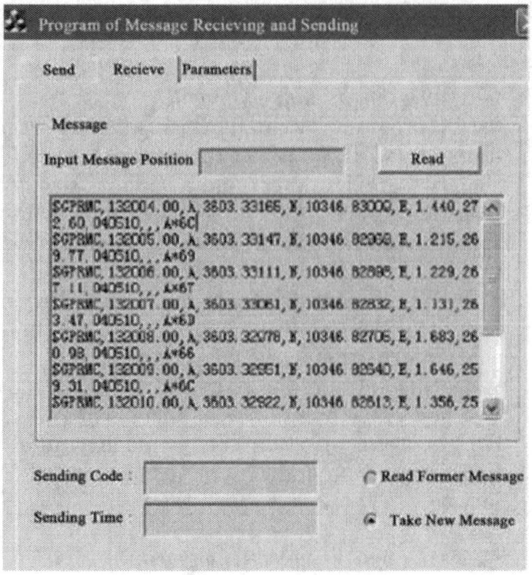

Figure 6. Window showing signal sending and receiving.

inserted into a phone card with the data traffic service. The work status of the module is revealed by the lights of D5 and D6. When D5 keeps shining and D6 flashes slowly, it means that the module has found a connection to the net. The debugging of the GPRS module is operated like this: link the serial port to the serial assistant, send AT instructions from the serial port by using the serial assistant, and observe the status of the GPRS from the serial assistant. AT instructions are sent to SIM900 A with the serial connection to MSP430. After sending the instructions, the serial port RXD starts to wait for the return-ing instruction from the receiving module. The instruction sending is judged to be successful by receiving the responding instructions. If the responding instructions are not received, the AT instructions would be sent again to ensure suc-cessful communication (see Figure 6).

7 CONCLUSIONS

The remote display system uses GPS to position refrigerated vehicles and transports the data of each end-node to the remote management plat-form through GPRS. This system also has the functions of displaying the driving states of refrig-erated vehicles in real-time, thereby recording posi-tions and enabling the dynamic replaying of the driving routes. To ensure the information safety of the refrigerated vehicles, the system permission is designed in user management and the functions such as add, delete, record, and modify of vehicles are also enabled in vehicle management.

ACKNOWLEDGMENT

This work was supported by the Special Fund for Agro-scientific Research in the Public Interest (No. 201303075).

REFERENCES

Huang Hengbo, Design and Implementation of Distribution and Monitoring System based on Internet of Thing. Xiamen University, 25–32 (2013).

Shen Yue, Chai Yanwei, Daily activity space of suburban mega-community residents in Beijing based on GPS data, ACTA GEOGRAPHICA SINICA, 04:50–51 (2013).

GuoChunyu, The locating and tracking monitoring center software design based on GSM/GPS sports car, Lanzhou University of technology, 7–9 (2011).

Yuan Guangsheng, Research on Vehicle Refrigerated Cargo Tracking System Based on RFID/GPS, Liaoning University of Technology, 40–43 (2016).

Jiang Xirui, The Design and Implementation of Position-ing System based on GSM/GPRS and GPS, Institute of Computing Technology Chinese Academy of Sciences, 22–34 (2006).

http://blog.sina.com.cn/s/blog_6721e3390102vlvz.html, 03–13(2015).

http://www.chinabgao.com/k/linglianwuliu/17808.html, 07–02(2015).

Advances in Energy Science and Equipment Engineering II – Zhou, Patty & Chen (Eds)
© 2017 Taylor & Francis Group, London, ISBN 978-1-138-71798-5

Pattern analysis of the square array antenna based on Hadamard difference sets by using MATLAB

Binghao Hu & Deming Zhong
Beijing University of Aeronautics and Astronautics, Beijing, China

Jian Dong
Central South University, Hunan Sheng, China

ABSTRACT: Hadamard difference sets is a branch of difference sets algorithm, which is a new array antenna algorithm. The algorithm used in array antenna can regularize the position of array elements of the arrangement, which is more advanced than the general random search. It can suppress the side lobe well and reduce the side lobe level. Mathematical analysis based on Hadamard difference sets is currently used to study array antennas in application; however, it cannot show the visualized distribution of the radiation field in space, thereby making it difficult to analyze. In this paper, a drawing pattern of the square array antenna based on Hadamard difference sets by using MATLAB is proposed, which can intuitively show the distribution of the radiation field in space. A further analysis on different antenna array elements is provided to illustrate the pros and cons of Hadamard difference sets.

1 INTRODUCTION

In wireless communications, an antenna plays an irreplaceable role in transmitting and receiving electromagnetic waves (Jia, 2008). Identical incentives on antenna aligned by different rules will result in different patterns. Research on the different alignment rules can improve the antenna gain coefficient and enhance its directional effect, or the appropriate alignment rules can be selected based on the demand conditions of the antenna. The research on arrangement rules is about arranging the array elements' positions with different mathematical methods (Dan, 2006). The Hadamard difference sets combined with the antenna array theory algorithm to arrange the distribution of the positions of array elements is a new idea of the antenna array. Hadamard difference sets array antenna allows the antenna elements to be arranged in the position of the rules; it is more advanced than general random search because it reduces the side lobe level effectively (Kopilovich, 2008). If the study of this method stays in the level of research based on mathematical equations, it will be very difficult to illustrate the Hadmard difference sets' impact on the radiation characteristics of the antenna. MATLAB has excellent graphics and numerical ability; using MATLAB can enhance the understanding and cognition of the antenna radiation field distribution theory, and can help to obtain more effective and intuitive results.

2 USING A MATHEMATICAL FORMULA ANALYSIS ANTENNA ARRAY PATTERN IS COMPLICATED

Pattern synthesis is the process of designing antenna parameters such as the side lobe value and main lobe width in order to achieve synthesis effect and optimization. Generally, pattern synthesis is a non-linear optimization problem. With the development of intelligent antenna technology, many pattern synthesis methods are proposed such as the Fourier transform method, numerical optimization method, Chebyshev method. Take the Dolph-Chebyshev pattern analysis method for example.

2.1 *Chebyshev polynomials*

The solution T_p in the following equation is the Chebyshev polynomial:

$$(1-x^2)\frac{d^2T_p}{dx^2} - x\frac{dT_p}{dx} + P^pT_p = 0 \qquad (1)$$

The solution of the equation is as follows:

$$T_p(x) =$$

$$\begin{cases} \sum_{n=0}^{M}(-1)^{M-n}\dfrac{M}{M+n}\binom{M+n}{2n}(2x)^{2n}, P=2M \\ \sum_{n=1}^{M}(-1)^{M-n}\dfrac{2M-1}{2(M+n-1)}\binom{M+n-1}{2n-1}(2x)^{2n-1}, P=2M-1 \end{cases} \quad (2)$$

$\binom{r}{s}$ of the equation is the binomial coefficient, which is given as follows:

$$\binom{r}{s} = \frac{r!}{s!(r-s)!} \quad (3)$$

Several low-order polynomials are given as follows:

$$\left.\begin{array}{l} T_0(x)=1 \\ T_1(x)=x \\ T_2(x)=2x^2-1 \\ T_3(x)=4x^3-3x \\ T_4=8x^4-8x^2+1 \end{array}\right\} \quad (4)$$

A higher-order equation derived by the use of a recursive equation is given as follows:

$$T_{p+1}(x)=2xT_p(x)-T_{p-1}(x) \quad (5)$$

Thus, one can deduce the P-order Chebyshev polynomial highest power by using the above-mentioned equations. Therefore, the above equation can also be written as follows:

$$T_p(x)=\begin{cases} \cos(P\arccos x), -1\le x\le 1 \\ \cosh(P\operatorname{arccos}h.x), |x|\ge 1 \end{cases} \quad (6)$$

2.2 Chebyshev array function

We found many properties of the Chebyshev polynomial after carrying out a thorough study. By combining research results, we found that Chebyshev polynomials can satisfy the requirements of the same side lobe. If we can make the independent variable x in a transformation that is associated with real variable θ, we can make some polynomial elements of $T_p(x)$ and the antenna array of the array factor correspond to each other.

For an evenly graded array of symmetric excitation and equidistant phase, when the array element is even, consider that $u=kd(\cos\theta-\cos\theta_0)$. The array factor can be written as follows:

$$S(u)=2\sum_{n=1}^{M}\frac{I_n}{I_0}\cos\left[(2n-1)\frac{u}{2}\right] \quad (7)$$

When the number of array elements is odd, that is $N=2M+1$, the array factor can be expressed as follows:

$$S(u)=1+2\sum_{n=1}^{M}\frac{I_n}{I_0}\cos\left(2n\frac{u}{2}\right) \quad (8)$$

Make

$$x=\cos\frac{u}{2} \quad (9)$$

The above-mentioned two equations correspond to the P-order Chebyshev polynomial (P = N − 1), and

$$S(u)=T_{N-1}(x)=T_{N-1}\left(\cos\frac{u}{2}\right) \quad (10)$$

The θ range is from 0 degrees to −180 degrees and X is in the range of −1 to 1. Therefore, to ensure that the scope of change is greater than the range of −1 to 1, it is necessary to change the substitution of equation (10) and re-order it as follows:

$$x=x_0\cos\frac{u}{2}, (x_0>1) \quad (11)$$

$$S(u)=T_{N-1}(x)=T_{N-1}\left(x_0\cos\frac{u}{2}\right) \quad (12)$$

2.3 The steps of the Dolph—Chebyshev method

According to the above conclusion, with the use of the Schelkunoff unit circle method, you can summarize the steps of the Chebyshev synthesis:

1. The first step is to determine the array element number N, and then we know the order of Chebyshev polynomials.
2. The second step is to obtain x_0 according to the requirements of the side lobe level R_0, via $T_{N-1}(x_0)=R_0$.

$$x_0=\cosh\left(\frac{1}{N-1}\operatorname{arccos}hR_0\right) \quad (13)$$

Because $x_0>1$, the calculating equation is inconvenient and therefore, the hyperbolic function equation is used in the above-mentioned equation, which is as follows:

$$x_0 = \frac{1}{2}\left[(R_0 + \sqrt{R_0^2 - 1})^{1/N-1} + (R_0 - \sqrt{R_0^2 - 1})^{1/N-1}\right] \quad (14)$$

3. By using equation (10), the root of $T_{N-1}(x)$ can be obtained:

$$x_i = \pm \cos\left[(2i-1)\frac{\pi}{2(N-1)}\right] \quad (15)$$

4. According to equation (14), the obtained u_i value corresponds to each x_i value, and one can get the root w_i of each position on the unit circle. Thus, $S(w)$ is obtained in the form of factoring, to get the distribution of the array excitation after each factor is multiplied.

Thus, a study of the square array antenna, through the mathematical analysis method, based on Hadamard difference sets does not show its visual image of the radiation field distribution in space and it is not easy to analyze.

3 THE PATTERN BY MATLAB SIMULATION CAN VISUALLY ANALYZE THE PROS AND CONS OF THE ANTENNA ARRAY

3.1 *Hadamard difference sets array elements' positions*

By definition, a Hadamard difference set of (V, K, Λ) is a set of K points defined on an integer $V_x \times V_y$ grid of V elements (Turyn, 1965).

$$HS = \left\{(b_0, c_0), (b_1, c_1), ..., (b_{k-1}, c_{k-1})\right\} \quad (16)$$

With $0 \le b_j V_x - 1$ and $0 \le j \le V_y - 1$, for any grid point $(m, n), 0 \le m \le V_x - 1$ and $0 \le n \le V_y - 1$, there exist exactly Λ pairs $\{(b_i, c_i), (b_j, c_j)\}$ that satisfy the equations $m \equiv (b_i - b_j) \bmod V_x$ and $n \equiv (c_i - c_j) \bmod V_y$, in which mod V means that the difference must be taken as modulo V;
Either $V = 4N^2, K = 2N^2 - N, \Lambda = N^2 - N$ or $V = 4N^2, K = 2N^2 + N, \Lambda = N^2 + N$, N is equal to 2r or $3 \cdot 2^r, r > 0$ (Kopilovich, 2008).
It is possible to synthesize a normalized thinned array by placing the array elements at the Hadamard difference sets locations.

3.2 *Autocorrelation of Hadamard difference sets*

In a difference set D, we can establish a sequence or array of ones and zeros (David, 1999)

$$A_v = \{a_i\} \ i = 0, 1, ..., K-1 \quad (17)$$

If $j \in D a_j = 1$ or $j \notin D a_j = 0$, we can create an infinite array of ones and zeros.

$$A_I = \{..., a_{-2}, a_{-1}, a_0, a_1, a_2 ...\} \quad (18)$$

For example, if D = {1,2,4} V = 7, K = 3, $\wedge$ = 1, we can get Av = {0110100}
In this array, A_V is periodically repeated, and we can define the autocorrelation function of A_I

$$C_I(\tau) = \sum_{n=0}^{V-1} a_n a_n + \tau \quad (19)$$

It shows that if and only if A_I is a difference set (Golomb, 1964), then

$$C_I(\tau) = \begin{cases} K, & if \ \tau(\bmod V) = 0 \\ \Lambda, & otherwise \end{cases} \quad (20)$$

When the positions of the elements 1 and 0 are determined, the periodically repeating element order is placed by a particular decision of difference sets; all side lobe peak power patterns of the difference sets array are limited to a relatively constant level, and are less than the main lobe peak by 1/K times.
When the infinite sequence is truncated to a single fixed period, the fixed level value still exists, and occupies half of the elements in the power pattern. The PSL is determined by the remaining points.

3.3 *MATLAB simulation results and analysis*

In this paper, the superiority of the Hadmard difference set in the antenna side lobe control is mainly studied. In order to compare the antenna array side lobe control ability in different cases, we introduce three important performance parameters of the antenna: main lobe width (HPBW), Peak Side lobe Level (PSL), and Average Side lobe Level (ASL). In the process of antenna array optimization, the main lobe width shows the capability of the side lobe control and the extent of main lobe sharpening. In the antenna pattern, the narrower the main lobe width is, the sharper the main lobe is. It means that the lesser the overflowed form of the electromagnetic waves of transmitting and receiving is, the smaller is the work power of the antenna. PSL and ASL show the quantitative antenna side lobe control ability, the side lobe causes the radiation of noise, and thus it is necessary to reduce the number of side lobes in the process of optimizing the antenna. The comparison of the Hadmard difference set between different elements is provided in the following section.

Among them:
Figure 1.a shows the power pattern when K = 15, V = 6

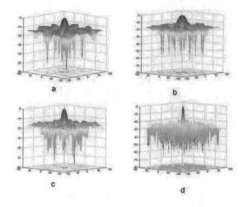

Figure 1.   Different array elements of Hadamard difference sets power patterns.

Table 1.   Four cases of the main lobe width (HPBW), the Peak Side lobe Level (PSL), and the Average Side lobe Level (ASL) parameter values.

	Main lobe width (HPBW)/rad	Peak Side lobe Level (PSL)/dB	Average Side lobe Level (ASL)/dB
K = 15	1.6057	−11.1243	−11.1243
K = 21	0.8378	−15.1414	−15.1414
K = 28	0.6283	−14.0503	−14.6910
K = 300	0.0698	−24.3510	−26.2085

Figure 1.b shows the power pattern when K = 21, V = 6

Figure 1.c shows the power pattern when K = 28, V = 8

Figure 1.d shows the power pattern when K = 300, V = 24

By comparing the simulation results shown above (Figure 1, Table 1), we can see that with an increase in array elements, the main lobe width of the Hadamard difference sets is smaller, which is the same as the Peak Side lobe Level and Average Side lobe Level, when K = 300 (Figure 1.d), compared with the three groups, is especially obvious. It can be seen, the more Hadamard difference sets array elements in the optimized antenna array, the effect better, it means that Hadamard difference set is suitable for large-scale sparse antenna array optimization. Of course, in practice, the selection should also consider the costs and other issues.

4   CONCLUSIONS

In conclusion, by using the mathematical method analysis, the antenna array obtained based on the Hadamard difference set has some limitations; the MATLAB simulation pattern has advantages in terms of analyzing the pros and cons of Hadamard difference sets of the antenna array, thereby clearly displaying the radiation characteristics of the antenna, making the concepts clear, and visualizing the theoretical results. Antenna technology involves a large number of mathematical algorithms and abstract concepts and therefore, drawing a pattern through MATLAB to describe the characteristics of the antenna is more concise than language.

Different array elements of the antenna array simulation pattern were analyzed, and we can see that the more array elements are present in the optimized antenna array, the better is the effect; but in practical applications, the antenna selection criteria are not only the optimization characteristics; other factors also influence the choice of antenna, such as cost and complexity.

ACKNOWLEDGMENTS

This research was supported by grants from the Civil Aviation Joint Funds Established by National Nature Science Foundation of China and Civil Aviation Administration of China (No. U1533201), a project of the Ministry of Industry and Information Technology of China (No. JSZL2015601C008), and the Major State Basic Research Development Program of China (973 Program) (No. 2014CB744904).

REFERENCES

Antenna Theory and Techniques (M) Xi'an University of Electronic Science and Technology Press Wanzheng lu 2004.
Antenna Theory and Techniques (M) Electronic Industry Press Shunshi zhong 2011.
Array Antenna Analysis and Synthesis (M) Beijing University of Aeronautics and Astronautics Press Zhenghui xue Weiming li Wu ren 2011 P108–P117.
Dandan zhu Optimization of Antenna Array (D) wuhan Huazhong University of Science and Technology 2006.
David G. Leeper, "Isophoric Arrays—Massively Thinned Phased Arrays with Well-Controlled Sidelobes", IEEE transactions on antennas and propagation, VOL. 47, pp. 1825–1835, DECEMBER 1999.
Golomb S.W. Digital Communications with Space Applications. EnglewoodCliffs, NJ: 1964.
Jia lao Studies of small ultra-wideband antennas and planar array antenna technology (D) shanghai Shanghai Jiaotong University 2008.
Kopilovich L.E. "Square array antennas based on Hadamard difference sets," IEEE Trans. Antennas Propag., vol. 56, pp. 263–266, 2008.
Modern Antenna Design (M) Second Edition Electronic Industry Press THOMAS A. MILLIGAN (Book) Yuchun guo (Translate) 2012.
Turyn R.J. "Character sums and difference sets," Pacific J. Math., vol.15, no. 1, pp. 319–346, 1965.

Advances in Energy Science and Equipment Engineering II – Zhou, Patty & Chen (Eds)
© 2017 Taylor & Francis Group, London, ISBN 978-1-138-71798-5

A study on the improved fuzzy C-means clustering algorithm

Yanrui Lei
Department of Network Engineering, Hainan College of Software Technology, Qionghai, Hainan, P.R. China

Laiquan Liu
Department of Digital Design, Hainan College of Software Technology, Qionghai, Hainan, P.R. China

Lei Bai
Department of Network Engineering, Hainan College of Software Technology, Qionghai, Hainan, P.R. China

Qingju Guo
Department of Information Management, Hainan College of Software Technology, Qionghai, Hainan, P.R. China

ABSTRACT: Fuzzy clustering is a kind of unsupervised clustering method for the important data analysis and modeling method. In this paper, an improved fuzzy FCM clustering algorithm is proposed to solve the problem of the Fuzzy C-Means (FCM) clustering algorithm due to the random selection of cluster centers that causes the algorithm converge to the local extremum. This algorithm takes complete consideration on the clustering center selection, makes full use of the idea of dividing sections, and combines with the division of the sampling algorithm and the standard clustering algorithm, so as to avoid the disadvantages of the Fuzzy C-Means (FCM) algorithm. The algorithm firstly divides the data set into each dimension to form each data area, and the data area is sampled according to the density of the data, so that the accuracy of the sampling can be improved; in the sample, the classification results and the optimal number of the F statistics are obtained by using the distance matrix and the fuzzy F statistics method. The next step is to obtain the center of the initial clustering; and then, run the FCM clustering algorithm to get the best results of the cluster based on the first stage of the cluster center. The experimental results show that the algorithm has certain advantages in accuracy and extreme value problem-solving when compared with other initial cluster centers.

1 INTRODUCTION

Clustering analysis is an important branch of unsupervised pattern recognition. It makes the sample sets, that don't have any signs, divide into several subsets (class) according to a certain rule, to make the detailed sample as far as possible in a class, and not a sample which is similar to a different class. The hard clustering divides each special sample to a certain class strictly, with the characteristics of either this or that. Fuzzy clustering is able to retain the uncertainty description of the sample, and can reflect the real world more objectively, and so it has become the main direction of the research of clustering analysis (Spragins, 2005).

Fuzzy clustering analysis has been applied in many fields, such as large scale data analysis, data mining, and graph processing. It has a very important application value (Yu, 2003). Fuzzy clustering analysis is widely used, in which the Fuzzy C-Means (FCM) clustering algorithm (Liu, 2004) is one of the most widely used cluster analysis methods.

The Fuzzy C-Means (FCM) clustering algorithm is a kind of typical dynamic clustering algorithm, which is based on the square of the error and fuzzy C-means algorithm. It makes an attempt to obtain the membership degree of each data sample point for all kinds of centers, through the optimization objective function, so as to determine the attribution of sample points to achieve the automatic classification of the sample data for the purpose of analysis (Asuhan, 1993). Because the FCM algorithm uses the gradient descent method to find the optimal solution, the initial value is more sensitive and is easy to fall into the local optimum. In order to improve the efficiency of the algorithm, it is not affected by the local minimum value, and the selection of initial value is very important. Choosing the proper initial value can not only improve the quality of the clustering results, but also speed up the convergence rate of the iterative process (Li, 2011).

In this paper, we first introduce the Fuzzy C-Means (FCM) clustering and data set classification and sampling, and focus on the traditional Fuzzy C-Means (FCM) clustering algorithm to improve the method (Jian,2010). An improved fuzzy C-means clustering algorithm based on partition sampling is proposed, by using a phased approach and combined with the sampling method, the fuzzy F statistics method, and the standard Fuzzy C-Means (FCM) clustering algorithm to solve the problem, and so the Fuzzy C-Means (FCM) clustering algorithm is avoided to converge to the local extremum problem, and a good result is achieved (Liu, 2008).

2 FUZZY C-MEANS (FCM) CLUSTERING

The dividing line of the common set theory is clear, definite, and unambiguous; however, with the continuous improvement of human cognition and the need of practical application, it is found that the traditional classification method cannot completely solve the problem of fuzzy classification in practical application. For example, when people are classified according to height, it is said that they are tall or short, very high, relatively high, very short, relatively short, etc. Similarly, the classification of the problem cannot have clear boundaries, and so there is the concept of fuzzy sets. With the emergence of the concept of fuzzy set, the concept of fuzzy division (Zhang, 2006) is proposed, and the method of the fuzzy set is used to deal with this kind of clustering problem, and it is called as the fuzzy clustering analysis.

The basic idea of the fuzzy C-means clustering algorithm is as follows: for a sample data $X = (x_1, x_2, \ldots x_n)$ set with n elements, the clustering process will be divided into a set of fuzzy ($2 \leq c \leq n$), and for each of the cluster centers $c_j (j = 1, 2, \ldots, c)$, to minimize the objective function.

The objective function is defined as follows:

$$J_c = \sum_{j=1}^{c} \sum_{i=1}^{n} \mu_{ij}^{\propto} \left\| x_i - c_j \right\|^2 \quad 1 \leq \propto \leq \infty \tag{1}$$

And satisfies the following equation:

$$\sum_{j=1}^{c} \mu_{ij} = 1 \quad \forall i = 1, 2, \ldots, n \tag{2}$$

$\mu_{ij} \in [0,1]$ represents that the first i data point is belonging to the first j membership degree of cluster centers;

C_j is the first j fuzzy clustering center, and the initial value is randomly selected in the training sample set; $\propto$ is the fuzzy degree, with an increase in $\propto$, the fuzzy clustering of the center increases; in this work, $\propto = 2$.

Fuzzy clustering is achieved through the iterative optimization of objective function J_c, which is a process of optimization. The fuzzy membership degree μ_{ij} and the cluster center c_j are obtained by using the following equation:

$$\mu_{ij} = \frac{1}{\sum_{k=1}^{c} \left(\frac{\left\| x_i - c_j \right\|^2}{\left\| x_i - c_k \right\|^2} \right)^{2(\alpha-1)}} \tag{3}$$

$$c_j = \frac{\sum_{i=1}^{n} \mu_{ij}^{\alpha} x_i}{\sum_{i=1}^{n} \mu_{ij}^{\alpha}} \tag{4}$$

In this process, the local minimum or saddle point in J_c is convergent. This process starts from a random cluster center, by searching the minimum point of the objective function, and constantly adjusts the fuzzy membership degree of the cluster center and each sample to determine the process of the sample class. This method relies heavily on the selection of cluster centers, and it is very easy for the fuzzy C-means clustering algorithm to fall into local minima if the initial classification is seriously deviated from the global optimal classification.

3 THE DIVISION AND SAMPLING OF DATA SETS

In a clustering data set $X = (x_1, x_2, \ldots x_n)$, D_n represents the n-dimensional space samples of the maximum and minimum value of the component difference, the number of clusters design for K. The data set is divided into the first step in the first division; the division of the idea is in each dimension, D_n all $d(d \geq k)$ are equal parts, and then the formation of $(d)^n$ area $Z_m (0 \leq m \leq (d)^n)$ occurs (LI, 2011).

Figure 1. shows a division sketch of the data set, among which d (d ≥ k).

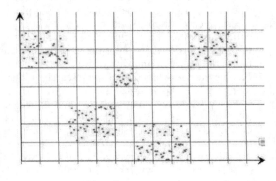

Figure 1. Partition diagram.

After the data set is divided into $Z_m (0 \leq m \leq (d)^n)$ regions, the number of data points in each region is divided into regions, and the data points are excluded for 0 of the data area, and the data points in each area are sampled according to the data points N_m in each area. Sampling idea: according to the number of data points from the region of the smallest, and within each region, sampled data points N_m follow the requirements: $M_m = N_m / n \times S$ (where S indicates each sampling point total) and n indicates the whole data set of total points.

4 IMPROVED FUZZY C-MEANS CLUSTERING ALGORITHM

The improved Fuzzy C-Means clustering algorithm is defined as the S-FCM algorithm, which is a clustering algorithm for determining the optimal number of clusters. Based on the sampling phase, the combination of the dynamic direct clustering algorithm and the standard clustering algorithm based on the sampling partition and the fuzzy F statistic method (WU, 2003), and by using the fuzzy F statistics method to obtain the advantage of the accuracy of the dynamic direct clustering, as far as possible to overcome the lack of sensitivity of the initial value of the algorithm, the algorithm has achieved good results.

Fuzzy C-Means clustering algorithm (FCM) is sensitive to the initial value, and it is easy for the deviation of the initial cluster center selection to affect the clustering results. Aiming at this problem, the S-FCM algorithm can be divided into two stages: the first stage of the original data set to use the method of division sampling (XIONG,2015) to take a part of the sample, by using the distance matrix and the fuzzy F statistics method, the classification results and the optimal number of the F statistics are obtained, and the next step is to obtain the center of the initial clustering; The second stage is set by using the first stage of the cluster center as the basis and running the FCM clustering algorithm, whereby the optimal clustering results are obtained. An S-FCM algorithm uses the dynamic clustering method, which can determine an optimal cluster number and the initial cluster center, and finally obtain the optimal solution. This algorithm has very good stability.

The overall idea of the improved algorithm is as follows:

First implementation phase: the original database according to the data set division and sampling method of the sample and in the sample data set based on, randomly selected the fuzzy clustering center, executive, which is based on the distance matrix and the fuzzy F statistic method of the clustering method:

Input: data set
Output result: cluster center and cluster number

1. First, the whole data set is divided into M and M_j sub space, $j = 1, 2, \ldots, M$
2. A random sampling is performed over the entire data set[5], followed by the $S_j, j = 1, 2, \ldots, J$ formation of a subset of the data. All of the S_j merged into a sample data set C, cluster $C_j, j = 1, 2, \ldots, J$ as the initial value respectively;
3. The Euclidean distance is used to calculate the distance between the sample points of the sampled data sets C, and the corresponding distance matrix is generated.
4. Hierarchical clustering tree based on the distance between classes is obtained;
5. The range of the cluster number is set up, and the segmentation results of different samples are generated according to the distance between classes;
6. Calculation of the different clustering results of the corresponding F statistics, F statistics, the larger the corresponding segmentation of the optimal results, in order to obtain the corresponding clustering results and the optimal number of clusters.

The second implementation stage: the first implementation phase is used to obtain the best

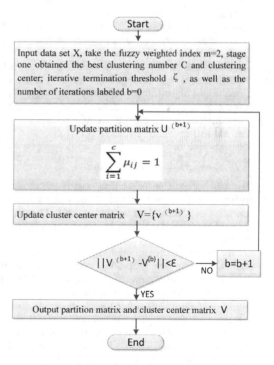

Figure 2.   Stage 2 execution processes.

number of clusters in this use of the standard FCM algorithm to calculate all the experimental data, which is the final result of the clustering.

Input: iteration termination condition ε, (experiments showed that, generally taking 2), experimental data sets, and first stage of implementation of the optimal cluster number.

Output result: fuzzy partition matrix U and cluster center matrix V

This algorithm improves the traditional FCM algorithm to blindly choose the cluster center and the number of the cluster, which is a kind of initial cluster centers and the number of cluster-oriented computing processes, to reduce the second phase of the algorithm to perform the FCM algorithm convergence time; the disadvantage is that the sampling data and the sampling data of the clustering process also need to spend a certain time, but can improve the accuracy of clustering and fall into the problem of extreme value.

5 EXPERIMENTAL PROCESS

5.1 Description of the experimental environment

A data set from UCI (California University, Irvine) is used in the machine learning database, from which the 3 sets of data sets must be picked (Xiong, 2015), by using the selected data to test the algorithm. Experimental data, such as Table 1, show that the number of selected data sets range from small to large, followed by the balance scale data sets, the nursery, and adult data sets.

In this paper, the program realized the use of Java language and achieved the test of the traditional FCM algorithm and the improved algorithm S-FCM; the experimental data of the same group are clustered and the experimental results of the two algorithms are analyzed respectively.

5.2 Comparison of cluster validity

By using the FCM algorithm and S-FCM algorithm in this paper, balance scale data sets, the nursery, and adult data sets are tested. The accuracy of the two algorithms is compared, as shown in Tables 1–3.

By analyzing the data in the above-mentioned tables, we can see that when the total number of sample points is more, the improved S-FCM algorithm is better. On the other hand, when the number of samples is small, the clustering accuracy of the improved algorithm is better than that of the FCM algorithm. And so, the improved S-FCM algorithm is more suitable for the data set of the sample point number because of the limitation of the sampling.

Table 1. Comparison of the balance scale data sets clustering accuracy.

Algorithm name	Total number of sample points	Correct number of clusters	Wrong number of clusters	Correct rate (%)
FCM	671	594	77	88.52
S-FCM	671	546	125	81.37

Table 2. Comparison of the nursery data sets clustering accuracy.

Algorithm name	Total number of sample points	Correct number of clusters	Wrong number of clusters	Correct rate (%)
FCM	12 960	8671	4289	66.90
S-FCM	12 960	9808	3152	75.67

Table 3. Comparison of the adult data sets clustering accuracy.

Algorithm name	Total number of sample points	Correct number of clusters	Wrong number of clusters	Correct rate (%)
FCM	48 842	35695	13 147	73.08
S-FCM	48 842	42156	6 686	86.31

5.3 Comparison of iteration times

For any clustering algorithm, owing to its always after an iterative clustering nature, it requires not only the clustering correctness, but also need to consider the clustering algorithm's objective function of the number of iterations, which is one of the standard algorithms. Especially when the amount of data is particularly large, the number of iterations is more able to reflect the efficiency of the algorithm.

The S-FCM clustering algorithm and FCM clustering algorithm in the convergence rate of the experimental comparison, the experiment using the scale balance data sets, nursery data sets, and adult data sets for the 20 iteration tests and take the average. The number of iteration times of each algorithm and its objective function value is shown in Figures 3–5.

As can be seen from the above example, the two algorithms are relatively close to the objective function at the end of the convergence, the S-FCM algorithm is used to predict the best cluster number and cluster center in advance, and so the convergence rate of the S-FCM algorithm is better than that of the FCM algorithm, and the number of iterations has been reduced.

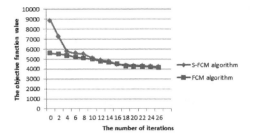

Figure 3. The convergence comparison of the S-FCM algorithm and FCM algorithm in the scale balance data set.

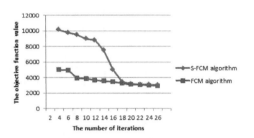

Figure 4. The convergence comparison of S-FCM algorithm and FCM algorithm in Nursery data set.

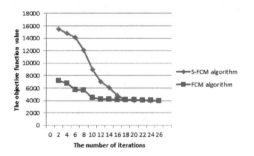

Figure 5. The convergence comparison of the S-FCM algorithm and FCM algorithm in an adult data set.

6 CONCLUSIONS

S-FCM algorithm improvement is performed for the random problems of the FCM algorithm in the selection of the initial clustering center, and it takes a complete consideration on the clustering center selection, makes full use of the idea of dividing sections, and combines with the division of the sampling algorithm and the standard clustering algorithm, so as to avoid the disadvantages of the Fuzzy C-Means (FCM) algorithm. According to the sampled data, the FCM algorithm is implemented to determine the optimal number of clusters and the center of the cluster, which makes the clustering result more stable. The experimental results of stage 2 show that the S-FCM algorithm has a better clustering effect than the traditional algorithm when the number of the sample data set is larger, and the execution time is lower than that of the traditional FCM algorithm. But when the number of the sample data set is smaller, the advantages of the improved algorithm cannot be seen.

ACKNOWLEDGMENT

This work is supported by the Natural Science Foundation of Hainan Province of China (No. 20156232).

REFERENCES

Asuhan K S, Selim S.A global algorithm for the fuzzy clustering problem[J]. Pattem Recognition, 1993, 26 (9): 1357~1361.

Jiang Lun, Ding Huafu. Improvement of the Fuzzy C-Means Clustering Algorithm [J]. Computer & Digital Engineering, 2010. 2(38): 5~6

Li Wei, Xue Hui-feng, Zhan Hai-liang. The Initial Cluster Center Algorithm Based on Segmenting and Sampling [J]. Journal of Taiyuan University of Technology, 2011, 4(42): 334~335.

Liu Xiaofang, Zeng Huanglin, Lv Bingchao. Clustering Analysis of Dot Desity FunctionWeighted Fuzzy C-Means Algorithm[J]. Computer Engineering and Applications, 2004, 24(4): 64~65.

Liurui-Jie, Zhang Jin-bo, LIURu. Fuzzy c-Means Clustering Algorithm [J]. Journal of Chongqing Institute of Technology(Natural Science), 2008, 2(22): 140~141

Mihong, Zhang Xi-wen, CHENWen-sheng. Application ofFuzzy C lustering A lgorithm on Data Mining Platform [J]. Journal of Xiamen University of Technology, 2006, 14(2): 17~21

Spragins J. Learning without a teacher[J]. IEEE Transactions of Information Theory, 2005, 23(6): 223~230.

Wu Cheng-mao, Fan Jiu-lun. Fuzzy F statistic and its application [J]. Journal of Xi'an University of Post and Telecommunications, 2003, 7(8): 57~59

Xiong Yongjun, Liu Weiguo, Ou Pengjie. New optimized fuzzy C-means clustering algorithm. [J]. Computer Engineering and applications, 2015, 51(11): 124–128.

Yu Jian. On the Fuzziness Index of the FCM Algorithms [J]. Chinese Journal of Computers, 2003, 26 (8): 968~973.

Advances in Energy Science and Equipment Engineering II – Zhou, Patty & Chen (Eds)
© 2017 Taylor & Francis Group, London, ISBN 978-1-138-71798-5

A correntropy-based direction-finding algorithm for cyclostationary signals in an impulsive noise environment

Sen Li, Xiaojing Chen & Rongxi He
College of Information and Science Technology, Dalian Maritime University, Dalian, China

ABSTRACT: In this paper, the problem of the Direction of Arrival (DOA) estimation for cyclostationary signals in impulsive noise environments, which is modeled as a stable distribution, is dealt with. Since the DOA estimation based on the cyclic correlation matrix degrades seriously in a stable distribution noise environment, we define a novel robust cyclic correlation matrix by using the robust statistics property of the correntropy technique, and apply this matrix with MUSIC algorithm to estimate the DOA in the presence of impulsive noise for cyclostationary signals. The simulation results demonstrate that the proposed algorithm outperforms the existing cyclic correlation matrix and the fractional lower order cyclic correlation matrix-based MUSIC algorithms.

1 INTRODUCTION

An array of sensors, such as radio antennas can be used to detect the presence of propagating signals, estimate their Directions of Arrival (DOAs) and other parameters. Application areas include radar, sonar, biomedical signal processing, communication systems, and others. Conventional array processing methods basically exploit the spatial properties of the signals impinging on the array of sensors, but many man-made signals encountered in radar, sonar, or telecommunications exhibit a cyclostationary (Gardner,2006) property, which can be exploited to cancel interference and background noise. Cyclostationarity was first introduced into array signal processing by Gardner (Gardner, 1988) due to its immunity to interference and noise and so, by incorporation of this property into the signal processing algorithm, the direction-finding performance of existing methods can be improved. Several algorithms can be found in the literature that exploit cyclostationarity to output conventional methods (Zeng, 2009; Du, 2015; Yan, 2007). Instead of using the correlation matrix as in the conventional methods, these algorithms require the estimation of the cyclic correlation matrix that reflects the cyclostationarity of incoming signals, assuming that they have baud rates and/or are carrier-modulated signals, as they would be in used in radar and radio communication applications.

The conventional methods and the cyclostationarity methods assume that the ambient noise is Gaussian-distributed and are based on second-order statistics. However, the noise in practice often exhibits non-Gaussian properties, and is sometimes accompanied by strong impulsiveness (Mahmood, 2012). For example, atmospheric noise (thunderstroms), car ignitions, microwave ovens, and other types of naturally occurring or man-made signal sources can result in aggregating noise components that may exhibit high amplitudes for small duration time intervals. Under investigation, it is found that the α-stable distribution is a suitable noise model to describe this type of noise (Nikias,1995). It can be considered as the greatest potential distribution to characterize various impulsive noises as different exponent parameters such as α are selected.

Since the α-stable distribution has no finite Second-Order Statistics (SOS), the DOA estimate algorithms for cyclostationary signals based on the cyclic correlation may be highly unreliable. Thus, the Fractional Lower-Order Cyclic Statistics (FLOCS) is proposed to cover this issue (You, 2013; Zhu, 2012; Song, 2013; Liu, 2015), such as the Fractional Lower Order Cyclic Correlation (FLOCC) (You, 2013; Zhu, 2012), the Phased Fractional Lower Order Cyclic Correlation (PFLOCC) (Song, 2013), and the fractional lower-order cyclic cross-ambiguity function (Liu, 2015). However, FLOCS requires a priori knowledge of the α-stable distribution, which is difficult to estimate in some practical applications. To solve this problem, a new operator referred to as the correntropy-based correlation (CECO) was proposed (Zhang, 2014; Qiu, 2012) and was applied to obtain DOA estimation. To deal with cyclostationary signals, Luan et al. (Luan, 2016) defined cyclic correntropy and the cyclic correntropy spectrum, and applied these in solving the time delay estimation problem in an impulsive noise environment.

In this paper, we introduce a novel operator based on CECO for cyclostationary signals, namely the correntropy-based cyclic correlation (CECCO), which can be used with MUSIC for cyclostationary signals in an impulsive noise environment. The derived DOA estimation algorithm is obtained by combining the CECCO matrix with MUSIC, namely the CECCO_MUSIC algorithm. Computer simulation experiments are provided to illustrate the performance superiority of the proposed CECCO_MUSIC algorithm over the FLOCC-based MUSIC algorithm (FLOCC_MUSIC) and cyclic correlation-based MUSIC algorithm (CCO_MUSIC) in an impulsive noise environment.

2 SIGNAL AND NOISE MODEL

In this paper, we consider a Uniform Linear Array (ULA) of L antennas. Suppose, there are K electromagnetic waves impinging on the array from angular directions $\theta_k, k = 1, \cdots, K$. The incident waves are assumed to be far-field narrowband point sources. In our study, we assume that K_a Signals Of Interest (SOIs) exhibit cyclostationary signals with cycle frequency ε $(with\ K_a \le K)$, and all of the remaining $L - K_a$ signals either have different cycle frequencies or are cyclically correlated with SOIs, and they are referenced as Signals Not Of Interest (SNOIs); the noise is an i.i.d random variable and is not correlated with signals. By using this assumption, the array output at the lth sensor can be expressed as follows:

$$x_l(t) = \sum_{k=1}^{K_a} A_{lk} s_k(t) + n_l(t), \quad l = 1, 2, \cdots, L \qquad (1)$$

where A_{lk} is the response of the lth sensor with respect to the kth SOI, $s_k(t)$ is the SOI emitted by the kth source that has a cycle frequency ε, and $n_l(t)$ represents all SNOIs plus noise received by the lth sensor.

Define the received data vector $X(t)$ as $X(t) = [x_1(t), \cdots, x_L(t)]^T$, and then it can be expressed as follows:

$$X(t) = A(\theta)S(t) + N(t) \qquad (2)$$

where $S(t) = [s_1(t), \ldots, s_{K_a}(t)]^T$ contains the SOIs, while the vector $N(t)$ represents the SNOIs and noise. The matrix $A(\theta) = \{A_{lk}\}_{L \times K_a} = [a(\theta_1), \ldots, a(\theta_{K_a})]$ contains the steering vectors of the impinging SOIs and the steering vector $a(\theta_k)$, $k = 1, \cdots, K_a$ can be expressed as follows:

$$a(\theta_k) = \left[1, \quad e^{-j\frac{2\pi}{\lambda}d\sin\theta_k} \quad, \cdots, \quad e^{-j\frac{2\pi}{\lambda}(L-1)d\sin\theta_k} \right]^T \qquad (3)$$

where, λ is the carrier wavelength of the signal and d is the interspacing.

The i.i.d random noises are assumed to be Symmetry α-Stable distributed (S α S) (Nikias, 1995), which was described by the characteristic exponent α, the dispersion parameter γ, and the location parameter a (a is usually set to zero). When $\alpha = 2$, the S α S distribution becomes a Gaussian distribution. An important difference between the Gaussian and the S α S distribution is that only moments of order less than α exist for the S α S distribution.

3 REVIEW OF THE CCO_MUSIC AND FLOCC_MUSIC ALGORITHMS

3.1 CCO_MUSIC algorithm

For the cyclic frequency ε and some lag parameter τ, the cyclic correlation (CCO) matrix of the receive data vector $X(t)$ is defined as follows:

$$R_{XX}(\varepsilon, \tau) = E[X(t)X^H(t+\tau)e^{-j2\pi\varepsilon t}] \qquad (4)$$

Instead of using the EigenValue Decomposition (EVD), the DOA estimation algorithm based on the Singular Value Decomposition (SVD) of the CCO matrix (4) is given as follows:

$$\begin{bmatrix} E_s & E_n \end{bmatrix} \begin{bmatrix} \Sigma_s & 0 \\ 0 & \Sigma_n \end{bmatrix} \begin{bmatrix} V_s & V_n \end{bmatrix}^H \qquad (5)$$

where the subscripts s and n stand for signal- and noise-subspace, respectively, $\begin{bmatrix} E_s & E_n \end{bmatrix}$ and $\begin{bmatrix} V_s & V_n \end{bmatrix}$ are unitary matrices, and the diagonal elements of the diagonal matrices Σ_s and Σ_n are arranged in the decreasing order. Σ_n tends to zero as the number of samples tends to infinity. Thus, the CCO_MUSIC algorithm can achieve the DOAs' estimation by searching for the peaks of the following spatial spectrum:

$$P(\theta) = \frac{1}{a^H(\theta)E_n E_n^H a(\theta)} \qquad (6)$$

3.2 FLOCC_MUSIC algorithm

The CCO_MUSIC algorithm plays an essential role in high-resolution direction-finding under Gaussian noise assumption. However, due to the lack of finite variances, the cyclic correlation does not exist under the SαS impulsive noise. Instead, the FLOCC_MUSIC algorithm was proposed (You, 2013; Qiu, 2012). In the FLOCC_MUSIC algorithm, the CCO matrix was replaced by the FLOCC matrix, which can be defined as follows:

$$R_{XX}^p(\varepsilon, \tau) = E\left[X(t)[X^T(t+\tau)]^{<p-1>} e^{-j2\pi\varepsilon t} \right] \qquad (7)$$

where, p is the order of the fractional lower order moment, and $1 < p < \alpha \leq 2$. For a complex process, $x^{<p>} = |x|^{p-1} x^*$. It was pointed out that such an operation is mainly used to suppress the amplitude of impulsive noise rather than the period. And so, the cyclic frequency for CCO is also suitable for FLOCC.

4 THE PROPOSED ALGORITHM

In this paper, we define a new cyclic statistics based on CECO, which can be applied in α-stable impulsive noise, and that would make it as an effective substitute for the conventional CCO.

For the array received signal vector $X(t)$, a CECO matrix was defined (Zhang, 2014), whose (i,l)th entry R_{il} is given by the following equation:

$$R_{il} = E\left[\exp\left(-\frac{|x_i(t) - \mu x_l^*(t)|^2}{2\sigma^2} \right) x_i(t) x_l^*(t) \right], \quad \mu \neq 1 \tag{8}$$

where μ is a given positive constant. It is stated in equation (12) that the CECO matrix would behave as a correlation matrix in Gaussian noise situations and a robust M-estimation correlation matrix in the impulsive noise environments.

Generalize the CECO matrix for cyclostationary signals, and we can define the CECCO matrix $R_{XX}^{ce}(\varepsilon, \tau)$. The (i,l)th element of it can be defined as follows:

$$\left[R_{XX}^{ce}(\varepsilon, \tau) \right]_{il} =$$
$$E\left[\exp\left(-\frac{|x_i(t) - \mu x_l^*(t+\tau)|^2}{2\sigma^2} \right) x_i(t) x_l^*(t+\tau) e^{-j2\pi\varepsilon t} \right],$$
$$\mu \neq 1 \tag{9}$$

Similarly, we can conclude that the CECCO matrix would behave as a cyclic correlation matrix in Gaussian noise situations and a robust M-estimation cyclic correlation matrix in the impulsive noise environments. Therefore, we can conduct a SVD on the matrix $R_{XX}^{ce}(\varepsilon, \tau)$ and formulate the corresponding spatial spectrum to obtain the DOA estimates of SOIs, and we define this as the CECCO_MUSIC algorithm.

5 SIMULATION RESULTS

In this section, we present some simulation results to compare the relative performance of the

proposed CECCO_MUSIC method with those of the CCO_MUSIC and FLOCC_MUSIC algorithms in an $S \alpha S$ impulsive noise environment. As the characteristic of the α-stable distribution makes the use of the standard SNR meaningless, a new SNR measure, named as the Generalized Signal-to-Noise Ratio (GSNR), is defined as $GSNR = 10 \log_{10} (\sigma_s^2/\gamma)^{[7]}$, where σ_s is the variance of the signal and γ is the dispersion parameter of the noise.

Two performance criteria are used to assess the performance of algorithms. The first one is the probability of resolution. To perform the resolution analysis of the algorithm, a popular resolution criterion is defined by using the following threshold equation (Li, 2014):

$$P(\theta_m) - \frac{1}{2}\{P(\theta_1) + P(\theta_2)\} > 0 \tag{10}$$

where θ_1 and θ_2 are the angles of arrival of the two SOIs and $\theta_m = (\theta_1 + \theta_2)/2$ is the mid-range between them. The two SOIs are said to be resolvable if inequality equation (10) holds. Two hundred independent Monte Carlo experiments are performed; the number that two incident angles can be resolved is denoted as N_{ok}, and then the probability of resolution is defined as $N_{ok}/200$.

In the case where two SOIs can be resolved, set $\bar{\theta}_i(n), i = 1,2$ is defined as the estimation of θ_l for the nth Monte Carlo experiment, the average Mean Square Error (MSE) of the DOA estimation is defined as follows:

$$\text{MSE} = \frac{1}{2N_{ok}} \sum_{n=1}^{N_{ok}} (\bar{\theta}_1(n) - \theta_1)^2 + \frac{1}{2N_{ok}} \sum_{n=1}^{N_{ok}} (\bar{\theta}_2(n) - \theta_2)^2 \tag{11}$$

Suppose, the ULA, which is made of ten sensors with an interspacing of half a wavelength, is used. The incoming signals are uncorrelated Binary Phase-Shift Keying (BPSK)-modulated sources. The sample frequency is $f_s = 900KHz$, and the carrier frequency of the BPSK SOIs is $f_1 = 100KHz$. Other signals are considered as SNOIs with a carrier frequency of $f_2 = 70KHz$. Depending on the type of modulation used, the cycle frequency is usually twice the carrier frequency or multiples of the baud rate or combinations of these. In this paper, the proposed algorithm and the CCO_MUSIC and FLOCC_MUSIC algorithms are simulated with the cycle frequency $\varepsilon = 2f_1$.

Simulation 1: suppose that two BPSK SOIs come from 30° and 40°, respectively, and one SNOI arrives from 50°. The noise is modeled as $S\alpha S$ distributed with $\alpha = 1.4$. The number of snapshots available to the algorithms is 600. The influence of the GSNR on the performance of the algorithms is shown in Figure 1. It can be seen that the performance of the CECCO_MUSIC algorithm is much better than that of FLOCC_MUSIC and CCO_MUSIC algorithms, especially when the GSNR values are low. The probability of the resolution of the proposed CECCO_MUSIC algorithms can be more than 90% when $GSNR \geq 10dB$, while the CCO_MUSIC algorithm almost fails when $GSNR < 14dB$.

Simulation 2: the influence of the characteristic exponent of the $S\alpha S$-stable impulsive noise on the performance of the algorithms is shown in Figure 2. The angles of arrival for the SOIs are 30° and 40°, and the DOA of SNOI is 50°. The GSNR is $GSNR = 14\ dB$. The number of snapshots available to the algorithms is 600. In Figure 2, we can see that the CECCO_MUSIC algorithm shows a significant performance improvement over FLOCC_MUSIC and CCO_MUSIC algorithms in terms of both probability of resolution and MSE.

Simulation 3: the influence of the number of snapshots on the performance of the algorithms is shown in Figure 3. The angles of arrival for the SOIs are 30° and 30° and the DOA of SNOI is 50°. The characteristic exponent is $\alpha = 1.4$ and GSNR is $GSNR = 14\ dB$. It indicates that the performance of all three methods becomes better as the

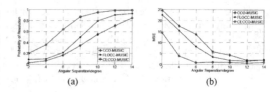

(a)   (b)

Figure 4. Probability of resolution (a) and MSE (b) versus angular separation.

number of snapshots increases, but for the same number of snapshots, the proposed methods have lower MSEs and a higher probability of resolution than the other two methods.

Simulation 4: the influence of the angular separation on the performance of the algorithms is shown in Figure 4. The angles of arrival for the SOIs are 30° and 30° + δ, where δ is the angular separation of the two SOIs, which varies from 2° to 14° in steps of 2°, and the DOA of SNOI is 50°. The characteristic exponent is $\alpha = 1.4$ and the GSNR is $GSNR = 14\ dB$. As we can see, both performances of the proposed algorithms are superior to that of the other two algorithms, and the probability of resolution of the proposed algorithms can reach above 80%, while the other two methods are no more than 50%, when the angular separation is 8°.

6 CONCLUSION

In this paper, a new operator referred to as the correntropy-based cyclic correlation (CECCO) is proposed. We formulate the CECCO-based matrix of the array outputs and apply it with MUSIC algorithm to obtain DOA estimates of SOIs. Experimental results illustrate that the CECCO_MUSIC can outperform the CCO and FLOCC-based MUSIC algorithms in impulsive noise environments.

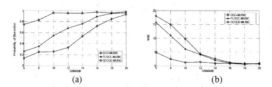

(a)   (b)

Figure 1. Probability of resolution (a) and MSE (b) versus GSNR.

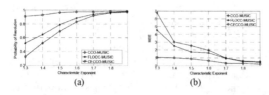

(a)   (b)

Figure 2. Probability of resolution (a) and MSE (b) versus characteristic exponent.

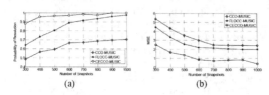

(a)   (b)

Figure 3. Probability of resolution (a) and MSE (b) versus number of snapshots.

ACKNOWLEDGMENTS

The authors acknowledge the National Natural Science Foundation of China (Grant number: 61301228) and the Fundamental Research Funds for the Central Universities (Grant number: 3132016318) for financial support.

REFERENCES

Du, C. L., H. C. Zeng, W.L.Lou, andY. Hou, "On Cyclostationary Analysis of WiFi Signals for Direction Estimation," *IEEE International Conference on Communications*, 3557–3561, (2015).

Gardner, W.A., "Simplification of MUSIC and ESPRIT by exploitation of cyclostationarity," IEEE Trans. Communnications, **76**, 845–847, (1988).

Gardner, W.A., A. Napolitano, and L. Paura, "Cyclostationarity: half a century of research," Signal Processing, **86**, 639–697, (2006).

Li, S., Z.X. Liang, "DOA Estimation of the Coherent Sources Based on SVD and the Fractional Lower Order Statistics", *Sixth International Conference on Wireless Communications and Signal Processing*, (2014).

Liu, Y., T. S. Qiu, and J. Li, "Joint estimation of time difference of arrival and frequency difference of arrival for cyclostationary signals under impulsive noise," Digital Signal Processing, **46**, 68–80, (2015).

Luan, S.Y., T.S. Qiu, Y.J.Zhu, L.Yu, "Cyclic correntropy and its spectrum in frequency estimation in the presence of impulsive noise," Signal Processing, **120**, 503–508, (2016).

Mahmood, A.,"PSK communication with passband additive symmetric α-stable noise," IEEE Trans. Communications, **60**, 2990–3000, (2012).

Nikias and M.Shao, C.L., "Signal processing with Alpha-stable distributions and applications," *New York, NY, USA: Wiely*, 1995.

Qiu, T.S., J.F Zhang, A.M.Song,and H. Tang, "The generalized correntropy-analogous statistics based direction of arrival estimation in impulsive noise environments," Journal of Signal Processing, **28**, 463–466, (2012).

Yan, H., H.H. Fan, "Signal-selective DOA tracking for wideband cyclostationary sources," IEEE Trans. Signal Processing, **55**, 2007–2015, (2007).

You, G. H., T.S. Qiu, A.M. Song, "Novel direction findings for cyclostationary signals in impulsive noise Environments," CSSP, **32**, 2939–2956, (2013).

You, G. H., T.S. Qiu, J. Yang, "A novel DOA estimation algorithm of cyclostationary signal based on UCA in impulsive noise," AEUE, **67**, 491–499, (2013).

You, G. H., T.S. Qiu, and Y. Zhu, "A novel extended fractional lower order cyclic MUSIC algorithm in impulsive noise," ICIC Express Letters, **6**, 2371–2376, (2012).

Zeng, W. J., X. L. Li, X. D. Zhang, and X. J. Jiang, "An improved signal-selective direction finding algorithm using second-order cyclic statistics," *IEEE International Conference on Acoustics, Speech and Signal Processing*, 2141–2144, (2009).

Zhang, J.F, T.S. Qiu, A.M. Song, and H. Tang, "A novel correntropy based DOA estimation algorithm in impulsive noise environments," Signal Processing, **104**, (2014).

Advances in Energy Science and Equipment Engineering II – Zhou, Patty & Chen (Eds)
© 2017 Taylor & Francis Group, London, ISBN 978-1-138-71798-5

The simulation and practical evaluation of shipping VHF communication equipment

Yuqin Wen & Yan Chen
Transportation Management College, Dalian Maritime University, Dalian, China

Hongxiang Ren
Navigation College, Dalian Maritime University, Dalian, China

ABSTRACT: In this paper, the technology of the development of the VHF simulator, the system architecture, the design of each module, and the development process as well as the practical evaluation function are mainly expounded. Among these, the steps assessment method and analytics hierarchy process method were used in the practical operation of the evaluation system. The weight of each evaluation element was determined based on the characteristics of each question. And based on assessing elements, the evaluation algorithm was designed from whether the task was completed, manipulation time, and manipulation steps. And then, the assessment result was calculated. Its implementation is based on Microsoft Visual 2010 development tools and the MFC development environment, thereby realizing the simulation of marine VHF communication equipment and the function of automatic assessment of the practical operation. The user can realize the training and evaluation for equipment operation through the terminal emulator.

1 INTRODUCTION

In accordance with SOLAS (Safety Of Life At Sea) and amendments of States Parties' Assembly to IMO (International Maritime Organization) and relevant provisions of the 1978 Seafarers' Training, Certification and Watchkeeping international contracts (STCW78/95) in 1995, operators of GMDSS (Global Maritime Distress and Safety System) must be trained, assessed, and must obtain certification. Owing to the high cost of GMDSS equipment, low efficiency of training and being prone to false alarms and other issues (Wang, 2008), the use of the GMDSS simulator for crew training and evaluation of the practical operation has been recognized by relevant organizations and conventions and it has gradually become a trend.

VHF (Very High Frequency) equipment can not only be used to achieve communication between the ship and the shore stations or communication between the ship and the ship in the near field, but also communication between the slipway and the public network users by transfer between land stations, and it is one of the indispensable GMDSS equipment (Tzannatos, 2004). Therefore, VHF simulator development and practical evaluation function are the actual needs of the GMDSS simulator and can promote the ability to enhance the practical operation of the crew.

JRC's JHS770S/780D VHF shares are higher in the domestic market, and therefore, we chose

this type of VHF as the research subject. Based on Microsoft Visual 2010 development tools and MFC development environments, we have developed a simulation system of marine VHF communications equipment, while achieving the automatic evaluation of the practical operation of the device, and integrating it into the Dalian Maritime University GMDSS simulator.

2 ACHIEVEMENT OF HUMAN–COMPUTER INTERACTION

Users' experience to manipulate the simulation equipment VHF should be consistent with the real VHF equipment (Ding, 2009); for this reason, it conducted a study of human–computer interaction simulation equipment function from the aspects of interface simulation and manipulation realization in this paper.

2.1 *Interface simulation*

VHF functions are achieved by using the keys on the device and knobs and handle in the real machine device.

In order to be consistent with the actual operation, the VHF operator interface need to be simulated in the simulation-based graphical interface in a WINDOWS environment (Jin, 2002). Therefore, we took pictures of the real machine equipment and used Photoshop and other tools for processing the formation of the bitmap resources; and then, by using double buffering and defining several canvas, we made the processed bitmap image and displayed it in canvas to achieve the simulation interface in the end (as shown in Figure 1).

2.2 *Manipulation realization*

On the main interface of emulation devices, various buttons, knobs, and the handle operations are achieved by picking up the mouse, namely the use of a mouse MFC message response mechanism (Sun, 2006), thereby adding OnLButton-Down(), OnLButtonUp() and OnMouseMove() response functions, which are performed in order to achieve key down and up, the knobs rotation, and so on.

2.3 *Editing of letters and characters*

At the time of editing, the directory name, numbers, letters, and special characters are required, but the JRC_JHS770S VHF models can only feed digital input by keys and inputting letters and special characters must be achieved through the knob. Therefore, in the development of a simulation device, you need to number each character, thereby changing the character number by using the knob to achieve the change of outputting characters on the interface. Taking the VHF Directory as an example, VHF Directory is divided into four classes: Coast station list, Ship station list, calling group list, and PSTN number list, of which all four can store multiple numbers which have names composed of 16 characters at the most; therefore, we use a three-dimensional array to store the integer number of characters in the device development. Among these, the first dimension is the representation of directory classification, the second dimension represents the number of each type in the number, and the third dimension represents the number of bits per character. As shown in Figure 2, we can change the first letter D of the number one name into H by using the enter knob.

Figure 1. Picture showing the operation interface of the VHF simulator.

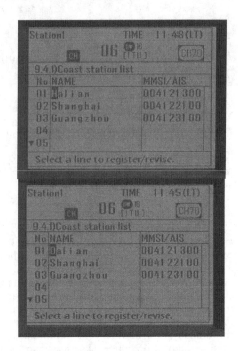

Figure 2. Picture showing the editing of the directory.

3 SIMULATION OF DEVICE FUNCTIONS

The VHF of JRC_JHS770S models has many functions, including DSC (Digital Security Controls) routine calling, DSC emergency calling, channel operations, equipment testing, device settings, etc. (Tzannatos, 2002); the simulation system achieves these functions through two major modules, as shown in Figure 3.

3.1 *Call simulation module*

The module is mainly used to achieve a single call, group call, all calls, and emergency call, simulation of communication between ocean transport vessels and other vessels or signals for help to other ship berths when in distress. When the system simulated a single call, the main process that occurs is shown in Figure 4. Process steps of group calls and general call are mainly consistent with the single call. The difference is that the user chooses GROUP CALLING menu for group calling, while ALL SHIPS menu choice for a full call (Shao. 2007). An emergency call system is used by holding down the emergency call button of the DISTRESS system for five seconds to automatically send a ship in distress and calling frequency and ship location information.

3.2 *Basic menu module*

The main achievement of this module is the DSC basic functions, such as channel scan, channel double

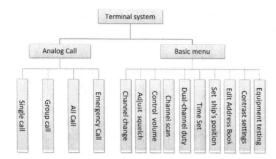

Figure 3. System structure diagram.

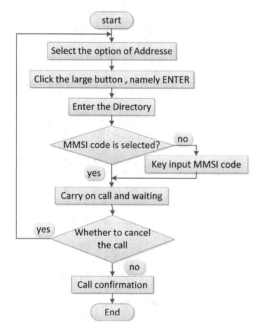

Figure 4. Flowchart of the calling process.

duty, clock settings, equipment testing, and simulation of VHF basic settings menu. The system function according to the sub-menu is divided into three parts when simulating functions of scan, as shown in Table 1. Of which, All CH scan is used to scan all channels of device, Memory CH is used to scan stored channel scanning, and Select CH scans manually the input scan's start channel and end channel.

4 AUTOMATIC ASSESSMENT OF PRACTICAL OPERATION

GMDSS's assessment of the practical operation is still of human judgment and it results in heavy workload while doing the assessment and it is difficult to ensure the objectivity and impartiality (Xiao, 2010). An automatic assessment of the practical operation is of the

Table 1. Scan menu.

Scan menu	Sub-menu	Illustration
Scan	All CH scan	Scans all CH
	Memory CH scan	Scans all memory CH
	Select CH scan	Scan specified CH

real-time recording the process of operation when the crew manipulates this simulator. After the operating result, in accordance with relevant state parameters and crew operating procedures, the systems automatically judge them. According to the seafarers' competence assessment standard, we have established an automatic assessment model of practical operation for the reasonableness of the crew operation assessment, which designs an evaluation algorithm based on three points of whether the task is completed, time of manipulation, and steps of manipulation.

The total score the VHF assessment is calculated as follows:

$$ER = (MT \cdot MTW + MS \cdot MSW) \cdot f(T_\alpha) \qquad (1)$$

$$MTW + MSW = 1 \qquad (2)$$

Among these, ER is the total score of the evaluation (Evaluation Result), MT is the manipulation time score of VHF evaluation (Manipulation Time), and MS is the score of manipulation steps obtained by evaluation (Manipulation Step). T_α is the actual using time of manipulation, $f(T)$ is the function used to determine whether the task is completed, MTW is the weight compared by the manipulation time score with the score of manipulation steps, and MSW is the weight compared by the score of manipulation steps with the manipulation time score. The function used to determine whether the task is completed is given as follows:

$$f(t) \begin{cases} 1, 0 < T_\alpha < T_m. \text{ finished} \\ 0, T_\alpha \geq T_m. \text{ unfinished} \end{cases} \qquad (3)$$

Among these, T_m is the maximum time limit for completion of the task and T_α is the time of manipulating actually.

$$MT = B_m \cdot [1 - (T_\alpha - T_m)/T_m], 0 < T_\alpha < T_m \qquad (4)$$

B_m is the basis score of the evaluation manipulation time and T_α is the time of manipulating actually. The equation for obtained scores of evaluation manipulating steps is given as follows:

$$MS = x_i + \sum_{j=1}^{S_e} D_j \qquad (5)$$

Among these, x_i is the reference score of the step i of the evaluation operation, Se is the total number of errors steps in VHF evaluation manipulation, and D_j is the deduction of the step j of evaluation operation because of mistakes. The specific process to the algorithm of the practical operation of VHF evaluation is shown in Figure 5.

Figure 6 shows the total interface of the VHF evaluation system, thereby selecting CH13 and setting the low-power and adjusts the squelch and starting speaker. After the crew answers these, the system automatically provides the results of the assessment.

In this evaluation test, the crew has completed the task and calculated based on the assessment of the total score, and the results are as follows:

$$f(t) = 1, MT \cdot MTW = 47.0, MS \cdot MSW = 50.0,$$
$$total\ ER = 97.0.$$

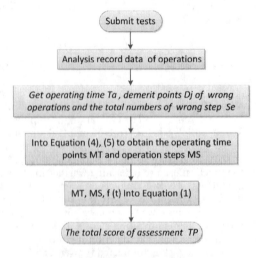

Figure 5. Flowchart of the evaluation algorithm.

Figure 6. Picture of the assessment interface Window.

5 CONCLUSION

This system realizes the marine VHF communications equipment realistic simulation of basic functions, such as the DSC routine call, DSC emergency call, channel operations, address book editor, call record storage and display, set the time and the ship's position, brightness and contrast adjustment etc. Meanwhile, it achieves the automatic evaluation of the crew manipulation and the purpose of crew training and simulation of real-time evaluation. The system has been integrated into the Dalian Maritime University GMDSS simulator and initially applied. In future work, we will improve the operational evaluation algorithm, so that the assessment function is more stable and evaluation results are more objective, thereby enabling the simulation training system to be more perfect.

ACKNOWLEDGMENTS

This work was supported by the National High Technology Research and Development Program of China (863 Program, No. 2015 AA010504), Science and Technology Planning Project of Ministry of Transport (No. 2015329225240), and Fundamental Research Funds for the Central Universities (No. 3132016324).

REFERENCES

Ding, F., "The develop1ing of VHF simu1ator of vesse1s", Ocean University of China, vol. 2, pp. 23–24, 2009.

Harbour Supervision, "International convention on standards of training, certification and watchkeeping for seafarers 78/95", China Science and technology Press, 1997.

Jin, Y. C., Y. Yin, "STCW convention and the development of marine simulator", Journal of Dalian Maritime University, vol. 3, pp. 51–55, 2002.

Shao, D. H., "The design of GMDSS digital selective calling simulation system and its implementation", Soochow University, 2007.

Sun, X., "VC++ in-depth and detailed explanation", Publishing House of Electronics Industry, 2006.

Tzannatos, E. S., "GMDSS false alerts: A persistent problem for the safety of navigation at sea", Journal of Navigation, vol. 1, pp. 153–159, 2004.

Tzannatos, E., "GMDSS operability: The operator-equipment interface", Journal of Navigation, vol. 55, pp. 74–83, 2002.

Wang, H. M., "GMDSS principle and integrated services", Dalian Maritime University press, 2008.

Xiao, F. B., "Evaluation system for GMDSS simulation training", Journal of Dalian Maritime University, vol. 2, pp. 62–66, 2010.

Advances in Energy Science and Equipment Engineering II – Zhou, Patty & Chen (Eds)
© 2017 Taylor & Francis Group, London, ISBN 978-1-138-71798-5

Design of a grounding wire management system based on a wireless sensor network

Wen Li, Yan Ma, Ningbo Liu, Youwei Wang & Quanhu Fei

State Grid Ningxia Electric Power Company Shizuishan Power Supply Company, Shizuishan, China

ABSTRACT: In order to regulate the management of temporary grounding wires, a controller composed of a MC9S12XS128MAA processor, 3G/WCDMA wireless module, WIFI wireless module, and RS-485 communication interface is designed. A grounding wire integrated management system is composed of the controller, server, hand-held devices, and client software. With the help of embedded technology and Internet technology, the client could monitor and control the whole process of temporary grounding wire utilization. The overall scheme of the system is presented and the implementation method of the hardware and software is introduced in detail. The practical application shows that the system works steadily and it is effective in improving the safety management level.

1 INTRODUCTION

The temporary grounding wire is a kind of safety equipment for an electrical circuit or equipment outage maintenance work, and it is the lifeline of personnel to ensure the safety of the operation. The operation of the grounding wire is dangerous, and once an accident has happened, the consequence is very serious. Therefore, the standard grounding wire management is an important work in the substation safety and error control system. At present, the management of temporary grounding wires is based on the microcomputer five protection system. Microcomputer five protection plays an important role in preventing man-made electrical accidents. But the current five protection system cannot solve the problem of temporary grounding (Xu, 2008). In order to avoid accidents of the hanging ground with power or closing the switch with grounding, many of the measures and auxiliary devices are introduced to temporary grounding wire management. These methods and apparatuses used in the grounding anti-misoperation play a certain role, but most of them rely on the system error prevention and human judgment, or use the independent operation, and the system background and five protection system are not combined, thereby leading to not achieving the total information of cross-platform sharing; These are still unable to meet the requirements of the current development of the construction of smart grid and digital substation technology.

In order to reduce the error operation effectively, such as leakage hang, removing the drain, shifting without permission, and so on, combined with the management status of our power supply company, in this work, embedded technology, wireless communication technology, and Internet technology have been used to develop a set of the grounding wire management system. The system can realize the whole process of the temporary grounding line, real-time monitoring and management, and effectively improve the safety management level of the grounding line.

2 SYSTEM STRUCTURE AND WORKING PRINCIPLE

2.1 System structure

The grounding wire management system structure is shown in Figure 1. The system is composed of a

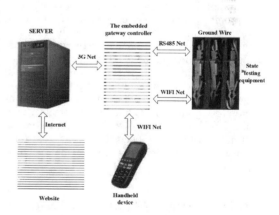

Figure 1. Schematic of the overall structure of the system.

ground wire state detection device, embedded gateway controller, server, client terminal, and hand-held terminal equipment.

The grounding wire state detection device can collect the information of the ground state and transmit it to the embedded gateway. The embedded gateway controller is used as a protocol converter to realize the data communication of the WIFI network, RS-485 network, and 3G/WCDMA network. Hand-held equipment mainly complete ground state display and related voice prompt function. The server releases the information to the external network. The administrator can achieve the monitoring and management of the docking ground wire of the operation or management personnel through the login integrated management system website.

2.2 *System working principle*

Each grounding clamp tail is provided with a set of ground wire state detection devices, and each ground wire can be connected with the embedded gateway controller by using WIFI network communication and RS-485 communication. In the work site, the embedded gateway controller automatically switches to WIFI network communication mode and starts the wireless access point WIFI devices, while trying to connect with the WIFI terminal equipment. When the grounding wire state detection device detects the grounding state of the grounding line, the WIFI device is switched on, and the bidirectional transmission of data can be carried out after the connection of the WIFI terminal and the wireless access point WIFI device. When the operation is completed, the controller will change into RS-485 communication mode automatically, and manage the system controller when the RS-485 detection ground wires are homing and migration is correct and the corresponding voice is prompted and displayed. Finally, the embedded gateway controller transmits the received data to the server through the 3G/WCDMA wireless module and is stored in the database through a preset protocol. Management personnel can be connected to the grounding line through the integrated management system to carry out the whole process of monitoring and controlling. A hand-held device also has the function of WIFI network communication with the embedded gateway controller, which is used for the operation of the operation personnel on-site.

3 SYSTEM HARDWARE IMPLEMENTATION

The hardware of the integrated management system is mainly composed of the grounding line state detection device, the controller, and the hand held terminal.

3.1 *Design of the embedded gateway controller*

An embedded gateway controller is the most critical part of the whole system, and it is the key point of the 3G/WCDMA network, WIFI network, and RS-485 network.

The embedded gateway controller is mainly composed of the host controller, WIFI module, 3G/WCDMA module, and some external devices. The host controller provides two standard UART communication interfaces, which are connected with the 3G module and the WIFI module, respectively. The 3G/WCDMA module mainly provides wireless Internet solutions, and the main controller operates through the AT instruction to achieve the 3G/WCDMA module configuration, dial-up Internet, and other related operations. The WIFI module mainly provides wireless data transmission solutions, the host controller operates through the AT command to configure the WIFI module and carry out the final implementation of the work site data transmission. The hardware structure of the system is shown in Figure 2.

The ground wire management system controller uses the MC9S12XS128MAA high performance sixteen bit single chip microcomputer as the host controller. The host controller integrates two UART interfaces; the main controller bus frequency works in 80 MHz in the system and the working voltage is 5V. The 3G/WCDMA module uses the ZTE MF210V2 module and the module supports AT instructions. The main controller sends AT instructions through the UART interface to complete the 3G/WCDMA module configuration and Internet access. As the input voltage limit of I/O port is 2.6 V for VIN, in order to achieve the compatibility between the host controller I/O port and the voltage of chip NLSX5014MUTAG. The WIFI module uses the production HF-A11 embedded module manufactured by the Shanghai Hanfeng Electronic Technology Co., Ltd. and this module can convert

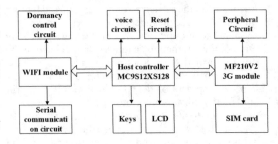

Figure 2. Hardware structure diagram of the embedded gateway controller.

the serial port and WIFI signal, and support AT commands, and the AP/STA mode. The server TCP, which supports multi-TCP link connection, supports 32 Client TCPs at the most, configures management system controller in the AP mode in the WIFI module, and acts as the Server TCP, and realize the connection with Client TCP.

3.2 *Design of the ground wire state detection device*

The ground wire state detection device is mainly composed of a controller, a lithium battery-charging circuit, a WIFI module, a ground wire state detection circuit, and a protection circuit. The main controller uses a C8051F340 microcontroller which is a Silicon company production, with a small size, low power consumption, with two UART interfaces, and which is very suitable for the small size of the embedded system. The TP4506 power management chip is the core of lithium battery-charging circuit, and has the function of the charging status detection, indication function, and adjustable charging current. The WIFI module in the ground wire state detection device is configured in the AP mode, and used as a Client TCP. What is more, the module is connected with the WIFI module of the server TCP, which is arranged in the embedded gateway controller. The ground wire state detection device provides a standard RS-485 interface, which is used as an embedded gateway controller of the RS-485 network. The SCM provides two UART ports, respectively as the RS-485 network and WIFI network interface. The hardware structure of the grounding wire state detection device is shown in Figure 3.

3.3 *Hand-held terminal design*

The hand-held terminal is mainly composed of a controller, a lithium battery-charging circuit, a WIFI module, a LCD display circuit, and a voice circuit. The main controller uses a C8051F340 microcontroller, which is a Silicon company production,

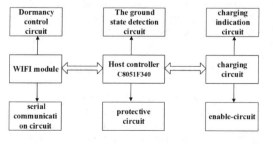

Figure 3.   Hardware structure diagram of the ground state detecting device.

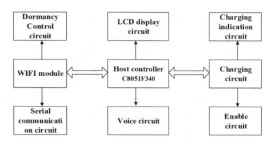

Figure 4.   Hardware structure of the hand-held terminal.

with a small size, low power consumption, and which is very suitable for hand-held devices. The lithium battery-charging circuit and WiFi module are described above. The LCD display circuit uses a Nokia lcd5110 LCD screen as the core and realizes the show of the ground state information and the realization of the speech circuit of the voice prompt function. The hardware structure of the hand-held terminal is shown in Figure 4.

4   THE IMPLEMENTATION OF SOFTWARE

4.1 *Controller state machine*

The working process of the ground line management system controller is divided into six states: the initia`l state, work mode, RS-485 mode, WIFI mode of operation, login server, and data communication. The specific points are given as follows:

a. Initialization: after the system is started, the necessary initialization of the controller is in the initial state [6].

b. Working mode judgment: need to judge the working mode of the controller according to the requirement of the field operation, when meeting the conditions (RS485Flag == 1&&WIFIFlag == 0) and configure into the RS-485 mode. While meeting the conditions (RS485Flag == 0&&WIFIFlag == 1), configure into the WiFi mode.

c. RS-485 working mode: Close the WIFI module, connect the host controller UART interface with the chip MAX485 UART interface through the electronic switch, and communicate between the integrated management system of the ground wire controller and the ground wire through the RS-485 network.

d. WIFI working mode: through the electronic switch, connect the integrated management system controller UART interface and WIFI module UART interface, complete the necessary configuration by starting the WIFI module, such as working mode, IP address, and ensure that the WIFI module is online.

e. Log in server: the module of the 3G/WCDMA is configured necessarily, such as the baud rate, IP address, and then makes a continuous connection between the specified IP address and port number server actively and realizes the communication until the login is successful.

f. Data communication: after logging in to the server, the controller will upload the connection data to the server according to the agreed protocol, and the server at the same time can issue instructions to achieve the docking grounding management function. After the system is reset, the controller working state returns to the initial state. The controller state machine model is shown in Figure 5.

4.2 *State machine of the ground wire state detection device*

The working process of the ground wire state detection device is divided into five states: the initial state, the resting state, the RS-485 mode, the WIFI mode, and the wake state. The specific points are given as follows:

a. Initialization: after the system is started, the necessary initialization of the ground line state detection device is in the initial state (Mo, 2015).

b. Sleep state: The ground wire state detection device starts a WIFI module every 15 minutes and tries to connect with the controller WIFI; if the connection is successful, enter the WIFI mode of operation, or on the contrary, close the WIFI module to enter the sleep state.

c. RS-485 working mode: when the single-chip microcomputer checks and finds that the conditions (SlaveRS485-Flag == 1&&SlaveWIFIFlag == 0) are satisfied, enter the RS485 working mode and make a communication.

d. WIFI working mode: when the single-chip microcomputer checks and finds that the conditions

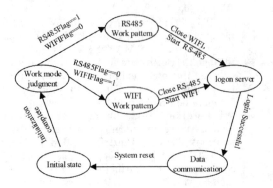

Figure 5.   The controller state machine model.

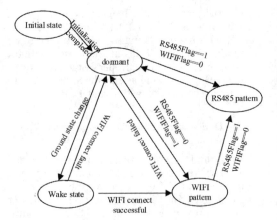

Figure 6.   Ground state detection device state machine model.

(SlaveRS485Flag == 0&&SlaveWIFIFlag == 1) are satisfied, enter the WIFI mode of operation and make a communication.

e. Wake state: when the state of the ground wire is changed, the WIFI module is started, and an attempt is made to connect to the WIFI module in the integrated management system controller.

The state machine model of the ground wire state detection device is shown in Figure 6.

4.3 *Implementation of ground homing detection*

In order to regulate the storage management of the ground wire, the system gives a unique ID number to each ground wire and stores it in EEPROM to achieve the binding between the ground wire and the storage cabinet. When the operation is completed, workers must place the ground wire into the corresponding storage cabinet for charge, at this time, thereby realizing canning communication between the controller and the ground wire, and displaying return information on the LCD. When it does not belong to the storage cabinet grounding wire homing, a voice prompt can be used. In this system, phase A, phase B, and phase C are stored in each storage cabinet, and every phase is connected to the ground.

5   SYSTEM TEST AND APPLICATION

The system has been tested in the power supply company, and six sets of equipment (a total of 18 ground wire) were put into operation at the same time and the operation of a detailed record and analysis was performed. Running results show that the temporary ground integrated management system monitor the grounding working station timely and accurately, and at the same time, the operators can configure query and manage the ground through the ground-integrated management system website.

Running results show that temporary ground-integrated management system timely and accurately carries out docking grounding work state monitoring, and at the same time, the operators through connected ground integrated management system website can be realized for arbitrary ground configuration, query, and management.

After analysis, one of the reasons for the problem is that the 3G/WCDMA network signal quality is not stable. The other is that the working site electromagnetic environment is relatively poor, the electromagnetic interference makes the grounding wire state detection device in the circuit work occasionally abnormally. To solve the problem, the author who worked with the software did further upgradation and increased the 3G/WCDMA module between the server and the timing of the handshake communication function, thereby ensuring the 3G/WCDMA module's on-line stability; to solve the second problem, the author working in the hardware circuit of the combined switch detected and isolated circuits and optimized error detection significantly reduced.

6 SUMMARY

The ground wire management system designed in this paper can realize the real-time monitoring of the grounding system. The operating personnel obtain ground wire monitoring and management by logging into the ground wire management system website, to ensure the timely and accurate operation of the operator to grasp the working status of the ground wire, which effectively improves the level of personnel to prevent errors. Through the network control of each ground wire, the storage of ground wires is more standardized. The authors elaborated the system design thought and the realization method from two aspects of the system hardware and the software, and verified the feasibility of the management system through practical application tests. This program has advantages such as low cost and good scalability for the grounding line number management and better application prospects.

REFERENCES

Jiang Jing, Yu Tao, Huang Jianbin, etal, Development of safety monitor for power distribution grounding wire [J], 2013, 50(6): 93–96.

Lin Yong, Yang Chunlai, Chang Ximao, Research and implementation of remote monitoring and control system based on GPRS [J]. Chemical automation and instrumentation, 2011, 38(11): 1367–1370.

Liu Jia-jun, Miao Jun. Research on New Type of Electric Power Overhaul Ground Wire Device [J]. Power System Protection and Control, 2009, 37(23): 119–121.

Mo Guo. Design and implementation of grounding line management system for overhead transmission line [D]. University of Electronic Science and technology, 2015.

Peng Bin, Yu Hao, Lv Xiaojun. Development and application of a new generation of online monitoring and management system for temporary grounding wire [J]. Automation of electric power system, 2015, 24: 110–114.

Wu Wencong, Jiang Jing, Yu tao. Study on detection method of temporary ground wire in distribution network [J]. Power system protection and control, 2012, 23: 151–155.

Xu Jianyuan, Dou Wenjun, Wang Aihong. Design of on line monitoring system of ground wire [J]. Electric power automation equipment, 2008, 07: 111–113.

Advances in Energy Science and Equipment Engineering II – Zhou, Patty & Chen (Eds)
© *2017 Taylor & Francis Group, London, ISBN 978-1-138-71798-5*

An analysis of information influence and communication based on user behavior

Chang Su
Telecommunication Engineering, Nanjing University of Posts and Telecommunications, Nanjing, China

ABSTRACT: User behavior can be reasonably predicted and one can achieve a certain degree of accuracy through big data analysis and optimized arithmetics. Finding what is required and desired by the information users is rising to become a hot spot. In this article, the interaction between the information impact and dissemination based on user behaviors is analyzed. The analysis in this paper is based on the big data sets of micro-blog, combined with insights on information impact and information dissemination in the micro-blog environment, and analysis developed through consideration of the relationship between the various factors of user behavior. Considering the influence of nodes and dissemination mechanism, the model of information transmission is established.

1 INTRODUCTION

With the rapid development of Internet, a variety of information and related applications are expanding at the speed of geometric progression geometrically. The boosting industry can portend an era of information explosion, or as it is commonly referred to, the Web 3.0 era. The contents the user generated has, on one hand, brought wealth to the user in the form of information; yet, on the other hand, produced a serious problem—information overload (Chen, 2008). A large amount of information far exceeds the range an average user can handle. The problems brought about by information overload is that users are unable to cope with the data upon time; what is worse than that is, the phenomena of the increasing dross content has exposed the users to the cacophony of information. Development of search engines, such as Weibo, that are based on social networking applications is badly affected and hindered by their inaccuracy. Figure 1 shows the proportion of the contents Weibo users forward on a daily basis. It reveals the contents that users are mainly concerned about. News on personal interests and social lives have taken up the first two places of the heated searching on Weibo.

While getting information through the Internet, people are overwhelmed by a mass amount of information at the same time. Therefore, how to turn the passive acceptance into an active search will become the key to solve the existing problems (Lin, 2007). Therefore, it is necessary to analyze users' behaviors. Giving users the information they want and service they need will be the key to solve

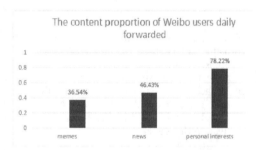

Figure 1. Chart showing the daily forwarded content proportion of Weibo users.

problems in social networking applications to survive the Web 3.0 era.

An individual user's behavior change is in accordance with the social group one belongs to. It is fairly impossible to understand the group features simply by analyzing the individual situation (Moore, 1995). Individuals belonging to the same group have similar interests over their action of searching. In order to decrease the number of irrelevant information and optimize the product, a personalized recommendation system comes into being (Liu, 2009).

The principle of user behavior analysis is based on the influence of social impact, which means the phenomenon of personal changing about thoughts, emotion and behavior under the effect of others. Social impact about social networking sites has been studied for a long time; varied scholars have put forward a number of representative theories. A team from Lazarsfeld presented the two-level communication theorem; it discussed the process of transmitting information from the media to the audience (Lazarsfeld, 1968). Grano-vetter proposed the theory of The Strength of Weak Ties, in which the degree of overlapping for two individuals' friendship networks is proportional to the strength of their tie to one another (Granovetter, 1973). Also, there are strong joint and several advantage theory, structural hole theory, and so on and these theories together have set a good basis for future research and development.

User behavior is determined by the communication of information to a great extent. Currently, there are more than 500 million registered users on Weibo and 1 million micro-blog posts are released every day. A social network represented by using a micro-blog has stood out to navigating the trend for the mass media. Therefore, a study on the user behavior and information transmission rule can help the network company grasp user preferences more accurately, and make the content with higher relativity that can be pushed to the users. Accuracy in message promoting could also assist the government with scientific control and guidance in controlling public sentiments through prediction over transmitted messages and views.

Based on these, in this article, the impact of social networking sites on user behavior in terms of impact and dissemination of information is discussed. The effect of the interaction between users of social networks will be included in the further analysis in this paper.

2 RELATED WORK OF INFLUENCING FACTORS

2.1 Information impact

The impact of user information in the network can be reflected through the interaction between the users. By using micro-blog information as an example, the user impact factors were mainly reflected in three aspects—user, user content, and user relevance. We can analyze these through the specific reference factors, as shown in Figure 2.

Previous studies have found significant relevance between the factors concerned with features of user activity on social networking sites. For example, Cha et al. compared the effects in terms of three different factors—forwarded times, mentioned times, and the number of followers. They also analyzed the time-dependent law of impact and arrived at the conclusion that the users with a large number of followers are not necessarily able to raise more forwarding or mentioned behaviors. According to the findings by Cha et al., the level of impact to which the social network users are exposed can be measured through the network

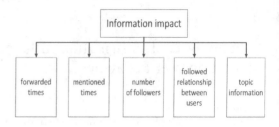

Figure 2. The reference factors of information impact.

typology, while this calculation should involve the time factors as they may influence the results of the measurement to a specific extent.

Based on the following relationship between users and topic information, Weng et al. analyze the type of user impact in terms of specific subjects and gradually evaluate the recommendation ranking system for users by using an algorithm (Weng, 2010). Weng's result has exhibited itself with significance.

Combining with these research results, in this article, the information dissemination with the impact model mentioned in prior studies will be introduced.

2.2 Information dissemination

The model analysis about the micro-blog information, concerning the relation between the layouts of information dissemination, can be distinguished as single key point type, chain type, and multiple key points type, as shown in Figure 3.

Single key point type: detailed information circulates quickly in this model; most of users are hot users with a strong orientation of public opinion, and they have the profound effect on information dissemination. In this type, the information dissemination is less affected by outside interference, and its communication range is limited to friends.

Chain type: the type of users within this model does not have a strong impact. It has many intermediate nodes; hence, its dissemination speed is slower than the single key point type. It is always used for forwarding e-mails.

Multiple key point type: information among this model carries with itself fast speed of dissemination, wider range, and strong dissemination and diffusion, which is commonly used in current hot spots and so on.

Concerned achievements in the study of information dissemination should be mentioned in accordance with the theme of this paper.

Leskovec J et al., based on the SIS epidemic model, established an information dissemination model in the network and found that the popularity of the specific micro-blog followed the rule of power law distribution (Leskovec, 2007).

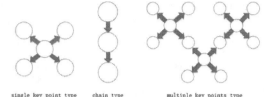

single key point type   chain type   multiple key points type

Figure 3. Schematic of the main types of information dissemination.

Zaman *et al.* used Twitter as an example and concluded that the dissemination of information included three different parties functioning simultaneously—the publisher, the follower, and the Weibo text itself, and they also established the forecast model based on the probabilistic collaborative filtering model (Zaman, 2010).

H. Kwak et al. analyzed the impact of time on forwarding posts and found that followed times have a direct relationship with nodes, but not the number of followers.

3 ANALYSIS OF INFORMATION IMPACT AND DISSEMINATION

By using the micro-blog as an example, the user behavior is directly reflected in two aspects—the information impact and dissemination. By combining with these two factors, the following models are established:

3.1 *The relationship between the information impact and the visual factors*

Information impact is directly reflected in the posted times and forwarded times. In this article, the posted times are used as a reference base and the ratio between the posted times, forwarded times, and the number of forwarded posts is analyzed and then the user's information impact is obtained.

I:

$$I = \ln \frac{R \cdot M}{S} (S \neq 0) \tag{1}$$

where S represents posted times, M represents forwarded times, and R represents the number of forwarded posts. S is always greater than R.

The greater the value of I is, the wider impact it represents within a certain range.

Concerning the impact of information dissemination on time delay and extension, in this article, impact is defined as $I(\Delta l, \Delta t)$, which has some relation with delayed time Δt and extended time Δl.

3.2 *The relationship between information dissemination and nodes*

The user's micro-blog information constitutes a network, and the user who is sending and forwarding becomes a network node. It can measure the importance of network nodes by using the point center method (Lv, 2013). In this paper, a model is cited, which is as follows: a node P_i in the network has a direct relation which can be expressed as C_{Di} and the node's point of the center can be expressed as follows:

$$C_{Di} = d \tag{2}$$

Two nodes, v_j and $v_k (j \neq k, v_j, v_k \in V)$ have the shortest path g_{jk}, and so the information between these two nodes has a relation with the shortest path $g_{jk}(v_i)$. The longer the shortest path is, the more impact is felt on clear interaction between the nodes.

$$C_{Bi} = \sum_{j}^{n} \sum_{k}^{n} g_{jk}(v_i) / g_{jk}, j \neq k \neq i, j < k \tag{3}$$

In an n-node network, the shortest path length between nodes v_j and v_k is represented as d_{ij} and closeness centrality of this node represented as C_{Ci} can be expressed as follows:

$$C_{Ci} = \left(\sum_{j=1}^{n} d_{ij} \right)^{-1} \tag{4}$$

The larger the value of the closeness centrality is, the closer this node is to the network center, and the greater the impact is.

4 SUMMARY

Besides the rise of Weibo, the studied example of social networking sites for this research work has provided an ideal experimental platform on the information impact phenomenon. In this article, it is found that the user posted times, forwarded times, the number of forwarded posts, and such factors will affect the user's information impact. While, the information impact, to some extent, would have a certain influence on the process of information dissemination. Cyclically, the change in information dissemination will gradually affect the users' posting habits.

Based on these findings, in this article, a method is proposed by combining with the relevant factors, which can measure the user impact, alongside with the measuring means of information dissemination through nodes.

1783

Actually, as a kind of social medium, the micro-blog plays a more important role as a medium rather than a social occasion; it is a chance as that users in the micro-blog just focus on celebrities and public media. To a big extent, the micro-blog does not have the spirit of equal communication on the Internet; the majority of its information is a one-way dissemination. The key point is controlled by the big V (verified micro-blog users who have more than 500,000 followers), while ordinary people can hardly show their views. Users realize that the contents they posted cannot get the positive or effective feedback; gradually they will lose their passion and decide to follow the big V indeed. The identity and mentality of people gradually turn from a content creator into the listener. Micro-blog almost makes the same function as the traditional newspaper media.

Because of this, more and more people begin to choose Wechat's moments instead of the micro-blog. In the latter, your voice will drown in the noise of social media, but in the former, your voice is very easy to be followed and forwarded by others. This phenomenon is worthy of our thinking.

In the next work, we will proceed from three aspects to further improve our method:

Firstly, the spam on the Web has a wide impact and special mode of transmission. The ways of getting rid of the spams by using the characteristics of the information dissemination would be otherwise a reminder of the points for discussion.

Secondly, the time delay and timeliness of information dissemination and information impact would be involved for study.

Last but not least, by screening the detailed data, people can carry out further studies on the emotional choices and precisely the range of information dissemination.

REFERENCES

Burke R. Hybrid Recommender Systems: Survey and Experiments. User Modeling and User-Adapted Interaction, 12(4):331–370 (2002).

Cha M, Haddi H, Benevenuto F, et al. Measuring user influence in twitter. Icwsm 10: International AAAI Conference on Weblogs & Social. (2015).

Fi Lin, Y Liu. The research on information overload. Information Studies: Theory & Application, 30 (05): 710–714.[3]Moore D C.(2007).

Frias-Martinez E, Chen S Y, Liu X. Investigation of behavior and perception of digital library users: A cognitive style perspective (International Journal of Information Management, 28(5):355–365.2008).

Granovetter M. The Strength of Weak Ties. American Journal of Sociology, 78(6,): 1360–1380 (1973).

Hong Yu,Xian Yang. Studying on the node's influence and propagation path modes in microblogging. Journal on Communications, 33(Z1):96–102.(2012).

Jianguo Liu, Tao Zhou, Binghong Wang. The research progress in personalized recommendation system [J]. Progress in Natural Science, personalized recommendation system. 19 (1): 1–15. (2009).

Jiemin Chen, Yung Tang, Jianguo Li, et al. Research on personalized recommendation algorithm. Journal of South China Normal University (natural science edition), 2014 (05): 8–15.

Lazarsfeld P F, Berelson B, Gaudet H. The peoples choice: how the voter makes up his mind in a presidential campaign. New York Columbia University Press, 77(2):177–186. (1968).

Leskovec J, Mcglohon M, Faloutsos C, et al. Cascading Behavior in Large Blog Graphs. Sdm, 15(1):9:3–9:56. (2007).

Lijuan Liu. Internet Users Data Mining and Behavior Analysis [D]. Beijing Jiaotong University. 8–9(2014).

Lin Yang. Analysis and prediction of user behavior based on social network. Xi'an University of post and Telecommunications (2013).

Moore, David Chioni. Colonial Discourse and Post-Colonial Theory: A Reader, by Patrick Williams; Laura Chrisman; The Post-Colonial Studies Reader, by Bill Ashcroft; Gareth Griffiths; Helen Tiffin. South Atlantic Review 60.4(1995).

Shaochen Lv. Information Dissemination Model Based on User Behavior in Microblog [D]. Huazhong University of Science and Technology. 2013,11–12.

Weng J, Lim E P, Jiang J, et al. Twitter Rank: finding topic-sensitive influential twitterers. ACM International Conference on Web Search and Data Mining. ACM, 261–270 (2010).

Xianjun Meng. Research on Web Text Clustering and Retrieval Technology. Harbin Institute of Technology (2009).

Zaman T R, Herbrich R, Gael J V, et al. Predicting Information Spreading in Twitter. Computational Social Science & the Wisdom of Crowds Workshop, (2010).

Advances in Energy Science and Equipment Engineering II – Zhou, Patty & Chen (Eds)
© 2017 Taylor & Francis Group, London, ISBN 978-1-138-71798-5

Research and analysis of the methods of color image noise reduction of living trees based on single-channel processing

Yingpeng Dai, Yutan Wang, Jingbo Fan, Chenghao Ma, Yaoyao Gao & Bohan Liu
School of Mechanical Engineering, Ningxia University, Ningxia, China

ABSTRACT: In this paper, we study the methods of image noise reduction on the basis of HSV color space to improve visual effect in the process of image noise reduction. The size of the main effect of various noises and the size of the interaction effects are estimated by the full factorial experiment. According to the characteristics of color components' independence in HSV color space, the color image is transformed from RGB color space to HSV color space. Different methods of noise reduction are used for different color channels to achieve better noise reduction effect. It shows the influence degree of the algorithm among the noises through the full factorial experiment so as to get the best performance of the algorithm to deal with the noise. In this paper, the full factorial experiment is introduced to the data processing of image noise, which is a new attempt. Meanwhile, it also plays a guiding role in the improved algorithm to achieve the best noise reduction effect.

1 INTRODUCTION

The vision, the main way to get information from external environment, accounts for 75–85% of information acquisition according to statistics (Jie, 2014). Therefore, the data of image form are particularly important. From the perspective of human vision, the true color image is beneficial to comprehension to human and includes a lot of information. However, various noises pollute the image to effect the quality of image in the process of image acquisition and transition.

Aiming at noise reduction of the true color image, several scholars put forward different effective methods, for example, teacher Jian-wei Wang from Northeast Forestry University (Wang, 2006), doctor Song-tao Liu from Dalian University of Technology (Liu, 2011), and professor Yan-quan Deng from Harbin Engineering University (Deng, 2009). The algorithms put forward by the above people are based on RGB color space. The association of color components has a great influence on noise reduction. On the basis of analyzing the differences between RGB color space and HSI color space, doctor Yu-tan Wang from Ningxia University put forward the method of noise reduction based on color space transformation. This algorithm combines vector algorithm and scalar algorithm in HSI color space, which has a significant effect on noise reduction (Wang, 2014). Although the algorithm, which is limited by RGB color space, deals with noise image in HSI color space, which has little association in color channels, it neglects random

distribution of a variety of noise in three channels of HSI color space. Therefore, the algorithm put forward by doctor Wang uses the same technique to each channel rather than dealing with image according to the distribution of mixed noise in each color channel. On the basis of HSV color space transformation, this paper put forward an algorithm that selects the best method in each channel by judging the optimal effect of different techniques in three channels of HSV color space in the process of dealing with mixed noise image. Then, the algorithm is used to design full factorial experiment to estimate the interaction effect among different noises.

2 TECHNICAL ROUTE

Three channels that belong to RGB color space are mutually interacted. Noises will exist in three channels when the image is polluted. As noise reduction, it will deal with three channels with same measures simultaneously rather than dealing with three channels with different measures separately. The color space with mutually independent channels should be found to process each channel separately. The model of HSV color space has independent components: (1) channel H represents hue, which is expressed by wavelength of color; (2) channel S represents saturation, which is expressed by depth of color; (3) channel V represents value. The above channels are independent and contain color information of RGB color space. Not only does HSV reserve color information, but also it avoids

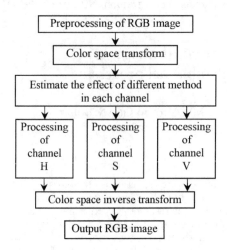

Figure 1.   Road map.

Figure 2.   Original image.

interaction with other channels. If each channel of HSV is processed with the same method for images containing various noises, which exist in every channel in different proportions, the effect of noise reduction is not optimal, as shown in Figure 1.

3 MATERIALS AND METHODS

3.1 *Image acquisition of living trees*

Hardware equipment of image acquisition includes a Levono mobile phone (P700, 8 million pixels) and an HP desktop computer (HP Pavilion g series, Inter(R) Core(TM) i3-2310M CPU @ 2.10 GHz 2.10 GHz, RAM 2.00, Windows 7 ultimate × 32). The image captured under sunny environment is saved as a .jpg file, which is convenient for experiment as the image can be cut to 900 × 1200 resolution, as shown in Figure 2.

3.2 *HSV color space*

HSV model, also called hexagonal cone model, visually describes the characteristics of color space. It could describe color information of RGB color space because each channel of HSV contains color information of RGB. Meanwhile, channels H, S, and V are mutually independent and hence do not affect each other.

The parameters of HSV include hue, saturation, and value, as shown in Figure 3.

3.3 *Experimental verification*

The three channels in HSV color space are processed, respectively, by mean filtering, median filtering, and adaptive filtering in experiment. Among them, the best method is used as the final

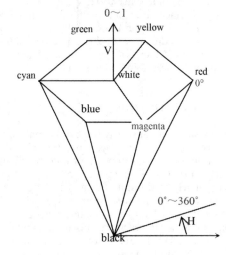

Figure 3.   HSV color space model.

treatment technology for the corresponding channel to reach the purpose of obtaining optimal quality of image. Finally, noise reduction effect is estimated by MSE or PSNR.

The image will be polluted by various noises in the process of image acquisition. Therefore, it uses multiple noises to simulate the actual situation in this experiment. The kinds of noise added are as follows:

Gaussian noise: mean is 0, variance is 0.02
Speckle noise: mean is 0, variance is 0.02
Salt-pepper noise: density is 0.03

First, the original image is added to the mixed noise. Then, the RGB color space is converted to HSV color space. Each channel of HSV is processed separately by mean filtering, median filtering, and adaptive filtering. The results are as shown in Table 1.

It is evident from the table that adaptive filtering is used in channel H and median filtering is used in channels S and V, as shown in Figure 4 and Table 2.

1786

Table 1. PSNR value corresponding to different filters in channels.

Processing method	PSNR of channel H/dB	PSNR of channel S/dB	PSNR of channel V/dB
Polluted image	56.9124	59.4966	62.7161
Mean filtering	70.2982	72.1389	74.7406
Median filtering	70.8488	74.8972	78.1662
Adaptive filtering	72.7381	71.3218	72.7741

a. Original image   b. Image added to mixed noise

c. Median filtering   d. Algorithm proposed in this paper

Figure 4. Comparison of median filtering and the algorithm proposed in this paper.

Table 2. The corresponding PSNR in different stage.

	PSNR/dB
Original image	
Polluted image	60.3159
Median filtering	72.0625
New algorithm	77.0354

Key algorithm steps:

Step 1: The original image is added to the mixed noise;

Step 2: Color space transformation, the RGB color space is converted to HSV color space;

Step 3: Find optimal method in each channel;

Step 4: Deal with channels with different technology.

It can be learned that not only median filtering but also the algorithm proposed in this paper can improve the quality of the polluted image. For the image added to mixed noise, compared with mean filtering, the PSNR value increased to about 9–12 dB, whereas the increase of that of the new algorithm was about 6–7 dB higher.

In addition, the image processed by new algorithm shows slight color difference because the transformation model takes approximate transform.

3.4 Analysis of the effect of different noises on the algorithm

It can be learned that the algorithm proposed in this paper has obvious effects on reducing mixed noise. It helps to design a full factorial experiment to estimate the interaction effect among different noises because different noise components in mixed noise produce different results.

3.4.1 Influence parameters and response
Influence parameter:

1. Variance of Gaussian noise
2. Density of slat-pepper noise
3. Variance of speckle noise
Response: PSNR

3.4.2 Determination of influence parameters
Variance of Gaussian noise:
  Low = 0.01 high = 0.03

Density of slat-pepper noise:
  Low = 0.01 high = 0.05

Variance of speckle noise:
  Low = 0.01 high = 0.03

3.4.3 Full factorial experiment
Test design and the response of the test are as shown in Tables 3–5 and Figure 5.

3.4.4 Result analysis
The main reaction responses of Gaussian, speckle, and salt-pepper are −1.197, −6.03, and −0.2831, respectively. It can be known that the noise reduction ability of this algorithm is gradually weakened with the increase of noise density or variance. In addition, the absolute value of the main response of the speckle is the smallest and that of the salt-pepper is the largest. Therefore, the processing ability of the three kinds of noise at different levels decreases in the following order: speckle noise > Gaussian noise > salt-pepper noise.

Table 3. Experimental response of full factorial experiment.

Test	Gaussian	Salt-pepper	Speckle	PSNR
1	0.01	0.01	0.01	79.0308
2	0.03	0.01	0.01	77.6565
3	0.01	0.05	0.01	78.2662
4	0.03	0.05	0.01	77.1222
5	0.01	0.01	0.03	78.6426
6	0.03	0.01	0.03	77.3856
7	0.01	0.05	0.03	77.9513
8	0.03	0.05	0.03	76.9638

Table 4. The value of main reaction response.

Main reaction response	Value
Gaussian	−1.1907
Salt-pepper	−6.03
Speckle	−0.2831

Table 5. Response value of interaction.

Interaction of noises	Value
Speckle and salt-pepper	0.04645
Salt-pepper and Gaussian	0.12495
Gaussian and speckle	0.06845
Gaussian speckle and salt-pepper	0.0098

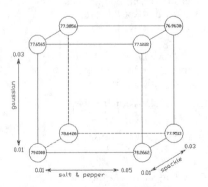

Figure 5. Full factorial experiment and its response.

The values of interactions between speckle and salt-pepper, between salt-pepper and Gaussian, between Gaussian and speckle, and among speckle, salt-pepper, and Gaussian are 0.04645, 0.12495, 0.06848, and 0.0098, respectively. It will be learned from the algorithm proposed in this paper by analysis of above data that:

1. There are interactions among Gaussian, speckle, and salt-pepper. That is, the noises are independent.
2. The larger the value of response, the higher is the effect of algorithm, and vice versa.

4 DISCUSSIONS AND CONCLUSIONS

In general, the methods of component processing, model transformation, and vector method are adopted for noise reduction of true color image (Koschan, 2009). In this paper, we convert RGB color space into HSV color space on the basis of color space transform. However, the color information will suffer certain impact in the process because RGB color space is a cub and HSV color space is a cone, which exist in the process of transformation. The effect depends on the precise degree of transfer function. In addition, it is also likely to cause larger noise to make the image bad with color space transformation. In addition to mean filtering, median filtering and adaptive filtering have better algorithms, which aim at the single channel to achieve the optimal result. This provides the idea for optimal processing of a single channel.

On the basis of single processing channel, this paper puts forward the noise reduction method for the color image of living trees. Compared with the classical vector median filtering method, this paper has a significant effect on the noise reduction and achieves the expected goal. Through the full factorial experiment to estimate the main response and the interaction among different noises, it guides the improved algorithm to have optimal effect.

ACKNOWLEDGMENT

This work was supported by the Natural Science Funds, Ningxia University (ZR1409).

REFERENCES

Deng Yanquan, Wang Xiaochen. New fuzzy vector median filtering algorithm for color images [J]. Computer Engineering and Applications, 2009, 45(1): 191–194.

Koschan A, Abidi M. Digital Color Image Processing [M]. [S. 1.]: John Wiley & Sons Inc. 2009.

Liu Songtao, Ma Linpo, Yin Fuliang. A color image vector median filtering algorithm based on noise estimation and double weighted spatial distance and magnitude value [J]. Journal of Optoelectronics Laser, 2011, 22(1): 131–135.

Wang Jianwei. Modification and application of median filtering algorithm for color image filter [J]. Journal of Harbin University of Commerce (Natural Sciences Edition). 2006, 22(4): 67–69.

Wang Yutan. Research on Methods of Lingwu Long Jujubes'Localization and Maturity Recognition Based on Machine Vision [D]. Bejing Forestry University. 2014.

Yang Jie. Shuzi Tuxiang Chuli Ji MATLAB Shixian [M]. Beijing: Publishing House of Electronics Industry. 2014.M. Ben Rabha, M.F. Boujmil, M. Saadoun, B. Bessaïs, Eur. Phys. J. Appl. Phys. (to be published).

Advances in Energy Science and Equipment Engineering II – Zhou, Patty & Chen (Eds)
© *2017 Taylor & Francis Group, London, ISBN 978-1-138-71798-5*

Change detection for remote sensing images with multiple thresholds based on particle swarm optimization

Limin Wu
Kunming Surveying and Mapping Management Center, Kunming, China
Faculty of Land Resource Engineering Kunming University of Science and Technology, Kunming, China

Liang Huang & Mingjia Wang
Faculty of Land Resource Engineering Kunming University of Science and Technology, Kunming, China

ABSTRACT: In this paper, we propose a change detection method for remote sensing images with multiple thresholds based on Particle Swarm Optimization (PSO). On the basis of this, the authors have attempted to solve the problems in the selection of varied thresholds in the change detection for multitemporal remote sensing images, such as large computation amount and long time for computation. To begin with, remote sensing images with multitemporal difference are constructed using the image difference method. Then, the principles of three-dimensional (3D) Otsu method are analyzed. Then, by introducing the PSO algorithm and investigating the principles and optimization process of the PSO algorithm, this paper presents a PSO-based 3D Otsu method with multiple thresholds for image segmentation. Subsequently, this method is applied in the binary segmentation of difference images. Finally, this method is utilized to carry out change detection for the selected experimental data, the results of which are then compared to those of one-dimensional Otsu algorithm and 3D Otsu exhaustive method. Experimental results show that the presented change detection method shows an accuracy of 94.86%, which indicates the effectiveness and reliability of this method for remote sensing images.

1 INTRODUCTION

Owing to the development activities of human and frequently occurring natural disasters, the land cover changes every day. To evaluate the effect of these changes on ecological environment and social economy, it is necessary to dynamically monitor the land cover. The traditional methods, represented by artificial regional survey and statistical methods, cannot satisfy the requirement of currently dynamic monitoring any more, as they require a lot of manpower and resources. Moreover, they need a long monitoring period and can only be used in a small area. Remote sensing images, characterized by high temporal resolution, a large real-time monitoring range, and easy to obtained, have been extensively applied in the dynamic monitoring of global land cover. Thereinto, the change detection of multitemporal remote sensing images is the key to realize the remote sensing dynamic monitoring. By using the remote sensing images obtained at different temporal stages, this technology quantitatively analyzes and identifies the characteristics, processes, and regions of land cover changes.

In the dynamic monitoring for the macroscopic change of the earth's surface, the remote sensing images with medium and low spatial resolutions obtained by satellites represented by LANDSAT 4–8 and MODIS are still the main data sources. Nowadays, dynamic monitoring for the macroscopic change of the earth's surface using the multitemporal remote sensing images with medium and low spatial resolutions has been utilized in various fields. They include urban expansion (Huang, 2016), present situation of land use or land cover change (Chen, 2006; Malmir, 2015), temporal and spatial evolution of rivers and lakes (Chan, 2013), land rocky desertification (Li, 2015), and so on.

Over the last three decades, many effective means have been put forward aiming at the change detection of multitemporal remote sensing images with medium and low spatial resolutions. These means raised are mainly based on two ideas: post-classification comparison and direct comparison (Huang, 2013). The latter refers to analyzing two remote sensing images of a same geographic area taken at different temporal stages pixel-by-pixel, thus obtaining reliable images concerning the change detection (Mishra, 2012). As it can be easily understood and achieved, it has been widely employed in the change detection of remote sensing images with medium and low spatial resolutions.

For the direct comparison method, the key is to construct difference images and select varied thresholds. Currently, many methods can be used to select varied thresholds (Rosin, 2010), including Ridler and Calvard (1978) algorithm, Tsai (1984) algorithm, Otsu (1979) algorithm, and Rosin (2001) algorithm. Among them, the Otsu algorithm and its improved version, two-Dimensional (2D) Otsu algorithm (Chen, 2012), have been utilized in the change detection of remote sensing images owing to their excellent effect in image segmentation (Huang, 2015). The 2D Otsu algorithm makes use of not only the gray-level distribution information of pixels but also the relevant information of pixels in neighborhood spaces, thus acquiring favorable segmentation effects. However, compared with other images, remote sensing images are more complicated as they have many peak patterns, such as single, double, three, and even more peaks. Therefore, if merely using the 2D Otsu algorithm, we possibly cannot obtain satisfactory results. To solve this problem, a three-Dimensional (3D) Otsu algorithm is investigated (Sthitpattanapongsa, 2011). However, with the increase of the dimensions of the Otsu algorithm, the amount of calculation gradually rises and therefore more time is taken to search the optimal threshold. Thus, finding an algorithm that can be used to rapidly and effectively select the optimal threshold of the 3D Otsu algorithm is the key of this segmentation method.

Particle swarm optimization (PSO) (Kennedy, 1995) is an intelligent optimization algorithm proposed by Kennedy et al. in 1995. It is realized by simulating the foraging behavior of birds based on the collective cooperation so as to achieve the expected goals of groups. PSO, as a kind of optimization algorithm, has been widely utilized. To reduce the long time consumed in searching the optimal threshold and large amounts of calculation of the 3D Otsu algorithm, this research adopts the PSO to find the optimal threshold of the 3D Otsu algorithm. The obtained thresholds are used for the binary segmentation of difference images, thus identifying the change and nonchange areas in the difference images. The experimental results suggest that this algorithm can be utilized to precisely segment images to obtain the change areas. Therefore, it is an effective and feasible method for change detection.

2 3D OTSU ALGORITHM

Suppose that an original image with the gray level valued as $0, 1, ..., L-1$ shows a size of $M \times N$ and the gray level of its pixels is represented by $f(x, y)$. Let $g(x, y)$ denote the average gray level in a pixel domain of $k \times k$ at (x, y) with the gradient being $G[f(x, y)]$. In this study, Sobel operators, which can be used to preferably process images with gradually changed gray level, much noise are selected in the gradient algorithm. $g(m, n)$ is defined as follows:

$$g(x, y) = \frac{1}{k^2} \sum_{m=-k/2}^{k/2} \sum_{n=-k/2}^{k/2} f(x+m, y+n) \qquad (1)$$

The triple (i, j, z) composed of $f(x, y)$, $g(x, y)$, and $G[f(x, y)]$ is defined as a 3D histogram in a cubic area of $L \times L \times L$. The three coordinates of the histogram represent the gray level, mean value, and gradient of the pixel of the image, as illustrated in Figure 1(a). The value of the histogram at any point is defined as p_{ijz}, indicating the occurrence frequency of the vector (i, j, z), where p_{ijz} is represented by:

$$p_{ijz} = \frac{r(i, j, z)}{M \times N} \qquad (2)$$

where $r(i, j, z)$ is the occurrence frequency of the triple $(i, j, z)(0 \le i, j, z \le L-1)$ and $\sum_{i=0}^{L-1} \sum_{j=0}^{L-1} \sum_{k=0}^{L-1} p_{ijz} = 1$.

On the basis of the definition of the 3D histogram mentioned above, the 3D histogram is divided into eight cuboid areas using cross lines at the threshold vector (p, s, t), as demonstrated

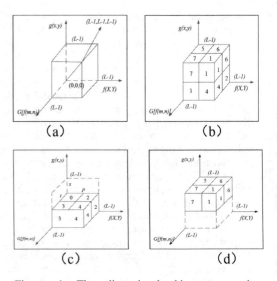

(a)   (b)

(c)   (d)

Figure 1. Three-dimensional histogram region: (a) Three-dimensional histogram domain of definition; (b) The division of three-dimensional histogram region; (c) The division of regional 0, 2, 3, 4; (d) The division of regional 1, 5, 6, 7.

in Figure 1(b). The specific division of areas is shown in Figure 1(c) and 1(d). In the 3D histogram, the target and background of the image are supposed to be located in the areas 0 and 1, respectively, while the noise and edges are expected to lie in other six areas. Because the pixels concerning the background and target of the image are much more than those relating to the boundary area, it is supposed that the sum of the probabilities of the pixels in areas 2–7 is approximately equal to 0. Under such assumption, the 3D histogram is more applicable than 2D histograms and the 3D Otsu algorithm shows a higher segmentation accuracy for the images with low contrast and a low signal-to-noise ratio.

Then, the occurrence probabilities of background and target areas are shown as follows:

$$\omega_0 = p_r(C_0) = \sum_{i=0}^{p}\sum_{j=0}^{s}\sum_{z=0}^{t} p_{ijz} = \omega_0(p,s,t) \qquad (3)$$

$$\omega_1 = p_r(C_1) = \sum_{i=p+1}^{L-1}\sum_{j=s+1}^{L-1}\sum_{z=t+1} p_{ijz} = \omega_1(p,s,t) \qquad (4)$$

where the occurrence probabilities of background and target areas are ω_0 and ω_1, respectively. The corresponding mean value of the background and target areas is expressed as follows:

$$\mu_0 = (\mu_{0i},\mu_{0j},\mu_{0z})^T = ip_{ijz}/\omega_0(p,s,t),$$
$$\left(\sum_{i=0}^{p}\sum_{j=0}^{s}\sum_{z=0}^{t}\sum_{i=0}^{p}\sum_{j=0}^{s}\sum_{z=0}^{t} jp_{ijz}/\omega_0(p,s,t),\right.$$
$$\left.\sum_{i=0}^{p}\sum_{j=0}^{s}\sum_{z=0}^{t} zp_{ijz}/\omega_0(p,s,t)\right)^T \qquad (5)$$

$$\mu_1 = (\mu_{1i},\mu_{1j},\mu_{1z})^T = ip_{ijz}/\omega_1(p,s,t),$$
$$\left(\sum_{i=p+1}^{L-1}\sum_{j=s+1}^{L-1}\sum_{z=t+1}^{L-1}\sum_{i=p+1}^{L-1}\sum_{j=s+1}^{L-1}\sum_{z=t+1}^{L-1} jp_{ijz}/\omega_1(p,s,t),\right.$$
$$\left.\sum_{i=p+1}^{L-1}\sum_{j=s+1}^{L-1}\sum_{z=t+1}^{L-1} zp_{ijz}/\omega_1(p,s,t)\right)^T \qquad (6)$$

The overall mean value of the 3D histogram is:

$$\mu_T = (\mu_{Ti},\mu_{Tj},\mu_{Tz})^T$$
$$= \left(\sum_{i=0}^{L-1}\sum_{j=0}^{L-1}\sum_{z=0}^{L-1} ip_{ijz}, \sum_{i=0}^{L-1}\sum_{j=0}^{L-1}\sum_{z=0}^{L-1} jp_{ijz}, \sum_{i=0}^{L-1}\sum_{j=0}^{L-1}\sum_{z=0}^{L-1} zp_{ijz}\right)^T$$
$$(7)$$

Assuming that the sum of the probabilities of the pixels in areas 2–7 are approximately equal to 0, we obtain:

$$\omega_0 + \omega_1 \approx 1, \mu_T \approx \omega_0\mu_0 + \omega_1\mu_1 \qquad (8)$$

A within-class scatter matrix is defined between the background and the target areas as follows:

$$\sigma_B = \omega_0\left[(\mu_0 - \mu_T)(\mu_0 - \mu_T)^T\right]$$
$$+ \omega_0\left[(\mu_1 - \mu_T)(\mu_1 - \mu_T)^T\right] \qquad (9)$$

By using the trace of σ_B to measure the discrete degree among classes, we get:

$$t_r\sigma_B(p,s,t) =$$
$$\omega_0\left[(\mu_{0i} - \mu_{Ti})^2 + (\mu_{0j} - \mu_{Tj})^2 + (\mu_{0z} - \mu_{Tz})^2\right]$$
$$+ \omega_1\left[(\mu_{1i} - \mu_{Ti})^2 + (\mu_{1j} - \mu_{Tj})^2 + (\mu_{1z} - \mu_{Tz})^2\right]$$
$$(10)$$

3 PSO

PSO, proposed by Kennedy and Eberhart, originates from a simple social simulation phenomenon. Initially, it was raised on the basis of the simulation of the foraging behavior of birds. Later, it is gradually used as an optimization tool in extensive applications. Similar to other optimization algorithms, PSO refers to moving the individuals in a group to excellent regions according to their fitness to the environment by combining the parameters to be optimized as a group. What is different from other algorithms is that when describing individuals, PSO regards individuals as a particle (point) not referred in a D-dimensional space. Therefore, in each iteration, by combining the historical optimal positions of particles (individual extrema p_{best}) with those of the group (global extrema g_{best}), the velocity and position of each particle are updated:

$$x_{k+1}^i = x_k^i + v_{k+1}^i \qquad (11)$$

$$v_{k+1}^i = w * x_k^i + c_1 r_1(p_{best}^i - x_k^i) + c_2 r_2(g_{best} - x_k^i) \qquad (12)$$

where k is the times of iteration; x^i and v^i are the position and velocity of the ith random particle, respectively; c_1 and c_2 denote the learning factors and generally $c_1 = c_2 = 2$; r_1 and r_2 indicate the random numbers in a range of 0–1; and w denotes weight.

With a simple concept, PSO can be easily realized and used in a large range at a high searching rate. Therefore, compared with other optimization algorithms, it shows more outstanding advantages.

4 CHANGE DETECTION BASED ON THE PSO AND 3D OTSU ALGORITHM

Assuming that A_1 and A_2 are two remote sensing images of the same scene taken at different moments through geometric registration with the pixel being

$M \times N$, $A_1 = \{A_1(x, y), 1 \le x \le M, 1 \le y \le N\}$ and $A_2 = \{A_2(x, y), 1 \le x \le M, 1 \le y \le N\}$. $A_1(x,y)$ and $A_2(x,y)$ show the gray levels of the i_{th} row and the j_{th} column in the image, respectively. To achieve the change detection based on PSO and 3D Otsu algorithms, the following steps are performed.

1. Construction of difference images. Difference images are constructed using the difference method in this research. The difference method performs subtraction operation to the corresponding pixel values of the remote sensing image at the temporal stages T_1 and T_2.

2. Extraction of change areas. Automatic threshold segmentation is carried out to the generated difference images so as to extract the change areas. The value of fitness is the basis for the selection of the individual extrema and global extrema using PSO. Therefore, different fitness functions need to be designed under different conditions. Under the condition that the target function is not complicated, we can directly use the target function as the fitness function. In this study, function $t_r\sigma_B(p,s,t)$ is served as the fitness function and the maximum vector (p_0, s_0, t_0) of the value of the within-class scatter function $t_r\sigma_B$ is acted as the optimal threshold of the 3D Otsu algorithm to solve its maximum value, that is:

$$(p_0, s_0, t_0) = \arg\max_{0 \le s, t, q \le L-1}\{t_r\sigma_B(p, s, t)\} \qquad (13)$$

The PSO-based 3D Otsu algorithm is used to solve the threshold p_0, s_0, t_0 for determining whether the pixels change. Then, the difference images are segmented to obtain the change areas.

3. Evaluation of the precision of change detection.

5 EXPERIMENTAL RESULTS AND ANALYSIS

5.1 Experimental data

In this study, a set of remote sensing images of the same scene taken at different temporal stages are used to verify the feasibility and effectiveness of the method proposed. The experimental data used are obtained from the remote sensing images with the spatial resolution of 30 m acquired using the satellite Landsat-5 in July, 1989 (Figure 2(a)), and January, 2010 (Figure 2(b)), in a local area of the Poyang Lake in China. The data set is provided by International Scientific & Technical Data Mirror Site, Computer Network Information Center, and Chinese Academy of Sciences. (http://www.gscloud.cn). These images all present the pixels of

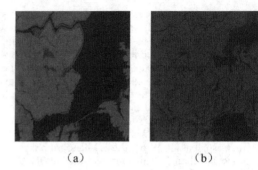

(a)   (b)

Figure 2. Multitemporal remote sensing images: (a) 15 July 1989; (b) 14 January 2010.

510×510, and the reference images of the change areas represented in black are demonstrated in Figure 3(a). Thereinto, the change areas show the pixels of 69,413, while the nonchange areas present the pixels of 190,687. The reference images are mainly acquired by interpreting the existing Geographic Information System (GIS) data.

5.2 Evaluation of the accuracy

To quantitatively and qualitatively detect the effectiveness of the method raised, this study evaluates the total wrong detection accuracy, false detection rate, and miss rate of the change detection (Bovolo, 2006). That is, total wrong detection accuracy is $Total = \frac{LH+LK}{S} \times 100\%$; while the false detection rate and miss rate are $P_E = \frac{LH}{SH} \times 100\%$ and $P_L = \frac{LK}{SS} \times 100\%$, respectively.

Here, LH represents the number of nonchange pixels incorrectly detected as change ones; LK shows the number of change pixels undetected; S denotes the sum of the pixels in the remote sensing image; SH indicates the sum of the nonchange pixels in the image; and SS stands for the sum of the change pixels.

By comparing the method presented in this study with the change detection methods based on PSO and one-dimensional Otsu algorithm with double thresholds, this research validates the advantages and disadvantages of this method.

5.3 Analysis of experimental results

Figure 3(b) shows the results of change detection obtained using the PSO-based one-dimensional Otsu algorithm with double thresholds, while Figure 3(c) shows the results acquired using the method developed in this paper. By evaluating the accuracy of the results obtained using these two methods, we obtain the following results, as illustrated in Table 1. As can be seen in Table 1, by using the method proposed in this study, 7,462 pixels of

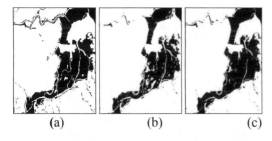

| (a) | (b) | (c) |

Figure 3. Results of change detection: (a) Reference map; (b) Result of one-dimensional Otsu algorithm with double thresholds; (c) Result of the proposed method.

Table 1. False alarms, missed alarms, and total errors for two methods in Poyang Lake.

Change detection methods	False alarms		Missed alarms		Total errors	
	Pixels	%	Pixels	%	Pixels	%
Otsu algorithm with double thresholds and PSO	14115	7.40%	2399	3.46%	16514	6.35%
Proposed Method	7462	3.91%	5918	8.53%	13380	5.14%

change areas are missed, showing a miss rate of 3.91%; while the change areas falsely detected show the pixel of 5,918 with the miss rate being 8.53%. Therefore, the total wrong detection rate is 5.14%. From the perspective of longitudinal comparison, compared with the traditional one-dimensional Otsu algorithm with double thresholds, 3,519 more pixels are falsely detected, while 6,653 less pixels are missed by using the method developed in this study. On the whole, the pixels wrongly detected are 3,143 less than the traditional Otsu algorithm. By comparing the three images in Figure 2, it can be found that in the change detection using the PSO-based one-dimensional Otsu algorithm with double thresholds, most change areas are missed as nonchange ones. Therefore, it favorably reduces the false detection rate while increases the wrong detection rate. In comparison, the method presented in this study accurately identifies most areas and shows the superior miss rate and total wrong detection accuracy to those of the PSO-based one-dimensional Otsu algorithm with double thresholds. The experimental results reveal that the change detection method for remote sensing images based on the PSO and 3D-Otsu algorithm is effective and feasible for change detection.

In terms of the operation time, because we adopt the PSO algorithm, the operation speed of the algorithm is effectively accelerated, thus ensuring the real time of the algorithm and shortening the computation time. The proposed method takes 1.5572s, while 47.8471s is used for the selection of thresholds with the application of the 3D Otsu exhaustive algorithm. This suggests that the method proposed in this paper can significantly improve the operation speed.

6 CONCLUSIONS

To solve the problems of the 3D Otsu algorithm in the change detection, that is, large calculation quantity and long computation time cost for each pixel, this study adopts the PSO to calculate the optimal threshold of the 3D Otsu algorithm. By studying the remote sensing images taken in a local area of Poyang Lake at two different temporal stages, this study verifies the accuracy of the proposed method. The experimental results indicate that the results of change detection obtained using the method presented show high accuracy (reaching 94.86%), which is superior to that of the PSO-based one-dimensional Otsu algorithm with double thresholds. Meanwhile, compared with the exhaustive method, the proposed method shows significantly increased operation efficiency. However, local optimum frequently occurs in the application of the PSO and the thresholds searched under low probabilities are not optimal ones, which influence the final results of change detection. Therefore, improving the PSO is the key point in the future.

ACKNOWLEDGMENTS

This work was supported by the Talent Cultivation Foundation of Kunming University of Science and Technology (No. KKSY201521040) and the foundation of Yunnan Educational Committee (No. 2016ZZX051). The authors would like to thank the anonymous reviewer for the constructive comments.

REFERENCES

Bovolo, F. and Bruzzone L., A detail-preserving scale-driven approach to change detection in multitemporal SAR images, IEEE T Geosci Remote, 43, pp. 2963–2972 (2006).

Chan, K.K.Y. and B. Xu., Perspective on remote sensing change detection of Poyang Lake wetland, Ann Gis, 19, pp. 231–243 (2013).

Chen, Q., L. Zhao, J. Lu, et al., Modified two-dimensional Otsu image segmentation algorithm and fast realization, IET Image Process, 6, pp. 426–433, (2012).

Chen, X.L., H.M. Zhao, P.X. Li, et al., Remote sensing image-based analysis of the relationship between urban heat island and land use/cover changes, Remote Sens Environ, 104, pp.133–146 (2006).

Huang, L., X.Q. Zuo and X.Q. Yu, Review on Change Detection Methods of Remote Sensing Images, Sci Surv Map, 38, pp. 203–206 (2013).

Huang, L., Y.M. Fang, X.Q. Zuo, et al., Automatic Change Detection Method of Multitemporal Remote Sensing Images Based on 2D-Otsu Algorithm Improved by Firefly Algorithm, J Sensors, 2015, pp. 1–8 (2015).

Huang, X., A. Schneider, and M.A. Friedl., Mapping sub-pixel urban expansion in China using MODIS and DMSP/OLS nighttime lights, Remote Sens Environ, 175, pp. 92–108 (2016).

Kennedy, J. and R. Eberhart., Particle swarm optimization, *IEEE International Conference on Neural Networks*, 4, pp. 1942–1948 (1995).

Li, S. and X. Luo., Study on Monitoring of Karst Rock Desertification Using Multi-temporal Remote Sensing, Chinese Agr Sci B, 31, pp. 262–267 (2015).

Malmir, M., M.M.K. Zarkesh, S.M. Monavari, et al., Urban development change detection based on Multi-Temporal Satellite Images as a fast tracking approach-a case study of Ahwaz County, southwestern Iran. Environ Monit Assess, 187, pp. 1–10 (2015).

Mishra, N.S., S. Ghosh and A. Ghosh., Fuzzy Clustering Algorithms Incorporating Local Information for Change Detection in Remotely Sensed Images, Appl Soft Comput, 12, pp. 2683–2692 (2012).

Rosin, P.L. and E. Ioannidis., Evaluation of Global Image Thresholding for Change Detection, Pattern Recogn Lett, 24, pp. 2345–2356 (2010).

Sthitpattanapongsa, P. and T. Srinark., An Equivalent 3D Otsu's Thresholding Method, *Lect Notes Comput Sci*, 7087, pp. 358–369 (2011).

An improved multivariate synchronization index method for SSVEP signal processing

Kun Chen
School of Mechanical and Electronic Engineering, Wuhan University of Technology, Wuhan, China
Key Laboratory of Fiber Optic Sensing Technology and Information Processing, Ministry of Education, Wuhan, China

Haojie Liu
School of Information Engineering, Wuhan University of Technology, Wuhan, China

Qingsong Ai
Key Laboratory of Fiber Optic Sensing Technology and Information Processing, Ministry of Education, Wuhan, China
School of Information Engineering, Wuhan University of Technology, Wuhan, China

ABSTRACT: Steady State Visual Evoked Potentials (SSVEPs) are widely used in Brain Computer Interfaces (BCIs) because of higher signal to noise ratio and greater information transfer rate. In this paper, an improved method based on Multivariate Synchronization Index (MSI) was proposed for multi-dimensional SSVEP signal processing. It combines Multi-set Canonical Correlation Analysis (MsetCCA) and MSI, which is named MsetCCA-MSI. Different experiments were conducted to compare the recognition accuracy for these three methods: CCA, MSI and MsetCCA-MSI. The results prove that the proposed MsetCCA-MSI method outperforms the other two no matter how the window length of data segments changes.

1 INTRODUCTION

A Brain Computer Interface (BCI) is a communication system that does not depend on the brain's normal output pathways of peripheral nerves and muscles (J. R. Wolpaw, 2000). It aims to create a specialized interface that allows an individual with severe motor disabilities to have effective control of devices such as computers and prostheses (S. G. Mason, 2003). Electroencephalography (EEG) is most commonly used for BCIs due to its portability and ease of use. Among different BCI paradigms, a Steady State Visual Evoked Potential (SSVEP) is a periodic response to a visual stimulus modulated at a frequency of higher than 4 Hz (D. Regan, 1989). It can be recorded from the scalp as a nearly sinusoidal oscillatory waveform with the same fundamental frequency as the stimulus, and often includes some higher harmonics, as seen in Figure 1. SSVEPs have the characteristics of higher signal to noise ratio and faster information transfer rate compared with other EEG signals.

In recent years, many methods have been developed for SSVEP signal processing (Q. Liu, 2014). Most researchers used Fourier-based methods for

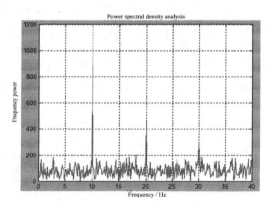

Figure 1. Power spectral density analysis of SSVEPs when stimulus frequency is 10 Hz.

Power Spectrum Density Analysis (PSDA). It computes the accumulative power at the stimulus frequencies and their harmonics, or the average power centred on the stimulus frequency (G. R. Muller-Putz, 2008). The advantage of Fourier-based transforms is their simplicity and small computation time.

However, these methods were originally designed for linear and stationary signals. SSVEPs are non-linear and no-stationary. Thus, methods like the wavelet transform or Hilbert Huang transform was employed (Y. Bian, 2011; Z. Zhang & L. Zhao, 2010). Other typical methods like Minimum Energy Combination (Mec), Maximum Contrast Combination (MCC) and canonical correlation analysis (CCA) were used for multi-channel SSVEP signal processing (O. Friman, 2007; H. Cecotti H, 2010 & Z.L. Lin, 2006).

Yangsong Zhang proposed a novel Multivariate Synchronization Index (MSI) for SSVEP frequency recognition (Y. Zhang, 2013). The experimental results showed it outperformed the widely used CCA and MEC methods, especially for short data length and few channels. Combining with the multiset CCA (MsetCCA) method, we made improvement to MSI and proposed a new method named MsetCCA-MSI for SSVEP signal processing. The data analysis showed this new method obtains better recognition accuracy than MSI.

The paper is organized as follows: experimental methods are described in Section 2. The results and discussion are illustrated in Section 3. The last section is conclusions.

2 EXPERIMENTAL METHODS

2.1 Experimental setup

A Symtop UE-16B EEG amplifier was used in our experiments as seen in Figure 2. It has 16 channels and a USB interface. The maximum sampling frequency can be adjusted to 1000 Hz. A low-pass filter and a notch filter are developed in the amplifier. In experiments, the sampling frequency was set to 200 Hz. Data from channels O1, O2, P3 and P4 were used for analysis.

2.2 Data pre-processing

The stimulus frequencies used in experiments ranged from 6 Hz to 12 Hz. Considering the highest second harmonic component of corresponding

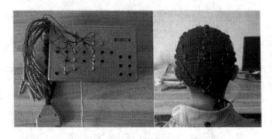

Figure 2. EEG acquisition device and EEG cap.

SSVEPs is 24 Hz, a band-pass filter with the bandwidth from 6 Hz to 30 Hz was employed.

EEG signal are easily contaminated by other bio-signals. Here, FastICA (fast independent component analysis) was used to remove these artefacts. The principle of FastICA will not be discussed in detail in this paper. We summarize the whole SSVEP pre-processing procedure as follows:

1. The raw EEG data were band-pass filtered;
2. The filtered data were centred and whitened;
3. FastICA was used to get the separating matrix and the Independent Components (ICs);
4. Set the ICs related to noise to zero;
5. Use the mixing matrix and new ICs to get the reconstructed signal.

2.3 CCA and MSI

2.3.1 Canonical correlation analysis

Canonical correlation analysis is typically used to compute the correlation coefficient between two multi-dimensional variables. If there are two variables $X = [x_1, x_2, ..., x_p]^T$ and $Y = [y_1, y_2, ..., y_q]^T$, a linear combination is applied to both X and Y to get two new variables $u = w_{x_1} x_1 + ... + w_{x_p} x_p = w_x^T X$ and $v = w_{y_1} y_1 + ... + w_{y_q} y_q = w_y^T Y$. When the correlation coefficient between u and v has the largest value, w_x and w_y can be obtained as follows:

$$\max_{w_x, w_y} \rho(u,v) = \max_{w_x, w_y} \frac{cov(u,v)}{\sqrt{var(u)var(v)}}$$
$$= \max_{w_x, w_y} \frac{w_x^T C_{xy} w_y}{\sqrt{\left(w_x^T C_{xx} w_x\right)\left(w_y^T C_{yy} w_y\right)}} \quad (1)$$

If the number of stimuli is K, the corresponding reference signals can be constructed as:

$$Y_k = \begin{pmatrix} \sin(2\pi f_k 1/f_s) & ... & \sin(2\pi f_k N/f_s) \\ \cos(2\pi f_k 1/f_s) & ... & \cos(2\pi f_k N/f_s) \\ \vdots & \vdots & \vdots \\ \sin(2\pi H f_k 1/f_s) & ... & \sin(2\pi H f_k N/f_s) \\ \cos(2\pi H f_k 1/f_s) & ... & \cos(2\pi H f_k N/f_s) \end{pmatrix} \quad (2)$$

f_s is the sampling frequency. The correlation coefficient ρ_k between the raw EEG data X and the reference signals Y_k is calculated. Finally, the target k is confirmed as $\max_k \rho_k$.

2.3.2 Multivariate synchronization index

The principle (Y. Zhang, 2013) of MSI is introduced here. The same as CCA, a reference signal Y with size $2N_h \times M$ is created. N is the number of channels. M is the number of sampling points, and N_h is the number of harmonics. X and Y are firstly normalized to have a zero mean and unitary variance. The correlation matrix between X and Y is calculated as:

$$C = \begin{bmatrix} C_{11} & C_{12} \\ C_{21} & C_{22} \end{bmatrix} \tag{3}$$

Here, $C_{11} = \frac{1}{M}XX^T$, $C_{22} = \frac{1}{M}YY^T$, $C_{12} = \frac{1}{M}XY^T$ and $C_{21} = \frac{1}{M}YX^T$. The matrix C contains both the autocorrelation and cross-correlation of X and Y. Then a linear transform is used:

$$U = \begin{bmatrix} C_{11}^{-0.5} & 0 \\ 0 & C_{22}^{-0.5} \end{bmatrix} \tag{4}$$

The transformed correlation matrix R is represented as:

$$R = UCU^T = \begin{bmatrix} I_{N\times N} & C_{11}^{-0.5}C_{12}C_{22}^{-0.5} \\ C_{22}^{-0.5}C_{21}C_{11}^{-0.5} & I_{2N_h \times 2N_h} \end{bmatrix} \tag{5}$$

where, $I_{N\times N}$ and $I_{2N_h \times 2N_h}$ are identity matrices. The normalized eigenvalues of R is calculated as:

$$\lambda_i = \frac{\lambda_i}{\sum_{i=1}^{P} \lambda_i} = \frac{\lambda_i}{tr(R)} \tag{6}$$

And, $P = N + 2N_h$.

The synchronization index between two multi-dimensional variables is calculates as:

$$S = 1 + \frac{\sum_{i=1}^{P} \lambda_i \log(\lambda_i)}{\log(P)} \tag{7}$$

If the number of stimuli is K, calculate the synchronization index between raw EEG data and the reference signal to get $S_1, S_2, ..., S_K$. The target stimulus T is recognized as:

$$T = \max_i S_i, \, i = 1, 2, ..., K \tag{8}$$

2.4 MsetCCA-MSI

Both CCA and MSI methods need to create reference signals with sine and cosine components corresponding to stimuli frequencies. However, the reference signal might lack in some components existing in real EEG data. Thus, a Multi-set CCA (MsetCCA) method was proposed by Zhang Yu (Y. Zhang, 2014). Inspired by this idea, in this paper, a multi-set CCA based multiple synchronization index (Mset-CCA MSI) method was suggested for SSVEP signal processing. The detailed procedure is described as follows:

Assume there are L sets of raw EEG data $X_i \in R^{M\times N}$ (M is the number of channels, N is the number of sampling points, $i = 1, 2, ..., L$). All data are firstly normalized. The objective function is to maximize the overall correlation among all canonical variates:

$$\max_{w_1, ..., w_L} \rho = \sum_{i\neq j}^{L} w_i^T X_i X_j^T w_j \tag{9}$$

where, $\frac{1}{L}\sum_{i=1}^{L} w_i^T X_i X_i^T w_i = 1$.

With Lagrange multipliers, the equation can be transformed into the following expression:

$$(R - S)w = \rho S w \tag{10}$$

where,

$$R = \begin{bmatrix} X_1 X_1^T & \cdots & X_1 X_L^T \\ \vdots & \ddots & \vdots \\ X_L X_1^T & \cdots & X_L X_L^T \end{bmatrix} \tag{11}$$

$$S = \begin{bmatrix} X_1 X_1^T & \cdots & 0 \\ \vdots & \ddots & \vdots \\ 0 & \cdots & X_L X_L^T \end{bmatrix} \tag{12}$$

$$w = \begin{bmatrix} w_1 \\ \vdots \\ w_L \end{bmatrix} \tag{13}$$

So the linear transforms $w_1, w_2, ..., w_L$ resulting in the largest overall canonical correlation among all canonical variates are given by eigenvectors corresponding to the largest generalized eigenvalus. The canonical variate is expressed as:

$$\tilde{z}_i = w_i^T X_i \tag{14}$$

If the stimulus frequency is fm, and there are L sets of EEG data $X_{1,m}, X_{2,m}, ..., X_{L,m} \in R^{M\times N}$ corresponding to stimulus fm, then try to find linear transforms $w_{1,m}, w_{2,m}, ..., w_{L,m}$ to get the maximum correlation between all canonical variates. The reference signal for the stimulus fm is created as:

$$\dot{Y}_m = \begin{bmatrix} \tilde{z}_{1,m}^T, \tilde{z}_{2,m}^T, ..., \tilde{z}_{L,m}^T \end{bmatrix}^T \tag{15}$$

Reference signals for other stimuli are created in the same way. And finally, the raw EEG data and those reference signal are processed with the MSI method.

The whole procedure for SSVEP signal processing is summarized as seen in Figure 3.

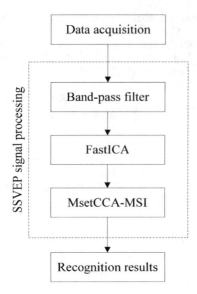

Figure 3.   SSVEP signal processing.

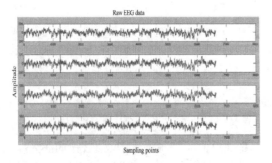

Figure 4.   Raw EEG data obtained from channels P3, P4, O1 and O2.

3   RESULTS AND DISCUSSION

Four stimulus frequencies (6 Hz, 6.5 Hz, 7 Hz, 7.5 Hz) were used to evoke SSVEPs. The sampling frequency is 200 Hz. There are 20 trials for each stimulus frequency and every trial lasts for 5 seconds (1000 sampling points). The window length of data segments is set to 1 second, 2 seconds, 3 seconds and 4 seconds to test the recognition accuracy, respectively.

Figure 4 illustrates raw EEG data obtained from channels P3, P4, O1 and O2. It can be seen that no obvious characteristics are readable in the time domain for EEG data. Thus, these data should be analyzed with other methods.

The CCA and MSI methods were used to compare the recognition results with the proposed MsetCCA-MSI method. Standard sine and cosine waves were created as the reference signals for CCA

Table 1.   Recognition accuracy for CCA, MSI and MsetCCA-MSI methods.

Window length	1s	2s	3s	4s
CCA	52.50%	81.25%	86.25%	85.00%
MSI	58.75%	78.75%	86.25%	87.50%
MsetCCA-MSI	66.25%	91.25%	98.75%	97.50%

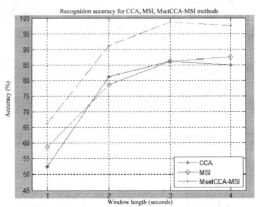

Figure 5.   Comparison of three methods for SSVEP recognition.

and MSI. The experimental results is illustrated in Table 1.

Figure 5 represents the recognition results of these three methods more obviously.

It can be seen from Figure 5 that the recognition accuracy of CCA and MSI methods are almost the same while the window length of data segments changes. However, the MsetCCA-MSI outperforms both CCA and MSI obviously no matter how the window length of data segments varies.

4   CONCLUSIONS

SSVEP-based BCIs have great potential in real world applications, and the signal processing algorithm is of great importance. In this paper, an improved method based on MSI, named MsetCCA-MSI was proposed for SSVEP recognition. Different experiments were conducted to prove the feasibility and efficiency of this proposed method. Compared with the CCA and MSI methods, MsetCCA-MSI obtains better recognition accuracy when the window length of data segments changes.

The future work is to implement this proposed method for online experiments and develop real SSVEP-based applications considering design of the stimuli, subject comfort and other factors.

ACKNOWLEDGEMENTS

This work was supported by the National Science Foundation (Grant No. 51475342) and the Fundamental Research Funds for the Central Universities (Grant No. 2016-III-019).

REFERENCES

Wolpaw, J.R., N. Birbaumer, W.J. Heetderks, D.J. McFarland, P.H. Peckham, G. Schalk, E. Donchin, L.A. Quatrano, C.J. Robinson and T.M. Vaughan, Brain-computer interface technology: a review of the first international meeting, IEEE Trans. Rehabil. Eng., 8: 164–173 (2000).

Mason S.G., and G.E. Birch, A general framework for brain-computer interface design, IEEE Trans. Neural Syst. Rehabil. Eng., 11: 70–85 (2003).

Regan, D., Human brain electrophysiology: evoked potentials and evoked magnetic fields in science and medicine, J. Clin. Neurophysiol., 7: 450–451 (1989).

Liu, Q., K. Chen, Q.S. Ai, S.Q. Xie, Review: Recent Development of Signal Processing Algorithms for SSVEP-based Brain Computer Interfaces, J. Med. Biol. Eng. 34:299–309 (2014).

Muller-Putz, G.R., G. Pfurtscheller, Control of an electrical prosthesis with an SSVEP-based BCI, IEEE Trans. Biomed. Eng, 55:361–364 (2008).

Bian, Y., H.W. Li, L. Zhao, G.H. Yang and L.Q. Geng, Research on steady state visual evoked potentials based on wavelet packet technology for brain-computer interface, Proc. Eng., 15: 2629–2633, (2011).

Zhang, Z., X. Li and Z. Deng, A CWT-based SSVEP classification method for brain-computer interface system, International Conference on Intelligent Control and Information Processing, 43–48 (2010).

Zhao, L., P.X. Yuan, L.T. Xiao, Q.G. Meng, D.F. Hu and H. Shen, Research on SSVEP feature extraction based on HHT, 7th Internatoinal Conference on Fuzzy Systems and Knowledge Discovery, 2220–2223 (2010).

Friman, O., T. Luth, I. Volosyak, A. Graser A, Spelling with steady-state visual evoked potentials. 3rd International IEEE/EMBS Conference on Neural Engineering, 54–357 (2007).

Cecotti H., H, A self-paced and calibration-less SSVEP-based brain-computer interface speller. IEEE Trans. Neural Syst. Rehabil. Eng. 18:127–133 (2010).

Lin, Z.L., C.S. Zhang, W. Wu, X.R. Gao, Frequency recognition based on canonical correlation analysis for SSVEP-based BCIs, IEEE Trans. Biomed. Eng. 53:2610–2614 (2006).

Zhang, Y., P. Xu, K. Cheng, D. Yao, Multivariate synchronization index for frequency recognition of ssvep-based brain–computer interface, J. Neurosci. Methods, 221:32–40 (2013).

Zhang, Y., G. Zhou, J. Jin, X. Wang, A. Cichocki. Frequency recognition in SSVEP-based BCI using multiset canonical correlation analysis. Int. J. Neural. Syst., 24 (2014).

Advances in Energy Science and Equipment Engineering II – Zhou, Patty & Chen (Eds)
© *2017 Taylor & Francis Group, London, ISBN 978-1-138-71798-5*

Reliability design of a humanoid robot

Yuling Shen
Northwestern University, Evanston, USA

ABSTRACT: Compared with conventional robots, a humanoid robot has a smaller volume, higher center of mass, less ground support, and more surrounding contacts. The stability of humanoid body posture can be easily affected because these robots are vulnerable to interferences of electromagnetic signal (EMI). In order to tackle with the aforementioned issues, reliability design of humanoid robot is conducted in this paper, which includes hardware electromagnetic compatibility (EMI) design, software reliability design, and impedance control design.

1 INTRODUCTION

Humanoid robot has become a long-term research hot spot in robot research field and its development level can represent the cutting-edge technology and exploration frontier of its associated field. Mr Shuuji Kajita, the pioneer of humanoid robot and staff of the National Institute of Advanced Industrial Science and Technology, Tokyo, Japan, has summarized three characters of humanoid robot: they can function under the environment as humans do, they can employ the tools humans use, and they have the shape and appearance similar to human beings. These advantages provide large utilization potentiality of humanoid robots in our society. Investigations on humanoid robot have been conducted within institutions and enterprises in the United States, Japan, South Korea, and Germany, with the support and guidance from their governments. Different types of humanoid robots have been developed; breakthroughs and industrialization promotions on humanoid robot key technology and core components have been realized, which promote the theoretical research study and engineering application of humanoid robot. Successes in humanoid robot developments of these countries have led to their domestic robot industrial technology upgrade and competitiveness enhancement. With subsidization from the National Science and Technology Development Program and guidance of relevant national policies, continuous breakthroughs and industrialization on robot and its relevant technologies have taken place. Considerable progress on robots have been achieved, which enables our command humanoid robot-related basic theories and its system integration technologies; however, the high-performance functional components of Chinese humanoid robots are still imported. The dynamic and controlling performance of Chinese home-made humanoid robot are not ideal, thus curbing the robot technology development and application prospect.

Different from conventional robots, humanoid robot has raised its unique research difficulty in posture balancing control. Traditionally, industrial robots are immobilized in their working conditions by mechanism; wheeled robots either distribute most of their mass on base plate or are well supported at their wheeled region. Therefore, for conventional robots such as industrial robot and wheeled robot, the posture influence from the surroundings can be neglected. Compared with conventional robots, humanoid robot has higher center of mass and less ground support. As a result, both accidental contact and external force disturbances will affect the robot posture, leading to fall over and bringing about serious damage to the hardware. The posture control issue of humanoid robot has become a must-solve research topic when under the practical operation.

The early studies on humanoid robot balance control can date back to the 1970s. Researchers such as Vukobratovic (Vukobratovic, 1972) and Golliday (Golliday, 1976) are representatives of their respective periods and they have accumulated rich experience on the robot balance control. With the development of technology and social perception, new challenges, specifically the miniaturization requirement from industrialization and constant working condition deterioration of robots, have been raised on humanoid robots. In order to guarantee humanoid robot's adaptability and smooth operation under complex working conditions, hardware circuit design on EMC, software design on system function expansibility, and posture control design by impedance control method, aiming at maintaining the robot stability during its waking movement, are elaborated in this

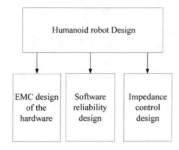

Figure 1. Humaniod robot design.

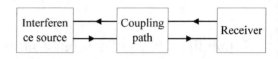

Figure 2. Three basic compositions of EMC.

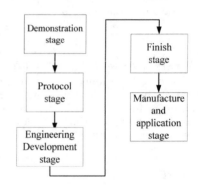

Figure 3. Design flow of humanoid robot electromagnetic compatibility.

paper. The overall structure of humaniod robot design can be seen in Fig. 1.

2 HARDWARE EMC DESIGN

Human civilization has suffered from unprecedentedly rapid technical innovations over the past decade. Three technological revolutions have elevated humans from the stage of adapting to nature to the stage of reshaping nature. Being a double-edged blade, modern industry, science, and technology, which have brought tremendous changes to our lives, could also destroy us. Chemical industry, for example, provides us with a myriad of chemical products, but also produces deadly toxicants that can potentially kill people; explosive, being an ideal assistant for us especially in mining, is also a key component in weapons production. Nuclear power, with the ability to generate enormous clean electricity, however, may cause long-term devastating harm to human beings and the environment by nuclear radio-contamination. Besides, terrorist attacks and regional wars around the world relying on modern technology bring people in conflict regions with irretrievable losses and sorrow.

Humanoid robot can provide its unique contribution to conduct sampling and explorations in places with chemical contamination, nuclear radio-contamination, and implement detection in battlefield and ruins after earthquake, where human beings can hardly access. The environments of these occasions are extremely complex and with all kinds of uncertainties. Therefore, how to guarantee the humanoid robot to function well under strong electromagnetic occasions and in places with intense radiation matters a lot to its performance.

By properly designing its EMC, the control board and drive board of humanoid robot can be without electromagnetic interferences, thereby fulfilling their original functions. Also, whether the humanoid robot has a qualified electromagnetic compatibility is also a good representative for its adaptability under complex industrial environment and complex working conditions.

The basic hardware Electromagnetic Compatibility (EMC) design (Matej, 2016) compositions are interference source, coupling trajectory, and receiver, as shown in Fig. 2. It can be seen that the humanoid robot is likely to be affected when under complex working conditions, thus electromagnetic compatibility design on humanoid robot through circuit design and shielding are needed. The overall electromagnetic compatibility design can be categorized into five steps: demonstration stage, protocol stage, engineering development stage, finish stage, and manufacture and application stage, which are shown in Fig. 3.

3 SOFTWARE RELIABILITY DESIGN

The well-designed hardware makes the robot reliable, while the well-designed software makes it consistent in logic. With the constant enrichment and consolidation of drivers, sensors, and computation, the controlling software structure has become complicated and redundant. Therefore, how to improve its controlling software system reliability, maintenance convenience, and expandability through structured design are urgent issues to be solved during the humanoid robot development process.

Intense coupling relationship exists between the robot hardware system design and its inherent software structure. For most robot prototypes, special hardware and corresponding matched controlling software are required to establish their designed functions, thus intense coupling also exists between the control algorithm and control hardware circuit. With the increasing demand on robot functions and the increasing complexity of robot actuating logic, the design mode has degraded the code readability and lead to code redundancy, thereby making the maintenance more difficult. By properly designing software framework into interfaces and modules, which is the mainstream thinking of robot software design, the direct code coupling between hardware and software can be tremendously decreased. With gradual clearing of classification on task- and structure-based robots, researchers have started to conduct effective abstraction and summarization on software level and proposed many robot design frameworks, which can be adapted to different types of robots. And researchers in this field have tried to formulate a unified platform standard for robot control software, thus making it with proper regulations.

Currently issued robot software include widely used closed source platforms such as CyberBotics Webots, White Box Roboticsml, and Evolution Robotics ERSPml; software platforms that aim at a specific special hardware such as Lego MindstormsH and CotsBotS (Bergbreiter, 2003) of LEg GOdt; open-sourced platforms such as OROCOSml (Smits, 2001), TeamBots, CARMEN[50], Pyro (Montemerlo, 2003), and ROS (Douglas, 2004).

An important element to guarantee the humanoid robot performance is its control program. The overall structure of Evolving Mind is adopted for the software design of humanoid robot in this paper. Evolving Mind is a control system programmed in C++, developed for robot Kong-II. It functions with a single process and multithread, and has fully considered researchers practical needs, system expansibility as well as maintainability, thereby making itself possible for researchers to add or modify software and hardware modules during the system development, which makes it convenient for researchers to enrich or perfect the robot performance. The implementation of Evolving Mind can largely decrease the cost of development in case the system requires any software or hardware upgrades in the future.

Evolving Mind is a layered structure system and its structure is shown in Fig. 4. According to the closeness of hardware, Evolving Mind can be categorized into four layers, including hardware interaction layer, data buffer layer, motion control layer, and decision supervision layer, stacking from the bottom to the top. Varying degrees of abstraction on control issues can be conducted independently at each layer. These abstractions can communicate with each other through predefined abstract function interfaces within this layered system; thus, it will decrease the code interdependence among

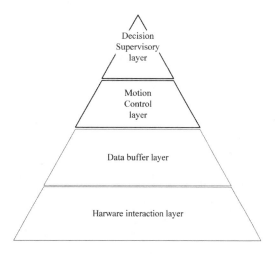

Figure 4. Evolving mind layered structure diagram.

different layers and provide the most convenience to its future upgrade. Also because of its independence, not every layer must be modified if any modifications are required in future implementations.

4 IMPEDANCE CONTROL DESIGN

The walking movement posture stability control is a long-term research hot spot and a challenge for humanoid robot motion control. Under practical operations, due to the interference of outer disturbance and system movement errors, actual movements of robot may differ from the expected movement trajectories. This disturbance may influence robot integral stability, in addition lead to the robot's falling over or capsizing, a malfunction that damages the robot itself or even jeopardizes surroundings. Therefore, robot kinematic error elimination is of a great significance to its posture stability and performance precision.

Humanoid robot kinematic errors are mainly caused by the following three factors: errors between the simplified yet inaccurate model, which is applied by motion planning algorithm and the practical kinematic model of humanoid robot, errors from every joint of humanoid robot when executing trajectory commands, and errors from unknown disturbances or uncertainties existing in its surroundings. Errors from movements are inevitable when the robot walks or functions; therefore, the movement itself may bring about unstable factors to the robot. In order to reduce these errors generated during the robot walking movement, impedance control theory (Gribovskaya, 2011) based on walking posture adjustment is proposed in this paper to conduct humanoid robot posture

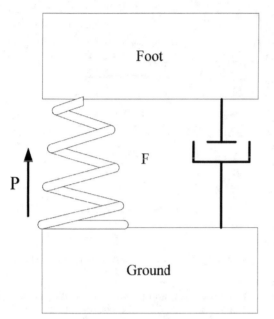

Figure 5. Spring damping system of a robot leg when touching the ground.

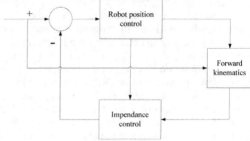

Figure 6. Impedance control method.

The system block diagram of impedance control-based humanoid posture stability control is shown in Fig. 6. Its main function is to best relieve the negative effect caused by ground-reactive force during switching period from single-leg support to double-leg support.

algorithm optimization. In common conditions, at least two objects shall be involved during robot operations. For instance, robot legs and the ground have been involved when the robot is walking. The most important characteristic of impedance control effort between the dynamic interaction of these two physical objects is that they two must be relatively complemented, that is, if one is impedance, the other must be admittance, and vice versa.

The main aim of impedance control-based robot leg movement trajectory adjustment is to decrease the touchdown impact on robot posture stability and to control robot touchdown process, taking the robot swing leg and ground as research objects. Fig. 5 introduces an equivalent system and a suppositional spring damping system to illustrate the connection between the robot swing leg and the center of robot hip. When the force between robot and ground is only enough to support its stability, deformation of the spring damped system is zero; when the force between robot and ground is larger or smaller than the needed force to support its stability, deformation on the spring damping system may occur, whose deformation scale equals the adjustment amount of the end of robot swing leg. By adopting impedance control-based robot leg movement trajectory adjustment (Jafari, 2012), the robot stability during its walking movement can be best guaranteed, therefore malfunctions such as fall over can be prevented.

4 CONCLUSION

Humanoid robot has similar body structures and athletic ability as human being. It can liberate human from manual labor and machinery repetitive operation. Moreover, they can play the roles of human being in many dangerous circumstances, which makes robot technologies and humanoid robot development become a key element to improve the development of our society in a considerably long time. In this paper, three aspects, hardware design which focused on EMC design; software design which focused on system function expansibility as well as maintainability; and impendence control method, which focused on ground reactive force relief, are conducted for the reliability design of humanoid robot. The composition of designs proves that proper design can guarantee the normal operation of humanoid robot under complex working conditions.

REFERENCES

10. Teambots. Teambots. http://www.teambots.org.
Bergbreiter S, Pister K S J. CotsBots: An off-the-shelf platfom for distributed robotics. Proceedings of 2003 IEEE Intemational Conference on Intelligent Robots and Systems, 2003,2:1632–1637.
Cyberbotics. Webots. Cyberbotics. http://www.cyber-botics.com/.
Douglas B, Kitmar D, Meeden L, Yanco H. Pyro: A python-based versatile programming environment for teaching robotics. Journal on Educational Resources in Computing, 2004,4 (3):3.
Evolution Robotics. Evolution Robotics ERSP. http://www.evolution.com/products/ersp/.

Golliday C, Hemami H. Postural stability of the two-degree-of-freedom biped by general linear feedback. IEEE Transactions on Automaitc Control, 1976. 21(1): 74–79.

Gribovskaya E, Kheddar A, Billard A. Motion learning and adaptive impedance for robot control during physical interaction with humans. Robotics and Automation (ICRA), 2011 IEEE International Conference on, 2011: 4326–4332.

Jafari A, Rezaei M, Talebi A, et al. An Adaptive Hybrid Force/Motion Control Design for Robot Manipulators Interacting in Constrained Motion With Unknown Non-Rigid Environments. ASME 2012 International Mechanical Engineering Congress and Exposition, 2012: 1063–1069.

Lego. Lego MindStorm http://www.1ego.com/eng/education/mindstorms/default.asp.

Matej Kučera; Milan Šebök: Electromagnetic compatibility analysing of electrical equipment. 2016 Diagnostic of Electrical Machines and Insulating Systems in Electrical Engineering (DEMISEE). 2016: 104–109

Montemerlo D, Rciy N, Thrun S. Perspectivs on standardization in mobile robot programming: The carnegie mellon navigation (CARMEN) toolkit. Proceedings of 2003 IEEE International Conference on Intelligent Robots and Systems. 2003, 3:2436–2441.

Smits R, Laett D, Claes K, Soetens P, Schutter J D, Bruyninckxh. Open robot control software: the OROCOS project. Proceedings of 2001 IEEE International Conference on Robotics and Automation. 2001.3:2523–2528.

Vukobratovic M, Stepaenko Y. On the stability of anthropomorphic systems. Mathematical Biosciences, 1972. 15:1–37.

White Box Robotics. White BOX Robotics. http://whiteboxrobotics.com/.

Advances in Energy Science and Equipment Engineering II – Zhou, Patty & Chen (Eds)
© 2017 Taylor & Francis Group, London, ISBN 978-1-138-71798-5

A novel monitoring algorithm for metric temporal logic

Xiang Qu, Han Li & Huijian Zhuang
State Key Laboratory of Rail Traffic Control and Safety, Beijing Jiaotong University, Beijing, China

ABSTRACT: Metric Temporal Logic (MTL) is a time-extension logic of Linear Temporal Logic (LTL), which has a more expression on monitoring real-time system. The program execution traces can be monitored by MTL efficiently. At the same time, MTL can reduce the complexity of the monitoring process. In this paper, we first describe the syntax and semantics of MTL on finite time-traces. Then, we present this monitoring algorithm on the basis of the rewriting logic of formula. Using our algorithm, we test an example of Vehicle On-Board Controller (VOBC) from the Chinese Train Control System (CTCS).

1 INTRODUCTION

Runtime verification (Clarke, 1999; Leucker, 2009; Barringer, 2009; DAmorim, 2005; Alur, 1994) is a simple dynamic technique of formal verification, which detects whether the current running traces of the system satisfy the desired properties, to judge the security situation of the system's operation. On the one hand, this method does not need to establish a system model and hence avoids the introduction of additional errors and the possibility of space explosion to the process of the complex system modeling. On the other hand, this method can monitor the whole system lifecycle state to avoid the test's incompleteness. Therefore, runtime verification plays an important role in guaranteeing the security of train control system.

The monitor's construct is the core of runtime verification. The monitor can be a piece of code or a hardware device, which is used to observe the running traces of the monitored system and to verify whether the traces satisfy the desired properties. The Linear Temporal Logic (LTL) (Havelund, 2002; Havelund, 2004; Bauer, 2007) is used to describe the system properties generally. The automata (Gastin, 2001; Bauer, 2007) and the formula rewriting (Havelund, 2001; Rosu, 2004) are two runtime verification methods based on the LTL to construct monitor. The former constructs monitor by transforming the LTL formula to automata (such as Büchi automata). The latter builds a rewriting rule to verify the execution traces based on the LTL.

The train control system is a typical real-time system with several time constraint properties. For example, the on-board equipment needs to connect with the RBC within 40 s after disconnecting. In this case, the monitor needs to monitor the temporal logic relation and the time relation among the system actions. Nevertheless, the LTL cannot describe the above property and hence cannot satisfy the monitoring needs of the train control system. In order to solve this problem, the Metric Temporal Logic (MTL) (Thati, 2005; Basin, 2012; Ouaknine, 2005; Ouaknine, 2008; Koymans, 1990) based on the LTL could be introduced to describe time-relevant properties.

The MTL is a real-time logic that adds time constraint properties to temporal operators on the basis of the LTL. Its model, which has a strong ability to express the time-related properties of the real-time system, is time-stamped state sequence, and can address the linear model inspection issue too. Therefore, the MTL can check the real-time system properties.

The Time Propositional Temporal Logic (TPTL) (Chai, 2013) is another extension mode with the storage of quantifiers based on the LTL, while it has a higher computational complexity. The Real-Time LTL (RTLTL) (Alur, 1990) adds temporal logic to the time operator, but has a fixed unit interval in execution trace, so RTLTL is inconvenient in verifying the real-time property. Furthermore, the typical runtime monitoring results have only two values (true and false), while the reference (Bauer, 2006) proposed a new LTL syntax and semantics based on three-value logic, which adds the inconclusive logical value to the typical logical value, and can improve the verification accuracy.

In order to improve the ability to express the runtime verification, this paper uses the verification method of the MTL to check time constraint properties, which is based on the real-time property of the verification system. This paper describes the syntax and semantics of the MTL on the basis of linear temporal logic and constructs the monitor (including the MTL formulas and formula rewriting algorithm) based on rewriting logic. The monitor

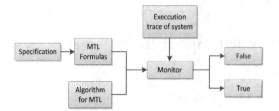

Figure 1. Verification of the runtime system.

checks the execution traces of the real-time system and determines whether the traces are satisfied with the given formulas by referring to CTCS-3 train control system overall technical proposal. Figure 1 shows the overall monitoring schematic.

At the end of this paper, combining the train control system, taking vehicle on-board control as an example, we demonstrate the practicability of this verification algorithm.

2 METRIC TEMPORAL LOGIC

By monitoring the execution traces of real-time system, runtime verification judges whether the running of the system is safety. The focus of this method is the finite prefix of infinite sequence. First, we present the definition of MTL based on finite trace.

Given a finite set of atomic propositions AP and a finite alphabet $\sum = 2^{AP}$, the formulas of MTL are defined as follows.

Definition 1. Syntax for MTL

$$\varphi := true \,|\, p \,|\, \neg\varphi \,|\, \varphi_1 \wedge \varphi_2 \,|\, \varphi_1 \, U_{[t_1,t_2]} \, \varphi_2 \quad (1)$$

where $p \in AP$, $t_1, t_2 \in R^{\geq 0}$ and $t_1 \leq t_2$, $t_2 > 0$. Especially, $\varphi_1 U_{[t_1,t_2]} \varphi_2$ can be equivalent to $\varphi_1 U_{\leq t_2} \varphi_2$ when $t_1 = 0$.

There are some other MTL equivalent formulas:

$$\varphi_1 \vee \varphi_2 = \neg(\neg\varphi_1 \wedge \neg\varphi_2) \quad (2)$$

$$\varphi_1 \rightarrow \varphi_2 = \neg\varphi_1 \vee \varphi_2 \quad (3)$$

$$\Diamond_{[t_1,t_2]}\varphi = true \, U_{[t_1,t_2]}\varphi \quad (4)$$

$$\Box_{[t_1,t_2]}\varphi = \neg\left(true \, U_{[t_1,t_2]}(\neg\varphi)\right) \quad (5)$$

The logical operators $\neg$ (negation), $\vee$ (or), $\wedge$ (and), $\rightarrow$ (implication) are Boolean operators. And the temporal operators are introduced as follows:

$\Diamond_{[t_1,t_2]}$ (Future): The execution trace will satisfy the MTL formula at least with an event in the $[t_1, t_2]$ interval.

$\Box_{[t_1,t_2]}$ (Globally): The execution trace satisfies the MTL formula with all events in the $[t_1, t_2]$ interval.

$U_{[t_1,t_2]}$ (Until): The execution trace will satisfy the MTL formula with an event and always satisfies another formula before the event in the $[t_1, t_2]$ interval.

Given a finite set of atomic propositions AP, the time-event defined on AP is denoted by (s,t). A finite time-trace ρ is a sequence of time-events, that is, $\rho = (s_0, t_0)(s_1, t_1)...(s_n, t_n)$.

Definition 2. Semantics for MTL

Given the time-trace ρ, a nonnegative integer I, and any MTL formulas φ_1, φ_2, the satisfaction relation $(\rho, i) \vdash \varphi$ is defined as follows:

$(\rho, i) \vdash true$

$(\rho, i) \vdash p$ iff $p \in s_i$

$(\rho, i) \vdash \neg\varphi$ iff $(\rho, i) \nvdash \varphi$

$(\rho, i) \vdash \varphi_1 \wedge \varphi_2$ iff $(\rho, i) \vdash \varphi_1$ and $(\rho, i) \vdash \varphi_2$

$(\rho, i) \vdash \varphi_1 U_{[t_1,t_2]} \varphi_2$ iff for some $j \geq i : (\rho, j) \vdash \varphi_2$ and $t_j \in [t_i + t_1, t_i + t_2]$ with for all $i \leq k < j : (\rho, k) \vdash \varphi_1$.

The definition of judgment concept is that for any time-trace (ρ, i) and MTL formula φ, we can get the result about whether (ρ, i) satisfies φ in a reasonable time.

There are some examples about the MTL having a strong expression ability in monitoring temporal properties of real-time system.

1. Sustainability: $\Box_{[t_1,t_2]}p$ says that the event p always occurs in the $[t_1, t_2]$ interval.
2. Eventually happen: $\Diamond_{[t_1,t_2]}p$ says that the event p will occur in the $[t_1, t_2]$ interval.
3. Bounded response property: $\Box(p \rightarrow (\Diamond_{[t_1,t_2]}q))$ says if the event p has occurred, the event q must occur in the $[t_1, t_2]$ interval.
4. Frequency limitation: $\Box(p \rightarrow (\Box_{\leq t_2} \neg p))$ says that two of the same event not less than the time interval t_2.

The atomic propositions p and q in the above examples can be any formulas of MTL. From the above description, we can get the result that MTL can describe better in the time constraints temporal properties of real-time system.

3 REWRITING THE ALGORITHM FOR MTL

3.1 Fundamentals of the algorithm

In runtime verification, the rewriting temporal logic is used to monitor whether the execution traces of system satisfy the given properties. The main idea of formula rewriting is to eliminate the first event about the system execution trace by using the rewriting rules and rewrite the formula, until all the events in the trace have been eliminated.

Then, we can get the verification verdict. For instance, the time-trace $\rho = (s_0, t_0)(s_1, t_1)\ldots(s_n, t_n)$ and the MTL formula φ, φ will be combined with the first event (s_0, t_0) and consume it. Then, we get a new formula φ' and a new time-trace $\rho' = (s_1, t_1)\ldots\ldots(s_n, t_n)$. After that, φ' will be combined with the time-trace $\rho' = (s_1, t_1)\ldots\ldots(s_n, t_n)$, consume the (s_1, t_1) event, and so on, until all the events have been consumed.

In general, there are two types of rewriting algorithm method on the formula rewriting. The first method is a simple rewriting, that is, the result formula of each rewriting will be applied directly to the next rewrite. However, when dealing with a long trace, the execution efficiency will be decreased significantly. The second method is called event consumption rewriting, that is, the result formula of each rewriting will be applied to the next rewriting after simplification. This method can greatly improve the efficiency of monitoring. The monitoring algorithm presented in this paper is based on the second method.

According to the description about semantics of MTL in the above section, we can develop the operation monitoring algorithm based on efficient rewriting. First, given the basic data types of this algorithm, *Int, Atom, Event, TimeWord, Trace,* and *Formula, Int* represents integer, that is, the time constraint of MTL formulas, *Atom* is the atomic proposition, *Event* means the event or the state, *TimeWord* is composed of event and time, *Trace* is the sequence of the system, and *Formula* is the MTL formula. In addition, there is another Boolean logic, that is, *true* and *false*.

The process of formula rewriting is that MTL formula consumes the first state of the trace and is rewritten to a new formula. In order to describe this process, we define _{_}(operator), that is, *Formula × TimeWord → Formula*. We take $A\{A'E : ts\}$ as an example to describe the specific meaning, A and A' are atomic propositions, and $A'E : ts$ is an event containing A'. First, we determine whether $A = A'$, if $A = A'$, it returns true; otherwise, it consumes A' and returns $A\{E : ts\}$.

Now given the relational operator $\_\vdash\_$ and the realization of the algorithm:

$$E : ts, T \vdash \varphi = T \vdash \varphi\{E : ts\} \tag{6}$$

where $E : ts, T$ is a trace with first event $E : ts$. The meaning of this formula is if the trace $E : ts, T$ satisfies the formula φ, then the remaining trace T satisfies the formula $\varphi\{E : ts\}$. This process will continue until the trace is consumed.

In addition, the operator priority of algorithm is presented by [prec]; the smaller is the number after [prec] the higher of priority. Table 1 shows the basic data types and operations.

Table 1. Basic data types and operations.

Variables
$A, A' : Atom$
$E : Event$
$ts : Int$
$TW : TimeWord$
$T : Trace$
Operation
$\_\{\_\} : Formula \times TimeWord \to Formula$
Case
$true\{TW\}$: **return** true
$false\{TW\}$: **return** false
$A\{nil\}$: **return** false
$A\{TW, T\}$: **return** $A\{TW\}$
$A\{A' : ts\}$: if $A = A'$ **return** true **else return** false
$A\{A' E : ts\}$: if $A = A'$ **return** true **else return** $A\{E : ts\}$

3.2 Rewriting the algorithm in Maude

Subsequently, we develop an algorithm for checking whether the traces satisfy the MTL formulas. Our algorithm is based on Maude (Clavel, 2002; Clavel, 2007), which is a high-performance system for model checking. It is convenient in runtime verification implementation of Maude support *system module* (mod) and *functional module* (fmod). In this algorithm, we use the functional module, whose pattern is as follows:

$$fmod \langle name \rangle is \langle body \rangle endfm \tag{7}$$

where *fmod*, *is* and *endfm* are the keyword of Maude. $\langle name \rangle$ is the module name defined by the user. $\langle body \rangle$ is the main part of the module.

If we want to use the algorithm in Maude, we need to define the necessary data types, which we have described in the above sections, including *Atom, Trace,* and so on. Now we take the global temporal operator (□) as an example and give the rewriting algorithm directly, which is the core of this algorithm. The rewriting algorithm is shown in Table 2.

The first line gives the definition of the global temporal operator. In order to describe the time constraint, we introduce some related operators like subtraction and conditional equation (ceq). The rest of the algorithm describes the rewriting rule of global temporal operator, where n, m, postion1, and postion2 are natural numbers; X and Y are formulas; T is a trace; E1 and E2 are events; TW is a TimeWord; and A and A' are atomic propositions.

We described the syntax and semantics of MTL, rewriting rules, and the implementation of this algorithm in Maude. Now we can verify the real-time constraint properties in Maude. Table 3 shows the operating monitor about some basic formulas in Maude.

Table 2. Rewriting the algorithm for global temporal operator.

Fmod MTL is
op []___ : Int Int Formula -> Formula [prec 11] .
ceq ([] n m X) = false if (m < 0) .
ceq ([] n m X) = ([] 0 m X) if (m > = 0) ∧ (n < 0) .
eq [] 0 0 X = X .
ceq (E1: postion1, E2: postion2, T) |- ([] n m X) = (E2: postion2, T) |- [] (n - (postion2 - postion1)) (m - (postion2 - postion1)) X
if (n > 0) and not ((postion2 - postion1) > m) .
ceq (E1: postion1, E2: postion2, T) |- ([] n m X) = false
if (n > 0) and ((postion2 - postion1) > m) .
ceq (E1: postion1, E2: postion2, T) |- ([] n m X) = (E1: postion1, E2: postion2, T) |- X and (E2: postion2, T) |- [] (n - (postion2 - postion1)) (m - (postion2 - postion1)) X
if (n = = 0) and not ((postion2 - postion1) > m) .
ceq (E1: postion1, E2: postion2, T) |- ([] n m X) = (E1: postion1, E2: postion2, T) |- X
if (n = = 0) and ((postion2 - postion1) > m) .
ceq (E1: postion1, E2: postion2) |- ([] n m X) = E1: postion1 |- X and E2: postion2 |- X
if (n = = 0) and ((postion2 - postion1) < = m) .
ceq (E1: postion1, E2: postion2) |- ([] n m X) = E1: postion1 |- X
if (n = = 0) and ((postion2 - postion1) > m) .
ceq (E1: postion1, E2: postion2) |- ([] n m X) = E2: postion2 |- X
if (n > 0) and ((postion2 - postion1) < = m) and ((postion2 - postion1) > = n) .
ceq (E1: postion1, E2: postion2) |- ([] n m X) = false
if (n > 0) and ((postion2 - postion1) > m) or ((postion2 - postion1) < n) .
ceq (E1: postion1) |- ([] n m X) = false
if (n > 0) .
ceq (E1: postion1) |- ([] n m X) = (E1: postion1) |- X
if (n = = 0) .
endfm

Table 3. Result of monitoring.

Formula result	Trace
□ 0 1 2 a	a: 0, a b: 3, b c: 6, b d: 9, a b e: 15 false
◇ 2 9 (b ∧ c)	a: 0, a b: 3, b c: 6, b d: 9, a b e: 15 true
a U 6 1 0 (d ∧ e)	a: 0, a b: 3, b c: 6, b d: 9, a b e: 15 false

---

4 CASE: VEHICLE ON-BOARD CONTROL

Vehicle On-Board Control (VOBC) is one of the most important equipment in CBTC, with all safety-related operations completed by VOBC. Therefore, realizing VOBC monitoring is crucial for monitoring the CBTC system. VOBC contains Automatic Train Protection (ATP), Automatic Train Operation (ATO), and some other subsystems. ATP provides operational safety of the train and driving information for ATO, and ATO achieves automatic driving of the train.

The functions of the VOBC system are described as follows:

1. It measures the speed of the train by using speed sensors, and according to the initial position and the acceleration of the train, it calculates the real-time location.
2. It provides continuous communication for trackside and center equipment by equipping with a two-way radio communication system.
3. It receives the Moving Authority (MA) from ZC and simultaneously sends the train's status information to ZC.
4. According to the operation mode of the train, it provides the recommended speed to the train and controls the state of the door.

Obviously, VOBC is a system with complex functions. Thus, it has a great significance for CBTC system to monitor it. In this paper, we apply our MTL runtime verification implementation to the concrete example of VOBC.

We consider some properties as follows:

Property 1: When outputting the braking command, the tractive command must last at least 40 s.

Property 2: When the running speed is greater than the recommended speed, it must be output the braking command within 60 s.

Property 3: After VOBC command, it must receive the feedback command from the train within 50 s.

We can describe the above properties using MTL formulas (the meanings of atomic propositions are shown in Table 4):

$$\varphi_1 = \square\, 0\, 40\, (TC \wedge \neg BC) \tag{8}$$

$$\varphi_2 = \square (OS \rightarrow (\diamond\, 0\, 60\, BC)) \tag{9}$$

$$\varphi_3 = \square (VOC \rightarrow (\diamond\, 0\, 50\, FC)) \tag{10}$$

The execution trace is extracted from the real-time system in the monitoring of the real system. However, for convenience, we construct some traces to test and verify the feasibility of our algorithm:

t_1 = VOC: 0, TC: 18, VOC: 29, FC: 42, TC: 68, FC: 79, BC: 94, OS: 124, VOC: 145, BC: 166, FC:

Table 4. Definitions of some atomic propositions.

Atomic propositions	Definition
TC	Tractive command
BC	Braking command
OS	Overspeed
VOC	Command from VOBC
FC	Feedback command from the train

190, VOC: 200, TC: 218, VOC: 229, FC: 242, TC: 268, FC: 279, BC: 294, OS: 324, VOC: 345, BC: 366, FC: 390, VOC: 400, TC: 418, VOC: 429, FC: 442, TC: 468, FC: 479, BC: 494, OS: 524, VOC: 545, BC: 566, FC: 590, VOC: 600, TC: 618, VOC: 629, FC: 642, TC: 668, FC: 679, BC: 694, OS: 724, VOC: 745, BC: 766, FC: 790

The results of $t_1 \vdash \varphi_1$, $t_1 \vdash \varphi_2$, and $t_1 \vdash \varphi_3$ in Maude are all true. It means that this execution trace satisfies all the properties. In order to verify that the algorithm can detect the error, we introduce some errors in the trace:

t_1 = VOC: 0, TC: 18, VOC: 29, FC: 42, TC: 68, FC: 79, BC: 94, OS: 124, VOC: 145, BC: 166, FC: 190, VOC: 200, TC: 218, VOC: 229, FC: 242, BC: 252, FC: 279, BC: 294, OS: 324, VOC: 345, BC: 366, FC: 390, VOC: 400, TC: 418, VOC: 429, FC: 442, TC: 468, FC: 479, BC: 494, OS: 524, VOC: 545, BC: 588, FC: 590, VOC: 600, TC: 618, VOC: 629, FC: 642, TC: 668, FC: 679, BC: 694, OS: 724, VOC: 745, BC: 766, FC: 800

The results of $t_1 \vdash \varphi_1$, $t_1 \vdash \varphi_2$, and $t_1 \vdash \varphi_3$ in Maude are all false. It means that this execution trace violates the properties. The above example shows the validity and feasibility of our runtime verification algorithm.

5 CONCLUSION

In this paper, we proposed a runtime verification method of MTL. We presented the implementation of the running monitoring algorithm based on formula rewriting, and verified this algorithm in Maude. We showed that this algorithm greatly improves the efficiency of verification in monitoring a long time-trace and hence it can be used in a complex system. Furthermore, we have presented a case about VOBC from the Chinese Train Control System. The results prove the validity and feasibility of our runtime verification algorithm.

ACKNOWLEDGEMENTS

This work was partially supported by the National Basic Research Program of China (no. 2014CB340703), the Foundation of State Key Laboratory of Rail Traffic Control and Safety (no. RCS2015ZT002), and the Fundamental Research Funds for the Central Universities (no. 2015 JBM113).

REFERENCES

Alur R., T, A. Henzinger. A Really Temporal Logic. Journal of the ACM (JACM) 1994, 41, 181–203.

Alur R., T, A. Henzinger. Real-Time Logics: Complexity and Expressiveness, [J]. Information and Computation, 1990, 104(1): 390–401.

Barringer H., K. Havelund, D. Rydeheard. A Groce. Rule Systems for Runtime Verification [C]: A Short Tutorial. In: Runtime Verification, Springer, 2009, 5779: 1–24.

Basin D., F. Klaedtke and E. Zalinescu. Algorithms for Monitoring Real-time Properties [C]. International Conference on Runtime Verification, 2012, 7186: 260–275.

Bauer A., M. Leucker and C. Schallhart. Runtime Verification for LTL and TLTL. Technical Report TUM-10724. Technische Universitat Munchen, 2007.

Bauer A., M. Leucker and C. Schallhart. The Good,the Bad,andthe Ugly-But How Ugly is Ugly [C]. Proceedings of International Workshop on Runtime Verification, 2007: 126–138.

Bauer A., M. Leucker. Model-Based Runtime Analysis of Distributed Reactive Systems [C]. Australian Software Engineering Conference, 2006: 243–252.

Chai M., H. Schlingloff. A Rewriting Based Monitoring Algorithm for TPTL [J]. Informatik Hu.[J],2013.

Clarke E., O. Grumberg, D. Peled. Model Checking [M]. MIT Press, 1999.

Clavel M., F. Duran and S. Eker. All About Maude-A High-Performance Logical Framework [M]. Berlin: Springer, 2007.

Clavel M., F. Eker, S. Lincoln, P. Marti-Oliet, N. Meseguer, J. Quesada, J.F. Maude: Specification and Programming in Rewriting Logic [C]. Theoretical Computer Science 285, 187–243, 2002.

DAmorim M., G. Rosu. Efficient Monitoring of Omega-Languages [C]. Proceedings of International Conference on Computer Aided Verification, 2005: 364–378.

Gastin P., D. Oddoux. Fast LTL to Büchi Automata Translation [C]. Computer Aided Verification, Springer, 2001, pp. 53–65.

Havelund K., G. Rosu. Efficient monitoring of safety properties [J]. J. Soflw. Tools Techn01. Transfer (STTT) 2004.

Havelund K., G. Rosu. Monitoring Programs Using Rewriting [C]. Proceedings of International Conference on Automated Software Engineering, 2001: 135–143.

Havelund K., G. Rosu. Synthesizing monitors for safety properties [J]. Tools and Algorithms for Construction and Analysis of Systems, 2002: 342–356.

Koymans R. Specifying Real-Time Properties with Metric Temporal Logic. Real-Time Systems, 1990, 2(4): 255–299.

Leucker M., C. Schallhart. A Brief Account of Runtime Verification [J]. Journal of Logic and Algebraic Programming, 2009, 78(5): 293–303.

Ouaknine J., J. Worrell. On the Decidability of Metric Temporal Logic [C]. Proceedings-Symposium on Logic in Computer Science, 2005: 188–197.

Ouaknine J., J. Worrell. Some Recent Results in Metric Temporal Logic [C]. Formal Modeling & Analysis of Timed Systems, Internatinal Conference, Formats, Sanit Malo, France. 2008, 5215: 1–13.

Rosu G., G. Havelund. Rewriting-Based Techniques for Runtime Verification [J]. Journal of Automated Software Engineering, 2004, 12(2): 151–197.

Thati P., G. Rosu. Monitoring Algorithms for Metric Temporal Logic Specifications [C]. Electronic Notes in Theoretical Computer Science, 2005, 113: 145–162.

Advances in Energy Science and Equipment Engineering II – Zhou, Patty & Chen (Eds)
© 2017 Taylor & Francis Group, London, ISBN 978-1-138-71798-5

Review of studies on big data

Wenjing Cai, Dong Wang & Guanghui Xue
China University of Mining and Technology, Beijing, China

ABSTRACT: In the recent years, the concept of "big data" has attracted much research interest. It is used to describe and definite tremendous amount of data created during information explosion. Gradually, big data begins to affect every industry, which marks the coming of big data time. In this paper, the essential conception and typical characters are introduced. The general steps of big data processing and some key technologies, including Hadoop and MapReduce, are analyzed. Finally, some challenges faced by us in big data time are summarized.

1 INTRODUCTION

Along with the economic globalization and global informatization, social development and human demands have been diversified, and cloud computing and IOT (Internet of Things) have been used frequently. Therefore, "big data" begins to attract more and more attention of common people, academia, industries, and even government agencies.

A special issue of *Nature*, "Big Data," was punished in 2008, which introduced the challenges brought by tremendous amount of data from Internet technology, Internet economy, supercomputing, environmental science, biological medicine, and other aspects (Bigdata, 2008). And in 2011, a special column named "Dealing with data" was published on *Science*, which analyzed the opportunities brought by Datadeluge. The United Nations also issued a white book, "Challenges and Opportunities with Big Data"(Agrawal, 2012), which introduced the birth of big data from the aspect of academy and analyzed the steps of big data processing as well as challenges faced by people in big data time. In March 2012, the Obama Government issued "Big data research and development initiative," and initiated "Plan of big data development." China put forward the guiding principle to set up the infrastructure of big data management. *The 12th Five-Year-Plan for National Economic and Social Development* pointed out: "China must focus on the study of theories and ways to process and find tremendous amount of information…" In 2013, Ministry of Science and Technology has officially initiated the program of 863-"targeting for advanced storage structure and key technology of big data" and started five projects of big data.

Therefore, the development trend of big data cannot be stopped. And it has become an important project for the whole world to realize objectively and take the best use of functions of big data and to push key industries, academia, and other industries to study and apply big data technology. All of these works have strategic importance.

2 DEFINITION OF BIG DATA

The definition of big data put forward by Mckinsey is: Big data is a collection of data, beyond other traditional databases in getting, saving, managing, and analyzing information. And not all big data is beyond a specific terabyte (Manyika, 2011).

According to the report of IDC published in 2011, big data is defined as that describing a new time of technology and system, and it is designed to get the value of data through its technology of getting, finding, and analyzing with high speed (Li, 2015).

The definition of big data from Wikimedia is: "Tremendous amount of information, which is also called big data, means the amount is too gigantic that common software and tools cannot get, manage, handle and clear up in affordable time, so as to help entrepreneurs to make a decision."

3 FEATURES OF BIG DATA

3.1 *Volume (big volume)*

The scale of big data is tremendous, which is about 10TB. However, in the real life, several big data are always used together, and the scale has to be measured in PB. That is, the scale increases from TB to PB, or even EB (1TB = 1024GB, 1PB = 1024TB).

Finally, only big data can deal with the tremendous amount of data effectively.

3.2 *Variety (a lot of varieties)*

Along with the rapid development of smart equipment, social communication technology, and sensor, the types and formats of data become richer and richer. The source of data also breaks up the former limited structured data, including Internet logs, RFID sensors, nonstructured social Internet data, and streaming video and audio. At the same time, the most important sources of many big data are relatively new, and the formats of data are various.

3.3 *Velocity (high velocity)*

The processing of big data follows the rule of one-second, which means acquiring the analysis results in several seconds. The speed of processing big data is high. As for the high speed, there are two meanings. One is the frequency of data generation, such as 1 ms, 1 min, 1 month, or 1 year, is fast. The other one is the frequency of data processing is fast. Some data have to be processed at once, whereas some can be processed when needed. It is also the most obvious difference of finding information between big data and traditional data that the former can acquire more valuable information in different types of data.

3.4 *Value (high authenticity and low density of the value of data)*

New source of data generates, and the limit of traditional source of data is broken up. Therefore, more effective information is needed to ensure the authenticity of the source of data. The density of the value of data is inversely proportional to the amount of data. Along with the increase of the amount of data, the amount of effective data does not increase. For example, as for a video surveillance with 1 h, only data with 1 or 2 s is effective.

To some extent, big data is the cutting-edge technology of data analysis. The four essential features of big data show that the data collected with high difficulty can be used easily. Through constant innovation, big data will create more and more value.

4 STEPS OF BIG DATA PROCESSING

There is not obvious difference of basic processing steps between big data and traditional data. The main difference is that big data needs to process a great amount of half-structured or nonstructured data. And there are four steps in processing data: data acquisition, data preprocessing, statistics and analysis, and data mining.

4.1 *Data acquisition*

Through RFID radio frequency data, sensor data, social network interaction data, and mobile Internet data, big data can acquire a great amount of structured, half-structured, and nonstructured data. The acquisition of data needs tremendous database, and sometimes, many databases will collect big data at the same time, which is the main feature during the process of data acquisition. A large number of concurrent and common methods of data acquisition are available, including sensor, log file, and Web crawler (Li, 2015).

4.2 *Data preprocessing*

In order to carry out effective analysis of massive amount of data, the data from many databases have to be imported to a large data path and chosen roughly. Some clear-up work is also needed on the basis of importation. This is called data processing. During data preprocessing, works such as extraction and clear-up are needed. Through extraction, multistructured or multityped data can be transmitted to simple structure or single type. And as not all big data is valuable, clear-up is needed to remove the identifiable error of data.

4.3 *Statistics and analysis*

Some tools of big data analysis can be used to classify different types of data according to their features, laying a basis for next processing. In general, DDB (Distributed Data Base) is used to analyze massive data and then summarize the classification.

4.4 *Data mining*

Data mining means extracting implicit, potential, and useful information and knowledge, which people do not know from large, incomplete, noisy, fuzzy, random, and practical data. Compared with statistics and analysis, there is no prearranged theme in data mining. On the basis of current data, computing is made and the data analyzing of higher class is realized (Yan, 2013).

5 RELATIVE TECHNOLOGIES OF BIG DATA

Cloud computing is the core principle of analysis and processing technology of big data as well as the basis of the application and analysis of big data (Chen, 2009). As for Google, a lot of technologies of big data processing and application platform are based on cloud-computing. The most typical two are the technology of big data processing with representatives of GFS—distributed file system, MapReduce—batch processing technology, and BigTable-DDB and Hadoop—open-source data processing platform, which is established on the basis of the technology of big data processing (Liu, 2014).

5.1 *MapReduce*

MapReduce is a programming model for parallel computing of large-scale data sets. Through assigning the amount of computation to different computer groups, MapReduce can deal with most analysis problems relating to big data. In general, MapReduce is used to finish extraction and retrieval of data. As for MapReduce, two typical functions, map function and reduce function, are combined. Map means operating every target in the collection and Reduce means traversal of the elements in the collection so as to produce comprehensive result. Massive data are divided into several parts, which are processed in different processors. And then the processing results are collected. Now MapReduce has become a main way of parallel data processing.

5.2 *Distributed file system*

Distributed system means several computers are networked and work collaboratively, just like a single computer dealing with some problems. Distributed file system is a subset of distributed systems, which is an inventory system in many computers. And the data on distributed file system will be distributed in different panel points. There are many distributed file systems, including GFS, HDFS, TFS, and Facebook Haystack. GFS is a system saving massive data designed by Google in early years. HDFS can be seemed as a simplified edition.

5.3 *Distributed Database (DDB)*

When processing massive data, relational database has been gradually given up. There are some limits of relational database, such as stiffness of data model and the lack of capability to process half-structured and nonstructured data. Nosql-non-relational database has gradually become a good method in big data. For example, BigTable, which is put forward by Google, is a distributed system, which is designed for managing large-scale structured data. And it can be expanded to the data of PB-class and thousands of servers (Chang, 2008).

5.4 *Hadoop*

Hadoop is an open-source and parallel-computing programming tool and distributed file system. It realizes the open source of MapReduce, becoming the first choice for big data processing because of its features of open source and easy to use (Tao, 2013).

The most common core designs of Hadoop are HDFS and MapReduce. HDFS enables the saving of massive data and MapReduce realizes computing of the data.

Hadoop Framework and MapReduce, the core technologies of cloud computing, are used to realize computing and saving of data, and HDFS-distributed file system and HBase-distributed database can be worked effectively in cloud computing framework. Therefore, distributed computing and saving and parallel computing and saving of cloud computing can be realized. Furthermore, the capability to process tremendous amount of data can be improved.

6 CHALLENGES FACED BY BIG DATA

At present, big data has developed rapidly and has been used in many fields. However, it still faces many challenges. Many of the challenges are brought by big data itself.

Although big data analysis provides us with tools to extract and use these data, security of data and problems of privacy also arise. Therefore, it has become an important challenge faced by big data to protect privacy of people and entrepreneurs and strengthen Internet security at the same time of acquiring valuable data and apply and share the benefits of big data.

Furthermore, there are some difficulties and challenges in the application of big data because of variety of data and complexity of data structure and model. The main technical problems are redundancy and noise reduction of big data; saving of big data with high efficiency and low cost; mining, analyzing tools, and developing environment of big data, which are adapted to different industries; and some new technologies of reducing energy consumption of data processing, saving, and communication. All of these problems are difficult to deal with.

Every circles of big data construction has to depend on professional peoples. Therefore, talents of big data analysis, management, and technical support are needed very much. Only the professional talents in big data are equipped with the skill of developing the application model of prediction analysis. At the same time, the lack of talents also hinders the development of big data market.

7 CONCLUSION

This paper expounds the basic concept of big data and provides a detailed introduction of its four characteristics: huge amount of data, various kinds of data, high processing speed, and high data reliability and low density. This paper also summarizes the general processing of big data, including data collection, pretreatment, statistics, and analysis as well as data mining. The key technologies applied in big data processing, such as Hadoop platform, distributed database, and MapReduce, are introduced briefly. However, current study on big data is still plain, and many basic problems have to be addressed. If these basic problems can be addressed properly and big data can be made full use of, more value will be created, which should be our vision.

REFERENCES

Agrawal, Divyakant, et al. "Challenges and Opportunities with Big Data. A community white paper developed by leading researchers across the United States."*Computing Research Association, Washington* (2012).

Bigdata, Nature, **455**(7209): 1–136 (2008).

Chang F, Dean J, Ghemawat S, et al. Bigtable: A distributed storage system for structured data, ACM Transactions on Computer Systems (TOCS), **26**(2):4(2008).

Chen, K., W.M. Zheng, Cloud Computing: System Example and Research Status, Journal of Software, **1**(20)(2009).

Dealing with data, Science, **3331**(6018): 639–806 (2011).

House, White. "Big data across the federal government."http://www.whitehouse.gov/sites/default/files/microsites/ostp/big_data_fact_sheet_final. Pdf

Li, X.L., H.G. Gong, Systematic Review of Big Data. China's Science: Information Science, **1**:1–44(2015).

Liu, Z.H., A.L. Zhang. Review of Studies on Big Data, Journal of Zhejiang University (Engineering Science Edition), **48**(6): 957–972(2014).

Manyika, James, et al. "Big data: The next frontier for innovation, competition, and productivity." (2011).

Tao, X.J., X.F. Hu, L. Liu. Review of Studies on Big Data, Journal of system simulation, **8**:(2013).

Tianjin University, 863 Program"Advanced storage structure and key technology for big data", kick—off Meeting.http:\\cs.tju.edu.cn/xwzx/xwdt/20130401103608433ETH.shtml

Yan, X.F., D.X. Zhang, Study on Big Data. Computer Science and its Development, **4**(32):4(2013).

Advances in Energy Science and Equipment Engineering II – Zhou, Patty & Chen (Eds)
© 2017 Taylor & Francis Group, London, ISBN 978-1-138-71798-5

Micro-blog short text and its processing technology: A review

Qi Fu
Jiangxi Science and Technology Normal University, NanChang, China

Jun Tan
Jiangxi University of Finance and Economics, NanChang, China

ABSTRACT: On the basis of the review of the extensive literature, this paper summarizes and classifies the concept, characteristics, and research status of micro-blog short text. Furthermore, it points out the limitations of the current related research and further improvements. Finally, the future research direction of micro-blog short text is prospected.

1 INTRODUCTION

Short text is an important product for the core of the social networking service era with client relationship. It comes mainly from media such as social networks, mobile terminal network, and instant communication tools, and has become indispensable in people's communication and information acquisition. Moreover, it has a great impact on people's production and life. Micro-blog, as a new social media network, is one of the most popular social applications as well as an important place for the short text. However, the update frequency and explosive growth trend of micro-blog short text not only increases the difficulty of micro-blog short text processing and research, but also causes great trouble to the users.

In recent years, there have been more and more studies on micro-blog short text in important journals and conferences worldwide. At present, the research on micro-blog short text mainly has two aspects: one is the processing technique of micro-blog short text, that is to say, the original data processing method of compression, screening, representation, and selection of a series of process, providing a good environment for the further research of micro-blog data short text learning and application. The other is by analyzing the characteristics of the micro-blog data to construct, classify, cluster, and use other learning model, in order to fully find the internal relations between the short text content, to help users find the potential rules hidden in the micro-blog data or resolve the practical problems in life, such as hot topic discovery, opinion leaders identification, network content monitoring, and detection of negative network public opinion. On the basis of the literature review of micro-blog short text, this paper

summarizes the research status of short text and hot topic of micro-blog and illustrated the existing problems in the current research of micro-blog short text to further explore the prospects.

2 CONCEPT AND CHARACTERISTICS OF MICRO-BLOG SHORT TEXT

Short text is referred to a short form of text generated by the users in the network, such as in micro-blog and Twitter, whose size is generally around 140 words. Micro-blog short text is different from other forms of short text in the following aspects: it has huge data size, contains a lot of hidden information in high value, and has some distinctive characteristics (Jing, 2012; Ellen, 2011).

1. The micro-blog text length is short, the information content is small, but the quantity is huge. These features result in severe data scarcity and high dimensionality of feature space.
2. Micro-blog short text has the characteristics of real-time update and dynamic change. Users often produce a large amount of short text in the network interaction, and the increase in the amount of data makes the short text preprocessing and learning process more complicated.
3. The micro-blog short text is not normative, its grammar is usually informal, and its language tends to colloquial and life, often with abbreviations, spelling errors, nonstandard terminology, noise, and expression symbols, increasing the difficulty of the user's understanding of the information and the incident.
4. Micro-blog short text often contains significant personal intention and obvious individualism emotional attitude, easy-to-generate new

themes in the current theme, forms theme drift or cross, and presents a long tail phenomenon, resulting in a serious imbalance in the distribution of data.

5. Micro-blog short text has semi-structured characteristics. In addition to the text content, it contains some metadata, such as the release time, the quantity of collection, forwarding quantity, comments, and other information. Compared with other traditional media short text, it will be more conducive to research.

3 MICRO-BLOG SHORT TEXT PROCESSING

There are many text processing technologies related to micro-blog short text. Here, we mainly introduce three typical topics: short text classification and clustering, information extraction, and topic detection.

3.1 Short text classification and clustering

The so-called micro-blog short text classification and clustering is based on the different themes of micro-blog. It will describe a class of topics micro-blog short text together to facilitate the user to read and refer. However, because of less word count of the micro-blog short text, as opposing to plain text, in the time of using machine learning methods for classification or clustering, it often produces serious data sparse problem thereby affecting the performance.

Thus, researchers have tried to solve the problem of sparse data. For example, Sriram showed that the characteristics of micro-blog short text were different from those of the ordinary text, and selected eight types of feature: author's information, publishing time, signs, and so on (Sriram, 2010). After the addition of these features, the classification performance was improved significantly, and the problem of sparse data was improved. And Liu suggested selection of the vocabulary of rich part of speech in micro-blog text as the initial feature during feature selection to consider part of speech and then adopted the HowNet semantic knowledge base. These words would be extended to the semantically related words, so as to achieve feature expansion. Eventually, it overcame sparse problem of the data in blog text (Liu, 2010). Experiments showed that the method was improved to a certain extent.

In addition, there are some research methods not limited to the micro-blog short text feature selection, but to the use of certain phenomena in the data to improve the effect of classification or clustering. For example, through experimental analysis, Peng Ze-ying found the micro-blog data categories with the phenomenon of long tail and thus put forward incomplete clustering information. It could effectively improve the information clustering performance (Peng, 2011). According to the social relationship of micro-blog users, A. Churchill clustered the user first and then combined with the Bayes classification algorithm, using the results of user clustering to improve the classification performance (Churchill, 2010). Similarly, M. Yoshida classified micro-blog's query word first, and then classified again according to the results of the query on the micro-blog search (Yoshida, 2010), and also achieved some improvement.

These studies are just a part of the characteristics and phenomena of micro-blog's text. If we can use much micro-blog text features as possible, it will be much possible to extract features in the process of feature selection, so as to improve the effect of micro-blog text classification or clustering.

3.2 Information extraction

Information extraction of micro-blog text is similar to the general text of the information extraction. That is, because of the shorter length of the micro-blog text, in the time of processing, it usually clusters the same topic of a number of micro-blog texts together first and then extracts the information needed by the user. For example, B. Sharifi proposed a summary of the method from a topic related to a number of micro-blog texts in the automatic extraction first (Sharifi, 2010). Phrase reinforcement algorithm is used in this paper to find phrases with the highest number of occurrences. Then, B. Sharifi applied the above method to the Twitter website and tapped into the specific domain of micro-blog resources (Sharifi, 2010). The experimental results showed that the effect of the system was very similar to that of the artificial intelligence.

Moreover, information extraction of micro-blog text is also related to natural language processing technology. For example, S.Petrovi'c integrated the event detection technology into micro-blog's text, and the processing speed of the method proposed by the article is better than most of the current event detection systems (Petrovi'c, 2010). With the help of user behavior characteristics, Sakaki made real-time monitoring of the network micro-blog text so as to let the user understand the recent occurrence of hot events at the first time (Sakaki, 2010). Experimental results showed that the real-time event detection system was effective. W.Zhao carried on the related keyword extraction to the Twitter text (Zhao, 2011). They proposed an algorithm of PageRank-based context, according to the correlation degree and sorting the keywords of the related topic, and finally extracted keywords. Of course, micro-blog itself has a lot of interesting information, which can be extracted. It needs to be further explored and thought by researchers.

3.3 *Topic detection*

Hot topic refers to the extensively concerned argument, topics, or information in a period of time. It is usually caused by some reasons or conditions, occurred at a particular time and place involving some of the objects, such as people or matters, and may be accompanied by some of the inevitable results. On the one hand, traditional topic detection is the event through the analysis of the structure of feature matrix between vocabulary and text, but the nature of the short text of micro-blog will lead to sparse feature matrix, so that the accuracy of the results is difficult to be satisfied. On the other hand, rich social information, super text data, emoticons, unique forwarding and comment data for the event that provides richer data base in the micro-blog data, and the traditional method are not very well integrated into the above kinds of data.

At present, the research results of micro-blog hot topic detection are mainly based on the following three kinds of analysis methods. First is the method of statistical analysis. On the micro-blog platform, hot events often attract the attention of a lot of people in a short time, resulting in a large number of comments and forwarding information. According to this feature, scholars believe whether monitoring occurrence frequency for the keywords of a given event can suddenly surge in a given time segment; if yes, it corresponds to the occurrence an event and if no, do not think. Through the changes of emotional keywords to find hot events micro-blog and emotional language distribution model, Liang Yang (2012) achieved finding of hot events by analyzing differences between the emotional-distribution language models in the adjacent time. Fei-ran Zheng (2012) detected a large number of emergent theme keywords online in micro-Bo, clustering them to find hot news event or events. Similarly, C. Lee (2011 & 2012) defined BursT weighted formula for keyword and introduced the sliding window, in order to achieve real-time monitoring of the occurrence of top events and discrimination.

Second is the method based on learning model analysis. According to topic keywords by four benchmarks chosen, Long Rui (2011) established graph model for topic keywords to cluster, and then through the clustering results found hot

micro-blog events. Moreover, putting forward graph model based on wavelet analysis in the literature (Weng, 2011) and time and space model based on probability in the literature (Sakaki, 2010), they also achieved effective detection of micro-blog events using their models. In addition, on the basis of the traditional LDA (latent Dirichlet allocation) model, and through the micro-blog data associated with their association characteristics, many scholars improved LDA model. For example, by considering contact relationship and text association of micro-blog, Chen-Yi Zhang (2012) proposed a micro-blog generational model-based LDA, that is, MB-LDA, to assist theme and event mining of micro-blog. On the basis of the relationship between the micro-blog post and the expanding LDA model, Li Jin (2012) proposed a method of hot topic discovery based on a special-topic space model.

Third is the similarity measure method obtained by improvement of the previous method. S. Phuvipadawat (2010) converted text to vector space model with TF-IDF method first and proposed the improved TF-IDF method based on named entities weighting. The method measured the similarity of information by adjusting the feature weight to find micro-blog events more accurately. And on the basis of considering the diverse characteristics of micro-blog data, Wei Tong (2013) proposed an event-discovery algorithm based on semantic similarity of Chinese text feature, time similarity, and social similarity. Under the premise of full extraction of user characteristics, blog features, and event characteristics, M. Gupta (2012) determined the relationship between features to achieve richer and more complete data to create an event detection model.

Analysis of these three methods can improve discovery or detection effect of micro-blog hot events. However, several analytical methods do not take into account the dynamic characteristics of the event's propagation and the specific characteristics of the micro-blog data on the impact of finding hot events. Then, we need to focus on dynamic detection methods of micro-blog hot events.

4 FUTURE DIRECTIONS FOR IMPROVEMENT

Although micro-blog short text research has made some progress, there are still many issues worthy of discussion and research.

1. At present, the research of short text mostly considers the sparse problem of short text, and the research about the imbalance of the distribution of micro-blog's short text is still rare. Imbalance

is a common problem in real life, and it is also a hot research topic in recent years. Imbalanced data research is aimed at improving the recognition rate of rare class data and mining the valuable information hidden in the small class of data. Imbalance research of micro-blog short text is conducive to discover the potential rules and trend of massive data, for example, the different between unexpected events and common event. In the initial stage, there is often only a small amount of Twitter users to participate in, and the produced short text only form a small sample of data set. Through the establishment of unbalanced clustering learning model it is possible to monitor the occurrence and change process of emergency. How to effectively improve the existing imbalance data learning methods and apply them to the micro-blog short text processing and learning is the focus on the issue of research. To solve this problem, in addition to considering the learning algorithm for the problem of bias, a new cost-sensitive learning method was proposed, in order to eliminate the impact of the imbalance of data distribution on learning, based on analyzing the topical characteristics of micro-blog and the degree of unbalanced data.

2. Micro-blog hot topic detection relies mainly on the results of keyword statistics and the study effect of topic model, but the openness and freedom of user participation will lead to the establishment of the semantic relationship between the topics of micro-blog short text, which makes the user not get the whole point of view and the direction of public opinion. To obtain the valuable information from the massive micro-blog data, it is needed to explore the hidden semantic information in short text, including vocabulary semantic mining, emotion semantic mining, and semantic mining of irregular text. To solve this problem, it requires in-depth understanding and analysis of the micro-blog short text features and grammatical rules, as well as it should make full use of them to improve the natural language processing technology and build a relatively systematic and complete semantic knowledge base of micro-blog short text.

3. The hierarchical structure of the micro-blog theme tends to lose a lot of important information, which leads to the difficulty of micro-blog users to analyze and forecast the development trend of the topic. And the organic integration of the multilevel clustering and the stream data clustering method is helpful for the study of topic tracking and prediction of micro-blog. Because of the massive nature and real-time features of micro-blog's short text, micro-blog theme tracking and prediction will become one of the hot spots in the future.

5 CONCLUSION

At present, micro-blog short text research shows the trend of diversification. To connect with international mainstream research and realize the localization of foreign frontier theory, active thinking about the main issues of the current domestic academic is necessary. In short, micro-blog short text research has still a long way to go before the actual commercial application. Furthermore, micro-blog short text and its application research have a broad research space.

ACKNOWLEDGMENT

This work was financially supported by the subject of Jiangxi "Twelfth Five-Year" plan for Social Science (Project No. 15TQ07).

REFERENCES

Churchill, A.L., E.G. Liodakis, S.H. Ye. Twitter Relevance Filtering via Joint Bayes Classifiers from User Clustering[R]. University of Stanford, Dec.12, 2010.

Ellen J. All about microtext-A working definition and a survey of current microtext research within artificial intelligence and natural language processing [C]// Proceedings of the 3th International Conference on Agents and Artificial Intelligence. Rome: Springer, 2011:329–336.

Gupta M, Zhao Peixiang, Han Jiawei. Evaluating event credibility on Twitter[OL].[2012–02–07]. http: //www. cs. uiuc.edu/hanj/pdf/sd12_mgupta.pdf.

Jing Wang, Ke Zhu, Binqiang Wang. A review of micro-blog research based on information data analysis[J]. computer application, 2012,32(7):2027–2029. In chinese.

Lee C, Wu C, Chien T. Burs T: A dynamic term weighting scheme for mining micro-blog messages [C] // Proceedings of the 8th International Symposium on Neural Networks. Guilin, 2011:548–557.

Lee C. Mining spatio-temporal information on micro-blog streams using a density-based online clustering method[J]. Expert Systems with Applications, 2012, 39(10): 9623–9641.

Li Jin, Zhang Hua, Wu Haoxiong. Chinese micro-blog hot topic mining system based on specific domain—BtopicMiner [J]. Computer application, 2012, 32(8): 2346–2349. In chinese.

Liu, Z., W. Yu, W. Chen, et al. Short Text Feature Selection and Classification for Micro Blog Mining [C]// Proceedings of CiSE' 2010:1–4, 2010.

Long Rui, Wang Haofeng, Chen Yuqiang, et al. Towards effective event detection, tracking and summarization on microblog data [C]//Proceedings of the 12th International Conference on Web-Age Information Management. Berlin: Springer-Verlag, 2011: 652–663.

Petrovi'c, S., M. Osborne, V. Lavrenko. Streaming First Story Detection with application to Twitter [C]//Proceedings of HLT-NAACL' 2010:181–189.

Phuvipadawat S, Murata T. Breaking news detection and tracking in Twitter[C]//Proceedings of the 2010 International Conference on Web Intelligence and Intelligent Agent Technology.Toronto: 2010:120–130.

Sakaki T, Okazaki M, Matsuo Y. Earthquake shakes Twitter users: Real-time event detection by social sensors[C]//Proceedings of the 19th International Conference on World Wide Web. Raleigh: ACM, 2010: 851–860.

Sakaki, T., M. Okazaki, Y. Matsuo. Earthquake Shakes Twitter Users: Real-time Event Detection by Social Sensors [C]//WWW 2010, Raleigh, North Carolina, 2010.

Sharifi, B., M.-A. Hutton, J. Kalita. Experiments in Micro-blog Summarization [C]//Proceedings of NAACL-HIT' 2010.

Sharifi, B., M.-A. Hutton, J. Kalita. Summarizing Micro-blogs Automatically [C]//Proceedings of NAACL-HIT' 2010:685–688.

Sriram, B., David Fuhry, Engin Demir, et al. Short Text Classification in Twitter to improve information Filtering [C]//Proceedings of SIGIR'10, Geneva, Switzerland, 2010.

Tong Wei, Chen Wei, Meng Xiaofeng. EDM: Micro-blog event detectional gorithm [J/OL]. [2013–02–26]. http://www.cnki.net/kcms/detail/11.5602.TP.201210 19.1017.001.html. In chinese.

Weng J, Lee B. Event detection in Twitter [C]//Proceedings of the 5th International AAAI Conference on Weblogs and Social Media. Barcelona: AAAI,2011: 401–408.

Yang Liang, Lin Yuan, Lin Hongfei. Micro-blog hot events discovery based on affective distribution[J]. Chinese Journal of information,2012, 26(1):84–90. In chinese.

Yoshida, M., S. Matsushima, S. Ono, etal. ITCUT: Tweet Categorization by Query Categorization for On-line Reputation Management[R]. University of Tokyo, 2010.

Zeying Peng, Xiaoming Yu, Hongbo Xu, et al. The incomplete clustering of large scale short texts [J]. Chinese Journal of information science, 2011, 25(1): 54–59.

Zhang Chenyi, Sun Jianling, Ding Yiqun. Micro-blog theme mining based on MB-LDA model [J]. Computer research and development, 2012, 48(10): 1795–1802. In chinese.

Zhao, W., J. Jiang, J. He, et al. Topical Key phrase Extraction from Twitter [C]//Proceedings of the 49th Annual Meeting of the Association for Computational Linguistics, Portland, Oregon, 2011:379–388.

Zheng Feiran, Miao Duoqian, Zhang Zhifei. A method of Chinese micro-blog news topic detection [J]. computer science, 2012, 39(1):138–141. In chinese.

Advances in Energy Science and Equipment Engineering II – Zhou, Patty & Chen (Eds)
© 2017 Taylor & Francis Group, London, ISBN 978-1-138-71798-5

Design of the direct sequence system in satellite communication system

Rong Li
College of Communication and Information Engineering, Xi'an University of Science and Technology, Xi'an, China

ABSTRACT: Spread spectrum technique is the key technique of CDMA (Code Division Multiple Access) in satellite communication system. This paper introduces the real-time simulation experiment system for direct sequence spread spectrum communication based on MATLAB/Simulink, and realizes four experiments about the transmitters and receivers. The Results show that this system has the features such as a good interactive interface, easy operation and high visualization. All of those are helpful to master the fundamental principles of satellite communication.

1 INTRODUCTION

Multi-access technique is one of the key techniques in satellite communication system, which includes FDMA, TDMA, CDMA and so on (Yao, 2011). Spread spectrum technique, being an important technique of CDMA, has many features such as strong theory and rich relative techniques, which includes anti interference technology, modulation and demodulation technology and multiple access technology. The paper introduces the real time simulation system for spread spectrum communication based on MATLAB/Simulink. In terms of the experiment system, the technique of spread spectrum communication can be understood more easily.

2 COMPOSITION OF THE EXPERIMENTAL SYSTEM

In the paper, the direct sequence spread spectrum system is simulated. The system includes two modules: the transmitting terminal and receiving terminal. The experimental system covers the most knowledge of the spread spectrum communication course and can be demonstrated for the students in class. It can help the students to remember and comprehend the theory of the course.

3 DESIGN OF THE EXPERIMENTAL SYSTEM

3.1 *Module of the transmitting end*

The parameters of the module of the transmitting terminal are listed as follows: 1 kb/s data rate and 127 kb/s spread spectrum code rate. The m sequence is used as spread spectrum code, whose generating polynomial is $f(x) = 1 + x^6 + x^7$ and cycle is 127. The Bpsk modulation technology is utilized with 500 KHz carrier frequency. The simulation block of the transmitting end is shown in Figure 1.

The data of the transmitting terminal are produced by the Data_GEN module which applies the Function S Data_GEN to produce a random "1" or "0" sequence. The process is controlled by the time module Pulse_Generator. The spread spectrum code is produced by the PN_GEN module which applies the Function S PN_GEN to produce a m sequence whose cycle is 127. The process is controlled by clock module Pulse_Generator1. The negative logic mapping is accomplished by the shaping module. In the course of mapping the binary "1" is changed to "1" level and "0" is changed to "–1" level. The spread spectrum modulation is completed by Product module. Besides, the Radio Frequency (RF) modulation is accomplished by

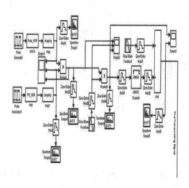

Figure1. Simulation block of the transmitting end.

Produce1 module and the magnitude of sine carrier is 1 which is produced by Sine Wave Function module. The modulated RF signal passes through the additive white noise channel and the interference signal is the single frequency sine carrier produced by Sine Wave Function1 module. In the course of simulation the signal to noise ratio is assumed to be 20 dB.

3.2 *Module of the receiving end*

Firstly, the frequency of the received signal is converted down to 250 KHz by heterodyne correlation dispreading device. Then the carrier is extracted and the data are regained by the Costas ring. The simulation block of the receiving end is shown in Figure 2.

A local carrier with frequency 250 KHz is produced by Sine Wave Function 2 module, which is modulated by the synchronized pseudo random code. Then the modulated signal is correlation processed with the received signal and the correlation processor is composed of Product3 and Digital Filter Design1 module. The Digital Filter Design1 module is a narrowband filter with 2 KHz bands whose center frequency is 250 KHz. The dispread signal is sent into the Costas ring to extract the medium frequency carrier and demodulate after amplified by the Gain module. The two-phase Costas ring is constructed because of Bpsk modulation technology. Besides, the static frequency of continuous time voltage controlled oscillator is 250 KHz, whose sensitivity of output frequency is 1 Hz/V. The filter is composed of Gain1, Gain3, Add2 and delay unit. Finally, the useful information is recovered after a simple judgment module.

4 REALIZATION OF THE EXPERIMENTAL SYSTEM

4.1 *The experiment of the transmitting end*

The simulation results of spread spectrum and modulation are shown in Figure 3. Among these figures, (a) is the waveform after spread spectrum and (b) includes the waveforms before modulation and after modulation. (c) is the spectrum waveform of the baseband signal before spread spectrum and (d) shows the spectrum waveform of the signal after spread spectrum. (e) is the spectrum waveform after radio frequency modulation. The experimental results correspond to those of the theory.

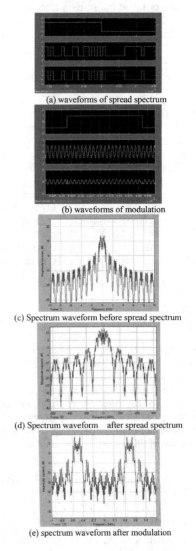

(a) waveforms of spread spectrum

(b) waveforms of modulation

(c) Spectrum waveform before spread spectrum

(d) Spectrum waveform after spread spectrum

(e) spectrum waveform after modulation

Figure 3. The simulation results of spread spectrum and modulation.

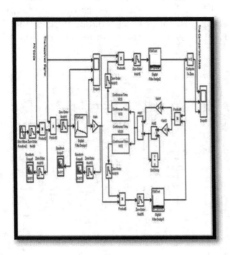

Figure 2. Simulation block of the receiving end.

4.1 *The experiment of the transmitting end*

The simulation waveforms of the receiving end are shown in Figure 4. The waveforms before and after correlation dispreading are illustrated in Figure (a). (b) and (c) include the spectrum waveforms before and after correlation dispreading. (d) is the comparison waveform between the demodulated signal and the original signal. All the experimental results are correct.

5 CONCLUSIONS

The experimental results obtained by the real time simulation system are correct and the output values will be changed with the change of the input parameters. Thus, the experimental system can be helpful to master the technique of spread spectrum. In a word, the simulation system of direct sequence can be applied to study on satellite communication.

ACKNOWLEDGMENT

This work is sponsored by the Natural Science Foundation of Shaanxi Province (2015 JM6341).

REFERENCES

Ding Yi-nong. "Simulink and Signal Processing" [M] (in Chinese). Beihang University Press, 2014.
Yao Jun. "Microwave and Satellite Communication" [M] (in Chinese). Beijing University of Posts and Telecommunications Press, 2011.Zhao Guo-sheng. "The Complete Learning Manual of Matlab" [M] (in Chinese). Tsinghua University Press, 2014.

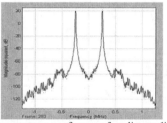

(a) waveforms of dispreading

(b) spectrum waveform before dispreading

(c) spectrum waveform   after dispreading

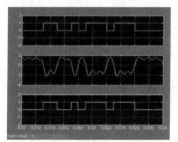

(d) waveforms of demodulation

Figure 4.   The simulation results of dispreading and demodulation.

Advances in Energy Science and Equipment Engineering II – Zhou, Patty & Chen (Eds)
© 2017 Taylor & Francis Group, London, ISBN 978-1-138-71798-5

Research of the data integration method for drilling data warehouse based on Hadoop

Shijing Xiang & Ming Fang
School of Computer Science, Xi'an Shiyou University, Shaanxi, Xi'an, China

ABSTRACT: The drilling information is wide distribution, large amounts of data, various kinds and difficult integration. That is not conducive to support the drilling decision analysis. To solve the problems, this paper constructs the general structure of the drilling data warehouse system based on Hadoop by using Hive, a data warehouse tool, and proposes a drilling data integration method by using Kettle, an ETL tool, and discusses in detail the technology scheme of loading historical data and incremental data from different kinds of relational database of each drilling unit to Hive. This research will provide the effective way and the new reference idea for the organization and integration of data in drilling data warehouse under the big data environment.

1 INTRODUCTION

Oil-gas drilling engineering is a systematic engineering with high investment and high risk, which involves multiple disciplines fields. Each decision analysis is related to the efficiency and success or failure of drilling engineering directly. In order to support the drilling decision-making process, the drilling data needs to be integrated, managed and utilized effectively. At the same time, it is necessary to establish the corresponding data warehouse. The existing drilling data warehouse systems (Zhang, 2004; Jing, 2012; Zhang, 2009) are mainly oriented to support a process or a phase in the decision-making, and they are applicable to the general, single and localized drilling data statistics and decision analysis. With the development of oil and gas drilling business, the drilling operation region is becoming wider and wider, and drilling data information is increasing massively. So, the key problems to be solved of improving the efficiency of drilling decision-making are how to construct a reasonable drilling data warehouse system and how to integrate the drilling data information effectively. With the theory of big data, the general structure of the drilling data warehouse system based on Hadoop is constructed by using Hive and a drilling data integration method is proposed by using Kettle. Consequently, this research will provide a fast and efficient approach for the organization and integration of data in drilling data warehouse under the big data environment.

2 THE GENERAL STRUCTURE OF THE DRILLING DATA WAREHOUSE SYSTEM BASED ON HADOOP

Drilling engineering decision-making involves a great amount of drilling data, and those data are dynamic, diversity, dispersion, mobility and timeliness. Therefore, distributed storage management of drilling data and the realization of parallel access control, data real-time analysis, data mining and utilization for drilling decision-making, are the foundation and the key to construct the drilling data warehouse and realize the parallel analysis under the big data environment. Hive (Liu, 2011) is an open source data warehouse infrastructure tool of being builted on Hadoop. Hive can map the structured data to form a database table, at the same time, it provides a standard SQL function and runs by the way of converting the SQL statement to the MapReduce task. Hive has a better support for data analysis and data mining, and it can be seen as a Hadoop client because it is independent of the cluster (Fan, 2015).

The drilling data has these characteristics of distribution widely, large amounts of data and variety kinds. Aiming on these characteristics, we apply the theory of big data and combine the technology of the Hive and Hadoop/MapReduce for supporting drilling decision analysis effectively. At the end, the drilling data warehouse system architecture based on Hadoop is constructed, which is composed of data layer, control layer and application layer, as shown in Figure 1.

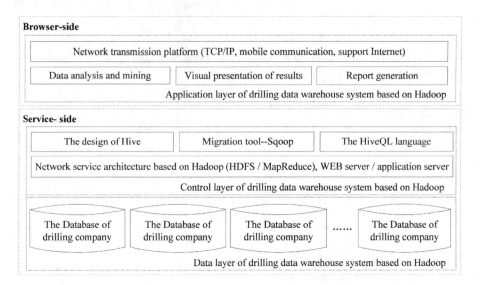

Figure 1. The general structure of the drilling data warehouse system based on Hadoop.

Among them, the data layer is the basic object of data analysis, it stores the drilling data information of each drilling company. The control layer is the core of the drilling data warehouse under the big data environment, it's role is to summarize, organize, harmonize and integrate the drilling data of the data layer by using ETL (Xu, 2011) tools. The application layer is mainly used to support drilling decision analysis, it encapsulates the functions, such as inquiry, analysis, decision-making and display. In addition, the application layer provides the easily accessible and friendly interface for drilling decision-making.

Firstly, the drilling data warehouse system based on Hadoop extracts the data from distributed and heterogeneous drilling data sources, and load those data to the data buffer by using ETL tools. Secondly, loading those data which are been cleaned, transformed, integrated to the auxiliary temporary database by using ETL tools. Thirdly, we migrate the data from the auxiliary temporary database to Hive by using Sqoop, analyse the drilling data information from multiple dimension and load the results to MySQL. Finally, the analysis results will be displayed to the user. The working process is shown in Figure 2.

The drilling data warehouse system based on Hadoop can store massive drilling data information in the form of HDFS distributed file, which can realize the storage of various types of data. System through a large number of built-in user function UDF to operate time, string and other data mining tools, enabling users to finish the operation which can't achieve easily. System translates the SQL query into the MapReduce job for executing

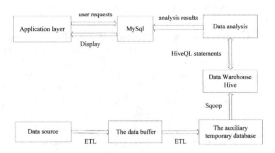

Figure 2. The working process of drilling data warehouse system based on Hadoop.

on the Hadoop cluster in the way of SQL-like query, it avoids the development of MapReduce programs, thus reducing the learning cost. In brief, it's ideal for statistical analysis of the drilling data warehouse.

3 THE DATA INTEGRATION METHOD OF DRILLING DATA WAREHOUSE BASED ON KETTLE

A data warehouse is an integrated, subject-oriented, time-variant, nonvolatile, consistent database, constructed from multiple sources, and made available to support decision-making in a business context (Wang, 2014). The establishment of data warehouse is to further explore the data resources for decision-making analysis. The quality of data in data warehouse determines the accuracy of decision-making. Due to the data source of data warehouse is not uniform, data format confusion, and various types of "dirty" data have increased the difficulty of data integration, so it is necessary to provide a complete solution for solving the problems of data consistency and integration, thus protecting the maneuverability of the data mining and analysis in later period.

Kettle (Casters, 2010; Liu, 2015; Wu, 2016) is an open source software of being written in Java, an ETL tool, which can run on multiple operating system platforms and is more effective and stable in data extraction. Kettle supports multiple input and output formats, which has a good support for the data warehouse. And, Kettle as an open source software, on the support of Hadoop is better than most of the ETL software. In this paper, we use the ETL tool, Kettle. When constructing the drilling data warehouse system, the data warehouse system needs to access the data from each drilling company business system continuously. Because of the huge amounts of data, it is necessary to establish a data buffer between the drilling data warehouse system and the drilling company business system. After the Kettle extracts the data from the drilling company business system to the data buffer, we transform the data. The way avoids drilling data warehouse system accesses the drilling company business system frequently, and prevents from a large number of works of integration and calculation in the business system. On the whole, the way reduces the impact on the performance of the business system and solves the heterogeneous data integration issues.

The data integration work in this data integration scheme is divided into three parts. In the first part, is to use the Kettle to extract and load historical data and incremental data from each drilling company business system to the data buffer. The part includes the synchronization of the historical data in the business system and the integration of the incremental data in the business system. In the second part, is to use the Kettle to extract, transform, clean and load data from the data buffer to the auxiliary temporary database. In the third part, is to use the Sqoop to migrate the data from the auxiliary temporary database to Hive, and those data include historical data and incremental data in drilling company business system which are cleaned and transformed.

3.1 The data integration from the drilling company business system to the data buffer

The drilling data warehouse system accesses the data in drilling company business system continuously, but the data is very large, it is necessary to establish a data buffer between the drilling data warehouse system and the drilling company business system.

The data in data buffer is integrated by the data in drilling company business system by using Kettle, and the data format is consistent with the source data format of drilling company business system. The importance of the data buffer is that it avoids a lot of integration, calculation and other work in the business system, and it reduces the impact on the performance of the business system. The data integration from the drilling business system to the data buffer includes the historical data integration and the incremental data integration (He, 2009).

The integration process of historical data is illustrated in Figure 3, the loading process does not contain any data processing work. In the loading process, the data in drilling company business system is treated as data input, then, the data is output to the data buffer directly.

The integration process of incremental data is illustrated in Figure 4, the role of the transformation is to synchronize the incremental data from the drilling company business system to the data buffer, for ensuring the data format is consistent with the source data format of drilling company business system. Due to the lack of time stamp fields in the source data in business system, the incremental data integration adopts the data comparison method (Casters, 2010) to synchronize the data between different data sources in this scheme. That is to say, compare the historical data with the new data one by one, then, according to the comparison result to operate the data in the data buffer. The data synchronization uses the "merge rows" component of Kettle in the transformation.

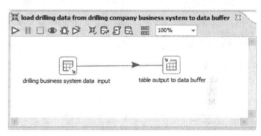

Figure 3. The integration process of loading historical data into the data buffer.

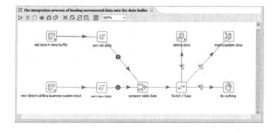

Figure 4. The integration process of loading incremental data into the data buffer.

In addition, by adding FLAG field in the data stream, the component compares the incremental data of the sub units (including the data generated by the delete, update and add operation) and the old data in the data buffer. According to the settings of Kettle for the "merge rows" component, FLAG field content can only be deleted, new, changed and identical, which indicates the change operation that the record has been experienced in the data flow. Among them, "deleted" indicates that the record has been deleted, "new" indicates that the record is new, "changed" indicates that the record has been updated, "identical" indicates that the record has not changed. In the "merge rows" component, the output data stream contains not only the field information in the data buffer, but also the FLAG field information, Figure 5 is the component configuration diagram of "merge rows" in the incremental data extraction and trans-formation. In the synchronization transformation of the incremental data, the "switch/case" components use the FLAG field information in data stream to make the data of having different FLAG information to point to the operation components of different database, the component configuration is shown in Figure 6. "Deleted data" component deletes the "deleted" data rows from the old data of the data buffer according to the unique identifier ID. "Insert/update data" component synchronizes the "new" and "changed" data rows to the data buffer by the way of inserting or updating. "Do nothing" component deals with the "identical" data rows, and the component does not do any operations on the database.

Above the data integration method, it not only ensures the synchronization of the historical data and the data in the data buffer, but also ensures that the incremental data in each drilling company business system can be loaded and updated in time. In addition, the method makes the data in Hive and the data in drilling company business system to maintain the latest status, and solves the problem of that drilling information is difficult to integrate and manage.

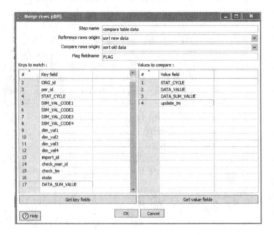

Figure 5. The component configuration diagram of "merge rows" in the incremental data extraction and transformation.

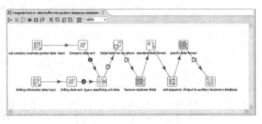

Figure 7. The transformation of the drilling data from the data buffer to the auxiliary temporary database.

Figure 6. The transformation of the "switch/case" component configuration diagram.

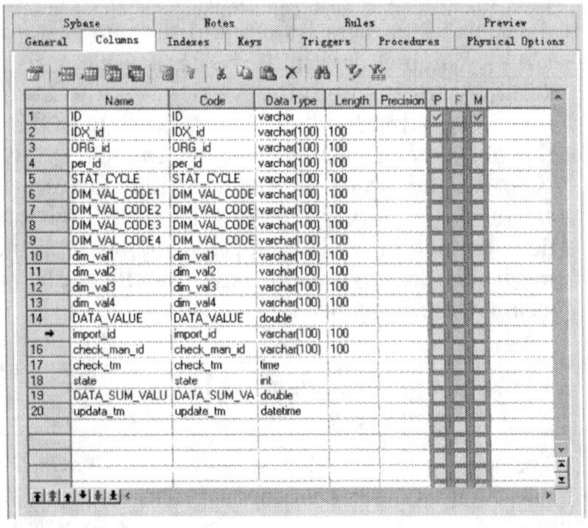

Figure 8. The structure of the data table of the auxiliary temporary database.

1830

Figure 9. Configuration of "shell components" in drilling data integration.

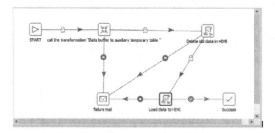

Figure 10. The complete workflow of loading the drilling data from the data buffer to Hive.

Figure 11. The component configuration diagram of "delete the historical data in Hive".

3.2 *The data integration from the data buffer to the auxiliary temporary database*

Sqoop can only load data directly from a relational database to Hive, so the data is stored temporarily in the auxiliary temporary database before loading it to Hive. The auxiliary temporary database is used to store the transformed data from the data buffer. The transformation process from the data buffer to the auxiliary temporary database is the process of cleaning the data from the data buffer by using Kettle. This transformation eliminates the differences of the data formats from different sources, eliminating all kinds of "dirty" data, so as to ensure that the data is accurate and uniform to be loaded into Hive.

Figure 7 is the transformation of the drilling data from the data buffer to the auxiliary temporary database. In this transformation includes the Kettle transformation components of "table input", "value mapper", "calculator" and "unique rows". The use of these components is to eliminate the differences of the data formats in the different sources, and eliminate all kinds of "dirty" data or duplicate data. Figure 8 is the structure of the data table of the auxiliary temporary database.

3.3 *The data integration from the auxiliary temporary database to Hive*

Loading the drilling data from the auxiliary temporary database to Hive requires the use of the Shell components of Kettle. The Shell components call Sqoop through the Shell scripts to load the data from a relational database to Hive. The Shell Scripts configuration is illustrated in Figure 9. The operational processes of loading the drilling data from the data buffer to Hive, as illustrated in Figure 10. Call the transformation of "the data from the data buffer to the auxiliary temporary database" in the operation process, which ensures that the latest data can store in auxiliary temporary database synchronously. And then delete all the old data in Hive, the component configuration of "delete the old data", as illustrated in Figure 11. At last, synchronize the data from the auxiliary temporary database to Hive.

4 CONCLUSIONS

The drilling data warehouse system based on Hadoop makes full use of the advantage of Hive, and it is depended on designing the ETL process based on Kettle. In addition, the system realizes the integration of the drilling data in the sub units from relational database to Hive, and makes full use of the scheduling plug-ins of Kettle. It not only achieves the automatic scheduling of data transfer of ETL process from business system to Hive, but also solves the heterogeneous data integration issues. Furthermore, the data integration method for drilling data warehouse based on Kettle perfects data transformation in ETL process, improves data quality, and optimizes the ETL process design to ensure the validity of data updates. Largely, this research will provide a strong support for drilling decision analysis in big data environment.

REFERENCES

Casters, Matt, R. Bouman, and J.V. Dongen, JOHN WILEY & SONS INC, Pentaho Kettle Solutions: Building Open Source ETL Solutions with Pentaho Data Integration, 23–73, (2010)

Fan, D.L., Beijing: Posts & Telecom Press, *Hadoop massive data processing technology and project*, 129–163, (2015)

He, T., Policy & Management, *Using the ETL tool KETTLE to realize the data integration of the library alliance information system*, (23), 47–48, (2009)

Hive, http://baike.sogou.com/v64982582.htm

Jing, N., H.H. Fan, Y.H. Zhai, et al, Netinfo Security, *Research on data warehouse solutions for drilling engineering*, (6), 93–96, (2012)

Liu, Y., Electronic Design Engineering, Research on integration of college multi-source heterogeneous data Based on KETTLE, **23**(10), 24–26, (2015)

Liu, Y.Z., X.J. Zhang, X.Y. Li, Journal of Guangxi University(Natural Science Edition), *Design of Web log analysis system based on Hadoop/Hive*, **36**(S1), 314–317, (2011)

Sqoop, http://baike.baidu.com/view/9452375.htm

Wang, F., G.F. Liu, Beijing: China Mechine Press, Business Intelligence in the era of big data architecture planning and case, 11–62, (2014)

Wu, Z.T., P.H. Jiang, J.G. Li, Database and information management, *Research on integration of college heterogeneous data Based on KETTLE,* 61–63, (2016)

Xu, J.G., Y. Pei, Computer Science, *Overview of Extraction, Transformation and Loading*, **38**(4), 15–20, (2011)

Zhang, H., Xi'an Shiyou University, The research of drilling tool requirement analysis system of drilling ERP based on data warehouse, (2009)

Zhang, Y., N.S. Zhang, Q. Liu, et al, Journal of Shengli Oilfield Teachers College, *Research and application of data warehouse in drilling information management system*, **18**(4), 80–82, (2004)

Advances in Energy Science and Equipment Engineering II – Zhou, Patty & Chen (Eds)
© 2017 Taylor & Francis Group, London, ISBN 978-1-138-71798-5

Predicting long non-coding RNAs based on openstack platform

Lei Sun, Xiaobin Zhang & Yun Li
School of Information Engineering, Yangzhou University, Jiangsu, China

ABSTRACT: Functional long non-coding RNAs (lncRNAs) can be identified from high-throughput sequencing data using bioinformatic tools. However, the increasingly expanded data for lncRNA prediction may lead to serious problems, including big data storage, highly intensive computation and high dimensional analysis, which are not trivial to manipulate in traditional way. In this paper, a cloud computing system based on OpenStack platform is proposed to deal with these problems. Specifically, OpenStack Swift was used to store the lncRNA dataset. A computational pipeline was constructed on Linux operating system for lncRNA prediction. To share the pipeline within the community, all programs and dependencies for lncRNA prediction were packaged in a Linux image. Experiment results represent that even though the cloud computing system takes longer computing time than a local workstation, it can be more flexible for resource allocation, big data storage, and method sharing, which can help solve the big data problem during the procedure of lncRNA prediction.

1 INTRODUCTION

Long non-coding RNAs (lncRNAs) belong to a category of non-coding RNAs with length larger than 200 nucleotides (nt), and it has been revealed that many of them can be functional in several cellular processes, such as Embryonic Stem Cell (ESC) pluripotency and diseases (Loewer, 2010). In recent years, high-throughput sequencing technology, e.g. RNA-seq (Trapnell, 2010), has captured tens of thousands of lncRNAs. And corresponding bioinformatic tools are used to predict the lncRNAs from the sequencing data. However, the increasingly expanded lncRNA data may lead to serious big data problems, such as data storage, highly intensive computation and high dimensional analysis, especially for lowly configured Personal Computers (PCs) (Sun, 2014). Specifically, the volume of sequencing data containing lncRNA information might be at least 5 Gigabyte (GB) per biological sample, and the total data volume of a single laboratory could be more than one Terabyte (TB). Second, the lncRNA prediction procedure may involve several computation-intensive steps, such as read mapping or transcriptome assembly, which are challenging for computer application. In general, a typical pipeline for biological data processing may require high-performance computing environment, which can provide multi-core processors, adequate Random Access Memory (RAM) and disks. And in the computing procedure there may output a large number of intermediate data files, which would aggravate the data storage problem. Last but not least, further functional analysis

on the lncRNAs predicted may need high dimensional computation and optimization. Therefore, current studies on lncRNA prediction as well as further study can be hampered by limited computing environment. Fortunately, such biological big data problem can be solved by cloud computing (Sun, 2014), which can provide users with scalable storage and computing resources through high-speed network (Krampis, 2012). In this paper, a cloud computing system based on OpenStack platform is proposed to tackle these problems for predicting the lncRNAs efficiently.

2 SYSTEM FRAMEWORK

The lncRNA prediction system based on OpenStack is mainly composed of four parts, namely Virtual Machine (VM) layer, OpenStack layer, dataset and Linux image (Figure 1).

2.1 *Virtual machine layer*

As the basic component of cloud computing platform, a VM layer is used to generate VMs by integrating various resources, including physical hardware and user software, which can be managed and allocated through hypervisors, and then responds to service requests from the hypervisors by a consistent and transparent way. For constructing a useful VM, the layer should at least allocate adequate resources including processors, RAMs, network, storage, system image and user software, as well as providing necessary configure

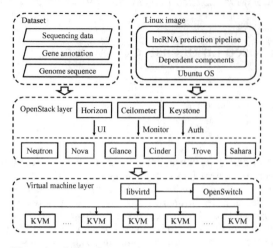

Figure 1. System framework.

information of VM, especially several unique tags in global environment, e.g. Media Access Control (MAC) address. As the development of computer technology, there appear amount of software tools providing the resource allocation service mentioned above. Specifically, Kernel-based Virtual Machine (KVM) or Quick Emulator (QEMU) provides basic hardware services, which can be used to create VMs having CPUs, RAMs and Network Interface Card (NIC) by simulating the computer. OpenSwitch acts as a switch networking the VMs. Disk files can be mapped to block devices by Virtual File System (VFS), which can be used to produce storage devices of VMs from files, e.g. Google File System (GFS) and Network File System (NFS) can provide storage service for the VMs, which prevents the VMs from manipulating physical storage directly, otherwise may affect the migration of VMs. As the core service of the VM layer, libvirtd can receive requests from the upper cloud management layer, call and integrate the services according to configure information, and generate VMs in line with user requirements, which will be return to the user as a response.

2.2 OpenStack layer

The OpenStack layer is used to collect and allocate the cloud computing resources. Firstly, this layer builds multiple resource pools collecting various types of resources of the entire computing cluster. Secondly, after receiving a user request, the layer can generate special user configure files according to the user requirements, meanwhile it sends resource allocation request to the lower layer, namely the VM layer, and then allocates the VMs packaging required resources to the user. OpenStack is composed a group of interrelated modules,

such as Compute (Nova), Block Storage (Cinder), Image Service (Glance), Networking (Neutron), Database (Trove), Elastic Map Reduce (Sahara), Dashboard (Horizon), Identity Service (Keystone) and Telemetry (Ceilometer).

- Nova is a cloud computing controller for managing and automating pools of computer resources.
- Cinder provides persistent block-level storage devices for use with OpenStack compute instances.
- Glance provides discovery, registration, and delivery services for disk and server images.
- Neutron is a system for managing networks and IP addresses and it ensures the network is not a bottleneck or limiting factor in a cloud deployment, and gives users self-service ability, even over network configurations.
- Trove is a database-as-a-service provisioning relational and non-relational database engine.
- Sahara provides users with simple means to provision Hadoop clusters by specifying several parameters, such as Hadoop version, cluster topology and nodes hardware details.
- Horizon provides administrators and users a graphical interface to access, provision, and automate cloud-based resources.
- Keystone provides a central directory of users mapped to the OpenStack services they can access.
- Ceilometer provides a single point of contact for billing systems, providing all the counters they need to establish customer billing, across all current and future OpenStack components.

Horizon can help users manage the cloud graphically without understanding details of the other components.

2.3 Dataset

The dataset for lncRNA prediction is mainly composed of sequencing data in Sequence Read Archive (SRA) or FASTQ format, gene annotation in Gene Transfer Format (GTF) and genome sequence in FASTA format. In general, researchers can download high-throughput sequencing data in SRA from Gene Expression Omnibus (GEO) or other sources, which are then transferred to FASTQ format using SRA toolkit. The dataset for lncRNA prediction can be stored in Swift through Cinder. In addition, other data needs to be stored when running some bioinformatic tools. For example, PhastCons scores are required for running lncRScan-SVM (Sun, 2015).

2.4 Linux image

A bioinformatics pipeline can be packaged in a Linux Operating System (OS) or image to

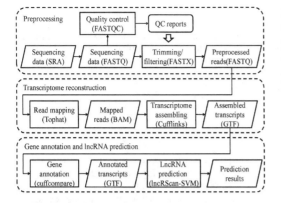

Figure 2. LncRNA prediction pipeline.

share with researchers in the community, which improves the reproducibility of computational experiments (Sun, 2014; Dudley, 2010). A lncRNA prediction pipeline (Figure 2) is mainly composed of three parts, namely 'Preprocessing', 'Transcriptome reconstruction' and 'Gene annotation and lncRNA prediction'. The 'Preprocessing' part is used to transfer data format and to conduct Quality Control (QC) on the sequencing data. For example, if we detect low-quality reads using QC tools such as FASTQC, we can filter or trim these reads using tools such as FASTX. The 'Preprocessing' part might not be processed automatically since the filtering or trimming strategy need to be conducted manually according to the QC reports. In the 'Transcriptome reconstruction' part, transcripts can be assembled from the preprocessed reads using a genome-guided strategy, e.g. Tophat-Cufflinks workflow (Trapnell, 2010; Trapnell, 2009). In the 'Gene annotation and lncRNA prediction' part, the assembled transcripts can be first classified into subclasses using cuffcompare, which can provide information of the novel assemblies. Then lncRScan-SVM (Sun, 2015) can be used to predict the lncRNAs. In addition to the prediction pipeline, various dependent components and software need to be installed and packaged in the OS. After that, the Linux image can be run on Nova through Horizon. The pre-configured Linux image containing lncRNA prediction programs and components can be run on any available VM, which is easy to manipulate.

3 EXPERIMENT DESIGN

In this section, we designed an experiment to evaluate the performance of lncRNA prediction on the OpenStack-based system.

3.1 Dataset preparation and storing

The original RNA-Seq data for testing were downloaded from European Bioinformatics Institute (EBI) (accession no. ERP000550) (Ren, 2012). According to the species of the sequencing data, we downloaded the genome sequences of human (GRCh37/hg19) from University of California Santa Cruz (UCSC) genome browser and human gene annotation (Version 19) from GENCODE (Harrow, 2012). The sizes of the data sets are listed in Table 1.

3.2 Resource allocation

The resources for storing the dataset and running the OS image were allocated through the Horizon module of the OpenStack layer (Table 2).

3.3 Program versions and parameters

Table 3 lists programs and parameters of the lncRNA prediction pipeline, which were packaged

Table 1. Dataset summary.

Type	Format	Source	Size (GB)
Sequencing data	FASTQ	EBI	16.7
Genome sequences	FASTA	UCSC genome browser	2.1
Gene annotation	GTF	GENCODE	1.1
Conservation (PhastCons)	BigWig	UCSC genome browser	5.4

Table 2. Resource allocation.

Resource	Number/Size
CPU	8
RAM	16GB
Disk	100GB

Table 3. Programs and parameters.

Stage	Program	Version	Parameters
Preprocessing	SRA Toolkit	2.5.2	Default
	FASTQC	0.11.3	Default
	FASTQX Toolkit	0.0.13	Default
Transcriptome reconstruction	Tophat	1.3.0	–p 8 –r 50
	Cufflinks	2.1.0	–p 8
lncRNA prediction	lncR Scan-SVM	1.0.1	Default

Table 4.   Executive scripts.

Script	Usage	Function
fastq-dump	fastq-dump test.sra	Convert SRA to FASTQ
fastqc	fastqc test.fastq	Quality control
fastx_trimmer	Fastx_trimmer—l 100 –i test.fastq	Trim reads
trans_construct. sh	sh trans_construct.sh	Transcriptome reconstruction
lncr_pred.sh	sh lncr_pred.sh	lncRNA prediction

Table 5.   Running time.

Transcriptome reconstruction	OpenStack (hour)	Workstation (hour)
Read mapping	7	4.5
Transcriptome assembling	3	1.5

in a Linux image (Ubuntu 12.04 64bit server). For creating an available image for lncRNA prediction, all programs and parameters used in the pipeline processing were chosen carefully. In addition, users can change these settings in the script.

3.4   Starting pipeline

After entering the OS, users do not need to install any programs or dependencies, which have been packaged by our developers. They can just start the pipeline using a set of scripts or programs as shown in Table 4.

4   RESULTS AND ANALYSIS

Since the reads passed several QC tests, we did not conduct trimming or filtering options on the original reads. To evaluate the system for lncRNA prediction, the pipeline was run on the cloud, and several information including running time, resource allocation, utility and data storage were recorded, in comparing with that resulting from a local workstation (24 CPUs, 32GB RAM, 64bit CentOS) with the same configuration.

4.1   Flexible resource allocation

The OpenStack layer can be used to allocate cloud resources more flexibly. We can allocate adequate virtual disks for storing the big biological dataset, necessary CPUs and RAMs for processing the data, and flexible network for accessing and transmitting the data, which would be serious for the workstation as data increases.

4.2   Running time

Since the 'Transcriptome reconstruction' step takes most of the running time, it was timed and

compared to that run on the workstation (Table 5). As a result, the running time on OpenStack is nearly two times than that on the workstation. This is due to the fact that a VM started on OpensStack may behave slower than a standalone workstation when running programs. However, this problem could be solved by parallel computing arranged on the Open-Stack, which will be investigated in our future work.

4.3   Utility

The results indicate that the cloud computing can efficiently improve the utility of system resources. In this experiment, we conducted computational prediction using the completely virtualized VMs, which can be freely migrated across heterogeneous platforms without being limited by the hardware of host computer. However, the computing efficiency of such VMs can be slightly decreased (See running time comparison of the results). A running system may have multiple computing jobs, each of which may occupy considerably variable resources. On one hand, in the beginning of computing, each job may request resources, part of which can be occupied throughout the computing procedure. On the other hand, the computing resources demanded by each job could be dynamic. For the latter one, we adopted a technique named Virtio Balloon, which makes each job obtain the demanded resources, thereby improving the system throughput. Finally, the pipeline predicted 67628 candidate lncRNAs from the total 73775 assembled transcripts.

4.4   Big data storage

The results indicate that the cloud computing platform has several advantages for data storage. For biological big data computing, a large number of storage resources are required. The system storage resources are centralized on disk arrays or distributed on physical machines. However, it is difficult to directly use the storage distributed using traditional methods. In this experiment, we used OpenStack to centralize the scattered storage fragments distributed on the physical machines, and to build storage pools by integrating the distributed resources and the disk arrays of the cluster. As a result, the data volume was greatly improved.

5 CONCLUSIONS

A system for prediction lncRNAs based on cloud computing (OpenStack) platform was proposed in the paper. This system presents advantages in solving several biological big data problems, such as data storage and computing-intensive analysis. First, we can efficiently allocate adequate computing resources for the prediction job on the system. Specifically, the data storage and computing problems were solved by Swift and Nova respectively. Second, the prediction pipeline and corresponding dependencies can be packaged in a Linux image, which is easy to run on the cloud as well as sharing with others in the community. This Whole System Snapshot Exchange (WSSE) concept can improve the reproducibility of experiments efficiently (Dudley, 2010). Meanwhile, the prediction on the cloud would take longer time than on the workstation, which can be solved by constructing a parallel computing environment in the future. Therefore, the cloud computing such as the OpenStack framework will be useful for the lncRNA prediction study as well as broader bioinformatic research during the coming biomedical big data era.

REFERENCES

Dudley J.T. and A.J. Butte, Nat biotechnology, **28**, 1181–1185 (2010).

Harrow, J., A. Frankish, J. M. Gonzalez, E. Tapanari, M. Diekhans, F. Kokocinski, B. L. Aken, D. Barrell, A. Zadissa, and S. Searle, Genome research, **22**, 1760–1774 (2012).

Krampis, K., T. Booth, B. Chapman, B. Tiwari, M. Bicak, D. Field, and K. E. Nelson, BMC bioinformatics, **13**, 42 (2012).

Loewer, S., M. N. Cabili, M. Guttman, Y.-H. Loh, K. Thomas, I. H. Park, M. Garber, M. Curran, T. Onder, S. Agarwal, P. D. Manos, S. Datta, E. S. Lander, T. M. Schlaeger, G. Q. Daley, and J. L. Rinn, Nat Genetics, **42**, 1113–1117 (2010).

Ren, S. Z. Peng, J.-H. Mao, Y. Yu, C. Yin, X. Gao, Z. Cui, J. Zhang, K. Yi, W. Xu, C. Chen, F. Wang, X. Guo, J. Lu, J. Yang, M. Wei, Z. Tian, Y. Guan, L. Tang, C. Xu, L. Wang, X. Gao, W. Tian, J. Wang, H. Yang, J. Wang, and Y. Sun, Cell Res, **22**, 806–821 (2012).

Sun, L., H. Liu, L. Zhang, and J. Meng, PLoS ONE, **10**, e0139654 (2015).

Sun, L., X. Hu, X. Zhang, and Y. Li, Journal of Electronic Measurement and Instrumentation, **28**, 1190–1197 (2014).

Trapnell, C., B. A. Williams, G. Pertea, A. Mortazavi, G. Kwan, M. J. van Baren, S. L. Salzberg, B. J. Wold, and L. Pachter, Nat Biotech, **28**, 511–515 (2010).

Trapnell, C., L. Pachter, and S. L. Salzberg, Bioinformatics, **25**, 1105–1111 (2009).

Advances in Energy Science and Equipment Engineering II – Zhou, Patty & Chen (Eds)
© 2017 Taylor & Francis Group, London, ISBN 978-1-138-71798-5

QoS-aware adaptive load-balancing algorithm for virtual clusters

Weihua Huang, Zhong Ma, Xinfa Dai, Mingdi Xu & Yi Gao
Wuhan Digital Engineering Institute, Wuhan, China

ABSTRACT: The load-balancing mechanism for virtual cluster can cause the degradation of Quality of Service (QoS). To solve this problem, a QoS-aware Adaptive Load Balancing Algorithm (QALBA) consisting of two control loops is proposed. The QoS control loop, as the internal control loop, triggers system resource allocation at the request of QoS requirements, to ensure that the performance of application layer services within the virtual machine is not affected, whereas the load-balancing control loop, as the external control loop, manages the migration of virtual machines within the domain according to the resource utilization of the host machine, so as to realize the macroscopic load balancing of the whole virtual cluster system. The collaboration of those two control loops can simultaneously handle QoS and load-balancing issues. The experimental results show that the proposed algorithm can guarantee QoS and balance system loads.

1 INTRODUCTION

As a new computing architecture, virtualization decouples the upper software from the underlying hardware through virtualization layer. This technology provides cluster system with elastic computing and dynamic resource allocation ability. Thus, it becomes crucial for virtual cluster to coordinate loads of computing nodes and maintain efficient resource utilization rate. Relevant studies demonstrate that load balancing can effectively improve the performance of virtual cluster systems. With the wide deployment of virtual cluster, Quality of Service (QoS) is drawing more attention. As a consequence, QoS guarantee poses new challenges to load-balancing algorithms.

Traditional load-balancing algorithms (Riyazuddin, 2015; Garima, 2015; Moumita, 2015; Zhao, 2016; Jain, 2016) fail to take QoS into full consideration. The migration of virtual cluster often leads to lower QoS during the load-balancing process for virtual cluster. Recently, some algorithms and strategies (Dai, 2015; Sun, 2014; Mohamed, 2013; Ye, 2013; Heiner, 2011) for load balancing have been put forward to consider the QoS issues. These algorithms drive the migration of virtual machines, schedule tasks, and adjust resource allocation on the basis of QoS. However, the above-mentioned algorithms have limits in evaluating the architecture of virtual cluster. For example, virtual machine migration and resource allocation are two different methods for regulating QoS and act on host machines and virtual machines, respectively. Their impacts on QoS are distinct in regulating granularity and if treated equally, they will cause unstable QoS.

The purpose of this paper is to design a load-balancing schedule that guarantees QoS for virtual cluster. By incorporating control theory, this paper presents a QoS-aware Adaptive Load Balancing Algorithm (QALBA). This algorithm has two layers of control loops: the external load-balancing control loop triggers the migration of virtual machines while taking QoS into consideration, so that appropriate host machine is selected for virtual machine. Thus, QoS of virtual machines is regulated in a coarse-grained way; the internal QoS control loop regulates QoS of virtual machines in a fine-grained way through dynamically assigning resources of virtual machines. The objective of the collaboration of these two control loops is to realize load balancing on the premise of guaranteeing the QoS for application-layer services within the virtual machine.

The rest of the paper is organized as follows. In Section II, we formulate the algorithm framework. The system modeling and controller design are discussed in Section III. Section IV describes the implementation of the algorithm. Experiments are performed in Section V. Finally, we conclude the paper in Section VI.

2 ALGORITHM FRAMEWORK

The framework of QALBA algorithm proposed in this paper contains two control loops: load balancing control loop and QoS control loop (Fig. 1).

The internal QoS control loop adjusts the CPU allocation of virtual machines on the same physical node on the premise of satisfying QoS. Meanwhile, the external load-balancing control loop

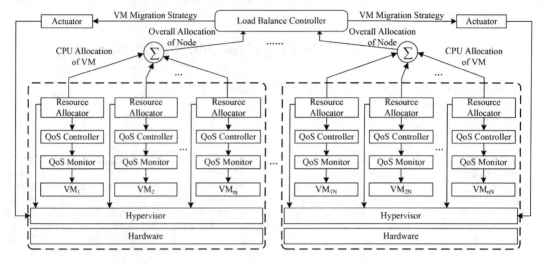

Figure 1. Framework of QALBA.

obtains output of internal QoS loop, makes scheduling decisions based on CPU allocation of virtual machines, and triggers virtual machine migration to realize load balancing for virtual cluster.

2.1 QoS control loop

Every virtual machine is controlled by a QoS control loop that optimizes QoS of application-layer services through adjusting CPU allocation of virtual machines. To facilitate the formation of control loop, this paper chooses the most important indicator of QoS, response time, as the performance index, according to Wang (2011). On the basis of this index, QoS control loop dynamically adjusts CPU allocation of virtual machines. The QoS control loop contains three components: QoS monitor, QoS controller, and resource allocator. This control loop completes the following tasks in the k-th control cycle:

Step 1. QoS monitor captures the response time of i-th virtual machine VM_i to all task requests at the k-th cycle and then calculates the average response time $t_i(k)$;

Step 2. Send $t_i(k)$ to QoS monitor as the controlled variable;

Step 3. QoS controller calculates the required CPU allocation $\vartheta_i(k+1)$ in the $(k+1)$-th control cycle according to the difference between QoS index and response time of i-th virtual machine VM_i. Meanwhile, it sends the value of $\vartheta_i(k+1)$ to resource allocator.

Step 4. Resource allocator adjusts the task scheduling parameter of virtual machine monitor on the basis of the load-balancing decisions that the QoS controller makes.

2.2 Load-balancing control loop

Aiming at controlling multiple physical nodes, load-balancing control loop is essentially a Multiple-Input and Multiple-Output (MIMO) control system. The purpose of this control loop is to guarantee that every physical node within the virtual cluster has almost the same load.

Definition 1. For physical node $Node_i$, suppose the upper limit of its CPU utilization rate is $\max_i$ and the actual utilization rate at k-th moment is $\eta_i(k)$. The relative load of CPU, $\eta_r(k)$, on node $Node_i$ is defined as:

$$\eta_r(k) = \eta_i(k)/\max_i \qquad (1)$$

The core components of this control loop consisted of load monitor, Model Predictive Controller (MPC), and actuator. The following tasks are completed at the k-th control cycle.

Step 1. The load-balancing monitor running on physical node $Node_i$ captures $\eta_i(k)$ and then calculates $\eta_r(k)$.

Step 2. On the basis of the value of $\eta_r(k)$, MPC calculates the number of virtual machines, $n_i(k+1)$, on every physical node $Node_i$ within the virtual cluster during $(k+1)$-th control cycle.

Step 3. According to the calculation results given by MPC, actuator calls the virtual machine monitor to migrate virtual machines.

2.3 Collaboration of two control loops

As both system load balancing and QoS guarantee will result in the allocation and scheduling of system resources, it is necessary to handle the coordination and decoupling relationship between these two control loops when designing the algorithm. Considering that load-balancing strategy has a greater impact on the performance of virtual machines than CPU allocation does, the proposed algorithm regards load-balancing control loop as the external loop and QoS control loop as the internal control loop. The load-balancing control loop migrates the virtual machines while the QoS control loop fine tunes the QoS through adjusting resource allocation for virtual machines.

According to control theories (Yang, 2012; Rubaai, 2011; Chan, 2012), stable state must be achieved for QoS control loop within the control cycle of load-balancing control loop, that is, the setting time of QoS control loop must be shorter than the control cycle of load-balancing control loop. Thus, the control cycle of load-balancing control loop can be determined by the setting time of QoS control loop. Besides, the control cycle of QoS control loop must ensure that virtual machines can process several task requests within one cycle.

3 SYSTEM MODELING AND CONTROLLER DESIGN

3.1 Modeling of controlled objects

The controlled variable of QoS control loop is the response time of application-layer services within the virtual cluster. Therefore, it is important to build performance model for controlled objects before designing QoS controller.

Let T_Q be the control cycle of QoS control loop and $t_i(k)$ be the average response time of the i-th virtual machine VM_i in the k-th control cycle. $t_i(k)$ is treated as a controlled variable, and its reference value is set to T_s, a response time that QoS requires. In the k-th control cycle, the CPU allocation assigned to VM_i is $\vartheta_i(k)$, which is treated as a control variable and is mapped to the internal variable of virtual machine monitor. Considering that the virtual cluster discussed here uses Xen Hypervisor as virtualization software, $\vartheta_i(k)$ is mapped to the internal task scheduling variable cap of Xen Hypervisor, aiming at guaranteeing QoS through adjusting this variable and optimizing task scheduling.

The model of controlled object interprets the relation between control variable $\vartheta_i(k)$ and controlled variable $t_i(k)$. In real practice, there is a nonlinear relationship between $\vartheta_i(k)$ and $t_i(k)$. To eliminate this nonlinear relationship, instead of using $\vartheta_i(k)$ and $t_i(k)$ directly, this algorithm uses difference values to their reference points. The reference point for $\vartheta_i(k)$ is the intermediate value

of CPU allocation, ϑ_i, and thus treating ϑ_i as system input, t_i, the reference point for $t_i(k)$ can be obtained. Let the model variables $\Delta \vartheta_i(k)$ and $\Delta t_i(k)$ be expressed as:

$$\Delta \vartheta_i(k) = \vartheta_i(k) - \vartheta_i \qquad (2)$$

$$\Delta t_i(k) = t_i(k) - t_i \qquad (3)$$

After linearization, the nonlinear relationship between model variables $\vartheta_i(k)$ and $t_i(k)$ is mapped to the linear relationship between $\Delta \vartheta_i(k)$ and $\Delta t_i(k)$. Therefore, the system model can be described by the following difference equation:

$$\Delta t_i(k) = \sum_{i=1}^{m_1} b_i \times \Delta t_i(k-i) + \sum_{i=1}^{m_2} c_i \times \Delta \vartheta_i(k-i) \quad (4)$$

where b_i and c_i are model variables that can be determined by system identification (Wang, 2011). Model orders m_1 and m_2 should be determined by considering both model precision and control overhead. In engineering applications, the order is usually not more than three. Load balancing is a slow control process. As an attempt to incorporate control theory into load balancing, the research adopts first-order model to facilitate program implementation. The parameters of system model can be determined by off-the-shelf business software, such as MATLAB. After obtaining model parameters, performing Z-transform for Equation (4) can deduce the following model expression:

$$G(Z) = c_1 /(Z - b_1) \qquad (5)$$

3.2 Design of QoS controller

The QoS controller is designed to ensure zero stable-state error and minimize setting time. In terms of control overhead, as every virtual machine is corresponding to one QoS control loop, there is a linear relationship between control overhead and the number of virtual machines. QoS controller is designed based on the Proportional Integral Derivative (PID) control strategy and its Z-domain transfer function is as follows:

$$F(Z) = (K_1 Z^2 - K_2 Z + K_3)/Z(Z - 1) \qquad (6)$$

Under the constraints of stability and zero stable-state error, parameters $K_1, K_2,$ and K_3 of controller can be determined using extreme value analysis. Fig. 2 expresses the QoS close control loop. Let the transfer functions of QoS controller and controlled subject be $F(Z)$ and $G(Z)$, respectively, then the closed-loop transfer function is expressed as:

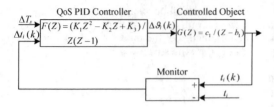

Figure 2. Close control loop of QoS.

$$H(Z) = F(Z)G(Z)/(1+F(Z)G(Z))$$
$$= (-c_1 K_1 Z^2 + c_1 K_2 Z - c_1 K_3)$$
$$/ (Z^3 - (c_1 K_1 + b_1 + 1)Z^2 \qquad (7)$$
$$+ (c_1 K_2 + b_1)Z - c_1 K_3)$$

After obtaining the CPU allocation increment, $\Delta \vartheta_i(k)$, for i-th virtual machine, VM_i, in the k-th control cycle, we can calculate the sum of CPU allocation, $\vartheta_{aj}(k)$, for all the virtual machines on the physical nodes, $Node_j$. $\vartheta_{aj}(k)$ can be expressed as:

$$\vartheta_{aj}(k) = \sum_{i=1}^{n_j} \vartheta_i(k) \qquad (8)$$

where $\vartheta_{aj}(k)$ is treated as a control input for the external load-balancing control loop. Through this method, the number of virtual machines running on every physical node is adjusted.

3.3 Design of load-balancing control loop

3.3.1 Problem description
Assume that the virtual cluster consists of N physical nodes and there are $n_j(k)$ virtual machines running on the j-th physical node, $Node_j$, during the k-th control cycle. The CPU allocation, $\vartheta_{aj}(k)$, for node $Node_j$ can be obtained referring to QoS control loop. Let m_j be the upper limit of CPU allocation on the physical node $Node_j$.
  Definition 2. Define the CPU relative allocation $\vartheta_{rj}(k)$ as:

$$\vartheta_{rj}(k) = \vartheta_j(k)/m_j \qquad (9)$$

Definition 3. Define the average value of CPU relative allocation $\vartheta_r(k)$ for all the physical nodes during the k-th control cycle as:

$$\vartheta_r(k) = \left[\sum_{j=1}^{N} \vartheta_{rj}(k) \right] / N \qquad (10)$$

On the basis of the above definition, we can calculate the control error $e_j(k)$, which is:

$$e_j(k) = \vartheta_{rj}(k) - \vartheta_r(k) \qquad (11)$$

where $e_j(k)$ is treated as controlled variable and its reference value is set to zero. To linearize the control process, control output is defined as $\Delta n_j(k)$, the difference between the number of virtual machines on the current nodes and the number of virtual machines satisfying load-balancing demands:

$$\Delta n_j(k) = n_j(k) - n \qquad (12)$$

where n is selected on the basis that every physical node has almost the similar CPU relative allocation.
  Based on $e_j(k)$, the control error vector $e(k)$ is:

$$e(k) = [e_1(k), \cdots, e_N(k)]^{\mathrm{T}} \qquad (13)$$

Based on $\Delta n_j(k)$, the vector denoting the changing number of virtual machines, $e(k)$, can be formed:

$$\Delta n(k) = [\Delta n_1(k), \cdots, \Delta n_N(k)]^{\mathrm{T}} \qquad (14)$$

The control purpose for load-balancing control loop is to determine vector $\Delta n_j(k)$ during the k-th control cycle, so that the sum of control error for all the physical nodes $Node_j$ during the $(k+1)$-th control cycle can be minimized, which can be denoted by:

$$\min_{\{\Delta n_l(k)|1 \le l \le N\}} \sum_{j=1}^{N} (e_j(k+1))^2 \qquad (15)$$

3.3.2 Controller design
On the basis of the above analysis, load-balancing control loop can be expressed by the following difference equation:

$$e(k) = \sum_{i=1}^{m_1} A_i * e(k-i) + \sum_{i=1}^{m_2} B_i * \Delta n(k-i) \qquad (16)$$

where m_1 and m_2 are the orders of control loop and A_i and B_i are parameter metrics that need to be determined through system identification.
  Load-balancing control loop is an MIMO control system that can process N physical nodes. This paper designs a controller based on LQR for the following consideration: linear quadratic regulator is suitable for handling decoupling MIMO control issues; and LQR can easily obtain the optimum control patterns for status feedback to form closed-loop optimum control, and the calculation amount is moderate.
  Before designing the controller, it is essential to transform the established model into state space model. The concept of accumulative error is

further introduced into the algorithm to enhance control performance.

Definition 4. Define accumulative error vector $q(k)$ as:

$$q(k) = \lambda * q(k-1) + e(k-1) \tag{17}$$

where λ represents forgetting factor. It is used to designate the speed at which the controller forgets control errors in the previous cycle.

According to engineering debugging experience, we assume $\lambda = 0.85$ and $m_1 = 2$ and $m_2 = 1$. Combining Equations (16) and (17), we can obtain the following expression for the model:

$$\begin{bmatrix} e(k) \\ e(k-1) \\ q(k) \end{bmatrix} = \begin{bmatrix} A_1 & A_2 & 0 \\ I & 0 & 0 \\ I & 0 & \lambda * I \end{bmatrix} * \begin{bmatrix} e(k-1) \\ e(k-2) \\ q(k-1) \end{bmatrix}$$
$$+ \begin{bmatrix} B_1 \\ 0 \\ 0 \end{bmatrix} * \Delta n(k-1) \tag{18}$$

Load-balancing control loop is a dynamic optimization process. The control goal of LQR controller for state space model (18) is to seek stable-state gain matrix F so that objective function L can be minimized:

$$L = \sum_{k=1}^{\infty} [e^T(k) \quad e^T(k-1) \quad q^T(k)] * Q$$
$$* \begin{bmatrix} e(k) \\ e(k-1) \\ e(k) \end{bmatrix} + \sum_{k=1}^{\infty} \Delta n^T(k) * R * \Delta n(k) \tag{19}$$
$$s.t. \quad \Delta n(k) = -F * [e(k) \quad e(k-1) \quad q(k)]^T$$

In Equation (19), both weight matrices Q and R are adopted to balance control performance and control overhead. At this point, we have transformed the design of load-balancing control loop into a solvable optimization problem that can be calculated and optimized using off-the-shelf business software packages.

4 ALGORITHM IMPLEMENTATION

QoS monitor runs like a demon in the virtual machine. It observes and calculates $t_i(k)$, the average response time of application layer services on the virtual machine during k-th control cycle, and sends $t_i(k)$ to QoS controller. Then, the QoS monitor sets the control cycle of QoS controller to T_Q and by using global macro defines the permissible maximum response time and reference value of response time for application layer services.

The QoS controller in every virtual machine executes QoS control algorithm according to response time and calculates corresponding $\Delta \vartheta_i(k)$. To improve efficiency, this control algorithm is programmed using C language. Finally, $\Delta \vartheta_i(k)$ is mapped to the increment of task scheduling variable *cap* of Xen Hypervisor. The scheduling algorithm for Xen Hypervisor is set as credit algorithm that can adjust the value of *cap* to regulate the task scheduling and guarantee QoS for application layer services.

For load-balancing control loop, a system-level service Daemon-S is implemented and created in every physical node. The functionality of this service is to call QoS controller to calculate the overall CPU allocation $\vartheta_{aj}(k)$ for all the virtual machines on every node. To facilitate implementation, this service is programmed in Python language. Then, Daemon-S sends $\vartheta_{aj}(k)$ as a control input to the load-balancing control loop. The load-balancing controller performs LQR control algorithm and calculates $\Delta n(k)$ according to $\vartheta_{aj}(k)$ sent from every physical node $Node_j$ during k-th control cycle. Every component $\Delta n_j(k)$ in vector $\Delta n(k)$ indicates the changes in the numbers of virtual machines on every node $Node_j$. On the basis of $\Delta n_j(k)$, Xen Hypervisor migrates virtual machines and assigns loads to appropriate physical nodes to realize load balancing.

5 EXPERIMENTS

5.1 *Experimental environment*

The virtual cluster is configured with four physical servers, three of which constitutes computing nodes $Node_1 \sim Node_3$ and the rest is used as network file system to store virtual machine mirror. Below is the hardware configuration for the computing nodes: Intel Xeon E5-2650-based Super Micro server, 16GB memory, CentOS 6.6 operating system, and virtualization hypervisor Xen 4.5.1. The virtual machine is configured with Apache server to respond to HTTP requests from the client, while the client uses Apache HTTP Server Benchmark as a tool for generating loads.

5.2 *Experiments for QoS controller*

To verify the effectiveness of the dynamic distribution of resources by the QoS controller according to the changing loads of virtual machines, two virtual machines running on node $Node_1$ are tested during multiple control cycles. In order to concisely describe the experimental process without loss of generality, we set the experimental process to 180 control cycles. In experiments, the load of one virtual machine is dynamically changing, while the load of the other is essentially unchanged. The two virtual machines are marked as Virtual Machine 1 (VM1) and Virtual Machine 2 (VM2). Apache Benchmark is used to stimulate load changes of VM1 to test whether QoS controller can adjust adaptively the CPU allocation assigned to virtual machine and satisfy QoS for application-layer services.

In Figs. 3 and 4, to show the variation trend of response time more clearly, we drew one response time every 10 control cycles. The following figures also use the same method.

CPU allocation in Fig. 4 is dimensionless because it is mapped to the scheduling variable *cap* of Xen Hypervisor.

As shown in Fig. 3, within the first 60 control cycles, those two virtual machines have almost the same response times to concurrent requests, meaning that the two virtual machines have similar QoS. During the 60th to 120th control cycles, the concurrent requests sent to VM1 are doubled, in order to imitate increased load. From Fig. 3, it can be observed that at the beginning of increased load, the response time for VM1 temporarily exceeds 600 ms before falling

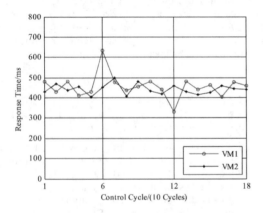

Figure 3.　Response time of VM1 and VM2.

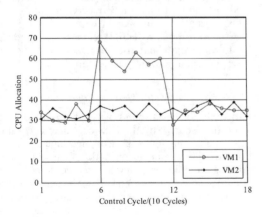

Figure 4.　CPU allocation of VM1 and VM2.

down. From Fig. 4, it can be observed that from the 60th to 120th control cycles, VM1 has higher CPU allocation than VM2. This is because QoS controller increases CPU allocation of VM1 so that VM1 can still satisfy QoS under heavy loads. With the loads of VM1 declining to the normal value starting from the 120th control cycle, its CPU allocation also restores to the previous value. Meanwhile, the CPU allocation of VM2 remains almost the same.

The experiments prove that QoS controller can distribute resources according to the loads of virtual machines and guarantee the QoS of virtual machines.

5.3　Collaboration experiments of QoS and load-balancing control loops

In this session, QoS control loop and load-balancing control loop are co-working to verify the control effect. Meanwhile, the experiment is set the same as in session 4.2 and it lasts 180 control cycles. From the outset of the experiment, the number of virtual machines running on node $Node_1 \sim Node_3$ is 4, 7, and 9, respectively. The concurrent requests are changed using Apache software to simulate load change of virtual machines.

To simulate the scenario where loads of virtual machines increase or decline, two adjustments are performed in the experiment.

First adjustment: Double the loads of all the virtual machines running on node $Node_2$ and node $Node_3$ starting from 60th control cycle.

Second adjustment: Reduce the loads of all virtual machines running on node $Node_1$ from 120th control cycle.

Figs. 5 and 6 show the changes in loads of physical nodes within virtual cluster and the shifting trend of virtual machines. Fig. 7 shows the CPU relative allocation of nodes.

For the first adjustment, the comparison between Figs. 5 and 6 can demonstrate that after increasing the loads for virtual machines on physical nodes $Node_2$ and $Node_3$, load-balancing algorithm has accordingly reduced the number of virtual machines on nodes $Node_2$ and $Node_3$ through virtual machine migration and some virtual machines are transferred to node $Node_1$ with lower loads.

For the second adjustment, only decline the loads of virtual machines on node $Node_1$. From Fig. 5, it can be observed that this adjustment automatically triggers the changes of loads in $Node_2$ and $Node_3$. The loads on $Node_1$ rapidly rises again after a temporary decline, and at the same time, the loads on $Node_2$ and $Node_3$ are declining. This is because load-balancing algorithm migrates part of virtual machines to $Node_1$, which can be shown in Fig. 6.

From Fig. 7, during the whole experiments, CPU relative allocation for nodes within the cluster is almost the same, except when the loads of virtual

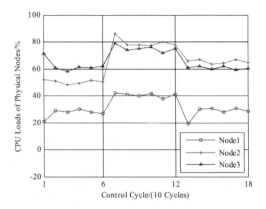

Figure 5. Changing process in loads of physical nodes $Node_1 \sim Node_3$.

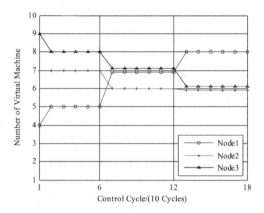

Figure 6. Number changing of virtual machines on physical nodes $Node_1 \sim Node_3$.

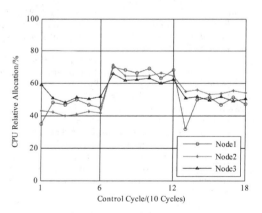

Figure 7. CPU relative allocation of physical nodes $Node_1 \sim Node_3$.

machines change dramatically. This means that during experiment, every physical node has almost the same relative load. The proposed QALBA algorithm realizes load balancing and guarantees QoS

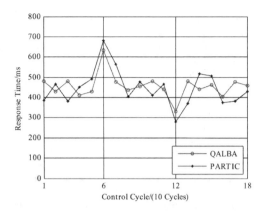

Figure 8. Response time comparison between QALBA and PARTIC.

for application-layer services within the virtual cluster through two methods: the external control loop transfers virtual machines and the internal control loop dynamically regulates resource allocation for every virtual machine.

5.4 Performance comparison

In this section, the proposed QALBA algorithm is compared with PARTIC algorithm presented in Wang (2011) in terms of performance. PARTIC is a classical algorithm that guarantees QoS using negative feedback control mechanism and it regards response time as the most important indicator just as our proposed algorithm does. The control cycle is still set to 180 during experiment. For those two algorithms, we use load generation software in Section 4.1 to double the loads from the 60th to 120th control cycles. Fig. 8 shows the data of response time during the experiment.

It can be observed from Fig. 8 that when the load changes, the response time of PARTIC algorithm has a wider variation than that of QALBA; but when the load is stable, the response time of PARTIC algorithm has a higher amplitude of fluctuation than that of QALBA algorithm. It can be proved that QALBA algorithm can better guarantee the stability of response time.

The above experiments show that QALBA algorithm can reasonably regulate CPU allocation at the requests of QoS and migrate virtual machines based on the macroscopic loads of all the physical nodes within virtual cluster. In this way, the proposed algorithm realizes load balancing on the premise of satisfying QoS.

6 CONCLUSION

With the rapid development of virtualization technologies, virtual cluster has become an important cluster architecture. This paper presents a QoS adaptive load-balancing algorithm that takes QoS issues of virtual cluster into consideration. The algorithm implements internal and external control loops: the QoS control loop triggers resource allocation for virtual machines, and the load-balancing control loop migrates virtual machines in the domain to achieve macroscopic load balancing. Experimental results validate the effectiveness of the proposed algorithm. Compared with existing algorithms, QALBA algorithm can not only realize load balancing, but also guarantee QoS for application-layer services within virtual cluster. As an attempt to introduce control theory into load-balancing algorithm for virtual cluster, this system model uses a simplified first-order model, and part of the parameters are defined according to engineering debugging experience. However, much work remains to be done on building more precise higher-order model and providing theoretical guidance for setting model parameters.

ACKNOWLEDGMENT

This work was financially supported by the National Science Foundation of China (Grant No. 61502438).

REFERENCES

Chan, K.Y., C. Yiu, T. S. Dillan, S. Nordholm, S. H. Ling, Enhancement of Speech Recognitions for Control Automation Using an Intelligent Particle Swarm Optimization, IEEE Transactions on Industrial Informatics, **8**, 4, 869–879(2012).

Chang, P., J. Jung, A Systematic Method for Gain Selection of Robust PID Control for Nonlinear Plants of Second-order Controller Canonical Form, IEEE Transactions on Control Systems Technology, **17**, 2, 473–483(2008).

Dai, X.M., Y. Wang, J. Wang, B, Bensaou, Energy-efficient Planning of QoS-constrained Virtual-cluster Embedding in Data Centers, *2015 IEEE 4th International Conference on Cloud Networking*, 267–272(2015).

Fukushima, M., O. Yamashita, A. Kanemura, S. Ishii, M. Kawato, et al. A State-space Modelling Approach for Localization of Focal Current Sources from MEG, IEEE Transactions on Biomedical Engineering, **59**, 6, 1561–1571(2012).

Garima, R., S. Rama, Analytical Literature Survey on Existing Load Balancing Schemes in Cloud Computing, *2015 International Conference on Green Computing and Internet of Things*, 1506–1510(2015).

Gomez, F., G. Indalecio, A. Garcia, T. Pena, A Flexible Cluster System for the Management of Virtual Clusters in the Cloud, *2015 IEEE 7th International Symposium on High Performance Computing and Communications*, 1693–1698(2015).

Heiner A., A, F. Simon, H Martin, S. Marcel, Distributed algorithms for QoS load balancing, Distributed Computing, **23**, 5, 321–330(2011).

Jain, A., R. Kumar, A Multi Stage Load Balancing Technique for Cloud Environment, *2016 International Conference on Information Communication and Embedded Systems*, 1–7(2016).

Khan, Z., R. Singh, J. Alam, and R. Kumar, Performance Analysis of Dynamic Load Balancing Techniques for Parallel and Distributed Systems, International Journal of Computer and Network Security, **2**, 2, 61–69(2010).

Lei, X., D. Fu, D. Zhu, C. Su, A Novel High-trans Conductance Operational Amplifier with Fast Setting Time, *Proc of the International Conference on Solid-State and Integrated Circuit Technology*, 500–502(2010).

Mohamed, M., S. Yahya, S. Mbaye, Load Balancing Approach for QoS Management of Multi-instance Applications in Clouds, *2013 International Conference on Cloud Computing and Big Data*, 119–126(2013).

Moumita, C., S. Setua, A New Clustered load Balancing Approach for Distributed Systems, *2015 Third International Conference on Computer, Communication, Control and Information Technology*, 1–7(2015).

Priess, M.C., R. Conway, J. Choi, J. M. Popovich, C. Radcliffe, Solutions to the Inverse LQR Problem with Application to Biological Systems Analysis, IEEE Transactions on Control Systems Technology, **23**, 2, 770–777(2014).

Riyazuddin, K., H. Mohd, H. Shahid, Different Technique of Load Balancing in Distributed System: a Review Paper, *2015 Global Conference on Communication Technologies*, 371–375(2015).

Rubaai, A., J. Jerry, S. T. Smith, Performance Evaluation of Fuzzy Switching Position Controller for Automation and Process Industry Control, IEEE Transactions on Industry Applications, **47**, 5, 2274–2282(2011).

Schmid, A., A. Pandey, T. Nguyen, Robust Pole Pplacement with Moore's Algorithm, IEEE Transactions on Automatic Control, **59**, 2, 500–505(2012).

Sun, H.L., T. Zhao, Y. Tang, X. D. Liu, A QoS-aware Load Balancing Policy in Multi-tenancy Environment, *2014 IEEE 8th International Symposium on Service Oriented System Engineering*, 140–147(2014).

Wang Y., Y.X. Wang, M. Chen, X. Zhu, PARTIC: Power-aware Response Time Control for Virtualized Web Servers, IEEE Trans Parallel and Distributed Systems, **22**, 2, 323–335(2011).

Wang, D., X. Zeng, J. Keane, An Output Constrained Clustering Approach for the Identification of Fuzzy Systems and Fuzzy Granular Systems, IEEE Transactions on Fuzzy Systems, **19**, 6, 1127–1140(2011).

Wang, J., K. Prakashan, and A Ilkay, A Physical and Virtual Compute Cluster Resource Load Balancing Approach to Data-parallel Scientific Workflow Scheduling, *2011 IEEE World Congress on Services*, 212–219(2011).

Yang, Y., Q. Zhu, M. Maasoumy, A. Sangiovanni, Development of Building Automation and Control Systems, IEEE Design & Test of Computers, **29**, 4, 45–55(2012).

Ye, F., Z. J. Wang, F. Xu, Y. C. Zhou, F. C. Zhou, et al. A Novel Cloud Load Balancing, Mechanism in Premise of Ensuring QoS, *2013 International Conference on Intelligent Automation and Soft Computing*, 151–163(2013).

Zhao, J., K. Yang, XH Wei, Y. Ding, L. Hu, et al. A Heuristic Clustering-based Task Deployment Approach for Load Balancing Using Bayes Theorem in Cloud Environment, IEEE Transactions on Parallel and Distributed Systems, **27**, 2, 305–316(2016).

Advances in Energy Science and Equipment Engineering II – Zhou, Patty & Chen (Eds)
© *2017 Taylor & Francis Group, London, ISBN 978-1-138-71798-5*

Research of a classification model based on complex network relationships

Weiwen He

*School of Mechatronic and Information Engineering, Guangzhou City Construction College,
Guangzhou, Guangdong, China*
College of Software, South China University of Technology, Guangzhou, Guangdong, China

ABSTRACT: Some machine learning algorithms are used to solve the problem of supervised learning. This idea is introduced into the relation classification issues based on the supervised learning; but when features are introduced into these models, neither the diversity between such features is taken into consideration, nor feature selection algorithm is utilized for treatment. To solve the above problems, this paper establishes a link classification model based on the feature selection, and this model proposes to use the modified RReliefF algorithm based on the local and global character information in order to get more identifiable character, so that the diversity of the characters is ensured and the accuracy of classifier is improved. The experimental result also shows that the classification model is effectiveness, good scalability and robustness. In addition, this model has a certain reference value for the study of complex network relationships.

1 INTRODUCTION

Complex network is a network structure formed jointly by a huge number of nodes and the complex relations there between. Now, the research on the principle and application of complex network has become a significant research direction for current interdisciplinary field. Link prediction is a very hot research focus, which is mainly to identify between which node pairs there are links through digging the known rich information of the complex network. Its numerous fields demonstrate its vast application prospects, among which one hot application is to analyze the networks of the transnational organizations engaged in drug trade and terrorism (Zhang, 2010).

As the main research direction of the data mining and the complex network analysis, link prediction (Kleinberg, 2007) 0 has been carried out to some extent. One mainstream research direction is the link prediction based on the supervised learning (classification model), which is also called the relation classification. The classification issue is a classical one in the machine learning field. Under the classification model, the input data is called training data, each group of which has a specific presentation class. The classification learning establishes a learning process in which there is a comparison between the actual results of the predict outcomes of the "training data", so as to achieve an accurate effect. Combining application scenarios of the link prediction for complex

networks, a complex network $G = (V, E)$ is given, in which V is the node collection, $E \in V \times V$ represents the collection of links between nodes, and $N(u) = \{u \in V \mid (u, v) \in E\}$ is defined as the adjacency node collection of node u. G can be decomposed as the training subgraph $G^{train} = (V, E^{train})$ and the prediction subgraph $G^{predict} = (V, E^{predict})$, so $G = G^{train} \cup G^{predict}$. In G^{train}, every possible link $e(u, v)$ is given a characteristic set $A(u, v)$ and a category $\delta(u, v) \rightarrow [0, 1]$, and there is a need to build a feature selection model and the classification function introduced in the machine learning field based on the training set G^{train} so as to determine the category of each link in $G^{predict}$:

$$f\left(G^{train}, G^{predict}\right) \rightarrow [0, 1]$$

A series of machine learning algorithms are designed to supervise the learning issues, for instance, the most classic Naive Bayesian, logistic regression and SVM for solving classification issues. This idea is introduced into the relation classification issues based on the supervised learning, for example, the non-parametric learning (Sarukkai, 2000) and the classification by maximal separation distance (Zhu, 2012); but these models neither take the diversity between features into consideration when introducing them, nor process them by utilizing the feature selection algorithm. As thus, this paper establishes a relation classification model based on RReliefF feature selection algorithm(Robnik-Šikonja, 2003), and this model proposes to select features by introducing the modified RReliefF algorithm based on the local and global character information. In the artificial datasets and the real datasets, PLS (Partial Least-Squares) algorithm (Sjöström, 2001) is applied to predict the classification and verify the validity of the feature selection. Generally, the contributions made herein are as follows:

- Introducing RReliefF algorithm to process features;
- Analyzing the contribution degree of different features to the classification effect;
- Applying the classical classification algorithm to carried out experiment to verifying its validity.

2 UNSUPERVISED CHARACTER CONSTRUCTION BASED ON LOCAL AND GLOBAL INFORMATION

The link prediction algorithm based on the similarity of the nodes, according to the characteristics of the network topology structure, calculates the information value of similarity between two nodes, and then sequences $\frac{N(N-1)}{2}$ unknown links. The higher the similarity is, the higher the likelihood of a link will be.

2.1 Local (node) character

Local characters generally refer to the character information built based on the common adjacent node collection $\{\omega_1, \omega_2 \ldots \omega_n\}$ of node v and node u. Due to the numerous quantity of characters, only a few of them are listed here (Zhou, 2010)[0]:

- Common Neighbors (CN), CN(u,v) = N(u) ∩ N(v), which is the number of the common adjacent node collection $\{\omega_1, \omega_2 \ldots \omega_n\}$ of the node v and node u. The more the adjacent nodes is, the larger the likelihood of a link will be.
- Resource Allocation (RA), RA(u,v) = $\sum\omega \in$ CN (u,v) $\frac{1}{|N(\omega)|}$, which distinguishes the role and contribution of the common adjacent nodes via the degree of each common adjacent node, and the high degree of common adjacent nodes contribute less than the small degree of common adjacent nodes.
- Jaccard, Jaccard, (u,v) = $\sum\omega \in$ CN(u,v) $\frac{|N(u) \cap N(v)|}{|N(u) \cup N(v)|}$, which not only considers the number of common adjacent nodes of the prediction node pairs, but also the ratio of common adjacent nodes in the collection of adjacent nodes of two nodes;

2.2 Global (path) character

These indexes of local character, based on the local topology property, is suitable for the network with a large number of network nodes or that requiring

fast response due to little required information, simple algorithm, low time complexity and high prediction accuracy. However, such property cannot deeply reflect the intrinsic property of the two nodes, and meanwhile, cannot reflect the impact of the entire network on the two nodes with respect to the smaller networks. Because of these two reasons, the similarity feature indexes based on the global (path) feature are also constructed:

- Katz, which is a characteristic index that, according to the integration of all paths, directly adds the collection on the path and gives higher weight to the shorter path, with the realization formula:

$$\mathrm{Katz}(u,v) = \alpha D_{u,v} + \alpha^2(D^2)_{u,v} + \alpha^3(D^3)_{u,v} + \ldots..$$

Among them, α is the parameter controlling the weight, D is figure matrix, $D_{u,v}$ represents one-step path of the node to u, v, $(D^2)_{u,v}$ is a two-step path, and so on.

- LHN2, which is a variant of Katz index, which is based on the assumption that if the nearest neighbor of the two nodes is similar, they are similar, with the realization formula:

$$\mathrm{LHN2}(u,v) = \varphi(I + \sigma D + \sigma^2 D^2 + \sigma^3 D^3 + \ldots..)$$

Φ and σ are parameters controlling the weight, and I is unit matrix.

- LP, a characteristic index used for describing local path information is defined as the collection of path from node u to node v:

$$\mathrm{LP}(u,v) = D^2 + \vartheta D^3 + \vartheta^2 D^4 + \ldots + \vartheta^{n-2} D^n$$

- SimRank, which, similar to LHN2, is based on the assumption that two nodes are similar if they are connected to the similar nodes, and however, is defined in recursion:

$$\mathrm{SimRank}(u,v) = C \cdot \frac{\sum_{w \in N(u)} \sum_{w' \in N(u)} \mathrm{SimRank}(w, w')}{|N(u)| \times |N(v)|}$$

It is based on the following theory: if two nodes are similar, the adjacent node of one node will be similar to the other node.

3 FEATURE SELECTION AND CLASSIFICATION MODEL

3.1 Relief feature selection algorithm

Feature Selection (FS), also called Feature Subset Selection (FSS), or attribute selection, refers that a feature subset is selected from all the features for better modeling. Feature selection can eliminate the irrelevant or redundant features, so as to reach the purpose of reducing the number of features, improving the accuracy of the model and reducing the run time. Moreover, the model is simplified by selecting the actually related features to facilitate researchers' comprehension on the process of data generation.

Relief (Robnik-Šikonja, 2003) algorithm is a feature selection algorithm based on feature weighting. Given a randomly selected instance I, Relief searches for its two nearest neighbors: one from the same class, called nearest hit H, and the other from the different class, called nearest miss M. On the basis of the values of the instances I, H and M, the feature weight W[A] can be calculated as for the attribute A. The larger weight the feature has, the stronger classification capacity the feature has, or otherwise, weaker.

The attribute A difference of different instances I_1 and I_2 is as follows:

$$\text{dif}(A, I_1, I_2) = \begin{cases} 0; \text{value}(A, I_1) = \text{value}(A, I_2) \\ 1; \text{value}(A, I_1) \neq \text{value}(A, I_2) \end{cases}$$

The function *dif* is used for calculating the distance between instances to find the nearest neighbors. W[A] can be given by the approximate probability distribution:

$$W[A] = P\left(\text{dif}(A, I_1, I_2) \mid \text{nearest inst. from diff.class}\right) - P(\text{dif}(A, I_1, I_2) \mid \text{nearest inst. from same class}) \tag{1}$$

If $A[I] = A[J]$, it means that the nearest instance is in the same class for the attribute A.

Kononenko extends the Relief to enable it more powerful so that can process the incomplete and noisy data and then obtains regressional ReliefF algorithm [RReliefF]. In regression problems the predicted value is continuous, therefore (nearest) hits and misses cannot be used. To solve this difficulty, instead of the require of whether two instances belong to the same class or not, a kind of probability that the predicted values of two instances are different is introduced.

The probability is defined as:

$$P_{\text{difA}} = P(\text{dif value}(A) \mid \text{nearest instances})$$

$$P_{\text{difC}} = P(\text{dif prediction} \mid \text{nearest instances})$$

According to the conditional probability:

$$P_{\text{difC}|\text{difA}} = P(\text{dif.prediction} \mid \text{dif. value}(A), \text{nearest instances})$$

Combined with Equation (1), the following can be acquired:

$$W[A] = \frac{P_{\text{difC}|\text{difA}} \times P_{\text{difA}}}{P_{\text{difC}}} - \frac{\left(1 - P_{\text{difC}|\text{difA}}\right) \times P_{\text{difA}}}{1 - P_{\text{difC}}} \tag{2}$$

Repeat the above process for k times, then the W[A] can be acquired, and then use W[A] for feature selection.

3.2 Classification model

In order to prove the effectiveness of feature selection, Partial Least-Squares (PLS) (Sjöström, 2001) is used for links classification. It is a new type of analysis method for multivariate statistics, which was firstly presented by S. World and C. Albano. By adopting the method of decomposing variables X and Y, the components (usually called factors) are extracted form variables X and Y at the same time, and the factors are arranged in descending order according to their correlation.

Combining with what mentioned above, the overall thoughts of model algorithm are as shown in Table 1:

In this model, different predicted value, different property, different predicted value & attribute

Table 1. Model procedure flowchart.

Algorithm: Prediction model for linking classification based on RReliefF Feature Selection.

Input: G

Output: Classification Accuracy

Initialize:

{in G^{train}, getting every link $e(u, v)$ character collection- $A(u, v)$ and the probability of classification $\tau(\theta)$; initializing: W[A] and other parameters are zero.}

Begin:

// Randomly select a link $[e_i(u, v)]$, and select k sides $[e_j(u, v)]$, which are adjacent to e_i.

for i=1 **to** m **do**

  for j: =1 **to** k **do** // for adjacent instance traversal

    $N_{dC} = N_{dC} + \text{dif}(\tau(\theta), e_i(u, v), e_j(u, v)) \cdot d(i, j)$

    for A:=1 **to** a **do** // for character collection traversal

      $N_{dA}[A] = N_{dA}[A] + \text{dif}(A, e_i(u, v), e_j(u, v)) \cdot d(i, j)$

      $N_{dC\&dA}[A] = N_{dC\&dA}[A] + \text{dif}(\tau(\theta), e_i(u, v), e_j(u, v)) \cdot$

        $\text{dif}(A, e_i(u, v), e_j(u, v)) \cdot d(i, j)$

    end

  end

end

for A:=1 **to** a **do**

  $W[A] = N_{dC\&dA}[A]/N_{dC}\text{-}(N_{dA}[A] - N_{dC\&dA}[A])/(m\text{-}N_{dC})$

end

for t=1,......, T **do** // implement T times training

  G is divided into $G^{\text{train}}, G^{\text{predict}}$

  Classifier ← Train(G^{train}, A, W[A]);

  Classification Accuracy ←Classifier.Predict(G^{predict});

End

are separately determined by N_{dC}, $N_{dA}[A]$ and $N_{dC\&dA}[A]$, and $d(i,j)$ signifies the distance between instances:

$$d(i,j) = \frac{d_1(i,j)}{\sum_{l=1}^{k} d_1(i,j)}; d_1(i,j) = e^{-\left(\frac{rank(I,I_j)}{\sigma}\right)^2},$$

where $rank(I,I_j)$ is the rank of the instance I_j in a sequence of instances ordered by the distance from I and σ is a user defined parameter controlling the influence of the distance.

4 EXPERIMENT AND ANALYSIS

4.1 Experiment setting

Matlab language is used for the development of the whole procedure. AUC evaluation index is adopted in experiment evaluation. AUC randomly selects two relations from the collections of existent links and inexistent links for n times, and separately calculate the similarity of the two relations in the model. If the former similarity is bigger than the latter for a times, and equal for b times, then:

$$AUC = \frac{a + 0.5b}{n}$$

4.2 Dataset

In order to verify the effectiveness of the model, artificial dataset and real dataset are applied in the experiment. First, the first manual network model is based on the thoughts of scale-free network model (Barabsi, 1999) by Barabasi and Albert, and such network is called BA3. The second artificial network introduces Master Equation Growth Model (Dorogovtsev, 2002) to generate a MEGM network of the node 200 and side 3028. The actual network uses the open data from several classical network data platforms, and the properties are as shown in Table 2:

4.3 Experimental results and analysis

4.3.1 Experiment results

The characters obtained are classified as Global (path) characters (Global), Local characters (Local) and All-Feature characters (All-F), and the results of datasets based on RReliefF feature selection algorithm and datasets without being processed with feature selection algorithm are listed in Table 3. According to the results, in all cases, characters being processed are easier to be distinguished, and their effectiveness is 2% to 12% higher than the original characters. This indicates that the RReliefF algorithm can effectively process the redundant information among the characters, thus selecting better character information for classification. Besides, it could be discovered that the Global character is more effective than the Local character (approximately 8% to 26%). This result is consistent with the unsupervised link prediction result, which proves that the Local character is only based on partial adjacent node structure and could not reflect the deep information of the links. This may cause low discrimination in the process of classification. On the contrary, the Global character can be more capable in discovering the global properties in complex networks, and at the same time, distinguishing different links effectively to obtain better classification. Therefore, it is a more practical method to introduce the Global character to the complex network links classification. For the record, experiments mentioned in this paper have been implemented with significance testing,

Table 2. Attributes of Network Structure. |V| and |E| are nodes and links number of the network. C stands for clustering coefficient, and AD stands for average node degree.

Network		|V|	|E|	C	AD
Artificial network	BA3	80	960	0.3969	48
	MEGM	200	3028	0.1112	30.28
real network	Yeast	2375	11693	0.388	9.85
	Jazz	198	5484	0.6334	55.39
	Facebook	333	5083	0.6678	28.55

Table 3. Results of Various Datasets (AUC).

	ALL-F		Local		Global	
	NOFS	FS	NOFS	FS	NOFS	FS
BA_3	0.687	0.7018	0.6927	0.6986	0.9706	0.9697
MEGM	0.7704	0.7824	0.7712	0.8779	0.9922	0.9923
Yeast	0.9826	0.9861	0.9828	0.991	0.9969	0.9986
Jazz	0.7779	0.8297	0.7768	0.8984	0.9559	0.9637
Facebook	0.9133	0.9296	0.9117	0.9207	0.9696	0.9797

according to which the test results prove that the model proposed herein is tenable.

4.3.2 Experiment analysis

The experiment of this section is to verify the expandability and robustness of the model. The method of modifying the training set scale and the k-folder cross-validation strategy are applied in the verification respectively. At first, according to different proportion: $trainRatio \in [0.5, 0.9]$, we divide G into G^{train} and $G^{predict}$, and then collect the experiment results, among which Jazz-FS and Jazz-NOFS separately stand for the results of Jazz datasets being introduced with feature selection and the results of direct classification. Fig. 1 and Fig. 2 show the corresponding effects and it could be clearly figured that the FS line is always above the NOFS line no matter in artificial datasets or in real datasets, and this indicates that the effect of FS is better. The k-folder divides G into k

subsets, and obtains each of the subsets a test set. The rests are training sets, with k times of cross tests repeatedly, and then the average value of the k times of test results is selected as the final result. The experiment sets the k as [5,10], and the results are listed in Fig. 3 and Fig. 4, which are consistent with the results above. The effects with RReliefF feature selection algorithm are always better in different situations, and this proves that the stability and scalability of the algorithm proposed herein are better.

4.3.3 Case analysis

In this section, Jazz dataset is taken as an example, and the Area Under the ROC Curve is introduced, which could be used to evaluate the average performance of feature selection models. If a model

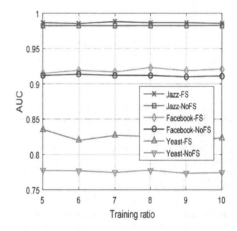

Figure 3.   Real datasets (K-folder).

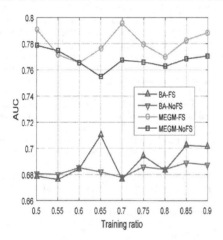

Figure 1.   Real-datasets.

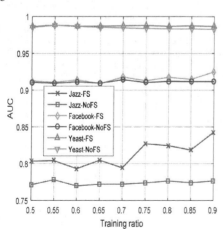

Figure 2.   Artificial-datasets.

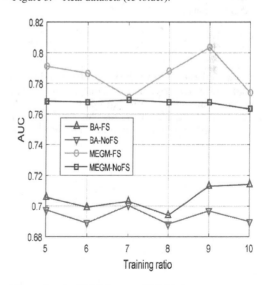

Figure 4.   Artificial datasets (K-folder).

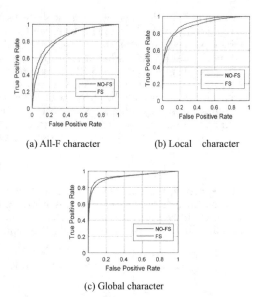

(a) All-F character   (b) Local character

(c) Global character

Figure 5.   ROC curve of Jazz dataset.

has a higher accuracy, its evaluation would be closer to 1, but if one model is better than another, then the area under the curve would be larger. The experiment outputs ROC curve charts of All-F character, Global character and Local character displayed in Fig. 5. From the curves, we can see that result from FS (the red solid line) is not only the closest to the top left corner, but also takes the largest area at the lower right corner. This indicates that it has the best effect, and also proves the effectiveness of RReliefF algorithm in relation classification scenes.

5   CONCLUSION

This paper analyzes the link classification based on complex network characters, and accordingly proposes a link classification model based on RReliefF FS in allusion to the problems of high noise information and big redundancy. This model firstly obtains local and global character information, and then adopts the improved RReliefF algorithm to get more identifiable character by analyzing the character, so that the diversity of the characters is ensured and the accuracy of classifier is improved. In addition, this paper creatively and deeply analyzes the different contributions of Local characters and Global characters to the classification effects, through which it is discovered that Global characters are more identifiable.

Later, it will be the focus at next research stage to apply this classification model on larger scale of dataset for accurate classification.

REFERENCES

Barabsi, A.-L., R. Albert, Emergence of scaling in random networks, science 286 (5439) (1999) 509–512.

Dorogovtsev, S. N., J. F. Mendes, Evolution of networks, Advances in physics 51 (4) (2002) 1079–1187.

Hasan, Mohammad A. Link Prediction using Supervised Learning [J]. Proc of Sdm Workshop on Link Analysis Counterterrorism & Security, 2005.

Liben Nowell D, Kleinberg J. The link prediction problem for social networks [J]. Journal of the American society for information science and technology, 2007, 58(7): 1019–1031.

Lü L, Jin C H, Zhou T. Similarity index based on local paths for link prediction of complex networks [J]. Physical Review E, 2009, 80(4): 046122.

Lü L, Zhou T. Link prediction in complex networks: A survey [J]. Physical A Statistical Mechanics & Its Applications, 2010, 390(6):1150–1170.

Robnik-Šikonja M, Kononenko I. Theoretical and Empirical Analysis of ReliefF and RReliefF [J]. Machine Learning, 2003, 53(1–2):23–69.

Sarukkai R R. Link prediction and path analysis using Markov chains [J]. Computer Networks, 2000, 33(1): 377–386.

Wold S, Sjöström M, Eriksson L. PLS-regression: a basic tool of chemometrics [J]. Chemometrics & Intelligent Laboratory Systems, 2001, 58(2):109–130.

Zhang Hai. Research on hidden network of terrorist organizations based on social network theory [D]. National University of Defense Technology. 2010.

Zhu J. Max-margin nonparametric latent feature models for link prediction [J]. arXiv preprint arXiv: 1206.4659, 2012.

Advances in Energy Science and Equipment Engineering II – Zhou, Patty & Chen (Eds)
© *2017 Taylor & Francis Group, London, ISBN 978-1-138-71798-5*

Data-driven distributed fault diagnosis strategy based on partial information fusion about a novel energy system

Lin Zhu, Yuntian Shen & Yupu Yang
Department of Automation, Shanghai Jiao Tong University, Shanghai, China

ABSTRACT: This paper addresses a novel combined partial information fusion and data-based learning method to achieve fault detection and location of the series-connected process. First, the serially connected system is analyzed to obtain the partition method of the whole system and provide the conditions to construct the subsystems. Then, the distributed Principle Component Analysis (PCA), which can extract the information from the real measurement of the subsystems is carried out for the modeling of the TS type of fuzzy inference. The constructional residuals are generated by comparing output signals of the TS models and the real measurements of each subsystem. The evaluation of the residuals examines the fault occurrence with the location information. Finally, the feasibility and efficiency of the method are evaluated by the Solid Oxide Fuel cells (SOFC), a new-energy power system.

1 INTRODUCTION

With the development of the modern industry, the demand of the assurance of safety during the productive process is growing while the structure of the plate is becoming complex. Early fault detection, diagnosis, and location could make the plate robust to avoid accidents of the production line in practice, which may cause a great economic loss. For plantwide process monitoring, the traditional concentrated monitoring strategy manages the whole production data by the dimensionality reduction and feature extraction (Kruger, 2001; Wang, 2004; Qin, 2003). However, the plantwide data reduction may reduce the location information of the system and can only determine the fault but not its location. Besides, the relationships among different parts of the plantwide process are also difficult to characterize (Simoglou, 2005). Thus, how to develop efficient monitoring methods for plant-wide processes has been a significant challenge in this area.

Because data-based process monitoring methods do not require the process model and the associated expert knowledge, they can be applied even when the models and expert knowledge of some complex industrial processes/systems are difficult to build and obtain in practice. Thus, they have become more and more popular in recent years. Those data contain the major process information and then can be used for modeling, monitoring, and control. In the past years, a significant progress has been made in the data-mining and processing area, which can provide new technologies for the utilization of process monitoring. The traditional multivariate statistics-based methods such as the Principal Component Analysis (PCA) (Demetriou, 1998; Cho, 2009; Ge, 2014; Cherry, 2006; Ghazzawi, 2008), independent component analysis (ICA) methods (Kruger, 2008; Li, 2002; Kano, 2003; Lee, 2004; Albazzaz, 2007), and the combination of ICA and PCA (Ge, 2007; Zhao, 2008) have several inherent limitations such as they can be carried out only by collecting a lot of data. Recently, a well-known one-class classification method Support Vector Data Description (SVDD) has been introduced for process monitoring (Liu, 2008; Ge, 2009).

As a benchmark simulation, the Tennessee Eastman process was considered as a representative plantwide process (Wang, 2004; Qin, 2003; Simoglou, 2005; Demetriou, 1998; Cho, 2009; Ge, 2014; Cherry, 2006; Ghazzawi, 2008; Kruger, 2008; Li, 2002; Kano, 2003; Lee, 2004; Albazzaz, 2007; Ge, 2007; Zhao, 2008; Liu, 2008; Ge, 2009), which has been widely used to test the performance of various monitoring approaches. However, as a representative new-energy resource and the new field of chemical industry, the fuel cell system has not been reported in existing works. Therefore, this article takes the solid oxide fuel cell to provide a reference to investigate the set of systems that are serially connected and have more variables than TE.

The remainder of this paper is organized as follows. In Section 2, the serially connected system is introduced and the sufficient condition of a serially connected subsystem is provided. In Section 3, the distributed PCA method is employed to substract the principle information of the subsystem.

Then, in Section 4, the two-dimensional Bayesian-based fuzzy model is deduced both offline and online forms of system identification and modeling method. Section 5 provides the fault detection and diagnosis technique. Finally, in Section 6, the monitoring method is verified by the solid oxide fuel cells system. Finally, conclusions and some discussions are made.

2 NEIGHBORHOOD–NEIGHBORHOOD INDUSTRY SYSTEM

In industry production line, such as chemical engineering, petrochemical engineering, and electric power industry, there always exists a series connected structure (Fig. 1). The serially connected system is composed of n subsystems. This class of cascade processes involves many subprocesses placed one after another in such a way that each subprocess is connected with dynamic control input coupling between its neighbor subprocesses. As a consequence, if malfunction happens from one subsystem, all the downstream subsystems would be out of order. Therefore, specialties of the serially connected system can be utilized to detect and locate the fault at the same time.

The model of the system can be described as:

$$\begin{cases} x(k+1) = Ax(k) + Bu(k) \\ y(k) = Cx(k) \end{cases} \tag{1}$$

$$\begin{cases} x_i(k+1) = A_{ii}x_i(k) + B_{ii}u_i(k) + \sum_{\substack{j=1 \\ j \neq i}}^{n} A_{ij}x_j(k) \\ y_i(k) = C_{ii}x_i(k) + \sum_{\substack{j=1 \\ j \neq i}} C_{ij}x_i(k) \end{cases} \tag{2}$$

The process system can be divided into several subsystems, as shown in Fig. 1, on the basis of the location or logic and treat all the correlation between the different subsystems as new inputs or disturbance. If the correlation between two subsystems is much higher than the self-correlation in one single system, then the two subsystems should be treated as one subsystem. When the correlation of the subsystem is particularly weak or impacts other subparts with a large time delay, those subsystems could be handled separately.

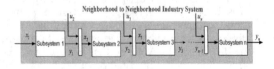

Figure 1. Serially connected process.

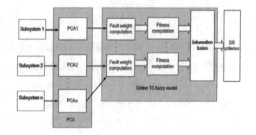

Figure 2. Distributed PCA method.

3 DISTRIBUTED PCA METHOD

Data preprocessing is also a critical step for data-based process monitoring. The aim of this step is to transform the original data in a more appropriate manner, which can be efficiently used for modeling. Although the traditional PCA analysis method is very efficiency to detect the fault, the variables of the whole system are reduced, the collinearity is eliminated, and the location information of the system will also be erased.

In this section, a partial dimension reduction method is proposed to obtain the subfeature space of each subsystem. The main critical data characteristics in the process industrial are analyzed in the PCA method. The series connected construction of industrial system is quite common in the plantwide process. The conventional PCA method can diagnose the occurrence of fault on the basis of the feature extraction of the system states. However, the PCA method has several inherent limitations, which are made under the assumption that the process data are Gaussian-distributed. Also, they require the process variables to be linearly correlated, and the process is operated under a single stationary condition, as shown in Fig. 2.

The initial PCA decomposition carried out upon the whole process variables can decrease the dimension of the system, but the information of the whole process is reduced. To identify the different patterns in the training database, the traditional PCA is modified to obtain the feature space of the model data. Instead of reducing the whole X of the system, the new method reduces the subsystem inputs to get the feature space of each subsystem. The improved PCA-based method is named part dimensionality reduction, which conserves the location information of the connected subsystem.

4 TS FUZZY MODEL

Fuzzy modeling often follows the approach of encoding expert knowledge expressed in a verbal form in a collection of if–then rules, creating a model structure.

The contribution of the ith TS fuzzy rules to the system was expressed in the form of "If...Then" statement as follows:

$R^i : If \, x_1(k) \, is \, A_1^i \, \cdots and \, x_n(k) \, is \, A_n^i$
$Then \, y^i(k+1) = p_0^i + p_1^i x_1 + \cdots + p_n^i x_n; \quad i = 1,2 \cdots c$

where c is the number of fuzzy rules, n is the input variables number of the T-S fuzzy model, $x_1(k), x_2(k), \cdots x_n(k)$ are the regressive variables consisting of output and input data at the kth instance and before, $x(k) = [x_1(k), x_2(k), \cdots x_n(k)]$ is the input vector of the T-S fuzzy model, $A_1^i, A_2^i, \cdots, A_n^i$ are the membership functions associated with the ith rule, and $p_0^i, p_1^i, \cdots, p_n^i$ are the consequent parameters of the submodel (fuzzy rules) i.

The fitness grade of the submodel i is denoted by β_i, thus the model output $y(k+1)$ at the $(k+1)$th instance can be calculated as follows [25]:

First, the model of the TS can be rewritten as:

$$y(k) = \sum_{i=1}^{c} \beta_i y^i(k) = \sum_{i=1}^{c} \beta_i (p_0^i + \cdots + p_n^i x_n(k))$$
$$= \sum_{i=1}^{c} (p_0^i + \cdots + p_n^i)(\beta_i + \cdots + \beta_i x_n(k))^T \quad (3)$$

$$\begin{cases} \Theta(k) = [\theta_1, \theta_2, \cdots, \theta_r]^T \\ \quad = [p_{10}, p_{20}, \cdots, p_{c0}, p_{11}, p_{21}, \cdots, p_{c1}, \cdots, p_{cn}]^T; \\ \Phi(k) = [\beta_1, \cdots, \beta_c, \beta_1 x_1(k), \cdots, \beta_c x_1(k), \cdots, \\ \quad \beta_1 x_n(k), \cdots, \beta_c x_n(k)]^T; \end{cases} \quad (4)$$

The outputs can be written as:

$$y(k+1) = \Phi(k)^T \cdot \Theta(k) \quad (5)$$

5 DISTRIBUTED FAULT DETECTION AND LOCATION

The traditional dimensionality reduction would put all the input into one matrix X and could get a complete dimension reduced data. However, in that way, some important information may be omitted, such as subsystem relationship and individuality. Hence, in this part, an innovative distributed fault detected strategy is proposed to address these problems.

The TS fuzzy model calculates the outputs during the working of the system. In each sampling period, the TS fuzzy model outputs the model-calculated state of each subsystem and compares it with the real system:

$$Y - \hat{Y} = [y_1, y_2, \cdots y_n] - [\hat{y}_1, \hat{y}_2, \cdots \hat{y}_n]$$
$$= [\Delta y_1, \Delta y_2, \cdots, \Delta y_n] \quad (6)$$

The $[\Delta y_1, \Delta y_2, \cdots \Delta y_n]$ would obey the normal distribution $N(0, \sigma_1)$ to $N(0, \sigma_n)$ if the fault did not occur. If some of the variables in the $[\Delta y_1, \Delta y_2, \cdots \Delta y_n]$ distribute outside of the 3σ, then the fault occurred from the location of the first deviation.

Offline form:

1. Use C cluster algorithm to recognize the subordinate degree function u_i of $x(k)$,
2. When the input equals $x(k)$, the fitness of the ith regulation to the system output can be calculated by:

$$\beta_i = \sum_{j=1}^{c} (\frac{u_i}{u_j}), \quad i = 1,2,\cdots,c \quad (7)$$

Its vector can be calculated by:

$$\Phi(k) = [\beta_1, \cdots, \beta_c, \beta_1 x_1(k), \cdots, \beta_c x_1(k), \\ \cdots, \beta_1 x_n(k), \cdots, \beta_c x_n(k)]^T \quad (8)$$

3. As both $y(k+1)$ and $\Phi(k)$ are known quantities, according to formula $y(k+1) = \Phi(k)\Theta(k)$, we can get:

$$\Theta(k) = (\Phi^T \Phi)^{-1} \Phi^T \cdot y(k+1) \quad (9)$$

4. Decode the fault-containing matrix and fault location.

The TS fuzzy model calculation is compared with the production system. Then, on the basis of the residual of the two types of data, the fault can be judged. Finally, the fault location is determined on the basis of the error start location in the fault-containing matrix.

6 CASE STUDY

Solid Oxide Fuel Cells (SOFC) is a new kind of power device (Fig. 3) that has recently attracted the attention of both academy and industry. The system structure of the SOFC is neighborhood-to-neighborhood topology structure. Over recent decades, the Solid Oxide Fuel Cell (SOFC) has attracted considerable attention because of its unique characteristics, including high efficiency, fuel flexibility, and environmentally friendly properties, when adopting a steam reformer to produce hydrogen at high conversion ratio.

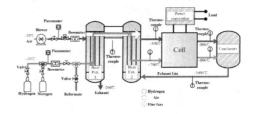

Figure 3. Structure of the SOFC system.

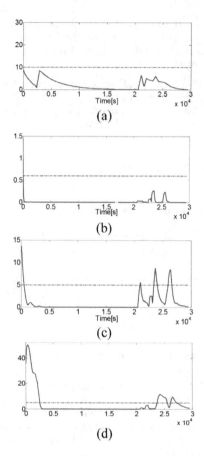

(a)

(b)

(c)

(d)

Figure 4. Residuals of the subsystems.

The system can work in two conditions: the constant-power condition and the time-variant power condition. The constant-power condition indicates that the power termination works in a single power rating. The SOFC system just supplies a constant power until needed. The time-variant power condition is more complex than the former one, which implies that the power provided by the fuel cell is non-constant; as a consequence, the fault detection would be very difficult. In this paper, we just diagnose the fault occurrence in the constant-power condition and provide the TS fuzzy learning of the SOFC states of the time-variant power condition but not the fault.

The simulation is put into effect in the MATLAB/Simulink condition with a computer $3GHz$ and $12G$ memory. The results are shown in Fig. 4. Because of the TS fuzzy model trace, the temperature increases during the normal operation of the variable. When the system deviates from the normal working condition, the TS output and the system output are not running at the same trajectory. In the first 4000s, the TS fuzzy model adjusts its parameters and learning error occurs. After the training

period, the TS model could follow the trajectory of the system. Figure 4 shows the R^2_i results after the smooth filter. It is evident from the figure that the first two subsystems work in a normal condition, as shown in Fig. 4.a-b, whereas the last two subsystems deviate from normal operation condition. The above section indicates the occurrence of fault in subsystem 3, which is the SOFC stack. In that way, the fault of the whole system can be detected when the location of the fault is obtained.

7 CONCLUSION

In this work, a fault detection and location method is presented by distributed PCA and dimensional Bayesian-based fuzzy model, and the SOFC system is employed to verify the method. The serially connected system is analyzed to obtain the partition method of the whole system and provide the conditions to construct subsystems. Then, the distributed principle component analysis is carried out to extract information from the real measurement of the subsystems for modeling of the TS type of fuzzy inference. Constructional residuals are generated by comparing output signals of the TS models and the real measurements of each subsystem. The evaluation of the residuals examines the fault occurrence with the location information. The simulation results show that this method can diagnose the system fault and accurately determine the location of the same.

REFERENCES

Albazzaz, H., X.Z. Wang, Chem. Eng., **94** (2007)
Cherry, G.A., S.J. Qin, IEEE Trans. Semiconductor Manufacturing, **19** (2006)
Cho, H.W. Expert Syst. Appl, **36** (2009)
Demetriou, M.A., M.M. Polycarpou, IEEE Transactions on Automatic Control, **43** (1998)
Ge, Z.Q., L. Xie, Z.H. Song, Ind. Eng. Chem. Res., **48** (2009)
Ge, Z.Q., Z.H. Song, Ind. Eng. Chem. Res., **46** (2007)
Ge, Z.Q., Z.H. Song, Ind. Eng. Chem. Res., **53** (2014)
Ghazzawi, A., B. Lennox, Control. Eng. Pract., **16** (2008)
Kano, M., S. Tanaka, S. Hasebe, I. Hashimoto, H. Ohno, AIChE J., **49** (2003)
Kruger, U., G. Dimitriadis, AIChE J., **54** (2008)
Kruger, U., Q. Chen, D.J. Sandoz, AIChE. J. E **47** (2001)
Lee, J.M., C.K. Yoo, I.B. Lee, J. Process Control, **14** (2004)
Li, R.F., X.Z. Wang, Comput. Chem. Eng., **26** (2002)
Liu, X.Q., L. Xie, U. Kruger, T. Littler, S.Q. Wang, AIChE J., **54** (2008)
Qin, S.J., J. Chemom, **17** (2003)
Simoglou, A., P. Georgieva, E.B. Martin, A.J. Morris, F. Azevedo, Comput. Chem. Eng., **29** (2005)
Wang, X.Z., S. Medasani, F. Marhoon, H. Albazzaz, Ind. Eng. Chem. Res. **43** (2004)
Zhao, C.H., F.L. Wang, F.R. Gao, Y.W. Zhang, Ind. Eng. Chem. Res., **47** (2008)

Advances in Energy Science and Equipment Engineering II – Zhou, Patty & Chen (Eds)
© *2017 Taylor & Francis Group, London, ISBN 978-1-138-71798-5*

Study on configuration optimization of metro fare collection facilities using AnyLogic

Xiao Liang
MOE Key Laboratory for Urban Transportation Complex Systems Theory and Technology,
Beijing Jiaotong University, Beijing, China

Meiquan Xie
School of Traffic and Transportation Engineering, Central South University, Changsha, China
Department of Civil Engineering, California State Polytechnic University, Pomona, CA, USA

Chaoyun Ma
MOE Key Laboratory for Urban Transportation Complex Systems Theory and Technology,
Beijing Jiaotong University, Beijing, China

Fei Wu
Hangzhou Metro, Hangzhou, China

ABSTRACT: Fare collection system is the main service facility for passengers in metro station, the configuration of which has a direct effect on station capacity and level of service. This paper focuses mainly on the stations served for commuting and seldom considers the stations with high recreational passenger flow. It considers fully stations served for recreation and tourism, taking Longxiangqiao station of Hangzhou Metro as an example. The capacity and configuration of fare collection facilities are analyzed through field investigation, and some optimization measures have been proposed to deal with the inappropriate facility layout. A simulation model based on AnyLogic software is established to verify the improved solutions. The results show that the service of each fare collection facility has reached level A, and the average walking time of passengers has decreased by around 20% after optimization. The approach could be helpful and referential to the design and reconstruction of metro stations.

1 INTRODUCTION

At present, tourism in China has entered a rapid development period. For many tourist cities, recreational passenger flow has become an important part of the metro passenger flow, especially at some stations mostly served for tourism, and the number of passengers on holidays is much more than that in weekdays (four times). This is in great contrast with the stations severed for commuting. Meanwhile, because the forecasting passenger volume at the stage of metro planning and design is far fewer than the actual number, the current fare collection facilities do not meet the passenger needs yet. Besides, the inappropriate facilities configuration would easily lead to passenger flow conflict and imbalanced usage of station facilities. There may be a large number of queuing passengers, gathering in the areas of Ticket Vending Machines (TVMs), fare gates, or other places, which reduces their walking speed in the stations, and even causes passenger congestion and stampede. Therefore, fare collection system has become the bottleneck of passenger

evacuation, and furthermore, affects the capacity and the Level of Service (LOS) of the whole station. Thus, it becomes crucial to analyze and optimize the configuration of fare collection facilities, without changing the structure of existing stations.

Fare collection system is one of the important metro infrastructures, including ticket office, TVMs, security scanners, and fare gates. There are many studies on fare collection facility configuration and optimization. Wu et al. divided the fare collection area into nine functional areas and established a microscopic simulation model to evaluate the adaptability of the facility configuration, as well as to obtain the optimum number of fare gates (Wu, 2011; Wu, 2014). Chen et al. developed a simulation model to optimize the hall facilities' layout of Beijing Zoo station using AnyLogic (Chen, 2013). Jiang and Luo optimized the configuration of the automatic fare collection by establishing an AnyLogic simulation model, considering the randomness in service time and passenger arrival (Jiang, 2013). Jiang et al. abstracted automatic ticketing process into a queuing system of a single service desk and used AnyLogic software to design the simulation experiments of the queuing system to obtain the optimal number of TVMs (Jiang, 2014). Xu et al. also used queuing theory to analyze subway station capacity (Xu, 2014). Bao et al. set up a microscopic simulation model of subway station by AnyLogic to analyze ticketing equipment layout and configuration. Disaggregate Logit model was used to depict passengers' behavior on the choice between TVM and manual ticket vending (Bao, 2014). Chen and Li modeled and simulated passengers' average queuing length, average waiting time, area density, and density map through AnyLogic and proposed facility layout optimized solution of Tianfu square metro station by simulating the operational data (Chen, 2016). Li et al. developed queuing simulation software with R language to determine the appropriate number of fare collection equipment (Li, 2016). Other scholars analyzed and evaluated the design of subway station using a simulation model based on hybrid Petri nets (Kaakai, 2007).

Most of the current studies paid more attention to metro stations, which mainly served for commuting, while there was seldom literature considering stations with high recreational passenger flow. Besides, scholars tended to use field survey data and simulation model by AnyLogic software to analyze capacity and LOS of fare collection facilities. The optimum number or layout of fare collection facilities was obtained through simulation optimization. However, many studies focused only on the configuration of a single facility, such as TVMs and fare gates, and did not take into account the fare collection facilities as a whole. Meanwhile, the optimization of ticket office and guardrail layout was likely to be overlooked in current studies. As a result, this paper takes Longxiangqiao station of Hangzhou Metro as an example, which serves for West Lake tourism with a large number of recreational passengers. Then, it investigates the passenger flow during peak hours and analyzes the configuration of the whole fare collection facilities. Some improved measures are proposed to optimize the facilities layout, and a simulation model based on AnyLogic software was developed to verify the optimization solutions. The results could provide a reference to designing the facilities of metro stations and reconstructing the existing facilities.

2 ANALYSIS OF THE CURRENT FARE COLLECTION FACILITIES

2.1 The situation of Longxiangqiao station

Longxiangqiao station of Hangzhou Metro is at a distance of 390 m from the West Lake, and it is the nearest metro station to the tourism areas. The station is underground and has two floors with island platform. In addition, it has four entrances and exits, namely A, B, C, and D. In the station hall, there are three sets of TVMs (15 machines), two ticket offices (six service desks), five sets of fare gates (two for exit with 16 service machines and three for entrance with 15 service machines), three security scanners, two escalators, and one lift, as shown in Figure 1.

As the West Lake is free for people, more and more tourists tend to alight from metro train at Longxiangqiao station, rather than Fengqi station, which is designed originally to serve for the passengers visiting the West Lake. This makes Longxiangqiao station the largest passenger volume station in Hangzhou Metro Line 1, with the total volume reaching 60000 passengers per day during holidays. By contrast, the predicting passenger volume was only 15500 when the station was planned and designed at the beginning. Therefore, the station hall area is small and lacks fare collection facilities, which could not satisfy the current passenger demand. In such situation, it is not realistic to rebuild the station or change the main structure, while optimizing the fare collection facilities' configuration provides an effective approach to improve the station capacity and LOS.

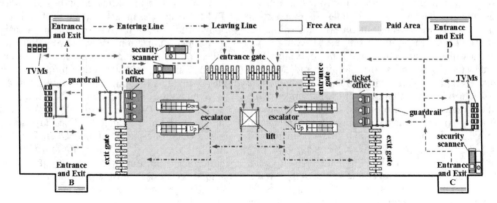

Figure 1.   Current configuration of the fare collection facilities.

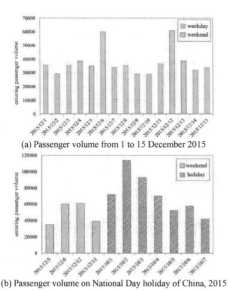

(a) Passenger volume from 1 to 15 December 2015

(b) Passenger volume on National Day holiday of China, 2015

Figure 2. Passenger volume entering Longxiangqiao station.

2.2 Passenger flow characteristics

There are many business and recreational facilities around Longxiangqiao Station. Through field investigation, it is found that passengers for shopping and tourism make up a large portion of the total passenger flow, while commuting passengers take only a small proportion. Furthermore, the passenger volume would reach a peak on holidays, which is about four times more than that in weekdays. Figure 2 shows the passenger volume entering the station. Figure 2(a) shows the passenger volume from 1 to 15 December 2015 and Figure 2(b) shows passenger volume during the National Day holiday of China from 1 to 7 October 2015.

It is noted from Figure 2 that passenger volume on weekends (e.g. Dec.6 and Dec. 12) has reached about 60000 passengers per day, while the value is about 40000 on weekdays. The passenger flow has increased by 50% on weekends than weekdays. Moreover, the passenger flow during holidays increases dramatically comparing to weekdays. From the above, it is concluded that, for metro stations served mainly for recreation and tourism such as Longxiangqiao station, the passenger volume has a great increase on holidays and weekends, which presents typical tourist flow characteristics.

2.3 Capacity of fare collection facilities

The passenger flow during peak hours (14:00 to 17:00) of the Ching Ming Festival (4 April 2016) was observed. At that time, the passenger flow was expected to be maximum at Longxiangqiao station.

Videos were recorded to obtain the data about passengers waiting in line for security and ticket checks. The capacity of fare collection facilities was assessed and the problems that blocked passenger flows were identified. Below is a summary of the assessment.

2.3.1 Ticketing facilities

The ticketing facilities at Longxiangqiao station are well under the capacity. A total of 6559 passengers enter the station in each observed hour. However, the station only has six ticket offices and 15 TVMs. The service rate of each ticket office and TVM is observed to be 144 passengers/h and 84 passengers/h, respectively. Given the fact that 21% of the passengers choose the TVMs while 20% get their tickets from the ticket offices, the total serving capacity of the ticketing facilities was 2124 passengers/h (or 1260 TVM passengers/h and 864 ticket office passengers/h).

The service rates of ticket offices and TVMs are low at Longxiangqiao station as it has a large number of tourist or nondomestic passengers who are not familiar with the ticketing process and spend more time on purchasing their tickets. This is in great contrast with Xizhimen station of Beijing subway, where commuting passengers dominate the passenger flow (i.e, up to 84% of passengers use transportation passes to check in and board a train, whereas only 16% of passengers buy the tickets from ticket offices), with the ticket office and TVM service rates recorded to be 327 and 300 passengers/h, respectively. Precisely, the ticketing facilities are used more frequently at Longxiangqiao station than the stations serving predominately commute passengers, and it requires more ticking facilities (ticket offices and TVMs).

2.3.2 Security check facilities

During the survey, 4204 passengers entered the station from Entrance A & B zone in each peak hour, while 2355 passengers entered from Entrance C. Two security scanners, each with a capacity of 2100 passengers/h, are operated at Entrance A & B zone. They serve the entering passengers well, while the scanner at Entrance C zone cannot meet the demand. An analysis of the scanner at Entrance C zone will be provided in later sections.

2.3.3 Entrance and exit gates

At the station, there are 15 entrance gates, each with a design capacity of 860 passengers per hour. These entrance gates performed well in handling 9404 passengers/h coming into the station. Meanwhile, 16 exit gates (each with a design capacity of 593 passengers) are able to handle 8256 passengers/h leaving the station. In this study, every passenger is assumed to spend 3–5 s to pass through a gate. The software

SPSS is used for curve fitting and statistical tests. The test results indicate that the passing time follows a Poisson distribution with the average of 3.55 s.

2.4 *Configuration issues*

According to the layout and capacity of the fare collection facilities, the station has the following four main shortcomings in passenger flow organization:

1. Six ticket offices with the design capacity of 864 passengers per hour cannot meet the demand of 1312 passengers/h (the total 6559 passengers multiplied by 20%), which often leads to long queues and congestion.
2. At Entrances A and B, the orientation of the security scanners is perpendicular to the passenger flow, which leads to unbalanced usage of the facilities. At Entrance C, there is only one scanner whose capacity is low for serving passengers. As a result, passengers are forced to form long queue(s) preparing for the security check. The rule "security check first and ticket purchase second" leads to a big issue when a short zone is provided for security check. More seriously, it is not safe when the entrance and exit passageways are blocked because of the long queue(s).
3. The orientation of entrance gates is unreasonable, which leads to conflicts among passenger flows and unbalanced use of the fare gates. Some gates are used more frequently than others. At Entrance D, three of nine entrance gates are oriented along their passenger flows, while the rest are at the right angle of the passenger flows. The ratio of the passengers passing through these two groups of gates is 3:2. It is found from the observations that the passengers prefer to use the parallel gates, which may cause long queues in front of these gates.
4. At Entrance A & B zone, passengers, after ticket transactions, walk to the security scanner, which is oriented perpendicular to the passenger flow. The passengers usually prefer to use the closest scanner for security check, which often leads to jams and long line in front of the scanner. In return, the jams could further affect the passengers entering the security check region. Besides, at Entrance and Exit C zone, passengers who leave from the exit gates interfere with passengers entering the station.

3 SIMULATION OPTIMIZATION OF THE FARE COLLECTION FACILITIES

AnyLogic software provides a virtual environment in which passenger flows can be simulated to evaluate the effect of security and ticketing facilities on passengers. In this paper, a simulation model is developed using AnyLogic to visualize the patterns of peak passenger flows regulated by security check and ticketing facilities during peak hours. The purpose of this model is to analyze the capacity and LOS of the facilities and provide recommendations for the improvement of facility configuration.

In this paper, queue density and walking time of passengers are used to measure the capacity and LOS. The queue density is defined as the average number of passengers waiting in line per unit time and unit area. The higher the density, the more crowded the region is and the lower the passengers' walking speed is. The walking time is the difference between the moment when a passenger arrives at the boarding platform from an entrance and the moment when the passenger arrives at the entrance.

When the passenger density is below 0.83 passenger/m^2, the LOS is defined as level A, from 0.83 to 1.11 passenger/m^2 as B, from 1.11 to 1.67 passenger/m^2 as C, from 1.67 to 3.33 passenger/m^2 as D, from 3.33 to 5.0 passenger/m^2 as E, and when the passenger density is above 5.0 passenger/m^2, the LOS is defined as level F.

3.1 *Model development*

The simulation model is developed through a set of steps. First, the Computer-Aided Design (CAD) layout of the Longxiangqiao station is developed and imported into AnyLogic. Second, the facilities (such as walls, TVMs, security scanners, entrance and exit gates, stairs, escalators, and guardrails) within the station are modeled using the space mark function of the AnyLogic pedestrian library. Finally, the logic module representing the interactions of the facilities and the passengers is constructed. The model parameters are adopted from the field survey data, as described in Section 2.2.

3.2 *Optimization measures*

The configuration of the security check and ticketing facilities is optimized to resolve the issues described in Section 2.4 Through a trial-and-error process, the optimal configuration consists of the following improvements, as shown in Figure 3.

1. Add three temporary ticket service desks and make the number of two groups of TVM the same at Entrance A & B zone. The current two groups of TVM are too close to each other. Passengers are expected to use them with almost equal probability. However, the field survey and simulation indicates that the LOS is much different at the A & B zone. In order to improve the performance of ticketing facilities, this study has provided the same number of VTMs at the two locations. Furthermore, the ticketing facilities always reserve a certain area,

which results in more passengers waiting in line. A part of the guardrails has been removed, which made the queues develop longitudinally on condition that the capacity of the ticketing facilities meets the passenger demand. At the same time, the ticket office should be helped with guardrails, as shown in the left zone of Figure 3. By doing so, the passenger flow is regulated. Passengers, after the ticket traction, can join the flow for security check.

2. Change the layout of the entrance gates at Entrance D zone and arrange the gates progressively at 45° against the passenger flow, as shown in the middle area of Figure 3. Through this, the unbalanced use of gates can be reduced.

3. Add a security scanner at Entrance C zone or move the existing scanner to the outside area. Also set up guardrails at certain locations (such as at the junctions of exit gates, entrance passageways, and exit passageways) to regulate passenger flows properly and then decrease conflicts among different passenger flows, as shown in the right zone of Figure 3.

3.3 Analysis of the simulation results

Figure 4 shows the simulation results of the passenger flows at the security check and ticketing facilities zone. The red areas represent the locations with the highest passenger density, where congestion is likely to occur.

It is noted that the passenger density is higher at the areas of ticket offices and TVMs at Exit A & B zone, entrance and exit gates, and the passageway at Exit C zone. The high density may cause a long queue for passengers and a low throughput. Meanwhile, passengers have to purchase their tickets first and then go through the security scanner

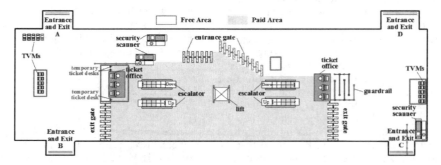

Figure 3. Optimized configuration of the fare collection facilities.

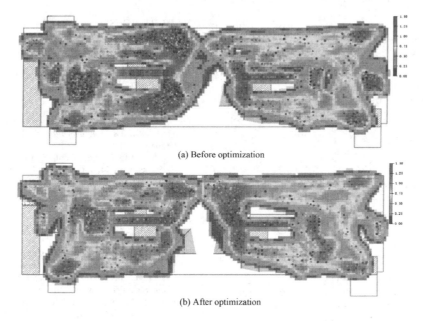

(a) Before optimization

(b) After optimization

Figure 4. Passenger flow density comparison at fare collection facilities in Longxiangqiao station.

at Exit A & B zone. With the accumulation of passengers entering the station constantly, congestion could be formed near the security scanner, especially during the peak hours on holidays. The security scanner capacity could not meet the passenger demand. As a result, an optimal arrangement of the facilities is required. Figure 2(b) shows the simulation result after optimization. Obviously, the red areas get smaller and the passenger density is relatively lower.

Furthermore, the LOS comparisons of the facilities are described in Table 1. After optimization, the LOS of all the facilities reaches level A.

In order to calibrate the model, 1080 passengers were used as samples to verify the walking time of passengers. Table 2 shows the average walking time of passengers entering from Entrance & Exit A, B, C, and D.

It is indicated from Table 2 that the passengers' walking time after optimization is decreased, especially for Entrances A, C, and D. The decreased rate is around 20%. The additional ticket office with guardrail at Exit A zone relieves the congestion and shortens the walking distance and walking time of passengers. After improvement of the entrance gate at Exit C and D zone, the passengers' walking time is dropped clearly. At Exit B zone, the optimization helps to reduce the time of passengers waiting in line. The walking distance seems to be a bit increased, and the decline of walking time for the passengers is not too obvious. It is noted that the decrease of walking time means the improvement of capacity in the station. That is, the optimization described in this paper is realistic and effective.

Table 1. LOS comparison of fare collection facilities.

Fare collection facility	Before optimization	After optimization
TVM at Exit A&B zone	C	A
Ticket office at Exit A&B zone	D	A
TVM at Exit C zone	B	A
Ticket office at Exit C zone	B	A
Entrance gates at Exit C zone	C	A
Passage of Exit C	D	A

Table 2. Average walking time for different entrances.

Entrance & Exit	Before optimization	After optimization	Decreased time	Decreased proportion
A	75.09 s	59.38 s	15.71 s	20.92%
B	85.01 s	80.72 s	4.29 s	5.05%
C	89.85 s	74.42 s	15.43 s	17.17%
D	63.05 s	50.62 s	12.43 s	19.71%

4 CONCLUSIONS

As tourism passengers have dominated the passenger flow in some metro stations, and the lack of consideration about the station facilities in the original design and construction, the current fare collection facilities could not satisfy the growing passenger demands sufficiently. This paper chooses Longxiangqiao station of Hangzhou Metro as a study case and analyzes the capacity and configuration of the fare collection facilities by field survey. The corresponding improved measures are proposed to optimize the existing facility configuration. Moreover, the paper develops a simulation model to evaluate the impact of fare collection facilities on passengers using AnyLogic software. The results show that the passenger volume has dramatically increased on holidays and weekends for stations served mainly for recreational passengers; the passenger density around fare collection facilities has been obviously reduced and LOS of each facility has been greatly improved after optimization. The method used in this paper is effective and could make a reference for metro station design and construction.

ACKNOWLEDGMENTS

This work was supported by the National Natural Science Foundation of China (71401006), the Fundamental Research Funds for the Central Universities (2014JBM057), and Hunan Provincial Natural Science Foundation of China (14JJ3030).

REFERENCES

Bao, G., L. Lu, H. Zhao, J. Next Gener. Inf. Technol **5**, 126–135 (2014)
Chen, L., R. Song, Z. Li, T. Li, J. Transp. Inf. Saf **31**, 19–24 (2013) (in Chinese)
Chen, W., Z. Li, J. Transp. Eng. Inf **14**, 110–115 (2016) (in Chinese)
Jiang, Y., N. Luo, *Fourth International Conference on Transportation Engineering*, 1142–1149 (Chengdu, China, 2013)
Jiang, Y., X. Lin, D. An, J. Zhu, L. Hu, J. Southeast Univ. (Nat. Sci. Ed.) **44**, 430–435 (2014) (in Chinese)
Kaakai, F., S. Hayat, A.E. Moudni, Simul. Modell. Pract. Theory **15**, 935–969 (2007)
Li, G., M. Xi, L. Ni, *Proceedings of the 2015 International Conference on Electrical and Information Technologies for Rail Transportation*, 429–436 (Springer Berlin Heidelberg, 2016)
Wu, X., Z. Yuan, Z.Z. Tian, *Transportation Research Board 90th Annual Meeting*, 11–0668 (Washington, DC, 2011)
Wu, X., Z. Yuan, *the 2nd International Symposium on Rail Transit Comprehensive Development (ISRTCD) Proceedings*, 67–74 (Springer Berlin Heidelberg, 2014)
Xu, X., J. Liu, H. Li, J. Hu, Transp. Res. Part C. Emerging Technol **38**, 28–43 (2014)

Research on session heterogeneity in the structured P2P cloud computing system

Dao-de He
School of Information Engineering, Guizhou University of Engineering Science, Bijie, China

Shan-xiong Chen
College of Computer and Information Science, Southwest University, Chongqing, China

Li Wang
School of Information Engineering, Guizhou University of Engineering Science, Bijie, China

Feng-yang Zhang
College of Computer and Information Science, Southwest University, Chongqing, China

ABSTRACT: This paper presents the design and evaluation of CHPC, a scalable, stable and price moderate network for P2P-Cloud computing. The structured P2P network Pastry is used to build the underlying overlay network and the P2P nodes to provide cloud services. Aimed at the problem of node session heterogeneity, the nodes in network are divided into cloud nodes and ordinary P2P nodes by the hierarchical way. The cloud nodes construct the network according to the Pastry structure. The ordinary P2P nodes are clustered around the cloud nodes. The simulation results show that this network can reduce the rate of joining and exiting, and improve network stability. Moreover, this research has important significance for breaking the enterprise monopoly of cloud services and promoting its popularization.

1 INTRODUCTION

At present, the resources of the cloud computing are provided and controlled by the CSP (Cloud Service Provider). So the cloud system has good reliability and controllability. However, according to the definition of the cloud computing (Wikipedia, 2015), it is a computing method of sharing software and hardware resources based on the Internet. The cloud system is monopolized by a few service providers, so it is bound to make the scalability and cost of the cloud computing severely restricted. Resources of P2P computing are provided by ordinary users in the Internet. The relationships of these users are point to point, so that resources can be fully utilized, thus the network scalability and resource cost can be ensured. But the user's session is heterogeneous, which results in unreliability of network, so that the service becomes uncontrollable and unreliable. We can draw the conclusion that the cloud computing and the P2P computing are two complementary technologies from the above analysis. In order to make full use of these two technologies, many scholars have combined the technology of P2P computing and cloud computing to form the P2P-Cloud computing technology in order to realize the sharing

of resources in the Internet (Wu, 2011; Ge, 2013; Wahid, 2015; Lopez-Fuentes, 2016; Rocha, 2016; Xu, 2012). These technologies are based on the use of P2P computing to construct overlay networks. And the nodes in this network are stable cloud nodes so then it can avoid the session heterogeneity of P2P nodes. Therefore, these technologies make the network have advantages of cloud computing and P2P computing. But the stable network's nodes will make it lose the excellent characteristics of low cost and scalability etc. Based on this, this paper analyses the session heterogeneity of P2P network, and improves the network topology so as to be suitable for cloud computing. Then it will make the ordinary Internet users become the provider of cloud services, so as to interpret the definition of cloud computing which is defined by Wikipedia.

2 RELATED RESEARCHES

In this paper, the network topology is based on Pastry (Rowstron, 2001). Pastry is a structured P2P system which is constructed by DHT (Distributed Hash Table) (Boon, 2003). Each node in the system has a 128 bit *nodeId*. When the node joins the system, this ID is obtained by using the correlation parameter V (Value, which can be IP address, public key and other content) of the node through the hash function. When resources need to be located, this value can be got by the hash function handling a parameter K (key, which can be the resource's name etc) of the resources. Then based on this value and Pastry's routing rules, the resources can be located.

2.1 The basic structure of Pastry

In Pastry system, each node has three status tables, the leaf set (L, leaf node set), the routing table (R, routing table) and the neighborhood set (N, neighbor node set). These status tables are used to maintain the topology information of the network. Figure 1 shows the status set of the node whose *nodeId* is 10233102 in the literature [8].

In Figure 1, the *nodeId* is 2^b-scale (e.g. node 10233102 is four-scale). The node's routing table (R) is organized with 2^b-1 entries for each row. The first n digits of the current node's *nodeId* are the same as the node which is in the nth row (the n's value starts from 0). And the *nodeId*'s format is the same prefix + the next digit + different residual bits. For example, there are n bits same prefix in the nth row, and the next digit (the (n+1)th digit) is the column number which is also numbered from zero. And this entry will be null when the number is the same as the present node's number. The leaf set (L) stores the routing information of the nodes whose *nodeId* is similar to local node. It contains two parts: the first half of node's *nodeId* is slightly larger than present node, and the latter is slightly smaller than this. Neighbor node set (N) is used to store the information of the nodes nearest the local node. This set is not used in normal routing process generally. Its main role is used to maintain the routing of the locality (He, 2009).

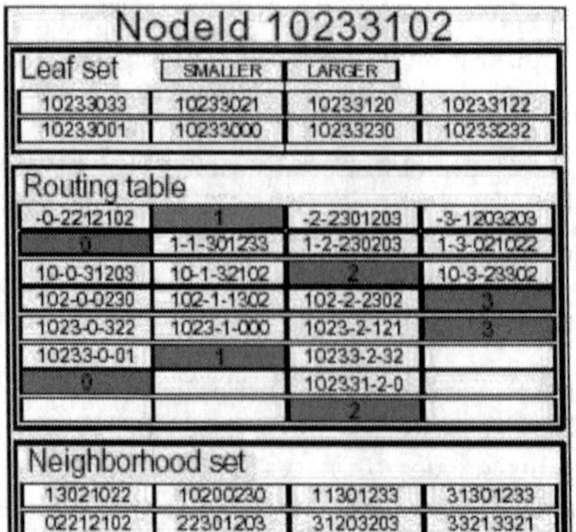

Figure 1. The status set of the node 10233102.

2.2 The joining and exiting principle of the Pastry node

When a new node needs to be added to the Pastry system, it must initialize its state table at first, and notify the other nodes which have been added to the system. Before the node X joins the system, it needs to search for the nearest node A, and submits admission request to this node. Next, node A sends a routing message to any node in the system (this routing message search key is the node identifier of the node X), and this message will be routed to the node Z whose *nodeId* is closest the node X's *nodeId*. Thirdly, they send their own status tables to the nodes which are the Node Z, A, as well as all nodes on the routing path. Then the node X uses this information to initialize its own status table, and notify other nodes that they have joined the Pastry system.

As for the problem of dealing with node failure or exit, the Pastry system uses the method that the node detects communication between itself with its adjacent node to deal with the problem of node failure. If the node A detects that a node B fails in its leaf set, it requests the leaf set of node C whose *nodeId* is the maximum or minimum in A's leaf set (if the B's *nodeId* is larger than A's, the C's *nodeId* is the maximum, otherwise, it is the minimum). Then, the node A selects a node which is not in its leaf set but in C's to replace the failed node B. If the node A detects that the node D fails which is in A's routing table, the node A selects the node E which is in the same row of this entry to replace the failure node. And if there is no other node in the row which has the failed node D's entry, the node A searches for a node from the next row to send the request, this process will be executed to find the corresponding table entry or traverse the entire routing table. In addition, the node A also checks whether there is a failure node in neighborhood set. If A's neighbor F is failure, then select a node from the neighborhood set of A's other neighbor H to replace this failure node.

From the above Pastry node joining and exiting mechanism, we can get that the system initializes or maintains the system status table to ensure that the resources can be released normally and positioned. But the system doesn't consider session heterogeneity of the nodes, so that the service terminating or the frequent access of the nodes can result in the instability of the system. And these can not support that the cloud computing provides the stable and continuous service to the users. Therefore, based on the existing structured P2P network system Pastry, this paper overcomes the session heterogeneity of its network nodes, and transforms it into a network system that can provide cloud services—CHPC (Conversation Heterogeneous P2P Cloud).

3 P2P CLOUD ARCHITECTURE DESIGN BASED ON SESSION HETEROGENEITY

In this section, the topology of CHPC is designed. There are two type nodes in the system, the ordinary P2P node and the cloud node. In order to distinguish between ordinary P2P node and cloud node, the following definitions are designed:

Definition 1: *CLOUD AREA*, which is constituted by the Pastry network topology. And the node in this region provides cloud services. This area is a stable region of CHPC, and the other is not stable.

Definition 2: *Cloud Node*, the node which provides cloud services in CHPC system is called cloud node. The Pastry network topology and the cloud node are used to build the underlying overlay network.

Definition 3: *Ordinary P2P Node*, when the node is initialized to join the system, it is an ordinary P2P node, and searches for a cloud node to as father node and be clustered around it. Ordinary P2P node can only use cloud services in the case of payment, cloud services can not be provided by this. And it can only get the cloud services from its father node from cloud area.

Definition 4: Online threshold (*ThrOnliT*), when the ordinary P2P node's online time exceeds this threshold value, this node can become a cloud node.

Definition 5: Performance threshold (*ThrP*), when the comprehensive performance of the ordinary P2P node (such as storage space, CPU, etc.) is beyond this threshold value, this node is possible to join the cloud network, provides cloud services.

According to the above definition, the topological structure of CHPC is composed of two parts. The network has been constructed by all cloud nodes

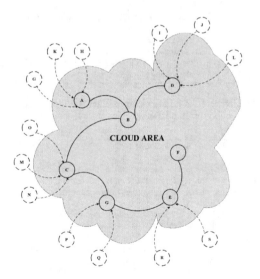

Figure 2. Network topology of CHPC.

in accordance with the Pastry system, and then the cluster is constructed by that the ordinary P2P nodes are clustered around the cloud node, thus the two-layer topology is formed. As shown in Figure 2:

In Figure 2, nodes (*A*, *B*, *C*, *D*, *E*, *F*, and *G*) are located in the cloud region (*CLOUD AREA*), they are the cloud nodes, and the structure of Pastry is used to construct the cloud area.

It is the ordinary P2P node which is located outside the cloud area. And this kind of node is clustered around the cloud node. For example, nodes (*O*, *M*, and *N*) are clustered around the node *C*.

4 NETWORK TOPOLOGY MAINTENANCE MECHANISM OF CHPC

In the third section, the CHPC network topology is designed to divide the nodes into ordinary P2P nodes and cloud nodes, and then it is divided into two layers of structure. Based on the above basis, in this section, we design the node joining and exiting (failure) mechanism of the CHPC, in order to achieve the network topology maintenance.

4.1 *Relevant definition*

In order to more clearly describe the network topology of CHPC and its maintenance mechanism, we need to give some precise definitions on the basis of the definition given in the third section, including the online time value of the node, the comprehensive performance value of the node, the maximum number of child nodes, the child node table etc.

Definition 6: Online Time of the Node (*T*), this value is used to record the online time of the node. And when $T >= ThrOnliT$, the node is eligible to join the cloud area.

Definition 7: Comprehensive Performance of the Node (*P*), this value is used to record the overall performance of the node. It is calculated according to the factors such as the size of the storage space and the performance of CPU etc. When $P >= ThrP$, the node is eligible to join the cloud area.

Definition 8: Maximum Number of Child Nodes (*MaxChildNumber*), this value is used to limit the size of the cluster. The cluster's size should be less than or equal to this value.

Definition 9: Child Node Table (*ChildTable*), each cloud node has this table, which is used to record the information about its child nodes.

Definition 10: Father Node Domain (*FatherNode*), the ordinary P2P node has this domain, which is used to record the related information of the cloud node which is clustered by this ordinary P2P node.

Definition 11: Flag of the Cloud Node Type (*IsCloudNode*), used to mark nodes as a cloud node, and if this value is 0, it is an ordinary P2P node; if this value is 1, it is represented as a cloud node.

Definition 12: Flag of the Candidate Node (*IsCandidate*), if the value of the node's comprehensive performance is more than *ThrP*, but the online time is less than *ThrOnliT*, the value of this flag is 1, which indicates that the node has the potential to become a cloud node.

Definition 13: *Candidate Cloud Node*, it is a special kind of ordinary P2P node, its *IsCandidate* value is 1, and it is possible to become a cloud node, provides cloud services.

Definition 14: Failure Detection Period ($T_{invalid}$), it is a timer, and when the detection time value of each node is equal to this value, this node sends a message to its related nodes (father node or child node or neighbor node etc) to determine whether the relevant nodes are invalid.

4.2 The joining algorithm of CHPC

When the node A is added to the CHPC network, it is the first to determine whether the node's online time value (T) and the node's comprehensive performance value (P) are equal to the threshold value of the cloud node (*ThrOnliT*, *ThrP*). If $T>=ThrOnliT$ and $P>=ThrP$, the node is added to the cloud area by Pastry joining algorithm. Otherwise, it applies to join the cluster which has been

---

1) *A.IsCloudNode*=0; /*When the network is initialized, the node A is an ordinary P2P node.*/

2) *A.IsCandidate*=0; /*The initial value of the candidate cloud node flag is 0.*/

3) while (1) {

4)   if   ($A.T>=ThrOnliT\&\&A.P>=ThrP$){/*Determine whether the online time value and the comprehensive performance value of the node A can reach the threshold value of the cloud node.*/

5) A.IsCloudNode=1; /*After setting this to 1, the node A is determined to be a cloud node.*/

6) joinPastry(A); /*The node A is added to the cloud area by using pastry joining algorithm.*/

7) return 1; /*The node A joins CHPC as a cloud node. */

8) } else {/*The case is that A is an ordinary P2P node. */

9) if($A.P>=ThrP\&\&A.IsCandidate==0$) {*IsCandidate*=1;}

10) *A.addCluster* (*A.nodeId*, *A.IsCandidate*, *A.T*) to *CLOUD AREA*; /* Node A puts in for joining the cluster which has been established by the cloud area's node. */

11) if(*B.ChildTable.Length<MaxChildNumber*) { /* When the any node B in cloud area receives the adding request and detects that its cluster is not full, it allows the node A to join its cluster. */

12) *B.ChildTable.add* (*A.nodeId*, *A.IsCandidate*, *A.T*);

13) *B.ackAdd* (*B.nodeId*) to A;

14) *A.FatherNode=B.nodeId*;

15) return 0; /* The node A join CHPC as an ordinary P2P node. */ }}}

Figure 3.   Initial joining in CHPC network algorithm.

established by the node B in cloud area, and if the number of the node B's child are less than *Max-ChildNumber*, then it is successful that the node A is added to the cluster of node B. If the above case fails, the node A continues to search for the cloud node's cluster until it can be add to a cluster or join the cloud area. The CHPC's node joining algorithm is shown in pseudo code form as shown in Figure 3.

In addition, when the online time value (T) of the candidate cloud node O is greater than or equal to *ThrOnliT* and its comprehensive performance (P) is greater than or equal to *ThrP*, then the node is separated from the cluster and applies to join the cloud area. Specifics are shown in Figure 4.

4.3 The exiting and failure algorithm of CHPC

Because the CHPC network is composed by two layers, cloud area and non-cloud area, to ensure the stability of the cloud services, we design the exiting and failure algorithm for the different types of nodes.

1. The ordinary P2P node exits from the network
When the ordinary P2P node A exits from the network, it only needs to exit from the cluster which is

---

1) if(*O.IsCandidate*=1&&*O.T>=ThrOnliT*){ /*When the candidate cloud node online time more than *ThrOnliT*, it exits from its father cluster and joins the cloud area. */

2) *O.reqOutCluster* (*O.nodeId*) to *O.FaterNode*;

3) Node *FatherNode=O.FatherNode*;

4) for (int i=0;i<*FaterNode.ChildTable.Length*; i++){ /* Empty related items which is the information of node O in the *ChildTable*. */

5) if(*FatherNode.ChildTable*[i].*nodeId==O.nodeId*){

6) *FatherNode.ChildTable*[i].*nodeId*=null;

7) return 1;}}

8) *O.FaterhNode*=null; /*Empty the father domain of the node O. */

9) *O.IsCandidate*=0;

10) *O.IsCloudNode*=1;

11) joinPastry (*O*); /* The node O is added to the cloud area by using pastry joining algorithm. */}

Figure 4.   The algorithm of the candidate cloud node joining the cloud area.

---

1) Node B=new Node ( );

2) B=*A*.fatherNode;

3) *A.withdrawCluster* (*A.nodeId*) to *B.nodeId*;

4) for (int i=0;i<*B.ChildTable.Length*; i++){/*Empty related items which is the information of node A in B's *ChildTable*. */

5) if(*B.ChildTable*[i].*nodeId==A.nodeId*){

6) *B.ChildTable*[i]=null;

7) return 1;}}

8) *A.FatherNode*=null; /*Empty the father domain. */

9) *A*.destroy ( ); /*The node A exits from the network. */

Figure 5.   The ordinary P2P node exiting algorithm.

established by its father node *B* and disconnect the corresponding connection. The ordinary P2P node exiting procedure is shown in pseudo code form as in Figure 5.

2. The cloud node exits from the network

The cloud node needs to be stable to provide cloud services, and should ensure that its service can not be interrupted. And it also needs to manage the cluster. So, when the cloud node *C* exits from the network, it must select a candidate cloud node in its *ChildTable* to replace it for continuing to provide cloud services and management cluster. Specific is shown in Figure 6.

---

1) int hasSuccessor=0; /*This variable is used to record whether a successor node is found. */

2) for (int i=0; i<*C.ChildTable.Length*;i++){ /*Select a node from its *ChildTable* to replace it. */

3) if (*C.ChildTable*[i].*IsCandidate*==1){

4) Node *D*=new Node ( );

5) *D*=*C.ChildTable*[i]; /*The node *D* is inherited node. */

6) hasSuccessor=1;

7) break; /* When the inherited node is found, the loop terminates. */}}

8) if(hasSuccessor==1){ /*After finding the successor node *D*, this node is used to replaced the node *C* for providing the cloud services and management cluster. */

9) *C.isMySuccesor*(*C*) to *D.nodeId*; /*The node *C* tells the node *D* that it is the inherited node. */

10) *D.ChildTable*=*C.ChildTable* ( ); /*The node *D* gets the node *C*'s *ChildTable*. */

11) for(int j=0; j<*D.ChildTable.Length*; j++){

12) if(*D.ChildTable*[j].*nodeId*==*D.nodeId*) {

13) *D.ChildTable*[j] =null; /*Clear the *D*'s information in its *ChildTable*. */

14) break;}}

15) *D.NodeId*=*C.NodeId*; /* The node *D* has the same ID with the node *C*, that is to say, it replaces *C* as a cloud node , and needs provide cloud services. */

16) *D.cloudService*=*C.cloudService*; /* *D* takes over *C*'s cloud services. */

17) *D.FaterhNode*=null; /*Because the node *D* is no longer a child node, so it empties its father node domain. */

18) *D.IsCandidate*=0;

19) *D.IsCloudNode*=1;

20) *C.* destroy ( ); /*The node *C* exits from network. */

21) return 1;

22) }else{ /* If the inheritance node is not found, the node *C* informs its child node to dissolve the cluster, and then exits from the network according to the Pastry's exiting algorithm. */

23) Node *childNode*=new Node ( );

24) for(int k=0; k<*C.ChildTable.Length*; k++){

25) *childNode*=*ChildTable*[k];

26) *C.dissolveCluster* (*C.nodeId*) to *childNode.nodeId*; /*The node *C* notifies its child nodes of that the cluster has been dissolved. */

27) *childNode.FatherNode*=null;

28) *childNode.inputCluster*(*childNode.nodeId*, *childNode.IsCandidate*, *childNode.T*) to *CLOUD AREA*; /*Its child nodes are added to the cluster which is built by other cloud nodes in the cloud area. */}

29) *withdrawPastry*(*C*); return 0; /*The node *C* exits from the network according to the Pastry's exiting algorithm. */}

Figure 6.   The cloud node exiting algorithm.

3. The node failure processing algorithm

In CHPC network, through the communication information with other nodes, their failure or not

---

1) Set the value of the failure detection cycle to $T_{invalid}$;

2) if(t_2-t_1==$T_{invalid}$){ /* When the failure detection cycle is reached, do the following. t_1 is the last failure detection time, and t_2 is the current time. */

3) Set the timer *t*; /* This timer is a time limit value to be used to judge whether child node or the neighbor node fails. If the response message is not received within this limit, the failure of a neighbor node or child node is determined. And this time value is less than the failure detection cycle. */

4) while(*t*){

5) if(*A.IsCloudNode* ==1){ /*If the node *A* is a cloud node, then execute the following code. */

6) for(int i=0;i< *A.ChildTable.Length*;i++) /* The *life* domain of all child nodes is initialized to 0. */

7) *A. ChildTable*[i].*life*=0;

8) for(i=0;i< *A.NeighborhoodSet.Length*;i++) /*The *life* domain of all neighborhood nodes is initialized to 0. */

9) *A. NeighborhoodSet*[i].*life*=0;

10) *A.broadcastInvalidChild* (*A.nodeId*) to *A.ChildTable*; /*The node *A* sends node failure detection message to each child node. */

11) Any child node *C* receives the *broadcastInvalidChild* message from the node *A*;

12) *C.isLife* (*C.nodeId*) to *A*;

13) The node *A* receives the *isLife* message from its child node *C*;

14) for(i=0;i< *A.ChildTable.Length*;i++)

15) if(*A.ChildTable*[i].*nodeId*==*C.nodeId*)

16) *A. ChildTable*[i].*life*=1; /* Set the *life* domain of the node *C* to 1. */

17) *A.broadcastInvalidNeighbor* (*A.nodeId*) to *A.NeighborhoodSet*; /*The node *A* sends node failure detection message to *NeighborhoodSet*. */

18) Any neighborhood node *N* receives the *broadcastInvalidNeighbor* message from the node *A*.

19) *N.isNeighborLife* (*N.nodeId*) to *A*;

20) The node *A* receives the *isNeighborLife* message from its neighborhood node *N*;

21) for(i=0;i< *A. NeighborhoodSet.Length*;i++)

22) if(*A. NeighborhoodSet*[i].*nodeId*==*N.nodeId*) *A. NeighborhoodSet*[i].*life*=1; /*Set the *life* domain of the node *N* to1. */}

23) else{ /*If the node *A* is a ordinary P2P cloud node, then execute the following code. */

24) *A.fatherLife*=0; /*The *life* domain of *A*'s father node is initialized to 0. */

25) *A.messageLife* (*A.nodeId*) to *A.FatherNode*; /*The node *A* sends message to its father node. */

26) if(The node *A* receives the response message from its father node) *A.fatherLife*=1;}}

27) if(*A.IsCloudNode* ==1){

28) The node *A* uses the ordinary P2P node exiting algorithm to let the failure node in the *ChildTable* exit from the system;

29) Traversal node A's *NeighborhoodSet*. And these failing neighborhood nodes exit from the network by using the Pastry node failure algorithm and the method of cloud node exiting;}

30) else{

31) if(*A*.fatherLife==0) { The node *A* uses the algorithm of ordinary P2P node joining CHPC to deal with the failure of its father node;}}}

Figure 7.   The node failure processing algorithm.

are judged. Then the corresponding failure mechanism is implemented. When the ordinary P2P node detects its father node failure, it searches for another cloud node to add itself into the cluster which established by this cloud node. When the cloud node detects that its child node is disabled, it only needs to reset relevant information in its *ChildTable*. When the cloud node can not communicate with another node in the cloud area, the Pastry's failure rules will be used. The CHPC's node failure algorithm is shown in pseudo code form in Figure 7.

5 ASSESSMENT OF SIMULATION

In order to verify that the CHPC protocol can ensure the stability of P2P cloud network in the environment of session heterogeneity, the simulation experiment is designed. Based on the FreePastry[11], 2000 network nodes are used for this experiment. In the test process, one node is added to the network every 5 seconds. In addition, we set the Online threshold (*ThrOnliT*) in the range of [0,1000] seconds, and the Performance threshold (*ThrP*) is fixed to 1000 units, and the maximum number of child nodes (*MaxChildNumber*) range is set in the range of [0,50]. When the value of *ThrOnliT* and *MaxChildNumber* is 0, the network degenerate into Pastry system, which is not suitable for providing cloud services.

The *ThrOnliT* is a limit of the online time of the node, which is an important parameter to determine whether the node is a stable node. When the *ThrOnliT* is 0, the system allows any node to join the cloud region, so that the system is degraded into Pastry network. In this simulation, firstly, the relationship between the *ThrOnliT* and the rate of node joining and exiting (*R*) is tested to determine an optimal *ThrOnliT* value. The value of *MaxChildNumber* is fixed to 50 which is the upper limit of this value. Figure 8 is the simulation result.

The Figure 8 shows the relationship between the *ThrOnliT* and the *R*. The X-axis shows the *ThrOnliT*(s). The Y-axis shows the rate of the joining and exiting (*pcs/min*). From this figure, we can get that: with the increase of *ThrOnliT* value, the rate of joining and exiting is decreasing, but when the *ThrOnliT* value is greater than 500 seconds, the decline rate tends to be gentle. There are three factors for the above phenomenon. Firstly, when the *ThrOnliT* value is 0, it makes the rate of joining and exiting equivalent to the Pastry that all the nodes can become cloud nodes to join the cloud network. Secondly, with the increasing of *ThrOnliT* value, the nodes with low online time are excluded from the network to improve the stability of the network. Thirdly, when the *ThrOnliT* value reaches

a certain value, the threshold value is increased to a limit, so that the threshold of the system is too high, so that *R*'s decline rate tends to be gentle.

Next, we test the relationship between *MaxChildNumber* and *R* which is another important factor for determining the stability of the network. The *ThrOnliT* value is fixed to the upper limit of 1000 seconds. The simulation results are shown in Figure 9.

This figure shows the relation of the maximum number of child nodes (*MaxChildNumber*) with the rate of node joining and exiting (*R*). The horizontal coordinate shows the upper bound of the cluster size (*pcs*). The vertical coordinate shows rate of joining and exiting (*pcs/min*). It can be seen from the figure that: with the increase of the size of the cluster, the rate of joining and exiting has a sharp decline, and when the size of the cluster is more than 20 nodes, the decline rate of *R* tends to be slow. The factor is as this, when the *X* is 0, the

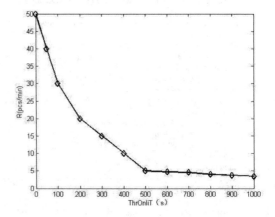

Figure 8. The relationship between the stability of CHPC system and *ThrOnliT*.

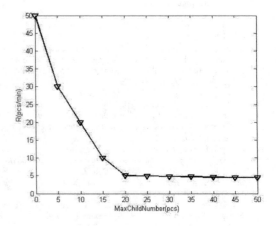

Figure 9. The relationship between the stability of CHPC system and *MaxChildNumber*.

system is degraded into Pastry system, and when the X reaches a certain limit, the system tends to be stable, so that the R decrease tends to be gentle.

6 CONCLUSIONS AND FUTURE WORK

The CHPC network proposed in this paper is a cloud computing platform based on P2P network. It uses P2P computing technology to integrate distributed resources in the Internet to construct the lower layer cloud platform. There is important significance for combining the advantages of P2P and cloud computing to constitute more in-depth sharing of Internet resources. The focus of the future research is that we will devote to the study of CHPC network data layer and application layer, as well as the security measures of the network.

ACKNOWLEDGMENTS

This study is supported by Fundamental Research Funds for the Central Universities (XDJK2014C002); The National Natural Science Foundation of China (61170192, 41271292); Chinese Postdoctoral Science Foundation (2015M580765); Joint funds for the Science and Technology Department of Guizhou Province and Bijie Municipal Science and Technology Bureau and Guizhou University of Engineering Science (KEHE LH word [2014]7530, [2016]7053 and KEHE J word LKB [2013]14 in Guizhou); Project of key disciplines of Guizhou Province (XUEWEIHE word ZDXK [2014]26 in Guizhou); Science and technology project of Bijie (KEHE word [2014]30 in Bijie).

REFERENCES

Boon Thau Loo, Scott Shenker, Ion Stoica. Complex Queries in DHT-based Peer-to-Peer Networks [C]. UC Berkeley, 2003: 1–6.

FreePastry [Z]. (2009-03-13). http://freepastry.org/FreePastry.

Ge Jun-wei, Wang Yan-feng, Fang Yi-qiu. Research and improvement of Chord algorithm in cloud computing [J]. Computer engineering and design, 2013, 34(10): 3413–3416 (in Chinese).

He Dao-de, Deng Xiao-heng. P2P Routing Algorithm Based on Physical Location and Access Locality [J]. Computer enginnering, 2009, 35(8): 146–149 (in Chinese).

Lopez-Fuentes F D A, Garcia-Rodriguez G. Collaborative Cloud Computing Based on P2P Networks [C]// International Conference on Advanced Information Networking and Applications Workshops. IEEE Computer Society, 2016: 209–213.

Rocha V, Kon F, Cobe R, et al. A hybrid cloud-P2P architecture for multimedia information retrieval on VoD services [J]. Computing, 2016, 98(1–2): 73–92.

Rowstron A, Druschel P. Pastry: Scalable, Decentralized Object Location and Routing for Large-scale Peer-to-Peer Systems [C]// Proc. of Int'l Conf. on Distributed Systems Platforms (Middle-ware). Heiderberg, Germany: [s. n.], 2001.

Wahid N W B A, Jenni K, Mandala S, et al. Review on Cloud Computing Application in P2P Video Streaming [J]. Procedia Computer Science, 2015, 50: 185–190.

Wikipedia. Cloud computing. Retrieved from https://en.wikipedia.org/wiki/Cloud_computing, 2015.

Wu Ji-yi. Study on DHT based Open P2P Cloud Storage Services Systems [D]. Hangzhou, China: Zhejiang University, 2011.

Xu Xiao-long, Wu Jia-xing, Yang Geng. Malicious code united-defense network based on Cloud-P2P model [J]. Application research of computers, 2012, 29(6): 2215–2217, 2257 (in Chinese).

Advances in Energy Science and Equipment Engineering II – Zhou, Patty & Chen (Eds)
© 2017 Taylor & Francis Group, London, ISBN 978-1-138-71798-5

Vehicle terminal system of intelligent logistics based on STM32F103 and BDS/GPS technologies

Fangyun Zhao
School of Information Engineering, Guizhou University of Engineering Science, Guizhou, Bijie, China

Mingfu Zhang
Southwest University, School of Computer and Information, Chongqing, China

Wei Sun
School of Software, BeiHang University, Beijing, Haidian, China

ABSTRACT: The application of the US GPS positioning system based on Internet of Things in the vehicle terminal system is influenced by unidirectional reception, multipath, radio interference, and adjustable pseudo-distance, leading to the inaccurate terminal positioning, poor communication effect, and unsafe information. 32-bit STM32F103 chips are taken as the host controller of vehicle terminal positioning system, BDS/GPS dual-mode system UM220-III N modules with the central node-type positioning characteristics and two-way communication function are used at the same time so as to design a set of vehicle terminal positioning systems with comprehensive positioning, digital short message communication, and timing function. The system test of data transmission and Baidu map moving track shows that the design has realized terminal positioning offset in 1–8 m, with the functions of high data transmission rate and two-way communication, meeting the needs of users, so it can fully achieve the real-time positioning and tracking communication and scheduling control of special logistics vehicles. Furthermore, it has solved the problems of unsmooth communication and unsafe information in the process of transporting special vehicles, thereby reducing transport costs and improving economic efficiency.

1 INTRODUCTION

"Internet of Things" has ensured that all items in the world can exchange information via the Internet. This trend has promoted the wide application of sensor technology, embedded technology, nanotechnology, and network technology (Yang, 2010). The vehicle terminal positioning technology has been introduced into various comprehensive technology applications with the development of US GPS, Russian Global, European Galileo, the Chinese BDS, and Indian IRNSS (Yang, 2015). Special logistics vehicles (such as cash transport, ammunition transport, initiating explosive devices transport, military goods transport, etc.), are needed not only to quickly locate and to have two-way communication and timing functions, but also to meet the requirements of all intelligent confidential information. And the electromagnetic wave of satellite navigation system positioning signal is susceptible to interference of obstacles, ionosphere, troposphere, the multipath effect, and other factors (Jie, 2014). From the aspects of national security and economic strategy, there are some problems such as unsmooth communication and unsafe information in the traditional transport positioning based on GPS positioning system used for special logistics vehicles.

Some researchers specifically simulated the position dilution of precision (GDOP), vertical dilution of precision (BDOP), horizontal dilution of precision (VDOP), and the Minimum Detectable Bias (MDB) for GPS, BDS, and GPS/BDS hybrid system. It was found that GPS/BDS had the smallest value of the DOP and MDB, namely at the same time observation period, BDS/GPS had the most uniform plane distribution in the observation area and the highest reliability (Yang, 2015). Through the analysis of GPS and BDS system monitoring of the on-orbit signal and collecting data quality, it was found that there was no big functional difference between GPS and BDS (Ouyang, 2013; Yang, 2011). It is predicted that by 2020, BDS will have the ability to provide global service and totally replace GPS, becoming a mainstream technology (Yang, 2015). Currently, the initial service of Beidou Satellite Positioning II has been involved in mapping, water conservancy, transportation, and national security (Cao, 2013).

In the past, some researchers had mostly selected AT 89S51, MSP430F 149, and other simple structures, with low cost, low power consumption, and manageable chip for system development, but these chips had slower computing and poor resource, which are not conducive to expansion and application; on the contrary, the Flash memory with only 64 K bytes enhanced low-power 51-type STC15 L2 K 60S 2 chip selected as the host controller (Li, 2014), which has some improvements in stability and cost of the system; however, as far as the new-generation free operating system-embedded ARM microprocessor STM32F103 with frequency up to 72 MHz, built-in Flash up to 128 k bytes is concerned, STC15 L2 K 60S2 and other chips have obviously deficiency in cost performance.

The selection of STM32F103 as the host controller has realized MCU functions, provided a development platform with low cost and high efficiency. The chip is small and has small pin number, low power consumption, great computing performance, and quick interrupt response. At the same time, the selection is based on UM220 III N BDS/GPS dual-mode navigation system, solving the data transmission rate, real-time two-way communication, and information security for the system.

2 COMPONENTS AND FUNCTIONS OF THE SYSTEM

2.1 Components of the system

The system is based on BDS/GPS dual-mode mobile positioning technology, GPRS general packet radio service technology, and STM32 MCU control technology, and it controls each module through the STM32 chip host controller. The module architecture of the system includes STM32 host controller, BDS module, GPRS communication module, a power supply circuit, and other auxiliary devices. The design and development tools of the circuit selected the powerful PADS VX.0, and it completed the overall design, layout, alignment for the system hardware circuit according to the PCB board and component function parameters (Yao, 2014). The circuit has the advantages of simple welding, low cost, small signal interference, stable performance, and reliability. The architecture of the system is shown in Figure 1.

2.2 Principles and module function of the system

The system used STM32F103 as the host controller and Beidou Satellite Positioning UM 22 0-III N module to receive position information of the vehicles sent by satellite. Through network, SIM900A performs data communication with control center, and OLED is used to display the position infor-

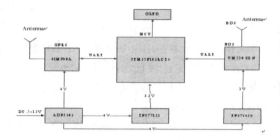

Figure 1. Architecture of the system.

mation of vehicles. The UAR T interface of STM32F103 communicates with U N M220—III and receives the position data sent from Beidou module UN220-III N. It is parsed to get a position data frame, further completing positioning data obtained from data analysis, and it realized real-time transmission of the position data through GPRS data communication function of SIM900A module, and the communication between the TCP/IP protocol and the monitoring center.

3 HARDWARE DESIGN

3.1 Design of STM32F103 module

STM32F103 module is an embedded microprocessor chip based on the Cortex-M 3 core, and it is the first RISC instruction processor based on the ARMv7-M architecture 32-bit standard. It has an external circuit requiring power, reset clock, and other devices. Its cost performance is much higher than that of the 51-series simple chip microcomputer.

In the circuit design, the STM32F103 chip with the crystals of 8 MHz, 32.768 kHz frequency rates, and the 8 MHz frequency rate crystal can be taken as master clock to provide frequency for the vehicle terminals. When the input end of the vehicle terminal system is connected to the circuit, the output end PDOOSC_OUT of STM32F103 chip internal oscillation circuit crystals provides frequency drive; the output end of the vehicle terminal system is connected to the end of internal oscillation circuit crystals PDOOSC_IN of the STM32F103 chip; the output of the crystals PC15-OSC32_OUT of the internal oscillation circuit of STM32F103 chip can provide a 32.768 kHz crystal vibration clock source for the real-time clock of the vehicle terminals. The pin allocation table is shown in Table 1.

3.2 Design of BDS/GPS modules

BDS/GPS based on UM220-III L are dual-mode positioning modules, including navigation and monitoring application. UM220 III-L is the third-generation product of UM220 III-NL series, with

No.	Pin name of the STM32	Corresponding module pin	Function
1	PC7	SIM900A-PWR	Start and restart to control SIM900A
2	PC8	SIM900A-STATUS	Get SIM900A working status
3	PA9	SIM900A-RX	Send data port to the SIM900A
4	PA10	SIM900A-TX	Receive data port from the SIM900A
5	PC10	BDS-RX	Send data port to the BDS
6	PC11	BDS-TX	Receive data port from the BDS

the smallest size, high integration, and easy application. It is very suitable for the high-cost GNSS-scale application. The module receives real-time data signals from Beidou II Satellite, which are amplified, filtered, and processed by the intermediate frequency and sent to the STM32 host controller through the UART serial ports, realizing the tracking, measuring, and locking of the data signals, and other processes

The circuit design of BDS/GPS modules: the circuit of BDS/GPS modules completes the signal settlement and analysis of the navigation satellite (Joseph, 2007) and selects UM220 III N chips with the features of high frequency and high performance of BDS/GPS double system modules from Star Core Technology Company, which can support the two frequency points of BDS B1 and GPS L1. The chip is of small volume, high performance, low cost, and low power consumption.

In the circuit design, it is only needed to provide 2.7–3.6V operating voltage, while the design of the circuit needed to be matched with the antenna 50Ω impedance; otherwise, BDS positioning module will not accept the signals sent from the Beidou Satellite.

3.3 Design of GPRS module

The design uses SIM900A chip of SMT package dual-band GSM/GPRS module, which is small and consumes less power, and it is 10 times faster than that sent by GSM high-speed transmission in the aspects of voice, SMS, data, and fax messages. SIM900A follows the embedded TCP/UDP protocol. GPRS module uses packet switching mode for transmission and receiving data sent from the UM 220 III N analysis and sends them to monitoring center through the Internet network at the same time.

UN220-III N module is initialized; STM32F103 can receive data frames sent from UN220-III N, including longitude, latitude, time, speed, and other positioning information. Only after the STM32F103 received a complete data frame, UN220-III N can determine whether to parse the positioning data frame and package the data frame format in accordance with the server agreement and then send them to the monitoring center server by SIM900A module, thus completing the process of positioning data sending, parsing, and transmission.

In the circuit design, it is needed to protect the SIM card, and in the circuit, the SMF05C can be taken conduct electrostatic protection so as to avoid burning of the SIM card. At the same time, the circuit design needs to match the 22Ω resistor to have a certain impedance between the module and the SIM card.

3.4 Circuit design of multi-channel power converters

This system consists of multiple hardware modules, with each module having different requirements for power supply, according to the different performance specifications of BDS/GPS, STM32F103, GPRS, OLED display, and it is needed to supply different power values for different power supply demands in the design.

Power supply index is known from working handbook for each module. The normal working voltage (Vi) of each module and the normal working current value (Ai) are shown in Table 2:

Therefore, it is needed to design a special variable power circuit, which is used as the power supply for each module. The power supply value provided by the common vehicle terminal is generally 5–20V. This system takes the buck DC converter ADP2303 as the primary power converter.

Buck DC converter TPS 76430 and TPS 77133 are taken as a secondary branch power converter; three chips are taken convert 4V/3 A, 3.3V/150 mA, 3V/150 mA, and other power supplies, as shown in Figure 2.

The ADP 2303 module in Figure P2 is a circuit converter, which can convert the power from +5V to + 12 V into a multivalue voltage source to

Table 2. Normal operating power value of components.

Name of components	Voltage	Current
BDS/GPS modules	2.7–3.6V	<150 mA
STM32F103 microprocessor	2.0–3.6V	Total current through VDD/VDDA <150 mA
GPRS module	3.3–4V	<10MA
OLED display	2.8–3.3V	<1 A

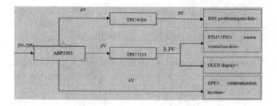

Figure 2. Circuit design of multiple power convertors.

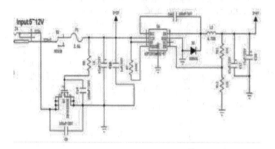

Figure 3. Circuit converter of ADP 2303 module.

provide +3V to +20V, so APD 2303 is suitable for the design of multi-input voltage, and the circuit design of ADP2303 power conversion multivalue is as shown in Figure 3.

In the design of circuit, it is needed to pay attention to the circuit safety and add the field effect tube SM4833NSK. This chip has high input impedance, low noise, good thermal stability, simple manufacturing process, and so on. The end of the SW pin is connected a variable resistance so as to conduct bypass restrictions of peak flow and prevent the circuit board burning when the power supply is connected reversely. F pin is connected in series with a pull-up resistor to regulate the output voltage; if V00 is the output voltage value, then the relationship between it and the pull-up resistor is shown in Formula (1, 1):

$$V_{OO} = 0.800 \times \left(1 + \frac{R_1}{R_2}\right) \quad (1)$$

Simulation debugging is conducted on the circuit to regulate the values of the pull-up resistors R1 and R2 and then the corresponding output voltage value VOO is obtained.

4 SOFTWARE DESIGN

4.1 Tools of system development

In this system, J-Link emulator PC is selected to connect with STM32F103 development board and Keil uVision5 is used to program STM 32 software as the development platform for system.

Keil uVision5 development tool is a C language software development platform of German Keil Software Company 51 series compatible microcontroller (Shi, 2014). Keil can provide debugging tools including C compilers, macro-assemblers, linker, library management, and a powerful simulation debugger, and it is based on the IDE development environment as well. Keil uVision5 program STM32 realizes the command control of STM32 over BDS module and GPRS modules.

4.2 Working process of host controller STM32

Host controller STM32 STM32 conducts effective judgment for the positioning data frames sent from the BDS, namely it determines whether the information received is in line with the rules of "MEA-0183" agreement (whether marked with "frame head") so as to determine to process or send the information. If the positioning data of Beidou Satellite is incomplete, STM32 will not process or send the following information. Only after the STM32 receives a full $GNRMC frame, can it send the data.

4.3 Reception and parsing of BDS positioning data

UM200-III N positioning module outputs positioning information from UART interface, and the information follows "NEMA-0183" specifications (Wang, 2013), that is, the frame information contains the "frame head" initially marked at the "$GPRMC" frame data, and it obtains the positioning information sent from BDS Satellite according to "NEMA-0183" agreement. UM200 III N module conducts intermediate frequency amplification, filtering, signal processing, and data parsing for the positioning information received and sends them to STM32F103 control module for positioning data processing through the UART serial port so as to obtain the accurate positioning information.

After the BDS/GPS dual-module system is normally connected to the electricity, it will automatically start searching satellite positioning, and send "NMEA-0183" specification information to the STM 32 F 103 control module through the UART serial port. All characters in "NMEA-0183" format statement are ASCII values, all statements start with the "$" signs, and the standard format is "$XXYYY". In the specific application, it is needed to eliminate unwanted data redundancy, to possibly concise positioning information in combination with the RMC statements of dynamic key information of actual vehicles selected (such as longitude, latitude, height, time, and speed).

For instance, in accordance with the basic structure of $GPRM frame: $GNRMC, hhmmss, sss, A, dddmm, mmmm, N, DDDMM, MMMM, E, X,

Y, ddmmyy, E, A* (CR)(LF), a set of data received by the BDS satellite are selected: $GNRMC, 054659.123, A, 6589.0227, N, 12270.2016, E, 0.027, 280.52, 030852, −2.3, E, A* 10 < CR><LF>, these statements are parsed one by one, such as: "$GPRMC" represents the initial mark of frames data, GN represents the GPS/BDS dual-mode system; "ddd mm, mmmm" represents the latitude; "N" represents the north latitude; "E" represents the east longitude; in "X" represents the driving speed; "Y" represents the direction angle; and "CR" and "LF" represent the ending mark of a positioning data frame.

4.4 Configuration and data sending of SIM900 A

"Transparent transmission mode" referred to as "transparent mode," is currently one of the communication modes most used in vehicle terminals. This design uses the mode, then inputs "AT + CIP-MODE = 1" instruction, and if the connection is successful, it starts the "transparent transmission mode." The system is initialized, dual-frequency GSM/GPRS communication module SIM900A is in normal operation, and starts network connections, sets the SIM900A operation state to be in TCP/IP "client mode", and starta the "transparent transmission mode." The SIM900A is connected to the designated port of the server and STM32F103 can conduct the two-way communication with the server by SIM900A. The specific operation is as follows: the system receives data from RX interfaces of SIM900A, which will be transferred to the server, and the BDS module receives the latitude, longitude, and speed of the logistic vehicle and sends them to STM32F103 host controller for processing through UART interface, thereby obtaining accurate positioning information of vehicle terminals. The host controller sends the information to GPRS module for packaging, and finally sends it to the specified IP interface or monitoring center through the Internet.

5 OVERALL TEST OF THE SYSTEM

At present, China has made a wide range of applications of Precise Point Positioning (PPP) technology used in high-precision measurement, such as low-orbit satellite determination and carrier and ground deformation monitoring areas (Zhang, 2015). Instead of using the modeling algorithm and analysis algorithm, the design directly uses UM220-III N, which has the dual-mode BDS/GPS module functions of navigation, monitoring, and positioning in all. It connects the UART port of MCU with the UART module of UM220 III-N module, sets up the corresponding parameters, starts the vehicle terminal system,

and starts the navigation and positioning test along a test route.

Line AB beginning with A and ending with B is selected to conduct the test of a vehicle terminal. UM220-III N will obtain the data transmitted from the Beidou Satellite positioning system and automatically conduct positioning calculation, again, in the form of NEMA-0183 format. It outputs geographical information (including longitude and latitude) of some test point from the UART interface, and is displayed on the screen of the terminal OLND.

The terminal OLND screen is turned on, and it is observed that the screen data are in real-time update with the information data such as the longitude, latitude, and speed of the vehicle on Line X. After comparing the corresponding records at real time (Lat: N 27.3025522, Lon: E 105.307717, Date Time) with the time coordinates of positioning information in the database table converted by the addtime, it is found that the difference between them is about 1–2 s, which proved that the system data receiving rate and the delay time are in normal operation.

The terminal positioning information is sent to the monitoring center by calling Baidu Map API through the Web interface (Zhang, 2015), the coordinate information converted in a database is dotted or lined on the Baidu Map so as to draw the X roadmap, and the test results are obtained.

Through many positioning tests for the same test line, the positioning data of some test point are obtained, and the mean value is used to obtain the average deviation of the test point to be 1–8 m.

6 CONCLUSION

Through on-site tests of the vehicles, the information contrast of Baidu map moving path showed that: the use of a new STM32F103 chip to conduct communication control over UM220 III N module had high efficiency, low cost, low power consumption, and high stability, realizing rapid positioning, accurate tracking, and two-way communication of transfer terminals, thereby improving the monitoring ability for special vehicles, so it is more transparent and highly efficient for on-road vehicles. The system has great application value in full range, and it is of intelligent management of the special logistics vehicles.

ACKNOWLEDGMENTS

The authors acknowledge the following basic research funds for this work: Southwestern University: Research on Logistics Monitoring and management system based on the technology of BDS and Internet of things (XDJK2014C109).

Guizhou Science and Technology Fund Project: Research on the construction of intelligent logistics system based on cloud computing and Internet of things technology, Department of Guizhou LH words ([2014]7536).

Jiont funds for the Science and Technology Department of Guizhou Province and Bijie Municipal Science and Technology Bureau and Guizhou University of Engineering Science (KEHE LH [2016]7053 in Guizhou).

REFERENCES

Cao Ling. Intelligent Public Transport Query System Based on Baidou Satellite Navigation System [D]. Huazhong Normal University, 2013.

Jie Qikui. Research and Imitation of IOT Terminal Position System Based on Beidou Navigation [D]. Beijing Jiaotong University, 2014.

Joseph Yiu. Getting Started with the KEIL Real View Microcontroller Development Kit. The Definitive Guide to the ARM Cortex-M3, 2007, Pages 289–313

Li Lidong, Gao Xuejiang, Mao Liqi, Chen Yourong, Zhou Ying. Intelligent Housing System Based on Wireless Sensor Networks. [J]. Journal of Zhejiang Shuren University. 2014.9:14–3

Ouyang Xiaofeng, Xu Chengtao, Liu Wenxiang et.al. Monitoring and Analysis of on-orbit signal data quality of Beidou satellite navigation systems [J] Global Positioning System, 2013.38(4)3:2–37.

Shi Shunyu Intelligent Vehicle Terminal Design Based on Beidou Satellite Navigation System [D]. Ocean University of China, 20 14.5.

Wang Li. Intelligent Streetlight Monitoring System Base on IOT and GPRS [D]. Hangzhou University of Electronic Science and Technology. 2013.

Yang Gang, Shen Peiyi, Zheng Chunhong. Theory and Technology of IOT [M]. Beijing: Science Press, 2010.

Yang Ling, Li Bofeng, Shen Yunzhong. Reliability Analysis of GPS and BDS Hybrid Position Model [J]. Journal of Navigation and Positioning, 2015.9.

Yang Ling, Li Bofeng, Shen Yunzhong. Reliability Analysis of GPS and BDS Hybrid Position Model [J]. Journal of Navigation and Positioning. 2015 3(3):27–34.

Yang Xinchun, Li Zhenghang, Wu Yun. Constellation of Beidou Satellite Navigation System and Analysis of XPL Performance. [J]. Journal of Surveying and Mapping. 2011 40:68–72

Yao Xianwei. Research and Design of Infrared control system Of Intelligent Housing System based on STM32. [D]. Yanshan University, 2014.5.

Zhang Chao, Gu Jihua. [J]. Design of Mobile Targeting Tracking System Based on GPS / BDS [J]. Modern Electronic Technology. 2015.91:121–123.

Zhang Xiaohong, Zou Xiang, Li Pan, Pan Yuming. Comparison between Positioning Convergence Time and Positioning Accuracy of BDS / GPS Precise Point. [J]. Journal of Surveying and Mapping. March, 2015. Vol.44, No.3.

Advances in Energy Science and Equipment Engineering II – Zhou, Patty & Chen (Eds)
© *2017 Taylor & Francis Group, London, ISBN 978-1-138-71798-5*

SWOT analysis of modern railway logistics and its development strategies

Lincan Li
China Academy of Railway Sciences, Beijing, China

Jichang Zhu
Transportation and Economics Research Institute, China Academy of Railway Sciences, Beijing, China

ABSTRACT: With the development of market economy and transformation of logistic mode, logistics forms represented by highway transportation have been greatly improved. However, the market shares of railway freight have been reducing. Therefore, the problems of how to locate the railway freight, how to analyze the level of development at the present stage, and how to improve the development of the railway modern logistics have become important for the long-term survival and better development of railway freight. In this paper, we summarize the improvement and reform measures to accelerate the development of railway modern logistics from three aspects: transport organization, marketing and service quality and information platform, and achievements in recent years. We use the SWOT strategic analysis method to analyze the internal and external environment of the modern railway logistics development. It summarizes the railway logistics strengths from three aspects: road network structure, consumer groups, and managerial and administrative expertise. We also analyze the railway logistics weaknesses from four aspects: the capability of total logistics service, railway rate mechanism, container transportation volume, and railway logistics professionals. Then, we make an outlook of the railway logistics on the basis of the development tendency of railway logistics, policy support of the country, and development process of the dedicated passenger traffic lines and the high-speed railway. Finally, we give the problems faced by the railway logistics from the two aspects: economic restructuring and logistics market competition. On the basis of the above analyses, we clearly locate the position of railway logistics development and study and put forward the corresponding development strategy.

1 INTRODUCTION

With the rapid development of high-speed railway in China, the defect of the capability of existing lines transport increases and the process of railway transportation to modern logistics development gradually improved. In April 2015, for the further development of modern logistics, China Railway Corporation decided to restore and rebuild the railway freight business ideas, business process, and management mechanism on the basis of the existing freight reform. In the aspect of the transportation organization, rail freight organization carries out the express train within the same railway bureau and different bureaus. To optimize marketing and service quality, China Railway Corporation and railway bureaus change ideas, strengthen the marketing consciousness, open service supervision telephone and a variety of service channels, disclose the public transport wagon resources, and ensure transportation to

the customer demand to achieve "promptness". In the aspect of information platform, the freight transportation aims to the comprehensive Internet and promotes China Railway 95306 website. The commodity service category expands to 25 categories, sugar, alcohol, oil, cotton, vegetables to name a few. At the same time, it opens regional service market, new special auction, trading, and other many types of trade. Through a series of reform measures, railway freight, no matter from the internal organization structure, operation mode, service quality, or the external social image, has greatly improved, but still has a long way to go to be a good modern logistics enterprise. Therefore, this paper uses the SWOT analysis method to analyze the competitive strengths, competitive weaknesses, and opportunities and threats of railway modern logistics to combine developing strategy and internal and external conditions and promote the development of railway modern logistics.

2 MODERN RAILWAY LOGISTICS DEVELOPMENT ANALYSIS

2.1 Strengths analysis

1. Complete network structure

Through long-term development and the support of national policy, the railway transport network has very strong integrity. In the aspect of inland transport, in recent years, the rapid development of high-speed rail business and the separation of passenger and freight traffic greatly improve the carrying capacity. At the same time, the construction of heavy haul railway, such as Daqin (Datong–Qinhuangdao) and Shuohuang (Shuozhou–Huanghuagang), further enhances the ability of railway transportation development space. According to "Long-term Railway Network Planning", jointly issued by the National Development and Reform Commission, Ministry of Transport of the People's Republic of China, (Zhang, 2015) China Railway Corporation in July 2016, China will have more perfect railway transportation network. According to international transportation, the normalized operation of China–Europe Block Train expands the railway network. According to the development of multimodal transport system, railway has port stations and port station networks, which almost cover the whole country. And railway has become the essential transport carrier to develop the multimodal transport logistics.

2. Stable customer group

Railway transportation, with features such as high loading, low freight rates, and safety, has a large number of stable customer groups in many years. At the same time, many major source clients have special railway lines, which connected the main railway directly. This part of the customers tend to give preference to railway transportation when transport demand produce. From the China Railway 95306 website statistics, the number of registered users has broken through 310000 until April 2016, and the number of online trading was more than 60000. A total of 95306 sites are gradually performing as a railway freight transporter, commodity marketer, and an online trading platform.

3. The improving management level

Compared with the former railway freight acceptance mode, rail freight now offers a variety of processing channels: 95306 nets, such as, "I want to deliver," 95306 customer service calls and administrations to accept the service, freight station business hall to handle, door-to-door service. Relied on the development of e-commerce, customers also can query the goods information in time. At the same time, the national railway bureau has its own railway freight station located in the center of the town area, with a large number of warehouses, freight yard, loading and unloading equipment, equipped with storage, warehousing, and "door-to-door" transportation conditions.

2.2 Weaknesses analysis

1. Insufficient marketing capabilities of the whole logistics

Progressive railway reform gradually advances the development of the railway modern logistics, but as a large transportation system, there is still a larger gap in the aspect of the business philosophy, product idea, and service consciousness, compared with the excellent logistics company. On the acceptance mode, Shunfeng, Zhongtong logistics companies will arrange pickup member to contact shipper directly and offer visiting service. However, for railway, when 95306 or freight administration are called, there is a choice of the ways, using vehicle logistics or less-than-truckload transportation. In this situation, it often requires the customer to call other service personnel to consult. In the aspect of delivering business, railway cooperates with other transportation companies, the cooperation transport company offers the visiting service, and the railway offers the transportation service. Therefore, when the customer puts forward transport demand, railway needs to contact the transportation company, and only when the company can have free car could it offer visiting service, so the real-time transport and "door-to-door" service cannot be guaranteed. On the service consciousness, at present, there still exists some deficiency in the quality of service among part of the business servers. For instance, when receiving customer calls, they are not familiar with logistics services, and cannot recommend the right product of transportation for the needs of customers (Liu, 2014; Zhao, 2014), so the customer is very difficult to enjoy the coordination with other logistics company comprehensive services.

2. The rigid rate mechanism

Railway freight rate lacks flexibility. In recent years, railway has taken a buyout price on freight charges. The rate is divided into two parts: transit charges and miscellaneous charges. Transit charges mainly include freight rate scale, mileage, and chargeable weight. The vehicle with less-than-one carload determines the freight rate scale by the category of goods. The container defines the freight rate scale by container type, miscellaneous charges depended on the actual project and the rate, but the actual freight pricing condition is complex. For vehicle transportation, vehicle goods respectively apply # 1–7 freight, the transit charges increase with the increase of freight number (except #7) (Liu, 2014), the highest rate #6 is 82% higher than that of #1. It often appears that although the property of the goods tends to be the same or similar, the same category of goods (Guo, 2010) is applied to different freight number, which leads to difference charges. At the same time, in terms of market supply and demand situation, highway and waterway use difference pricing, but the pricing flexibility of railway is insufficient, making the railway transport dominated on freight competition in transport demand.

3. Lack of attention to improve container volume

Container can be used to organize small batch of goods. At the same time, because the handling container is convenient, it is also an important component of the multimodal transport. China Railway Corporation has relaxed restriction on the limit of goods applied to container in recent years. At the same time, for small container used in scattered conditions on the express train, no container fees should be charged. Compared to other modes of transportation using container (Athanasios Ballis, 2002; Michael Michael, 2001), railway container transport has cost advantage. However, in recent years, container volume has slow growth, and the railway container is still used only as a single mode of transport organization, and mainly carries bulk cargo instead of a way of containerization transport. In addition, limited by weight and transport time, railway container itself shares less of the market. Compared with highways and waterways, it attracts less supply of goods and the freight volume growth rate is low.

4. Lack of the railway logistics professionals

Modern logistics management is a well-known practical activity. Throughout the whole process of supply chain, railway modern logistics relates to transportation, economy, management, information, and other fields. Furthermore, it has high-quality request for railway management personnel, and for customer requirements, it demands a relatively comprehensive solution, and can deal with abnormal changes. The current railway freight center is generally separated from the locomotive depot, and the staff is shipping. Their market consciousness and service consciousness are not strong, and there still exists the saying of "asking for transiting" when transportation demand happens. In the development of railway to the modern logistics, the introduction and training of logistics professionals will become the key factor affecting the development of logistics enterprises.

2.3 *Opportunity analysis*

1. The trend of the rapid development of modern logistics

Although in recent years logistics enterprises have developed rapidly in China, but compared with the developing level of logistics in the United States and Germany, Chinese logistics is still in the beginning stages of development, for it has only more than 10 years of logistics industry till now. From the point of view of railway transportation, the development level of the e-commerce logistics is suitable for railway transport joining the logistics environment. The railway should use its own advantages and take the initiative to increase the transportation market share.

2. The national policy support

The State Council issued "The Medium and Long-term Development of Logistics Industry Planning," making the logistics industry as the basis of the national economic development and strategic industries to new heights, providing a good opportunity for the railway to join modern logistics. The National Development and Reform Commission issued the "Internet + efficient logistics implementation opinion". It requires constructing logistics information Internet sharing system, promoting the traditional logistics activities to develop in the direction of informatization, digital (Jie, 2010; Dulin, 2010), and promoting logistics information, especially the government logistics information open sharing to improve the convenience for further integration for modern railway logistics and market.

3. The development of passenger rail line and high-speed railway

Upon the completion of a large number of passenger rail lines and high-speed railway, main busy lines will realize the separation of passenger and cargo, which will greatly increase the capacity of railway, thereby increasing the running speed of bulk goods trains. High-speed express business may also expand extensively, and the competitiveness of railway in freight market will be further enhanced.

2.4 *Threat analysis*

1. The adjustment of the economic structure leads to decrease of bulk cargo volume

Since the 18th National Congress of the Communist Party of China, the adjustment of economic structure has accelerated. China eliminates a series of backward production capacity, rectifies the related industries that not conform to the sustainable development of enterprises, and intensifies protection of the environment pollution. The bulk cargo, such as coal, steel, fertilizer, power, and so on, are mainly affected. For instance, as for coal, the coal consumption ability of power generation enterprises drops and the coal stocks increase, so the demand for coal further reduced. In terms of steel and iron, the six ministries and commissions, such as the State Environmental Protection Department, the National Development and Reform Commission, jointly issued the "Action Plan Implementation Guidelines for Prevention and Control of Atmospheric Pollution in Beijing-Tianjin-Hebei and surrounding areas," which clearly put forward that in 2017, Hebei Province needs to reduce 600 million tons of steel production capacity, which means one-third of Hebei iron and steel production capacity should be reduced. And two-thirds of Tangshan iron and steel production capacity will be cut, about 40 million tons. Influenced by the decrease of iron and steel and reduction of raw material inventory, upstream raw materials such as iron ore, coking coal, and steel billet, the demand for steel is flourishing. Railway bulk cargo generally showed a sharp decline, which has a great influence on the volume (Shigeru Yurimoto, 2002; Suzuki T, 2012; William, 2002).

2. The logistics market competition

Highway, air transport enterprises take advantage of its own to establish logistics enterprise and to occupy domestic transportation market share. At the same time, until August 2016, domestic oil prices suffered "two losses," which leads to highway, water, and air transport enterprise transportation cost reduction, so more supply of goods will flow to the highway, water, and air transport enterprises. From the point of international logistics enterprises, DH (DHL) in China, Fedex, Deppon Express, and other foreign logistics companies relied on the highway and air transport and maintained a sustained capital growth momentum into China. And they have relatively high-quality logistics resources, software and hardware facilities, and technical support, which will increase the competition of domestic transportation market.

3 STRATEGIES

1. Stabilize bulk cargo volume and gradually promote the railway modern logistics transformation process

Railway should keep the advantage of bulk cargo transportation and use the constantly improving transportation network to increase the proportion of direct transport, accelerate the delivery, set out customized services, mine potential customers, further stabilize railway bulk cargo volume, and on the basis of taking advantages of railway transport, draw lessons from Deppon Express and Shunfeng and other logistics enterprise to optimize transportation production, process, operation, and management optimization, setting up railway logistics center node on the mode of modern logistics to implement goods storage, processing,

distribution, and other functions in the process of delivery, which will result in actively moving closer to the modern logistics enterprises.

2. Developing railway logistics information network

Spreading railway 95306 net brand, solving the poor user experience, and friendly interface is not enough. Lack of integration with offline and other problems in current 95306 net, making full use of cloud computing, Internet of things, big data technology to build 95306 net as a railway electric business logistics platform with online trading, logistics service, information interaction, industry information, material purchasing and investment, and other functions to realize the business flow, logistics, cash flow, information flow of highly centralized, unified, and harmonious systems, and to build "Internet +" business model promote the development of online services and offline entity economic integration. At the same time, under the background of the development of transportation logistics, accelerating the establishment of a railway freight information synchronous transmission system and railway good electronic commerce information query system, forming data package that covers every order of goods quantity, type, station, the consignee detailed information, and realizing the packet before or synchronous transport of goods in the database provide real-time monitoring goods information for the customer and broaden transport demand.

3. Innovating the goods product, creating the express products under the thought of entire logistics

Excellent domestic and overseas logistics enterprises exclusively adapt to express logistics market brand, such as Deppon Express launches "VIP Privilege," "Precise Transition," and "Accurate Transition." [14], Shunfeng launches "Deliver Today," "Standard Transition," and "Special Price Transition." To fit the needs of logistics market customers, railway needs to strengthen the idea of taking the customer as the leading design express products under the idea of the whole logistics. At present, the Shanghai railway bureau has established express, fast, and ordinary trains to central Europe and central Asia, using the molten iron intermodal trains "3+3" transport product system, based on three speed levels: 160, 120, and 80 km/h. It has also launched long triangle scattered express train and batch bulk cargo express to strengthen the region's competitiveness of railway transportation.

4. The multimode of transport cooperation

Railway can give full play to the main line advantage and make effective integration and utilization of highways, waterways, and aviation logistics resources to extend railway logistics services and break the bottleneck of railway logistics and social comprehensive transportation system integration. At the same time, the railway can jointly produce logistics and social logistics public information platform and establish large radiation area to realize the railways, highways, waterways, and other logistics information and data, to make social logistics demand find the most suitable logistics service integrators and solution in the shortest possible time and finally enhance social logistics operational efficiency.

5. Product designing under the thought of the entire logistics

Learning advanced theory based on the idea of the entire logistics to establish the modern logistics management mode. In the aspect of the railway freight transportation organization, railway traffic organization should combine with its own capacity with the logistics demand and schedule car to coordinate the logistics needs of customers. According to different kinds of traffic flow, railway dispatching departments should determine reasonable and effective traffic organization scheme and provide information for tracking, refrigerated anti-staling value-added services. Besides, railway can provide the logistics financial, demand management, information consultation, and other value-added services.

6. Set up a rate system that matches the market effectively

In the current railway freight system, the vehicle charges depend on the freight rate scale, freight rate, and weight; container charges depend on the types; less-than-one carload depends on the freight rate scale, and the freight rate is based on highway. Therefore, railway should establish freight rate system leading by railway and based on cost–benefit analysis and effective matching [15]. That is, on pricing principles, highlighting railways leading the internal traffic of fate and social goods to match the best benefit of cost, its bottom pricing line (floor) is variable, and ceiling (ceiling price) is the guide formulated by the state. On price control, railway can use mathematical model, that is, on the basis of the railway traffic flow, it carries on "scheduled train" management, and through the role of price lever, it makes each of the direction of the goods to meet the needs of the traffic as far as possible. Running direct train and through the role of price lever to realize the optimization of the national railway traffic organization.

7. Promoting freight unitized

Consummating the unitized transport supporting policies and making comprehensive arrangement of multimodal transport and manufacturing enterprise demands. Establishing the goods packaging, unitized equipment, loading and unloading machinery configuration, stacking, loading and unloading of goods craft standard system, developing the use of unitized equipment to realize containerization are the primary functions. Second, studying using freight preferential measures to encourage customers unitized transport. At the same time, according to the scattered cargo express, use of 1.5t small container in scattered express station accelerates the organization of container express transportation.

8. Introduction of logistics professionals strengthens personnel training

Excellent enterprise logistics talents can not only bring the advanced concept of logistics, logistics technology, and logistics operation mode, but also use their personnel logistics relationship to develop the logistics market of the enterprise. Thus, under the situation of lack of advanced logistics talented personnel, railway logistics enterprises should take positive and effective measures to set up the mechanism of attracting talent, from the introduction of various logistics management talents, advanced logistics technology talents, advanced education talented person, through that railway can establish effective logistics management system in a relatively short time. In addition, China has universities such as Tsinghua University, Tongji University, Beijing Jiaotong University, and Southwest Jiaotong University; logistics engineering; and scientific research institutes such as the Chinese Academy of Sciences and China Academy of Railway Sciences, which have logistics engineering units. These colleges, universities, and research institutes have trained a large number of logistics talents, which provides talent reserve for the railway modern logistics development.

4 CONCLUSIONS

With the rapid development of the modern logistics, the period of railway logistics transition has extended, but with a bright future. If constant progress wants to be reached, it is necessary to have a clear understanding of the reality, feasible development strategies, and right force points. This dissertation has a certain reference value and guides significance for the transformation from railway freight to modern logistics in the electronic commerce environment and for the increase in the market share of railway freight.

ACKNOWLEDGMENT

This work was financially supported by China Academy of Railway Sciences (2014YJ099 & 2015X006-C).

REFERENCES

Athanasios Ballis, John Golias. Comparative Evaluation of Existing and innovative Rail-road Freight Transport Terminals. Transportation Research Part A 36(2002): 593–611.

Dulin. Rail freight logistics modernization reform research [J]. Science and technology exchange, 2010 (6): 46 and 47.

Guo Yuhua. Rail freight development of modern logistics research [J]. Journal of railway transportation economy, 2010 (2): 10 to 13.

Jie Dongdong, Gai yu-xian, Li Cheng, etc. The SWOT analysis of the rail freight logistics development in our country [J]. Journal of Lanzhou Jiaotong University, 2010, 29 (3): 71–74, 78. DOI: 10.3969/ji SSN. 1001–4373.2010.03.015.

Liu Pengxin. Based on the railway logistics bulk cargo transportation organization optimization research [D]. Southwest Jiaotong University, 2014.

Liu Pengxin. Railway of bulk cargo transportation logistics SWOT analysis and countermeasures study [J]. Journal of Heilongjiang science and technology information, 2014, (6): 90–90. The DOI: 10.3969/ji SSN. 1673–1328.2014.06.097.

Michael J. Bauer, Charles C. Pokier, Lawrence Lapide, PhD, and John Bermudez. E-Business: The Strategic Impact on Supply Chain and Logistics[M]. 2001:58–256.

Shigeru Yurimoto, Naoto Katayama. A model for the optimal number and locations of public distribution centers and its application to the Tokyo metropolitan area[J]. International Journal of Industrial Engineering, 2002, 9(4): 363–371.

SuzukiT, LiGQ. An Analysis on the Railway-Based Intermodal Freight Transport in Japan Regarding The Effect of Disasters[J]. 8th International Conference on Transport Transportation Studies(ICTTS), Beijing. PEOPLES R CHINA, Procedia Social and Behavioral Sciences, 2012(43): 111–118.

Wang Haibo. The railway goods transportation logistics measures analysis [J]. China's collective economy, 2007 (5): 48.

Wang Zhande, Wang Yao. Analysis of the concept and characteristics of railway logistics [J]. Journal of logistics technology, 2008, 27 (2): 16–19.

Wang lei, China railway freight reform strategy orientation research [D]. Beijing Jiaotong University, 2007.

William C. Vantuono. KCS Moves to Scheduled Operations. Railway Age. 2002, (11).

Zhang Changjun. Under the "Real. Freight" rail freight logistics development research [J]. Logistics technology. 2015, 38(2):151–153. DOI:10.3969/j.issn. 1002–3100.2015.02.047.

Zhao Haikuan, Wang Tao, Song Wenkai, etc. Accelerate the reform of railway freight yard logistics will enhance the capacity and quality of railway logistics [J]. Journal of China railway, 2014, (9): 1–5. DOI: 10.3969/j.i SSN. 1001–683 - x. 2014.09.001.

Advances in Energy Science and Equipment Engineering II – Zhou, Patty & Chen (Eds)
© 2017 Taylor & Francis Group, London, ISBN 978-1-138-71798-5

Query optimization design based on cloud computing

Shanxiong Chen
School of Information Engineering, Guizhou University of Engineering Science, Bijie, China
College of Computer and Information Science, Southwest University, Chongqing, China

Li Wang, Qirong Zhang & Daode He
College of Computer and Information Science, Southwest University, Chongqing, China

ABSTRACT: Cloud computing, because of its advantages in large data processing and resource sharing, is being used in several industries. Data storage and retrieval mode in cloud computing are different from the traditional database system, and the data query method of traditional database cannot be directly migrated to a cloud platform. In this paper, we use suffix tree to build an algorithm of cloud storage and discuss cloud query model based on suffix tree. The query model can be embedded into existing data query system, achieving migration from traditional query platform to cloud platform.

1 INTRODUCTION

With the rapid development of information technology, increasingly more enterprises and institutions have been benefitted by its advantages, but the management and analysis of large amounts of data followed really make many industries, such as medical treatment, communications, transportation, finance, and Internet, complicated (Li, 2014; Lai, 2012). Traditional data processing methods and means cannot deal with the management of such large-scale data, and related hardware and software as well as the high cost of maintenance make most users awkward. Cloud computing is an emerging computing model, which conceals the computing resources and the implementation process of computing, and users only need to submit computing tasks or service requests through a browser or application interface, without having to consider how to build computing architecture and how to organize and schedule computing resources. More and more organizations are more willing to transfer the data center from expensive high-performance computing clusters to public or private cloud environments (Winterberg, 2011).

As cloud computing is a data storage and computing model built on the basis of distributed resources storage and equipment sharing, traditional database technology cannot be directly migrated to a cloud computing platform. Therefore, a key technology to build a cloud computing platform is to establish database service of cloud storage, which is a challenging task. In this paper, we mainly discuss the query optimization technology of cloud computing to provide a viable method

for achieving fast retrieval and manipulation of cloud platform data.

2 QUERY TECHNOLOGY OF CLOUD COMPUTING

With the needs of big data processing, more and more application service and data processing transform from high-performance server to public or private cloud systems. How to provide data processing service and conduct efficient data management has become one of the most critical tasks in cloud computing system. As the data storage and management mode of cloud computing system are completely different from the management mode of traditional relational database, existing database computing cannot be directly migrated to cloud computing system (Chen, 2011). Furthermore, cloud computing system requires that data management capability can provide good scalability and fast and accurate data access capability, and has efficient solutions for clustered data analysis and high-density concurrency transaction processing. Similar to existing database systems, query processing and optimization is the key technology of data management in cloud computing system. Data retrieval capability is an important safeguard for cloud computing system to provide rapid response service. Under the three major cloud computing service models, framework service, platform service, and software service, query technology is an important technical aspect and also an important function that both the user and the system will apply. Index technology can effectively improve the quality of query in data management system, and the index is used to reduce CPU time used in the query, disk reading, and other operations, in order to improve query performance. Constructing effective index in cloud computing environment can also improve query processing performance (Song, 2014; Zhang, 2014). In this paper, we presents a fast retrieval technique of suffix tree to achieve fast data query in cloud computing system.

3 INDEX TECHNOLOGY BASED ON SUFFIX DATA

The index technology currently used by cloud computing is divided into two categories (Loh, 2011; Barsky, 2011): centralized index and distributed index. The centralized index is to divide file into a number of fixed-size data blocks and centrally store data index in the central management node, in order to ensure access efficiency of metadata. The distributed index evenly stores data to each node in the cloud, and data query only requires the positioning of the node route in the entire cloud system. The suffix tree index proposed in this paper can conduct centralized index and distributed index (Rasheed, 2010; Dorohonceanu, 2000). It can conduct centralized index in small-scale private cloud, to improve management efficiency, and can conduct distributed index in large-scale cloud platform, to relieve system pressure (Koschke, 2014).

3.1 Suffix tree

The query of a path expression can be calculated through the connection between the tables of element names and attribute names on corresponding paths. For example, the query of e1/eZ/e3 can be decomposed into el/e2 and eZ/e3, and then connecting their results once can get search result, but there will be more connection times for longer path, which would cause high processing cost and then affect query efficiency.

Aggregating the data nodes with the same semanteme into a node and being able to query the node within limited steps are the basic ideas of constructing the suffix tree of cloud computing query path: a polymerization suffix tree that contains all suffix strings of each node's semantic path string (namely, suffix tree sharing semantic path) is constructed, and the suffix tree is called the suffix index tree of corresponding data figure (SuffIndex), as is shown in Figure 1. It is also corresponding suffix index tree based on cloud-structure data, and data nodes contained in each node in index tree is called an extended set of the node. As is shown in Figure 2, the expansion set of the

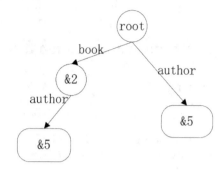

Figure 1.   SuffIndex of book.&2.author.&5.

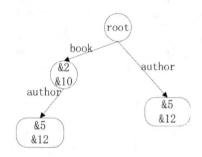

Figure 2.   SuffIndex of book.&10.author.&12 enters.

corresponding index node in path book is $\{\& 2, \& 10\}$. The definition of SuffIndex is given below:

Let $\sigma(T_0) = \{s_1, s_2, \ldots s_L\}$ be the collection of all nodes in V_0 on the data path string of V_0, and for all $s_i(i = 1, 2, \ldots L)$, adopt their semantic path strings to construct suffix tree $T_{\sigma(T_0)} = \{V_{suff}, root_{suff}, T_{suff}, E_{suff}, \Sigma_0, F\}$, in which $V_{suff}, root_{suff}, T_{suff}, E_{suff}$ are node set, root node, leaf node, and edge set of the tree, respectively, and Σ_0 is from T_0. $F(v') = \{v \mid v \in V_0 \wedge v' \in V_{suff} \wedge v_{path} = v'_{path}\}$, v_{suff}, v'_{path} are, respectively, the semantic paths of node v and v'.

While constructing SuffIndex, because the path of all internal nodes is a substring of leaf node path, it is only necessary to deal with all leaf node paths in Figure 2. The parameter $\sigma(T_0)$ only needs to contain all leaf node paths, namely, $\sigma(T_0) = \{s_1, s_2 \ldots s_l\}$, and l is the number of leaf nodes.

3.2 Query design of suffix tree

Let the current suffix tree be T_{i-1} and root node be root, then the algorithm of the process of constructing $s_i = l_{l_1} d_{l_1} l_{l_2} d_{l_2} \wedge l_{l_m} d_{l_m}$ suffix tree in $T_{i-1}T$ is as follows.

Process of constructing $s_i = l_{l_1} d_{l_1} l_{l_2} d_{l_2} \wedge l_{l_m} d_{l_m}$ suffix tree in $T_{i-1}T$:

1. Take semantic path $p \leftarrow l_{i_1}l_{i_2} \wedge l_{i_m}$, which is also the suffix of p, and suppose data string $q \leftarrow d_{i_1}d_{i_2} \wedge d_{i_m}$
2. Search whether there is an edge P [1] from root, and if so, integrate q [1] into the expansion set of nodes it points to in $T_{i\_}1$.
3. Continue the search process of Step 2 until no matching edge can be found. Suppose the final matching edge is P [j], the node it points to is M, $k \leftarrow j$, and move to Step 4.
4. $k \leftarrow k + 1$, if $k \leq |p|$, move to Step 5, and if not, move to Step 6.
5. Create a new node N, integrate d[k] into the expansion set of N, edge p[k] points from M to N, and move to Step 4.
6. If $|p| \leq 1$, move to Step 7, and if $p \leftarrow p[2k \,|\, p \,|]$, $q \leftarrow [2k \,|\, q \,|]$, move to Step 2.
7. End.

There may be multiple document models in cloud database, so several different SuffIndex trees can be built. Assuming that they have a common virtual root, a collection of SuffIndex tree formed by virtual roots can be added. Therefore, we can dynamically create SuffIndex tree containing multiple document models. Path navigation is a very important issue in structured data query of cloud computing. Because suffix tree can solve a lot of problems in linear time, taking the path in document as a character string to encode, using the suffix tree to index these paths, and merging the same index paths, we can find whether this path exists in the database or not within the linear time of drawing the path pattern length. Furthermore, example of each element can be grouped according to the root node of document mode it is located and its path pattern.

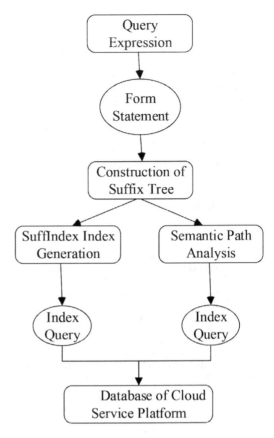

Figure 3. Cloud query process based on suffix tree.

then assign to each node, and for centralized query mode, unified management of index database is conducted.

4 CLOUD QUERY MODEL BASED ON SUFFIX TREE

This section gives a clear flow structure diagram based on the query process in cloud computing, as is shown in Figure 3. The query process can be easily integrated and embedded into the query of other distributed systems. The query system uses the index method of characteristic value to initialize the query sentence set to generate the index table of feature value, which deals with the query index newly added from time to time in the operation and calculates the characteristic value of single index. If the characteristic value of the query sentence hits an item in the index table of characteristic value, the public part of the query sentence newly added and the original query sentence set can be reused. Suffix tree query of cloud computing is to use suffix tree to establish query index and

5 CONCLUSION

The application of cloud computing in various industries is bound to bring rapid development in industry information, especially the cloud platform with low cost and high efficiency, making resource sharing, distributed processing of information, and big data processing become the key technology of industry information building. This paper explores the query optimization technique in cloud computing and proposes a query method of suffix data. The method is adaptable to both centralized and distributed queries, and its cloud query method based on suffix tree can be embedded in the existing database management system, and it realizes the migration from traditional database technology to cloud computing data management technology.

ACKNOWLEDGMENTS

This study was supported by the National Natural Science Foundation of China (61170192), the Fundamental Research Funds for the Central Universities (XDJK2014C039, XDJK2016C045), Doctoral Fund of Southwestern University (swu114033), and Joint funds for the Science and Technology Department of Guizhou Province and Bijie Municipal Science and Technology Bureau and Guizhou University of Engineering Science (KEHE LH [2016]7053 in Guizhou).

REFERENCES

Barsky, M., U. Stege, and A. Thomo, "Suffix trees for inputs larger than main memory," *Information Systems,* vol. 36, pp. 644–654, May 2011.

Chen J.G. and T.T. He, "Query-focused Multi-document Summarization Using Cloud Model," *Information-an International Interdisciplinary Journal,* vol. 14, pp. 951–956, Mar 2011.

Dorohonceanu B. and C. Nevill-Manning, "A practical suffix-tree implementation for string searches," *Dr Dobbs Journal,* vol 25, pp. 133–+, Jul 2000.

Koschke, R. "Large-scale inter-system clone detection using suffix trees and hashing", *Journal of Software-Evolution and Process,* vol. 26, pp. 747–769, Aug 2014.

Lai K.C. and Y.F. Yu, "A scalable multi-attribute hybrid overlay for range queries on the cloud," *Information Systems Frontiers,* vol. 14, pp. 895–908, Sep 2012.

Li, R.X., Z.Y. Xu, W.S. Kang, K.C. Yow, and C.Z. Xu, "Efficient multi-keyword ranked query over encrypted data in cloud computing," *Future Generation Computer Systems-the International Journal of Grid Computing and Escience,* vol. 30, pp. 179–190, Jan 2014.

Loh W.K. and H. Ahn, "A Storage-Efficient Suffix Tree Construction Algorithm for Human Genome Sequences," *Ieice Transactions on Information and Systems,* vol. E94d, pp. 2557–2560, Dec 2011.

Rasheed F. and R. Alhajj, "STNR: A suffix tree based noise resilient algorithm for periodicity detection in time series databases," *Applied Intelligence,* vol. 32, pp. 267–278, Jun 2010.

Song, W., Z.Y. Peng, Q. Wang, F.Q. Cheng, X.X. Wu, and Y.H. Cui, "Efficient privacy-preserved data query over ciphertext in cloud computing," *Security and Communication Networks,* vol. 7, pp. 1049–1065, Jun 2014.

Winterberg, F. "The clouds of physics and Einstein's last query: Can quantum mechanics be derived from general relativity?," *Physics Essays,* vol. 24, pp. 169–174, Jun 2011.

Zhang, Y.G., S. Su, Y.L. Wang, W.F. Chen, and F.C. Yang, "Privacy-assured substructure similarity query over encrypted graph-structured data in cloud," *Security and Communication Networks,* vol. 7, pp. 1933–1944, Nov 2014.

Advances in Energy Science and Equipment Engineering II – Zhou, Patty & Chen (Eds)
© 2017 Taylor & Francis Group, London, ISBN 978-1-138-71798-5

A study on shift strategy of AMT for pure electric vehicles based on driving intentions

Jun Zhang & Xiao-bing Wang
National Engineering Laboratory for Electric Vehicles, Beijing Institute of Technology, Beijing, China

ABSTRACT: In order to make the shift timing of electric vehicles better meet the driver's driving experience, taking pure electric vehicle equipped with two-gear Automated Mechanical Transmission (AMT) as the research platform, shift control strategy was proposed, and the MATLAB/Simulink simulation model of the whole vehicle was established. The simulation results show that this system can identify the driver's different driving intentions accurately, and the corresponding shift strategy given by it improves the adaptability and practicality of the shift control strategy.

1 INTRODUCTION

In order to improve the safety, comfort, and driving convenience of the vehicle, shift operate automation is an inevitable trend during the development of electric vehicles (Xi, 2010). Shift control strategy is an important factor that affects the performance of automatic transmission. In the past, shift schedule often required the driver to choose dynamic or economic buttons, which greatly affects the driver's driving experience. Some shift schedules were divided into different areas according to the speed of the vehicle or combined with the size of the throttle opening to choose a compromise schedule between the economic and dynamic shift schedule (Wang, 2005). With the development trend of intelligent vehicles, the identification of driving intentions and vehicle running environment has become a hot research topic in recent years (Liu, 2005).

In this paper, we studied the AMT shift control strategy based on driving intentions recognition of pure electric vehicles. This transmission system has no clutch. By using the control characteristics of the electric vehicle's drive motor, when ready to shift, the drive motor is turned to free mode, namely the output torque of the motor turned to zero. By this way, the traditional automobile power interruption process is achieved, thus completing the shift operation.

2 ANALYSIS AND EXPERIENCE SUMMARY OF DRIVING INTENTIONS

During the driving process of the vehicle, driver can operate the vehicle according to the vehicle's state and its external environment. Driving intention recognition is based on the driver's operation on accelerator pedal, brake pedal, and other parts of the car.

According to the interviews conducted on experienced drivers, driving intentions are classified into urgent acceleration, acceleration, maintain, sliding, braking, and urgent braking.

Urgent acceleration is the intention that the driver quickly steps on the accelerator pedal to the end or greater opening degree. By this operation, the driver wants to get a larger acceleration to complete overtaking or emergency movement intent. In this case, in order to obtain a larger driving torque, it should be delayed to shift up. If it is staying in the high gear, then it should be shifted to low gear ahead of time.

Acceleration intention is that the driver steps on the accelerator pedal slowly; the desired acceleration of the vehicle is small. In this case, it should be appropriately delayed to shift up, timely shift down.

Maintain intention is that the driver maintains the accelerator pedal in current position or its change range is very small. By that operation, the driver wants to maintain the current state of motion, so that the basic shift schedule is used in this case.

Sliding intention results when the driver releases the accelerator pedal and the brake pedal so that the vehicle can be driven by inertia. This case uses the basic shift schedule as well.

Braking intention is the driver releases the accelerator pedal and brakes slowly to decelerate the vehicle. In this case, it should be limited to shift up, timely shift down.

Urgent braking intention is that the driver releases the accelerator pedal and steps on the

Table 1. Shift operating principles under different driving intentions.

Driving intention	Shift operating principle
Urgent acceleration	Delay to shift up, ahead to shift down
Acceleration	Delay to shift up, timely shift down
Maintenance	Basic shift schedule
Sliding	Basic shift schedule
Braking	Limited to shift up, timely shift down
Urgent braking	No shift

brake pedal hardly so that the vehicle slows down very quickly. The shift in this case, as the vehicle slows down very quickly, resulting in synchronous speed calculated by the motor varies greatly with the rotational speed of the output shaft of the gearbox when engaging to the aimed gear, which can result in big shifting impact, even fail to complete the shift operation. In order to avoid this situation, no shift operation would engage. After the brake pedal is lifted, the gear to shift is given according to the current speed and gear. Table 1 lists the shift operating principles under different driving intentions.

3 DRIVING INTENTION RECOGNITION

As drivers operate, their intention is difficult to be described accurately with mathematical model, an empirical model is employed. Fuzzy control has been widely studied and applied to the problem of dealing with empirical models (Liu, 2013). At present, the identification methods of driving intentions are based on the opening degree and change rate of the accelerator pedal (Yang, 2009). However, this cannot fully reflect the true intentions of the driver. Taking urgent acceleration intention as an example, driver operation characteristics is stepping on the accelerator pedal to the end and remaining for a period of time. If we only use variables of the change rate of accelerate pedal and its opening degree and not consider the vehicle body acceleration, we can only identify instantaneous intention when the driver steps on the pedal, but acute accelerated process will continue for a period of time and the driving intention will as well. The other intentions are the same.

For fuzzy reasoning, the more the input, the more complex is reasoning. Too many inputs will result in a very large fuzzy rule base, which also exerts a huge load on storage. The choosing of membership degree function must also consider the simplified calculation. Triangular and trapezoidal function is simple in calculation, and the

Figure 1. Fuzzy recognition model of driver's intention.

Table 2. Input, output variable domain and fuzzy set list.

Variable name	Discourse domain	Fuzzy set
Accelerator pedal opening (%)	[0,100]	ZO PS PM PB
The change rate of accelerator pedal opening (%/s)	[−100,100]	NB NM NS ZO PS PM PB
Acceleration (m/s²)	[−6.8,3.3]	NB NS ZO PS PB
Driving intention	3	Urgent acceleration
	2	Acceleration
	1	Maintain
	0	Sliding
	−1	Braking
	−2	Urgent braking

performance difference is very small compared to that of the complex curve membership degree functions (Shi, 2011).

On the basis of the above considerations, this paper takes accelerator pedal opening degree, change rate of accelerator pedal, and body acceleration as input and the quantified driving intentions as output. Combined with the driver's driving experience, a three-input single-output driver intention fuzzy recognizer is established, whose composition principle is shown in Figure 1.

This is a Multi-Input Single-Output (MISO) fuzzy reasoning model. The input—output variable domain and fuzzy sets are shown in Table 2. The domain of accelerator pedal opening is [0, 100] and its four fuzzy sets are Zero (ZO), Positive Small (PS), Positive Middle (PM), and Positive Big (PB); the domain of the accelerator pedal

Table 3. Fuzzy rule base for driving intention recognition.

Accelerator pedal opening	Change rate of accelerator pedal opening	Acceleration	Driving intention
ZO	ZO	NS	Braking
ZO	ZO	ZO	Sliding
ZO	ZO	NB	Urgent braking
/	NM	NS	Braking
/	NB	NS	Braking
PS	PS	ZO	Maintain
PS	ZO	ZO	Maintain
PS	NS	ZO	Maintain
PM	NS	ZO	Maintain
PM	PS	ZO	Maintain
PM	ZO	ZO	Maintain
PB	PS	PS	Acceleration
PM	PM	PS	Acceleration
PM	PB	PS	Acceleration
PS	PS	PS	Acceleration
PM	PS	PS	acceleration
PS	PB	PB	Urgent acceleration
PM	PB	PB	Urgent acceleration
PB	ZO	PB	Urgent acceleration
PB	PS	PB	Urgent acceleration
PB	PM	PB	Urgent acceleration
PB	PB	PB	Urgent acceleration

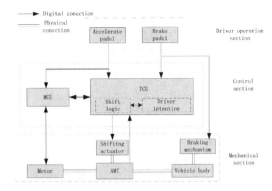

Figure 2. Schematic diagram of AMT system simulation.

Table 4. Parameters of the mini pure electric vehicle.

Parameters	
Total mass (kg)	670/850
Tire radius (m)	0.247
Drag coefficient	0.38
Frontal area (m²)	2.418
Dimensions (mm)	2765*1560*1550
Coefficient of rolling resistance	0.015
Low gear ratio	8.166
High gear ratio	4.935
Final ratio	1.0

opening change rate is [−100, 100] and its seven fuzzy sets are Negative Big (NB), Negative Middle (NM), Negative Small (NS), Zero (ZO), Positive Small (PS), Positive Middle (PM), and Positive Big (PB); the domain of acceleration is [−6.8, 3.3] and its five fuzzy sets are Negative Big (NB), Negative Small (NS), Zero (ZO), Positive Small (PS), and Positive Big (PB); the domain of driver's intention is a single value, where 3 represents urgent acceleration, 2 is acceleration, 1 is maintain, 0 is sliding, −1 is braking, and −2 represents urgent braking.

According to the driving experience and the characteristics of the vehicle, the rules are formulated in Table 3.

4 SHIFT STRATEGY SIMULATION BASED ON DRIVING INTENTION RECOGNITION

The fuzzy recognition model is generated by MATLAB/Simulink Fuzzy logic toolbox, combined with the controller model and the vehicle model built by SimDriveline and Stateflow, and the simulation of the driver's driving intentions identification is completed. SimDriveline is a physical modeling module; the transmission system it built takes characteristics of stiffness, inertia, and so on into account. Therefore, it can be more close to real reactions (Inc, 2012). Compared with traditional digital modules, it has obvious advantages. The simulation structure of the AMT system studied in this paper is shown in Figure 2.

The simulation model is composed of three parts: the driver's operation part, control part, and the vehicle transmission system part.

Parameters of the mini pure electric vehicle studied in this paper are shown in Table 4.

During the shift operation, vehicles will unload and pick off to neutral gear. This operation will disturb the process of driving intentions identification. Therefore, when shift operation is actuated, driving intention remains unchanged and starts again until the end of the shift process.

The simulation tests are actuated in the following order: start the car with big accelerator opening, gradually reduce the accelerator, lightly brake, accelerate with a small accelerator pedal opening,

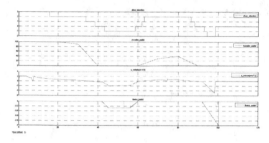

Figure 3.   Driving intentions, vehicle body acceleration, accelerator, and brake pedal.

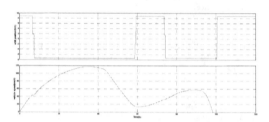

Figure 4.   Gear box position and speed curve.

and then emergency braking. The simulation result is shown in Figures 3 and 4. The vehicle acceleration is large when start with a big throttle, and the recognized driving intention is urgent acceleration. With its speed increasing, vehicle body acceleration decreases gradually, the driving intention is recognized as acceleration. Subsequently, the throttle is reduced and the automobile acceleration is reduced further, driving intentions are identified to maintain sliding. Then, the brake pedal is lightly stepped, the car slows down gradually, and the driving intention is recognized as braking intention; after the accelerator pedal is slowly stepped on, automobile acceleration increases gradually, and the identification of driving intention is followed by to sliding, maintain, and acceleration; finally, the accelerator pedal is lifted and the brake pedal is stepped to the end, automobile acceleration is negative, and the driving intention is recognized as urge braking.

The simulation results show that the fuzzy recognizer can accurately identify the driving intentions and complete the shift operation on time, according to the driving intention. As shown in Figures 3 and 4, the shift-up point under urgent acceleration intention is significantly delayed compared with under acceleration intention. When braking intentions shift down as basic schedule, under urgent braking intention, the gear does not move. And after braking, it shifts back to the low gear.

5   CONCLUSIONS

Driving intentions recognition is one of the key technologies of intelligent gear shifting. Fuzzy control has been widely applied due to its advantages on empirical models. For fuzzy controller, different input, same input, different domain, and fuzzy rules will cause huge difference in the recognition results. In this paper, we took the throttle open degree and its change rate and acceleration of the vehicle as the input, quantified driver intention as output, and made a three-input single-output fuzzy identifier. The identification results are applied to the selection and correction of the shift rules. Simulation results illustrate its correctness and applicability. This paper lays the foundation of future research on the integrated shifting strategy, which combines driving intentions and driving environment of the vehicle.

REFERENCES

Liu Jin-kun, Shen Xiao-rong, Zhao Long, System identification theory and MATLAB simulation. Publishing House of Electronics Industry (2013).
Liu Zhen-jun, Doctoral thesis, Chongqing University (2005).
Shi Jun-wu, Doctoral thesis, Shanghai Jiao Tong University (2011).
The MathWorks Inc., SimDriveline User's Guide (2012).
Wang Yu-hai, Song Jian, Li Xing-kun, Journal of Highway and Transportation Research and Development. **22**, 113–118 (2005).
Xi Jun-qiang, Wang Lei, Fu Wen-qing. Transactions of Beijing Institute of Technology. **30**, 42–45 (2010).
Yang Wei-bin, Chen Quan-shi, Wu Guang-qiang, Jounal of Mechanical Engineering, **45**, 206–210 (2009).

Advances in Energy Science and Equipment Engineering II – Zhou, Patty & Chen (Eds)
© 2017 Taylor & Francis Group, London, ISBN 978-1-138-71798-5

Measuring characteristics and performance of HTTP video streaming in China

Hui Tang & Liang Chen
College of Information Engineering, Shenzhen University, Guangdong Sheng, China

Yue Wang
Department of Information Management, Central University of Finance and Economics, Beijing, China

Na Wang
College of Information Engineering, Shenzhen University, Guangdong Sheng, China

ABSTRACT: Nowadays, MSPs witness an increasing network traffic generated by online video services. Facing the intense mutual competition, it is necessary for MSPs to improve quality of media content and performance of online video services to seize market share. In this study, we aim to evaluate the performance of online video services for MSPs. First, we detect media content delivery and characteristics of technology (e.g., playback rate, video slice thresholds, and slice durations) of MSPs by using network sniffer. Second, we evaluate performance of MSPs with start-up delay, number, and total time of video stalling. Our difference from previous works is that we use automated Web application to record video playback timestamps automatically, which is necessary to obtain start-up delay, number, and total time of video stalling. Experiments show advantages and disadvantages for MSPs in China.

1 INTRODUCTION

With the rapid development of the Internet, users are shifting from cable TV to online video, especially on mobile devices. Over 70% of data traffic is contributed by wireless online video services (Cisco, 2014). Media Service Providers (MSPs) are facing crucial challenges of improving the quality of media content and performance of online video services for users. Media content delivery based on HTTP not only improves performance but also makes configuration easier than that of RTSP. Both HTTP Progressive Download (HPD) and HTTP Adaptive Streaming (HAS) are often applied to online video services. Thus, media content delivery based on HTTP gradually occupies the leading role.

In this study, we aimed to evaluate the performance of online video services for MSPs in China. However, the media content delivery and characteristics of technology adopted by MSPs are different. We use network general sniffer named Wireshark (2002) to sniff video streaming and automated Web browsers application named Selenium (2005) to record various timestamps of video playback (i.e. start timestamp, end timestamp, stalling timestamp, and playback resuming timestamp). As a result, we determine the media content delivery,

average playback rate, slice size, and slice threshold under different video definitions for MSPs in China. Moreover, we evaluate the performance of online video services with start-up delay, number, and total time of stalling for a video.

In previous studies, Wang et al. (2003) proposed various metrics to evaluate the performance of UDP-based video streaming under different network conditions. Authors (Pessemier, 2013; Huynh-Thu, 2009) proposed metrics that rely on subjective human responses. Mok et al. (2011) investigated how the Quality of Service (QoS) affects the Quality of Experience (QoE) of HTTP video streaming. Our work considers start-up delay and stalling as performance metrics for TCP-based video streaming. Cidon (1996) and Xu (2011) calculated the packet loss probability in terms of the recursive approach. Furthermore, Xu et al. (2011) proposed a fluid model to analyze the probability of video stalling on the file level. Luan et al. (Luan, 2010) analyzed the probability of video stalling, considering video size and playback buffer size. However, they do not provide insights into the exact number and total time of stalling for a video.

The rest of the paper is organized as follows. The background is introduced in Section 2. Section 3 presents the measurement methods with Wireshark and Selenium. Section 4 presents experiments and results for MSP services. Section 5 concludes this paper and discusses the future work.

2 BACKGROUND

2.1 *Transport protocols, start-up delay, and slicing*

The transport protocols of media content delivery mainly derive from two classes: RTSP, which is based on RTP/RTCP and is stateful, and HTTP which is based on TCP and is stateless.

When the transport protocol is RTSP, and a user makes a request for the video to an RTSP server, the server transmits packets of video frames over RTP as a continuous stream until the user is disconnected. The advantage of RTSP is that the server monitors the client state over RTCP during the connection and controls the transmission rate effectively. However, the shortcoming of media streaming over RTSP is serious and significant. First, RTP uses UDP to transmit packets, and it can lose packets severely due to volatile network condition, traffic congestion, or buffer limit. Second, it requires a high workload to deploy RTSP servers on Content Delivery Network (CDN). Therefore, RTSP sees a drop in popularity of the transmission of the online video stream.

Fortunately, we can avoid these two problems by HTTP. When a user makes a request to an HTTP server, the server responds once and transmits packets over TCP. Media content delivery over HTTP mainly includes HPD and HAS, which provides firewall-friendly and reliable transmission. The shortcomings of HPD are as follows: First, bandwidth will be wasted if a user stops watching the content or switch to another content after the content is downloaded. Second, due to latency, it generally does not support live video services. Nevertheless, HPD still plays an important role in media content delivery. HAS is proposed by different companies to improve HPD, such as Microsoft IIS Smooth Streaming (MS IIS), HTTP Living Streaming (HLS), and Dynamic Adaptive Streaming over HTTP (DASH). HAS dynamically transmits video frame of appropriate definition in terms of network condition and client platform. The features (Stockhammer, 2011) of different HAS are shown in Table 1. Nowadays, almost all MSPs adopt the media content delivery over HTTP.

When an online video is requested, the media server first transmits the packets of an advertisement, which is a major source of income for video companies. Notably, the packets of advertisement are transmitted by a different server rather than the same video server. The packets of the requested video start transmitting after the advertisement are downloaded. The start-up delay is closely related

Table 1. HTTP adaptive steaming comparison.

Feature	MS IIS	HLS	DASH
Streaming protocol	HTTP	HTTP	HTTP
Supported platforms	Silverlight	IOS	General
Media container	Mp4	Ts	Mp4
Live streaming	Yes	Yes	Yes
Default slice duration	2 s	10 s	Flexible
End-to-end latency	>1.5 s	30 s	Flexible

Table 2. Parameters of video definitions.

Definition	Pixels	Byte rate	Frames rate
LSD	320×240	≥ 50 KB/s	≥ 20f/s
SD	480×360	≥ 200 KB/s	≥ 20f/s
MD	640×480	≥ 600 KB/s	≥ 20f/s
HD	1280×720	≥ 1.2MB/s	≥ 30f/s
VHD	1920×1080	≥ 2MB/s	≥ 30f/s

to video stalling as it determines the initial length of playback buffer.

A video is cut into several slices when its length exceeds a certain threshold. A slice is a several-minutes video fragment, whose length depends on each MSP, and it is transmitted by a TCP connection. The advantage of video slicing is that it can prevent the attacker from downloading the whole video easily.

2.2 *Video definition*

The definitions of online video mainly are the Low-Standard Definition (LSD), Standard Definition (SD), Medium Definition (MD), High Definition (HD), and Very High Definition (VHD), whose parameters are shown in Table 2.

The video playback is composed of a sequence of Group of Pictures (GoP). The most direct way to estimate the video playback rate is by considering the frame rate and the size of a GoP consisting of I, P, and B frames. The size of GoP is determined by the number of I, P, and B frames, which have different compression ratios. MSPs usually adopt Variable Byte Rate (VBR) to encode a video, which results in the dynamic size per GOP. However, it is difficult to measure the size of per GoP during video playback. We use average byte rate as metric to measure playback rate of a video, but different videos may be played under different definitions.

2.3 *Measurement tools*

In this paper, we use Wireshark and Selenium to sniff network traffic and record timestamp of video playback. Wireshark is the most popular network protocol analyzer, which lets us know what is happening on our network at a microscopic level by capturing packets information (e.g., arrival timestamp and protocol header).

Selenium is an automated Web application, which can simulate a user to operate Web browser according to a script. Moreover, it provides rich API interfaces and supports multiple languages.

3 MEASUREMENT METHODS

In this section, we introduce how to use Wireshark to capture video streaming and code a script to record the timestamp of video playback on the local machine.

Wireshark capture the video streaming from MSP servers by setting host IP, which belongs to MSPs. When a user makes an HTTP request for a video, server transmits packets over TCP. Some header information is necessary to be obtained. First, both HTTP and TCP packets contain the same IP addresses of server and client, which maps HTTP and TCP-streaming for a requested video. Second, we download video files from URL contained in HTTP header and obtain the duration of video slice by playing it on the local media player. Third, stream index and sequence number in TCP header denote the index of a TCP-streaming from the server to client and total bytes of ack packets for a video streaming, respectively.

Selenium can record corresponding timestamps of video playback automatically by monitoring the changes of labels in clients' HTML pages. Details about the Selenium script are as follows. First, Selenium needs to initialize a browser engine. Then, selector locates <video> label or Adobe's Flash Player, and monitors sublabel (e.g., play button). During the playback, Selenium will save timestamps and states of video in the database when the sublabel is changed. Comparing the timestamp recorded by Selenium with that captured from Wireshark, we obtain the state of video and bytes of ack packets at a given timestamp.

4 EXPERIMENTS AND RESULTS

In this section, we introduce our method to achieve media content delivery and obtain the characteristics of technology (e.g., playback rate of definition, duration and threshold of video slice, start-up delay, and duration of video stalling) for MSPs in China, such as Tencent, Iqiyi, Youku, and Letv.

4.1 Media content-transmitted protocols

By downloading all of the video files from URL contained in HTTP header, we find that the types

of file are different for MSPs. We can determine the media content delivery for an MSP according to the types of files. The media content delivery is HPD if the video files include media files only and is HAS if the video files include both the described file and media file. Table 3 shows types of the video file of MSPs.

4.2 Video playback rate

Duration of a video can be obtained by the difference between end timestamp and start timestamp, and the number of bytes of the video is a sum of bytes of video slices. We have:

$$r_p = \frac{\sum_{i=1}^{n} s_i}{t} \tag{1}$$

where s_i denotes the size of ith video slice, t denotes the duration of a video, and r_p denotes the playback rate.

Table 4 shows our measured video playback rates of the main definitions popularly used in China. We find that the video playback rate of the same definition depends on each MSP.

Table 3. Types of video file of MSPs.

Feature	Described file	Media file
Tencent	None	Mp4 & Flv
Iqiyi	None	Mp4 & F4v
Youku	None	Mp4 & Flv
Letv	M3u8	Ts

Table 4. Video playback rates of the main definitions of each MSP.

MSP	SD	MD	HD
Tencent	34–39 KB/s	110–120 KB/s	195–200 KB/s
Iqiyi	44–52 KB/s	76–84 KB/s	188–200 KB/s
Youku	50–53 KB/s	112–120 KB/s	195–200 KB/s
Letv	58–65 KB/s	88–118 KB/s	200–220 KB/s

Table 5. Characteristics of slice.

MSP	Slicing threshold	Duration of slice
Tencent	10 min	4–5 min
Iqiyi	9 min	6–7 min
Youku	7 min	4–7 min
Letv	None	10–20 s

Table 6. Video specification and the measure result.

MSP	Definition	Length	Start-up delay	Number of stalling	Total time of stalling	Feature
Tencent	SD	893 s	17 s	0	0 s	128 kB/s
Iqiyi	SD	814 s	13 s	0	0 s	128 kB/s
Youku	SD	799 s	9 s	0	0 s	128 kB/s
Letv	SD	872 s	11 s	0	0 s	128 kB/s
Tencent	MD	893 s	14 s	1	11 s	128 kB/s
Iqiyi	MD	814 s	12 s	1	7 s	128 kB/s
Youku	MD	799 s	13 s	7	17 s	128 kB/s
Letv	MD	872 s	8 s	9	13 s	128 kB/s
Tencent	HD	893 s	10 s	1	6 s	256 kB/s
Iqiyi	HD	814 s	12 s	1	7 s	256 kB/s
Youku	HD	799 s	10 s	4	8 s	256 kB/s
Letv	HD	872 s	12 s	4	3 s	256 kB/s

4.3 Video slicing

By downloading a mass of various videos, we find the video slice thresholds and slice durations of four MSPs in China, which are shown in Table 5. For a given MSP, the durations of slices are almost the same for different video definitions.

4.4 Performance and evaluation

In this section, we consider start-up delay, number, and duration of video stalling as performance metrics. Usually, an advertisement is downloaded and played before video playback. The video begins downloading during the playback of the advertisement. The video's start-up delay is the difference between the duration of the video's playback start timestamp and the duration of the video's downloading start timestamp. Video occurs stalling when playback buffer is empty and resumes after playback buffer exceeds the buffer threshold. The duration of video stalling is the difference between the video's playback resuming timestamp and the video's stalling timestamp.

We measure the performance of wired online videos with campus network of Shenzhen University. Especially, we measure the performance of videos in the same network condition and period of time for each MSP. Table 6 shows video specification and our measurement result. We obtained several conclusions as follows. First, start-up delay for each MSP is similar and the duration of the advertisement may cause long start-up delay. Second, even though the downloading rate is greater than the playback rate of video (in Table 4), the video may occur stalling due to network jitter and latency of the next video slice. Third, Tencent and Iqiyi keep long waiting time during video stalling. Fourth, Youku and Letv cause more frequent video stalling than the others, although each video stalling lasts a short time.

5 CONCLUSION AND FUTURE

In this work, we evaluated the performance of video service according to start-up delay, number, and total time of video stalling. Our work is divided into two parts. First, we detected media content delivery and characteristics of technology including playback rate, video slice thresholds, and slice durations. Second, we evaluated the performance of video service with our measured results. It is necessary for MSPs to balance frequency with the duration of video stalling.

Currently, we study the performance of MSPs in China. In the future, we will study to reconstruct video playback with network traffic data only.

ACKNOWLEDGMENTS

This work was supported by the Natural Science Foundation of China (No. 61502315, No. 61309030), the Natural Science Foundation of Guangdong Province (No. 2015A030310366), the Technological Innovation of Shenzhen city (No. JCYJ20160422112909302), and the Fundamental Research Funds for the Shenzhen University (201558). Liang Chen is the first corresponding author (lchen@szu.edu.cn), Yue Wang is the second corresponding author (yueLwang@163.com), and Na Wang is the third corresponding author (wangna@szu.edu.cn).

REFERENCES

Cisco, C.V.N.I. "Global mobile data traffic forecast update, 2013–2018," white paper, 2014.

Cidon I, Khamisy A, Sidi M. Analysis of packet loss processes in high-speed networks [J]. IEEE Transactions on Information Theory, 1996, **39**(1):98–108.

Huynh-Thu Q, Ghanbari M. No-reference temporal quality metric for video impaired by frame freezing artefacts [C]//IEEE International Conference on Image Processing. IEEE, 2009:2221–2224.

Luan T H, Cai L X, Shen X. Impact of Network Dynamics on User's Video Quality: Analytical Framework and QoS Provision [J]. IEEE Transactions on Multimedia, 2010, 12(1):64–78.

Mok, R.K.P., Chan, E.W.W., Chang, R.K.C. Measuring the quality of experience of HTTP video streaming. [C]//Ifip/ieee International Symposium on Integrated Network Management, IM 2011, Dublin, Ireland, 23–27 May 2011:485–492.

Pessemier T D, Moor K D, Joseph W, et al. Quantifying the Influence of Rebuffering Interruptions on the User's Quality of Experience During Mobile Video Watching [J]. IEEE Transactions on Broadcasting, 2013, 59(1):47–61.

Selenium, https://www.seleniumhq.org/.

Stockhammer, T. "Dynamic adaptive streaming over http-: standards and design principles," in ACM Sigmm Conference on Multimedia Systems, Mmsys 2011, Santa Clara, Ca, Usa, February, 2011, pp. 133–144.

Wang Z, Banerjee S, Jamin S. Studying streaming video quality: from an application point of view[C]// Proceedings of the eleventh ACM international conference on Multimedia. ACM, 2003: 327–330.

Wireshark, https://www.wireshark.org/.

Xu Y, Altman E, El-Azouzi R, et al. Probabilistic analysis of buffer starvation in Markovian queues [J]. Proceedings—IEEE INFOCOM, 2011, 131(5):1826–1834.

Robust optimization research for the bi-objective vehicle routing problem with time windows

Pei Xu, Shan Lu & Quansheng Lei
Beijing University of Posts and Telecommunications, Beijing, China

ABSTRACT: This paper presents a bi-objective model to design a transport cost–wait time minimization vehicle routing problem with time windows. Our model shows that some uncertainty parameters, such as the demands of customers, travel time, and service time, belong to an uncertainty polytope. Then, we introduce a robust optimization approach to address the VRPTW under uncertainty (RVRPTW). We compare the numerical experiments of the deterministic VRPTW and RVRPTW and demonstrate the benefits of our robust optimization approach.

1 INTRODUCTION

In the fields of transportation system and logistics distribution, how to meet the delivery of goods or service for customers is becoming increasingly important. The research on the vehicle routing problem or vehicle scheduling problem is of utmost importance.

Vehicle Routing Problem (VRP) is a type of combinatorial optimization problems, which was first studied by Dantzig and Ramser (Dantzig, 1959) in 1959. Because of the development of VRP, the problem has been extended to many variations. We consider that there is a time restriction for all customers so that they must be served within a time window, that is, Vehicle Routing Problem with Time Windows (VRPTW). It is an extended form of the VRP with capacity constraints, which is an NP-hard problem.

In the real world of VRPTW, not all the objective factors can be known in advance. On the contrary, there are many uncertain factors, which perform in the following situations: (1) the uncertainty of demands, (2) the uncertainty of travel time, and (3) the uncertainty of service time. We consider the vehicle routing problems with time windows under uncertainty environment and establish a robust optimization model called RVRPTW. Robustness generally refers to the insensitivity of parameters perturbations for a logistics system. Robust optimization is an effective approach to deal with decision problems with uncertainty.

The contributions of this paper are as follows.

1. A bi-objective model is developed, which achieves minimization of the transport costs and wait time.
2. We consider that the uncertain factors may arise in the model, including demands of customers, travel time, and service time and propose a robust optimization model for VRPTW under those uncertainties.

The structure of this paper is as follows. The next section introduces the relevant literature of VRPTW and robust optimization method. In Section 3, we present an uncertainty set for linear programming and introduce two different robust optimization approaches. Section 4 provides a deterministic bi-objective model of VRPTW. In Section 5, we use robust optimization approach to solve the bi-objective VRPTW under uncertainty. We present the numerical results of our formulations in Section 6 and compare the robust solution against the deterministic solution. Finally, we conclude the paper in Section 7.

Notation: In order to facilitate the calculation, the letter symbols appearing in this paper are described as follows. Lowercase bold (b) represents the vector, whereas uppercase boldface (A) denotes matrices. Tilde ($\tilde{a}_{ij}$) denotes nominal values and check ($\hat{a}_{ij}$) represents uncertainty range for the given entry.

2 LITERATURE REVIEW

Solving vehicle routing problem is a scientific challenge as it is NP-hard. The problem was first studied by Dantzig and Ramser (Dantzig, 1959) in 1959 to describe the truck dispatching problem. In 1987, Solomon (Solomon, 1987) added the constraint of time windows to the vehicle routing problem for the first time and obtained very good results through an insertion-type heuristic. In recent years, many accurate algorithm and heuristic algorithm have been established to solve the problem. Even so, uncertainty is often ignored in most of the vehicle routing problems. Bertsimas and Simchi-Levi

(Bertsimas, 1996) considered the uncertainties into the vehicle routing problem. They supposed that uncertain parameters belong to a distribution function, where the combination of probabilistic and combinatorial optimization model is used to generate an approximate optimal solution.

In recent years, the robust optimization method is a new way to solve the optimization problem under uncertainties, which has been concerned and studied by many scholars in many fields. In the robust optimization method, we hope to find a suboptimal solution so as to ensure that the solution is feasible and close to the optimal solution when the data are changed. Soyster (Soyster, 1973) proposed a linear optimization model to obtain a solution that is feasible for all uncertain data that belong to a convex set. Sungur et al. (Sungur, 2008) considered a capacitated vehicle routing problem with demand uncertainty and constructed a robust vehicle routing problem. However, they assumed that all the travel time reach a maximum value in the optimal solution so that the results will be too conservative. Ben-Ta and Nemirovski (Ben-Tal, 1998) presented the convex optimal problem for the uncertain data belonging to a uncertain set $\mathcal{U}$.

3 ROBUST OPTIMIZATION APPROACH UNDER UNCERTAINTY

Consider a typical linear programming problem:

$$(\text{LP})\ min\ cx$$

$$s.t. \quad Ax \le b \tag{1}$$

$$x \in \mathbb{R}^n$$

where A is a matrix of $n \times m$, which is equivalent to $[a_{ij}]^{n \times m}$, $c = (c_1, c_2, ..., c_m)$, $b = (b_1, b_2, ..., b_n)^T$, $i = 1, 2, ..., n$, $j = 1, 2, ..., m$. For a typical model of deterministic VRPTW, we assume that the values of A, c, and b are known.

3.1 Uncertainty factors and uncertainty set for linear programming

Without loss of generality, we assume that the data uncertainty only affects the entry of matrix A, without affecting the c of objective function. If uncertainty data appear in the objective vector c, we can introduce a maximum z and add a new constraint $z - cx \le 0$. By this way, the new constraint will be included in $Ax \le b$.

We assume that A belongs to an uncertainty set $U \subset \mathbb{R}^{n \times m}$ and the problem is subject to uncertainty only in matrix A. The robust optimization model (LP^U) of LP is:

$$(\text{LP}^U)\ min\ cx$$

$$s.t. \quad Ax \le b \quad A \in U \tag{2}$$

$$x \in \mathbb{R}^n$$

where the n linear constraint in (1) must be satisfied for $A \in U$. Hence, the finite n constraint has been replaced by infinite linear constraint of (2). In order to facilitate the calculation, constraint (2) can be rewritten as the following inequality:

$$a_i^T x \le b_i,\ a_i \in U, i = 1, 2, \cdots, n \tag{3}$$

where a_i^T is the i-th row vector of A, $a_i^T = (a_{i1}, a_{i2}, \cdots, a_{im})$, $i = 1, 2, \cdots, n$. Then, we can further simplify the constraint (3), which is equivalent to $max_{a_i \in U} a_i^T x \le b_i$, $i = 1, 2, \cdots, n$.

Thus, the robust optimization model (LP^U) can be further expressed by the following formulation:

$$(\text{LP}^U)\ min\ cx$$

$$s.t. \quad max_{a_i \in U}\ a_i^T x \le b_i, \quad i = 1, 2, \cdots, n \tag{4}$$

$$x \in \mathbb{R}^n$$

We now consider i-th row of the coefficient matrix A. Let J_i denote the set of coefficients in row i that are subject to uncertainty. Each entry a_{ij} takes values in $\left[\tilde{a}_{ij} - \hat{a}_{ij}, \tilde{a}_{ij} + \hat{a}_{ij} \right]$, $j \in J_i$ (Ben-Tal and Nemirovski [34]), where $\tilde{a}_{ij}$ is the nominal value of a_{ij} and $\hat{a}_{ij}$ represents the uncertainty range for the given entry. Moreover, we define a random variable $\varepsilon = \left(a_{ij} - \tilde{a}_{ij} \right) / \hat{a}_{ij}$, and the range of ε is $[-1,1]$. Thus, the constraint (4) can be expressed as $max_{a_i \in U}\ a_i^T x \le b_i, i = 1, 2, \cdots, n$.

$$max_{a_i \in U} \left(a_i^T x \right) = max\left\{ \sum\nolimits_{j \in J_i} a_{ij} x_j + \sum\nolimits_{j \notin J_i} a_{ij} x_j \right\}$$
$$= max\left\{ \sum\nolimits_{j \in J_i} a_{ij} x_j + \sum\nolimits_{j \notin J_i} \tilde{a}_{ij} x_j \right\}$$
$$= max\left\{ \sum\nolimits_{j} \tilde{a}_{ij} x_j + \sum\nolimits_{j \in J_i} \left(a_{ij} - \tilde{a}_{ij} \right) x_j \right\} \tag{5}$$

3.2 The robust optimization approach of Soyster

As we mentioned in Section 2, Soyster (Soyster, 1973) proposed a box uncertainty set for linear programming, which is defined as a box of form U^S:

$$U^S = Box_1 = \left\{ a_{ij} \in R^{n \times m} : \left| a_{ij} - \tilde{a}_{ij} \right| \le \varepsilon \hat{a}_{ij}, \forall j \in J_i, \|\varepsilon\|_\infty \le 1 \right\}$$

Hence, the constraint (5) can be stated further as:

$$max \left\{ \sum\nolimits_{j} \tilde{a}_{ij} x_j + \sum\nolimits_{j \in J_i} \left(a_{ij} - \tilde{a}_{ij} \right) x_j \right\}$$
$$= \sum\nolimits_{j} \tilde{a}_{ij} x_j + \sum\nolimits_{j \in J_i} \varepsilon \hat{a}_{ij} x_j \tag{6}$$

And the robust optimization of LP is:

$$\left(LP^{US}\right) min\ cx$$
$$s.t.\ \sum_j \tilde{a}_{ij} x_j + \sum_{j \in J_i} \varepsilon \hat{a}_{ij} y_j \le b_i \qquad (7)$$
$$-y_j \le x_j \le y_j, \quad \forall i$$
$$x \in \mathbb{R}^n$$

Let x^* be the optimal solution of model $\left(LP^{US}\right)$; obviously, the objective formulation can obtain optimal solution when $y_j = |x_j^*|$. Therefore, the constraint (7) is equivalent to:

$$\sum_j \tilde{a}_{ij} x_j^* + \sum_{j \in J_i} \hat{a}_{ij} |x_j^*| \le b_i, \quad \forall i$$

For every possible uncertain data a_{ij}, as $\sum_j a_{ij} x_j^* = \sum_j \tilde{a}_{ij} x_j^* + \sum_{j \in J_i} \varepsilon \hat{a}_{ij} x_j^* \le \sum_j \tilde{a}_{ij} x_j^* + \sum_{j \in J_i} \hat{a}_{ij} |x_j^*| \le b_i$, the solution is always feasible, that is, the solution is "robust." Although the solution obtained by the Soyster can satisfy all the uncertainties, the results are more conservative, and in the sense, the objective function value of the robust solution is far worse than the optimal value of the nominal linear programming.

3.3 The robust optimization approach of Ben-Tal and Nemirovski

In order to address the above conservatism, Ben-Tal and Nemirovski (Ben-Tal, 1998) proposed a new robust optimization method that the data uncertainty belongs to an ellipsoidal uncertainty set. We assume that there is an ellipsoidal uncertainty set U^B, which can be seen as a sphere of radius Ω, and defined as:

$$U^B = Ball_\Omega = \left\{ a_{ij} \in R^{n \times m} : \left(a_{ij} - \tilde{a}_{ij}\right)^T \Lambda^{-1} \left(a_{ij} - \tilde{a}_{ij}\right) \le \Omega^2 \right\}$$

where Λ is a positive definite matrix of $|J_i| \times |J_i|$ and Ω is the safety parameter of uncertainty. By using a process of affine transformation, U^B can also be expressed as: $U^B = \left\{ a_{ij} \in R^{n \times m} : a_{ij} = \tilde{a}_{ij} + \Delta \xi, \|\xi\|_2 \le \Omega, \Delta \in R^{m \times m}, \Omega \in R \right\}$, where $\Delta = \Lambda^{\frac{1}{2}}$.

Then, the constraint (4) is equivalent to:

$$max\left\{ \sum_j \tilde{a}_{ij} x_j + \sum_{j \in J_i} \left(a_{ij} - \tilde{a}_{ij}\right) x_j \right\} \le b_i, a_{ij} \in U^B,$$
$$i = 1, 2, \cdots, n \qquad (8)$$

As the value of a_{ij} is equivalent to the nominal value $\tilde{a}_{ij}$ when $j \notin J_i$, we only consider the part of $j \in J_i$. That is, $max\{\sum_{j \in J_i} (a_{ij} - \tilde{a}_{ij}) x_j\} \le b_i, a_{ij} \in U^B$, $i = 1, 2, \cdots, n$. It is equivalent to:

$$max_{j \in J_i} \left\{ \left(a_i - \tilde{a}_i\right) X : \left(a_i - \tilde{a}_i\right)^T \Lambda^{-1} \left(a_i - \tilde{a}_i\right) \le \Omega^2 \right\} \le b_i \quad (9)$$

It is obvious that the left of the inequality (7) is a convex problem. According to Lagrange multiplier and KKT (Karush–Kuhn–Tucker) conditions, inequality (9) can be expressed as:

$$min f\left(a_i^*\right) = -\left(a_i - \tilde{a}_i\right) X$$
$$s.t.\ g\left(a_i^*\right) = \left(a_i - \tilde{a}_i\right)^T \Lambda^{-1} \left(a_i - \tilde{a}_i\right) - \Omega^2 \le 0, j \in J_i \qquad (10)$$

where a_i^* is the optimal solution of the above formulation. Let u^* be the Lagrange multiplier and z^* be the optimal objective value. According to the KKT condition, the formulation can be rewritten as:

$$L\left(a_i^*, u^*\right) = -\left(a_i - \tilde{a}_i\right) X + u^* [\left(a_i - \tilde{a}_i\right)^T \Lambda^{-1} \left(a_i - \tilde{a}_i\right) - \Omega^2]$$
$$s.t.\ - X + 2u^* \Lambda^{-1} \left(a_i - \tilde{a}_i\right) = 0 \qquad (11)$$

$$u^* [\left(a_i - \tilde{a}_i\right)^T \Lambda^{-1} \left(a_i - \tilde{a}_i\right) - \Omega^2] = 0 \qquad (12)$$

$$u^* \ge 0 \qquad (13)$$

Then, we can obtain that:

$$a_i - \tilde{a}_i = \frac{\Omega \Lambda X}{\sqrt{X^T \Lambda X}} \qquad (14)$$

Thus, the optimal objective value z^* is:

$$z^* = z^{*T} = X^T \left(a_i - \tilde{a}_i\right) = \Omega \sqrt{X^T \Lambda X} \qquad (15)$$

Finally, the inequality (5) can be expressed as:

$$\sum_j \tilde{a}_{ij} x_j + \Omega \sqrt{X^T \Lambda X} \le b_i$$
$$= \tilde{a}_i X + \Omega \sqrt{X^T \Lambda X} \le b_i$$

Ben-Tal and Nemirovski also showed that under the model of data uncertainty, the probability that the constraint i is violated is at most $exp - (-\Omega^2/2)$.

4 THE DETERMINISTIC MODEL OF VRPTW

In this section, we first introduce the model of VRPTW and establish a deterministic bi-objective model of VRPTW. Then, we consider the impact of uncertainties of the model and establish the corresponding robust optimization model for VRPTW.

VRPTW is a generalization of the VRP involving the time windows, which is a combinatorial optimization and integer programming problem. This problem poses the following question: Given a distribution center and a list of customers whose demands are different from each other, what is the optimal set of route for a fleet of vehicles starting from the depot to serve the sets of customers without violating the vehicle capacity and time window constraints and finally return to the depot.

4.1 Model assumptions

To simplify the problem and to facilitate the calculation, we make the following reasonable assumptions before modeling the problem:

1. The capacities of the set of vehicle are the same and known in advance;
2. The distance from the depot to each customer point is known in advance;
3. The transportation cost per unit distance from the depot to each customer point is known in advance;
4. The time for each vehicle starting from the depot is 0;
5. A customer's location and its parameters (e.g., demands, time windows, and service time) are independent of each other.

4.2 Symbolic representation

To model the problem, the following representation is used throughout the whole paper.

Sets:

N – set of customers, indexed by I, $I \in N$

N_0 – set of customers and depot O, $i \in N$, $N_0 = N \cup \{O\}$

K – set of vehicles departure from the depot, indexed by k, $k \in K$

A – set of arc between customer i and j, $(i, j) \in A$

Parameters:

μ_i – the demand of customer i, $i \in N$

c – unit transportation cost

d_{ij} – distance between customer i and j, $i, j \in N$

t_{ij} – travel time between customer i and j, $(i, j) \in A$

$[e_i, I_i]$ – the pickup of customer i must occur in the time window between ties e_i and I_i

r_i^k – arrive time when vehicle k arrive at customer i, $i \in N$, $k \in K$

s_i – the pickup of customer i takes s_i units of time by vehicle k, $i \in N$

w_i^k – wait time between the arrive time and service time by vehicle k at customer i, $i \in N$, $k \in K$

Decision variables:

$$x_{ij}^k = \begin{cases} 1, & \text{if vehicle } k \text{ travels from customer } i \text{ to } j, \\ 0, & \text{otherwise.} \end{cases}$$

$$y_i^k = \begin{cases} 1, & \text{if customer } i \text{ is assigned to the vehicle } k, \\ 0, & \text{otherwise.} \end{cases}$$

4.3 The deterministic model

The deterministic bi-objective model that minimizes total transport costs and wait time of vehicle can be expressed in the following form:

$$(VRPTW) \min \sum_{k \in K} \sum_{(i,j) \in A} d_{ij} c x_{ij}^k \tag{16}$$

$$\min \sum_{k \in K} \sum_{i \in N_0} w_i^k y_i^k \tag{17}$$

$$s.t. \sum_{k \in K} \sum_{j \in \Delta^+(i)} x_{ij}^k = 1, \quad \forall i \in N \tag{18}$$

$$\sum_{j \in \Delta^+(i)} x_{0j}^k = 1, \quad \forall k \in K \tag{19}$$

$$\sum_{i \in \Delta^-(j)} x_{ij}^k - \sum_{i \in \Delta^+(j)} x_{ji}^k = 0, \quad \forall k \in K, j \in N \tag{20}$$

$$\sum_{i \in \Delta^-(n+1)} x_{i,n+1}^k = 1, \quad \forall k \in K \tag{21}$$

$$\sum_{i \in N} \mu_i \sum_{j \in \Delta^+(i)} x_{ij}^k \leq C, \quad \forall k \in K \tag{22}$$

$$e_i \leq r_i^k \leq l_i, \quad \forall k \in K, i \in N_0 \tag{23}$$

$$x_{ij}^k \left(r_i^k + s_i + t_{ij} - r_j^k \right) \leq 0, \quad \forall k \in K, (i,j) \in A \tag{24}$$

$$w_i^k = \max\{0, e_i - r_i^k\}, \quad \forall i \in N \tag{25}$$

$$y_i^k, x_{ij}^k \in \{0,1\}, \quad \forall k \in K, (i,j) \in A \tag{26}$$

The objectives (16) and (17) are to minimize the sum of the transport costs and wait time, respectively. Constraint (18) restricts that each customer must be served by exactly a single vehicle; that is, the demand of a customer cannot be split over several vehicles. Constrains (19)–(21) characterize the flow on the route to be followed by vehicle k. Constraint (22) is the capacity constraint for the vehicles, whereas constrains (23) and (24) guarantee time windows feasibility with respect to each customer. Constraint (25) is the calculation of wait time for customer i served by vehicle k. Finally, constraint (26) imposes binary conditions on the decision variables.

The following figures are the three different situations for the pickup of customer j by vehicle k.

5 THE ROBUST FORMULATION OF VRPTW

In this section, we transform the deterministic bi-objective model into its robust formulation by replacing the constraints that have uncertain coefficients with other constraints, which consider the uncertainty sets. We will therefore only take into account the uncertainty coefficients in constraints (22) and (25).

5.1 Uncertainty in demand

We consider that the demand of each customer is uncertain and belongs to an ellipsoidal uncertainty set U_{D1}. By using affine transformation, we obtain:

$$U_{D1} = \left\{ \begin{array}{l} U \in R^{|I|} : U = \tilde{U} + \Delta \xi, \|\xi\| \le \theta, \tilde{U} \in R^{|I|}, \\ \Delta \in R^{|I| \times |I|}, \theta \in R \end{array} \right\}$$

where $U = \begin{bmatrix} \mu_1 \\ \mu_2 \\ \vdots \\ \mu_n \end{bmatrix}, \tilde{U} = \begin{bmatrix} \tilde{\mu}_1 \\ \tilde{\mu}_2 \\ \vdots \\ \tilde{\mu}_n \end{bmatrix}, n = |I|, \tilde{U}$ is the nominal

value of $U. X_{.j}^k = \begin{bmatrix} x_{1j}^k \\ x_{2j}^k \\ \vdots \\ x_{nj}^k \end{bmatrix}, \Delta = \Sigma^{\frac{1}{2}}, \Sigma$ is a positive

definite matrix of uncertain demands.

The formulation (22) can be stated further as follows:

$$\tilde{U}^T X_{.j}^j + \theta \sqrt{X_{.j}^{k^T} \Sigma X_{.j}^k} \le C \tag{27}$$

5.2 Uncertainty in travel time and service time

We use the method of Ben-Tal and Nemirovski (1998) to formulate our bi-objective model. First, we assume that the uncertainty of travel time and service time belong to an ellipsoidal uncertainty set, which is defined as affine transformation. Travel time t_{ij} belongs to a set U_{T1}, while service time s_i of each customer belongs to a set S_{E1}:

$$W_{T1} = \left\{ W \in R^{|I|} : W = \tilde{W} + \Lambda \varsigma, \|\varsigma\| \le \gamma, \tilde{W} \in R^{|I|}, \right.$$
$$\left. \Lambda \in R^{|I| \times |I|}, \gamma \in R \right\}$$

$$S_{E1} = \left\{ S \in R^{|I|} : S = \tilde{S} + \Xi \eta, \|\eta\| \le \delta, \tilde{S} \in R^{|I|}, \right.$$
$$\left. \Xi \in R^{|I| \times |I|}, \delta \in R \right\}$$

Similarly, $W = \begin{bmatrix} t_{1j} \\ t_{2j} \\ \vdots \\ t_{nj} \end{bmatrix}, \tilde{W} = \begin{bmatrix} \tilde{t}_{1j} \\ \tilde{t}_{2j} \\ \vdots \\ \tilde{t}_{nj} \end{bmatrix}, j = 1, 2, \cdots, n,$

Figure 1 (a). The situation where the arrival time r_j^k is later than the earliest pickup time e_j.

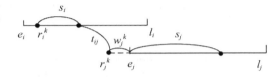

Figure 1 (b). The situation where the arrival time r_j^k is earlier than the earliest pickup time e_j.

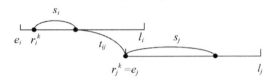

Figure 1 (c). The situation where the arrival time r_j^k is equal to the earliest pickup time e_j.

$$S = \begin{bmatrix} s_1 \\ s_2 \\ \vdots \\ s_n \end{bmatrix}, \tilde{S} = \begin{bmatrix} \tilde{s}_1 \\ \tilde{s}_2 \\ \vdots \\ \tilde{s}_n \end{bmatrix}, n = |I|.$$

And $\tilde{W}$ is the nominal value of W, $\tilde{S}$ is the nominal value of S. $\Lambda = \Psi^{\frac{1}{2}}, \Xi = \Gamma^{\frac{1}{2}}$. ψ and Γ are positive definite matrices of uncertain travel time and service time, respectively.

The robust formulation of constraint (24) can be transformed into:

$$\left(\tilde{W}^T + \tilde{S}^T \right) X_{.j}^k + \gamma \sqrt{X_{.j}^{k^T} \Psi X_{.j}^k} + \delta \sqrt{X_{.j}^{k^T} \Gamma X_{.j}^k}$$
$$\le \left(w_i^{k^T} - w_j^{k^T} \right) X_{.j}^k \tag{28}$$

In practice, solving robust formulation of VRPTW is likely more difficult than solving the deterministic model.

6 NUMERICAL EXPERIMENTS

In this section, we will combine the robust and deterministic VRPTW models to calculate the numerical experiments. We first specify the parameters for the calculation. Then, we will present the performance measure, which is used to solve the formulation and receive the computational results.

Our models consist of 100 customers, 20 vehicles, and one depot. The distances between any two customers are Euclidean. The capacity C of each vehicle is 700, while unit transportation cost c is 2; average speed of the vehicles v is 2. The service time per customer is set to 90.

In this study, we use genetic algorithm to solve the deterministic and robust problems. All models and algorithm are coded using C++ programming language based on Microsoft Visual Studio 2010. Moreover, a time limit of 1 h is set for all instances. We consider the situations when the values of safety parameters θ, γ, and δ take different values, and assume that $\theta = \gamma = \delta$. In each situation, we obtain the computational solutions through 10 iterations.

Next, we demonstrate the trade-off between the two objective functions in Section 4. Let W be a weight to perform the trade-off between objective functions (16) and (17), and the bi-objective function can be written as:

$$min \sum_{k \in K} \sum_{(i,j) \in A} d_{ij} c x_{ij}^k + W \sum_{k \in K} \sum_{i \in N_0} w_i^k y_i^k$$

Because minimizing the transport cost of vehicle routes is of a higher priority than minimizing the wait time, weight W is set to 10%, 20%, and 50%. Both the position and demands of customers came from historical data of the major distribution districts in Beijing, China. In addition, ρ is defined as $\rho = \frac{z_r - z_d}{z_d}$, where z_r and z_d are the optimal objective functional values of the robust and deterministic VRPTW, respectively. Obviously, the smaller ρ means better performance of the model.

Tables 1, 2, and 3 show the results based on the performance measure ρ under the different safety

Table 1. Robust and deterministic optimal solution comparison when safety parameters $\theta = \gamma = \delta = 0.002$.

$\theta = \gamma = \delta = 0.002$	$W = 10\%$			$W = 20\%$			$W = 50\%$		
NO	z_d	z_r	$\rho = \frac{z_r - z_d}{z_d}$	z_d	z_r	$\rho = \frac{z_r - z_d}{z_d}$	z_d	z_r	$\rho = \frac{z_r - z_d}{z_d}$
1	776.92	825.04	0.0619	833.54	885.38	0.0622	1003.40	1066.40	0.0628
2	776.90	833.78	0.0732	831.50	894.66	0.0760	995.30	1077.30	0.0824
3	775.82	811.65	0.0462	831.84	870.70	0.0467	999.90	1047.85	0.0480
4	752.43	835.48	0.1104	809.76	895.66	0.1061	981.75	1076.20	0.0962
5	763.63	832.71	0.0905	818.76	894.02	0.0919	984.15	1077.95	0.0953
6	756.27	817.34	0.0808	813.94	877.28	0.0778	986.95	1057.10	0.0711
7	753.04	830.66	0.1031	811.18	891.72	0.0993	985.60	1074.90	0.0906
8	760.13	809.80	0.0653	815.46	868.80	0.0654	981.45	1045.80	0.0656
9	778.88	828.10	0.0632	833.96	887.30	0.0640	999.20	1064.90	0.0658
10	771.14	830.09	0.0764	825.38	890.18	0.0785	988.10	1070.45	0.0833
AVERAGE	766.52	825.47	0.0771	822.53	885.57	0.0768	990.58	1065.89	0.0761

Table 2. Robust and deterministic optimal solution comparison when safety parameters $\theta = \gamma = \delta = 0.005$.

$\theta = \gamma = \delta = 0.005$	$W = 10\%$			$W = 20\%$			$W = 50\%$		
NO	z_d	z_r	$\rho = \frac{z_r - z_d}{z_d}$	z_d	z_r	$\rho = \frac{z_r - z_d}{z_d}$	z_d	z_r	$\rho = \frac{z_r - z_d}{z_d}$
1	783.41	847.10	0.0813	841.32	907.60	0.0788	1015.05	1089.10	0.0730
2	810.51	844.27	0.0417	869.72	905.04	0.0406	1047.35	1087.35	0.0382
3	808.46	813.83	0.0066	866.02	874.46	0.0097	1038.70	1056.35	0.0170
4	789.51	829.56	0.0507	846.92	893.42	0.0549	1019.15	1085.00	0.0646
5	777.08	843.72	0.0858	834.16	905.04	0.0850	1005.40	1089.00	0.0832
6	814.97	835.78	0.0255	872.54	899.06	0.0304	1045.25	1088.90	0.0418
7	770.23	842.65	0.0940	828.06	904.40	0.0922	1001.55	1089.65	0.0880
8	782.60	835.27	0.0673	839.50	898.24	0.0700	1010.20	1087.15	0.0762
9	785.22	848.75	0.0809	844.24	909.10	0.0768	1021.30	1090.15	0.0674
10	765.35	847.60	0.1075	824.20	911.80	0.1063	1000.75	1104.40	0.1036
AVERAGE	788.73	838.85	0.0641	846.67	900.82	0.0645	1020.47	1086.71	0.0653

Table 3. Robust and deterministic optimal solution comparison when safety parameters $\theta = \gamma = \delta = 0.01$.

$\theta = \gamma = \delta = 0.0$	$W = 10\%$			$W = 20\%$			$W = 50\%$		
NO	z_d	z_r	$\rho = \dfrac{z_r - z_d}{z_d}$	z_d	z_r	$\rho = \dfrac{z_r - z_d}{z_d}$	z_d	z_r	$\rho = \dfrac{z_r - z_d}{z_d}$
1	840.72	855.38	0.0174	903.94	920.76	0.0186	1093.60	1116.90	0.0213
2	818.42	854.93	0.0446	879.14	916.16	0.0421	1061.30	1099.85	0.0363
3	844.76	881.79	0.0438	905.62	945.08	0.0436	1088.20	1134.95	0.0430
4	849.19	879.34	0.0355	911.08	940.78	0.0326	1096.75	1125.10	0.0258
5	806.34	852.30	0.0570	867.08	917.20	0.0578	1049.30	1111.90	0.0597
6	829.98	888.45	0.0704	894.36	954.60	0.0674	1087.50	1153.05	0.0603
7	795.51	863.51	0.0855	854.72	926.92	0.0845	1032.35	1117.15	0.0821
8	814.29	885.93	0.0880	873.58	950.26	0.0878	1051.45	1143.25	0.0873
9	833.34	848.77	0.0185	892.68	910.44	0.0199	1070.70	1095.45	0.0231
10	837.20	869.83	0.0390	901.20	930.16	0.0321	1093.20	1111.15	0.0164
AVERAGE	826.98	868.02	0.0500	888.34	931.24	0.0486	1072.44	1120.88	0.0455

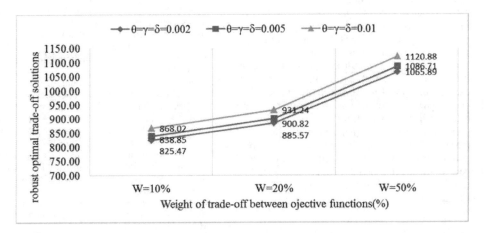

Figure 2. Robust optimal trade-off solutions between cost and wait time for VRPTW under different values of safety parameters.

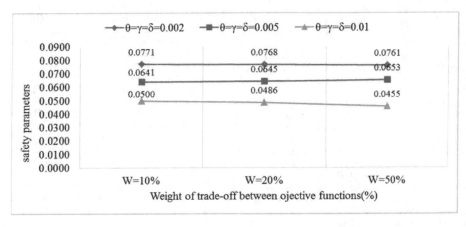

Figure 3. Ratio ρ for the solution of robust with respect to deterministic model under different values of safety parameters.

parameters to compare the robust and deterministic optimal solutions of VRPTW. Figures 2 and 3 illustrate robust optimal cost and ratio ρ for VRPTW under different safety parameters, respectively.

By analyzing the tables and figures, we can obtain the following conclusions. The higher the values of safety parameters, the smaller is performance measure ρ and the better is the cost and wait time performance. The higher is the weight W, the more the robust optimal trade-off solutions between cost and wait time, but the change is not obvious when the safety parameters are varied.

7 CONCLUSION

The main objective of this paper is to develop a bi-objective model of VRPTW and apply the robust optimal model to cope with the VRPTW under conditions of uncertainty. First, we present a bi-objective mode of VRPTW, where the transport costs and wait time are taken into account. We then consider that the uncertainty appeared in demands, travel time, and service time, and build a robust optimization function. Finally, we analyze the performance of robust and deterministic solutions under different situations.

There are several further prospects for our future work. For instance, it is meaningful to consider the unmet demands or unmet service time, or we can introduce a penalty function to face the situation of arrive time of vehicles beyond the time windows.

REFERENCES

Ben-Tal A, El Ghaoui L, Nemirovski A. *Robust optimization* [M]. Princeton University Press, 2009.

Ben-Tal A, Nemirovski A. Robust convex optimization [J]. Mathematics of operations research, **23**(4): 769–805(1998).

Ben-Tal A, Nemirovski A. Robust solutions of linear programming problems contaminated with uncertain data [J]. Mathematical programming, **88**(3): 411–424(2000).

Bertsimas D J, Simchi-Levi D. A new generation of vehicle routing research: robust algorithms, addressing uncertainty [J]. Operations Research, **44**(2): 286–304(1996).

Bertsimas D, Sim M. The price of robustness [J]. Operations research, **52**(1): 35–53(2004).

Christofides N, Mingozzi A, Toth P. Exact algorithms for the vehicle routing problem, based on spanning tree and shortest path relaxations [J]. Mathematical programming, **20**(1): 255–282(1981).

Dantzig G. B, Ramser J. H. The truck dispatching problem [J]. Management science, **6**(1): 80–91(1959).

Desrochers M, Desrosiers J, Solomon M. A new optimization algorithm for the vehicle routing problem with time windows [J]. Operations research, **40**(2): 342–354(1992).

Eiichi Taniguchi, Russell G Thompson, Tadashi Yamada and Ron van Duin. *City Logistics Network Modelling and Intelligent Transport Systems* [M]. Beijing: Publishing House of Electronics Industry, 2011: 93–112.

Gillett B E, Miller L R. A heuristic algorithm for the vehicle-dispatch problem [J]. Operations research, **22**(2): 340–349(1974).

Julien Bramel and David Simchi-Levi. *The Logic of Logistics: Theory, Algorithms, and Applications for Logistics Management* [M]. Berlin: Springer, 1997.

Kohl N, Madsen O B G. An optimization algorithm for the vehicle routing problem with time windows based on lagrangian relaxation [J]. Operations Research, **45**(3): 395–406(1997).

Kolen, AWJ, Kan AHG Rinnooy and HWIM Trienekens. Vehicle Routing With Time Windows [J].Operations Research, 1987(35): 266–273.

Laporte G, Mercure H, Nobert Y. An exact algorithm for the asymmetrical capacitated vehicle routing problem [J]. Networks, **16**(1): 33–46(1986).

Lenstra J.K. and A.H.G. Rinnooy Kan. Complexity of Vehicle Routing and Scheduling Problems [J]. Networks, **11**: 221–227(1981).

Madsen, O.B.G. Optimal scheduling of trucks—A routing problem with tight due times for delivery. In H. Strobel, R. Genser, and M. Etschmaier, editors, Optimization applied to transportation systems, IIASA, International Institute for Applied System Analysis, Laxenburgh, Austria, pp.126–136(1976).

Marius M. Solomon. Algorithms for the Vehicle Routing and Scheduling Problems with Time Window Constraints [J]. Operations Research, **35**(2): 254–265(1987).

Pullen H. and M. Webb. A computer application to a transport scheduling problem [J]. Computer Journal, **10**: 10–13(1967).

Reinhard Diestel. *Graph Theory* [M]. Beijing: Higher Education Press (2013).

Solomon M. M. Algorithms for the vehicle routing and scheduling problems with time window constraints [J]. Operations research, **35**(2): 254–265(1987).

Soyster, A.L. Convex programming with set-inclusive constraints and applications to inexact linear programming [J]. Operations research, **21**,1154–1157(1973).

Sungur I, Ordóñez F, Dessouky M. A robust optimization approach for the capacitated vehicle routing problem with demand uncertainty[J]. IIE Transactions, **40**(5): 509–523(2008).

Toth P, Vigo D. *The vehicle routing problem* [M]. Society for Industrial and Applied Mathematics, (2001).

William J. Cook, William H. Cunningham, William R. Pulleyblak and Alexander Schrijver. *Combinatorial Optimization* [M]. Beijing: Higher Education Press (2011).

Zhang Z H, Jiang H. A robust counterpart approach to the bi-objective emergency medical service design problem [J]. Applied Mathematical Modelling, **38**(3): 1033–1040(2014).

A medical cloud storage strategy based on discrete PSO algorithm

Xihua Peng
School of Computer Science and Information, Southwest University, Chongqing, China

Shanxiong Chen
School of Computer Science and Information, Southwest University, Chongqing, China
School of Information Engineering, Guizhou University of Engineering Science, Bijie, China

Maoling Peng
Chongqing City Management College, Chongqing, China

ABSTRACT: In the process of implementing medical cloud services, people usually pay more attention to data security, load balancing of the system and other issues. One of the key technologies to solve these problems is to optimize medical cloud storage deployment plan. In this paper, the related technologies of the medical cloud storage are abstracted to multi-objective optimization problems, and then the discrete PSO algorithm is used to introduce the custom updating method of speed and position, as well as re-optimized judgment rules. Finally, the IDPSO algorithm could be used to solve the problems of storage and optimization in cloud computing. Then simulation tests are carried out in the simulation platform CloudSim Toolkit and the results show that this method is effective.

1 INTRODUCTION

With the rapid development of the standardization of international medical information, as well as the electronic processes, the regional medical information platform has become the basic infrastructure of hospitals, and HIS, PACS and other applications have gradually become the core of information technology (Kim, 2015; Elit, 2015). Due to the increasing data volume in these applications, users have to suffer risks caused by the data loss or the downtime of the system, for example, a standard pathological picture may be close to 5GB, which demonstrates the value of the data. Therefore, effectively protecting and building the most efficient information system for the hospital has become a priority for all.

Currently, the cloud storage system consists of multiple components, which can be used to collect, store and process the underlying data, and then implement other cloud services on this basis. However, the existing network storage technologies, such as Network-Attached Storage (NAS), Storage Area Networks (SAN), etc., fail to meet the core needs of cloud storage services due to their own technical restrictions. The Object-Based Storage (OBS) method may contribute to the development of cloud storage(Niu, 2010; Zhang, 2011). In the implementation process, there is a certain coupling between performance parameters of the internal nodes, such as the memory capacity, processing speed and so on, so it is important to optimize the deployment in a dynamic environment, and better configure virtual nodes and allocate resources. To solve this problem, many solutions have been proposed, especially those heuristic swarm intelligence algorithms (Kaveshgar, 2015) such as particle swarm algorithm or genetic algorithm. But there are still some limitations of these programs in speed and quality. Therefore, this paper, on the basis of OBS cloud storage model, uses the Improved Discrete Particle Swarm Optimization (IDPSO) algorithm to achieve the optimization solution of the deployment of the cloud storage of medical data.

2 MEDICAL CLOUD STORAGE SYSTEM MODEL

2.1 System modelling

The main purpose of cloud storage services is to allocate the files which need to be saved to the data storage nodes in the cloud platform according to the reasonable number of copies. This could meet the data security requirements of users, accelerate the execution of tasks in the cloud services system and make the load of the system achieve a relatively balance. Figure 1 describes the system model of medical cloud storage.

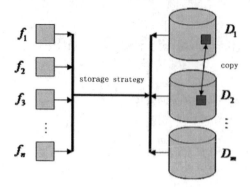

Figure 1.   The system model of Medical cloud storage.

To simplify the question but keep its generality, this paper makes the following assumptions: 1) in literature (Li, 2013) we find that when users access to f_n, they carry out sequential read most of the time in the existing cloud file system, so we will focus on this file operation. 2) as a medical cloud services platform, most of the data we are facing, such as electronic medical records, CT images, and pathological picture and other documents has the feature of "Write-Once Read-Many", the storage policy in this paper does not consider the mechanism of data consistency.

Therefore, the medical cloud storage deployment issues in this paper will be transformed into a multi-objective optimization problem, that is, n files will be deployed to m data storage nodes and thus associated system performance can be optimized.

2.2 Optimization goals

In the cloud storage optimization process, the following performance parameters are mainly considered.

1. Average delay: Minimizing the system delay is important for any cloud storage platform. As mentioned above, since the medical data has a feature of "Write-Once Read-Many", this paper mainly considers the delay of file read. As there may be multiple copies of each file, the average delay D_n of file f_n is shown as follows:

$$\overline{D_i} = \frac{1}{r_i} \times \sum_{j=1}^{m}\left(\Gamma(i,j) \times \frac{s_i}{B_j} \times A(i,j)\right) \qquad (1)$$

In the above formula, $A(i,j)$ represents the ratio of incoming file read request of file f_i on the data node D_j to the total requests, B_j is the bandwidth data node, $\Gamma(i,j)$ is a decision variable, when there are files on data nodes, $\Gamma(i,j) = 1$ and oth-

erwise it is 0. s_i is the number of file in data node r_i denotes the number of data node.

2. The storage cost. It is calculated mainly based on the cost of the data node where the file is stored. In general, the node with strong processing power costs more, so the system should minimize storage costs under the premise of ensuring small system delay.

$$\overline{C_i} = \sum_{j=1}^{m}\left(\Gamma(i,j) \times C(i,j)\right) \qquad (2)$$

In the above formula, $\Gamma(i,j)$ plays the same role as it does in formula (1), $C(i,j)$ is the cost function of file i which is stored on data node j.

3. Load balancing: this paper convert the system load to the sum of requests of all the stored objects on the storage node. Thus every storage node can be expressed as

$$V(j) = \sum_{i=1}^{n}\left(\Gamma(i,j) \times R(i)\right) \qquad (3)$$

$R(i)$ is the number of requests of file i. The average load can be approximately expressed as:

$$V_{avg} = \sum_{j=1}^{m} V(j) \bigg/ M \qquad (4)$$

When the difference between the load of storage nodes and V_{avg} approaches 0, the system load is close to equilibrium.

3 CLOUD STORAGE STRATEGY BASED ON IDPSO

Particle Swarm Optimization (PSO) algorithm is swarm intelligence evolutionary algorithm proposed by Kennedy et al (2011). Since the standard PSO algorithm could only be used to solve continuous domain problems, a binary version of PSO algorithm was proposed in literature (Unler, 2010) to optimize the discrete space. To apply discrete PSO algorithm to cloud storage, the first problem to be solved is to establish a suitable mapping between the particles and the solution space. This paper adopts the integer encoding mechanism to correspond solution space, shown as the following formula:

$$\begin{cases} (f_{1,1}, D_1) & (f_{1,2}, D_2) & \cdots & (f_{1,z}, D_m) \\ (f_{2,1}, D_1) & (f_{2,1}, D_2) & \cdots & (f_{2,z}, D_m) \\ \cdots & \cdots & \cdots & \cdots \\ (f_{n,1}, D_1) & (f_{n,2}, D_2) & \cdots & (f_{n,z}, D_m) \end{cases}$$

The line number $f_{1,1}$ is a copy of f_1, D_m is the number of storage nodes, $(f_{n,2}, D_2)$ indicates whether or not there is a copy of file on the storage node.

3.1 Fitness function

Taking into account the multiple optimization goals of cloud storage deployment modeling, this paper adopts an efficient measurement approach, namely, introduce the weighting factors, construct structural fitness function and achieve multi-objective optimization. Combining formula (1)–(3), this paper converts the fitness function into the following form:

$$F_{fit} = a_1 * \overline{D_c} + a_2 * f\left(\overline{V_{avg}}\right) + a_3 * \overline{L_c} \qquad (5)$$

a_1, a_2 and a_3 are the weight coefficients and could meet $a_1 + a_2 + a_3 = 1$, $f(\overline{V_{avg}})$ means the load balancing function of the whole system is calculated by the formula (4). As each property has different units in the above formula, this paper uses min-max standardized data to conduct normalization.

3.2 Update methods of particle velocity and position

The essence of particle swarm algorithm is to use the information gap between individual extreme and global extreme of the current particles to adjust the velocity of particles, and thus change the position of the next iteration of particles. Based on the calculation formula of speed and position in literature, the crossover operation in the genetic algorithm is introduced, shown as follows:

1. the transformation operation of matrix position is defined as ch (a_{11}, a_{22}), indicating the exchange between a_{11}th bit and a_{22}th bit of matrix, and the sequence consisting of a plurality of switching operations is referred to exchange sequences, denoted by $CH = (ch_1, ch_2 \cdots ch_m)$.
2. The basic switching sequence set: subtract two matrices to produce a switching sequence, which means the matrix A' can be obtained after sequence transformation of matrix A. As different exchanging orders could produce the same transformation from matrix A to matrix A', the exchange sequence with the least number of transformations is called the basic switching sequence set. According to the above method, the speed and position formula of the discrete PSO algorithm is modified as follows:

$$V_{id}^{k+1} = V_{id}^k + \alpha \otimes \left(pBest_{id}^k - p_{id}^k\right) + \beta \otimes \left(gBest_{id}^k - p_{id}^k\right) \qquad (6)$$

In the above formula, $\otimes$ is the matrix multiplication, α, β are random numbers of $(0,1)$, $\alpha \otimes \left(pBest_{id}^k - p_{id}^k\right)$ means to keep the commutators with the probability α and exchange X_{id}. So the new position formula will be generated by the following formula.

$$X_{id}' = X_{id} \oplus CH \qquad (7)$$

In the above formula, $\oplus$ means the matrixes are exchanged by a basic switching sequence.

Due to the prematurity of the standard discrete PSO algorithm, this paper draws the design idea from literature and introduces a mechanism to determine the population re-optimization: If a certain percentage of particles in the current population could meet the requirement, the algorithm will re-optimize the population. The judgment condition is that if the number of elements is less than μ in the basic switching sequence of particle velocity, and the solution which meets the requirement has not been determined, the re-optimization will be carried out.

4 PERFORMANCE EVALUATION

The simulation tests are conducted in the platform CloudSim Toolkit and the improved three performance indicators are simulated including service delay, expense cost, and load balancing of medical cloud storage under the influence of the new storage policy. Then a comparison is carried out with Genetic Algorithm (GA).

When searching for the optimal feasible solution of memory scheduling with IDPSO, the population size is 50 particles and the maximum allowable number of iterations is 300. The related parameters of the standard discrete PSO algorithm and some parameters of IDPSO algorithm are set according to literature. For all simulation experiments in this paper, for each problem, the independent experiments will be repeated 10 times and then average them.

In Figure 2, the total number of files is used to represent the load of the system. Generally speaking, they are in direct proportion. Compared with HDFS and GA, IDPSO algorithm could reduce the average length of service. As can be seen from Figure 2, IDPSO algorithm could achieve better results when the total number of files varies.

Figure 3 is a node load after the optimization algorithm. The ordinate represents the load value corresponding to each node in the system. As can be seen from Figure 3, the load varies on each node in HDFS system, however, with the increase of the number of iterations, the load situation stabilizes, and gradually tends to be in a stable level. Affected by other factors, the load values of each node

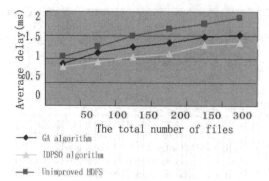

problems, the discrete PSO algorithm is used to introduce the custom updating method of speed and position, as well as re-optimized judgment rules. Finally the IDPSO algorithm which could be used to solve the problems of storage and optimization in the cloud computing. Then simulation tests are carried out in the simulation platform CloudSim Toolkit and the results show that this method is effective.

Figure 2.   The average delay of reading file.

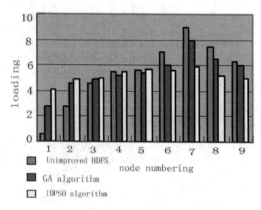

Figure 3.   The node loading on the termination of the algorithm.

cannot be completely equal, but they are so close, which means the load balancing has been basically achieved.

5   CONCLUSION

In the process of implementing medical cloud services, people usually pay more attention to data security, load balancing of the system and other issues. This paper discusses the optimization algorithm of storage deployment, aiming to improve the performance of the entire medical cloud services by adjusting the mapping between file and storage nodes. Then after the medical cloud storage are abstracted to multi-objective optimization

ACKNOWLEDGEMENT

This study was supported by the National Natural Science Foundation of China (61170192), the Fundamental Research Funds for the Central Universities (XDJK2014C039, XDJK2016C045), Doctoral Fund of Southwestern University (swu1114033)

REFERENCES

Elit L.M., C. Charles, A. Gafni, J. Ranford, S. Tedford-Gold, and I. Gold, "How oncologists communicate information to women with recurrent ovarian cancer in the context of treatment decision making in the medical encounter," *Health Expectations,* vol. 18, pp. 1066–1080, Oct 2015.

Kaveshgar N. and N. Huynh, "A genetic algorithm heuristic for solving the quay crane scheduling problem with time windows," *Maritime Economics & Logistics,* vol. 17, pp. 515–537, Dec 2015.

Kim S.H. and K.Y. Chung, "Medical information service system based on human 3D anatomical model," *Multimedia Tools and Applications,* vol. 74, pp. 8939–8950, Oct 2015.

Li R., D. Yang, H.B. Hu, J. Xie, and L. Fu, "Scalable Rdf Graph Querying Using Cloud Computing," *Journal of Web Engineering,* vol. 12, pp. 159–180, Feb 2013.

Niu Z.Y., K. Zhou, D. Feng, and T.M. Yang, "Access Control Lists for Object-Based Storage Systems," *Chinese Journal of Electronics,* vol. 19, pp. 431–436, Jul 2010.

Unler A. and A. Murat, "A discrete particle swarm optimization method for feature selection in binary classification problems," *European Journal of Operational Research,* vol. 206, pp. 528–539, Nov 1 2010.

Zhang Y.H., H.Y. Wang, D.S. Wang, and W.M. Zheng, "Employing Object-Based Storage Devices to Embed File Access Control in Storage," *Intelligent Automation and Soft Computing,* vol. 17, pp. 1–11, 2011.

Zhang H.B., D.D. Kennedy, G.P. Rangaiah, and A. Bonilla-Petriciolet, "Novel bare-bones particle swarm optimization and its performance for modeling vapor-liquid equilibrium data," *Fluid Phase Equilibria,* vol. 301, pp. 33–45, Feb 15 2011.

Advances in Energy Science and Equipment Engineering II – Zhou, Patty & Chen (Eds)
© *2017 Taylor & Francis Group, London, ISBN 978-1-138-71798-5*

A survey on cyber security of connected vehicles

Liang Bai & Zhen Zhang
*The National Computer Network Emergency Response Technical Team/Coordination Center of China,
Beijing, P.R. China*

Haoyuan Huo
Science Department, Beijing University of Posts and Telecommunications, Beijing, China

Yueying He
*The National Computer Network Emergency Response Technical Team/Coordination Center of China,
Beijing, P.R. China*

ABSTRACT: The Internet tidal wave has a great influence on the vehicle industry, making vehicles more frequently exchange information with the outside world through network. Although it brings great advancements in efficiency, it also provides the entry point for cyber-attacks. In this paper, a survey on cyber security of connected vehicles is made. We first enumerate the potential attack interfaces of modern vehicles and their security vulnerabilities. Then, the attack approaches, indicating the ways to externally control over the vehicle system, are described. Finally, the protection mechanism and the future development against vehicle cyber-attacks are presented.

1 INTRODUCTION

In-vehicle network system consists of automotive sensors and Electric Controls Units (ECUs) (Othmane, 2015). ECUs could achieve a broad range of functionality, including the drivetrain, brakes, lighting, etc. They are connected to the sensors through an in-vehicle network, so that they can receive information about the environment in real time. In addition to the in-vehicle network, modern vehicles also use a variety of wireless technology, such as cellular network (Zhang, 2005), WiFi, Bluetooth, NFC and IR, to communicate with other vehicles or devices, so that they can exchange the information of location, speed, direction, road condition, etc. It is called "connected vehicles", which could allow the driver to remotely control the various components of the vehicle, such as head lights, interior lights, windshield wipers, heat, air conditioner, etc., or even control the engine by a smart phone (Mahmud, 2006).

Obviously, the driving efficiency is improved by connected vehicles. However, since the cyber security issue is scarcely taken into account in designing, lots of security vulnerabilities exist in connected vehicles, which make it easy to remotely get access to the in-vehicle network and control over the key ECUs. So far, lots of research has been made on this topic, which shows that some

security vulnerabilities will give rise to privacy disclosure, and more seriously, even endanger people's lives and property. Since the Global Positioning System (GPS) could remotely control vehicles in Texas, more and more security vulnerabilities have been exposed (He, 2016). The way on how to unlock Subaru backdoors by using text messages, which are sent over the phone links to wireless devices in the vehicles, was presented at DefCon2011 (Greene, 2011). At DefCon2013, a hacker disclosed how to remotely control the steering wheel, brake, throttle of Ford Escape and Toyota Prius via OBD-II. Qihoo360 indicated the vulnerabilities in Tesla vehicles in April 2014 (Rosenblatt, 2014). Charlie Miller and Chris Valasek (Miller, 2015) indicated how to get access to steering wheels, brakes, engines and doors to demonstrate the feasibility of remote control over vehicles, which led to the recall of 1.4 million vehicles by Fiat Chrysler. Therefore, all the cases show that the cyber security of connected vehicles should be emphasized.

In this paper, the potential cyber-attack interfaces of connected vehicles and their security vulnerabilities are enumerated. Then, we describe the attack approaches in three categories. Finally, the current research on protection mechanism and the future development for the security of connected vehicles are presented.

2 POTENTIAL CYBER SECURITY RISKS

In this section, we talk about the attack interfaces and vulnerabilities of connected vehicles, which indicate the potential cyber security risks.

- **On-Board Diagnostic (OBD).** The main design target of in-vehicle networks (e.g. CAN bus) is going to achieve reliability and efficiency rather than cyber security, which makes vehicles in lack of protection mechanism against cyber-attacks. As all of the ECUs are connected to CAN bus inside vehicles, they communicate with each other via CAN data packets. From a cyber-security perspective, there are three ways to use the CAN bus (Paret, 2007):
  - Firstly, the data can be read and sent on CAN bus via OBD-II interfaces.
  - Secondly, the data can only be read but cannot be sent on the bus through OBD interfaces.
  - Thirdly, the diagnostic data be sent and feedback can be received only by J1939 protocol.

  As described above, attackers could monitor and crack CAN data, and attack vehicles with OBD-II interfaces.
- **Key fobs.** The working principle of this technology is based on Radio Frequency Identification (RFID). The rolling code, a kind of pseudo random code, is stored in key fobs (Tieman, 2009). When communication initiates, the key fobs send both rolling code and function code (the code represents different functions, such as unlocking the doors, etc.). Then the vehicle makes a check. Only when the rolling code from key fobs matches its own code, the vehicle performs the operations indicated by the function code. Key fobs and the vehicle keep holding synchronization of the rolling code at all times. Obviously, attackers can crack the key fobs and control vehicles when he obtains the signals or algorithms of the rolling code.
- **Bluetooth key.** It connects to in-vehicle network via Bluetooth, and could control vehicles by the smartphone applications. For instance, the driver can unlock the vehicle body doors and engine cover without the real key but use the control signals from a smartphone. It is very dangerous that attackers can send malicious command to CAN-bus easily when he gets into the network of Bluetooth. Since the smartphone applications of Bluetooth key is lack of security reinforce, attackers could obtain the method to control the vehicle by reverse analysis.
- **Vehicular services.** The connected vehicles install some applications for services (Lu, 2014), such as firmware updates, in-vehicle multimedia entertainment, vehicular social networking, and location-based services, etc. These applications are allowed to send or receive data through Internet, so attackers may research on the vulnerabilities of them by reverse analysis and fuzzy test, and then exploit to control the connected vehicles
- **Ultrasonic radar.** The ultrasonic radar receives the sound waves as input signals to the vehicle system (Alonge, 2009). The cost of an attack to ultrasonic radar is no more than one hundred dollars. At present, ultrasonic sensor is often used by driverless vehicle to detect the obstacles nearby. An attacker may confuse the vehicles by controlling an ultrasonic emitter to send the waves with same period and frequency, and make the obstacles around the vehicles become 'invisible'.
- **Millimetre-wave radar.** Currently, the most advanced autonomous vehicles are using millimetre wave radars (Okai, 2001). The detection range of long-distance millimetre wave radar is almost 150 m and the frequency band is between 77 and 79 GHz. Meanwhile, the detection range of short-distance millimetre wave radars often installed on two sides of the vehicle is 50 m and the frequency band is 24 GHz. The safety of the autonomous vehicles is improved by these two kinds of radars. However, it is also possible to cheat the radars by sending the millimetre wave with the same period and frequency, which could stop the vehicles or make them deviate from the original lane.
- **High Definition Digital Camera (HDDC).** the HDDC is the eye of smart connected vehicles, which helps vehicles identify the lane and road sign, or judge the speed of vehicles and walkers nearby. An easy way to attack the HDDC is to blind the cameras by strong light (Pouliot, 2002). Attackers can also construct a special video image to make the camera disabled. However, the second way is more difficult to implement because the video image can only be constructed after reverse analysis of algorithms of HDDC.
- **Light Detection and Ranging (LIDAR).** Replay attacks can be implied because the light signals of LIDAR (Reutebuch, 2005) are not encoded and enciphered. When the searching signals are sent by LIDAR, attackers could create some pseudo images of obstacles and feed them back to LIDAR on time. Then, the LIDAR will 'see' illusory objects surrounding them. Having deeply analysed the characteristics of LIDAR signals, attackers may forge many obstacles, such as vehicles, walls, walkers etc., and create the illusion of moving objects. Besides, attackers can also launch denial of service attack to LIDAR.
- **802.11P.** Wireless transmission is the key technology of vehicular communication system. It integrates each single vehicle as a whole,

which will better support for moving environment and strengthen the safety (Lochert, 2008; Anon,2013). IEEE 802.11p, an extension of IEEE 802.11a/g standard, is also called Wireless Access in the Vehicular Environment (WAVE). It is mainly used for wireless communication of in-vehicle electronics, and support for vehicle-infrastructure communication and vehicle-vehicle communication. From a technical point of view, WAVE makes lots of improvements for connected vehicles, such as support for the switching of hotspot, the strengthening of identity authentication, etc. 802.11p is based on IEEE 802.11 protocol family, thus the cyber risk of wireless communication is inevitably introduced, although the functions of identification authentication and data transmission encryption are added on application layer

- **Internet of Vehicles (IoV) service platform.** The IoV service platform provides a set of powerful functions including navigation, entertainment, security protection, SNS and remote maintenance (Song 2014), which is usually deployed in the cloud, so the platform also faces the cyber threats to the cloud infrastructure. For example, attackers can get access to the host by the technique of virtual machine escape, and further obtain the interfaces, secret keys, certificates and other important information of connected vehicles. Therefore, the IoV service platform deployed in the cloud not only depends on its own self-security, but also considers other security elements, such as anti-DOS, traditional IT protection and security management, etc.

3 THE ATTACK APPROACHES

There is no doubt that connected vehicles will face a variety of cyber-attacks, as it is a mobile information platform based on network. The information exchange between connected vehicles and external network is becoming more frequent by means of wireless communication. There can be a variety of interfaces to a vehicle, such as a smart mobile phone connected to the vehicle, or even the ECU node connected to the in-vehicle network. In this section, we will describe three ways used to attack the vehicle system.

3.1 *Attacks by physical connection*

This kind of attack is launched by connecting a device to the physical interfaces, such as OBD, the charging port, etc. A method was presented to control the brake and wiper of a vehicle by setting a special instrument on the OBD port and attacking the vehicle system through it (Lochert, 2008).

The vehicle can be easily controlled because it has no mechanism to analyse and verify the communication data, and set no access limits to the internal communication address. Miller and Valasek (Valasek, 2015) physically connected their laptop to in-vehicle network, which could read and write via the CAN bus. Then, by analysing the CAN bus protocol, they further achieved several functionality, including the drivetrain, brakes, lighting, etc. There is a detailed introduction of the source code, software and hardware environment they used to attack vehicle, which is recorded in their white paper.

3.2 *Attacks through peripheral products*

Nowadays, more entertainment devices and applications are installed in vehicles. For instance, smart phone, as one kind of peripheral devices, can be achieved as the interface to the in-vehicle network. The other side, there are many kinds of malicious software or applications in smart phones, which will also have serious threats on the vehicle cyber security. Woo attacked the ECU node through the mobile applications which connect to OBD interface using Bluetooth or WiFi. With this method, Woo could change the dashboard and close the engine. Francillon recorded and analysed the relay messages between a vehicle and the key fob. Then, he used relay attack to get access to the internal network of the vehicle and launched it.

3.3 *Attacks from external network*

Tire Pressure Monitoring System (TPMS) and vehicle-road communication use short-distance wireless communication, which exists the threats of signal interception and interference. Due to the analysis of the wireless communication protocol, Rouf pointed out that the messages sent by TPMS sensor are not encrypted and can be intercepted easily. The TPMS sensor has a unique 32-bit ID, and it can wirelessly communicate with other equipments within 40 meters, which exists the leakage risk of location information and the TPMS alarm messages can be forged to launch replay attacks. Moreover, the telematics system and GPS may also face the threats of being attacked from external network.

4 THE PROTECTION TECHNOLOGY

There is no system or technology which could defend against all kinds of cyber-attacks. The research in this section focuses on some kinds of security protection technology and solutions, which are classified into three categories.

4.1 In-vehicle network

Kleberger (Kleberger, 2014) did some research on the security threats of in-vehicle network and discussed the solutions. Han (Han, 2014) analysed the security needs of integrating a target device, such as a phone, into vehicles. He defined a security objective and put forward the attack scenarios, and designed a three-step authentication mechanism to estimate the integration security index of the target device. Yang put forward a vehicular keyless access system against relay attack and a new agreement of distance boundary, which can help detect the malicious access from attackers.

4.2 Vehicle-Vehicle and Vehicle-Infrastructure network

As one of Mobile Ad-hoc NETworks (MANET), Vehicular Ad-hoc NETworks (VANET) has made a great development in recent years, especially the Vehicle-to-Vehicle (V2V) and Vehicle-to-Infrastructure (V2I) model. However, their security problems have drawn common attention. Due to the self-organizing characteristics of VANET, the running status of entire network is decided by the terminals which use wireless communication. At present most cyber security protection techniques are designed based on the traditional wired network. The trust of wireless communication on both sides can't be guaranteed in V2V and V2I, so whether the information is secure also cannot be guaranteed. Therefore, the cyber security protection technology based on wired network is not suitable for VANET. Freudiger (Freudiger, 2011) used a method of random identify label to guarantee the location information security of VANET.

4.3 Vehicle-Internet network

Some researchers presented a vehicular gateway authentication system based on OPENSSL and IPV6, which uses digital certificate to verify the identity to ensure the secure connection between the vehicle terminals and the server on the Internet. Some scholars are studying the secure way of accessing the embedded system and some research has done on the encryption protocols for vehicle-Internet communication. Raya (Raya, 2005) presented a method of distributing a large number of encrypted public and private key pair, and matched public key certifications for each vehicle and server. Besides, SNEP protocol is suggested to be applied in the vehicle gateway to make the communication secure, whose corresponding encryption algorithm is also being studied.

5 THE FUTURE DEVELOPMENT

The cyber security problem of vehicles is becoming the first issue when the vehicle manufacturers apply new technology and make new products. Toyota General Motors are applying communication data encryption module and access authentication module. Tesla, Audi and BMW will found the professional Computer Emergency Response Teams (CERT). Google and Baidu are researching to build a new security framework for connected vehicles. European, American and Japanese organizations related are actively funding research programs of vehicular cyber security, such as OVERSEE, EVITA, etc. Japanese Information-technology Promotion Agency (IPA) is defining various cyber security measures according to different stages of vehicle lifecycle, including design stage, development stage, usage stage, scrap stage and management stage. For all stages, there are detail descriptions about security measures that would be used.

6 CONCLUSION

The development of vehicular network technology, intelligent transportation system, driver assistance system, self-driving technology and cloud service has greatly enhanced the function and application of vehicles. However, more and more cyber security risks against modern vehicles have risen up at the same time. In this paper, we list the potential attack interfaces and the vulnerabilities existing in connected vehicles, and describe the cyber-attack approaches which are classified into three categories. Finally, the introduction to the protection mechanism and the future development against vehicle cyber-attacks is presented.

REFERENCES

Anon, http://www.en.wikipedia.org/wiki/Vehicular_adhoc network (2013).
Alonge, F., M. Branciforte, F. Motta, IEEE Transactions on Instrumentation & Measurement, 58(2), 318–329 (2009).
Freudiger, C.W., W. Min, G.R. Holtom, Nature Photonics, 5(2), 103–109 (2011).
Greene. T., Black Hat and Defcon (2011).
Han. M., US 8725310 B2 (2014).
He, C.K., J. Song, W. Zhou, Industrial Technology Innovation, 3–2, 108–114 (2016).
IEEE P802.11pTM/D1.1. Draft amendment (2005).
IEEE P802.11pTM/D10.0. Draft standard (2010).

IEEE P802.11pTM/D10.0. IEEE 802.11 working group of the IEEE 802 committee (2010).

IEEE P802.11pTM/D3.0. Draft standard (2007).

International Standard Organization, **ISO 26262–1** (2011).

International Standard Organization, ISO-11898-1(CAN) (2003).

Kleberger, P., G. Moulin. Chalmers University of Technology, 202–205 (2014).

Lochert, C., B. Scheuermann, C. Wewetzer, Proceedings of the Fifth ACM International Workshop on Vehicular Internet working, 58–65 (2008).

Lu, N., N. Cheng, N. Zhang, Internet of Things Journal, 1(4), 289–299 (2014)

Mahmud, S., S. Shanker, IEEE Transactions on Vehicular Technology, 55, 1051–1061(2006).

Miller, C., C. Valasek. *Black Hat* (2015).

Okai, F., K. Hanawa, K. Takano, US20010026237A1 (2001).

Othmane, L.B., H. Weffers, M.M. Mohamad, and M. Wolf, *Norwell, MA: Springer* (2015).

Paret, D., R. Riesco, *Chichester: John Wiley &Sons Ltd*, 14–23 (2007).

Pouliot, D.A., D.J. King, F.W. Bell, Remote Sensing of Environment, **82(2–3)**, 322–334 (2002).

Raya, M., J.P. Hubaux, ACM Workshop on Security of Ad Hoc and Sensor Networks, 11–21 (2005).

Reutebuch, S.E., H.E. Andersen, R.J. Mcgaughey, Journal of Forestry, 103(6), 286–292 (2005).

Rosenblatt, S., https://www.cnet.com/news/chinese-hackers-take-command-of-tesla-model-s/ (2014).

Song, J., Y.H. Cheng, X.Z. Zhang, Applied Mechanics & Materials, **548–549**, 1844–1847 (2014).

Tieman, C.A., A.C. Nascimento, V.L. Brooks, US20090243821 (2009).

Valasek, C., C. Miller, US20150113638 (2015).

Zhang, J., I. Stojmenovic, *Wiley*, **I-2**, 654–663 (2005).

Advances in Energy Science and Equipment Engineering II – Zhou, Patty & Chen (Eds)
© 2017 Taylor & Francis Group, London, ISBN 978-1-138-71798-5

Aerodynamic heating calculation by lattice Boltzmann equation

Qing Shao
College of Astronautics, Northwestern Polytechnical University, Xi'an, China
Shanghai Electro-Mechanical Engineering Institute, Shanghai, China

Fu-ting Bao
College of Astronautics, Northwestern Polytechnical University, Xi'an, China

Chao-feng Liu
Shanghai Electro-Mechanical Engineering Institute, Shanghai, China

ABSTRACT: In this paper, an innovative and practicable scheme combining Euler equation with lattice Boltzmann method is proposed to enhance potential applications in the aerodynamic heating prediction of hypersonic fight vehicles. In order to suit the complex three-dimensional configurations, CFD simulation based on commercial software is applied to obtain the aerodynamic parameters outside the boundary layer. To improve the accuracy of reference enthalpy method and to reduce the high computational cost of direct solution of Navier–Stokes equation, a compressible thermal lattice Boltzmann algorithm is put forward to calculate the viscous effect inside the boundary layer by simulating the stream and collision of nonequilibrium particles. The thermal LBM is improved by updating density in each process of stream and collision to couple the velocity and temperature calculation. Meanwhile, an analogy between laminar flow and turbulent flow is drawn to expand the application scope of this scheme. The feasibility of the new scheme is validated by experimental data of a hypersonic vehicle, and the advantages and disadvantages are compared to those of other aerodynamic heating prediction methods.

1 INTRODUCTION

High aerodynamic heating leads to a series of problems, such as surface ablation, weakening of structural properties, and abnormal working temperature environment of the instruments. Therefore, it is important to accurately predict aerodynamic heating environment in the design of hypersonic vehicles. For a hypersonic vehicle with wide-range trajectory, both flight test and ground-based test, which experience vastly deferent flow regimes, are expensive and technically challenging. Therefore, numerical modeling and computation become increasingly important.

Nowadays, some engineering algorithms such as Reference Enthalpy Method (REM) and Tracing Streamline Method (TSM) are widely applied in vehicle design (Dejarnette, 1987). With the development of Computational Fluid Dynamics (CFD), the Direct solution of Navier–Stokes equation (DNS) and some simplified models based on Navier–Stokes (N-S) solution, such as Viscous Shock Layer method (VSL), Parabolic Navier–Stokes method (PNS), are increasingly employed to calculate the thermal environment of hypersonic vehicles. The computational results of

aerodynamic heating by CFD methods strongly depended on trajectory, shape, Mach number, Reynolds number, Knudsen number, mesh resolution, and so on (Orszag, 2000). Although the existing CFD capability provides considerable accuracy for aerodynamic force calculation, it is still far from the same accuracy for the calculation of aerodynamic heating calculation. In general, the simple engineering algorithms are restricted from complex configuration and the pure CFD methods are faulted for huge computational cost.

In a microscopic view, there are great gradients in molecular or macroscopic aerodynamic properties for a hypersonic flow. It means that the Maxwell equilibrium distribution is not necessarily conformed to and it requires a relaxation time for the gas to come to equilibrium. To improve the prediction's accuracy of thermal environment under the condition of nonequilibrium gas and complicated boundaries, the microscopic Direct Simulation Monte–Carlo (DSMC) method was employed in the calculation of aerodynamic heating. Moss and Bird (Moss, 1997; Bird, 1984; Glass, 2001; Moss, 2005) simulated aerodynamic heating of shuttle in low-density air with DSMC. Lofthouse (Lofthouse, 2008) compared

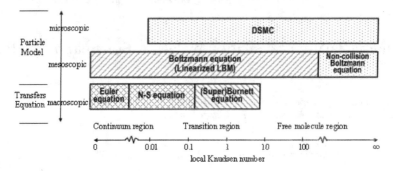

Figure 1. Application scopes of CFD techniques.

nonequilibrium hypersonic aerodynamic heating by N-S equation to that by DSMC method and obtained appropriate accuracy for hypersonic flight. However, the huge computational cost of DSMC simulation prevents these methods from being widely applied in engineering design of hypersonic vehicles.

In the past decades, the Lattice Boltzmann Method (LBM) has attracted great attention in computational fluid dynamics and compressible thermal fluid simulations (Qian, 1992; Fang, 2002; Chen, 1991; Zou, 1995; Ansumali, 2005; Karlin, 2009; Bruce, 2000; Sun, 2000; Lallemand, 2003; Li, 2007; Fu, 2010). Bruce LBM is a mesoscopic theory, which is constructed based on kinetic theory of gas. It is different from traditional macroscopic and microscopic methods such as the Finite Difference Method (FDM), Finite-Element Method (FEM), and DSMC simulation. LBM is an explicit solving scheme of linearized collision term of the discrete Boltzmann equation. In LBM scheme, the flow is simulated by the stream and collision of microscopic particles; therefore, the range of application of LBM is larger than that of N-S equation, as shown in Fig. 1. As a particle method, LBM has the advantages in handling complicated boundary conditions, and essentially goes beyond the limitation of the continuum assumption. As a linearized explicit numerical scheme, LBM obtains a controllable and parallel computational efficiency over DSMC and DNS solving.

In consideration of these advantages of LBM, this work attempts to construct a practicable scheme combining Euler equation with lattice Boltzmann method for the predicting heating environment of hypersonic vehicles. In the following paragraphs, the assumptions of the scheme are explained and the modeling steps are described. Then, a numerical simulation compared with experimental data is carried out to validate the scheme. Finally, the advantages and disadvantages of this scheme are discussed.

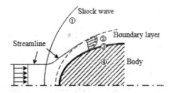

Figure 2. Four levels of aerodynamic heating.

2 ASSUMPTIONS AND GENERAL SCHEME DESIGN

To calculate the aerodynamic heating along the trajectory, it is necessary to understand the mechanism of aerodynamic heating inside and outside the boundary layer (Boyd, 2003; Sun, 2002; Santos, 2004). When air flows through object's surface, tremendous kinetic energy is changed into heat energy by the effects of compression, friction, and block. In a macroscopic view, the changing of density and temperature are the result of compression effect and viscosity effect, which are characterized by Mach number and Reynolds number. In a microscopic view, it is the result of the normal and tangential interactions of gas molecules, which can be simulated by the processes of collision and stream in particle methods. The levels of heat transfer and exchange between gas and solid wall are shown in Fig. 2. The first level is thermal convection between free flow and outside boundary layer, including the heat changes initiated by shock wave and expansion wave. The second level is the heat transfer and convection between gas molecules inside boundary layer, in which the thermal convection is caused by density gradient initiated by temperature gradient in boundary layer. In close vicinity of object's surface, a great amount of heat is generated by the great velocity gradient. The third level is the heat transfer between gas molecules and wall. The last level is the heat transfer and thermal radiation of object's surface and inner body.

On the basis of the above understanding of the mechanism of aerodynamic heating, some assumptions are taken in this work as follows:

1. The irrelevant assumption of shock wave and boundary layer. Despite the shock wave layer becomes thinner for high-speed flow, most boundary layers are still supposed to be kept from the shock wave line, as the front/nose of most hypersonic vehicles is designed using a bluff body, where a detached shock wave is created. On the contrary, the knowledge about the growth and separation of turbulent flow in the boundary layer under the effect of shock wave is not yet sufficient. To avoid rigorous requirements for numerical method to simulate these complicated phenomena, the irrelevant assumption is taken in this work.
2. The Prandtl boundary layer assumption. The high velocity gradient inside the boundary layer is the main source of the aerodynamic heating and the viscosity effect is ignored outside the boundary layer. Because designers of vehicles only ultimately concern the heat flux on the surface in a thermal environment prediction, the thickness of the boundary layer will be dealt with in a normalized thickness and in a uniform format.
3. The assumption of small deviation from thermodynamics equilibrium. We suppose the non-equilibrium process is not very large inside the boundary layer for the two conveniences. One is to use linear collision term when solving Boltzmann equation. The other is to simplify the relation between macroscopic transport coefficients and microscopic relaxation time.
4. Because aerodynamic heating involves many influence factors and the work cannot cover all of the conditions, the effects of the material ablation, particle erosion, mass injection, surface roughness, and so on are ignored.

On the basis of the above assumptions, a new scheme combining Euler equation, LBM, and heat conduction equation is proposed for aerodynamic heating prediction of hypersonic vehicles in complicated geometries. The methodology of the combined method is shown in Fig. 3. Outside the boundary layer, Euler equation without a fine boundary layer is solved, which reflects the compressible effect of complex configurations. Inside the boundary layer, viscous LBM is solved instead of the engineering method such as the reference enthalpy method in order to obtain higher accuracy. In inner solid body, heat conduction equation is solved to obtain the temperature distribution of the structure (for a thin-walled metal structure, one-dimensional heat conduction equation meets accuracy requirements). The grid of the above layers is respectively three-dimensional, two-dimensional, and one-dimensional, so we call the above scheme as "3+2+1" scheme.

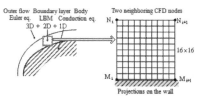

Figure 3. Methodology of "3+2+1" scheme.

Outside the boundary layer, statistical parameters are directly given from Euler equation; inside the boundary layer, microscopic parameters of LBM are used for the particles. The transform of the two types of parameters is illustrated by the equilibrium distribution functions of the velocity and temperature of the particles. The scheme of the interface of the boundary layer is as follows: For each macroscopic there-dimensional surface node (or cell) in the CFD network, a uniform 16×16 LBM network is constructed by the D2Q12 (two dimensions and 12 speeds) model to simulate the inner flow in vicinity of the node. The collision time and particle distribution functions are derived from macroscopic conservation of energy. For the fact of zero normal velocity near the wall for a laminar flow, the flow inside the boundary layer of each CFD nodes is simplified into a viscous Couette flow with given pressure and velocity condition in the thermal LBM. Inside the solid body, the one-dimensional heat conduction equation is solved in the corresponding nodes to get the temperature boundary condition of the flow. The core of the method is the modeling of thermal LBM inside the boundary layer.

3 THERMAL LBM MODEL INSIDE THE BOUNDARY LAYER

When calculating the outer flow with CFD method on the basis of Euler equation, the aerodynamic parameters of the nearest node are used. According to the boundary layer theory, viscosity is not effective on press field in the case of large number of Reynolds. Therefore, the aerodynamic parameters of pressure, velocity, and density outside the boundary layer, which are the inputs of temperature calculation inside the boundary layer, are solved by Euler equation instead of N-S equation. Inside the boundary layer, the linear Boltzmann equation is applied to simulate the heat flow, in which the relaxation process with velocity and energy express as τ. The Boltzmann equation without an external force is written as:

$$\frac{\partial f}{\partial t} + \mathbf{v} \cdot \nabla f = \Omega(f) \tag{1}$$

where $f = f(\mathbf{x}, \mathbf{v}, t)$ is the density distribution function. Macroscopic pressure ρ, velocity $\mathbf{u}$, and internal energy e are defined as:

$$\begin{cases} \rho = \int f \, d\mathbf{v} \\ \rho \mathbf{u} = \int \mathbf{v} \, f \, d\mathbf{v} \\ \rho e = \dfrac{DRT}{2} = \dfrac{1}{2}\int (\mathbf{v} - \mathbf{u})^2 f \, d\mathbf{v} \end{cases} \quad (2)$$

The flow of boundary layer is simulated by the coupled double-distribution-function model of LBM, and a total energy distribution function when considering the pressure of work, and the viscosity heat dissipation is introduced as:

$$h(\mathbf{x},\mathbf{v},t) = \frac{v^2}{2} f(\mathbf{x},\mathbf{v},t) \quad (3)$$

The total energy is calculated by:

$$\rho E = \rho e + \rho \frac{u^2}{2} = \int h d\mathbf{v} \quad (4)$$

The total energy distribution function satisfies:

$$\frac{\partial h}{\partial t} + \mathbf{v} \cdot \nabla h = \frac{v^2}{2}\Omega(f) = \Omega(h) \quad (5)$$

where the collision operator of and velocity distribution and energy distribution can be calculated by the Bhatnagar–Gross–Krook collision term as the function of the relaxation time (Bhatnagar, 1954)

$$\begin{cases} \Omega(f) = -\dfrac{1}{\tau_f}(f - f^{(eq)}) \\ \Omega(h) = -\dfrac{1}{\tau_h}(h - h^{(eq)}) \end{cases} \quad (6)$$

The velocity equilibrium distribution and the energy equilibrium distribution satisfy the Maxwell distribution:

$$\begin{cases} f^{(eq)} = \dfrac{\rho}{(2\pi RT)^{D/2}} \exp\left[-\dfrac{(\mathbf{v}-\mathbf{u})^2}{2RT}\right] \\ h^{(eq)} = \dfrac{\rho v^2}{2(2\pi RT)^{D/2}} \exp\left[-\dfrac{(\mathbf{v}-\mathbf{u})^2}{2RT}\right] \end{cases} \quad (7)$$

where τ_f and τ_h are the relaxation times of velocity and energy respectively, which are describing the process of particles from nonequilibrium state to equilibrium state. Using Chapman–Enskog expansion, macroscopic N-S equation and heat conduction equation can be derived from Boltzmann equation of Eqs. (1–7) (Qian, 1992).

Some important factors leading to the nonequilibrium characteristics, such as the high velocity gradient caused by high speed, the oscillate energy mode caused by high temperature, and the long relaxation time caused by high altitude and low density. Considering the nonequilibrium flow characteristics in the slip region and the transition region, the Knudsen number is used to denote the dispersion degree of the gas molecules:

$$Kn = \frac{\lambda}{L} \quad (8)$$

where the molecular free path λ is calculated by the macroscopic viscosity coefficient μ:

$$\lambda = \frac{\sqrt{\pi}\mu}{2P}\sqrt{\frac{2\sigma T}{m}} \quad (9)$$

Furthermore, the relation between the microscopic Knudsen number and the macroscopic parameters is deduced as (Liu, 2009; Rao, 2009; Ni, 2009; Yang, 2011):

$$\begin{cases} Kn = \dfrac{\mu\sqrt{\pi RT}}{\sqrt{2}PL} \\ \tau_f = \dfrac{\pi\mu}{4P}, \tau_h = \dfrac{\tau_f}{Pr} \end{cases} \quad (10)$$

where Pr is the Prandtl number. For a perfect gas, the pressure is calculated by:

$$p = \rho RT \quad (11)$$

In the existing double distributing thermal LBM method, the temperature in Eq. (5) is not coupled back into Eq. (1) (Bruce, 2000; Sun, 2000; Lallemand, 2003; Li, 2007). To obtain the real temperature and density, we updated the density in each process of stream and collision by Eq. (12), recalculated the velocity distributing function as Eq. (13), and then realized the double-direction coupling between temperature and velocity. The formula of gas density of high temperature is:

$$\begin{cases} \rho = 0.213833\left(\dfrac{P}{P_0}\right)\left(\dfrac{P}{1755.52}\right)^{-a}, P_0 = 101325 \text{ Pa}, \\ a = 0.972, \text{when } 167.5\,\text{kJ}/\text{kg} < \text{H} \le 1755.52\,\text{kJ}/\text{kg} \\ a = 0.715 + 1.38974 \times 10^{-2}\ln(P/P_0), \\ \qquad \text{when } 1755.52\,\text{kJ}/\text{kg} < \text{H} \le 35026\,\text{kJ}/\text{kg} \end{cases}$$
$$(12)$$

To discretize the Boltzmann equation and its distribution functions, a D2Q12 model (Li, 2007) (as seen in Fig. 4) is utilized in this research, which ensures the sixth-rank lattice tensor isotropic and the second-order Taylor expansion of equilibrium distribution function. On the basis of the D2Q12 model, the discrete Boltzmann equation is written as:

$$\begin{cases} \dfrac{\partial f_\alpha}{\partial t} + (\mathbf{e}_\alpha \cdot \nabla)f_\alpha = -\dfrac{1}{\tau_f}\left(f_\alpha - f_\alpha^{eq}\right) \\ \dfrac{\partial h_\alpha}{\partial t} + (\mathbf{e}_\alpha \cdot \nabla)h_\alpha = -\dfrac{1}{\tau_h}\left(h_\alpha - h_\alpha^{eq}\right) + \dfrac{1}{\tau_{hf}}\left(\mathbf{e}_\alpha \cdot \mathbf{u} - \dfrac{v^2}{2}\right)\left(f_\alpha - f_\alpha^{eq}\right) \\ \tau_{hf} = \tau_h \tau_f \big/ \left(\tau_f - \tau_h\right), \alpha = 1\cdots 12 \end{cases} \tag{13}$$

By the second-order Taylor expansion, the discrete equilibrium distribution functions corresponding to Eqs. (7) are deduced as (Li, 2007):

$$\begin{cases} f_{1,2,3,4}^{eq} = \dfrac{\rho}{12}\left[(5-4p)+(8-12p)(\mathbf{e}_\alpha\cdot\mathbf{u})-2u^2+2(\mathbf{e}_\alpha\cdot\mathbf{u})^2-6(\mathbf{e}_\alpha\cdot\mathbf{u})u^2+4(\mathbf{e}_\alpha\cdot\mathbf{u})^3\right] \\ f_{5,6,7,8}^{eq} = \dfrac{\rho}{8}\left[(2p-1)+2p(\mathbf{e}_\alpha\cdot\mathbf{u})-u^2+(\mathbf{e}_\alpha\cdot\mathbf{u})^2-(\mathbf{e}_\alpha\cdot\mathbf{u})u^2+(\mathbf{e}_\alpha\cdot\mathbf{u})^3\right] \\ f_{9,10,11,12}^{eq} = \dfrac{\rho}{48}\left[2(2p-1)+2(3p-1)(\mathbf{e}_\alpha\cdot\mathbf{u})+2u^2+(\mathbf{e}_\alpha\cdot\mathbf{u})^2+0.5(\mathbf{e}_\alpha\cdot\mathbf{u})^3\right] \\ h_{1,2,3,4}^{eq} = \dfrac{5\rho E}{12}-\dfrac{p}{2c^2}(E+RT)+\dfrac{1}{6}\left[(\rho E+p)(\mathbf{e}_\alpha\cdot\mathbf{u})+(pE+2p)((\mathbf{e}_\alpha\cdot\mathbf{u})^2-u^2))\right] \\ h_{5,6,7,8}^{eq} = -\dfrac{\rho E}{8}+\dfrac{p}{4c^2}(E+RT)+\dfrac{1}{8}\left[(\rho E+p)(\mathbf{e}_\alpha\cdot\mathbf{u})+(pE+2p)((\mathbf{e}_\alpha\cdot\mathbf{u})^2-u^2))\right] \\ h_{9,10,11,12}^{eq} = -\dfrac{\rho E}{24}+\dfrac{p}{12c^2}(E+RT)+\dfrac{1}{48}\left[(\rho E+p)(\mathbf{e}_\alpha\cdot\mathbf{u})+(pE+2p)((\mathbf{e}_\alpha\cdot\mathbf{u})^2+2u^2))\right] \end{cases} \tag{14}$$

Therefore, the temperature of each node is:

$$T = \frac{2}{bR}\left(\sum_\alpha h_\alpha - \frac{u^2}{2}\right) \tag{15}$$

where b is the molecule freedom degree and R is the gas constant. For a laminar flow, the heating flux on the solid wall is calculated by:

$$q_w = \kappa(T_{aw} - T_w) \tag{16}$$

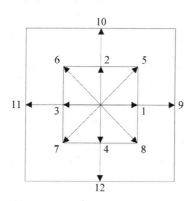

Figure 4. D2Q12 model for thermal LBM.

where $\kappa = \mu c_p / \mathrm{Pr} = \tau_h \mu c_p / \tau_f$ is the thermal conductivity coefficient and c_p is the ratio of specific heats of gas.

The heating flux of Eq. (16) is only appropriate to a laminar flow. When the state of flow changes from laminar to turbulent, the heating flux increases greatly. Some simplified models can calculate the heat flux models for turbulent flows. In this scheme, the Reynolds analogy relation is utilized to calculate the heat flux for turbulent flows:

$$\mathrm{Re}_{CR} = 10^{5.37+0.2325Ma-0.004015Ma^2} \tag{17}$$

when $\mathrm{Re}_x \geq \mathrm{Re}_{CR}$ is the heat flux for a turbulent flow:

$$q_{w,turb} = q_w\left(\frac{0.0296}{0.332}\times \mathrm{Re}_x^{3/10}\rho_x^{3/10}\mu_x^{-3/10}\right) \tag{18}$$

The treatment of the boundary conditions is very important in LBM. For a particle model, the incidence-rebound boundary condition is a natural advantage, which can be used in any complex boundary. The method of the wall boundary condition can be found elsewhere (Liu, 2009; Cornubert, 1991; Guo, 2002; Li, 2013).

4 NUMERICAL EXAMPLES AND COMMENTS

To validate the above scheme, a numerical simulation of aerodynamic heating of blunt cone is taken. This example is derived from the report of NASA TN D-5450 (Cleary, 1969), in which a large number of experiments and detailed studies were carried out and the data were believable. The geometry of the blunt cone is shown in Fig. 5, in which the radius of head $R_n = 0.0095$, angle of half cone $\theta_c = 15°$, and total length $L = 0.5687$ m. The computational conditions of flow are: Mach number $M_1 = 10.6$, pressure $P_1 = 132$ Pa, temperature $T_1 = 47.34$ K, velocity $V_1 = 1461.92$ m/s, and temperature of wall $T_w = 294.44$ K. In the example, the temperature of wall is constant, so that the "3+2+1" scheme of aerodynamic heating reduces to a "3+2" scheme in the absence of heat conduction of solid body.

In the computation of gas parameters outside the boundary layer, the modeling and meshing of FLUENT software is shown in Fig. 6, in which 1608576 nodes are involved. To solve the steady flow field, the density based on viscid solver, the explicit formulation, and the Roe-FDs solution of the first-order upwind discretization are chosen. The distribution of pressure and temperature outside the boundary layer of the cone are shown in Figs. 7 and 8. The maximum temperature is 1105.6 K (seen in Fig. 8), which has an error of only 0.5% compared with the result of 1111.2 K by Eq. (19) of the stagnation point. The pressure and velocity distribution curves outside the boundary layer are shown in Figs. 9 and 10, in which great gradients of pressure and velocity at the nose of the cone can be seen:

$$T_0 = T_\infty \left(1 + \frac{\gamma - 1}{2} Ma_\infty^2 \right) \qquad (19)$$

After the outer aerodynamic parameters are obtained, the thermal LBM is applied to simulate the flow inside the boundary layer in a 16×16 network. Figure 11 shows the velocity profile of the boundary layer at a certain node at $x_s/L = 0.5$ by the thermal LBM, where u/u_0 is the dimensionless velocity, xs/L is the dimensionless length, and y/D is the dimensionless thickness of the boundary layer. The dimensionless factor u_0 is the velocity

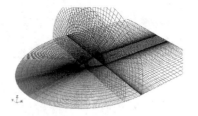

Figure 6.   Modeling and meshing of CFD.

Figure 7.   Pressure outside the boundary layer.

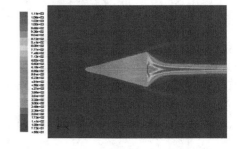

Figure 8.   Temperature outside the boundary layer.

outside the boundary layer, $y = 1 \cdots 16, D = 16$ and L is defined in Fig. 5.

In the simulation, $\theta = 0°$ denotes the leeward side of the cone and $\theta = 180°$ denotes the windward side of the cone. The Cleary table in the report of NASA TN D-5450 had listed the experimental data of thermal laminar flow for the cone of $\theta_c = 15°$. The experimental data involved 10 points but lacked the data at the nose of the cone. Figure 12–14 show the heat flux calculated by thermal LBM at angles of attack $\alpha = 0°$, $10°$, $20°$. For the convenience of comparison with the experimental results, the location and the heat flux of wall are normalized, where the dimensionless factor $q_0 = 215.764$ kW/m² is the heat flux at the stagnation point of a calibration model.[30] It can be seen from the figures that numerical results at front of the cone are a little less than experimental results while numerical results at leeward side of the cone are a little larger than experimental results. The deviations between numerical results and experimental results on the windward side are less than those on the leeward side. The main

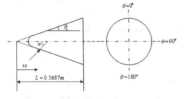

Figure 5.   Geometry of the blunt cone of 15°.

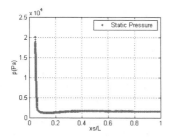

Figure 9. Pressure distribution along the wall ($\theta = 0°$ and $\theta = 180°$).

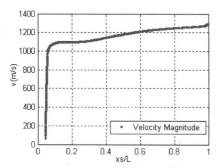

Figure 10. Velocity distribution along the wall ($\theta = 0°$ and $\theta = 180°$).

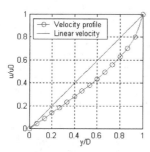

Figure 11. Velocity profile of the boundary layer at $x_s/L = 0.5$.

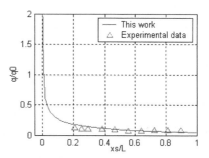

Figure 12. Heating flux of the wall when angle of attack $\alpha = 0°$.

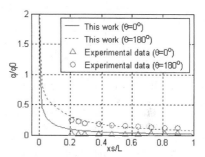

Figure 13. Heating flux of the wall when angle of attack $\alpha = 10°$.

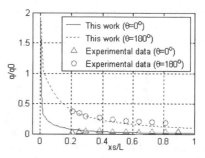

Figure 14. Heating flux of the wall when angle of attack $\alpha = 20°$.

reason lies in the fact that the boundary layer on the leeward side is larger and the fluid structure is more complicated on the leeward side. Overall, the heat fluxes of three angles of attack on the windward side and the leeward side are in agreement with the experimental data.

The above example shows the feasibility of this scheme in aerodynamic heating calculation. For an effective engineering application, there is a balance between computational accuracy and computational efficiency. The feature of the scheme is the mesoscopic thermal LBM based on the kinetic theory of gas inside the boundary layer, in which viscosity is a function of space and time, with the background of the conventional incompressible methods such as the reference enthalpy method. Therefore, the scheme obtains some computational accuracy and loses some computational efficiency. Table 1 compares the existing methods of aerodynamic heating. As discussed in Part 2, this scheme has its hypothesis of application and has still shortages to be improved. For example, the outer flow field may be solved by the LBM of Euler equation, which requires more capability of adaptive mesh refinement (the capability of nested grid) of LBM software. Meanwhile, the inner flow field is possibly to be solved directly by the LBM of turbulent flow, and it also depends on the development of LBM theory and application in the future.

Table 1. Comparison of several computational methods.

Method	Accuracy	Efficiency
Reference enthalpy	Low	Very fast
Euler+Engineering	Low	Fast
Euler+LBM (this work)	Moderate	Moderate
Euler+DSMC	Moderate	Slow
DNS	High	Very slow
DSMC	High	Very slow

5 CONCLUSIONS

A new scheme of aerodynamic heating combining Euler equation and thermal LBM is proposed in this paper. The technique of this scheme uses CFD simulation to obtain the aerodynamic parameters of outside the boundary layer and applies the thermal LBM to simulate the nonequilibrium particles' stream and collision inside the boundary layer. Furthermore, the thermal LBM is improved on two aspects. One is the coupling calculation of velocity and temperature by updating the density in each process of stream and collision. The other is an analogy between laminar flow and turbulent flow, which is drawn to expand the application scope of the thermal LBM. The scheme joins the computational accuracy and computational efficiency against the existing methods of aerodynamic heating. Finally, it is proved to be practicable by comparing with experimental data of a hypersonic vehicle.

REFERENCES

Ansumali, S, IV Karlin, CE Frouzakis, KB Boulouchos, Entropic lattice Boltzmann method for microflows [J]. Physica A—Statistical Mechanics and Its Applications, 359, 289, 2009.

Ansumali, S, and IV Karlin. Consistent lattice Boltzmann method. Physical Review Letters. 95, 260605, 2005.

Bhatnagar, PL, EP Gross, M Krook. A model collision processes in gases [J]. Physical Review, 94, 511, 1954.

Boyd, ID, JF Padilla. Simulation of sharp leading edge aerothermodynamics[R]. AIAA Paper, 7062, 2003.

Bruce, JP, RR David. Lattice Boltzmann algorithm for simulating thermal flow in compressible fluids [J]. Journal of Computational Physics. 161(1), 1, 2000.

Chen, SY, HD Chen, D Martinez, W Matthaeus. Lattice Boltzmann model for simulation of magnetohydrodynamics [J]. Physical Review Letters. 67, 3776, 1991.

Cleary, JW. Effects of attack and bluntness on laminar heating-rate distributions of a 15o cone at a Mach number of 10.6 [R]. NASA Technical Note. NASA TN D-5450, 1969.

Cornubert, R, D d'Humieres, D Levermore. A Knudsen layer theory for lattice gases [J]. Physica D, 47(1), 241, 1991.

Dejarnette, FR, FM Cheatwood, HH Hamilton, KJ Weilmuenster. A Review of Some Approximate Method Used in Aerodynamic Heating Analyses [J]. Journal of Thermophysics and Heat transfer, 1(1), 5, 1987.

Fang, HP, ZW Wang, ZF Lin, MR Liu. Lattice Boltzmann method for simulating the viscous flow in large distensible blood vessels. Physical Review E, 65, 051925, 2002.

Glass, CE, JN Moss. Aerothermodynamic characteristics in the hypersonic continuum-rarefied transitional regime [R]. AIAA Paper, 2962, 2001.

Guo, ZL, CG Zheng, BC Shi. Non-equilibrium extrapolation method for velocity and pressure boundary conditions in the lattice Boltzmann method. Chinese Physics, 11(4), 366, 2002.

Lallemand, P, LS Luo. Theory of the lattice Boltzmann method: Acoustic and thermal properties in two and three dimensions [J]. Physical Review E, 68, 036706, 2003.

Li, LK, RW Mei, JF Klausner. Boundary conditions for thermal lattice Boltzmann equation method. Journal of Computational Physics, 237(15), 366, 2013.

Li, Q, YL He, Y Wang, WQ Tao. Coupled double-distribution-function lattice Boltzmann method for the compressible Navier-Stokes equations [J]. Physical Review E, 76, 056705, 2007.

Liu, CF, JD Yang, YS Ni. A multiplicative decomposition of Poiseuille number on rarefaction and roughness by lattice Boltzmann simulation [J]. Computers and Mathematics with Applications, 61(12) 3528, 2011.

Liu, CF, YS Ni, The fractal roughness effect of micro Poiseuille flows using the lattice Boltzmann method [J], International Journal of Engineering Science, 47(5), 660, 2009.

Liu, CF, YS Ni, Y Rao. Roughness effect of different geometries on micro gas flows by Lattice Boltzmann simulation [J], International Journal of Modern Physics C, 20(6), 953, 2009.

Liu, CF, YS Ni. The effect of surface roughness on rarefied gas flows by lattice Boltzmann method [J], Chinese Physics B, 17(12), 4554, 2008.

Lofthouse, AJ. Hypersonic Aerothermodynamics using the Direct Simulation Monte Carlo and Navier-Stokes models, University of Michigan, Dissertation of Doctor, 2008.

Moss, JN, GA Bird. Direct Simulation Monte Carlo simulations of hypersonic flows with shock interactions [J]. AIAA Journal, 43(12), 2565, 2005.

Moss, JN, GA Bird. Direct simulation of transitional flow for hypersonic re-entry conditions [J]. AIAA Paper, 0223, 1984.

Moss, JN, RG Wilmoth, JM Price. DSMC simulations of blunt body flows for Mars entries: Mars pathfinder and Mars microprobe capsules [R]. AIAA Paper, 2508, 1997.

Orszag, SA, I Staroselsky. CFD: Progress and problems [J], Computer Physics Communications, 127(1), 165, 2000.

Qian, YH, D d'Humieres, P Lallemand. Lattice BGK models for Navier-Stokes equation [J]. Europhysics Letter, 17(6), 479, 1992.

Santos, WFN, MJ Lewis. The effect of compressibility on rarefied hypersonic flow over power law leading edges[R]. AIAA Paper, 1181, 2004.

So, RMC, SC Fu, RCK Leung. Finite difference lattice Boltzmann method for compressible thermal fluids [J]. AIAA Journal, 48, 1059, 2010.

Sun, CH. Thermal lattice Boltzmann model for compressible fluid [J]. Chinese Physics Letters, 17(3), 209, 2000.

Sun, QH, ID Boyd. Numerical simulation of gas flow over microscale airfoils [J]. Journal of thermophysics and Heat transfer, 16(2), 171, 2002.

Zou, Q, S Hou, S Chen, GD Doolen. An improved incompressible lattice Boltzmann model for time-independent flows [J]. Journal of Statistical Physics, 81(1), 35, 1995.

Advances in Energy Science and Equipment Engineering II – Zhou, Patty & Chen (Eds)
© *2017 Taylor & Francis Group, London, ISBN 978-1-138-71798-5*

Mixed marriage in honey bees optimization algorithm and its convergence analysis

Chenguang Yang
Chinese Electronic Equipment System Corporation Institute, Beijing, P.R. China

Qiaoge Liu
Agricultural Bank of China, Beijing, P.R. China

ABSTRACT: Marriage in Honey Bees Optimization (MBO) is a new swarm intelligence method; however, it is limited by low speed and complex computation process. By changing the structure of MBO and utilizing a local search method to perform the local characteristic, we propose a new optimization algorithm. The global convergence characteristic of the proposed algorithm is proved by using the Markov Chain theory. And then some simulations are done on Traveling Salesman Problem (TSP) and several public evaluation functions. The simulation results obtained by comparing the proposed algorithm with MBO and genetic algorithm show that the proposed algorithm has better convergence.

1 INTRODUCTION

Marriage in Honey Bees Optimization (MBO) is a swarm intelligence method proposed by Jason Teo and Hussein A. Abbass (Abbass, 2001) and updated by Jason Teo, Hussein A. Abbass (Teo, 2001), and Omid Bozorg Haddad et al. (Omid, 2006; Hyeong, 2006).

The behavior of honey bees shows many features like cooperation and communication, and hence they have aroused great interests in modeling intelligent behavior these years. In Marriage in Honey Bees Optimization (MBO), mating behavior of honey bees is considered as a typical swarm-based optimization approach. The behavior of honey bees is related to the product of their genetic potentiality, ecological and physiological environments, the social conditions of the colony, and various prior and ongoing interactions among these three (Abbass, 2001).

The five main processes of MBO are: (a) the mating-flight of the queen bees with drones encounters some probability, (b) creating new broods by the queen bees, (c) improving the broods' fitness by workers, (d) updating the workers' fitness, and (e) replacing the least-fit queen(s) with the fittest brood(s).

The objective of this paper is to improve the performance of MBO. By combining the interior-reflective Newton method, an improved algorithm is proposed and its convergence is analyzed on the basis of the theory of Markov Chain.

2 MIXED MARRIAGE IN HONEY BEES OPTIMIZATION (MMBO)

2.1 *Interior-reflective Newton method*

The interior-reflective Newton method is a commonly used nonlinear optimization algorithm proposed by Coleman, T.F. (1996), and Y. Li (1994). It is a trust region approach for minimizing a nonlinear function subject to simple bounds. It defines a solution to a trust region subproblem by minimizing a quadratic function subject only to an ellipsoidal constraint. The iterations generated by these methods are always strictly feasible. This method reduces to a standard trust region approach for the unconstrained problem when there are no upper or lower bounds on the variables.

The framework of the method subject to bounds is shown in Coleman (1996).

Δ_k is the trust region size.

$$D(x) \overset{def}{=} diag(|v(x)|^{\frac{1}{2}})$$

$$H(x) \overset{def}{=} \nabla^2 f(x)$$

$$g(x) \overset{def}{=} \nabla f(x)$$

$$\psi_k(s) \overset{def}{=} s^T g_k + \frac{1}{2} s^T M_k s$$

$$C(x) \overset{def}{=} D(x) diag(g(x)) J^v(x) D(x)$$

$$M(x) \overset{def}{=} B(x) + C(x)$$

$$x_0 \in \text{int}(F)$$

For $k = 0,1,\ldots\ldots$
1. Compute $f(x_k), g_k, H_k$, and C_k; Define the quadratic model:

$$\psi_k(s) = g_k^T s + \frac{1}{2} s^T (H_k + C_k)s.$$

2. Compute s_k using

$$\min_s \left\{ \psi_k(s) : \|D_k s\|_2 \le \Delta_k \right\},$$

such that $x_k + s_k \in \text{int}(F)$.
3. Compute
$$\rho_k^f = \frac{f(x_k + s_k) - f(x_k) + \frac{1}{2}s_k^T C(x_k)s_k}{\psi_k(s_k)}$$

4. If $\rho_k^f > \mu$, then set $x_{k+1} = x_k + s_k$.
Otherwise, set $x_{k+1} = x_k$.
5. Update the model ψ_k, scaling matrix D_k, and Δ_k as specified.

2.2 MMBO

Therefore, we utilize the local search ability and replace the worker of MBO algorithm by the interior-reflective Newton method.

Some studies related to MBO have been carried out in our research. One of them is to increase the convergence speed. Here we provide some information about it, as the main work in this paper will be based on such improved MBO algorithm.

In MBO algorithm, the probability of a drone mates with a queen is defined by the annealing function (Abbass, 2001). Not only the calculation of probability is complex, but also its calculation participants are complicated. Therefore, the whole process has a large computation burden.

On the contrary, we have found that MBO with low speed needs enough iteration times to approach optimization result. However, several variables in MBO, such as energy and speed, cannot ensure this. As the process proceeds, the mating probability becomes lower, which neither help the calculation process put up, nor help converge globally. Therefore, on the basis of the original MBO algorithm, we have done some improvement on the original MBO algorithm. That is, by randomly initializing drones and restricting the condition of iteration, the computation process will become easier. Details about this improvement have been discussed elsewhere.

Here we further our research to improve the performance of MBO and propose an algorithm of Mixed Marriage in Honey Bees Optimization

(MMBO) by taking the interior-reflective Newton method as the worker.

Details of MMBO are shown in Figure 1.

In MMBO, we define three operators: crossover, mutation, and heuristic. Crossover and mutate are same as that in GA. However, the heuristic operator is a new one proposed in MMBO.

Crossover and mutation are the same operators as in the GA. Heuristic operator improves a set of broods. It help conduct local search on broods. For the good local convergence performance, we use interior-reflective Newton method as the heuristic operator.

3 CONVERGENCE ANALYSIS OF MMBO ALGORITHM

3.1 Markov chain

The probabilities of a Markov chain are usually entered into a transition matrix, indicating which state or symbol follows which other state or symbol.

Definition 1 (Rudolph, 1994): A square matrix is $A = \left[a_{ij} \right]_{n \times n}$

(a) if $\forall i, j \in \{1,\ldots n\} : a_{ij} > 0$, A is positive ($A > 0$);
(b) if $\forall i, j \in \{1,\ldots n\} : a_{ij} \ge 0$, A is nonnegative ($A \ge 0$);
(c) if $A \ge 0$ and $\exists m \in N : A^m > 0$, A is primitive;
(d) if $A \ge 0$ and $\forall i \in \{1,\ldots n\} : \sum_{j=1}^{n} a_{ij} = 1$, A is stochastic.

(1)

```
Define Q: the number of queens
       D: the number of drones
       M: the sperm theca size
Initialize each worker with a unique heuristic
Initialize each queen's genotype at random
Apply interior-reflective Newton method to improve the
queen's genotype
While the stopping criteria is not satisfied (cycle times bigger
than max cycle number or result is good enough)
   for queen = 1 to Q
     for i=1 to M
       generate a drone randomly
       add spermatozoa to the queen's sperm theca
       generate a brood by crossovering the queen's genome with
the selected sperm,
       mutate the generated brood's genotype
       use interior-reflective Newton method to improve the
drone's genotype
         if the new brood is better than the worst queen
           replace the least-fit queen with the new brood
           refresh the queen list depending on fitness
         end if
     end for
   end for
end while
```

Figure 1. Mixed Marriage in Honey Bees Optimization (MMBO) algorithm.

Definition 2 (Rudolph, 1994): If the state space S is finite ($|S| = n$) and the transition probability $p_{ij}(t)$ is independent of t:

$$\exists i,j \in S, \exists u,v \in \mathbf{N}, p_{ij}(u) = p_{ij}(v) \quad (2)$$

the Markov chain is said to be finite and homogeneous. $p_{ij}(t)$ is the probability of transitioning from state $i \in S$ to state $j \in S$ at step t.

Theorem 1 (Rudolph, 1994): For a homogeneous finite Markov chain, with the transition matrix $P = (p_{ij})$, if:

$$\exists m \in \mathbf{N}: P^m > 0 \quad (3)$$

then this Markov chain is ergodic and with finite distribution. $\lim_{t \to \infty} p_{ij}(t) = \bar{p}_j, i,j \in S$ is the steady distribution of the homogeneous finite Markov chain.

Theorem 2 (Rudolph, 1994) (The basic limit theorem of Markov chain): If P is a primitive homogeneous Markov chain's transition matrix, then

(a) $\exists! \omega^T > 0: \omega^T P = \omega^T$, ω. a probability vector.
(b) $\forall \varphi_i \in S$ (φ_i is the start state and it's probability vector is g_i^T): $\lim_{k \to \infty} g_i^k P^k = \omega^T$
(c) From $\lim_{k \to \infty} P^k = \bar{P}$, we can get a limit probability matrix $\bar{P}$, it is a $n \times n$ matrix and it's all rows are same to ω^T.

$$(4)$$

Theorem 3 (Rudolph, 1994): Let P be a reducible stochastic matrix, where $C: m \times m$ is a primitive stochastic matrix and $R \neq 0, T \neq 0$. Then:

$$P^\infty = \lim_{k \to \infty} P^k = \lim_{k \to \infty} \begin{pmatrix} C^k & 0 \\ \sum_{i=0}^{k-1} T^i R C^{k-i} & T^k \end{pmatrix} = \begin{pmatrix} C^\infty & 0 \\ R^\infty & 0 \end{pmatrix} \quad (5)$$

is a stable stochastic matrix.

3.2 Convergence analysis

There are only three ways to change from one generation to another: crossover, mutate, and heuristic. These operators depend only on the inputs and not restricted to time. Then, we can get the following theorems.

Definition 3: The state space of MMBO is:

$$X = \left\{ x = [t_1, t_2, \ldots, t_N] \middle| t_i \in \{0,1\}, i = 1, \ldots, N \right\} \quad (6)$$

where $[t_1, t_2, \ldots, t_N]$ is the binary bit cluster listed in turn.

Define $f(x)$ as the fitness function based on x, and y is the fitness. Therefore, the fitness aggregate Y is:

$$Y = \left\{ y \middle| y = f(x), x \in X \right\} \quad (7)$$

It is easy to see:

$$\forall x \in X, y > 0 \quad (8)$$

Defining $g = |Y|$, we can get an ordered aggregate:

$$\{y_1, y_2, \ldots, y_g\}, y_1 > y_2 > \ldots > y_g \quad (9)$$

Crossover, mutate, and heuristic operators lead to probable transition in the state space. And we use three transition matrices: C, M, and H to describe their infections respectively. Finally, we can get:

$$Tr = C \cdot M \cdot H \quad (10)$$

where Tr is the transition matrix of the Markov chain of the MMBO algorithm.

Theorem 4: The Markov Chain of MMBO is finite and homogeneous.

Proof:

The aggregate $\{x_1, x_2, \ldots, x_M\}$ is finite. So the Markov chain composed of $\{x_1, x_2, \ldots, x_M\}$ is finite. This finite space can also be said as a state space X.

With $\rho_i, \rho_j \in X$, the probability of transformation from the state ρ_i to the state ρ_j at step t only depends on ρ_i and is independent of time. Therefore, the Markov chain of the MMBO algorithm is homogeneous.

End.

Theorem 5: The transition matrices of the crossover probability (C) and heuristic probability (H) in the MMBO algorithm are all stochastic.

Proof:

The square matrix is $C = [c_{ij}]_{n \times n}$. Then:

$$\forall i,j \in \{1, \ldots n\}: c_{ij} \geq 0 \ and \ \forall i \in \{1, \ldots n\}: \sum_{j=1}^{n} c_{ij} = 1 \quad (11)$$

Therefore, C is stochastic.

The square matrix H is $H = [h_{ij}]_{n \times n}$. Then:

$$\forall i,j \in \{1, \ldots n\}: h_{ij} \geq 0 \ and \ \forall i \in \{1, \ldots n\}: \sum_{j=1}^{n} h_{ij} = 1 \quad (12)$$

Therefore, H is stochastic.

End

Theorem 6: The transition matrix of the MMBO with mutation probability (M) is stochastic and positive.

Proof:

$M = [m_{ij}]_{n \times n}$ is a square matrix. Then:

$$\forall i,j \in \{1,...n\} : m_{ij} \geq 0 \text{ and } \forall i \in \{1,...n\} : \sum_{j=1}^{n} m_{ij} = 1 \quad (13)$$

So, M is stochastic.

And the mutation has an influence on every position of a state vector. We can easily know $\forall x_i, x_j \in X$. Each position of x_i can mutate to the value of x_j. Therefore, the probability of x_i mutate to x_j is positive. So, M is positive.

End

Theorem 7: The Markov chain of the MMBO (Tr) is ergodic and with finite distribution: $\lim_{t \to \infty} tr_{ij}(t) = \overline{tr_j} > 0, i,j \in X$

Proof:

According to Theorem 5, Theorem 6, and (10), Tr is positive. And according to Theorem 1, this proposition is proved.

End

Definition 4: The fitness of one generation is the largest one of the individuals in this generation:

$$f(\{x_1, x_2 ... x_K\}) = \max_{i=1,2,...,K} \{f(x_i)\} \quad (14)$$

Define $X_i = \{x_1, x_2,...,x_K \mid f(\{x_1, x_2,...,x_K\}) = y_i, x_1, x_2,...,x_K \in X\}$, y_i is defined in (9), that is, the fitness of all the individuals in X_i is equal to y_i.

Definition 5: For an arbitrary initial generation $X(0)$, y_1 has the largest fitness:

$$\lim_{t \to \infty} \Pr(f(X(t)) = y_1) = 1 \quad (15)$$

Then, the algorithm is a global convergence.

Theorem 8: The MMBO converges to the global optimum.

Proof:

We can define

$$TX = \{X_i \mid i \in N\} \quad (16)$$

For Definition 4 and Theorem 4, TX is a Markov chain. In the same time, we define:

$$P(X_i) = P\{iX \in X_i\} \quad (17)$$

iX is defined in (12).

We can see that $P(X_i) > 0$ and $\sum_{i=1}^{n} P(X_i) = 1$.

Defining $P(X_i, X_j)$ as the probability state X_i, we can get

$$P(X_i, X_j) = \sum_{ni=1}^{N_i} \sum_{nj=1}^{N_j} P(x_{ni}, x_{nj}), x_{ni} \in X_i, x_{nj} \in X_j \quad (18)$$

Because MMBO saves the best individual at every generation, $P(X_i, X_j) = 0, i < j$. And the transition matrix of TX's Markov chain can be written as follows:

$$P = \begin{bmatrix} P(X_1, X_1) & \cdots & P(X_1, X_n) \\ \vdots & \cdots & \vdots \\ P(X_n, X_1) & \cdots & P(X_n, X_n) \end{bmatrix}$$

$$= \begin{bmatrix} 1 & 0 & \cdots & 0 \\ P(X_2, X_1) & P(X_2, X_2) & \ddots & \vdots \\ \vdots & \vdots & \ddots & 0 \\ P(X_n, X_1) & \cdots & \cdots & P(X_n, X_n) \end{bmatrix} \quad (19)$$

For Theorem 3,

$$C = 1, T = \begin{bmatrix} P(X_2, X_2) & \cdots & 0 \\ \vdots & \ddots & \vdots \\ P(X_n, X_2) & \cdots & P(X_n, X_n) \end{bmatrix},$$

$$R = \begin{bmatrix} P(X_2, X_1) \\ \vdots \\ P(X_n, X_1) \end{bmatrix} \quad (20)$$

$$P^{\infty} = \lim_{k \to \infty} P^k = \lim_{k \to \infty} \begin{pmatrix} C^k & 0 \\ \sum_{i=0}^{k-1} T^i R C^{k-i} & T^k \end{pmatrix} = \begin{pmatrix} C^{\infty} & 0 \\ R^{\infty} & 0 \end{pmatrix} \quad (21)$$

For Theorems 7 and 1, P^{∞} is a stable random matrix, so $R^{\infty} = 1$. That is:

$$R^{\infty} = \lim_{k \to \infty} R^k = \begin{bmatrix} \lim_{k \to \infty} (P(X_2, X_1))^k \\ \vdots \\ \lim_{k \to \infty} (P(X_n, X_1))^k \end{bmatrix} = \begin{bmatrix} 1 \\ \vdots \\ 1 \end{bmatrix} \quad (22)$$

So every state in TX will go to X_1, if the iteration number is big enough, this proposition is proved.

End

4 SIMULATION

To test the convergence performance of MMBO, we choose original MBO algorithm and genetic algorithm for comparison. We did the simulation on two parts: one using some popular complex evaluation functions and the other using Traveling Salesman Problem (TSP).

4.1 Comparison on evaluation functions

The initial value is generated randomly, and each figure shows the average results of 20 times simulation with one evaluation function.

- (Figure 2) Evaluation Function 1: Schwefel's Problem 1:

$$f(s) = \sum_{i=1}^{30} |x_i| + \prod_{i=1}^{30} |x_i|, \quad |x_i| \leq 10 \quad (23)$$

- (Figure 3) Evaluation Function 2: Schwefel's Problem 2:

$$f(x) = \sum_{i=1}^{i} (\sum_{j=1}^{i} x_j)^2, \quad |x_i| \leq 100 \quad (24)$$

- (Figure 4) Evaluation Function 3: Generalized Rosenbrock's Function:

$$f(x) = \sum_{i=1} [100(x_{i+1} - x_i^2)^2 + (1 - x_i)^2], \quad |x_i| \leq 30 \quad (25)$$

- (Figure 5) Evaluation Function 4: Generalized Rastrigin's Function:

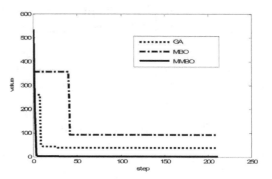

Figure 4. Results of MMBO, MBO, and GA with Evaluation Function 3.

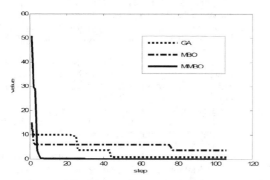

Figure 2. Results of MMBO, MBO, and GA with Evaluation Function 1.

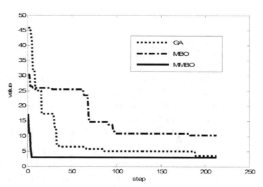

Figure 5. Results of MMBO, MBO, and GA with Evaluation Function 4.

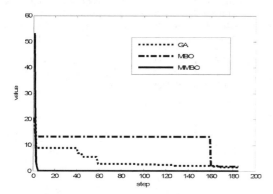

Figure 3. Results of MMBO, MBO, and GA with Evaluation Function 2.

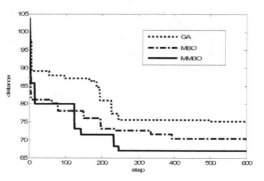

Figure 6. TSP with 16 nodes solved by MMBO, MBO, and GA.

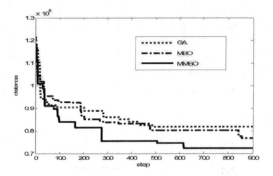

Figure 7. TSP with 48 nodes solved by MMBO, MBO, and GA.

$$f(x) = \sum_{i=1} \left[x_i^2 - 10 \cdot \cos(2\pi x_i) + 10 \right], \quad |x_i| \le 5.12$$

(26)

4.2 Traveling Salesman Problem

Here, TSP based on the data form TSPLIB is solved by MMBO, MBO, and GA, as shown in Figures 6–7.

5 CONCLUSIONS

From the above, we can see that the MMBO shows better performance than MBO and GA, not only to solve TSP but also to optimize evaluation functions, and can keep converge faster with different node's number. The simulation results show that:

- MMBO is convergent and keeps good performance for all these test functions, although these test functions are more complex than the normal ones and may have many local optimization points.
- MMBO performs better than MBO and GA. MMBO converges more quickly, especially at the initial part. Particularly, even if the initial condition is worse than MBO and GA, MMBO can show finer result.
- As for MBO, because of the process of choosing drones with some probability, MBO's performance is not always high. It often keeps staying at a value for some time and then drops dramatically at some step. Sometimes MBO is better than GA, but sometimes not.

In this paper, we proposed an algorithm of Mixed Marriage in Honey Bees Optimization (MMBO) to overcome the slowness of the original MBO. MBO has a set of parameters to coordinate and much of calculation time is cost. MMBO avoids such complex process and also can reach the desired result. It generates a drone part randomly each time and mates with a finite quantity of queens. Therefore, MMBO can not only avoid the local optimum, but also increase the speed. And the global convergence is preserved for optimization. Simulating with complex evaluation functions and TSP, MMBO shows better performance than MBO and GA.

REFERENCES

Abbass, H.A., "A Single Queen Single Worker Honey-Bees Approach to 3-SAT", Proceedings of the Genetic and Evolutionary Computation Conference, GECCO2001, San Francisco, USA, 2001, pp. 807–814.

Abbass, H.A., "Marriage in Honey Bees Optimization (MBO): A Haplometrosis Polygynous Swarming Approach", Congress on Evolutionary Computation, CEC2001, Seoul, Korea, 2001, pp. 207–214.

Coleman, T.F., Y. Li, "An Interior, Trust Region Approach for Nonlinear Minimization Subject to Bounds", SIAM Journal on Optimization, 6, 1996, pp. 418–445.

Coleman, T.F., Y. Li, "On the Convergence of Reflective Newton Methods for Large-Scale Nonlinear Minimization Subject to Bounds", Mathematical Programming, 67(2), 1994, pp. 189–224.

Hyeong Soo Chang, "Converging Marriage in Honey-Bees Optimization and Application to Stochastic Dynamic Programming", Journal of Global Optimization, 35, 2006, pp. 423–441.

Jason Teo, Hussein A. Abbass, "An Annealing Approach to the Mating-Flight Trajectories in the Marriage in Honey Bees Optimization Algorithm", Technical Report CS04/01, School of Computer Science, University of New South Wales at ADFA, 2001.

Omid Bozorg Haddad, Abbas Afshar and Miguel A. Marino, "Honey-Bees Mating Optimization (HBMO) Algorithm: A New Heuristic Approach for Water Resources Optimization", Water Resources Management, 20, 2006, pp. 661–680.

Rudolph G., "Convergence Analysis of Canonical Genetic Algorithms", IEEE Transaction Neural Networks. 5(1), 1994, pp. 96–101.

Advances in Energy Science and Equipment Engineering II – Zhou, Patty & Chen (Eds)
© 2017 Taylor & Francis Group, London, ISBN 978-1-138-71798-5

Complexity analysis of logistics service supply chain network structure

Guangsheng Zhang
Logistics Department, Pinghu Campus of Jiaxing University, Pinghu, China

ABSTRACT: In this paper, we review the existing literature on LSSC structure, and describe the general state of LSSC structure based on the basic mechanism of logistics service supply chain. We introduce statistical description of complex network theory and analyze the whole macro behavior and inherent law of supply chain system from the perspective of complexity. Finally, complex networks theory was applied to LSSC structural study. LSSC network complex topological structure and local evolution growth mechanism were described. This provides a good foundation for analyzing evolution mechanism of LSSC network, exploring coordination mechanism between network nodes and optimizing supply chain network. The research of LSSC network structure based on complex network theory is helpful to study the inner mechanism from a new perspective.

1 INTRODUCTION

In recent years, the service industry has gradually grown into a new core of national economy and become the most vital industry of them all. Traditional vertical integration product supply chain of logistics industry as an important service industry has been unable to adapt to pressures, such as sharp changes of market demand, shortened delivery time, quality improvement, cost reduction, and service improvement. Logistics service promotion is the most important measure to reduce manufacturing cost and improve comprehensive competitive power of enterprises. Especially in the case of small scale and weak service of logistics enterprises in our country, LSSC research will be beneficial to solve problems of logistics service industry development. Logistics Service Supply Chain (LSSC) is a traditional subsidiary supply chain, enterprise logistics network as basic unit, and value creation mechanism and law of LSSC in operation process are studied from the perspective of value stream. LSSC relies on integrated logistics services, can optimize service processes, integrate logistics resources, balance service capabilities, and has become the trend of modern supply chain management. The existing literature of supply chain mainly focuses on industry, manufacturing, and supply chain, with supply chain services attracting the least attention. Scholars mainly focus on connotation, service outsourcing, supply chain coordination, and cooperation. The number of studies conducted on the LSSC structure mode is less. Goran Persson (2001) classified logistics service providers on the basis of different attributes, listed examples of various logistics provider, and

illustrated the relationship between mutual cooperation and supply of providers. Ellram et al. (2004) compared adaptability of GSCF, SCOR, and Hewlett-Packard products supply chain model and constructed LSSC management framework on the basis of operation mode of logistics service supply chain. Choy K L et al. (2007) believed that logistics service supply chain was a structure model of "functional logistics provider, integrated logistics supplier, customer," and put forward solution of LMIS. Yu Tian (2003) explored LSSC process code and believed that LSSC was "supplier of integrated logistics service providers and manufacturing and retail enterprises" oriented. Xiuxia Yan (2005) proposed that LSSC operation is based on supply chain model of logistics service integrator operation and integrated logistics service providers. Weihua Liu (2007) showed network structure of multifunctional logistics services. Literature review showed that compared with studies on logistics product supply chain, current studies on LSSC are still growing, lacking depth, and not considering characteristics of logistics services, and they have just been introduced into mature product supply chain. Most studies of LSSC structure has analyzed microscopic behavior problems of service supply chain and failed to explain macroscopic behavior and inherent law. LSSC is a typical complex dynamic supply and demand network, composed of functional logistics providers and influenced by internal and external impact. The address on LSSC as a simple two-level or multilevel structure is unreasonable. As for these disadvantages, in order to expand research breadth and depth of LSSC network structure, complex network theory is introduced into LSSC structure research. It can reveal important properties and inherent law of LSSC and have important reference value for operation management and scientific decision making of LSSC. In addition, complex network theory is the research front of systematic science. Because the complexity of supply and demand network service is higher than general network system and literature is little on complexity of LSSC structure, the study has not only academic meanings but also practical value on complexity of LSSC structure, decision-making relations, and research direction.

2 LSSC NETWORK STRUCTURE

Existing LSSC literature mainly studies two structure frameworks, composed of functional service provider and service integrators. Logistics service integrator is limited to the function of single function or regional limited functional logistics service provider to become a member of the service network, providing professional logistics services for the logistics service supply chain. The basic structure is shown in Figure 1.

Figure 1 shows the basic structure of LSSC. In practice, on the one hand, functional logistics service providers provide services not only for some integrators but also for multiple logistics integrators. A large number of LSSC cross structures form a network, thereby increasing the difficulty of coordination management. On the other hand, logistics service integrator has its advantages to meet the needs of logistics services in many business networks. At the same time, individual growth, market competition, and other factors of network structure affect the constant entry and departure of nodes. Its structure is evolving, the complex logistics service supply, and customer demand network make the LSSC present a complex multilevel network structure. Therefore, multilevel complex networks have the general LSSC

structure. LSSC network structure is based on the demand for logistics services as the network pull power, in accordance with logistics service direction of arrangement of various service chain nodes enterprises show the network structure relationship between various levels of the provider and the demand side.

3 COMPLEX NETWORK THEORY

Complex network node is the individual of complex system, and the edge between nodes is a kind of relationship that is naturally formed or artificially constructed according to some rules, mainly presenting a complex system. Statistical indicators of complex networks are as follows: Average distance (L) refers to average value of all nodes in the network, which reflects the degree of separation between nodes in the network. Efficiency (E) refers to average value of reciprocal between two nodes, which indicates the degree of ease. Degree is defined as the number of adjacent edges of the nodes, which is the most important description of the statistical properties of the nodes, directed graph $G \leq V, E \geq$ degree of v is the starting point of the number of sides, denoted by $d(v)^+$, v is the degree for the end to have a number of sides, the degree of penetration is denoted by $d(v)^-$, and the degree is $d(v)$. Degree distribution P(k) is the degree of any node, just as the probability of K. The nonscaling property is a complex network with power law distribution, which is an intrinsic property of a large number of complex systems. Cluster coefficient describes the ratio of network nodes, and network clustering coefficient is the average value of clustering coefficient nodes. Small-world experiment on the average distance between any two points of network (L) showed a logarithmic growth with the increase in the number of network nodes (N), L~lnN. And the local structure of network still has more obvious characteristics. LSSC network is a

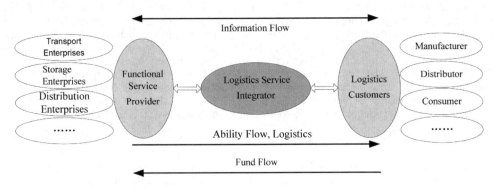

Figure 1.   Basic structure of supply chain logistics services.

complex network structure formed by the complex network structure, and enterprises constantly enter and leave with continuous evolution of structure. We can build realistic network model using growth and evolution model of complex network. Because of the regionalism of LSSC network, we select the local-world evolution model to establish growth evolution model of LSSC network. A specific model can be described as: (1) the initial network has M_0 nodes and E_0 edges; (2) m nodes are selected from the existing network nodes, as that of local-world model; (3) adding a new node and m sides and choosing M nodes of local world according to preferential attachment probability. The corresponding formula is as follows:

$$\Pi_{local}(k_i) = \frac{\Pi'(i \in Local - World * k_i)}{\sum_{j} {}_{local} k_j}$$
$$= \frac{M}{m_0 + t} * \frac{k_i}{\sum_{j} {}_{local} k_j} (M \geq m)$$

When $M = m$, a new node is connected to all nodes of the local world, and degree distribution follows the exponential function. When $M = m_0 + t$, each node of the local world is entire network, which grows with time. When $m < M < m_0 + t$, degree distribution follows the distribution between exponential distribution and power law distribution.

4 COMPLEXITY ANALYSIS OF LSSC NETWORK

4.1 Feature analysis of LSSC complex network

Because of connections of more applications of information technology and Internet media, links between nodes are more close and frequent, and LSSC networks exhibit complex topological features. From the perspective of complex network topology, LSSC network system can be represented by the graph, participation enterprises are network "apex," according to the location of the node enterprises in the supply chain to be classified, such as suppliers, integrators, and customers. The link between vertex enterprises can be expressed as "edge." The adaptive capacity of vertex enterprises is represented as node degree, which is mainly shown as comprehensive competitiveness of enterprises. Number of fixed time period is defined as the weight of the corresponding enterprise node. Ultimate functional logistics service providers of LSSC network sell service products through logistics service integrators and direct selling. A certain business contact exits. On the basis of the two-level LSSC network, flow process of

multilayer complex LSSC network is analyzed. Network enterprises are composed of node sets and subnodes $V = \{VC, VI, VF\}$, and each subnode set includes a number of nodes. VC is the logistics service demander, $VC = \{c_1, c_2,...,c_l\}$, c_1, $c_2,..., c_L$ represents market client node set. VI is a logistics integrator node, and $VI = \{i_1, i_2,...,i_g\}$, i_1, $i_2,..., ig$ represents logistics integrator node set. VF is a logistics service provider node, $F = \{f_1, f_2,...,f_h\}$, f_1, $f_2,..., f_H$ represents functional logistics service provider node set, $F1 = \{f_{11}, f_{12},...,f_{1n}\}$, $f_{11}, f_{12},...,f_{1n}$ represents logistics function to provide enterprise node set, e_{i-j-k} represents competition and cooperation relationship $E(G)$, i represents the first-tier supplier, j represents the second-tier logistics integrators, and k represents logistics demand side. For example, $e_{f11-0-c1}$ represents that transport function supplier F11 provides service products for logistics demand side. Node set V and edge set $E(G)$ constitute LSSC complex network structure $G(V, E)$, as shown in Figure 2.

Based on LSSC characteristics and basic concepts of complex networks, statistical descriptions of LSSC network are mainly as follows:

1. e_{i-j-k} of LSSC network edge set $E(G)$ is the relationship between enterprise nodes, and starting point and end point forms directed network. Figure 2 is the LSSC-directed network graph. The ordered pair $<f_{11}, c_1>$ represents directed connection between the first transport logistics provider and the first customer node. Node degree $d(v)$ is the sum of $d(v)^+$ and $d(v)^-$, proportional relation of $d(v)^+$ and $d(v)^-$ produces preferred connection principle. When a new node appears in the network, the node edge will be connected with the existing m network nodes. If the new node is starting point, probability W of $d(v)^-$ of node vi is the basis for the selection of endpoint vi, $W(d(v_i)^-) = d(v_i)^- / \sum d(v_j)^-$. If the new node is terminal point, probability W of $d(v)^-$ of node vi is the basis for the selection of initial point vi, $W(d(v_i)^+) = d(v_i)^+ / \sum d(v_j)^+$. Superior node enterprises of network have large in-degree and out-degree. These advantages make the enterprise more likely to become a new partner to the entire enterprise.

2. LSSC networks can show characteristics of small average path length L and high clustering coefficient and reflect the small-world aggregation characteristics. Average distance length (L) indicates service time length. With enhancement of customer requirement of service efficiency and logistics capability of node enterprises, how to efficiently complete logistics demand has become the key strategic issue that node enterprises must deal with. LSSC network clustering coefficient is defined as the average of

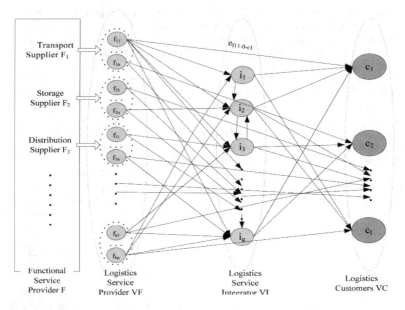

Figure 2. LSSC complex network diagram.

all node clustering coefficients; average clustering coefficient reflects high network clustering coefficient with the increase of communication between node enterprises, and especially, more and more logistics enterprises using advanced LMIS prompted the node enterprises more closely linked.

Growth mechanism of LSSC network node is presented as characteristic of scale-free network. A small number of nodes in dynamic evolution system have a large number of links, and a large number of nodes have a small number of links, and node growth is subject to random distribution. The law distribution is not the same with different growth. Most of LSSC network node degrees are very low, only a small amount of central node degree is higher. One or more logistics integrators exist in LSSC network, namely the LSSC core enterprise. Logistics integration establishes a close contact with other levels of enterprise through LMIS, establishes logistics service supply system, and embodies nonscaling. In recent years, the fourth reflected LSSC network nonscaling has played the role of logistics integrator. It plays a pivotal role in LSSC contact exchange through optimal allocation of resources, integrates LSSC network systems, and provides efficient supply chain optimization solutions.

4.2 Evolution analysis of LSSC local area network

LSSC have typical characteristics of complex network and can build a complex network evolution

model. Because of the strong regional characteristic of LSSC network, local-world evolution model was selected and used to construct LSSC network growth model. Behavior force of network connections and weights of evolutionary model is the choice of node degrees, reflecting the phenomenon of "the fittest rich," namely the new nodes connected to the higher node degrees. Because of market competition, survival environment uncertainty, and other factors, original enterprise exits network continuously. Because of transformation and bankruptcy, a new enterprise will constantly enrich the network system, whose structure is evolving. Therefore, LSSC network evolution process is a complex dynamic process. Complexity analysis of LSSC network structure is a new perspective.

Different from the traditional product supply chain network, the influence factors of the connection of LSSC network node enterprises are various, so it is not enough to only consider the connection between nodes. Connection strength and connection nature of LSSC network enterprise are also very important, and connection strength has a decisive significance. On the basis of local evolution model, the concept of network node correlation is added, local-world selection rules are improved, and evolution mode of local scale is established as the growth model of LSSC network.

The closer the position of the nodes, the greater the degree of correlation. Node a_i is assigned to the "position" parameter value i, using position parameters to measure degree of similarity between the node and other nodes in the network. Correlation degree between node i and node j is reciprocal of Euclidean distance between two points, that is, $h_{ij} = 1/|a_i - a_j|$ shows that the closer the relationship between two nodes, the greater the degree of correlation. Because each node enterprise of LSSC network has different roles and marketing force, such as functional supplier, logistics integrator, and customers, we can evaluate the correlation strength of enterprises cooperation. According to the correlation principle of node enterprises, the greater the correlation degree, the position is more important, the greater the probability of new connections. As a rule, the node degree of preference is connected to the local world, which has a large correlation with the new nodes, and the number of nodes in the local world is increased.

5 CONCLUSIONS

In this paper, the general structure of LSSC was described on the basis of complex characteristics of LSSC network structure and studies of service supply chain. Whole macro-behavior and inherent law of service supply chain system were analyzed from the perspective of complexity. Finally, complex networks theory was applied to LSSC structural study. LSSC network complex topological structure and local evolution growth mechanism were described. The conclusions provide a good foundation for analyzing the growth and evolution mechanism of LSSC network, exploring coordination mechanism between network nodes, and optimizing supply chain network.

REFERENCES

Choy, K.L., Chung-Lun Li, Stuart C K So, Henry Lau, et al, "Managing uncertainty in logistics service supply chain", International Journal of Risk Assessment and Manage- ment. Geneva, vol. 7, no. 1, (2007).

Goran Persson, Helge Virum, "Growth strategies for logistics service providers, A case study", International Journal of Logistics Management, vol. 12, no. 1, (2001).

Lisa M. Ellram, Wendy L.T., Corey B, "Understanding and Managing the Service Supply Chain", Journal of Supply Chain Management, (2004).

Liu Weihua, "Coordination of logistics service supply chain capabilities cooperation", Shanghai Jiaotong University, (2007).

Tian Yu, "Logistics Service Supply Chain Constructing Supplier Selection", Systems Engineering Theory and Practice, vol. 5, (2003).

Xiuxia Yan, Sun Lin Yan, Wang Kan Chang, "Logistics Service Supply Chain Performance Evaluation and Characteristics Study", China Mechanical Engineering, vol. 16, no. 11, (2005).

Advances in Energy Science and Equipment Engineering II – Zhou, Patty & Chen (Eds)
© *2017 Taylor & Francis Group, London, ISBN 978-1-138-71798-5*

Research on adaptive learning model based on distributed cognition

Qishen Zhu & Yinghua Zha
School of Computer Science and Software, Nanjing Vocational Institute of Industry Technology, Nanjing, China

ABSTRACT: Online learning has become a new way for students. On the basis of the theory of distributed cognition, the adaptive learning model is established for students' professional learning. This model makes the students more efficient by using online learning system, according to their own interests and professional foundation, and builds a better professional knowledge structure, thereby achieving personal learning. This paper takes computer science as an example to analyze and explore the online adaptive learning system model. A personalized and adaptive online learning system model DC-AHAM is proposed. The function and system structure of this model are discussed, and the application using the model for a course as an example is designed.

1 INTRODUCTION

Nowadays, the Internet has been used in many aspects of human life. Cell phones and PCs have played a special role in people's education, using which they can visit BBS, Weibo, WeChat, and many other learning websites, which are normal information resources of learning. More and more universities, enterprises, affiliations, and persons will be converted to online educational platforms. The proportion of traditional education is declining. Overall, the Internet has made an important change in student learning.

In this paper, a personal adaptive online learning mode is studied. Using this system, students can select their own directions of learning on the basis of their characteristics and interests. Then, the system can push related knowledge and some knowledge need to be further acquired by students on the basis of the feedback of study, which can guide them to construct their own specialty system. This system will improve learning interests and efficiency of students, and will help them to meet social requirements. An adaptive learning model based on distributed cognition is proposed, which can recommend specialty knowledge on the basis of learners' characteristics, hobbies, and learning results.

2 ONLINE LEARNING

The Internet supplies mass information including abundant learning resources, and hence, online learning is formed with multiple media interactions and multiple open platforms. Using the Internet technology, a new system of learning and teaching with student center has been a direction of online learning study. This system can increase learning chances, improve study efficiency, and then help students to construct knowledge system. However, there are differences from classroom centered learning; online learning, especially mobile learning, can support learning resources whenever and wherever possibly as well as learning chances. Therefore, students can make good use of fragmented time, which is an important complementary and extension.

By the online learning platform, especially mobile learning, students can manage contents of learning, processing of learning progress, and process by themselves. Special personal study environment is established and then the personal study is realized. In order to achieve the initial learning aims, students can communicate with teachers and themselves conveniently and learn with cooperation.

Features of online learning:

1. Diversity of learning resources: such as technical websites, technical special subjects, Wiki, Weibo, and education platforms of schools. Students can select the resources that suit them on the basis of their own hobbies and interests.
2. Autonomous learning: As the online platform supplies a relatively independent environment for personal learning, students should have self-constraints, independent thinking, and self-learning and testing, which are important to form the personal learning methods and habits.
3. Interactive learning: there are two important factors for the learning procedure, such as interacting and communication. It includes students and

learning platforms, students and students, as well as students and tutors. This may increase students learning interests during the procedure of learning, creating, and sharing, so that the potential talents and creative abilities are dug. It also has the functions of guiding students and constrains.

4. Real-time learning: the online learning platform, especially mobile learning, will make student learning whenever and wherever possible, as well as thinking and creating. If there are any questions during learning, they will be solved by communicating with others. Therefore, this can increase students using the fragment time to improve learning efficiency.

Online learning is an important way for digital learning, using which students can accomplish knowledge studying. Online learning has many advantages, and hence many companies and schools have begun to construct various online learning platforms. At the same time, online learning has become an important research topic.

3 DISTRIBUTED COGNITIVE THEORIES

In the 1980s, Hutchins et al. of University of California put forward the concept of "Distributed Cognition," and thought that it would be a new way of recognition (Hutchins E, 1991).

3.1 *Definition of distributed cognition*

Distributed cognition is a system comprising the cognitive subject and environment. And it is a new unit of analysis involving all things that participated in cognition. It is the information processing process of internal and external representation (Chuah J, 1999). Distributed cognition is regarded as a kind of system which contains many subjects and many kinds of tools and techniques that can coordinate the internal and external representation. Furthermore, it is helpful to provide a dynamic information processing system. Hutchins believes that the distributed cognitive analysis unit is not an individual, but a cognitive process based on the functional relationship between the elements of the cognitive processing (Hutchins E, 1991). Distributed cognition is a process of interaction between individual and cognitive elements. They are independent as well as interactive. It is a progressive process of mutual promotion and spiral (Salomon, 1993).

3.2 *Distributed cognitive learning*

Distributed cognition theory guides the students to learn, during which learning resources is in a dominant position. Resource information representation and propagation is conducive for students learning cognition, increases interests in learning, and makes learning and interaction with exchange information easy. Students must be actively involved in the resources of learning and develop the habit of self-study. Their professional ability and interest continue to be improved and optimized in interactive activities.

Traditional cognition emphasizes on individual cognition, but distributed cognition takes into account all the factors involved in cognitive activities. Distributed cognition uses the individual mutual interaction and the interaction between individual and distributed resources for learning activities (Zhang, 2010), with both sides being interdependent. At the same time, the interests of the students and the learning model can be returned to the resources platform. Thus, the knowledge of the resources is expanded and the effectiveness of the students is improved.

3.3 *Distributed cognition and online learning*

Online learning offers anytime learning opportunities for students. Distributed cognition theory provides theoretical support for the design and planning of the learning platform. Through the online learning platform, students can make the individual learning environment and realize personalized learning and collaboration learning between students and teachers or between students. The distributed cognition of online learning has the following main features:

1. The students are the main part of cognition: during the learning process, students should play dominant position, and the teaching resources should be designed around the subject. According to the step-by-step learning of students, the teaching resources can make them initiatively learn and improve their interests.
2. Interactive cognitive learning: each student has his/her own interest, hobby, and professional foundation, so communicating with each other between students and between student and teacher can solve the problem from different angles. And everyone can benefit from it.
3. Refine design for the learning resources: according to the characteristics of online learning, especially mobile learning, the learning resources should be short and fine, and have various expression. And the resources can be presented by video, audio, pictures, texts, and other forms, and released in multiple platforms. While the resources provide students' feedback, questions and test enable students to understand their own learning, so as to make them initiatively complete their own cognitive goals.

Distributed learning becomes a kind of new teaching model; it is a strong complement for traditional teaching methods. And students' learning can be independent of time and place. The learning resources are various, and teaching content can distribute in different noncentric positions. Online learning, especially mobile learning, is an important way of distributed learning.

4 ADAPTIVE LEARNING SYSTEMS

Adaptive learning is studied from the intelligent system. It explores how the learning system can better meet the different needs of the students. Thus, there are corresponding teaching strategies according to the various characteristics and behavioral tendencies, such as learning goals, preferences, and cognitive level. Personalized learning path and learning resources are recommended, and students' individualized learning environment should be formed (Tseng, 2008; Gao, 2012).

With the research of machine learning and recommendation algorithms, adaptive learning has been widely used. Through the rich resources of the Internet and computer intelligent algorithm, and according to the professional backgrounds of the students, interest, hobby, study progress, and the learning platform can recommend related professional knowledge, transfer learning content, and teach students according to their aptitude. Through online learning, students continuously deepen their knowledge and improve their professional architecture. As the learners, in the adaptive learning environment, students are no longer passive recipients of knowledge, but they can initiatively find knowledge. Therefore, the process is a spiral one.

5 DISTRIBUTED ADAPTIVE MODEL (DC-AHAM)

According to the adaptive learning system, many scholars put forward some better reference models. In 1996, Debra explicitly proposed reference model AHAM (Adaptive Hypermedia Architecture). The model includes four tuples, such as the domain model, adaptive model, student model, and adaptive engine (Paul, 1999). On the basis of this model, the distributed adaptive model DC-AHAM is improved and optimized from resource adaptive aspects and learning environment construction based on students' major interest, hobby, learning habits, and personalization.

5.1 DC-AHAM model structure

Distributed adaptive model DC-AHAM is shown in six-dimensional space: characteristics, process, analysis, resource, recommendation, and search.

The relationship between six-dimensional models is shown in Figure 1:

Characteristic is the student character model, which includes two parts: static and dynamic students' information. The static information includes age, gender, grade, and professional interest. The dynamic information includes behavioral information, such as browsing, attention, and preferences.

Process is learning model, including knowledge of curriculum, curriculum set, and knowledge points.

Analysis is the model of result of the test, including basic knowledge, knowledge application, and knowledge points of individual evaluation.

Resource includes online learning resources, including curriculum knowledge point, test set, knowledge application, and extended knowledge points set.

Recommendation is the recommended technologies. According to the students' interests and the knowledge learned, with the combination of student characteristics recommendation and collaborative filtering recommendation, it can push different learning materials and expand new knowledge for the students. At the same time it pushes related social needs about their learning, so students can adjust their professional development direction according to their own learning.

Search is a resource search technologies set, which can search the resources needed by the students and expand system resources.

In this model, learning characteristic similarity can be analyzed according to the behavior of students, learning test schedule, and test results, meanwhile the results are categorized and compared with learning resources. The shortest-path probability and the mean value of the length of the shortest path are obtained as the basis for the recommendation algorithm. And the recommendation

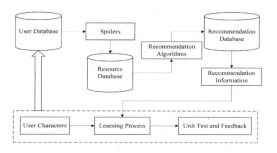

Figure 1.   Six-dimensional model.

information can be pushed to the students. The test results as the main basis to modify the characteristics of the students, and feed it back to characteristics model of the students and modify the student's interests. At the same time, the results can be as analytic basis of the recommended resources to realize personalized learning recommendation system.

The student characteristics model can also be a basis for resource search system. Through the automatic Web crawler or manual looking for corresponding resources, new knowledge can be increased for the resources of this system. The adaptive updated process of system resources is achieved. Through the analysis of feedbacks of student learning behaviors, the students' interesting resources are recommended, which can guide them in the professional learning for knowledge expansion and migration. This can help them to form a complete professional knowledge system.

The purpose of DC-AHAM model is to provide students with learning content, knowledge transfer, and knowledge expansion, which are the most consistent with their characteristics.

5.2 "Cold start" problem

The "cold start" problem needs to be solved in practical application of DC-AHAM model. When the students use at the first time or access to new learning process, there is no individual leading knowledge content of the student in the system, so that the adaptive resource cannot be effectively formed. The simple and efficient solution is that the basic knowledge during the learning process is the learning part. According to the results of students' learning, distributed cognition adaptive learning process is formed.

Online learning system, which is constructed by DC-AHAM model, uses student characteristics and collaborative filtering algorithm to finish student learning resource adaptive recommendation by increasing the feedback. When students' personalized learning environment is built, DC-AHAM uses students' learning feedback information to adjust the characteristics of students. According to the characteristics of students, learning resources, and search for related resources, this model is more suitable for online professional learning.

6 DC-AHAM MODEL APPLICATIONS

Adaptive process is a continuous approximation to the target. When students use the online platform system, their professional level is constantly improved. And using the system can change students' interests and preferences. When students'

characteristics are changed, DC-AHAM models refine the recommendation information and finish the system adaptability.

6.1 DC-AHAM structural framework

The framework of distributed adaptive model (DC-AHAM) is shown in Figure 2. The framework is composed of three parts: user interface, adaptive engine, and data storage.

User interface mainly includes user module, courses module, learning process, information push module, and test module. The user module is about students' characteristics, which records the basic information, professional interests, and hobbies. Students choose their own interests and preferences as the basis for recommended information through the online platform. Student behaviors include browsing behavior and preference. The courses module is mainly about the current curriculum knowledge; in this module, teachers set related knowledge points and fragment knowledge to facilitate students' online learning. Learning process records the students' learning progress and learning characteristics. Test module is about the test of the students' knowledge and save the test results. Information push module is mainly for adaptive engine to push information.

Logical layer is adaptive engine with three parts: user characters, collaborative filtering recommendation, and resources. User characters analysis mainly formed characteristic information of students based on student characteristics, behavior information, learning progress. It saves the characteristic information into the feature library as an important data source of recommendation algorithm. Resource classification is resources similarity matching according to the division of different specialty. Through characteristics recommendation and resource coordination filtering algorithm and according to different professional knowledge, social needs, and technical requirements, the recommended information is

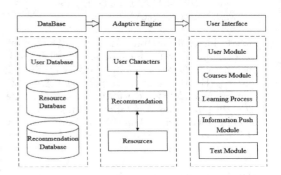

Figure 2. Distributed adaptive model (DC-AHAM) framework.

formed and is pushed to the display layer for helping students to learn and improve their learning interests and preferences. Meanwhile, through relevant knowledge, students can build their own knowledge structure and personal learning environment. The characteristics of the students, learning resources, behavior of students, and other information are persisted by data storage.

6.2 DC-AHAM curriculum design

DC-AHAM model is used for online learning system design, for example, "operating system" course. According to the DC-AHAM model, the professional skills knowledge is subdivided into small knowledge. The curriculum resources include various forms of text, voice, images, simple animation, Internet links, and other hypermedia. Through micro-classes, forums, application skills, problems solving, unit testing, interest expansion projects, and learning communities are established, so that the students can learn online, communicate and collaborate each other, timely push the new resources and information, and change their passive learning to adaptive autonomous learning to meet individual learning needs.

DC-AHAM model recommends the relative learning resources with the results of the student's unit tests, which is the key to guide students to learn the professional knowledge. Online test questions are designed for different professional direction; the test result data are the main basis for measuring the professional interests of student. The "operating system" course is the basis of program design, Internet management, embedded design, Linux driver design, and so on. The specialty directions are represented by vector $M := (l_1, ..., l_n)$, the unit test questions are presented by vector $T := (l_1, ..., l_n)$, and the professional direction similarity data matrix is shown as:

$$Sim_{i,j} = \begin{Bmatrix} sim_{1,1} & \cdots & sim_{1,j} \\ \vdots & \vdots & \vdots \\ sim_{i,1} & \cdots & sim_{i,j} \end{Bmatrix} \quad (1)$$

where $sim_{i,j}$ is the correlation of unit test i and specialty j, the scores of the tests are summed by different specialty, the score matrix of professional similarity is generated and defined in Eq. (1), and the test scores vector is $\vec{S} := (s_{1,j}, ..., s_{m,j})$. This algorithm is described as follows:

$$Total_j = \sum_{i=1}^{m} sim_{i,j} \quad (2)$$

The pseudo-code of this algorithm is described as follows:

```
for j < n do
    weight_j = relative_j · score_j
    degree_j = total/weight_j
end for
```

where $relative_j$ is the weight of specialty j, $score_j$ is the score of specialty j, $total_j$ is the sum of similarity of all test T_i and professional direction j, $weight_j$ is the test score of all tests with the professional direction, and $degree_j$ is degree that the student mastered in the specialty direction j, which is an important indicator of measuring the interests and hobbies of different students. Compared with the specialty directions represented by the first two maximum values of $degree_j$ with learning resources, Euclidean distance of specialty characteristics of the cluster center is obtained, the shortest path probability and the length of the shortest path are obtained, and the recommended resources are pushed on the basis of this for the students' best interest. The two minimum values of $degree_j$ indicate the weak specialty directions of the students. Accordingly, the relevant basic specialty knowledge is recommended to improve the students' ability. This can guide students to build their own professional system and build a personalized learning environment.

DC-AHAM model uses collaborative filtering recommendation algorithm based on integration of students' characteristics and learning contents to finish recommendation. With questionnaire, test results and the abilities of knowledge application, knowledge transfer, and students' interests, it can find their new hobbies and correct their characteristic value to improve the recommendation algorithm, and more effectively recommend their professional knowledge. It has better matching accuracy, timeliness, and student satisfaction.

6.3 Adaptive design

According to the characteristics of the students, such as interests, hobbies, and professional background, it makes comprehensive analysis and recommends related knowledge to students. It can not only solve the cold start problem in recommendation algorithm but also guide the students to independent thinking, construct their own knowledge system, and improve the interest of learning.

Because of the different foundation of students, the expected learning goals of the system recommended have 80% accuracy. When the accuracy of students' test results is <70% threshold, the system will send related knowledge for review; when it is >70%, interest direction of specialty knowledge is pushed, and the next step of learning resources is specified to form the effective adaptive feedback.

According to the students' learning feedback, the module can grasp and update the relevant professional information and social demand, and the information is timely pushed to the students. Students can obtain the relevant resources and information of the professional direction.

7 CONCLUSIONS

In this paper, according to the online learning characteristics, distributed cognition theory, and adaptive learning, an adaptive learning model DC-AHAM based on distributed cognition is presented. The system structure of the model is designed: one computer professional course model is designed using this model. The design, development, and implementation of online learning system are studied. According to the DC-AHAM model, the learning community is helpful for students to establish a personalized learning environment, to construct their own professional learning direction, and to finish adaptive online learning. In the practical application, the model is important for online learning. In the future, we will work on the optimization of collaborative filtering algorithm, improvement of the diversity of students' learning feedback, and the accuracy of professional recommendations.

ACKNOWLEDGMENT

This work was financially supported by the research project of Higher Education Reform of 2015 in Jiangsu Province, "Research on adaptive learning model based on mobile-Internet environment" (item number: 2015 JSJG364).

REFERENCES

Chuah J, Zhang J, Johnson t R. distributed cognition of a navigational instrument display task[C] proceedings of the twenty first annual conference of the cognitive science Society. Mahweh, new Jersey: Lawrence-Erlbaumm. 1999 to here.

Gao Huzi, Zhou Dongdai. Adaptive learning system learning style model of the research status and prospect [J]. Audio visual education research, 2012.2:32–38.

Hutchins E. The social organization of distributed cognition[J]. Biosemiotics, 1991.

Paul De Bra, Licia Calvi. AHA! an open Adaptive Hypermedia Architecture[J]. New Review in Hypermedia & Multimedia, 1999, 4(1):115–140.

Salomon G. No distribution without individuals cognition: A dynamic interactional view[J]. Distributed cognitions: Psychological and educational considerations 1993: 111–138.

Tseng J C R, Chu H C, Hwang G J, et al. Development of an adaptive learning system with two sources of personalization information ☆[J]. Computers & Education, 2008, 51(2):776–786.

Zhang Wei, Chen Lin, Ding Yan. Mobile learning age learning view: Based on the perspective of distributed cognition theory [J]. The Chinese audio-visual education, 2010 (4): 21–25.

Advances in Energy Science and Equipment Engineering II – Zhou, Patty & Chen (Eds)
© 2017 Taylor & Francis Group, London, ISBN 978-1-138-71798-5

From a literature review to a theoretical framework for cloud-based internet of things and optimization enabled perishable food cold chain management

Sichao Lu & Xifu Wang
School of Traffic and Transportation, Beijing Jiaotong University, Beijing, China

ABSTRACT: This paper introduces a conceptual framework for perishable food cold chain management which is enabled by the cloud-based internet of things and optimization techniques. This framework consists of five components. The first one is about leveraging the cloud-based internet of things platform to collect and share critical data throughout the distribution process. Component 2 mainly copes with the optimization of allocating perishable food products with various temperature requirements for storage and mutual affected biochemical characteristics to a cold store with different cabins. Optimizing routes for long-haul transportation and the load plan for point-to-point short-haul transportation are themes of the component 3 and the component 4 respectively. The last component is about optimization for multi-temperature joint distribution.

1 INTRODUCTION

For transporting and storing perishable food products, a proper environmental condition can effectively reduce the risk of food spoilage and maintain food freshness. Therefore, the cold store, the refrigerated vehicle, and the refrigerated container has been vastly used by logistics providers. Furthermore, there has been an increasing number of IT solution providers interested in utilizing cutting-edge technologies such as Internet of Things (IoT), cloud computing, and evolutionary algorithms to improve the perishable food cold chain management under uncertainty. In the meantime, academic interest in this field has also risen significantly in recent years.

Following this trend, there exist a lot of academic papers relate to perishable food cold chain management. Among those papers, many of them focus on how to leverage the internet of things to monitor perishable food products. For example, Aung et al. presented a cold chain management system that is implemented by Radio Frequency Identification (RFID) and Wireless Sensor Network (WSN) (Aung, 2011). Verdouw et al. analyzed the concept of virtual food supply chains from an IoT perspective and proposed an architecture, which was applied to a case study of a fish supply chain (Verdouw, 2016).

Optimizing routes for perishable food transportation is the other key research area for perishable food cold chain management. Many vehicle routing problems based models have been proposed for this type of optimization problems. For example, Song and Ko (2016) proposed a vehicle routing problem based model that encompasses both refrigerated—and general-type of vehicles for multi-commodity perishable food products delivery and solved it by a priority-based heuristic. Amorim and Almada-Lobo (2014) proposed the Multi-Objective Vehicle Routing Problem with Time Windows dealing with Perishability (MO-VRPTW-P), which aims at minimizing total routing cost and maximizing the average freshness of all requests. They used the ε-constraint method to solve small-sized problems and two versions of a multi-objective evolutionary algorithm to solve instances with 100 customers. Given that many parameters or variables are vague, some studies have taken the concept of fuzzy set theory into consideration. Brito et al. (2012) proposed a mathematical model of a vehicle routing for distribution of frozen food with fuzzy travel times and solved it by a hybrid algorithm based on greedy randomized adaptive search procedure and Variable Neighborhood Search. Chakraborty et al. (2016) formulated the model of multi-objective multi-item intuitionistic fuzzy solid transportation problem and used the expected value of intuitionistic fuzzy numbers in credibility measures to defuzzify it.

However, few papers refer to the theoretical framework for managing perishable food cold chain by leveraging cutting-edge technologies. Therefore, this paper is devoted to remedying this problem.

2 CLOUD-BASED IOT ENABLED ASSET MANAGEMENT

As the ubiquitous computing has become the trend of business environment (Lee, 2007), the IoT and the cloud computing have become popular technologies in the logistics industry. Through these advanced technologies, logistics information can be exchanged seamlessly to enhance the collaboration of perishable food cold chain. Based on various commercial applications and academic studies, Lu and Wang proposed a cloud-based IoT platform for perishable food transportation management (Lu, 2016). In (Lu, 2016) they extended the platform by adding the cold storage warehouse management system and the food traceability system. Kuo and Chen pointed out that RFID and temperature sensors can be incorporated in the mobile information system for Multi-Temperature Joint Distribution (MTJD) management (Kuo, 2010). Furthermore, IoT techniques can also be applied in the processes of agricultural field monitoring and food processing. The featured service offerings and their key enabling technologies are shown in Table 1.

In practice, common functions such as logistics finance (Li, 2010) and order management are usually incorporated in a logistics platform. Furthermore, a cloud-based harvest management system (Ampatzidis, 2016) could also be incorporated for handling fresh fruit and vegetables.

As a national standard of China, the Ecode can be used to identify any product across the whole life cycle. End consumers can trace the information in the perishable food cold chain by reading the two-dimensional Ecode via mobile applications or other intelligent devices. In terms of the WSN technologies, experiments show that 6 LoWPAN (Wang, 2010; Miao, 2014) can also be adopted as a promising technology for managing perishable

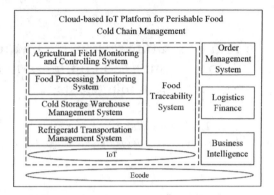

Figure 1. Overview of the cloud-based IoT platform for perishable food cold chain management.

food cold chain. The overview of this platform is briefly presented in Figure 1.

He et al. proposed seven business value drivers of the IoT-enabled supply chain (He, 2010). Lu and Wang categorized the business value elements enabled by the cloud based IoT platform for IT-enabled perishable food transportation into four groups (Lu, 2016). In general, economic losses relate to food spoilage can be prevented and efficiency can be increased by leveraging the IoT while IT investments and labor costs can be reduced to a large extent within the cloud computing environment.

3 OPTIMIZATION FOR COLD STORAGE MANAGEMENT UNDER FUZZY ENVIRONMENT

It is common that some perishable food products should be preserved in the cold store during many business processes. However, there exist few studies relate to the mathematical model and algorithm for optimizing perishable food storage location assignment. One of the related problems is the generalized segregated storage problem which aims at minimizing the cost of storing goods for allocating a certain number of goods to compartments subject to segregation constraints (Barbucha, 2004). Lu and Wang stated that the cabin allocation optimization for perishable food storage and that for durable goods storage differ in terms of three points (Lu, 2016): (1) Different temperature and humidity demands among perishable foods should be taken into consideration. (2) Due to problems such as ethylene sensitivity and odor contamination, incompatible perishable foods cannot be put together. (3) Prices of perishable food products can be taken as fuzzy numbers because they cannot be determined precisely in advance. Furthermore,

Table 1. Featured service offerings of the cloud-based IoT platform for perishable food cold chain management.

Service offering	Key enabling technologies	Representative literature
Master data management	Cloud computing	(Zhang, 2016)
Virtual perishable food cold chain monitoring	RFID, WSN, GSM/3G/4G, WiFi, GPS, GIS	(Aung, 2011; Wang, 2015; Guo, 2011; Wang, 2011; L, 2014; Chandra, 2014)
Food traceability	RFID, EPCIS	(Thakur, 2011; J, 2013)

they proposed a single-objective fuzzy cold storage problem to deal with the situation when perishable food products are too many to be stored in a refrigerated warehouse. Given that this model is NP-hard, they designed a discrete firefly algorithm based on the MCFA (Lu, 2015). This algorithm consists of five parts that are k-preference integration based defuzzification, population initialization, movement of fireflies by mutual attraction, a four-phase repair operation, and movement of the brightest firefly.

Based on this problem, not only can the food quality be guaranteed for next business processes, but also some business models can be supported. For example, uncertain profitability caused by seasonal price variation can be improved by storing the seasonal fruit for a certain period and then selling them to satisfy uneven demands throughout the year.

4 OPTIMIZATION FOR LINE HAUL PERISHABLE FOOD TRANSPORTATION

Numerous studies have been carried about line haul transportation, while few papers relate to the field of cold chain. Therefore, two components of the theoretical framework are proposed in this section.

4.1 Optimization for long-haul perishable food transportation under fuzzy environment

For point-to-point long haul, the above mentioned mathematical model of fuzzy cold storage problem can be applied by regarding each cabin of a cold store as a refrigerated container. For long haul through a transportation network, the net optimization models can be leveraged. Qian built two maximum flow based mathematical models which depict the agricultural products transportation from Taiwan to Sichuan and solved them with the EXCEL (H, 2012). Shih and Lee (1999) took the cost parameter and capacity constraints of the fuzzy multi-level minimum cost flow problem as fuzzy numbers.

Chakraborty et al. proposed a multi-objective, multi-item intuitionistic fuzzy solid transportation problem for damageable items, which uses a parameter to denote the percentage of unit damage items (Chakraborty, 2016). Based on this idea, Lu and Wang proposed a fuzzy minimum cost flow problem for damageable items transportation, which can be applied in some scenarios of the long-haul perishable food transportation (Lu, 2016). They defuzzified that model by three techniques respectively and solved them with Wolfram Mathematica 9. We think parameters such as the demand quantity and the supply quantity could also be taken as fuzzy numbers.

4.2 Load planning for point-to-point short-haul perishable food transportation

Load planning plays a pivotal role to the point-to-point transportation when there are too many products to be loaded by given vehicles. In a comprehensive review of related literature, few studies have fully considered the particularity of perishable foods. Xing et al. proposed a mathematical model of 3-dimensional bin packing problem for agricultural products which aims to orthogonally pack all the rectangular-shaped products into a three-dimensional container at the lowest cost (B, 2011). However, the relation between the temperature demands of perishable food products and the compartment temperature provided by refrigerated vehicles is neglected by most studies. Furthermore, multi-objective optimization with more than two objectives are recommended because decision makers tend to set multiple goals for shipments sometimes.

5 OPTIMIZATION FOR MULTI-TEMPERATURE JOINT DISTRIBUTION

Through a comprehensive review of the literature, many mathematical models which deal with the food distribution problem are proposed based on the model of Vehicle Routing Problem with Time Windows (VRPTW). However, those models are based on the assumption that perishable food products with different temperature demands can be put together in a vehicle with single temperature zone. Therefore, they are not applicable to the route optimization of MTJD.

In recent years, due to rapid advances in e-business and web technologies, consumers can purchase perishable food products such as vegetables and ice creams through Internet shopping conveniently. As a result, the cold distribution market and the home delivery market continues to expand, which contributes to the development of MTJD. Hsu and Liu (2011) divided the MTJD into two types. The first type is mechanical refrigerated compartment division. It uses a technique which can divide a single vehicle compartment into different temperature zones. The other type is replaceable cold accumulation and insulated box. In this type, standardized cold insulated boxes and cabinets are put in regular vehicles to store perishable food products of varying temperatures. When there is only one temperature zone or all of the insulated boxes have the same temperatures, the MTJD become the single-temperature joint distribution.

Studies relate to optimization for MTJD are also quite few. Hsu and Liu constructed a mathematical model to determine optimal multi-temperature logistic techniques and food handling volume (Hsu, 2011). Wang and Zhao concluded that the MTJD can bring economic benefits and contribute to the low-carbon economy (Wang, 2012). Cho and Li built the mathematical model of the Multi-Temperature Refrigerated Container Vehicle Routing Problem (MRCVRP) and designed a two-stage heuristic method which consists of the modified savings algorithm and four interchange-based heuristics (Cho, 2005). Wang and Sun proposed a model of MTJD with stochastic demands and time limit. Furthermore, they designed an optimization approach which blends the K-means clustering algorithm, the ant colony algorithm, and a stochastic dynamic programming algorithm to solve it (Wang, 2016). Sun proposed a mathematical model for MTJD with carbon emission constraints, which aims to minimize total cost of food cold chain and total carbon emissions (Sun, 2015).

Based on the above mentioned papers, we suggest taking travel times of the MTJD as fuzzy numbers. Furthermore, the relation between customer satisfaction and arrival time and be expressed by a Z-shaped curve. According to the consumer behaviour, a quantified customer satisfaction value tends to decrease slowly when the shipment of perishable food products is just in transit. As the real arrival time passes the expected arrival time of the consignee, the satisfaction value drops significantly. After a certain period of time, the satisfaction value drops slowly and then it may equal to a small number in the end. The MATLAB fuzzy toolbox includes a Z-shaped built-in membership function, which can be used to depict the customer satisfaction. A similar proposal has been presented to depict the relation between the profit and the degree of food freshness by Chen and Fang (Chen, 2013).

6 TOWARD A THEORETICAL FRAMEWORK

Based on the components mentioned above, we propose a theoretical framework for cloud-based IoT and optimization enabled perishable food cold chain management, which is shown in Figure 2.

As illustrated in Figure 2, it clearly shows that the theoretical framework we proposed consists of five components: (1) the cloud-based internet of things enabled asset management; (2) the optimization enabled cold storage management under fuzzy environment; (3) the optimization enabled long haul through a transportation network under fuzzy environment; (4) the optimization enabled

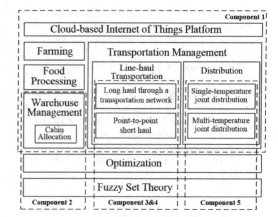

Figure 2. A theoretical framework for cloud-based IoT and optimization enabled perishable food cold chain management.

point-to-point short haul under fuzzy environment; (5) the optimization enabled MTJD under fuzzy environment.

7 CONCLUSIONS

In this paper, we proposed a theoretical framework for perishable food cold chain management which consists of five components. Component 1 is about leveraging the cloud-based internet of things platform to collect and share critical data of perishable food products throughout the complete food cold chain. When the perishable food products are too more to be put in a cold store with different cabins, an optimal plan for allocating these products among cabins with several restrictions should be made by using the component 2. The component 3 is used to make an optimal plan of transporting perishable foods from multiple sources to multiple destinations through a transport network for long-haul transportation. For point-to-point short-haul transportation with limited load capacities and multiple goals, the component 4 aims to generate an optimized load plan. When perishable food products with different temperature demands are to be distributed to end customers in one vehicle within a service area, the component 5 can be adopted to generate optimized routes.

The theoretical framework we proposed can enable visibility and intelligence across the entire food cold chain, which contributes to the cost reduction and improved customer satisfaction. Furthermore, environmental and social issues of this theoretical framework such as reduced food spoilage and food poisoning as well as improved resource utilization can contribute to a better sustainable development (Seuring, 2008).

Given that business models vary across different industries, sub-sectors of the foodstuff industry, and even different enterprises with same roles in a business ecosystem, it is definite that there exist various theoretical frameworks. Therefore, it is hoped that the theoretical framework we proposed can enable propositions of other frameworks. Furthermore, a specific theory or business model for cloud-based Internet of Things and optimization enabled perishable food cold chain management can be further developed based on this framework.

REFERENCES

Amorim, P., B. Almada-Lobo, Ind. Eng. **67** 223 (2014).

Ampatzidis, Y., L. Tan, R. Haley, M.D. Whiting, Comput. Electron. Agr. **122** 161 (2016).

Aung, M.M., F.W. Liang, Y.S. Chang, C. Makatsoris, J. Chang, Int. J. Manuf. Res. **6** 291 (2011).

Barbucha, D., Eur. J. Oper. Res. **156** 54 (2004).

Brito, J., F.J. Martinez, J.A. Moreno, J.L. Verdegay, Fuzzy Optim. Decis. Making, **11** 337 (2012).

Chakraborty, D., D.K. Jana, T.K. Roy, J. Intell. Fuzzy Syst. **30** 1109 (2016).

Chandra A.A., and S.R. Lee, Int. J. Multimedia Ubiquit. Eng. **9** 145 (2014).

Chen, L. Fang, J. Guangxi, U.: Nat Sci Ed, **38** 729 (2013).

Cho, Y.J., C.C. Li, J. East. Asia Soc. Transp. Stud. **6** 2749 (2005).

Feng, J., Z. Fu, Z. Wang, M, Xu, X. Zhang, Food Control **31** 314 (2013).

General Administration of Quality Supervision, Inspection and Quarantine of the People's Republic of China and Standardization Administration of the People's Republic of China, *GB/T 31866-2015 Identification system for internet of things—Entity code, National Standards of the People's Republic of China* (Standards Press of China, Beijing, 2015).

Guo, B., J. Qian, T. Zhang, X. Yang, T. Chinese Soc. Agr. Eng. **27** 208 (2011).

He, M., C. Ren, Q. Wang, B. Shao, J. Dong, Proc of 2010 IEEE 7th International Conference on e-Business Engineering, IEEE Press, 326 (2010).

Hsu, C.I., K.P. Liu, Food Control **22** 1873 (2011).

Kuo, J.C., M.C. Chen, Food Control **21** 559 (2010).

Lee, H.J., S. Kim, Int. J. Technol. Manage. **38** 424 (2007).

Li, Y.X., S.Y. Wang, and G.Z. Feng, Syst. Eng. Theory Pract. **30** 1 (2010).

Lu, S., X. Wang, Adv. J. Food Sci. and Technol. **11** 190 (2016).

Lu, S., X. Wang, J. Intell. Fuzzy Syst. **31** 2431 (2016).

Lu, S., X. Wang, *Proc. of 2015 IEEE 6th International Conference on Software Engineering and Service Science*, IEEE Press, 1003 (2015).

Lu, S., X. Wang, *Proc. of 2016 IEEE 7th International Conference on Software Engineering and Service Science*, IEEE Press, 852 (2016).

Lu, S., X. Wang, *Proc. of Fuzzy System and Data Mining 2016, Frontiers in Artificial Intelligence and Applications* (IOS Press, Amsterdam, 2016). (to be published).

Miao, J., *Research and Implementation on 6LoWPAN Based Cold Chain Environment Monitoring System*, MSc Dissertation, Nanjing University of Posts and Telecommunications (2014).

Qi, L., M. Xu, Z. Fu, T. Mira, X. Zhang, Food Control **38** 19 (2014).

Qian, H. Log. Technol. **34** 47 (2012).

Seuring, S., M. Müller, J. Clean. Prod. **16** 1699 (2008).

Shih, H.S., E.S. Lee, Fuzzy Sets Syst. **107** 59 (1999).

Song, B.D., Y.D. Ko, J. Food Eng. **169** 61 (2016).

Sun, H., *Proc. of International Conference on Logistics Engineering, Management and Computer Science*, Atlantis Press, 1022 (2015).

Thakur, M., C.F. Sørensen, F.O. Bjørnson, E. Forås, C.R. Hurburgh, J. Food Eng. **103** 417 (2011).

Verdouw, C.N., J. Wolfert, A.J.M. Beulens, A. Rialland, J. Food Eng. **176** 128 (2016).

Wang, J., H, Wang, J. He, L, Li, M. Shen, et al., Comput. Electron. Agr. **110** 196 (2015).

Wang, S., H. Sun, Ind. Eng. Manage. **21** 49 (2016).

Wang, S., M. Zhao, J. Highw. Transp. Res. Dev. **29** 144 (2012).

Wang, T., X. Zhang, W. Chen, Z. Fu, Z. Peng, T. Chinese Soc. Agr. Eng. **27** 141 (2011).

Wang, X., X. Yin, T. Chinese Soc. Agr. Eng. **26** 224 (2010).

Xing, B., X. Yang, J. Qian, F. Wang, T. Chinese Soc. Agr. Eng. **27** 237 (2011).

Zhang, C., X. Zhang, D. Wang, Y. Wang, T. Chinese Soc. Agr. Eng. **32** 2739 (2016).

Advances in Energy Science and Equipment Engineering II – Zhou, Patty & Chen (Eds)
© 2017 Taylor & Francis Group, London, ISBN 978-1-138-71798-5

Research on the detecting technology of DC crosstalk without impedance

Hai-lun Han, Qiang Yin, Qi-liang Zhao, Xiao-dan Ren, Dong-yao Li & Yan-yan Wei
XJ Power Co. Ltd., State Grid Corporation of China, Xuchang, Henan, China

ABSTRACT: In view of the problem of DC crosstalk without impedance in substation, a new detecting method is proposed based on DC crosstalk without impedance. The insulation monitoring devices for DC power system is designed. The kind of this grounding fault is prevented. A prototype test shows that it is able to monitor the DC crosstalk without impedance in real time, send out the alarm information and improve the reliability of the DC system, which has the advantages of simple operation, accurate measurement and good stability.

1 INTRODUCTION

DC system has a very important position in substation and power plant. DC system is power equipment for the signal equipment, the protection automation equipments, the emergency lighting, the emergency power and the operating power supply. DC system should give service to the users of the hydraulic power plant, the thermal power plant, the substation and the other users using DC power equipment.

The two independent storage batteries are chosen in 110 kV substation and 220 kV substation in view of the importance and the demand of relay protection and circuit breaker tripping mechanism based on the reliability of DC power supply in the technical code for designing DC system of power projects.

In case of DC crosstalk appears in substation, many adverse effects and harm will be introduced. It will bring the double misoperation, cause related equipment refuse operation, affect the precision positioning of the grounding fault point, result in the feed burning or fire in DC system, shorten the service life of the battery, and affect the safety of production (REN, 2011; LUO, 2008; XU, 2008). Therefore, DC crosstalk is an urgent problem need to be resolved.

It is pointed out that the feed network should use the radial power supply mode instead of the annular power supply mode in the DC system of substation. When DC crosstalk appears in DC system, the devices should be able to send out the DC crosstalk alarm information and can choose the fault slip road.

Therefore, this paper presents the detecting technology of DC crosstalk and designs the insulation monitoring devices for DC power system, which can monitor the situation of DC crosstalk in real time and can send out the alarm information of DC crosstalk. A prototype tests that the validity and feasibility of the devices are verified.

2 THE DETECTING PRINCIPLE OF DC CROSSTALK

The equivalent circuit diagram of DC system is shown in Fig. 1. Where, R1 and R2 are the resistors in equalization bridge, R3 is the resistor in inspection bridge, S1 is the switch in inspection bridge, R+ and R− are the resistors of the positive and negative electrode to the grounding, C+ and C− are the capacitances of the positive and negative electrode to the grounding, and the imaginary frame is the insulation monitoring devices for DC power system.

$$R_{\Sigma+} = R1 // R+ // C+ \tag{1}$$

$$R_{\Sigma-} = R1 // R- // C- \tag{2}$$

$$E = (V+)-(V-) \tag{3}$$

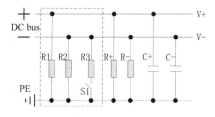

Figure 1. The equivalent circuit diagram of DC system.

where, $R_{\Sigma+}$ is the insulation resistor of the positive electrode to the grounding, $R_{\Sigma-}$ is the insulation resistor of the negative electrode to the grounding, and E is the voltage between the positive bus and the negative bus.

2.1 The homopolarity interconnection

The homopolarity interconnection is the connection mode in the positive or negative connection operation in the two stage DC system, even the positive and negative connection operation in the two stage DC system. It is the reason for such interconnecton that the parasitic loop is existence of the secondary loop, resulting in the homopolarity interconnection operation. The mathematical model of the homopolarity interconnection is shown in Fig. 2. The homopolarity interconnection mode between the positive electrode and the positive electrode is named A mode shown in Fig. 2(a), and the homopolarity interconnection mode between the negative electrode and the negative electrode is named B mode shown in Fig. 2(b).

While $R_{1\Sigma} \neq R_2\Sigma$ and E1 = E2, E1+ = E2+ and E1− = E2− are shown in Fig. 2(a), and then the output current of the battery is bigger in the lager insulation resistor. At the same time, and E1+ = E2+ and E1− = E2− are shown in Fig. 2(b), and then the output current of the battery is bigger in the small insulation resistor. While $R_{1\Sigma} \neq R_{2\Sigma}$ and E1 ≠ E2, E1+ = E2+, E1− = (E1) − E1+ and E2− = (E2) − E2+ are shown in Fig. 2(a). At the same time, E1− = E2−, E1+ = (E1) − E1− and E2+ = (E2) − E2− are shown in Fig. 2(b).

When the monitoring situation appears which is described above the paper in the operation of the insulation monitoring devices for DC power system, the inspection bridge of any set will be open belonging to the two sets of devices through the automatic mode. The voltage is equal between the positive electrode to the grounding and the positive electrode to the grounding, and then, between the negative electrode to the grounding and the negative electrode to the grounding, when the inspection bridge enters into the system, which can be used to judge the existence of interconnection systems. The grounding point of the equalization

bridge is off in the insulation monitoring devices for DC power system, The voltage equal relationship has not appear between the positive electrode to the grounding and the positive electrode to the grounding, and then, between the negative electrode to the grounding and the negative electrode to the grounding, which can be used to completely judge the existence of interconnection systems. At this point, if the value of E1+, E1−, E2+ and E2− have changed in the numerical value, and the interrelation between them is still hold, the system shows the existence of DC crosstalk, otherwise, the system does not exist the DC crosstalk.

2.2 The heteropolarity interconnection

The heteropolarity interconnection is the connection mode in such operation between the positive electrode of I stage DC and the negative electrode of II stage DC or between the negative electrode of I stage DC and the positive electrode of II stage DC. The mathematical model of the heteropolarity interconnection is shown in Fig. 3. The heteropolarity interconnection mode between the positive electrode and the negative electrode is named C mode shown in Fig. 3(a), and the heteropolarity interconnection mode between the negative electrode and the positive electrode is named D mode shown in Fig. 3(b).

While $R_{1\Sigma} \neq R_{2\Sigma}$ and E1 = E2, E1− = E2+ and (E1+) + E2− = 2E1 = 2E2 are shown in Fig. 3(a), and then, the electrode voltage of the lager insulation resistor will exceed the power supply voltage. At the same time, E1+ = E2− and (E1−) + E2+ = 2E1 = 2E2 are shown in Fig. 3(b), and then, the electrode voltage of the lager insulation resistor will exceed the power supply voltage. While $R_{1\Sigma} \neq R_{2\Sigma}$ and E1 ≠ E2, E1− = E2+ and (E1+) + E2− = E1 + E2 are shown in Fig. 3(a), and then, the electrode voltage of the lager insulation resistor will exceed the power supply voltage. At the same time, E1+ = E2− and (E1−) + E2+ = E1 + E2 are shown in Fig. 3(b), and then, the electrode voltage of the lager insulation resistor will exceed the power supply voltage.

When the monitoring situation appears which is described above the paper in the operation of

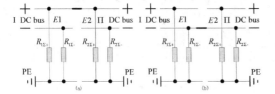

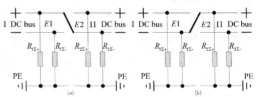

Figure 2. The mathematical model of the homopolarity interconnection.

Figure 3. The mathematical model of the heteropolarity interconnection.

the insulation monitoring devices for DC power system, the grounding point of the equalization bridge is off or the detecting bridge is on in the two sets of the insulation monitoring devices for DC power system though the automatic and alternant mode. If the value of E1+, E1−, E2+ and E2− have changed in the numerical value, and the interrelation between them is still hold, the system has the existence of DC crosstalk, otherwise, the system does not exist the DC crosstalk.

3 THE DESIGN OF DC CROSSTALK

The voltage of DC system, the voltage of the positive electrode to the grounding and the voltage of the negative electrode to the grounding were sampled, divided and converted through the signal acquisition unit, and then the signal is sent to the single chip microcomputer for analyzing and processing, and then, the processing signal is matched with the DC crosstalk fault models of the devices with built-in fault models. So it can be used to accurately judge the existence of interconnection systems and the DC crosstalk fault models, and then send out the alarm information of DC crosstalk. It can significantly reduce the DC crosstalk fault operation time and reduce the DC crosstalk found difficulty with the two stage DC power system in substation or power plant(ZHAO, 2013; HUANG, 2014).

Hardware design is mainly divided into two parts. The first part is composed of the main control unit and the external device of the main control unit, the other part is composed of the sampling unit, the conditioning unit, and the alarm controll unit based on the DC bus and branch. The hardware block diagram is shown in Fig. 4.

The hardware detection circuit of DC crosstalk is shown in Fig. 5. Where, Network labelled Samx can represent the sampling points V+, the sampling points V− and the sampling points grounding. They form the Sam_ADx signals into AD pin of the controller through the conditioning circuit, and then, the sampling value of the real-time detection is realized through the software operation.

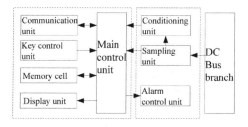

Figure 4.   The block diagram of the hardware.

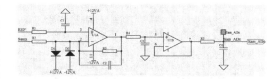

Figure 5.   The hardware detection circuit of DC crosstalk.

4 EXPERIMENT

In order to verify the validity and reliability of the detection technology, the insulation monitoring devices for DC power system is designed. Test instruments have the MX30-3PI-400-LF-SNK of the programmable power supply and the FLUKE15B of the multimeter.

Where, (a) is the page of interconnection type, (b) is the page of the alarm information of interconnection, (c) is the page of the voltage of the positive electrode to the grounding, and (c) is the page of the negative electrode to the grounding.

The diagram of the homopolarity interconnection between the positive electrode of I stage DC and the positive electrode of II stage DC based on E1 ≠ E2 is shown in Fig. 6. And then, it can be seen that the display status is given in the device displaying unit, and the conclusion is verified above the paper.

The diagram of the homopolarity interconnection between the negative electrode of I stage DC and the negative electrode of II stage DC based on E1 ≠ E2 is shown in Fig. 7. And then, it can be seen that the display status is given in the device displaying unit, and the conclusion is verified above the paper.

The diagram of the heteropolarity interconnection between the positive electrode of I stage DC and the negative electrode of II stage DC based on E1 ≠ E2 is shown in Fig. 8. And then, it can be seen that the display status is given in the device displaying unit, and the conclusion is verified above the paper.

The diagram of the heteropolarity interconnection between the negative electrode of I stage DC and the positive electrode of II stage DC based on E1 ≠ E2 is shown in Fig. 9. And then, it can be seen that the display status is given in the device displaying unit, and the conclusion is verified above the paper.

In view of the operation mode, the grounding point of the equalization bridge are interconnected in the two sets of the insulation monitoring devices for DC power system, so the $R_{1\Sigma+}$ and $R_{2\Sigma+}$ are paralleled, or the $R_{1\Sigma-}$ and $R_{2\Sigma-}$ are paralleled. The paralleled resistor value are series among the source negative electrode, $R_{1\Sigma-}$ and $R_{2\Sigma-}$ or among the source negative electrode, $R_{1\Sigma+}$ and $R2_{\Sigma+}$. What-

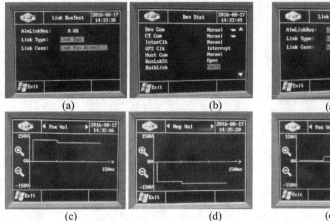

(a)   (b)

(c)   (d)

Figure 6. The A mode on E1 ≠ E2.

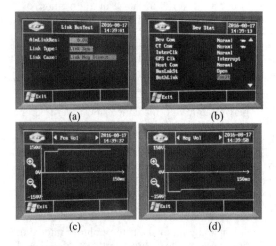

(a)   (b)

(c)   (d)

Figure 7. The B mode on E1 ≠ E2.

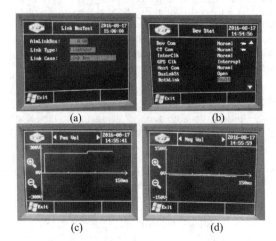

(a)   (b)

(c)   (d)

Figure 8. The C mode on E1 ≠ E2.

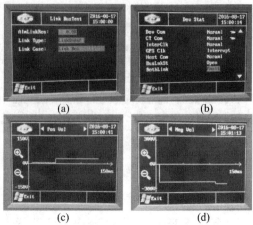

(a)   (b)

(c)   (d)

Figure 9. The D mode on E1 ≠ E2.

ever the grounding point of the equalization bridge is off or the detecting bridge is on in the two sets of the insulation monitoring devices for DC power system though the automatic and alternant mode, and the interrelation between them (E1+ = E2+ and E1− = E2−)is still hold in the A mode and the B mode, but the value E1+ = E2+ and E1− = E2−have changed in the numerical value in the A mode and the B mode, or the interrelation between them (E1− = E2+ and E1+ = E2−) is still hold in the C mode and the D mode, but the value E1− = E2+ and E1+ = E2− have changed in the numerical value in the C mode and the D mode. So, the system has the existence of DC crosstalk, otherwise, the system does not exist the DC crosstalk.

The processing method of the DC crosstalk has several suggestions:

1. The switch is off which can be resolved the interconnection for the same load in the two stage DC bus power system.
2. The DC crosstalk point is that the interconnection signal enters into the DC bus of another stage by finding the direction point from the interconnection branches to the load. The load will be back to the bus that belongs to this bus.
3. The parasitic circuit can be removed, which has been find out.

5 CONCLUSION

In this paper, a detection method of DC crosstalk is presented. The insulation monitoring devices for DC power system is designed. A prototype test that it is able to monitor DC crosstalk in real time, send out the alarm information of DC crosstalk. This device has been running for more than half a year in the field of substation, with the advantages of reliable, accurate, stable and innovative, at the same time, with strong market competitiveness.

ACKNOWLEDGEMENTS

This work was financially supported by Science and Technology Project of State Grid Corporation of China(Research on the intelligent on-line monitoring and maintaining technology for AC-DC integration power in substation, and then, Research on DC electrical source supervisor and the decision support system of operation and maintenance in substation).

REFERENCES

DL/T 1392–2014 Technical specification of insulation monitoring devices for DC Power system[S].

DL/T 5044–2004 Technical code for designing DC system of power projects[S].

HUANG Dong-shan, ZHOU Wei, YANG Li-cai, et al. Study on insulation failure and loop fault in the DC system[J]. Electronic Test, 2014. 24:130–132.

LUO Zhi-ping, XIONG Di, XIE Zhi-hao, etal. Looped network problems and their solutions in DC system in substations[J]. Relay, 2008. 36(3):71–74.

REN Dong-hong, FAN Shu-gen. Analysis and measures for protection misoperation caused by grounding of DC power supply system[J]. Distribution & Utilization, 2011. 28(3): 44–48.

The 18 items of grid major anti accident measures of SGCC[Z]. Beijing: China Electric Power Press. 2011.

XU Yu-feng, YUN Chang-an, YIN Xing-guang. Hazards and Handling of Ring Faults in DC Systems[J]. GUANG-DONG ELECTRIC POWER, 2008. 36(3):71–74.

ZHAO Ying-chun, JIANG Xiao-hong, LIU Qing, etal. Design of DC ring alarm and loop find device[J]. Automation Application, 2013. 10:65–66.

Study of the influence factors and calculation method of saturation flow rate at signalized intersection based on measured data

Weiming Luo
Traffic Management Research Institute of Ministry of Public Security Ministry, Wuxi, Jiangsu, China

Jianhua Yuan
Traffic Management Research Institute of Ministry of Public Security Ministry, Wuxi, Jiangsu, China
Jiangsu Province Collaborative Innovation Center of Modern Urban Traffic Technologies, Nanjing, Jiangsu, China

Min Wang, Lin Xu & Chengsheng Liu
Traffic Management Research Institute of Ministry of Public Security Ministry, Wuxi, Jiangsu, China

ABSTRACT: In this paper, we study the effects of uncertainty factors of traffic and present a new calculation method based on Grubbs test for calculating the saturation flow rate at signalized intersection. The calculation results are based on the practical measured data by cameras at the signalized intersections of Dachi Road and Qianrong Road in Wuxi, China. By analyzing the statistics, this paper shows that a portion of the samples containing a large number of the measured traffic data affected by the uncertainty factors should be removed. The developed method is used to calculate the saturation flow rate with different conditions. The calculation results illustrate that the proposed method can overcome the effects of uncertainty factors and reflect the laws of saturation flow rate under different conditions.

1 INTRODUCTION

Statistics from the Ministry of Public Security, shows that China is the world's second largest vehicles consumption market after the United States. The growing number of cars puts huge pressure on the urban transport system and makes for enhanced demand on the reliability and controllability of urban traffic control system. Traffic capacity is the maximum number of cars passing a given point with prevailing urban road, traffic, and control in a certain time period. Capacity is an important parameter of traffic signal timing, which can measure the dispersion assignment ability of traffic. The capacities differ for different roads (Leutzbach, 2011). Yang et al. (Yang, 2014) proposed that the saturation flow rate method is the dominant way for calculating signal intersection capacity based on the study of the relevant results of signalized intersection capacity.

Saturation flow rate at signalized intersection is one of the important parameters of saturation flow method for calculating the vehicle capacity. Saturation flow rate can usually be calculated by modeling or measurement. Model-based calculation methods need to be corrected by various factors which are difficult to be quantified or measured,

but these factors influence the accuracy of saturation flow rate (Tarko, 2000). Therefore, saturation flow rate is calculated with the field-measured data of headway in engineering applications. The time headway is defined as the interval between the time point when the head of consecutive cars start crossing the stop line or any other reference line. After the traffic signal turns green, it should exclude the negative influence by various uncertainty factors. The saturation flow rate equals the reciprocal of the average headway times 3600 (Jin, 2009). One of the most widely used traffic detectors is coil detector, which is sensitive, tunable, adaptable, and cheap. However, a single coil detector cannot measure the speed of vehicles, and the measured data, which are the time of vehicles arriving and passing the coil, cannot distinguish the types of cars (Zhao, 2008).

This paper makes an effort to calculate the saturation flow rate with measured data in the field. It also discusses the effects of uncertainty factors on the traffic by statistical analysis. The developed method based on Grubbs test is used for calculating the saturation flow rate at signalized intersection. It will be shown through calculation results that the proposed method can overcome the effects of uncertainty factors and reflect the laws of saturation flow rate under different conditions.

2 ANALYSIS OF INFLUENTIAL FACTORS FOR SATURATION FLOW AT SIGNALIZED INTERSECTION

In this section, the various uncertainty factors are analyzed, which influence the accuracy calculation of saturation flow rate based on the measured data in the field. It is a complex task to calculate the saturation flow rate with practically measured data, because of the influences of various factors (Shao, 2011). The measured data that are influenced by various factors are referred to as influenced headway. The values of influenced data are higher than the headway under normal conditions (Akçelik, 2002). The main influential factors are weather, vehicle type, road condition, and psychological state of the driver (Shao, 2011). Figs. 1–4 show the ratios of influenced headway in every cycle under different weather conditions.

Figs. 1 and 2 show the ratios of influenced headway in every cycle at the through lanes and the left turn lanes on sunny day, respectively. The number of cycles with the ratios of the influenced headway less than 30% account for 86.4% of the total cycle at the through lanes. Similarly, at the left turn lanes, the number of cycles with the ratios of the influenced headway less than 30% account for 88% of the total cycle. However, the ratios of influenced headway in every cycle at the left turn lanes are mostly between 20% and 30%.

Fig. 3 shows the ratios of the influenced headway in every cycle at the through lanes on rainy day. It is evident from the figure that the ratios of the influenced headway on rainy day are higher than the ratios in every cycle at the through lanes on sunny day. Fig. 4 shows the ratios of the influenced headway in every cycle at the left turn lanes on rainy day. It is evident from the figure that the ratios of the influenced headway are higher than those in every cycle at the left turn lanes on sunny day.

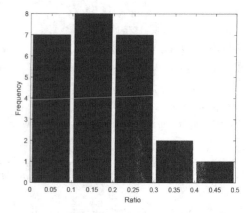

Figure 2. Histogram of the ratio for influenced headway in every cycle at the left turn lanes on sunny days.

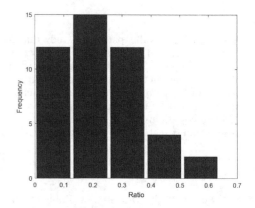

Figure 3. Histogram of the ratio for influenced headway in every cycle at the through lanes on rainy days.

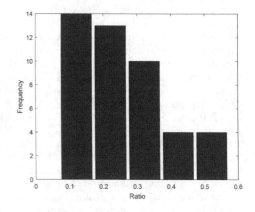

Figure 4. Histogram of the ratio for influenced headway in every cycle at the left turn lanes on rainy days.

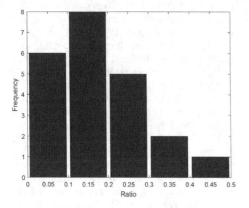

Figure 1. Histogram of the ratio for influenced headway in every cycle at the through lanes on sunny days.

Combined with the four figures, the comparative results show that different lanes and weather have large impacts to the headway. This paper provides the percentage of time headway influenced

Table 1. Ratio of headway influenced by different factors.

Weather	Lane	Driver psychology	Car types	Merging	Congestion
Rain day	Through	51.43%	37.14%	2.86%	8.57%
	Left	53.13%	31.24%	3.13%	12.50%
Sunny day	Through	27.27%	40.91%	22.73%	9.09%
	Left	25.64%	46.16%	12.82%	15.38%

by different factors with analyzing the video data. The analysis results are shown in Table 1.

Because more drivers frequently cut across the road without notice at the through lane, the ratio of headway influenced by driver psychology is 27.27% more than the ratio 25.64% at the left turn lane on sunny day. The traffic jam of the left lane is worse than the traffic on the through lane based on the statistical data. As shown in Table 1, it is obvious that the ratio of abnormal headway influenced by driver psychological under sunny day is less than that on rainy day. The reason is that drivers of car-following, who are affected by the slippery roads and the stadia, choose longer distance and lower speed.

3 GRUBBS-BASED CALCULATION METHOD OF THE SATURATED FLOW RATE

In this section, a new computational procedure is developed for the calculation of the saturated flow rate at signalized intersection. This method is based on Grubbs test method, which is optimal for small sample size problem. Supposing the uncertainty is influenced by various factors, the phenomenon called stochastic noise, is normal distribution (Wood, 1998). The time series of noise can be written as: $w_i \sim N(0, \sigma^2), i = 1, 2, ..., n$. Then, the measured data can be written as:

$$t_{si} = t_s + w_i \qquad (1)$$

where t_{si} is the i th measured queue discharge headway in the field and t_s is the ideal saturated headway.

The mean and variance of measured queue headway is:

$$\bar{t}_s = \frac{1}{n}\sum_{i=1}^{n} t_{si} \qquad (2)$$

$$\bar{s} = \sqrt{\frac{1}{n-1}\sum_{i=1}^{n}(t_{si} - \bar{t}_s)^2} \qquad (3)$$

The constructed discrimination statistic is:

$$G_i = \frac{\max\limits_{1 \le i \le n}|t_{si} - \bar{t}_s|}{\bar{s}} \qquad (4)$$

The discrimination statistic (4) can be transformed into t distribution with degree of freedom n–2,

$$\frac{G_i}{\sqrt{(n-1-G_i^2)/(n-2)}} \sim t_{\alpha i}(n-2) \qquad (5)$$

where α is the level of test. The discrimination statistic $G_{\alpha i}$, which is used to determine the influenced headway, can be written as:

$$G_{\alpha i} = \frac{t_{\alpha i}\sqrt{n-1}}{\sqrt{n-2-t_{\alpha i}^2}} \qquad (6)$$

If $G_i > G_{\alpha i}$, in other words, when $P(G_i > G_{\alpha i}) = \alpha$, the measured headway t_{si} is influenced by some factors and needs to be rejected, that is, the number of the measured data minus 1. Repeat above mentioned process until $G_i < G_{\alpha i}, i = 1, 2, ..., n_1, n_1 \le n$.

There are many ratios of headway that are influenced by various factors above 10%. However, the Grubbs test method is the best when the ratio of the outlier is less than 10% (Chen, 2009). To solve this problem, it is supposed to reduce the influence of factors by rejecting a portion of the samples, which contain a significant number of abnormal values.

The data $t_s^j(1), t_s^j(2), ..., t_s^j(n)$, which are the measured headway in j th cycle, are sorted in ascending order:

$$\hat{n} = \begin{cases} [0.7n], & \text{if sunny} \\ [0.6n], & \text{if rainy} \end{cases} \qquad (7)$$

where n is the number of measured queue discharge headway, $\hat{n}$ is the number of the remaining data which contain a small amount of measured data influenced by various factors, and $[\cdot]$ is integer-valued function.

However, there are still some situations where the ratio of the influenced headway is too higher to reject on the basis of the above process during the rush hour. To solve this problem, the above process must be repeated for m cycles and the remaining data should be merged in ascending order. Then, one-third of the remaining data are selected for the calculation of average of measured queue discharge headways:

$$\bar{t}_s = \frac{1}{\hat{n}_m}\sum_{i=1}^{\hat{n}_m} \hat{t}_{si} \qquad (8)$$

where $\bar{t}_s$ is the average of measured queue discharge headways. When $n \rightarrow \infty$, $\bar{t}_s$ is equal to the saturated headway, $\hat{t}_{si}$ is the remaining data, and $\hat{n}_m$ is the number of remaining data during m cycles.

Thus, the saturated flow rate can be obtained as follows:

$$S = \frac{3600}{\bar{t}_s} \qquad (9)$$

4 CALCULATION RESULTS OF SATURATION FLOW RATE AT SIGNALIZED INTERSECTION

In this section, the calculation results of saturation flow rate will be presented to verify the effectiveness of the proposed calculation scheme based on Grubbs test. For this purpose, we consider the measured traffic data, which are surveyed at the signalized intersection of Dachi Road and Qianrong Road in Wuxi, are collected by cameras. Both Dachi Road and Qianrong Road are important sections of urban road in Wuxi. There are many vehicles at the intersections during peak periods. The intersection is a traffic complex area.

Table 2 gives the calculation values of saturation flow rate. From lane function, it is obvious from the table that the volume of the left turn lane is less than the traffic flow of through lane. Because most drivers choose to drive carefully when they turn left, the measured headway becomes larger than the ideal value and the saturation flow rate of left turn lane is relatively smaller. By viewing different periods, the number of queuing vehicles is uncertain during off-peak periods, the maximum and minimum of traffic volume are both less than the volume during peaking periods. The main reasons are that the number of queuing vehicles is uncertain and the traffic flow is not saturated. In this case, the saturation flow rate gets lower than that in saturated case. Weather data show that the

Table 2. Results of saturation flow rate calculated by the Grubbs-based calculation method (pch/h).

Weather	Period	Lane	Min.	Max.	Mean
Sunny Day	Morning peak	Through	2347	2606	2496
		Left	2189	2500	2285
	Off peak	Through	2204	2554	2424
		Left	1999	2400	2202
	Evening peak	Through	2426	2769	2600
		Left	2170	2342	2247
Rainy Day	Morning peak	Through	1992	2322	2284
		Left	1862	2226	2038
	Evening peak	Through	2025	2170	2113
		Left	1763	1975	1906

volume during peaking periods on a rainy day is less than that on a sunny day. Consequently, the proposed calculation scheme based on Grubbs test can eliminate the effects of uncertainty factors and reflect the laws of saturation flow rate under different conditions.

5 CONCLUSION

In this paper, a new calculation method based on Grubbs test has been proposed for calculating the saturation flow rate at signalized intersection. Considering the effect of various uncertainty factors, this paper presents that a portion of the samples containing a significant number of the measured headway, which affects the main uncertainty factors, should be removed. The calculation results based on practically measured data present that the developed method can eliminate the effects of uncertainty factors and reflect the laws of saturation flow rate under different conditions.

REFERENCES

Akçelik, R., M. Besley. Queue discharge flow and speed models for signalised intersections. In Transportation and Traffic Theory in the 21st Century, Proceedings of the 15th International Symposium on Transportation and Traffic Theory. 99–118(2002).

Chen, R., S. Zhou, The application of improved Grubbs' criterion for inspecting the count of radon concentration. Nuclear Electronics & Detection Technology, 1: 027(2009).

Jin, X., Y. Zhang, F. Wang et al. Departure headway at signalized intersection: A log-normal distribution model approach. Transportation research part C: emerging technologies,17(3), 318–327(2009).

Leutzbach, W. Introduction to the theory of traffic flow. Spriner, Berlin.

MHC2000, Transportation research board. National Research Council, Washington, DC, 113(2000).

Shao, C.Q., J. Rong, L. Zhao, Developing adjustment factors of saturation flow rates at signalized intersections. Beijing Gongye Daxue Xuebao(Journal of Beijing University of Technology), 37(10), 1505–1510(2011).

Shao, C.Q., J. Rong, X.M. Liu, Study on the saturation flow rate and its influence factors at signalized intersection in China. Procedia-Social and Behavioural Sciences, 16, 504–514(2011).

Tarko, A.P., M. Tracz, Uncertainty in saturation flow predictions. Red, 1: P2(2000).

Wood, S.N. Data driven statistical method. Biometrics, 54(4), 1678(1998).

Yang, X., J. Zhao, W. Ma et al, Review on calculation method for signalized intersection capacity, China Journal of Highway and Transport, 05: 148–157(2014).

Zhao, L., C.Q. Shao, J. Rong, Analysis of traffic composition affecting saturation flow at signalized intersection. Highway Engineering, 4, 023(2008).

Advances in Energy Science and Equipment Engineering II – Zhou, Patty & Chen (Eds)
© *2017 Taylor & Francis Group, London, ISBN 978-1-138-71798-5*

A new method of designing a fuzzy classification system based on Michigan code

Zhipeng Hu & Ming Ma
Information Technology and Media College of Beihua University, Jilin, China

ABSTRACT: In order to construct the fuzzy rule set and reduce the number of fuzzy rules and improve the interpretation of the fuzzy classification system, this paper presents a design method of the improved genetic fuzzy classification system based on the Michigan code. New method improves the original algorithm in the premise of the explanation of the fuzzy classification system. In the genetic operation, the introduction of irrelevant items is identified, which is identified as the attribute of unrelated items without additional calculation of membership degree. Reduce the time complexity of the algorithm, and improve the running speed of the system. Simulation results show that the method can improve the accuracy of the system while maintaining the stability of the system.

1 INTRODUCTION

Fuzzy classification system has the ability to deal with uncertain information, therefore, the theory of fuzzy classification has been widely used in all fields (Ye, 2016; Fabio, 2016). Fuzzy classification system can be based on the relevant data by the domain experts to build, but in most cases, either the relevant data is not complete, or the field experts do not exist. How to construct the fuzzy classification system from the limited data rapidly and accurately has become a hot topic in recent years (Rocco, 2016) There are three main research directions of the fuzzy classification system: Based on fuzzy clustering, genetic algorithm and fuzzy neural network (Wu, 2016).

This paper presents a design method of fuzzy classification system based on improved genetic encoding of Michigan, various properties of the new algorithm based on the original algorithm for fuzzy rules to add do not care to identification, reduces the time complexity of fuzzy classification system, and improve the operation speed of the system. The algorithm uses a data sample to generate a complete set of rules, and the rule set is optimized by genetic algorithm. In the process of genetic algorithm coding, the Michigan code is used in the selection operation, and the speed of the sample identification is introduced in the process of mutation operation. The effectiveness of the proposed method is verified by simulation experiments.

2 RULE CONSTRUCTION AND CODING METHOD

2.1 Constructing fuzzy rule antecedent

In the training sample, the data of the known sample points are divided by 5 triangle membership functions, and the triangle membership function is expressed as follows:

$$f(x; a, b, c) = \max\left(0, \min\left(\frac{x-a}{b-a}, \frac{c-x}{c-b}\right)\right) \quad (1)$$

a, b, c: three vertices of the triangle membership function, x: data sample.

Taking into account the integrity of the fuzzy classification system, in the initial stage of the algorithm, the sample points are all set as the initial rule set:

$$R_{set} = \begin{pmatrix} a_{11}, b_{11}, c_{11}, a_{12}, b_{12}, c_{12}, \ldots, a_{1n}, b_{1n}, c_{1n} \\ \ldots \\ a_{m1}, b_{m1}, c_{m1}, a_{m2}, b_{m2}, c_{m2}, \ldots, a_{mn}, b_{mn}, c_{mn} \end{pmatrix}$$

$$(2)$$

m: initial population size; n: the number of attributes in the data sample.

2.2 Fuzzy rule set

The excitation intensity of each rule is calculated separately:

$$u_j(x) = \prod_{i=1}^{n} u_{ji}(x_i) \qquad (3)$$

n: the number of attributes of the current training sample, then, that is, the value of membership function. And then calculate the sum of the incentive intensity of each rule in each category:

$$\beta_{class}(R_j) = \sum_{x \in c} (u_j(x)) \qquad (4)$$

In the formula, $u_j(x)$ represents the incentive intensity of the j rule on the current classification of the sample point X, and we classify the $\beta_{class}(R_j)$ value maximum as the post of the rule. An initial fuzzy rule set is obtained by combining the fuzzy rule antecedent and the fuzzy rule.

2.3 Adjust fuzzy rule set

The fuzzy rule set is established on the basis of considering the integrity of the fuzzy classification system. The number of rules is large, and the size of the fuzzy rule set is 5^n. In order to improve the interpretation of the fuzzy classification system, the training samples are used to adjust the initial rule set, remove the rule of fuzzy rule set value less than 0, and the fitness function is as follows:

$$fitness(R_j) = CP(R_j) - \omega MP(R_j) \qquad (5)$$

$CP(R_j)$ is the correct classification of all the training samples, $MP(R_j)$ is the number of correct classification of all the training samples, ω is the number of the wrong classification of all the training samples, and the penalty coefficient of the error classification is generally set between [0.2,0.5].

2.4 Coding method of fuzzy classification system

Fuzzy classification system is mainly used in the generation of fuzzy rules set by the training samples, the main use of two learning methods: one is the 1978 proposed by approach Michigan; one is proposed in 1980 approach Pittsburgh.

The fuzzy classification system of Michigan type was first designed by Holland in 1978. The main feature of Michigan encoding is that each chromosome in genetic algorithm corresponding to the fuzzy classification system a fuzzy rule in the classifier according to the individual incentive intensity selection, reproduction, crossover and mutation operations, the algorithm will get a set of optimal rules(Assis, 2013). The Michigan classification system in the choice of operation choose higher fitness individuals, this will result in some

samples for individual fitness smaller individuals are ignored, causing premature lower system diversity(Wesley, 2014).

The fuzzy classification system of Pittsburgh type was first designed by Smith in 1980. Different from Michigan code, Pittsburgh code is the main feature of the genetic algorithm in each chromosome is a fuzzy classification system of the complete set of rules. Although the algorithm avoids the conflict between the single rule, but because each generation requires simultaneous evaluation of multiple rule sets, Pittsburgh classification system is huge in computation. Rule set corresponding exchange in the crossover operation, rules, which also makes the performance of a single rule change is blocked, thereby reducing the efficiency of evolution(Upshaw, 2015).

3 ALGORITHM DESCRIPTION

The fuzzy rule sets are optimized by genetic algorithms, selection of outstanding individual genetic to the next generation, to mark the speed of the algorithm is introduced to enhance the selection, the variation of expanding the search space into words calculation of lifting system diversity.

3.1 Code operation

In this paper, we use Michigan code, each of the rules of fuzzy classification system is a chromosome. Each chromosome is composed of regular and regular parts, in which the rules are composed of triangular membership function parameters. In order to reduce the time complexity of the algorithm and improve the algorithm speed, while encoding, adding independent item identification mark flag. each iteration independent random values 0 or 1,0 shows the current property is irrelevant attributes for each attribute of each chromosome, do not need to participate in the membership calculation, 1 shows the properties for the relevant attribute, need to participate in the calculation of membership.

3.2 Select operation

In order to calculate the initial fitness value of each individual in the population, according to the fitness values in descending order of individuals in the initial population. In this paper, the "roulette wheel method" is used to calculate the probability of the bet. Finally, according to the selection of the current

Figure 1. The coding operation.

round of the gambling wheel pointer variable area selection of the current probability of the elite of the elite of the individual to the next generation. Due to the random direction of the roulette wheel pointer, so in the choice of individual, may appear in a classification of the situation can not choose any rule. In order to avoid this kind of situation, the paper proposed a method of deleting similar individuals in the selection strategy of (Wu,1999). The method needs to calculate the similarity of the parameters of each chromosome, and then randomly delete the rules. Although the algorithm to a certain extent, to ensure the diversity of the rule set, but it is also possible to adapt to a larger degree of elite individuals to delete. In order to ensure that random selection of offspring in each category have fuzzy rules, this paper proposes a "remedial strategy" scheme, namely each genetic selection of offspring before, must be directly labeled as elite individual genetic to the next generation of the rules to each category, the rest of the roulette wheel selection rules. In this way, it is guaranteed that the genetic descendants have the fuzzy rule of the elite in each category, which can ensure the accuracy of the system, but also have small probability to select the edge rule with little adaptation degree, and keep the diversity of the system.

3.3 Crossover operation

According to the sequence of roulette wheel selection, the random properties of each pair of individuals in accordance with the crossover probability of crossover probability to expand the search space algorithm. If the number of selected individuals is odd, then the last one does not carry out the crossover operation directly to the next generation.

3.4 Mutation operation

According to the probability of mutation, a rule is found in the population and the parameters of all the triangular membership functions of a property are randomly variant.

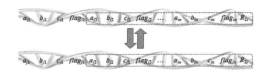

Figure 2. The crossover operation.

Figure 3. The mutation operation.

The introduction of words in mutation operation calculation of lifting system diversity, random selection for the triangular membership function parameters a_{in}, b_{in}, c_{in} were as follows: in the specified range of random mutation operator δ, a new triangular membership function is transformed vertices a'_{in} and c'_{in} are as follows:

$$a'_{in} = \max\{b_{in} - \delta(b_{in} - a_{in}), x_{\min}\} \qquad (6)$$

$$c'_{in} = \min\{b_{in} + \delta(b_{in} - a_{in}), x_{\max}\} \qquad (7)$$

The fuzzy rules which are obtained by the algorithm have some duplicate rules, and the fuzzy rule sets are sorted.

4 SIMULATION EXPERIMENT

In this paper, the Wine data set is used to test the system. Wine system is a typical classification system, and it has been used as the evaluation standard of classification algorithms for many years. Wine data set contains a total of 178 records from 3 different origins of wine, 13 properties are 13 kinds of chemical composition of wine, the origin of wine can be inferred through chemical analysis. It is worth mentioning that all attribute variables are continuous variables.

The simulation experiment uses Michigan code, the crossover probability is 0.9, the probability of mutation is 0.1, the iteration number is 800, the data sample is Wine data set, 178 sample points, 13 attributes, and 3 categories. Among them, 119 training samples, 59 test samples.

The new algorithm keeps the fuzzy classification system with good accuracy, and it can shorten the length of the fuzzy rules. The simulation results show that the new algorithm can shorten the average length of the fuzzy rules effectively compared with the article (Wang, 2014).

Table 1. 100 Experiments on Wine data.

	Number of fuzzy rules	Accuracy (%)
Article (Shi,1999)	3	96.2
This Paper	3	97.1
Article (Ishibuchi, 2001)	4	98
Article (Upshaw, 2015)	5	98

Table 2. Non-dominated rule set on Wine data.

	Average rule length	Accuracy (%)
Article (Wang, 2014)	1.33	96.1
This Paper	0.75	97.1

The experimental results show that the algorithm can obtain a better explanation in the case of a little loss of system accuracy.

5 CONCLUSION

An improved genetic fuzzy classification system based on Michigan code is introduced to reduce the complexity of the sample identification, and the operation speed of the algorithm is improved. Unlike other fuzzy classification systems, the proposed algorithm can improve the understanding of the accuracy of the algorithm in consideration of the accuracy of the algorithm to a certain extent. The experimental results verify the effectiveness of the proposed method. The new algorithm is fast and the system is explained, but the accuracy is decreased. How to improve the accuracy of the fuzzy system, this will be the next step of the work.

ACKNOWLEDGMENTS

This work is supported by the Science and Technology Research Project of the Education Department of Jilin Province (No. 2012126) and Science and Technology Department of Jilin Province Natural Science Foundation (20140101185JC).

REFERENCES

Assis, M, J L Jacobsen, I Jensen et al. The hard hexagon partition function for complex fugacity[J]. Journal of Physics A: Mathematical and Theoretical, 2013, 46(44).

Fabio Grandi. Dynamic class hierarchy management for multi-version ontology-based personalization [J]. Journal of Computer and System Sciences, 2016, 82(1).

Guohua Wu, Rammohan Mallipeddi, P.N. Suganthan et al.. Differential evolution with multi-population based ensemble of mutation strategies[J]. Information Sciences, 2016, 329.

Ishibuchi H, Nakashmia T, Murata T. Three-objective genetics-based machine learning for linguistic rule extraction[J]. Information Sciences, 2001, 136(1–4): 109–133.

Juan Ye, Graeme Stevenson, Simon Dobson. Detecting abnormal events on binary sensors in smart home environments[J]. Pervasive and Mobile Computing, 2016.

Rocco Trombetti, Yue Zhou. On the 2-ranks of a class of unitals[J]. Finite Fields and Their Applications, 2016.

Shi Y, Eberhart R, Chen Y. Implementation of evolutionary fuzzy system[J]. IEEE Trans. Fuzzy Systems, 1999, 7(2): 109–119.

Upshaw CR, Rhodes J D, Webber M E. Modeling peak load reduction and energy consumption enabled by an integrated thermal energy and water storage system for residential air conditioning systems in Austin, Texas. Energy and Build. 2015.

Wang D, Jia H, Wang C, et al. Performance evaluation of controlling thermostatically controlled appliances as virtual generators using comfort-constrained state-queueing models. IET Generation, Transmission & Distribution. 2014.

Wesley J. Cole, Joshua D. Rhodes, William Gorman, Krystian X. Perez, Michael E. Webber, Thomas F. Edgar. Community-scale residential air conditioning control for effective grid management [J]. Applied Energy. 2014.

Wu TP, Chen SM. A new method for constructing membership functions and fuzzy rules from training examples [J]. IEEE Trans System Man Cybernetic Part B. 1999, 29(1): 25–40.

Author index

Liu, K. 1277
Liu, K.-W. 461, 603
Liu, L. 535
Liu, L. 977
Liu, L. 1739
Liu, L.-J. 455
Liu, L.Q. 1505, 1759
Liu, N.B. 1775
Liu, P. 1511
Liu, Q. 95
Liu, Q. 125
Liu, Q. 343
Liu, Q. 881
Liu, Q.F. 585
Liu, Q.G. 1923
Liu, Q.Y. 1163
Liu, R. 907
Liu, S. 1273
Liu, S.M. 129, 785
Liu, S.Y. 211, 1691
Liu, S.Z. 585
Liu, T. 227, 1537
Liu, W. 1673
Liu, W.J. 19
Liu, W.Q. 1375
Liu, X. 1407
Liu, X.B. 969
Liu, X.-C. 1695
Liu, X.C. 973
Liu, X.C. 1431
Liu, X.G. 1709
Liu, X.J. 739, 755
Liu, X.L. 195
Liu, X.Y. 1703
Liu, Y. 407
Liu, Y. 531
Liu, Y. 1297
Liu, Y. 1471
Liu, Y. 1751
Liu, Y.B. 609
Liu, Y.K. 803
Liu, Y.-L. 1187
Liu, Z.G. 1283
Liu, Z.W. 551
Liu, Z.W. 945
Long, N.B. 919
Lu, J.H. 75
Lu, L. 1331
Lu, R.T. 1551, 1587
Lu, S. 1897
Lu, S.C. 1941
Lu, X.M. 369
Lu, Y.-L. 1471
Lü, Y.-Q. 1385
Lu, Z.H. 65
Luo, H.T. 1537
Luo, M. 1183
Luo, Q. 31

Luo, W.M. 1953
Luo, X.-F. 793
Luo, Y. 517
Luo, Z. 1613
Luo, Z.Q. 1119
Lv, D. 747
Lv, H. 747
Lv, K. 993
Lv, P. 473
Lv, T.Z. 311
Lv, Y.Z. 887

Ma, C.H. 1785
Ma, C.Y. 1857
Ma, F.-Y. 45
Ma, L. 897
Ma, L.L. 863
Ma, L.T. 1173
Ma, M. 1957
Ma, M.J. 1355
Ma, S.J. 1511
Ma, S.L. 57
Ma, Y. 1775
Ma, Y.J. 1619
Ma, Z. 1839
Ma, Z.J. 999
Mao, Y.Y. 281
Mei, X.C. 499
Meng, J. 1273
Meng, S.-S. 1385
Ming, T.F. 1261
Mohamad Zamri, S.F. 1335
Mohamed, O. 1061
Muhd Zailani, N.A. 1335

Ni, K. 1249
Ni, W.K. 495, 579, 1067
Ni, Y.P. 221
Nian, H.F. 1033, 1177
Nie, L. 1277
Nie, P. 681
Ning, N. 999
Niu, C. 1477
Niu, Q.L. 1327

Ou, L. 973

Paik, S.M. 615
Pan, F.Y. 1277
Pan, J.-W. 915
Pan, Z.X. 281
Pang, H. 301
Pang, H.X. 159
Pang, Y.H. 19
Pang, Z.Y. 359
Pei, D.J. 1145
Pei, G.-Q. 301
Pei, W.B. 1157

Peng, D. 79
Peng, H.T. 1163
Peng, M.L. 1905
Peng, Q.-C. 1311
Peng, R.M. 125
Peng, X.H. 1905
Peng, Y.F. 891
Pu, S.Y. 1091
Pu, Z.H. 859

Qi, B. 887
Qi, S.B. 761
Qian, C.X. 49
Qian, J. 1203
Qian, W.X. 1399
Qin, K. 959
Qin, S.L. 1445
Qin, W. 499
Qiu, J.L. 467
Qiu, T. 1023
Qu, W.J. 999
Qu, X. 1807
Quan, R.K. 1575
Quan, Y.K. 1033

Ren, H.X. 1771
Ren, X.-D. 1947
Ren, Y. 1571
Ren, Z. 1297
Rong, M. 815
Ruan, Z. 215
Rusbintardjo, G. 1193

Sang, G.C. 703
Sha, S. 477
Shan, B.L. 887
Shan, P.F. 65
Shao, Q. 1915
Shek, P.-N. 595
Shen, J. 1725
Shen, J.-R. 603
Shen, L.M. 1627, 1633
Shen, P. 1639
Shen, Q. 1725
Shen, R.C. 1709
Shen, Y.L. 1801
Shen, Y.T. 1853
Shi, D.P. 363
Shi, H.P. 1399
Shi, J.F. 79
Shi, S.Z. 203
Shi, W.L. 129
Shi, X. 1079
Shi, Z.-X. 713
Shu, D.K. 1355
Shuai, X.-X. 45
Si, W. 1371
Song, J. 1119

Yue, M.C. 709
Yue, X.J. 65

Zang, D.J. 1113
Zang, H.X. 1677
Zang, J.L. 1703
Zang, Z.X. 499
Zeng, M. 13
Zeng, M.Z. 1685
Zeng, W.J. 1327
Zeng, X. 1421
Zeng, X.X. 521
Zeng, Y.N. 1751
Zha, B.-L. 897
Zha, Y.H. 1935
Zhan, X.P. 285, 1739
Zhang, A.L. 1431
Zhang, B. 259
Zhang, B. 659
Zhang, C. 1103
Zhang, C. 1733
Zhang, D.B. 1283, 1291
Zhang, D.L. 349
Zhang, D.W. 327
Zhang, E.Y. 321
Zhang, F.-Y. 1863
Zhang, G.L. 1399
Zhang, G.S. 1929
Zhang, G.X. 477, 1439
Zhang, H. 547
Zhang, H. 869
Zhang, H. 1511
Zhang, H. 1709
Zhang, H.H. 1661
Zhang, H.R. 381
Zhang, H.X. 1451
Zhang, J. 333
Zhang, J. 969
Zhang, J. 1493, 1497
Zhang, J. 1639
Zhang, J. 1887
Zhang, J.B. 531
Zhang, J.H. 1163
Zhang, J.-W. 599
Zhang, J.W. 1253
Zhang, L.C. 285
Zhang, L.L. 249
Zhang, L.L. 1145
Zhang, M. 45
Zhang, M.F. 1871
Zhang, Q. 1177
Zhang, Q. 1451
Zhang, Q.R. 1883
Zhang, Q.S. 525
Zhang, R. 1055
Zhang, R.F. 919
Zhang, R.M. 1745

Zhang, R.P. 387
Zhang, S. 1163
Zhang, S.P. 881
Zhang, T.X. 977
Zhang, W. 625
Zhang, W.-B. 85
Zhang, W.S. 619
Zhang, W.W. 99
Zhang, W.X. 969
Zhang, X. 447
Zhang, X. 1407
Zhang, X. 1471
Zhang, X.B. 1833
Zhang, X.H. 1297
Zhang, X.L. 19
Zhang, X.M. 687
Zhang, X.P. 1163
Zhang, X.R. 321
Zhang, X.-S. 1055
Zhang, X.Y. 807
Zhang, X.Z. 901
Zhang, Y. 897
Zhang, Y. 1321
Zhang, Y. 1327
Zhang, Y.J. 1315
Zhang, Y.J. 1643
Zhang, Y.L. 579, 1067
Zhang, Y.S. 1527
Zhang, Y.T. 1157
Zhang, Y.Z. 65
Zhang, Z. 423
Zhang, Z. 1909
Zhang, Z.B. 1217
Zhang, Z.H. 733
Zhang, Z.H. 983
Zhang, Z.-Q. 1385
Zhang, Z.Y. 1681
Zhao, B. 285
Zhao, B.Y. 353
Zhao, C.H. 65
Zhao, F.Y. 1871
Zhao, H.W. 359
Zhao, H.Y. 203
Zhao, J. 387
Zhao, J.J. 1079
Zhao, J.J. 1487
Zhao, J.S. 95
Zhao, K. 945
Zhao, Q. 1163
Zhao, Q.-L. 1947
Zhao, S. 1301
Zhao, S.G. 1005
Zhao, S.Q. 1733
Zhao, T.X. 1355
Zhao, W.G. 1451
Zhao, Y. 483
Zhao, Y. 811

Zhao, Y. 929, 937
Zhao, Y.B. 729
Zhao, Y.F. 1113
Zhao, Y.F. 1227
Zhao, Y.H. 363
Zhao, Y.K. 13
Zhao, Z.D. 407
Zhao, Z.Q. 1145
Zheng, B. 349
Zheng, C.M. 965
Zheng, H.J. 215
Zheng, L.P. 559, 573
Zheng, R.L. 441
Zheng, Y.B. 1551, 1587
Zheng, Y.D. 733
Zheng, Z.-S. 1695
Zhong, D.M. 1755
Zhong, H. 187
Zhong, S. 1199
Zhou, D. 1167
Zhou, F. 1135, 1141
Zhou, F. 1471
Zhou, F.X. 327
Zhou, G.J. 1561
Zhou, H.-K. 1461
Zhou, J.L. 681
Zhou, J.L. 1681
Zhou, J.M. 1091
Zhou, J.Q. 1017
Zhou, J.Z. 381
Zhou, K.Q. 1575
Zhou, L.J. 19
Zhou, X. 1725
Zhou, Y.M. 1515
Zhuo, Z. 1481
Zhou, Z.Q. 525
Zhu, C. 659
Zhu, F. 1667
Zhu, H.B. 521
Zhu, H.Y. 1211
Zhu, J.C. 1877
Zhu, J.W. 1515
Zhu, K.P. 1315
Zhu, L. 1853
Zhu, L.P. 799
Zhu, Q.Q. 1067
Zhu, Q.S. 1935
Zhu, Q.-T. 1385
Zhu, R. 1403
Zhu, T. 1381
Zhu, Y. 691
Zhu, Y.H. 211,
   1591, 1691
Zhu, Z.-Z. 451
Zhuang, H.J. 1807
Zhuang, W.L. 837
Zou, M.M. 969

Printed and bound by CPI Group (UK) Ltd, Croydon, CR0 4YY

24/10/2024

01778311-0001